MAR 2007
R
CH
cont

CHILTON®

ASIAN
MECHANICAL SERVICE
VOLUME 1
2006 EDITION
Acura
Honda
Hyundai
Isuzu
Kia

THOMSON ™
DELMAR LEARNING

Australia • Canada • Mexico • Singapore • Spain • United Kingdom • United States

THOMSON
DELMAR LEARNING

Chilton®

Asian
Mechanical Service
Volume I
Acura, Honda, Hyundai, Isuzu, Kia
2006 Edition

**Vice President,
Technology Professional Business Unit:**
Gregory L. Clayton

**Publisher,
Technology Professional Business Unit:**
David Koontz

Director of Marketing:
Beth A. Lutz

Production Director:
Patty Stephan

Production Manager:
Andrew Crouth

Marketing Specialist:
Brian McGrath

Marketing Coordinator:
Marissa Maiella

Marketing Assistant:
Jennifer Stall

Sr. Production Editor:
Elizabeth C. Hough

Editorial Assistant:
Christine Wade

Editors:
Nick D'Andrea
Thomas A. Mellon
Rich J. Rivelle
Jon Wallace

Cover Design:
Melinda Possinger

ISBN: 1-4180-0947-4

ISSN: 1548-0887

NOTICE TO THE READER

Table of Contents

Model Index

USING THIS INFORMATION

Organization

To find where a particular model section or procedure is located, look in the Table of Contents. Main topics are listed with the page number on which they may be found. Following the main topics is an alphabetical listing of all of the procedures within the section and their page numbers.

Manufacturer and Model Coverage

This product covers 2002–2006 Asian models that are produced in sufficient quantities to warrant coverage, and which have technical content available from the vehicle manufacturers before our publication date. Although this information is as complete as possible at the time of publication, some manufacturers may make changes which cannot be included here. While striving for total accuracy, the publisher cannot assume responsibility for any errors, changes, or omissions that may occur in the compilation of this data.

Part Numbers & Special Tools

Part numbers and special tools are recommended by the publisher and vehicle manufacturer to perform specific jobs. Before substituting any part or tool for the one recommended, you must be completely satisfied that neither your personal safety, nor the performance of the vehicle will be endangered.

ACKNOWLEDGEMENT

The publisher would like to express appreciation to the following vehicle manufacturers for their assistance in producing this publication. No further reproduction or distribution of the material in this manual is allowed without the expressed written permission of the vehicle manufacturers and the publisher. American Honda Motor Co., including Acura and Honda Divisions, Hyundai Group, including Hyundai and Kia Motor, Isuzu Motors of America, Inc.

PRECAUTIONS

Before servicing any vehicle, please be sure to read all of the following precautions, which deal with personal safety, prevention of component damage, and important points to take into consideration when servicing a motor vehicle:

• Always wear safety glasses or goggles when drilling, cutting, grinding or prying.

• Steel-toed work shoes should be worn when working with heavy parts. Pockets should not be used for carrying tools. A slip or fall can drive a screwdriver into your body.

• Work surfaces, including tools and the floor should be kept clean of grease, oil or other slippery material.

• When working around moving parts, don't wear loose clothing. Long hair should be tied back under a hat or cap, or in a hair net.

• Always use tools only for the purpose for which they were designed. Never pry with a screwdriver.

• Keep a fire extinguisher and first aid kit handy.

• Always properly support the vehicle with approved stands or lift.

• Always have adequate ventilation when working with chemicals or hazardous material.

• Carbon monoxide is colorless, odorless and dangerous. If it is necessary to operate the engine with vehicle in a closed area such as a garage, always use an exhaust collector to vent the exhaust gases outside the closed area.

• When draining coolant, keep in mind that small children and some pets are attracted by ethylene glycol antifreeze, and are quite likely to drink any left in an open container, or in puddles on the ground. This will prove fatal in sufficient quantity. Always drain the coolant into a sealable container.

• To avoid personal injury, do not remove the coolant pressure relief cap while the engine is operating or hot. The cooling system is under pressure; steam and hot liquid can come out forcefully when the cap is loosened slightly. Failure to follow these instructions may result in personal injury. The coolant must be recovered in a suitable, clean container for reuse. If the coolant is contaminated it must be recycled or disposed of correctly.

• When carrying out maintenance on the starting system be aware that heavy gauge leads are connected directly to the battery. Make sure the protective caps are in place when maintenance is completed. Failure to follow these instructions may result in personal injury.

• Do not remove any part of the engine emission control system. Operating the engine without the engine emission control system will reduce fuel economy and engine ventilation. This will weaken engine performance and shorten engine life. It is also a violation of Federal law.

• Due to environmental concerns, when the air conditioning system is drained, the refrigerant must be collected using refrigerant recovery/recycling equipment. Federal law requires that refrigerant be recovered into appropriate recovery equipment and the process be conducted by qualified technicians who have been certified by an approved organization, such as MACS, ASI, etc. Use of a recovery machine dedicated to the appropriate refrigerant is necessary to reduce the possibility of oil and refrigerant incompatibility concerns. Refer to the instructions provided by the equipment manufacturer when removing refrigerant from or charging the air conditioning system.

• Always disconnect the battery ground when working on or around the electrical system.

• Batteries contain sulfuric acid. Avoid contact with skin, eyes, or clothing. Also, shield your eyes when working near batteries to protect against possible splashing of the acid solution. In case of acid contact with skin or eyes, flush immediately with water for a minimum of 15 minutes and get prompt medical attention. If acid is swallowed, call a physician immediately. Failure to follow these instructions may result in personal injury.

• Batteries normally produce explosive gases. Therefore, do not allow flames, sparks or lighted substances to come near the battery. When charging or working near a battery, always shield your face and protect your eyes. Always provide ventilation. Failure to follow these instructions may result in personal injury.

• When lifting a battery, excessive pressure on the end walls could cause acid to spew through the vent caps, resulting in personal injury, damage to the vehicle or battery. Lift with a battery carrier or with your hands on opposite corners. Failure to follow these instructions may result in personal injury.

• Observe all applicable safety precautions when working around fuel. Whenever servicing the fuel system, always work in a well-ventilated area. Do not allow fuel spray or vapors to come in contact with a spark, open flame, or excessive heat (a hot drop light, for example). Keep a dry chemical fire extinguisher near the work area. Always keep fuel in a container specifically designed for fuel storage; also, always properly seal fuel containers to avoid the possibility of fire or explosion. Do not smoke or carry lighted tobacco or open flame of any type when working on or near any fuel-related components.

• Fuel injection systems often remain pressurized, even after the engine has been turned OFF. The fuel system pressure must be relieved before disconnecting any fuel lines. Failure to do so may result in fire and/or personal injury.

• The evaporative emissions system contains fuel vapor and condensed fuel vapor. Although not present in large quantities, it still presents the danger of explosion or fire. Disconnect the battery ground cable from the battery to minimize the possibility of an electrical spark occurring, possibly causing a fire or explosion if fuel vapor or liquid fuel is present in the area. Failure to follow these instructions can result in personal injury.

• The EPA warns that prolonged contact with used engine oil may cause a number of skin disorders, including cancer! You should make every effort to minimize your exposure to used engine oil. Protective gloves should be worn when changing oil. Wash your hands and any other exposed skin areas as soon as possible after exposure to used engine oil. Soap and water, or waterless hand cleaner should be used.

• Some vehicles are equipped with an air bag system, often referred to as a Supplemental Restraint System (SRS) or Supplemental Inflatable Restraint (SIR) system. The system must be disabled before performing service on or around system components, steering column, instrument panel components, wiring and sensors. Failure to follow safety and disabling procedures could result in accidental air bag deployment, possible personal injury and unnecessary system repairs.

• Always wear safety goggles when working with, or around, the air bag system. When carrying a non-deployed air bag, be sure the bag and trim cover are pointed away from your body. When placing a non-deployed air bag on a work surface, always face the bag and trim cover upward, away from the surface. This will reduce the motion of the module if it is accidentally deployed.

• Electronic modules are sensitive to electrical charges. The ABS module can be damaged if exposed to these charges.

• Brake pads and shoes may contain asbestos, which has been determined to be a cancer-causing agent. Never clean brake surfaces with compressed air. Avoid inhaling brake dust. Clean all brake surfaces with a commercially available brake cleaning fluid.

• When replacing brake pads, shoes, discs or drums, replace them as complete axle sets.

• When servicing drum brakes, disassemble and assemble one side at a time, leaving the remaining side intact for reference.

• Brake fluid often contains polyglycol ethers and polyglycols. Avoid contact with the eyes and wash your hands thoroughly after handling brake fluid. If you do get brake fluid in your eyes, flush your eyes with clean, running water for 15 minutes. If eye irritation persists, or if you have taken brake fluid internally, immediately seek medical assistance.

• Clean, high quality brake fluid from a sealed container is essential to the safe and proper operation of the brake system. You should always buy the correct type of brake fluid for your vehicle. If the brake fluid becomes contaminated, completely flush the system with new fluid. Never reuse any brake fluid. Any brake fluid that is removed from the system should be discarded. Also, do not allow any brake fluid to come in contact with a painted or plastic surface; it will damage the paint.

• Never operate the engine without the proper amount and type of engine oil; doing so will result in severe engine damage.

• Timing belt maintenance is extremely important! Many models utilize an interference-type, non-freewheeling engine. If the timing belt breaks, the valves in the cylinder head may strike the pistons, causing potentially serious (also time-consuming and expensive) engine damage.

• Disconnecting the negative battery cable on some vehicles may interfere with the functions of the on-board computer system(s) and may require the computer to undergo a relearning process once the negative battery cable is reconnected.

• Steering and suspension fasteners are critical parts because they affect performance of vital components and systems and their failure can result in major service expense. They must be replaced with the same grade or part number or an equivalent part if replacement is necessary. Do not use a replacement part of lesser quality or substitute design. Torque values must be used as specified during reassembly to ensure proper retention of these parts.

ACURA

3.2TL • 3.2CL • 3.5RL • RSX • TSX

SPECIFICATION AND MAINTENANCE CHARTS

ENGINE AND VEHICLE IDENTIFICATION

			Engine						Model Year	
Code	Liters (cc)	Cu. In.	Cyl.	Fuel Sys.	Engine Type	Eng. Mfg.		Code ①		Year
K20A2	2.0 (1999)	122	4	PGM-FI	DOHC	Honda		2		2002
K20A3	2.0 (1999)	122	4	PGM-FI	DOHC	Honda		3		2003
K20Z1	2.0 (1999)	122	4	PGM-FI	DOHC	Honda		4		2004
K24A4	2.4 (2354)	144	4	PGM-FI	DOHC	Honda		5		2005
J32A1	3.2 (3210)	196	6	PGM-FI	SOHC	Honda		6		2006
J32A2	3.2 (3210)	196	6	PGM-FI	SOHC	Honda				
J32A3	3.2 (3210)	196	6	PGM-FI	SOHC	Honda				
C35A1	3.5 (3474)	211	6	PGM-FI	SOHC	Honda				
J35A8	3.5 (3471)	222	6	PGM-FI	SOHC	Honda				

PGM-FI: Programmed Fuel Injection

DOHC: Double Overhead Camshaft

SOHC: Single Overhead Camshaft

① 10th digit of the Vehicle Identification Number (VIN)

09474_ACUCAR_C0001

GENERAL ENGINE SPECIFICATIONS

Year	Model	Engine Displacement Liters	Engine ID	Net Horsepower @ rpm	Net Torque @ rpm (ft. lbs.)	Bore x Stroke (in.)	Compression Ratio	Oil Pressure @ rpm
2002	RSX	2.0	K20A3	160@6500	141@4000	3.39x3.39	9.8:1	44@3000
	RSX ①	2.0	K20A2	200@7400	142@6000	3.39x3.39	11.0:1	44@3000
	3.2CL	3.2	J32A1	225@5600	217@4700	3.50x3.39	9.8:1	71@3000
	3.2CL	3.2	J32A2	260@6100	232@3500	3.50x3.39	10.5:1	71@3000
	3.2TL	3.2	J32A1	225@5500	216@5000	3.50x3.39	9.8:1	71@3000
	3.5RL	3.5	C35A1/KA9	210@5200	224@2800	3.54x3.58	9.6:1	50@3000
2003	RSX	2.0	K20A3	160@6500	141@4000	3.39x3.39	9.8:1	44@3000
	RSX ①	2.0	K20A2	200@7400	142@6000	3.39x3.39	11.0:1	44@3000
	3.2CL	3.2	J32A1	225@5600	217@4700	3.50x3.39	9.8:1	71@3000
	3.2CL	3.2	J32A2	260@6100	232@3500	3.50x3.39	10.5:1	71@3000
	3.2TL	3.2	J32A1	225@5500	216@5000	3.50x3.39	9.8:1	71@3000
	3.5RL	3.5	C35A1/KA9	210@5200	224@2800	3.54x3.58	9.6:1	50@3000
2004	RSX	2.0	K20A3	155@6500	139@4000	3.39x3.39	9.8:1	44@3000
	RSX ①	2.0	K20A2	201@7800	140@7000	3.39x3.39	11.0:1	44@3000
	TSX	2.4	K24A2	200@6800	166@4500	3.42X3.89	10.5:1	44@3000
	3.2TL	3.2	J32A3	258@6200	233@5000	3.50x3.39	11.0:1	71@3000
	3.5RL	3.5	C35A1/KA9	210@5200	224@2800	3.54x3.58	9.6:1	50@3000
2005-06	RSX	2.0	K20A3	155@6500	139@4000	3.39x3.39	9.8:1	44@3000
	RSX ①	2.0	K20Z1	201@7800	140@7000	3.39x3.39	11.0:1	44@3000
	TSX	2.4	K24A2	200@6800	166@4500	3.42x3.89	10.5:1	44@3000
	3.2TL	3.2	J32A3	258@6200	233@5000	3.50x3.39	11.0:1	71@3000
	3.5RL	3.5	J35A8	290@6200	256@5000	3.50x3.66	11.0:1	71@3000

PGM-FI: Programmed Fuel Injection

① Type-S

09474_ACUCAR_C0002

ENGINE TUNE-UP SPECIFICATIONS

Year	Engine Displacement Liters	Engine ID/VIN	Spark Plug Gap (in.)	Ignition Timing (deg.) MT	Ignition Timing (deg.) AT	Fuel Pump (psi)	Idle Speed (rpm) MT	Idle Speed (rpm) AT	Valve Clearance In.	Valve Clearance Ex.
2002	2.0	K20A3	0.039-0.043	8B	8B	47-54 ①	600-700	600-700	0.008-0.010	0.011-0.013
	2.0	K20A2	0.039-0.043	8B	8B	47-54 ①	650-750	650-750	0.008-0.010	0.011-0.013
	3.2	J32A1	0.039-0.043	—	8-10B	41-48 ①	—	630-730	0.008-0.009	0.011-0.013
	3.2	J32A2	0.039-0.043	—	8-10B	41-48 ①	—	700-800	0.008-0.009	0.011-0.013
	3.5	C35A1/KA9	0.039-0.043	—	15B	43-50 ①	—	600-700	HYD	HYD
2003	2.0	K20A3	0.039-0.043	8B	8B	47-54 ①	600-700	600-700	0.008-0.010	0.011-0.013
	2.0	K20A2	0.039-0.043	8B	8B	47-54 ①	650-750	650-750	0.008-0.010	0.011-0.013
	3.2	J32A1	0.039-0.043	—	8-10B	41-48 ①	—	630-730	0.008-0.009	0.011-0.013
	3.2	J32A2	0.039-0.043	—	8-10B	41-48 ①	—	700-800	0.008-0.009	0.011-0.013
	3.5	C35A1/KA9	0.039-0.043	—	15B	43-50 ①	—	600-700	HYD	HYD
2004	2.0	K20A3	0.039-0.043	6-10 B	6-10 B	47-54 ①	600-700	600-700	0.008-0.010	0.011-0.013
	2.0	K20A2	0.039-0.043	6-10 B	6-10 B	47-54 ①	650-750	650-750	0.008-0.010	0.011-0.013
	2.4	K24A2	0.039-0.043	6-10 B	6-10 B	48-55 ①	700-800	750-850	0.008-0.010	0.010-0.011
	3.2	J32A3	0.039-0.043	8-10B	8-10B	57-64 ①	—	700-800	0.008-0.009	0.011-0.013
	3.5	C35A1/KA9	0.039-0.043	—	15B	43-50 ①	—	600-700	HYD	HYD
2005-06	2.0	K20A3	0.039-0.043	6-10 B	6-10 B	47-54 ①	600-700	600-700	0.008-0.010	0.011-0.013
	2.0	K20Z1	0.039-0.043	6-10 B	6-10 B	47-54 ①	800-900	800-900	0.008-0.010	0.010-0.011
	2.4	K24A2	0.039-0.043	6-10 B	6-10 B	48-55 ①	700-800	750-850	0.008-0.010	0.010-0.011
	3.2	J32A3	0.039-0.043	8-10B	8-10B	57-64 ①	—	700-800	0.008-0.009	0.011-0.013
	3.5	J35A8	0.039-0.043	—	8-12B	55-63 ①	—	630-730	HYD	HYD

NOTE: The Vehicle Emission Control Information label reflects specification changes during production and must be used if they differ from this chart.

B: Before Top Dead Center

HYD: Hydraulic

① At idle, pressure regulator vacuum hose disconnected

CAPACITIES

Year	Model	Engine Displacement Liters	Engine ID/VIN	Engine Oil with Filter (qts.)	Transmission (pts.)			Transfer Case (pts.)	Drive Axle		Fuel Tank (gal.)	Cooling System (qts.)
					5-Spd	6-Spd	Auto.		Front (pts.)	Rear (pts.)		
2002	RSX	2.0	K20A3	4.4	3.2	—	5.8	—	—	—	13.2	①
	RSX ②	2.0	K20A2	5.0	—	3.2	—	—	—	—	13.2	5.4
	3.2CL	3.2	J32A1	4.6	—	—	7.6	—	—	—	17.1	5.9
	3.2CL	3.2	J32A2	4.6	—	—	7.6	—	—	—	17.1	5.9
	3.2TL	3.2	J32A1/UA5	4.6	—	—	6.2	—	—	—	17.1	5.9
	3.5RL	3.5	C35A1/KA9	4.9	—	—	6.4	—	2.2	—	18.0	6.4
2003	RSX	2.0	K20A3	4.4	3.2	—	5.8	—	—	—	13.2	①
	RSX ②	2.0	K20A2	5.0	—	3.2	—	—	—	—	13.2	5.4
	3.2CL	3.2	J32A1	4.6	—	—	7.6	—	—	—	17.1	5.9
	3.2CL	3.2	J32A2	4.6	—	—	7.6	—	—	—	17.1	5.9
	3.2TL	3.2	J32A1/UA5	4.6	—	—	6.2	—	—	—	17.1	5.9
	3.5RL	3.5	C35A1/KA9	4.9	—	—	6.4	—	2.2	—	18.0	6.4
2004	RSX	2.0	K20A3	4.4	3.2	—	6.2	—	—	—	13.2	①
	RSX ②	2.0	K20A2	5.0	—	3.2	—	—	—	—	13.2	5.4
	TSX	2.4	K24A2	4.4	—	4.2	6.0	—	—	—	17.1	③
	3.2TL	3.2	J32A3	4.5	—	4.6	6.4	—	—	—	17.1	5.9
	3.5RL	3.5	C35A1/KA9	4.9	—	—	6.4	—	2.2	—	18.0	6.4
2005-06	RSX	2.0	K20A3	4.4	3.2	—	6.2	—	—	—	13.2	①
	RSX ②	2.0	K20Z1	5.0	—	3.2	—	—	—	—	13.2	5.4
	TSX	2.4	K24A2	4.4	—	4.2	6.0	—	—	—	17.1	③
	3.2TL	3.2	J32A3	4.5	—	4.6	6.4	—	—	—	17.1	5.9
	3.5RL	3.5	J35A8	4.5	—	—	5.8	—	—	1.5	19.4	6.4

NOTE: All capacities are approximate. Add fluid gradually and ensure a proper fluid level is obtained.

NOTE: Capacities given are service, not overhaul capacities

① Automatic transmission: 5.3
   Manual Transmission: 5.4

② Type-S

③ Automatic transmission: 5.6
   Manual Transmission: 5.7

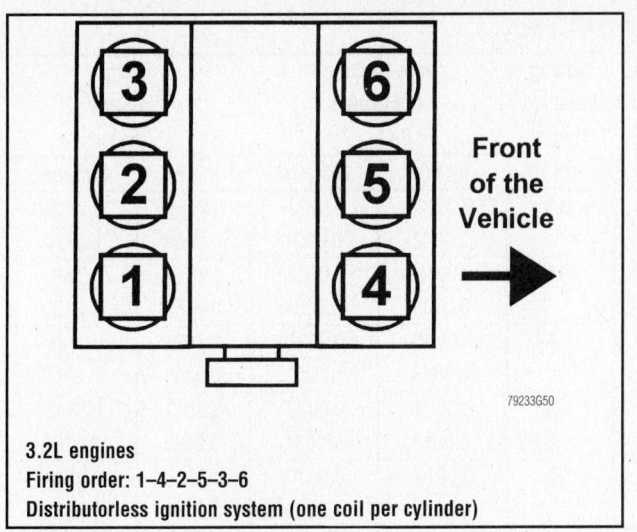

3.2L engines
Firing order: 1–4–2–5–3–6
Distributorless ignition system (one coil per cylinder)

79233G50

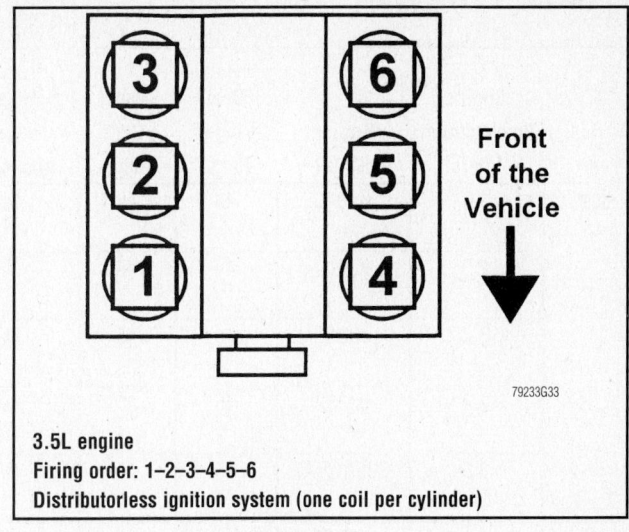

3.5L engine
Firing order: 1–2–3–4–5–6
Distributorless ignition system (one coil per cylinder)

79233G33

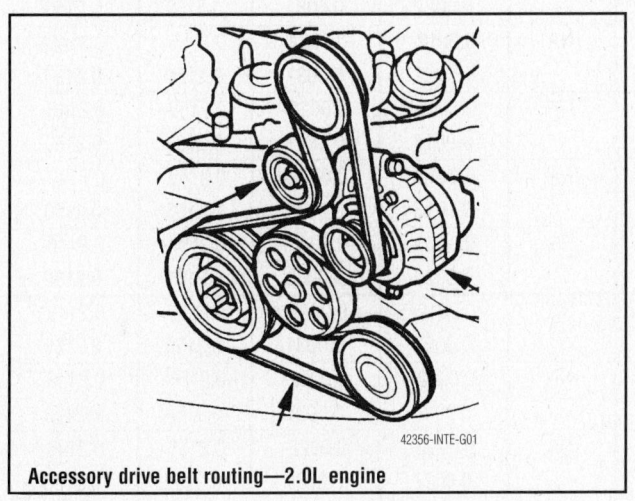

42356-INTE-G01

Accessory drive belt routing—2.0L engine

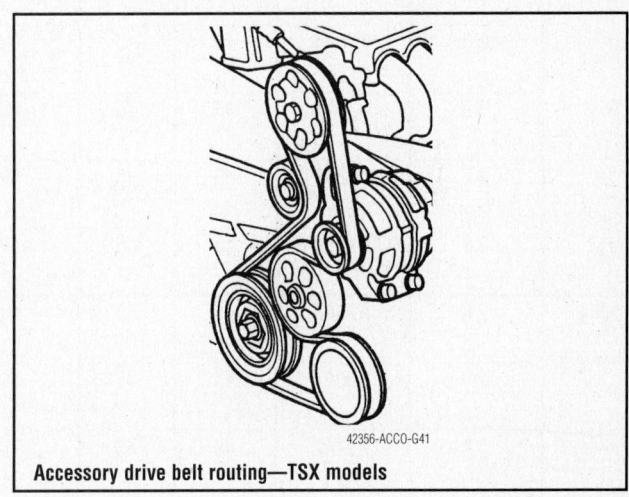

42356-ACCO-G41

Accessory drive belt routing—TSX models

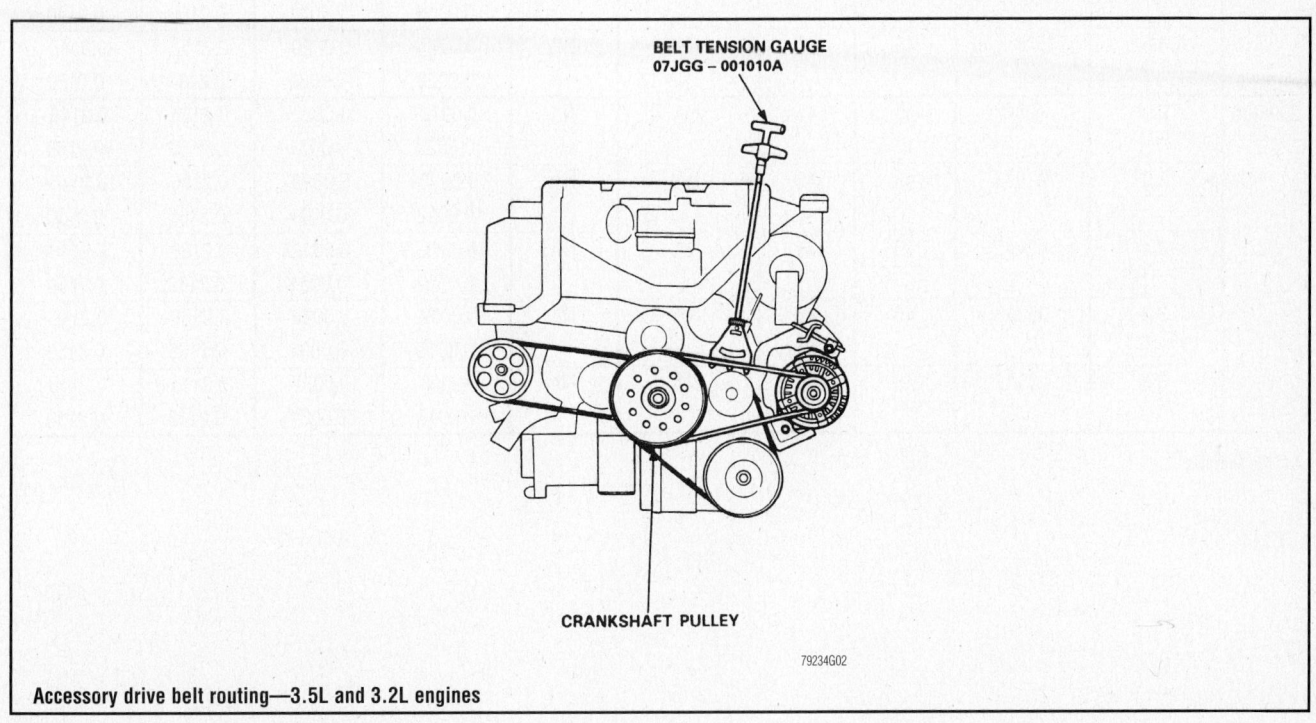

BELT TENSION GAUGE
07JGG – 001010A

CRANKSHAFT PULLEY

79234G02

Accessory drive belt routing—3.5L and 3.2L engines

VALVE SPECIFICATIONS

Year	Engine Displacement Liters	Engine ID/VIN	Seat Angle (deg.)	Face Angle (deg.)	Spring Test Pressure (lbs. @ in.)	Spring Installed Height (in.)	Stem-to-Guide Clearance (in.) Intake	Stem-to-Guide Clearance (in.) Exhaust	Stem Diameter (in.) Intake	Stem Diameter (in.) Exhaust
2002	2.0	K20A3	45	45	NA	NA	0.0012-0.0022	0.0022-0.0031	0.2156-0.2159	0.2146-0.2150
	2.0	K20A4	45	45	NA	NA	0.0012-0.0022	0.0022-0.0031	0.2156-0.2159	0.2146-0.2150
	3.2	J32A1	45	45	NA	NA	0.0008-0.0018	0.0022-0.0031	0.2159-0.2163	0.2146-0.2150
	3.2	J32A2	45	45	NA	NA	0.0008-0.0018	0.0022-0.0031	0.2159-0.2163	0.2146-0.2150
	3.5	C35A1/KA9	45	45	NA	NA	0.0010-0.0020	0.0020-0.0030	0.2157-0.2161	0.2146-0.2150
2003	2.0	K20A3	45	45	NA	NA	0.0012-0.0022	0.0022-0.0031	0.2156-0.2159	0.2146-0.2150
	2.0	K20A4	45	45	NA	NA	0.0012-0.0022	0.0022-0.0031	0.2156-0.2159	0.2146-0.2150
	3.2	J32A1	45	45	NA	NA	0.0008-0.0018	0.0022-0.0031	0.2159-0.2163	0.2146-0.2150
	3.2	J32A2	45	45	NA	NA	0.0008-0.0018	0.0022-0.0031	0.2159-0.2163	0.2146-0.2150
	3.5	C35A1/KA9	45	45	NA	NA	0.0010-0.0020	0.0020-0.0030	0.2157-0.2161	0.2146-0.2150
2004	2.0	K20A3	45	45	NA	NA	0.0012-0.0022	0.0022-0.0031	0.2156-0.2159	0.2146-0.2150
	2.0	K20A4	45	45	NA	NA	0.0012-0.0022	0.0022-0.0031	0.2156-0.2159	0.2146-0.2150
	2.4	K24A2	45	45	NA	NA	0.0012-0.0022	0.0022-0.0031	0.2156-0.2159	0.2146-0.2150
	3.2	J32A3	45	45	NA	NA	0.0008-0.0018	0.0022-0.0031	0.2159-0.2163	0.2146-0.2150
	3.5	C35A1/KA9	45	45	NA	NA	0.0010-0.0020	0.0020-0.0030	0.2157-0.2161	0.2146-0.2150
2005-06	2.0	K20A3	45	45	NA	NA	0.0012-0.0022	0.0022-0.0031	0.2156-0.2159	0.2146-0.2150
	2.0	K20Z1	45	45	NA	NA	0.0012-0.0022	0.0022-0.0031	0.2156-0.2159	0.2146-0.2150
	2.4	K24A2	45	45	NA	NA	0.0012-0.0022	0.0022-0.0031	0.2156-0.2159	0.2146-0.2150
	3.2	J32A3	45	45	NA	NA	0.0008-0.0018	0.0022-0.0031	0.2159-0.2163	0.2146-0.2150
	3.5	J35A8	45	45	NA	NA	0.0008-0.0018	0.0022-0.0031	0.2159-0.2163	0.2146-0.2150

NA: Not Available

CRANKSHAFT AND CONNECTING ROD SPECIFICATIONS

All measurements are given in inches.

Year	Engine Displacement Liters	Engine ID/VIN	Crankshaft Main Brg. Journal Dia.	Crankshaft Main Brg. Oil Clearance	Crankshaft Shaft End-play	Thrust on No.	Connecting Rod Journal Diameter	Connecting Rod Oil Clearance	Connecting Rod Side Clearance
2002	2.0	K20A3	2.1648-2.1657	①	0.0040-0.0140	4	1.7707-1.7717	0.0008-0.0019	0.0060-0.0160
	2.0	K20A2	2.1646-2.1655	①	0.0040-0.0140	4	1.8888-1.8898	0.0013-0.0024	0.0060-0.0160
	3.2	J32A1	2.8337-2.8346	0.0008-0.0017	0.0040-0.0140	3	2.1644-2.1654	0.0008-0.0017	0.0060-0.0140
	3.2	J32A2	2.8337-2.8346	0.0008-0.0017	0.0040-0.0140	3	2.1644-2.1654	0.0008-0.0017	0.0060-0.0140
	3.5	C35A1/KA9	2.6762-2.6772	0.0008-0.0017	0.0040-0.0110	3	2.1248-2.1257	0.0009-0.0018	0.0060-0.0120
2003	2.0	K20A3	2.1648-2.1657	①	0.0040-0.0140	4	1.7707-1.7717	0.0008-0.0019	0.0060-0.0160
	2.0	K20A2	2.1646-2.1655	①	0.0040-0.0140	4	1.8888-1.8898	0.0013-0.0024	0.0060-0.0160
	3.2	J32A1	2.8337-2.8346	0.0008-0.0017	0.0040-0.0140	3	2.1644-2.1654	0.0008-0.0017	0.0060-0.0140
	3.2	J32A2	2.8337-2.8346	0.0008-0.0017	0.0040-0.0140	3	2.1644-2.1654	0.0008-0.0017	0.0060-0.0140
	3.5	C35A1/KA9	2.6762-2.6772	0.0008-0.0017	0.0040-0.0110	3	2.1248-2.1257	0.0009-0.0018	0.0060-0.0120
2004	2.0	K20A3	2.1648-2.1657	①	0.0040-0.0140	4	1.7707-1.7717	0.0008-0.0019	0.0060-0.0160
	2.0	K20A2	2.1646-2.1655	①	0.0040-0.0140	4	1.8888-1.8898	0.0013-0.0024	0.0060-0.0160
	2.4	K24A4	②	①	0.0040-0.0140	4	1.8888-1.8898	0.0013-0.0026	0.0060-0.0160
	3.2	J32A3	2.8337-2.8346	0.0008-0.0017	0.0040-0.0140	3	2.1644-2.1654	0.0008-0.0017	0.0060-0.0140
	3.5	C35A1/KA9	2.6762-2.6772	0.0008-0.0017	0.0040-0.0110	3	2.1248-2.1257	0.0009-0.0018	0.0060-0.0120
2005-06	2.0	K20A3	2.1648-2.1657	①	0.0040-0.0140	4	1.7707-1.7717	0.0008-0.0019	0.0060-0.0160
	2.0	K20Z1	2.1646-2.1655	①	0.0040-0.0140	4	1.8888-1.8898	0.0013-0.0024	0.0060-0.0160
	2.4	K24A4	②	①	0.0040-0.0140	4	1.8888-1.8898	0.0013-0.0026	0.0060-0.0160
	3.2	J32A3	2.8337-2.8346	0.0008-0.0017	0.0040-0.0140	3	2.1644-2.1654	0.0008-0.0017	0.0060-0.0140
	3.5	J35A8	2.8337-2.8346	0.0008-0.0017	0.0040-0.0140	3	2.1644-2.1654	0.0008-0.0017	0.0060-0.0140

① Nos. 1, 2, 4 and 5: 0.0007-0.0016
No. 3: 0.0010-0.0019
② Nos. 1, 2, 4 and 5: 2.1648-2.1657
No. 3: 2.1644-2.1654

PISTON AND RING SPECIFICATIONS
All measurements are given in inches

Year	Engine Displacement Liters	Engine ID/VIN	Piston Clearance	Ring Gap			Ring Side Clearance		
				Top Compression	Bottom Compression	Oil Control	Top Compression	Bottom Compression	Oil Control
2002	2.0	K20A3	0.0008-0.0016	0.0080-0.0014	0.0160-0.0220	0.0100-0.0260	0.0014-0.0024	0.0012-0.0022	NA
	2.0	K20A2	0.0008-0.0016	0.0080-0.0014	0.0200-0.0260	0.0080-0.0280	0.0016-0.0026	0.0018-0.0028	NA
	3.2	J32A1	0.0006-0.00160	0.0080-0.0140	0.0160-0.0220	0.0080-0.0280	0.0014-0.0024	0.0012-0.0022	NA
	3.2	J32A2	0.0006-0.00160	0.0080-0.0140	0.0160-0.0220	0.0080-0.0280	0.0022-0.0031	0.0012-0.0022	NA
	3.5	C35A1/KA	0.0010-0.0020	0.0100-0.0160	0.0160-0.0220	①	0.0022-0.0031	0.0012-0.0022	NA
2003	2.0	K20A3	0.0008-0.0016	0.0080-0.0014	0.0160-0.0220	0.0100-0.0260	0.0014-0.0024	0.0012-0.0022	NA
	2.0	K20A2	0.0008-0.0016	0.0080-0.0014	0.0200-0.0260	0.0080-0.0280	0.0016-0.0026	0.0018-0.0028	NA
	3.2	J32A1	0.0006-0.00160	0.0080-0.0140	0.0160-0.0220	0.0080-0.0280	0.0014-0.0024	0.0012-0.0022	NA
	3.2	J32A2	0.0006-0.00160	0.0080-0.0140	0.0160-0.0220	0.0080-0.0280	0.0022-0.0031	0.0012-0.0022	NA
	3.5	C35A1/KA	0.0010-0.0020	0.0100-0.0160	0.0160-0.0220	①	0.0022-0.0031	0.0012-0.0022	NA
2004	2.0	K20A3	0.0008-0.0016	0.0080-0.0014	0.0160-0.0220	0.0100-0.0260	0.0014-0.0024	0.0012-0.0022	NA
	2.0	K20A2	0.0008-0.0016	0.0080-0.0014	0.0200-0.0260	0.0080-0.0280	0.0016-0.0026	0.0018-0.0028	NA
	2.4	K24A2	0.0008-0.0016	0.0080-0.0014	0.0200-0.0260	0.0080-0.0280	0.0018-0.0028	0.0016-0.0026	NA
	3.2	J32A3	0.0006-0.00160	0.0080-0.0140	0.0160-0.0220	0.0080-0.0280	0.0022-0.0031	0.0012-0.0022	NA
	3.5	C35A1/KA	0.0010-0.0020	0.0100-0.0160	0.0160-0.0220	③	0.0022-0.0031	0.0012-0.0022	NA
2005-06	2.0	K20A3	0.0008-0.0016	0.0080-0.0014	0.0160-0.0220	0.0100-0.0260	0.0014-0.0024	0.0012-0.0022	NA
	2.0	K20Z1	0.0008-0.0016	0.0080-0.0014	0.0200-0.0260	0.0080-0.0280	0.0016-0.0026	0.0018-0.0028	NA
	2.4	K24A2	0.0008-0.0016	0.0080-0.0014	0.0200-0.0260	0.0080-0.0280	0.0018-0.0028	0.0016-0.0026	NA
	3.2	J32A3	0.0006-0.00160	0.0080-0.0140	0.0160-0.0220	0.0080-0.0280	0.0022-0.0031	0.0012-0.0022	NA
	3.5	J35A8	0.0006-0.00160	0.0080-0.0140	0.0160-0.0220	0.0080-0.0280	0.0022-0.0031	0.0012-0.0022	NA

NA; Not Applicable

① RIKEN: 0.0080-0.0280 inches
   TEIKOKU: 0.0080-0.0200 inches

TORQUE SPECIFICATIONS
All readings in ft. lbs.

Year	Engine Displacement Liters	Engine ID/VIN	Cylinder Head Bolts	Main Bearing Bolts	Rod Bearing Bolts	Crankshaft Damper Bolts	Flywheel Bolts	Manifold Intake	Manifold Exhaust	Spark Plugs	Oil Pan Drain Plug
2002	2.0	K20A3	①	②	③	④	⑤	16	33	⑥	33
	2.0	K20A2	①	②	⑦	④	⑤	16	33	⑥	29
	3.2	J32A1	⑧	⑨	⑩	181	54	16	23	13	29
	3.2	J32A2	⑧	⑨	⑩	181	54	16	23	13	29
	3.5	C35A1/KA9	⑧	⑨	⑩	④	54	16	23	13	33
2003	2.0	K20A3	①	②	③	④	⑤	16	33	⑥	33
	2.0	K20A2	①	②	⑦	④	⑤	16	33	⑥	29
	3.2	J32A1	⑧	⑨	⑩	181	54	16	23	13	29
	3.2	J32A2	⑧	⑨	⑩	181	54	16	23	13	29
	3.5	C35A1/KA9	⑧	⑨	⑩	④	54	16	23	13	33
2004	2.0	K20A3	①	②	③	④	⑤	16	33	⑥	33
	2.0	K20A2	①	②	⑦	④	⑤	16	33	⑥	29
	2.4	K24A2	①	②	④	④	76	16	33	13	33
	3.2	J32A3	⑧	⑨	⑩	⑪	⑫	16	23	13	29
	3.5	C35A1/KA9	⑧	⑨	⑩	④	54	16	23	13	33
2005-06	2.0	K20A3	①	②	③	④	⑤	16	33	⑥	33
	2.0	K20Z1	①	②	⑦	④	⑤	16	33	⑥	29
	2.4	K24A2	①	②	④	④	76	16	33	13	33
	3.2	J32A3	⑧	⑨	⑩	⑪	⑫	16	23	13	29
	3.5	J35A8	⑬	⑭	⑩	④	54	16	23	13	29

① Step 1: 29 ft. lbs.
Step 2: Rotate 90 degrees
Step 3: Rotate an additional 90 degrees
Step 4 (new bolts only): additional 90 degrees

② Step 1: 22 ft. lbs.
Step 2: Rotate 56 degrees

③ 14 ft. lbs. Plus 90 degrees

④ Step 1: 36 ft. lbs.
Step 2: Rotate 90 degrees

⑤ Manual trans 6 speed: 90 ft. lbs.
Manual trans 5 speed: 76 ft. lbs.
Auto trans.: 54 ft. lbs.

⑥ NGK type IFRG-11KS & Denso type SK22PRM11S: 18 ft. lbs.
All others: 13 ft. lbs.

⑦ 22 ft. lbs. Plus 90 degrees

⑧ Step 1: 29 ft. lbs.
Step 2: 51 ft. lbs.
Step 3: 72.3 ft. lbs.

⑨ Step 1: Cap bolts 56 ft. lbs.
Step 2: Side bolts 36 ft. lbs.

⑩ Step 1: 14 ft. lbs.
Step 2: Rotate 90 degrees

⑪ Step 1: 14 ft. lbs.
Step 2: Rotate 90 degrees

⑫ Manual tran: 76 ft. lbs.
Auto trans.: 54 ft. lbs.

⑬ 6 point bolts:
Step 1: 29 ft. lbs.
Step 2: 51 ft. lbs.
Step 3: 72 ft. lbs.

12 point bolts:
Step 1: 29 ft. lbs.
Step 2: Rotate 90 degrees
Step 3: Rotate an additional 90 degrees
Step 4 (new bolts only): additional 90 degrees

⑭ Step 1: Cap bolts 54 ft. lbs.
Step 2: Side bolts 36 ft. lbs.

WHEEL ALIGNMENT

Year	Model		Caster Range (+/-Deg.)	Caster Preferred Setting (Deg.)	Camber Range (+/-Deg.)	Camber Preferred Setting (Deg.)	Toe-in (in.)
2002	3.2 TL	F	1.00	+2.80	1.00	0	0 +/- 0.06
		R	—	—	0.50	-0.50	0.06 +/- 0.06
	3.2 CL	F	1.00	+2.80	1.00	0	0 +/- 0.06
		R	—	—	1.00	-0.50	0.06 +/- 0.06
	3.5 RL	F	1.00	+2.81	1.00	0	0 +/- 0.06
		R	—	—	1.00	-0.50	0.06 +/- 0.06
	RSX ①	F	1.00	+1.16	1.00	+0.16	0 +/- 0.16
		R	—	—	1.00	+0.75	0.16 +/- 0.16
	RSX ②	F	1.00	+1.16	1.00	+0.16	0 +/- 0.06
		R	—	—	1.00	+0.50	0.06 +/- 0.06
2003	3.2 TL	F	1.00	+2.80	1.00	0	0 +/- 0.06
		R	—	—	0.50	-0.50	0.06 +/- 0.06
	3.2 CL	F	1.00	+2.80	1.00	0	0 +/- 0.06
		R	—	—	1.00	-0.50	0.06 +/- 0.06
	3.5 RL	F	1.00	+2.81	1.00	0	0 +/- 0.06
		R	—	—	1.00	-0.50	0.06 +/- 0.06
	RSX ①	F	1.00	+1.16	1.00	+0.16	0 +/- 0.16
		R	—	—	1.00	+0.75	0.16 +/- 0.16
	RSX ②	F	1.00	+1.16	1.00	+0.16	0 +/- 0.06
		R	—	—	1.00	+0.50	0.06 +/- 0.06
2004	3.2 TL	F	1.00	+2.80	1.00	0	0 +/- 0.06
		R	—	—	0.50	-0.50	0.06 +/- 0.06
	3.5 RL	F	1.00	+2.81	1.00	0	0 +/- 0.06
		R	—	—	1.00	-0.50	0.06 +/- 0.06
	RSX ①	F	1.00	+1.16	1.00	+0.16	0 +/- 0.16
		R	—	—	1.00	+0.75	0.16 +/- 0.16
	RSX ②	F	1.00	+1.16	1.00	+0.16	0 +/- 0.06
		R	—	—	1.00	+0.50	0.06 +/- 0.06
	TSX	F	0.45	+3.13	0.45	0	0 +/- 0.08
		R	—	—	0.30	-1.00	0.08 +/- 0.08
2005-06	3.2 TL	F	1.00	+2.80	1.00	0	0 +/- 0.06
		R	—	—	0.50	-0.50	0.06 +/- 0.06
	3.5 RL	F	1.00	+2.81	1.00	0	0 +/- 0.06
		R	—	—	1.00	-0.50	0.06 +/- 0.06
	RSX ①	F	1.00	+1.16	1.00	+0.16	0 +/- 0.16
		R	—	—	1.00	+0.75	0.16 +/- 0.16
	RSX ②	F	1.00	+1.16	1.00	+0.16	0 +/- 0.06
		R	—	—	1.00	+0.50	0.06 +/- 0.06
	TSX	F	0.45	+3.13	0.45	0	0 +/- 0.08
		R	—	—	0.30	-1.00	0.08 +/- 0.08

① Except Type S

② Type S

TIRE, WHEEL AND BALL JOINT SPECIFICATIONS

| Year | Model | OEM Tires | | Tire Pressures (psi) | | Wheel | Ball Joint | Lug Nut |
		Standard	Optional	Front	Rear	Size	Inspection	(ft. lbs.)
2002	RSX	P205/55R16	None	32	30	6-JJ	NS	80
	3.2TL	P205/60VR16	None	29	29	6-JJ	NS	80
	3.2CL	P205/60VR16	None	29	29	6-JJ	NS	80
	3.5RL	P225/55/R16	None	32	29	6-JJ	NS	80
2003	RSX	P205/55R16	None	32	30	6-JJ	NS	80
	3.2TL	P205/60VR16	None	29	29	6-JJ	NS	80
	3.2CL	P205/60VR16	None	29	29	6-JJ	NS	80
	3.5RL	P225/55/R16	None	32	29	6-JJ	NS	80
2004	RSX	P205/55R16	None	32	30	6-JJ	NS	80
	3.2TL	P235/45VR17	None	29	29	6-JJ	NS	80
	3.5RL	P225/55/R16	None	32	29	6-JJ	NS	80
	TSX	P215/50/R17	None	32	30	6-JJ	NS	80
2005-06	RSX	P205/55R16	None	32	30	6-JJ	NS	80
	3.2TL	P235/45VR17	None	29	29	6-JJ	NS	80
	3.5RL	P245/50/R17	None	32	29	6-JJ	NS	80
	TSX	P215/50/R17	None	32	30	6-JJ	NS	94

OEM: Original Equipment Manufacturer

PSI: Pounds Per Square Inch

STD: Standard

OPT: Optional

NS: Not Specified by manufacturer

① Front tires: P215/40R17

Rear tires: P255/40R17

09474_ACUCAR_C0010

BRAKE SPECIFICATIONS
All measurements in inches unless noted

| Year | Model | | Brake Disc | | | Brake Drum Diameter | | | Minimum Lining Thickness | | Brake Caliper | |
			Original Thickness	Minimum Thickness	Maximum Runout	Original Inside Diameter	Max. Wear Limit	Maximum Machine Diameter	Front	Rear	Bracket Bolts (ft. lbs.)	Mounting Bolts (ft. lbs.)
2002	3.5RL	F	0.910	0.830	0.004	—	—	—	0.06	—	80	36
		R	0.350	0.300	0.004	—	—	—	—	0.06	28	17
	3.2TL	F	1.100	1.020	0.004	—	—	—	0.06	—	80	36
		R	0.350	0.310	0.004	①	6.693 ②	②	—	③	41	17
	3.2CL	F	1.100	1.020	0.004	—	—	—	0.06	—	80	36
		R	0.350	0.300	0.004	①	6.693 ②	②	—	③	41	17
	RSX	F	0.830	0.750	0.004	—	—	—	0.06	—	80	④
		R	0.350	0.310	0.004	—	—	—	—	0.06	41	16
	RSX Type-S	F	0.980	0.910	0.004	—	—	—	0.06	—	80	16
		R	0.350	0.310	0.004	—	—	—	—	0.06	41	16
2003	3.5RL	F	0.910	0.830	0.004	—	—	—	0.06	—	80	36
		R	0.350	0.300	0.004	—	—	—	—	0.06	28	17
	3.2TL	F	1.100	1.020	0.004	—	—	—	0.06	—	80	36
		R	0.350	0.310	0.004	①	6.693 ②	②	—	③	41	17
	3.2CL	F	1.100	1.020	0.004	—	—	—	0.06	—	80	36
		R	0.350	0.300	0.004	①	6.693 ②	②	—	③	41	17
	RSX	F	0.830	0.750	0.004	—	—	—	0.06	—	80	④
		R	0.350	0.310	0.004	—	—	—	—	0.06	41	16
	RSX Type-S	F	0.980	0.910	0.004	—	—	—	0.06	—	80	16
		R	0.350	0.310	0.004	—	—	—	—	0.06	41	16
2004	3.5RL	F	0.910	0.830	0.004	—	—	—	0.06	—	80	36
		R	0.350	0.300	0.004	—	—	—	—	0.06	28	17
	3.2TL	F	1.100	1.020	0.004	—	—	—	0.06	—	80	37
		R	0.990	0.910	0.004	—	—	—	—	0.06	41	17
	RSX	F	0.830	0.750	0.004	—	—	—	0.06	—	80	④
		R	0.350	0.310	0.004	—	—	—	—	0.06	41	16
	RSX Type-S	F	0.980	0.910	0.004	—	—	—	0.06	—	80	16
		R	0.350	0.310	0.004	—	—	—	—	0.06	41	16
	TSX	F	0.980	0.910	0.004	—	—	—	0.06	—	80	36
		R	0.390	0.310	0.004	—	—	—	—	0.06	28	27
2005-06	3.5RL	F	1.10-1.11	1.020	0.004	—	—	—	0.06	—	—	58
		R	0.625-0.634	0.550	0.006	—	—	—	—	0.06	79	17
	3.2TL	F	1.100	1.020	0.004	—	—	—	0.06	—	80	37
		R	0.990	0.910	0.004	—	—	—	—	0.06	41	17
	RSX	F	0.830	0.750	0.004	—	—	—	0.06	—	80	④
		R	0.350	0.310	0.004	—	—	—	—	0.06	41	16
	RSX Type-S	F	0.980	0.910	0.004	—	—	—	0.06	—	80	16
		R	0.350	0.310	0.004	—	—	—	—	0.06	41	16
	TSX	F	0.980	0.910	0.004	—	—	—	0.06	—	80	36
		R	0.390	0.310	0.004	—	—	—	—	0.06	28	27

NA: Not Available

F: Front

R: Rear

① Rear parking brake drum: 6.689 inches

② Rear parking brake drum maximum diameter: 6.732 inches

③ Rear pad: 0.06 inches

   Rear parking brake shoes: 0.04 inches

④ 14 inch model 16 ft. lbs.

   16 inch model 24 ft. lbs.

SCHEDULED MAINTENANCE INTERVALS
ACURA—3.2CL, 3.2TL, 3.5RL, RSX, NSX & TSX

TO BE SERVICED	TYPE OF SERVICE	VEHICLE MILEAGE INTERVAL (x1000)												
		7.5	15	22.5	30	37.5	45	52.5	60	67.5	75	82.5	90	97.5
Engine oil	R	✓	✓	✓	✓	✓	✓	✓	✓	✓	✓	✓	✓	✓
Rear brake discs, calipers & pads	S/I		✓		✓		✓		✓		✓		✓	
Rotate tires	S/I	✓	✓	✓	✓	✓	✓	✓	✓	✓	✓	✓	✓	✓
A/C filter	R		✓		✓		✓		✓		✓		✓	
A/C filter (3.5RL)	R				✓				✓					
Brake hoses & lines (including ABS)	S/I		✓		✓		✓		✓		✓		✓	
Cooling system hoses & connections	S/I		✓		✓		✓		✓		✓		✓	
Driveshaft boots	S/I		✓		✓		✓		✓		✓		✓	
Exhaust system	S/I		✓		✓		✓		✓		✓		✓	
Front brake discs & calipers	S/I		✓		✓		✓		✓		✓		✓	
Fuel pipes, hoses & connections	S/I		✓		✓		✓		✓		✓		✓	
Suspension components	S/I		✓		✓		✓		✓		✓		✓	
Suspension mounting bolts	S/I		✓		✓		✓		✓		✓		✓	
Tie rods, steering gear box & boots	S/I		✓		✓		✓		✓		✓		✓	
Steering operation, tie rod ends, steering gearbox & boots	S/I		✓		✓				✓				✓	
Parking brake	S/I		✓		✓		✓		✓		✓		✓	
Air cleaner element	R				✓				✓				✓	
Automatic transmission fluid	R				✓				✓				✓	
Brake fluid (including ABS) (Integra & RSX)	R				✓				✓				✓	
Brake fluid (including ABS) (3.5RL)	R								✓				✓	
Brake fluid (including ABS) (3.2TL & NSX)	R						✓				✓			
Front differential fluid (3.2TL & 3.5RL)	R								✓				✓	
Manual transmission fluid	R				✓				✓				✓	
ABS operation	S/I				✓				✓				✓	
Drive belt(s)	S/I				✓				✓				✓	
Spark plugs (RSX exc. Type S)	R				✓				✓				✓	
Spark plugs (3.2TL)	R								✓				✓	
Spark plugs (3.5.RL)	R								✓					
Engine coolant	R					✓					✓			
Fuel filter	R								✓					
PCV valve	S/I								✓					

SCHEDULED MAINTENANCE INTERVALS
ACURA—3.2CL, 3.2TL, 3.5RL, RSX, NSX & TSX

TO BE SERVICED	TYPE OF SERVICE	VEHICLE MILEAGE INTERVAL (x1000)												
		7.5	15	22.5	30	37.5	45	52.5	60	67.5	75	82.5	90	97.5
Timing belt (except as noted below)	R												✓	
Timing belt & timing balancer belt (3.5RL) ①	R													
Transmission fluid	R												✓	
Idle speed (3.2TL)	S/I								✓					
Idle speed (3.5RL) ②	S/I		✓		✓		✓		✓		✓		✓	
Ignition wires	S/I		✓		✓				✓				✓	
TWC converter heat shield	S/I		✓		✓		✓		✓		✓		✓	
Water pump	S/I				✓				✓				✓	
Water pump (3.5RL) ②	S/I		✓		✓		✓		✓		✓		✓	

R: Replace   S/I: Service or Inspect

① Replace at 105,000 miles.

② Service or inspect at 105,000 miles.

FREQUENT OPERATION MAINTENANCE (SEVERE SERVICE)

If a vehicle is operated under any of the following conditions it is considered severe service:

- Extremely dusty areas.

-50% or more of the vehicle operation is in 32°C (90°F) or higher temperatures, or constant operation in temperatures below 0°C (32°F).

-Prolonged idling (vehicle operation in stop and go traffic).

-Frequent short running periods (engine does not warm to normal operating temperatures).

-Police, taxi, delivery usage or trailer towing usage.

Oil & oil filter: change every 3750 miles.

Brake hoses & lines (including ABS) (3.5RL): service or inspect every 7500 miles.

Cooling system hoses & connections (3.5RL): service or inspect every 7500 miles.

Driveshaft boots (3.2TL & 3.5RL): check every 7500 miles.

Exhaust system (3.5RL): check every 7500 miles.

Brake discs, calipers & pads: service or inspect every 7500 miles.

Fuel pipes, hoses & connections (3.5RL): check every 7500 miles.

Power steering system: service or inspect every 7500 miles.

Suspension components: service or inspect every 7500 miles.

Tie rod ends, steering gear box & boots (3.2TL & 3.5RL): service or inspect every 7500 miles.

Front differential fluid (3.2TL, 3.5RL): replace every 15,000 miles.

Transmission fluid (3.2TL & 3.5RL): replace every 30,000 miles

Timing belt (3.2TL): replace every 60,000 miles.

Water pump (3.2TL): service or inspect every 60,000 miles.

ENGINE REPAIR

➡**Disconnecting the negative battery cable on some vehicles may interfere with the functions of the on board computer system. The computer may undergo a relearning process once the negative battery cable is reconnected.**

Alternator

REMOVAL & INSTALLATION

CL Model

1. Before servicing the vehicle, refer to the precautions section.
2. Remove or disconnect the following:
   • Both battery cables
   • Adjusting bolts
   • Accessory drive belt
   • Mounting bolts
3. Turn the alternator 90° in a counter–clockwise direction.
   • Alternator
   • 4 prong connector
   • Harness clip and bracket assembly
   • Black wire from the terminal
   • Alternator from the vehicle

To install:
4. Install or connect the following:
   • Alternator
   • Black wire
   • Harness clip and bracket assembly and t tighten the bolt to 104 inch lbs. (12 Nm)
   • 4 prong connector
   • Mounting bolts and tighten the 10 x 1.25mm bolts to 33 ft. lbs. (44 Nm), the 8 x 1.25mm locknut/lock bolt(s) to 16 ft. lbs. (22 Nm) and the 6 x 1.0mm 104 inch lbs. (12 Nm)
   • Accessory drive belt

➡**There is no belt tension adjustment due to the use of an automatic tensioner.**

5. The Powertrain Control Module (PCM) idle memory must be reset after reconnecting the battery. Start the engine and hold it at 3000 rpm until the cooling fan comes on. Then allow the engine to idle for about 5 minutes with all accessories OFF and with the transmission in Park or Neutral.
6. Connect the positive, then the negative battery cable.

2002–04 RL Model

1. Before servicing the vehicle, refer to the precautions section.

2. Remove or disconnect the following:
   • Both battery cables
   • 4 prong connector and the black wire from the rear of the alternator
   • Alternator adjusting bolt(s)
   • Mounting bolt(s)
   • Alternator belt
   • Alternator assembly

To install:
3. Install or connect the following:
   • Alternator assembly
   • Mounting bolts and tighten them to 33 ft. lbs. (44 Nm)
   • Alternator adjusting bolt, hand tight
   • Alternator belt
4. Adjust alternator belt to tension.
5. Tighten the 8 x 1.25mm locknut/lock bolt(s) to 16 ft. lbs. (22 Nm).
6. Install or connect the following:
   • 4 prong connector and the black wire to the rear of the alternator
   • Both battery cables

✳✳ WARNING

Be sure to adjust the alternator belt to the proper tension or alternator bearing failure may occur.

➡**The Powertrain Control Module (PCM) idle memory must be reset after reconnecting the battery. Start the engine and hold it at 3000 rpm until the cooling fan comes on. Then allow the engine to idle for about 5 minutes with all accessories OFF and with the transmission in Park or Neutral.**

2005–06 RL Model

Make sure you have the anti-theft codes for the radio and the navigation system, then write down the customer's audio presets. Make sure the ignition switch is OFF.
1. Before servicing the vehicle, refer to the precautions section.
2. Remove the right upper fender trim, battery trim, left upper fender trim, then remove the upper grille cover.
3. Disconnect the negative cable from the battery, then disconnect the positive cable.
4. Remove the splash shield.
5. Remove the harness clamps and connector from the A/C condenser fan shroud.
6. Loosen the two bolts securing the A/C condenser fan shroud.
7. Disconnect the fan motor connector, and remove the reserve tank .
8. Remove the two bolts, then remove the A/C condenser fan shroud.

9. Remove the drive belt.
10. Disconnect the alternator connector and BLK wire from the alternator.
11. Remove the bolt securing the harness holder.
12. Remove the mounting bolt and alternator bracket mounting bolt, then remove the alternator

To install:
13. Installation is the reverse of removal. Tighten the upper bolt to 16 ft. lbs. (22 Nm) and the lower bolt to 33 ft. lbs. (44 Nm).

RSX Models

1. Before servicing the vehicle, refer to the precautions section.
2. Remove or disconnect the following:
   • Both battery cables
   • Drive belt
   • Auto tensioner
   • Connector and the black wire from the rear of the alternator
   • Mounting bolt(s)
   • Alternator assembly

To install:
3. Install or connect the following:
   • Alternator assembly
   • Mounting bolts and tighten them to 16 ft. lbs. (22 Nm)
   • Connector and the black wire to the rear of the alternator
   • Auto tensioner
   • Drive belt
   • Both battery cables

✳✳ WARNING

Be sure to adjust the alternator belt to the proper tension or alternator bearing failure may occur.

➡**The Powertrain Control Module (PCM) idle memory must be reset after reconnecting the battery. Start the engine and hold it at 3000 rpm until the cooling fan comes on. Then allow the engine to idle for about 5 minutes with all accessories OFF and with the transmission in Park or Neutral.**

2002–03 TL Models

1. Before servicing the vehicle, refer to the precautions section.
2. Remove or disconnect the following:
   • Both battery cables
   • Adjusting bolts
   • Accessory drive belt
   • Mounting bolts

3. Turn the alternator 90° in a counter-clockwise direction.
   - Alternator
   - 4 prong connector
   - Harness clip and bracket assembly
   - Black wire from the terminal
   - Alternator from the vehicle

To install:

4. Install or connect the following:
   - Alternator
   - Black wire
   - Harness clip and bracket assembly and tighten the bolt to 104 inch lbs. (12 Nm)
   - 4 prong connector
   - Mounting bolts and tighten the 10 x 1.25mm bolts to 33 ft. lbs. (44 Nm), the 8 x 1.25mm locknut/lock bolt(s) to 16 ft. lbs. (22 Nm) and the 6 x 1.0mm 104 inch lbs. (12 Nm)
   - Accessory drive belt

➡**There is no belt tension adjustment due to the use of an automatic tensioner.**

5. The Powertrain Control Module (PCM) idle memory must be reset after reconnecting the battery. Start the engine and hold it at 3000 rpm until the cooling fan comes on. Then allow the engine to idle for about 5 minutes with all accessories OFF and with the transmission in Park or Neutral.

6. Connect the positive, then the negative battery cable.

2004–06 TL Models

1. Before servicing the vehicle, refer to the precautions section.

➡**Make sure you have the anti-theft code for the radio, and the navigation system, then write down the radio channel presets. Make sure the ignition switch is OFF.**
Disconnect the negative cable from the battery first, then disconnect the positive cable.

2. Disconnect the A/C condenser fan motor connector and the A/C compressor clutch connector, then remove the reserve tank.

3. Remove the two bolts, loosen bolt, and then remove the A/C condenser fan shroud.

4. Remove the drive belt.

5. Disconnect the alternator connector and the BLK wire from the alternator.

6. Remove the bolt retaining the harness bracket

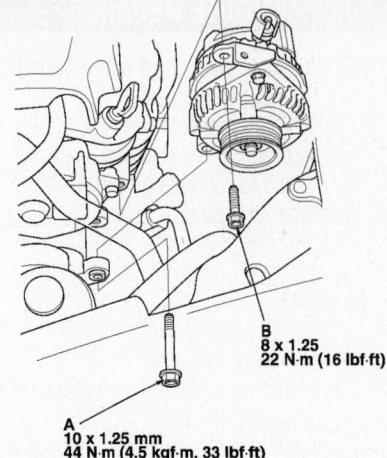

B
8 x 1.25
22 N·m (16 lbf·ft)

A
10 x 1.25 mm
44 N·m (4.5 kgf·m, 33 lbf·ft)

09474_ACUCAR_G0001

Alternator mounting—2004–06 TL model

7. Remove the mounting bolt, alternator bracket mounting bolt and the remove the alternator.

To install:

8. Installation is the reverse of removal. Tighten the mounting bolt to 33 ft. lbs. (44 Nm), the bracket bolt to 16 ft. lbs. (22 Nm) and the harness bolt to 9 ft. lbs. (12 Nm).

TSX Models

1. Before servicing the vehicle, refer to the precautions section.

2. Note the radio security code and the radio presets.

3. Remove or disconnect the following:
   - Negative battery cable, then the positive
   - Drive belt
   - Auto-tensioner
   - Connectors from the alternator
   - 3 alternator mounting bolts and the alternator

To install:
   - Alternator and 3 mounting bolts. Torque the bolts to 16 ft. lbs. (22 Nm).

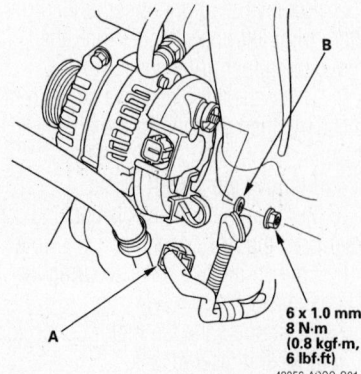

B

6 x 1.0 mm
8 N·m
(0.8 kgf·m,
6 lbf·ft)

A

42356-ACCO-G01

Alternator mounting—TSX models

   - Electrical connectors
   - Auto-tensioner
   - Drive belt
   - Positive, then negative battery cables

4. Enter the security code and radio presets

Ignition Timing

ADJUSTMENT

CL Model

➡**The ignition timing is controlled by the Powertrain Control Module (PCM) and can be checked for diagnostic purposes. If the timing is out of specification, all mechanical and electrical systems should checked for proper operation before replacing the PCM.**

1. Before servicing the vehicle, refer to the precautions section.

2. Check the idle speed and adjust if necessary.

3. To check the ignition timing, start the engine and allow it to fast idle at 3000 rpm with all electrical accessories off and the transmission in **N** or **P**. Allow the engine to warm up and reach normal operating temperature. The engine cooling fan should cycle at least 1 time.

4. Connect a timing light to the No. 1 plug wire. With the engine idling at normal operating temperature point the timing light toward the pointer on the timing belt cover.

5. Inspect the ignition timing. The specifications is 8–12 degrees Before Top Dead Center (BTDC) at idle.

➡**All mechanical and electrical systems should checked for proper operation before replacing the PCM.**

6. If the ignition timing is incorrect, replace the PCM. Only replace the PCM as a last resort.

7. Remove the timing light.

2002–04 RL Model

➡**The ignition timing is controlled by the Powertrain Control Module (PCM) and can be checked for diagnostic purposes. If the timing is out of specification, all mechanical and electrical systems should checked for proper operation before replacing the PCM.**

1. Before servicing the vehicle, refer to the precautions section.

2. To check the ignition timing, start the engine and allow it to fast idle at 3000 rpm with all electrical accessories off and the

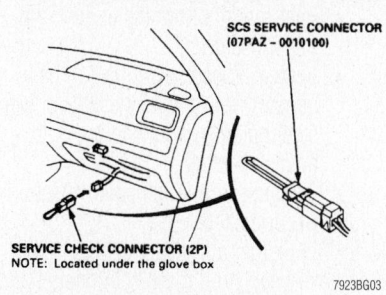

Service check connector—3.5RL

transmission in **N** or **P**. Allow the engine to warm up and reach normal operating temperature. The engine cooling fan should cycle at least 1 time.

3. Locate the Service Check (SCS) connector under the glove box and connect the service connector tool part number 07PAZ–0010100 to it.

4. Check the idle speed and adjust if necessary.

5. Connect a timing light to the No. 1 plug wire. With the engine idling at normal operating temperature point the timing light toward the pointer on the timing belt cover.

6. Inspect the ignition timing. The specifications are 13–17 degrees BTDC at 700–800 rpm

➡**All mechanical and electrical systems should checked for proper operation before replacing the PCM.**

7. If the ignition timing is incorrect, replace the PCM. Only replace the PCM as a last resort.

8. Remove the timing light.

9. Disconnect the special tool (SCS service connector) from the service check connector.

2004–06 TL and 2005–06 RL

1. Before servicing the vehicle, refer to the precautions section.

2. Connect the Honda Diagnostic System (HDS) to the Data Link Connector (DLC), and check for DTCs.

3. If a DTC is present, diagnose and repair the cause before inspecting the ignition timing.

4. Start the engine. Hold the engine at 3,000 rpm with no load (in Neutral) until the radiator fan comes on, then let it idle.

5. Check the idle speed.

6. Select "SCS" mode using the HDS.

7. Remove the right side engine compartment cover.

8. Connect the timing light to the service loop.

9. Aim the light toward the pointer on the timing belt cover. Check the ignition

timing under a no load condition. Headlights, blower fan, rear window defogger, and air conditioner are turned off.

10. The timing should be 8–12degrees Before Top Dead Center (TDC) in park or neutral.

11. If the ignition timing differs from the specification, check the cam timing.

12. Disconnect the HDS and the timing light.

RSX Models

➡**The ignition timing is controlled by the Powertrain Control Module (PCM) and can be checked for diagnostic purposes. If the timing is out of specification, all mechanical and electrical systems should be checked for proper operation before replacing the PCM.**

1. Before servicing the vehicle, refer to the precautions section.

2. To check the ignition timing, start the engine and allow it to fast idle at 3000 rpm with all electrical accessories off and the transmission in **N** or **P**. Allow the engine to warm up and reach normal operating temperature. The engine cooling fan should cycle at least 1 time.

3. Locate the Service Check (SCS) connector under the glove box. Connect the service connector tool part number 07PAZ–0010100 to the SCS terminals.

4. Check the idle speed and adjust if necessary.

5. Connect a timing light to the No. 1 plug wire. While engine idles, point the light toward the pointer on the timing belt cover.

6. Inspect the ignition timing at idle. The specification is 6–10 degrees Before Top Dead Center (BTDC) at idle.

➡**All mechanical and electrical systems should checked for proper operation before replacing the PCM.**

7. If the ignition timing is incorrect, replace the PCM.

8. Remove the service connector.

2002–03 TL Models

➡**The ignition timing is controlled by the Powertrain Control Module (PCM) and can be checked for diagnostic purposes. If the timing is out of specification, all mechanical and electrical systems should checked for proper operation before replacing the PCM.**

1. Before servicing the vehicle, refer to the precautions section.

2. To check the ignition timing, start the engine and allow it to fast idle at 3000 rpm

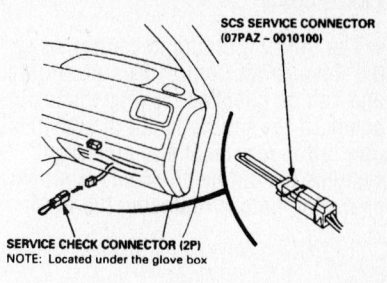

Service check connector—2002-03 3.2TL

Timing light attachment—2002-03 3.2TL

with all electrical accessories off and the transmission in **N** or **P**. Allow the engine to warm up and reach normal operating temperature. The engine cooling fan should cycle at least 1 time.

3. Locate the Service Check (SCS) connector under the glove box and connect the service connector tool part number 07PAZ–0010100 to it.

4. Check the idle speed and adjust if necessary.

5. Connect a timing light to the No. 1 plug wire. With the engine idling at normal operating temperature point the timing light toward the pointer on the timing belt cover.

6. Inspect the ignition timing. The specifications is 13–17 degrees Before Top Dead Center (BTDC) at 590–690 rpm

➡**All mechanical and electrical systems should checked for proper operation before replacing the PCM.**

7. If the ignition timing is incorrect, replace the PCM. Only replace the PCM as a last resort.

8. Remove the timing light.

9. Disconnect the special tool (SCS service connector) from the service check connector.

TSX Models

➡ **The ignition timing is controlled by the Powertrain Control Module (PCM) and can be checked for diagnostic purposes. If the timing is out of specification, all mechanical and electrical systems should be checked for proper operation before replacing the PCM.**

1. Before servicing the vehicle, refer to the precautions section.

2. To check the ignition timing, start the engine and allow it to fast idle at 3000 rpm with all electrical accessories off and the transmission in **N** or **P**. Allow the engine to warm up and reach normal operating temperature. The engine cooling fan should cycle at least 1 time.

3. Locate the Service Check (SCS) connector under the glove box. Connect the service connector tool part number 07PA7–0010100 to the SCS terminals.

4. Check the idle speed and adjust if necessary.

5. Connect a timing light to the No. 1 plug wire. While engine idles, point the light toward the pointer on the timing belt cover.

6. Inspect the ignition timing at idle. The specification is 6–10 degrees Before Top Dead Center (BTDC) at idle.

➡ **All mechanical and electrical systems should checked for proper operation before replacing the PCM.**

7. If the ignition timing is incorrect, replace the PCM.

8. Remove the service connector.

Engine Assembly

REMOVAL & INSTALLATION

CL Model

1. Before servicing the vehicle, refer to the precautions section.

2. Do not remove the hood. Disconnect the hood support strut and reconnect it to hold the hood in a vertical position.

3. Properly relieve the fuel pressure.

4. Drain the engine oil, coolant, transmission fluid and the differential fluid.

5. Remove or disconnect the following:
- Negative battery cable, then the positive battery cable
- Intake manifold cover
- Water bypass hoses
- Traction Control System (TCS) control valve actuator connector
- TCS control valve angel sensor connector
- EVAP canister hose
- Vacuum hose and breather pipe and the air intake duct
- Battery, battery tray and tray bracket
- Starter and ground cables
- Throttle cable and cruise control cable from the throttle and bracket
- Left side engine wiring harness connectors
- Fuel feed and return hoses
- Brake booster vacuum hose
- Engine wire clamps
- Vacuum hose and the Heated Oxygen (HO2S) sensor connector
- Battery cables from the under–hood fuse/relay box
- Under–hood fuse/relay box
- Powertrain Control Module (PCM) electrical connectors
- Grommet mounting nuts and pull out the wiring harness
- Accessory drive belts
- Power steering pump without disconnecting the hoses
- Front wheels
- Lower splash shield
- Front exhaust pipe from the manifold
- Shift cable cover
- Shift cable with the control lever from the transaxle
- Front damper forks
- Lower ball joints from the steering knuckles
- Halfshafts from the differential and the intermediate shaft
- Power steering hose clamps
- Upper and lower radiator hoses
- Heater hoses
- Power steering hose clamps at the rear beam
- Radiator

6. Attach a suitable chain hoist to the engine lifting hooks and support the engine.

➡ **The engine and transmission assembly is removed by lowering it from the vehicle. Be sure the vehicle is in a position that will allow the engine and transmission assembly enough clearance to be moved from the vehicle once it is lowered away from the vehicle.**

- Side, rear and front engine mount support fasteners
- Ensure the hoist brackets are properly positioned and raise the hoist to its full height
- Front suspension radius rod bolts

7. Make alignment marks on the front beam and remove the front beam.

- A/C compressor with the hoses still attached.
- Rear mounts from the engine and transmission
- Vehicle Speed sensor (VSS) connector, then remove the VSS/power steering sensor leaving the fluid hoses attached
- Transmission fluid cooler hoses
- Ground cable

8. Check that all hoses, cables and wires have been properly disconnected and slowly lower the engine about 6 inches (150 mm). Recheck that all hoses, cables and wires have been properly disconnected.

9. Carefully lower the engine/transmission assembly from the vehicle.

To install:

10. Installation is the reverse of the removal procedure, while using the following torque values:

- Transmission rear mount bolts: 28 ft. lbs. (38 Nm)
- Air conditioning compressor bolts: 16 ft. lbs. (22 Nm)
- Front beam nuts: 28 ft. lbs. (38 Nm)
- Front beam bolts: 76 ft. lbs. (103 Nm)
- Rear bolts: 28 ft. lbs. (38 Nm)
- Front suspension radius rod bolts: 119 ft. lbs. (162 Nm)
- Front engine mount nut: 40 ft. lbs. (54 Nm)
- Rear engine mount nut: 40 ft. lbs. (54 Nm)
- Rear engine mount bolt: 28 ft. lbs. (38 Nm)
- Side engine mount bracket bolts: 33 ft. lbs. (44 Nm)
- Side engine mount through bolt: 40 ft. lbs. (54 Nm)
- Lower ball joint nuts: 40 ft. lbs. (54 Nm)
- Damper fork bolts: 47 ft. lbs. (64 Nm)
- Front exhaust pipe nuts: 40 ft. lbs. (54 Nm)
- Shift cable and control lever bolts: 10 ft. lbs. (14 Nm)
- Power steering pump bolts: 16 ft. lbs. (22 Nm)

11. Fill the engine with oil and the transmission with fluid.

12. Fill and bleed (if necessary) the air from the cooling system.

13. Connect the positive, then the negative battery cable and enter the radio security code.

14. Switch the ignition **ON** but do not engage the starter. The fuel pump should run for approximately 2 seconds, building pressure within the lines. Switch the igni-

tion **OFF**, then **ON** 2 or 3 more times to build full system pressure. Check for fuel leaks.

15. Start the engine, allowing it to idle. Check the hoses and lines carefully for any sign of leakage.

16. Check the timing and idle speed.

17. After the engine has warmed up fully and the fan(s) have come on at least once, recheck the engine for fluid leaks. Switch the engine OFF.

18. Adjust the belts and throttle cable as necessary.

2002–04 RL Model

1. Before servicing the vehicle, refer to the precautions section.

2. Move the front passenger's seat forward.

3. Relieve the fuel system pressure.

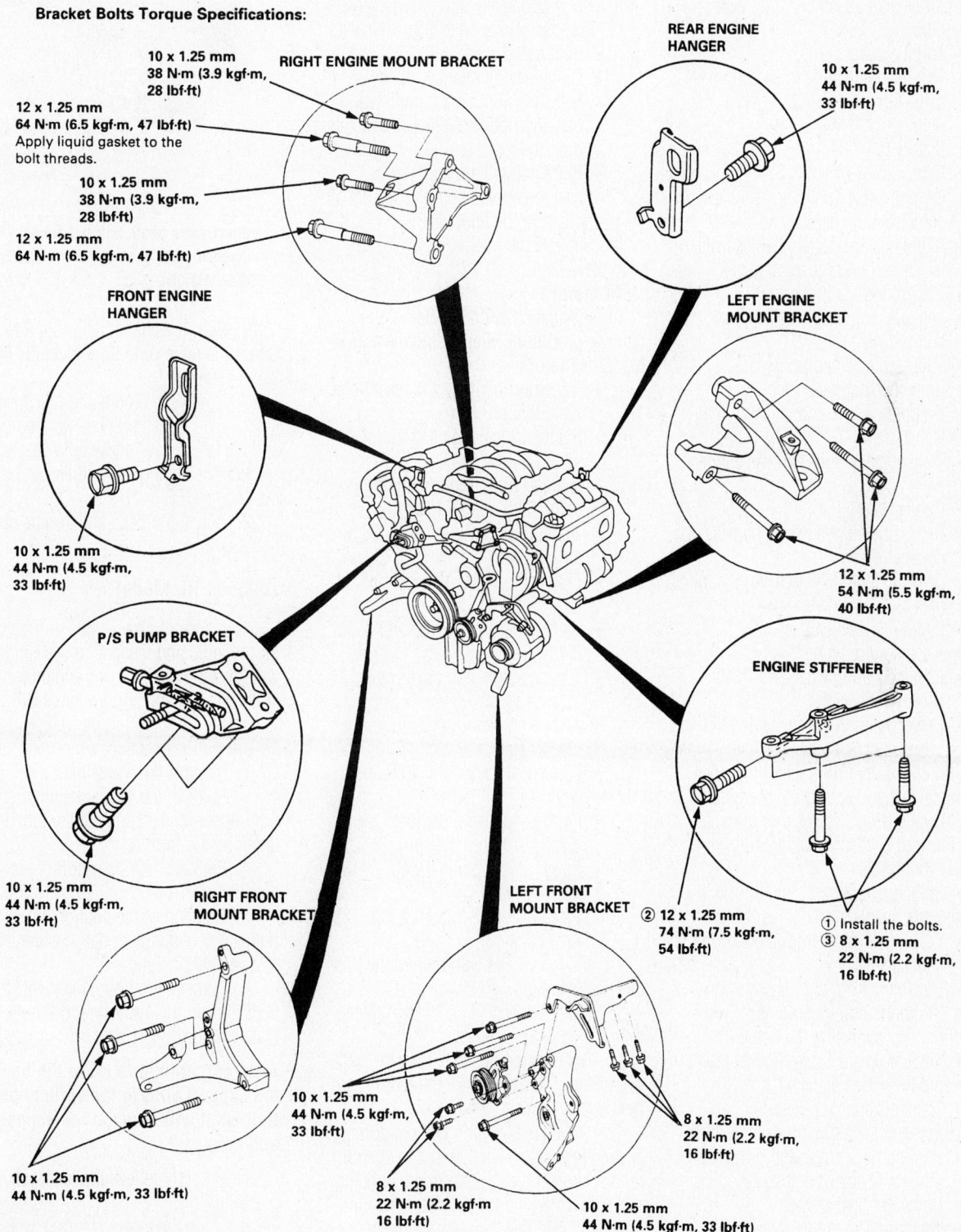

Bracket Bolts Torque Specifications:

10 x 1.25 mm
38 N·m (3.9 kgf·m,
28 lbf·ft)

RIGHT ENGINE MOUNT BRACKET

12 x 1.25 mm
64 N·m (6.5 kgf·m, 47 lbf·ft)
Apply liquid gasket to the bolt threads.

10 x 1.25 mm
38 N·m (3.9 kgf·m,
28 lbf·ft)

12 x 1.25 mm
64 N·m (6.5 kgf·m, 47 lbf·ft)

REAR ENGINE HANGER

10 x 1.25 mm
44 N·m (4.5 kgf·m,
33 lbf·ft)

FRONT ENGINE HANGER

10 x 1.25 mm
44 N·m (4.5 kgf·m,
33 lbf·ft)

LEFT ENGINE MOUNT BRACKET

12 x 1.25 mm
54 N·m (5.5 kgf·m,
40 lbf·ft)

P/S PUMP BRACKET

10 x 1.25 mm
44 N·m (4.5 kgf·m,
33 lbf·ft)

ENGINE STIFFENER

② 12 x 1.25 mm
74 N·m (7.5 kgf·m,
54 lbf·ft)

① Install the bolts.
③ 8 x 1.25 mm
22 N·m (2.2 kgf·m,
16 lbf·ft)

RIGHT FRONT MOUNT BRACKET

LEFT FRONT MOUNT BRACKET

10 x 1.25 mm
44 N·m (4.5 kgf·m,
33 lbf·ft)

10 x 1.25 mm
44 N·m (4.5 kgf·m, 33 lbf·ft)

8 x 1.25 mm
22 N·m (2.2 kgf·m
16 lbf·ft)

8 x 1.25 mm
22 N·m (2.2 kgf·m,
16 lbf·ft)

10 x 1.25 mm
44 N·m (4.5 kgf·m, 33 lbf·ft)

7923BG77

View of the engine mounting bracket showing torque specifications—2002–04 3.5RL

4. Drain the engine oil, coolant, transmission fluid and differential fluid.

5. Remove or disconnect the following:
- Hood
- Negative battery cable, then the positive battery cable
- Strut brace
- Engine cover
- Air cleaner assembly and intake duct
- Throttle cover
- Throttle cable and cruise control cable from the throttle and bracket
- Battery
- Battery tray
- Relay box
- Ground cable and wiring harness clips from the firewall
- Alternator and battery cables from the under–hood fuse/relay box
- Underhood fuse/relay box
- Left side engine wiring harness connectors
- Fuel feed and return hoses
- Brake booster vacuum hose
- Evaporative emissions (EVAP) canister hose
- Transmission sub–harness connector
- Control box
- Right side engine wiring harness connectors
- Spark plug voltage detection module
- Engine ground cables
- Accessory drive belts
- Power Steering Pressure (PSP) switch connector
- Power steering pump leaving the hoses attached

6. Pull the carpet back under the front passengers seat and detach the secondary Heated Oxygen (HO2S) sensor connector.

7. Remove or disconnect the following:
- Front wheels
- Splash shield
- Front suspension damper forks
- Lower ball joints from the steering knuckles
- Halfshafts from the transmission
- Air conditioning compressor without disconnecting the hoses
- Vehicle Speed Sensor (VSS) leaving the hoses attached
- Transmission stop collars
- Front exhaust pipe from the vehicle
- Wire harness cover and grommet
- Three way catalytic converter
- Converter heat shield
- Transmission fluid cooler lines
- Shift cable cover
- Shift cable from the transmission

- Control lever from the control shaft
- Upper and lower radiator hoses
- Radiator
- Heater hoses

8. Loosen the locknut on the fuel pressure regulator and rotate it 180 degrees.

9. Attach a chain hoist to the engine lifting eyelets.

10. Raise and safely support the vehicle.

11. Remove or disconnect the following:
- Shift cable guide
- Transmission beam
- Transmission mount and bracket
- Left and right front mount brackets from the mounts
- Right and left engine mounts

12. Raise the engine slightly, be sure all connections have be removed.

13. Remove the engine/transmission from the vehicle.

To install:

14. Installation is the reverse of the removal procedure, while using the following torque values:
- Transmission mount bracket bolts: 28 ft. lbs. (38 Nm)
- Right and left engine mount nuts: 47 ft. lbs. (64 Nm)
- Left and right front mount through bolts: 52 ft. lbs. (74 Nm)
- Transmission mount bolts: 28 ft. lbs. (38 Nm)
- Shift cable bolts: 108 inch lbs. (12 Nm)
- Front exhaust pipe nuts: 40 ft. lbs. (54 Nm)
- Transmission stop collar bolts: 28 ft. lbs. (38 Nm)
- A/C compressor bolts: 16 ft. lbs. (22 Nm)
- Lower ball joint nuts: 40 ft. lbs. (54 Nm)
- Front suspension damper fork bolts: 41 ft. lbs. (69 Nm)
- Power steering pump bolt: 33 ft. lbs. (44 Nm)
- Power steering nut: 16 ft. lbs. (22 Nm)

15. Fill the engine with oil and the transmission with fluid.

16. Fill and bleed (if necessary) the air from the cooling system.

17. Connect the positive, then the negative battery cable and enter the radio security code.

18. Switch the ignition **ON** but do not engage the starter. The fuel pump should run for approximately 2 seconds, building pressure within the lines. Switch the ignition **OFF**, then **ON** 2 or 3 more times to build full system pressure. Check for fuel leaks.

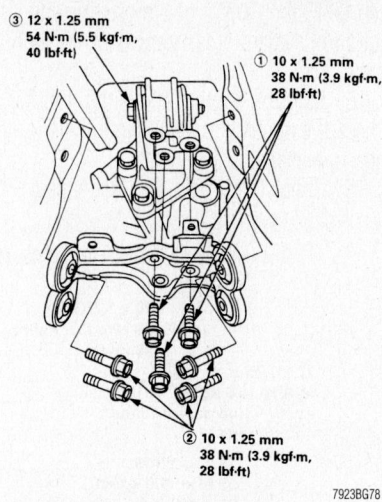

Transmission beam bolt tightening sequence and torque specifications—2002–04 3.5RL

19. Start the engine, allowing it to idle. Check the hoses and lines carefully for any sign of leakage.

20. Check the timing and idle speed.

21. After the engine has warmed up fully and the fan(s) have come on at least once, recheck the engine for fluid leaks. Switch the engine OFF.

22. Adjust the belts and throttle cable as necessary.

2005–06 RL Model

Make sure you have the anti-theft codes for the radio and navigation system, then write down the customer's audio presets. Make sure the ignition switch is OFF.

1. Before servicing the vehicle, refer to the precautions section.

2. Remove the windshield wiper arms.

3. Remove the intake manifold cover, right upper fender trim, battery trim and left upper fender trim.

4. Remove the upper grille cover and cowl cover.

5. Disconnect the support struts from both sides of the pivot ball. Secure the hood in a vertical position.

6. Remove the right side pivot ball and install it into the lower threaded hole, then reattach the support strut.

➡**Do not attempt to close the hood with the support strut in the vertical position, as it will damage the support strut and hood.**

7. Properly relieve the fuel system pressure.

8. Drain the power steering system fluid, then plug the fluid reservoir and return hose.

9. Disconnect the negative cable from the battery first, then disconnect the positive cable.

10. Remove the battery.

11. Remove the air cleaner.

12. Remove the air intake duct cover

13. Remove the harness clamps, starter cable and ground cable

14. Remove the battery base.

15. Remove the battery cable from the under-hood fuse/relay box.

16. Remove the harness clamp, then disconnect the harness connector.

17. Remove the two bolts securing the under-hood fuse/relay box, then remove the under-hood fuse/relay box.

18. Remove the harness clamp and disconnect the engine wire harness connectors on the left side of the engine compartment.

19. Remove the connector and transfer breather hose, then remove the strut brace.

20. Remove the quick-connect fitting cover, then disconnect the fuel feed hose.

21. Remove the brake booster vacuum hose and Evaporative Emission canister hose.

22. Remove the shift cable.

23. Remove the drive belt.

24. Remove the Power Steering (P/S) pump outlet line from the P/S pump, then plug the outlet line and P/S pump.

25. Remove the P/S hose from the clamp.

26. Remove the power steering system fluid reservoir from the clamp.

27. Remove the heat shield.

28. Remove the steering wheel.

29. Remove the steering joint cover.

30. Make a reference mark across the steering joint and steering gearbox pinion shaft. Remove the steering joint bolt and disconnect the steering joint from the steering gearbox pinion shaft.

31. Remove the radiator cap.

32. Raise the vehicle on the hoist to full height.

33. Remove the front wheels.

34. Remove the splash shield.

35. Drain and recycle the engine coolant.

36. Drain the automatic transmission fluid.

37. Drain the engine oil.

38. Disconnect the stabilizer links.

39. Remove the damper fork.

40. Separate the tie-rod end ball joints from the knuckles.

41. Separate the knuckles from the lower arms.

42. Remove the driveshafts.

43. Remove the exhaust pipe assembly.

44. Remove the propeller shaft .

45. Remove the transfer assembly.

46. Remove the ATF cooler hoses from the transmission.

47. Remove the P/S hose, then plug the line and hose.

48. Disconnect the power steering pressure switch connector.

49. Disconnect the connector from the steering gearbox.

50. Remove the bolts securing the transmission lower mount.

51. Lower the vehicle on the hoist

52. Remove the radiator hoses.

53. Remove the heater hoses.

54. Remove the radiator.

55. Remove the four bolts securing the A/C compressor.

56. Remove the connector bracket from the front cylinder head; use the bracket bolt hole to attach engine balancer bar front arm.

57. Remove the engine mount control solenoid valve bracket from the rear cylinder head; use the bracket bolt hole to attach engine balancer bar rear arm.

58. Lift and support the engine with engine hanger and engine balancer bar. Attach the front arm to the front cylinder head with a spacer and the 10 x 1.25 mm bolt. Attach the rear arm to the rear cylinder head with the 8 x 1.25 mm bolt.

59. Remove the vacuum hose and front mount stop, then remove the front mount bolt.

60. Remove the vacuum hose and rear mount stop, then remove the rear mount bolt.

61. Remove the bolt securing the shift cable bracket.

62. Raise the vehicle on the hoist to full height.

63. Attach the tool to the subframe by hanging the belt over the front of the subframe, then secure the belt with its stop.

64. Raise the jack and line up the slots in the special tool arms with the bolt holes on the jack base, then securely attach them with four bolts.

65. Remove the subframe.

66. Lower the vehicle and attach a chain hoist to the engine hook and transmission hook. Lift up on the engine/transmission assembly until it's securely supported by the chain hoist, and remove the engine support hanger from the engine and vehicle.

67. Remove the two bolts securing the upper bracket.

68. Check that the engine/transmission is completely free of vacuum hoses, fuel and coolant hoses, and electrical wiring.

69. Slowly lower the engine/transmission assembly about 150 mm (6 in.). Check once again that all hoses and electrical wiring are disconnected and free from the

engine/transmission, then lower it all the way.

70. Disconnect the chain hoist from the engine/transmission assembly.

71. Raise the vehicle all the way, and remove the engine/transmission assembly from under the vehicle.

To install:

72. Position the engine/transmission assembly under the vehicle. Be sure that they are properly aligned. Carefully lower the vehicle until the engine and transmission are properly positioned in the engine compartment.

73. Make sure the vehicle is not resting on any part of the engine or transmission. Lift and support the engine with a chain hoist and carefully raise the engine/transmission assembly into place.

➡**Reinstall the mounting bolts/support nuts in the sequence given in the following steps. Failure to follow this sequence may cause excessive noise and vibration, and reduce engine mount life.**

74. Install the two bolts securing the upper bracket. Tighten the bolts 40 ft. lbs. (54 Nm).

75. Install the engine balancer bar # VSB02C000019; attach the front arm to the front cylinder head with a spacer and 10 x 1.25 mm bolt, attach the rear arm to the rear cylinder head with 8 x 1.25 mm bolt.

76. Install the engine support hanger # AAR-T-12566 to the vehicle, and attach the hook to the slotted hole in the engine balancer bar. Tighten the wing nut by hand to lift and support the engine/transmission assembly.

77. Remove the chain hoist, then raise the vehicle on the hoist to full height.

78. Attach the front subframe adapter to the subframe by hanging the belt over the front of the subframe, then secure the belt with its stop. Raise the subframe up to the body with a jack

79. Loosely install the four subframe mounting bolts, four 10 x 1.25 mm bolts and two 12 x 1.25 mm bolts with the stiffeners.

80. Insert the tool through the positioning slot on the right rear stiffener, through the positioning hole on the subframe, and into the positioning hole on the body, then loosely tighten the subframe right rear mounting bolt.

81. Insert the tool through the positioning slot on the left rear stiffener, through the positioning hole on the subframe, and into the positioning hole on the body, then loosely tighten the subframe left rear mounting bolt.

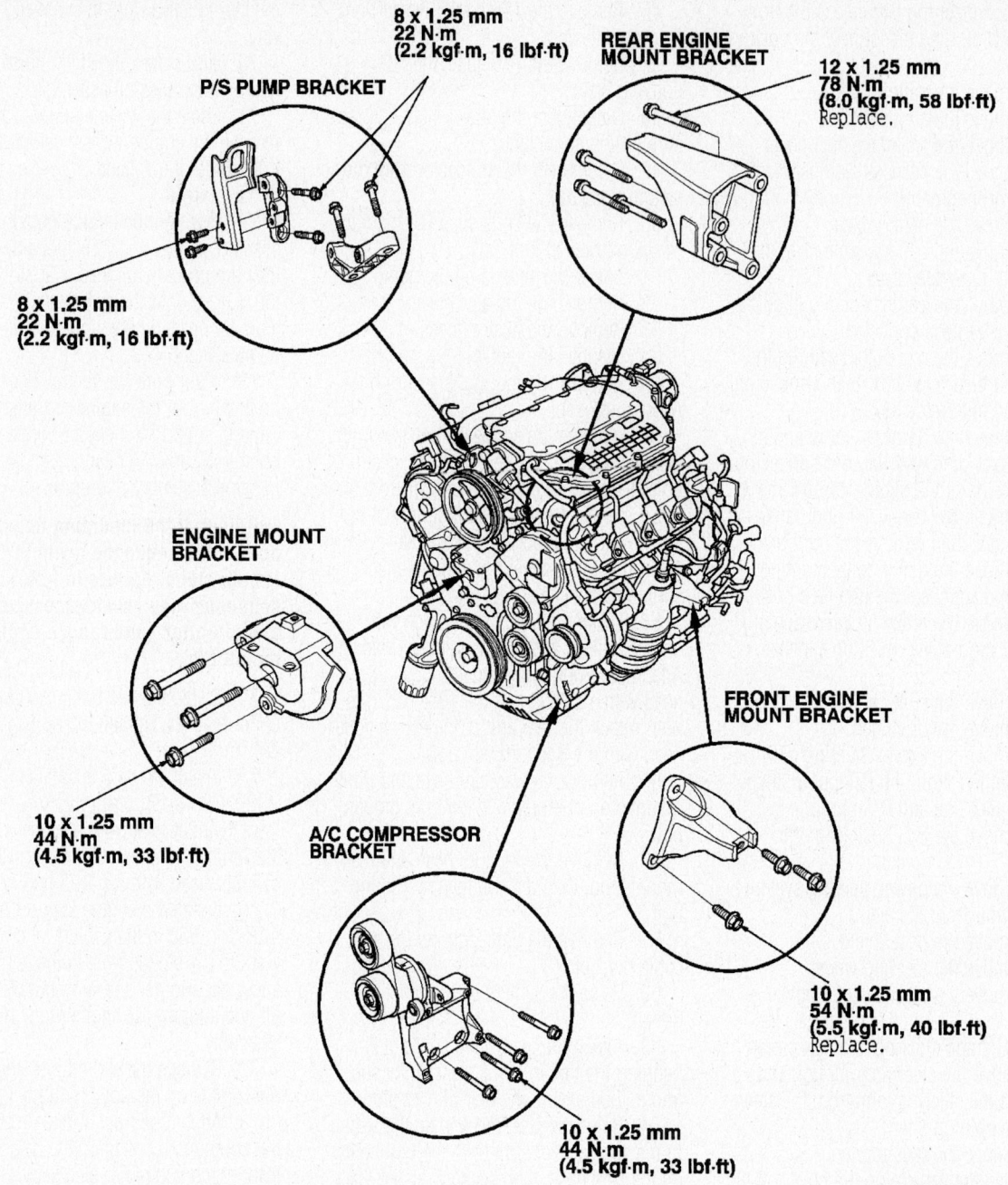

8 x 1.25 mm
22 N·m
(2.2 kgf·m, 16 lbf·ft)

P/S PUMP BRACKET

REAR ENGINE MOUNT BRACKET

12 x 1.25 mm
78 N·m
(8.0 kgf·m, 58 lbf·ft)
Replace.

8 x 1.25 mm
22 N·m
(2.2 kgf·m, 16 lbf·ft)

ENGINE MOUNT BRACKET

10 x 1.25 mm
44 N·m
(4.5 kgf·m, 33 lbf·ft)

A/C COMPRESSOR BRACKET

FRONT ENGINE MOUNT BRACKET

10 x 1.25 mm
54 N·m
(5.5 kgf·m, 40 lbf·ft)
Replace.

10 x 1.25 mm
44 N·m
(4.5 kgf·m, 33 lbf·ft)

09474_ACUCAR_G0124

Exploded view of the engine mounting—2005–06 RL model

82. Loosely tighten the subframe right/left front mounting bolts.

83. Loosen the subframe right rear mounting bolt, then retighten the subframe right rear mounting bolt to the torque specified in the accompanying illustration.

84. Tighten the subframe left rear mounting bolt to the torque specified in the accompanying illustration.

85. Tighten the subframe right/left front mounting bolts to the torque specified in the accompanying illustration.

86. Check that the positioning slots on the right/left rear stiffener, the positioning holes on the subframe, and the positioning holes on the body are aligned using the special tool.

87. Tighten the stiffener mounting bolts to the torque specified in the accompanying illustration.

88. Remove the jack and front subframe adapter.

89. Tighten the bolts securing the transmission lower mount to 33 ft. lbs. (44 Nm).

90. Lower the vehicle on the hoist, then remove the engine hanger and balancer bar.

91. Install the shift cable and tighten the bolt securing the shift cable bracket.

92. Install the vacuum hose.

93. Tighten the rear mount bolt to 40 ft. lbs. (54 Nm), then install the rear mount stop to 47 ft. lbs. (64 Nm)

94. Install the vacuum hose.

95. Tighten the front mount bolt to 40 ft. lbs. (54 Nm), then install the front mount stop to 47 ft. lbs. (64 Nm).

96. Loosen the mounting bolts for the upper half of the side engine mount bracket, then retighten them to 40 ft. lbs. (54 Nm).

97. Install the remaining components in the reverse order of removal. Keeping in mind the following torque specifications:

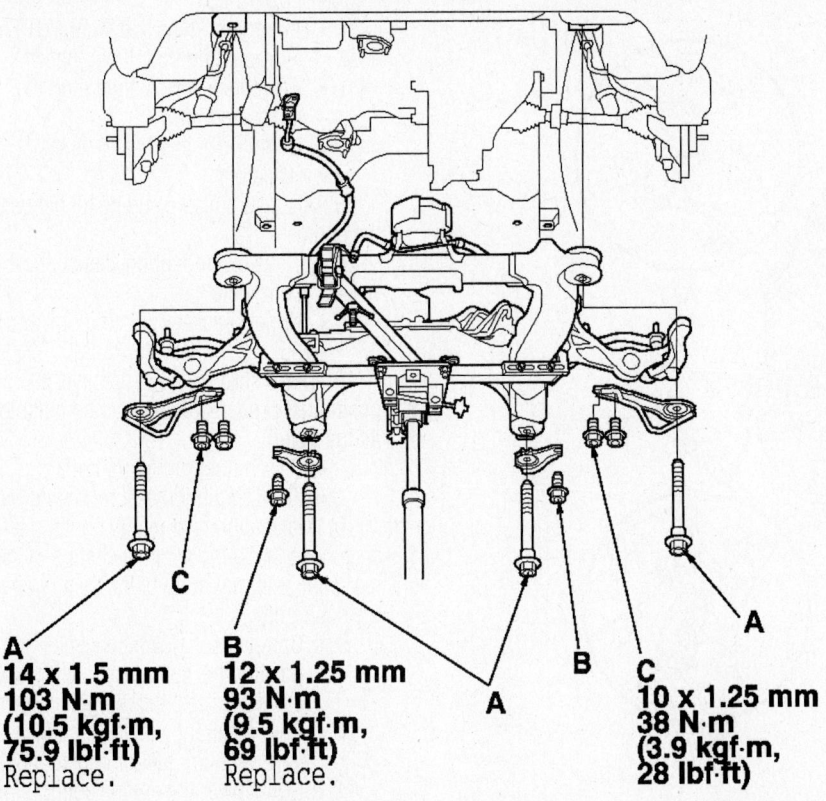

A
14 x 1.5 mm
103 N·m
(10.5 kgf·m,
75.9 lbf·ft)
Replace.

B
12 x 1.25 mm
93 N·m
(9.5 kgf·m,
69 lbf·ft)
Replace.

C
10 x 1.25 mm
38 N·m
(3.9 kgf·m,
28 lbf·ft)

09474_ACUCAR_G0119

Sub-frame bolt locations and torque specifications—2005–06 RL model

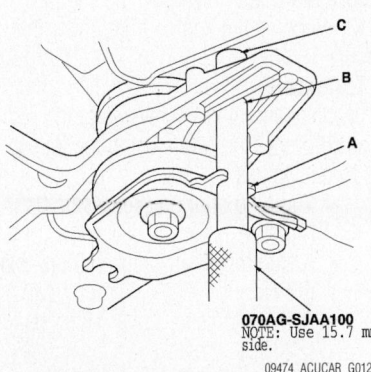

070AG-SJAA100
NOTE: Use 15.7 mm side.

09474_ACUCAR_G0120

Insert the special tool through the positioning slot (B) on the rear stiffener, through the positioning hole (C) on the subframe—2005–06 RL model

a. Tighten the A/C compressor bolts to 16 ft. lbs. (22 Nm).

b. Tighten the connector bracket to the front cylinder head bolts to 16 ft. lbs. (22 Nm).

c. Tighten the engine mount control solenoid valve bracket to the rear cylinder head bolts to 16 ft. lbs. (22 Nm).

d. Tighten the steering joint bolt to 21 ft. lbs. (28 Nm).

98. Refill the engine with engine oil.

99. Refill the transmission with fluid

100. Refill the radiator with engine coolant, and bleed the air from the cooling system with the heater valve open.

101. Refill the power steering system fluid.

102. Do the engine control module (ECM)/powertrain control module (PCM) reset procedure.

103. Do the Crankshaft Position (CKP) pattern clear/CKP pattern learn procedure.

104. Inspect the idle speed.

105. Inspect the ignition timing.

106. Enter the anti-theft codes for the radio and the navigation system, then enter the customer's radio channel presets.

RSX Models

1. Before servicing the vehicle, refer to the precautions section.

2. Obtain the anti–theft code for the radio.

3. Drain the engine oil, coolant, and transmission oil (or fluid) into sealable containers and carefully reinstall the drain plugs using new sealing washers.

4. Properly relieve the fuel system pressure.

5. Remove or disconnect the following:

• Negative and positive battery cables

• Battery

• Intake manifold cover

• Intake Air Temperature (IAT) sensor connector

• Breather hose

• Air cleaner housing

• Intake air duct

• Battery cable from the underhood fuse/relay box, then the harness clamps and ground cable

• Throttle cover, if equipped

6. Fully open the throttle link and cruise control link by hand, then remove the cables from the links.

7. Loosen the locknuts and remove the cables from the bracket.

• Engine Control Module (ECM)/Powertrain Control Module (PCM) connectors and main wire harness connector

• Harness clamps and grommet, then pull the engine wire harness through the bulkhead

• Fuel feed hose

• Evaporative Emission (EVAP) canister hose and brake booster vacuum hose

• Clutch slave cylinder and clutch line bracket mounting bolt, if equipped with a manual transaxle

• Shift cable and select cable, on manual transaxles

• Drive belt

• Power steering pump and position aside without disconnecting the hoses

• Bolt holding the power steering hose bracket

• Radiator cap

• Front wheels and tires

• Splash shield

• Air Fuel (A/F) ratio sensor connector

• Secondary Heated Oxygen (HO$_2$) sensor connector

• Three way catalytic converter assembly

• Lower ball joints and stabilizer links

• Halfshafts. Coat all machined surfaces with clean engine oil, then tie plastic bags over the halfshaft ends to protect them.

• Shift cable holder and the shift cable cover. To avoid damaging the control lever joint, be sure to remove the bolts holding the shift cable cover, on automatic transaxles.

• Spring clip and control pin, then the shift cable from the control lever, on automatic transaxles

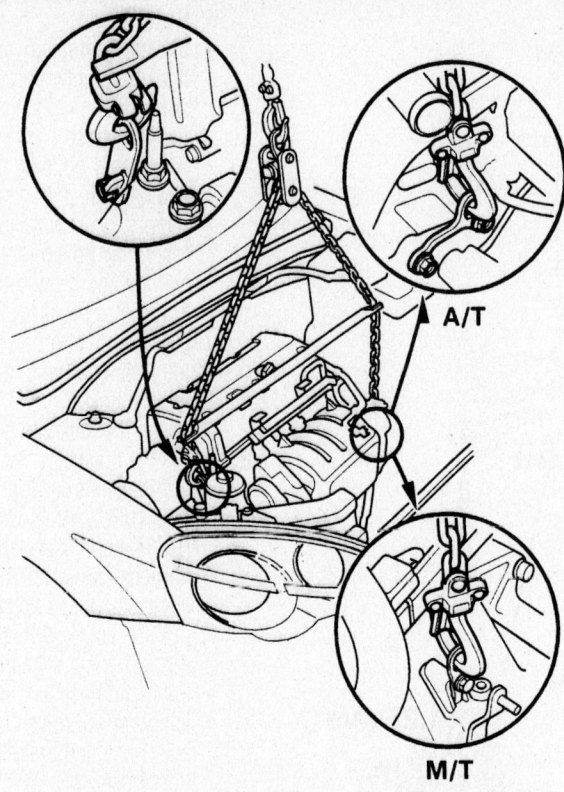

Attach a suitable engine tilt hanger to the engine–RSX models

42356-INTE-G02

- Lower hose
- Automatic Transaxle Fluid (ATF) filter mounting bolt, if equipped
- ATF cooler hoses, then plug the hoses and lines
- Upper hose and heater hoses
- Radiator

8. Attach a suitable engine tilt hanger to the engine.

- Transaxle mount bracket support bolts/nuts
- Upper bracket mounting bolt and nut

9. Make sure the hoist brackets are positioned properly, then raise the hoist to its full height.

- Rear mount mounting bolts
- Front mount bracket mounting bolt

10. Make alignment marks on the reference lines that align with the centers of the rear subframe mounting bolts, then remove the subframe.

- Compressor clutch connector and position the compressor aside without disconnect the refrigerant lines
- Any remaining electrical connectors, vacuum, fuel or coolant hoses

11. Slowly lower the engine about 6 in. (15 cm). Check that all hoses and wires are disconnected from the engine/transaxle.

12. Lower the engine all the way, then remove the chain hoist from the engine.

13. Remove the engine from under the vehicle.

To install:

14. Installation is the reverse of the removal procedure, while using the following torque values:

- Engine mount bracket: 33 ft. lbs. (44 Nm)
- A/C compressor bracket: 33 ft. lbs. (44 Nm)
- A/C compressor bolts: 16 ft. lbs. (22 Nm)
- New subframe bolts: 76 ft. lbs. (103 Nm)
- Rear mount mounting bolts: 43 ft. lbs. (59 Nm)
- Upper bracket mounting bolt: 40 ft. lbs. (54 Nm)
- Support bolts/nuts: 40 ft. lbs. (54 Nm)
- Front engine mount bracket mounting bolts: 47 ft. lbs. (64 Nm)
- A/T shift control cable: 7.2 ft. lbs. (9.8 Nm)
- New catalytic converter self–locking nuts: 25 ft. lbs. (33 Nm)
- Catalytic converter bolt: 16 ft. lbs. (22 Nm)
- ATF filter mounting bolt: 104 inch lbs. (12 Nm)

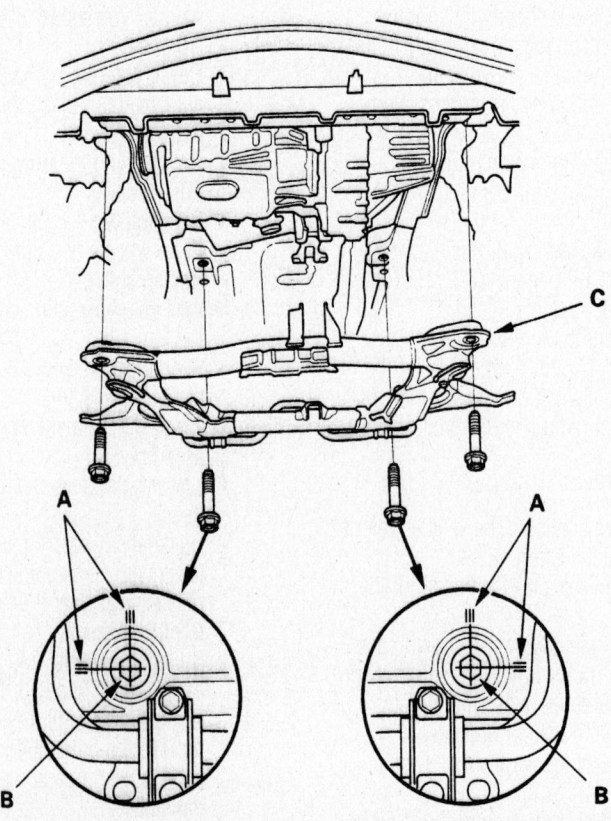

42356-INTE-G03

Make alignment marks on the reference lines (A) that align with the centers of the rear subframe mounting bolts (B), then remove the front subframe (C)–RSX models

- Power steering pump bolts: 16 ft. lbs. (22 Nm)
- Power steering hose bracket bolt: 7.2 ft. lbs. (9.8 Nm)
- M/T cable bolts: 7.2 ft. lbs. (9.8 Nm)
- Clutch slave cylinder mounting bolts: 17 ft. lbs. (24 Nm)
- Clutch line bracket mounting bolt: 7.2 ft. lbs. (9.8 Nm)
- Battery cable harness clamp bolt: 104 inch lbs. (12 Nm)
- Air cleaner housing bolts: 104 inch lbs. (12 Nm)
- Intake manifold cover bolts: 104 inch lbs. (12 Nm)

15. Fill the engine with oil and the transaxle with fluid.

16. Fill and bleed (if necessary) the air from the cooling system.

17. Connect the positive, then the negative battery cable and enter the radio security code.

18. Switch the ignition **ON** but do not engage the starter. The fuel pump should run for approximately 2 seconds, building pressure within the lines. Switch the ignition **OFF**, then **ON** 2 or 3 more times to build full system pressure. Check for fuel leaks.

19. Start the engine, allowing it to idle. Check the hoses and lines carefully for any sign of leakage.

20. Check the timing and idle speed.

21. After the engine has warmed up fully and the fan(s) have come on at least once, recheck the engine for fluid leaks. Switch the engine OFF.

22. Adjust the belts and throttle cable as necessary.

2002–03 TL Models

1. Before servicing the vehicle, refer to the precautions section.

2. Do not remove the hood. Disconnect the hood support strut and reconnect it to hold the hood in a vertical position.

3. Properly relieve the fuel pressure.

4. Drain the engine oil, coolant, transmission fluid and the differential fluid.

5. Remove or disconnect the following:
- Negative battery cable, then the positive battery cable
- battery and battery tray
- Air cleaner assembly
- Throttle cable and cruise control cable from the throttle and bracket
- Left side engine wiring harness connectors
- Fuel feed and return hoses
- Brake booster vacuum hose

- Battery cables from the under–hood fuse/relay box
- Under–hood fuse/relay box
- Powertrain Control Module (PCM) electrical connectors
- Accessory drive belts
- Power steering pump without disconnecting the hoses
- Vehicle Speed sensor (VSS) connector, then remove the VSS/power steering sensor leaving the fluid hoses attached
- Front wheels
- Lower splash shield
- Heated Oxygen (HO2S) sensor connector
- Front exhaust pipe from the manifold
- Front damper forks
- Lower ball joints from the steering knuckles
- Halfshafts from the differential and the intermediate shaft
- Shift cable cover
- Shift cable with the control lever from the transaxle
- Power steering hose clamps and the engine mount control vacuum hose
- Upper and lower radiator hoses
- Heater hoses
- Transmission fluid cooler hoses
- Ground cable
- Power steering hose clamp from the rear beam assembly

6. Attach a suitable chain hoist to the engine lifting hooks and support the engine.

➡**The engine and transmission assembly is removed by lowering it from the vehicle. Be sure the vehicle is in a position that will allow the engine and transmission assembly enough clearance to be moved from the vehicle once it is lowered away from the vehicle.**

7. Remove or disconnect the following:
- Side, rear and front engine mount support fasteners
- Front suspension radius rod bolts

8. Make alignment marks on the front beam and remove the front beam.

9. Remove or disconnect the following:
- Air conditioning compressor leaving the hoses attached
- Rear mounts from the engine and transmission

10. Check that all hoses, cables and wires have been properly disconnected and slowly lower the engine about 6 inches (150 mm). Recheck that all hoses, cables and wires have been properly disconnected.

11. Carefully lower the engine/transmission assembly from the vehicle.

To install:

12. Installation is the reverse of the removal procedure, while using the following torque values:
- Transmission rear mount bolts: 28 ft. lbs. (38 Nm)
- Air conditioning compressor bolts: 16 ft. lbs. (22 Nm)
- Front beam nuts: 28 ft. lbs. (38 Nm)
- Front beam bolts: 76 ft. lbs. (103 Nm)
- Rear bolts: 28 ft. lbs. (38 Nm)
- Front suspension radius rod bolts: 119 ft. lbs. (162 Nm)
- Front engine mount nut: 40 ft. lbs. (54 Nm)
- Rear engine mount nut: 40 ft. lbs. (54 Nm)
- Rear engine mount bolt: 28 ft. lbs. (38 Nm)
- Side engine mount bracket bolts: 33 ft. lbs. (44 Nm)
- Side engine mount through bolt: 40 ft. lbs. (54 Nm)
- Lower ball joint nuts: 40 ft. lbs. (54 Nm)
- Damper fork bolts: 47 ft. lbs. (64 Nm)
- Front exhaust pipe nuts: 40 ft. lbs. (54 Nm)
- Shift cable and control lever bolts: 10 ft. lbs. (14 Nm)
- Power steering pump bolts: 16 ft. lbs. (22 Nm)

13. Fill the engine with oil and the transmission with fluid.

14. Fill and bleed (if neccssary) the air from the cooling system.

15. Connect the positive, then the negative battery cable and enter the radio security code.

16. Switch the ignition **ON** but do not engage the starter. The fuel pump should run for approximately 2 seconds, building pressure within the lines. Switch the ignition **OFF**, then **ON** 2 or 3 more times to build full system pressure. Check for fuel leaks.

17. Start the engine, allowing it to idle. Check the hoses and lines carefully for any sign of leakage.

18. Check the timing and idle speed.

19. After the engine has warmed up fully and the fan(s) have come on at least once, recheck the engine for fluid leaks. Switch the engine OFF.

20. Adjust the belts and throttle cable as necessary.

2004–06 TL Models

➡Be sure you have the anti-theft codes for the radio and navigation system, then write down the audio channel presets. Make sure the ignition switch is OFF.

1. Before servicing the vehicle, refer to the precautions section.

2. Disconnect the support struts from both sides of the pivot ball (bolted to the hood).

3. Secure the hood in a vertical position. Remove the right side pivot ball and install it into the lower threaded hole, then reattach the support strut.

➡Do not attempt to close the hood with the support strut in the vertical position, as it will damage the support strut and hood.

4. Remove the left side engine compartment cover and the left rear engine compartment cover.

5. Remove the right side engine compartment cover, then remove the right rear engine compartment cover.

6. Remove the intake manifold cover.

7. Drain the power steering system fluid, then plug the fluid reservoir and return hose.

8. Relieve the fuel system pressure.

9. Disconnect the negative cable from the battery first, then disconnect the positive cable.

10. Remove the battery.

11. Remove the air cleaner housing.

12. Remove the harness clamp.

13. Remove the four bolts, then remove the battery base.

14. Remove the battery cables from the under-hood fuse/relay box, then disconnect the harness connector.

15. Remove the under-hood fuse/relay box from the body.

16. Remove the harness clamp, and remove the two 6 mm bolts, then remove the strut brace.

17. Remove the brake booster vacuum hose and evaporative emission (EVAP) canister hose.

18. Remove the harness clamp, and disconnect the engine wire harness connectors on the left side of the engine compartment.

19. Remove the three bolts retaining the shift cable holder, then remove the shift cable and select cable.

➡Do not bend the cables excessively if equipped with a manual transmission.

20. Remove the clutch slave cylinder and clutch line bracket mounting bolt.

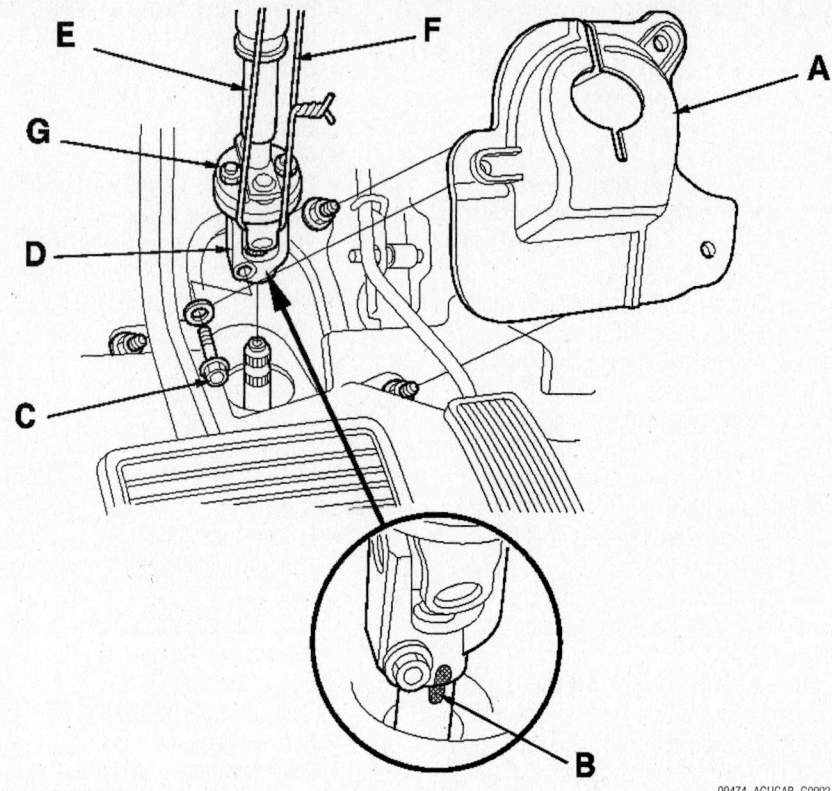

Make a reference mark across the steering joint and steering gearbox pinion shaft—2004–06 TL model

➡Do not operate the clutch pedal once the slave cylinder has been removed if equipped with a manual transmission.

21. Remove the drive belt.

22. Remove the Power Steering (PS) pump outlet line from the PS pump, then plug the outlet line and the PS pump.

23. Remove the PS hose from the clamp.

24. Remove the PS fluid reservoir from the clamp.

25. Remove the steering joint cover.

26. Lock the steering wheel. Make a reference mark across the steering joint and steering gearbox pinion shaft.

27. Remove the steering joint bolt, and disconnect the steering joint from the steering gearbox pinion shaft. To prevent damage to the cable reel, do not turn the steering wheel once the steering joint has been removed.

28. Disconnect the A/C condenser fan motor connector and A/C compressor clutch connector, and remove the reserve tank. Wipe up any spilled engine coolant immediately.

29. Remove the two bolts, loosen the bottom bolt, and then remove the A/C condenser fan shroud.

30. Remove the four bolts retaining the A/C compressor.

31. Remove the radiator cap.

32. Raise the vehicle on the hoist to full height.

33. Remove the front tires/wheels

34. Remove the engine under cover.

35. Remove the splash shield.

36. Loosen the drain plug in the radiator, and drain the engine coolant.

37. Drain the transmission fluid:

38. Drain the engine oil.

39. Disconnect the stabilizer links.

40. Remove the damper fork.

41. Separate the tie-rod end ball joints from the knuckles.

42. Separate the knuckles from the lower arms.

43. Remove the driveshafts and coat all precision-finished surfaces with clean engine oil. Tie plastic bags over the driveshaft ends.

44. Remove the shift cable. Do not bend the shift cable excessively if equipped with an automatic transmission.

45. Remove exhaust pipe.

46. Remove the PS hose, then plug the line and the hose.

47. Disconnect the power steering pressure switch connector.

48. Remove the nuts retaining the transmission lower front mount and transmission lower rear mount.

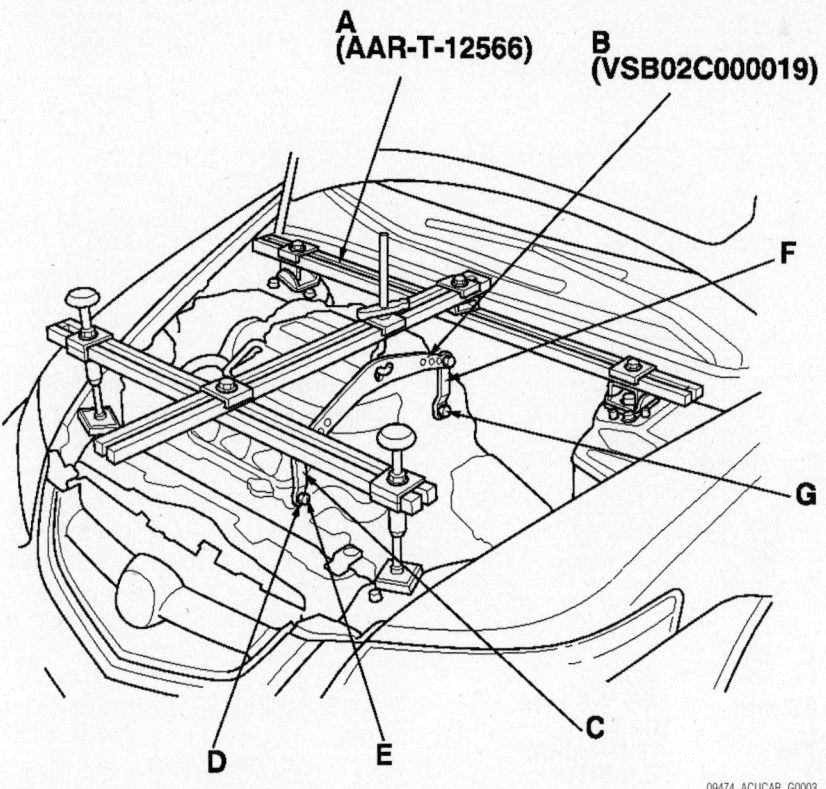

A
(AAR-T-12566)

B
(VSB02C000019)

09474_ACUCAR_G0003

Lift and support the engine with engine hanger and engine balancer bar—2004–06 TL model

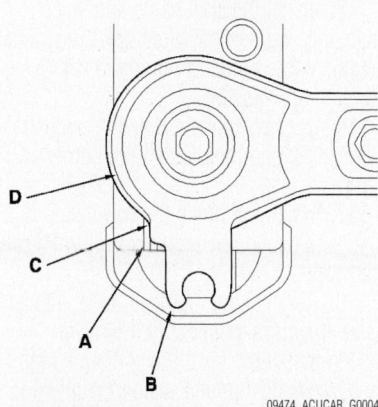

09474_ACUCAR_G0004

Make the appropriate reference lines at both ends of the sub-frame that line up—2004–06 TL model

49. Remove the upper radiator hose and lower radiator hose.

50. Remove the heater hoses

51. Remove the two bolts retaining the side engine mount bracket

52. Remove the right fender trim, left fender trim, and front bulkhead cover.

53. Lift and support the engine with engine hanger and engine balancer bar as follows:

    a. Attach the front arm to the front cylinder head with a spacer and the connector bracket bolt (10 x 1.25 mm).

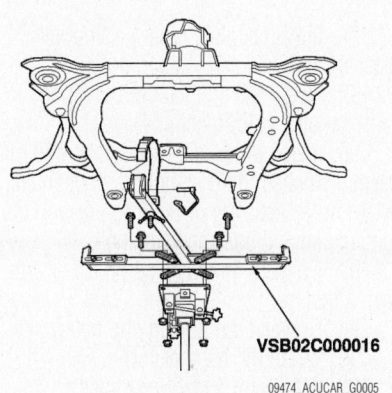

VSB02C000016

09474_ACUCAR_G0005

Attach an engine sub-frame adapter tool to the sub-frame—2004–06 TL model

    b. Attach the rear arm to the rear cylinder head with the harness clamp bracket bolt (8 x 1.25 mm).

54. Remove the ground cable.

55. Remove the transmission upper mount bracket, and remove the vacuum hose.

56. Remove the front mount stop, then remove the front mount bolt.

57. Remove the two bolts retaining the rear engine damper on models with a manual transmission.

58. Remove the rear mount stop, then remove the rear mount bolt.

59. Remove the vacuum hose.

60. Remove the two bolts retaining the shift cable bracket if equipped with a manual transmission.

61. Make sure the hoist brackets are positioned properly.

62. Make the appropriate reference lines at both ends of the sub-frame that line up.

63. Attach an engine sub-frame adapter tool to the sub-frame by hanging the belt over the front of the sub-frame, then secure the belt with its stop.

64. Raise the jack and line up the slots in the engine sub-frame adapter arms with the bolt holes on the jack base, then securely attach them with four bolts.

65. Remove the sub-frame mid-mounts

66. Remove the sub-frame.

67. Check that the engine/transmission is completely free of vacuum hoses, fuel and coolant hoses, and electrical wiring.

68. Lower the vehicle and securely support the engine and transmission assembly.

69. When the engine and transmission are securely supported, and there is no tension on the engine support hanger, remove the engine hanger from the engine.

70. Slowly raise the vehicle about 150 mm (6 in.). Check once again that all hoses and wires are disconnected from the engine/transmission assembly.

71. Raise the vehicle all the way.

72. Remove the engine/transmission assembly from under the vehicle.

To install:

73. Position the engine/transmission assembly under the vehicle.

74. Lift and support the engine with engine hanger and engine balancer bar.

75. Attach the front arm to the front cylinder head with a spacer and the connector bracket bolt (10 x 1.25 mm).

76. Attach the rear arm to the rear cylinder head with the harness clamp bracket bolt (8 x 1.25 mm).

77. Lift the engine into position in the vehicle.

78. Reinstall all mounting bolts/support nuts in the sequences given. Failure to follow this may cause excessive noise and vibration, and reduce bushing life

79. Attach the engine sub-frame adapter tool to the sub-frame by hanging the belt over the front of the sub-frame, then secure the belt with its stop. Raise the sub-frame up to the body with a jack

80. Loosely install the four sub-frame mounting bolts and four 12 x 1.25 mm bolts, with the stiffeners.

81. Loosely install the sub-frame mid mounts, then remove the tool.

82. Align the reference marks with edge

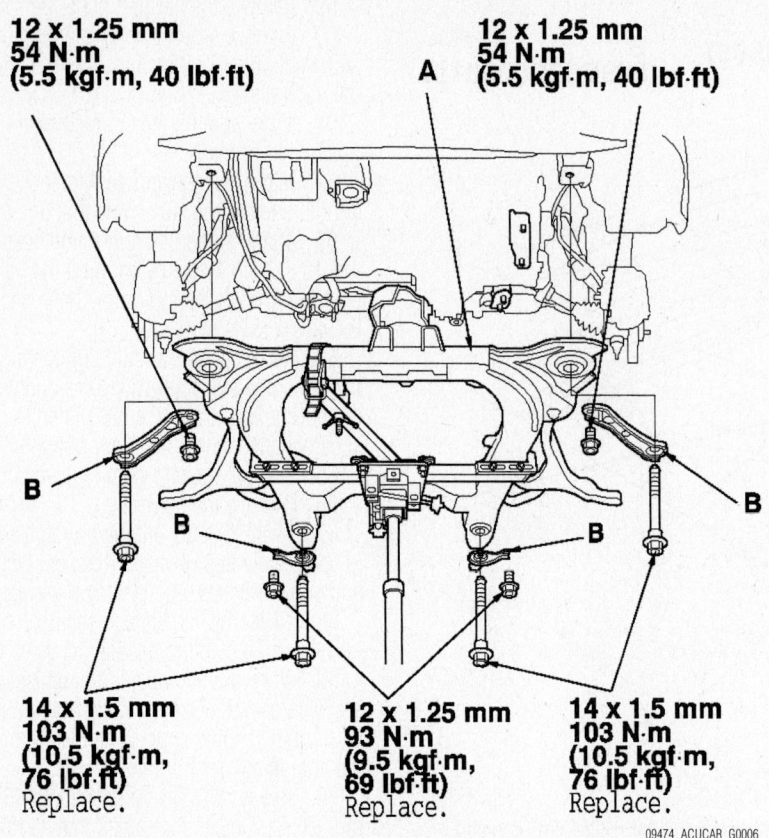

12 x 1.25 mm
54 N·m
(5.5 kgf·m, 40 lbf·ft)

12 x 1.25 mm
54 N·m
(5.5 kgf·m, 40 lbf·ft)

14 x 1.5 mm
103 N·m
(10.5 kgf·m,
76 lbf·ft)
Replace.

12 x 1.25 mm
93 N·m
(9.5 kgf·m,
69 lbf·ft)
Replace.

14 x 1.5 mm
103 N·m
(10.5 kgf·m,
76 lbf·ft)
Replace.

09474_ACUCAR_G0006

Sub-frame bolt locations and torque specifications—2004–06 TL model

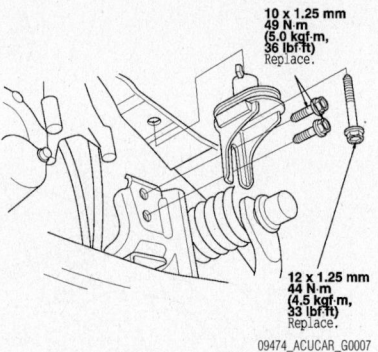

10 x 1.25 mm
49 N·m
(5.0 kgf·m,
36 lbf·ft)
Replace.

12 x 1.25 mm
44 N·m
(4.5 kgf·m,
33 lbf·ft)
Replace.

09474_ACUCAR_G0007

Sub-frame mid-mounts bolt specifications—2004–06 TL model

of both rear stiffener, and tighten the rear sub-frame mounting bolts, then front bolts, and tighten the stiffener bolts to the specified torque shown in the accompanying illustration.

83. Tighten the bolts retaining the sub-frame mid-mounts to the specified torque shown in the accompanying illustration.

84. Tighten the nuts retaining the transmission lower front mount and transmission lower rear mount to 33 ft. lbs. (44 Nm).

85. Remove the jack and front sub-frame adapter.

86. Install the shift cable, and tighten the two bolts retaining the shift cable bracket if equipped with a manual transmission.

87. Tighten the two bolts retaining the rear engine damper if equipped with a manual transmission.

88. Tighten the front mount bolt, then install the front mount stop.

89. Tighten the two bolts retaining the side engine mount bracket.

90. Install the transmission upper mount bracket, then tighten the bolts in the numbered sequence shown.

91. Install the vacuum hose.

92. Install the ground cable.

93. Connect the power steering pressure switch connector.

94. Install exhaust pipe A using new gaskets and new self locking nuts.

95. Install the shift cable if equipped with an automatic transmission.

96. Install a new set ring on the end of each driveshaft, then install the driveshafts. Make sure each ring clicks into place in the differential and intermediate shaft.

97. Connect the suspension lower arm ball joints.

98. Connect the tie-rod end ball joints.

99. Install the damper fork.

100. Connect the stabilizer links

101. Install the splash shield

102. Install the engine under cover.

103. Install the front tires/wheels.

→On models with a manual transmission, be careful not to damage or chip the paint of the brake calipers when installing the wheels.

104. Lower the vehicle on the hoist.

105. Install the heater hoses.

106. Install the upper radiator hose and lower radiator hose.

107. Install the A/C compressor

108. Install the A/C condenser fan shroud and coolant reserve tank, then connect the A/C condenser fan motor connector and A/C compressor clutch connector.

109. Align the reference mark on the steering joint and steering gearbox pinion shaft. Connect the steering joint to the steering gearbox pinion shaft. Tighten the steering joint bolt to 21 ft. lbs. (28 Nm).

110. Install the PS pump outlet hose with a new O-ring.

111. Install the PS hose to the clamp.

112. Install the PS fluid reservoir to the clamp.

113. Install the drive belt.

114. Install the clutch slave cylinder and clutch line bracket mounting bolt if equipped with a manual transmission.

115. Install the shift cable and select cable, then tighten the three bolts retaining the shift cable holder if equipped with a manual transmission.

116. Connect the engine wire harness connectors, and then install the harness clamp.

117. Install the brake booster vacuum hose, and evaporative emission (EVAP) canister hose.

118. Connect the fuel feed hose, then install the quick-connect fitting cover.

119. Install the strut brace, then install the harness clamp, and install the two 6 mm bolts.

120. Install the under-hood fuse/relay box, and connect the harness connector.

121. Install the battery cables.

122. Install the battery base, then install the harness clamp.

123. Install the air cleaner housing.

124. Install the resonator.

125. Install the intake manifold cover.

126. Install the battery.

127. Move the shift lever to each gear, and verify that the automatic transmission gear position indicator follows the transmission range switch if equipped with an automatic transmission.

128. Check that the transmission shifts into gear smoothly if equipped with a manual transmission.

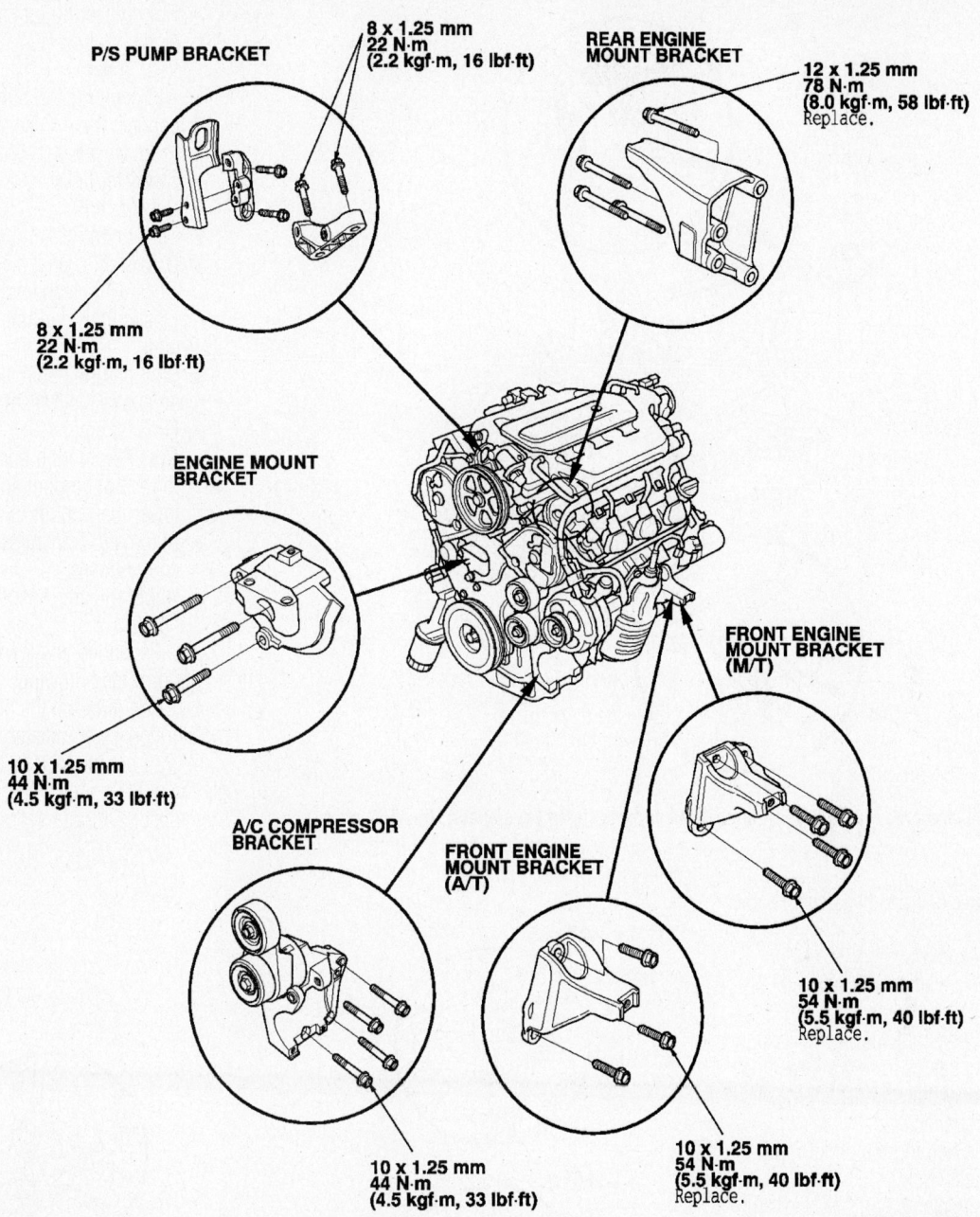

P/S PUMP BRACKET

8 x 1.25 mm
22 N·m
(2.2 kgf·m, 16 lbf·ft)

8 x 1.25 mm
22 N·m
(2.2 kgf·m, 16 lbf·ft)

**REAR ENGINE
MOUNT BRACKET**

12 x 1.25 mm
78 N·m
(8.0 kgf·m, 58 lbf·ft)
Replace.

**ENGINE MOUNT
BRACKET**

10 x 1.25 mm
44 N·m
(4.5 kgf·m, 33 lbf·ft)

**A/C COMPRESSOR
BRACKET**

**FRONT ENGINE
MOUNT BRACKET
(A/T)**

**FRONT ENGINE
MOUNT BRACKET
(M/T)**

10 x 1.25 mm
54 N·m
(5.5 kgf·m, 40 lbf·ft)
Replace.

10 x 1.25 mm
44 N·m
(4.5 kgf·m, 33 lbf·ft)

10 x 1.25 mm
54 N·m
(5.5 kgf·m, 40 lbf·ft)
Replace.

09474_ACUCAR_G0008

Exploded view of the engine mounts and their bolt torque specifications—2004–06 TL model

129. Turn the ignition switch ON (II) (do not operate the starter) so the fuel pump runs for about 2 seconds and pressurizes the fuel line. Repeat this operation two or three times, then check for fuel leakage at any point in the fuel line.

130. Refill the engine with engine oil.

131. Refill the transmission with fluid

132. Refill the radiator with engine coolant, and bleed the air from the cooling system with the heater valve open.

133. Refill the power steering system fluid.

134. Do the engine control module (ECM)/powertrain control module (PCM) reset procedure.

135. Do the Crankshaft Position (CKP) pattern clear/CKP pattern learn procedure.

136. Inspect the idle speed.

137. Inspect the ignition timing.

138. Install the right rear engine compartment cover, then install the right side engine compartment cover.

139. Install the left rear engine compartment cover and the left side engine compartment cover.

140. Enter the anti-theft codes for the radio and the navigation system, then enter the customer's radio channel presets.

TSX Models

1. Before servicing the vehicle, refer to the precautions section.

2. Obtain the anti–theft code for the radio.

3. Remove the splash shield in order to drain the vehicle's fluids.

4. Drain the engine oil, coolant, and transmission oil (or fluid) into sealable containers and carefully reinstall the drain plugs using new sealing washers.

5. Properly relieve the fuel system pressure.

6. Remove or disconnect the following:

- Power steering pump and the hose clamp, but do NOT disconnect the fluid lines
- A/C compressor, but do not disconnect the A/C hoses
- Front wheels and tires
- Stabilizer links
- Damper fork
- Lower control arm ball joints
- Driveshafts. Coat all machined surfaces with clean engine oil, then tie plastic bags over the driveshaft ends.
- Shift control cable, if equipped with an automatic transmission
- Exhaust pipe "A"
- Nuts holding the transmission lower front mount and transmission lower rear mount
- Automatic Transmission Fluid cooler hoses
- Upper radiator hose and heater hoses

7. To disconnect the lower radiator hose, perform the following:

   a. Clean dirt from the quick-connector, radiator and lower radiator hose.

   b. Pull out the lock by hand, and wiggle the quick-connector to remove it from the radiator. Do not use tools to remove the quick-connector, do it by hand.

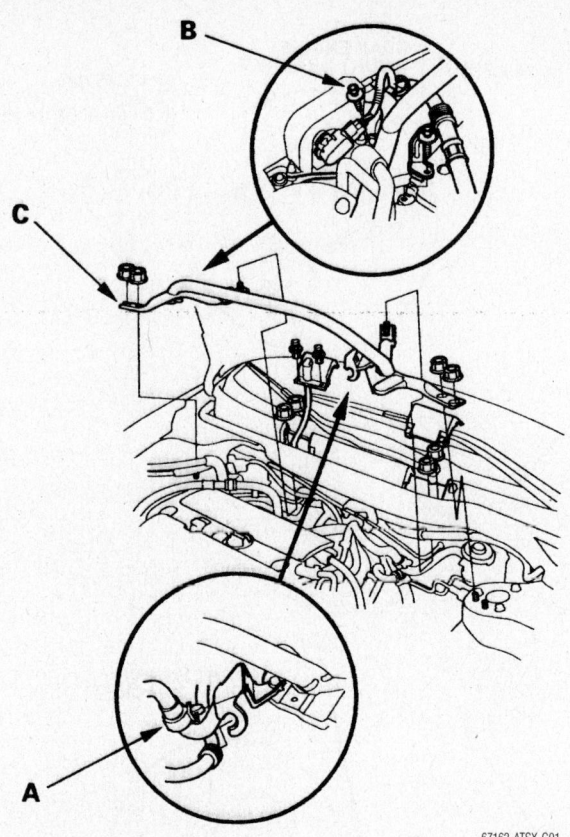

67162-ATSX-G01

To remove the strut brace (C), you must remove the harness clamp (A) and bolt (B)—TSX models

- Negative and positive battery cables
- Battery
- Air cleaner housing
- Harness clamps and harness bracket
- Battery base retainers and base
- Battery cables from the underhood fuel/relay box and detach the electrical connector
- 2 bolts holding the underhood fuse/relay box
- Harness clamp and bolt and strut brace
- Quick-connect fitting cover and fuel feed hose
- Evaporative emission (EVAP) canister hose and brake booster vacuum hose
- Engine Control Module/Powertrain Control Module (ECM/PCM) connectors and main wire harness
- Accelerator Pedal Position (APP) sensor connector
- Harness clamps and grommet, then pull the engine wire harness through the bulkhead
- Clutch slave cylinder, shift cable and select cable, if equipped with a manual transmission
- Drive belt

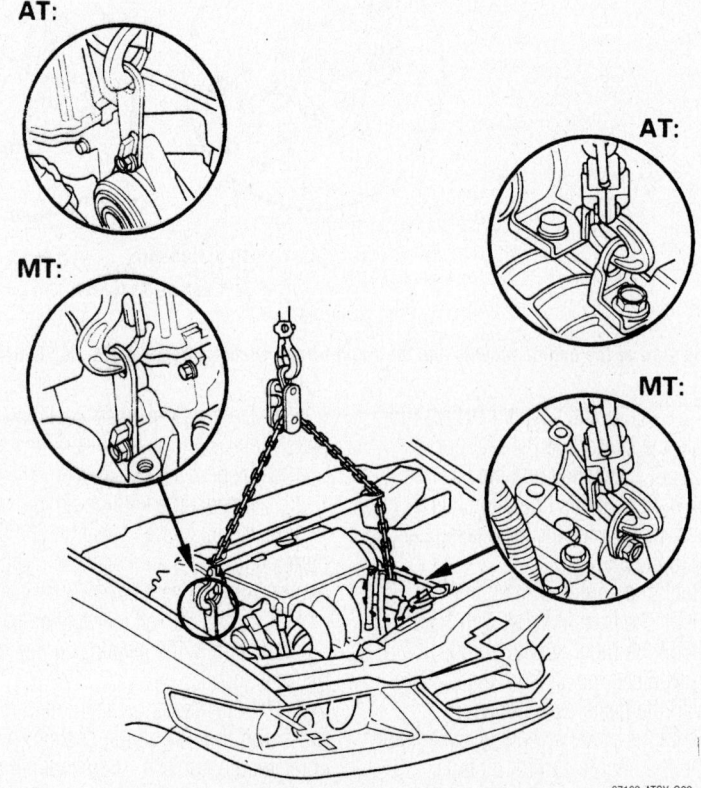

Make sure the chain hoist is properly attached to the engine—TSX models

67162-ATSX-G02

8. Remove or disconnect the following:
- Ground cable and upper bracket

9. Proper attach a chain hoist to the engine.

10. For vehicles with a equipped with a manual transmission, remove the ground cable, transmission upper mount/bracket assembly and clutch line clamp bracket

11. For vehicles with an equipped with an automatic transmission, remove the ground cable, then remove the bracket plate, transmission upper bracket and transmission upper mount.

12. Remove or disconnect the following:
- Vacuum hose from the vacuum line
- Front mount stop and vacuum hose clamp bracket and the front mount bolt
- Rear mount stop and the rear mount bolt

13. Make sure that the engine/transmission assembly is totally free of all vacuum lines, wiring and fuel and coolant hoses.

14. Slowly raise the engine about 6 in. (150mm), then recheck that all wires/hoses are disconnected from the engine/transmission assembly.

15. Raise the engine/transmission assembly all the way and remove it from the vehicle.

To install:

16. Install the accessory brackets to the engine and tighten to the following torque specifications:
- Side engine mount bracket: 33 ft. lbs. (45 Nm)
- Rear mount bracket (use NEW bolts): 65 ft. lbs. (88 Nm)
- Front mount bracket (use NEW bolts): 47 ft. lbs. (64 Nm)
- A/C compressor bracket: 33 ft. lbs. (45 Nm)

➡**You must tighten the mounting bolts/support nuts in the order listed below, following the tightening sequences when given. Failure to follow the proper sequence may cause excessive noise and vibration and may also reduce the life of the bushings.**

17. Installation is the reverse of the removal procedure, while using the following torque values:
- Front mount bolt (use a NEW bolt): 40 ft. lbs. (54 Nm)
- Front mount stop nuts: 58 ft. lbs. (78 Nm).
- Rear mount bolt (use a NEW bolt): 40 ft. lbs. (54 Nm)
- Rear mount stop nuts: 51 ft. lbs. (69 Nm)

- 2 top upper bracket bolts: 40 ft. lbs. (54 Nm)
- Side upper bracket bolt: 47 ft. lbs. (64 Nm)
- Ground cable bolt: 104 inch lbs. (12 Nm)

18. If equipped with a equipped with a manual transmission, loosen the transmission upper mount bolt, then install the transmission upper mount/bracket assembly, clutch line clamp bracket and ground cable.

19. If equipped with an equipped with an automatic transmission, install the transmission upper mount, transmission upper bracket and ground cable, then loosely install the transmission upper bracket plate.

20. On equipped with a manual transmission equipped vehicles, tighten a NEW transmission upper mount bolt to 40 ft. lbs. (54 Nm). On equipped with an automatic transmission equipped vehicles, tighten NEW transmission upper mount bracket plate mounting bolts to 40 ft. lbs. (54 Nm) and NEW transmission upper mount bolt to 40 ft. lbs. (54 Nm).

21. The remainder of installation is the reverse of the removal procedure. Note the following tightening specifications:
- Transmission lower front and rear mounts: 33 ft. lbs. (44 Nm)
- Exhaust pipe "A" self-locking nuts (use NEW nuts): 25 ft. lbs.
- Exhaust pipe "A" self-locking bolt (use a NEW bolt): 16 ft. lbs. (22 Nm)
- A/C compressor mounting bolts: 16 ft. lbs. (22 Nm)
- Power steering pump mounting bolts: 16 ft. lbs. (22 Nm)
- Power steering hose clamp bolt: 104 inch lbs. (12 Nm)
- Strut brace nuts: 16 ft. lbs. (22 Nm)
- Harness clamp bolt: 104 inch lbs. (12 Nm)
- Underhood fuse/relay box bolts: 104 inch lbs. (12 Nm)
- Battery base bolts: 16 ft. lbs. (22 Nm)
- Harness bracket bolt: 104 inch lbs. (12 Nm)

22. Fill the engine with oil and the transmission with fluid.

23. Fill and bleed (if necessary) the air from the cooling system.

24. Connect the positive, then the negative battery cable and enter the radio security code.

25. Switch the ignition **ON** but do not engage the starter. The fuel pump should run for approximately 2 seconds, building pressure within the lines. Switch the ignition

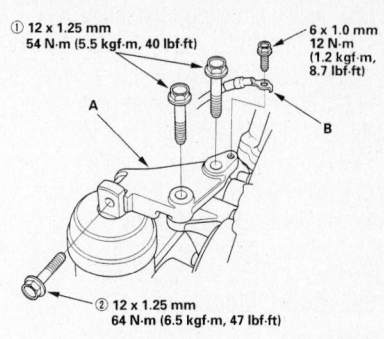

① 12 x 1.25 mm
54 N·m (5.5 kgf·m, 40 lbf·ft)

6 x 1.0 mm
12 N·m (1.2 kgf·m, 8.7 lbf·ft)

A

B

② 12 x 1.25 mm
64 N·m (6.5 kgf·m, 47 lbf·ft)

67162-ATSX-G03

Proper tightening sequence for the upper bracket bolts—TSX models

OFF, then **ON** 2 or 3 more times to build full system pressure. Check for fuel leaks.

26. Start the engine, allowing it to idle. Check the hoses and lines carefully for any sign of leakage.

27. Check the timing and idle speed.

28. After the engine has warmed up fully and the fan(s) have come on at least once, recheck the engine for fluid leaks. Switch the engine OFF.

29. Adjust the belts and throttle cable as necessary.

Water Pump

REMOVAL & INSTALLATION

CL Model

➡**Perform this service operation with the engine cold.**

1. Before servicing the vehicle, refer to the precautions section.

2. Disconnect the negative battery cable.

3. Remove the front splash panel.

4. Drain the cooling system.

5. Remove the timing belt. Inspect the timing belt for any signs of damage or oil and coolant contamination. Replace the timing belt if there is any doubt about its condition.

6. Remove the timing belt tensioner.

7. Remove the water pump bolts. Then, remove the water pump and sprocket assembly from the engine block.

To install:

8. Install the water pump with a new O–ring. Use new bolts and tighten the mounting bolts evenly to 104 inch lbs. (12 Nm.

9. If removed, install the timing belt rear cover and camshaft pulley.

10. If removed, install the timing belt tensioner.

11. Install or connect the following:
   • Timing belt and timing belt covers
   • Accessory drive belts

12. Close the cooling system drain plug. Refill and bleed the cooling system.

13. Connect the negative battery cable.

14. Start the engine, allow it to reach normal operating temperature, check for leaks, and top off as necessary.

2002–04 RL Model

➡**Perform this service operation with the engine cold.**

1. Before servicing the vehicle, refer to the precautions section.

2. Disconnect the negative battery cable.

3. Remove the front splash panel.

4. Drain the cooling system.

5. Remove the timing belt. Inspect the timing belt for any signs of damage or oil and coolant contamination. Replace the timing belt if there is any doubt about its condition.

6. Remove the left camshaft pulley and back cover.

7. Remove the water pump bolts. Then, remove the water pump and sprocket assembly from the engine block.

To install:

8. Install the water pump with a new O–ring. Use new bolts and tighten the 6mm mounting bolts evenly to 104 inch lbs. (12 Nm) and the 8mm bolts to 16 ft. lbs. (22 Nm).

9. If removed, install the timing belt rear cover and camshaft pulley.

10. If removed, install the timing belt tensioner.

11. Install or connect the following:
   • Timing belt and timing belt covers
   • Accessory drive belts

12. Close the cooling system drain plug. Refill and bleed the cooling system.

13. Connect the negative battery cable.

14. Start the engine, allow it to reach normal operating temperature, check for leaks, and top off as necessary.

2004–06 TL and 2005–06 RL

1. Before servicing the vehicle, refer to the precautions section.

2. Disconnect the negative battery cable.

3. Remove the timing belt.

4. Remove the timing belt adjuster.

5. Remove the water pump by removing the five bolts.

6. Inspect and clean the O-ring groove and the mating surface of the engine block.

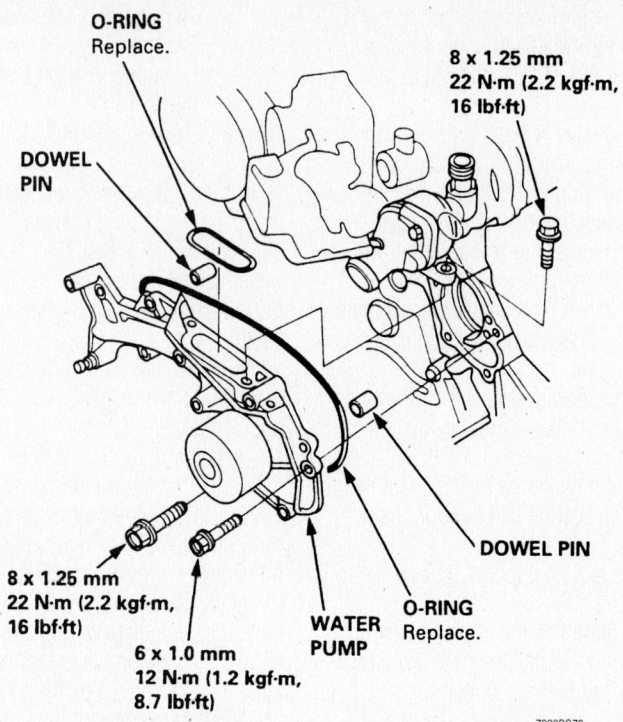

O-RING
Replace.

8 x 1.25 mm
22 N·m (2.2 kgf·m, 16 lbf·ft)

DOWEL PIN

DOWEL PIN

8 x 1.25 mm
22 N·m (2.2 kgf·m, 16 lbf·ft)

6 x 1.0 mm
12 N·m (1.2 kgf·m, 8.7 lbf·ft)

WATER PUMP

O-RING
Replace.

7923BG79

Water pump mounting and bolt torque specifications—2002–04 RL models

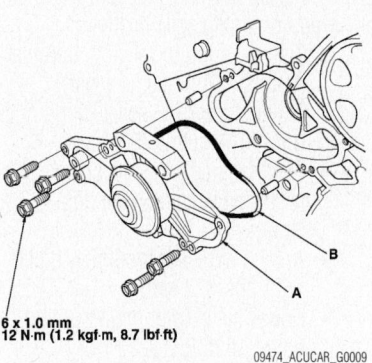

6 x 1.0 mm
12 N·m (1.2 kgf·m, 8.7 lbf·ft)

B

A

09474_ACUCAR_G0009

Water pump mounting—2004–06 TL and 2005–06 RL models

To install:

7. Install the water pump with a new O-ring in the reverse order of removal. Tighten the bolts to 8.7 ft. lbs. (12 Nm).

8. Install the timing belt adjuster.

9. Install the timing belt.

10. Refill the radiator with engine coolant, then bleed the air from the cooling system.

RSX Models

1. Before servicing the vehicle, refer to the precautions section.

2. Disconnect the negative battery cable.

3. Drain the engine coolant.

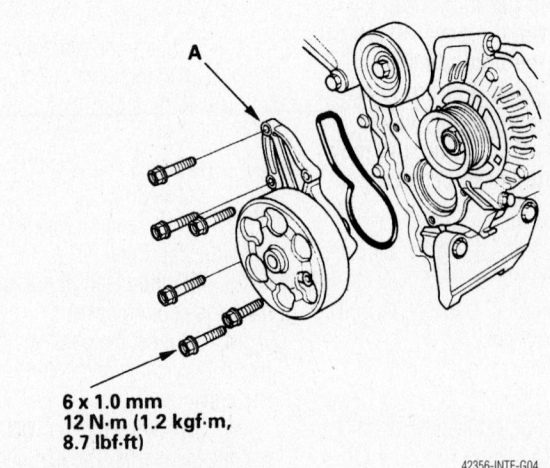

A

6 x 1.0 mm
12 N·m (1.2 kgf·m, 8.7 lbf·ft)

42356-INTE-G04

Water pump mounting—2.0L (K20A3) engine

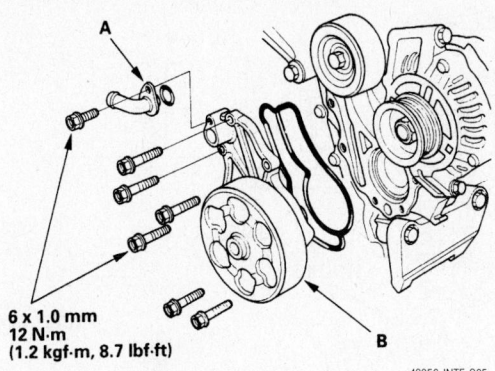

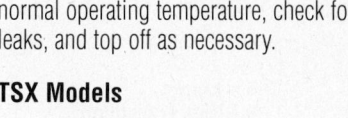

6 x 1.0 mm
12 N·m
(1.2 kgf·m, 8.7 lbf·ft)

42356-INTE-G05

Water pump mounting—2.0L (K20A2/K20Z1) engine

4. Remove or disconnect the following:
   • Drive belt
   • Crankshaft pulley
   • For the K20A3 (base) engine, 6 bolts securing the water pump
   • For the K20A2 and K20Z1 (Type–S) engine, oil cooler joint pipe and the 7 bolts securing the water pump
   • Water pump
5. Clean and inspect the O–ring groove and mating surfaces.

To install:
6. Install or connect the following:
   • Water pump with a new O–ring. Tighten the mounting bolts to 104 inch lbs. (12 Nm).
   • Crankshaft pulley
   • Drive belt
7. Fill the engine with coolant and bleed the air from the cooling system.
8. Connect the negative battery cable and enter the radio security code.
9. Run the engine and check for cooling system leaks.

2002–03 TL Models

➡**Perform this service operation with the engine cold.**

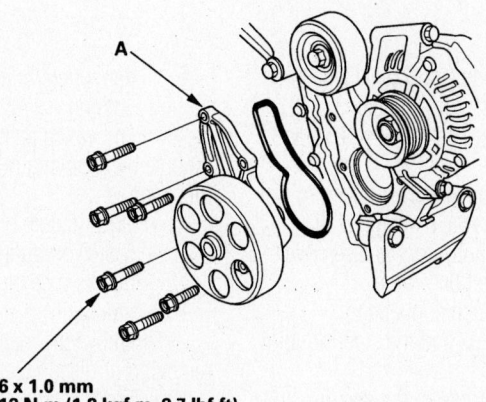

A

6 x 1.0 mm
12 N·m (1.2 kgf·m, 8.7 lbf·ft)

42356-ACCO-G02

Water pump mounting—TSX models

1. Before servicing the vehicle, refer to the precautions section.
2. Disconnect the negative battery cable.
3. Remove the front splash panel.
4. Drain the cooling system.
5. Remove the timing belt. Inspect the timing belt for any signs of damage or oil and coolant contamination. Replace the timing belt if there is any doubt about its condition.
6. Remove the timing belt tensioner.
7. Remove the water pump bolts. Then, remove the water pump and sprocket assembly from the engine block.

To install:
8. Install the water pump with a new O–ring. Use new bolts and tighten the 6mm mounting bolts evenly to 104 inch lbs. (12 Nm) and the 8mm bolts to 16 ft. lbs. (22 Nm).
9. If removed, install the timing belt rear cover and camshaft pulley.
10. If removed, install the timing belt tensioner.
11. Install or connect the following:
    • Timing belt and timing belt covers
    • Accessory drive belts
12. Close the cooling system drain plug. Refill and bleed the cooling system.
13. Connect the negative battery cable.

14. Start the engine, allow it to reach normal operating temperature, check for leaks, and top off as necessary.

TSX Models

1. Before servicing the vehicle, refer to the precautions section.
2. Drain the cooling system.
3. Remove or disconnect the following:
   • Crankshaft pulley
   • 6 water pump mounting bolts, then the pump and O-ring seal
4. Clean the seal groove and mating surfaces.
5. Install or connect the following:
   • Water pump, with a new seal. Tighten the bolts to 104 inch lbs. (12 Nm).
   • Crankshaft pulley
6. Fill the cooling system.
7. Start the engine and check for leaks.

Heater Core

REMOVAL AND INSTALLATION

CL Model

➡**Be sure to acquire the anti-theft code for the radio; then, write down the frequencies for the preset buttons.**

1. Before servicing the vehicle, refer to the precautions section.
2. Disconnect the negative battery cable.

✳✳ CAUTION

After disconnecting the negative battery cable, wait for at least 3 minutes for the SRS module to deplete its energy.

3. Drain the cooling system into a clean container for reuse.
4. Remove or disconnect the following:
   • Heater hoses from the heater core
   • 2 heater housing-to-cowl nuts, located in the engine compartment
5. Place the front wheel in the straight-ahead position.
6. Remove the SRS module and the steering wheel by removing or disconnecting the following:
   • Access panel-to-steering wheel screws and the panel
   • SRS module electrical connector
   • SRS module-to-steering wheel covers from both sides of the steering wheel
   • Both SRS module-to-steering wheel bolts, using a T30 Torx® bit

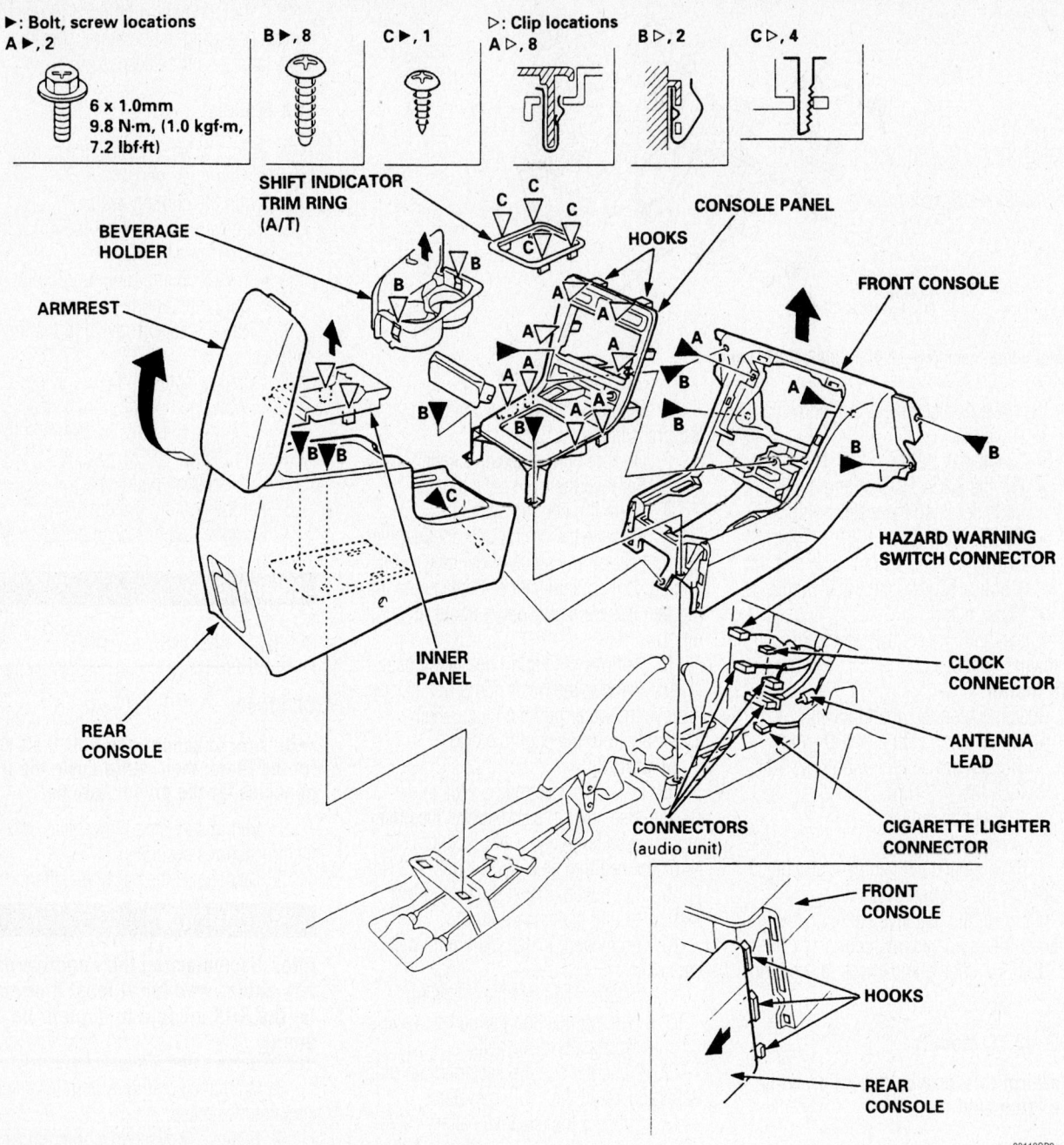

►: Bolt, screw locations
A ►, 2

6 x 1.0mm
9.8 N·m, (1.0 kgf·m,
7.2 lbf·ft)

B ►, 8

C ►, 1

▷: Clip locations
A ▷, 8

B ▷, 2

C ▷, 4

View of the front and rear consoles and related components—3.2CL

- SRS module from the steering wheel
- Electrical connectors from the steering wheel
- Steering wheel from the steering column

7. Remove the steering column by removing or disconnecting the following:
- Coin pocket, the lower instrument panel-to-dash screw and the lower instrument panel from the driver's side
- Electrical connector and the air hose at the knee bolster; then,

remove the knee bolster-to-dash bolts and the knee bolster
- Steering column cover screws and the covers
- Combination switch-to-steering column screws, disconnect the electrical connectors and remove the combination switch
- Ignition switch connectors
- Clamps, clips and the steering joint cover
- Steering joint from the steering column shaft
- Steering column-to-instrument

panel nuts/bolts and the steering column

8. Remove the passenger's side SRS module by removing or disconnecting the following:
- SRS module electrical connector
- 5 SRS module-to-instrument panel nuts and carefully remove the SRS module

9. Remove the instrument panel by removing or disconnecting the following:
- Front and rear consoles
- Cruise control master switch and the panel brightness controller;

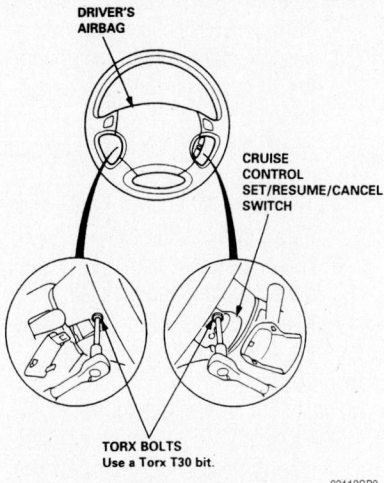

Removing the driver's side SRS module-to-steering wheel bolts—3.2CL

then, disconnect the electrical connectors
- Instrument cluster-to-instrument panel screws and pry out the instrument cluster and disconnect the electrical connectors
- Glove box-to-damper screw
- 2 glove box-to-dash screws and the glove box
- Audio unit-to-instrument panel fasteners and the audio unit
- 3 passenger's dashboard panel screws (located on the passengers side); then, carefully pry out the dashboard panel and the climate control unit as an assembly
- Side defogger trim from both sides of the instrument panel
- Instrument panel electrical connectors, the recirculation control motor connector and the resistor connector
- Instrument panel-to-chassis bolts and the instrument panel
- Steering hanger beam-to-chassis nuts, bolts and the beam

10. Discharge and recover the air conditioning system refrigerant.

11. Remove the evaporator housing by removing or disconnecting the following:
- Refrigerant lines from the evaporator core. Discard the O-rings and plug the openings to prevent contamination.
- Thermostat electrical connector at the evaporator housing
- Wiring harness clips from the evaporator housing
- Evaporator housing-to-chassis screws, the nut and bolt
- Drain hose
- Evaporator housing

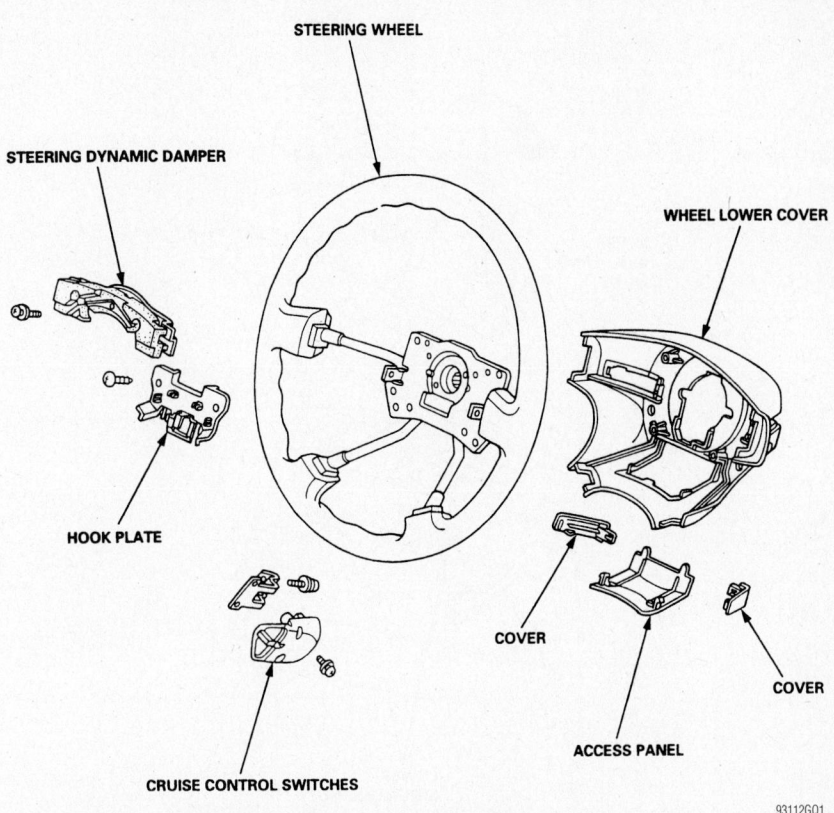

Exploded view of the steering wheel and related components—3.2CL

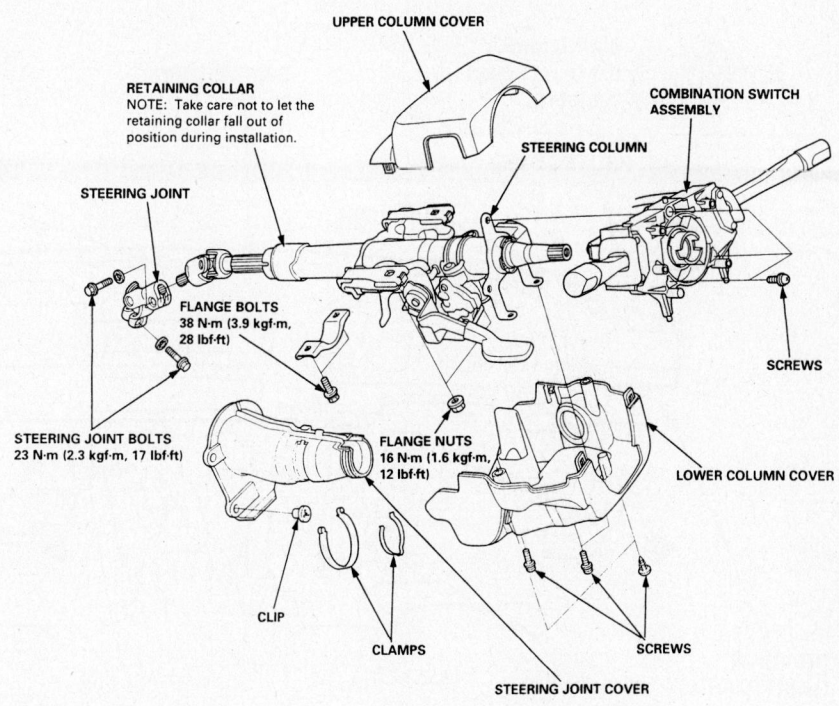

Exploded view of the steering wheel column and related components—3.2CL

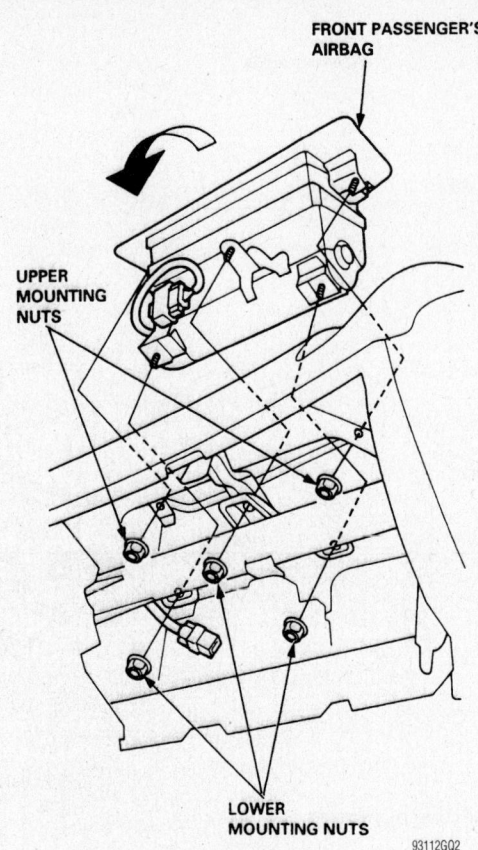

Exploded view of the passenger's side SRS module—3.2CL

12. Remove or disconnect the following:
- Wiring harness clip; then, disconnect the mode control motor and the air mix control motor connectors
- Heater housing-to-chassis bolt and the heater housing
- Vent/defroster duct-to-heater housing screws and the duct
- Heater core pipe clamp-to-heater housing screw and the clamp
- Heater core clamp-to-heater housing screw and the heater core

To install:

13. Install or connect the following:
- Heater core clamp and the heater core-to-heater housing screw
- Heater core pipe clamp and the clamp-to-heater housing screw
- Vent/defroster duct and the duct-to-heater housing screws
- Heater housing and the heater housing-to-chassis bolt
- Wiring harness clip; then, connect the mode control motor and the air mix control motor connectors

14. Install the evaporator housing by installing or connecting the following:
- Evaporator housing
- Drain hose

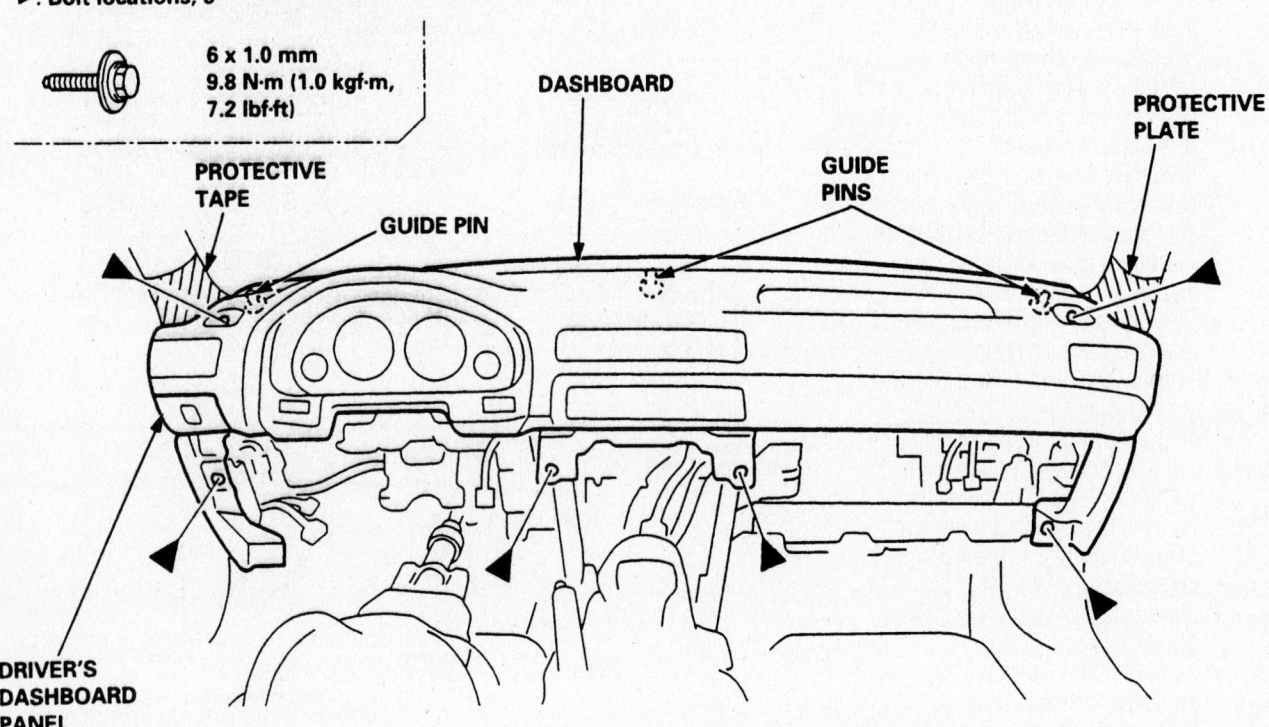

View of the dashboard bolt locations—3.2CL

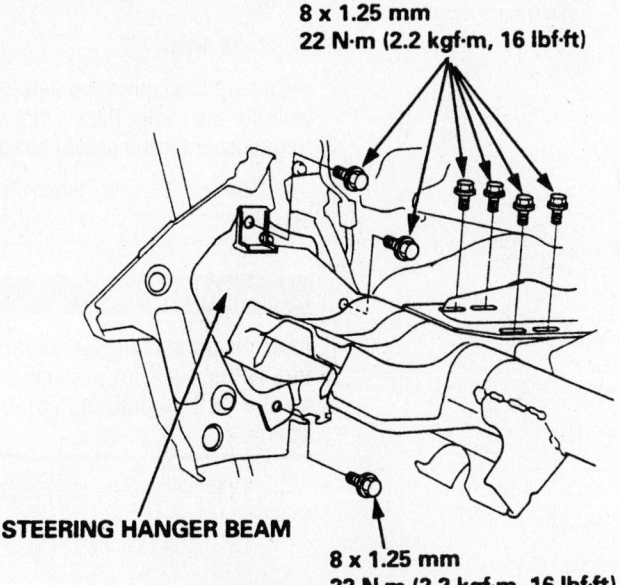

8 x 1.25 mm
22 N·m (2.2 kgf·m, 16 lbf·ft)

STEERING HANGER BEAM

8 x 1.25 mm
22 N·m (2.2 kgf·m, 16 lbf·ft)

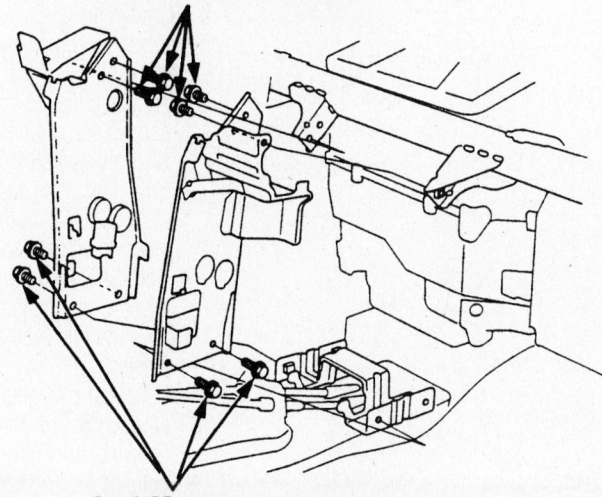

8 x 1.25 mm
22 N·m (2.2 kgf·m, 16 lbf·ft)

8 x 1.25 mm
22 N·m (2.2 kgf·m, 16 lbf·ft)

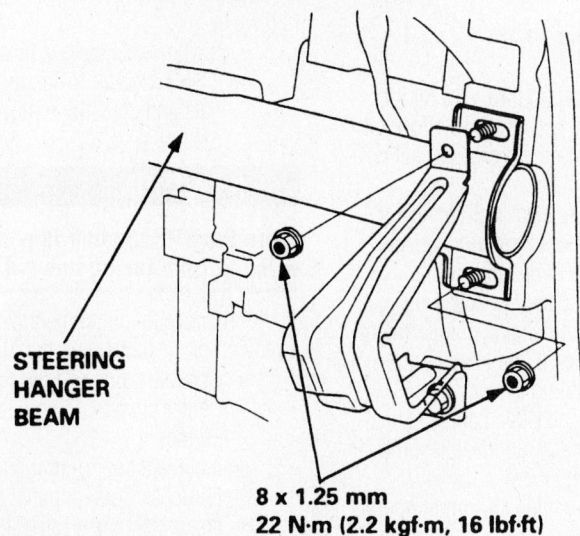

**STEERING
HANGER
BEAM**

8 x 1.25 mm
22 N·m (2.2 kgf·m, 16 lbf·ft)

93112GQ5

View of the steering hanger beam and related components—3.2CL

- Evaporator housing-to-chassis screws, the nut and bolt
- Wiring harness clips to the evaporator housing
- Thermostat electrical connector, at the evaporator housing
- Refrigerant lines to the evaporator core, make sure to use new O–rings
- Steering hanger beam and the beam-to-chassis nuts and bolts

15. Install the instrument panel by installing or connecting the following:

- Instrument panel and the instrument panel-to-chassis bolts
- Instrument panel electrical connectors, the recirculation control motor connector and the resistor connector
- Side defogger trim at both sides of the instrument panel
- Dashboard panel and the climate control unit as an assembly; then, install the 3 passenger's dashboard panel screws at the passenger's side
- Audio unit and the audio unit-to-instrument panel fasteners
- Glove box and the 2 glove box-to-dash screws
- Glove box-to-damper screw
- Instrument cluster and the instrument cluster-to-instrument panel screws; then, connect the electrical connectors
- Cruise control master switch and the panel brightness controller; then, connect the electrical connectors
- Front and rear consoles

16. Install the passenger's side SRS module by installing or connecting the following:

- SRS module and the 5 SRS module-to-instrument panel nuts and torque to 84 inch lbs. (9.8 Nm)
- SRS module electrical connector

17. Install the steering column by installing or connecting the following:

- Steering column and the steering column-to-instrument panel nuts/bolts
- Steering joint to the steering column shaft
- Clamps, clips and the steering joint cover
- Ignition switch connectors
- Combination switch, connect the electrical connectors and install the combination switch-to-steering column screws
- Steering column covers and the cover screws

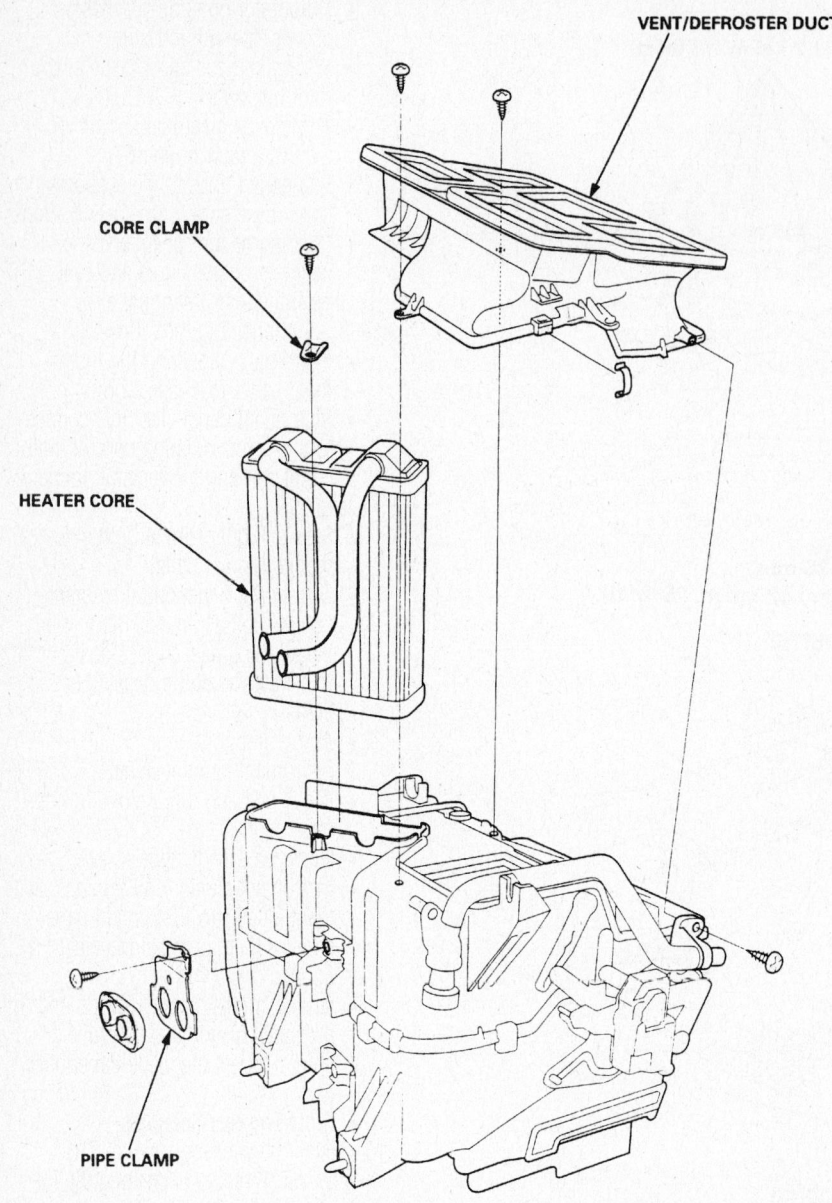

VENT/DEFROSTER DUCT

CORE CLAMP

HEATER CORE

PIPE CLAMP

93112GQ6

Exploded view of the heater core, heater housing and related components—3.2CL

- Knee bolster and the knee bolster-to-dash bolts; then, connect the electrical connector and the air hose.
- Lower instrument panel, the lower instrument panel-to-dash screw and the coin pocket

18. Install the SRS module and the steering wheel by installing or connecting the following:
- Steering wheel to the steering column and torque the nut to 28 ft. lbs. (38 Nm)
- Electrical connectors to the steering wheel
- SRS module to the steering wheel
- Both SRS module-to-steering

wheel bolts and torque to 84 inch lbs. (9.8 Nm) using a T30 Torx bit
- SRS module-to-steering wheel covers
- SRS module electrical connector
- Access panel-to-steering wheel screws and the panel.
- 2 heater housing-to-cowl nuts located in the engine compartment
- Heater hoses to the heater core

19. Refill the cooling system.
20. Connect the negative battery cable.
21. Evacuate, charge and leak test the air conditioning system.
22. Operate the engine to normal operating temperatures; then, check the climate control operation and check for leaks.

3.5RL

2002–04 MODELS

➡Be sure to acquire the anti–theft code for the radio; then, write down the frequencies for the preset buttons.

1. Before servicing the vehicle, refer to the precautions section.
2. Disconnect the negative battery cable.

✳✳ CAUTION

After disconnecting the negative battery cable, wait for at least 3 minutes for the SRS module to deplete its energy.

3. Drain the cooling system into a clean container for reuse.
4. Disconnect the heater hoses from the heater core.
5. Place the front wheel in the straight–ahead position.
6. Remove the SRS module and the steering wheel by removing or disconnecting the following:
- Access panel-to-steering wheel screws and the panel
- SRS module electrical connector
- SRS module-to-steering wheel covers, located at both sides of the steering wheel
- Both SRS module-to-steering wheel bolts using a T30 Torx bit
- SRS module from the steering wheel
- Electrical connectors from the steering wheel
- Steering wheel from the steering column

7. Remove the passenger's side SRS module by removing or disconnecting the following:
- SRS module electrical connector
- 2 SRS module-to-instrument panel nuts and carefully remove the SRS module

✳✳ CAUTION

Store the SRS module in a safe place with the front facing upward.

8. Remove the instrument panel by removing or disconnecting the following:
- Console panel and the rear console
- Center air vent using a suitable prytool
- 4 climate control unit/audio assembly-to-instrument panel bolts; then, disconnect the electrical connectors and remove the climate control unit/audio assembly

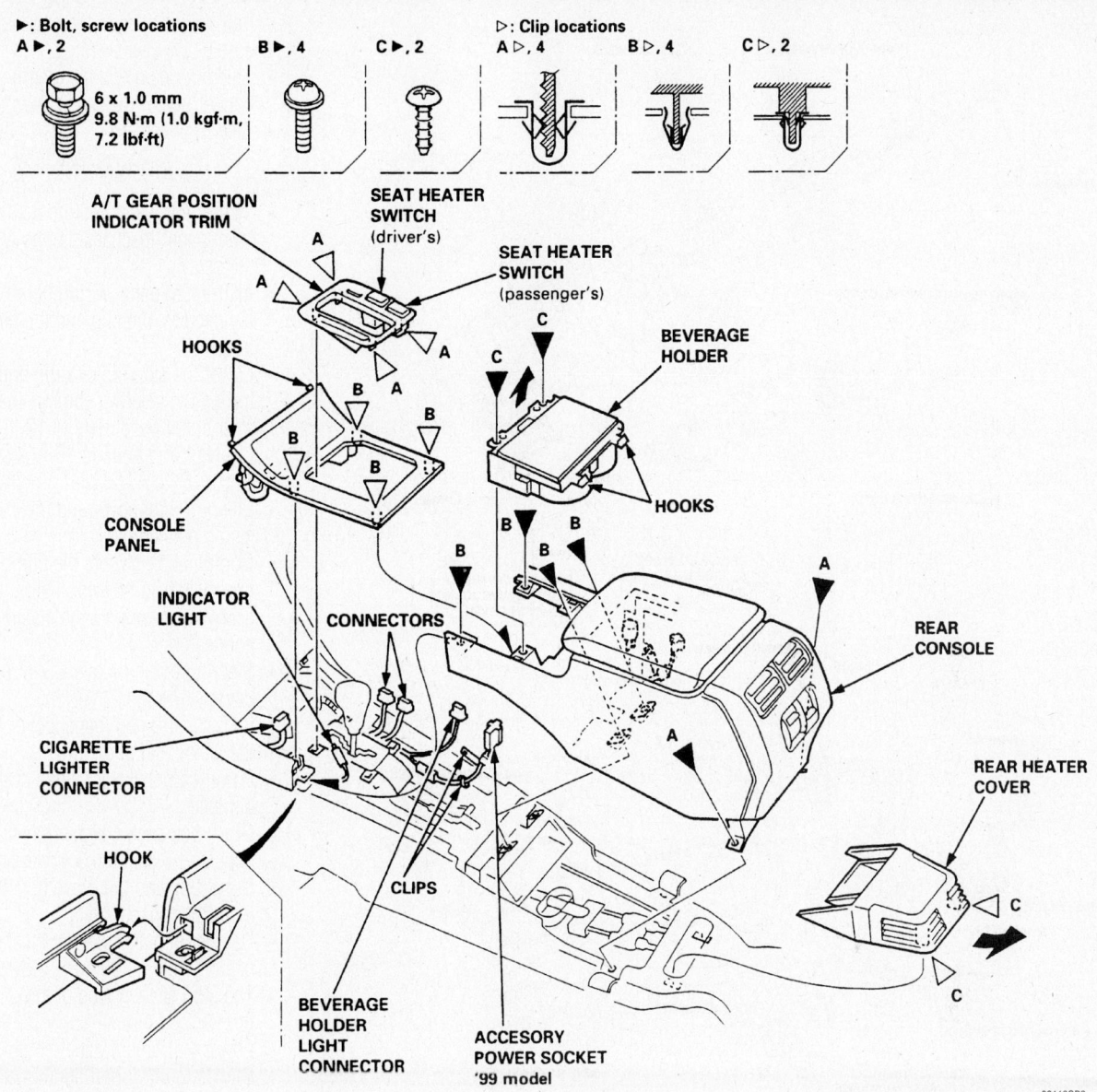

▶: Bolt, screw locations

A ▶, 2   B ▶, 4   C ▶, 2

6 x 1.0 mm
9.8 N·m (1.0 kgf·m,
7.2 lbf·ft)

▷: Clip locations

A ▷, 4   B ▷, 4   C ▷, 2

A/T GEAR POSITION
INDICATOR TRIM

SEAT HEATER
SWITCH
(driver's)

SEAT HEATER
SWITCH
(passenger's)

BEVERAGE
HOLDER

HOOKS

CONSOLE
PANEL

HOOKS

INDICATOR
LIGHT

CONNECTORS

REAR
CONSOLE

CIGARETTE
LIGHTER
CONNECTOR

REAR HEATER
COVER

HOOK

CLIPS

BEVERAGE
HOLDER
LIGHT
CONNECTOR

ACCESORY
POWER SOCKET
'99 model

93112GR2

Exploded view of the console panel, rear console and related components—2002–04 3.5RL

- Lower instrument panel cover at the passenger's side
- The stop located at each side of the glove box
- Damper clip and the electrical connector while holding the glove box
- Glove box-to-instrument panel bolts and the glove box
- 3 glove box back cover screws, disconnect the clips and remove the cover
- Lower carpet at both sides of the console
- Lower instrument panel cover and the kick panel at the driver's side
- Steering column cover screws and the covers
- Combination switch-to-steering column screws, the combination

switch and disconnect the electrical connectors
- Steering joint cover clamp, screws and the cover
- Steering column-to-instrument panel nuts and bolts; then, lower the steering column
- Instrument panel wiring harness connectors, the clip and the air hose
- Instrument panel-to-chassis bolts, the screws and the instrument panel
- Steering beam-to-chassis bolts at the left side
- Steering beam-to-chassis nuts, the bolt and the steering hanger beam at the right side

9. Discharge and recover the air conditioning system refrigerant.

10. Remove the evaporator/blower housing by removing or disconnecting the following:
- Refrigerant lines from the evaporator core
- Electrical connectors from the evaporator/blower housing
- Evaporator/blower housing-to-chassis fasteners and the evaporator/blower housing
- Clips and the floor heater duct

11. At the left side of the heater housing, remove the cruise control unit -to-heater housing bolt and the cruise control unit.

12. Remove or disconnect the following:
- Mode control motor and the air mix control motor connectors
- Wiring harness clips and the wiring harness

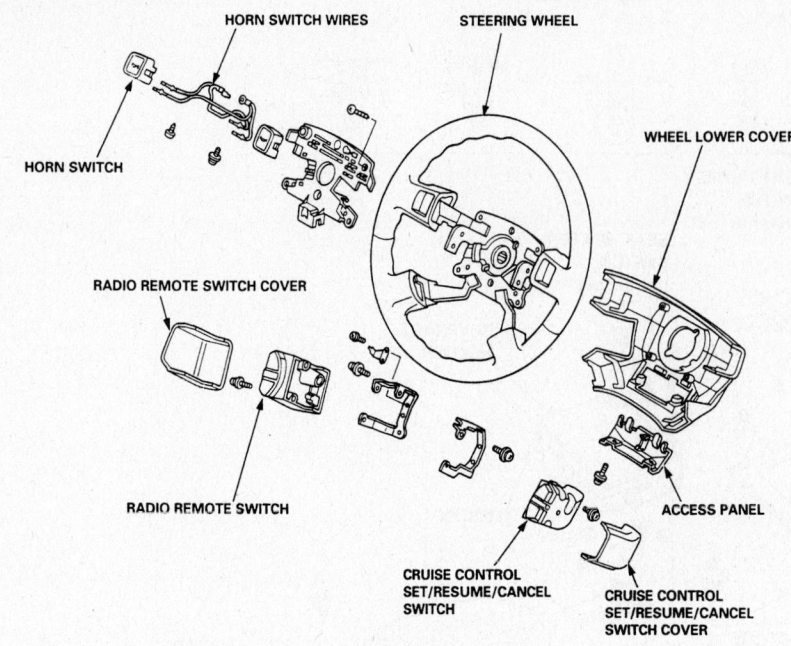

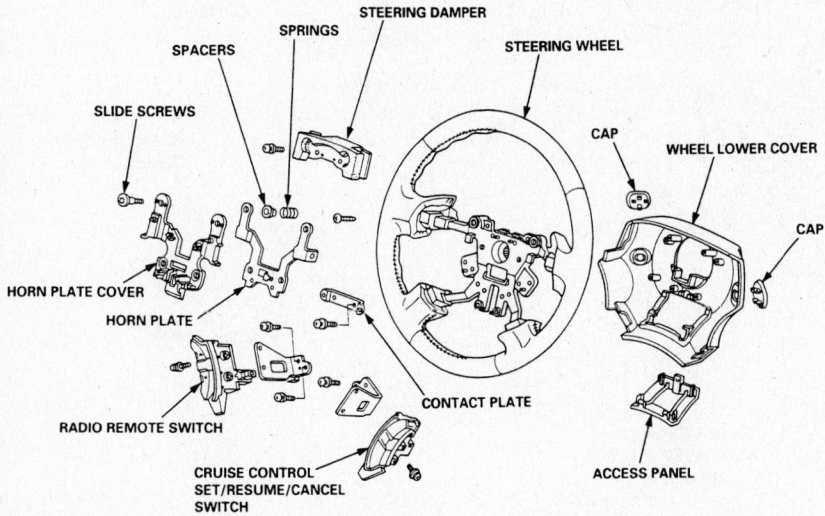

Exploded view of the SRS module, the steering wheel and related components—2002–04

- Heater housing-to-chassis nuts, the bolt and the heater housing
- Vent/defroster duct-to-heater housing screws and the duct
- Heater core-to-heater housing pipe clamp screws and the clamps
- Heater core-to-heater housing clamp screws and the clamp
- Heater core from the heater housing

To install:

13. Install or connect the following:
- Heater core to the heater housing
- Heater core-to-heater housing clamp and the clamp screws
- Heater core-to-heater housing pipe clamps and the clamp screws
- Vent/defroster duct and the duct-to-heater housing screws

- Heater housing and the heater housing-to-chassis nuts and bolt
- Wiring harness and the wiring harness clips
- Mode control motor and the air mix control motor connectors
- Cruise control unit and the cruise control unit-to-heater housing bolt at the left side of the heater housing
- Floor heater duct and the clips

14. Install the evaporator/blower housing by installing or connecting the following:
- Evaporator/blower housing and the evaporator/blower housing-to-chassis fasteners
- Electrical connectors to the evaporator/blower housing

- Refrigerant lines to the evaporator core using new O–rings
- Steering beam and the steering hanger beam-to-chassis nuts
- Steering beam-to-chassis bolts

15. Install the instrument panel by installing or connecting the following:
- Instrument panel and the instrument panel-to-chassis bolts and the screws
- Instrument panel wiring harness connectors, the clip and the air hose
- Lower the steering column and torque the steering column-to-instrument panel nuts to 12 ft. lbs. (16 Nm) and bolts to 16 ft. lbs. (22 Nm)
- Steering joint cover and the cover clamp and screws
- Combination switch, the combination switch-to-steering column screws and connect the electrical connectors
- Steering column covers and the cover screws
- Lower instrument panel cover and the kick panel at the driver's side
- Lower carpet at both sides of the console
- Glove box back cover, connect the clips and install the 3 cover screws
- Glove box and the glove box-to-instrument panel bolts
- Damper clip and the electrical connector while holding the glove box
- The stop at each side of the glove box
- Lower instrument panel cover
- Climate control unit/audio assembly; then, connect the electrical connectors and install the 4 climate control unit/audio assembly-to-instrument panel bolts
- Center air vent.
- Console panel and the rear console

16. Install the passenger's side SRS module by installing or connecting the following:
- SRS module and the 2 SRS module-to-instrument panel nuts; then, torque the nuts to 84 inch lbs. (9.8 Nm)
- SRS module electrical connector

17. Install the SRS module and the steering wheel by installing or connecting the following:
- Steering wheel to the steering column and torque the nut to 36 ft. lbs. (49 Nm)
- Electrical connectors to the steering wheel

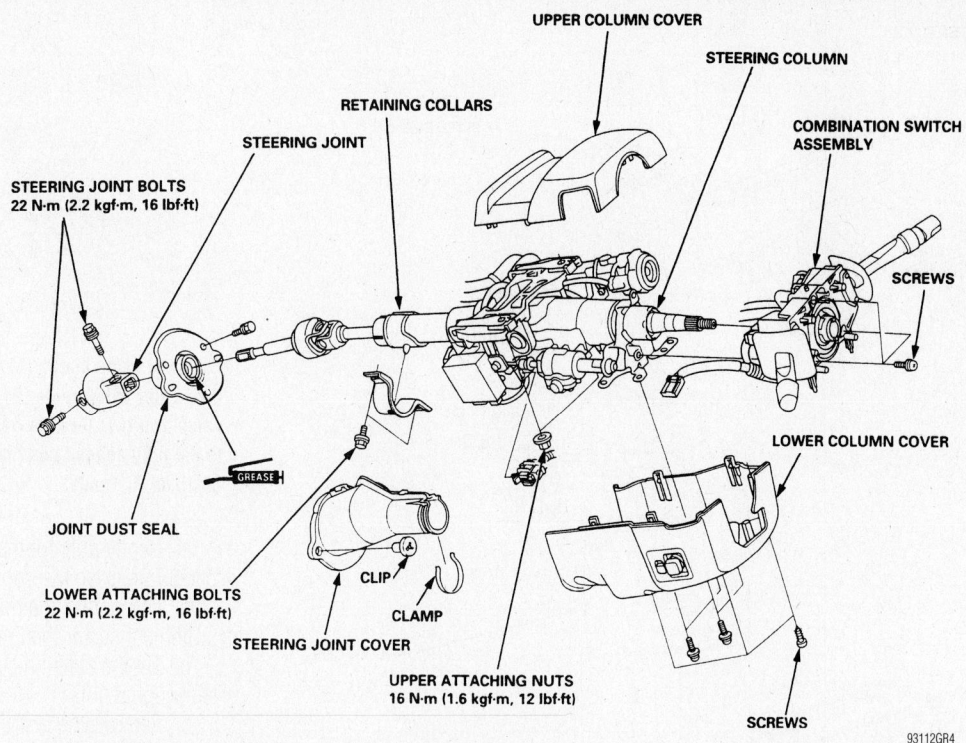

UPPER COLUMN COVER

STEERING COLUMN

RETAINING COLLARS

COMBINATION SWITCH
ASSEMBLY

STEERING JOINT

STEERING JOINT BOLTS
22 N·m (2.2 kgf·m, 16 lbf·ft)

SCREWS

JOINT DUST SEAL

LOWER COLUMN COVER

LOWER ATTACHING BOLTS
22 N·m (2.2 kgf·m, 16 lbf·ft)

CLIP

CLAMP

STEERING JOINT COVER

UPPER ATTACHING NUTS
16 N·m (1.6 kgf·m, 12 lbf·ft)

SCREWS

93112GR4

Exploded view of the steering column and related components—2002–04 3.5RL

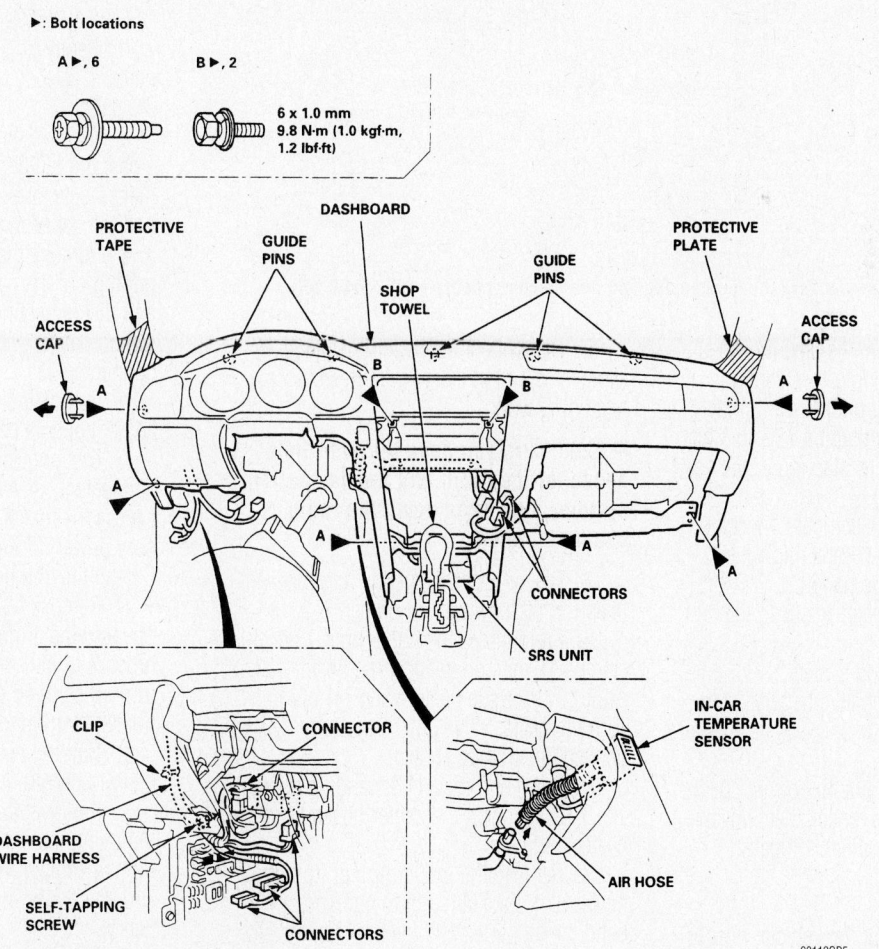

▶: Bolt locations

A ▶, 6   B ▶, 2

6 x 1.0 mm
9.8 N·m (1.0 kgf·m,
1.2 lbf·ft)

PROTECTIVE
TAPE

DASHBOARD

PROTECTIVE
PLATE

GUIDE
PINS

SHOP
TOWEL

GUIDE
PINS

ACCESS
CAP

A

B

B

A

ACCESS
CAP

A

A

A

CONNECTORS

SRS UNIT

CLIP

CONNECTOR

IN-CAR
TEMPERATURE
SENSOR

DASHBOARD
WIRE HARNESS

SELF-TAPPING
SCREW

CONNECTORS

AIR HOSE

93112GR5

View of the instrument panel bolt locations—2002–04 3.5RL

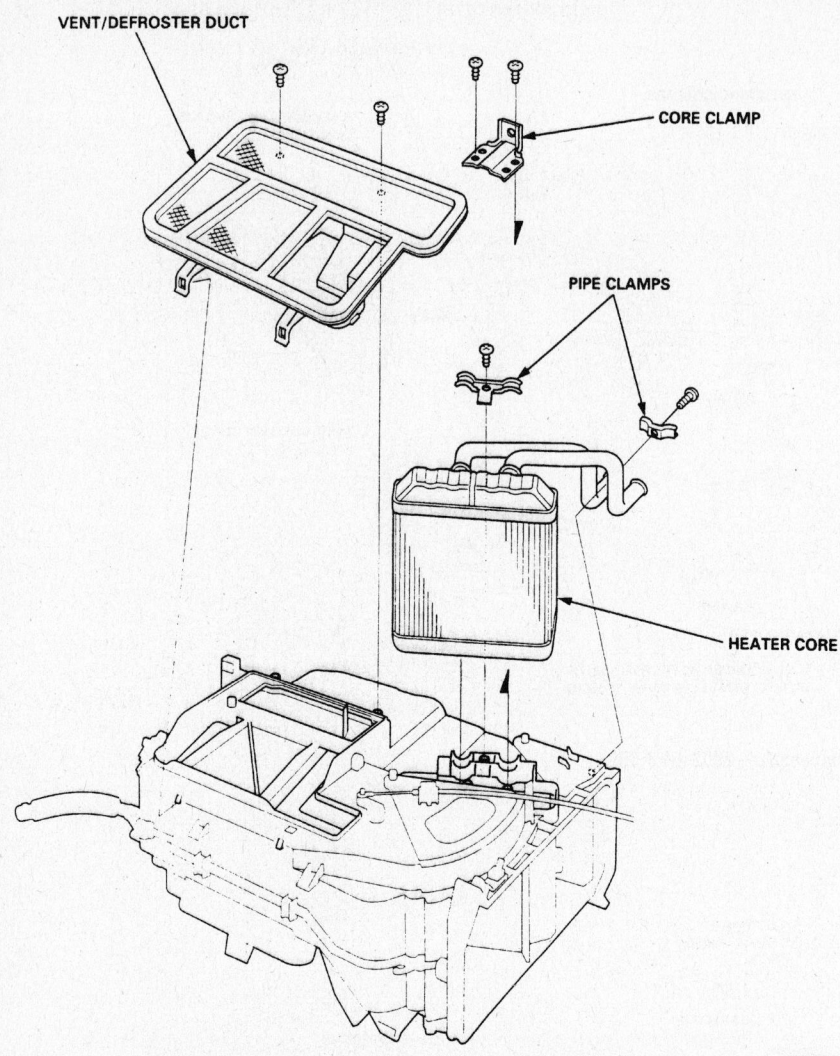

VENT/DEFROSTER DUCT

CORE CLAMP

PIPE CLAMPS

HEATER CORE

93112GR6

Exploded view of the heater core, heater housing and related components—2002–04 3.5RL

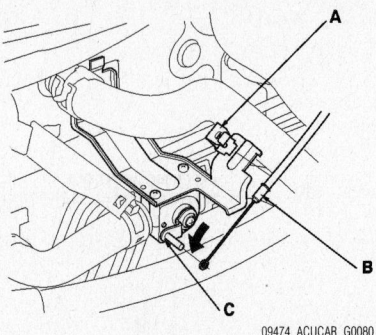

09474_ACUCAR_G0080

Open the cable clamp (A), then disconnect the heater valve cable (B) from the heater valve arm (C). Turn the heater valve arm to the fully opened position as shown— 2004–06 RL model

- SRS module to the steering wheel
- Both SRS module-to-steering wheel bolts and torque to 84 inch lbs. (9.8 Nm) using a T30 Torx bit
- SRS module-to-steering wheel covers
- SRS module electrical connector
- Access panel-to-steering wheel screws and the panel

18. Connect the heater hoses to the heater core.

19. Refill the cooling system.

20. Connect the negative battery cable.

21. Evacuate, charge and leak test the air conditioning system.

22. Operate the engine to normal operating temperatures; then, check the climate control operation and check for leaks.

2005–06 MODELS

1. Before servicing the vehicle, refer to the precautions section.

2. Disconnect the battery negative cable, and wait at least 3 minutes before beginning work.

➡**Make sure you have the anti-theft codes for the radio and the navigation system, then write down the radio channel presets.**

3. Disconnect the suction and receiver lines from the evaporator core

4. From under the hood, open the cable clamp (A), then disconnect the heater valve cable (B) from the heater valve arm (C). Turn the heater valve arm to the fully opened position as shown.

5. Drain and recycle the engine coolant.

6. Disconnect the heater hoses from the heater unit.

7. Remove the mounting nut from the heater unit. Take care not to damage or bend the fuel lines and the brake lines, etc.

8. Remove the dashboard as follows:

a. Gently pull out the driver's switch panel to release the hooks and detach the clips. Disconnect the power mirror switch connector then remove the panel.

b. Tilt the steering column down, and telescope it out.

c. Remove the instrument fascia by pulling out the bottom to release the clips gently and disconnecting the illumination switch connector and CMS OFF/VSA OFF/AFS OFF switch connector

d. Gently pull out the driver's vent panel to release the clips, then remove the panel.

e. Remove the center trim on both sides, the utility pocket housing, instrument fascia, audio-HVAC-control module and the passenger's vent panel.

f. Remove the glove box by gently pulling out the panel to release the clips and disconnecting the hazard warning switch/passenger airbag cutoff indicator connector.

g. Detach the clips by pulling the center console panel up.

h. Disconnect the seat heater switch connectors, or AVS switch connectors, accessory socket connector and accessory power socket light connector. Detach the harness clip.

i. Remove the center console trim by opening the armrest, open the beverage holder lid, pull the front portion of the trim up by hand to detach the clips and hooks, then remove it.

j. Pull the beverage holder up by hand to detach the clips then remove it

k. Remove the accessory trim by removing the console box mat, unfastening the screws and removing the trim and disconnecting the power socket connector.

l. Detach the clips by pulling the driver's front console cover and passenger's front console cover out, and pull them backward to release their pins from the heater unit, then remove them.

m. Remove the center console by first disconnecting the sub harness connector, and detach it from the vent duct. Remove the bolts and pull the console backwards to remove it

n. Adjust the steering column upward.

o. Remove the driver's dashboard under cover.

p. Release the clip and pull out the driver's dashboard lower cover to detach the clips and hooks, then remove the cover.

q. Remove the bolts securing the front of the center console, then pull out both sides as needed.

r. Remove the center trim.

s. Gently pull out along the top of the utility pocket housing to release the clips, then lower the front side down to release the pins from the select lever bracket.

t. Tilt the steering column down, and telescope it out.

u. Remove the instrument fascia.

v. Remove the self tapping screws, then pull out the audio unit.

w. Disconnect the connectors and remove the audio unit.

x. Remove the self-tapping screws and the climate control unit.

y. Discharge the static electricity (which accumulated on you when you removed the climate control unit) by touching the door striker or other body parts

z. Remove the dashboard upper visor.

aa. Remove the three screws, then remove the display unit. The forward screw can be seen through the windshield or with a mirror.

bb. Remove the front door panel.

cc. Pull the speaker straight out, just enough to release the upper clips. Then lift the speaker straight up to release the lower clips. If you pull the speaker out too far from the door, you will damage the lower clips.

dd. Remove the sunlight sensor from the dashboard, then disconnect the 2P connector. Be careful not to damage the sensor and the dashboard.

ee. Remove the in-car temperature sensor from the driver's dashboard lower cover, then disconnect the 4P connector.

ff. Remove the A-pillar trim from both sides.

gg. Remove the access panel from the steering wheel, then disconnect the

driver's airbag 4P connector from the cable reel.

hh. Remove the two Torx bolts using a Torx T30 bit.

ii. Disconnect the horn switch connector, then remove the driver's airbag.

jj. Disconnect the cable reel connector.

kk. Loosen the steering wheel nut.

ll. Install a steering wheel puller on the steering wheel. Free the steering wheel from the steering column shaft by turning the pressure bolt of the puller.

➡**These items when removing the steering wheel:**

- Do not tap on the steering wheel or the steering column shaft when removing the steering wheel.
- If you thread the puller bolts into the wheel hub more than five threads, the bolts will hit the cable reel and damage it. To prevent this, install a pair of jam nuts five threads up on each puller bolt.

mm. Remove the steering wheel puller, then remove the steering wheel nut and steering wheel from the steering column

nn. Remove the steering joint cover.

oo. Remove the upper column cover and disconnect the connectors from the tilt/telescopic switch on the lower column cover.

pp. Disconnect the wire harness connectors from the combination switch assembly.

qq. Remove the combination switch assembly from the steering column shaft by removing the screws.

rr. Disconnect the connectors from the ignition switch, and release the wire harness clips from the steering column.

ss. Disconnect the connectors from the tilt/telescopic motors.

tt. Disconnect the connectors from the tilt/telescopic control unit.

uu. Remove the steering joint bolts, then disconnect the steering joint from the pinion shaft and the steering column shaft.

vv. Remove the steering column by removing the attaching nuts and bolts on the passenger side

ww. Remove the screw, then remove the passenger's joint duct.

xx. Disconnect and detach the passenger's airbag connector from the steering hanger beam and detach the steering hanger beam wire harness clip. Remove the passenger's airbag mounting nuts.

yy. Disconnect the RF unit connector and detach the harness clip from the unit bracket. Tie a string to the sunlight sensor harness connector then pull the connector out through the hole in the dashboard. Remove the bolts, then remove the RF unit.

zz. From the front of the dashboard, remove the screws and bolts securing the dashboard/steering hanger beam.

aa. Lift up on the dashboard to release it from the guide pins then carefully remove the dashboard through the front door opening.

➡**Lay the dashboard on its front or back. Do not rest it on the lower console opening or you may damage it.**

9. Disconnect the connectors from the power tilt/telescopic steering control unit, the adaptive front lighting control unit, the power steering control unit, and the daytime running lights control unit.

10. Remove the relay and the power tilt/telescopic steering control unit, then remove the self-tapping screws and the bracket.

11. Disconnect the connectors from the driver's cool vent control motor, the driver's air mix control motor, the mode control motor, and the evaporator temperature sensor, then remove the wire harness clips.

12. Disconnect the connectors from the blower motor, the power transistor, the control motor relay, the throttle actuator control module sub harness, and dashboard wire harnesses.

13. Remove the connector clip and the wire harness clips.

14. Disconnect the connectors from the recirculation control motor, the passenger's cool vent control motor, the passenger's air mix control motor, and the rear vent control motor, and then remove the wire harness clip and the wire harness.

15. Remove the clips, the mounting nuts, and the blower-heater unit.

16. Remove the self-tapping screws and the passenger's heater duct and the expansion valve cover, then remove the bolts and the expansion valve.

17. Remove the self-tapping screws the joint duct and the heater core cover.

18. Remove the self-tapping screws the heater pipe brackets and the grommets and carefully pull out the heater core.

To install:

19. Installation is the reverse of removal. refer to the illustrations accompanying the removal procedure for component locations and torque specifications.

RSX Models

➡ **Be sure to acquire the anti–theft code for the radio; then, write down the frequencies for the preset buttons.**

1. Before servicing the vehicle, refer to the precautions section.
2. Disconnect the negative battery cable.

✳✳ CAUTION

After disconnecting the negative battery cable, wait for at least 3 minutes for the SRS module to deplete its energy.

3. Drain the cooling system into a clean container for reuse.
4. Remove or disconnect the following:
   - Heater hoscs from the heater core
   - Heater housing-to-cowl nut, located in the engine compartment
5. Remove the instrument panel by removing or disconnecting the following:
   - Front seats
   - Front and rear consoles
   - Lower dashboard-to-instrument panel screws and the lower dashboard, located on the driver's side
   - Knee bolster
   - Glove box
   - 4 glove box frame-to-instrument panel bolts and the frame
   - Clock
   - Moon roof switch
   - Stereo radio/cassette
   - Lower steering column cover clamps and the cover
   - Wrap a shop towel around the steering column to prevent damage to the column.
   - Steering column-to-instrument panel nuts/bolts and lower the steering column
   - SRS module-to-instrument panel nuts (located on the driver's side); then, carefully, remove the SRS module and disconnect the electrical connector
   - Air mix control cable and electrical connectors, located at the center of the instrument panel
   - Antenna lead
   - Electrical connectors, located at the under–dash fuse/relay box
   - Pry out the access panels, from both sides of the instrument panel
   - Instrument panel-to-chassis bolts and remove the instrument panel
   - Wiring harness clips; then, remove the 4 heater housing-to-blower motor duct screws and the duct,

(models not equipped with air conditioning only)

6. If equipped with air conditioning, remove the evaporator housing removing or disconnecting the following:
   - Discharge and recover the air conditioning system refrigerant
   - Refrigerant lines from the evaporator core, discard the O–rings and plug the openings to prevent contamination
   - Cut the insulation pad (at the firewall) at the lower evaporator housing-to-chassis location
   - Thermostat electrical connector, located at the evaporator housing
   - Wiring harness clips from the evaporator housing
   - Evaporator housing-to-chassis screws, the nut and bolt
   - Drain hose
   - Evaporator housing, carefully
   - 2 SRS support beam bolts, nut and the SRS beam, located at the passenger's side
   - Mode control motor connector and the wiring harness clip from the heater housing
   - Heater housing-to-chassis nuts and the heater housing

7. Remove the damper arm cover-to-heater housing screw and the cover.
8. Remove or disconnect the following:
   - Damper arm link; then, remove the damper arm-to-heater housing screw and the arm
   - 2 heater core cover-to-heater housing screws and the cover
   - Pipe clamp screw and the clamp
   - Heater core from the heater housing

To install:

9. Install or connect the following:
   - Heater core to the heater housing
   - Pipe clamp screw and the clamp
   - Heater core cover and the 2 cover-to-heater housing screws
   - Damper arm-to-heater housing screw and the arm; then, connect the damper arm link
   - Damper arm cover and the cover-to-heater housing screw
   - Heater housing and the heater housing-to-chassis nuts
   - Mode control motor connector and the wiring harness clip to the heater housing
   - SRS support beam, nut and the 2 SRS beam bolts on the passengers side

10. If equipped with air conditioning, install the evaporator housing by installing or connecting the following:

   - Evaporator housing
   - Drain hose
   - 4 evaporator housing-to-chassis screws, the nut and bolt
   - Wiring harness clips to the evaporator housing
   - Thermostat electrical connector at the evaporator housing
   - Refrigerant lines to the evaporator core making sure to use new O–rings
   - Heater housing-to-blower motor duct and the 4 duct screws; then, connect the wiring harness clips (models not equipped with air conditioning only)

11. Install the instrument panel by installing or connecting the following:
   - Instrument panel and the instrument panel-to-chassis bolts panels on both sides of the instrument panel
   - Electrical connectors at the under-dash fuse/relay box
   - Antenna lead
   - Air mix control cable and electrical connectors at the center of the instrument panel
   - SRS module electrical connector (located on the passenger's side) and torque the SRS module-to-instrument panel nuts to 84 inch lbs. (9.8 Nm)
   - Steering column and torque the steering column-to-instrument panel nuts to 108 ft. lbs. (13 Nm) and the bolts to 16 ft. lbs. (22 Nm)
   - Lower steering column cover and the cover clamps
   - Stereo radio/cassette
   - Moon roof switch
   - Clock
   - Glove box frame and the 4 frame-to-instrument panel bolts
   - Glove box
   - Knee bolster
   - Lower dashboard and the lower dashboard-to-instrument panel screws on the drivers side
   - Front and rear consoles
   - Front seats
   - Heater housing-to-cowl nut (located in the engine compartment) and torque to 16 ft. lbs. (22 Nm)
   - Heater hoses to the heater core

12. Refill the cooling system.
13. Connect the negative battery cable.
14. Evacuate, charge and leak test the air conditioning system refrigerant.
15. Operate the engine to normal operating temperatures; then, check the climate control operation and check for leaks.

TL

2002–03 MODELS

➡**Be sure to acquire the anti–theft code for the radio; then, write down the frequencies for the preset buttons.**

1. Before servicing the vehicle, refer to the precautions section.
2. Disconnect the negative battery cable.

❋❋ CAUTION

After disconnecting the negative battery cable, wait for at least 3 minutes for the SRS module to deplete its energy.

3. Drain the cooling system into a clean container for reuse.
4. Disconnect the heater hoses from the heater core.
5. In the engine compartment, remove the heater housing-to-cowl nut.
6. Remove the instrument panel by removing or disconnecting the following:

- Clips and remove the dashboard side cover, located at the driver's side
- Lower dashboard cover screw, detach the clips and remove the lower dashboard cover
- Console panel screws and the console panel
- Glove box-to-instrument panel screw; then, remove the damper from the glove box
- Glove box stop located at each side while holding the glove box
- Both glove box-to-instrument panel screws and the glove box
- Clips and remove the dashboard side cover located at the passenger's side
- Glove box cover-to-instrument panel screws, disconnect the electrical connectors and remove the glove box cover
- Rear console and the front console cover
- 4 audio unit-to-instrument panel screws, disconnect the electrical connectors and remove the audio unit
- Pry out the front pillar trim at both sides
- Bolts at the rear vent duct, disconnect the clips and the OBD–II connector; then, detach the tabs and remove the rear vent duct
- Steering column cover screws and the covers

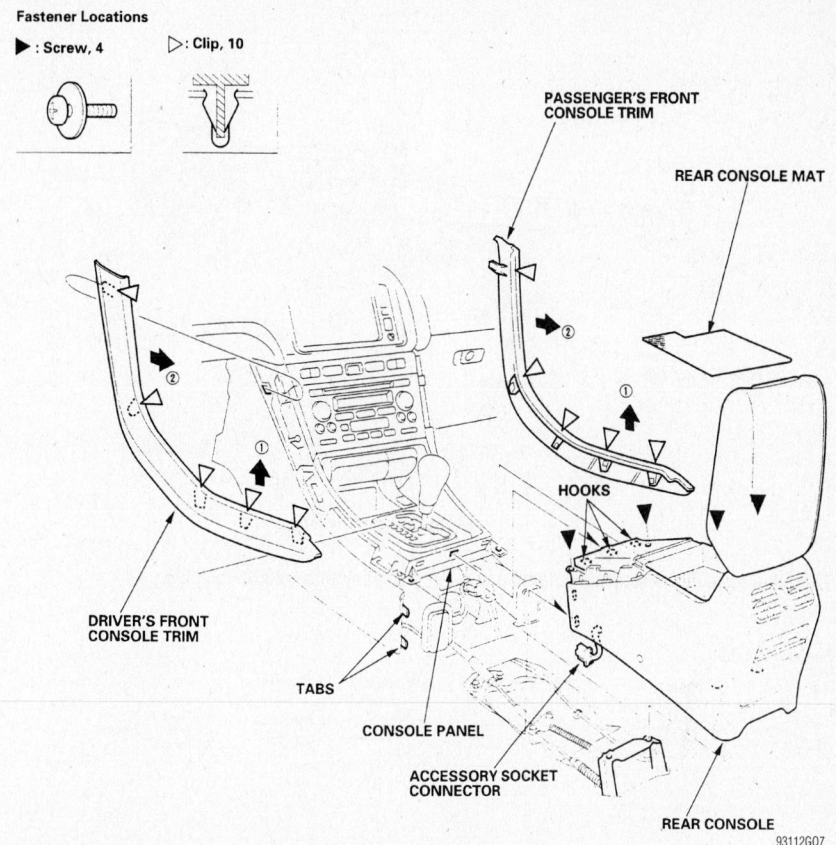

Exploded view of the rear console and related components—2002–03 3.2TL

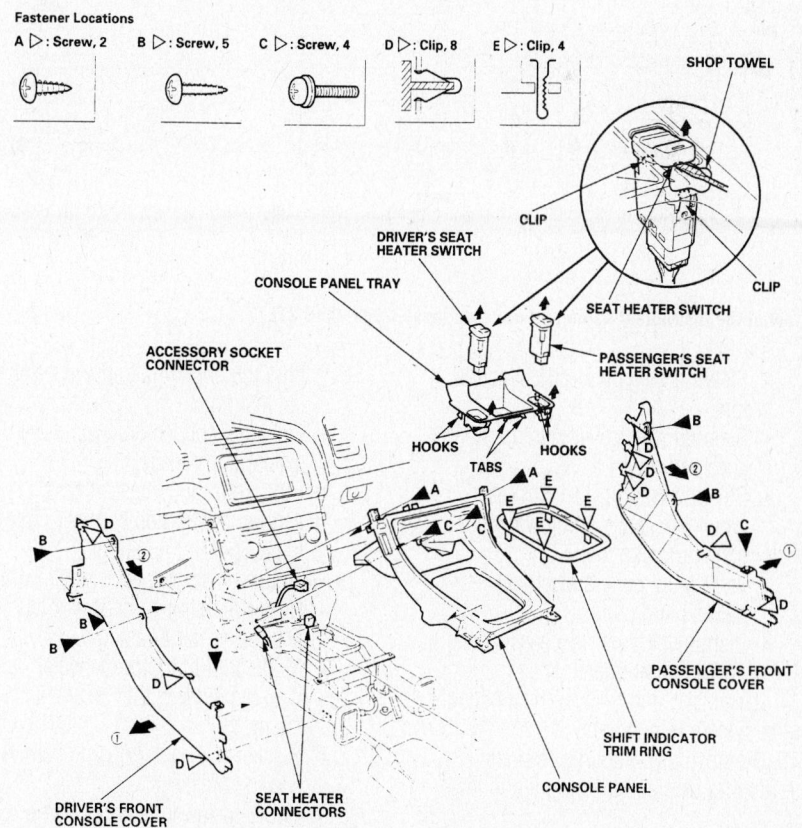

Exploded view of the console panel and related components—2002–03 3.2TL

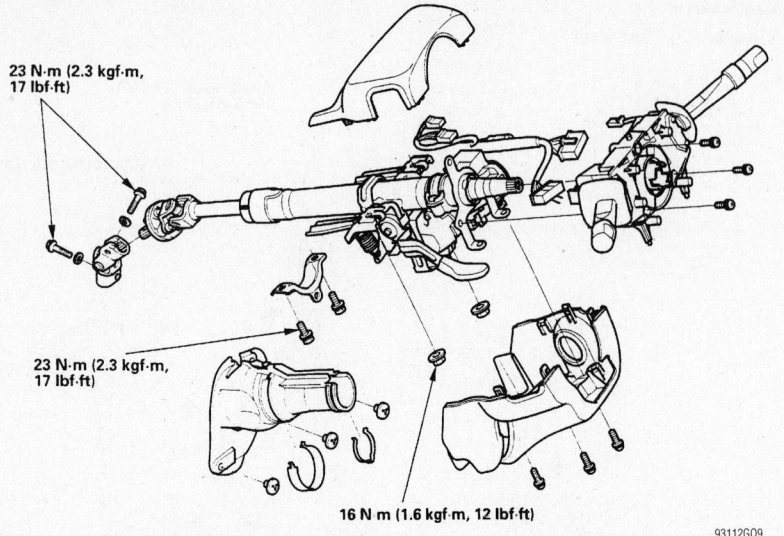

23 N·m (2.3 kgf·m, 17 lbf·ft)

23 N·m (2.3 kgf·m, 17 lbf·ft)

16 N·m (1.6 kgf·m, 12 lbf·ft)

93112GQ9

Exploded view of the steering column and related components—2002–03 3.2TL

Fastener Locations

B ▶ : Bolt, 6   C ▶ : Bolt, 3   D ▶ : Bolt, 1   E ▶ : Bolt, 3   F ▶ : Bolt, 1

8 x 1.25 mm
22 N·m (2.2 kgf·m, 16 lbf·ft)

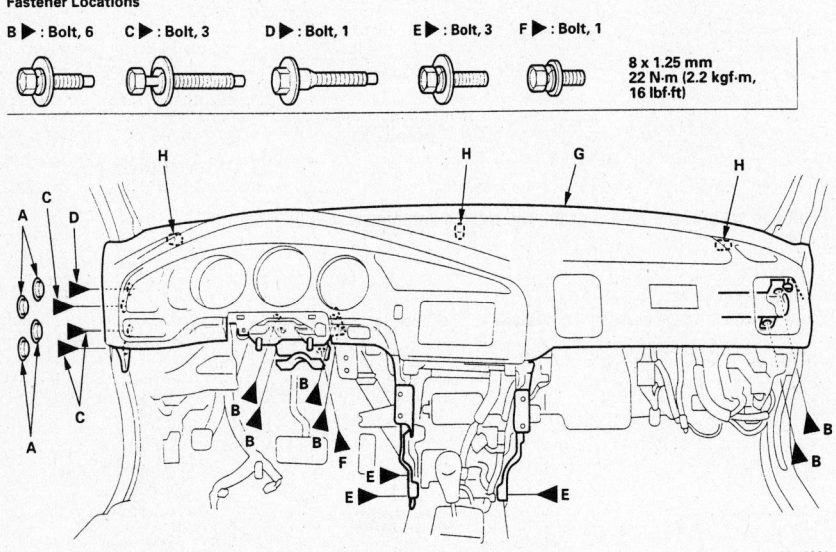

93112GQ0

View of the instrument panel and bolt locations—2002–03 3.2TL

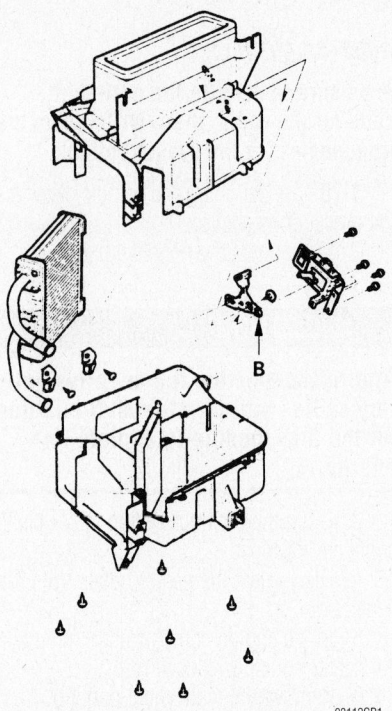

93112GR1

Exploded view of the heater core, heater housing and related components—2002–03 3.2TL

- Steering column electrical connectors
- Steering joint cover clamp, screws and the cover
- Steering column-to-instrument panel nuts/bolts and lower the steering column
- Instrument panel's electrical connectors and clips
- Instrument panel-to-chassis bolts and the instrument panel

7. Discharge and recover the air conditioning system refrigerant.

8. Remove the evaporator housing by removing or disconnecting the following:

- Refrigerant lines from the evaporator core. Discard the O–rings and plug the openings to prevent contamination.
- Electrical connectors from the evaporator housing
- Evaporator housing-to-chassis screws, nut and bolts; then, remove the evaporator housing

9. Remove or disconnect the following:

- Heater housing-to-chassis bolts and the heater housing
- Mode control motor-to-heater housing screws, the linkage and the motor
- Bracket-to-heater housing screws and the brackets
- Upper-to-lower heater housing screws and separate the housings
- Heater core

To install:

10. Install or connect the following:

- Heater core
- Upper-to-lower heater housing and install the heater housing screws
- Brackets and the bracket-to-heater housing screws
- Mode control motor, the linkage and the motor-to-heater housing screws
- Heater housing and the heater housing-to-chassis bolts

11. Install the evaporator housing by installing or connecting the following:

- Evaporator housing and the evaporator housing-to-chassis screws, nut and bolts
- Electrical connectors to the evaporator housing
- Refrigerant lines to the evaporator core, make sure to use new O–rings

12. Install the instrument panel by installing or connecting the following:

- Instrument panel and the instrument panel-to-chassis bolts
- Instrument panel's electrical connectors and clips
- Steering column and torque the steering column-to-instrument panel nuts to 12 ft. lbs. (16 Nm) and bolts to 17 ft. lbs. (23 Nm)

- Steering joint cover and the cover clamp and screws
- Steering column electrical connectors
- Steering column covers and the cover screws
- Rear vent duct, connect the clips and the OBD–II connector, attach the tabs and install the bolts
- Front pillar trim. On both sides.
- Audio unit, connect the electrical connectors and install the 4 audio unit-to-instrument panel screws
- Rear console and the front console cover
- Glove box cover, connect the electrical connectors and install the glove box cover-to-instrument panel screws
- Dashboard side cover and attach the clips at the passenger's side
- Both glove box and the glove box-to-instrument panel screws
- Glove box stop (located at each side), while holding the glove box
- Damper to the glove box and the glove box-to-instrument panel screw
- Console panel and the console panel screws
- Lower dashboard cover, attach the clips and install the lower dashboard cover screw
- Dashboard side cover and attach the clips at the driver's side
- Heater housing-to-cowl nut located in the engine compartment
- Heater hoses to the heater core

13. Refill the cooling system.
14. Connect the negative battery cable.
15. Evacuate, charge and leak test the air conditioning system.
16. Operate the engine to normal operating temperatures; then, check the climate control operation and check for leaks.

2004–06 MODELS

1. Before servicing the vehicle, refer to the precautions section.
2. Disconnect the battery negative cable, and wait at least 3 minutes before beginning work.

➡**Make sure you have the anti-theft codes for the radio and the navigation system, then write down the radio channel presets.**

3. Disconnect the suction and receiver lines from the evaporator core
4. From under the hood, open the cable clamp (A), then disconnect the heater valve cable (B) from the heater valve arm (C).

Turn the heater valve arm to the fully opened position as shown.
5. Drain and recycle the engine coolant.
6. Disconnect the heater hoses from the heater unit.
7. Remove the mounting nut from the heater unit. Take care not to damage or bend the fuel lines and the brake lines, etc.
8. Remove the dashboard as follows:
   a. Remove the driver' and passenger seats.
   b. If equipped with a manual transmission, lower the shift lever boot to release the hooks from the boot, then remove the shift knob.
   c. Detach the clips and release the hooks by carefully inserting a trim tool between the HVAC panel and the center console trim and prying gently on both sides.
   d. Open the console box lid then remove the console mat and screws.
   e. Slide the rear section of the center console rearward to release the hooks.
   f. Lift up the rear of the console.
   g. Release the harness retainer clip from the air duct, and disconnect the accessory power socket connector and the light bulb socket connector, and then remove the rear section of the center console.
   h. Remove the drivers side lower cover, by adjusting the steering column upward.
   i. Remove the clip, and disconnect the in-car temperature sensor connector and air hose. Gently pull out the bottom of the dashboard lower cover to detach the clips and release the pin and hooks.
   j. Disconnect the VSA OFF switch connector, power mirror switch connector, and trunk lid opener switch connector.
   k. To remove the glove box, release the clips by gently pulling the bottom of the glove box housing out and disconnect the trunk lid opener main switch connector and light bulb connector.
   l. Remove the kick panels.
   m. Remove the A-pillar trim
   n. To remove the steering column, align the front wheels straight ahead position
   o. Remove the access panel from the steering wheel, then disconnect the driver's airbag 4P connector from the cable reel.
   p. Remove the two Torx bolts using a Torx T30 bit.
   q. Disconnect the horn switch connector, then remove the driver's airbag.
   r. Disconnect the cruise control set/resume switch, audio remote switch

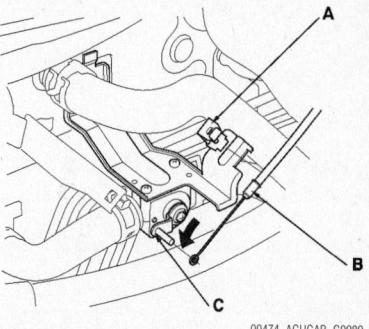

Open the cable clamp (A), then disconnect the heater valve cable (B) from the heater valve arm (C). Turn the heater valve arm to the fully opened position as shown— 2004–06 TL model

and HFL/HFL-voice control switch connectors (if equipped).
   s. Loosen the steering wheel bolt.
   t. Install a commercially available steering wheel puller on the steering and remove the wheel.

✳✳ CAUTION

Do not tap on the steering wheel or the steering column shaft when removing the steering wheel. If you thread the puller bolts into the wheel hub more than five threads, the bolts will hit the cable reel and damage it. To prevent this, install a pair of jam nuts five threads up on each puller bolt.

   u. Remove the column covers.
   v. Remove the steering joint cover.
   w. Release the tilt/telescopic lever, and adjust the steering column to full tilt up position, and to the full telescopic in position. Tighten the tilt/telescopic lever.

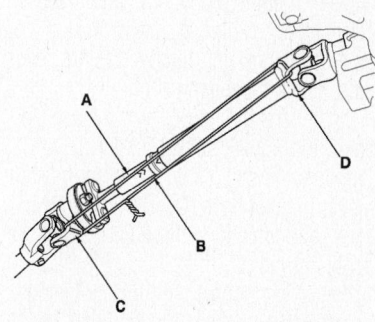

Hold the slider shaft (A) on the column with a piece of wire (B) between the joint yoke (C) of the slider shaft and joint yoke (D) of the upper shaft to prevent the slider shaft from pulling out—2004–06 TL model

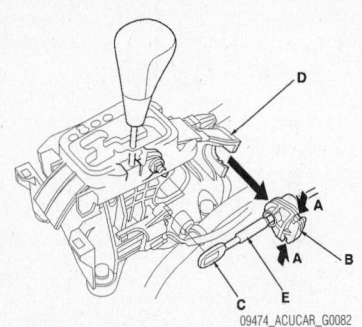

09474_ACUCAR_G0082

Press the holder lock release (A), and pull out the socket holder (B) to remove the shift cable (C) from the shift lever bracket base (D). Do not remove the shift cable by pulling on the shift cable guide (E)—2004–06 TL model

x. Hold the slider shaft (A) on the column with a piece of wire (B) between the joint yoke (C) of the slider shaft and joint yoke (D) of the upper shaft to prevent the slider shaft from pulling out.

y. Release the tilt/telescopic lever, and adjust the steering column to the full telescopic out position, then tighten the tilt/telescopic lever.

z. Disconnect the wire harness connectors from the combination switch assembly.

aa. Remove the combination switch assembly from the steering column shaft by removing the screws.

bb. Disconnect the connectors from the ignition switch, and release the wire harness clips from the steering column.

cc. Remove the steering joint bolt, then disconnect the steering joint from the pinion shaft.

dd. Remove the steering column by removing the attaching nuts and bolts. If the lower slide shaft is removed, slip it into the upper shaft by aligning the marks.

ee. If equipped with an automatic transmission, shift the transmission into the R position and remove the nut securing the shift cable end.

ff. Press the holder lock release (A), and pull out the socket holder (B) to remove the shift cable (C) from the shift lever bracket base (D). Do not remove the shift cable by pulling on the shift cable guide (E).

gg. Disconnect shift lock solenoid connector and transmission gear selection switch/park pin switch connector.

hh. Remove the harness clamps.

ii. Remove the shift lever assembly.

jj. Detach the clips, then remove the rear vent duct

Fastener Locations

A ▶ : Bolt, 6

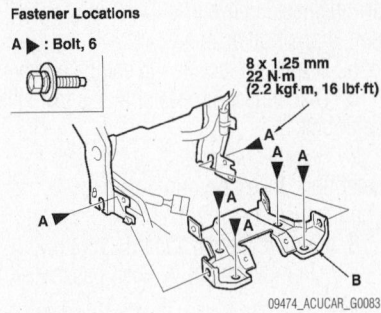

8 x 1.25 mm
22 N·m
(2.2 kgf·m, 16 lbf·ft)

09474_ACUCAR_G0083

Remove the center bracket—2004–06 TL model

Fastener Locations

B ▶ : Bolt, 3   C ▶ : Bolt, 6   D ▷ : Clip, 2

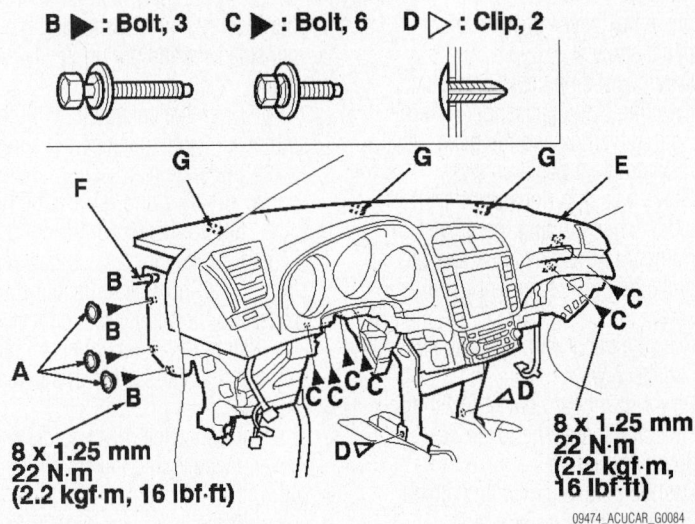

8 x 1.25 mm
22 N·m
(2.2 kgf·m, 16 lbf·ft)

8 x 1.25 mm
22 N·m
(2.2 kgf·m, 16 lbf·ft)

09474_ACUCAR_G0084

Dashboard mounting—2004–06 TL model

kk. Remove the clips, then pull back the carpet.

ll. Remove the rear joint vent ducts on both sides.

mm. Disconnect the connector and remove the bolts, then remove the bracket.

nn. Remove the heater ducts.

oo. Disconnect the SRS connectors and remove the Torx bolts, then pull out the SRS unit.

pp. From under the dash, disconnect any wiring harness connectors that would interfere with dashboard removal.

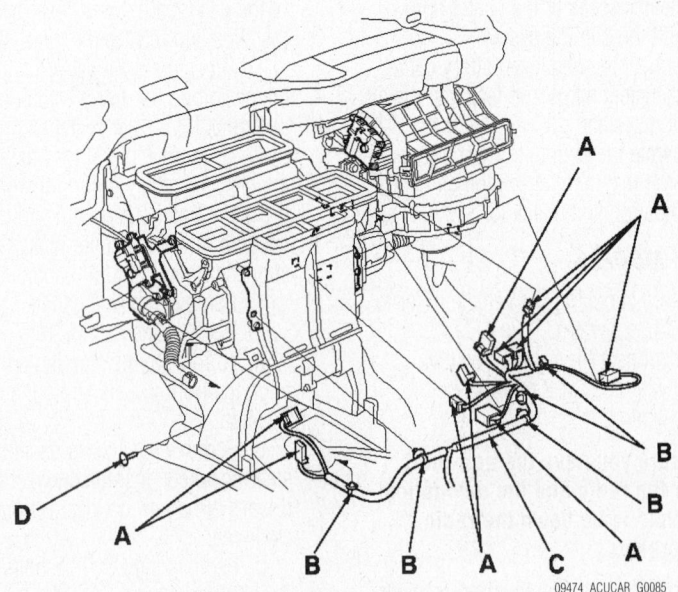

09474_ACUCAR_G0085

Unplug the connectors (A) from the driver's mode control motor, driver's air mix control motor, evaporator temperature sensor, power transistor, passenger's mode control motor, passenger's air mix control motor, recirculation control motor, wire harness clips (B), wire harness (C) and carpet clip (D)—2004–06 TL model

6 x 1.0 mm
9.8 N·m
(1.0 kgf·m, 7.2 lbf·ft)

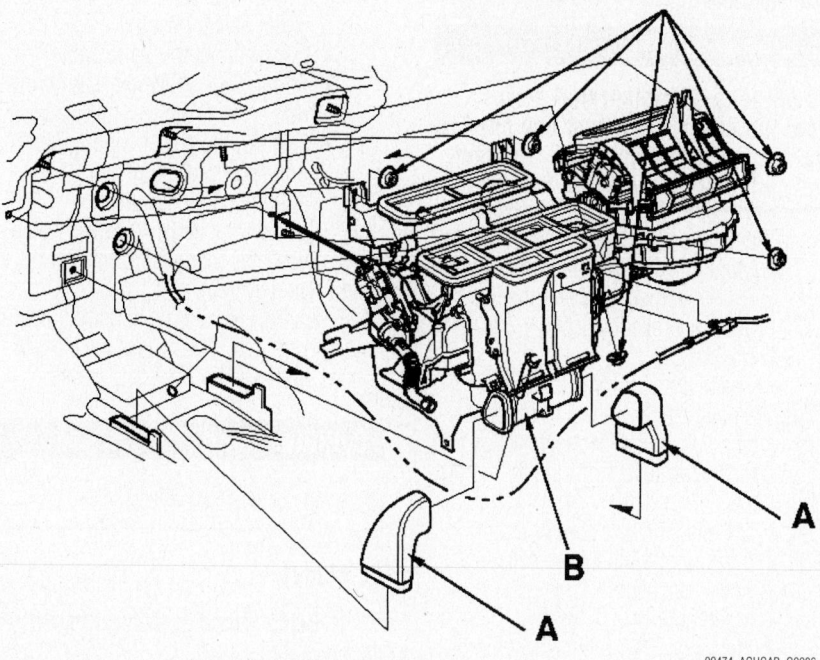

09474_ACUCAR_G0086

Remove the blower/heater unit—2004–06 TL model

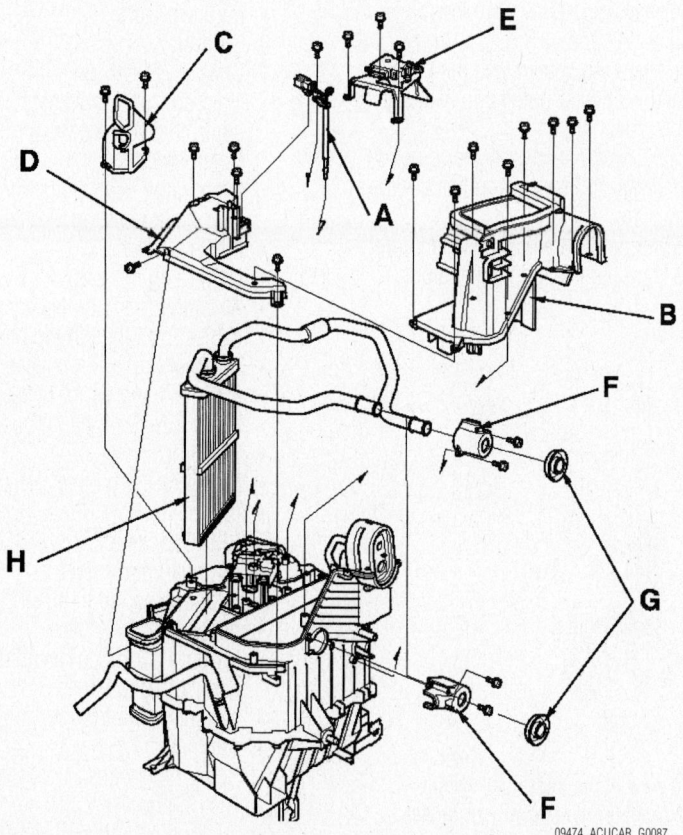

09474_ACUCAR_G0087

Exploded view of the blower/heater unit components—2004–06 TL model

➡ **Lift the white wire harness connector locks before trying to remove the connectors from the fuse box.**

qq. Disconnect the parking brake switch connector.

rr. Using a T30 Torx bit, remove the ground bolt.

ss. If equipped with an automatic transmission, disconnect the park pin switch connector.

tt. From under the dash, disconnect the ECM/PCM connector, antenna lead, engine wire harness connector and A/C sub harness connector.

uu. If equipped with an automatic transmission, disconnect the shift lock solenoid connector

vv. From under the dash, disconnect the passenger's door wire harness connectors, the three floor wire harness connectors (if the vehicle has a navigation system) or the one floor wire harness connector (if the vehicle does not have a navigation system), stereo amplifier connectors, and ETC unit connector.

ww. Detach all of the harness and connector clips.

xx. Remove bolts, then remove the brake pedal support member.

yy. Unbolt the parking brake handle assembly and the center console bracket. Do not disconnect the parking brake cable; move the assembly to the side of the shifter.

zz. Remove the bolts, then remove the center bracket.

aaa. From outside the driver's door, remove the caps, then remove the bolts and clips, then lift up on the dashboard to release it from the guide pins.

➡ **Before removing the dashboard, make sure all the harnesses have been disconnected.**

bbb. Carefully remove the dashboard through the front door opening.

9. Disconnect the connectors (A) from the driver's mode control motor, the driver's air mix control motor, the evaporator temperature sensor, the power transistor, the passenger's mode control motor, passenger's air mix control motor, and the recirculation control motor, then remove the wire harness clips (B), the wire harness (C), and the carpet clip (D). Refer to the illustration for component locations.

10. Remove the heater ducts. then remove the mounting nuts and the blower-heater unit.

11. Remove the self-tapping screws, the evaporator temperature sensor (A) and the joint duct (B). Remove the self-tapping

screws, then remove the passenger's heater outlet (C), and the heater core cover (D). Remove the self-tapping screws and the passenger's air mix control motor (E). Remove the self-tapping screws, the heater pipe brackets (F), the grommets (G) and carefully pull out the heater core (H) so you don't bend the inlet and outlet pipes. Refer to the illustration for component locations.

To install:

12. Installation is the reverse of removal. refer to the illustrations accompanying the removal procedure for component locations and torque specifications.

TSX Models

➡Be sure to acquire the anti–theft code for the radio; then, write down the frequencies for the preset buttons.

1. Before servicing the vehicle, refer to the precautions section.

2. Disconnect the negative battery cable.

✳✳ CAUTION

After disconnecting the negative battery cable, wait for at least 3 minutes for the SRS module to deplete its energy.

3. Properly drain the cooling system. Properly discharge the air conditioning system using an approved refrigerant recover/recycling system.

4. Remove or disconnect the following:
   • Suction and receiver lines from the evaporator core

5. From under the hood, open the cable clamp and disconnect the heater valve cable from the heater valve arm. Turn the heater valve arm to the fully opened position.

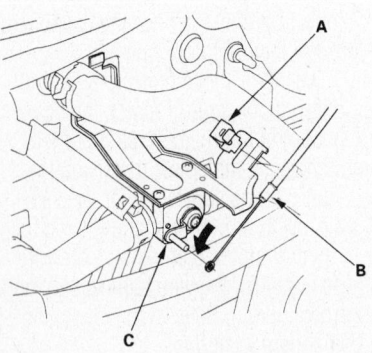

67162-ATSX-G04

Open the heater valve cable clamp (A), then disconnect the cable (B) from the heater valve arm. Turn the heater valve arm (C) to the fully opened position—TSX models

6. Position a drain pan under the hoses. Slide the hose clamps back Remove the bolt and water valve bracket, then disconnect the inlet and outlet heater hoses from the heater unit.

✳✳ WARNING

Do not let any coolant that drains from the hoses to contact any painted surfaces or electrical parts. Immediately wipe up and coolant that spills.

7. Remove or disconnect the following:
   • Mounting nut from the heater unit. Do not damage or bend the fuel lines or brake lines
   • Dashboard
   • Footrest and footrest bracket

8. Remove or disconnect the connectors, wire harness clips, connector clips and wire harness, after disconnecting the following:
   • Driver's side mix control motor
   • Evaporator temperature sensor
   • Power transistor
   • Mode control
   • Passenger's air mix control motor

9. Remove or disconnect the following:
   • Heater ducts
   • Mounting nuts and blower/heater unit
   • Self-tapping screws and the joint ducts "A" and "B"
   • Self-tapping screws and the passenger's heater outlet and heater core cover
   • Self-tapping screws, heater pipe brackets, grommets and heater pipe brackets
   • Heater core, carefully, being sure not to bend the inlet and outlet pipes.

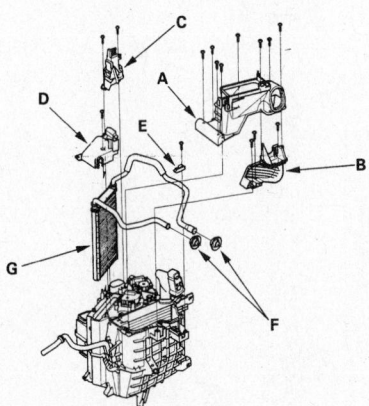

67162-ATSX-G05

Exploded view of the passenger heater outlet (C), heater core cover (D), heater pipe brackets (E), grommets (F) and heater core (G)—TSX models

To install:

10. Installation is the reverse of the removal procedure, noting the following points:
   • Do not interchange the inlet and outlet heater hoses
   • Adjust the heater valve cable
   • Perform the PCM idle learn procedure
   • Perform the power window control unit reset procedure

11. Refill the cooling system.

12. Connect the negative battery cable.

13. Evacuate, charge and leak test the air conditioning system.

14. Operate the engine to normal operating temperatures; then, check the climate control operation and check for leaks.

Cylinder Head

REMOVAL & INSTALLATION

CL Model

1. Before servicing the vehicle, refer to the precautions section.

2. Disconnect the negative battery cable.

3. Drain the coolant.

4. Relieve the fuel system pressure.

5. Remove or disconnect the following:
   • Engine covers
   • Strut brace
   • Water bypass hose
   • Traction Control System (TCS) control valve from the throttle body (if equipped)
   • Evaporative Emissions (EVAP) canister hose from the throttle body
   • Intake air duct
   • Upper engine covers
   • Accelerator and cruise control cables from the throttle body
   • Breather and water bypass hoses
   • Fuel feed and return hoses
   • Brake booster vacuum hose
   • PCV hose
   • Power steering hose clamp
   • Intake Manifold Runner Control (IMRC) actuator
   • Wire harness holder
   • Side engine mount bracket
   • Accessory drive belts
   • Power steering pump without disconnecting the lines
   • Ground cable from the engine
   • Alternator
   • Spark plug wires
   • Distributor
   • Intake Air Temperature (IAT) sensor connector

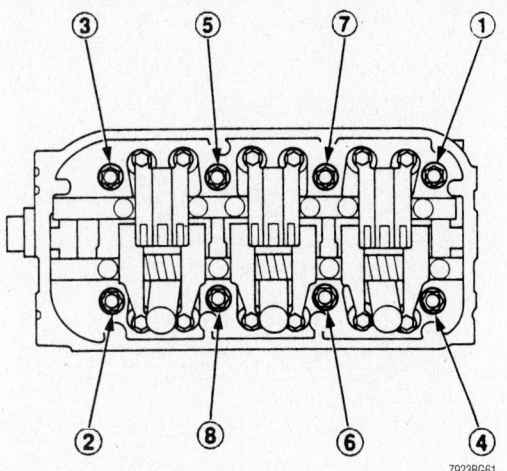

Loosen the cylinder head bolts in the sequence shown—CL models

- Idle Air Control (IAC) valve connector
- Throttle Position (TPS) sensor connector
- Manifold Absolute Pressure (MAP) sensor connector
- Engine Coolant Temperature (ECT) sensor connector
- Radiator fan switch connectors
- Crankshaft Position (CKP) sensor connector
- Top Dead Center (TDC) sensor connector
- Exhaust Gas Recirculation (EGR) valve connector
- Engine oil pressure switch connector
- Ignition coils
- Intake manifold
- Fuel injector connectors
- Fuel rails
- Intake Air Bypass (IAB) control valve vacuum hoses
- Heater hoses
- Upper and lower radiator hoses
- Exhaust manifolds
- Water passage assembly
- Crankshaft pulley
- Front timing belt cover

6. Set the engine to TDC by aligning the marks on the crankshaft and camshaft pulleys.

- Timing belt
- Camshaft pulleys
- Rear timing belt covers

7. Loosen each cylinder head bolt ⅓ turn at a time in the reverse order of the tightening sequence.

8. Remove the cylinder heads and the oil control orifices.

To install:

9. Install the oil control orifices using new O–rings.

10. If removed, install the dowel pins.

11. Position new cylinder head gaskets on the cylinder block.

12. If moved, set the crankshaft and camshaft pulleys to TDC by aligning the marks on the pulley and oil pump.

13. Carefully position the cylinder heads on the engine.

14. Lubricate the cylinder head bolts with clean engine oil.

15. Tighten the cylinder head bolts in 3 steps. Be sure to follow the tightening torque sequence:
   a. Step 1: 29 ft. lbs. (39 Nm).
   b. Step 2: 51 ft. lbs. (69 Nm).
   c. Step 3: 72 ft. lbs. (98 Nm).

16. Install or connect the following:
- Exhaust manifolds with new gaskets and torque the nuts to 23 ft. lbs. (31 Nm)
- Rear timing belt covers and torque the bolts to 16 ft. lbs. (22 Nm)

17. Install the camshaft pulleys and torque to 67 ft. lbs. (90 Nm).

- Timing belt
- Front timing belt cover and torque the bolts to 108 inch lbs. (12 Nm)

18. Check and adjust the valve clearance.

- Crankshaft pulley and torque the bolt to 181 ft. lbs. (245 Nm)
- Water passage assembly and torque the bolts to 16 ft. lbs. (22 Nm)
- Intake manifold with new gaskets and O–rings and tighten the bolts to 16 ft. lbs. (22 Nm)
- Cylinder head cover and tighten the bolts to 108 inch lbs. (12 Nm)
- Ignition coils and torque the bolts to 108 inch lbs. (12 Nm)
- Exhaust manifolds and torque the bolts to 23 ft. lbs. (31 Nm)
- Upper and lower radiator hoses
- Heater hoses
- IAB vacuum hoses
- Fuel rails and torque the bolts to 84 inch lbs. (9.5 Nm)
- Fuel injector connectors
- Intake manifold and torque the bolts to 16 ft. lbs. (22 Nm)
- Engine oil pressure switch connector
- EGR valve connector
- TDC sensor connector
- CKP sensor connector
- Radiator fan switch connectors
- MAP sensor connector
- ECT sensor connector
- TPS sensor connector
- IAC sensor connector
- IAT sensor connector
- Spark plug wires
- Distributor
- Alternator and torque the upper bolt to 16 ft. lbs. (22 Nm) and the lower bolt to 33 ft. lbs. (44 Nm)

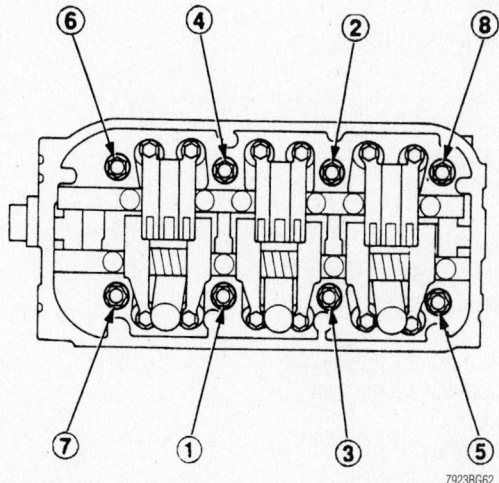

Cylinder head torque sequence—CL model

- Ground cable
- Power steering pump and torque the bolts to 17 ft. lbs. (24 Nm)
- Accessory drive belts
- Side engine mount and torque the mounting bolts to 33 ft. lbs. (44 Nm) and the through bolt to 40 ft. lbs. (54 Nm)
- Wire harness holder
- IMRC actuator
- PCV hose
- Brake booster vacuum hose
- Fuel feed and return lines and torque the fitting to 16 ft. lbs. (22 Nm)
- Accelerator and cruise control cables
- Engine covers
- Intake air duct
- Fvaporative Emissions (EVAP) canister hose from the throttle body
- TCS control valve
- Strut brace and torque the bolts to 16 ft. lbs. (22 Nm)
- Engine covers and torque the bolts to 108 inch lbs. (12 Nm)

19. Change the engine oil and filter.
20. Fill and bleed the cooling system.
21. Connect the negative battery cable.
22. Bring the engine to operating temperature and inspect for any fluid leaks. Top off all fluid levels as necessary.
23. Enter the security code for the radio.

➡The PCM idle memory must be reset after reconnecting the battery. Start the engine and hold it at 3000 rpm until the cooling fan comes on. Then allow the engine to idle for about 5 minutes with all accessories OFF and with the transmission in Park or Neutral.

2002–04 RL Model

1. Before servicing the vehicle, refer to the precautions section.
2. Disconnect the negative battery cable.
3. Drain the coolant.
4. Relieve the fuel system pressure.
5. Remove or disconnect the following:
- Engine covers
- Strut brace
- Water bypass hose
- Traction Control System (TCS) control valve from the throttle body (if equipped)
- Evaporative Emissions (EVAP) canister hose from the throttle body
- Intake air duct
- Upper engine covers
- Accelerator and cruise control cables from the throttle body

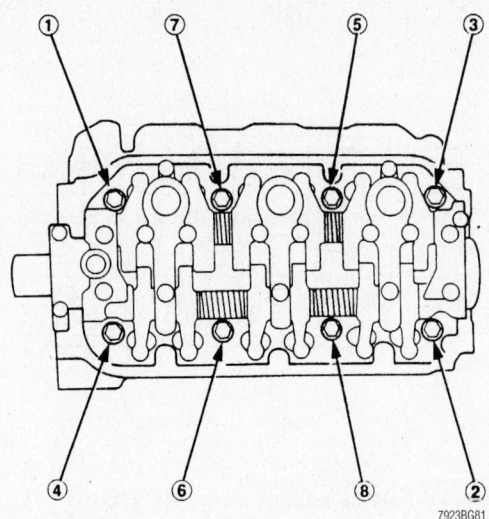

Loosen the cylinder head bolts in the sequence shown—2002–04 RL model

- Fuel feed and return hoses
- Brake booster vacuum hose
- PCV hose
- Intake Manifold Runner Control (IMRC) actuator
- Wire harness holder
- Side engine mount bracket
- Accessory drive belts
- Power steering pump without disconnecting the lines
- Ground cable from the engine
- Alternator
- Spark plug wires
- Distributor
- Intake Air Temperature (IAT) sensor connector
- Idle Air Control (IAC) valve connector
- Throttle Position (TPS) sensor connector
- Manifold Absolute Pressure (MAP) sensor connector
- Engine Coolant Temperature (ECT) sensor connector
- Radiator fan switch connectors
- Crankshaft Position (CKP) sensor connector
- Top Dead Center (TDC) sensor connector
- Exhaust Gas Recirculation (EGR) valve connector
- Engine oil pressure switch connector
- Ignition coils
- Intake manifold
- Fuel injector connectors

CYLINDER HEAD BOLTS TORQUE SEQUENCE

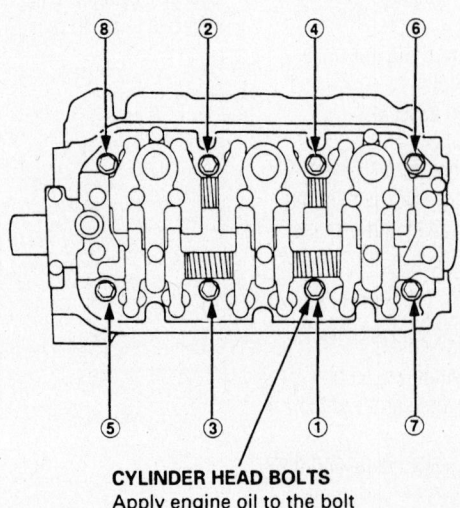

CYLINDER HEAD BOLTS
Apply engine oil to the bolt threads.

Cylinder head torque sequence—2002–04 RL model

- Fuel rails
- Intake Air Bypass (IAB) control valve vacuum hoses
- Heater hoses
- Upper and lower radiator hoses
- Exhaust manifolds
- Water passage assembly
- Crankshaft pulley
- Front timing belt cover

6. Set the engine to TDC by aligning the marks on the crankshaft and camshaft pulleys.

- Timing belt
- Camshaft pulleys
- Rear timing belt covers

7. Loosen each cylinder head bolt ⅓ turn at a time in the reverse order of the tightening sequence.

8. Remove the cylinder heads and the oil control orifices.

To install:

9. Install the oil control orifices using new O–rings.

10. If removed, install the dowel pins.

11. Position new cylinder head gaskets on the cylinder block.

12. If moved, set the crankshaft and camshaft pulleys to TDC by aligning the marks on the pulley and oil pump.

13. Carefully position the cylinder heads on the engine.

14. Lubricate the cylinder head bolts with clean engine oil.

15. Tighten the cylinder head bolts in 3 steps. Be sure to follow the tightening torque sequence:

   a. Step 1: 29 ft. lbs. (39 Nm).
   b. Step 2: 51 ft. lbs. (69 Nm).
   c. Step 3: 72 ft. lbs. (98 Nm).

16. Install or connect the following:

- Exhaust manifolds with new gaskets and torque the nuts to 23 ft. lbs. (31 Nm)
- Rear timing belt covers and torque the bolts to 16 ft. lbs. (22 Nm)

17. Install the camshaft pulleys and torque to 23 ft. lbs. (31 Nm).

18. Install or connect the following:

- Timing belt
- Front timing belt cover and torque the bolts to 108 inch lbs. (12 Nm)

19. Check and adjust the valve clearance.

- Crankshaft pulley and torque the bolt to 181 ft. lbs. (245 Nm)
- Water passage assembly and torque the bolts to 16 ft. lbs. (22 Nm)
- Intake manifold with new gaskets and O–rings and tighten the bolts to 16 ft. lbs. (22 Nm)
- Cylinder head cover and tighten the bolts to 108 inch lbs. (12 Nm)

- Ignition coils and torque the bolts to 108 inch lbs. (12 Nm)
- Exhaust manifolds and torque the bolts to 23 ft. lbs. (31 Nm)
- Upper and lower radiator hoses
- Heater hoses
- IAB vacuum hoses
- Fuel rails and torque the bolts to 84 inch lbs. (9.5 Nm)
- Fuel injector connectors
- Intake manifold and torque the bolts to 16 ft. lbs. (22 Nm)
- Engine oil pressure switch connector
- EGR valve connector
- TDC sensor connector
- CKP sensor connector
- Radiator fan switch connectors
- MAP sensor connector
- ECT sensor connector
- TPS sensor connector
- IAC sensor connector
- IAT sensor connector
- Spark plug wires
- Distributor
- Alternator and torque the upper bolt to 16 ft. lbs. (22 Nm) and the lower bolt to 33 ft. lbs. (44 Nm)
- Ground cable
- Power steering pump and torque the bolts to 17 ft. lbs. (24 Nm)
- Accessory drive belts
- Side engine mount and torque the mounting bolts to 33 ft. lbs. (44 Nm) and the through bolt to 40 ft. lbs. (54 Nm)
- Wire harness holder
- IMRC actuator
- PCV hose
- Brake booster vacuum hose
- Fuel feed and return lines and torque the fitting to 16 ft. lbs. (22 Nm)
- Accelerator and cruise control cables
- Engine covers
- Intake air duct
- Evaporative Emissions (EVAP) canister hose from the throttle body
- TCS control valve
- Strut brace and torque the bolts to 16 ft. lbs. (22 Nm)
- Engine covers and torque the bolts to 108 inch lbs. (12 Nm)

20. Change the engine oil and filter.

21. Fill and bleed the cooling system.

22. Connect the negative battery cable.

23. Bring the engine to operating temperature and inspect for any fluid leaks. Top off all fluid levels as necessary.

24. Enter the security code for the radio.

➡ **The PCM idle memory must be reset after reconnecting the battery. Start the engine and hold it at 3000 rpm until the cooling fan comes on. Then allow the engine to idle for about 5 minutes with all accessories OFF and with the transmission in Park or Neutral.**

2005–06 RL Model

Make sure you have the anti-theft codes for the radio and the navigation system, then write down the customer's audio presets. Make sure the ignition switch is OFF.

1. Before servicing the vehicle, refer to the precautions section.

2. Relieve the fuel pressure.

3. Disconnect the negative cable from the battery.

4. Drain the engine coolant.

5. Remove the air cleaner.

6. Remove the drive belt.

7. Remove the timing belt.

8. Remove the Power Steering (P/S) pump and P/S hose clamp.

9. Remove the alternator.

10. Remove the intake manifold.

11. Remove the six ignition coils.

12. Remove the following engine wire harness connectors and wire harness clamps from the cylinder head:

- Six injector connectors
- Engine coolant temperature (ECT)
- Engine Coolant Temperature (ECT)
- Crankshaft Position (CKP) sensor

13. Exhaust Gas Recirculation (EGR) valve

14. Rocker arm oil control solenoid connector

15. Rocker arm oil pressure switch connector

16. Oil pressure switch connector

17. Two Air Fuel ratio (A/F) sensor connectors

18. Two secondary Heated Oxygen (O2s) sensor connectors

19. Remove the front warm up three way catalytic converter and rear warm up three way catalytic converter.

20. Remove the quick-connect fitting cover then disconnect the fuel feed hose.

21. Remove the two nuts securing the purge joint.

22. Remove the radiator hoses.

23. Remove the heater hoses and water bypass hose.

24. Remove the two bolts securing the harness holder.

25. Remove the two bolts securing the vacuum line.

26. Remove the harness clamp.

27. Remove the connector bracket from the front cylinder head.

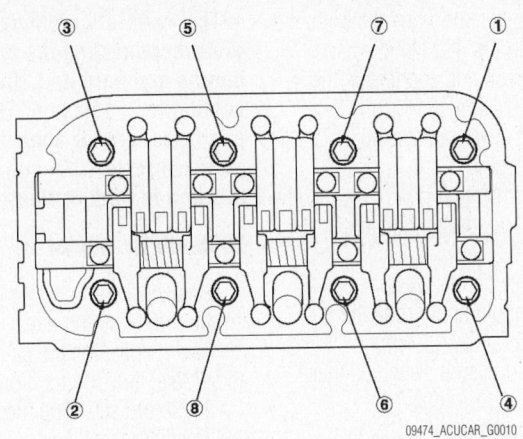

Remove the cylinder head bolts in the sequence shown using several passes—2005–06 RL model

28. Remove the engine mount control solenoid valve bracket from the rear cylinder head.

29. Remove the fuel rails.

30. Remove the ground cable then remove the water passage and connecting pipe.

31. Remove the front and rear camshaft pulleys and front and rear back covers.

32. Remove the intake manifold.

33. Remove the six ignition coils.

34. Remove the three bolts securing the harness holder, and remove the dipstick.

35. Remove the bolt securing the power steering hose clamp.

36. Remove the two bolts securing the harness holder.

37. Remove the breather hose.

38. Remove the cylinder head covers.

39. Remove the cylinder head bolts. To prevent warpage, unscrew the bolts in sequence 1/3 turn at a time and repeat the sequence until all bolts are loosened.

40. Remove the cylinder heads.

To install:

41. Clean the cylinder head and block surface.

42. Clean and install the oil control orifices with new O-rings.

43. Install the dowel pins and new cylinder head gaskets.

44. Put the cylinder head onto the engine block.

45. Clean the timing belt pulleys, timing belt guide plate, and the upper and lower covers.

46. Set the timing belt drive pulley to Top Dead Center (TDC) by aligning the TDC mark on the tooth of the timing belt drive pulley with the pointer on the oil pump.

47. Set the camshaft pulleys to TDC by aligning the TDC marks on the camshaft pulleys with the pointers on the back covers.

48. Apply new engine oil to the threads and flanges of the cylinder head bolts

49. Tighten the 6 point cylinder head bolts sequentially in three steps:

    a. Step 1: 29 ft. lbs. (39 Nm).

    b. Step 2: 51 ft. lbs. (69 Nm).

    c. Step 3: 72 ft. lbs. (98 Nm).

50. Tighten the 12 point cylinder head bolts sequentially in three steps:

    a. Step 1: 29 ft. lbs. (39 Nm).

    b. Step 2: plus 90 degrees

    c. Step 3: plus 90 degrees

    d. Step 4: if using new bolts: plus 90 degrees

➡**Use a beam-type torque wrench. When using a preset-type torque wrench, be sure to tighten slowly and not to overtighten. If a bolt makes any noise while you are torquing it, loosen the bolt, and retighten it from the 1st step.**

51. Install the remaining components in the reverse order of removal.

RSX Models

1. Before servicing the vehicle, refer to the precautions section.

2. Disconnect the negative battery cable.

3. Drain the coolant.

4. Relieve the fuel system pressure.

5. Remove or disconnect the following:
- Fuel feed hose, K20A3 engine only
- Drive belt
- Intake manifold
- Water bypass hose
- Exhaust manifold
- Timing chain

6. Disconnect the following engine wire harness connectors and wire harness clamps from the cylinder head:
- Fuel injector connectors, K20A3 engine
- Engine Coolant Temperature (ECT) sensor connector
- Exhaust and intake Camshaft Position (CMP) sensor connectors

7. Remove or disconnect the following:
- Upper radiator hose and heater hose
- Harness holder from the bracket
- Connecting pipe mounting bolt and water bypass line mounting bolts

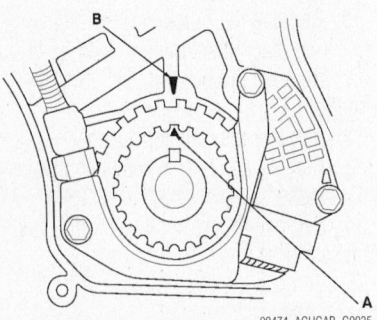

Align the TDC mark on the tooth of the timing belt drive pulley with the pointer on the oil pump —200–06 RL model

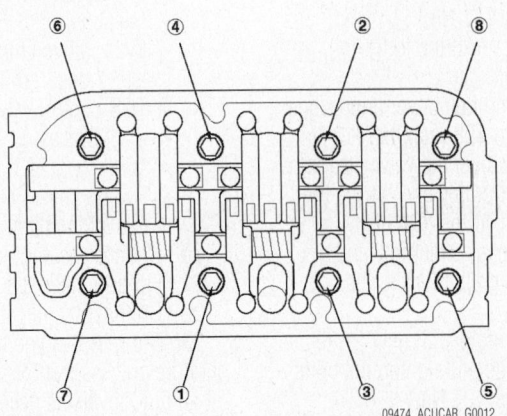

Tighten the cylinder head bolts in the sequence shown—2005–06 RL model

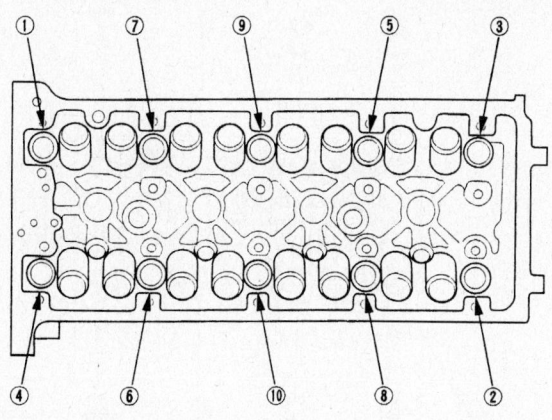

Cylinder head bolt loosening sequence—RSX models

- Water bypass hose
- Rocker arm assembly
- Cylinder head bolts, loosening in sequence, ⅓ turn at a time until removed
- Cylinder head

To install:

8. Clean and inspect the cylinder head and mating surfaces.

9. Install a new cylinder head gasket and dowel pins on the cylinder block

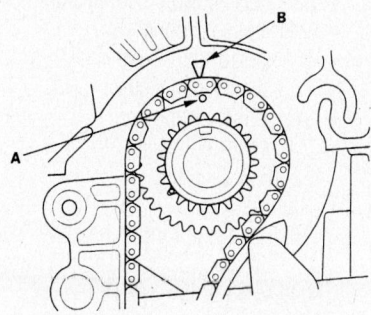

Align the TDC mark on the crankshaft sprocket (A) with the pointer (B) on the cylinder block—2 RSX models

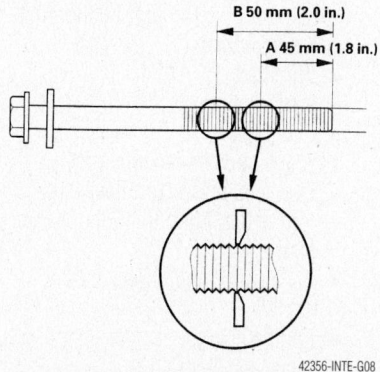

Measure the diameter of the head bolts to determine if they are reusable—RSX models

10. Set the crankshaft to Top Dead Center (TDC). Align the TDC mark on the crankshaft sprocket with the pointer on the cylinder block.

11. Measure the diameter of each cylinder head bolt at point A and point B, as shown in the illustration. If either specification is less than 0.42 in. (10.6mm), you must replace the cylinder head bolt.

12. Carefully position the cylinder heads on the engine.

13. Lubricate the cylinder head bolts with clean engine oil.

14. Tighten the cylinder head bolts in 3 steps. Be sure to follow the tightening torque sequence:

   a. Step 1: 29 ft. lbs. (39 Nm).
   b. Step 2: 90 degrees.
   c. Step 3: 90 degrees.
   d. Step 4 (new bolts only): 90 degrees.

15. Install or connect the following:

- Rocker arm assembly
- Water bypass hose
- Connecting pipe mounting bolt and water bypass line mounting bolts
- Harness holder on the bracket
- Upper radiator hose and heater hose
- Water bypass hose
- Intake manifold
- Exhaust manifold
- Timing chain
- Fuel feed hose, K20A3 engine only

16. Adjust the valve clearance.

- Drive belt
- Negative battery cable and enter the radio security code.

17. After installation, check to see that all hoses and wires are installed correctly.

18. Fill and bleed the air from the cooling system.

19. Attach the negative battery cable.

20. Enter the radio security code.

2002–03 TL Models

1. Before servicing the vehicle, refer to the precautions section.

2. Disconnect the negative battery cable.

3. Drain the coolant.

4. Relieve the fuel system pressure.

5. Remove or disconnect the following:

- Engine covers
- Strut brace
- Water bypass hose
- Traction Control System (TCS) control valve from the throttle body (if equipped)
- Evaporative Emissions (EVAP) canister hose from the throttle body
- Intake air duct
- Upper engine covers
- Accelerator and cruise control cables from the throttle body
- Fuel feed and return hoses
- Brake booster vacuum hose
- PCV hose
- Intake Manifold Runner Control (IMRC) actuator
- Wire harness holder
- Side engine mount bracket

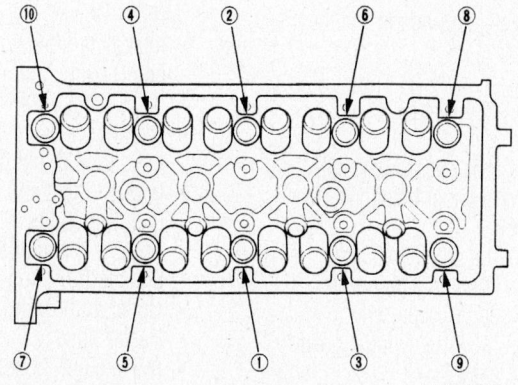

Cylinder head bolt tightening sequence—RSX models

- Accessory drive belts
- Power steering pump without disconnecting the lines
- Ground cable from the engine
- Alternator
- Spark plug wires
- Distributor
- Intake Air Temperature (IAT) sensor connector
- Idle Air Control (IAC) valve connector
- Throttle Position (TPS) sensor connector
- Manifold Absolute Pressure (MAP) sensor connector
- Engine Coolant Temperature (ECT) sensor connector
- Radiator fan switch connectors
- Crankshaft Position (CKP) sensor connector
- Top Dead Center (TDC) sensor connector
- Exhaust Gas Recirculation (EGR) valve connector
- Engine oil pressure switch connector
- Ignition coils
- Intake manifold
- Fuel injector connectors
- Fuel rails
- Intake Air Bypass (IAB) control valve vacuum hoses
- Heater hoses
- Upper and lower radiator hoses
- Exhaust manifolds
- Water passage assembly
- Crankshaft pulley
- Front timing belt cover

6. Set the engine to TDC by aligning the marks on the crankshaft and camshaft pulleys.

- Timing belt
- Camshaft pulleys
- Rear timing belt covers

7. Loosen each cylinder head bolt ⅓ turn at a time in the reverse order of the tightening sequence.

8. Remove the cylinder heads and the oil control orifices.

To install:

9. Install the oil control orifices using new O–rings.

10. If removed, install the dowel pins.

11. Position new cylinder head gaskets on the cylinder block.

12. If moved, set the crankshaft and camshaft pulleys to TDC by aligning the marks on the pulley and oil pump.

13. Carefully position the cylinder heads on the engine.

14. Lubricate the cylinder head bolts with clean engine oil.

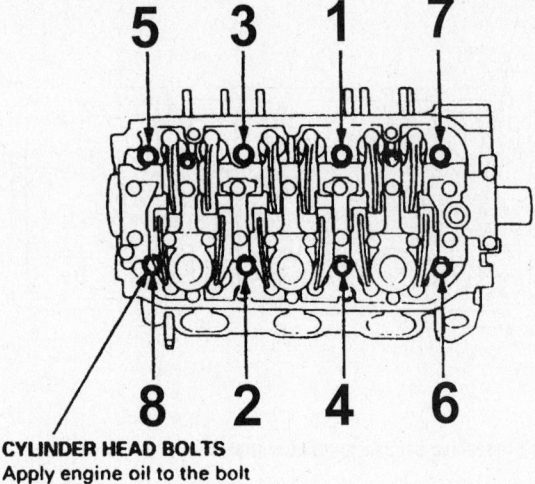

CYLINDER HEAD BOLTS
Apply engine oil to the bolt threads.

7923BG12

Cylinder head torque sequence—2002–03 3.2L engine

15. Tighten the cylinder head bolts in 3 steps. Be sure to follow the tightening torque sequence:
 a. Step 1: 29 ft. lbs. (39 Nm).
 b. Step 2: 51 ft. lbs. (69 Nm).
 c. Step 3: 72 ft. lbs. (98 Nm).
16. Install or connect the following:
 - Exhaust manifolds with new gaskets and torque the nuts to 23 ft. lbs. (31 Nm)
 - Rear timing belt covers and torque the bolts to 16 ft. lbs. (22 Nm)
17. Install the camshaft pulleys and torque to:
 a. 3.2L engines: 67 ft. lbs. (90 Nm).
 b. 3.5L engines: 23 ft. lbs. (31 Nm).
18. Install or connect the following:
 - Timing belt
 - Front timing belt cover and torque the bolts to 108 inch lbs. (12 Nm)
19. Check and adjust the valve clearance.
 - Crankshaft pulley and torque the bolt to 181 ft. lbs. (245 Nm)
 - Water passage assembly and torque the bolts to 16 ft. lbs. (22 Nm)
 - Intake manifold with new gaskets and O–rings and tighten the bolts to 16 ft. lbs. (22 Nm)
 - Cylinder head cover and tighten the bolts to 108 inch lbs. (12 Nm)
 - Ignition coils and torque the bolts to 108 inch lbs. (12 Nm)
 - Exhaust manifolds and torque the bolts to 23 ft. lbs. (31 Nm)
 - Upper and lower radiator hoses
 - Heater hoses
 - IAB vacuum hoses
 - Fuel rails and torque the bolts to 84 inch lbs. (9.5 Nm)

- Fuel injector connectors
- Intake manifold and torque the bolts to 16 ft. lbs. (22 Nm)
- Engine oil pressure switch connector
- EGR valve connector
- TDC sensor connector
- CKP sensor connector
- Radiator fan switch connectors
- MAP sensor connector
- ECT sensor connector
- TPS sensor connector
- IAC sensor connector
- IAT sensor connector
- Spark plug wires
- Distributor
- Alternator and torque the upper bolt to 16 ft. lbs. (22 Nm) and the lower bolt to 33 ft. lbs. (44 Nm)
- Ground cable
- Power steering pump and torque the bolts to 17 ft. lbs. (24 Nm)
- Accessory drive belts
- Side engine mount and torque the mounting bolts to 33 ft. lbs. (44 Nm) and the through bolt to 40 ft. lbs. (54 Nm)
- Wire harness holder
- IMRC actuator
- PCV hose
- Brake booster vacuum hose
- Fuel feed and return lines and torque the fitting to 16 ft. lbs. (22 Nm)
- Accelerator and cruise control cables
- Engine covers
- Intake air duct
- Evaporative Emissions (EVAP) canister hose from the throttle body
- TCS control valve

- Strut brace and torque the bolts to 16 ft. lbs. (22 Nm)
- Engine covers and torque the bolts to 108 inch lbs. (12 Nm)

20. Change the engine oil and filter.
21. Fill and bleed the cooling system.
22. Connect the negative battery cable.
23. Bring the engine to operating temperature and inspect for any fluid leaks. Top off all fluid levels as necessary.
24. Enter the security code for the radio.

➡The PCM idle memory must be reset after reconnecting the battery. Start the engine and hold it at 3000 rpm until the cooling fan comes on. Then allow the engine to idle for about 5 minutes with all accessories OFF and with the transmission in Park or Neutral.

2004–06 TL Models

➡Make sure you have the anti-theft codes for the radio and the navigation system, then write down the radio channel presets. Make sure the ignition switch is OFF.

1. Before servicing the vehicle, refer to the precautions section.
2. Relieve the fuel system pressure.
3. Disconnect the battery negative cable from the battery.
4. Drain the engine coolant.
5. Remove the front warm up three way catalytic converter (front WU-TWC) and rear warm up three way catalytic converter (rear WU-TWC).
6. Remove the drive belt.
7. Remove the timing belt.
8. Remove the Power Steering (PS) pump, and PS hose bracket.
9. Remove the alternator.
10. Remove the intake manifold.
11. Remove the six ignition coils.
12. Remove the engine wire harness connectors and wire harness clamps from the cylinder head as follows:
   - Six injector connectors
   - Engine Coolant Temperature (ECT) sensor connector
   - Crankshaft Position (CKP) sensor A and B connector
   - Exhaust Gas Recirculation (EGR) valve connector
   - VTEC solenoid valve connector
   - VTEC oil pressure switch connector
   - Oil pressure switch connector
   - Two air fuel ratio (A/F) sensor connectors
   - Two secondary heated oxygen sensor (secondary HO2S) connectors
13. Remove the upper radiator hose and lower radiator hose.

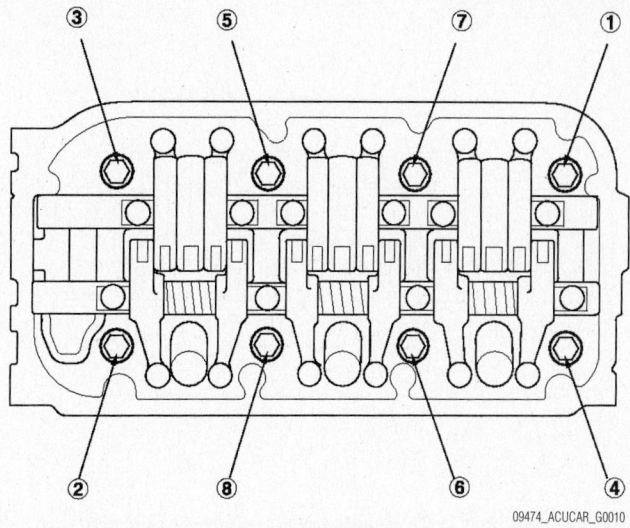

09474_ACUCAR_G0010

Remove the cylinder head bolts in the sequence shown using several passes—2004-06 TL model

14. Remove the heater hose and water bypass hose(s).
15. Remove the bolt retaining the harness holder.
16. Remove the bolt retaining the harness bracket.
17. Remove the fuel rails
18. Remove the bolt retaining the harness bracket.
19. Remove the water passage
20. Remove the front and rear camshaft pulleys and front and rear back covers.
21. Remove the cylinder head covers.
22. Remove the cylinder head bolts. To prevent warpage, unscrew the bolts in sequence 1/3 turn at a time; repeat the sequence until all bolts are loosened.
23. Remove the cylinder heads.
   To install:
24. Clean the cylinder head and block surface.
25. Clean and install the oil control orifices with new O-rings.
26. Install the dowel pins and new cylinder head gaskets.
27. Put the cylinder head onto the engine block.
28. Clean the timing belt pulleys, timing belt guide plate, and the upper and lower covers.
29. Set the timing belt drive pulley to Top Dead Center (TDC) by aligning the TDC mark on the tooth of the timing belt drive pulley with the pointer on the oil pump.
30. Set the camshaft pulleys to TDC by aligning the TDC marks on the camshaft pulleys with the pointers on the back covers.
31. Apply new engine oil to the threads and flanges of the cylinder head bolts

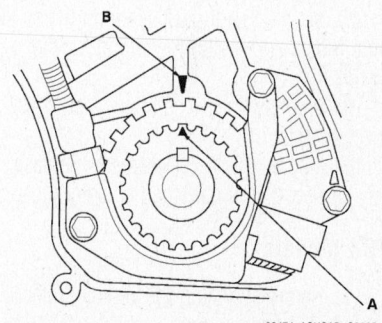

09474_ACUCAR_G0035

Align the TDC mark on the tooth of the timing belt drive pulley with the pointer on the oil pump —2004-06 TL model

➡Perform each step twice.

32. Tighten the 6 point cylinder head bolts sequentially in three steps:
   a. Step 1: 29 ft. lbs. (39 Nm).
   b. Step 2: 51 ft. lbs. (69 Nm).
   c. Step 3: 72 ft. lbs. (98 Nm).
33. Tighten the 12 point cylinder head bolts sequentially in three steps:
   a. Step 1: 29 ft. lbs. (39 Nm).
   b. Step 2: plus 90 degrees
   c. Step 3: plus 90 degrees
   d. Step 4: if using new bolts: plus 90 degrees

➡Use a beam-type torque wrench. When using a preset-type torque wrench, be sure to tighten slowly and not to overtighten. If a bolt makes any noise while you are torquing it, loosen the bolt, and retighten it from the 1st step.

34. Adjust the valve clearance.
35. Install the cylinder head covers.

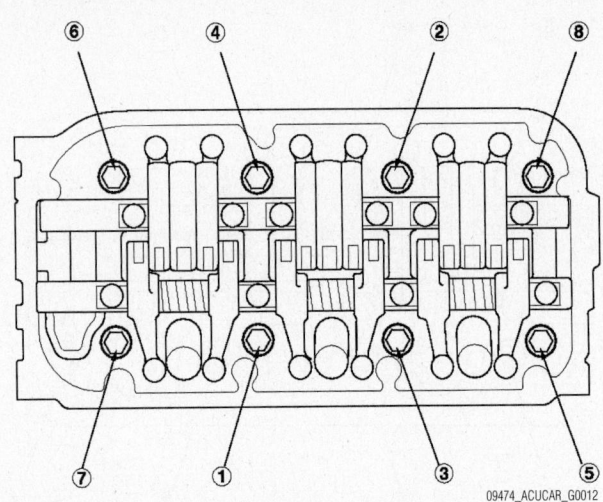

Tighten the cylinder head bolts in the sequence shown—2004–06 TL model

09474_ACUCAR_G0012

36. Install the timing belt.

37. Clean the head cover contacting surfaces with a shop towel.

38. Set the spark plug seals on the spark plug tubes, and install the cylinder head covers.

39. Visually check the spark plug seals for damage.

40. Inspect the cover washers. Replace any washer that is damaged or deteriorated.

TSX Models

1. Before servicing the vehicle, refer to the precautions section.

2. Obtain the security code for the radio.

3. Disconnect the negative battery cable.

4. Drain the coolant.

5. Remove or disconnect the following:
   - Intake manifold cover
   - 4 ignition coils
   - 2 bolts securing the vacuum line
   - Bolt securing the power steering hose bracket
   - Dipstick and breather hose
   - Retainers and cylinder head cover
   - Fuel line
   - Drive belt
   - Intake Air Temperature (IAT) sensor connector
   - Vacuum hose (B) and breather pipe (C), then the intake air duct (D)
   - Bolt securing the connecting pipe
   - Evaporative emission (EVAP) canister hose and brake booster vacuum hose
   - Intake manifold
   - Exhaust manifold
   - Positive Crankcase Ventilation (PCV) hose, vacuum hose and ground cable
   - Upper radiator hose, heater hoses and water bypass hose
   - Engine wire harness connectors and wire harness clamps from the cylinder head

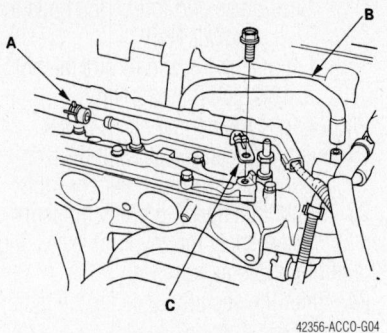

Disconnect the PCV hose (A), vacuum hose (B) and ground cable (C)—TSX models

   - 4 injector connectors
   - Engine Coolant Temperature (ECT) sensor connector
   - Camshaft Position (CMP) sensor connectors
   - Exhaust Gas Recirculation (EGR) valve connector
   - VTEC solenoid valve connector
   - Engine Oil Pressure (EOP) sensor connector
   - 2 bolts securing the EVAP canister purge valve bracket and the bolt securing the harness bracket
   - Timing chain
   - Rocker arm assembly
   - Cylinder head bolts, in sequence, ⅓ turn at a time until completely loosened
   - Cylinder head. Discard the gasket.

To install:

6. Be sure all cylinder head and block gasket surfaces are clean. Check the cylinder head for warpage. If warpage is less than 0.002 in. (0.05mm), cylinder head resurfacing is not required. Maximum resurface limit is 0.008 in. (0.2mm) based on a cylinder head height of 3.94 in. (100mm).

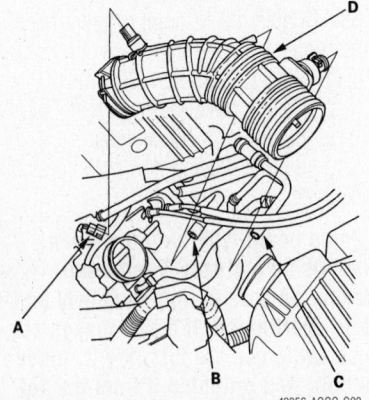

Remove the vacuum hose (B), breather pipe (C), then remove the intake air duct (D)—TSX models

42356-ACCO-G03

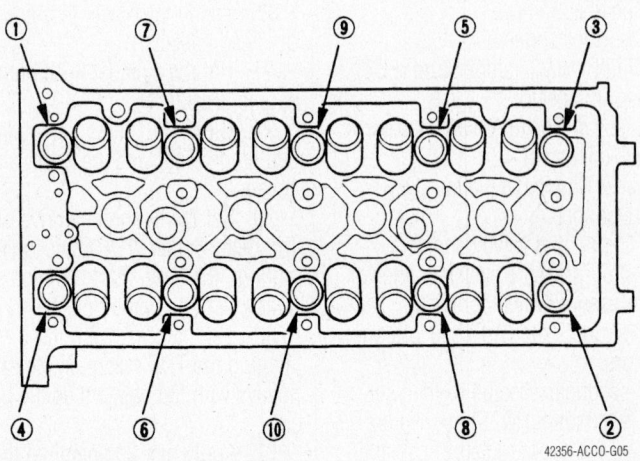

Cylinder head bolt loosening sequence—TSX models

42356-ACCO-G05

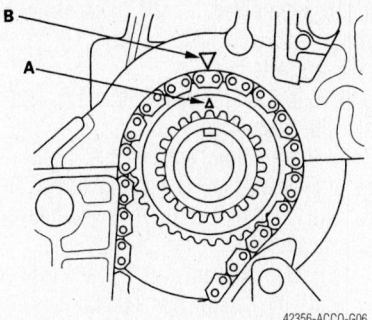

Set the crankshaft to TDC by aligning the mark (A) on the crankshaft sprocket with the pointer (B) on the cylinder block—TSX models

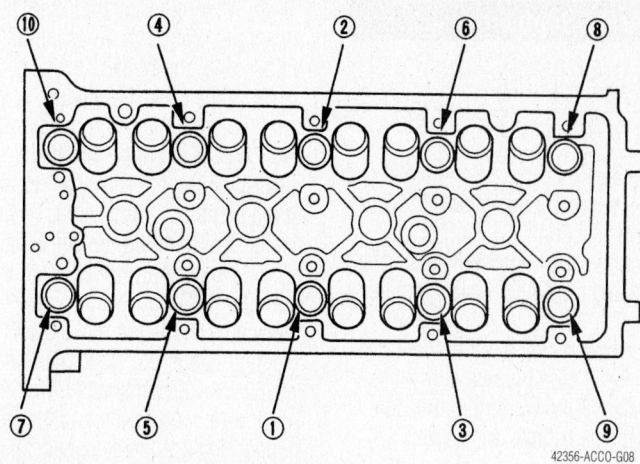

Cylinder head bolt tightening sequence—TSX models

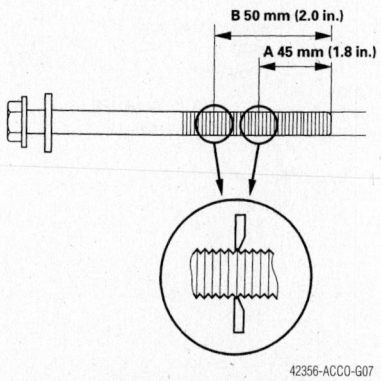

You must measure the cylinder head bolts to see if they can be reused or need to be replaced—TSX models

7. Install or connect the following:
   • New gasket and dowel pins on the cylinder block
8. Set the crankshaft to Top Dead Center (TDC). Align the TDC mark (A) on the crankshaft sprocket with the pointer (B) on the cylinder block.
9. Measure the diameter of each cylinder head bolt at points A & B, as shown in the illustration. If either diameter is less than 0.42 in. (10.6mm), replace the head bolt
10. Apply engine oil to the threads and under the bolt heads of all of the bolts.
11. Install the cylinder head. Tighten the bolts in sequence as follows:
   a. Step 1: 29 ft. lbs. (39 Nm).
   b. Step 2: Plus 90 degrees.
   c. Step 3: Plus 90 degrees.
   d. Step 4: If using new cylinder head bolts, add an additional 90 degrees.
12. Install or connect the following:
   • Rocker arm assembly
   • Timing chain
   • 2 bolts securing the EVAP canister purge valve bracket and tighten to 16 ft. lbs. (22 Nm)

• Bolt securing the harness bracket and tighten to 104 ft. lbs. (12 Nm)
• Upper radiator hose, heater hoses and water bypass hose
• PCV hose, vacuum hose and ground cable
• Exhaust manifold
• Intake manifold
• EVAP canister hose and brake booster vacuum hose
• Fuel line
• Bolt securing the connecting pipe and tighten to 16 ft. lbs. (22 Nm)
• Intake air duct, IAT sensor connector, vacuum hose and breather pipe
• Cylinder head cover gasket in the groove of the cylinder head cover
• Apply liquid gasket P/N 08718-0009 or equivalent on the chain cover and No. 5 rocker shaft holder mating areas. The parts must be installed within 5 minutes of applying liquid gasket.

• Spark plug seals on the spark plug tubes
• Cylinder head cover on the cylinder head, then slide the cover back and forth gently to seat it
• Cover washers
• Cylinder head cover bolts. Torque, in sequence, in 2 or 3 steps to 104 inch lbs. (12 Nm).
• Dipstick and breather hose
• Bolt securing the power steering hose bracket
• 2 bolts securing the vacuum line
• 4 ignition coils
• Intake manifold cover, and tighten the retainers to 104 inch lbs. (12 Nm)
• All of the remaining hoses, tubes, and connectors are installed correctly.
13. Fill the cooling system.
14. Connect the negative battery cable and enter the radio security code.
15. Start the engine and check carefully for any leaks.

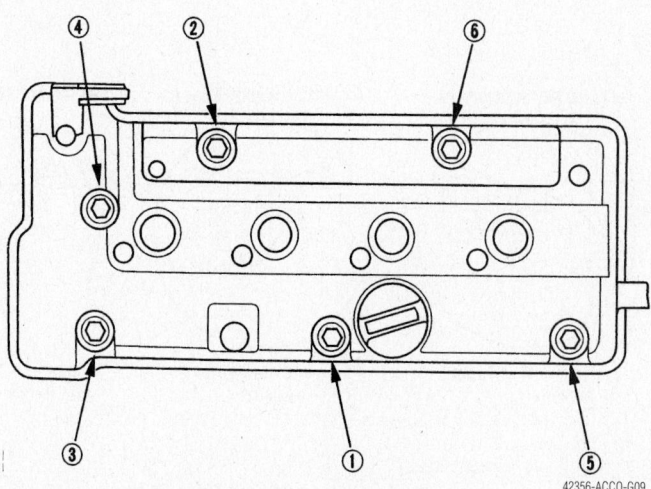

Cylinder head cover bolt tightening sequence—TSX models

Rocker Arms/Shafts

REMOVAL & INSTALLATION

CL Model

1. Before servicing the vehicle, refer to the precautions section.

2. Disconnect the negative battery cable.

3. Remove the cylinder head from the vehicle.

4. Loosen the rocker shaft holder bolts 1 turn at a time in the opposite of the installation sequence. Following this procedure will prevent the camshafts and rocker assemblies from warping.

5. After all bolts are loose, remove the rocker arm shafts as an assembly with the bolts still in the holders.

6. If the rocker shafts are to be disassembled, note that each rocker arm has a letter **A** or **B** stamped into the side. Before disassembling the rocker arms, make a note of the position of each letter so the arms can be reassembled the same way.

7. Do not remove the hydraulic tappets from the rocker arms unless they are to be replaced. Handle the rocker arms carefully so the oil does not drain out of the tappets.

8. Lift the camshafts from the cylinder head, wipe them clean and inspect the lift ramps. Replace the camshafts and rockers if the lobes are pitted, scored, or excessively worn.

To install:

9. Lubricate the camshaft and its journals with fresh engine oil.

10. Place a new camshaft seal on the end of the camshaft. The spring side of the seal must face in. Lubricate the journals and set the camshaft in place on the head.

11. Install the camshaft onto the cylinder head with the keyway pointed up.

12. Apply liquid gasket to the mounting surfaces of the camshaft end holders.

13. Set the rocker arm assemblies in place and start all the cam holder bolts. Be sure the rocker arms are properly positioned

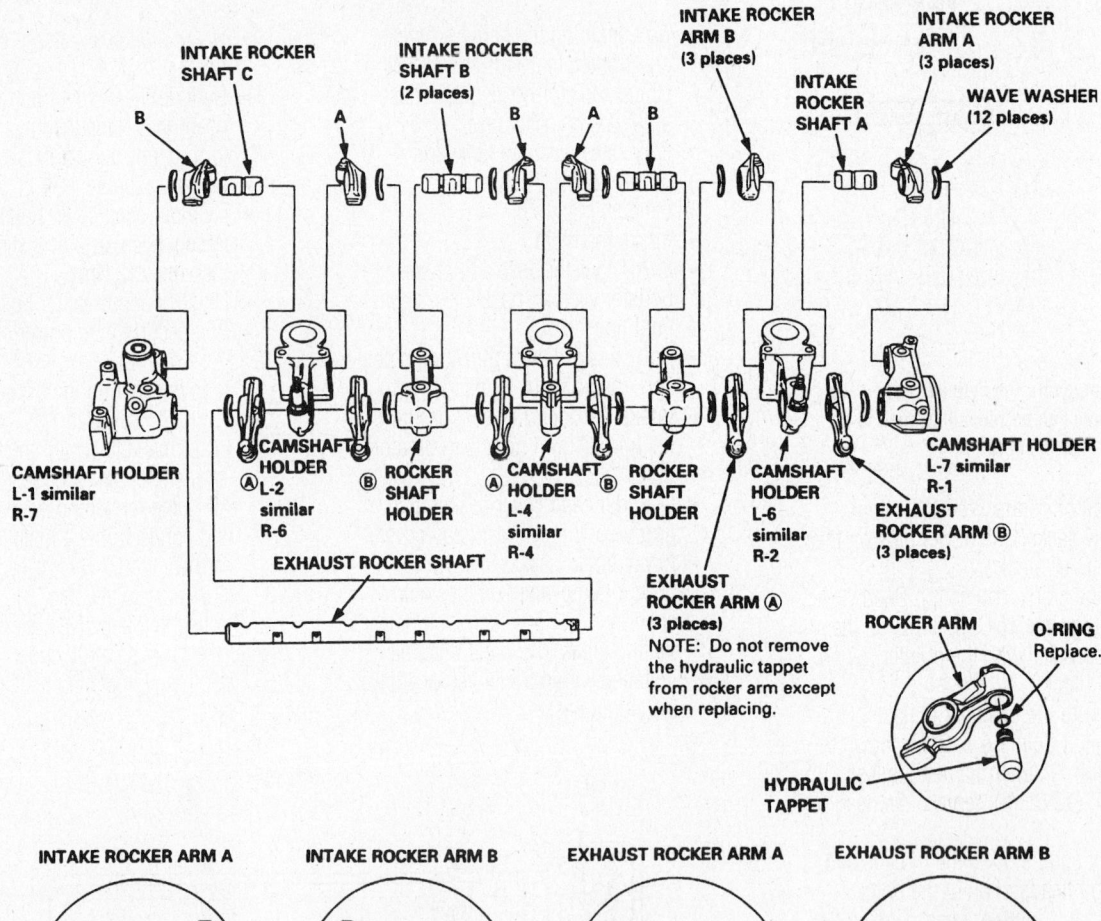

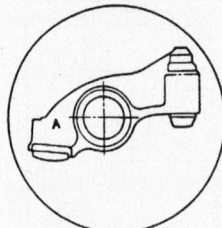

INTAKE ROCKER ARM A

Letter "A" is stamped on rocker arm.

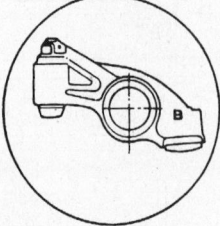

INTAKE ROCKER ARM B

Letter "B" is stamped on rocker arm.

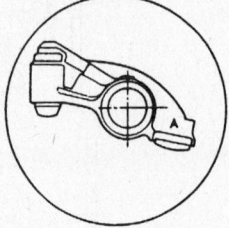

EXHAUST ROCKER ARM A

Letter "A" is stamped on rocker arm.

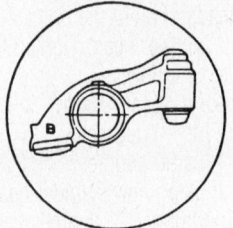

EXHAUST ROCKER ARM B

Letter "B" is stamped on rocker arm.

L: Left
R: Right

Exploded view of the rocker arms and related components—CL with 3.2L engine

7923BG84

and turn each bolt in sequence 2 turns at a time until the holders are seated on the head. Follow this procedure to avoid damaging the camshaft and rocker assemblies.

14. When all the camshaft and rocker holders are seated, tighten the bolts in the same sequence. Tighten the 8mm bolts to 16 ft. lbs. (22 Nm) and the 6mm bolts to 104 inch lbs. (12 Nm).

15. Install or connect the following:
- Cylinder head
- Negative battery cable

16. Check for proper engine and valve train operation.

2002–04 RL Model

1. Before servicing the vehicle, refer to the precautions section.
2. Disconnect the negative battery cable.
3. Remove the cylinder head from the vehicle.
4. Loosen the rocker shaft holder bolts 1 turn at a time in the opposite of the installation sequence. Following this procedure

will prevent the camshafts and rocker assemblies from warping.

5. After all bolts are loose, remove the rocker arm shafts as an assembly with the bolts still in the holders.

6. If the rocker shafts are to be disassembled, note that each rocker arm has a letter **A** or **B** stamped into the side. Before disassembling the rocker arms, make a note of the position of each letter so the arms can be reassembled the same way.

7. Do not remove the hydraulic tappets

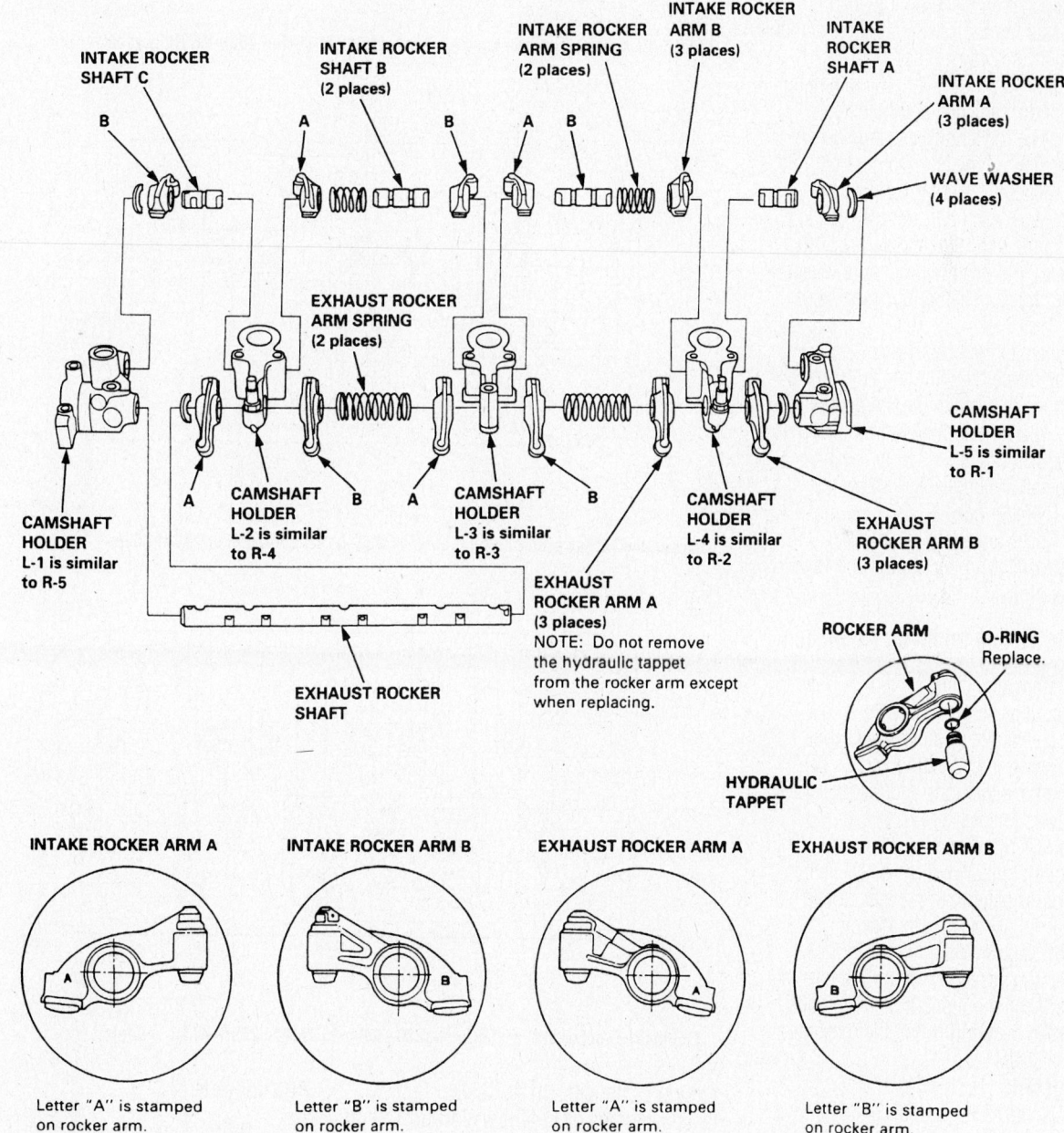

Exploded view of the rocker arms and related components—2002–04 3.5L engine

from the rocker arms unless they are to be replaced. Handle the rocker arms carefully so the oil does not drain out of the tappets.

8. Lift the camshafts from the cylinder head, wipe them clean and inspect the lift ramps. Replace the camshafts and rockers if the lobes are pitted, scored, or excessively worn.

To install:

9. Lubricate the camshaft and its journals with fresh engine oil.

10. Place a new camshaft seal on the end of the camshaft. The spring side of the seal must face in. Lubricate the journals and set the camshaft in place on the head.

11. Install the camshaft onto the cylinder head with the keyway pointed up.

12. Apply liquid gasket to the mounting surfaces of the camshaft end holders.

13. Set the rocker arm assemblies in place and start all the cam holder bolts. Be sure the rocker arms are properly positioned and turn each bolt in sequence 2 turns at a time until the holders are seated on the head. Follow this procedure to avoid damaging the camshaft and rocker assemblies.

14. When all the camshaft and rocker holders are seated, tighten the bolts in the same sequence. Tighten the 8mm bolts to 16 ft. lbs. (22 Nm) and the 6mm bolts to 104 inch lbs. (12 Nm).

15. Install or connect the following:
- Cylinder head
- Negative battery cable

16. Check for proper engine and valve train operation.

2004–06 TL and 2005–06 RL Models

1. Before servicing the vehicle, refer to the precautions section.

2. Remove the cylinder head cover.

3. Loosen the adjusting screws (A).

4. Remove the bolts and the rocker arm assembly as follows:

a. Step 1: Unscrew the rocker shaft mounting bolts two turns at a time, in a crisscross pattern, to prevent damaging the valves or rocker arm assembly.

b. Step 2: When removing the rocker arm assembly, do not remove the rocker shaft mounting bolts. The bolts will keep the springs and the rocker arms on the shafts

To install:

5. Set the rocker arm assembly in place, and loosely install the bolts. Make sure that the rocker arms are properly positioned on the valve stems.

6. Tighten each bolt two turns at a time in the sequence shown to ensure that the

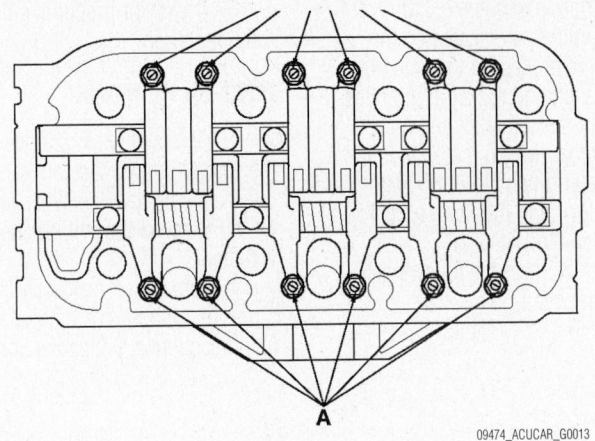

Loosen the rocker arm adjusting screws—2004–06 TL and 2005–06 RL models

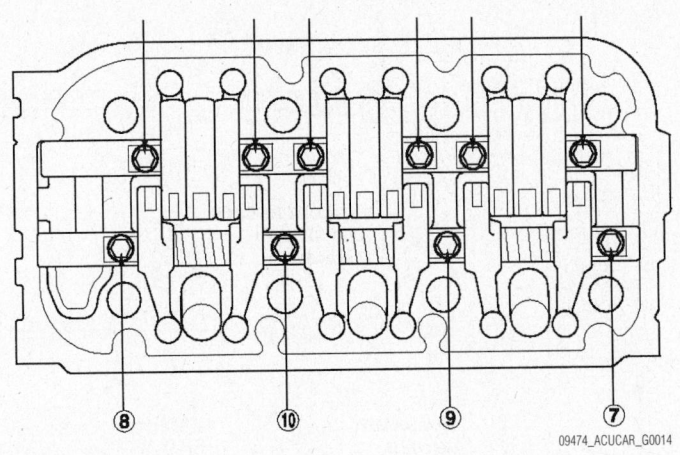

Loosen the rocker arm mounting bolts—2004–06 TL and 2005–06 RL models

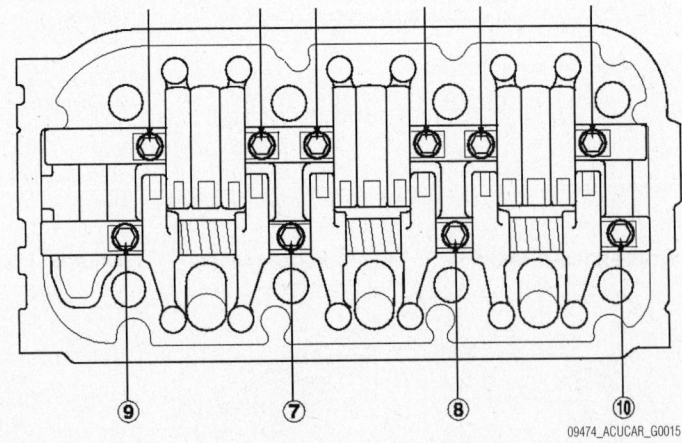

Tighten the rocker arm mounting bolts—2004–06 TL and 2005–06 RL models

rockers do not bind on the valves. Tighten to a final torque of 17 ft. lbs. (24 Nm).

➡**Apply new engine oil to the threads and flange of the exhaust rocker shaft mounting bolts.**

7. Install the cylinder head cover.

RSX Models

1. Before servicing the vehicle, refer to the precautions section.

2. Remove or disconnect the following:
- Negative battery cable
- Cylinder head from the vehicle

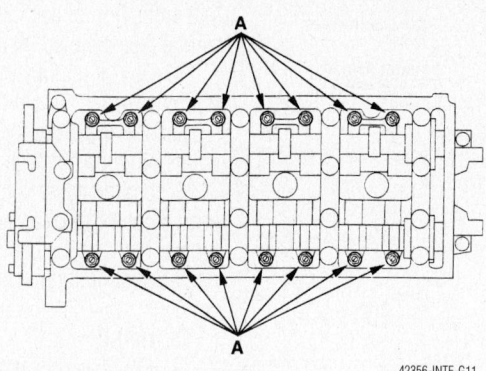

Loosen the rocker arm adjusting screws (A)—RSX models

42356-INTE-G11

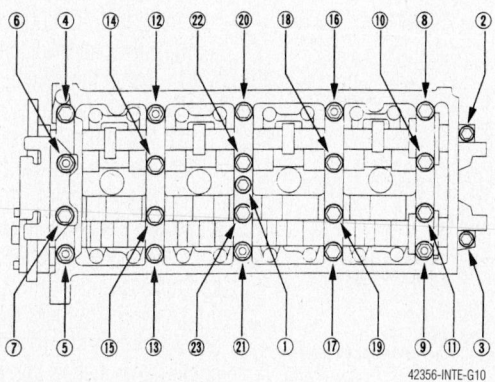

Camshaft holder bolt loosening sequence—RSX models

42356-INTE-G10

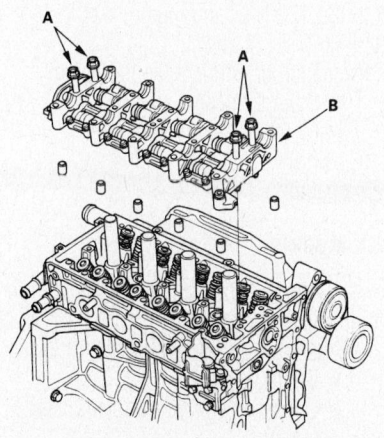

Insert the bolts (A) into the rocker shaft holder, then remove the rocker arm assembly (B)—RSX models

42356-INTE-G12

- Loosen the rocker arm adjusting screws
- Camshaft holder bolts, 2 turns at a time in a criss-cross pattern
- Timing chain guide B, camshaft holders and camshafts

3. Insert the bolts into the rocker shaft holder, then remove the rocker arm assembly.

4. Disassemble the rocker shafts as necessary.

To install:

5. Clean and dry the No. 5 rocker shaft holder mating surface.

6. Apply liquid gasket part No. 08718–0009 evenly to the cylinder head mating surface of the No. 5 rocker shaft holder and install within 5 minutes.

7. Reassemble the rocker arm assembly.

8. Insert the bolts into the rocker shaft holder, then install the rocker arm assembly on the cylinder head. Remove the bolts from the rocker shaft holder.

9. Make sure the punch marks on the VTC actuator and exhaust camshaft sprocket are facing up, then set the camshafts in the holder.

10. Set the camshaft holders and timing chain guide B in place.

11. Tighten the bolts, in sequence, to the following specifications:
- 8mm bolts: 16 ft. lbs. (22 Nm)
- 6mm bolts (nos. 21, 22 & 23): 104 inch lbs. (12 Nm)

12. Install or connect the following:
- Timing chain, then adjust the valve clearance
- Cylinder head
- Negative battery cable

13. Check for proper engine and valve train operation.

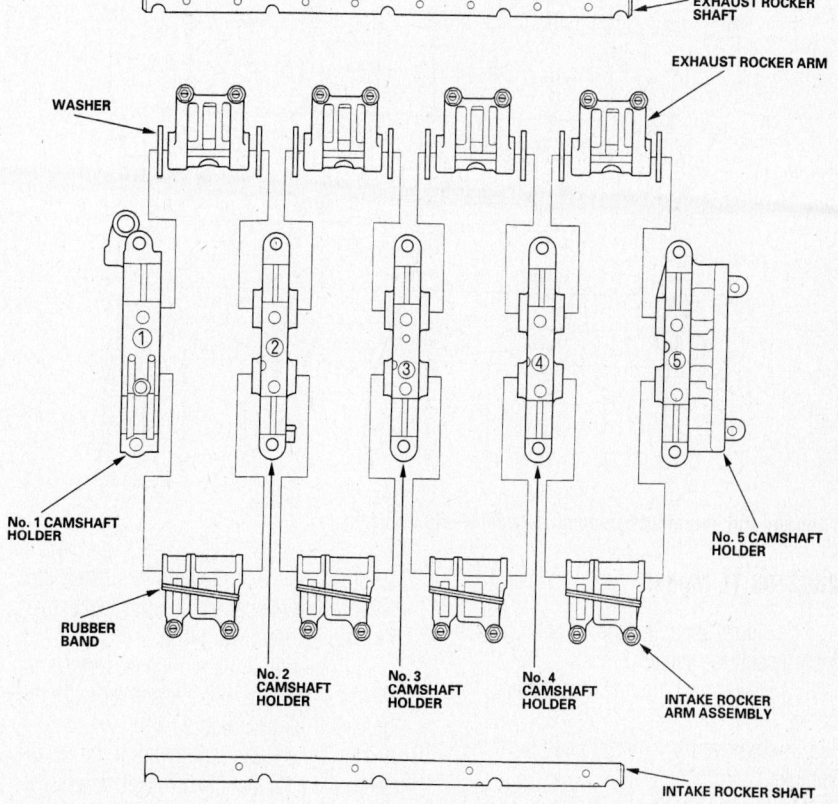

Rocker arm disassembly—2.0L (K20A3) engine

42356-INTE-G13

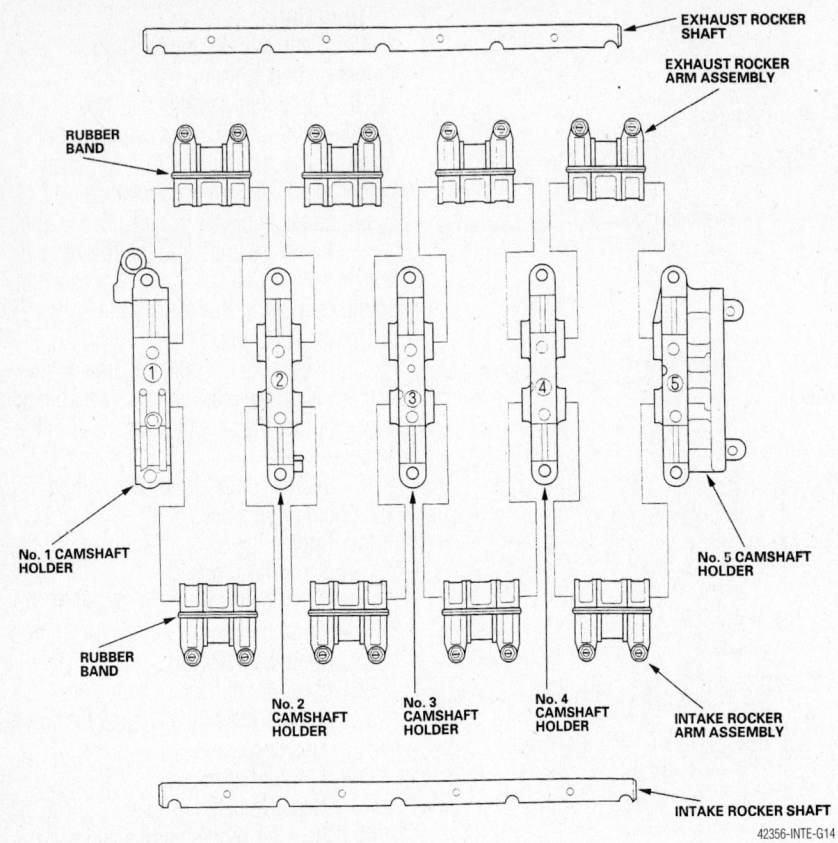

Rocker arm disassembly—2.0L (K20A2/K20Z1) engines

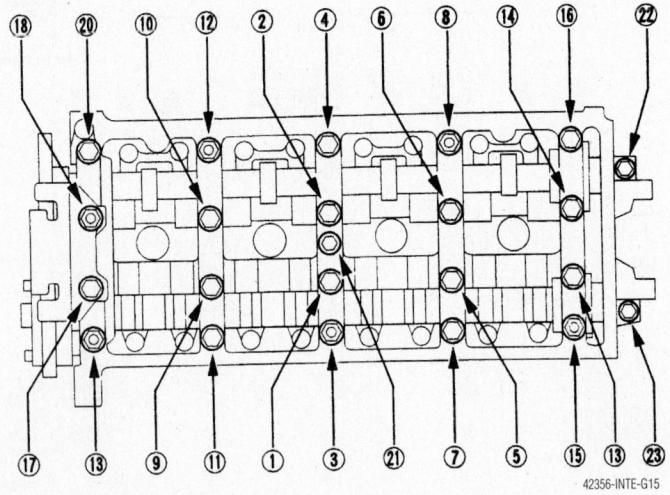

Rocker arm assembly tightening sequence—RSX models

2002–03 TL Models

1. Before servicing the vehicle, refer to the precautions section.

2. Disconnect the negative battery cable.

3. Remove the cylinder head from the vehicle.

4. Loosen the rocker shaft holder bolts 1 turn at a time in the opposite of the installation sequence. Following this procedure will prevent the camshafts and rocker assemblies from warping.

5. After all bolts are loose, remove the rocker arm shafts as an assembly with the bolts still in the holders.

6. If the rocker shafts are to be disassembled, note that each rocker arm has a letter **A** or **B** stamped into the side. Before disassembling the rocker arms, make a note of the position of each letter so the arms can be reassembled the same way.

7. Do not remove the hydraulic tappets from the rocker arms unless they are to be replaced. Handle the rocker arms carefully so the oil does not drain out of the tappets.

8. Lift the camshafts from the cylinder head, wipe them clean and inspect the lift ramps. Replace the camshafts and rockers if the lobes are pitted, scored, or excessively worn.

To install:

9. Lubricate the camshaft and its journals with fresh engine oil.

10. Place a new camshaft seal on the end of the camshaft. The spring side of the seal must face in. Lubricate the journals and set the camshaft in place on the head.

11. Install the camshaft onto the cylinder head with the keyway pointed up.

12. Apply liquid gasket to the mounting surfaces of the camshaft end holders.

13. Set the rocker arm assemblies in place and start all the cam holder bolts. Be sure the rocker arms are properly positioned and turn each bolt in sequence 2 turns at a time until the holders are seated on the head. Follow this procedure to avoid damaging the camshaft and rocker assemblies.

14. When all the camshaft and rocker holders are seated, tighten the bolts in the same sequence. Tighten the 8mm bolts to 16 ft. lbs. (22 Nm) and the 6mm bolts to 104 inch lbs. (12 Nm).

15. Install or connect the following:
   • Cylinder head
   • Negative battery cable

16. Check for proper engine and valve train operation.

TSX Models

1. Before servicing the vehicle, refer to the precautions section.

2. Remove or disconnect the following:
   • Timing chain
   • Loosen the rocker arm adjusting screws
   • Camshaft holder bolts, two turns at a time in sequence
   • Timing chain guide (B), camshaft and camshafts

3. Insert the bolts (A) into the rocker shaft holder, then remove the rocker arm assembly (B)

To install:

4. Clean and dry the No. 5 rocker shaft holding mating surface.

5. Apply a suitable liquid gasket P/N 08718-0009, or equivalent, evenly to the cylinder head mating surface of the No. 5 rocker shaft holder.

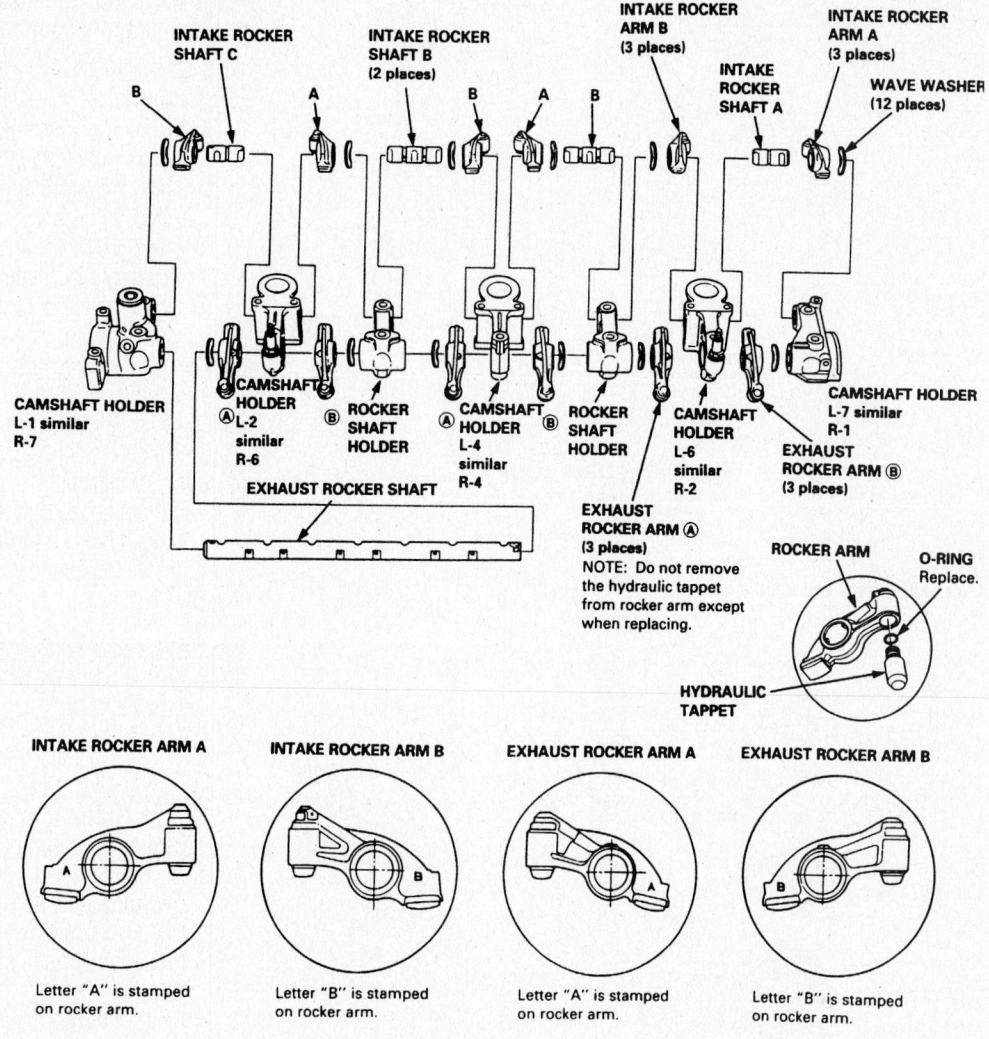

Exploded view of the rocker arms and related components—2002–03 3.2L engine

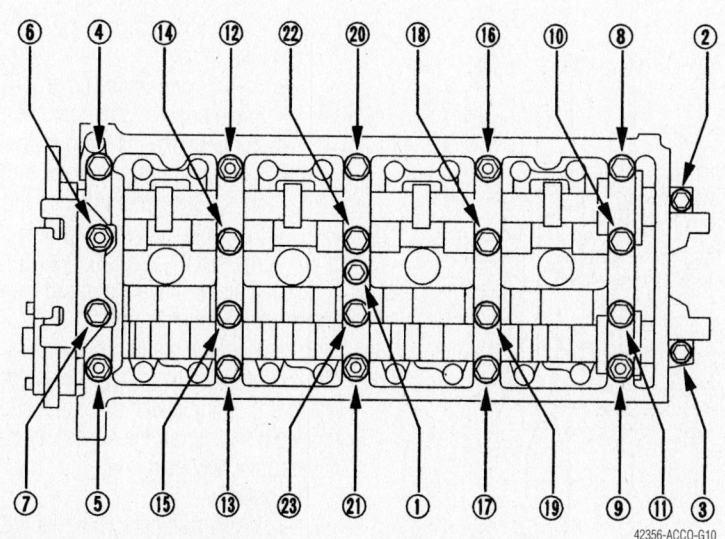

Camshaft holder bolt loosening sequence—TSX models

42356-ACCO-G10

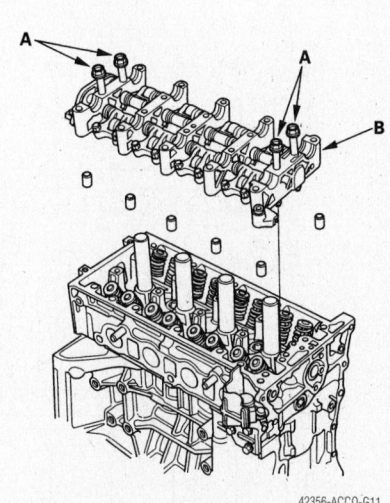

42356-ACCO-G11

Insert the bolts (A) into the rocker shaft holder, then remove the rocker arm assembly (B)—TSX models

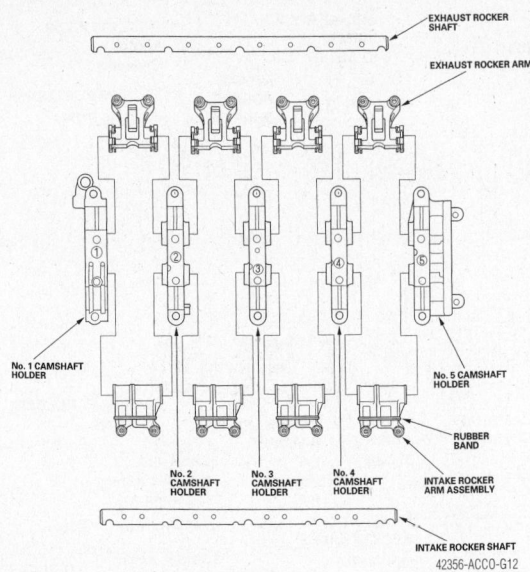

Exploded view of the rocker arms and related components—TSX models

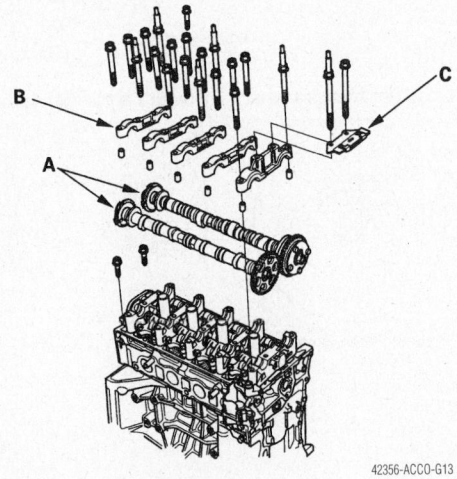

When installing the camshafts (A) make sure the punch marks on the VTC actuator and exhaust cam sprockets are facing up—TSX models

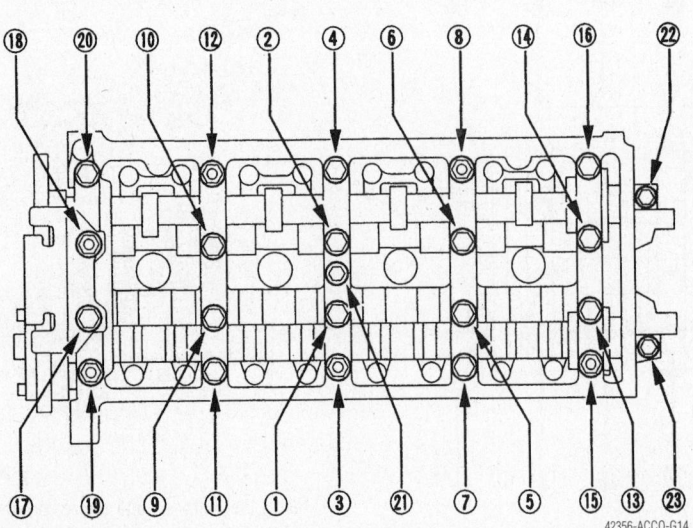

Rocker arm assembly bolt tightening sequence—TSX models

➡ **The parts must be installed within 5 minutes of applying the liquid gasket.**

6. Reassemble the rocker arm assembly, as necessary.

7. Install or connect the following
- Bolts (A) into the rocker shaft holder, then the rocker arm assembly on the cylinder head. Remove the bolts from the rocker shaft holder.

8. Make sure the punch marks on the variable valve timing control (VTC) actuator and exhaust camshaft sprocket are facing up, then set the camshafts (A) in the holder

9. Set the camshaft holders (B) and timing chain guide B (C) in place.

10. Tighten the bolts, in sequence, to the following specification:
  a. 8mm bolts: 16 ft. lbs. (22 Nm)
  b. 6mm bolts: 104 inch lbs. (12 Nm)

11. Install the timing chain and adjust the valve lash.

Intake Manifold

REMOVAL & INSTALLATION

CL Model

1. Before servicing the vehicle, refer to the precautions section.
2. Disconnect the negative battery cable.
3. Drain the cooling system.
4. Properly relieve the fuel system pressure.
5. Remove or disconnect the following:
- Intake air duct
- Strut brace (if equipped)
- Intake manifold cover
- Water bypass hose
- Traction Control System (TCS) actuator (if equipped)
- Evaporative Emission (EVAP) canister hose
- Throttle and cruise control cables
- Brake booster vacuum hose
- Upper intake manifold cover
- Intake manifold nuts and bolts in a crisscross pattern, beginning from the center and moving out

6. Verify that all vacuum lines are disconnected and remove the intake manifold and throttle body as a unit.

7. Inspect the manifold for cracks, flatness, or other damage; replace any damaged parts. If the intake manifold is to be replaced, transfer all the necessary components to the new manifold.

To install:
- Intake fasteners:16 ft. lbs. (22 Nm)
- Upper intake manifold cover fasteners: 108 inch lbs. (12 Nm)

- TCS actuator bolts: 16 ft. lbs. (22 Nm)
- Strut brace bolts: 16 ft. lbs. (22 Nm)

2002–04 RL Model

1. Before servicing the vehicle, refer to the precautions section.
2. Disconnect the negative battery cable.
3. Drain the cooling system.
4. Properly relieve the fuel system pressure.
5. Remove or disconnect the following:
   - Intake air duct
   - Strut brace (if equipped)
   - Intake manifold cover
   - Water bypass hose
   - Traction Control System (TCS) actuator (if equipped)
   - Evaporative Emission (EVAP) canister hose
   - Throttle and cruise control cables
   - Brake booster vacuum hose
   - Upper intake manifold cover
   - Intake manifold nuts and bolts in a crisscross pattern, beginning from the center and moving out
6. Verify that all vacuum lines are disconnected and remove the intake manifold and throttle body as a unit.
7. Inspect the manifold for cracks, flatness, or other damage; replace any damaged parts. If the intake manifold is to be replaced, transfer all the necessary components to the new manifold.

To install:

- Intake fasteners: 16 ft. lbs. (22 Nm)
- Upper intake manifold cover fasteners: 108 inch lbs. (12 Nm)
- TCS actuator bolts: 16 ft. lbs. (22 Nm)
- Strut brace bolts: 16 ft. lbs. (22 Nm)

2004–06 TL and 2005–06 RL Models

1. Before servicing the vehicle, refer to the precautions section.
2. Remove the intake manifold cover.
3. Remove the air intake duct.
4. Remove the engine mount control solenoid valve, Positive Crankcase Ventilation (PCV) hose, brake booster vacuum hose, and vacuum hose.
5. Remove the evaporative emission (EVAP) canister purge hose and water bypass hoses, then plug the water bypass hoses.
6. Remove the following engine wire harness connectors and wire harness clamps from the intake manifold:
   - Intake Air Temperature (IAT) sensor connector

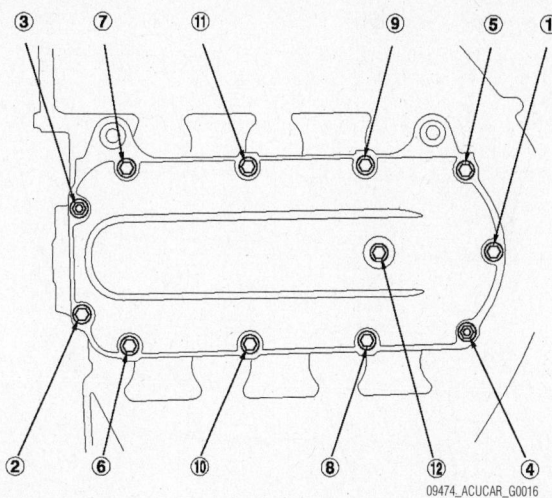

Remove the upper cover mounting bolts and nuts sequentially in two or three steps—2004–06 TL and 2005–06 RL models

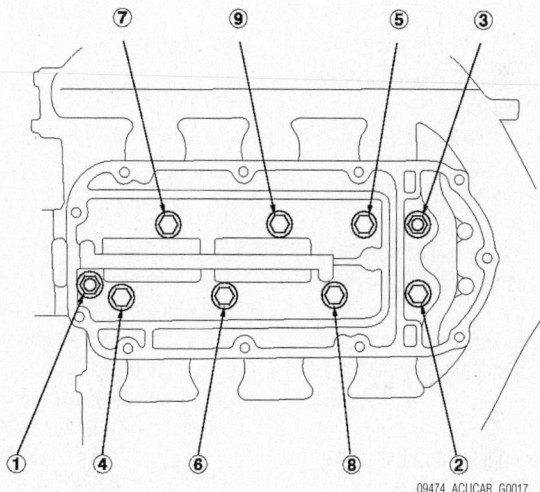

Remove the intake manifold mounting bolts and nuts sequentially in two or three steps—2004–06 TL and 2005–06 RL models

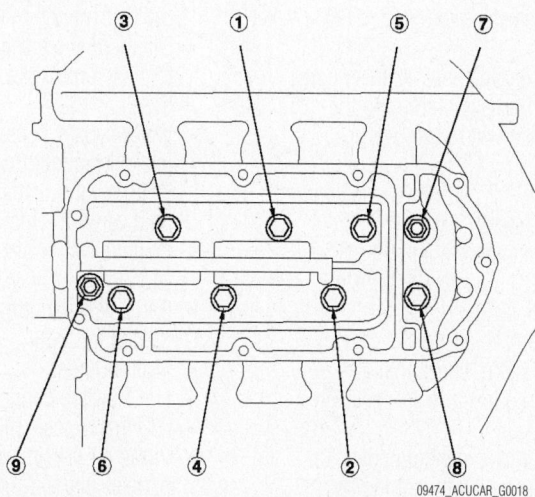

Tighten the intake manifold mounting bolts and nuts sequentially in two or three steps—2004–06 TL and 2005–06 RL models

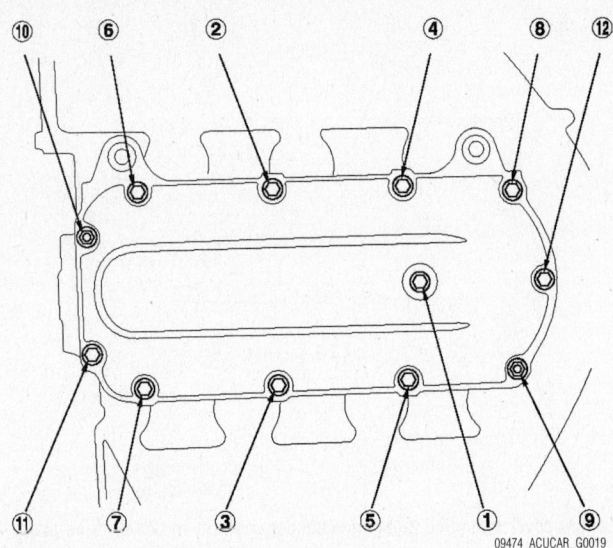

09474_ACUCAR_G0019

Tighten the upper cover mounting bolts and nuts sequentially in two or three steps—2004–06 TL and 2005–06 RL models

- Throttle actuator connector
- Manifold Absolute Pressure (MAP) sensor connector
- Evaporative Emission (EVAP) canister purge valve connector
- Intake Manifold Tuning (IMT) valve connector

7. Remove the upper cover mounting bolts and nuts sequentially in two or three steps.

8. Remove the intake manifold mounting bolts and nuts sequentially in two or three steps.

To install:

9. Install a new gasket and the intake manifold. Tighten the bolts and nuts sequentially in two or three steps to 16 ft. lbs. (22 Nm).

10. Install a new gasket and the upper cover. Tighten the bolts and nuts sequentially in two or three steps to 9 ft. lbs. (12 Nm).

11. Install the engine wire harness connectors and wire harness clamps to the intake manifold that where removed earlier.

12. Install the EVAP canister purge hose and water bypass hoses.

13. Install the engine mount control solenoid valve, PCV hose, brake booster vacuum hose, and vacuum hose.

14. Install the intake air duct.

15. After installation, check that all tubes, hoses, and connectors are installed correctly.

16. Install the intake manifold cover.

17. Refill the radiator with engine coolant, then bleed air from the cooling system with the heater valve open.

RSX Models

2.0L (K20A3) ENGINES

1. Before servicing the vehicle, refer to the precautions section.

2. Drain the cooling system into a sealable container.

3. Remove or disconnect the following:
- Negative battery cable
- Intake manifold cover
- Intake Air Temperature (IAT) sensor connector
- Breather hose
- Air cleaner housing
- Throttle cover. Fully open the throttle link and cruise control link by hand, then remove the throttle cable and cruise control cable from the links
- Locknut, loosen only, then the cables from the bracket
- Evaporative emission (EVAP) canister hose and brake booster vacuum hose
- Water bypass hoses, then plug them
- Intake Manifold Runner Control (IMRC) valve actuator control solenoid valve connector
- Positive Crankcase Ventilation (PCV) hose
- IMRC valve control solenoid valve mounting bolt
- Front bumper
- Hood switch connector
- A/C line bracket mounting bolt
- Intake air duct mounting bolt and harness clamps
- Upper bracket and cushion mounting bolts, then the bulkhead

- Idle Air Control (IAC) valve connector
- Throttle Position (TP) sensor connector
- Manifold Absolute Pressure (MAP) sensor connector
- Evaporative Emission (EVAP) canister purge valve connector
- Intake Manifold Runner Control (IMRC) valve position sensor connector
- Intake manifold and O–rings

To install:

4. Install or connect the following:
- Intake manifold, with new O–rings. Tighten the bolts/nuts in a criss–cross pattern in 2 or 3 steps to 16 ft. lbs. (22 Nm).
- Bulkhead, and upper bracket and cushion mounting bolts
- Hood switch connector
- A/C line bracket mounting bolt and tighten to 104 inch lbs. (10 Nm)
- Intake air duct mounting bolt and tighten to 104 inch lbs. (10 Nm)
- Harness clamps
- Front bumper
- IMRC valve actuator control solenoid valve connector
- PCV hose and IMRC valve actuator control solenoid valve mounting bolt
- Water bypass hoses
- EVAP canister hose and brake booster vacuum hose
- Throttle and cruise control cables. Adjust the cables.
- Air cleaner housing
- IAT sensor connector
- Breather hose
- Intake manifold cover and tighten to 104 inch lbs. (10 Nm)
- Negative battery cable

5. Fill the engine with coolant, then start the engine and check for leaks.

2.0L K20A2 AND K20Z1 ENGINES

1. Before servicing the vehicle, refer to the precautions section.

2. Drain the cooling system into a sealable container.

3. Remove or disconnect the following:
- Negative battery cable
- Intake manifold cover
- Evaporative emission (EVAP) canister hose and brake booster vacuum hose
- Intake Air Temperature (IAT) sensor connector
- Breather hose
- Air cleaner housing

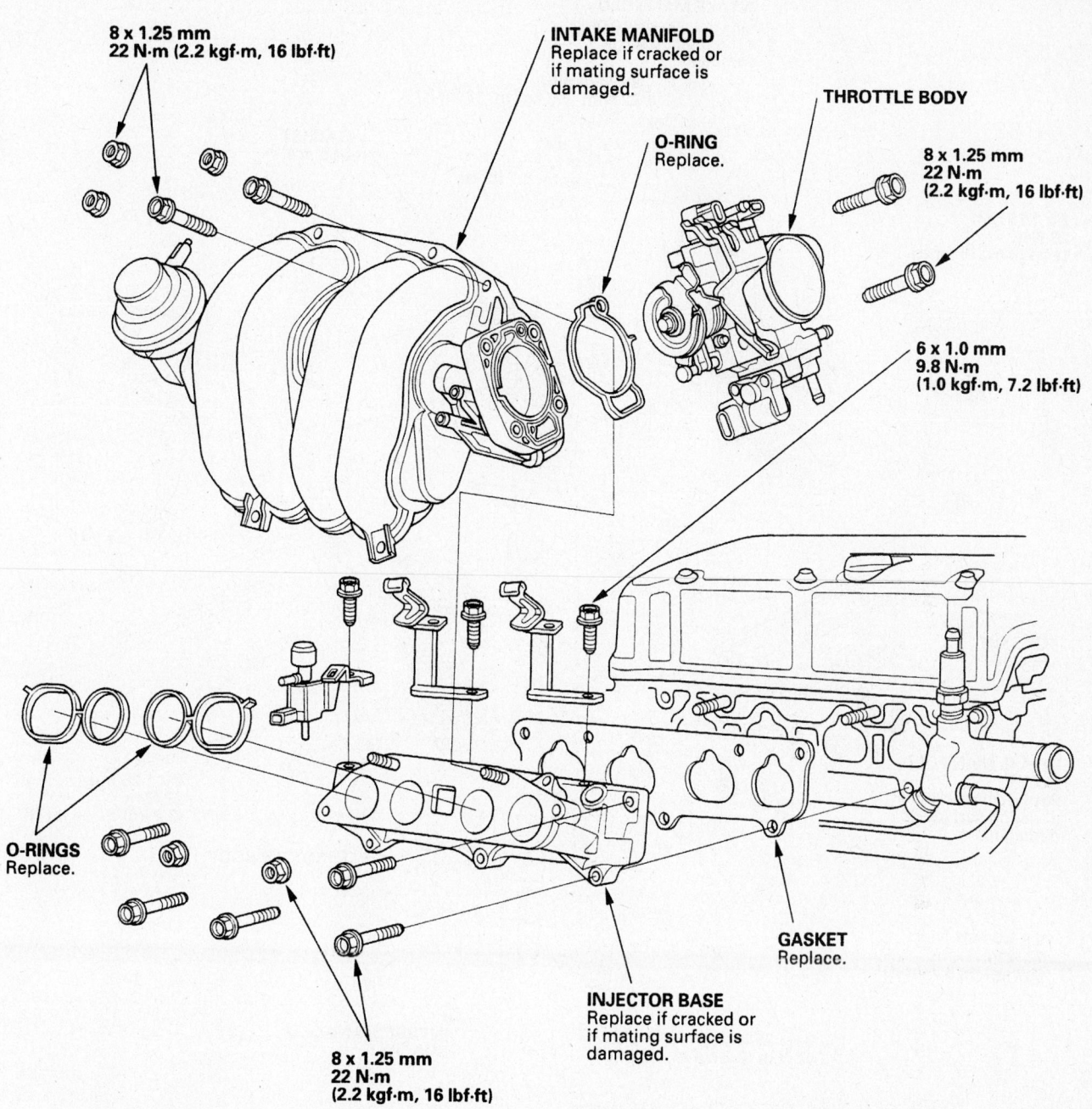

8 x 1.25 mm
22 N·m (2.2 kgf·m, 16 lbf·ft)

INTAKE MANIFOLD
Replace if cracked or
if mating surface is
damaged.

THROTTLE BODY

O-RING
Replace.

8 x 1.25 mm
22 N·m
(2.2 kgf·m, 16 lbf·ft)

6 x 1.0 mm
9.8 N·m
(1.0 kgf·m, 7.2 lbf·ft)

O-RINGS
Replace.

GASKET
Replace.

INJECTOR BASE
Replace if cracked or
if mating surface is
damaged.

8 x 1.25 mm
22 N·m
(2.2 kgf·m, 16 lbf·ft)

42356-INTE-G16

Exploded view of the intake manifold—2.0L (K20A3) engine

- Throttle cover. Fully open the throttle link and cruise control link by hand, then remove the throttle cable and cruise control cable from the links.
- Locknut, loosen only, then the cables from the bracket
- Water bypass hoses, then plug them
4. Relieve the fuel system pressure.
  - Fuel feed hose
  - Positive Crankcase Ventilation

(PCV) hose, harness holder mounting bolt and harness clamp mounting bolt
- Idle Air Control (IAC) valve connector
- Throttle Position (TP) sensor connector
- Manifold Absolute Pressure (MAP) sensor connector
- Evaporative Emission (EVAP) canister purge valve connector
- Intake Manifold Runner Control

(IMRC) valve position sensor connector
- 2 bolts securing the intake manifold and brackets
- Intake manifold mounting bolts and nuts
- 2 stud bolts and the intake manifold. Discard the gasket(s).

To install:
5. Install or connect the following:
  - Intake manifold with a new gasket, then tighten the 2 stud bolts to 16

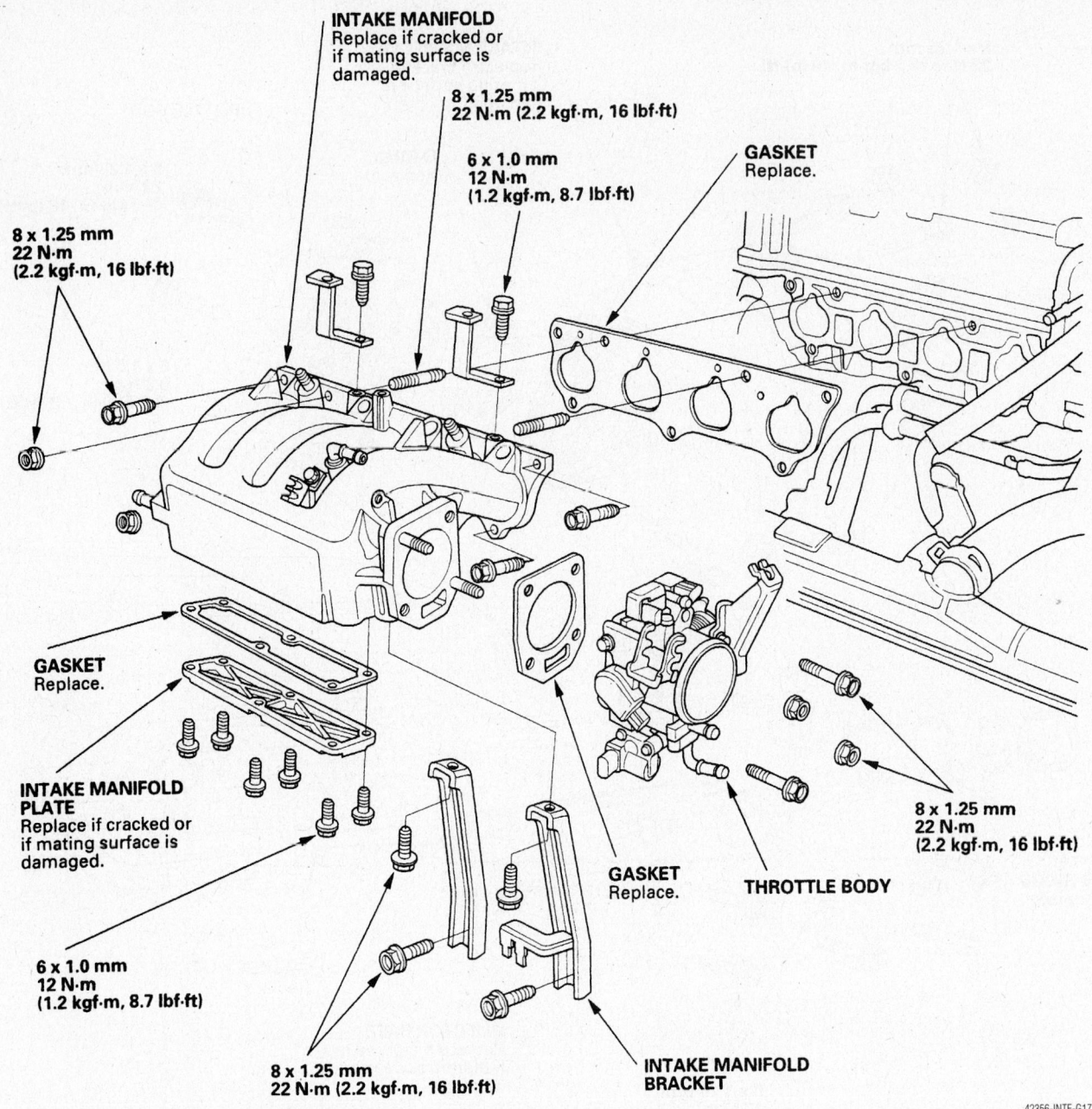

INTAKE MANIFOLD
Replace if cracked or if mating surface is damaged.

8 x 1.25 mm
22 N·m (2.2 kgf·m, 16 lbf·ft)

6 x 1.0 mm
12 N·m
(1.2 kgf·m, 8.7 lbf·ft)

GASKET
Replace.

8 x 1.25 mm
22 N·m
(2.2 kgf·m, 16 lbf·ft)

GASKET
Replace.

INTAKE MANIFOLD PLATE
Replace if cracked or if mating surface is damaged.

6 x 1.0 mm
12 N·m
(1.2 kgf·m, 8.7 lbf·ft)

8 x 1.25 mm
22 N·m (2.2 kgf·m, 16 lbf·ft)

GASKET
Replace.

THROTTLE BODY

8 x 1.25 mm
22 N·m
(2.2 kgf·m, 16 lbf·ft)

INTAKE MANIFOLD BRACKET

42356-INTE-G17

Exploded view of the intake manifold—2.0L K20A3 and K20Z1 engines

ft. lbs. (22 Nm). Tighten the intake manifold bolts/nuts in a criss-cross pattern, beginning with the inner bolt to 16 ft. lbs. (22 Nm).
- 2 bolts securing the intake manifold brackets and tighten to 16 ft. lbs. (22 Nm)
- PCV hose, harness holder mounting bolt and harness clamp mounting bolt. Tighten the bolts to 104 inch lbs. (10 Nm).
- Fuel feed hose
- Water bypass hoses
- Throttle and cruise control cables. Adjust the cables.

- Air cleaner housing
- IAT sensor connector
- Breather hose
- EVAP canister hose, brake booster vacuum hose and vacuum hoses
- Intake manifold cover and tighten to 104 inch lbs. (10 Nm)
- Negative battery cable

6. Fill the engine with coolant, then start the engine and check for leaks.

2002–03 TL Models

1. Before servicing the vehicle, refer to the precautions section.

2. Disconnect the negative battery cable.
3. Drain the cooling system.
4. Properly relieve the fuel system pressure.
5. Remove or disconnect the following:
- Intake air duct
- Strut brace (if equipped)
- Intake manifold cover
- Water bypass hose
- Traction Control System (TCS) actuator (if equipped)
- Evaporative Emission (EVAP) canister hose
- Throttle and cruise control cables
- Brake booster vacuum hose

- Upper intake manifold cover
- Intake manifold nuts and bolts in a crisscross pattern, beginning from the center and moving out

6. Verify that all vacuum lines are disconnected and remove the intake manifold and throttle body as a unit.

7. Inspect the manifold for cracks, flatness, or other damage; replace any damaged parts. If the intake manifold is to be replaced, transfer all the necessary components to the new manifold.

To install:

- Intake fasteners:16 ft. lbs. (22 Nm)
- Upper intake manifold cover fasteners: 108 inch lbs. (12 Nm)
- TCS actuator bolts: 16 ft. lbs. (22 Nm)
- Strut brace bolts: 16 ft. lbs. (22 Nm)

TSX Models

1. Before servicing the vehicle, refer to the precautions section.
2. Disconnect the negative battery cable.
3. Drain the engine coolant into a sealable container.
4. Remove or disconnect the following:
   - Intake Air Temperature (IAT) sensor electrical connector
   - Vacuum hose and breather pipe and the air intake duct
   - Intake manifold cover
   - Throttle and cruise control cables by loosening the locknuts, then slipping the cable ends out of the accelerator linkage.

➡**Do not bend the cables during removal. Always replace any throttle or cruise control cables that get kinked during removal.**

   - Evaporative emission (EVAP) canister hose and brake booster vacuum hose
   - Idle Air Control (IAC) valve connectors
   - Throttle Position (TP) sensor connector
   - Manifold Absolute Pressure (MAP) sensor connector
   - Necessary engine wire harness connectors and wire harness clamps from the intake manifold
   - Bolt securing the harness holder and remove the harness clamps
   - Water bypass hoses, then plug them
   - Harness clamp and harness connector from the intake manifold bracket

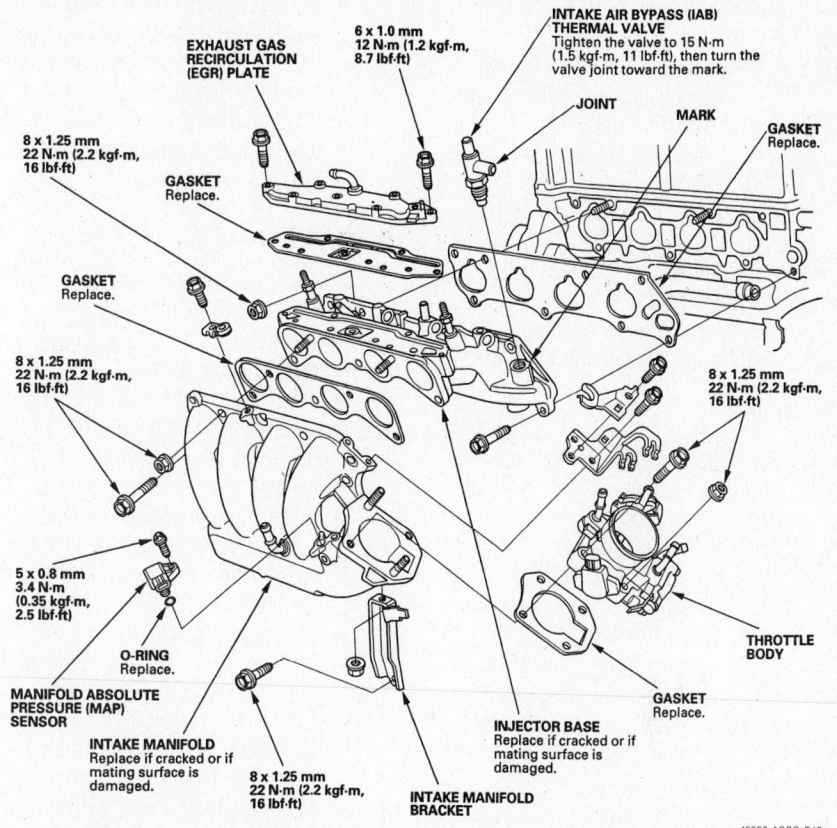

Exploded view of the intake manifold and related components—TSX models

42356-ACCO-G15

   - Intake manifold bracket
   - If equipped, automatic transmission vacuum hose
   - Retainer and intake manifold

To install:

5. Clean the mounting surfaces.
6. Install or connect the following:
   - New gasket
   - Intake manifold. Tighten the bolts, in a criss-cross pattern beginning with the inner bolt, to 16 ft. lbs. (22 Nm).
   - If equipped, automatic transmission vacuum hose
   - Intake manifold bracket
   - Harness clamp and connector to the intake manifold bracket
   - Water bypass hoses
   - Bolt securing the harness holder and tighten to 104 inch lbs. (12 Nm)
   - Harness clamps
   - EVAP canister hose and brake booster vacuum hose
   - Throttle and cruise control cables
   - Intake manifold cover
   - Intake air duct
   - IAT sensor connector, vacuum hose and breather pipe

7. Refill the cooling system.

8. Connect the negative battery cable, start the engine, and check for leaks.

Exhaust Manifold

REMOVAL & INSTALLATION

CL Model

1. Before servicing the vehicle, refer to the precautions section.
2. Remove or disconnect the following:
   - Negative battery cable
   - Exhaust manifold covers
   - Small heat shields from the cylinder heads (if equipped)
   - Exhaust pipe from the manifold
   - Heated Oxygen (HO$_2$S) sensors
   - Exhaust manifold nuts in a crisscross pattern starting from the center of the manifold
   - Exhaust manifold

To install:

3. Install or connect the following:
   - Exhaust manifold with a new gasket and new nuts and tighten the nuts in a crisscross pattern starting from the center to 22 ft. lbs. (30 Nm)
   - Small heat shields and tighten the

attaching bolts to 16 ft. lbs. (22 Nm) (if equipped)
- Exhaust pipe to the manifold with a new gasket and tighten the nuts to 40 ft. lbs. (55 Nm)
- HO$_2$S sensor and tighten it to 33 ft. lbs. (45 Nm)
- Manifold covers and tighten the bolts to 16 ft. lbs. (22 Nm)

4. Verify that all vacuum lines and wiring are properly connected.

5. Reconnect the negative battery cable.

6. Start the engine and check for leaks.

RL Models

1. Before servicing the vehicle, refer to the precautions section.

2. Remove or disconnect the following:
- Negative battery cable
- Exhaust manifold covers
- Small heat shields from the cylinder heads (if equipped)
- Exhaust pipe from the manifold
- Heated Oxygen (HO$_2$S) sensors
- Exhaust manifold nuts in a criss-cross pattern starting from the center of the manifold
- Exhaust manifold

To install:

3. Install or connect the following:
- Exhaust manifold with a new gasket and new nuts and tighten the nuts in a crisscross pattern starting from the center to 22 ft. lbs. (30 Nm)
- Small heat shields and tighten the attaching bolts to 16 ft. lbs. (22 Nm) (if equipped)
- Exhaust pipe to the manifold with a new gasket and tighten the nuts to 40 ft. lbs. (55 Nm)
- HO$_2$S sensor and tighten it to 33 ft. lbs. (45 Nm)
- Manifold covers and tighten the bolts to 16 ft. lbs. (22 Nm)

4. Verify that all vacuum lines and wiring are properly connected.

5. Reconnect the negative battery cable.

6. Start the engine and check for leaks.

RSX Models

1. Before servicing the vehicle, refer to the precautions section.

2. Remove or disconnect the following:
- VTEC solenoid valve
- Intermediate shaft cover
- Cover and exhaust manifold bracket
- Mounting nuts and bolts and exhaust manifold. Discard the gasket.

To install:

3. Install or connect the following:

- Exhaust manifold with a new gasket. Tighten the retainers to 33 ft. lbs. (45 Nm), in a criss–cross pattern starting at the inner bolt.
- Cover and exhaust manifold bracket
- Intermediate shaft cover
- VTEC solenoid valve

TL Models

1. Before servicing the vehicle, refer to the precautions section.

2. Remove or disconnect the following:
- Negative battery cable
- Exhaust manifold covers
- Small heat shields from the cylinder heads (if equipped)
- Exhaust pipe from the manifold
- Heated Oxygen (HO$_2$S) sensors
- Exhaust manifold nuts in a criss-cross pattern starting from the center of the manifold
- Exhaust manifold

To install:

3. Install or connect the following:
- Exhaust manifold with a new gasket and new nuts and tighten the nuts in a crisscross pattern starting from the center to 22 ft. lbs. (30 Nm)
- Small heat shields and tighten the

---

attaching bolts to 16 ft. lbs. (22 Nm) (if equipped)
- Exhaust pipe to the manifold with a new gasket and tighten the nuts to 40 ft. lbs. (55 Nm)
- HO$_2$S sensor and tighten it to 33 ft. lbs. (45 Nm)
- Manifold covers and tighten the bolts to 16 ft. lbs. (22 Nm)

4. Verify that all vacuum lines and wiring are properly connected.

5. Reconnect the negative battery cable.

6. Start the engine and check for leaks.

TSX Models

1. Before servicing the vehicle, refer to the precautions section.

2. Raise and safely support the vehicle.

3. Remove or disconnect the following:
- VTEC solenoid valve
- Driveshaft heat shield
- Cover and exhaust manifold bracket
- Exhaust manifold

To install:

4. Clean the mounting surfaces.

5. Install or connect the following:
- New gasket on the cylinder head
- Exhaust manifold. Tighten the nuts, in a criss-cross pattern starting

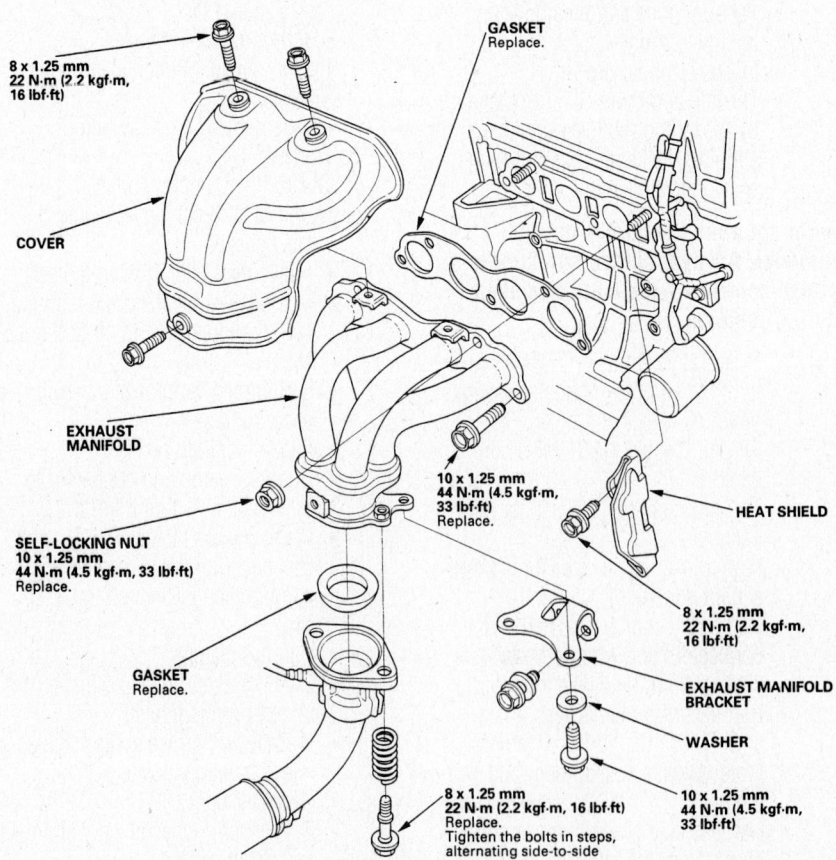

GASKET Replace.

COVER

EXHAUST MANIFOLD

SELF-LOCKING NUT 10 x 1.25 mm 44 N·m (4.5 kgf·m, 33 lbf·ft) Replace.

GASKET Replace.

8 x 1.25 mm 22 N·m (2.2 kgf·m, 16 lbf·ft)

10 x 1.25 mm 44 N·m (4.5 kgf·m, 33 lbf·ft) Replace.

HEAT SHIELD

8 x 1.25 mm 22 N·m (2.2 kgf·m, 16 lbf·ft)

EXHAUST MANIFOLD BRACKET

WASHER

8 x 1.25 mm 22 N·m (2.2 kgf·m, 16 lbf·ft) Replace. Tighten the bolts in steps, alternating side-to-side

10 x 1.25 mm 44 N·m (4.5 kgf·m, 33 lbf·ft)

42356-ACCO-G16

Exploded view of the exhaust manifold and related components—TSX models

with the inner nut, to 33 ft. lbs. (45 Nm).
- Exhaust manifold bracket and cover
- Driveshaft heat shield
- VTEC solenoid valve

Front Crankshaft Seal

REMOVAL & INSTALLATION

CL, RSX, TSX, 2002–04 RL and 2002–03 TL Models

1. Before servicing the vehicle, refer to the precautions section.
2. Disconnect negative cable at the battery.
3. Raise and safely support the vehicle. Drain the engine oil and properly dispose of it.
4. Be sure the crankshaft is at TDC on No. 1 cylinder by aligning the white mark on the crankshaft pulley with the pointer on the lower timing belt cover.
5. Remove or disconnect the following:
- Crankshaft pulley
- Cylinder head cover
- Timing belt cover
- Timing belt
- Crankshaft Speed Fluctuation (CKF) sensor (if equipped)
- Timing belt drive gear from the crankshaft
6. Using a suitable prytool, carefully remove the seal.

To install:
7. Apply a light coat of oil to the seal lip.
8. Position the seal, then using a seal driver, install the seal into the housing.
9. Install or connect the following:
- Timing belt drive gear
- Timing belt
- Timing belt cover
- Cylinder head cover
- CKF sensor and tighten the attach-

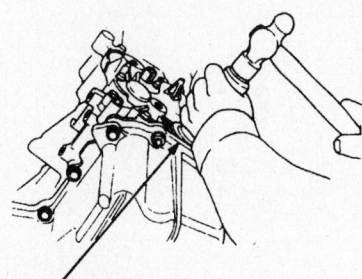

SEAL DRIVER
Install seal with the part number side facing out.

7923BG19

Installing the seal

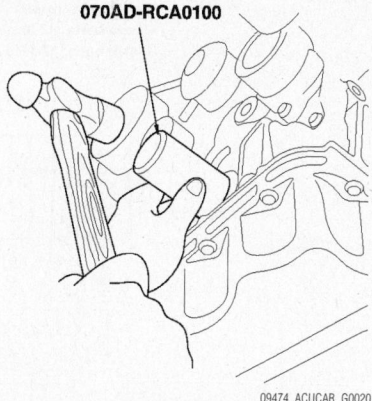

070AD-RCA0100

09474_ACUCAR_G0020

Use seal driver 070AD-RCA0100 to install the crankshaft oil seal—2004–06 TL and 2005–06 RL models

ing bolts to 96 inch lbs. (11 Nm) (if equipped)
- Crankshaft pulley
10. Lower the vehicle and check and fill the engine with oil as necessary.
11. Connect the negative battery cable and enter the radio security code.
12. Run the engine and check for leaks.
13. Turn off engine and check the oil level. Top off the oil level if necessary.

2004–06 TL and 2005–06 RL

1. Before servicing the vehicle, refer to the precautions section.

2. Remove the Crankshaft Position (CKP) sensor A and B, timing belt, and timing belt drive pulley.
3. Remove the pulley end crankshaft oil seal.
4. Clean and dry the crankshaft oil seal housing.
5. Apply a light coat of multipurpose grease to the crankshaft and to the lip of the seal.
6. Using the seal driver 070AD-RCA0100, drive in the crankshaft oil seal until the driver bottoms against the oil pump.
7. When the seal is in place, clean any excess grease off the crankshaft, and check that the oil seal lip is not distorted.
8. Install the timing belt drive pulley, CKP sensor A and B, and timing belt.

Camshaft and Valve Lifters

REMOVAL & INSTALLATION

CL Model

➡**The radio may have a coded theft protection circuit. Obtain the code from the owner before disconnecting the battery, removing the radio fuse, or removing the radio.**

1. Before servicing the vehicle, refer to the precautions section.

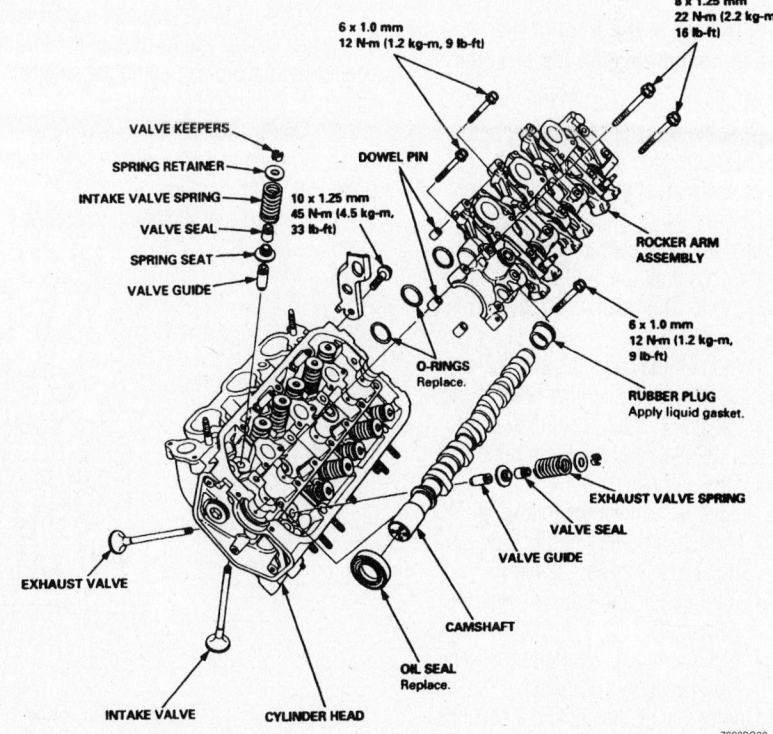

7923BG22

Camshaft and rocker arm assembly—CL model

2. Remove or disconnect the following:
- Negative battery cable
- Timing belt covers and cylinder head covers

3. Rotate the crankshaft to Top Dead Center (TDC) for the No. 1 piston and remove the timing belt.

4. Remove the camshaft sprockets.

5. Loosen the rocker shaft holder bolts 1 turn at a time in the reverse of the torque sequence to avoid damaging the valves, camshafts, or rocker assemblies.

6. After all bolts are loose, remove the rocker arm shafts as an assembly with the bolts still in the holders.

7. If the rocker shafts are to be disassembled, note that each rocker arm has a letter **A** or **B** stamped into the side. Before disassembling the rocker arms, make a note of the position of each letter so that the arms can be reassembled in the same position.

8. Do not remove the hydraulic tappets from the rocker arms unless they are to be replaced. Handle the rocker arms carefully so the oil does not drain out of the tappets.

9. Lift the camshafts from the cylinder head, wipe them clean and inspect the lift ramps. Replace the camshafts and rockers if the lobes are pitted, scored, or excessively worn.

To install:

10. Place a new seal on the end of the camshaft, lubricate the journals and set the camshaft in place on the head.

➡**The pin hole in the front of the camshaft designates the top position.**

11. Apply liquid gasket to the mounting surfaces of the camshaft end holders.

12. Set the rocker arm assemblies in place and start all of the camshaft holder bolts. Be sure the rocker arms are properly positioned and turn each bolt in sequence 2 turns at a time until the holders are seated on the head to avoid damaging the valves or rocker assemblies.

13. When all the camshaft and rocker holders are seated, tighten the bolts in the same sequence. Tighten the 8mm bolts to 16 ft. lbs. (22 Nm) and the 6mm bolts to 104 inch lbs. (12 Nm).

14. Install or connect the following:
- Camshaft pulleys and tighten the bolts to 23 ft. lbs. (32 Nm)
- Timing belt and pour oil over the camshafts
- Cylinder head cover and reassemble accessory components

15. Verify that all electrical connections and vacuum lines are connected.

16. Reconnect the negative battery cable.

17. Run the engine and check for leaks and proper operation.

2002–04 RL Model

1. Before servicing the vehicle, refer to the precautions section.

2. Remove or disconnect the following:
- Negative battery cable
- Timing belt covers and cylinder head covers

3. Rotate the crankshaft to Top Dead Center (TDC) for the No. 1 piston and remove the timing belt.

4. Remove the camshaft sprockets.

5. Loosen the rocker shaft holder bolts 1 turn at a time in the reverse of the torque sequence to avoid damaging the valves, camshafts, or rocker assemblies.

6. After all bolts are loose, remove the rocker arm shafts as an assembly with the bolts still in the holders.

7. If the rocker shafts are to be disassembled, note that each rocker arm has a letter **A** or **B** stamped into the side. Before disassembling the rocker arms, make a note of the position of each letter so that the arms can be reassembled in the same position.

8. Do not remove the hydraulic tappets from the rocker arms unless they are to be replaced. Handle the rocker arms carefully so the oil does not drain out of the tappets.

9. Lift the camshafts from the cylinder head, wipe them clean and inspect the lift ramps. Replace the camshafts and rockers if the lobes are pitted, scored, or excessively worn.

To install:

10. Place a new seal on the end of the camshaft, lubricate the journals and set the camshaft in place on the head.

➡**The pin hole in the front of the camshaft designates the top position.**

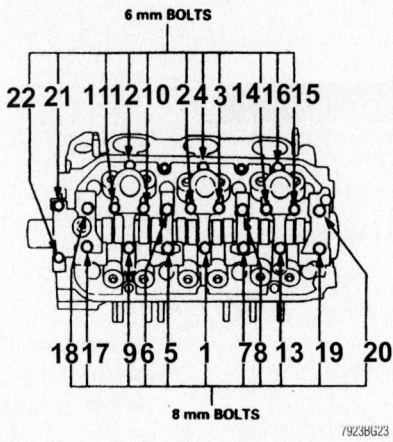

Specified torque:
8 mm bolts: 22 N·m (2.2 kg-m, 16 lb-ft)
6 mm bolts: 12 N·m (1.2 kg-m, 9 lb-ft)

Camshaft holder bolt tightening sequence—CL model

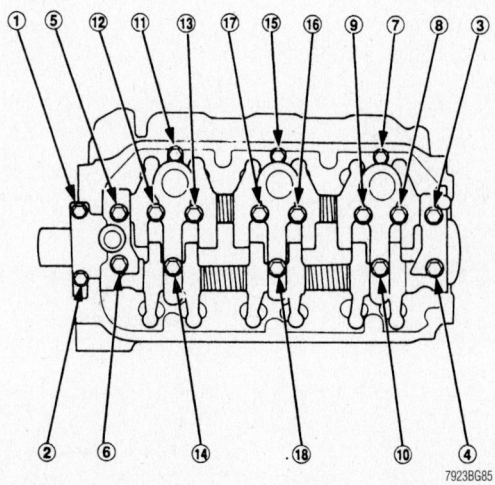

Loosen the camshaft holder bolts in the specified sequence—2002–04 RL models

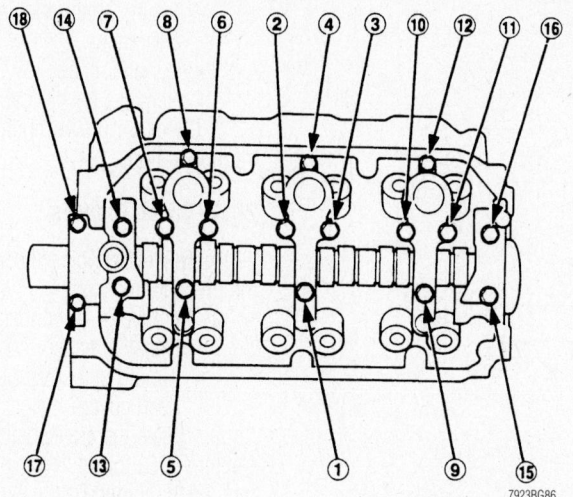

Tighten the camshaft holder bolts in the specified sequence—2002–04 RL models

7923BG86

11. Apply liquid gasket to the mounting surfaces of the camshaft end holders.

12. Set the rocker arm assemblies in place and start all of the camshaft holder bolts. Be sure the rocker arms are properly positioned and turn each bolt in sequence 2 turns at a time until the holders are seated on the head to avoid damaging the valves or rocker assemblies.

13. When all the camshaft and rocker holders are seated, tighten the bolts in the same sequence. Tighten the 8mm bolts to 16 ft. lbs. (22 Nm) and the 6mm bolts to 104 inch lbs. (12 Nm).

14. Install or connect the following:
- Camshaft pulleys and tighten the bolts to 23 ft. lbs. (32 Nm)
- Timing belt and pour oil over the camshafts
- Cylinder head cover and reassemble accessory components

15. Verify that all electrical connections and vacuum lines are connected.

16. Reconnect the negative battery cable.

17. Run the engine and check for leaks and proper operation.

2005–06 RL Model

FRONT

➡**Make sure you have the anti-theft codes for the radio and the navigation system, then write down the radio channel presets. Make sure the ignition switch is OFF.**

1. Before servicing the vehicle, refer to the precautions section.

2. Remove battery trim and the left upper fender trim.

3. Disconnect the negative cable from the battery first, then disconnect the positive cable.

4. Remove the battery.

5. Drain the engine coolant.

6. Remove the radiator hoses.

7. Remove the Exhaust Gas Recirculation (EGR) valve.

8. Remove the timing belt.

9. Remove the rocker arm assembly.

10. Remove the front camshaft pulley.

11. Remove the thrust cover then remove the front camshaft.

To install:

12. Install the front camshaft in the reverse order of removal. Always use a new O-ring and apply new engine oil to the journals and camshaft lobes.

13. Tighten the camshaft thrust plate to 16 ft. lbs. (22 Nm).

14. Apply new engine oil to the threads of the camshaft pulley mounting bolt, then install the front camshaft pulley. Tighten the bolt to 67 ft. lbs. (90 Nm).

15. Install the rocker arm assembly.

16. Install the timing belt.

17. Adjust the valve clearance.

18. Install the battery.

19. Fill the radiator with engine coolant and bleed the air out.

20. Enter the anti-theft codes for the radio and the navigation system, then enter the customer's radio channel presets.

REAR

➡**Make sure you have the anti-theft codes for the radio and the navigation system, then write down the radio channel presets. Make sure the ignition switch is OFF.**

1. Before servicing the vehicle, refer to the precautions section.

2. Relieve the fuel system pressure.

3. Disconnect the negative battery cable.

4. Drain the cooling system.

5. Remove the timing belt.

6. Remove the rocker arm assembly.

7. Remove the rear camshaft pulley.

8. Remove the two nuts securing the purge joint.

9. Remove the thrust cover, then remove the rear camshaft.

To install:

10. Install the rear camshaft in the reverse order of removal. Always use a new O-ring and apply new engine oil to the journals and camshaft lobes.

11. Tighten the camshaft thrust plate to 16 ft. lbs. (22 Nm).

12. Apply new engine oil to the threads of the camshaft pulley mounting bolt, then install the rear camshaft pulley. Tighten the bolt to 67 ft. lbs. (90 Nm).

13. Install the rocker arm assembly.

14. Install the timing belt.

15. Adjust the valve clearance.

16. Install the battery.

17. Fill the radiator with engine coolant and bleed the air out.

18. Enter the anti-theft codes for the radio and the navigation system, then enter the customer's radio channel presets.

RSX Models

1. Before servicing the vehicle, refer to the precautions section.

➡**The radio may have a coded theft protection circuit. Obtain the code from the owner before disconnecting the battery, removing the radio fuse, or removing the radio.**

2. Remove or disconnect the following:
- Negative battery cable
- Cylinder head from the vehicle
- Loosen the rocker arm adjusting screws
- Camshaft holder bolts, 2 turns at a time in a criss–cross pattern
- Timing chain guide B, camshaft holders and camshafts

To install:

3. Make sure the punch marks on the VTC actuator and exhaust camshaft sprocket are facing up, then set the camshafts in the holder.

4. Set the camshaft holders and timing chain guide B in place.

5. Tighten the bolts, in sequence, to the following specifications:
- 8mm bolts: 16 ft. lbs. (22 Nm)
- 6mm bolts (nos. 21, 22 & 23): 104 inch lbs. (12 Nm)

6. Install or connect the following:

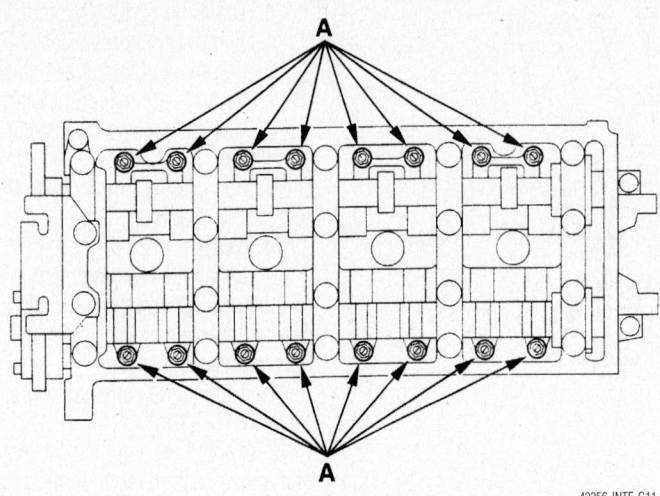

Loosen the rocker arm adjusting screws (A)—RSX models

42356-INTE-G11

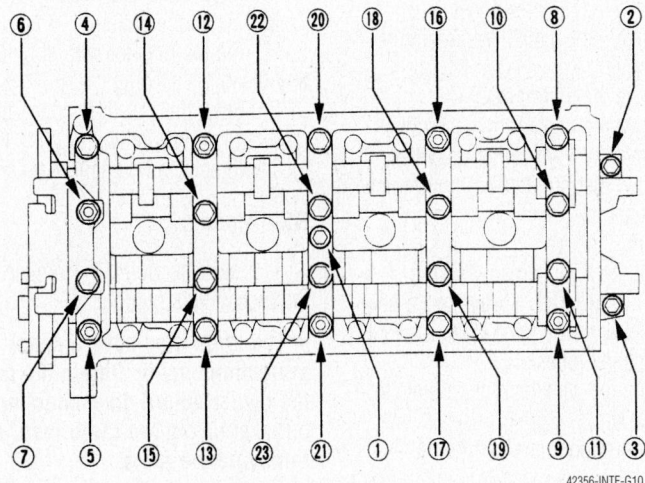

Camshaft holder bolt loosening sequence—RSX models

42356-INTE-G10

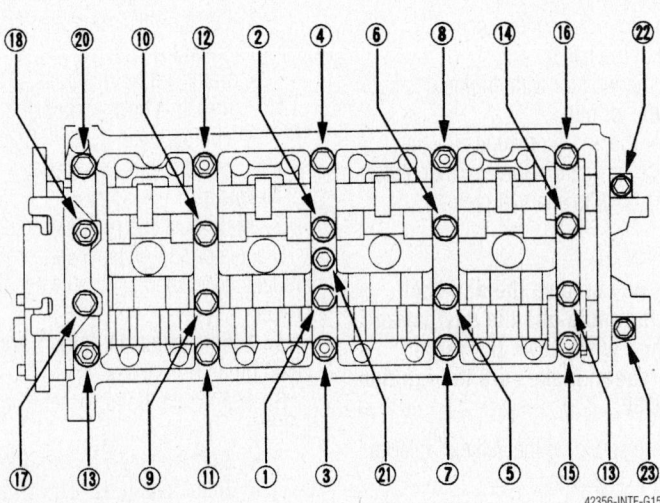

Rocker arm assembly tightening sequence—RSX models

42356-INTE-G15

- Timing chain, then adjust the valve clearance
- Cylinder head
- Negative battery cable

7. Check for proper engine and valve train operation.

2002–03 TL Models

1. Before servicing the vehicle, refer to the precautions section.

2. Remove or disconnect the following:
- Negative battery cable
- Timing belt covers and cylinder head covers

3. Rotate the crankshaft to Top Dead Center (TDC) for the No. 1 piston and remove the timing belt.

4. Remove the camshaft sprockets.

5. Loosen the rocker shaft holder bolts 1 turn at a time in the reverse of the torque sequence to avoid damaging the valves, camshafts, or rocker assemblies.

6. After all bolts are loose, remove the rocker arm shafts as an assembly with the bolts still in the holders.

7. If the rocker shafts are to be disassembled, note that each rocker arm has a letter **A** or **B** stamped into the side. Before disassembling the rocker arms, make a note of the position of each letter so that the arms can be reassembled in the same position.

8. Do not remove the hydraulic tappets from the rocker arms unless they are to be replaced. Handle the rocker arms carefully so the oil does not drain out of the tappets.

9. Lift the camshafts from the cylinder head, wipe them clean and inspect the lift ramps. Replace the camshafts and rockers if the lobes are pitted, scored, or excessively worn.

To install:

10. Place a new seal on the end of the camshaft, lubricate the journals and set the camshaft in place on the head.

➡The pin hole in the front of the camshaft designates the top position.

11. Apply liquid gasket to the mounting surfaces of the camshaft end holders.

12. Set the rocker arm assemblies in place and start all of the camshaft holder bolts. Be sure the rocker arms are properly positioned and turn each bolt in sequence 2 turns at a time until the holders are seated on the head to avoid damaging the valves or rocker assemblies.

13. When all the camshaft and rocker holders are seated, tighten the bolts in the same sequence. Tighten the 8mm bolts to

VALVE KEEPERS
SPRING RETAINER
INTAKE VALVE SPRING
VALVE SEAL
SPRING SEAT
VALVE GUIDE

6 x 1.0 mm
12 N·m (1.2 kg-m, 9 lb-ft)

8 x 1.25 mm
22 N·m (2.2 kg-m, 16 lb-ft)

DOWEL PIN

10 x 1.25 mm
45 N·m (4.5 kg-m, 33 lb-ft)

ROCKER ARM
ASSEMBLY

6 x 1.0 mm
12 N·m (1.2 kg-m, 9 lb-ft)

RUBBER PLUG
Apply liquid gasket.

O-RINGS
Replace.

EXHAUST VALVE SPRING
VALVE SEAL
VALVE GUIDE

EXHAUST VALVE

CAMSHAFT

OIL SEAL
Replace.

INTAKE VALVE   CYLINDER HEAD

7923BG22

Camshaft and rocker arm assembly—2002–03 3.2L engines

Specified torque:
8 mm bolts: 22 N·m (2.2 kg-m, 16 lb-ft)
6 mm bolts: 12 N·m (1.2 kg-m, 9 lb-ft)

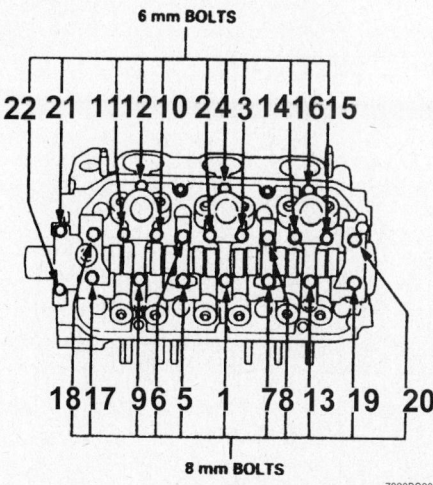

6 mm BOLTS

22 21 11 12 10 2 4 3 14 16 15

18 17 9 6 5 1 7 8 13 19 20

8 mm BOLTS

7923BG23

Camshaft holder bolt tightening sequence—2002–03 3.2L engine

16 ft. lbs. (22 Nm) and the 6mm bolts to 104 inch lbs. (12 Nm).
14. Install or connect the following:
   • Camshaft pulleys and tighten the bolts to 23 ft. lbs. (32 Nm)
   • Timing belt and pour oil over the camshafts

   • Cylinder head cover and reassemble accessory components
15. Verify that all electrical connections and vacuum lines are connected.
16. Reconnect the negative battery cable.
17. Run the engine and check for leaks and proper operation.

2004–06 TL Models

FRONT

➡ **Make sure you have the anti-theft codes for the radio and the navigation system, then write down the radio channel presets. Make sure the ignition switch is OFF.**

1. Before servicing the vehicle, refer to the precautions section.
2. Remove the left side engine compartment cover.
3. Disconnect the negative cable from the battery first, then disconnect the positive cable.
4. Remove the battery.
5. Drain the engine coolant.
6. Remove the upper radiator hose.
7. Remove the Exhaust Gas Recirculation (EGR) valve.
8. Remove the timing belt.
9. Remove the rocker arm assembly.
10. Remove the front camshaft pulley.
11. Remove the thrust cover, then remove the front camshaft.

To install:

12. Install the front camshaft in the reverse order of removal. Always use a new O-ring and apply new engine oil to the journals and camshaft lobes.
13. Tighten the camshaft thrust plate to 16 ft. lbs. (22 Nm).
14. Apply new engine oil to the threads of the camshaft pulley mounting bolt, then install the front camshaft pulley. Tighten the bolt to 67 ft. lbs. (90 Nm).
15. Install the rocker arm assembly.
16. Install the timing belt.
17. Adjust the valve clearance.
18. Install the battery.
19. Fill the radiator with engine coolant and bleed the air out.
20. Enter the anti-theft codes for the radio and the navigation system, then enter the customer's radio channel presets.

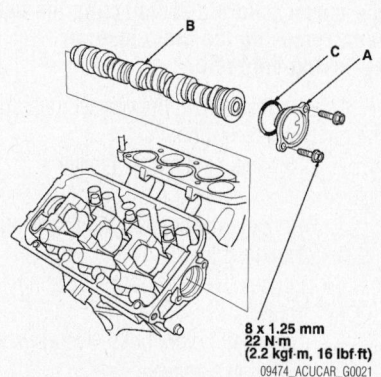

8 x 1.25 mm
22 N·m
(2.2 kgf·m, 16 lbf·ft)
09474_ACUCAR_G0021

Typical camshaft mounting—2004–06 TL and 2005–06 RL models

REAR

➥Make sure you have the anti-theft codes for the radio and the navigation system, then write down the radio channel presets. Make sure the ignition switch is OFF.

1. Before servicing the vehicle, refer to the precautions section.
2. Relieve the fuel system pressure.
3. Disconnect the negative battery cable.
4. Remove the fuse/relay box from under the hood.
5. Drain the cooling system.
6. Remove the heater hose and two nuts retaining the fuel line.
7. Remove the timing belt.
8. Remove the rocker arm assembly.
9. Remove the rear camshaft pulley.
10. Remove the thrust cover, then remove the rear camshaft.

To install:

11. Install the rear camshaft in the reverse order of removal. Always use a new O-ring and apply new engine oil to the journals and camshaft lobes.
12. Tighten the camshaft thrust plate to 16 ft. lbs. (22 Nm).
13. Apply new engine oil to the threads of the camshaft pulley mounting bolt, then install the rear camshaft pulley. Tighten the bolt to 67 ft. lbs. (90 Nm).
14. Install the rocker arm assembly.
15. Install the timing belt.
16. Adjust the valve clearance.
17. Install the battery.
18. Fill the radiator with engine coolant and bleed the air out.
19. Enter the anti-theft codes for the radio and the navigation system, then enter the customer's radio channel presets.

TSX Models

➥The radio may have a coded theft protection circuit. Obtain the code from the owner before disconnecting the battery, removing the radio fuse, or removing the radio.

1. Before servicing the vehicle, refer to the precautions section.
2. Disconnect the negative battery cable.
3. Remove or disconnect the following:
   • Timing chain
   • Loosen the rocker arm adjusting screws
   • Camshaft holder bolts, two turns at a time in sequence
   • Timing chain guide (B), camshaft and camshafts

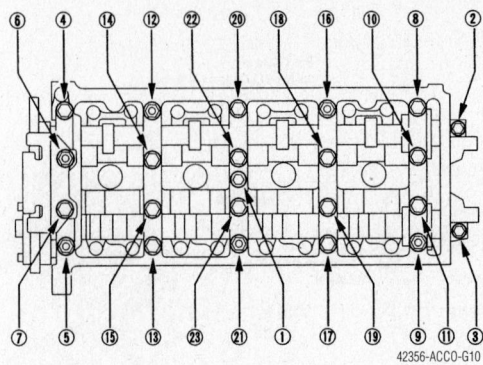

Camshaft holder bolt loosening sequence—TSX models

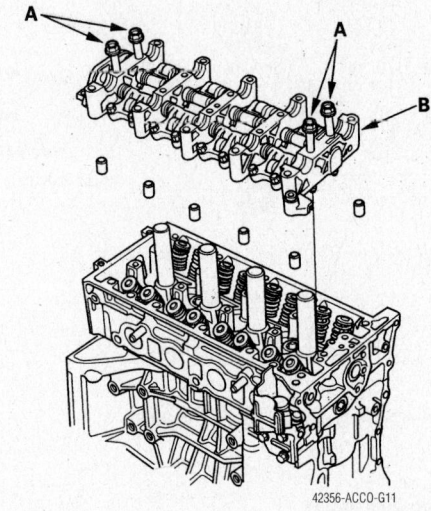

Insert the bolts (A) into the rocker shaft holder, then remove the rocker arm assembly (B)—TSX models

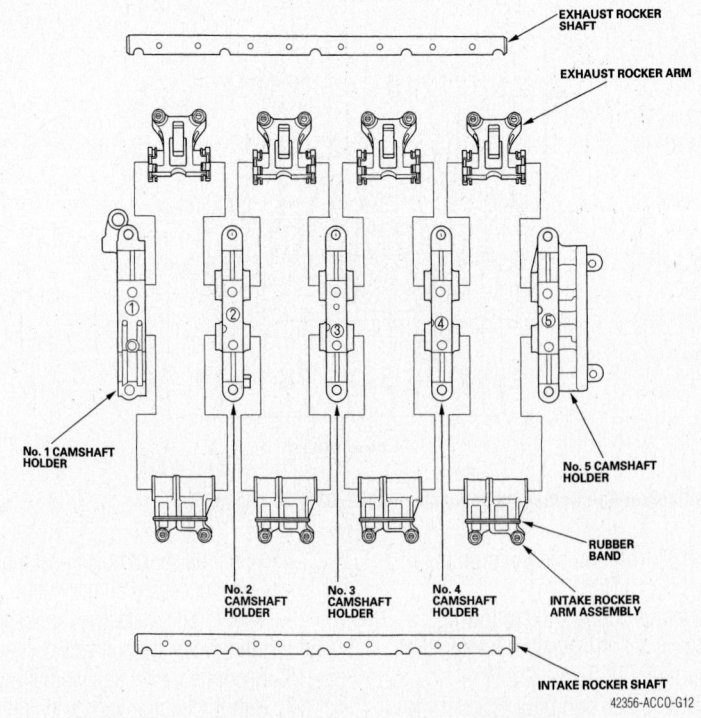

Exploded view of the rocker arms and related components—TSX models

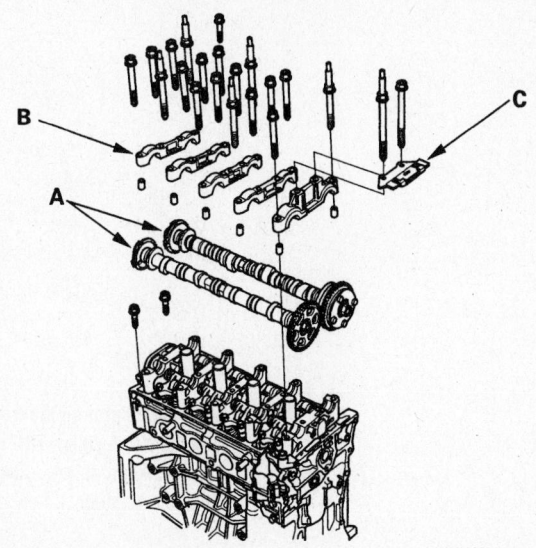

When installing the camshafts (A) make sure the punch marks on the VTC actuator and exhaust cam sprockets are facing up—TSX models

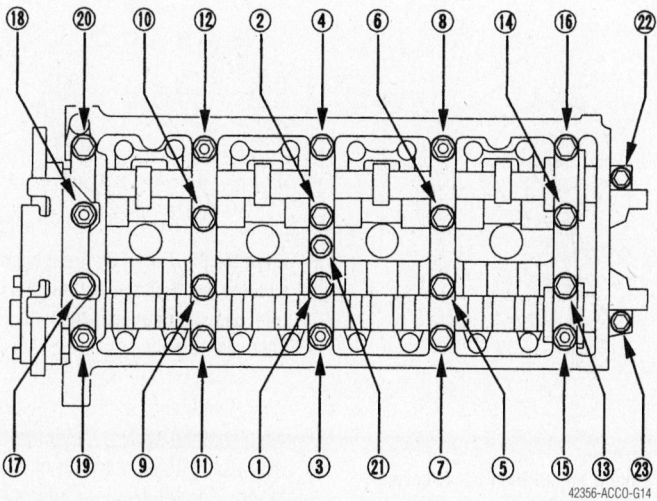

Rocker arm assembly bolt tightening sequence—TSX models

4. Insert the bolts (A) into the rocker shaft holder, then remove the rocker arm assembly (B)

- Camshafts by carefully lifting them out of the cylinder head

To install:

5. Clean and dry the No. 5 rocker shaft holding mating surface.

6. Apply a suitable liquid gasket P/N 08718-0009, or equivalent, evenly to the cylinder head mating surface of the No. 5 rocker shaft holder.

➡**The parts must be installed within 5 minutes of applying the liquid gasket.**

7. Reassemble the rocker arm assembly, as necessary.

8. Install or connect the following

- Bolts (A) into the rocker shaft holder, then the rocker arm assembly on the cylinder head. Remove the bolts from the rocker shaft holder.

9. Make sure the punch marks on the variable valve timing control (VTC) actuator and exhaust camshaft sprocket are facing up, then set the camshafts (A) in the holder

10. Set the camshaft holders (B) and timing chain guide B (C) in place.

11. Tighten the bolts, in sequence, to the following specification:

   a. 8mm bolts: 16 ft. lbs. (22 Nm)
   b. 6mm bolts: 104 inch lbs. (12 Nm)

12. Install the timing chain and adjust the valve lash.

Valve Lash

ADJUSTMENT

CL, 2002–04 RL and 2002–03 TL Models

These engines are equipped with hydraulic valve lash adjusters on the rocker arms. No valve clearance adjustments are possible or necessary.

2004–06 TL and 2005–06 RL Models

➡**Adjust the valves only when the cylinder head temperature is less than 100 ßF (38 ßC).**

1. Before servicing the vehicle, refer to the precautions section.

2. Remove the right side engine compartment cover.

3. Remove the cylinder head covers.

4. Set the No. 1 piston at Top Dead Center (TDC).

5. Align the pointer on the front upper cover with the No. 1 piston TDC mark on the front camshaft pulley.

6. Select the correct thickness feeler gauge for the valves you're going to check.

7. Valve clearance on the intake valves is 0.008–0.009 inch (0.20–0.24 mm) and on the exhaust side it is 0.011–0.013 inch (0.28–0.32 mm).

8. Insert the feeler gauge between the adjusting screw and the end of the valve stem on No. 1 cylinder and slide it back and forth; you should feel a slight amount of drag.

9. If you feel too much or too little drag, loosen the locknut, and turn the adjusting screw until the drag on the feeler gauge is correct.

10. Tighten the locknut to 14 ft. lbs. (20 Nm) and recheck the clearance.

11. Repeat the adjustment, if necessary.

12. Rotate the crankshaft clockwise. Align the pointer on the front upper cover with the No. 4 piston TDC mark on the front camshaft pulley.

13. Check and, if necessary, adjust the valve clearance on No. 4 cylinder.

14. Rotate the crankshaft clockwise. Align the pointer on the front upper cover with the No. 2 piston TDC mark on the front camshaft pulley.

15. Check and, if necessary, adjust the valve clearance on No. 2 cylinder

16. Rotate the crankshaft clockwise. Align the pointer on the front upper cover with the No. 5 piston TDC mark on the front camshaft pulley.

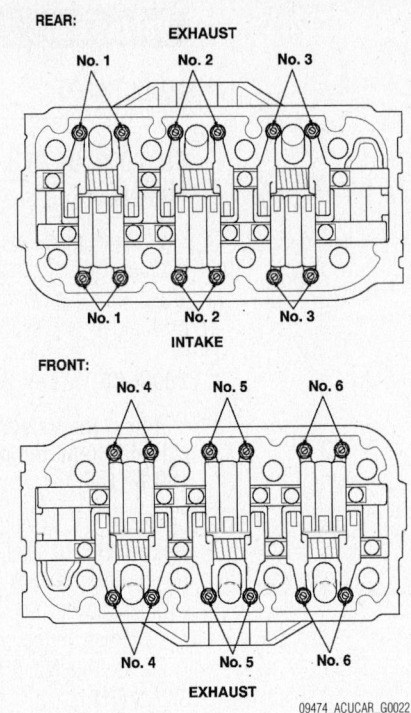

Valve adjusting screw locations—2004–06 TL and 2005–06 RL models

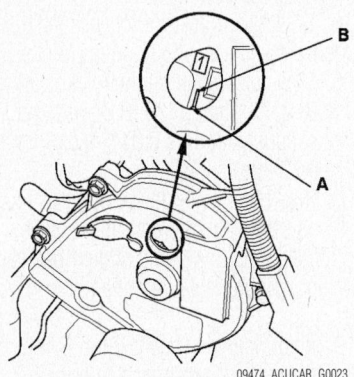

Align the pointer on the front upper cover with the No. 1 piston TDC mark on the front camshaft pulley—2004–06 TL and 2005–06 RL models

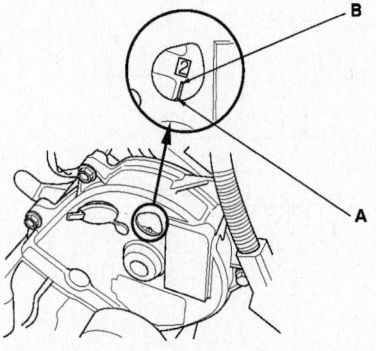

Align the pointer on the front upper cover with the No. 2 piston TDC mark on the front camshaft pulley—2004–06 TL and 2005–06 RL models

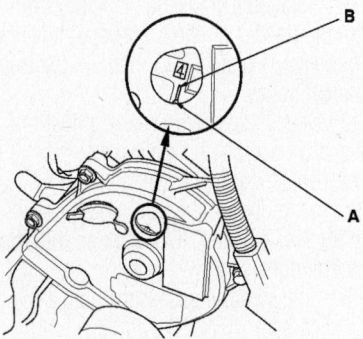

Align the pointer on the front upper cover with the No. 4 piston TDC mark on the front camshaft pulley—2004–06 TL and 2005–06 RL models

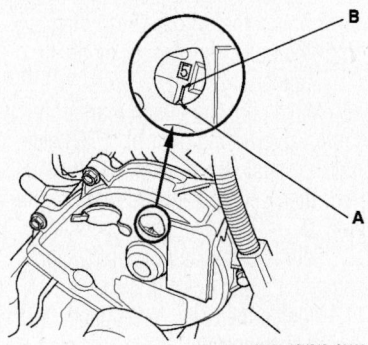

Align the pointer on the front upper cover with the No. 5 piston TDC mark on the front camshaft pulley—2004–06 TL and 2005–06 RL models

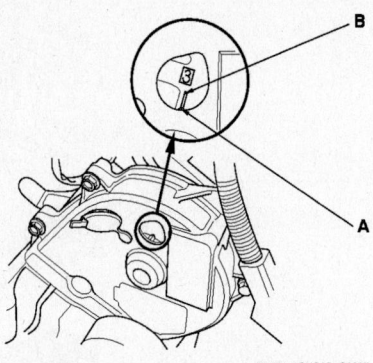

Align the pointer on the front upper cover with the No. 3 piston TDC mark on the front camshaft pulley—2004–06 TL and 2005–06 RL models

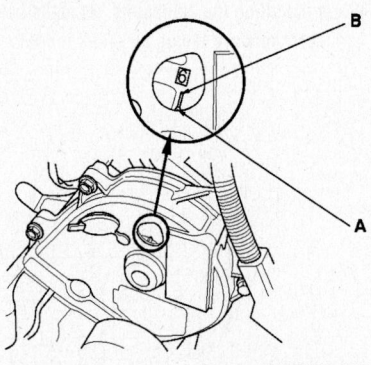

Align the pointer on the front upper cover with the No. 6 piston TDC mark on the front camshaft pulley—2004–06 TL and 2005–06 RL models

17. Check and, if necessary, adjust the valve clearance on No. 5 cylinder

18. Rotate the crankshaft clockwise. Align the pointer on the front upper cover with the No. 3 piston TDC mark on the front camshaft pulley.

Check and, if necessary, adjust the valve clearance on No. 3 cylinder

19. Rotate the crankshaft clockwise. Align the pointer on the front upper cover with the No. 6 piston TDC mark on the front camshaft pulley.

20. Check and, if necessary, adjust the valve clearance on No. 6 cylinder.

21. Install the cylinder head covers.

22. Install the right side engine compartment cover.

RSX Models

➡While all valve adjustments must be as accurate as possible, it is better to have the valve adjustment slightly loose rather than too tight. Burned valves may result from overly tight

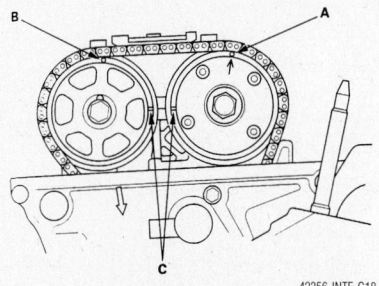

42356-INTE-G18

The punch mark (A) on the VTC actuator and the punch mark (B) on the exhaust camshaft sprockets should be at the top. Align the TDC marks (C) on the VTC actuator and exhaust camshaft sprocket—RSX models

EXHAUST

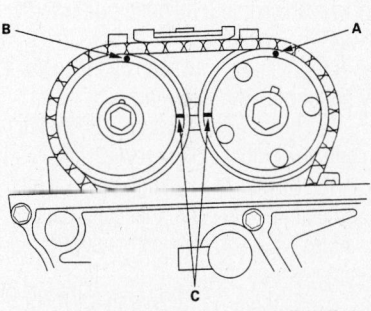

42356-INTE-G19

Valve arrangement—RSX models

adjustments. **Perform the valve adjustment for each cylinder in the same sequence as the firing order: 1–3–4–2.**

1. Before servicing the vehicle, refer to the precautions section.

2. Be sure the engine is cold; cylinder head temperature must be below 100° F (38° C). Overnight cold is best.

3. Remove the cylinder head cover.

4. Set the No. 1 cylinder to Top Dead Center (TDC).

a. The punch mark (arrow) on the Variable Valve Timing Control (VTC) actuator and the punch mark on the exhaust camshaft sprockets should be at the top. Align the TDC marks on the VTC actuator and exhaust camshaft sprocket.

5. Valve clearances are:

- K20A3 engine: Intake— 0.008–0.010 in. (0.21–0.25mm) Exhaust—0.011–0.013 in. (0.28–0.32mm)
- K20A2 and K20Z1 engines: Intake—0.008–0.010 in. (0.21–0.25mm) Exhaust— 0.010–0.011 in. (0.25–0.29mm)

6. With the No. 1 cylinder at TDC, adjust the valves of the No. 1 cylinder by performing the following procedures:

a. Hold the rocker arm against the valve and place the feeler gauge between the rocker arm and the camshaft lobe. There should be a slight drag on the feeler gauge.

b. If adjustment is required, loosen the valve adjusting the screw locknut.

c. Turn the adjusting screw to obtain the proper clearance.

d. Hold the adjusting screw and tighten the locknut(s). Tighten the locknut to 14 ft. lbs. (20 Nm) for all except K20A3 exhaust locknuts. For the K20A3 exhaust, tighten the locknut to 10 ft. lbs. (14 Nm).

e. Recheck the clearance.

7. Turn the crankshaft 180 degrees counterclockwise; the cam pulley will turn 90 degrees. With the No. 3 cylinder at TDC, the **UP** marks should be at the exhaust side. Adjust the valves on the No. 3 cylinder.

8. Turn the crankshaft 180 degrees counterclockwise; the cam pulley will turn 90 degrees. With the No. 4 cylinder at TDC, both **UP** marks should be at the bottom. Adjust the valves on the No. 4 cylinder.

9. Turn the crankshaft 180 degrees counterclockwise. The No. 2 cylinder will now be on TDC and the **UP** marks should be at the intake side. Adjust the valves on the No. 2 cylinder.

10. Install the cylinder head cover and upper timing belt cover.

TSX Models

➡ **The valve clearance should be adjusted when the engine is cold, the cylinder head temperature should be less than 100°F (38°C).**

➡ **The radio may contain a coded theft protection circuit. Always obtain the code number before disconnecting the battery.**

1. Before servicing the vehicle, refer to the precautions section.

2. Remove or disconnect the following:
- Negative battery cable

➡ **Label the wires before disconnecting them.**

- Spark plug wires from the spark plugs
- Positive Crankcase Ventilation (PCV) hose
- Cylinder head cover. Replace the rubber seals if damaged or deteriorated.

The punch mark (A) on the VTC actuator and the punch mark (B) on the exhaust cam sprocket should be at the top. Align the TDC marks (C) on the VTC actuator and exhaust cam sprockets—TSX models

3. Turn the engine to align the timing marks and set cylinder No.1 to TDC. The white mark on the crankshaft pulley should align with the pointer on the timing belt cover. The words **UP** embossed on the camshaft pulley should be aligned in the upward position. The marks on the edge of the pulley should be aligned with the cylinder head or the back cover upper edge.

4. Adjust the valves on cylinder No. 1 by performing the following:

a. Insert a feeler gauge in between the camshaft lobe and the rocker arm.

➡ **The intake valve clearance specification is 0.008–0.010 in. (0.21–0.25mm). The exhaust valve clearance specification is 0.010–0.011 in. (0.25–0.29mm).**

b. Loosen the locknut and turn the adjusting screw until the feeler gauge slides back and forth with a slight amount of drag.

c. Tighten the locknut to 14 ft. lbs. (20 Nm) and recheck the valve clear-

67162-ATSX-G06

ance. Repeat the valve adjustment if necessary.

5. Rotate the crankshaft 180° counterclockwise (the camshaft pulleys will turn 90°) The **UP** arrow marks should be pointing to the exhaust side of the cylinder head.

6. Adjust the valves on cylinder No. 3 by performing the following:

a. Insert a feeler gauge in between the camshaft lobe and the rocker arm.

b. Loosen the locknut and turn the adjusting screw until the feeler gauge slides back and forth with a slight amount of drag.

c. Tighten the locknut to 14 ft. lbs. (20 Nm) and recheck the valve clearance. Repeat the valve adjustment if necessary.

7. Rotate the crankshaft 180° counterclockwise (the camshaft pulleys will turn 90°) to bring No. 4 piston to TDC. The **UP** arrow marks should be pointing down, toward the crankshaft.

8. Adjust the valves on cylinder No. 4 by performing the following:

a. Insert a feeler gauge in between the camshaft lobe and the rocker arm.

b. Loosen the locknut and turn the adjusting screw until the feeler gauge slides back and forth with a slight amount of drag.

c. Tighten the locknut to 14 ft. lbs. (20 Nm) and recheck the valve clearance. Repeat the valve adjustment if necessary.

9. Rotate the crankshaft 180° counterclockwise (the camshaft pulleys will turn 90°) to bring piston No. 2 to TDC. The **UP** arrow marks should be pointing to the intake side of the cylinder head.

10. Adjust the valves on cylinder No. 2 by performing the following:

a. Insert a feeler gauge in between the camshaft lobe and the rocker arm.

b. Loosen the locknut and turn the adjusting screw until the feeler gauge slides back and forth with a slight amount of drag.

c. Tighten the locknut to 14 ft. lbs. (20 Nm) and recheck the valve clearance. Repeat the valve adjustment if necessary.

11. Install the cylinder head cover gasket cover to the groove of the cylinder head cover. Before installing the gasket, thoroughly clean the seal and the groove. Seat the recesses for the camshaft first, then work it into the groove around the outside edges. Be sure the gasket is seated securely in the corners of the recesses.

12. Apply liquid gasket to the 4 corners of the recesses of the cylinder head cover gasket. Do not install the parts if 5 minutes or more have elapsed since applying liquid

Valve clearance adjusting screw locations—TSX models

gasket. After assembly, wait at least 20 minutes before filling the engine with oil.

13. Install or connect the following:
- Cylinder head (valve) cover. Tighten the bolts attaching to 84 inch lbs. (10 Nm).
- Spark plug wires to the correct spark plugs.
- Positive, then the negative battery cable and enter the radio security code.

Starter Motor

REMOVAL & INSTALLATION

CL Model

1. Before servicing the vehicle, refer to the precautions section.
2. Disconnect the negative battery cable.
3. Remove or disconnect the following:
- Starter electrical connectors
- 2 starter mounting bolts
- Starter motor

To install:

4. Installation is the reverse of the removal procedure. Tighten the starter motor bolts to 33 ft. lbs. (44 Nm).

➡**When installing the starter cable, be sure to place the closed loop connector over the stud on the starter with the crimped side of the connector facing up. This is to ensure a proper fit against the stud.**

2002–04 RL Model

➡**This procedure requires the use of an engine hoist to lift the engine slightly.**

1. Before servicing the vehicle, refer to the precautions section.
2. Remove or disconnect the following:

- Both battery cables
- Battery and tray
- Alternator and belt
- Left exhaust manifold cover
- Left damper fork
- Left lower ball joint from the suspension
- Left drive shaft
- Transmission stop collar
- Exhaust system Y–pipe
- Front mounting bolts

3. Attach a suitable engine hoist and slightly lift the engine.
- Left engine mount bracket
- Starter electrical connectors
- Starter motor

To install:

4. Install or connect the following:
- Starter motor and tighten the bolts to 33 ft. lbs. (44 Nm)
- Starter electrical connectors

➡**Upon installation of the starter cable and the black/white wire, make sure that the crimped side of the connector is facing up.**

5. Lower the engine onto the motor mount and tighten the nut to 47 ft. lbs. (64 Nm) and the bolts to 40 ft. lbs. (54 Nm)

6. Install or connect the following:
- Exhaust system Y–pipe with new gaskets and tighten the 10mm bolts and nuts to 40 ft. lbs. (54 Nm) and the 8mm nuts to 16 ft. lbs. (22 Nm)
- Transmission stop collar and tighten the bolts to 28 ft. lbs. (38 Nm)
- Left drive shaft
- Damper fork
- Left lower ball joint
- Left exhaust manifold cover
- Alternator and belt
- Battery tray
- Battery and battery cables

2005–06 RL Model

Make sure you have the anti-theft codes for the radio and the navigation system, then write down the customer's radio presets. Make sure the ignition switch is OFF.

1. Before servicing the vehicle, refer to the precautions section.

2. Remove the battery trim.

3. Disconnect the negative cable from the battery first, then disconnect the positive cable.

4. Remove the intake manifold cover

5. Remove the vacuum hose and transmission dipstick.

6. Remove the harness clamp.

7. Disconnect the starter cable from the B terminal, then disconnect the BLK/WHT wire from the S terminal.

8. Remove the two bolts holding the starter, then remove the starter.

To install:

9. Install in the reverse order of removal. Tighten the starter bolts to 33 ft. lbs. (44 Nm).

10. Make sure the crimped side of the ring terminal is facing out.

11. Connect the battery positive cable to the battery first, then connect the negative cable.

12. Start the engine to make sure the starter works properly.

13. Enter the anti-theft codes for the radio and the navigation system, then enter the customer's radio channel presets.

RSX Models

1. Before servicing the vehicle, refer to the precautions section.

2. Remove or disconnect the following:
- Negative, then the positive battery cables
- Knock Sensor (KS) connector
- Bolt securing the harness bracket, K20A3 engine
- Bolt securing the harness bracket, then the intake manifold brackets, K20A2 engine
- Starter cable from the B terminal on the solenoid
- Black and white wire from the S terminal
- 2 starter mounting bolts and starter

To install:

3. Installation is the reverse of the removal procedure, noting the following:
- Make sure the crimped side of the B terminal connector ring terminal is facing out
- Tighten the top starter mounting bolt to 33 ft. lbs. (44 Nm) and the bottom bolt to 47 ft. lbs. (64 Nm)

2002–03 TL Models

1. Before servicing the vehicle, refer to the precautions section.

2. Disconnect the negative battery cable.

3. Remove or disconnect the following:
- Starter electrical connectors
- 2 starter mounting bolts
- Starter motor

To install:

4. Installation is the reverse of the removal procedure. Tighten the starter motor bolts to 33 ft. lbs. (44 Nm).

➡**When installing the starter cable, be sure to place the closed loop connector over the stud on the starter with the crimped side of the connector facing up. This is to ensure a proper fit against the stud.**

2004–06 TL Models

➡**Make sure you have the anti-theft code for the radio, and the navigation system, then write down the radio channel presets. Make sure the ignition switch is OFF.**

1. Before servicing the vehicle, refer to the precautions section.

2. Remove the left side engine compartment cover.

3. Disconnect the negative cable and positive cable.

4. Remove the battery hold-down bracket, then remove the battery and battery tray.

5. Remove the harness clamp.

6. Disconnect the starter cable from the B terminal, then disconnect the BLK/WHT wire from the S terminal.

7. Remove the two bolts holding the starter, then remove the starter.

To install:

8. Install in the reverse order of removal. Tighten the starter bolts to 47 ft. lbs. (64 Nm).

9. Make sure the crimped side of the ring terminal is facing out.

10. Connect the battery positive cable to the battery first, then connect the negative cable.

11. Start the engine to make sure the starter works properly.

12. Enter the anti-theft codes for the radio and the navigation system, then enter the customer's radio channel presets.

TSX Models

➡**The factory sound system has a coded theft protection system. It is recommended that you know your reset code before you begin.**

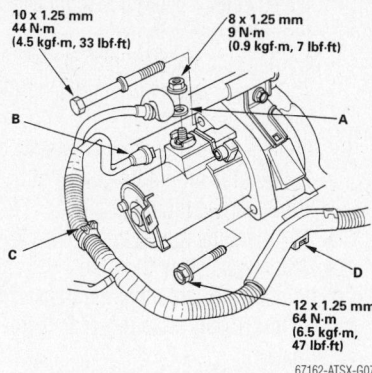

10 x 1.25 mm
44 N·m
(4.5 kgf·m, 33 lbf·ft)

8 x 1.25 mm
9 N·m
(0.9 kgf·m, 7 lbf·ft)

12 x 1.25 mm
64 N·m
(6.5 kgf·m, 47 lbf·ft)

67162-ATSX-G07

Starter mounting—TSX models

1. Before servicing the vehicle, refer to the precautions section.

2. Remove or disconnect the following:
- Negative then the positive battery cables
- Intake manifold
- Starter cable from the B terminal
- Black/white wire from the S (solenoid) terminal
- Harness clamp and holder
- Two bolts that mount the starter to the transaxle assembly
- Starter

To install:

3. Install in the reverse order of removal.

➡**When installing the heavy gauge starter cable, make sure the crimped side of the terminal end is facing out.**

4. Enter the anti-theft code and radio presets.

Oil Pan

REMOVAL & INSTALLATION

CL Model

1. Before servicing the vehicle, refer to the precautions section.

2. Drain the engine oil.

3. Drain the differential oil.

4. Remove or disconnect the following:
- Negative battery cable
- Accessory drive belts
- Power steering pump without disconnecting the lines
- Exhaust manifold covers
- Front wheels
- Splash shield
- Strut forks
- Lower ball joints
- Halfshafts from the differential
- Intermediate shaft
- Vehicle speed/power steering

speed sensor without disconnecting the hoses
- Lower plate from the rear beam
- A/C compressor without disconnecting the lines

5. Attach a chain hoist to the engine.
- Left engine mount bracket
- Right engine mount bracket
- Right engine mount
- 36mm sealing bolt on the transaxle. Ensure that the transaxle is in 1st gear (manual) or **P** (automatic).
- Extension shaft from the differential with an extension shaft puller
- Differential mounting bolts and the 26mm shim, then remove the differential
- Rear engine stiffener
- Flywheel cover or the torque converter covers
- Oil pan

➡️**Do not lose the dowel pins from the oil pan**

To install:

6. Apply liquid gasket to the cylinder block. Be sure that the mating surfaces are clean and dry before installing the liquid gasket. Do not apply liquid gasket to the O–ring grooves.

7. Install or connect the following:
- Oil pan with new O–rings and torque the bolts to 16 ft. lbs. (22 Nm)

➡️**Be sure the dowel pins are still in place.**

- Flywheel or torque converter cover and torque the bolts to 108 inch lbs. (12 Nm)
- Rear engine stiffener and tighten the bolt attaching the engine stiffener to the transaxle first, to 47 ft.

lbs. (64 Nm), then tighten the bolts to the engine block to 16 ft. lbs. (22 Nm)
- Differential to the engine and torque the bolts to 47 ft. lbs. (64 Nm)

➡️**Be sure to install the shim in the original location**

- Air conditioning compressor to the engine block and tighten the mounting bolts to 16 ft. lbs. (22 Nm)
- New set ring to the extension shaft—Using an extension shaft installer tool, install the shaft to the differential.
- Extension shaft with a new set ring

8. Fill the secondary gear with super high temperature grease. Applying sealer to the threads of the 36mm sealing bolt, then install the bolt and tighten it to 58 ft. lbs. (78 Nm).
- Right engine mount and torque the bolts to 28 ft. lbs. (38 Nm)
- Right engine mount bracket and torque the nut and bolts to 47 ft. lbs. (64 Nm)
- Left engine mount bracket and torque the bolts to 40 ft. lbs. (54 Nm)
- A/C compressor and torque the bolts to 16 ft. lbs. (22 Nm)
- Lower plate and torque the bolts to 28 ft. lbs. (38 Nm)
- Vehicle speed/power steering speed sensor and torque the bolts to 108 inch lbs. (12 Nm)
- Intermediate shaft and the half-shafts
- Lower ball joints and tighten the nuts to 40 ft. lbs. (54 Nm)
- Strut forks and torque the bolts to 51 ft. lbs. (69 Nm)

- Engine splash shield and tighten the bolts to 84 inch lbs. (9.5 Nm)
- Front wheels
- Exhaust manifold covers and torque the bolts to 16 ft. lbs. (22 Nm)
- Power steering pump and torque the bolt to 33 ft. lbs. (44 Nm) and the nut to 16 ft. lbs. (22 Nm)
- Accessory drive belts
- Negative battery cable

9. Fill the engine with oil.
10. Fill the differential with oil.
11. Run the engine and check for leaks.
12. Check the front wheel alignment.

2002–04 RL Model

1. Before servicing the vehicle, refer to the precautions section.
2. Drain the engine oil.
3. Drain the differential oil.
4. Remove or disconnect the following:
- Negative battery cable
- Accessory drive belts
- Power steering pump without disconnecting the lines
- Exhaust manifold covers
- Front wheels
- Splash shield
- Strut forks
- Lower ball joints
- Halfshafts from the differential
- Intermediate shaft
- Vehicle speed/power steering speed sensor without disconnecting the hoses
- Lower plate from the rear beam
- A/C compressor without disconnecting the lines

5. Attach a chain hoist to the engine.
- Left engine mount bracket
- Right engine mount bracket
- Right engine mount
- 36mm sealing bolt on the transaxle. Ensure that the transaxle is in 1st gear (manual) or **P** (automatic).
- Extension shaft from the differential with an extension shaft puller
- Differential mounting bolts and the 26mm shim, then remove the differential
- Rear engine stiffener
- Flywheel cover or the torque converter covers
- Oil pan

➡️**Do not lose the dowel pins from the oil pan**

To install:

6. Apply liquid gasket to the cylinder block. Be sure that the mating surfaces are

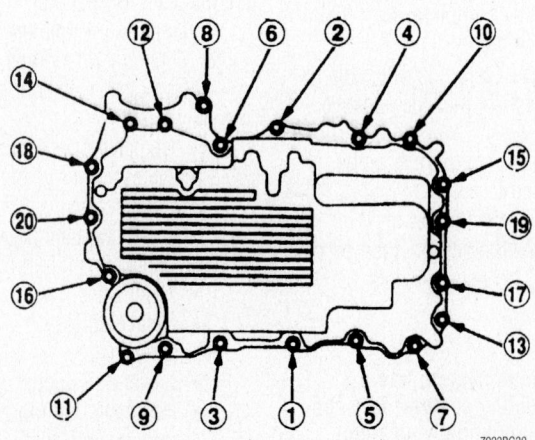

7923BG30

Be sure to tighten the oil pan bolts in the sequence shown—CL model

clean and dry before installing the liquid gasket. Do not apply liquid gasket to the O–ring grooves.

7. Install or connect the following:
- Oil pan with new O–rings and torque the bolts to 16 ft. lbs. (22 Nm)

➡**Be sure the dowel pins are still in place.**

- Flywheel or torque converter cover and torque the bolts to 108 inch lbs. (12 Nm)
- Rear engine stiffener and tighten the bolt attaching the engine stiffener to the transaxle first, to 47 ft. lbs. (64 Nm), then tighten the bolts to the engine block to 16 ft. lbs. (22 Nm)
- Differential to the engine and torque the bolts to 47 ft. lbs. (64 Nm)

➡**Be sure to install the shim in the original location**

- Air conditioning compressor to the engine block and tighten the mounting bolts to 16 ft. lbs. (22 Nm)
- New set ring to the extension shaft—Using an extension shaft installer tool, install the shaft to the differential.
- Extension shaft with a new set ring

8. Fill the secondary gear with super high temperature grease. Applying sealer to the threads of the 36mm sealing bolt, then install the bolt and tighten it to 58 ft. lbs. (78 Nm).
- Right engine mount and torque the bolts to 28 ft. lbs. (38 Nm)
- Right engine mount bracket and torque the nut and bolts to 47 ft. lbs. (64 Nm)
- Left engine mount bracket and torque the bolts to 40 ft. lbs. (54 Nm)
- A/C compressor and torque the bolts to 16 ft. lbs. (22 Nm)
- Lower plate and torque the bolts to 28 ft. lbs. (38 Nm)
- Vehicle speed/power steering speed sensor and torque the bolts to 108 inch lbs. (12 Nm)
- Intermediate shaft and the half-shafts
- Lower ball joints and tighten the nuts to 40 ft. lbs. (54 Nm)
- Strut forks and torque the bolts to 51 ft. lbs. (69 Nm)
- Engine splash shield and tighten the bolts to 84 inch lbs. (9.5 Nm)

- Front wheels
- Exhaust manifold covers and torque the bolts to 16 ft. lbs. (22 Nm)
- Power steering pump and torque the bolt to 33 ft. lbs. (44 Nm) and the nut to 16 ft. lbs. (22 Nm)
- Accessory drive belts
- Negative battery cable

9. Fill the engine with oil.
10. Fill the differential with oil.
11. Run the engine and check for leaks.
12. Check the front wheel alignment.

2005–06 RL Model

1. Before servicing the vehicle, refer to the precautions section.
2. Drain the engine oil.
3. Remove the splash shield.
4. Remove the under cover.
5. Remove exhaust pipe A.
6. Remove the rear warm up three way catalytic converter bracket.
7. Remove the torque converter cover and the bolts retaining the transmission.
8. Remove the bolts retaining the oil pan.
9. Using a flat blade screwdriver, separate the oil pan from the block in the places shown.
10. Remove the oil pan.

To install:

11. Remove any old liquid gasket from the oil pan mating surfaces, bolts, and bolt holes.
12. Clean and dry the oil pan mating surfaces.
13. Apply liquid gasket, P/N 08717-0004, 08718-0001, 08718-0003, or 08718-0009, evenly to the oil pan mating surface of the engine block.

➡**Do not install the parts if 4 minutes or more have elapsed since applying liquid gasket. Instead, re-apply liquid gasket after removing the old residue.**

14. Install the oil pan on the engine block. Tighten the bolts in two or three steps. In the final step, tighten all bolts, in sequence, to 9 ft. lbs. (12 Nm).

➡**After assembly, wait at least 30 minutes before filling the engine with oil.**

15. Tighten the bolts retaining the transmission to 47 ft. lbs. (64 Nm), then install the torque converter cover.
16. Remove the rear warm up three way catalytic converter bracket. Tighten to 16 ft. lbs. (22 Nm)
17. Install the remaining components in the reverse order of removal, after 30 minutes has elapsed, fill the engine with new oil.

RSX Models

K20A3 ENGINE

1. Before servicing the vehicle, refer to the precautions section.
2. Raise and safely support the vehicle.
3. Drain the engine oil.
4. Attach a chain hoist to the engine.
5. Remove or disconnect the following:
- Lower ball joints
- Rear mount mounting bolts
- Front mount mounting bolts
- Automatic Transaxle Fluid (ATF) filter mounting bolt, if equipped with an automatic transaxle

6. Make alignment marks on the reference lines that align with the centers of the rear subframe mounting bolts.
- Front subframe
- Bolts/nuts securing the oil pan
- Oil pan by driving an oil pan seal cutter between the oil pan and cylinder block, then cutting the oil pan seal by striking the side of the cutter to slide the cutter along the oil pan

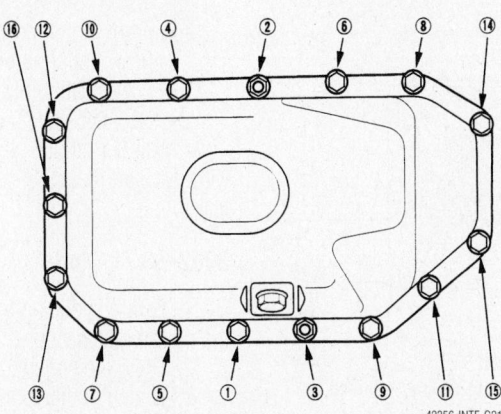

42356-INTE-G21

Oil pan tightening sequence—2.0L (K20A3) engine

To install:

7. Thoroughly clean the mating surfaces, bolts and bolt holes. Apply liquid gasket part no. 08718–0009 evenly to the cylinder block mating surface of the oil pan and inner threads of the bolt holes.

➡**Make sure to install the oil pan within 5 minutes of applying the liquid gasket.**

8. Install or connect the following:
   • Oil pan
   • Oil pan mounting bolts. Tighten, in sequence, in 2 or 3 steps to 104 inch lbs. (12 Nm).
   • Subframe. Align the reference lines on the subframe with the bolt head center, tighten the bolts to 76 ft. lbs. (103 Nm).
   • ATF filter mounting bolt
   • Front mounting bolt
   • Rear mounting bolts
   • Lower ball joints
   • Negative battery cable

9. Wait 30 minutes, then fill the engine with oil. Start the engine and check for leaks.

K20A2 AND K20Z1 ENGINES

1. Before servicing the vehicle, refer to the precautions section.
2. Raise and safely support the vehicle.
3. Drain the engine oil.
4. Attach a chain hoist to the engine.
5. Remove or disconnect the following:
   • Lower ball joints
   • Rear mount mounting bolts
   • Front mount mounting bolts

6. Make alignment marks on the reference lines that align with the centers of the rear subframe mounting bolts.
   • Front subframe
   • Clutch cover and 2 bolts securing the transaxle
   • Bolts/nuts securing the oil pan

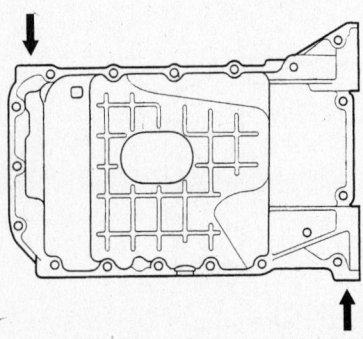

42356-INTE-G20

Insert a flat tip prytool where indicated by the arrows to carefully separate the oil pan from the block—2.0 (K20A2) engine

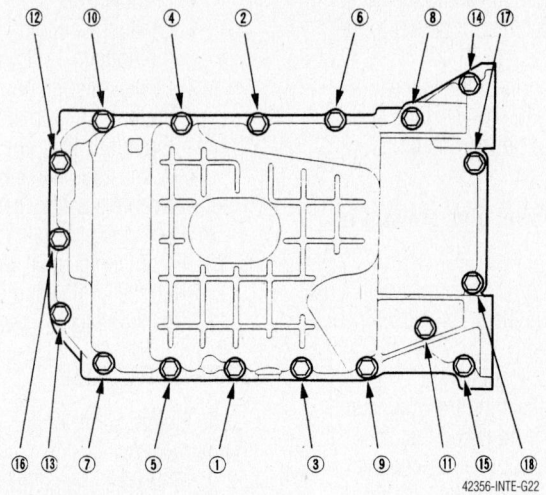

42356-INTE-G22

Oil pan tightening sequence—2.0L (K20A2) engine

   • Oil pan by inserting a flat tip prytool where shown in the illustration and separate the oil pan from the block

To install:

7. Thoroughly clean the mating surfaces, bolts and bolt holes. Apply liquid gasket part no. 08718–0009 evenly to the cylinder block mating surface of the oil pan and inner threads of the bolt holes.

➡**Make sure to install the oil pan within 5 minutes of applying the liquid gasket.**

8. Install or connect the following:
   • Oil pan
   • Oil pan mounting bolts. Tighten, in sequence, in 2 or 3 steps to 104 inch lbs. (12 Nm).
   • Clutch cover. Tighten the bolts to 104 inch lbs. (12 Nm).
   • 2 bolts securing the transaxle and tighten to 47 ft. lbs. (64 Nm)
   • Subframe. Align the reference lines on the subframe with the bolt head center, tighten the bolts to 76 ft. lbs. (103 Nm).
   • Front mounting bolt
   • Rear mounting bolts
   • Lower ball joints
   • Negative battery cable

9. Wait 30 minutes, then fill the engine with oil. Start the engine and check for leaks.

2002–03 TL Models

1. Before servicing the vehicle, refer to the precautions section.
2. Drain the engine oil.
3. Drain the differential oil.
4. Remove or disconnect the following:

   • Negative battery cable
   • Accessory drive belts
   • Power steering pump without disconnecting the lines
   • Exhaust manifold covers
   • Front wheels
   • Splash shield
   • Strut forks
   • Lower ball joints
   • Halfshafts from the differential
   • Intermediate shaft
   • Vehicle speed/power steering speed sensor without disconnecting the hoses
   • Lower plate from the rear beam
   • A/C compressor without disconnecting the lines

5. Attach a chain hoist to the engine.
   • Left engine mount bracket
   • Right engine mount bracket
   • Right engine mount
   • 36mm sealing bolt on the transaxle. Ensure that the transaxle is in 1st gear (manual) or **P** (automatic).
   • Extension shaft from the differential with an extension shaft puller
   • Differential mounting bolts and the 26mm shim, then remove the differential
   • Rear engine stiffener
   • Flywheel cover or the torque converter covers
   • Oil pan

➡**Do not lose the dowel pins from the oil pan**

To install:

6. Apply liquid gasket to the cylinder block. Be sure that the mating surfaces are clean and dry before installing the liquid gasket. Do not apply liquid gasket to the O–ring grooves.

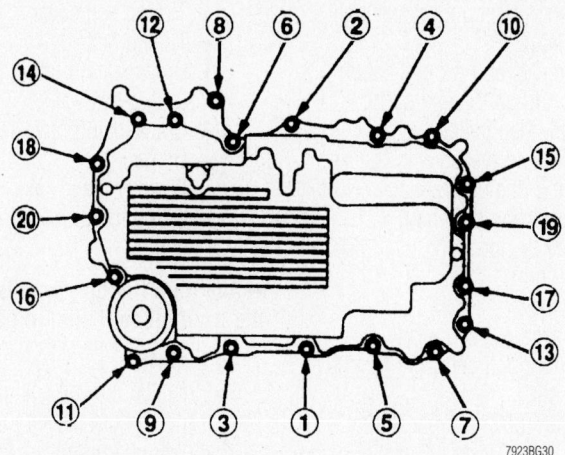

Be sure to tighten the oil pan bolts in the sequence shown—2002–03 TL models

7923BG30

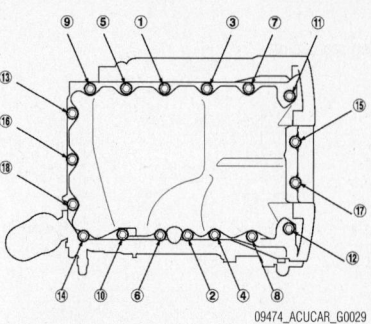

09474_ACUCAR_G0029

Oil pan bolt torque sequence—2004–06 TL and 2005–06 RL models

7. Install or connect the following:
- Oil pan with new O–rings and torque the bolts to 16 ft. lbs. (22 Nm)

➥**Be sure the dowel pins are still in place.**

- Flywheel or torque converter cover and torque the bolts to 108 inch lbs. (12 Nm)
- Rear engine stiffener and tighten the bolt attaching the engine stiffener to the transaxle first, to 47 ft. lbs. (64 Nm), then tighten the bolts to the engine block to 16 ft. lbs. (22 Nm)
- Differential to the engine and torque the bolts to 47 ft. lbs. (64 Nm)

➥**Be sure to install the shim in the original location**

- Air conditioning compressor to the engine block and tighten the mounting bolts to 16 ft. lbs. (22 Nm)
- New set ring to the extension shaft—Using an extension shaft installer tool, install the shaft to the differential.
- Extension shaft with a new set ring
8. Fill the secondary gear with super high temperature grease. Applying sealer to the threads of the 36mm sealing bolt, then install the bolt and tighten it to 58 ft. lbs. (78 Nm).
- Right engine mount and torque the bolts to 28 ft. lbs. (38 Nm)
- Right engine mount bracket and torque the nut and bolts to 47 ft. lbs. (64 Nm)
- Left engine mount bracket and torque the bolts to 40 ft. lbs. (54 Nm)

- A/C compressor and torque the bolts to 16 ft. lbs. (22 Nm)
- Lower plate and torque the bolts to 28 ft. lbs. (38 Nm)
- Vehicle speed/power steering speed sensor and torque the bolts to 108 inch lbs. (12 Nm)
- Intermediate shaft and the half-shafts
- Lower ball joints and tighten the nuts to 40 ft. lbs. (54 Nm)
- Strut forks and torque the bolts to 51 ft. lbs. (69 Nm)
- Engine splash shield and tighten the bolts to 84 inch lbs. (9.5 Nm)
- Front wheels
- Exhaust manifold covers and torque the bolts to 16 ft. lbs. (22 Nm)
- Power steering pump and torque the bolt to 33 ft. lbs. (44 Nm) and the nut to 16 ft. lbs. (22 Nm)
- Accessory drive belts
- Negative battery cable
9. Fill the engine with oil.
10. Fill the differential with oil.
11. Run the engine and check for leaks.
12. Check the front wheel alignment.

2004–06 TL Models

1. Before servicing the vehicle, refer to the precautions section.
2. Drain the engine oil.
3. Remove the splash shield.
4. Remove the under cover.
5. Remove exhaust pipe A.
6. Remove the torque converter cover and the two bolts retaining the transmission.
7. Remove the bolts retaining the oil pan.
8. Using a flat blade screwdriver, separate the oil pan from the block in the places shown.
9. Remove the oil pan.

To install:

10. Remove any old liquid gasket from the oil pan mating surfaces, bolts, and bolt holes.
11. Clean and dry the oil pan mating surfaces.
12. Apply liquid gasket, P/N 08717-0004, 08718-0001, 08718-0003, or 08718-0009, evenly to the oil pan mating surface of the engine block.

➥**Do not install the parts if 4 minutes or more have elapsed since applying liquid gasket. Instead, re-apply liquid gasket after removing the old residue.**

13. Install the oil pan on the engine block. Tighten the bolts in two or three steps. In the final step, tighten all bolts, in sequence, to 9 ft. lbs. (12 Nm).

➥**After assembly, wait at least 30 minutes before filling the engine with oil.**

14. Tighten the two bolts retaining the transmission to 28 ft. lbs. (38 Nm), then install the torque converter cover.
15. Install the remaining components in the reverse order of removal, after 30 minutes has elapsed, fill the engine with new oil.

TSX Models

1. Before servicing the vehicle, refer to the precautions section.
2. Remove or disconnect the following:
- Negative battery cable
- Engine oil
- Front tire and wheel assemblies
- Splash shield
- Stabilizer links
- Right side damper fork
- Right side lower ball joint
- Right side halfshaft. Coat all machined surfaces with clean engine oil and secure a plastic bag over the end of the halfshaft.
3. From the engine compartment, remove the front mount stop and front mount bolt

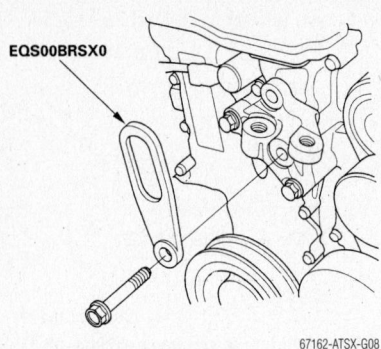

EQS00BRSX0

67162-ATSX-G08

Proper installation of Engine Hanger Plate EQS00BRSX0—TSX models

- Rear mount stop and rear mount bolt
- Ground cable and upper bracket
- Bolt holding the side engine mount bracket and attach Engine Hanger Plate EQS00BRSX0, or equivalent.

4. Use a jack to lift the engine 1.2–2.4 in. (30–60mm).

- Stiffener
- Oil pan bolts and nuts

5. Hammer a seal cutter between the engine block and oil pan to break the seal.

6. Remove the oil pan.

To install:

7. Clean the oil pan flange and engine block mounting surface.

8. Install or connect the following:

- Sealant to the oil pan flange. Be sure to apply sealant toward the inside of the bolt holes.
- Oil pan on the engine. Tighten the bolts in sequence, in 2 or 3 steps, to 104 inch lbs. (12 Nm).
- Stiffener. Torque the retaining bolts to 33 ft. lbs. (45 Nm)

9. Remove the engine hanger tool and

---

tighten the side engine mount bracket bolt to 33 ft. lbs. (45 Nm).

- Upper bracket and ground cable
- New set ring on the end of the driveshaft, then install the driveshaft. Make sure each ring "clicks" into plate in the differential.
- Right side lower ball joint
- Right side damper fork
- Stabilizer links
- Splash shield
- Tire and wheel assemblies
- Front mount bolt and front mount stop
- Rear mount bolt and rear mount stop

✳✳ WARNING

Operating the engine without the proper amount and type of engine oil will result in severe engine damage.

10. After 30 minutes, fill the engine with the correct amount of oil.
11. Connect the negative battery cable.
12. Start the engine and check for leaks.

Oil Pump

REMOVAL & INSTALLATION

CL Model

1. Before servicing the vehicle, refer to the precautions section.
2. Drain the engine oil.
3. Be sure the crankshaft is at Top Dead Center (TDC) on the No. 1 cylinder.
4. Remove or disconnect the following:

- Negative battery cable
- Cylinder head cover

- Timing belt cover
- Timing belt
- Crankshaft Position (CKP) sensor (if necessary)
- Crankshaft timing belt gear
- Oil pan
- Pickup screen
- Oil pump from the front of the engine

➡ **Any time the oil pump is removed, the front oil seal should be replaced.**

To install:

5. Install a new oil seal in the oil pump.
6. Apply liquid gasket to the mounting surface of the oil pump.
7. Install the oil pump, using new O–rings. Tighten the bolts to 108 inch lbs. (12 Nm).
8. Install or connect the following:

- Oil pump pickup screen
- Oil pan and tighten the bolts to 108 inch lbs. (12 Nm)
- Crankshaft timing belt gear
- Timing belt
- Crankshaft Position (CKP) sensor and torque the bolt to 108 inch lbs. (12 Nm)
- Timing belt cover
- Cylinder head cover
- Negative battery cable

9. Wait at least 30 minutes after completion of procedure before refilling the engine with oil. The waiting period is to allow the silicone sealant (liquid gasket) to cure.
10. Refill the engine with oil.
11. Start the engine and check the engine for leaks.
12. Turn off engine and check the oil level. Top off the oil level if necessary.

2002–04 RL Model

1. Before servicing the vehicle, refer to the precautions section.
2. Drain the engine oil.
3. Be sure the crankshaft is at Top Dead Center (TDC) on the No. 1 cylinder.
4. Remove or disconnect the following:

- Negative battery cable
- Cylinder head cover
- Timing belt cover
- Timing belt
- Crankshaft Position (CKP) sensor (if necessary)
- Crankshaft timing belt gear
- Oil pan
- Pickup screen
- Oil pump from the front of the engine

➡ **Any time the oil pump is removed, the front oil seal should be replaced.**

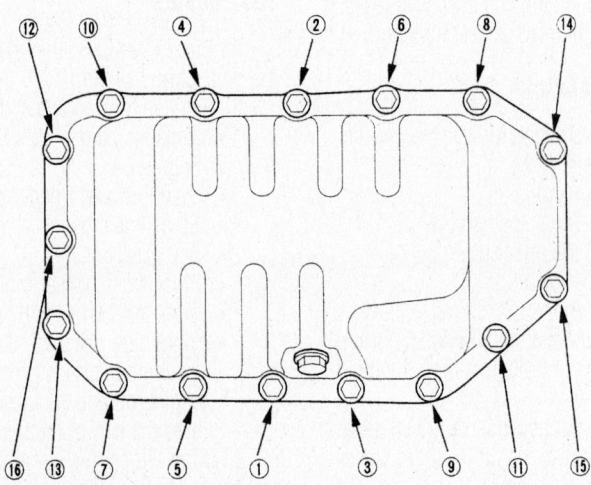

67162-ATSX-G09

Oil pan bolt tightening sequence—TSX models

To install:

5. Install a new oil seal in the oil pump.

6. Apply liquid gasket to the mounting surface of the oil pump.

7. Install the oil pump, using new O–rings. Tighten the 6mm bolts to 108 inch lbs. (12 Nm) and the 8mm bolts to 16 ft. lbs. (22 Nm).

8. Install or connect the following:
   • Oil pump pickup screen
   • Oil pan and tighten the bolts to 108 inch lbs. (12 Nm)
   • Crankshaft timing belt gear
   • Timing belt
   • Crankshaft Position (CKP) sensor and torque the bolt to 108 inch lbs. (12 Nm)
   • Timing belt cover
   • Cylinder head cover
   • Negative battery cable

9. Wait at least 30 minutes after completion of procedure before refilling the engine with oil. The waiting period is to allow the silicone sealant (liquid gasket) to cure.

10. Refill the engine with oil.

11. Start the engine and check the engine for leaks.

12. Turn off engine and check the oil level. Top off the oil level if necessary.

NOTE:
• Use new O-rings when reassembling.
• Apply oil to O-rings before installation.
• Use liquid gasket, Part No. 08718 – 0001 or 08718 – 0003.
• Clean the oil control orifice before installing.
• Remove the balancer shaft

OIL CONTROL ORIFICE (HYDRAULIC TAPPET)
Remove with 6 x 1.0 mm bolt and clean.

OIL CONTROL ORIFICES (CAMSHAFT JOURNAL)
Remove with 6 x 1.0 mm bolt and clean.

OIL CONTROL ORIFICE (HYDRAULIC TAPPET)
Remove with 6 x 1.0 mm bolt and clean.

O-RING
Replace.

OIL PUMP

O-RINGS
Replace.

O-RINGS
Replace.

6 x 1.0 mm
12 N·m (1.2 kgf·m, 8.7 lbf·ft)

BAFFLE PLATE

O-RING
Replace.

6 x 1.0 mm
12 N·m (1.2 kgf·m, 8.7 lbf·ft)

8 x 1.25 mm
22 N·m (2.2 kgf·m, 16 lbf·ft)

O-RING
Replace.

6 x 1.0 mm
12 N·m (1.2 kgf·m, 8.7 lbf·ft)

LOWER BAFFLE PLATE

DOWEL PIN

OIL PAN
Apply liquid gasket to mating surface.

OIL SCREEN

O-RING
Replace.

8 x 1.25 mm
22 N·m (2.2 kgf·m, 16 lbf·ft)

O-RINGS
Replace.

SNAP RING
Install with open side facing up.

WASHER
Replace.

OIL FILTER

DRAIN BOLT
14 x 1.5 mm
44 N·m (4.5 kgf·m, 33 lbf·ft)
Do not overtighten.

OIL PAN INNER PIPE
Clean when installing.

7923BG89

Exploded view of the lubrication system—2002–04 RL models

2005–06 RL Model

1. Before servicing the vehicle, refer to the precautions section.
2. Drain the engine oil.
3. Remove the windshield wiper arms.
4. Remove the intake manifold cover, right upper fender trim, battery trim, and left upper fender trim, then remove the upper grille cover and cowl cover.
5. Remove the under-hood fuse/relay box, connector, and transfer breather hose from the strut brace, then remove the strut brace.
6. Remove the connector bracket from the front cylinder head; use the bracket bolt hole to attach engine balancer bar front arm.
7. Remove the engine mount control solenoid valve bracket from the rear cylinder head; use the bracket bolt hole to attach engine balancer bar rear arm.
8. Lift and support the engine with engine hanger and engine balancer bar Attach the front arm to the front cylinder head with a spacer and the 10 x 1.25 mm bolt. Attach the rear arm to the rear cylinder head with the 8 x 1.25 mm bolt.
9. Remove the timing belt.
10. Remove the Crankshaft Position (CKP) sensor.
11. Remove the rocker arm oil control solenoid/oil filter assembly.
12. Remove the oil pan.
13. Remove the oil screen then remove the oil pump.

To install:

14. Remove the old oil seal from the oil pump.
15. Gently tap in the new oil seal until the seal driver bottoms on the pump.
16. Remove any old liquid gasket from the oil pump mating surfaces, bolts, and bolt holes.
17. Clean and dry the oil pump mating surfaces.

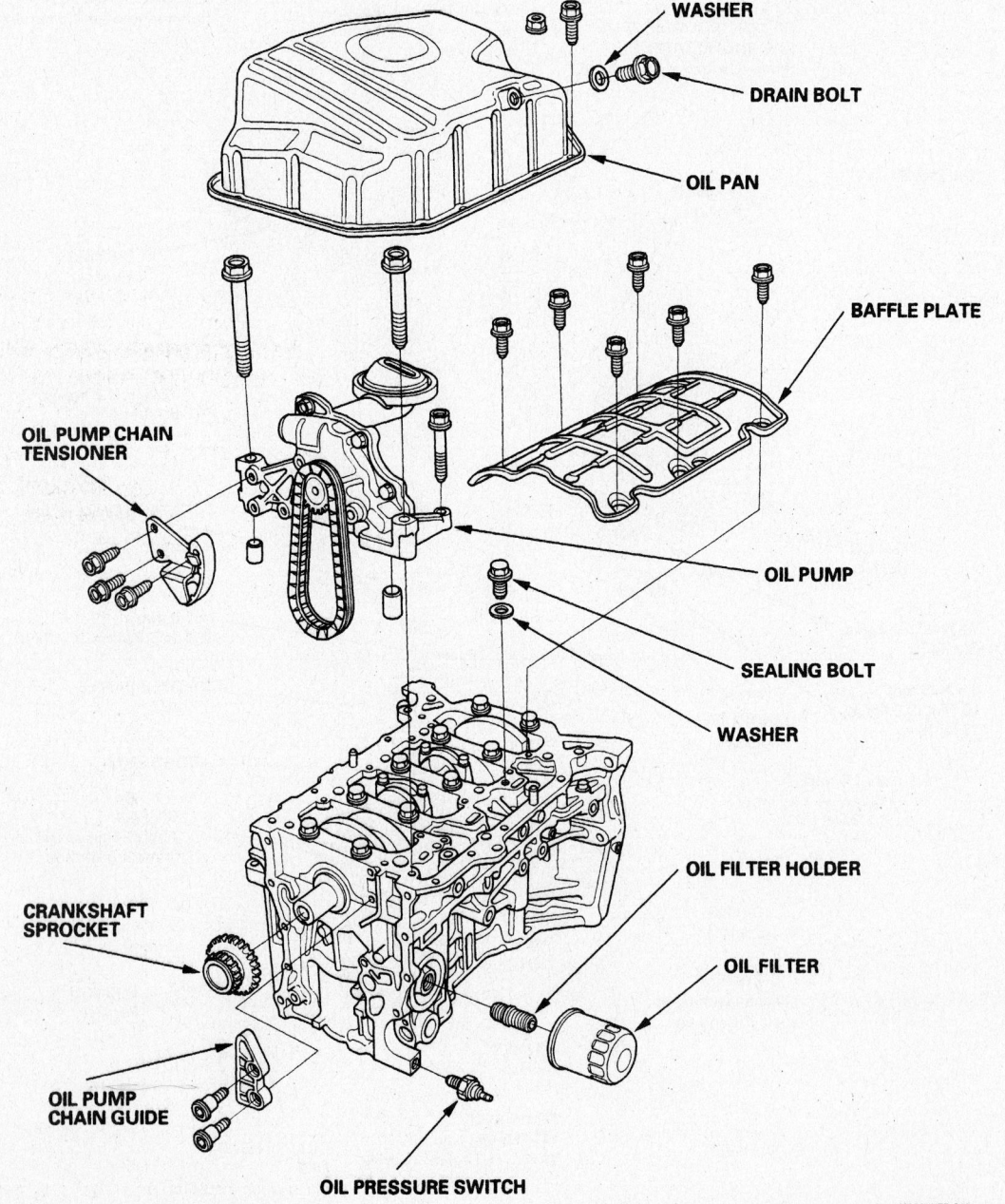

Exploded view of the lubrication system—2.0L (K20A3) engine shown

42356-INTE-G23

18. Apply liquid gasket evenly to the engine block mating surface of the oil pump.

➡**Do not install the parts if 4 minutes or more have elapsed since applying liquid gasket. Instead, re-apply liquid gasket after removing the old residue.**

19. Grease the lip of the oil seal, and apply oil to the new O-ring.

20. Install the dowel pins, then align the inner rotor with the crankshaft and install the oil pump. Tighten the bolts to 9 ft. lbs. (12 Nm).

21. Clean the excess grease off the crankshaft, and check the seal for distortion.

22. Apply oil to the new O-ring, install the oil screen.

23. Install the oil pan.

24. Install the remaining components in the reverse order of removal.

25. After assembly, wait at least 30 minutes before filling the engine with oil.

RSX Models

1. Before servicing the vehicle, refer to the precautions section.
2. Drain the engine oil.
3. Remove or disconnect the following:
   - Oil pan
   - Oil pump chain tensioner
   - Oil pump

To install:
4. Install or connect the following:
   - Oil pump
   - Oil pump chain tensioner
   - Oil pan
5. Wait 30 minutes, then fill the engine with oil. Start the engine and check for leaks.

2002–03 TL Models

1. Before servicing the vehicle, refer to the precautions section.
2. Drain the engine oil.
3. Be sure the crankshaft is at Top Dead Center (TDC) on the No. 1 cylinder.
4. Remove or disconnect the following:
   - Negative battery cable
   - Cylinder head cover
   - Timing belt cover
   - Timing belt
   - Crankshaft Position (CKP) sensor (if necessary)
   - Crankshaft timing belt gear
   - Oil pan
   - Pickup screen
   - Oil pump from the front of the engine

➡**Any time the oil pump is removed, the front oil seal should be replaced.**

To install:
5. Install a new oil seal in the oil pump.
6. Apply liquid gasket to the mounting surface of the oil pump.
7. Install the oil pump, using new O-rings. Tighten the 6mm bolts to 108 inch lbs. (12 Nm) and the 8mm bolts to 16 ft. lbs. (22 Nm).
8. Install or connect the following:
   - Oil pump pickup screen
   - Oil pan and tighten the bolts to 108 inch lbs. (12 Nm)
   - Crankshaft timing belt gear
   - Timing belt
   - Crankshaft Position (CKP) sensor and torque the bolt to 108 inch lbs. (12 Nm)
   - Timing belt cover
   - Cylinder head cover
   - Negative battery cable
9. Wait at least 30 minutes after completion of procedure before refilling the engine with oil. The waiting period is to allow the silicone sealant (liquid gasket) to cure.
10. Refill the engine with oil.
11. Start the engine and check the engine for leaks.
12. Turn off engine and check the oil level. Top off the oil level if necessary.

2004–06 TL Models

1. Before servicing the vehicle, refer to the precautions section.
2. Drain the engine oil.
3. Remove the timing belt.
4. Remove the Crankshaft Position (CKP) sensor A/B.
5. Remove the left side engine compartment cover and left rear engine compartment cover.
6. Remove the right fender trim and right side engine compartment cover.
7. Remove the right rear engine compartment cover, left fender trim and front bulkhead cover.
8. Remove the intake manifold cover.
9. Remove the strut brace.
10. Lift and support the engine with an engine support tool such as AAR-T-12566.
11. Remove the VTEC solenoid valve and oil filter assembly.
12. Remove the oil pan.
13. Remove the oil screen.
14. Remove the mounting bolts and the oil pump assembly.

To install:
15. Remove the old oil seal from the oil pump.
16. Gently tap in the new oil seal until the seal driver bottoms on the pump.

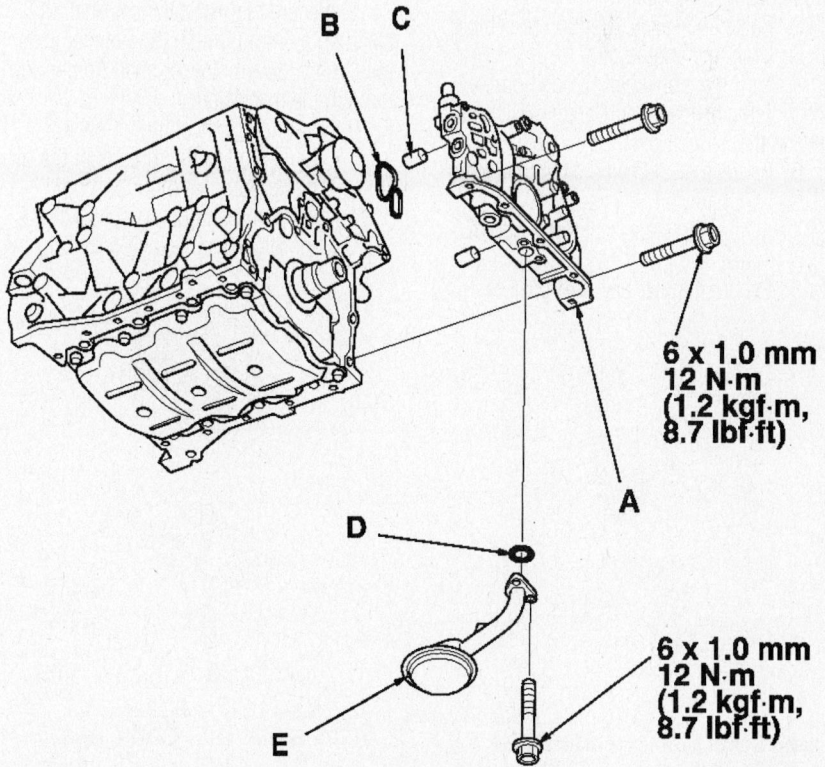

6 x 1.0 mm
12 N·m
(1.2 kgf·m, 8.7 lbf·ft)

6 x 1.0 mm
12 N·m
(1.2 kgf·m, 8.7 lbf·ft)

09474_ACUCAR_G0030

Oil pump assembly mounting—2004–06 TL and 2005–06 RL models

17. Remove any old liquid gasket from the oil pump mating surfaces, bolts, and bolt holes.

18. Clean and dry the oil pump mating surfaces.

19. Apply liquid gasket, P/N 08717-0004, 08718-0001, 08718-0003, or 08718-0009, evenly to the engine block mating surface of the oil pump.

➡**Do not install the parts if 4 minutes or more have elapsed since applying liquid gasket. Instead, re-apply liquid gasket after removing the old residue.**

20. Grease the lip of the oil seal, and apply oil to the new O-ring.

21. Install the dowel pins, then align the inner rotor with the crankshaft and install the oil pump. Tighten the bolts to 9 ft. lbs. (12 Nm).

22. Clean the excess grease off the crankshaft, and check the seal for distortion.

23. Apply oil to the new O-ring, install the oil screen.

24. Install the oil pan.

25. Install the remaining components in the reverse order of removal.

26. After assembly, wait at least 30 minutes before filling the engine with oil.

TSX Models

1. Before servicing the vehicle, refer to the precautions section.

2. Raise and safely support the vehicle.

3. Drain the engine oil.

4. Turn the crankshaft to position the No. 1 piston at Top Dead Center (TDC) on the compression stroke.

5. Remove or disconnect the following:
   • Oil pan
   • Oil pump chain tensioner, and discard

6. Secure the rear balancer shaft by

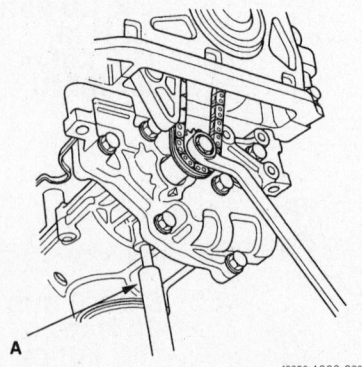

Insert a 6mm pin into the maintenance hole in the lower balancer shaft holder and through the rear balancer shaft—TSX models

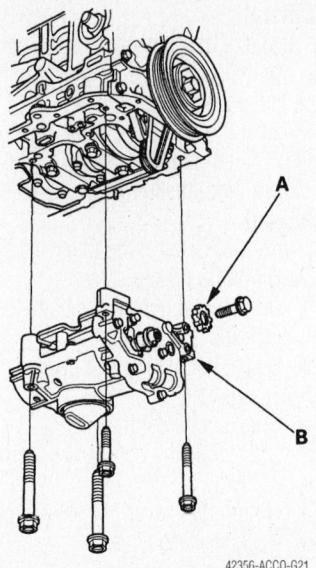

42356-ACCO-G21

Exploded view of the oil pump sprocket (A) and oil pump (B)—TSX models

inserting a 6mm pin into the maintenance hole in the lower balancer shaft holder and through the rear balancer shaft
   • Oil pump sprocket mounting bolt
   • Oil pump sprocket and oil pump

To install:

7. Make sure the No. 1 piston is still at TDC.

8. Align the dowel pin on the rear balancer shaft wit the mark on the oil pump

9. Secure the rear balancer shaft by inserting a 6mm pin into the maintenance hole in the lower balancer shaft holder and through the rear balancer shaft

10. Install or connect the following:
   • Engine oil to the threads of the oil pump sprocket mounting bolt
   • Oil pump and pump sprocket, loosely. Remove the 6mm pin.

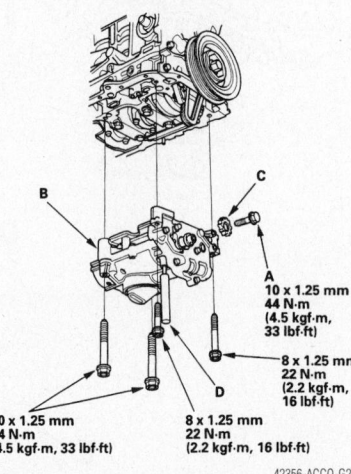

A
10 x 1.25 mm
44 N·m
(4.5 kgf·m,
33 lbf·ft)

8 x 1.25 mm
22 N·m
(2.2 kgf·m,
16 lbf·ft)

10 x 1.25 mm
44 N·m
(4.5 kgf·m, 33 lbf·ft)

8 x 1.25 mm
22 N·m
(2.2 kgf·m, 16 lbf·ft)

42356-ACCO-G22

Oil pump tightening specifications—TSX models

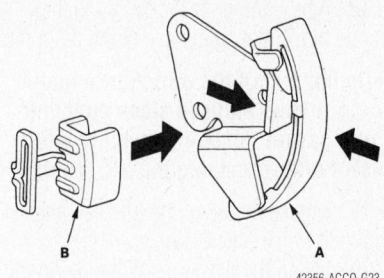

42356-ACCO-G23

Squeeze a new oil pump chain tensioner (A) then install the set clip (B) on it—TSX models

   • Oil pump mounting bolts and tighten as shown in the illustration

11. Squeeze a new oil pump chain tensioner then install the set clip on it. The set clip is provided with the oil pump chain tensioner
   • New oil pump chain tensioner and tighten the bolts to 8.7 ft. lbs. (12 Nm). Remove the set clip from the tensioner.
   • Oil pan

12. Fill the crankcase with the proper amount of new engine oil.

Rear Main Seal

REMOVAL & INSTALLATION

1. Before servicing the vehicle, refer to the precautions section.

2. Remove or disconnect the following:
   • Transaxle
   • Clutch (if equipped)
   • Flexplate
   • Crankshaft seal by prying it out of the retainer

To install:

3. Install or connect the following:
   • Clean engine oil to the lip of the new seal
   • New seal into the retainer using an appropriate seal driver
   • Flywheel
   • Clutch (if equipped)
   • Transmission

Timing Belt

REMOVAL & INSTALLATION

CL Model

➡**The radio may have a coded theft protection circuit. Obtain the code before disconnecting the battery, removing the radio fuse, or removing the radio.**

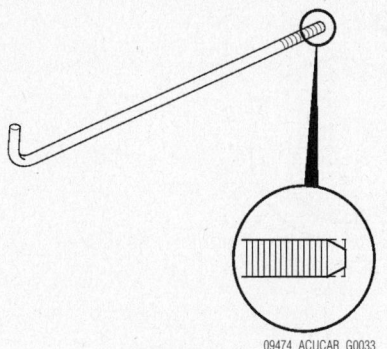

09474_ACUCAR_G0033

Grind the battery tray clamp bolt as shown—CL model

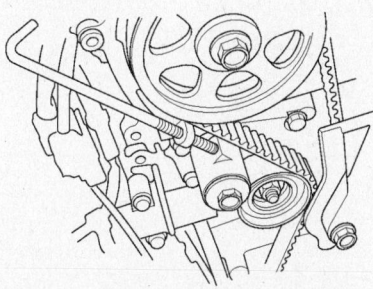

09474_ACUCAR_G0034

Screw the battery clamp bolt in as shown to hold the timing belt adjuster in its current position—CL model

1. Before servicing the vehicle, refer to the precautions section.

2. Check that the No. 1 piston Top Dead Center (TDC) mark on the front camshaft pulley and the pointer on the front upper cover are aligned.

3. Remove the front wheel.

4. Remove the splash shield.

5. Remove the drive belts.

6. Support the engine with a jack and wood block under the oil pan.

7. Remove the side engine mount bracket.

8. Remove the dipstick tube.

9. Remove the crankshaft pulley.

10. Remove the front upper cover and rear upper cover.

11. Remove one of the battery clamp bolts from the battery tray, and grind the end of it as shown.

12. Screw the battery clamp bolt in as shown to hold the timing belt adjuster in its current position. Only tighten it by hand, do not use a wrench.

13. Remove the engine mount bracket

14. Remove the idler pulley bolt and idler pulley, then remove the timing belt.

To install:

15. If reusing the old belt, perform the following steps.

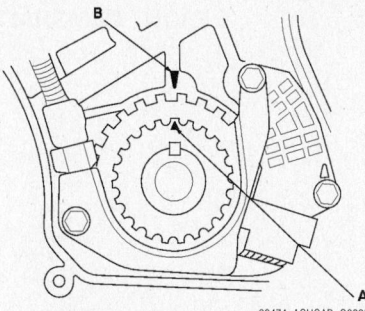

09474_ACUCAR_G0035

Set the timing belt drive pulley to TDC by aligning the TDC mark on the tooth of the timing belt drive pulley with the pointer on the oil pump—CL model

a. Clean the timing belt pulleys, timing belt guide plate, and the upper and lower covers.

b. Set the timing belt drive pulley to TDC by aligning the TDC mark on the tooth of the timing belt drive pulley with the pointer on the oil pump.

c. Set the camshaft pulleys to TDC by aligning the TDC marks on the camshaft pulleys with the pointers on the back covers.

d. Apply liquid thread lock (P/N 08713-0001) to the idler pulley bolt, then loosely install the idler pulley bolt as the pulley can move and does not come off.

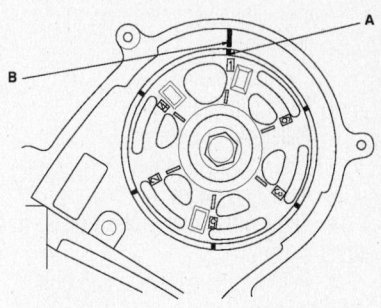

09474_ACUCAR_G0036

Align the front camshaft pulley to TDC as shown—CL model

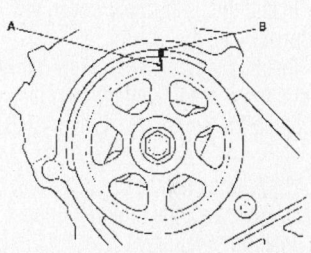

Install the timing belt in a counterclockwise sequence starting with the drive pulley.

09474_ACUCAR_G0037

Align the rear camshaft pulley to TDC as shown—CL model

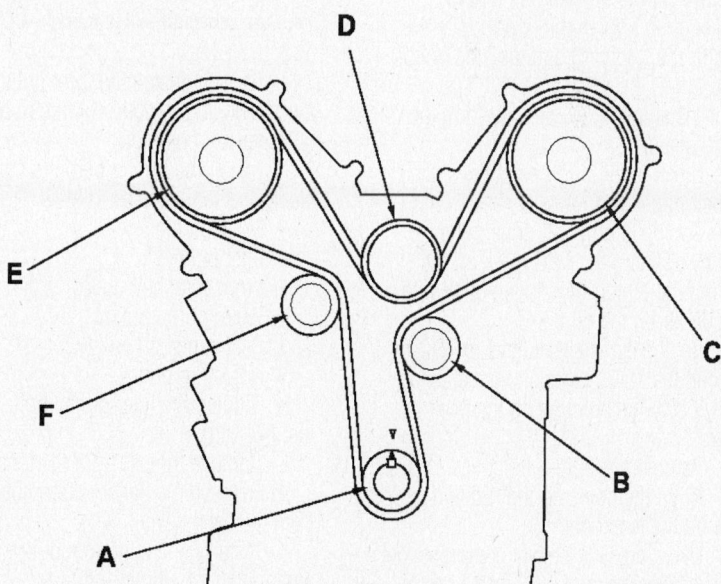

A. Drive pulley
B. Idler pulley
C. Front camshaft pulley
D. Water pump pulley
E. Rear camshaft pulley
F. Adjusting pulley

09474_ACUCAR_G0038

Timing belt routing—CL model

e. Install the timing belt in a counter-clockwise sequence as follows:

- Drive pulley
- Idler pulley
- Front camshaft pulley
- Water pump pulley
- Rear camshaft pulley
- Adjusting pulley

f. Tighten the idler pulley bolt 33 ft. lbs. (44 Nm).

g. Remove the battery clamp bolt from the back cover.

h. Install the engine mount bracket and tighten the upper bolt to 9 ft. lbs. (12 Nm) and the lower bolts to 33 ft. lbs. (44 Nm).

i. Install the lower cover.

j. Install the front upper cover and rear upper cover.

k. Install the crankshaft pulley, do not use an impact wrench. Hold the pulley and tighten the bolt to 181 ft. lbs. (245 Nm).

l. Rotate the crankshaft pulley six turns clockwise so the timing belt positions itself on the pulleys.

m. Turn the crankshaft pulley so its white mark lines up with the pointer.

n. Check the camshaft pulley marks, they should be as illustrated.

16. Install the dipstick tube.

17. Install the drive belts.

a. Install the side engine mount bracket, then tighten the upper bolts first to 33 ft. lbs. (44 Nm) and the side bolts last to 40 ft. lbs. (54 Nm).

18. If installing a new belt, perform the following steps.

a. Clean the timing belt pulleys, timing belt upper and lower covers.

b. Set the timing belt drive pulley to TDC by aligning the TDC mark on the tooth of the timing belt drive pulley with the pointer on the oil pump.

c. Set the camshaft pulleys to TDC by aligning the TDC marks on the camshaft pulleys with the pointers on the back covers.

d. Remove the auto-tensioner.

e. Align the holes on the rod and body on the tensioner.

f. Use a press to slowly compress the tensioner and insert a 0.08 inch (2 mm) pin through the body and rod. Make sure the pressure to compress the tensioner does not exceed 2,200 lbs. ft. (9,800 N).

g. Install the tensioner with the pin still in place.

h. Screw the battery clamp bolt in as shown to hold the timing belt adjuster in its current position. Only tighten it by hand, do not use a wrench.

FRONT CAMSHAFT PULLEY:

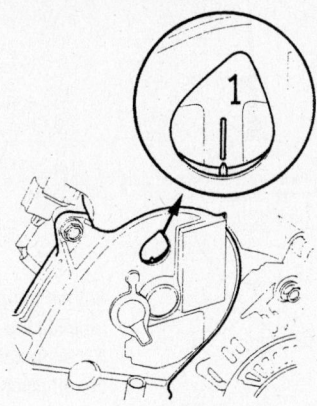

REAR CAMSHAFT PULLEY:

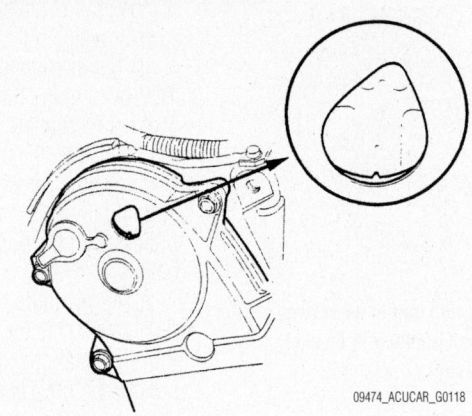

09474_ACUCAR_G0118

Check the camshaft pulley marks—CL model

i. Loosely install the idler pulley with a new bolt so the pulley is free to move but does not come off.

j. Install the timing belt in a counter-clockwise sequence as follows:

- Drive pulley
- Idler pulley
- Front camshaft pulley
- Water pump pulley
- Rear camshaft pulley
- Adjusting pulley

k. Tighten the idler pulley bolt 33 ft. lbs. (44 Nm).

l. Remove the pin from the tensioner.

m. Remove the battery clamp bolt from the back cover.

n. Install the engine mount bracket and tighten the upper bolt to 9 ft. lbs. (12 Nm) and the lower bolts to 33 ft. lbs. (44 Nm).

o. Install the lower cover.

p. Install the front upper cover and rear upper cover.

q. Install the crankshaft pulley, do not use an impact wrench. Hold the pulley and tighten the bolt to 181 ft. lbs. (245 Nm).

r. Rotate the crankshaft pulley six turns clockwise so the timing belt positions itself on the pulleys.

s. Turn the crankshaft pulley so its white mark lines up with the pointer.

t. Check the camshaft pulley marks, they should be as illustrated.

19. Install the dipstick tube.

20. Install the drive belts.

a. Install the side engine mount bracket, then tighten the upper bolts first to 33 ft. lbs. (44 Nm) and the side bolts last to 40 ft. lbs. (54 Nm).

21. Install the remaining components in the reverse order of removal.

2002–04 RL Model

➡**Under normal driving conditions, the timing belt and timing balancer belt are to be replaced at 105,000 miles (168,000 km).**

1. Before servicing the vehicle, refer to the precautions section.

2. Be sure to acquire the anti–theft code for the radio and the frequencies for the radio's preset buttons.

3. Remove or disconnect the following:
- Negative battery cable
- Strut brace located at the top rear of the engine, if necessary

4. Rotate the crankshaft pulley so that the No. 1 piston is at Top Dead Center (TDC) of its compression stroke.
- Top engine cover-to-engine bolts and the cover
- Air intake duct and air cleaner housing
- Alternator and A/C compressor drive belts
- TCS control valve upper and lower brackets
- Power steering belt
- TCS throttle sensor connector, the TCS throttle actuator connector and the Throttle Position (TP) sensor
- TCS control valve assembly.
- Vehicle Speed Sensor (VSS) harness connector and remove the wire harness holder
- Breather and vacuum hoses
- Ignition Control Module (ICM) bracket from the right timing belt cover
- Idler pulley bracket from the left side of the crankshaft pulley
- Dipstick and the dipstick tube-to-engine bolt; then, pull the tube from its O-ring mount. Discard the O-ring.

5. Using the Holder Handle and the 50mm Offset Holder Attachment tool 07MAB-PY3010A and a 19mm socket, secure the crankshaft pulley and remove the crankshaft pulley bolt.
- Washer and pull the crankshaft pulley from the engine
- Upper and lower timing belt cover

6. Loosen the balancer belt adjusting

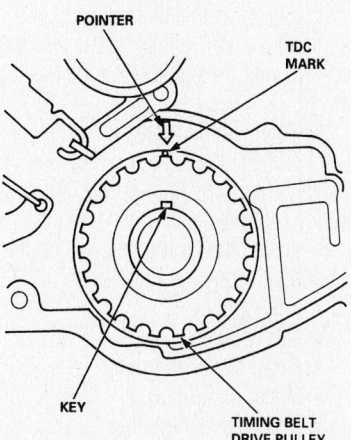

Crankshaft timing belt pulley alignment mark locations—2002–04 RL models

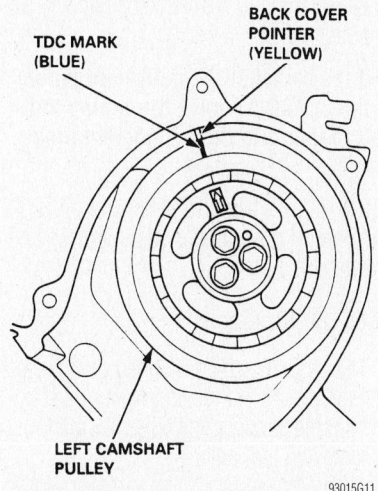

Left camshaft timing belt pulley alignment mark locations—2002–04 RL models

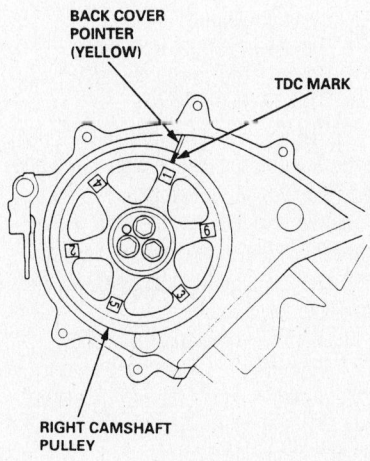

Right camshaft timing belt pulley alignment mark locations—2002–04 RL models

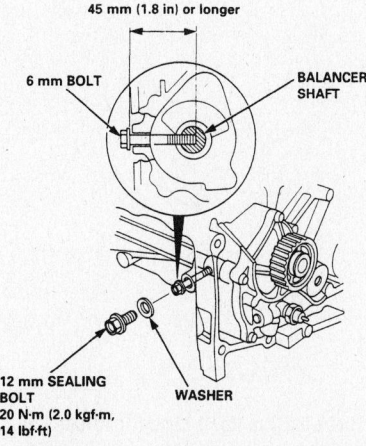

Securing the balancer shaft—2002–04 RL models

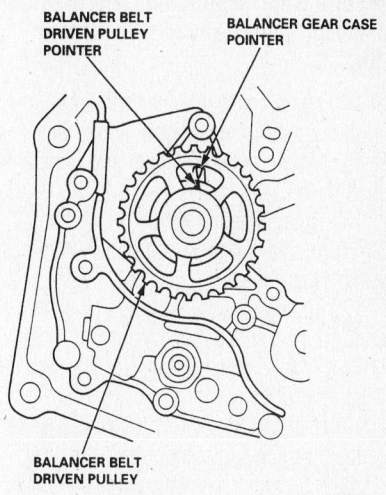

Balancer shaft alignment mark locations—2002–04 RL models

nut 180° (½) turn. Push the tensioner to relieve the tension from the balancer belt; then, retighten the adjusting bolt. Remove the balancer belt.

7. Loosen the timing belt adjusting nut 180° (½) turn. Push the tensioner to relieve the tension from the timing belt; then, retighten the adjusting bolt. Remove the timing belt.

To install:

8. Remove the spark plugs.

9. Remove the balancer belt drive pulley and the timing belt guide plate from the crankshaft.

10. Clean the upper and lower timing belt covers.

11. Position the timing belt pulley so the No. 1 piston is at TDC of its compression stroke. Align the mark on the pulley (near keyway) with the pointer mark on the oil pump.

12. Adjust the camshaft pulley so that the No. 1 piston is the TDC of the compression stroke. Align the TDC marks on the pulley with the upper surface of the back cover; the arrow mark on the left camshaft pulley and the "1" on the right camshaft pulley should be facing the back cover pointer.

13. Install the timing belt in the following sequence:
   a. Crankshaft timing belt pulley sprocket.
   b. Adjusting pulley.
   c. Left camshaft pulley.
   d. Water pump pulley.
   e. Right camshaft pulley.

❊❊ WARNING

Make sure that the camshaft and crankshaft pulleys are at TDC.

➡️**For easier installation, turn the right camshaft pulley about ½ tooth from TDC.**

14. Loosen and retighten the timing belt adjusting bolt to tension the timing belt.

15. Install the lower cover and the crankshaft pulley.

16. Rotate the crankshaft pulley about 5–6 turns clockwise to position the timing belt on the pulleys.

17. To adjust the timing belt tension, perform the following procedure:

   a. Make sure that the No. 1 piston is at TDC of its compression stroke.

   b. Rotate the crankshaft clockwise ten teeth on the camshaft pulley; the blue mark on the crankshaft pulley should align with the lower cover pointer.

   c. Loosen the adjusting nut 180° (½ turn).

   d. Tighten the adjusting nut to 31 ft. lbs. (42 Nm).

18. Remove the crankshaft pulley and the lower cover; then, install the timing belt guide plate and the balancer belt drive pulley.

19. Align the balancer shaft pulley by performing the following procedure:

   a. Using a 6 x 45mm bolt, insert it into the maintenance hole and the balancer shaft.

   b. Align the pointer on the balancer belt pulley with the pointer on the balancer gear case.

20. Adjust the timing belt drive pulley so that the No. 1 piston is at TDC of the compression stroke.

21. Install the balancer belt drive pulley and the balancer belt.

22. Loosen and retighten the balancer adjuster bolt to place tension on the balancer belt.

23. Remove the 6mm bolt and install the sealing bolt in the maintenance hole using a new washer.

24. Install the crankshaft pulley. Rotate the crankshaft pulley about 5–6 turns clockwise to position the timing belt on the pulleys.

25. Loosen the balancer belt adjuster bolt 180° (½ turn) and retighten the bolt to 33 ft. lbs. (44 Nm).

26. Remove the crankshaft pulley.

27. Install the upper and lower timing belt covers and the crankshaft pulley.

28. Install the crankshaft pulley and finger-tighten the bolt and washer. Using the Holder Handle and the 50mm Offset Holder Attachment tool 07MAB-PY3010A and a 19mm socket with a torque wrench, tighten the crankshaft pulley bolt to 181 ft. lbs. (245 Nm).

29. Make sure the crankshaft and camshaft pulleys are at TDC.

➡️**If the camshaft or crankshaft pulley is not at TDC, remove the timing belt and re-perform the adjustment procedure.**

30. To complete the installation, reverse the removal procedures. Adjust the tension of the drive belts.

2005–06 RL Model

The radio may have a coded theft protection circuit. Obtain the code before disconnecting the battery, removing the radio fuse, or removing the radio.

1. Before servicing the vehicle, refer to the precautions section.

2. Remove the right upper fender trim.

3. Check that the No. 1 piston Top Dead Center (TDC) mark on the front camshaft pulley and the pointer on the front upper cover are aligned.

4. Remove the right front wheel.

5. Remove the splash shield.

6. Remove the drive belt.

7. Remove the drive belt auto-tensioner.

8. Support the engine with a jack and wood block under the oil pan.

9. Remove the ground cable, then remove the side engine mount bracket.

10. Remove the front upper cover and rear upper cover.

11. Remove the crankshaft pulley.

12. Remove the lower cover.

13. Remove one of the battery clamp bolts from the battery tray, and grind the end of it as shown.

14. Screw the battery clamp bolt in as shown to hold the timing belt adjuster in its current position. Only tighten it by hand, do not use a wrench.

15. Remove the engine mount bracket

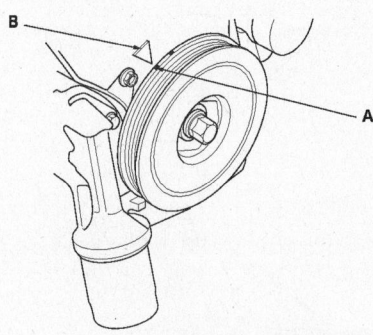

09474_ACUCAR_G0032

Check that the No. 1 piston Top Dead Center (TDC) mark on the front camshaft pulley and the pointer on the front upper cover are aligned—2005–06 RL model

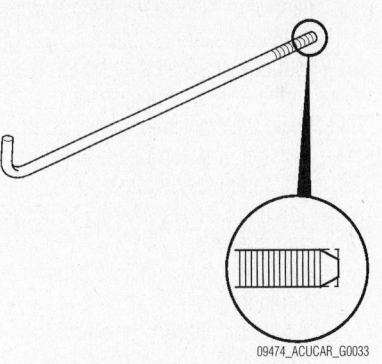

09474_ACUCAR_G0033

Grind the battery tray clamp bolt as shown—2005–06 RL model

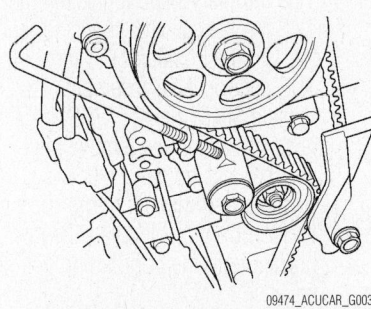

09474_ACUCAR_G0034

Screw the battery clamp bolt in as shown to hold the timing belt adjuster in its current position—2005–06 RL model

16. Remove the idler pulley bolt and idler pulley, then remove the timing belt.

To install:

17. If reusing the old belt, perform the following steps.

   a. Clean the timing belt pulleys, timing belt guide plate, and the upper and lower covers.

   b. Set the timing belt drive pulley to TDC by aligning the TDC mark on the tooth of the timing belt drive pulley with the pointer on the oil pump.

   c. Set the camshaft pulleys to TDC by aligning the TDC marks on the camshaft pulleys with the pointers on the back covers.

   d. Apply liquid thread lock (P/N 08713-0001) to the idler pulley bolt, then loosely install the idler pulley bolt as the pulley can move and does not come off.

   e. Install the timing belt in a counterclockwise sequence as follows:

- Drive pulley
- Idler pulley
- Front camshaft pulley
- Water pump pulley
- Rear camshaft pulley
- Adjusting pulley

   f. Tighten the idler pulley bolt 33 ft. lbs. (44 Nm).

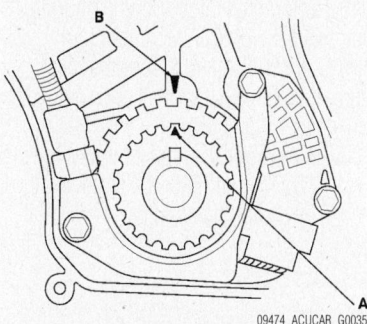

09474_ACUCAR_G0035

Set the timing belt drive pulley to TDC by aligning the TDC mark on the tooth of the timing belt drive pulley with the pointer on the oil pump—2005–06 RL model

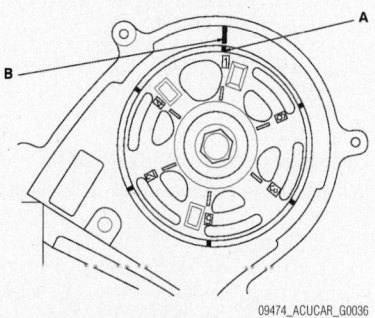

09474_ACUCAR_G0036

Align the front camshaft pulley to TDC as shown—2005–06 RL model

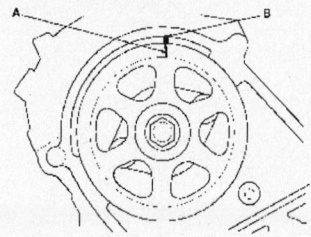

Install the timing belt in a counterclockwise sequence starting with the drive pulley.

09474_ACUCAR_G0037

Align the rear camshaft pulley to TDC as shown—2005–06 RL model

g. Remove the battery clamp bolt from the back cover.

h. Install the engine mount bracket and tighten the upper bolt to 9 ft. lbs. (12 Nm) and the lower bolts to 33 ft. lbs. (44 Nm).

i. Install the timing belt guide plate as shown.

j. Install the crankshaft pulley, do not use an impact wrench. Hold the pulley and tighten the bolt to 47 ft. lbs. (64 Nm) plus turn an additional 60 degrees.

k. Install the front upper cover and rear upper cover.

l. Rotate the crankshaft pulley six turns clockwise so the timing belt positions itself on the pulleys.

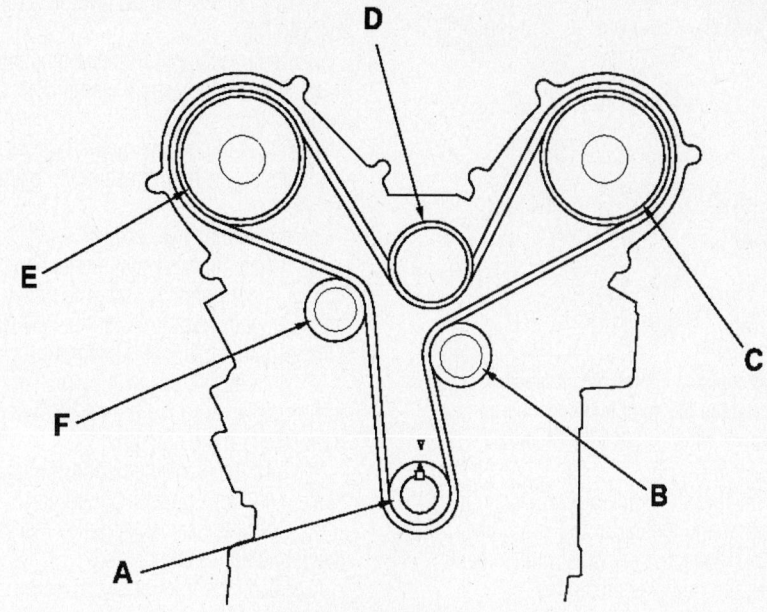

A. Drive pulley
B. Idler pulley
C. Front camshaft pulley
D. Water pump pulley
E. Rear camshaft pulley
F. Adjusting pulley

09474_ACUCAR_G0038

Timing belt routing—2005–06 RL model

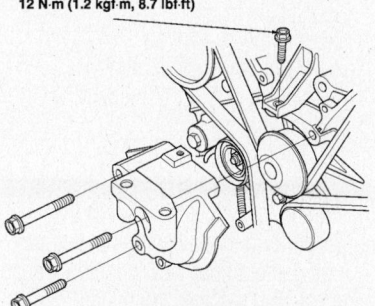

6 x 1.0 mm
12 N·m (1.2 kgf·m, 8.7 lbf·ft)

10 x 1.25 mm
44 N·m (4.5 kgf·m, 33 lbf·ft)

09474_ACUCAR_G0039

Install the engine mount bracket—2005–06 RL model

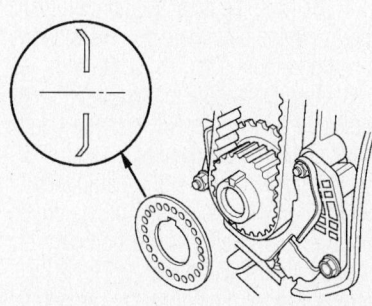

09474_ACUCAR_G0040

Install the timing belt guide plate—2005–06 RL model

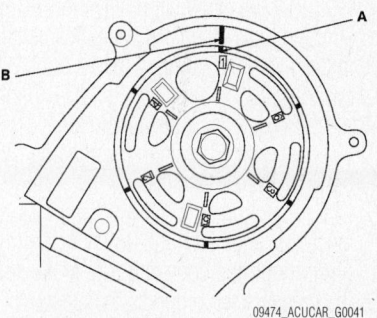

09474_ACUCAR_G0041

Check the front camshaft pulley marks—2005–06 RL model

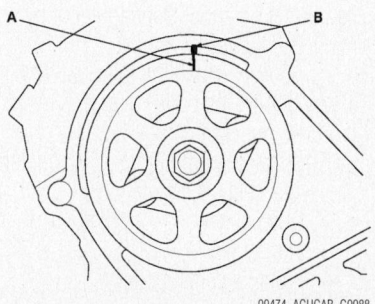

09474_ACUCAR_G0088

Check the rear camshaft pulley marks—2005–06 RL model

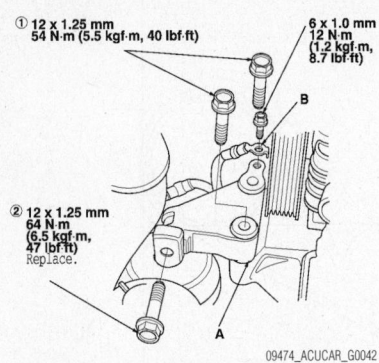

Install the side engine mount bracket, then tighten the bolts in the numbered sequence shown—2005–06 RL model

m. Turn the crankshaft pulley so its white mark lines up with the pointer.

n. Check the camshaft pulley marks, they should be as illustrated.

o. Install the side engine mount bracket, then tighten the bolts in the numbered sequence shown.

18. If installing a new belt, perform the following steps.

a. Clean the timing belt pulleys, timing belt guide plate, and the upper and lower covers.

b. Set the timing belt drive pulley to TDC by aligning the TDC mark on the tooth of the timing belt drive pulley with the pointer on the oil pump.

c. Set the camshaft pulleys to TDC by aligning the TDC marks on the camshaft pulleys with the pointers on the back covers.

d. Remove the auto-tensioner.

e. Align the holes on the rod and body on the tensioner.

f. Use a press to slowly compress the tensioner and insert a 0.08 inch (2 mm) pin through the body and rod. Make sure the pressure to compress the tensioner does not exceed 2,200 lbs. ft. (9,800 N).

g. Install the tensioner with the pin still in place.

h. Screw the battery clamp bolt in as shown to hold the timing belt adjuster in its current position. Only tighten it by hand, do not use a wrench.

i. Loosely install the idler pulley with a new bolt so the pulley is free to move but does not come off.

j. Install the timing belt in a counter-clockwise sequence as follows:
- Drive pulley
- Idler pulley
- Front camshaft pulley
- Water pump pulley
- Rear camshaft pulley
- Adjusting pulley

k. Tighten the idler pulley bolt 33 ft. lbs. (44 Nm).

l. Remove the pin from the tensioner.

m. Remove the battery clamp bolt from the back cover.

n. Install the engine mount bracket.

o. Install the timing belt guide plate as shown.

p. Install the lower cover.

q. Install the crankshaft pulley, do not use an impact wrench. Hold the pulley and tighten the bolt to 47 ft. lbs. (64 Nm) plus turn an additional 60 degrees.

r. Rotate the crankshaft pulley six turns clockwise so the timing belt positions itself on the pulleys.

s. Turn the crankshaft pulley so its white mark lines up with the pointer.

t. Check the camshaft pulley marks, they should be as illustrated.

u. Install the front upper cover and rear upper cover.

v. Install the side engine mount bracket, then tighten the bolts in the numbered sequence shown.

19. Install the remaining components in the reverse order of removal.

2002–03 TL Models

1. Before servicing the vehicle, refer to the precautions section.

2. Remove or disconnect the following:
- Negative battery cable
- Ignition coil cover
- Front tire/wheel assemblies
- Splash shield from under the vehicle
- Drive belts

3. To remove the engine-to-chassis side mount, located at the front of the engine, perform the following procedure:

a. Using a floor jack, position a cushion between the jack and the oil pan.

b. Raise the engine slightly to take the weight off of the side mount.

c. Remove the side mount-to-engine bracket bolt, the side mount-to-chassis bolts and the side mount.

4. Remove the dipstick and the dipstick tube-to-engine bolt; then, pull the tube from its O–ring mount. Discard the O–ring.

5. Turn the engine to align the timing marks and set cylinder No. 1 to Top Dead Center (TDC) on the compression stroke. The white mark on the crankshaft pulley should align with the pointer on the timing belt cover. Remove the inspection caps on the upper timing belt covers to check the alignment of the timing marks. The pointers for the camshaft pulleys should align with the marks on rear upper cover mark.

6. Using the Holder Handle and the 50mm Offset Holder Attachment tool 07MAB–PY3010A and a 19mm socket, secure the crankshaft pulley and remove the crankshaft pulley bolt.

7. Remove or disconnect the following:
- Washer and pull the crankshaft pulley from the engine

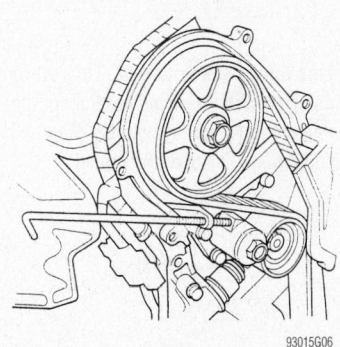

Using battery clamp bolt to hold timing belt adjuster in position—2002–TL models

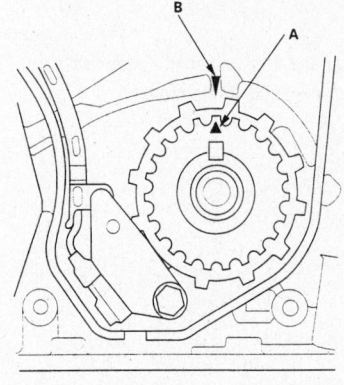

Crankshaft sprocket alignment mark positioning for TDC—2002–TL models

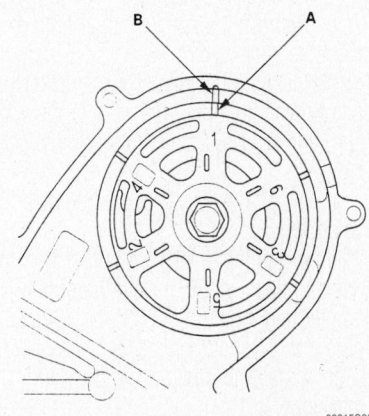

Left camshaft sprocket alignment mark positioning for TDC—2002–TL models

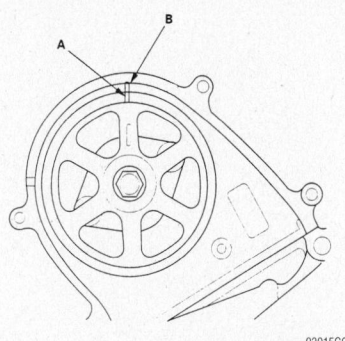

Right camshaft sprocket alignment mark positioning for TDC—2002–TL models

- All necessary components for access to the timing belt covers
- Upper and lower timing belt covers. Clean any dirt, oil or grease from the covers. Do not use the covers for storing removed items.
- One of the battery clamp bolts and grind a 45° bevel on the threaded end. Screw the battery clamp bolt into hole provided at the base of the right cylinder head to hold the timing belt adjuster in it's current position; tighten the bolt by hand. Do NOT use a wrench.
- Engine mount bracket

8. At the base of the left cylinder head, loosen the idler pulley bolt about 5–6 turns; then, remove the timing belt.

To install:

❋❋ CAUTION

Do not rotate the crankshaft pulley or camshaft pulleys with the timing belt removed. The pistons may hit the valves and cause damage.

9. Remove the spark plugs.
10. Set the timing belt drive (crankshaft) sprocket so that the No. 1 piston is at Top Dead Center (TDC). Align the TDC mark on the tooth of the timing belt drive sprocket with the pointer on the oil pump.
11. Set the camshaft pulleys so that the No. 1 piston is at TDC. Align the TDC mark on the camshaft pulleys to the pointers on the back covers.
12. Remove the battery clamp bolt from the back cover. Remove the auto-tensioner.
13. Service the auto-tensioner by performing the following procedure:
   a. Position the auto-tensioner in a soft jawed vise with the maintenance bolt facing upward. DO NOT grip the body of the auto-tensioner.
   b. Remove the maintenance bolt.
   c. Be careful not to spill the oil from

inside the tensioner. If oil is spilled, replenish it; the total capacity is 0.22 fl. oz. (6.5 ml).
   d. Using Stopper tool 14540–P8A–A01, position it on the auto-tensioner while turning the internal screw.
   e. Insert a flat–blade screwdriver into the maintenance hole and turn it clockwise to compress the bottom.

❋❋ WARNING

Be careful not to damage the threads or the gasket contact surface with the screwdriver.

   f. Using a new gasket, reinstall the maintenance bolt and torque it to 6 ft. lbs. (8 Nm).
   g. Make sure that no oil is leaking around the maintenance bolt and install the auto-tensioner; torque the bolts 33 ft. lbs. (44 Nm).

➡**Make sure that the Stopper tool 14540–P8A–A01 stays in place.**

14. Install the timing belt on the sprockets in the following sequence: drive sprocket (crankshaft), idler pulley, left camshaft sprocket, water pump pulley, right camshaft sprocket and adjusting pulley.
15. Torque the idler pulley bolt to 33 ft. lbs. (44 Nm).
16. Remove the Stopper tool from the auto-tensioner.

17. Install or connect the following:
- Engine mount-to-engine and torque the No. 10 bolts to 33 ft. lbs. (44 Nm) and the No. 6 bolt to 104 inch lbs. (12 Nm)
- Lower and upper timing belt covers
- Crankshaft pulley and finger–tighten the bolt and washer. Using the Holder Handle and the 50mm Offset Holder Attachment tool 07MAB–PY3010A and a 19mm socket with a torque wrench, tighten the crankshaft pulley bolt to 181 ft. lbs. (245 Nm).

18. Rotate the crankshaft 5–6 turns clockwise so that the timing belt positions itself properly on the sprockets.
19. Set cylinder No. 1 to TDC by aligning the timing marks. If the timing marks do not align, remove the timing belt, then adjust the components and reinstall the timing belt.
20. Install all applicable components.
21. To complete the installation, reverse the removal procedures. Adjust the tension of the drive belts.

2004–06 TL Models

➡**The radio may have a coded theft protection circuit. Obtain the code before disconnecting the battery, removing the radio fuse, or removing the radio.**

1. Before servicing the vehicle, refer to the precautions section.

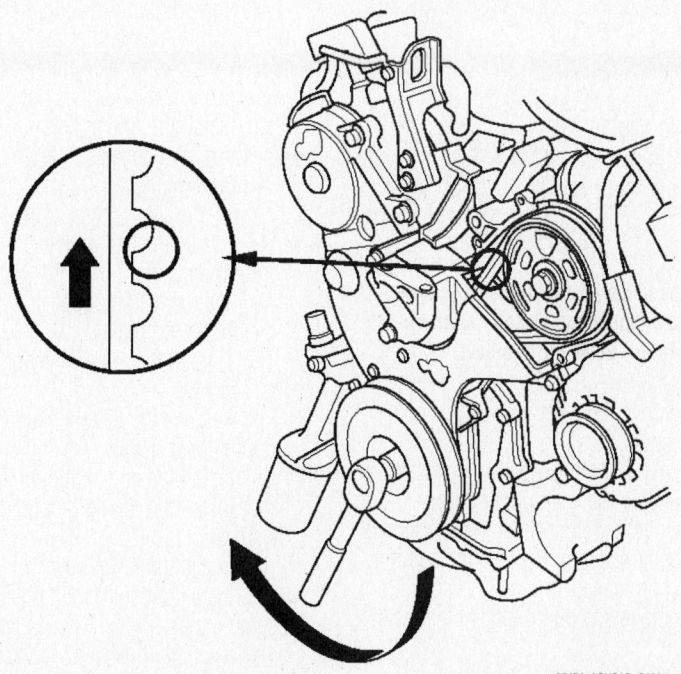

Inspect the timing belt for cracks and oil or coolant soaking—2004–06 TL model

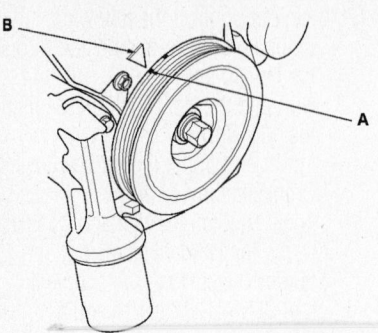

09474_ACUCAR_G32

Check that the No. 1 piston Top Dead Center (TDC) mark on the front camshaft pulley and the pointer on the front upper cover are aligned—2004–06 TL model

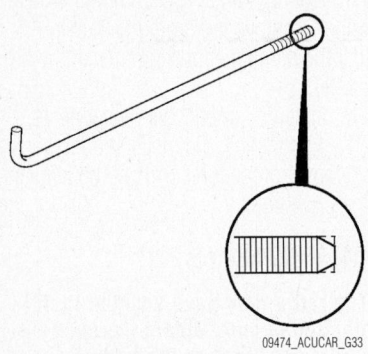

09474_ACUCAR_G33

Grind the battery tray clamp bolt as shown—2004–06 TL model

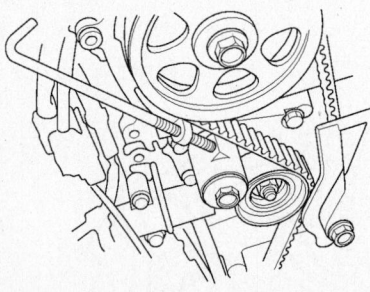

09474_ACUCAR_G0034

Screw the battery clamp bolt in as shown to hold the timing belt adjuster in its current position—2004–06 TL model

2. Remove the right side engine compartment cover.

3. Remove the drive belt.

4. Remove the front upper cover.

5. Inspect the timing belt for cracks and oil or coolant soaking. Replace the belt if it is cracked or soaked.

6. Remove any oil or solvent that gets on the belt.

7. Check that the No. 1 piston Top Dead Center (TDC) mark on the front

camshaft pulley and the pointer on the front upper cover are aligned.

8. Remove the right front wheel.

9. Remove the splash shield.

10. Remove the drive belt.

11. Support the engine with a jack and wood block under the oil pan.

12. Remove the ground cable, then remove the side engine mount bracket.

13. Remove the front upper cover and rear upper cover.

14. Remove the crankshaft pulley.

15. Remove the lower cover.

16. Remove one of the battery clamp bolts from the battery tray, and grind the end of it as shown.

17. Screw the battery clamp bolt in as shown to hold the timing belt adjuster in its current position. Only tighten it by hand, do not use a wrench.

18. Remove the engine mount bracket

19. Remove the idler pulley bolt and idler pulley, then remove the timing belt.

To install:

20. If reusing the old belt, perform the following steps.

   a. Clean the timing belt pulleys, timing belt guide plate, and the upper and lower covers.

   b. Set the timing belt drive pulley to TDC by aligning the TDC mark on the tooth of the timing belt drive pulley with the pointer on the oil pump.

   c. Set the camshaft pulleys to TDC by aligning the TDC marks on the camshaft pulleys with the pointers on the back covers.

   d. Apply liquid thread lock (P/N 08713-0001) to the idler pulley bolt, then loosely install the idler pulley bolt as the pulley can move and does not come off.

   e. Install the timing belt in a counter-clockwise sequence as follows:

    • Drive pulley
    • Idler pulley
    • Front camshaft pulley
    • Water pump pulley
    • Rear camshaft pulley
    • Adjusting pulley

   f. Tighten the idler pulley bolt 33 ft. lbs. (44 Nm).

   g. Remove the battery clamp bolt from the back cover.

   h. Install the engine mount bracket.

   i. Install the timing belt guide plate as shown.

   j. Install the lower cover.

   k. Install the crankshaft pulley, do not use an impact wrench. Hold the pulley and tighten the bolt to 47 ft. lbs. (64 Nm) plus turn an additional 60 degrees.

   l. Rotate the crankshaft pulley six

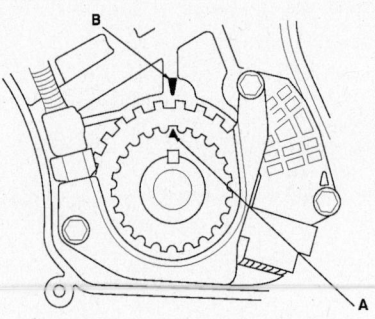

09474_ACUCAR_G0035

Set the timing belt drive pulley to TDC by aligning the TDC mark on the tooth of the timing belt drive pulley with the pointer on the oil pump—2004–06 TL model

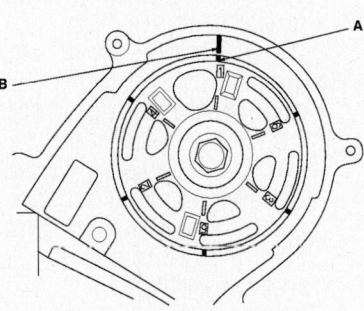

09474_ACUCAR_G0036

Align the front camshaft pulley to TDC as shown—2004–06 TL model

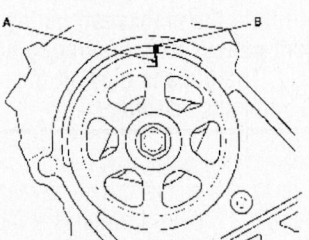

Install the timing belt in a counterclockwise sequence starting with the drive pulley.

09474_ACUCAR_G0037

Align the rear camshaft pulley to TDC as shown—2004–06 TL model

turns clockwise so the timing belt positions itself on the pulleys.

   m. Turn the crankshaft pulley so its white mark lines up with the pointer.

   n. Check the camshaft pulley marks, they should be as illustrated.

   o. Install the front upper cover and rear upper cover.

   p. Install the side engine mount bracket, then tighten the bolts in the numbered sequence shown.

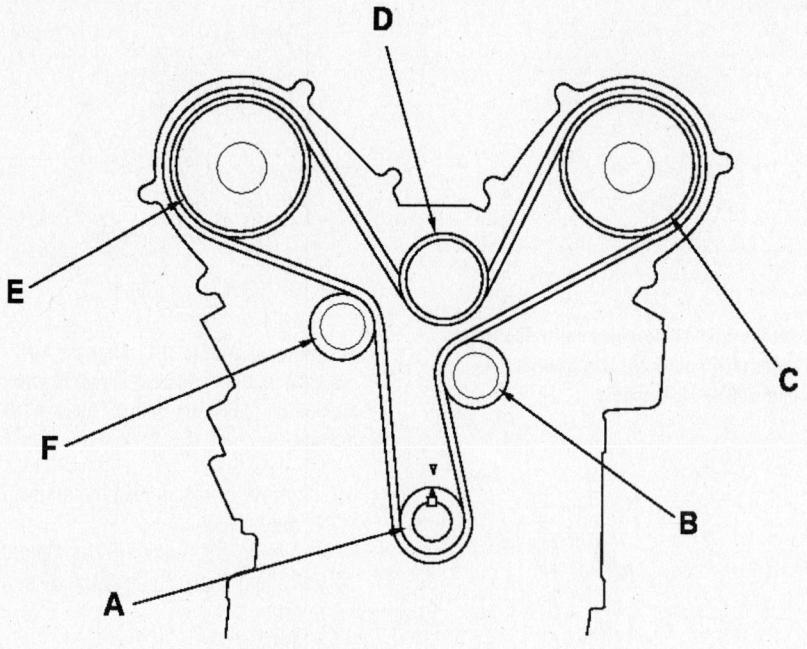

A. Drive pulley
B. Idler pulley
C. Front camshaft pulley
D. Water pump pulley
E. Rear camshaft pulley
F. Adjusting pulley

09474_ACUCAR_G0038

Timing belt routing—2004–06 TL model

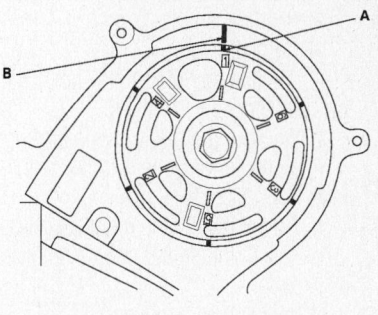

09474_ACUCAR_G0041

Check the front camshaft pulley marks—2004–06 TL model

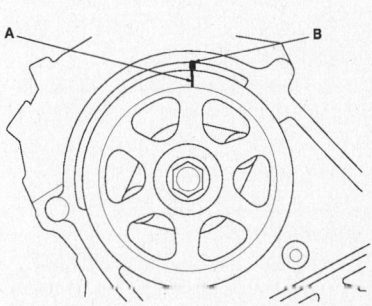

09474_ACUCAR_G0088

Check the rear camshaft pulley marks—2004–06 TL model

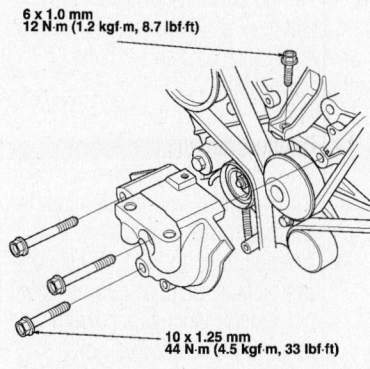

6 x 1.0 mm
12 N·m (1.2 kgf·m, 8.7 lbf·ft)

10 x 1.25 mm
44 N·m (4.5 kgf·m, 33 lbf·ft)

09474_ACUCAR_G0039

Install the engine mount bracket—2004–06 TL model

21. If installing a new belt, perform the following steps.

a. Clean the timing belt pulleys, timing belt guide plate, and the upper and lower covers.

b. Set the timing belt drive pulley to TDC by aligning the TDC mark on the tooth of the timing belt drive pulley with the pointer on the oil pump.

c. Set the camshaft pulleys to TDC by

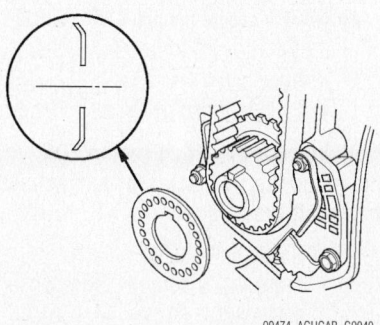

09474_ACUCAR_G0040

Install the timing belt guide plate—2004–06 TL model

aligning the TDC marks on the camshaft pulleys with the pointers on the back covers.

d. Remove the auto-tensioner.

e. Align the holes on the rod and body on the tensioner.

f. Use a press to slowly compress the tensioner and insert a 0.08 inch (2 mm) pin through the body and rod. Make sure the pressure to compress the tensioner does not exceed 2,200 lbs. ft. (9,800 N).

g. Install the tensioner with the pin still in place.

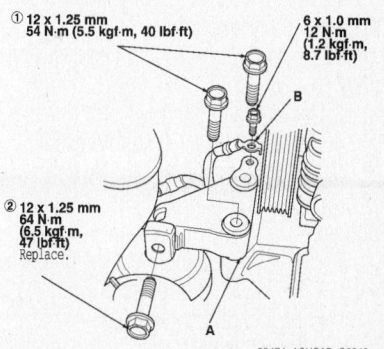

① 12 x 1.25 mm
54 N·m (5.5 kgf·m, 40 lbf·ft)

6 x 1.0 mm
12 N·m
(1.2 kgf·m,
8.7 lbf·ft)

② 12 x 1.25 mm
64 N·m
(6.5 kgf·m,
47 lbf·ft)
Replace.

09474_ACUCAR_G0042

Install the side engine mount bracket, then tighten the bolts in the numbered sequence shown—2004–06 TL model

h. Screw the battery clamp bolt in as shown to hold the timing belt adjuster in its current position. Only tighten it by hand, do not use a wrench.

i. Loosely install the idler pulley with a new bolt so the pulley is free to move but does not come off.

j. Install the timing belt in a counter-clockwise sequence as follows:

- Drive pulley
- Idler pulley
- Front camshaft pulley

- Water pump pulley
- Rear camshaft pulley
- Adjusting pulley

k. Tighten the idler pulley bolt 33 ft. lbs. (44 Nm).

l. Remove the pin from the tensioner.

m. Remove the battery clamp bolt from the back cover.

n. Install the engine mount bracket.

o. Install the timing belt guide plate as shown.

p. Install the lower cover.

q. Install the crankshaft pulley, do not use an impact wrench. Hold the pulley and tighten the bolt to 47 ft. lbs. (64 Nm) plus turn an additional 60 degrees.

r. Rotate the crankshaft pulley six turns clockwise so the timing belt positions itself on the pulleys.

s. Turn the crankshaft pulley so its white mark lines up with the pointer.

t. Check the camshaft pulley marks, they should be as illustrated.

u. Install the front upper cover and rear upper cover.

v. Install the side engine mount bracket, then tighten the bolts in the numbered sequence shown.

22. Install the remaining components in the reverse order of removal.

TSX Models

1. Before servicing the vehicle, refer to the precautions section.

2. Set the engine to Top Dead Center (TDC).

3. Drain the cooling system.

4. Relieve the fuel system pressure.

5. Turn the crankshaft pulley so its Top Dead Center (TDC) mark lines up with the pointer.

6. Remove or disconnect the following:
- Negative battery cable
- Front tires and wheels
- Splash shield

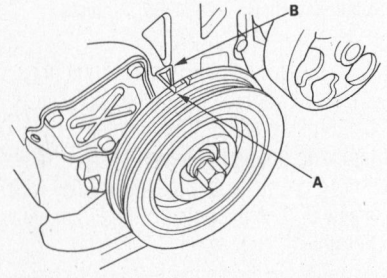

42356-ACCO-G24

Turn the crankshaft pulley so the TDC mark (A) is aligned with the pointer (B)— TSX models

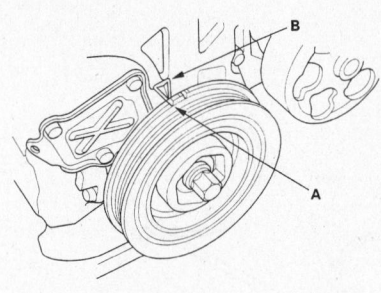

67162-ATSX-G10

Turn the crankshaft pulley so its Top Dead Center (TDC) mark (A) line up with the pointer (B)—TSX models

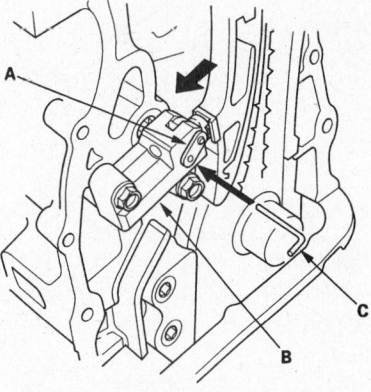

42356-ACCO-G25

Align the holes on the lock (A) and the auto-tensioner (B), then place a 1.5mm pin into the holes. Turn the crankshaft clockwise to secure the pin—TSX models

- Drive belt
- Cylinder head cover

➡**Make sure the No. 1 piston TDC marks on the VTC actuator and exhaust camshaft are aligned.**

- Crankshaft pulley
- Crankshaft Position (CKP) sensor connector
- Variable Valve Timing Control (VTC) oil control solenoid valve connector
- VTC oil control solenoid valve

7. Support the engine with a suitable jack with a wooden block under the oil pan.
- Ground cable and upper bracket
- Side engine mount bracket
- Chain cover/case

8. Loosely install the crankshaft pulley. Turn the crankshaft counterclockwise to compress the auto-tensioner.

9. Align the holes on the lock (A) and the auto-tensioner (B), then place a 1.5mm pin into the holes. Turn the crankshaft clockwise to secure the pin.

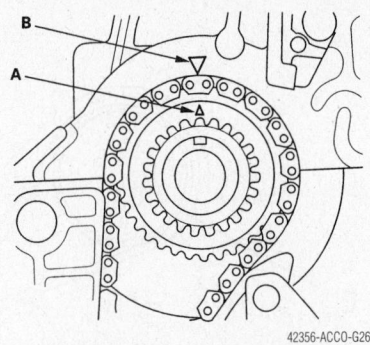

42356-ACCO-G26

Set the crankshaft to TDC. Align the TDC mark (A) on the crankshaft sprocket with the pointer (B) on the cylinder block—TSX models

10. Remove or disconnect the following:
- Auto-tensioner
- Timing chain guide B (top guide)
- Timing chain guide A and tensioner arm
- Timing chain

✳✳ WARNING

Do not place the timing chain near any magnetic fields.

To install:

11. Set the crankshaft to TDC. Align the TDC mark (A) on the crankshaft sprocket with the pointer (B) on the cylinder block.

12. Set the camshafts to TDC. The punch mark (A) on the VTC actuator and the punch mark (B) on the exhaust camshaft (C) should be at the top. Align the TDC marks (C) on the VTC actuator and exhaust camshaft sprockets.

13. Install or connect the following:
- Timing chain the crankshaft sprocket with the colored link of the chain aligned with the mark on the crank sprocket
- Timing chain on the VTC actuator and exhaust camshaft sprocket with the punch marks aligned with the center of the 2 colored links
- Timing chain guide A and tensioner arm. Tighten the guide bolts to 104 inch lbs. (12 Nm) and the tensioner arm retainer to 16 ft. lbs. (22 Nm).
- Auto-tensioner and tighten the bolts to 104 inch lbs. (12 Nm)
- Timing chain guide B and tighten the retainers to 16 ft. lbs. (22 Nm)

14. Remove the pin from the auto-tensioner.

15. Inspect the chain cover seal for damage and replace if necessary. Clean and dry the chain cover mating surfaces.

16. Install or connect the following:
- Liquid gasket, P/N 08718-0009

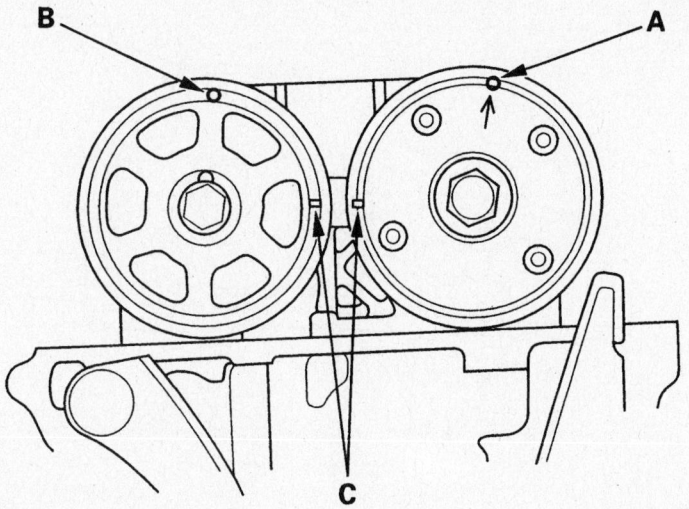

42356-ACCO-G27

The mark (A) on the VTC actuator and the mark (B) on the exhaust cam (C) should be at the top. Align the TDC marks (C) on the VTC actuator and exhaust cam sprockets—TSX models

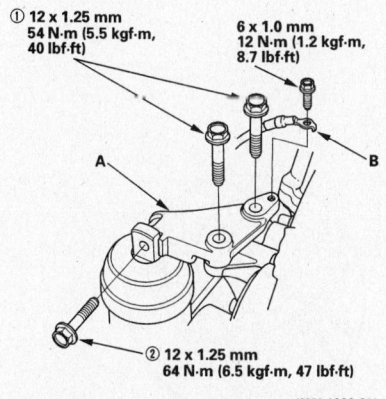

① 12 x 1.25 mm
54 N·m (5.5 kgf·m, 40 lbf·ft)

6 x 1.0 mm
12 N·m (1.2 kgf·m, 8.7 lbf·ft)

A

B

② 12 x 1.25 mm
64 N·m (6.5 kgf·m, 47 lbf·ft)

42356-ACCO-G28

Tighten the upper bracket upper bolt/nuts in the proper order to the correct specification—TSX models

- Side engine mounting bracket and tighten the retainers to 33 ft. lbs. (44 Nm)
- Upper bracket, then tighten the bolts/nuts as shown in the illustration
- Ground cable
- VTC oil control solenoid valve
- CKP sensor and VTC oil control solenoid valve connectors
- Crankshaft pulley
- Cylinder head cover
- Drive belt
- Splash shield
- Wheels and tires

17. Fill the engine cooling system and connect the negative battery cable.

evenly to the cylinder block mating surface of the timing chain cover and the inner threads of the holes
- Liquid gasket to the cylinder block upper surface contact areas on the chain cover and the oil pan mating surface of the chain cover in the inner threads of the holes

➡ **Make sure to install the components within 5 minutes of applying the sealer.**

- New O-ring the timing chain cover. Set the edge of the cover to the edge of the oil pan, then install the cover on the engine block. Tighten the retainers to 104 inch lbs. (12 Nm).

➡ **When installing the chain case, do not slide the bottom surface on the oil pan mounting surface.**

Timing Chain

REMOVAL & INSTALLATION

RSX Models

1. Before servicing the vehicle, refer to the precautions section.
2. Rotate the crankshaft pulley so that the No. 1 piston is at Top Dead Center (TDC) of its compression stroke.
3. Remove or disconnect the following:
- Front wheels and tires
- Splash shield
- Drive belt
- Cylinder head cover
4. Hold the crankshaft pulley with holder handle 07JAB–001020A and holder attachment 07NAB–001040A, then remove the pulley bolt with a 19mm socket and breaker bar.
- Oil cooler hose joint pipe from the water pump, K20A2 engines
- Crankshaft Position (CKP) sensor connector and Variable Valve Timing Control (VTC) oil control solenoid valve connector
- VTC oil control solenoid valve
5. Support the engine under the oil pan with a jack and block of wood.
- Ground cable and upper bracket
- Side engine mounting bracket
- Retaining bolts and timing chain case
6. Loosely install the crankshaft pulley.
7. Turn the crankshaft counterclockwise to compress the auto-tensioner.
8. Align the holes on the lock and auto-tensioner, then insert a 0.06 in. (1.5mm) pin

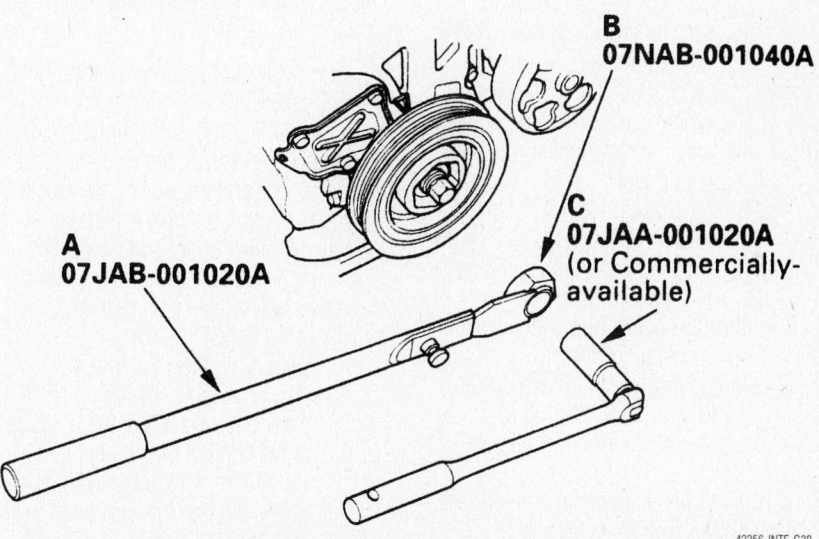

B
07NAB-001040A

A
07JAB-001020A

C
07JAA-001020A
(or Commercially-available)

42356-INTE-G38

Hold the crankshaft pulley with holder handle (A) and holder attachment (B)—RSX models

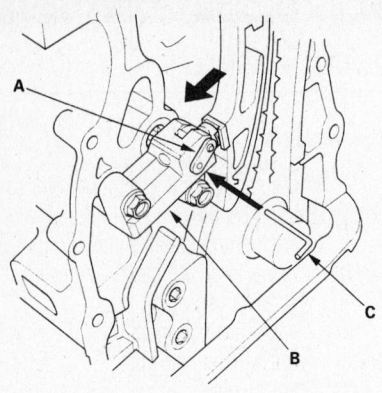

42356-INTE-G37

Align the holes on the lock (A) and auto-tensioner (B), then insert a 0.06 in. (1.5mm) pin (C) into the hole–RSX models

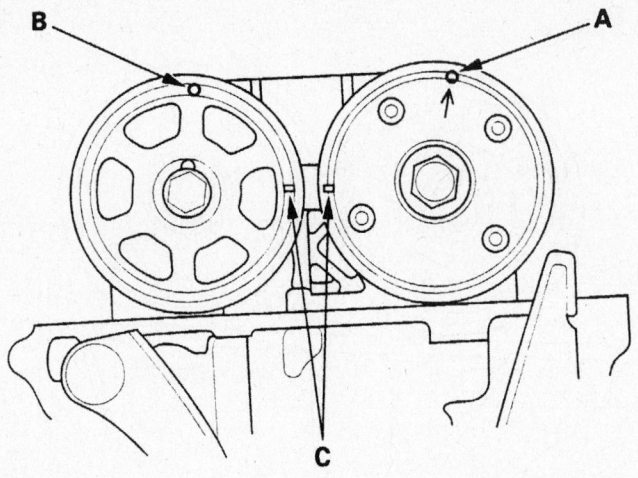

42356-INTE-G24

The punch mark (arrow) (A) on the VTC actuator and punch mark (B) on the exhaust camshaft sprocket should be at the top. Align the TDC marks (C)on the VTC actuator and exhaust camshaft sprocket–RSX models

into the hole. Turn the crankshaft clockwise to hold the pin.

9. Auto-tensioner
   • Timing chain guide B
   • Timing chain guide A and tensioner arm
   • Timing chain

➡ **Do not let the timing chain near any magnetic fields.**

To install:

10. Set the crankshaft to TDC. Align the TDC mark on the crankshaft sprocket with the pointer on the cylinder block.

11. Set the camshafts to TDC. The punch mark (arrow) on the VTC actuator and punch mark on the exhaust camshaft sprocket should be at the top. Align the TDC marks on the VTC actuator and exhaust camshaft sprocket.

12. Install or connect the following:
   • Timing chain on the crankshaft sprocket with the colored piece aligned with the punch mark on the crankshaft sprocket
   • Timing chain on the VTC actuator and exhaust camshaft sprocket with the punch marks aligned with the 2 colored pieces
   • Timing chain guide A and tighten the bolts to 104 inch lbs. (12 Nm)
   • Tensioner arm and tighten the bolts to 16 ft. lbs. (22 Nm)
   • Auto-tensioner. Tighten the bolts to 104 inch lbs. (12 Nm).
   • Timing chain guide B and tighten the bolts to 16 ft. lbs. (22 Nm)

13. Remove the pin from the auto-tensioner.

14. Check the timing chain case oil seal for damage and replace if necessary.

15. Remove the oil liquid gasket material

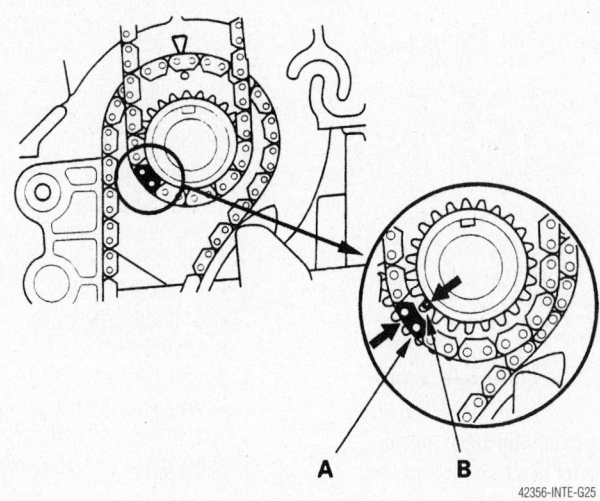

42356-INTE-G25

Install the timing chain on the crankshaft sprocket with the colored piece aligned with the punch mark on the crankshaft sprocket—RSX models

from the chain case mating surfaces, bolts and bolt holes.

16. Apply liquid gasket part no. 08718–0009, evenly to the cylinder block mating surface chain case and the inner threads of the holes. Apply liquid gasket to the cylinder block upper surface contact areas on the chain case and to the oil pan mating surface of the chain case and inner threads of the holes.

17. Install or connect the following:
   • New O–ring on the timing chain case. Set the edge of the chain case to the edge of the oil pan, then install the case on the cylinder block. Tighten the case bolts to 104 inch lbs. (12 Nm).

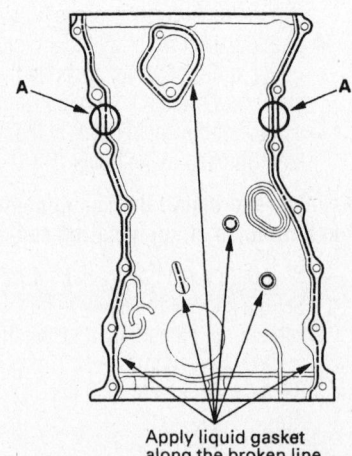

Apply liquid gasket along the broken line.

42356-INTE-G26

Apply liquid gasket to the points shown—RSX models

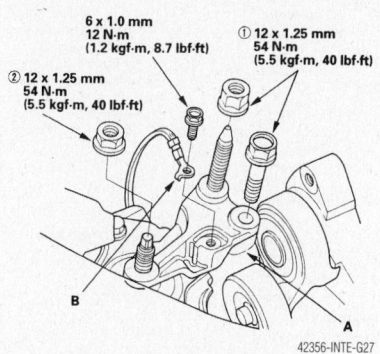

Upper bracket tightening sequence and specifications—RSX models

✷✷ WARNING

When installing the chain case, do not slide the bottom surface on the oil pan mounting surface.

- Side engine mount bracket and tighten to 33 ft. lbs. (44 m)
- Upper bracket, then tighten the bolts/nuts in sequence as shown in the illustration
- Ground cable
- VTC coil control solenoid valve
- CKP sensor connector and VTC oil control solenoid valve connector

- Oil cooler hose using a new O–ring
18. Clean the crankshaft pulley and pulley bolt, then apply lubrication to the pulley bolt and washer.
- Crankshaft pulley and hold the pulley with the holder handle and holder attachment. Using a 19mm socket and torque wrench, tighten the pulley bolt to 181 ft. lbs. (245 Nm).
- Cylinder head cover
- Drive belt
- Splash shield
- Front wheels and tires

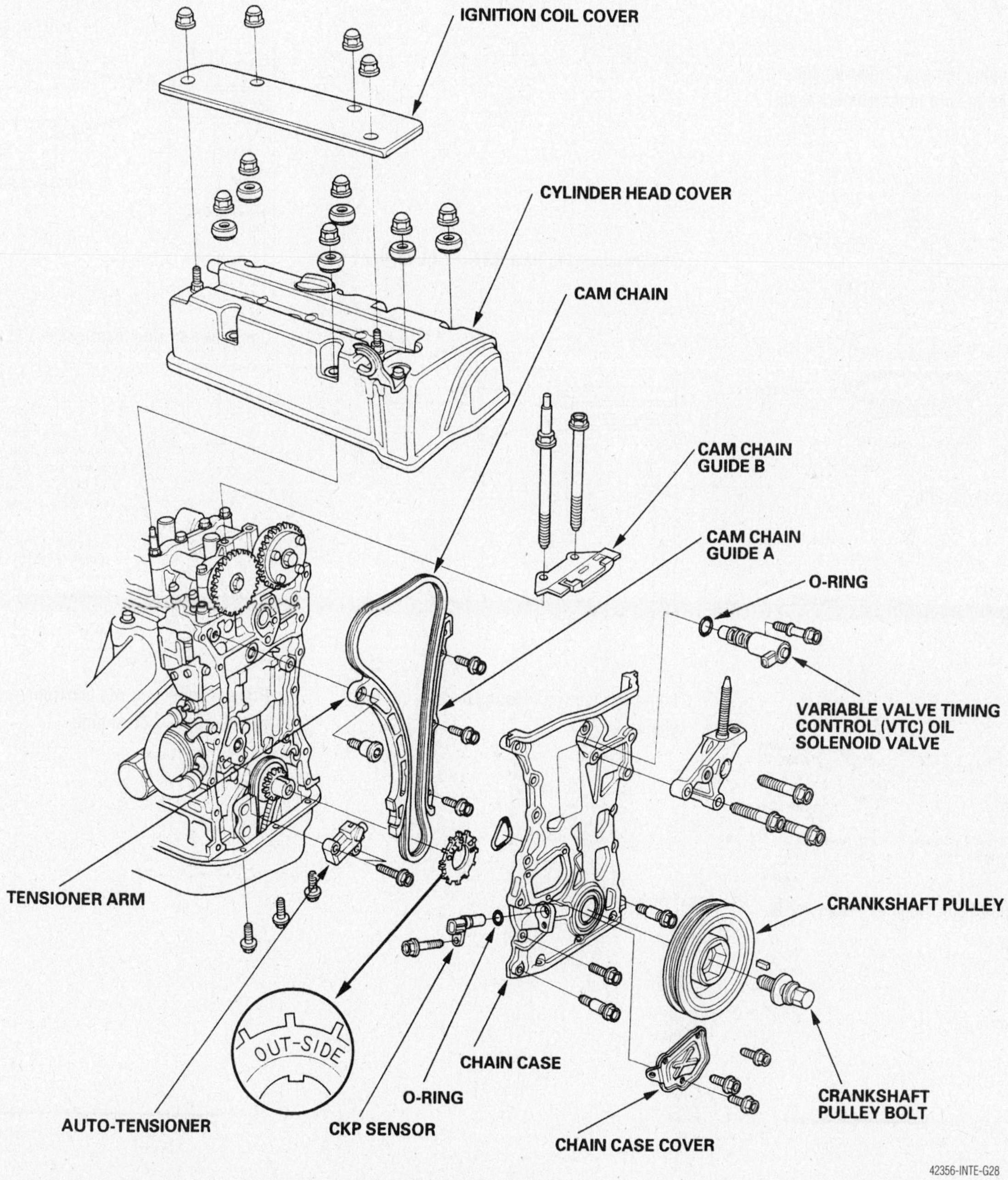

IGNITION COIL COVER

CYLINDER HEAD COVER

CAM CHAIN

CAM CHAIN GUIDE B

CAM CHAIN GUIDE A

O-RING

VARIABLE VALVE TIMING CONTROL (VTC) OIL SOLENOID VALVE

TENSIONER ARM

OUT-SIDE

CRANKSHAFT PULLEY

CRANKSHAFT PULLEY BOLT

AUTO-TENSIONER

O-RING

CKP SENSOR

CHAIN CASE

CHAIN CASE COVER

42356-INTE-G28

Exploded view of the timing (cam) chain and related components—RSX models

Piston and Ring

POSITIONING

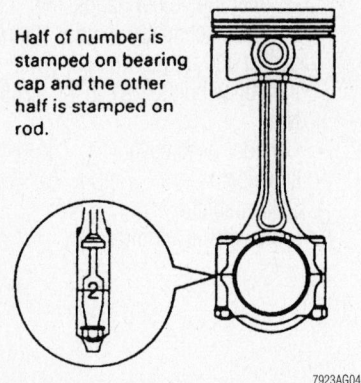

Half of number is stamped on bearing cap and the other half is stamped on rod.

7923AG04

Before removing the caps from the connecting rods, be sure to matchmark them as shown

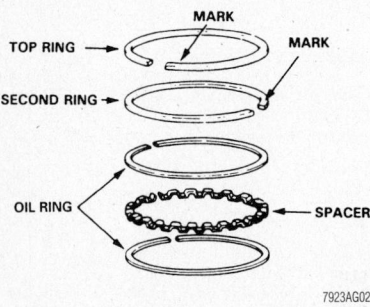

7923AG02

Piston ring positioning

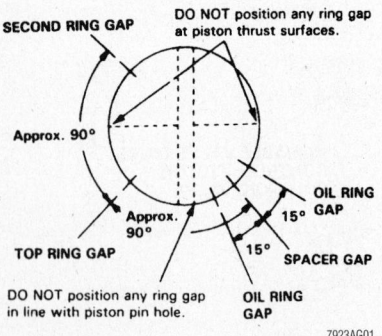

7923AG01

Piston ring end-gap spacing—except 2.0L engines

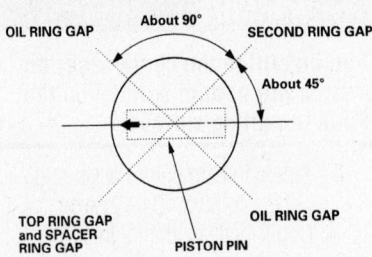

42356-INTE-G30

Piston ring end-gap spacing—2.0L engines

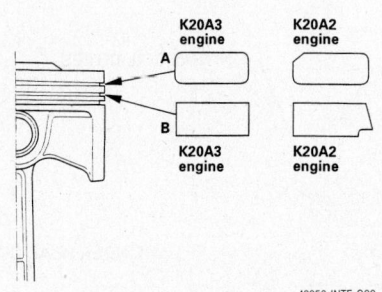

42356-INTE-G29

Compression ring locations—2.0L engines

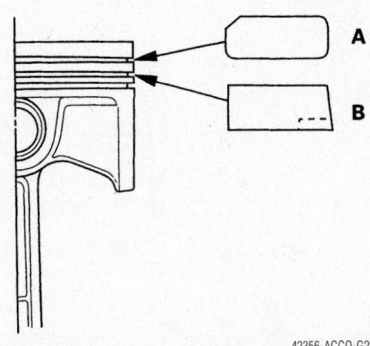

42356-ACCO-G29

Compression ring locations—2.4L (K24A2) engines

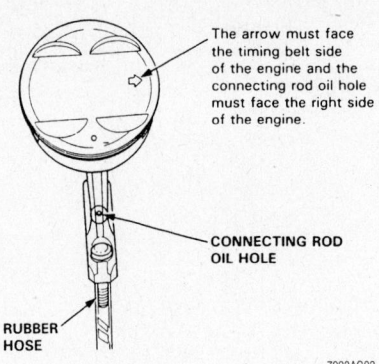

The arrow must face the timing belt side of the engine and the connecting rod oil hole must face the right side of the engine.

CONNECTING ROD OIL HOLE

RUBBER HOSE

7923AG03

Piston/connecting rod assembly-to-engine orientation—TL models

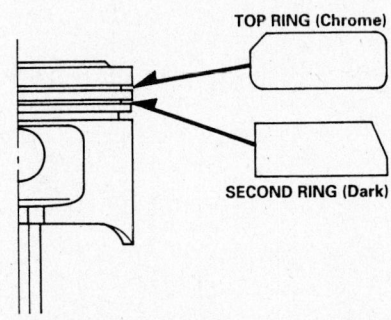

TOP RING (Chrome)

SECOND RING (Dark)

7923AG16

Compression ring locations—3.5L engine

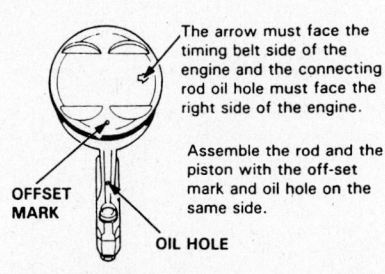

The arrow must face the timing belt side of the engine and the connecting rod oil hole must face the right side of the engine.

Assemble the rod and the piston with the off-set mark and oil hole on the same side.

OFFSET MARK

OIL HOLE

7923AG15

Piston/connecting rod assembly-to-engine orientation—3.5L engine

FUEL SYSTEM

Fuel System Service Precautions

Safety is the most important factor when performing not only fuel system maintenance but also any type of maintenance. Failure to conduct maintenance and repairs in a safe manner may result in serious personal injury or death. Maintenance and testing of the vehicle's fuel system components can be accomplished safely and effectively by adhering to the following rules and guidelines.

• To avoid the possibility of fire and personal injury, always disconnect the negative battery cable unless the repair or test procedure requires that battery voltage be applied.

• Always relieve the fuel system pressure prior to disconnecting any fuel system component (injector, fuel rail, pressure regulator, etc.), fitting or fuel line connection. Exercise extreme caution whenever relieving fuel system pressure, to avoid exposing skin, face and eyes to fuel spray. Please be advised that fuel under pressure may penetrate the skin or any part of the body that it contacts.

• Always place a shop towel or cloth around the fitting or connection prior to loosening to absorb any excess fuel due to spillage. Ensure that all fuel spillage (should it occur) is quickly removed from engine surfaces. Ensure that all fuel soaked cloths or towels are deposited into a suitable waste container.

• Always keep a dry chemical (Class B) fire extinguisher near the work area.

• Do not allow fuel spray or fuel vapors to come into contact with a spark or open flame.

• Always use a back–up wrench when loosening and tightening fuel line connection fittings. This will prevent unnecessary stress and torsion to fuel line piping. Always follow the proper torque specifications.

• Always replace worn fuel fitting O–rings with new. Do not substitute fuel hose or equivalent, where fuel pipe is installed.

Fuel System Pressure

RELIEVING

CL Model

✳✳ CAUTION

The fuel injection system remains under pressure after the engine has

been turned OFF. Properly relieve fuel pressure before disconnecting any fuel lines. Failure to do so may result in fire or personal injury.

➡**The radio may contain a coded theft protection circuit. Always obtain the code number before disconnecting the battery. If the vehicle is equipped with 4WS, the steering control unit is shut down when the battery is disconnected. After connecting the battery, turn the steering wheel lock-to-lock to reset the steering control unit.**

1. Before servicing the vehicle, refer to the precautions section.
2. Disconnect the negative battery cable.
3. Remove the fuel filler cap.
4. Use a box wrench on the 6mm service bolt or fuel pulsation damper, as applicable, on the fuel rail while holding the special banjo bolt with another wrench.
5. Place a rag or shop towel over the bolt/pulsation damper.
6. Slowly loosen the bolt/damper 1 complete turn.
7. After work is completed, tighten the bolt to 16 ft. lbs. (22 Nm). Don't overtighten the service bolts, their threads may strip and cause leaks.
8. Reconnect the negative battery cable.
9. Turn the ignition **ON**, but don't start the engine. Repeat this 2 or 3 times to pressurize the fuel system. Check for fuel leaks.
10. Enter the radio security code.

RL Model

✳✳ CAUTION

The fuel injection system remains under pressure after the engine has been turned OFF. Properly relieve fuel pressure before disconnecting any fuel lines. Failure to do so may result in fire or personal injury.

➡**The radio may contain a coded theft protection circuit. Always obtain the code number before disconnecting the battery. If the vehicle is equipped with 4WS, the steering control unit is shut down when the battery is disconnected. After connecting the battery, turn the steering wheel lock-to-lock to reset the steering control unit.**

1. Before servicing the vehicle, refer to the precautions section.

2. Disconnect the negative battery cable.
3. Remove the kick panel, then remove the PGM-FI main relay (FUEL PUMP) from the under-dash fuse/relay box. Start the engine and let it run until it stalls.
4. Remove the fuel filler cap.
5. On vehicles without quick connect fittings:
   a. Use a box wrench on the 6mm service bolt or fuel pulsation damper, as applicable, on the fuel rail while holding the special banjo bolt with another wrench.
   b. Place a rag or shop towel over the bolt/pulsation damper.
   c. Slowly loosen the bolt/damper 1 complete turn.
   d. On vehicles with quick-connect fittings:
   e. Remove the quick-connect fitting cover.
   f. Clean any dirt from the quick-connect fitting.
   g. Place a rag or shop towel over quick-connect fitting.
   h. Detach the quick-connect fitting by holding the connector with one hand, then squeeze the retainer tabs with the other hand to release them from the locking pawls. Pull the connector off.

✳✳ CAUTION

Do not allow fuel spray or fuel vapors to come in contact with a spark or open flame. Keep a dry chemical fire extinguisher nearby. Never store fuel in an open container due to risk of fire or explosion.

➡**A fuel pressure gauge may be attached at the 6mm service bolt/pulsation damper or quick-connect location. Always replace the washer between the service bolt and the banjo bolt whenever the service bolt is loosened.**

6. If equipped, remove the service bolt/damper and install a new washer. Tighten the 6mm service bolt to 104 inch lbs. (12 Nm). If equipped with a fuel pulsation damper, tighten to 16 ft. lbs. (22 Nm). Don't overtighten the service bolts, their threads may strip and cause leaks.
7. If equipped with a quick-connect fitting, connect the fitting, making sure the locking pawls are properly engaged,
8. Clean up any fuel spilled on the engine and intake manifold.
9. Install the fuel pump relay to the

under dash fuel/relay box and install the kick-panel cover.

10. Install the fuel filler cap.

11. Properly dispose of the fuel soaked rag or shop towel.

12. Reconnect the negative battery cable.

13. Turn the ignition **ON**, but don't start the engine. Repeat this 2 or 3 times to pressurize the fuel system. Check for fuel leaks.

14. Enter the radio security code.

RSX Models

❋❋ CAUTION

The fuel injection system remains under pressure after the engine has been turned OFF. Properly relieve fuel pressure before disconnecting any fuel lines. Failure to do so may result in fire or personal injury.

➡The radio may contain a coded theft protection circuit. Always obtain the code number before disconnecting the battery. If the vehicle is equipped with 4WS, the steering control unit is shut down when the battery is disconnected. After connecting the battery, turn the steering wheel lock-to-lock to reset the steering control unit.

1. Before servicing the vehicle, refer to the precautions section.

2. Disconnect the negative battery cable.

3. Remove the kick panel, then remove the PGM-FI main relay (FUEL PUMP) from the under-dash fuse/relay box. Start the engine and let it run until it stalls.

4. Remove the fuel filler cap.

5. On vehicles without quick connect fittings:

   a. Use a box wrench on the 6mm service bolt or fuel pulsation damper, as applicable, on the fuel rail while holding the special banjo bolt with another wrench.

   b. Place a rag or shop towel over the bolt/pulsation damper.

   c. Slowly loosen the bolt/damper 1 complete turn.

   d. On vehicles with quick-connect fittings:

   e. Remove the quick-connect fitting cover.

   f. Clean any dirt from the quick-connect fitting.

   g. Place a rag or shop towel over quick-connect fitting.

   h. Detach the quick-connect fitting by holding the connector with one hand, then squeeze the retainer tabs with the

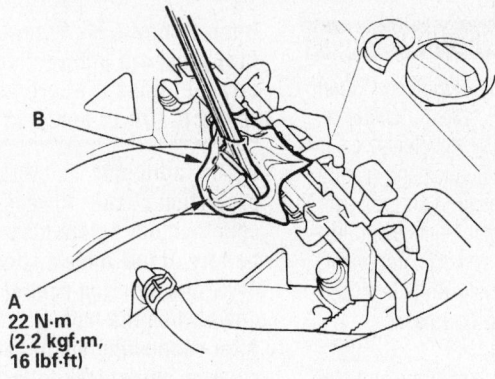

K20A2 engine

B

A
**22 N·m
(2.2 kgf·m,
16 lbf·ft)**

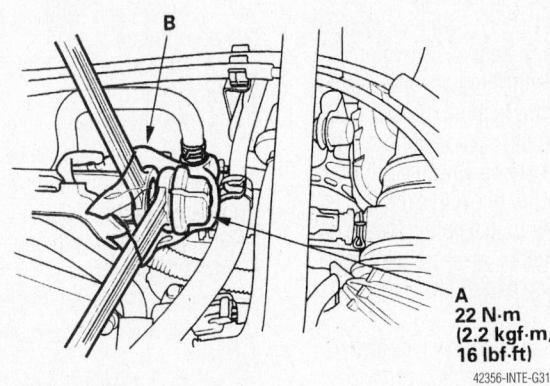

K20A3 engine

B

A
**22 N·m
(2.2 kgf·m,
16 lbf·ft)**

42356-INTE-G31

View of the fuel pulsation damper (A)—RSX models

other hand to release them from the locking pawls. Pull the connector off.

❋❋ CAUTION

Do not allow fuel spray or fuel vapors to come in contact with a spark or open flame. Keep a dry chemical fire extinguisher nearby. Never store fuel in an open container due to risk of fire or explosion.

➡A fuel pressure gauge may be attached at the 6mm service bolt/pulsation damper or quick-connect location. Always replace the washer between the service bolt and the banjo bolt whenever the service bolt is loosened.

6. If equipped, remove the service bolt/damper and install a new washer. Tighten the 6mm service bolt to 104 inch lbs. (12 Nm). If equipped with a fuel pulsation damper, tighten to 16 ft. lbs. (22 Nm). Don't overtighten the service bolts, their threads may strip and cause leaks.

7. If equipped with a quick-connect fitting, connect the fitting, making sure the locking pawls are properly engaged,

8. Clean up any fuel spilled on the engine and intake manifold.

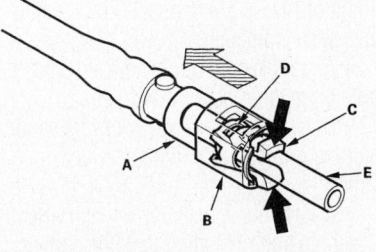

42356-ACCO-G30

Hold the quick-connect (A) connector (B) with one hand, then squeeze the retainer tabs (C) with the other hand to release them from the locking pawls (D)

9. Install the fuel pump relay to the under dash fuel/relay box and install the kick-panel cover.

10. Install the fuel filler cap.

11. Properly dispose of the fuel soaked rag or shop towel.

12. Reconnect the negative battery cable.

13. Turn the ignition **ON**, but don't start the engine. Repeat this 2 or 3 times to pressurize the fuel system. Check for fuel leaks.

14. Enter the radio security code.

2002–03 TL Models

✳✳ CAUTION

The fuel injection system remains under pressure after the engine has been turned OFF. Properly relieve fuel pressure before disconnecting any fuel lines. Failure to do so may result in fire or personal injury.

➡The radio may contain a coded theft protection circuit. Always obtain the code number before disconnecting the battery. If the vehicle is equipped with 4WS, the steering control unit is shut down when the battery is disconnected. After connecting the battery, turn the steering wheel lock-to-lock to reset the steering control unit.

1. Before servicing the vehicle, refer to the precautions section.
2. Disconnect the negative battery cable.
3. Remove the kick panel, then remove the PGM-FI main relay (FUEL PUMP) from the under-dash fuse/relay box. Start the engine and let it run until it stalls.
4. Remove the fuel filler cap.
5. On vehicles without quick connect fittings:
   a. Use a box wrench on the 6mm service bolt or fuel pulsation damper, as applicable, on the fuel rail while holding the special banjo bolt with another wrench.
   b. Place a rag or shop towel over the bolt/pulsation damper.
   c. Slowly loosen the bolt/damper 1 complete turn.
   d. On vehicles with quick-connect fittings:
   e. Remove the quick-connect fitting cover.
   f. Clean any dirt from the quick-connect fitting.
   g. Place a rag or shop towel over quick-connect fitting.
   h. Detach the quick-connect fitting by holding the connector with one hand, then squeeze the retainer tabs with the other hand to release them from the locking pawls. Pull the connector off.

✳✳ CAUTION

Do not allow fuel spray or fuel vapors to come in contact with a spark or open flame. Keep a dry chemical fire extinguisher nearby. Never store fuel in an open container due to risk of fire or explosion.

➡A fuel pressure gauge may be attached at the 6mm service bolt/pulsation damper or quick-connect location. Always replace the washer between the service bolt and the banjo bolt whenever the service bolt is loosened.

6. If equipped, remove the service bolt/damper and install a new washer. Tighten the 6mm service bolt to 104 inch lbs. (12 Nm). If equipped with a fuel pulsation damper, tighten to 16 ft. lbs. (22 Nm). Don't overtighten the service bolts, their threads may strip and cause leaks.
7. If equipped with a quick-connect fitting, connect the fitting, making sure the locking pawls are properly engaged,
8. Clean up any fuel spilled on the engine and intake manifold.
9. Install the fuel pump relay to the under dash fuel/relay box and install the kick-panel cover.
10. Install the fuel filler cap.
11. Properly dispose of the fuel soaked rag or shop towel.
12. Reconnect the negative battery cable.
13. Turn the ignition **ON**, but don't start the engine. Repeat this 2 or 3 times to pressurize the fuel system. Check for fuel leaks.
14. Enter the radio security code.

2004–06 TL Models

➡The radio may have a coded theft protection circuit. Obtain the code before disconnecting the battery, removing the radio fuse, or removing the radio.

1. Before servicing the vehicle, refer to the precautions section.
2. Remove the left kick panel, then remove PGM-FI main relay 2 (FUEL PUMP) from the under-dash fuse/relay box.
3. Start the engine, and let it idle until it stalls.

➡If any DTCs are stored, clear and ignore them.

4. Turn the ignition switch OFF.
5. Remove the fuel fill cap, and relieve the pressure in the fuel tank.
6. Make sure the ignition switch is OFF, then disconnect the negative cable from the battery.
7. Remove the quick-connect fitting cover.
8. Check the fuel quick-connect fitting for dirt, and clean it if needed.
9. Place a rag or shop towel over the quick-connect fitting.
10. Disconnect the quick-connect fitting by holding the connector with one hand, and squeezing the retainer tabs with the

other hand to release them from the locking tabs, then pull the connector off.

➡To prevent the remaining fuel in the fuel feed line or hose from flowing out, use a rag or shop towel.

11. If the connector does not move, keep the retainer tabs pressed down, and alternately pull and push the connector until it comes off easily.
12. Do not remove the retainer from the line; once removed, the retainer must be replaced with a new one.
13. After disconnecting the quick-connect fitting, check it for dirt or damage.

TSX Models

✳✳ CAUTION

The fuel injection system remains under pressure after the engine has been turned OFF. Properly relieve fuel pressure before disconnecting any fuel lines. Failure to do so may result in fire or personal injury.

➡The radio may contain a coded theft protection circuit. Always obtain the code number before disconnecting the battery. If the vehicle is equipped with 4WS, the steering control unit is shut down when the battery is disconnected. After connecting the battery, turn the steering wheel lock-to-lock to reset the steering control unit.

1. Before servicing the vehicle, refer to the precautions section.
2. Disconnect the negative battery cable.
3. Remove the kick panel, then remove the PGM-FI main relay (FUEL PUMP) from the under-dash fuse/relay box. Start the engine and let it run until it stalls.
4. Remove the fuel filler cap.
5. On vehicles without quick connect fittings:
   a. Use a box wrench on the 6mm service bolt or fuel pulsation damper, as applicable, on the fuel rail while holding the special banjo bolt with another wrench.
   b. Place a rag or shop towel over the bolt/pulsation damper.
   c. Slowly loosen the bolt/damper 1 complete turn.
   d. On vehicles with quick-connect fittings:
   e. Remove the quick-connect fitting cover.
   f. Clean any dirt from the quick-connect fitting.

g. Place a rag or shop towel over quick-connect fitting.

h. Detach the quick-connect fitting by holding the connector with one hand, then squeeze the retainer tabs with the other hand to release them from the locking pawls. Pull the connector off.

✳✳ CAUTION

Do not allow fuel spray or fuel vapors to come in contact with a spark or open flame. Keep a dry chemical fire extinguisher nearby. Never store fuel in an open container due to risk of fire or explosion.

➡A fuel pressure gauge may be attached at the 6mm service bolt/pulsation damper or quick-connect location. Always replace the washer between the service bolt and the banjo bolt whenever the service bolt is loosened.

6. If equipped, remove the service bolt/damper and install a new washer. Tighten the 6mm service bolt to 104 inch lbs. (12 Nm). If equipped with a fuel pulsation damper, tighten to 16 ft. lbs. (22 Nm). Don't overtighten the service bolts, their threads may strip and cause leaks.

7. If equipped with a quick-connect fitting, connect the fitting, making sure the locking pawls are properly engaged,

8. Clean up any fuel spilled on the engine and intake manifold.

9. Install the fuel pump relay to the under dash fuel/relay box and install the kick-panel cover.

10. Install the fuel filler cap.

11. Properly dispose of the fuel soaked rag or shop towel.

12. Reconnect the negative battery cable.

13. Turn the ignition **ON**, but don't start the engine. Repeat this 2 or 3 times to pressurize the fuel system. Check for fuel leaks.

14. Enter the radio security code.

Fuel Filter

REMOVAL & INSTALLATION

Except Fuel Pump-Mounted Filters

1. Before servicing the vehicle, refer to the precautions section.

2. Disconnect the negative battery cable.

3. Properly relieve the fuel pressure.

4. Wrap a shop towel around the filter fittings. Use a properly sized wrench to slowly loosen the fuel line fittings.

5. Remove or disconnect the following:
  • Fuel pipes from the fuel filter

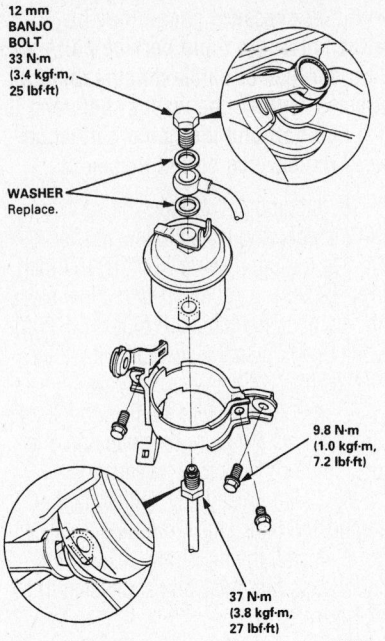

12 mm BANJO BOLT
33 N·m (3.4 kgf·m, 25 lbf·ft)

WASHER Replace.

9.8 N·m (1.0 kgf·m, 7.2 lbf·ft)

37 N·m (3.8 kgf·m, 27 lbf·ft)

7923BG90

Fuel filter assembly

  • Fuel filter clamp
  • Filter from the vehicle

✳✳ WARNING

It is very important that ALL of the fuel line banjo bolt washers be replaced every time the banjo bolts are loosened. If the washers are not replaced, the fuel lines will leak pressurized fuel, causing the risk of fire or explosion.

To install:

6. Install or connect the following:
  • New filter in position and tighten the clamp mounting bolt to 89 inch lbs. (10 Nm)
  • Banjo bolt with new washers and tighten it to 25 ft. lbs. (33 Nm)
  • Fuel feed line and tighten the fitting to 27 ft. lbs. (37 Nm)
  • Negative battery cable and enter the radio security code.

7. Start the vehicle and check for leaks.

Fuel Pump-Mounted Filter

1. Before servicing the vehicle, refer to the precautions section.

2. Properly relieve the fuel pressure.

3. Remove or disconnect the following:
  • Negative battery cable
  • Fuel pump
  • Fuel filter set

To install:

4. Install or connect the following:
  • Fuel filter set, using a new base gasket and new O–rings

• Fuel pump. When installing the fuel gauge sending unit, make sure the is secure and the connector is locked into place.

• Negative battery cable

Fuel Pump

REMOVAL & INSTALLATION

CL Model

1. Before servicing the vehicle, refer to the precautions section.

2. Properly relieve the fuel system pressure.

3. Remove or disconnect the following:
  • Negative battery cable
  • Spare tire lid
  • Floor access panel
  • Pump electrical connectors
  • Quick connect fittings
  • Bolts and the pump

To install:

4. Installation is the reverse of the removal procedure.

RL and RSX Models

1. Before servicing the vehicle, refer to the precautions section.

2. Properly relieve the fuel system pressure.

3. Remove or disconnect the following:
  • Rear seat cushion
  • Access panel from the floor
  • Fuel line and wiring from the fuel pump assembly
  • Mounting nuts and the fuel pump from the fuel tank

To install:

4. Installation is the reverse of the removal procedure. Tighten the fuel pump mounting nuts to 53 inch lbs. (6 Nm).

2002–03 TL Models

1. Before servicing the vehicle, refer to the precautions section.

2. Remove or disconnect the following:
  • Negative battery cable
  • Left rear wheel
  • Tank drain bolt and drain the fuel into an approved container
  • Pump and float wiring connectors located under the trunk floor
  • Fuel hose and pipe covers from the inside of the quarter panel

3. Support the tank with a transmission jack, remove the straps and lower the tank out of the vehicle. If it sticks on the under-

coating, carefully pry it free using a blunt or wooden instrument as a lever.

4. Disconnect the fuel line by removing the banjo bolt or uncoupling the quick–connect fittings.

5. Remove the fuel pump mounting nuts. Remove the fuel pump from the fuel tank.

To install:

6. Installation is the reverse of the removal procedure, while using the following torque values:

- Fuel pump mounting nuts: 48 inch lbs. (6 Nm)
- Fuel tank strap bolts: 28 ft. lbs. (38 Nm)
- Fuel tank drain bolt: 36 ft. lbs. (49–50 Nm)

2004–06 TL Models

1. Before servicing the vehicle, refer to the precautions section.

2. Properly relieve the fuel pressure.

3. Remove the fuel fill cap.

4. Remove the trunk floor.

5. Remove the access panel from the floor.

6. Disconnect fuel pump 5P connector.

7. Disconnect the quick-connect fitting from the fuel tank unit.

8. Using Fuel sender wrench 07AAA-S0XA100, loosen the locknut.

9. Remove the locknut and the fuel tank unit.

10. Remove the fuel filter, the fuel gauge sending unit, the case, the wire harness, and the fuel pressure regulator.

To install:

11. Installation is the reverse of removal, use a new base gasket, new O-rings, and a new locknut

➡**After installation, make sure the base gasket is not pinched.**

TSX Models

1. Before servicing the vehicle, refer to the precautions section.

2. Properly relieve the fuel pressure.

3. Remove or disconnect the following:

- Negative battery cable, if not already done
- Fuel fill cap
- Trunk floor
- Access panel from the floor
- Fuel pump connector
- Quick-connect fitting from the fuel tank unit
- Fuel pump/sender assembly locknut using tool 07XAA-001010A
- Locknut and fuel pump/sender assembly

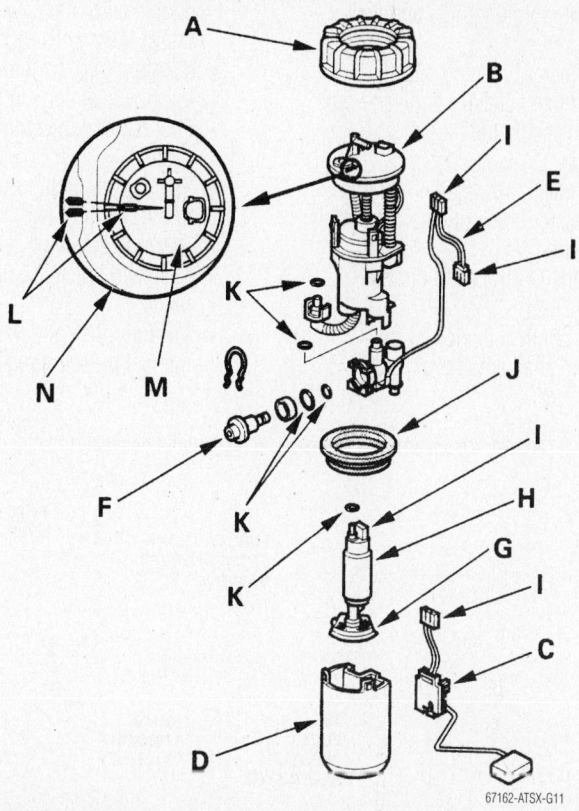

Exploded view of the fuel filter (B), fuel gauge sending unit (C), case (D), wire harness (E), fuel pressure regulator (F) and related components—TSX models

67162-ATSX-G11

- Fuel filter, fuel gauge sending unit, case, wire harness and fuel pressure regulator

To install:

4. Installation is the reverse of the removal procedure, noting the following points:

a. When connecting the wire harness, make sure the connection is secure and the connectors are firmly locked into place. Do not bend or twist the fuel gauge sending unit more than necessary.

b. Install the pump assembly components in the reverse order of removal, with a new base gasket and new O-rings. Tighten the fuel pump/sender locknut to 69 ft. lbs. (93 Nm) with special tool 07XAA-001010A.

c. When installing the fuel tank unit, align the marks on the unit and the fuel tank.

Fuel Injector

REMOVAL & INSTALLATION

✹✹ CAUTION

Fuel injection systems remain under pressure, even after the engine has

been turned OFF. The fuel system pressure must be relieved before disconnecting any fuel lines. Failure to do so may result in fire and/or personal injury. Observe all applicable safety precautions when working around fuel. Whenever servicing the fuel system, always work in a well-ventilated area. Do not allow fuel spray or vapors to come in contact with a spark or open flame. Keep a dry chemical fire extinguisher near the work area. Always keep fuel in a container specifically designed for fuel storage; also, always properly seal fuel containers to avoid the possibility of fire or explosion.

CL Model

1. Before servicing the vehicle, refer to the precautions section.

2. Disconnect the negative battery cable.

3. Relieve the fuel system pressure.

4. Remove or disconnect the following:

- Intake manifold cover, if equipped
- Fuel injector electrical connectors
- Fuel line quick-connect fittings
- Fuel feed line from the fuel rail
- Vacuum hose and fuel return line from the fuel pressure regulator

- Fuel rail mounting nuts and fuel rail
- Fuel injector retaining clip

5. Grasp the fuel injectors body and pull up while gently rocking the fuel injector from side to side.

6. Once removed, inspect the fuel injector cap and body for signs of deterioration. Replace as required.

7. Remove the O–rings and discard.

To install:

8. Install or connect the following:
- Apply a small amount of clean engine oil to the new O–rings and install them onto each injector
- Injectors into the fuel rail
- Injector retaining clips
- Fuel rail and injectors into injector base
- Fuel rail mounting nuts and tighten to 7 ft. lbs. (10 Nm)
- Ground cable bolt, if removed
- Fuel feed line and quick-connect fittings
- Vacuum hose and fuel return line to the fuel pressure regulator

- Fuel injector electrical connectors
- Intake manifold cover, if equipped
- Negative battery cable

9. Run the engine at idle for 2 minutes, then turn the engine **OFF** and check for fuel leaks and proper operation.

2002–04 RL Model

1. Before servicing the vehicle, refer to the precautions section.

2. Disconnect the negative battery cable.

3. Relieve the fuel system pressure.

4. Remove or disconnect the following:

NOTE: Check all hose clamps and retighten if necessary.

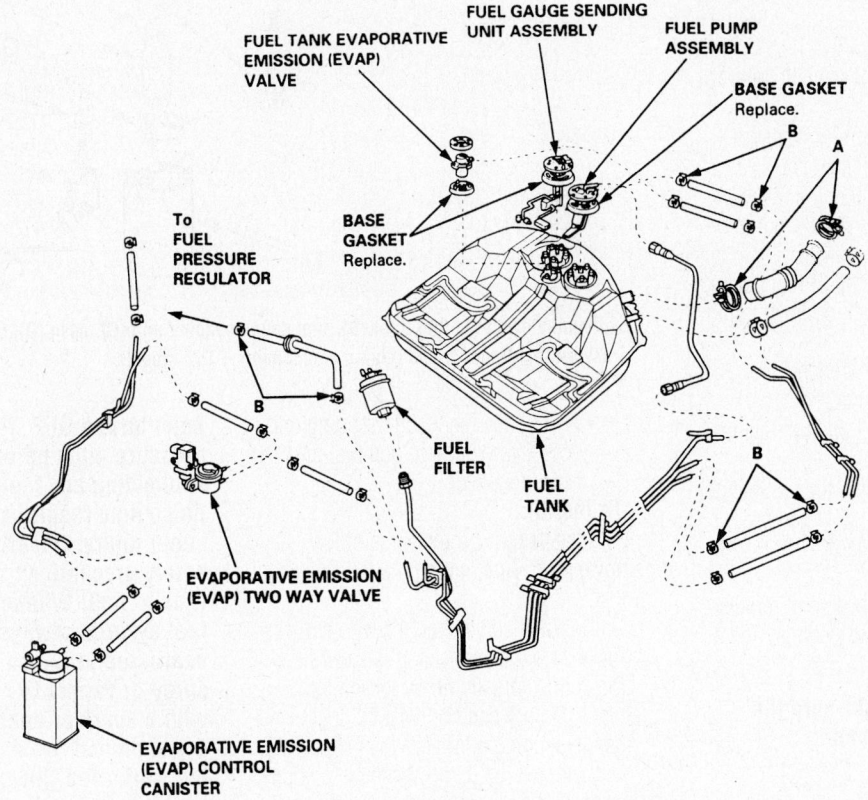

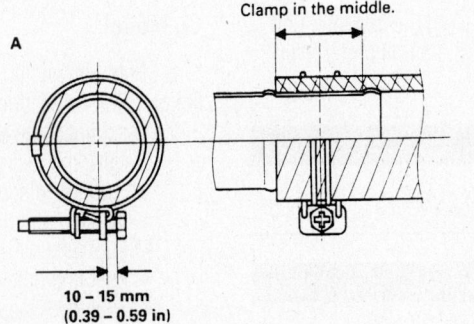

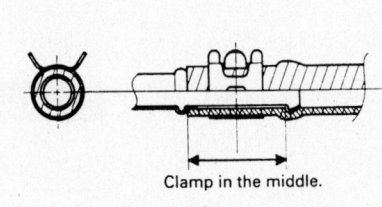

7923BG91

Exploded view of the fuel line routing—2002–04 RL models

- Intake manifold cover, if equipped
- Fuel injector electrical connectors
- Ground cable bolt, if necessary
- Fuel line quick-connect fittings
- Fuel feed line from the fuel rail
- Vacuum hose and fuel return line from the fuel pressure regulator
- Fuel rail mounting nuts and fuel rail
- Fuel injector retaining clip

5. Grasp the fuel injectors body and pull up while gently rocking the fuel injector from side to side.

6. Once removed, inspect the fuel injector cap and body for signs of deterioration. Replace as required.

7. Remove the O–rings and discard.

To install:

8. Install or connect the following:
- Apply a small amount of clean engine oil to the new O–rings and install them onto each injector
- Injectors into the fuel rail
- Injector retaining clips
- Fuel rail and injectors into injector base
- Fuel rail mounting nuts and tighten to 16 ft. lbs. (22 Nm)
- Ground cable bolt, if removed
- Fuel feed line and quick-connect fittings
- Vacuum hose and fuel return line to the fuel pressure regulator
- Fuel injector electrical connectors
- Intake manifold cover, if equipped
- Negative battery cable

9. Run the engine at idle for 2 minutes, then turn the engine **OFF** and check for fuel leaks and proper operation.

2004–06 TL and 2005–06 RL Models

1. Before servicing the vehicle, refer to the precautions section.

2. Properly relieve the fuel pressure.

3. Remove the intake manifold.

4. Disconnect the connectors from the injectors.

5. Disconnect the quick-connect fittings.

6. Remove the fuel rail mounting bolts from the fuel rail.

7. Remove the injector clip from the injector.

8. Remove the injector from the fuel rail.

To install:

9. Coat the new O–rings with clean engine oil, and insert the injectors into the fuel rail.

10. Install the injector clip.

11. Coat the new injector O-rings with clean engine oil.

12. Install the injectors in the injector base.

13. Install the fuel rail mounting nuts.

14. Connect the connectors on the injectors.

15. Connect the quick-connect fittings.

16. Turn the ignition switch ON, but do not operate the starter. After the fuel pump runs for about 2 seconds, the fuel rail will be pressurized. Repeat this two or three times, then check for fuel leakage.

17. Install the intake manifold.

RSX Models

1. Before servicing the vehicle, refer to the precautions section.

2. Disconnect the negative battery cable.

3. Relieve the fuel system pressure.

4. Remove or disconnect the following:
- Intake manifold cover, if equipped
- Fuel injector electrical connectors
- Ground cable bolt, if necessary
- Fuel line quick-connect fittings
- Fuel feed line from the fuel rail
- Vacuum hose and fuel return line from the fuel pressure regulator
- Fuel rail mounting nuts and fuel rail
- Fuel injector retaining clip

5. Grasp the fuel injectors body and pull up while gently rocking the fuel injector from side to side.

6. Once removed, inspect the fuel injector cap and body for signs of deterioration. Replace as required.

7. Remove the O–rings and discard.

To install:

8. Install or connect the following:
- Apply a small amount of clean engine oil to the new O–rings and install them onto each injector
- Injectors into the fuel rail
- Injector retaining clips
- Fuel rail and injectors into injector base
- Fuel rail mounting nuts and tighten to 16 ft. lbs. (22 Nm)
- Ground cable bolt, if removed
- Fuel feed line and quick-connect fittings
- Vacuum hose and fuel return line to the fuel pressure regulator
- Fuel injector electrical connectors
- Intake manifold cover, if equipped
- Negative battery cable

9. Run the engine at idle for 2 minutes, then turn the engine **OFF** and check for fuel leaks and proper operation.

2002–03 TL Models

> **✳✳ CAUTION**
>
> **Fuel injection systems remain under pressure, even after the engine has been turned OFF. The fuel system pressure must be relieved before disconnecting any fuel lines. Failure to do so may result in fire and/or personal injury. Observe all applicable safety precautions when working around fuel. Whenever servicing the fuel system, always work in a well-ventilated area. Do not allow fuel spray or vapors to come in contact with a spark or open flame. Keep a dry chemical fire extinguisher near the work area. Always keep fuel in a container specifically designed for fuel storage; also, always properly seal fuel containers to avoid the possibility of fire or explosion.**

1. Before servicing the vehicle, refer to the precautions section.

2. Disconnect the negative battery cable.

3. Relieve the fuel system pressure.

4. Remove or disconnect the following:
- Intake manifold cover, if equipped
- Fuel injector electrical connectors
- Ground cable bolt, if necessary
- Fuel line quick-connect fittings
- Fuel feed line from the fuel rail
- Vacuum hose and fuel return line from the fuel pressure regulator
- Fuel rail mounting nuts and fuel rail
- Fuel injector retaining clip

5. Grasp the fuel injectors body and pull up while gently rocking the fuel injector from side to side.

6. Once removed, inspect the fuel injector cap and body for signs of deterioration. Replace as required.

7. Remove the O–rings and discard.

To install:

8. Install or connect the following:
- Apply a small amount of clean engine oil to the new O–rings and install them onto each injector
- Injectors into the fuel rail
- Injector retaining clips
- Fuel rail and injectors into injector base
- Fuel rail mounting nuts and tighten to 16 ft. lbs. (22 Nm)
- Ground cable bolt, if removed
- Fuel feed line and quick-connect fittings
- Vacuum hose and fuel return line to the fuel pressure regulator

- Fuel injector electrical connectors
- Intake manifold cover, if equipped
- Negative battery cable

9. Run the engine at idle for 2 minutes, then turn the engine **OFF** and check for fuel leaks and proper operation.

TSX Models

☀ CAUTION

Fuel injection systems remain under pressure, even after the engine has been turned OFF. The fuel system pressure must be relieved before disconnecting any fuel lines. Failure to do so may result in fire and/or personal injury. Observe all applicable safety precautions when working around fuel. Whenever servicing the fuel system, always work in a well-ventilated area. Do not allow fuel spray or vapors to come in contact with a spark or open flame. Keep a dry chemical fire extinguisher near the work area. Always keep fuel in a container specifically designed for fuel storage; also, always properly seal fuel containers to avoid the possibility of fire or explosion.

1. Disconnect the negative battery cable.
2. Relieve the fuel system pressure.
3. Remove or disconnect the following:
   - Intake manifold cover, if equipped
   - Fuel injector electrical connectors
   - Ground cable bolt, if necessary
   - Fuel line quick-connect fittings
   - Fuel feed line from the fuel rail
   - Vacuum hose and fuel return line from the fuel pressure regulator
   - Fuel rail mounting nuts and fuel rail
   - Fuel injector retaining clip
4. Grasp the fuel injectors body and pull

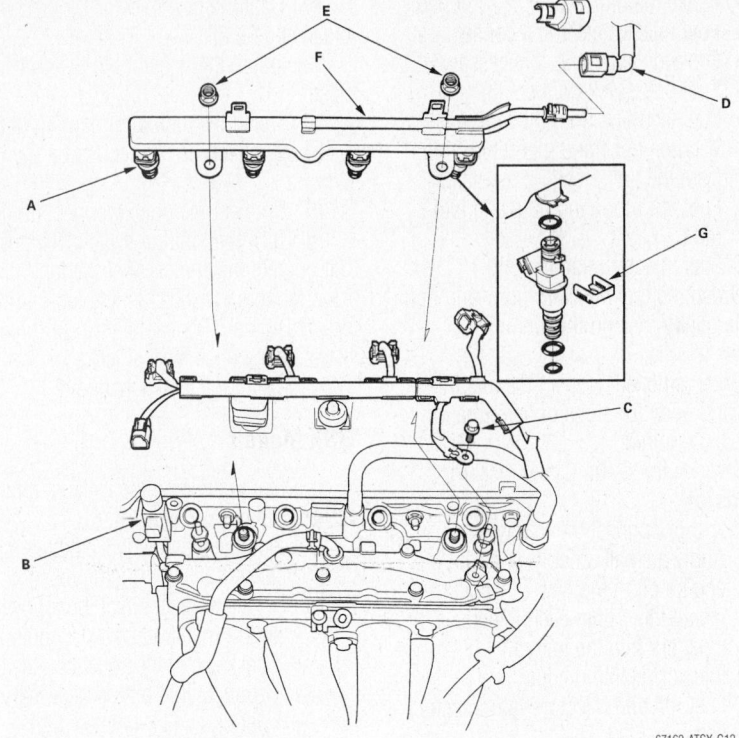

Exploded view of the fuel rail and injectors—TSX models

67162-ATSX-G12

up while gently rocking the fuel injector from side to side.

5. Once removed, inspect the fuel injector cap and body for signs of deterioration. Replace as required.

6. Remove the O–rings and discard.

To install:

7. Install or connect the following:
   - Apply a small amount of clean engine oil to the new O–rings and install them onto each injector
   - Injectors into the fuel rail
   - Injector retaining clips
   - Fuel rail and injectors into injector base

- Fuel rail mounting nuts and tighten to 16 ft. lbs. (22 Nm)
- Ground cable bolt, if removed
- Fuel feed line and quick-connect fittings
- Vacuum hose and fuel return line to the fuel pressure regulator
- Fuel injector electrical connectors
- Intake manifold cover, if equipped
- Negative battery cable

8. Run the engine at idle for 2 minutes, then turn the engine **OFF** and check for fuel leaks and proper operation.

DRIVE TRAIN

Transaxle Assembly

REMOVAL & INSTALLATION

Manual

RSX MODELS

1. Before servicing the vehicle, refer to the precautions section.

2. Disconnect the negative battery cable, then the positive battery cable.

3. Drain the transaxle oil. Install the drain plug with a new washer.

4. Remove or disconnect the following:

- Intake manifold cover
- Air cleaner housing
- Intake air duct
- Battery tray
- Transmission ground cable
- Back–up light switch connector
- Vehicle Speed Sensor (VSS) connector
- Reverse lockout solenoid connector
- Cable bracket and cable from the top of the transaxle. Remove the cables and bracket together to avoid bending the cables.
- Harness clips
- Slave cylinder, being careful not to

bend the clutch line. Do not depress the clutch pedal after the slave cylinder has been removed.
- Engine wire harness cover by lifting up on the lock tab, then sliding the harness forward off the air cleaner housing mounting bracket
- Water pipe mounting bolt and lower the pipe slightly
- Loosen the air cleaner housing bracket mounting bolt, then the mounting bolt
- Brake booster and Evaporative emission (EVAP) line brake mounting bolts and attach special tool

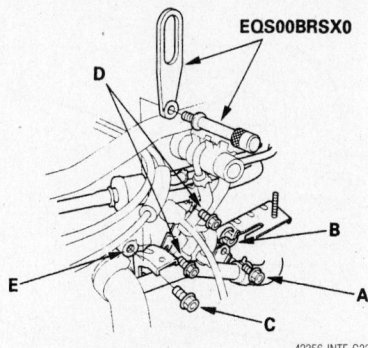

Attach the special tool to the threaded hole (E) in the cylinder head —RSX

42356-INTE-G33

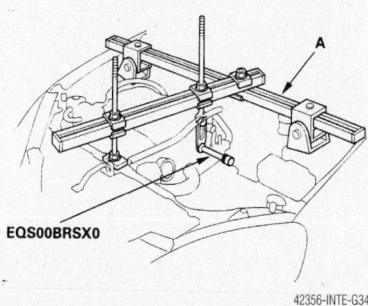

Install the engine support hanger (A) to the vehicle and attach the hook to the special tool—RSX

42356-INTE-G34

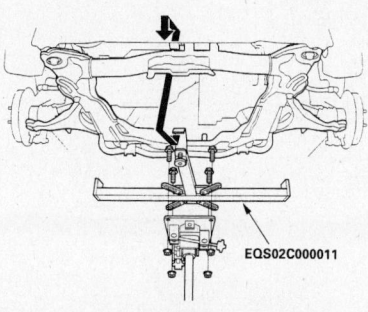

Support the subframe with the subframe adapter and a jack—RSX

42356-INTE-G35

EQS00BRSX0 to the threaded hole in the cylinder head

5. Install the engine support hanger to the vehicle and attach the hook to the special tool.

- 2 upper transmission mounting bolts
- Transmission mount bracket and transmission mounting bolt
- Air cleaner bracket
- Splash shield
- Three–way catalytic converter
- Halfshafts
- Intermediate shaft
- Front engine mount bracket mounting bolt

- 3 bolts securing the transaxle rear mount

6. Support the subframe with the subframe adapter and a jack.

7. Make reference marks of the installed position on the front suspension subframe and mounting bolts, then remove the subframe.

8. Remove or disconnect the following:
- Clutch cover. The cover will have 2 bolts on 5–speed transaxles and 3 bolts on 6–speed transaxles.
- Front engine mount
- Transaxle mounting bolts, after placing a jack under the transaxle. 5–speed models will have 4 lower bolts and 6–speed models will have 2 rear and 2 lower bolts.
- Harness bracket and intake manifold bracket, 6–speed transaxles only
- 2 front transaxle mounting bolts, 5–speed transaxles only

9. Pull the transaxle away from the engine until the mainshaft clears the clutch pressure plate, then lower the transmission on the jack.
- Transaxle rear mount and rear mount bracket
- Boot, release fork and release bearing from the transaxle

10. Installation is the reverse of the removal procedure, while using the following torque values:
- Transaxle rear mount bracket bolts: 47 ft. lbs. (64 Nm)
- 2 front transaxle mounting bolts (5–speed only): 47 ft. lbs. (64 Nm)
- 4 lower transaxle mounting bolts (5–speed only): 47 ft. lbs. (64 Nm)
- 2 rear transaxle mounting bolts (6–speed only): 47 ft. lbs. (64 Nm)
- 2 lower transaxle mounting bolts (6–speed only): 33 ft. lbs. (44 Nm)
- Front engine mount bolts: 47 ft. lbs. (64 Nm)
- Clutch cover bolts: 9 ft. lbs. (12 Nm)
- Subframe mounting bolts: 76 ft. lbs. (98 Nm)
- 3 transaxle rear mount mounting bolts: 43 ft. lbs. (59 Nm)
- Transaxle mount bracket and transmission mounting bolt: 40 ft. lbs. (54 Nm)
- Air cleaner bracket bolts: 9 ft. lbs. (12 Nm)
- 2 upper transaxle mounting bolts: 47 ft. lbs. (64 Nm)
- 2 Slave cylinder mounting bolts: 16 ft. lbs. (22 Nm)
- Transmission ground cable: 16 ft. lbs. (22 Nm)

2004–06 TL MODELS

The following tools are required to perform this procedure:
- Engine support hanger, A and Reds AAR-T-12566
- Engine hanger balance bar VSB02C000019
- Front sub-frame adapter VSB02C000016

1. Before servicing the vehicle, refer to the precautions section.

2. Set the wheels in the straight ahead position.

3. Lock the steering wheel.

4. Disconnect the support struts from both sides of the pivot ball (bolted to the hood). Secure the hood in a vertical position. Remove the right side pivot ball, and install it into the lower threaded hole, then reattach the support strut.

➡**Do not attempt to close the hood with the support strut in the vertical position; it will damage the support strut and hood.**

5. Remove the left rear engine compartment cover and the left side engine compartment cover.

6. Remove the right side engine compartment cover and the right rear engine compartment cover.

7. Remove the intake manifold cover and the bulkhead cover.

8. Make sure you have the anti-theft codes for the radio and the navigation system, then write down the customer's radio channel presets. Make sure the ignition switch is OFF.

9. Disconnect the negative cable from the battery first, then disconnect the positive cable.

10. Remove the battery.

11. Remove the air cleaner housing and resonator chamber.

12. Remove the battery base.

13. Remove the under-hood fuse/relay box.

14. Remove the strut bar.

15. Disconnect the back-up light switch connector and reverse lockout solenoid connector, then remove the harness clips.

16. Disconnect the input shaft speed sensor connector and output shaft speed sensor connector.

17. Remove the harness clips, then disconnect the vacuum hose.

18. Disconnect the starter cable, then remove the starter motor.

19. Remove the harness bracket.

20. Remove the cable bracket, then disconnect the cables from the top of the trans-

mission housing. Carefully remove both cables and the bracket together to avoid bending the cables.

21. Disconnect the vacuum hose.

22. Carefully remove the slave cylinder without bending the clutch line. Do not press the clutch pedal once the slave cylinder has been removed.

23. Remove the engine front mount mounting nuts, engine front mount stop, and the engine front mount mounting bolt.

24. Remove the engine rear mount mounting nuts and engine rear mount stop.

25. Remove the transmission upper mounting bolts.

26. Remove and plug the return hose from the power steering fluid reservoir. Wipe off any spilled fluid at once.

27. Remove the power steering pump outlet line from the power steering pump, and remove the hose from its clamp.

28. Lift and support the engine with the engine support hanger and the engine hanger balance bar:

    a. Attach the front arm to the front cylinder head with a spacer and the 10 x 1.25 mm bolt.

    b. Attach the rear arm to the rear cylinder head with the 8 x 1.25 mm bolt.

29. Remove the steering joint cover.

30. Make a reference mark across the steering joint and steering gearbox pinion shaft. Remove the steering joint bolt, and disconnect the steering joint by removing the steering joint toward the steering column. Hold the slider shaft on the column with a piece of wire between the joint yoke on the slider shaft to the joint yoke on the upper shaft.

31. Remove the transmission mount bracket and ground cable.

32. Remove the under cover.

33. Drain the transmission fluid. Install the drain bolt with a new washer.

34. Remove the exhaust pipe.

35. Separate the front stabilizer link.

36. Remove the damper fork.

37. Separate the knuckle from the lower arm.

38. Separate the tie-rod ball joint.

39. Remove the left and right driveshaft inboard joints.

40. Remove the intermediate shaft.

41. Remove the rear engine lower mount mounting bolts

42. Disconnect the power steering pressure switch connector.

43. Remove the transmission lower mount mounting nuts.

44. Remove the middle sub-frame mounting bolts.

45. Make the appropriate reference lines

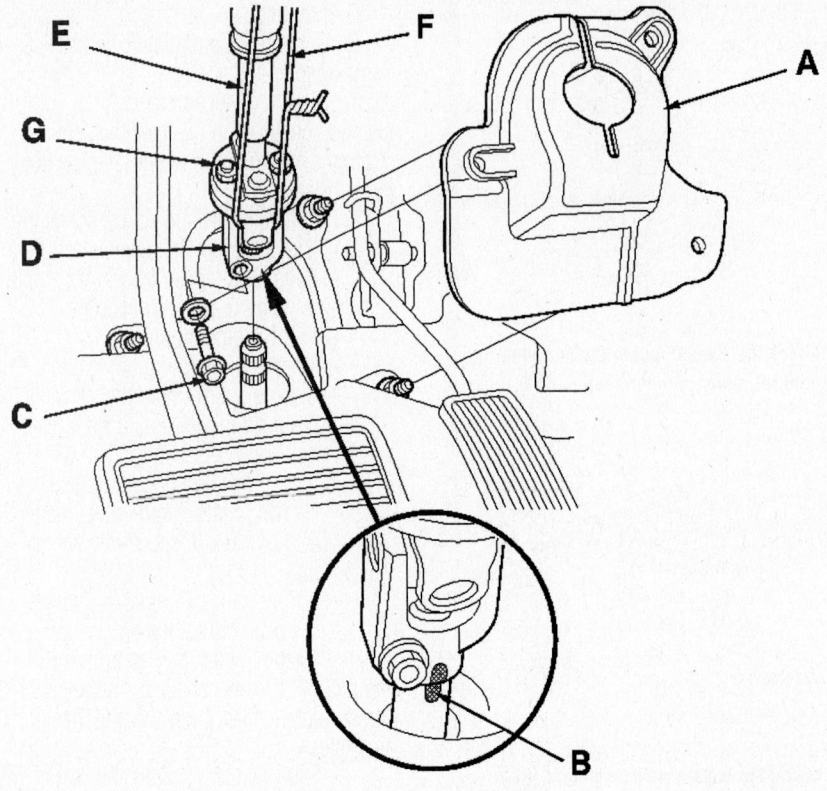

Make a reference mark across the steering joint and steering gearbox pinion shaft —2004–06 TL model

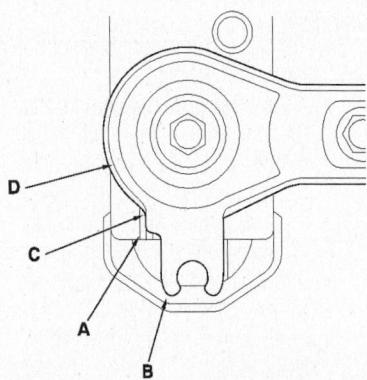

Make the appropriate reference lines at both ends of the sub-frame that line up— 2004–06 TL model

at both ends of the sub-frame that line up with the edge of the stiffeners.

46. Support the sub-frame with the front sub-frame adapter and a jack

47. Remove the front suspension sub-frame stays and front suspension sub-frame.

48. Remove the front engine mount upper bracket.

49. Remove the transmission lower front mount and the transmission lower rear mount.

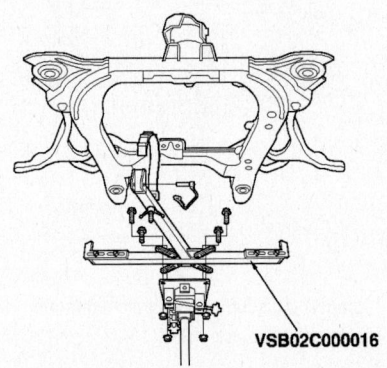

Attach an engine sub-frame adapter tool to the sub-frame—2004–06 TL model

50. Remove the clutch cover.

51. Place the transmission jack under the transmission, and remove the transmission lower mounting bolts.

52. Pull the transmission away from the engine until the transmission mainshaft clears the clutch pressure plate, then lower the transmission on the transmission jack.

To install:

53. Check that the two dowel pins are installed in the clutch housing.

54. Apply super high temp urea grease

to the release fork and the release bearing. Install the release fork, the release bearing, and the boot.

55. Place the transmission on the transmission jack, and raise it to the engine level.

56. Install the transmission lower mounting bolts.

57. Install the clutch cover

58. Install the front engine mount upper bracket.

59. Support the subframe with the front subframe adapter and a jack

60. Install the front suspension subframe (A) and front suspension subframe stays (B).

61. Align the reference marks (A) with edge (B) of both rear stiffer (C), and tighten the rear subframe mounting bolts, then the front bolts, and tighten the stiffener bolts to the specified torque.

62. Install the middle subframe mounting bolts.

63. Install the transmission lower front mount (A) and the transmission lower rear mount (B).

64. Install the transmission lower mount mounting nuts

65. Install the rear engine lower mount mounting bolts.

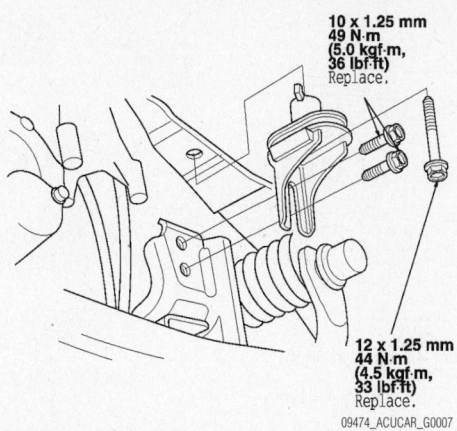

10 x 1.25 mm
49 N·m
(5.0 kgf·m,
36 lbf·ft)
Replace.

12 x 1.25 mm
44 N·m
(4.5 kgf·m,
33 lbf·ft)
Replace.

09474_ACUCAR_G0007

Sub-frame mid-mounts bolt specifications—2004–06 TL model with a manual transmission

66. Connect the power steering pressure switch connector.

67. Connect the return line to the power steering fluid reservoir.

68. Install the tie-rod ball joint.

69. Install the front wheels, and set them in the straight ahead position.

70. Connect the steering joint to the steering gearbox pinion shaft by aligning the reference mark, and remove the wire from the joint yoke.

71. Install the steering joint cover.

72. Install the intermediate shaft.

73. Install the left and right driveshaft inboard joints.

74. Install the knuckle onto the lower arm.

75. Install the damper fork.

76. Connect the front stabilizer

77. Install the exhaust pipe A and new gaskets (B).

78. Install the splash shield

79. Install the under cover

80. Lay the fender cover over the fender, then install the transmission mount bracket (A) and ground cable (B).

81. Remove the engine hanger and tool illustrated from the engine.

82. Install the transmission upper mounting bolts.

83. Install the engine rear mount stop.

84. Install the engine front mount mounting bolt (A) and the engine front mount stop (B).

85. Apply super high temp urea grease to the end of the cylinder rod. Install the slave cylinder. Be careful not to bend the clutch line.

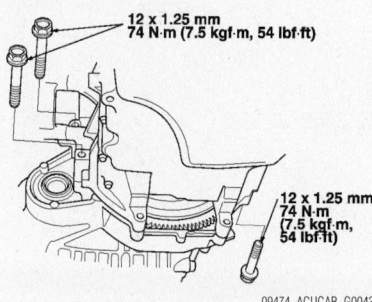

12 x 1.25 mm
74 N·m (7.5 kgf·m, 54 lbf·ft)

12 x 1.25 mm
74 N·m
(7.5 kgf·m,
54 lbf·ft)

09474_ACUCAR_G0043

Install the transmission lower bolts— 2004–06 TL model with a manual transmission

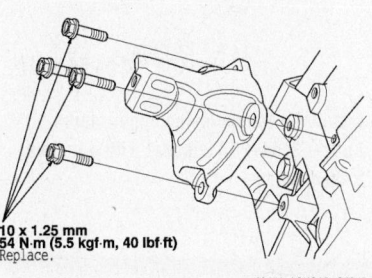

10 x 1.25 mm
54 N·m (5.5 kgf·m, 40 lbf·ft)
Replace.

09474_ACUCAR_G0045

Install the front engine mount bolts— 2004–06 TL model with a manual transmission

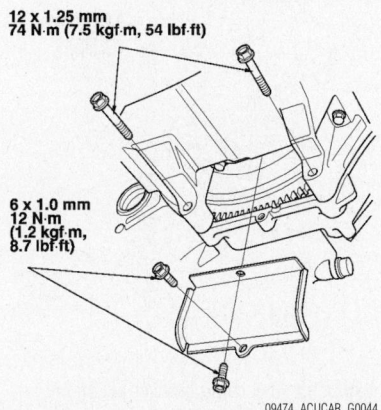

12 x 1.25 mm
74 N·m (7.5 kgf·m, 54 lbf·ft)

6 x 1.0 mm
12 N·m
(1.2 kgf·m,
8.7 lbf·ft)

09474_ACUCAR_G0044

Install the clutch cover bolts—2004–06 TL model with a manual transmission

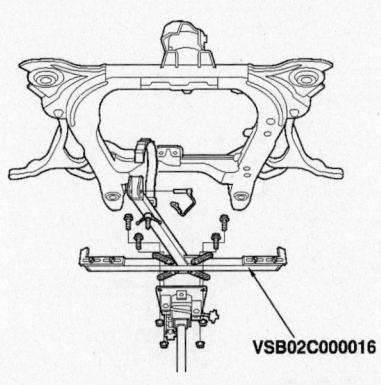

VSB02C000016

09474_ACUCAR_G0005

Attach an engine sub-frame adapter tool to the sub-frame—2004–06 TL model with a manual transmission

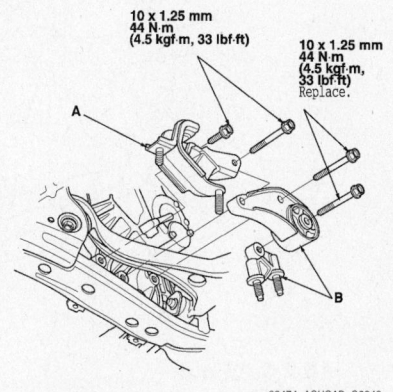

10 x 1.25 mm
44 N·m
(4.5 kgf·m, 33 lbf·ft)

10 x 1.25 mm
44 N·m
(4.5 kgf·m,
33 lbf·ft)
Replace.

09474_ACUCAR_G0046

Install the transmission lower front mount (A) and transmission lower rear mount bolts—2004–06 TL model with a manual transmission

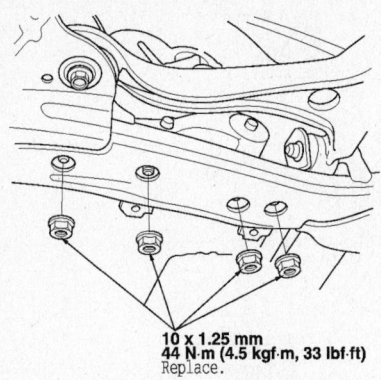

10 x 1.25 mm
44 N·m (4.5 kgf·m, 33 lbf·ft)
Replace.

09474_ACUCAR_G0047

Install the transmission lower mount nuts—2004–06 TL model

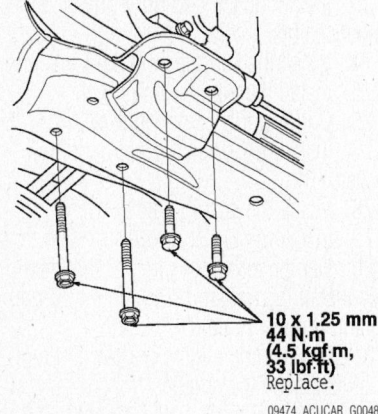

10 x 1.25 mm
44 N·m
(4.5 kgf·m,
33 lbf·ft)
Replace.

09474_ACUCAR_G0048

Install rear engine lower mount bolts—2004–06 TL model with a manual transmission

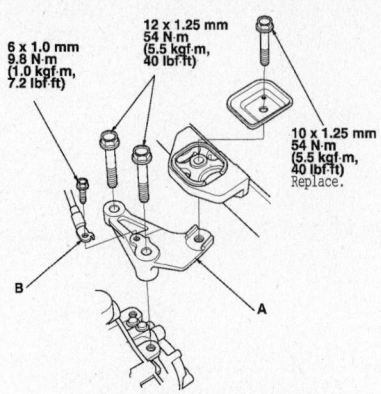

6 x 1.0 mm
9.8 N·m
(1.0 kgf·m,
7.2 lbf·ft)

12 x 1.25 mm
54 N·m
(5.5 kgf·m,
40 lbf·ft)

10 x 1.25 mm
54 N·m
(5.5 kgf·m,
40 lbf·ft)
Replace.

B
A

09474_ACUCAR_G0050

Transmission mount bracket (A) and ground cable (B) torque specs—2004–06 TL model with a manual transmission

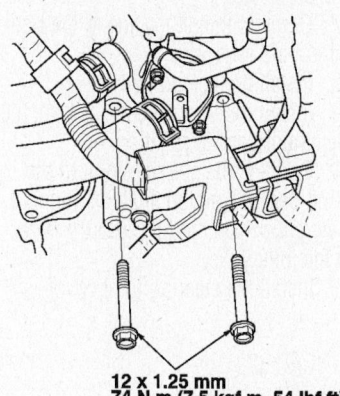

12 x 1.25 mm
74 N·m (7.5 kgf·m, 54 lbf·ft)

09474_ACUCAR_G0051

Install the transmission upper mount bolts—2004–06 TL model with a manual transmission

86. Connect the vacuum hose.
87. Install the cable bracket and cables.
88. Apply a light coat of super high temp urea grease to the cable ends, and install new cotter pins.
89. Install the harness bracket.

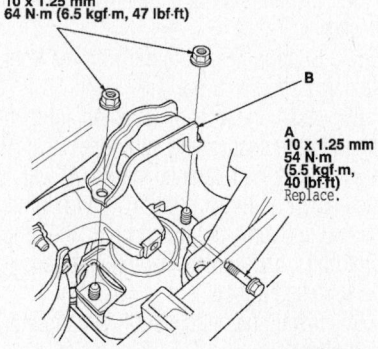

10 x 1.25 mm
64 N·m (6.5 kgf·m, 47 lbf·ft)

B

A

10 x 1.25 mm
54 N·m
(5.5 kgf·m,
40 lbf·ft)
Replace.

09474_ACUCAR_G0052

Install the front engine mount bolts (A) the front mount stop (B) and the mounting nuts (C)—2004–06 TL model with a manual transmission

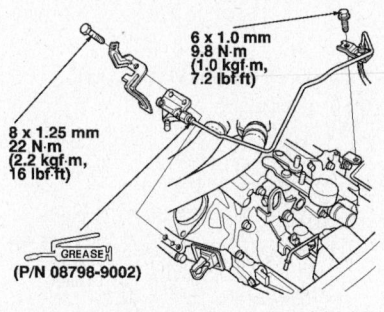

6 x 1.0 mm
9.8 N·m
(1.0 kgf·m,
7.2 lbf·ft)

8 x 1.25 mm
22 N·m
(2.2 kgf·m,
16 lbf·ft)

GREASE
(P/N 08798-9002)

09474_ACUCAR_G0053

Slave cylinder mounting—2004–06 TL model with a manual transmission

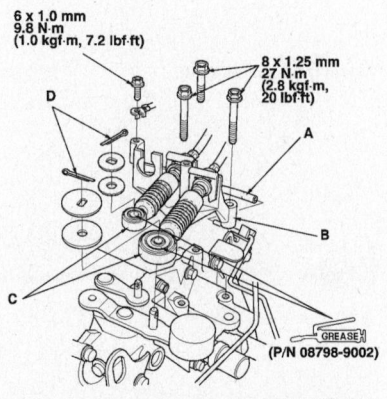

6 x 1.0 mm
9.8 N·m
(1.0 kgf·m, 7.2 lbf·ft)

8 x 1.25 mm
27 N·m
(2.8 kgf·m,
20 lbf·ft)

D
A
B
C

GREASE
(P/N 08798-9002)

09474_ACUCAR_G0054

Exploded view cable bracket assembly, cables and the retainer torque values—2004–06 TL model with a manual transmission

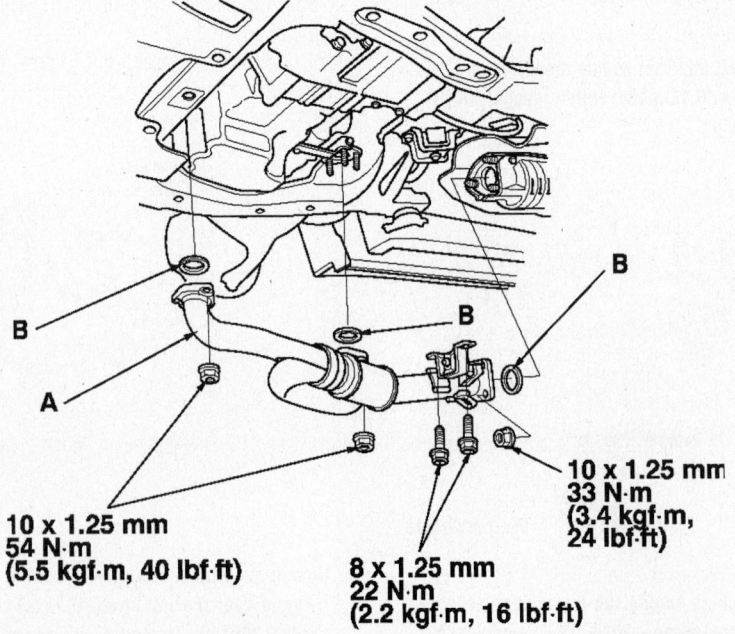

B
B
B
A

10 x 1.25 mm
33 N·m
(3.4 kgf·m,
24 lbf·ft)

10 x 1.25 mm
54 N·m
(5.5 kgf·m, 40 lbf·ft)

8 x 1.25 mm
22 N·m
(2.2 kgf·m, 16 lbf·ft)

09474_ACUCAR_G0049

Exhaust pipe torque specifications—2004–06 TL model with a manual transmission

TSX MODELS

➡ This procedure requires the use of the following special tools, or their equivalents: Engine hanger/adapter VSB02C000015, Engine support hanger A & Reds AAR-T-12566 and Subframe adapter VSB02C000016.

1. Before servicing the vehicle, refer to the precautions section.

2. Note the radio security code and the radio presets.

3. Remove or disconnect the following:

4. Drain the transaxle fluid.

5. Remove or disconnect the following:
   - Negative battery cable, then the positive battery cable
   - Air cleaner housing assembly
   - Battery base
   - Clutch slave cylinder, but do NOT bend the clutch line. Do NOT depress the clutch pedal once the slave cylinder is removed.
   - Cable bracket and cables from the top of the transaxle housing
   - Countershaft speed sensor connector
   - Back-up light switch connector
   - Reverse solenoid lockout connector
   - Secondary Heater Oxygen Sensor (HO2S) connector and the bracket
   - Engine front mount stop and the engine mount front mounting bolt
   - Engine rear mount stop
   - Right and left side steering gearbox mounting bolts
   - Transaxle upper mounting bolts

6. Attach Engine hanger/adapter VSB02C000015, or equivalent to the threaded holes in the cylinder head.

7. Install the engine support hanger to the vehicle, then attach the hook to the special tool.

8. Remove or disconnect the following:
   - Transaxle mount bracket and ground cable. Do NOT drop the mount collar.
   - Splash shield
   - Exhaust pipe "A"
   - Front stabilizer link
   - Damper fork
   - Knuckle from the lower control arm
   - Inboard joint of the driveshaft
   - Intermediate shaft
   - Heat shield
   - Power steering line bracket mounting bolt
   - Power steering line from the subframe. There are 2 clamps and 1 mounting bracket bolt.
   - Rear engine mount lower mounting bolts

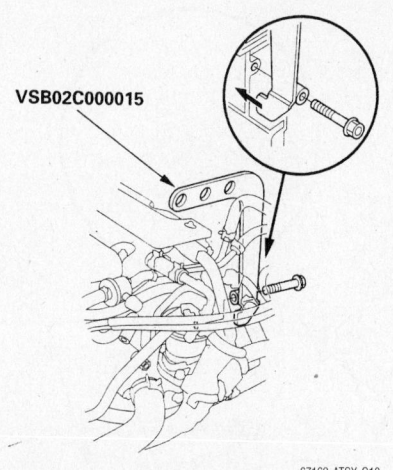

67162-ATSX-G13

Attach Engine hanger/adapter VSB02C000015, or equivalent to the threaded holes in the cylinder head—TSX models

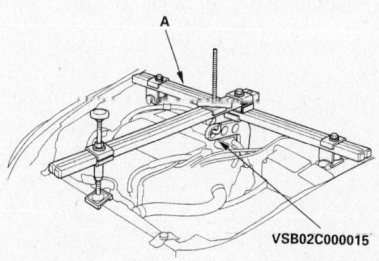

67162-ATSX-G14

Install the engine support hanger to the vehicle, then attach the hook to the special tool—TSX models

   - Transaxle lower mount mounting nuts
   - Middle subframe mounting bolts

9. Matchmark the ends of the subframe to the edge of the stiffeners.

10. Support the subframe with the subframe adapter tool and a suitable jack.

11. Suspend the steering gearbox with a suitable support, then remove the front suspension subframe stays and subframe.

12. Remove or disconnect the following:
   - Front engine mount upper bracket
   - Shift cable bracket, rear engine mount upper mounting bolt and rear engine mount
   - Rear engine mount upper bracket
   - Clutch cover

13. Place a transmission jack under the transaxle, then remove the lower transaxle mounting bolts.

14. Pull the transaxle away from the engine until the mainshaft clears the clutch pressure plate, then carefully lower the transaxle using the jack.

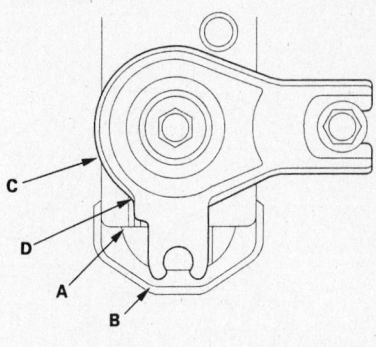

67162-ATSX-G15

Matchmark (A) the ends of the subframe (B) to the edge (C) of the stiffeners (D)—TSX models

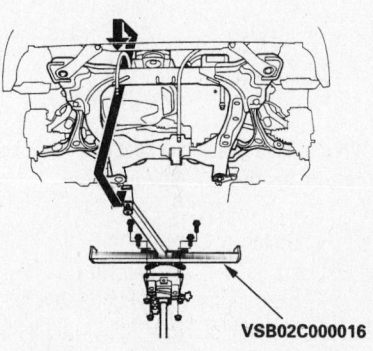

VSB02C000016

67162-ATSX-G16

Support the subframe with the subframe adapter tool and a suitable jack —TSX models

15. Remove or disconnect the following:
   - Transmission lower front mount and lower rear mount
   - Release fork boot from the clutch housing
   - Release fork from the clutch housing by squeezing the release fork set spring with pliers
   - Release bearing

To install:

16. Make sure the 2 dowel pins are installed in the clutch housing.

17. Apply super high temp urea grease (P/N 087989002) to the release fork and release bearing, then install the release fork, release bearing and boot.

18. Make sure the mount bracket collars are installed in the transaxle, then install the transmission lower front and rear mounts. Tighten the front mount bolts to 43 ft. lbs. (59 Nm) and the lower rear mount bolts to 33 ft. lbs. (45 Nm).

19. Place the transaxle on the transaxle jack and raise it to engine level.

20. Install or connect the following:
   - Transmission mounting lower bolts and tighten to 47 ft. lbs. (64 Nm)

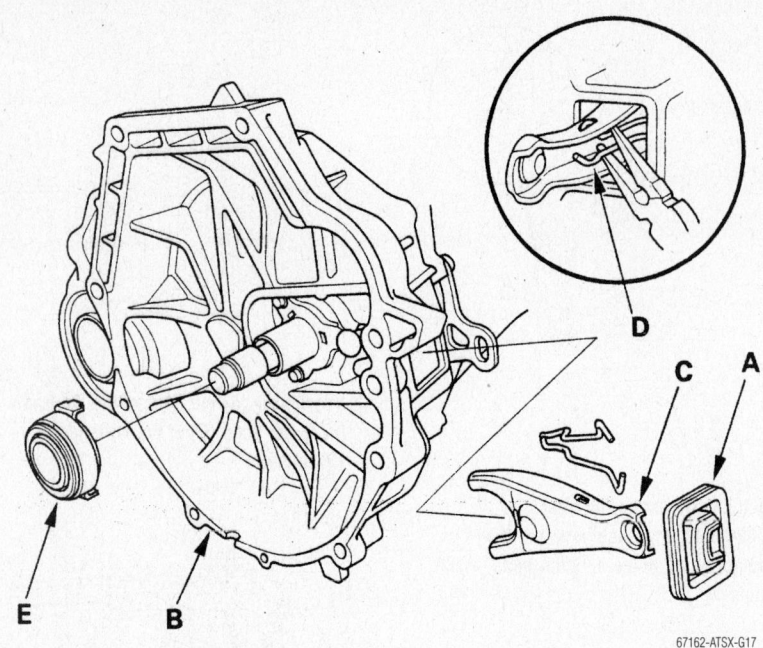

67162-ATSX-G17

View of the release fork boot (A), clutch housing (B), release fork (C), set spring (D) and release bearing (E)—TSX models

- Clutch cover and tighten the retainers to 33 ft. lbs. (45 Nm)
- Rear engine mount upper bracket and tighten NEW bolts to 65 ft. lbs. (90 Nm)
- Rear engine mount
- New rear engine mount upper mounting bolt and tighten to 40 ft. lbs. (54 Nm)
- Shift cable bracket and tighten the bolt to 16 ft. lbs. (22 Nm)
- Front engine mount upper bracket and tighten NEW bolts to 47 ft. lbs. (64 Nm)

21. Support the subframe with the subframe adapter tool and a jack.

22. Install the front suspension subframe and subframe stays.

23. Align the matchmarks with the edge of both rear stiffeners and tighten the rear subframe mounting bolts, then the front bolts, as shown in the accompanying figure.

24. Install or connect the following:
- New middle subframe mounting bolts. Torque the side (short) bolts to 36 ft. lbs. (49 Nm) and the bottom (long) bolt to 33 ft. lbs. (44 Nm).
- Transaxle lower mount mounting nuts and tighten to 33 ft. lbs. (45 Nm)
- New rear engine mount lower mounting bolts and tighten to 33 ft. lbs. (45 Nm)
- Power steering line to the subframe

with the 2 clamps and bolt. Torque the bolt to 7.2 ft. lbs. (9.8 Nm).
- Power steering line bracket mounting bolt and tighten to 7.2 ft. lbs. (9.8 Nm)
- Heat shield. Tighten the retainers to 7.2 ft. lbs. (9.8 Nm).
- Intermediate shaft

- Driveshaft inboard joint
- Knuckle onto the lower arm
- Damper fork
- Front stabilizer
- Exhaust pipe "A" with new gaskets. Torque the bolt to 16 ft. lbs. (22 Nm) and the nut to 25 ft. lbs. (33 Nm).
- Splash shield

25. Make sure the mount bracket collars are installed in the transaxle, then install the transaxle mount bracket and ground cable. Torque the mount bracket bolts to 40 ft. lbs. (54 Nm) and the ground cable bolt to 7.2 ft. lbs. (9.8 Nm).

26. Install or connect the following:
- Transmission upper mounting bolts and tighten to 47 ft. lbs. (64 Nm)
- New steering gearbox mounting bolts. Tighten the right side bolts to 28 ft. lbs. (38 Nm) and the left side bolts to 43 ft. lbs. (59 Nm).
- Engine rear mount stop and tighten the nuts to 51 ft. lbs. (69 Nm)
- New engine mount front mounting bolt and tighten to 40 ft. lbs. (54 Nm)
- Engine front mount stop and tighten the nuts to 58 ft. lbs. (78 Nm)
- Bracket and secondary HO2S sensor connector

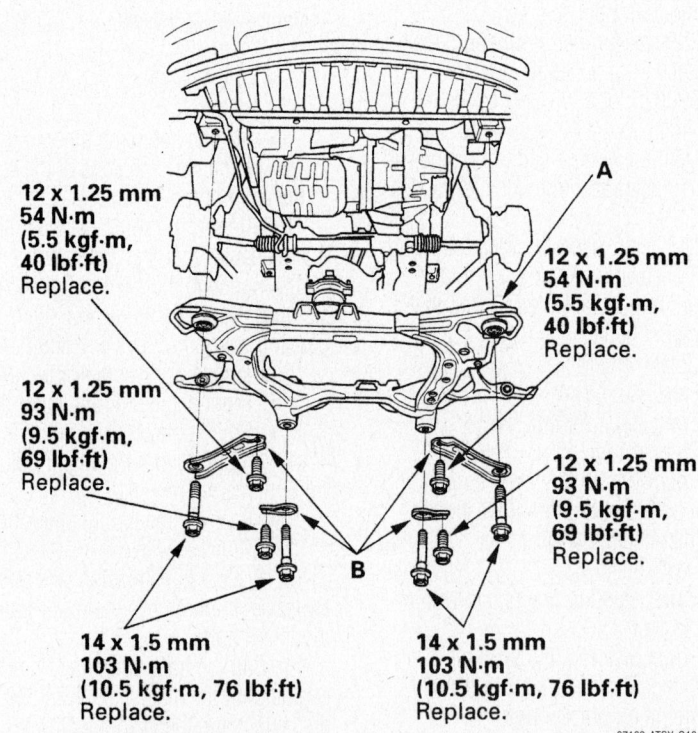

12 x 1.25 mm
54 N·m
(5.5 kgf·m, 40 lbf·ft)
Replace.

12 x 1.25 mm
54 N·m
(5.5 kgf·m, 40 lbf·ft)
Replace.

12 x 1.25 mm
93 N·m
(9.5 kgf·m, 69 lbf·ft)
Replace.

12 x 1.25 mm
93 N·m
(9.5 kgf·m, 69 lbf·ft)
Replace.

14 x 1.5 mm
103 N·m
(10.5 kgf·m, 76 lbf·ft)
Replace.

14 x 1.5 mm
103 N·m
(10.5 kgf·m, 76 lbf·ft)
Replace.

Subframe bolt tightening specifications—TSX models

67162-ATSX-G18

- Reverse lockout solenoid, back-up light switch and countershaft speed sensor connectors
- Cable bracket and cables. Tighten the bolts to 20 ft. lbs. (27 Nm). Apply a light coat of super high temp grease to the cable ends, then install new cotter pins.

27. Apply super high temp urea grease to the end of the cylinder rod, then install the slave cylinder. Tighten the slave cylinder mounting bolts to 16 ft. lbs. (22 Nm) and the fluid line bolt to 7.2 ft. lbs. (9.8 Nm). Do not bend the clutch line.

- Battery base
- Air cleaner housing
- Battery

28. Fill the transaxle with fluid.

29. Connect the positive and negative battery cables.

30. Program the station presets, if necessary.

31. Test drive and check clutch operation and front wheel alignment.

Automatic

CL MODEL

1. Before servicing the vehicle, refer to the precautions section.
2. Shift the transmission into **P**.
3. Drain the transmission fluid.
4. Remove or disconnect the following:
   - Both battery cables
   - Battery and the tray
   - Intake air duct
   - Transmission oil cooler hoses
   - Starter motor
   - Transmission ground cable
   - Shift control solenoid valve connectors
   - Clutch pressure switch electrical connector
   - Mainshaft speed sensor electrical connector
   - Clutch pressure control solenoid valve electrical connector
   - Lock–up control solenoid valve electrical connector
   - Countershaft speed sensor electrical connector
   - Gear position switch connector
   - Vehicle Speed (VSS) sensor without disconnecting the hoses
   - Front mount nut
   - Engine cover
   - Splash shield
   - Strut forks
   - Ball joints from the control arms
   - Radius rods from the control arms
   - Both halfshafts
   - Front beam

5. Raise the transmission with a jack to take the pressure off of the mounts.
   - Lower rear mount
   - Intermediate shaft
   - Shift cable holder
   - Shift cable cover
   - Shift cable with the control lever
   - Torque converter cover
   - Drive plate bolts
   - Engine stiffener
   - Front mount bracket
   - Transmission-to-engine bolts
   - Transmission from the vehicle

To install:

6. Installation is the reverse of the removal procedure, while using the following torque values:
   - Transmission mounting bolts: 47 ft. lbs. (64 Nm)
   - Front mount bracket bolts: 28 ft. lbs. (38 Nm)
   - Engine stiffener bolts: 28 ft. lbs. (38 Nm)
   - Drive plate bolts: 20 ft. lbs. (26 Nm)
   - Torque converter cover bolts: 108 inch lbs. (12 Nm)
   - Control lever bolt: 10 ft. lbs. (14 Nm)
   - Cable cover bolts: 16 ft. lbs. (22 Nm)
   - Shift cable holder bolts: 84 inch lbs. (9.5 Nm)
   - Intermediate shaft bolts: 30 ft. lbs. (39 Nm)
   - Lower transmission mount bolts: 28 ft. lbs. (38 Nm)
   - Front beam 10mm bolts: 28 ft. lbs. (38 Nm)
   - Front beam 12mm bolts: 47 ft. lbs. (64 Nm)
   - Front beam 14mm bolts: 76 ft. lbs. (103 Nm)
   - Lower transmission mount nuts: 28 ft. lbs. (38 Nm)
   - Strut fork pinch bolts: 32 ft. lbs. (43 Nm)
   - Strut fork through bolts to 47 ft. lbs. (64 Nm)
   - Ball joint nuts: 36–43 ft. lbs. (49–59 Nm)
   - Radius rod bolts: 132 ft. lbs. (179 Nm)
   - Front mount nut: 40 ft. lbs. (54 Nm)
   - VSS sensor bolts: 20 ft. lbs. (26 Nm)
   - Transmission ground cable bolt: 20 ft. lbs. (26 Nm)
   - Starter motor bolts: 33 ft. lbs. (44 Nm)

2002–04 RL MODELS

1. Before servicing the vehicle, refer to the precautions section.
2. Disconnect the negative, then the positive battery cables.
3. Shift the transaxle into **P**.
4. Drain the transmission.
5. Remove or disconnect the following:
   - Strut brace
   - Control box
   - Transaxle sub–harness connectors, and remove the sub–harness clamp
   - 3 bolts securing the transaxle dipstick pipe bracket
   - Upper transaxle mounting bolts

6. Pull the carpet back under the passenger seat to expose the secondary Heated Oxygen (HO$_2$S) sensor connector. Detach the connector and push it out from the inside of the vehicle.

7. Remove or disconnect the following:
   - Transmission stop collars
   - Front exhaust pipe from the vehicle
   - HO$_2$S sensor wiring harness cover and grommet
   - Catalytic converter
   - Exhaust heat shield from the floor of the vehicle
   - Transaxle oil cooler hoses
   - Shift solenoid valve electrical connector
   - Shift cable cover from the transaxle
   - Shift cable holder from the holder base
   - Control lever from the control shaft
   - Transaxle dipstick pipe from the torque converter housing
   - Range switch connector
   - Lower plate from under the steering gear, then re–install the 2 steering gear mounting bolts
   - Shift cable guide bracket from the transaxle beam

8. Raise the transmission slightly to take the weight off of the mounts.
   - Transaxle beam
   - Rear transaxle mount bracket and the mount
   - Exhaust pipe bracket
   - Sealing bolt from the differential
   - Extension shaft from the differential
   - Transmission-to-engine bolts
   - Engine stiffener
   - Torque converter covers
   - Drive plate bolts
   - Transmission from the vehicle

To install:

9. Installation is the reverse of the removal procedure, while using the following torque values:

- Drive plate bolts: Step 1: 108 inch lbs. (12 Nm). Step 2: 20 ft. lbs. (26 Nm)
- Torque converter cover bolts: 108 inch lbs. (12 Nm)
- Transaxle housing 8mm bolts: 16 ft. lbs. (22 Nm)
- Transaxle housing 12mm bolts: 47 ft. lbs. (64 Nm)
- Engine stiffener bolts: 16 ft. lbs. (22 Nm)
- Transaxle beam bolts: 28 ft. lbs. (38 Nm)
- Rear transaxle mount bracket bolts: 28 ft. lbs. (38 Nm)
- Rear transaxle mount bolts: 40 ft. lbs. (54 Nm)
- Shift cable guide bolt: 84 inch lbs. (9.5 Nm)
- Differential sealing bolt: 58 ft. lbs. (78 Nm)
- Lower plate bolts: 28 ft. lbs. (38 Nm)
- Steering gear bolts: 43 ft. lbs. (59 Nm)
- Shift control lever nut: 12 ft. lbs. (16 Nm)
- Shift cable holder bolts: 108 inch lbs. (12 Nm)
- Shift cable cover bolts: 108 inch lbs. (12 Nm)
- Exhaust heat shield bolts: 84 inch lbs. (9.5 Nm)
- Catalytic converter nuts: 25 ft. lbs. (33 Nm)
- Front exhaust pipe manifold nuts: 40 ft. lbs. (54 Nm)
- Transmission stop collar bolts: 28 ft. lbs. (38 Nm)
- Control box mounting bolts: 108 inch lbs. (12 Nm)
- Strut brace bolts: 16 ft. lbs. (22 Nm)

2005–06 RL MODELS

Make sure you have the customer's radio and navigation anti-theft codes, and write down the radio presets.

1. Before servicing the vehicle, refer to the precautions section.

2. Remove the battery trim, left upper fender trim, cowl cover upper trim, right upper fender trim, upper grille cover and intake manifold cover.

3. Remove the windshield wiper arms and cowl cover.

4. Disconnect the support strut from the pivot ball on both sides.

5. Raise and secure the hood in a vertical position. Remove the right side pivot ball and install it into the lower threaded hole, then reattach the support strut.

➡**Do not attempt to close the hood with the support strut in the vertical position; it will damage the support strut and the hood.**

6. Set the wheels in the straight ahead position, and lock the steering wheel.

7. Drain the power steering system fluid from the reservoir.

8. Make sure the ignition switch is OFF. Disconnect the negative terminal from the battery, then disconnect the positive terminal.

9. Remove the battery hold-down bracket, and remove the battery cover, battery and battery tray.

10. Remove the air intake duct and air cleaner housing.

11. Remove the battery base.

12. Remove the splash shield.

13. Drain the automatic transmission fluid (ATF).

14. Reinstall the drain plug and a new sealing washer.

15. Remove the strut brace.

16. Disconnect the steering joint

17. Remove the power steering pump outlet line from the power steering pump and remove the hose from its clamp.

18. Disconnect the power steering pressure switch connector.

19. Remove the starter.

20. Disconnect the solenoid harness connector.

21. Remove the bolt securing the radiator hose clamp.

22. Disconnect the vacuum hose from the vacuum line and remove the vacuum line bolt.

23. Remove the nuts securing the shift cable bracket.

24. Remove the spring clip/washer and the control pin, then separate the shift cable end from the control lever.

25. Disconnect the A/T clutch pressure control solenoid valve A connector, A/T clutch pressure control solenoid valve B connector, and 4th clutch transmission fluid pressure switch connectors, and remove the harness cover mounting bolt.

26. Remove the harness cover mounting bolt, and remove the harness clamps from the clamp brackets.

27. Disconnect the A/T clutch pressure control solenoid valve C connector, and remove the connector bracket and the harness clamp bracket.

28. Disconnect the transmission range switch connector and remove the harness clamp from the clamp bracket.

29. Remove the transmission ground cable.

30. Disconnect the output shaft speed sensor connector, input shaft speed sensor connector, 3rd clutch transmission fluid pressure switch connector and ATF temperature sensor connector.

31. Disconnect the 2nd clutch transmission fluid pressure switch connector and remove the vacuum line bolt.

32. Remove the ATF cooler hoses from the ATF cooler lines. Turn the ends of the cooler hoses up to prevent ATF from flowing out, then plug the cooler hoses and lines.

33. Remove the connector bracket from the engine front cylinder head, use the bracket bolt hole to attach the engine balancer bar front arm.

34. Disconnect the solenoid connector, disconnect the vacuum tube from the joint, then remove the bracket from the engine rear cylinder head; use the bracket bolt hole to attach the engine balancer bar rear arm.

35. Install the engine balancer bar # VSB02C000019, attach the front arm to the front cylinder head with the spacer and the 10 mm bolt, and attach the rear arm to the rear cylinder head with the 8 mm bolt.

36. Install the engine support hanger # AAR-T-12566 to the vehicle, and attach the hook to the engine balancer bar slot. Tighten the wing nut by hand, and lift and support the engine.

37. Remove the front mount stop and the front mount bolt.

38. Remove the exhaust pipe.

39. Make a reference mark across the propeller shaft and the transfer companion flange.

40. Separate the propeller shaft from the transfer companion flange.

41. Remove the propeller shaft.

42. Insert a 6 mm Allen wrench in the top of the ball joint pin and remove the nuts, then separate the stabilizer link from the lower arms.

43. Remove the damper pinch bolts, damper fork bolts and self-locking nuts, knuckle holder bolts and nuts. Remove the damper forks.

44. Remove the cotter pins and nuts and separate the steering tie-rod end ball joints from the knuckle.

45. Remove the bolt securing the transfer assembly breather tube bracket and disconnect the breather tube from the breather pipe on the transfer assembly.

46. Remove the transfer assembly from the transmission.

47. Remove the steering gearbox heat shield.

48. Loosen the hose clamp bolt, then disconnect the power steering fluid hose from the line at the right front of the subframe.

49. Remove the ATF cooler hose from the ATF cooler. Turn the end of the cooler hose up to prevent ATF from flowing out, then plug the hose.

50. Remove the torque converter cover and remove the drive plate bolts while rotating the crankshaft pulley.

51. Remove the transmission lower mount bolts.

52. Disconnect the power steering angle sensor connector.

53. Remove the shift cable bracket on the steering gearbox stiffener.

54. Remove the rear mount stop, and remove the rear mount bolt.

55. Attach the front subframe adapter # VSB02C000016 to the subframe by looping the strap over the front of the subframe, then secure the strap.

56. Raise the jack and line up the slots in the arms with the bolt holes on the corner of the jack base, then tighten the bolts.

57. Remove the four bolts securing the stiffeners and the four bolts securing the front subframe, then lower the front subframe.

58. Remove the transmission lower mount.

59. Pry out the driveshafts from the differential and the intermediate shaft.

60. Remove the exhaust manifold bracket and the heat shield.

61. Remove the intermediate shaft.

62. Remove the upper transmission housing mounting bolts.

63. Remove the front mount bracket.

64. Remove the sensor harness from the harness clamp.

65. Remove the transmission housing mounting bolt using a socket 22 mm in length.

66. Remove the rear transmission housing mounting bolts.

67. Lower the transmission by loosening the wing nut of the engine support hanger, and tilt the engine just enough for the transmission to clear the side frame.

68. Place a jack under the transmission

69. Remove the lower transmission housing mounting bolts.

70. Slide the transmission away from the engine to remove it from the vehicle.

To install:

71. Place the transmission on the jack, and raise it to engine level.

72. Attach the transmission to the engine, and install the lower transmission housing mounting bolts. Tighten to 47 ft. lbs. (64 Nm).

73. Install the rear and upper transmission housing mounting bolts. Tighten to 47 ft. lbs. (64 Nm).

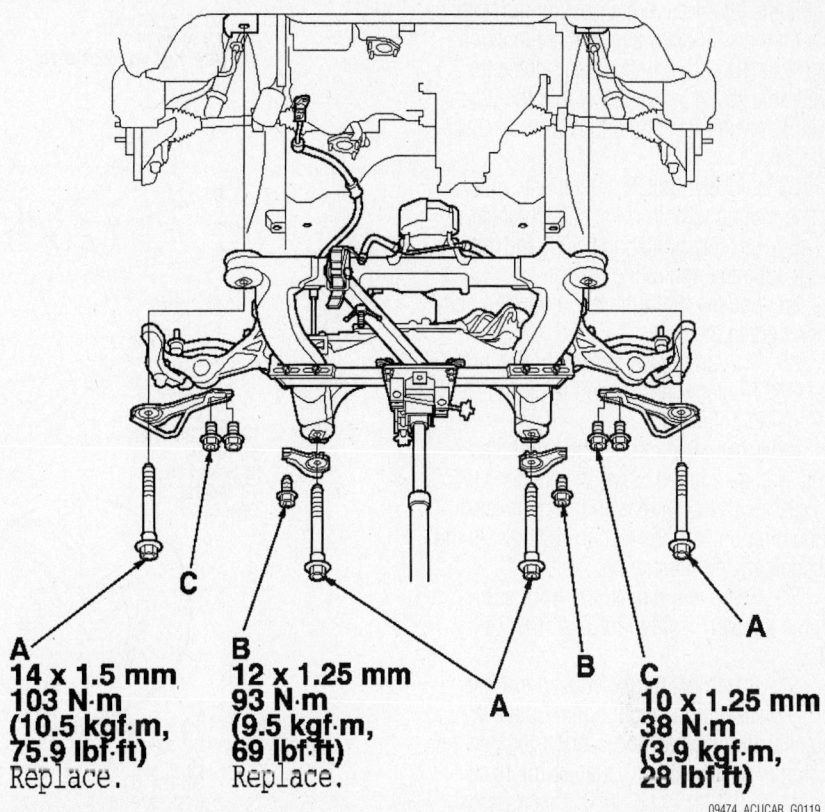

A
**14 x 1.5 mm
103 N·m
(10.5 kgf·m,
75.9 lbf·ft)
Replace.**

B
**12 x 1.25 mm
93 N·m
(9.5 kgf·m,
69 lbf·ft)
Replace.**

C
**10 x 1.25 mm
38 N·m
(3.9 kgf·m,
28 lbf·ft)**

09474_ACUCAR_G0119

Sub-frame bolt locations and torque specifications—2005–06 RL model

74. Install the transmission housing mounting bolt. Tighten to 47 ft. lbs. (64 Nm).

75. Install the sensor harness clamps on the clamp bracket.

76. Install the front mount bracket with the new mounting bolts. Tighten to 40 ft. lbs. (54 Nm).

77. Install the new set ring on the intermediate shaft and install the intermediate shaft in the differential.

➡**While installing the intermediate shaft in the differential, be sure not to allow dust or other foreign particles to enter the transmission.**

78. Install the exhaust manifold bracket and heat shield.

79. Install the new set ring on the left driveshaft, then install the left driveshaft in the differential. While installing the driveshaft in the differential, be sure not to allow dust or other foreign particles to enter the transmission. Install the right driveshaft over the intermediate shaft.

➡**Turn the right and left steering knuckle fully outward, and slide the driveshaft and intermediate shaft into the differential and intermediate shaft until you feel its set ring engage the side gear.**

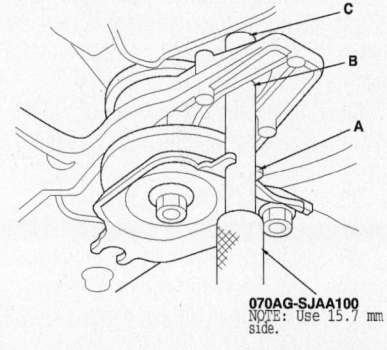

070AG-SJAA100
NOTE: Use 15.7 mm side.

09474_ACUCAR_G0120

Insert the special tool through the positioning slot (B) on the rear stiffener, through the positioning hole (C) on the subframe—2005–06 RL model

80. Support the front subframe with the front subframe adapter and a jack, and lift it up to the body.

81. Loosely install the new subframe mounting bolts (A), new rear stiffener mounting bolts (B), front stiffener mounting bolts (C), and front and rear stiffeners. Refer to the illustration for bolt location and its torque specification.

82. Loosely tighten the right rear subframe mounting bolt (A); insert the special tool through the positioning slot (B) on the

rear stiffener, through the positioning hole (C) on the subframe, and into the positioning hole (D) on the body, then tighten the subframe mounting bolt. Refer to the illustration for bolt location and its torque specification.

83. Loosely tighten the left rear subframe mounting bolt in the same manner.

84. Loosely tighten the right and left front subframe mounting bolts.

85. Loosen the right rear mounting bolt, then tighten the bolt.

86. Tighten the left rear mounting bolt.

87. Tighten the right and left front mounting bolts.

88. Check that the positioning holes and slot are aligned using the special tool.

89. Tighten the rear and front stiffener mounting bolts to the torque specified in the accompanying illustration.

90. Remove the jack and front subframe adapter. Tighten to 33 ft. lbs. (44 Nm).

91. Attach the torque converter to the drive plate with the eight bolts. Rotate the crankshaft pulley as necessary to tighten the bolt to 4 ft. lbs. (6 Nm), then to the final torque of 9 ft. lbs. (12 Nm), in a crisscross pattern. After tightening the last bolt, check that the crankshaft rotates freely.

92. Install the torque converter cover.

93. Connect the power steering fluid hose to the line at the right front of the front subframe, and secure the hose with its hose clamp.

94. Connect the ATF cooler hose to the ATF cooler, and secure the hose with the clip.

95. Install the steering gearbox heat shield

96. Install the dowel pin in the transmission, and install the transfer assembly on the transmission. Tighten the bolts to 38 ft. lbs. (51 Nm).

97. Secure the transfer assembly breather tube bracket on the transfer assembly with the bolt, and install the breather tube over the breather pipe. If the breather tube was removed from the clamp, install the tube at the dot on the clamp.

98. Install the propeller shaft.

99. Install the propeller shaft to the transfer companion flange by aligning the previously made reference mark. Tighten the bolts to 54 ft. lbs. (74 Nm).

100. Install exhaust pipe with the new self-locking nuts and the new gaskets.

101. Install the damper forks (A) with the damper pinch bolts, the new damper fork bolts, and the new self-locking nuts. Attach the knuckle holder (B) to the knuckle, and secure the knuckle holder with the bolt and

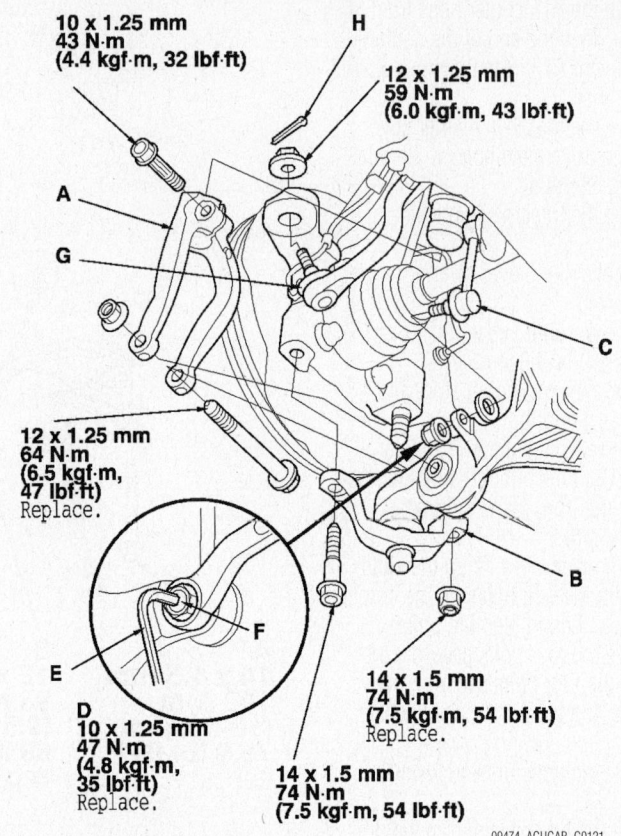

Exploded view of the damper fork assembly and related components—2005–06 RL model

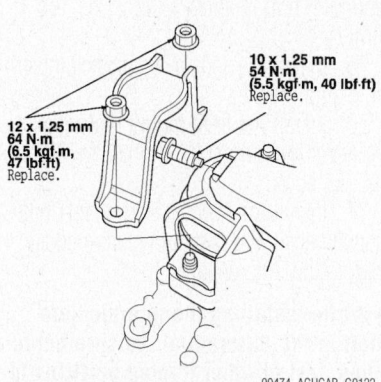

09474_ACUCAR_G0122

Exploded view of the rear mount—2005–06 RL model

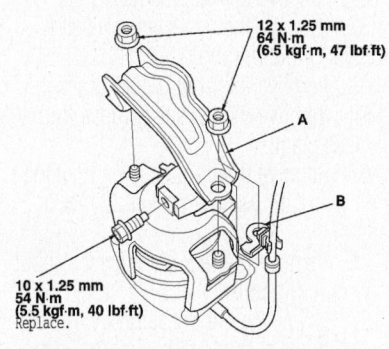

09474_ACUCAR_G0123

Exploded view of the front mount—2005–06 RL model

new nut. Refer to the illustration for bolt location and its torque specification.

102. Install the stabilizer links (C) to the lower arms, and install the nuts (D). Insert a 6 mm Allen wrench (E) in the ball joint pin (F), and tighten the nuts. Refer to the illustration for bolt location and its torque specification.

103. Install the tie-rod end ball joints (G) to each knuckle with the nuts and the new cotter pins (H). Refer to the illustration for bolt location and its torque specification.

104. Install the new rear mount bolt, and

install the rear mount stop with new nuts. Refer to the illustration for bolt location and its torque specification.

105. Install the new front mount bolt, and install the front mount stop and vacuum hose clamp. Refer to the illustration for bolt location and its torque specification.

106. Remove the engine support hanger and engine hanger balancer bar.

107. Install the connector bracket on the engine front cylinder head. Tighten the bolt to 33 ft. lbs. (44 Nm).

108. Install the clamp brackets on the

engine rear cylinder head, and connect the vacuum tube and solenoid connector. Tighten the bolt to 20 ft. lbs. (26 Nm).

109. Install the remaining components in the reverse order of removal.

RSX MODELS

1. Before servicing the vehicle, refer to the precautions section.
2. Drain the transaxle fluid.
3. Remove or disconnect the following:
   - Negative battery cable, then the positive battery cable
   - Air cleaner housing assembly with intake air duct
   - Starter motor cables and holder
   - Transaxle ground cable from the transaxle hanger
   - Lockup control solenoid valve connector
   - Shift control solenoid valve connector
   - Harness clamp on the lockup control solenoid harness from the harness stay
   - Vehicle Speed (VSS) sensor electrical connector
   - Main shaft speed sensor electrical connector
   - Counter shaft speed sensor electrical connector
   - Upper transaxle mounting bolts
   - Splash shield
   - Front wheels
   - Ball joints from the lower control arms
   - Right strut fork
   - Both halfshafts from the vehicle
   - Heated Oxygen (HO2S) sensor connector
   - Front exhaust pipe from the vehicle
   - Intermediate shaft
   - Shift cable cover
   - Shift cable by removing the control lever

✲✲ WARNING

Do not bend the shift control cable when removing it.

   - Right front mount/bracket
   - End of the throttle control cable from the throttle control drum
   - Transmission oil cooler hoses from the joint pipes
   - Engine stiffener
   - Torque converter cover
   - 8 drive plate bolts
4. Support the transaxle.
   - Transaxle mounting bolts and rear engine mounting bolts

   - Transaxle from the engine
   - Starter motor from the transaxle (if necessary)

To install:

5. Installation is the reverse of the removal procedure, while using the following torque values:
   - Starter motor bolts: 33 ft. lbs. (45 Nm)
   - Transaxle mounting bolts: 47 ft. lbs. (64 Nm)
   - Rear engine mount bolts: 87 ft. lbs. (118 Nm)
   - Transmission mount bolt: 54 ft. lbs. (74 Nm)
   - Transmission mount nuts: 47 ft. lbs. (64 Nm)
   - Drive plate bolts: 108 inch lbs. (12 Nm)
   - Torque converter cover 6mm bolts: 108 inch lbs. (12 Nm)
   - Torque converter cover 10mm bolt: 33 ft. lbs. (44 Nm)
   - Engine stiffener transaxle bolt: 32 ft. lbs. (43 Nm)
   - Engine stiffener engine bolts: 17 ft. lbs. (24 Nm)
   - Right front mount/bracket 12mm bolts: 47 ft. lbs. (64 Nm)
   - Right front mount/bracket 10mm bolts: 33 ft. lbs. (44 Nm)
   - Control lever bolt: 10 ft. lbs. (14 Nm)
   - Intermediate shaft mounting bolts: 29 ft. lbs. (39 Nm)
   - Front exhaust pipe manifold nuts: 40 ft. lbs. (54 Nm)
   - Catalytic converter nuts: 16 ft. lbs. (22 Nm)
   - Strut fork pinch bolt: 32 ft. lbs. (43 Nm)
   - Strut fork lower bolt: 47 ft. lbs. (64 Nm)
   - Splash shield bolts: 108 inch lbs. (12 Nm)

2002–03 TL MODELS

1. Before servicing the vehicle, refer to the precautions section.
2. Shift the transmission into **P**.
3. Drain the transmission fluid.
4. Remove or disconnect the following:
   - Both battery cables
   - Battery and the tray
   - Intake air duct
   - Transmission oil cooler hoses
   - Starter motor
   - Transmission ground cable
   - Shift control solenoid valve connectors
   - Clutch pressure switch electrical connector

   - Mainshaft speed sensor electrical connector
   - Clutch pressure control solenoid valve electrical connector
   - Lock–up control solenoid valve electrical connector
   - Countershaft speed sensor electrical connector
   - Gear position switch connector
   - Vehicle Speed (VSS) sensor without disconnecting the hoses
   - Front mount nut
   - Engine cover
   - Splash shield
   - Strut forks
   - Ball joints from the control arms
   - Radius rods from the control arms
   - Both halfshafts
   - Front beam

5. Raise the transmission with a jack to take the pressure off of the mounts.
   - Lower rear mount
   - Intermediate shaft
   - Shift cable holder
   - Shift cable cover
   - Shift cable with the control lever
   - Torque converter cover
   - Drive plate bolts
   - Engine stiffener
   - Front mount bracket
   - Transmission-to-engine bolts
   - Transmission from the vehicle

To install:

6. Installation is the reverse of the removal procedure, while using the following torque values:
   - Transmission mounting bolts: 47 ft. lbs. (64 Nm)
   - Front mount bracket bolts: 28 ft. lbs. (38 Nm)
   - Engine stiffener bolts: 28 ft. lbs. (38 Nm)
   - Drive plate bolts: 20 ft. lbs. (26 Nm)
   - Torque converter cover bolts: 108 inch lbs. (12 Nm)
   - Control lever bolt: 10 ft. lbs. (14 Nm)
   - Cable cover bolts: 16 ft. lbs. (22 Nm)
   - Shift cable holder bolts: 84 inch lbs. (9.5 Nm)
   - Intermediate shaft bolts: 30 ft. lbs. (39 Nm)
   - Lower transmission mount bolts: 28 ft. lbs. (38 Nm)
   - Front beam 10mm bolts: 28 ft. lbs. (38 Nm)
   - Front beam 12mm bolts: 47 ft. lbs. (64 Nm)
   - Front beam 14mm bolts: 76 ft. lbs. (103 Nm)

- Lower transmission mount nuts: 28 ft. lbs. (38 Nm)
- Strut fork pinch bolts: 32 ft. lbs. (43 Nm)
- Strut fork through bolts to 47 ft. lbs. (64 Nm)
- Ball joint nuts: 36–43 ft. lbs. (49–59 Nm)
- Radius rod bolts: 132 ft. lbs. (179 Nm)
- Front mount nut: 40 ft. lbs. (54 Nm)
- VSS sensor bolts: 20 ft. lbs. (26 Nm)
- Transmission ground cable bolt: 20 ft. lbs. (26 Nm)
- Starter motor bolts: 33 ft. lbs. (44 Nm)

2004–06 TL MODELS

The following tools are required to perform this procedure:
- Engine support hanger, A and Reds AAR-T-12566
- Engine hanger balance bar VSB02C000019
- Front sub-frame adapter VSB02C000016

1. Before servicing the vehicle, refer to the precautions section.

2. Set the wheels in the straight ahead position.

3. Lock the steering wheel.

4. Disconnect the support strut from both sides of the pivot ball (bolted to the hood). Secure the hood in a vertical position. Remove the right side pivot ball and install it into the lower threaded hole, then reattach the support strut.

➡Do not attempt to close the hood with the support strut in the vertical position; it will damage the support strut and the hood.

5. Set the wheels in the straight ahead position.

6. Lock the steering wheel.

7. Drain the power steering system fluid from the reservoir.

8. Remove the steering joint cover.

9. Make a reference mark across the steering joint and steering gearbox pinion shaft.

10. Remove the steering joint bolt, and disconnect the steering joint by removing the steering joint toward the steering column.

11. Hold the slider shaft on the column with a peace of wire between the joint yoke on the slider shaft to the joint yoke on the upper shaft.

12. Remove the covers in the following order:

- Left side engine compartment cover
- Left rear engine compartment cover
- Right fender trim
- Right side engine compartment cover
- Right rear engine compartment cover
- Left fender trim
- Front bulkhead cover
- Intake manifold cover

13. Remove the harness clamp, two 6 mm bolts and the strut brace.

14. Remove the power steering pump outlet line from the pump, and remove the hose from its clamp.

15. Remove the transmission under cover.

16. Remove the splash shield.

17. Drain the transmission fluid.

18. Reinstall the drain plug with a new sealing washer.

19. Disconnect the battery cables, negative cable first.

20. Remove the battery hold-down bracket, the battery and battery tray.

21. Remove the resonator cover and resonator.

22. Remove the intake air duct and air cleaner housing.

23. Remove the bolts retaining the battery base from under the vehicle, and in the engine compartment, then remove the base.

24. Remove the starter.

25. Remove the transmission upper mount bracket and bracket plate.

26. Remove the transmission sub-harness connector from its bracket and disconnect it.

27. Disconnect the input shaft speed sensor connector and 4th clutch transmission fluid pressure switch connector.

28. Disconnect the vacuum hose from the vacuum line.

29. Disconnect shift solenoid valve connectors and the automatic transmission clutch pressure control solenoid valve connector, then remove the harness clamps from the clamp brackets.

30. Disconnect the transmission clutch pressure control solenoid valve connectors, then remove the harness clamp from the clamp bracket.

31. Disconnect the connectors from the torque converter clutch solenoid valve and shift solenoid valve.

32. Remove the transmission range switch connector from its bracket, and disconnect it.

33. Disconnect the output shaft speed sensor connector, and remove the harness clamps from the brackets.

34. Disconnect the vacuum hose from the vacuum line.

35. Remove the harness clamps from the brackets.

36. Remove the harness cover from the bracket.

37. Remove the bolt retaining the bracket.

38. Remove the transmission ground cable, and disconnect the breather tube.

39. Unfasten the bolts retaining the ATF warmer and the bracket.

40. Remove the ATF warmer from the transmission housing. Cover the fluid passages on the transmission and ATF warmer with tape. Do not disconnect the hoses.

41. Remove the connector bracket from the engine front cylinder head; use the bracket bolt to attach engine balancer bar front arm

42. Remove the harness clamp bracket from the engine rear cylinder head; use the bracket bolt to attach engine balancer bar rear arm

43. Lift and support the engine with engine hanger and engine balancer bar. Attach the front arm to the front cylinder head with a spacer and the connector bracket 10 x 1.25 mm bolt. Attach the rear arm to the rear cylinder head with the harness clamp bracket 8 x 1.25 mm bolt.

44. Remove the front mount stop, and remove the front mount bolt.

45. Insert a 6 mm Allen wrench in the top of the ball joint pin, remove the nuts and separate the stabilizer link from the lower arms.

46. Separate the ball joints from the lower arms.

47. Separate the tie-rod end ball joints from the knuckles.

48. Remove exhaust pipe and its mount.

49. Remove the steering gearbox heat shield.

50. Remove the power steering fluid hose from its line on the front sub-frame.

51. Disconnect the power steering pressure switch connector.

52. Remove the torque converter cover and the drive plate bolts while rotating the crankshaft pulley.

53. Remove the engine-to-torque converter housing mounting bolts.

54. Remove the bolts attaching the shift cable holder and the shift cable cover.

➡To prevent damage to the control lever joint, remove the bolts retaining the holder before removing the bolts retaining the cover.

55. Remove the lock bolt retaining the selector control lever, then remove the shift cable and the control lever. Do not bend the shift cable excessively

56. Install a 6 x 1.0 x 14 mm bolt and nut on the shift cable cover, then reinstall the shift cable cover to the torque converter housing. If you do not perform this step, the bolt head of the cable cover may prevent you from removing the torque converter during transmission removal

57. Remove the rear mount base bracket bolts.

58. Remove the transmission lower mount nuts.

59. Remove the both mid-mounts.

60. Make the appropriate reference lines at both ends of the sub-frame that line up with the edge of the stiffeners.

61. Attach the tool VSB02C000016to the sub-frame by hanging the strap of the tool over the front of the sub-frame, then secure the strap with its stop.

62. Raise the jack and line up the slots in the arms with the bolt holes on the corner of the jack base, then attach them with bolts securely

63. Remove the four bolts retaining the stiffeners, the four bolts retaining the front sub-frame and lower the front sub-frame

64. Remove the transmission lower mounts.

65. Remove the driveshafts from the differential and intermediate shaft

66. Remove the exhaust manifold bracket and heat shield. Remove the intermediate shaft.

67. Coat all precision finished surfaces with clean engine oil, then tie plastic bags over both ends of driveshaft and intermediate shaft.

68. Place a jack under the transmission

69. Remove the transmission housing mounting bolts

70. Remove the bolt retaining the harness clamp bracket, and remove the front mount bracket.

71. Remove the transmission housing mounting bolts.

72. Slide the transmission away from the engine to remove it from the vehicle.

To install:

73. Place the transmission on the jack, and raise the transmission to the engine level.

74. Attach the transmission to the engine, then install the transmission housing mounting bolts.

75. Install the transmission housing mounting bolts and tighten to 47 ft. lbs. (64 Nm).

76. Remove the jack from the transmission

77. Install the front mount bracket with the new bolts, tighten the bolts to 40 ft. lbs. (54 Nm).

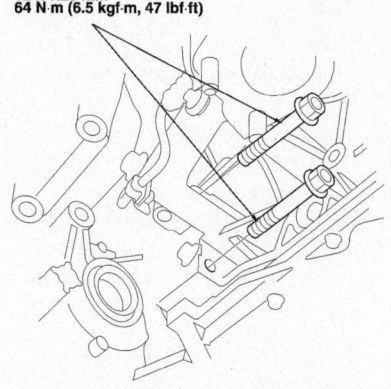

12 x 1.25 mm
64 N·m (6.5 kgf·m, 47 lbf·ft)

09474_ACUCAR_G0055

Rear transmission housing mounting bolt torque specifications—2004–06 TL model with an automatic transmission

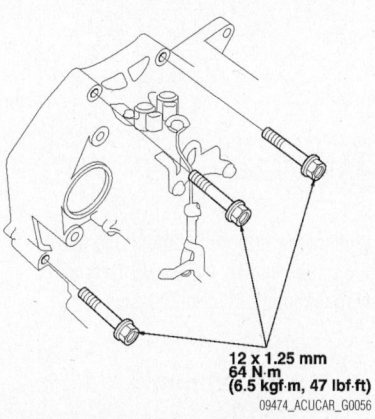

12 x 1.25 mm
64 N·m
(6.5 kgf·m, 47 lbf·ft)

09474_ACUCAR_G0056

Upper and front transmission housing mounting bolt torque specifications—2004–06 TL model with an automatic transmission

78. Install the harness clamp on the mount bracket and tighten to 16 ft. lbs. (22 Nm).

79. Install the engine-to-torque converter housing mounting bolts and tighten to 47 ft. lbs. (64 Nm).

80. Attach the torque converter to the drive plate with the eight bolts. Rotate the crankshaft pulley as necessary to tighten the bolts to 4.5 ft. lbs. (6 Nm), then to the final torque in a crisscross pattern to 9 ft. lbs. (12 Nm). After tightening the last bolt, check that the crankshaft rotates freely.

81. Install the torque converter cover.

82. Install the new set ring on the intermediate shaft.

83. Install the exhaust manifold bracket tighten the bolts to 20 ft. lbs. (26 Nm). Install the heat shield tighten the bolts to 20 ft. lbs. (26 Nm).

84. Install the new set ring on the left driveshaft, then install the left driveshaft in

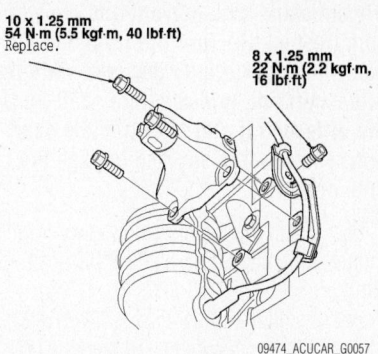

10 x 1.25 mm
54 N·m (5.5 kgf·m, 40 lbf·ft)
Replace.

8 x 1.25 mm
22 N·m (2.2 kgf·m, 16 lbf·ft)

09474_ACUCAR_G0057

Front mount bracket and harness clamp bolt torque specifications—2004–06 TL model with an automatic transmission

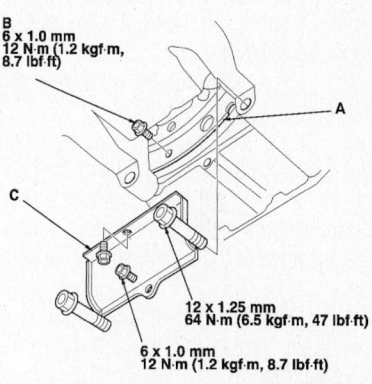

B
6 x 1.0 mm
12 N·m (1.2 kgf·m, 8.7 lbf·ft)

A

C

12 x 1.25 mm
64 N·m (6.5 kgf·m, 47 lbf·ft)

6 x 1.0 mm
12 N·m (1.2 kgf·m, 8.7 lbf·ft)

09474_ACUCAR_G0058

Engine-to-torque converter housing mounting bolt torque specifications—2004–06 TL model with an automatic transmission

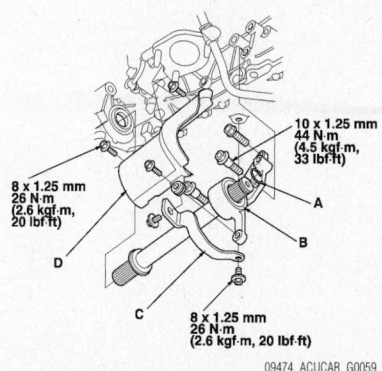

10 x 1.25 mm
44 N·m (4.5 kgf·m, 33 lbf·ft)

8 x 1.25 mm
26 N·m (2.6 kgf·m, 20 lbf·ft)

A

B

D

C

8 x 1.25 mm
26 N·m (2.6 kgf·m, 20 lbf·ft)

09474_ACUCAR_G0059

Install the new set ring (A) on the shaft (B) and tighten the exhaust manifold bracket (C) and heat shield (D) bolts—2004–06 TL model with an automatic transmission

the differential. While installing the driveshaft in the differential, make sure not to allow dirt or other particles to enter the transmission. Install the left driveshaft over the intermediate shaft.

➡**Clean the areas where the driveshaft and intermediate shaft contact the** transmission with solvent and dry with compressed air. Turn the right and left steering knuckle fully out, and slide the driveshaft and intermediate shaft into the differential and intermediate shaft until you feel the set ring engage the side gear.

85. Install the transmission lower front mount tighten the bolts to 33 ft. lbs. (44 Nm).

86. Install the transmission lower rear mount with the new bolts. Tighten the bolts to 33 ft. lbs. (44 Nm).

87. Support the front sub-frame with the tool and a jack, and lift it up to body.

88. Loosely install the new sub-frame mounting bolts, and new rear stiffener mounting bolts, and the front stiffener mounting bolts.

89. Loosely install both of the new both mid-mount mounting bolts.

90. Align the reference marks with edge of both rear stiffeners, and tighten the rear sub-frame mounting bolts, the front bolts and the stiffener bolts in that order to the torque shown in the accompanying illustration.

91. Tighten the mid-mount mounting bolts to the torque shown in the accompanying illustration.

92. Install the new rear mount bracket bolts to the torque shown in the accompanying illustration.

93. Install the new transmission lower mount nuts to the torque shown in the accompanying illustration.

94. Connect the power steering pressure switch connector.

95. Connect the power steering hose to the power steering line.

96. Install selector control lever on the

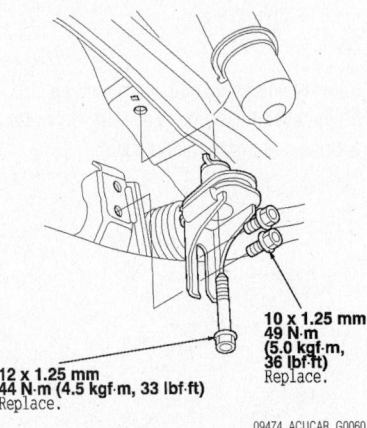

**12 x 1.25 mm
44 N·m (4.5 kgf·m, 33 lbf·ft)**
Replace.

**10 x 1.25 mm
49 N·m
(5.0 kgf·m,
36 lbf·ft)**
Replace.

09474_ACUCAR_G0060

Tighten the mid-mount mounting bolts to the torque shown—2004–06 TL model with an automatic transmission

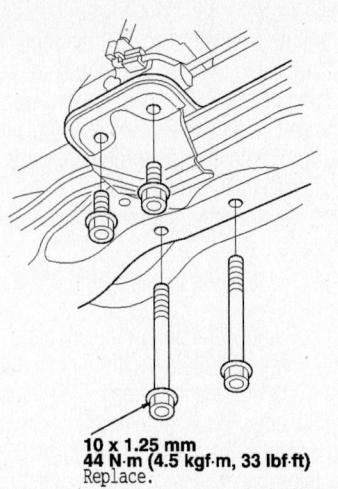

**10 x 1.25 mm
44 N·m (4.5 kgf·m, 33 lbf·ft)**
Replace.

09474_ACUCAR_G0061

Tighten the new rear mount bracket bolts to the torque shown—2004–06 TL model with an automatic transmission

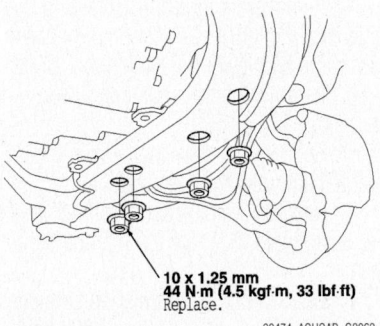

**10 x 1.25 mm
44 N·m (4.5 kgf·m, 33 lbf·ft)**
Replace.

09474_ACUCAR_G0062

Tighten the new transmission lower mount nuts to the specifications shown—2004–06 TL model with an automatic transmission

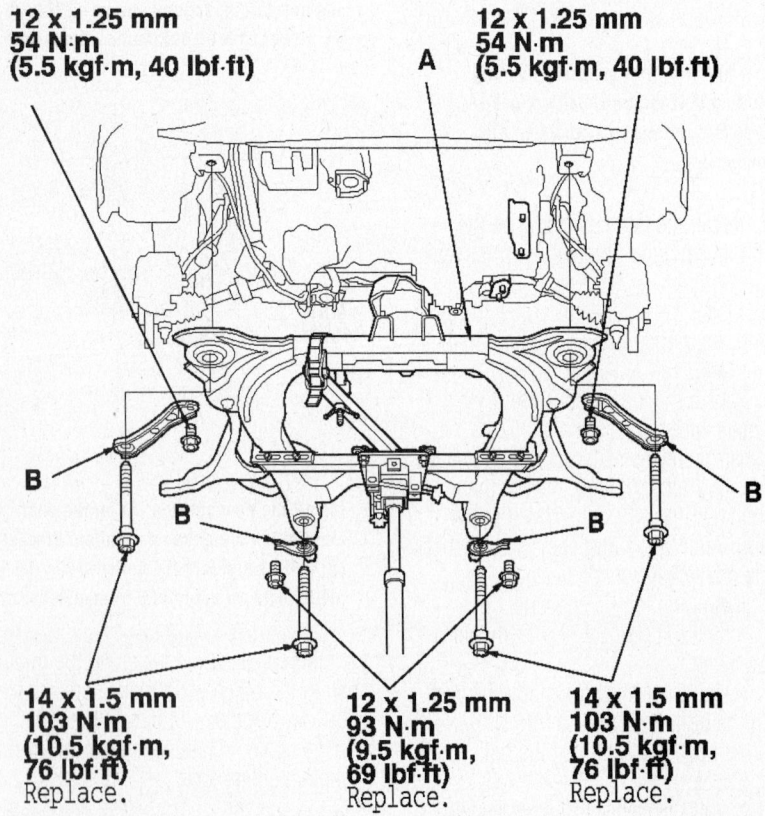

**12 x 1.25 mm
54 N·m
(5.5 kgf·m, 40 lbf·ft)**

A

**12 x 1.25 mm
54 N·m
(5.5 kgf·m, 40 lbf·ft)**

B   B

B   B

**14 x 1.5 mm
103 N·m
(10.5 kgf·m,
76 lbf·ft)**
Replace.

**12 x 1.25 mm
93 N·m
(9.5 kgf·m,
69 lbf·ft)**
Replace.

**14 x 1.5 mm
103 N·m
(10.5 kgf·m,
76 lbf·ft)**
Replace.

09474_ACUCAR_G0006

Sub-frame bolt locations and torque specifications—2004–06 TL model

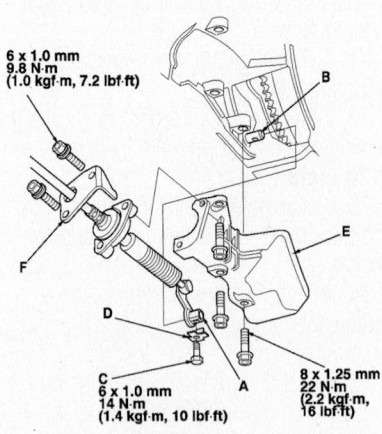

**6 x 1.0 mm
9.8 N·m
(1.0 kgf·m, 7.2 lbf·ft)**

B

**6 x 1.0 mm
14 N·m
(1.4 kgf·m, 10 lbf·ft)**

**8 x 1.25 mm
22 N·m
(2.2 kgf·m,
16 lbf·ft)**

E

F   D   C   A

09474_ACUCAR_G0063

Install selector control lever on the selector control shaft. Tighten all fasteners to the specifications shown—2004–06 TL model with an automatic transmission

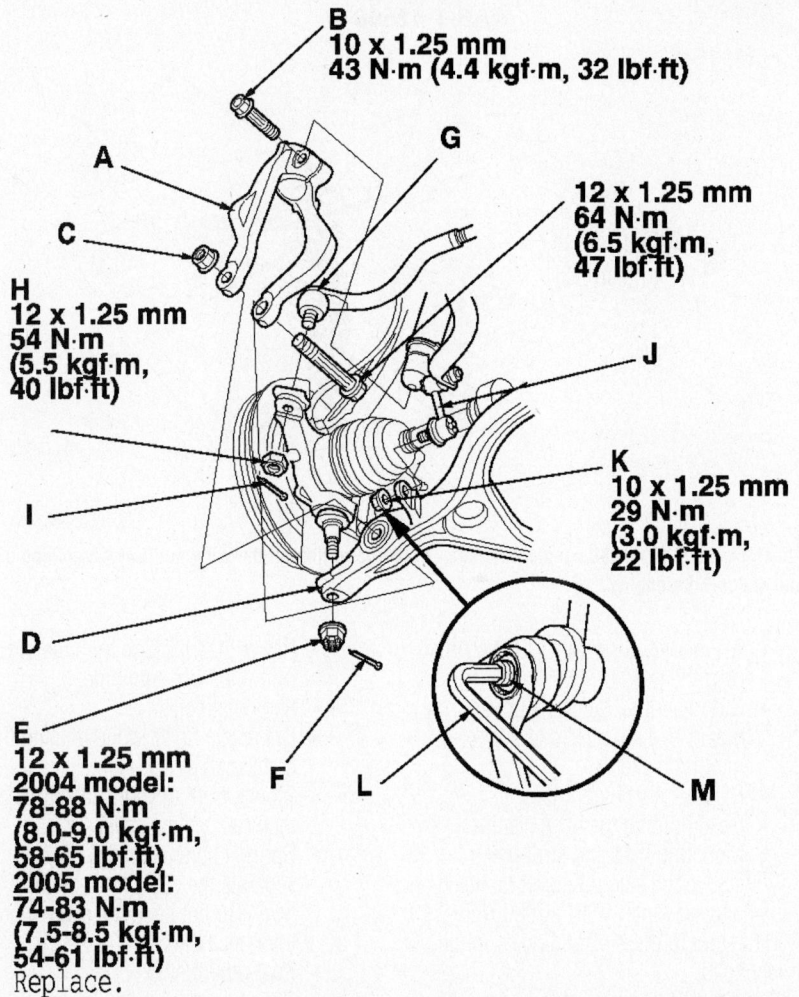

B
10 x 1.25 mm
43 N·m (4.4 kgf·m, 32 lbf·ft)

G

12 x 1.25 mm
64 N·m
(6.5 kgf·m,
47 lbf·ft)

A

C

H
12 x 1.25 mm
54 N·m
(5.5 kgf·m,
40 lbf·ft)

J

K
10 x 1.25 mm
29 N·m
(3.0 kgf·m,
22 lbf·ft)

I

D

E
12 x 1.25 mm
2004 model:
78-88 N·m
(8.0-9.0 kgf·m,
58-65 lbf·ft)
2005 model:
74-83 N·m
(7.5-8.5 kgf·m,
54-61 lbf·ft)
Replace.

F **L** **M**

09474_ACUCAR_G0064

Exploded view of the damper fork assembly and related components—2004–06 TL model with an automatic transmission

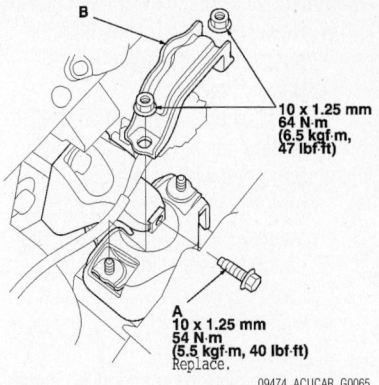

B

10 x 1.25 mm
64 N·m
(6.5 kgf·m,
47 lbf·ft)

A
10 x 1.25 mm
54 N·m
(5.5 kgf·m, 40 lbf·ft)
Replace.

09474_ACUCAR_G0065

Tighten the new front mount bolts (A) and the mount stop (B) to the specifications shown—2004–06 TL model with an automatic transmission

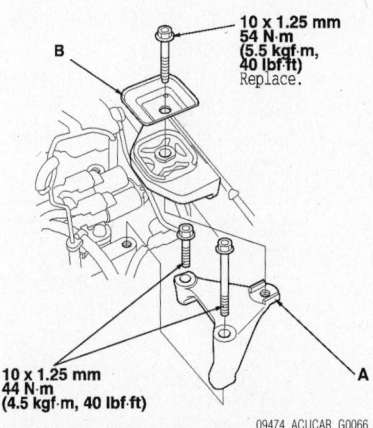

10 x 1.25 mm
54 N·m
(5.5 kgf·m,
40 lbf·ft)
Replace.

B

10 x 1.25 mm
44 N·m
(4.5 kgf·m, 40 lbf·ft)

A

09474_ACUCAR_G0066

Tighten the front mount stop transmission upper mount bracket (A) and new upper mount bolt and bracket (B) to the torque shown —2004–06 TL model with an automatic transmission

selector control shaft. Do not bend the shift cable excessively.

97. Install the lock bolt with a new lock washer, then bend the lock washer tab against the bolt head.

98. Install the shift cable cover, then secure the shift cable holder on the cover with the bolts.

➡**To prevent damage to the control lever joint, be sure install the shift cable holder after installing the shift cable cover to the torque converter housing.**

99. Install the remaining components in the reverse order of removal keeping in mind the following component torque specifications:

100. Install the damper forks (A) with damper pinch bolts (B) and the new self-locking nuts (C), then install the ball joints on the lower arms (D) with the ball joint nuts (E) and new cotter pins (F). Install the

tie-rod end ball joints (G) to each knuckle with the nuts (H) and new cotter pins (I). Install the stabilizer links (J) to the lower arms, and install the nuts (K). Insert a 6 mm Allen wrench (L) in the ball joint (M), and tighten the nuts. Refer to the accompanying illustration for component location and torque value.

101. Install the new front mount bolt to the torque shown in the accompanying illustration.

102. Install the front mount stop to the torque shown in the accompanying illustration.

103. Install the transmission upper mount bracket on the transmission with two bolts to the torque shown in the accompanying illustration.

104. Install the new upper mount bolt and bracket plate to the torque shown in the accompanying illustration.

105. Connect the steering joint to the steering gearbox pinion shaft by aligning

the reference mark, and remove the wire from the joint yoke. Tighten the bolt to 20 ft. lbs. (26 Nm).

TSX MODELS

➡**This procedure requires the use of the following special tools, or their equivalents: Engine hanger/adapter VSB02C000015, Engine support hanger A & Reds AAR-T-12566 and Front sub-frame adapter EQS02C00011.**

1. Before servicing the vehicle, refer to the precautions section.

2. Note the radio security code and the radio presets.

3. Remove or disconnect the following:

4. Remove the splash shield.

5. Drain the transaxle fluid. Reinstall the drain plug with a new washer and tighten the plug to 36 ft. lbs. (49 Nm).

6. Remove or disconnect the following:
- Negative battery cable, then the positive battery cable
- Battery tray
- Air cleaner housing assembly
- Battery base
- Equipped with an automatic transmission clutch pressure control solenoid valve A connector, 2nd clutch transmission fluid pressure switch connector
- Harness clams from the clamp brackets
- Transmission range switch connector from its bracket and disconnect it
- AF sensor connector from its bracket and disconnect it
- Mainshaft speed sensor connector
- Countershaft speed sensor connector and harness clamps from the clamp brackets
- 3rd clutch transmission fluid pressure switch connector
- Harness clamp from the clamp bracket
- Shift solenoid harness connector equipped with an automatic transmission clutch pressure control solenoid valve B and C connectors
- Harness clamp from the clamp bracket
- ATF cooler hoses from the lines. Plug the hoses and lines to prevent fluid from leaking out. Check for signs of leakage at the hose joints.
- ATF cooler hose from the hose clamp

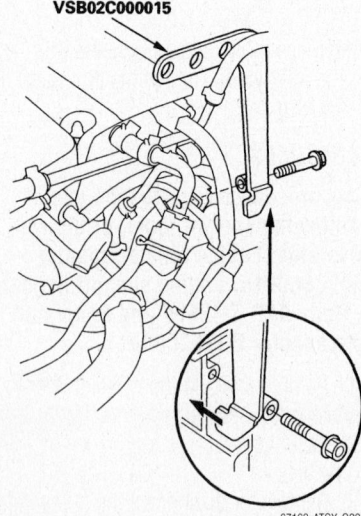

VSB02C000015

Attach Engine hanger/adapter VSB02C000015, or equivalent to the threaded holes in the cylinder head—TSX models

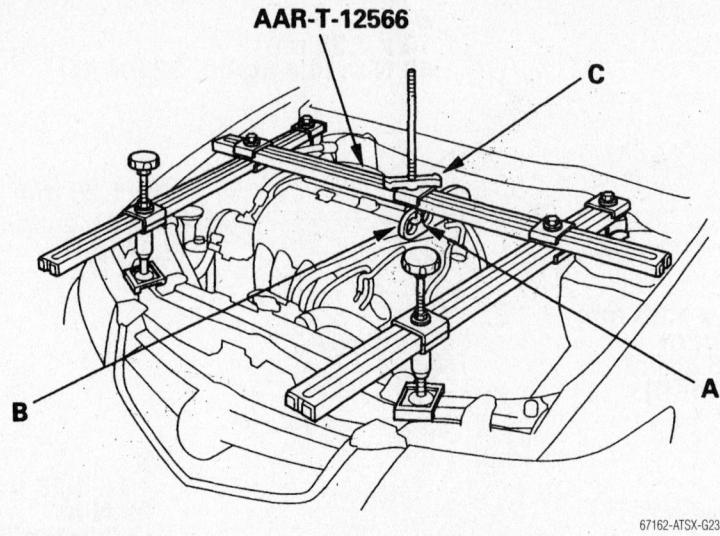

AAR-T-12566

67162-ATSX-G23

Attach the hook (A) to the special tool adapter (B), then tighten the wing nut (C) by hand and lift and support the engine

- ATF cooler hose from the line and plug the hose
- Ground cable, transaxle upper mount bracket plate and upper mount bracket
- Harness clamp and hose
- Bolts holding the hose clamps
- Mounting nuts and strut brace

7. Remove the clamp bracket to attach Engine hanger/adapter VSB02C000015. Attach the tool to the threaded holes in the cylinder head.

8. Install the engine support hanger to the vehicle, then attach the hook to the special tool.

9. Install the Engine support hanger AAR-T-12566, or equivalent to the vehicle. Attach the hook to the special tool adapter as shown in the accompanying figure. Tighten the wing nut by hand and lift and support the engine.

10. Remove or disconnect the following:
- Vacuum hose from the clamp and hose from the vacuum line
- Front mount stop and clamp bracket
- Front mount bolt

11. Insert a 6mm hex wrench in the top of the ball joint pin, remove the nuts, then separate the stabilizer link from the lower control arms.
- Cotter pins, castle nuts, damper pinch bolt, self-locking nut, bolt and damper forks, then separate the ball joints from the lower control arms
- Exhaust pipe "A" and its mount
- Steering gearbox heat shield and bolt securing the power steering fluid line bracket

- Power steering line from the clips from the front subframe
- Engine stiffener
- Driveplate bolts, while rotating the crankshaft pulley
- 3 bolts from the shift cable holder, then remove the cover
- Spring clip and control pin and separate the cable from the control lever. Do not bend the cable more than necessary.
- Rear mount stop
- Power steering fluid line bracket from the front subframe
- Steering gearbox mounting bolts and stiffener
- Steering gearbox mounting bracket bolts
- Damper and rear mount base bracket bolts
- Transmission lower mount nuts
- Both mid mounts

12. Matchmark the ends of the subframe to the edge of the stiffeners.

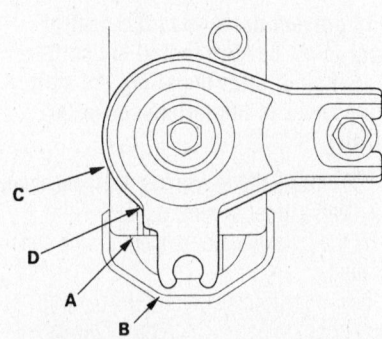

67162-ATSX-G15

Matchmark (A) the ends of the subframe (B) to the edge (C) of the stiffeners (D)—TSX models

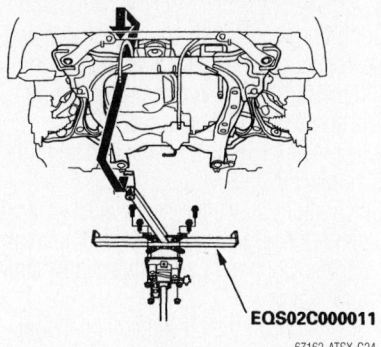

EQS02C000011

67162-ATSX-G24

Support the subframe with the subframe adapter tool and a suitable jack —TSX models

13. Attach the special tool to the subframe with hanging the hook of the special tool over the front of the subframe, then tighten the special tool screw.

14. Raise the jack and line up the slots in the arms with the bolt holes on the corner of the jack base, then attach them with the bolts securely.

15. Remove the 4 bolts holding the stiffeners and 4 bolts holding the front subframe. Lower the front subframe while sliding the steering gearbox out to clear the gearbox mounting bracket on the subframe.

16. Securely suspend the gearbox from the body with a piece of wire or rope.

17. Remove or disconnect the following:
- Transaxle lower mounts
- Driveshafts from the differential and intermediate shaft by prying them out
- Intermediate shaft

➡ **Coat all machined surfaces with clean engine oil, then put plastic bags over the driveshaft and intermediate shaft ends to protect them.**

- Rear mount base/bracket
- Rear mount bracket

18. Put a jack under the transmission.
- Transmission housing mounting bolts
- Front mount bracket
- Transmission housing mounting bolts from the front and rear of the transmission

19. Slide the transmission away from the engine to remove it from the vehicle.

20. Remove the torque converter. Check the drive plate and replace if necessary.

21. Inspect the transaxle lower front mount and lower rear mount. If the mount rubber is worn or damage, it must be replaced.

To install:

22. Installation is the reverse of the removal procedure, while using the following torque values:

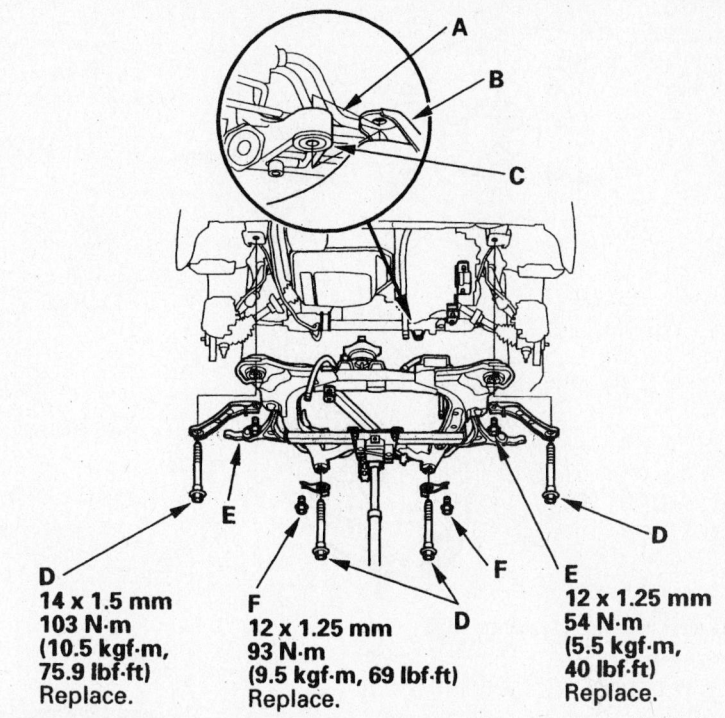

D
14 x 1.5 mm
103 N·m
(10.5 kgf·m,
75.9 lbf·ft)
Replace.

F
12 x 1.25 mm
93 N·m
(9.5 kgf·m, 69 lbf·ft)
Replace.

E
12 x 1.25 mm
54 N·m
(5.5 kgf·m,
40 lbf·ft)
Replace.

67162-ATSX-G25

Stiffener (E & F) and subframe (D) mounting bolts—TSX models

- Drive plate mounting bolts: 54 ft. lbs. (74 Nm)
- Transmission housing mounting bolts: 47 ft. lbs. (64 Nm)
- New front mount bracket bolts: 47 ft. lbs. (64 Nm)
- New rear mount bracket A bolts: 65 ft. lbs. (88 Nm)
- New rear mount brace bracket bolts: 40 ft. lbs. (54 Nm)
- Intermediate shaft bolts: 29 ft. lbs. (39 Nm)
- Transmission lower mount bolts: 33 ft. lbs. (45 Nm)
- New subframe mounting bolts: 75.9 ft. lbs. (103 Nm)
- Stiffener mounting bolts (after aligning matchmarks made during removal): Refer to accompanying figure
- Mid mount short bolts: 36 ft. lbs. (49 Nm)
- Mid mount long bolt 33 ft. lbs. (44 Nm)
- Rear mount base bracket short bolts: 16 ft. lbs. (22 Nm)
- Rear mount base bracket long bolts: 36 ft. lbs. (49 Nm)
- Transaxle lower mount nuts: 33 ft. lbs. (44 Nm)
- Steering gearbox mounting bolts: 43 ft. lbs. (60 Nm)
- Steering gearbox mounting bracket bolts: 28 ft. lbs. (38 Nm)

- Rear mount stop: 51 ft. lbs. (69 Nm)
- Torque converter-to-drive plate bolts (in 2 steps): 8.7 ft. lbs. (12 Nm)
- Engine stiffener: 33 ft. lbs. (45 Nm)
- Transaxle upper mount bracket bolts: 40 ft. lbs. (54 Nm)
- Transmission upper mount bracket plate: 40 ft. lbs. (54 Nm)

Clutch

REMOVAL & INSTALLATION

RSX Models

1. Before servicing the vehicle, refer to the precautions section.

2. Disconnect the negative battery cable.

3. Remove the transaxle assembly from the vehicle.

4. Insert a clutch alignment tool. Use a feeler gauge and measure the clearance between the pressure plate spring fingers and the clutch alignment disc. There should be a maximum of 0.02 in. (0.6mm) of clearance for a new pressure plate with 0.03 in. (0.8mm) limit for a used pressure plate. If the height is more than the service limit, replace the pressure plate.

5. Remove the clutch alignment disc.

6. Install a flywheel holder to aid in the

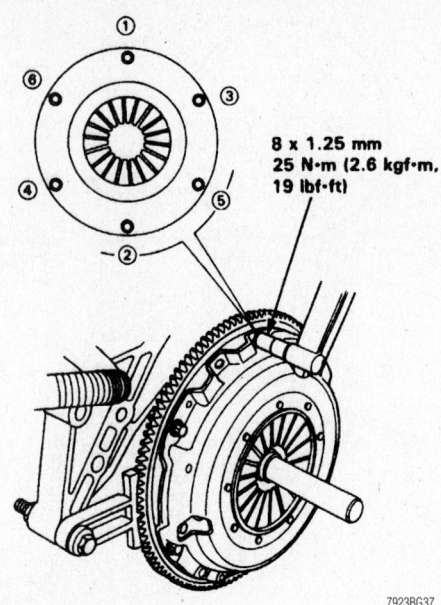

8 x 1.25 mm
25 N·m (2.6 kgf·m,
19 lbf·ft)

7923BG37

Pressure plate bolt torque sequence—RSX

removal of the pressure plate and clutch disc.

7. Matchmark the flywheel and pressure plate for easy reassembly. Remove the pressure plate bolts in a crisscross pattern 2 turns at a time to prevent warping the plate.

8. Remove the pressure plate, then the clutch disc with the alignment shaft.

To install:

9. Installation is the reverse of the removal procedure, while using the following torque values:

- Flywheel mounting bolts: 76 ft. lbs. (103 Nm) for 5–speed transaxles or 90 ft. lbs. (122 Nm) for 6–speed transaxles
- Pressure plate bolts: 19 ft. lbs. (25 Nm)

2004–06 TL Models

1. Before servicing the vehicle, refer to the precautions section.

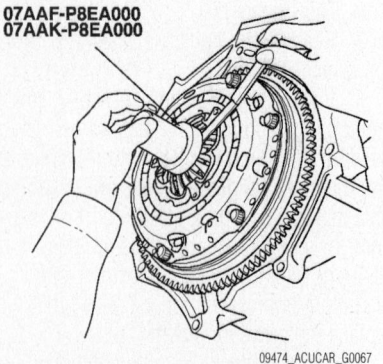

07AAF-P8EA000
07AAK-P8EA000

09474_ACUCAR_G0067

Checking the height of the diaphragm spring fingers—2004–06 TL model

2. Check the height of the diaphragm spring fingers using the tool illustrated and a feeler gauge. If the height is more than 0.00 inch (0.8 mm), replace the pressure plate and clutch disc as a set.

3. Install the tools shown and perform the following steps:

a. Turn the center screw clockwise by hand to apply pressure on the diaphragm spring. Continue turning the center screw until it stops.

b. Loosen the pressure plate mounting bolts in a star pattern in several steps, then remove the bolts.

c. Turn the center screw on the pressure plate compressor counterclockwise by hand to release the pressure, then install two pressure plate mounting bolts, hand-tight, to hold the pressure plate.

d. Remove the tools and the pressure plate.

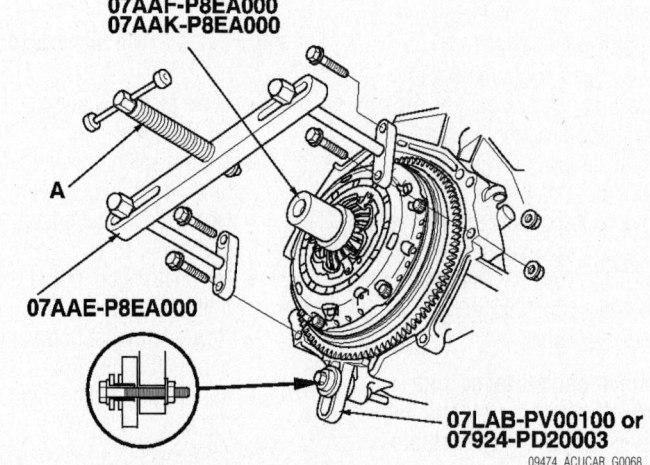

07AAF-P8EA000
07AAK-P8EA000

A

07AAE-P8EA000

07LAB-PV00100 or
07924-PD20003

09474_ACUCAR_G0068

The tools illustrated are required to remove the clutch assembly—2004–06 TL model

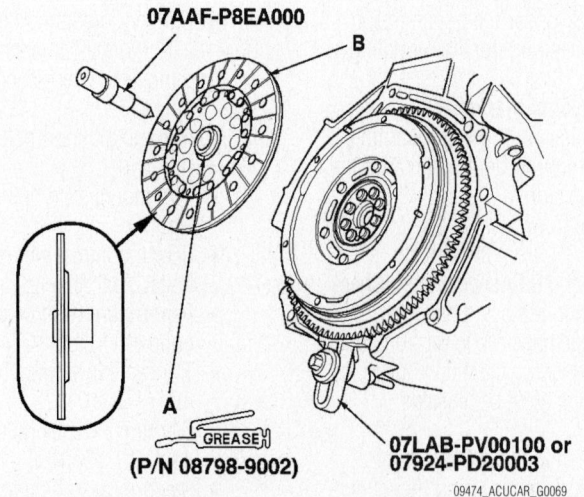

07AAF-P8EA000

B

A

GREASE
(P/N 08798-9002)

07LAB-PV00100 or
07924-PD20003

09474_ACUCAR_G0069

Remove the tools and the pressure plate—2004–06 TL model

➡ **The clutch disc and pressure plate are a matched set and must be replaced together.**

To install:

4. Temporarily install the clutch disc onto the splines of the transmission mainshaft. Make sure the clutch disc slides freely on the mainshaft.

5. Install the ring gear holder.

6. Apply super high temp urea grease to the splines of the clutch disc, then install the clutch disc using the tools

7. Align a point mark across the pressure plate and flywheel.

8. Install the pressure plate and the mounting bolts, finger-tight.

9. Install the tools shown and perform the following steps:

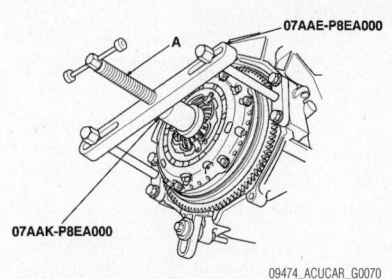

07AAE-P8EA000

07AAK-P8EA000

09474_ACUCAR_G0070

Turn the center screw clockwise by hand to apply pressure on the diaphragm spring—2004–06 TL model

a. Turn the center screw clockwise by hand to apply pressure on the diaphragm spring. Continue turning the center screw until it stops.

10. Be careful not to damage the pressure plate. Tighten the pressure plate mounting bolts in a star pattern to 19 ft. lbs. (25 Nm) in several steps.

a. Turn the center screw on the pressure plate compressor counterclockwise by hand to release the pressure, then remove the tool illustrated.

b. Make sure the diaphragm spring fingers are all the same height.

TSX Models

1. Before servicing the vehicle, refer to the precautions section.

2. Disconnect the negative battery cable.

3. Remove the transaxle assembly from the vehicle.

4. Insert a clutch alignment tool. Use a

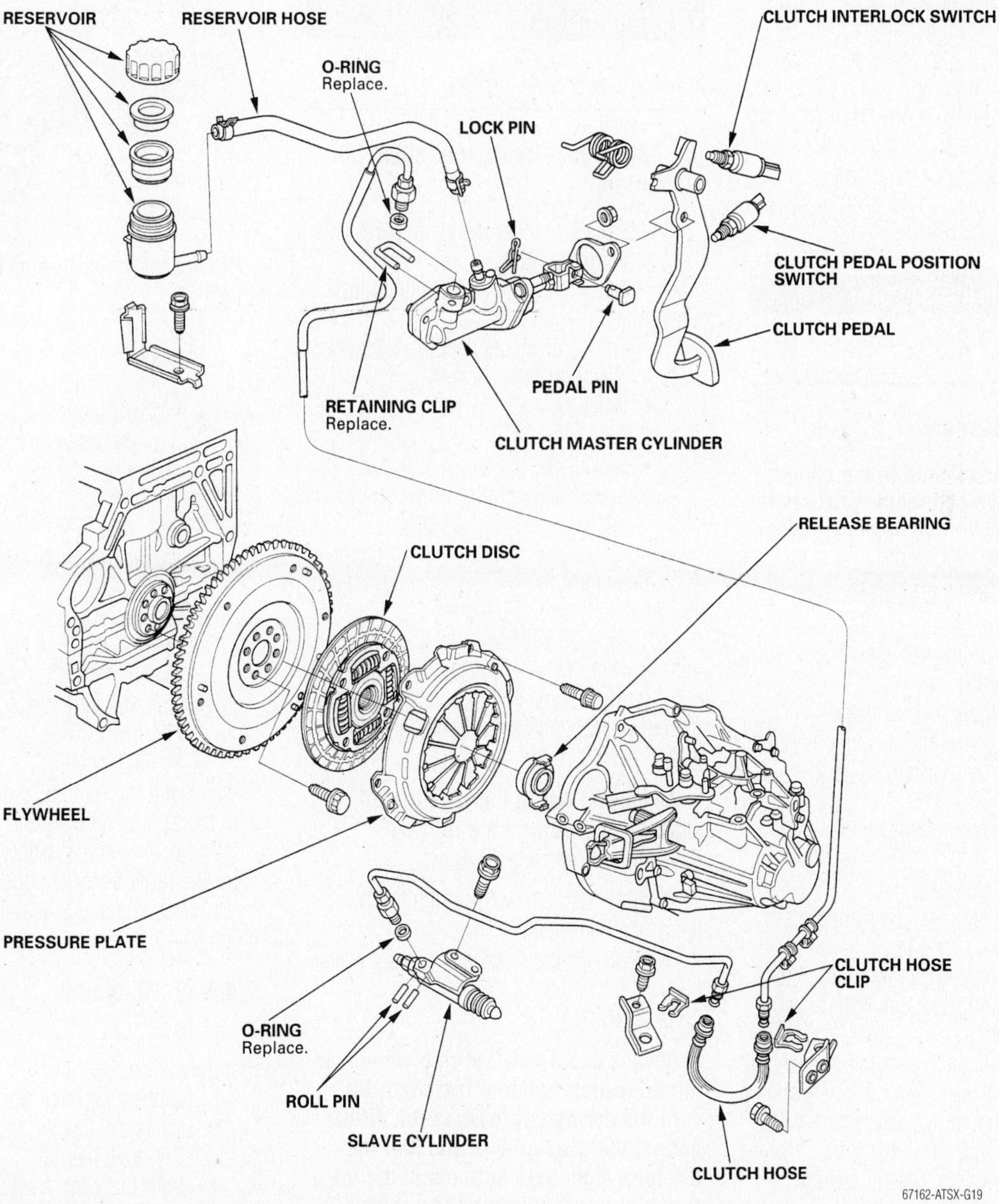

RESERVOIR

RESERVOIR HOSE

O-RING
Replace.

LOCK PIN

CLUTCH INTERLOCK SWITCH

CLUTCH PEDAL POSITION SWITCH

CLUTCH PEDAL

RETAINING CLIP
Replace.

PEDAL PIN

CLUTCH MASTER CYLINDER

RELEASE BEARING

CLUTCH DISC

FLYWHEEL

PRESSURE PLATE

O-RING
Replace.

ROLL PIN

SLAVE CYLINDER

CLUTCH HOSE CLIP

CLUTCH HOSE

67162-ATSX-G19

Exploded view of the clutch system components—TSX models

feeler gauge and measure the clearance between the pressure plate spring fingers and the clutch alignment disc. There should be a maximum of 0.02 in. (0.6mm) of clearance for a new pressure plate with 0.03 in. (0.8mm) limit for a used pressure plate. If the height is more than the service limit, replace the pressure plate.

5. Remove the clutch alignment disc.

6. Install a flywheel holder to aid in the removal of the pressure plate and clutch disc.

7. Matchmark the flywheel and pressure plate for easy reassembly. Remove the pressure plate bolts in a crisscross pattern 2 turns at a time to prevent warping the plate.

8. Remove the pressure plate, then the clutch disc with the alignment shaft.

To install:

9. Installation is the reverse of the removal procedure, while using the following torque values:

- Flywheel mounting bolts: 76 ft. lbs. (103 Nm) for 5–speed transaxles or 90 ft. lbs. (122 Nm) for 6–speed transaxles
- Pressure plate bolts: 19 ft. lbs. (25 Nm)

Hydraulic Clutch System

BLEEDING

RSX and TSX Models

➡**Use DOT 3 brake fluid in the clutch master and slave cylinders. Brake fluid will damage the vehicle's paint. Immediately clean up any spills.**

1. Before servicing the vehicle, refer to the precautions section.

2. Fit a flare or box end wrench onto the slave cylinder bleeder screw.

3. Attach a rubber tube to the slave cylinder bleeder screw and suspend it into a clear drain container partially filled with brake fluid.

4. Fill the clutch master cylinder with brake fluid.

5. Proceed as follows:
    a. Open the bleeder screw and press the clutch pedal to the floor.

6. Close the bleeder screw. Tighten to 70 inch lbs. (6 Nm). Be careful not to overtighten the screw.

7. Release the clutch pedal and recheck the reservoir fluid level. Top off if necessary.

8. Continue the above procedure until no more bubbles appear in the tube.

9. Top off the clutch master cylinder reservoir with brake fluid.

2004–06 TL Models

1. Before servicing the vehicle, refer to the precautions section.

2. Attach a hose to the bleeder screw, and suspend the hose in a container of brake fluid.

3. Make sure there is an adequate supply of fluid in the clutch master cylinder, then slowly pump the clutch pedal until no more bubbles appear at the bleeder hose.

4. It may be necessary to limit the movement of the release fork with a block of wood to remove all the air from the system.

5. Tighten the bleeder screw.

6. Refill the clutch master cylinder with fluid when done.

7. Confirm clutch operation, and check for leaking fluid.

Halfshaft

REMOVAL & INSTALLATION

CL, TSX, 2002–04 RL and 2002–03 TL Models

1. Before servicing the vehicle, refer to the precautions section.

2. Drain the differential or transmission lubricant.

3. Remove or disconnect the following:
    - Negative battery cable
    - Wheel(s)
    - Axle nut
    - Strut fork
    - Lower ball joint from the control arm

4. Pull the knuckle outward and remove the halfshaft outboard CV–joint from the knuckle using a plastic mallet.

5. Using a small pry bar carefully pry out the inboard CV–joint approximately ½ in. (13mm) in order to force the spring clip out of the groove in the differential side gears.

➡**Be careful not to damage the oil seal. Do not pull on the inboard CV–joint, it may come apart.**

6. Pull the halfshaft out of the differential or the intermediate shaft.

7. Remove the halfshaft from the wheel hub.

To install:

➡**Always use a new set ring whenever the driveshaft is being installed. Be sure the driveshaft locks in the differential side gear groove and that the CV–joint stub–axle bottoms in the differential or the intermediate shaft.**

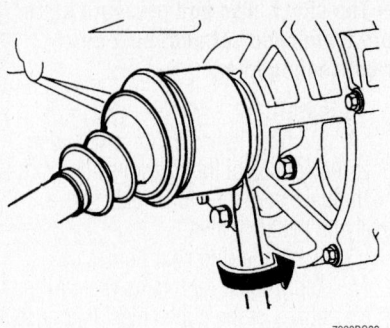

Carefully pry the inboard joint from the transaxle

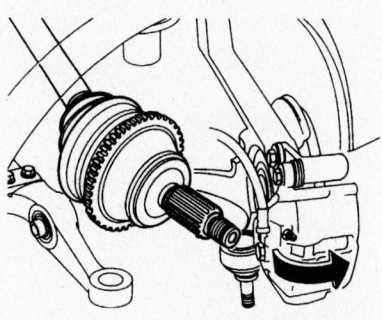

Pull the hub assembly from the outboard joint

8. Install or connect the following:
    - Outboard joint to the wheel hub
    - Inboard joint with a new set ring into the differential or intermediate shaft until the set ring locks in the groove
    - Ball joint to the control arm and torque the nut to 36–43 ft. lbs. (49–59 Nm)
    - Strut fork and torque the pinch bolt to 32 ft. lbs. (43 Nm) and the lower bolt to 47 ft. lbs. (64 Nm)
    - Axle nut and torque it to 181 ft. lbs. (245 Nm)
    - Wheel(s)

9. Refill the transmission or differential with the correct amount and type of fluid.

10. Reconnect the battery cable and enter the radio security code.

11. Measure and adjust the wheel alignment.

2005–06 RL Model

1. Before servicing the vehicle, refer to the precautions section.

2. Remove the front wheels.

3. Lift up the locking tab on the spindle nut, then remove the nut.

4. Drain the transmission fluid, then reinstall the drain plug using a new washer.

5. Remove exhaust pipe.

6. Hold the stabilizer ball joint pin with a hex wrench, and remove the flange nut. Separate the front stabilizer link from the lower arm.

7. Remove the self-locking nut, 12 mm flange bolt, and 10 mm flange bolt, then remove the damper fork.

8. Remove the knuckle holder bolt and nut.

9. Pull the knuckle outward, and remove the outboard joint from the front wheel hub using a plastic hammer.

10. Remove the exhaust pipe.

11. On the left driveshaft, pry the inboard joint from the differential case with a prybar.

12. On the right driveshaft, drive the inboard joint off of the intermediate shaft with a drift and hammer.

13. Remove the driveshaft as an assembly.

➡**Do not pull on the driveshaft, because the inboard joint may come apart. Pull the driveshaft straight out to avoid damaging the oil seal.**

To install:

14. Install a new set ring in the set ring groove of the left driveshaft.

15. Apply 0.02-0.04 ounces (0.5-1 gram) of grease to the whole splined surface of the right driveshaft.

16. After applying the grease, remove the grease from the splined grooves at intervals of 2-3 splines and from the set ring groove so that air can bleed from the intermediate shaft.

17. Clean the areas where the driveshaft contacts the differential thoroughly with solvent and dry with compressed air.

18. Install the outboard joint into the front hub.

19. Install the exhaust pipe.

20. Install the knuckle holder bolt and nut. Tighten to 54 ft. lbs. (74 Nm).

21. Insert the inboard end of the driveshaft into the differential or intermediate shaft until the new set ring locks in the groove.

22. Install the damper fork over the driveshaft and onto the lower arm. Install the damper in the damper fork so the aligning tab is aligned with the slot in the damper fork. Loosely install the flange bolt.

23. Loosely install the flange bolt and a new self-locking nut.

24. Connect the front stabilizer link to the lower arm. Hold the stabilizer link ball joint pin with a hex wrench, and tighten the new flange nut to 40 ft. lbs. (54 Nm).

25. Install a new spindle nut, then tighten the nut to 242 ft. lbs. (349 Nm). After tightening, use a drift to stake the spindle nut shoulder against the driveshaft.

26. Install the front wheel.

27. Turn the front wheel by hand, and make sure there is no interference between the driveshaft and related parts.

28. Tighten the flange bolt and the self-locking nut with the vehicle's weight on the damper.

29. Refill the transmission with recommended transmission fluid:

30. Check the front wheel alignment, and adjust it if necessary

RSX Models

1. Before servicing the vehicle, refer to the precautions section.

2. Drain the differential or transmission lubricant.

3. Remove or disconnect the following:
- Negative battery cable
- Wheel(s)
- Axle nut
- Strut fork
- Lower ball joint from the control arm

4. Pull the knuckle outward and remove the halfshaft outboard CV–joint from the knuckle using a plastic mallet.

5. Using a small pry bar carefully pry out the inboard CV–joint approximately ½ in. (13mm) in order to force the spring clip out of the groove in the differential side gears.

➡**Be careful not to damage the oil seal. Do not pull on the inboard CV–joint, it may come apart.**

6. Pull the halfshaft out of the differential or the intermediate shaft.

7. Remove the halfshaft from the wheel hub.

To install:

➡**Always use a new set ring whenever the driveshaft is being installed. Be sure the driveshaft locks in the differential side gear groove and that the CV–joint stub–axle bottoms in the differential or the intermediate shaft.**

8. Install or connect the following:
- Outboard joint to the wheel hub
- Inboard joint with a new set ring into the differential or intermediate shaft until the set ring locks in the groove
- Ball joint to the control arm and torque the nut to 36–43 ft. lbs. (49–59 Nm)
- Strut fork and torque the pinch bolt

to 32 ft. lbs. (43 Nm) and the lower bolt to 47 ft. lbs. (64 Nm)
- Axle nut and torque it to 181 ft. lbs. (245 Nm)
- Wheel(s)

9. Refill the transmission or differential with the correct amount and type of fluid.

10. Reconnect the battery cable and enter the radio security code.

11. Measure and adjust the wheel alignment.

2004–06 TL Models

1. Before servicing the vehicle, refer to the precautions section.

2. Remove the wheel nuts and front wheels.

3. Lift up the locking tab on the spindle nut, then remove the nut.

4. Drain the transmission fluid, then reinstall the drain plug using a new washer:

5. Remove exhaust pipe.

6. Hold the stabilizer ball joint pin with a hex wrench, and remove the flange nut. Separate the front stabilizer link from the lower arm.

7. Remove the self-locking nut, 12 mm flange bolt, and 10 mm flange bolt, then remove the damper fork.

8. Remove the cotter pin from the lower arm ball joint, and remove the nut.

➡**To avoid damaging the ball joint, install the tool illustrated on the threads of the ball joint. Be careful not to damage the ball joint boot when installing the remover. Do not force or hammer on the lower arm, or pry between the lower arm and the knuckle. You could damage the ball joint.**

9. Disconnect the lower ball joint from the lower arm using the tool illustrated.

➡**The collar on the lower arm is removed with ball joint, the lower arm must be replaced.**

10. Pull the knuckle outward, and remove the outboard joint from the front wheel hub using a plastic hammer

11. On the left driveshaft, pry the inboard joint from the differential case with a prybar.

12. On the right driveshaft, drive the inboard joint off of the intermediate shaft with a drift and hammer.

13. Remove the driveshaft as an assembly.

➡**Do not pull on the driveshaft, because the inboard joint may come apart. Pull the driveshaft straight out to avoid damaging the oil seal.**

To install:

14. Install a new set ring in the set ring groove of the left driveshaft.

15. Apply 0.02-0.04 ounces (0.5-1 gram) of grease to the whole splined surface of the right driveshaft.

16. After applying the grease, remove the grease from the splined grooves at intervals of 2-3 splines and from the set ring groove so that air can bleed from the intermediate shaft.

17. Clean the areas where the driveshaft contacts the differential thoroughly with solvent and dry with compressed air.

18. Insert the inboard end of the driveshaft into the differential or intermediate shaft until the new set ring locks in the groove.

19. Install the outboard joint into the front hub.

20. Clean off any grease from the ball joint tapered section and threads, then install the knuckle onto the lower arm. Torque the new castle nut to the 58 ft. lbs. (78 Nm) on 2004 models or 54 ft. lbs. (74 Nm) on 2005–06 models, then tighten it only far enough to align the slot with the ball joint pin hole. Do not align the nut by loosening it.

➡**Make sure the ball joint boot is not damaged or cracked.**

21. Install the new cotter pin into the ball joint pin hole, and bend the cotter pin.

22. Install the damper fork over the

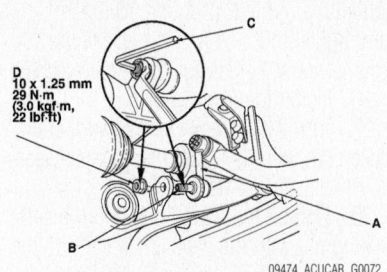

09474_ACUCAR_G0072

Loosely install the flange bolt and a new self-locking nut—2004–06 TL model

driveshaft and onto the lower arm. Install the damper in the damper fork so the aligning tab is aligned with the slot in the damper fork. Loosely install the flange bolt.

23. Loosely install the flange bolt and a new self-locking nut.

24. Connect the front stabilizer link to the lower arm. Hold the stabilizer link ball joint pin with a hex wrench, and tighten the new flange nut.

25. Install the exhaust pipe.

26. Install a new spindle nut, then tighten the nut to 181 ft. lbs. (245 Nm). After tightening, use a drift to stake the spindle nut shoulder against the driveshaft.

27. Install the front wheel.

28. Turn the front wheel by hand, and make sure there is no interference between the driveshaft and related parts.

29. Tighten the flange bolt and the self-locking nut with the vehicle's weight on the damper.

30. Refill the transmission with recommended transmission fluid:

31. Check the front wheel alignment, and adjust it if necessary

CV–Joints

OVERHAUL

CL Model

1. Before servicing the vehicle, refer to the precautions section.

2. Remove or disconnect the following:
   - Negative battery cable
   - Halfshaft from the vehicle
   - Boot bands
   - Circlip from the shaft (outer joint only)
   - Joint from the shaft after match-marking the rollers and the joint for re–installation

➡**There is a spring located under the outer joint.**

   - Rollers from the spider
   - Set ring
   - Spider using a bearing puller
   - Stop ring

To install:

3. Wrap the splines of the halfshaft with vinyl tape to protect the boots from damage.

4. Install or connect the following:
   - CV boot, then remove the tape
   - Stop ring
   - Spider
   - Set ring
   - Rollers onto the spider
   - Pack the outer joint with the grease provided (approx. 6.2 oz. for outer joint or 4.4 oz. for the inner)
   - Spring and cap (outer joint only)
   - CV joint onto the shaft, matching the marks made earlier
   - Circlip into the outer joint inner groove

5. Adjust the length of the halfshaft as shown in the illustration.

6. Adjust the boots to halfway between full compression and full extension.

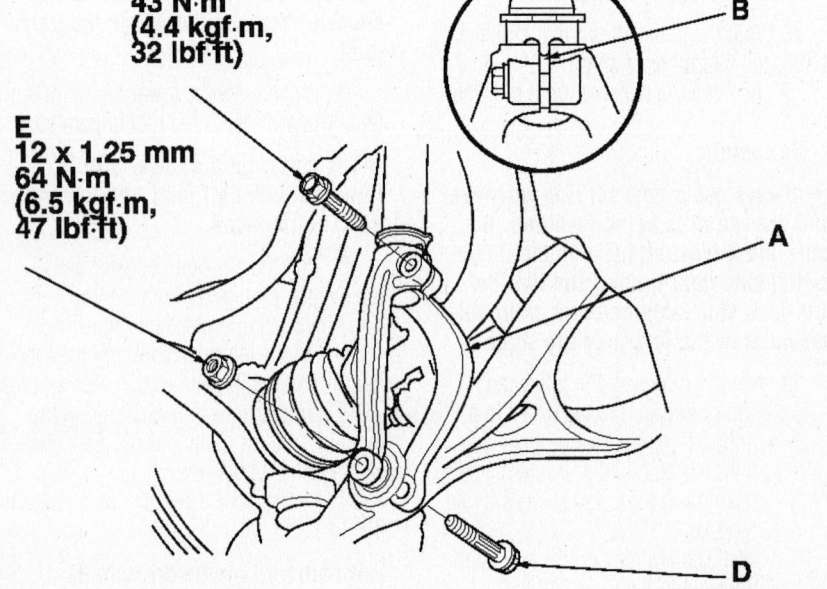

C
**10 x 1.25 mm
43 N·m
(4.4 kgf·m,
32 lbf·ft)**

E
**12 x 1.25 mm
64 N·m
(6.5 kgf·m,
47 lbf·ft)**

B

A

D

09474_ACUCAR_G0071

Install the damper fork over the driveshaft and onto the lower arm—2004–06 TL model, the 2005–06 RL model is similar

Left Driveshaft: 554 – 559 mm (21.8 – 22.0 in.)
Right Driveshaft: 544 – 549 mm (21.4 – 21.6 in.)

09474_ACUCAR_G0140

Adjust the length of the halfshaft—CL model

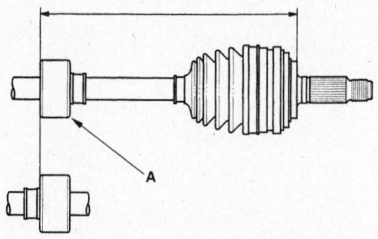

276.3 – 280.3 mm (10.88 – 11.04 in.)

09474_ACUCAR_G0141

Position the dynamic damper as shown— CL model

7. Install new boot bands on the boots and bend both sets of locking tabs down.

8. Install a new set ring in the halfshaft groove.

9. Install the halfshaft in the vehicle.

2002–04 RL Model

➡**The outer joint is not serviceable, if wear or excessive play is found the joint must be replace.**

1. Before servicing the vehicle, refer to the precautions section.

2. Remove or disconnect the following:
- Negative battery cable
- Halfshaft from the vehicle
- Set ring from the inner joint
- Inner joint boot bands
- Inboard joint after marking relationship of joint-to-rollers for later installation
- Rollers from the spider
- Circlip from the shaft
- Spider using a bearing remover
- Stop ring
- Inboard boot
- Dynamic damper band and damper (if equipped)
- Outer joint boot bands and boot (if necessary)

To install:

3. Wrap the splines with vinyl tape to prevent damage to the boots.

4. Install or connect the following:
- Outer joint boot
- Dynamic damper
- Inner joint boot then remove the tape

5. Pack the outer joint boot with grease included with the kit (approx. 4.7 oz.)
- Stop ring into the halfshaft groove
- Spider on the halfshaft
- Circlip into the halfshaft groove
- Rollers onto the spider

6. Pack the inner joint with the grease included with the kit (approx. 4.7 oz.)
- Inner joint on the halfshaft matching the marks made earlier

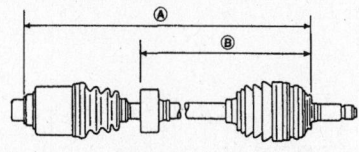

Left driveshaft:
- Ⓐ: 580 – 585 mm (22.8 - 23.0 in)
- Ⓑ: 315.6 – 319.6 mm (12.4 - 12.6 in)

Right driveshaft: 512 - 517 mm (20.2 - 20.4 in)

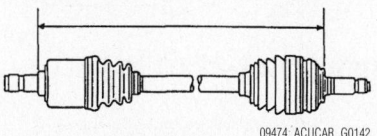

09474_ACUCAR_G0142

Adjust the length of the halfshaft— 2002–04 RL model

7. Adjust the length of the halfshaft to the specification shown in the illustration.

8. Adjust the boots to halfway between full compression and full extension.

9. Position the dynamic damper (if equipped) as shown in the illustration.

10. Install a new dynamic damper band and bend down both locking tabs.

11. Install new inner and outer joint boot bands.

12. Install a new set ring in the halfshaft groove.

13. Install the halfshaft in the vehicle.

2004–06 TL and 2005–06 RL Models

INBOARD JOINT SIDE

1. Before servicing the vehicle, refer to the precautions section.

2. Remove the set ring from the inboard joint.

3. Remove the boot bands. Be careful not to damage the boot and dynamic damper.

4. If the boot band is a welded type, cut the boot band.

5. If the boot band is a double loop type, lift up the band end, then push it into the clip.

6. If the boot band is a low profile type, pinch the boot band using commercially available boot band pincers.

7. Make a mark on each roller and inboard joint to identify the locations of rollers and grooves in the inboard joint. Then remove the inboard joint on the shop towel. Be careful not to drop the rollers when separating them from the inboard joint.

8. Make a mark on the rollers and spider to identify the locations of the rollers on the spider, then remove the rollers.

9. Remove the circlip.

10. Mark the spider and driveshaft to identify the position of the spider on the shaft.

11. Remove the spider

12. Wrap the splines on the driveshaft with vinyl tape to prevent damage to the boot.

13. Remove the inboard boot. Be careful not to damage the boot.

14. Remove the vinyl tape

To install:

15. Wrap the splines with vinyl tape to prevent damage to the inboard boot.

16. Install the inboard boot onto the driveshaft, then remove the vinyl tape. Be careful not to damage the inboard boot

17. Install the spider onto the driveshaft by aligning the marks on the spider and the end of the driveshaft.

18. Fit the circlip into the driveshaft groove. Rotate the circlip in its groove to make sure it is fully seated.

19. Fit the rollers onto the spider with their high shoulders facing outward, and note these items:

20. Reinstall the rollers in their original positions on the spider by aligning the marks.

21. Hold the driveshaft pointed up to prevent the rollers from falling off

22. Pack the inboard joint with the joint grease included in the new driveshaft set.

23. Reinstall the inboard joint onto the driveshaft by aligning the marks on the inboard joint and the rollers.

24. Hold the driveshaft so the inboard joint is pointing up to prevent it from falling off.

25. Adjust the inboard joint so the rollers are in the middle of the joint, then check the driveshaft length measurement to those shown in the illustration.

26. Install the boot bands.

27. Install the new low profile band onto the boot, then hook the tab of the band.

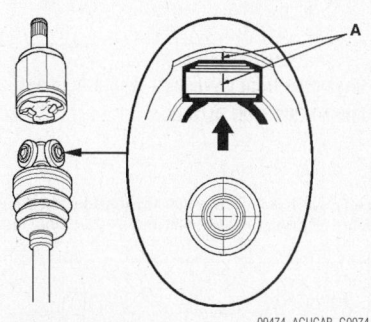

09474_ACUCAR_G0074

Reinstall the inboard joint onto the driveshaft by aligning the marks on the inboard joint and the rollers—2004–06 TL and 2005–06 RL models

Right driveshaft: 512-517 mm (20.16-20.35 in.)

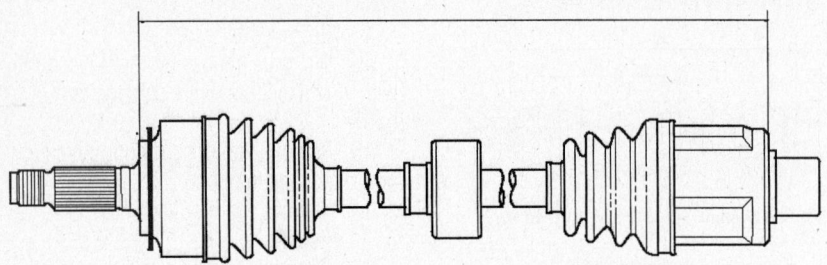

Left driveshaft: 554-559 mm (21.81-22.01 in.)

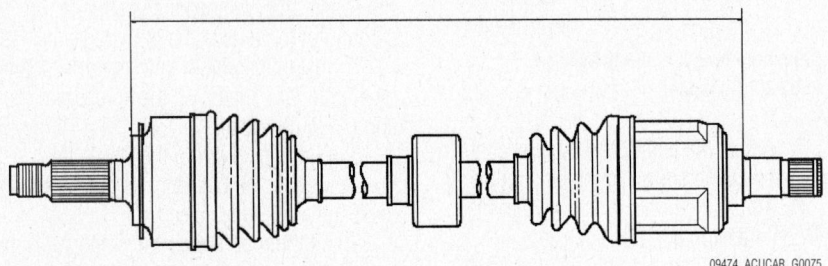

09474_ACUCAR_G0075

Check the driveshaft length measurement—2004–06 TL model

28. Close the hook portion of the band with a commercially available boot band pincers, then hook the tabs of the band.

29. Install the boot band on the other end of the boot.

30. Fit the boot ends onto the driveshaft and the inboard joint, then install the new double loop band onto the boot.

31. Pull up the slack in the band by hand

Right driveshaft: 532–537 mm (20.94–21.14 in.)

Left driveshaft: 549–554 mm (21.61–21.81 in.)

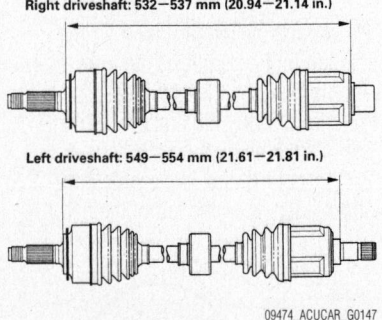

09474_ACUCAR_G0147

Check the front driveshaft length measurement—2005–06 RL model

Left driveshaft: 480–485 mm (18.90–19.09 in.)
Right driveshaft: 515–520 mm (20.28–20.47 in.)

09474_ACUCAR_G0148

Check the rear driveshaft length measurement—2005–06 RL model

32. Mark a position on the band 0.4–0.6 inches (10–14 mm) from the clip.

33. Thread the free end of the band through the nose section of a commercially available boot band tool KD-3191 or equivalent, and into the slot on the winding mandrel.

34. Place a wrench on the winding mandrel of the boot band tool, and tighten the band until the marked spot on the band meets the edge of the clip.

35. Lift up the boot band tool to bend

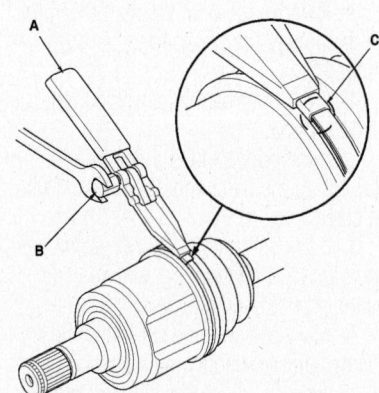

09474_ACUCAR_G0076

Thread the free end of the band through the nose section of a commercially available boot band tool KD-3191 or equivalent, and into the slot on the winding mandrel when reassembling the inboard joint—2004–06 TL model and 2005–06 RL models

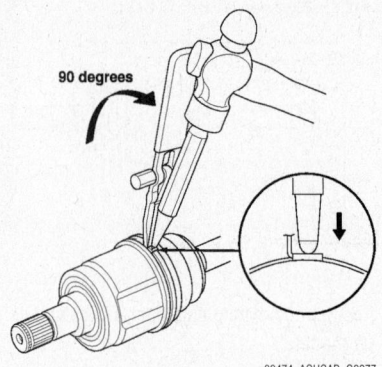

09474_ACUCAR_G0077

Center-punch the clip, then fold over the remaining tail onto the clip when reassembling the inboard joint—2004–06 TL model and 2005–06 RL models

the free end of the band 90 degrees to the clip. Center-punch the clip, then fold over the remaining tail onto the clip.

36. Unwind the boot band tool, and cut off the excess free end of the band to leave a 5-10 mm (0.2-0.4 in.) tail protruding from the clip.

37. Bend the band end by tapping it down with a hammer.

➡**Make sure the band and clip do not interfere with anything on the vehicle and the band does not move.**

38. Remove any grease remaining on the surrounding surfaces

39. Install the new set ring.

OUTBOARD JOINT SIDE

1. Before servicing the vehicle, refer to the precautions section.

2. Remove the boot bands. Be careful not to damage the boot and dynamic damper.

3. If the boot band is an ear clamp type, lift up the three tabs with a screwdriver.

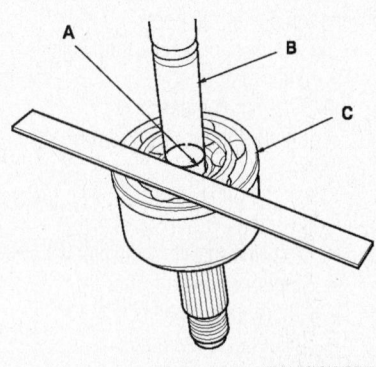

09474_ACUCAR_G0073

Make a mark on the driveshaft at the same position of the outboard joint end—2004–06 TL model and 2005–06 RL models

4. Slide the outboard boot partially to the inboard joint side. Be careful not to damage the boot.

5. Wipe off the grease to expose the driveshaft and the outboard joint inner race.

6. Make a mark on the driveshaft at the same position of the outboard joint end.

7. Carefully clamp the driveshaft in a vise.

8. Remove the outboard joint using the tool threaded adapter, 26 x 1.5 mm 07XAC-001030A and a commercially available 5/8" x 18 UNF slide hammer.

9. Remove the driveshaft from the vise.

10. Remove the stop ring from the drive-shaft.

11. Wrap the splines on the driveshaft with vinyl tape to prevent damage to the boot.

12. Remove the outboard boot. Be careful not to damage the boot.

13. Remove the vinyl tape.

To install:

14. Wrap the splines with vinyl tape to prevent damage to the outboard boot.

15. Install new ear clamp bands and outboard boot, then remove the vinyl tape. Be careful not to damage the outboard boot.

16. Install the new stop ring in the drive-shaft groove.

17. Insert the driveshaft into the outboard joint until the stop ring is closed.

18. To completely seat the outboard joint, pick up the driveshaft and joint, and tap them on a hard surface.

➡**Do not use a hammer as excessive force may damage the driveshaft. Be careful not to damage the threaded section of the outboard joint.**

19. Check the alignment of the paint mark with the outboard joint end.

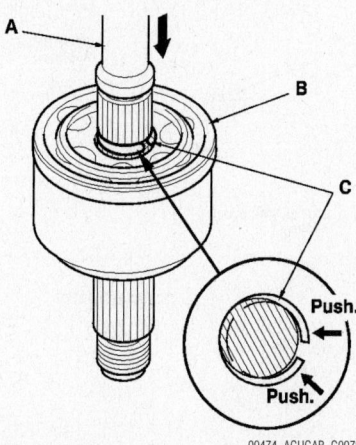

Insert the driveshaft into the outboard joint until the stop ring is closed—2004–06 TL model and 2005–06 RL models

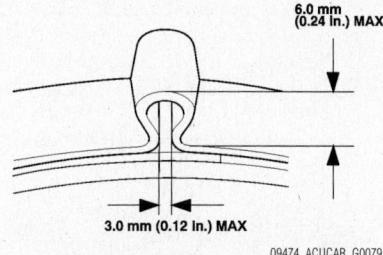

09474_ACUCAR_G0079

Check the clearance between the closed ear portion of the band. If the clearance is not within the standard, close the ear portion of the band farther—2004–06 TL model and 2005–06 RL models

20. Pack the outboard joint with the joint grease included in the new joint boot set.

21. Adjust the length of the driveshafts to those shown in the illustration, then adjust the boots to halfway between full compression and full extension. Make sure the ends of the boots seat in the groove of the driveshaft and joint.

22. Fit the boot ends onto the driveshaft and outboard joint.

23. Close the ear portion of the band with commercially available boot band pincers Kent-Moore J-35910 or equivalent.

24. Check the clearance between the closed ear portion of the band. If the clearance is not within the standard, close the ear portion of the band farther.

RSX Models

➡**The outer joint is not serviceable, if wear or excessive play is found the joint must be replace.**

1. Before servicing the vehicle, refer to the precautions section.

2. Remove or disconnect the following:

- Negative battery cable
- Halfshaft from the vehicle
- Set ring from the inner joint
- Inner joint boot bands
- Inboard joint after marking relationship of joint-to-rollers for later installation
- Rollers from the spider
- Circlip from the shaft
- Spider using a bearing remover
- Stop ring
- Inboard boot
- Dynamic damper band and damper (if equipped)
- Outer joint boot bands and boot (if necessary)

To install:

3. Wrap the splines with vinyl tape to prevent damage to the boots.

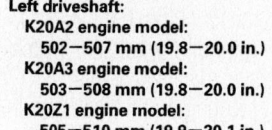

Left driveshaft:
K20A2 engine model:
  502—507 mm (19.8—20.0 in.)
K20A3 engine model:
  503—508 mm (19.8—20.0 in.)
K20Z1 engine model:
  505—510 mm (19.9—20.1 in.)

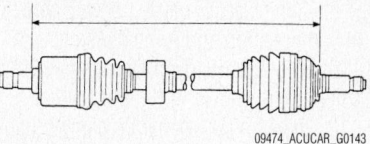

09474_ACUCAR_G0143

Adjust the length of the left halfshaft— RSX model

Right driveshaft:
K20A2 engine model:
  481—486 mm (18.9—19.1 in.)
K20A3 engine model:
  485—490 mm (19.1—19.3 in.)
K20Z1 engine model:
  484—489 mm (19.1—19.3 in.)

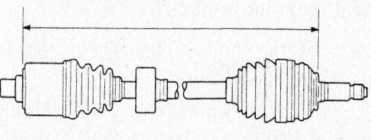

09474_ACUCAR_G0144

Adjust the length of the right halfshaft— RSX model

Left driveshaft:
K20A2 engine model:
  287—291 mm (11.3—11.5 in.)
K20A3 engine model:
  271—275 mm (10.7—10.8 in.)
K20Z1 engine model:
  270—274 mm (10.6—10.7 in.)
Right driveshaft:
  271—275 mm (10.7—10.8 in.)

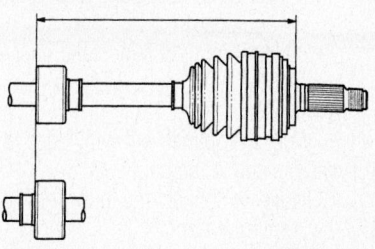

09474_ACUCAR_G0145

Position the dynamic damper as shown— CL model

4. Install or connect the following:
- Outer joint boot
- Dynamic damper
- Inner joint boot then remove the tape

5. Pack the outer joint boot with grease included with the kit (approx. 4.7 oz.)
- Stop ring into the halfshaft groove
- Spider on the halfshaft
- Circlip into the halfshaft groove
- Rollers onto the spider

6. Pack the inner joint with the grease included with the kit (approx. 4.7 oz.)

- Inner joint on the halfshaft matching the marks made earlier

7. Adjust the length of the halfshaft to the specification shown in the illustration.

8. Adjust the boots to halfway between full compression and full extension.

9. Position the dynamic damper (if equipped) as shown in the illustration.

10. Install a new dynamic damper band and bend down both locking tabs.

11. Install new inner and outer joint boot bands.

12. Install a new set ring in the halfshaft groove.

13. Install the halfshaft in the vehicle.

2002–03 TL Models

➡The outer joint is not serviceable, if wear or excessive play is found the joint must be replace.

1. Before servicing the vehicle, refer to the precautions section.

2. Remove or disconnect the following:
- Negative battery cable
- Halfshaft from the vehicle
- Set ring from the inner joint
- Inner joint boot bands
- Inboard joint after marking relationship of joint-to-rollers for later installation
- Rollers from the spider
- Circlip from the shaft
- Spider using a bearing remover
- Stop ring
- Inboard boot
- Dynamic damper band and damper (if equipped)
- Outer joint boot bands and boot (if necessary)

To install:

3. Wrap the splines with vinyl tape to prevent damage to the boots.

4. Install or connect the following:
- Outer joint boot
- Dynamic damper
- Inner joint boot then remove the tape

5. Pack the outer joint boot with grease included with the kit (approx. 4.7 oz.)
- Stop ring into the halfshaft groove
- Spider on the halfshaft
- Circlip into the halfshaft groove
- Rollers onto the spider

6. Pack the inner joint with the grease included with the kit (approx. 4.7 oz.)
- Inner joint on the halfshaft matching the marks made earlier

7. Adjust the length of the halfshaft to the specification shown in the illustration.

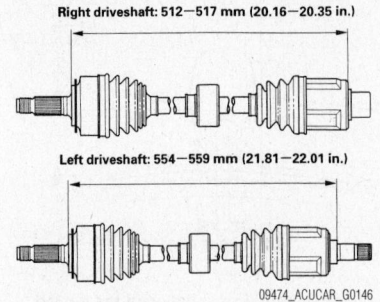

Right driveshaft: 512—517 mm (20.16—20.35 in.)

Left driveshaft: 554—559 mm (21.81—22.01 in.)

09474_ACUCAR_G0146

Adjust the length of the halfshafts— 2002–03 TL models

8. Adjust the boots to halfway between full compression and full extension.

9. Position the dynamic damper (if equipped) as shown in the illustration.

10. Install a new dynamic damper band and bend down both locking tabs.

11. Install new inner and outer joint boot bands.

12. Install a new set ring in the halfshaft groove.

13. Install the halfshaft in the vehicle.

TSX Models

➡The outer joint is not serviceable, if wear or excessive play is found the joint must be replace.

1. Before servicing the vehicle, refer to the precautions section.

2. Remove or disconnect the following:

- Negative battery cable
- Halfshaft from the vehicle
- Set ring from the inner joint
- Inner joint boot bands
- Inboard joint after marking relationship of joint-to-rollers for later installation
- Rollers from the spider
- Circlip from the shaft
- Spider using a bearing remover
- Stop ring
- Inboard boot
- Dynamic damper band and damper (if equipped)
- Outer joint boot bands and boot (if necessary)

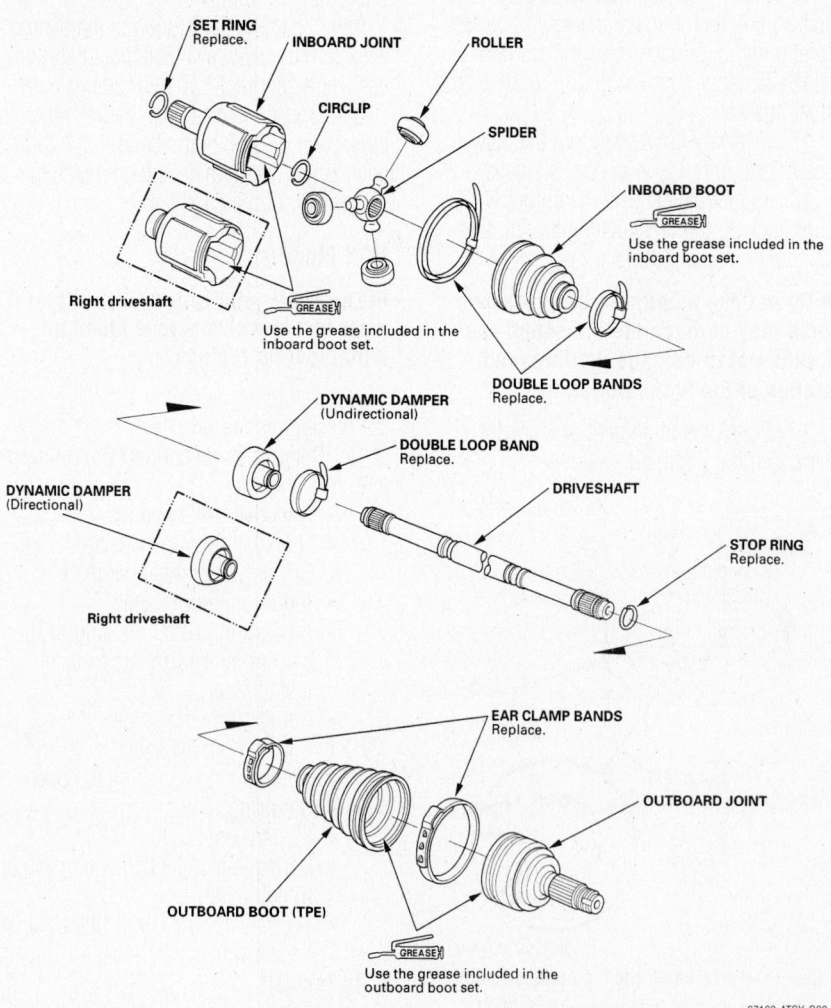

Exploded view of the CV-joint—TSX models

67162-ATSX-G20

To install:

3. Wrap the splines with vinyl tape to prevent damage to the boots.

4. Install or connect the following:
- Outer joint boot
- Dynamic damper
- Inner joint boot then remove the tape

5. Pack the outer joint boot with grease included with the kit (approx. 4.7 oz.)
- Stop ring into the halfshaft groove
- Spider on the halfshaft
- Circlip into the halfshaft groove
- Rollers onto the spider

6. Pack the inner joint with the grease included with the kit (approx. 4.7 oz.)

- Inner joint on the halfshaft matching the marks made earlier

7. Adjust the length of the halfshaft to the specification shown in the illustration.

8. Adjust the boots to halfway between full compression and full extension.

9. Position the dynamic damper (if equipped) as shown in the illustration.

10. Install a new dynamic damper band and bend down both locking tabs.

11. Install new inner and outer joint boot bands.

12. Install a new set ring in the halfshaft groove.

13. Install the halfshaft in the vehicle.

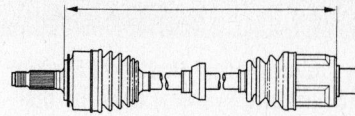

Right driveshaft: 452—457 mm (17.80—17.99 in.)

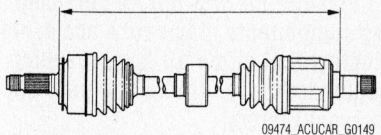

Left driveshaft: 531—536 mm (20.91—21.10 in.)

09474_ACUCAR_G0149

Adjust the length of the halfshafts—TSX models

STEERING AND SUSPENSION

Air Bag

✳✳ CAUTION

Some vehicles are equipped with an air bag system, also known as the Supplemental Restraint System (SRS). The system must be disabled before performing service on or around system components, steering column, instrument panel components, wiring and sensors. Failure to follow safety and disabling procedures could result in accidental air bag deployment, possible personal injury and unnecessary system repairs.

PRECAUTIONS

Several precautions must be observed when handling the inflator module to avoid accidental deployment and possible personal injury.

- Never carry the inflator module by the wires or connector on the underside of the module.
- When carrying a live inflator module, hold securely with both hands, and ensure that the bag and trim cover are pointed away.
- Place the inflator module on a bench or other surface with the bag and trim cover facing up.
- With the inflator module on the bench, never place anything on or close to the module which may be thrown in the event of an accidental deployment.

DISARMING

CL Model

✳✳ CAUTION

The SRS must be disarmed before any of its components are disconnected or removed. Failing to disable the SRS before servicing its components may cause accidental deployment of the air bag, resulting in unnecessary repairs and possible personal injury.

1. Before servicing the vehicle, refer to the precautions section.

2. Turn the ignition switch to the **LOCK** position. Remove the key.

3. Disconnect the negative and positive battery cables.

4. Always wait at least 3 minutes after disconnecting the battery before working around the air bags.

5. Remove or disconnect the following:
- Steering wheel lower access cover
- Clip securing the air bag module/cable reel connection to the steering column
- Air bag and cable reel connection—Immediately install the red shorting connector onto the air bag module connector.

➡**The driver's side air bag connection contains a spring–contact self–disabling device. A shorting connector doesn't need to be installed on the driver's air bag connector.**

6. After servicing has been completed, couple the air bag and cable reel connectors.

7. Install or connect the following:
- Clip securing the air bag/cable reel connection to the steering column
- Access cover

To enable:

8. Reconnect the positive and negative battery cables.

9. Turn the ignition switch to the **ON** position, but don't start the engine. The SRS indicator light should turn on for 6 seconds, then turn off. If the SRS indicator light doesn't come on, or stays on longer than 6 seconds, the system fault must be diagnosed.

10. Enter the radio security code.

RL Model

1. Before servicing the vehicle, refer to the precautions section.

2. Disconnect the negative battery cable, then the positive cable.

3. Wait 3 minutes for the air bag reserve power to discharge before preceding with work.

To enable:

4. Reconnect the positive and negative battery cables.

5. Turn the ignition switch to the **ON** position, but don't start the engine. The SRS indicator light should turn on for 6 seconds, then turn off. If the SRS indicator light doesn't come on, or stays on longer than 6 seconds, the system fault must be diagnosed.

6. Enter the radio security code.

RSX Models

※ CAUTION

The Supplemental Restraint System (SRS, air bag system) must be disarmed before any of its components are disconnected or removed. Failing to disable the SRS before servicing its components may cause accidental deployment of the air bag, resulting in unnecessary repairs and possible personal injury.

1. Before servicing the vehicle, refer to the precautions section.
2. Turn the ignition switch **OFF**.
3. Disconnect the negative battery cable, then wait 3 minutes.
4. For the driver air bag:
    a. Remove the access panel from the steering wheel and disconnect the driver's airbag 4P connector from the cable reel.
5. For the passenger air bag:
    a. Remove the glove box.
    b. Disconnect the front passenger's 4P connector from the dashboard wire harness.
6. For the side airbag:
    a. Disconnect both side airbag 2P connectors from the floor wire harness.
7. For the seat belt tensioners:
    a. Remove the B–pillar trim panels.
    b. Disconnect both seat belt tensioner 2P connectors from the floor wire harness.
8. For the seat belt buckle tensioners:
    a. Disconnect both seat belt buckle tensioner 4P connectors.

To enable:
9. For the seat belt buckle tensioners:
    a. Connect both seat belt buckle tensioner 4P connectors.
10. For the seat belt tensioners:
    a. Connect both seat belt tensioner 2P connectors to the floor wire harness.
    b. Install the B–pillar trim panels.
11. For the side airbag:
    a. Connect both side airbag 2P connectors to the floor wire harness.
12. For the passenger air bag:
    a. Connect the front passenger's 4P connector to the dashboard wire harness.
    b. Install the glove box.
13. For the driver air bag:
    a. Connect the driver's airbag 4P connector to the cable reel. Install the access panel to the steering wheel.
14. Reconnect the negative battery cable.
15. Turn the ignition switch to the **ON** position, but don't start the engine. The SRS

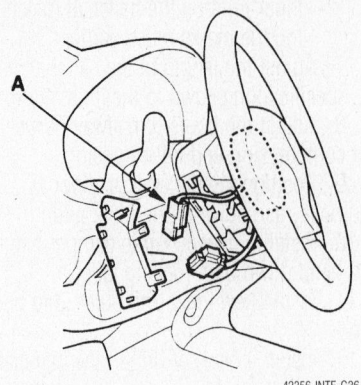

42356-INTE-G36

Remove the access panel from the steering wheel and disconnect the driver's airbag 4P connector from the cable reel— RSX

indicator light should turn on for 6 seconds, then turn off. If the SRS indicator light doesn't come on, or stays on longer than 6 seconds, the system fault must be diagnosed.
16. Enter the radio security code.

2002–03 TL Models

※ CAUTION

The SRS must be disarmed before any of its components are disconnected or removed. Failing to disable the SRS before servicing its components may cause accidental deployment of the air bag, resulting in unnecessary repairs and possible personal injury.

1. Before servicing the vehicle, refer to the precautions section.
2. Turn the ignition switch to the **LOCK** position. Remove the key.
3. Disconnect the negative and positive battery cables.
4. Always wait at least 3 minutes after disconnecting the battery before working around the air bags.
5. Remove or disconnect the following:
    • Steering wheel lower access cover
    • Clip securing the air bag module/cable reel connection to the steering column
    • Air bag and cable reel connection—Immediately install the red shorting connector onto the air bag module connector.

➡**The driver's side air bag connection contains a spring–contact self–disabling device. A shorting connector doesn't need to be installed on the driver's air bag connector.**

6. After servicing has been completed, couple the air bag and cable reel connectors.
7. Install or connect the following:
    • Clip securing the air bag/cable reel connection to the steering column
    • Access cover

To enable:
8. Reconnect the positive and negative battery cables.
9. Turn the ignition switch to the **ON** position, but don't start the engine. The SRS indicator light should turn on for 6 seconds, then turn off. If the SRS indicator light doesn't come on, or stays on longer than 6 seconds, the system fault must be diagnosed.
10. Enter the radio security code.

2004–06 TL and 2005–06 RL Models

1. Disconnect the battery negative cable, and wait at least 3 minutes before beginning work.
    a. Remove the access panel from the steering wheel, then disconnect the driver's airbag 4P connector from the cable reel.

TSX Models

※ CAUTION

The Supplemental Restraint System (SRS, air bag system) must be disarmed before any of its components are disconnected or removed. Failing to disable the SRS before servicing its components may cause accidental deployment of the air bag, resulting in unnecessary repairs and possible personal injury.

1. Before servicing the vehicle, refer to the precautions section.
2. Turn the ignition switch **OFF**.
3. Disconnect the negative battery cable, then wait 3 minutes.
4. For the driver air bag:
    a. Remove the access panel from the steering wheel and disconnect the driver's airbag 4P connector from the cable reel.
5. For the passenger air bag:
    a. Remove the glove box.
    b. Disconnect the front passenger's 4P connector from the dashboard wire harness.
6. For the side airbag:
    a. Disconnect both side airbag 2P connectors from the floor wire harness.
7. For the side curtain air bag:
    a. Disconnect the side curtain airbag

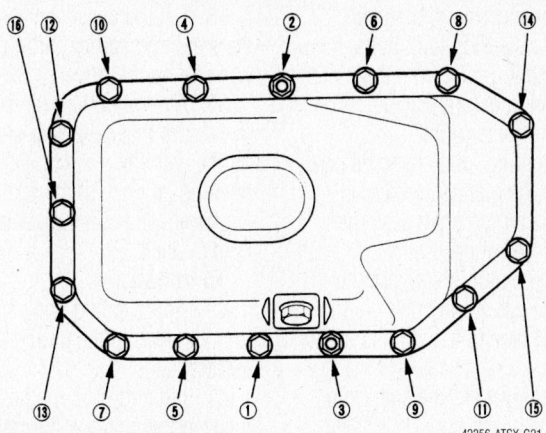

Locations of the SRS electrical connections—TSX models

42356-ATSX-G21

2P connector from the side curtain airbag sub harness.

8. For the seat belt tensioners:

a. Disconnect both seat belt tensioner 2P connectors from the floor wire harness.

To enable:

9. For the seat belt tensioners:

a. Connect both seat belt tensioner 2P connectors to the floor wire harness.

b. Install the B–pillar trim panels.

10. For the side curtain air bag:

a. Disconnect the side curtain airbag 2P connector from the side curtain airbag sub harness.

11. For the side airbag:

a. Connect both side airbag 2P connectors to the floor wire harness.

12. For the passenger air bag:

a. Connect the front passenger's 4P connector to the dashboard wire harness.

b. Install the glove box.

13. For the driver air bag:

a. Connect the driver's airbag 4P connector to the cable reel. Install the access panel to the steering wheel.

14. Reconnect the negative battery cable.

15. Turn the ignition switch to the **ON** position, but don't start the engine. The SRS indicator light should turn on for 6 seconds, then turn off. If the SRS indicator light doesn't come on, or stays on longer than 6 seconds, the system fault must be diagnosed.

16. Enter the radio security code.

Rack and Pinion Steering Gear

REMOVAL & INSTALLATION

CL Model

1. Before servicing the vehicle, refer to the precautions section.

2. Note the radio security code and the radio presets.

3. Drain the power steering fluid.

4. Remove or disconnect the following:
- Both battery cables
- Wheels
- Steering wheel lower access panel
- Supplemental Inflatable Restraint (SIR) electrical connector
- Steering wheel side panels
- Air bag
- Horn electrical connector
- Cruise control electrical connector
- Audio remote switch and navigation guide switch, if equipped
- Steering wheel
- Steering joint cover
- Steering joint lower bolt and pull the joint toward the column
- Tie rods from the steering knuckles
- Shift linkage (if equipped with a manual transmission)
- Heated Oxygen (HO$_2$S) sensor electrical connector
- Catalytic converter from the vehicle
- Return line from the steering gear
- Rear beam brace rod
- Left tie rod end, then slide the steering gear all of the way to the right
- 2 lines from the valve body unit on the steering gear

✳✳ CAUTION

After disconnecting the hose and pipe, plug or cap the hose and pipe to prevent foreign materials from entering the valve body unit.

➡**Do not loosen the cylinder pipes between the valve body unit and the cylinder.**

5. Remove the steering gear mounting bolts.

6. Pull the steering gear all the way down to clear the pinion shaft from the bulkhead, and remove the pinion shaft grommet.

7. Slide the rack all of the way to the right, then place the left rack end below the rear beam.

8. Move the steering gear assembly to the left, and tilt the left side down to remove it from the vehicle.

To install:

9. Installation is the reverse of the removal procedure, while using the following torque values:
- Left steering gear mounting bolts: 28 ft. lbs. (38 Nm)
- Right steering gear mounting bolts: 43 ft. lbs. (58 Nm)

➡**After installing the steering gear, check the air hose connections for interference with adjacent parts.**

- Intermediate shaft pinch bolt: 16 ft. lbs. (22 Nm)
- Rear beam brace rod bolts: 28 ft. lbs. (38 Nm)
- Catalytic converter nuts: 25 ft. lbs. (33 Nm)
- Tie rod end nuts: 33 ft. lbs. (44 Nm)
- Steering wheel nut: 36 ft. lbs. (49 Nm)
- Air bag bolts: 84 inch lbs. (9.5 Nm)

2002–04 RL Model

1. Before servicing the vehicle, refer to the precautions section.

2. Note the radio security code and the radio presets.

3. Drain the power steering fluid.

4. Remove or disconnect the following:
- Negative then the positive battery cables
- Steering wheel lower access panel
- Supplemental Inflatable Restraint (SIR) electrical connector
- Steering wheel side panels
- Air bag
- Horn electrical connector
- Cruise control electrical connector
- Audio remote switch and navigation guide switch, if equipped
- Steering wheel
- Steering joint bolts then move the joint toward the column
- Front wheels
- Tie rods from the steering knuckles
- Splash guard
- Feed line from the valve body
- Line mounting clamps
- Feed line from the line mounting cushions

- Sensor inlet line and both return lines from the hoses
- Steering gear mounting brackets
- Steering gear from the vehicle

To install:

5. Installation is the reverse of the removal procedure, while using the following torque values:

- Line mounting clamps bolts: 84 inch lbs. (9.5 Nm)
- Feed line bolts: 96 inch lbs. (11 Nm)
- Mounting bracket bolts: 28 ft. lbs. (38 Nm)
- Splash shield short bolts: 28 ft. lbs. (38 Nm)
- Splash shield long bolts: 43 ft. lbs. (59 Nm)
- Steering joint pinch bolt: 16 ft. lbs. (22 Nm)
- Steering wheel nut: 36 ft. lbs. (49 Nm)
- Air bag bolts: 84 inch lbs. (9.5 Nm)

2005–06 RL Model

1. Before servicing the vehicle, refer to the precautions section.
2. Remove the right and left front fender trim.
3. Drain the power steering fluid.
4. Remove the relay box and release the wire harness clips.
5. Remove the front strut brace.
6. Remove the access panel from the steering wheel, then disconnect the driver's airbag 4P connector from the cable reel.
7. Remove the two Torx bolts using a Torx T30 bit.
8. Disconnect the horn switch connector, then remove the driver's airbag.
9. Disconnect the cable reel connector.
10. Loosen the steering wheel nut.
11. Install a steering wheel puller on the steering wheel. Free the steering wheel from the steering column shaft by turning the pressure bolt of the puller.

➡**These items when removing the steering wheel:**

- Do not tap on the steering wheel or the steering column shaft when removing the steering wheel.
- If you thread the puller bolts into the wheel hub more than five threads, the bolts will hit the cable reel and damage it. To prevent this, install a pair of jam nuts five threads up on each puller bolt.

12. Remove the steering wheel puller, then remove the steering wheel nut and steering wheel from the steering column

13. Remove the steering joint cover.
14. Remove the steering joint bolts, then disconnect the steering joint from the pinion shaft and the steering column shaft.
15. Remove the center guide.
16. Remove steering joint cover. Be careful not to damage the mating surface on joint cover and the pinion shaft grommet. Replace the cover seal if necessary.
17. Remove the pinion shaft grommet from the top of the valve body unit.
18. Apply vinyl tape to the splines on the pinion shaft.
19. Remove and discard the cotter pin from the tie-rod ball joint nut and loosen the nut.
20. Separate the tie-rod ball joint and knuckle using the ball joint remover
21. Remove the nuts from the rear engine mount.
22. Remove the rear engine mount and the engine mount pressure hose holder mounting bolt on the gearbox stiffener.
23. Disconnect the engine mount pressure hose from the rear engine mount.
24. Remove the feed line holder mounting bolt and A/T wire line holder mounting bolt on the gearbox stiffener.
25. Disconnect the return hose from the clamp.
26. Remove the feed line holder mounting bolt and return line holder mounting bolt on the gearbox.
27. Disconnect the stepping valve motor connector.
28. Place shop towels under the line connections, and cover the gearbox mounting part to protect it from the power steering fluid. Loosen the adjustable hose clamp and disconnect the return hose.
29. Loosen the 14 mm flare nut and disconnect the feed line.
30. Remove the return line joint from the steering gearbox.
31. Remove the P/S heat baffle plate.
32. Remove the gearbox stiffener.
33. Remove the dynamic damper between the gearbox stiffener and front suspension subframe.
34. Remove the steering gearbox mounting bolts and washers on the right, then left gearbox mount.
35. Remove the 4-way brake line joint bolts.
36. Move the steering gearbox to the passenger's side, and rotate it so the pinion shaft points toward the front of the vehicle.
37. Carefully move the steering gearbox as an assembly toward the left side of the vehicle until the pinion shaft clears the wheel well opening. Be careful not to damage the brake lines with the pinion shaft.

38. Lift the passenger's side, and remove the steering gearbox through the wheel well opening on the driver's side.
39. After removing the steering gearbox, make sure that no power steering fluid gets on the gearbox mount cushions, gearbox housing, surface of the front suspension subframe, and stiffener. Wipe off any spilled fluid at once.

To install:

40. Apply a mild soap and water solution to both sides of the mount cushion mating surfaces (A).
41. Lift and pass the passenger's side of the steering gearbox through the wheel well opening on the driver's side.
42. Carefully move the steering gearbox toward the passenger's side until the pinion shaft clears the wheel well opening on the body.
43. Rotate the steering gearbox so the pinion shaft points upward, then move the steering gearbox to the driver's side.
44. Install the 4-way brake line joint.
45. Install the steering gearbox mounting

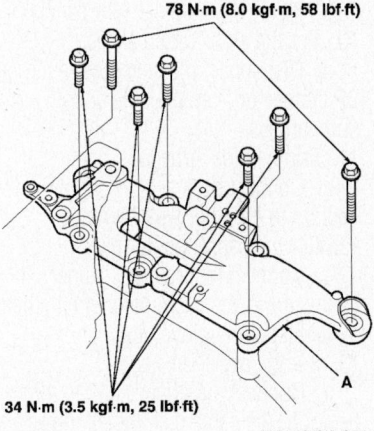

78 N·m (8.0 kgf·m, 58 lbf·ft)

34 N·m (3.5 kgf·m, 25 lbf·ft)

09474_ACUCAR_G0126

Tighten the gearbox stiffener bolts as shown—2005–06 RL model

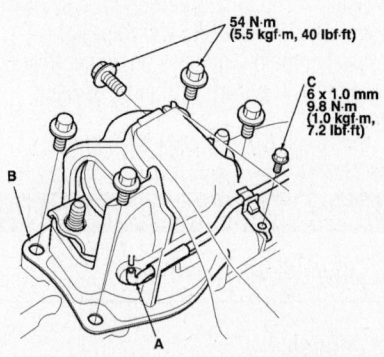

54 N·m (5.5 kgf·m, 40 lbf·ft)

C
6 x 1.0 mm
9.8 N·m
(1.0 kgf·m, 7.2 lbf·ft)

09474_ACUCAR_G0128

Tighten the rear engine mount bolts as shown—2005–06 RL model

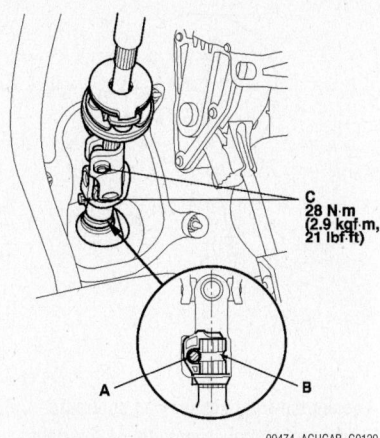

09474_ACUCAR_G0129

Align the bolt hole on the steering joint with the groove around the pinion shaft—2005–06 RL model

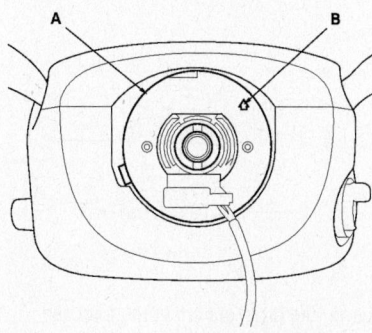

09474_ACUCAR_G0130

Center the cable reel (A), the arrow mark (B) on the cable reel label point should point straight up —2005–06 RL model

bolts and washers on the mounts, then tighten them to 58 ft. lbs. (78 Nm).

46. Install the gearbox stiffener, then tighten the mounting bolts to the torque shown in the illustration.

47. Install a dynamic damper in the reverse order of removal.

48. Install the remaining components keeping in mind the following:

a. Tighten the rear engine mount bolt to the specifications shown in the illustration.

b. Install the nuts on the rear engine mount, then tighten them to 47 ft. lbs. (64 Nm).

c. Install the tie-rod end ball joint nut, and tighten it 43 ft. lbs. (59 Nm). Install the new cotter pin. Install the pinion shaft grommet. Align the slot in the pinion shaft grommet with the lug portion on the valve housing.

d. Align the bolt hole on the steering joint with the groove around the pinion shaft, and loosely install the joint bolts. Be sure that the joint bolt is securely in

the groove in the pinion shaft. Pull on the steering joint to make sure that the steering joint is fully seated. Tighten the steering joint bolt to 21 ft. lbs. (28 Nm).

e. Before installing the steering wheel, make sure the front wheels are aligned straight ahead, then center the cable reel (A). Do this by first rotating the cable reel clockwise until it stops. Then rotate it counterclockwise about three full turns. The arrow mark (B) on the cable reel label point should point straight up. Refer to the illustration for locations.

f. Position the two tabs of the turn signal canceling sleeve. Route the airbag connector through the steering wheel and install the steering wheel on to the steering column shaft, making sure the steering wheel hub engages the pins of the cable reel and tabs of the turn signal canceling sleeve. Do not tap on the steering wheel or steering column shaft when installing the steering wheel.

g. Install the steering wheel nut and tighten it to 36 ft. lbs. (49 Nm). Connect the cable reel connector.

h. Make sure the wire harness is routed and fastened properly.

i. Fill the system with power steering fluid, and bleed air from the system.

j. Start the engine, allow it to idle, and turn the steering wheel from lock-to-lock several times to warm up to the fluid. Check the gearbox for leaks.

RSX Models

1. Before servicing the vehicle, refer to the precautions section.

2. Note the radio security code and the radio presets.

3. Drain the power steering fluid.

4. Remove or disconnect the following:
- Both battery cables
- Wheels
- Steering wheel lower access panel
- Supplemental Inflatable Restraint (SIR) electrical connector
- Steering wheel side panels
- Air bag
- Horn electrical connector
- Cruise control electrical connector
- Audio remote switch and navigation guide switch, if equipped
- Steering wheel
- Steering joint cover
- Steering joint lower bolt and pull the joint toward the column
- Tie rods from the steering knuckles
- Shift linkage (if equipped with a manual transmission)

- Heated Oxygen (HO2S) sensor electrical connector
- Catalytic converter from the vehicle
- Return line from the steering gear
- Rear beam brace rod
- Left tie rod end, then slide the steering gear all of the way to the right
- 2 lines from the valve body unit on the steering gear

✳✳ CAUTION

After disconnecting the hose and pipe, plug or cap the hose and pipe to prevent foreign materials from entering the valve body unit.

➡**Do not loosen the cylinder pipes between the valve body unit and the cylinder.**

5. Remove the steering gear mounting bolts.

6. Pull the steering gear all the way down to clear the pinion shaft from the bulkhead, and remove the pinion shaft grommet.

7. Slide the rack all of the way to the right, then place the left rack end below the rear beam.

8. Move the steering gear assembly to the left, and tilt the left side down to remove it from the vehicle.

To install:

9. Installation is the reverse of the removal procedure, while using the following torque values:
- Left steering gear mounting bolts: 28 ft. lbs. (38 Nm)
- Right steering gear mounting bolts: 43 ft. lbs. (58 Nm)

➡**After installing the steering gear, check the air hose connections for interference with adjacent parts.**

- Intermediate shaft pinch bolt: 16 ft. lbs. (22 Nm)
- Rear beam brace rod bolts: 28 ft. lbs. (38 Nm)
- Catalytic converter nuts: 25 ft. lbs. (33 Nm)
- Tie rod end nuts: 33 ft. lbs. (44 Nm)
- Steering wheel nut: 36 ft. lbs. (49 Nm)
- Air bag bolts: 84 inch lbs. (9.5 Nm)

2002–03 TL Models

1. Before servicing the vehicle, refer to the precautions section.

2. Note the radio security code and the radio presets.

3. Drain the power steering fluid.

4. Remove or disconnect the following:
- Negative then the positive battery cables
- Steering wheel lower access panel
- Supplemental Inflatable Restraint (SIR) electrical connector
- Steering wheel side panels
- Air bag
- Horn electrical connector
- Cruise control electrical connector
- Audio remote switch and navigation guide switch, if equipped
- Steering wheel
- Steering joint bolts then move the joint toward the column
- Front wheels
- Tie rods from the steering knuckles
- Splash guard
- Feed line from the valve body
- Line mounting clamps
- Feed line from the line mounting cushions
- Sensor inlet line and both return lines from the hoses
- Steering gear mounting brackets
- Steering gear from the vehicle

To install:

5. Installation is the reverse of the removal procedure, while using the following torque values:
- Line mounting clamps bolts: 84 inch lbs. (9.5 Nm)
- Feed line bolts: 96 inch lbs. (11 Nm)
- Mounting bracket bolts: 28 ft. lbs. (38 Nm)
- Splash shield short bolts: 28 ft. lbs. (38 Nm)
- Splash shield long bolts: 43 ft. lbs. (59 Nm)
- Steering joint pinch bolt: 16 ft. lbs. (22 Nm)
- Steering wheel nut: 36 ft. lbs. (49 Nm)
- Air bag bolts: 84 inch lbs. (9.5 Nm)

2004–06 TL Models

1. Before servicing the vehicle, refer to the precautions section.

2. Remove the right side engine component cover, right rear engine component cover and left rear engine component cover.

3. Drain the power steering fluid.

4. Remove the front wheels.

5. Remove the access panel from the steering wheel, then disconnect the driver's airbag 4P connector from the cable reel.

6. Remove the two Torx bolts using a Torx T30 bit.

7. Disconnect the horn switch connector, then remove the driver's airbag.

8. Disconnect the cruise control set/resume switch, audio remote switch and HFL/HFL-voice control switch connectors (if equipped).

9. Loosen the steering wheel bolt.

10. Install a commercially available steering wheel puller on the steering and remove the wheel.

✳✳ CAUTION

Do not tap on the steering wheel or the steering column shaft when removing the steering wheel.

11. If you thread the puller bolts into the wheel hub more than five threads, the bolts will hit the cable reel and damage it. To prevent this, install a pair of jam nuts five threads up on each puller bolt.

12. Remove the steering joint cover.

13. Remove the steering joint bolt, and disconnect the steering joint by moving the steering joint toward the column. Hold the slider shaft on the column with a piece of wire between the joint yoke on the slider shaft to the joint yoke on the upper shaft.

14. Remove the center guide and discard it.

15. Remove the steering joint cover. Be careful not to damage the mating surface on the joint cover and pinion shaft grommet. Replace the cover seal if necessary.

16. Remove and discard the cotter pin from the tie-rod ball joint nut, and loosen the nut.

17. Separate the tie-rod ball joint from the knuckle.

18. Disconnect the power steering pressure switch connector.

19. Remove the splash shield.

20. Remove the engine under cover.

21. Attach the tool illustrated to the front suspension subframe by hanging the hook of the tool over the front of the subframe, then tighten the screw.

22. Raise the jack and line up the slots in the arms with the bolt holes on the corner of the jack base, then attach them with bolts securely.

23. Remove the front suspension subframe right middle mounting bolts.

24. Remove the front suspension subframe left middle mounting.

25. Loosen the front suspension subframe front bracket mounting bolts on the right and left of the vehicle so they are about 1 ³⁄₁₆ inch (30 mm) from the mounting surface.

26. Remove the front suspension sub-

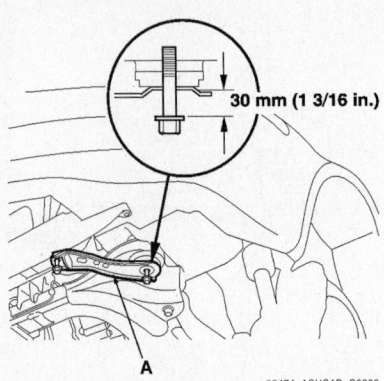

30 mm (1 3/16 in.)

09474_ACUCAR_G0089

Loosen the front suspension subframe front bracket mounting bolts on the right and left of the vehicle so they are about 1 ³⁄₁₆ inch (30 mm) from the mounting surface—2004–06 TL model

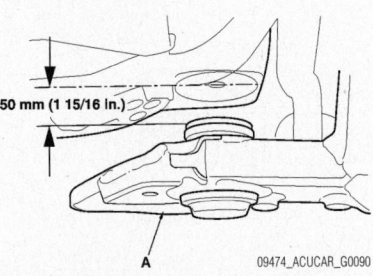

50 mm (1 15/16 in.)

09474_ACUCAR_G0090

Lower the jack supporting the front suspension subframe with the tool illustrated slowly until the front suspension subframe has dropped about 1 ¹⁵⁄₁₆ inch (50 mm)—2004–06 TL model

frame rear bracket on the right and left of the vehicle from the front suspension subframe.

27. Lower the jack supporting the front suspension subframe with the tool illustrated slowly until the front suspension subframe has dropped about 1 ¹⁵⁄₁₆ inch (50 mm).

28. Remove the P/S heat baffle plate.

29. Remove the feed line holder mounting bolt on the front suspension subframe.

30. Remove the feed line holder mounting bolt and return hose from the gearbox mounting bracket.

31. Place several shop towels under the line connections, and cover the gearbox mounting part to protect it from the power steering fluid. Loosen the flare nut and disconnect the feed line.

32. Loosen the flare nut and disconnect the return line on 2004 models or the pipe on 2005–06 models.

33. After disconnecting the lines or pipe, plug or seal them with a piece of tape or equivalent to prevent foreign materials from entering.

➡**Do not loosen the cylinder line A and B between the valve body unit and the cylinder.**

34. Remove the steering gearbox mounting bolts and washers on the left gearbox mount.

35. Remove the two flange bolts from the right side of the gearbox, then remove the gearbox mounting bracket and cushion.

36. Move the steering gearbox toward the front, and remove the pinion shaft grommet from the top of the valve body unit.

37. Apply vinyl tape to the splines on the pinion shaft.

38. Apply vinyl tape or equivalent material to brake lines to protect if from the pinion shaft.

39. Move the steering gearbox to the driver's side, and rotate it so the pinion shaft points toward the front of the vehicle.

40. Carefully move the steering gearbox as an assembly toward the left side of the vehicle until the pinion shaft clears the wheel well opening. Be careful not to damage the brake lines with the pinion shaft.

To install:

41. Before installing the steering gearbox, make sure that no power steering fluid is on the mating surface of the gearbox and front suspension subframe. To prevent the gearbox mounting bolts from loosening after the installation, remove any power steering fluid from the mount cushions and bolt holes.

42. Apply a mild soap and water solution to both sides of the mount cushion mating surfaces.

43. Pass the cylinder of the steering gearbox through the wheel well opening on the driver's side.

44. Carefully move the steering gearbox toward the passenger's side until the pinion shaft clears the wheel well opening on the body.

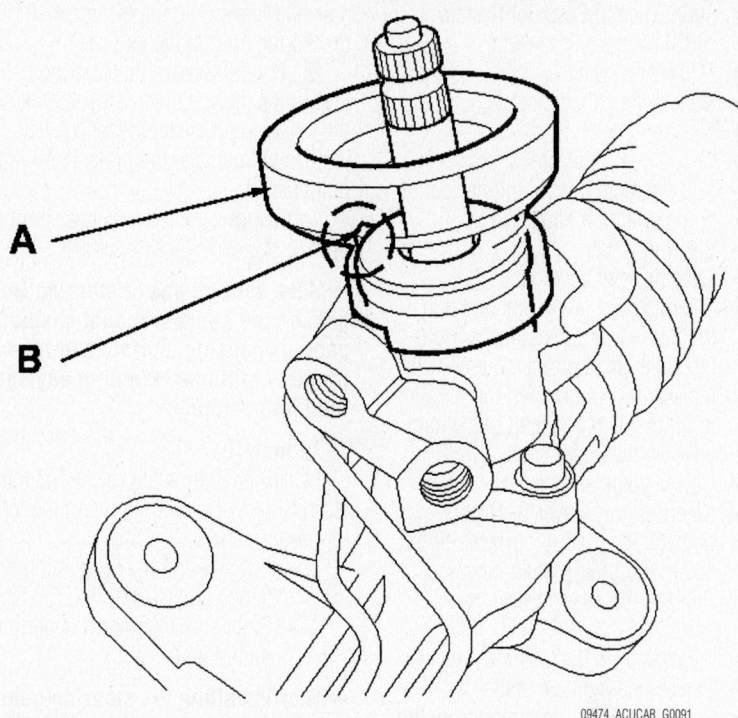

09474_ACUCAR_G0091

Install the pinion shaft grommet. Align the slot in the pinion shaft grommet with the lug portion on the valve housing—2004–06 TL model

45. Rotate the steering gearbox so the pinion shaft points upward.

46. Continue moving the gearbox toward the passenger's side until the steering gearbox is in position.

47. Make sure the power steering return line and feed line are routed above the gearbox

48. Remove the vinyl tape from the pinion shaft, and install the pinion shaft grommet. Align the slot in the pinion shaft grommet with the lug portion on the valve housing.

49. Position the cutout on the mounting cushion as shown, and install it on the cylinder of the gearbox securely.

50. Install the new gearbox mounting

bracket over the mounting cushion, and loosely install the two flange bolts.

51. Install the new gearbox mounting bolts and washers on the left side of the gearbox, then tighten them to the 28 ft. lbs. (38 Nm).

52. Tighten the flange bolts on the right side of the gearbox to 43 ft. lbs. (59 Nm) alternately in two or more steps.

53. Loosely connect the return line/pipe and feed line by hand.

54. Install the feed line holder and return hose on the gearbox mounting bracket.

55. Install the feed line holder on the front suspension subframe. Make sure that there is no interference between the feed and return lines and any other parts.

56. Install the remaining components in the reverse order of removal.

TSX Models

1. Before servicing the vehicle, refer to the precautions section.

2. Note the radio security code and the radio presets.

3. Disconnect the negative battery cable and wait at least 3 minutes.

4. Make sure the front wheels are in the straight ahead position.

5. Drain the power steering fluid.

6. Remove or disconnect the following:
   • Wheels
   • Steering wheel access panel

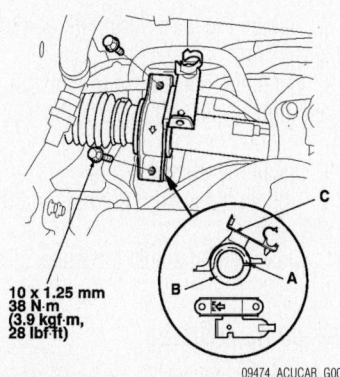

10 x 1.25 mm
38 N·m
(3.9 kgf·m,
28 lbf·ft)

09474_ACUCAR_G0092

Install the new gearbox mounting bolts and washers on the left side of the gearbox—2004–06 TL model

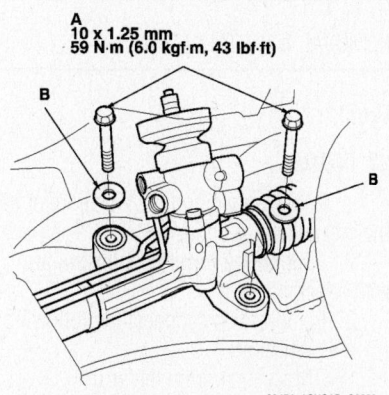
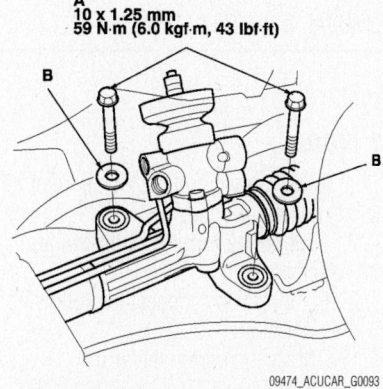

A
10 x 1.25 mm
59 N·m (6.0 kgf·m, 43 lbf·ft)

09474_ACUCAR_G0093

Tighten the flange bolts on the right side of the gearbox—2004–06 TL model

- Supplemental Restraint System (SRS) electrical connector
- 2 Torx® screws
- Horn electrical connector
- Air bag
- Cruise control electrical connector
- Audio remote switch and navigation guide switch, if equipped
- Steering wheel
- Steering joint cover A
- Steering joint lower bolt and pull the joint toward the column. Hold the slider shaft on the column with a piece of wire between the yoke joint on the slider shaft to the joint yoke on the upper shaft.
- Center guide and discard
- Steering joint cover B. Do not damage the mating surfaces of the cover and pinion shaft grommet. Replace the cover seal if necessary.
- Tie rods from the steering knuckles
- Power steering heat baffle
- Feed line holder mounting from the front suspension subframe
- Feed line holder mounting bolt and return hose from the gearbox mounting bracket
- Shift linkage (if equipped with a manual transmission)

7. Place some rags under the fluid line connections and cover the gearbox mounting part to protect it from power steering fluid. Loosen the flare nut and disconnect the fluid and return lines.

✻✻ CAUTION

After disconnecting the fluid and return lines, plug or cap the hose and pipe to prevent foreign materials from entering the valve body unit.

➥**Do not loosen the cylinder pipes between the valve body unit and the cylinder.**

8. Remove or disconnect the following:
- Front suspension subframe left mid mount
- Steering gear mounting bolts from the left gearbox mount, then remove the steering stiffener plate
- 2 flange bolts from the right side of the gearbox, then remove the gearbox mounting bracket, and cushion

9. Move the steering gear toward the front, and remove the pinion shaft grommet from the top of the valve body unit.

10. Put vinyl tape on the brake lines to protect them from the pinion shaft.

11. Move the steering gear to the driver's side and rotate it so the pinion shaft points toward the front of the vehicle.

12. Carefully move the steering gearbox toward the driver's side of the vehicle until the pinion shaft clears the wheel well opening. Do not damage the brake lines with the pinion shaft.

13. Remove the steering gear from the driver's side wheel well opening.

➥**Make sure no power steering fluid gets on the gearbox mount cushions, gearbox housing, surfaces of the subframe or stiffener. Wipe up any spilled fluid immediately.**

To install:

14. Installation is the reverse of the removal procedure, while using the following torque values:
- Left steering gear mounting bracket flange bolts: 28 ft. lbs. (38 Nm)
- Right steering gear mounting bolts: 43 ft. lbs. (58 Nm)

➥**After installing the steering gear, check the air hose connections for interference with adjacent parts.**

- Return line flare nut: 21 ft. lbs. (28 Nm)
- Feed line flare nut: 31 ft. lbs. (42 Nm)
- Front suspension subframe mid mount bottom (long) bolt: 33 ft. lbs. (44 Nm)
- Front suspension subframe mid mount side (short) bolts: 36 ft. lbs. (49 Nm)
- Steering joint to pinion shaft bolt: 21 ft. lbs. (29 Nm)
- Tie rod end nuts: 32 ft. lbs. (43 Nm)
- Steering wheel nut: 29 ft. lbs. (39 Nm)
- Air bag bolts: 86 inch lbs. (9.8 Nm)

Strut

REMOVAL & INSTALLATION

Front

CL MODEL

1. Before servicing the vehicle, refer to the precautions section.

2. Support the lower suspension arm with a jack.

3. Remove or disconnect the following:
- Wheel(s)
- Brake hose from the strut
- Strut fork
- Upper strut mounting nuts
- Strut from the vehicle

To install:

4. Install the strut and loosely install the mounting nuts

5. Install the strut fork, do not tighten the bolts at this time

➥**All suspension nuts and bolts should be tightened with the vehicle on the ground, or with a floor jack supporting the vehicle's weight.**

6. Tighten the pinch bolt to 32 ft. lbs. (43 Nm).

7. Tighten the lower strut fork bolt to 47 ft. lbs. (64 Nm).

8. Install the brake hose to the strut fork and tighten the bolt to 16 ft. lbs. (22 Nm).

9. Install the wheels.

10. Tighten the upper strut nuts to 37 ft. lbs. (50 Nm)

11. Measure and adjust the wheel alignment.

2002–04 RL MODELS

1. Before servicing the vehicle, refer to the precautions section.

2. Support the lower suspension arm with a jack.

3. Remove or disconnect the following:
- Wheel(s)
- Brake hose from the strut
- Strut fork
- Upper strut mounting nuts
- Strut from the vehicle

To install:

4. Install the strut and loosely install the mounting nuts

5. Install the strut fork, do not tighten the bolts at this time

➥**All suspension nuts and bolts should be tightened with the vehicle on the ground, or with a floor jack supporting the vehicle's weight.**

6. Tighten the pinch bolt to 32 ft. lbs. (43 Nm).

7. Tighten the lower strut fork bolt to 47 ft. lbs. (64 Nm).

8. Install the brake hose to the strut fork and tighten the bolt to 16 ft. lbs. (22 Nm).

9. Install the wheels.

10. Tighten the upper strut nuts to 28 ft. lbs. (38 Nm).

11. Measure and adjust the wheel alignment.

2005–06 RL MODELS

1. Before servicing the vehicle, refer to the precautions section.

2. Remove covers from the engine compartment cover.

3. Remove the strut tower bar with the vehicle on the ground.

4. Remove the front wheel.

5. Remove the damper fork from the damper and lower arm.

6. Remove the flange nuts from the top of the damper, then remove the damper assembly.

To install:

7. Position the damper assembly in the body with the aligning tab (B) facing inside, then loosely install the flange nuts.

8. Loosely install the damper pinch bolt into the damper fork.

9. Install the flange bolt to the damper fork and lower arm, then lightly tighten the new damper fork mounting nut.

➡**Use a new damper fork mounting nut on reassembly.**

10. Place the floor jack under the lower arm, and raise the suspension to load it with the vehicle's weight.

11. Tighten the flange nuts on the top of the damper to 25 ft. lbs. (34 Nm) on the 8 mm nuts and 42 ft. lbs. (55 Nm) on the 10 mm nuts.

12. Tighten the damper pinch bolts to

the torque specified in the accompanying illustration.

13. Tighten the flange nut on the damper fork to the torque specified in the accompanying illustration.

14. Install the strut tower bar with the vehicle on the ground.

15. Install the engine compartment covers.

16. Install the front wheel.

RSX MODELS

1. Before servicing the vehicle, refer to the precautions section.

CAUTION:
● Replace the self-locking nuts after removal.
● The vehicle should be on the ground before any bolts or nuts connected to rubber mounts or bushings are tightened.
● Torque the castle nut to the lower torque specification, then tighten it only far enough to align the slot with the pin hole. Do not align the nut by loosening.

NOTE: Wipe off the grease before tightening the nut at the ball joint.

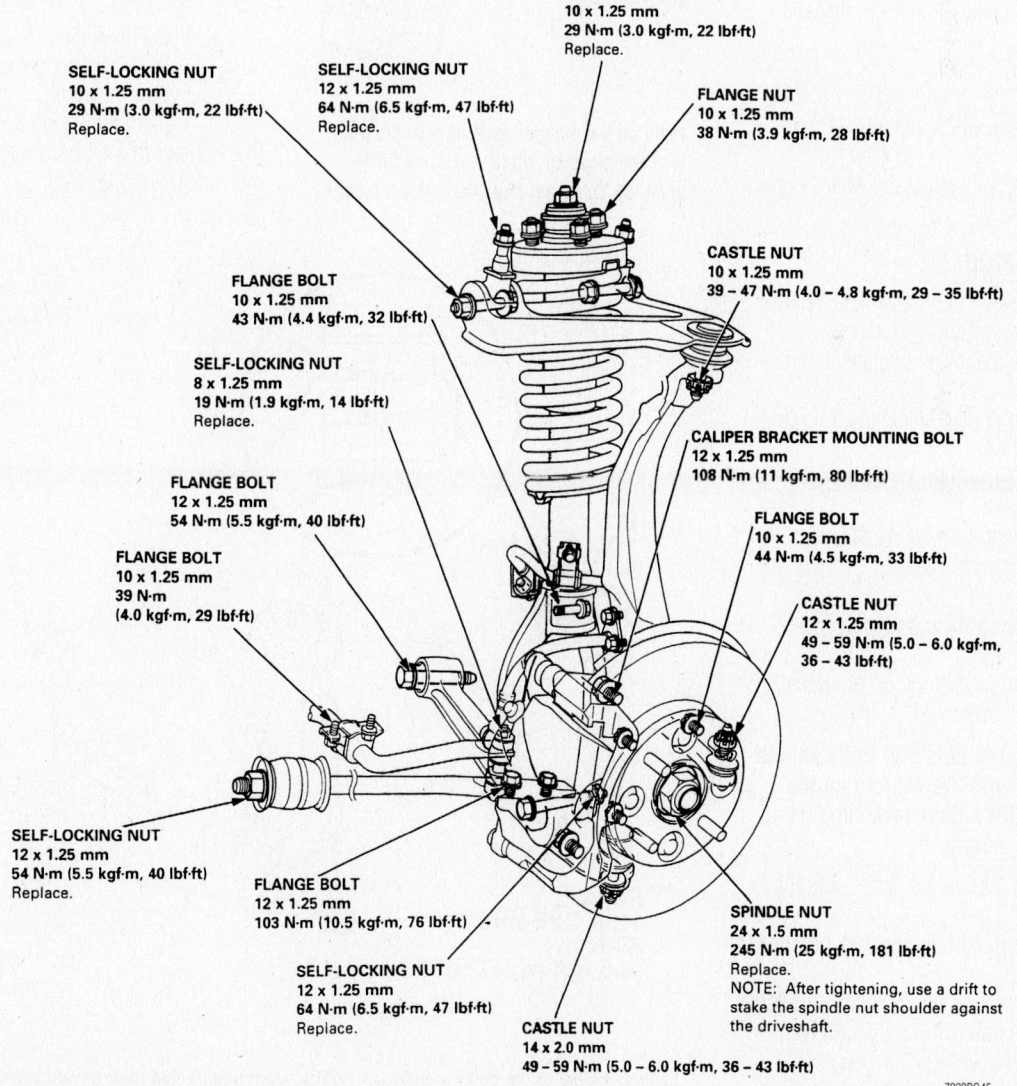

Front suspension showing the torque specifications

7923BG45

2. Support the lower suspension arm with a jack.

3. Remove or disconnect the following:
- Wheel(s)
- Brake hose from the strut
- Strut fork
- Upper strut mounting nuts
- Strut from the vehicle

To install:

4. Install the strut and loosely install the mounting nuts

5. Install the strut fork, do not tighten the bolts at this time

➡**All suspension nuts and bolts should be tightened with the vehicle on the ground, or with a floor jack supporting the vehicle's weight.**

6. Tighten the pinch bolt to 32 ft. lbs. (43 Nm).

7. Tighten the lower strut fork bolt to 47 ft. lbs. (64 Nm).

8. Install the brake hose to the strut fork and tighten the bolt to 16 ft. lbs. (22 Nm).

9. Install the wheels.

10. Tighten the upper strut nuts to 28 ft. lbs. (38 Nm) (except 3.2 models 37 ft. lbs. [50 Nm])

11. Measure and adjust the wheel alignment.

2002–03 TL MODELS

1. Before servicing the vehicle, refer to the precautions section.

2. Support the lower suspension arm with a jack.

3. Remove or disconnect the following:
- Wheel(s)
- Brake hose from the strut
- Strut fork
- Upper strut mounting nuts
- Strut from the vehicle

To install:

4. Install the strut and loosely install the mounting nuts

5. Install the strut fork, do not tighten the bolts at this time

➡**All suspension nuts and bolts should be tightened with the vehicle on the ground, or with a floor jack supporting the vehicle's weight.**

6. Tighten the pinch bolt to 32 ft. lbs. (43 Nm).

7. Tighten the lower strut fork bolt to 47 ft. lbs. (64 Nm).

8. Install the brake hose to the strut fork and tighten the bolt to 16 ft. lbs. (22 Nm).

9. Install the wheels.

10. Tighten the upper strut nuts to 37 ft. lbs. 50 Nm).

11. Measure and adjust the wheel alignment.

2004–06 TL MODELS

1. Before servicing the vehicle, refer to the precautions section.

2. Remove the right side engine compartment cover, right rear engine compartment cover and left rear engine compartment cover.

3. Remove the strut tower bar with the vehicle on the ground.

4. Remove the front wheel.

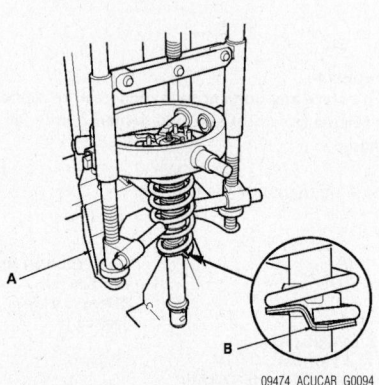

09474_ACUCAR_G0094

Position the damper assembly in the body with the aligning tab (B) facing inside—2004–06 TL model and 2005–06 RL model

5. Remove the damper fork from the damper and lower arm.

6. Remove the flange nuts from the top of the damper, then remove the damper assembly.

To install:

7. Position the damper assembly in the body with the aligning tab (B) facing inside, then loosely install the flange nuts.

8. Install the damper fork (A) over the driveshaft and onto the lower arm. Install the front damper in the damper fork so the aligning tab (B) is aligned with the slot (C) in the damper fork. Refer to the illustration for component locations and torque specifications.

9. Loosely install the damper pinch bolt into the damper fork.

10. Install the flange bolt to the damper fork and lower arm, then lightly tighten the new damper fork mounting nut.

➡**Use a new damper fork mounting nut on reassembly.**

11. Place the floor jack under the lower arm, and raise the suspension to load it with the vehicle's weight.

12. Tighten the flange nuts on the top of the damper to the torque specified in the accompanying illustration.

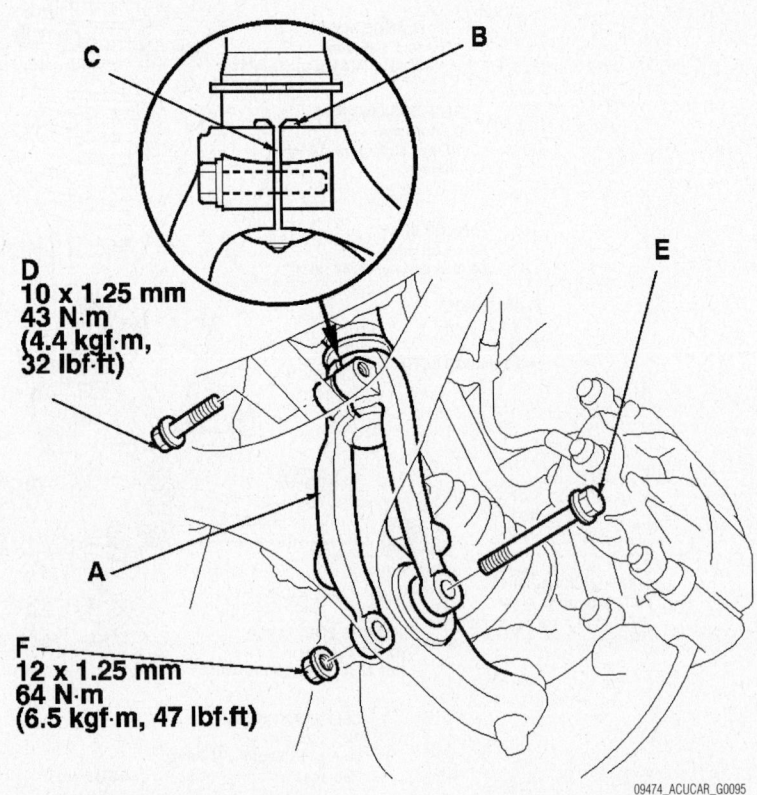

D
10 x 1.25 mm
43 N·m
(4.4 kgf·m, 32 lbf·ft)

F
12 x 1.25 mm
64 N·m
(6.5 kgf·m, 47 lbf·ft)

09474_ACUCAR_G0095

Exploded view of the damper fork and related components and their torque specifications—2004–06 TL model, 2005–06 RL similar

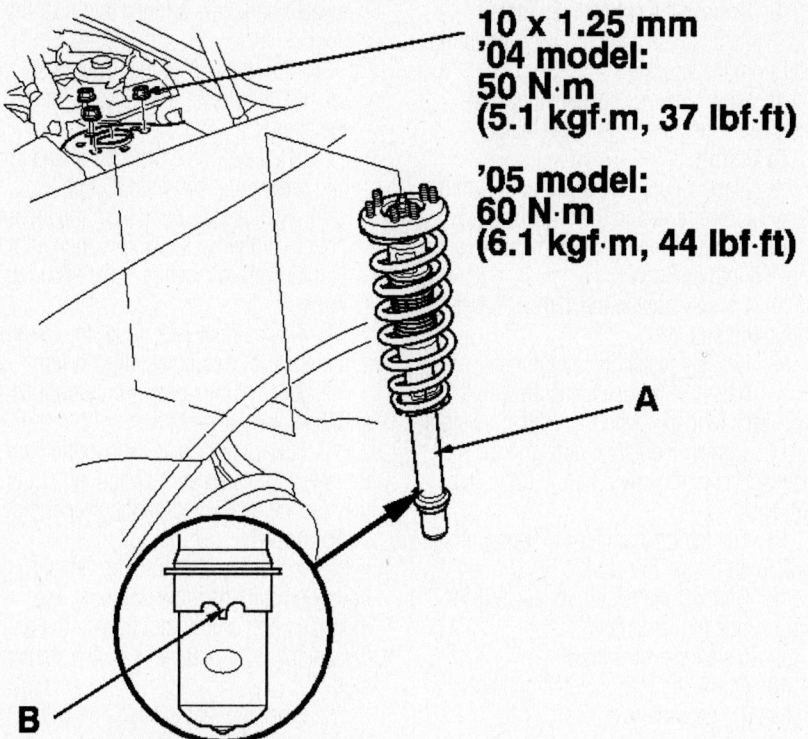

10 x 1.25 mm
'04 model:
50 N·m
(5.1 kgf·m, 37 lbf·ft)

'05 model:
60 N·m
(6.1 kgf·m, 44 lbf·ft)

A

B

09474_ACUCAR_G0096

View of the upper strut mounting—2004–06 TL model

13. Tighten the damper pinch bolts to the torque specified in the accompanying illustration.

14. Tighten the flange nut on the damper fork to the torque specified in the accompanying illustration.

15. Install the strut tower bar with the vehicle on the ground.

16. Install the engine compartment covers.

17. Install the front wheel.

TSX MODELS

1. Before servicing the vehicle, refer to the precautions section.

2. Raise and safely support the vehicle.

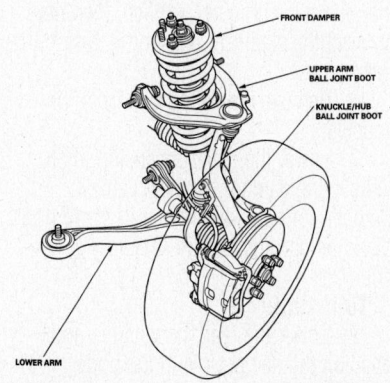

FRONT DAMPER

UPPER ARM
BALL JOINT BOOT

KNUCKLE/HUB
BALL JOINT BOOT

LOWER ARM

42356-ACCO-G35

Front suspension components—TSX models

3. Remove or disconnect the following:
• Front wheels
• Damper fork bolts, then the damper fork from the damper and lower arm
• 2 8mm flange nuts and 3 10mm flange nuts
• Strut (damper assembly) from the vehicle

To install:

➡**Use new self-locking bolts when installing the struts and assembling the damper forks.**

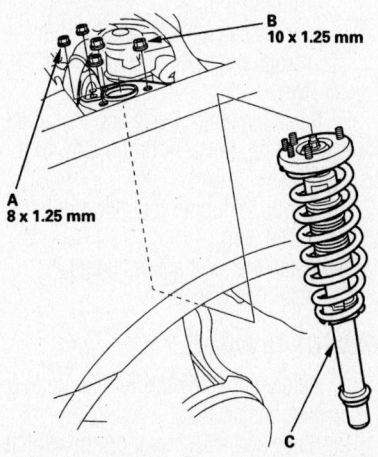

B
10 x 1.25 mm

A
8 x 1.25 mm

C

42356-ACCO-G36

Front strut (damper) mounting—TSX models

4. Install or connect the following:
• Strut into the vehicle with the aligning tab facing inside, if equipped. Hand-tighten the mounting nuts.
• Strut into the damper fork. The alignment mark on the strut tube fits into the slot on the damper fork.
• Pinch bolt and damper fork bolt/nut. Only hand-tighten these bolts.
• Front wheels and lower the vehicle.

5. With all 4 of the vehicle's wheels on the ground, tighten the damper fork nut to 47 ft. lbs. (65 Nm) while holding the damper fork bolt. Tighten the damper fork pinch bolt to 32 ft. lbs. (44 Nm). Tighten the damper assembly 8mm bolts to 16 ft. lbs. (22 Nm) and the 10mm bolts to 37 ft. lbs. (50 Nm).

6. Tighten the wheel nuts to 80 ft. lbs. (110 Nm).

7. Check and adjust the vehicle's front end alignment.

Rear

CL MODEL

1. Before servicing the vehicle, refer to the precautions section.

2. Remove the rear wheel.

3. Remove the rear shelf.

4. Remove the two flange nuts.

5. Remove the flange nut while holding the joint pin with a hex wrench, then disconnect the stabilizer link from the stabilizer link bracket.

6. Remove the flange bolt from the knuckle.

7. Lower the rear suspension, then remove the damper from the vehicle. Damper springs are different, left and right. Mark the springs L and R before you continue

To install:

8. Lower the rear suspension, and position the damper (A) in the body.

9. Loosely install the flange nuts onto the top of the damper.

10. Loosely install the flange bolt on the bottom of the damper. Connect the stabilizer link on the bracket then loosely install the flange nut.

11. Raise the rear suspension with a floor jack to load the vehicle weight, and tighten the flange bolt to the torque specified in the accompanying illustration.

12. Tighten the flange nut to 29 ft. lbs. (39 Nm) and the bolt to 43 ft. lbs. (59 Nm) while holding the joint pin with a hex wrench.

13. Tighten the two flange nuts on top of the damper to 37 ft. lbs. (50 Nm).

14. Install the rear shelf.

15. Install the rear wheel.

16. Check the rear wheel alignment, and adjust if necessary.

2002–04 RL MODELS

1. Before servicing the vehicle, refer to the precautions section.

2. Raise and safely support the vehicle and remove the rear wheels.

3. Remove or disconnect the following:
   - Upper strut mount cover from the rear panel, just below the speaker—On sedans, remove the trunk side panel.
   - Trim cover, then remove the upper mount nuts
   - Wheel sensor wire brackets, on cars with ABS—do not disconnect the wheel sensor connector
   - Lower strut mounting bolt

4. Lower the rear suspension and remove the strut assembly from the vehicle.

5. If necessary, use a spring compressor to remove the spring from the strut assembly.

To install:

6. Reassemble the strut and coil spring assembly. Tighten the strut self–locking nut to 22 ft. lbs. (30 Nm).

7. Lower the rear suspension and position the strut assembly in the vehicle. The nut welded to the lower strut mounting should face the front of the vehicle.

8. Loosely install the upper mounting nuts.

9. Install the strut lower mounting bolt.

10. Raise the vehicle until the vehicle just lifts off the safety stand and tighten the lower strut bolt and lower control arm bolt. Tighten the lower strut mounting bolt to 76 ft. lbs. (103 Nm).

11. Install the wheel sensor wire bracket on cars with ABS.

12. Tighten the upper mounting nuts to 36 ft. lbs. (49 Nm).

13. Install the rear wheels, then lower the vehicle.

14. Install the trim panel, or the trunk side panel.

15. Check the vehicle's alignment.

2005–06 RL MODELS

1. Before servicing the vehicle, refer to the precautions section.

2. Remove the rear wheel.

3. Place a floor jack at the connecting point of lower arm and the stabilizer link to support them.

4. Disconnect the headlight leveling sensor linkage from the damper.

5. Remove the rear shelf.

6. Remove the two flange nuts.

7. Remove the damper lower mounting bolt from the knuckle

8. Lower the rear suspension, then remove the damper from the vehicle

To install:

9. Place a floor jack at the connecting point of lower arm and the stabilizer link.

10. Compress the damper by hand, and move it into position.

11. Loosely tighten the damper lower mounting bolt.

12. Loosely install the flange nuts.

13. Raise the rear suspension with a floor jack to load it with the vehicle's weight.

14. Tighten the flange nuts and the damper lower mounting bolt to 28 ft. lbs. (38 Nm).

15. Tighten the damper lower mounting bolt to 43 ft. lbs. (59 Nm).

16. Connect the headlight leveling sensor linkage to the damper.

17. Install the rear wheel.

18. Check the rear wheel alignment, and adjust it if necessary.

RSX MODELS

1. Before servicing the vehicle, refer to the precautions section.

2. Raise and safely support the vehicle.

3. Support the control arm with a jack.

4. Remove or disconnect the following:
   - Rear wheels
   - Strut access panel
   - Upper mounting nuts
   - Wheel sensor wire brackets (if equipped)
   - Lower mounting bolt

5. Lower the jack and remove the strut from the vehicle.

To install:

6. Install or connect the following:
   - Strut and hand tighten the upper mounting nuts
   - Wheel sensor wire bracket and torque the bolts to 84 inch lbs. (9.5 Nm)

7. Raise the control arm and install the lower mounting bolt and torque it to 40 ft. lbs. (54 Nm).

8. Torque the upper mounting nuts to 36 ft. lbs. (49 Nm).

9. Install the strut access panel.

10. Install the wheels.

2002–03 TL MODELS

1. Before servicing the vehicle, refer to the precautions section.

2. Raise and safely support the vehicle and remove the rear wheels.

3. Remove the rear seat:
   a. Remove the lower cushion bolts located under the armrest and near the floor.

   b. Pull the rear of the lower cushion up and lift it forward to release it from the clips.

   c. Pull down the trunk bulkhead trim and release the armrest lid clips.

   d. Remove the bolts from the top and bottom of the back cushion, then lift it up and forward to disengage the securing hooks.

4. Place a floor jack under the lower arm and slightly compress the spring.

5. Remove the upper mounting nuts and the lower flange bolt.

6. Lower the jack to remove the strut. Be sure and mark the right and left struts so they can be reinstalled on the proper sides.

To install:

7. Install the strut into the vehicle. Loosely install the mounting nuts and mounting bolt, but do not tighten them until the weight of the vehicle is on the suspension.

8. Raise the rear suspension with a floor jack until the weight of the vehicle is on the strut. Tighten the upper mounting nuts to 28 ft. lbs. (39 Nm), then tighten the lower mounting bolts to 40 ft. lbs. (55 Nm). Be careful not to pinch the ABS speed sensor wire between the strut and bracket.

9. Install the rear wheels and lower the vehicle.

10. Install the rear seat back and torque the bolts to 84 inch lbs. (9.5 Nm).

11. Install the armrest lid.

12. Install the lower seat cushion and torque the bolts to 84 inch lbs. (9.5 Nm).

2004–06 TL MODELS

1. Before servicing the vehicle, refer to the precautions section.

2. Remove the rear wheel.

3. Remove the rear shelf.

4. Remove the two flange nuts.

5. Remove the flange nut while holding the joint pin with a hex wrench, then disconnect the stabilizer link from the stabilizer link bracket.

6. Remove the flange bolt from the knuckle.

7. Lower the rear suspension, then remove the damper from the vehicle. Damper springs are different, left and right. Mark the springs L and R before you continue

To install:

8. Lower the rear suspension, and position the damper (A) in the body.

9. Loosely install the flange nuts onto the top of the damper.

10. Loosely install the flange bolt (A) on

CAUTION:
- Replace the self-locking nuts after removal.
- The vehicle should be on the ground before any bolts or nuts connected to rubber mounts or bushings are tightened.
- Torque the castle nut to the lower torque specification, then tighten it only far enough to align the slot with the pin hole. Do not align the nut by loosening.

NOTE: Wipe off the grease before tightening the nut at the ball joint.

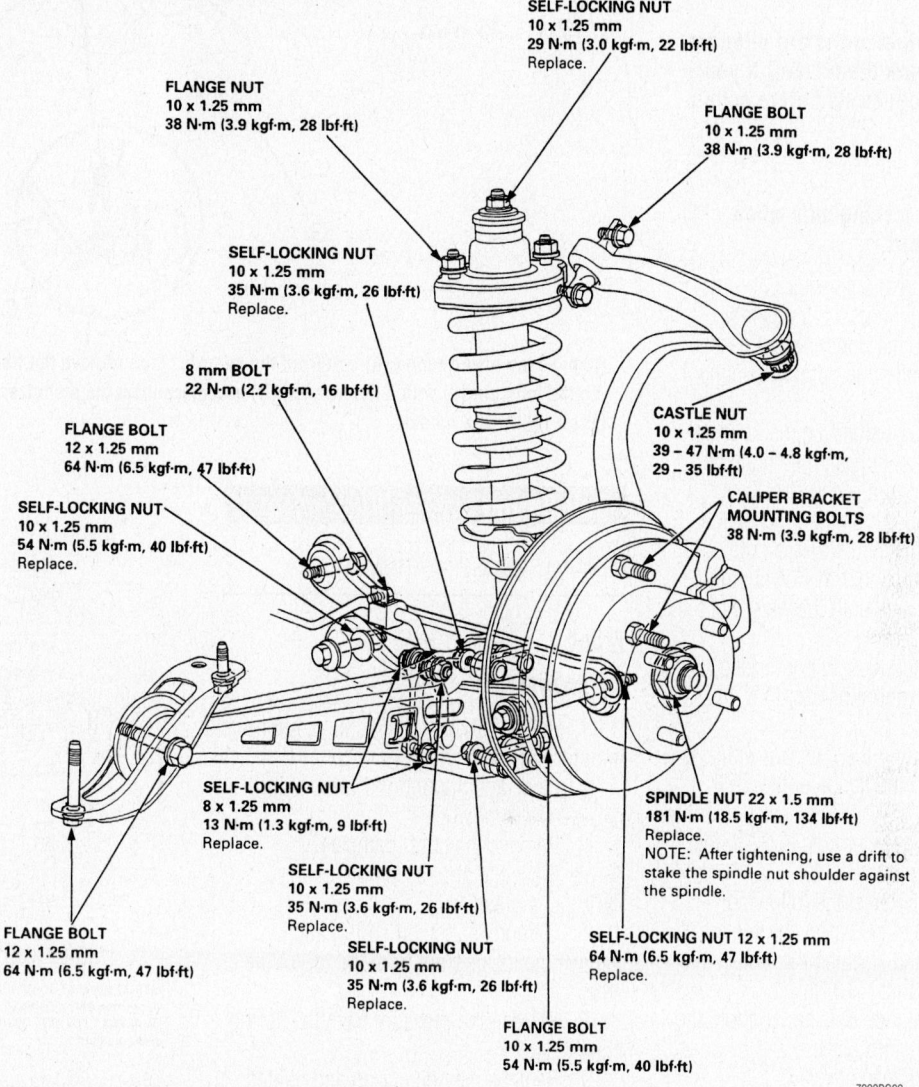

SELF-LOCKING NUT
10 x 1.25 mm
29 N·m (3.0 kgf·m, 22 lbf·ft)
Replace.

FLANGE NUT
10 x 1.25 mm
38 N·m (3.9 kgf·m, 28 lbf·ft)

FLANGE BOLT
10 x 1.25 mm
38 N·m (3.9 kgf·m, 28 lbf·ft)

SELF-LOCKING NUT
10 x 1.25 mm
35 N·m (3.6 kgf·m, 26 lbf·ft)
Replace.

8 mm BOLT
22 N·m (2.2 kgf·m, 16 lbf·ft)

FLANGE BOLT
12 x 1.25 mm
64 N·m (6.5 kgf·m, 47 lbf·ft)

CASTLE NUT
10 x 1.25 mm
39 – 47 N·m (4.0 – 4.8 kgf·m,
29 – 35 lbf·ft)

SELF-LOCKING NUT
10 x 1.25 mm
54 N·m (5.5 kgf·m, 40 lbf·ft)
Replace.

CALIPER BRACKET
MOUNTING BOLTS
38 N·m (3.9 kgf·m, 28 lbf·ft)

SELF-LOCKING NUT
8 x 1.25 mm
13 N·m (1.3 kgf·m, 9 lbf·ft)
Replace.

SELF-LOCKING NUT
10 x 1.25 mm
35 N·m (3.6 kgf·m, 26 lbf·ft)
Replace.

SPINDLE NUT 22 x 1.5 mm
181 N·m (18.5 kgf·m, 134 lbf·ft)
Replace.
NOTE: After tightening, use a drift to
stake the spindle nut shoulder against
the spindle.

SELF-LOCKING NUT 12 x 1.25 mm
64 N·m (6.5 kgf·m, 47 lbf·ft)
Replace.

FLANGE BOLT
12 x 1.25 mm
64 N·m (6.5 kgf·m, 47 lbf·ft)

SELF-LOCKING NUT
10 x 1.25 mm
35 N·m (3.6 kgf·m, 26 lbf·ft)
Replace.

FLANGE BOLT
10 x 1.25 mm
54 N·m (5.5 kgf·m, 40 lbf·ft)

7923BG92

Typical early model rear suspension components showing the torque specifications

the bottom of the damper. Connect the stabilizer link (B) on the bracket (C), then loosely install the flange nut (D).

11. Raise the rear suspension with a floor jack to load the vehicle weight, and tighten the flange bolt to the torque specified in the accompanying illustration.

12. Tighten the flange nut while holding the joint pin with a hex wrench.

13. Tighten the two flange nuts on top of the damper to 37 ft. lbs. (50 Nm).

14. Install the rear shelf.

15. Install the rear wheel.

16. Check the rear wheel alignment, and adjust if necessary.

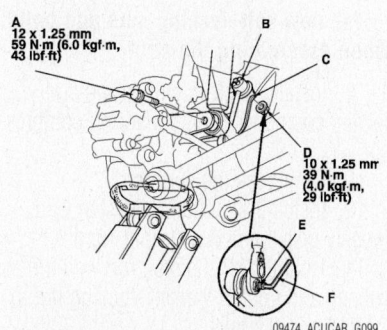

A
12 x 1.25 mm
59 N·m (6.0 kgf·m,
43 lbf·ft)

D
10 x 1.25 mm
39 N·m
(4.0 kgf·m,
29 lbf·ft)

09474_ACUCAR_G099

Install the flange bolt (A) on the bottom of the damper. Connect the stabilizer link (B) on the bracket (C), then loosely install the flange nut (D)—2004–06 TL model

TSX MODELS

1. Before servicing the vehicle, refer to the precautions section.

2. Fold the rear seat forward.

3. Remove or disconnect the following:
- Rear shelf cover
- Seat side bolster cushions. The side bolster cushions are secured by 1 screw at the bottom, and 2 clips at the top.
- Strut mount cap, if equipped, and upper strut mounting nuts

4. Raise and safely support the vehicle.
- Rear wheels, then support the knuckle with a floor jack

- Strut lower flange bolt from the knuckle
- Strut flange nut while holding the joint pin with a hex wrench
- Stabilizer link from the bracket
- Strut, while lowering rear suspension

➡ **The left and right struts are different, so be sure to mark them L & R if you are removing both struts before continuing.**

To install:

➡ **Use new self-locking nuts when installing the strut.**

5. Lower the rear suspension.
6. Install or connect the following:
   - Strut into the upper mount. Only hand-tighten the upper mounting nuts.
   - Strut into position on the knuckle, then loosely install the flange bolt on the bottom of the strut
   - Stabilizer link on the bracket and loosely install the flange nut
7. Place a jack under the lower strut mount. Raise the jack until the weight of the vehicle is on the jack.
8. With the suspension under load, tighten the lower mount bolt to 43 ft. lbs. (59 Nm).
9. While holding the joint pin with a cotter pin, tighten the flange nut to 29 ft. lbs. (39 Nm).
10. Tighten the upper nuts to 37 ft. lbs. (50 Nm).
11. Install or connect the following:
    - Rear wheel(s). Lower the vehicle to the ground. Tighten the wheel nuts to 80 ft. lbs. (110 Nm).
    - Rear seat side bolsters and fold the seat back into place.
    - Rear bulkhead cover
12. Check and adjust the vehicle's rear wheel alignment.

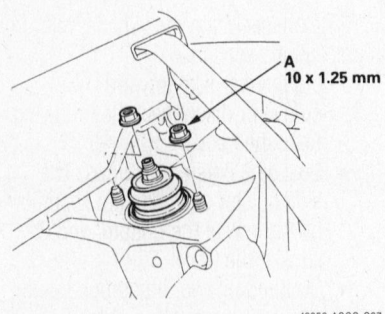

Rear strut upper mounting nut locations— TSX models

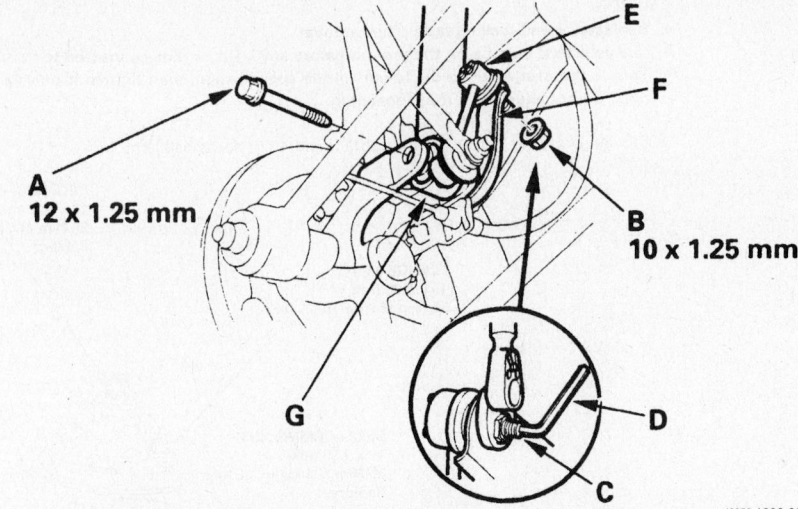

Remove the lower flange (A) bolt from the knuckle, then remove the flange nut (B) while holding the joint pin (C) with a hex wrench (D) and disconnect the stabilizer link (E) from the bracket (F)—TSX models

Coil Spring

REMOVAL & INSTALLATION

Front

CL MODEL

1. Before servicing the vehicle, refer to the precautions section.
2. Raise and support the vehicle and remove the front wheels.
3. Remove the strut (damper).
4. Place the strut assembly in a coil spring compressor.
5. Compress the coil spring and remove the locking nut from the top of the strut.
6. Release the pressure from the spring compressor.
7. Remove the coil spring and related pieces from the strut.

To install:

➡ **Use new self–locking nuts and bolts when assembling the strut.**

8. Install the strut, coil spring and related components on the spring compressor.
9. Compress the spring.
10. Install the mounting washer, and loosely install a new self–locking nut.
11. Hold the strut piston rod with a hex wrench and tighten the self–locking nut to 22 ft. lbs. (30 Nm).
12. Install the strut in the vehicle.
13. Check and adjust the vehicle's front wheel alignment.

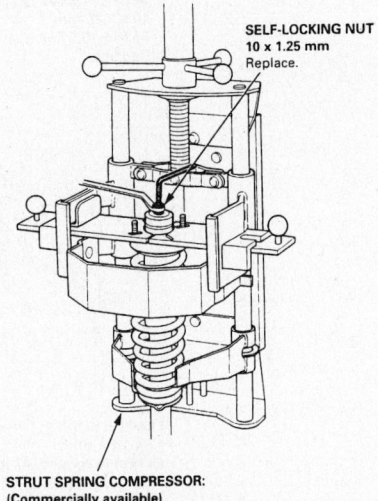

Compress the coil spring until the spring moves away from the seat and use a hex wrench to hold the piston rod while removing the nut

2002–04 RL MODELS

1. Before servicing the vehicle, refer to the precautions section.
2. Raise and support the vehicle and remove the front wheels.
3. Remove the strut (damper).
4. Place the strut assembly in a coil spring compressor.
5. Compress the coil spring and remove the locking nut from the top of the strut.

6. Release the pressure from the spring compressor.

7. Remove the coil spring and related pieces from the strut.

To install:

➡ **Use new self–locking nuts and bolts when assembling the strut.**

8. Install the strut, coil spring and related components on the spring compressor.

9. Compress the spring.

10. Install the mounting washer, and loosely install a new self–locking nut.

11. Hold the strut piston rod with a hex wrench and tighten the self–locking nut to 22 ft. lbs. (30 Nm).

12. Install the strut in the vehicle.

13. Check and adjust the vehicle's front wheel alignment.

RSX MODELS

1. Before servicing the vehicle, refer to the precautions section.

2. Raise and support the vehicle and remove the front wheels.

3. Remove the strut (damper).

4. Place the strut assembly in a coil spring compressor.

5. Compress the coil spring and remove the locking nut from the top of the strut.

6. Release the pressure from the spring compressor.

7. Remove the coil spring and related pieces from the strut.

To install:

➡ **Use new self–locking nuts and bolts when assembling the strut.**

8. Install the strut, coil spring and related components on the spring compressor.

9. Compress the spring.

10. Install the mounting washer, and loosely install a new self–locking nut.

11. Hold the strut piston rod with a hex wrench and tighten the self–locking nut to 22 ft. lbs. (30 Nm).

12. Install the strut in the vehicle.

13. Check and adjust the vehicle's front wheel alignment.

2002–03 TL MODELS

1. Before servicing the vehicle, refer to the precautions section.

2. Raise and support the vehicle and remove the front wheels.

3. Remove the strut (damper).

4. Place the strut assembly in a coil spring compressor.

5. Compress the coil spring and

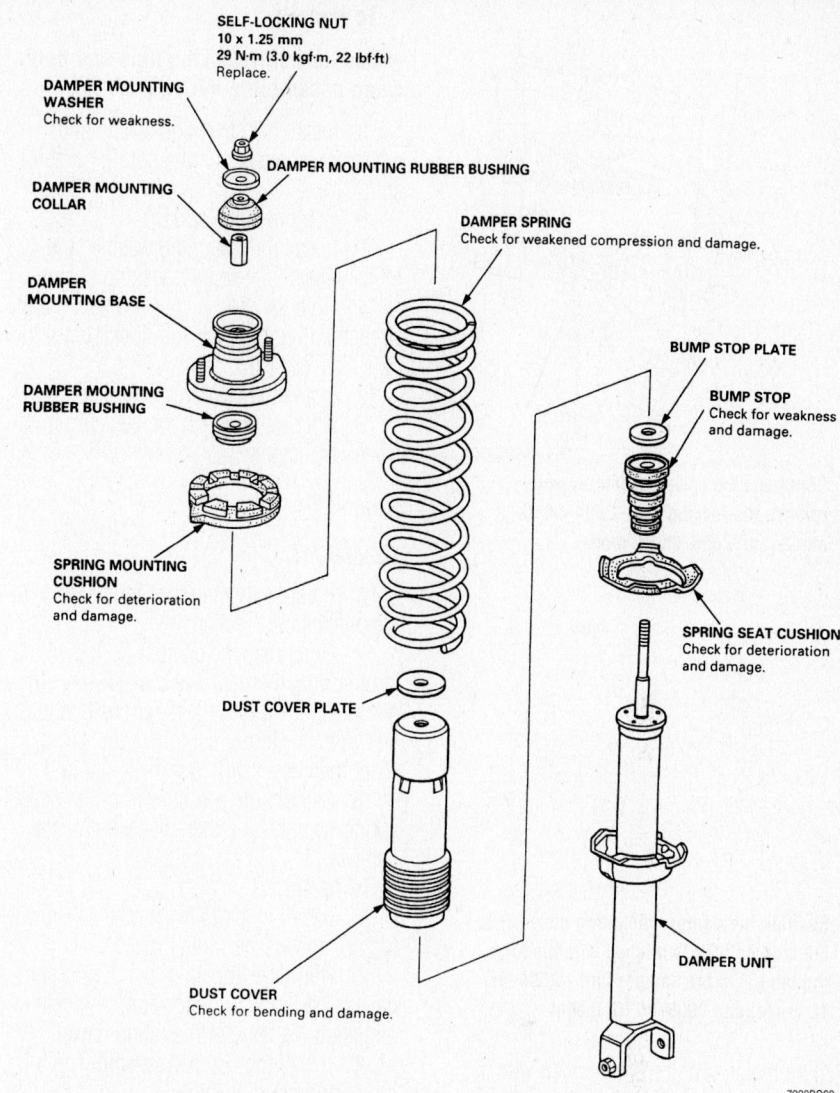

SELF-LOCKING NUT
10 x 1.25 mm
29 N·m (3.0 kgf·m, 22 lbf·ft)
Replace.

DAMPER MOUNTING WASHER
Check for weakness.

DAMPER MOUNTING COLLAR

DAMPER MOUNTING RUBBER BUSHING

DAMPER MOUNTING BASE

DAMPER MOUNTING RUBBER BUSHING

DAMPER SPRING
Check for weakened compression and damage.

SPRING MOUNTING CUSHION
Check for deterioration and damage.

BUMP STOP PLATE

BUMP STOP
Check for weakness and damage.

SPRING SEAT CUSHION
Check for deterioration and damage.

DUST COVER PLATE

DUST COVER
Check for bending and damage.

DAMPER UNIT

7923BG93

Exploded view of the rear strut (damper) on early model vehicles

remove the locking nut from the top of the strut.

6. Release the pressure from the spring compressor.

7. Remove the coil spring and related pieces from the strut.

To install:

➡ **Use new self–locking nuts and bolts when assembling the strut.**

8. Install the strut, coil spring and related components on the spring compressor.

9. Compress the spring.

10. Install the mounting washer, and loosely install a new self–locking nut.

11. Hold the strut piston rod with a hex wrench and tighten the self–locking nut to 22 ft. lbs. (30 Nm).

12. Install the strut in the vehicle.

13. Check and adjust the vehicle's front wheel alignment.

2004–06 TL AND 2005–06 RL

1. Before servicing the vehicle, refer to the precautions section.

2. Compress the damper spring with a strut spring compressor then remove the self-locking nut while holding the damper shaft with a hex wrench. Do not compress the spring more than necessary to remove the nut.

3. Release the pressure from the strut spring compressor, then disassemble the damper.

To install:

4. Reassemble the damper except for the washer and self-locking nut.

Install the damper assembly on a strut spring compressor and compress the spring lightly.

5. Align the bottom of the spring and the stepped part of the lower spring seat.

6. Position the damper mounting base

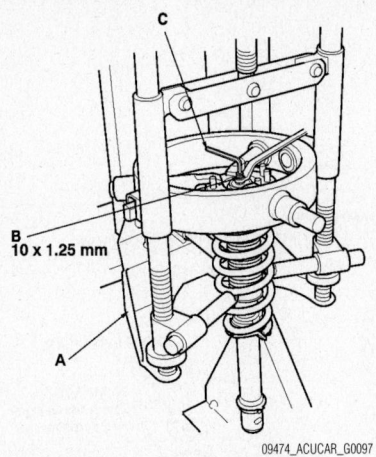

Compress the spring assembly and remove the locking nut—2004–06 TL model and 2005–06 RL model

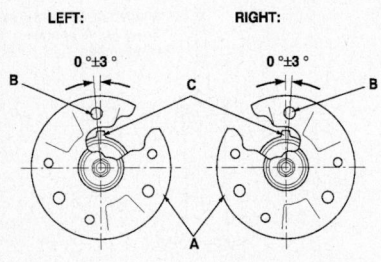

Position the damper mounting base (A) so the stud bolt (B) is aligned with the aligning tab (C) in the damper unit—2004–06 TL model and 2005–06 RL model

(A) so the stud bolt (B) is aligned with the aligning tab (C) in the damper unit.

7. Compress the damper spring. Do not compress the spring excessively.

8. Install the washer and a new 10 mm self-locking nut. Hold the damper shaft with a hex wench and tighten the 10 mm self-locking nut to the 22 ft. lbs. (29 Nm).

9. Remove the damper assembly from the strut spring compressor.

TSX MODELS

1. Before servicing the vehicle, refer to the precautions section.

2. Raise and support the vehicle and remove the front wheels.

3. Remove the strut (damper).

4. Place the strut assembly in a coil spring compressor.

5. Compress the coil spring and remove the locking nut from the top of the strut.

6. Release the pressure from the spring compressor.

7. Remove the coil spring and related pieces from the strut.

To install:

➡**Use new self–locking nuts and bolts when assembling the strut.**

8. Install the strut, coil spring and related components on the spring compressor.

9. Compress the spring.

10. Install the mounting washer, and loosely install a new self–locking nut.

11. Hold the strut piston rod with a hex wrench and tighten the self–locking nut to 22 ft. lbs. (30 Nm).

12. Install the strut in the vehicle.

13. Check and adjust the vehicle's front wheel alignment.

Rear

CL MODEL

1. Before servicing the vehicle, refer to the precautions section.

2. Compress the damper spring with a strut spring, then remove the self-locking nut while holding the damper shaft with a hex wrench. Do not compress the spring more than necessary to remove the nut.

3. Release the pressure from the strut spring compressor, then disassemble the damper.

To install:

4. Install all parts except the self-locking nut and washer onto the.

5. Align the bottom of the spring and the stepped part of the lower spring seat, and align the damper mounting base.

6. Install the damper assembly on a strut spring compressor.

7. Compress the damper spring with the spring compressor.

8. Install the washer and a new self-locking nut. Hold the damper shaft with a hex wench and tighten the 10 mm self-locking nut to the 22 ft. lbs. (29 Nm).

9. Remove the damper assembly from the strut spring compressor.

RL MODELS

1. Before servicing the vehicle, refer to the precautions section.

2. Remove the rear wheel.

3. Remove the self-locking nut and the washer while holding the joint pin with hex wrench and disconnect the stabilizer link from lower arm.

4. Place a floor jack at the connecting point of lower arm and the stabilizer link.

5. Remove the cotter pin from the lower arm ball joint, and loosen the nut.

➡**During installation, insert the new cotter pin into the ball joint pin hole from the rear to the front of vehicle, and bend its end as shown.**

6. Disconnect the lower arm ball joint from the knuckle using the ball joint remover.

7. Lower the floor jack gradually.

8. Remove the spring, spring mounting cushion and lower spring seat.

To install:

9. Install the spring mounting cushion and spring.

10. Align the bottom of the spring, the stepped part of the lower spring seat and lower arm.

11. Place the floor jack at the connecting point of lower arm and the stabilizer link.

12. Raise the jack slowly until you can align the bolt hole of lower arm and the knuckle ball joint pin, then loosely install the castle nut.

13. Install the stabilizer link on the lower arm with the washer and the new self-locking nut, and lightly tighten them.

14. Raise the rear suspension with a floor jack to load it with the vehicle's weight.

15. Tighten the castle nut and self-locking nut to the 54–61 ft. lbs. (74–83 Nm).

➡**Torque the castle nut to the lower torque specification, then tightens if only far enough to align the slot with the ball joint pin hole. Do not align the castle nut by loosening it.**

16. Insert a new cotter pin into the ball joint pin from the rear to the front of the vehicle, and bend its end.

17. Install the rear wheel.

18. Check the rear wheel alignment, and adjust it if necessary.

2004–06 TL MODELS

1. Before servicing the vehicle, refer to the precautions section.

2. Compress the damper spring with a strut spring, then remove the self-locking nut while holding the damper shaft with a hex wrench. Do not compress the spring more than necessary to remove the nut.

3. Release the pressure from the strut spring compressor, then disassemble the damper.

To install:

4. Install all parts except the self-locking nut and washer onto the.

5. Align the bottom of the spring (A) and the stepped part of the lower spring seat (B), and align the damper mounting base (C).

6. Install the damper assembly on a strut spring compressor.

7. Compress the damper spring with the spring compressor.

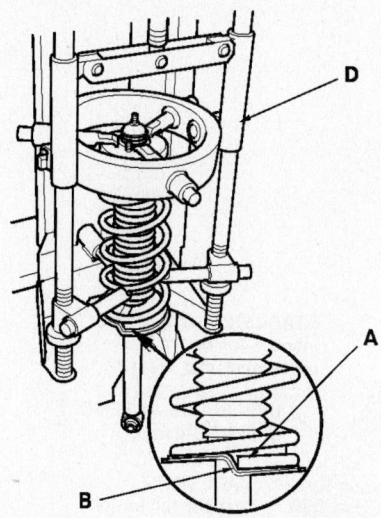

09474_ACUCAR_G0100

Align the bottom of the spring (A) and the stepped part of the lower spring seat (B), and align the damper mounting base (C)—2004–06 TL and TSX models

8. Install the washer and a new self-locking nut. Hold the damper shaft with a hex wench and tighten the 10 mm self-locking nut to the 22 ft. lbs. (29 Nm).

9. Remove the damper assembly from the strut spring compressor.

TSX MODELS

1. Before servicing the vehicle, refer to the precautions section.

2. Compress the damper spring with a strut spring, then remove the self-locking nut while holding the damper shaft with a hex wrench. Do not compress the spring more than necessary to remove the nut.

3. Release the pressure from the strut spring compressor, then disassemble the damper.

To install:

4. Install all parts except the self-locking nut and washer onto the.

5. Align the bottom of the spring (A) and the stepped part of the lower spring seat (B), and align the damper mounting base (C).

6. Install the damper assembly on a strut spring compressor.

7. Compress the damper spring with the spring compressor.

8. Install the washer and a new self-locking nut. Hold the damper shaft with a hex wench and tighten the 10 mm self-locking nut to the 22 ft. lbs. (29 Nm).

9. Remove the damper assembly from the strut spring compressor.

Upper Ball Joint

REMOVAL & INSTALLATION

➡**The upper ball joint cannot be removed from the control arm. If the ball joint is damaged, the upper arm assembly must be replaced.**

Lower Ball Joint

REMOVAL & INSTALLATION

➡**The lower ball joint cannot be removed from the steering knuckle.**

1. Before servicing the vehicle, refer to the precautions section.

2. Raise and safely support the vehicle

3. Remove or disconnect the following:
   - Wheels
   - Axle nut
   - Brake hose from the knuckle
   - Brake caliper mounting from the knuckle
   - Wheel sensor wire bracket and the sensor from the knuckle
   - Tie rod end from the knuckle
   - Lower ball joint from the control arm
   - Upper ball joint from the knuckle
   - Knuckle and hub by sliding the assembly off the halfshaft. Tap the end of the halfshaft with a plastic mallet to release it from the knuckle.
   - Hub and rotor assembly from the knuckle
   - Splash guard from the knuckle

To install:

4. Install or connect the following:
   - Splash guard and torque the bolts to 84 inch lbs. (9.5 Nm)
   - Hub assembly and tighten the self–locking bolts to 33 ft. lbs. (45 Nm)

➡**Be sure that all the hub bolts are properly tightened to avoid warpage of the brake disc.**

   - Knuckle and hub assembly onto the halfshaft
   - Tie rod and torque the nut to 36–43 ft. lbs. (49–59 Nm)
   - Upper ball joint and torque the nut to 29–35 ft. lbs. (39–47 Nm)
   - Lower ball joint and torque the nut to 36–43 ft. lbs. (49–59 Nm)
   - Wheel sensor and torque the bolts to 16 ft. lbs. (22 Nm)
   - Wheel sensor wire and torque the bolts to 84 inch lbs. (9.5 Nm)

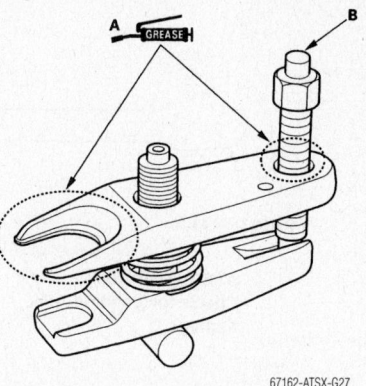

67162-ATSX-G27

Apply grease to the A & B areas on the ball joint separator tool—TSX models

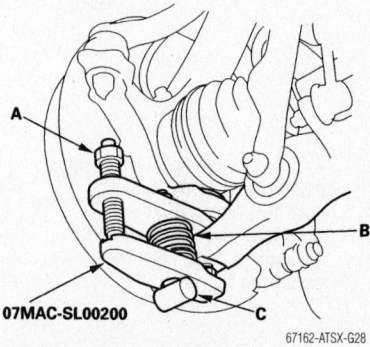

07MAC-SL00200   67162-ATSX-G28

Pressure bolt (A), adjusting bolt (B) and proper position of the head of the adjusting bolt (C)—TSX models

   - Brake caliper and torque the bolts to 80 ft. lbs. (108 Nm)
   - Brake hoses and torque the bolts to 84 inch lbs. (9.5 Nm)
   - New axle nut and torque it to 181 ft. lbs. (245 Nm)
   - Wheel

5. Measure and adjust the wheel alignment.

Upper Control Arm

REMOVAL & INSTALLATION

Front

CL MODEL

1. Before servicing the vehicle, refer to the precautions section.

2. Raise and safely support the vehicle.

3. Remove or disconnect the following:
   - Front wheel
   - Front damper/strut
   - Wheel speed sensor bracket from the upper control arm
   - Cotter pin from the upper ball joint and loosen the nut

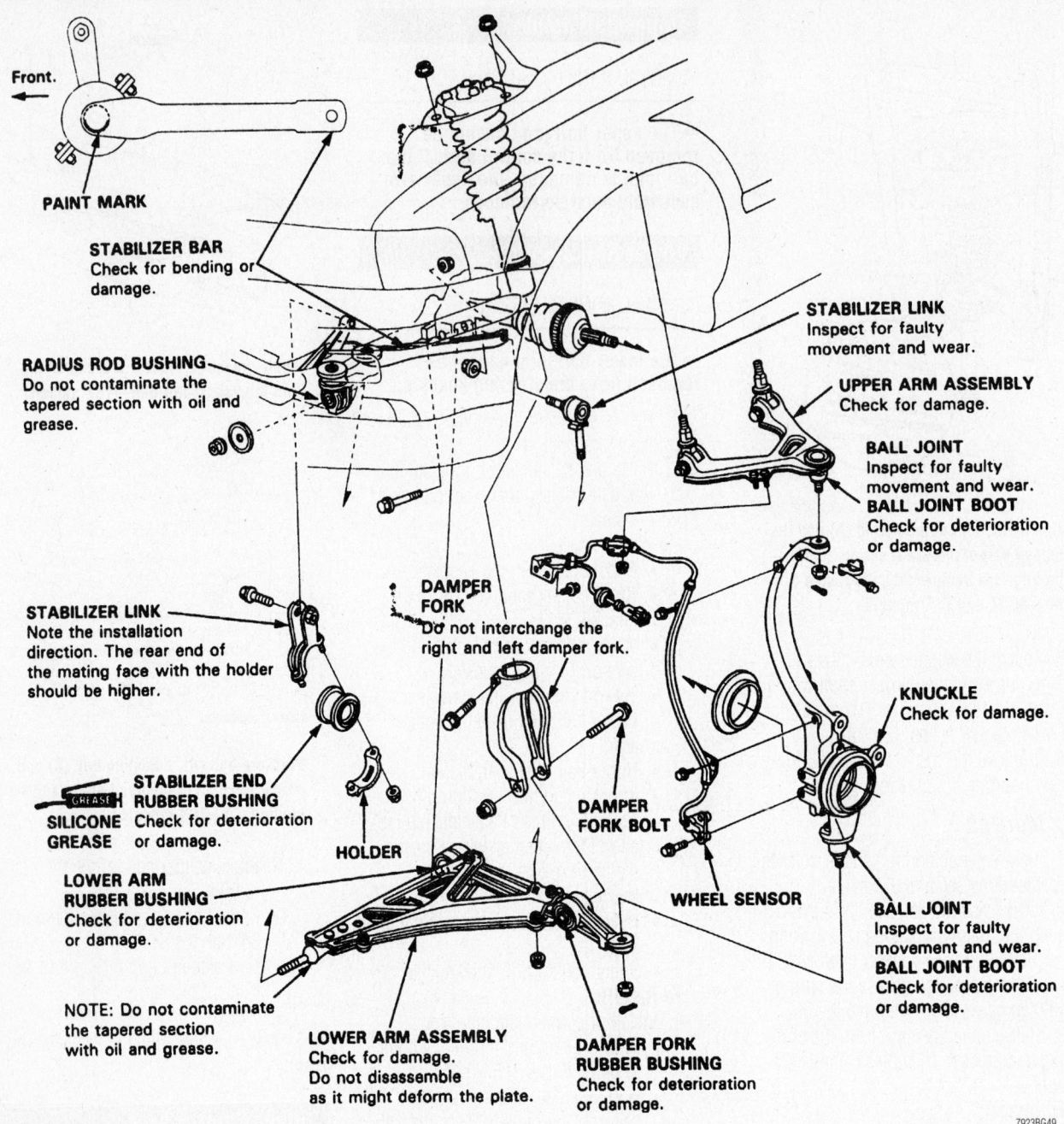

Front.

PAINT MARK

STABILIZER BAR
Check for bending or
damage.

RADIUS ROD BUSHING
Do not contaminate the
tapered section with oil and
grease.

STABILIZER LINK
Inspect for faulty
movement and wear.

UPPER ARM ASSEMBLY
Check for damage.

BALL JOINT
Inspect for faulty
movement and wear.

BALL JOINT BOOT
Check for deterioration
or damage.

STABILIZER LINK
Note the installation
direction. The rear end of
the mating face with the holder
should be higher.

DAMPER FORK
Do not interchange the
right and left damper fork.

KNUCKLE
Check for damage.

GREASE
SILICONE GREASE

STABILIZER END RUBBER BUSHING
Check for deterioration
or damage.

HOLDER

DAMPER FORK BOLT

WHEEL SENSOR

BALL JOINT
Inspect for faulty
movement and wear.

BALL JOINT BOOT
Check for deterioration
or damage.

LOWER ARM RUBBER BUSHING
Check for deterioration
or damage.

NOTE: Do not contaminate
the tapered section
with oil and grease.

LOWER ARM ASSEMBLY
Check for damage.
Do not disassemble
as it might deform the plate.

DAMPER FORK RUBBER BUSHING
Check for deterioration
or damage.

7923BG49

A common upper control arm and ball joint assembly—early model vehicle

- Upper ball joint from the knuckle, using a suitable separator tool
- Upper control arm bolts or nuts, as applicable
- Upper control arm from the vehicle

To install:
4. Install or connect the following:
- Upper control arm
- Upper control arm bolts: 23 ft. lbs. (31 Nm)
- Upper control arm-to-chassis nuts and torque them to 47 ft. lbs. (64 Nm)

- Ball joint to the steering knuckle and torque the nut to 29–35 ft. lbs. (39–47 Nm)
- Front wheel
5. Check the Front wheel alignment and adjust if necessary.

2002–04 RL MODEL

1. Before servicing the vehicle, refer to the precautions section.
2. Raise and safely support the vehicle.
3. Remove or disconnect the following:
- Front wheel
- Front damper/strut

- Wheel speed sensor bracket from the upper control arm
- Cotter pin from the upper ball joint and loosen the nut
- Upper ball joint from the knuckle, using a suitable separator tool
- Upper control arm bolts or nuts, as applicable
- Upper control arm from the vehicle

To install:
4. Install or connect the following:
- Upper control arm
- Upper control arm bolts: 23 ft. lbs. (31 Nm)

- Upper control arm-to-chassis nuts and torque them to 47 ft. lbs. (64 Nm)
- Ball joint to the steering knuckle and torque the nut to 29–35 ft. lbs. (39–47 Nm)
- Front wheel

5. Check the front wheel alignment and adjust if necessary.

2005–06 RL MODEL

1. Before servicing the vehicle, refer to the precautions section.
2. Remove the front strut assembly.
3. Remove the wheel sensor bracket from the upper arm.
4. Remove the cotter pin from the upper arm ball joint, then loosen the nut.

➡**During installation, insert the new cotter pin into the ball joint pin hole from the front to the rear of vehicle, and bend its end.**

5. Disconnect the upper arm ball joint from the knuckle using a suitable ball joint separator tool.
6. Remove the upper arm mounting bolts then remove the upper arm.

To install:

7. Install the upper arm by inserting a 6mm diameter 300mm long rod into the positioning holes and place the upper arm on the rod to position it before tightening the upper arm mounting bolts.
8. Install the remaining parts in the reverse order of removal, keeping in mind the following:

a. Be careful not to damage the ball joint boot when installing the knuckle.

b. First install all the components and lightly tighten the bolts and nuts, then raise the suspension to load it with the vehicle's weight before fully tightening to the specified torque values. Do not place the jack against the ball joint pin of the knuckle. Tighten the upper arm bolts to 23 ft. lbs. (31 Nm). Tighten ball joint nut to 43–51 ft. lbs. (59–69 Nm) and install a new cotter pin.

RSX MODELS

1. Before servicing the vehicle, refer to the precautions section.
2. Raise and safely support the vehicle.
3. Remove or disconnect the following:

- Front wheel
- Front damper/strut
- Wheel speed sensor bracket from the upper control arm
- Cotter pin from the upper ball joint and loosen the nut

- Upper ball joint from the knuckle, using a suitable separator tool
- Upper control arm bolts or nuts, as applicable
- Upper control arm from the vehicle

To install:

4. Install or connect the following:

- Upper control arm
- Upper control arm bolts: 23 ft. lbs. (31 Nm)
- Upper control arm-to-chassis nuts and torque them to 47 ft. lbs. (64 Nm)
- Ball joint to the steering knuckle and torque the nut to 29–35 ft. lbs. (39–47 Nm)
- Front wheel

5. Check the front wheel alignment and adjust if necessary.

2002–03 TL MODELS

1. Before servicing the vehicle, refer to the precautions section.
2. Raise and safely support the vehicle.
3. Remove or disconnect the following:

- Front wheel
- Front damper/strut
- Wheel speed sensor bracket from the upper control arm
- Cotter pin from the upper ball joint and loosen the nut
- Upper ball joint from the knuckle, using a suitable separator tool
- Upper control arm bolts or nuts, as applicable
- Upper control arm from the vehicle

To install:

4. Install or connect the following:

- Upper control arm
- Upper control arm bolts: 23 ft. lbs. (31 Nm)
- Upper control arm-to-chassis nuts and torque them to 47 ft. lbs. (64 Nm)
- Ball joint to the steering knuckle and torque the nut to 29–35 ft. lbs. (39–47 Nm)
- Front wheel

5. Check the front wheel alignment and adjust if necessary.

2004–06 TL MODELS

1. Before servicing the vehicle, refer to the precautions section.
2. Remove the front strut assembly.
3. Remove the wheel sensor bracket from the upper arm.
4. Remove the cotter pin from the upper arm ball joint, then loosen the nut.

➡**During installation, insert the new cotter pin into the ball joint pin hole**

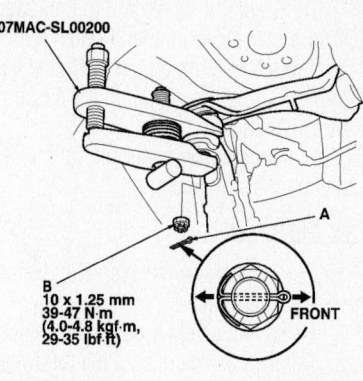

07MAC-SL00200

10 x 1.25 mm
39-47 N·m
(4.0-4.8 kgf·m,
29-35 lbf·ft)

09474_ACUCAR_G0101

Use the following tool to separate the ball joint and tighten the nut as shown— 2004–06 TL model

from the front to the rear of vehicle, and bend its end as shown.

5. Disconnect the upper arm ball joint from the knuckle using the tool illustrated.
6. Remove the upper arm mounting bolts then remove the upper arm.

To install:

7. Install the upper arm by inserting a 6mm diameter 300mm long rod into the positioning holes and place the upper arm on the rod to position it before tightening the upper arm mounting bolts.
8. Install the remaining parts in the reverse order of removal, keeping in mind the following:

a. Be careful not to damage the ball joint boot when installing the knuckle.

b. First install all the components and lightly tighten the bolts and nuts, then raise the suspension to load it with the vehicle's weight before fully tightening to the specified torque values. Do not place the jack against the ball joint pin of the knuckle.

c. Tighten all mounting hardware to specified torque values.

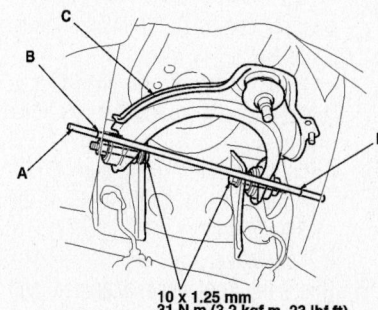

10 x 1.25 mm
31 N·m (3.2 kgf·m, 23 lbf·ft)

09474_ACUCAR_G0102

Insert a 6mm diameter 300mm long rod into the positioning holes and place the upper arm on the rod to position it before tightening the upper arm mounting bolts— 2004–06 TL model

d. Torque the castle nut to the lower torque specification, then tighten it only far enough to align the slot with the ball joint pin hole. Do not align the castle nut by loosening it.

e. Install a new cotter pin on the castle nut after tightening.

TSX MODELS

1. Before servicing the vehicle, refer to the precautions section.
2. Raise and safely support the vehicle.
3. Remove or disconnect the following:
   - Front wheel
   - Front damper/strut
   - Wheel speed sensor bracket from the upper control arm
   - Cotter pin from the upper ball joint and loosen the nut
   - Upper ball joint from the knuckle, using a suitable separator tool
   - Upper control arm bolts or nuts, as applicable
   - Upper control arm from the vehicle

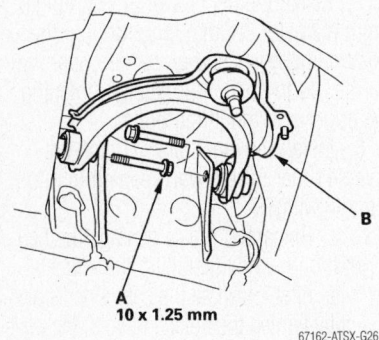

10 x 1.25 mm

67162-ATSX-G26

Upper control arm (B) and bolt (A) mounting—TSX model

To install:

4. Install or connect the following:
   - Upper control arm
   - Upper control arm bolts: 23 ft. lbs. (31 Nm)
   - Upper control arm-to-chassis nuts and torque them to 47 ft. lbs. (64 Nm)
   - Ball joint to the steering knuckle and torque the nut to 29–35 ft. lbs. (39–47 Nm)
   - Front wheel
5. Check the front wheel alignment and adjust if necessary.

Rear

2005–06 RL MODEL

1. Before servicing the vehicle, refer to the precautions section.

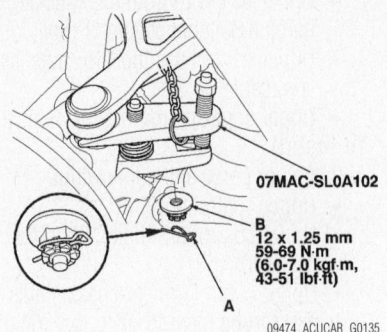

07MAC-SL0A102

B
12 x 1.25 mm
59-69 N·m
(6.0-7.0 kgf·m,
43-51 lbf·ft)

A

09474_ACUCAR_G0135

Ball joint removal/installation, rear upper arm—2005–06 RL model

2. Remove the rear wheel.
3. Remove the control arm mounting nut) and washer from the knuckle side.
4. Place a floor jack under the trailing arm, and support the suspension.
5. Remove the wheel sensor harness bracket from the upper arm.

➡ **Use a new bracket on reassembly.**

6. Remove the lock pin from the upper arm ball joint, and loosen the nut.

➡ **During installation, insert the new lock pin as shown after tightening the nut.**

7. Disconnect the upper arm ball joint from the knuckle using the ball joint remover.
8. Remove the flange bolts and nuts and remove the upper arm.

To install:

9. Install the upper arm in the reverse order of removal, and please note the following:

   a. First install all the suspension components and lightly tighten the bolts and nuts, then raise the suspension to load it with the vehicle's weight before fully tightening the bolts and nuts to the specified torque.

   b. Tighten all the mounting hardware to the specified torque.

   c. Be careful not to damage the ball joint boot when installing the knuckle.

   d. Before connecting the ball joint to the knuckle, degrease the threaded section and tapered portion of the ball joint pin, the connecting hole, and the threaded section and mating surface of the castle nut.

   e. Torque the castle nut to the lower torque specification, then tighten it only far enough to align the slot with the ball joint pin hole. Do not align the castle nut by loosening it.

   f. Check the wheel alignment, and adjust it if necessary.

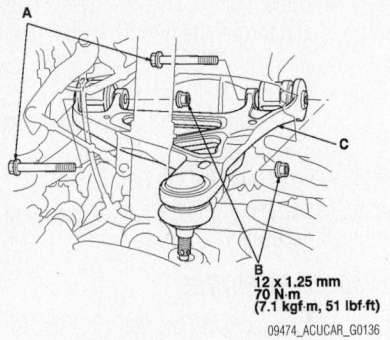

A

C

B
12 x 1.25 mm
70 N·m
(7.1 kgf·m, 51 lbf·ft)

09474_ACUCAR_G0136

Exploded view of the rear upper arm mounting—2005–06 RL model

2002–03 TL MODELS

1. Before servicing the vehicle, refer to the precautions section.
2. Raise and safely support the vehicle.
3. Remove or disconnect the following:
   - Front wheels
   - Lower damper fork bolt
   - Stabilizer bar from the arm
   - Lower ball joint from the steering knuckle
   - Radius rod from the lower control arm
   - Lower control arm mounting bolts
   - Lower arm from the vehicle

To install:

4. Install or connect the following:
   - Lower control arm and torque the bolts to 40 ft lbs. (54 Nm)
   - Lower ball joint to the steering knuckle and torque the nut to 36–43 ft. lbs. (49–59 Nm)
   - Stabilizer bar and torque the bolts to 16 ft. lbs. (22 Nm)
   - Lower damper fork bolt and torque it to 47 ft. lbs. (64 Nm)
   - Radius rod and torque the bolts to 76 ft. lbs. (103 Nm)
   - Front wheels
5. Measure and adjust the wheel alignment.

2004–06 TL MODELS

1. Before servicing the vehicle, refer to the precautions section.
2. Remove the rear wheel.
3. Remove the control arm mounting nut) and washer from the knuckle side.
4. Mark the cam positions of the adjusting bolt and adjusting cam, then remove the self-locking nut, adjusting cam, and adjusting bolt. Discard the self-locking nut and control arm mounting nut.
5. Remove the control arm.

To install:

6. Installation is the reverse of removal, keep in mind the following:

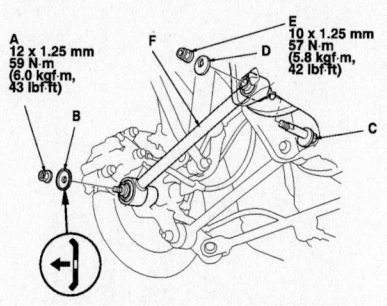

Exploded view of the rear upper control arm assembly—2004-06 TL model

a. Align the cam positions of the adjusting bolt and adjusting cam with the marked positions when tightening.

b. Use a new self-locking nut and control arm mounting nut on reassembly.

c. Tighten all mounting hardware to the torque specified in the accompanying illustration.

d. First install all the components and lightly tighten the bolts and nuts, then raise the suspension to load it with the vehicle's weight before fully tightening to the final torque specs.

e. Check the wheel alignment, and adjust if necessary.

CONTROL ARM BUSHING REPLACEMENT

➡ **The bushings are an integral part of the control arm and are not replaceable. If they are damaged the control arm should be replaced.**

Lower Control Arm

REMOVAL & INSTALLATION

Front

CL MODEL

1. Before servicing the vehicle, refer to the precautions section.
2. Raise and safely support the vehicle.
3. Remove or disconnect the following:
   - Front wheels
   - Lower damper fork bolt
   - Stabilizer bar from the arm
   - Lower ball joint from the steering knuckle
   - Radius rod from the lower control arm
   - Lower control arm mounting bolts
   - Lower arm from the vehicle

To install:
4. Install or connect the following:
   - Lower control arm and torque the bolts to 40 ft lbs. (54 Nm)

- Lower ball joint to the steering knuckle and torque the nut to 36–43 ft. lbs. (49–59 Nm)
- Stabilizer bar and torque the bolts to 16 ft. lbs. (22 Nm)
- Lower damper fork bolt and torque it to 47 ft. lbs. (64 Nm)
- Radius rod and torque the bolts to 76 ft. lbs. (103 Nm)
- Front wheels

5. Measure and adjust the wheel alignment.

2002-04 RL MODEL

1. Before servicing the vehicle, refer to the precautions section.
2. Raise and safely support the vehicle.
3. Remove or disconnect the following:
   - Front wheels
   - Lower damper fork bolt
   - Stabilizer bar from the arm
   - Lower ball joint from the steering knuckle
   - Radius rod from the lower control arm
   - Lower control arm mounting bolts
   - Lower arm from the vehicle

To install:
4. Install or connect the following:
   - Lower control arm and torque the bolts to 40 ft lbs. (54 Nm)
   - Lower ball joint to the steering knuckle and torque the nut to 36–43 ft. lbs. (49–59 Nm)
   - Stabilizer bar and torque the bolts to 16 ft. lbs. (22 Nm)
   - Lower damper fork bolt and torque it to 47 ft. lbs. (64 Nm)
   - Radius rod and torque the bolts to 76 ft. lbs. (103 Nm)
   - Front wheels

5. Measure and adjust the wheel alignment.

2005-06 RL MODEL

1. Remove the front wheels.
2. Remove the holder from the knuckle.
3. Remove the cotter pin from the lower ball joint castle nut and then remove the nut.

➡ **To avoid damaging the ball joint, install a hex nut on the threads of the ball joint. Be careful not to damage the ball joint boot when installing the remover. Do not force or hammer on the lower arm, or pry between the lower arm and knuckle. You could damage the ball joint. Insert the new cotter pin into the ball joint pin hole from the rear to the front of vehicle, and bend its end as shown.**

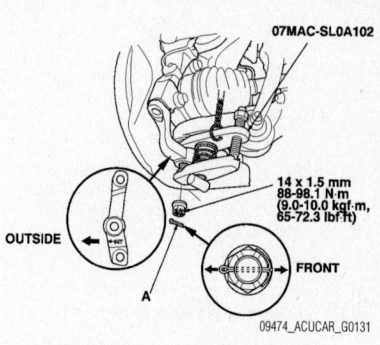

Ball joint removal/installation, front lower arm—2005-06 RL model

4. Disconnect the lower arm ball joint from the holder using the ball joint remover.
5. Remove the self-locking nut and washer, then remove the damper fork mounting nut and bolt.
6. Remove the flange bolts and self-locking nut, then remove the lower arm.

To install:
7. In case of loosening lower arm bracket, follow the steps to retighten its bolts, refer to the illustration for bolt locations:
   a. Lightly tighten the three bolts.
   b. Align the center of the cam (B) to the edge (C).
   c. Tighten the adjusting bolt (D).
   d. Tighten the adjusting bolt (E).
   e. Tighten the adjusting bolt (F).
8. Install the lower arm in the reverse order of removal, and note these items:
9. Be careful not to damage the ball joint boot when installing the knuckle.
10. Before connecting the lower ball joint to the holder, degrease the threaded section and tapered portion of the ball joint pin, the holder connecting hole, the threaded section and mating surface of the castle nut.
11. Use a new self-locking nut on reassembly.

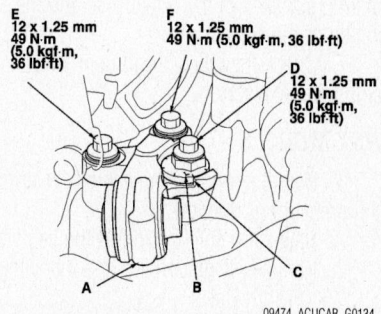

In case of loosening the front lower arm bracket, follow the steps to retighten its bolts—2005-06 RL model

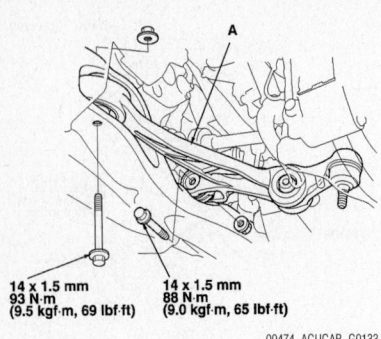

Exploded view of the front lower arm and the flange bolts—2005–06 RL model

14 x 1.5 mm
93 N·m
(9.5 kgf·m, 69 lbf·ft)

14 x 1.5 mm
88 N·m
(9.0 kgf·m, 65 lbf·ft)

09474_ACUCAR_G0133

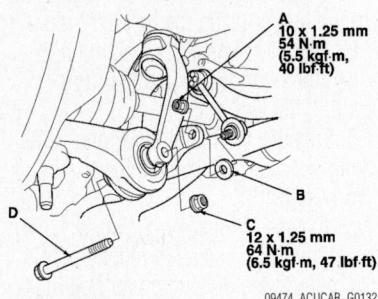

A
10 x 1.25 mm
54 N·m
(5.5 kgf·m, 40 lbf·ft)

C
12 x 1.25 mm
64 N·m
(6.5 kgf·m, 47 lbf·ft)

09474_ACUCAR_G0132

Exploded view of the self-locking nut and washer and damper fork mounting nut and bolt—2005–06 RL model

12. First install all the components and lightly tighten the bolts and nuts, then raise the suspension to load it with the vehicle's weight before fully tightening to the specified torque values. Do not place the jack against the ball joint pin of the knuckle.

13. Tighten all mounting hardware to the specified torque values.

14. Torque the castle nut to the lower torque specification, then tighten it only far enough to align the slot with the ball joint pin hole. Do not align the castle nut by loosening it.

15. Install a new cotter pin on the castle nut after torquing.

16. Before installing the wheel, clean the mating surfaces of the brake disc and the inside of the wheel.

17. Check the wheel alignment, and adjust if necessary.

RSX MODELS

1. Before servicing the vehicle, refer to the precautions section.
2. Raise and safely support the vehicle.
3. Remove or disconnect the following:
   - Front wheels
   - Lower damper fork bolt
   - Stabilizer bar from the arm
   - Lower ball joint from the steering knuckle

- Radius rod from the lower control arm
- Lower control arm mounting bolts
- Lower arm from the vehicle

To install:
4. Install or connect the following:
   - Lower control arm and torque the bolts to 40 ft lbs. (54 Nm)
   - Lower ball joint to the steering knuckle and torque the nut to 36–43 ft. lbs. (49–59 Nm)
   - Stabilizer bar and torque the bolts to 16 ft. lbs. (22 Nm)
   - Lower damper fork bolt and torque it to 47 ft. lbs. (64 Nm)
   - Radius rod and torque the bolts to 76 ft. lbs. (103 Nm)
   - Front wheels
5. Measure and adjust the wheel alignment.

2002–03 TL MODELS

1. Before servicing the vehicle, refer to the precautions section.
2. Raise and safely support the vehicle.
3. Remove or disconnect the following:
   - Front wheels
   - Lower damper fork bolt
   - Stabilizer bar from the arm
   - Lower ball joint from the steering knuckle
   - Radius rod from the lower control arm
   - Lower control arm mounting bolts
   - Lower arm from the vehicle

To install:
4. Install or connect the following:
   - Lower control arm and torque the bolts to 40 ft lbs. (54 Nm)
   - Lower ball joint to the steering knuckle and torque the nut to 36–43 ft. lbs. (49–59 Nm)
   - Stabilizer bar and torque the bolts to 16 ft. lbs. (22 Nm)
   - Lower damper fork bolt and torque it to 47 ft. lbs. (64 Nm)
   - Radius rod and torque the bolts to 76 ft. lbs. (103 Nm)
   - Front wheels
5. Measure and adjust the wheel alignment.

2004–06 TL MODELS

1. Before servicing the vehicle, refer to the precautions section.
2. Remove the front wheels.
3. Remove the damper fork from the damper and lower arm.

➡**During installation, insert the damper fork into the damper lower end so the aligning tab (D) is aligned with**

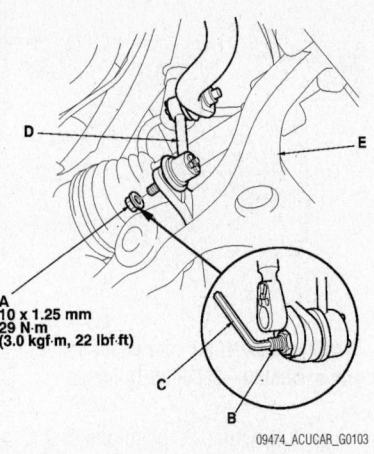

A
10 x 1.25 mm
29 N·m
(3.0 kgf·m, 22 lbf·ft)

09474_ACUCAR_G0103

Remove the flange nut while holding the joint pin with a hex wrench—2004–06 TL model

the slot (E) in the damper fork. Replace the damper fork mounting nut (F) with a new one.

4. Remove the flange nut while holding the joint pin with a hex wrench then disconnect the stabilizer link from the lower arm.

5. Remove the cotter pin from the lower ball joint castle nut, then remove the nut.

➡**To avoid damaging the ball joint, install the tool illustrated on the threads of the ball joint.**

6. Be careful not to damage the ball joint boot when installing the remover.

7. Do not force or hammer on the lower arm, or pry between the lower arm and knuckle. You could damage the ball joint.

8. Disconnect the lower arm ball joint from the knuckle using the tools shown.

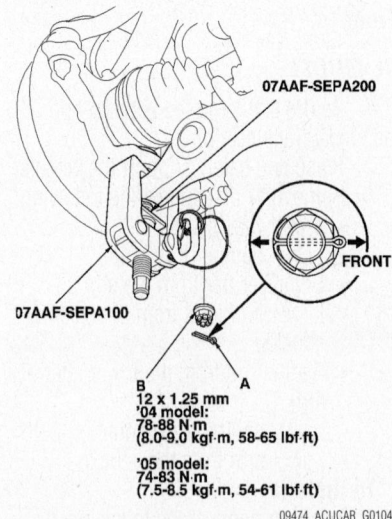

07AAF-SEPA200

07AAF-SEPA100

FRONT

B
12 x 1.25 mm
'04 model:
78–88 N·m
(8.0–9.0 kgf·m, 58–65 lbf·ft)

'05 model:
74–83 N·m
(7.5–8.5 kgf·m, 54–61 lbf·ft)

09474_ACUCAR_G0104

Removing/installing the lower ball joint—2004–06 TL model

vehicle's weight before fully tightening to the specified torque values. Do not place the jack against the ball joint pin of the knuckle.

f. Torque the castle nut to the lower torque specification, then tighten it only far enough to align the slot with the ball joint pin hole. Do not align the castle nut by loosening it.

g. Install a new cotter pin on the castle nut after torquing.

h. Check the wheel alignment, and adjust if necessary.

TSX MODELS

1. Before servicing the vehicle, refer to the precautions section.
2. Raise and safely support the vehicle.
3. Remove or disconnect the following:
   - Front wheels
   - Damper fork from the damper and lower control arm
   - Flange nut, while holding the joint pin with a hex wrench
   - Stabilizer link from the lower control arm
   - Cotter pin and nut from the lower ball joint
   - Lower control arm from the knuckle using a suitable separator tool
   - Flange bolts and lower control arm

To install:

4. Installation is the reverse of the removal procedure, noting the following steps and torque specifications:

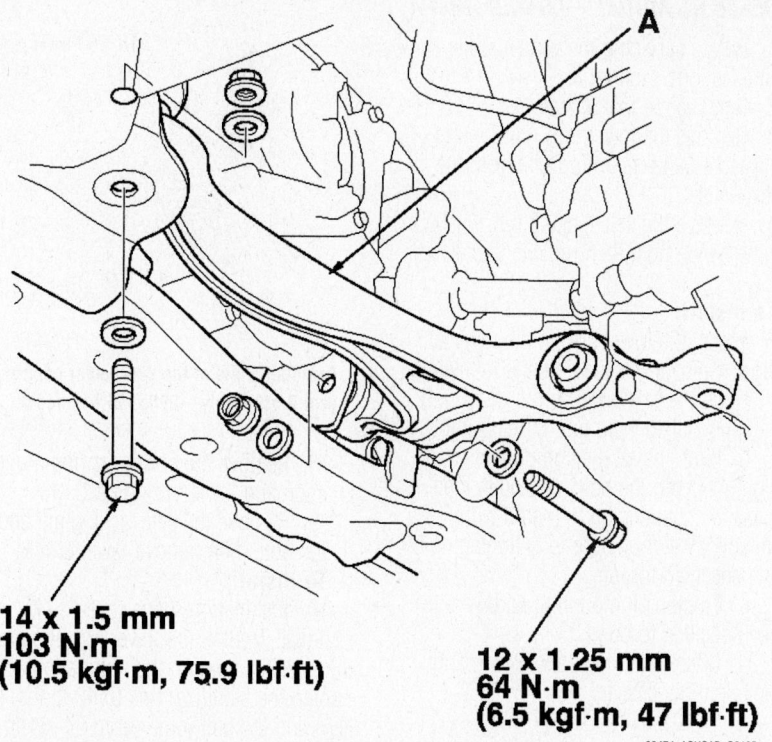

14 x 1.5 mm
103 N·m
(10.5 kgf·m, 75.9 lbf·ft)

12 x 1.25 mm
64 N·m
(6.5 kgf·m, 47 lbf·ft)

09474_ACUCAR_G0105

Removing/installing the flange bolts, nuts, washers and lower arm—2004–06 TL model

9. Remove the flange bolts, nuts and washers, then remove the lower arm.

To install:

10. Install the lower arm in the reverse order of removal, and note these items:

a. Check the collar sleeve (A) on the lower arm (B). Replace the lower arm if the collar/sleeve is loose or damaged.

b. Be careful not to damage the ball joint boot when installing the knuckle.

c. Tighten all mounting hardware to the torque specified in the accompanying illustrations.

d. Before connecting the lower ball joint to the lower arm, degrease the threaded section and tapered portion of

the ball joint pin, the lower arm connecting hole, the threaded section and mating surface of the castle nut.

e. First install all the components and lightly tighten the bolts and nuts, then raise the suspension to load it with the

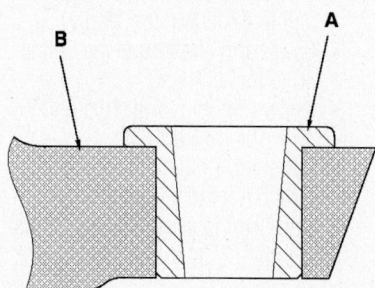

09474_ACUCAR_G0106

Check the collar sleeve (A) on the lower arm (B). Replace the lower arm if the collar/sleeve is loose or damaged—2004–06 TL model

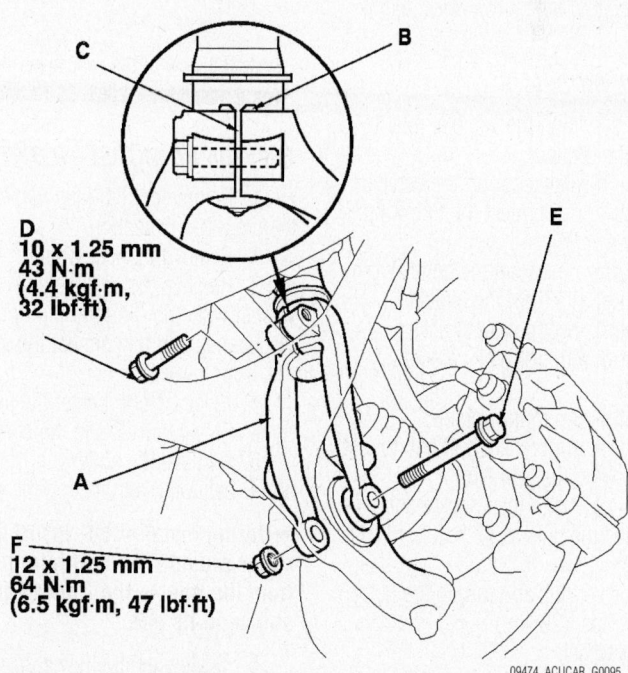

D
10 x 1.25 mm
43 N·m
(4.4 kgf·m, 32 lbf·ft)

F
12 x 1.25 mm
64 N·m
(6.5 kgf·m, 47 lbf·ft)

09474_ACUCAR_G0095

Exploded view of the damper fork and related components and their torque specifications—2004–06 TL model

- Lower control arm flange bolts: 14x1.5mm bolt to 61 ft. lbs. (83 Nm) and 12x1.25mm bolt to 47 ft. lbs. (64 Nm)
- Lower ball joint castle nut: 58–65 ft. lbs. (78–88 Nm)
- Stabilizer link-to-control arm flange nut: 22 ft. lbs. (29 Nm)
- Insert the damper fork into the damper lower end so the aligning tab is aligned with the slot in the damper fork. Use a new damper fork mounting nut
- Lower damper fork mounting bolt: 47 ft. lbs. (64 Nm)
- Upper damper fork mounting bolt: 32 ft. lbs. (44 Nm)
- Check and adjust the front wheel alignment

Rear

CL MODEL

1. Before servicing the vehicle, refer to the precautions section.
2. Raise and safely support the vehicle.
3. Remove or disconnect the following:
   - Rear wheels
   - Control arm from the knuckle
   - Control arm from the subframe
   - Control arm from the vehicle

To install:

4. Install the control arm to the vehicle and torque the subframe nut to 40 ft. lbs. (54 Nm) and the knuckle nut to 43 ft. lbs. (59 Nm).
5. Measure and adjust the wheel alignment.

2002–04 RL MODEL

1. Before servicing the vehicle, refer to the precautions section.
2. Raise and safely support the vehicle.
3. Remove or disconnect the following:
   - Rear wheels
   - Stabilizer link from the control arm
   - Control arm from the knuckle
   - Control arm from the bracket
   - Control arm from the vehicle

To install:

- Control arm to the vehicle
- Control arm bracket bolt and torque it to 47 ft. lbs. (64 Nm)
- Control arm to the knuckle and torque the nuts to 47 ft. lbs. (64 Nm)
- Stabilizer link and torque the nut to 22 ft. lbs. (29 Nm)
- Rear wheels

4. Measure and adjust the wheel alignment.

2005–06 RL MODEL—LOWER ARM A

1. Before servicing the vehicle, refer to the precautions section.
2. Remove the rear wheel.
3. Remove the lower arm A mounting nut, washers and mounting bolt from the knuckle side.
4. Remove the self-locking nut, washers and the flange bolt, then remove the lower arm A.

To install:

5. Install the lower arm in the reverse order of removal, please note the following:
   a. Use a new lower arm A mounting nut and self-locking nut on reassembly.
   b. First, install the components and lightly tighten the bolts and nuts, then raise the suspension to load it with the vehicle's weight before fully tightening to the specified torque.
   c. Tighten all mounting hardware to the specified torque.
   d. Check the wheel alignment, and adjust it if necessary.

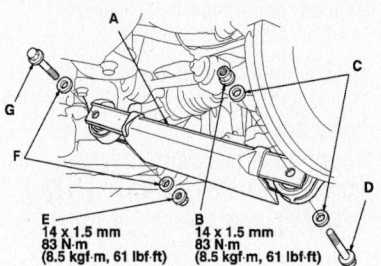

09474_ACUCAR_G0137

Exploded view of the rear lower control arm A assembly—2005–06 RL model

2005–06 RL MODEL—LOWER ARM B

1. Before servicing the vehicle, refer to the precautions section.
2. Remove the rear wheel.
3. Remove the self-locking nut and the washer while holding the joint pin with a hex wrench and disconnect the stabilizer link from lower arm B.
4. Place a floor jack at the connecting point of lower arm B and the stabilizer link.
5. Remove the cotter pin from the lower arm B ball joint, and loosen the nut.

➡**During installation, insert the new cotter pin into the ball joint pin hole from the rear to the front of vehicle, and bend its end.**

6. Disconnect the lower arm ball joint from the knuckle using a separator tool.
7. Lower the floor jack gradually.

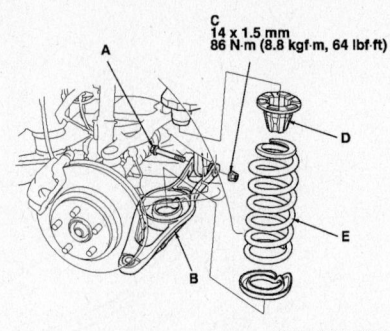

09474_ACUCAR_G0138

Exploded view of the rear lower control arm B assembly—2005–06 RL model

8. Remove the spring, spring mounting cushion and lower spring seat.
9. Remove the self-locking nut and flange bolt, then remove lower arm B.

To install:

10. Installation is the reverse of removal. Tighten the lower arm B bolts to the specifications in the illustration and tighten the stabilizer link to 40 ft. lbs. (54 Nm) and the ball joint nut to 54–61 ft. lbs. (74–83 Nm).

RSX MODELS

1. Before servicing the vehicle, refer to the precautions section.
2. Raise and safely support the vehicle.
3. Remove or disconnect the following:
   - Rear wheels
   - Hub/bearing assembly
   - Splash guard
   - Lower strut mounting bolt
   - Upper arm from the control arm
   - Lower arm from the control arm
   - Compensator arm
   - Control arm mounting bolts
   - Control arm from the vehicle

To install:

4. Install or connect the following:
   - Control arm and torque the bolts to 47 ft. lbs. (64 Nm)
   - Compensator arm and torque the bolt to 47 ft. lbs. (64 Nm)
   - Lower arm and torque the bolt to 40 ft. lbs. (54 Nm)
   - Upper arm and torque the bolt to 40 ft. lbs. (54 Nm)
   - Lower strut mounting bolt and torque it to 40 ft. lbs. (54 Nm)
   - Hub/bearing assembly and torque the nut to 134 ft. lbs. (181 Nm)

2002–03 TL MODELS

1. Before servicing the vehicle, refer to the precautions section.
2. Raise and safely support the vehicle.
3. Remove or disconnect the following:

- Rear wheels
- Control arm from the knuckle
- Control arm from the subframe
- Control arm from the vehicle

To install:

4. Install the control arm to the vehicle and torque the subframe nut to 40 ft. lbs. (54 Nm) and the knuckle nut to 43 ft. lbs. (59 Nm).

5. Measure and adjust the wheel alignment.

2004–06 TL MODELS

1. Before servicing the vehicle, refer to the precautions section.

2. Remove the rear wheel.

3. Remove the lower arm mounting nut and mounting bolt from the knuckle side.

4. Remove the flange bolt and the lower arm.

To install:

5. Installation is the reverse of removal, keep in mind the following:

   a. Align the cam positions of the adjusting bolt and adjusting cam with the marked positions when tightening.

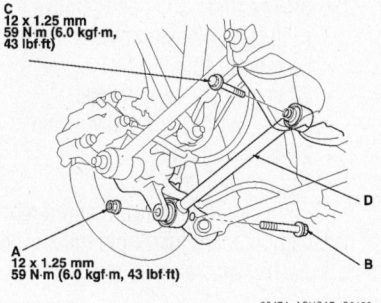

C
12 x 1.25 mm
59 N·m (6.0 kgf·m, 43 lbf·ft)

A
12 x 1.25 mm
59 N·m (6.0 kgf·m, 43 lbf·ft)

09474_ACUCAR_G0108

Exploded view of the rear lower control arm assembly—2004–06 TL model

   b. Use a new lower arm mounting nut on reassembly.

   c. Tighten all mounting hardware to the torque specified in the accompanying illustration.

   d. First install all the components and lightly tighten the bolts and nuts, then raise the suspension to load it with the vehicle's weight before fully tightening to the final torque specs.

   e. Check the wheel alignment, and adjust if necessary.

TSX—LOWER ARM A

1. Before servicing the vehicle, refer to the precautions section.

2. Raise and safely support the vehicle.

3. Remove or disconnect the following:

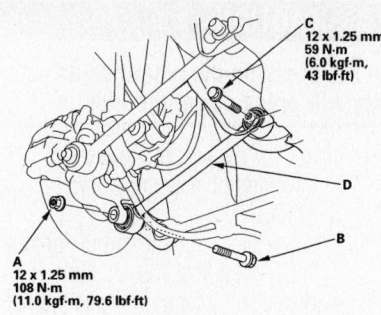

C
12 x 1.25 mm
59 N·m
(6.0 kgf·m,
43 lbf·ft)

D

B

A
12 x 1.25 mm
108 N·m
(11.0 kgf·m, 79.6 lbf·ft)

67162-ATSX-G29

Lower arm mounting A—TSX models

- Rear wheels
- Lower arm mounting nut and mounting bolt from the knuckle side
- Flange bolt
- Lower arm

4. Installation is the reverse of the removal procedure. Tighten the flange bolt to 43 ft. lbs. (59 Nm) and the mounting nut to 79.6 ft. lbs. (108 Nm).

TSX—LOWER CONTROL ARM B

1. Before servicing the vehicle, refer to the precautions section.

2. Raise and safely support the vehicle.

3. Remove or disconnect the following:

- Rear wheels
- Control arm mounting nut and washer from the knuckle side

4. Matchmark the cam positions of the adjusting bolt and adjusting cam, then remove the self-locking nut and adjusting cam and adjusting bolt. Discard the self-locking nut and control arm mounting nut.

- Lower control arm

5. Installation is the reverse of the removal procedure. Tighten the self-locking nut to 40 ft. lbs. (54 Nm) and the mounting nut to 51 ft. lbs. (69 Nm).

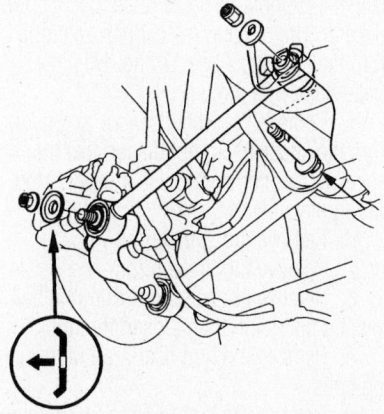

67162-ATSX-G30

Control arm mounting B—TSX

CONTROL ARM BUSHING REPLACEMENT

➡**The bushings are an integral part of the control arm and are not replaceable. If they are damaged the control arm should be replaced.**

Wheel Bearings

ADJUSTMENT

The front and rear wheel bearings are not adjustable or repairable and should be replaced if found defective.

REMOVAL & INSTALLATION

Front

CL MODEL

1. Before servicing the vehicle, refer to the precautions section.

2. Raise and safely support the vehicle.

3. Remove or disconnect the following:

- Wheel
- Axle nut
- Wheel sensor and wire brackets from the knuckle
- Brake caliper from the knuckle
- Brake rotor from the knuckle
- Tie rod from the knuckle
- Lower control arm from the knuckle
- Upper arm from the knuckle
- Knuckle/hub assembly from the halfshaft

4. Clamp the knuckle in a vise and secure a slide hammer to the wheel studs to separate the hub from the knuckle.

5. Remove the splash guard.

6. Remove the snapring from the knuckle.

7. Support the knuckle and press the bearing out towards the wheel side.

8. If the inner bearing race stayed on the hub, use a puller to remove it.

To install:

9. Press a new inner race on the hub.

10. Press a new bearing into the knuckle.

11. Install the outer snapring.

12. Install the splash guard and torque the bolts to 4 ft. lbs. (4.9 Nm).

13. Properly support the knuckle and press the hub into the bearing.

❉❉ CAUTION

Do not press on the wheel studs or they will press out of the hub.

14. Install or connect the following:
- Knuckle/hub assembly onto the halfshaft
- Lower ball joint and torque the nut to 51–58 ft. lbs. (69–78 Nm)
- Upper ball joint and torque the nut to 29–35 ft. lbs. (40–48 Nm)
- Tie rod end and torque the nut to 36–43 ft. lbs. (50–60 Nm)
- Brake rotor and torque the bolts to 84 inch lbs. (9.5 Nm)
- Brake caliper and torque the bolts to 80 ft. lbs. (108 Nm)
- Brake hose brackets and torque the bolts to 16 ft. lbs. (22 Nm)
- Wheel sensor and torque the bolts to 84 inch lbs. (9.5 Nm)
- Wheel sensor wire brackets and torque the bolts to 84 inch lbs. (9.5 Nm)
- New axle nut and torque it to 181 ft. lbs. (245 Nm)
- Wheel

15. Measure and adjust the wheel alignment

2002–04 RL MODEL

1. Before servicing the vehicle, refer to the precautions section.
2. Raise and safely support the vehicle.
3. Remove or disconnect the following:
- Wheel
- Axle nut
- Wheel sensor and wire brackets from the knuckle
- Brake caliper from the knuckle
- Brake rotor from the knuckle
- Tie rod from the knuckle
- Lower control arm from the knuckle
- Upper arm from the knuckle
- Knuckle/hub assembly from the halfshaft

4. Clamp the knuckle in a vise and secure a slide hammer to the wheel studs to separate the hub from the knuckle.
5. Remove the splash guard.
6. Remove the snapring from the knuckle.
7. Support the knuckle and press the bearing out towards the wheel side.
8. If the inner bearing race stayed on the hub, use a puller to remove it.

To install:
9. Press a new inner race on the hub.
10. Press a new bearing into the knuckle.
11. Install the outer snapring.
12. Install the splash guard and torque the bolts to 4 ft. lbs. (4.9 Nm).
13. Properly support the knuckle and press the hub into the bearing.

✳✳ CAUTION

Do not press on the wheel studs or they will press out of the hub.

14. Install or connect the following:
- Knuckle/hub assembly onto the halfshaft
- Lower ball joint and torque the nut to 51–58 ft. lbs. (69–78 Nm)
- Upper ball joint and torque the nut to 29–35 ft. lbs. (40–48 Nm)
- Tie rod end and torque the nut to 36–43 ft. lbs. (50–60 Nm)
- Brake rotor and torque the bolts to 84 inch lbs. (9.5 Nm)
- Brake caliper and torque the bolts to 80 ft. lbs. (108 Nm)
- Brake hose brackets and torque the bolts to 16 ft. lbs. (22 Nm)
- Wheel sensor and torque the bolts to 84 inch lbs. (9.5 Nm)
- Wheel sensor wire brackets and torque the bolts to 84 inch lbs. (9.5 Nm)
- New axle nut and torque it to 181 ft. lbs. (245 Nm)
- Wheel

15. Measure and adjust the wheel alignment

2005–06 RL MODEL

➡The front wheel bearings on these vehicles are part of the wheel hub and are not serviceable. If the bearing is bad the wheel hub must be replaced.

1. Before servicing the vehicle, refer to the precautions section.
2. Remove the front wheels.
3. Remove the brake hose mounting bracket.
4. Remove the brake caliper bracket mounting bolts and caliper assembly from the knuckle. To prevent damage to the caliper assembly or brake hose, use a short piece of wire to hang the caliper assembly from the undercarriage. Do not twist the brake hose with force.
5. Remove the wheel sensor and o-ring from the knuckle. Do not disconnect the wheel sensor connector. use a new O-ring during installation.
6. Remove the spindle nut.
7. Remove the brake rotor.
8. Remove the cotter pin from the tie-rod end ball joint, then loosen the nut.
9. Remove the hub bearing unit mounting bolts.
10. Remove the hub bearing unit by tapping the driveshaft end with plastic hammer while drawing the hub bearing unit outward.

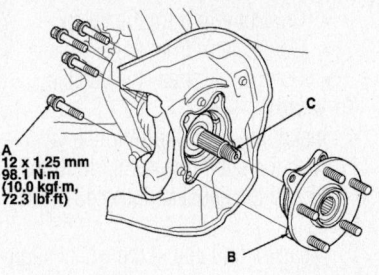

09474_ACUCAR_G0139

Exploded view of the front wheel bearing assembly—2005–06 RL model

To install:
11. Installation is the reverse of removal. Tighten the mounting bolts to 72 ft. lbs. (93 Nm).

RSX MODELS

1. Before servicing the vehicle, refer to the precautions section.
2. Raise and safely support the vehicle.
3. Remove or disconnect the following:
- Front wheels
- Axle nut
- Brake hose mounting bolts
- Brake caliper bolts and remove the caliper from the knuckle
- Disc brake rotor
- Wheel sensor wire bracket
- Wheel sensor from the knuckle
- Lower ball joint
- Upper ball joint

4. Pull the knuckle outward and remove the halfshaft outboard joint from the knuckle using a plastic hammer, then remove the knuckle.
5. Place the knuckle on a base. Insert a disassembly tool into the hub, then using a press, remove the hub from the knuckle.
6. Remove the circlip and the splash guard from the knuckle.
7. Place the knuckle on the disassembly base and install a driver to the bearing. Using a press, remove the bearing from the knuckle.
8. Press the wheel bearing inner race from the hub using the hub disassembly tool and a bearing separator.

To install:
9. Press a new inner race and wheel bearing into the knuckle with a suitable driver.
10. Install or connect the following:
- Splash guard and tighten the screws to 43 inch lbs. (5 Nm)
- Circlip securely in the knuckle groove

11. Use a suitable driver to press the knuckle onto the hub.

12. Install or connect the following:
- Knuckle/hub assembly onto the halfshaft
- Lower ball joint to the knuckle and torque the nut to 36–43 ft. lbs. (49–59 Nm)
- Tie rod and tighten the nut to 29–35 ft. lbs. (39–47 Nm)
- Upper ball joint and tighten the nut to 29–35 ft. lbs. (39–47 Nm)
- Wheel sensor and torque the bolts to 84 inch lbs. (9.5 Nm)
- Wheel sensor wire bracket and torque the bolts to 84 inch lbs. (9.5 Nm)
- Brake rotor and tighten the screws to 84 inch lbs. (10 Nm)
- Brake caliper and torque the bolts to 80 ft. lbs. (108 Nm)
- Brake hose mounting bolts and torque them to 84 inch lbs. (9.5 Nm)
- New axle nut and tighten it to 134 ft. lbs. (181 Nm)
- Wheels

13. Measure and adjust the wheel alignment.

2002–03 TL MODELS

1. Before servicing the vehicle, refer to the precautions section.
2. Raise and safely support the vehicle.
3. Remove or disconnect the following:
- Wheel
- Axle nut
- Wheel sensor and wire brackets from the knuckle
- Brake caliper from the knuckle
- Brake rotor from the knuckle
- Tie rod from the knuckle
- Lower control arm from the knuckle
- Upper arm from the knuckle
- Knuckle/hub assembly from the halfshaft
4. Clamp the knuckle in a vise and

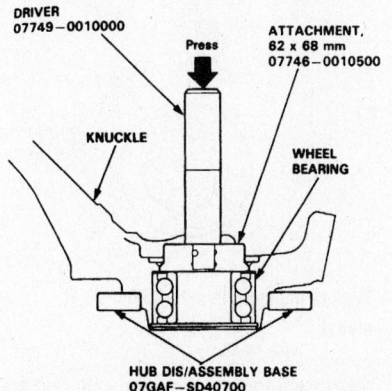

DRIVER
07749–0010000

Press

ATTACHMENT,
62 x 68 mm
07746–0010500

KNUCKLE

WHEEL
BEARING

HUB DIS/ASSEMBLY BASE
07GAF–SD40700

7923BG53

Pressing out the wheel bearing—2002–03 3.2TL shown, other older models similar

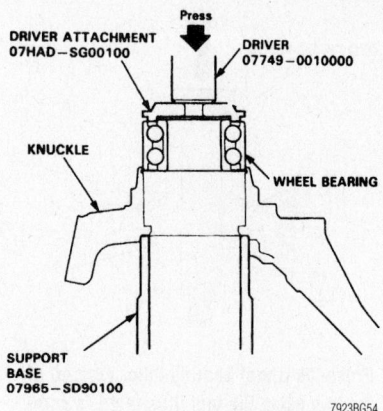

Press

DRIVER ATTACHMENT
07HAD–SG00100

DRIVER
07749–0010000

KNUCKLE

WHEEL BEARING

SUPPORT
BASE
07965–SD90100

7923BG54

Installing the wheel bearing—2002–03 3.2TL shown, other older models similar

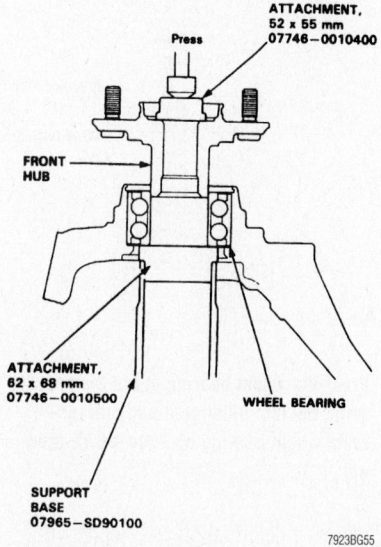

ATTACHMENT,
52 x 55 mm
07746–0010400

Press

FRONT
HUB

ATTACHMENT,
62 x 68 mm
07746–0010500

WHEEL BEARING

SUPPORT
BASE
07965–SD90100

7923BG55

Hub installation—2002–03 3.2TL models

secure a slide hammer to the wheel studs to separate the hub from the knuckle.

5. Remove the splash guard.
6. Remove the snapring from the knuckle.
7. Support the knuckle and press the bearing out towards the wheel side.
8. If the inner bearing race stayed on the hub, use a puller to remove it.

To install:

9. Press a new inner race on the hub.
10. Press a new bearing into the knuckle.
11. Install the outer snapring.
12. Install the splash guard and torque the bolts to 4 ft. lbs. (4.9 Nm).
13. Properly support the knuckle and press the hub into the bearing.

✳✳ CAUTION

Do not press on the wheel studs or they will press out of the hub.

14. Install or connect the following:
- Knuckle/hub assembly onto the halfshaft
- Lower ball joint and torque the nut to 51–58 ft. lbs. (69–78 Nm)
- Upper ball joint and torque the nut to 29–35 ft. lbs. (40–48 Nm)
- Tie rod end and torque the nut to 36–43 ft. lbs. (50–60 Nm)
- Brake rotor and torque the bolts to 84 inch lbs. (9.5 Nm)
- Brake caliper and torque the bolts to 80 ft. lbs. (108 Nm)
- Brake hose brackets and torque the bolts to 16 ft. lbs. (22 Nm)
- Wheel sensor and torque the bolts to 84 inch lbs. (9.5 Nm)
- Wheel sensor wire brackets and torque the bolts to 84 inch lbs. (9.5 Nm)
- New axle nut and torque it to 181 ft. lbs. (245 Nm)
- Wheel

15. Measure and adjust the wheel alignment

2004–06 TL MODELS

1. Before servicing the vehicle, refer to the precautions section.
2. Remove the front wheels.
3. Remove the brake hose mounting bracket.
4. Remove the brake caliper bracket mounting bolts and caliper assembly from the knuckle. To prevent damage to the caliper assembly or brake hose, use a short piece of wire to hang the caliper assembly from the undercarriage. Do not twist the brake hose with force.
5. Remove the washers on models with a manual transmission.
6. Remove the wheel sensor from the knuckle. Do not disconnect the wheel sensor connector.
7. Remove the spindle nut.
8. Remove the brake rotor.
9. Remove the cotter pin from the tie-rod end ball joint, then loosen the nut.

➡**During installation, install the new cotter pin after tightening the nut, and bend its end as shown.**

10. Disconnect the tie-rod end ball joint from the knuckle.
11. Remove the cotter pin from the lower arm ball joint, and remove the nut.

➡**To avoid damaging the ball joint, install the tool shown in the control arm procedures on the threads of the ball joint.**

➡️Be careful not to damage the ball joint boot when installing the remover. Do not force or hammer on the lower arm, or pry between the lower arm and the knuckle. You could damage the ball joint.

12. Disconnect the lower ball joint from the lower arm using the tool illustrated.

➡️If the collar/sleeve on the lower arm is removed with ball joint, the lower arm must be replaced.

13. Remove the cotter pin from the upper arm ball joint, then loosen the nut.

➡️During installation, insert the new cotter pin into the ball joint pin from the front to the rear of the vehicle, and bend its end as shown.

14. Disconnect the upper arm ball joint from the knuckle.

15. Remove the driveshaft outboard joint from the, then remove the knuckle.

➡️Do not pull the driveshaft end outward. The inner driveshaft joint may come apart.

16. Separate the hub from the knuckle using the tool illustrated and a hydraulic press. Hold the knuckle with the attachment of the hydraulic press or equivalent tool. Be careful not to deform the splash guard.

17. Hold onto the hub to keep it from falling when pressed clear

18. Press the wheel bearing inner race off of the hub using the tool illustrated, a commercially available bearing separator, and a press.

19. Remove the snap ring and the splash guard from the knuckle.

20. Check the front knuckle ring for

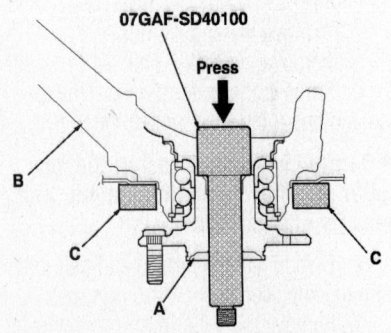

Separate the hub from the knuckle using the tool illustrated and a hydraulic press. Hold the knuckle with the attachment of the hydraulic press or equivalent tool—Front wheel bearing on 2004–06 TL models

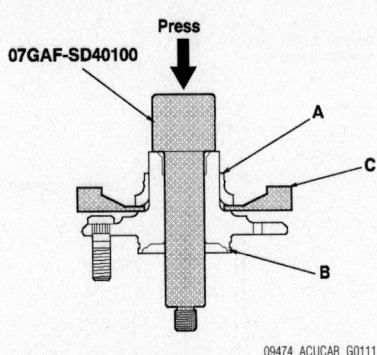

Press the wheel bearing inner race off of the hub using the tool illustrated, a commercially available bearing separator, and a press—Front wheel bearing on 2004–06 TL models

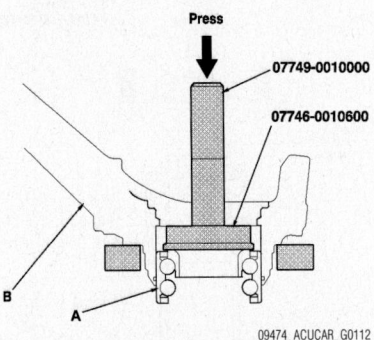

Press the wheel bearing out of the knuckle using the tool illustrated and a press—Front wheel bearing on 2004–06 TL models

damage or deformation, and replace it if necessary.

21. Press the wheel bearing out of the knuckle using the tool illustrated and a press.

To install:

22. Wash the knuckle and hub thoroughly in high flash point solvent before reassembly.

23. Press a new wheel bearing into the knuckle using the old bearing, a steel plate, the tool illustrated, and a press.

➡️Install the wheel bearing with the wheel sensor magnetic encoder (brown color), toward the inside of the knuckle.

24. Remove any oil, grease, dust, metal debris, and other foreign material from the encoder surface.

25. Keep all magnetic tools away from the encoder surface.

➡️Be careful not to damage the encoder surface when you insert the wheel bearing.

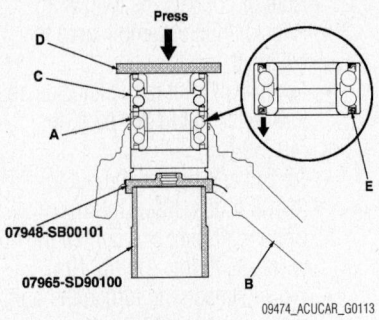

Press a new wheel bearing into the knuckle using the old bearing, a steel plate, the tool illustrated, and a press—Front wheel bearing on 2004–06 TL models

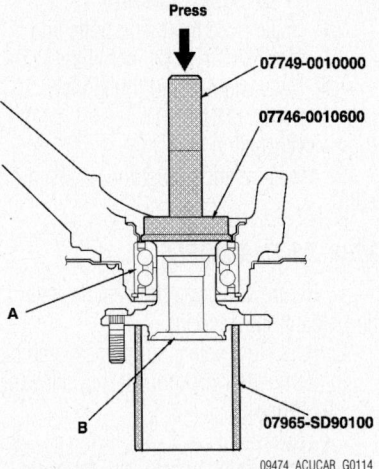

Press the wheel bearing onto the hub using the tool illustrated and a press—Front wheel bearing on 2004–06 TL models

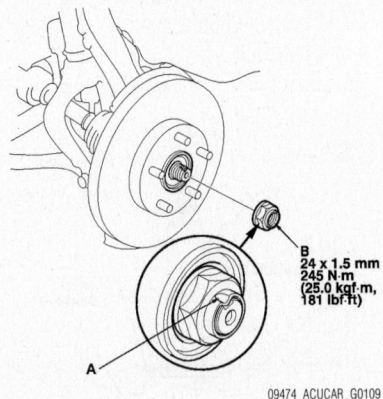

Tighten the spindle nut—2004–06 TL model

26. Install the new front knuckle ring on the inside of the knuckle by aligning the cutout portion on the ring with the wheel sensor hole in the knuckle. Be careful not to

damage or deform the ring when installing it.

27. Install the snap ring securely in the knuckle.

28. Install the splash guard and tighten the screws.

29. Press the wheel bearing onto the hub using the tool illustrated and a press.

30. Install the remaining components in the reverse order of removal.

TSX MODELS

1. Before servicing the vehicle, refer to the precautions section.

2. Raise and safely support the vehicle.

3. Remove or disconnect the following:
- Wheel
- Axle nut
- Wheel sensor and wire brackets from the knuckle
- Brake caliper from the knuckle
- Brake rotor from the knuckle
- Tie rod from the knuckle
- Lower control arm from the knuckle
- Upper arm from the knuckle
- Knuckle/hub assembly from the halfshaft

4. Clamp the knuckle in a vise and secure a slide hammer to the wheel studs to separate the hub from the knuckle.

5. Remove the splash guard.

6. Remove the snapring from the knuckle.

7. Support the knuckle and press the bearing out towards the wheel side.

8. If the inner bearing race stayed on the hub, use a puller to remove it.

To install:

9. Press a new inner race on the hub.

10. Press a new bearing into the knuckle.

11. Install the outer snapring.

12. Install the splash guard and torque the bolts to 4 ft. lbs. (4.9 Nm).

13. Properly support the knuckle and press the hub into the bearing.

✳✳ CAUTION

Do not press on the wheel studs or they will press out of the hub.

14. Install or connect the following:
- Knuckle/hub assembly onto the halfshaft
- Lower ball joint and torque the nut to 51–58 ft. lbs. (69–78 Nm)
- Upper ball joint and torque the nut to 29–35 ft. lbs. (40–48 Nm)
- Tie rod end and torque the nut to 36–43 ft. lbs. (50–60 Nm)
- Brake rotor and torque the bolts to 84 inch lbs. (9.5 Nm)
- Brake caliper and torque the bolts to 80 ft. lbs. (108 Nm)
- Brake hose brackets and torque the bolts to 16 ft. lbs. (22 Nm)
- Wheel sensor and torque the bolts to 84 inch lbs. (9.5 Nm)
- Wheel sensor wire brackets and torque the bolts to 84 inch lbs. (9.5 Nm)
- New axle nut and torque it to 181 ft. lbs. (245 Nm)
- Wheel

15. Measure and adjust the wheel alignment

Rear

CL MODEL

➡The rear wheel bearings on these vehicles are part of the wheel hub and are not serviceable. If the bearing is bad the wheel hub must be replaced.

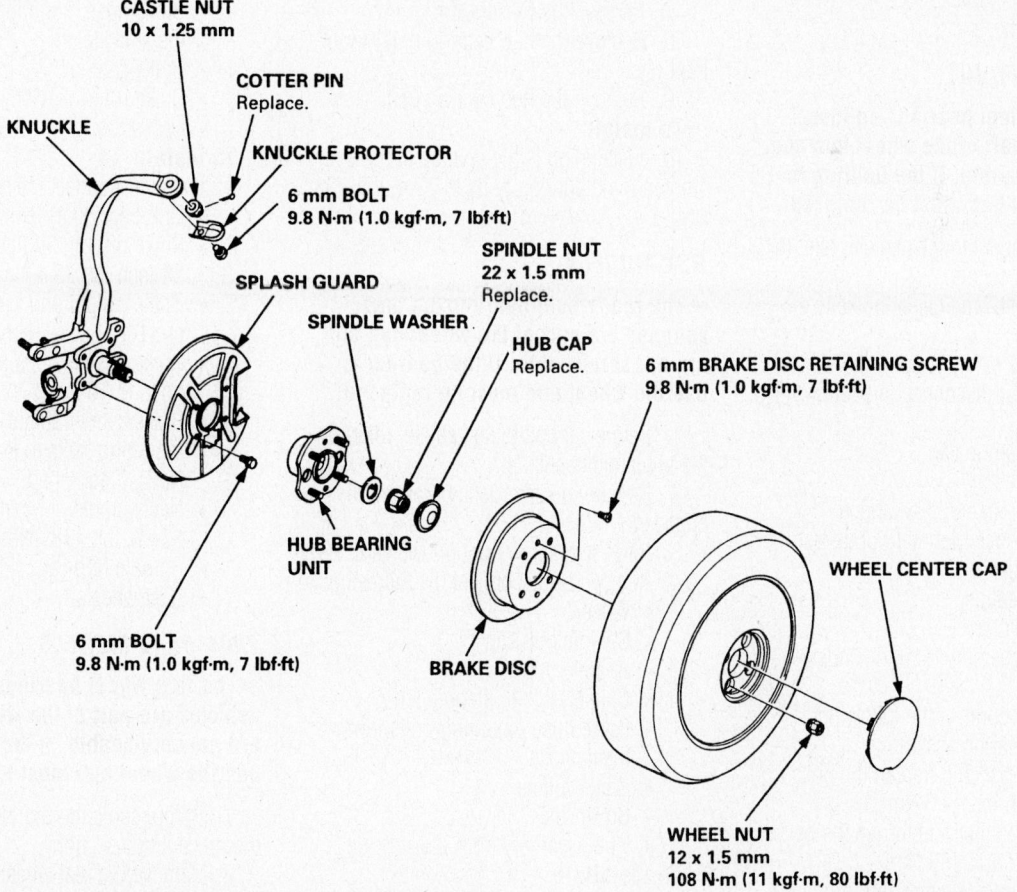

CASTLE NUT
10 x 1.25 mm

COTTER PIN
Replace.

KNUCKLE

KNUCKLE PROTECTOR

6 mm BOLT
9.8 N·m (1.0 kgf·m, 7 lbf·ft)

SPLASH GUARD

SPINDLE NUT
22 x 1.5 mm
Replace.

SPINDLE WASHER

HUB CAP
Replace.

6 mm BRAKE DISC RETAINING SCREW
9.8 N·m (1.0 kgf·m, 7 lbf·ft)

HUB BEARING UNIT

6 mm BOLT
9.8 N·m (1.0 kgf·m, 7 lbf·ft)

BRAKE DISC

WHEEL CENTER CAP

WHEEL NUT
12 x 1.5 mm
108 N·m (11 kgf·m, 80 lbf·ft)

7923BGA3

Exploded view of the rear bearing and related components—3.2CL

1. Before servicing the vehicle, refer to the precautions section.
2. Be sure the emergency brake is disengaged.
3. Raise and safely support the vehicle.
4. Remove or disconnect the following:
   - Wheel
   - Spindle hub cap
   - Spindle nut
   - Caliper shield, if equipped
   - Brake hose mounting bolts from the knuckle
   - Brake caliper
   - Brake rotor
   - Hub assembly from the knuckle

To install:

5. Install or connect the following:
   - Hub/bearing assembly
   - Brake disc and tighten the bolts to 84 inch lbs. (9.5 Nm)
   - Brake caliper and torque the bolts to 41 ft. lbs. (56 Nm)
   - Brake hose clamps and torque the bolts to 16 ft. lbs. (22 Nm)
   - Brake caliper shield and torque the mounting bolts to 84 inch lbs. (9.5 Nm),, if equipped
   - New spindle nut and torque it to 134 ft. lbs. (181 Nm)
   - Spindle hub cap
   - Rear wheels

2002–04 RL MODEL

➡**The rear wheel bearings on these vehicles are part of the wheel hub and are not serviceable. If the bearing is bad the wheel hub must be replaced.**

1. Before servicing the vehicle, refer to the precautions section.
2. Be sure the emergency brake is disengaged.
3. Raise and safely support the vehicle.
4. Remove or disconnect the following:
   - Wheel
   - Spindle hub cap
   - Spindle nut
   - Caliper shield, if equipped
   - Brake hose mounting bolts from the knuckle
   - Brake caliper
   - Brake rotor
   - Hub assembly from the knuckle

To install:

5. Install or connect the following:
   - Hub/bearing assembly
   - Brake disc and tighten the bolts to 84 inch lbs. (9.5 Nm)
   - Brake caliper and torque the bolts to 41 ft. lbs. (56 Nm)
   - Brake hose clamps and torque the bolts to 16 ft. lbs. (22 Nm)

- Brake caliper shield and torque the mounting bolts to 84 inch lbs. (9.5 Nm),, if equipped
- New spindle nut and torque it to 134 ft. lbs. (181 Nm)
- Spindle hub cap
- Rear wheels

2005–06 RL MODEL

➡**The rear wheel bearings on these vehicles are part of the wheel hub and are not serviceable. If the bearing is bad the wheel hub must be replaced.**

1. Before servicing the vehicle, refer to the precautions section.
2. Remove the rear wheels.
3. Remove the brake caliper bracket mounting bolts and caliper assembly from the knuckle. To prevent damage to the caliper assembly or brake hose, use a short piece of wire to hang the caliper assembly from the undercarriage. Do not twist the brake hose with force.
4. Remove the 2 washers.
5. Remove the wheel sensor and o-ring from the knuckle. Do not disconnect the wheel sensor connector. use a new O-ring during installation.
6. Remove the spindle nut.
7. Remove the brake rotor.
8. Remove the hub bearing unit mounting bolts.
9. Remove the hub bearing unit.

To install:

10. Installation is the reverse of removal. Tighten the mounting bolts to 72 ft. lbs. (93 Nm).

RSX MODELS

➡**The rear wheel bearings on these vehicles are part of the wheel hub and are not serviceable. If the bearing is bad the wheel hub must be replaced.**

1. Before servicing the vehicle, refer to the precautions section.
2. Be sure the emergency brake is disengaged.
3. Raise and safely support the vehicle.
4. Remove or disconnect the following:
   - Wheel
   - Spindle hub cap
   - Spindle nut
   - Caliper shield, if equipped
   - Brake hose mounting bolts from the knuckle
   - Brake caliper
   - Brake rotor
   - Hub assembly from the knuckle

To install:

5. Install or connect the following:
   - Hub/bearing assembly

- Brake disc and tighten the bolts to 84 inch lbs. (9.5 Nm)
- Brake caliper and torque the bolts to 41 ft. lbs. (56 Nm)
- Brake hose clamps and torque the bolts to 16 ft. lbs. (22 Nm)
- Brake caliper shield and torque the mounting bolts to 84 inch lbs. (9.5 Nm),, if equipped
- New spindle nut and torque it to 134 ft. lbs. (181 Nm)
- Spindle hub cap
- Rear wheels

2002–03 TL MODELS

➡**The rear wheel bearings on these vehicles are part of the wheel hub and are not serviceable. If the bearing is bad the wheel hub must be replaced.**

1. Before servicing the vehicle, refer to the precautions section.
2. Be sure the emergency brake is disengaged.
3. Raise and safely support the vehicle.
4. Remove or disconnect the following:
   - Wheel
   - Spindle hub cap
   - Spindle nut
   - Caliper shield, if equipped
   - Brake hose mounting bolts from the knuckle
   - Brake caliper
   - Brake rotor
   - Hub assembly from the knuckle

To install:

5. Install or connect the following:
   - Hub/bearing assembly
   - Brake disc and tighten the bolts to 84 inch lbs. (9.5 Nm)
   - Brake caliper and torque the bolts to 41 ft. lbs. (56 Nm)
   - Brake hose clamps and torque the bolts to 16 ft. lbs. (22 Nm)
   - Brake caliper shield and torque the mounting bolts to 84 inch lbs. (9.5 Nm),, if equipped
   - New spindle nut and torque it to 134 ft. lbs. (181 Nm)
   - Spindle hub cap
   - Rear wheels

2004–06 TL MODELS

➡**The rear wheel bearings on these vehicles are part of the wheel hub and are not serviceable. If the bearing is bad the wheel hub must be replaced.**

1. Before servicing the vehicle, refer to the precautions section.
2. Remove the rear wheels.
3. Remove the brake hose mounting bracket.

4. Remove the brake caliper bracket mounting bolts and caliper assembly from the knuckle. To prevent damage to the caliper assembly or brake hose, use a short piece of wire to hang the caliper assembly from the undercarriage. Do not twist the brake hose with force.

5. Remove the hub cap (A).

6. Remove the spindle nut.

7. Remove the brake disc.

8. Remove the hub bearing unit.

9. Installation is the reverse of removal. Tighten the hub bearing unit nuts to 28 ft. lbs. (38 Nm).

TSX MODELS

➡ **The rear wheel bearings on these vehicles are part of the wheel hub and** are not serviceable. If the bearing is bad the wheel hub must be replaced.

1. Before servicing the vehicle, refer to the precautions section.

2. Be sure the emergency brake is disengaged.

3. Raise and safely support the vehicle.

4. Remove or disconnect the following:
    *   Wheel
    *   Spindle hub cap
    *   Spindle nut
    *   Caliper shield, if equipped
    *   Brake hose mounting bolts from the knuckle
    *   Brake caliper
    *   Brake rotor
    *   Hub assembly from the knuckle

To install:

5. Install or connect the following:
    *   Hub/bearing assembly
    *   Brake disc and tighten the bolts to 84 inch lbs. (9.5 Nm)
    *   Brake caliper and torque the bolts to 41 ft. lbs. (56 Nm)
    *   Brake hose clamps and torque the bolts to 16 ft. lbs. (22 Nm)
    *   Brake caliper shield and torque the mounting bolts to 84 inch lbs. (9.5 Nm),, if equipped
    *   New spindle nut and torque it to 134 ft. lbs. (181 Nm)
    *   Spindle hub cap
    *   Rear wheels

BRAKES

Brake Caliper

REMOVAL & INSTALLATION

Front

CL MODEL

1. Before servicing the vehicle, refer to the precautions section.

2. Remove or disconnect the following:
    *   Wheels
    *   Banjo bolt and disconnect the brake line from the caliper
    *   Caliper mounting bolts
    *   Caliper
    *   Brake pads and shims
    *   Pad spring from the caliper body, if equipped
    *   Caliper bracket mounting bolts and bracket

To install:

3. Install or connect the following:
    *   Bracket and torque the bolts to 80 ft. lbs. (109 Nm)
    *   Pad spring, brake pads, shims, caliper and slide mounting bolts
    *   Caliper slide mounting bolt and torque to 37 ft. lbs. (49 Nm)
    *   Brake line and the banjo bolt. Replace the crush washers and torque the banjo bolt to 25 ft. lbs. (35 Nm).

4. Bleed the brake system.
    *   Front wheels

2002–03 RL MODEL

1. Before servicing the vehicle, refer to the precautions section.

2. Remove or disconnect the following:
    *   Wheels
    *   Banjo bolt and disconnect the brake line from the caliper
    *   Caliper mounting bolts
    *   Caliper
    *   Brake pads and shims
    *   Pad spring from the caliper body, if equipped
    *   Caliper bracket mounting bolts and bracket

To install:

3. Install or connect the following:
    *   Bracket and torque the bolts to 80 ft. lbs. (109 Nm)
    *   Pad spring, brake pads, shims, caliper and slide mounting bolts
    *   Caliper slide mounting bolt and torque to 37 ft. lbs. (49 Nm)
    *   Brake line and the banjo bolt. Replace the crush washers and torque the banjo bolt to 25 ft. lbs. (35 Nm).

4. Bleed the brake system.
    *   Front wheels

2005–06 RL MODEL

1. Before servicing the vehicle, refer to the precautions in the beginning of this section.

2. Remove or disconnect the following:
    *   Wheels
    *   Brake line fitting and disconnect the brake line from the caliper
    *   Caliper mounting bolts
    *   Caliper
    *   Brake pads and shims

To install:

3. Install or connect the following:
    *   Brake pads and shims
    *   Caliper bolts to 58 ft. lbs. (78 Nm)

4. Install or connect the following:
    *   Brake line and fitting. Torque the fitting to 11 ft. lbs. (15 Nm).

5. Bleed the brake system. Torque the bleed screws to 6 ft. lbs. (8 Nm).
    *   Front wheels

RSX MODELS

1. Before servicing the vehicle, refer to the precautions section.

2. Remove or disconnect the following:
    *   Wheels
    *   Banjo bolt and disconnect the brake line from the caliper
    *   Caliper mounting bolts
    *   Caliper
    *   Brake pads and shims
    *   Pad spring from the caliper body, if equipped
    *   Caliper bracket mounting bolts and bracket

To install:

3. Install or connect the following:
    *   Bracket and torque the bolts to 80 ft. lbs. (109 Nm)
    *   Pad spring, brake pads, shims, caliper and slide mounting bolts
    *   Caliper slide mounting bolt. Torque the bolts to 16 ft. lbs. (22 Nm) on the 14 inch model, or 24 ft. lbs. (32 Nm) on the 16 inch model.
    *   Brake line and the banjo bolt. Replace the crush washers and torque the banjo bolt to 25 ft. lbs. (35 Nm).

4. Bleed the brake system.
    *   Front wheels

2002–03 TL MODELS

1. Before servicing the vehicle, refer to the precautions section.

2. Remove or disconnect the following:
- Wheels
- Banjo bolt and disconnect the brake line from the caliper
- Caliper mounting bolts
- Caliper
- Brake pads and shims
- Pad spring from the caliper body, if equipped
- Caliper bracket mounting bolts and bracket

To install:

3. Install or connect the following:
- Bracket and torque the bolts to 80 ft. lbs. (109 Nm)
- Pad spring, brake pads, shims, caliper and slide mounting bolts
- Caliper slide mounting bolt and torque to 37 ft. lbs. (49 Nm)
- Brake line and the banjo bolt. Replace the crush washers and torque the banjo bolt to 25 ft. lbs. (35 Nm).

4. Bleed the brake system.
- Front wheels

2004–06 TL MODELS

1. Before servicing the vehicle, refer to the precautions in the beginning of this section.

2. Remove or disconnect the following:
- Wheels
- Banjo bolt and disconnect the brake line from the caliper
- Caliper mounting bolts
- Caliper
- Brake pads and shims
- Pad spring from the caliper body, if equipped
- Caliper bracket mounting bolts and bracket

To install:

3. Install or connect the following:
- Bracket and torque the bolts to 80 ft. lbs. (109 Nm)
- Pad spring, brake pads, shims, caliper and slide mounting bolts

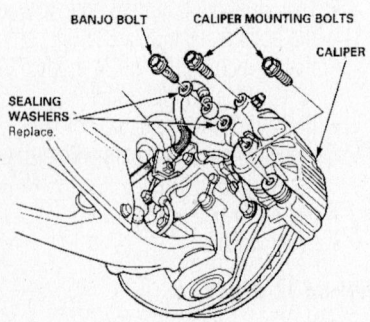

BANJO BOLT   CALIPER MOUNTING BOLTS

CALIPER

SEALING WASHERS
Replace.

93016G01

Front caliper mounting

- Caliper slide mounting bolt to 37 ft. lbs. (49 Nm)

4. Install or connect the following:
- Brake line and the banjo bolt. Replace the crush washers and torque the banjo bolt to 25 ft. lbs. (35 Nm).

5. Bleed the brake system. Torque the bleed screws to 84 inch lbs. (9 Nm).
- Front wheels and tighten the wheel nuts to 80 ft. lbs. (109 Nm)

TSX MODELS

1. Before servicing the vehicle, refer to the precautions section.

2. Remove or disconnect the following:
- Wheels
- Banjo bolt and disconnect the brake line from the caliper
- Caliper mounting bolts
- Caliper
- Brake pads and shims
- Pad spring from the caliper body, if equipped
- Caliper bracket mounting bolts and bracket

To install:

3. Install or connect the following:
- Bracket and torque the bolts to 80 ft. lbs. (109 Nm)
- Pad spring, brake pads, shims, caliper and slide mounting bolts
- Caliper slide mounting bolt and torque to 36 ft. lbs. (49 Nm)
- Brake line and the banjo bolt. Replace the crush washers and torque the banjo bolt to 25 ft. lbs. (35 Nm).

4. Bleed the brake system.
- Front wheels

Rear

CL MODEL

1. Before servicing the vehicle, refer to the precautions section.

2. Remove or disconnect the following:
- Wheels
- Caliper dust shield
- Parking brake cable from the caliper arm, if equipped
- Brake line from the caliper
- Caliper mounting bolts and pull the caliper off the bracket
- Pads, shim, and pad retainer spring.
- Caliper bracket mounting bolts
- Bracket from the rotor

To install:

3. Install or connect the following:
- Caliper bracket. Torque the mounting bolts to 41 ft. lbs. (55 Nm).

- Pads, shims, and pad retainer springs
- Caliper. Torque the mounting bolts to 17 ft. lbs. (23 Nm).
- Brake hose with new crush washers and torque the banjo bolt to 25 ft. lbs. (35 Nm).
- Parking brake cable, if equipped

4. Bleed the brake system.
- Rear wheels

2002–03 RL MODEL

1. Before servicing the vehicle, refer to the precautions section.

2. Remove or disconnect the following:
- Wheels
- Caliper dust shield
- Parking brake cable from the caliper arm, if equipped
- Brake line from the caliper
- Caliper mounting bolts and pull the caliper off the bracket
- Pads, shim, and pad retainer spring.
- Caliper bracket mounting bolts
- Bracket from the rotor

To install:

3. Install or connect the following:
- Caliper bracket. Torque the mounting bolts to 28 ft. lbs. (39 Nm).
- Pads, shims, and pad retainer springs
- Caliper. Torque the mounting bolts to 41 ft. lbs. (55 Nm).
- Brake hose with new crush washers and torque the banjo bolt to 25 ft. lbs. (35 Nm).
- Parking brake cable, if equipped

4. Bleed the brake system.
- Rear wheels

2005–06 RL MODEL

1. Before servicing the vehicle, refer to the precautions in the beginning of this section.

2. Remove or disconnect the following:
- Wheels
- Parking brake cable from the caliper arm, if equipped
- Brake line from the caliper
- Caliper mounting bolts and pull the caliper off the bracket
- Pads, shim, and pad retainer spring.
- Caliper bracket mounting bolts
- Bracket from the rotor

To install:

3. Install or connect the following:
- Caliper bracket. Torque the mounting bolts to 79 ft. lbs. (108 Nm).
- Pads, shims, and pad retainer springs

- Caliper. Torque the mounting bolts to 17 ft. lbs. (23 Nm).
- Brake hose.
- Parking brake cable, if equipped
4. Bleed the brake system.
  - Rear wheels

RSX MODELS

1. Before servicing the vehicle, refer to the precautions section.
2. Remove or disconnect the following:
   - Wheels
   - Caliper dust shield
   - Parking brake cable from the caliper arm, if equipped
   - Brake line from the caliper
   - Caliper mounting bolts and pull the caliper off the bracket
   - Pads, shim, and pad retainer spring.
   - Caliper bracket mounting bolts
   - Bracket from the rotor
To install:
3. Install or connect the following:
   - Caliper bracket. Torque the mounting bolts to 55 ft. lbs. (41 Nm).
   - Pads, shims, and pad retainer springs
   - Caliper. Torque the mounting bolts to 16 ft. lbs. (22 Nm).
   - Brake hose with new crush washers and torque the banjo bolt to 25 ft. lbs. (35 Nm).
   - Parking brake cable, if equipped
4. Bleed the brake system.
   - Rear wheels

2002–03 TL MODELS

1. Before servicing the vehicle, refer to the precautions section.
2. Remove or disconnect the following:
   - Wheels
   - Caliper dust shield
   - Parking brake cable from the caliper arm, if equipped
   - Brake line from the caliper
   - Caliper mounting bolts and pull the caliper off the bracket
   - Pads, shim, and pad retainer spring.
   - Caliper bracket mounting bolts
   - Bracket from the rotor
To install:
3. Install or connect the following:
   - Caliper bracket. Torque the mounting bolts to 28 ft. lbs. (39 Nm).
   - Pads, shims, and pad retainer springs
   - Caliper. Torque the mounting bolts to 41 ft. lbs. (55 Nm).
   - Brake hose with new crush washers

and torque the banjo bolt to 25 ft. lbs. (35 Nm).
   - Parking brake cable, if equipped
4. Bleed the brake system.
   - Rear wheels

2004–06 TL MODELS

1. Before servicing the vehicle, refer to the precautions in the beginning of this section.
2. Remove or disconnect the following:
   - Wheels
   - Caliper dust shield
   - Parking brake cable from the caliper arm, if equipped
   - Brake line from the caliper
   - Caliper mounting bolts and pull the caliper off the bracket
   - Pads, shim, and pad retainer spring.
   - Caliper bracket mounting bolts
   - Bracket from the rotor
To install:
3. Install or connect the following:
   - Caliper bracket. Torque the mounting bolts to 28 ft. lbs. (39 Nm).
   - Pads, shims, and pad retainer springs
   - Caliper. Torque the mounting bolts to 17 ft. lbs. (23 Nm).
   - Brake hose with new crush washers and torque the banjo bolt to 25 ft. lbs. (35 Nm).
   - Parking brake cable, if equipped
4. Bleed the brake system.
   - Caliper dust shield and tighten the bolts to 84 inch lbs. (10 Nm)
   - Rear wheels and torque the wheel nuts to 80 ft. lbs. (109 Nm)

TSX MODELS

1. Before servicing the vehicle, refer to the precautions section.
2. Remove or disconnect the following:
   - Wheels
   - Caliper dust shield
   - Parking brake cable from the caliper arm, if equipped
   - Brake line from the caliper
   - Caliper mounting bolts and pull the caliper off the bracket
   - Pads, shim, and pad retainer spring.
   - Caliper bracket mounting bolts
   - Bracket from the rotor
To install:
3. Install or connect the following:
   - Caliper bracket. Torque the mounting bolts to 28 ft. lbs. (39 Nm).
   - Pads, shims, and pad retainer springs
   - Caliper. Torque the mounting bolts to 17 ft. lbs. (23 Nm).

- Brake hose with new crush washers and torque the banjo bolt to 25 ft. lbs. (35 Nm).
- Parking brake cable, if equipped
4. Bleed the brake system.
  - Rear wheels

Disc Brake Pads

REMOVAL & INSTALLATION

Front

CL MODEL

1. Before servicing the vehicle, refer to the precautions section.
2. Remove or disconnect the following:
   - Wheels
   - Lower caliper bolt and pivot the caliper up and away from the rotor
   - Pads, shims, and pad retainer springs
To install:
3. Clean all points where the pads and shims touch the caliper and mount. Apply a thin film of silicone grease to the cleaned areas.
4. Install or connect the following:
   - Pad retainers in position on the caliper bracket
   - High temperature brake grease to the back side of the pads and both sides of shims and wipe off the excess.
   - Pads and shims
   - Inner brake pad with the wear indicator facing upward
5. Loosen the bleed screw slightly and push in the caliper piston to allow mounting of the caliper over the rotor. Torque the bleed screw to 84 inch lbs. (9 Nm).
6. Pivot the caliper down over the rotor and install the caliper bolts. Torque the bolts to 36 ft. lbs. (49 Nm)
7. If disconnected, install the brake pad wear indicator connector. Install the wheels.

2002–03 RL MODEL

1. Before servicing the vehicle, refer to the precautions section.
2. Remove or disconnect the following:
   - Wheels
   - Lower caliper bolt and pivot the caliper up and away from the rotor
   - Pads, shims, and pad retainer springs
To install:
3. Clean all points where the pads and shims touch the caliper and mount. Apply a thin film of silicone grease to the cleaned areas.

4. Install or connect the following:
- Pad retainers in position on the caliper bracket
- High temperature brake grease to the back side of the pads and both sides of shims and wipe off the excess.
- Pads and shims
- Inner brake pad with the wear indicator facing upward

5. Loosen the bleed screw slightly and push in the caliper piston to allow mounting of the caliper over the rotor. Torque the bleed screw to 84 inch lbs. (9 Nm).

6. Pivot the caliper down over the rotor and install the caliper bolts. Torque the bolts to 36 ft. lbs. (50 Nm).

7. If disconnected, install the brake pad wear indicator connector. Install the wheels.

2005–06 RL MODEL

1. Before servicing the vehicle, refer to the precautions in the beginning of this section.

2. Remove the front wheels.

3. Turn and twist out the clip from the caliper hole and pull the clip out from the pad pins.

4. Remove the pad pins and the pad spring.

5. Remove the pad.

6. Clean the caliper thoroughly; remove any rust, and check for grooves and cracks.

7. Check the brake disc for damage and cracks.

To install:

8. Install the brake caliper piston compressor on the caliper.

9. Press in the piston with the brake caliper piston compressor 07AAE-SEPA101 so that the caliper will fit over the pads. Make sure the piston boot is in position to prevent damaging it.

➡**Be careful when in pressing the piston the brake fluid might overflow from the master cylinder's reservoir.**

10. Remove the brake caliper piston compressor.

11. Apply M-77 assembly paste to both sides of the pad shim, back of the brake pads and the other contact areas. Wipe excess assembly paste off the pads. Contaminated brake discs or pads reduce stopping ability.

12. Install the brake pads correctly. Install the brake pad with the wear indicator on the inside.

13. If you are reusing the brake pads, always reinstall the brake pads in their original positions to prevent a momentary loss of braking efficiency

14. Install the pad spring. Hold the pad spring and install the pad pins into the caliper from the outside to the inside of vehicle.

15. First insert the clip ends to the pad pins and then twist the clip into the caliper hole to stabilize.

16. Press the brake pedal several times to make sure the brakes work.

➡**The brakes may require a greater pedal stroke immediately after the brake pads have been replaced as a set. Several applications of the brake pedal will restore the normal pedal stroke.**

17. After installation, check for leaks at the brake hose and line joints or connections, and retighten if necessary.

18. Reinstall the front wheels, then test-drive the vehicle.

19. Check for leaks.

RSX MODELS

1. Before servicing the vehicle, refer to the precautions section.

2. Remove or disconnect the following:
- Wheels
- Lower caliper bolt and pivot the caliper up and away from the rotor
- Pads, shims, and pad retainer springs

To install:

3. Clean all points where the pads and shims touch the caliper and mount. Apply a thin film of silicone grease to the cleaned areas.

4. Install or connect the following:
- Pad retainers in position on the caliper bracket
- High temperature brake grease to the back side of the pads and both sides of shims and wipe off the excess.
- Pads and shims
- Inner brake pad with the wear indicator facing upward

5. Loosen the bleed screw slightly and push in the caliper piston to allow mounting of the caliper over the rotor. Torque the bleed screw to 84 inch lbs. (9 Nm).

6. Pivot the caliper down over the rotor and install the caliper bolts. Torque the bolts to 16 ft. lbs. (22 Nm) on the 14 inch model, or 24 ft. lbs. (32 Nm) on the 16 inch model.

7. If disconnected, install the brake pad wear indicator connector. Install the wheels.

2002–03 TL MODELS

1. Before servicing the vehicle, refer to the precautions section.

2. Remove or disconnect the following:
- Wheels
- Lower caliper bolt and pivot the caliper up and away from the rotor
- Pads, shims, and pad retainer springs

To install:

3. Clean all points where the pads and shims touch the caliper and mount. Apply a thin film of silicone grease to the cleaned areas.

4. Install or connect the following:
- Pad retainers in position on the caliper bracket
- High temperature brake grease to the back side of the pads and both sides of shims and wipe off the excess.
- Pads and shims
- Inner brake pad with the wear indicator facing upward

5. Loosen the bleed screw slightly and push in the caliper piston to allow mounting of the caliper over the rotor. Torque the bleed screw to 84 inch lbs. (9 Nm).

6. Pivot the caliper down over the rotor and install the caliper bolts. Torque the bolts to 36 ft. lbs. (50 Nm)

7. If disconnected, install the brake pad wear indicator connector. Install the wheels.

2004–06 TL MODELS WITH AN AUTOMATIC TRANSMISSION

1. Before servicing the vehicle, refer to the precautions in the beginning of this section.

2. Remove or disconnect the following:
- Wheels
- Lower caliper bolt and pivot the caliper up and away from the rotor
- Pads, shims, and pad retainer springs

To install:

3. Clean all points where the pads and shims touch the caliper and mount. Apply a thin film of silicone grease to the cleaned areas.

4. Install or connect the following:
- Pad retainers in position on the caliper bracket
- High temperature brake grease to the back side of the pads and both sides of shims and wipe off the excess.
- Pads and shims
- Inner brake pad with the wear indicator facing upward

5. Loosen the bleed screw slightly and push in the caliper piston to allow mounting of the caliper over the rotor. Tighten the bleed screw to 84 inch lbs. (9 Nm).

6. Pivot the caliper down over the rotor

and install the caliper bolts. Torque the bolts to 37 ft. lbs. (50 Nm)

7. If disconnected, install the brake pad wear indicator connector. Install the wheels.

2004–06 TL MODELS WITH A MANUAL TRANSMISSION

1. Before servicing the vehicle, refer to the precautions in the beginning of this section.

2. Remove the pad pins using a $\frac{5}{32}$ inch pin punch from the outside to the inside of vehicle, and remove the pad spring.

3. Remove the brake pads.

4. Clean the caliper.

5. Check the brake disc for damage and cracks.

6. Mount the piston compressor tool on the caliper.

7. Press in the outside and inside pistons with the brake caliper wrench so the caliper will fit over the pads.

8. Remove the tool.

9. Apply Molykote CU-7439 PLUS PASTE to the back of the brake pads. Wipe excess grease off the pads.

✳✳ CAUTION

Contaminated brake discs or brake pads reduce stopping ability. Keep grease off the brake disc and pads.

10. Install the brake pad with the wear indicator on the outside bottom.

11. If you are reusing the brake pads, always reinstall the brake pads in their original positions to prevent a momentary loss of braking efficiency.

12. Install the pad spring.

13. Hold the pad spring, and install the pad pins into the caliper from the inside to the outside of vehicle using the tool.

14. Install the front wheels.

15. Press the brake pedal several times to make sure the brakes work.

➥**Engagement of the brakes may require a greater pedal stroke immediately after the brake pads have been replaced as a set. Several applications of the brake pedal will restore the normal pedal stroke.**

16. After installation, check for leaks at the brake hose and line joints or connections, and retighten if necessary.

TSX MODELS

1. Before servicing the vehicle, refer to the precautions section.

2. Remove or disconnect the following:

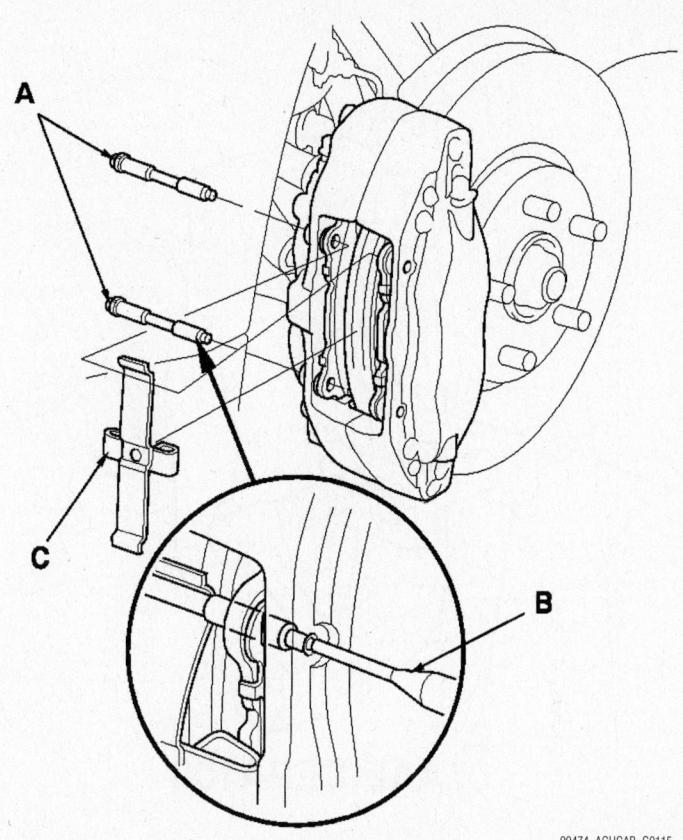

09474_ACUCAR_G0115

Remove the pad pins using a $\frac{5}{32}$ inch pin punch—2004–06 TL model with a manual transmission

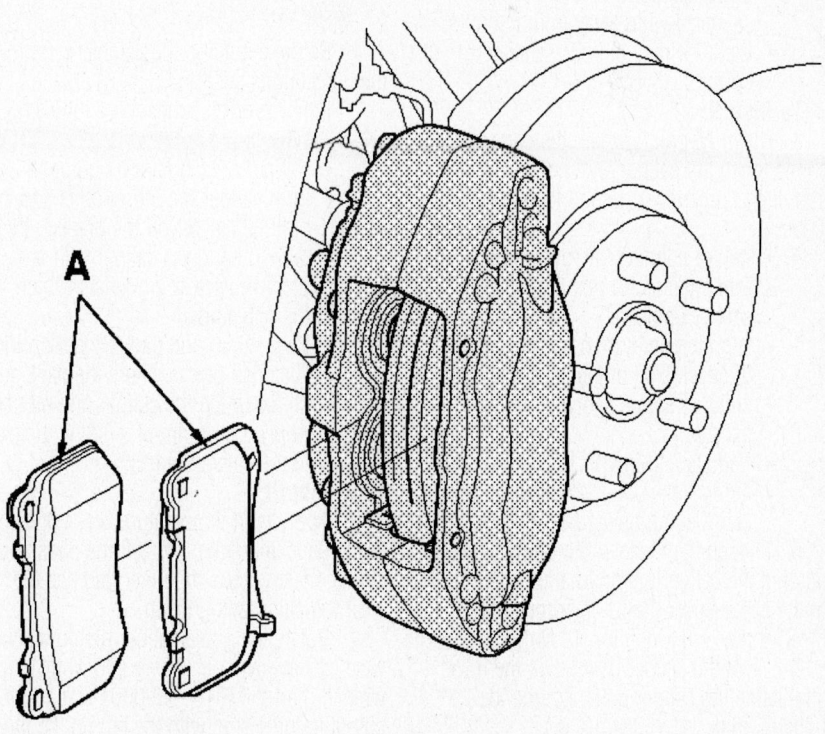

09474_ACUCAR_G0116

Remove the pads—2004–06 TL model with a manual transmission

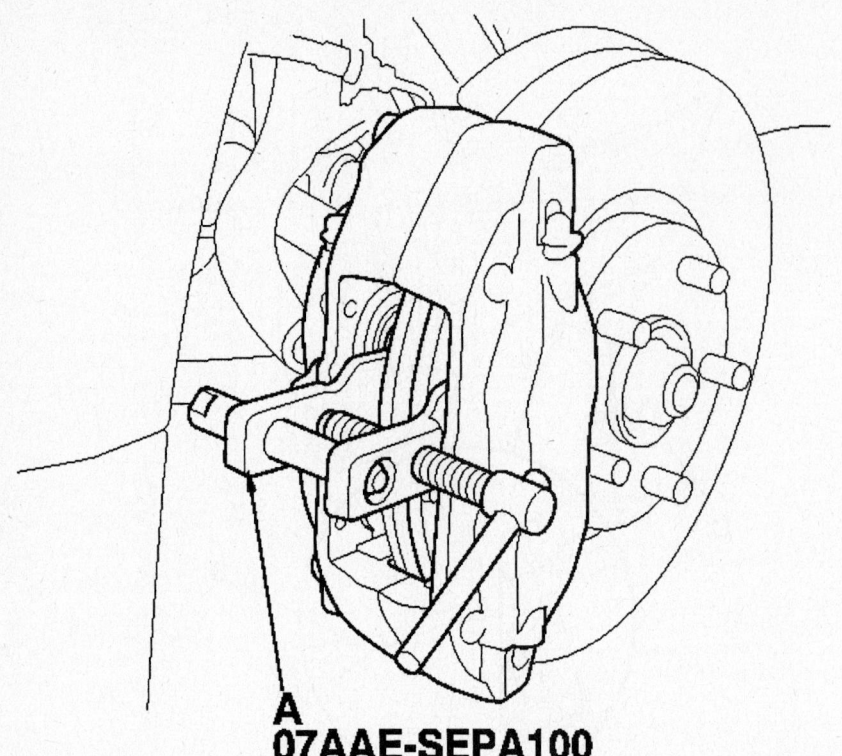

A
07AAE-SEPA100

09474_ACUCAR_G0117

Mount the piston compressor tool on the caliper—2004–06 TL model with a manual transmission

- Wheels
- Lower caliper bolt and pivot the caliper up and away from the rotor
- Pads, shims, and pad retainer springs

To install:

3. Clean all points where the pads and shims touch the caliper and mount. Apply a thin film of silicone grease to the cleaned areas.

4. Install or connect the following:
- Pad retainers in position on the caliper bracket
- High temperature brake grease to the back side of the pads and both sides of shims and wipe off the excess.
- Pads and shims
- Inner brake pad with the wear indicator facing upward

5. Loosen the bleed screw slightly and push in the caliper piston to allow mounting of the caliper over the rotor. Torque the bleed screw to 84 inch lbs. (9 Nm).

6. Pivot the caliper down over the rotor and install the caliper bolts. Torque the bolts to 36 ft. lbs. (50 Nm).

7. If disconnected, install the brake pad wear indicator connector. Install the wheels.

Rear

CL MODEL

1. Before servicing the vehicle, refer to the precautions section.

2. Remove or disconnect the following:
- Wheels
- Caliper dust shield, if equipped
- Both caliper mounting bolts and pull the caliper off the bracket. Be sure to hang the caliper with a piece of wire so no tension is on the brake line.
- Pads, shim and pad retainer spring. Clean all points where the pads and shims touch the caliper and mount. Apply a thin film of silicone grease to the cleaned areas.

To install:

3. Apply high temperature brake grease to the pads and shims. Install the pads and shims, making sure the inner pad has the wear indicator facing down.

4. Rotate the brake caliper piston clockwise into the cylinder, using a locknut wrench (part # 07916–6390001). Align the cutout in the piston with the tab on the inner pad by turning the piston back.

✳✳ WARNING

Lubricate the piston boot with grease to avoid twisting the piston boot. If the piston boot is twisted, back the piston out so it sits properly.

5. Install or connect the following:
- Caliper on the bracket and torque the bolts to 17 ft. lbs. (23 Nm)
- Parking brake cable and the dust shield, if removed
- Wheels and lower the vehicle

2002–03 RL MODEL

1. Before servicing the vehicle, refer to the precautions section.

2. Remove or disconnect the following:
- Wheels
- Caliper dust shield, if equipped
- Both caliper mounting bolts and pull the caliper off the bracket. Be sure to hang the caliper with a piece of wire so no tension is on the brake line.
- Pads, shim and pad retainer spring. Clean all points where the pads and shims touch the caliper and mount. Apply a thin film of silicone grease to the cleaned areas.

To install:

3. Apply high temperature brake grease to the pads and shims. Install the pads and shims, making sure the inner pad has the wear indicator facing down.

4. Rotate the brake caliper piston clockwise into the cylinder, using a locknut wrench (part # 07916–6390001). Align the cutout in the piston with the tab on the inner pad by turning the piston back.

✳✳ WARNING

Lubricate the piston boot with grease to avoid twisting the piston boot. If the piston boot is twisted, back the piston out so it sits properly.

5. Install or connect the following:
- Caliper on the bracket and torque the bolts to 17 ft. lbs. (23 Nm)
- Parking brake cable and the dust shield, if removed
- Wheels and lower the vehicle

2005–06 RL MODEL

1. Before servicing the vehicle, refer to the precautions in the beginning of this section.

2. Remove the rear wheels.

3. Release the parking brake.

4. Remove the caliper bracket mounting

bolts while holding the caliper pins) with a wrench being careful not to damage the pin boot, and remove the caliper. Check the hose and pin boots for damage and deterioration. Thoroughly clean the outside of the caliper to prevent dust and dirt from entering inside.

5. Support the caliper with a piece of wire so it does not hang from the brake hose.

6. Remove the pad shims and brake pads.

7. Remove the pad retainers.

8. Clean the caliper thoroughly; remove any rust and check for grooves and cracks.

9. Check the brake disc/drum for damage and cracks.

To install:

10. Apply M-77 assembly paste to the retainers on their mating surfaces against the caliper bracket.

11. Install the pad retainers. Wipe excess assembly paste off the retainers. Contaminated brake discs and pads reduce stopping ability. Keep assembly paste off the discs and pads.

12. Install the pad retainers.

13. Apply M-77 assembly paste to the pad side of the shims and back of the brake pads and the all other contact. Wipe excess assembly paste off the pad shims and brake pads. Contaminated brake discs or pads reduce stopping ability.

14. Install the brake pads and pad shims on the caliper bracket. Install the inner brake pad with its wear indicator facing on top.

15. If you are reusing the brake pads, always reinstall the brake pads in their original positions to prevent a momentary loss of braking efficiency

Push in the piston (A) so the caliper will fit over the brake pads. Make sure the piston boot is in position to prevent damaging it when installing the caliper.

➡️**Be careful when pushing in the caliper, brake fluid might overflow from the master cylinder's reservoir.**

16. Apply M-77 assembly paste to the piston edges on their mating surfaces against the inner pad shim.

17. Install the brake caliper.

18. Install the caliper bolts, and torque them to 17 ft. lbs. (23 Nm) while holding the caliper pins with a wrench being and careful not to damage the pin boot.

19. Press the brake pedal several times to make sure the brakes work, then road-test the vehicle.

➡️**The brake may require a greater pedal stroke immediately after the brake pads have been replaced as a set. Several applications of the brake pedal will restore the normal pedal stroke.**

20. After installation, check for leaks at the hose and line joints and connections, and retighten if necessary.

RSX MODELS

1. Before servicing the vehicle, refer to the precautions section.

2. Remove or disconnect the following:
   • Wheels
   • Caliper dust shield, if equipped
   • Both caliper mounting bolts and pull the caliper off the bracket. Be sure to hang the caliper with a piece of wire so no tension is on the brake line.
   • Pads, shim and pad retainer spring. Clean all points where the pads and shims touch the caliper and mount. Apply a thin film of silicone grease to the cleaned areas.

To install:

3. Apply high temperature brake grease to the pads and shims. Install the pads and shims, making sure the inner pad has the wear indicator facing down.

4. Rotate the brake caliper piston clockwise into the cylinder, using a locknut wrench (part # 07916–6390001). Align the cutout in the piston with the tab on the inner pad by turning the piston back.

✳✳ WARNING

Lubricate the piston boot with grease to avoid twisting the piston boot. If the piston boot is twisted, back the piston out so it sits properly.

5. Install or connect the following:
   • Caliper on the bracket and torque the bolts to 16 ft. lbs. (22 Nm)
   • Parking brake cable and the dust shield, if removed
   • Wheels and lower the vehicle

2002–03 TL MODELS

1. Before servicing the vehicle, refer to the precautions section.

2. Remove or disconnect the following:
   • Wheels
   • Caliper dust shield, if equipped
   • Both caliper mounting bolts and pull the caliper off the bracket. Be sure to hang the caliper with a

piece of wire so no tension is on the brake line.
   • Pads, shim and pad retainer spring. Clean all points where the pads and shims touch the caliper and mount. Apply a thin film of silicone grease to the cleaned areas.

To install:

3. Apply high temperature brake grease to the pads and shims. Install the pads and shims, making sure the inner pad has the wear indicator facing down.

4. Rotate the brake caliper piston clockwise into the cylinder, using a locknut wrench (part # 07916–6390001). Align the cutout in the piston with the tab on the inner pad by turning the piston back.

✳✳ WARNING

Lubricate the piston boot with grease to avoid twisting the piston boot. If the piston boot is twisted, back the piston out so it sits properly.

5. Install or connect the following:
   • Caliper on the bracket and torque the bolts to 17 ft. lbs. (23 Nm)
   • Parking brake cable and the dust shield, if removed
   • Wheels and lower the vehicle

2004–06 TL MODELS

1. Before servicing the vehicle, refer to the precautions in the beginning of this section.

2. Remove or disconnect the following:
   • Wheels
   • Caliper dust shield, if equipped
   • Both caliper mounting bolts and pull the caliper off the bracket. Be sure to hang the caliper with a piece of wire so no tension is on the brake line.
   • Pads, shim and pad retainer spring. Clean all points where the pads and shims touch the caliper and mount. Apply a thin film of silicone grease to the cleaned areas.

To install:

3. Apply high temperature brake grease to the pads and shims. Install the pads and shims, making sure the inner pad has the wear indicator facing down.

4. Compress the brake caliper.

✳✳ WARNING

Lubricate the piston boot with grease to avoid twisting the piston boot. If the piston boot is twisted, back the piston out so it sits properly.

5. Install or connect the following:
  - Caliper on the bracket and torque the bolts to 17 ft. lbs. (23 Nm)
  - Parking brake cable and the dust shield, if removed
  - Wheels and lower the vehicle

TSX MODELS

1. Before servicing the vehicle, refer to the precautions section.

2. Remove or disconnect the following:

  - Wheels
  - Caliper dust shield, if equipped
  - Both caliper mounting bolts and pull the caliper off the bracket. Be sure to hang the caliper with a piece of wire so no tension is on the brake line.
  - Pads, shim and pad retainer spring. Clean all points where the pads and shims touch the caliper and mount. Apply a thin film of silicone grease to the cleaned areas.

To install:

3. Apply high temperature brake grease to the pads and shims. Install the pads and shims, making sure the inner pad has the wear indicator facing down.

4. Rotate the brake caliper piston clockwise into the cylinder, using a locknut wrench (part # 07916–6390001). Align the cutout in the piston with the tab on the inner pad by turning the piston back.

❊❊ WARNING

Lubricate the piston boot with grease to avoid twisting the piston boot. If the piston boot is twisted, back the piston out so it sits properly.

5. Install or connect the following:
  - Caliper on the bracket and torque the bolts to 17 ft. lbs. (23 Nm)
  - Parking brake cable and the dust shield, if removed
  - Wheels and lower the vehicle

ACURA

MDX

SPECIFICATION AND MAINTENANCE CHARTS

ENGINE AND VEHICLE IDENTIFICATION CHART

		Engine Code						Model Year	
Code	Liters (cc)	Cu. In.	Cyl.	Fuel Sys.	Engine Type	Eng. Mfg.	Code ①		Year
J35A3	3.5 (3471)	222	6	SMFI	SOHC	Honda	2		2002
							3		2003
							4		2004
							5		2005
							6		2006

SOHC: Single Overhead Cam

SMFI: Sequential Multi-port Fuel Injection

① 10th position of VIN

09474_AMDX_C0001

GENERAL ENGINE SPECIFICATIONS

Year	Model	Engine Displacement Liters (VIN)	Net Horsepower @ rpm	Net Torque @ rpm (ft. lbs.)	Bore x Stroke (in.)	Com- pression Ratio	Oil Pressure @ rpm
2002	MDX	3.5 (J35A3)	210@5200	229@4300	3.50x3.66	9.4:1	71@3000
2003	MDX	3.5 (J35A3)	210@5200	229@4300	3.50x3.66	9.4:1	71@3000
2004	MDX	3.5 (J35A3)	210@5200	229@4300	3.50x3.66	9.4:1	71@3000
2005-06	MDX	3.5 (J35A3)	265@5800	253@3500	3.50x3.66	10.0:1	71@3000

SMFI: Sequential Multi-port Fuel Injection

09474_AMDX_C0002

ENGINE TUNE-UP SPECIFICATIONS

Year	Engine Displacement Liters (VIN)	Spark Plug Gap (in.)	Ignition Timing (deg.) MT	AT	Fuel Pump (psi)	Idle Speed (rpm) MT	AT	Valve Clearance (in.) In.	Ex.
2002	3.5 (J35A1)	0.039-0.043	—	8-12 ①	57-64	—	680-780	0.008-0.009	0.011-0.013
2003	3.5 (J35A1)	0.039-0.043	—	8-12 ①	57-64	—	680-780	0.008-0.009	0.011-0.013
2004	3.5 (J35A1)	0.039-0.043	—	8-12 ①	57-64	—	680-780	0.008-0.009	0.011-0.013
2005-06	3.5 (J35A1)	0.039-0.043	—	8-12 ①	57-64	—	680-780	0.008-0.009	0.011-0.013

NOTE: The Vehicle Emission Control Information label often reflects changes made during production and must be used if they differ from this chart.

NOTE: The fuel pressure readings are given with the vacuum hose connected to the regulator and the engine running

① Before top dead center

09474_AMDX_C0003

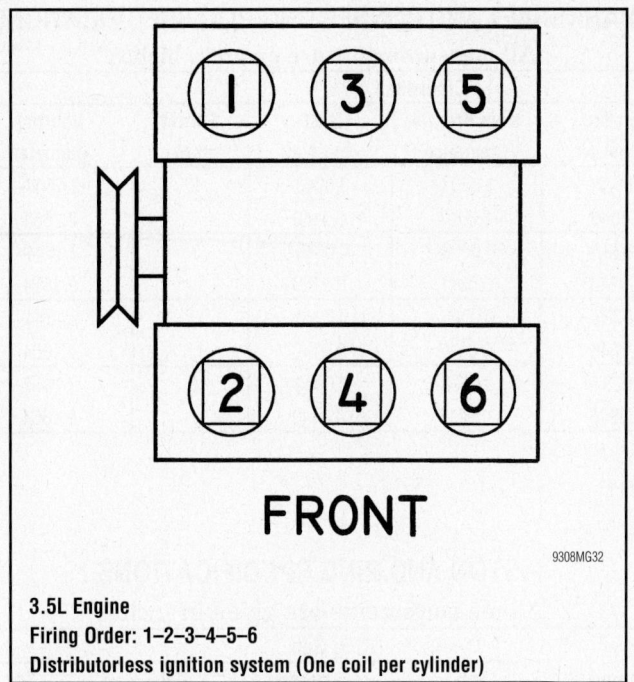

FRONT

9308MG32

3.5L Engine
Firing Order: 1–2–3–4–5–6
Distributorless ignition system (One coil per cylinder)

CAPACITIES

Year	Model	Engine Displacement Liters (VIN)	Engine Oil with Filter (qts.)	Transmission (qts.) 5-Spd	Transmission (qts.) Auto.	Transfer Case (pts.)	Drive Axle Front (pts.)	Drive Axle Rear (pts.)	Fuel Tank (gal.)	Cooling System (qts.)
2002	MDX	3.5 (J35A3)	4.5	—	7.7	—	—	—	19.2	8.0
2003	MDX	3.5 (J35A3)	4.5	—	7.7	—	—	—	19.2	8.0
2004	MDX	3.5 (J35A3)	4.5	—	7.7	—	—	—	19.2	8.0
2005-06	MDX	3.5 (J35A3)	4.5	—	7.7	—	—	—	19.2	8.0

NOTE: All capacities are approximate. Add fluid gradually and check to be sure a proper fluid level is obtained.

09474_AMDX_C0004

VALVE SPECIFICATIONS

Year	Engine Displacement Liters (cc)	Seat Angle (deg.)	Face Angle (deg.)	Spring Test Pressure (lbs. @ in.)	Spring Installed Height (in.)	Stem-to-Guide Clearance (in.) Intake	Stem-to-Guide Clearance (in.) Exhaust	Stem Diameter (in.) Intake	Stem Diameter (in.) Exhaust
2002	3.5 (J35A3)	45	45	NA	①	0.0008-0.0018	0.0022-0.0031	0.2159-0.2163	0.2146-0.2150
2003	3.5 (J35A3)	45	45	NA	①	0.0008-0.0018	0.0022-0.0031	0.2159-0.2163	0.2146-0.2150
2004	3.5 (J35A3)	45	45	NA	①	0.0008-0.0018	0.0022-0.0031	0.2159-0.2163	0.2146-0.2150
2005-06	3.5 (J35A3)	45	45	NA	①	0.0008-0.0018	0.0022-0.0031	0.2159-0.2163	0.2146-0.2150

NA: Not Available

① Valve spring free length:
  Intake: 2.029 in.
  Exhaust: 2.010 in.

09474_AMDX_C0005

CRANKSHAFT AND CONNECTING ROD SPECIFICATIONS
All measurements are given in inches

Year	Engine Displacement Liters (cc)	Crankshaft				Connecting Rod		
		Main Brg. Journal Dia.	Main Brg. Oil Clearance	Shaft End-play	Thrust on No.	Journal Diameter	Oil Clearance	Side Clearance
2002	3.5 (J35A3)	2.8337-2.8346	0.0008-0.0017	0.0040-0.0140	3	2.1644-2.1654	0.0008-0.0017	0.0060-0.0140
2003	3.5 (J35A3)	2.8337-2.8346	0.0008-0.0017	0.0040-0.0140	3	2.1644-2.1654	0.0008-0.0017	0.0060-0.0140
2004	3.5 (J35A3)	2.8337-2.8346	0.0008-0.0017	0.0040-0.0140	3	2.1644-2.1654	0.0008-0.0017	0.0060-0.0140
2005-06	3.5 (J35A3)	2.8337-2.8346	0.0008-0.0017	0.0040-0.0140	3	2.1644-2.1654	0.0008-0.0017	0.0060-0.0140

09474_AMDX_C0006

PISTON AND RING SPECIFICATIONS
All measurements are given in inches

Year	Engine Displacement Liters (cc)	Piston Clearance	Ring Gap			Ring Side Clearance		
			Top Compression	Bottom Compression	Oil Control	Top Compression	Bottom Compression	Oil Control
2002	3.5 (J35A3)	0.0006-0.0016	0.0080-0.0140	0.0160-0.0220	0.0080-0.0280	0.0022-0.0031	0.0012-0.0022	NA
2003	3.5 (J35A3)	0.0006-0.0016	0.0080-0.0140	0.0160-0.0220	0.0080-0.0280	0.0022-0.0031	0.0012-0.0022	NA
2004	3.5 (J35A3)	0.0006-0.0016	0.0080-0.0140	0.0160-0.0220	0.0080-0.0280	0.0022-0.0031	0.0012-0.0022	NA
2005-06	3.5 (J35A3)	0.0006-0.0016	0.0080-0.0140	0.0160-0.0220	0.0080-0.0280	0.0022-0.0031	0.0012-0.0022	NA

NA: Not Available

09474_AMDX_C0007

TORQUE SPECIFICATIONS
All readings in ft. lbs.

Year	Engine Displacement Liters (VIN)	Cylinder Head Bolts	Main Bearing Bolts	Rod Bearing Bolts	Crankshaft Damper Bolts	Flywheel Bolts	Manifold		Spark Plugs	Oil Pan Drain Plug
							Intake	Exhaust		
2002	3.5 (J35A1)	①	②	③	181	54	16	23	13	29
2003	3.5 (J35A1)	①	②	③	④	54	16	23	13	29
2004	3.5 (J35A1)	①	②	③	④	54	16	23	13	29
2005-06	3.5 (J35A1)	①	②	③	④	54	16	23	13	29

NOTE: Dip main bearing bolts and crankshaft damper bolt in clean engine oil prior to tightening.

① Step 1: 29 ft. lbs.
   Step 2: 51 ft. lbs.
   Step 3: 72 ft. lbs.

② 11mm bolt 54 ft. lbs.
   10mm bolt 36 ft. lbs.

③ Step 1: 14 ft. lbs.
   Step 2: 90 degrees

④ Step 1: 47 ft. lbs.
   Step 2: plus 60 degrees

09474_AMDX_C0008

WHEEL ALIGNMENT

Year	Model		Caster Range (+/-Deg.)	Caster Preferred Setting (Deg.)	Camber Range (+/-Deg.)	Camber Preferred Setting (Deg.)	Toe-in (in.)
2002	MDX	F	1.00	1.00	1.00	30	0+/-1/16
		R	—	—	0	30	0+/-1/16
2003	MDX	F	1.00	1.00	1.00	30	0+/-1/16
		R	—	—	0	30	0+/-1/16
2004	MDX	F	1.00	1.00	1.00	30	0+/-1/16
		R	—	—	0	30	0+/-1/16
2005-06	MDX	F	1.00	1.00	1.00	30	0+/-1/16
		R	—	—	0	30	0+/-1/16

09474_AMDX_C0009

TIRE, WHEEL AND BALL JOINT SPECIFICATIONS

Year	Model	OEM Tires Standard	OEM Tires Optional	Tire Pressures (psi) Front	Tire Pressures (psi) Rear	Wheel Size	Ball Joint Inspection	Lug Nut
2002	MDX	P235/65R17	None	32	32	R17	NS	80
2003	MDX	P235/65R17	None	32	32	R17	NS	80
2004	MDX	P235/65R17	None	32	32	R17	NS	80
2005-06	MDX	P235/65R17	None	32	32	R17	NS	80

OEM: Original Equipment Manufacturer

PSI: Pounds Per Square Inch

NS: Not specified by manufacturer

09474_AMDX_C0010

BRAKE SPECIFICATIONS
Acura MDX
All measurements in inches unless noted

Year	Model		Brake Disc Original Thickness	Brake Disc Minimum Thickness	Brake Disc Maximum Runout	Brake Drum Diameter Original Inside Diameter	Brake Drum Diameter Max. Wear Limit	Brake Drum Diameter Maximum Machine Diameter	Minimum Lining Thickness Front	Minimum Lining Thickness Rear	Brake Caliper Bracket Bolts (ft. lbs.)	Brake Caliper Mounting Bolts (ft. lbs.)
2002	MDX	F	1.100	1.020	0.004	—	—	—	0.060	—	101	27
		R	0.430	0.350	0.004	—	—	—	—	0.41	41	27
2003	MDX	F	1.100	1.020	0.004	—	—	—	0.060	—	101	27
		R	0.430	0.350	0.004	—	—	—	—	0.41	41	27
2004	MDX	F	1.100	1.020	0.004	—	—	—	0.060	—	101	27
		R	0.430	0.350	0.004	—	—	—	—	0.41	41	27
2005-06	MDX	F	1.100	1.020	0.004	—	—	—	0.060	—	101	27
		R	0.430	0.350	0.004	—	—	—	—	0.41	41	27

F: Front

R: Rear

09474_AMDX_C0011

SCHEDULED MAINTENANCE INTERVALS
ACURA—MDX

TO BE SERVICED	TYPE OF SERVICE	VEHICLE MILEAGE INTERVAL (x1000)															
		7.5	15	22.5	30	37.5	45	52.5	60	67.5	75	82.5	90	97.5	105	112.5	120
Accessory drive belts	I & A				✓				✓				✓				✓
Air cleaner element	R				✓				✓				✓				✓
Brake fluid	R	Every 3 years															
Brake hoses & lines (including ABS)	I		✓		✓		✓		✓		✓		✓		✓		✓
Cooling system hoses & connections	I		✓		✓		✓		✓		✓		✓		✓		✓
Engine coolant ①	R						✓						✓				
Engine oil	R	✓	✓	✓	✓	✓	✓	✓	✓	✓	✓	✓	✓	✓	✓	✓	✓
Engine oil and coolant levels	I	Inspect at each fuel stop															
Engine oil filter	R		✓		✓		✓		✓		✓		✓		✓		✓
Exhaust system	I		✓		✓		✓		✓		✓		✓		✓		✓
Fluid levels and condition	I		✓		✓		✓		✓		✓		✓		✓		✓
Front and rear brakes	I		✓		✓		✓		✓		✓		✓		✓		✓
Fuel lines & connection	I		✓		✓		✓		✓		✓		✓		✓		✓
Halfshaft boots	I		✓		✓		✓		✓		✓		✓		✓		✓
Idle speed	I & A														✓		
Parking brake system	I & A		✓		✓		✓		✓		✓		✓		✓		✓
Rear differential fluid	R	✓			✓		✓		✓				✓				✓
Rotate and inspect tires	I	✓	✓	✓	✓	✓	✓	✓	✓	✓	✓	✓	✓	✓	✓	✓	✓
Spark plugs	R														✓		
Supplemental Restrain system (SRS)	I	Inspect the SRS 10 years after production															
Suspension components	I		✓		✓		✓		✓		✓		✓		✓		✓
Tie rod ends, steering gear box & boots	I		✓		✓		✓		✓		✓		✓		✓		✓
Timing belt	R														✓		
Transmission fluid	R						✓				✓				✓		
Valve clearance	I	Adjust if valves are noisy															
Water pump	S/I														✓		

R: Replace   I: Inspect   A: Adjust

① Every 12,000 miles or 10 years, then every 60,000 miles or 5 years

FREQUENT OPERATION MAINTENANCE (SEVERE SERVICE)

If a vehicle is operated under any of the following conditions it is considered severe service:

- Towing a trailer or using a camper or car-top carrier.
- Repeated short trips of less than 5 miles in temperatures below freezing, or trips of less than 10 miles in any temperature.
- Extensive idling or low-speed driving for long distances as in heavy commercial use, such as delivery, taxi or police cars.
- Operating on rough, muddy or salt-covered roads.
- Operating on unpaved or dusty roads.
- Driving in extremely hot (over 90°) conditions.

Air cleaner element: replace every 15,000 miles

Engine oil and filter: replace every 3750 miles or 6 months, whichever occurs first.

Timing belt: replace every 60,000 miles if the vehicle is regularly driven in temperatures above 110°F or below -20°F, or if frequently towing a trailer.

Transmission fluid: replace every 30,000 miles.

Rear differential fluid: replace every 60,000 miles.

Front and rear brakes: inspect every 7500 miles or 6 months, whichever occurs first.

Locks and hinges: lubricate every 15,000 miles.

Tie rods, steering gear box, boots: inspect every 7500 miles or 6 months, whichever occurs first.

Suspension components: inspect every 7500 miles or 6 months, whichever occurs first.

Halfshaft boots: inspect every 7500 miles or 6 months, whichever occurs first.

ENGINE REPAIR

➡️Disconnecting the negative battery cable on some vehicles may interfere with the functions of the on board computer system. The computer may undergo a relearning process once the negative battery cable is reconnected.

Distributor

The MDX is equipped with a Distributorless Ignition System (DIS).

Alternator

REMOVAL

1. Before servicing the vehicle, refer to the precautions section.
2. Remove or disconnect the following:
   • Negative battery cable
   • Intake manifold and ignition coil covers
   • Accessory drive belt
   • Alternator wiring harness connectors
   • Alternator mounting bolts
   • Wiring harness clamp
   • Alternator

INSTALLATION

1. Install or connect the following:
   • Alternator
   • Wiring harness clamp. Tighten the bolt to 105 inch lbs. (12 Nm).
   • Alternator mounting bolts. Tighten the 10mm bolt to 33 ft. lbs. (44 Nm) and the 8mm bolt to 16 ft. lbs. (22 Nm).
   • Alternator wiring harness connectors. Tighten the battery terminal nut to 105 inch lbs. (12 Nm).
   • Accessory drive belt
   • Intake manifold and ignition coil covers
   • Negative battery cable

Ignition Timing

ADJUSTMENT

The MDX is equipped with a Distributorless Ignition System (DIS). The ignition timing is controlled by the Powertrain Control module (PCM). No adjustment is necessary.

Engine Assembly

REMOVAL & INSTALLATION

➡️The engine and transaxle are removed from the vehicle as a unit.

1. Before servicing the vehicle, refer to the precautions section.
2. Remove the two clips, pull the insulator and then remove the support struts from the hood, then move the hood to a vertical position and reinstall the support strut.
3. Drain the cooling system.
4. Drain the power steering system.
5. Drain the transaxle fluid.
6. Drain the engine oil.
7. Relieve fuel system pressure.
8. Remove or disconnect the following:
   • Negative battery cable
   • Battery and tray
   • Intake and ignition coil covers
   • Electrical connections and the air intake duct
   • Left engine wire harness connectors
   • Relay bracket
   • Starter cable and harness clamp
   • Shift control cable
   • Accelerator cable
   • Cruise control cable
   • Fuel lines
   • Brake booster EVAP canister hoses
9. Remove the drivers side center console lower panel and pull back the carpet to access steering joint cover.
   • Steering joint cover
   • Steering joint bolt
   • Powertrain Control Module (PCM) connectors
   • Alternator cables
   • Drive belt
   • Power steering hose's clamps and clips
   • Heated Oxygen (HO$_2$S) sensor connector and grommet. Pull the PCM harness through the firewall.
   • Fuse/Relay box battery cable
   • Front wheels
   • Splash shield
   • Front sub-frame stiffener
   • Exhaust front pipe
   • Propeller shaft
   • Transfer assembly
   • Stabilizer bar links
   • Ball joints
   • Knuckles from the lower arms

   • Driveshafts
   • Power steering hose and pressure switch connector
   • Transaxle lower front mount
   • Transaxle lower rear mount
   • Rear engine mount bolts
   • A/C compressor and move aside and attach with wire to keep it out of the way during engine removal
   • Heater hoses
   • Radiator hoses
   • Ground cable
   • Transaxle oil cooler lines
   • Radiator
10. Attach a hoist to the engine lifting eyes and support the powertrain weight.
11. Remove or disconnect the following:
    • Side engine mount bracket
    • Front mount bracket support nut
12. Matchmark the front subframe to the mounting points.
13. Remove or disconnect the following:
    • Front subframe
    • All remaining hoses and electrical connections
14. Lower the powertrain away from the vehicle.
To install:
15. Raise the powertrain into position.
16. Installation is the reverse of removal but please note the following steps:
    • A/C compressor bolts to 16 ft. lbs. (22 Nm)
    • Front subframe. Use new bolts and tighten the 14mm bolts to 76 ft. lbs. (103 Nm). Tighten the front brace bolts to 54 ft. lbs. (74 Nm) and the rear brace bolts to 86 ft. lbs. (117 Nm).
    • Transaxle lower front mount nuts to 28 ft. lbs. (38 Nm)
    • Transaxle lower rear mount bolts to 28 ft. lbs. (38 Nm)
    • Front mount bracket support nut to 40 ft. lbs. (54 Nm)
    • Side engine mount bracket bolts to 33 ft. lbs. (44 Nm) and the through bolt to 40 ft. lbs. (54 Nm)
17. Fill the engine crankcase to the correct level.
18. Fill the transaxle to the correct level.
19. Fill the cooling system.
20. Fill the power steering system.
21. Start the engine and check for leaks.
22. Check the wheel alignment and adjust as necessary.

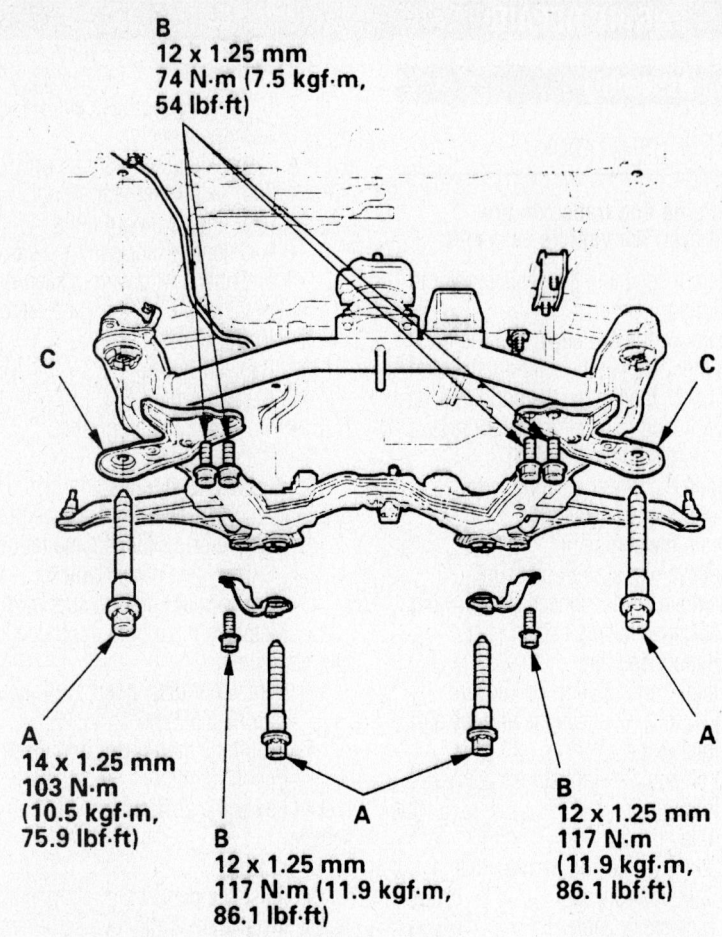

B
12 x 1.25 mm
74 N·m (7.5 kgf·m,
54 lbf·ft)

C

C

A
14 x 1.25 mm
103 N·m
(10.5 kgf·m,
75.9 lbf·ft)

A

A

B
12 x 1.25 mm
117 N·m (11.9 kgf·m,
86.1 lbf·ft)

B
12 x 1.25 mm
117 N·m
(11.9 kgf·m,
86.1 lbf·ft)

A

9302MG69

Sub-frame fastener locations and tightening torque—MDX

Water Pump

REMOVAL & INSTALLATION

1. Before servicing the vehicle, refer to the precautions section.
2. Drain the cooling system.
3. Remove or disconnect the following:
   - Negative battery cable
   - Accessory drive belts
   - Front cover
   - Timing belt
   - Timing belt tensioner
   - Water pump

To install:

4. Install or connect the following:
   - Water pump. Use a new O-ring seal and tighten the bolts to 105 inch lbs. (12 Nm).
   - Timing belt tensioner
   - Timing belt
   - Front cover
   - Accessory drive belts
   - Negative battery cable
5. Fill the cooling system.
6. Start the engine and check for leaks.

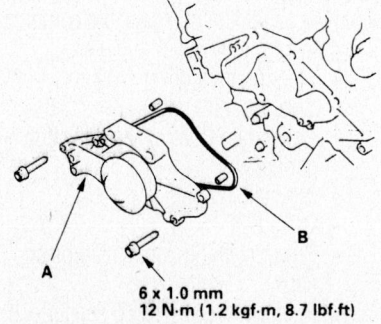

A

B

6 x 1.0 mm
12 N·m (1.2 kgf·m, 8.7 lbf·ft)

93552G01

Exploded view of the water pump mounting

Heater Core

REMOVAL & INSTALLATION

1. Before servicing the vehicle, refer to the precautions section.
2. Drain the cooling system.
3. Remove or disconnect the following:
   - Negative battery cable
4. Recover the refrigerant using approved equipment.

- Heater valve cable from the valve arm. Turn the valve arm to the fully opened position.
- Heater hoses from the heater unit
- Mounting nut from the heater unit. Be careful not to bend or damage fuel or brake lines.

5. Remove the dashboard as follows:

   a. Remove the center console by unlatching the clips.

   b. Remove the dashboard lower cover screw, gently pull down on the cover to disengage the clips and disconnect the electrical connections.

   c. Remove the dashboard side cover by gently pulling and turning to unfasten the clips.

   d. While holding the glove box, remove the box stop from each side, then disconnect the lock from the damper.

   e. Remove the glove ox bolts and the glove box.

   f. Remove the shift lever assembly.

   g. Remove the front door trim, lick panel and A-pillar trim from both sides.

   h. Remove the cap from the front pillar corner trim. Unfasten the screw, slide the trim upward along the pillar and remove it. Remove the remaining clips from the body.

   i. On the drivers side, remove the fuel/relay box nut and pull out the box.

   j. Remove the steering column

   k. On the passenger side remove the fuse/relay bolt and pull out the box.

   l. Disconnect all electrical connections from the dashboard.

   m. If equipped with a navigation system, pull back the carpet, remove the harness cushions and then pull out the GPS harness.

   n. Remove all harness and connector clips.

   o. Remove all the bolts and lift up on the dashboard to release the dashboard and steering hanger beam from the guide pins.

   p. Remove the dashboard through the door.

6. Remove the evaporator as follows:

   a. Disconnect the receiver and suction lines from the evaporator.

   b. Remove the mounting nuts and plug the lines to avoid system contamination.

   c. Remove the plastic brace and glove box frame.

   d. Disconnect the wire harness and evaporator temperature sensor connector.

   e. Remove the self-tapping screws, the nuts and the evaporator.

7. Remove or disconnect the following:

- Mounting bolts and the heater unit
- Self-tapping screws and the clamp, then pull the heater core from the case being careful not to bend the pipes

To install:

8. Install or connect the following:
- Heater core in the case
- Clamp and the screws
- Heater unit and tighten the bolts to 7 ft. lbs. (10 Nm)
- Evaporator in the reverse order of removal. Tighten all the retainers to 7 ft. lbs. (10 Nm) .

9. Install the dashboard in the reverse order of removal keeping in mind the following points:

a. Make sure the dashboard is seated properly and that the wiring harness and steering hanger beam wire harness are not pinched.

b. Referring to the accompanying illustration, tighten bolts **(A)** to 7 ft. lbs. (10 Nm). Tighten all the other bolts to 16 ft. lbs. (22 Nm). Apply thread lock to the **B** bolts before installation.

c. Ensure that all electrical connectors are properly connected.

10. Install or connect the following:

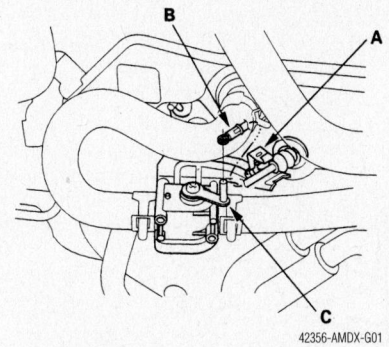

42356-AMDX-G01

In the engine compartment, open the cable clamp (A), then disconnect the heater valve cable (B) from the valve arm (C)

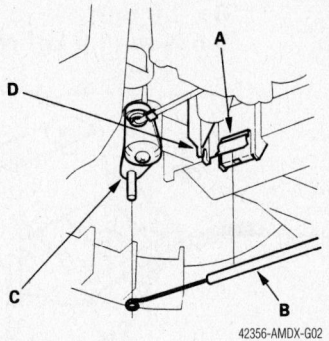

42356-AMDX-G02

Under the dashboard, disconnect the valve cable housing from the clamp (A) and the cable (B) from the mix control linkage (C)

- Mounting nut to the heater unit and tighten to 9 ft. lbs. (13 Nm)
- Heater hoses

11. Connect the heater valve cable and adjust as follows:

a. In the engine compartment, open the cable clamp (A), then disconnect the heater valve cable (B) from the valve arm (C).

b. Under the dashboard, disconnect the valve cable housing from the clamp (A) and the cable (B) from the mix control linkage (C).

c. Set the temperature control button

to the MAX COOL position with the ignition switch in the on position.

d. Attach the valve cable (B) to the mix control linkage (C) as shown in the illustration, hold the end of the cable housing against the stop, then snap the cable housing into the clamp.

Sin the engine compartment, turn the valve arm (C) to the fully closed position as shown in the accompanying illustration and hold it there. Attach the cable (B) to the vale arm and pull gently on the cable housing to take up the slack, then

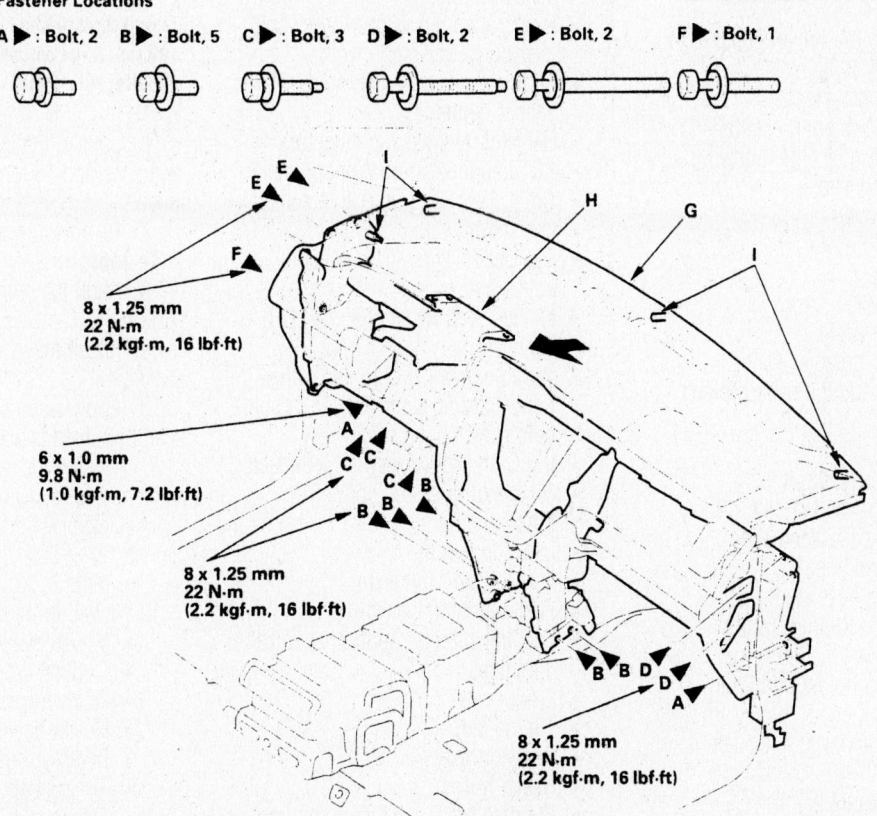

Fastener Locations

A ▶ : Bolt, 2   B ▶ : Bolt, 5   C ▶ : Bolt, 3   D ▶ : Bolt, 2   E ▶ : Bolt, 2   F ▶ : Bolt, 1

8 x 1.25 mm
22 N·m
(2.2 kgf·m, 16 lbf·ft)

6 x 1.0 mm
9.8 N·m
(1.0 kgf·m, 7.2 lbf·ft)

8 x 1.25 mm
22 N·m
(2.2 kgf·m, 16 lbf·ft)

8 x 1.25 mm
22 N·m
(2.2 kgf·m, 16 lbf·ft)

93552G91

Exploded view of the dashboard mounting—Acura MDX

6 x 1.0 mm
9.8 N·m (1.0 kgf·m, 7.2 lbf·ft)

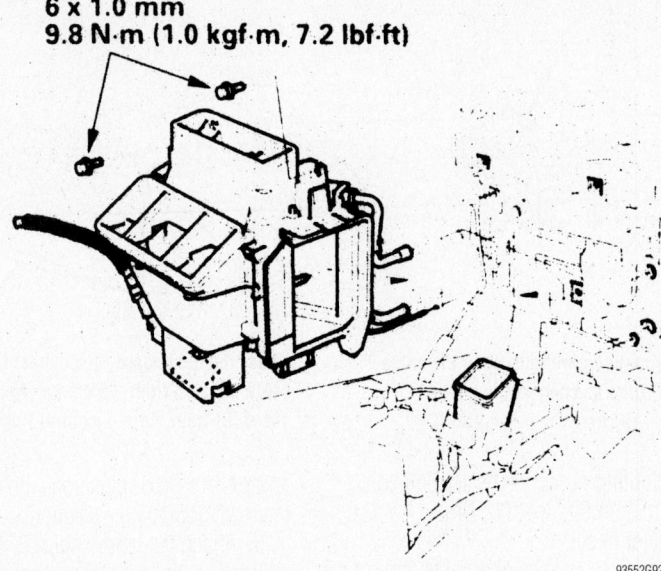

93552G92

Exploded view of the evaporator mounting—Acura MDX

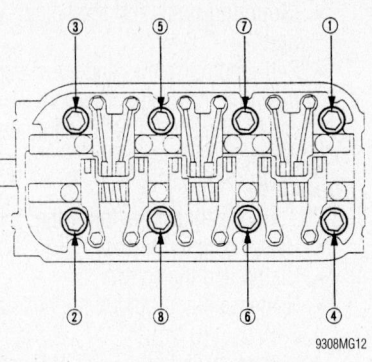

9308MG12

Cylinder head bolt loosening sequence—MDX

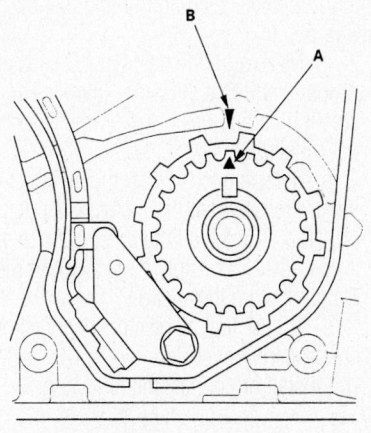

9302MG74

Crankshaft timing belt sprocket TDC marks. Align sprocket mark (A) with pointer (B)—MDX

install the cable housing into the clamp (A).

12. Fill the cooling system
13. Connect the battery cable.

Cylinder Head

REMOVAL & INSTALLATION

1. Before servicing the vehicle, refer to the precautions section.
2. Drain the cooling system.
3. Relieve the fuel system pressure.
4. Remove or disconnect the following:
   - Negative battery cable
   - Ignition coil covers
   - Intake manifold cover
   - Air intake tube
   - Accelerator cable
   - Cruise control cable
   - EVAP canister, breather, water bypass hoses and the bypass pipe bolt
   - Fuel lines
   - Brake booster vacuum line
   - Intake manifold vacuum line
   - Positive Crankcase Ventilation (PCV) valve and hose
   - Power steering hose clamp
   - Intake Manifold Runner Control (IMRC) actuator
   - Wiring harness holder and joint connector
5. Support the engine with a jack and a block of wood.
   - Side engine mount bracket
   - Accessory drive belts
   - Power steering pump

- Alternator
- Intake Air Temperature (IAT) sensor connector
- Idle Air Control (IAC) valve connector
- Throttle Position (TP) sensor connector
- Manifold Absolute Pressure (MAP) sensor connector
- Evaporative Emission (EVAP) canister purge valve connector
- Engine Coolant Temperature (ECT) sensor connector
- Radiator fan switch connectors
- ECT gauge sending unit connector
- Crankshaft Position (CKP) sensor connector
- Top Dead Center (TDC) sensor connector
- Exhaust Gas Recirculation (EGR) connector
- Variable Valve Timing and Valve Lift Electronic Control (VTEC) solenoid valve connector
- VTEC oil pressure switch connector
- Oil pressure switch connector
- Ignition coils
- Intake manifold
- Fuel injector connectors
- Fuel supply manifold
- Fuel injection air control valve vacuum lines
- Front cover
- Timing belt
- Radiator hoses
- Heater hoses
- Front and rear exhaust manifolds
- Coolant cross-over pipe
- Valve covers

6. Loosen the cylinder head bolts in sequence and 1/3 turns until all bolts are loose.
7. Remove the cylinder head.

To install:

8. Align the crankshaft and camshaft sprocket TDC marks as shown.
9. Install the cylinder heads with new gaskets.
10. Apply clean engine oil to the cylinder head bolt threads and flanges.
11. Tighten the cylinder head bolts in sequence as follows:
    a. Step 1: 29 ft. lbs. (39 Nm).
    b. Step 2: 51 ft. lbs. (69 Nm).
    c. Step 3: 72 ft. lbs. (98 Nm).
12. Install or connect the following:
    - Valve covers
    - Coolant cross-over pipe
    - Front and rear exhaust manifolds
    - Heater hoses
    - Radiator hoses
    - Timing belt
    - Front cover
    - Fuel injection air control valve vacuum lines

FRONT:

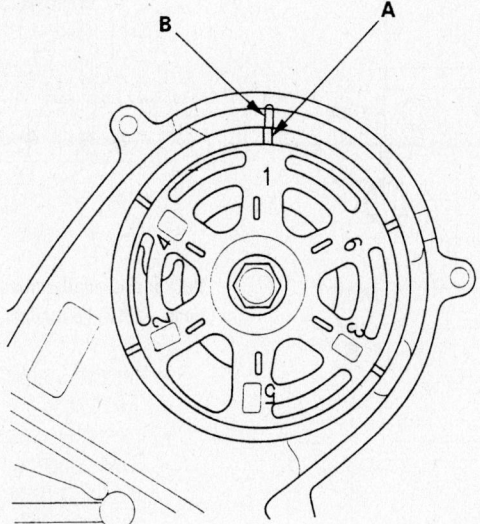

REAR:

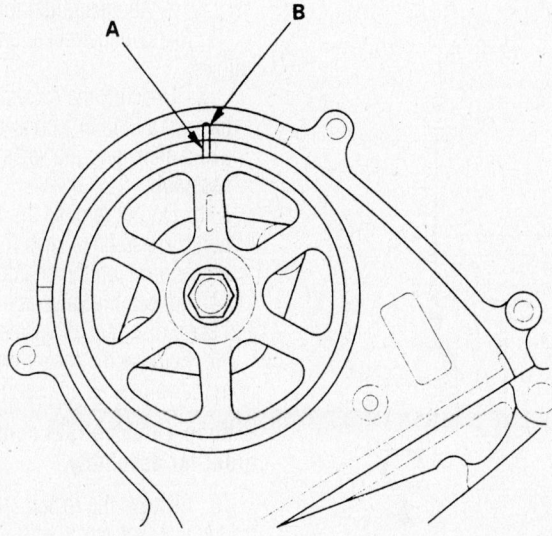

9302MG85

Camshaft TDC marks. Align sprocket mark (A) with the back cover pointer (B)—MDX

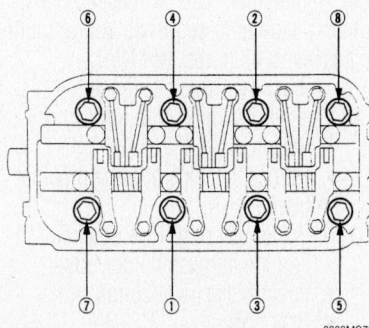

9302MG70

Cylinder head bolt tightening sequence—MDX

- Fuel supply manifold
- Fuel injector connectors
- Intake manifold
- Ignition coils
- Oil pressure switch connector
- VTEC oil pressure switch connector
- VTEC solenoid valve connector
- EGR connector
- TDC sensor connector
- CKP sensor connector
- ECT gauge sending unit connector
- Radiator fan switch connectors
- ECT sensor connector
- MAP sensor connector
- TP sensor connector

- IAC valve connector
- IAT sensor connector
- Alternator
- Power steering hose clamp
- Power steering pump
- Side engine mount bracket
- PCV valve and hose
- Intake manifold vacuum line
- Brake booster vacuum line
- Fuel lines
- Cruise control cable
- Accelerator cable
- Intake manifold cover
- Ignition coil covers
- Accessory drive belts
- Air intake tube
- EVAP control canister hose and vacuum hose
- Negative battery cable

13. Fill the cooling system.
14. Start the engine and check for leaks.

Rocker Arms/Shafts

REMOVAL & INSTALLATION

2002 Models

1. Before servicing the vehicle, refer to the precautions section.
2. Remove or disconnect the following:
   - Negative battery cable
   - Air intake tube
   - Ignition coil covers
   - Intake manifold cover
   - Intake manifold
   - Valve cover
3. Loosen the valve adjuster locknuts and screws so that all valves are closed.
4. Loosen the rocker arm shaft bolts evenly in sequence.
5. Remove the rocker arms and shafts from the vehicle as an assembly.

➡**Keep all valvetrain components in order for assembly.**

6. Remove the rocker arms and springs from the rocker arm shafts.

To install:

7. Assemble the rocker arms and springs to the rocker arm shafts in their original positions.
8. Install the rocker arm assemblies. Tighten the bolts in sequence and in multiple passes to 17 ft. lbs. (24 Nm).
9. Adjust the valve clearance.
10. Install or connect the following:
    - Valve covers
    - Intake manifold
    - Intake manifold cover
    - Ignition coil covers

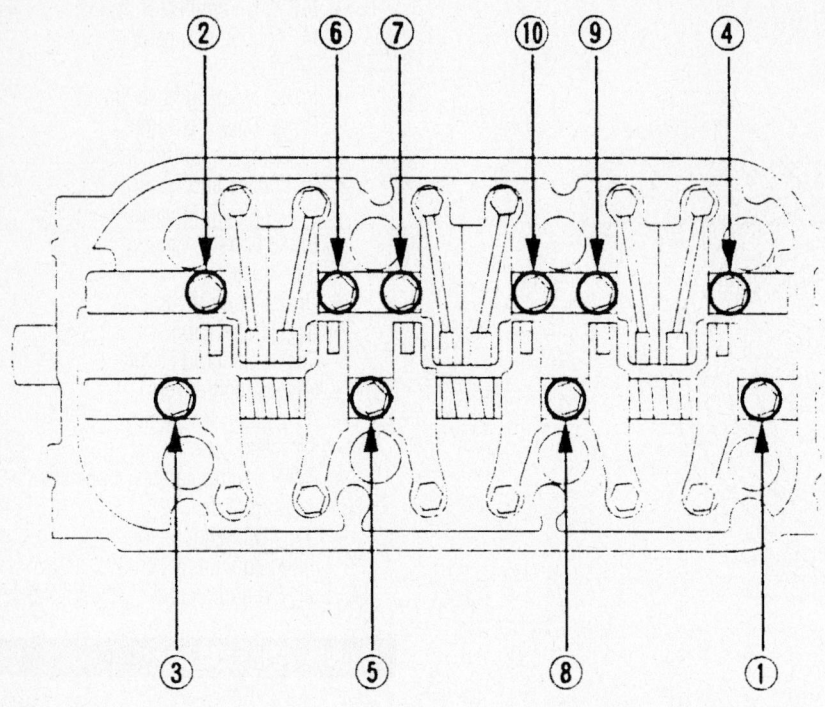

Rocker arm shaft loosening sequence—2002 models

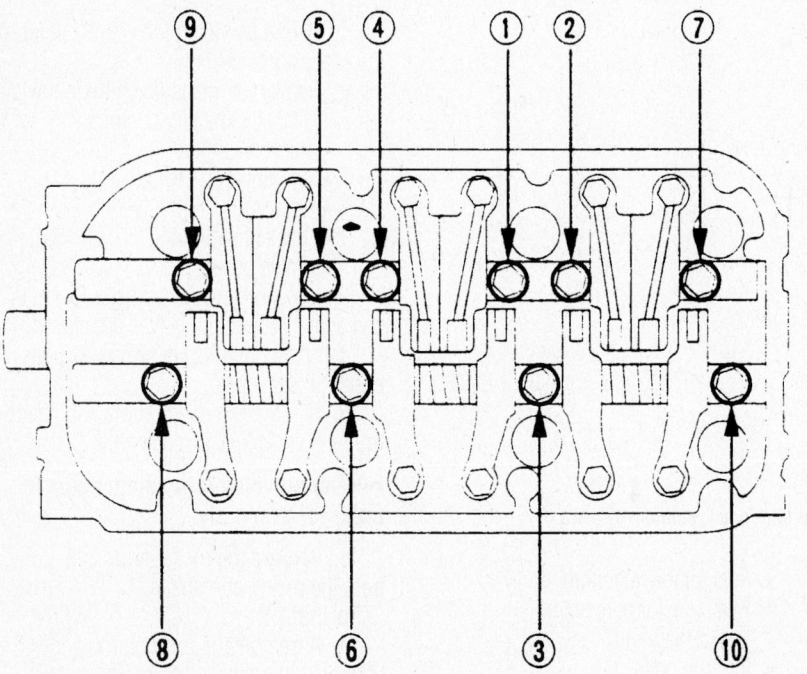

Rocker shaft tightening sequence—2002 models

- Air intake tube
- Negative battery cable

11. Start the engine and check for leaks.

2003–06 Models

1. Before servicing the vehicle, refer to the precautions section.

2. Remove or disconnect the following:
- Negative battery cable
- Intake manifold cover
- Ignition coil covers
- Ignition coils
- Dipstick
- 2 bolts attaching the harness holder

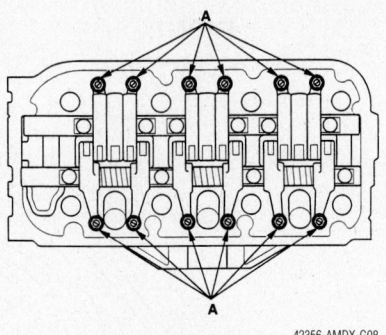

Rocker arm shaft adjusting screw locations—2003–06 models

- Positive Crankcase ventilation (PCV) hose
- Injector electrical connections
- Power steering hose bracket bolt and the harness holder bolts
- Harness clamps and the breather hose
- Valve cover
- Rocker arm adjusting screws. refer to the illustration for location.

3. Remove the rocker arm assembly as follows:

a. Unscrew the rocker shaft bolts 2 turns at a time in a criss-cross pattern to avoid damaging the vales or rocker assembly.

b. Do not remove the rocker shaft bolts. These bolts keep the springs and rocker arms on the shafts.

4. Loosen the valve adjuster locknuts and screws so that all valves are closed.

5. Remove the rocker arms and shafts from the vehicle as an assembly.

➡Keep all valvetrain components in order for assembly.

6. Remove the rocker arms and springs from the rocker arm shafts.

To install:

7. Assemble the rocker arms and springs to the rocker arm shafts in their original positions.

8. Install the rocker arm assemblies. Tighten the bolts in sequence and in multiple passes to 17 ft. lbs. (24 Nm).

9. Adjust the valve clearance.

10. Install or connect the following:
- Valve covers
- Harness clamps and the breather hose
- Power steering hose bracket bolt and the harness holder bolts
- Injector electrical connections
- PCV hose
- 2 bolts attaching the harness holder
- Dipstick

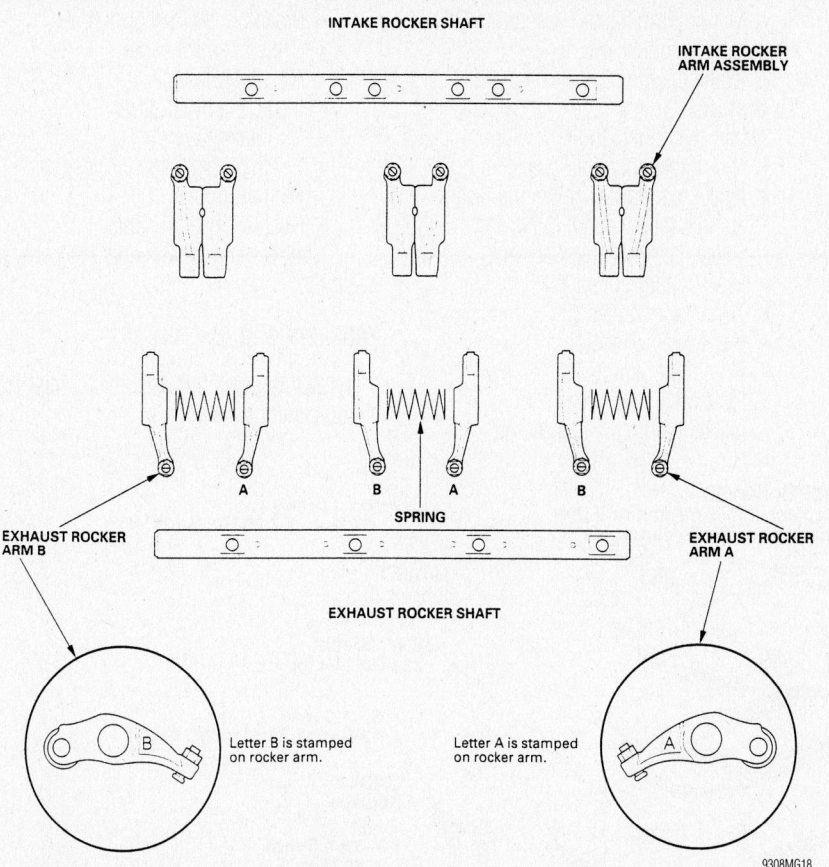

Exploded view of the rocker arms and shafts—MDX

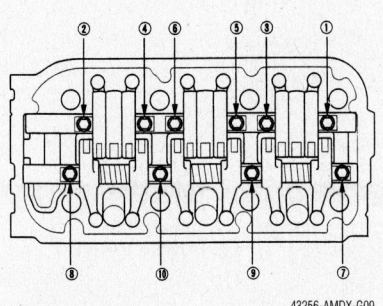

43256-AMDX-G09

Rocker arm shaft loosening sequence—2003–06 models

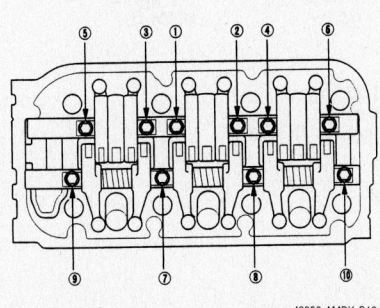

42356-AMDX-G10

Rocker shaft tightening sequence—2003–06 models

- Ignition coils and torque the retainers to 9 ft. lbs. (12 Nm)
- Ignition coil covers
- Intake manifold cover
- Negative battery cable

Intake Manifold

REMOVAL & INSTALLATION

2002 Models

1. Before servicing the vehicle, refer to the precautions section.
2. Remove or disconnect the following:
   - Negative battery cable
   - Evaporative Emissions (EVAP) control canister hose and vacuum hose
   - Air intake tube
   - Intake manifold cover
   - Accelerator cable
   - Cruise control cable
   - Brake booster vacuum line
   - Intake manifold vacuum line
   - Positive Crankcase Ventilation (PCV) valve and hose
   - Intake Air Temperature (IAT) sensor connector
   - Idle Air Control (IAC) valve connector
   - Throttle Position (TP) sensor connector

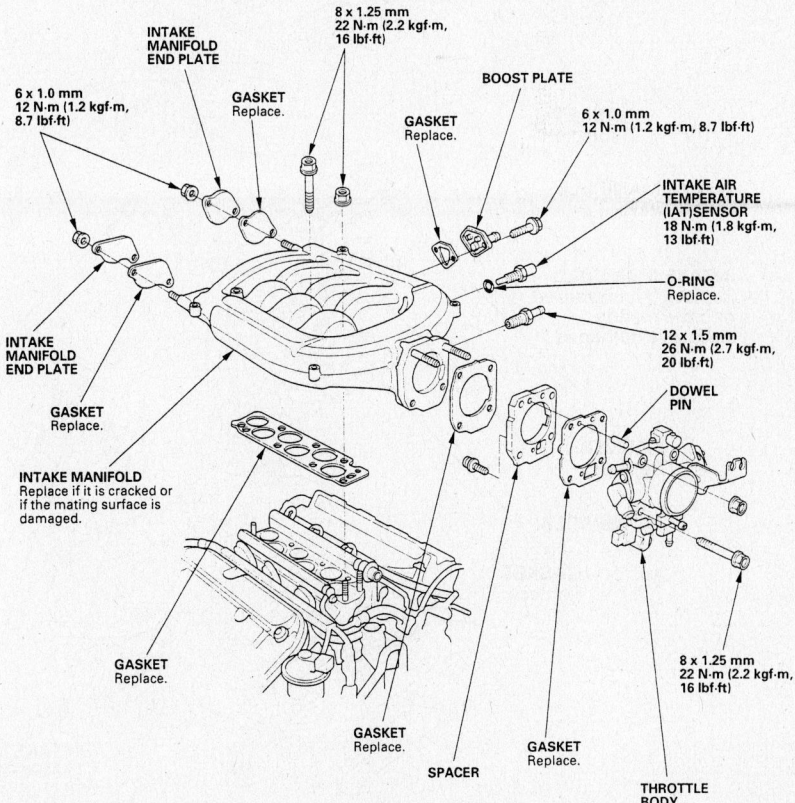

Exploded view of the intake manifold—2002 models

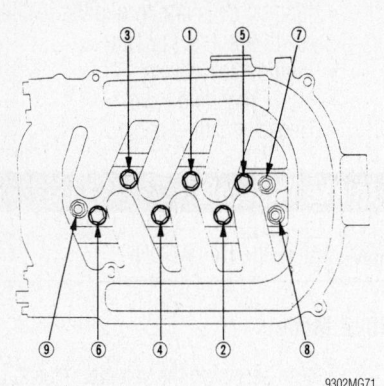

Intake manifold torque sequence—MDX

9302MG71

- Manifold Absolute Pressure (MAP) sensor connector
- Intake manifold

To install:

3. Install or connect the following:
   - New intake manifold gasket
   - Intake manifold. Tighten the fasteners in sequence and in several passes to 16 ft. lbs. (22 Nm).
   - MAP sensor connector
   - TP sensor connector
   - IAC valve connector
   - IAT sensor connector
   - PCV valve and hose
   - Intake manifold vacuum line

- Brake booster vacuum line
- Cruise control cable
- Accelerator cable
- Intake manifold cover
- Air intake tube
- EVAP control canister hose and vacuum hose
- Negative battery cable

4. Start the engine and check for proper operation.

2003–2006 Models

1. Before servicing the vehicle, refer to the precautions section.

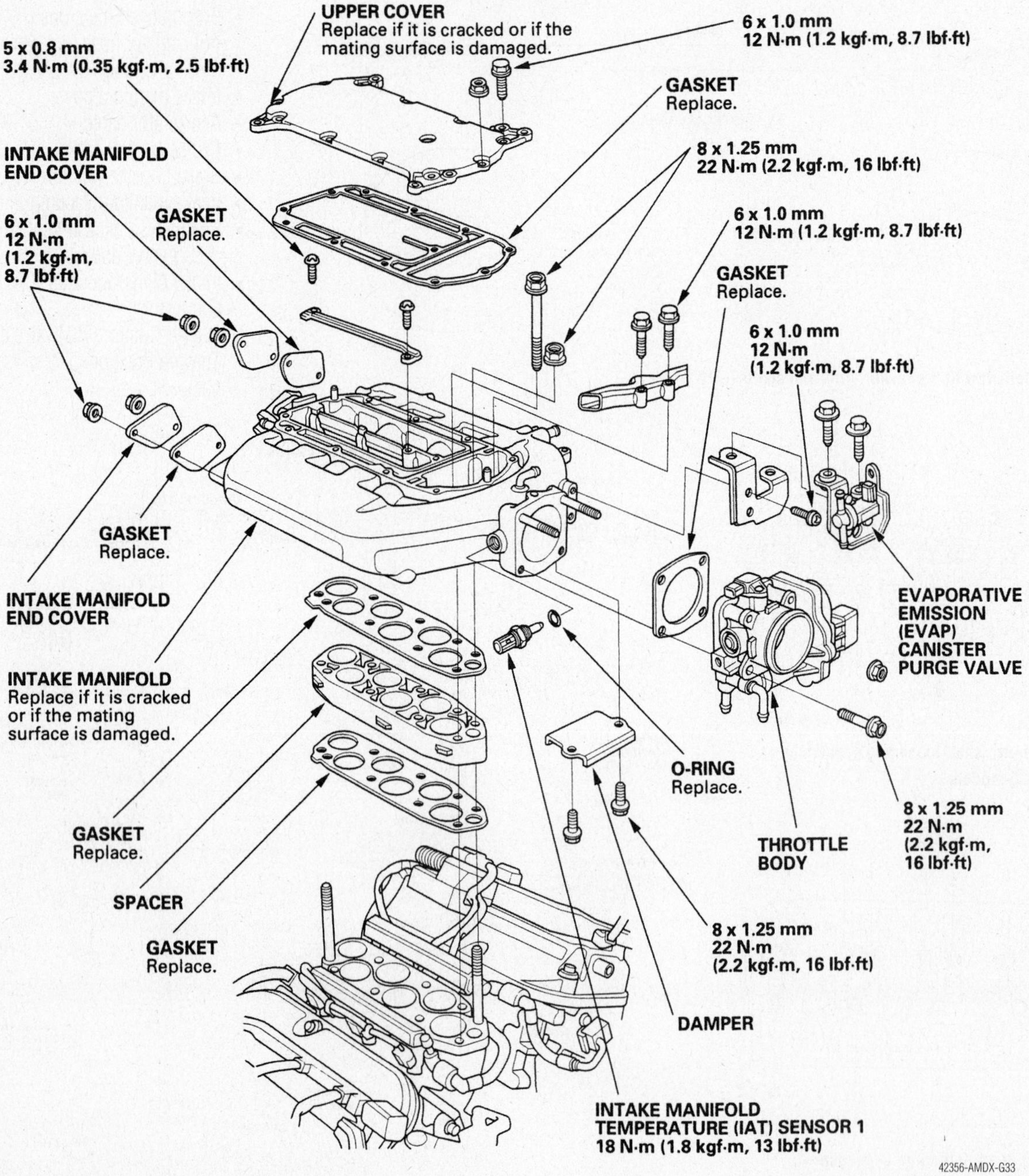

UPPER COVER
Replace if it is cracked or if the mating surface is damaged.

5 x 0.8 mm
3.4 N·m (0.35 kgf·m, 2.5 lbf·ft)

6 x 1.0 mm
12 N·m (1.2 kgf·m, 8.7 lbf·ft)

GASKET
Replace.

INTAKE MANIFOLD END COVER

6 x 1.0 mm
12 N·m
(1.2 kgf·m, 8.7 lbf·ft)

GASKET
Replace.

8 x 1.25 mm
22 N·m (2.2 kgf·m, 16 lbf·ft)

6 x 1.0 mm
12 N·m (1.2 kgf·m, 8.7 lbf·ft)

GASKET
Replace.

6 x 1.0 mm
12 N·m
(1.2 kgf·m, 8.7 lbf·ft)

GASKET
Replace.

INTAKE MANIFOLD END COVER

INTAKE MANIFOLD
Replace if it is cracked or if the mating surface is damaged.

GASKET
Replace.

SPACER

GASKET
Replace.

O-RING
Replace.

THROTTLE BODY

EVAPORATIVE EMISSION (EVAP) CANISTER PURGE VALVE

8 x 1.25 mm
22 N·m
(2.2 kgf·m, 16 lbf·ft)

8 x 1.25 mm
22 N·m
(2.2 kgf·m, 16 lbf·ft)

DAMPER

INTAKE MANIFOLD TEMPERATURE (IAT) SENSOR 1
18 N·m (1.8 kgf·m, 13 lbf·ft)

42356-AMDX-G33

Exploded view of the intake manifold—2003–06 models

2. Remove or disconnect the following:
- Negative battery cable
- Intake manifold cover
- Intake Air Temperature (IAT) sensor 2 connector
- Air intake tube
- Positive Crankcase Ventilation (PCV) hose
- Brake booster vacuum line
- Intake manifold vacuum line
- Evaporative Emissions (EVAP) control canister hose and clamp bracket
- Water bypass hoses from the throttle body and plug the hoses

3. Remove the following electrical connections and clamps from the manifold:
- Intake Air Temperature (IAT) sensor 1 connector
- Throttle Position (TP) sensor connector
- Manifold Absolute Pressure (MAP) sensor connector
- EVAP canister purge valve connector
- Intake Manifold Runner Control (IMRC) actuator connector

4. Remove or disconnect the following:
- Upper cover bolts and nuts in the sequence illustrated using two or three passes

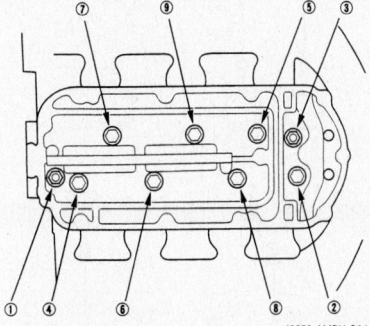

Intake manifold loosening sequence—2003–06 models

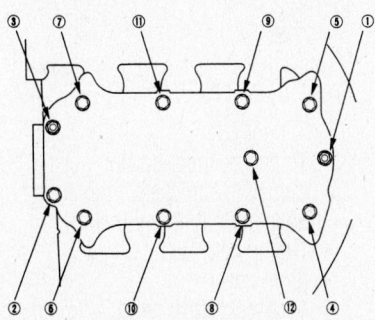

Upper cover loosening sequence—2003–05 models

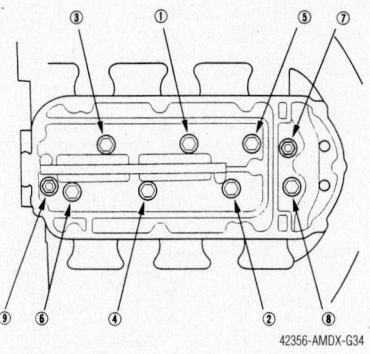

Intake manifold torque sequence—2003–05 models

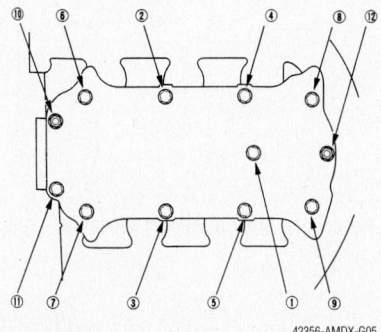

Upper cover torque sequence—2003–06 models

- Intake manifold bolts in the sequence illustrated
- Intake manifold and spacer

To install:

5. Install or connect the following:
- New intake manifold gasket and spacer
- Intake manifold. Tighten the fasteners in sequence and in several passes to 16 ft. lbs. (22 Nm).
- Upper cover bolts and nuts in the sequence illustrated using two or three passes to 9 ft. lbs. (12 Nm)

6. Connect the following electrical connections and clamps to the manifold:
- Intake Manifold Runner Control (IMRC) actuator connector
- EVAP canister purge valve connector
- MAP sensor connector
- TP sensor connector
- IAT sensor 1 connector
- Water bypass hoses to the throttle body
- EVAP control canister hose and clamp bracket
- Intake manifold vacuum line
- Brake booster vacuum line
- PCV hose
- Air intake tube

- IAT sensor 2 connector
- Intake manifold cover
- Negative battery cable

7. Start the engine and check for proper operation.

Exhaust Manifold

REMOVAL & INSTALLATION

1. Before servicing the vehicle, refer to the precautions section.
2. Remove or disconnect the following:
- Negative battery cable
- Exhaust manifold heat shield
- Heated Oxygen (HO2S) sensor connector
- Exhaust front pipe
- Exhaust manifold bracket, if equipped
- Exhaust manifold

To install:

3. Install or connect the following:
- Exhaust manifold. Tighten the fasteners to 23 ft. lbs. (31 Nm).
- Exhaust manifold bracket, if equipped. Tighten the bolts to 33 ft. lbs. (44 Nm).
- Exhaust front pipe. Tighten the nuts to 40 ft. lbs. (55 Nm).
- Heated Oxygen (HO2S) sensor connector
- Exhaust manifold heat shield
- Negative battery cable

Front Crankshaft Seal

REMOVAL & INSTALLATION

1. Before servicing the vehicle, refer to the precautions section.
2. Remove or disconnect the following:
- Negative battery cable
- Accessory drive belts
- Side engine mount
- Valve cover
- Crankshaft pulley
- Front cover
- Balance shaft belt, if equipped
- Timing belt
- Top Dead Center (TDC) sensor, if equipped
- Crankshaft timing sprocket
- Front crankshaft seal

To install:

3. Lubricate the crankshaft seal lip with grease prior to installation.
4. Install the front crankshaft seal so that it is flush with the surface of the oil pump housing.

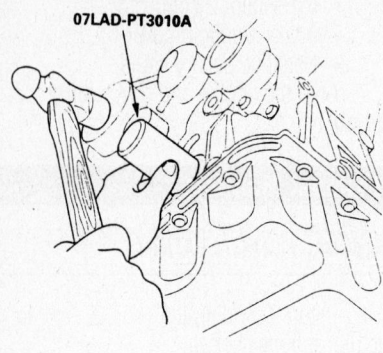

07LAD-PT3010A

93352G02

Front crankshaft seal installation

5. Install or connect the following:
- Crankshaft timing sprocket
- Top Dead Center (TDC) sensor, if equipped
- Timing belt
- Balance shaft belt, if equipped
- Front cover
- Crankshaft pulley. Tighten the bolt to 181 ft. lbs. (245 Nm) on 2002 models, or 47 ft. lbs. (64 Nm) plus an additional 60 degrees on 2003–06 models.
- Valve cover
- Side engine mount
- Accessory drive belts
- Negative battery cable
6. Check the engine oil level and add if necessary.
7. Start the engine and check for leaks.

Camshaft

REMOVAL & INSTALLATION

2002 Models

1. Before servicing the vehicle, refer to the precautions section.
2. Remove or disconnect the following:
- Negative battery cable
- Air intake tube
- Accessory drive belts
- Front cover
- Timing belt
- Camshaft sprockets
- Timing belt rear covers
- Ignition coil covers
- Intake manifold cover
- Intake manifold
- Valve cover
- Rocker arms and shaft assembly
- Camshaft thrust cover
- Camshaft

To install:

➡**Use new O-rings, seals and gaskets when installing the camshaft.**

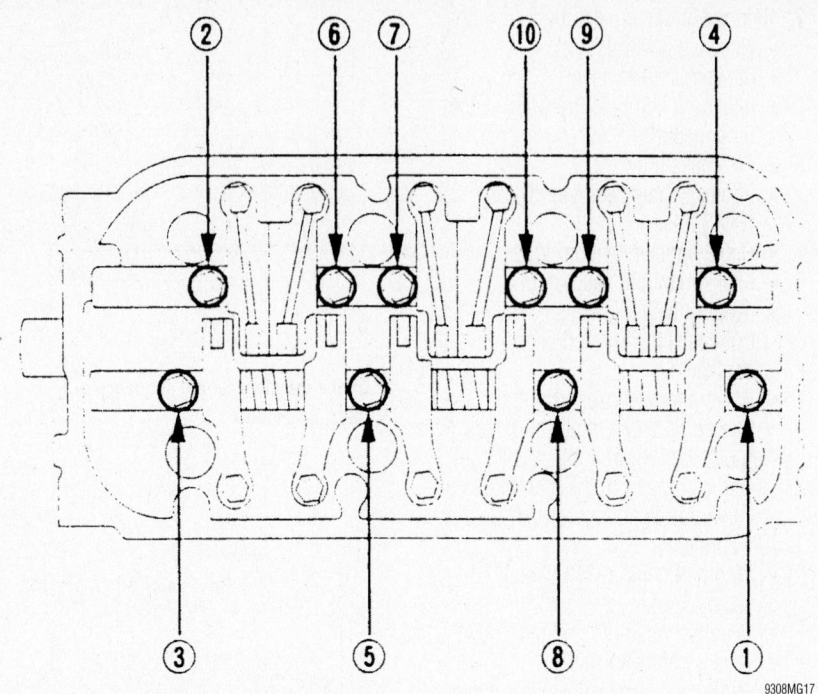

9308MG17

Rocker arm shaft loosening sequence—2002 models

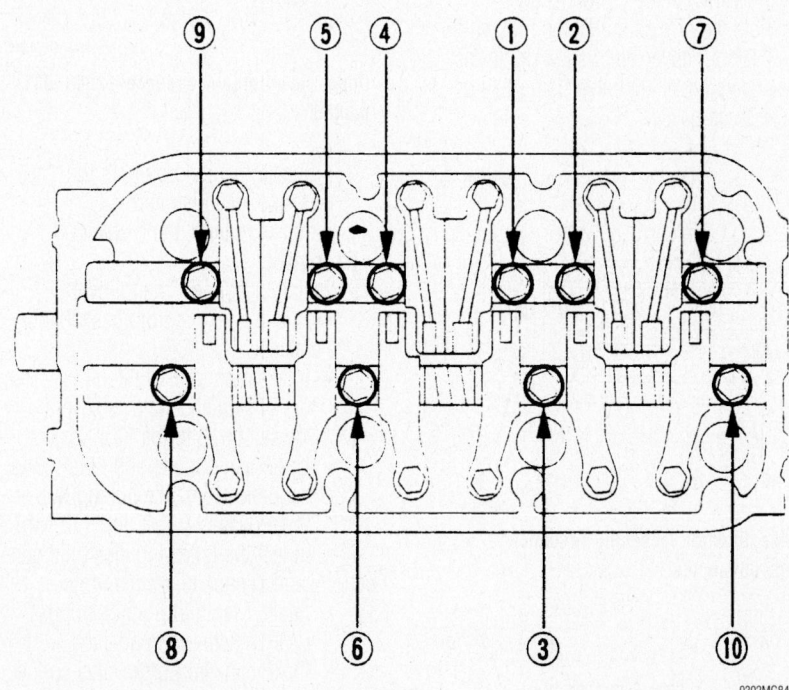

9302MG84

Rocker shaft tightening sequence—2002 models

3. Install or connect the following:
- Camshaft
- Camshaft thrust cover. Tighten the bolts to 16 ft. lbs. (22 Nm).
- Rocker arms and shaft assembly
- Valve cover
- Intake manifold
- Intake manifold cover
- Ignition coil covers
- Timing belt rear covers
- Camshaft sprockets. Tighten the bolts to 67 ft. lbs. (90 Nm).
- Timing belt
- Front cover
- Accessory drive belts
- Air intake tube
- Negative battery cable
4. Start the engine and check for leaks.

2003–06 Models

FRONT

1. Before servicing the vehicle, refer to the precautions section.
2. Remove or disconnect the following:
   - Negative and positive battery cables
   - Battery
3. Drain the coolant.
   - Upper radiator hose
   - Exhaust Gas Recirculation (EGR) vale
   - Timing belt
   - Rocker arm assembly
   - Front camshaft pulley
   - Thrust plate and camshaft

To install:

4. Install or connect the following:
   - Camshaft using a new O-ring. Tighten the thrust plate to 16 ft. lbs. (22 Nm).
   - Front camshaft pulley
   - Rocker arm assembly
   - Timing belt
   - Exhaust Gas Recirculation (EGR) vale
   - Battery
   - Positive, then negative battery cables

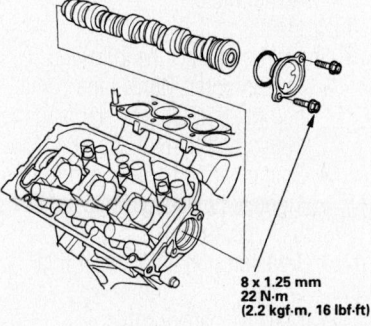

**8 x 1.25 mm
22 N·m
(2.2 kgf·m, 16 lbf·ft)**

42356-AMDX-G06

Front camshaft assembly—2003–06 models

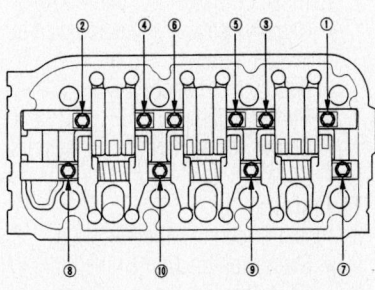

42356-AMDX-G09

Rocker arm shaft loosening sequence—2003–06 models

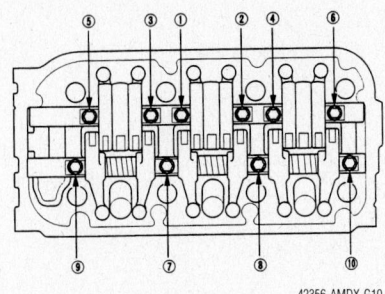

42356-AMDX-G10

Rocker shaft tightening sequence—2003–05 models

5. Fill the cooling system.
   - Camshaft
6. Start the engine and check for leaks.

REAR

1. Before servicing the vehicle, refer to the precautions section.
2. Drain the cooling system.
3. Relieve the fuel system pressure.
4. Remove or disconnect the following:
   - Negative battery cable
   - Under-hood fuse box
   - Fuel feed hose
   - Nuts securing the fuel line
   - Brake lines from the master cylinder
   - Timing belt
   - Rocker arm assembly
   - Rear camshaft pulley
   - Thrust plate and camshaft

To install:

5. Install or connect the following:
   - Camshaft using a new O-ring. Tighten the thrust plate to 16 ft. lbs. (22 Nm).
   - Rear camshaft pulley
   - Rocker arm assembly
   - Timing belt
   - Brake lines to the master cylinder
   - Nuts securing the fuel line
   - Fuel feed hose
   - Under-hood fuse box
   - Negative battery cable

Valve Lash

ADJUSTMENT

Adjust the valves only when the cylinder head temperature is less than 100°F (38°C).

1. Before servicing the vehicle, refer to the precautions section.
2. Remove or disconnect the following:
   - Negative battery cable
   - Air intake tube
   - Intake manifold
   - Valve cover
3. Rotate the crankshaft so that the valves to be adjusted are closed and the rocker arm is contacting the camshaft lobe base circle.
4. Measure the valve clearance. If adjustment is necessary, loosen the locknut

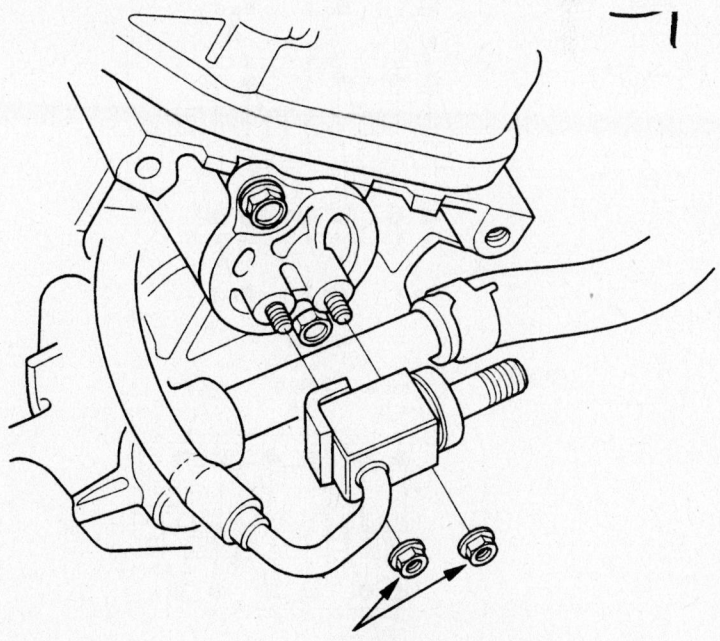

**6 x 1.0 mm
12 N·m (1.2 kgf·m, 8.7 lbf·ft)**

42356-AMDX-G07

Remove the nuts attaching the fuel line when removing the rear camshaft—2003–06 models

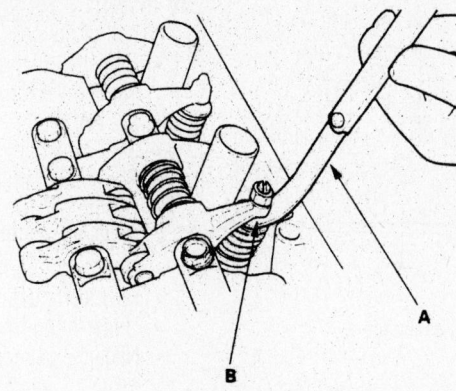

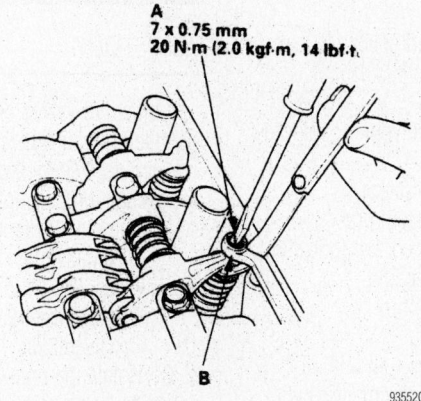

A
7 x 0.75 mm
20 N·m (2.0 kgf·m, 14 lbf·t.)

93552G04

Inspect the valve clearance, adjust to specification and tighten the retainer to specification

Adjusting screw locations:

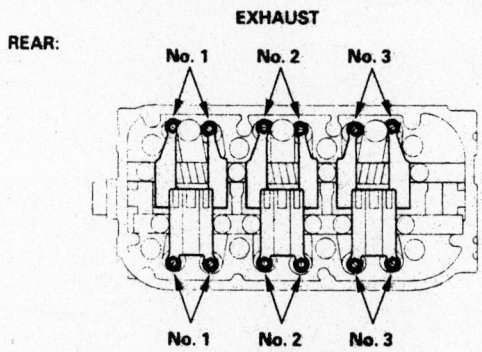

Valve adjusting retainer locations

and turn the adjusting screw as necessary to achieve the correct valve clearance.

  5. The correct valve clearance is:
- Intake valves: 0.008–0.009 inches (0.20–0.24mm)
- Exhaust valves: 0.011–0.013 inches (0.28–0.32mm)

  6. After adjustment, tighten the locknuts to 14 ft. lbs. (20 Nm).

  7. Install or connect the following:
- Valve cover
- Intake manifold
- Air intake tube
- Negative battery cable

  8. Start the engine and check for proper operation.

Starter Motor

REMOVAL & INSTALLATION

2002 Models

  1. Before servicing the vehicle, refer to the precautions section.

  2. Remove or disconnect the following:
- Negative battery cable and wait at least 3 minutes
- Transmission fluid cooler line clamp
- Starter wiring harness connectors
- Starter motor

To install:

  3. Install or connect the following:
- Starter motor. Tighten the 10mm bolt to 33 ft. lbs. (44 Nm) and the 12mm bolt to 47 ft. lbs. (64 Nm).
- Starter wiring harness connectors. Tighten the battery cable nut to 79 inch lbs. (9 Nm).
- Transmission fluid cooler line clamp
- Negative battery cable

2003–06 models

  1. Before servicing the vehicle, refer to the precautions section.

  2. Remove or disconnect the following:
- Negative battery cable and wait at least 3 minutes
- Battery and battery tray
- Harness clamps
- Starter wiring harness connectors
- Starter motor

To install:

  3. Install or connect the following:
- Starter motor. Tighten the bolts to 54 ft. lbs. (74 Nm).
- Starter wiring harness connectors. Tighten the battery cable nut to 79 inch lbs. (9 Nm).

- Harness clamps
- Battery tray
- Negative battery cable

Oil Pan

REMOVAL & INSTALLATION

1. Before servicing the vehicle, refer to the precautions section.
2. Drain the engine oil.
3. Remove or disconnect the following:
   - Negative battery cable
   - Front splash shield
   - Heated Oxygen (HO_2S) sensor connector
   - Exhaust front pipe
   - Torque converter cover, if equipped with an automatic transaxle
   - Oil pan

To install:

4. Install the oil pan. Apply liquid gasket as shown.
5. Tighten the bolts in sequence to 105 inch lbs. (12 Nm) using several passes.
6. Wait at least 30 minutes before adding oil to the engine.
7. Install or connect the following:
   - Torque converter cover, if removed
   - Exhaust front pipe

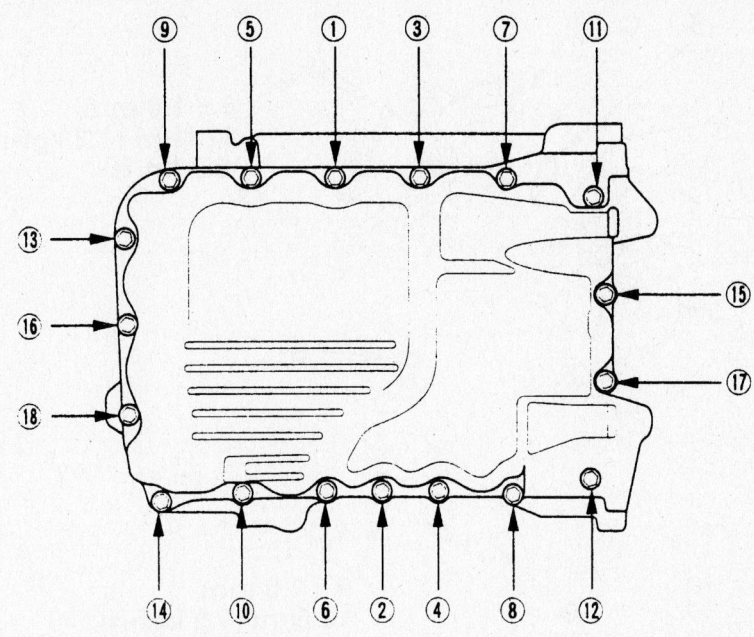

Oil pan tightening sequence—MDX

- Subframe center beam, if removed
- HO_2S sensor connector
- Front splash shield
- Negative battery cable

Oil Pump

REMOVAL & INSTALLATION

1. Before servicing the vehicle, refer to the precautions section.
2. Drain the engine oil.
3. Remove or disconnect the following:
   - Negative battery cable
   - Accessory drive belts
   - Front cover
   - Timing belt
   - Timing belt idler pulley
   - Crankshaft Position (CKP) sensor
   - Crankshaft timing sprocket
   - Variable Valve Timing and Valve Lift Electronic Control (VTEC) solenoid valve connector
   - Oil filter adapter
   - Oil pan
   - Oil pump pickup tube
   - Oil pump

To install:

➡**Use new gaskets and O-ring seals for assembly.**

4. Apply liquid gasket to the oil pump and to the bolt hole threads.
5. Install or connect the following:
   - Oil pump. Tighten the bolts to 105 inch lbs. (12 Nm).
   - Oil pump pickup tube. Tighten the bolts to 105 inch lbs. (12 Nm).
   - Oil pan
   - Oil filter adapter

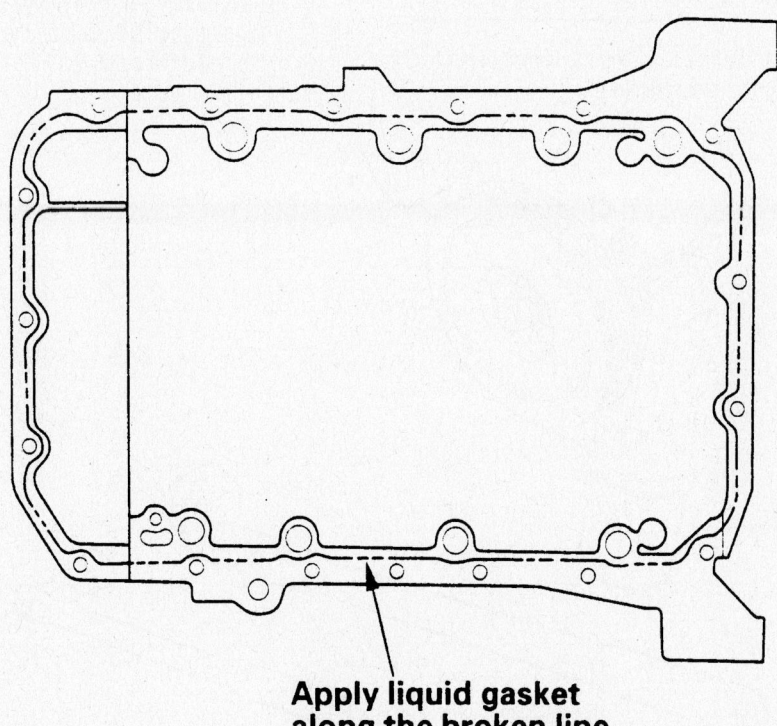

Apply liquid gasket along the broken line.

Apply liquid gasket to the inner threads of the bolt holes and the engine block along the area indicated by the broken line—MDX

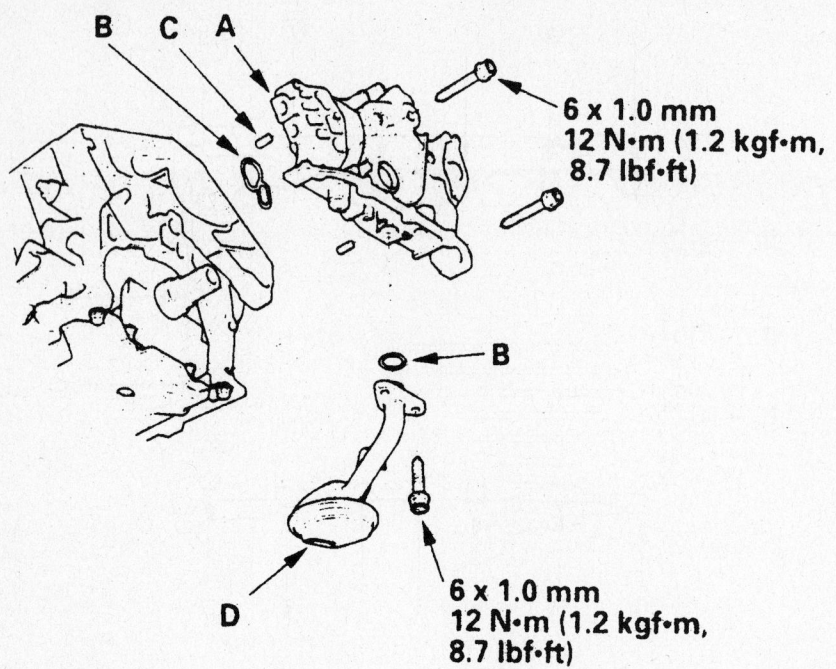

6 x 1.0 mm
12 N•m (1.2 kgf•m,
8.7 lbf•ft)

6 x 1.0 mm
12 N•m (1.2 kgf•m,
8.7 lbf•ft)

93552G05

Exploded view of the oil pump assembly

- VTEC solenoid valve connector
- Crankshaft timing sprocket
- CKP sensor
- Timing belt idler pulley
- Timing belt
- Front cover
- Accessory drive belts
- Negative battery cable
6. Fill the crankcase to the correct level.
7. Start the engine and check for leaks.

Rear Main Seal

REMOVAL & INSTALLATION

1. Before servicing the vehicle, refer to the precautions section.
2. Remove or disconnect the following:
- Transaxle
- Clutch pressure plate and disc, if equipped
- Flywheel
- Oil seal
To install:
3. Install or connect the following:
- Oil seal. Drive the seal square into the seal case.
- Flywheel. Tighten the bolts in a crossing pattern to 54 ft. lbs. (73 Nm).
- Clutch pressure plate and disc, if equipped
- Transaxle

4. Check the fluid levels.
5. Start the engine and check for leaks.

Timing Belt

REMOVAL & INSTALLATION

1. Before servicing the vehicle, refer to the precautions section.

2. Turn the crankshaft so the white mark aligns with the pointer.
3. Make sure the number 1 piston is at Top Dead center (TDC).
4. Remove or disconnect the following:
- Negative battery cable
- Wheels and splash shield
- Drive belt
5. Support the engine with a block of wood and a jack under the oil pan.
- Upper side engine mount
6. Remove the crankshaft pulley using holder tool shown in the accompanying illustration and a breaker and socket, loosen the 19mm bolt and remove the pulley.
- Front upper cover, rear upper cover and the lower cover
- One of the battery clamp bolts and grind the end as illustrated
7. Screw the battery clamp bolt as illustrated to hold the belt adjuster in position. Do not use a wrench, hand tighten only.
- Lower side engine mount
- Idler pulley bolt and the pulley
- Timing belt
To install:
8. If installing a new belt, perform the following steps:
a. Clean the pulleys, belt guide plate and the upper and lower covers.
b. Set the timing belt drive pulley to TDC by aligning the TDC mark on the tooth of the belt drive pulley with the pointer on the oil pump.

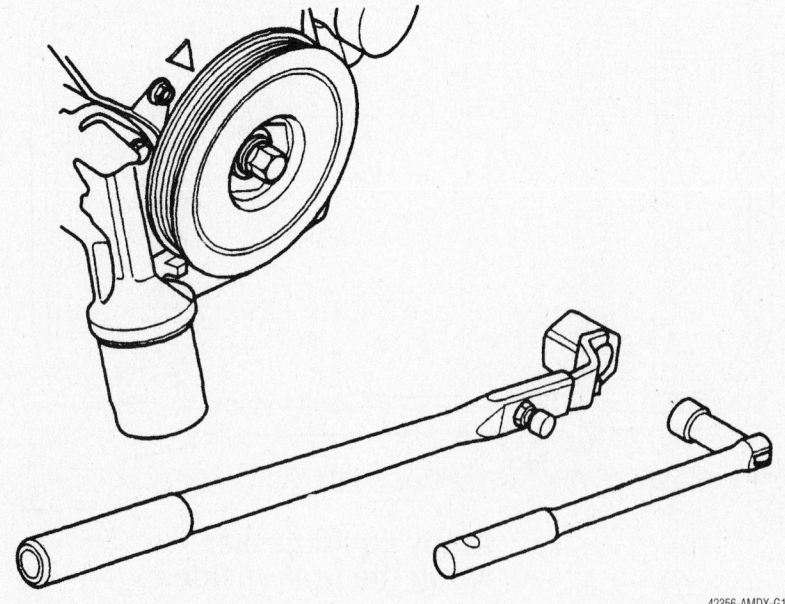

42356-AMDX-G15

Remove the crankshaft pulley using holder tool shown

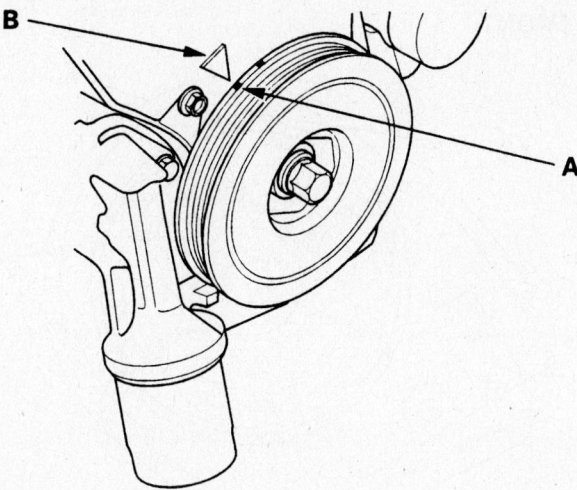

42356-AMDX-G11

Turn the crankshaft so the white mark (A) aligns with the pointer (B)

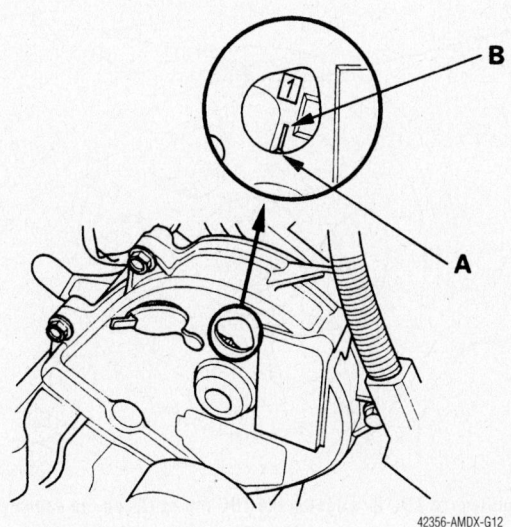

42356-AMDX-G12

Make sure the number 1 piston is at top dead center (A) on the front camshaft pulley and pointer (B)

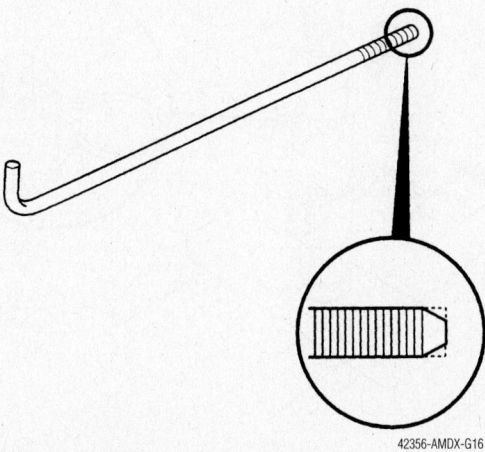

42356-AMDX-G16

Remove a battery clamp bolt and grind the end as shown

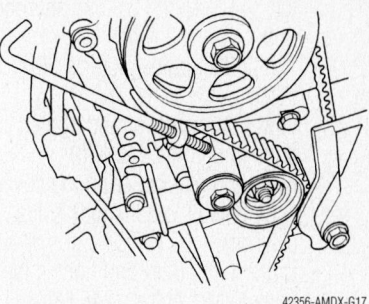

42356-AMDX-G17

Install the battery clamp bolt as shown to hold the belt adjuster in position

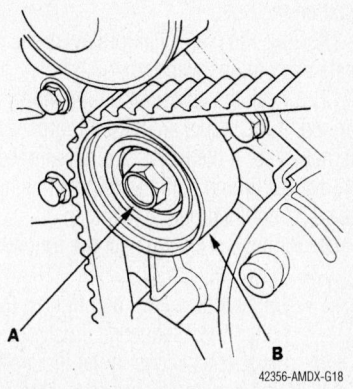

42356-AMDX-G18

Remove the idler pulley bolt (A), pulley (B) and the timing belt

c. Set the camshaft pulleys to TDC by aligning the TDC marks on the camshaft pulleys with the pointers on the back covers.

d. Remove the battery clamp bolt.

e. Remove the belt tensioner.

f. Align the holes on the rod and housing of the tensioner.

g. Using a press or other suitable device, slowly compress the tensioner and insert a 0.08 inch (2mm) pin through the housing and rod.

h. Install the tensioner making sure the pin is still installed.

i. Apply thread locker to idler pulley bolt then hand tighten the bolt.

j. Install the belt over the pulleys in this sequence; drive pulley, idler pulley, front camshaft pulley, water pump pulley, rear camshaft pulley and adjusting pulley.

k. Tighten the idler pulley bolt to 33 ft. lbs. (44 Nm).

l. Remove the pin from the tensioner.

9. Install or connect the following:
  • Lower half of the side mount and tighten the 3 long bolts to 33 ft.

lbs. (44 Nm) and the one short bolt to 9 ft. lbs. (12 Nm)

- Timing belt guide plate as illustrated
- Lower timing cover and tighten the bolts to 9 ft. lbs. (12 Nm)
- Front and rear upper timing covers and tighten the bolts to 9 ft. lbs. (12 Nm)
- Crankshaft pulley and tighten the bolts to 181 ft. lbs. (245 Nm), using the holding tool to prevent the unit from turning

10. Rotate the crankshaft pulley about 5 or 6 degrees clockwise so the belt positions itself on the pulleys.

11. Turn the crankshaft pulley so the white mark aligns with the pointer.

12. Check the camshaft pulley marks are aligned. If the marks are aligned, proceed to the next step. If the marks are not aligned, remove the timing belt and reinstall using the steps outlined before this step.

13. Remove or disconnect the following:
- Drive belt
- Upper side mount and tighten the bolts in the sequence illustrated to the specifications in the illustration

14. Using a suitable scan tool, perform the Powertrain Control Module (PCM) reset and the Crankshaft position (CKP) pattern clear/learn procedures, following the scan tool manufactures instructions.

15. If installing the old belt, perform the following steps:

    a. Clean the pulleys, belt guide plate and the upper and lower covers.

    b. Set the timing belt drive pulley to TDC by aligning the TDC mark on the tooth of the belt drive pulley with the pointer on the oil pump.

    c. Set the camshaft pulleys to TDC by aligning the TDC marks on the camshaft

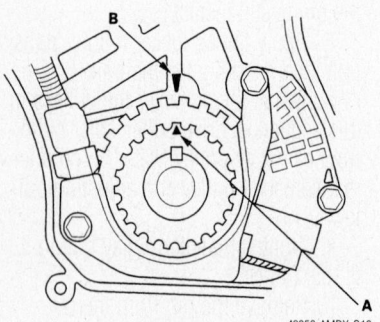

42356-AMDX-G19

Set the timing belt pulley to TDC by aligning the TDC mark (A) on the tooth of the belt pulley with the pointer (B) on the oil pump

FRONT:

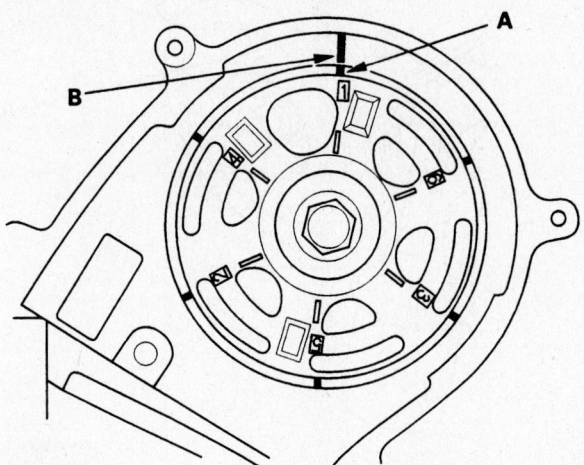

REAR:

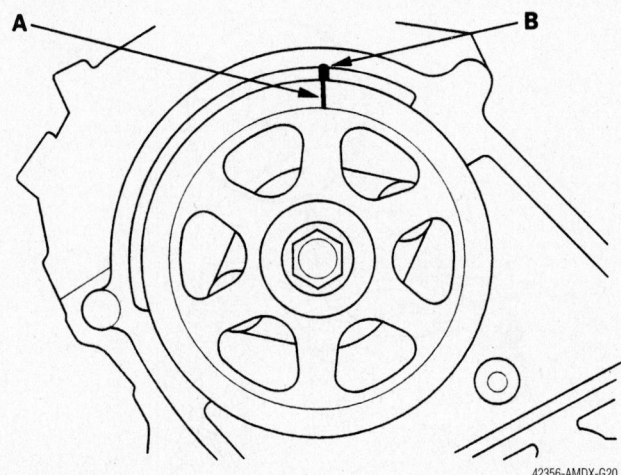

42356-AMDX-G20

Set the camshaft pulleys to TDC by aligning the TDC marks (A) on the camshaft pulleys with the pointers (B) on the back covers

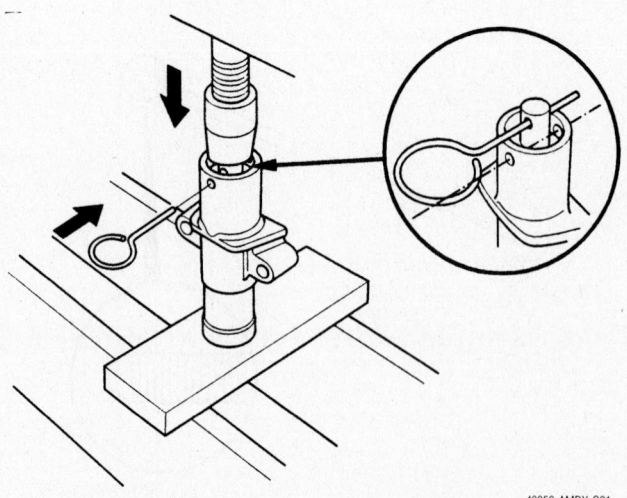

42356-AMDX-G21

Insert a 0.08 inch (2mm) pin through the tensioner housing and rod

-1 Drive pulley (A).
-2 Idler pulley (B).
-3 Front camshaft pulley (C).
-4 Water pump pulley (D).
-5 Rear camshaft pulley (E).
-6 Adjusting pulley (F).

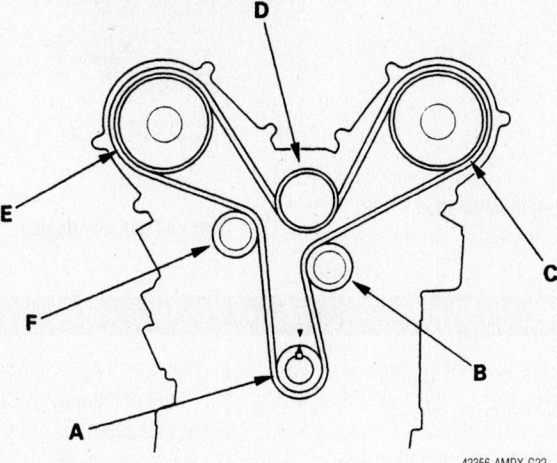

42356-AMDX-G22

Route the belt as shown in the sequence listed

- Timing belt guide plate as illustrated
- Lower timing cover and tighten the bolts to 9 ft. lbs. (12 Nm)
- Front and rear upper timing covers and tighten the bolts to 9 ft. lbs. (12 Nm)
- Crankshaft pulley and tighten the bolts to 181 ft. lbs. (245 Nm), using the holding tool to prevent the

18. Rotate the crankshaft pulley about 5 or 6 degrees clockwise so the belt positions itself on the pulleys.

19. Turn the crankshaft pulley so the white mark aligns with the pointer.

20. Check the camshaft pulley marks are aligned. If the marks are aligned, proceed to the next step. If the marks are not aligned, remove the timing belt and reinstall using the steps outlined before this step.

21. Remove or disconnect the following:
- Drive belt
- Upper side mount and tighten the

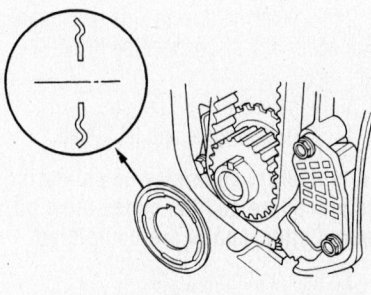

42356-AMDX-G23

Install the timing belt guide plate as shown

pulleys with the pointers on the back covers.

d. Apply thread locker to idler pulley bolt then hand tighten the bolt.

16. If the tensioner was extended and the belt cannot be installed, perform the steps above for the new belt installation.

a. Install the belt over the pulleys in this sequence; drive pulley, idler pulley, front camshaft pulley, water pump pulley, rear camshaft pulley and adjusting pulley.

b. Tighten the idler pulley bolt to 33 ft. lbs. (44 Nm).

c. Remove the battery clamp bolt.

17. Install or connect the following:
- Lower half of the side mount and tighten the 3 long bolts to 33 ft. lbs. (44 Nm) and the one short bolt to 9 ft. lbs. (12 Nm)

FRONT CAMSHAFT PULLEY:

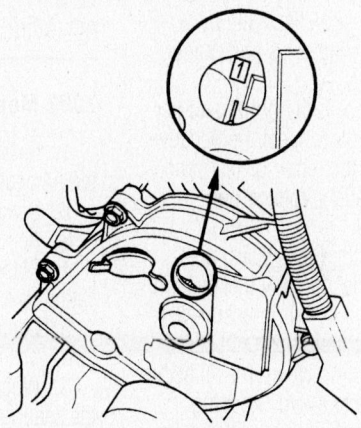

REAR CAMSHAFT PULLEY:

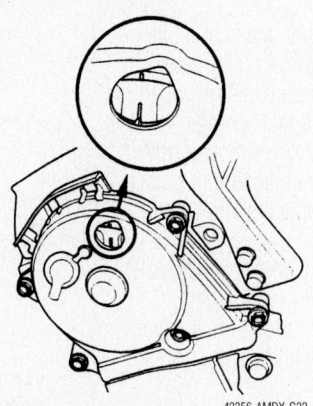

42356-AMDX-G32

Check that the camshaft pulley marks are aligned as shown

bolts in the sequence illustrated to the specifications in the illustration

22. Using a suitable scan tool, perform the Powertrain Control Module (PCM) reset and the Crankshaft position (CKP) pattern clear/learn procedures, following the scan tool manufactures instructions.

Piston and Ring

POSITIONING

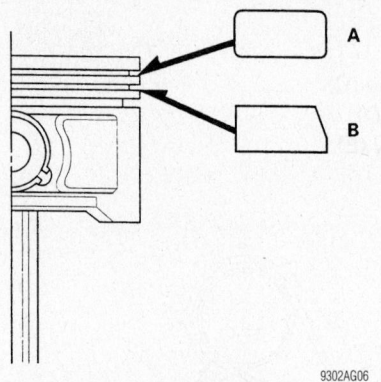

Compression ring identification—3.5L engine

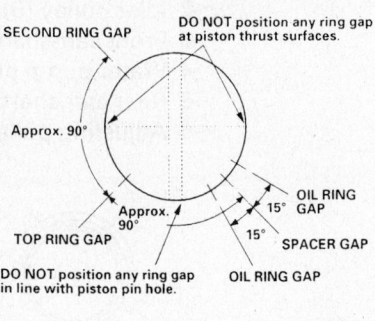

Ring end gap positioning

FUEL SYSTEM

Fuel System Service Precautions

Safety is the most important factor when performing not only fuel system maintenance, but any type of maintenance. Failure to conduct maintenance and repairs in a safe manner may result in serious personal injury or death. Maintenance and testing of the vehicle's fuel system components can be accomplished safely and effectively by adhering to the following rules and guidelines:

• To avoid the possibility of fire and personal injury, always disconnect the negative battery cable unless the repair or test procedure requires that battery voltage be applied.

• Always relieve the fuel system pressure prior to disconnecting any fuel system component (injector, fuel rail, pressure regulator, etc.), fitting or fuel line connection. Exercise extreme caution whenever relieving fuel system pressure to avoid exposing skin, face and eyes to fuel spray. Please be advised that fuel under pressure may penetrate the skin or any part of the body that it contacts.

• Always place a shop towel or cloth around the fitting or connection prior to loosening to absorb any excess fuel due to spillage. Ensure that all fuel spillage (should it occur) is quickly removed from engine surfaces. Ensure that all fuel soaked cloths or towels are deposited into a suitable waste container.

• Always keep a dry chemical (Class B) fire extinguisher near the work area.

• Do not allow fuel spray or fuel vapors to come into contact with a spark or open flame.

• Always use a backup wrench when loosening and tightening fuel line connection fittings. This will prevent unnecessary stress and torsion to fuel line piping. Always follow the proper torque specifications.

• Always replace worn fuel fitting O-rings with new. Do not substitute fuel hose or equivalent, where fuel pipe is installed.

Fuel System Pressure

RELIEVING

2002 Models

1. Before servicing the vehicle, refer to the precautions section.
2. Disconnect the negative battery cable.
3. Remove the fuel filler cap.
4. Place a shop towel over the fuel pulsation damper.
5. Loosen the fuel pulsation damper 1 turn.
6. When service is completed, replace the sealing washer and tighten the pulsation damper to 16 ft. lbs. (22 Nm).

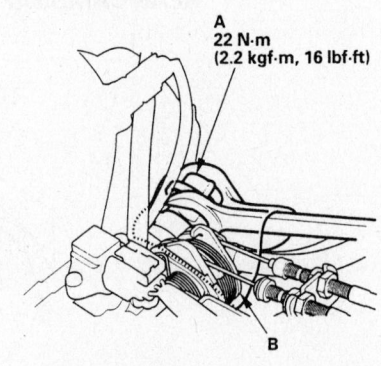

Use a wrench on the fuel pulsation damper (A). Place a rag over the damper (B) when relieving residual fuel pressure—2002 MDX

7. Replace the fuel filler cap.
8. Connect the negative battery cable.
9. Start the engine and check for leaks.

2003–06 Models

1. Before servicing the vehicle, refer to the precautions section.
2. Remove the drivers side dashboard lower cover and disconnect the PGM-FI main relay.
3. Remove the fuel filler cap.
4. Start the engine and let it stall.

➡A temporary code may be set during this procedure and the codes must be cleared after repairs are completed.

5. Turn the ignition off.
6. Disconnect the negative battery cable.
7. Remove the quick connect fitting cover.
8. Place a shop towel over the quick connect fitting and disconnect the fitting.
9. After the pressure is release, reconnect the fitting and install the cover.
10. After repairs are complete install the relay and connect the negative battery cable.

Fuel Filter

REMOVAL & INSTALLATION

1. Before servicing the vehicle, refer to the precautions section.
2. Relieve the fuel system pressure.
3. Remove or disconnect the following:
   • Negative battery cable
   • Driver's side second row seat and cut the carpet along the dotted line. Be careful not to cut the wiring harness under the carpet.

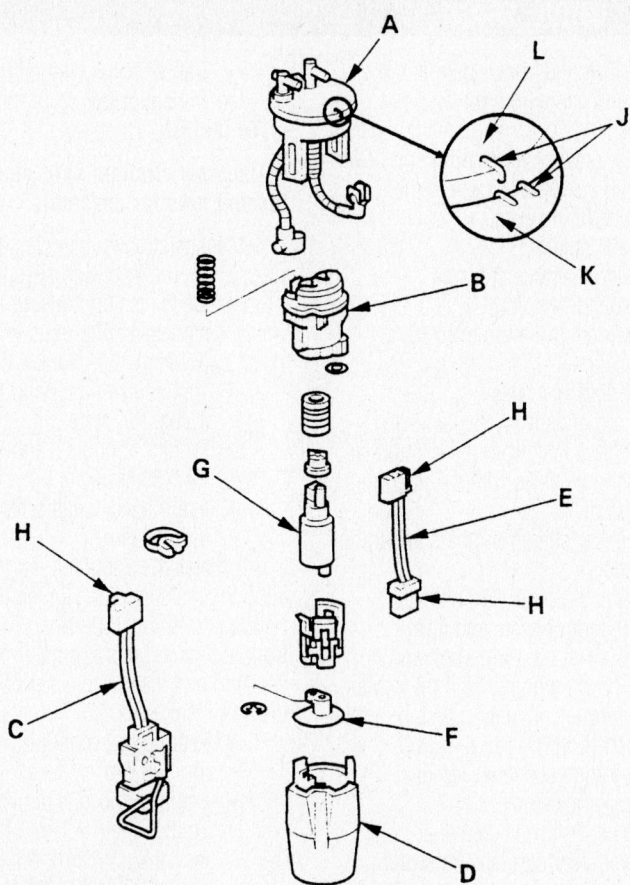

A. Bracket
B. Fuel filter
C. Fuel gauge sender
D. Case
E. Wire harness
F. Suction filter
G. Fuel pump
H. Connectors
J. Alignment marks
K. Fuel tank
L. Fuel pump module

9308MG26

Exploded view of the fuel pump module—MDX

- Access panel
- Fuel pump module

4. Disassemble the fuel pump module and remove the fuel filter.

To install:

5. Install the fuel filter and assemble the fuel pump module.

6. Install or connect the following:
- Fuel pump module
- Access panel
- Carpet and seat
- Negative battery cable

7. Start the engine and check for leaks.

Fuel Pump

REMOVAL & INSTALLATION

1. Before servicing the vehicle, refer to the precautions section.
2. Relieve the fuel system pressure.
3. Remove or disconnect the following:
- Negative battery cable
- Driver's side second row seat and cut the carpet along the dotted line. Be careful not to cut the wiring harness under the carpet.

- Access panel
- Fuel pump module wiring connector
- Fuel supply and return lines
- Fuel pump locknut using tool 07AAA-S0XA100
- Fuel pump module

To install:

4. Install or connect the following:
- Fuel pump module. Use a new seal and align the matchmarks.
- Fuel pump locknut and tighten to 69 ft. lbs. (93 Nm).
- Fuel supply and return lines
- Fuel pump module wiring connector
- Access panel
- Carpet and seat
- Negative battery cable

5. Start the engine and check for leaks.

Fuel Injector

REMOVAL & INSTALLATION

1. Before servicing the vehicle, refer to the precautions section.
2. Relieve the fuel system pressure.
3. Remove or disconnect the following:
- Negative battery cable
- Intake manifold
- Fuel lines
- Fuel injector connectors
- Fuel pressure regulator vacuum line
- Fuel supply manifold

4. Separate the fuel injectors from the fuel supply manifold.

To install:

5. Install the fuel injectors to the fuel supply manifold with new cushion rings and O-rings. Coat the injectors with clean engine oil prior to removal.

6. Install new seal rings to the intake manifold.

7. Install or connect the following:
- Fuel supply manifold and injector assembly. Tighten the bolts to 86 inch lbs. (10 Nm).
- Fuel pressure regulator vacuum line
- Fuel injector connectors
- Fuel lines
- Intake manifold
- Negative battery cable

8. Start the engine and check for leaks.

DRIVE TRAIN

Transaxle Assembly

REMOVAL & INSTALLATION

1. Before servicing the vehicle, refer to the precautions section.
2. Drain the transaxle.
3. Drain the power steering system.
4. Remove the engine appearance covers.
5. Remove the drivers side center console lower panel and pull back the cover to access steering joint cover.
6. Remove or disconnect the following:

- Steering joint bolt
- Steering joint from the steering gearbox pinion shaft
- Air intake assembly
- Battery
- Battery tray
- Power steering pump hose and the clamp bolt
- Transmission breather tube
- Cooler hose from the clamp on the starter
- Transaxle oil cooler lines
- Starter motor
- Shift control solenoid valve connectors
- Transaxle ground cable
- 8P connector from the bracket and the connector
- Clutch pressure switch connectors
- Joint connector and transmission range switch connector from the brackets
- Countershaft speed sensor connector
- Heated Oxygen (HO2S) sensor connectors
- Transmission housing mounting bolts

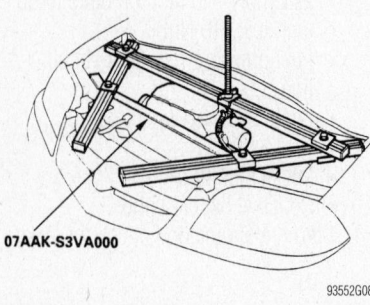

07AAK-S3VA000

93552G08

Support the engine while removing the transaxle—MDX

- Nut from the front mount and the ground cable from the engine
- Bulkhead cover, windshield wiper arms, cowl cover sealing and cover
- Install a support fixture to the engine lifting eyes.
- Splash shield
- Front sub-frame stiffener
- Exhaust front pipe
- Lower control arms from the knuckle
- Stabilizer bar links
- Tie rod ends from the knuckle
- Left driveshaft from the differential
- Right driveshaft from the intermediate shaft
- Propeller shaft from the companion flange
- Shift cable cover and holder
- Shift control cable and lever

7. Install a 6 x 1 x 14mm bolt and nut on the cable cover, then reinstall the cable cover to the torque converter housing. If this is not done, the bolt head of the cable cover may prevent torque converter removal.

- Transfer assembly
- Engine-to-torque converter bolts
- Power steering pressure switch connection
- Power steering hose clamp, then the hose from the pipe at the sub-frame
- Transmission lower mount nuts

8. Matchmark the front subframe to the vehicle body.

- Rear mount bracket bolts

9. Support the sub-frame with a 4 x 4 x 50 inch piece of wood and a jack.

- Sub-frame
- Transaxle lower mounts
- Driveshafts from the differential and intermediate shaft
- Intermediate shaft
- Transmission front mount bracket

93552G09

Support the sub-frame with a 4 x 4 x 50 inch piece of wood and a jack

- Transmission flange bolts
- Transmission

To install:

➡ **Use new circlips, split pins and self-locking nuts for assembly.**

10. Installation is the reverse of removal. Please note the following specifications:

- Transmission housing bolts and harness clamp bolts to 47 ft. lbs. (64 Nm)
- Transmission housing bolts to 47 ft. lbs. (64 Nm)
- Front mount bracket bolts to 28 ft. lbs. (38 Nm)
- Intermediate shaft bolts to 29 ft. lbs. (39 Nm)

11. Raise the subframe into position and align the matchmarks. Tighten the subframe bolts to 76 ft. lbs. (103 Nm). Tighten the front subframe bracket bolts to 54 ft. lbs. (74 Nm) and the rear bracket bolts to 86 ft. lbs. (117 Nm).

- Rear engine mount bolts to 28 ft. lbs. (38 Nm)
- Engine-to-torque converter bolts. Tighten the 6 x 1 mm bolts to 105 inch lbs. (12 Nm), 10 x 1.25mm bolt to 28 ft. lbs. (38 Nm).
- Front motor mount nut to 40 ft. lbs. (54 Nm)

12. Fill the transaxle to the correct level.
13. Start the engine and check for leaks.
14. Check the wheel alignment and adjust as necessary.

Transfer Assembly

REMOVAL & INSTALLATION

1. Before servicing the vehicle, refer to the precautions section.
2. Drain the transmission fluid.
3. Remove or disconnect the following:

- Negative battery cable
- Front sub-frame stiffener
- Heated Oxygen (HO2S) sensor connectors
- Exhaust front pipe
- Breather tube bracket bolt, then the tube from the breather pipe
- Propeller shaft from the transfer assembly
- Transfer assembly bolts and the assembly

To install:

4. Install or connect the following:

- New O-ring on the transfer cover

- Dowel pin on the assembly
- Transfer assembly and tighten the bolts to 33 ft. lbs. (44 Nm) on 2002 models and 38 ft. lbs. (51 Nm) on 2003–06 models
- Propeller shaft and tighten the bolts to 53 ft. lbs. (72 Nm)
- Breather tube bracket, attach the tube with the dot facing outwards and tighten the bolt to 9 ft. lbs. (12 Nm)
- Exhaust front pipe
- Front sub-frame stiffener and tighten the bolts to 40 ft. lbs. (54 Nm)
- Heated Oxygen (HO$_2$S) sensor connectors
- Negative battery cable

Halfshaft

REMOVAL & INSTALLATION

Front

1. Before servicing the vehicle, refer to the precautions section.
2. Drain the transaxle if removing the left halfshaft. If is not necessary to drain the fluid if removing the right halfshaft.
3. Remove or disconnect the following:
   - Negative battery cable
   - Front wheels
   - Spindle nut
   - Stabilizer bar link
   - Lower ball joint
4. Pry the inboard joint from the transaxle or intermediate shaft.
5. Remove the outer CV-joint stub shaft from the hub by tapping the stub shaft with a plastic hammer.
   To install:

➡**Use new circlips, split pins and self-locking nuts for assembly.**

6. Install the outer CV-joint stub shaft into the hub.
7. Install the inner CV-joint to the transaxle or intermediate shaft until the circlip locks in the retaining groove.
8. Install or connect the following:
   - Lower ball joint. Tighten the nut to 43–51 ft. lbs. (59–69 Nm).
   - Stabilizer bar link. Tighten the nut to 58 ft. lbs. (78 Nm).
   - Spindle nut. Tighten the nut to 181 ft. lbs. (245 Nm) on 2002 models, or 210 ft. lbs. (285 Nm) on 2003–06 models.
   - Front wheels

- Negative battery cable
9. Fill the transaxle to the correct level and check for leaks.

Rear

1. Before servicing the vehicle, refer to the precautions section.
2. Drain the differential fluid.
3. Remove the wheels.
4. Remove and discard the nut.
5. Remove the upper arm bolt.
6. Remove the rear shock absorber.
7. Remove the lower arm flange bolt
8. Remove the wheel sensor bracket.
9. Pull the knuckle outward and disconnect the rear driveshaft joint from the hub using a plastic mallet.
10. Disconnect the driveshaft from the hub.
11. Pry the inboard joint from the differential. Do not use a hammer just a prying tool.
12. Remove the driveshaft.
    To install:
13. Clean all mating surfaces. Apply 0.05–0.07 Ounces (1.5–2.0 grams) of grease to the whole splined surface.
14. After applying the grease, remove the grease from the splined grooves at intervals of 2–3 splines and from the set ring groove so that air can bleed from the differential.
15. Place a new set ring in the set ring groove of the differential.
16. Clean the areas where the driveshaft contacts the differential with solvent and dry completely.
17. Insert the inboard end of the driveshaft into the differential until the set ring locks in the groove.
18. Install the outboard end into the hub.
19. Install the wheel sensor bracket.
20. Install the lower arm flange bolt and tighten to 54 ft. lbs. (74 Nm).
21. Install the upper arm bolt and tighten to 47 ft. lbs. (64 Nm).
22. Install a new spindle nut and tighten to 181 ft. lbs. (245 Nm). Use a drift to stake the spindle nut shoulder once the nut is tightened to specification.
23. Install the wheel
24. Turn the wheel to make sure there is no binding between the driveshaft and wheel.
25. Refill the differential until the fluid level is at the bottom of the fill hole. A complete oil change would require 2.79 quarts of VTM–4 differential fluid. Install the plug and tighten to 35 ft. lbs. (47 Nm).
26. Check and adjust the wheel alignment.

CV-Joint

OVERHAUL

Front

OUTBOARD JOINT

1. Before servicing the vehicle, refer to the precautions section.
2. Remove or disconnect the following:
   - Axle halfshaft from the vehicle and place it in a vise
   - Outboard joint boot clamps and push the boot back
   - Outboard joint by driving it off the axle shaft with a brass drift and hammer
   - Outboard joint boot
   To install:

➡**Use new circlips and boot clamps for assembly.**

3. Install the outboard joint boot and clamps to the axle shaft.
4. Fill the outboard joint with grease. Install the outboard joint to the axle shaft. Tap the stub shaft with a brass hammer to seat the circlip.
5. Fill the outboard joint boot with grease and install the boot clamps.
6. Install the axle halfshaft to the vehicle.

INBOARD JOINT

1. Before servicing the vehicle, refer to the precautions section.
2. Remove or disconnect the following:
   - Axle halfshaft from the vehicle.
   - Inboard joint boot clamps and push the boot back
   - Inboard joint housing from the axle
   - Rollers from the spider
   - Snapring and the spider from the axle shaft
   - Inboard joint boot
   To install:

➡**Use new circlips and boot clamps for assembly.**

3. Install or connect the following:
   - Inboard joint boot and clamps to the axle shaft
   - Spider with a new snapring
   - Rollers to the spider
4. Fill the joint housing with grease and install it.
5. Fill the inboard joint boot with grease and install the boot clamps.
6. Install the axle halfshaft to the vehicle.

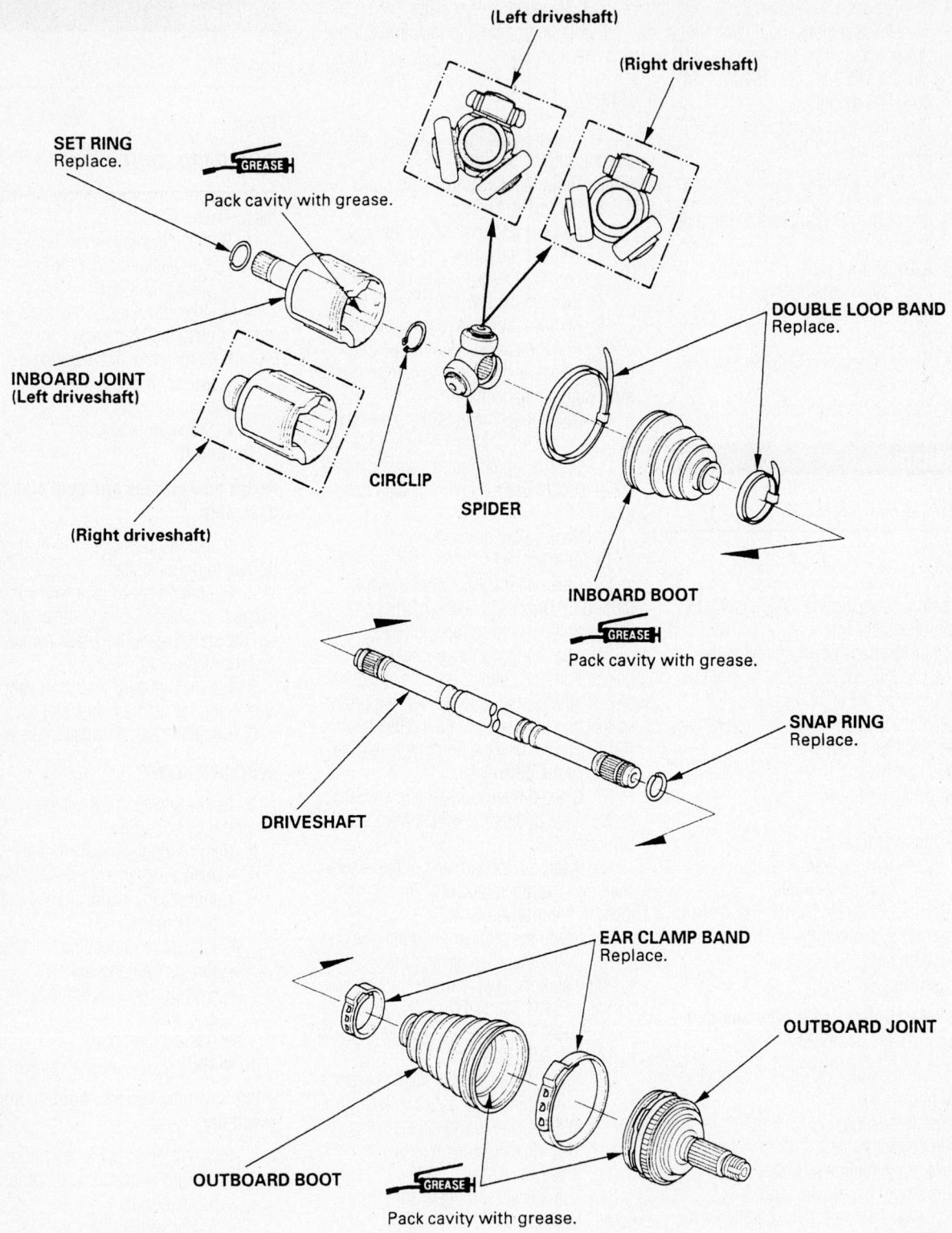

SET RING
Replace.

GREASE

Pack cavity with grease.

(Left driveshaft)

(Right driveshaft)

INBOARD JOINT
(Left driveshaft)

DOUBLE LOOP BAND
Replace.

CIRCLIP

(Right driveshaft)

SPIDER

INBOARD BOOT

GREASE

Pack cavity with grease.

SNAP RING
Replace.

DRIVESHAFT

EAR CLAMP BAND
Replace.

OUTBOARD JOINT

OUTBOARD BOOT

GREASE

Pack cavity with grease.

9308MG29

Front axle exploded view—MDX

STEERING AND SUSPENSION

Air Bag

✳✳ CAUTION

Some vehicles are equipped with an air bag system. The system must be disarmed before performing service on, or around, system components, the steering column, instrument panel components, wiring and sensors. Failure to follow the safety precautions and the disarming procedure could result in accidental air bag deployment, possible injury and unnecessary system repairs.

PRECAUTIONS

Several precautions must be observed when handling the inflator module to avoid accidental deployment and possible personal injury.

• Never carry the inflator module by the wires or connector on the underside of the module.

• When carrying a live inflator module, hold securely with both hands, and ensure that the bag and trim cover are pointed away.

• Place the inflator module on a bench or other surface with the bag and trim cover facing up.

• With the inflator module on the bench, never place anything on or close to the module which may be thrown in the event of an accidental deployment.

Before servicing the vehicle, also make sure to refer to the precautions in the beginning of this section as well.

DISARMING

Disconnect and isolate the negative battery cable. Wait 3 minutes for the system capacitor to discharge before performing any service.

Power Rack and Pinion Steering Gear

REMOVAL & INSTALLATION

✳✳ WARNING

Do not permit the steering wheel to turn whenever the steering gear is disconnected from the steering column. Damage to the air bag wiring can result.

1. Before servicing the vehicle, refer to the precautions section.
2. Center the steering wheel and lock it in position.
3. Attach a support fixture to the engine lifting eyes.
4. Remove or disconnect the following:
   • Negative battery cable
   • Air bag and steering wheel
   • Steering joint cover
   • Steering flexible joint
   • Power steering fluid lines
   • 10mm bolt on the engine side mount bracket
   • Front wheels
   • Outer tie rod ends
   • Sub-frame stiffener
   • Heated Oxygen (HO_2S) sensor connectors
   • 3 way catalytic converter from the mufflers
   • Flange bolts from the exhaust rubber mount
   • Power steering pressure switch connector
   • Propeller shaft protector
   • Splash shield
5. Support the front subframe with a jack and support the transmission with a second jack.
6. Loosen the 14mm subframe bolts.
7. Lower the subframe about 1 3/16 inches (30mm).
8. Remove or disconnect the following:
   • Two 12mm and two 14 stiffener plate bolts
9. Support the transfer case by raising the transmission jack and remove the two 12mm bolts.
   • Two 14mm bolts and the rear stiffener plats from the sub-frame
10. Lower the transmission jack until the front subframe has dropped about 1 15/16 inch (50mm).
    • Power steering line brackets
    • Feed line
    • Return hose
    • Two 10mm bolts from the right side gearbox
    • Mounting bracket and cushion
    • Two 10mm bolts from the left side gearbox
11. Lower the transmission jack until the front subframe has dropped about 3 15/16 inch (100mm).
    • Gearbox stiffener bracket
12. Slide the gearbox between the body and front sub-frame towards the left and from the vehicle.

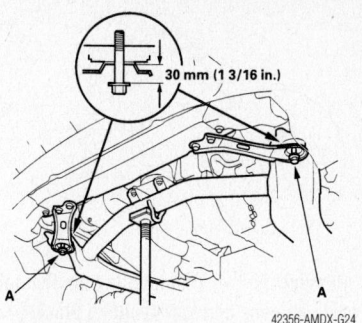

42356-AMDX-G24

Loosen the 14mm subframe bolts and lower the subframe about 1 3/16 inches (30mm)

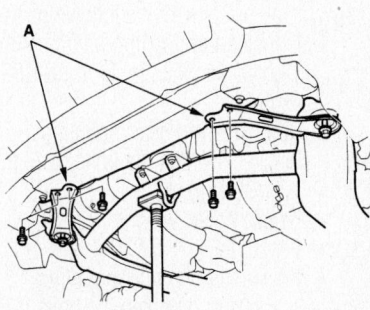

42356-AMDX-G25

Remove the four 12mm stiffener plate bolts

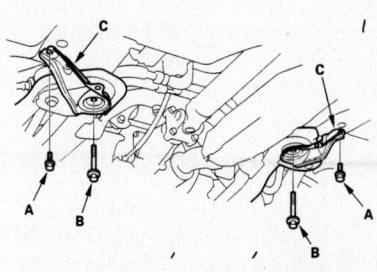

42356-AMDX-G26

Remove the two 12mm bolts (A), then the 14mm bolts (B) and the rear stiffener plates (C) from the sub-frame

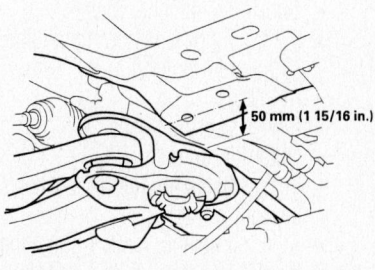

42356-AMDX-G27

Lower the transmission jack until the front subframe has dropped about 1 15/16 inch (50mm)

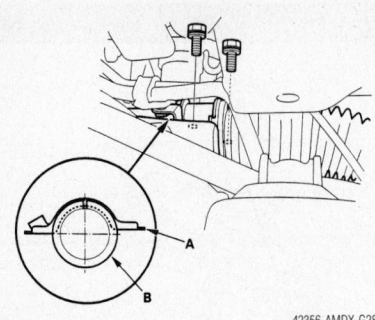

42356-AMDX-G28

Remove the two 10mm bolts from the right side gearbox and the mounting bracket and cushion

To install:

13. Position the steering gear in the vehicle.

14. Install or connect the following:
- Left steering gear mounting bolts. Tighten the bolts to 43 ft. lbs. (58 Nm).
- Right steering gear mounting bracket. Tighten the bolts to 29 ft. lbs. (39 Nm).
- Return hose
- Feed line
- Power steering line mounting brackets and tighten the bolts to 7 ft. lbs. (10 Nm)

15. Raise the subframe into position. Tighten the 14mm bolts to 76 ft. lbs. (103 Nm) and the 12mm bolts to 86 ft. lbs. (117 Nm).

16. Install or connect the following:
- Front stiffener plates. Tighten the 14mm bolts to 76 ft. lbs. (103 Nm) and the 12mm bolts to 54 ft. lbs. (74 Nm).
- Splash shield
- Propeller shaft protector
- Power steering pressure switch
- 3 way catalytic converter and mufflers. Tighten the nuts to 25 ft. lbs. (33 Nm)
- Rubber exhaust mount and tighten the bolts to 28 ft. lbs. (38 Nm)
- HO_2S sensor connectors
- Sub-frame stiffener plate
- 10mm flange bolts on the engine side mount bracket to 33 ft. lbs. (44 Nm)
- Power steering hoses
- Outer tie rod ends
- Front wheels. Position the wheels straight-ahead.
- Steering flexible joint. Tighten the pinch bolts to 16 ft. lbs. (22 Nm).
- Steering joint cover
- Negative battery cable

17. Fill the power steering system.

18. Check the wheel alignment and adjust as necessary.

Strut

REMOVAL & INSTALLATION

Front

1. Before servicing the vehicle, refer to the precautions section.

2. Remove or disconnect the following:
- Front wheel
- Wheel speed sensor wiring bracket
- Brake hose bracket
- Stabilizer bar link
- Strut pinch bolts
- Upper mount nuts
- Strut

To install:

3. Install or connect the following:
- Strut. Tighten the upper mount nuts to 43 ft. lbs. (59 Nm).
- Strut pinch bolts. Tighten the nuts to 116 ft. lbs. (157 Nm).
- Stabilizer bar link. Tighten the nut to 58 ft. lbs. (78 Nm).

- Brake hose bracket
- Wheel speed sensor wiring bracket
- Front wheel

4. Check the wheel alignment and adjust as necessary.

Shock Absorber

REMOVAL & INSTALLATION

Rear

1. Before servicing the vehicle, refer to the precautions section.

2. Support the vehicle under the lower control arm.

3. Remove or disconnect the following:
- Rear wheel
- Upper shock absorber flange bolt
- Lower shock absorber nut
- Shock absorber

To install:

4. Install or connect the following:
- Shock absorber. Tighten the fasteners to 47 ft. lbs. (64 Nm).
- Rear wheel

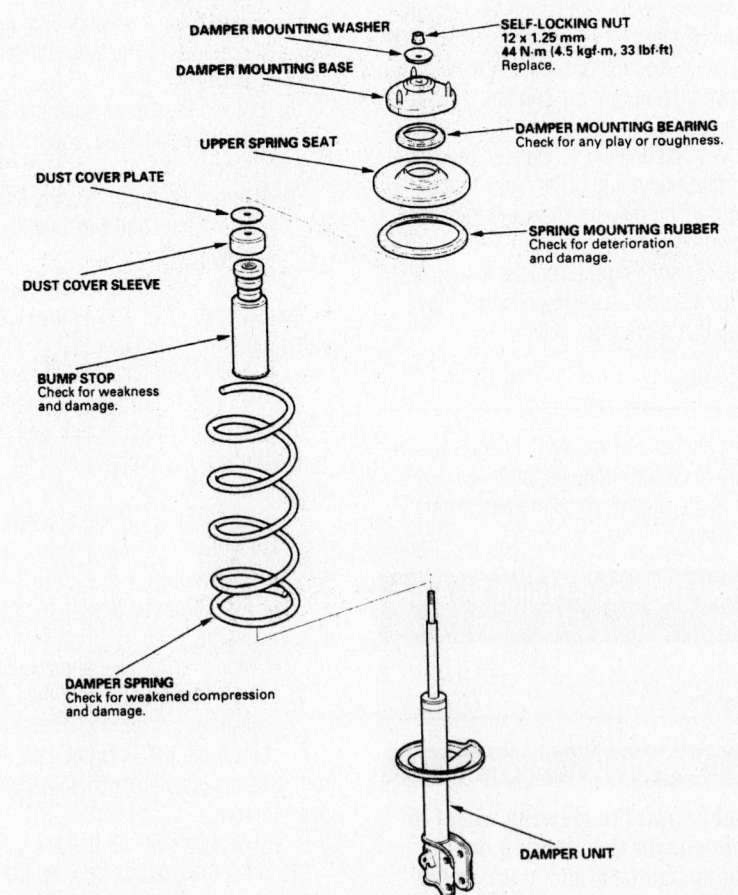

Exploded view of the front strut assembly

SELF-LOCKING NUT
12 x 1.25 mm
44 N·m (4.5 kgf·m, 33 lbf·ft)
Replace.

DAMPER MOUNTING WASHER

DAMPER MOUNTING BASE

DAMPER MOUNTING BEARING
Check for any play or roughness.

UPPER SPRING SEAT

DUST COVER PLATE

SPRING MOUNTING RUBBER
Check for deterioration and damage.

DUST COVER SLEEVE

BUMP STOP
Check for weakness and damage.

DAMPER SPRING
Check for weakened compression and damage.

DAMPER UNIT

93552G11

Coil Spring

REMOVAL & INSTALLATION

Front

1. Before servicing the vehicle, refer to the precautions section.
2. Remove the strut from the vehicle and install in a strut spring compressor. Compress the spring until the end of the spring comes away from the spring seat.
3. Remove the upper strut mount, spring seat and related components.
4. Remove the coil spring from the strut spring compressor.

To install:

➡**Use a new self-locking nut.**

5. Compress the spring and position the strut so that the end of the spring aligns with the notch in the spring seat.
6. Install the upper strut mounting components and tighten the nut to 33 ft. lbs. (44 Nm).
7. Install the strut to the vehicle.
8. Check the wheel alignment and adjust as necessary.

Rear

1. Before servicing the vehicle, refer to the precautions section.

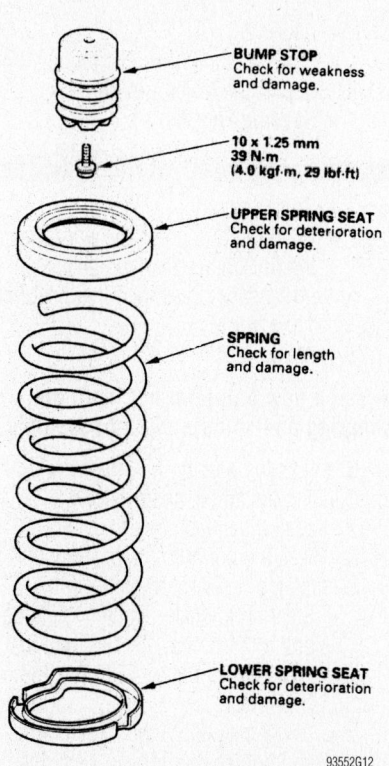

BUMP STOP
Check for weakness and damage.

10 x 1.25 mm
39 N·m
(4.0 kgf·m, 29 lbf·ft)

UPPER SPRING SEAT
Check for deterioration and damage.

SPRING
Check for length and damage.

LOWER SPRING SEAT
Check for deterioration and damage.

93552G12

Exploded view of the rear spring assembly

2. Support the vehicle under the lower control arm.
3. Remove or disconnect the following:
   - Rear wheel
   - Stabilizer link from the lower arm
   - Wheel speed sensor wiring harness from the lower arm. Do not disconnect the connector.
   - Upper shock absorber flange bolt
   - Lower control arm bolts
4. Lower the floor jack and remove the coil spring and spring seats.

To install:

➡**Use new self-locking nuts for assembly.**

5. Place the coil spring and spring seats on the lower control arm and raise into position. Tighten the inboard bolt to 61 ft. lbs. and the outer bolt to 54 ft. lbs. (74 Nm).
6. Install or connect the following:
   - Rear wheel

Ball Joint

REMOVAL & INSTALLATION

The lower ball joints are replaced with the control arms as an assembly.

Upper Control Arm

REMOVAL & INSTALLATION

Rear

1. Before servicing the vehicle, refer to the precautions section.

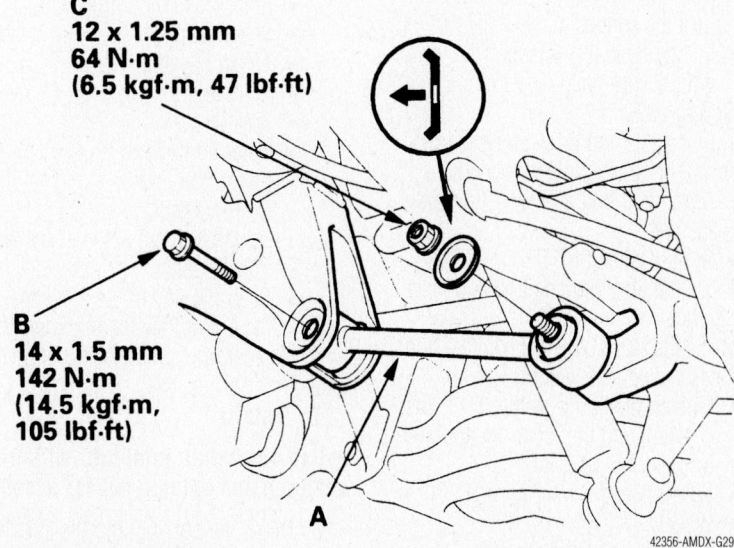

C
12 x 1.25 mm
64 N·m
(6.5 kgf·m, 47 lbf·ft)

B
14 x 1.5 mm
142 N·m
(14.5 kgf·m,
105 lbf·ft)

A

42356-AMDX-G29

Rear lower arm (A) mounting

2. Support the control arm at the knuckle.
3. Remove or disconnect the following:
   - Wheel
   - Upper ball joint from the knuckle
   - Upper arm bolt and the arm
4. Installation is the reverse of removal. Tighten the arm bolt to 47 ft. lbs. (64 Nm) and the ball joint nut to 36–43 ft. lbs. (49–59 Nm).

Lower Control Arm

REMOVAL & INSTALLATION

Front

1. Before servicing the vehicle, refer to the precautions section.
2. Remove or disconnect the following:
   - Front wheel
   - Lower ball joint
   - Front inner flange bolt
   - Rear inner flange bolt
   - Lower control arm

To install:

➡**Use a new split pin for assembly.**

3. Install or connect the following:
   - Lower control arm. Tighten the inner flange bolts to 69 ft. lbs. (93 Nm).
   - Lower ball joint. Tighten the nut to 43–51 ft. lbs. (59–69 Nm).
   - Front wheel
4. Check the wheel alignment and adjust as necessary.

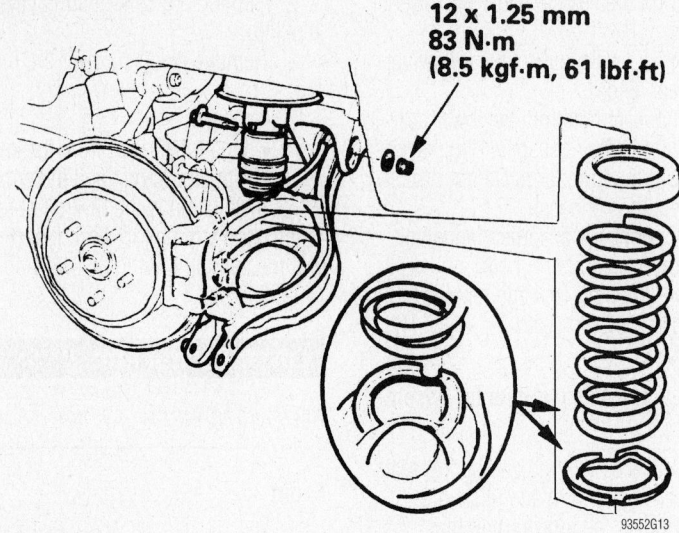

12 x 1.25 mm
83 N·m
(8.5 kgf·m, 61 lbf·ft)

93552G13

Rear lower arm (B) mounting

Rear

LOWER ARM (A)

1. Before servicing the vehicle, refer to the precautions section.
2. Remove or disconnect the following:
   - Lower arm mounting bolt and nut
   - Lower arm
3. Installation is the reverse of removal. Tighten the bolt to 105 ft. lbs. (142 Nm) and the nut to 47 ft. lbs. (64 Nm).

LOWER ARM (B)

1. Before servicing the vehicle, refer to the precautions section.
2. Support the control arm with a jack.
3. Remove or disconnect the following:
   - Wheel
   - Stabilizer link from the lower arm
   - Wheel speed sensor wiring harness from the lower arm. Do not disconnect the connector.
   - Flange bolts that attaches the lower arm to the knuckle
4. Spring assembly
   - Inner nuts and bolts and the arm
5. Install or connect the following:
   - Arm, inner bolt and loosely install the nut
   - Spring assembly
6. Raise the arm into position and install the flange bolt.
7. Raise the rear suspension with a floor jack to load the vehicle weight.
   - Tighten the flange bolt to 54 ft. lbs. (74 Nm) and the inner nut and bolt to 61 ft. lbs. (83 Nm).
   - Wheel speed sensor harness
   - Wheel
8. Check the vehicle alignment.

CONTROL ARM BUSHING REPLACEMENT

The control arm bushings are serviced with the control arms as an assembly.

Wheel Bearings

ADJUSTMENT

The wheel bearings are sealed units and are not adjustable.

REMOVAL & INSTALLATION

Front

1. Before servicing the vehicle, refer to the precautions section.
2. Remove or disconnect the following:
   - Front wheel
   - Brake hose mounting bolt
   - Brake caliper
   - Wheel speed sensor
   - Spindle nut
   - Brake rotor
   - Outer tie rod end
   - Lower ball joint
   - Steering knuckle
3. Press the hub out of the wheel bearing.
   - Splash guard
   - Snapring and press the wheel bearing out of the steering knuckle
4. If necessary, press the inner bearing race off of the hub.

To install:

➡**Use a new ball joint nut, split pin, snapring and spindle nut for assembly.**

5. Press the bearing into the steering knuckle and install the snapring.

6. Install the splash guard.
7. Press the hub into the bearing.
8. Install or connect the following:
   - Steering knuckle. Tighten the ball joint nut to 43–51 ft. lbs. (59–69 Nm) and the damper flange bolts to 116 ft. lbs. (157 Nm).
   - Outer tie rod end. Tighten the nut to 40 ft. lbs. (54 Nm).
   - Wheel speed sensor, if equipped
   - Brake caliper and rotor
   - Brake hose
   - Spindle nut. Tighten the nut to 181 ft. lbs. (245 Nm) on 2002 models. On 2003–06 models, torque to 210 ft. lbs. (285 Nm).
   - Front wheel
9. Check the wheel alignment and adjust as necessary.

Rear

1. Before servicing the vehicle, refer to the precautions section.
2. Remove or disconnect the following:
   - Rear wheel
   - Brake hose bracket mounting bolts from the trailing arm and the knuckle
   - Brake caliper
   - Wheel speed sensor
   - Spindle nut
   - Brake rotor
   - Upper ball joint
   - Lower arm (A)
   - Lower arm (B) from the trailing arm
3. Support the lower arm (B)
   - Steering knuckle
4. Press the hub out of the wheel bearing.
   - Splash guard
   - Snapring and press the wheel bearing out of the steering knuckle
5. If necessary, press the inner bearing race off of the hub.

To install:

➡**Use a new ball joint nut, split pin, snapring and spindle nut for assembly.**

6. Press the bearing into the steering knuckle and install the snapring.
7. Install the splash guard.
8. Press the hub into the bearing.
9. Install or connect the following:
   - Steering knuckle. Tighten the flange bolt to 54 ft. lbs. (74 Nm) and the lower shock nut to 47 ft. lbs. (64 Nm)
   - Lower arm (B) to the trailing arm and tighten the bolts to 47 ft. lbs. (64 Nm)

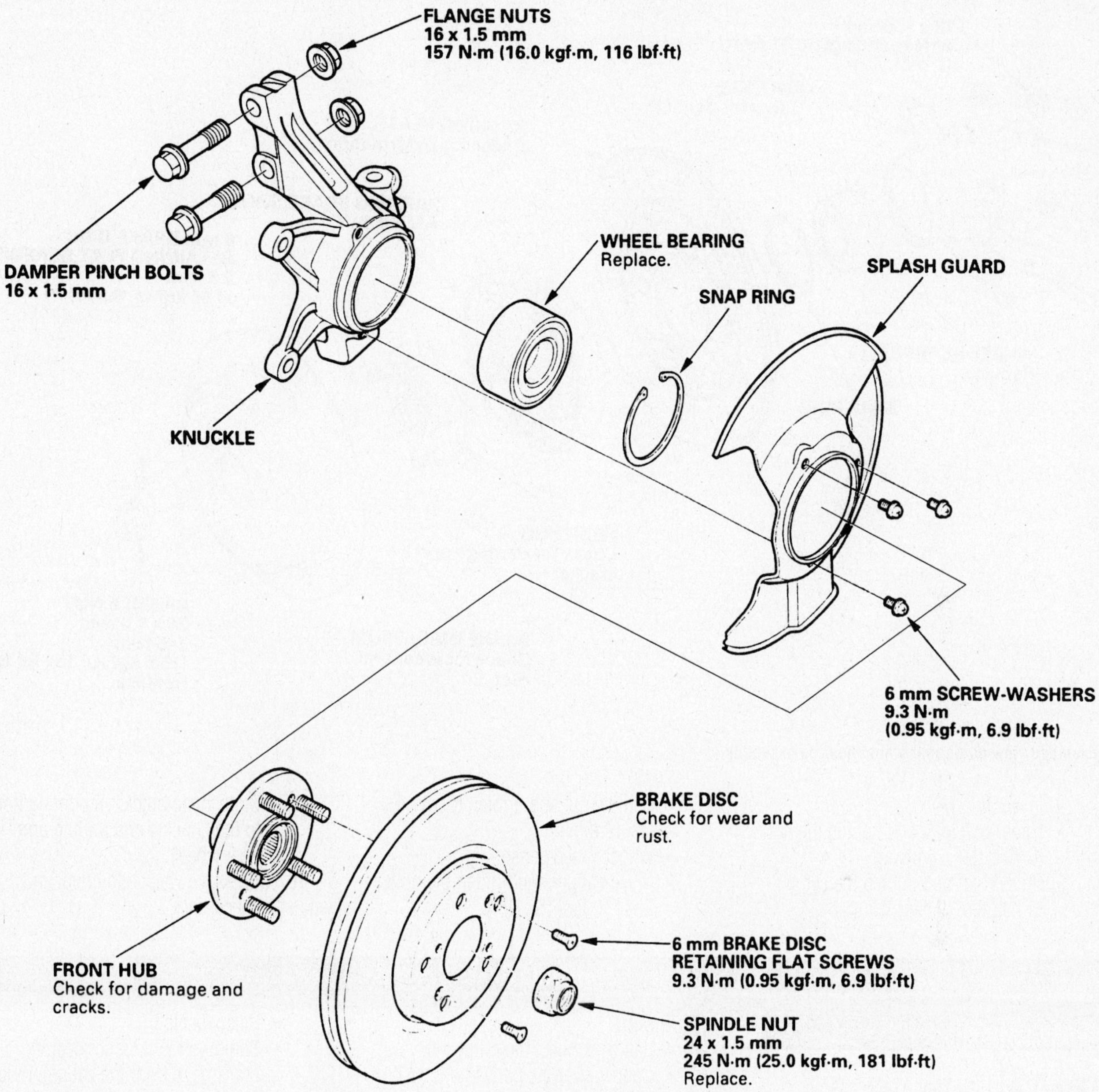

FLANGE NUTS
16 x 1.5 mm
157 N·m (16.0 kgf·m, 116 lbf·ft)
WHEEL BEARING
Replace.
SPLASH GUARD
SNAP RING
DAMPER PINCH BOLTS
16 x 1.5 mm
KNUCKLE
6 mm SCREW-WASHERS
9.3 N·m
(0.95 kgf·m, 6.9 lbf·ft)
BRAKE DISC
Check for wear and rust.
FRONT HUB
Check for damage and cracks.
6 mm BRAKE DISC
RETAINING FLAT SCREWS
9.3 N·m (0.95 kgf·m, 6.9 lbf·ft)
SPINDLE NUT
24 x 1.5 mm
245 N·m (25.0 kgf·m, 181 lbf·ft)
Replace.
93552G16

Front wheel bearing assembly, spindle nut torque shown is for 2002, on 2003–06 torque is 210 ft. lbs. (285 Nm)

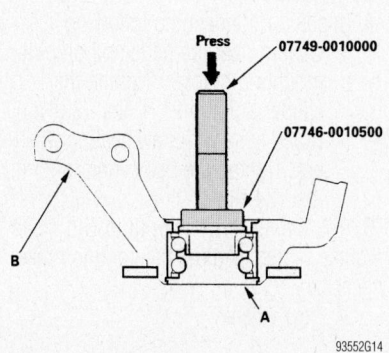

Press
07749-0010000
07746-0010500
B
A
93552G14

Press the wheel bearing out of the knuckle

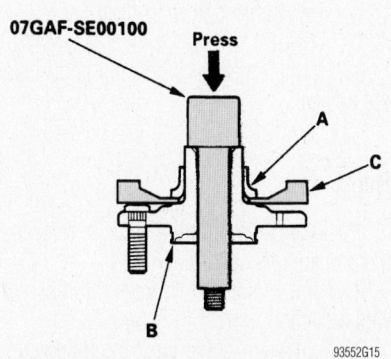

07GAF-SE00100
Press
A
C
B
93552G15

Press the wheel bearing inner race from the hub

Press
07749-0010000
07746-0010500
B
C
A
07965-SD90100
42356-AMDX-G31

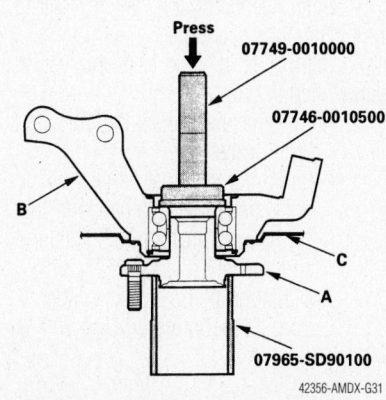

Press the wheel bearing into the knuckle

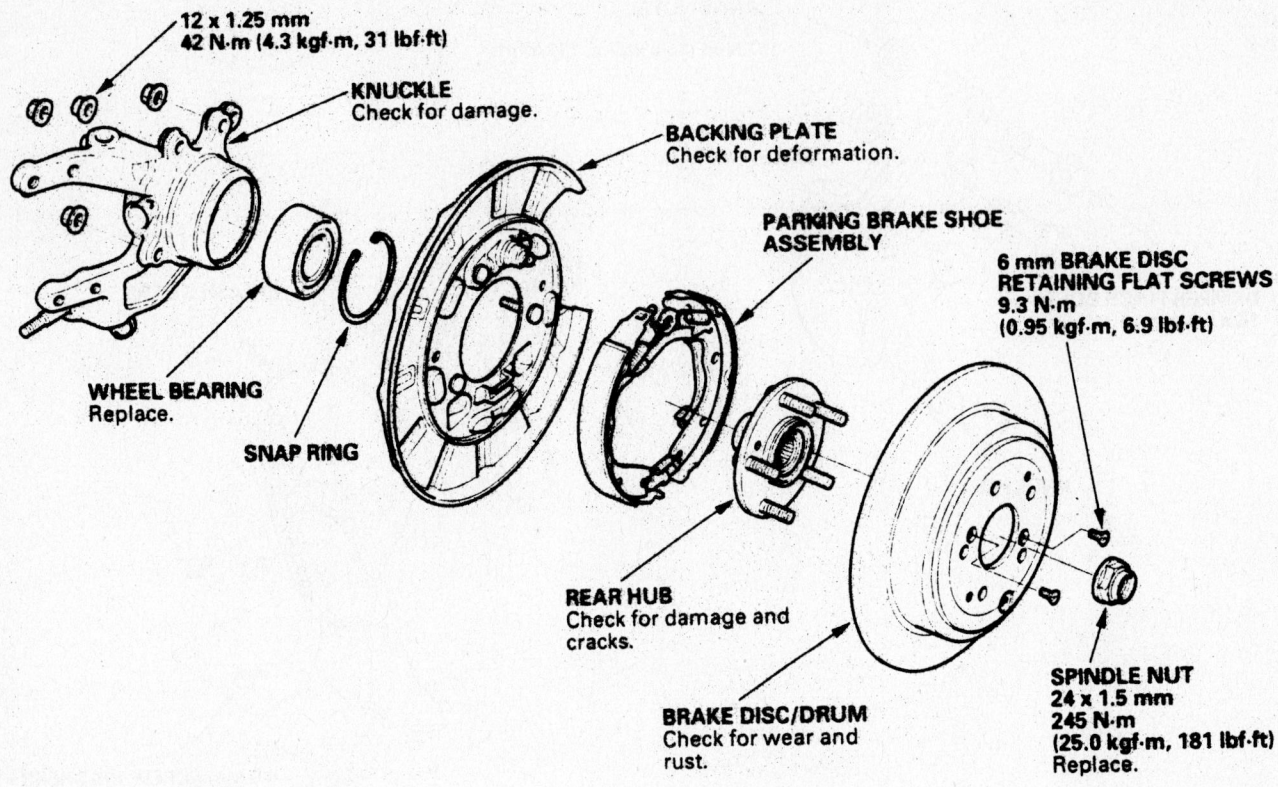

12 x 1.25 mm
42 N·m (4.3 kgf·m, 31 lbf·ft)

KNUCKLE
Check for damage.

BACKING PLATE
Check for deformation.

PARKING BRAKE SHOE ASSEMBLY

6 mm BRAKE DISC RETAINING FLAT SCREWS
9.3 N·m
(0.95 kgf·m, 6.9 lbf·ft)

WHEEL BEARING
Replace.

SNAP RING

REAR HUB
Check for damage and cracks.

BRAKE DISC/DRUM
Check for wear and rust.

SPINDLE NUT
24 x 1.5 mm
245 N·m
(25.0 kgf·m, 181 lbf·ft)
Replace.

93552G17

Exploded view of the rear wheel bearing assembly

- Lower arm (A)
- Upper ball joint and tighten the nut to 40 ft. lbs. (54 Nm)
- Brake rotor and tighten the screws to 7 ft. lbs. (9 Nm)
- Spindle nut and tighten to 181 ft, lbs. (245 Nm)
- Wheel speed sensor
- Brake caliper and tighten the bolts to 41 ft. lbs. (55 Nm)
- Brake hose bracket mounting bolts to the knuckle and trailing arm
- Rear wheel

10. Check the wheel alignment and adjust as necessary.

BRAKES

Brake Caliper

REMOVAL AND INSTALLATION

Front

1. Before servicing the vehicle, refer to the precautions section.
2. Remove some fluid from the reservoir with a suction pump.
3. Remove or disconnect the following:
   - Front wheels
   - Banjo bolt and disconnect the brake hose from the caliper. Plug the hose to prevent fluid loss and contamination.
   - Mounting bolts and the caliper from its mounting bracket

To Install:
4. Install or connect the following:
   - Caliper over the pads and onto its mounting bracket. Torque both caliper bolts to 27 ft. lbs. (36 Nm).
   - Brake hose to the caliper using new sealing washers. Carefully torque the banjo bolt to 25 ft. lbs. (34 Nm).
5. Fill the reservoir with fluid and bleed the brakes.
   - Front wheels

Rear

1. Before servicing the vehicle, refer to the precautions section.
2. Remove some fluid from the reservoir with a suction pump.
3. Remove or disconnect the following:

- Rear wheels
- Banjo bolt and disconnect the brake hose from the caliper. Plug the hose to prevent fluid loss and contamination.
- 2 caliper mounting bolts and the caliper from its mounting bracket

To Install:
4. Install or connect the following:
   - Caliper over the pads and onto its mounting bracket. Tighten the caliper bolts to 27 ft. lbs. (37 Nm).
   - Brake hose with new sealing washers. Tighten the banjo bolt to 25 ft. lbs. (34 Nm).
5. Fill the reservoir with fluid and bleed the brake system. Adjust the parking brake if necessary.
   - Rear wheels

Front Brake Caliper Overhaul

⚠**CAUTION**

Frequent inhalation of brake pad dust, regardless of material composition, could be hazardous to your health.
- Avoid breathing dust particles.
- Never use an air hose or brush to clean brake assemblies. Use an OSHA-approved vacuum cleaner.

Remove, disassemble, inspect, reassemble, and install the caliper, and note these items:

- Do not spill brake fluid on the vehicle; it may damage the paint; if brake fluid gets on the paint, wash it off immediately with water.
- To prevent dripping brake fluid, cover disconnected hose joints with rags or shop towels.
- Clean all parts in brake fluid and air dry; blow out all passages with compressed air.
- Before reassembling, check that all parts are free of dirt and other foreign particles.
- Replace parts with new ones as specified in the illustration.
- Make sure no dirt or other foreign matter gets in the brake fluid.
- Make sure no grease or oil gets on the brake discs or pads.
- When reusing pads, always reinstall them in their original positions to prevent loss of braking efficiency.
- Do not reuse drained brake fluid. Use only clean Genuine Honda DOT 3 Brake Fluid. Non-Honda brake fluid can cause corrosion and shorten the life of the system.
- Coat the piston, piston seal groove, and caliper bore with clean brake fluid.
- Replace all rubber parts with new ones.
- After installing the caliper, check the brake hose and line for leaks, interference, and twisting.

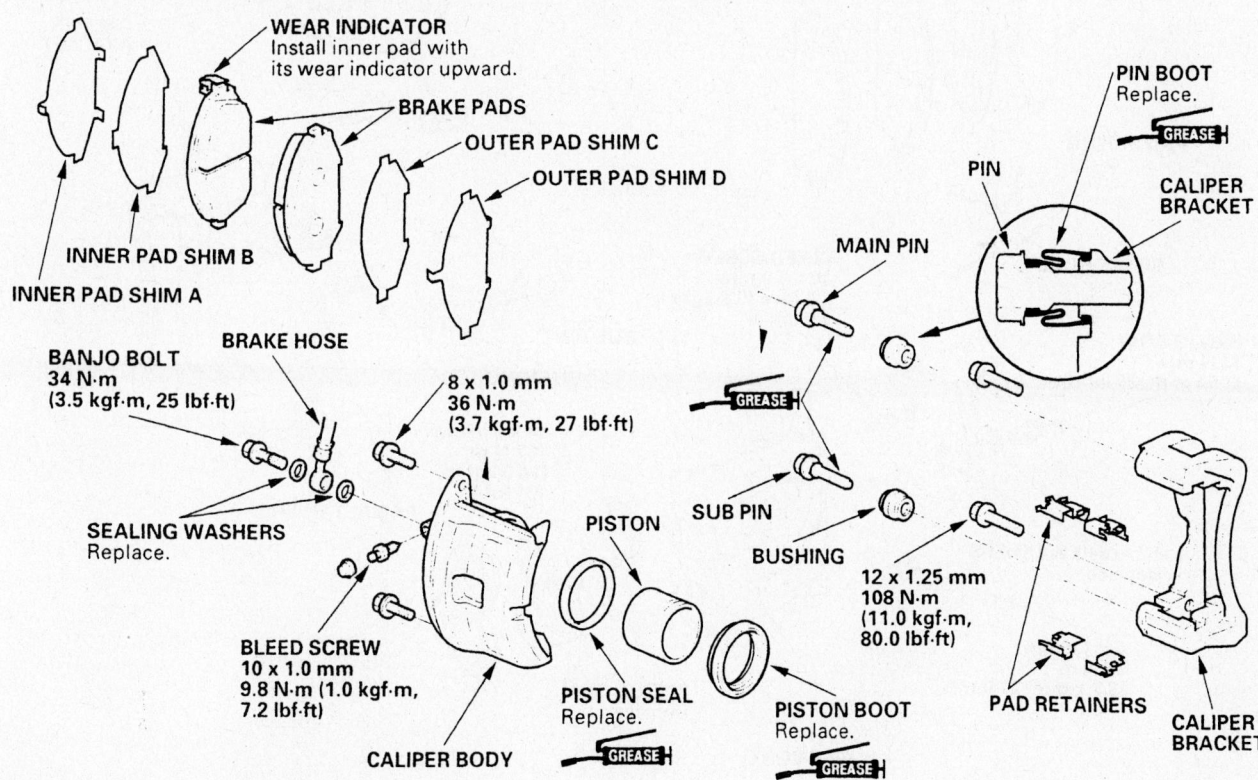

Exploded view of the front caliper components—2002 models

⚠CAUTION

Frequent inhalation of brake pad dust, regardless of material composition, could be hazardous to your health.
• Avoid breathing dust particles.
• Never use an air hose or brush to clean brake assemblies. Use an OSHA-approved vacuum cleaner.

Remove, disassemble, inspect, reassemble, and install the caliper, and note these items:

• Do not spill brake fluid on the vehicle; it may damage the paint; if brake fluid gets on the paint, wash it off immediately with water.
• To prevent dripping brake fluid, cover disconnected hose joints with rags or shop towels.
• Clean all parts in brake fluid and air dry; blow out all passages with compressed air.
• Before reassembling, check that all parts are free of dirt and other foreign particles.
• Replace parts with new ones as specified in the illustration.
• Make sure no dirt or other foreign matter gets in the brake fluid.
• Make sure no grease or oil gets on the brake discs or pads.
• When reusing pads, always reinstall them in their original positions to prevent loss of braking efficiency.
• Do not reuse drained brake fluid. Use only clean Honda DOT 3 Brake Fluid. Non-Honda brake fluid can cause corrosion and shorten the life of the system.
• Coat the piston, piston seal groove, and caliper bore with clean brake fluid.
• Replace all rubber parts with new ones.
• After installing the caliper, check the brake hose and line for leaks, interference, and twisting.

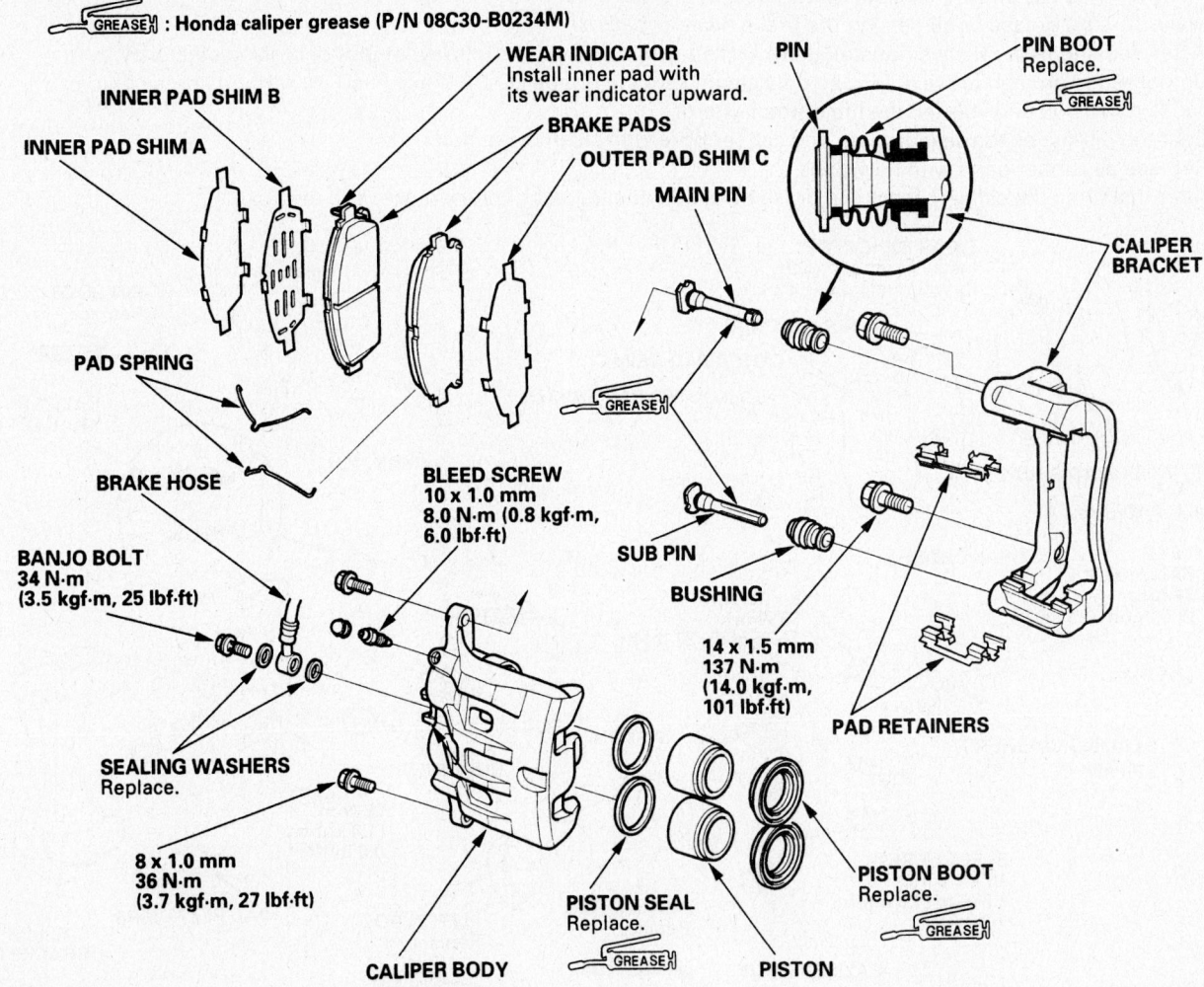

Exploded view of the front caliper components—2003–06 models

42356-AMDX-G30

Rear Brake Caliper Overhaul

> ### ⚠ CAUTION
>
> Frequent inhalation of brake pad dust, regardless of material composition, could be hazardous to your health.
> • Avoid breathing dust particles.
> • Never use an air hose or brush to clean brake assemblies. Use an OSHA-approved vacuum cleaner.

Remove, disassemble, inspect, reassemble, and install the caliper, and note these items:

• Do not spill brake fluid on the vehicle; It may damage the paint; If brake fluid gets on the paint, wash it off immediately with water.
• To prevent dripping brake fluid, cover disconnected hose joints with rags or shop towels.
• Clean all parts in brake fluid and air dry; blow out all passages with compressed air.
• Before reassembling, check that all parts are free of dirt and other foreign particles.
• Replace parts with new ones as specified in the illustration.
• Make sure no dirt or other foreign matter gets into the brake fluid.
• Make sure no grease or oil gets on the brake discs or pads.
• When reusing pads, always reinstall them in their original positions to prevent loss of braking efficiency.
• Do not reuse drained brake fluid. Use only clean Genuine Honda DOT 3 Brake Fluid. Non-Honda brake fluid can cause corrosion and shorten the life of the system.
• Coat the piston, piston seal groove, and caliper bore with clean brake fluid.
• Replace all rubber parts with new ones.
• After installing the caliper, check the brake hose and line for leaks, interference, and twisting.

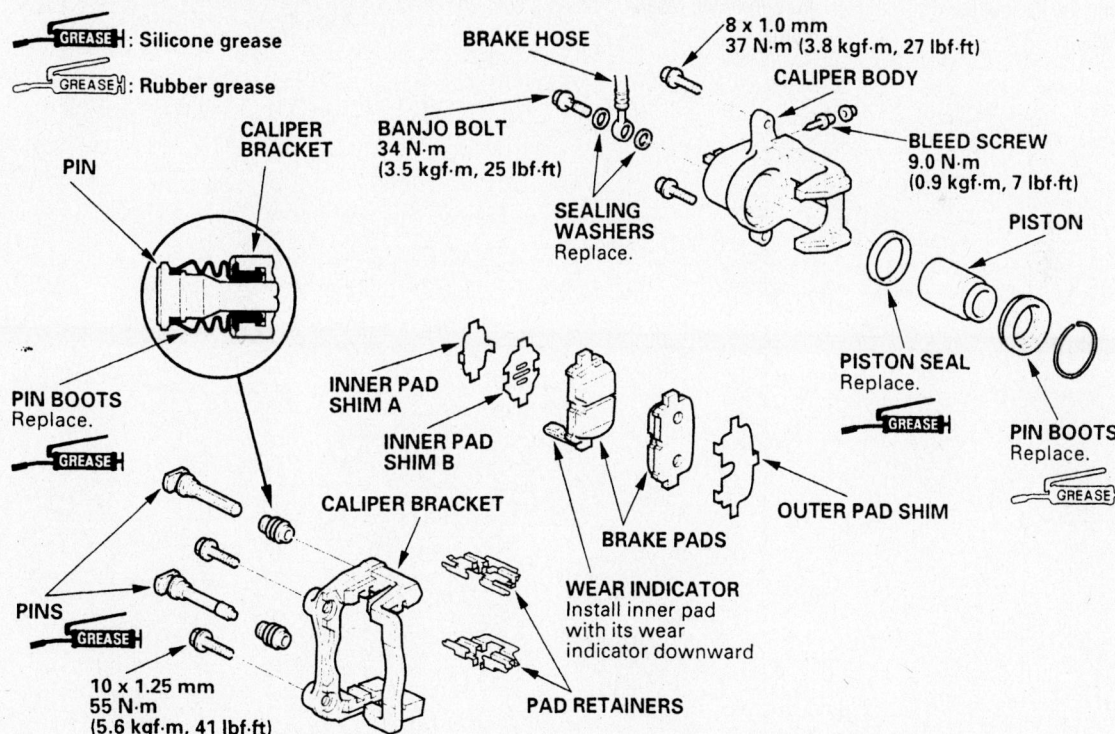

Exploded view of the rear caliper components—MDX

93552GZB

Disc Brake Pads

REMOVAL AND INSTALLATION

Front

1. Before servicing the vehicle, refer to the precautions section.

2. Remove or disconnect the following:
   - Front wheels

3. Remove a small amount of brake fluid from the reservoir using a suction pump.
   - Brake hose clamp from the knuckle by unfastening the retaining bolts
   - Lower caliper retaining bolt and pivot the caliper upward, off of the pads
   - Pad shim and pad retainers
   - Disc brake pads from the caliper

To install:

4. Clean the caliper thoroughly; remove any rust from the lip of the disc or rotor. Check the brake rotor for grooves or cracks. If any heavy scoring is present, the rotor must be replaced.

5. Install or connect the following:
   - Pad retainers. Apply molybdenum brake grease to both surfaces of the shims and the back of the disc brake pads.

- Pads and shims. The pad with the wear indicator goes in the inboard position.

6. Push in the caliper piston so the caliper will fit over the pads. This is most easily accomplished with a pad spreader or large C-clamp.
   - Caliper down into position and tighten the mounting bolt to 27 ft. lbs. (37 Nm)
   - Brake hose to the knuckle, if removed
   - Wheels

7. Add brake fluid to the master cylinder reservoir and install the cap.

8. Depress the brake pedal several times and make sure that the movement feels normal. The first brake pedal application may result in a very long pedal action due to the pistons being retracted. Always make several brake applications before starting the vehicle. Bleed the system if necessary.

Rear

1. Before servicing the vehicle, refer to the precautions section.

2. Remove a small amount of brake fluid from the reservoir using a suction pump.

3. Remove or disconnect the following:
   - Rear wheels
   - 2 caliper mounting bolts and the caliper from the bracket
   - pads, shims, and pad retainers

To install:

4. Clean the caliper thoroughly; remove any dirt or dust. Check the brake rotor for grooves or cracks and machine or replace, as necessary.

5. Install or connect the following:
   - Pad retainers. Apply molybdenum brake grease to both surfaces of the shims and the back of the disc brake pads.
   - Pads and shims. The wear retainer on the inboard pad faces down.

6. Use a suitable tool to push caliper piston into its bore and enable the caliper to fit over the pads. Lubricate the piston boot with silicon grease. Avoid twisting the boot.
   - Brake caliper and tighten the mounting bolts to 27 ft. lbs. (37 Nm)
   - Rear wheels

7. Add brake fluid to the master cylinder reservoir. Depress the brake pedal several times to seat the pads. Bleed the brakes if necessary.

HONDA

Accord • Civic • Prelude • S2000

3

SPECIFICATIONS AND MAINTENANCE CHARTS

ENGINE AND VEHICLE IDENTIFICATION

Engine							Model Year	
Code	Liters	Cu. In. (cc)	Cyl.	Fuel Sys.	Eng. Mfg.		Code	Year
D17A1	1.7	101.7 (1668)	4	PGM-FI	Honda		2	2002
D17A2	1.7	101.7 (1668)	4	PGM-FI	Honda		3	2003
D17A6	1.7	101.7 (1668)	4	PGM-FI	Honda		4	2004
R18A1	1.8	110.0 (1798)	4	PGM-FI	Honda		5	2005
K20Z3	2.0	121.9 (1997)	4	PGM-FI	Honda		6	2006
F20C1	2.0	121.9 (1997)	4	PGM-FI	Honda			
D17A7	1.7	101.7 (1668)	4	PGM-FI	Honda			
K20A3	2.0	121.9 (1997)	4	PGM-FI	Honda			
F22C1	2.2	132.0 (2157)	4	PGM-FI	Honda			
F23A1	2.3	137.0 (2254)	4	PGM-FI	Honda			
F23A4	2.3	137.0 (2254)	4	PGM-FI	Honda			
F23A5	2.3	137.0 (2254)	4	PGM-FI	Honda			
K24A4	2.4	144.0 (2354)	4	PGM-FI	Honda			
K24A8	2.4	144.0 (2354)	4	PGM-FI	Honda			
J30A5	3.0	183.0 (2997)	6	PGM-FI	Honda			
J30A1	3.0	183.0 (2997)	6	PGM-FI	Honda			
J30A4	3.0	183.0 (2997)	6	PGM-FI	Honda			

PGM-FI: Programmed Fuel Injection

09474_ACCO_C0001

GENERAL ENGINE SPECIFICATIONS

Year	Model	Engine Displacement Liters	Engine ID/VIN	Net Horsepower @ rpm	Net Torque @ rpm (ft. lbs.)	Bore X Stroke (in.)	Compression Ratio	Oil Pressure @ rpm
2002	Civic	1.7	D17A1	115@6100	110@4500	2.95X3.72	9.5:1	50@3000
	Civic	1.7	D17A2	127@6100	117@4500	2.95X3.72	9.9:1	50@3000
	Civic	1.7	D17A6	127@6100	117@4500	2.95X3.72	9.9:1	50@3000
	Civic	1.7	D17A7	NA	NA	2.95X3.72	12.5:1	50@3000
	Civic	2.0	K20A3	NA	NA	3.38x3.39	9.8:1	44@3000
	Accord	2.3	F23A1	150@5700	152@4900	3.39X3.82	9.3:1	50@3000
	Accord	2.3	F23A4	150@5700	152@4900	3.39X3.82	9.3:1	50@3000
	Accord	2.3	F23A5	150@5700	152@4900	3.39X3.82	9.3:1	50@3000
	Accord	3.0	J30A1	200@5500	195@4700	3.39X3.39	9.4:1	50@3000
	S2000	2.0	F20C1	240@8300	153@7500	3.43X3.31	11:01	85@3000
2003	Civic	1.7	D17A1	115@6100	110@4500	2.95X3.72	9.5:1	50@3000
	Civic	1.7	D17A2	127@6100	117@4500	2.95X3.72	9.9:1	50@3000
	Civic	1.7	D17A6	127@6100	117@4500	2.95X3.72	9.9:1	50@3000
	Civic	1.7	D17A7	NA	NA	2.95X3.72	12.5:1	50@3000
	Civic	2.0	K20A3	NA	NA	3.38x3.39	9.8:1	44@3000
	Accord	2.4	K24A4	160@5500	161@4500	3.42X3.89	9.7:1	50@3000
	Accord	3.0	J30A4	240@6250	212@5000	3.38X3.38	10.0:1	50@3000
	S2000	2.0	F20C1	240@8300	153@7500	3.43X3.31	11:01	85@3000
2004	Civic	1.7	D17A1	115@6100	110@4500	2.95X3.72	9.5:1	50@3000
	Civic	1.7	D17A2	127@6100	117@4500	2.95X3.72	9.9:1	50@3000
	Civic	1.7	D17A6	127@6100	117@4500	2.95X3.72	9.9:1	50@3000
	Civic	2.0	K20A3	NA	NA	3.38x3.39	9.8:1	44@3000
	Accord	2.4	K24A4	160@5500	161@4500	3.42X3.89	9.7:1	50@3000
	Accord	3.0	J30A4	240@6250	212@5000	3.38X3.38	10.0:1	50@3000
	S2000	2.2	F22C1	NA	NA	3.43X3.57	11:01	85@3000
2005	Civic	1.7	D17A1	115@6100	110@4500	2.95X3.72	9.5:1	50@3000
	Civic	1.7	D17A2	127@6100	117@4500	2.95X3.72	9.9:1	50@3000
	Civic	1.7	D17A6	127@6100	117@4500	2.95X3.72	9.9:1	50@3000
	Civic	2.0	K20A3	NA	NA	3.38x3.39	9.8:1	44@3000
	Accord	2.4	K24A4	160@5500	161@4500	3.42X3.89	9.7:1	50@3000
	Accord	3.0	J30A4	240@6250	212@5000	3.38X3.38	10.0:1	50@3000
	S2000	2.2	F22C1	NA	NA	3.43X3.57	11:01	85@3000
2006	Civic	1.8	R18A1	140@6300	128@4800	3.19x3.44	10.5:1	50@3000
	Civic	2.0	K20Z3	197@7800	139@6100	3.39x3.39	9.8:1	44@3000
	Accord	2.4	K24A8	166@5800	160@4000	3.43x3.90	9.7:1	44@3000
	Accord	3.0	J30A5	244@6250	211@5000	3.39x3.39	10.0:1	71@3000
	S2000	2.2	F22C1	237@7800	162@6800	3.43X3.57	11:01	85@3000

NA: Information not available

ENGINE TUNE-UP SPECIFICATIONS

Year	Engine Displacement Liters	Engine ID	Spark Plugs Gap (in.)	Ignition Timing (deg.) MT	Ignition Timing (deg.) AT	Fuel Pump (psi)	Idle Speed (rpm) MT	Idle Speed (rpm) AT	Valve Clearance In.	Valve Clearance Ex.
2002	1.7	D17A1	0.039-0.043	8B	8B	40-47	650-750	650-750	0.007-0.009	0.009-0.011
	1.7	D17A2	0.039-0.043	8B	8B	40-47	650-750	650-750	0.007-0.009	0.009-0.011
	1.7	D17A6	0.039-0.043	8B	8B	40-47	650-750	650-750	0.007-0.009	0.009-0.011
	1.7	D17A7	0.028-0.031	8B	8B	NA	650-750	650-750	0.007-0.009	0.009-0.011
	2.0	K20A3	0.039-0.043	8B	8B	47-52	650-800	650-800	0.008-0.010	0.011-0.013
	2.0	F20C1	0.039-0.043	5B	—	47-54	750-850	—	0.008-0.010	0.010-0.011
	2.3	F23A1	0.039-0.043	12B	12B	40-47	650-750	650-750	0.009-0.011	0.011-0.013
	2.3	F23A4	0.039-0.043	12B	12B	40-47	650-750	650-750	0.009-0.011	0.011-0.013
	2.3	F23A5	0.039-0.043	12B	12B	40-47	650-750	650-750	0.009-0.011	0.011-0.013
	3.0	J30A1	0.039-0.043	—	10B	41-48	700-800	740-840	0.008-0.009	0.011-0.013
2003	1.7	D17A1	0.039-0.043	8B	8B	40-47	650-750	650-750	0.007-0.009	0.009-0.011
	1.7	D17A2	0.039-0.043	8B	8B	40-47	650-750	650-750	0.007-0.009	0.009-0.011
	1.7	D17A6	0.039-0.043	8B	8B	40-47	650-750	650-750	0.007-0.009	0.009-0.011
	1.7	D17A7	0.028-0.031	8B	8B	NA	650-750	650-750	0.007-0.009	0.009-0.011
	2.0	K20A3	0.039-0.043	8B	8B	47-52	650-800	650-800	0.008-0.010	0.011-0.013
	2.0	F20C1	0.039-0.043	5B	—	47-54	750-850	—	0.008-0.010	0.010-0.011
	2.4	K24A4	①	8B	8B	②	650-750	750-850	0.008-0.010	0.011-0.013
	3.0	J30A4	0.039-0.043	—	10B	48-55	700-800	740-840	0.008-0.009	0.011-0.013
2004	1.7	D17A1	0.039-0.043	8B	8B	40-47	650-750	650-750	0.007-0.009	0.009-0.011
	1.7	D17A2	0.039-0.043	8B	8B	40-47	650-750	650-750	0.007-0.009	0.009-0.011
	1.7	D17A6	0.039-0.043	8B	8B	40-47	650-750	650-750	0.007-0.009	0.009-0.011
	2.0	K20A3	0.039-0.043	8B	8B	47-52	650-800	650-800	0.008-0.010	0.011-0.013
	2.2	F22C1	0.039-0.043	5B	—	40-47	750-850	—	0.008-0.010	0.010-0.011
	2.4	K24A4	①	8B	8B	②	650-750	750-850	0.008-0.010	0.011-0.013

ENGINE TUNE-UP SPECIFICATIONS

Year	Engine Displacement Liters	Engine ID	Spark Plugs Gap (in.)	Ignition Timing (deg.) MT	AT	Fuel Pump (psi)	Idle Speed (rpm) MT	AT	Valve Clearance In.	Ex.
2004 cont.	3.0	J30A4	0.039-0.043	—	10B	48-55	700-800	740-840	0.008-0.009	0.011-0.013
2005	1.7	D17A1	0.039-0.043	8B	8B	40-47	650-750	650-750	0.007-0.009	0.009-0.011
	1.7	D17A2	0.039-0.043	8B	8B	40-47	650-750	650-750	0.007-0.009	0.009-0.011
	1.7	D17A6	0.039-0.043	8B	8B	40-47	650-750	650-750	0.007-0.009	0.009-0.011
	2.0	K20A3	0.039-0.043	8B	8B	47-52	650-800	650-800	0.008-0.010	0.011-0.013
	2.2	F22C1	0.039-0.043	5B	—	40-47	750-850	—	0.008-0.010	0.010-0.011
	2.4	K24A4	①	8B	8B	②	620-770	620-770	0.008-0.010	0.011-0.013
	3.0	J30A4	0.039-0.043	—	10B	55-63	700-800	740-840	0.008-0.009	0.011-0.013
2006	1.8	R18A1	0.038-0.043	8B	8B	55-63	620-720	620-720	0.007-0.009	0.009-0.011
	2.0	K20Z3	0.039-0.043	8B	8B	48-55	700-800	700-800	④	⑤
	1.7	D17A6	0.039-0.043	8B	8B	40-47	650-750	650-750	0.007-0.009	0.009-0.011
	2.2	F22C1	0.039-0.043	5B	—	40-47	850-950	—	0.008-0.010	0.010-0.011
	2.4	K24A8	①	8B	8B	③	670-770	750-850	0.008-0.010	0.011-0.013
	3.0	J30A5	0.039-0.043	10B	10B	55-63	700-800	740-840	0.008-0.009	0.011-0.013

NOTE: The Vehicle Emission Control Information label often reflects specification changes made during production.

The label figures must be used if they differ from those in this chart

B: Before Top Dead Center

NA: Information not available

① Except SULEV: 0.039-0.043
   SULEV: 0.047-0.051

② Except SULEV: 48-55 psi
   SULEV: 47-54

③ Except SULEV: 47-55 psi
   SULEV: 48-55 psi

④ 0.009 +/- 0.0008

⑤ 0.010 +/- 0.0008

09474_ACCO_C0004

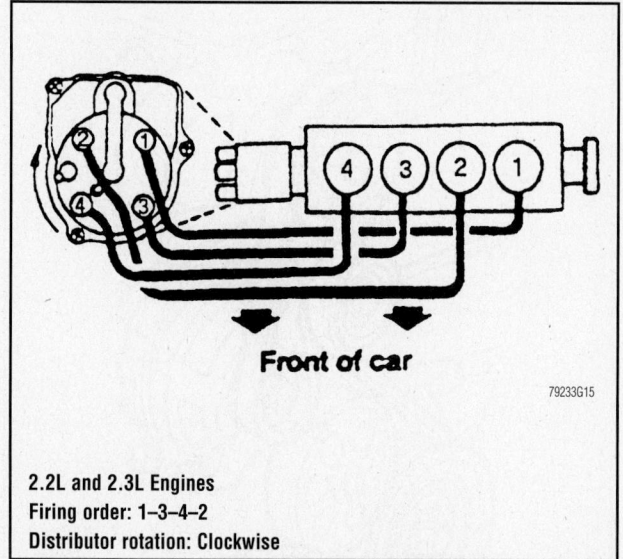

2.2L and 2.3L Engines
Firing order: 1–3–4–2
Distributor rotation: Clockwise

79233G15

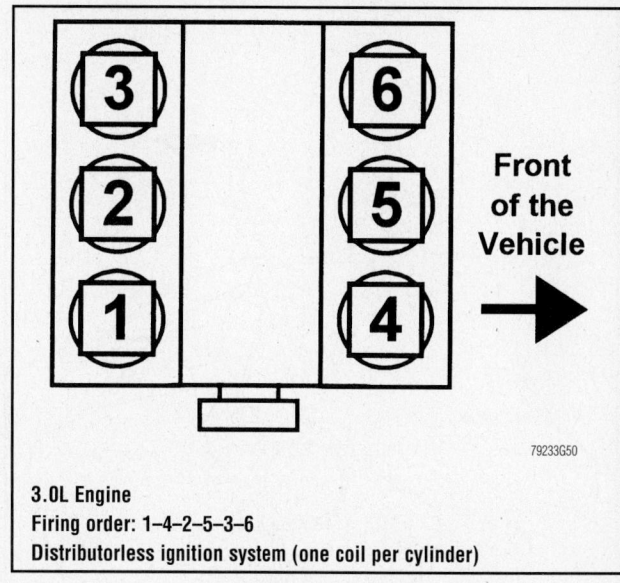

Front of the Vehicle →

3.0L Engine
Firing order: 1–4–2–5–3–6
Distributorless ignition system (one coil per cylinder)

79233G50

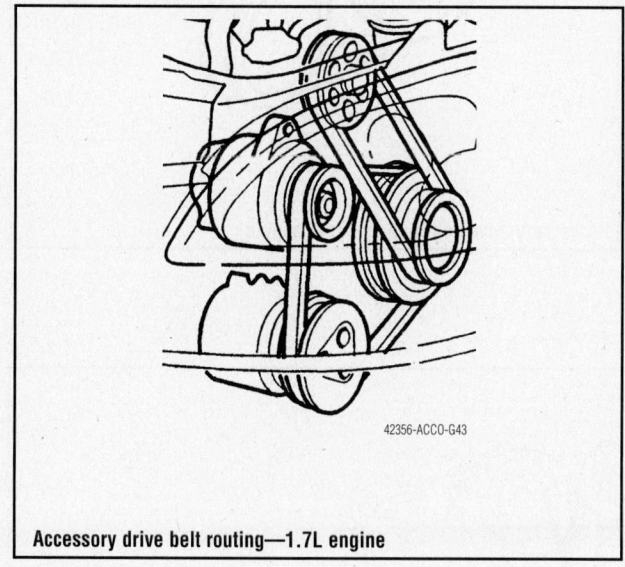

42356-ACCO-G43

Accessory drive belt routing—1.7L engine

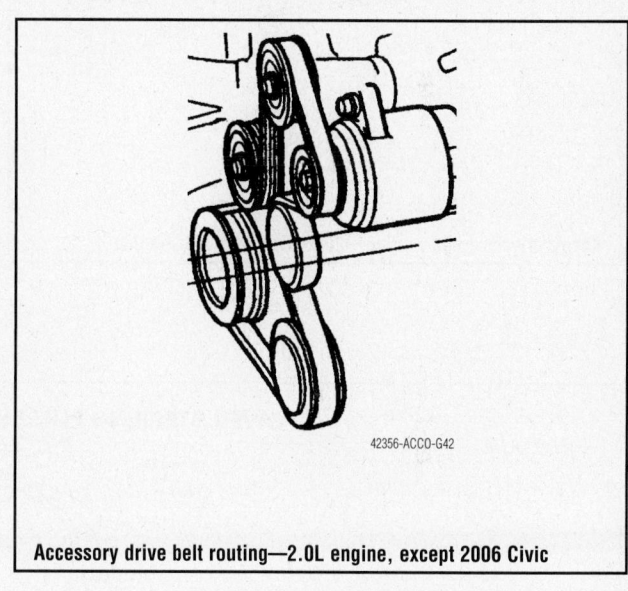

42356-ACCO-G42

Accessory drive belt routing—2.0L engine, except 2006 Civic

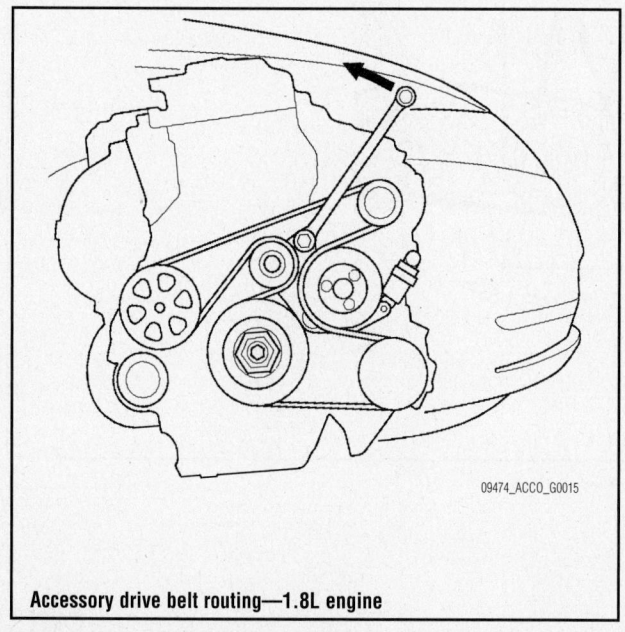

09474_ACCO_G0015

Accessory drive belt routing—1.8L engine

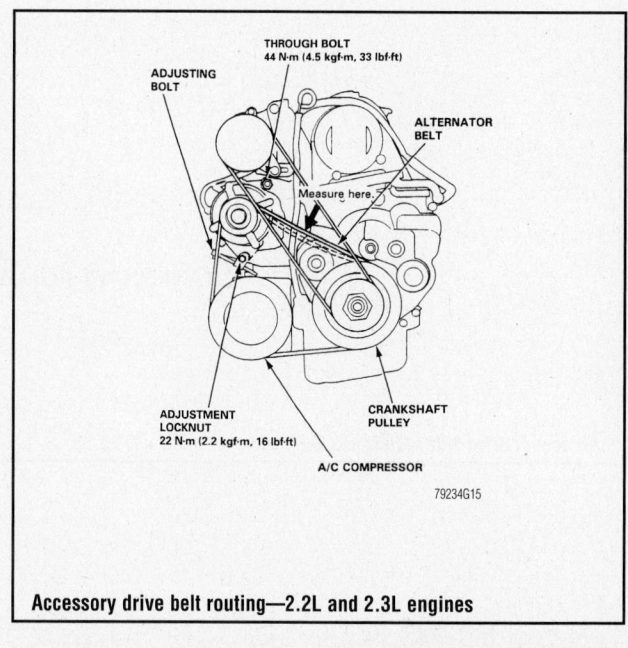

THROUGH BOLT
44 N·m (4.5 kgf·m, 33 lbf·ft)

ADJUSTING BOLT

ALTERNATOR BELT

Measure here.

ADJUSTMENT LOCKNUT
22 N·m (2.2 kgf·m, 16 lbf·ft)

CRANKSHAFT PULLEY

A/C COMPRESSOR

79234G15

Accessory drive belt routing—2.2L and 2.3L engines

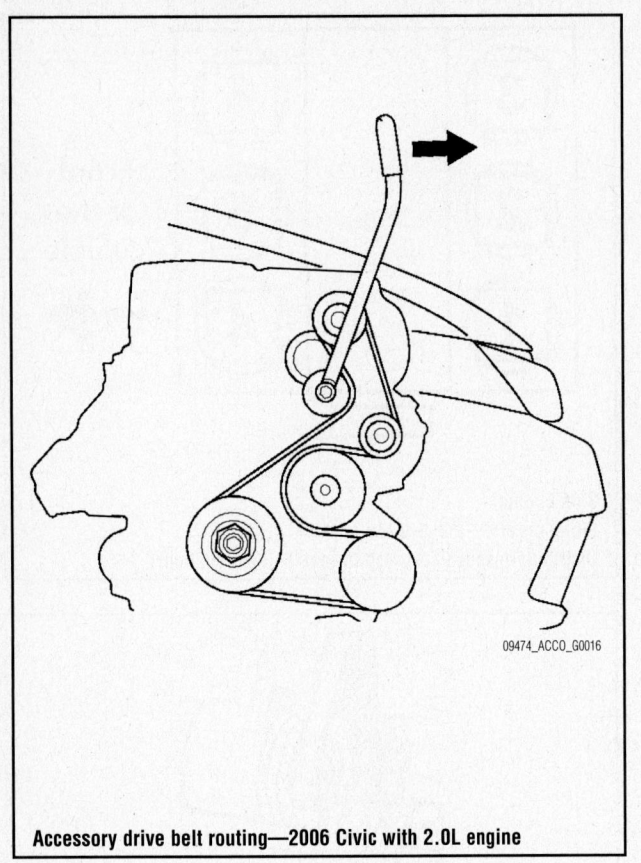

Accessory drive belt routing—2006 Civic with 2.0L engine

09474_ACCO_G0016

Accessory drive belt routing—2.4L engine

42356-ACCO-G41

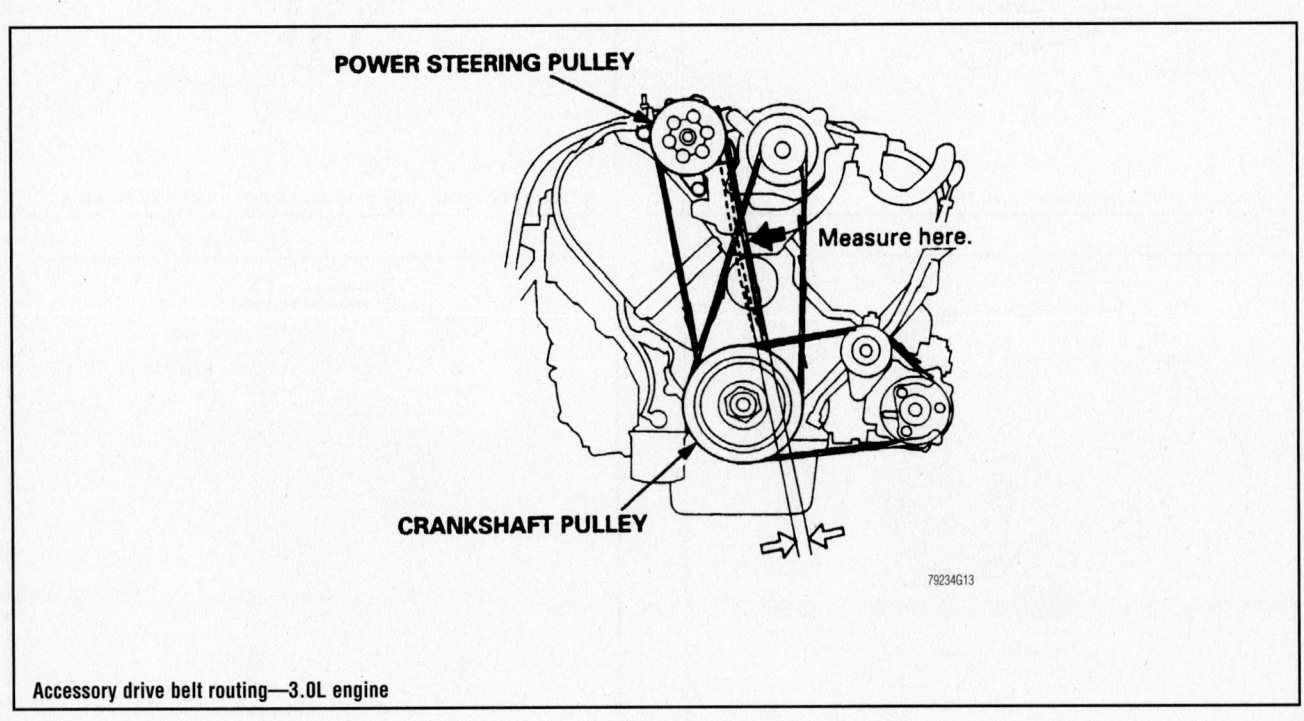

POWER STEERING PULLEY

Measure here.

CRANKSHAFT PULLEY

79234G13

Accessory drive belt routing—3.0L engine

CAPACITIES

Year	Model	Engine Displacement Liters	Engine ID	Engine Oil with Filter	Transmission (pts.) 5-Spd	Transmission (pts.) Auto.	Drive Axle Front (pts.)	Drive Axle Rear (pts.)	Fuel Tank (gal.)	Cooling System (qts.)
2002	Accord	2.3	F23A1	4.0	4.0	5.0	—	—	17.0	①
	Accord	2.3	F23A4	4.5	4.0	5.0	—	—	17.0	①
	Accord	2.3	F23A5	4.5	4.0	5.2	—	—	17.0	①
	Accord	3.0	J30A1	4.6	—	6.2	—	—	17.1	5.9
	Civic	1.7	D17A1	3.7	2.5	5.8	—	—	13.2	4.1
	Civic	1.7	D17A2	3.7	2.5	5.8	—	—	13.2	4.1
	Civic	1.7	D17A6	3.7	2.5	5.8	—	—	13.2	4.1
	Civic	1.7	D17A7	3.7	—	6.8	—	—	NA	4.1
	Civic	2.0	K20A3	4.4	2.5	—	—	—	13.2	4.1
	S2000	2.0	F20C1	5.1	③	—	—	—	13.2	6.9
2003	Accord	2.4	K24A4	4.4	4.0	6.0	—	—	17.1	②
	Accord	3.0	J30A4	4.5	④	6.0	—	—	17.1	7.1
	Civic	1.7	D17A1	3.4	2.5	5.8	—	—	13.2	4.1
	Civic	1.7	D17A2	3.7	2.5	5.8	—	—	13.2	4.1
	Civic	1.7	D17A6	3.7	2.5	5.8	—	—	13.2	4.1
	Civic	1.7	D17A7	3.7	—	6.8	—	—	NA	4.1
	Civic	2.0	K20A3	4.4	2.5	—	—	—	13.2	4.1
	S2000	2.0	F20C1	5.1	③	—	—	—	13.2	6.9
2004	Accord	2.4	K24A4	4.4	4.0	6.0	—	—	17.1	②
	Accord	3.0	J30A4	4.5	④	6.0	—	—	17.1	7.1
	Civic	1.7	D17A1	3.4	2.5	5.8	—	—	13.2	4.1
	Civic	1.7	D17A2	3.7	2.5	5.8	—	—	13.2	4.1
	Civic	1.7	D17A6	3.7	2.5	5.8	—	—	13.2	4.1
	Civic	2.0	K20A3	4.4	2.5	—	—	—	13.2	4.1
	S2000	2.2	F22C1	5.1	③	—	—	—	13.2	6.9
2005	Accord	2.4	K24A4	4.4	4.0	6.0	—	—	17.1	②
	Accord	3.0	J30A4	4.5	④	6.0	—	—	17.1	7.1
	Civic	1.7	D17A1	3.4	2.5	5.8	—	—	13.2	4.1
	Civic	1.7	D17A2	3.7	2.5	5.8	—	—	13.2	4.1
	Civic	1.7	D17A6	3.7	2.5	5.8	—	—	13.2	4.1
	Civic	2.0	K20A3	4.4	2.5	—	—	—	13.2	4.1
	S2000	2.2	F22C1	5.1	③	—	—	—	13.2	6.9
2006	Accord	2.4	K24A8	4.4	4.0	6.0	—	—	17.1	②
	Accord	3.0	J30A5	4.5	④	6.0	—	—	17.1	7.1
	Civic	1.8	R18A1	4.0	3.0	5.0	—	—	13.2	⑤
	Civic	2.0	K20Z3	4.5	③	—	—	—	13.2	4.1
	S2000	2.2	F22C1	5.1	③	—	—	—	13.2	6.9

NOTE: All capacities are approximate. Add fluid gradually and ensure a proper fluid level is obtained.

NOTE: Capacities given are service, not overhaul capacities

NA: Information not available

① Automatic Transaxle: 5.7
 Manual Transaxle: 5.8

② Automatic Transaxle: 5.3
 Manual Transaxle: 5.4

③ 6-speed 3.1 pts.

④ 6-speed 4.2 pts.

⑤ Manual Transaxle 4 door: 5.2
 Automatic transaxle 4 door: 5.8
 Manual transaxle 2 door: 5.8
 Automatic transaxle 4 door: 5.9

VALVE SPECIFICATIONS

Year	Engine Displacement Liters	Engine ID	Seat Angle (deg.)	Face Angle (deg.)	Spring Test Pressure (lbs. @ in.)	Spring Installed Height (in.)	Stem-to-Guide Clearance (in.)		Stem Diameter (in.)	
							Intake	Exhaust	Intake	Exhaust
2002	1.7	D17A1	45	45	NA	NA	0.0008-0.0020	0.0020-0.0031	0.2157-0.2161	0.2146-0.2150
	1.7	D17A2	45	45	NA	NA	0.0008-0.0020	0.0020-0.0031	0.2157-0.2161	0.2146-0.2150
	1.7	D17A6	45	45	NA	NA	0.0008-0.0020	0.0020-0.0031	0.2157-0.2161	0.2146-0.2150
	1.7	D17A7	45	45	NA	NA	0.0008-0.0020	0.0020-0.0031	0.2157-0.2161	0.2146-0.2150
	2.0	K20A3	45	45	NA	NA	0.0012-0.0022	0.0022-0.0031	0.2156-0.2159	0.2146-0.2150
	2.0	F20C1	45	45	NA	NA	0.0010-0.0020	0.0020-0.0030	0.2157-0.2162	0.2146-0.2150
	2.3	F23A1	45	45	NA	NA	0.0008-0.0018	0.0022-0.0031	0.2159-0.2163	0.2146-0.2150
	2.3	F23A4	45	45	NA	NA	0.0008-0.0018	0.0022-0.0031	0.2159-0.2163	0.2146-0.2150
	2.3	F23A5	45	45	NA	NA	0.0008-0.0018	0.0022-0.0031	0.2159-0.2163	0.2146-0.2150
	3.0	J30A1	45	45	NA	NA	0.0008-0.0018	0.0022-0.0031	0.2159-0.2163	0.2146-0.2150
2003	1.7	D17A1	45	45	NA	NA	0.0008-0.0020	0.0020-0.0031	0.2157-0.2161	0.2146-0.2150
	1.7	D17A2	45	45	NA	NA	0.0008-0.0020	0.0020-0.0031	0.2157-0.2161	0.2146-0.2150
	1.7	D17A6	45	45	NA	NA	0.0008-0.0020	0.0020-0.0031	0.2157-0.2161	0.2146-0.2150
	1.7	D17A7	45	45	NA	NA	0.0008-0.0020	0.0020-0.0031	0.2157-0.2161	0.2146-0.2150
	2.0	K20A3	45	45	NA	NA	0.0012-0.0022	0.0022-0.0031	0.2156-0.2159	0.2146-0.2150
	2.0	F20C1	45	45	NA	NA	0.0010-0.0020	0.0020-0.0030	0.2157-0.2162	0.2146-0.2150
	2.4	K24A4	45	45	NA	NA	0.0012-0.0022	0.0022-0.0031	0.2156-0.2159	0.2146-0.2150
	3.0	J30A4	45	45	NA	NA	0.0008-0.0018	0.0022-0.0031	0.2159-0.2163	0.2146-0.2150
2004	1.7	D17A1	45	45	NA	NA	0.0008-0.0020	0.0020-0.0031	0.2157-0.2161	0.2146-0.2150
	1.7	D17A2	45	45	NA	NA	0.0008-0.0020	0.0020-0.0031	0.2157-0.2161	0.2146-0.2150
	1.7	D17A6	45	45	NA	NA	0.0008-0.0020	0.0020-0.0031	0.2157-0.2161	0.2146-0.2150
	2.0	K20A3	45	45	NA	NA	0.0012-0.0022	0.0022-0.0031	0.2156-0.2159	0.2146-0.2150
	2.2	F22C1	45	45	NA	NA	0.0010-0.0020	0.0020-0.0030	0.2157-0.2162	0.2146-0.2150
	2.4	K24A4	45	45	NA	NA	0.0012-0.0022	0.0022-0.0031	0.2156-0.2159	0.2146-0.2150

VALVE SPECIFICATIONS

Year	Engine Displacement Liters	Engine ID	Seat Angle (deg.)	Face Angle (deg.)	Spring Test Pressure (lbs. @ in.)	Spring Installed Height (in.)	Stem-to-Guide Clearance (in.)		Stem Diameter (in.)	
							Intake	Exhaust	Intake	Exhaust
2004 cont.	3.0	J30A4	45	45	NA	NA	0.0008-0.0018	0.0022-0.0031	0.2159-0.2163	0.2146-0.2150
2005	1.7	D17A1	45	45	NA	NA	0.0008-0.0020	0.0020-0.0031	0.2157-0.2161	0.2146-0.2150
	1.7	D17A2	45	45	NA	NA	0.0008-0.0020	0.0020-0.0031	0.2157-0.2161	0.2146-0.2150
	1.7	D17A6	45	45	NA	NA	0.0008-0.0020	0.0020-0.0031	0.2157-0.2161	0.2146-0.2150
	2.0	K20A3	45	45	NA	NA	0.0012-0.0022	0.0022-0.0031	0.2156-0.2159	0.2146-0.2150
	2.2	F22C1	45	45	NA	NA	0.0010-0.0020	0.0020-0.0030	0.2157-0.2162	0.2146-0.2150
	2.4	K24A4	45	45	NA	NA	0.0012-0.0022	0.0022-0.0031	0.2156-0.2159	0.2146-0.2150
	3.0	J30A4	45	45	NA	NA	0.0008-0.0018	0.0022-0.0031	0.2159-0.2163	0.2146-0.2150
2006	1.8	D17A1	45	45	NA	NA	0.0008-0.0020	0.0020-0.0031	0.2157-0.2161	0.2146-0.2150
	2.0	D17A2	45	45	NA	NA	0.0012-0.0022	0.0022-0.0031	0.2156-0.2159	0.2146-0.2150
	2.2	F22C1	45	45	NA	NA	0.0010-0.0020	0.0020-0.0030	0.2157-0.2162	0.2146-0.2150
	2.4	K24A8	45	45	NA	NA	0.0012-0.0022	0.0022-0.0031	0.2156-0.2159	0.2146-0.2150
	3.0	J30A5	45	45	NA	NA	0.0008-0.0018	0.0022-0.0031	0.2159-0.2163	0.2146-0.2150

NA: Information not available

09474_ACCO_C0007

CRANKSHAFT AND CONNECTING ROD SPECIFICATIONS

All measurements are given in inches.

Year	Engine Displacement Liters	Engine ID	Crankshaft				Connecting Rod		
			Main Brg. Journal Dia.	Main Brg. Oil Clearance	Shaft End-play	Thrust on No.	Journal Diameter	Oil Clearance	Side Clearance
2002	1.7	D17A1	2.1644-2.1654	①	0.004-0.014	4	1.7707-1.7717	0.0009-0.0017	0.0006-0.0016
	1.7	D17A2	2.1644-2.1654	①	0.004-0.014	4	1.7707-1.7717	0.0009-0.0017	0.0006-0.0016
	1.7	D17A6	2.1644-2.1654	①	0.004-0.014	4	1.7707-1.7717	0.0009-0.0017	0.0006-0.0016
	1.7	D17A7	2.1644-2.1654	①	0.004-0.014	4	1.7707-1.7717	0.0009-0.0017	0.0006-0.0016
	2.0	K20A3	②	③	0.004-0.014	4	1.7707-1.7717	0.0008-0.0019	0.0006-0.0016
	2.0	F20C1	2.1644-2.1654	0.0007-0.0016	0.004-0.014	4	1.8888-1.8898	0.0012-0.0021	0.0006-0.0016
	2.3	F23A1	④	⑤	0.004-0.018	4	1.7707-1.7717	0.0008-0.0024	0.0006-0.0016
	2.3	F23A4	④	⑤	0.004-0.018	4	1.7707-1.7717	0.0008-0.0024	0.0006-0.0018
	2.3	F23A5	④	⑤	0.004-0.018	4	1.7707-1.7717	0.0008-0.0024	0.0006-0.0016
	3.0	J30A1	2.8337-2.8346	0.0008-0.0020	0.004-0.018	3	2.0857-2.0866	0.0008-0.0020	0.0006-0.0018
2003	1.7	D17A1	2.1644-2.1654	①	0.004-0.014	4	1.7707-1.7717	0.0009-0.0017	0.0006-0.0016
	1.7	D17A2	2.1644-2.1654	①	0.004-0.014	4	1.7707-1.7717	0.0009-0.0017	0.0006-0.0016
	1.7	D17A6	2.1644-2.1654	①	0.004-0.014	4	1.7707-1.7717	0.0009-0.0017	0.0006-0.0016
	1.7	D17A7	2.1644-2.1654	①	0.004-0.014	4	1.7707-1.7717	0.0009-0.0017	0.0006-0.0016
	2.0	K20A3	②	③	0.004-0.014	4	1.7707-1.7717	0.0008-0.0019	0.0006-0.0016
	2.0	F20C1	2.1644-2.1654	0.0007-0.0016	0.004-0.014	4	1.8888-1.8898	0.0012-0.0021	0.0006-0.0016
	2.4	K24A4	②	③	0.004-0.014	4	1.8888-1.8898	0.0002-0.0006	0.0006-0.0016
	3.0	J30A4	2.8337-2.8346	0.0008-0.0017	0.004-0.014	3	2.0857-2.0866	0.0008-0.0017	0.0006-0.0018
2004	1.7	D17A1	2.1644-2.1654	①	0.004-0.014	4	1.7707-1.7717	0.0009-0.0017	0.0006-0.0016
	1.7	D17A2	2.1644-2.1654	①	0.004-0.014	4	1.7707-1.7717	0.0009-0.0017	0.0006-0.0016
	1.7	D17A6	2.1644-2.1654	①	0.004-0.014	4	1.7707-1.7717	0.0009-0.0017	0.0006-0.0016
	2.0	K20A3	②	③	0.004-0.014	4	1.7707-1.7717	0.0008-0.0019	0.0006-0.0016
	2.2	F22C1	2.1644-2.1654	0.0007-0.0016	0.004-0.014	4	1.8888-1.8898	0.0012-0.0021	0.0006-0.0016
	2.4	K24A4	②	③	0.004-0.014	4	1.8888-1.8898	0.0002-0.0006	0.0006-0.0016

CRANKSHAFT AND CONNECTING ROD SPECIFICATIONS

All measurements are given in inches.

Year	Engine Displacement Liters	Engine ID	Crankshaft				Connecting Rod		
			Main Brg. Journal Dia.	Main Brg. Oil Clearance	Shaft End-play	Thrust on No.	Journal Diameter	Oil Clearance	Side Clearance
2004 cont.	3.0	J30A4	2.8337-2.8346	0.0008-0.0017	0.004-0.014	3	2.0857-2.0866	0.0008-0.0017	0.0006-0.0018
2005	1.7	D17A1	2.1644-2.1654	①	0.004-0.014	4	1.7707-1.7717	0.0009-0.0017	0.0006-0.0016
	1.7	D17A2	2.1644-2.1654	①	0.004-0.014	4	1.7707-1.7717	0.0009-0.0017	0.0006-0.0016
	1.7	D17A6	2.1644-2.1654	①	0.004-0.014	4	1.7707-1.7717	0.0009-0.0017	0.0006-0.0016
	2.0	K20A3	②	③	0.004-0.014	4	1.7707-1.7717	0.0008-0.0019	0.0006-0.0016
	2.2	F22C1	2.1644-2.1654	0.0007-0.0016	0.004-0.014	4	1.8888-1.8898	0.0012-0.0021	0.0006-0.0016
	2.4	K24A4	②	③	0.004-0.014	4	1.8888-1.8898	0.0002-0.0006	0.0006-0.0016
	3.0	J30A4	2.8337-2.8346	0.0008-0.0017	0.004-0.014	3	2.0857-2.0866	0.0008-0.0017	0.0006-0.0018
2006	1.8	R18A1	2.1644-2.1654	0.0007-0.0014	0.004-0.014	4	1.7707-1.7716	0.0009-0.0017	0.0006-0.0018
	2.0	K20Z3	②	③	0.004-0.014	4	1.7707-1.7717	0.0013-0.0026	0.0006-0.0016
	1.7	D17A6	2.1644-2.1654	①	0.004-0.014	4	1.7707-1.7717	0.0009-0.0017	0.0006-0.0016
	2.2	F22C1	2.1644-2.1654	0.0007-0.0016	0.004-0.014	4	1.8888-1.8898	0.0012-0.0021	0.0006-0.0016
	2.4	K24A8	②	③	0.004-0.014	4	1.8888-1.8898	0.0002-0.0006	0.0006-0.0016
	3.0	J30A5	2.8337-2.8346	0.0008-0.0017	0.004-0.014	3	2.0857-2.0866	0.0008-0.0017	0.0006-0.0018

① Journals 1 and 5: 0.0007-0.0014
  Journals 2, 3 and 4: 0.0009 - 0.0017

② Journals 1, 2 4 and 5: 2.1648-2.1657
  Journal 3: 2.1644-2.1654

③ Journals 1, 2 4 and 5: 0.0007-0.0016
  Journal 3: 0.0010-0.0019

④ Journals 1, 2 and 4: 2.1646 - 2.1655
  Journal 3: 2.1644 - 2.1654
  Journal 5: 2.1650 - 2.1660

⑤ Journals 1, 2 and 4: 0.0008 - 0.0020
  Journal 3: 0.0010 - 0.0022
  Journal 5: 0.0004 - 0.0016

PISTON AND RING SPECIFICATIONS

All measurements are given in inches.

| Year | Engine Displacement Liters | Engine ID | Piston Clearance | Ring Gap | | | Ring Side Clearance | | |
				Top Compression	Bottom Compression	Oil Control	Top Compression	Bottom Compression	Oil Control
2002	1.7	D17A1	0.0004-0.0016	0.006-0.024	0.012-0.024	0.008-0.031	0.0014-0.0024	0.0012-0.0022	NA
	1.7	D17A2	0.0004-0.0016	0.006-0.024	0.012-0.024	0.008-0.031	0.0014-0.0024	0.0012-0.0022	NA
	1.7	D17A6	0.0004-0.0016	0.006-0.024	0.012-0.024	0.008-0.031	0.0014-0.0024	0.0012-0.0022	NA
	1.7	D17A7	0.0004-0.0016	0.006-0.024	0.012-0.024	0.008-0.031	0.0014-0.0024	0.0012-0.0022	NA
	2.0	K20A3	0.0008-0.0016	0.008-0.024	0.016-0.022	①	0.0014-0.0024	②	NA
	2.0	F20C1	0.0002-0.0011	0.010-0.024	0.024-0.035	0.008-0.031	0.0018-0.0035	0.0016-0.0028	NA
	2.3	F23A1	0.0008-0.0020	0.008-0.024	0.016-0.028	0.008-0.031	0.0014-0.0050	0.0012-0.0050	NA
	2.3	F23A4	0.0008-0.0020	0.008-0.024	0.016-0.028	0.008-0.031	0.0014-0.0050	0.0012-0.0050	NA
	2.3	F23A5	0.0008-0.0020	0.008-0.024	0.016-0.028	0.008-0.031	0.0014-0.0050	0.0012-0.0050	NA
	3.0	J30A1	0.0006-0.0030	0.008-0.024	0.016-0.028	0.008-0.031	0.0014-0.0050	0.0012-0.0050	NA
2003	1.7	D17A1	0.0004-0.0016	0.006-0.024	0.012-0.024	0.008-0.031	0.0014-0.0024	0.0012-0.0022	NA
	1.7	D17A2	0.0004-0.0016	0.006-0.024	0.012-0.024	0.008-0.031	0.0014-0.0024	0.0012-0.0022	NA
	1.7	D17A6	0.0004-0.0016	0.006-0.024	0.012-0.024	0.008-0.031	0.0014-0.0024	0.0012-0.0022	NA
	1.7	D17A7	0.0004-0.0016	0.006-0.024	0.012-0.024	0.008-0.031	0.0014-0.0024	0.0012-0.0022	NA
	2.0	K20A3	0.0008-0.0016	0.008-0.024	0.016-0.022	①	0.0014-0.0024	②	NA
	2.0	F20C1	0.0002-0.0011	0.010-0.024	0.024-0.035	0.008-0.031	0.0018-0.0035	0.0016-0.0028	NA
	2.4	K24A4	0.0008-0.0016	0.008-0.0024	0.016-0.028	0.008-0.031	0.0018-0.0028	0.0020-0.0030	NA
	3.0	J30A4	0.0006-0.0030	0.008-0.024	0.016-0.028	0.008-0.031	0.0022-0.0060	0.0012-0.0050	NA
2004	1.7	D17A1	0.0004-0.0016	0.006-0.024	0.012-0.024	0.008-0.031	0.0014-0.0024	0.0012-0.0022	NA
	1.7	D17A2	0.0004-0.0016	0.006-0.024	0.012-0.024	0.008-0.031	0.0014-0.0024	0.0012-0.0022	NA
	1.7	D17A6	0.0004-0.0016	0.006-0.024	0.012-0.024	0.008-0.031	0.0014-0.0024	0.0012-0.0022	NA
	2.0	K20A3	0.0008-0.0016	0.008-0.024	0.016-0.022	①	0.0014-0.0024	②	NA
	2.2	F22C1	0.0002-0.0011	0.010-0.024	0.024-0.035	0.008-0.031	0.0018-0.0035	0.0016-0.0028	NA
	2.4	K24A4	0.0008-0.0016	0.008-0.0024	0.016-0.028	0.008-0.031	0.0018-0.0028	0.0020-0.0030	NA

PISTON AND RING SPECIFICATIONS
All measurements are given in inches.

Year	Engine Displacement Liters	Engine ID	Piston Clearance	Ring Gap			Ring Side Clearance		
				Top Compression	Bottom Compression	Oil Control	Top Compression	Bottom Compression	Oil Control
2004 cont.	3.0	J30A4	0.0006-0.0030	0.008-0.024	0.016-0.028	0.008-0.031	0.0022-0.0060	0.0012-0.0050	NA
2005	1.7	D17A1	0.0004-0.0016	0.006-0.024	0.012-0.024	0.008-0.031	0.0014-0.0024	0.0012-0.0022	NA
	1.7	D17A2	0.0004-0.0016	0.006-0.024	0.012-0.024	0.008-0.031	0.0014-0.0024	0.0012-0.0022	NA
	1.7	D17A6	0.0004-0.0016	0.006-0.024	0.012-0.024	0.008-0.031	0.0014-0.0024	0.0012-0.0022	NA
	2.0	K20A3	0.0008-0.0016	0.008-0.024	0.016-0.022	①	0.0014-0.0024	②	NA
	2.2	F22C1	0.0002-0.0011	0.010-0.024	0.024-0.035	0.008-0.031	0.0018-0.0035	0.0016-0.0028	NA
	2.4	K24A4	0.0008-0.0016	0.008-0.0024	0.016-0.028	0.008-0.031	0.0018-0.0028	0.0020-0.0030	NA
	3.0	J30A4	0.0006-0.0030	0.008-0.024	0.016-0.028	0.008-0.031	0.0022-0.0060	0.0012-0.0050	NA
2006	1.8	R18A1	0.0004-0.0016	0.008-0.024	0.016-0.028	0.008-0.031	0.0018-0.0028	0.0014-0.0024	NA
	2.0	K20Z3	0.0008-0.0016	0.008-0.0024	0.020-0.030	0.008-0.031	0.0018-0.0028	0.0016-0.0026	NA
	2.2	F22C1	0.0002-0.0011	0.010-0.024	0.024-0.035	0.008-0.031	0.0018-0.0035	0.0016-0.0028	NA
	2.4	K24A8	0.0008-0.0016	0.008-0.0024	0.016-0.028	0.008-0.031	0.0018-0.0028	0.0020-0.0030	NA
	3.0	J30A5	0.0006-0.0030	0.008-0.024	0.016-0.028	0.008-0.031	0.0022-0.0060	0.0012-0.0050	NA

NA: Information not available

① Reken: 0.010-0.030
Federal Mogul: 0.008-0.031

② Reken: 0.012-0.0022
Federal Mogul: 0.0010-0.0024

TORQUE SPECIFICATIONS
All readings in ft. lbs.

Year	Engine Displacement Liters	Engine ID	Cylinder Head Bolts	Main Bearing Bolts	Rod Bearing Bolts	Crankshaft Damper Bolts	Flywheel Bolts	Manifold Intake	Manifold Exhaust	Spark Plugs	Oil Pan Drain Plug
2002	1.7	D17A1	①	②	24	⑨	⑭	16	23	13	33
	1.7	D17A2	①	②	24	⑨	⑭	16	23	13	29
	1.7	D17A6	①	②	24	⑨	⑭	16	23	13	29
	1.7	D17A7	①	②	24	⑨	⑭	16	23	13	29
	2.0	K20A3	⑪	⑫	⑮	⑬	76	16	23	13	33
	2.0	F20C1	④	⑤	⑥	192	94	16	23	13	29
	2.3	F23A1	④	⑧	⑨	181	③	16	23	13	33
	2.3	F23A4	④	⑧	⑨	181	③	16	23	13	33
	2.3	F23A5	④	⑧	⑨	181	③	16	23	13	33
	3.0	J30A1	⑦	⑩	⑨	181	③	16	23	13	29
2003	1.7	D17A1	①	②	24	③	⑭	16	23	13	33
	1.7	D17A2	①	②	24	⑨	⑭	16	23	13	29
	1.7	D17A6	①	②	24	⑨	⑭	16	23	13	29
	1.7	D17A7	①	②	24	⑨	⑭	16	23	13	29
	2.0	K20A3	⑪	⑫	⑮	⑬	76	16	23	13	33
	2.0	F20C1	④	⑤	⑥	192	94	16	23	13	29
	2.4	K24A4	⑪	⑫	⑮	⑬	③	16	33	13	33
	3.0	J30A4	⑦	⑩	⑨	181	③	16	23	13	29
2004	1.7	D17A1	①	②	24	③	⑭	16	23	13	33
	1.7	D17A2	①	②	24	⑨	⑭	16	23	13	29
	1.7	D17A6	①	②	24	⑨	⑭	16	23	13	29
	2.0	K20A3	⑪	⑫	⑮	⑬	76	16	23	13	33
	2.2	F22C1	④	⑤	⑥	192	94	16	23	13	29
	2.4	K24A4	⑪	⑫	⑮	⑬	③	16	33	13	33
	3.0	J30A4	⑦	⑩	⑨	181	③	16	23	13	29
2005	1.7	D17A1	①	②	24	③	⑭	16	23	13	33
	1.7	D17A2	①	②	24	⑨	⑭	16	23	13	29
	1.7	D17A6	①	②	24	⑨	⑭	16	23	13	29
	2.0	K20A3	⑪	⑫	⑮	⑯	76	16	23	13	33
	2.2	F22C1	④	⑤	⑥	192	94	16	23	13	29
	2.4	K24A4	⑪	⑫	⑮	⑬	③	16	33	13	33
	3.0	J30A4	⑦	⑩	⑨	181	③	16	23	13	29
2006	1.8	R18A1	⑰	⑲	⑮	⑯	③	17	23	13	30
	2.0	K20Z3	⑰	⑫	⑱	⑬	③	16	23	13	30
	2.2	F22C1	④	⑤	⑥	192	94	16	23	13	29
	2.4	K24A8	⑪	⑫	⑮	⑬	③	16	33	13	33
	3.0	J30A5	⑦	⑩	⑨	181	③	16	23	13	29

① Step 1: 14 ft. lbs.
   Step 2: 36 ft. lbs.
   Step 3: 49 ft. lbs.
② Step 1: 18 ft. lbs.
   Step 2: 38 ft. lbs.
③ Automatic transaxle: 54 ft. lbs.
   Manual transaxle: 76 ft. lbs.

④ Step 1: 22 ft. lbs.
   Step 2: Rotate 90 degrees
   Step 3: Rotate 90 degrees
   Step 4: If new bolt rotate
   additional 90 degrees
⑤ Step 1: 22 ft. lbs.
   Step 2: Rotate 60 degrees
   Step 3: 8mm bolts to 16 ft. lbs.

⑥ Step 1: 18 ft. lbs.
   Step 2: Rotate 90 degrees
⑦ Step 1: 29 ft. lbs.
   Step 2: 51 ft. lbs.
   Step 3: 72.3 ft. lbs.
⑧ Step 1: 11mm bolts, 29 ft. lbs.
   Step 2: 11mm bolts, 58 ft. lbs.
   Step 3: 6mm bolts, 8.7 ft. lbs.

09474_ACCO_C0012

Torque Specifications Chart
Footnotes Continued

⑨ Step 1: 14.5 ft. lbs.

Step 2: Rotate 90 degrees

⑩ Step 1: Cap bolts, 56 ft. lbs.

Step 2: Side bolts, 36 ft. lbs.

⑪ Step 1: 29 ft. lbs.

Step 2: Rotate 90 degrees

Step 3: Rotate 90 degrees

Step 4: If new bolt rotate

additional 90 degrees

⑫ Step 1: 22 ft. lbs.

Step 2: Rotate 56 degrees

⑬ Step 1: 36 ft. lbs.

Step 2: Rotate 90 degrees

⑭ Automatic transaxle: 54 ft. lbs.

Manual transaxle: 87 ft. lbs.

⑮ Step 1: 14 ft. lbs.

Step 2: Rotate 90 degrees

⑯ Step 1: 51 ft. lbs.

Step 2: Rotate 90 degrees

⑰ Step 1: 29 ft. lbs.

Step 2: Rotate 90 degrees

Step 3: Rotate 90 degrees

Step 4: If new bolt rotate

additional 90 degrees

⑱ Step 1: 22 ft. lbs.

Step 2: Rotate 90 degrees

⑲ Step 1: 18 ft. lbs.

Step 2: Rotate 57 degrees

09474_ACCO_C0013

WHEEL ALIGNMENT

Year	Model		Caster Range (+/-Deg.)	Caster Preferred Setting (Deg.)	Camber Range (+/-Deg.)	Camber Preferred Setting (Deg.)	Toe-in (in.)
2002	Accord	F	1.00	+2.80	1.00	+0.06	0 +/- 0.03
		R	—	—	0.50	-0.50	0.03 +/- 0.03
	Civic-exc Hatchback	F	—	+1.33	—	0 00'	0
		R	—	—	—	-0 45'	1/16"
	Civic- Hatchback	F	—	1 03'	—	0 00' +/- 1 45'	0 +/- 12"
		R	—	—	—	-0 45' +/- 45'	0.08
	S2000	F	—	6.00'	—	-0.30'	0
		R	—	—	—	-1.30'	0.25 +/- 0.08
2003	Accord	F	—	+3 10'	—	+0.06	0
		R	—	—	0.50	—	1/16"
	Civic-exc Hatchback	F	—	+1.33	—	0 00'	0
		R	—	—	—	-0 45'	1/16"
	Civic- Hatchback	F	—	1 03'	—	0 00' +/- 1 45'	0 +/- 12"
		R	—	—	—	-0 45' +/- 45'	0.08
	S2000	F	0.75	6.00'	—	-0.30'	0
		R	—	—	—	-1.30'	0.25 +/- 0.08
2004	Accord	F	1.00	+3 10'	—	+0.06	0
		R	—	—	0.50	—	1/16"
	Civic-exc Hatchback	F	—	+1.33	—	0 00'	0
		R	—	—	—	-0 45'	1/16"
	Civic- Hatchback	F	—	1 36' +/- 1	—	-0 09" +/- 45'	0 +/- 12"
		R	—	—	—	-0 51' +/- 45'	0.08
	S2000	F	0.75	6.00'	—	-0.30'	0
		R	—	—	—	-1.30'	0.14 +/- 0.08
2005	Accord	F	1.00	+3 10'	—	+0.06	0
		R	—	—	0.50	—	1/16"
	Civic-exc. Hatchback	F	—	+1.33	—	0 00'	0
		R	—	—	—	-0 45'	1/16"
	Civic- Hatchback	F	—	1 36' +/- 1	—	-0 09' +/- 45'	0 +/- 12"
		R	—	—	—	-0 51' +/- 45'	0.08
	S2000	F	0.75	6.00'	—	-0.30'	0
		R	—	—	—	-1.30'	0.14 +/- 0.08
2006	Accord	F	—	+3 15"	—	0 00'	0
		R	—	—	0.50	-1 00'	1/16"
	Civic	F	—	7 00'	—	-1. 30'	0
		R	—	—	—	-0 45'	1/16"
	S2000	F	0.75	6.00'	—	-0.30'	0
		R	—	—	—	-1.30'	0.14 +/- 0.08

09474_ACCO_C0014

TIRE, WHEEL AND BALL JOINT SPECIFICATIONS

Year	Model	OEM Tires		Tire Pressures (psi)		Wheel Size	Ball Joint Inspection	Lug Nut (ft. lbs.)
		Standard	Optional	Front	Rear			
2002	Civic	①	P195/55VR15	②	②	5-J	NA	80
	Accord DX	P195/70SR14	None	32	32	5-J	NA	80
	Accord EX, LX 4-cyl.	P195/65HR15	None	32	32	6.5	NA	80
	Accord LX V6	P205/65VR15	None	30	30	5-J	NA	80
	S2000	③	None	32	32	NA	NA	80
2003	Civic	①	P195/55VR15	②	②	5-J	NA	80
	Accord DX	P195/65R15	None	32	30	5-J	NA	80
	Accord LX 4-cyl.	P205/65R15	None	32	32	6.5	NA	80
	Accord EX 4-cyl.	P205/60R16	None	32	32	6.5	NA	80
	Accord V6	④	None	30	30	5-J	NA	80
	S2000	③	None	32	32	NA	NA	80
2004	Civic	⑦	P195/55VR15	30	30	5-J	NA	80
	Accord DX	P195/65R15	None	32	30	5-J	NA	80
	Accord LX 4-cyl.	P205/65R15	None	30	29	6.5	NA	80
	Accord EX 4-cyl.	P205/60R16	None	30	29	6.5	NA	80
	Accord V6	④	None	32	⑤	5-J	NA	80
	S2000	⑥	None	32	32	NA	NA	80
2005	Civic	⑦	P195/55VR15	30	30	5-J	NA	80
	Accord DX	P195/65R15	None	32	30	5-J	NA	80
	Accord LX 4-cyl.	P205/65R15	None	30	29	6.5	NA	80
	Accord EX 4-cyl.	P205/60R16	None	30	29	6.5	NA	80
	Accord V6	④	None	32	⑤	5-J	NA	80
	S2000	⑥	None	32	32	NA	NA	80
2006	Civic- exc SI	⑩	None	32	30	NA	NA	80
	Civic- SI	P215/45R17	215/45R17	32	30	NA	NA	80
	Accord DX	P205/60R16	None	32	30	5-J	NA	80
	Accord LX 4-cyl.	P205/65R15	None	⑨	⑨	6.5	NA	80
	Accord EX 4-cyl.	⑧	None	⑨	⑨	6.5	NA	80
	Accord V6	T215/50R17	None	32	29	5-J	NA	80
	S2000	⑥	None	32	32	NA	NA	80

OEM: Original Equipment Manufacturer

PSI: Pounds Per Square Inch

NA: Information not available

① Except EX and Hatchback: P185/70R14
   EX: P185/65R15
   Hatchback: P195/60R15

② Except Hatchback: Front 30, Rear 30
   Hatchback: Front 33, Rear 30

③ Front: P205/55R16
   Rear P225/50R16

④ Except EX w/manual transaxle: P205/60R16
   EX w/manual transaxle: P215/50R17

⑤ LX and EX 4 door: 32
   LX and EX 2 door: 30
   EX w/manual transaxle: 29

⑥ Front: P215/45R17
   Rear P245/40R17

⑦ Except EX and Hatchback: P185/70R14 LX 4 door: P205/65R15
   EX: P185/65R15
   Hatchback: P205/55R16

⑧ LX 4 door: P205/65R15
   LX 2 door: P205/60R16
   LX 4 door: P205/60R16

⑨ LX 2 door: Front 32, Rear 30
   LX 4 door: Front 30, Rear 29

⑩ Except DX: R205/55R16
   DX: R195/65R15

09474_ACCO_C0015

BRAKE SPECIFICATIONS
All measurements in inches unless noted

Year	Model		Brake Disc Original Thickness	Brake Disc Minimum Thickness	Brake Disc Maximum Runout	Brake Drum Diameter Original Inside Diameter	Brake Drum Diameter Max. Wear Limit	Brake Drum Diameter Maximum Machine Diameter	Minimum Lining Thickness Front	Minimum Lining Thickness Rear	Brake Caliper Bracket Bolts (ft. lbs.)	Brake Caliper Mounting Bolts (ft. lbs.)
2002	Accord	F	0.910	0.830	0.004	—	—	—	0.060	—	80	①
		R	0.360	0.310	0.004	8.66	8.70	8.70	—	②	41	17
	Civic	F	0.860	0.750	0.004	—	—	—	0.060	—	80	25
		R	0.397	0.350	0.004	7.87	7.91	7.91	—	②	41	17
	S2000	F	0.990	0.910	0.004	—	—	—	0.060	—	80	24
		R	0.476	0.390	0.004	—	—	—	—	0.060	41	17
2003	Accord	F	③	④	0.004	—	—	—	0.060	—	80	⑥
		R	⑤	0.310	0.004	8.66	8.70	8.70	—	0.060	41	17
	Civic	F	0.860	0.750	0.004	—	—	—	0.060	—	80	25
		R	0.397	0.350	0.004	7.87	7.91	7.91	—	②	41	17
	S2000	F	0.990	0.910	0.004	—	—	—	0.060	—	80	24
		R	0.476	0.390	0.004	—	—	—	—	0.060	41	17
2004	Accord	F	③	④	0.004	—	—	—	0.060	—	80	⑥
		R	⑤	0.310	0.004	8.66	8.70	8.70	—	0.060	41	17
	Civic	F	0.860	0.750	0.004	—	—	—	0.060	—	80	⑦
		R	0.397	0.350	0.004	7.87	7.91	7.91	—	②	41	17
	S2000	F	0.990	0.910	0.004	—	—	—	0.060	—	84	24
		R	0.476	0.390	0.004	—	—	—	—	0.060	41	17
2005	Accord	F	③	④	0.004	—	—	—	0.060	—	80	⑥
		R	⑤	0.310	0.004	8.66	8.70	8.70	—	0.060	41	17
	Civic	F	0.860	0.750	0.004	—	—	—	0.060	—	80	⑦
		R	0.397	0.350	0.004	7.87	7.91	7.91	—	②	41	17
	S2000	F	0.990	0.910	0.004	—	—	—	0.060	—	84	24
		R	0.476	0.390	0.004	—	—	—	—	0.060	41	17
2006	Accord	F	③	④	0.004	—	—	—	0.060	—	80	⑥
		R	⑤	0.310	0.004	8.66	8.70	8.70	—	0.060	41	17
	Civic	F	⑧	⑨	0.004	—	—	—	0.060	—	80	25
		R	0.360	0.310	0.004	7.87	7.91	7.91	—	②	41	17
	S2000	F	0.990	0.910	0.004	—	—	—	0.060	—	84	24
		R	0.476	0.390	0.004	—	—	—	—	0.060	41	17

NA: Information not available

F: Front

R: Rear

① Calipers with long pins beyond bolt threads: 54 ft. lbs.
Calipers with no pin beyond threads: 20 ft. lbs.

② Drum brakes: 0.080
Disk Brake 0.060

③ Except V6 w/manual transaxle: 0.910
V6 w/automatic transaxle: 1.110

④ Except V6 w/manual transaxle: 0.830
V6 w/automatic transaxle: 1.020

⑤ Except V6 w/manual transaxle: 0.360
V6 w/automatic transaxle: 0.040

⑥ V6: 37 ft. lbs.
4 cyl. 26 ft. lbs.

⑦ Except 2004-05 Hatchback: 25 ft. lbs.
2004-05 Hatchback 16 ft. lbs.

⑧ 4 cyl (R18A1) engine: 0.83
4 cyl (K20Z3) engine: 0.99

⑨ 4 cyl (R18A1) engine: 0.75
4 cyl (K20Z3) engine: 0.91

09474_ACCO_C0016

SCHEDULED MAINTENANCE INTERVALS
2002-05 Honda—Civic, Accord & S2000

TO BE SERVICED	TYPE OF SERVICE	VEHICLE MILEAGE INTERVAL (x1000)												
		7.5	15	22.5	30	37.5	45	52.5	60	67.5	75	82.5	90	97.5
Engine oil & filter	R	✓	✓	✓	✓	✓	✓	✓	✓	✓	✓	✓	✓	✓
Front brake pads	S/I	✓	✓	✓	✓	✓	✓	✓	✓	✓	✓	✓	✓	✓
Rotate tires	S/I	✓	✓	✓	✓	✓	✓	✓	✓	✓	✓	✓	✓	✓
Cooling system, hoses & connections	S/I		✓		✓		✓		✓		✓		✓	
Driveshaft boots	S/I		✓		✓		✓		✓		✓		✓	
Exhaust system	S/I		✓		✓		✓		✓		✓		✓	
Front brake discs & calipers	S/I		✓		✓		✓		✓		✓		✓	
Front wheel alignment	S/I		✓		✓		✓		✓		✓		✓	
Fuel pipes, hoses & connections	S/I		✓		✓		✓		✓		✓		✓	
Parking brake adjustment	S/I		✓		✓		✓		✓		✓		✓	
Power steering system	S/I		✓		✓		✓		✓		✓		✓	
Rear brake discs, calipers & pads	S/I		✓		✓		✓		✓		✓		✓	
Suspension components	S/I		✓		✓		✓		✓		✓		✓	
Suspension mounting bolts	S/I		✓		✓		✓		✓		✓		✓	
Tie rods, steering gear box & boots	S/I		✓		✓		✓		✓		✓		✓	
Valve clearance	S/I				✓				✓				✓	
Parking brake	S/I		✓		✓				✓				✓	
Air cleaner element	R				✓				✓				✓	
Transmission fluid (Civic CVT)	R				✓		✓		✓		✓		✓	
Transmission fluid (A/T or M/T) (except as noted below)	R				✓				✓				✓	
Brake fluid (including ABS) (Accord V6)	R				✓				✓				✓	
Brake fluid (including ABS) (Accord L4 & Civic)	R						✓						✓	
Spark plugs (non-VTEC)	R				✓				✓				✓	
Spark plugs (VTEC) ①	R								✓				✓	
ABS operation	S/I				✓				✓				✓	
Alternator drive belt	S/I				✓				✓					
Power steering pump belt	S/I				✓				✓				✓	
Rear brake drums, wheel cylinders & linings	S/I				✓				✓				✓	
Engine coolant	R						✓				✓			
ABS high pressure hose	R								✓					
Fuel filter	R								✓					
Timing belt	R												✓	
Timing balancer belt	R												✓	
Distributor, ignition cap & rotor	S/I								✓					
Idle speed	S/I								✓					

09474_ACCO_C0017

SCHEDULED MAINTENANCE INTERVALS
2002-05 Honda—Civic, Accord & S2000

TO BE SERVICED	TYPE OF SERVICE	VEHICLE MILEAGE INTERVAL (x1000)												
		7.5	15	22.5	30	37.5	45	52.5	60	67.5	75	82.5	90	97.5
Ignition wires	S/I								✔					
PCV valve	S/I								✔					
TWC converter heat shield	S/I								✔					
Water pump	S/I												✔	

R: Replace   S/I: Service or Inspect

① S2000: 105,000 miles

FREQUENT OPERATION MAINTENANCE (SEVERE SERVICE)

If a vehicle is operated under any of the following conditions it is considered severe service:

- Extremely dusty areas.

- 50% or more of the vehicle operation is in 32°C (90°F) or higher temperatures, or constant operation in temperatures below 0°C (32°F).

- Prolonged idling (vehicle operation in stop and go traffic).

- Frequent short running periods (engine does not warm to normal operating temperatures).

- Police, taxi, delivery usage or trailer towing usage.

Oil & oil filter: change every 3750 miles.

Driveshaft boots: service or inspect every 7500 miles.

Front brake discs & calipers, & rear brake discs, calipers & pads: service or inspect every 7500 miles.

Power steering system: service or inspect every 7500 miles.

Suspension components: service or inspect every 7500 miles.

Tie rods, steering gear box & boots: service or inspect every 7500 miles.

Air cleaner element: service or inspect every 15,000 miles.

Transmission fluid (Accord V6 & Civic CVT): replace every 15,000 miles.

Transmission fluid (Accord L4 & Civic): replace every 30,000 miles.

Timing balancer belt: replace every 60,000 miles.

Timing belt: replace every 60,000 miles.

Water pump: service or inspect every 60,000 miles.

09474_ACCO_C0018

MAINTENANCE MINDER SCHEDULE
2006 Honda—Civic, Accord & S2000

All 2006 Honda's displays engine oil life and maintenance service items in the information display to indicate when to perform maintenance service. If the engine oil life is 15% or less, based on the onboard computer's caluculations, you will see SERVICE DUE SOON in the information display every time the ignition key is turned to ON. The maintenance minder indicator will also come on and the maintenance code(s) for other scheduled maintenance items needing service will be displayed below the message.

Symbol	Item	Service
A	Engine oil ①	Change
B	Engine oil and filter	Change
	Tires	Rotate
	Brakes	Inspect
	Parking brake adjustment	Check
	Steering gear and linkage	Inspect
	Suspension components	Inspect
	Driveshaft boots	Inspect
	Brake hoses and lines	Inspect
	Exhaust system	Inspect
	Fuel lines and connections	Inspect
1	Tires	Rotate
2	Engine air filter ②	Replace
	Dust and pollen filter ③	Replace
	Accessory drive belt	Inspect
3	Transmission fluid ④	Replace
	Transfer case fluid ④	Replace
4	Spark plugs	Replace
	Timing belt ⑤	Replace
	Water pump	Inspect
	Valve clearance ⑥	Inspect
5	Engine coolant	Replace
6	VTM-4 rear differential fluid	Replace

① If the message SERVICE DUE NOW does not appear more than 12 months after the display is reset, change every year.

② If driven in dusty conditions, replace every 15,000 miles.

③ If driven in urban areas that have a high concentration of soot from industry and diesel, replace every 15,000 miles

④ If regularly driven in mountainous areas at very low speed or trailer towing, change the fluid every 30,000 miles.

⑤ If driven regularly in temperatures over 110 deg.F or below -20 deg.F, or towing a trailer, replace every 60,000 miles.

⑥ Adjust if necessary.

Additionally, replace the brake fluid every 3 years, and inspect the idle speed every 160,000 miles.
To reset the Engine Oil Life Display:
1. Turn the ignition switch to ON.
2. Press the SELECT button repeatedly until the engine oil life display or the service message is displayed.
3. Press the RESET button for about 10 seconds. You will see a MAINT RESET message.
4. Select the appropriate answer, MAINT RESET >N (NO) or MAINT RESET > y (YES) by pressing the SELECT button repeatedly.
   >N or >Y is displayed on the outside temperature >N or >Y is displayed on the outside temperature display.
5. Select the MAINT RESET > Y (YES), and press and hold the RESET button again to reset the engine oil life to 100%.

ENGINE REPAIR

➡**Disconnecting the negative battery cable on some vehicles may interfere with the functions of the on board computer systems and may require the computer to undergo a relearning process, once the negative battery cable is reconnected.**

Distributor

REMOVAL & INSTALLATION

1.7L Engine

1. Before servicing the vehicle, refer to the Precautions Section.
2. Disconnect the negative battery cable. Disconnect the electrical connectors from the distributor.
3. Label and disconnect the spark plug wires.
4. Remove the distributor mounting bolts. Remove the distributor from its mounting.

To install:

5. Position the engine at TDC on the compression stroke.
6. Coat a new oil ring with clean engine oil. Install it in position. Position the distributor in its mounting.

➡**The lugs on the end of the distributor and their mating grooves in the camshaft end are offset to eliminate the possibility of installing the distributor 180 degrees out of time.**

7. Install the mounting bolts. Install the ignition wires.
8. Connect the electrical connector. Check the ignition timing

Alternator

REMOVAL & INSTALLATION

Civic 1.7L Engine

1. Before servicing the vehicle, refer to the Precautions Section.
2. Remove or disconnect the following:

- Negative battery cable
- Accessory drive belts
- Power steering pump
- 4P connector and battery terminal wire
- Alternator bolts
- Alternator

To install:

- Alternator and tighten the bolts to 33 ft. lbs. (44 Nm)
- 4P connector and battery terminal wire. Tighten the battery terminal wire nut to 108 inch lbs. (12 Nm).
- Power steering pump
- Accessory drive belts
- Negative battery cable

3. Reprogram the ECM/PCM engine idle characteristics. Be sure all electrical items are OFF.
4. Start the engine. Hold the idle speed at 3000 RPM's in park or neutral until the radiator fan comes on or the temperature reached 194 degrees.
5. Let the engine idle for about five minutes with the throttle fully closed.
6. If the radiator fan comes on during the five minutes, do not count this toward the five minute programming time.
7. Set the clock.

Civic 1.8L Engine

1. Before servicing the vehicle, refer to the Precautions Section.
2. Remove or disconnect the following:

- Negative battery cable, then the positive
- Accessory drive belt
- 4P connector and battery terminal wire
- Alternator bolts
- Alternator

To install:

3. Installation is the reverse of removal.
4. Adjust the alternator belt tension.
5. Connect the negative battery cable, then the positive cable.
6. Reprogram the power window control unit as follows.
7. Turn the ignition switch ON. Lower the window, all the way down. Open the driver's side door.
8. Turn the ignition switch OFF. Push and hold the drivers window DOWN switch. Turn the ignition switch ON. Release the driver's window DOWN switch. This must be done within five seconds.
9. Repeat the above step three more times. Wait one second.
10. Confirm that the AUTO UP and AUTO DOWN do not work. If they work repeat the procedure, paying close attention to the five second time limit
11. Move the driver's window all the way down by holding the drivers window DOWN switch to the AUTO DOWN position.

12. Pull up and hold the driver's window UP switch to the AUTO UP position until the window reaches the fully closed position. Hold the switch for one second.
13. Confirm that the power window master switch has been reset by using the driver's window AUTO UP and DOWN function.
14. If the AUTO UP and DOWN feature is still not working, repeat the complete procedure, paying close attention to the five second time limit.
15. Set the clock.

Civic 2.0L Engine

2002–2005

1. Before servicing the vehicle, refer to the Precautions Section.
2. Remove or disconnect the following:

- Negative battery cable, then the positive
- Front bumper
- Right side headlight
- Reserve tank from its mounting bracket
- Accessory drive belt
- Alternator bolts
- Electrical connectors
- Alternator

To install:

3. Installation is the reverse of removal.
4. Adjust the alternator belt tension.
5. Reprogram the ECM engine idle characteristics. Be sure all electrical items are OFF.
6. Start the engine. Hold the idle speed at 3000 RPM's in neutral until the radiator fan comes on or the temperature reached 194 degrees.
7. Let the engine idle for about five minutes with the throttle fully closed.
8. If the radiator fan comes on during the five minutes, do not count this toward the five minute programming time.
9. Set the clock.

2006

1. Before servicing the vehicle, refer to the Precautions Section.
2. Disconnect the negative battery cable, then the positive cable.
3. Remove the drive belt.
4. Remove the front grille cover. Disconnect the fan motor connector, hood switch connector. Remove the harness clamps.
5. Remove the reservoir hose, radiator

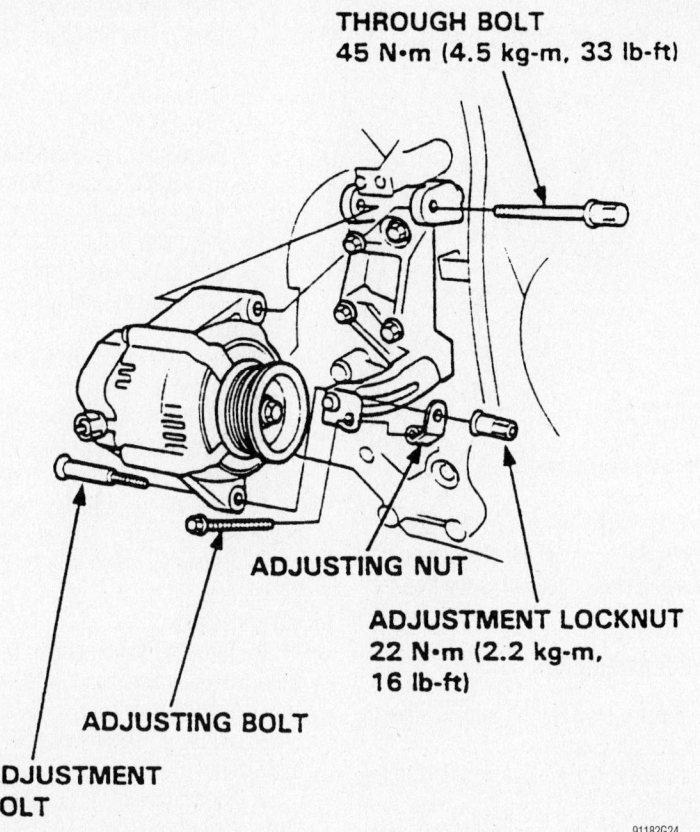

THROUGH BOLT
45 N•m (4.5 kg-m, 33 lb-ft)

ADJUSTING NUT

ADJUSTMENT LOCKNUT
22 N•m (2.2 kg-m,
16 lb-ft)

ADJUSTING BOLT

ADJUSTMENT
BOLT

91182G24

Alternator mounting bolt locations

cap mounting bolts and upper radiator brackets.

6. Remove the condenser mounting bolts. Remove the bulkhead.

7. Remove the alternator retaining bolts. Disconnect the electrical connections. Remove the alternator from its mounting.

To install:

8. Installation is the reverse of removal.

9. Adjust the alternator belt tension.

10. Reprogram the power window control unit as follows.

11. Turn the ignition switch ON. Lower the window, all the way down. Open the driver's side door.

12. Turn the ignition switch OFF. Push and hold the drivers window DOWN switch. Turn the ignition switch ON. Release the driver's window DOWN switch. This must be done within five seconds.

13. Repeat the above step three more times. Wait one second.

14. Confirm that the AUTO UP and AUTO DOWN do not work. If they work repeat the procedure, paying close attention to the five second time limit

15. Move the driver's window all the way down by holding the drivers window DOWN switch to the AUTO DOWN position.

16. Pull up and hold the driver's window

UP switch to the AUTO UP position until the window reaches the fully closed position. Hold the switch for one second.

17. Confirm that the power window master switch has been reset by using the driver's window AUTO UP and DOWN function.

18. If the AUTO UP and DOWN feature is still not working, repeat the complete procedure, paying close attention to the five second time limit.

19. Set the clock.

S2000

1. Before servicing the vehicle, refer to the Precautions Section.

2. Remove or disconnect the following:
   - Negative battery cable, then the positive
   - Accessory drive belt
   - 4P connector and battery terminal wire
   - Alternator bolts
   - Alternator

To install:
   - Alternator and tighten the bolts to 33 ft. lbs. (44 Nm)
   - 4P connector and battery terminal wire. Tighten the battery terminal wire nut to 108 inch lbs. (12 Nm).

   - Accessory drive belt
   - Negative battery cable, then the positive

3. On 2002–05 vehicles, reprogram the ECM engine idle characteristics. Be sure all electrical items are OFF.

4. Start the engine. Hold the idle speed at 3000 RPM's in neutral until the radiator fan comes on or the temperature reached 176 degrees.

5. Let the engine idle for about five minutes with the throttle fully closed.

6. If the radiator fan comes on during the five minutes, do not count this toward the five minute programming time.

7. Set the clock.

Accord 2.3L Engine

1. Before servicing the vehicle, refer to the Precautions Section.

2. Note the radio security code and the radio presets.

3. Remove or disconnect the following:
   - Negative battery cable, then the positive
   - 4P connector and battery terminal wire from the alternator
   - Adjusting bolt, locknut and the mounting bolt
   - Alternator belt
   - Alternator from the bracket

To install:

4. Installation is the reverse of removal.

5. Adjust the alternator belt tension.

6. Enter the anti-theft code for the radio.

Accord 2.4L Engine

1. Before servicing the vehicle, refer to the Precautions Section.

2. Note the radio security code and the radio presets.

3. Remove or disconnect the following:
   - Negative battery cable, then the positive

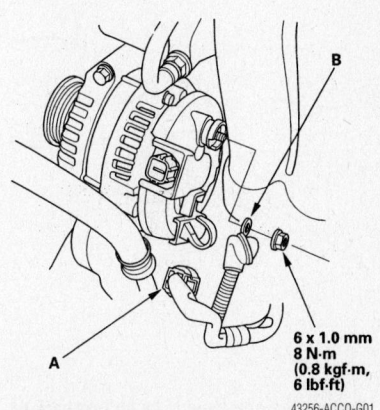

B

A

6 x 1.0 mm
8 N•m
(0.8 kgf·m,
6 lbf·ft)

43256-ACCO-G01

Alternator mounting—2.4L engine

- Drive belt
- Auto-tensioner
- Connectors from the alternator
- 3 alternator mounting bolts and the alternator

To install:

- Alternator and 3 mounting bolts. Torque the bolts to 16 ft. lbs. (22 Nm).
- Electrical connectors
- Auto-tensioner
- Drive belt
- Positive, then negative battery cables

4. Enter the security code and radio presets.

5. Reprogram the power window control unit as follows.

6. Turn the ignition switch ON. Lower the window, all the way down. Open the driver's side door.

7. Turn the ignition switch OFF. Push and hold the drivers window DOWN switch. Turn the ignition switch ON. Release the driver's window DOWN switch. This must be done within five seconds.

8. Repeat the above step three more times. Wait one second.

9. Confirm that the AUTO UP and AUTO DOWN do not work. If they work repeat the procedure, paying close attention to the five second time limit

10. Move the driver's window all the way down by holding the drivers window DOWN switch to the AUTO DOWN position.

11. Pull up and hold the driver's window UP switch to the AUTO UP position until the window reaches the fully closed position. Hold the switch for one second.

12. Confirm that the power window master switch has been reset by using the driver's window AUTO UP and DOWN function.

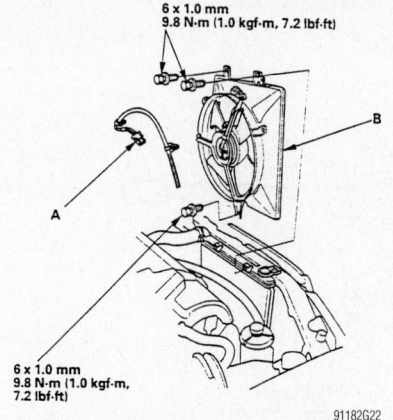

6 x 1.0 mm
9.8 N·m (1.0 kgf·m, 7.2 lbf·ft)

6 x 1.0 mm
9.8 N·m (1.0 kgf·m, 7.2 lbf·ft)

91182G22

Remove the condenser fan—3.0L engine

13. If the AUTO UP and DOWN feature is still not working, repeat the complete procedure, paying close attention to the five second time limit.

Accord 3.0L Engine

1. Before servicing the vehicle, refer to the Precautions Section.

2. Note the radio security code and the radio presets.

3. Remove or disconnect the following:

- Negative battery cable, then the positive

- Alternator belt tension by pulling back on the adjuster and then remove the belt
- Condenser fan motor connector from the shroud
- Condenser fan assembly
- Four prong connector from the rear of the alternator
- Alternator mounting bolts
- Wiring harness clamp
- Alternator assembly

To install:

4. Alternator installation is the reverse of the removal procedure.

5. Connect the positive battery cable, then the negative battery cable. Enter the radio security code and station presets.

6. On 2003–06 vehicles, reprogram the power window control unit as follows.

7. Turn the ignition switch ON. Lower the window, all the way down. Open the driver's side door.

8. Turn the ignition switch OFF. Push and hold the drivers window DOWN switch. Turn the ignition switch ON. Release the driver's window DOWN switch. This must be done within five seconds.

9. Repeat the above step three more times. Wait one second.

10. Confirm that the AUTO UP and AUTO DOWN do not work. If they work

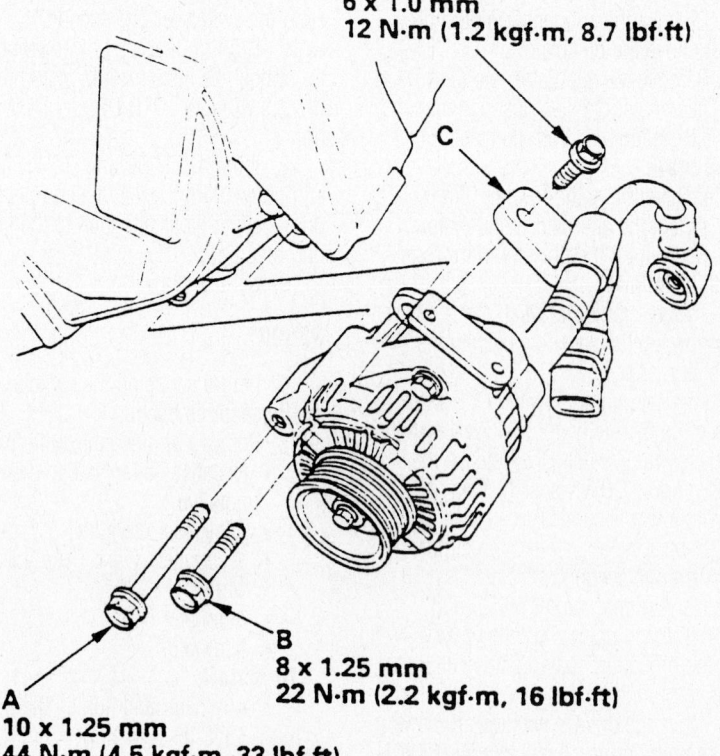

6 x 1.0 mm
12 N·m (1.2 kgf·m, 8.7 lbf·ft)

C

8 x 1.25 mm
22 N·m (2.2 kgf·m, 16 lbf·ft)

A

B

10 x 1.25 mm
44 N·m (4.5 kgf·m, 33 lbf·ft)

91182G23

Torque the alternator bolts to the specs shown—3.0L engine

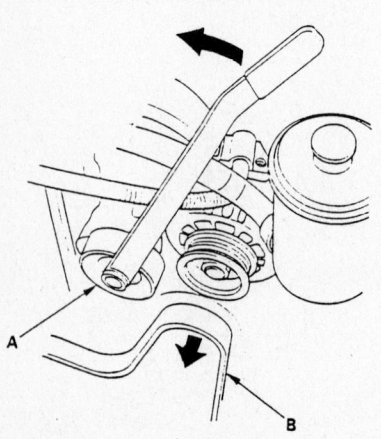

A

B

91182G21

Relieve the belt tension by pulling back on the tensioner—3.0L engine

repeat the procedure, paying close attention to the five second time limit

11. Move the driver's window all the way down by holding the drivers window DOWN switch to the AUTO DOWN position.

12. Pull up and hold the driver's window UP switch to the AUTO UP position until the window reaches the fully closed position. Hold the switch for one second.

13. Confirm that the power window master switch has been reset by using the driver's window AUTO UP and DOWN function.

14. If the AUTO UP and DOWN feature is still not working, repeat the complete procedure, paying close attention to the five second time limit.

15. Set the clock.

❊❊ WARNING

Be sure to adjust the alternator belt to the proper tension or alternator bearing failure may occur.

Ignition Timing

ADJUSTMENT

Except 2.3L Engine

All engines are equipped with Distributorless Ignition Systems (DIS). No adjustment is necessary.

2.3L Engine

This engine is equipped with a distributor; however ignition timing is not adjustable.

CHECKING

The ignition timing is not adjustable, but the ignition base timing can be checked by performing the following:

1. Before servicing the vehicle, refer to the Precautions Section.

2. Connect a PGM tester (scan tool) to the data link connector.

3. Connect a timing light to the No. 1 ignition cable.

4. Start the engine and allow it to warm up until the electric fan comes on.

5. Be sure to turn off all accessories.

6. Verify that the idle speed is set to specification.

7. Point the light at the timing belt cover near the crankshaft pulley and read the timing. Verify that the ignition timing is within specification.

Engine Assembly

REMOVAL & INSTALLATION

➡ **The original radio contains a coded anti-theft circuit. Obtain the security code number before disconnecting the battery cables.**

Civic

2002–2005

1. Before servicing the vehicle, refer to the Precautions Section.

2. Disconnect the negative and positive battery cables. Wait at least 3 minutes before working around the air bags.

➡ **The engine and transaxle are removed from the vehicle as 1 unit.**

3. Support the hood as far open as possible. If the hood is to be removed, first matchmark the hinge plates with a felt-tipped marker.

4. Remove the battery from the vehicle. Unbolt and remove the battery tray.

5. Disconnect the battery and alternator cables from the underhood fuse and relay box on the right shock tower.

6. Remove the lower right kick panel to expose the Powertrain Control Module (PCM).

7. Label and disconnect the 5 wiring harness connections from the PCM.

8. Unbolt the main wiring harness retainer from the rear of the fuse and relay box on the right side of the bulkhead. Carefully pull the grommet out of its bulkhead opening. Next, pull the PCM harness and connectors through the opening. Be careful not to damage the wiring, insulation, or connectors.

9. Relieve the fuel pressure:
   a. Loosen the fuel filler cap.
   b. Use a box-end wrench and a flare nut wrench to hold the fuel filter banjo fitting.
   c. Place a shop towel over the fuel filter to catch the fuel spray.
   d. Slowly loosen the fuel filter service bolt 1 full turn.
   e. Clean up any spilled fuel.

10. Remove or disconnect the following:
   • Intake air duct and air cleaner
   • Intake Air Temperature (IAT) sensor connector from the air cleaner case, if equipped
   • Fuel feed hose from the fuel filter
   • Fuel return hose from the fuel rail
   • Intake manifold/throttle body vacuum hoses
   • Brake booster vacuum hose

   • Evaporative emissions (EVAP) canister vacuum hose
   • Power Steering Pressure (PSP) switch and detach its clamp from the bracket below the brake booster
   • Transaxle ground cable
   • Radiator hose bracket

11. Loosen the throttle cable locknut, then disconnect the cable from the throttle body linkage. Move the cable aside without kinking it.

12. Loosen the power steering pump mounting bolts. Slip power steering belt off its pulleys. Unbolt the steering pump and move it out of the work area. Don't disconnect the hydraulic hoses.

➡ **Label the connectors before detaching them.**

13. Remove or disconnect the following:
   • Engine wiring harness connectors at the left side of the engine compartment

14. Drain the coolant from the radiator and engine block.

15. Remove or disconnect the following:
   • Upper and lower radiator hoses
   • Heater hoses from the cylinder head

16. If equipped with a CVT transaxle, loosen the shift cable locknut. Remove the spring clip and washers and disconnect the shift cable from its linkage. Be careful not to kink the cable or damage its boot.

17. Remove or disconnect the following:
   • Hydraulic line brackets from the top of the transaxle case, if equipped with a manual transaxle

18. Attach a chain hoist to the engine lifting brackets. Don't raise the hoist to lift the engine yet.

19. Raise the vehicle and support it safely. Remove the front wheels.

20. Remove or disconnect the following:
   • Engine splash shield

21. Drain the engine oil.

22. Drain the fluid from the transaxle.

23. Remove or disconnect the following:
   • Left front engine mount bracket from the shock tower, if equipped with air conditioning

24. Loosen the compressor idler pulley and adjusting bolt. Slip the belt around the engine mount stud to remove it.

25. Remove or disconnect the following:
   • Compressor mounting bolts to separate the compressor from its mounting plate. Move the compressor out of the work area. Do not disconnect the air conditioning refrigerant lines.

26. Remove or disconnect the following:

- ATF cooler lines, if equipped. Plug the cooler lines to prevent fluid leakage and contamination
- Slave cylinder from the transaxle case without disconnecting its hydraulic line, if equipped with manual transaxle
- Front exhaust pipe from the exhaust manifold and catalytic converter. Unbolt its hanger bracket and remove the exhaust pipe.
- Shift cable from the transaxle control shaft, if equipped with automatic transaxle
- Shift rod and extension rod from the transaxle, if equipped with manual transaxle
- Strut damper fork
- Steering knuckle ball joint from the lower control arm using a ball joint separator

27. Pry the inboard CV-joints from the transaxle. Then, move the halfshafts away from the transaxle and wire them to the undercarriage of the vehicle. Tie plastic bags over the inboard CV-joints to prevent damage to the boots and splined shafts.

28. Raise the hoist slightly to take up the weight of the engine and transaxle assembly.

29. Disconnect the engine mounts in the following order:

a. Unbolt and remove the left front engine mount.

b. Unbolt and remove the right front engine mount and bracket assembly.

c. Remove the rear engine mount through-bolt. Then, unbolt the rear mount bracket from the engine block.

30. If necessary, lower the vehicle slightly to gain access to the side engine and transaxle mounts. Do not release the tension on the chain hoist. The engine must be securely supported.

31. Unbolt the side engine mount bracket from the engine block bracket and mount damper.

32. Unbolt the transaxle mount bracket from the transaxle case. Then, unbolt the mount from the shock tower.

33. Raise the chain hoist to lift the engine a few inches off of its mounts.

34. Verify that all electrical, vacuum, and fuel lines have been disconnected.

35. Raise the engine and transaxle assembly and remove it from the vehicle.

To install:

➡Use new self-locking nuts and gaskets when installing the front exhaust pipe and when assembling the front suspension. Use new set rings on the inboard CV-joint splined shafts.

36. Lower the engine and transaxle assembly into the vehicle.

37. Install and connect the engine and transaxle mounts and brackets. Use new self-locking nuts and color-coded bolts. At this point, only tighten the mounting nuts and bolts by hand.

38. Before installing the left front engine mount, fit the air conditioning compressor back into place and install the compressor belt. Tighten the compressor bolts to 17 ft. lbs. (24 Nm).

➡Failure to tighten the bolts in the proper sequence can cause excessive noise and vibration and reduce bushing life. Be sure to check that the bushings are not twisted or offset.

39. The engine and transaxle mount and bracket fasteners must be tightened in the proper sequence with the weight of the engine resting upon them. This step is important for engine mount pre-loading. Tighten the engine mount bolts in the following sequence:

a. Transaxle mount bolts: 47 ft. lbs. (64 Nm); or 28 ft. lbs. (38 Nm) for CVT-equipped vehicles.

b. Side engine mount bracket nuts: 54 ft. lbs. (74 Nm).

c. Rear mount bracket bolts: 61 ft. lbs. (83 Nm); or 43 ft. lbs. (59 Nm) for CVT-equipped vehicles.

d. Rear mount through-bolt: 43 ft. lbs. (59 Nm).

e. Transaxle mount bracket nuts or bolts: 47 ft. lbs. (64 Nm).

f. Transaxle mount through-bolt: 54 ft. lbs. (74 Nm).

g. Right front mount bracket bolts: 33 ft. lbs. (44 Nm).

h. Right front mount carrier bolts: 33 ft. lbs. (44 Nm).

i. Left front mount: stud: 61 ft. lbs. (85 Nm); carrier bolts: 33 ft. lbs. (44 Nm); nut: 43 ft. lbs. (59 Nm).

40. Remove the chain hoist from the engine lifting hooks.

41. Install or connect the following:

- New set rings on the inboard splined shafts of each halfshaft. Check that the set ring on each inboard CV-joint clicks into place when the halfshafts are installed into the transaxle.
- Damper fork and reconnect the lower ball joint. When the weight of the vehicle is resting on its suspension, tighten the pinch bolt to 32 ft. lbs. (44 Nm) and the fork bolt to 47 ft. lbs. (65 Nm). Tighten the ball joint castle nut to 36–43 ft. lbs.

(50–60 Nm). Next, tighten the castle nut only enough to install a new cotter pin.

- Slave cylinder, if equipped. Tighten the slave cylinder mounting bolts to 16 ft. lbs. (22 Nm). If the clutch hydraulic line was disconnected, air must be bled from the system.
- Transaxle shift and extension rods to the linkage at the transaxle case, if equipped. Install a new 8mm spring pin into the shift rod linkage. Then, install the retainer clip and boot. Tighten the extension rod bolt to 16 ft. lbs. (22 Nm).
- Shift cable to the control shaft, if equipped with an automatic transaxle. Use a new lockwasher and tighten the lockbolt to 10 ft. lbs. (14 Nm). Tighten the shift cable cover bolts to 16 ft. lbs. (22 Nm). Install the shift cable cover and tighten its bolts to 16 ft. lbs. (22 Nm).

42. Install the front exhaust pipe using new self-locking nuts:

a. Tighten the exhaust manifold nuts to 40 ft. lbs. (55 Nm).

43. Install or connect the following:

- Tighten the exhaust flange bolts to 16 ft. lbs. (22 Nm).
- ATF cooler lines. If the rubber cooler lines are cracked or stressed, they must be replaced.
- Engine splash shield

44. Refill the engine with fresh oil.

45. Refill the transaxle with the proper fluid.

46. Lower the vehicle.

47. If equipped, fit the clutch hydraulic line brackets back into place. Tighten the 8mm bolts to 17 ft. lbs. (24 Nm). Tighten the 6mm bolts to 96 inch lbs. (11 Nm).

48. If equipped with a CVT transaxle, reconnect the shift cable to the linkage. Use new plastic washers and a new spring clip. Tighten the locknut to 22 ft. lbs. (29 Nm).

49. Adjust the alternator and air conditioning compressor belt tensions.

50. Install or connect the following:

- Upper and lower radiator hoses and the heater hoses
- Power steering pump into its mounts. Adjust the pump belt tension, then tighten the mounting bolts to 17 ft. lbs. (24 Nm).
- PSP switch connector and attach its harness clamp
- Intake manifold/throttle body vacuum hoses
- Brake booster vacuum hose
- EVAP canister vacuum hose

- Fuel line fittings to the fuel filter and fuel rail. Use new sealing washers. Tighten the banjo fittings to 25 ft. lbs. (33 Nm), and the service bolts to 11 ft. lbs. (15 Nm). Don't over tighten the fittings
- Throttle cable and adjust its deflection to 10–12mm (0.39–0.47 in.).

51. Feed the PCM harness through the hole in the bulkhead. Apply sealant to the grommet, then install the retainer.

52. Install or connect the following:

- Engine wiring harness and ground cables that were disconnected during removal. Be sure the grounds are free of corrosion to ensure good contact.
- Fuse and relay box back into position
- Battery and alternator cables
- Air cleaner case and air intake duct
- IAT connector
- 5 PCM connectors and kick panel
- Battery tray and the battery

53. Verify that all wiring harnesses and grounds, vacuum lines, fuel lines have been reconnected.

54. Refill the radiator with fresh coolant.

55. If it was removed, install the hood. Reconnect the windshield washer tubing. After installation, check to be sure that the hood, fender, and grille panel gaps are equal.

56. Reconnect the positive and negative battery cables.

57. Turn the ignition switch to the **ON** position, but don't start the engine. Then, turn the ignition **OFF**. Repeat this procedure 2 or 3 times and check for fuel leaks.

58. Start the engine and allow it to warm up to its normal operating temperature.

59. Reprogram the ECM engine idle characteristics. Be sure all electrical items are OFF.

60. Start the engine. Hold the idle speed at 3000 RPM's in neutral until the radiator fan comes on or the temperature reached 194 degrees.

61. Let the engine idle for about five minutes with the throttle fully closed.

62. If the radiator fan comes on during the five minutes, do not count this toward the five minute programming time.

63. Bleed the air from the cooling system with the heater valve open.

64. Check the throttle cable deflection and operation.

65. Check and adjust the ignition timing.

66. Shut the engine off and check the drive belt adjustments.

67. Check all fluid levels and top up as necessary.

68. Check and adjust the front wheel alignment.

69. Road test the vehicle.

2006 1.8L ENGINE

1. Before servicing the vehicle, refer to the Precautions Section.

2. Note the radio security code and the radio presets.

3. Properly relieve the fuel system pressure.

4. Disconnect the negative battery cable, than the positive cable. Remove the hood support from the driver's side of the vehicle. Lock the hood in the full up position.

5. Remove the battery. Remove the air cleaner assembly. Remove the cowl cover and undercover panel.

6. Remove the fuel line quick connect fitting cover. Disconnect the fuel feed hose. Remove the evaporative emission canister hose, power brake vacuum booster hoses and power steering hose clamp.

7. Remove the battery cables from the under hood fuse/relay box. Remove the harness clamps. Remove the ECM /PCM control module cover. Remove the assembly retaining bolts. Disconnect the electrical connectors. Disconnect the engine wiring harness connector. Remove the harness clamps.

8. If equipped with manual transaxle, remove the bolts retaining the harness holder. Remove the harness clamp. Remove the shift cable. Remove the clutch slave cylinder and line bracket mounting bolt.

9. Remove the drive belt. Remove the harness cover. Disconnect the ignition coil connectors. Remove the four retaining bolts holding the harness holders in place. Remove the front harness holder.

10. Remove the breather pipe, air cleaner bracket and air harness bracket. Remove the harness holder bracket.

11. Attach the engine support tool (AAR-T-12566) to the vehicle. Remove the radiator cap.

12. Raise and support the vehicle safely. Remove the front tires. Remove the splash shield. Drain the radiator. Drain the engine oil. Drain the transaxle fluid.

13. Remove the exhaust pipe. If equipped with automatic transaxle, remove the shift cable.

14. Separate the stabilizer links. Separate the knuckles from the lower arms. Remove the driveshafts.

15. Remove the power steering pump retaining bolts and position it to the side. It is not necessary to disconnect the hoses from the pump.

16. Unclamp the power steering line clamp from the subframe. Remove the power steering line mounting bolts.

17. Remove the gearbox mounting bolt. Remove the steering gearbox stiffener

18. Lower the vehicle to the ground. Remove the radiator.

19. Disconnect the air conditioning compressor clutch connector and remove the harness clamp. Remove the compressor assembly, without disconnecting the refrigerant lines.

20. Remove the heater hoses. Raise and support the vehicle safely. Remove the lower torque rod.

21. Matchmark the subframe with the edge of the subframe stiffner. Attach special tool VSB02C000016 by hanging the belt over the front of the subframe. Secure the belt to the stop.

22. Raise the jack, line up the slots in the tool arms with the bolt holes on the corner of the jack base. Attach the four bolts.

23. Remove the subframe mid mount. Remove the subframe. Lower the vehicle to the ground. Remove the ground cable.

24. Attach a lifting device. Raise the engine/transaxle assembly up slightly. Remove the engine support tool.

25. Remove the side engine mount bracket. Remove the transaxle mounting bolt and nuts. Remove the ground cable.

26. Be sure that the engine/transaxle assembly is free from all hoses, vacuum lines and electrical wires and connectors.

27. Remove the engine/transaxle assembly from under the vehicle.

To install:

28. Installation is the reverse of the removal procedure. Be sure to check and adjust all required fluid levels.

29. Reprogram the ECM engine idle characteristics. Be sure all electrical items are OFF.

30. Start the engine. Hold the idle speed at 3000 RPM's in neutral until the radiator fan comes on or the temperature reached 194 degrees.

31. Let the engine idle for about five minutes with the throttle fully closed.

32. If the radiator fan comes on during the five minutes, do not count this toward the five minute programming time.

33. Reprogram the crankshaft position (CKP) pattern. Run the engine until the operating temperature reaches 176 degree. With the engine stopped clear the CKP pattern. Turn the ignition switch OFF. Turn the ignition switch ON and wait thirty seconds.

34. Roadtest the vehicle on a level surface. Decelerate the engine speed of 2500 rpm to 1000 rpm. If equipped with auto-

matic transaxle use two Drive positions. If equipped with manual transaxle use first gear.

35. Stop the vehicle, but keep the engine running.

36. Check PULSAR F/B LEARN in the data list with the HDS. If not complete repeat the procedure. If complete, roadtest the vehicle on a level surface. Decelerate the engine speed of 5000 rpm to 3000 rpm. If equipped with automatic transaxle use two Drive positions. If equipped with manual transaxle use first gear.

37. Stop the vehicle, but keep the engine running.

38. Check PULSAR F/B LEARN in the data list with the HDS. If not complete repeat the procedure.

39. If completed, turn the ignition switch OFF. Turn the ignition switch ON, wait thirty seconds. The learning procedure is now complete.

40. Enter the antitheft codes for the radio and the navigation system. Set the clock.

41. Reprogram the power window control unit as follows.

42. Turn the ignition switch ON. Lower the window, all the way down. Open the driver's side door.

43. Turn the ignition switch OFF. Push and hold the drivers window DOWN switch. Turn the ignition switch ON. Release the driver's window DOWN switch. This must be done within five seconds.

44. Repeat the above step three more times. Wait one second.

45. Confirm that the AUTO UP and AUTO DOWN do not work. If they work repeat the procedure, paying close attention to the five second time limit

46. Move the driver's window all the way down by holding the drivers window DOWN switch to the AUTO DOWN position.

47. Pull up and hold the driver's window UP switch to the AUTO UP position until the window reaches the fully closed position. Hold the switch for one second.

48. Confirm that the power window master switch has been reset by using the driver's window AUTO UP and DOWN function.

49. If the AUTO UP and DOWN feature is still not working, repeat the complete procedure, paying close attention to the five second time limit.

2006 2.0L ENGINE

1. Before servicing the vehicle, refer to the Precautions Section.

2. Note the radio security code and the radio presets.

3. Properly relieve the fuel system pressure.

4. Disconnect the negative battery cable, than the positive cable. Remove the hood support from the driver's side of the vehicle. Lock the hood in the full up position.

5. Remove the battery. Remove the air cleaner assembly. Remove the cowl cover and undercover panel.

6. Disconnect the harness clamp. Remove the resonator unit.

7. Remove the battery cables from the under hood fuse/relay box. Remove the harness clamps. Remove the ECM /PCM control module cover. Remove the assembly retaining bolts. Disconnect the electrical connectors. Disconnect the engine wiring harness connector. Remove the harness clamps.

8. Remove the intake manifold cover. Remove the fuel line quick connect fitting cover. Disconnect the fuel feed hose. Remove the evaporative emission canister hose, power brake vacuum booster hoses and power steering hose clamp.

9. If equipped with manual transaxle, remove the bolts retaining the harness holder. Remove the harness clamp. Remove the shift cable. Remove the clutch slave cylinder and line bracket mounting bolt.

10. Remove the air cleaner housing bracket. Remove the drive belt. Remove the idler pulley base. Remove the radiator cap.

11. Raise and safely support the vehicle. Remove the front tires. Remove the splash shield. Drain the radiator. Drain the engine oil. Drain the transaxle fluid.

12. Disconnect the air fuel ratio (A/F) sensor connector. Remove the grommet and disconnect the oxygen sensor connector. Remove the three way catalytic converter.

13. Separate the stabilizer links. Separate the knuckles from the lower arms. Remove the driveshafts.

14. Remove the steering gearbox bracket. Remove the steering gearbox mounting bolt, stiffner mounting bolt and stiffner. Remove the harness clamp from the subframe.

15. Disconnect the air conditioning compressor clutch connector and remove the harness clamp. Remove the compressor assembly, without disconnecting the refrigerant lines. Lower the vehicle to the ground.

16. Remove the radiator. Remove the heater hoses. Attach the engine support tool (AAR-T-12566) to the vehicle.

17. Raise and support the vehicle safely. Remove the lower torque rod. Remove the lower radiator hose from the clamp. Remove the front mount retaining bolt.

18. Matchmark the subframe with the edge of the subframe stiffner. Attach special

tool VSB02C000016 by hanging the belt over the front of the subframe. Secure the belt to the stop.

19. Raise the jack, line up the slots in the tool arms with the bolt holes on the corner of the jack base. Attach the four bolts.

20. Remove the subframe. Lower the vehicle to the ground. Remove the side engine mount bracket retaining bolt and nut. Remove the transaxle mount bracket and retaining bolt and nuts.

21. Attach a lifting device. Raise the engine/transaxle assembly up slightly. Remove the engine support tool.

22. Remove the side engine mount bracket. Remove the transaxle mounting bolt and nuts. Remove the ground cable.

23. Be sure that the engine/transaxle assembly is free from all hoses, vacuum lines and electrical wires and connectors.

24. Remove the engine/transaxle assembly from under the vehicle.

To install:

25. Installation is the reverse of the removal procedure. Be sure to check and adjust all required fluid levels.

26. Reprogram the ECM engine idle characteristics. Be sure all electrical items are OFF.

27. Start the engine. Hold the idle speed at 3000 RPM's in neutral until the radiator fan comes on or the temperature reached 194 degrees.

28. Let the engine idle for about five minutes with the throttle fully closed.

29. If the radiator fan comes on during the five minutes, do not count this toward the five minute programming time.

30. Reprogram the crankshaft position (CKP) pattern. Run the engine until the operating temperature reaches 176 degree. With the engine stopped clear the CKP pattern. Turn the ignition switch OFF. Turn the ignition switch ON and wait thirty seconds.

31. Roadtest the vehicle on a level surface. Decelerate the engine speed of 2500 rpm to 1000 rpm. If equipped with automatic transaxle use two Drive positions. If equipped with manual transaxle use first gear.

32. Stop the vehicle, but keep the engine running.

33. Check PULSAR F/B LEARN in the data list with the HDS. If not complete repeat the procedure. If complete, roadtest the vehicle on a level surface. Decelerate the engine speed of 5000 rpm to 3000 rpm. If equipped with automatic transaxle use two Drive positions. If equipped with manual transaxle use first gear.

34. Stop the vehicle, but keep the engine running.

35. Check PULSAR F/B LEARN in the data list with the HDS. If not complete repeat the procedure.

36. If completed, turn the ignition switch OFF. Turn the ignition switch ON, wait thirty seconds. The learning procedure is now complete.

37. Enter the antitheft codes for the radio and the navigation system. Set the clock.

38. Reprogram the power window control unit as follows.

39. Turn the ignition switch ON. Lower the window, all the way down. Open the driver's side door.

40. Turn the ignition switch OFF. Push and hold the drivers window DOWN switch. Turn the ignition switch ON. Release the driver's window DOWN switch. This must be done within five seconds.

41. Repeat the above step three more times. Wait one second.

42. Confirm that the AUTO UP and AUTO DOWN do not work. If they work repeat the procedure, paying close attention to the five second time limit

43. Move the driver's window all the way down by holding the drivers window DOWN switch to the AUTO DOWN position.

44. Pull up and hold the driver's window UP switch to the AUTO UP position until the window reaches the fully closed position. Hold the switch for one second.

45. Confirm that the power window master switch has been reset by using the driver's window AUTO UP and DOWN function.

46. If the AUTO UP and DOWN feature is still not working, repeat the complete procedure, paying close attention to the five second time limit.

S2000

1. Before servicing the vehicle, refer to the Precautions Section.

2. Note the radio security code and the radio presets.

3. Properly relieve the fuel system pressure.

4. Disconnect the negative battery cable, than the positive cable. Remove the hood support from the driver's side of the vehicle. Lock the hood in the full up position.

5. Remove the battery.

6. Raise and support the vehicle safely. Drain the engine oil.

7. Remove the transmission. Lower the vehicle.

8. On 2002–2005 vehicles, disconnect the electronic control module (ECM) connectors from the ECM.

9. On 2006 vehicles, disconnect the dashboard wiring harness connector.

10. On 2002–2005 vehicles, Remove the vacuum tank and throttle cable assembly.

11. Remove the electrical power steering (EPS) retaining bolts. Remove the control unit from its mounting.

12. Remove the battery cable from the main under hood fuse/relay box. Remove the harness clamps.

13. Remove the battery cable from the auxiliary under hood fuse box. Remove the ground cable and harness clamps.

14. Remove the harness retaining grommet from its mounting and pull out the ECM connectors.

15. Remove the radiator cap. Raise and safely support the vehicle. Drain the engine coolant.

16. Remove the engine stop bracket cushion and stop bracket. Lower the vehicle.

17. Disconnect and remove the heater hoses. Remove the lower radiator hose. Remove the upper radiator hose.

18. On 2002–2005 vehicles, remove the fuel feed hose, fuel return hose and brake booster vacuum hose.

19. On 2006 vehicles, Remove the intake manifold cover. Remove the quick connect fitting cover, then disconnect the fuel feed hose and brake booster vacuum hose.

20. Disconnect the evaporative emission (EVAP) canister hose.

21. Properly attach the engine lifting device.

22. Remove the support nut from the left side engine mount bracket. Remove the support nut and the four mounting bolts. Remove the right side engine mount bracket.

23. Raise the engine. Check to insure that no wires, hoses etc are attached. Remove the engine from the vehicle.

To install:

24. Installation is the reverse of the removal procedure, while using the following torque values:

- Right motor mount bracket bolts: 28 ft. lbs. (38 Nm)
- Motor mount nuts: 40 ft. lbs. (54 Nm)
- Front motor mount bolts: 16 ft. lbs. (22 Nm)

25. Reprogram the ECM engine idle characteristics. Be sure all electrical items are OFF.

26. Start the engine. Hold the idle speed at 3000 RPM's in neutral until the radiator fan comes on or the temperature reached 176 degrees.

27. Let the engine idle for about five minutes with the throttle fully closed.

28. If the radiator fan comes on during the five minutes, do not count this toward the five minute programming time.

29. Set the clock.

Accord

2002

1. Before servicing the vehicle, refer to the Precautions Section.

2. Obtain the anti-theft code for the radio, then disconnect the battery cables. Be sure to disconnect the negative cable first.

3. Remove the air intake duct.

4. Secure the hood in the open position with a long prop rod such as P/N 74145-S84-A00.

5. Remove or disconnect the following:
- Both battery cables and the connector from the underhood relay box
- Battery and tray, on the 3.0L engine
- Bolt securing the relay box to the body
- Accelerator and cruise control cables from the throttle body and bracket

6. Properly relieve the fuel system pressure.
- Fuel hoses from the fuel rail
- Brake booster vacuum and evaporative emissions (EVAP) canister hoses
- Vacuum hose from the canister
- Hose securing the power steering hose on the engine
- Power steering pump belt, then remove the pump and position it out of the way. Use wire if necessary.

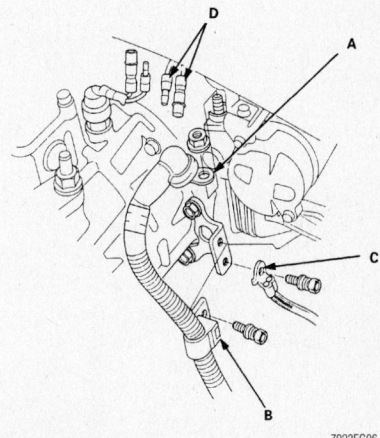

Starter cable, clamp, ground cable and back-up light switch connector locations— 2.3L engine

7923FG06

- Powertrain Control Module (PCM) connectors from the control module. Remove the grommet and pull the connectors through.
- Wiring harness connectors at the right side of the engine compartment for 2.3L engine and on the left side for the 3.0L engine.

7. On the 2.3L engine, remove the starter cable (A) and clamp (B). Remove the ground cable (C) and back-up light switch connectors (D). On the 3.0L engine, remove the starter wiring from the engine compartment attaching points.

8. On vehicles with a manual transaxle, disconnect the shift and select cables from the transaxle. Remove the slave cylinder mounting bolts and position the cylinder out of the way. Be sure not to bend the line.

9. Remove or disconnect the following:
- Rear engine mount through-bolt and stiffener
- Front engine mount bracket mounting bolts and loosen the through-bolt
- Radiator cap

10. Raise and safely support the vehicle.

11. Remove or disconnect the following:
- Front tires
- Engine under cover

12. Loosen the radiator drain plug and drain the coolant.

13. Drain the transaxle oil or fluid, and then reinstall the plug using a new washer.

14. Drain the engine oil, and then reinstall the plug using a new washer.

15. Lower the vehicle and remove the upper and lower radiator hoses and heater hoses from the engine.

16. On vehicles with an automatic transaxle, disconnect the ATF fluid cooler lines.

17. Remove the air conditioning com-

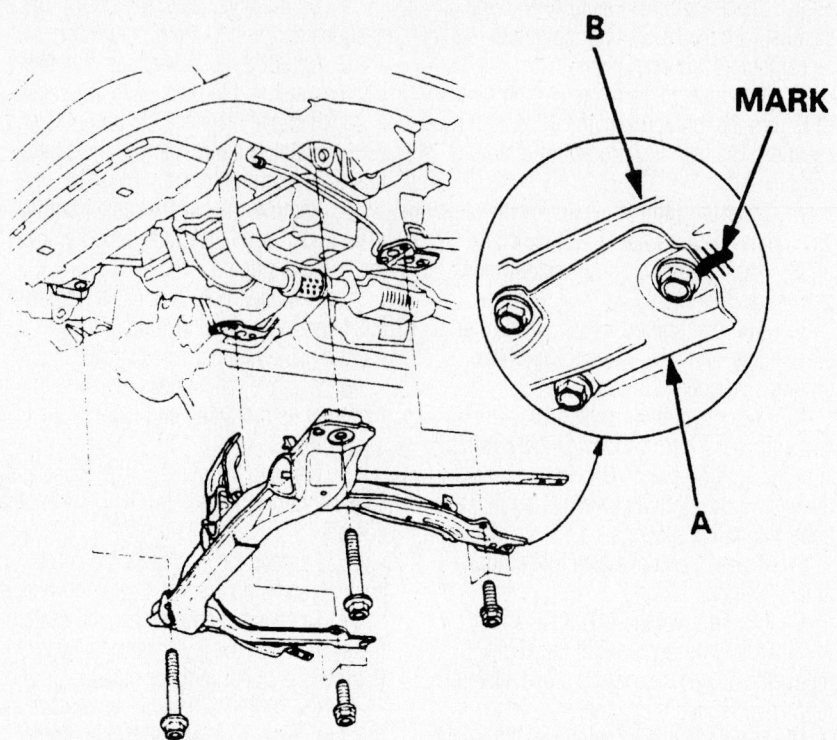

Be sure to mark the location of the front beams (A) on the rear beams (B) before removing the subframe—2.3L engine

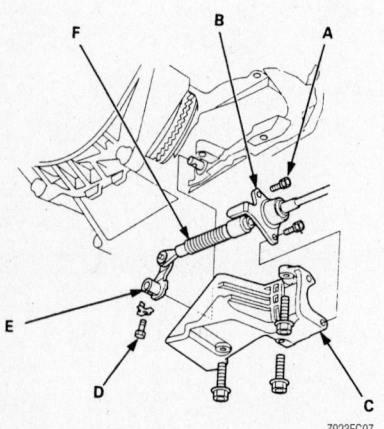

Automatic transaxle linkage components—2.3L engine

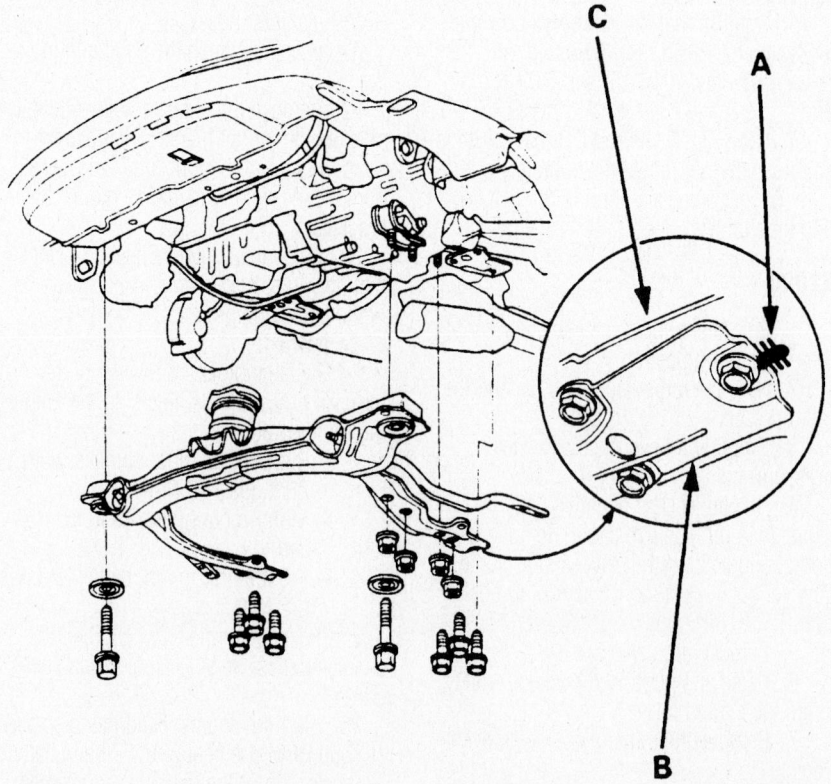

Mark the location of the front beams (A) on the rear beams (B) before removing the subframe—3.0L engine

pressor from the engine and position it to the side without disconnecting the hoses.

18. Raise the vehicle and remove the front exhaust pipe.

19. On vehicles with automatic transaxle, remove the 2 bolts (A) for the shift cable holder (B), then remove the shift cable cover (C). To prevent damage to the linkage, be sure to remove the shift cable holder before removing the bolts for the cover.

20. Remove or disconnect the following:
- Lockbolt (D) from the control lever (E), then the shift cable (F) with the control lever
- Through-bolt securing the bottom of the shock absorber to the control arm
- Halfshafts
- Rear engine mounting bracket
- 2 flange bolts from each of the radius rods

21. Mark the location of the front beams (A) on the rear beams (B). Remove the 4 bolts and the subframe.

22. Lower the vehicle about half way and attach a chain hoist to the engine lifting points as shown. Apply slight upward pressure to the engine/transaxle assembly.

23. Remove the remaining engine and transaxle mounting brackets.

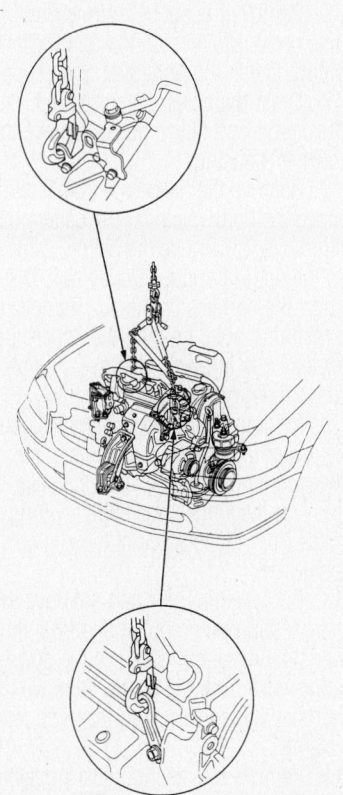

7923FG09

Engine lifting points—2.3L engine shown, 2.4L engine similar

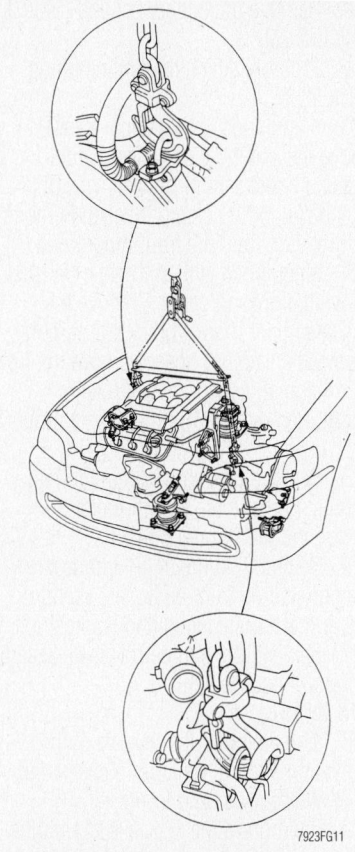

7923FG11

Engine lifting points—3.0L engine

24. Lower the engine about 6 in. (150mm) and check that the engine/transaxle is free of any hoses, cables or wiring.

25. Lower the assembly completely and remove it from under the vehicle.

To install:

26. Lift the engine into position and install the engine mounting brackets. Tighten the retainers as follows:

    a. On the 2.3L engine, tighten the engine mounting bolts and nuts to 40 ft. lbs. (54 Nm).

    b. On the 3.0L, tighten the bolts to 28 ft. lbs. (38 Nm).

27. On the 3.0L engine, install the air conditioning compressor. Tighten the bolts to 16 ft. lbs. (22 Nm).

28. Install the transaxle mounting bracket, and tighten the retainers as follows:

    a. On the 2.3L engine, tighten the nuts to 28 ft. lbs. (38 Nm) and the through-bolt to 40 ft. lbs. (54 Nm).

    b. On the 3.0L engine, tighten the bolts to 28 ft. lbs. (38 Nm).

29. Install the sub-frame in its original position, and tighten the retainers as follows:

    a. On the 2.3L engine, tighten the rear bolts to 47 ft. lbs. (64 Nm) and the front bolts to 76 ft. lbs. (103 Nm).

    b. On the 3.0L engine, tighten the rear bolts to 40 ft. lbs. (54 Nm), front bolts to 76 ft. lbs. (103 Nm) and the nuts to 28 ft. lbs. (38 Nm).

30. On the 2.3L engine, install or connect the following:
- Radius rod bolts: 119 ft. lbs. (162 Nm)
- Rear mount bracket: 40 ft. lbs. (54 Nm)
- Stiffener. Tighten the through-bolt to 47 ft. lbs. (64 Nm) for manual transaxles or the nut and bolt to 28 ft. lbs. (38 Nm) for automatic transaxles.
- 3 front mounting bracket bolts: 28 ft. lbs. (38 Nm). Then, tighten the through-bolt to 47 ft. lbs. (64 Nm).
- Air conditioning compressor: 16 ft. lbs. (22 Nm)

31. On the 3.0L engine, install or connect the following:
- Radius rod bolts: 119 ft. lbs. (162 Nm)
- Front mounting bracket support nut: 40 ft. lbs. (54 Nm)
- Rear mounting bracket nut and bolt. Tighten the nut to 40 ft. lbs. (54 Nm) and the bolt to 28 ft. lbs. (38 Nm).
- Side mounting bracket. Tighten the bolts to 40 ft. lbs. (54 Nm) and the through-bolt to 40 ft. lbs. (54 Nm).
- Exhaust system
- Shift linkage, if equipped with an automatic transaxle

32. The remainder of the installation is the reverse of the removal.

33. Refill and bleed the cooling system.

❋❋ WARNING

Operating the engine without the proper amount and type of engine oil will result in severe engine damage.

34. Fill the engine with the correct amount of oil.

35. Install the battery if removed. Start the engine and check for leaks.

2003–2006 2.4L ENGINE

1. Before servicing the vehicle, refer to the Precautions Section.

2. Note the radio security code and the radio presets.

3. Properly relieve the fuel system pressure.

4. Disconnect the negative battery cable, than the positive cable. Remove the hood support from the driver's side of the vehicle. Lock the hood in the full up position.

5. Remove the battery.

6. Disconnect the air intake sensor. Remove the vacuum hose, and breather pipe then remove the intake air duct.

7. Remove the air cleaner housing. Remove the harness clamp.

8. Remove the harness terminal wires. Remove the bolts retaining the battery base and remove it. Remove the battery cables from the under hood fuse/relay box. Disconnect the harness connector. Remove the two retaining bolts holding the under hood fuse/relay box in place. Remove the box.

9. On 2003–2005 vehicles, remove the throttle cable and cruise control actuator cable.

10. On 2006 vehicles, remove the harness clamp and the strut brace.

11. Remove the quick connect fitting cover. Disconnect the fuel feed hose. Remove the evaporative emission canister hose and brake booster vacuum hose.

12. On 2003–2005 vehicles, remove the harness clamp then disconnect the engine wiring harness connectors on the left side of the engine compartment.

13. On 2006 vehicles, disconnect the ECM/PCM connectors and the main wire harness connectors. Disconnect the accelerator pedal position (APP) sensor connector. Remove the harness clamps and grommet. Pull the wire harness thru the bulkhead.

14. If equipped with manual transaxle, remove the clutch slave cylinder and clutch line bracket mounting bolts. Remove the shift and select cables.

15. Disconnect the air fuel sensor connector. Remove the drive belt.

16. Remove the power steering pump. Remove the air conditioning compressor without disconnecting the air conditioning hoses. Position the compressor to the side.

17. Remove the radiator cap. Raise and support the vehicle safely. Remove the front tires. Drain the radiator. Drain the transaxle. Drain the engine oil.

18. Disconnect the stabilizer links. Remove the damper fork. Separate the knuckles from the lower arm ball joints. Remove the driveshafts. Remove the exhaust pipe.

19. On automatic transaxle remove the bolts securing the shift cable holder. Remove the shift cable cover. Remove the spring clip and control pin then separate the shift cable from the control lever.

20. Remove the transaxle lower front and lower rear mounts. Lower the vehicle. On automatic transaxle models, remove the fluid cooler lines.

21. Remove the upper radiator hose and heater hoses. Remove the lower radiator

hoses. Remove the battery ground cable and upper bracket.

22. Properly attach the engine lifting device.

23. If equipped with manual transaxle, remove the ground cable. Remove the transaxle upper mount/bracket assembly and the clutch line clamp bracket. Remove the front stop then the front mount bolt.

24. If equipped with automatic transaxle, remove the ground cable. Remove the transaxle upper mount/bracket assembly. Remove the vacuum hose. Remove the front mount stop and vacuum hose clamp bracket. Remove the front transaxle mount bolt.

25. Remove the rear mount stop retaining nuts. Remove the rear mount stop. Remove the rear mount bolt.

26. Be sure that the engine/transaxle assembly is free from all hoses, vacuum lines and electrical wires and connectors.

27. Remove the engine/transaxle assembly from under the vehicle.

To install:

28. Installation is the reverse of the removal procedure. Be sure to check and adjust all required fluid levels.

29. Reprogram the ECM/PCM with the HDS, turn the ignition switch the OFF position. Turn the ignition switch to the ON position. Wait thirty seconds. Turn the ignition switch OFF and disconnect the HDS from the DLC.

30. Reprogram the ECM engine idle characteristics. Be sure all electrical items are OFF.

31. Start the engine. Hold the idle speed at 3000 RPM's in neutral until the radiator fan comes on or the temperature reached 176 degrees.

32. Let the engine idle for about five minutes with the throttle fully closed.

33. If the radiator fan comes on during the five minutes, do not count this toward the five minute programming time.

34. Enter the antitheft codes for the radio and the navigation system. Set the clock.

35. Reprogram the power window control unit as follows.

36. Turn the ignition switch ON. Lower the window, all the way down. Open the driver's side door.

37. Turn the ignition switch OFF. Push and hold the drivers window DOWN switch. Turn the ignition switch ON. Release the driver's window DOWN switch. This must be done within five seconds.

38. Repeat the above step three more times. Wait one second.

39. Confirm that the AUTO UP and AUTO DOWN do not work. If they work

repeat the procedure, paying close attention to the five second time limit

40. Move the driver's window all the way down by holding the drivers window DOWN switch to the AUTO DOWN position.

41. Pull up and hold the driver's window UP switch to the AUTO UP position until the window reaches the fully closed position. Hold the switch for one second.

42. Confirm that the power window master switch has been reset by using the driver's window AUTO UP and DOWN function.

43. If the AUTO UP and DOWN feature is still not working, repeat the complete procedure, paying close attention to the five second time limit.

2003–2006 3.0L ENGINE

1. Before servicing the vehicle, refer to the Precautions Section.

2. Note the radio security code and the radio presets.

3. Properly relieve the fuel system pressure.

4. Disconnect the negative battery cable, than the positive cable. Remove the battery.

5. Remove the windshield wiper arms. Remove the cowl cover. Remove the bulkhead cover.

6. Remove the support struts from the engine hood. Move the hood to a vertical position. Install the right side hood support.

7. Drain the power steering fluid. Plug the reservoir and return hose. Remove the air cleaner housing.

8. Remove the harness terminal wires. Remove the bolts retaining the battery base and remove it. Remove the battery cables from the under hood fuse/relay box. Disconnect the harness connector. Remove the two retaining bolts holding the under hood fuse/relay box in place. Remove the box. Remove the ground cable.

9. Remove the brake booster vacuum hose, evaporative emission canister hose and vacuum hose. Remove the harness clamp and disconnect the engine wiring harness from the left side of the engine compartment.

10. Remove the drive belt. Remove the power steering pump outlet line from the power steering pump. Remove the power steering hose from the clamp. Remove the power steering system fluid reservoir from the clamp.

11. Remove the steering gear protective cover. Lock the steering wheel in position. Make a reference mark across the steering joint and steering gear box pinion shaft. Remove the steering joint bolt. Disconnect

the steering joint from the steering gearbox pinion shaft.

➡**To prevent damage to the cable reel, do not turn the steering wheel once the steering joint has been removed.**

12. Disconnect the fan motor connector. Disconnect the compressor clutch connector. Remove the coolant tank.

13. Remove the condenser fan shroud retaining bolts. Remove the fan shroud. Remove the four bolts retaining the air conditioning compressor.

14. If equipped with manual transaxle, remove the ground cable from the shift cable holder. Remove the bolts retaining the shift cable holder. Remove the shift cable and the select cable. Remove the clutch slave cylinder mounting bolt. Remove the slave cylinder.

➡**Do not operate the clutch pedal once the slave cylinder has been removed.**

15. Remove the radiator cap. Raise and safely support the vehicle. Remove the front tires. Remove the splash pan.

16. Drain the radiator. Drain the engine oil. Drain the transaxle.

17. Disconnect the stabilizer links. Remove the damper fork. Separate the tie rod end ball joints from the knuckles. Separate the knuckles from the lower arms.

18. Remove the driveshafts. Remove the exhaust pipe.

19. On automatic transaxle, Remove the shift cable cover retaining bolt. Remove the cover. Remove the lock bolt retaining then control lever, remove the lever assembly.

➡**To prevent damage to the control lever joint, be sure to remove the bolts retaining the shift cable holder before removing the bolts retaining the shift cable cover.**

20. Remove and plug the power steering line hose. Disconnect the power steering pressure switch connector.

21. Remove the nuts retaining the transaxle lower front mount and lower rear mount. Lower the vehicle to the ground.

22. Remove the radiator hoses. Remove the heater hoses. Remove the connector clamp from the front and rear cylinder head.

➡**Removal of these bolts is necessary if using engine support tool AAR-T-12566 and VSB02C000919.**

23. Lift and support the engine, using the engine support tool.

24. If equipped with manual transaxle, remove the ground cable. If equipped with

automatic transaxle, remove the transaxle upper mount bracket and vacuum hose.

25. Remove the front mount stop. Remove the front mount bolt. Remove the rear mount stop. Remove the rear mount bolt.

26. If equipped with manual transaxle, remove the two bolts retaining the chift cable bracket. Raise and safely support the vehicle.

27. Matchmark the subframe with the edge of the subframe stiffner. Attach special tool VSB02C000016 by hanging the belt over the front of the subframe. Secure the belt to the stop.

28. Raise the jack, line up the slots in the tool arms with the bolt holes on the corner of the jack base. Attach the four bolts.

29. Remove the subframe mid mount. Remove the subframe. Lower the vehicle to the ground.

30. Attach a lifting device. Raise the engine/transaxle assembly up slightly. Remove the engine support tool.

31. Remove the two bolts retaining the side engine mount bracket.

32. Be sure that the engine/transaxle assembly is free from all hoses, vacuum lines and electrical wires and connectors.

33. Remove the engine/transaxle assembly from under the vehicle.

To install:

34. Installation is the reverse of the removal procedure. Be sure to check and adjust all required fluid levels.

35. Reprogram the ECM/PCM with the HDS, turn the ignition switch the OFF position. Turn the ignition switch to the ON position. Wait thirty seconds. Turn the ignition switch OFF and disconnect the HDS from the DLC.

36. Reprogram the crankshaft position (CKP) pattern. Run the engine until the operating temperature reaches 176 degree. With the engine stopped clear the CKP pattern. Turn the ignition switch OFF. Turn the ignition switch ON and wait thirty seconds.

37. Roadtest the vehicle on a level surface. Decelerate the engine speed of 2500 rpm to 1000 rpm. If equipped with automatic transaxle use two Drive positions. If equipped with manual transaxle use first gear.

38. Stop the vehicle, but keep the engine running.

39. Check PULSAR F/B LEARN in the data list with the HDS. If not complete repeat the procedure. If complete, roadtest the vehicle on a level surface. Decelerate the engine speed of 5000 rpm to 3000 rpm. If equipped with automatic transaxle use two Drive positions. If equipped with manual transaxle use first gear.

40. Stop the vehicle, but keep the engine running.

41. Check PULSAR F/B LEARN in the data list with the HDS. If not complete repeat the procedure.

42. If completed, turn the ignition switch OFF. Turn the ignition switch ON, wait thirty seconds. The learning procedure is now complete.

43. Enter the antitheft codes for the radio and the navigation system. Set the clock.

44. Reprogram the power window control unit as follows.

45. Turn the ignition switch ON. Lower the window, all the way down. Open the driver's side door.

46. Turn the ignition switch OFF. Push and hold the drivers window DOWN switch. Turn the ignition switch ON. Release the driver's window DOWN switch. This must be done within five seconds.

47. Repeat the above step three more times. Wait one second.

48. Confirm that the AUTO UP and AUTO DOWN do not work. If they work repeat the procedure, paying close attention to the five second time limit

49. Move the driver's window all the way down by holding the drivers window DOWN switch to the AUTO DOWN position.

50. Pull up and hold the driver's window UP switch to the AUTO UP position until the window reaches the fully closed position. Hold the switch for one second.

51. Confirm that the power window master switch has been reset by using the driver's window AUTO UP and DOWN function.

52. If the AUTO UP and DOWN feature is still not working, repeat the complete procedure, paying close attention to the five second time limit.

Water Pump

REMOVAL & INSTALLATION

1.7 Engine

➡**The original radio contains a coded anti-theft circuit. Obtain the security code number before disconnecting the battery cables.**

1. Before servicing the vehicle, refer to the Precautions Section.

2. Drain the cooling system.

3. Remove the timing belt

4. Remove the water pump retaining bolts. Remove the water pump from the engine.

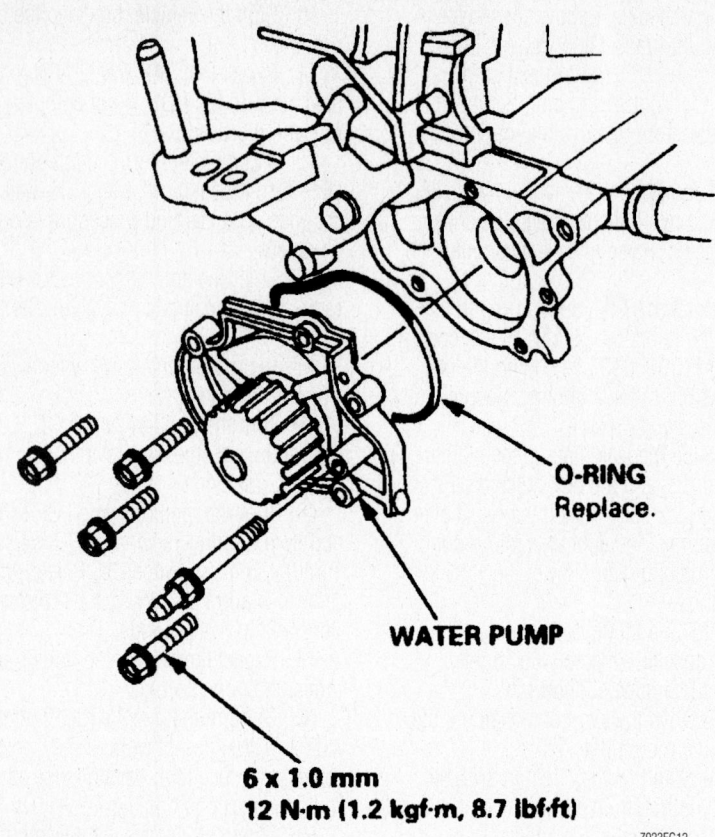

O-RING
Replace.

WATER PUMP

6 x 1.0 mm
12 N·m (1.2 kgf·m, 8.7 lbf·ft)

7923FG12

Water pump—1.7L and 2.3L engines

To install:

5. Clean the water pump and O-ring mating surfaces before installation.

6. Install the water pump to the engine using a new O-ring. Install the timing belt.

7. Refill the radiator. Start the engine and check for leaks.

1.8L, 2.0L, 2.2L and 2.4L Engines

1. Before servicing the vehicle, refer to the Precautions Section.

2. Drain the cooling system.

3. Remove the drive belt.

4. On Hatchback and 2006 Civic with 2.0L engine, remove the crankshaft pulley.

5. On S2000, remove the water pump pulley.

6. On Accord and 2006 Civic with 2.0L engine, remove the drive belt automatic tensioner.

➡It may first be necessary to remove the power steering pump from its mounting, without disconnecting the fluid hoses.

7. Remove the water pump retaining bolts. Remove the water pump from the engine.

To install:

8. Clean the water pump and O-ring mating surfaces before installation.

9. Install the water pump to the engine using a new O-ring.

10. Continue the installation in the reverse order of the removal procedure.

11. Refill the radiator. Start the engine and check for leaks.

2.3L Engine

➡**The original radio contains a coded anti-theft circuit. Obtain the security code number before disconnecting the battery cables.**

1. Before servicing the vehicle, refer to the Precautions Section.

2. Remove or disconnect the following:
   • Negative battery cable

3. Drain the cooling system.

4. Remove or disconnect the following:
   • Accessory drive belts, the valve cover, and the upper timing belt cover

5. Set the timing at Top Dead Center (TDC)/compression for No. 1 piston.

6. Remove or disconnect the following:
   • Crankshaft pulley and lower timing belt cover
   • Timing belt. Replace the timing belt if it is contaminated with oil or coolant or shows any signs of wear and damage.

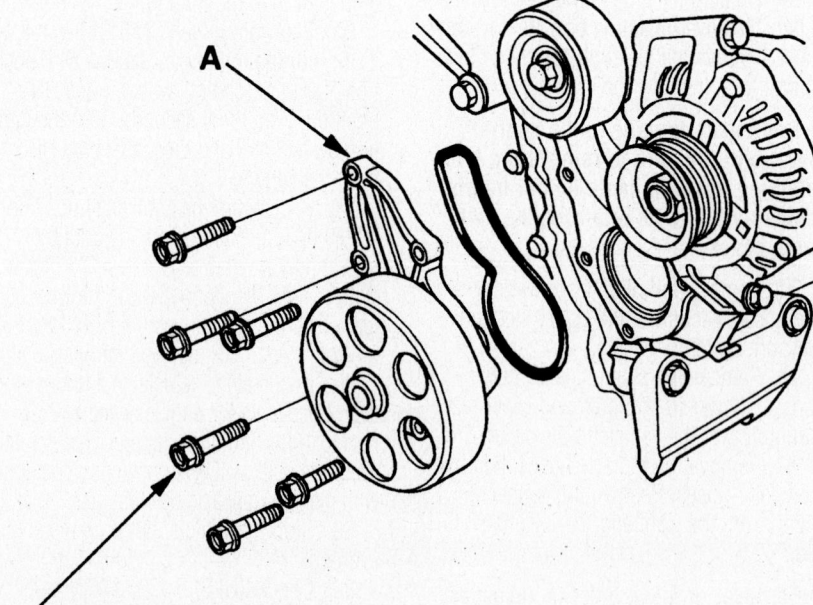

A

6 x 1.0 mm
12 N·m (1.2 kgf·m, 8.7 lbf·ft)

43256-ACCO-G02

Water pump mounting—except 1.7L, 2.3L and 3.0L engines

- Crankshaft Speed Fluctuation (CKF) sensor bracket and move the sensor out of the way, if equipped. Cover the sensor with a shop towel to keep coolant off of it.
- Water pump from the engine block.

To install:

7. Clean the water pump and O-ring mating surfaces before installation.

8. Install or connect the following:

- Water pump with a new O-ring. Coat only the bolt threads with liquid gasket and tighten them to 108 inch lbs. (12 Nm).
- Timing belt. Be sure it is fitted and adjusted properly.
- CKF sensor, if equipped, and tighten the bracket bolts to 108 inch lbs. (12 Nm).
- Lower belt cover and crankshaft pulley
- Upper timing belt cover, the valve cover, and the accessory drive belts

9. Be sure the cooling system drain plug is closed. Refill and bleed the cooling system.

10. Connect the negative battery cable and enter the radio security code.

11. Start the engine, allow it to reach normal operating temperature, and check for coolant leaks.

3.0L Engine

1. Before servicing the vehicle, refer to the Precautions Section.
2. Drain the cooling system.
3. Remove the timing belt.
4. Remove the water pump retaining bolts. Remove the water pump from the engine.

To install:

5. Clean the water pump and O-ring mating surfaces before installation.
6. Install the water pump to the engine using a new O-ring. Install the timing belt.
7. Refill the radiator. Start the engine and check for leaks.

Heater Core

REMOVAL AND INSTALLATION

Civic

2002–2005

➡**Make sure to acquire the anti-theft code from the radio and write down the frequencies for the radio's preset buttons.**

1. Disconnect the negative battery cable.

✱✱ CAUTION

After disconnecting the negative battery cable, wait for at least 3 minutes for the air bag module to deplete its energy before working the on the instrument panel or steering wheel.

2. Discharge the air conditioning system. Disconnect the suction and the receiver lines from the evaporator core.
3. Drain the engine coolant.
4. From under the hood, open the cable clamp. Disconnect the heater control cable from the valve. Turn the valve to the fully opened position. Disconnect the heater hoses from heater core assembly.
5. Remove the heater control valve mounting bolt and remove the valve. Remove the heater unit mounting nut. Take care not to bend or damage fuel and brake lines.
6. Remove the center console as follows.

- If equipped with manual transaxle, remove the shifter knob
- Detach the center console retaining clips and pull the assembly upward
- If equipped with heated seats, detach the heat switch connectors
- Pry up on the right side of the console pocket, detach the clips and remove the pocket
- Remove the screws, detach the clips and disconnect the front power accessory connector
- Open the console lid and remove the console mat
- If equipped, disconnect the rear accessory power socket electrical connector
- Remove the console retaining screws and remove the console from the vehicle

7. Remove the drivers side lower dashboard cover.
8. Remove the passengers side lower dashboard cover.
9. Remove the center pocket assembly.
10. Remove the drivers and passengers kick panels. Remove the A-pillar trim panels.
11. Remove the driver's side air bag. Remove the steering wheel and cable reel. Remove the steering column covers.
12. Remove the combination switch assembly from the column shaft by disconnecting the connectors and removing the screws. Disconnect the ignition switch connectors from the under dash fuse/relay box.
13. Remove the steering joint bolt, at the

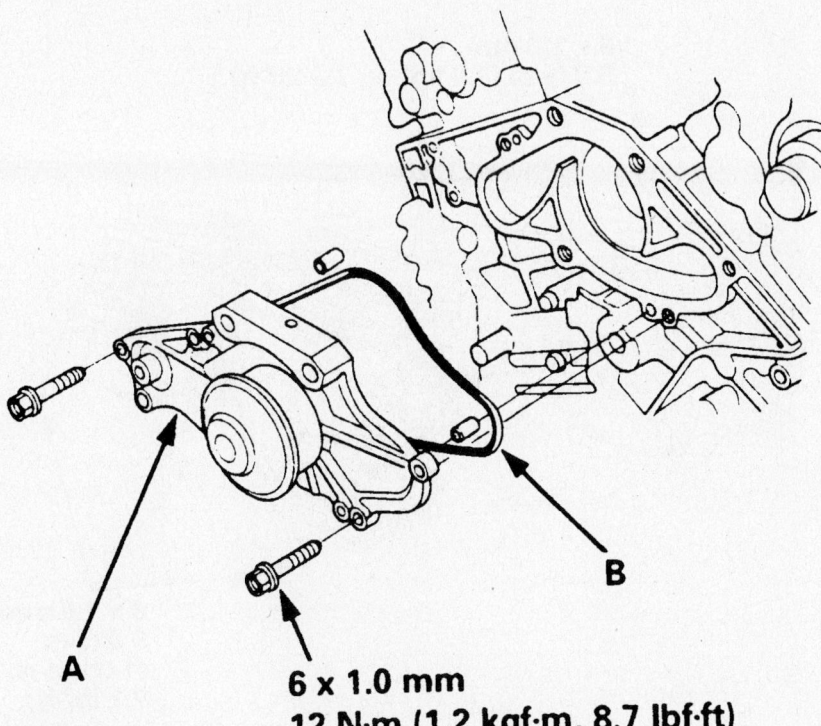

6 x 1.0 mm
12 N·m (1.2 kgf·m, 8.7 lbf·ft)

7923FG14

Exploded view of the water pump mounting—3.0L engine

base of the column. Remove the steering column retaining bolts and remove the column.

14. If equipped with manual transaxle, remove the rear vent duct. Reposition the shift lever for clearance.

15. From under the center dash, remove the rear vent duct. Open the passenger's side door and pull outward on the bottom of the dashboard side cover to release the lower clip. Remove the cover.

16. On the driver's side disconnect all electrical connectors, air hoses and harness that interfere with the removal of the dashboard.

17. On vehicles equipped with automatic transaxle, disconnect the parking pin shift connector and the shift lock solenoid connector.

18. In the middle section of the dashboard disconnect all electrical connectors, air hoses and harness that interfere with the removal of the dashboard.

19. On the passenger's side disconnect all electrical connectors, air hoses and harness that interfere with the removal of the dashboard. Disconnect the ECM/PCM connector.

20. Open the driver's side door remove the bolts and clips after remove their protective caps. Lift upward on the dashboard/steering hanger beam to release if from the guide pins. Be sure all electrical connectors are unplugged.

21. Carefully remove the dashboard assembly from the vehicle.

22. Remove the passengers side lower dashboard cover, the right kick panel and the glovebox.

23. Remove the dash brace center bolt. Cut the plastic brace in the glove box opening. Discard the plastic piece.

24. Remove the relays. Remove the bolts and the glovebox frame. Remove the ECM/PCM.

25. Disconnect the connectors from the air mix control motor, evaporator temperature sensor, power transistor and recirculation control motor.

26. Remove the wiring harness clips, connector clip and wire harness. Remove the heater ducts.

27. Fold back the carpet and remove the mounting bolts. Disconnect the drain hose. Remove the blower unit from the vehicle.

28. Remove the self taping screws and the expansion valve cover. Pry out the evaporator core.

29. Remove the self taping screws, remove the grommet and carefully pry out the heater core.

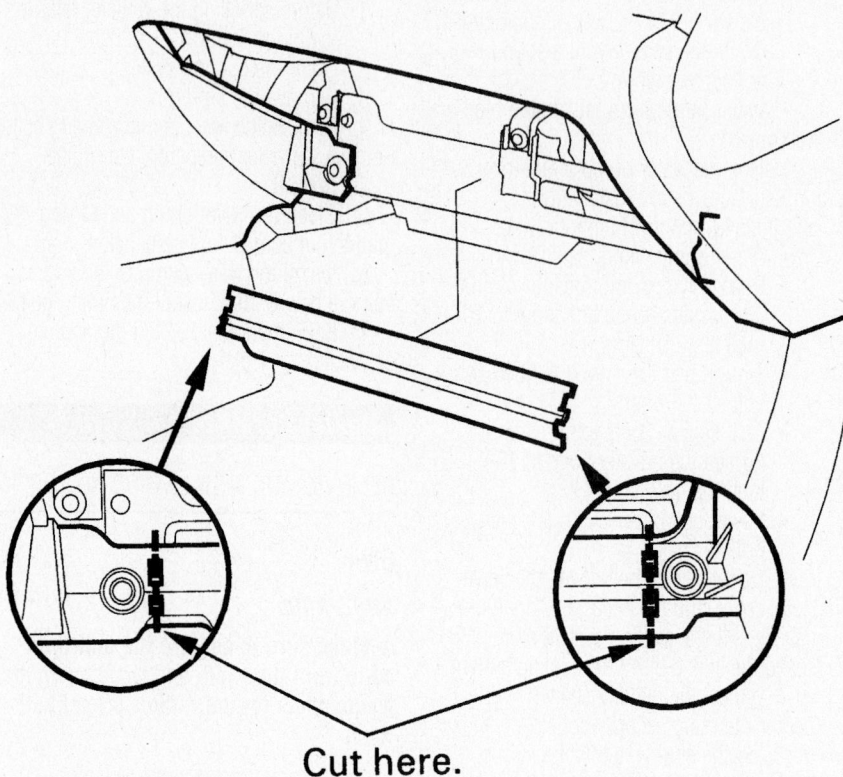

Cut here.

09474_ACCO_G0005

Glovebox cut out location—2002—2005 Civic

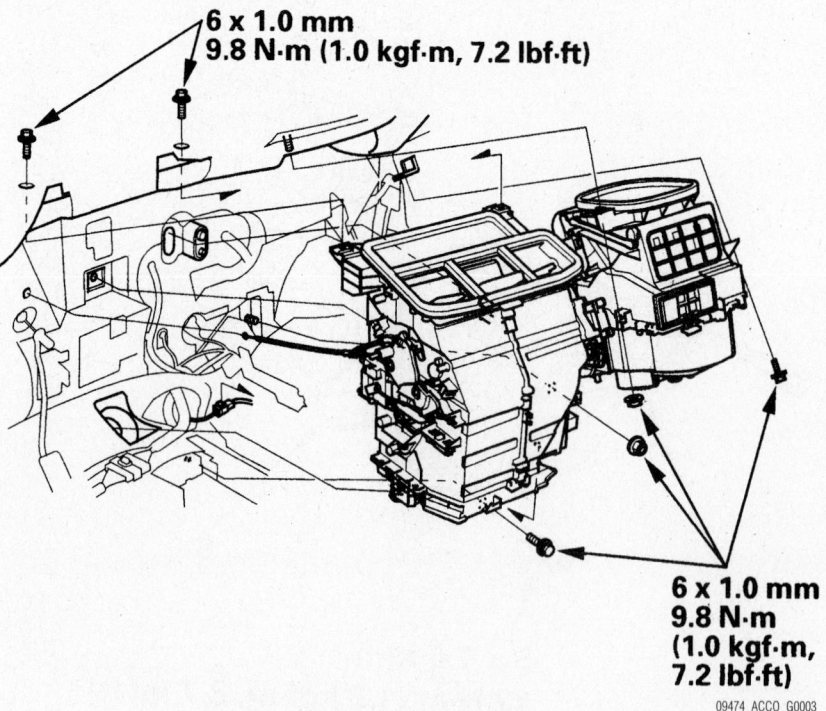

6 x 1.0 mm
9.8 N·m (1.0 kgf·m, 7.2 lbf·ft)

6 x 1.0 mm
9.8 N·m
(1.0 kgf·m,
7.2 lbf·ft)

09474_ACCO_G0003

View of the heater housing, evaporator housing and related components—2002—2005 Civic

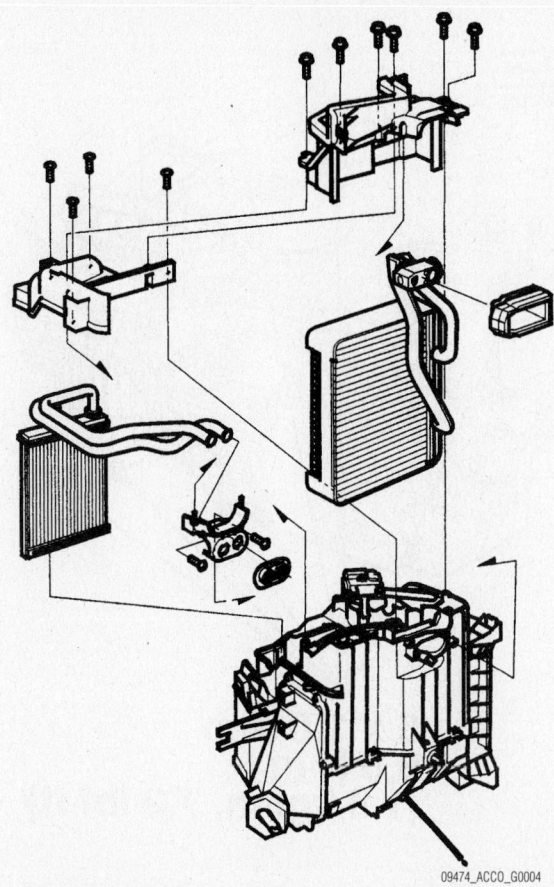

09474_ACCO_G0004

Exploded view of the heater core—2002—2005 Civic

To install:

30. Installation is the reverse of the removal procedure.

31. Evacuate, charge and leak test the air conditioning system refrigerant.

32. Operate the engine to normal operating temperatures; then, check the climate control operation and check for leaks.

33. Enter the antitheft codes for the radio and the navigation system.

34. Reprogram the ECM/PCM engine idle characteristics. Be sure all electrical items are OFF.

35. Start the engine. Hold the idle speed at 3000 RPM's in park or neutral until the radiator fan comes on or the temperature reached 194 degrees.

36. Let the engine idle for about five minutes with the throttle fully closed.

37. If the radiator fan comes on during the five minutes, do not count this toward the five minute programming time.

38. Set the clock.

2006

➡**Make sure to acquire the anti-theft code from the radio and write down the frequencies for the radio's preset buttons.**

1. Disconnect the negative battery cable.

✳✳ CAUTION

After disconnecting the negative battery cable, wait for at least 3 minutes for the air bag module to deplete its energy before working the on the instrument panel or steering wheel.

2. Discharge the air conditioning system.

3. Remove the air cleaner assembly. Disconnect the heater hoses. Remove the mounting nut from the heater unit.

4. Remove the subdisplay visor. Remove the navigation system, if equipped. On vehicles without navigation system remove the radio.

5. Remove the dashboard retaining screws. Detach the retaining clips along the lower edge of the instrument panel.

6. Detach the retaining clips along the upper edge of the instrument panel. Gently pull forward to release the hooks from the holder of the gauge module.

7. Remove the instrument panel.

8. Disconnect the electrical connector from the blower motor. Remove the wire harness clip. Disconnect the connector from the recirculation control motor.

9. Disconnect the connectors from the mode control motor, the evaporator temperature sensor and the power transistor. Remove the wire clip harness.

10. Disconnect the connectors from the air mix control motor and the air conditioning wire harness. Remove the connector clip and the wire harness clips.

11. Remove the mounting bolt, mounting nuts and the heater/blower unit from the vehicle.

12. Remove the self taping screws, remove the grommet and carefully pry out the heater core.

To install:

13. Installation is the reverse of the removal procedure.

14. Evacuate, charge and leak test the air conditioning system refrigerant, as required.

15. Operate the engine to normal operating temperatures; then, check the climate control operation and check for leaks.

16. Enter the antitheft codes for the radio and the navigation system.

17. Reprogram the power window control unit as follows.

18. Turn the ignition switch ON. Lower the window, all the way down. Open the driver's side door.

19. Turn the ignition switch OFF. Push and hold the drivers window DOWN switch. Turn the ignition switch ON. Release the driver's window DOWN switch. This must be done within five seconds.

20. Repeat the above step three more times. Wait one second.

21. Confirm that the AUTO UP and AUTO DOWN do not work. If they work repeat the procedure, paying close attention to the five second time limit

22. Move the driver's window all the way down by holding the drivers window DOWN switch to the AUTO DOWN position.

23. Pull up and hold the driver's window UP switch to the AUTO UP position until the window reaches the fully closed position. Hold the switch for one second.

24. Confirm that the power window master switch has been reset by using the driver's window AUTO UP and DOWN function.

25. If the AUTO UP and DOWN feature is still not working, repeat the complete procedure, paying close attention to the five second time limit.

26. Set the clock.

S2000

➡**Make sure to acquire the anti-theft code from the radio and write down the frequencies for the radio's preset buttons.**

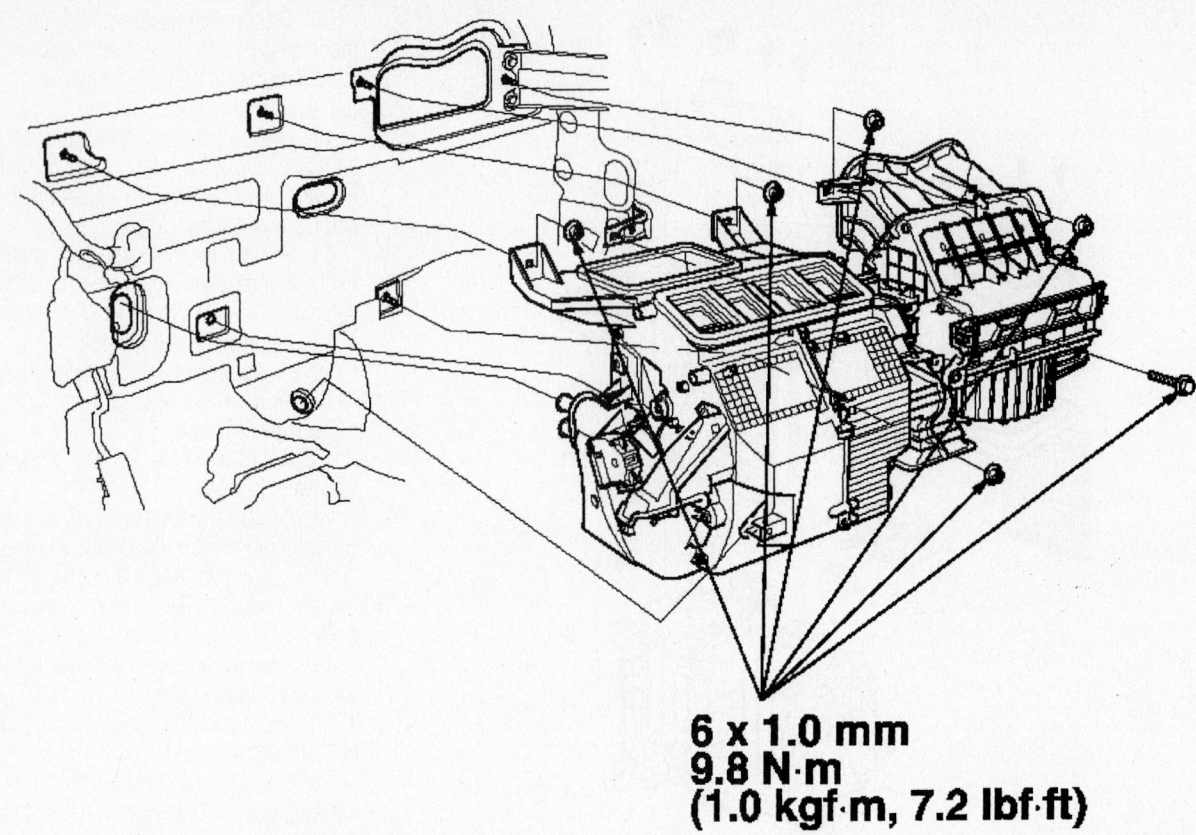

**6 x 1.0 mm
9.8 N·m
(1.0 kgf·m, 7.2 lbf·ft)**

09474_ACCO_G0006

View of the heater housing, evaporator housing and related components—2006 Civic

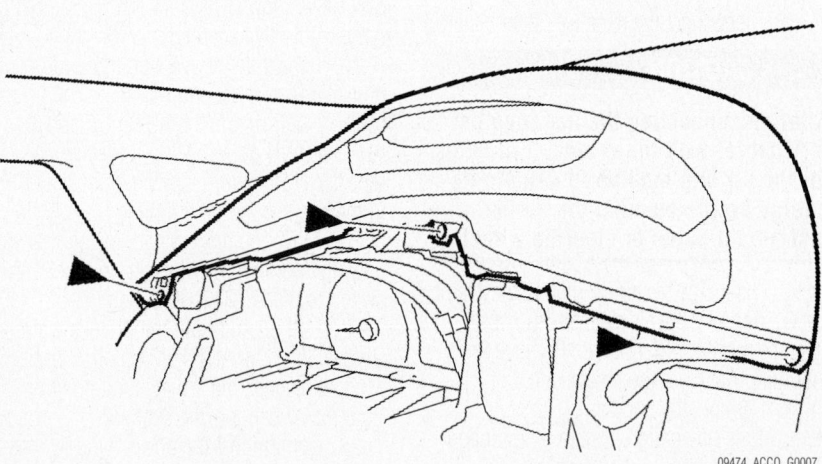

09474_ACCO_G0007

Dash pad retaining screw locations—2006 Civic

1. Disconnect the negative battery cable.

2. Drain the engine coolant. Disconnect the heat shield from the exhaust manifold. Remove the battery.

3. Discharge the air conditioning system. Disconnect the suction and the receiver lines from the evaporator core.

4. From under the hood, open the cable clamp. Disconnect the heater control cable from the valve. Turn the valve arm to the fully open position.

5. Remove the mounting bolt from the heater valve. Disconnect the heater hoses from the heater unit. Remove the mounting nut from the heater unit. Disconnect the air conditioning lines from the evaporator housing.

6. Remove the radio panel retaining clips and remove the panel. Remove the radio retaining screws and pull the unit forward. Disconnect the antenna lead and electrical connectors and remove the radio.

7. Remove the driver's side air bag. Remove the steering wheel and cable reel. Remove the steering column covers.

8. Remove the combination switch assembly from the column shaft by disconnecting the connectors and removing the screws. Disconnect the ignition switch connectors from the under dash fuse/relay box.

9. Remove the steering joint bolt, at the base of the column. Remove the steering column retaining bolts and remove the column.

10. Remove the drivers side lower dash cover. Remove the passenger's side lower dash cover. Remove the drivers and passengers side front console cover.

11. Remove the passenger's side air bag assembly.

12. Remove the kick panels and A-pilar trim from both the drivers and passenger's side.

13. Disconnect all electrical connectors, air hoses and harness that interfere with the removal of the dashboard. Remove the ground bolts.

14. Open the driver's side door remove the bolts and clips after remove their protective caps. Lift upward on the dashboard/steering hanger beam to release if from the guide pins. Be sure all electrical connectors are unplugged.

15. Remove the blower/evaporator housing from the vehicle.

16. Remove the mounting bolts, the center brackets and the radio mounting brackets.

17. Remove the SRS unit. Remove the defroster outlet and wire harness clips.

18. Disconnect the connectors from the mode control motor and the air mix control motor. Remove the wire harness clip.

19. Remove the mounting nuts, the mounting bolts and remove the heater unit from the vehicle.

20. Remove the heater core cover. Remove the heater core.

To install:

21. Installation is the reverse of the removal procedure.

22. Evacuate, charge and leak test the air conditioning system refrigerant.

23. Operate the engine to normal operating temperatures; then, check the climate control operation and check for leaks.

24. Enter the antitheft codes for the radio and the navigation system. Set the clock.

25. On 2002–05 vehicles, reprogram the ECM engine idle characteristics. Be sure all electrical items are OFF.

26. Start the engine. Hold the idle speed at 3000 RPM's in neutral until the radiator

fan comes on or the temperature reached 176 degrees.

27. Let the engine idle for about five minutes with the throttle fully closed.

28. If the radiator fan comes on during the five minutes, do not count this toward the five minute programming time.

29. On 2006 vehicles, reprogram the ECM engine idle characteristics. Be sure all electrical items are OFF.

30. Reset the ECM with the HDS. Turn the ignition switch to the ON position. Wait two seconds.

31. Start the engine. Hold the idle speed at 3000 RPM's in neutral until the radiator fan comes on or the temperature reached 194 degrees.

32. Let the engine idle for about five minutes with the throttle fully closed.

33. If the radiator fan comes on during

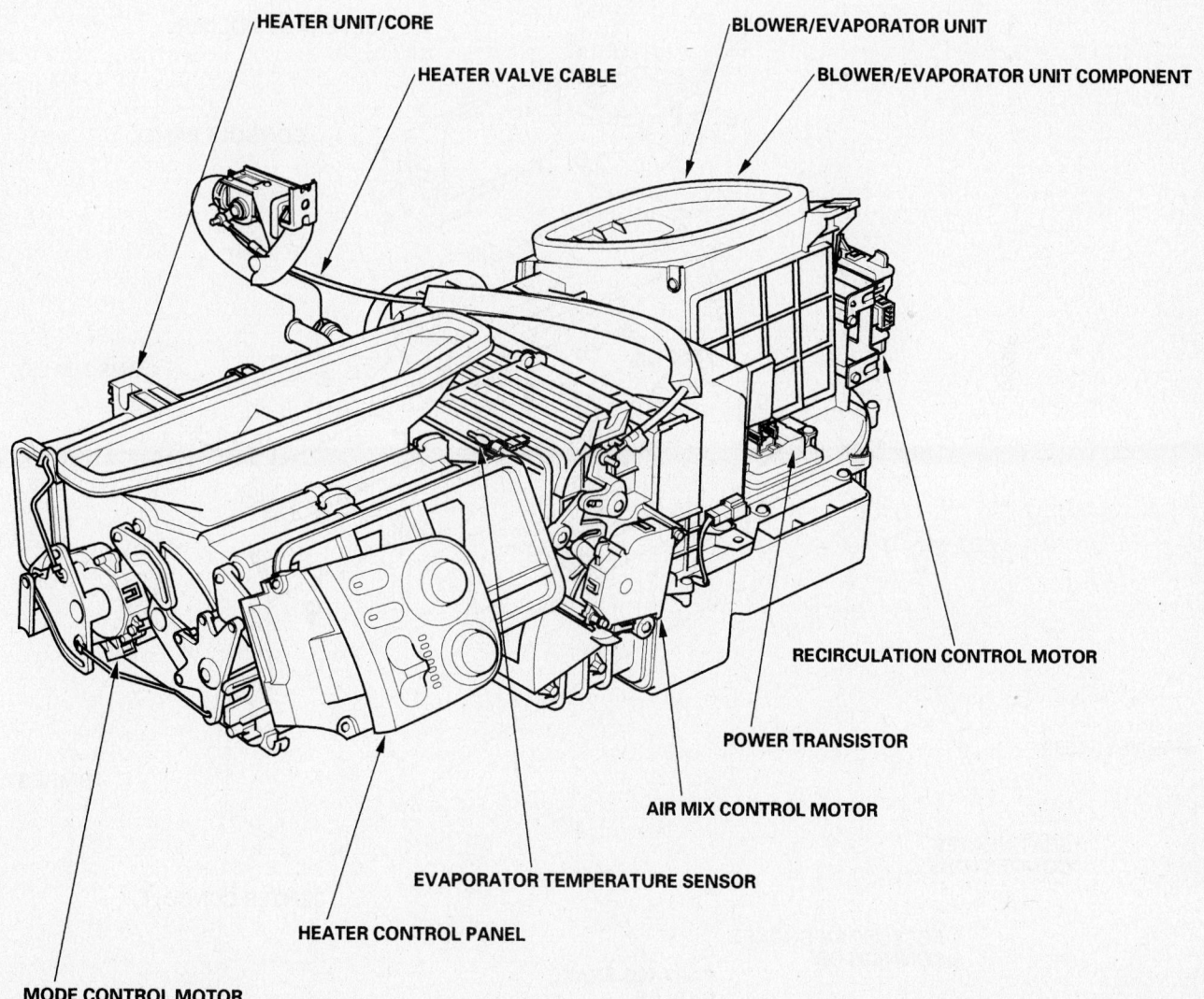

View of the heater housing, evaporator housing and related components—S2000

the five minutes, do not count this toward the five minute programming time.

Accord

➡ Make sure to acquire the anti-theft code from the radio and write down the frequencies for the radio's preset buttons.

1. Disconnect the negative battery cable.

❊❊ CAUTION

After disconnecting the negative battery cable, wait for at least 3 minutes for the air bag module to deplete its energy before working the on the instrument panel or steering wheel.

2. Discharge the air conditioning system. Disconnect the suction and the receiver lines from the evaporator core.

3. Drain the engine coolant.

4. From under the hood, open the cable clamp. Disconnect the heater control cable from the valve. Turn the valve to the fully opened position. Disconnect the heater hoses from heater core assembly. Remove the mounting nut from the heater unit.

5. Remove the center console as follows.

• If equipped with manual transaxle, remove the shifter knob

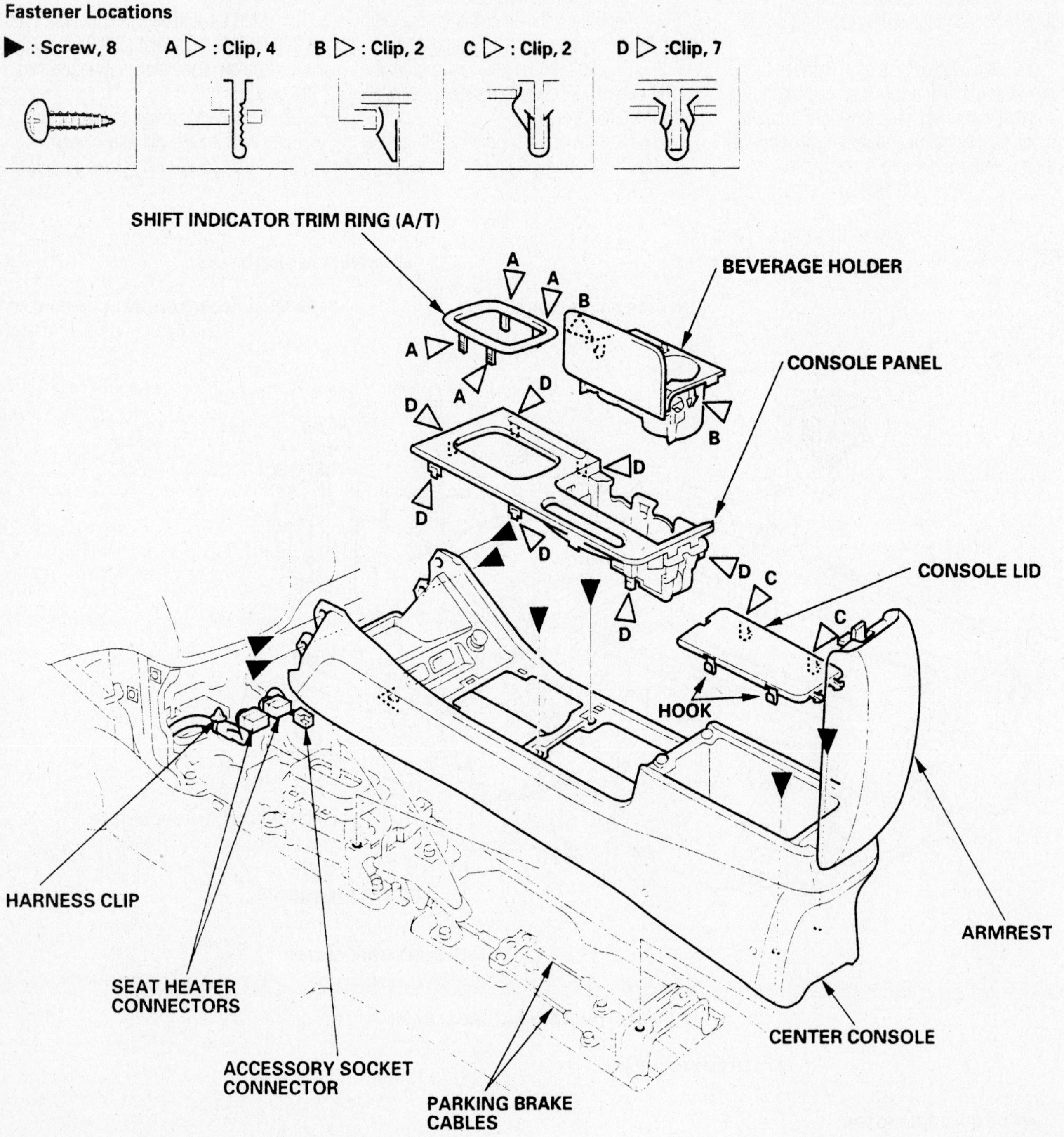

Fastener Locations

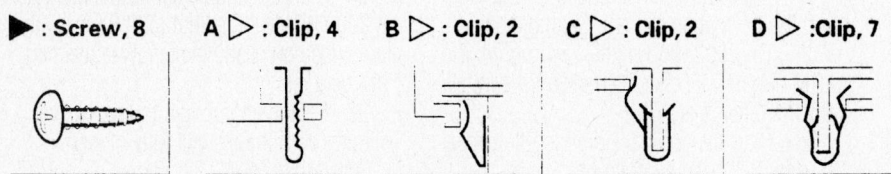

▶ : Screw, 8   A ▷ : Clip, 4   B ▷ : Clip, 2   C ▷ : Clip, 2   D ▷ : Clip, 7

Exploded view of the center console and related components—Accord

93112GJ7

- Detach the center console retaining clips and pull the assembly upward
- If equipped with heated seats, detach the heat switch connectors
- Pry up on the right side of the console pocket, detach the clips and remove the pocket
- Remove the screws, detach the clips and disconnect the front power accessory connector
- Open the console lid and remove the console mat
- If equipped, disconnect the rear accessory power socket electrical connector
- Remove the console retaining screws and remove the console from the vehicle

6. Remove the drivers side lower dashboard cover.

7. Remove the passengers side lower dashboard cover.

8. Remove the center pocket assembly. Remove the glovebox.

9. On 2004–06 vehicles remove the

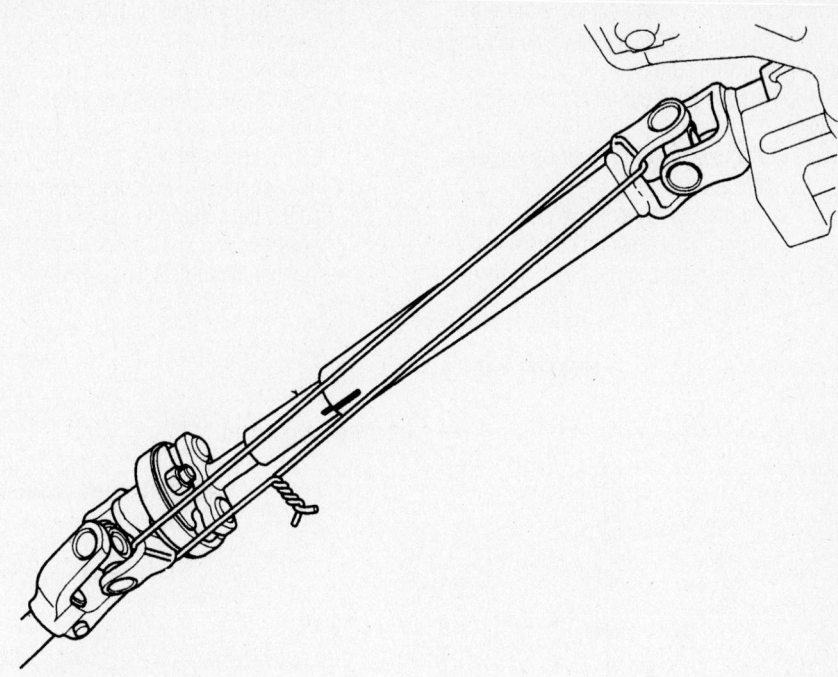

09474_ACCO_G0001

Slider shaft wire positioning—Accord

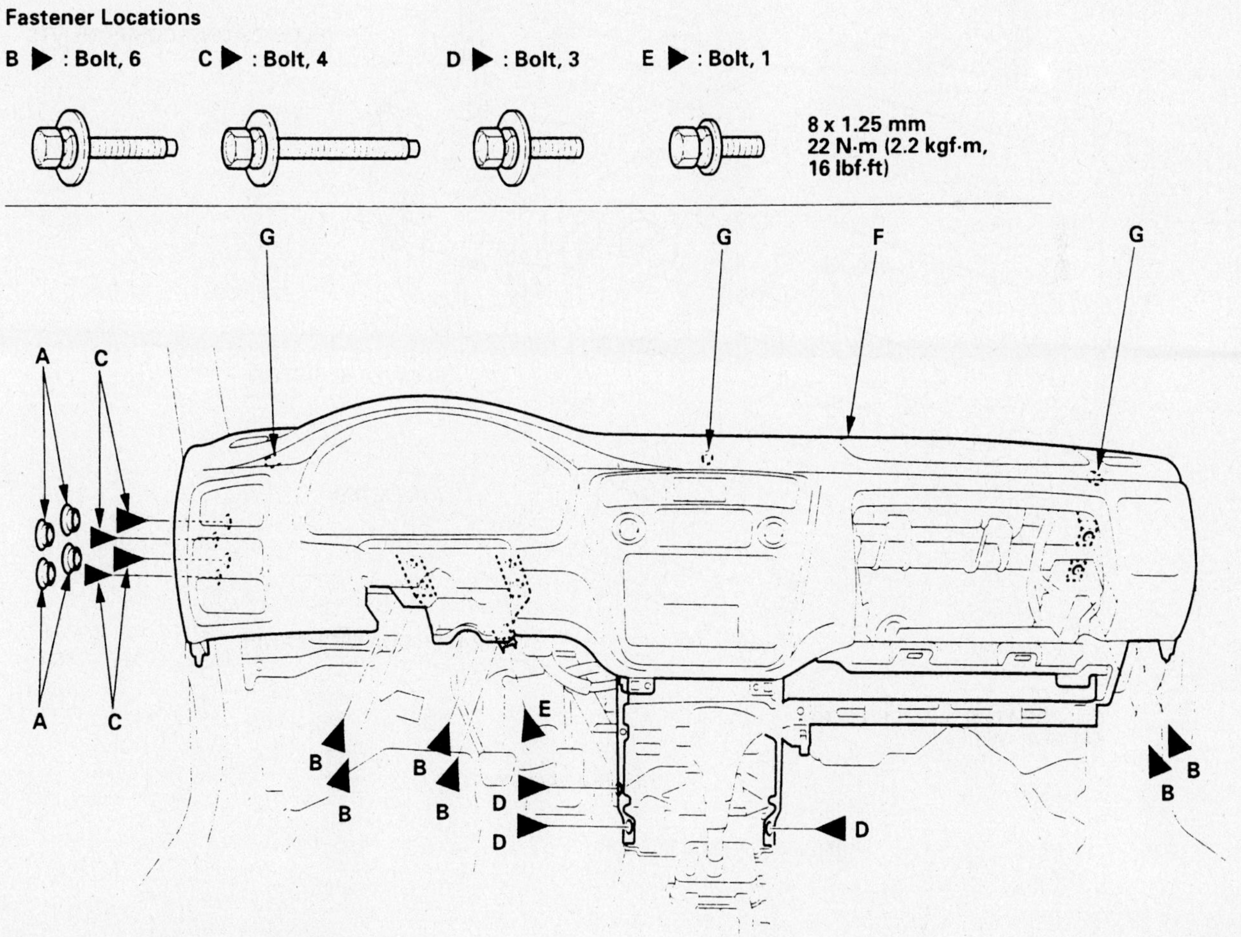

Fastener Locations

B ▶ : Bolt, 6   C ▶ : Bolt, 4   D ▶ : Bolt, 3   E ▶ : Bolt, 1

8 x 1.25 mm
22 N·m (2.2 kgf·m, 16 lbf·ft)

View of the instrument panel bolt locations—Accord

93112GJ8

drivers side under cover panel. Remove the drivers and passengers kick panels. Remove the A-pillar trim panels.

10. Remove the steering column as follows.

- Remove the driver's air bag assembly
- Remove the steering wheel
- Remove the steering column covers
- Remove the steering joint cover

- Release the tilt/telescopic lever and adjust the column in the full up position
- Tighten the tilt/telescopic lever
- Hold the slider shaft on the column using a piece of wire
- Release the tilt/telescopic lever and adjust the column to the full up position
- Tighten the tilt/telescopic lever

- Disconnect the wire harness from the combination switch and remove the switch
- Disconnect the connectors from the ignition switch and release the wire harness clips from the column
- Remove the steering joint bolt and disconnect the steering joint from the pinion shaft
- Remove the steering column retain-

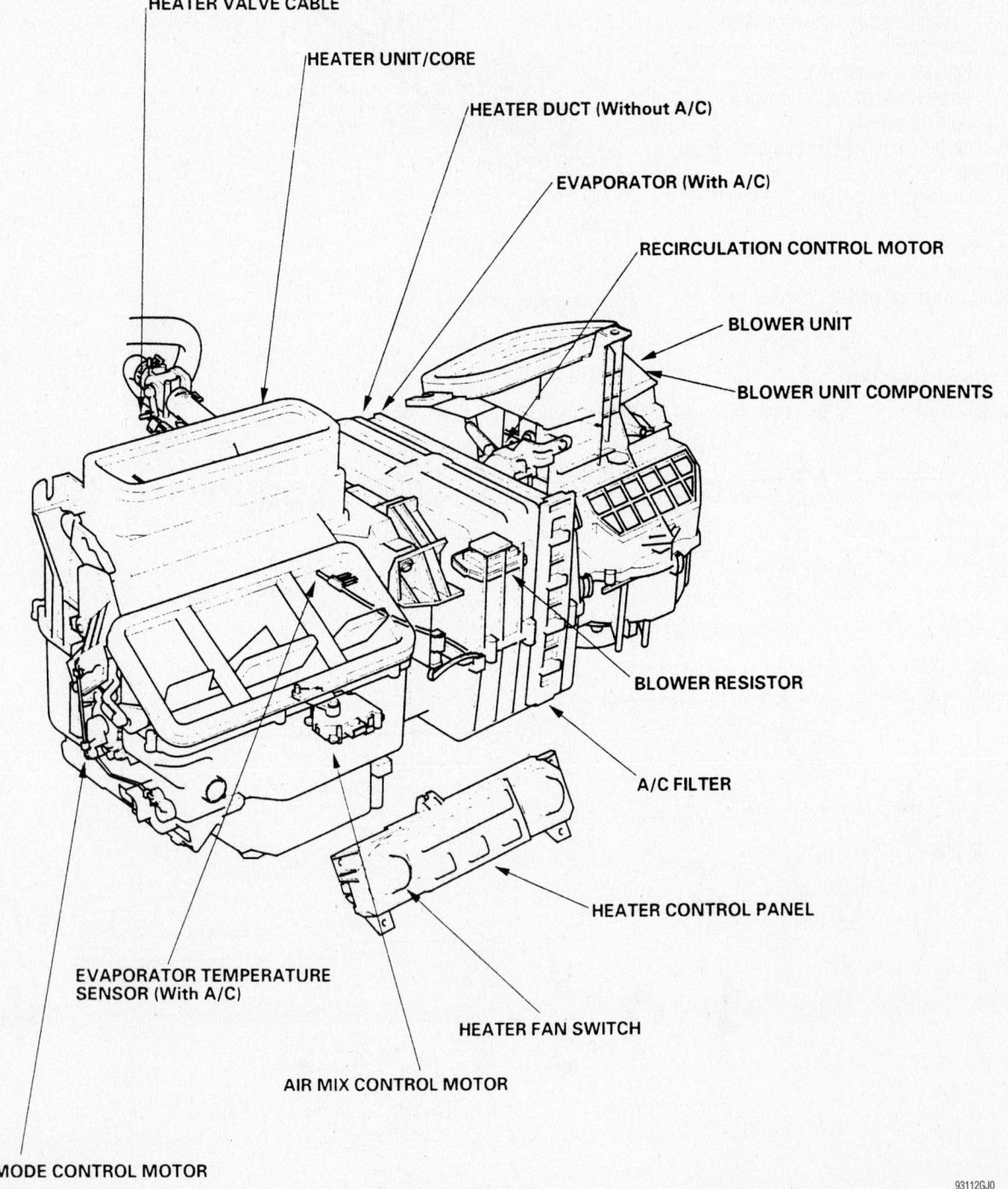

HEATER VALVE CABLE

HEATER UNIT/CORE

HEATER DUCT (Without A/C)

EVAPORATOR (With A/C)

RECIRCULATION CONTROL MOTOR

BLOWER UNIT

BLOWER UNIT COMPONENTS

BLOWER RESISTOR

A/C FILTER

HEATER CONTROL PANEL

HEATER FAN SWITCH

AIR MIX CONTROL MOTOR

EVAPORATOR TEMPERATURE SENSOR (With A/C)

MODE CONTROL MOTOR

93112GJ0

View of the heater housing, evaporator housing and related components—Accord

ing nuts and remove the column from the vehicle

11. Remove the driver's side center lower cover. Remove the passenger's side center lower cover and rear duct vent. Remove the SRS control unit.

12. On the driver's side disconnect all electrical connectors, air hoses and harness that interfere with the removal of the dashboard.

13. In the middle section of the dashboard disconnect the parking brake switch connector, radio antenna connector and lead. Disconnect all electrical connectors, air hoses and harness that interfere with the removal of the dashboard. Disconnect the ECM/PCM connector.

14. On the passenger's side disconnect all electrical connectors, air hoses and harness that interfere with the removal of the dashboard.

15. Remove the brake pedal support member.

16. Open the driver's side door remove the bolts and clips after remove their protective caps. Lift upward on the dashboard/steering hanger beam to release if from the guide pins. Be sure all electrical connectors are unplugged.

17. Carefully remove the dashboard assembly from the vehicle.

18. Disconnect the connectors from the air mix control motor, evaporator temperature sensor, power transistor and recirculation control motor.

19. Remove the wiring harness clips, connector clip and wire harness. Remove the heater ducts.

20. Remove the mounting nuts and remove the blower-heater unit from the vehicle.

21. Remove the heater core cover. Remove the heater core.

To install:

22. Installation is the reverse of the removal procedure.

23. Evacuate, charge and leak test the air conditioning system refrigerant.

24. Operate the engine to normal operating temperatures; then, check the climate control operation and check for leaks.

25. Enter the antitheft codes for the radio and the navigation system. Set the clock.

26. Reprogram the power window control unit as follows.

27. Turn the ignition switch ON. Lower the window, all the way down. Open the driver's side door.

28. Turn the ignition switch OFF. Push and hold the drivers window DOWN switch. Turn the ignition switch ON. Release the driver's window DOWN switch. This must be done within five seconds.

29. Repeat the above step three more times. Wait one second.

30. Confirm that the AUTO UP and AUTO DOWN do not work. If they work repeat the procedure, paying close attention to the five second time limit

31. Move the driver's window all the way down by holding the drivers window DOWN switch to the AUTO DOWN position.

32. Pull up and hold the driver's window UP switch to the AUTO UP position until the window reaches the fully closed position. Hold the switch for one second.

33. Confirm that the power window master switch has been reset by using the driver's window AUTO UP and DOWN function.

34. If the AUTO UP and DOWN feature is still not working, repeat the complete procedure, paying close attention to the five second time limit.

Cylinder Head

REMOVAL & INSTALLATION

➡The radio may contain a coded theft protection circuit. Always obtain the code number before disconnecting the battery.

Civic

2002–2005 1.7L ENGINE

1. Before servicing the vehicle, refer to the Precautions Section.

2. Be sure the cylinder head is cool to the touch before beginning the removal procedure. The coolant temperature must be below 100°F (38°C).

3. Relieve the fuel system pressure. Disconnect the negative battery cable. Drain the engine coolant.

4. Remove the throttle cable and the cruise control actuator cable. Remove the intake resonator.

5. Disconnect the intake air temperature (IAT) sensor. Remove the breather hose and the air cleaner assembly. Remove the brake booster vacuum hose and the PCV valve hose.

6. Remove the evaporative emission (EVAP) canister hose. Remove the ground cable. Remove the upper and lower radiator hoses. Remove the heater hoses and the water bypass hose.

7. Remove the two bolts securing the connecting pipe and remove the pipe from the water passage.

8. Remove the adjusting plate mounting bolt, locknut and mounting bolt. Remove the power steering pump belt. Remove the

pump and position it to the side without disconnecting the hoses.

9. Remove the alternator. Remove the air conditioning hose bracket. Remove the power steering hose bracket.

10. Remove the engine wiring harness connectors and wire harness clamps from the intake manifold.

11. Remove the idle air control valve connector, throttle position sensor connector, evaporative emission control canister connector and the engine coolant temperature sensor connector.

12. Remove the radiator fan switch connector, crankshaft position sensor connector, camshaft position sensor connector and the camshaft position TDC sensor connector.

13. Remove the EGR connectors, VTEC solenoid valve connector and VETC oil pressure switch connector, except Civic GL. Remove the oil pressure sensor connector.

14. Properly support the engine under the oil pan. Remove the upper bracket. Remove the exhaust manifold cover. Remove the exhaust manifold.

15. Remove the intake manifold retaining bolts and remove the intake manifold.

16. Remove the four ignition coils. Remove the throttle cable clamps and harness holder from the cylinder head cover.

17. Remove the cylinder head cover retaining bolts. Remove the cylinder head cover.

18. Remove the timing belt. Remove the camshaft pulley and back cover from the cylinder head.

19. Remove the cylinder head retaining bolts.

20. To prevent warpage, loosen the cylinder head bolts in a 3-step crisscross pattern in the reverse order of the tightening sequence.

21. Remove the cylinder head from the engine.

To install:

➡Use new O-ring, seals, and gaskets when installing the cylinder head and its components.

22. Be sure the cylinder head and the engine block surfaces are clean, level, and straight.

23. Be sure the cylinder head dowel pins and control orifice are aligned. Clean the oil control orifice and reinstall it with a new O-ring.

24. Set the crankshaft to TDC. Align the TDC mark on the timing belt drive pulley with the pointer on the oil pump.

25. Set the camshaft pulley to TDC. The

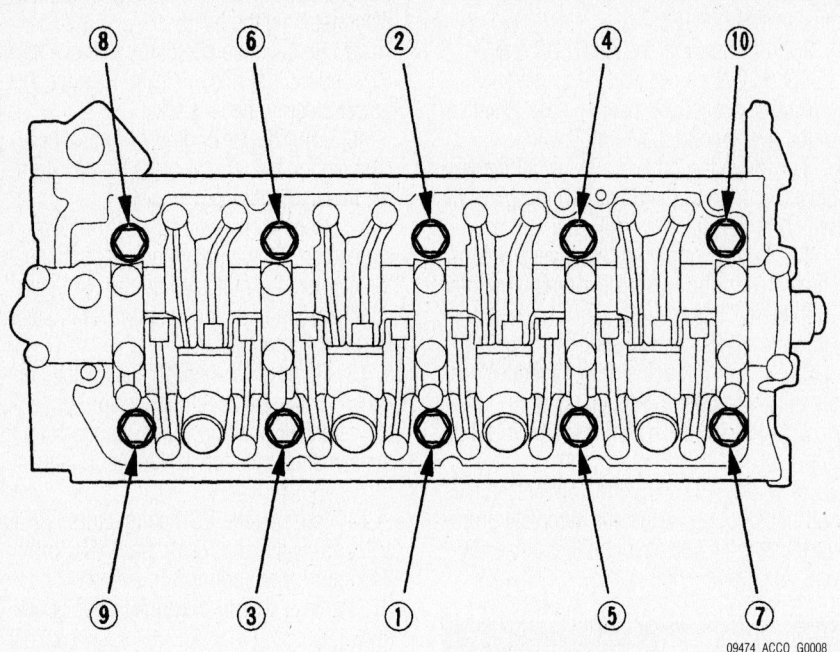

09474_ACCO_G0008

Cylinder head torque sequence—1.7L engine

"UP" mark on the camshaft pulley should be at the top. Align the TDC marks on the camshaft pulley with the top edge of the cylinder head.

26. Position the cylinder head on the engine. Coat the threads of the cylinder head retaining bolts with clean engine oil. Tighten the cylinder head bolts sequentially to the proper torque.

27. Continue the installation in the reverse order of the removal procedure.

28. Enter the antitheft codes for the radio and the navigation system.

29. Reprogram the ECM/PCM engine idle characteristics. Be sure all electrical items are OFF.

30. Start the engine. Hold the idle speed at 3000 RPM's in park or neutral until the radiator fan comes on or the temperature reached 194 degrees.

31. Let the engine idle for about five minutes with the throttle fully closed.

32. If the radiator fan comes on during the five minutes, do not count this toward the five minute programming time.

33. Set the clock.

2006 1.8L ENGINE

1. Before servicing the vehicle, refer to the Precautions Section.

2. Be sure the cylinder head is cool to the touch before beginning the removal procedure. The coolant temperature must be below 100°F (38°C).

3. Relieve the fuel system pressure.

Disconnect the negative battery cable. Drain the engine coolant.

4. Remove the drive belt. Remove the intake manifold.

5. Remove the harness clamps. Remove the PCV valve hose from the clamp.

Remove the air cleaner housing bracket and the harness holder from the cylinder head.

6. If equipped with manual transaxle, remove the upper radiator hose and heater hoses.

7. If equipped with automatic transaxle, remove the upper radiator hose, heater hoses and water bypass hose.

8. Remove the engine wiring harness connectors and the wiring harness clamps from the cylinder head.

9. Remove the injector connectors, coolant temperature connector, air fuel ratio sensor connector and the secondary heated oxygen sensor connector.

10. Disconnect the EGR valve connector, rocker arm oil control solenoid sensor and the rocker arm oil pressure switch connector.

11. Remove the four ignition coils. Remove the three way catalytic converter. Remove the thermostat housing.

12. Remove the cam chain. Remove the cylinder head retaining bolts.

13. To prevent warpage, loosen the cylinder head bolts in a 3-step crisscross pattern in the reverse order of the tightening sequence.

14. Remove the cylinder head from the engine.

To install:

➡**Use new O-ring, seals, and gaskets when installing the cylinder head and its components.**

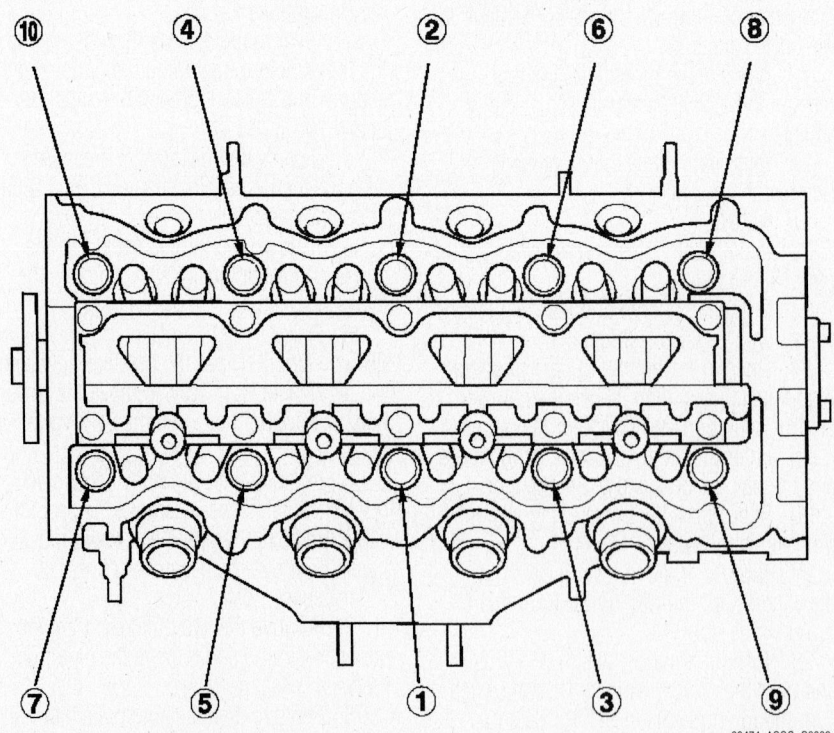

09474_ACCO_G0009

Cylinder head torque sequence—1.8L engine

15. Be sure the cylinder head and the engine block surfaces are clean, level, and straight.

16. Be sure the cylinder head dowel pins and control orifice are aligned. Clean the oil control orifice and reinstall it with a new O-ring.

17. Set the crankshaft to TDC. Align the TDC mark on the crankshaft sprocket with the pointer on the engine block.

18. Set the camshaft to TDC. The "UP" mark on the camshaft sprocket should be at the top and the TDC grooves on the camshaft sprocket should line up with the top edge of the cylinder head.

19. Position the cylinder head on the engine.

20. Measure the diameter of each cylinder head bolt between point A and B. If either diameter is less than 0.42 inch, replace the bolt.

21. Coat the threads of the cylinder head retaining bolts with clean engine oil. Tighten the cylinder head bolts sequentially to the proper torque.

22. Continue the installation in the reverse order of the removal procedure.

23. Reprogram the crankshaft position (CKP) pattern. Run the engine until the operating temperature reaches 176 degree. With the engine stopped clear the CKP pattern. Turn the ignition switch OFF. Turn the ignition switch ON and wait thirty seconds.

24. Roadtest the vehicle on a level surface. Decelerate the engine speed of 2500 rpm to 1000 rpm. If equipped with automatic transaxle use two Drive positions. If equipped with manual transaxle use first gear.

25. Stop the vehicle, but keep the engine running.

26. Check PULSAR F/B LEARN in the data list with the HDS. If not complete repeat the procedure. If complete, roadtest the vehicle on a level surface. Decelerate the engine speed of 5000 rpm to 3000 rpm. If equipped with automatic transaxle use two Drive positions. If equipped with manual transaxle use first gear.

27. Stop the vehicle, but keep the engine running.

28. Check PULSAR F/B LEARN in the data list with the HDS. If not complete repeat the procedure.

29. If completed, turn the ignition switch OFF. Turn the ignition switch ON, wait thirty seconds. The learning procedure is now complete.

30. Enter the antitheft codes for the radio and the navigation system. Set the clock.

31. Reprogram the power window control unit as follows.

32. Turn the ignition switch ON. Lower the window, all the way down. Open the driver's side door.

33. Turn the ignition switch OFF. Push and hold the drivers window DOWN switch. Turn the ignition switch ON. Release the driver's window DOWN switch. This must be done within five seconds.

34. Repeat the above step three more times. Wait one second.

35. Confirm that the AUTO UP and AUTO DOWN do not work. If they work repeat the procedure, paying close attention to the five second time limit

36. Move the driver's window all the way down by holding the drivers window DOWN switch to the AUTO DOWN position.

37. Pull up and hold the driver's window UP switch to the AUTO UP position until the window reaches the fully closed position. Hold the switch for one second.

38. Confirm that the power window master switch has been reset by using the driver's window AUTO UP and DOWN function.

39. If the AUTO UP and DOWN feature is still not working, repeat the complete procedure, paying close attention to the five second time limit.

2006 2.0L ENGINE

1. Before servicing the vehicle, refer to the Precautions Section.

2. Be sure the cylinder head is cool to the touch before beginning the removal procedure. The coolant temperature must be below 100°F (38°C).

3. Relieve the fuel system pressure. Disconnect the negative battery cable. Drain the engine coolant.

4. Remove the air cleaner housing. Remove the drive belt. Remove the intake manifold.

5. Remove the exhaust manifold. Remove the evaporative emission canister hose and the brake booster vacuum hose.

6. Remove the quick-connect fitting cover. Disconnect the fuel feed hose. Remove the harness holder from the bracket then remove the harness holder bracket.

7. Remove the upper radiator hose and the heater hoses. Remove the bolt securing the connecting pipe.

8. Remove the fuel injector connectors, the engine coolant temperature sensor connector, the camshaft position intake sensor connector and the camshaft position exhaust sensor connector.

9. Remove the rocker arm oil control solenoid connector, the rocker arm oil pressure switch connector and the EVAP canister purge valve connector.

10. Remove the cam chain. Remove the rocker arm cover. Remove the rocker arm assembly. Remove the cylinder head retaining bolts.

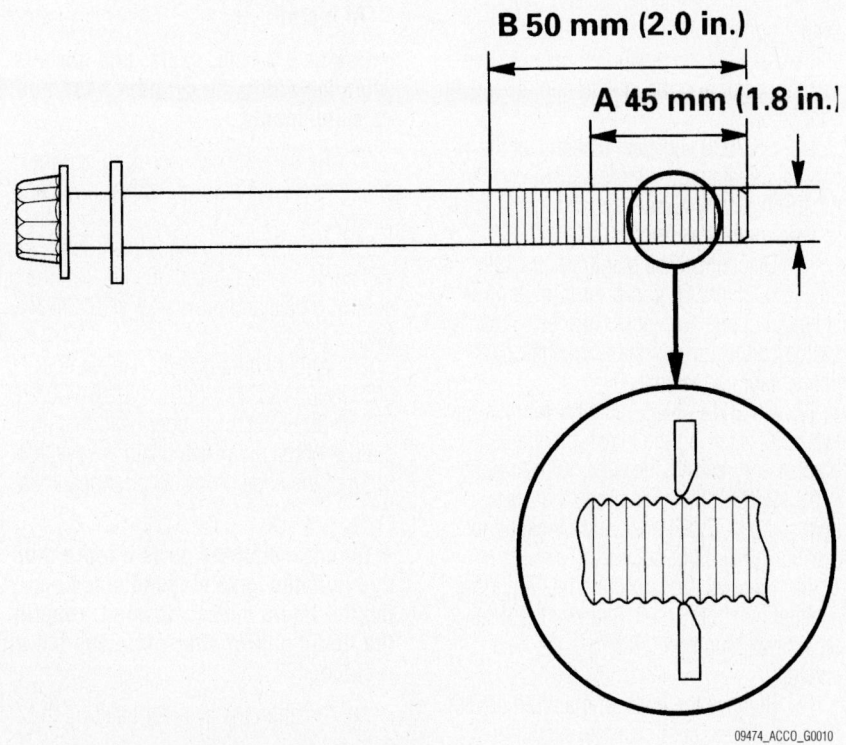

B 50 mm (2.0 in.)

A 45 mm (1.8 in.)

09474_ACCO_G0010

Cylinder head bolt measurement reference points

11. To prevent warpage, loosen the cylinder head bolts in a 3-step crisscross pattern in the reverse order of the tightening sequence.

12. Remove the cylinder head from the engine.

To install:

➡ **Use new O-ring, seals, and gaskets when installing the cylinder head and its components.**

13. Be sure the cylinder head and the engine block surfaces are clean, level, and straight.

14. Be sure the cylinder head dowel pins and control orifice are aligned. Clean the oil control orifice and reinstall it with a new O-ring.

15. Set the crankshaft to TDC. Align the TDC mark on the crankshaft sprocket with the pointer on the engine block.

16. Position the cylinder head on the engine.

17. Measure the diameter of each cylinder head bolt between point A and B. If either diameter is less than 0.42 inch, replace the bolt.

18. Coat the threads of the cylinder head retaining bolts with clean engine oil. Tighten the cylinder head bolts sequentially to the proper torque.

19. Continue the installation in the reverse order of the removal procedure.

20. Enter the antitheft codes for the radio and the navigation system. Set the clock.

21. Reprogram the power window control unit as follows.

22. Turn the ignition switch ON. Lower the window, all the way down. Open the driver's side door.

23. Turn the ignition switch OFF. Push and hold the drivers window DOWN switch. Turn the ignition switch ON. Release the driver's window DOWN switch. This must be done within five seconds.

24. Repeat the above step three more times. Wait one second.

25. Confirm that the AUTO UP and AUTO DOWN do not work. If they work repeat the procedure, paying close attention to the five second time limit

26. Move the driver's window all the way down by holding the drivers window DOWN switch to the AUTO DOWN position.

27. Pull up and hold the driver's window UP switch to the AUTO UP position until the window reaches the fully closed position. Hold the switch for one second.

28. Confirm that the power window master switch has been reset by using the driver's window AUTO UP and DOWN function.

29. If the AUTO UP and DOWN feature is still not working, repeat the complete procedure, paying close attention to the five second time limit.

S2000

1. Before servicing the vehicle, refer to the Precautions Section.

2. Be sure the cylinder head is cool to the touch before beginning the removal procedure. The coolant temperature must be below 100°F (38°C).

3. Relieve the fuel system pressure. Disconnect the negative battery cable. Drain the engine coolant. Drain the engine oil.

4. Remove the air cleaner assembly. Remove the intake air cleaner housing. Remove the drive belt.

5. On 2002–2005 vehicles remove the throttle cable, fuel feed hose, fuel return hose and power brake booster hose.

6. On 2006 vehicles, remove the intake manifold cover, power brake booster hose and the quick-connect fitting cover, then disconnect the fuel feed hose.

7. Disconnect the evaporative emission canister hose. Remove the intake manifold bracket retaining bolt. Remove the water outlet cover.

8. Remove the engine wire harness connectors and the wire harness clamps from the cylinder head and intake manifold. Remove the four fuel injector connectors

9. On 2002–2005 vehicles, disconnect the IAT sensor connector, heated oxygen sensor connector, throttle position sensor connector, engine coolant temperature sensor connector and the idle air control sensor connector.

10. On 2006 vehicles, disconnect the engine coolant temperature sensor connector, throttle body connector, and the air/fuel sensor connector.

11. Disconnect the manifold absolute pressure connector, rocker arm oil control solenoid connector, rocker arm oil pressure switch connector and the crankshaft position sensor connector.

12. On 2002–2005 vehicles, remove the two bolts securing the intake manifold bracket and remove the water bypass hose.

13. 2006 vehicles, remove the water bypass hose, EVAP hose and bracket and the intake manifold bracket.

14. Remove the four bolts retaining the exhaust manifold cover. Remove the heat shield retaining bolts. Remove the heat shield.

15. Remove the exhaust manifold cover. Remove the exhaust manifold retaining bolts and remove the exhaust manifold.

16. On 2002–2005 vehicles, remove the injectors.

17. Remove the intake manifold bracket clips. Remove the intake manifold retaining bolts. Remove the intake manifold.

18. Remove the cylinder head cover retaining bolts. Remove the cylinder head cover.

19. Position the No. 1 piston at TDC. The TDC marks on the cam chain sprocket should align with the cylinder head surface.

20. Remove the end cover and nozzle from the cam chain auto tensioner. Thread a nut onto a 5x0.8 mm bolt at is at least 40 mm long. Thread the bolt into the maintenance hole in the cam chain auto tensioner.

21. Turn the bolt clockwise to compress the cam chain auto tensioner and lock it in place with the nut. Remove the cam chain auto tensioner.

22. Loosen the rocker arm adjusting screws. Remove the camshaft holders and camshafts.

23. Insert the bolts into the rocker shaft holder and remove the rocker arm assembly.

24. Remove the idler gear/cam chain sprocket assembly, idler gear collar and washer.

25. Remove the cylinder head retaining bolts. To prevent warpage, loosen the cylinder head bolts in a 3-step crisscross pattern in the reverse order of the tightening sequence.

26. Remove the cylinder head from the engine.

To install:

➡ **Use new O-ring, seals, and gaskets when installing the cylinder head and its components.**

27. Be sure the cylinder head and the engine block surfaces are clean, level, and straight.

28. Be sure the cylinder head dowel pins and control orifice are aligned. Clean the oil control orifice and reinstall it with a new O-ring.

29. Apply liquid gasket, part number 08717-0004, 08718-0001, 08718-0003 or 08718-0009 to the cylinder head mating surface of the block and chain case within 5mm of the edge of the cylinder head gasket.

➡ **Do not install the parts if more than five minutes have elapsed since applying the liquid gasket. Instead, reapply the liquid gasket after removing the old residue.**

30. Position the cylinder head on the engine.

31. Coat the threads of the cylinder head

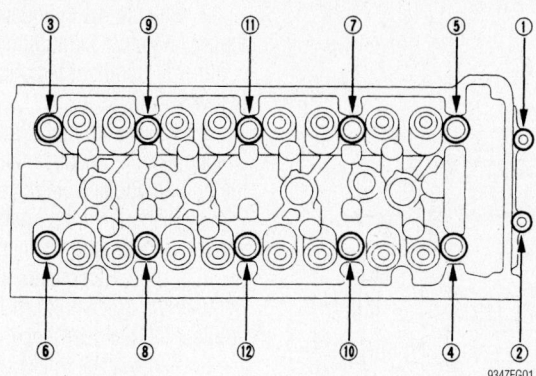

Cylinder head loosening sequence—2.0L and 2.2L engines

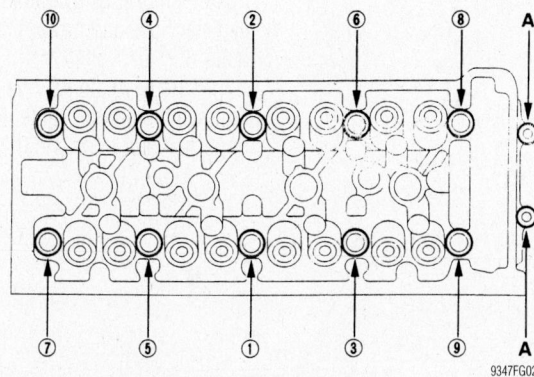

Cylinder head torque sequence—2.0L and 2.2L engines

retaining bolts with clean engine oil. Tighten the cylinder head bolts sequentially to the proper torque.

32. Continue the installation in the reverse order of the removal procedure.

33. On 2002–05 vehicles, reprogram the ECM engine idle characteristics. Be sure all electrical items are OFF.

34. Start the engine. Hold the idle speed at 3000 RPM's in neutral until the radiator fan comes on or the temperature reached 176 degrees.

35. Let the engine idle for about five minutes with the throttle fully closed.

36. If the radiator fan comes on during the five minutes, do not count this toward the five minute programming time.

37. On 2006 vehicles, reprogram the crankshaft position (CKP) pattern. Run the engine until the operating temperature reaches 176 degree. With the engine stopped clear the CKP pattern. Turn the ignition switch OFF. Turn the ignition switch ON and wait thirty seconds.

38. Roadtest the vehicle on a level surface. Decelerate the engine speed of 2500 rpm to 1000 rpm. If equipped with automatic transaxle use two Drive positions. If equipped with manual transaxle use first gear.

39. Stop the vehicle, but keep the engine running.

40. Check PULSAR F/B LEARN in the data list with the HDS. If not complete repeat the procedure. If complete, roadtest the vehicle on a level surface. Decelerate the engine speed of 5000 rpm to 3000 rpm. If equipped with automatic transaxle use two Drive positions. If equipped with manual transaxle use first gear.

41. Stop the vehicle, but keep the engine running.

42. Check PULSAR F/B LEARN in the data list with the HDS. If not complete repeat the procedure.

43. If completed, turn the ignition switch OFF. Turn the ignition switch ON, wait thirty seconds. The learning procedure is now complete.

44. Enter the antitheft codes for the radio and the navigation system. Set the clock.

45. Reprogram the power window control unit as follows.

46. Turn the ignition switch ON. Lower the window, all the way down. Open the driver's side door.

47. Turn the ignition switch OFF. Push and hold the drivers window DOWN switch. Turn the ignition switch ON. Release the

driver's window DOWN switch. This must be done within five seconds.

48. Repeat the above step three more times. Wait one second.

49. Confirm that the AUTO UP and AUTO DOWN do not work. If they work repeat the procedure, paying close attention to the five second time limit

50. Move the driver's window all the way down by holding the drivers window DOWN switch to the AUTO DOWN position.

51. Pull up and hold the driver's window UP switch to the AUTO UP position until the window reaches the fully closed position. Hold the switch for one second.

52. Confirm that the power window master switch has been reset by using the driver's window AUTO UP and DOWN function.

53. If the AUTO UP and DOWN feature is still not working, repeat the complete procedure, paying close attention to the five second time limit.

Accord

2.3L ENGINE

1. Before servicing the vehicle, refer to the Precautions Section.

2. Drain the cooling system.

3. Relieve the fuel system pressure.

4. Remove or disconnect the following:
- Negative battery cable
- Air intake duct
- Throttle and cruise control cables
- Positive Crankcase Ventilation (PCV) valve and hose
- Fuel lines
- Brake booster vacuum hose
- Evaporative Emissions (EVAP) canister hoses
- Water bypass hoses
- Accessory drive belts
- Power steering pump
- Alternator wiring harness
- Radiator hoses
- Heater hoses
- Fuel injector connectors
- Intake Air Temperature (IAT) sensor connector
- Idle Air Control (IAC) valve connector
- Throttle Position (TP) sensor connector
- Manifold Absolute Pressure (MAP) sensor connector
- Heated Oxygen (HO$_2$S) sensor connector (F23A1, F23A5 engines)
- Air/Fuel ratio sensor connector (F23A4 engine)
- Engine Coolant Temperature (ECT) sensor connector

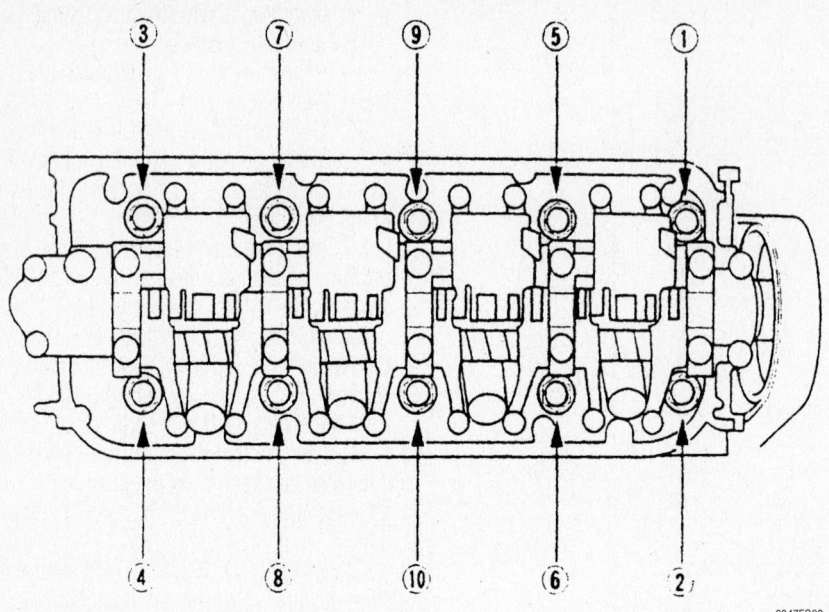

Cylinder head loosening sequence—2.3L engine

- Radiator fan switch connector
- Coolant temperature gauge sender connector
- Exhaust Gas Recirculation (EGR) valve connector
- Crankshaft Position (CKP) sensor connector
- VTEC solenoid valve connector (F23A1, F23A4 engines)
- VTEC oil pressure switch connector (F23A1, F23A4 engines)
- Distributor
- Front motor mount bracket
- Valve cover
- Timing belt
- Camshaft pulley and back cover
- Intake manifold
- Exhaust manifold
- Cylinder head. Loosen the bolts in sequence and in 1/3 turns.

To install:

5. Set the crankshaft pulley to Top Dead Center (TDC).

6. Set the camshaft pulley to TDC.

7. Install the cylinder head with a new gasket. Tighten the bolts in sequence as follows:

    a. Step 1: 22 ft. lbs. (29 Nm).

    b. Step 2: Plus 90 degrees.

    c. Step 3: Plus 90 degrees.

    d. Step 4: If using new cylinder head bolts, add an additional 90 degrees.

8. The remainder of the installation is the reverse of the removal procedure.

2.4L ENGINE

1. Before servicing the vehicle, refer to the Precautions Section.

2. Be sure the cylinder head is cool to the touch before beginning the removal procedure. The coolant temperature must be below 100°F (38°C).

3. Relieve the fuel system pressure. Disconnect the negative battery cable. Drain the engine coolant. Remove the drive belt.

4. Disconnect the intake air temperature sensor. Remove the vacuum hose and breather pipe. Remove the intake air duct. Remove the quick connect fitting cover. Disconnect the fuel feed hose.

5. Remove the intake manifold. Remove the exhaust manifold.

6. Remove the bolt securing the connecting pipe. Remove the water bypass hose. Remove the evaporative emission canister hose and power brake vacuum hose.

7. Remove the PCV valve hose. Remove the ground cable. Remove the upper radiator hose and the heater hoses.

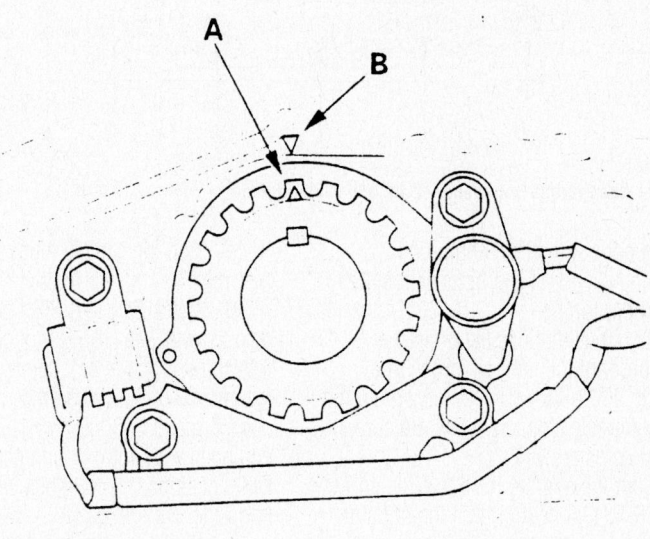

Set the crankshaft to TDC by aligning pointers A and B—2.3L engine

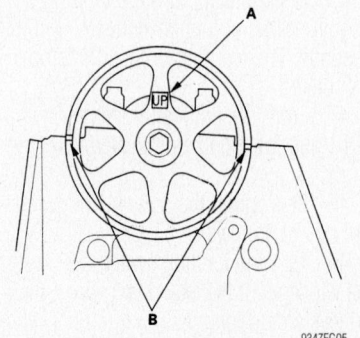

Align the camshaft pulley as shown prior to cylinder head installation—2.3L engine

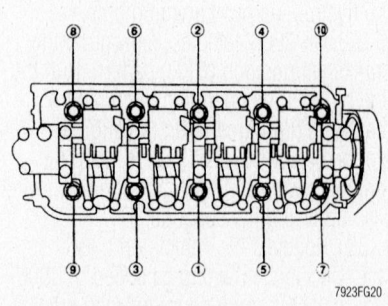

Cylinder head torque sequence—2.3L engine

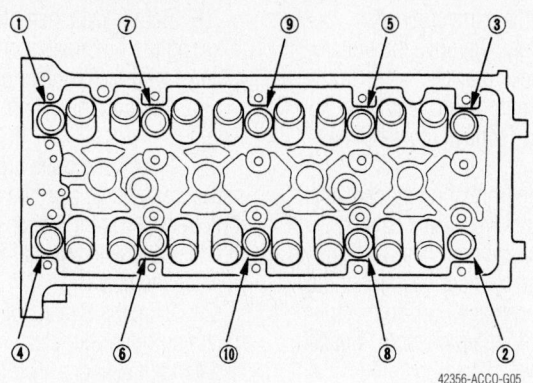

Cylinder head loosening sequence—2.4L engines

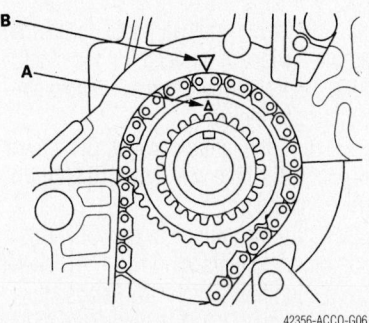

Set the crankshaft to TDC by aligning the mark (A) on the crankshaft sprocket with the pointer (B) on the cylinder block—2.4L engine

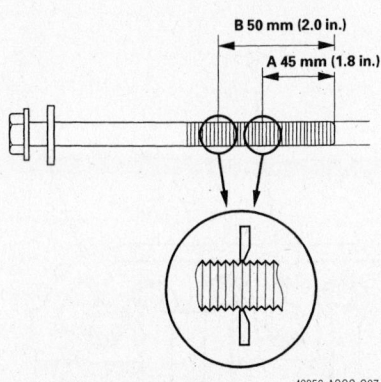

You must measure the cylinder head bolts to see if they can be reused or need to be replaced—2.4L engine

8. Remove the engine wiring harness connectors and the wire harness clamps from the cylinder head.

9. Disconnect the fuel injector connectors, engine coolant temperature sensor connector, camshaft position sensor connectors and the EGR valve connector.

10. Disconnect the rocker arm oil control solenoid connector and engine oil pressure sensor connector.

11. Remove the two bolts securing the

EVAP canister purge valve bracket and remove the bolt securing the harness bracket.

12. Remove the cam chain. Remove the rocker arm assembly.

13. Remove the cylinder head retaining bolts. To prevent warpage, loosen the cylinder head bolts in a 3-step crisscross pattern in the reverse order of the tightening sequence.

14. Remove the cylinder head from the engine.

To install:

➡ Use new O-ring, seals, and gaskets when installing the cylinder head and its components.

15. Be sure the cylinder head and the engine block surfaces are clean, level, and straight.

16. Be sure the cylinder head dowel pins and control orifice are aligned. Clean the oil control orifice and reinstall it with a new O-ring.

17. Set the crankshaft to TDC. Align the TDC mark on the crankshaft sprocket with the pointer on the engine block.

18. Position the cylinder head on the engine.

19. Measure the diameter of each cylinder head bolt at points A & B, as shown in the illustration. If either diameter is less than 0.42 in. (10.6mm), replace the head bolt

20. Coat the threads of the cylinder head retaining bolts with clean engine oil. Tighten the cylinder head bolts sequentially to the proper torque.

21. Continue the installation in the reverse order of the removal procedure.

22. Reprogram the ECM/PCM with the HDS, turn the ignition switch the OFF position. Turn the ignition switch to the ON position. Wait thirty seconds. Turn the ignition switch OFF and disconnect the HDS from the DLC.

23. Reprogram the ECM engine idle characteristics. Be sure all electrical items are OFF.

24. Start the engine. Hold the idle speed at 3000 RPM's in neutral until the radiator fan comes on or the temperature reached 176 degrees.

25. Let the engine idle for about five minutes with the throttle fully closed.

26. If the radiator fan comes on during the five minutes, do not count this toward the five minute programming time.

27. Enter the antitheft codes for the radio and the navigation system. Set the clock.

28. Reprogram the power window control unit as follows.

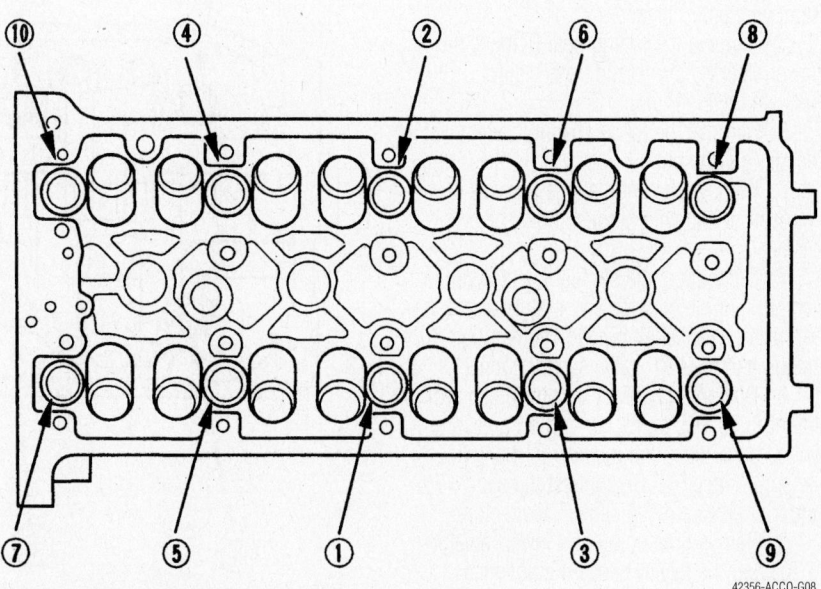

Cylinder head bolt tightening sequence—2.4L engine

29. Turn the ignition switch ON. Lower the window, all the way down. Open the driver's side door.

30. Turn the ignition switch OFF. Push and hold the drivers window DOWN switch. Turn the ignition switch ON. Release the driver's window DOWN switch. This must be done within five seconds.

31. Repeat the above step three more times. Wait one second.

32. Confirm that the AUTO UP and AUTO DOWN do not work. If they work repeat the procedure, paying close attention to the five second time limit

33. Move the driver's window all the way down by holding the drivers window DOWN switch to the AUTO DOWN position.

34. Pull up and hold the driver's window UP switch to the AUTO UP position until the window reaches the fully closed position. Hold the switch for one second.

35. Confirm that the power window master switch has been reset by using the driver's window AUTO UP and DOWN function.

36. If the AUTO UP and DOWN feature is still not working, repeat the complete procedure, paying close attention to the five second time limit.

3.0L ENGINE

1. Before servicing the vehicle, refer to the Precautions Section.

2. Be sure the cylinder head is cool to the touch before beginning the removal procedure. The coolant temperature must be below 100°F (38°C).

3. Relieve the fuel system pressure. Disconnect the negative battery cable. Drain the engine coolant. Remove the drive belt. Remove the air cleaner.

4. Remove the timing belt. Remove the power steering pump and hose clamp. Remove the alternator.

5. Remove the intake manifold. Remove the six ignition coils.

6. Remove the engine wire harness connectors and wire harness from the cylinder head.

7. Disconnect the six fuel injector connectors, engine coolant temperature sensor connector, crankshaft position sensor connector, and the EGR valve sensor connector.

8. Disconnect the rocker arm oil control solenoid connector, rocker arm oil pressure switch connector, and the two air fuel ratio sensor connectors. Disconnect the secondary heated oxygen sensor connectors.

9. Remove the upper and lower radiator hoses and the heater hoses. Remove the water bypass hose.

10. Remove the ground cable. Remove the bolt securing the harness bracket. Remove the fuel rails. Remove the bolt securing the harness bracket.

11. Remove the front and rear three way catalytic converter. Remove the water passage.

12. Remove the front and rear camshaft pulleys and front and rear back covers.

13. Remove the cylinder head covers.

14. Remove the cylinder head retaining bolts. To prevent warpage, loosen the cylinder head bolts in a 3-step crisscross pattern in the reverse order of the tightening sequence.

15. Remove the cylinder head from the engine.

To install:

➡ Use new O-ring, seals, and gaskets when installing the cylinder head and its components.

16. Be sure the cylinder head and the engine block surfaces are clean, level, and straight.

17. Be sure the cylinder head dowel pins and control orifice are aligned. Clean the oil control orifice and reinstall it with a new O-ring.

18. Set the timing belt drive pulley to TDC by aligning the TDC mark on the tooth of the timing belt drive pulley with the pointer on the oil pump.

19. Set the camshaft pulleys to TDC by aligning the TDC marks on the camshaft pulleys with the pointers on the back covers

20. Position the cylinder head on the engine.

21. If the head bolts are Dodecagon type, measure the diameter of each cylinder head bolt between point A and B, see illustration. If either diameter is less than 0.42 inch, replace the bolt.

22. Coat the threads of the cylinder head retaining bolts with clean engine oil. Tighten the cylinder head bolts sequentially to the proper torque.

23. Continue the installation in the reverse order of the removal procedure.

24. Enter the antitheft codes for the radio and the navigation system. Set the clock.

25. Reprogram the power window control unit as follows.

26. Turn the ignition switch ON. Lower the window, all the way down. Open the driver's side door.

27. Turn the ignition switch OFF. Push and hold the drivers window DOWN switch. Turn the ignition switch ON. Release the driver's window DOWN switch. This must be done within five seconds.

28. Repeat the above step three more times. Wait one second.

29. Confirm that the AUTO UP and AUTO DOWN do not work. If they work

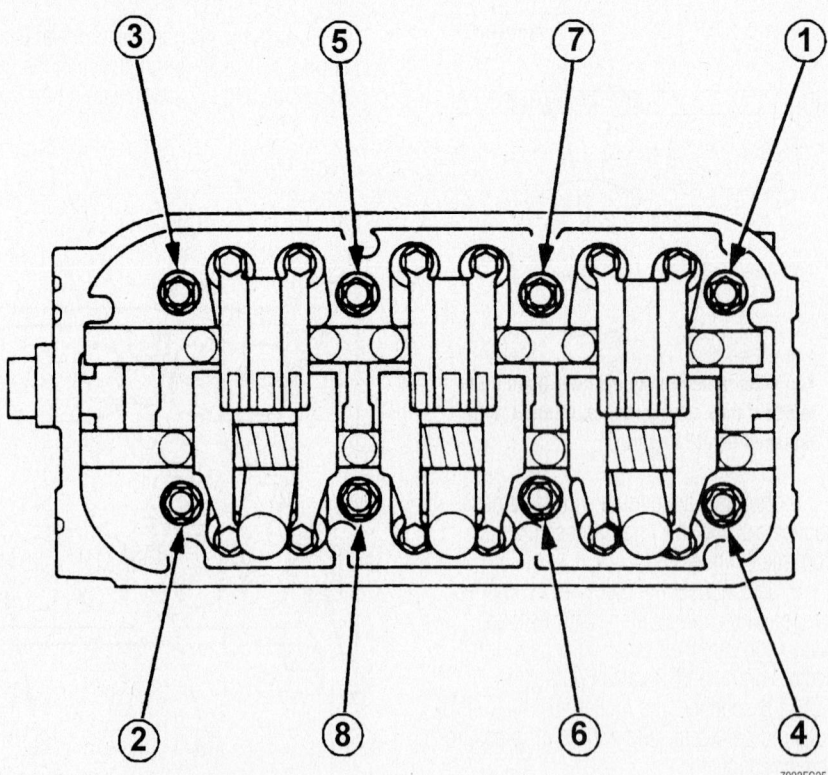

Cylinder head loosening sequence—3.0L engine

7923FG25

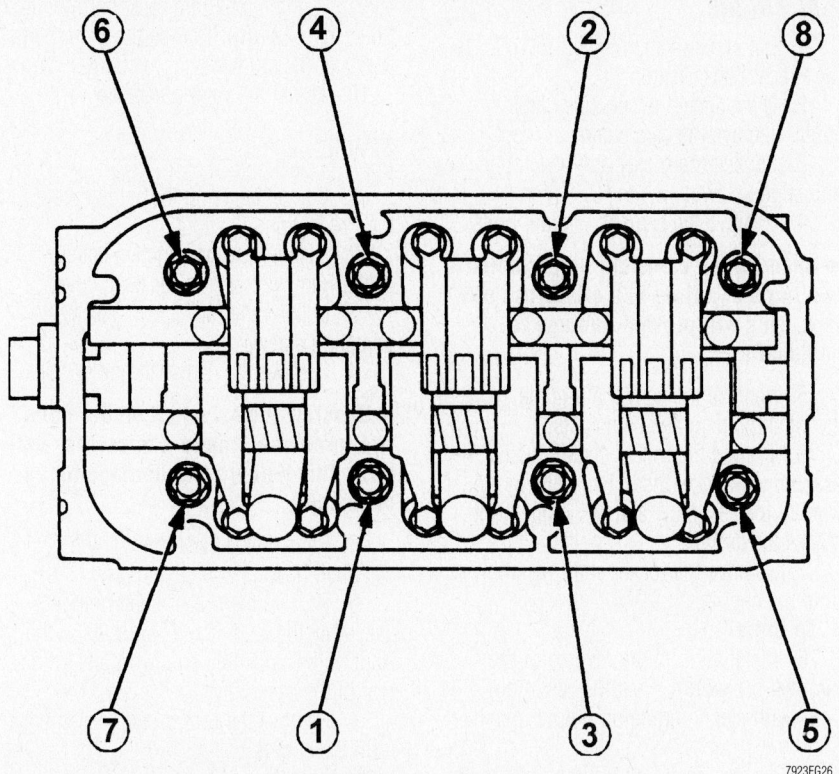

Cylinder head torque sequence—3.0L engine

2. Disconnect the negative battery cable. Remove the timing belt.

3. Loosen the rocker arm assembly adjusting screws.

4. On 1.8L engine, remove the lost motion holder bolts. Remove the lost motion holder and the lost motion assemblies.

5. Remove the rocker arm assembly retaining bolts. Remove the rocker arm assembly from the engine.

➡Unscrew the camshaft holder bolts two turns at a time in a crisscross pattern. This will prevent damage to the valves or rocker arm assembly. When removing the assembly, do not remove the camshaft holder bolts. These bolts will keep the camshaft holders, springs and rocker arms on the shafts.

To install:

6. Installation is the reverse of the removal procedure. Coat the retaining bolts with clean engine oil before installation.

7. Tighten each bolt 2 turns at a time in the crisscross sequence so that the rockers

repeat the procedure, paying close attention to the five second time limit

30. Move the driver's window all the way down by holding the drivers window DOWN switch to the AUTO DOWN position.

31. Pull up and hold the driver's window UP switch to the AUTO UP position until the window reaches the fully closed position. Hold the switch for one second.

32. Confirm that the power window master switch has been reset by using the driver's window AUTO UP and DOWN function.

33. If the AUTO UP and DOWN feature is still not working, repeat the complete procedure, paying close attention to the five second time limit.

Rocker Arms/Shafts

REMOVAL & INSTALLATION

➡**The radio may contain a coded theft protection circuit. Always obtain the code number before disconnecting the battery.**

Civic

EXCEPT 2.0L ENGINE

1. Before servicing the vehicle, refer to the Precautions Section.

Specified Torque
8 mm Bolts: 20 N·m (2.0 kgf·m, 14 lbf·ft)
6 mm Bolts: 12 N·m (1.2 kgf·m, 8.7 lbf·ft)
6 mm Bolts: ⑪, ⑫, ⑬, ⑭

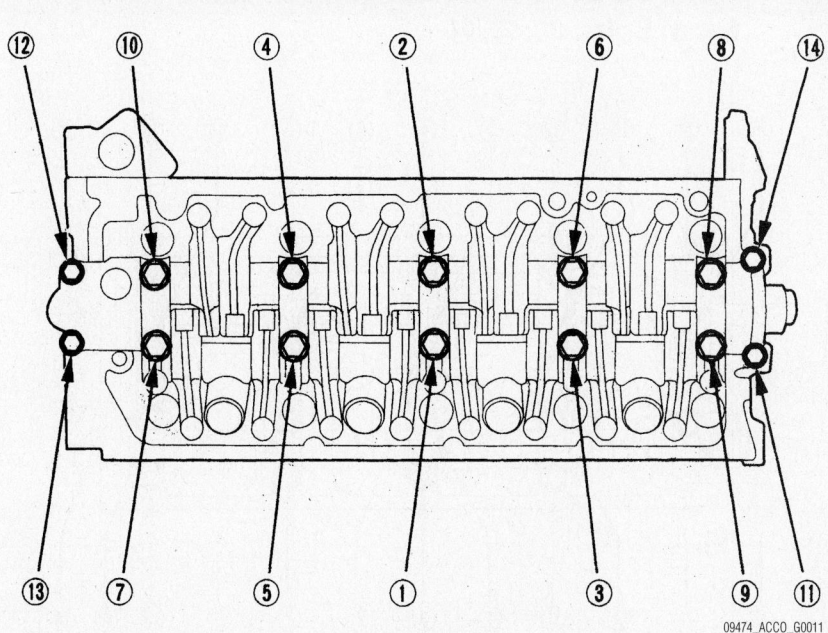

Rocker arm assembly bolt tightening sequence—1.7L and 1.8L engines

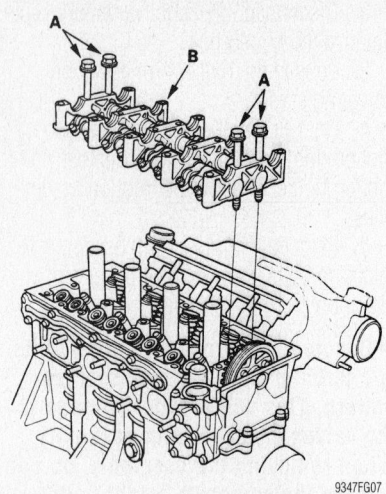

Use bolts (A) to hold the rocker arm assembly (B) together while removing or replacing the assembly—2.0L and 2.2L engines

are evenly tightened and don't bind on the valves. Tighten the 8mm rocker arm bolts to 14 ft. lbs. (20 Nm). Tighten the 6mm bolts to 8.7 ft. lbs (12 Nm).

8. Torque the lost motion holder retaining bolts to 11 ft. lbs (15 Nm).

9. Adjust the valve clearance.

NOTE: If the engine does not have bolt ㉑, skip it and continue the torque sequence.

Specified Torque
8 mm Bolts: 22 N·m (2.2 kgf·m, 16 lbf·ft)
6 mm Bolts: 12 N·m (1.2 kgf·m, 8.7 lbf·ft)
6 mm Bolts: ㉑, ㉒, ㉓

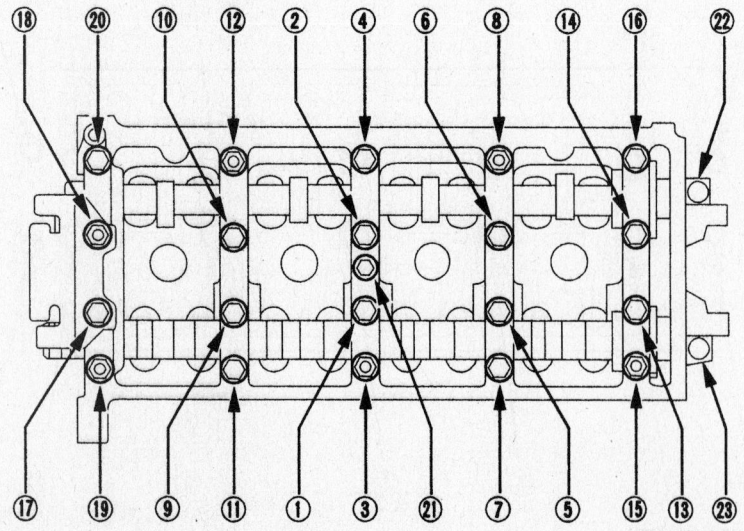

Rocker arm assembly bolt tightening sequence—Civic 2.0L engine

2.0L ENGINE

1. Before servicing the vehicle, refer to the Precautions Section.

2. Disconnect the negative battery cable. Remove the cam chain.

3. Loosen the rocker arm assembly adjusting screws.

4. Remove the camshaft holder bolts.

➡Unscrew the camshaft holder bolts two turns at a time in a crisscross pattern. This will prevent damage to camshafts.

5. Remove the cam chain guide, camshaft holders and camshafts.

6. To hold the rocker shaft holder assembly in place, insert two bolts in the front of the assembly and two bolts in the rear of the assembly.

7. Remove the rocker shaft assembly from the engine.

To install:

8. Installation is the reverse of the removal procedure. Coat the retaining bolts with clean engine oil before installation.

9. Tighten each bolt 2 turns at a time in the crisscross sequence so that the rockers are evenly tightened and don't bind on the valves. Tighten the 8mm rocker arm bolts to 16 ft. lbs. (22 Nm). Tighten the 6mm bolts to 8.7 ft. lbs (12 Nm).

10. Adjust the valve clearance.

S2000

1. Before servicing the vehicle, refer to the Precautions Section.

2. Disconnect the negative battery cable. Remove the cylinder head cover.

3. Loosen the rocker arm assembly adjusting screws.

4. Remove the camshaft holder bolts.

➡Unscrew the camshaft holder bolts two turns at a time in a crisscross pattern. This will prevent damage to camshafts.

5. Remove the cam chain guide, camshaft holders and camshafts.

6. To hold the rocker shaft holder assembly in place, insert two bolts in the front of the assembly and two bolts in the rear of the assembly.

7. Remove the rocker shaft assembly from the engine.

To install:

8. Be sure the crankshaft pulley is at TDC. Align the TDC mark on the pulley with the pointer on the chain case.

9. Check the alignment of the TDC marks on the cam chain sprocket with the cylinder head surface. Align the TDC marks on the intake camshaft gear and the exhaust camshaft gear.

10. Install the camshafts.

11. Continue the installation is the reverse order of the removal procedure. Coat the retaining bolts with clean engine oil before installation.

12. Tighten each bolt 2 turns at a time in the crisscross sequence so that the rockers are evenly tightened and don't bind on the valves. Tighten all bolts to 16 ft. lbs. (22 Nm).

13. Adjust the valve clearance.

Accord

2.3L ENGINE

1. Before servicing the vehicle, refer to the Precautions Section.

2. Disconnect the negative battery cable.

3. Turn the crankshaft so the No. 1 piston is at Top Dead Center (TDC).

➡The No. 1 piston is at top dead center when the pointer on the block aligns with the white painted mark on the flywheel (manual transaxle) or driveplate (automatic transaxle).

Specified torque:
8 x 1.25 mm
22 N·m (2.2 kgf·m, 16 lbf·ft)

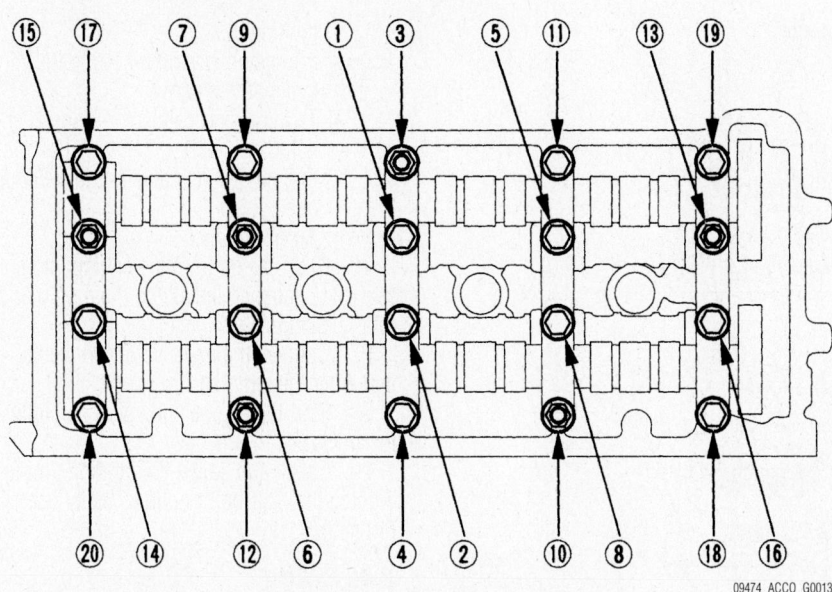

09474_ACCO_G0013

Rocker arm assembly bolt tightening sequence—2.2L and 2.2L engines

4. Remove or disconnect the following:
   - Air intake duct
   - Engine ground cable from the cylinder head cover
   - Connector and the terminal from the alternator
   - Engine wiring harness from the valve cover
   - Ignition coil

➡**Label all electrical connectors before detaching them.**

   - Electrical connectors from the distributor and the spark plug wires from the spark plugs.

➡**Matchmark the installed position if the distributor before removal.**

   - Distributor from the cylinder head
   - Ignition coil wire from the distributor
   - Positive Crankcase Ventilation (PCV) hose
   - Cylinder head cover. Replace the rubber seals if damaged or deteriorated.
   - Timing belt middle cover

5. Ensure the words **UP** embossed on the camshaft pulleys are aligned in the upward position.

6. Mark the rotation of the timing belt if it is to be used again. Loosen the timing belt adjusting nut ½ turn, then release the tension on the timing belt. Push the ten-

sioner to release tension from the belt, then tighten the adjusting nut.

7. Remove the timing belt from the camshaft sprockets.

⁂ WARNING

Do not crimp or bend the timing belt more than 90°, or less than 1 inch (25mm) in diameter

8. Insert a 5.0mm pin punch in each of the camshaft caps, nearest to the sprockets, through the holes provided. Remove the camshaft sprocket attaching bolts, then remove the sprockets. Do not lose the sprocket keys.

9. Remove or disconnect the following:
   - Side engine mount bracket B, then the timing belt back cover from behind the camshaft sprockets.
   - Rocker arm adjusting screws, then the pin punches from the camshaft caps

➡**Note the camshaft holder's locations for ease of installation. Loosen the bolts in the reverse order of the holder bolts torque sequence.**

   - Camshaft holders
   - Camshafts from the cylinder head, then discard the camshaft seals.
   - Rubber cap from the head, located at the end of the intake camshaft

   - Rocker arms from the cylinder head. Note the locations of the rocker arms.

➡**The rocker arms have to be installed to their original locations if being reused.**

   To install:

➡**Lubricate the rocker arms with clean oil before installation.**

10. Install or connect the following:
    - Rocker arms on the pivot bolts and the valve stems. If the rocker arms are being reused, install them to their original locations. The locknuts and adjustment screws should be loosened before installing the rocker arms.

11. Lubricate the camshafts with clean oil.

12. Install or connect the following:
    - Camshaft seals to the end of the camshafts that the timing belt sprockets attach to. The open side (spring) should be facing into the cylinder head when installed.

➡**Be sure the keyways on the camshafts are facing up and install the camshafts to the cylinder head.**

    - Rubber plug to the cylinder head at the end of the intake camshaft

13. Apply liquid gasket to the head mating surfaces of the No. 1 and No. 6 camshaft holders, then install them along with No. 2, 3, 4 and 5. **I** or **E** marks are stamped on the camshaft holders to identify them as Intake or Exhaust side holders. The arrows stamped on the holders should point toward the timing belt.

14. Snug the camshaft holders in place.

15. Press the camshaft seals securely into place.

16. Tighten the camshaft holder bolts in 2 steps, following the proper sequence, to ensure that the rockers do not bind on the valves. Tighten all the bolts, except the 4 studs, to 84 inch lbs. (10 Nm). Tighten the studs (number 5 and 7 bolts in the correct sequence) to 108 inch lbs. (12 Nm).

17. Install or connect the following:
    - Timing belt back cover.
    - Side engine mount bracket B. Tighten the bolt attaching the bracket to the cylinder head to 33 ft. lbs. (45 Nm). Tighten the bolts attaching the bracket to the side engine mount to 16 ft. lbs. (22 Nm).

18. Insert a 5.0mm pin punch in each of the camshaft caps, nearest to the pulleys,

Specified torque:
Except Intake ⑤, ⑦, Exhaust ⑥, ⑧:
　　　　　10 N·m (1.0 kg-m, 7 lb-ft)
Intake ⑤, ⑦, Exhaust ⑥, ⑧:
　　　　　12 N·m (1.2 kg-m, 9 lb-ft)

TIGHTENING SEQUENCE

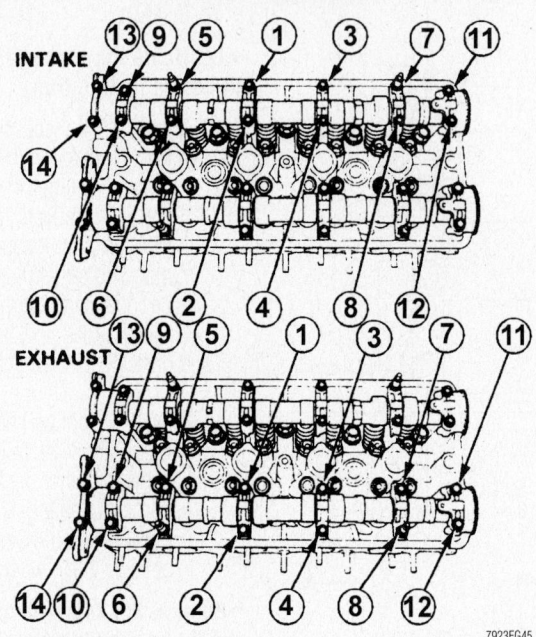

Camshaft holders torque sequence—2.3L engine

through the holes provided. Install the keys into the camshaft grooves.

19. Push the camshaft sprockets onto the camshafts, then tighten the retaining bolts to 27 ft. lbs. (38 Nm).

20. Ensure the words **UP** embossed on the camshaft pulleys are aligned in the upward position. Install the timing belt to the camshaft sprockets, then remove the 2, 5.0mm pin punches from the camshaft bearing caps.

21. Loosen, then tighten the timing belt adjuster nut.

22. Turn the crankshaft counterclockwise until the cam pulley has moved 3 teeth; this creates tension on the timing belt. Loosen, then tighten the adjusting nut and tighten it to 33 ft. lbs. (45 Nm).

23. Adjust the valves.

24. Tighten the crankshaft pulley bolt to 181 ft. lbs. (250 Nm).

25. Install or connect the following:
- Middle timing belt cover and tighten the attaching bolts to 108 inch lbs. (12 Nm).
- Cylinder head cover and tighten the cap nuts to 84 inch lbs. (10 Nm).
- PCV hose to the cylinder head cover
- Distributor. Snug the attaching

bolts until the timing has been checked and adjusted.
- Spark plug wires, then the distributor electrical connectors
- Ignition coil wire to the distributor.
- Ignition coil
- Alternator wiring harness to the cylinder head cover
- Terminal and connector to the alternator
- Engine ground cable to the cylinder head cover
- Air intake duct

26. Drain the oil from the engine into a sealable container. Install the drain plug and refill the engine with clean oil.

27. Connect the negative battery cable and enter the radio security code.

28. Start the engine and check carefully for any leaks.

29. Check and adjust the ignition timing. Tighten the distributor bolts to 13 ft. lbs. (18 Nm).

2.4L ENGINE

1. Before servicing the vehicle, refer to the Precautions Section.

2. Disconnect the negative battery cable. Remove the cylinder head cover.

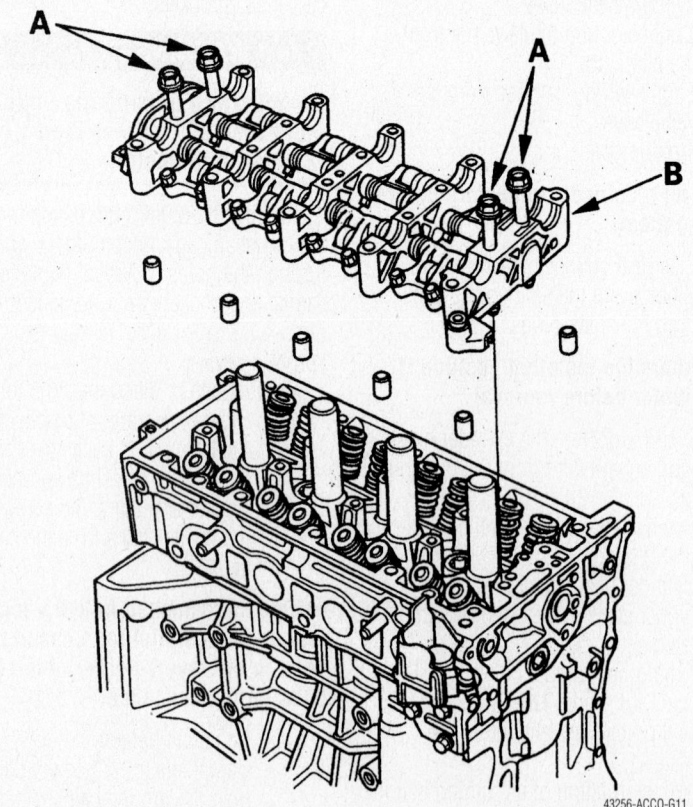

Insert the bolts (A) into the rocker shaft holder, then remove the rocker arm assembly (B)—2.4L engine

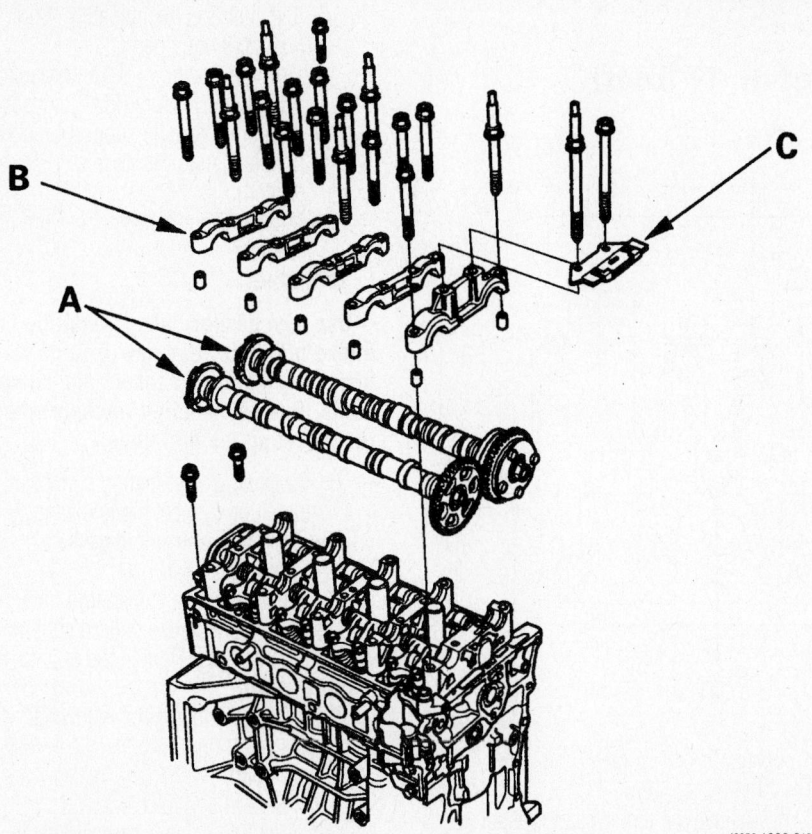

When installing the camshafts (A) make sure the punch marks on the VTC actuator and exhaust cam sprockets are facing up—2.4L engine

42356-ACCO-G13

3. Loosen the rocker arm assembly adjusting screws.

4. Remove the camshaft holder bolts.

➡**Unscrew the camshaft holder bolts two turns at a time in a crisscross pattern. This will prevent damage to camshafts.**

5. Remove the cam chain guide, camshaft holders and camshafts.

6. To hold the rocker shaft holder assembly in place, insert two bolts in the front of the assembly and two bolts in the rear of the assembly.

7. Remove the rocker shaft assembly from the engine.

To install:

8. Clean and dry the No. 5 rocker shaft holding mating surface.

9. Apply a suitable liquid gasket P/N 08718-0009, or equivalent, evenly to the cylinder head mating surface of the No. 5 rocker shaft holder.

➡**The parts must be installed within 5 minutes of applying the liquid gasket.**

10. Reassemble the rocker arm assembly, as necessary.

11. Install or connect the following

• Bolts (A) into the rocker shaft holder, then the rocker arm assembly on the cylinder head. Remove the bolts from the rocker shaft holder.

12. Make sure the punch marks on the variable valve timing control (VTC) actuator and exhaust camshaft sprocket are facing up, then set the camshafts (A) in the holder

13. Set the camshaft holders (B) and timing chain guide B (C) in place.

14. Tighten the bolts, in sequence, to the following specification:

a. 8mm bolts: 16 ft. lbs. (22 Nm)

b. 6mm bolts: 8.7 ft. lbs. (12 Nm)

15. Continue the installation in the reverse order of the removal procedure.

16. Adjust the valve lash.

3.0L ENGINE

1. Before servicing the vehicle, refer to the Precautions Section.

2. Disconnect the negative battery cable. Remove the timing belt.

3. Loosen the rocker arm assembly adjusting screws.

4. Remove the rocker arm assembly retaining bolts. Remove the rocker arm assembly from the engine.

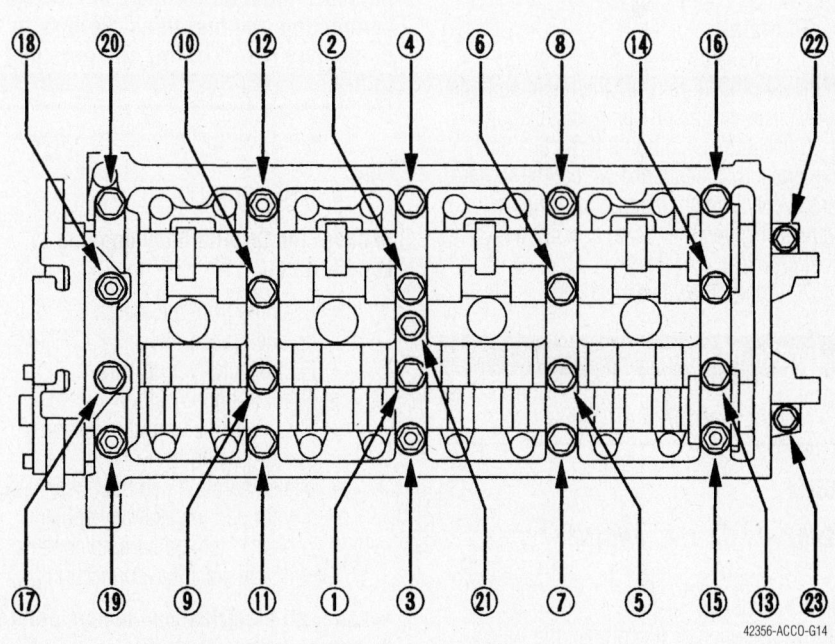

42356-ACCO-G14

Rocker arm assembly bolt tightening sequence—2.4L engine

Specified torque
8 x 1.25 mm: 24 N·m (2.4 kgf·m, 17 lbf·ft)

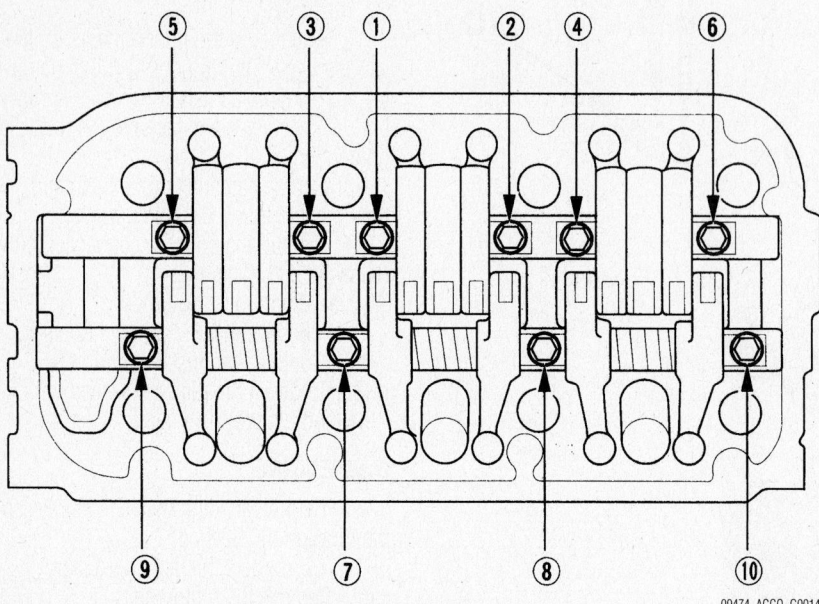

Rocker arm assembly bolt tightening sequence—3.0L engine

09474_ACCO_G0014

➡Unscrew the camshaft holder bolts two turns at a time in a crisscross pattern. This will prevent damage to the valves or rocker arm assembly. When removing the assembly, do not remove the camshaft holder bolts. These bolts will keep the camshaft holders, springs and rocker arms on the shafts.

To install:

5. Installation is the reverse of the removal procedure. Coat the retaining bolts with clean engine oil before installation.

6. Tighten each bolt 2 turns at a time in the crisscross sequence so that the rockers are evenly tightened and don't bind on the valves. Tighten the rocker arm bolts to 17 ft. lbs. (24 Nm).

7. Adjust the valve clearance.

Intake Manifold

REMOVAL & INSTALLATION

Civic

2002–03 GX 1.7L ENGINE

1. Before servicing the vehicle, refer to the Precautions Section.

2. Remove or disconnect the following:

• Negative battery cable

3. Drain the cooling system to a level below the upper radiator hose.

4. Relieve the fuel system pressure by loosening the fuel filter service bolt.

⁂ CAUTION

The fuel injection system remains under pressure even after the engine has been turned off. The fuel system pressure must be relieved before disconnecting any fuel lines. Failure to do so may result in fire and personal injury.

5. Remove or disconnect the following:

• Intake air duct

➡Cover the throttle body opening to keep dirt out.

• Fuel line from the fuel rail. Clean up any spilled fuel.
• Fuel injector wiring harnesses
• Fuel rail and injectors
• Throttle cable from the linkage at the throttle body
• Intake manifold cooling hoses. Use a drain pan to catch any spilled coolant. Also, be sure no coolant spills on electrical connections.

➡Label all electrical connectors before detaching them.

• Engine wiring harness connectors from the intake manifold sensors
• Idle Air Control (IAC) valve

• Exhaust Gas Recirculation (EGR valve), if equipped
• Throttle Position (TP) and Manifold Absolute Pressure (MAP) sensor
• Manifold from its support bracket
• Intake manifold nuts in a crisscross pattern.
• Intake manifold assembly from the vehicle.

To install:

➡Use new gaskets when installing the intake manifold. Use new O-rings when installing manifold sensors and components. Use new sealing washers when reconnecting the fuel lines.

6. Clean all gasket mating surfaces.
7. Install or connect the following:

• New intake manifold gaskets
• Intake manifold

8. Tighten the intake manifold nuts in 2 or 3 steps in a crisscross pattern starting with the inside nuts. Tighten the nuts to 16 ft. lbs. (22 Nm).

9. Install or connect the following:

• Support bracket bolts: 17 ft. lbs. (24 Nm)
• Fuel rail and injectors
• Fuel line using new washers
• EGR valve and tighten its nuts to 15 ft. lbs. (21 Nm).
• IAC valve. Tighten its mounting bolts to 16 ft. lbs. (22 Nm).
• Fuel injector wiring harnesses
• Intake manifold wiring harnesses
• Intake manifold cooling hoses
• Throttle cable
• Intake air duct and air cleaner assembly

10. Refill and bleed the cooling system.
11. Connect the negative battery cable.
12. Verify that all sensors, valves, and vacuum lines are installed and connected properly. Be sure there are no loose electrical connections.
13. Turn the ignition on and off several times without starting the engine to pressurize the fuel system. Run the engine and check for proper operation. Check for vacuum leaks.
14. After the engine has warmed up, check the operation of the throttle cable and adjust it if necessary.

2002–03 GX 1.7L ENGINE

➡Make sure to acquire the anti-theft code from the radio and write down the frequencies for the radio's preset buttons.

1. Before servicing the vehicle, refer to the Precautions Section.

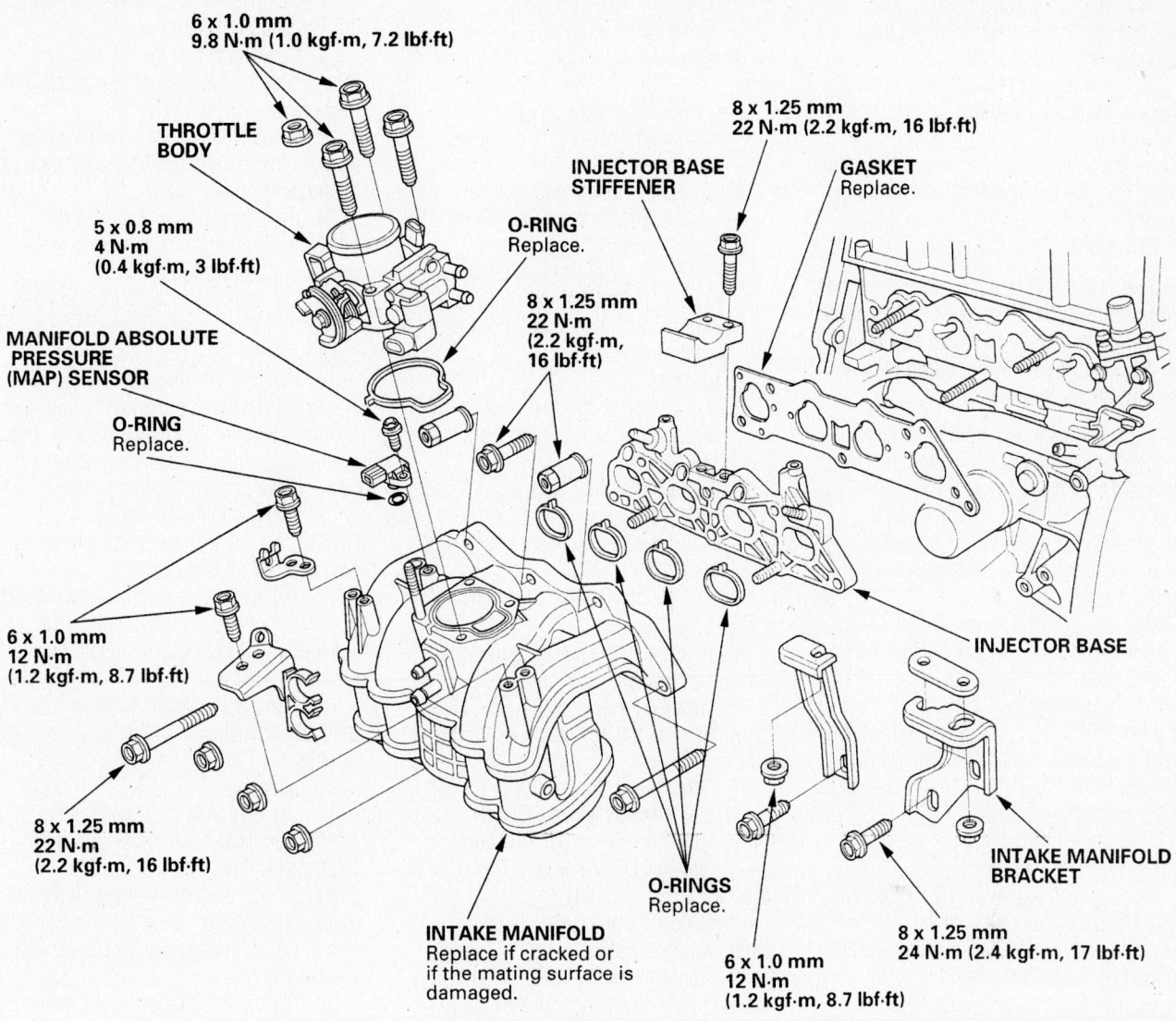

6 x 1.0 mm
9.8 N·m (1.0 kgf·m, 7.2 lbf·ft)

**THROTTLE
BODY**

8 x 1.25 mm
22 N·m (2.2 kgf·m, 16 lbf·ft)

**INJECTOR BASE
STIFFENER**

GASKET
Replace.

5 x 0.8 mm
4 N·m
(0.4 kgf·m, 3 lbf·ft)

O-RING
Replace.

8 x 1.25 mm
22 N·m
(2.2 kgf·m,
16 lbf·ft)

**MANIFOLD ABSOLUTE
PRESSURE
(MAP) SENSOR**

O-RING
Replace.

INJECTOR BASE

6 x 1.0 mm
12 N·m
(1.2 kgf·m, 8.7 lbf·ft)

8 x 1.25 mm
22 N·m
(2.2 kgf·m, 16 lbf·ft)

O-RINGS
Replace.

**INTAKE MANIFOLD
BRACKET**

INTAKE MANIFOLD
Replace if cracked or
if the mating surface is
damaged.

6 x 1.0 mm
12 N·m
(1.2 kgf·m, 8.7 lbf·ft)

8 x 1.25 mm
24 N·m (2.4 kgf·m, 17 lbf·ft)

67162-ACCO-G01

Exploded view of the intake manifold and related components—1.7L engine except 2002–03 GX

2. Turn off the manual shut off valve. To reduce pressure in the lines start the engine and run it until it stalls.

3. Disconnect the negative battery cable. Drain the cooling system.

4. Remove the throttle cable. Remove the cruise control cable.

5. Remove the intake resonator. Remove the air cleaner. Remove the water bypass hoses.

6. Disconnect the power brake vacuum hose and the four way joint valve.

7. Remove the fuel injectors. Remove the fuel feed pipe. Disconnect the fuel feed hose. Remove the air cleaner housing bracket. Remove the thermo valve bolt and intake manifold bracket bolt.

8. Remove the intake manifold retaining bolts. Remove the intake manifold from the engine.

To install:

➡Use new gaskets when installing the intake manifold. Use new O-rings when installing manifold sensors and components. Use new sealing washers when reconnecting the fuel lines.

9. Clean all gasket mating surfaces.

10. Torque the intake manifold retaining nuts to 16 ft. lbs. (22 Nm).

11. Continue the installation in reverse order of the removal procedure.

1.8L ENGINE

➡Make sure to acquire the anti-theft code from the radio and write down the frequencies for the radio's preset buttons.

1. Before servicing the vehicle, refer to the Precautions Section.

2. Disconnect the negative battery cable. Drain the cooling system.

3. Remove the cowl cover and the under cowl panel. Remove the air cleaner housing assembly. Remove the air intake duct.

4. Remove the injector cover. Remove the evaporative emission canister hose, power brake booster hose and the power steering pump hose clamp.

5. Remove the quick connect fitting cover. Disconnect the fuel feed hose.

6. Remove the engine wire harness connectors and the wire harness clamps from the intake manifold.

7. Disconnect the throttle actuator connector, the manifold absolute pressure connector, the evaporative emission control canister purge connector and the intake manifold tuning valve actuator connector.

8. Remove the water bypass hoses. Remove the throttle body. Remove the heater hose clamp bracket.

9. Raise and safely support the vehicle. Remove the intake manifold bracket. Lower the vehicle.

10. Remove the intake manifold retaining bolts. Remove the intake manifold from the engine.

To install:

➡ **Use new gaskets when installing the intake manifold. Use new O-rings when installing manifold sensors and components. Use new sealing washers when reconnecting the fuel lines.**

11. Clean all gasket mating surfaces.

12. Torque the intake manifold retaining nuts to 17 ft. lbs. (24 Nm).

13. Continue the installation in the reverse order of the removal procedure.

14. Reprogram the power window control unit as follows.

15. Turn the ignition switch ON. Lower the window, all the way down. Open the driver's side door.

16. Turn the ignition switch OFF. Push and hold the drivers window DOWN switch. Turn the ignition switch ON. Release the driver's window DOWN switch. This must be done within five seconds.

17. Repeat the above step three more times. Wait one second.

18. Confirm that the AUTO UP and AUTO DOWN do not work. If they work repeat the procedure, paying close attention to the five second time limit

19. Move the driver's window all the way down by holding the drivers window DOWN switch to the AUTO DOWN position.

20. Pull up and hold the driver's window UP switch to the AUTO UP position until the window reaches the fully closed position. Hold the switch for one second.

21. Confirm that the power window master switch has been reset by using the driver's window AUTO UP and DOWN function.

22. If the AUTO UP and DOWN feature is still not working, repeat the complete procedure, paying close attention to the five second time limit.

2002–05 2.0L ENGINE

➡ **Make sure to acquire the anti-theft code from the radio and write down the frequencies for the radio's preset buttons.**

1. Before servicing the vehicle, refer to the Precautions Section.

2. Drain the cooling system.

3. Relieve the fuel system pressure.

4. Remove or disconnect the following:

- Negative battery cable
- Cooling hoses from the intake manifold
- Vacuum hoses and electrical connectors from the manifold and throttle body
- Throttle cable from the throttle body
- Fuel rail and fuel injectors
- Intake manifold support brackets
- Intake manifold

To install:

5. Installation is the reverse of the removal procedure, while using the following torque values:

- Intake manifold fasteners: 16 ft. lbs. (22 Nm)
- Throttle body fasteners: 16 ft. lbs. (22 Nm)
- Intake manifold bracket bolts: 33 ft. lbs. (44 Nm)

6. Reprogram the power window control unit as follows.

7. Turn the ignition switch ON. Lower the window, all the way down. Open the driver's side door.

8. Turn the ignition switch OFF. Push and hold the drivers window DOWN switch. Turn the ignition switch ON. Release the driver's window DOWN switch. This must be done within five seconds.

9. Repeat the above step three more times. Wait one second.

10. Confirm that the AUTO UP and AUTO DOWN do not work. If they work repeat the procedure, paying close attention to the five second time limit

11. Move the driver's window all the way down by holding the drivers window DOWN switch to the AUTO DOWN position.

12. Pull up and hold the driver's window UP switch to the AUTO UP position until the window reaches the fully closed position. Hold the switch for one second.

13. Confirm that the power window master switch has been reset by using the driver's window AUTO UP and DOWN function.

14. If the AUTO UP and DOWN feature is still not working, repeat the complete procedure, paying close attention to the five second time limit.

2006 2.0L ENGINE

➡ **Make sure to acquire the anti-theft code from the radio and write down the frequencies for the radio's preset buttons.**

1. Before servicing the vehicle, refer to the Precautions Section.

2. Relieve the fuel system pressure. Disconnect the negative battery cable. Drain the cooling system.

3. Remove the intake manifold cover. Remove the vacuum hose, breather pipe and the intake air duct.

4. Remove the engine wire harness connectors and the wire harness clamps from the intake manifold.

5. Disconnect the four fuel injector connectors, the manifold absolute pressure connector and the throttle connector.

6. Remove the ground cable, harness clamp bracket and the harness holder from its mounting. Remove the PCV valve hose, evaporative emission canister hose and the power brake booster vacuum hose.

7. Remove the water bypass hoses. Remove the quick connect fitting cover. Disconnect the fuel feed hose.

8. Raise and support the vehicle safely. Remove the intake manifold connector cover. Remove the intake manifold bracket. Lower the vehicle.

9. Remove the intake manifold retaining bolts. Remove the intake manifold from the vehicle.

To install:

10. Installation is the reverse of the removal procedure, while using the following torque values:

11. Torque the manifold retaining bolts and nut to specification in a crisscross pattern in two or three steps, beginning with the inner bolt.

12. Reprogram the power window control unit as follows.

13. Turn the ignition switch ON. Lower the window, all the way down. Open the driver's side door.

14. Turn the ignition switch OFF. Push and hold the drivers window DOWN switch. Turn the ignition switch ON. Release the driver's window DOWN switch. This must be done within five seconds.

15. Repeat the above step three more times. Wait one second.

16. Confirm that the AUTO UP and AUTO DOWN do not work. If they work repeat the procedure, paying close attention to the five second time limit

17. Move the driver's window all the way down by holding the drivers window DOWN switch to the AUTO DOWN position.

18. Pull up and hold the driver's window UP switch to the AUTO UP position until the window reaches the fully closed position. Hold the switch for one second.

19. Confirm that the power window mas-

ter switch has been reset by using the driver's window AUTO UP and DOWN function.

20. If the AUTO UP and DOWN feature is still not working, repeat the complete procedure, paying close attention to the five second time limit.

S2000

➡**Make sure to acquire the anti-theft code from the radio and write down the frequencies for the radio's preset buttons.**

1. Before servicing the vehicle, refer to the Precautions Section.

2. Drain the cooling system.
3. Relieve the fuel system pressure.
4. Remove or disconnect the following:
   • Negative battery cable
   • Cooling hoses from the intake manifold
   • Vacuum hoses and electrical con-

NOTE: Use new O-rings and gaskets when reassembling.

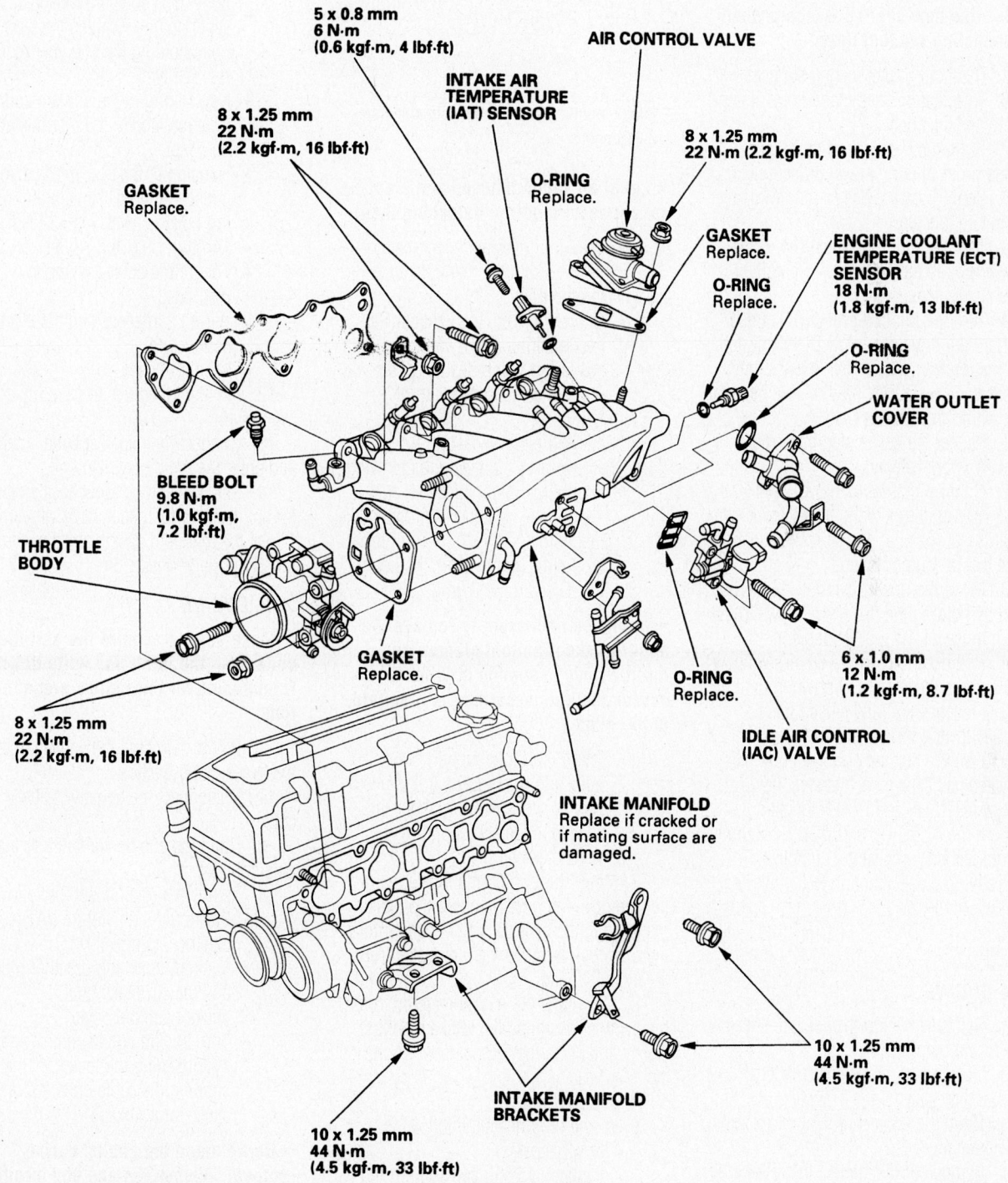

Exploded view of the intake manifold and related components—S2000 with 2.0L engine

9347FG09

nectors from the manifold and throttle body
- Throttle cable from the throttle body
- Fuel rail and fuel injectors
- Intake manifold support brackets
- Intake manifold

To install:

➡**Use new gaskets when installing the intake manifold. Use new O-rings when installing manifold sensors and components. Use new sealing washers when reconnecting the fuel lines.**

5. Clean all gasket mating surfaces.
6. Torque the intake manifold retaining nuts to specification.
7. Continue the installation in the reverse order of the removal procedure.
8. Reprogram the power window control unit as follows.
9. Turn the ignition switch ON. Lower the window, all the way down. Open the driver's side door.
10. Turn the ignition switch OFF. Push and hold the drivers window DOWN switch. Turn the ignition switch ON. Release the driver's window DOWN switch. This must be done within five seconds.
11. Repeat the above step three more times. Wait one second.
12. Confirm that the AUTO UP and AUTO DOWN do not work. If they work repeat the procedure, paying close attention to the five second time limit
13. Move the driver's window all the way down by holding the drivers window DOWN switch to the AUTO DOWN position.
14. Pull up and hold the driver's window UP switch to the AUTO UP position until the window reaches the fully closed position. Hold the switch for one second.
15. Confirm that the power window master switch has been reset by using the driver's window AUTO UP and DOWN function.
16. If the AUTO UP and DOWN feature is still not working, repeat the complete procedure, paying close attention to the five second time limit.

Accord

2.3L ENGINE

1. Before servicing the vehicle, refer to the Precautions Section.
2. Disconnect the negative battery cable.
3. Drain the engine coolant into a sealable container.
4. Remove or disconnect the following:
- Cooling hoses from the intake manifold

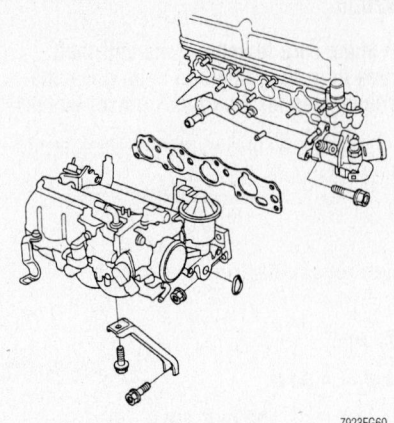

7923FG60

Intake manifold and related components— 2.3L engine

➡**Label all vacuum hoses and electrical connectors before detaching them.**

- Vacuum hoses and electrical connectors from the manifold and throttle body
- Connector from the Exhaust Gas Recirculation (EGR) valve. Position the wiring harnesses out of the way.
- Throttle cable from the throttle body

5. Relieve the fuel pressure.
6. Remove or disconnect the following:
- Fuel rail and fuel injectors
- Thermostat housing from the intake manifold and the connecting pipe by pulling and twisting the housing. Discard the O-rings.

➡**It may be necessary to remove the upper intake manifold plenum and throttle body assembly in order to access the nuts securing the manifold to the head.**

- Intake manifold support bracket bolts and the bracket. If it is necessary to access it from under the vehicle; raise and support the vehicle safely.

7. While supporting the intake manifold, remove the nuts attaching the intake manifold to the cylinder head, then remove the manifold. Remove the old gasket from the cylinder head.
8. Clean any old gasket material from the cylinder head and the intake manifold. Check and clean the FIA chamber on the cylinder head.

To install:

9. Install or connect the following:
- New gasket
- Intake manifold, and support the manifold.
- Support bracket to the manifold.

Tighten the retaining bolt to 16 ft. lbs. (22 Nm).
10. Starting with the inner or center nuts, tighten the nuts, in a crisscross pattern, to the correct torque. The tension must be even across the entire face of the manifold if leaks are to be prevented. Correct torque is 16 ft. lbs. (22 Nm).
11. Install or connect the following:
- New gasket
- Upper intake manifold and throttle body assembly, if removed as a separate unit. Tighten the nuts and bolts holding the chamber to 16 ft. lbs. (22 Nm).
- New O-ring to the coolant connecting pipe, and to the thermostat housing.
- Housing to the coolant pipe and the intake manifold. Tighten the mounting bolts to 16 ft. lbs. (22 Nm).
- Throttle cable and adjust.
- Fuel rail/injector assembly
- Fuel lines
- Wiring harnesses and the electrical connectors
- Vacuum hoses

12. Fill and bleed the air from the cooling system.
13. Connect the negative battery cable and enter the radio security code.
14. Start the engine and check carefully for any leaks of fuel, coolant or vacuum. Check the manifold gasket areas carefully for any leakage of vacuum.

2.4L ENGINE

➡**Make sure to acquire the anti-theft code from the radio and write down the frequencies for the radio's preset buttons.**

1. Before servicing the vehicle, refer to the Precautions Section.
2. Disconnect the negative battery cable.
3. Drain the engine coolant into a sealable container.
4. Remove or disconnect the following:
- Intake Air Temperature (IAT) sensor electrical connector
- Vacuum hose and breather pipe and the air intake duct
- Intake manifold cover
- Throttle and cruise control cables by loosening the locknuts, then slipping the cable ends out of the accelerator linkage.

➡**Do not bend the cables during removal. Always replace any throttle or cruise control cables that get kinked during removal.**

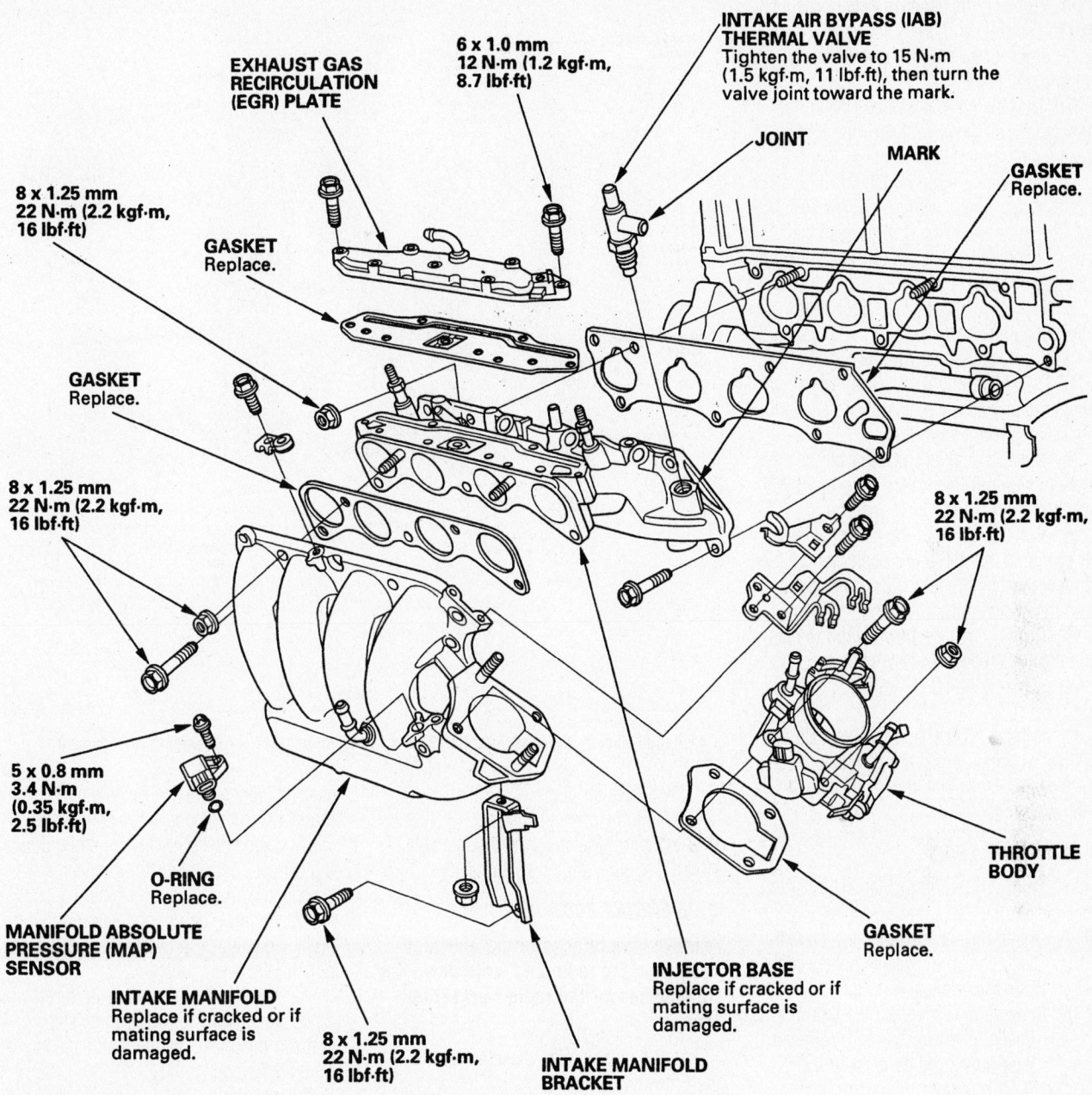

EXHAUST GAS RECIRCULATION (EGR) PLATE

**6 x 1.0 mm
12 N·m (1.2 kgf·m, 8.7 lbf·ft)**

INTAKE AIR BYPASS (IAB) THERMAL VALVE
Tighten the valve to 15 N·m (1.5 kgf·m, 11 lbf·ft), then turn the valve joint toward the mark.

JOINT

MARK

GASKET
Replace.

**8 x 1.25 mm
22 N·m (2.2 kgf·m, 16 lbf·ft)**

GASKET
Replace.

GASKET
Replace.

**8 x 1.25 mm
22 N·m (2.2 kgf·m, 16 lbf·ft)**

**8 x 1.25 mm
22 N·m (2.2 kgf·m, 16 lbf·ft)**

**5 x 0.8 mm
3.4 N·m
(0.35 kgf·m, 2.5 lbf·ft)**

O-RING
Replace.

THROTTLE BODY

MANIFOLD ABSOLUTE PRESSURE (MAP) SENSOR

GASKET
Replace.

INTAKE MANIFOLD
Replace if cracked or if mating surface is damaged.

**8 x 1.25 mm
22 N·m (2.2 kgf·m, 16 lbf·ft)**

INTAKE MANIFOLD BRACKET

INJECTOR BASE
Replace if cracked or if mating surface is damaged.

42356-ACCO-G15

Exploded view of the intake manifold and related components—2.4L engine

- Evaporative emission (EVAP) canister hose and brake booster vacuum hose
- Idle Air Control (IAC) valve connectors
- Throttle Position (TP) sensor connector
- Manifold Absolute Pressure (MAP) sensor connector
- Necessary engine wire harness connectors and wire harness clamps from the intake manifold
- Bolt securing the harness holder and remove the harness clamps
- Water bypass hoses, then plug them

- Harness clamp and harness connector from the intake manifold bracket
- Intake manifold bracket
- A/T vacuum hose
- Retainer and intake manifold

To install:
5. Clean the mounting surfaces.
6. Install or connect the following:
   - New gasket
   - Intake manifold. Tighten the bolts, in a crisscross pattern beginning with the inner bolt, to 16 ft. lbs. (22 Nm).
   - A/T vacuum hose

- Intake manifold bracket
- Harness clamp and connector to the intake manifold bracket
- Water bypass hoses
- Bolt securing the harness holder and tighten to 8.7 ft. lbs. (12 Nm)
- Harness clamps
- EVAP canister hose and brake booster vacuum hose
- Throttle and cruise control cables
- Intake manifold cover
- Intake air duct
- IAT sensor connector, vacuum hose and breather pipe
7. Refill the cooling system.

8. Connect the negative battery cable, start the engine, and check for leaks.

9. Reprogram the power window control unit as follows.

10. Turn the ignition switch ON. Lower the window, all the way down. Open the driver's side door.

11. Turn the ignition switch OFF. Push and hold the drivers window DOWN switch. Turn the ignition switch ON. Release the driver's window DOWN switch. This must be done within five seconds.

12. Repeat the above step three more times. Wait one second.

13. Confirm that the AUTO UP and AUTO DOWN do not work. If they work repeat the procedure, paying close attention to the five second time limit

14. Move the driver's window all the way down by holding the drivers window DOWN switch to the AUTO DOWN position.

15. Pull up and hold the driver's window UP switch to the AUTO UP position until the window reaches the fully closed position. Hold the switch for one second.

16. Confirm that the power window master switch has been reset by using the driver's window AUTO UP and DOWN function.

17. If the AUTO UP and DOWN feature is still not working, repeat the complete procedure, paying close attention to the five second time limit.

3.0L ENGINE 2002

1. Before servicing the vehicle, refer to the Precautions Section.
2. Obtain the security code for the radio.
3. Disconnect the negative battery cable.
4. Drain the coolant.
5. Remove or disconnect the following:
   • Evaporative emissions (EVAP) canister hose from the throttle body.
   • Air intake duct
   • Upper engine covers
   • Accelerator and cruise control cables from the throttle body.

➡**Ensure that all components have been removed from the intake manifold.**

   • Intake manifold
   To install:
6. Clean the mounting surfaces.
7. Install or connect the following:
   • New gasket
   • Intake manifold. Tighten the bolts to 16 ft. lbs. (22 Nm).
   • All removed hoses and wiring on the intake manifold and throttle body.
   • Engine covers

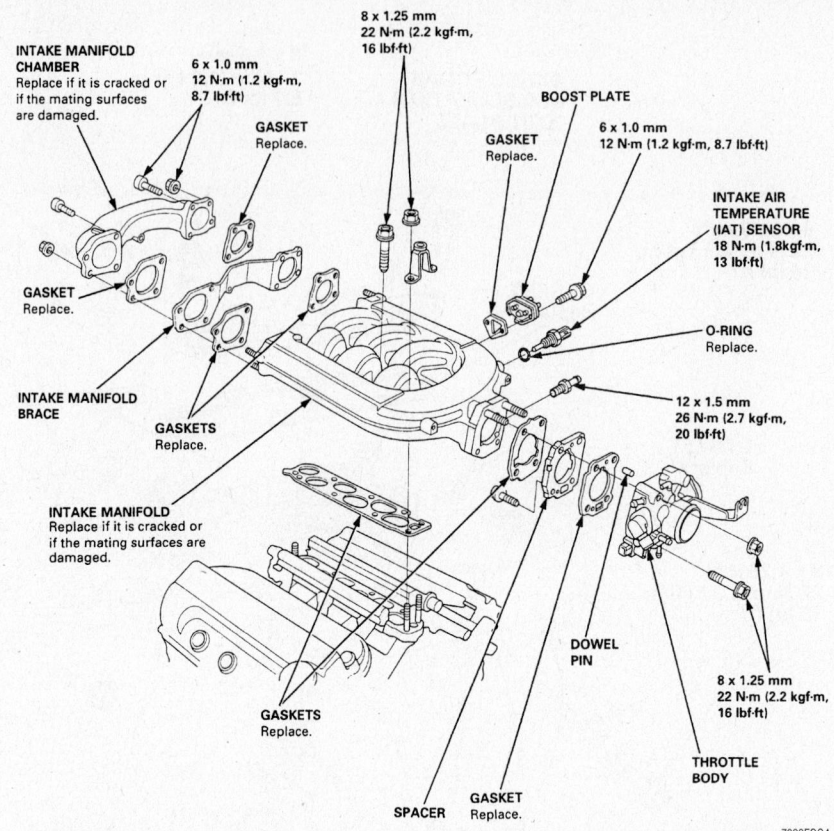

Exploded view of the intake manifold and related components—3.0L engine 2002 Accord

7923FGC4

   • Intake air duct
8. Refill the cooling system.
9. Connect the negative battery cable, start the engine, and check for leaks.

3.0L ENGINE 2003–06

➡**Make sure to acquire the anti-theft code from the radio and write down the frequencies for the radio's preset buttons.**

1. Before servicing the vehicle, refer to the Precautions Section.
2. Relieve the fuel system pressure. Disconnect the negative battery cable. Drain the cooling system.
3. Remove the ignition coil cover. Remove the breather pipe and the air intake duct.
4. Remove the PCV valve hose, power brake booster hose, evaporative emission canister hose and the vacuum hose.
5. Remove the water bypass hoses. Remove the engine wire harness connectors and the wire harness clamps from the intake manifold
6. Disconnect the intake air temperature sensor connector, throttle actuator connector and the manifold absolute sensor connector.
7. Disconnect the evaporative emission

canister purge valve connector and the engine mount control solenoid valve connector.
8. Disconnect the intake manifold tuning runner control actuator connector, if equipped with manual transaxle.
9. Remove the upper cover mounting bolts sequentially in two or three steps and remove the upper cover.
10. Remove the intake manifold retaining bolts sequentially in two or three steps. Remove the intake manifold and spacer from the engine.

To install:

➡**Use new gaskets when installing the intake manifold. Use new O-rings when installing manifold sensors and components. Use new sealing washers when reconnecting the fuel lines.**

11. Clean all gasket mating surfaces.
12. Torque the intake manifold retaining nuts to specification, sequentially in two or three steps.
13. Torque the upper cover retaining bolts to 8.7 ft. lbs. (12 Nm), sequentially in two or three steps.
14. Continue the installation in the reverse order of the removal procedure.
15. Reprogram the power window control unit as follows.

16. Turn the ignition switch ON. Lower the window, all the way down. Open the driver's side door.

17. Turn the ignition switch OFF. Push and hold the drivers window DOWN switch. Turn the ignition switch ON. Release the driver's window DOWN switch. This must be done within five seconds.

18. Repeat the above step three more times. Wait one second.

19. Confirm that the AUTO UP and AUTO DOWN do not work. If they work repeat the procedure, paying close attention to the five second time limit

20. Move the driver's window all the way down by holding the drivers window DOWN switch to the AUTO DOWN position.

21. Pull up and hold the driver's window UP switch to the AUTO UP position until the window reaches the fully closed position. Hold the switch for one second.

22. Confirm that the power window master switch has been reset by using the driver's window AUTO UP and DOWN function.

23. If the AUTO UP and DOWN feature is still not working, repeat the complete procedure, paying close attention to the five second time limit.

Exhaust Manifold

REMOVAL & INSTALLATION

1.7L Engine

1. Before servicing the vehicle, refer to the Precautions Section.

2. Disconnect the negative battery cable.

3. Raise and support the front of the vehicle and block the rear wheels.

4. Remove or disconnect the following:
- Front exhaust pipe from the exhaust manifold/catalytic converter
- Exhaust manifold support brackets if their bolts are accessible from this angle. The splash shield may be removed for better access.

5. Lower the vehicle.

➡**Remove any rust or dirt from the exhaust manifold before removal. This will prevent dirt from entering the exhaust pipes.**

6. Remove or disconnect the following:
- Manifold heat shield
- Heated Oxygen Sensor (HO2S) harness
- HO2S, using an oxygen sensor

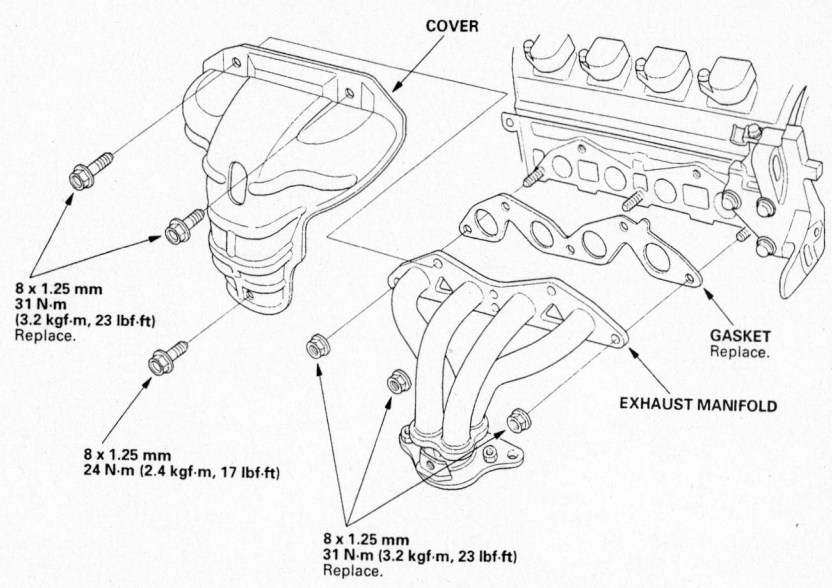

COVER

8 x 1.25 mm
31 N·m
(3.2 kgf·m, 23 lbf·ft)
Replace.

8 x 1.25 mm
24 N·m (2.4 kgf·m, 17 lbf·ft)

8 x 1.25 mm
31 N·m (3.2 kgf·m, 23 lbf·ft)
Replace.

GASKET
Replace.

EXHAUST MANIFOLD

67162-ACCO-G02

Exploded view of the exhaust manifold—1.7L (D17A2) engine

socket or box end wrench to unscrew the sensor from the manifold. Handle the sensor carefully.
- Exhaust manifold brackets
- Exhaust manifold and separate it from the cylinder head
- Exhaust manifold and gasket

To install:

➡**Use new gaskets and self-locking nuts when installing the exhaust manifold.**

7. Clean the gasket mating surfaces of the manifold and cylinder head ports.

8. Install or connect the following:
- New gasket onto the cylinder head
- New gaskets onto the exhaust pipe flange
- Exhaust manifold. Apply anti-seize paste to the studs. Tighten the self-locking nuts to 23 ft. lbs. (32 Nm) in a crisscross pattern starting in the center of the manifold and working outward.
- Manifold brackets and tighten their bolts to 33 ft. lbs. (45 Nm) for all other engines

9. Carefully coat only the threads of the oxygen sensor body with anti-seize paste. Don't get any anti-seize on the sensor probe.

10. Install or connect the following:
- HO2S and carefully tighten it to 33 ft. lbs. (45 Nm)
- Heat shield and tighten the bolts to 16 ft. lbs. (22 Nm)
- HO2S connector

11. Raise and support the front of the vehicle and block the rear wheels.

12. Install or connect the following:
- Front exhaust pipe and the exhaust manifold/catalytic converter. Tighten the self-locking nuts to 40 ft. lbs. (55 Nm), if the converter is not attached to the manifold. If the converter is attached, tighten to 25 ft. lbs., (34 Nm). Install any manifold brackets and tighten them to 33 ft. lbs. (45 Nm).
- Splash shield if it was removed

13. Lower the vehicle and connect the negative battery cable.

14. Run the engine and check for exhaust leaks.

2.0L Engine

1. Before servicing the vehicle, refer to the Precautions Section.

2. On 2006 Civic, drain the cooling system and relieve the fuel system pressure.

3. Remove or disconnect the following:
- Negative battery cable
- Catalytic converter
- Exhaust manifold heat shields
- Heated Oxygen (HO2S) sensor connector
- Exhaust manifold bracket
- Exhaust manifold

To install:

4. Installation is the reverse of the removal procedure.

5. As required, reprogram the power window control unit as follows.

6. Turn the ignition switch ON. Lower the window, all the way down. Open the driver's side door.

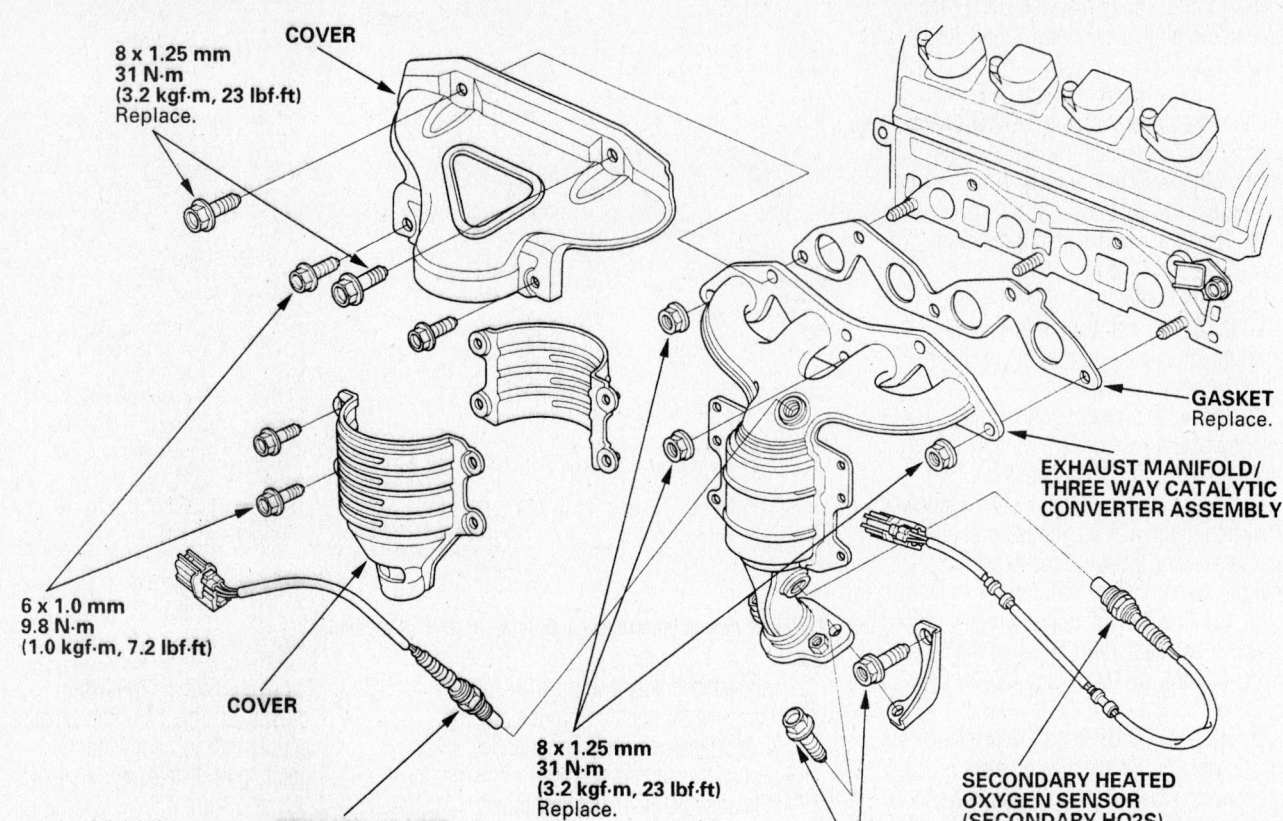

COVER

8 x 1.25 mm
31 N·m
(3.2 kgf·m, 23 lbf·ft)
Replace.

6 x 1.0 mm
9.8 N·m
(1.0 kgf·m, 7.2 lbf·ft)

COVER

**PRIMARY HEATED
OXYGEN SENSOR
(PRIMARY HO2S)**
44 N·m (4.5 kgf·m, 33 lbf·ft)

8 x 1.25 mm
31 N·m
(3.2 kgf·m, 23 lbf·ft)
Replace.

10 x 1.25 mm
44 N·m (4.5 kgf·m, 33 lbf·ft)

GASKET
Replace.

**EXHAUST MANIFOLD/
THREE WAY CATALYTIC
CONVERTER ASSEMBLY**

**SECONDARY HEATED
OXYGEN SENSOR
(SECONDARY HO2S)**
44 N·m (4.5 kgf·m, 33 lbf·ft)

67162-ACCO-G03

Exploded view of the exhaust manifold —1.7L (D17A1, D17A6 & D17A7) engines

7. Turn the ignition switch OFF. Push and hold the drivers window DOWN switch. Turn the ignition switch ON. Release the driver's window DOWN switch. This must be done within five seconds.

8. Repeat the above step three more times. Wait one second.

9. Confirm that the AUTO UP and AUTO DOWN do not work. If they work repeat the procedure, paying close attention to the five second time limit

10. Move the driver's window all the way down by holding the drivers window DOWN switch to the AUTO DOWN position.

11. Pull up and hold the driver's window UP switch to the AUTO UP position until the window reaches the fully closed position. Hold the switch for one second.

12. Confirm that the power window master switch has been reset by using the driver's window AUTO UP and DOWN function.

13. If the AUTO UP and DOWN feature is still not working, repeat the complete procedure, paying close attention to the five second time limit.

2.3L Engine

1. Before servicing the vehicle, refer to the Precautions Section.

2. Remove or disconnect the following:

- Negative battery cable

3. Safely raise and support the vehicle.

4. Remove or disconnect the following:

- Oxygen Sensor (O$_2$S) connector, if it is located in the manifold.
- Exhaust manifold upper cover
- Heat insulator from the manifold, if equipped with air conditioning.
- Nuts attaching the exhaust manifold to the front exhaust pipe.
- Pipe from the manifold and discard the gasket. Support the pipe with wire; do not allow it to hang by itself.
- Exhaust manifold bracket(s) bolts and bracket(s).
- Exhaust manifold attaching nuts, using a crisscross pattern (starting from the center).
- Manifold and discard the gasket.

Clean the manifold and cylinder head mating surfaces.

- Lower manifold cover from the manifold, if equipped.

To install:

5. Use new gaskets when installing the exhaust manifold.

6. Clean all gasket mating surfaces.

7. Starting with the manifold inner or center nuts, tighten the nuts in a crisscross pattern to the correct torque. The tension must be even across the entire face of the manifold if leaks are to be prevented.

8. Continue the installation in the reverse order of the removal procedure.

9. As required, reprogram the power window control unit as follows.

10. Turn the ignition switch ON. Lower the window, all the way down. Open the driver's side door.

11. Turn the ignition switch OFF. Push and hold the drivers window DOWN switch. Turn the ignition switch ON. Release the driver's window DOWN switch. This must be done within five seconds.

NOTE: Use new gaskets and self-locking nuts when reassembling.

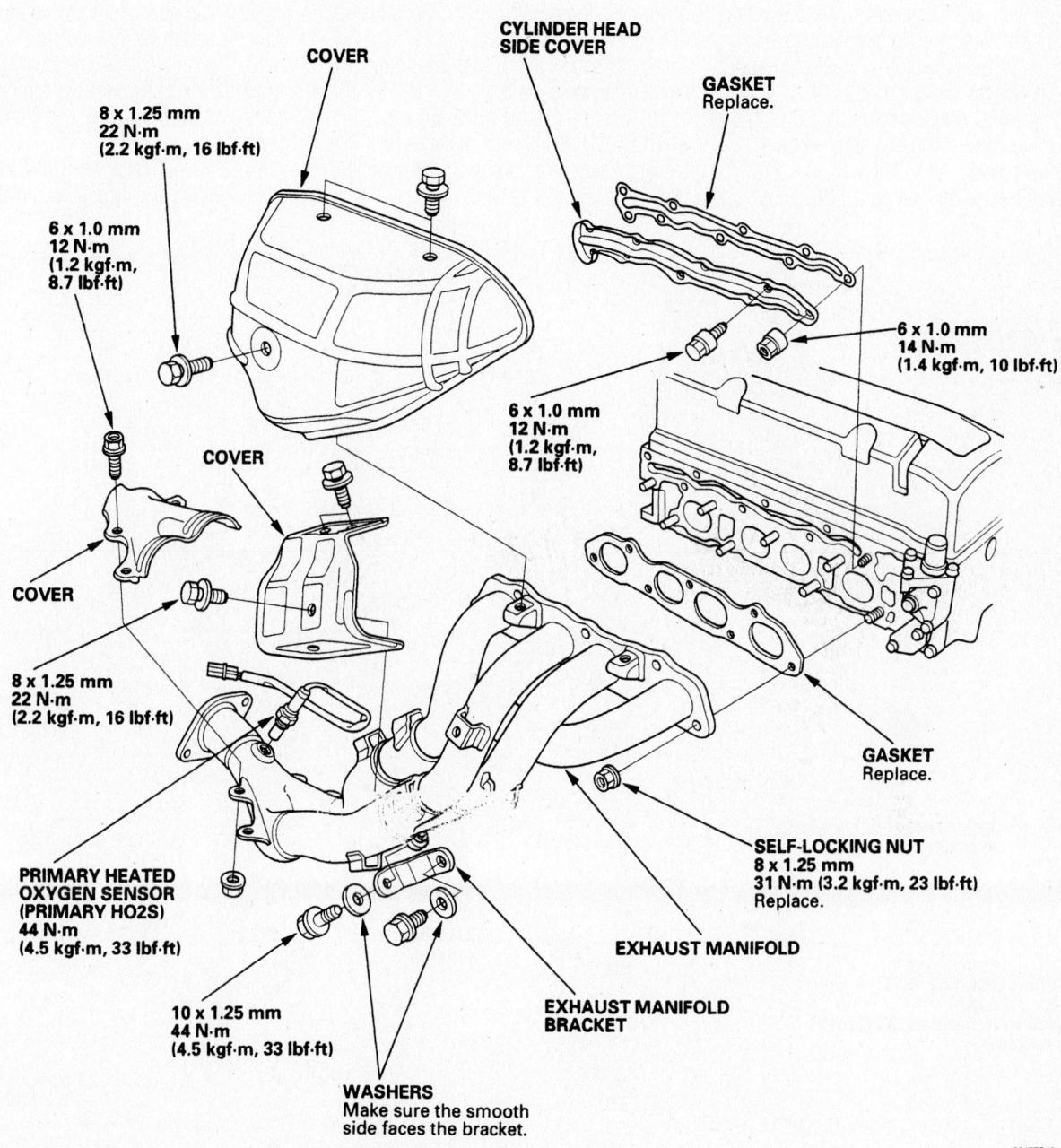

COVER

CYLINDER HEAD SIDE COVER

GASKET
Replace.

8 x 1.25 mm
22 N·m
(2.2 kgf·m, 16 lbf·ft)

6 x 1.0 mm
12 N·m
(1.2 kgf·m,
8.7 lbf·ft)

6 x 1.0 mm
14 N·m
(1.4 kgf·m, 10 lbf·ft)

6 x 1.0 mm
12 N·m
(1.2 kgf·m,
8.7 lbf·ft)

COVER

COVER

8 x 1.25 mm
22 N·m
(2.2 kgf·m, 16 lbf·ft)

GASKET
Replace.

SELF-LOCKING NUT
8 x 1.25 mm
31 N·m (3.2 kgf·m, 23 lbf·ft)
Replace.

**PRIMARY HEATED
OXYGEN SENSOR
(PRIMARY HO2S)
44 N·m
(4.5 kgf·m, 33 lbf·ft)**

EXHAUST MANIFOLD

**EXHAUST MANIFOLD
BRACKET**

10 x 1.25 mm
44 N·m
(4.5 kgf·m, 33 lbf·ft)

WASHERS
Make sure the smooth
side faces the bracket.

9347FG10

Exploded view of the exhaust manifold—S2000 with 2.0L and 2.2L engines

12. Repeat the above step three more times. Wait one second.

13. Confirm that the AUTO UP and AUTO DOWN do not work. If they work repeat the procedure, paying close attention to the five second time limit

14. Move the driver's window all the way down by holding the drivers window DOWN switch to the AUTO DOWN position.

15. Pull up and hold the driver's window

UP switch to the AUTO UP position until the window reaches the fully closed position. Hold the switch for one second.

16. Confirm that the power window master switch has been reset by using the driver's window AUTO UP and DOWN function.

17. If the AUTO UP and DOWN feature is still not working, repeat the complete procedure, paying close attention to the five second time limit.

2.4L Engine

1. Before servicing the vehicle, refer to the Precautions Section.

2. Raise and safely support the vehicle.

3. Remove or disconnect the following:
- VTEC solenoid valve
- Driveshaft heat shield
- Cover and exhaust manifold bracket
- Exhaust manifold

To install:

4. Clean the mounting surfaces. Be sure to use a new gasket.

5. Install or connect the following:
- Exhaust manifold. Tighten the nuts, in a crisscross pattern starting with the inner nut, to specification.
- Exhaust manifold bracket and cover
- Driveshaft heat shield
- VTEC solenoid valve

6. As required, reprogram the power window control unit as follows.

7. Turn the ignition switch ON. Lower the window, all the way down. Open the driver's side door.

8. Turn the ignition switch OFF. Push and hold the drivers window DOWN switch. Turn the ignition switch ON. Release the driver's window DOWN switch. This must be done within five seconds.

9. Repeat the above step three more times. Wait one second.

10. Confirm that the AUTO UP and AUTO DOWN do not work. If they work repeat the procedure, paying close attention to the five second time limit

11. Move the driver's window all the way down by holding the drivers window DOWN switch to the AUTO DOWN position.

12. Pull up and hold the driver's window UP switch to the AUTO UP position until the window reaches the fully closed position. Hold the switch for one second.

13. Confirm that the power window master switch has been reset by using the driver's window AUTO UP and DOWN function.

14. If the AUTO UP and DOWN feature is still not working, repeat the complete proce-

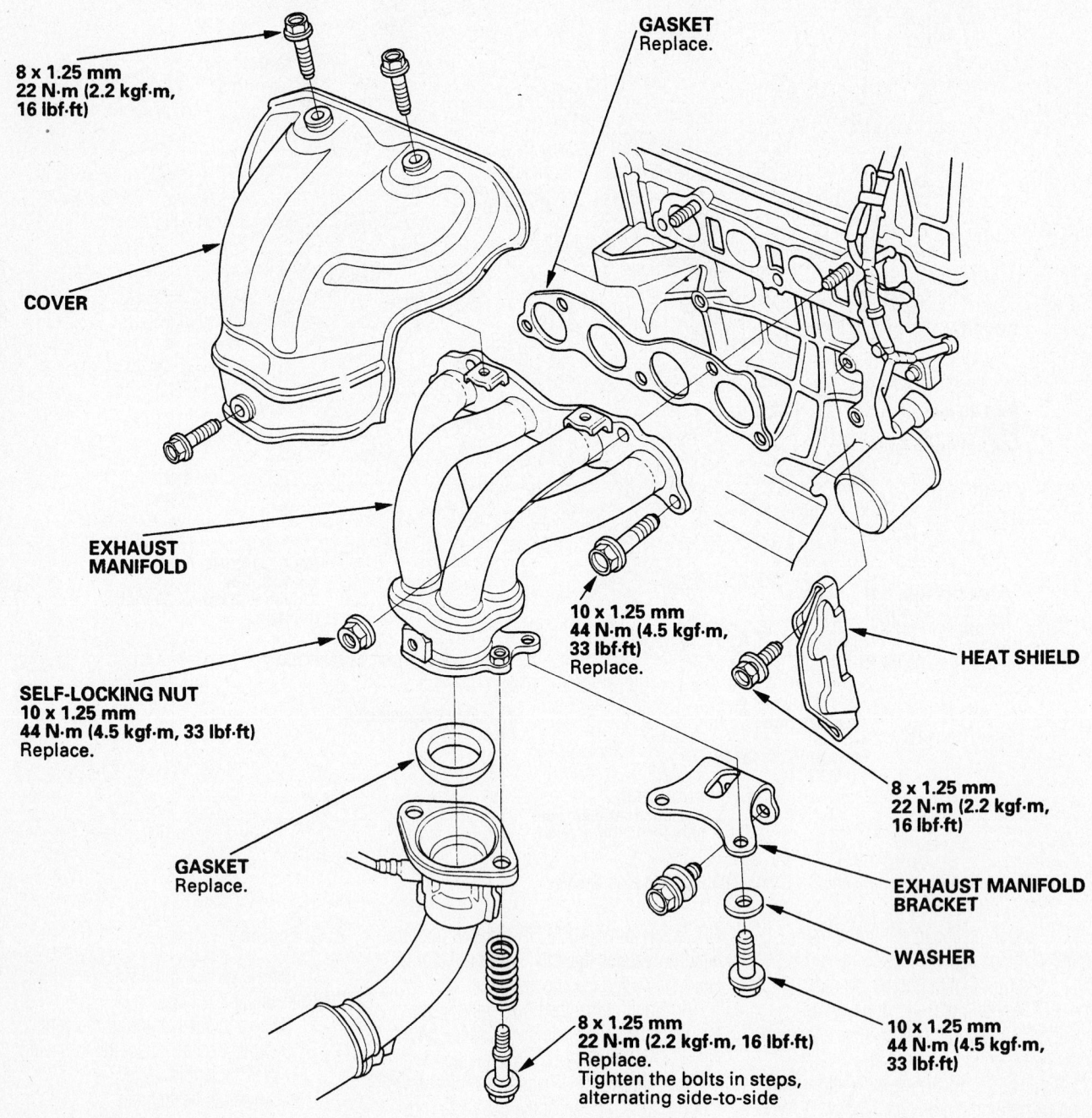

Exploded view of the exhaust manifold—2.4L engine

dure, paying close attention to the five second time limit.

3.0L Engine

1. Before servicing the vehicle, refer to the Precautions Section.
2. Raise and safely support the vehicle.
3. Remove or disconnect the following:
   - Engine undercover
   - Exhaust pipe from the manifold to be removed
4. Lower the vehicle.
5. Remove or disconnect the following:
   - Exhaust manifold heat shield
   - Mounting nuts and the exhaust manifold.

To install:
6. Use new gaskets when installing the exhaust manifold.
7. Clean all gasket mating surfaces.
8. Install or connect the following:
   - Exhaust manifold. Tighten the nuts to specification.
   - Heat shield. Tighten the bolts to 16 ft. lbs. (22 Nm).
9. Raise the vehicle and connect the exhaust pipe to the manifold using a new gasket. Tighten the nuts to 40 ft. lbs. (54 Nm).
10. As required, reprogram the power window control unit as follows.
11. Turn the ignition switch ON. Lower the window, all the way down. Open the driver's side door.
12. Turn the ignition switch OFF. Push and hold the drivers window DOWN switch. Turn the ignition switch ON. Release the

driver's window DOWN switch. This must be done within five seconds.
13. Repeat the above step three more times. Wait one second.
14. Confirm that the AUTO UP and AUTO DOWN do not work. If they work repeat the procedure, paying close attention to the five second time limit
15. Move the driver's window all the way down by holding the drivers window DOWN switch to the AUTO DOWN position.
16. Pull up and hold the driver's window UP switch to the AUTO UP position until the window reaches the fully closed position. Hold the switch for one second.
17. Confirm that the power window master switch has been reset by using the driver's window AUTO UP and DOWN function.
18. If the AUTO UP and DOWN feature is still not working, repeat the complete procedure, paying close attention to the five second time limit.

Front Crankshaft Seal

REMOVAL & INSTALLATION

➡️**The original radio may contain a coded anti-theft circuit. Obtain the security code number before disconnecting the battery cables.**

1. Before servicing the vehicle, refer to the Precautions Section.
2. Disconnect the negative battery cable.
3. Safely raise and support the vehicle.

4. Remove or disconnect the following:
   - Splash shield
   - Engine accessory drive belts
5. Turn the engine to align the timing marks and set cylinder No.1 to TDC. The white mark on the crankshaft pulley should align with the pointer on the timing belt cover. Remove the inspection caps on the upper timing belt covers to check the alignment of the timing marks. The pointers for the camshafts should align with the green marks on the camshaft sprockets.
6. Remove or disconnect the following:
   - Upper timing belt covers and crankshaft pulley
   - Lower timing belt cover

➡️**Mark the direction of the timing belt rotation if it is to be reinstalled.**

   - Timing belt
   - Crankshaft Position (CKP) sensor from the oil pump, if equipped
   - Stopper plate
   - Timing belt sprocket from the crankshaft. Do not lose the sprocket key.
   - Seal from the front of the engine, using a suitable seal removal tool

To install:
7. Clean the seal mounting surfaces on the engine block.
8. Apply a thin coat of grease on the crankshaft and seal lips.
9. Install or connect the following:
   - Seal with the part number facing out. Use a seal driver to seat the seal against the oil pump. Clean any excess grease off the crankshaft and be sure the seal lip is not distorted.
   - Timing belt sprocket and key to the crankshaft
   - Stopper plate and if equipped, the CKP sensor to the oil pump. Tighten the stopper plate and sensor mounting bolts to 108 inch lbs. (12 Nm).

➡️**Verify that the engine is at Top Dead Center (TDC) for the no. 1 cylinder on the compression stroke.**

   - Timing belt
   - Timing belt covers and crankshaft pulley. Tighten the crankshaft pulley bolt to 181 ft. lbs. (245 Nm), with the aid of a crank pulley holder.
   - Accessory drive belts, then adjust.

➡️**Verify that all engine components that may have been removed have been reinstalled correctly.**

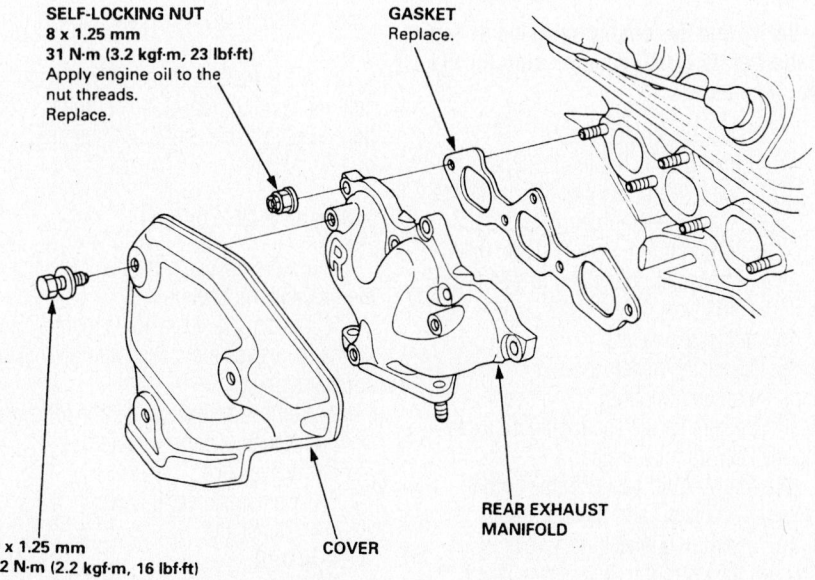

SELF-LOCKING NUT
8 x 1.25 mm
31 N·m (3.2 kgf·m, 23 lbf·ft)
Apply engine oil to the nut threads.
Replace.

GASKET
Replace.

8 x 1.25 mm
22 N·m (2.2 kgf·m, 16 lbf·ft)

COVER

REAR EXHAUST MANIFOLD

7923FG93

Exploded view of the exhaust manifold —3.0L engine

- Splash shield and lower the vehicle.
- Negative battery cable
10. Top up the engine oil if necessary.
11. Run the engine and check for leaks.

Camshaft

REMOVAL & INSTALLATION

1.7L and 1.8L Engines

1. Before servicing the vehicle, refer to the Precautions Section.
2. Remove or disconnect the following components:
- Negative battery cable
- Ignition wires
- Valve cover and the upper timing belt cover.
3. Rotate the crankshaft to set the No. 1 cylinder at Top Dead Center (TDC) for the compression stroke. Once the engine is in this position, it shouldn't be disturbed.
4. Remove or disconnect the following components:
- Timing belt. If the timing belt is contaminated with oil or coolant, it must be replaced. If the timing belt is to be reused, mark its direction of rotation.
- Distributor, if equipped
- Camshaft sprocket and its key. Remove the upper rear timing cover.
- On the 1.8L engine remove the lost motion holder retaining bolts.
- Rocker arm locknuts and the valve adjusting screws.
- Camshaft holder bolts in a 2-step crisscross sequence, starting at the edges and working toward the center of the cylinder head.
- Rocker arm and shaft assembly. Leave the camshaft holder bolts in the camshaft holders to hold the rocker arm and shaft assembly together.
5. Wrap rubber bands around the VTEC rocker arm assemblies so that they do not separate.
6. Store the rocker arm and shaft assembly away from your work area. Cover the assembly with shop towels or a sheet of plastic to protect it from dust.
7. Lift the camshaft from the cylinder head. Remove the camshaft seal.
8. Inspect the camshaft journals and lobes for signs of scoring or other damage.

To install:

9. Remove the oil control orifice. Thoroughly clean it and reinstall it with a new O-ring.

10. Clean and inspect the camshaft bearing caps in the cylinder head.
11. Lubricate the lobes and journals of the camshaft prior to installation.
12. Install or connect the following:
- Camshaft with the keyway facing up so that the camshaft will be at Top Dead Center (TDC)/compression for the No. 1 cylinder.
- New, lightly lubricated, camshaft seal
13. Install the rocker arm and shaft assembly as follows:
a. Remove the rubber bands from the VTEC rocker arms.
b. Lubricate the rocker arm contact surfaces.
c. Apply liquid gasket to the head mating surfaces of the No. 1 and No. 5 camshaft holders. Don't allow the sealant to cure before installing the rocker arm assembly.
d. Set the rocker arm and shaft assembly in place. If equipped, install the lost motion assembly holder.
e. Coat the threads of the camshaft holder bolts with clean oil and loosely install them.
f. Tighten each bolt 2 turns at a time in the crisscross sequence to ensure that the rockers and camshaft holder do not bind on the camshaft journals.
g. Tighten the 8mm camshaft holder bolts to 14 ft. lbs. (20 Nm), and the 6mm camshaft holder bolts to 108 inch lbs. (12 Nm).
14. Install or connect the following:
- Camshaft sprocket and key. Tighten the retaining bolt to 27 ft. lbs. (38 Nm).

➡**Verify that the engine remains at Top Dead Center (TDC)/compression for the No. 1 cylinder.**

- Distributor, if equipped
- Timing belt. Tighten the tensioner bolt to 33 ft. lbs. (44 Nm) once the belt has been properly tensioned.
- Lower timing cover. Tighten the crankshaft pulley bolt to 134 ft. lbs. (181 Nm).
15. Adjust the valves.
16. Manually inspect the VTEC rocker arms for smooth motion.
17. Be sure all the spark plug tube sealing gaskets are fully seated.
18. Apply liquid gasket to the corner recesses of a new valve cover gasket.
19. Install or connect the following:
- Gasket to the valve cover. Don't allow the sealant to cure before installation.

- Valve cover. Gently wiggle the valve cover to be sure it is fully seated. Tighten the valve cover bolts in a crisscross pattern to 84 inch lbs. (10 Nm).
- Ignition wires
20. Refill the engine with fresh oil and install a new filter.
21. Reconnect the battery cable.
22. Warm the engine up to normal operating temperature. Check for oil leaks.
23. If equipped with a distributor, check the ignition timing and adjust it if necessary. Then, if equipped, tighten the distributor mounting bolts to 17 ft. lbs. (24 Nm).
24. As required, reprogram the power window control unit as follows.
25. Turn the ignition switch ON. Lower the window, all the way down. Open the driver's side door.
26. Turn the ignition switch OFF. Push and hold the drivers window DOWN switch. Turn the ignition switch ON. Release the driver's window DOWN switch. This must be done within five seconds.
27. Repeat the above step three more times. Wait one second.
28. Confirm that the AUTO UP and AUTO DOWN do not work. If they work repeat the procedure, paying close attention to the five second time limit
29. Move the driver's window all the way down by holding the drivers window DOWN switch to the AUTO DOWN position.
30. Pull up and hold the driver's window UP switch to the AUTO UP position until the window reaches the fully closed position. Hold the switch for one second.
31. Confirm that the power window master switch has been reset by using the driver's window AUTO UP and DOWN function.
32. If the AUTO UP and DOWN feature is still not working, repeat the complete procedure, paying close attention to the five second time limit.

2.0L and 2.2L Engine

1. Before servicing the vehicle, refer to the Precautions Section.
2. Loosen the valve adjustment screws so that all valves are closed and all rocker arms are loose.
3. Remove or disconnect the following:
- Negative battery cable
- Valve cover
- Camshaft bearing caps
- Camshafts

To install:

4. Set the engine to Top Dead Center (TDC) so that the timing chain sprocket tim-

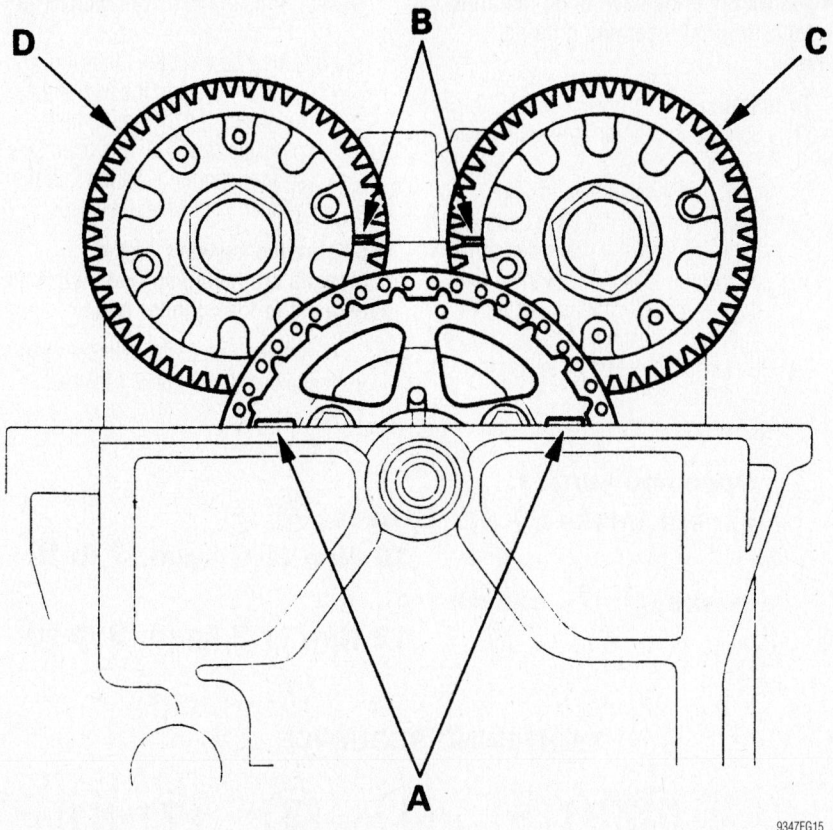

Timing chain sprocket alignment marks (A) and camshaft sprocket alignment marks (B)—2.0L and 2.2L engine

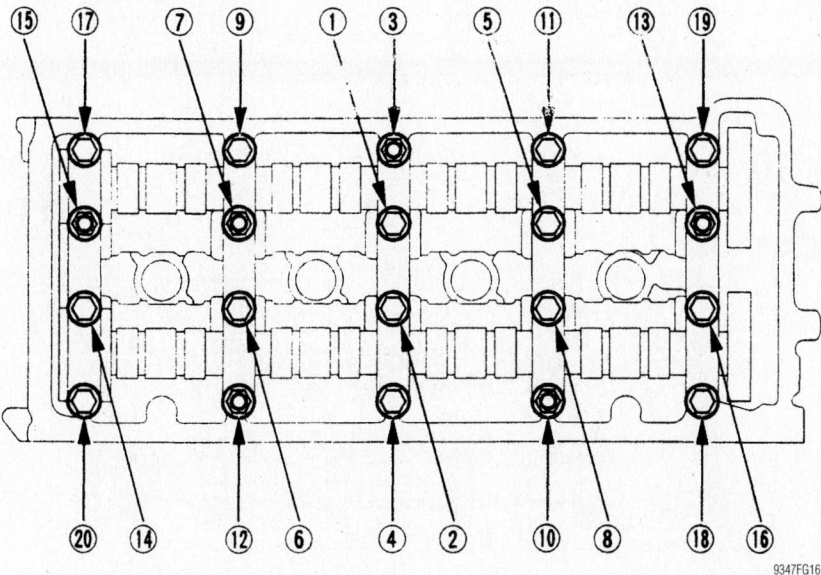

Camshaft bearing cap torque sequence—2.0L and 2.2L engine

ing marks are aligned with the cylinder head surface as shown.

5. Install or connect the following:
- Camshafts with the sprocket timing marks aligned as shown
- Camshaft bearing caps and tighten the bolts in sequence to 16 ft. lbs. (22 Nm). Adjust the valve clearance.
- Valve cover
- Negative battery cable

6. As required, reprogram the power window control unit as follows.

7. Turn the ignition switch ON. Lower the window, all the way down. Open the driver's side door.

8. Turn the ignition switch OFF. Push and hold the drivers window DOWN switch. Turn the ignition switch ON. Release the driver's window DOWN switch. This must be done within five seconds.

9. Repeat the above step three more times. Wait one second.

10. Confirm that the AUTO UP and AUTO DOWN do not work. If they work repeat the procedure, paying close attention to the five second time limit

11. Move the driver's window all the way down by holding the drivers window DOWN switch to the AUTO DOWN position.

12. Pull up and hold the driver's window UP switch to the AUTO UP position until the window reaches the fully closed position. Hold the switch for one second.

13. Confirm that the power window master switch has been reset by using the driver's window AUTO UP and DOWN function.

14. If the AUTO UP and DOWN feature is still not working, repeat the complete procedure, paying close attention to the five second time limit.

2.3L Engine

1. Before servicing the vehicle, refer to the Precautions Section.

2. Disconnect the negative battery cable.

3. Turn the crankshaft so the No. 1 piston is at Top Dead Center (TDC).

➡**The No. 1 piston is at Top Dead Center (TDC) when the pointer on the block aligns with the white painted mark on the flywheel (manual transaxle) or driveplate (automatic transaxle).**

4. Remove or disconnect the following:
- Air intake duct
- Engine ground cable from the cylinder head cover
- Connector and the terminal from the alternator

- Engine wiring harness from the valve cover
- Ignition coil

➡ **Tag all electrical connectors before disconnecting them.**

- Electrical connectors from the distributor
- Spark plug wires from the spark plugs
- Distributor from the cylinder head
- Ignition coil wire from the distributor
- Positive Crankcase Ventilation (PCV) hose
- Cylinder head cover. Replace the rubber seals if damaged or deteriorated.
- Timing belt middle cover

5. Ensure the words **UP** embossed on the camshaft pulleys are aligned in the upward position.

6. Mark the rotation of the timing belt if it is to be used again. Loosen the timing belt adjusting nut ½ turn, then release the tension on the timing belt. Push the tensioner to release tension from the belt, then tighten the adjusting nut.

7. Remove or disconnect the following:
- Timing belt from the camshaft sprockets

✳✳ WARNING

Do not crimp or bend the timing belt more than 90°, or less than 1 inch (25mm) in diameter

8. Insert a 5.0mm pin punch in each of the camshaft caps nearest to the sprockets through the holes provided.

9. Remove or disconnect the following:
- Camshaft sprocket attaching bolts, then the sprockets. Do not lose the sprocket keys.
- Side engine mount bracket B, then the timing belt back cover from behind the camshaft sprockets
- Rocker arm adjusting screws, then the pin punches from the camshaft caps
- Camshaft holders, note the holders locations for ease of installation. Loosen the bolts in the reverse order of the installation.
- Camshafts from the cylinder head, then discard the camshaft seals
- Rubber cap from the head, located at the end of the intake camshaft

10. Remove the rocker arms from the cylinder head. Note the locations of the rocker arms.

➡ **The rocker arms have to be installed to their original locations if being reused.**

To install:

11. Lubricate the rocker arms with clean oil.

12. Install or connect the following:
- Rocker arms on the pivot bolts and the valve stems. If the rocker arms are being reused, install them to their original locations. The locknuts and adjustment screws should be loosened before installing the rocker arms.

13. Lubricate the camshafts with clean oil.

14. Install or connect the following:
- Camshaft seals to the end of the camshafts that the timing belt sprockets attach to. The open side (spring) should be facing into the cylinder head when installed.

➡ **Be sure the keyways on the camshafts are facing up and install the camshafts to the cylinder head.**

- Rubber plug to the cylinder head at the end of the intake camshaft

Specified torque:
Except Intake ⑤, ⑦. Exhaust ⑥, ⑧:
10 N·m (1.0 kg-m, 7 lb-ft)

Intake ⑤, ⑦. Exhaust ⑥, ⑧:
12 N·m (1.2 kg-m, 9 lb-ft)

TIGHTENING SEQUENCE

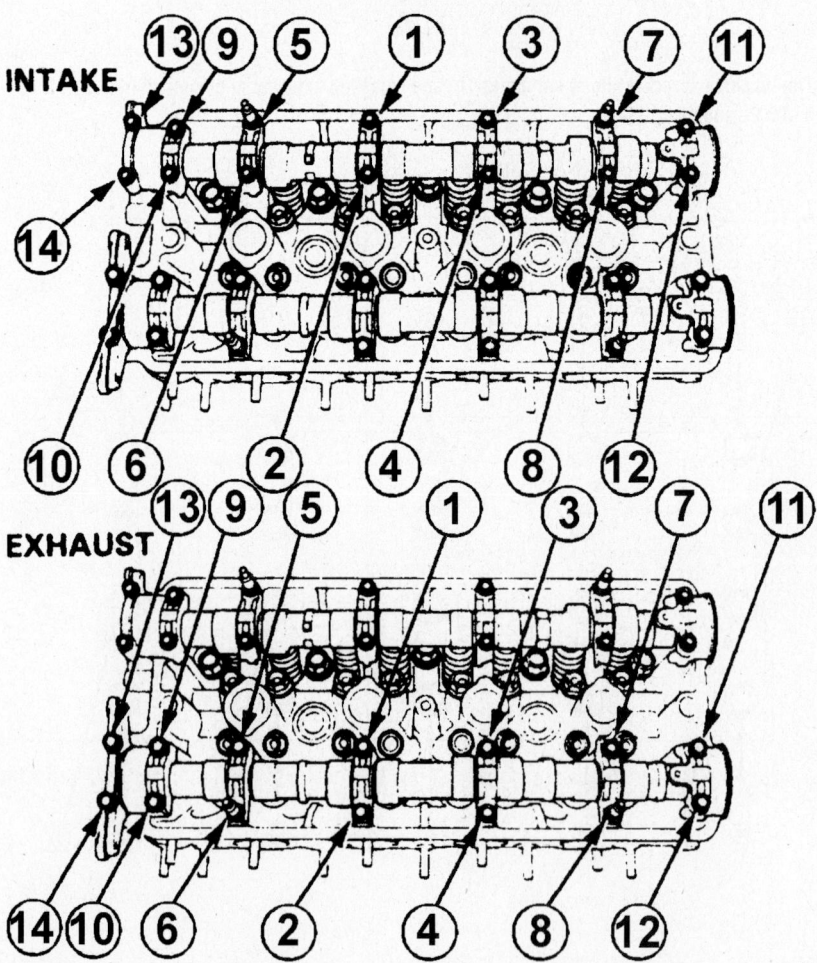

Camshaft holders torque sequence—2.3L engine

7923FG67

15. Apply liquid gasket to the head mating surfaces of the No. 1 and No. 6 camshaft holders, then install them along with No. 2, 3, 4 and 5. **I** or **E** marks are stamped on the camshaft holders to identify them as Intake or Exhaust side holders. The arrows stamped on the holders should point toward the timing belt.

16. Snug the camshaft holders in place.

17. Press the camshaft seals securely into place.

18. Tighten the camshaft holder bolts in 2 steps, following the proper sequence, to ensure that the rockers do not bind on the valves. Tighten all the bolts, except the 4 studs, to 84 inch lbs. (10 Nm). Tighten the studs (number 5 and 7 bolts in the correct sequence) to 108 inch lbs. (12 Nm).

19. Install or connect the following:
- Timing belt back cover
- Side engine mount bracket B. Tighten the bolt attaching the bracket to the cylinder head to 33 ft. lbs. (45 Nm). Tighten the bolts attaching the bracket to the side engine mount to 16 ft. lbs. (22 Nm).
- 5.0mm pin punch in each of the camshaft caps, nearest to the pulleys, through the holes provided.
- Keys into the camshaft grooves

20. Push the camshaft sprockets onto the camshafts, then tighten the retaining bolts to 27 ft. lbs. (38 Nm).

21. Ensure the words **UP** embossed on the camshaft pulleys are aligned in the upward position.

22. Install or connect the following:
- Timing belt to the camshaft sprockets, then remove the 2, 5.0mm pin punches from the camshaft bearing caps

23. Loosen, then tighten the timing belt adjuster nut.

24. Turn the crankshaft counterclockwise until the cam pulley has moved 3 teeth; this creates tension on the timing belt. Loosen the adjusting nut, then tighten it to 33 ft. lbs. (45 Nm).

25. Adjust the valves.

26. Tighten the crankshaft pulley bolt to 181 ft. lbs. (250 Nm).

27. Install or connect the following:
- Middle timing belt cover and tighten the attaching bolts to 108 inch lbs. (12 Nm).
- Cylinder head cover and tighten the cap nuts to 84 inch lbs. (10 Nm)
- PCV hose to the cylinder head cover
- Distributor to the cylinder head, snug the attaching bolts until the timing has been checked and adjusted
- Spark plug wires and distributor electrical connectors
- Ignition coil wire to the distributor
- Ignition coil
- Alternator wiring harness to the cylinder head cover
- Terminal and connector to the alternator
- Engine ground cable to the cylinder head cover
- Air intake duct

28. Drain the oil from the engine into a sealable container. Install the drain plug and refill the engine with clean oil.

29. Connect the negative battery cable and enter the radio security code.

30. Start the engine and check carefully for any leaks.

31. Check and adjust the ignition timing.

Tighten the distributor bolts to 13 ft. lbs. (18 Nm).

2.4L Engine

1. Before servicing the vehicle, refer to the Precautions Section.

2. Disconnect the negative battery cable.

3. Remove or disconnect the following:
- Timing chain
- Loosen the rocker arm adjusting screws
- Camshaft holder bolts, two turns at a time in sequence
- Timing chain guide (B), camshaft and camshafts

4. Insert the bolts (A) into the rocker shaft holder, then remove the rocker arm assembly (B)
- Camshafts by carefully lifting them out of the cylinder head

To install:

5. Clean and dry the No. 5 rocker shaft holding mating surface.

6. Apply a suitable liquid gasket P/N 08718-0009, or equivalent, evenly to the cylinder head mating surface of the No. 5 rocker shaft holder.

➡**The parts must be installed within 5 minutes of applying the liquid gasket.**

7. Reassemble the rocker arm assembly, as necessary.

8. Install or connect the following
- Bolts (A) into the rocker shaft holder, then the rocker arm assembly on the cylinder head. Remove

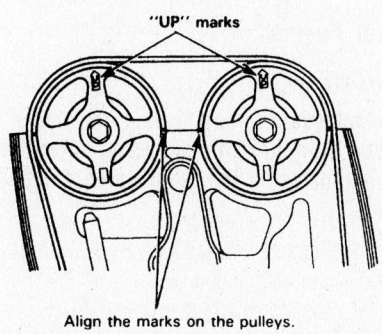

"UP" marks

Align the marks on the pulleys.

7923FG68

Camshaft sprockets alignment—2.3L engine

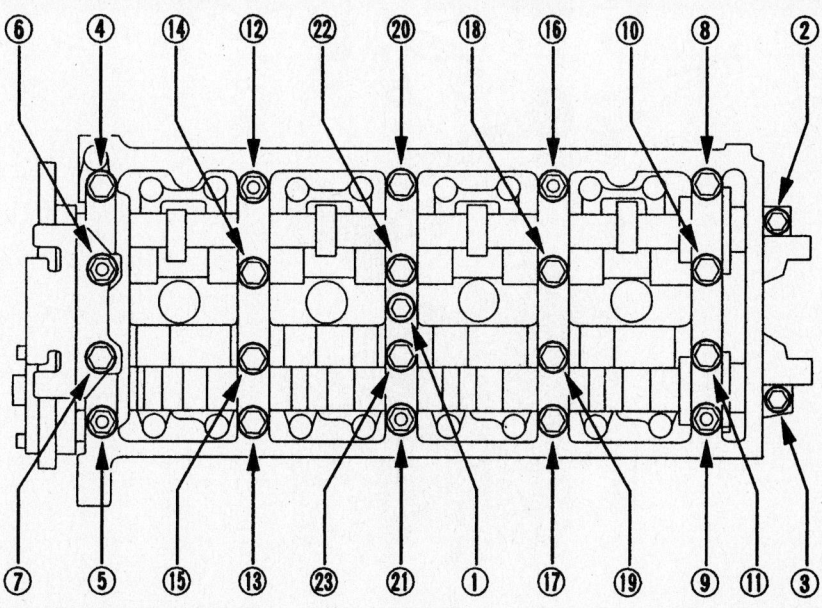

42356-ACCO-G10

Camshaft holder bolt loosening sequence—2.4L engine

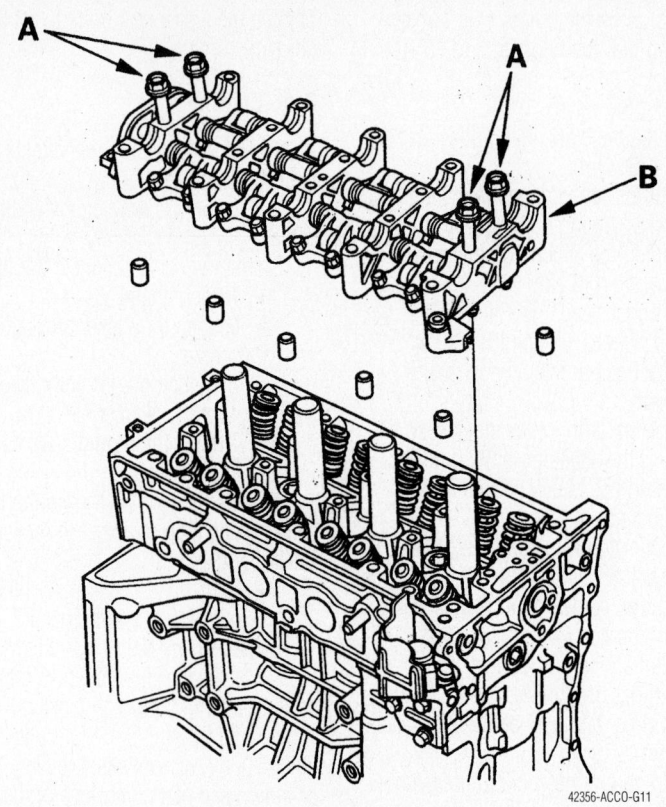

Insert the bolts (A) into the rocker shaft holder, then remove the rocker arm assembly (B)—2.4L engine

the bolts from the rocker shaft holder.

9. Make sure the punch marks on the variable valve timing control (VTC) actuator and exhaust camshaft sprocket are facing up, then set the camshafts (A) in the holder

10. Set the camshaft holders (B) and timing chain guide B (C) in place.

11. Tighten the bolts, in sequence, to the following specification:

a. 8mm bolts: 16 ft. lbs. (22 Nm)
b. 6mm bolts: 8.7 ft. lbs. (12 Nm)

12. Install the timing chain and adjust the valve lash.

13. Continue the installation in the reverse order of the removal procedure.

14. As required, reprogram the power window control unit as follows.

15. Turn the ignition switch ON. Lower the window, all the way down. Open the driver's side door.

16. Turn the ignition switch OFF. Push and hold the drivers window DOWN switch. Turn the ignition switch ON. Release the driver's window DOWN switch. This must be done within five seconds.

17. Repeat the above step three more times. Wait one second.

18. Confirm that the AUTO UP and AUTO DOWN do not work. If they work repeat the procedure, paying close attention to the five second time limit

19. Move the driver's window all the way down by holding the drivers window DOWN switch to the AUTO DOWN position.

20. Pull up and hold the driver's window UP switch to the AUTO UP position until the window reaches the fully closed position. Hold the switch for one second.

21. Confirm that the power window master switch has been reset by using the driver's window AUTO UP and DOWN function.

22. If the AUTO UP and DOWN feature is still not working, repeat the complete procedure, paying close attention to the five second time limit.

3.0L Engine

FRONT

➡Make sure to acquire the anti-theft code from the radio and write down the frequencies for the radio's preset buttons.

1. Before servicing the vehicle, refer to the Precautions Section.

2. Disconnect the negative battery cable. Remove the battery from the vehicle.

3. Drain the cooling system. Remove the upper radiator hose.

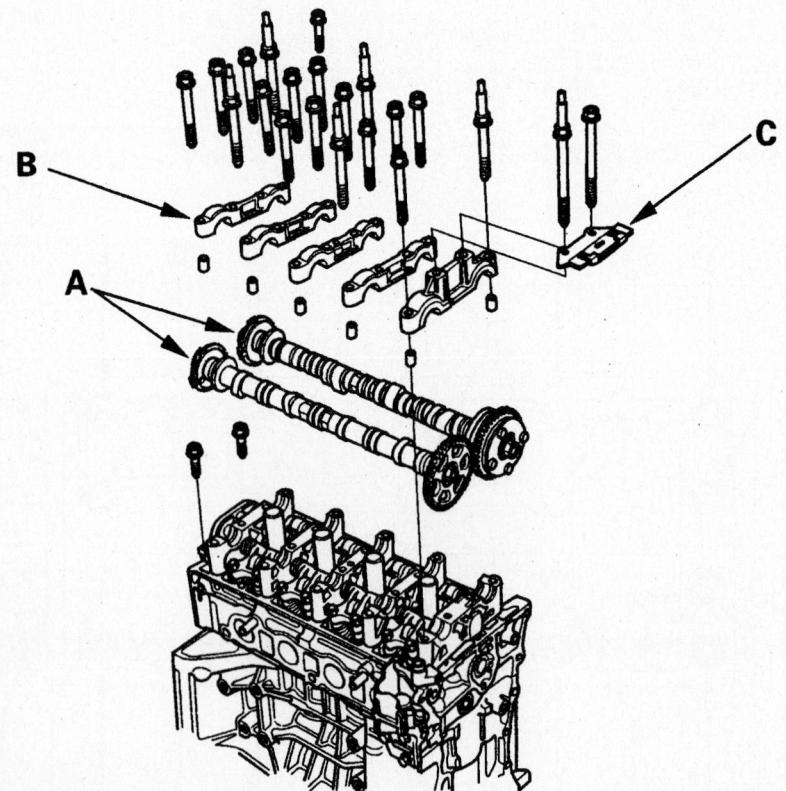

When installing the camshafts (A) make sure the punch marks on the VTC actuator and exhaust cam sprockets are facing up—2.4L engine

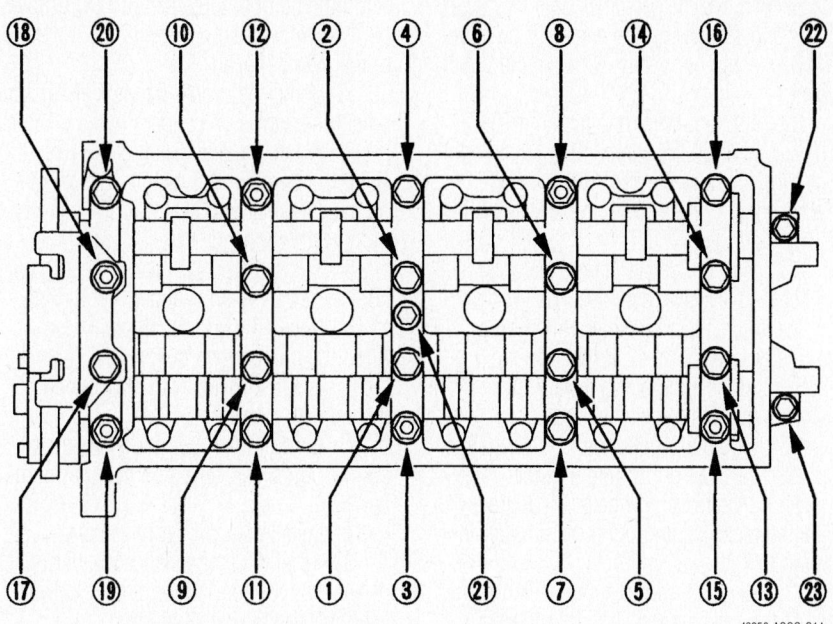

Rocker arm assembly bolt tightening sequence—2.4L engine

42356-ACCO-G14

4. Remove the EGR valve. Remove the timing belt.

5. Remove the rocker arm assembly. Remove the camshaft pulley.

6. Remove the thrust cover. Remove the camshaft.

To install:

7. Installation is the reverse of the removal procedure. Be sure to use new O-rings and gaskets. Coat the camshaft with clean engine oil prior to installation.

8. Coat the camshaft pulley mounting bolts with clean engine oil prior to installation and torque to 16 ft. lbs. (22 Nm).

9. Adjust the valve clearance.

10. Reprogram the crankshaft position (CKP) pattern. Run the engine until the operating temperature reaches 176 degree. With the engine stopped clear the CKP pattern. Turn the ignition switch OFF. Turn the ignition switch ON and wait thirty seconds.

11. Roadtest the vehicle on a level surface. Decelerate the engine speed of 2500 rpm to 1000 rpm. If equipped with automatic transaxle use two Drive positions. If equipped with manual transaxle use first gear.

12. Stop the vehicle, but keep the engine running.

13. Check PULSAR F/B LEARN in the data list with the HDS. If not complete repeat the procedure. If complete, roadtest the vehicle on a level surface. Decelerate the engine speed of 5000 rpm to 3000 rpm. If equipped with automatic transaxle use two Drive positions. If equipped with manual transaxle use first gear.

14. Stop the vehicle, but keep the engine running.

15. Check PULSAR F/B LEARN in the data list with the HDS. If not complete repeat the procedure.

16. If completed, turn the ignition switch OFF. Turn the ignition switch ON, wait thirty seconds. The learning procedure is now complete.

17. Enter the antitheft codes for the radio and the navigation system. Set the clock.

18. Reprogram the power window control unit as follows.

19. Turn the ignition switch ON. Lower the window, all the way down. Open the driver's side door.

20. Turn the ignition switch OFF. Push and hold the drivers window DOWN switch. Turn the ignition switch ON. Release the driver's window DOWN switch. This must be done within five seconds.

21. Repeat the above step three more times. Wait one second.

22. Confirm that the AUTO UP and AUTO DOWN do not work. If they work repeat the procedure, paying close attention to the five second time limit

23. Move the driver's window all the way down by holding the drivers window DOWN switch to the AUTO DOWN position.

24. Pull up and hold the driver's window UP switch to the AUTO UP position until the window reaches the fully closed position. Hold the switch for one second.

25. Confirm that the power window master switch has been reset by using the driver's window AUTO UP and DOWN function.

26. If the AUTO UP and DOWN feature is still not working, repeat the complete procedure, paying close attention to the five second time limit.

REANS **REAR**

➡ **Make sure to acquire the anti-theft code from the radio and write down the frequencies for the radio's preset buttons.**

1. Before servicing the vehicle, refer to the Precautions Section.

2. Disconnect the negative battery cable. Relieve the fuel system pressure.

3. Remove the under hood fuse/relay box.

4. Drain the cooling system. Remove the upper radiator hose. Remove the heater hose and the nuts retaining the fuel line hose.

5. Remove the timing belt. Remove the rocker arm assembly. Remove the camshaft pulley.

6. Remove the thrust cover. Remove the camshaft.

To install:

7. Installation is the reverse of the removal procedure. Be sure to use new O-rings and gaskets. Coat the camshaft with clean engine oil prior to installation.

8. Coat the camshaft pulley mounting bolts with clean engine oil prior to installation and torque to 16 ft. lbs. (22 Nm).

9. Adjust the valve clearance.

10. Reprogram the crankshaft position (CKP) pattern. Run the engine until the operating temperature reaches 176 degree. With the engine stopped clear the CKP pattern. Turn the ignition switch OFF. Turn the ignition switch ON and wait thirty seconds.

11. Roadtest the vehicle on a level surface. Decelerate the engine speed of 2500 rpm to 1000 rpm. If equipped with automatic transaxle use two Drive positions. If equipped with manual transaxle use first gear.

12. Stop the vehicle, but keep the engine running.

13. Check PULSAR F/B LEARN in the data list with the HDS. If not complete repeat the procedure. If complete, roadtest the vehicle on a level surface. Decelerate the engine speed of 5000 rpm to 3000 rpm. If equipped with automatic transaxle use two Drive positions. If equipped with manual transaxle use first gear.

14. Stop the vehicle, but keep the engine running.

15. Check PULSAR F/B LEARN in the data list with the HDS. If not complete repeat the procedure.

16. If completed, turn the ignition switch OFF. Turn the ignition switch ON, wait thirty seconds. The learning procedure is now complete.

17. Enter the antitheft codes for the radio and the navigation system. Set the clock.

18. Reprogram the power window control unit as follows.

19. Turn the ignition switch ON. Lower the window, all the way down. Open the driver's side door.

20. Turn the ignition switch OFF. Push and hold the drivers window DOWN switch. Turn the ignition switch ON. Release the driver's window DOWN switch. This must be done within five seconds.

21. Repeat the above step three more times. Wait one second.

22. Confirm that the AUTO UP and AUTO DOWN do not work. If they work repeat the procedure, paying close attention to the five second time limit

23. Move the driver's window all the way down by holding the drivers window DOWN switch to the AUTO DOWN position.

24. Pull up and hold the driver's window UP switch to the AUTO UP position until the window reaches the fully closed position. Hold the switch for one second.

25. Confirm that the power window master switch has been reset by using the driver's window AUTO UP and DOWN function.

26. If the AUTO UP and DOWN feature is still not working, repeat the complete procedure, paying close attention to the five second time limit.

Valve Clearance

ADJUSTMENT

➡ **Adjust valves only when the cylinder head temperature is less than 100 degrees F.**

Civic

1.7L ENGINE

1. Before servicing the vehicle, refer to the Precautions Section. Disconnect the negative battery cable.

2. Note the radio security code and the radio presets.

3. Remove the ignition coil cover. Remove the ignition coils. Remove the throttle cable clamps and harness holder from the cylinder head cover.

4. Remove the cylinder head cover retaining bolts. Remove the cylinder head cover from the engine.

5. Remove the grommet from the upper cover and disconnect the camshaft position (CMO) sensor connector. Remove the upper cover.

6. Set the number one piston at TDC. The UP mark on the camshaft pulley should be at the top, and the TDC marks on the pulley should line up with the top edge of the cylinder head.

7. Using the proper gauge feeler gauge, adjust the valves on cylinder number one.

8. Rotate the crankshaft 180 degrees counterclockwise. The UP mark on the camshaft pulley should be toward the exhaust side of the cylinder head.

9. Using the proper gauge feeler gauge, adjust the valves on cylinder number three.

10. Rotate the crankshaft 180 degrees counterclockwise and bring the number four piston to TDC.

11. Using the proper gauge feeler gauge, adjust the valves on cylinder number four.

12. Rotate the crankshaft 180 degrees counterclockwise. The UP mark on the camshaft pulley should be toward the intake side of the cylinder head.

13. Using the proper gauge feeler gauge, adjust the valves on cylinder number two.

14. Install the cylinder head cover.

15. As required, reprogram the ECM/PCM engine idle characteristics. Be sure all electrical items are OFF.

16. Start the engine. Hold the idle speed at 3000 RPM's in park or neutral until the radiator fan comes on or the temperature reached 194 degrees.

17. Let the engine idle for about five minutes with the throttle fully closed.

18. If the radiator fan comes on during the five minutes, do not count this toward the five minute programming time.

19. Set the clock.

1.8L ENGINE

1. Before servicing the vehicle, refer to the Precautions Section. Disconnect the negative battery cable.

2. Note the radio security code and the radio presets.

3. Remove the cylinder head cover retaining bolts. Remove the cylinder head cover from the engine.

4. Set the number one piston at TDC. The UP mark on the camshaft sprocket should be at the top, and the TDC grooves on the camshaft sprocket should line up with the top edge of the cylinder head.

5. Using the proper gauge feeler gauge, adjust the valves on cylinder number one.

6. Rotate the crankshaft clockwise.

Align the number three piston TDC groove on the camshaft sprocket with the top edge of the cylinder head.

7. Using the proper gauge feeler gauge, adjust the valves on cylinder number three.

8. Rotate the crankshaft clockwise. Align the number four piston TDC groove on the camshaft sprocket with the top edge of the cylinder head.

9. Using the proper gauge feeler gauge, adjust the valves on cylinder number four.

10. Rotate the crankshaft clockwise. Align the number two piston TDC groove on the camshaft sprocket with the top edge of the cylinder head.

11. Using the proper gauge feeler gauge, adjust the valves on cylinder number two.

12. Install the cylinder head cover.

13. Reset the maintenance minder light.

14. As required, reprogram the power window control unit as follows.

15. Turn the ignition switch ON. Lower the window, all the way down. Open the driver's side door.

16. Turn the ignition switch OFF. Push and hold the drivers window DOWN switch. Turn the ignition switch ON. Release the driver's window DOWN switch. This must be done within five seconds.

17. Repeat the above step three more times. Wait one second.

18. Confirm that the AUTO UP and AUTO DOWN do not work. If they work repeat the procedure, paying close attention to the five second time limit

19. Move the driver's window all the way down by holding the drivers window DOWN switch to the AUTO DOWN position.

20. Pull up and hold the driver's window UP switch to the AUTO UP position until the window reaches the fully closed position. Hold the switch for one second.

21. Confirm that the power window master switch has been reset by using the driver's window AUTO UP and DOWN function.

22. If the AUTO UP and DOWN feature is still not working, repeat the complete procedure, paying close attention to the five second time limit.

23. Set the clock.

2.0L ENGINE

1. Before servicing the vehicle, refer to the Precautions Section. Disconnect the negative battery cable.

2. Note the radio security code and the radio presets.

3. Remove the cylinder head cover retaining bolts. Remove the cylinder head cover from the engine.

4. Set the number one piston at TDC. The punch mark on the variable timing control (VTC) actuator and the punch mark on the exhaust camshaft sprocket should be at the top. Align the TDC marks on the VTC actuator and exhaust camshaft sprocket.

5. Using the proper gauge feeler gauge, adjust the valves on cylinder number one.

6. Rotate the crankshaft 180 degrees. Using the proper gauge feeler gauge, adjust the valves on cylinder number three.

7. Rotate the crankshaft 180 degrees. Using the proper gauge feeler gauge, adjust the valves on cylinder number four.

8. Rotate the crankshaft 180 degrees. Using the proper gauge feeler gauge, adjust the valves on cylinder number two.

9. Install the cylinder head cover.

10. As required, on 2002–2005 vehicles, reprogram the ECM engine idle characteristics. Be sure all electrical items are OFF.

11. Start the engine. Hold the idle speed at 3000 RPM's in neutral until the radiator fan comes on or the temperature reached 194 degrees.

12. Let the engine idle for about five minutes with the throttle fully closed.

13. If the radiator fan comes on during the five minutes, do not count this toward the five minute programming time.

14. Set the clock.

15. As required, reprogram the power window control unit as follows.

16. Turn the ignition switch ON. Lower the window, all the way down. Open the driver's side door.

17. Turn the ignition switch OFF. Push and hold the drivers window DOWN switch. Turn the ignition switch ON. Release the driver's window DOWN switch. This must be done within five seconds.

18. Repeat the above step three more times. Wait one second.

19. Confirm that the AUTO UP and AUTO DOWN do not work. If they work repeat the procedure, paying close attention to the five second time limit

20. Move the driver's window all the way down by holding the drivers window DOWN switch to the AUTO DOWN position.

21. Pull up and hold the driver's window UP switch to the AUTO UP position until the window reaches the fully closed position. Hold the switch for one second.

22. Confirm that the power window master switch has been reset by using the driver's window AUTO UP and DOWN function.

23. If the AUTO UP and DOWN feature is still not working, repeat the complete procedure, paying close attention to the five second time limit.

S2000

2.0L AND 2.2L ENGINES

1. Before servicing the vehicle, refer to the Precautions Section. Disconnect the negative battery cable.

2. Note the radio security code and the radio presets.

3. Remove the cylinder head cover retaining bolts. Remove the cylinder head cover from the engine.

4. Set the number one piston at TDC. The TDC marks on the cam chain sprocket should align with the cylinder head surface.

5. Using the proper gauge feeler gauge, adjust the valves on cylinder number one.

6. Rotate the crankshaft 180 degrees. Using the proper gauge feeler gauge, adjust the valves on cylinder number three.

7. Rotate the crankshaft 180 degrees. Using the proper gauge feeler gauge, adjust the valves on cylinder number four.

8. Rotate the crankshaft 180 degrees. Using the proper gauge feeler gauge, adjust the valves on cylinder number two.

9. Install the cylinder head cover.

10. As required, on 2002–05 vehicles, reprogram the ECM engine idle characteristics. Be sure all electrical items are OFF.

11. Start the engine. Hold the idle speed at 3000 RPM's in neutral until the radiator fan comes on or the temperature reached 176 degrees.

12. Let the engine idle for about five minutes with the throttle fully closed.

13. If the radiator fan comes on during the five minutes, do not count this toward the five minute programming time.

14. Set the clock.

15. As required, reprogram the power window control unit as follows.

16. Turn the ignition switch ON. Lower the window, all the way down. Open the driver's side door.

17. Turn the ignition switch OFF. Push and hold the drivers window DOWN switch. Turn the ignition switch ON. Release the driver's window DOWN switch. This must be done within five seconds.

18. Repeat the above step three more times. Wait one second.

19. Confirm that the AUTO UP and AUTO DOWN do not work. If they work repeat the procedure, paying close attention to the five second time limit

20. Move the driver's window all the way down by holding the drivers window DOWN switch to the AUTO DOWN position.

21. Pull up and hold the driver's window UP switch to the AUTO UP position until the window reaches the fully closed position. Hold the switch for one second.

22. Confirm that the power window master switch has been reset by using the driver's window AUTO UP and DOWN function.

23. If the AUTO UP and DOWN feature is still not working, repeat the complete procedure, paying close attention to the five second time limit.

Accord

2.3L ENGINE

1. Before servicing the vehicle, refer to the Precautions Section. Disconnect the negative battery cable.

2. Note the radio security code and the radio presets.

3. Remove the cylinder head cover retaining bolts. Remove the cylinder head cover from the engine.

4. Set the number one piston at TDC. The UP mark on the camshaft pulley should be at the top, and the TDC marks on the pulley should line up with the top edge of the cylinder head.

5. Using the proper gauge feeler gauge, adjust the valves on cylinder number one.

6. Rotate the crankshaft 180 degrees counterclockwise. The UP mark on the camshaft pulley should be toward the exhaust side of the cylinder head.

7. Using the proper gauge feeler gauge, adjust the valves on cylinder number three.

8. Rotate the crankshaft 180 degrees counterclockwise and bring the number four piston to TDC.

9. Using the proper gauge feeler gauge, adjust the valves on cylinder number four.

10. Rotate the crankshaft 180 degrees counterclockwise. The UP mark on the camshaft pulley should be toward the intake side of the cylinder head.

11. Using the proper gauge feeler gauge, adjust the valves on cylinder number two.

12. Install the cylinder head cover.

2.4L ENGINE

1. Before servicing the vehicle, refer to the Precautions Section. Disconnect the negative battery cable.

2. Note the radio security code and the radio presets.

3. Remove the cylinder head cover retaining bolts. Remove the cylinder head cover from the engine.

4. Set the number one piston at TDC. The punch mark on the variable timing control (VTC) actuator and the punch mark on the exhaust camshaft sprocket should be at the top. Align the TDC marks on the VTC actuator and exhaust camshaft sprocket.

5. Using the proper gauge feeler gauge, adjust the valves on cylinder number one.

6. Rotate the crankshaft 180 degrees. Using the proper gauge feeler gauge, adjust the valves on cylinder number three.

7. Rotate the crankshaft 180 degrees. Using the proper gauge feeler gauge, adjust the valves on cylinder number four.

8. Rotate the crankshaft 180 degrees. Using the proper gauge feeler gauge, adjust the valves on cylinder number two.

9. Install the cylinder head cover.

10. As required, reprogram the power window control unit as follows.

11. Turn the ignition switch ON. Lower the window, all the way down. Open the driver's side door.

12. Turn the ignition switch OFF. Push and hold the drivers window DOWN switch. Turn the ignition switch ON. Release the driver's window DOWN switch. This must be done within five seconds.

13. Repeat the above step three more times. Wait one second.

14. Confirm that the AUTO UP and AUTO DOWN do not work. If they work repeat the procedure, paying close attention to the five second time limit

15. Move the driver's window all the way down by holding the drivers window DOWN switch to the AUTO DOWN position.

16. Pull up and hold the driver's window UP switch to the AUTO UP position until the window reaches the fully closed position. Hold the switch for one second.

17. Confirm that the power window master switch has been reset by using the driver's window AUTO UP and DOWN function.

18. If the AUTO UP and DOWN feature is still not working, repeat the complete procedure, paying close attention to the five second time limit.

19. Set the clock.

3.0L ENGINE

1. Before servicing the vehicle, refer to the Precautions Section. Disconnect the negative battery cable.

2. Note the radio security code and the radio presets.

3. Remove the cylinder head cover retaining bolts. Remove the cylinder head cover from the engine.

4. Set the number one piston at TDC. Align the pointer on the front of the upper cover with the number one piston TDC mark on the front camshaft pulley.

5. Using the proper gauge feeler gauge, adjust the valves on cylinder number one.

6. Rotate the crankshaft clockwise. Align the pointer on the front of the upper cover with the number four piston TDC mark on the front camshaft pulley.

7. Using the proper gauge feeler gauge, adjust the valves on cylinder number four.

8. Rotate the crankshaft clockwise. Align the pointer on the front of the upper cover with the number two piston TDC mark on the front camshaft pulley.

9. Using the proper gauge feeler gauge, adjust the valves on cylinder number two.

10. Rotate the crankshaft clockwise. Align the pointer on the front of the upper cover with the number five piston TDC mark on the front camshaft pulley.

11. Using the proper gauge feeler gauge, adjust the valves on cylinder number five.

12. Rotate the crankshaft clockwise. Align the pointer on the front of the upper cover with the number three piston TDC mark on the front camshaft pulley.

13. Using the proper gauge feeler gauge, adjust the valves on cylinder number three.

14. Rotate the crankshaft clockwise. Align the pointer on the front of the upper cover with the number six piston TDC mark on the front camshaft pulley.

15. Using the proper gauge feeler gauge, adjust the valves on cylinder number six.

16. Install the cylinder head cover.

17. As required, reprogram the crankshaft position (CKP) pattern. Run the engine until the operating temperature reaches 176 degree. With the engine stopped clear the CKP pattern. Turn the ignition switch OFF. Turn the ignition switch ON and wait thirty seconds.

18. Roadtest the vehicle on a level surface. Decelerate the engine speed of 2500 rpm to 1000 rpm. If equipped with automatic transaxle use two Drive positions. If equipped with manual transaxle use first gear.

19. Stop the vehicle, but keep the engine running.

20. Check PULSAR F/B LEARN in the data list with the HDS. If not complete repeat the procedure. If complete, roadtest the vehicle on a level surface. Decelerate the engine speed of 5000 rpm to 3000 rpm. If equipped with automatic transaxle use two Drive positions. If equipped with manual transaxle use first gear.

21. Stop the vehicle, but keep the engine running.

22. Check PULSAR F/B LEARN in the data list with the HDS. If not complete repeat the procedure.

23. If completed, turn the ignition switch OFF. Turn the ignition switch ON, wait thirty seconds. The learning procedure is now complete.

24. Enter the antitheft codes for the radio and the navigation system. Set the clock.

25. As required, reprogram the power window control unit as follows.

26. Turn the ignition switch ON. Lower the window, all the way down. Open the driver's side door.

27. Turn the ignition switch OFF. Push and hold the drivers window DOWN switch. Turn the ignition switch ON. Release the driver's window DOWN switch. This must be done within five seconds.

28. Repeat the above step three more times. Wait one second.

29. Confirm that the AUTO UP and AUTO DOWN do not work. If they work repeat the procedure, paying close attention to the five second time limit

30. Move the driver's window all the way down by holding the drivers window DOWN switch to the AUTO DOWN position.

31. Pull up and hold the driver's window UP switch to the AUTO UP position until the window reaches the fully closed position. Hold the switch for one second.

32. Confirm that the power window master switch has been reset by using the driver's window AUTO UP and DOWN function.

33. If the AUTO UP and DOWN feature is still not working, repeat the complete procedure, paying close attention to the five second time limit.

34. Set the clock.

Starter Motor

REMOVAL & INSTALLATION

Civic

1.7L ENGINE

1. Before servicing the vehicle, refer to the Precautions Section.

2. Note the radio security code and the radio presets.

3. Remove or disconnect the following:
- Negative battery cable
- Air intake resonator
- Starter motor wiring harness
- Starter motor

To install:

4. Install or connect the following:
- Starter motor and tighten the bolts to 33 ft. lbs. (44 Nm)
- Starter motor wiring harness and tighten the battery cable terminal bolt to 84 inch lbs. (9 Nm)
- Negative battery cable

5. As required, reprogram the ECM/PCM engine idle characteristics. Be sure all electrical items are OFF.

6. Start the engine. Hold the idle speed at 3000 RPM's in park or neutral until the radiator fan comes on or the temperature reached 194 degrees.

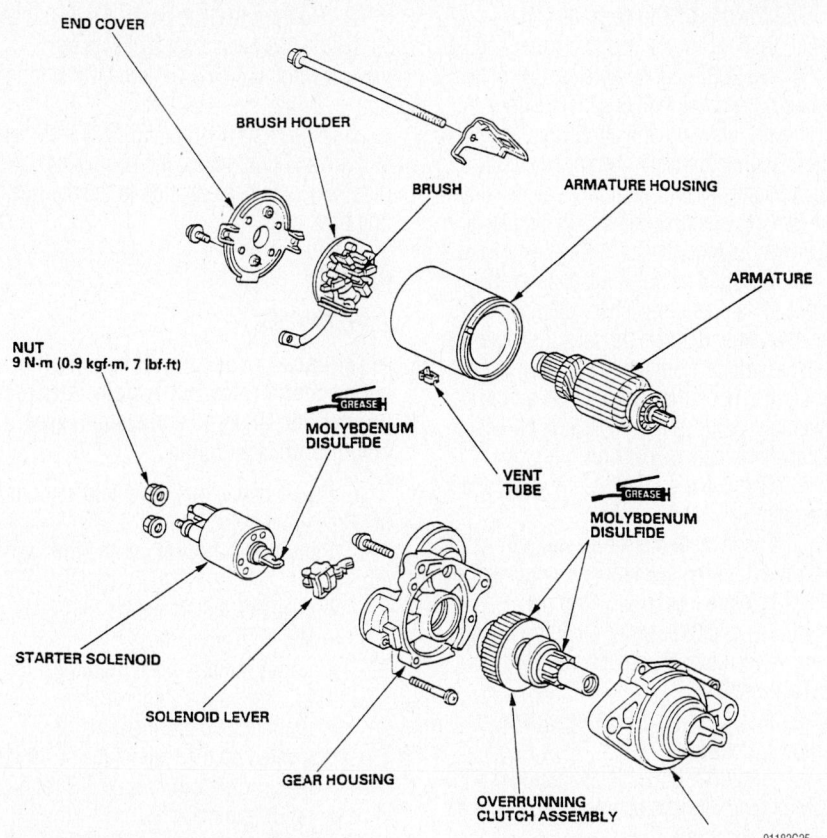

END COVER

BRUSH HOLDER

BRUSH

ARMATURE HOUSING

ARMATURE

NUT
9 N·m (0.9 kgf·m, 7 lbf·ft)

GREASE
MOLYBDENUM DISULFIDE

VENT TUBE

GREASE
MOLYBDENUM DISULFIDE

STARTER SOLENOID

SOLENOID LEVER

GEAR HOUSING

OVERRUNNING CLUTCH ASSEMBLY

91182G25

Exploded view of a typical Honda Starter

7. Let the engine idle for about five minutes with the throttle fully closed.

8. If the radiator fan comes on during the five minutes, do not count this toward the five minute programming time.

9. Set the clock.

1.8L ENGINE

1. Before servicing the vehicle, refer to the Precautions Section.

2. Note the radio security code and the radio presets.

3. Disconnect the negative battery cable. Disconnect the positive battery cable.

4. Raise and support the vehicle safely. Remove the exhaust pipe. Remove the intake manifold bracket.

5. Remove the harness clamps and harness connector from each clamp.

6. Remove the starter retaining bolts. Remove the starter from the engine.

7. Disconnect the starter cable from the terminal. Disconnect the connector from the S terminal. Remove the starter from the vehicle.

To install:

8. Installation is the reverse of the removal procedure.

9. Reprogram the power window control unit as follows.

10. Turn the ignition switch ON. Lower the window, all the way down. Open the driver's side door.

11. Turn the ignition switch OFF. Push and hold the drivers window DOWN switch. Turn the ignition switch ON. Release the driver's window DOWN switch. This must be done within five seconds.

12. Repeat the above step three more times. Wait one second.

13. Confirm that the AUTO UP and AUTO DOWN do not work. If they work repeat the procedure, paying close attention to the five second time limit

14. Move the driver's window all the way down by holding the drivers window DOWN switch to the AUTO DOWN position.

15. Pull up and hold the driver's window UP switch to the AUTO UP position until the window reaches the fully closed position. Hold the switch for one second.

16. Confirm that the power window master switch has been reset by using the driver's window AUTO UP and DOWN function.

17. If the AUTO UP and DOWN feature is still not working, repeat the complete procedure, paying close attention to the five second time limit.

18. Set the clock.

2002–2005 2.0L ENGINE

1. Before servicing the vehicle, refer to the Precautions Section.

2. Note the radio security code and the radio presets.

3. Disconnect the negative battery cable. Disconnect the positive battery cable.

4. Disconnect the knock sensor connector. Remove the bolt securing the harness bracket. Remove the intake manifold bracket.

5. Disconnect the starter cable from the B terminal on the solenoid. Disconnect the BLK/WHT wire from the S terminal.

6. Remove the starter retaining bolts. Remove the started from the vehicle.

To install:

7. Installation is the reverse of the removal procedure.

8. As required, reprogram the ECM engine idle characteristics. Be sure all electrical items are OFF.

9. Start the engine. Hold the idle speed at 3000 RPM's in neutral until the radiator fan comes on or the temperature reached 194 degrees.

10. Let the engine idle for about five minutes with the throttle fully closed.

11. If the radiator fan comes on during the five minutes, do not count this toward the five minute programming time.

12. Set the clock.

13. As required, reprogram the power window control unit as follows.

14. Turn the ignition switch ON. Lower the window, all the way down. Open the driver's side door.

15. Turn the ignition switch OFF. Push and hold the drivers window DOWN switch. Turn the ignition switch ON. Release the driver's window DOWN switch. This must be done within five seconds.

16. Repeat the above step three more times. Wait one second.

17. Confirm that the AUTO UP and AUTO DOWN do not work. If they work repeat the procedure, paying close attention to the five second time limit

18. Move the driver's window all the way down by holding the drivers window DOWN switch to the AUTO DOWN position.

19. Pull up and hold the driver's window UP switch to the AUTO UP position until the window reaches the fully closed position. Hold the switch for one second.

20. Confirm that the power window master switch has been reset by using the driver's window AUTO UP and DOWN function.

21. If the AUTO UP and DOWN feature is still not working, repeat the complete proce-

dure, paying close attention to the five second time limit.

2006 2.0L ENGINE

1. Before servicing the vehicle, refer to the Precautions Section.

2. Note the radio security code and the radio presets.

3. Disconnect the negative battery cable. Disconnect the positive battery cable.

4. Raise and support the vehicle safely. Remove the splash shield.

5. Remove the harness clamp. Remove the starter retaining bolts. Remove the starter from the engine.

6. Disconnect the starter cable from the B terminal on the solenoid. Disconnect the BLK/WHT wire from the S terminal.

7. Remove the starter retaining bolts. Remove the started from the vehicle.

To install:

8. Installation is the reverse of the removal procedure.

9. Reprogram the power window control unit as follows.

10. Turn the ignition switch ON. Lower the window, all the way down. Open the driver's side door.

11. Turn the ignition switch OFF. Push and hold the drivers window DOWN switch. Turn the ignition switch ON. Release the driver's window DOWN switch. This must be done within five seconds.

12. Repeat the above step three more times. Wait one second.

13. Confirm that the AUTO UP and AUTO DOWN do not work. If they work repeat the procedure, paying close attention to the five second time limit

14. Move the driver's window all the way down by holding the drivers window DOWN switch to the AUTO DOWN position.

15. Pull up and hold the driver's window UP switch to the AUTO UP position until the window reaches the fully closed position. Hold the switch for one second.

16. Confirm that the power window master switch has been reset by using the driver's window AUTO UP and DOWN function.

17. If the AUTO UP and DOWN feature is still not working, repeat the complete procedure, paying close attention to the five second time limit.

18. Set the clock.

S2000

2.0L AND 2.2L ENGINES

1. Before servicing the vehicle, refer to the Precautions Section.

2. Note the radio security code and the radio presets.

3. Disconnect the negative battery cable. Disconnect the positive battery cable.

4. On 2002–2005 vehicles, disconnect the air hose and breather pipe. Remove the air cleaner housing and air cleaner assembly. Disconnect the air control solenoid valve connector and vacuum hose. Remove the harness clamps. Remove the intake air cleaner housing.

5. On 2006 vehicles, disconnect the intake air temperature sensor connector and breather pipe. Remove the manifold absolute sensor harness from the holder. Remove the air cleaner housing cover and the air cleaner assembly. Remove the IAT sensor harness clamps and the intake air housing.

6. Remove the drive belt. Remove the alternator.

7. Disconnect the starter cable from the B terminal on the solenoid. Disconnect the BLK/WHT wire from the S terminal.

8. Remove the starter retaining bolts. Remove the started from the vehicle.

To install:

9. Installation is the reverse of the removal procedure.

10. On 2002–05 vehicles, reprogram the ECM engine idle characteristics. Be sure all electrical items are OFF.

11. Start the engine. Hold the idle speed at 3000 RPM's in neutral until the radiator fan comes on or the temperature reached 176 degrees.

12. Let the engine idle for about five minutes with the throttle fully closed.

13. If the radiator fan comes on during the five minutes, do not count this toward the five minute programming time.

14. Set the clock.

15. As required, reprogram the power window control unit as follows.

16. Turn the ignition switch ON. Lower the window, all the way down. Open the driver's side door.

17. Turn the ignition switch OFF. Push and hold the drivers window DOWN switch. Turn the ignition switch ON. Release the driver's window DOWN switch. This must be done within five seconds.

18. Repeat the above step three more times. Wait one second.

19. Confirm that the AUTO UP and AUTO DOWN do not work. If they work repeat the procedure, paying close attention to the five second time limit

20. Move the driver's window all the way down by holding the drivers window DOWN switch to the AUTO DOWN position.

21. Pull up and hold the driver's window UP switch to the AUTO UP position until the window reaches the fully closed position. Hold the switch for one second.

22. Confirm that the power window master switch has been reset by using the driver's window AUTO UP and DOWN function.

23. If the AUTO UP and DOWN feature is still not working, repeat the complete procedure, paying close attention to the five second time limit.

Accord

2.3L ENGINE

➡The factory sound system has a coded theft protection system. It is recommended that you know your reset code before you begin.

1. Before servicing the vehicle, refer to the Precautions Section.

2. Remove or disconnect the following:

- Negative battery cable
- Wiring from harness
- Lower radiator hose from the bracket on the starter motor
- Starter cable from terminal B located on the back of the solenoid
- Black/white wire from the S (solenoid) terminal
- Two bolts that mount the starter to the transaxle assembly
- Starter

To install:

3. Install in the reverse order of removal.

➡When installing the heavy gauge starter cable, make sure the crimped side of the terminal end is facing out.

4. Enter the anti-theft code and radio presets.

2.4L ENGINE

1. Before servicing the vehicle, refer to the Precautions Section.

2. Note the radio security code and the radio presets.

3. Disconnect the negative battery cable. Disconnect the positive battery cable.

4. Drain the cooling system. Remove the intake manifold.

5. Disconnect the starter cable from the B terminal on the solenoid. Disconnect the BLK/WHT wire from the S terminal.

6. Remove the harness clamp. Remove the harness holder.

7. Remove the starter retaining bolts. Remove the started from the vehicle.

To install:

8. Installation is the reverse of the removal procedure.

9. Reprogram the power window control unit as follows.

M/T:

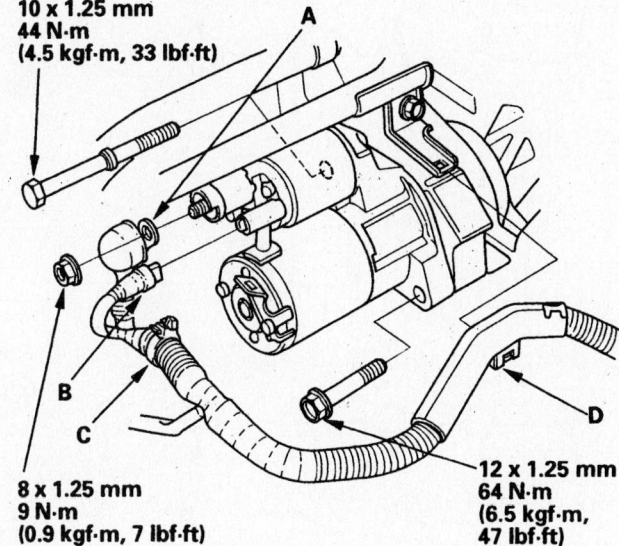

10 x 1.25 mm
44 N·m
(4.5 kgf·m, 33 lbf·ft)

A

8 x 1.25 mm
9 N·m
(0.9 kgf·m, 7 lbf·ft)

B

C

D

12 x 1.25 mm
64 N·m
(6.5 kgf·m,
47 lbf·ft)

A/T:

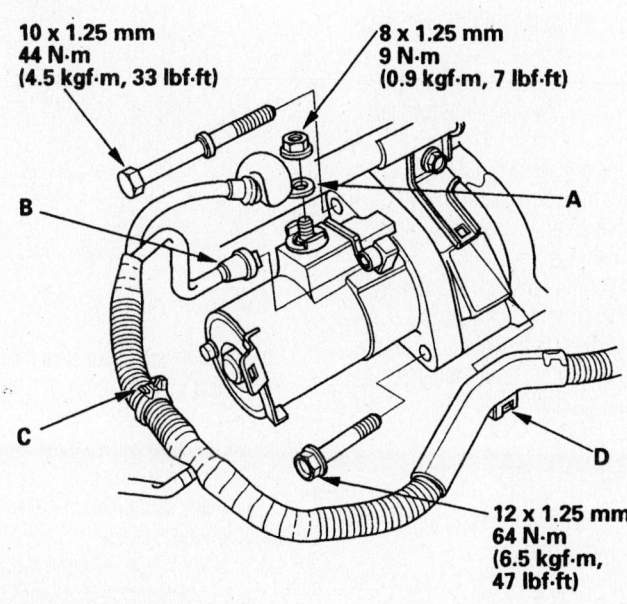

10 x 1.25 mm
44 N·m
(4.5 kgf·m, 33 lbf·ft)

8 x 1.25 mm
9 N·m
(0.9 kgf·m, 7 lbf·ft)

B

A

C

D

12 x 1.25 mm
64 N·m
(6.5 kgf·m,
47 lbf·ft)

42356-ACCO-G18

Starter mounting—2.4L engine

10. Turn the ignition switch ON. Lower the window, all the way down. Open the driver's side door.

11. Turn the ignition switch OFF. Push and hold the drivers window DOWN switch. Turn the ignition switch ON. Release the driver's window DOWN switch. This must be done within five seconds.

12. Repeat the above step three more times. Wait one second.

13. Confirm that the AUTO UP and AUTO DOWN do not work. If they work repeat the procedure, paying close attention to the five second time limit

14. Move the driver's window all the way down by holding the drivers window DOWN switch to the AUTO DOWN position.

15. Pull up and hold the driver's window UP switch to the AUTO UP position until the window reaches the fully closed position. Hold the switch for one second.

16. Confirm that the power window master switch has been reset by using the driver's window AUTO UP and DOWN function.

17. If the AUTO UP and DOWN feature is still not working, repeat the complete procedure, paying close attention to the five second time limit.

18. Set the clock.

3.0L ENGINE

1. Before servicing the vehicle, refer to the Precautions Section.

2. Note the radio security code and the radio presets.

3. Disconnect the negative battery cable. Disconnect the positive battery cable.

4. Remove the battery. Remove the harness clamp.

5. Disconnect the starter cable from the B terminal on the solenoid. Disconnect the BLK/WHT wire from the S terminal.

6. Remove the starter retaining bolts. Remove the started from the vehicle.

To install:

7. Installation is the reverse of the removal procedure.

8. Reprogram the power window control unit as follows.

9. Turn the ignition switch ON. Lower the window, all the way down. Open the driver's side door.

10. Turn the ignition switch OFF. Push and hold the drivers window DOWN switch. Turn the ignition switch ON. Release the driver's window DOWN switch. This must be done within five seconds.

11. Repeat the above step three more times. Wait one second.

12. Confirm that the AUTO UP and AUTO DOWN do not work. If they work repeat the procedure, paying close attention to the five second time limit

13. Move the driver's window all the way down by holding the drivers window DOWN switch to the AUTO DOWN position.

14. Pull up and hold the driver's window UP switch to the AUTO UP position until the window reaches the fully closed position. Hold the switch for one second.

15. Confirm that the power window master switch has been reset by using the driver's window AUTO UP and DOWN function.

16. If the AUTO UP and DOWN feature is still not working, repeat the complete proce-

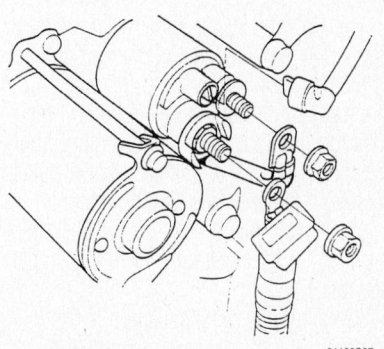

91182G27

Location of starter wiring—3.0L engine

dure, paying close attention to the five second time limit.

17. Set the clock.

Oil Pan

REMOVAL & INSTALLATION

Civic

1.7L ENGINE

1. Before servicing the vehicle, refer to the Precautions Section.

2. Note the radio security code and the radio presets. Disconnect the negative battery cable.

3. Drain the engine oil. Raise and support the vehicle safely.

4. On the D17A2 engine, disconnect the primary heated oxygen sensor connector and the secondary heated oxygen sensor connector. Remove the exhaust pipe/three way catalytic converter.

5. On the D17A6 engine, disconnect the third heated oxygen sensor connector. Remove the exhaust pipe/three way catalytic converter.

6. On the D17A1 engine, remove the exhaust pipe/muffler assembly.

7. Remove the oil pan retaining bolts. Remove the oil pan from the vehicle.

To install:

8. Remove any old gasket from the mating surfaces. Be sure these surfaces are clean and dry.

9. Apply liquid gasket, part number 08717-004, 08718-0001, 08718-0003 or 08718-0009 on the oil pump and engine block end cover mating surfaces (steel pan) and on the pan mating surface of the block and the inner threads of the bolt holes (aluminum pan).

➡**Do not install the parts if more than four minutes have elapsed since applying the liquid gasket. Instead, reapply after removing the previous coating material.**

10. On vehicles equipped with the steel pan, apply liquid gasket at the four corners of the pan recesses, prior to installation. Once again, do not install the parts if more than four minutes have elapsed since applying the liquid gasket. Instead, reapply after removing the previous coating material.

11. Install the oil pan. Torque the retaining bolts to specification and in the proper sequence.

12. Continue the installation in the reverse order of the removal procedure.

13. After assembly, wait at least thirty

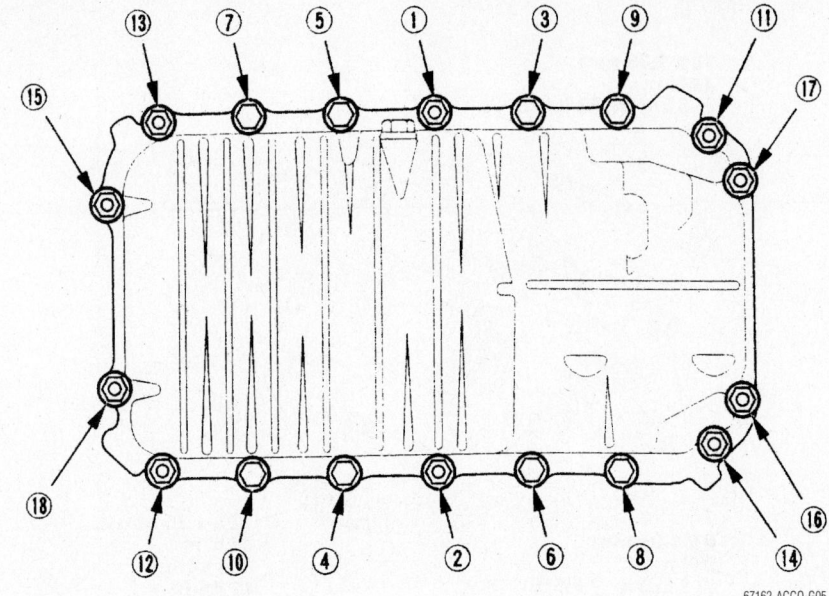

Oil pan bolt tightening sequence—Civic with aluminum oil pan

minutes before filling the engine with clean engine oil.

14. Reprogram the ECM/PCM engine idle characteristics. Be sure all electrical items are OFF.

15. Start the engine. Hold the idle speed at 3000 RPM's in park or neutral until the radiator fan comes on or the temperature reached 194 degrees.

16. Let the engine idle for about five minutes with the throttle fully closed.

17. If the radiator fan comes on during the five minutes, do not count this toward the five minute programming time.

18. Set the clock.

1.8L ENGINE

1. Before servicing the vehicle, refer to the Precautions Section.

2. Note the radio security code and the radio presets. Disconnect the negative battery cable.

3. Remove the drive belt. Remove the air conditioning condenser fan shroud.

4. Disconnect the compressor electrical connectors. Remove the compressor retaining bolts, and position it to the side without discharging the system.

5. Raise and support the vehicle safely. Remove the splash shield. Drain the engine oil.

6. Remove the exhaust pipe. Properly support the oil pan. Remove the lower torque rod. Remove the oil pan support tool.

7. Remove the lower torque rod bracket. Remove the air conditioning compressor bracket.

8. If equipped with automatic transaxle,

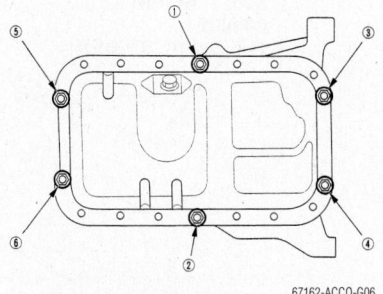

Oil pan bolt tightening sequence—Civic with steel oil pan

remove the shift cable cover. Remove the torque converter cover.

9. Remove the clutch cover if equipped with manual transaxle.

10. Remove the oil pan retaining bolts. Using a flat bladed tool, carefully separate the oil pan from the engine block. Remove the oil pan from the vehicle.

To install:

11. Remove any old gasket from the mating surfaces. Be sure these surfaces are clean and dry.

12. Apply liquid gasket, part number 08717-004, 08718-0001, 08718-0003 or 08718-0009 evenly to the engine block mating surface of the oil pan.

➡**Do not install the parts if more than five minutes have elapsed since applying the liquid gasket. Instead, reapply after removing the previous coating material.**

13. Install the dowel pins, using new O-rings.

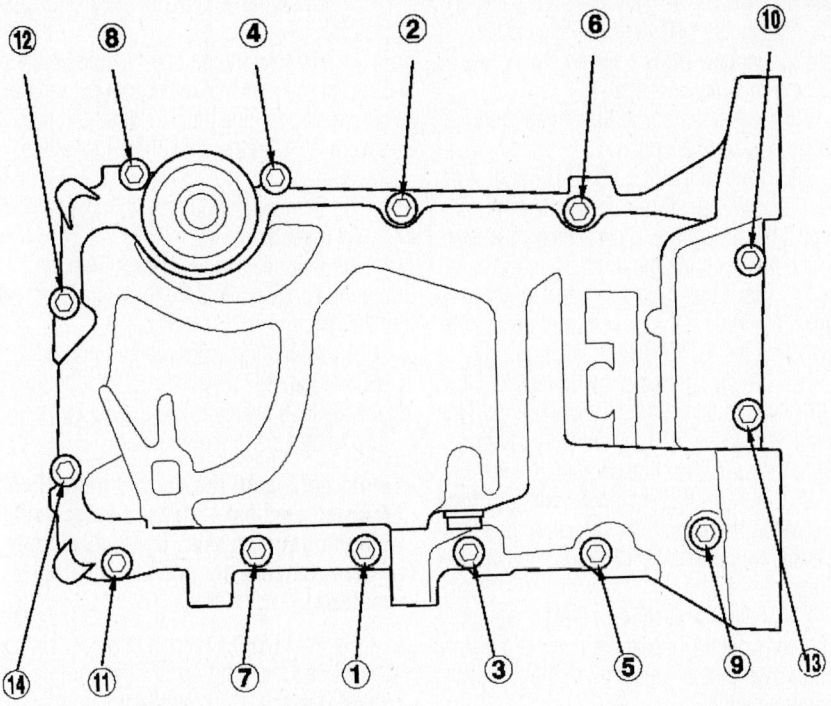

Oil pan bolt tightening sequence—1.8L engine

09474_ACCO_G0017

14. Position the oil pan in place. Tighten the oil pan retaining bolts in two or three steps to specification and in the proper sequence.

15. Continue the installation in the reverse order of the removal procedure.

16. After assembly, wait at least thirty minutes before filling the engine with clean engine oil.

17. Do not run the engine for at least three hours after installing the oil pan.

18. As required, reprogram the power window control unit as follows.

19. Turn the ignition switch ON. Lower the window, all the way down. Open the driver's side door.

20. Turn the ignition switch OFF. Push and hold the drivers window DOWN switch. Turn the ignition switch ON. Release the driver's window DOWN switch. This must be done within five seconds.

21. Repeat the above step three more times. Wait one second.

22. Confirm that the AUTO UP and AUTO DOWN do not work. If they work repeat the procedure, paying close attention to the five second time limit

23. Move the driver's window all the way down by holding the drivers window DOWN switch to the AUTO DOWN position.

24. Pull up and hold the driver's window UP switch to the AUTO UP position until the window reaches the fully closed position. Hold the switch for one second.

25. Confirm that the power window master switch has been reset by using the driver's window AUTO UP and DOWN function.

26. If the AUTO UP and DOWN feature is still not working, repeat the complete procedure, paying close attention to the five second time limit.

27. Set the clock.

2002–2005 2.0L ENGINE

1. Before servicing the vehicle, refer to the Precautions Section.

2. Note the radio security code and the radio presets. Disconnect the negative battery cable.

3. Remove the subframe. Position the engine support tool, VSB02C000051, on the engine and radiator support bracket.

4. Raise and support the vehicle safely. Drain the engine oil.

5. Disconnect the suspension lower arm ball joints. Remove the rear mount mounting bolts. Remove the front mount mounting bolt.

6. Use a marker and make alignment marks on the reference lines that align with the centers of the rear subframe mounting bolts. Remove the front subframe.

7. Remove the oil pan retaining bolts. Drive an oil pan seal cutter tool between the oil pan and engine block. Cut the oil pan seal by striking the side of the tool along the oil pan

8. Remove the oil pan from the vehicle.

To install:

9. Remove any old gasket from the mating surfaces. Be sure these surfaces are clean and dry.

10. Apply liquid gasket, part number 08717-004, 08718-0001, 08718-0003 or 08718-0009 evenly to the engine block mating surface of the oil pan.

➡**Do not install the parts if more than five minutes have elapsed since applying the liquid gasket. Instead, reapply after removing the previous coating material.**

11. Position the oil pan in place. Tighten the oil pan retaining bolts in two or three steps to specification and in the proper sequence.

12. Continue the installation in the reverse order of the removal procedure.

13. After assembly, wait at least thirty minutes before filling the engine with clean engine oil.

14. Do not run the engine for at least three hours after installing the oil pan.

15. As required, reprogram the power window control unit as follows.

16. Turn the ignition switch ON. Lower the window, all the way down. Open the driver's side door.

17. Turn the ignition switch OFF. Push and hold the drivers window DOWN switch. Turn the ignition switch ON. Release the driver's window DOWN switch. This must be done within five seconds.

18. Repeat the above step three more times. Wait one second.

19. Confirm that the AUTO UP and AUTO DOWN do not work. If they work repeat the procedure, paying close attention to the five second time limit

20. Move the driver's window all the way down by holding the drivers window DOWN switch to the AUTO DOWN position.

21. Pull up and hold the driver's window UP switch to the AUTO UP position until the window reaches the fully closed position. Hold the switch for one second.

22. Confirm that the power window master switch has been reset by using the driver's window AUTO UP and DOWN function.

23. If the AUTO UP and DOWN feature is still not working, repeat the complete procedure, paying close attention to the five second time limit.

24. Set the clock.

2006 2.0L ENGINE

1. Before servicing the vehicle, refer to the Precautions Section.

2. Note the radio security code and the

radio presets. Disconnect the negative battery cable.

3. Raise and support the vehicle safely. Drain the engine oil. Remove the front tires.

4. Remove the splash shield. Separate the stabilizer links. Separate the knuckles from the lower arms.

5. Remove the steering gearbox bracket. Remove the steering gearbox mounting bolt, stiffner mounting bolt and stiffner.

6. Remove the gearbox mounting bolt, stiffner mounting bolt and stiffner. Remove the harness clamp from the subframe

7. Remove the lower torque rod. Remove the front mount mounting bolt.

8. Use a marker and make alignment marks on the reference lines that align with the centers of the rear subframe mounting bolts.

9. Loosen the mid-stiffner mounting bolts, on both sides. Support the subframe using the proper support tool.

10. Remove the front subframe. Remove the lower torque rod bracket.

11. Remove the clutch cover and the transaxle mounting bolts.

12. Remove the oil pan retaining bolts. Using a flat bladed tool, carefully separate the oil pan from the engine block. Remove the oil pan from the vehicle.

To install:

13. Remove any old gasket from the mating surfaces. Be sure these surfaces are clean and dry.

14. Apply liquid gasket, part number 08717-004, 08718-0001, 08718-0003 or 08718-0009 evenly to the engine block mating surface of the oil pan.

➡**Do not install the parts if more than five minutes have elapsed since applying the liquid gasket. Instead, reapply after removing the previous coating material.**

15. Position the oil pan in place. Tighten the oil pan retaining bolts in two or three steps to specification and in the proper sequence.

16. Continue the installation in the reverse order of the removal procedure.

17. After assembly, wait at least thirty minutes before filling the engine with clean engine oil.

18. Do not run the engine for at least three hours after installing the oil pan.

19. As required, reprogram the power window control unit as follows.

20. Turn the ignition switch ON. Lower the window, all the way down. Open the driver's side door.

21. Turn the ignition switch OFF. Push

and hold the drivers window DOWN switch. Turn the ignition switch ON. Release the driver's window DOWN switch. This must be done within five seconds.

22. Repeat the above step three more times. Wait one second.

23. Confirm that the AUTO UP and AUTO DOWN do not work. If they work repeat the procedure, paying close attention to the five second time limit

24. Move the driver's window all the way down by holding the drivers window DOWN switch to the AUTO DOWN position.

25. Pull up and hold the driver's window UP switch to the AUTO UP position until the window reaches the fully closed position. Hold the switch for one second.

26. Confirm that the power window master switch has been reset by using the driver's window AUTO UP and DOWN function.

27. If the AUTO UP and DOWN feature is still not working, repeat the complete procedure, paying close attention to the five second time limit.

28. Set the clock.

S2000

2.0L AND 2.2L ENGINES

1. Before servicing the vehicle, refer to the Precautions Section.

2. Note the radio security code and the radio presets. Disconnect the negative battery cable.

3. Drain the engine oil. Raise and support the vehicle safely.

4. Remove the oil pan retaining bolts. Drive an oil pan seal cutter tool between the oil pan and engine block. Cut the oil pan seal by striking the side of the tool along the oil pan

5. Remove the oil pan from the vehicle.

To install:

6. Remove any old gasket from the mating surfaces. Be sure these surfaces are clean and dry.

7. Apply liquid gasket, part number 08717-004, 08718-0001, 08718-0003 or 08718-0009 evenly to the engine block mating surface of the oil pan.

➡**Do not install the parts if more than four minutes have elapsed since applying the liquid gasket. Instead, reapply after removing the previous coating material.**

8. Position the oil pan in place. Tighten the oil pan retaining bolts in two or three steps to specification and in the proper sequence.

9. Continue the installation in the reverse order of the removal procedure.

10. After assembly, wait at least thirty minutes before filling the engine with clean engine oil.

11. Do not run the engine for at least three hours after installing the oil pan.

12. As required, reprogram the power window control unit as follows.

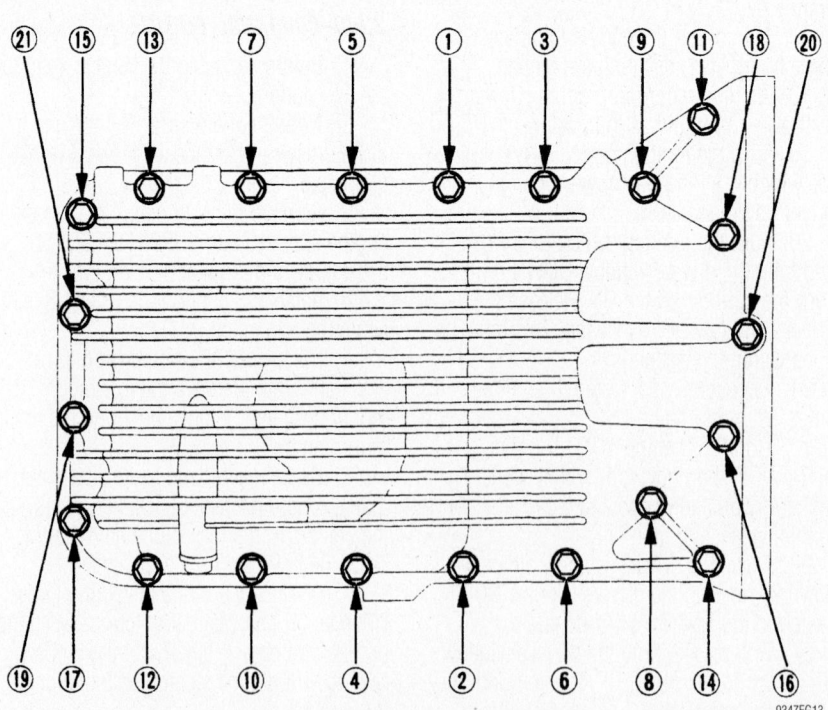

Oil pan torque sequence—S2000 with 2.0L and 2.2L engines

9347FG13

13. Turn the ignition switch ON. Lower the window, all the way down. Open the driver's side door.

14. Turn the ignition switch OFF. Push and hold the drivers window DOWN switch. Turn the ignition switch ON. Release the driver's window DOWN switch. This must be done within five seconds.

15. Repeat the above step three more times. Wait one second.

16. Confirm that the AUTO UP and AUTO DOWN do not work. If they work repeat the procedure, paying close attention to the five second time limit

17. Move the driver's window all the way down by holding the drivers window DOWN switch to the AUTO DOWN position.

18. Pull up and hold the driver's window UP switch to the AUTO UP position until the window reaches the fully closed position. Hold the switch for one second.

19. Confirm that the power window master switch has been reset by using the driver's window AUTO UP and DOWN function.

20. If the AUTO UP and DOWN feature is still not working, repeat the complete procedure, paying close attention to the five second time limit.

21. Set the clock.

Accord

2.3L ENGINE

1. Before servicing the vehicle, refer to the Precautions Section.

2. Disconnect the negative battery cable.

3. Raise and safely support the vehicle.

4. Drain the engine oil into a sealable container. Install the drain bolt with a new gasket and tighten to 33 ft. lbs. (44 Nm).

5. Remove or disconnect the following:

- Front wheels and the splash shield
- Center beam
- Oxygen (O_2S) sensor electrical connector
- Bolts from the support bracket on the exhaust pipe
- Nuts attaching the exhaust pipe to the exhaust manifold and the catalytic converter
- Exhaust pipe and discard the gaskets
- Converter cover, if equipped with an automatic transaxle
- Clutch cover, if equipped with a manual transaxle
- Oil pan nuts and bolts (in a crisscross pattern) and the oil pan; if necessary, use a mallet to tap the

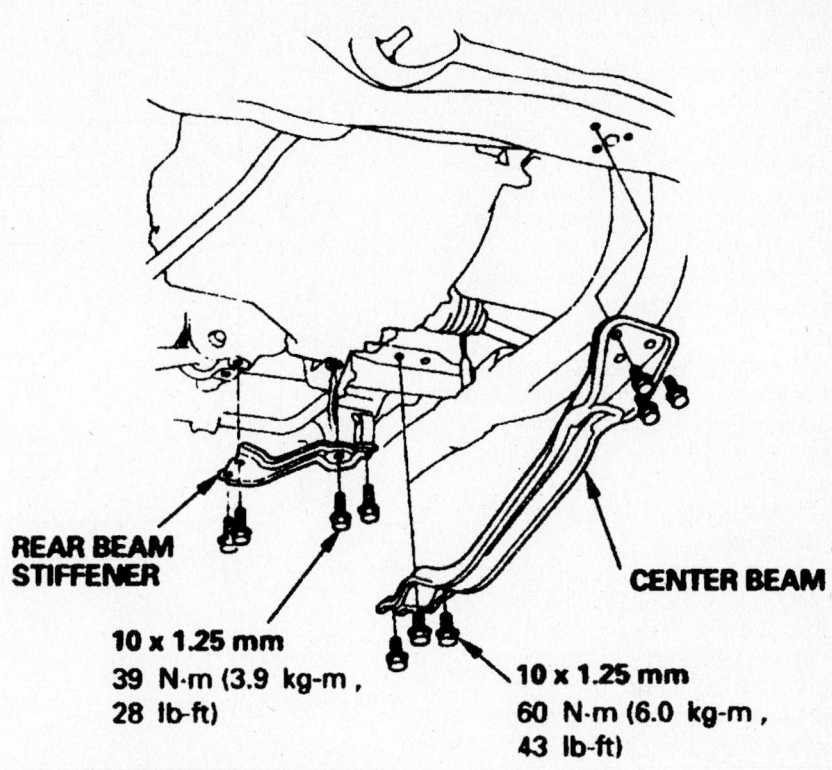

REAR BEAM STIFFENER

10 x 1.25 mm
39 N·m (3.9 kg-m, 28 lb-ft)

CENTER BEAM

10 x 1.25 mm
60 N·m (6.0 kg-m, 43 lb-ft)

7923FG84

To gain access to the oil pan, remove the center beam—2.3L engine

corners of the oil pan. DO NOT pry on the pan to get it loose

6. Clean the oil pan mounting surface of old gasket material and engine oil.

To install:

7. Install or connect the following:

- New oil pan gasket to the oil pan. Apply liquid gasket to the corners of the curved section of the gasket.
- Oil pan to the engine
- Oil pan nuts and bolts and tighten the nuts and bolts in sequence. Tighten the nuts and bolts in 2 steps to 10 ft. lbs. (14 Nm).
- Torque converter cover or clutch cover, as applicable. Tighten the bolts to 108 inch lbs. (12 Nm).
- Exhaust pipe with new gaskets and new locknuts. Tighten the nuts attaching the exhaust pipe to the manifold to 40 ft. lbs. (54 Nm) and tighten the nuts attaching the exhaust pipe to the catalytic converter to 25 ft. lbs. (33 Nm).
- Bolts to the exhaust pipe support bracket and tighten to 13 ft. lbs. (18 Nm)
- O_2S electrical connector
- Center beam and tighten the mounting bolts to 43 ft. lbs. (60 Nm)

- Splash shield and tighten the mounting bolts to 84 inch lbs. (10 Nm)
- Front wheels

8. Lower the vehicle and fill the engine with oil.

9. Connect the negative battery cable and enter the radio security code.

10. Start the engine and check for leaks.

2.4L ENGINE

1. Before servicing the vehicle, refer to the Precautions Section.

2. Note the radio security code and the radio presets. Disconnect the negative battery cable.

3. Disconnect the battery positive cable. Remove the battery. Remove the air cleaner housing. Remove the battery base.

4. If equipped with manual transaxle remove the clutch slave cylinder and clutch line bracket mounting bolt.

5. Remove the ground cable. Remove the transaxle upper mount/bracket assembly.

6. Remove the front mount stop. Remove the front mount bolt.

7. Remove the rear mount stop. Remove the rear mount bolt.

8. Raise and support the vehicle

safely. Drain the engine oil. Remove the front tires.

9. Disconnect the stabilizer links. Remove the left side damper fork. Disconnect the left side suspension lower arm ball joint.

10. Remove the left side driveshaft. Remove the nuts securing the transaxle lower front mount and transaxle lower rear mount.

11. Use a transaxle jack and support the transaxle assembly. Remove the stiffener.

12. Remove the oil pan retaining bolts. Drive an oil pan seal cutter tool between the oil pan and engine block. Cut the oil pan seal by striking the side of the tool along the oil pan

13. Remove the oil pan from the vehicle.

To install:

14. Remove any old gasket from the mating surfaces. Be sure these surfaces are clean and dry.

15. Apply liquid gasket, part number 08717-004, 08718-0001, 08718-0003 or 08718-0009 evenly to the engine block mating surface of the oil pan.

➡**Do not install the parts if more than four minutes have elapsed since applying the liquid gasket. Instead, reapply after removing the previous coating material.**

16. Position the oil pan in place. Tighten the oil pan retaining bolts in two or three steps to specification and in the proper sequence.

17. Continue the installation in the reverse order of the removal procedure.

18. After assembly, wait at least thirty minutes before filling the engine with clean engine oil.

19. Do not run the engine for at least three hours after installing the oil pan.

20. As required, reprogram the power window control unit as follows.

21. Turn the ignition switch ON. Lower

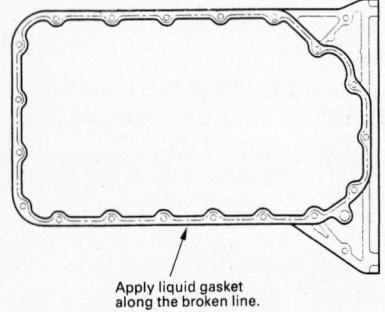

Apply liquid gasket
along the broken line.

9347FG12

**Apply liquid gasket along the broken line—
2.0L (2006 Civic) engine and 2.4L engine**

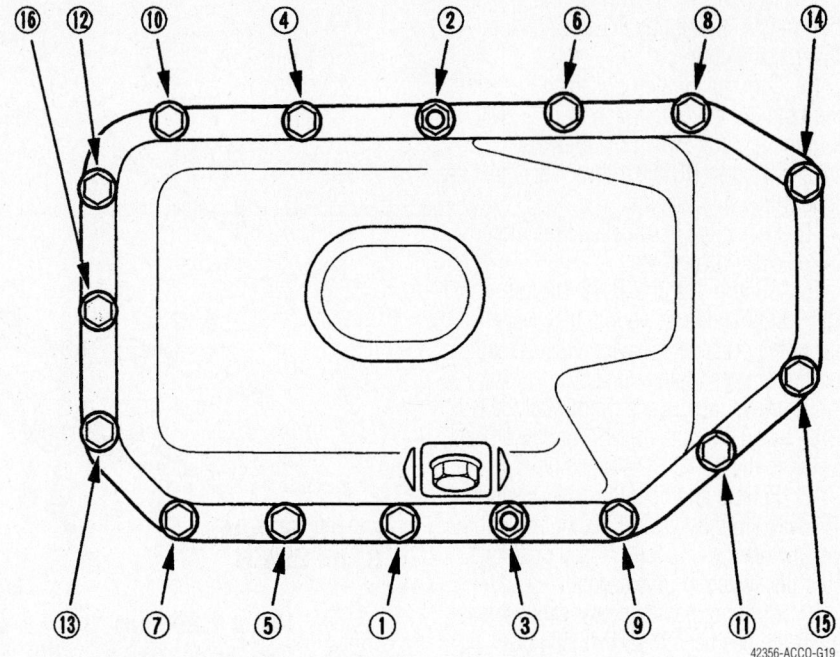

Oil pan bolt tightening sequence—2.0L (2006 Civic) engine and 2.4L engine

42356-ACCO-G19

the window, all the way down. Open the driver's side door.

22. Turn the ignition switch OFF. Push and hold the drivers window DOWN switch. Turn the ignition switch ON. Release the driver's window DOWN switch. This must be done within five seconds.

23. Repeat the above step three more times. Wait one second.

24. Confirm that the AUTO UP and AUTO DOWN do not work. If they work repeat the procedure, paying close attention to the five second time limit

25. Move the driver's window all the way down by holding the drivers window DOWN switch to the AUTO DOWN position.

26. Pull up and hold the driver's window UP switch to the AUTO UP position until the window reaches the fully closed position. Hold the switch for one second.

27. Confirm that the power window master switch has been reset by using the driver's window AUTO UP and DOWN function.

28. If the AUTO UP and DOWN feature is still not working, repeat the complete procedure, paying close attention to the five second time limit.

29. Set the clock.

3.0L ENGINE

1. Before servicing the vehicle, refer to the Precautions Section.

2. Note the radio security code and the radio presets. Disconnect the negative battery cable.

3. Raise and support the vehicle safely. Drain the engine oil.

4. Remove the exhaust pipe. Remove the torque converter cover. Remove the two lower transaxle retaining bolts.

5. Remove the oil pan retaining bolts. Using a flat bladed tool, carefully separate the oil pan from the engine block. Remove the oil pan from the vehicle.

To install:

6. Remove any old gasket from the mating surfaces. Be sure these surfaces are clean and dry.

7. Apply liquid gasket, part number 08717-004, 08718-0001, 08718-0003 or 08718-0009 evenly to the engine block mating surface of the oil pan.

➡**Do not install the parts if more than four minutes have elapsed since applying the liquid gasket. Instead, reapply after removing the previous coating material.**

8. Position the oil pan in place. Tighten the oil pan retaining bolts in two or three steps to specification and in the proper sequence.

9. Continue the installation in the reverse order of the removal procedure.

10. After assembly, wait at least thirty minutes before filling the engine with clean engine oil.

11. Do not run the engine for at least three hours after installing the oil pan.

12. As required, reprogram the power window control unit as follows.

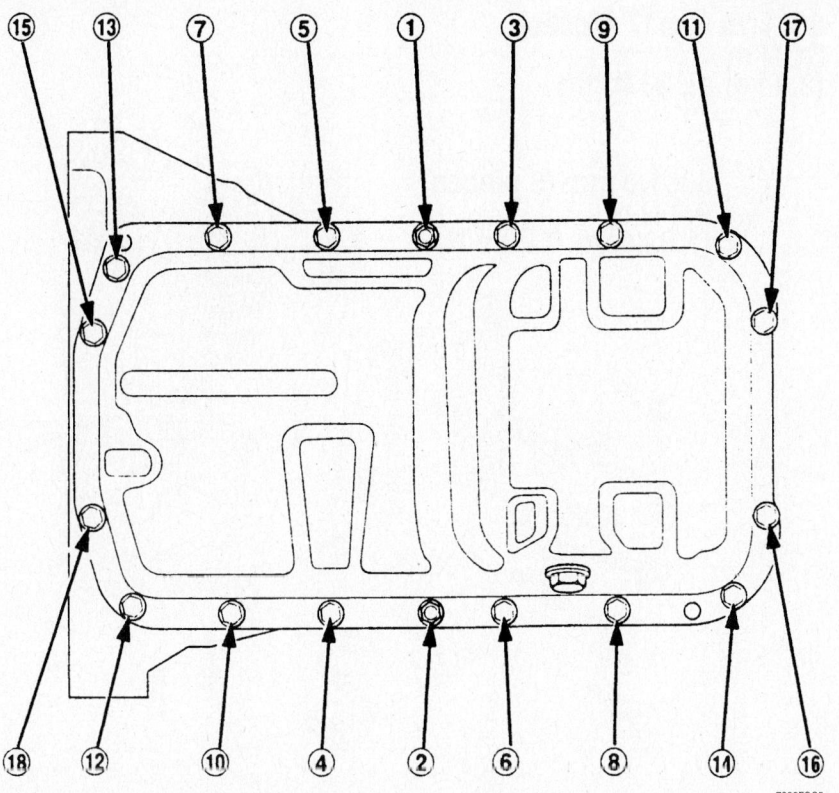

Oil pan mounting bolt tightening sequence—3.0L engine

7923FGC5

Oil Pump

REMOVAL & INSTALLATION

Civic

1.7L ENGINE

1. Before servicing the vehicle, refer to the Precautions Section.

2. Note the radio security code and the radio presets. Disconnect the negative battery cable.

3. Remove the timing belt. Remove the oil pan.

4. Remove the oil screen. Remove the oil pump mounting bolts. Remove the oil pump.

5. Remove the screws from the pump housing and separate the housing and cover.

To install:

6. Remove any old gasket from the mating surfaces. Be sure these surfaces are clean and dry.

7. Apply liquid gasket, part number 00717-004, 08718-0001, 08718-0003 or 08718-0009 evenly to the block mating surface of the oil pump.

13. Turn the ignition switch ON. Lower the window, all the way down. Open the driver's side door.

14. Turn the ignition switch OFF. Push and hold the drivers window DOWN switch. Turn the ignition switch ON. Release the driver's window DOWN switch. This must be done within five seconds.

15. Repeat the above step three more times. Wait one second.

16. Confirm that the AUTO UP and AUTO DOWN do not work. If they work repeat the procedure, paying close attention to the five second time limit

17. Move the driver's window all the way down by holding the drivers window DOWN switch to the AUTO DOWN position.

18. Pull up and hold the driver's window UP switch to the AUTO UP position until the window reaches the fully closed position. Hold the switch for one second.

19. Confirm that the power window master switch has been reset by using the driver's window AUTO UP and DOWN function.

20. If the AUTO UP and DOWN feature is still not working, repeat the complete procedure, paying close attention to the five second time limit.

21. Set the clock.

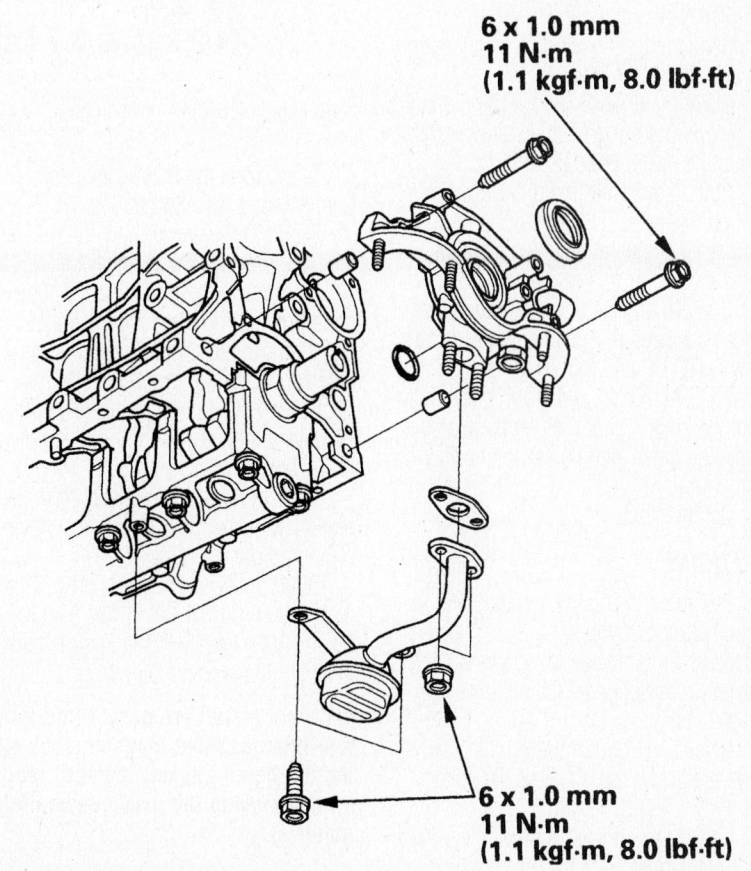

6 x 1.0 mm
11 N·m
(1.1 kgf·m, 8.0 lbf·ft)

6 x 1.0 mm
11 N·m
(1.1 kgf·m, 8.0 lbf·ft)

Oil pump mounting—1.7L engine

09474_ACCO_G0019

➡Do not install the parts if more than four minutes have elapsed since applying the liquid gasket. Instead, reapply after removing the previous coating material.

8. Grease the lips on the oil seal and apply oil to the new O-ring.

9. Install the dowel pins. Align the inner rotor with the crankshaft and install the oil pump.

10. Continue the installation in the reverse order of the removal procedure.

11. After assembly, wait at least thirty minutes before filling the engine with clean engine oil.

12. Do not run the engine for at least three hours after installing the oil pan.

13. As required, reprogram the power window control unit as follows.

14. Turn the ignition switch ON. Lower the window, all the way down. Open the driver's side door.

15. Turn the ignition switch OFF. Push and hold the drivers window DOWN switch. Turn the ignition switch ON. Release the driver's window DOWN switch. This must be done within five seconds.

16. Repeat the above step three more times. Wait one second.

17. Confirm that the AUTO UP and AUTO DOWN do not work. If they work repeat the procedure, paying close attention to the five second time limit

18. Move the driver's window all the way down by holding the drivers window DOWN switch to the AUTO DOWN position.

19. Pull up and hold the driver's window UP switch to the AUTO UP position until the window reaches the fully closed position. Hold the switch for one second.

20. Confirm that the power window master switch has been reset by using the driver's window AUTO UP and DOWN function.

21. If the AUTO UP and DOWN feature is still not working, repeat the complete procedure, paying close attention to the five second time limit.

22. Set the clock.

1.8L ENGINE

1. Before servicing the vehicle, refer to the Precautions Section.

2. Note the radio security code and the radio presets. Disconnect the negative battery cable.

3. Raise and support the vehicle safely. Remove the front tires. Remove the splash shield.

4. Lower the vehicle. Remove the drive belt auto tensioner. Remove the cylinder head cover.

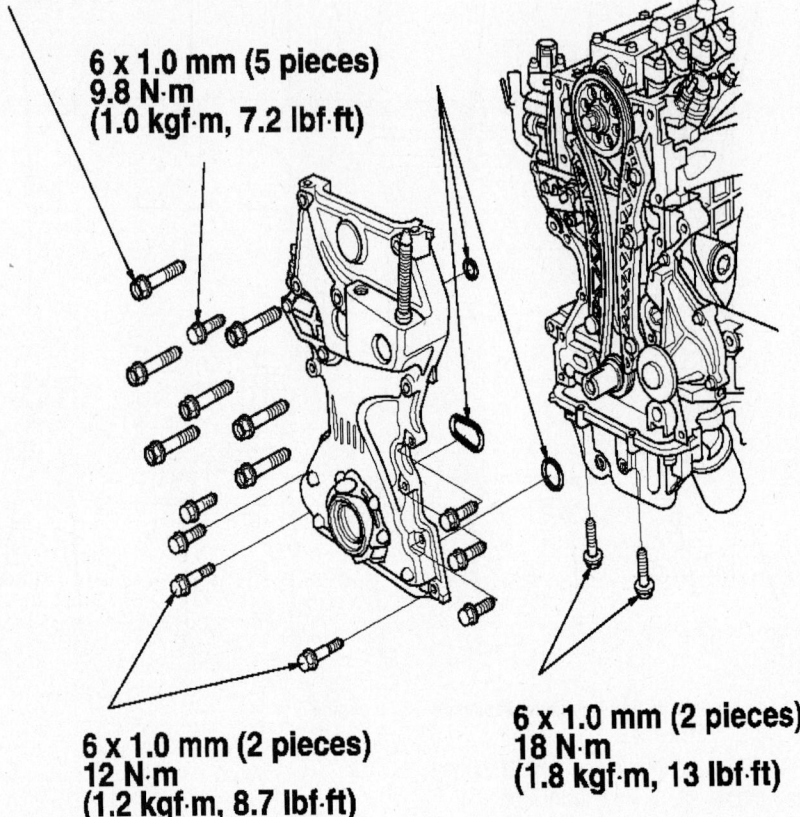

8 x 1.25 mm (7 pieces)
31 N·m
(3.2 kgf·m, 23 lbf·ft)

6 x 1.0 mm (5 pieces)
9.8 N·m
(1.0 kgf·m, 7.2 lbf·ft)

6 x 1.0 mm (2 pieces)
12 N·m
(1.2 kgf·m, 8.7 lbf·ft)

6 x 1.0 mm (2 pieces)
18 N·m
(1.8 kgf·m, 13 lbf·ft)

09474_ACCO_G0020

Oil pump mounting—1.8L engine

5. Remove the PCV valve hose. Remove the crankshaft pulley.

6. Properly support the engine, using a jack and block of wood under the oil pan.

7. Remove the bolt securing the air conditioning line. Remove the upper torque rod. Remove the ground cable. Remove the side engine mount bracket.

8. Remove the oil pump retaining bolts. Remove the oil pump from the engine.

To install:

9. Remove any old gasket from the mating surfaces. Be sure these surfaces are clean and dry.

10. Apply liquid gasket, part number 08717-004, 08718-0001, 08718-0003 or 08718-0009 evenly to the engine block mating surface of the oil pump.

➡Do not install the parts if more than four minutes have elapsed since applying the liquid gasket. Instead, reapply after removing the previous coating material.

11. Apply liquid gasket to the engine block upper surface contact areas on the oil pump, lower block upper surface contact areas of the oil pump.

12. Install a new O-ring on the oil pump. Set the edge of the oil pump on the edge of the oil pan. Install the oil pump on the engine block.

13. Loosely install the dowel bolts, then tighten the 8mm bolts. Tighten the 6mm bolts and the dowel bolts.

14. Continue the installation in the reverse order of the removal procedure.

15. After assembly, wait at least thirty minutes before filling the engine with clean engine oil.

16. Do not run the engine for at least three hours after installing the oil pan.

17. As required, reprogram the power window control unit as follows.

18. Turn the ignition switch ON. Lower the window, all the way down. Open the driver's side door.

19. Turn the ignition switch OFF. Push and hold the drivers window DOWN switch. Turn the ignition switch ON. Release the driver's window DOWN switch. This must be done within five seconds.

14. Continue the installation in the reverse order of the removal procedure.

15. After assembly, wait at least thirty minutes before filling the engine with clean engine oil.

16. Do not run the engine for at least three hours after installing the oil pan.

17. As required, reprogram the power window control unit as follows.

18. Turn the ignition switch ON. Lower the window, all the way down. Open the driver's side door.

19. Turn the ignition switch OFF. Push and hold the drivers window DOWN switch. Turn the ignition switch ON. Release the driver's window DOWN switch. This must be done within five seconds.

20. Repeat the above step three more times. Wait one second.

21. Confirm that the AUTO UP and AUTO DOWN do not work. If they work repeat the procedure, paying close attention to the five second time limit

22. Move the driver's window all the way down by holding the drivers window DOWN switch to the AUTO DOWN position.

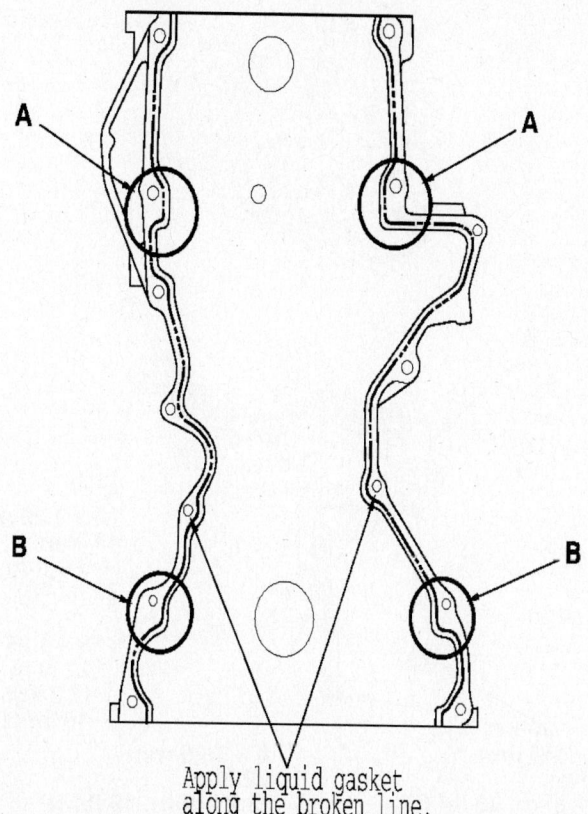

Apply liquid gasket along the broken line.

09474_ACCO_G0021

Apply liquid gasket to the engine block upper surface contact areas (A) on the oil pump, lower block upper surface contact areas (B) of the oil pump—1.8L engine

20. Repeat the above step three more times. Wait one second.

21. Confirm that the AUTO UP and AUTO DOWN do not work. If they work repeat the procedure, paying close attention to the five second time limit

22. Move the driver's window all the way down by holding the drivers window DOWN switch to the AUTO DOWN position.

23. Pull up and hold the driver's window UP switch to the AUTO UP position until the window reaches the fully closed position. Hold the switch for one second.

24. Confirm that the power window master switch has been reset by using the driver's window AUTO UP and DOWN function.

25. If the AUTO UP and DOWN feature is still not working, repeat the complete procedure, paying close attention to the five second time limit.

26. Set the clock.

2.0L ENGINE

1. Before servicing the vehicle, refer to the Precautions Section.

2. Note the radio security code and the radio presets. Disconnect the negative battery cable.

3. Position the number one piston at TDC. Remove the oil pan.

4. Remove and discard the oil pump chain tensioner.

5. To hold the rear balancer shaft, insert a 6mm pin driver into the maintenance hole in the lower balancer shaft holder and through the rear balancer shaft.

6. Loosen the oil pump sprocket mounting bolt.

7. Remove the oil pump sprocket. Remove the oil pump.

To install:

8. Remove any old gasket from the mating surfaces. Be sure these surfaces are clean and dry.

9. Apply clean engine oil to the threads of the oil pump sprocket mounting bolt.

10. Loosely install the oil pump, and then install the oil pump sprocket. Remove the pin driver.

11. Tighten the pump retaining bolts to specification.

12. Squeeze the new oil pump chain tensioner and then install the set clip.

13. Install the oil pump chain tensioner. Remove the set clip from the pump chain tensioner.

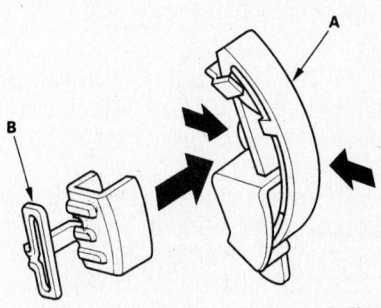

9347FG14

Compress the oil pump chain tensioner (A) and install the retaining clip (B)—2.0L (2006 Civic) engine and 2.4L engine

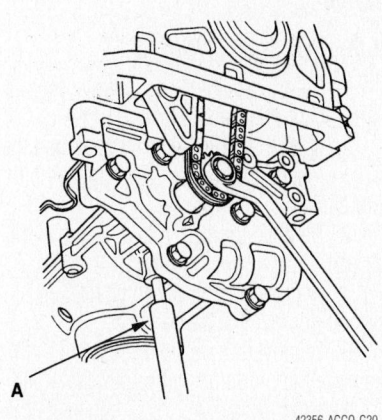

42356-ACCO-G20

Insert a 6mm pin into the maintenance hole in the lower balancer shaft holder and through the rear balancer shaft—2.0L (2006 Civic) engine and 2.4L engine

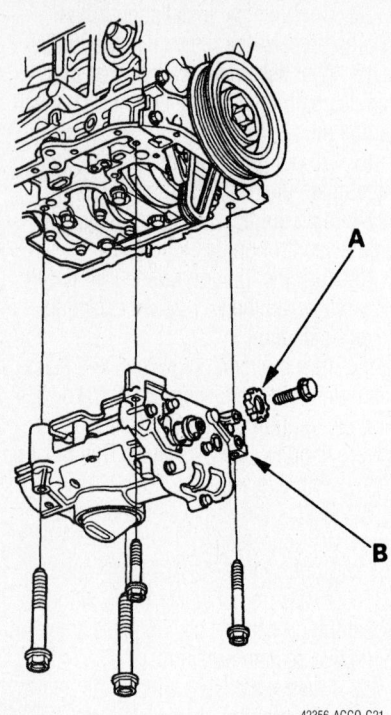

42356-ACCO-G21

Exploded view of the oil pump sprocket (A) and oil pump (B)—2.0L (2006 Civic) engine and 2.4L engine

23. Pull up and hold the driver's window UP switch to the AUTO UP position until the window reaches the fully closed position. Hold the switch for one second.

24. Confirm that the power window master switch has been reset by using the driver's window AUTO UP and DOWN function.

25. If the AUTO UP and DOWN feature is still not working, repeat the complete procedure, paying close attention to the five second time limit.

26. Set the clock.

S2000

2.0L AND 2.2L ENGINES

1. Before servicing the vehicle, refer to the Precautions Section.

2. Note the radio security code and the radio presets. Disconnect the negative battery cable.

3. Remove the cam chain. Remove the oil pan.

4. Remove the oil pump chain tensioner. Remove the baffle plate.

5. Remove the oil pump retaining bolts. Remove the oil pump, oil pump chain and crankshaft sprocket.

To install:

6. Remove any old gasket from the mating surfaces. Be sure these surfaces are clean and dry.

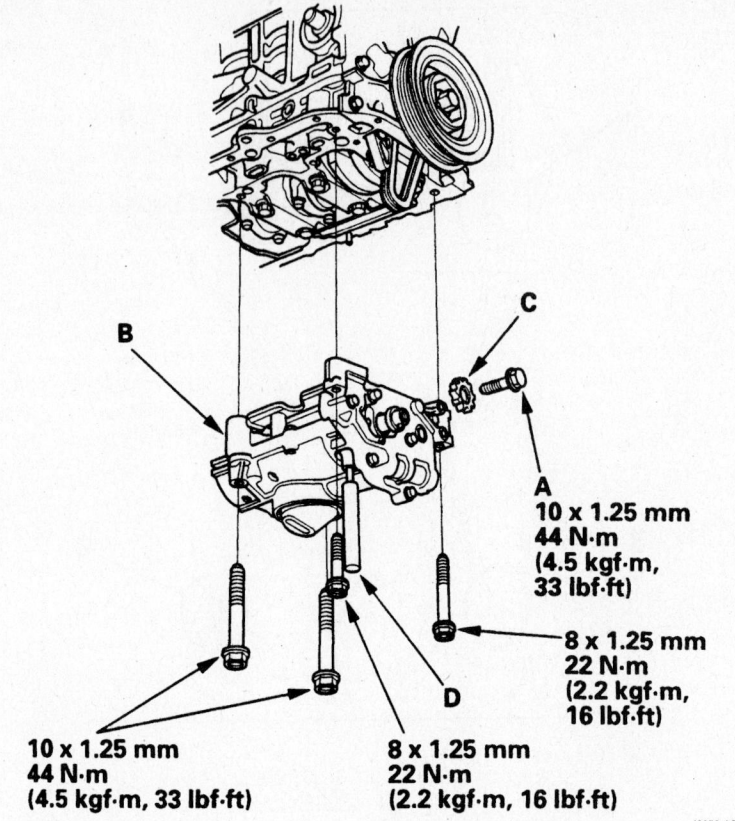

A
10 x 1.25 mm
44 N·m
(4.5 kgf·m,
33 lbf·ft)

8 x 1.25 mm
22 N·m
(2.2 kgf·m,
16 lbf·ft)

10 x 1.25 mm
44 N·m
(4.5 kgf·m, 33 lbf·ft)

8 x 1.25 mm
22 N·m
(2.2 kgf·m, 16 lbf·ft)

42356-ACCO-G22

Oil pump tightening specifications—2.0L (2006 Civic) engine and 2.4L engine

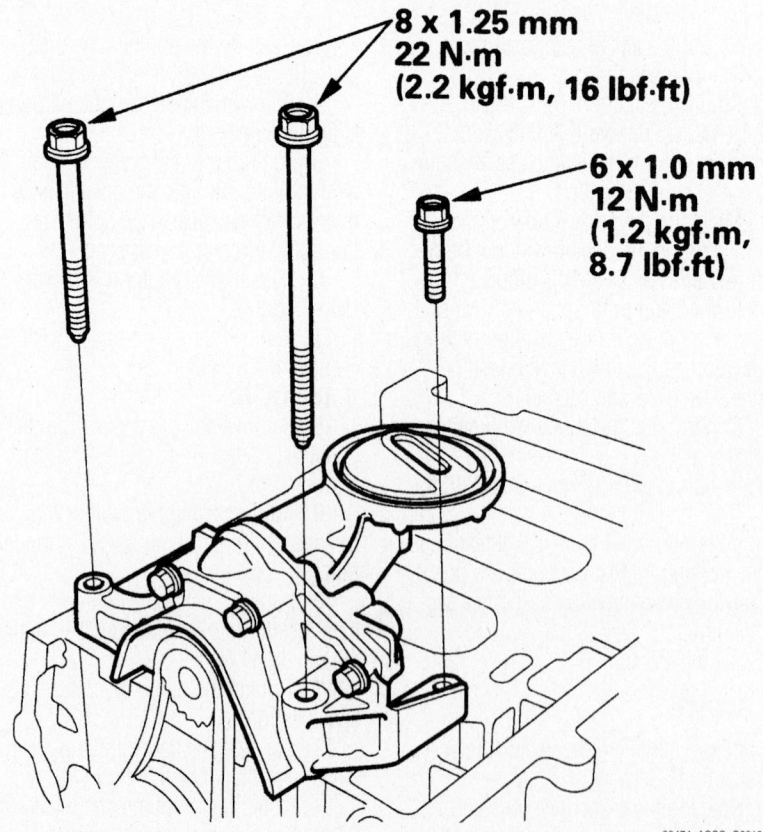

8 x 1.25 mm
22 N·m
(2.2 kgf·m, 16 lbf·ft)

6 x 1.0 mm
12 N·m
(1.2 kgf·m,
8.7 lbf·ft)

09474_ACCO_G0018

Oil pump mounting—S2000 with 2.0L and 2.2L engines

7. Squeeze the new oil pump chain tensioner and then install the set clip.

8. Install the crankshaft sprocket, oil pump chain and oil pump. Install the baffle plate.

9. Set the crankshaft sprocket so that the number one piston is at TDC. Align the key on the sprocket and crankshaft with the pointer on the engine block.

10. Move the cam chain so that the colored piece aligns with the punched mark on the crankshaft.

11. Install the oil pump chain guide and the oil pump chain tensioner with the seat clip.

12. Remove the set clip from the oil pump chain tensioner.

13. Continue the installation in the reverse order of the removal procedure.

14. After assembly, wait at least thirty minutes before filling the engine with clean engine oil.

15. Do not run the engine for at least three hours after installing the oil pan.

16. As required, reprogram the power window control unit as follows.

17. Turn the ignition switch ON. Lower the window, all the way down. Open the driver's side door.

18. Turn the ignition switch OFF. Push and hold the drivers window DOWN switch. Turn the ignition switch ON. Release the driver's window DOWN switch. This must be done within five seconds.

19. Repeat the above step three more times. Wait one second.

20. Confirm that the AUTO UP and AUTO DOWN do not work. If they work repeat the procedure, paying close attention to the five second time limit

21. Move the driver's window all the way down by holding the drivers window DOWN switch to the AUTO DOWN position.

22. Pull up and hold the driver's window UP switch to the AUTO UP position until the window reaches the fully closed position. Hold the switch for one second.

23. Confirm that the power window master switch has been reset by using the driver's window AUTO UP and DOWN function.

24. If the AUTO UP and DOWN feature is still not working, repeat the complete procedure, paying close attention to the five second time limit.

25. Set the clock.

Accord

2.3L ENGINE

1. Before servicing the vehicle, refer to the Precautions Section.

2. Disconnect the negative battery cable.

3. Drain the engine oil into a sealable container.

4. Turn the engine to align the timing marks and set cylinder No.1 to TDC. The white mark on the crankshaft pulley should align with the pointer on the timing belt cover.

5. Remove or disconnect the following:
    *   Valve cover and upper timing belt cover
    *   Power steering pump belt and the alternator belt, also the air conditioning belt if so equipped
    *   Crankshaft pulley and the lower timing belt cover
    *   Balancer belt and the timing belt. Be sure to mark the rotation of the timing belt if it is going to be reused
    *   Timing belt and balancer belt tensioners
    *   Crankshaft Position (CKP) sensor, if equipped
    *   Timing belt drive pulley and key from the crankshaft
    *   Balancer driven pulley, by inserting a suitable tool into the maintenance hole in the front balancer shaft.

6. Align the rear timing balancer pulley using a 6 x 100mm bolt or rod. Mark the bolt or rod at a point 2.9 in. (74mm) from the end. Remove the bolt from the maintenance hole on the side of the block; insert the bolt/rod into the hole. Align the 74mm mark with the face of the hole. This pin will hold the shaft in place.

7. Remove or disconnect the following:
    *   Balancer gear case and the dowel pins. Discard the O-ring.
    *   Balancer driven gear attaching bolt and the balancer driven gear
    *   Oil pan and the oil screen. Discard the screen gasket.
    *   Oil pump mounting bolts and oil pump assembly
    *   Dowel pins from the engine and clean the oil pump mating surfaces of old gasket material and oil. Discard the O-rings.

To install:

8. Install the 2 dowel pins and new O-rings to the cylinder block.

9. Be sure that the mating surfaces are clean and dry. Apply a liquid gasket evenly in a narrow bead, centered on the mating surface. Once the sealant is applied, do not wait longer than 20 minutes to install the parts; the sealant will become ineffective. After final assembly, wait at least 30 minutes before adding oil to the engine, giving the sealant time to set. To prevent leakage of

oil, apply a suitable thread sealer to the inner threads of the bolt holes.

10. Install or connect the following:
    *   Oil pump to the engine block. Tighten the mounting bolts to 108 inch lbs. (12 Nm).
    *   Oil screen. Tighten the screen mounting bolts and nuts to 108 inch lbs. (12 Nm).
    *   Oil pan
    *   Balancer driven pulley to the front balancer belt, hold the balancer shaft in place with a suitable tool. Tighten the attaching bolt to 22 ft. lbs. (29 Nm).
    *   Balancer driven gear to the rear balancer shaft. Tighten the bolt to 18 ft. lbs. (25 Nm).

➡ **Before installing the balancer driven gear and the gear case, apply molybdenum disulfide (lithium grease) to the thrust surfaces of the balancer gears.**

11. Align the groove on the pulley edge to the pointer on the balancer gear case.

12. Install or connect the following:
    *   Balancer gear case to the engine and the mounting bolts and nut. The rear balancer shaft is being held in place with a 6 x 100mm bolt. Tighten the mounting bolts and nut to 18 ft. lbs. (25 Nm).

13. Check the alignment of the pointer on the balancer pulley to the pointer on the oil pump.

14. Install or connect the following:
    *   Drive pulley to the crankshaft
    *   CKP sensor. Tighten the mounting bolts to 108 inch lbs. (12 Nm).
    *   Timing belt tensioners
    *   Timing belt and the balancer belt
    *   Crankshaft pulley and the lower timing belt cover
    *   Drive belts for the alternator, power steering, and air conditioning compressor. Tension the belts properly.
    *   Valve cover and upper timing belt cover

15. Refill the engine with clean, fresh oil.

16. Connect the negative battery cable and enter the radio security code.

2.4L ENGINE

1. Before servicing the vehicle, refer to the Precautions Section.

2. Note the radio security code and the radio presets. Disconnect the negative battery cable.

3. Position the number one piston at TDC. Remove the oil pan.

4. Remove and discard the oil pump chain tensioner.

5. To hold the rear balancer shaft, insert a 6mm pin driver into the maintenance hole in the lower balancer shaft holder and through the rear balancer shaft.

6. Loosen the oil pump sprocket mounting bolt.

7. Remove the oil pump sprocket. Remove the oil pump.

To install:

8. Remove any old gasket from the mating surfaces. Be sure these surfaces are clean and dry.

9. Apply clean engine oil to the threads of the oil pump sprocket mounting bolt.

10. loosely install the oil pump, and then install the oil pump sprocket. Remove the pin driver.

11. Tighten the pump retaining bolts to specification.

12. Squeeze the new oil pump chain tensioner and then install the set clip.

13. Install the oil pump chain tensioner. Remove the set clip from the pump chain tensioner.

14. Continue the installation in the reverse order of the removal procedure.

15. After assembly, wait at least thirty minutes before filling the engine with clean engine oil.

16. Do not run the engine for at least three hours after installing the oil pan.

17. As required, reprogram the power window control unit as follows.

18. Turn the ignition switch ON. Lower the window, all the way down. Open the driver's side door.

19. Turn the ignition switch OFF. Push and hold the drivers window DOWN switch. Turn the ignition switch ON. Release the driver's window DOWN switch. This must be done within five seconds.

20. Repeat the above step three more times. Wait one second.

21. Confirm that the AUTO UP and AUTO DOWN do not work. If they work repeat the procedure, paying close attention to the five second time limit

22. Move the driver's window all the way down by holding the drivers window DOWN switch to the AUTO DOWN position.

23. Pull up and hold the driver's window UP switch to the AUTO UP position until the window reaches the fully closed position. Hold the switch for one second.

24. Confirm that the power window master switch has been reset by using the driver's window AUTO UP and DOWN function.

25. If the AUTO UP and DOWN feature is still not working, repeat the complete procedure, paying close attention to the five second time limit.

26. Set the clock.

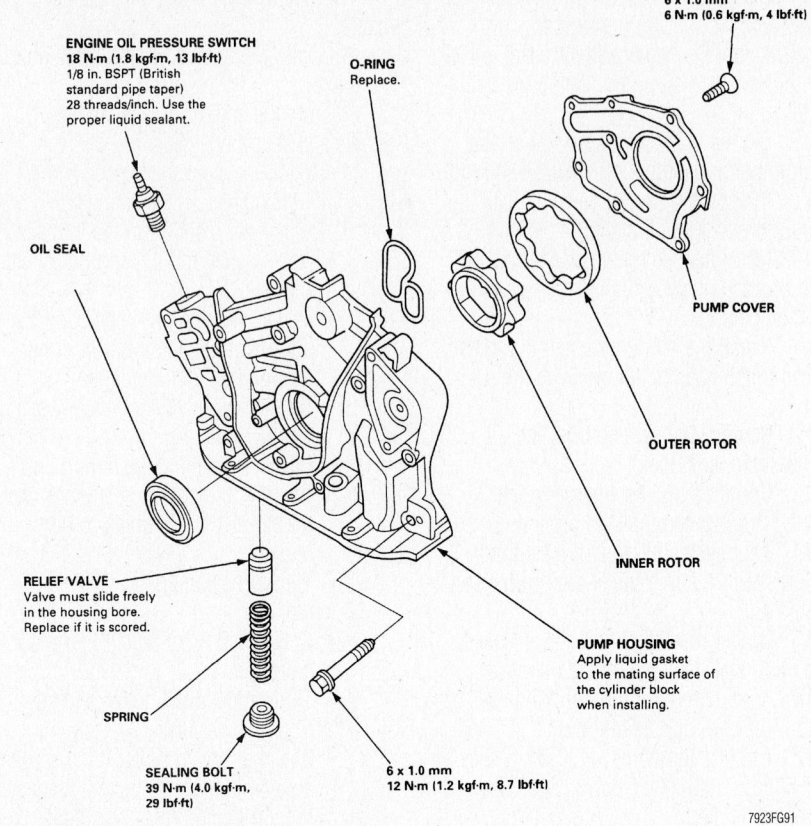

ENGINE OIL PRESSURE SWITCH
18 N·m (1.8 kgf·m, 13 lbf·ft)
1/8 in. BSPT (British standard pipe taper) 28 threads/inch. Use the proper liquid sealant.

O-RING
Replace.

6 x 1.0 mm
6 N·m (0.6 kgf·m, 4 lbf·ft)

OIL SEAL

PUMP COVER

OUTER ROTOR

INNER ROTOR

RELIEF VALVE
Valve must slide freely in the housing bore. Replace if it is scored.

SPRING

PUMP HOUSING
Apply liquid gasket to the mating surface of the cylinder block when installing.

SEALING BOLT
39 N·m (4.0 kgf·m, 29 lbf·ft)

6 x 1.0 mm
12 N·m (1.2 kgf·m, 8.7 lbf·ft)

7923FG91

Exploded view of the oil pump—3.0L engine

3.0L ENGINE

1. Before servicing the vehicle, refer to the Precautions Section.

2. Note the radio security code and the radio presets. Disconnect the negative battery cable.

3. Raise and support the vehicle safely. Drain the engine oil.

4. Remove the bulkhead cover. Lower the vehicle.

5. Remove the windshield wiper arms. Remove the cowl cover.

6. Remove the connector bracket from the front of the cylinder head. Remove the harness clamp bracket from the rear of the cylinder head. Attach engine support tool, VSB02C0000019. Lift and support the engine.

7. Remove the timing belt. Remove the crankshaft position sensor (CKP). Remove the rocker arm oil control solenoid and oil filter assembly.

8. Raise and support the vehicle safely. Remove the oil pan.

9. Remove the oil pump screen. Remove the oil pump retaining bolts. Remove the oil pump from the engine.

To install:

10. Remove any old gasket from the mating surfaces. Be sure these surfaces are clean and dry.

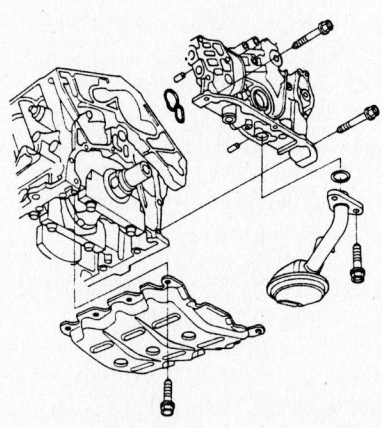

7923FG92

Oil pump mounting—3.0L engine

11. Apply liquid gasket, part number 08717-004, 08718-0001, 08718-0003 or 08718-0009 evenly to the block mating surface of the oil pump.

➡**Do not install the parts if more than four minutes have elapsed since applying the liquid gasket. Instead, reapply after removing the previous coating material.**

12. Grease the lip of the oil seal and apply clean oil to the new O-rings.

13. Install the dowel pins then align the

inner rotor with the crankshaft. Install the oil pump.

14. Continue the installation in the reverse order of the removal procedure.

15. After assembly, wait at least thirty minutes before filling the engine with clean engine oil.

16. Do not run the engine for at least three hours after installing the oil pan.

17. As required, reprogram the power window control unit as follows.

18. Turn the ignition switch ON. Lower the window, all the way down. Open the driver's side door.

19. Turn the ignition switch OFF. Push and hold the drivers window DOWN switch. Turn the ignition switch ON. Release the driver's window DOWN switch. This must be done within five seconds.

20. Repeat the above step three more times. Wait one second.

21. Confirm that the AUTO UP and AUTO DOWN do not work. If they work repeat the procedure, paying close attention to the five second time limit

22. Move the driver's window all the way down by holding the drivers window DOWN switch to the AUTO DOWN position.

23. Pull up and hold the driver's window UP switch to the AUTO UP position until the window reaches the fully closed position. Hold the switch for one second.

24. Confirm that the power window master switch has been reset by using the driver's window AUTO UP and DOWN function.

25. If the AUTO UP and DOWN feature is still not working, repeat the complete procedure, paying close attention to the five second time limit.

26. Set the clock.

Rear Main Seal

REMOVAL & INSTALLATION

1. Before servicing the vehicle, refer to the Precautions Section.

2. Remove the transaxle from the vehicle.

3. Remove the driveplate from the crankshaft.

4. Carefully pry the crankshaft seal out of the retainer.

To install:

5. Apply clean engine oil to the lip of the new seal.

6. Install the seal onto the crankshaft and into the retainer using the appropriate seal driver.

7. Install the driveplate and the transmission.

Timing Belt, Sprockets, Front Cover and Seal

REMOVAL & INSTALLATION

1.7L Engine

1. Before servicing the vehicle, refer to the Precautions Section.

2. Note the radio security code and the radio presets. Disconnect the negative battery cable.

3. Rotate the crankshaft to set the engine at Top Dead Center (TDC) on the compression stroke for the No. 1 piston. The white mark on the crankshaft pulley should align with the pointers on the timing cover. Once the engine is in this position, it must not be disturbed.

4. Raise and support the vehicle safely. Remove the front tires. Remove the splash shield. Lower the vehicle.

5. Remove the power steering pump and position it to the side without disconnecting the hoses. Remove the alternator.

6. Remove the ignition coil cover. Remove the four ignition coils. Remove the throttle cable clamps and harness holder mounting bolts. Remove the cylinder head cover. Remove the crankshaft pulley.

7. Properly support the engine using a block of wood under the oil pan. Remove the upper bracket.

8. Remove the grommet from the upper cover. Disconnect the camshaft position (CMP) top dead center (TDC) sensor connector.

9. Remove the side engine mount bracket.

10. Remove the crankshaft position (CKP) sensor from the oil pump, without disconnecting the connector.

11. Apply slight force to the timing belt at the midpoint between the camshaft pulley and the water pump pulley and check that the auto tensioner moves smoothly, if not replace it.

12. Move the auto tensioner to remove tension from the timing belt. Remove the timing belt.

To install:

13. Set the crankshaft to TDC. Align the TDC mark on the timing belt drive pulley with the pointer on the oil pump.

14. Set the camshaft to TDC. The UP mark on the camshaft pulley should be at the top. Align the TDC marks on the camshaft pulley with the top edge of the cylinder head.

15. Install the timing belt in a counter-

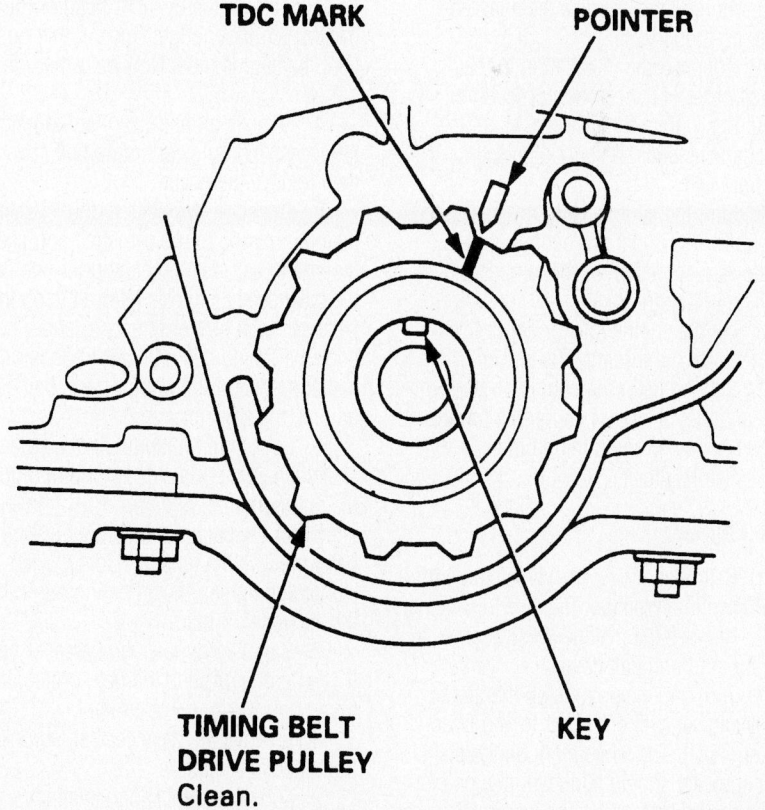

TDC alignment mark locations for the crankshaft sprocket—1.7L engine

79235G22

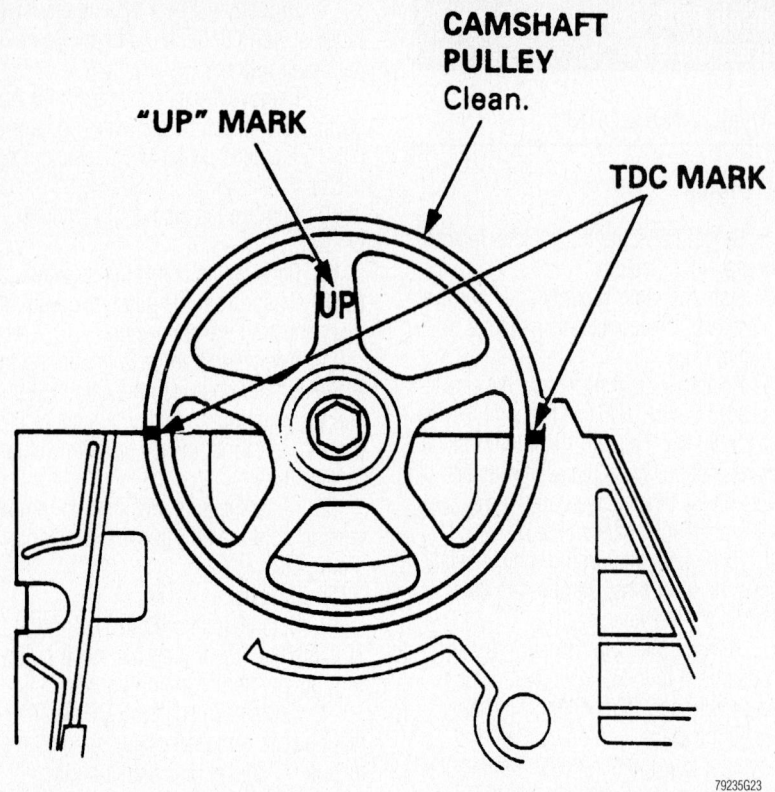

CAMSHAFT PULLEY
Clean.

"UP" MARK

TDC MARK

79235G23

Timing belt sprocket TDC mark positioning for timing belt installation—1.7L engine

clockwise sequence, starting with the drive pulley.

16. As required, replace the auto tensioner assembly.

17. Continue the installation in the reverse order of the removal procedure.

18. Reprogram the ECM/PCM engine idle characteristics. Be sure all electrical items are OFF.

19. Start the engine. Hold the idle speed at 3000 RPM's in park or neutral until the radiator fan comes on or the temperature reached 194 degrees.

20. Let the engine idle for about five minutes with the throttle fully closed.

21. If the radiator fan comes on during the five minutes, do not count this toward the five minute programming time.

22. Set the clock.

2.3L Engine

1. Before servicing the vehicle, refer to the Precautions Section.

2. Remove the cylinder head (valve) and upper timing belt covers.

3. Turn the engine to align the timing marks and set cylinder No. 1 to Top Dead Center (TDC). The white mark on the crankshaft sprocket should align with the pointer on the timing belt cover. The words **UP** embossed on the camshaft sprocket should

be aligned in the upward position. The marks on the edge of the sprocket should be aligned with the cylinder head or the back cover upper edge. Once in this position, the engine must NOT be turned or disturbed.

4. Remove all necessary components for access to the lower timing belt cover, then remove the cover.

5. There are two belts in this system; the one running to the camshaft sprocket is the timing belt. The other, shorter one drives the balance shaft and is referred to as the balancer shaft belt or timing balancer belt. Lock the timing belt adjuster in position, by installing one of the lower timing belt cover bolts to the adjuster arm.

6. Loosen the timing belt and balancer shafts tensioner adjuster nut, do not loosen the nut more than one turn. Push the tensioner for the balancer belt away from the belt to relieve the tension. Hold the tensioner and tighten the adjusting nut to hold the tensioner in place.

7. Carefully remove the balancer belt. Do not crimp or bend the belt; protect it from contact with oil or coolant.

8. Remove the balancer belt sprocket from the crankshaft.

9. Loosen the lockbolt installed to the timing belt adjuster and loosen the adjusting nut. Push the timing belt adjuster to

remove the tension on the timing belt, then tighten the adjuster nut.

10. Remove the timing belt by sliding it off the sprockets. Do not crimp or bend the belt; protect it from contact with oil or coolant.

11. If defective, remove the belt tensioners by performing the following:

a. Remove the springs from the balancer belt and the timing belt tensioners.

b. Remove the adjusting nut from the belt tensioners.

c. Remove the bolt from the balancer belt adjuster lever, and then remove the lever and the tensioner pulley.

d. Remove the lockbolt from the timing belt tensioner lever, then remove the tensioner pulley and lever from the engine.

12. This is an excellent time to check or replace the water pump. Even if the timing belt is only being replaced as part of a good maintenance schedule, consider replacing the pump at the same time.

To install:

13. If the water pump is to be replaced, install a new O-ring and make certain it is properly seated. Install the water pump and tighten the mounting bolts to 106 inch lbs. (12 Nm).

14. If the tensioners were removed, perform the following procedures:

a. Install the timing belt tensioner lever and the tensioner pulley.

b. Install the balancer belt pulley and adjuster lever.

c. Install the adjusting nut and the bolt to the balancer belt adjuster lever.

d. Install the springs to the tensioners.

e. Install the lockbolt to the timing belt tensioner, then move it its full deflection and tighten the lockbolt.

f. Move the balancer it's full deflection and tighten the adjusting nut to hold its position.

15. The pointer on the crankshaft sprocket should be aligned with the pointer on the oil pump; the camshaft sprocket must be aligned so that the word **UP** is at the top of the sprocket and the marks on the edge of the sprocket are aligned with the surfaces of the head or the back cover upper edge.

16. Install the timing belt on the sprockets in the following sequence: crankshaft sprocket, tensioner sprocket, water pump sprocket and camshaft sprocket.

17. Check the timing marks to be sure that they did not move.

18. Loosen, then retighten the timing belt adjusting nut; this will apply the proper amount of tension to the timing belt.

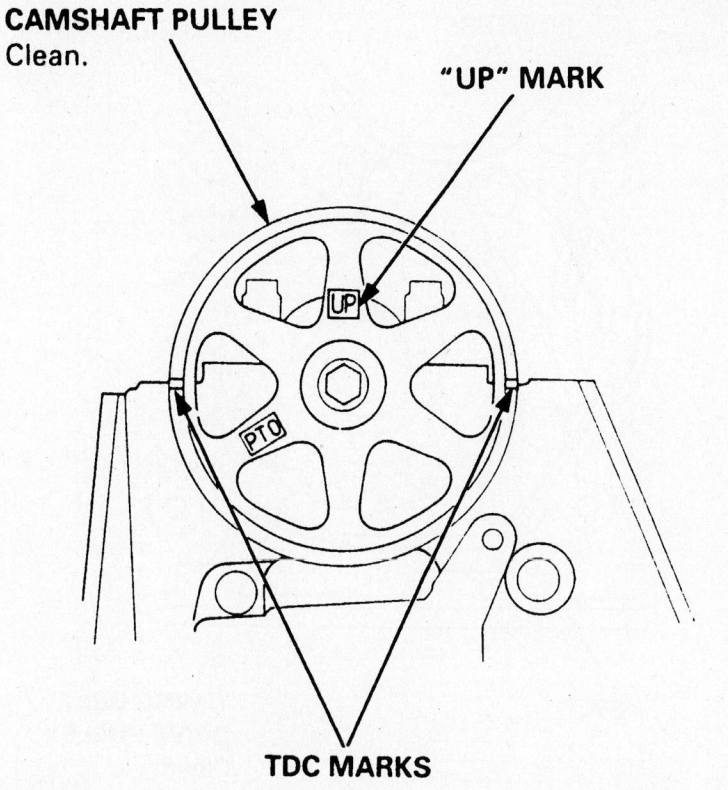

CAMSHAFT PULLEY
Clean.

"UP" MARK

TDC MARKS

79235G28

Position the camshaft sprocket as indicated for timing belt installation—2.3L engine

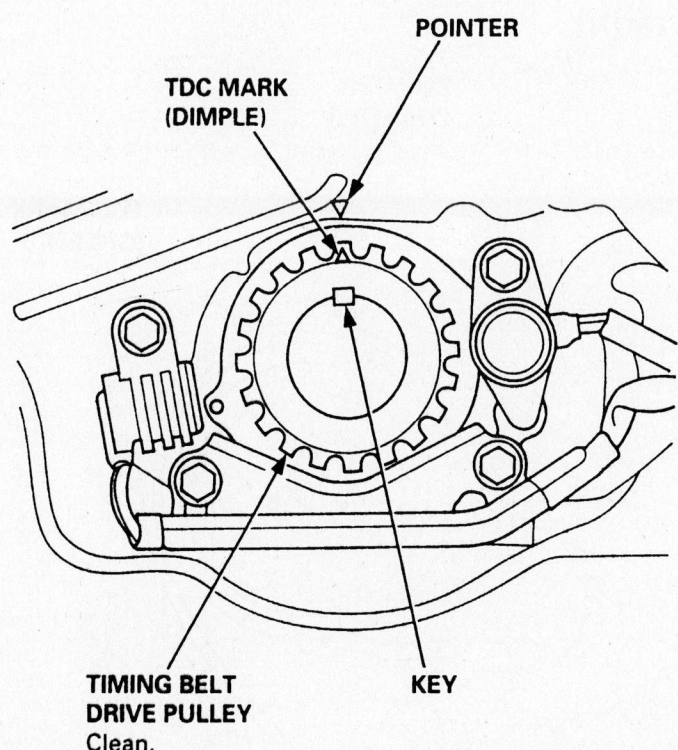

POINTER

TDC MARK
(DIMPLE)

TIMING BELT
DRIVE PULLEY
Clean.

KEY

79235G27

Before installing the timing belt, ensure the crankshaft sprocket marks are properly aligned—2.3L engine

19. Install the timing balancer belt drive sprocket and the lower timing belt cover.

20. Install the crankshaft pulley and bolt, tighten the bolt to 181 ft. lbs. (245 Nm). Rotate the crankshaft sprocket 5–6 turns to position the timing belt on the sprockets.

21. Set the No. 1 cylinder to TDC and loosen the timing belt adjusting nut one turn. Turn the crankshaft counterclockwise until the cam sprocket has moved 3 teeth; this creates tension on the timing belt.

22. Tighten the timing belt adjusting nut.

23. Set the crankshaft sprocket and the camshaft sprocket to TDC. If the sprockets do not align, remove the belt to realign the marks, then install the belt.

24. Remove the crankshaft pulley and the lower cover.

25. With the timing marks aligned, lock the timing belt adjuster in place with one of the lower cover mounting bolts.

26. Loosen the adjusting nut and ensure the timing balancer belt adjuster moves freely.

27. Align the rear timing balancer sprocket using a 6 x 100mm bolt or rod. Mark the bolt or rod at a point 2.9 in. (74mm) from the end. Remove the bolt from the maintenance hole on the side of the block; insert the bolt/rod into the hole and align the 2.9 in. (74mm) mark with the face of the hole. This will hold the shaft in place during installation.

28. Align the groove on the front balancer shaft sprocket with the pointer on the oil pump.

29. Install the balancer belt. Once the belts are in place, be sure that all the engine alignment marks are still correct. If not, remove the belts, realign the engine and reinstall the belts. Once the belts are properly installed, slowly loosen the adjusting nut, allowing the tensioner to move against the belt. Remove the bolt from the maintenance hole and reinstall the bolt and washer.

30. Install the crankshaft pulley, then turn the crankshaft sprocket 1 turn counterclockwise and tighten the timing belt adjusting nut to 33 ft. lbs. (45 Nm).

31. Remove the crankshaft pulley and the bolt locking the timing belt adjuster in place.

32. Install the lower and upper timing belt covers, and all applicable components. When installing the crankshaft pulley, coat the threads and seating face of the pulley bolt with engine oil, then install and tighten the bolt to 181 ft. lbs. (250 Nm).

33. Install the cylinder head cover gasket cover to the groove of the cylinder head cover. Before installing the gasket thor-

oughly clean the seal and the groove. Seat the recesses for the camshaft first, then work it into the groove around the outside edges. Be sure the gasket is seated securely in the corners of the recesses.

34. Apply liquid gasket to the four corners of the recesses of the cylinder head cover gasket. Do not install the parts if 5 minutes or more have elapsed since applying liquid gasket. After assembly, wait at least 20 minutes before filling the engine with oil.

35. Install the cylinder head (valve) cover and all other applicable components.

3.0L Engine

1. Before servicing the vehicle, refer to the Precautions Section.

2. Note the radio security code and the radio presets. Disconnect the negative battery cable.

3. Turn the crankshaft so that its white mark lines up with the pointer. Check to insure that number one piston is at TDC. Be sure that the mark on the front camshaft pulley and the pointer on the front upper cover are aligned.

4. Raise and support the vehicle safely. Remove the front tires. Remove the splash shield.

5. Remove the drive belt. Remove the drive belt auto tensioner.

6. Properly support the engine using a block of wood under the oil pan. Remove the ground cable. Remove the side engine mount bracket.

7. Remove the front upper cover. Remove the rear upper cover.

8. Remove the crankshaft pulley. Remove the lower cover.

9. Using the proper size bolt, screw it into the timing belt adjuster. Tighten by hand, do not use a wrench.

10. Remove the engine mount bracket.

11. Remove the idler pulley bolt and idler pulley. Remove the timing belt.

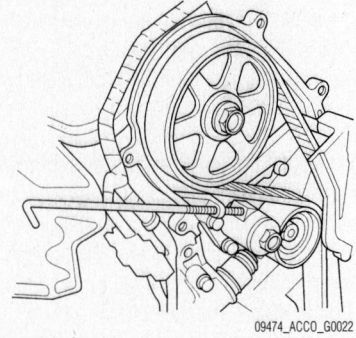

09474_ACCO_G0022

Timing belt adjuster bolt and installation location—3.0L engine

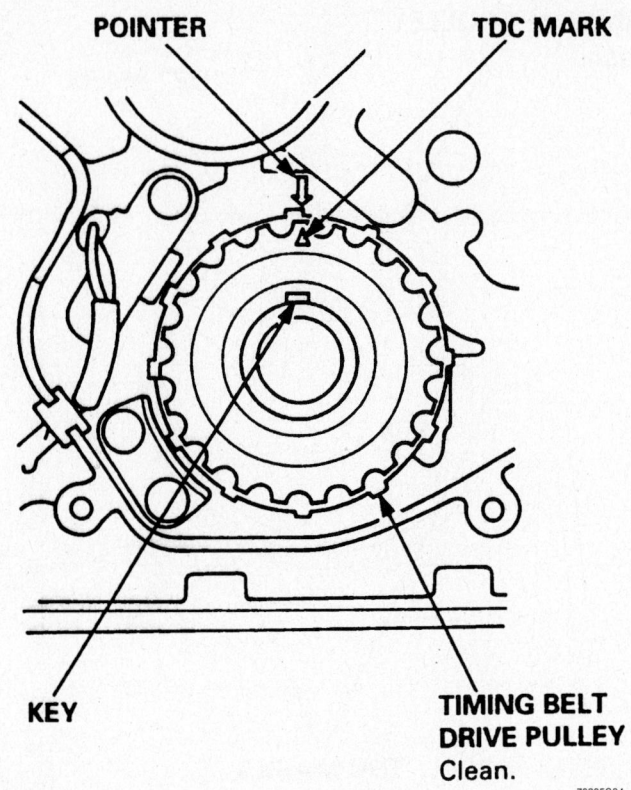

79235G24

Crankshaft timing belt sprocket alignment mark locations—3.0L engine

FRONT:

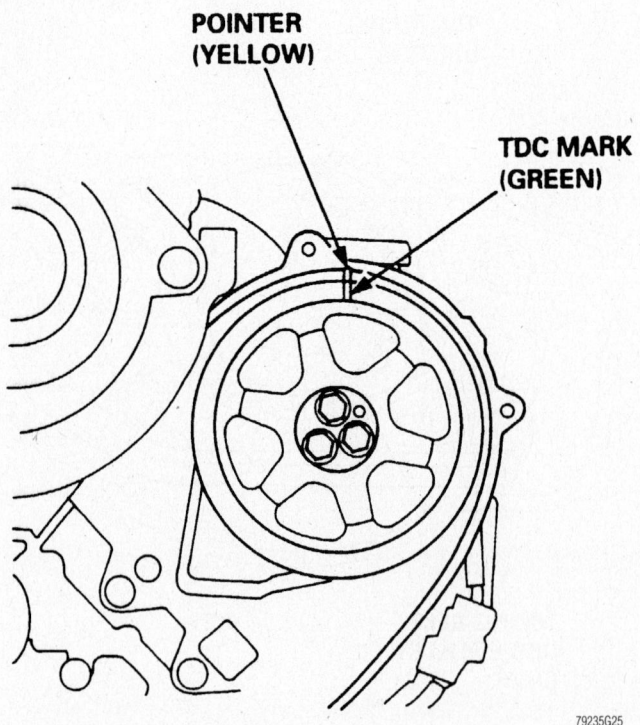

79235G25

Left camshaft timing belt sprocket alignment mark location—3.0L engine

REAR:

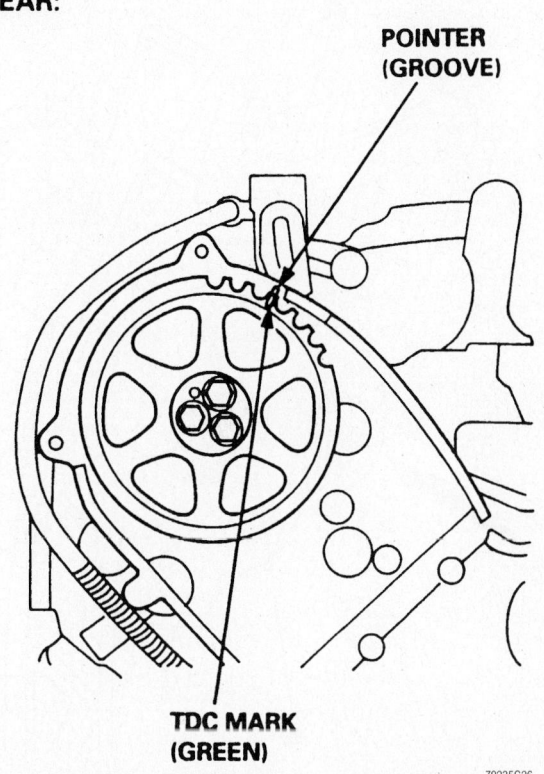

POINTER
(GROOVE)

TDC MARK
(GREEN)

79235G26

Rear camshaft timing belt sprocket alignment mark location—3.0L engine

To install:

12. Set the timing belt drive pulley to TDC by aligning the TDC mark on the tooth of the timing belt drive pulley with the pointer on the oil pump.

13. Set the camshaft pulleys to TDC by aligning the TDC marks on the camshaft pulleys with the pointers on the back covers.

14. Remove the auto tensioner. Align the holes on the rod and housing of the tensioner.

15. Using a hydraulic press to slowly compress the auto tensioner, insert a 0.08 inch pin through the housing and the rod.

16. Install the auto tensioner. Be sure that the pin stays in place.

17. By hand, screw down on the timing belt adjuster bolt to hold the timing belt adjuster.

18. Install the timing belt in a counter-clockwise sequence starting with the drive pulley.

19. Remove the pin from the auto tensioner. Remove the bolt from the timing belt adjuster.

20. Continue the installation in the reverse order of the removal procedure.

21. Reprogram the crankshaft position (CKP) pattern. Run the engine until the operating temperature reaches 176 degree. With the engine stopped clear the CKP pattern. Turn the ignition switch OFF. Turn the ignition switch ON and wait thirty seconds.

22. Roadtest the vehicle on a level surface. Decelerate the engine speed of 2500 rpm to 1000 rpm. If equipped with automatic transaxle use two Drive positions. If equipped with manual transaxle use first gear.

23. Stop the vehicle, but keep the engine running.

24. Check PULSAR F/B LEARN in the data list with the HDS. If not complete repeat the procedure. If complete, roadtest the vehicle on a level surface. Decelerate the engine speed of 5000 rpm to 3000 rpm. If equipped with automatic transaxle use two Drive positions. If equipped with manual transaxle use first gear.

25. Stop the vehicle, but keep the engine running.

26. Check PULSAR F/B LEARN in the data list with the HDS. If not complete repeat the procedure.

27. If completed, turn the ignition switch OFF. Turn the ignition switch ON, wait thirty seconds. The learning procedure is now complete.

28. As required, reprogram the power window control unit as follows.

29. Turn the ignition switch ON. Lower the window, all the way down. Open the driver's side door.

30. Turn the ignition switch OFF. Push and hold the drivers window DOWN switch. Turn the ignition switch ON. Release the driver's window DOWN switch. This must be done within five seconds.

31. Repeat the above step three more times. Wait one second.

32. Confirm that the AUTO UP and AUTO DOWN do not work. If they work repeat the procedure, paying close attention to the five second time limit

33. Move the driver's window all the way down by holding the drivers window DOWN switch to the AUTO DOWN position.

34. Pull up and hold the driver's window UP switch to the AUTO UP position until the window reaches the fully closed position. Hold the switch for one second.

35. Confirm that the power window master switch has been reset by using the driver's window AUTO UP and DOWN function.

36. If the AUTO UP and DOWN feature is still not working, repeat the complete procedure, paying close attention to the five second time limit.

37. Set the clock.

Timing Chain, Sprockets, Front Cover and Seal

REMOVAL & INSTALLATION

Civic

1.8L ENGINE

1. Before servicing the vehicle, refer to the Precautions Section.

2. Note the radio security code and the radio presets. Disconnect the negative battery cable.

3. Rotate the crankshaft to set the engine at Top Dead Center (TDC) on the compression stroke for the No. 1 piston. The UP mark on the camshaft sprocket should be at the top, and the TDC grooves on the camshaft sprocket should line up with the top edge of the cylinder head.

4. Raise and support the vehicle safely. Remove the front tires. Remove the splash shield. Lower the vehicle.

5. Remove the drive belt auto tensioner. Remove the cylinder head cover. Remove the PCV valve hose. Remove the crankshaft pulley.

6. Properly support the engine using a block of wood under the oil pan. Remove the bolt securing the air conditioning line

and remove the upper torque rod. Remove the ground cable. Remove the engine mount bracket.

7. Remove the oil pump.

8. Loosely install the crankshaft pulley. Turn the crankshaft counterclockwise to compress the auto tensioner. Align the holes on the lock and the auto tensioner. Insert a 0.04 inch diameter pin into the holes. Turn the crankshaft clockwise to secure the pin.

9. Remove the auto tensioner. Remove the crankshaft pulley.

10. Remove the cam chain guide and cam chain tensioner arm. Remove the cam chain from the engine.

To install:

11. Set the crankshaft to TDC. Align the TDC mark on the crankshaft sprocket with the pointer on the engine block.

12. Set the camshaft to TDC. The UP mark on the camshaft sprocket should be at the top and the TDC grooves on the camshaft sprocket should line up with the top edge of the cylinder head.

13. Install the cam chain on the crankshaft sprocket with the colored piece aligned with the mark on the crankshaft sprocket.

14. Install the cam chain on the camshaft sprocket with the colored link plate aligned with the mark on the crankshaft sprocket.

15. Install the cam chain guide and tensioner arm. Install the auto tensioner. Remove the pin from the auto tensioner.

16. Install the oil pump.

17. Continue the installation in the reverse order of the removal procedure.

18. Reprogram the crankshaft position (CKP) pattern. Run the engine until the operating temperature reaches 176 degree. With the engine stopped clear the CKP pattern. Turn the ignition switch OFF. Turn the ignition switch ON and wait thirty seconds.

19. Roadtest the vehicle on a level surface. Decelerate the engine speed of 2500 rpm to 1000 rpm. If equipped with automatic transaxle use two Drive positions. If equipped with manual transaxle use first gear.

20. Stop the vehicle, but keep the engine running.

21. Check PULSAR F/B LEARN in the data list with the HDS. If not complete repeat the procedure. If complete, roadtest the vehicle on a level surface. Decelerate the engine speed of 5000 rpm to 3000 rpm. If equipped with automatic transaxle use two Drive positions. If equipped with manual transaxle use first gear.

22. Stop the vehicle, but keep the engine running.

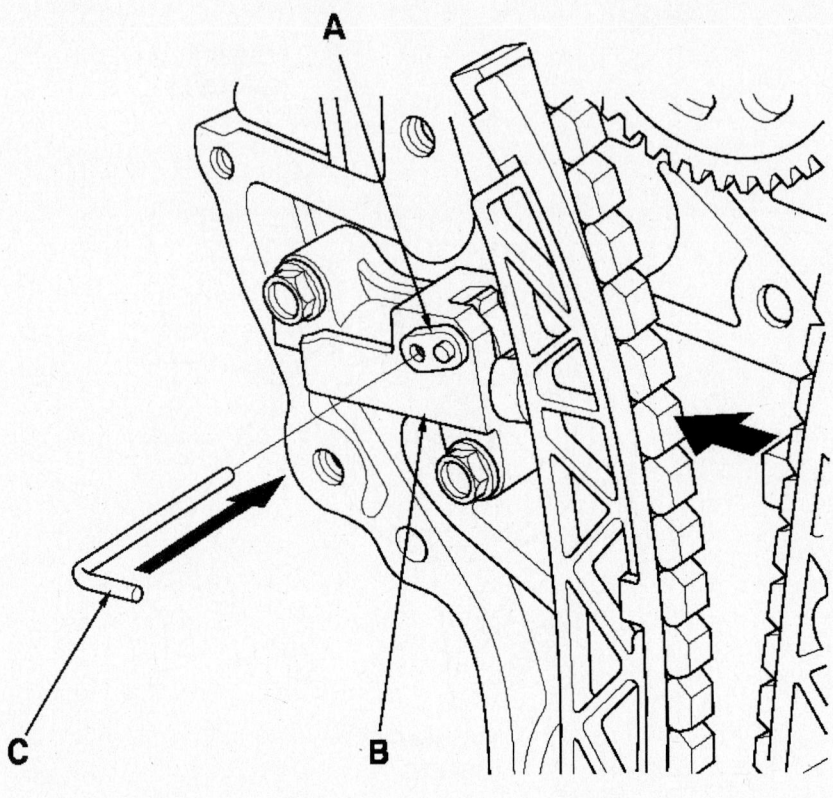

09474_ACCO_G0024

Locking auto tensioner: (A) alignment hole, (B) tensioner and (C) lock pin—1.8L engine

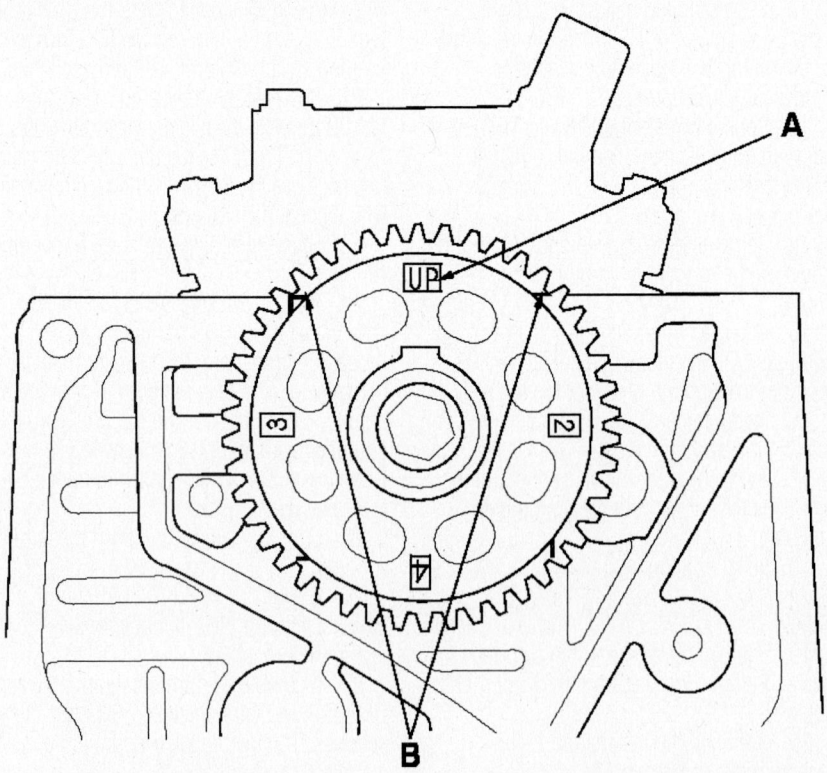

09474_ACCO_G0025

Camshaft TDC location—1.8L engine

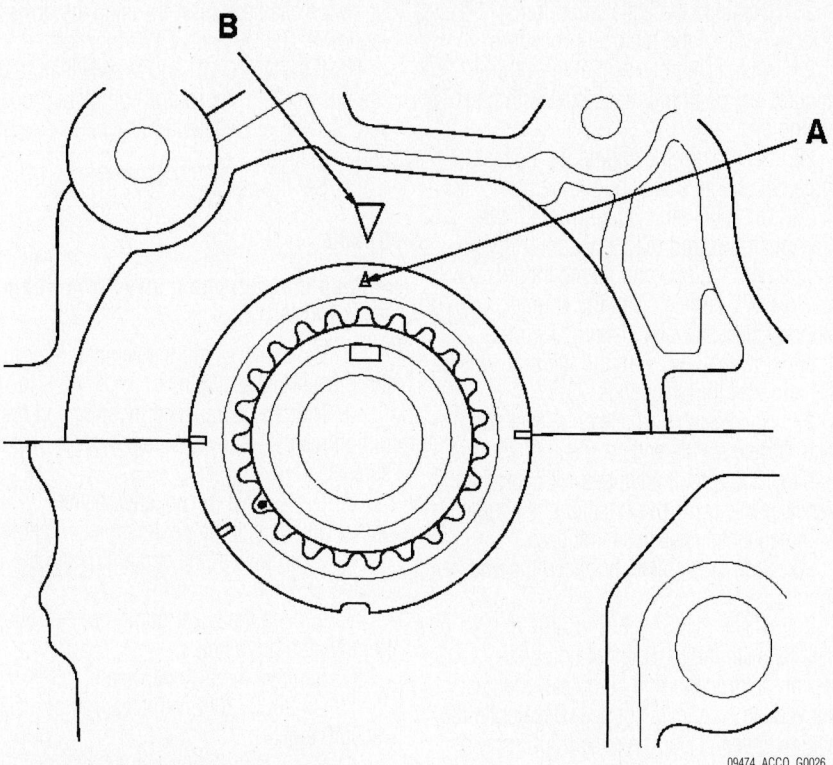

Crankshaft TDC location—1.8L engine

09474_ACCO_G0026

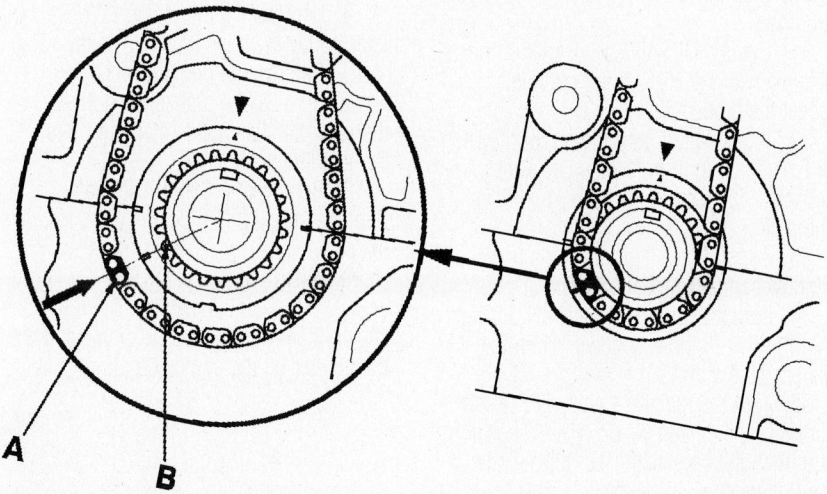

Timing chain marking (A) crankshaft position (B)—1.8L engine

09474_ACCO_G0027

23. Check PULSAR F/B LEARN in the data list with the HDS. If not complete repeat the procedure.

24. If completed, turn the ignition switch OFF. Turn the ignition switch ON, wait thirty seconds. The learning procedure is now complete.

25. Reprogram the power window control unit as follows.

26. Turn the ignition switch ON. Lower the window, all the way down. Open the driver's side door.

27. Turn the ignition switch OFF. Push and hold the drivers window DOWN switch. Turn the ignition switch ON. Release the driver's window DOWN switch. This must be done within five seconds.

28. Repeat the above step three more times. Wait one second.

29. Confirm that the AUTO UP and AUTO DOWN do not work. If they work repeat the procedure, paying close attention to the five second time limit

30. Move the driver's window all the way down by holding the drivers window DOWN switch to the AUTO DOWN position.

31. Pull up and hold the driver's window UP switch to the AUTO UP position until the window reaches the fully closed position. Hold the switch for one second.

32. Confirm that the power window master switch has been reset by using the driver's window AUTO UP and DOWN function.

33. If the AUTO UP and DOWN feature is still not working, repeat the complete procedure, paying close attention to the five second time limit.

34. Set the clock.

2.0L ENGINE

➡ **Keep the cam chain away from magnetic fields.**

1. Before servicing the vehicle, refer to the Precautions Section.

2. Note the radio security code and the radio presets. Disconnect the negative battery cable.

3. Rotate the crankshaft to set the engine at Top Dead Center (TDC) on the compression stroke for the No. 1 piston. The TDC mark on the crankshaft pulley should line up with the pointer.

4. Raise and support the vehicle safely. Remove the front tires. Remove the splash shield. Lower the vehicle.

5. Remove the drive belt. Remove the cylinder head cover. Remove the crankshaft pulley.

6. Disconnect the crankshaft position (CKP) sensor connector and the variable valve timing control (VTC) oil control solenoid valve connector.

7. Remove the VTC oil control solenoid valve.

8. Properly support the engine using a block of wood under the oil pan. Remove the ground cable. Remove the upper bracket. Remove the side engine mount bracket.

9. Remove the chain case cover.

10. Loosely install the crankshaft pulley. Turn the crankshaft counterclockwise to compress the auto tensioner. Align the holes on the lock and the auto tensioner. Insert a 0.05 inch diameter pin into the holes. Turn the crankshaft clockwise to secure the pin. Remove the auto tensioner.

11. Remove the cam chain guide. Remove the Remove the tensioner arm.

12. Remove the cam chain from the engine.

To install:

13. Set the crankshaft to TDC. Align the TDC mark on the crankshaft sprocket with the pointer on the engine block.

14. Set the camshafts to TDC. The punch mark on the VTC actuator and the punch mark on the exhaust camshaft sprocket should be at the top. Align the TDC marks on the VTC actuator and the exhaust camshaft sprocket.

15. Install the cam chain on the crankshaft sprocket with the colored piece aligned with the punch mark on the crankshaft sprocket.

16. Install the cam chain on the VTC actuator and exhaust camshaft sprocket with the punch marks aligned with the two colored pieces.

17. Install the cam chain guide and tensioner arm. Install the auto tensioner. Install the cam chain guide. Remove the pin from the auto tensioner.

18. Replace the chain case oil seal.

19. Remove any old gasket from the mating surfaces. Be sure these surfaces are clean and dry.

20. Apply liquid gasket, part number 08717-004, 08718-0001, 08718-0003 or 08718-0009 evenly to the cylinder block mating surface of the chain case and to the inner threads of the holes.

➡**Do not install the parts if more than four minutes have elapsed since applying the liquid gasket. Instead, reapply after removing the previous coating material.**

21. Apply liquid gasket to the cylinder block upper surface contact areas on the chain case.

22. Install a new O-ring on the chain case. Set the edge of the chain case on the edge of the oil pan. Install the chain case to the cylinder block.

➡**When installing the chain case, do not slide the bottom surface on the oil pan mounting surface.**

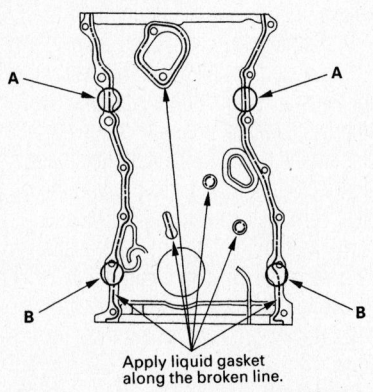

Apply liquid gasket along the broken line.

09474_ACCO_G0023

Apply liquid gasket at points (A) and points (B)—2.0L (2006 Civic) engine and 2.4L engine

23. Continue the installation in the reverse order of the removal procedure.

24. After assembly, wait at least thirty minutes before filling the engine with clean engine oil.

25. Do not run the engine for at least three hours after installing the oil pan.

26. On 2006 vehicles, reprogram the crankshaft position (CKP) pattern. Run the engine until the operating temperature reaches 176 degree. With the engine stopped clear the CKP pattern. Turn the ignition switch OFF. Turn the ignition switch ON and wait thirty seconds.

27. Roadtest the vehicle on a level surface. Decelerate the engine speed of 2500 rpm to 1000 rpm. If equipped with automatic transaxle use two Drive positions. If equipped with manual transaxle use first gear.

28. Stop the vehicle, but keep the engine running.

29. Check PULSAR F/B LEARN in the data list with the HDS. If not complete repeat the procedure. If complete, roadtest the vehicle on a level surface. Decelerate the engine speed of 5000 rpm to 3000 rpm. If equipped with automatic transaxle use two Drive positions. If equipped with manual transaxle use first gear.

30. Stop the vehicle, but keep the engine running.

31. Check PULSAR F/B LEARN in the data list with the HDS. If not complete repeat the procedure.

32. If completed, turn the ignition switch OFF. Turn the ignition switch ON, wait thirty seconds. The learning procedure is now complete.

33. As required, reprogram the power window control unit as follows.

34. Turn the ignition switch ON. Lower the window, all the way down. Open the driver's side door.

35. Turn the ignition switch OFF. Push and hold the drivers window DOWN switch. Turn the ignition switch ON. Release the driver's window DOWN switch. This must be done within five seconds.

36. Repeat the above step three more times. Wait one second.

37. Confirm that the AUTO UP and AUTO DOWN do not work. If they work repeat the procedure, paying close attention to the five second time limit

38. Move the driver's window all the way down by holding the drivers window DOWN switch to the AUTO DOWN position.

39. Pull up and hold the driver's window UP switch to the AUTO UP position until the window reaches the fully closed position. Hold the switch for one second.

40. Confirm that the power window mas-

ter switch has been reset by using the driver's window AUTO UP and DOWN function.

41. If the AUTO UP and DOWN feature is still not working, repeat the complete procedure, paying close attention to the five second time limit.

42. Set the clock.

S2000

➡**Keep the cam chain away from magnetic fields.**

1. Before servicing the vehicle, refer to the Precautions Section.

2. Note the radio security code and the radio presets. Relieve the fuel system pressure.

3. Disconnect the negative battery cable. Disconnect the positive battery cable.

4. Drain the engine coolant. Drain the engine oil.

5. Loosen the water pump pulley bolts. Remove the drive belt.

6. Remove the cylinder head.

7. On 2002–2005 vehicles, remove the vacuum tank.

8. Remove the water bypass hose. Remove the water bypass tube retaining bolts and tube. Remove the water pump pulley. Remove the auto tensioner.

9. Remove the alternator. Remove the idler pulley. Remove the idler pulley base.

10. Remove the oil pan.

11. Remove the crankshaft pulley. Remove the chain case retaining bolts. Remove the chain case.

12. Remove the CKP pulse plate. Remove the oil pump chain guide. Remove the cam chain.

To install:

13. Set the crankshaft to TDC. Align the key on the sprocket and the crankshaft with the pointer on the engine block.

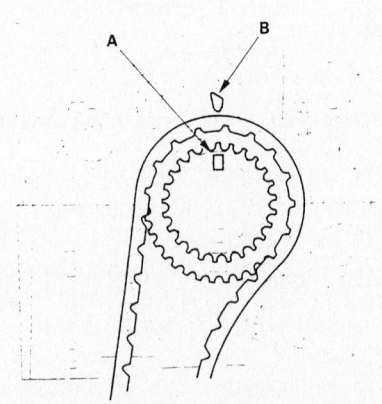

9347FG19

Align the sprocket key (A) with the cylinder block pointer (B) to set the engine to TDC—S2000

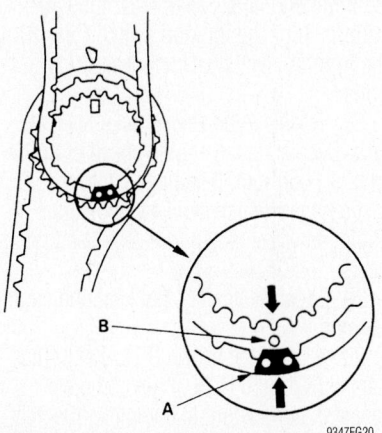

Install the timing chain with the colored link (A) aligned with the crankshaft sprocket punch mark (B)— S2000

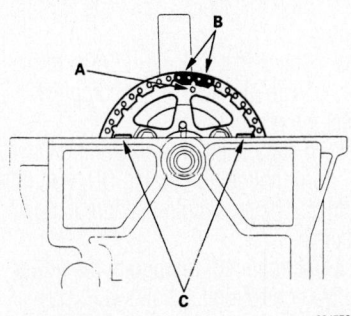

Timing chain idler sprocket punch mark (A), colored links (B) and TDC marks (C) in proper alignment— S2000

14. Install the cam chain with the colored piece aligned with the punch mark on the crankshaft sprocket.

15. Install the oil pump chain guide. Install the CKP pulse plate.

16. Replace the chain case oil seal.

17. Remove any old gasket from the mating surfaces. Be sure these surfaces are clean and dry.

18. Apply liquid gasket, part number 08717-004, 08718-0001, 08718-0003 or 08718-0009 evenly to the cylinder block mating surface of the chain case.

➡**Do not install the parts if more than four minutes have elapsed since applying the liquid gasket. Instead, reapply after removing the previous coating material.**

19. Install the dowel pins and the chain case using a new O-ring.

20. Continue the installation in the reverse order of the removal procedure.

21. After assembly, wait at least thirty minutes before filling the engine with clean engine oil.

22. Do not run the engine for at least three hours after installing the oil pan.

23. On 2002–2005 vehicles, reprogram the ECM/PCM engine idle characteristics. Be sure all electrical items are OFF.

24. Start the engine. Hold the idle speed at 3000 RPM's in neutral until the radiator fan comes on or the temperature reached 176 degrees.

25. Let the engine idle for about five minutes with the throttle fully closed.

26. If the radiator fan comes on during the five minutes, do not count this toward the five minute programming time.

27. On 2006 vehicles, reprogram the crankshaft position (CKP) pattern. Run the engine until the operating temperature reaches 176 degree. With the engine stopped clear the CKP pattern. Turn the ignition switch OFF. Turn the ignition switch ON and wait thirty seconds.

28. Roadtest the vehicle on a level surface. Decelerate the engine speed of 2500 rpm to 1000 rpm. If equipped with automatic transaxle use two Drive positions. If equipped with manual transaxle use first gear.

29. Stop the vehicle, but keep the engine running.

30. Check PULSAR F/B LEARN in the data list with the HDS. If not complete repeat the procedure. If complete, roadtest the vehicle on a level surface. Decelerate the engine speed of 5000 rpm to 3000 rpm. If equipped with automatic transaxle use two Drive positions. If equipped with manual transaxle use first gear.

31. Stop the vehicle, but keep the engine running.

32. Check PULSAR F/B LEARN in the data list with the HDS. If not complete repeat the procedure.

33. If completed, turn the ignition switch OFF. Turn the ignition switch ON, wait thirty seconds. The learning procedure is now complete.

34. As required, reprogram the power window control unit as follows.

35. Turn the ignition switch ON. Lower the window, all the way down. Open the driver's side door.

36. Turn the ignition switch OFF. Push and hold the drivers window DOWN switch. Turn the ignition switch ON. Release the driver's window DOWN switch. This must be done within five seconds.

37. Repeat the above step three more times. Wait one second.

38. Confirm that the AUTO UP and AUTO DOWN do not work. If they work repeat the procedure, paying close attention to the five second time limit

39. Move the driver's window all the way down by holding the drivers window DOWN switch to the AUTO DOWN position.

40. Pull up and hold the driver's window UP switch to the AUTO UP position until the window reaches the fully closed position. Hold the switch for one second.

41. Confirm that the power window master switch has been reset by using the driver's window AUTO UP and DOWN function.

42. If the AUTO UP and DOWN feature is still not working, repeat the complete procedure, paying close attention to the five second time limit.

43. Set the clock.

Accord

➡**Keep the cam chain away from magnetic fields.**

1. Before servicing the vehicle, refer to the Precautions Section.

2. Note the radio security code and the radio presets. Disconnect the negative battery cable.

3. Rotate the crankshaft to set the engine at Top Dead Center (TDC) on the compression stroke for the No. 1 piston. The TDC mark on the crankshaft pulley should line up with the pointer.

4. Raise and support the vehicle safely. Remove the front tires. Remove the splash shield. Lower the vehicle.

5. Remove the drive belt. Remove the cylinder head cover. Remove the crankshaft pulley.

6. Check that the number one piston TDC marks on the variable timing control (VTC) actuator and exhaust camshaft sprocket are aligned.

7. Disconnect the crankshaft position (CKP) sensor connector and the variable valve timing control (VTC) oil control solenoid valve connector.

8. Remove the VTC oil control solenoid valve.

9. Properly support the engine using a block of wood under the oil pan. Remove the ground cable. Remove the upper bracket. Remove the side engine mount bracket.

10. Remove the chain case cover.

11. Loosely install the crankshaft pulley. Turn the crankshaft counterclockwise to compress the auto tensioner. Align the holes on the lock and the auto tensioner. Insert a 0.05 inch diameter pin into the holes. Turn the crankshaft clockwise to secure the pin. Remove the auto tensioner.

12. Remove the cam chain guide. Remove the Remove the tensioner arm.

13. Remove the cam chain from the engine.

To install:

➡**Check that the VTC actuator is locked by turning the actuator counterclockwise. If not locked, turn the actuator clockwise until it stops. Recheck it. If it is still not locked, replace it.**

14. Set the crankshaft to TDC. Align the TDC mark on the crankshaft sprocket with the pointer on the engine block.

15. Set the camshafts to TDC. The punch mark on the VTC actuator and the punch mark on the exhaust camshaft sprocket should be at the top. Align the TDC marks on the VTC actuator and the exhaust camshaft sprocket.

16. Install the cam chain on the crankshaft sprocket with the colored piece aligned with the punch mark on the crankshaft sprocket.

17. Install the cam chain on the VTC actuator and exhaust camshaft sprocket with the punch marks aligned with the two colored pieces.

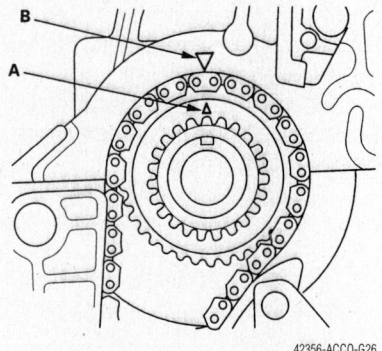

42356-ACCO-G26

Set the crankshaft to TDC. Align the TDC mark (A) on the crankshaft sprocket with the pointer (B) on the cylinder block—2.0L (2006 Civic) engine and 2.4L engine

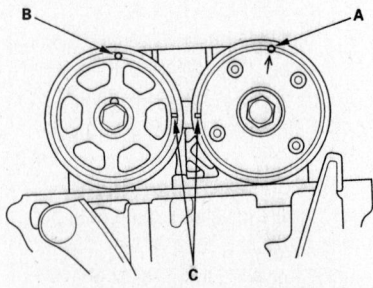

42356-ACCO-G27

The mark (A) on the VTC actuator and the mark (B) on the exhaust cam (C) should be at the top. Align the TDC marks (C) on the VTC actuator and exhaust cam sprockets—2.0L (2006 Civic) engine and 2.4L engine

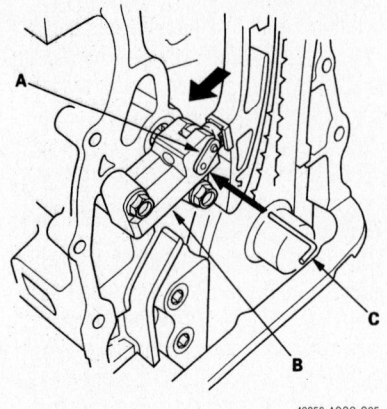

42356-ACCO-G25

Align the holes on the lock (A) and the auto-tensioner (B), then place a 1.5mm pin into the holes. Turn the crankshaft clockwise to secure the pin—2.0L (2006 Civic) engine and 2.4L engine

18. Install the cam chain guide and tensioner arm. Install the auto tensioner. Install the cam chain guide. Remove the pin from the auto tensioner.

19. Replace the chain case oil seal.

20. Remove any old gasket from the mating surfaces. Be sure these surfaces are clean and dry.

21. Apply liquid gasket, part number 08717-004, 08718-0001, 08718-0003 or 08718-0009 evenly to the cylinder block mating surface of the chain case and to the inner threads of the holes.

➡**Do not install the parts if more than four minutes have elapsed since applying the liquid gasket. Instead, reapply after removing the previous coating material.**

22. Apply liquid gasket to the cylinder block upper surface contact areas on the chain case.

23. Install a new O-ring on the chain case. Set the edge of the chain case on the edge of the oil pan. Install the chain case to the cylinder block.

➡**When installing the chain case, do not slide the bottom surface on the oil pan mounting surface.**

24. Continue the installation in the reverse order of the removal procedure.

25. After assembly, wait at least thirty minutes before filling the engine with clean engine oil.

26. Do not run the engine for at least three hours after installing the oil pan.

27. Reprogram the crankshaft position (CKP) pattern. Run the engine until the operating temperature reaches 176 degree.

With the engine stopped clear the CKP pattern. Turn the ignition switch OFF. Turn the ignition switch ON and wait thirty seconds.

28. Roadtest the vehicle on a level surface. Decelerate the engine speed of 2500 rpm to 1000 rpm. If equipped with automatic transaxle use two Drive positions. If equipped with manual transaxle use first gear.

29. Stop the vehicle, but keep the engine running.

30. Check PULSAR F/B LEARN in the data list with the HDS. If not complete repeat the procedure. If complete, roadtest the vehicle on a level surface. Decelerate the engine speed of 5000 rpm to 3000 rpm. If equipped with automatic transaxle use two Drive positions. If equipped with manual transaxle use first gear.

31. Stop the vehicle, but keep the engine running.

32. Check PULSAR F/B LEARN in the data list with the HDS. If not complete repeat the procedure.

33. If completed, turn the ignition switch OFF. Turn the ignition switch ON, wait thirty seconds. The learning procedure is now complete.

34. As required, reprogram the power window control unit as follows.

35. Turn the ignition switch ON. Lower the window, all the way down. Open the driver's side door.

36. Turn the ignition switch OFF. Push and hold the drivers window DOWN switch. Turn the ignition switch ON. Release the driver's window DOWN switch. This must be done within five seconds.

37. Repeat the above step three more times. Wait one second.

38. Confirm that the AUTO UP and AUTO DOWN do not work. If they work repeat the procedure, paying close attention to the five second time limit

39. Move the driver's window all the way down by holding the drivers window DOWN switch to the AUTO DOWN position.

40. Pull up and hold the driver's window UP switch to the AUTO UP position until the window reaches the fully closed position. Hold the switch for one second.

41. Confirm that the power window master switch has been reset by using the driver's window AUTO UP and DOWN function.

42. If the AUTO UP and DOWN feature is still not working, repeat the complete procedure, paying close attention to the five second time limit.

43. Set the clock.

Piston and Ring

POSITIONING

When assembling the pistons, piston rings and connecting rods, and when installing these assemblies into the engine block, it is vitally important to ensure that these three components are properly positioned with respect to each other. Often times the engine block is designed so that if a connecting rod or piston is installed backwards, or in the wrong bank of cylinders, internal engine damage may occur once the engine is started. The piston ring end-gap spacing that is recommended by the engine manufacturer is often with the purpose of increased compression pressures during the engine break-in period. Failure to properly space the piston ring end-gaps may lead to increased oil consumption and extended break-in time. Therefore, always be sure to position the pistons, rings and connecting rods as shown in the accompanying illustrations.

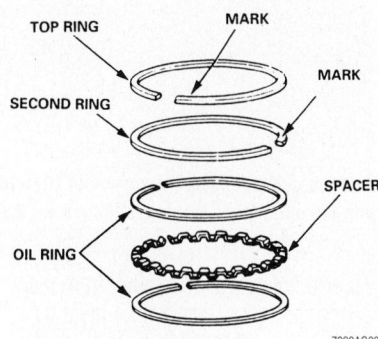

Piston ring positioning

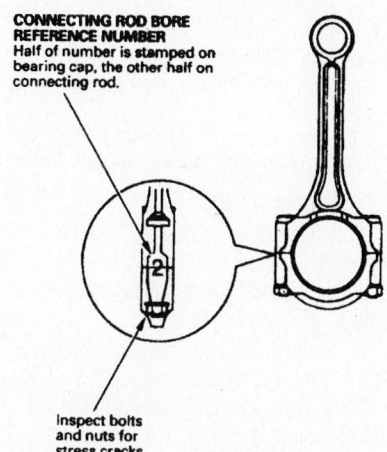

Before removing the caps from the connecting rods, be sure to matchmark them as shown

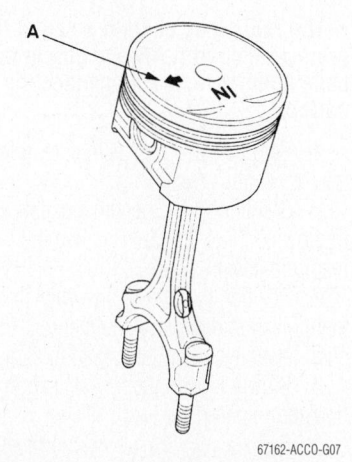

Piston/connecting rod assembly-to-engine orientation. The arrow (A) must face the timing belt side of the engine—1.7L engine

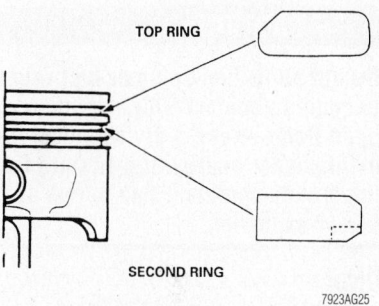

Compression ring location—2.3L engine

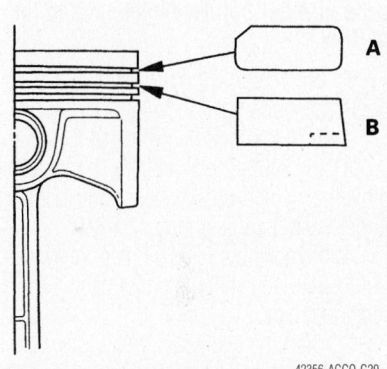

Compression ring location—2.4L engine

FUEL SYSTEM

Fuel System Service Precautions

Safety is the most important factor when performing not only fuel system maintenance but any type of maintenance. Failure to conduct maintenance and repairs in a safe manner may result in serious personal injury or death. Maintenance and testing of the vehicle's fuel system components can be accomplished safely and effectively by adhering to the following rules and guidelines.

• To avoid the possibility of fire and personal injury, always disconnect the negative battery cable unless the repair or test procedure requires that battery voltage be applied.

• Always relieve the fuel system pressure prior to disconnecting any fuel system component (injector, fuel rail, pressure regulator, etc.), fitting or fuel line connection. Exercise extreme caution whenever relieving fuel system pressure, to avoid exposing skin, face and eyes to fuel spray. Please be advised that fuel under pressure may penetrate the skin or any part of the body that it contacts.

• Always place a shop towel or cloth around the fitting or connection prior to loosening to absorb any excess fuel due to spillage. Ensure that all fuel spillage (should it occur) is quickly removed from engine surfaces. Ensure that all fuel soaked cloths or towels are deposited into a suitable waste container.

• Always keep a dry chemical (Class B) fire extinguisher near the work area.

• Do not allow fuel spray or fuel vapors to come into contact with a spark or open flame.

• Always use a back-up wrench when loosening and tightening fuel line connection fittings. This will prevent unnecessary stress and torsion to fuel line piping. Always follow the proper torque specifications.

• Always replace worn fuel fitting O-rings with new. Do not substitute fuel hose or equivalent, where fuel pipe is installed.

Fuel System Pressure

RELIEVING

> ※※ **CAUTION**
>
> **The fuel injection system remains under pressure after the engine has been turned OFF. Properly relieve fuel pressure before disconnecting any fuel lines. Failure to do so may result in fire or personal injury.**

> ※※ **CAUTION**
>
> **Do not allow fuel spray or fuel vapors to come in contact with a spark or open flame. Keep a dry chemical fire extinguisher nearby. Never store fuel in an open container due to risk of fire or explosion.**

Civic

➡The radio may contain a coded theft protection circuit. Always obtain the code number before disconnecting the battery.

1. Before servicing the vehicle, refer to the Precautions Section.
2. Turn the ignition switch OFF.
3. On 2002–2005 vehicles, remove the glove box. On 2006 vehicles, remove the under dash fuse/relay box.
4. Remove the PGM-FI main relay. On 2006 Civic, reinstall the under dash fuse/relay box.
5. Start the engine and let it idle until it stalls.

➡If any codes are stored, ignore them.

6. Turn the ignition OFF.
7. On Hatchback remove the engine cover.
8. Disconnect the negative battery cable.
9. Remove the fuel filler cap to relieve the fuel pressure in the tank.
10. On vehicles equipped with a quick connect cover, remove it. Clean any dirt from the quick-connect fitting. Place a rag or shop towel over quick-connect fitting.
11. Detach the quick-connect fitting by holding the connector with one hand, then squeeze the retainer tabs with the other hand to release them from the locking pawls. Pull the connector off.
12. If the connector does not move, keep the retainer tabs pressed down and alternately pull and push the connector until it comes off easily.

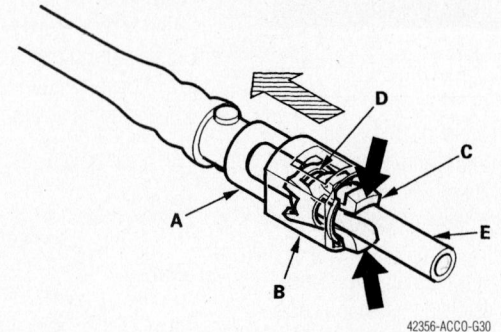

42356-ACCO-G30

Hold the quick-connect (A) connector (B) with one hand, then squeeze the retainer tabs (C) with the other hand to release them from the locking pawls (D)

13. Do not move the retainer from the line; once removed, the retainer must be replaced with a new one.

S2000

➡The radio may contain a coded theft protection circuit. Always obtain the code number before disconnecting the battery.

1. Before servicing the vehicle, refer to the Precautions Section.
2. Disconnect the negative battery cable. Remove the fuel fill cap. Remove the intake manifold cover.
3. Position the proper size wrench on the fuel pulsation damper at the fuel rail. Place a shop towel over the fuel pulsation damper.
4. Slowly loosen the fuel pulsation damper on complete turn.
5. Replace all washers whenever the fuel pulsation damper is loosened or removed.

Accord

2.3L ENGINE

➡The radio may contain a coded theft protection circuit. Always obtain the code number before disconnecting the battery.

1. Before servicing the vehicle, refer to the Precautions Section.
2. Disconnect the negative battery cable. Remove the fuel fill cap.
3. Position the proper size wrench on the fuel pulsation damper at the fuel rail. Place a shop towel over the fuel pulsation damper.
4. Slowly loosen the fuel pulsation damper on complete turn.
5. Replace all washers whenever the fuel pulsation damper is loosened or removed.

EXCEPT 2.3L ENGINE

➡The radio may contain a coded theft protection circuit. Always obtain the

code number before disconnecting the battery.

1. Before servicing the vehicle, refer to the Precautions Section.
2. Turn the ignition switch OFF.
3. Remove the left side kick panel. Remove the PGM-FI main relay.
4. Start the engine and let it idle until it stalls.

➡If any codes are stored, ignore them.

5. Turn the ignition OFF.
6. On Hatchback remove the engine cover.
7. Disconnect the negative battery cable.
8. Remove the fuel filler cap to relieve the fuel pressure in the tank.
9. On vehicles equipped with a quick connect cover, remove it. Clean any dirt from the quick-connect fitting. Place a rag or shop towel over quick-connect fitting.
10. Detach the quick-connect fitting by holding the connector with one hand, then squeeze the retainer tabs with the other hand to release them from the locking pawls. Pull the connector off.
11. If the connector does not move, keep the retainer tabs pressed down and alternately pull and push the connector until it comes off easily.
12. Do not move the retainer from the line; once removed, the retainer must be replaced with a new one.

Fuel Filter

REMOVAL & INSTALLATION

1. Before servicing the vehicle, refer to the Precautions Section.
2. Relieve the fuel system pressure.
3. Remove or disconnect the following:
   • Fuel pump from the tank
   • Fuel filter from the pump module

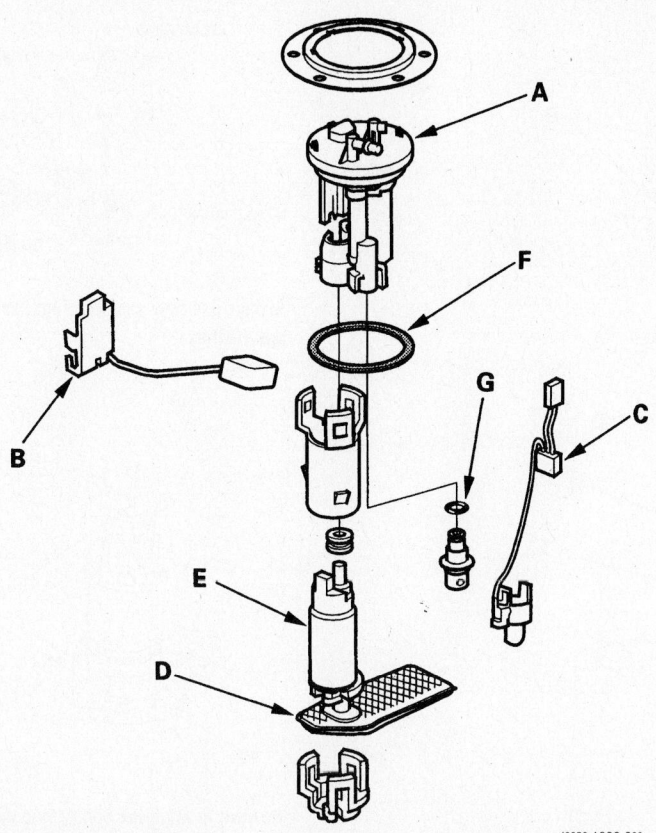

42356-ACCO-G32

Exploded view of the strainer case (A), fuel gauge sending unit (B), wire harness (C), suction filter (D) and fuel pump (E)—Accord shown

To install:

4. Install or connect the following:
   - New filter on the pump module
   - Fuel pump in the tank

Fuel Pump

REMOVAL & INSTALLATION

➡**The radio may contain a coded theft protection circuit. Always obtain the code number before disconnecting the battery.**

❈❈ CAUTION

The fuel injection system remains under pressure, even after the engine has been turned OFF. The fuel system pressure must be relieved before disconnecting any fuel lines. Failure to follow this procedure may result in fire, explosion, or personal injury.

Civic

1. Before servicing the vehicle, refer to the Precautions Section.
2. Disconnect the negative battery cable.
3. Relieve the fuel pressure.
4. Remove or disconnect the following:
   - Rear seat cushions
   - Remove the fuel filler cap
   - On 2006 Civic, remove the rear floor upper crossmember
   - Fuel pump access panel
   - Fuel pump electrical harness

➡**Clean the fuel line fittings before disconnecting them.**

5. Disconnect the quick connect fitting from the pump assembly.
6. Using tool 07AAA-S0XA100, or equivalent, loosen the fuel tank unit locknut. Remove the locknut and the fuel tank unit. Allow the fuel in the pump drain into the tank before removing the pump from the vehicle.

To install:

7. Be sure to use a new base gasket and new O-rings.
8. When installing the fuel tank unit, align the marks on the fuel tank and the fuel tank unit.
9. Make sure the electrical connections are secure and the connector firmly locked in place.
10. Make sure the fuel line connections

are secure and the connector firmly locked in place.
11. Continue the installation in the reverse order of the removal procedure.

S2000

1. Before servicing the vehicle, refer to the Precautions Section.
2. Relieve the fuel system pressure.
3. Remove or disconnect the following:
   - Negative battery cable
   - Remove the fuel tank filler cap
   - Rear package tray
   - Fuel pump access panel
   - Fuel pump electrical harness

➡**Clean the fuel line fittings before disconnecting them.**

4. Disconnect the quick connect fitting from the pump assembly.
5. Remove the fuel tank retaining bolts. Remove the fuel tank unit from the vehicle.

To install:

6. Be sure to use a new base gasket.
7. When installing the fuel tank unit, align the marks on the fuel tank and the fuel tank unit.
8. Make sure the electrical connections are secure and the connector firmly locked in place.
9. Make sure the fuel line connections are secure and the connector firmly locked in place.
10. Continue the installation in the reverse order of the removal procedure.

Accord

1. Before servicing the vehicle, refer to the Precautions Section.
2. Relieve the fuel pressure.
3. Remove or disconnect the following:

   - Negative battery cable
   - Remove the fuel filler cap
   - Spare tire cover, if equipped
   - Trunk floor
   - Access panel from the floor
   - electrical connector from the pump assembly.

➡**Clean the fuel line fittings before disconnecting them.**

4. Disconnect the quick connect fitting from the pump assembly.
5. Depending upon the year and engine, either use tool 07AAA-S0XA100, or equivalent, to loosen and remove the fuel tank unit locknut and the fuel tank unit, or remove the fuel tank retaining bolts and remove the fuel tank unit.

Except SULEV model:

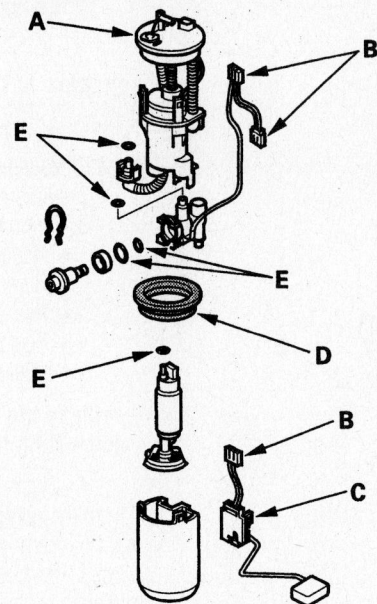

SULEV model:

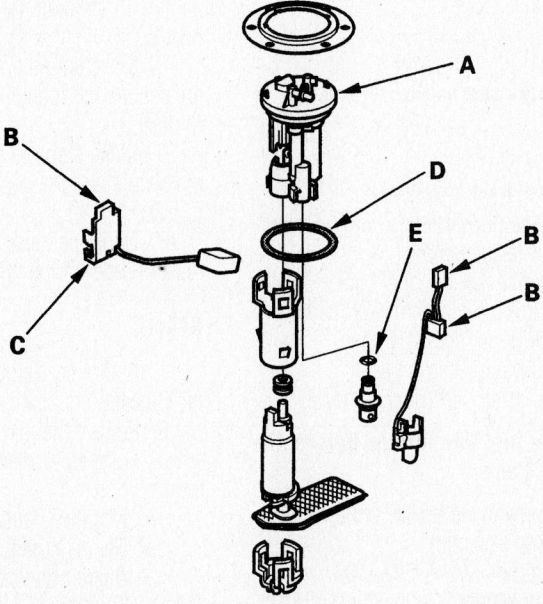

Fuel filter (A) and related components—2003–06 Accord shown

To install:

6. Be sure to use a new base gasket and new O-rings.

7. When installing the fuel tank unit, align the marks on the fuel tank and the fuel tank unit.

8. Make sure the electrical connections are secure and the connector firmly locked in place.

9. Make sure the fuel line connections are secure and the connector firmly locked in place.

10. Continue the installation in the reverse order of the removal procedure.

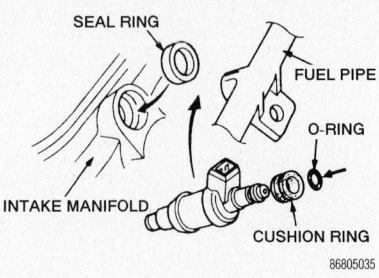

Always use new cushion rings, seal rings and O-rings

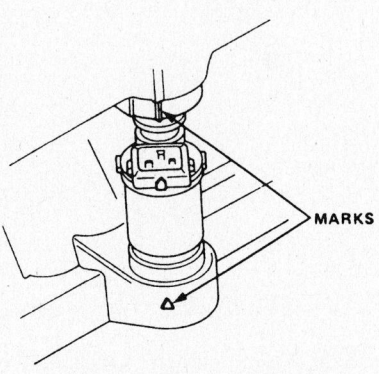

Be sure to align the center line on the injector with the mark on the fuel rail

Fuel Injector

REMOVAL & INSTALLATION

1. Disconnect the negative battery cable.

2. Relieve the fuel system pressure.

3. Remove the necessary components in order to gain access to the fuel rail assembly.

4. Remove the fuel rail assembly

5. Carefully pull the injectors from the intake manifold.

6. Discard the seal rings, cushion rings and O-rings.

To install:

7. Slide new cushion rings onto the injectors.

8. Coat new O-rings with clean engine oil and put them on the injectors.

9. Insert the injectors into the fuel rail. Be sure to align the center line on the injector with the mark on the fuel rail.

10. Coat new seal rings with clean engine oil and insert them into the intake manifold.

11. Install the fuel rail assembly.

DRIVE TRAIN

Transaxle Assembly

REMOVAL & INSTALLATION

Manual Transaxle

CIVIC 2002–05

❈❈ WARNING

Use only genuine Honda manual transaxle fluid (MTF)-it is specially formulated for use in Honda transaxles. If Honda MTF is not available, API SG/SJ 10W-30 or 10W-40 motor oil may be used as a temporary lubricant. However, motor oil will cause increased transaxle wear and shifting effort. Refill the transaxle with Honda MTF as soon as possible.

1. Before servicing the vehicle, refer to the Precautions Section.
2. Remove or disconnect the following:
   - Negative and positive battery cables
   - Resonator, the air cleaner box, and the air intake duct.
   - Starter cables and the transaxle ground cable.
   - Back-up light switch connection
   - Upper radiator hose out of its bracket.
   - Vehicle Speed Sensor (VSS) connector.
   - Clutch fluid line bracket
   - Slave cylinder. It isn't necessary to disconnect the clutch fluid line.
3. Raise and safely support the vehicle.
4. Drain the transaxle fluid.
5. Remove or disconnect the following:
   - Front wheels
   - Strut pinch bolt and fork bolt
   - Lower ball joint from the steering knuckle using a ball joint remover.
   - Halfshaft inboard joints out of the transaxle case. Swing the steering knuckles out to free the halfshafts from the transaxle.
6. Tie the halfshafts up and out of the way with wire so that the joints will not be stressed. Tie plastic bags over the inboard joints to prevent damage to the CV-boots and splined shafts.
7. Remove or disconnect the following:
   - Shift rod and extension rod from the transaxle case. Drive the shift rod retaining pin out with a pin punch.

- Front exhaust pipe
- Engine-to-transaxle stiffener brackets and the clutch cover plate
8. Attach a lifting chain to the engine and lift slightly to ease the tension on the mounts.
9. Remove or disconnect the following:
   - Splash shield from underneath the vehicle.
   - Right-front mount/bracket assembly
10. Place a jack under the transaxle to support its weight.
11. Remove or disconnect the following:
    - Transaxle side mount and bracket
    - Starter's lower mounting bolt and the upper 3 transaxle case bolts.
    - 3 rear transaxle mount bracket bolts, then the lower 3 transaxle case bolts.
    - Pull the transaxle away from the engine until it clears the mainshaft. Lower the transaxle out of the vehicle Be careful not to bend the clutch hydraulic line.

To install:

➡**Use new self-locking nuts and color-coded bolts when installing the transaxle and suspension components.**

12. Apply high temperature grease to the mainshaft splines, release fork contact points, and throw-out bearing. The manufacturer recommends part No. 08798-9002, Honda Super High temp Urea Grease.
13. Place the transaxle on a transaxle jack and raise it to the level of the engine.
14. Align the transaxle and engine. Be sure the transaxle case dowel pins are securely seated, and fit the transaxle onto the engine. Install the upper and lower transaxle case bolts and the 14mm rear mount bolts and washers. Only hand-tighten them at this time.
15. Raise the transaxle and install the side mount. Tighten the upper and lower transaxle case bolts to 47 ft. lbs. (64 Nm). Tighten the 14mm rear mount bracket bolts to 61 ft. lbs. (84 Nm).
16. First, tighten the transaxle side mount bracket nuts and bolt to 47 ft. lbs. (64 Nm) each. Next, tighten the mount bushing bolts to 47 ft. lbs. (64 Nm). Finally, tighten the through-bolt to 54 ft. lbs. (74 Nm).
17. Install or connect the following:
    - Right-front mount/bracket assembly. Use 3 new 12mm bolt and washers, and tighten them to 47 ft.

lbs. (64 Nm). Tighten the 2 10mm bolts to 33 ft. lbs. (45 Nm).
- Clutch cover
- Engine-to-transaxle stiffener brackets and tighten the 8mm bolts to 17 ft. lbs. (24 Nm). Tighten the 10mm bolts to 33 ft. lbs. (44 Nm).
18. Once the transaxle is bolted to the engine, and the transaxle mounts are securely tightened, the engine lifting chain may be removed.
19. Install or connect the following:
    - Shift rod with a new spring pin and clip. Then, fit the shift rod boot back into place. Connect the torque rod and tighten the bolt to 16 ft. lbs. (22 Nm).
    - Front exhaust pipe. Use new self-locking nuts and gaskets. Tighten the rear flange nuts to 16 ft. lbs. (22 Nm). Tighten the front flange nuts to 25 ft. lbs. (33Nm) for all others.
    - New set rings on the inboard CV-joint splines
    - Halfshafts into the transaxle case and intermediate shaft. The inboard joints must snap into place.
    - Lower ball joint and damper fork
    - Front wheels
    - Slave cylinder and clutch pipe stay. Coat the slave cylinder's tip with high temperature grease. Be sure it snaps into the release fork. Tighten the slave cylinder mounting bolts to 16 ft. lbs. (22 Nm).
    - VSS connector and back-up light switch connectors
    - Wiring harness clamps and starter cables.
    - Resonator, air cleaner box, and air intake duct. Fit the upper radiator hose back into its bracket.
20. Lower the vehicle and tighten the strut pinch bolts to 32 ft. lbs. (44 Nm). Tighten the fork bolts to 47 ft. lbs. (65 Nm). Tighten the ball joint castle nuts to 40 ft. lbs. (55 Nm), then tighten them only enough to install new cotter pins.
21. Turn the breather cap so that the arrow with the **F** mark points toward the front of the vehicle.
22. Refill the transaxle with the Honda MTF fluid.
23. Reconnect the positive and negative battery cables.
24. Bleed the clutch hydraulic system.
25. Check the clutch and transaxle for smooth operation.

26. Check and adjust the front wheel alignment.

CIVIC 2006

1. Before servicing the vehicle, refer to the Precautions Section.

2. The radio may contain a coded theft protection circuit. Always obtain the code number before disconnecting the battery.

3. Disconnect the negative battery cable. Disconnect the positive battery cable. Remove the battery.

4. Remove the cowl cover and under cowl panel.

5. Remove the air cleaner assembly. Remove the harness clips and the intake air duct. Remove the battery base with the reservoir tank.

6. Remove the clutch bracket and carefully remove the slave cylinder. Do not depress the clutch pedal once the slave cylinder has been removed.

7. Disconnect the back up switch electrical connector, the vehicle speed sensor, the reverse solenoid lockout connector and the harness clips.

8. Remove the cable bracket, then disconnect the cables from the top of the transaxle housing. Carefully remove the cables and the bracket.

9. Remove the harness clips from the clutch cable bracket and the harness bracket. Remove the engine wire harness cover. Slide the harness forward off of the air cleaner housing mounting bracket.

10. Attach tool VSP02C000015, or equivalent to the holes in the cylinder head. Install the front leg of the tool on to the engine hanger in the front of the vehicle. This operation will support the engine and transaxle assembly.

11. Remove the two upper transaxle upper mounting bolts. Remove the under hood fuse/relay box and position it to the side.

12. Remove the ECM stay and position it to the side. Remove the clutch line clamp. Disconnect the ground cable. Remove the transaxle mount bracket bolts and nuts. Remove the transaxle mount bracket.

13. Raise and safely support the vehicle. Drain the transaxle fluid.

14. Remove the splash shield. Separate the lower arm. Remove the stiffner plate and mounting bracket from the steering gearbox. Disconnect the exhaust mounting rubber.

15. Remove the stiffner plate and harness clip. Remove the front engine mount bracket mounting bolt and nut. Remove the lower radiator hose from the front mount bracket.

16. Remove the front engine mount

bracket from the transaxle and engine. Remove the rear engine mount bracket mounting bolt.

17. Remove the middle subframe mounting bolts. Install tool VSB02C000016, or equivalent, to the subframe. Raise the jack and line up the slots in the arms with the bolt holes on the corner of the jack base. Attach them securely.

18. Remove the front suspension subframe mounting bolts and front suspension subframe. Suspend the steering gearbox to the side.

19. Pry out the driveshafts inboard joint. Remove the intermediate shafts.

20. Remove the clutch cover. Position a transaxle jack under the transaxle assembly. Remove the transaxle mounting bolts. Carefully remove the transaxle from the vehicle.

To install:

21. Apply high temperature grease to the mainshaft splines, release fork contact points, and throw-out bearing. The manufacturer recommends part No. 08798-9002, Honda Super High temp Urea Grease.

22. Place the transaxle on a transaxle jack and raise it to the level of the engine.

23. Align the transaxle and engine. Be sure the transaxle case dowel pins are securely seated. Install the transaxle retaining bolts.

24. Continue the installation in the reverse order of the removal procedure.

25. Reprogram the power window control unit as follows.

26. Turn the ignition switch ON. Lower the window, all the way down. Open the driver's side door.

27. Turn the ignition switch OFF. Push and hold the drivers window DOWN switch. Turn the ignition switch ON. Release the driver's window DOWN switch. This must be done within five seconds.

28. Repeat the above step three more times. Wait one second.

29. Confirm that the AUTO UP and AUTO DOWN do not work. If they work repeat the procedure, paying close attention to the five second time limit

30. Move the driver's window all the way down by holding the drivers window DOWN switch to the AUTO DOWN position.

31. Pull up and hold the driver's window UP switch to the AUTO UP position until the window reaches the fully closed position. Hold the switch for one second.

32. Confirm that the power window master switch has been reset by using the driver's window AUTO UP and DOWN function.

33. If the AUTO UP and DOWN feature is still not working, repeat the complete proce-

dure, paying close attention to the five second time limit.

34. Set the clock.

S2000

1. Before servicing the vehicle, refer to the Precautions Section.

2. Remove or disconnect the following:
   - Battery cables and the battery
   - Shift lever knob
   - Center console
   - Shift lever boot
   - Shift lever
   - Air cleaner housing
   - Steering shaft from the steering gear box
   - Alternator
   - A/C compressor
   - Exhaust manifold heat shields
   - Upper starter mounting bolt
   - Upper intake manifold bracket mounting bolt
   - Suction valve hose
   - Camshaft Position (CMP) sensor connectors
   - Splash shield
   - Steering gear box electrical connector
   - Torque sensor connector
   - Intake manifold bracket
   - Heated Oxygen (HO$_2$S) sensor connectors
   - Catalytic converter
   - Exhaust manifold
   - Driveshaft
   - Shifter boot holder bolts
   - Clutch slave cylinder
   - Clutch release fork
   - Lower transmission flange bolts

3. Support the front subframe with a floor jack and remove the two center mounting bolts.

4. Loosen the four outer mounting bolts 3 inches (75 mm).

5. Lower the front subframe until it is supported by the loosened bolts.

6. Support the transmission with the floor jack.

7. Remove or disconnect the following:
   - Rear transmission mount
   - Speed sensor connector and wiring harness
   - Upper transmission flange bolts
   - Transmission

To install:

➡Use new subframe bolts for assembly.

8. Installation is the reverse of the removal procedure, while using the following torque values:
   - Transmission flange bolts: 47 ft. lbs. (64 Nm)

- Rear transmission mount bolts: 28 ft. lbs. (38 Nm)
- Subframe mounting bolts: 14mm bolts to 85 ft. lbs. (116 Nm) and 12mm bolts to 43 ft. lbs. (59 Nm)
- Clutch slave cylinder bolts: 16 ft. lbs. (22 Nm)
- A/C compressor bolts: 33 ft. lbs. (44 Nm)
- Steering shaft pinch bolt: 16 ft. lbs. (22 Nm)
- Shift lever bolts: 86 inch lbs. (10 Nm)

ACCORD

1. Before servicing the vehicle, refer to the Precautions Section.
2. Shift the transaxle into **R**.
3. Remove or disconnect the following:
   - Negative and positive battery cables and the battery.
   - Idle Air Control (IAC) solenoid connector
   - Intake duct, resonator, air cleaner case, and battery base.
   - Starter wires and starter
   - Transaxle ground cable and the back-up light switch wire.
   - Cable stay, then the cables from the top housing of the transaxle
   - Both cables and the stay together
   - Vehicle Speed Sensor (VSS). Leave the speed sensor hoses connected.
   - Shift cable bracket
   - Shift and select cables from the top of the transaxle case. Leave the cables and bracket together, and wire them out of the work area.
   - Mounting bolts and clutch slave cylinder with the clutch pipe and pushrod.
   - Mounting bolt and clutch hose joint with the clutch pipe and clutch hose.

➡**Do not depress the clutch pedal once the slave cylinder has been removed. Be careful not to kink the metal hydraulic lines.**

   - 2 upper transaxle case bolts
4. Raise and safely support the vehicle.
5. Remove or disconnect the following:
   - Front wheels
   - Engine splash shield
6. Drain the transaxle fluid.
7. Remove or disconnect the following:
   - Clutch damper bracket and raise it out of the way.
   - Subframe center beam
   - Cotter pins and lower arm ball joint nuts
   - Ball joints and lower arms using a press type ball joint tool.

- Right damper fork bolt
- Right damper pinch bolt, then separate the damper fork and damper.
- Radius rod bolts and nut, then the right radius rod.
- Right and left halfshafts from the differential and the intermediate shaft, using a suitable prytool.
- Left halfshaft
- Intermediate shaft from the differential by removing its 3 bearing shaft mounting bolts.

8. Swing the right halfshaft out and wire it up inside the right fender well. Tie plastic bags over the inboard CV-joints to protect the boots and splines from damage.
9. Remove or disconnect the following:
   - Engine stiffener and the clutch cover
   - Intake manifold bracket
   - Rear engine mount bracket
   - 3 rear engine mount bracket mounting bolts, then discard.
10. Place a transaxle jack under the transaxle. Raise the transaxle just enough to take the weight off the mounts.

➡**A chain hoist may be attached the transaxle lifting hooks to steady it and aid in lowering it from the vehicle.**

11. Remove or disconnect the following:
   - Transaxle housing mounting bolt on the engine side
   - Transaxle mount bolt and loosen the mount bracket nuts.
   - 3 transaxle housing mounting bolts, then the transaxle from the vehicle.

To install:

➡**Use new self-locking nuts when assembling the front suspension. Install new set rings onto the inboard CV-joints. Use new self-locking bolts when installing transaxle rear mount bracket (the bolts are color coded by type). New fasteners are available from a Honda dealer.**

12. Be sure the 2 dowel pins are installed into the transaxle case.
13. Apply heavy duty high temperature grease (use Honda part No. 08798–9002) to the release bearing, mainshaft splines, and the release fork pawls. Install the release fork and release bearing.
14. Raise the transaxle into position.
15. Install or connect the following:
   - 3 lower transaxle case bolts and tighten to 47 ft. lbs. (65 Nm).
   - Transaxle mount and mount bracket
   - Through-bolt and tighten temporarily. Be sure the engine is level and

tighten the 3 mount bracket nuts to 40 ft. lbs. (55 Nm). Tighten the through-bolt to 47 ft. lbs. (65 Nm).
   - Upper transaxle case bolts on the engine side and tighten to 47 ft. lbs. (65 Nm).
   - 3 new rear engine bracket mounting bolts and tighten to 40 ft. lbs. (55 Nm).
   - Intake manifold bracket and tighten the bolts to 16 ft. lbs. (22 Nm).
   - Clutch cover and tighten the bolts to 108 inch lbs. (12 Nm).
   - Subframe center beam with new self-locking bolts. Evenly tighten the bolts to 37 ft. lbs. (50 Nm).
   - Engine stiffener plate, if equipped. Loosely install the mounting bolts. Tighten the stiffener-to-transaxle case mounting bolt to 28 ft. lbs. (39 Nm), then tighten the 2 stiffener-to-engine block mounting bolts to 28 ft. lbs. (39 Nm) beginning with the bolt closest to the transaxle.
   - Radius rod and the damper fork. Hand-tighten all the fasteners.
   - Intermediate shaft and tighten its mounting bolts to 28 ft. lbs. (39 Nm).
   - New set ring on the end of each halfshaft.
   - Right and left halfshafts. Turn the right and left steering knuckle fully outward and slide the axle into the differential, until the set ring is felt engaging the differential side gear.
   - Lower control arm ball joints. Tighten the castle nuts to 40 ft. lbs. (50 Nm). Then, tighten them only enough to install a new cotter pin.
   - Clutch damper and tighten its mounting bolts to 16 ft. lbs. (22 Nm).
   - Front wheels. Lower the vehicle.
16. Place a floor jack under the right front knuckle, and raise the jack until it is supporting the vehicle's weight.
17. Tighten the radius rod mounting bolts to 76 ft. lbs. (105 Nm) and the radius rod nut to 32 ft. lbs. (44 Nm). Tighten the damper fork nut while holding the damper fork bolt to 40 ft. lbs. (55 Nm). Tighten the damper pinch bolt to 32 ft. lbs. (44 Nm).
18. Coat the tip of the slave cylinder with high temperature grease. Install the clutch hose joint and clutch slave cylinder to the transaxle housing. Be sure the slave cylinders tip snaps into the release fork. Tighten the slave cylinder mounting bolts to 16 ft. lbs. (22 Nm).
19. Install or connect the following:

- Speed sensor. Tighten the mounting bolt to 13 ft. lbs. (18 Nm).
- Shift cable and select cable to the shift arm lever. Tighten the cable bracket mounting bolts to 20 ft. lbs. (27 Nm).
- New cotter pins
- Back-up light switch connector
- Starter. Tighten the 10mm bolt to 32 ft. lbs. (45 Nm) and the 12mm bolt to 54 ft. lbs. (75 Nm).
- Starter wires
- Transaxle ground cable

20. Fill the transaxle with the proper type and quantity of oil.
21. Install or connect the following:
- Air cleaner case and the resonator, then the intake duct.
- Battery tray bracket and battery tray and tighten the bolts to 16 ft. lbs. (22 Nm).
- Battery and the battery cables.

22. Check the clutch pedal free-play.
23. Check and adjust the front wheel alignment.
24. Road test the vehicle and check the transaxle for smooth operation.
25. Loosen the 3 front engine mount bracket mounting bolts, then retighten them to 28 ft. lbs. (38 Nm).
26. Enter the radio security code.

Automatic Transaxle

CIVIC 2002–05

1. Before servicing the vehicle, refer to the Precautions Section.
2. Remove or disconnect the following:
- Negative and positive battery cables
- Resonator, the air cleaner box, and the air intake duct.
- Starter cables and the transaxle ground cable
- Engine wiring harness clip

➡**Label all electrical connectors before removing them.**

- Lock-up control solenoid connector
- Vehicle Speed Sensor and countershaft speed sensor connectors
- Upper transaxle case bolts and the rear engine mounting bolt

3. Raise and safely support the vehicle.
4. Remove or disconnect the following:
- Front wheels

5. Drain the automatic transaxle fluid. Then, install the drain plug with a new crush washer. Note the color, consistency, and odor of the drained fluid.
6. Remove or disconnect the following:
- Front splash shield

- Shift control and linear solenoid connectors
- Mainshaft speed sensor connector
- Strut pinch bolt and fork bolt
- Lower ball joint using a ball joint remover.

7. Pry the halfshaft inboard joints out of the transaxle case and intermediate shaft. Swing the steering knuckles out to free the halfshafts from the transaxle.
8. Tie the halfshafts up and out of the way with wire. Tie plastic bags over the inboard joints to prevent damage to the CV-boots and splined shafts.
9. Remove or disconnect the following:
- Front exhaust pipe
- Shift cable cover
- Shift cable from the transaxle control shaft. Move the shift cable out of the way, and tie it up with wire.
- Automatic Transaxle Fluid (ATF) cooler hoses from the cooler lines. Cap the lines to prevent fluid lose and contamination.
- Right-front mount and bracket assembly
- Engine stiffener and the torque converter cover plate.
- 8 torque converter-to-driveplate bolts 1 at a time by rotating the crankshaft pulley.

➡**There are no gear teeth on the driveplate; the starter motor engages a ring gear on the inner edge of the torque converter.**

10. After unbolting the torque converter from the driveplate, rotate the crankshaft to set the engine at Top Dead Center (TDC)/compression for the No. 1 cylinder.
11. Remove or disconnect the following:
- Ignition wires
- Distributor, if equipped

12. Attach a lifting chain to the engine and lift slightly to ease the tension on the mounts.
13. Place a transaxle jack under the transaxle and remove the transaxle side mount and bracket.
14. With the transaxle supported, remove the transaxle rear mount bracket bolts and transaxle case bolts.
15. Pull the transaxle away from the engine until it clears the locating dowel pins. Carefully lower the transaxle from the vehicle with the torque converter angled upward so it doesn't drop out of the transaxle.
16. Remove the torque converter from the transaxle. Inspect the ring gear teeth for breakage and inspect the converter's hub for burrs and scoring. Check the condition of

the converter fluid. Replace the torque converter if necessary.
17. Inspect the transaxles front oil pump bearing and seal for signs of leakage and scoring. Inspect the mainshaft for burrs, scoring, and roughness.
18. With the transaxle removed, carefully inspect the driveplate for stress cracks, enlarged bolt holes, and other defects. Replace it if necessary.

To install:

➡**Use new self-locking nuts and color-coded bolts when installing the transaxle and suspension components.**

19. Flush the transaxle cooler lines to remove any contaminated fluid and residual clutch material:
a. Use a pressurized flusher (Honda J38405-A or equivalent). Use only Honda flushing fluid (Honda J35944–20); other fluids may damage the system.
b. Fill the flusher with 21 ounces of fluid. Pressurize the flusher to 80–120 psi, following the procedure on the fluid container and flusher.
c. Clamp the discharge hose of the flusher to the cooler return line. Clamp the drain hose to the cooler inlet line and route it into a bucket or drain tank.
d. Connect the flusher to air and water lines. The air line use a water trap to keep excess moisture out.
e. Open the flusher water valve and flush the cooler for 10 seconds.
f. Depress the flusher trigger to mix flushing fluid with the water. Flush for 2 minutes, turning the air valve on and off for 5 seconds every 15–20 seconds to create a surging action.
g. After finishing 1 flushing cycle, reverse the hose and flush in the opposite direction.
h. Dry the cooler lines with compressed air for 2 minutes or longer to remove all excess moisture from the system.
20. Install or connect the following:
- Starter motor onto the transaxle case and tighten its mounting bolts to 33 ft. lbs. (45 Nm).
- Torque converter with a new hub O-ring.

21. Place the transaxle on a transaxle jack and raise it to the level of the engine.
22. Align the transaxle and engine. Install the transaxle case bolts. Install new 14mm rear mount bolts and washers.
23. Raise the transaxle and install the side mount. Tighten the case bolts to 47 ft. lbs. (64 Nm). Tighten all of the 14mm rear mount bolts to 61 ft. lbs. (85 Nm).

24. Install or connect the following:
- Transaxle side mount and bracket. Tighten the bracket nuts to 47 ft. lbs. (64 Nm). Tighten the mount through-bolt to 54 ft. lbs. (74 Nm).

25. Remove the transaxle jack.

26. Rotate the crankshaft and install the torque converter-to-driveplate bolts. Tighten the bolts to 108 inch lbs. (12 Nm) in a crisscross pattern. Tighten the bolts to the specification in 2 steps.

27. Rotate the crankshaft to reset the engine at Top Dead Center (TDC)/compression for the No. 1 cylinder. After the engine is set at Top Dead Center (TDC), it must not be disturbed until the distributor, if equipped, has been reinstalled.

28. Install or connect the following:
- Torque converter cover and tighten the bolts to 108 inch lbs. (12 Nm).
- Engine stiffener and tighten the 8mm bolts to 17 ft. lbs. (24 Nm). Tighten the 10mm bolts to 33 ft. lbs. (45 Nm).
- Right-front mount and bracket assembly. Tighten the 10mm bolt to 33 ft. lbs. (44 Nm), and the 12mm bolts to 47 ft. lbs. (64 Nm).

29. Remove the lifting chain and chain hooks.

30. Verify that the engine is at Top Dead Center (TDC)/compression for the No. 1 cylinder. Align the tabs on the distributor drive with the grooves on the end of the camshaft. Install the distributor and hand-tighten the mounting bolts. Reconnect the ignition wires.

31. Tighten the crankshaft pulley to 134 ft. lbs. (181 Nm).

32. Install or connect the following:
- Transaxle cooler lines
- New set rings on the inboard CV-joint splines
- Halfshafts into the transaxle case and intermediate shaft. The inboard joints must snap into place.
- Lower ball joint and damper fork
- Front wheels
- Shift cable linkage to the transaxle control shaft.
- New lockwasher and tighten the linkage bolt to 10 ft. lbs. (14 Nm).
- Shift cable cover and tighten its bolt to 16 ft. lbs. (22 Nm).
- Front exhaust pipe. Use new self-locking nuts and gaskets. Tighten the rear flange nuts to 16 ft. lbs. (22 Nm), and the front flange nuts to 47 ft. lbs. (64 Nm).
- VSS and countershaft speed sensor connectors
- Lock-up control solenoid connector.

- Shift control and linear solenoid connectors
- Mainshaft speed sensor connector
- Wiring harness clamps and starter cables
- Resonator, air cleaner box, and air intake duct
- Front splash shield

33. Lower the vehicle and tighten the strut pinch bolts to 32 ft. lbs. (44 Nm). Tighten the fork bolts to 47 ft. lbs. (65 Nm). Tighten the ball joint castle nuts to 40 ft. lbs. (55 Nm), then tighten them only enough to install new cotter pins.

34. Refill the transaxle with fresh ATF. Use only Honda Premium ATF or DEXRON®II or III ATF. Reconnect the positive and negative battery cables.
   a. Leave the flusher drain hose attached to the cooler return line.
   b. With the transaxle in park, run the engine for 30 seconds, or until approximately 1 quart of fluid is discharged. Immediately shut the engine off. This completes the cooler flushing process.
   c. Remove the drain hose and reconnect the cooler return line.
   d. Refill the transaxle to the proper level.

35. Check shift cable and throttle cable adjustments.

36. Check the ignition timing. Rotate the distributor counterclockwise to advance the timing, or clockwise to retard the timing. When the timing has been set, tighten the distributor mounting bolts to 13 ft. lbs. (18 Nm).

37. Start the engine and shift through all the gears 3 times.

38. Let the engine warm up to operating temperature and check the fluid level with the transaxle in the **P** or **N** position.

39. Check and adjust the front wheel alignment.

40. Road test the vehicle. Recheck the transaxle fluid level.

CIVIC–2006

1. Before servicing the vehicle, refer to the Precautions Section.

2. The radio may contain a coded theft protection circuit. Always obtain the code number before disconnecting the battery.

3. Remove the cowl cover and the under cowl panel. Remove the front grille cover.

4. Disconnect the negative battery cable. Disconnect the positive battery cable. Remove the battery.

5. Remove the air cleaner housing and air intake duct. Remove the battery base and resonator.

Raise and support the vehicle safely. Remove the splash shield. Drain the transaxle fluid.

6. Lower the vehicle. Position the hood in the vertical position.

7. Remove the mounting bolts securing the harness cover and remove the harness clamp.

8. Disconnect the transaxle pressure control switch solenoid valve, solenoid valve B connector and solenoid valve C connector.

9. Remove the harness clamp from its bracket. Remove the air cleaner housing mounting bracket. Disconnect the transaxle range switch connector.

10. Disconnect the output shaft speed sensor connector and remove the harness clamps from the clamp brackets.

11. Disconnect the input shaft speed sensor connector and the second clutch transaxle fluid pressure switch connector.

12. Disconnect the ATF warmer hoses from the transaxle fluid lines. Plug the hoses and lines to prevent fluid leakage.

13. Remove the bolts securing the ATF warmer. Do not disconnect the ATF hoses and water by pass hose from the ATF warmer.

14. Disconnect the shift solenoid harness connector and the third clutch transaxle fluid pressure switch connector. Remove the harness clamps from the clamp brackets.

15. Remove the harness clamp from its bracket. Remove the radiator hose from the clamp. Remove the air cleaner hose mounting bracket.

16. Install the support eyelet part number 07AAK-SNAA400, or equivalent behind the breather pipe and down to the threaded hole on the cylinder head. Attach another support eyelet to the cylinder head with the support bolt part number 07AAK-SNAA500, or equivalent. Hand tighten the bolt.

17. Install tool VSB02C000025, or equivalent onto the engine hanger. Carefully position the engine hanger on the vehicle and support the engine and transaxle assembly.

18. Remove the nuts and bolt securing the lower arm and ball joint. Separate the lower arms from the ball joints. Remove both body mount brackets. Remove the steering gearbox mounting bracket bolts.

19. Remove the rear steering gearbox mounting bolt, stiffner mounting bolt and stiffner. Remove the bolt securing the power steering fluid line clamp bracket.

20. Remove the steering gearbox mounting bolt, stiffner mounting bolt and stiffner. Remove the bolt securing the power steer-

ing fluid line bracket on the right of the front subframe. Remove the power steering line from the clamp. Remove the lower torque rod bolts.

21. Make reference marks on the front subframe that line up with the edge of the body. Attach the front subframe adapter tool, VSB02000016 or equivalent, to the subframe. Secure the strap with the stop and tighten the wing nut. Raise the jack and line up the slots in the arms with the bolt holes on the corner of the jack base. Tighten the bolts.

22. Remove the four bolts securing the front subframe. Lower the subframe. Hang the steering gearbox to the side, with rope or wire.

23. Remove the driveshafts from the differential. Remove the shift cable cover. Remove the three bolts securing the shift cable holder. Pry up on the lock tab of the lock washer. Remove the lock bolt and lock washer. Separate the shift cable from the control shaft.

24. Remove the shift cable holder bracket from the transaxle. Remove the torque converter cover. Remove the drive plate bolts. Remove the upper transaxle housing mounting bolts.

25. Remove the transaxle mount bracket bolts. Remove the front transaxle housing mounting bolts. Remove the rear transaxle housing mounting bolts.

26. Lower the transaxle assembly by loosening the wing nut of the engine support hanger. Lift the engine just enough for the transaxle to clear the side frame.

27. Position a jack under the transaxle. Remove the lower transaxle housing mounting bolts.

28. Slide the transaxle away from the engine and remove it from the vehicle.

29. Remove the torque converter and dowel pins.

To install:

30. Install the torque converter on the mainshaft, using a new O-ring.

31. Install the dowel pins in the torque converter housing. Place the transaxle on the transaxle jack. Raise the unit to engine level.

32. Install the lower transaxle housing mounting bolt part way in the bolt hole on the engine. Attach the transaxle to the engine.

33. Install the lower transaxle housing mounting bolt, tighten the bolts.

34. Continue the installation in the reverse order of the removal procedure.

35. Check and adjust the front wheel alignment

36. Properly fill and check the automatic transaxle fluid level.

37. Reprogram the power window control unit as follows.

38. Turn the ignition switch ON. Lower the window, all the way down. Open the driver's side door.

39. Turn the ignition switch OFF. Push and hold the drivers window DOWN switch. Turn the ignition switch ON. Release the driver's window DOWN switch. This must be done within five seconds.

40. Repeat the above step three more times. Wait one second.

41. Confirm that the AUTO UP and AUTO DOWN do not work. If they work repeat the procedure, paying close attention to the five second time limit

42. Move the driver's window all the way down by holding the drivers window DOWN switch to the AUTO DOWN position.

43. Pull up and hold the driver's window UP switch to the AUTO UP position until the window reaches the fully closed position. Hold the switch for one second.

44. Confirm that the power window master switch has been reset by using the driver's window AUTO UP and DOWN function.

45. If the AUTO UP and DOWN feature is still not working, repeat the complete procedure, paying close attention to the five second time limit.

46. Set the clock.
47. Roadtest the vehicle.

ACCORD–2002 EXCEPT 3.0L ENGINE

1. Before servicing the vehicle, refer to the Precautions Section.

2. Remove or disconnect the following:
   - Negative, then the positive battery cables
   - Battery

3. Shift the transaxle into **N**.

4. Remove or disconnect the following:
   - Air intake hose, air cleaner housing, and the resonator assembly.
   - Battery base and the base stay
   - Throttle cable from the throttle control lever.
   - Transaxle ground cable and the speed sensor connectors.
   - Solenoid valve connectors
   - Lock-up control solenoid valve and shift control solenoid valve connectors.
   - Transaxle cooler hoses from the joint pipes and plug the hoses.
   - Starter cables and starter
   - Countershaft Speed Sensor (CSS) and Vehicle Speed Sensor (VSS) connectors

5. Install a hoist to the engine.
6. Remove or disconnect the following:

   - 4 upper bolts attaching the transaxle to the engine block.
   - 3 bolts attaching the front engine mount bracket to the engine.
   - Transaxle mount

7. Raise and safely support the vehicle. Remove the front wheels.

8. Drain the transaxle fluid and reinstall the drain plug with a new washer.

9. Remove or disconnect the following:
   - Splash shield
   - Subframe center beam
   - Cotter pins and lower arm ball joint nuts, then separate the ball joints from the lower arms using a suitable tool.
   - Right damper pinch bolt, then separate the damper fork and damper.
   - Bolts and nut, then the right radius rod.

10. Using a small prying device, carefully pry the right and left halfshafts out of the differential. Remove the right and left halfshafts. Tie plastic bags over the halfshaft ends to prevent damage to the CV boots and splines.

11. Remove or disconnect the following:
   - Bolts mounting the intermediate shaft, then the intermediate shaft from the differential.
   - Torque converter cover and shift cable cover.
   - Shift control cable by removing the lock bolt.
   - Shift cable lever from the control shaft. Don't disconnect the control lever from the shift cable. Wire the shift cable out of the work area and be careful not to kink it.
   - 8 drive plate bolts, one at a time while rotating the crankshaft pulley.

12. Place a suitable jack under the transaxle and raise the jack just enough to take weight off of the mounts.

13. Remove or disconnect the following:
   - Intake manifold bracket.
   - Transaxle housing mounting bolts
   - Mounting bolts from the rear engine mount bracket.
   - 4 transaxle housing mounting bolts and 3 mount bracket nuts.

14. Pull the transaxle away from the engine until it clears the 14mm dowel pins, then lower it using the jack.

To install:

➡**Use new self-locking nuts when assembling the front suspension components. Install new set rings onto the halfshaft inboard joint splines. Replace any color-coded self-locking bolts.**

15. Flush the transaxle cooler lines before installing the transaxle. Use a pres-

surized flushing unit such as Honda J38405-A or equivalent. Use only Honda biodegradable flushing fluid, Honda J35944–20. Other fluids will damage the automatic transmission cooling system.

   a. Fill the flusher with 21 ounces of fluid. Pressurize the flusher to 80–120 psi, following the procedure on the fluid container and flusher.

   b. Clamp the discharge hose of the flusher to the cooler return line. Clamp the drain hose to the cooler inlet line and route it into a bucket or drain tank.

   c. Connect the flusher to air and water lines. Open the flusher water valve and flush the cooler for 10 seconds. The air line should be equipped with a water trap to keep the system dry.

   d. Depress the flusher trigger to mix flushing fluid with the water. Flush for 2 minutes, turning the air valve on and off for 5 seconds every 15–20 seconds to create a surging action.

   e. After finishing 1 flushing cycle, reverse the hose and flush in the opposite direction following the same steps.

   f. Dry the cooler lines with compressed air so that no moisture is left in the cooler system.

16. Be sure the 2, 14mm dowel pins are installed into the torque converter housing.

17. Install or connect the following:
- Torque converter onto the transaxle mainshaft with a new hub O-ring.
- Starter motor onto the transaxle case and tighten the mounting bolts to 33 ft. lbs. (44 Nm).
- Transaxle and transaxle housing mounting bolts: 47 ft. lbs. (65 Nm)
- Rear engine mounting bolts: 40 ft. lbs. (54 Nm)
- Intake manifold bracket and tighten the bolts to 16 ft. lbs. (22 Nm).
- Upper bolts attaching the transaxle to the engine: 47 ft. lbs. (64 Nm)
- Front engine mount bracket bolts: 28 ft. lbs. (38 Nm)
- Transaxle mount and the nuts and bolt that attach the mount. Tighten the nuts first to 28 ft. lbs. (38 Nm), then tighten the bolt to 47 ft. lbs. (64 Nm).

18. Remove the jack from the transaxle.
19. Install or connect the following:
- Torque converter to the drive plate with the 8 bolts. Tighten the bolts in 2 steps in a crisscross pattern: first to 54 inch lbs. (6 Nm), and finally to 108 inch lbs. (12 Nm). Check for free rotation after tightening the last bolt.

- Shift control cable and control cable holder. Tighten the shift cable lock bolt to 10 ft. lbs. (14 Nm). Tighten the shift cable cover bolts to 13 ft. lbs. (18 Nm).
- Torque converter cover and tighten the bolts to 108 inch lbs. (12 Nm).
20. Remove the engine hoist.
21. Install or connect the following:
- Radius rod and damper fork
- Intermediate shaft into the differential and tighten the mounting bolts to 28 ft. lbs. (38 Nm).
- New set ring on the end of each halfshaft.
22. Turn the right steering knuckle fully outward and slide the axle into the differential until the set ring snaps into the differential side gear. Repeat the procedure on the left side.
23. Install or connect the following:
- Damper fork bolts and ball joint nuts to the lower arms: 40 ft. lbs. (55 Nm) with a new cotter pin.
- Subframe center beam and tighten the center beam bolts to 28 ft. lbs. (39 Nm).
- Splash shield
- Front wheels and lower the vehicle.
- Speed sensor connector
24. Support the right front knuckle with a floor jack, until the weight of the vehicle is held by the jack. Tighten the damper fork pinch bolt to 32 ft. lbs. (44 Nm). Tighten the radius rod bolts to 76 ft. lbs. (105 Nm), and the radius rod nut to 32 ft. lbs. (44 Nm). Hold the damper fork bolt with a wrench, and tighten the nut to 40 ft. lbs. (55 Nm). Remove the floor jack.
25. Install or connect the following:
- Cables to the starter
- Throttle control cable
- Lock-up control solenoid valve and shift control solenoid valve connectors
- Speed sensor connectors and the transaxle ground cable
- Transaxle cooler inlet hose to the joint pipe. Attach a drain hose to the return line.
- Battery base stay and the battery base
26. Install the resonator assembly, the air cleaner assembly, and the air intake hose.

- Battery, positive then the negative battery cables
27. Refill the transaxle with ATF. Use only Honda Premium ATF or DEXRON®II ATF.

   a. With the flusher drain hose attached to the cooler return line.

   b. Place the transaxle in **P**, run the engine for 30 seconds, or until approximately 1 quart of fluid is discharged. Immediately shut off the engine. This completes the cooler flushing process.

   c. Remove the drain hose and reconnect the cooler return line.

   d. Refill the transaxle to the proper level with ATF.
28. Start the engine, set the parking brake, and shift the transaxle through all gears 3 times. Check for proper shift cable adjustment.
29. Let the engine reach operating temperature with the transaxle in **P** or **N**. Then, shut off the engine and check the fluid level.
30. Road test the vehicle.
31. After road testing the vehicle, loosen the front engine mount bracket bolts, then retighten them to 28 ft. lbs. (39 Nm).
32. Check and adjust the vehicle's front end alignment.
33. Enter the radio security code.

ACCORD–2002 3.0L ENGINE

1. Before servicing the vehicle, refer to the Precautions Section.
2. Remove or disconnect the following:
- Negative, then the positive battery cables
- Battery and tray
- Clamps securing the battery cables to the base.
- Intake air duct and the air cleaner assembly
3. Raise the vehicle and drain the transaxle fluid. Replace the drain plug with a new washer.
4. Remove or disconnect the following:
- Starter wiring and harness clamps, then the breather and radiator hoses from the retainer.
- Wiring connectors from the transaxle assembly
- Cooler lines; point them up to prevent fluid drainage.
- Bolt and nut, then the rear stiffener.
- Bolts securing the transaxle to the engine.
- Front mounting bracket bolts
- Engine under cover
- Lower shock absorber mounting and the lower ball joints from the control arms.
- Bolts securing the radius rods to the lower arms.
- Halfshafts. Keep the splined ends of the shafts clean.

➡**Matchmark the installed position of the sub-frame on the main-frame before removing it.**

- Sub-frame from the main-frame
- Engine brace from the rear of the engine
- Shift cable cover, bracket and cable
- 8 bolts securing the drive plate to the torque converter

5. Attach a chain hoist to the engine and raise it slightly.

6. Place a jack under the transaxle.

7. Remove or disconnect the following:
- Transaxle mount bracket
- Intake manifold support bracket
- Rear mount bracket

8. Pull the transaxle back slightly until it comes off the dowels and lower it from the vehicle. Do not let the torque converter fall out of the transaxle.

To install:

9. Install or connect the following:
- Torque converter using a new O-ring, if removed.
- Dowel pins in the torque converter housing.
- Transaxle to the engine and the rear mount bracket. Tighten the 8mm bolt to 16 ft. lbs. (22 Nm) and the 12mm bolts to 40 ft. lbs. (54 Nm).
- Transaxle-to-engine bolts: 47 ft. lbs. (64 Nm)
- Breather tube with the dot facing up
- Transaxle mount bracket. Tighten the nuts to 28 ft. lbs. (38 Nm) and the through-bolt to 40 ft. lbs. (54 Nm).
- Driveplate-to-torque converter bolts: 108 inch lbs. (12 Nm) in a crisscross pattern.
- Shift cable, bracket and cover
- Engine brace on the rear of the engine.
- Halfshafts
- Sub-frame after aligning the match-marks. Tighten the rear bolts to 47 ft. lbs. (64 Nm) and the front bolts to 76 ft. lbs. (103 Nm).
- Front mount. Tighten the bolts to 28 ft. lbs. (38 Nm).
- Shock absorbers and the radius rods to the lower control arms.
- Engine under cover
- All the wiring connectors
- Starter wiring and harness clamps
- Battery
- Air cleaner assembly and intake duct

10. Refill the transaxle with Genuine Honda® premium automatic transmission fluid.

ACCORD–2003–2006 EXCEPT 3.0L ENGINE

1. Before servicing the vehicle, refer to the Precautions Section.

2. The radio may contain a coded theft protection circuit. Always obtain the code number before disconnecting the battery.

3. Disconnect the negative battery cable. Disconnect the positive battery cable. Remove the battery.

4. Position the front wheels in the straight ahead position. Lock the steering column. Remove the steering joint cover.

5. Make a reference mark across the steering joint and steering gearbox pinion shaft. Remove the steering joint bolt. Disconnect the steering by removing the steering joint toward the steering column. Hold the slider shaft on the column using a piece of wire between the joint yoke on the slider shaft to the joint yoke on the upper shaft.

6. Remove and plug the return hose from the power steering pump reservoir. Remove the power steering pump outlet line. Remove the bolt securing the power steering hose clamp.

7. Raise and support the vehicle safely. Remove the splash shield. Drain the transaxle fluid.

8. Lower the vehicle. Remove the battery tray. Remove the air intake duct and the air cleaner assembly. Remove the battery base.

9. Disconnect the transaxle clutch pressure control solenoid, valve A connector, second clutch pressure switch connector and remove the harness clamps from the clamp brackets

10. Remove the transaxle range selector switch connector from its bracket. Disconnect it from its mounting.

11. Remove the AF sensor connector from its bracket and disconnect it.

12. Disconnect the input shaft speed sensor connector and the output shaft speed sensor. Remove the harness clamps from the brackets.

13. Disconnect the third clutch transaxle fluid pressure switch connector. Remove the harness clamp from the clamp bracket.

14. Disconnect connector C108, AT pressure control solenoid valve B connector and the solenoid valve C connector. Remove the harness clamp from the clamp bracket.

15. Remove the transaxle cooler lines. Plug the lines to prevent leakage. Remove the cooler hose from the clamp. Disconnect and plug the cooler hose from the cooler line.

16. Remove the ground cable. Transaxle upper mount bracket plate and transaxle upper mount bracket.

17. Attach the special adapter tool VSB02C000015, or equivalent, to the threaded hole in the cylinder head. Install the engine support tool AAR-T-12566 or

equivalent. Lift and support the engine and transaxle assembly.

18. Remove the vacuum hose from its clamp. Disconnect the hose from the vacuum line. Remove the front mount stop and clamp bracket. Remove the front mount bolt.

19. Remove the rear mount stop and remove the rear mount bolt.

20. Remove the exhaust pipe and its mount. Disconnect the power steering pressure switch connector.

21. Insert a 6mm allen wrench in the top of the ball joint pin. Remove the nuts. Separate the stabilizer link from the lower arms.

22. Remove the cotter pins, castle nuts, damper pinch bolt, self locking nut, bolt and damper forks. Separate the ball joints from the lower arms.

23. Remove the cotter pins and nuts. Separate the tie rod end ball joints from the knuckles.

24. Remove the engine stiffner and remove the drive plate bolts. Remove the shift cable holder and cover.

25. Remove the spring clip and control pin, then separate the shift cable from the selector control lever.

26. Remove the transaxle lower mount nuts. Remove both mid mounts.

27. Make a reference line at both ends of the subframe that line up with the edge of the stiffners.

28. Attach the front subframe adapter tool, VSB02000016 or equivalent, to the subframe. Secure the strap with the stop and tighten the wing nut. Raise the jack and line up the slots in the arms with the bolt holes on the corner of the jack base. Tighten the bolts.

29. Remove the four bolts securing the front subframe. Lower the subframe. Remove the transaxle lower mounts. Remove the driveshaft boot cover and bracket. Pry the driveshafts and remove them from the differential. Remove the rear mount bracket.

30. Position a jack under the transaxle. Remove the upper transaxle housing mounting bolts. Remove the front mount bracket. Remove the front transaxle housing mounting bolts.

31. Remove the rear transaxle housing mounting bolts.

32. Slide the transaxle away from the engine and remove it from the vehicle.

To install:

33. Install the torque converter on the mainshaft, using a new O-ring.

34. Install the dowel pins in the torque converter housing. Place the transaxle on the transaxle jack. Raise the unit to engine

level. Attach the transaxle to the engine. Install the retaining bolts.

35. Continue the installation in the reverse order of the removal procedure.

36. Check and adjust the front wheel alignment

37. Properly fill and check the automatic transaxle fluid level.

38. Reprogram the power window control unit as follows.

39. Turn the ignition switch ON. Lower the window, all the way down. Open the driver's side door.

40. Turn the ignition switch OFF. Push and hold the drivers window DOWN switch. Turn the ignition switch ON. Release the driver's window DOWN switch. This must be done within five seconds.

41. Repeat the above step three more times. Wait one second.

42. Confirm that the AUTO UP and AUTO DOWN do not work. If they work repeat the procedure, paying close attention to the five second time limit

43. Move the driver's window all the way down by holding the drivers window DOWN switch to the AUTO DOWN position.

44. Pull up and hold the driver's window UP switch to the AUTO UP position until the window reaches the fully closed position. Hold the switch for one second.

45. Confirm that the power window master switch has been reset by using the driver's window AUTO UP and DOWN function.

46. If the AUTO UP and DOWN feature is still not working, repeat the complete procedure, paying close attention to the five second time limit.

47. Set the clock.

48. Roadtest the vehicle.

ACCORD–2003–2006 3.0L ENGINE

1. Before servicing the vehicle, refer to the Precautions Section.

2. The radio may contain a coded theft protection circuit. Always obtain the code number before disconnecting the battery.

3. Disconnect the negative battery cable. Disconnect the positive battery cable. Remove the battery.

4. Position the front wheels in the straight ahead position. Lock the steering column. Remove the steering joint cover.

5. Make a reference mark across the steering joint and steering gearbox pinion shaft. Remove the steering joint bolt. Disconnect the steering by removing the steering joint toward the steering column. Hold the slider shaft on the column using a piece of wire between the joint yoke on the slider shaft to the joint yoke on the upper shaft.

6. Remove and plug the return hose from the power steering pump reservoir. Remove the power steering pump outlet line. Remove the bolt securing the power steering hose clamp.

7. Remove the bolts securing the support strut brackets on both sides of the hood. Secure the hood in a vertical position, and then install the right strut bracket by turning it over with its bolt in the lower position.

8. Raise and support the vehicle safely. Remove the splash shield. Remove the power steering pump outlet line from the power steering pump. Remove the hose from its clamp.

9. Drain the transaxle fluid. Remove the battery tray.

10. Remove the air intake duct and air cleaner assembly. Remove the battery base. Remove the starter.

11. Remove the transaxle upper mount bracket and bracket plate. Remove the transaxle sub harness connector from its bracket and disconnect it.

12. Disconnect the input shaft speed sensor connector, the fourth clutch transaxle fluid pressure switch connector and disconnect the vacuum hose from the vacuum line.

13. Disconnect the shift solenoid valve A connector, the A/T clutch pressure solenoid C connector. Remove the harness clamps from the clamp brackets.

14. Disconnect the A/T clutch pressure control solenoid valve A connector, the solenoid valve B connector. Remove the harness clamps from the clamp brackets.

15. Disconnect the connectors from the torque converter clutch solenoid valve and shift solenoid valve. Remove the transaxle range switch connector from its bracket and disconnect it.

16. Disconnect the output shaft speed sensor connector and remove the harness clamps from the brackets. Disconnect the vacuum hose from the vacuum line.

17. Remove the harness clamps from the brackets and remove the bolt securing the bracket. Remove the transaxle ground cable and disconnect the breather tube.

18. Remove the bolts securing the ATF warmer and the bracket. Remove the ATF warmer from the transaxle housing. Do not disconnect the hoses.

19. Remove the connector bracket from the engine front cylinder head, use the bracket bolt to attach the engine balancer bar front arm.

20. Remove the harness clamp bracket from the engine rear cylinder head, use the bracket bolt hole to attach the engine balancer bar rear arm.

21. Remove the front bulkhead cover. Lift and support the engine and transaxle assembly using tool VSB02C000019, or equivalent.

22. Remove the front stop mount. Remove the front mount.

23. Remove the rear stop mount. Remove the rear mount.

24. Insert a 6mm allen wrench in the top of the ball joint pin. Remove the nuts. Separate the stabilizer link from the lower arms.

25. Remove the cotter pins, castle nuts, damper pinch bolt, self locking nut, bolt and damper forks. Separate the ball joints from the lower arms.

26. Remove the cotter pins and nuts. Separate the tie rod end ball joints from the knuckles.

27. Remove the exhaust pipe and its mount. Remove the power steering fluid hose from its line on the front subframe. Disconnect the power steering pressure switch connector.

28. Remove the torque converter cover. Remove the drive plate bolts. Remove the engine to torque converter housing mounting bolts.

29. Remove the bolts securing the shift cable holder. Remove the shift cable cover. To prevent damage to the control lever joint, remove the bolts securing the holder before removing the bolts securing the cover.

30. Remove the lock bolt securing the control lever. Remove the shift cable and control lever.

➡**Install a 6x1.0-14mm bolt and nut on the shift cable cover. Reinstall the shift cable cover to the torque converter housing. If you do not do this, the bolt head of the cable cover may prevent you from removing the torque converter during transaxle removal.**

31. Remove the transaxle lower mount nuts. Remove both middle mounts.

32. Make a reference line at both ends of the subframe that line up with the edge of the stiffners.

33. Attach the front subframe adapter tool, VSB02000016 or equivalent, to the subframe. Secure the strap with the stop and tighten the wing nut. Raise the jack and line up the slots in the arms with the bolt holes on the corner of the jack base. Tighten the bolts.

34. Remove the four bolts securing the stiffners and the four bolts securing the front subframe. Lower the subframe. Remove the transaxle lower mounts.

35. Remove the driveshafts from the dif-

ferential and intermediate shaft. Remove the exhaust manifold bracket and heat shield. Remove the intermediate shaft.

36. Position a jack under the transaxle. Remove the upper transaxle housing mounting bolts. Remove the front mount bracket. Remove the front transaxle housing mounting bolts.

37. Remove the bolt securing the harness clamp bracket. Remove the front mount bracket. Remove the rear transaxle housing mounting bolts.

38. Slide the transaxle away from the engine and remove it from the vehicle.

To install:

39. Install the torque converter on the mainshaft, using a new O-ring.

40. Install the dowel pins in the torque converter housing. Place the transaxle on the transaxle jack. Raise the unit to engine level. Attach the transaxle to the engine. Install the retaining bolts.

41. Continue the installation in the reverse order of the removal procedure.

42. Check and adjust the front wheel alignment

43. Properly fill and check the automatic transaxle fluid level.

44. Reprogram the power window control unit as follows.

45. Turn the ignition switch ON. Lower the window, all the way down. Open the driver's side door.

46. Turn the ignition switch OFF. Push and hold the drivers window DOWN switch. Turn the ignition switch ON. Release the driver's window DOWN switch. This must be done within five seconds.

47. Repeat the above step three more times. Wait one second.

48. Confirm that the AUTO UP and AUTO DOWN do not work. If they work repeat the procedure, paying close attention to the five second time limit

49. Move the driver's window all the way down by holding the drivers window DOWN switch to the AUTO DOWN position.

50. Pull up and hold the driver's window UP switch to the AUTO UP position until the window reaches the fully closed position. Hold the switch for one second.

51. Confirm that the power window master switch has been reset by using the driver's window AUTO UP and DOWN function.

52. If the AUTO UP and DOWN feature is still not working, repeat the complete procedure, paying close attention to the five second time limit.

53. Set the clock.
54. Roadtest the vehicle.

Clutch

REMOVAL & INSTALLATION

1. Before servicing the vehicle, refer to the Precautions Section.
2. Remove or disconnect the following:
   • Negative battery cable
3. Raise and safely support the vehicle.
4. Remove or disconnect the following:
   • Transmission from the vehicle. Matchmark the flywheel and clutch for reassembly.
5. Use a flywheel ring-gear holder to lock the flywheel in position.
6. Remove or disconnect the following:
   • Pressure plate bolts, 2 turns at a time working in a crisscross pattern to prevent warping the pressure plate.
   • Pressure plate and clutch disc
7. Inspect the flywheel, disc, and pressure plate for wear, cracks, and warpage. Light scoring of the flywheel may be polished out; gouges, warpage, burn marks, cracks, or chipped teeth require replacement of the flywheel.

➡If the flywheel is to be removed, but is going to be reused, matchmark it to the engine block prior to removal. Aligning the matchmarks upon reassembly will preserve driveline balance.

8. Inspect the flywheel's ball bearing: turn the inner race of the bearing with your finger, and be sure it turns smoothly and quietly. If the bearing is loose or noisy, or exhibits rough motion, replace it.
9. Remove or disconnect the following:
   • Release fork boot. Squeeze the release fork retaining spring to disengage the fork from its pivot.
   • Release fork from the clutch housing.

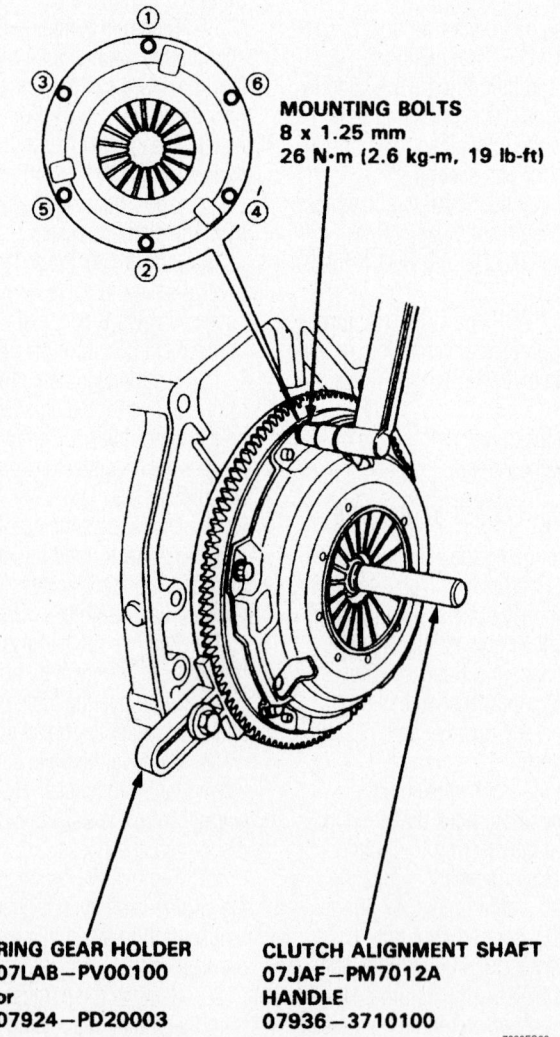

MOUNTING BOLTS
8 x 1.25 mm
26 N·m (2.6 kg-m, 19 lb-ft)

RING GEAR HOLDER
07LAB—PV00100
or
07924—PD20003

CLUTCH ALIGNMENT SHAFT
07JAF—PM7012A
HANDLE
07936—3710100

7923FG99

Clutch alignment tools and pressure plate torque sequence

- Release bearing. Spin the bearing by hand to check its degree of play. Replace the release bearing if it has excessive play or is leaking grease.

10. Inspect the rear main bearing oil seal for signs of leakage. If necessary, replace the seal to prevent oil leakage onto the clutch's friction surfaces.

To install:

11. If necessary, drive out the flywheel bearing, then use a suitably-sized bearing driver to install a new one. Use a crisscross pattern to tighten the flywheel mounting bolts in several steps.

12. Install or connect the following:
- Clutch disc and pressure plate by aligning the dowels on the flywheel with the dowel holes in the pressure plate. If a new pressure plate is not being installed, align the matchmarks that were made during removal.
- Pressure plate bolts, hand-tight

13. Insert a suitable clutch disc alignment tool into the splined hole in the clutch disc. Align the clutch and pressure plate.

14. Tighten the pressure plate bolts in a crisscross pattern 2 turns at a time to prevent warping the pressure plate. The final torque is 19 ft. lbs. (26 Nm).

15. Remove the alignment tool and ring gear holder.

16. Coat the mainshaft with heavy-duty high-temperature grease. The manufacturer recommends part No. 08798–9002, Honda super high-temp urea grease.

17. Coat the release fork pawls and the inner race of the release bearing with high temperature grease and install them into the clutch housing. Be sure the release fork retainer spring snaps into place on the pivot. The bearing and fork must fit together properly and slide back and forth smoothly.

18. Coat the tip of the slave cylinder with grease. Install the release fork boot.

19. Install or connect the following:
- Transmission, making sure the mainshaft is properly aligned with the clutch disc splines, and the transmission case dowels are properly aligned with the engine block.
- Transmission case bolts: 47 ft. lbs. (65 Nm), sequentially

20. Bleed the clutch hydraulic system.

21. Adjust the clutch pedal free-play.

22. Verify that all engine and transaxle components are installed and connected properly.

23. Reconnect the negative battery cable.

24. Road test the vehicle.

Hydraulic Clutch System

BLEEDING

1. Before servicing the vehicle, refer to the Precautions Section.

2. Fill the clutch master cylinder reservoir with clean DOT 3 or 4 brake fluid.

3. Attach a rubber tube to the clutch slave cylinder bleed screw. Route the tube into a container of clean brake fluid.

4. Loosen the bleed screw.

5. Slowly pump the clutch pedal until the fluid draining from the slave cylinder is free of air bubbles.

6. Tighten the bleed screw to 72–84 inch lbs. (8–10 Nm).

7. Refill the clutch master cylinder reservoir with brake fluid.

Halfshaft

REMOVAL & INSTALLATION

Civic–2002—2005

1. Before servicing the vehicle, refer to the Precautions Section.

2. Raise and support the vehicle safely. Remove the front tires.

3. Lift up the locking tab on the spindle nut. Remove the nut. Drain the transaxle fluid.

4. Hold the stabilizer ball joint pin with a hex wrench and remove the flange nut. Separate the front stabilizer link from the lower arm.

5. Remove the lock pin from the lower arm ball joint castle nut. Remove the nut.

6. Separate the ball joint from the lower arm. Pull the knuckle outward and remove the outboard joint from the front wheel hub, using a plastic hammer.

7. Pry the inboard joint, with a pry bar, and remove the halfshaft from the differential case as an assembly.

➡**Do not pull on the halfshaft because the inboard joint may come apart. Pull the halfshaft straight out to avoid damaging the oil seal.**

To install:

8. Installation is the reverse of the removal procedure.

9. After the front tire is installed, turn the front wheel by hand and make sure there is no interference between the halfshaft and surrounding parts.

10. As required, refill the transaxle using the proper grade and type fluid.

11. Check the front wheel alignment. Adjust if necessary.

Civic–2006

1. Before servicing the vehicle, refer to the Precautions Section.

2. Raise and support the vehicle safely. Remove the front tires.

3. Lift up the locking tab on the spindle nut. Remove the nut. Drain the transaxle fluid.

4. Remove the halfshaft outboard joint from the front wheel hub, using a plastic hammer.

5. Remove the nuts and bolts and separate the lower arm using a prybar.

6. Pull the knuckle outward, and remove the halfshaft outboard joint from the front wheel hub.

7. Left and right halfshaft, except manual transaxle right halfshaft, pry the inboard joint from the differential case with a prybar.

8. Right halfshaft on manual transaxle vehicles, drive the inboard joint off of the intermediate shaft with a drift and hammer. Remove the halfshaft as an assembly.

➡**Do not pull on the halfshaft because the inboard joint may come apart. Pull the halfshaft straight out to avoid damaging the oil seal.**

To install:

9. Installation is the reverse of the removal procedure.

10. After the front tire is installed, turn the front wheel by hand and make sure there is no interference between the halfshaft and surrounding parts.

11. As required, refill the transaxle using the proper grade and type fluid.

12. Check the front wheel alignment. Adjust if necessary.

S2000

1. Before servicing the vehicle, refer to the Precautions Section.

2. Raise and support the vehicle safely. Remove the rear tires.

3. Lift up the locking tab on the spindle nut. Remove the nut.

4. Remove the cotter pin from the lower arm ball joint castle nut. Remove the nut. Separate the ball joint from the lower arm.

5. Remove the wheel sensor harness from the upper arm.

6. Make reference marks across the inboard joint and the rear differential.

7. Remove the six inboard joint mounting bolts and nuts. Remove the inboard joint from the rear differential.

8. Pull the knuckle outward and remove the outboard joint from the wheel hub using a plastic hammer.

9. Remove the driveshaft.

To install:

10. Installation is the reverse of the removal procedure.

11. After the front tire is installed, turn the front wheel by hand and make sure there is no interference between the driveshaft and surrounding parts.

12. Check the rear wheel alignment. Adjust if necessary.

Accord–2002

1. Before servicing the vehicle, refer to the Precautions Section.

2. Loosen the front spindle nut.

3. Raise and safely support the vehicle.

4. Remove or disconnect the following:
   - Front wheels and the spindle nut

5. Drain the transaxle fluid and install the drain plug with a new washer. If the halfshaft to be removed is installed into the intermediate shaft, the transaxle fluid does not need to be drained.
   - Flange nut, while holding the stabilizer ball joint pin with a hex wrench
   - Front stabilizer link from the lower arm
   - Damper fork nut and damper pinch bolt(s)
   - Damper fork
   - Cotter pin and castle nut from the lower arm ball joint.

6. Install a hex nut flush onto the ball joint stud to prevent the ball joint tool from damaging the stud threads.

7. Using a ball joint tool, separate the lower arm from the knuckle.

8. Pull the knuckle outward.

9. Remove or disconnect the following:
   - Heat shield, if equipped
   - Halfshaft outboard joint from the hub by tapping it with a plastic hammer.
   - Inner CV-joint away from the transaxle case to force the halfshaft set ring out of the groove.
   - Halfshaft from the differential case or intermediate shaft by pulling on the inboard CV-joint

➡**Do not pull on the halfshaft as the CV-joint may come apart. Use care when prying out the assembly and pull it straight to avoid damaging the differential oil seal or intermediate shaft oil or dust seals.**

To install:

10. Replace the differential oil seal or intermediate shaft seal if either were damaged during removal.

11. Install or connect the following:

- New set rings on the ends of the halfshafts

12. Apply 0.02–0.04 oz. of grease to the whole splined surface of the right side halfshaft. Then, remove the grease from the splined grooves at intervals of 2–3 splines and from the set ring groove to let air bleed from the intermediate shaft.
   - Halfshafts and be sure the set ring locks in the differential gear groove and the halfshaft bottoms in the differential or intermediate shaft.
   - Outboard joint into the hub. Be sure the splines mesh together and the joint is fully seated into the hub.
   - Heat shield, if equipped
   - Ball joint stud into the lower control arm.

13. Torque the new castle nut to the lower of the following specifications, then tighten only enough to install a new cotter pin. Never loosen the nut to install the cotter pin.
   a. Accord: 36–43 ft. lbs. (49–59 Nm)

14. Install or connect the following:
   - Damper fork into position. Make sure the aligning tab is lined up with the slot in the damper fork. Tighten the retainers loosely
   - Front stabilizer link to the lower arm. Hold the stabilizer link ball joint pin with a hex wrench and tighten the new flange nut to 22 ft. lbs. (30 Nm).
   - Tighten the upper damper pinch bolt to 32 ft. lbs. (44 Nm) and the fork nut to 47 ft. lbs. (65 Nm).
   - Front wheels
   - New spindle nut; don't tighten it yet.

15. Lower the vehicle.

16. Tighten the spindle nut to 134 ft. lbs. (181 Nm) for the 4-cylinder Accord with A/T. For all other models, tighten the spindle nut to 181 ft. lbs. (245 Nm) and stake its tab. Tighten the wheel nuts to 80 ft. lbs. (110 Nm).

17. Fill the transaxle with the proper type and quantity of fluid.

18. Warm the engine up, check the transaxle fluid level, and road test the vehicle.

Accord–2003–2006

1. Before servicing the vehicle, refer to the Precautions Section.

2. Raise and support the vehicle safely. Remove the front tires.

3. Lift up the locking tab on the spindle nut. Remove the nut. Drain the transaxle fluid.

4. Hold the stabilizer ball joint pin with a hex wrench and remove the flange nut. Separate the front stabilizer link from the lower arm.

5. Remove the lock pin from the lower arm ball joint castle nut. Remove the nut.

6. Separate the ball joint from the lower arm. Pull the knuckle outward and remove the outboard joint from the front wheel hub, using a plastic hammer.

7. If equipped with automatic transaxle, remove the heat shield.

8. If removing the left halfshaft, pry the inboard joint from the differential case with a prybar.

9. If removing the right halfshaft on manual transaxle vehicles, drive the inboard joint off of the intermediate shaft with a drift hammer

10. If removing the right halfshaft on automatic transaxle vehicles, pry the inboard joint from the differential case with a prybar.

➡**Do not pull on the halfshaft because the inboard joint may come apart. Pull the halfshaft straight out to avoid damaging the oil seal.**

To install:

11. Installation is the reverse of the removal procedure.

12. After the front tire is installed, turn the front wheel by hand and make sure there is no interference between the halfshaft and surrounding parts.

13. As required, refill the transaxle using the proper grade and type fluid.

14. Check the front wheel alignment. Adjust if necessary.

CV-Joints

OVERHAUL

1. Before servicing the vehicle, refer to the Precautions Section.

2. Remove the halfshaft.

3. Remove the large retaining band from the inboard boot. Remove the smaller band from the inboard boot and slide the boot off the joint.

4. Carefully remove the stub end of the inboard joint. Check the splines for cracks, wear or damage. Check the inside bore for any sign of wear.

5. Remove and discard the snapring from the end of the halfshaft. This will allow removal of the spider assembly.

6. Mark the rollers, spider and the stub end of the axle so that all parts may be

reassembled in the same position. Remove the rollers from the spider.

7. Remove the second snapring from the shaft. Remove the joint boot. If equipped, remove the dynamic damper from the shaft.

8. If the outer joint's boot is to be replaced, remove the boot clamps and slide the boot off the joint, then off the shaft. Hold the outer joint and swivel the end. If the joint is noisy, it must be replaced. The replacement joint will come with a new shaft; the inner joint must be assembled onto the shaft.

9. Clean and inspect all disassembled parts. Any sign of wear requires replacement.

To install:

10. Thoroughly pack the inboard and outboard joints with moly grease. Use only moly grease; other lubricants will not last. Wrap the splines of the shaft in vinyl or electrical tape to protect the boots as they are installed.

11. Slide the boot for the outer joint over the shaft and onto the joint. Do not install the bands yet.

12. Slide the inner boot onto the shaft. Install the dynamic damper if it was removed.

Left driveshaft:
  Japan-produced models:
    M/T: 788—793 mm (31.0—31.2 in.)
    A/T: 792—797 mm (31.2—31.4 in.)
  U.S.-and Canada-produced models:
    M/T: 788.2—793.2 mm (31.0—31.2 in.)
    A/T: 791.7—796.7 mm (31.2—31.4 in.)

Right driveshaft:
  Japan-produced models:
    2001-2002 models: 502—507 mm
                              (19.8—20.0 in.)
    2003-2005 models M/T: 497—502 mm
                              (19.6—19.8 in.)
    2003-2005 models A/T: 502—507 mm
                              (19.8—20.0 in.)
  U.S.-and Canada-produced models:
    501.2—506.2 mm (19.7—19.9 in.)

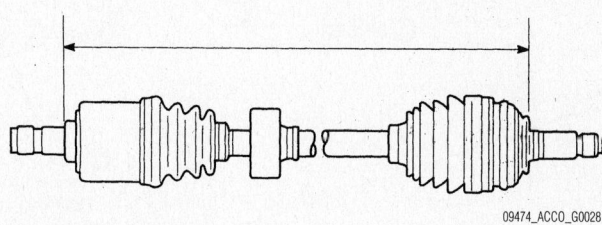

Halfshaft length specifications—Civic

LEFT DRIVESHAFT

RIGHT DRIVESHAFT

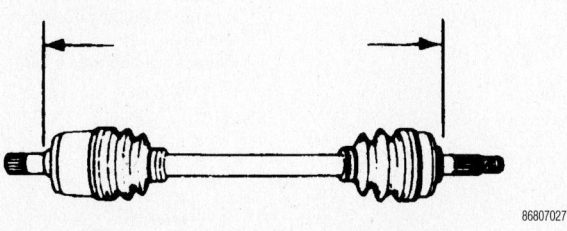

Halfshafts must be set to the correct length before installing boot bands

BOOT BANDS
Bend both sets of locking tabs.

LOCKING TABS

86807028

Always use new boot bands

13. Install the inboard snapring on the shaft. Install the rollers and bearing races on the spider shafts. Hold the shaft upright, then slide the spider assembly into the inboard shaft joint. Install the outer snapring.

14. Slide the boots over both joints. Position the small end of the boot so that the band will be centered between the locating humps on the shaft. Install the band; bend both sets of locking tabs. Once the band is in place, expand and compress the boots once or twice; allow the boots to return to their normal size and length.

15. Adjust the length of the halfshaft to

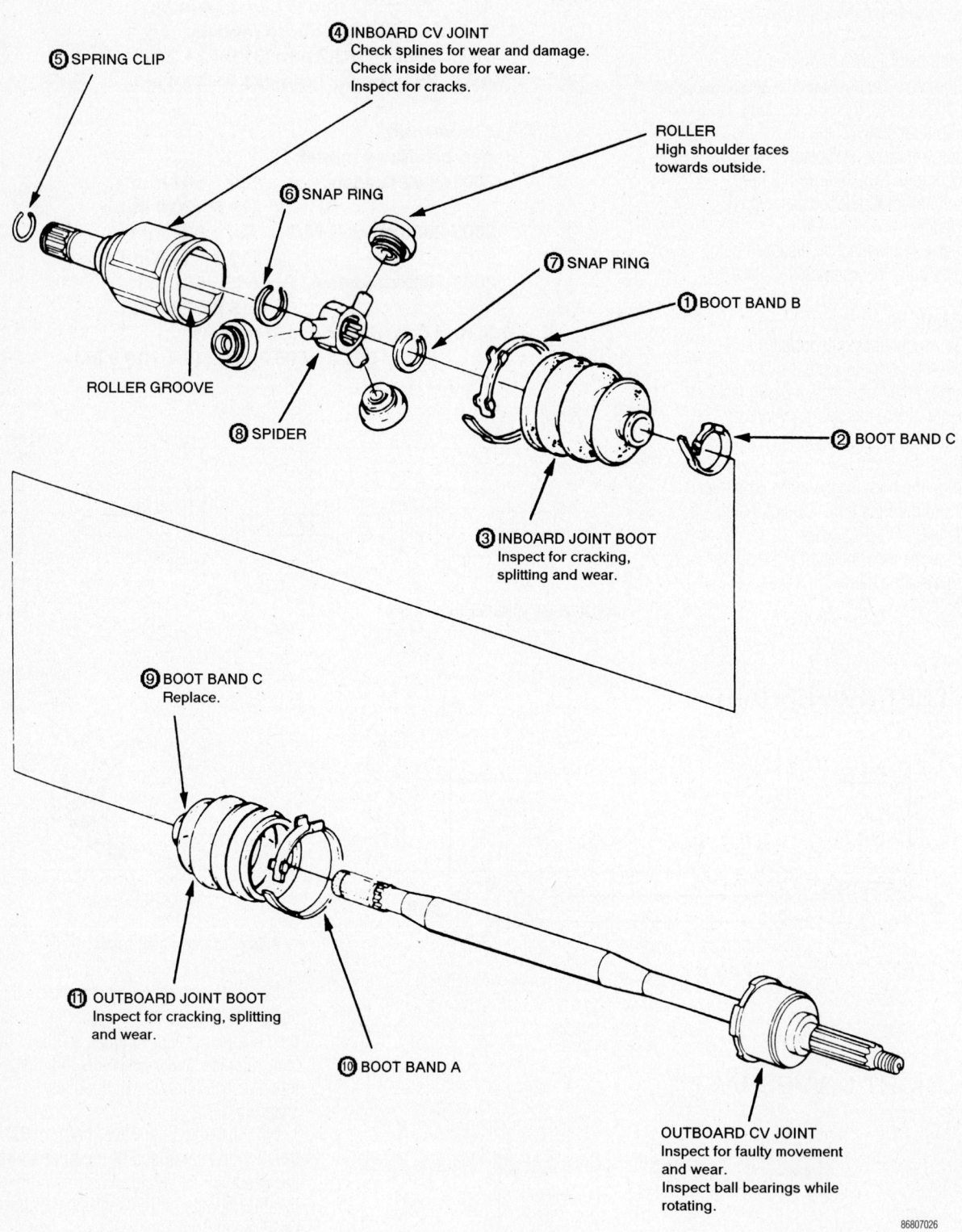

⑤ SPRING CLIP

④ INBOARD CV JOINT
Check splines for wear and damage.
Check inside bore for wear.
Inspect for cracks.

ROLLER
High shoulder faces
towards outside.

⑥ SNAP RING

⑦ SNAP RING

① BOOT BAND B

ROLLER GROOVE

⑧ SPIDER

② BOOT BAND C

③ INBOARD JOINT BOOT
Inspect for cracking,
splitting and wear.

⑨ BOOT BAND C
Replace.

⑪ OUTBOARD JOINT BOOT
Inspect for cracking, splitting
and wear.

⑩ BOOT BAND A

OUTBOARD CV JOINT
Inspect for faulty movement
and wear.
Inspect ball bearings while
rotating.

86807026

Exploded view of the halfshaft

the specifications listed below, then adjust the boots to halfway between full compression and full extension. Make sure the ends of the boots seat in the groove of the driveshaft and joint. Correct shaft lengths are:

- 2002 Accord with M/T—Left and right shafts: 19.1–19.3 in. (486–491mm).
- 2002 Accord with A/T—Left shaft: 33.3–33.5 in. (845–850mm). Right shaft: 19.1–19.9 in. (486–491mm).
- 2003–06 Accord (4 cyl.) with M/T—Left shaft: 21.42–21.61 in. (544–549mm).Right shaft: 18.62–18.82 in. (473–478mm).
- 2003–06 Accord (4 cyl.) with A/T—Left shaft: 21.61–21.81 in. (549–554mm). Right shaft: 33.43–33.62 in. (849–854mm).
- 2003–06 Accord (V6)—Left shaft: 21.81–22.01 in. (554–559mm). Right shaft: 20.16–20.35 in. (512–517mm)
- Civic—see accompanying illustration.
- S2000—Left halfshaft: 22.8–23 inches (579–584 mm). Right halfshaft—24.6–24.8 inches (624–629 mm).

16. Install new boot bands on the large ends of the boots. Be sure to bend both sets of locking tabs. Lightly tap the doubled-over portion of the band(s) to reduce the height. Do NOT hit the boot.
17. Position the dynamic damper so that it is 0.1–1.2 in. (3–7mm) from the CV-boot. Install a new retaining band in the same fashion as the boot bands.
18. Install a new snapring on the inboard end of the joint, then install the halfshaft.

Pinion Seal

REMOVAL & INSTALLATION

S2000

1. Before servicing the vehicle, refer to the Precautions Section.
2. Drain the axle housing fluid.
3. Remove or disconnect the following:

- Negative battery cable
- Rear wheels
- Driveshaft
- Brake calipers and pads

➡The brake calipers and pads must be removed so that there is no additional drag when measuring pinion bearing preload.

4. Use an inch lb. torque wrench and measure and record the amount of torque required to maintain pinion rotation through several revolutions.
5. Remove or disconnect the following:

- Pinion flange
- Pinion seal
- Pinion bearing
- Collapsible spacer

To install:

➡Use a new collapsible spacer and flange nut for assembly.

6. Install or connect the following:
- Collapsible spacer
- Pinion bearing
- Pinion seal
- Pinion flange
7. Rotate the pinion flange occasionally while tightening the flange nut to make sure the pinion bearings seat correctly.
8. Tighten the flange nut to 94 ft. lbs. (127 Nm) and then measure bearing preload torque.
9. Continue tightening the flange nut to achieve the bearing preload torque originally measured. Do not exceed 210 ft. lbs. (284 Nm) flange nut torque.
10. If using new pinion bearings, add 8–12 inch lbs. (0.88–1.37 Nm) to the originally measured bearing preload.

✳✳ CAUTION

Never loosen the pinion nut to reduce bearing preload. If it is necessary to reduce bearing preload, install a new collapsible spacer and pinion nut.

11. Install or connect the following:
- Driveshaft
- Brake calipers and pads
- Wheels
- Negative battery cable
12. Fill the differential with gear lubricant and check for leaks.

STEERING

Air Bag

The air bag modules must be disabled if they, or any other part of the SRS, must be serviced or disconnected. Failing to disable the SRS before servicing its components may cause accidental air bag deployment and possible personal injury.

PRECAUTIONS

Several precautions must be observed when handling the inflator module to avoid accidental deployment and possible personal injury.

- Never carry the inflator module by the wires or connector on the underside of the module.
- When carrying a live inflator module, hold securely with both hands, and ensure that the bag and trim cover are pointed away.

- Place the inflator module on a bench or other surface with the bag and trim cover facing up.
- With the inflator module on the bench, never place anything on or close to the module which may be thrown in the event of an accidental deployment.
- Except when performing electrical inspections, always turn the ignition switch OFF, disconnect the negative battery cable from the battery and wait at least 3 minutes before working around the air bag.
- Be aware that the SRS memory is not erased even if the ignition switch is turned OFF or the battery cables are disconnected.
- Before removing any SRS parts, including disconnection of connectors, always disconnect the SRS connector.
- The radio may contain a coded theft protection circuit. Always obtain the code number before disconnecting the battery.

- Misalignment of the cable reel could cause an open in the wiring, making the SRS system and the horn inoperative. Center the cable reel whenever the steering wheel, steering column or other steering column related procedures are performed.

DISCONNECTING SYSTEM CONNECTORS

➡Before removing the front airbag, side air bag, any SRS related devices, disconnecting connectors from related devices, removing the dashboard, removing the steering column, or disconnecting the airbag connectors or the side airbag connectors to prevent accidental deployment, turn the ignition switch OFF, disconnect the negative battery cable and wait at least three minutes for starting the repair operation.

- Before disconnecting SRS unit connector A (A) from the SRS unit, disconnect the driver's airbag 4P connector (C), the front passenger's airbag 4P connector (D), the driver's seat belt tensioner 2P connector (F), and the front passenger's seat belt tensioner 2P connector (G).
- Before disconnecting SRS unit connector B (I) from the SRS unit, disconnect both side airbag 2P connectors (L, M) and both seat belt buckle tensioner 4P connectors (J, K).
- Before disconnecting the cable reel 4P connector (B), disconnect the driver's airbag 4P connector (C).
- Before disconnecting the floor wire harness 4P connector (E), disconnect both seat belt tensioner 2P connectors (F, G).

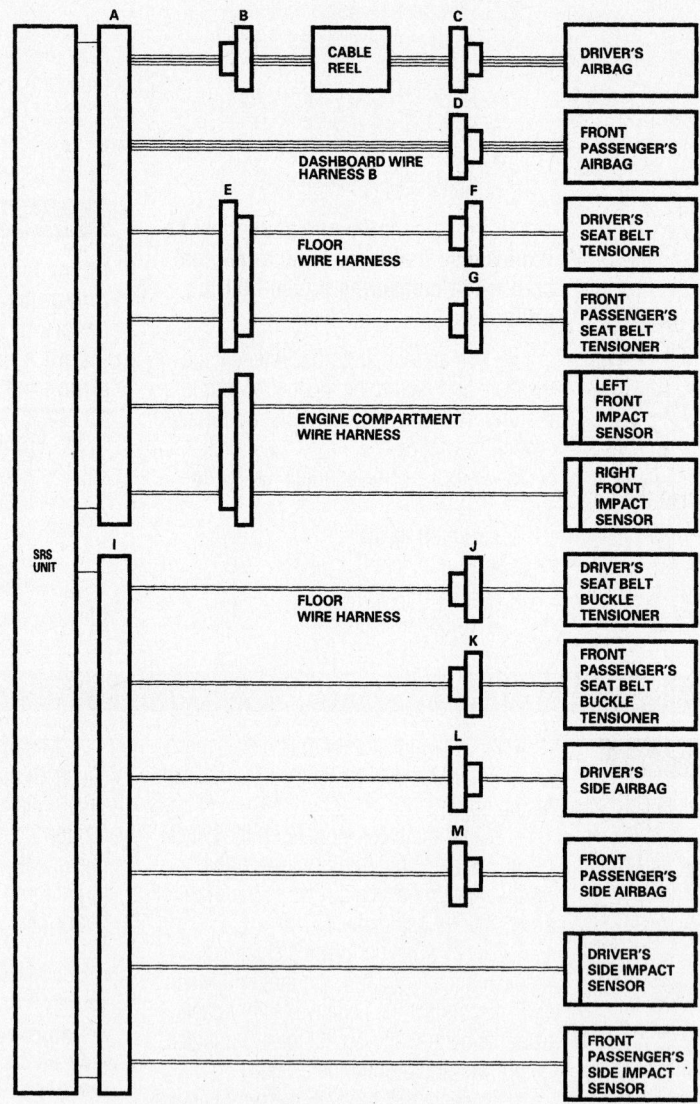

09474_ACCO_G0029

SRS connector location and related information—2002—2005 Civic except Hatchback

- Before disconnecting SRS unit connector A (1) from the SRS unit, disconnect the driver's airbag 4P connector (3), the front passenger's airbag 4P connector (4), the driver's seat belt tensioner 2P connector (6), and the front passenger's seat belt tensioner 2P connector (7).
- Before disconnecting SRS unit connector B (8) from the SRS unit, disconnect both side airbag 2P connectors (11, 12), and both seat belt buckle tensioner 4P connectors (9, 10).
- Before disconnecting the cable reel 4P connector (2), disconnect the driver's airbag 4P connector (3).
- Before disconnecting the floor wire harness 4P connector (5), disconnect both seat belt tensioner 2P connectors (6, 7).

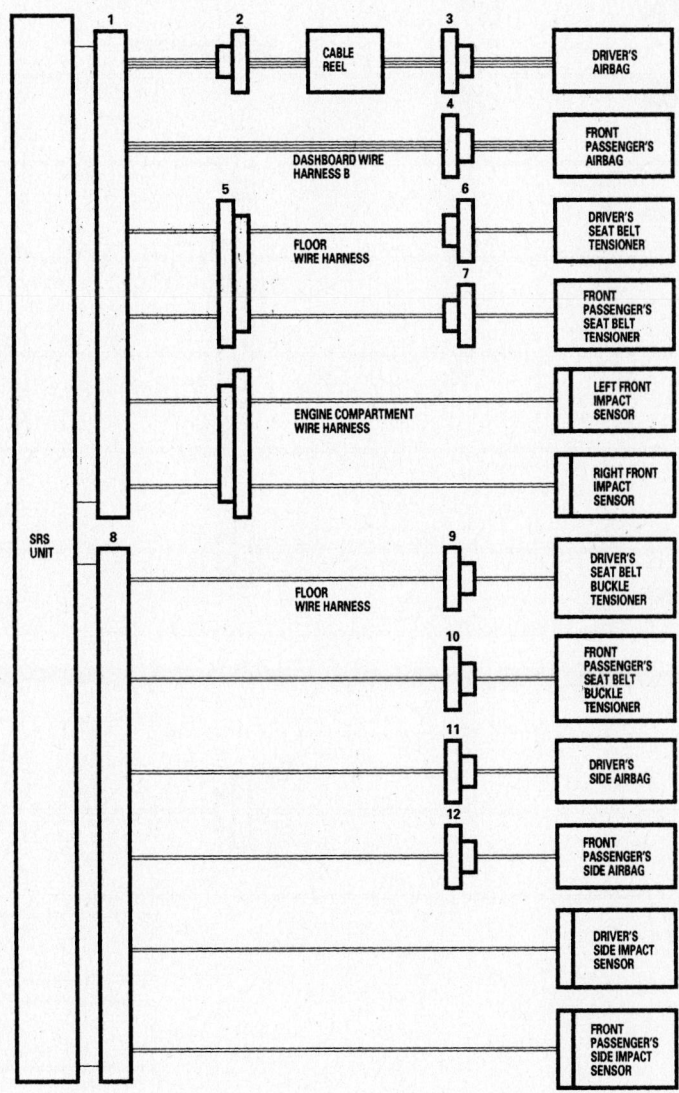

09474_ACCO_G0030

SRS connector location and related information—2002—2005 Hatchback

- Before disconnecting the cable reel 4P connector (1), disconnect the driver's airbag 4P connector (2).

- Before disconnecting SRS unit connector B from SRS unit, disconnect both seat belt tensioner 4P connectors and both seat belt buckle tensioner 4P connectors (3, 4, 5, 6).

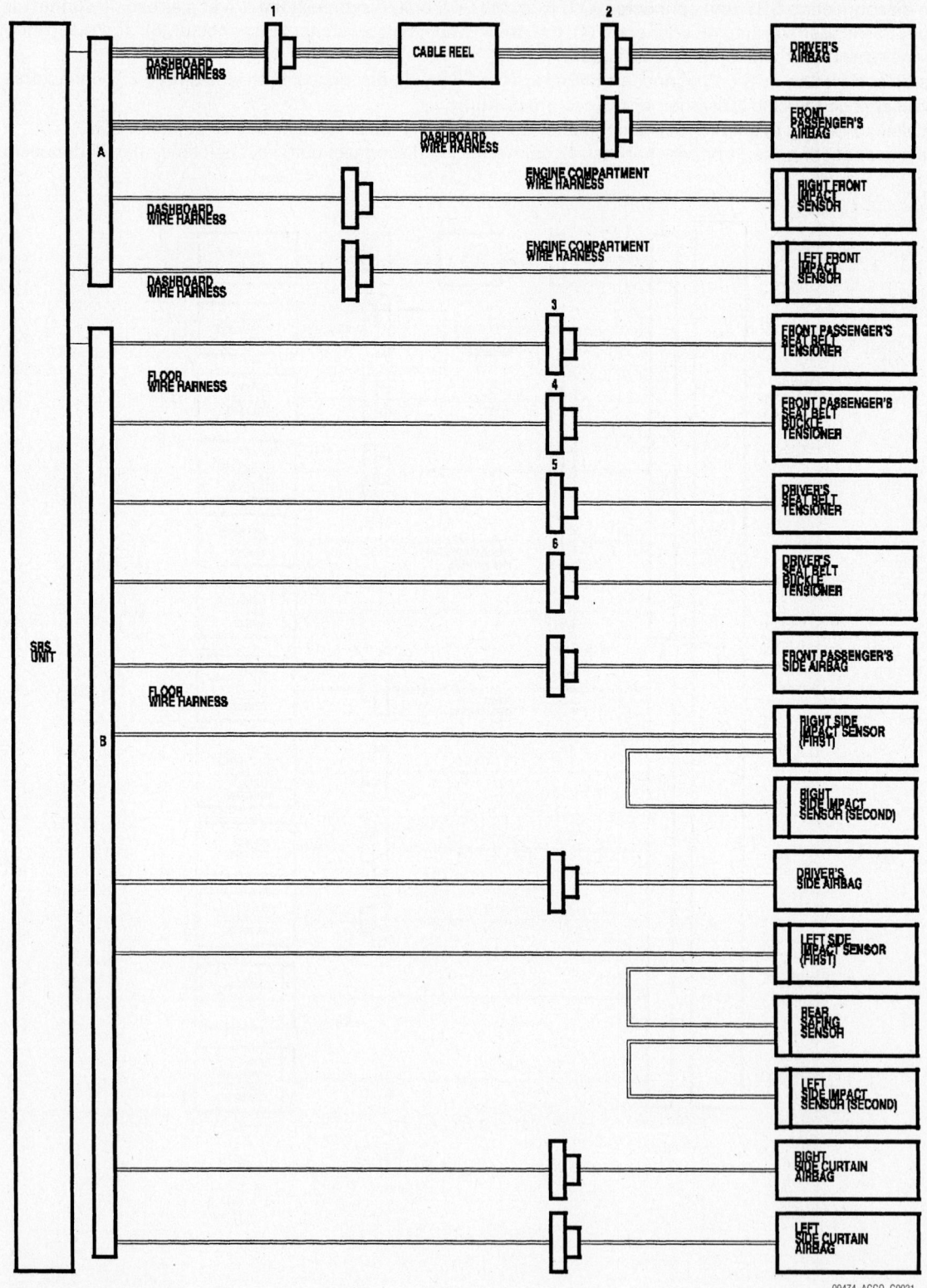

09474_ACCO_G0031

SRS connector location and related information—2006 Civic

- Before disconnecting the SRS unit connector A (18P) from the SRS unit, disconnect both airbag 2P connectors (1, 2) and both seat belt tensioner 2P connectors (3, 4).

- Before disconnecting the cable reel 2P connector (5), disconnect the driver's airbag 2P connector (1).

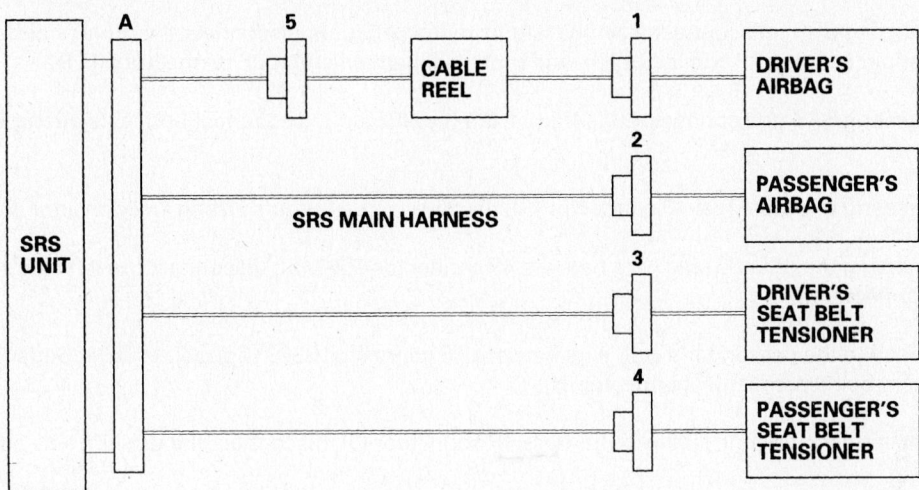

09474_ACCO_G0032

SRS connector location and related information—2002—2005 S2000

- Before disconnecting the cable reel 4P connector (1), disconnect the driver's airbag 4P connector (2).
- Before disconnecting SRS unit connector B (28P) from the SRS unit, disconnect both seat belt tensioner 2P connectors (3, 4).

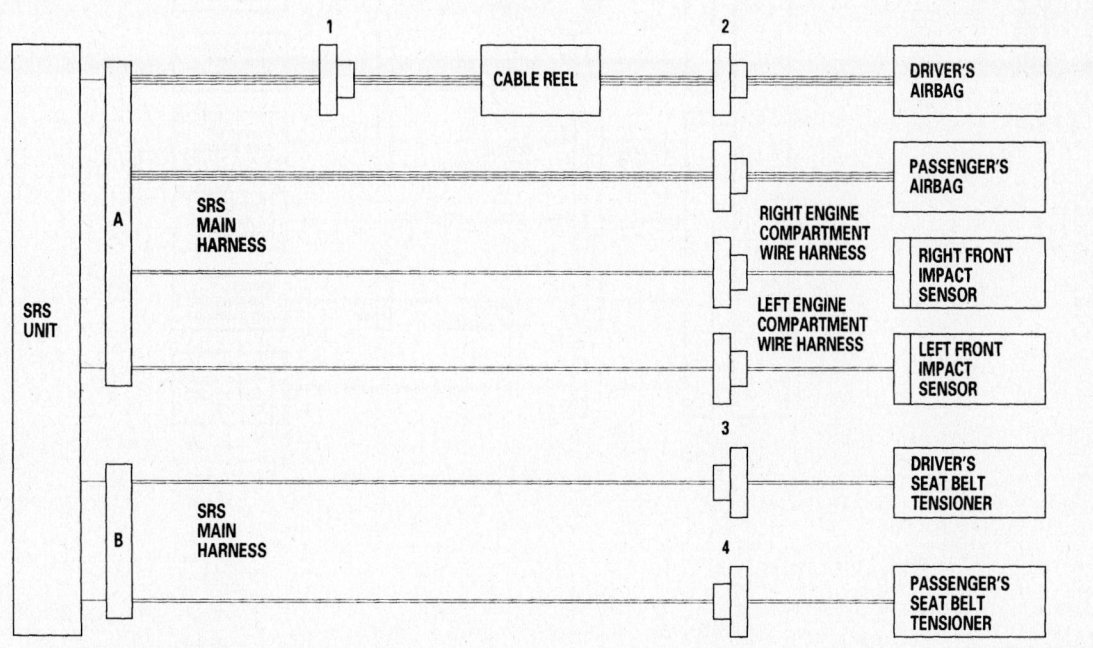

09474_ACCO_G0033

SRS connector location and related information—2006 S2000

- Before disconnecting SRS unit connector A (18P) (A) from the SRS unit, disconnect the driver's airbag 4P connector (C), front passenger's airbag 4P connector (D), and both seat belt tensioner 2P connectors (F, H).

- Before disconnecting SRS unit connector B (14P) (I) from the SRS unit, disconnect both side airbag 2P connectors (K, M).

- Before disconnecting the cable reel 4P connector (B), disconnect the driver's airbag 4P connector (C).

- Before disconnecting the driver's side wire harness 4P connector C557 (E), disconnect the left side seat belt tensioner 2P connector (F).

- Before disconnecting the passenger's side wire harness 4P connector C583 (Coupe) or C584 (Sedan) (G), disconnect the right side seat belt tensioner 2P connector (H).

- Before disconnecting the driver's side wire harness 4P connector (J), disconnect the driver's side airbag 2P connector (K).

- Before disconnecting the passenger's side wire harness 4P connector (L), disconnect the front passenger's side airbag 2P connector (M).

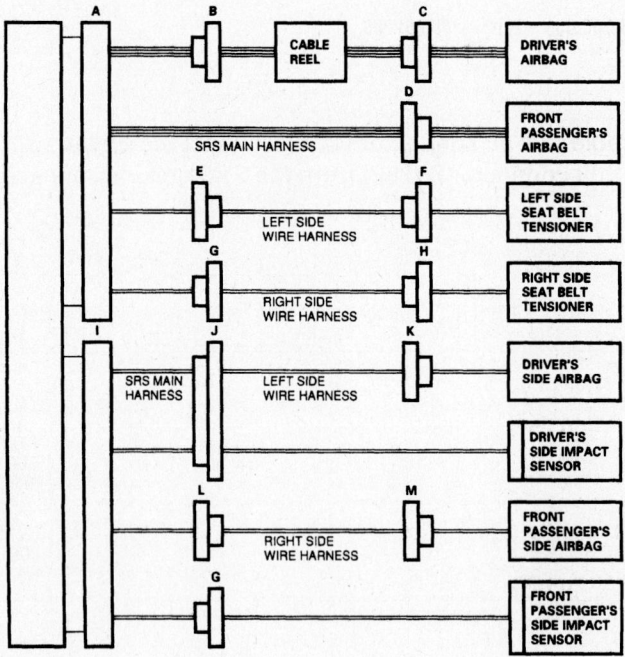

09474_ACCO_G0034

• Before disconnecting SRS unit connector B (3) from SRS unit, disconnect both seat belt tensioner 4P connectors (4, 5).
• Before disconnecting the cable reel 4P connector (1), disconnect the driver's airbag 4P connector (2).

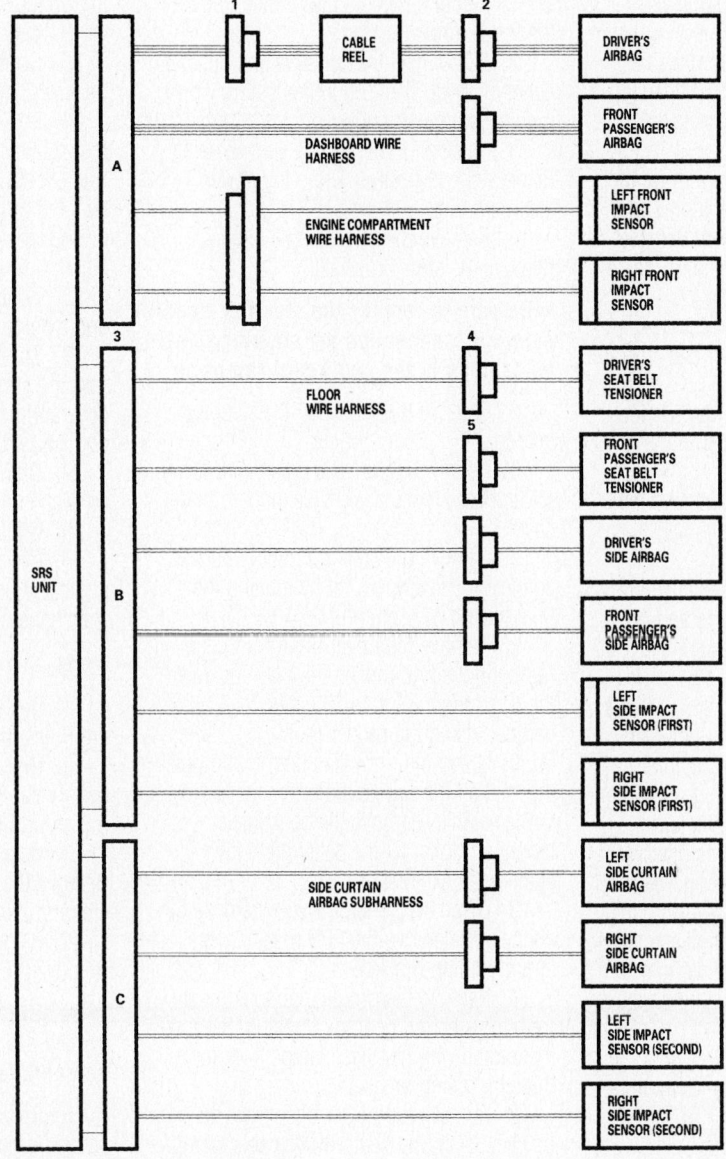

SRS connector location and related information—2003—2006 Accord

Power Steering Gear

REMOVAL & INSTALLATION

✳✳ CAUTION

The air bag must be disabled before removing the steering wheel to center the cable reel. Failure to disarm the air bag system may cause accidental air bag deployment, resulting in unnecessary air bag system repairs and the risk of personal injury.

➡**The radio may contain a coded theft protection circuit. Always obtain the code number before disconnecting the battery.**

Civic

2002–2005 EXCEPT HATCHBACK

1. Before servicing the vehicle, refer to the Precautions Section.

2. Disconnect the negative and positive battery cables. Wait 3 minutes before working around the air bags.

3. Raise and support the vehicle safely. Remove the front tires. Drain the power steering fluid.

4. Remove the driver's side airbag. Remove the steering wheel.

➡**Be sure to remove the steering wheel before disconnecting the steering joint as damage to the cable reel can occur.**

5. Remove the driver's dashboard lower cover. Remove the driver's dashboard under cover.

6. Remove the steering joint bolts. Disconnect the steering joint by moving the joint toward the column.

7. Remove the air cleaner and resonator.

8. Open the heater valve cable clamp.

Disconnect the heater valve cable. Remove the heater valve from the bulkhead. Move it to the side.

9. Above the gearbox, remove the feed line clamp. Disconnect the feed line and return line from the valve body.

10. Remove the cotter pin from the tie rod ball joint and loosen the nut. Separate the tie rod ball joint and the steering damper, using the proper tool.

11. From under the gearbox, remove the return hose clamp. Remove the engine wire harness from the harness bracket. Remove the steering rack guard clips, remove the guard.

12. On 2002—03 vehicles, remove the steering gearbox nut and washer and steering stiffner.

13. On 2004—05 vehicles, remove the steering gearbox nut and washer on the right gearbox mounting bracket.

14. On 2002—03 vehicles, remove the steering gearbox attachment bolts and washers and the steering stiffner.

15. On 2004—05 vehicles, remove the steering gearbox attachment bolts and washers on the left gearbox mounting bracket.

16. Pull on the steering gearbox to free it from the mounting stud, on the right gearbox mounting bracket. Remove the three mounting bolts and bracket.

17. Lower the steering gearbox and rotate it so the pinion shaft points upward.

18. Carefully remove the steering gearbox and tie rods as an assembly toward the passenger's side until the pinion shaft clears the heater hoses. Remove the steering gearbox through the wheel well opening on the passenger's side of the vehicle.

To install:

19. Position the assembly to it's mounting on the vehicle.

20. Install the mounting bolts.

21. Continue the installation in the reverse order of the removal procedure.

22. Be sure to center the cable reel by first rotating it clockwise until it stops. Then rotate it counterclockwise (approximately two and a half turns) until the arrow mark on the label points straight up. Install the steering wheel.

23. Fill the system with the proper grade and type power steering fluid.

24. Start the engine, allow it to idle, and turn the steering wheel from lock-to-lock several times to warm up the fluid. Check and adjust the fluid level.

25. Check the gearbox for fluid leaks, correct as necessary.

26. Check front end alignment.

27. Check the steering wheel spoke

angle. Adjust by turning the right and left tie rods equally, if necessary.

2002–2005 HATCHBACK

1. Before servicing the vehicle, refer to the Precautions Section.

2. Disconnect the negative and positive battery cables. Wait 3 minutes before working around the air bags.

3. Raise and support the vehicle safely. Remove the front tires. Drain the power steering fluid.

4. Remove the driver's side airbag. Remove the steering wheel.

➡**Be sure to remove the steering wheel before disconnecting the steering joint as damage to the cable reel can occur.**

5. Remove the motor on the steering gearbox.

6. Remove the driver's dashboard lower cover. Remove the driver's dashboard under cover.

7. Remove the steering joint bolts. Disconnect the steering joint by moving the joint toward the column.

8. Remove the cotter pin from the tie rod ball joint and loosen the nut. Separate the tie rod ball joint and the steering damper, using the proper tool.

9. Grasp the right side tie rod and pull the rack all the way to the passenger's side of the vehicle. Remove the heat shield mounting bolts, let the heat shield lay against the exhaust.

10. Disconnect the EPS wire harness 6P connector. Remove the EPS wire harness and mounting bracket.

11. Remove the ground cable terminal from the steering gearbox housing. Remove the engine wire harness clamps from the three mounting brackets.

12. Remove the heater hose from the bracket. Open the heater valve cable clamp. Disconnect the heater valve cable. Remove the heater valve from the bulkhead. Move it to the side.

13. Remove the side frame stiffner. Remove the steering stiffner from the left side of the steering gearbox. Remove the steering stiffner from the right side of the steering gearbox.

14. Lower the steering gearbox and rotate it so the pinion shaft points upward.

15. Remove the pinion shaft grommet from the top of the torque sensor.

16. Carefully remove the steering gearbox and tie rods as an assembly toward the wheel well opening on the passenger's side.

17. Carefully raise the driver's side (pinion side) of the steering gearbox and tie rod until it clears the master cylinder and

underhood fuse/relay box. Remove the assembly from the vehicle.

To install:

18. Position the assembly to it's mounting on the vehicle.

19. Install the mounting bolts.

20. Continue the installation in the reverse order of the removal procedure.

21. Be sure to center the cable reel by first rotating it clockwise until it stops. Then rotate it counterclockwise (approximately two and a half turns) until the arrow mark on the label points straight up. Install the steering wheel.

22. Fill the system with the proper grade and type power steering fluid.

23. Start the engine, allow it to idle, and turn the steering wheel from lock-to-lock several times to warm up the fluid. Check and adjust the fluid level.

24. Check the gearbox for fluid leaks, correct as necessary.

25. Check front end alignment.

26. Check the steering wheel spoke angle. Adjust by turning the right and left tie rods equally, if necessary.

2006

1. Before servicing the vehicle, refer to the Precautions Section.

2. Disconnect the negative and positive battery cables. Wait 3 minutes before working around the air bags.

3. Raise and support the vehicle safely. Remove the front tires. Drain the power steering fluid.

4. Remove the driver's side airbag. Remove the steering wheel.

➡**Be sure to remove the steering wheel before disconnecting the steering joint as damage to the cable reel can occur.**

5. Remove the driver's dashboard lower cover. Remove the driver's dashboard under cover.

6. Remove the steering joint bolts. Disconnect the steering joint by moving the joint toward the column. Hold the lower slide shaft on the column with a piece of wire between the joint yoke on the lower slide shaft to the joint yoke on the upper shaft.

7. Remove the center guide, if equipped and discard it. The center guide is for factory assembly only.

8. Remove the harness holder and the air cleaner housing bracket. Remove the breather pipe and the harness holder bracket.

9. Attach special tool 07AAK-SNAA120, or equivalent to the cylinder head. Install the engine support hanger, AAR-12566) or

equivalent. Attach the hook to the tool and tighten the wing nut. Lift and support the engine.

10. Remove the cotter pin from the tie rod ball joint and loosen the nut. Separate the tie rod ball joint and the knuckle, using the proper tool.

11. Remove the pump outlet hose clamp from the intake manifold. Remove the inlet line clamp bolt. Open the return line holder and remove the return line clamp bolt.

12. Loosen the adjustable hose clamp and disconnect the return hose. Loosen the flare nut and disconnect the inlet line.

13. Remove the splash shield. Remove the lower ball joint mounting bolt and flange nuts from the lower arm. Disconnect the lower arm from the lower ball housing.

14. Attach tool VSB02C000016 to the front subframe and a transaxle jack or powertrain lift tool. Make sure that the subframe is securely supported.

15. Remove the front subframe middle mount bolt from the left side. Remove the front subframe middle mount bolt from the right side. Remove the two 12mm flange bolts from the lower torque rod and bracket.

16. Remove the front subframe front mounting bolts from the right and left sides of the vehicle. Discard them.

17. Remove the front subframe rear mounting bolts from the right and left sides of the vehicle. Discard them.

18. Lower the front subframe and steering gearbox as an assembly. Remove the pinion shaft grommet from the top of the valve body unit.

19. Remove the two 10mm bolts from the right side of the steering gearbox. Remove the mounting bracket and mounting cushion.

20. Remove the four 10mm flange bolts from the left side of the steering gearbox. Remove the stiffner plates.

21. Remove the steering gearbox from the front subframe

To install:

22. Position the assembly to it's mounting on the subframe.

23. Loosely install the stiffner plates and gearbox mounting bolts on the left side of the steering gearbox.

24. Position the cutout on the mounting cushion and install it on the right side of the steering gearbox.

25. Install the gearbox mounting bracket over the mounting cushion and loosely install the two 10mm bolts.

26. Tighten the 10mm bolts on both sides of the steering gearbox to 40 ft. lbs. right side and 28 ft. lbs. left side, alternately in two or more steps.

**10 x 1.25 mm
38 N·m
(3.9 kgf·m, 28 lbf·ft)**

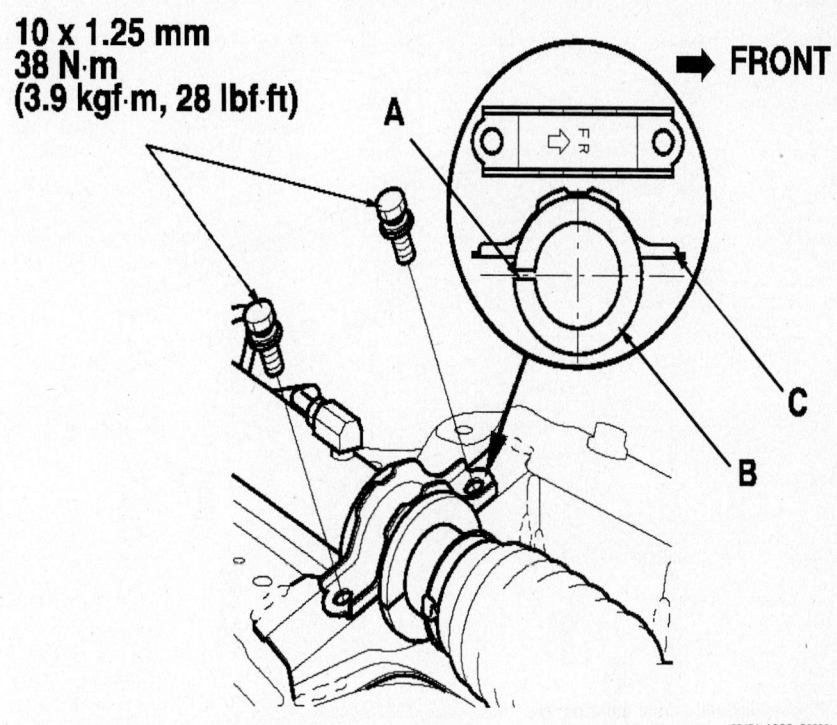

Cutout positioning on mounting cushion—2006 Civic

27. Install the pinion shaft grommet. Align the slot in the pinion shaft grommet with the lug portion on the valve housing. The grommet must not have a gap at the mating surface of the grommet and valve housing.

28. Carefully raise the front subframe in position.

29. Continue the installation in the reverse order of the removal procedure.

30. Be sure to center the cable reel by first rotating it clockwise until it stops. Then rotate it counterclockwise (approximately two and a half turns) until the arrow mark on the label points straight up. Install the steering wheel.

31. Fill the system with the proper grade and type power steering fluid.

32. Start the engine, allow it to idle, and turn the steering wheel from lock-to-lock several times to warm up the fluid. Check and adjust the fluid level.

33. Check the gearbox for fluid leaks, correct as necessary.

34. Check front end alignment.

35. Check the steering wheel spoke angle. Adjust by turning the right and left tie rods equally, if necessary.

S2000

1. Before servicing the vehicle, refer to the Precautions Section.

2. Remove or disconnect the following:

- Negative battery cable
- Front wheels
- Driver's air bag
- Steering wheel
- Steering coupler
- Outer tie rod ends
- Splash shield
- Stabilizer bar brackets
- Steering gear wiring connectors
- Steering gear mounting bolts

3. Move the steering gear forward and to the right to remove the steering gear.

To install:

4. Installation is the reverse of the removal procedure, while using the following torque values:

- Steering gear mounting bolts: 33 ft. lbs. (44 Nm)
- Steering gear ground cable bolt: 88 inch lbs. (10 Nm)
- Stabilizer bar bracket bolts: 61 ft. lbs. (83 Nm)
- Splash shield bolts: 88 inch lbs. (10 Nm)
- Outer tie rod end nuts: 40 ft. lbs. (54 Nm)
- Steering coupler pinch bolts: 16 ft. lbs. (22 Nm)

Accord

2002

1. Before servicing the vehicle, refer to the Precautions Section.

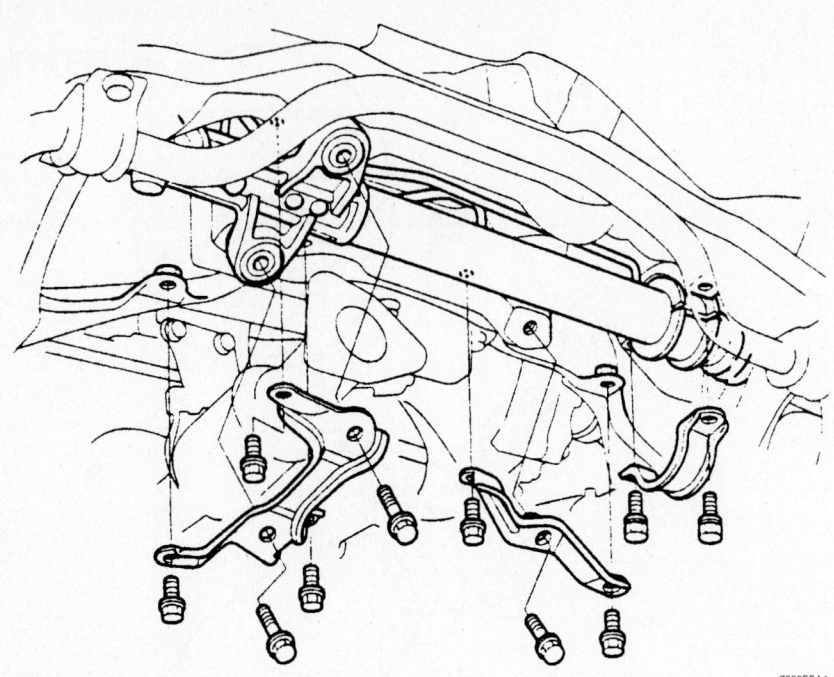

Power rack and pinion steering gear mounting—2002 Accord

2. Lift the power steering reservoir off of its mount and disconnect the inlet hose.

3. Insert a length of tubing into the inlet hose and route the tubing into a drain container.

4. With the engine running at idle, turn the steering wheel lock-to-lock several times until fluid stops running out of the hose. Immediately shut off the engine.

5. Position the front wheels straight ahead. Lock the steering column with the ignition key. Reconnect the reservoir inlet hose.

6. Remove or disconnect the following:
- Negative battery cable
- Steering joint cover and the upper and lower steering joint bolts.

7. Raise and support the vehicle safely.

8. Remove or disconnect the following:
- Front wheels
- Tie rod end cotter pins and castle nuts
- Tie rod ends from the steering knuckles, using a ball joint tool.
- Left tie rod end and slide the rack all the way to the right.
- Heated Oxygen Sensor (HO2S) connector
- Self-locking nuts, then separate the catalytic converter from the exhaust pipe.
- Catalytic converter
- Shift linkage from the transaxle case, if equipped with a manual transaxle.
- Shift cable cover and cable (wire it

up and out of the way), if equipped with an automatic transaxle.
- 2 hydraulic lines from the rack valve body, using a flare nut wrench. Plug the lines to keep dirt and moisture out. Carefully move the disconnected lines to the rear of the rack assembly so that they are not damaged when the rack is removed.
- Rack stiffener plate, then the steering rack mounting bolts.

9. Pull the steering rack down to release it from the pinion shaft.

10. Drop the steering rack far enough to permit the end of the pinion shaft to come out of the hole in the frame channel.

11. Slide the steering rack to the right until the left tie rod clears the subframe, then drop it down and out of the vehicle to the left.

To install:

➡Use new gaskets and self-locking nuts when installing the catalytic converter.

❊❊ WARNING

Use only genuine Honda power steering fluid. Any other type or brand of fluid will damage the power steering pump.

12. Before installing the rack & pinion, slide the ends all the way to the right.

13. Install or connect the following:
- Pinion shaft grommet. The lug on the pinion shaft grommet aligns with the slot on the valve body.
- Steering rack into position
- Pinion shaft grommet and insert the pinion through the hole in the bulkhead.
- Rack mounting bolts. Tighten the bracket bolts to 28 ft. lbs. (39 Nm). Tighten the stiffener plate mounting bolts to 32 ft. lbs. (43 Nm).

14. Center the rack ends within their steering strokes.

15. Center the air bag cable reel, as follows:

a. Turn the steering wheel left approximately 150°, to check the cable reel position with the indicator.

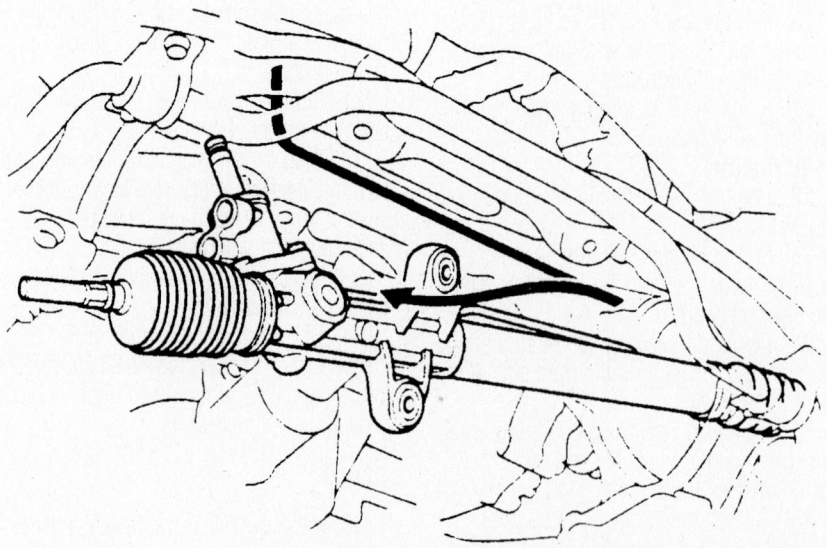

Move the steering rack to the right, then down and out of the vehicle—2002 Accord

b. If the cable reel is centered, the yellow gear tooth lines up with the alignment mark on the cover.

c. Return the steering wheel right approximately 150° to position the steering wheel in the straight-ahead position.

16. Line up the bolt hole in the steering joint with the groove in the pinion shaft. Slip the joint onto the pinion shaft. Pull the joint up and down to be sure the splines are fully seated. Tighten the joint bolts to 16 ft. lbs. (22 Nm).

➡**Connect the steering joint and pinion shaft with the cable reel and steering rack centered. Verify that the lower joint bolt is securely seated in the pinion shaft groove. If the steering wheel and rack are not centered, reposition the serrations at the lower end of the steering joint.**

17. Install or connect the following:
- Steering joint cover and the rack and pinion cover
- 2 hydraulic lines to the rack valve body. Carefully tighten the 14mm inlet fitting to 27 ft. lbs. (37 Nm) and the 16mm outlet fitting to 21 ft. lbs. (28 Nm).
- Shift cable and the select cable to the transaxle with new cotter pins, if equipped with a manual transaxle.
- Shift cable to the transaxle using a new lockwasher, if equipped with an automatic transaxle. Tighten the lockbolt to 10 ft. lbs. (14 Nm).
- Catalytic converter using new gaskets and self-locking nuts. Tighten the front nuts to 16 ft. lbs. (22 Nm), and the rear nuts to 25 ft. lbs. (34 Nm).
- HO_2S sensor connector
- Tie rod ends onto the rack ends
- Tie rod ends to the steering knuckles, then the castle nuts.

18. Tighten the ball joint castle nuts to 29–35 ft. lbs. (40–48 Nm). Then, tighten them only enough to install new cotter pins.

19. Install the front wheels.

20. Lower the vehicle.

21. Reconnect the negative battery cable.

22. Be sure the reservoir inlet line has been reconnected. Fill the reservoir to the upper line with Honda power steering fluid. Run the engine at idle and turn the steering wheel lock-to-lock several times to bleed air from the system and fill the rack valve body. Recheck the fluid level and add more if necessary.

23. Check the power steering system for leaks.

24. Check the front wheel alignment and steering wheel spoke angle. Make adjustments by turning the left and right tie rod ends equally.

25. Road test the vehicle.

2003–2006

1. Before servicing the vehicle, refer to the Precautions Section.

2. Disconnect the negative and positive battery cables. Wait 3 minutes before working around the air bags.

3. Raise and support the vehicle safely. Remove the front tires. Drain the power steering fluid.

4. Remove the driver's side airbag. Remove the steering wheel.

➡**Be sure to remove the steering wheel before disconnecting the steering joint as damage to the cable reel can occur.**

5. Remove the top half of the steering joint cover.

6. Remove the steering joint bolts. Disconnect the steering joint by moving the joint toward the column. Hold the lower slide shaft on the column with a piece of wire between the joint yoke on the lower slide shaft to the joint yoke on the upper shaft.

7. Remove the lower half of the steering joint cover.

8. Remove the cotter pin from the tie rod ball joint and loosen the nut. Separate the tie rod ball joint and the knuckle, using the proper tool.

9. Remove the power steering heat baffle plate. Remove the feed line holder mounting bolt on the front suspension subframe.

10. Remove the feed line holder mounting bolt and return hose from the gearbox mounting bracket.

11. Place shop towels under the fluid lines and disconnect them. Plug the lines to prevent dirt from entering the system.

➡**Do not loosen the cylinder line between the valve body unit and the cylinder.**

12. Attach tool EQS02C000016 to the front subframe.

13. Remove the front subframe right middle mount bolt. Remove the front subframe left middle mount bolt.

14. Remove the steering gearbox mounting bolts on the left gearbox mount. Remove the steering stiffner plate.

15. Remove the front suspension subframe rear bracket and mounting bolt.

16. Loosen the two front subframe rear mounting bolts on the right and left. Slowly lower the special tool supporting the subframe until the subframe has dropped 1 1/8 in.

17. Remove the two flange bolts from the right side of the gearbox. Remove the gearbox mounting bracket and cushion.

18. Move the steering gearbox toward the front and remove the pinion shaft grommet from the top of the valve body unit.

19. Apply vinyl tape to the splines on the pinion shaft. Tape the brake lines to protect them from the pinion shaft.

20. Move the steering gearbox to the driver's side, and rotate it so that the pinion shaft points toward the front of the vehicle.

21. Carefully move the steering gearbox as an assembly toward the driver's side of the vehicle until the pinion shaft clears the wheel well opening.

22. Remove the steering gearbox through the wheel well opening on the driver's side of the vehicle.

To install:

23. Position the assembly to it's mounting on the vehicle.

24. Install the mounting bolts.

25. Continue the installation in the reverse order of the removal procedure.

26. Be sure to center the cable reel by first rotating it clockwise until it stops. Then rotate it counterclockwise (approximately two and a half turns) until the arrow mark on the label points straight up. Install the steering wheel.

27. Fill the system with the proper grade and type power steering fluid.

28. Start the engine, allow it to idle, and turn the steering wheel from lock-to-lock several times to warm up the fluid. Check and adjust the fluid level.

29. Check the gearbox for fluid leaks, correct as necessary.

30. Check front end alignment.

31. Check the steering wheel spoke angle. Adjust by turning the right and left tie rods equally, if necessary.

FRONT SUSPENSION

Strut

REMOVAL & INSTALLATION

Civic

2002–2005

1. Before servicing the vehicle, refer to the Precautions Section.

2. Raise and support the vehicle safely. Remove the front tires.

3. Remove the cotter pin from the tie rod end ball joint. Remove the nut.

4. Disconnect the tie rod end from the steering arm on the damper.

5. Remove the wheel sensor harness bracket and brake hose bracket from the damper. Do not disconnect the wheel sensor connector.

6. Remove the damper pinch bolts from the bottom of the damper. Remove the flange nuts from the top of the damper.

7. Lower the lower arm and remove the damper assembly from the vehicle.

To install:

8. Position the assembly to it's mounting on the vehicle.

9. Install the mounting bolts.

10. Continue the installation in the reverse order of the removal procedure.

11. Check front end alignment.

2006

1. Before servicing the vehicle, refer to the Precautions Section.

2. Raise and support the vehicle safely. Remove the front tires.

3. Remove the wheel sensor harness bracket and brake hose bracket from the damper. Do not disconnect the wheel sensor connector.

4. Remove the damper pinch bolts from the bottom of the damper. Do not allow the knuckle to rotate too far outward as this may cause the inner CV joint bearing to unseat.

5. Turn the ignition switch to the ON position. Turn the windshield wipers on; turn the ignition switch off leaving the wipers near the A-pillars.

6. Remove the service cap and lid. Remove the three flange nuts at the top of the damper.

➡Damper springs are different, left and right. Mark the springs before continuing.

7. Remove the damper assembly from the vehicle.

To install:

8. Position the assembly to it's mounting on the vehicle.

9. Install the mounting bolts.

10. Continue the installation in the reverse order of the removal procedure.

11. Check front end alignment.

S2000

1. Before servicing the vehicle, refer to the Precautions Section.

2. Raise and support the vehicle safely. Remove the front tires.

3. Remove the flange bolt and brake hose mounting bracket from the damper.

4. Remove the cotter pin from the lower arm ball joint. Remove the castle nut. Remove the lower arm ball joint.

5. Remove the flange nuts from the top of the damper. Remove the flange bolt from the bottom of the damper.

6. Lower the lower arm and remove the damper assembly from the vehicle.

To install:

7. Position the assembly to it's mounting on the vehicle.

8. Install the mounting bolts.

9. Continue the installation in the reverse order of the removal procedure.

10. Check front end alignment.

Accord

1. Before servicing the vehicle, refer to the Precautions Section.

2. Raise and support the vehicle safely. Remove the front tires.

3. Remove the damper fork from the damper and lower arm.

4. Remove the flange nuts from the top of the damper assembly.

5. Remove the damper assembly from the vehicle.

To install:

6. Position the assembly to it's mounting on the vehicle.

7. Install the mounting bolts.

8. Continue the installation in the reverse order of the removal procedure.

9. Check front end alignment.

DISASSEMBLY

1. Before servicing the vehicle, refer to the Precautions Section.

2. Compress the damper spring with a commercially available strut spring compressor according to the manufacturer's instructions.

3. Remove the self locking nut while holding the damper shaft with a hex wrench. Do not compress the spring more than necessary to remove the nut.

4. Release the pressure from the strut spring compressor. Disassemble the damper assembly. Reassemble all parts, except the spring.

5. Compress the damper assembly by hand and check for smooth operation thru full stroke, both compression and extension.

6. The damper should extend smoothly

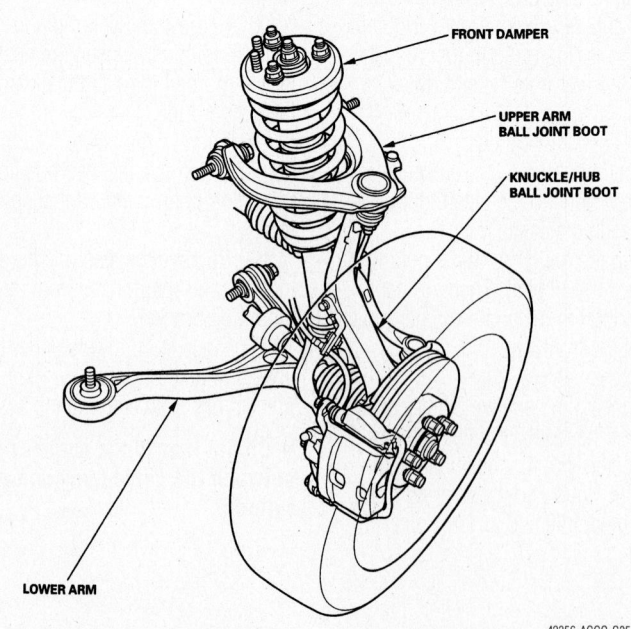

FRONT DAMPER

UPPER ARM
BALL JOINT BOOT

KNUCKLE/HUB
BALL JOINT BOOT

LOWER ARM

42356-ACCO-G35

Front suspension components—Accord

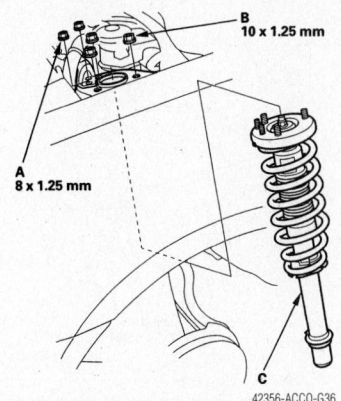

Front strut (damper) mounting—2003 Accord shown

and constantly when compression is released. If not, gas is leaking and the damper should be replaced.

ASSEMBLY

1. Install all parts except the self locking nut onto the damper unit.

2. Align the bottom of the spring and the stepped part of the lower spring seat. The hole in the upper spring seat and the arrow on the damper mounting base must point toward the knuckle mounting area.

3. Install the damper assembly on a commercially available strut spring compressor.

4. Compress the damper spring with the strut spring compressor. Install a new self locking nut on the damper shaft.

5. Hold the damper shaft with a hex wrench and tighten the self locking nut.

Stabilizer Bar

REMOVAL & INSTALLATION

Civic

2002–2005

1. Before servicing the vehicle, refer to the Precautions Section.

After removing the subframe mounting bolts, front suspension subframe middle mounting rubber mounting bolts, front suspension subframe rear bracket mounting bolts, and front suspension subframe rear damper front mounting bolt, be sure to replace them with new ones.

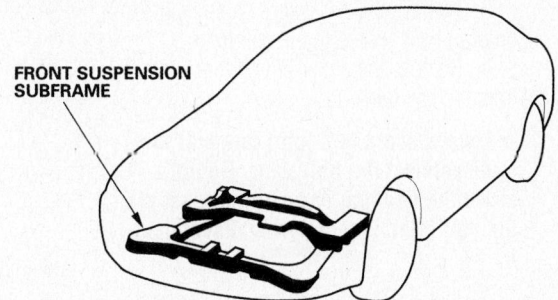

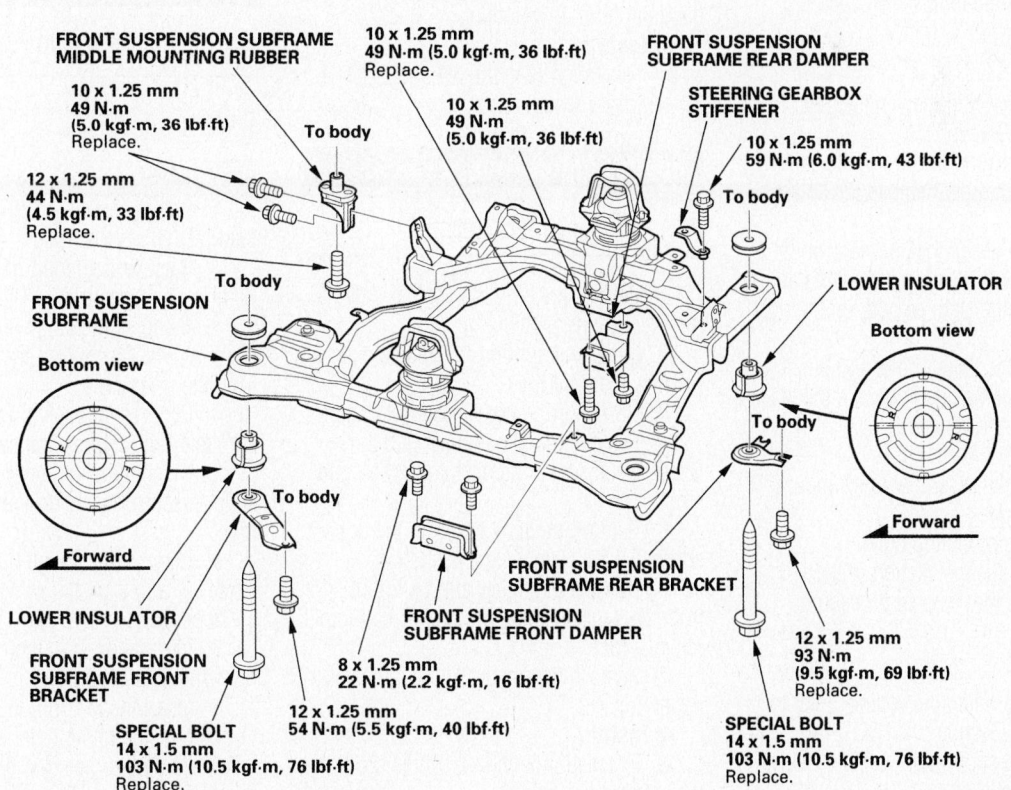

Front subframe and related components—2003—2006 Accord

09474_ACCO_G0037

2. Raise and support the vehicle safely. Remove the front tires.

3. Remove the self locking nuts while holding the joint pin. Disconnect the stabilizer links from the stabilizer bar on the right and left sides.

4. Remove the flange bolts and bushing holders. Remove the bushings and the stabilizer bar from the vehicle.

To install:

5. Position the assembly to it's mounting on the vehicle.

6. Install the mounting bolts.

7. Continue the installation in the reverse order of the removal procedure.

2006

1. Before servicing the vehicle, refer to the Precautions Section.

2. Raise and support the vehicle safely. Remove the front tires.

3. Disconnect both stabilizer links from the stabilizer bar.

4. Remove the flange bolts and the bushing holders. Remove the bushings and the stabilizer bar from the front suspension subframe.

To install:

5. Position the assembly to it's mounting on the vehicle.

6. Install the mounting bolts.

7. Continue the installation in the reverse order of the removal procedure.

S2000

1. Before servicing the vehicle, refer to the Precautions Section.

2. Raise and support the vehicle safely. Remove the front tires. Remove the splash shield.

3. Remove the self locking nuts while holding the joint pin. Disconnect the stabilizer links from the stabilizer bar on the right and left sides.

4. Remove the flange bolts and bushing holders. Remove the bushings and the stabilizer bar from the vehicle.

To install:

5. Position the assembly to it's mounting on the vehicle.

6. Install the mounting bolts.

7. Continue the installation in the reverse order of the removal procedure.

Accord

1. Before servicing the vehicle, refer to the Precautions Section.

2. Raise and support the vehicle safely. Remove the front tires.

3. Disconnect the stabilizer links from the stabilizer bar on the right and left sides.

4. Remove the front suspension subframe from the body.

5. Remove the flange bolts and bushing holders. Remove the bushings and the stabilizer bar from the vehicle.

To install:

6. Position the assembly to it's mounting on the vehicle.

7. Install the mounting bolts.

8. Continue the installation in the reverse order of the removal procedure.

Ball Joint

REMOVAL & INSTALLATION

1. Before servicing the vehicle, refer to the Precautions Section.

2. Raise and support the vehicle safely. Remove the front tires.

3. Remove the necessary components in order to gain access to the ball joint.

4. Remove the cotter pin and lock nut from the ball joint.

➡**Always use a ball joint removal tool to disconnect the ball joint. Do not strike the housing or any other part of the ball joint connection to disconnect it.**

5. Install the ball joint removal tool onto the threads of the ball joint pin. Remove the ball joint from its mounting.

To install:

6. Installation is in the reverse order of the removal procedure.

Upper Control Arm

REMOVAL & INSTALLATION

1. Before servicing the vehicle, refer to the Precautions Section.

2. Raise and support the vehicle safely. Remove the front tires.

3. On Accord, remove the front damper.

4. Remove the flange bolts and the wheel speed sensor harness from the upper arm.

5. Remove the lock pin from the upper arm ball joint. Remove the castle nut.

6. Remove the upper arm ball joint from the knuckle suing the proper ball joint removal tool.

7. Remove the flange bolts and the upper arm.

To install:

8. Installation is in the reverse order of the removal procedure.

9. On Accord, install the upper arm by inserting a rod (outside diameter 6mm,

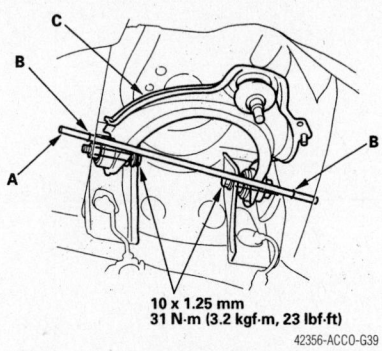

10 x 1.25 mm
31 N·m (3.2 kgf·m, 23 lbf·ft)

42356-ACCO-G39

Insert a rod (A) into the positioning holes (B) and place the upper control arm (C) on the rod to position it before tightening the bolts—Accord

length 300mm) into the positioning holes of the frame. Place the upper arm on the rod to position it before tightening the upper arm mounting bolts.

10. On Accord, torque the control arm retaining bolts to 23 ft. lbs.

11. On S2000, torque the control arm retaining bolts to 75.9 ft. lbs.

12. Be sure to use new flange bolts, castle nut and lock pin.

13. Check and adjust the wheel alignment, as necessary.

Lower Control Arm

REMOVAL & INSTALLATION

Civic

2002–2005

1. Before servicing the vehicle, refer to the Precautions Section.

2. Raise and support the vehicle safely. Remove the front tires.

3. Remove the flange nut while holding the joint pin. Disconnect the stabilizer link from the lower arm.

4. Remove the lock pin from the lower arm ball joint, remove the castle nut.

5. Disconnect the lower arm from the knuckle assembly, using the proper removal tool.

6. Remove the flange bolts and remove the lower arm from the vehicle.

To install:

7. Installation is in the reverse order of the removal procedure.

8. Insert the lock pin into the ball joint from the inside to the outside of the vehicle.

9. Be sure to use new flange bolts, castle nut and lock pin.

10. Check and adjust the wheel alignment, as necessary.

2006

1. Before servicing the vehicle, refer to the Precautions Section.

2. Raise and support the vehicle safely. Remove the front tires.

3. Remove the flange nut while holding the ball joint pin. Disconnect the stabilizer links from the lower arm.

4. Turn the stabilizer bar backward to gain access to the front side of the lower arm mounting bolt.

5. Remove the flange bolt and nuts from the lower arm.

6. Disconnect the lower ball joint from the lower arm.

7. Remove the lower arm mounting bolts. Remove the lower arm from the front suspension subframe.

To install:

8. Installation is in the reverse order of the removal procedure.

9. Be sure to use new flange bolts, castle nut and lock pin.

10. Check and adjust the wheel alignment, as necessary.

S2000

1. Before servicing the vehicle, refer to the Precautions Section.

2. Raise and support the vehicle safely. Remove the front tires.

3. Remove the flange nut while holding the ball joint pin. Disconnect the stabilizer links from the lower arm.

4. Remove the flange bolt and nuts from the lower arm.

5. Remove the cotter pin from the lower arm ball joint. Remove the castle nut.

6. Remove the lower arm ball joint from the knuckle, using the ball joint removal tool.

7. Remove the self locking nut and self locking cam nut. Remove the cam plate adjusting bolt, cam collar, flange bolt and the lower arm.

To install:

8. Installation is in the reverse order of the removal procedure.

9. Be sure to use new flange bolts, castle nut and lock pin.

10. Check and adjust the wheel alignment, as necessary.

Accord

1. Before servicing the vehicle, refer to the Precautions Section.

2. Raise and support the vehicle safely. Remove the front tires.

3. Remove the damper fork from the damper and the lower control arm.

4. Remove the flange nut while holding the ball joint pin. Disconnect the stabilizer links from the lower arm.

5. Remove the cotter pin from the lower arm ball joint. Remove the castle nut.

6. Remove the lower arm ball joint from the knuckle, using the ball joint removal tool.

7. Remove the flange bolts and remove the lower arm from the vehicle.

To install:

8. Installation is in the reverse order of the removal procedure.

9. Insert the damper fork into the damper lower end, so the aligning tab is aligned with the slot in the damper fork. Replace the damper fork retaining nut, with a new one.

10. Be sure to use new flange bolts, castle nut and lock pin.

11. Check and adjust the wheel alignment, as necessary.

Wheel Bearings

ADJUSTMENT

1. Before servicing the vehicle, refer to the Precautions Section.

2. Raise and support the vehicle safely. Remove the front tires.

3. Install suitable flat washers and wheel nuts, and tighten the nuts to 79.6 ft. lbs. (108 Nm), to hold the brake disc securely against the hub.

4. Attach a dial gauge against the hub flange. Bearing end play should be 0—0.002 inch (0—0.05mm).

5. On 2006 Civic and 2003—2006 Accord, measure the end play by moving the disc inward and outward.

6. If the bearing end play is not within specification, replace the wheel bearing.

REMOVAL & INSTALLATION

Civic

➡A hydraulic press and several bearing drivers and attachments are needed to remove and install the hub and bearing.

1. Before servicing the vehicle, refer to the Precautions Section.

2. Pry the spindle nut stake away from the spindle, then loosen the nut.

3. Raise and safely support the vehicle.

4. Remove or disconnect the following:
- Front wheel and the spindle nut
- Wheel sensor wire bracket from the knuckle, but don't disconnect it.
- Caliper mounting bolts and the caliper. Support the caliper out of the way with a length of wire. Do not let the caliper hang from the brake hose.
- 6mm brake disc retaining screws. Screw 2, 12mm bolts into the disc to push it away from the hub.
- Tie rod castle nut
- Tie rod ball joint using a suitable ball joint remover.
- Cotter pin and loosen the lower arm ball joint nut half the length of the joint threads.

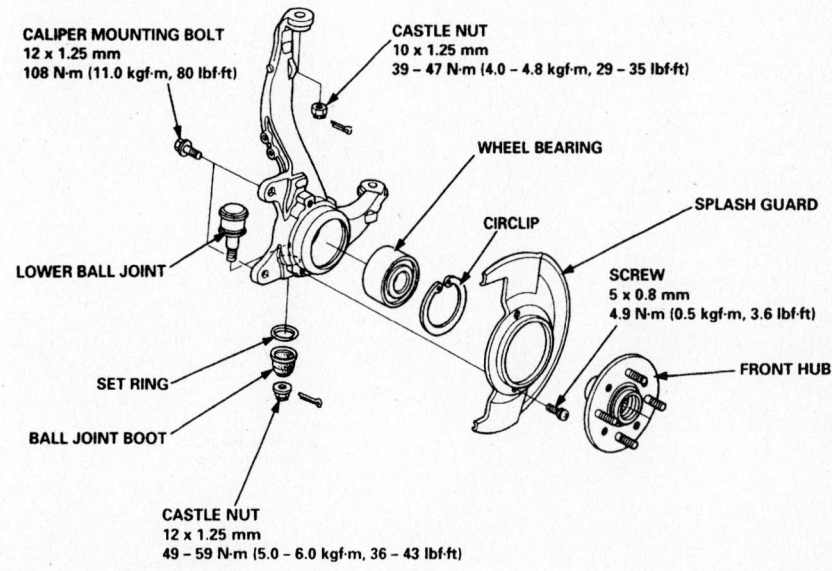

Knuckle components—Civic

- Ball joint and lower arm using a suitable puller with the pawls applied to the lower arm.

➡**Avoid damaging the ball joint boot. If necessary, apply penetrating type lubricant to loosen the ball joint.**

- Ball joint nut cover
- Cotter pin and the upper ball joint nut.
- Upper ball joint and knuckle using a ball joint remover.

5. Use a plastic mallet to free the half-shaft from the knuckle. Pull the knuckle out to remove it.

➡**A new wheel bearing must be used when the hub is removed.**

6. Place the knuckle in a press and use a base and pilot to press the hub assembly out of the wheel bearing.

7. Remove the knuckle ring seal and circlip. Remove the splash guard from the knuckle.

8. Press the wheel bearing out of the knuckle using a driving attachment.

To install:

9. Clean the knuckle and hub assembly and inspect them for damage.

10. Install or connect the following:
- New wheel bearing into the hub using a driving tool.
- Circlip in the outer groove of the knuckle.
- Splash guard
- Hub assembly into the steering knuckle using a base and a driving and guide tool.
- Knuckle ring seal
- Knuckle onto the spindle
- Knuckle onto the upper and lower ball joints and tighten the castle nuts.
- Tie rod ball joint onto the steering knuckle.

11. Tighten the upper ball joint nut and tie rod nut to 29–35 ft. lbs. (40–48 Nm) and the lower ball joint castle nut to 36–43 ft. lbs. (50–60 Nm).

12. Install or connect the following:
- Anti-lock Brake System (ABS) wheel sensor wire brackets onto the knuckle. Tighten the mounting bolts to 84 inch lbs. (10 Nm).
- Brake disc; use 2 lug nuts to evenly draw the disc onto the hub.
- Retainer screws: 84 inch lbs. (10 Nm)
- Spindle washer and nut. Don't tighten the nut until the vehicle is on the ground.
- Brake caliper and tighten the bolts to 80 ft. lbs. (110 Nm).
- Front wheels and lower the vehicle.

13. Tighten the spindle nut to 134 ft. lbs. (185 Nm), stake the nut, and install the grease cap.

14. Check and adjust the vehicle's front wheel alignment.

➡**Avoid damaging the ball joint boot. If necessary, apply penetrating-type lubricant to loosen the ball joint.**

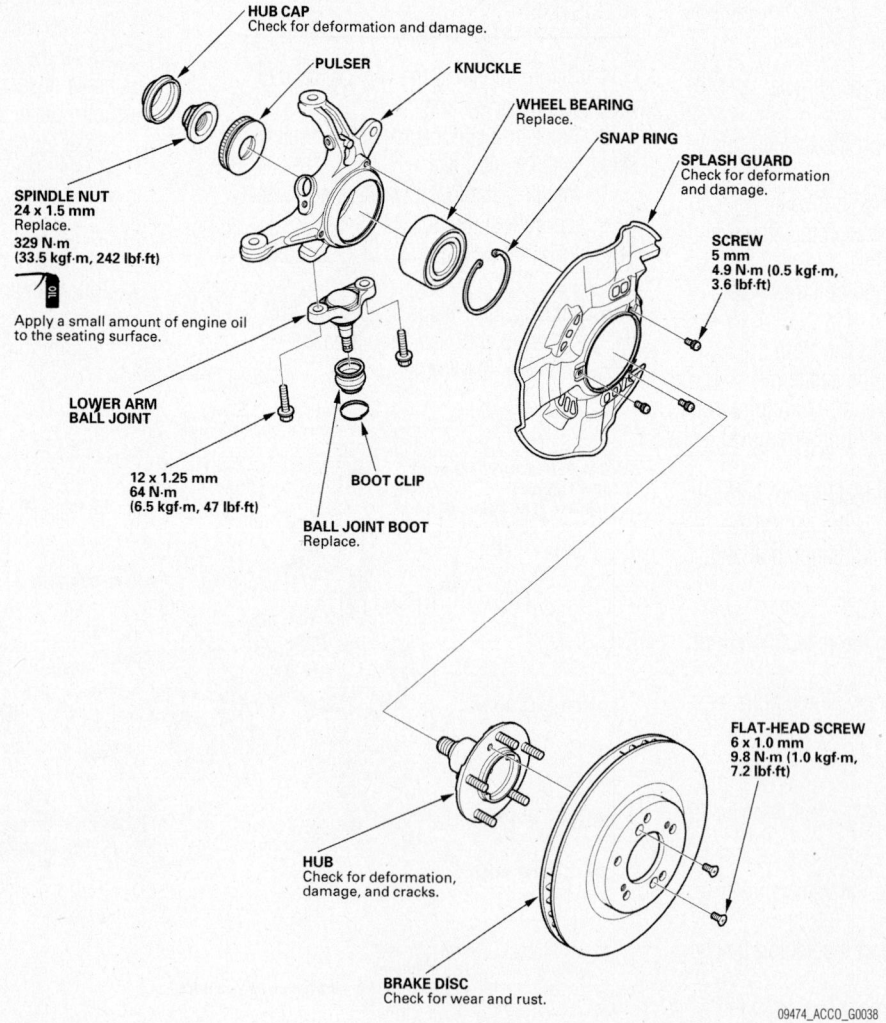

Knuckle components—S2000

09474_ACCO_G0038

S2000

➡ **A hydraulic press and several bearing drivers and attachments are needed to remove and install the hub and bearing.**

1. Before servicing the vehicle, refer to the Precautions Section.

2. Remove or disconnect the following:
   - Front wheel
   - Brake hose bracket mounting bolts
   - Brake caliper and caliper support
   - Wheel speed sensor
   - Brake rotor
   - Outer tie rod end
   - Upper and lower ball joints
   - Steering knuckle from the vehicle
   - Dust cover
   - Spindle nut
   - Wheel speed pulse ring

3. Mount the steering knuckle in a press and press the hub out of the wheel bearing.

4. Remove the splash guard and the wheel bearing snapring.

5. Press the wheel bearing out of the steering knuckle.

To install:

6. Installation is the reverse of the removal procedure, while using the following torque values:
   - Splash guard screws: 48 inch lbs. (5 Nm)
   - Spindle nut: 242 ft. lbs. (329 Nm)
   - Upper ball joint nut: 36–43 ft. lbs. (49–59 Nm)
   - Lower ball joint nut: 43–51 ft. lbs. (56–69 Nm)
   - Outer tie rod end nut: 40 ft. lbs. (54 Nm)
   - Brake caliper support bolts: 83 ft. lbs. (113 Nm)

Accord

➡ **A hydraulic press and several bearing drivers and attachments are needed to remove and install the hub and bearing.**

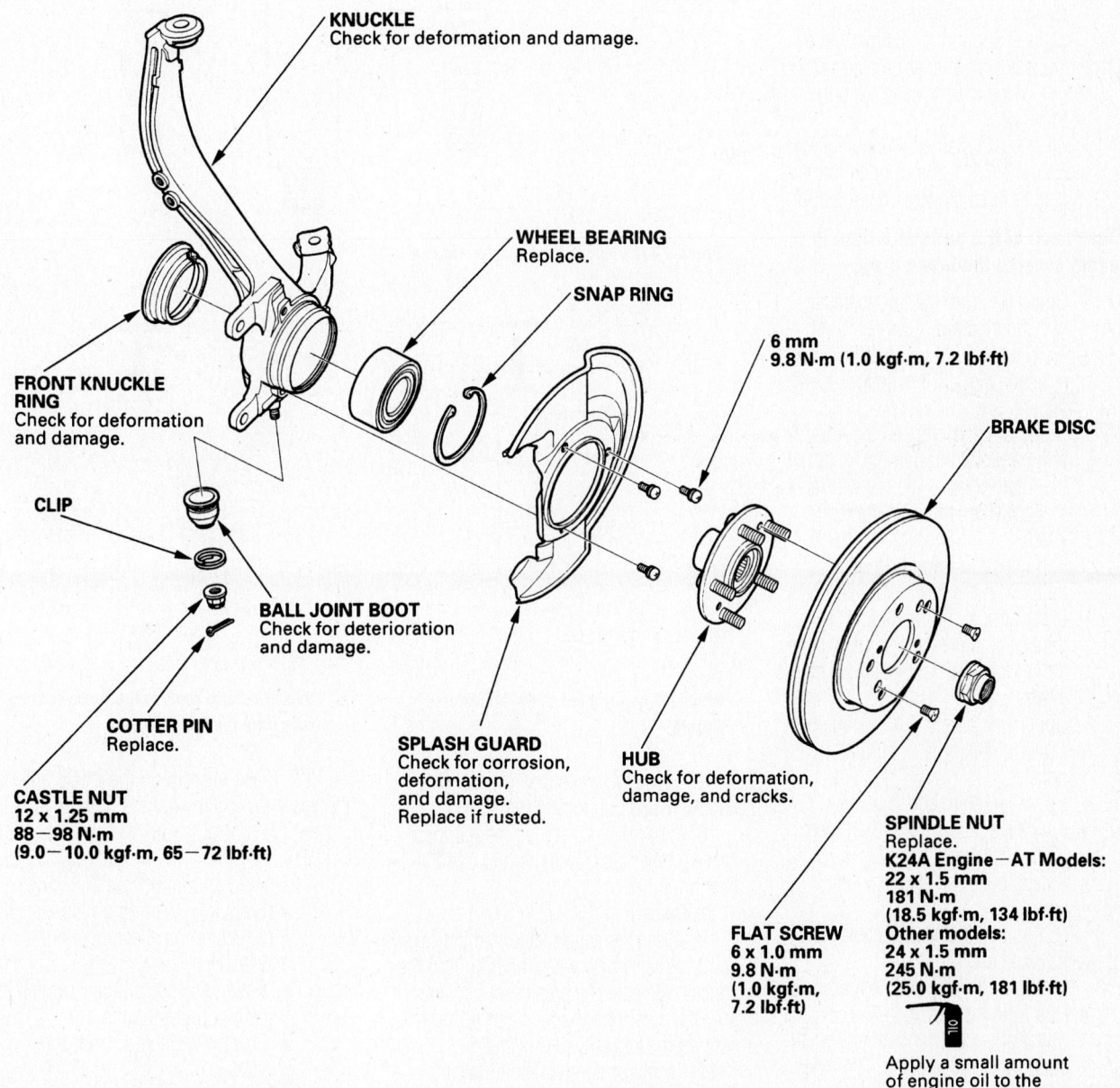

KNUCKLE
Check for deformation and damage.

WHEEL BEARING
Replace.

SNAP RING

6 mm
9.8 N·m (1.0 kgf·m, 7.2 lbf·ft)

BRAKE DISC

FRONT KNUCKLE RING
Check for deformation and damage.

CLIP

BALL JOINT BOOT
Check for deterioration and damage.

COTTER PIN
Replace.

CASTLE NUT
12 x 1.25 mm
88—98 N·m
(9.0—10.0 kgf·m, 65—72 lbf·ft)

SPLASH GUARD
Check for corrosion, deformation, and damage. Replace if rusted.

HUB
Check for deformation, damage, and cracks.

FLAT SCREW
6 x 1.0 mm
9.8 N·m
(1.0 kgf·m, 7.2 lbf·ft)

SPINDLE NUT
Replace.
K24A Engine—AT Models:
22 x 1.5 mm
181 N·m
(18.5 kgf·m, 134 lbf·ft)
Other models:
24 x 1.5 mm
245 N·m
(25.0 kgf·m, 181 lbf·ft)

Apply a small amount of engine oil to the surface of the nut.

Knuckle components—Accord

42356-ACCO-G40

➡**Once the hub has been removed, the wheel bearings must be replaced.**

A hydraulic press and bearing drivers must be used to remove and install the bearing.

1. Before servicing the vehicle, refer to the Precautions Section.

2. Pry the spindle nut stake away from the spindle and loosen the nut. Do not tighten or loosen a spindle nut unless the vehicle is sitting on all 4 wheels. The torque required is high enough to cause the vehicle to fall off the stands even when properly supported.

3. Raise and safely support the vehicle.

4. Remove or disconnect the following:
- Wheel and the spindle nut
- Caliper mounting bolts and the caliper. Support the caliper out of the way with a length of wire. Do not let the caliper hang from the brake hose.
- 6mm brake disc retaining screws. Screw 2, 8 x 1.25mm bolts into the disc to push it away from the hub.

➡**Turn each bolt 2 turns at a time to prevent cocking the brake disc.**

- Cotter pin from the tie rod castle nut, then the nut.
- Tie rod ball joint using a ball joint remover, then lift the tie rod out of the knuckle.
- Cotter pin, then loosen the lower arm ball joint nut half the length of the joint threads. The nut will retain the arm when the joint comes loose.
- Ball joint and lower arm using a puller with the pawls applied to the lower arm. Avoid damaging the ball joint boot. If necessary, apply penetrating lubricant to loosen the ball joint.
- Upper ball joint shield, if equipped.
- Cotter pin and the upper ball joint nut.
- Upper ball joint and knuckle
- Knuckle and hub by sliding them off the halfshaft.
- Splash guard screws from the knuckle.

5. Position the knuckle/hub assembly in a hydraulic press.

6. Remove or disconnect the following:
- Hub from the knuckle using a driver of the proper diameter while supporting the knuckle. The inner bearing race may stay on the hub.
- Splash guard and snapring from the knuckle.

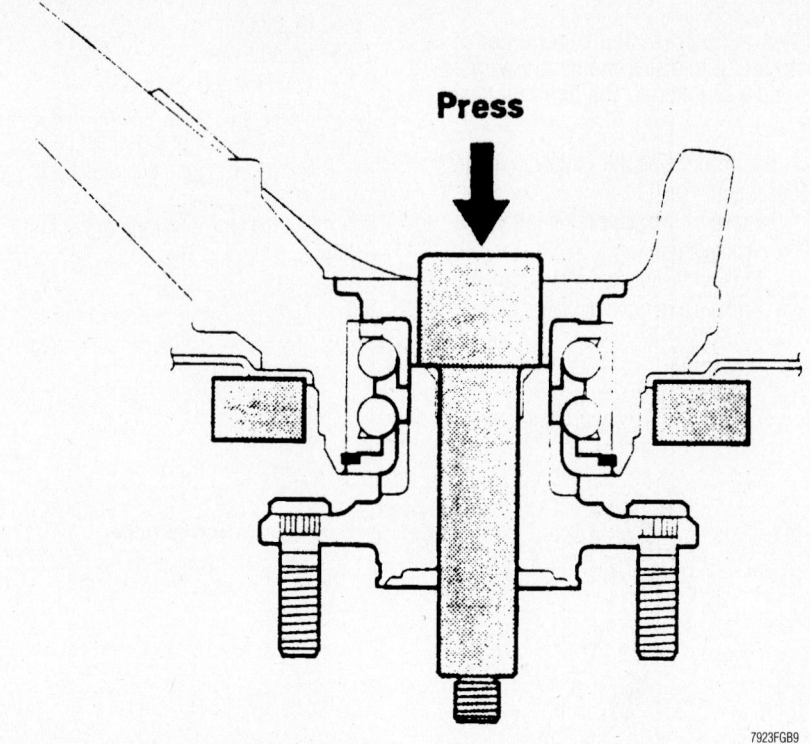

Press the hub out of the knuckle—Accord

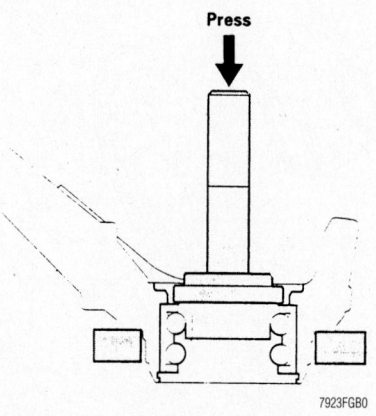

Press the bearing out of the knuckle—Accord

7. Press the wheel bearing out of the knuckle while supporting the knuckle.

8. If necessary, remove the outboard bearing inner race from the hub using a bearing puller.

To install:

9. Clean the knuckle and hub thoroughly.

10. Press a new wheel bearing into the knuckle. Be sure the press tool contacts only the outer bearing race and properly support the knuckle so it is stable.

11. Install or connect the following:
- Snapring
- Splash shield. Don't over tighten the screws.

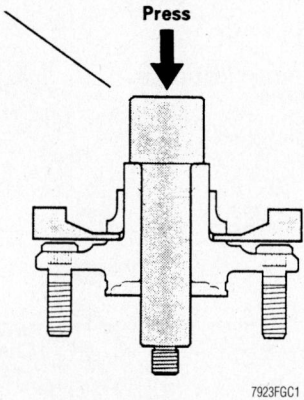

Use a press to remove the inner bearing race from the hub—Accord

12. Place the hub on the press table and press the knuckle onto the hub. Be sure the press tool contacts only the inner bearing race.

13. Install or connect the following:
- Front knuckle ring on the knuckle
- Knuckle/hub assembly on the vehicle.
- Brake disc and caliper. Tighten the caliper bracket bolts.
- Front wheels and lower the vehicle.

14. Tighten the spindle nut.

15. Check and adjust the vehicle's front wheel alignment.

REAR SUSPENSION

Strut

REMOVAL & INSTALLATION

Civic

2002–2005

1. Before servicing the vehicle, refer to the Precautions Section.

2. Raise and support the vehicle safely. Remove the rear tires.

3. Remove the flange bolt from the bottom of the damper.

4. Remove the flange nuts from the bottom of the damper.

5. Open the trunk and remove the access panel cover, if necessary.

6. Remove the flange nuts from the top of the damper

7. Remove the damper assembly from the vehicle.

To install:

8. Position the damper assembly in the vehicle.

➡ **Be aware of the direction of the damper mounting base so that the small hole dot on it is toward the front inside of the vehicle,**

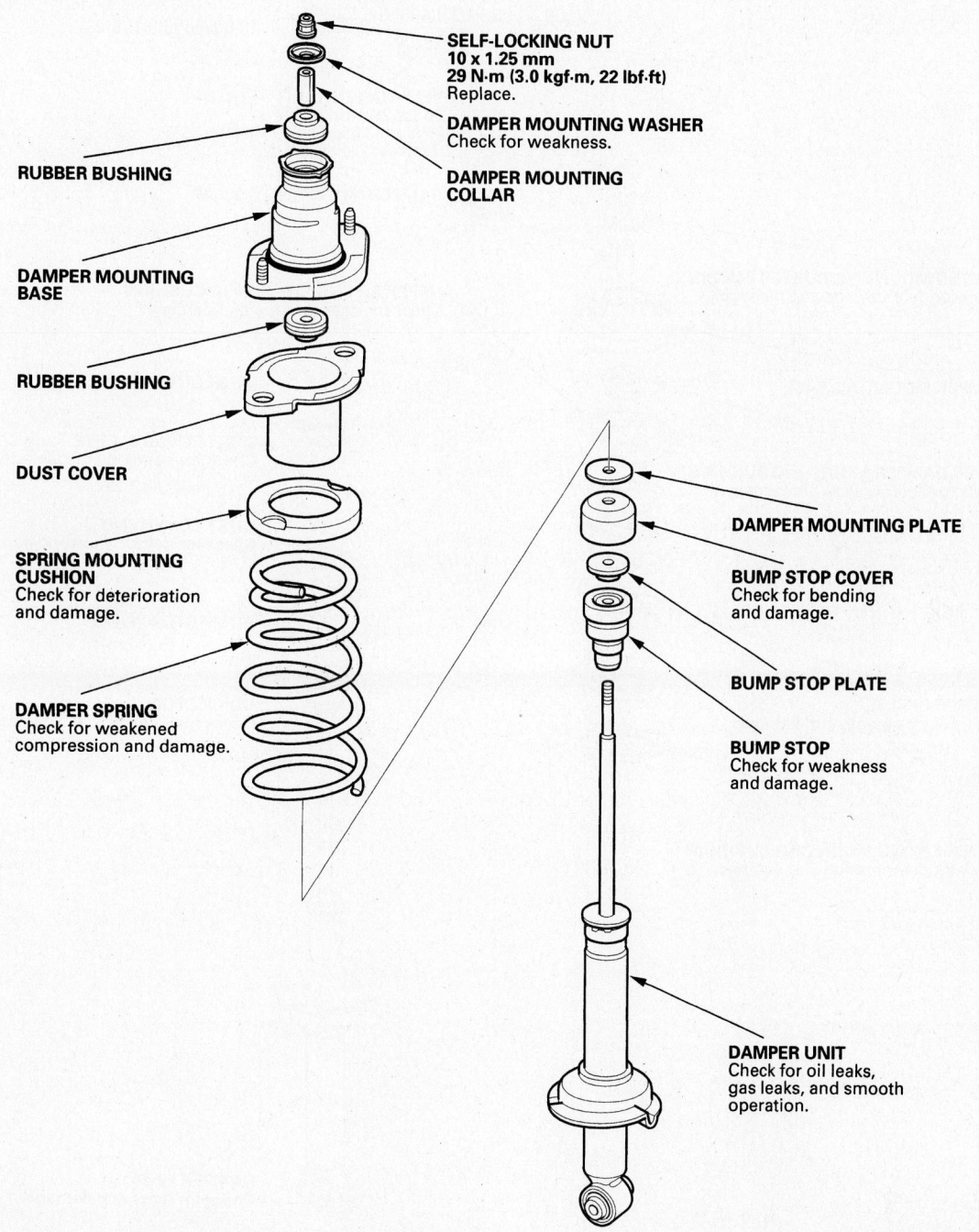

SELF-LOCKING NUT
10 x 1.25 mm
29 N·m (3.0 kgf·m, 22 lbf·ft)
Replace.

DAMPER MOUNTING WASHER
Check for weakness.

DAMPER MOUNTING COLLAR

RUBBER BUSHING

DAMPER MOUNTING BASE

RUBBER BUSHING

DUST COVER

SPRING MOUNTING CUSHION
Check for deterioration and damage.

DAMPER SPRING
Check for weakened compression and damage.

DAMPER MOUNTING PLATE

BUMP STOP COVER
Check for bending and damage.

BUMP STOP PLATE

BUMP STOP
Check for weakness and damage.

DAMPER UNIT
Check for oil leaks, gas leaks, and smooth operation.

09474_ACCO_G0039

Exploded view of the rear suspension strut—2002—2005 Civic

9. Loosely install the flange nuts on the top of the damper.

10. Loosely install the flange bolt on the bottom of the damper.

11. Raise the suspension with a floor jack to load it with the vehicles weight before tightening the nuts and bolt.

12. Continue the installation in the reverse order of the removal procedure.

2006

1. Before servicing the vehicle, refer to the Precautions Section.

2. Raise and support the vehicle safely. Remove the rear tires.

3. Position a floor jack at the connecting point of the trailing arm and the knuckle. Raise the floor jack until the suspension begins to compress.

4. Remove the flange bolt and discard it.

5. Remove the trunk side trim panel.

6. Remove the self locking nut while holding the damper shaft. Compress the damper unit, by hand, and remove it from the vehicle.

To install:

7. Position the damper assembly in the vehicle.

8. Position a floor jack under the trailing to support the suspension. Install a new damper mounting bolt.

9. Loosely tighten the damper mounting bolt. Raise the floor jack until the suspension begins to compress. Tighten the damper mounting bolt.

10. Continue the installation in the reverse order of the removal procedure.

S2000

1. Before servicing the vehicle, refer to the Precautions Section.

2. Remove the spare tire from the trunk.

3. Raise and support the vehicle safely. Remove the rear tires.

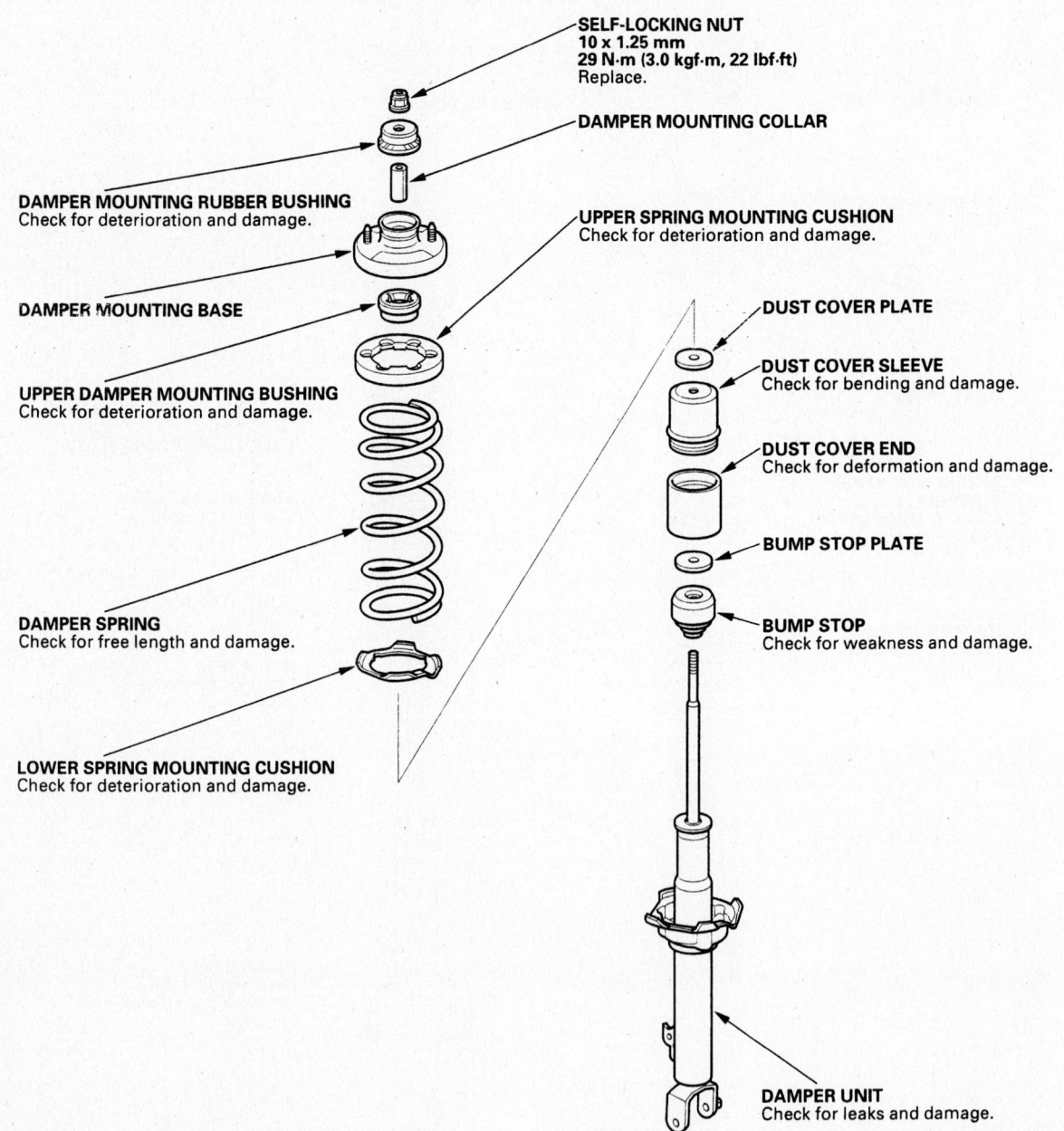

Exploded view of the strut and spring assembly—S2000

9347FG24

4. Remove the flange nuts from the top of the damper.

5. Remove the flange bolt from the bottom of the damper.

6. Lower the lower arm and remove the damper assembly.

To install:

7. Position the damper assembly in the vehicle.

8. Loosely install new flange nuts onto the damper studs.

9. Position the bottom of the damper assembly on the lower arm. Install a new flange bolt.

10. Raise the floor jack until the suspension begins to compress. Tighten the damper mounting bolt.

11. Continue the installation in the reverse order of the removal procedure.

Accord

2002

1. Before servicing the vehicle, refer to the Precautions Section.

2. Raise and support the vehicle safely. Remove the rear tires.

3. Remove the rear bulkhead cover. Remove the two flange nuts.

4. Remove the flange bolt from the knuckle. Remove the flange nut from the stabilizer link.

5. Lower the rear suspension. Remove the damper from the vehicle.

To install:

6. Position the damper assembly in the vehicle.

➡**Damper springs are different for the left and right side. Mark the springs prior to removal.**

7. Lower the rear suspension, position the damper and loosely install the two flange nuts.

8. Loosely install the flange bolt and nut.

9. Raise the floor jack until the suspension begins to compress. Tighten the bolts.

10. Continue the installation in the reverse order of the removal procedure.

2003–2006

1. Before servicing the vehicle, refer to the Precautions Section.

2. Raise and support the vehicle safely. Remove the rear tires.

3. Remove the rear bulkhead cover. Remove the seat side bolster.

4. Remove the two flange nuts, while holding the joint pin. Disconnect the stabilizer link from the stabilizer bracket.

5. Remove the flange bolt from the knuckle. Lower the rear suspension and remove the damper from the vehicle.

To install:

6. Position the damper assembly in the vehicle.

➡**Damper springs are different for the left and right side. Mark the springs prior to removal.**

7. Loosely install the flange nuts onto the top of the damper.

8. Loosely install the flange bolt on the bottom of the damper.

9. Connect the stabilizer link on the bracket and loosely install the flange nut.

10. Raise the floor jack until the suspension begins to compress. Tighten the bolts.

11. Continue the installation in the reverse order of the removal procedure.

DISASSEMBLY

1. Before servicing the vehicle, refer to the Precautions Section.

2. Compress the damper spring with a commercially available strut spring com-

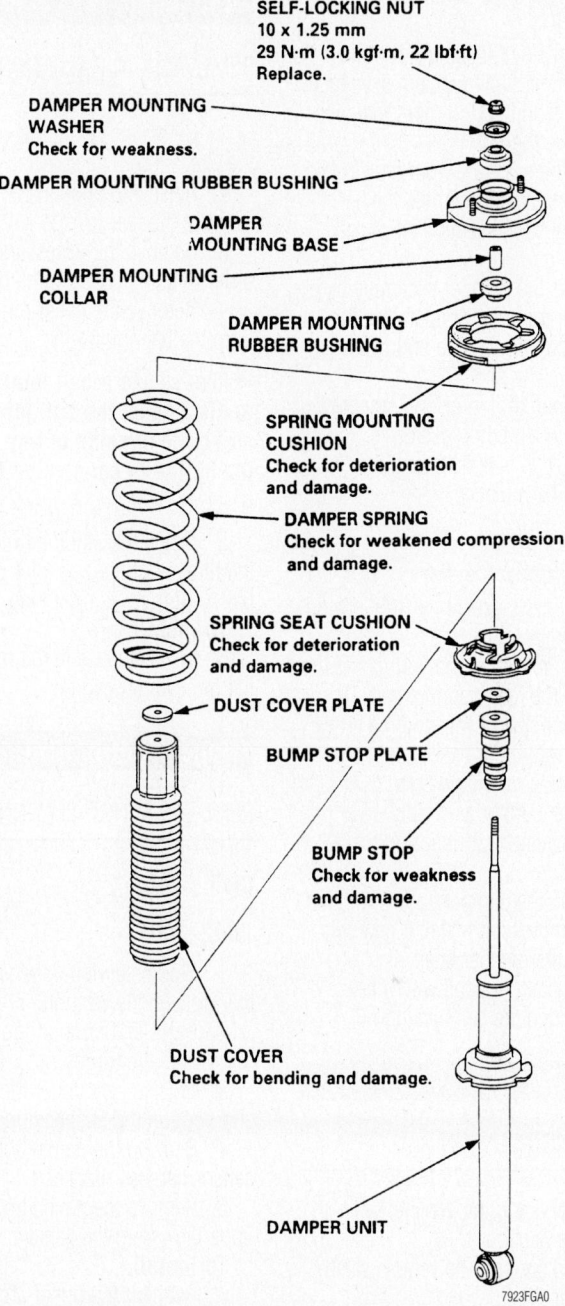

SELF-LOCKING NUT
10 x 1.25 mm
29 N·m (3.0 kgf·m, 22 lbf·ft)
Replace.

DAMPER MOUNTING WASHER
Check for weakness.

DAMPER MOUNTING RUBBER BUSHING

DAMPER MOUNTING BASE

DAMPER MOUNTING COLLAR

DAMPER MOUNTING RUBBER BUSHING

SPRING MOUNTING CUSHION
Check for deterioration and damage.

DAMPER SPRING
Check for weakened compression and damage.

SPRING SEAT CUSHION
Check for deterioration and damage.

DUST COVER PLATE

BUMP STOP PLATE

BUMP STOP
Check for weakness and damage.

DUST COVER
Check for bending and damage.

DAMPER UNIT

7923FGA0

Exploded view of the rear suspension strut assembly—Accord

pressor according to the manufacturer's instructions.

3. Remove the self locking nut while holding the damper shaft with a hex wrench. Do not compress the spring more than necessary to remove the nut.

4. Release the pressure from the strut spring compressor. Disassemble the damper assembly. Reassemble all parts, except the spring.

5. Compress the damper assembly by hand and check for smooth operation thru full stroke, both compression and extension.

6. The damper should extend smoothly and constantly when compression is released. If not, gas is leaking and the damper should be replaced.

ASSEMBLY

1. Install all parts except the self locking nut onto the damper unit.

2. Align the bottom of the spring and the stepped part of the lower spring seat. The hole in the upper spring seat and the arrow on the damper mounting base must point toward the knuckle mounting area.

3. Install the damper assembly on a commercially available strut spring compressor.

4. Compress the damper spring with the strut spring compressor. Install a new self locking nut on the damper shaft.

5. Hold the damper shaft with a hex wrench and tighten the self locking nut.

Stabilizer Bar

REMOVAL & INSTALLATION

1. Before servicing the vehicle, refer to the Precautions Section.

2. Raise and support the vehicle safely. Remove the rear tires.

3. Remove the self locking nuts, while holding the joint pins.

4. Disconnect the stabilizer links from the stabilizer bar on the right and left sides.

5. Remove the flange bolts and bushing holders. Remove the bushings and the stabilizer bar from the vehicle.

To install:

6. Position the stabilizer bar on the vehicle.

7. Use new self locking nuts.

8. Be sure the right and left ends of the stabilizer bar are installed on their respective sides of the vehicle.

Properly align the ends of the paint marks on the stabilizer bar with the bushings

Ball Joint

REMOVAL & INSTALLATION

1. Before servicing the vehicle, refer to the Precautions Section.

2. Raise and support the vehicle safely. Remove the rear tires.

3. Remove the necessary components in order to gain access to the ball joint.

4. Remove the cotter pin and lock nut from the ball joint.

➡**Always use a ball joint removal tool to disconnect the ball joint. Do not strike the housing or any other part of the ball joint connection to disconnect it.**

5. Install the ball joint removal tool onto the threads of the ball joint pin. Remove the ball joint from its mounting.

To install:

6. Installation is in the reverse order of the removal procedure.

Upper Control Arm

REMOVAL & INSTALLATION

Civic

2002–2005

1. Before servicing the vehicle, refer to the Precautions Section.

2. Raise and support the vehicle safely. Remove the rear tires.

3. Place a jack under the trailing arm and support the suspension.

4. Remove the flange bolt and wheel sensor harness bracket.

5. Remove the flange bolts. Remove the upper arm from the vehicle.

To install:

6. Position the upper arm on the vehicle. Be sure to use new retaining bolts and nuts.

7. Install all the suspension components and lightly tighten the bolts and nuts. Position a jack under the trailing arm and raise the suspension to load it with the vehicles weight, before fully tightening the bolts and nuts.

8. Continue the installation in the reverse order of the removal procedure.

9. Check and adjust the wheel alignment, as required.

2006

1. Before servicing the vehicle, refer to the Precautions Section.

2. Raise and support the vehicle safely. Remove the rear tires.

3. Place a jack under the trailing arm and the knuckle.

4. Remove the upper arm mounting bolts and the flange bolt. Remove the upper arm from the vehicle.

To install:

5. Position the upper arm on the vehicle. Be sure to use new retaining bolts and nuts.

6. Install all the suspension components and lightly tighten the bolts and nuts. Position a jack under the trailing arm and raise the suspension to load it with the vehicles weight, before fully tightening the bolts and nuts.

7. Continue the installation in the reverse order of the removal procedure.

8. Check and adjust the wheel alignment, as required.

S2000

1. Before servicing the vehicle, refer to the Precautions Section.

2. Raise and support the vehicle safely. Remove the rear tires.

3. Remove the flange bolts and wheel sensor harness from the upper arm.

4. Remove the lock pin from the upper arm ball joint, remove the castle nut.

5. Remove the upper arm ball joint from the knuckle using the ball joint removal tool.

6. Remove the flange bolts. Remove the upper arm from the vehicle.

To install:

7. Position the upper arm on the vehicle. Be sure to use new retaining bolts and nuts.

8. Install all the suspension components and lightly tighten the bolts and nuts. Position a jack under the trailing arm and raise the suspension to load it with the vehicles weight, before fully tightening the bolts and nuts.

9. Continue the installation in the reverse order of the removal procedure.

10. Check and adjust the wheel alignment, as required.

Accord

1. Before servicing the vehicle, refer to the Precautions Section.

2. Raise and support the vehicle safely. Remove the rear tires.

3. Remove the lock pin from the upper arm ball joint, and loosen the castle nut.

4. Disconnect the upper arm ball joint from the knuckle using the ball joint removal tool.

5. Remove the brake hose mounting bracket. Do not disconnect the brake line.

6. Remove the wheel sensor harness mounting bracket.

7. Remove the flange bolt. Remove the upper arm from the vehicle.

To install:

8. Position the upper arm on the vehicle. Be sure to use new retaining bolts and nuts.

9. Install all the suspension components and lightly tighten the bolts and nuts. Position a jack under the trailing arm and raise the suspension to load it with the vehicles weight, before fully tightening the bolts and nuts.

10. Continue the installation in the reverse order of the removal procedure.

11. Check and adjust the wheel alignment, as required.

Lower Control Arm

REMOVAL & INSTALLATION

S2000

1. Before servicing the vehicle, refer to the Precautions Section.

2. Raise and support the vehicle safely. Remove the rear tires.

3. Remove the flange nut while holding the joint pin. Disconnect the stabilizer link from the lower arm.

4. Remove the flange bolt and disconnect the damper from the lower arm.

Remove the cotter pin from the lower arm ball joint. Remove the castle nut.

5. Remove the lower arm ball joint from the knuckle, using a ball joint removal tool.

6. Remove the flange bolt, self locking nut, cam plate and adjusting nut. Remove the lower arm from the vehicle.

To install:

7. Position the lower arm on the vehicle. Be sure to use new retaining bolts and nuts.

8. Install all the suspension components and lightly tighten the bolts and nuts. Position a jack under the trailing arm and raise the suspension to load it with the vehicles weight, before fully tightening the bolts and nuts.

9. Continue the installation in the reverse order of the removal procedure.

10. Check and adjust the wheel alignment, as required.

Accord

1. Before servicing the vehicle, refer to the Precautions Section.

2. Raise and support the vehicle safely. Remove the rear tires.

3. Remove the lower arm mounting nut and mounting bolt from the knuckle side.

4. Remove the flange bolt. Remove the lower arm from the vehicle.

To install:

5. Position the lower arm on the vehicle. Be sure to use new retaining bolts and nuts.

6. Install all the suspension components and lightly tighten the bolts and nuts. Position a jack under the trailing arm and raise the suspension to load it with the vehicles weight, before fully tightening the bolts and nuts.

7. Continue the installation in the reverse order of the removal procedure.

8. Check and adjust the wheel alignment, as required.

Wheel Bearings

ADJUSTMENT

1. Before servicing the vehicle, refer to the Precautions Section.

2. Raise and support the vehicle safely. Remove the rear tires.

3. Install suitable flat washers and wheel nuts, and tighten the nuts to 79.6 ft. lbs.

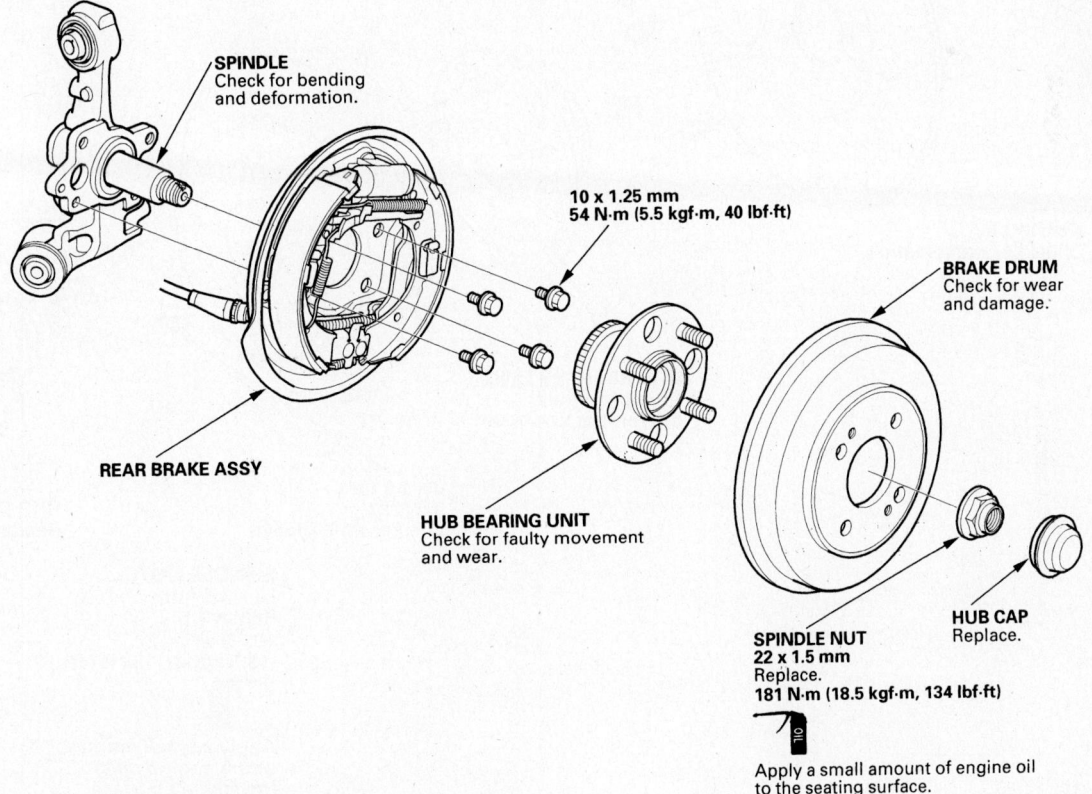

Knuckle and related components—2002—2005 Civic with drum brakes

(108 Nm), to hold the brake disc or drum securely against the hub.

4. Attach a dial gauge against the center of the hub cap. Bearing end play should be 0—0.002 inch (0—0.05mm).

5. On 2006 Civic, S2000 and 2003—2006 Accord, measure the end play by moving the disc or drum inward and outward.

6. If the bearing end play is not within specification, replace the wheel bearing.

REMOVAL & INSTALLATION

Civic

DRUM BRAKE

1. Before servicing the vehicle, refer to the Precautions Section.

2. Raise and support the vehicle safely. Remove the rear tires.

3. Remove the hub cap, raise the stake and remove the spindle nut.

4. Screw two 8x1.25mm bolts into the brake drum to push it away from the hub. Turn each bolt two turns at a time to prevent cocking the drum excessively. Remove the brake drum.

5. Remove the hub bearing unit from the spindle.

To install:

6. Position the hub bearing unit on the vehicle. Be sure to use new bolts and nuts, as necessary. Use a new hub cap.

7. Continue the installation in the reverse order of the removal procedure.

DISC BRAKE

1. Before servicing the vehicle, refer to the Precautions Section.

2. Raise and support the vehicle safely. Remove the rear tires.

3. Remove the brake hose bracket mounting bolt from the knuckle.

4. Remove the brake caliper mounting bolts. Remove the caliper assembly from the knuckle. Remove the two washers.

5. Remove the 6mm brake disc retaining screws. Remove the rotor from the hub bearing nut.

➡ If the brake disc has clung to the hub bearing unit, screw two 8x1.25MM bolts into the rotor to push it away from the hub bearing unit. Turn each bolt 90 degrees at a time to prevent cocking the rotor.

6. Remove the hub bearing unit and the O-ring.

To install:

7. Position the hub bearing unit on the vehicle. Be sure to use new bolts and nuts, as necessary. Use a new hub cap.

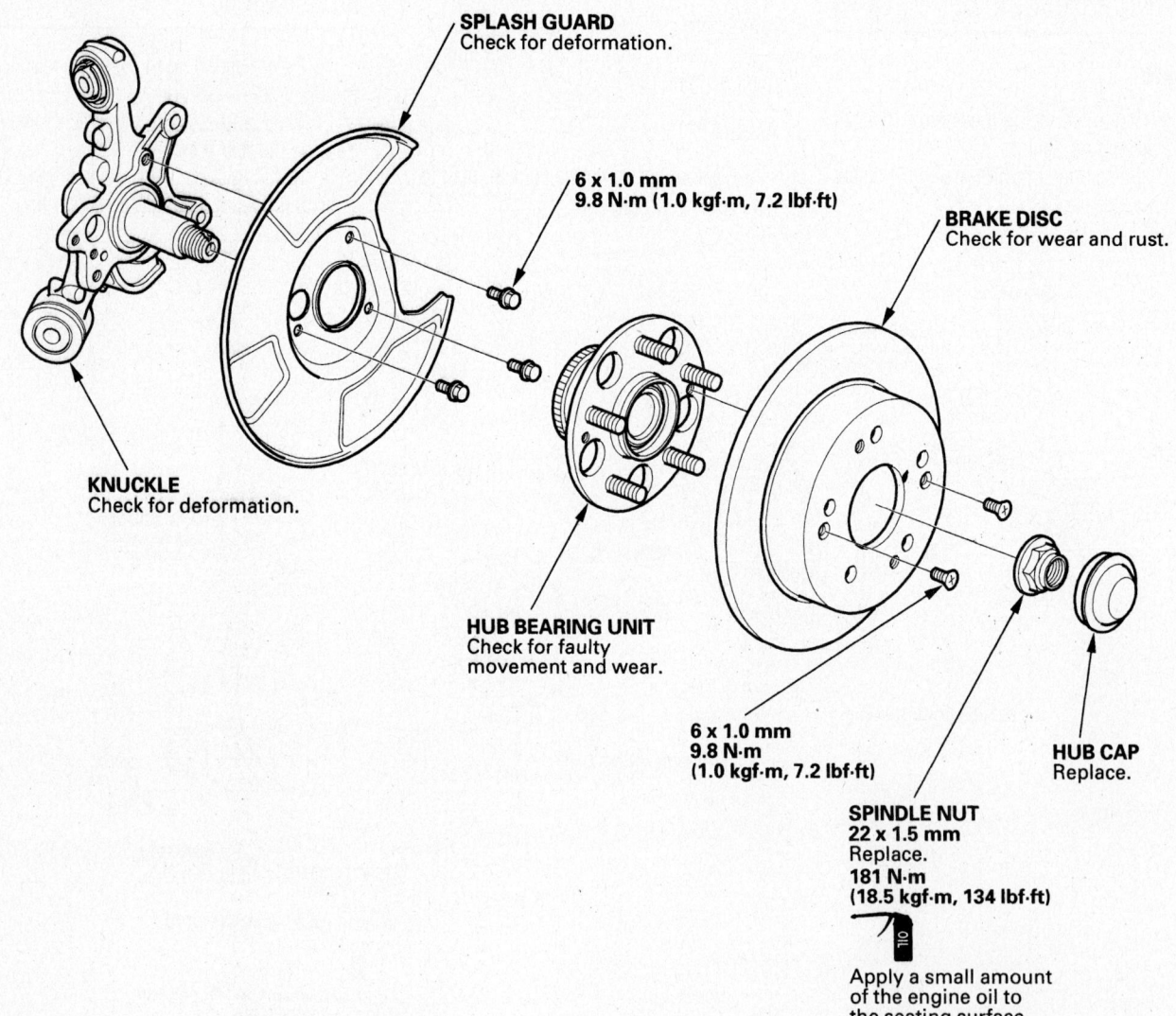

SPLASH GUARD
Check for deformation.

6 x 1.0 mm
9.8 N·m (1.0 kgf·m, 7.2 lbf·ft)

BRAKE DISC
Check for wear and rust.

KNUCKLE
Check for deformation.

HUB BEARING UNIT
Check for faulty movement and wear.

6 x 1.0 mm
9.8 N·m
(1.0 kgf·m, 7.2 lbf·ft)

HUB CAP
Replace.

SPINDLE NUT
22 x 1.5 mm
Replace.
181 N·m
(18.5 kgf·m, 134 lbf·ft)

Apply a small amount of the engine oil to the seating surface.

09474_ACCO_G0041

Knuckle and related components—2002—2005 Civic with disc brakes

8. Continue the installation in the reverse order of the removal procedure.

9. Check the wheel alignment, adjust as required.

S2000

1. Before servicing the vehicle, refer to the Precautions Section.

2. Remove or disconnect the following:
- Rear wheel
- Brake caliper support bracket
- Wheel speed sensor
- Spindle nut
- Brake rotor
- Control arm
- Upper and lower ball joints
- Spindle from the vehicle

3. Mount the steering knuckle in a press and press the hub out of the wheel bearing.

4. Remove the splash guard and the wheel bearing snapring.

5. Press the wheel bearing out of the steering knuckle.

To install:

6. Installation is the reverse of the removal procedure, while using the following torque values:
- Splash guard screws: 48 inch lbs. (5 Nm)
- Spindle nut: 181 ft. lbs. (245 Nm)
- Upper ball joint nut: 36–43 ft. lbs. (49–59 Nm)
- Lower ball joint nut: 51–58 ft. lbs. (68–78 Nm)
- Control arm ball joint nut: 36–43 ft. lbs. (49–59 Nm)
- Brake caliper support bolts: 41 ft. lbs. (55 Nm)

Accord

➡ **The rear wheel bearing and hub unit are replaced as a unit.**

1. Before servicing the vehicle, refer to the Precautions Section.

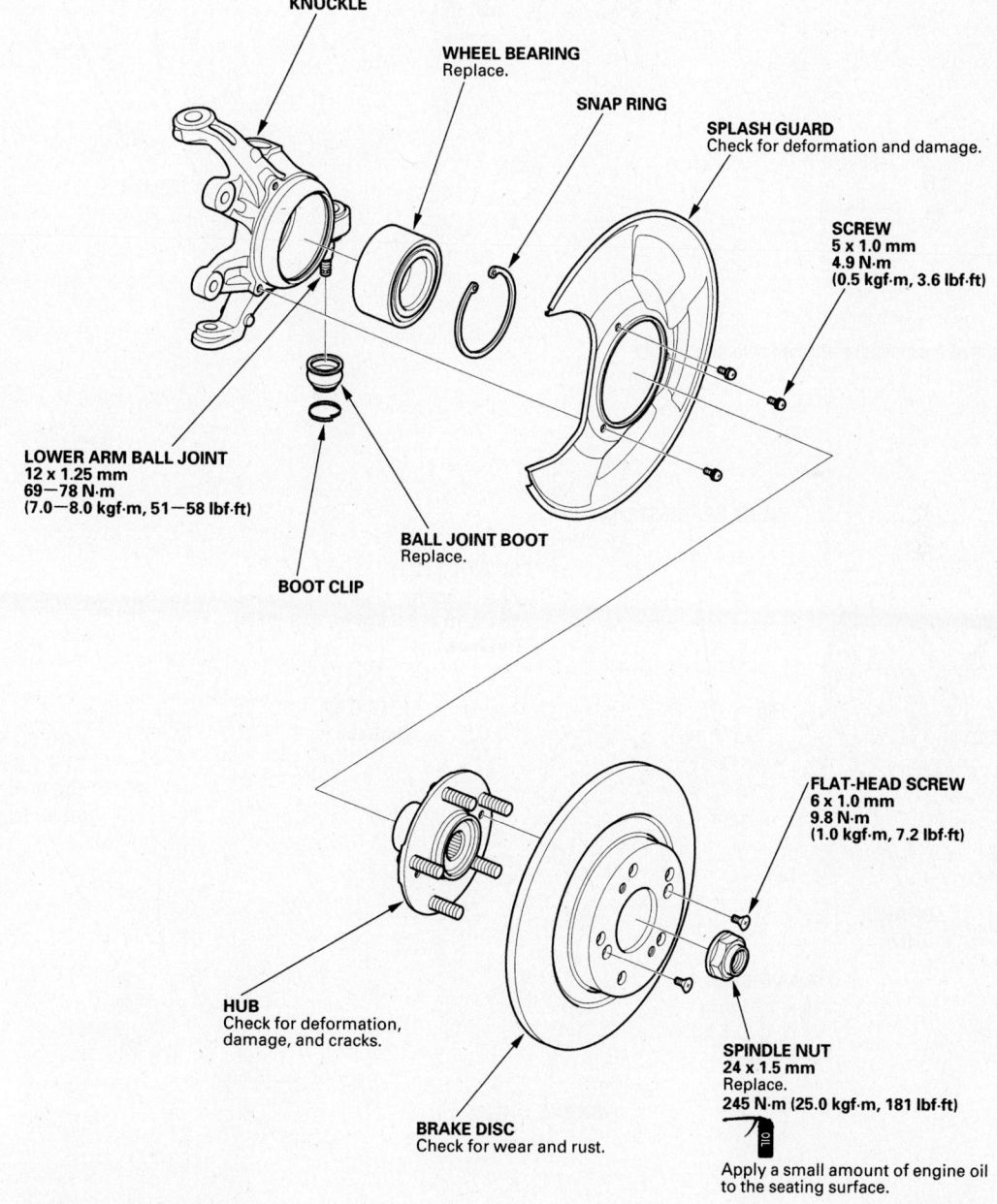

KNUCKLE

WHEEL BEARING
Replace.

SNAP RING

SPLASH GUARD
Check for deformation and damage.

SCREW
5 x 1.0 mm
4.9 N·m
(0.5 kgf·m, 3.6 lbf·ft)

LOWER ARM BALL JOINT
12 x 1.25 mm
69—78 N·m
(7.0—8.0 kgf·m, 51—58 lbf·ft)

BALL JOINT BOOT
Replace.

BOOT CLIP

FLAT-HEAD SCREW
6 x 1.0 mm
9.8 N·m
(1.0 kgf·m, 7.2 lbf·ft)

HUB
Check for deformation, damage, and cracks.

SPINDLE NUT
24 x 1.5 mm
Replace.
245 N·m (25.0 kgf·m, 181 lbf·ft)

Apply a small amount of engine oil
to the seating surface.

BRAKE DISC
Check for wear and rust.

09474_ACCO_G0040

Knuckle and related components—S2000

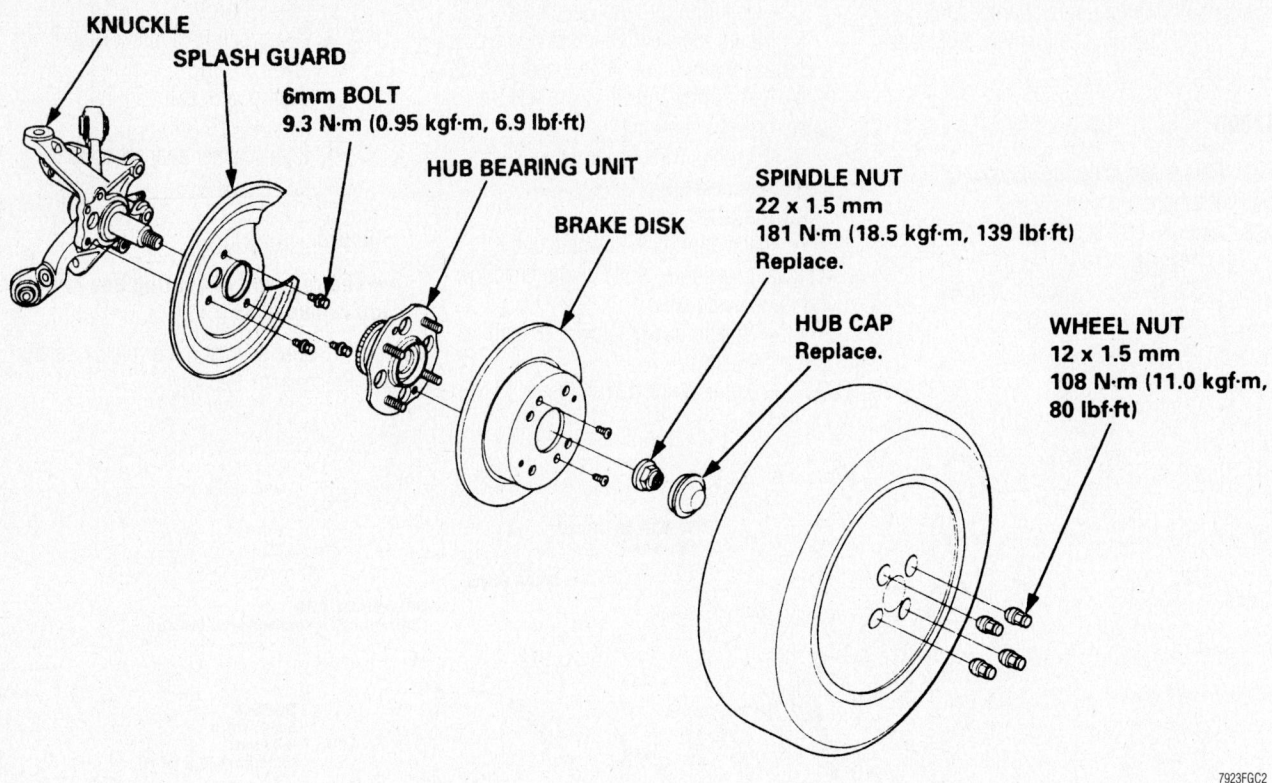

KNUCKLE

SPLASH GUARD

6mm BOLT
9.3 N·m (0.95 kgf·m, 6.9 lbf·ft)

HUB BEARING UNIT

BRAKE DISK

SPINDLE NUT
22 x 1.5 mm
181 N·m (18.5 kgf·m, 139 lbf·ft)
Replace.

HUB CAP
Replace.

WHEEL NUT
12 x 1.5 mm
108 N·m (11.0 kgf·m, 80 lbf·ft)

7923FGC2

Knuckle and related components—Accord with disc brakes

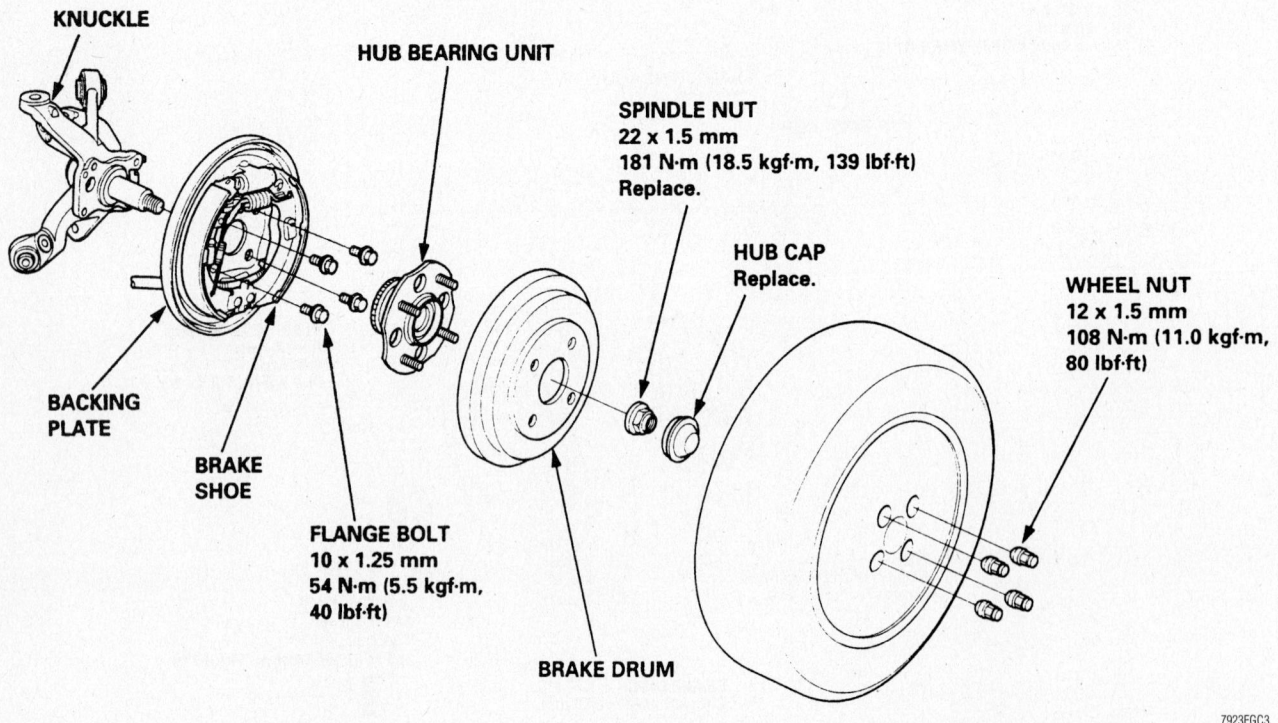

KNUCKLE

HUB BEARING UNIT

SPINDLE NUT
22 x 1.5 mm
181 N·m (18.5 kgf·m, 139 lbf·ft)
Replace.

HUB CAP
Replace.

WHEEL NUT
12 x 1.5 mm
108 N·m (11.0 kgf·m, 80 lbf·ft)

BACKING PLATE

BRAKE SHOE

FLANGE BOLT
10 x 1.25 mm
54 N·m (5.5 kgf·m, 40 lbf·ft)

BRAKE DRUM

7923FGC3

Knuckle and related components—Accord with drum brakes

2. Loosen the spindle nut.

3. Raise the vehicle and support it safely.

4. Remove or disconnect the following:
- Rear wheels
- Brake disc retaining screws
- Brake hose brackets from the knuckle
- Caliper bracket mounting bolts and hang the caliper out of the way with a piece of wire.
- Brake disc. If the disc is frozen on the hub, screw 2, 8 x 1.25mm bolts evenly into the disc to push it away from the hub.

- Spindle nut and pull the hub unit off of the spindle.

➡**Clean the backing plate and the mating surfaces of the brake disc and hub with brake cleaner. Clean the spindle, washer, and hub with solvent.**

To install:

5. Inspect the hub unit for signs of damage or wear. If the bearings are worn, the entire unit must be replaced.

6. Install or connect the following:
- Hub unit and spindle washer onto the spindle.
- Spindle nut but do not tighten it.

- Brake disc and tighten the retaining screws to 84 inch lbs. (10 Nm).
- Brake caliper and tighten the mounting bolts to 28 ft. lbs. (39 Nm).
- Brake hose brackets onto the knuckle and tighten the bolts to 16 ft. lbs. (22 Nm).
- Rear wheels and lower the vehicle.

7. With the vehicle on the ground, tighten the new spindle nut to 134 ft. lbs. (185 Nm), then stake the nut with a punch.

8. Tighten the wheel nuts to 80 ft. lbs. (110 Nm).

9. Test the operation of the brakes.

BRAKES

Brake Caliper

REMOVAL & INSTALLATION

Civic

FRONT

1. Before servicing the vehicle, refer to the Precautions Section.

2. Remove and discard brake fluid from the master cylinder, as necessary.

3. Raise and support the vehicle safely. Remove the front tires.

4. Disconnect and plug the brake line hose.

5. On 2004—05 Hatchback, remove the brake hose bracket mounting bolt.

6. Remove the caliper flange bolts.

7. Remove the brake pads and shims.

8. Remove the pad retainers and check the caliper pins for free movement.

To install:

9. Installation is the reverse of the removal procedure.

10. Apply a thin coat of M-77 assembly paste part number 08798-9010 to the brake pad sides of the pad shims and the back of the brake pads, wipe excess paste off the shim.

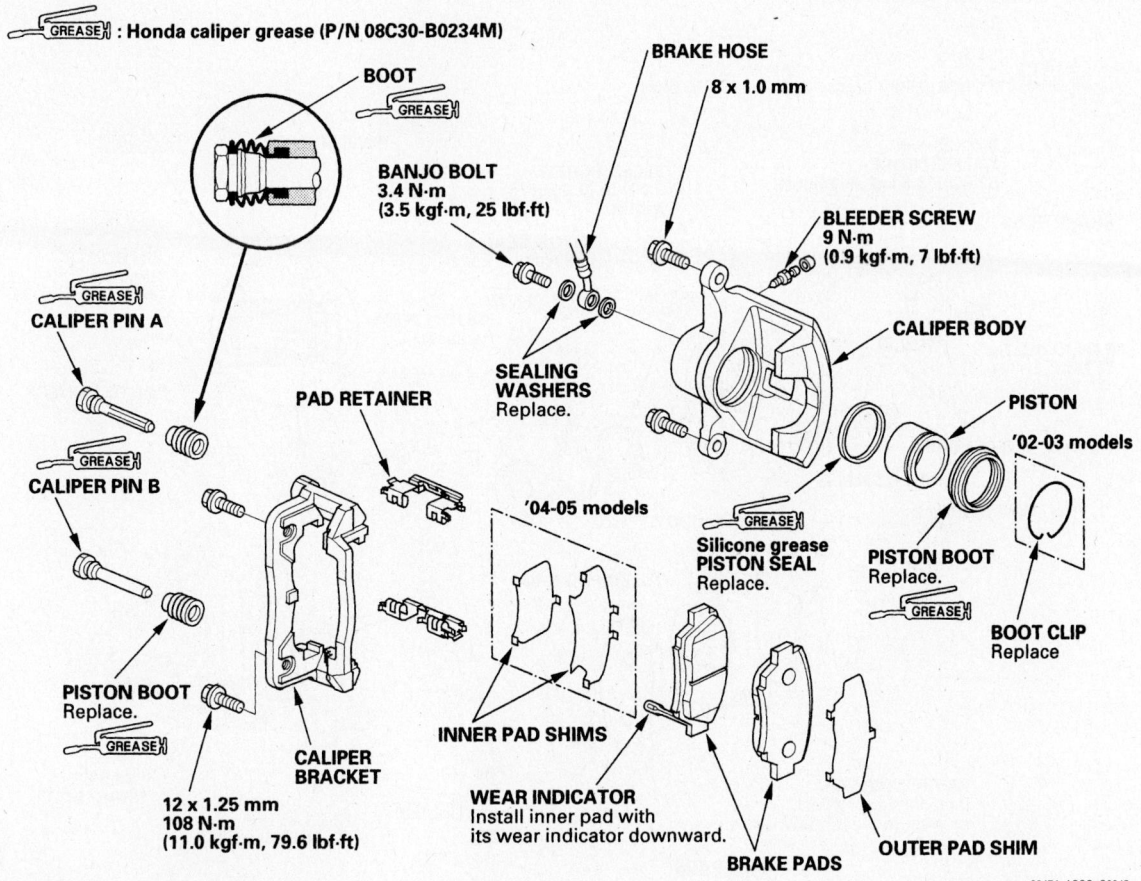

Exploded view of the front disc brakes—Civic

09474_ACCO_G0045

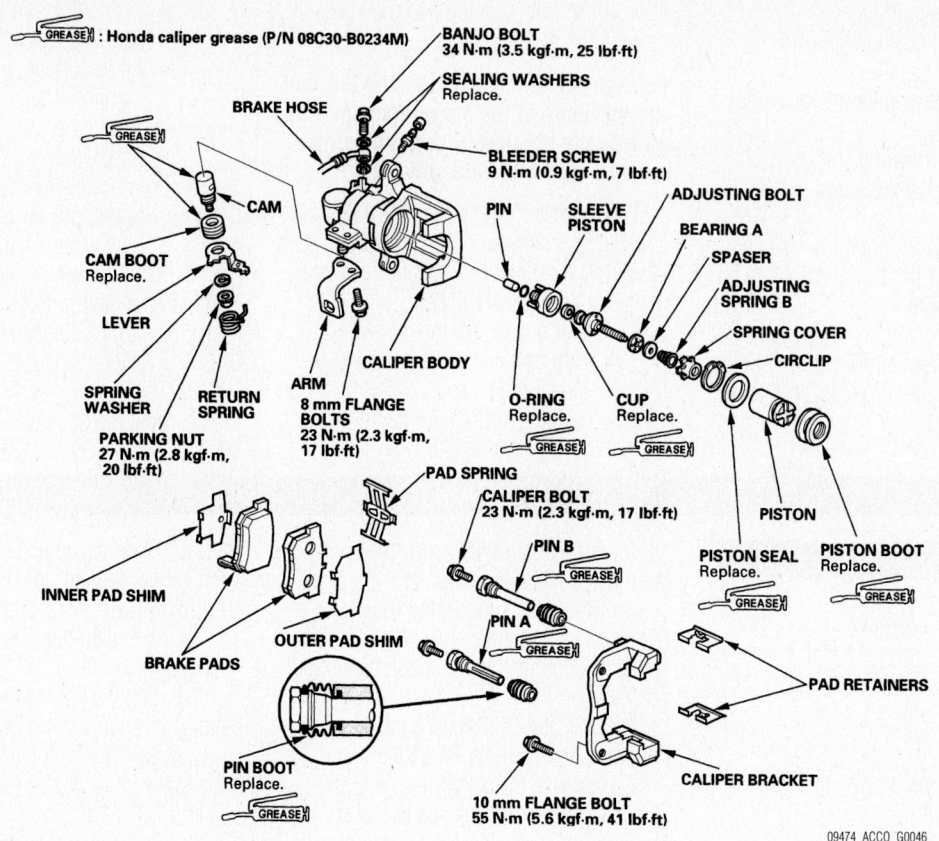

Exploded view of the rear disc brakes—Civic

09474_ACCO_G0046

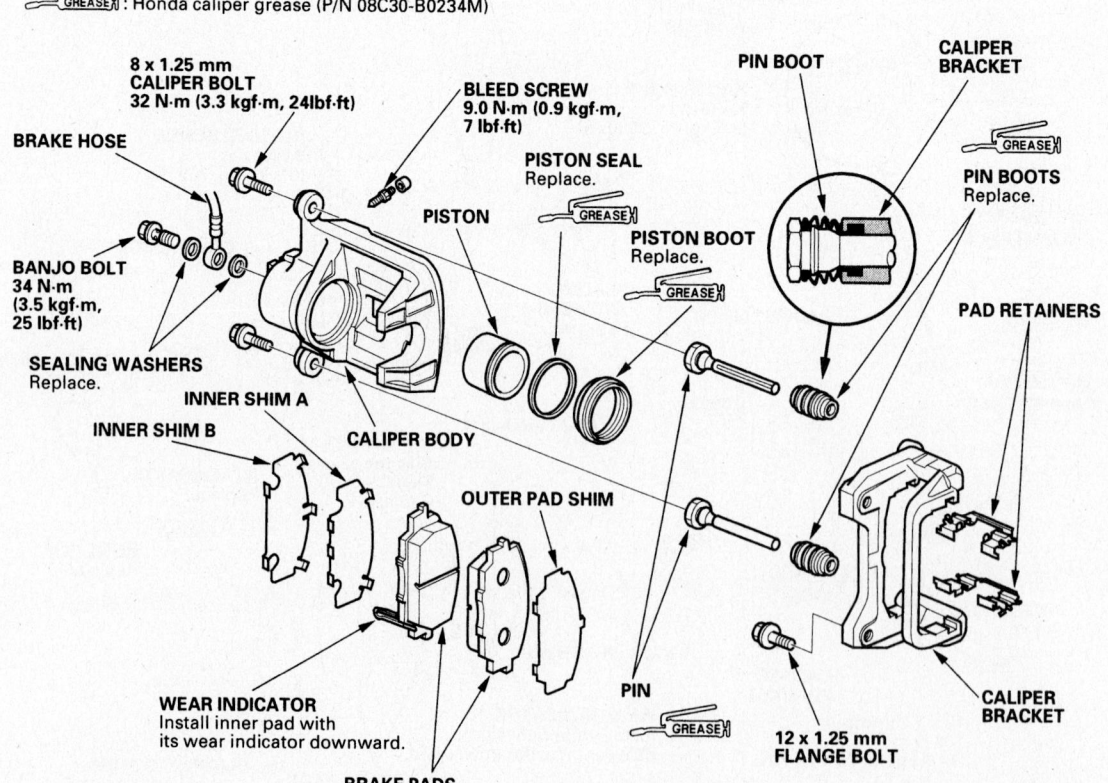

Exploded view of the front disc brakes—S2000

09474_ACCO_G0043

11. Check and refill the master cylinder, as necessary.

12. Roadtest the vehicle.

REAR

1. Before servicing the vehicle, refer to the Precautions Section.

2. Remove and discard brake fluid from the master cylinder, as necessary.

3. Raise and support the vehicle safely. Remove the rear tires.

4. Remove the bolt and brake hose from the mounting bracket.

5. Disconnect and plug the brake line hose.

6. Remove the caliper bolts. Remove the caliper from the caliper bracket.

7. Remove the pad shims and brake pads.

8. Remove the pad retainers and check the caliper pins for free movement.

To install:

9. Apply a thin coat of M-77 assembly paste part number o8798-9010 to the brake pad sides of the pad shims and the back of the brake pads, wipe excess paste off the shim.

10. Install the brake pads and shims.

11. Rotate the caliper piston clockwise into the cylinder, then align the cutout in the piston with the tab by turning the piston back so the caliper can be installed on the brake pad. Lubricate the boot with rubber grease to avoid twisting the piston boot.

12. Continue the installation in the reverse order of the removal procedure.

13. After installation depress the brake pedal several times to be sure the brake works. Roadtest the vehicle.

➡**Engagement of the brake may require a greater pedal stroke effort**

immediately after the brake pads have been replaced as a set.

14. Check and refill the master cylinder, as necessary.

15. Check the parking brake adjustment.

S2000

FRONT

1. Before servicing the vehicle, refer to the Precautions Section.

2. Remove and discard brake fluid from the master cylinder, as necessary.

3. Raise and support the vehicle safely. Remove the front tires.

4. Disconnect and plug the brake line hose.

5. Remove the brake hose bracket mounting bolt.

6. Remove the caliper flange bolts.

7. Remove the brake pads and shims.

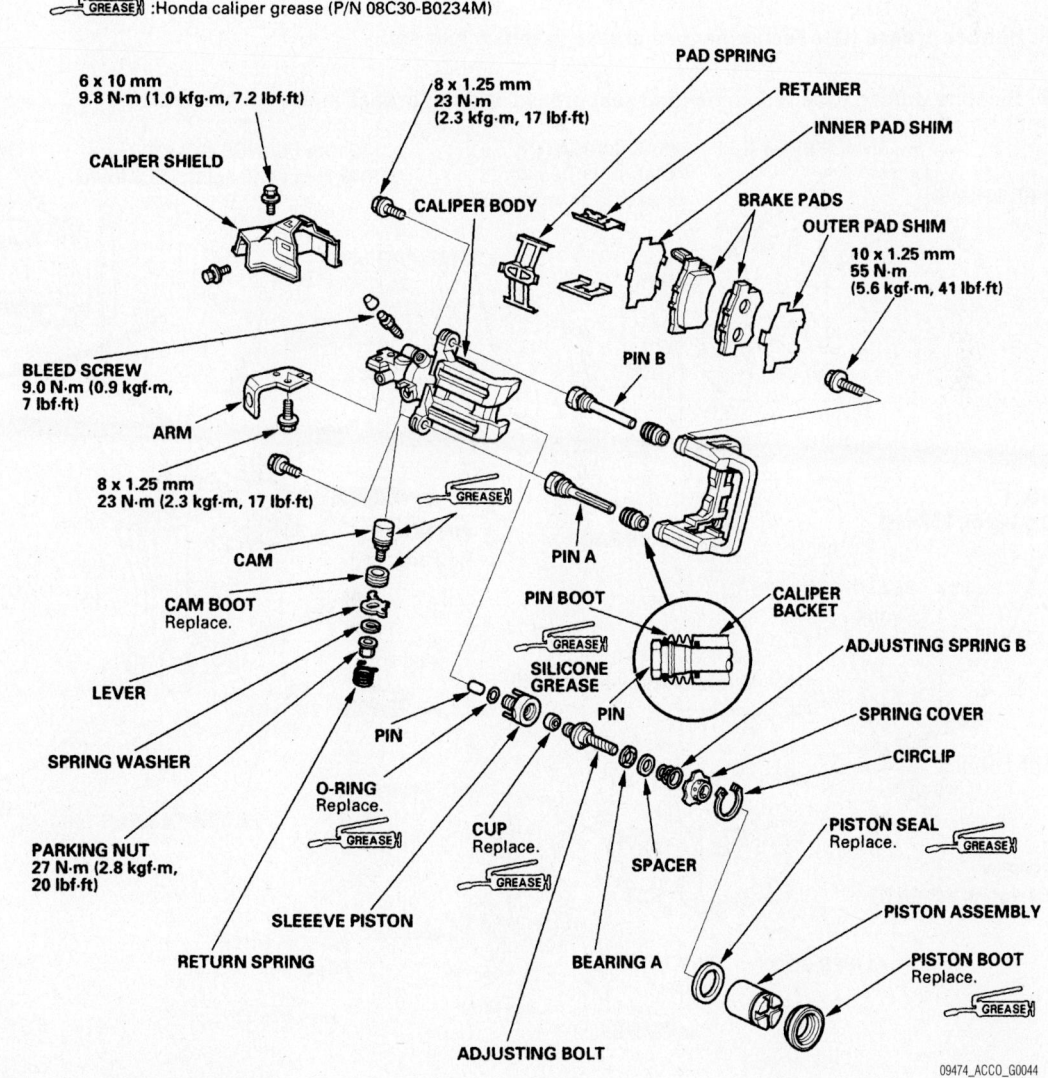

Exploded view of the rear disc brakes—S2000

09474_ACCO_G0044

8. Remove the pad retainers and check the caliper pins for free movement.

To install:

9. Apply a thin coat of M-77 assembly paste part number o8798-9010 to the brake pad sides of the pad shims and the back of the brake pads, wipe excess paste off the shim.

10. Continue the installation in the reverse order of the removal procedure.

11. Check and refill the master cylinder, as necessary.

12. Roadtest the vehicle.

REAR

1. Before servicing the vehicle, refer to the Precautions Section.

2. Remove and discard brake fluid from the master cylinder, as necessary.

3. Raise and support the vehicle safely. Remove the rear tires. Release the parking brake.

4. Disconnect and plug the brake line hose.

5. Remove the caliper bolts. Remove the caliper from the caliper bracket.

6. Remove the pad shims and brake pads.

7. Remove the pad retainers and check the caliper pins for free movement.

To install:

8. Apply a thin coat of M-77 assembly paste part number o8798-9010 to the brake pad sides of the pad shims and the back of the brake pads, wipe excess paste off the shim.

9. Install the brake pads and shims.

10. Rotate the caliper piston clockwise into the cylinder, then align the cutout in the piston with the tab on the inner pad by turning the piston back so the caliper can be installed on the brake pad. Lubricate the boot with rubber grease to avoid twisting the piston boot.

11. Continue the installation in the reverse order of the removal procedure.

12. After installation depress the brake pedal several times to be sure the brake works. Roadtest the vehicle.

➥**Engagement of the brake may require a greater pedal stroke effort immediately after the brake pads have been replaced as a set.**

13. Check the parking brake adjustment.

14. Check and refill the master cylinder, as necessary.

15. Roadtest the vehicle.

Accord

FRONT

1. Before servicing the vehicle, refer to the Precautions Section.

2. Remove and discard brake fluid from the master cylinder, as necessary.

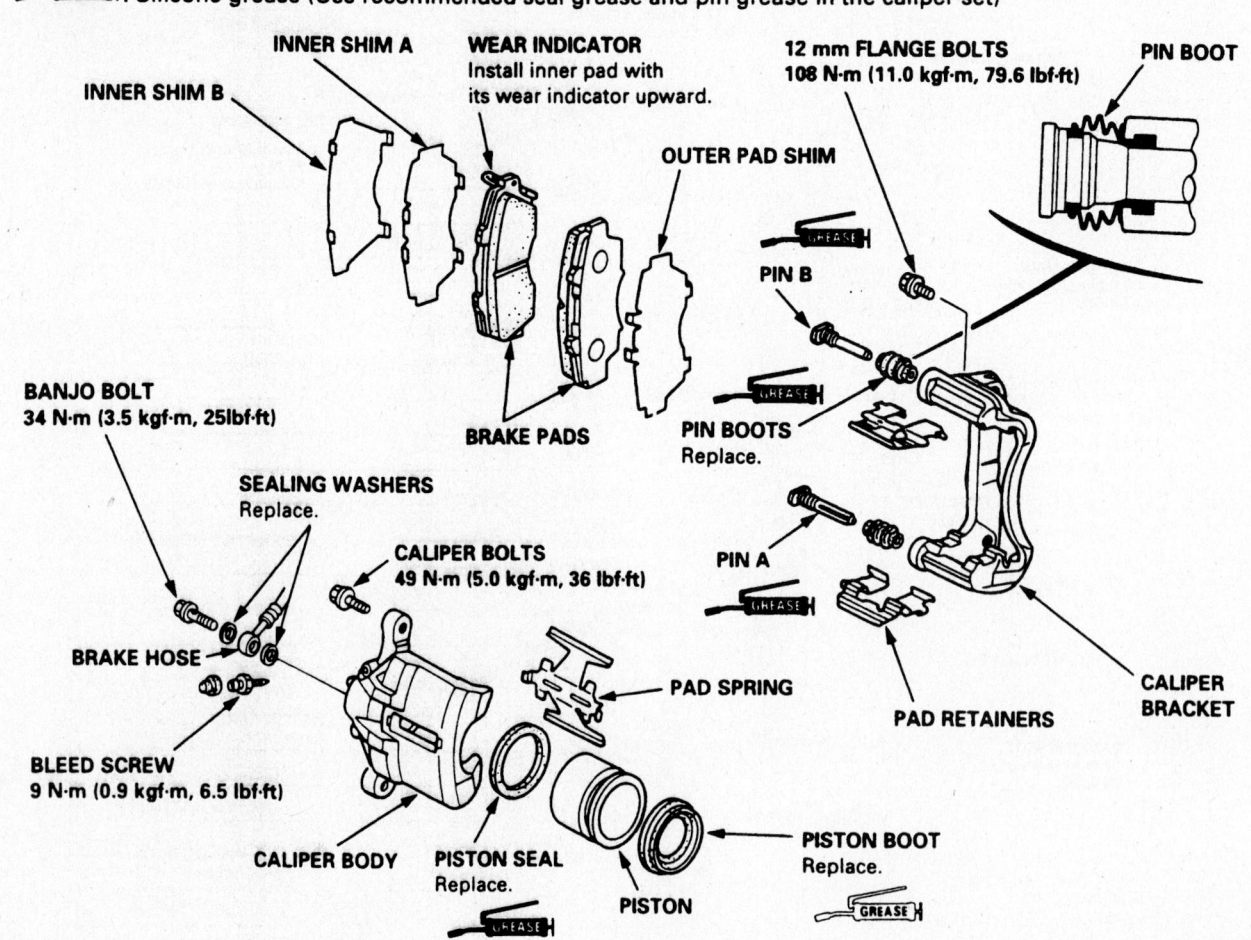

GREASE: Rubber grease (Use recommended grease in the caliper set)

GREASE: Silicone grease (Use recommended seal grease and pin grease in the caliper set)

Exploded view of the front disc brakes—Accord V6 shown

3. Raise and support the vehicle safely. Remove the front tires.

4. Disconnect and plug the brake line hose.

5. Remove the brake hose bracket mounting bolt.

6. Remove the caliper flange bolts.

7. Remove the brake pads and shims.

8. Remove the pad retainers and check the caliper pins for free movement.

To install:

9. Apply a thin coat of M-77 assembly paste part number o8798-9010 to the brake pad sides of the pad shims and the back of the brake pads, wipe excess paste off the shim.

10. Continue the installation in the reverse order of the removal procedure.

11. Check and refill the master cylinder, as necessary.

12. Roadtest the vehicle.

REAR

1. Before servicing the vehicle, refer to the Precautions Section.

2. Remove and discard brake fluid from the master cylinder, as necessary.

3. Raise and support the vehicle safely. Remove the rear tires. Release the parking brake.

4. Disconnect and plug the brake line hose.

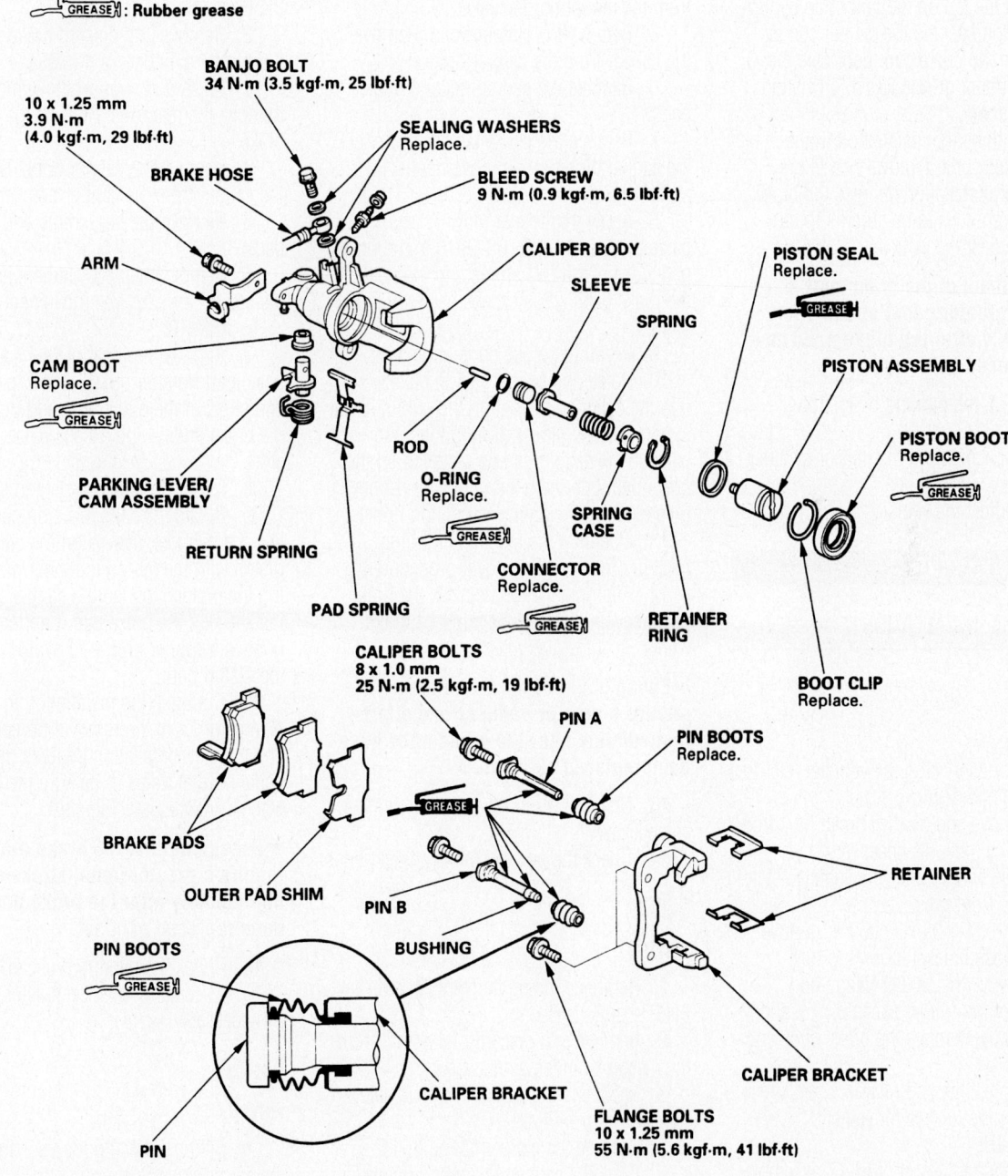

Sedan:

Exploded view of the rear disc brakes—Accord

93016G08

5. Remove the caliper bolts. Remove the caliper from the caliper bracket.

6. Remove the pad shims and brake pads.

7. Remove the pad retainers and check the caliper pins for free movement.

To install:

8. Apply a thin coat of M-77 assembly paste part number o8798-9010 to the brake pad sides of the pad shims and the back of the brake pads, wipe excess paste off the shim.

9. Install the brake pads and shims.

10. Rotate the caliper piston clockwise into the cylinder, then align the cutout in the piston with the tab on the inner pad by turning the piston back so the caliper can be installed on the brake pad. Lubricate the boot with rubber grease to avoid twisting the piston boot.

11. Continue the installation in the reverse order of the removal procedure.

12. After installation depress the brake pedal several times to be sure the brake works. Roadtest the vehicle.

➡**Engagement of the brake may require a greater pedal stroke effort immediately after the brake pads have been replaced as a set.**

13. Check the parking brake adjustment.

14. Check and refill the master cylinder, as necessary.

15. Roadtest the vehicle.

Disc Brake Pads

REMOVAL & INSTALLATION

Civic

FRONT

1. Before servicing the vehicle, refer to the Precautions Section.

2. Remove and discard brake fluid from the master cylinder, as necessary.

3. Raise and support the vehicle safely. Remove the front tires.

4. On 2004—05 Hatchback, remove the brake hose bracket mounting bolt.

5. Remove the caliper flange bolt.

6. Pivot the caliper assembly up and out of the way. Remove the brake pads and shims.

7. Remove the pad retainers and check the caliper pins for free movement.

To install:

8. Apply a thin coat of M-77 assembly paste part number o8798-9010 to the brake pad sides of the pad shims and the back of

the brake pads, wipe excess paste off the shim.

9. Continue the installation in the reverse order of the removal procedure.

10. Check and refill the master cylinder, as necessary.

REAR

1. Before servicing the vehicle, refer to the Precautions Section.

2. Remove and discard brake fluid from the master cylinder, as necessary.

3. Raise and support the vehicle safely. Remove the rear tires.

4. Remove the bolt and brake hose from the mounting bracket.

5. Remove the caliper bolts. Remove the caliper from the caliper bracket.

6. Remove the pad shims and brake pads.

7. Remove the pad retainers and check the caliper pins for free movement.

To install:

8. Apply a thin coat of M-77 assembly paste part number o8798-9010 to the brake pad sides of the pad shims and the back of the brake pads, wipe excess paste off the shim.

9. Install the brake pads and shims.

10. Rotate the caliper piston clockwise into the cylinder, then align the cutout in the piston with the tab by turning the piston back so the caliper can be installed on the brake pad. Lubricate the boot with rubber grease to avoid twisting the piston boot.

11. Continue the installation in the reverse order of the removal procedure.

12. After installation depress the brake pedal several times to be sure the brake works. Roadtest the vehicle.

➡**Engagement of the brake may require a greater pedal stroke effort immediately after the brake pads have been replaced as a set.**

13. Check and refill the master cylinder, as necessary.

14. Check the parking brake adjustment.

S2000

FRONT

1. Before servicing the vehicle, refer to the Precautions Section.

2. Remove and discard brake fluid from the master cylinder, as necessary.

3. Raise and support the vehicle safely. Remove the front tires.

4. Remove the caliper flange bolt.

5. Pivot the caliper assembly up and out of the way. Remove the brake pads and shims.

6. Remove the pad retainers and check the caliper pins for free movement.

To install:

7. Apply a thin coat of M-77 assembly paste part number o8798-9010 to the brake pad sides of the pad shims and the back of the brake pads, wipe excess paste off the shim.

8. Continue the installation in the reverse order of the removal procedure.

9. Check and refill the master cylinder, as necessary.

REAR

1. Before servicing the vehicle, refer to the Precautions Section.

2. Remove and discard brake fluid from the master cylinder, as necessary.

3. Raise and support the vehicle safely. Remove the rear tires. Release the parking brake.

4. Remove the caliper bolts. Remove the caliper from the caliper bracket.

5. Remove the pad shims and brake pads.

6. Remove the pad retainers and check the caliper pins for free movement.

To install:

7. Apply a thin coat of M-77 assembly paste part number o8798-9010 to the brake pad sides of the pad shims and the back of the brake pads, wipe excess paste off the shim.

8. Install the brake pads and shims.

9. Rotate the caliper piston clockwise into the cylinder, then align the cutout in the piston with the tab on the inner pad by turning the piston back so the caliper can be installed on the brake pad. Lubricate the boot with rubber grease to avoid twisting the piston boot.

10. Continue the installation in the reverse order of the removal procedure.

11. After installation depress the brake pedal several times to be sure the brake works. Roadtest the vehicle.

➡**Engagement of the brake may require a greater pedal stroke effort immediately after the brake pads have been replaced as a set.**

12. Check the parking brake adjustment.

13. Check and refill the master cylinder, as necessary.

Accord

FRONT

1. Before servicing the vehicle, refer to the Precautions Section.

2. Remove and discard brake fluid from the master cylinder, as necessary.

3. Raise and support the vehicle safely. Remove the front tires.

4. Remove the brake hose bracket mounting bolt.

5. Remove the caliper flange bolt.

6. Pivot the caliper assembly up and out of the way. Remove the brake pads and shims.

7. Remove the pad retainers and check the caliper pins for free movement.

To install:

8. Apply a thin coat of M-77 assembly paste part number 08798-9010 to the brake pad sides of the pad shims and the back of the brake pads, wipe excess paste off the shim.

9. Continue the installation in the reverse order of the removal procedure.

10. Check and refill the master cylinder, as necessary.

REAR

1. Before servicing the vehicle, refer to the Precautions Section.

2. Remove and discard brake fluid from the master cylinder, as necessary.

3. Raise and support the vehicle safely. Remove the rear tires. Release the parking brake.

4. Remove the caliper bolts. Remove the caliper from the caliper bracket.

5. Remove the pad shims and brake pads.

6. Remove the pad retainers and check the caliper pins for free movement.

To install:

7. Apply a thin coat of M-77 assembly paste part number 08798-9010 to the brake pad sides of the pad shims and the back of

the brake pads, wipe excess paste off the shim.

8. Install the brake pads and shims.

9. Rotate the caliper piston clockwise into the cylinder, then align the cutout in the piston with the tab on the inner pad by turning the piston back so the caliper can be installed on the brake pad. Lubricate the boot with rubber grease to avoid twisting the piston boot.

10. Continue the installation in the reverse order of the removal procedure.

11. After installation depress the brake pedal several times to be sure the brake works. Roadtest the vehicle.

➡**Engagement of the brake may require a greater pedal stroke effort immediately after the brake pads have been replaced as a set.**

12. Check the parking brake adjustment.

13. Check and refill the master cylinder, as necessary.

Brake Drums

REMOVAL & INSTALLATION

Civic and Accord

1. Before servicing the vehicle, refer to the Precautions Section.

2. Raise and support the vehicle safely. Remove the front tires.

3. Release the parking brake.

4. Remove the brake drum

To install:

5. Installation is the reverse of the removal procedure.

6. Make certain the brake shoes are adjusted to allow the drum clearance during installation.

Brake Shoes

REMOVAL & INSTALLATION

Accord and Civic

1. Before servicing the vehicle, refer to the Precautions Section.

2. Remove or disconnect the following:
   - Rear wheels and brake drums
   - Upper return spring from the brake shoes

3. Push the retainer springs and turn the tension pins to release the shoes from the backing plate.
   - Lower the brake shoe assembly and remove the lower return spring
   - Brake shoe assembly from the backing plate
   - Parking brake cable from the brake shoe lever
   - Upper return spring, self-adjuster lever, and self-adjuster spring. Separate the brake shoes.
   - Wave washer, parking brake lever, and pivot pin from the brake shoe by removing the U-clip.

To install:

4. Install or connect the following:
   - Brake cylinder grease to the sliding surface of the pivot pin and insert the pin into the brake shoe.
   - Parking brake lever and wave washer on the pivot pin and

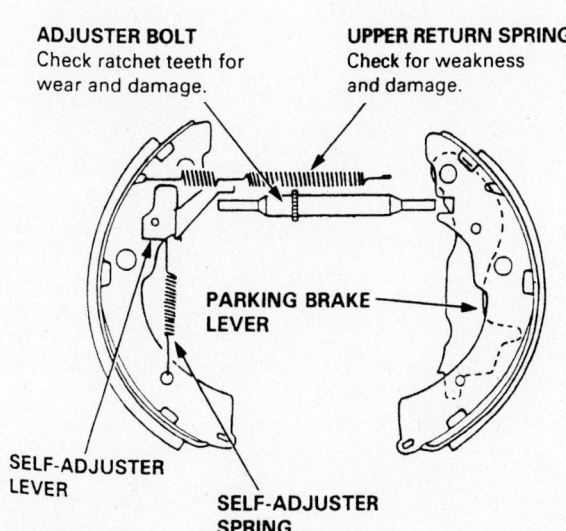

ADJUSTER BOLT
Check ratchet teeth for wear and damage.

UPPER RETURN SPRING
Check for weakness and damage.

PARKING BRAKE LEVER

SELF-ADJUSTER LEVER

SELF-ADJUSTER SPRING

Rear drum brakes—Civic shown

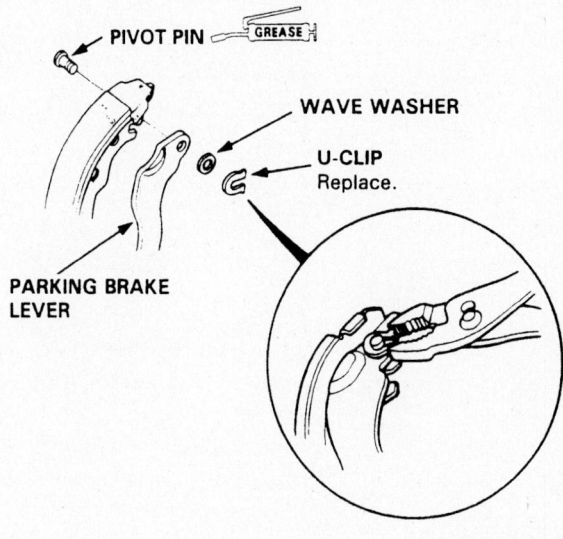

PIVOT PIN [GREASE]

WAVE WASHER

U-CLIP
Replace.

PARKING BRAKE LEVER

93016G09

pinch the U-clip with a pair of pliers to secure the pivot pin to the shoe

- Parking brake cable to the parking brake lever
- Adjuster spring to the adjuster lever first, then to the brake shoe
- Adjuster bolt/clevis assembly and the upper return spring
- Brake shoes to the backing plate
- Lower return spring, the tension pins and retaining springs
- Connect the upper return spring

5. Turn the adjuster bolt to force the brake shoes out until the brake drum will not easily go on. Back off the adjuster bolt just enough that the brake drum will go on and turn easily.

6. Install the wheels.

7. Depress the brake pedal several times to set the self-adjusting brake. Adjust the parking brake.

HONDA

CR-V

SPECIFICATIONS AND MAINTENANCE CHARTS

ENGINE AND VEHICLE IDENTIFICATION CHART

		Engine Code					Model Year	
Code	Liters (cc)	Cu. In.	Cyl.	Fuel Sys.	Engine Type	Eng. Mfg.	Code ①	Year
K24A1	2.4 (2354)	146	4	SMFI	DOHC	Honda	2	2002
DOHC: Double Overhead Cam							3	2003
SMFI: Sequential Multi-port Fuel Injection							4	2004
① 10th position of VIN							5	2005
							6	2006

09474_CRV_C0001

GENERAL ENGINE SPECIFICATIONS

Year	Model	Engine Displacement Liters	Engine ID	Net Horsepower @ rpm	Net Torque @ rpm (ft. lbs.)	Bore x Stroke (in.)	Compression Ratio	Oil Pressure @ rpm
2002	CR-V	2.4	K24A1	160@6000	162@3600	3.43x3.90	9.6:1	44@3000
2003	CR-V	2.4	K24A1	160@6000	162@3600	3.43x3.90	9.6:1	44@3000
2004	CR-V	2.4	K24A1	160@6000	162@3600	3.43x3.90	9.6:1	44@3000
2005	CR-V	2.4	K24A1	160@6000	162@3600	3.43x3.90	9.6:1	44@3000
2006	CR-V	2.4	K24A1	156@5900	160@3600	3.43x3.90	9.6:1	44@3000

09474_CRV_C0002

ENGINE TUNE-UP SPECIFICATIONS

Year	Engine Displacement Liters	Engine ID	Spark Plug Gap (in.)	Ignition Timing (deg.) MT	Ignition Timing (deg.) AT	Fuel Pump (psi)	Idle Speed (rpm) MT	Idle Speed (rpm) AT	Valve Clearance (in.) In.	Valve Clearance (in.) Ex.
2002	2.4	K24A1	0.039-0.043	6-10B	6-10B	50	600-700	600-700	0.008-0.010	0.011-0.013
2003	2.4	K24A1	0.039-0.043	6-10B	6-10B	50	600-700	600-700	0.008-0.010	0.011-0.013
2004	2.4	K24A1	0.039-0.043	6-10B	6-10B	50	600-700	600-700	0.008-0.010	0.011-0.013
2005	2.4	K24A1	0.039-0.043	6-10B	6-10B	50	600-700	600-700	0.008-0.010	0.011-0.013
2006	2.4	K24A1	0.039-0.043	6-10B	6-10B	50	600-700	600-700	0.008-0.010	0.011-0.013

NOTE: The Vehicle Emission Control Information label often reflects changes made during production and must be used if they differ from this chart.

NOTE: The fuel pressure readings are given with the vacuum hose connected to the regulator and the engine running

B: Before top dead center

HYD: Hydraulic

09474_CRV_C0003

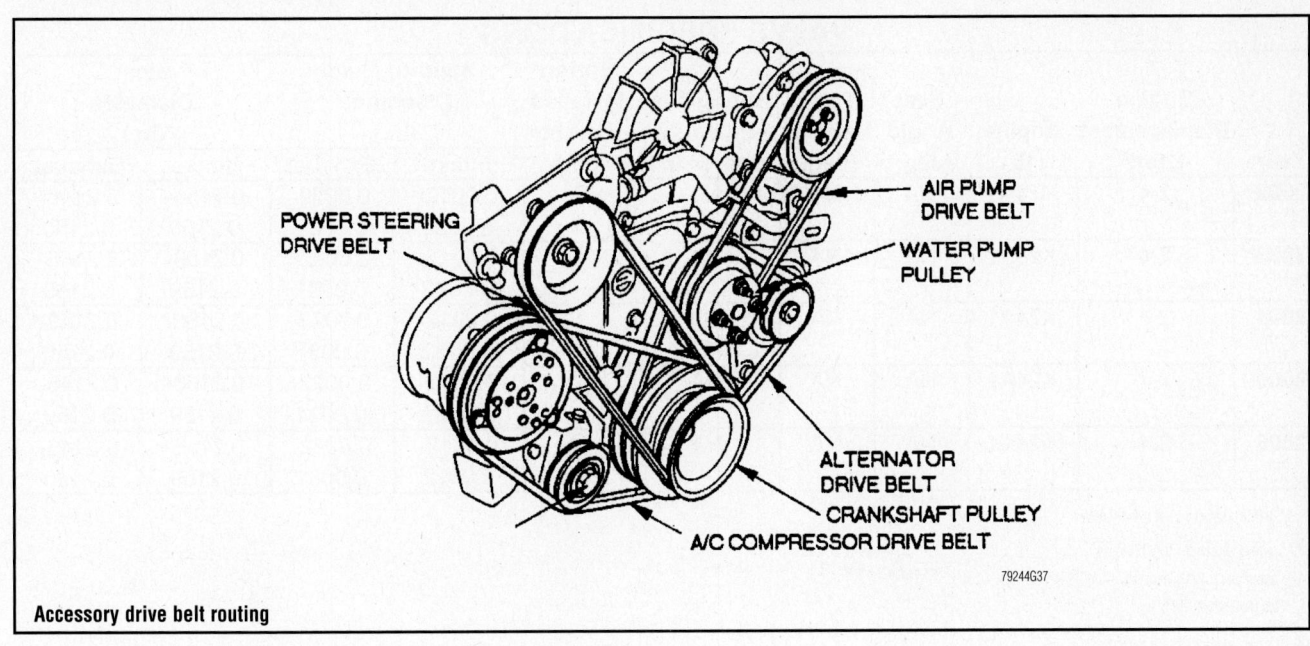

POWER STEERING
DRIVE BELT

AIR PUMP
DRIVE BELT

WATER PUMP
PULLEY

ALTERNATOR
DRIVE BELT

CRANKSHAFT PULLEY

A/C COMPRESSOR DRIVE BELT

79244G37

Accessory drive belt routing

CAPACITIES

Year	Model	Engine Displacement Liters	Engine ID	Engine Oil with Filter (qts.)	Transmission (pts.)* MT	Transmission (pts.)* AT	Transfer Case (pts.)	Drive Axle Front (pts.)	Drive Axle Rear (pts.)	Fuel Tank (gal.)	Cooling System (qts.)
2002	CR-V	2.4	K24A1	4.4	4.0	①	②	②	2.2	15.3	5.8
2003	CR-V	2.4	K24A1	4.4	4.0	①	②	②	2.2	15.3	5.8
2004	CR-V	2.4	K24A1	4.4	4.0	①	②	②	2.2	15.3	5.8
2005	CR-V	2.4	K24A1	4.4	4.0	①	②	②	2.2	15.3	5.8
2006	CR-V	2.4	K24A1	4.4	4.0	①	②	②	2.2	15.3	5.8

NOTE: All capacities are approximate. Add fluid gradually and check to be sure a proper fluid level is obtained.

* Fluid change

① 4WD: 6.6 pts.
  2WD: 6.2 pts.

② Included in transaxle refill figure

09474_CRV_C0004

VALVE SPECIFICATIONS

Year	Engine Displacement Liters	Engine ID	Seat Angle (deg.)	Face Angle (deg.)	Spring Test Pressure (lbs. @ in.)	Spring Installed Height (in.)	Stem-to-Guide Clearance (in.)		Stem Diameter (in.)	
							Intake	Exhaust	Intake	Exhaust
2002	2.4	K24A1	NA	NA	NA	①	0.0012-0.0022	0.0022-0.0031	0.2156-0.2159	0.2146-0.2150
2003	2.4	K24A1	NA	NA	NA	①	0.0012-0.0022	0.0022-0.0031	0.2156-0.2159	0.2146-0.2150
2004	2.4	K24A1	NA	NA	NA	①	0.0012-0.0022	0.0022-0.0031	0.2156-0.2159	0.2146-0.2150
2005	2.4	K24A1	NA	NA	NA	②	0.0012-0.0022	0.0022-0.0031	0.2156-0.2159	0.2146-0.2150
2006	2.4	K24A1	NA	NA	NA	②	0.0012-0.0022	0.0022-0.0031	0.2156-0.2159	0.2146-0.2150

NA: Information not available

① Valve spring free length:
  Intake: 1.668 in.
  Exhaust: 1.745 in.

② Valve spring free length:
  Intake: 1.874 in.
  Exhaust: 1.954 in.

CRANKSHAFT AND CONNECTING ROD SPECIFICATIONS

All measurements are given in inches

Year	Engine Displacement Liters	Engine ID	Crankshaft				Connecting Rod		
			Main Brg. Journal Dia.	Main Brg. Oil Clearance	Shaft End-play	Thrust on No.	Journal Diameter	Oil Clearance	Side Clearance
2002	2.4	K24A1	①	②	0.0040-0.0140	3	1.8888-1.8898	0.0008-0.0019	0.016
2003	2.4	K24A1	①	②	0.0040-0.0140	3	1.8888-1.8898	0.0008-0.0019	0.016
2004	2.4	K24A1	①	②	0.0040-0.0140	3	1.8888-1.8898	0.0008-0.0019	0.016
2005	2.4	K24A1	①	②	0.0040-0.0140	3	1.8888-1.8898	0.0008-0.0019	0.006-0.0120
2006	2.4	K24A1	①	②	0.0040-0.0140	3	1.8888-1.8898	0.0008-0.0019	0.006-0.0120

① Except No. 3: 2.1648-2.1657
  No. 3: 2.1644-2.1654

② Except No. 3: 0.0007-0.0016
  No. 3: 0.0010-0.0019

PISTON AND RING SPECIFICATIONS

All measurements are given in inches

Year	Engine Displacement Liters	Engine ID	Piston Clearance	Ring Gap			Ring Side Clearance		
				Top Compression	Bottom Compression	Oil Control	Top Compression	Bottom Compression	Oil Control
2002	2.4	K24A1	0.0008-0.0016	0.0080-0.0140	0.0160-0.0220	0.0080-0.0280	0.0018-0.0028	0.0020-0.0030	NA
2003	2.4	K24A1	0.0008-0.0016	0.0080-0.0140	0.0160-0.0220	0.0080-0.0280	0.0018-0.0028	0.0020-0.0030	NA
2004	2.4	K24A1	0.0008-0.0016	0.0080-0.0140	0.0160-0.0220	0.0080-0.0280	0.0018-0.0028	0.0020-0.0030	NA
2005	2.4	K24A1	0.0008-0.0016	0.0080-0.0140	0.0160-0.0220	0.0080-0.0280	0.0018-0.0028	0.0020-0.0030	NA
2006	2.4	K24A1	0.0008-0.0016	0.0080-0.0140	0.0160-0.0220	0.0080-0.0280	0.0018-0.0028	0.0020-0.0030	NA

NA: Information not available

09474_CRV_C0007

TORQUE SPECIFICATIONS

All readings in ft. lbs.

Year	Engine Displacement Liters	Engine ID	Cylinder Head Bolts	Main Bearing Bolts	Rod Bearing Bolts	Crankshaft Damper Bolts	Flywheel Bolts	Manifold		Spark Plugs	Oil Pan Drain Plug
								Intake	Exhaust		
2002	2.4	K24A1	①	②	③	181	90	16	33	13	33
2003	2.4	K24A1	①	②	③	181	90	16	33	13	33
2004	2.4	K24A1	①	②	③	④	90	16	33	13	33
2005	2.4	K24A1	①	②	③	④	90	16	33	13	33
2006	2.4	K24A1	①	②	③	④	90	16	33	13	33

NOTE: Dip main bearing bolts and crankshaft damper bolt in clean engine oil prior to tightening.

① Step 1: 29 ft. lbs.
  Step 2: +90 degrees
  Step 3: +90 degrees
  Step 4: NEW BOLT ONLY +90 degrees

② Bearing cap bolts in sequence:
  Step 1: 22 ft. lbs.
  Step 2: +56 degrees
  Tighten the 8mm bolts to 16 ft. lbs., in sequence

③ 14 ft. lbs. +90 degrees

④ 36 ft. lbs. +90 degrees

09474_CRV_C0008

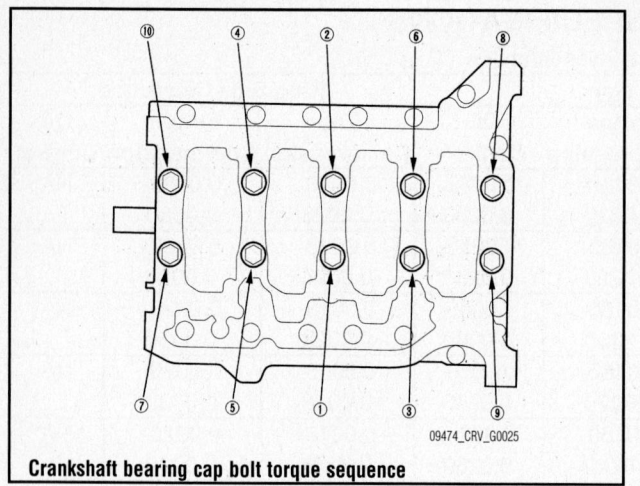

Crankshaft bearing cap bolt torque sequence

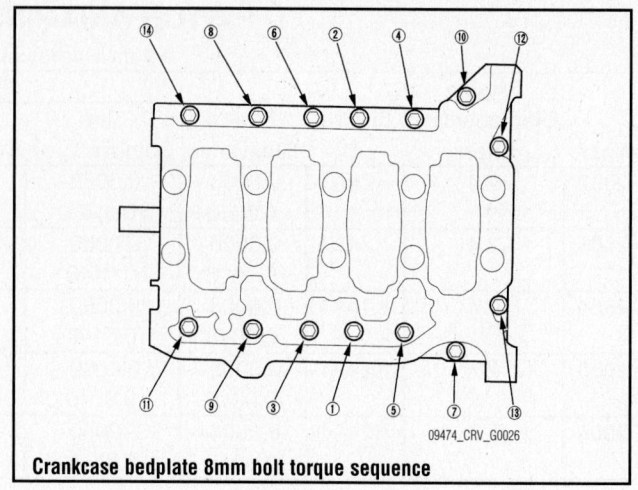

Crankcase bedplate 8mm bolt torque sequence

WHEEL ALIGNMENT

Year	Model		Caster Range (+/-Deg.)	Caster Preferred Setting (Deg.)	Camber Range (+/-Deg.)	Camber Preferred Setting (Deg.)	Toe-in (in.)
2002	CR-V	F	1.00	+1.75	0.75	0	0+/-0.08
		R	—	—	0.75	-1.00	0.08+/-0.08
2003	CR-V	F	1.00	+1.75	0.75	0	0+/-0.08
		R	—	—	0.75	-1.00	0.08+/-0.08
2004	CR-V	F	1.00	+1.75	0.75	0	0+/-0.08
		R	—	—	0.75	-1.00	0.08+/-0.08
2005	CR-V	F	1.00	+1.75	0.75	0	0+/-0.08
		R	—	—	0.75	-1.00	0.08+/-0.08
2006	CR-V	F	1.00	+1.75	0.75	0	0+/-0.08
		R	—	—	0.75	-1.00	0.08+/-0.08

09474_CRV_C0009

TIRE, WHEEL AND BALL JOINT SPECIFICATIONS

Year	Model	OEM Tires		Tire Pressures (psi)		Wheel Size	Ball Joint Inspection	Lugnut Torque (ft. lbs.)
		Standard	Optional	Front	Rear			
2002	CR-V	P205/70R15	None	26	26	6JJ	NS	80
2003	CR-V	P205/70R15	None	26	26	6JJ	NS	80
2004	CR-V	P205/70R15	None	26	26	6JJ	NS	80
2005	CR-V	P215/65R16	None	①	①	NA	NS	80
2006	CR-V	P215/65R16	None	①	①	NA	NS	80

NA: Information not available

OEM: Original Equipment Manufacturer

PSI: Pounds Per Square Inch

NS: Not specified by manufacturer

① See placard on vehicle

09474_CRV_C0010

BRAKE SPECIFICATIONS
All measurements in inches unless noted

Year	Model		Brake Disc			Minimum Lining Thickness	Brake Caliper	
			Original Thickness	Minimum Thickness	Maximum Runout		Bracket Bolts (ft. lbs.)	Mounting Bolts (ft. lbs.)
2002	CR-V	F	0.910	0.830	0.004	0.060	80	25
		R	0.350	0.280	0.004	0.040	41	16
2003	CR-V	F	0.910	0.830	0.004	0.060	80	25
		R	0.350	0.280	0.004	0.040	41	16
2004	CR-V	F	0.910	0.830	0.004	0.060	80	25
		R	0.350	0.280	0.004	0.040	41	16
2005	CR-V	F	0.985	0.910	0.004	0.060	80	25
		R	0.350	0.280	0.004	0.060	41	16
2006	CR-V	F	0.985	0.910	0.004	0.060	80	25
		R	0.350	0.280	0.004	0.060	41	16

F: Front

R: Rear

09474_CRV_C0011

SCHEDULED MAINTENANCE INTERVALS
NORMAL CONDITIONS

2002-03 HONDA CR-V

TO BE SERVICED	TYPE OF SERVICE	VEHICLE MILEAGE INTERVAL (x1000)											
		10	20	30	40	50	60	70	80	90	100	110	120
Accessory drive belts	I & A			✓			✓			✓			✓
Air cleaner element	R			✓			✓			✓			✓
Air conditioning filter	R			✓			✓			✓			✓
Brake fluid	R											✓	
Brake hoses & lines (including ABS)	I		✓		✓		✓		✓		✓		
Cooling system hoses & connections	I		✓		✓		✓		✓		✓		
Engine coolant	R												✓
Engine oil	R	✓	✓	✓	✓	✓	✓	✓	✓	✓	✓	✓	✓
Engine oil and coolant levels	I	Inspect at each fuel stop											
Engine oil filter	R		✓		✓		✓		✓		✓		
Exhaust system	I		✓		✓		✓		✓		✓		
Fluid levels and condition	I		✓		✓		✓		✓		✓		
Front and rear brakes	I		✓		✓		✓		✓		✓		
Fuel lines & connection	I		✓		✓		✓		✓		✓		
Halfshaft boots	I		✓		✓		✓		✓		✓		
Idle speed	I & A											✓	
Parking brake system	I & A		✓		✓		✓		✓		✓		
Rear differential fluid	R										✓		
Rotate and inspect tires	I	✓	✓	✓	✓	✓	✓	✓	✓	✓	✓	✓	✓
Spark plugs	R											✓	
Suspension components	I		✓		✓		✓		✓		✓		
Tie rod ends, steering gear box & boots	I		✓		✓		✓		✓		✓		
Transmission fluid	R												✓
Valve clearance	I											✓	

R: Replace   I: Inspect   A: Adjust

09474_CRV_C0012

SCHEDULED MAINTENANCE INTERVALS
NORMAL CONDITIONS

2004-05 HONDA CR-V

TO BE SERVICED	TYPE OF SERVICE	VEHICLE MILEAGE INTERVAL (x1000)											
		10	20	30	40	50	60	70	80	90	100	110	120
Accessory drive belts	I			✓			✓			✓			✓
Air cleaner element	R			✓			✓			✓			✓
Air conditioning filter	R			✓			✓			✓			✓
Brake fluid level	I		✓		✓		✓		✓		✓		✓
Brake fluid	R	Every 3 years											
Brake hoses & lines (including ABS)	I		✓		✓		✓		✓		✓		✓
Cooling system hoses & connections	I		✓		✓		✓		✓		✓		✓
Engine coolant	R	Every 120,000 miles or 10 years, then every 60,000 or 5 years											
Engine oil	R	✓	✓	✓	✓	✓	✓	✓	✓	✓	✓	✓	✓
Engine oil and coolant levels	I	Inspect at each fuel stop											
Engine oil filter	R		✓		✓		✓		✓		✓		✓
Exhaust system	I		✓		✓		✓		✓		✓		✓
Fluid levels and condition	I		✓		✓		✓		✓		✓		✓
Front and rear brakes	I		✓		✓		✓		✓		✓		✓
Fuel lines & connection	I		✓		✓		✓		✓		✓		✓
Halfshaft boots	I		✓		✓		✓		✓		✓		✓
Idle speed	I	Every 160,000 miles											
Manual transmission fluid	R	Every 120,000 miles or 6 years, whichever comes first											
Parking brake system	I & A		✓		✓		✓		✓		✓		✓
Power steering fluid level	I		✓		✓		✓		✓		✓		✓
Rear differential fluid level	I		✓		✓		✓		✓		✓		✓
Rear differential fluid	R									✓			
Rotate and inspect tires	I	✓	✓	✓	✓	✓	✓	✓	✓	✓	✓	✓	✓
Spark plugs	R											✓	
Suspension components	I		✓		✓		✓		✓		✓		✓
Tie rod ends, steering gear box & boots	I		✓		✓		✓		✓		✓		✓
Transmission fluid level	I		✓		✓		✓		✓		✓		✓
Automatic Transmission fluid	R												✓
Valve clearance	I/A											✓	

R: Replace   I: Inspect   A: Adjust

09474_CRV_C0013

SCHEDULED MAINTENANCE INTERVALS
NORMAL CONDITIONS

2006 HONDA CR-V

TO BE SERVICED	TYPE OF SERVICE	VEHICLE MILEAGE INTERVAL (x1000)											
		10	20	30	40	50	60	70	80	90	100	110	120
Accessory drive belts	I			✓			✓			✓			✓
Air cleaner element	R			✓			✓			✓			✓
Air conditioning filter	R			✓			✓			✓			✓
Automatic Transmission fluid	R												✓
Brake fluid level	I	Once a month											
Brake fluid	R	Every 3 years											
Brake hoses & lines (including ABS)	I		✓		✓		✓		✓		✓		✓
Cooling system hoses & connections	I		✓		✓		✓		✓		✓		✓
Engine coolant	R	Every 120,000 miles or 10 years, then every 60,000 or 5 years											
Engine oil	R	✓	✓	✓	✓	✓	✓	✓	✓	✓	✓	✓	✓
Engine oil and coolant levels	I	Inspect at each fuel stop											
Engine oil filter	R		✓		✓		✓		✓		✓		✓
Exhaust system	I		✓		✓		✓		✓		✓		✓
Fluid levels and condition	I		✓		✓		✓		✓		✓		✓
Front and rear brakes	I		✓		✓		✓		✓		✓		✓
Fuel lines & connection	I		✓		✓		✓		✓		✓		✓
Halfshaft boots	I		✓		✓		✓		✓		✓		✓
Idle speed	I	Every 160,000 miles											
Lights	I	Once a month											
Manual transmission fluid	R	Every 120,000 miles or 6 years, whichever comes first											
Parking brake system	I & A		✓		✓		✓		✓		✓		✓
Power steering fluid level	I		✓		✓		✓		✓		✓		✓
Rear differential fluid level	I		✓		✓		✓		✓		✓		✓
Rear differential fluid	R	Every 90,000 miles or 5 years, whichever comes first											
Rotate and inspect tires	I	✓	✓	✓	✓	✓	✓	✓	✓	✓	✓	✓	✓
Spark plugs	R											✓	
Suspension components	I		✓		✓		✓		✓		✓		✓
Tie rod ends, steering gear box & boots	I		✓		✓		✓		✓		✓		✓
Tire inflation & condition	I	Once a month											
Transmission fluid level	I		✓		✓		✓		✓		✓		✓
Valve clearance	I/A											✓	

R: Replace   I: Inspect   A: Adjust

SCHEDULED MAINTENANCE INTERVALS
SEVERE CONDITIONS

2002-05 HONDA CR-V

TO BE SERVICED	TYPE OF SERVICE	VEHICLE MILEAGE INTERVAL (x1000)											
		5	10	15	20	25	30	35	40	45	50	55	60
Accessory drive belts	I						✓						✓
Air cleaner element	R						✓						✓
Air conditioning filter	R			✓			✓			✓			✓
Brake fluid level	I				✓				✓				✓
Brake fluid	R	Every 3 years											
Brake hoses & lines (including ABS)	I		✓		✓		✓		✓		✓		✓
Cooling system hoses & connections	I		✓		✓		✓		✓		✓		✓
Dust and pollen filter	R						✓						✓
Engine coolant	R	At 120,000 miles or 10 years, then every 60,000 miles or 5 years											
Engine oil	R	✓	✓	✓	✓	✓	✓	✓	✓	✓	✓	✓	✓
Engine oil and coolant levels	I	Inspect at each fuel stop											
Engine oil filter	R		✓		✓		✓		✓		✓		✓
Exhaust system	I				✓				✓				✓
Fluid levels and condition	I				✓				✓				✓
Front and rear brakes	I		✓		✓		✓		✓		✓		✓
Fuel lines & connection	I		✓		✓		✓		✓		✓		✓
Halfshaft boots	I		✓		✓		✓		✓		✓		✓
Hinges, locks and latches	L		✓		✓		✓		✓		✓		✓
Idle speed	I	Every 160,000 miles											
Lights and controls	I				✓				✓				✓
Parking brake system	I & A				✓				✓				✓
Power steering fluid level	I				✓				✓				✓
Rear differential fluid level	I				✓				✓				✓
Rear differential fluid	R												✓
Spark plugs	R	Every 110,000 miles											
Suspension components	I		✓		✓		✓		✓		✓		✓
Tie rod ends, steering gear box & boots	I		✓		✓		✓		✓		✓		✓
Tire inflation & condition	I	Once a month											
Tires	Rotate		✓		✓		✓		✓		✓		✓
Transmission fluid level	I				✓				✓				✓
Transmission fluid	R												✓
Valve clearance	I	Every 110,000 miles											

R: Replace   I: Inspect   A: Adjust   L: Lubricate

Use this schedule if the vehicle is driven mainly in Canada or under any of the following conditions.
If used only occasionally under these condtions, use the Normal Schedule.

Driving fewer than 5 miles per trip.
Driving fewer than 10 miles per trip in freezing weather.
Extensive idling.
Long periods of stop-and-go driving.
Taxi or commercial use.
Trailer towing, driving with a car-top carrier, or driving in mountainous areas.
Driving on muddy, dusty or de-iced roads.

09474_CRV_C0015

SCHEDULED MAINTENANCE INTERVALS
SEVERE CONDITIONS

2006 HONDA CR-V

TO BE SERVICED	TYPE OF SERVICE	VEHICLE MILEAGE INTERVAL (x1000)											
		5	10	15	20	25	30	35	40	45	50	55	60
Accessory drive belts	I						✓						✓
Air cleaner element	R			✓			✓			✓			✓
Air conditioning filter	R						✓						✓
Automatic Transmission fluid	R												✓
Brake fluid level	I				✓				✓				✓
Brake fluid	R	Every 3 years											
Brake hoses & lines (including ABS)	I				✓				✓				✓
Cooling system hoses & connections	I		✓		✓		✓		✓		✓		✓
Dust and pollen filter	R						✓						✓
Engine coolant	R	At 120,000 miles or 10 years, then every 60,000 miles or 5 years											
Engine oil	R	✓	✓	✓	✓	✓	✓	✓	✓	✓	✓	✓	✓
Engine oil and coolant levels	I	Inspect at each fuel stop											
Engine oil filter	R		✓		✓		✓		✓		✓		✓
Exhaust system	I				✓				✓				✓
Fluid levels and condition	I				✓				✓				✓
Front and rear brakes	I		✓		✓		✓		✓		✓		✓
Fuel lines & connection	I				✓				✓				✓
Halfshaft boots	I		✓		✓		✓		✓		✓		✓
Hinges, locks and latches	L		✓		✓		✓		✓		✓		✓
Idle speed	I	Every 160,000 miles											
Lights and controls	I				✓				✓				✓
Manual transmission fluid	R	Every 60,000 miles or 3 years, whichever comes first											
Parking brake system	I & A				✓				✓				✓
Power steering fluid level	I				✓				✓				✓
Rear differential fluid level	I				✓				✓				✓
Rear differential fluid	R												✓
Spark plugs	R	Every 110,000 miles											
Suspension components	I		✓		✓		✓		✓		✓		✓
Tie rod ends, steering gear box & boots	I		✓		✓		✓		✓		✓		✓
Tire inflation & condition	I	Once a month											
Tires	Rotate		✓		✓		✓		✓		✓		✓
Transmission fluid level	I				✓				✓				✓
Valve clearance	I/A	Every 110,000 miles											

R: Replace   I: Inspect   A: Adjust   L: Lubricate

Use this schedule if the vehicle is driven MAINLY in Canada or under any of the following conditions.

Driving fewer than 5 miles per trip.

Driving fewer than 10 miles per trip in freezing weather.

Driving in temperatures over 90 deg.F (32 deg.C)

Extensive idling.

Long periods of stop-and-go driving.

Taxi or commercial use.

Trailer towing, driving with a car-top carrier, or driving in mountainous areas.

Driving on muddy, dusty or de-iced roads.

ENGINE REPAIR

➡**Disconnecting the negative battery cable on some vehicles may interfere with the functions of the on board computer system. The computer may undergo a relearning process once the negative battery cable is reconnected.**

Alternator

REMOVAL & INSTALLATION

1. Before servicing the vehicle, refer to the Precautions Section.
2. Remove or disconnect the following:
   - Negative battery cable
   - Front cover
   - Accessory drive belt
   - The 3 bolts holding the alternator
   - Alternator wiring harness connectors
   - Alternator

To install:

3. Install or connect the following:
   - Alternator. Tighten the bolts to 16 ft. lbs. (22 Nm).
   - Alternator wiring harness connectors. Tighten the battery terminal nut to 70 inch lbs. (8 Nm).
   - Accessory drive belts
   - Front cover
   - Negative battery cable

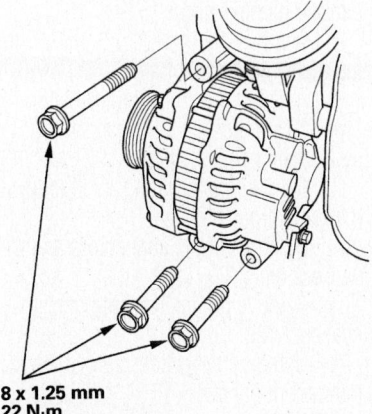

**8 x 1.25 mm
22 N·m
(2.2 kgf·m, 16 lbf·ft)**

09474_CRV_G0001

Alternator mounting

Engine Assembly

REMOVAL & INSTALLATION

➡**The engine and transaxle are removed from the vehicle as a unit.**

1. Before servicing the vehicle, refer to the Precautions Section.
2. Drain the cooling system.
3. Drain the transaxle fluid.
4. Drain the engine oil.
5. Relieve fuel system pressure.
6. Remove or disconnect the following:
   - Negative battery cable
   - Fuse/Relay box battery cables
   - Battery and tray
   - Intake manifold cover
   - IAT sensor connector
   - Breather hose
   - Intake duct
   - Cables from the power distribution center
   - Throttle and cruise cables
   - Powertrain Control Module (PCM) connectors and grommet. Pull the PCM harness through the firewall.
   - Fuel lines
   - EVAP canister
   - Brake booster vacuum line
   - Clutch slave cylinder
   - Clutch hose bracket
   - Shift cables
   - Drive belt
   - Power steering pump, leaving the hoses connected
7. Attach a hoist to the engine lifting eyes and support the powertrain weight.
8. Remove or disconnect the following:
   - Splash shield
   - Wheels
   - Catalytic converter
   - Rear driveshaft
   - Stabilizer links
   - Right damper fork
   - Halfshafts
   - Shift cable
   - Radiator
   - Upper bracket
   - Transaxle mount and bracket
   - Front mount bolt
   - Rear mount bracket bolts. Match-mark the sub-frame mounting bolt centers.

There is a special tool necessary for sub-frame removal. The Honda tool number is EQS02C000011. Attach the tool as explained in the tool instructions, attach a floor jack with adapter, remove the 4 sub-frame bolts and lower the sub-frame.

9. Remove or disconnect the following:
   - A/C compressor without disconnecting the hoses
10. Check that all hoses and wires are disconnected.

11. Lower the engine about 6 inches and recheck all clearances.
12. Lower the engine all the way.
13. Remove the chain hoist.

To install:

14. Installation is the reverse of removal. Observe the following torques:
   - Front engine mount bracket bolts: 33 ft. lbs. (44Nm)
   - A/C compressor bracket: 33 ft. lbs. (44Nm)
   - Stiffener 10mm bolts: 33 ft. lbs. (44Nm); 6mm bolts 9 ft. lbs. (12 Nm)
   - A/C compressor bolts: 33 ft. lbs. (44 Nm)
   - Subframe front bolt: 47 ft. lbs. (64 Nm)
   - Subframe rear bolts: 43 ft. lbs. (59 Nm)
   - Upper bracket bolt and nut: 40 ft. lbs. (54 Nm)
   - Transmission mount bracket support bolts/nuts: 40 ft. lbs. (54 Nm)
   - PS pump bolts: 16 ft. lbs. (22 Nm)
   - Intake manifold cover: 9 ft. lbs. (12 Nm)

➡**Use new self-locking nuts and color-coded self-locking bolts when installing the engine mounts and suspension components.**

➡**Do not tighten the engine or transaxle mount fasteners until instructed to do so.**

15. Lower the powertrain into position.
16. Install or connect the following:
   - Transaxle mount and bracket. Tighten the frame mounting bolts to 47 ft. lbs. (64 Nm).
   - Upper bracket. Tighten the nuts in sequence to 54 ft. lbs. (74 Nm).
   - Rear mount bracket through bolt
   - Right front mount and bracket
   - Left front mount and bracket
17. Tighten the remaining mount fasteners as follows:

   a. Transaxle mount fasteners to 47 ft. lbs. (64 Nm) and the through bolt to 54 ft. lbs. (74 Nm).

   b. Rear mount bracket through bolt to 43 ft. lbs. (59 Nm).

   c. Right front mount 12mm bolts to 47 ft. lbs. (64 Nm) and the 10mm bolts to 33 ft. lbs. (44 Nm).

   d. Left front mount 12mm stud bolt to 61 ft. lbs. (83 Nm), 10mm bolts to 33 ft. lbs. (44 Nm), and 12mm nut to 43 ft. lbs. (59 Nm).

e. Right front mount 12mm nut to 43 ft. lbs. (59 Nm).

18. Install or connect the following:
- Rear driveshaft, if equipped
- A/C compressor
- Radiator
- A/C hose clamp

19. If equipped with a manual transaxle, install or connect the following:
- Shift cables
- Transaxle ground cable
- Clutch hose bracket
- Clutch slave cylinder

20. If equipped with an automatic transaxle, install or connect the following:
- Transaxle fluid cooler lines
- Transaxle ground cable and hose clamp
- Shift cable
- Shift cable cover

21. For all vehicles, install or connect the following:
- Axle halfshafts
- Lower ball joints
- Right damper fork
- Exhaust front pipe
- HO2S connector
- Heater hoses
- Radiator hoses
- Splash shield
- PSP switch
- Accelerator cable
- Brake booster vacuum line
- Fuel lines
- A/C compressor drive belt
- Power steering pump and belt
- Cruise control actuator
- Left engine wire harness connectors
- PCM connectors and grommet
- Air intake assembly
- Battery and tray
- Fuse/Relay box battery cables
- Negative battery cable

22. Fill the engine crankcase to the correct level.
23. Fill the transaxle to the correct level.
24. Fill the cooling system.
25. Start the engine and check for leaks.
26. Check the wheel alignment and adjust as necessary.

Water Pump

REMOVAL & INSTALLATION

1. Before servicing the vehicle, refer to the Precautions Section.
2. Drain the cooling system.
3. Remove or disconnect the following:
- Negative battery cable

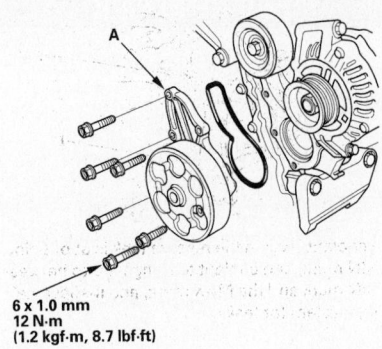

6 x 1.0 mm
12 N·m
(1.2 kgf·m, 8.7 lbf·ft)

09474_CRV_G0011

Water pump installation

- Accessory drive belt
- Crankshaft pulley
- Water pump (6 bolts)

Installation is the reverse of removal. Use new O-rings. Torque the bolts to 104 inch lbs. (12 Nm).

Heater Core

REMOVAL & INSTALLATION

1. Disconnect the negative battery cable.
2. Drain the cooling system into a clean container for reuse.
3. In the engine compartment, open the heater valve cable clamp and disconnect the cable from the heater valve arm. Then, turn the heater valve to the fully opened position.
4. Disconnect the heater hoses from the heater core.
5. Remove the heater housing-to-chassis nut.

➡ **When removing the heater housing nut, be careful not to damage or bend the fuel lines, the brake lines, etc.**

6. Remove the instrument panel by performing the following procedure:

a. Remove the driver's side lower instrument panel cover screws, disengage the clips and remove the lower cover.

b. Remove the knee bolster bolts and the knee bolster.

c. Remove the glove box stops from each side of the glove box.

d. Remove the glove box-to-instrument panel bolts and the glove box.

e. Remove the lower console cover by disengaging the 4 clips and removing the cover.

f. Remove the 6 center pocket-to-instrument panel screws; then, insert a flat tipped screwdriver at the upper right side corner of the center pocket, push

down on the top of the hook and remove the center pocket/beverage holder assembly.

g. Remove the center instrument panel lower cover screws and disengage the clips on the upper left side; then, disconnect the electrical connectors and remove the cover.

h. Gently, push the power window switch from the instrument panel's lower cover opening by hand. Disconnect the electrical connectors and remove the power window switch.

i. Close the driver's side air vent; then, gently, push out the clips and pull out the vent. Disconnect the electrical connectors and remove the vent.

j. Gently, push out the driver's side defogger trim; then, disconnect the electrical connector and remove the side defogger trim.

k. At the base of the steering wheel, remove the access panel and disconnect the air bag electrical connector.

l. Remove the steering column covers screws and the covers.

m. Remove the steering column-to-instrument panel nuts/bolts and lower the steering column.

n. Remove the instrument panel side covers.

o. Disconnect the wiring harness connector and remove the nuts.

p. Move the under-dash fuse/relay box.

q. Disconnect the antenna connector and the harness clips.

r. Remove the connector holder from the instrument panel frame.

s. Remove the control unit/relay bracket from behind the center of the instrument panel.

t. Remove the passenger's side lower instrument panel cover.

u. Disconnect the connectors and the harness clips.

v. Remove the instrument panel-to-chassis bolts.

w. Using an assistant, remove the instrument panel.

7. Remove the evaporator housing by performing the following procedure:

a. Discharge and recover the air conditioning system refrigerant.

b. In the engine compartment, remove the refrigerant lines-to-evaporator housing bolts.

c. Separate the lines, discard the grommets and plug the openings to prevent contamination.

d. Disconnect the evaporator housing's temperature sensor connector.

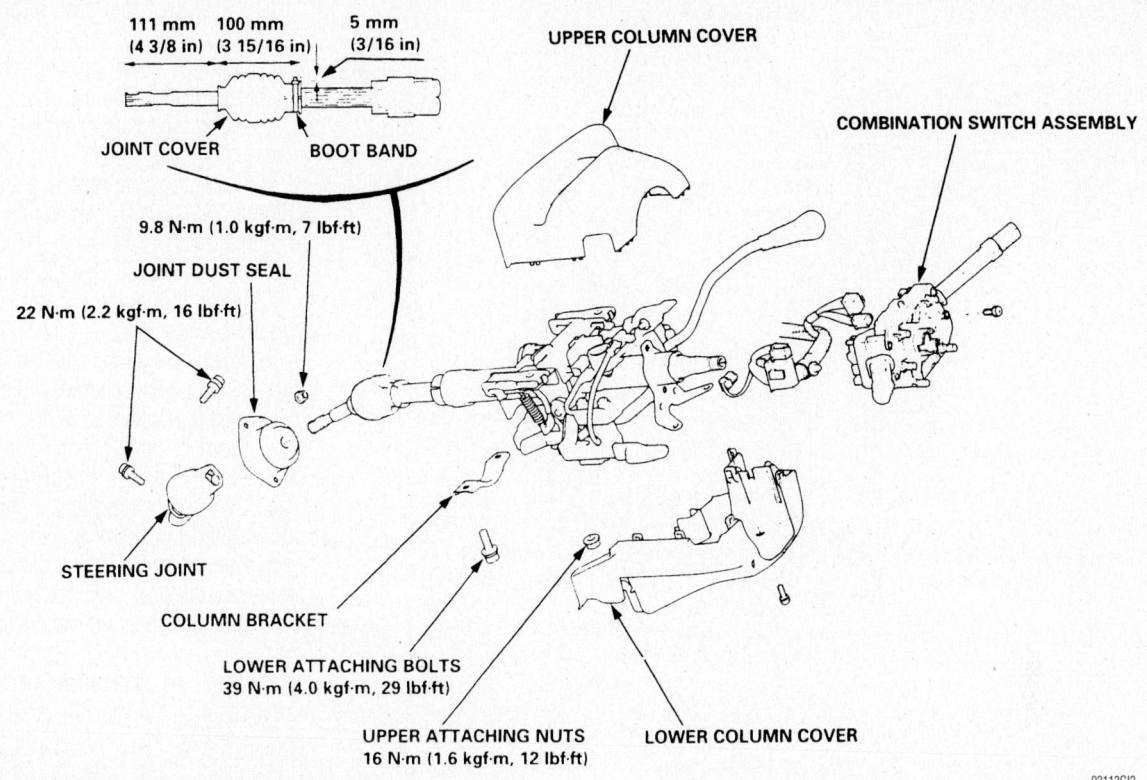

111 mm   100 mm   5 mm
(4 3/8 in) (3 15/16 in)   (3/16 in)

JOINT COVER   BOOT BAND

UPPER COLUMN COVER

COMBINATION SWITCH ASSEMBLY

9.8 N·m (1.0 kgf·m, 7 lbf·ft)

JOINT DUST SEAL

22 N·m (2.2 kgf·m, 16 lbf·ft)

STEERING JOINT

COLUMN BRACKET

LOWER ATTACHING BOLTS
39 N·m (4.0 kgf·m, 29 lbf·ft)

UPPER ATTACHING NUTS
16 N·m (1.6 kgf·m, 12 lbf·ft)

LOWER COLUMN COVER

93113GI2

Exploded view of the steering column and related components

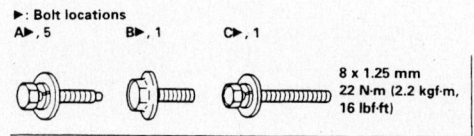

▶: Bolt locations
A▶, 5   B▶, 1   C▶, 1

8 x 1.25 mm
22 N·m (2.2 kgf·m,
16 lbf·ft)

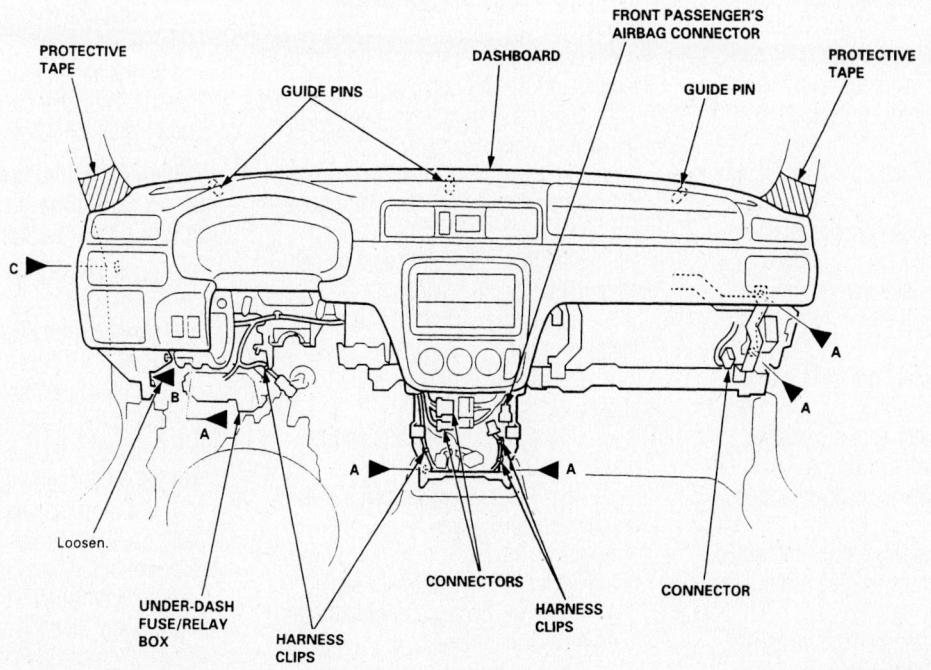

PROTECTIVE
TAPE

GUIDE PINS

DASHBOARD

FRONT PASSENGER'S
AIRBAG CONNECTOR

GUIDE PIN

PROTECTIVE
TAPE

C ▶

A

B

A

A ▶   ▶ A

Loosen.

UNDER-DASH
FUSE/RELAY
BOX

HARNESS
CLIPS

CONNECTORS

HARNESS
CLIPS

CONNECTOR

93113GI3

Exploded view of the instrument panel and related components

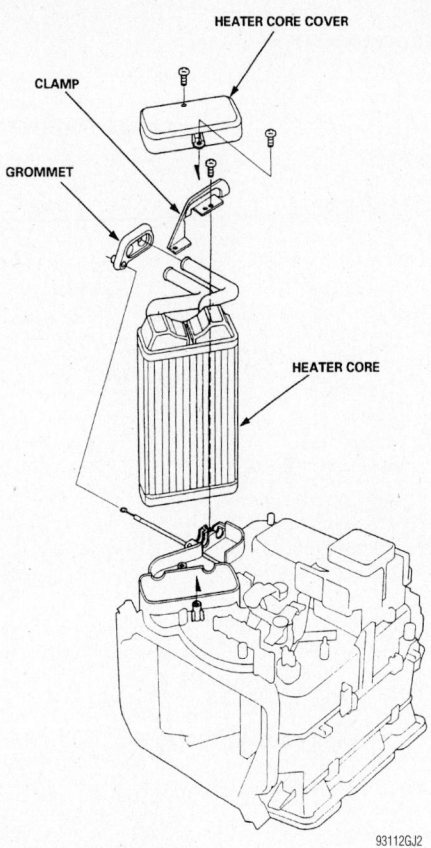

CLAMP

GROMMET

HEATER CORE COVER

HEATER CORE

93112GJ2

Exploded view of the heater core and housing

e. Remove the evaporator housing-to-chassis screws/nut and remove the evaporator housing.

8. Disconnect the mode control motor and the air mix control motor electrical connectors and remove the wiring harness clips and the wiring harness from the heater housing.

9. Remove the heater duct clip, the heater housing-to-chassis nuts and the heater housing.

10. Remove the heater core cover screws and the cover.

11. Remove the heater core pipe clamp screws and the clamp.

12. Remove the heater core from the heater housing.

To install:

13. Install the heater core in the heater housing.

14. Install the heater core pipe clamp and the clamp screws.

15. Install the heater core cover and the cover screws.

16. Install the heater housing, the heater housing-to-chassis nuts and the heater duct clip.

17. Install the wiring harness clips and the wiring harness to the heater housing and connect the mode control motor and

the air mix control motor electrical connectors.

18. Install the evaporator housing by performing the following procedure:

a. Install the evaporator housing and the evaporator housing-to-chassis screws/nut.

b. Connect the evaporator housing's temperature sensor connector.

c. Using new grommets, connect the refrigerant lines.

d. In the engine compartment, install the refrigerant lines-to-evaporator housing bolts.

19. Install the instrument panel by performing the following procedure:

a. Using an assistant, install the instrument panel.

b. Install the instrument panel-to-chassis bolts.

c. Connect the connectors and the harness clips.

d. Install the passenger's side lower instrument panel cover.

e. Install the control unit/relay bracket to the center of the instrument panel.

f. Install the connector holder to the instrument panel frame.

g. Connect the antenna connector and the harness clips.

h. Install the under-dash fuse/relay box.

i. Connect the wiring harness connector and install the nuts.

j. Install the instrument panel side covers.

k. Install the steering column and the column-to-instrument panel nuts/bolts. Torque the nuts to 12 ft. lbs. (16 Nm) and the bolts to 29 ft. lbs. (39 Nm).

l. Install the steering column covers and the cover screws.

m. At the base of the steering wheel, connect the air bag electrical connector and install the access panel.

n. Connect the electrical connector and install the driver's side defogger trim.

o. Connect the electrical connectors and install driver's side air vent.

p. Connect the electrical connectors and install the power window switch to the instrument panel's lower cover opening.

q. Install the center instrument panel lower cover and engage the clips on the upper left side; then, connect the electrical connectors and Install the cover screws.

r. Install the center pocket/beverage holder assembly and the 6 center pocket-to-instrument panel screws.

s. Install the lower console cover by engaging the 4 clips.

t. Install the glove box and the glove box-to-instrument panel bolts.

u. Install the glove box stops to each side of the glove box.

v. Install the knee bolster and the knee bolster bolts.

w. Install the driver's side lower instrument panel cover, engage the clips and install the lower cover screws.

➡**When installing the heater housing nut, be careful not to damage or bend the fuel lines, the brake lines or etc.**

20. Install the heater housing-to-chassis nut.

21. Connect the heater hoses to the heater core.

22. In the engine compartment, connect the cable to the heater valve arm and close the heater valve cable clamp.

23. Refill the cooling system.

24. Connect the negative battery cable.

25. Evacuate and charge and leak test the air conditioning system refrigerant.

26. Run the engine to normal operating temperatures; then, check the climate control operation and check for leaks.

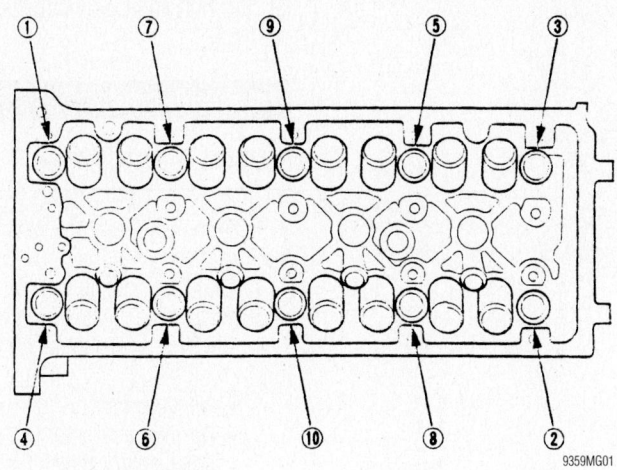

Cylinder head bolt loosening sequence

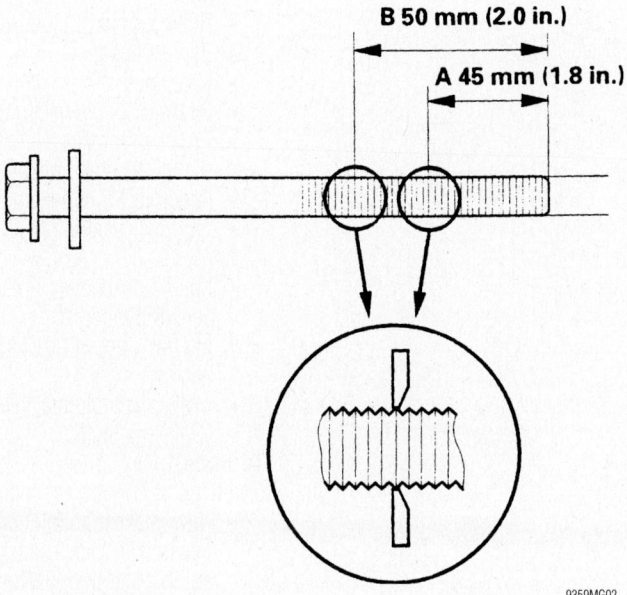

Cylinder head bolt inspection

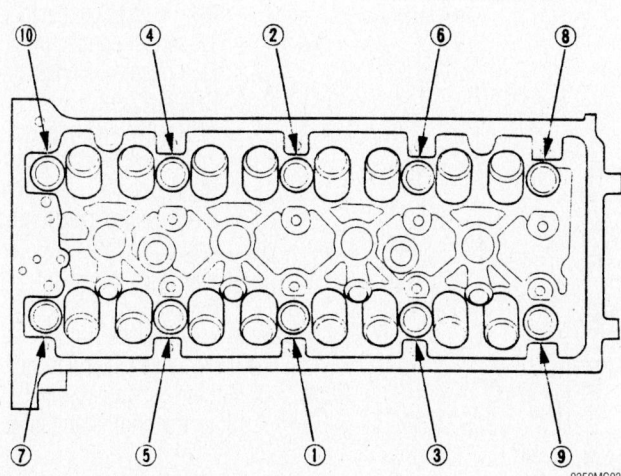

Cylinder head bolt torque sequence

Cylinder Head

REMOVAL & INSTALLATION

1. Before servicing the vehicle, refer to the Precautions Section.
2. Drain the cooling system.
3. Relieve the fuel system pressure.
4. Remove or disconnect the following:

- Negative battery cable
- Accessory drive belt
- Fuel lines
- Intake manifold
- Bypass hoses
- Exhaust manifold
- Cam chain
- Engine wiring harness connectors
- Upper radiator hose
- Heater hose
- Brake booster vacuum line
- Intake manifold vacuum line
- Rocker arms

5. Loosen the cylinder head bolts in sequence and ⅓ turns until all bolts are loose.
6. Remove the cylinder head.
7. Installation is the reverse of removal.

Rocker Arms/Shafts

REMOVAL & INSTALLATION

1. Remove the camshaft chain
2. Loosen the rocker adjusting screws
3. Remove the camshaft holder bolts 2 turns at a time in the sequence shown.
4. Remove the camshaft chain guide, camshaft holders and camshafts.
5. Remove the rocker arm assembly.
6. Installation is the reverse of removal. Prior to installation, clean the No.5 rocker shaft holder mating surface and apply RTV gasket sealer to the mounting point on the head. See the illustration for the torque sequence. Torque the 8mm bolts to 16 ft.

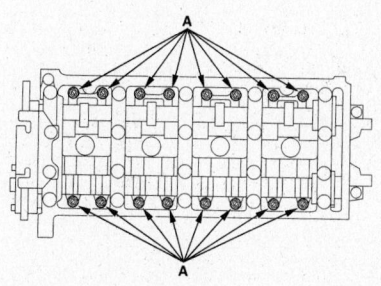

Rocker arm adjusting screws

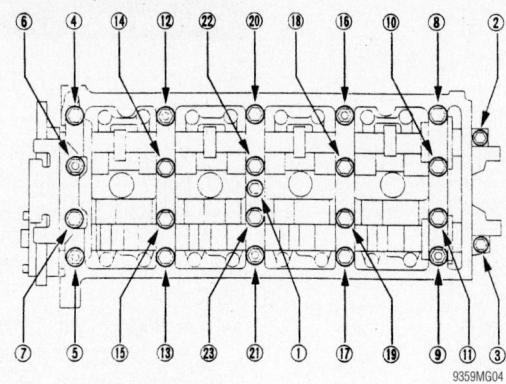

Camshaft holder bolt loosening sequence

9359MG04

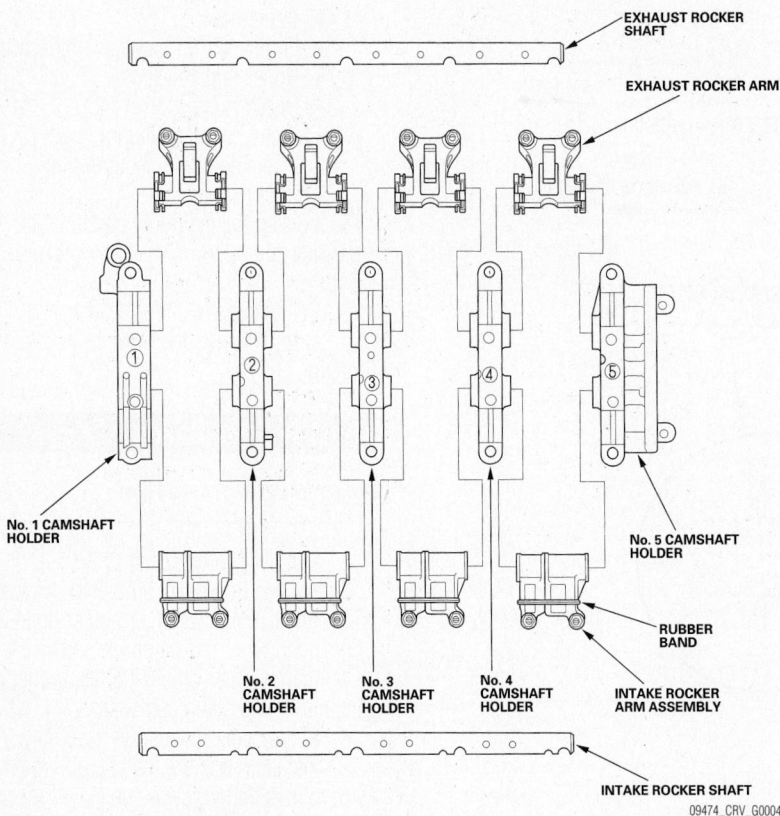

09474_CRV_G0004

Rocker arms/shafts exploded view

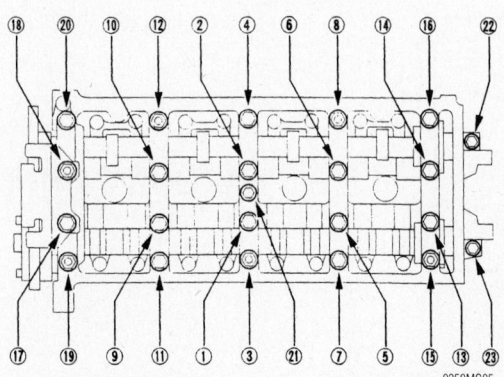

9359MG05

Camshaft holder bolt torque sequence

lbs. (22 Nm) and the 6mm bolts to 9 ft. lbs. (12 Nm).

Intake Manifold

REMOVAL & INSTALLATION

1. Before servicing the vehicle, refer to the Precautions Section.
2. Drain the cooling system.
3. Relieve the fuel system pressure.
4. Remove or disconnect the following:

- Negative battery cable
- Air intake assembly
- Accelerator cable
- Cruise control cable
- Evaporative Emissions (EVAP) control canister hose and vacuum hose
- Brake booster vacuum line
- Bypass hoses
- Fuel lines
- Intake manifold vacuum line
- Positive Crankcase Ventilation (PCV) valve and hose
- Fuel injector connectors
- Throttle Position (TP) sensor connector
- Manifold Absolute Pressure (MAP) sensor connector
- Idle Air Control (IAC) valve connector
- Intake manifold brackets
- Intake manifold

To install:

5. Install or connect the following:
- New intake manifold gasket
- Intake manifold. Tighten the fasteners to 16 ft. lbs. (22Nm).
- Intake manifold brackets
- Cruise control actuator
- IAC valve connector
- MAP sensor connector
- TP sensor connector
- Fuel injector connectors
- Bypass hoses
- PCV valve and hose
- Intake manifold vacuum line
- Brake booster vacuum line
- EVAP control canister hose and vacuum hose
- Fuel lines
- Accelerator cable
- Air intake assembly
- Negative battery cable

6. Fill the cooling system.
7. Start the engine and check for leaks.

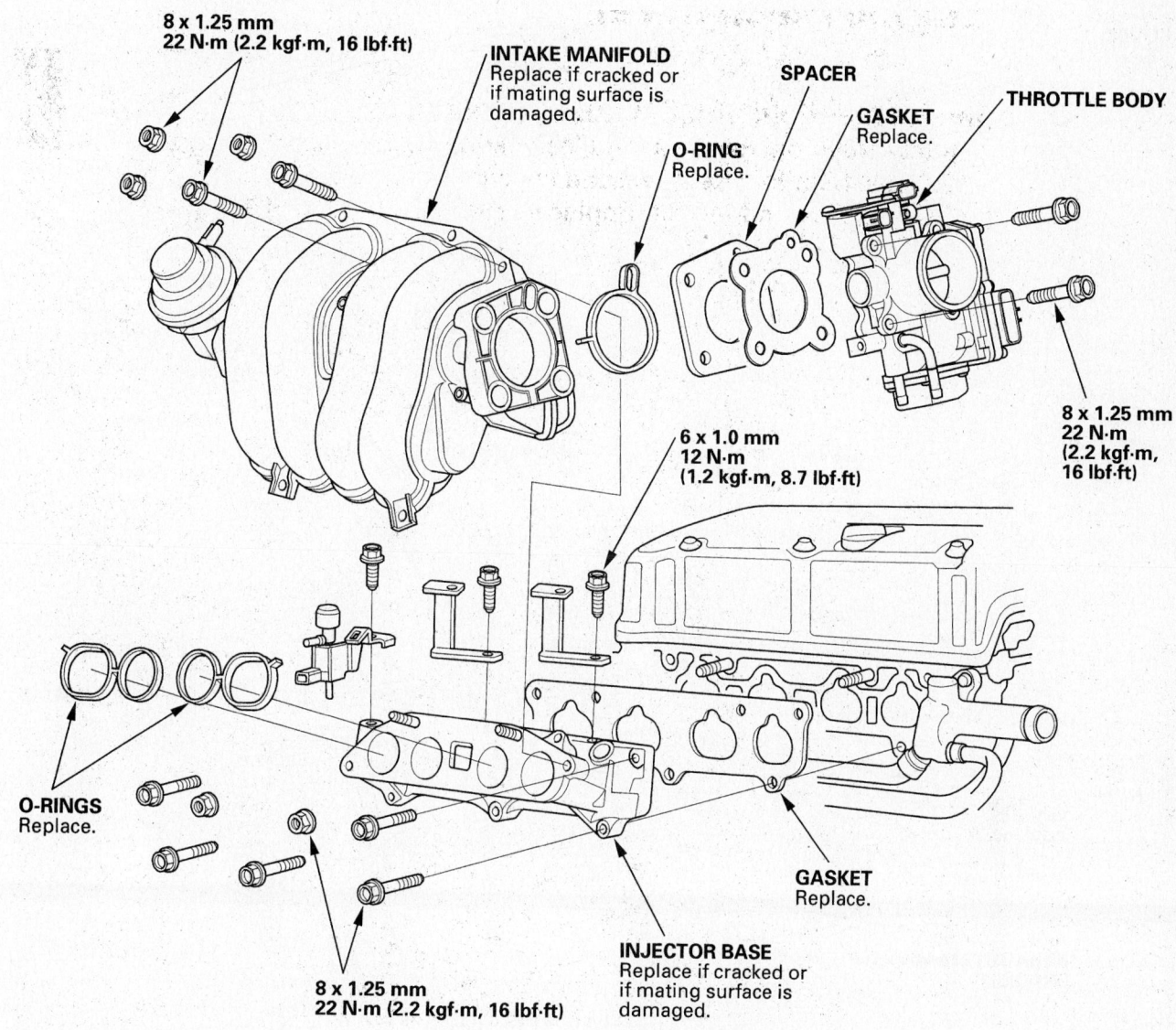

8 x 1.25 mm
22 N·m (2.2 kgf·m, 16 lbf·ft)

INTAKE MANIFOLD
Replace if cracked or
if mating surface is
damaged.

SPACER

GASKET
Replace.

THROTTLE BODY

O-RING
Replace.

8 x 1.25 mm
22 N·m
(2.2 kgf·m,
16 lbf·ft)

6 x 1.0 mm
12 N·m
(1.2 kgf·m, 8.7 lbf·ft)

O-RINGS
Replace.

GASKET
Replace.

INJECTOR BASE
Replace if cracked or
if mating surface is
damaged.

8 x 1.25 mm
22 N·m (2.2 kgf·m, 16 lbf·ft)

09474_CRV_G0009

Intake manifold and related parts

Exhaust Manifold

REMOVAL & INSTALLATION

1. Before servicing the vehicle, refer to
the Precautions Section.
2. Remove or disconnect the following:
- Negative battery cable
- VTEC solenoid valve

- Driveshaft heat shield
- Exhaust manifold heat shield
- Exhaust front pipe
- Exhaust manifold bracket, if
equipped
- Exhaust manifold

To install:
3. Install or connect the following:
- Exhaust manifold. Tighten the fasteners to 3 ft. lbs. (44m).

- Exhaust manifold bracket, if
equipped. Tighten the bolts to 33 ft.
lbs. (44 Nm).
- Exhaust front pipe. Tighten the nuts
to 16 ft. lbs. (22m).
- Exhaust manifold heat shield
- Driveshaft heat shield
- VTEC solenoid valve
- Negative battery cable

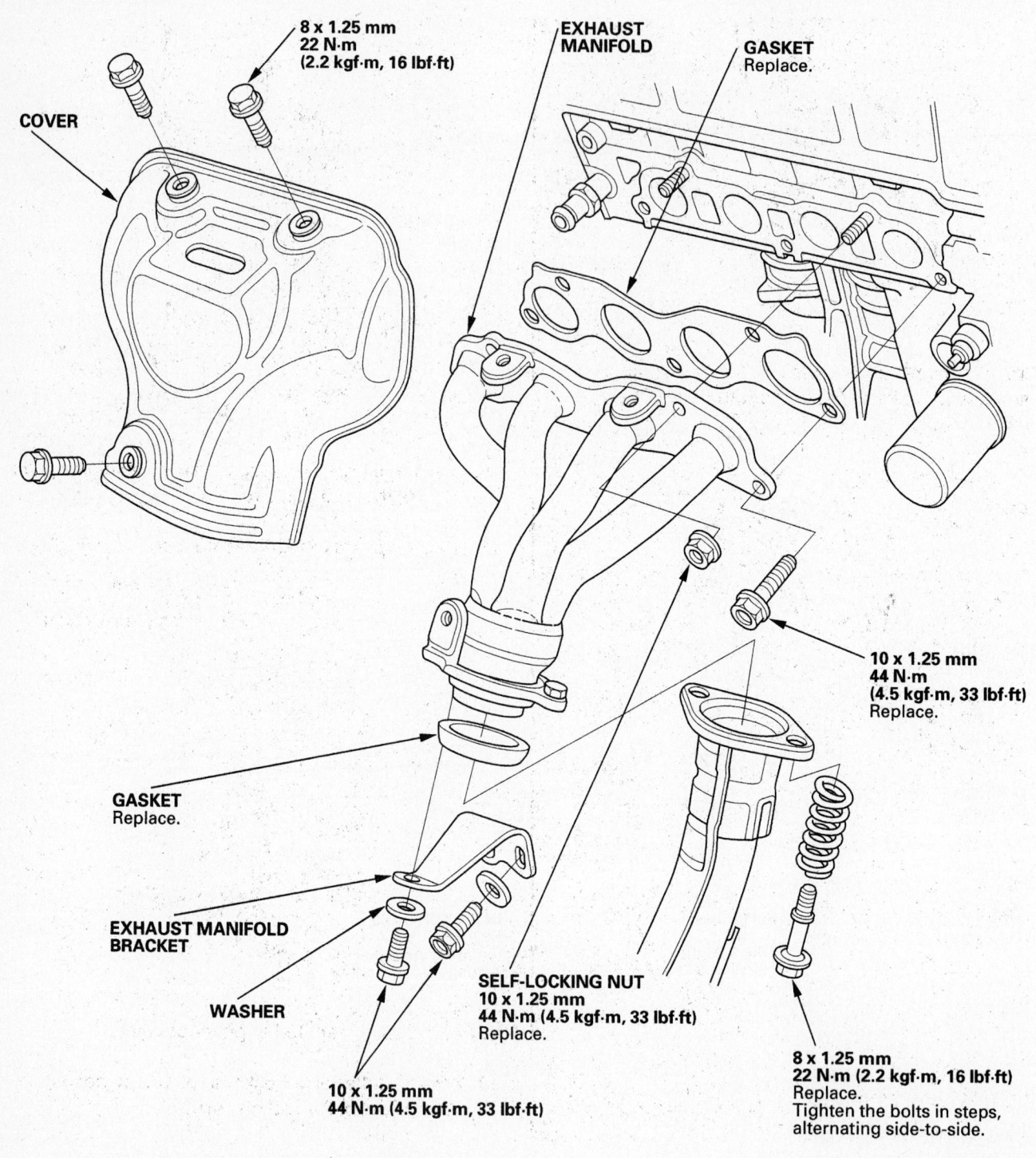

COVER

**8 x 1.25 mm
22 N·m
(2.2 kgf·m, 16 lbf·ft)**

EXHAUST MANIFOLD

GASKET
Replace.

GASKET
Replace.

EXHAUST MANIFOLD BRACKET

WASHER

**SELF-LOCKING NUT
10 x 1.25 mm
44 N·m (4.5 kgf·m, 33 lbf·ft)**
Replace.

**10 x 1.25 mm
44 N·m (4.5 kgf·m, 33 lbf·ft)**

**10 x 1.25 mm
44 N·m
(4.5 kgf·m, 33 lbf·ft)**
Replace.

**8 x 1.25 mm
22 N·m (2.2 kgf·m, 16 lbf·ft)**
Replace.
Tighten the bolts in steps,
alternating side-to-side.

09474_CRV_G0010

Exhaust manifold and related parts

Front Crankshaft Seal

REMOVAL & INSTALLATION

See the Timing Chain Removal & Installation procedure.

Camshafts

REMOVAL & INSTALLATION

See the Rocker Arm Shaft Removal & Installation procedure.

Valve Lash

ADJUSTMENT

Adjust the valves only when the cylinder head temperature is less than 100°F (38°C).

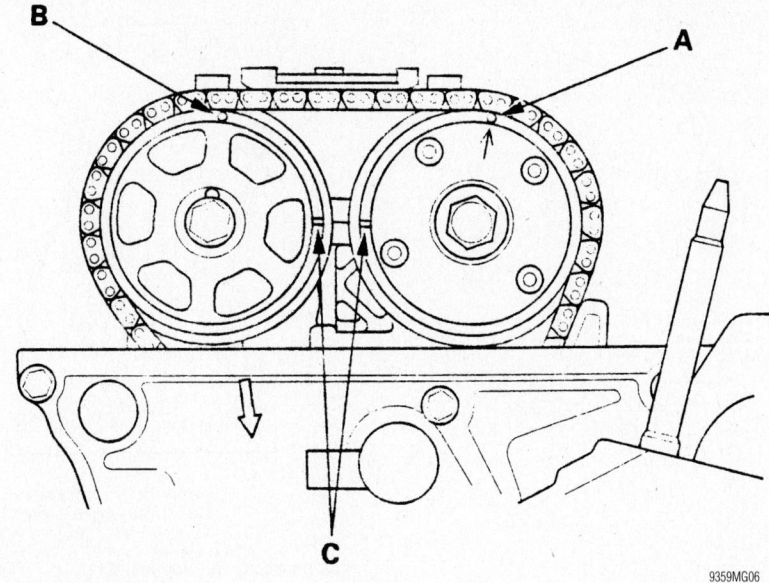

9359MG06

Align the timing marks

1. Before servicing the vehicle, refer to the Precautions Section.
2. Remove or disconnect the following:
   - Negative battery cable
   - cylinder head cover
3. Set the timing marks as shown in the illustration with N0.1 at TDC. Check all clearances. Intake should be 0.008–0.010 in.; exhaust should be 0.011–0.013 in. Intake locknut torque is 14 ft. lbs.; exhaust is 10 ft. lbs.
4. Rotate the crankshaft 180 degrees clockwise and recheck No.3.
5. Rotate the crankshaft 180 degrees clockwise and recheck No.4.
6. Rotate the crankshaft 180 degrees clockwise and recheck No.2.

Starter Motor

REMOVAL & INSTALLATION

1. Before servicing the vehicle, refer to the Precautions Section.
2. Remove or disconnect the following:
   - Negative battery cable
   - Knock sensor connector
   - Starter wiring harness connectors
   - Starter motor

To install:

3. Install or connect the following:
   - Starter motor. Tighten the upper bolt to 33 ft. lbs. (44 Nm); the lower bolt to 47 ft. lbs. (64 Nm).
   - Starter wiring harness connectors. Tighten the battery cable nut to 84 inch lbs. (9 Nm).
   - Knock sensor connector
   - Negative battery cable

8 x 1.25 mm
9 N·m (0.9 kgf·m, 7 lbf·ft)

10 x 1.25 mm
44 N·m
(4.5 kgf·m,
33 lbf·ft)

12 x 1.25 mm
64 N·m (6.5 kgf·m, 47 lbf·ft)

09474_CRV_G0002

Starter mounting

Timing Chain and Front Seal

REMOVAL & INSTALLATION

1. Before servicing the vehicle, refer to the Precautions Section.
2. Drain the engine oil.
3. Align the timing marks at TDC No.1.
4. Remove or disconnect the following:
   - Negative battery cable
   - Front splash shield
   - Drive belt
   - Cylinder head cover
   - Crankshaft pulley
   - CKP sensor
   - VTC oil control connector
   - VTC oil control solenoid valve
5. Support the engine with a block of wood and jack.
6. Remove or disconnect the following:
   - Ground cable
   - Upper support bracket
   - Side engine mount
   - Chain case
7. Loosely install the crank pulley. Turn the crankshaft counterclockwise to compress the auto-tensioner. Align the holes on the lock and auto-tensioner and insert a 1.5mm pin into the holes. Turn the crank clockwise to hold the pin.
8. Remove the auto-tensioner.
9. Remove the chain guides.
10. Remove the tensioner arm.
11. Remove the chain.
12. With the case removed, drive out the old seal and install a new one.

To install:

13. Set the crankshaft to TDC.

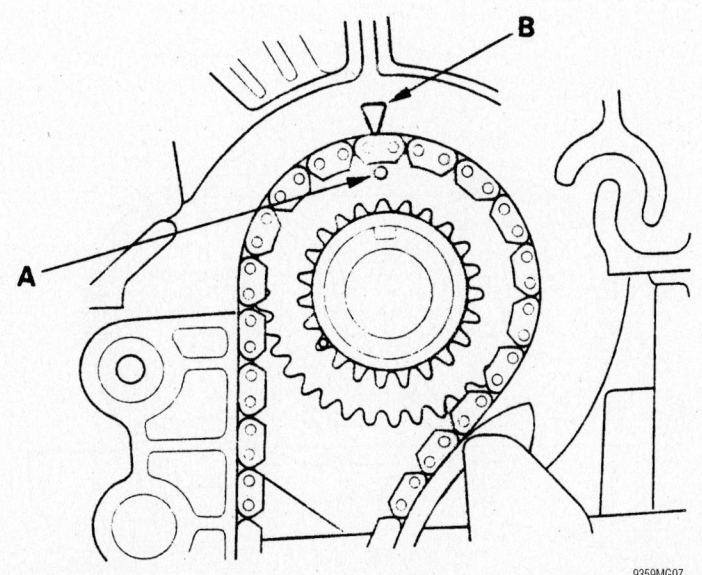

9359MG07

Align the crankshaft timing marks

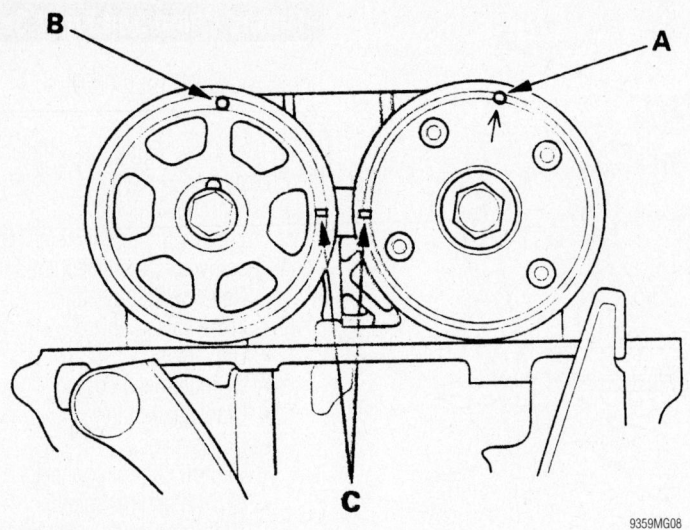

9359MG08

Align the camshaft timing marks

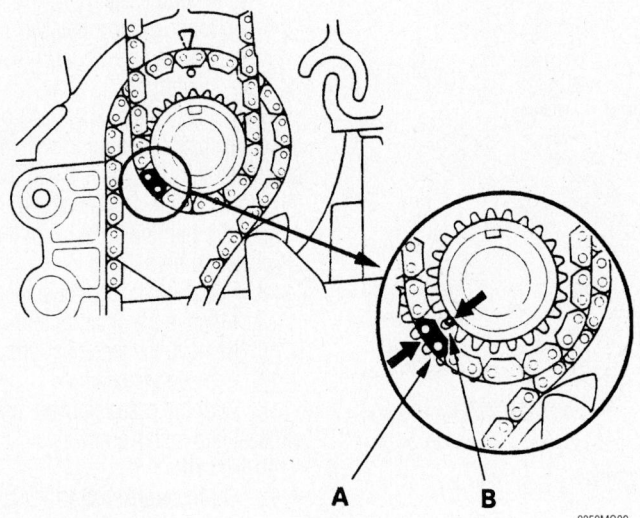

9359MG09

Chain installed on crankshaft

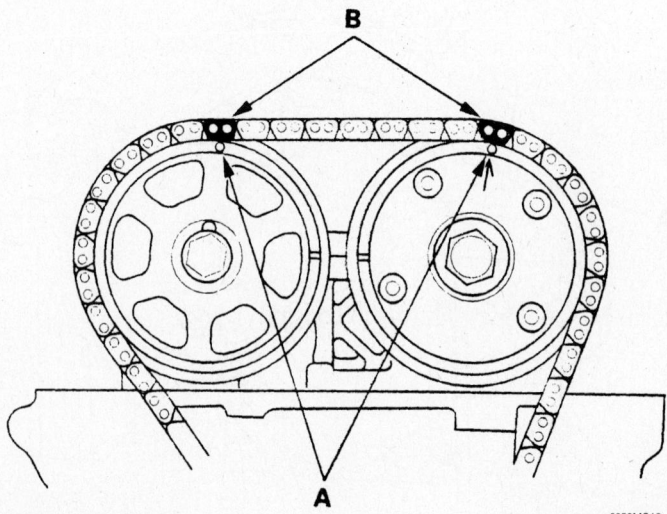

9359MG10

Chain installed on camshafts

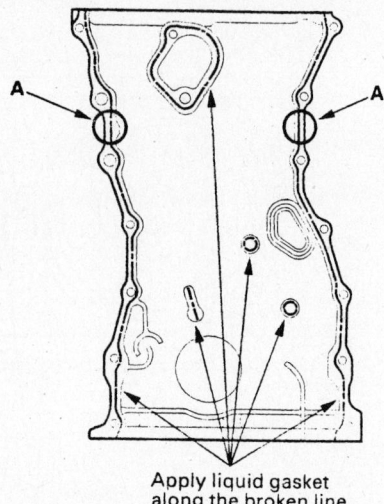

Apply liquid gasket along the broken line.

9359MG11

Chain case sealer application

14. Set the camshafts to TDC.

15. Install the chain on the sprocket with the colored piece A aligned with the punch mark B.

16. The remainder of installation is the reverse of removal. See the accompanying illustration for sealer application. Observe the following torques:

- Chain guide: 9 ft. lbs. (12 Nm)
- Tensioner arm: 16 ft. lbs. (22 Nm)
- Auto-tensioner: 9 ft. lbs. (12 Nm)
- Upper chain guide: 16 ft. lbs. (22 Nm)
- Case: 9 ft. lbs. (12 Nm)
- Side mount: 33 ft. lbs. (44 Nm)
- Upper bracket: 40 ft. lbs. (54 Nm)

Oil Pan

REMOVAL & INSTALLATION

1. Before servicing the vehicle, refer to the Precautions Section.
2. Drain the engine oil.

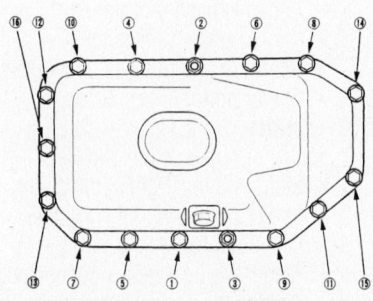

9359MG12

Oil pan fastener tightening sequence

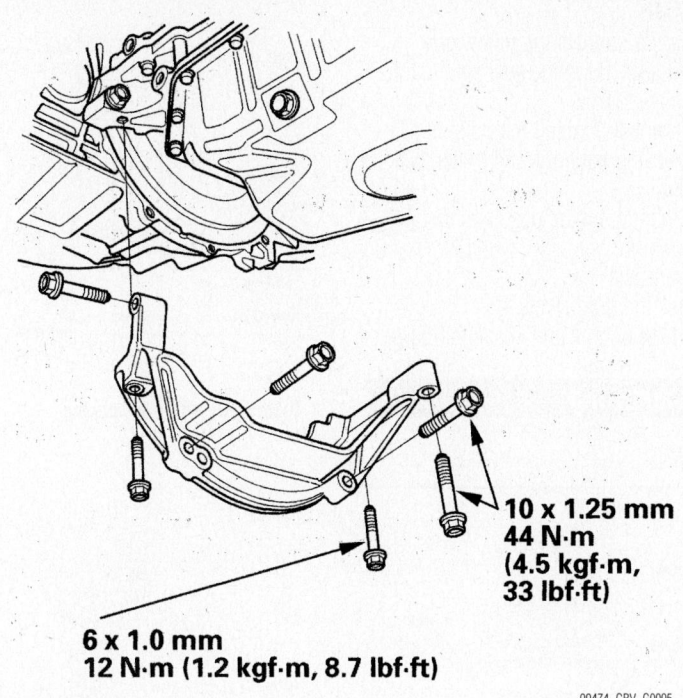

**10 x 1.25 mm
44 N·m
(4.5 kgf·m,
33 lbf·ft)**

**6 x 1.0 mm
12 N·m (1.2 kgf·m, 8.7 lbf·ft)**

09474_CRV_G0005

Stiffener installation

3. Remove or disconnect the following:
   • Subframe. See engine Removal and Installation.
   • With MT, the stiffener
   • Oil pan bolts
   • Oil pan. A gasket cutter will be needed.
4. Installation is the reverse of removal. Torque the bolts, in sequence, in 2 or 3 steps, to 9 ft. lbs. (12 Nm).

Oil Pump

REMOVAL & INSTALLATION

1. Before servicing the vehicle, refer to the Precautions Section.

2. Drain the engine oil.
3. Remove or disconnect the following:
   • Negative battery cable
   • Oil pan
   • Pump chain
   • Pump sprocket
   • Pump

To install:

4. Make sure that No.1 piston is at TDC.
5. Align the dowel pin on the rear balance shaft with the mark on the pump.
6. Install the pump and sprocket loosely.
7. Remove the balance shaft holding pin.
8. Torque the 10mm mounting bolts to 33 ft. lbs. (44 Nm); the 8mm bolts to 16 ft. lbs. (22 Nm).
9. Torque the pulley bolt to 33 ft. lbs. (44 Nm).
10. Torque the tensioner bolts to 9 ft. lbs. (12 nm).

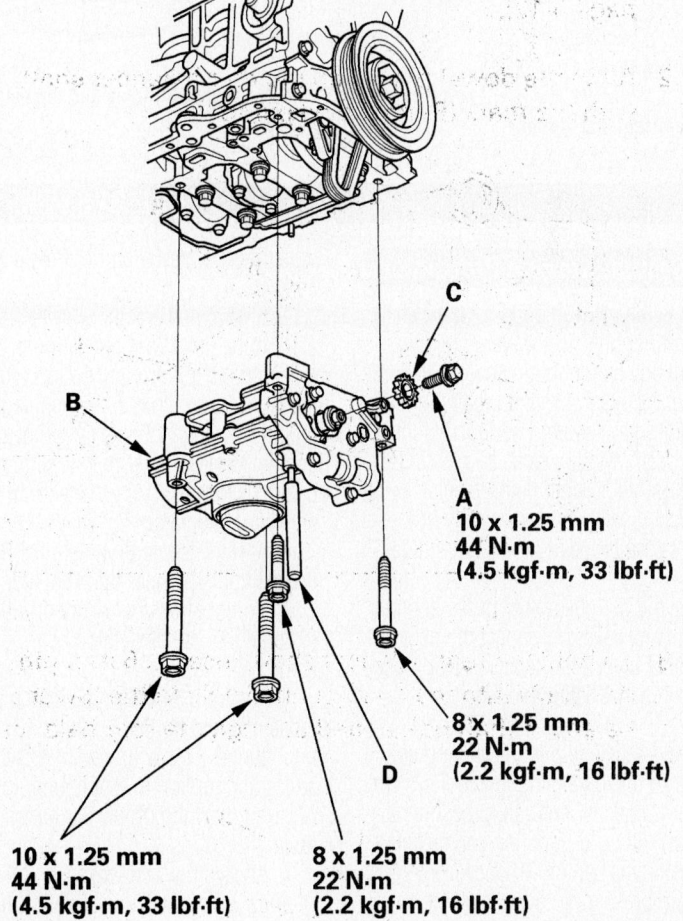

**A
10 x 1.25 mm
44 N·m
(4.5 kgf·m, 33 lbf·ft)**

**8 x 1.25 mm
22 N·m
(2.2 kgf·m, 16 lbf·ft)**

**10 x 1.25 mm
44 N·m
(4.5 kgf·m, 33 lbf·ft)**

**8 x 1.25 mm
22 N·m
(2.2 kgf·m, 16 lbf·ft)**

Oil pump installation

09474_CRV_G0008

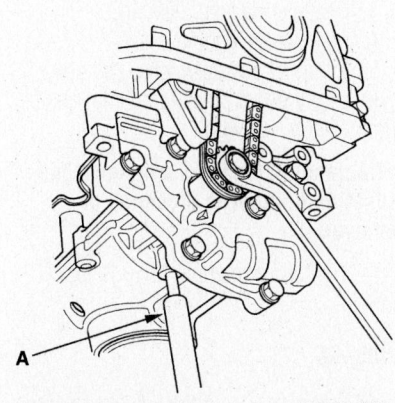

09474_CRV_G0007

Oil pump chain removal

Rear Main Seal

REMOVAL & INSTALLATION

1. Before servicing the vehicle, refer to the Precautions Section.
2. Remove or disconnect the following:
   - Transaxle
   - Clutch pressure plate and disc, if equipped
   - Flywheel
   - Oil seal

Oil Seal Installed Height: 5.5—6.5 mm
(0.22—0.26 in.)

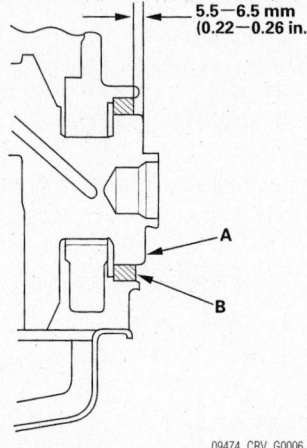

09474_CRV_G0006

Rear main seal installation

To install:

3. Install or connect the following:
   - Oil seal. Drive the seal square into the seal case.
   - Flywheel. Tighten the bolts in a crossing pattern to 54 ft. lbs. (73 Nm).
   - Clutch pressure plate and disc, if equipped
   - Transaxle
4. Check the fluid levels.
5. Start the engine and check for leaks.

Piston and Ring

POSITIONING

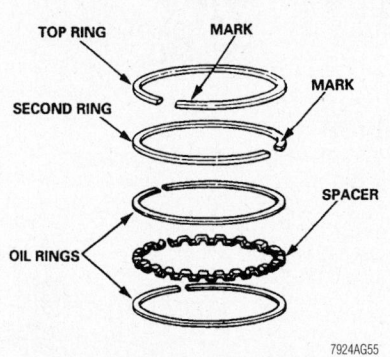

7924AG55

Piston ring positioning and top mark location

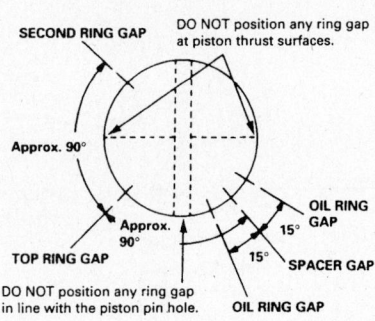

7924AG54

Piston ring end-gap spacing

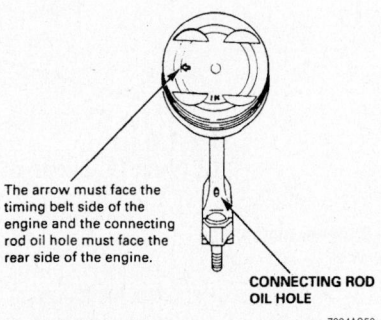

7924AG53

Piston and connecting rod assembly

FUEL SYSTEM

Fuel System Service Precautions

Safety is the most important factor when performing not only fuel system maintenance, but any type of maintenance. Failure to conduct maintenance and repairs in a safe manner may result in serious personal injury or death. Maintenance and testing of the vehicle's fuel system components can be accomplished safely and effectively by adhering to the following rules and guidelines:

- To avoid the possibility of fire and personal injury, always disconnect the negative battery cable unless the repair or test procedure requires that battery voltage be applied.
- Always relieve the fuel system pressure prior to disconnecting any fuel system component (injector, fuel rail, pressure regulator, etc.), fitting or fuel line connection. Exercise extreme caution whenever relieving fuel system pressure to avoid exposing skin, face and eyes to fuel spray. Please be advised that fuel under pressure may penetrate the skin or any part of the body that it contacts.

- Always place a shop towel or cloth around the fitting or connection prior to loosening to absorb any excess fuel due to spillage. Ensure that all fuel spillage (should it occur) is quickly removed from engine surfaces. Ensure that all fuel soaked cloths or towels are deposited into a suitable waste container.
- Always keep a dry chemical (Class B) fire extinguisher near the work area.
- Do not allow fuel spray or fuel vapors to come into contact with a spark or open flame.
- Always use a backup wrench when loosening and tightening fuel line connection fittings. This will prevent unnecessary stress and torsion to fuel line piping. Always follow the proper torque specifications.
- Always replace worn fuel fitting O-rings with new. Do not substitute fuel hose or equivalent, where fuel pipe is installed.

Fuel System Pressure

RELIEVING

2002–04

1. Before servicing the vehicle, refer to the Precautions Section.
2. Disconnect the negative battery cable.
3. Remove the fuel filler cap.
4. Remove the engine cover.
5. Using a back up wrench and shop towel, turn the fuel pulsation damper one complete turn, slowly.

➡**Replace all washers whenever the pulsation damper is loosened or removed.**

6. Tighten the damper to 16 ft. lbs. (22 Nm).

2005–06

1. Before servicing the vehicle, refer to the Precautions Section.

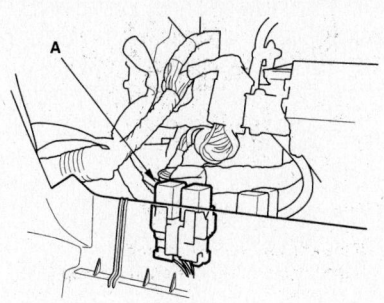

PGM-FI main relay

2. Remove the glove box.

3. Remove the PGM-FI main relay 2 (A).

4. Start the engine and let it run until it shuts off.

5. Turn the engine OFF.

6. Remove the fuel filler cap.

7. Disconnect the battery ground cable.

8. Remove the intake manifold cover and the quick-connect fitting cover.

9. Place a rag over the quick-connect fitting. Disconnect the fitting.

Fuel Filter

REMOVAL & INSTALLATION

➡**The fuel filter should be replaced whenever the fuel pressure drops below 48 psi, after making sure that the fuel pump and fuel pressure regulator are okay.**

1. Before servicing the vehicle, refer to the Precautions Section.

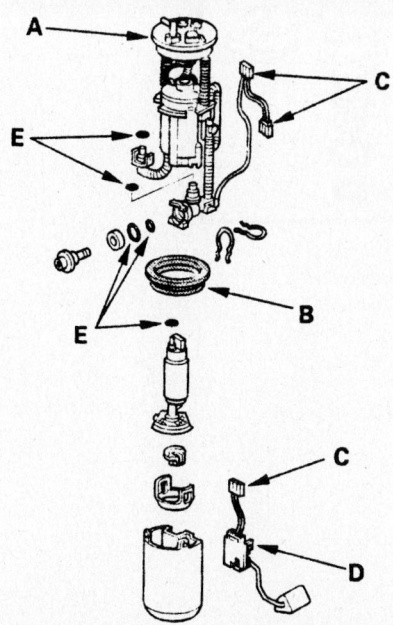

Exploded view of the fuel filter mounting

2. Relieve the fuel system pressure.

3. Remove or disconnect the following:
- Negative battery cable
- Fuel pump
- Fuel filter carrier (A)
- Fuel filter

To install:

4. Install or connect the following:
- Fuel filter
- Fuel lines
- New gasket (B)
- New O-rings (E)
- Connectors (C)
- Sending unit (D)

5. Start the engine and check for leaks.

Fuel Pump

REMOVAL & INSTALLATION

1. Before servicing the vehicle, refer to the Precautions Section.

A. Fuel filter
B. Gasket
C. O-rings
D. Terminal
E. Sending unit

Fuel module exploded view

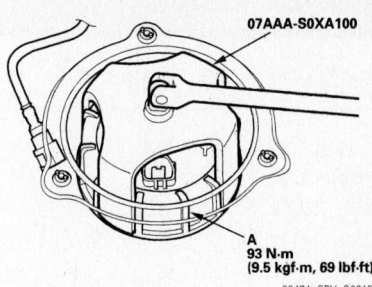

Lockring installation

2. Relieve the fuel system pressure.

3. Remove or disconnect the following:
- Negative battery cable
- Fuel filler cap
- Access panel, under the rear seats
- Fuel pump connector
- Fuel supply and return lines
- Fuel pump locknut
- Fuel pump module

4. Installation is the reverse of removal.

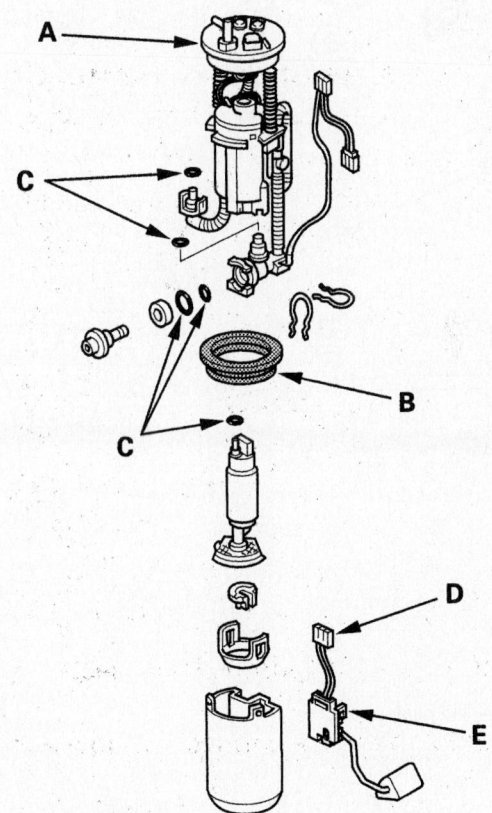

Fuel Injector

REMOVAL & INSTALLATION

1. Before servicing the vehicle, refer to the Precautions Section.
2. Relieve the fuel system pressure.
3. Remove or disconnect the following:

- Negative battery cable
- Injector connectors, ground cable and harness holder
- Fuel line
- Fuel supply manifold
- Fuel injectors

To install:

4. Install the fuel injectors to the fuel supply manifold with new O-rings coated with clean engine oil.

5. Install new seal rings, coated with clean engine oil, in the intake manifold.
6. Install or connect the following:

- Fuel supply manifold and injector assembly. Tighten the nuts to 16 ft. lbs. (22 Nm).
- Fuel lines
- Injector connectors
- Negative battery cable

7. Start the engine and check for leaks.

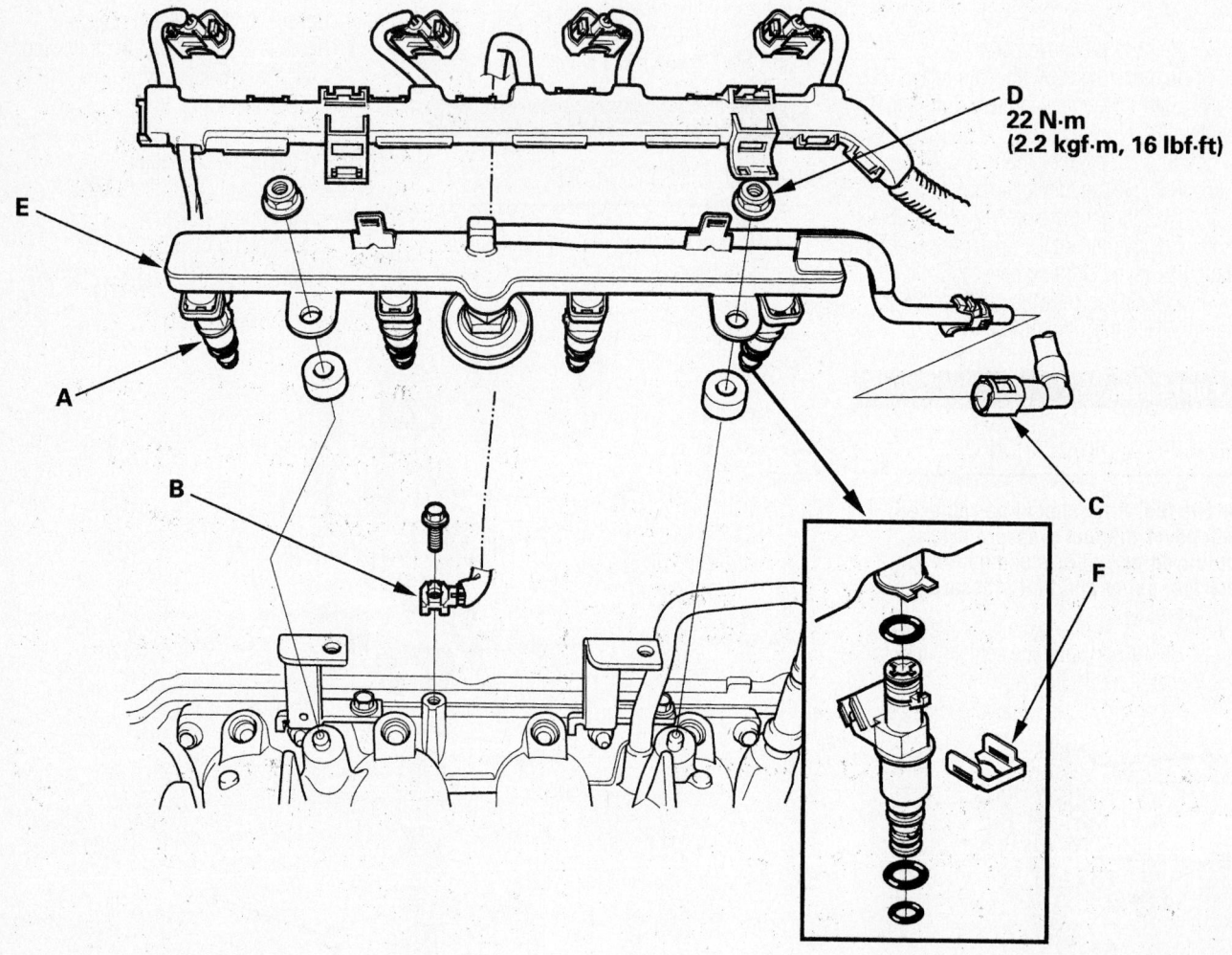

D
22 N·m
(2.2 kgf·m, 16 lbf·ft)

Fuel supply manifold and related parts

09474_CRV_G0012

DRIVE TRAIN

Transaxle Assembly

REMOVAL & INSTALLATION

Automatic Transaxle

1. Before servicing the vehicle, refer to the Precautions Section.
2. Drain the transaxle.
3. Remove or disconnect the following:
   - Battery
   - Battery tray
   - Air intake assembly
   - Splash shield
   - Transaxle ground cable
   - Starter motor
   - Clutch pressure control solenoid valve connector
   - Mainshaft speed sensor connector
   - Clutch pressure switch connectors
   - Shift control solenoid valve connectors
   - Lockup control solenoid connector
   - Countershaft speed sensor connector
   - Transaxle oil cooler lines
   - Engine wiring harness from the air cleaner bracket
   - Water pipe mounting bolt
   - Brake booster and EVAP line mounting bolts. Attach an engine support hanger to the head to support the weight of the engine.
   - Stabilizer link from the lower arm
   - Lower arms from the knuckles
   - Torque converter nuts
   - Shift cable
   - Front mount bolt and nut
   - Rear mount bolts
   - Subframe (see the Engine Removal and Installation procedure)
   - Rear driveshaft, if equipped
4. Separate the inner CV-joints from the transaxle and intermediate shaft and support the axle halfshafts out of the work area with safety wire.
   - Intermediate shaft
   - Engine stiffener
   - Transaxle mount bolts and nuts
   - Transaxle
5. Installation is the reverse of removal. Observe the following torques:
   - Air cleaner housing bracket bolt: 16 ft. lbs. (22 Nm)
   - Front mount bolts: 47 ft. lbs. (64 Nm)
   - Rear mount bracket bolts: 40 ft. lb. (54 Nm)

- Transmission-to-engine bolts: 47 ft. lbs. (64 Nm)
- Upper transmission mount bolt: 40 ft. lbs. (54 Nm)
- Upper transmission mount nuts: 40 ft. lbs. (54 Nm)
- Rear driveshaft bolts: 24 ft. lbs. (32 Nm)
- Subframe bolts: 76 ft. lbs. (103 Nm)

Manual Transaxle

1. Before servicing the vehicle, refer to the Precautions Section.
2. Remove or disconnect the following:
   - Negative battery cable
   - Air intake assembly
   - Transaxle ground cable
   - Vehicle Speed Sensor (VSS) connector
   - Splash shield
   - Shift cables and bracket
   - Clutch slave cylinder and hose bracket
   - Wire harness bracket
   - Water pipe mounting bolt
   - Brake booster and EVAP line mounting bolts. Attach an engine support hanger to the head to support the weight of the engine.
   - Upper transmission mounting bolts
   - Transaxle mount and bracket
3. Separate the inner CV-joints from the transaxle and intermediate shaft and support the axle halfshafts out of the work area with safety wire.
4. Remove or disconnect the following:
   - Intermediate shaft
   - Right front mount and bracket
   - Rear engine mounting bolts
   - Rear driveshaft
   - Subframe (see the Engine Removal and Installation procedure)
   - Clutch housing cover
   - Transaxle
5. Installation is the reverse of removal. Observe the following torques:
   - Transaxle rear mount and bracket. Tighten the bracket bolts to 40 ft. lbs. (54 Nm) and the through bolt to 47 ft. lbs. (64 Nm)
   - Transaxle. Tighten the flange bolts to 47 ft. lbs. (64 Nm)
   - Front mount and bracket. Tighten the bolts to 47 ft. lbs. (64 Nm).
   - Clutch housing cover. Tighten the bolts to 29 ft. lbs. (39 Nm)
   - Subframe: 76 ft. lbs. (98 Nm)

Clutch

ADJUSTMENTS

The CR-V is equipped with a hydraulic clutch system. No adjustment is necessary.

REMOVAL & INSTALLATION

1. Before servicing the vehicle, refer to the Precautions Section.
2. Remove or disconnect the following:
   - Negative battery cable
   - Transaxle
   - Pressure plate. Loosen the bolts evenly in a crossing pattern.
   - Clutch disc

To install:
3. Install the clutch disc and pressure plate. Tighten the pressure plate bolts in a crossing pattern and in several steps to 19 ft. lbs. (25 Nm).

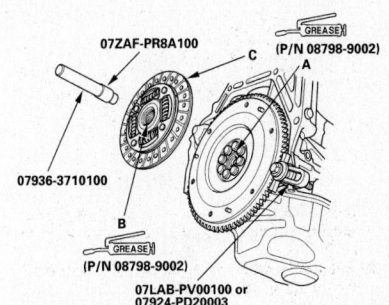

09474_CRV_G0016

Clutch disc installation

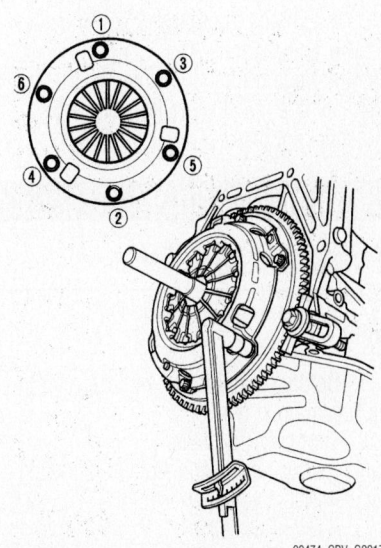

09474_CRV_G0017

Clutch pressure plate torque sequence

4. Install or connect the following:
- Transaxle
- Negative battery cable

HYDRAULIC CLUTCH SYSTEM BLEEDING

1. Before servicing the vehicle, refer to the Precautions Section.
2. Attach a hose to the bleeder screw and suspend the other end in a container of clean brake fluid.
3. Open the bleeder screw.
4. Slowly pump the clutch pedal until no more air bubbles appear at the bleeder hose.
5. Tighten the bleeder screw to 70 inch lbs. (8 Nm).
6. Refill the clutch master cylinder as necessary.
7. Check for leaks and proper clutch operation.

Transfer Assembly

REMOVAL & INSTALLATION

1. Before servicing the vehicle, refer to the Precautions Section.
2. Drain the transaxle fluid.
3. Remove or disconnect the following:
- Negative battery cable
- Rear driveshaft
- Transfer assembly and bracket

To install:

4. Install the transfer assembly and bracket with a new O-ring seal. Tighten the 10mm bolts to 33 ft. lbs. (44 Nm) and the 8mm bolts to 17 ft. lbs. (24 Nm).
5. Install or connect the following:
- Rear driveshaft. Tighten the bolts to 24 ft. lbs. (32 Nm).
- Exhaust front pipe
- HO2S connector
- Negative battery cable
6. Fill the transaxle to the correct level and check for leaks.

Halfshaft

REMOVAL & INSTALLATION

Front

1. Before servicing the vehicle, refer to the Precautions Section.
2. Drain the transaxle.
3. Remove or disconnect the following:
- Negative battery cable
- Front wheels
- Stabilizer bar

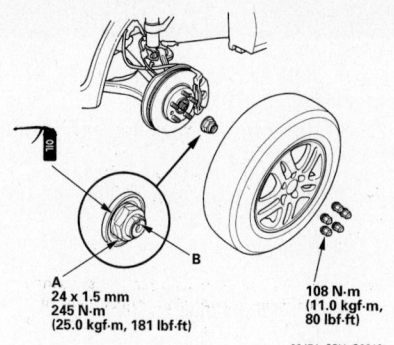

A
24 x 1.5 mm
245 N·m
(25.0 kgf·m, 181 lbf·ft)

B

108 N·m
(11.0 kgf·m,
80 lbf·ft)

09474_CRV_G0018

Front halfshaft locknut (A), shoulder (B)

- Lower ball joint
- Spindle nut
4. On the left side, pry the inboard joint from the case with a prybar.
5. On the right side, drive the inboard shaft off the intermediate shaft with a drift and hammer.
6. Installation is the reverse of removal. Grease the splined ends of the shaft. Insert the shaft into the differential until the set ring locks in the groove. Observe the following torques:
- Ball stud nut: 40 ft. lbs. (54 Nm)
- Stabilizer link nuts: 29 ft. lbs. (39 Nm)
- Spindle nut: 181 ft. lbs. (245 Nm)

Rear

1. Before servicing the vehicle, refer to the Precautions Section.
2. Drain the differential.
3. Remove or disconnect the following:
- Negative battery cable
- Rear wheels
- Spindle nut
4. Pry the inboard joint from the differential.
5. Remove the outer CV-joint stub shaft from the hub by tapping the stub shaft with a plastic hammer.

To install:

➡Use new circlips and self-locking nuts for assembly.

6. Install the outer CV-joint stub shaft into the hub.
7. Install the inner CV-joint to the differential until the circlip locks in the retaining groove.
8. Install or connect the following:
- Spindle nut. Tighten the nut to 134 ft. lbs. (181 Nm).
- Rear wheels
- Negative battery cable
9. Fill the differential to the correct level and check for leaks.

CV-Joint

OVERHAUL

Front

OUTBOARD JOINT

1. Before servicing the vehicle, refer to the Precautions Section.
2. Remove or disconnect the following:
- Axle halfshaft from the vehicle and place it in a vise
- Outboard joint boot clamps and push the boot back
- Outboard joint by driving it off the axle shaft with a brass drift and hammer
- Outboard joint boot

To install:

➡Use new circlips and boot clamps for assembly.

3. Install the outboard joint boot and clamps to the axle shaft.
4. Fill the outboard joint with grease. Install the outboard joint to the axle shaft. Tap the stub shaft with a brass hammer to seat the circlip.
5. Fill the outboard joint boot with grease and install the boot clamps.
6. Install the axle halfshaft to the vehicle.

INBOARD JOINT

1. Before servicing the vehicle, refer to the Precautions Section.
2. Remove or disconnect the following:
- Axle halfshaft from the vehicle.
- Inboard joint boot clamps and push the boot back
- Inboard joint housing from the axle
- Rollers from the spider
- Snapring and the spider from the axle shaft
- Inboard joint boot

To install:

➡Use new circlips and boot clamps for assembly.

3. Install or connect the following:
- Inboard joint boot and clamps to the axle shaft
- Spider with a new snapring
- Rollers to the spider
4. Fill the joint housing with grease and install it.
5. Fill the inboard joint boot with grease and install the boot clamps.
6. Install the axle halfshaft to the vehicle.

Rear

1. Before servicing the vehicle, refer to the Precautions Section.
2. Remove or disconnect the following:
   - Axle halfshaft from the vehicle
   - Joint boot clamps and push the boot back
   - Joint housing from the axle
   - Rollers from the spider
   - Snapring and the spider from the axle shaft
   - Joint boot

To install:

➡**Use new circlips and boot clamps for assembly.**

3. Install or connect the following:
   - Joint boot and clamps to the axle shaft
   - Spider with a new snapring
   - Rollers to the spider
4. Fill the joint housing with grease and install it.
5. Fill the joint boot with grease and install the boot clamps.
6. Install the axle halfshaft to the vehicle.

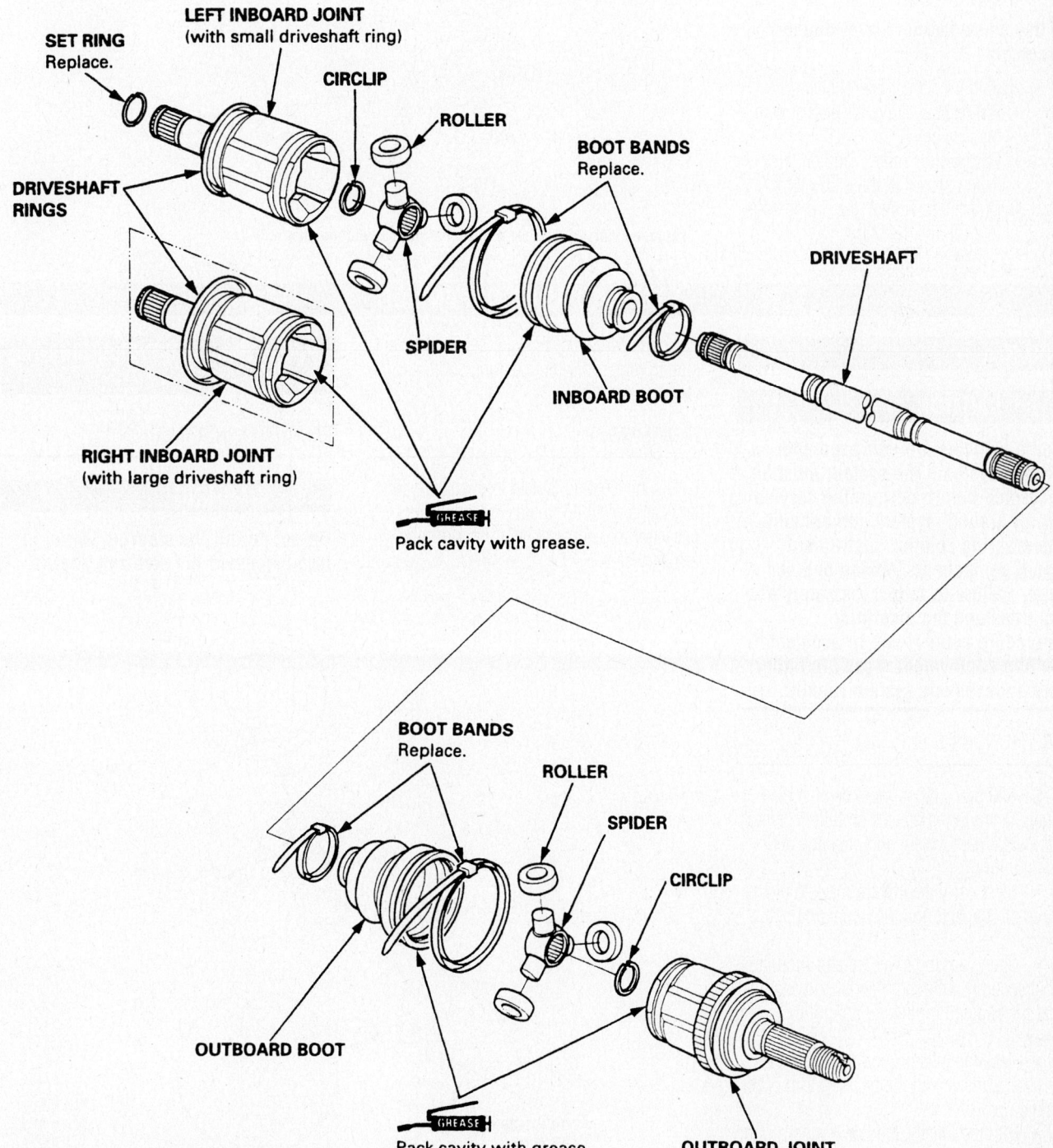

Exploded view of the rear axle—CR-V

9308MG30

Pinion Seal

REMOVAL & INSTALLATION

1. Before servicing the vehicle, refer to the Precautions Section.
2. Remove or disconnect the following:
   - Driveshaft
   - Companion flange
   - Pinion seal

To install:

➡ **Use a new locknut and O-ring for assembly.**

3. Install or connect the following:
   - Pinion seal. Drive the seal square into the bore.
   - Companion flange. Tighten the locknut to 87 ft. lbs. (118 Nm).
   - Driveshaft. Tighten the flange bolts to 24 ft. lbs. (32 Nm).

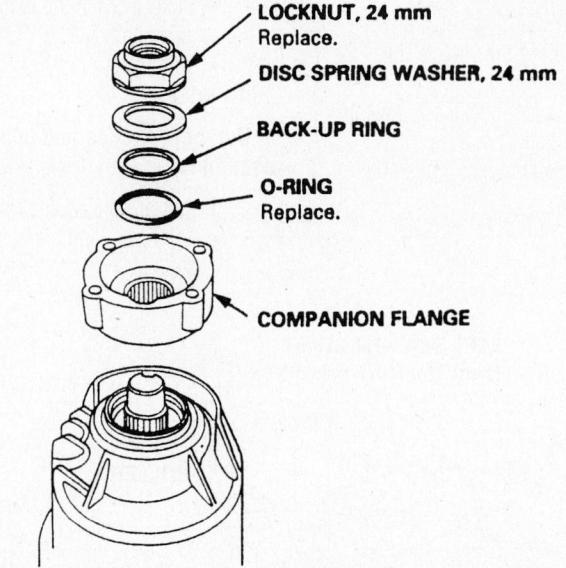

LOCKNUT, 24 mm
Replace.

DISC SPRING WASHER, 24 mm

BACK-UP RING

O-RING
Replace.

COMPANION FLANGE

9308MG31

Exploded view of the rear differential pinion components—CR-V

STEERING AND SUSPENSION

Air Bag

❊❊ CAUTION

Some vehicles are equipped with an air bag system. The system must be disarmed before performing service on, or around, system components, the steering column, instrument panel components, wiring and sensors. Failure to follow the safety precautions and the disarming procedure could result in accidental air bag deployment, possible injury and unnecessary system repairs.

PRECAUTIONS

Several precautions must be observed when handling the inflator module to avoid accidental deployment and possible personal injury.

- Never carry the inflator module by the wires or connector on the underside of the module.
- When carrying a live inflator module, hold securely with both hands, and ensure that the bag and trim cover are pointed away.
- Place the inflator module on a bench or other surface with the bag and trim cover facing up.
- With the inflator module on the bench, never place anything on or close to the module which may be thrown in the event of an accidental deployment.

Before servicing the vehicle, also make sure to refer to the Precautions Section as well.

DISARMING

Disconnect and isolate the negative battery cable. Wait 3 minutes for the system capacitor to discharge before performing any service.

Power Rack and Pinion Steering Gear

REMOVAL & INSTALLATION

❊❊ WARNING

Do not permit the steering wheel to turn whenever the steering gear is

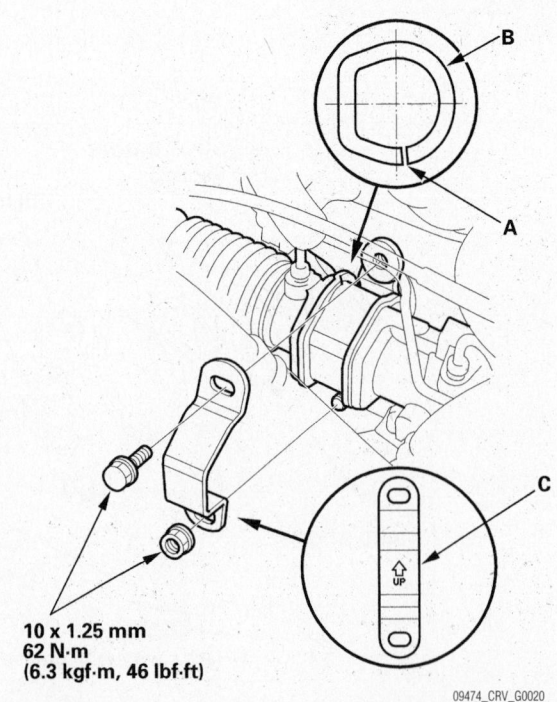

10 x 1.25 mm
62 N·m
(6.3 kgf·m, 46 lbf·ft)

09474_CRV_G0020

Steering gear installation

**disconnected from the steering col-
umn. Damage to the air bag wiring
can result.**

1.   Before servicing the vehicle, refer to the Precautions Section.

2.   Center the steering wheel and lock it in position.

3.   Remove or disconnect the follow-ing:

- Negative battery cable
- Air bag and steering wheel
- Front wheels
- Driver's side dashboard lower cover and undercover
- Air cleaner housing

- Steering joint bolts
- Tie rod ends
- Steering hoses
- Left side flange bolts
- Mounting brackets

4.   Lower the unit so the pinion shaft points outward. Remove the pinion shaft grommet. The steering gear is removed through the driver's side.

5.   Installation is the reverse of removal. Position the cutout (A) on the mounting cushion (B) as shown. Install the mounting bracket (C) over the cushion. Observe the following torques:

- Mounting bracket and side flange bolts: 46 ft. lbs. (62 Nm)

- Supply line flare nut: 27 ft. lbs. (37 Nm)
- Tie rod ball stud nuts: 32 ft. lbs. (43 Nm)
- Steering joint bolts: 21 ft. lbs. (28 Nm)

Strut

REMOVAL & INSTALLATION

Front

1.   Before servicing the vehicle, refer to the Precautions Section.

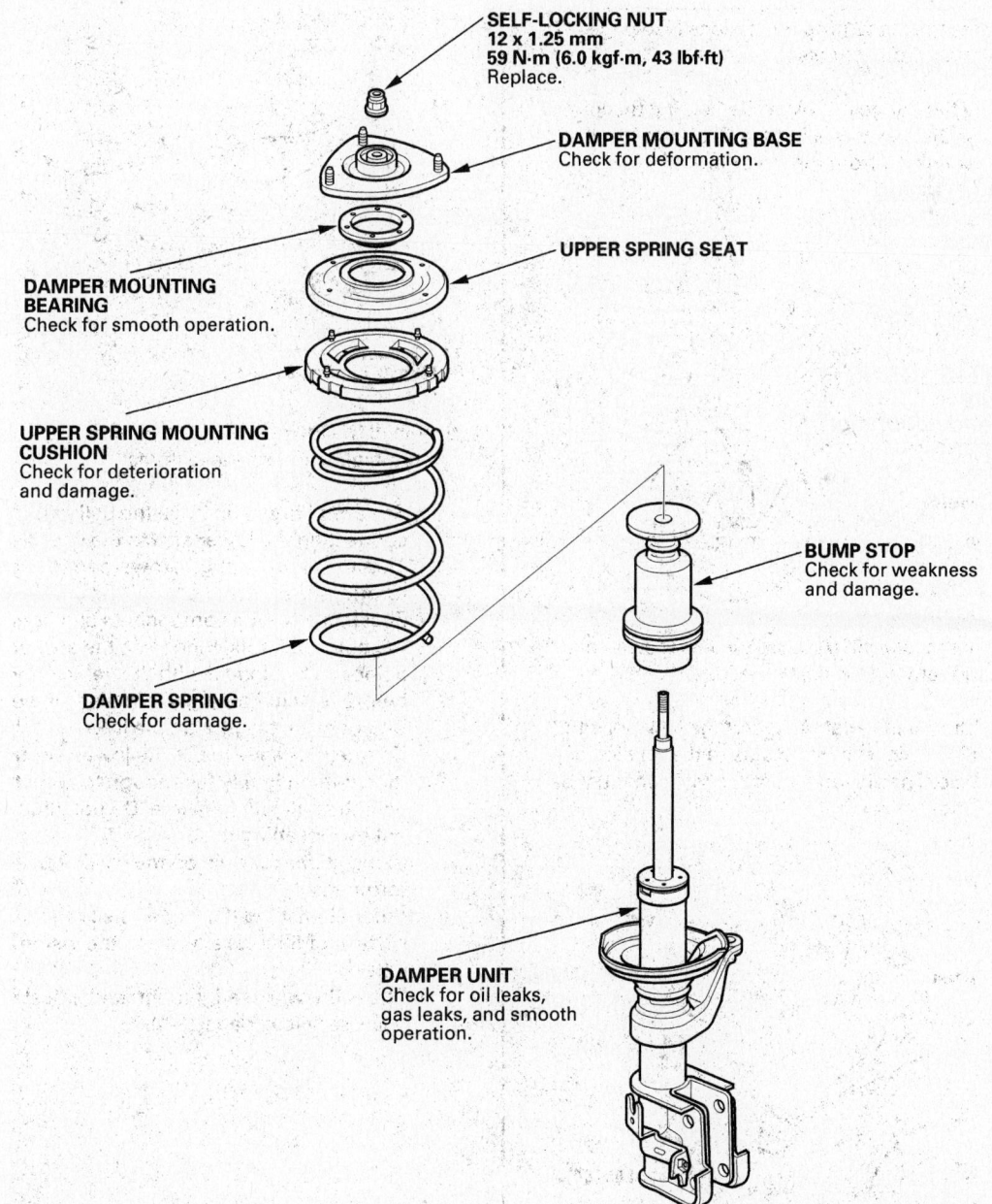

SELF-LOCKING NUT
12 x 1.25 mm
59 N·m (6.0 kgf·m, 43 lbf·ft)
Replace.

DAMPER MOUNTING BASE
Check for deformation.

UPPER SPRING SEAT

DAMPER MOUNTING
BEARING
Check for smooth operation.

UPPER SPRING MOUNTING
CUSHION
Check for deterioration
and damage.

BUMP STOP
Check for weakness
and damage.

DAMPER SPRING
Check for damage.

DAMPER UNIT
Check for oil leaks,
gas leaks, and smooth
operation.

09474_CRV_G0021

Front strut exploded view

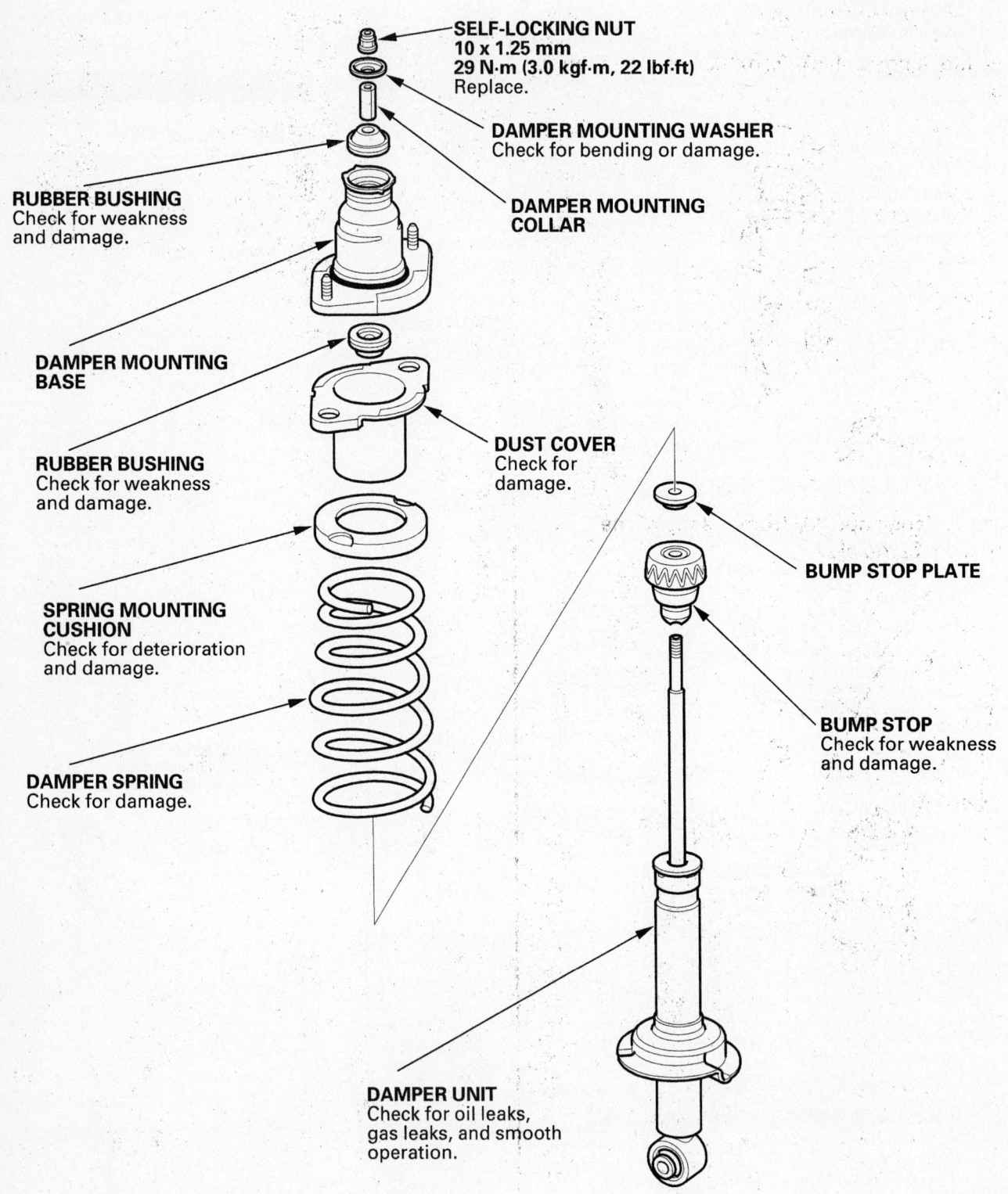

SELF-LOCKING NUT
10 x 1.25 mm
29 N·m (3.0 kgf·m, 22 lbf·ft)
Replace.

DAMPER MOUNTING WASHER
Check for bending or damage.

DAMPER MOUNTING COLLAR

RUBBER BUSHING
Check for weakness and damage.

DAMPER MOUNTING BASE

RUBBER BUSHING
Check for weakness and damage.

DUST COVER
Check for damage.

SPRING MOUNTING CUSHION
Check for deterioration and damage.

BUMP STOP PLATE

DAMPER SPRING
Check for damage.

BUMP STOP
Check for weakness and damage.

DAMPER UNIT
Check for oil leaks, gas leaks, and smooth operation.

09474_CRV_G0022

Rear strut exploded view

2. Remove or disconnect the following:
- Front wheel
- Tie rod end
- Brake hose retainer
- ABS sensor
- Strut

To install:

➡**Use new self-locking fasteners for assembly.**

3. Install or connect the following:
- Strut. Tighten the upper mounting nuts to 33 ft. lbs. (44 Nm).
- Tighten the pinch bolts to 116 ft. lbs. (157 Nm)
- ABS sensor
- Tie rod end
- Brake hose retainer
- Front wheel

Rear

1. Before servicing the vehicle, refer to the Precautions Section.
2. Support the vehicle under the lower control arm.
3. Remove or disconnect the following:
- Rear wheel
- Interior access panel
- Damper cap
- Upper strut mount nuts
- Lower strut flange bolt
- Strut

To install:

4. Install or connect the following:
- Strut. Tighten the nuts to 54 ft. lbs. (74 Nm); the bolt to 69 ft. lbs. (93 Nm).
- Damper cap
- Interior access panel
- Rear wheel

Coil Spring

REMOVAL & INSTALLATION

Front

1. Before servicing the vehicle, refer to the Precautions Section.
2. Remove the strut from the vehicle and install in a strut spring compressor. Compress the spring until the end of the spring comes away from the spring seat.
3. Remove the upper strut mount, spring seat and related components.
4. Remove the coil spring from the strut spring compressor.

To install:

➡**Use a new self-locking nut.**

5. Compress the spring and position the strut so that the end of the spring aligns with the notch in the spring seat.
6. Install the upper strut mounting components and tighten the nut to 43 ft. lbs. (59 Nm).
7. Install the strut to the vehicle.
8. Check the wheel alignment and adjust as necessary.

Rear

1. Before servicing the vehicle, refer to the Precautions Section.
2. Remove the strut from the vehicle and install in a strut spring compressor. Compress the spring until the end of the spring comes away from the spring seat.
3. Remove or disconnect the following:
- Upper strut mount, spring seat and related components
- Coil spring from the strut spring compressor

To install:

➡**Use a new self-locking nut.**

4. Compress the spring and position the strut so that the end of the spring aligns with the notch in the spring seat.
5. Install or connect the following:
- Upper strut mounting components and tighten the nut to 22 ft. lbs. (29 Nm).
- Strut to the vehicle
6. Check the wheel alignment and adjust as necessary.

Upper Ball Joint

REMOVAL & INSTALLATION

The upper ball joints are replaced with the upper control arms as an assembly.

Lower Ball Joint

REMOVAL & INSTALLATION

The ball joint is not replaceable.

Upper Control Arm

REMOVAL & INSTALLATION

1. Before servicing the vehicle, refer to the Precautions Section.
2. Support the lower control arm assembly with a floor jack.
3. Remove or disconnect the following:
- Upper ball joint
- Inner control arm flange bolts.
- Upper control arm

To install:

➡**Use new self-locking nuts for assembly.**

4. Install the upper control arm. Tighten the ball joint nut to 29–35 ft. lbs. (39–47 Nm) and the inner flange bolts to 40 ft. lbs. (54 Nm).

CONTROL ARM BUSHING REPLACEMENT

The upper control arm bushings are serviced with the upper control arm as an assembly.

Lower Control Arm

REMOVAL & INSTALLATION

1. Before servicing the vehicle, refer to the Precautions Section.
2. Remove or disconnect the following:
- Front wheel
- Stabilizer link
- Lower arm from the knuckle
- Lower arm
3. Installation is the reverse of removal. Observe the following torques:
- Lower arm bolts: 61 ft. lbs. (83 nm)
- Ball stud nut: 2002–04: 40 ft. lbs. (54 Nm); 2005–06 48 ft. lbs. (65 Nm)
- Stabilizer link: 29 ft. lbs. (39 Nm)

CONTROL ARM BUSHING REPLACEMENT

The lower control arm front inner bushing and the damper fork bushing are serviced with the control arm as an assembly.

Rear inner bushing

1. Before servicing the vehicle, refer to the Precautions Section.
2. Remove or disconnect the following:
- Front wheel
- Rear bushing bracket
- Rear bushing

To install:

➡**Use a new self-locking nut for assembly.**

3. Install or connect the following:
- Rear bushing. Tighten the nut to 61 ft. lbs. (83 Nm).
- Rear bushing bracket. Tighten the bolts to 66 ft. lbs. (89 Nm).
- Front wheel
4. Check the wheel alignment and adjust as necessary.

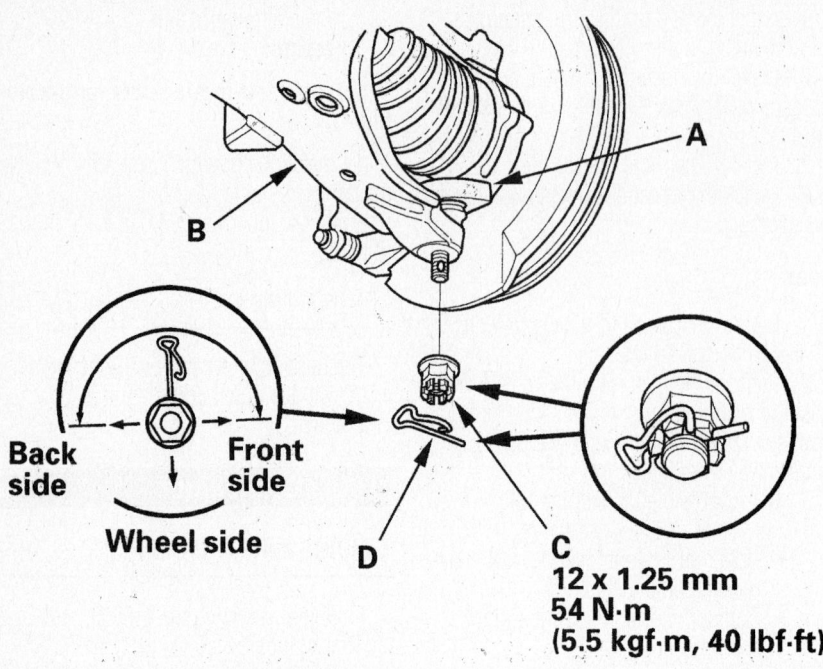

Back side

Front side

Wheel side

D

C
12 x 1.25 mm
54 N·m
(5.5 kgf·m, 40 lbf·ft)

09474_CRV_G0019

Lower ball stud nut installation—2002–04 shown

Wheel Bearings

ADJUSTMENT

The wheel bearings are not adjustable.

REMOVAL & INSTALLATION

Front

1. Before servicing the vehicle, refer to the Precautions Section.
2. Remove or disconnect the following:
   - Front wheel
   - Spindle nut
   - Brake caliper and rotor. Forcing screws are needed to remove the rotor.
   - Brake hose bracket
   - ABS sensor
   - Stabilizer link
   - Lower arm from the knuckle
   - Strut-to-knuckle bolts
   - Steering hub/knuckle assembly
3. Press the hub from the knuckle. The bearings and races can now be pressed out and replaced.

➡ **With ABS, install the bearing with the magnetic encoder (brown color) toward the inside of the knuckle.**

4. Observe the following torques:
   - Strut bolts: 116 ft. lbs. (157 Nm)
   - Ball stud nuts: 40 ft. lbs. (54 Nm)
   - Stabilizer bar link: 29 ft. lbs. (39 Nm)
   - Spindle nut: 181 ft. lbs. (245 Nm)

Rear

1. Before servicing the vehicle, refer to the Precautions Section.
2. Remove or disconnect the following:
   - Rear wheel
   - Brake caliper
   - Rotor
   - Spindle nut
   - Axle shaft (2wd)
   - Parking brake shoes
   - Parking brake cable
   - Wheel sensor, if equipped
3. Support the trailing arm.
4. Remove or disconnect the following:
   - Upper arm from the knuckle
5. Matchmark the trailing arm cam adjusting bolt and cam. Remove the bolt. Discard the nut.
6. Remove the flange bolt.
7. Remove the knuckle assembly.
8. Press the hub from the knuckle. The bearings and races can now be pressed out and replaced.

➡ **With ABS, install the bearing with the magnetic encoder (brown color) toward the inside of the knuckle.**

9. Observe the following torques:
   - Flange bolt: 69 ft. lbs. (93 Nm)
   - Cam bolts: 43 ft. lbs. (59 Nm)
   - Spindle nut: 134 ft. lbs. (181 Nm)
   - Caliper mounting bolts: 41 ft. lbs. (55 Nm)

BRAKES

Brake Caliper

REMOVAL & INSTALLATION

Front

1. Remove the upper and lower bolts.
2. Lift off the caliper.
3. Remove the pad springs.
4. Remove the pads and shims.
5. Remove the pad retainers.
6. Installation is the reverse of removal. Coat both sides of the shims and the backs of the pads with brake grease. Torque the bolts to 25 ft. lbs. (34 Nm). If the hose was disconnected, torque the banjo bolt to 25 ft. lbs. (34 Nm).

Rear

1. Remove the caliper pin bolts.
2. Lift off the caliper and suspend it safely.
3. Remove the pads and shims.
4. Remove the pad retainers.
5. Installation is the reverse of removal. Coat both sides of the shims and the backs of the pads with brake grease. Torque the bolts to 16 ft. lbs. (22 Nm). If the hose was disconnected, torque the banjo bolt to 16 ft. lbs. (22 Nm).

Disc Brake Pads

REMOVAL & INSTALLATION

Front

1. Remove the lower bolt.
2. Pivot the caliper up and hold the pads.

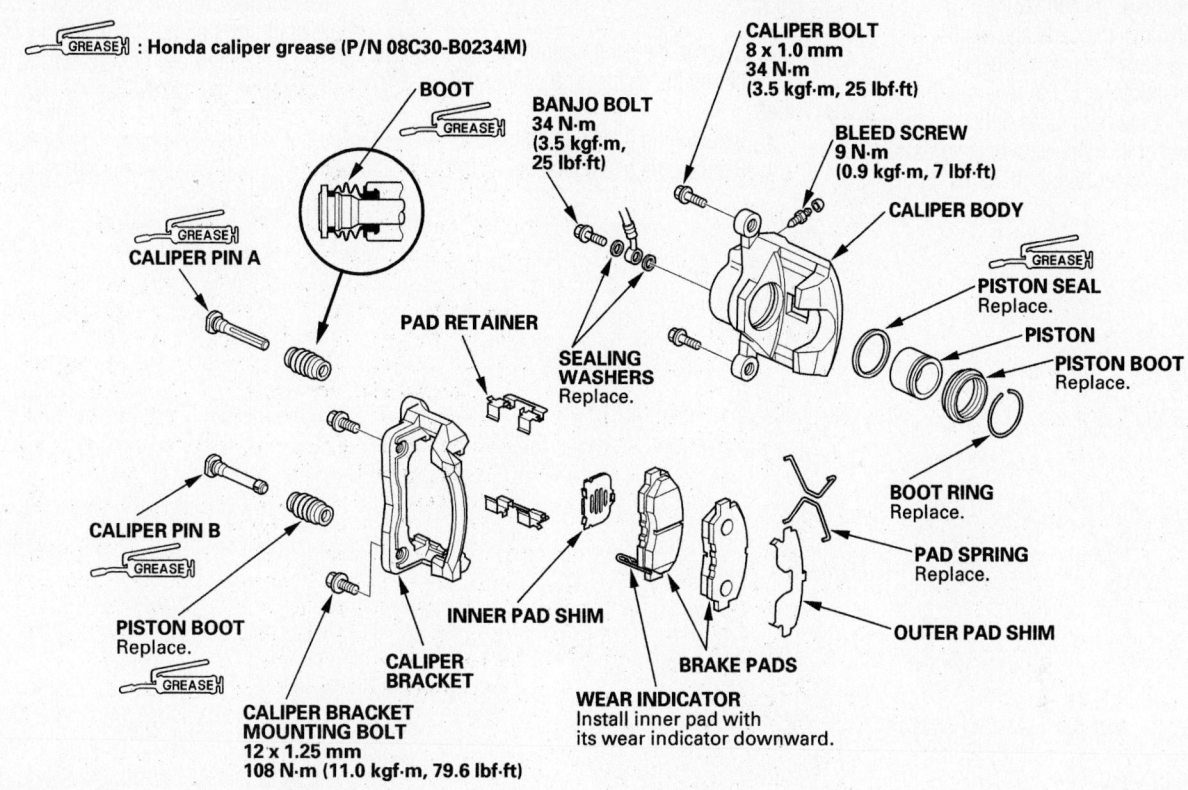

⟨GREASE⟩ : Honda caliper grease (P/N 08C30-B0234M)

BOOT

⟨GREASE⟩

CALIPER PIN A

⟨GREASE⟩

BANJO BOLT
34 N·m
(3.5 kgf·m,
25 lbf·ft)

CALIPER BOLT
8 x 1.0 mm
34 N·m
(3.5 kgf·m, 25 lbf·ft)

BLEED SCREW
9 N·m
(0.9 kgf·m, 7 lbf·ft)

CALIPER BODY

⟨GREASE⟩
PISTON SEAL
Replace.

PISTON

PISTON BOOT
Replace.

PAD RETAINER

**SEALING
WASHERS**
Replace.

BOOT RING
Replace.

CALIPER PIN B

⟨GREASE⟩

PISTON BOOT
Replace.

⟨GREASE⟩

**CALIPER BRACKET
MOUNTING BOLT**
12 x 1.25 mm
108 N·m (11.0 kgf·m, 79.6 lbf·ft)

**CALIPER
BRACKET**

INNER PAD SHIM

WEAR INDICATOR
Install inner pad with
its wear indicator downward.

BRAKE PADS

PAD SPRING
Replace.

OUTER PAD SHIM

09474_CRV_G0023

Front brake components—2005 models shown; others similar

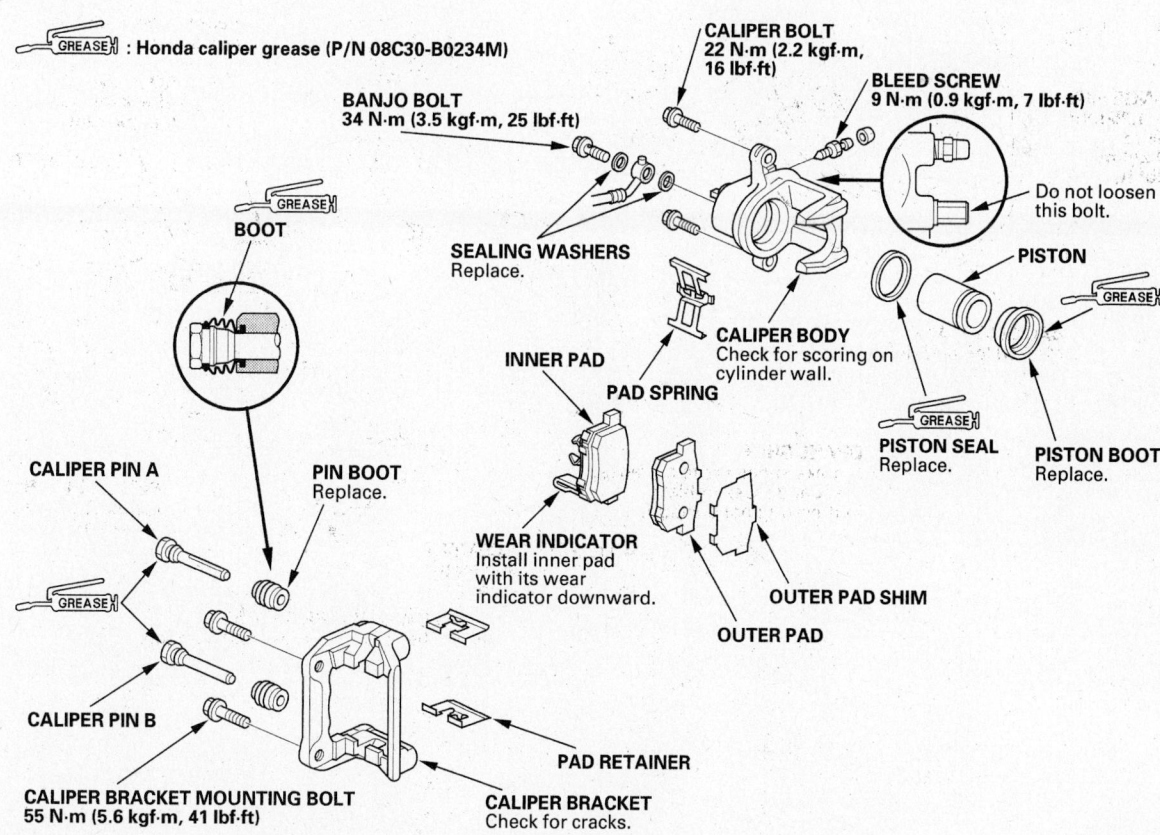

⟨GREASE⟩ : Honda caliper grease (P/N 08C30-B0234M)

CALIPER BOLT
22 N·m (2.2 kgf·m,
16 lbf·ft)

BLEED SCREW
9 N·m (0.9 kgf·m, 7 lbf·ft)

BANJO BOLT
34 N·m (3.5 kgf·m, 25 lbf·ft)

Do not loosen
this bolt.

⟨GREASE⟩

BOOT

SEALING WASHERS
Replace.

INNER PAD

PAD SPRING

CALIPER BODY
Check for scoring on
cylinder wall.

PISTON

⟨GREASE⟩

⟨GREASE⟩

PISTON SEAL
Replace.

PISTON BOOT
Replace.

CALIPER PIN A

⟨GREASE⟩

PIN BOOT
Replace.

WEAR INDICATOR
Install inner pad
with its wear
indicator downward.

OUTER PAD SHIM

OUTER PAD

CALIPER PIN B

CALIPER BRACKET MOUNTING BOLT
55 N·m (5.6 kgf·m, 41 lbf·ft)

PAD RETAINER

CALIPER BRACKET
Check for cracks.

09474_CRV_G0024

Rear brake components—2005 models shown; others similar

3. Remove the pad springs.
4. Remove the pads and shims.
5. Remove the pad retainers.
6. Installation is the reverse of removal. Coat both sides of the shims and the backs of the pads with brake grease. Torque the lower bolt to 25 ft. lbs. (34 Nm).

Rear

1. Remove the caliper pin bolts.
2. Lift off the caliper and suspend it safely.
3. Remove the pads and shims.
4. Remove the pad retainers.

5. Installation is the reverse of removal. Coat both sides of the shims and the backs of the pads with brake grease. Torque the bolts to 16 ft. lbs. (22 Nm).

HONDA

Element

SPECIFICATION AND MAINTENANCE CHARTS

ENGINE AND VEHICLE IDENTIFICATION CHART

Engine Code							Model Year		
Code	Liters (cc)	Cu. In.	Cyl.	Fuel Sys.	Engine Type	Eng. Mfg.	Code ①	Year
K24A4	2.4 (2354)	144	4	SMFI	DOHC	Honda	3	2003
DOHC: Double Overhead Cam							4	2004
SMFI: Sequential Multi-port Fuel Injection							5	2005
① 10th position of VIN							6	2006

09474_ELEM_C0001

GENERAL ENGINE SPECIFICATIONS

Year	Model	Engine Displacement Liters (VIN)	Net Horsepower @ rpm	Net Torque @ rpm (ft. lbs.)	Bore x Stroke (in.)	Com- pression Ratio	Oil Pressure @ rpm
2003	Element	2.4 (K24A4)	160@5500	161@4500	3.42x3.90	9.7:1	44@3000
2004	Element	2.4 (K24A4)	160@5500	161@4500	3.42x3.90	9.7:1	44@3000
2005-06	Element	2.4 (K24A4)	160@5500	161@4500	3.42x3.90	9.7:1	44@3000

SMFI: Sequential Multi-port Fuel Injection

09474_ELEM_C0002

ENGINE TUNE-UP SPECIFICATIONS

Year	Engine Displacement Liters (VIN)	Spark Plug Gap (in.)	Ignition Timing (deg.) MT	Ignition Timing (deg.) AT	Fuel Pump (psi)	Idle Speed (rpm) MT	Idle Speed (rpm) AT	Valve Clearance (in.) In.	Valve Clearance (in.) Ex.
2003	2.4 (K24A4)	0.039-0.043	6-10B	6-10B	48-55	650-750	650-750	0.008-0.010	0.011-0.013
2004	2.4 (K24A4)	0.039-0.043	6-10B	6-10B	48-55	650-750	650-750	0.008-0.010	0.011-0.013
2005-06	2.4 (K24A4)	0.039-0.043	6-10B	6-10B	48-55	650-750	650-750	0.008-0.010	0.011-0.013

NOTE: The Vehicle Emission Control Information label often reflects changes made during production and must be used if they differ from this chart.

NOTE: The fuel pressure readings are given with the vacuum hose connected to the regulator and the engine running

B: Before top dead center

HYD: Hydraulic

09474_ELEM_C0003

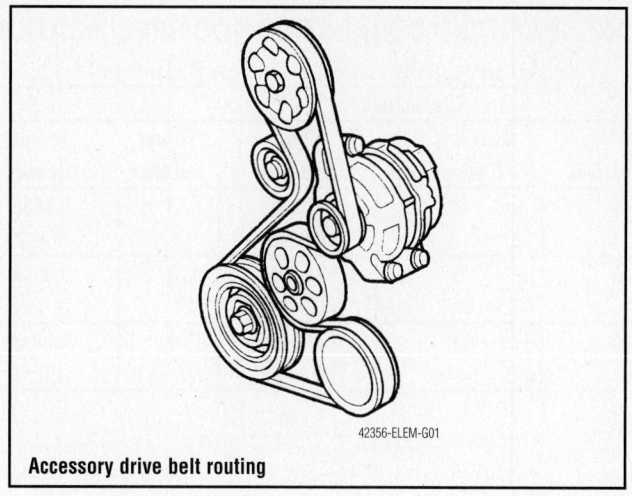

42356-ELEM-G01

Accessory drive belt routing

CAPACITIES

Year	Model	Engine Displacement Liters (VIN)	Engine Oil with Filter (qts.)	Transmission (pts.) 5-Spd	Transmission (pts.) Auto.	Transfer Case (pts.)	Drive Axle Front (pts.)	Drive Axle Rear (pts.)	Fuel Tank (gal.)	Cooling System (qts.)
2003	Element	2.4 (K24A4)	4.4	4.0	①	②	②	2.2	15.9	③
2004	Element	2.4 (K24A4)	4.4	4.0	①	②	②	2.2	15.9	③
2005-06	Element	2.4 (K24A4)	4.4	4.0	①	②	②	2.2	15.9	③

NOTE: All capacities are approximate. Add fluid gradually and check to be sure a proper fluid level is obtained.

① 2WD: 6.6 pts. for fluid change, 15.2 for overhaul

    4WD: 6.3 pts. for fluid change, 13.8 for overhaul

② Included in transaxle refill figure

③ Manual trans: 7.6 qts

    Auto trans: 7.5 qts.

09474_ELEM_C0004

VALVE SPECIFICATIONS

Year	Engine Displacement Liters (VIN)	Seat Angle (deg.)	Face Angle (deg.)	Spring Test Pressure (lbs. @ in.)	Spring Installed Height (in.)	Stem-to-Guide Clearance (in.) Intake	Stem-to-Guide Clearance (in.) Exhaust	Stem Diameter (in.) Intake	Stem Diameter (in.) Exhaust
2003	2.4 (K24A4)	NA	NA	NA	①	0.0012-0.0022	0.0022-0.0031	0.2156-0.2159	0.2146-0.2150
2004	2.4 (K24A4)	NA	NA	NA	①	0.0012-0.0022	0.0022-0.0031	0.2156-0.2159	0.2146-0.2150
2005-06	2.4 (K24A4)	NA	NA	NA	①	0.0012-0.0022	0.0022-0.0031	0.2156-0.2159	0.2146-0.2150

NA: Not Available

① Valve spring free length:

    Intake: 1.668 in.

    Exhaust: 1.745 in.

09474_ELEM_C0005

CRANKSHAFT AND CONNECTING ROD SPECIFICATIONS

All measurements are given in inches

Year	Engine Displacement Liters (VIN)	Crankshaft				Connecting Rod		
		Main Brg. Journal Dia.	Main Brg. Oil Clearance	Shaft End-play	Thrust on No.	Journal Diameter	Oil Clearance	Side Clearance
2003	2.4 (K24A4)	①	②	0.0040-0.0140	3	1.8888-1.8898	0.0008-0.0019	0.016
2004	2.4 (K24A4)	①	②	0.0040-0.0140	3	1.8888-1.8898	0.0008-0.0019	0.016
2005-06	2.4 (K24A4)	①	②	0.0040-0.0140	3	1.8888-1.8898	0.0008-0.0019	0.016

① Except No. 3: 2.1648-2.1657
   No. 3: 2.1644-2.1654

② Except No. 3: 0.0007-0.0016
   No. 3: 0.0010-0.0019

09474_ELEM_C0006

PISTON AND RING SPECIFICATIONS

All measurements are given in inches

Year	Engine Displacement Liters (VIN)	Piston Clearance	Ring Gap			Ring Side Clearance		
			Top Compression	Bottom Compression	Oil Control	Top Compression	Bottom Compression	Oil Control
2003	2.4 (K24A4)	0.0008-0.0016	0.0080-0.0140	0.0160-0.0220	0.0080-0.0280	0.0018-0.0028	0.0020-0.0030	NA
2004	2.4 (K24A4)	0.0008-0.0016	0.0080-0.0140	0.0160-0.0220	0.0080-0.0280	0.0018-0.0028	0.0020-0.0030	NA
2005-06	2.4 (K24A4)	0.0008-0.0016	0.0080-0.0140	0.0160-0.0220	0.0080-0.0280	0.0018-0.0028	0.0020-0.0030	NA

NA: Not Applicable

09474_ELEM_C0007

TORQUE SPECIFICATIONS

All readings in ft. lbs.

Year	Engine Displacement Liters (VIN)	Cylinder Head Bolts	Main Bearing Bolts	Rod Bearing Bolts	Crankshaft Damper Bolts	Flywheel Bolts	Manifold		Spark Plugs	Oil Pan Drain Plug
							Intake	Exhaust		
2003	2.4 (K24A4)	①	②	③	④	76	16	33	13	33
2004	2.4 (K24A4)	①	②	③	④	76	16	33	13	33
2005-06	2.4 (K24A4)	①	②	③	④	76	16	33	13	33

NOTE: Dip main bearing bolts and crankshaft damper bolt in clean engine oil prior to tightening.

① Step 1: 29 ft. lbs.
   Step 2: +90 degrees
   Step 3: +90 degrees
   Step 4: NEW BOLT ONLY +90 degrees

④ 36 ft. lbs. +90 degrees

② 22 ft. lbs. +56 degrees

③ 14 ft. lbs. +90 degrees

09474_ELEM_C0008

WHEEL ALIGNMENT

Year	Model		Caster Range (+/-Deg.)	Caster Preferred Setting (Deg.)	Camber Range (+/-Deg.)	Camber Preferred Setting (Deg.)	Toe-in (in.)
2003	Element	F	1.00	+1.75	0.75	0	0+/-0.08
		R	—	—	0.75	-1.00	0.08+/-0.08
2004	Element	F	1.00	+1.75	0.75	0	0+/-0.08
		R	—	—	0.75	-1.00	0.08+/-0.08
2004-06	Element	F	1.00	+1.75	0.75	0	0+/-0.08
		R	—	—	0.75	-1.00	0.08+/-0.08

09474_ELEM_C0009

TIRE, WHEEL AND BALL JOINT SPECIFICATIONS

Year	Model	OEM Tires Standard	OEM Tires Optional	Tire Pressures (psi) Front	Tire Pressures (psi) Rear	Wheel Size	Ball Joint Inspection	Lug Nut
2003	Element	P215/70R16	None	26	26	6JJ	NS	80
2004	Element	P215/70R16	None	26	26	6JJ	NS	80
2005-06	Element	P215/70R16	None	26	26	6JJ	NS	80

OEM: Original Equipment Manufacturer

PSI: Pounds Per Square Inch

NS: Not specified by manufacturer

09474_ELEM_C0010

BRAKE SPECIFICATIONS
HONDA ELEMENT
All measurements in inches unless noted

Year	Model		Brake Disc Original Thickness	Brake Disc Minimum Thickness	Brake Disc Maximum Runout	Brake Drum Diameter Original Inside Diameter	Max. Wear Limit	Maximum Machine Diameter	Minimum Lining Thickness Front	Minimum Lining Thickness Rear	Brake Caliper Bracket Bolts (ft. lbs.)	Brake Caliper Mounting Bolts (ft. lbs.)
2003	Element	F	0.910	0.830	0.004	—	—	—	0.060	—	80	25
		R	0.350	0.300	0.004	—	—	—	0.060	—	41	16
2004	Element	F	0.910	0.830	0.004	—	—	—	0.060	—	80	25
		R	0.350	0.300	0.004	—	—	—	0.060	—	41	16
2005-06	Element	F	0.910	0.830	0.004	—	—	—	0.060	—	80	25
		R	0.350	0.300	0.004	—	—	—	0.060	—	41	16

F: Front

R: Rear

09474_ELEM_C0011

SCHEDULED MAINTENANCE INTERVALS

HONDA—ELEMENT

TO BE SERVICED	TYPE OF SERVICE	VEHICLE MILEAGE INTERVAL (x1000)											
		10	20	30	40	50	60	70	80	90	100	110	120
Accessory drive belts	I & A			✓			✓			✓			✓
Air cleaner element	R			✓			✓			✓			✓
Air conditioning filter	R			✓			✓			✓			✓
Brake fluid	R											✓	
Brake hoses & lines (including ABS)	I		✓		✓		✓		✓		✓		
Cooling system hoses & connections	I		✓		✓		✓		✓		✓		
Engine coolant	R												✓
Engine oil	R	✓	✓	✓	✓	✓	✓	✓	✓	✓	✓	✓	✓
Engine oil and coolant levels	I	Inspect at each fuel stop											
Engine oil filter	R		✓		✓		✓		✓		✓		
Exhaust system	I		✓		✓		✓		✓		✓		
Fluid levels and condition	I		✓		✓		✓		✓		✓		
Front and rear brakes	I		✓		✓		✓		✓		✓		
Fuel lines & connection	I		✓		✓		✓		✓		✓		
Halfshaft boots	I		✓		✓		✓		✓		✓		
Idle speed	I & A											✓	
Parking brake system	I & A		✓		✓		✓		✓		✓		
Rear differential fluid	R										✓		
Rotate and inspect tires	I	✓	✓	✓	✓	✓	✓	✓	✓	✓	✓	✓	✓
Spark plugs	R											✓	
Suspension components	I		✓		✓		✓		✓		✓		
Tie rod ends, steering gear box & boots	I		✓		✓		✓		✓		✓		
Transmission fluid	R												✓
Valve clearance	I											✓	

R: Replace   I: Inspect   A: Adjust

FREQUENT OPERATION MAINTENANCE (SEVERE SERVICE)

If a vehicle is operated under any of the following conditions it is considered severe service:

- Towing a trailer or using a camper or car-top carrier.

- Repeated short trips of less than 5 miles in temperatures below freezing, or trips of less than 10 miles in any temperature.

- Extensive idling or low-speed driving for long distances as in heavy commercial use, such as delivery, taxi or police cars.

- Operating on rough, muddy or salt-covered roads.

- Operating on unpaved or dusty roads.

- Driving in extremely hot (over 90°) conditions.

Air cleaner element: replace every 15,000 miles

Engine oil and filter: replace every 3750 miles or 6 months, whichever occurs first.

Timing belt: replace every 60,000 miles if the vehicle is regularly driven in temperatures above 110°F or below -20°F.

Transmission fluid: replace every 30,000 miles.

Rear differential fluid: replace every 60,000 miles.

Front and rear brakes: inspect every 7500 miles or 6 months, whichever occurs first.

Locks and hinges: lubricate every 15,000 miles.

Tie rods, steering gear box, boots: inspect every 7500 miles or 6 months, whichever occurs first.

Suspension components: inspect every 7500 miles or 6 months, whichever occurs first.

Halfshaft boots: inspect every 7500 miles or 6 months, whichever occurs first.

ENGINE REPAIR

➡️**Disconnecting the negative battery cable on some vehicles may interfere with the functions of the on board computer system. The computer may undergo a relearning process once the negative battery cable is reconnected.**

Distributor

The 2.4L engine does not have a distributor.

Alternator

REMOVAL

1. Before servicing the vehicle, refer to the precautions section.
2. Remove or disconnect the following:
   - Negative, then the positive battery cables
   - Accessory drive belt
   - Auto-tensioner
   - Alternator wiring harness connectors and harness clamp
   - Positive Crankcase Ventilation (PCV) valve
   - 3 bolts holding the alternator
   - Alternator

INSTALLATION

1. Install or connect the following:
   - Alternator. Tighten the bolts to 16 ft. lbs. (22 Nm).
   - PCV valve
   - Alternator wiring harness connectors and harness clamp
   - Auto tensioner
   - Accessory drive belt
   - Negative battery cable

Ignition Timing

ADJUSTMENT

Adjustment is not possible on the 2.4L engine.

Engine Assembly

REMOVAL & INSTALLATION

➡️**The engine and transaxle are removed from the vehicle as a unit.**

1. Before servicing the vehicle, refer to the precautions section.

2. Drain the cooling system.
3. Drain the transaxle fluid.
4. Drain the engine oil.
5. Relieve fuel system pressure.
6. Remove or disconnect the following:
   - Negative, then the positive battery cables
   - IAT sensor connector
   - Intake Air Temperature (IAT) sensor connector
   - Breather and vacuum hoses
   - Intake duct
   - Battery
   - Air cleaner housing
   - Battery tray
   - Cables from the power distribution center
   - Ground cable
   - Throttle and cruise cables
   - Fuel lines
   - EVAP canister
   - Powertrain Control Module (PCM) connectors and grommet. Pull the PCM harness through the firewall.
   - Clutch slave cylinder and line mounting bracket if equipped with a manual transmission
   - Shift cable if equipped with a manual transmission
   - Drive belt
   - Power steering pump, leaving the hoses connected
   - Power steering hose from the bracket on the valve cover
   - Wheels
   - Splash shield
   - Air/Fuel ratio (A/F) sensor connector
   - Secondary Heated Oxygen (O_2s) sensor connector
   - Catalytic converter
   - Stabilizer links
   - Lower ball joints
   - Halfshafts
   - Shift cable on models with an automatic transmission
   - Propeller shaft on 4 wheel drive models
   - Transmission cooler lines and ATF filter mounting if equipped with an automatic transmission
   - Radiator and heater hoses
   - Bolt attaching the power steering line if equipped with an automatic transmission
7. Support the transmission with a jack and block of wood if equipped with an automatic transmission.
   - Transmission mount

8. Attach a hoist to the engine lifting eyes and support the powertrain weight.
9. Remove the jack and block of wood if equipped with an automatic transmission.
   - Transmission mount bracket
   - Upper engine mount bracket retainers
   - Rear mount bracket bolts.
   - Front mount bolt
10. Matchmark the sub-frame mounting bolt centers.

There is a special tool necessary for sub-frame removal. The Honda tool number is EQS02C000011. Attach the tool as explained in the tool instructions, attach a floor jack with adapter, remove the 4 sub-frame bolts and lower the sub-frame.

11. Remove or disconnect the following:
12. A/C compressor without disconnecting the hoses
13. Check that all hoses and wires are disconnected.
14. Lower the engine about 6 inches and recheck all clearances.
15. Lower the engine all the way.
16. Remove the chain hoist.

To install:

17. Installation is the reverse of removal. Observe the following torques:
    - Front engine mount bracket bolts: 33 ft. lbs. (44Nm)
    - A/C compressor bracket: 33 ft. lbs. (44Nm)
    - Stiffener 10mm bolts: 33 ft. lbs. (44Nm); 6mm bolts 9 ft. lbs. (12 Nm)
    - A/C compressor bolts: 16 ft. lbs. (22 Nm)
    - Subframe bolts: 76 ft. lbs. (98 Nm)
    - Upper bracket bolt and nut: 45 ft. lbs. (61 Nm)
    - Transmission mount bracket support bolts/nuts: 40 ft. lbs. (54 Nm)
    - PS pump bolts: 16 ft. lbs. (22 Nm)

➡️**Use new self-locking nuts and color-coded self-locking bolts when installing the engine mounts and suspension components.**

➡️**Do not tighten the engine or transaxle mount fasteners until instructed to do so.**

18. Lower the powertrain into position.
19. Install or connect the following:
    - Transaxle mount and bracket. Tighten the frame mounting bolts to 47 ft. lbs. (64 Nm).

- Upper bracket. Tighten the nuts in sequence to 54 ft. lbs. (74 Nm).
- Rear mount bracket through bolt
- Right front mount and bracket
- Left front mount and bracket

20. Tighten the remaining mount fasteners as follows:

   a. Transaxle mount fasteners to 47 ft. lbs. (64 Nm) and the through bolt to 54 ft. lbs. (74 Nm).

   b. Rear mount bracket through bolt to 43 ft. lbs. (59 Nm).

   c. Right front mount 12mm bolts to 47 ft. lbs. (64 Nm) and the 10mm bolts to 33 ft. lbs. (44 Nm).

   d. Left front mount 12mm stud bolt to 61 ft. lbs. (83 Nm), 10mm bolts to 33 ft. lbs. (44 Nm), and 12mm nut to 43 ft. lbs. (59 Nm).

   e. Right front mount 12mm nut to 43 ft. lbs. (59 Nm).

21. Install or connect the following:
- Propeller shaft, if equipped
- A/C compressor
- Radiator
- A/C hose clamp

22. If equipped with a manual transaxle, install or connect the following:
- Shift cables
- Transaxle ground cable
- Clutch hose bracket
- Clutch slave cylinder

23. If equipped with an automatic transaxle, install or connect the following:
- Transaxle fluid cooler lines
- Transaxle ground cable and hose clamp
- Shift cable
- Shift cable cover

24. For all vehicles, install or connect the following:
- Axle halfshafts
- Lower ball joints
- Right damper fork
- Exhaust front pipe
- HO2S connector
- Heater hoses
- Radiator hoses
- Splash shield
- PSP switch
- Accelerator cable
- Brake booster vacuum line
- Fuel lines
- A/C compressor drive belt
- Power steering pump and belt
- Cruise control actuator
- Left engine wire harness connectors
- PCM connectors and grommet
- Air intake assembly
- Battery and tray

- Fuse/Relay box battery cables
- Negative battery cable

25. Fill the engine crankcase to the correct level.

26. Fill the transaxle to the correct level.

27. Fill the cooling system.

28. Start the engine and check for leaks.

29. Check the wheel alignment and adjust as necessary.

Water Pump

REMOVAL & INSTALLATION

1. Before servicing the vehicle, refer to the precautions section.
2. Drain the cooling system.
3. Remove or disconnect the following:
- Negative battery cable
- Accessory drive belt
- Crankshaft pulley
- Water pump (6 bolts)

To install:

4. Clean the water pump mating surfaces.
5. Install or connect the following:
- Water pump with a new O-ring. Torque the bolts to 8.7 ft. lbs. (12 Nm).
- Crankshaft pulley
- Accessory drive belt
- Negative battery cable

6. Refill the engine cooling system.

Heater Core

REMOVAL & INSTALLATION

1. Before servicing the vehicle, refer to the precautions section.
2. Disconnect the negative battery cable.
3. Drain the cooling system into a clean container for reuse.
4. Disconnect the A/C lines from the evaporator core if equipped with A/C.
5. Open the cable clamp (A) and disconnect the heater valve cable (B) from the heater valve arm (C). Turn the heater valve arm to the full open position as illustrated.

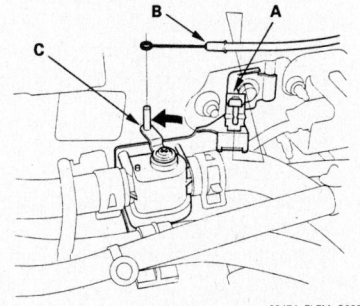

09474_ELEM_G0001

Open the cable clamp (A) and disconnect the heater valve cable (B) from the heater valve arm (C). Turn the heater valve arm to the full open position

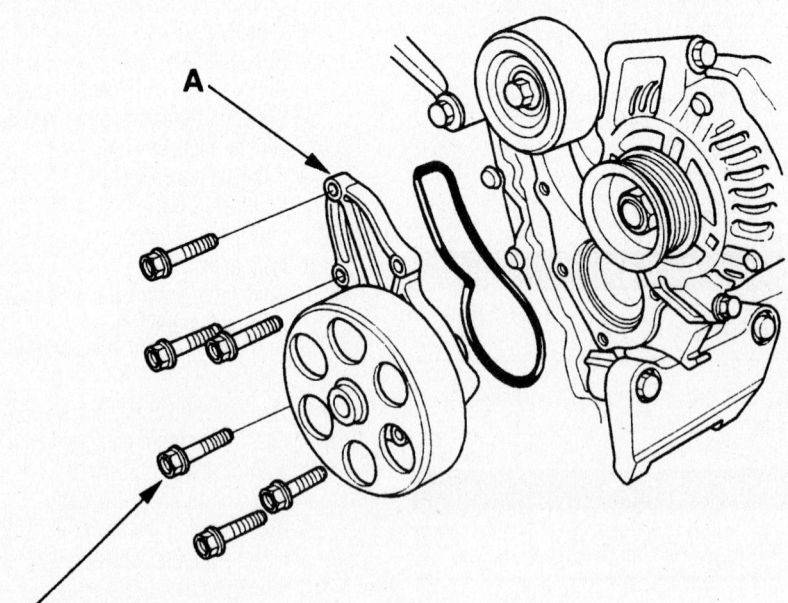

**6 x 1.0 mm
12 N·m (1.2 kgf·m, 8.7 lbf·ft)**

42356-ELEM-G02

Exploded view of the water pump mounting

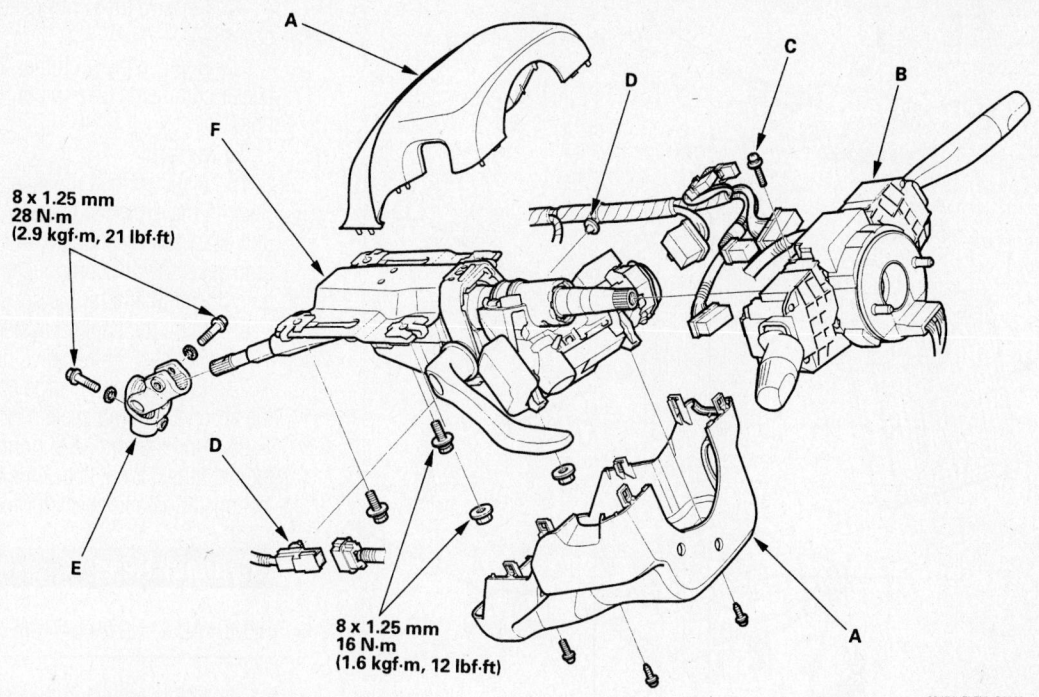

8 x 1.25 mm
28 N·m
(2.9 kgf·m, 21 lbf·ft)

8 x 1.25 mm
16 N·m
(1.6 kgf·m, 12 lbf·ft)

09474_ELEM_G0002

Exploded view of the steering column assembly

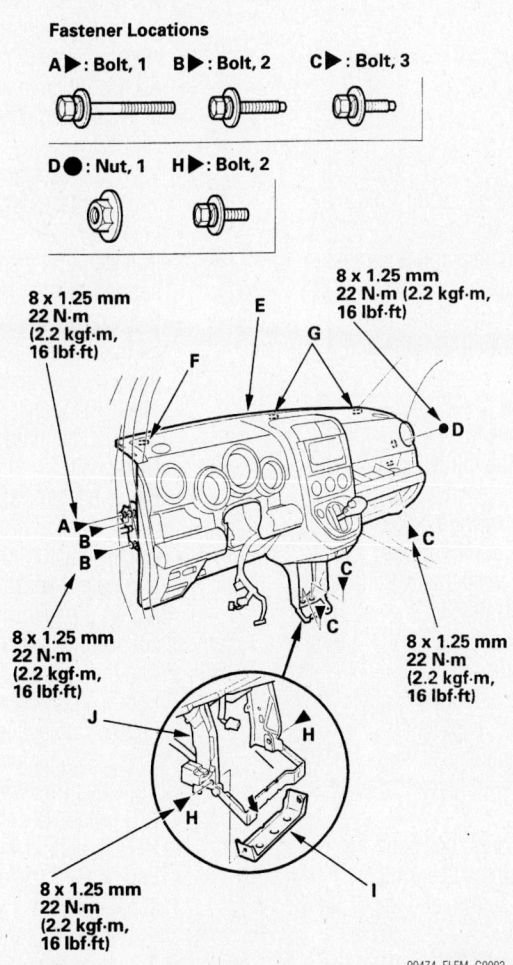

Fastener Locations

A ▶ : Bolt, 1   B ▶ : Bolt, 2   C ▶ : Bolt, 3

D ● : Nut, 1   H ▶ : Bolt, 2

8 x 1.25 mm
22 N·m
(2.2 kgf·m,
16 lbf·ft)

8 x 1.25 mm
22 N·m (2.2 kgf·m,
16 lbf·ft)

8 x 1.25 mm
22 N·m
(2.2 kgf·m,
16 lbf·ft)

8 x 1.25 mm
22 N·m
(2.2 kgf·m,
16 lbf·ft)

8 x 1.25 mm
22 N·m
(2.2 kgf·m,
16 lbf·ft)

09474_ELEM_G0003

Exploded view of the dashboard, retainers and the retainer torque specifications

6. Remove the heater hoses from the heater core

7. Remove the mounting bolt and heater valve. Remove the nut from the heater unit being careful of any lines, hoses and wiring in the vicinity.

8. Remove the dashboard as follows:

a. Remove the driver's side lower instrument panel cover clips and remove the lower cover.

b. Remove the glove box stops from each side of the glove box.

c. Remove the glove box-to-instrument panel bolts and the glove box.

d. Remove the passengers side lower instrument panel cover clips and remove the lower cover.

e. Remove the center lower cover clips and remove the lower cover.

f. Remove the passenger vent.

g. Remove the A–trim trim on both sides.

h. Remove the passenger side kick panel.

i. Remove the steering wheel.

j. Remove the steering column covers.

k. Disconnect the wiring from the combination switch and remove the assembly by removing the screw on top of the switch.

l. Disconnect the ignition switch connectors and release the wire harness clips from the column.

m. Disconnect the steering joint bolt and disconnect it from the column shaft.

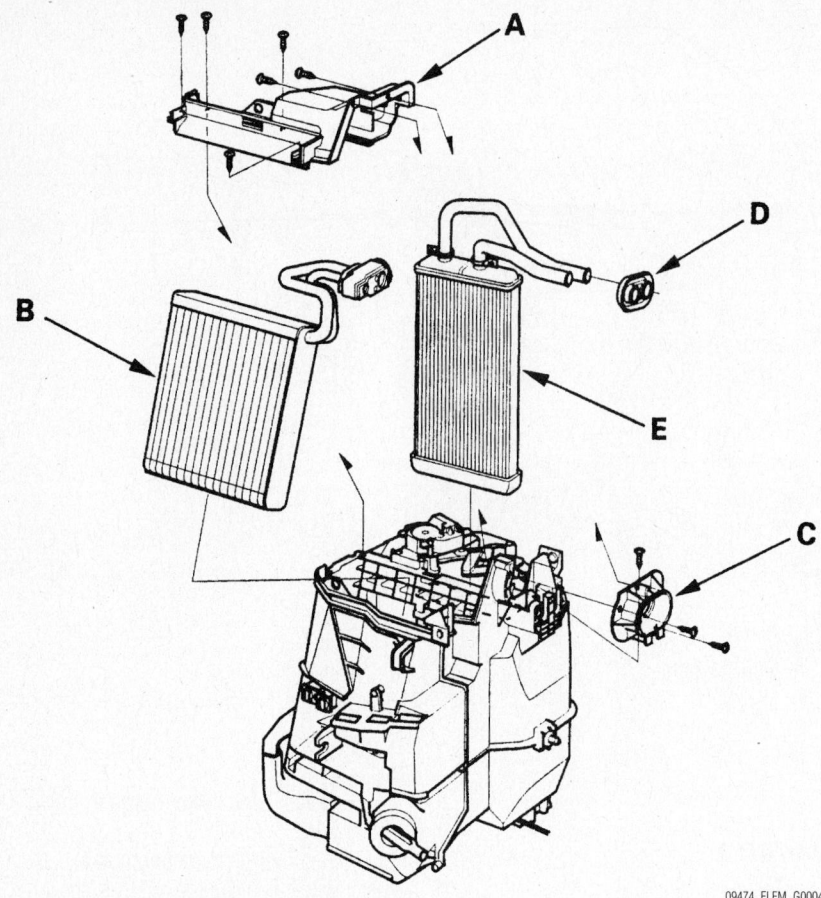

09474_ELEM_G0004

Exploded view of the heater unit assembly

n. Remove the steering column retainers and the column.

o. Control cable if equipped with an automatic transmission or shift cable if equipped with a manual transaxle.

p. Woofer, if equipped.

q. On the drivers side disconnect the following:
- Tweeter connector
- Drivers door wiring connectors
- Brake switch connector
- Clutch switch connector, if equipped
- Engine compartment harness connectors from the fuse/relay box

r. In the middle of the dashboard, disconnect the SRS control unit connector, floor harness connector and engine compartment harness connectors.

s. On the passenger side, disconnect the following:
- Passenger door wiring connectors
- Antenna lead
- PCM connectors
- Engine wire harness connectors
- Heater sub-harness connectors
- Passenger airbag connectors.
- Amplifier connectors

- Wire harness protector from the amplifier, if equipped

t. Remove any remaining harness and connector clips.

u. Dashboard bolts. Refer to the exploded view for bolt location and torque values.

v. Remove the dashboard.

9. Remove the PCM.

10. Disconnect the following connectors:
- Dashboard wiring harness
- Air mixture control motor
- Evaporator temperature sensor
- Power transistor
- Mode control motor
- Blower motor

11. Disconnect the following clips:
- Wire harness clips
- Connector clips

12. Remove the wire harness, heater duct and clip.

13. Remove the drain hose and the mounting nuts and the heater unit.

14. Remove the screws and the expansion valve cover (A).

15. If equipped with A/C, remove the evaporator core (B).

16. Remove the screws and the flange cover (C).

17. Remove the grommet (D) and the heater core being careful not to damage any lines.

To install:

18. Installation is the reverse of removal. Refer to the exploded views of the heater unit assembly, dashboard and steering column assembly for component location, fastener location and torque specifications.

19. Refill the cooling system.

20. Connect the negative battery cable.

21. Evacuate and charge and leak test the air conditioning system refrigerant.

22. Run the engine to normal operating temperatures; then, check the climate control operation and check for leaks.

Cylinder Head

REMOVAL & INSTALLATION

1. Before servicing the vehicle, refer to the precautions section.

2. Drain the cooling system.

3. Relieve the fuel system pressure.

4. Remove or disconnect the following:
- Negative battery cable
- Accessory drive belt
- Intake Air Temperature (IAT) sensor connector
- Vacuum hoses and breather pipe and air intake duct
- Fuel feed hose
- Bolt securing the connecting pipe support bracket to the engine block
- Evaporative emission (EVAP) canister hose and brake booster vacuum hose
- Intake manifold
- Exhaust manifold
- Cam chain
- Positive Crankcase Ventilation (PCV) hose and ground cable
- Upper radiator hose, heater hoses and water bypass hose

5. Remove the following engine wire harness connectors and wire harness clamps from the cylinder head:
- Four injector connector
- Engine Coolant Temperature (ECT) sensor connector
- Camshaft Position (CMP) sensor A & B (intake & exhaust) connectors
- VTEC solenoid valve connector
- Engine Oil Pressure (EOP) sensor connector

6. Remove or disconnect the following:
- 3 bolts holding the EVAP canister purge valve bracket and remove the

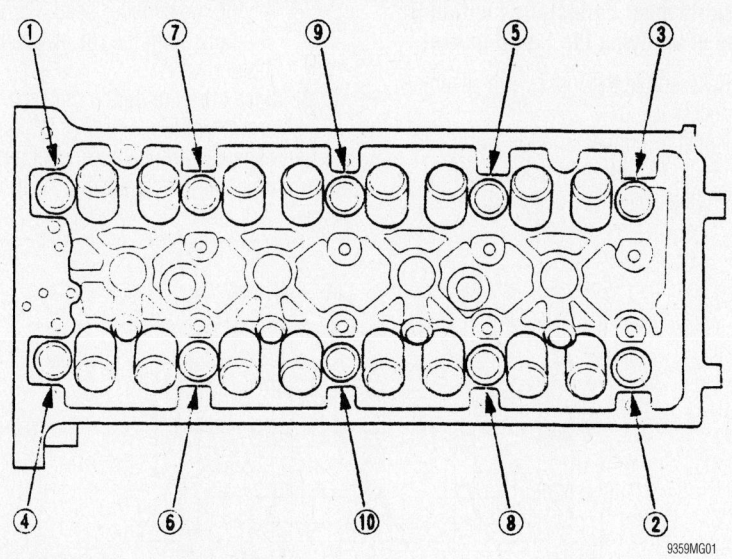

Cylinder head bolt loosening sequence

9359MG01

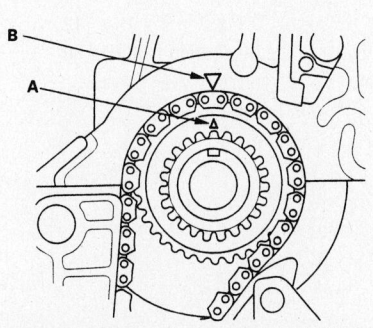

42356-ELEM-G03

Set the crankshaft to TDC by aligning the mark (A) on the crankshaft sprocket with the pointer (B) on the cylinder block

two bolts (B) securing the harness bracket
- Timing (cam) chain
- Rocker arm assembly

7. Loosen the cylinder head bolts in sequence and ⅓ turns until all bolts are loose.

8. Remove the cylinder head.

To install:

9. Be sure all cylinder head and block gasket surfaces are clean. Check the cylinder head for warpage. If warpage is less than 0.002 in. (0.05mm), cylinder head resurfacing is not required. Maximum resurface limit is 0.008 in. (0.2mm) based on a cylinder head height of 3.94 in. (100mm).

10. Install or connect the following:
- New gasket and dowel pins on the cylinder block

11. Set the crankshaft to Top Dead Center (TDC). Align the TDC mark (A) on the crankshaft sprocket with the pointer (B) on the cylinder block.

12. Measure the diameter of each cylinder head bolt at points A & B, as shown in the illustration. If either diameter is less than 0.42 in. (10.6mm), replace the head bolt

13. Apply engine oil to the threads and under the bolt heads of all of the bolts.

14. Install the cylinder head. Tighten the bolts in sequence as follows:
   a. Step 1: 29 ft. lbs. (39 Nm).
   b. Step 2: Plus 90 degrees.
   c. Step 3: Plus 90 degrees.
   d. Step 4: If using new cylinder head bolts, add an additional 90 degrees.

15. The remainder of installation is the reverse of removal.

16. Fill the cooling system.

17. Connect the negative battery cable and enter the radio security code.

18. Start the engine and check carefully for any leaks.

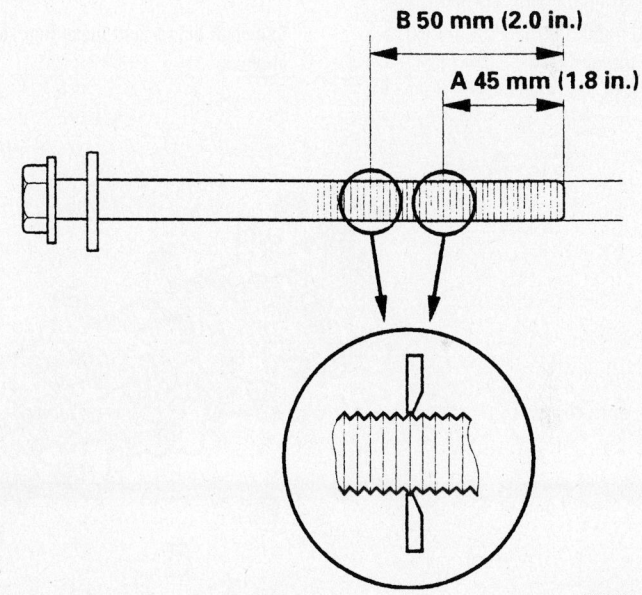

9359MG02

Cylinder head bolt inspection

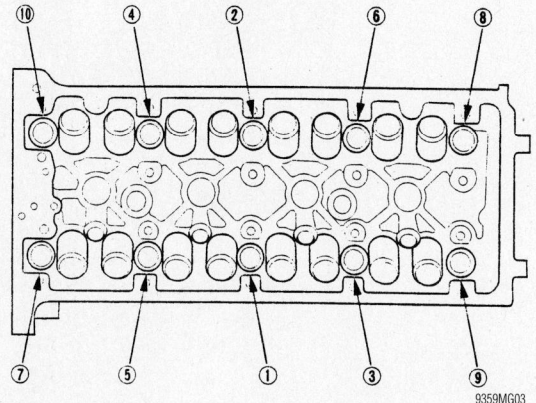

9359MG03

Cylinder head bolt torque sequence

Rocker Arms/Shafts

REMOVAL & INSTALLATION

1. Before servicing the vehicle, refer to the precautions section.
2. Remove or disconnect the following:
   - Timing (cam) chain
   - Loosen the rocker arm adjusting screws
   - Camshaft holder bolts, two turns at a time in sequence
   - Timing chain guide (B), camshaft holders and camshafts
3. Insert the bolts (A) into the rocker shaft holder, then remove the rocker arm assembly (B)

To install:

4. Clean and dry the No. 5 rocker shaft holding mating surface.
5. Apply a suitable liquid gasket P/N 08718-0009, or equivalent, evenly to the cylinder head mating surface of the No. 5 rocker shaft holder.

➥**The parts must be installed within 5 minutes of applying the liquid gasket.**

6. Reassemble the rocker arm assembly, as necessary.
7. Install or connect the following
   - Bolts (A) into the rocker shaft holder, then the rocker arm assembly on the cylinder head. Remove the bolts from the rocker shaft holder.
8. Make sure the punch marks on the variable valve timing control (VTC) actuator and exhaust camshaft sprocket are facing up, then set the camshafts (A) in the holder

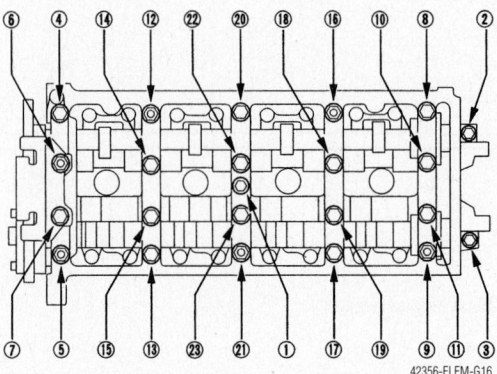

Camshaft holder bolt loosening sequence. Note that bolt 1 in the illustration is not on all engines.

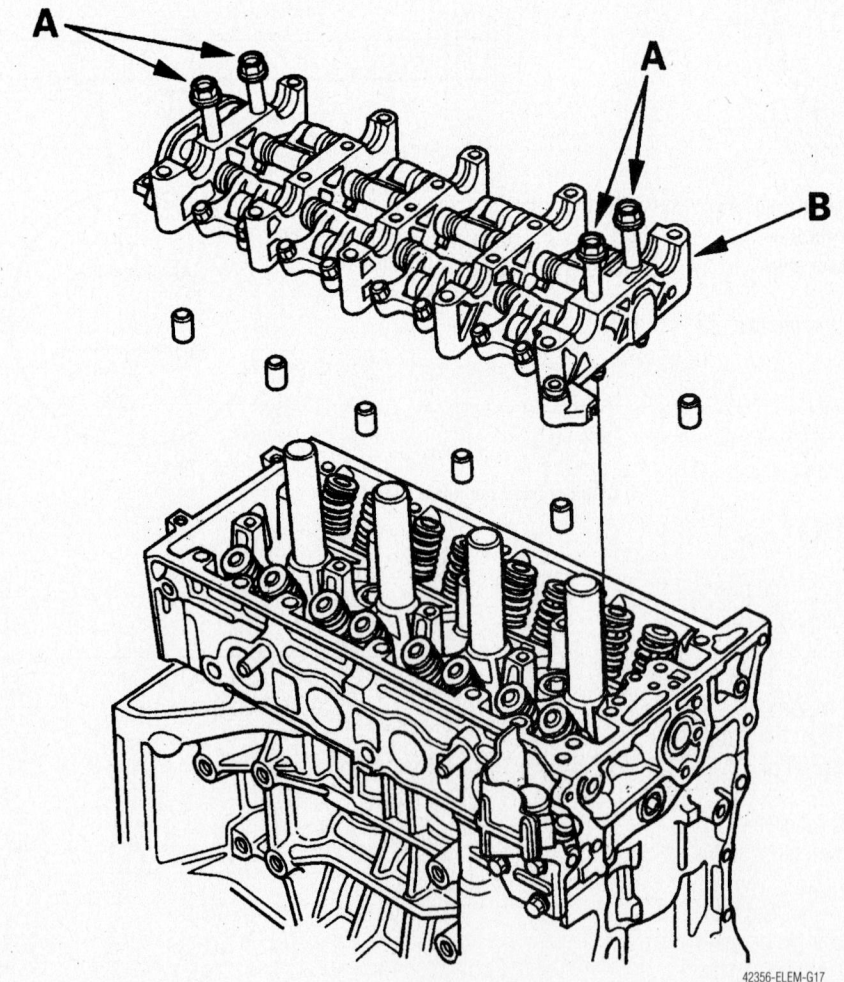

Insert the bolts (A) into the rocker shaft holder, then remove the rocker arm assembly (B)

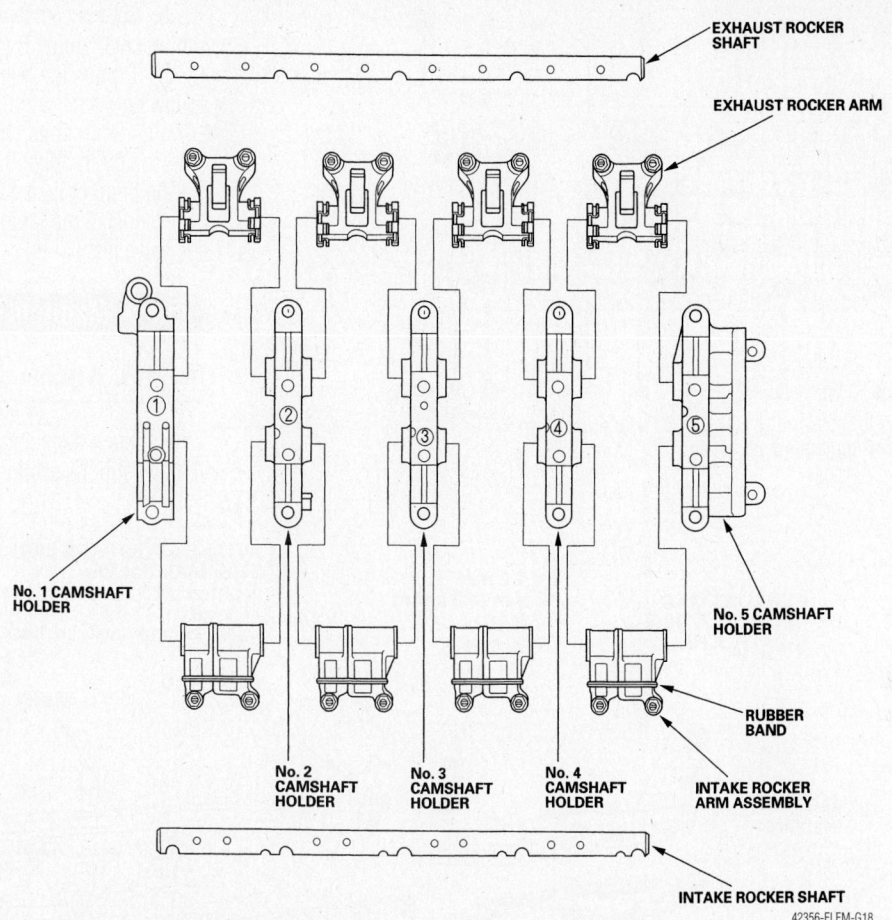

EXHAUST ROCKER SHAFT

EXHAUST ROCKER ARM

No. 1 CAMSHAFT HOLDER

No. 5 CAMSHAFT HOLDER

RUBBER BAND

No. 2 CAMSHAFT HOLDER

No. 3 CAMSHAFT HOLDER

No. 4 CAMSHAFT HOLDER

INTAKE ROCKER ARM ASSEMBLY

INTAKE ROCKER SHAFT

42356-ELEM-G18

Exploded view of the rocker arms and related components

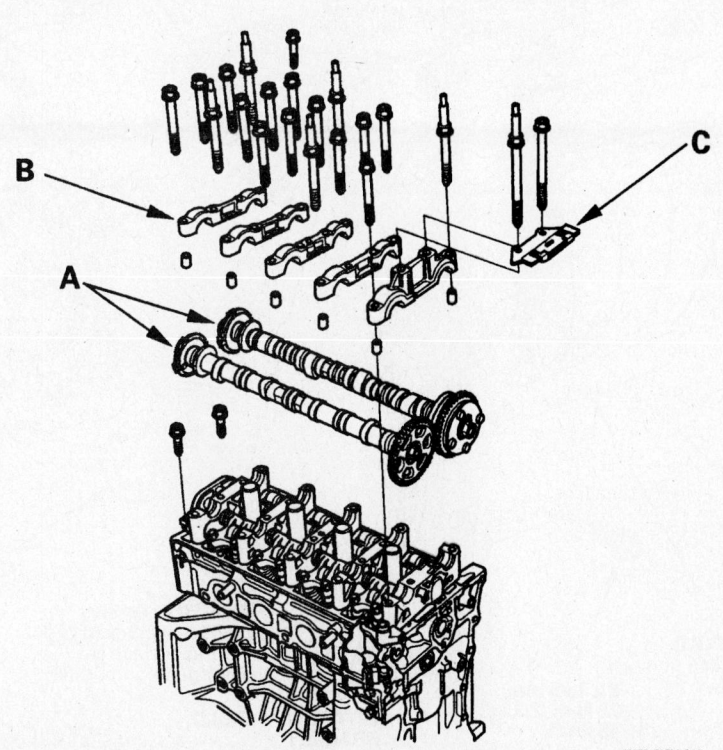

42356-ELEM-G19

When installing the camshafts (A) make sure the punch marks on the VTC actuator and exhaust cam sprockets are facing up

18 20 10 12 2 4 6 8 14 16 22
17 19 9 11 1 3 21 7 5 15 13 23
42356-ELEM-G20

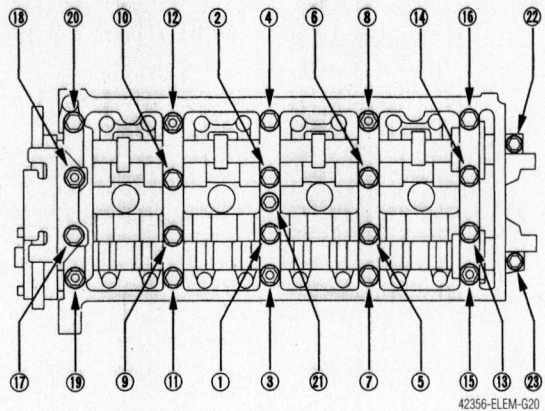

Rocker arm assembly bolt tightening sequence

9. Set the camshaft holders (B) and timing chain guide B (C) in place.

10. Tighten the bolts, in sequence, to the following specification:

a. 8mm bolts: 16 ft. lbs. (22 Nm).

b. 6mm bolts: 8.7 ft. lbs. (12 Nm).

The 6mm bolts are 21, 22 and 23.

11. Install the timing chain and adjust the valve lash.

Intake Manifold

REMOVAL & INSTALLATION

1. Before servicing the vehicle, refer to the precautions section.

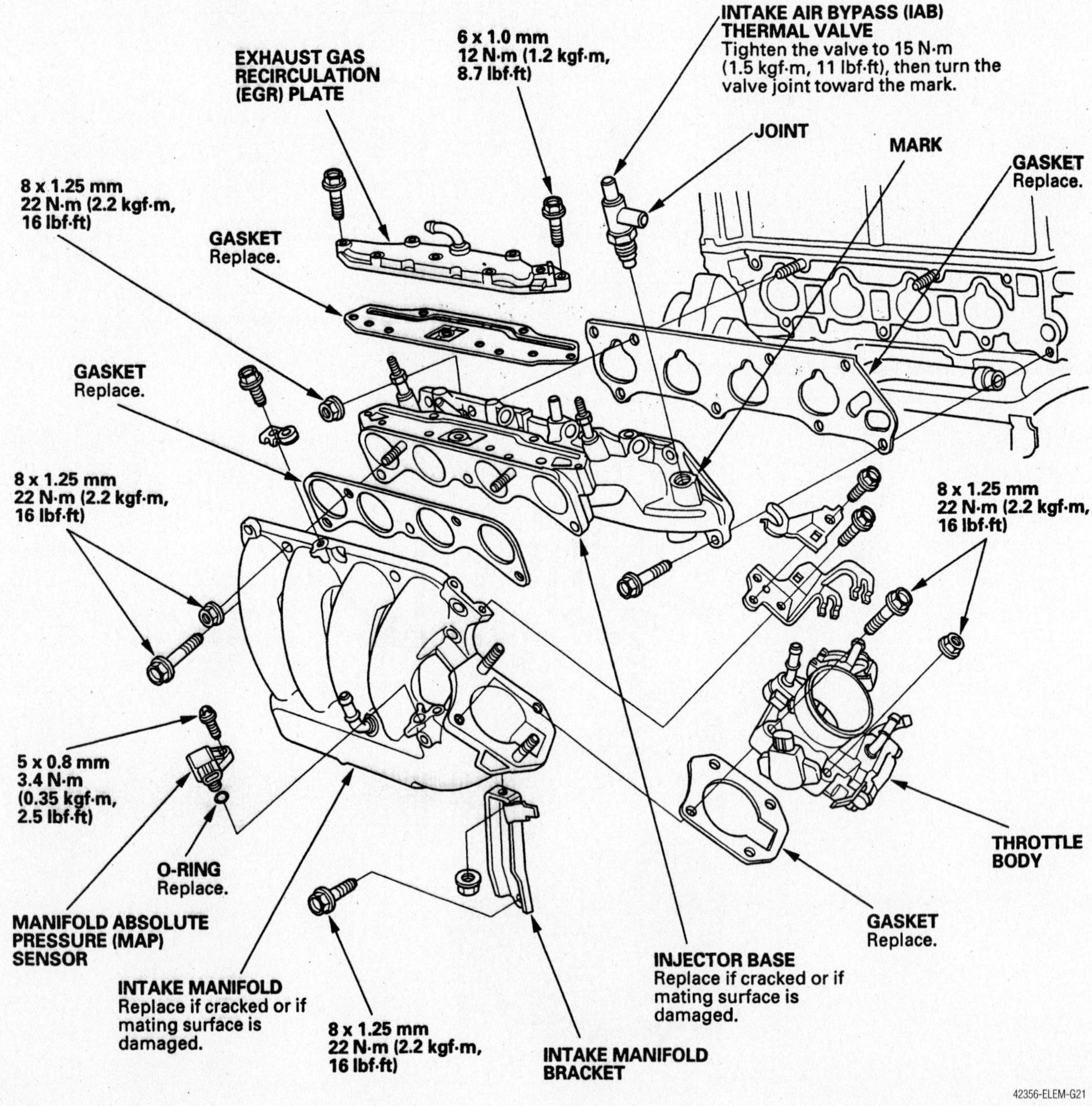

INTAKE AIR BYPASS (IAB) THERMAL VALVE
Tighten the valve to 15 N·m (1.5 kgf·m, 11 lbf·ft), then turn the valve joint toward the mark.
6 x 1.0 mm
12 N·m (1.2 kgf·m, 8.7 lbf·ft)
EXHAUST GAS RECIRCULATION (EGR) PLATE
JOINT
MARK
GASKET
Replace.
8 x 1.25 mm
22 N·m (2.2 kgf·m, 16 lbf·ft)
GASKET
Replace.
GASKET
Replace.
8 x 1.25 mm
22 N·m (2.2 kgf·m, 16 lbf·ft)
8 x 1.25 mm
22 N·m (2.2 kgf·m, 16 lbf·ft)
5 x 0.8 mm
3.4 N·m
(0.35 kgf·m, 2.5 lbf·ft)
O-RING
Replace.
THROTTLE BODY
MANIFOLD ABSOLUTE PRESSURE (MAP) SENSOR
GASKET
Replace.
INJECTOR BASE
Replace if cracked or if mating surface is damaged.
INTAKE MANIFOLD
Replace if cracked or if mating surface is damaged.
8 x 1.25 mm
22 N·m (2.2 kgf·m, 16 lbf·ft)
INTAKE MANIFOLD BRACKET
42356-ELEM-G21

Exploded view of the intake manifold and related components

2. Disconnect the negative battery cable.
3. Drain the engine coolant into a sealable container.
4. Remove or disconnect the following:
- Intake Air Temperature (IAT) sensor electrical connector
- Vacuum hose and breather pipe and the air intake duct
- Intake manifold cover
- Throttle and cruise control cables by loosening the locknuts, then slipping the cable ends out of the accelerator linkage.

➡**Do not bend the cables during removal. Always replace any throttle or cruise control cables that get kinked during removal.**

- Evaporative emission (EVAP) canister hose and brake booster vacuum hose
- Idle Air Control (IAC) valve connectors
- Throttle Position (TP) sensor connector

- Manifold Absolute Pressure (MAP) sensor connector
- Necessary engine wire harness connectors and wire harness clamps from the intake manifold
- Bolt securing the harness holder and remove the harness clamps
- Water bypass hoses, then plug them
- Harness clamp and harness connector from the intake manifold bracket
- Intake manifold bracket
- A/T vacuum hose
- Retainer and intake manifold

To install:
5. Clean the mounting surfaces.
6. Install or connect the following:
- New gasket
- Intake manifold. Tighten the bolts, in a criss-cross pattern beginning with the inner bolt, to 16 ft. lbs. (22 Nm).
- A/T vacuum hose
- Intake manifold bracket

- Harness clamp and connector to the intake manifold bracket
- Water bypass hoses
- Bolt securing the harness holder and tighten to 8.7 ft. lbs. (12 Nm)
- Harness clamps
- EVAP canister hose and brake booster vacuum hose
- Throttle and cruise control cables
- Intake manifold cover
- Intake air duct
- IAT sensor connector, vacuum hose and breather pipe
7. Refill the cooling system.
8. Connect the negative battery cable, start the engine, and check for leaks.

Exhaust Manifold

REMOVAL & INSTALLATION

1. Before servicing the vehicle, refer to the precautions section.
2. Raise and safely support the vehicle.

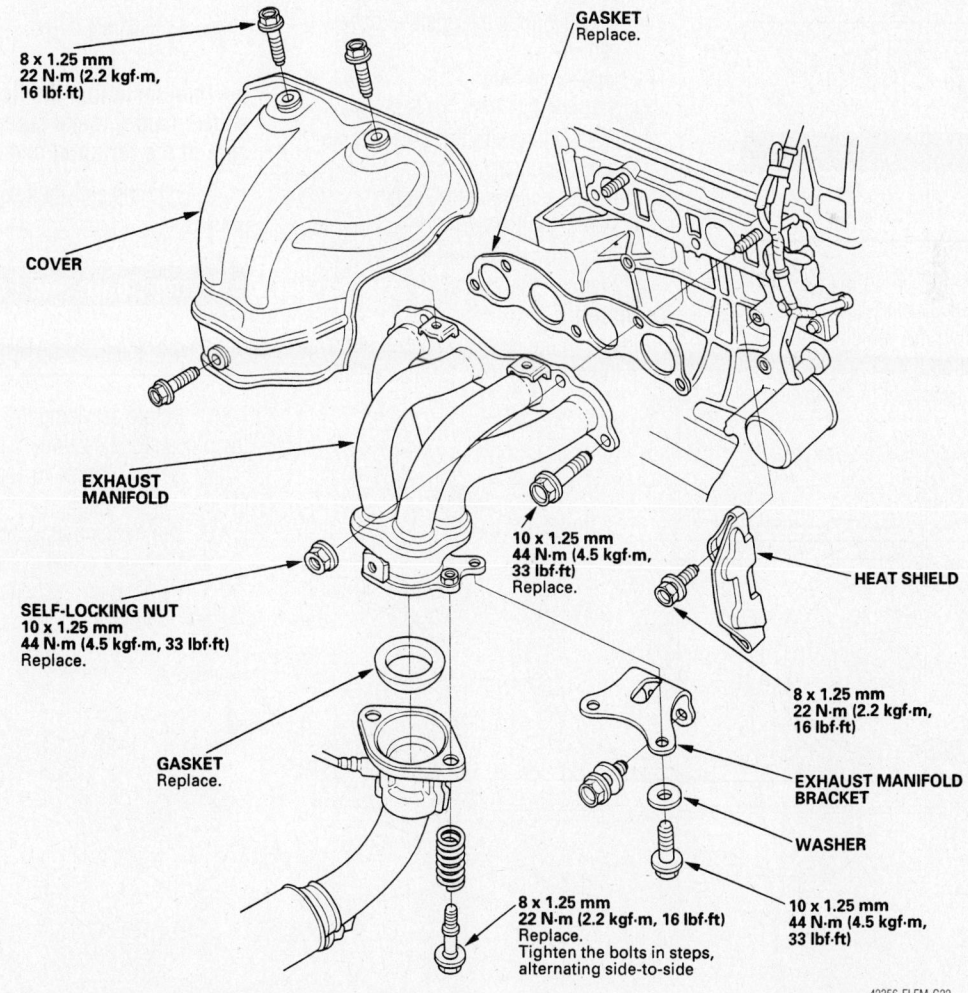

**8 x 1.25 mm
22 N·m (2.2 kgf·m, 16 lbf·ft)**

GASKET
Replace.

COVER

**EXHAUST
MANIFOLD**

**10 x 1.25 mm
44 N·m (4.5 kgf·m, 33 lbf·ft)
Replace.**

HEAT SHIELD

**SELF-LOCKING NUT
10 x 1.25 mm
44 N·m (4.5 kgf·m, 33 lbf·ft)
Replace.**

**8 x 1.25 mm
22 N·m (2.2 kgf·m, 16 lbf·ft)**

GASKET
Replace.

**EXHAUST MANIFOLD
BRACKET**

WASHER

**8 x 1.25 mm
22 N·m (2.2 kgf·m, 16 lbf·ft)
Replace.
Tighten the bolts in steps, alternating side-to-side.**

**10 x 1.25 mm
44 N·m (4.5 kgf·m, 33 lbf·ft)**

42356-ELEM-G22

Exploded view of the exhaust manifold and related components

3. Remove or disconnect the following:
- VTEC solenoid valve
- Intermediate shaft heat cover
- Cover and exhaust manifold bracket
- Exhaust manifold

To install:

4. Clean the mounting surfaces.
5. Install or connect the following:
- New gasket on the cylinder head
- Exhaust manifold. Tighten the nuts, in a criss-cross pattern starting with the inner nut, to 33 ft. lbs. (45 Nm).
- Exhaust manifold bracket and cover
- Intermediate shaft heat cover
- VTEC solenoid valve

Front Crankshaft Seal

REMOVAL & INSTALLATION

For the 2.4L engine, see the Timing Chain Removal & Installation procedure.

Camshaft

REMOVAL & INSTALLATION

See the Rocker Arm Shaft Removal & Installation procedure.

Valve Lash

ADJUSTMENT

Adjust the valves only when the cylinder head temperature is less than 100°F (38°C).

1. Before servicing the vehicle, refer to the precautions section.
2. Remove or disconnect the following:

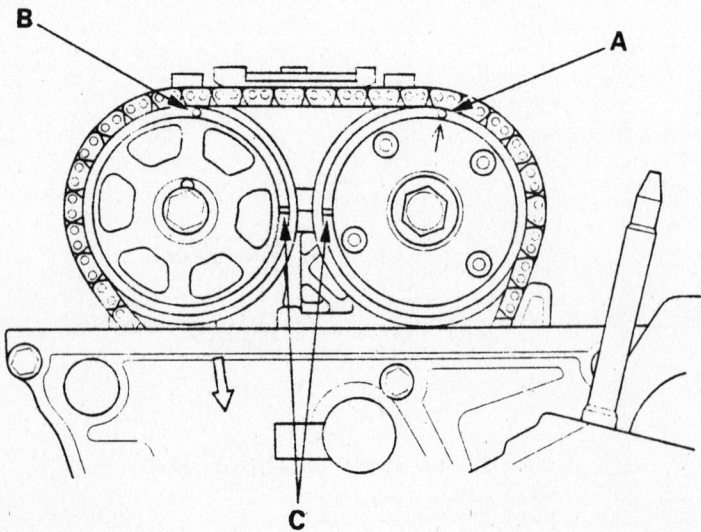

Align the timing marks

- Negative battery cable
- Cylinder head cover

3. Set the timing marks as shown in the illustration with N0.1 at TDC. Check all clearances. Intake should be 0.008–0.010 in.; exhaust should be 0.011–0.013 in. Intake locknut torque is 14 ft. lbs. (19 Nm); exhaust is 10 ft. lbs. (14 Nm).

4. Rotate the crankshaft 180 degrees clockwise and recheck No.3.
5. Rotate the crankshaft 180 degrees clockwise and recheck No.4.
6. Rotate the crankshaft 180 degrees clockwise and recheck No.2.

Starter Motor

REMOVAL & INSTALLATION

➡ **The factory sound system has a coded theft protection system. It is recommended that you know your reset code before you begin.**

1. Before servicing the vehicle, refer to the precautions section.
2. Remove or disconnect the following:
- Negative then the positive battery cables
- Intake manifold
- Starter cable from the B terminal
- Black/white wire from the S (solenoid) terminal
- Harness clamp and holder
- Two bolts that mount the starter to the transaxle assembly
- Starter

To install:

3. Install in the reverse order of removal. Refer to the illustration for torque specifications.

M/T:

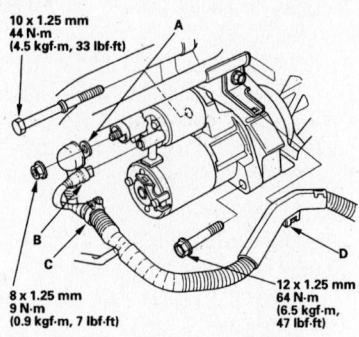

A/T:

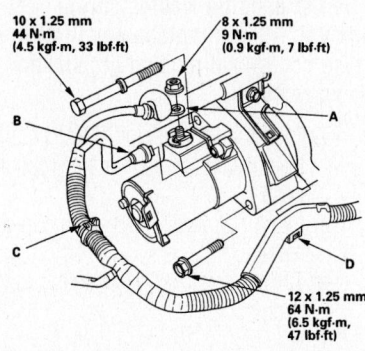

42356-ELEM-G23

Starter mounting

➡ **When installing the heavy gauge starter cable, make sure the crimped side of the terminal end is facing out.**

4. Enter the anti-theft code and radio presets.

Timing Chain and Front Seal

REMOVAL & INSTALLATION

1. Before servicing the vehicle, refer to the precautions section.
2. Set the engine to Top Dead Center (TDC).
3. Drain the cooling system.
4. Relieve the fuel system pressure.
5. Remove or disconnect the following:
- Negative battery cable

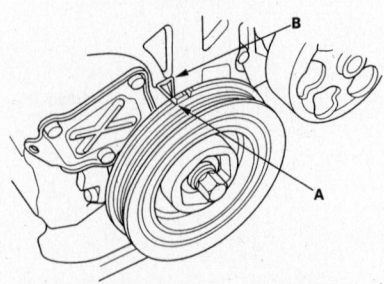

42356-ELEM-G24

Turn the crankshaft pulley so the TDC mark (A) is aligned with the pointer (B)

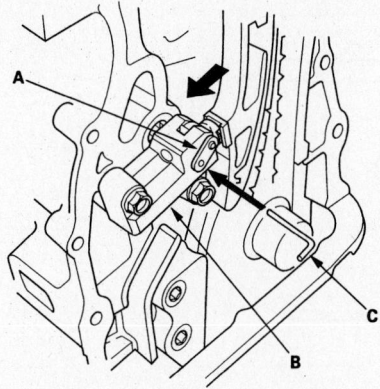

42356-ELEM-G25

Align the holes on the lock (A) and the auto-tensioner (B), then place a 1.5mm pin into the holes. Turn the crankshaft clockwise to secure the pin

- Front tires and wheels
- Splash shield
- Drive belt
- Cylinder head cover.

6. Check that the No. 1 piston TDC marks on the Variable Valve Timing Control (VTC) actuator and exhaust camshaft sprocket are aligned
- Crankshaft pulley
- Crankshaft Position (CKP) sensor connector
- VTC oil control solenoid valve connector
- VTC oil control solenoid valve

7. Support the engine with a suitable jack with a wooden block under the oil pan.
- Ground cable and upper engine mount bracket
- Side engine mount bracket
- Chain (case) cover

8. Loosely install the crankshaft pulley. Turn the crankshaft counterclockwise to compress the auto-tensioner.

9. Align the holes on the lock (A) and the auto-tensioner (B), then place a 1.5mm pin into the holes. Turn the crankshaft clockwise to secure the pin.

10. Remove or disconnect the following:
- Auto-tensioner
- Timing chain guide B (top guide)
- Timing chain guide A and tensioner arm
- Timing chain

✱✱ WARNING

Do not let the timing chain near any magnetic fields.

To install:

11. Set the crankshaft to TDC. Align the TDC mark (A) on the crankshaft sprocket with the pointer (B) on the cylinder block.

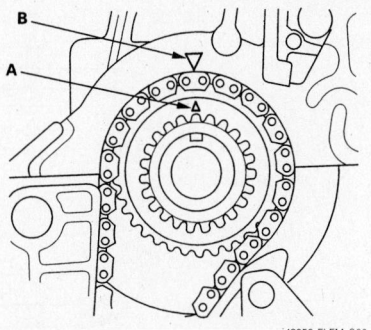

42356-ELEM-G26

Set the crankshaft to TDC. Align the TDC mark (A) on the crankshaft sprocket with the pointer (B) on the cylinder block

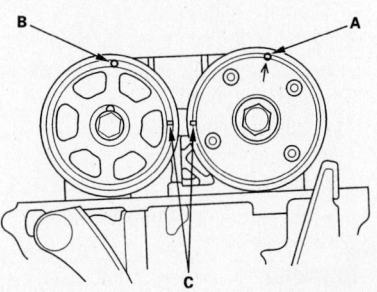

42356-ELEM-G27

The mark (A) on the VTC actuator and the mark (B) on the exhaust cam (C) should be at the top. Align the TDC marks (C) on the VTC actuator and exhaust cam sprockets

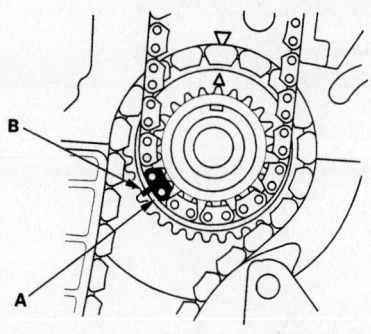

09474_ELEM_G0005

Install the timing chain on the crankshaft sprocket with the colored link of the chain aligned with the mark on the crank sprocket

12. Set the camshafts to TDC. The punch mark (A) on the VTC actuator and the punch mark (B) on the exhaust camshaft (C) should be at the top. Align the TDC marks (C) on the VTC actuator and exhaust camshaft sprockets.

13. Install or connect the following:
- Timing chain on the crankshaft sprocket with the colored link of the chain aligned with the mark on the crank sprocket
- Timing chain on the VTC actuator

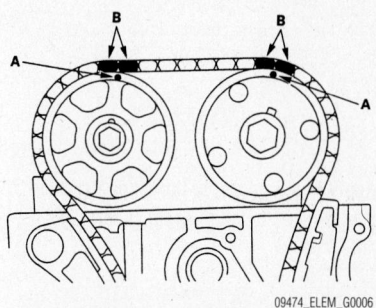

09474_ELEM_G0006

Install the timing chain on the VTC actuator and exhaust camshaft sprocket with the punch marks aligned with the center of the 2 colored links

and exhaust camshaft sprocket with the punch marks aligned with the center of the 2 colored links
- Timing chain guide A and tensioner arm. Tighten the guide bolts to 8.7 ft. lbs. (12 Nm) and the tensioner arm retainer to 16 ft. lbs. (22 Nm).
- Auto-tensioner and tighten the bolts to 8.7 ft. lbs. (12 Nm)
- Timing chain guide B and tighten the retainers to 16 ft. lbs. (22 Nm)

14. Remove the pin from the auto-tensioner.

15. Inspect the chain cover seal for damage and replace if necessary. Clean and dry the chain cover mating surfaces.

16. Install or connect the following:
- Liquid gasket, P/N 08718-0009 evenly to the cylinder block mating surface of the timing chain cover and the inner threads of the holes
- Liquid gasket to the cylinder block upper surface contact areas on the chain cover and the oil pan mating surface of the chain cover in the inner threads of the holes
- Liquid gasket to the oil pan surface where it contacts the chain cover

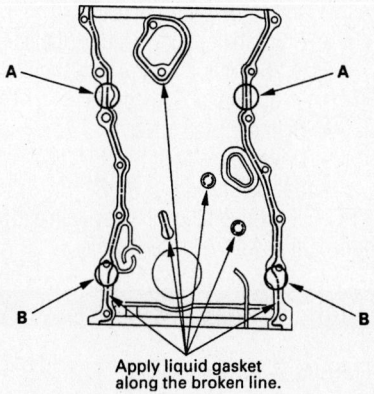

Apply liquid gasket along the broken line.

09474_ELEM_G0007

Apply liquid gasket to the chain cover locations illustrated

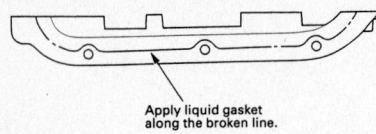

Apply liquid gasket along the broken line.

09474_ELEM_G0008

Apply liquid gasket to the oil pan surface where it contacts the chain cover

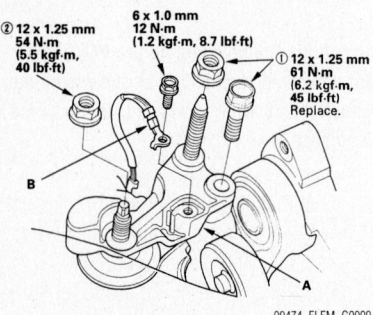

② 12 x 1.25 mm
54 N·m
(5.5 kgf·m,
40 lbf·ft)

6 x 1.0 mm
12 N·m
(1.2 kgf·m, 8.7 lbf·ft)

① 12 x 1.25 mm
61 N·m
(6.2 kgf·m,
45 lbf·ft)
Replace.

B
A

09474_ELEM_G0009

Tighten the upper bracket upper bolt/nuts in the proper order to the correct specification

➡**Make sure to install the components within 4 minutes of applying the sealer.**

- New O-ring the timing chain cover. Set the edge of the cover to the edge of the oil pan, then install the cover on the engine block. Tighten the retainers to 8.7 ft. lbs. (12 Nm).

➡**When installing the chain case, do not slide the bottom surface on the oil pan mounting surface.**

- Side engine mounting bracket and tighten the retainers to 33 ft. lbs. (44 Nm)
- Upper mount, then tighten the bolts/nuts as shown in the illustration
- Ground cable
- VTC oil control solenoid valve
- CKP sensor and VTC oil control solenoid valve connectors
- Crankshaft pulley. Tighten the bolt to 36 ft. lbs. (49 Nm), then tighten an additional 90 degrees.
- Cylinder head cover
- Drive belt
- Splash shield

17. Fill the engine cooling system and connect the negative battery cable.

Oil Pan

REMOVAL & INSTALLATION

1. Before servicing the vehicle, refer to the precautions section.

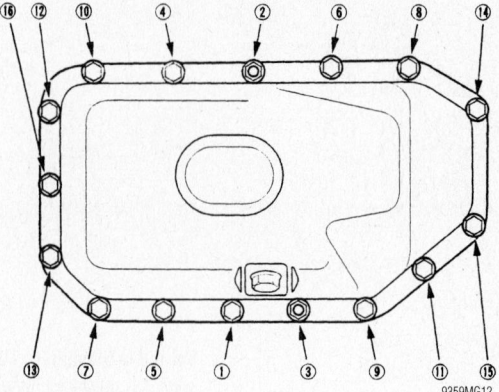

9359MG12

Oil pan fastener tightening sequence

2. Drain the engine oil.
3. Remove or disconnect the following:
   - Subframe. See engine Removal and Installation.
   - With a manual transmission, remove the stiffener
   - Oil pan bolts
   - Oil pan. A gasket cutter will be needed.

To install:

4. Apply a bead of liquid gasket to the oil pan mating surface. make sure to install the pan within 4 minutes of applying the gasket maker.

5. Installation is the reverse of removal. Torque the bolts, in sequence, in 2 or 3 steps, to 9 ft. lbs. (12 Nm).

Oil Pump

REMOVAL & INSTALLATION

1. Before servicing the vehicle, refer to the precautions section.

2. Drain the engine oil.

3. Set the No. 1 piston to Top Dead Center (TDC).

4. Remove or disconnect the following:
   - Negative battery cable
   - Oil pan
   - Oil pump chain tensioner and discard

5. Insert a 6mm pin driver into the maintenance hole in the lower balance shaft holder and through the rear balancer shaft to hold the rear balancer shaft.

6. Loosen the oil pump sprocket mounting bolt.
   - Oil pump sprocket
   - Oil pump

To install:

7. Make sure that No.1 piston is at TDC.

8. Align the dowel pin on the rear balance shaft with the mark on the pump.

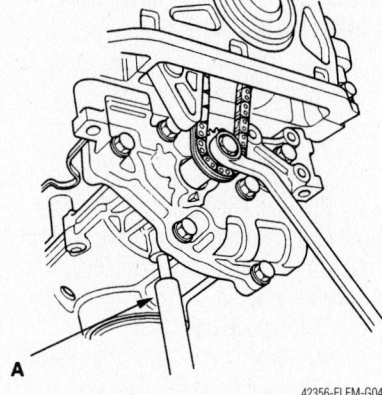

A

42356-ELEM-G04

Insert a 6mm pin into the maintenance hole in the lower balance shaft holder, through the rear balancer shaft to hold the shaft, then loosen the sprocket mounting bolt

9. Insert a 6mm pin into the maintenance hole in the lower balance shaft holder, through the rear balancer shaft to hold the shaft.

10. Install or connect the following:
   - Engine oil to the threads of the oil pump sprocket mounting bolt
   - Oil pump and sprocket loosely

11. Remove the balance shaft holding pin.

12. Torque the 10mm mounting bolts to 33 ft. lbs. (44 Nm); the 8mm bolts to 16 ft. lbs. (22 Nm).

13. Torque the pulley bolt to 33 ft. lbs. (44 Nm).

14. Squeeze the new oil pump chain tensioner then install the set clip on it as shown in the illustration.

15. Install or connect the following:
   - New oil pump chain tensioner and torque the bolts to 9 ft. lbs. (12 Nm). Remove the set clip from the tensioner.
   - Oil pan

16. Fill the engine with oil.

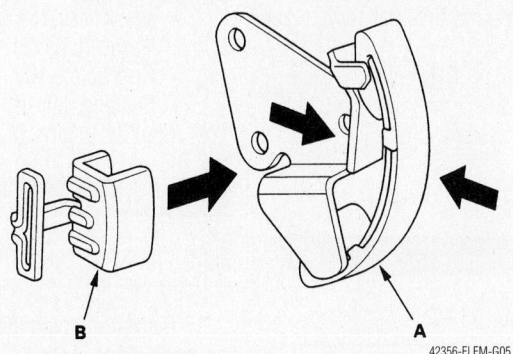

42356-ELEM-G05

Squeeze the new oil pump chain tensioner (A) then install the set clip (A) on it as shown. The clip is supplied with the new tensioner

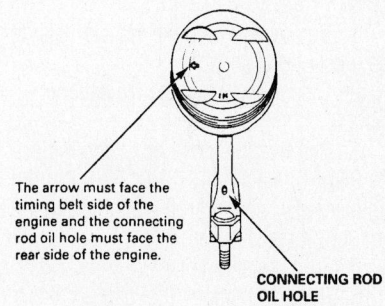

7924AG54

Piston ring end-gap spacing

Rear Main Seal

REMOVAL & INSTALLATION

1. Before servicing the vehicle, refer to the precautions section.
2. Remove or disconnect the following:
   - Transaxle
   - Clutch pressure plate and disc, if equipped
   - Flywheel
   - Oil seal

To install:

3. Install or connect the following:
   - Oil seal. Drive the seal square into the seal case.
   - Flywheel. Tighten the bolts in a crossing pattern to 76 ft. lbs. (103 Nm).
   - Clutch pressure plate and disc, if equipped
   - Transaxle

4. Check the fluid levels.
5. Start the engine and check for leaks.

Piston and Ring

POSITIONING

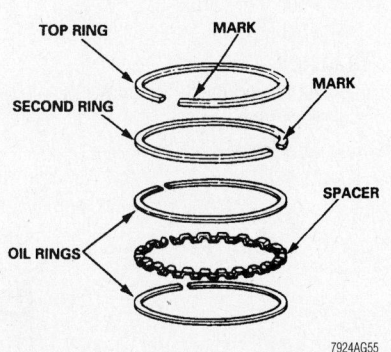

7924AG55

Piston ring positioning and top mark location

The arrow must face the timing belt side of the engine and the connecting rod oil hole must face the rear side of the engine.

7924AG53

Piston and connecting rod assembly

FUEL SYSTEM

Fuel System Service Precautions

Safety is the most important factor when performing not only fuel system maintenance, but any type of maintenance. Failure to conduct maintenance and repairs in a safe manner may result in serious personal injury or death. Maintenance and testing of the vehicle's fuel system components can be accomplished safely and effectively by adhering to the following rules and guidelines:

- To avoid the possibility of fire and personal injury, always disconnect the negative battery cable unless the repair or test procedure requires that battery voltage be applied.
- Always relieve the fuel system pressure prior to disconnecting any fuel system component (injector, fuel rail, pressure regulator, etc.), fitting or fuel line connection. Exercise extreme caution whenever relieving fuel system pressure to avoid exposing skin, face and eyes to fuel spray. Please be advised that fuel under pressure may penetrate the skin or any part of the body that it contacts.

- Always place a shop towel or cloth around the fitting or connection prior to loosening to absorb any excess fuel due to spillage. Ensure that all fuel spillage (should it occur) is quickly removed from engine surfaces. Ensure that all fuel soaked cloths or towels are deposited into a suitable waste container.
- Always keep a dry chemical (Class B) fire extinguisher near the work area.
- Do not allow fuel spray or fuel vapors to come into contact with a spark or open flame.
- Always use a backup wrench when loosening and tightening fuel line connection fittings. This will prevent unnecessary stress and torsion to fuel line piping. Always follow the proper torque specifications.
- Always replace worn fuel fitting O-rings with new. Do not substitute fuel hose or equivalent, where fuel pipe is installed.

Fuel System Pressure

RELIEVING

✳✳ CAUTION

The fuel injection system remains under pressure after the engine has been turned OFF. Properly relieve fuel pressure before disconnecting any fuel lines. Failure to do so may result in fire or personal injury.

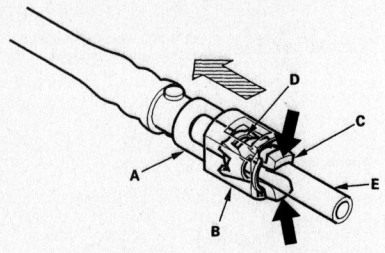

42356-ELEM-G29

Hold the quick-connect (A) connector (B) with one hand, then squeeze the retainer tabs (C) with the other hand to release them from the locking pawls (D)

➡**The radio may contain a coded theft protection circuit. Always obtain the code number before disconnecting the battery.**

1. Before servicing the vehicle, refer to the precautions section.
2. Disconnect the negative battery cable.
3. Remove the glove box, then remove the PGM-FI main relay (FUEL PUMP) from the fuse/relay box. Start the engine and let it run until it stalls.
4. Turn the engine OFF.
5. Remove the fuel filler cap.
6. Remove the quick-connect fitting cover.
7. Clean any dirt from the quick-connect fitting.
8. Place a rag or shop towel over quick-connect fitting.
9. Detach the quick-connect fitting by holding the connector with one hand, then squeeze the retainer tabs with the other hand to release them from the locking pawls. Pull the connector off.

✷✷ CAUTION

Do not allow fuel spray or fuel vapors to come in contact with a spark or open flame. Keep a dry chemical fire extinguisher nearby. Never store fuel in an open container due to risk of fire or explosion.

➡**A fuel pressure gauge may be attached at the quick-connect location.**

10. Connect the quick-connect fitting, making sure the locking pawls are properly engaged.
11. Clean up any fuel spilled on the engine and intake manifold.
12. Install the fuel pump relay to the underdash fuel/relay box and install the glove box.
13. Install the fuel filler cap.

14. Reconnect the negative battery cable.
15. Turn the ignition **ON**, but don't start the engine. Repeat this 2 or 3 times to pressurize the fuel system. Check for fuel leaks.
16. Enter the radio security code.

Fuel Filter

REMOVAL & INSTALLATION

➡**The fuel filter should be replaced whenever the fuel pressure drops below 48 psi, after making sure that the fuel pump and fuel pressure regulator are okay.**

1. Before servicing the vehicle, refer to the precautions section.
2. Relieve the fuel system pressure.
3. Remove or disconnect the following:
   - Negative battery cable
   - Fuel pump
   - Fuel filter carrier (A)
   - Fuel filter

To install:

4. Install or connect the following:
   - Fuel filter
   - Fuel lines

- New gasket (B)
- New o-rings (E)
- Connectors (C)
- Sending unit (D)

5. Start the engine and check for leaks.

Fuel Pump

REMOVAL & INSTALLATION

1. Before servicing the vehicle, refer to the precautions section.
2. Relieve the fuel system pressure.
3. Remove or disconnect the following:
   - Negative battery cable
   - Fuel filler cap
   - Center console, then both track floor covers and sill trims.
4. Fold back the floor covering until you can get to the access panel
   - Access panel from the floor
   - Fuel pump connector
   - Fuel supply and return line quick-connect fittings
   - Fuel pump locknut, using special tool No. 07XAA-001010A
   - Fuel pump sending assembly
5. Installation is the reverse of removal.

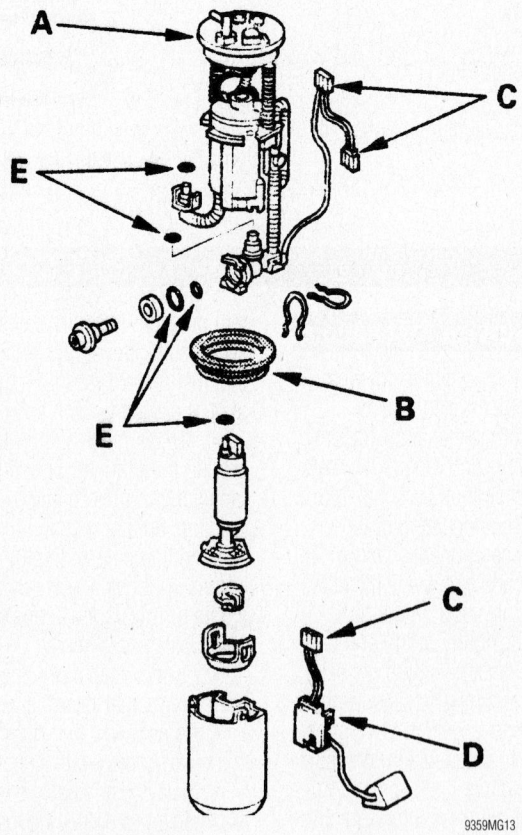

9359MG13

Exploded view of the fuel filter mounting

Fuel Injector

REMOVAL & INSTALLATION

1. Before servicing the vehicle, refer to the precautions section.
2. Relieve the fuel system pressure.
3. Remove or disconnect the following:
   - Negative battery cable
   - Engine cover
   - Injector connectors, ground cable and harness holder
   - Fuel line quick-connect fittings
   - Fuel rail mounting nuts
   - Injector clip(s) from the injector(s)
   - Fuel injectors from the fuel rail

To install:
4. Install or connect the following:
   - Injectors to the fuel rail with new O-rings coated with clean engine oil.
   - Injector clips
   - Injectors in the injector base
   - Fuel rail and injector assembly. Tighten the nuts to 16 ft. lbs. (22 Nm).
   - Ground cable bolt
   - Injector connectors
   - Fuel lines
   - Negative battery cable
5. Start the engine and check for leaks.

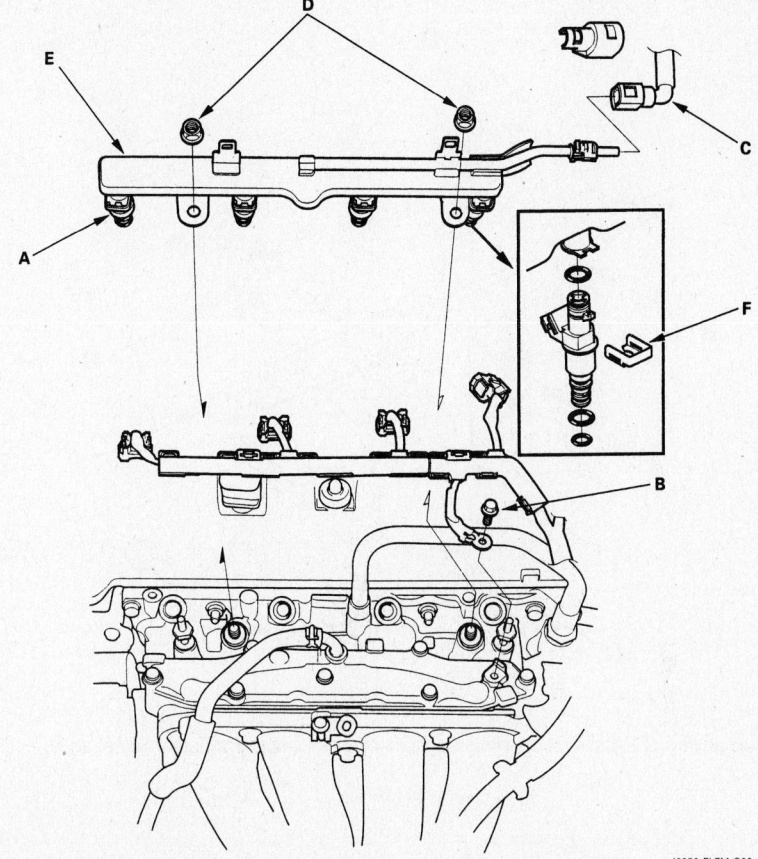

Exploded view of the fuel rail (E), injectors (A) and related components

42356-ELEM-G06

DRIVE TRAIN

Transaxle Assembly

REMOVAL & INSTALLATION

Automatic Transaxle

1. Before servicing the vehicle, refer to the precautions section.
2. Drain the transaxle.
3. Remove or disconnect the following:
   - Splash shield
   - Battery cables, negative cable first
   - Air intake assembly
   - Battery
   - Battery tray
   - Harness clamp from the batter base and remove the base
   - Transaxle ground cable
   - Clutch pressure switch connectors
   - 2nd clutch pressure control solenoid valve connector
   - Clutch pressure control valve connector
   - Harness clamps from the brackets
   - Countershaft speed sensor connector
   - Mainshaft speed sensor connector

- Transmission Range Switch (TRS) connector from the bracket and disconnect it
- 3rd clutch pressure control solenoid valve connector
- Shift control solenoid valve connectors
- Harness clamps from the brackets
- Transaxle oil cooler lines
- Harness clamp from the clamp bracket and harness cover from the bracket
- Water pipe mounting bolt and brackets

4. Attach engine hanger VSB02C000015 to the treaded holes in the cylinder head.
5. Install engine support hanger AAR-T-12566 to the engine and hanger VSB02C000015.
6. Install a 5mm hex wrench in the top ball joint pin and remove the nut. Separate the stabilizer link from the lower arm.
   - Lower arms from the knuckles
   - Torque converter cover and converter bolts
7. On four wheel drive models, refer to the illustration for location and perform the following:
   a. Remove the bolts (A) attaching the shift cable bracket (B) and remove the cable cover (C).
   b. Remove the spring clip (D) and control pin (E) and separate the cable (F) from the selector lever (G) being careful not to bend the cable too much.
   c. Remove the bolts (H) attaching the shift cable bracket (I) and remove the bracket from the cable.
8. On two wheel drive models, refer to the illustration for location and perform the following:
   a. Remove the shift cable cover (A).
   b. Remove the spring clip (B) and control pin (C) the separate the cable (D) from the selector lever (E).
   c. Remove the bolts attaching the shift cable bracket (F) and remove the bracket from the cable.
9. Remove or disconnect the following:
   - Transmission cooler hose from the cooler line and plug the hose
   - Front mount bolt and nut
   - Rear mount bolts

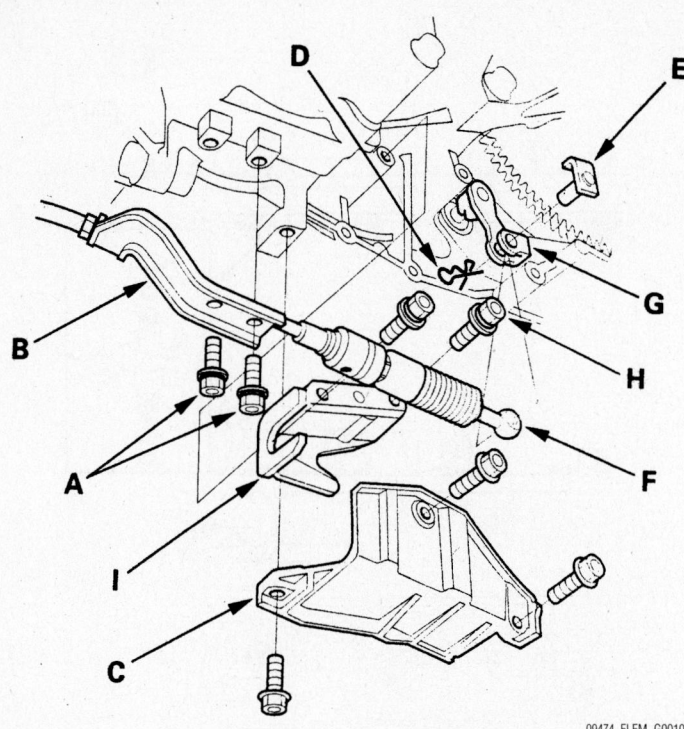

Exploded view of the shift cable components on models equipped with four wheel drive

09474_ELEM_G0010

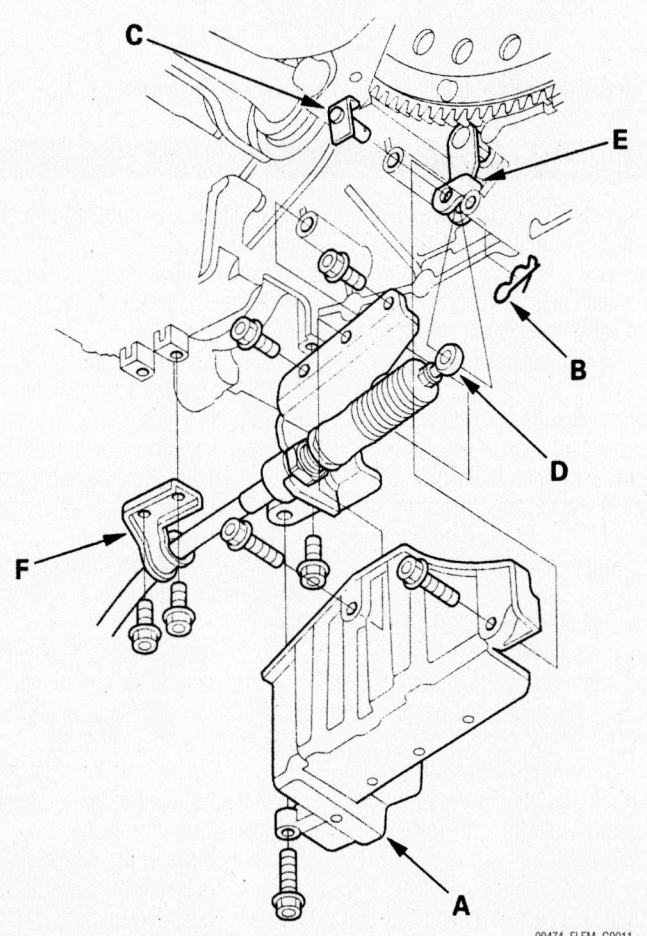

Exploded view of the shift cable components on models equipped with two wheel drive

09474_ELEM_G0011

10. Matchmark the sub-frame mounting bolt centers.

There is a special tool necessary for sub-frame removal. The Honda tool number is EQS02C000011.

11. Attach the tool as explained in the tool instructions, attach a floor jack with adapter, remove the 4 sub-frame bolts and lower the sub-frame.

- Rear driveshaft, if equipped

12. Separate the inner CV-joints from the transaxle and intermediate shaft and support the axle halfshafts out of the work area with safety wire.

- Intermediate shaft

13. Support the transmission with a suitable jack.

- Transmission mount bolts and nuts
- Transmission housing bolts
- Transaxle

To install:

14. Installation is the reverse of removal. Observe the following torques:

- Air cleaner housing bracket bolt: 16 ft. lbs. (22 Nm)
- Front mount bolts: 47 ft. lbs. (64 Nm)
- Rear mount bracket bolts: 40 ft. lb. (54 Nm)
- Transmission-to-engine bolts: 47 ft. lbs. (64 Nm)
- Upper transmission mount bolt: 40 ft. lbs. (54 Nm)
- Upper transmission mount nuts: 40 ft. lbs. (54 Nm)
- Rear driveshaft bolts: 24 ft. lbs. (32 Nm)
- Subframe bolts: 76 ft. lbs. (103 Nm)

Manual Transaxle

1. Before servicing the vehicle, refer to the precautions section.

2. Secure the hood in a vertical position.

3. Remove or disconnect the following:

- Air intake assembly
- Battery cables, battery and tray
- Transaxle ground cable
- Vehicle Speed Sensor (VSS) connector
- Back-up light switch connector
- Cable bracket and the cables from the top of the transmission housing. Be careful when removing the cables to avoid bending them.
- Clutch slave cylinder without bending the clutch line. Do not press the clutch pedal after the slave cylinder has been removed.
- Two upper transmission bolts

4. Attach engine hanger VSB02C000015 to the treaded holes in the cylinder head.

5. Install engine support hanger AAR-T-12566 to the engine and hanger VSB02C000015.
- Transmission mount bracket and bolt
- Air cleaner bracket

6. Drain the transmission fluid
- Splash shield
- Driveshafts
- Intermediate shaft
- Front engine mount bracket bolt
- 3 bolts attaching the rear transmission mount

7. Matchmark the sub-frame mounting bolt centers.

There is a special tool necessary for sub-frame removal. The Honda tool number is EQS02C000011.

8. Attach the tool as explained in the tool instructions, attach a floor jack with adapter, remove the 4 sub-frame bolts and lower the sub-frame.
- Clutch cover
- Front engine mount

9. Support the transmission with a suitable jack and remove the 4 lower mounting bolts.

10. Pull the transmission away from the engine until the mainshaft clears the pressure plate and lower the transmission.

To install:

11. Installation is the reverse of removal. Observe the following torques:
- Transaxle rear mount and bracket. Tighten the bracket bolts to 40 ft. lbs. (54 Nm) and the through bolt to 47 ft. lbs. (64 Nm)
- Transaxle. Tighten the flange bolts to 47 ft. lbs. (64 Nm)
- Front mount and bracket. Tighten the bolts to 47 ft. lbs. (64 Nm).

- Clutch housing cover. Tighten the bolts to 29 ft. lbs. (39 Nm)
- Subframe: 76 ft. lbs. (98 Nm)

12. Fill the transmission with the correct type and amount of fluid.

13. Road test the vehicle.

Clutch

ADJUSTMENTS

The Element is equipped with a hydraulic clutch system. No adjustment is necessary.

REMOVAL & INSTALLATION

1. Before servicing the vehicle, refer to the precautions section.

2. Remove or disconnect the following:
- Negative battery cable
- Transaxle

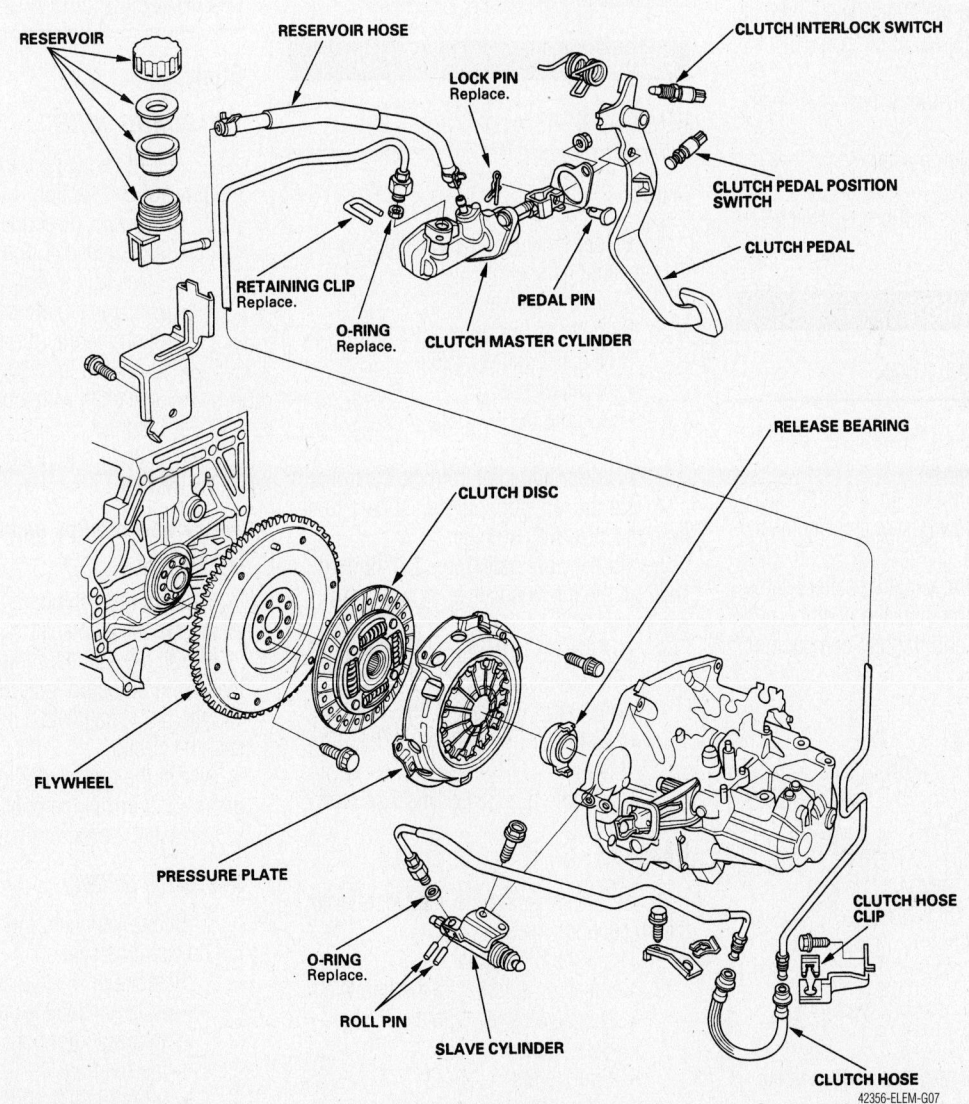

Exploded view of the clutch system components

- Pressure plate. Loosen the bolts evenly in a crossing pattern.
- Clutch disc

To install:

3. Install the clutch disc and pressure plate. Tighten the pressure plate bolts in a crisscross pattern, in several steps to 19 ft. lbs. (26 Nm).

4. Install or connect the following:
- Transaxle
- Negative battery cable

Hydraulic Clutch System

BLEEDING

1. Before servicing the vehicle, refer to the precautions section.

2. Attach a hose to the bleeder screw and suspend the other end in a container of clean brake fluid.

3. Open the bleeder screw.

4. Slowly pump the clutch pedal until no more air bubbles appear at the bleeder hose.

5. Tighten the bleeder screw to 70 inch lbs. (8 Nm).

6. Refill the clutch master cylinder as necessary.

7. Check for leaks and proper clutch operation.

Transfer Assembly

REMOVAL & INSTALLATION

1. Before servicing the vehicle, refer to the precautions section.

2. Drain the transaxle fluid. Install the drain plug with a new gasket and tighten to 36 ft. lbs. (49 Nm).

3. Disconnect the negative battery cable.

4. Matchmark the installed position of the propeller shaft and transfer companion flange.

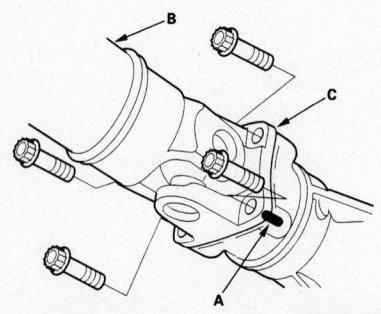

Matchmark (A) the installed position of the propeller shaft (B) and transfer companion flange (C)

42356-ELEM-G08

5. Remove or disconnect the following:
- Propeller shaft from the transfer assembly
- Mounting bolts and transfer assembly

To install:

6. Clean the transfer assembly mating surfaces, then apply clean transmission fluid to the mating surfaces.

7. Install or connect the following:
- New O-ring seal on the transfer assembly
- 4 bolts in the transfer housing, then the transfer assembly with the dowel pin. Tighten the 10mm bolts to 33 ft. lbs. (44 Nm).
- Propeller shaft to the transfer companion flange, aligning the mark made during removal. Tighten the 8mm bolts to 24 ft. lbs. (33 Nm).
- Negative battery cable

8. Fill the transaxle to the correct level and check for leaks.

Halfshaft

REMOVAL & INSTALLATION

Front

1. Before servicing the vehicle, refer to the precautions section.

2. Drain the transaxle.

3. Remove or disconnect the following:
- Negative battery cable
- Front wheels
- Spindle nut
- Stabilizer bar
- Lower ball joint from the control arm

4. On the left side, pry the inboard joint from the case with a prybar.

5. On the right side, drive the inboard shaft off the intermediate shaft with a drift and hammer.

6. Installation is the reverse of removal. Observe the following torques:
- Ball stud nuts: 40 ft. lbs. (54 Nm)
- Stabilizer link nuts: 29 ft. lbs. (39 Nm)
- Spindle nut: 181 ft. lbs. (245 Nm)

Rear

1. Before servicing the vehicle, refer to the precautions section.

2. Drain the differential.

3. Remove or disconnect the following:
- Negative battery cable
- Rear wheels
- Spindle nut

4. Pry the inboard joint from the differential.

5. Remove the outer CV-joint stub shaft from the hub by tapping the stub shaft with a plastic hammer.

To install:

➡**Use new circlips and self-locking nuts for assembly.**

6. Install the outer CV-joint stub shaft into the hub.

7. Install the inner CV-joint to the differential until the circlip locks in the retaining groove.

8. Install or connect the following:
- Spindle nut. Tighten the nut to 134 ft. lbs. (181 Nm).
- Rear wheels
- Negative battery cable

9. Fill the differential to the correct level and check for leaks.

CV-Joint

OVERHAUL

Front

OUTBOARD JOINT

1. Before servicing the vehicle, refer to the precautions section.

2. Remove or disconnect the following:
- Axle halfshaft from the vehicle and place it in a vise
- Outboard joint boot clamps and push the boot back
- Outboard joint by driving it off the axle shaft with a brass drift and hammer
- Outboard joint boot

To install:

➡**Use new circlips and boot clamps for assembly.**

3. Install the outboard joint boot and clamps to the axle shaft.

4. Fill the outboard joint with grease. Install the outboard joint to the axle shaft. Tap the stub shaft with a brass hammer to seat the circlip.

5. Fill the outboard joint boot with grease and install the boot clamps.

6. Install the axle halfshaft to the vehicle.

INBOARD JOINT

1. Before servicing the vehicle, refer to the precautions section.

2. Remove or disconnect the following:
- Axle halfshaft from the vehicle.
- Inboard joint boot clamps and push the boot back
- Inboard joint housing from the axle
- Rollers from the spider

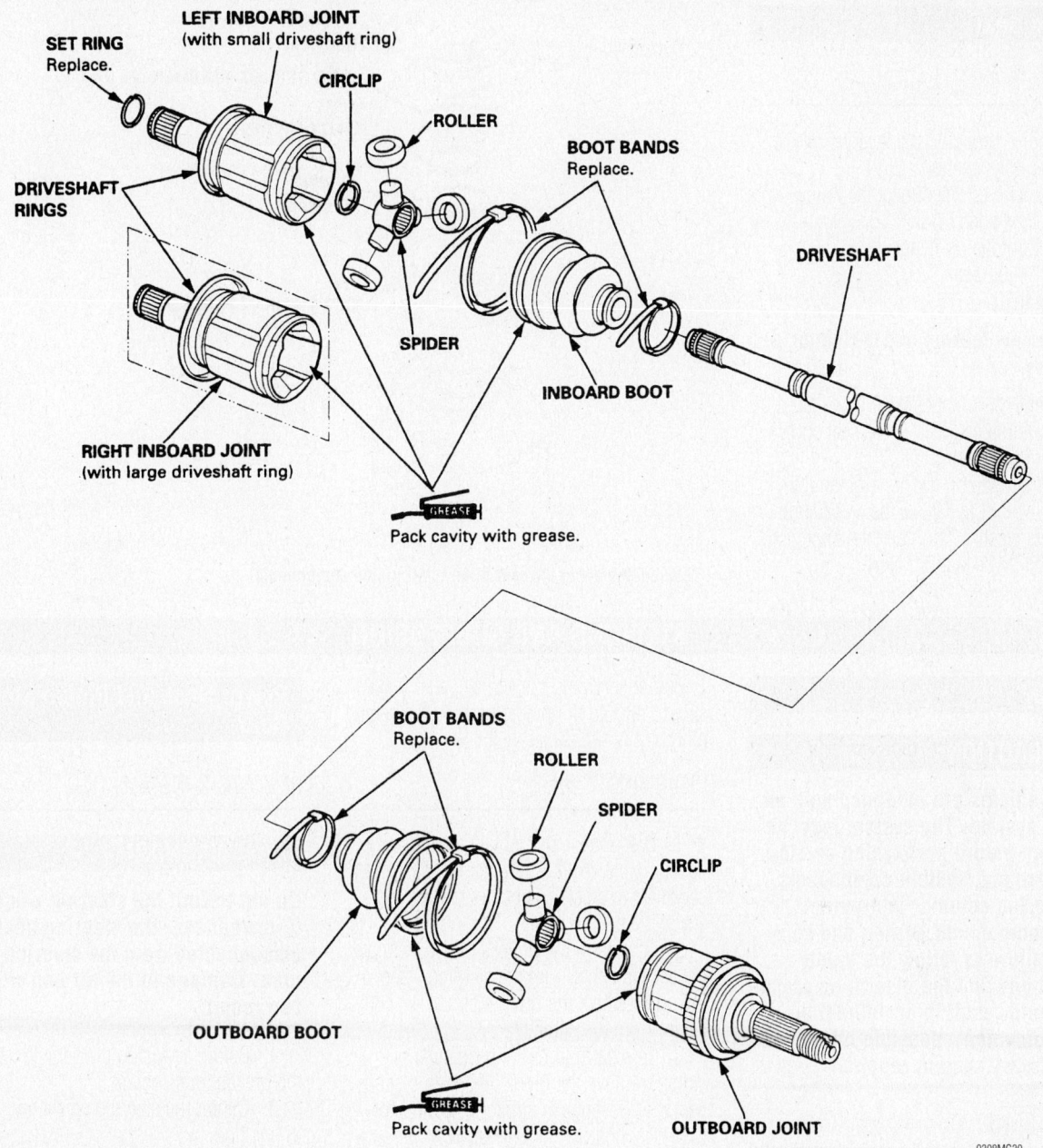

Exploded view of the rear axle

- Snapring and the spider from the axle shaft
- Inboard joint boot

To install:

➡ **Use new circlips and boot clamps for assembly.**

3. Install or connect the following:
  - Inboard joint boot and clamps to the axle shaft
  - Spider with a new snapring
  - Rollers to the spider
4. Fill the joint housing with grease and install it.
5. Fill the inboard joint boot with grease and install the boot clamps.

6. Install the axle halfshaft to the vehicle.

Rear

1. Before servicing the vehicle, refer to the precautions section.
2. Remove or disconnect the following:
  - Axle halfshaft from the vehicle
  - Joint boot clamps and push the boot back
  - Joint housing from the axle
  - Rollers from the spider
  - Snapring and the spider from the axle shaft
  - Joint boot

To install:

➡ **Use new circlips and boot clamps for assembly.**

3. Install or connect the following:
  - Joint boot and clamps to the axle shaft
  - Spider with a new snapring
  - Rollers to the spider
4. Fill the joint housing with grease and install it.
5. Fill the joint boot with grease and install the boot clamps.
6. Install the axle halfshaft to the vehicle.

Pinion Seal

REMOVAL & INSTALLATION

1. Before servicing the vehicle, refer to the precautions section.
2. Remove or disconnect the following:
   - Driveshaft
   - Companion flange
   - Pinion seal

To install:

➡ **Use a new locknut and O-ring for assembly.**

3. Install or connect the following:
   - Pinion seal. Drive the seal square into the bore.
   - Companion flange. Tighten the locknut to 108 ft. lbs. (147 Nm).
   - Driveshaft. Tighten the flange bolts to 24 ft. lbs. (32 Nm).

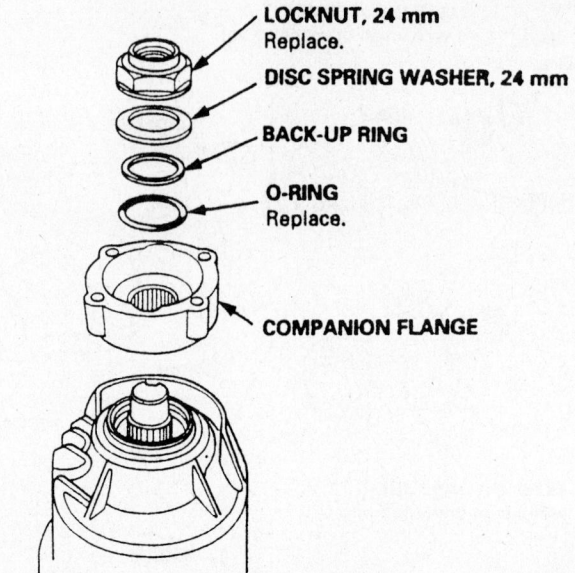

LOCKNUT, 24 mm
Replace.

DISC SPRING WASHER, 24 mm

BACK-UP RING

O-RING
Replace.

COMPANION FLANGE

9308MG31

Exploded view of the rear differential pinion components

STEERING AND SUSPENSION

Air Bag

✳✳ CAUTION

Some vehicles are equipped with an air bag system. The system must be disarmed before performing service on, or around, system components, the steering column, instrument panel components, wiring and sensors. Failure to follow the safety precautions and the disarming procedure could result in accidental air bag deployment, possible injury and unnecessary system repairs.

PRECAUTIONS

Several precautions must be observed when handling the inflator module to avoid accidental deployment and possible personal injury.

- Never carry the inflator module by the wires or connector on the underside of the module.
- When carrying a live inflator module, hold securely with both hands, and ensure that the bag and trim cover are pointed away.
- Place the inflator module on a bench or other surface with the bag and trim cover facing up.
- With the inflator module on the bench, never place anything on or close to the module which may be thrown in the event of an accidental deployment.

Before servicing the vehicle, also make sure to refer to the precautions in the beginning of this section as well.

DISARMING

1. Disconnect and isolate the negative battery cable. Wait 3 minutes for the system capacitor to discharge before performing any service.
2. To disarm the driver's airbag, remove the access panel from the steering wheel, then disconnect the driver's airbag 4P connector from the cable reel.
3. To disarm the front passenger's airbag, remove the glove box, then disconnect the passenger's airbag 4P connector from dashboard wire harness B.
4. To disarm the side airbag, disconnect the side airbag 2P connector from the floor wire harness.
5. To disarm the seat belt tensioner, disconnect the seat belt tensioner 2P connector from the rear door wire harness.
6. To disarm the seat belt buckle tensioner, disconnect the seat belt buckle tensioner 4P connector.
7. To disarm the SRS unit, disconnect the SRS unit connector A, B or C, as applicable.

REARMING

1. To rearm, connect the electrical connector(s) as necessary, then connect the negative battery cable.

Power Rack and Pinion Steering Gear

REMOVAL & INSTALLATION

✳✳ WARNING

Do not permit the steering wheel to turn whenever the steering gear is disconnected from the steering column. Damage to the air bag wiring can result.

1. Before servicing the vehicle, refer to the precautions section.
2. Center the steering wheel and lock it in position.
3. Remove or disconnect the following:
   - Negative battery cable and wait at least 3 minutes before continuing
   - Front wheels
4. Remove the air bag and steering wheel as follows:
   a. Align the front wheels in the straight ahead position
   b. Remove the access panel from the steering wheel and disconnect the drivers airbag 4P connector.
   c. Remove the two Torx® bolts using a T30 bit.
   d. Remove the airbag.
   e. Disconnect the cruise control connector and horn switch connector.
   f. Loosen the steering wheel bolt and using a suitable puller, free the steering wheel.

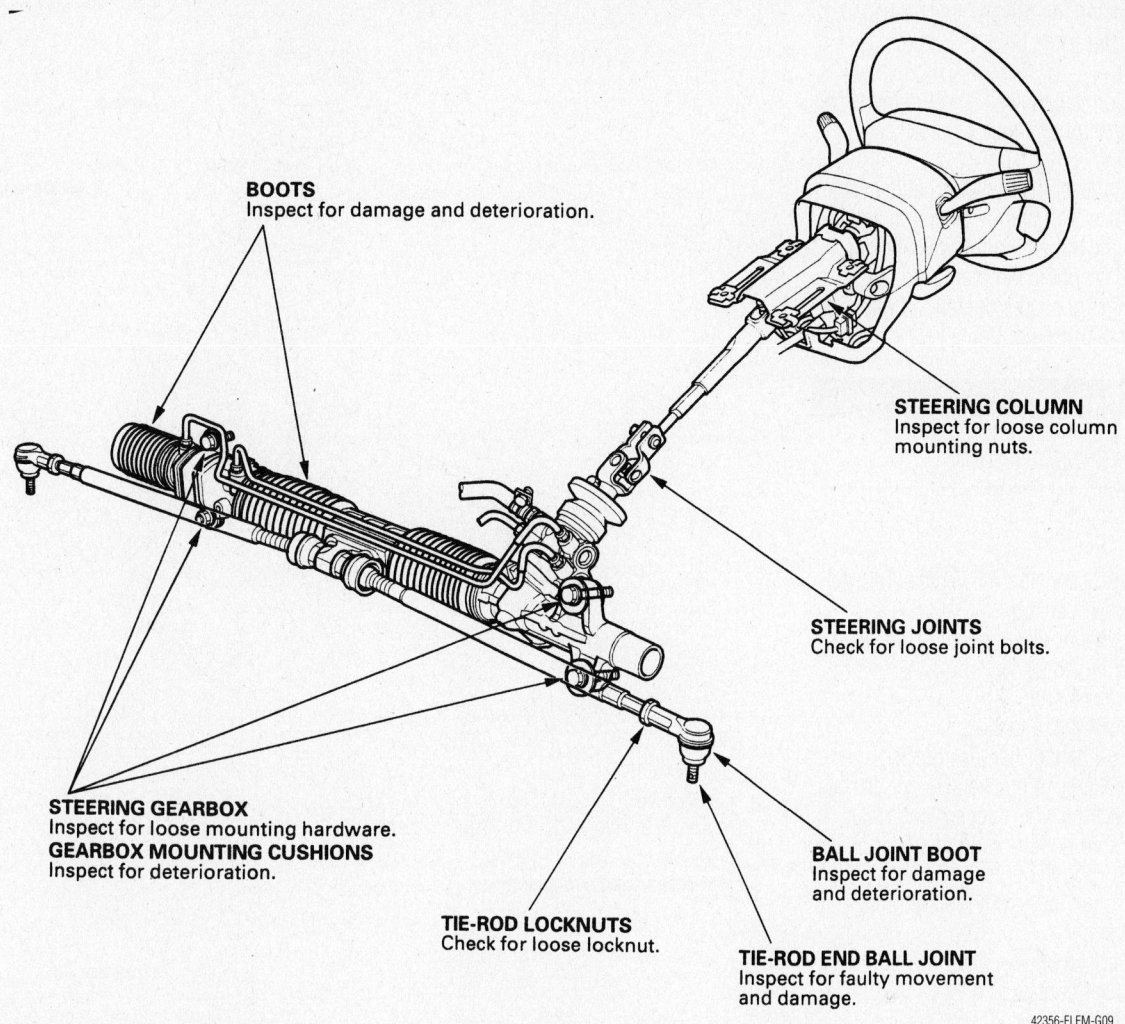

BOOTS
Inspect for damage and deterioration.

STEERING COLUMN
Inspect for loose column
mounting nuts.

STEERING JOINTS
Check for loose joint bolts.

STEERING GEARBOX
Inspect for loose mounting hardware.
GEARBOX MOUNTING CUSHIONS
Inspect for deterioration.

BALL JOINT BOOT
Inspect for damage
and deterioration.

TIE-ROD LOCKNUTS
Check for loose locknut.

TIE-ROD END BALL JOINT
Inspect for faulty movement
and damage.

42356-ELEM-G09

Power steering gear and related components

➡ **Do not tap on the steering wheel or column shaft during removal. If you thread the puller bolts more than 5 threads into the wheel hub you will hit the cable reel and damage it. To prevent damage, insert a pair of jam nuts 5 threads up on each puller bolt.**

   g. Remove the puller, steering wheel bolt and wheel.
  5. Remove or disconnect the following:
- Driver's side dashboard lower cover and undercover
- Steering joint bolts and disconnect the joint by moving the joint towards the column
- Center pin from the top of the pinion shaft, if equipped and discard the pin
- Tie rod ends
- Power steering heat baffle plate
- Engine wiring harness clamp and clip from their brackets
- Loosen the adjustable hose clamp and disconnect the return hose

- Loosen the 14mm flare nut and disconnect the feed line
- Open the hose holders on the return hose and remove the clamp
- Power steering pressure switch connector
- Feed line on the power steering line mounting bracket and set it aside
- Body stiffener
- Left, then right side flange bolts and washers
- Mounting brackets

  6. Lower the unit so the pinion shaft points upward. Remove the pinion shaft grommet. The steering gear is removed through the driver's side.
  7. Installation of the steering gear is the reverse of removal. Observe the following torques:
- Mounting bracket and side flange bolts: 46 ft. lbs. (62 Nm)
- Supply line flare nut: 27 ft. lbs. (37 Nm)

- Tie rod ball stud nuts: 40 ft. lbs. (54 Nm)
- Steering joint bolts: 21 ft. lbs. (28 Nm)

  8. Install the steering wheel and air bag as follows:
   a. First make sure the front wheels are aligned straight ahead. Center the cable reel by rotating the cable reel clockwise until it stops, then rotate it counterclockwise about 2 ½ turns. The arrow mark on the cable reel should point straight up.
   b. Position the tabs on the turn signal canceling sleeve, install the steering wheel and make sure the wheel hub engages the pins of the cable reel and tabs of the canceling sleeve. Do not tap on the wheel or column.
   c. Install the steering wheel bolt and tighten to 29 ft. lbs. (39 Nm). Connect the horn switch, cruise control switch and ensure the wiring is routed correctly and properly secured.

d. Install the drivers side air bag and tighten the Torx® bolts to 7 ft. lbs. (9 Nm).

e. Connect the cable reel to the airbag 4P connector and install the access panel.

f. Connect the negative battery cable.

g. Turn the ignition switch on and ensure the airbag light illuminates for about 6 seconds and then goes out.

h. Ensure proper operation of the horn and cruise control.

Strut (Damper)

REMOVAL & INSTALLATION

Front

1. Before servicing the vehicle, refer to the precautions section.
2. Remove or disconnect the following:
   - Front wheel
   - Tie rod end
   - Brake hose retainer
   - ABS sensor harness bracket and brake hose bracket. Do not disconnect the wheel sensor connector.
   - Pinch bolts from the damper, while holding the nuts
   - Flange nuts from the top of the damper
   - Strut (damper), after lowering the lower control arm

To install:

➡**Use new self-locking fasteners for assembly.**

3. Install or connect the following:
   - Strut (damper). Tighten the upper mounting nuts to 33 ft. lbs. (44 Nm).
   - Tighten the pinch bolts to 116 ft. lbs. (157 Nm)
   - ABS sensor
   - Tie rod end
   - Brake hose retainer
   - Front wheel

Rear

1. Before servicing the vehicle, refer to the precautions section.
2. Support the vehicle under the lower control arm.
3. Remove or disconnect the following:
   - Rear wheel
   - Flange bolt from the bottom of the damper (strut)
   - Evaporative emission (EVAP) canister bolts, and loosen the EVAP canister mounting (left side only)

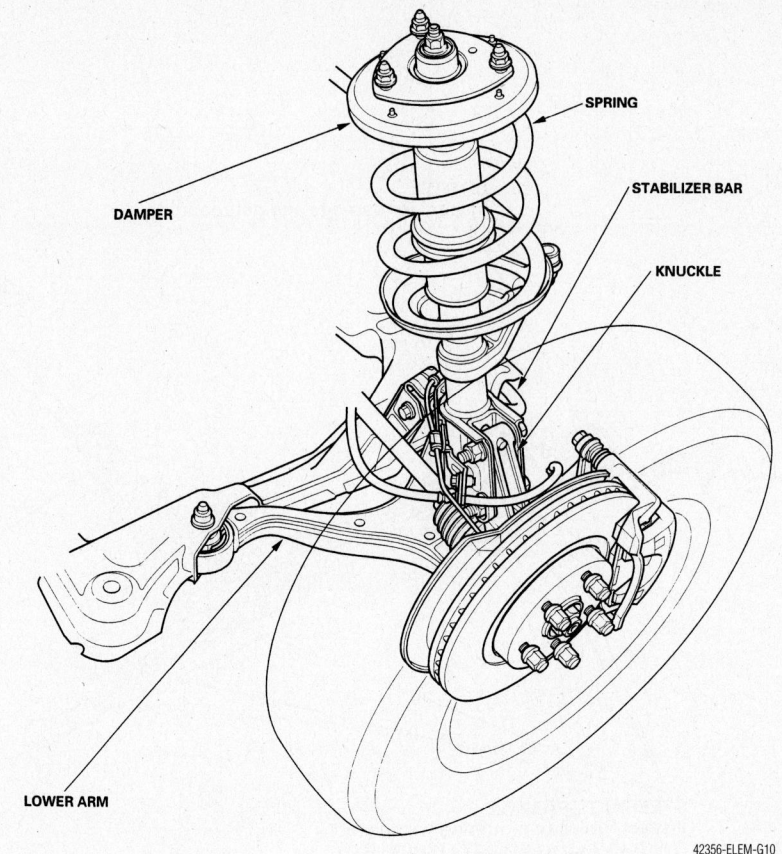

Front suspension components

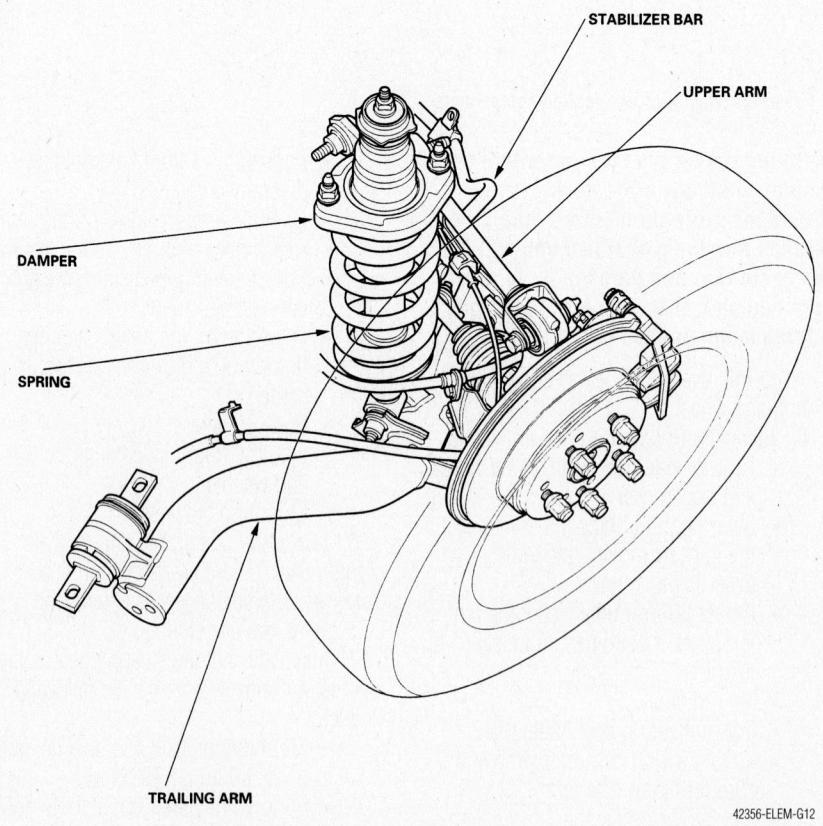

Rear suspension components

- Interior access panel, if necessary
- Flange nuts from the top of the damper in the cargo area
- Strut

To install:

4. Install or connect the following:
- Strut. Position the damper mounting base so the indent mark is toward the inside of the vehicle,
- Upper flange nuts, hand-tight only
- Bottom flange bolt, hand-tight only

5. With the suspension raised with a jack to load it with the vehicles weight, tighten the bottom bolt to 69 ft. lbs. and the top nuts to 54 ft. lbs. (74 Nm).

- Interior access panel, if necessary
- EVAP canister mounting bolts
- Rear wheel

Coil Spring

REMOVAL & INSTALLATION

Front

1. Before servicing the vehicle, refer to the precautions section.

2. Remove the strut from the vehicle and install in a strut spring compressor. Compress the spring until the end of the spring comes away from the spring seat.

3. Remove the upper strut mount, spring seat and related components.

4. Remove the coil spring from the strut spring compressor.

To install:

➡**Use a new self-locking nut.**

5. Compress the spring and position the strut so that the end of the spring aligns with the notch in the spring seat.

6. Install the upper strut mounting components and tighten the nut to 33 ft. lbs. (44 Nm).

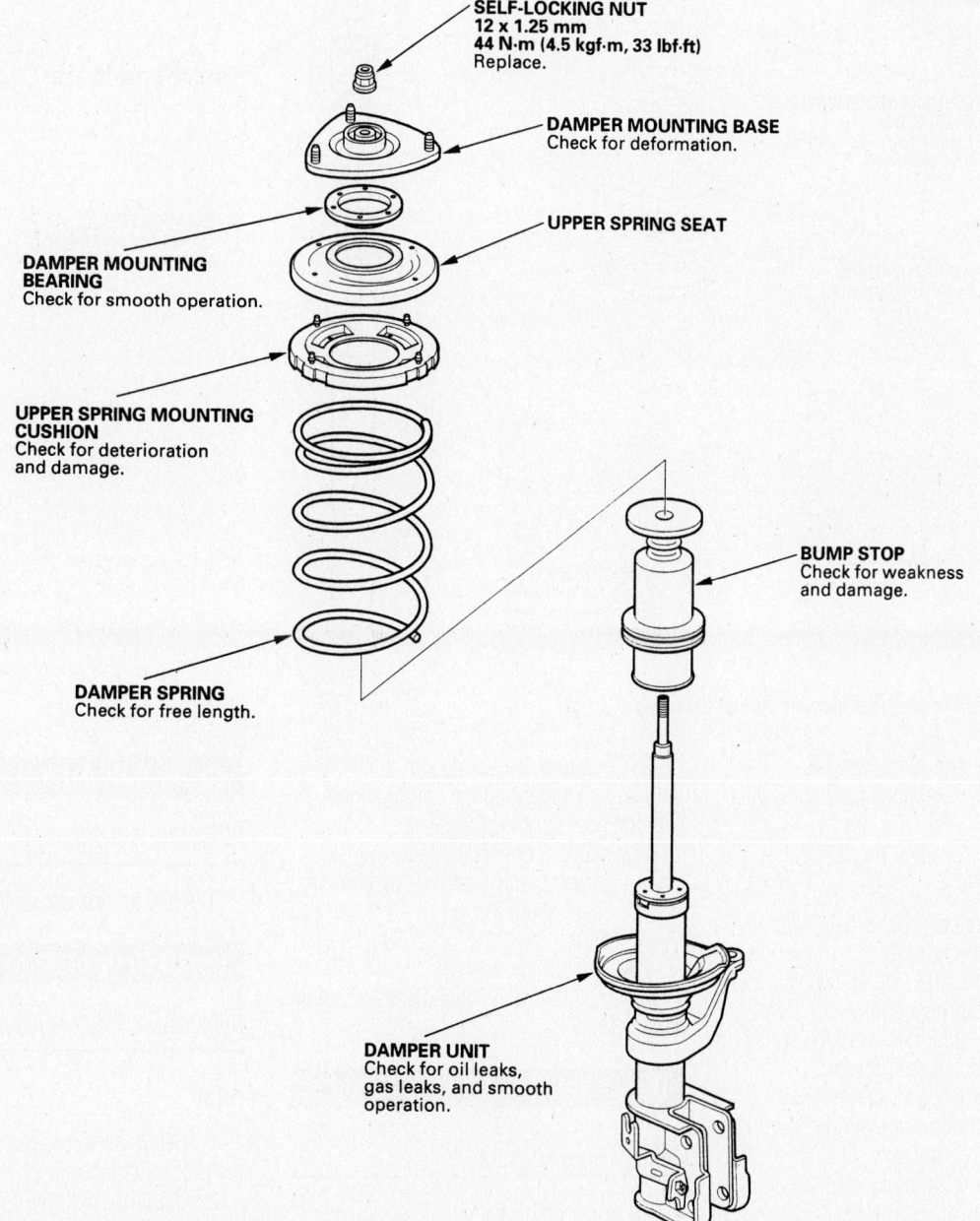

SELF-LOCKING NUT
12 x 1.25 mm
44 N·m (4.5 kgf·m, 33 lbf·ft)
Replace.

DAMPER MOUNTING BASE
Check for deformation.

UPPER SPRING SEAT

DAMPER MOUNTING BEARING
Check for smooth operation.

UPPER SPRING MOUNTING CUSHION
Check for deterioration and damage.

DAMPER SPRING
Check for free length.

BUMP STOP
Check for weakness and damage.

DAMPER UNIT
Check for oil leaks, gas leaks, and smooth operation.

42356-ELEM-G11

Exploded view of the front strut (damper and spring) assembly

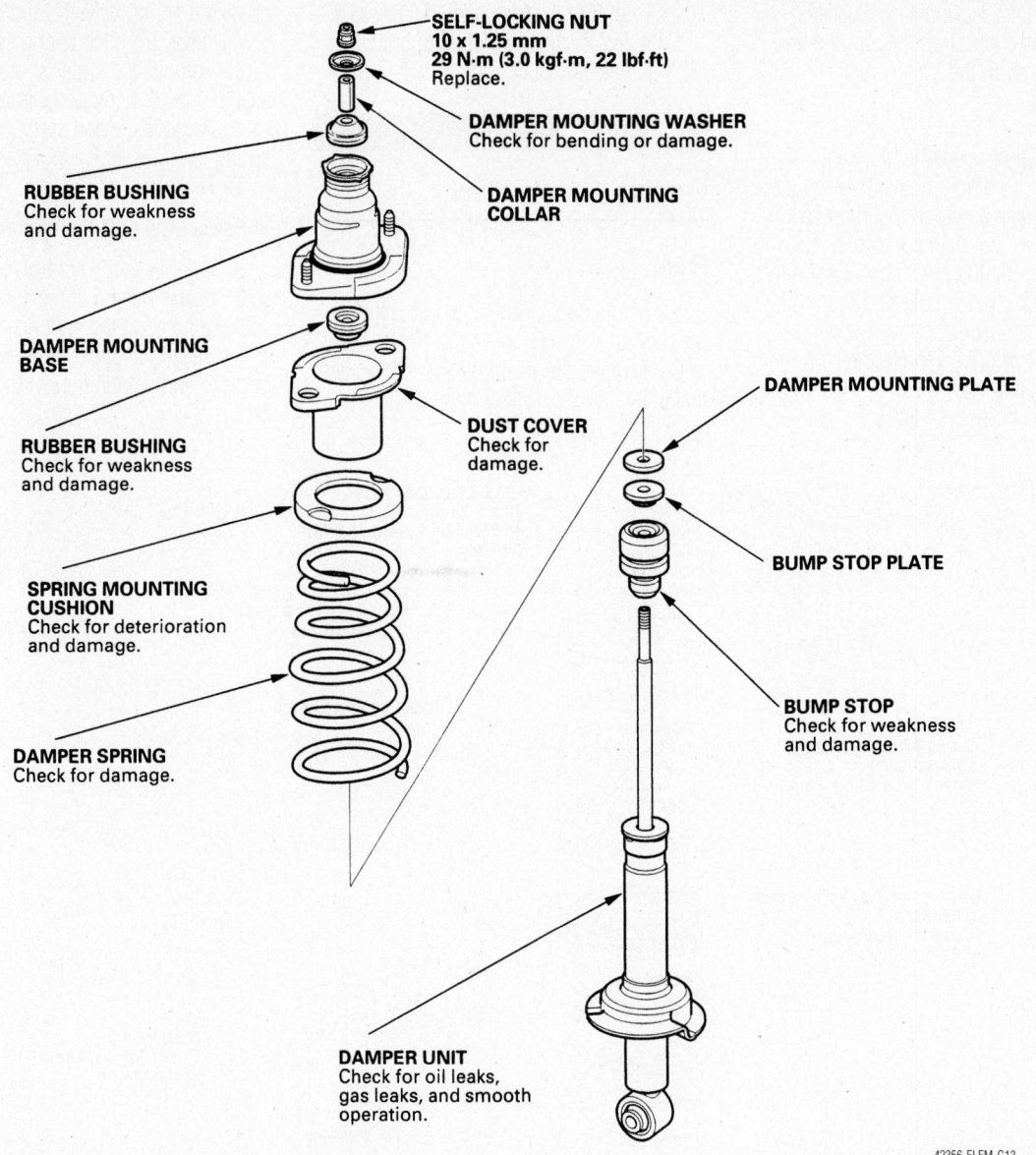

SELF-LOCKING NUT
10 x 1.25 mm
29 N·m (3.0 kgf·m, 22 lbf·ft)
Replace.

DAMPER MOUNTING WASHER
Check for bending or damage.

DAMPER MOUNTING COLLAR

RUBBER BUSHING
Check for weakness and damage.

DAMPER MOUNTING BASE

RUBBER BUSHING
Check for weakness and damage.

DUST COVER
Check for damage.

DAMPER MOUNTING PLATE

SPRING MOUNTING CUSHION
Check for deterioration and damage.

BUMP STOP PLATE

DAMPER SPRING
Check for damage.

BUMP STOP
Check for weakness and damage.

DAMPER UNIT
Check for oil leaks, gas leaks, and smooth operation.

42356-ELEM-G13

Exploded view of the strut (damper and spring) assembly

7. Install the strut to the vehicle.

8. Check the wheel alignment and adjust as necessary.

Rear

1. Before servicing the vehicle, refer to the precautions section.

2. Remove the strut from the vehicle and install in a strut spring compressor. Compress the spring until the end of the spring comes away from the spring seat.

3. Remove or disconnect the following:
- Upper strut mount, spring seat and related components
- Coil spring from the strut spring compressor

To install:

➡**Use a new self-locking nut.**

4. Compress the spring and position the strut so that the end of the spring aligns with the notch in the spring seat.

5. Install or connect the following:
- Upper strut mounting components and tighten the nut to 22 ft. lbs. (29 Nm).
- Strut to the vehicle

6. Check the wheel alignment and adjust as necessary.

Upper Ball Joint

REMOVAL & INSTALLATION

The upper ball joints are replaced with the upper control arms as an assembly.

Lower Ball Joint

REMOVAL & INSTALLATION

The ball joint is not replaceable.

Upper Control Arm

REMOVAL & INSTALLATION

Rear

1. Before servicing the vehicle, refer to the precautions section.

2. Support the lower control arm assembly with a floor jack.

3. Remove or disconnect the following:

- Wheel speed sensor harness bracket, if equipped
- Flange bolts and the control arm

To install:

4. Install all suspension components and fasteners and hand tighten them.. Place a jack under the trailing arm, raise the suspension with the jack and load the jack with the vehicle weight.

5. Tighten the upper control arm flange bolts to 69 ft. lbs. (93 Nm).

6. Clean the wheel, mating surface of the brake disc or drum and inside of the wheel.

7. Check and adjust the wheel alignment as needed.

CONTROL ARM BUSHING REPLACEMENT

The upper control arm bushings are serviced with the upper control arm as an assembly.

Lower Control Arm

REMOVAL & INSTALLATION

Front

1. Before servicing the vehicle, refer to the precautions section.

2. Remove or disconnect the following:
- Front wheel
- Stabilizer link
- Lower arm from the knuckle
- Lower arm

To install:

3. Install all suspension components and fasteners and hand tighten them.. Place a jack under the suspension, raise the suspension with the jack and load the jack with the vehicle weight.

4. Installation is the reverse of removal. Observe the following torques:
- Lower arm bolts: 61 ft. lbs. (83 Nm)
- Ball stud nut: 51 ft. lbs. (69 Nm). Install the cotter pin into the ball joint from the inside to the outside of the vehicle.
- Stabilizer link: 29 ft. lbs. (39 Nm)

CONTROL ARM BUSHING REPLACEMENT

The lower control arm front inner bushing and the damper fork bushing are serviced with the control arm as an assembly.

REAR INNER BUSHING

1. Before servicing the vehicle, refer to the precautions section.

2. Remove or disconnect the following:
- Front wheel
- Rear bushing bracket
- Rear bushing

To install:

➡ **Use a new self-locking nut for assembly.**

3. Install or connect the following:
- Rear bushing. Tighten the nut to 61 ft. lbs. (83 Nm).
- Rear bushing bracket. Tighten the bolts to 66 ft. lbs. (89 Nm).
- Front wheel

4. Check the wheel alignment and adjust as necessary.

Wheel Bearings

ADJUSTMENT

The wheel bearings are sealed units and are not adjustable.

REMOVAL & INSTALLATION

Front

1. Before servicing the vehicle, refer to the precautions section.

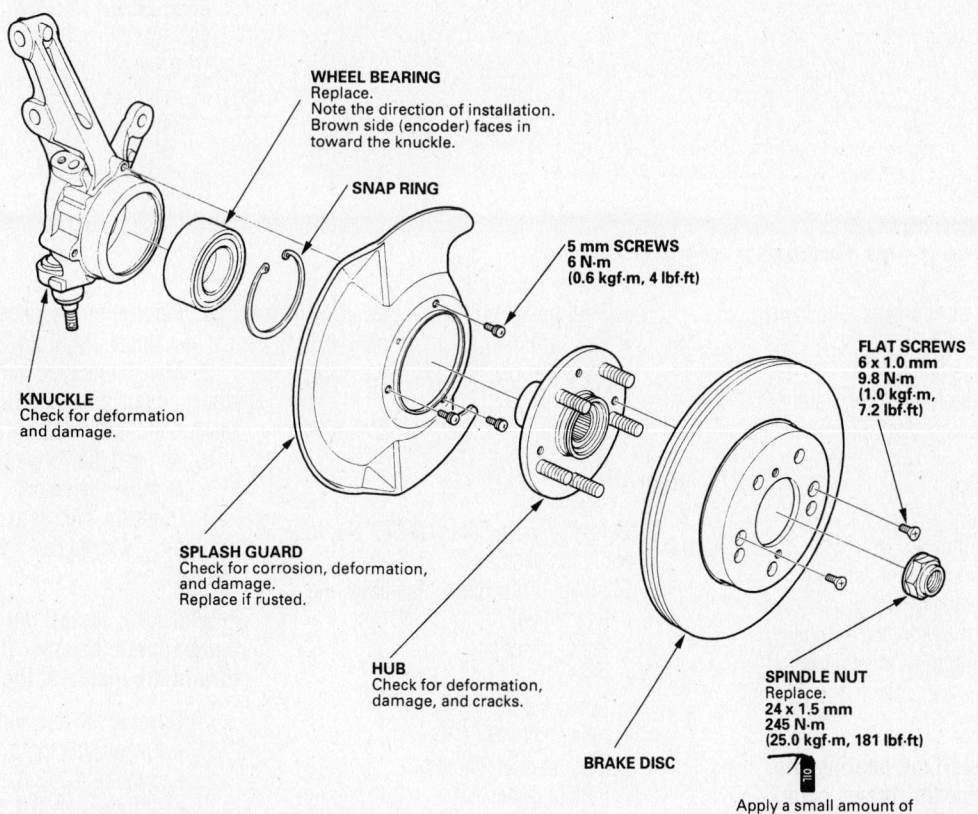

WHEEL BEARING
Replace.
Note the direction of installation.
Brown side (encoder) faces in toward the knuckle.

SNAP RING

5 mm SCREWS
6 N·m
(0.6 kgf·m, 4 lbf·ft)

FLAT SCREWS
6 x 1.0 mm
9.8 N·m
(1.0 kgf·m, 7.2 lbf·ft)

KNUCKLE
Check for deformation and damage.

SPLASH GUARD
Check for corrosion, deformation, and damage.
Replace if rusted.

HUB
Check for deformation, damage, and cracks.

SPINDLE NUT
Replace.
24 x 1.5 mm
245 N·m
(25.0 kgf·m, 181 lbf·ft)

Apply a small amount of engine oil to the seating surface.

BRAKE DISC

42356-ELEM-G14

Exploded view of the front hub, wheel bearing and related components

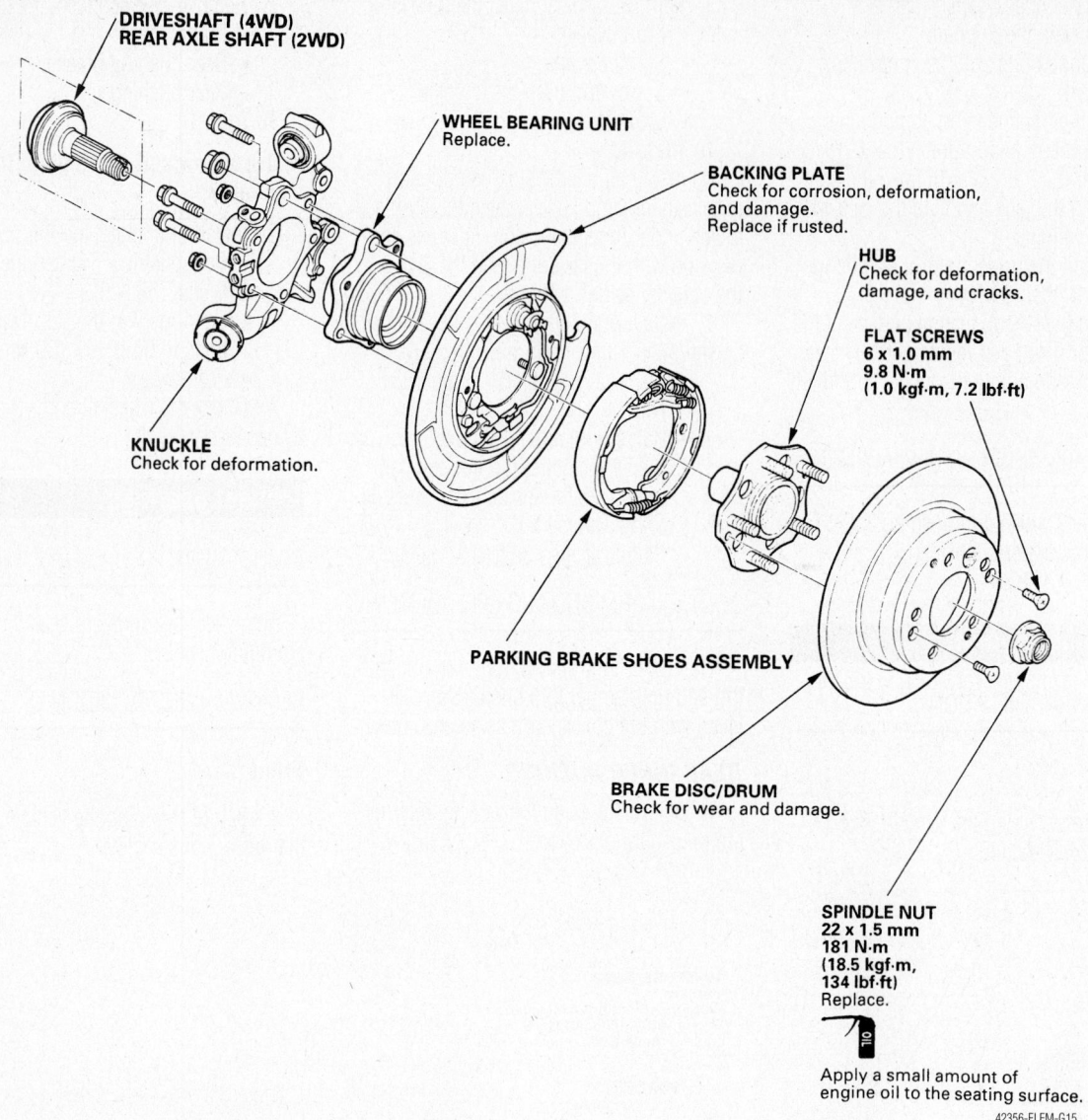

DRIVESHAFT (4WD)
REAR AXLE SHAFT (2WD)

WHEEL BEARING UNIT
Replace.

BACKING PLATE
Check for corrosion, deformation,
and damage.
Replace if rusted.

HUB
Check for deformation,
damage, and cracks.

FLAT SCREWS
6 x 1.0 mm
9.8 N·m
(1.0 kgf·m, 7.2 lbf·ft)

KNUCKLE
Check for deformation.

PARKING BRAKE SHOES ASSEMBLY

BRAKE DISC/DRUM
Check for wear and damage.

SPINDLE NUT
22 x 1.5 mm
181 N·m
(18.5 kgf·m,
134 lbf·ft)
Replace.

Apply a small amount of
engine oil to the seating surface.

42356-ELEM-G15

Exploded view of the rear hub, wheel bearing and related components

2. Remove or disconnect the following:
- Front wheel
- Brake hose bracket
- Brake caliper and rotor. Forcing screws are needed to remove the rotor.
- Spindle nut
- ABS sensor
- Stabilizer link
- Lower arm from the knuckle
- Strut-to-knuckle bolts
- Steering hub/knuckle assembly

3. Press the hub from the knuckle. The bearings and races can now be pressed out and replaced.

➡**With ABS, install the bearing with the magnetic encoder (brown color) toward the inside of the knuckle.**

4. Observe the following torques:

- Strut bolts: 116 ft. lbs. (157 Nm)
- Ball stud nuts: 51 ft. lbs. (69 Nm)
- Stabilizer bar link: 29 ft. lbs. (39 Nm)
- Spindle nut: 181 ft. lbs. (245 Nm)

Rear

1. Before servicing the vehicle, refer to the precautions section.

2. Remove or disconnect the following:
- Rear wheel
- Brake caliper
- Rotor
- Spindle nut
- Axle shaft (4wd)
- Parking brake shoes
- Parking brake cable
- Wheel sensor, if equipped

3. Support the trailing arm.

4. Remove or disconnect the following:
- Upper arm from the knuckle

5. Matchmark the trailing arm cam adjusting bolt and cam. Remove the bolt. Discard the nut.

6. Remove the flange bolt.

7. Remove the knuckle assembly.

8. Press the hub from the knuckle. The bearings and races can now be pressed out and replaced.

➡**With ABS, install the bearing with the magnetic encoder (brown color) toward the inside of the knuckle.**

9. Observe the following torques:
- Flange bolt: 69 ft. lbs. (93 Nm)
- Cam bolts: 43 ft. lbs. (59 Nm)
- Spindle nut: 134 ft. lbs. (181 Nm)
- Caliper mounting bolts: 41 ft. lbs. (55 Nm)

BRAKES

Brake Caliper

REMOVAL & INSTALLATION

Front

1. Remove the wheel
2. Remove the brake hose banjo bolt and washers. Discard the washers.
3. Remove the upper and lower bolts.
4. Lift off the caliper.
5. Remove the pad springs.
6. Remove the pads and shims.
7. Remove the pad retainers.
8. Installation is the reverse of removal. Coat both sides of the shims and the backs of the pads with brake grease. Torque the bolts to 25 ft. lbs. (34 Nm). Install new washers and torque the banjo bolt to 25 ft. lbs. (34 Nm).

Rear

1. Remove the wheel
2. Remove the brake hose banjo bolt and washers. Discard the washers.
3. Remove the caliper pin bolts.
4. Lift off the caliper and suspend it safely.
5. Remove the pads and shims.
6. Remove the pad retainers.
7. Installation is the reverse of removal. Coat both sides of the shims and the backs of the pads with brake grease. Torque the bolts to 16 ft. lbs. (22 Nm). Install new washers and torque the banjo bolt to 25 ft. lbs. (34 Nm).

Disc Brake Pads

REMOVAL & INSTALLATION

Front

1. Remove the lower bolt.
2. Pivot the caliper up and hold the pads.
3. Remove the pad springs.
4. Remove the pads and shims.
5. Remove the pad retainers.
6. Installation is the reverse of removal. Coat both sides of the shims and the backs of the pads with brake grease. Torque the lower bolt to 25 ft. lbs. (34 Nm).

Rear

1. Remove the caliper pin bolts.
2. Lift off the caliper and suspend it safely.

3. Remove the pads and shims.
4. Remove the pad retainers.
5. Installation is the reverse of removal. Coat both sides of the shims and the backs of the pads with brake grease. Torque the bolts to 16 ft. lbs. (22 Nm).

Brake Drums

REMOVAL & INSTALLATION

1. Raise and safely support the vehicle. Release the parking brake.
2. Remove the rear wheels.
3. Use chalk to mark the brake drum to one of the wheel studs as an index mark for reinstallation.
4. Remove the retaining screw that holds the brake drum to the axle flange.
5. Pull the brake drum from the axle flange.

To install:
6. Align the index mark and install the brake drum to the axle flange.
7. Install the retaining screw to secure the brake drum to the axle flange.
8. Install the rear wheels.

Rear Brake Shoes

REMOVAL & INSTALLATION

1. Raise and safely support the vehicle.
2. Remove the rear wheels.
3. Remove the brake drums.
4. Remove the brake return springs.
5. Remove the leading shoe holding pin and spring, and then the leading shoe.
6. Remove the self-adjuster and the adjuster lever.
7. Remove the trailing shoe holding pin and spring.
8. Disconnect the parking brake cable from the trailing shoe and remove the trailing shoe. Remove the parking brake lever from the trailing shoe.

To install:
9. Attach the parking brake lever to the trailing shoe.
10. Connect the parking brake cable to the parking brake lever.
11. Apply a thin coat of high temperature grease to the shoe contact points on the brake backing plate contact surface (B), and self-adjuster (D).

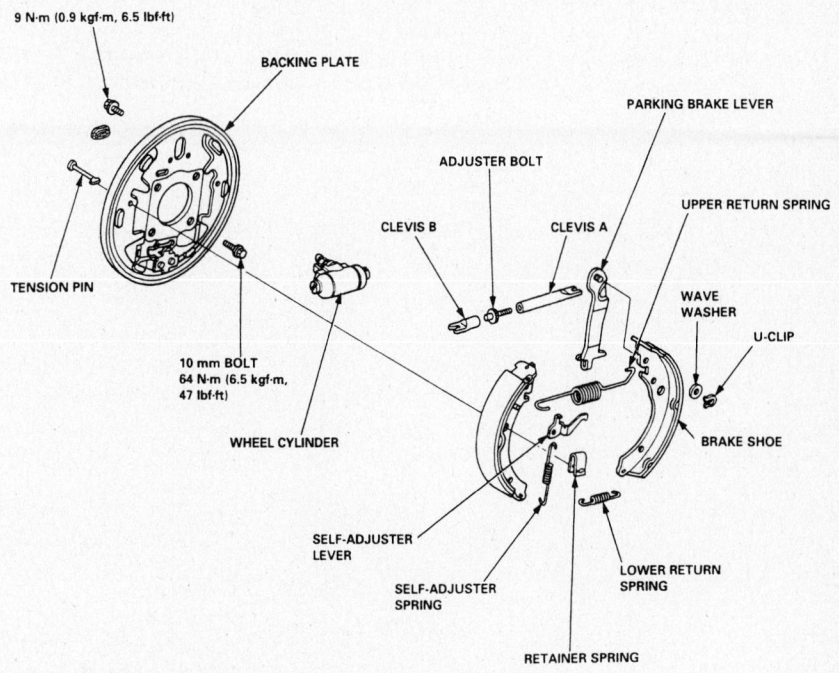

Exploded view of the rear drum brakes

93026G59

12. Position the trailing shoe on the backing plate and install the hold-down pin, spring, and retainer. Be careful not to stretch the return spring when fitting the shoes onto the backing plate.

13. Connect the upper return spring and the leading shoe to the trailing shoe and position the leading brake shoe on the backing plate.

14. Install the adjuster assembly and the hold-down pin, spring, and retainer.

15. Use a brake spring tool to install the lower return spring.

16. Install the self-adjuster lever and adjuster spring.

17. Adjust the shoe-to-drum clearance to 0.0098–0.0157 in. (0.25–0.40mm) and install the brake drum.

18. Check the brake drum for scoring or other wear. Machine or replace as necessary. Check the maximum brake drum diameter specification when machining.

19. Install the rear wheels. Lower the vehicle.

20. Road-test the vehicle.

HONDA

Insight

6

SPECIFICATIONS AND MAINTENANCE CHARTS

ENGINE AND VEHICLE IDENTIFICATION CHART

		Engine Code						Model Year	
Code	Liters (cc)	Cu. In.	Cyl.	Fuel Sys.	Engine Type	Eng. Mfg.		Code ①	Year
ECA1	1.0 (999)	61	3	SMFI	SOHC	Honda		2	2002
								3	2003
SOHC: Single Overhead Cam								4	2004
SMFI: Sequential Multi-port Fuel Injection								5	2005
① 10th position of VIN								6	2006

09474_INSI_C0001

GENERAL ENGINE SPECIFICATIONS

Year	Model	Engine Displacement Liters	Engine ID/VIN	Net Horsepower @ rpm	Net Torque @ rpm (ft. lbs.)	Bore x Stroke (in.)	Compression Ratio	Oil Pressure @ rpm
2002	Insight	1.0	ECA1	71@5700	89@2000	2.83x3.21	①	50@3000
2003	Insight	1.0	ECA1	71@5700	89@2000	2.83x3.21	①	50@3000
2004	Insight	1.0	ECA1	71@5700	89@2000	2.83x3.21	①	50@3000
2005	Insight	1.0	ECA1	71@5700	89@2000	2.83x3.21	①	50@3000
2006	Insight	1.0	ECA1	71@5700	89@2000	2.83x3.21	①	50@3000

① Man. Trans. 10.8:1
Auto. Trans. 10.3:1

09474_INSI_C0002

ENGINE TUNE-UP SPECIFICATIONS

Year	Engine Displacement Liters	Engine ID/VIN	Spark Plug Gap (in.)	Ignition Timing (deg.) MT	Ignition Timing (deg.) AT	Fuel Pump (psi)	Idle Speed (rpm) MT	Idle Speed (rpm) AT	Valve Clearance (in.) In.	Valve Clearance (in.) Ex.
2002	1.0	ECA1	0.039-0.043	10-14B	10-14B	40-47	850-950	850-950	0.007-0.009	0.008-0.010
2003	1.0	ECA1	0.039-0.043	10-14B	10-14B	40-47	850-950	850-950	0.007-0.009	0.008-0.010
2004	1.0	ECA1	0.039-0.043	10-14B	10-14B	40-47	850-950	850-950	0.007-0.009	0.008-0.010
2005	1.0	ECA1	0.039-0.043	10-14B	10-14B	40-47	850-950	850-950	0.007-0.009	0.008-0.010
2006	1.0	ECA1	0.039-0.043	10-14B	10-14B	40-47	850-950	850-950	0.007-0.009	0.008-0.010

NOTE: The Vehicle Emission Control Information label often reflects changes made during production and must be used if they differ from this chart.

NOTE: The fuel pressure readings are given with the vacuum hose connected to the regulator and the engine running

B: Before top dead center

09474_INSI_C0003

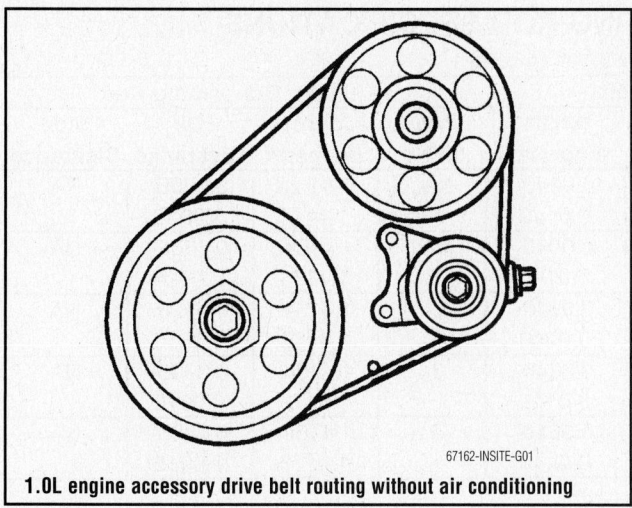

67162-INSITE-G01

1.0L engine accessory drive belt routing without air conditioning

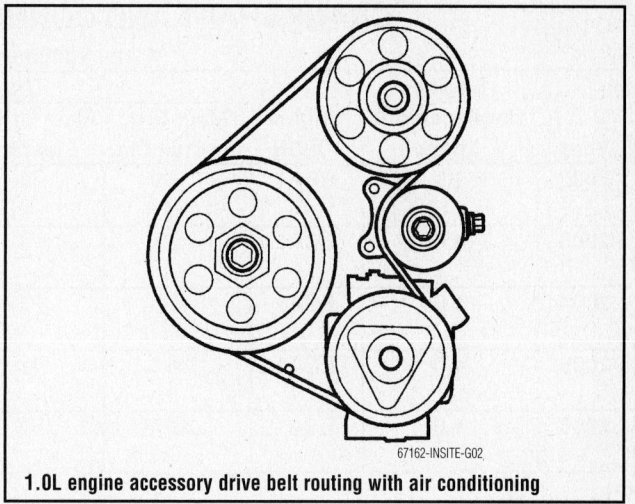

67162-INSITE-G02

1.0L engine accessory drive belt routing with air conditioning

CAPACITIES

Year	Model	Engine Displacement Liters	Engine ID/VIN	Engine Oil with Filter (qts.)	Transmission (pts.)		Drive Axle		Fuel Tank (gal.)	Cooling System (qts.)
					5-Spd	Auto.	Front (pts.)	Rear (pts.)		
2002	Insight	1.0	ECA1	2.6	3.2	7.0	—	—	10.4	4.2
2003	Insight	1.0	ECA1	2.6	3.2	7.0	—	—	10.4	4.2
2004	Insight	1.0	ECA1	2.6	3.2	7.0	—	—	10.4	4.1
2005	Insight	1.0	ECA1	2.6	3.2	7.0	—	—	10.4	4.1
2006	Insight	1.0	ECA1	2.6	3.2	7.0	—	—	10.4	4.1

NOTE: All capacities are approximate. Add fluid gradually and check to be sure a proper fluid level is obtained.

09474_INSI_C0004

VALVE SPECIFICATIONS

Year	Engine Displacement Liters	Engine ID/VIN	Seat Angle (deg.)	Face Angle (deg.)	Spring Test Pressure (lbs. @ in.)	Spring Installed Height (in.)	Stem-to-Guide Clearance (in.)		Stem Diameter (in.)	
							Intake	Exhaust	Intake	Exhaust
2002	1.0	ECA1	45	45	NA	①	0.0010-0.0020	0.0020-0.0031	0.2157-0.2161	0.2146-0.2150
2003	1.0	ECA1	NA	NA	NA	①	0.0010-0.0020	0.0020-0.0031	0.2157-0.2161	0.2146-0.2150
2004	1.0	ECA1	NA	NA	NA	①	0.0010-0.0020	0.0020-0.0031	0.2157-0.2161	0.2146-0.2150
2005	1.0	ECA1	NA	NA	NA	①	0.0010-0.0020	0.0020-0.0030	0.2157-0.2161	0.2146-0.2150
2006	1.0	ECA1	NA	NA	NA	①	0.0010-0.0020	0.0020-0.0030	0.2157-0.2161	0.2146-0.2150

NA: Not Available

① Valve spring free length:
   Intake: 2.689 in.
   Exhaust: 2.800 in.

09474_INSI_C0005

CRANKSHAFT AND CONNECTING ROD SPECIFICATIONS

All measurements are given in inches

Year	Engine Displacement Liters	Engine ID/VIN	Crankshaft				Connecting Rod		
			Main Brg. Journal Dia.	Main Brg. Oil Clearance	Shaft End-play	Thrust on No.	Journal Diameter	Oil Clearance	Side Clearance
2002	1.0	ECA1	①	②	0.0040-0.0140	NA	1.4164-1.4173	0.0008-0.0015	NA
2003	1.0	ECA1	①	②	0.0040-0.0140	NA	1.4164-1.4173	0.0008-0.0015	NA
2004	1.0	ECA1	①	②	0.0040-0.0140	NA	1.4164-1.4173	0.0008-0.0015	NA
2005	1.0	ECA1	①	②	0.0040-0.0140	3	1.4164-1.4173	0.0008-0.0015	NA
2006	1.0	ECA1	①	②	0.0040-0.0140	3	1.4164-1.4173	0.0008-0.0015	NA

NA: Not available

① Nos. 1 and 4: 1.5741-1.5750
   Nos. 2 and 3: 1.5739-1.5748

② Nos. 1 and 4: 0.0006-0.0013
   Nos. 2 and 3: 0.0008-0.0015

09474_INSI_C0006

PISTON AND RING SPECIFICATIONS

All measurements are given in inches

Year	Engine Displacement Liters	Engine ID/VIN	Piston Clearance	Ring Gap			Ring Side Clearance		
				Top Compression	Bottom Compression	Oil Control	Top Compression	Bottom Compression	Oil Control
2002	1.0	ECA1	0.0002-0.0017	0.0060-0.0120	0.0140-0.0200	0.0080-0.0200	0.0022-0.0031	0.0012-0.0022	NA
2003	1.0	ECA1	0.0002-0.0017	0.0060-0.0120	0.0140-0.0200	0.0080-0.0280	0.0022-0.0031	0.0012-0.0022	NA
2004	1.0	ECA1	0.0002-0.0017	0.0060-0.0120	0.0140-0.0200	0.0080-0.0280	0.0022-0.0031	0.0012-0.0022	NA
2005	1.0	ECA1	0.0002-0.0017	0.0060-0.0120	0.0140-0.0200	0.0080-0.0280	0.0022-0.0031	0.0012-0.0022	NA
2006	1.0	ECA1	0.0002-0.0017	0.0060-0.0120	0.0140-0.0200	0.0080-0.0280	0.0022-0.0031	0.0012-0.0022	NA

NA: Not Available

09474_INSI_C0007

TORQUE SPECIFICATIONS
All readings in ft. lbs.

Year	Engine Displacement Liters	Engine ID/VIN	Cylinder Head Bolts	Main Bearing Bolts	Rod Bearing Bolts	Crankshaft Damper Bolts	Flywheel Bolts	Manifold		Spark Plugs	Oil Pan Drain Plug
								Intake	Exhaust		
2002	1.0	ECA1	①	②	③	④	33	16	17	17	29
2003	1.0	ECA1	①	②	③	④	33	16	17	17	29
2004	1.0	ECA1	①	②	③	④	33	16	17	17	29
2005	1.0	ECA1	①	②	③	④	33	16	17	17	29
2006	1.0	ECA1	①	②	③	④	33	16	17	17	29

NOTE: Dip connecting rod and main bearing bolts and crankshaft damper bolt in clean engine oil prior to tightening.

① Step 1: 29 ft. lbs.
Step 2: +90 degrees
Step 3: 6 mm bolts to 106 inch lbs.

② Step 1: 18 ft. lbs.
Step 2: +60 degrees

③ Step 1: 87 inch lbs.
Step 2: +90 degrees

④ Step 1: 14 ft. lbs.
Step 2: +90 degrees

09474_INSI_C0008

WHEEL ALIGNMENT

Year	Model		Caster		Camber		Toe-in (in.)
			Range (+/-Deg.)	Preferred Setting (Deg.)	Range (+/-Deg.)	Preferred Setting (Deg.)	
2002	Insight	F	1.00	+2.00	1.00	0	0+/- 0.08
		R	—	—	1.00	-1.00	0.12+/- 0.12
2003	Insight	F	1.00	+2.00	1.00	0	0+/- 0.08
		R	—	—	1.00	-1.00	0.12+/- 0.12
2004	Insight	F	1.00	+2.00	1.00	0	0+/- 0.08
		R	—	—	1.00	-1.00	0.12+/- 0.12
2005	Insight	F	1.00	+2.00	1.00	0	0+/- 0.08
		R	—	—	1.00	-1.00	0.12+/- 0.12
2005	Insight	F	1.00	+2.00	1.00	0	0+/- 0.08
		R	—	—	1.00	-1.00	0.12+/- 0.12

09474_INSI_C0009

TIRE, WHEEL AND BALL JOINT SPECIFICATIONS

Year	Model	OEM Tires		Tire Pressures (psi)		Wheel Size	Ball Joint Inspection	Lug Nuts
		Standard	Optional	Front	Rear			
2002	Insight	P165/65R14 78S	None	38	35	NS	NS	80
2003	Insight	P165/65R14 78S	None	38	35	NS	NS	80
2004	Insight	P165/65R14 78S	None	38	35	NS	NS	80
2005	Insight	P165/65R14 78S	None	38	35	NS	NS	80
2006	Insight	P165/65R14 78S	None	38	35	NS	NS	80

OEM: Original Equipment Manufacturer

PSI: Pounds Per Square Inch

NS: Not specified by manufacturer

09474_INSI_C0010

BRAKE SPECIFICATIONS
All measurements in inches unless noted

Year	Model		Brake Disc			Brake Drum Diameter			Minimum Lining Thickness		Brake Caliper	
			Original Thickness	Minimum Thickness	Maximum Runout	Original Inside Diameter	Max. Wear Limit	Maximum Machine Diameter	Front	Rear	Bracket Bolts (ft. lbs.)	Mounting Bolts (ft. lbs.)
2002	Insight	F	0.670	0.590	0.002	—	—	—	0.080	—	80	36
		R	—	—	—	7.08	7.12	7.12	—	0.040	—	—
2003	Insight	F	0.670	0.590	0.002	—	—	—	0.080	—	80	36
		R	—	—	—	7.08	7.12	7.12	—	0.040	—	—
2004	Insight	F	0.670	0.590	0.002	—	—	—	0.080	—	80	36
		R	—	—	—	7.08	7.12	7.12	—	0.040	—	—
2005	Insight	F	0.665-0.673	0.590	0.002	—	—	—	0.060	—	23	16
		R	—	—	—	7.08	7.126	7.126	—	0.040	—	—
2006	Insight	F	0.665-0.673	0.590	0.002	—	—	—	0.060	—	23	16
		R	—	—	—	7.08	7.126	7.126	—	0.040	—	—

F: Front

R: Rear

09474_INSI_C0011

SCHEDULED MAINTENANCE INTERVALS

2002-03 HONDA—Insight

TO BE SERVICED	TYPE OF SERVICE	VEHICLE MILEAGE INTERVAL (x1000)															
		7.5	15	22.5	30	37.5	45	52.5	60	67.5	75	82.5	90	97.5	105	112.5	120
Accessory drive belts	I & A				✓				✓				✓				✓
Air cleaner element	R				✓				✓				✓				✓
Air conditioning filter	R				✓				✓				✓				✓
Brake fluid	R	Replace every 3 years															
Brake hoses & lines	I		✓		✓		✓		✓		✓		✓		✓		✓
Cooling system	I		✓		✓		✓		✓		✓		✓		✓		✓
Engine coolant	R														✓		
Engine oil	R	✓	✓	✓	✓	✓	✓	✓	✓	✓	✓	✓	✓	✓	✓	✓	✓
Engine oil and coolant levels	I	Inspect at each fuel stop															
Engine oil filter	R		✓		✓		✓		✓		✓		✓		✓		✓
Exhaust system	I		✓		✓		✓		✓		✓		✓		✓		✓
Fluid levels and condition	I		✓		✓		✓		✓		✓		✓		✓		✓
Front and rear brakes	I		✓		✓		✓		✓		✓		✓		✓		✓
Fuel lines & connection	I		✓		✓		✓		✓		✓		✓		✓		✓
Halfshaft boots	I		✓		✓		✓		✓		✓		✓		✓		✓
Idle speed	I & A														✓		
Parking brake system	I & A		✓		✓		✓		✓		✓		✓		✓		✓
Rotate and inspect tires	I	✓	✓	✓	✓	✓	✓	✓	✓	✓	✓	✓	✓	✓	✓	✓	✓
Spark plugs	R				✓				✓				✓				✓
Supplemental Restrain system	I	Inspect the SRS 10 years after production															
Suspension components	I		✓		✓		✓		✓		✓		✓		✓		✓
Tie rod ends, steering gear box & boots	I		✓		✓		✓		✓		✓		✓		✓		✓
Automatic transmission fluid	R				✓				✓				✓				✓
Manual transmission fluid	R																✓
Valve clearance	I				✓				✓				✓				✓

R: Replace   I: Inspect   A: Adjust

FREQUENT OPERATION MAINTENANCE (SEVERE SERVICE)

If a vehicle is operated under any of the following conditions it is considered severe service:

- Towing a trailer or using a camper or car-top carrier.
- Repeated short trips of less than 5 miles in temperatures below freezing, or trips of less than 10 miles in any temperature.
- Extensive idling or low-speed driving for long distances as in heavy commercial use, such as delivery, taxi or police cars.
- Operating on rough, muddy or salt-covered roads.
- Operating on unpaved or dusty roads.
- Driving in extremely hot (over 90°) conditions.

Air cleaner element: replace every 15,000 miles

Engine oil and filter: replace every 3750 miles or 6 months, whichever occurs first.

Automatic transmission fluid: replace every 15,000 miles.

Front and rear brakes: inspect every 7500 miles or 6 months, whichever occurs first.

Locks and hinges: lubricate every 15,000 miles.

Tie rods, steering gear box, boots: inspect every 7500 miles or 6 months, whichever occurs first.

Suspension components: inspect every 7500 miles or 6 months, whichever occurs first.

Halfshaft boots: inspect every 7500 miles or 6 months, whichever occurs first.

09474_INSI_C0012

SCHEDULED MAINTENANCE INTERVALS

2004-06 HONDA—Insight

TO BE SERVICED	TYPE OF SERVICE	VEHICLE MILEAGE INTERVAL (x1000)															
		7.5	15	22.5	30	37.5	45	52.5	60	67.5	75	82.5	90	97.5	105	112.5	120
Accessory drive belts	I & A				✓				✓				✓				✓
Air cleaner element	R				✓				✓				✓				✓
Air conditioning filter	R				✓				✓				✓				✓
Brake fluid	R	Replace every 3 years															
Brake hoses & lines	I		✓		✓		✓		✓		✓		✓		✓		✓
Cooling system	I		✓		✓		✓		✓		✓		✓		✓		✓
Engine coolant	R								✓								
Engine oil	R	✓	✓	✓	✓	✓	✓	✓	✓	✓	✓	✓	✓	✓	✓	✓	✓
Engine oil and coolant levels	I	Inspect at each fuel stop															
Engine oil filter	R		✓		✓		✓		✓		✓		✓		✓		✓
Exhaust system	I		✓		✓		✓		✓		✓		✓		✓		✓
Fluid levels and condition	I		✓		✓		✓		✓		✓		✓		✓		✓
Front and rear brakes	I		✓		✓		✓		✓		✓		✓		✓		✓
Fuel lines & connection	I		✓		✓		✓		✓		✓		✓		✓		✓
Halfshaft boots	I		✓		✓		✓		✓		✓		✓		✓		✓
Idle speed	I & A													✓			
Parking brake system	I & A		✓		✓		✓		✓		✓		✓		✓		✓
Rotate and inspect tires	I	✓	✓	✓	✓	✓	✓	✓	✓	✓	✓	✓	✓	✓	✓	✓	✓
Spark plugs	R																✓
Supplemental Restrain system	I	Inspect the SRS 10 years after production															
Suspension components	I		✓		✓		✓		✓		✓		✓		✓		✓
Tie rod ends, steering gear box & boots	I		✓		✓		✓		✓		✓		✓		✓		✓
Automatic transmission fluid	R				✓				✓				✓				✓
Manual transmission fluid	R																✓
Valve clearance	I																✓

R: Replace   I: Inspect   A: Adjust

FREQUENT OPERATION MAINTENANCE (SEVERE SERVICE)

If a vehicle is operated under any of the following conditions it is considered severe service:

- Towing a trailer or using a camper or car-top carrier.
- Repeated short trips of less than 5 miles in temperatures below freezing, or trips of less than 10 miles in any temperature.
- Extensive idling or low-speed driving for long distances as in heavy commercial use, such as delivery, taxi or police cars.
- Operating on rough, muddy or salt-covered roads.
- Operating on unpaved or dusty roads.
- Driving in extremely hot (over 90°) conditions.

Air cleaner element: replace every 15,000 miles

Engine oil and filter: replace every 3750 miles or 6 months, whichever occurs first.

Automatic transmission fluid: replace every 15,000 miles.

Front and rear brakes: inspect every 7500 miles or 6 months, whichever occurs first.

Locks and hinges: lubricate every 15,000 miles.

Tie rods, steering gear box, boots: inspect every 7500 miles or 6 months, whichever occurs first.

Suspension components: inspect every 7500 miles or 6 months, whichever occurs first.

Halfshaft boots: inspect every 7500 miles or 6 months, whichever occurs first.

ENGINE REPAIR

➡Disconnecting the negative battery cable on some vehicles may interfere with the functions of the on board computer system. The computer may undergo a relearning process once the negative battery cable is reconnected.

Distributor

The 1.0L engine does not use a distributor.

Alternator

REMOVAL

The 1.0L engine has a DC electric converter and therefore does not use a standard alternator.

Ignition Timing

ADJUSTMENT

Adjustment is not possible on the 1.0L engine. Use the following procedure to check the timing.

CHECKING

1. Connect the Honda Diagnostic System (HDS) to the Data Link Connector (DLC) and check for Diagnostic Trouble Codes (DTC's). If any DTC's are present, diagnose and repair the cause.
2. Start the engine and hold it at 3000 rpm until the cooling fan comes on, then return to idle.
3. Place the HDS tester in the "SCS" mode.
4. Connect a timing light to the no. 1 ignition coil wire.
5. With all accessories off, check the ignition timing on the oil pump pointer at the crankshaft pulley. Timing should be 10–14 degrees BTDC.
6. If the timing is off, check the cam timing. If the cam timing is okay, check the Electronic Control Module (ECM) for the latest software and update, or substitute a known good ECM and check the timing again.
7. If the timing is okay with a substitute ECM, replace the original ECM.

Engine Assembly

REMOVAL & INSTALLATION

➡The engine, electric motor and transaxle are removed from the vehicle as a unit.

1. Before servicing the vehicle, refer to the Precautions Section. Also see the IMA system disabling before proceeding with any procedure.
2. Turn the battery module switch off and measure the voltage.
3. Disable the IMA system.
4. Drain the cooling system, engine oil and transmission fluids.
5. Relieve the fuel system pressure.
6. Disconnect the battery cables and remove the battery.
7. Remove the battery box.
8. Remove the engine appearance cover.
9. Remove the breather pipe and brake booster vacuum line bracket, then remove the air cleaner and intake air duct.
10. Disconnect the throttle cable linkage.
11. Remove or disconnect the following:
    - Brake booster vacuum hose
    - EVAP canister hose
    - Starter ground and positive cables
12. Mark the location and disconnect the "U", "V" and "W" phase power cables from the electric motor.
13. Remove or disconnect the following:
    - Electronic Control Module (ECM) and the main wiring harness connectors
    - 2002–2005 Models: Fuel feed and return lines
    - 2006 Model: Fuel feed line
    - Accessory drive belts
    - Clutch slave cylinder, if equipped

Power Cables
Cover

9.8 N·m
(1.0 kgf·m, 7.2 lbf·ft)

67162-INSITE-G03

Electric motor power cable locations—Insight

- Transmission shift cables
- Electric motor power cable clamps
14. Raise and support the vehicle.
15. Remove or disconnect the following:
    - Engine splash shields and brackets
    - Lower control arm ball joints
    - Axle halfshafts
    - A/C compressor without disconnecting refrigerant lines, if equipped
    - Oxygen sensor and catalytic converter
    - Motor power cable and cable holder
    - Radiator and heater hoses
    - Shift cables
    - Transmission oil cooler lines, if equipped
16. Attach a suitable engine chain hoist device to the engine.
17. Remove the rear and side engine mounting nuts and bolts.
18. Remove the transaxle mount nuts and bolts.
19. Ensure that all vacuum hoses, wiring harnesses and fuel lines are free from interfering with the engine removal.
20. Place a suitable rolling jack under the engine assembly, and lower the chain hoist until the engine assembly is on the jack.
21. Remove the engine/electric motor/transmission from the vehicle.
To install:

➡Use new self-locking nuts and color-coded self-locking bolts when installing the engine mounts and suspension components.

➡Do not tighten the engine or transaxle mount fasteners until instructed to do so.

22. Raise the powertrain into position.
23. Install or connect the following:
    - Transaxle mount and bracket. Tighten the bolts and nut to 41 ft. lbs. (64 Nm) on manual transmission, or 76 ft. lbs. (103 Nm) on automatic transmission. Loosen the mount through bolt.
    - Side engine mount. Tighten the bolt and nuts to 38 ft. lbs. (52 Nm). Loosen the mount through bolt
    - Upper bracket. Tighten the nuts in sequence to 54 ft. lbs. (74 Nm).
    - Raise the vehicle
    - Rear mount. Tighten the bolts to 54 ft. lbs. (74 Nm). Loosen the through bolt.

- Lower the vehicle and remove the chain hoist.
- Tighten the transaxle mount through bolt to 40 ft. lbs. (54 Nm).
- Tighten the side mount through bolt to 40 ft. lbs. (54 Nm).
- Tighten the rear mount through bolt to 66 ft. lbs. (89 Nm).
- Motor power cable and cable holder
- Left engine undercover
- Oxygen sensor and catalytic converter
- A/C compressor
- Axle halfshafts
- Lower control arm ball joints
- Engine splash shields and brackets
- Lower the vehicle
- Electric motor power cable clamps
- Transmission oil cooler lines
- Shift cables
- Accessory drive belts
- Radiator and heater hoses
- ECM and main wiring harness connectors
- Motor power cables
- 2002–2005 Models: Fuel feed and return lines
- 2006 Model: Fuel feed line
- Starter ground and positive cables
- EVAP canister hose
- Brake booster vacuum hose
- Throttle cable linkage
- Clutch slave cylinder
- Air cleaner and intake duct
- Breather pipe and brake booster vacuum line bracket
- Battery box
- Battery cables
- Fill the cooling system, engine oil and transmission fluids
- Engine appearance cover

24. Check that the IMA battery level indicator reads full.

25. If the battery level indicator displays no segments, start the engine and hold between 3500-4000 rpm with no load until the IMA battery level indicator displays at least three segments.

26. Check the engine idle.

Water Pump

REMOVAL & INSTALLATION

1. Before servicing the vehicle, refer to the Precautions Section. Also see the IMA system disabling before proceeding with any procedure.

2. Drain the cooling system.

3. Remove or disconnect the following:

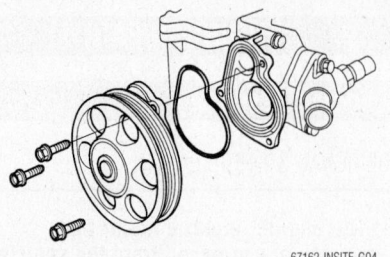

67162-INSITE-G04

Exploded view of the water pump mounting—Insight

- Negative battery cable
- Accessory drive belt
- Water pump

To install:

4. Install or connect the following:
- Water pump. Use a new seal and tighten the bolts to 105 inch lbs. (12 Nm).
- Accessory drive belt
- Negative battery cable

5. Fill the cooling system.

6. Start the engine and check for leaks.

Heater Core

REMOVAL & INSTALLATION

1. Before servicing the vehicle, refer to the Precautions Section. Also see the IMA system disabling before proceeding with any procedure.

2. Disconnect the negative battery cable and wait for 3 minutes.

3. Place the front wheels in the straight ahead position.

4. Drain the cooling system into a clean container for reuse.

5. In the engine compartment, open the heater valve cable clamp and disconnect the cable from the heater valve arm. Then, turn the heater valve to the fully opened position.

6. Disconnect the heater hoses from the heater core.

7. Remove the heater housing-to-chassis nut.

➡**When removing the heater housing nut, be careful not to damage or bend the fuel lines, the brake lines, etc.**

8. Remove the front center floor console.

9. Remove the glove box.

10. Remove the center lower cover.

11. Remove the radio.

12. Remove the steering wheel access cover and disconnect the driver air bag connector.

13. Remove the steering wheel cover and remove the driver air bag.

14. Loosen the steering wheel nut, then use a puller and remove the steering wheel.

15. Remove the steering column covers.

16. Disconnect the cable wheel connectors and remove the cable reel.

17. Disconnect the combination switch and ignition switch connectors.

18. Remove the steering joint pinch bolts.

19. Remove the steering column attaching nuts and remove the steering column.

20. Disconnect the passenger air bag connector.

21. Remove the 3 air bag mounting nuts, then pry carefully with a screwdriver and lift the air bag out of the dashboard.

22. Remove the A pillar trim on both sides.

23. From under and behind the dashboard, disconnect and label all electrical connectors.

24. Remove the driver coin tray.

25. Remove the dashboard attaching bolts (A) and screws (B) as shown.

26. Carefully lift up on the dashboard and disengage it from the holders (C) and (D).

27. Remove the dashboard through the door opening.

28. Discharge and recover the air conditioning system refrigerant.

29. In the engine compartment, remove the refrigerant lines-to-evaporator housing bolts.

30. Separate the lines, discard the grommets and plug the openings to prevent contamination.

31. Disconnect the evaporator housing's temperature sensor connector.

32. Remove the evaporator housing-to-chassis screws/nut and remove the evaporator housing.

33. Disconnect the mode control motor and the air mix control motor electrical connectors and remove the wiring harness clips and the wiring harness from the heater housing.

34. Remove the heater duct clip, the heater housing-to-chassis nuts and the heater housing.

35. Remove the heater core cover screws and the cover.

36. Remove the heater core pipe clamp screws and the clamp.

37. Remove the heater core from the heater housing.

To install:

38. Install the heater core in the heater housing.

39. Install the heater core pipe clamp and the clamp screws.

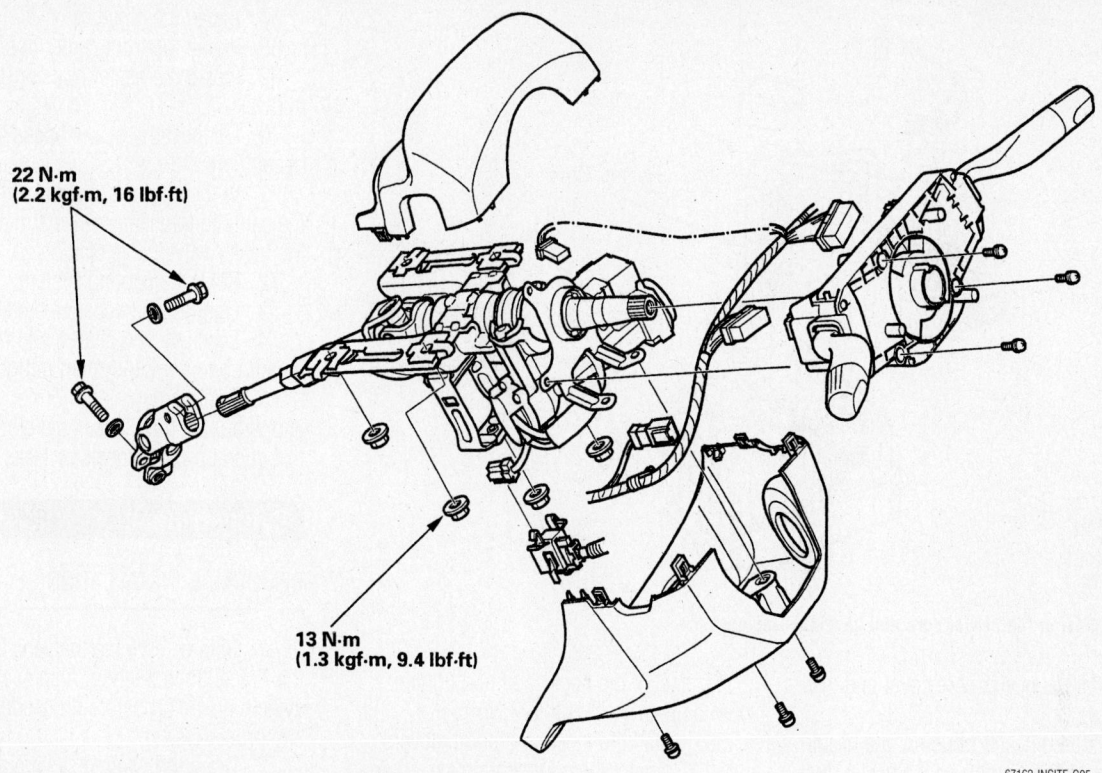

22 N·m
(2.2 kgf·m, 16 lbf·ft)

13 N·m
(1.3 kgf·m, 9.4 lbf·ft)

67162-INSITE-G05

Exploded view of the steering column and related components—Insight

Fastener Locations

A ▶ : Bolt, 7   B ▶ : Screw, 2

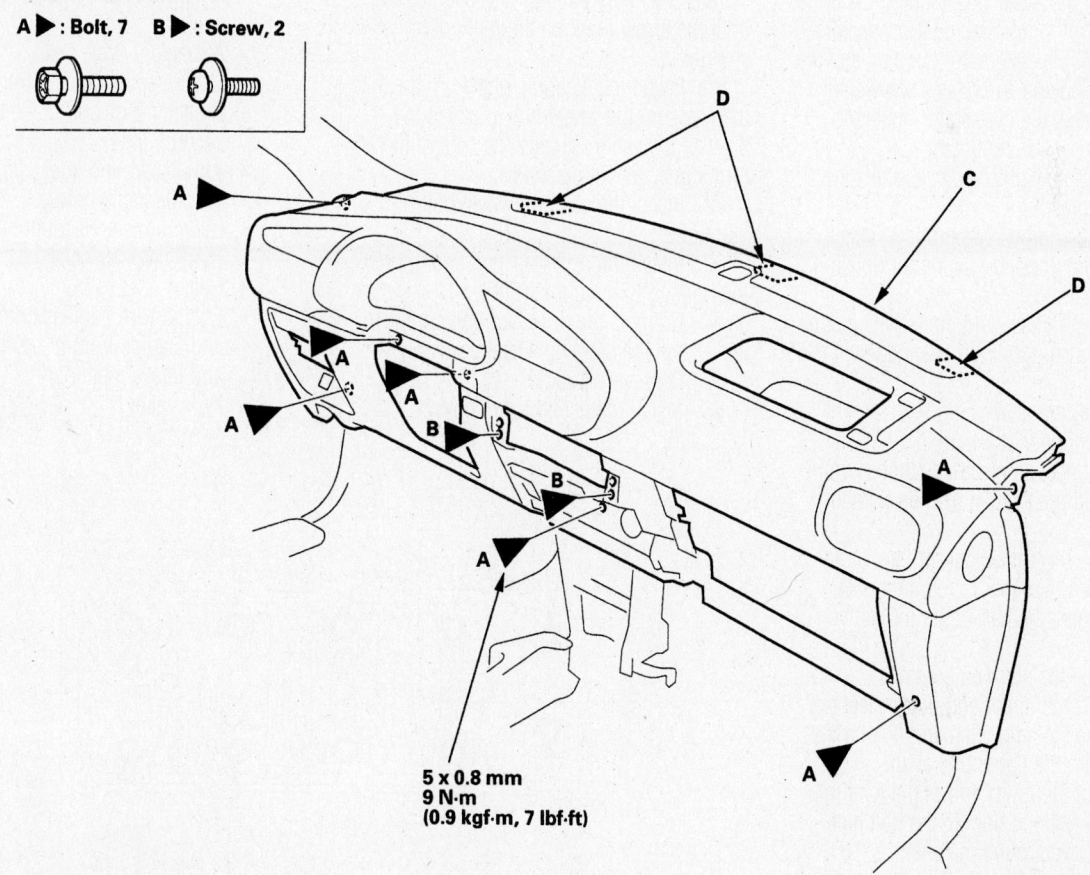

5 x 0.8 mm
9 N·m
(0.9 kgf·m, 7 lbf·ft)

67162-INSITE-G06

Exploded view of the instrument panel—Insight

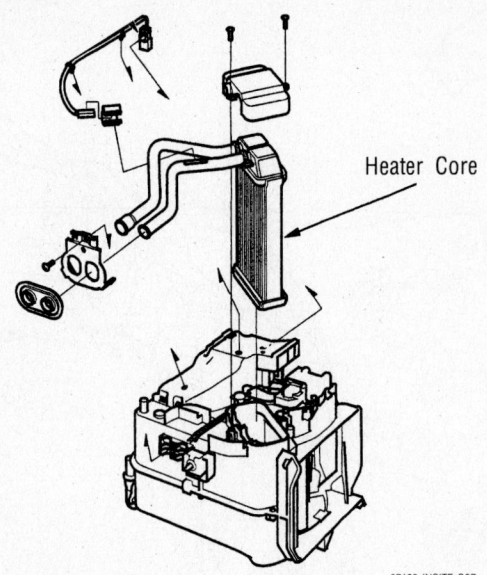

Heater Core

67162-INSITE-G07

Exploded view of the heater core and housing—Insight

40. Install the heater core cover and the cover screws.

41. Install the heater housing, the heater housing-to-chassis nuts and the heater duct clip.

42. Install the wiring harness clips and the wiring harness to the heater housing and connect the mode control motor and the air mix control motor electrical connectors.

43. Install the evaporator housing and the evaporator housing-to-chassis screws/nut.

44. Connect the evaporator housing's temperature sensor connector.

45. Using new grommets, connect the refrigerant lines.

46. In the engine compartment, install the refrigerant lines-to-evaporator housing bolts.

47. Using an assistant, install the instrument panel making sure it is engaged in the holders.

48. Install the driver coin tray.

49. From under and behind the dashboard, reconnect all electrical connectors.

50. Install the A pillar trim on both sides.

51. Place the passenger air bag in the dashboard and tighten 3 mounting nuts.

52. Connect the passenger air bag connector.

53. Install the steering column.

54. Insert the upper end of the steering joint into the steering shaft and line up the bolt hole with the flat on the shaft.

55. Insert the lower end of the joint into the pinion shaft and line up the bolt hole with the groove around the shaft.

56. Install the lower bolt and pull up on the joint to ensure it is securely seated.

57. Install the upper bolt and tighten both bolts to 16 ft. lbs. (22 Nm).

58. Install the steering column covers.

59. Connect the combination switch and ignition switch connectors.

60. Install the cable and center it by rotating it clockwise until it stops. Rotate it counterclockwise about 2 and one half turns until the arrow mark on the reel point straight up.

61. Install the steering wheel on the shaft making sure the wheel hub engages the pins on the cable reel and the tabs of the turn signal canceling sleeve.

62. Install the steering wheel bolt nut and tighten it to 29 ft. lbs. (39 Nm).

63. Install the driver air bag and tighten the bolts to 87 inch lbs. (9.8 Nm).

64. Connect the driver air bag connector and install the steering wheel access cover.

65. Remove the radio.

66. Install the center lower cover.

67. Install the glove box

68. Install the front center floor console.

69. Install the heater housing-to-chassis nut.

70. Connect the heater hoses to the heater core.

71. In the engine compartment, connect the cable to the heater valve arm and close the heater valve cable clamp.

72. Refill the cooling system.

73. Connect the negative battery cable.

74. Evacuate and charge and leak test the air conditioning system refrigerant.

75. Run the engine to normal operating temperatures; then, check the climate control operation and check for leaks.

Cylinder Head

REMOVAL & INSTALLATION

1. Before servicing the vehicle, refer to the Precautions Section. Also see the IMA system disabling before proceeding with any procedure.

2. Drain the cooling system and engine oil.

3. Relieve the fuel system pressure.

4. Remove or disconnect the following:
- Negative and then positive battery cables
- Accessory drive belt
- Engine appearance cover
- Breather pipe and brake booster vacuum line
- Air cleaner and intake air duct
- Throttle cable linkage
- Brake booster vacuum hose
- EVAP canister hose
- Heater hoses
- Engine wire harness connectors and clamps from the cylinder head and intake manifold
- PCV valve

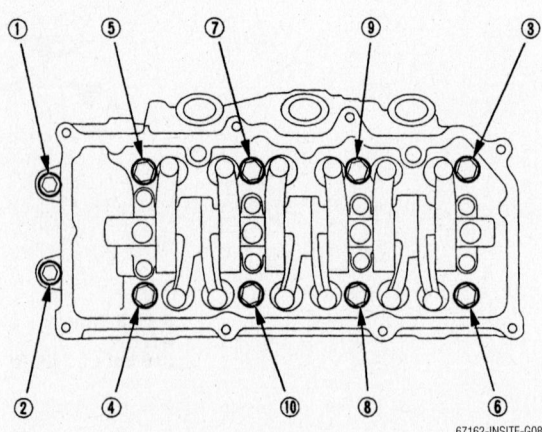

67162-INSITE-G08

Cylinder head bolt loosening sequence—Insight

- Fuel lines
- Cylinder head cover
- Intake manifold
- Radiator hoses
- Catalytic converter
- Dipstick
- Cam chain tensioner by loosening the bolt in sequence one turn at a time.
- Cylinder head plug
- Timing chain

5. Hold the camshaft with an open end wrench, then remove the camshaft sprocket bolt.

6. Remove the camshaft sprocket.

7. Remove the engine mount bracket bolt on the cylinder head side.

8. Loosen the cylinder head bolts in sequence and 1/3 turns until all bolts are loose.

9. Remove the cylinder head.

To install:

10. Clean the cylinder head and block mating surfaces.

11. Ensure the crankshaft pulley is at TDC.

12. Apply liquid gasket to the areas of the engine block as shown.

13. Install the cylinder head.

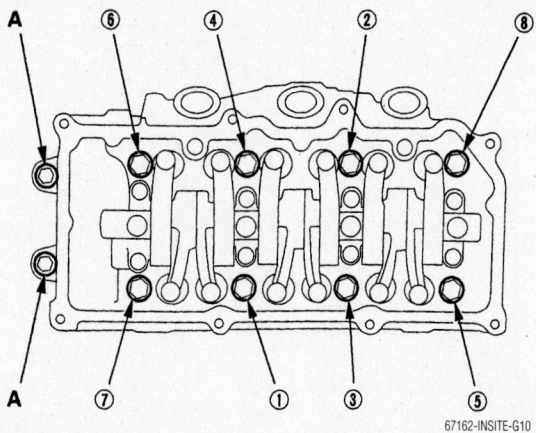

Cylinder head bolt tightening sequence—Insight

14. Coat the cylinder head bolt threads with clean engine oil.

15. Tighten the cylinder head bolts in the sequence shown to 29 ft. lbs. (39 Nm), plus an additional 90 degrees.

16. Tighten the 6 mm bolts (A) to 105 inch lbs. (12 Nm).

17. Install the timing chain on the camshaft sprocket and place the sprocket on the camshaft.

18. Turn the camshaft sprocket counterclockwise to relieve the free play and check that the TDC mark on the sprocket lines up with the cylinder head surface.

19. If the mark is not aligned correctly, remove the sprocket from the camshaft and reposition the timing chain until the mark is aligned.

20. Hold the camshaft with an open end wrench and tighten the sprocket bolt to 41 ft. lbs. (56 Nm).

21. Install a new cylinder head plug.

22. Press the rod to pump the oil out of the timing chain tensioner.

23. Install a new o-ring to the tensioner spacer and a new gasket on the tensioner and then tighten the bolts and nuts equally

to 105 inch lbs. (12 Nm), while pressing the tensioner against the head.

24. Adjust the valve clearance.

25. The remainder of the installation is the reverse of the removal procedure.

26. Start the engine and hold at 3500 rpm with no load for 10 minutes.

27. Check that the IMA battery level indicator reads full.

28. Check the engine idle.

Rocker Arms/Shafts

REMOVAL & INSTALLATION

1. Before servicing the vehicle, refer to the Precautions Section. Also see the IMA system disabling before proceeding with any procedure.

2. Remove the valve cover.

3. Loosen the rocker adjusting screws

4. Remove the rocker shaft bolts 2 turns at a time in the sequence shown.

5. Remove the rocker arm assembly.

6. Installation is the reverse of removal. Tighten the bolts in sequence as shown to 16 ft. lbs. (22 Nm).

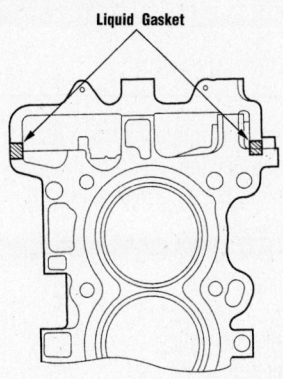

Liquid Gasket

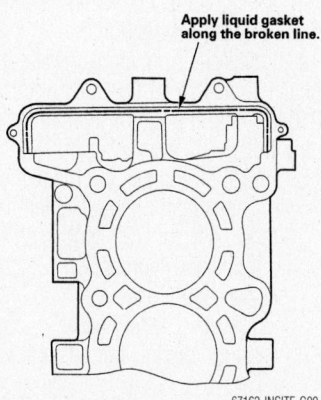

Engine block sealant application areas—Insight

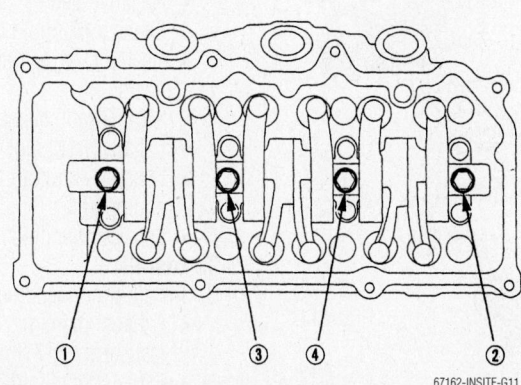

Rocker arm shaft bolt loosening sequence—Insight

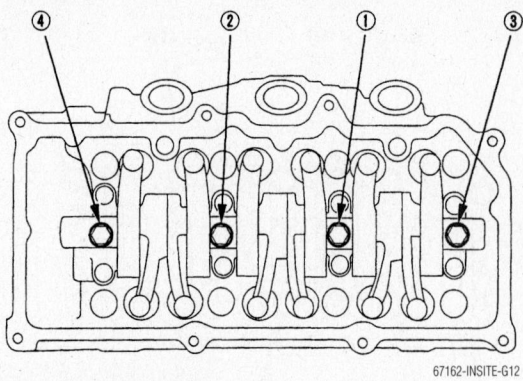

Rocker arm shaft bolt tightening sequence—Insight

Intake Manifold

REMOVAL & INSTALLATION

1. Before servicing the vehicle, refer to the Precautions Section. Also see the IMA system disabling before proceeding with any procedure.
2. Drain the cooling system and engine oil.
3. Relieve the fuel system pressure.
4. Remove or disconnect the following:
   - Negative and then positive battery cables
   - Accessory drive belt
   - Engine appearance cover
   - Breather pipe and brake booster vacuum line
   - Air cleaner and intake air duct
   - Throttle cable linkage
   - Brake booster vacuum hose
   - EVAP canister hose
   - Heater hoses
5. Disconnect and label all the electrical connectors that would interfere with the valve cover removal, then remove the valve cover.
6. Remove the intake manifold.

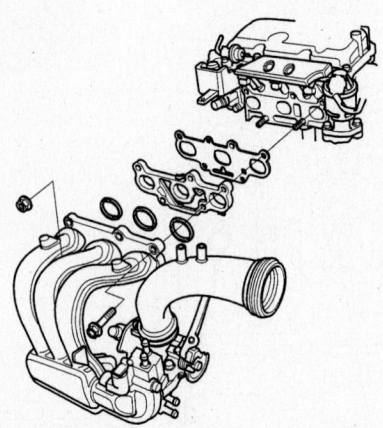

Intake manifold exploded view—Insight

To install:

7. Install or connect the following:
   - New intake manifold gasket and o-rings
   - Intake manifold. Tighten the fasteners to 16 ft. lbs. (22 Nm).
   - Valve cover
   - Electrical connectors
   - Heater hoses
   - EVAP canister hose
   - Brake booster vacuum hose
   - Throttle cable linkage
   - Air cleaner and intake air duct
   - Breather pipe and brake booster vacuum line
   - Engine appearance cover
   - Accessory drive belt
   - Negative and then positive battery cables
8. Fill the cooling system.
9. Start the engine and check for leaks.

Exhaust Manifold

REMOVAL & INSTALLATION

1. Before servicing the vehicle, refer to the Precautions Section. Also see the IMA system disabling before proceeding with any procedure.
2. Remove or disconnect the following:
   - Negative battery cable
   - Exhaust manifold heat shield
   - Heated Oxygen Sensor (HO2S) connector
   - Air fuel ratio sensor connector
   - Exhaust front pipe
   - Exhaust manifold bracket, if equipped
   - Exhaust manifold

To install:

3. Install or connect the following:
   - Exhaust manifold. Tighten the new fasteners to 17 ft. lbs. (24 Nm).
   - Exhaust manifold bracket, if

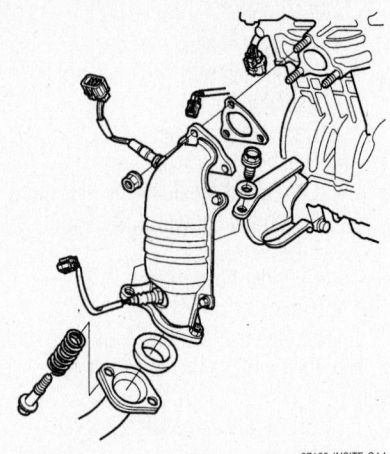

Exploded view of the exhaust system— Insight

equipped. Tighten the bolts to 33 ft. lbs. (44 Nm).
   - Exhaust front pipe. Tighten the bolts alternately to 16 ft. lbs. (22 Nm).
   - Heated Oxygen Sensor (HO2S) connector
   - Air fuel ratio sensor connector
   - Exhaust manifold heat shield
   - Negative battery cable

Front Crankshaft Seal

REMOVAL & INSTALLATION

1. Before servicing the vehicle, refer to the Precautions Section. Also see the IMA system disabling before proceeding with any procedure.
2. Remove or disconnect the following:
   - Negative battery cable
   - Accessory drive belt
   - Crankshaft pulley
   - Front crankshaft seal

To install:

3. Lubricate the crankshaft seal lip with grease prior to installation.
4. Install the front crankshaft seal using Special Tools 07749-0010000 and 07746-0010400 until it bottoms against the oil pump housing.
5. Install or connect the following:
   - Crankshaft pulley. Tighten the bolt to 14 ft. lbs. (20 Nm), plus an additional 90 degrees.
   - Accessory drive belt
   - Negative battery cable
6. Start the engine and check for leaks.

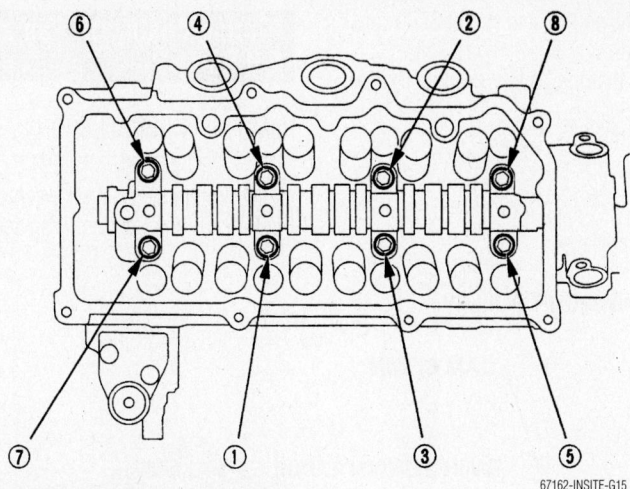

Camshaft holder tightening sequence—Insight

67162-INSITE-G15

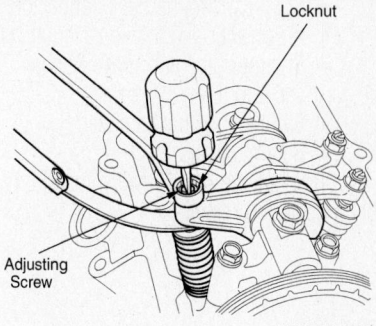

09474_INSI_G0001

Loosening the locknut to turn the adjusting screw—Insight

Camshaft

REMOVAL & INSTALLATION

1. Before servicing the vehicle, refer to the Precautions Section. Also see the IMA system disabling before proceeding with any procedure.
2. Remove or disconnect the following:
   - Negative battery cable
   - Accessory drive belt
   - Valve cover
   - Camshaft sprocket
   - Rocker arms and shaft

➡**Keep all valvetrain components in order for assembly.**

3. Remove the camshaft holder bolts in reverse of the tightening sequence.
4. Remove the camshaft.
To install:
   - Camshafts. Tighten the holder bolts in sequence to 105 inch lbs. (12 Nm).
   - Rocker arms and shaft
   - Camshaft sprocket
   - Valve cover
   - Accessory drive belt
   - Negative battery cable
5. Adjust the valve clearance.
6. Start the engine and check for leaks.

Valve Lash

ADJUSTMENT

Adjust the valves only when the cylinder head temperature is less than 100°F (38°C).
1. Before servicing the vehicle, refer to the Precautions Section. Also see the IMA system disabling before proceeding with any procedure.

2. Remove or disconnect the following:
   - Negative battery cable
   - Valve cover
3. Set the No. 1 piston at TDC. The TDC mark on the camshaft sprocket should line up with the cylinder head surface.
4. Check the clearance on No. 1 cylinder. Intake should be 0.007–0.009 in.; exhaust should be 0.008–0.010 in.
5. If too much or too little drag is present, loosen the locknut and turn the adjusting screw until the drag on the feeler gauge is correct. Tighten the locknut and recheck the clearance.
6. Rotate the crankshaft 240 degrees clockwise and check No.3.
7. Rotate the crankshaft 240 degrees clockwise and check No. 2.

Starter Motor

REMOVAL & INSTALLATION

1. Before servicing the vehicle, refer to the Precautions Section. Also see the IMA system disabling before proceeding with any procedure.
2. Remove or disconnect the following:
   - Negative and then positive battery cable
   - Engine appearance cover
   - Breather pipe and brake booster vacuum line
   - Air cleaner and intake air duct
   - Starter wiring harness connectors
   - Starter motor
To install:
3. Install or connect the following:
   - Starter motor. Tighten the bolts to 33 ft. lbs. (44 Nm).

Adjusting screw location:

EXHAUST

No. 1   No. 2   No. 3

No. 1   No. 2   No. 3

INTAKE

67162-INSITE-G16

Valve adjustment locations—Insight

- Starter wiring harness connectors
- Air cleaner and intake air duct
- Breather pipe and brake booster vacuum line
- Engine appearance cover
- Battery cables

4. Remove the No. 15 (40A) fuse from the under hood relay box.

5. Start the engine and run at 3500 rpm for 10 minutes.

6. Check that the battery module indicator reads full.

7. Install the No. 15 fuse.

Timing Chain, Oil Pan and Oil Pump

REMOVAL & INSTALLATION

1. Before servicing the vehicle, refer to the Precautions Section. Also see the IMA

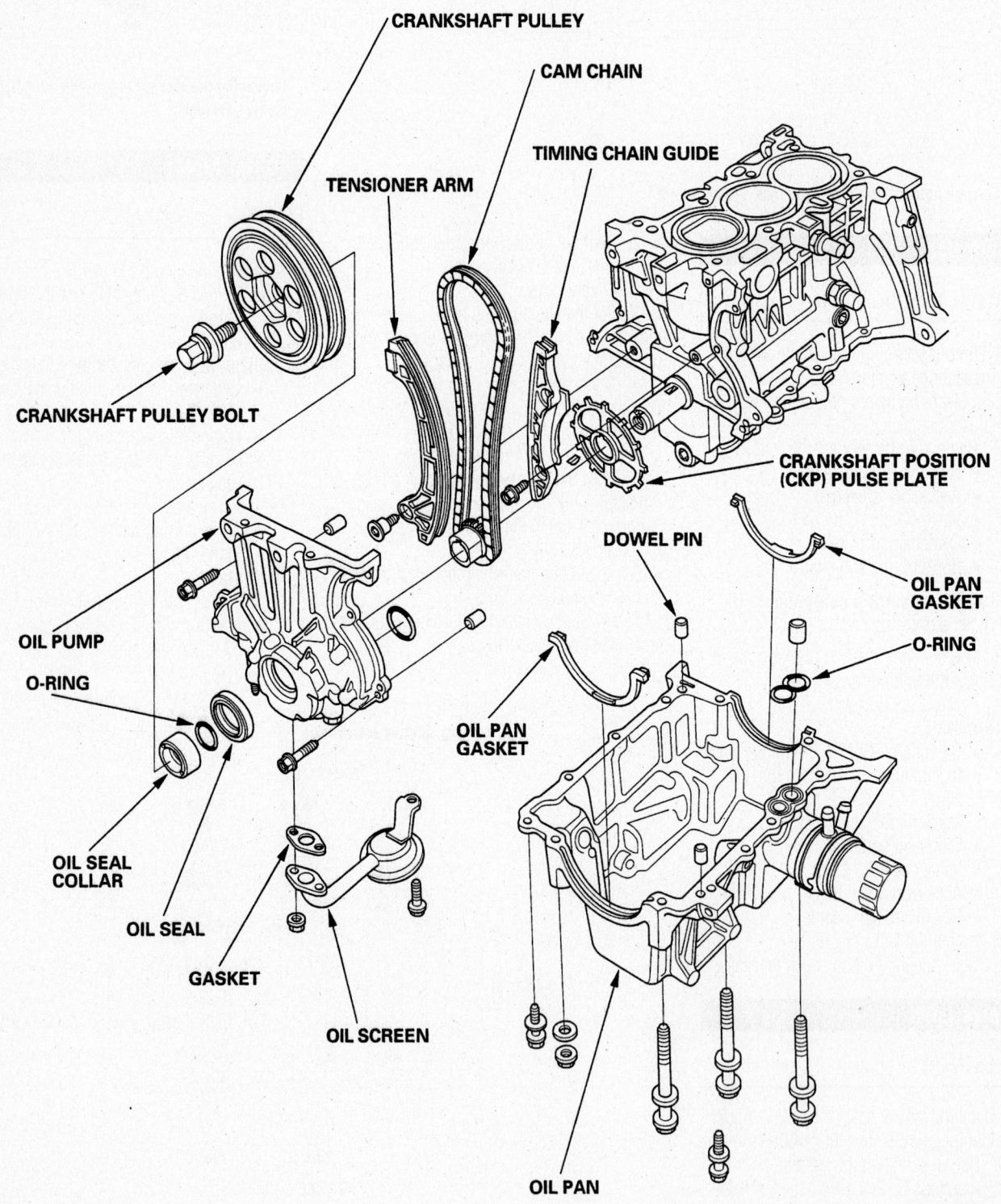

CRANKSHAFT PULLEY

CAM CHAIN

TIMING CHAIN GUIDE

TENSIONER ARM

CRANKSHAFT PULLEY BOLT

CRANKSHAFT POSITION (CKP) PULSE PLATE

OIL PUMP

O-RING

DOWEL PIN

OIL PAN GASKET

O-RING

OIL PAN GASKET

OIL SEAL COLLAR

OIL SEAL

GASKET

OIL SCREEN

OIL PAN

Exploded view of the timing chain components—Insight

system disabling before proceeding with any procedure.

2. Disconnect the negative battery cable.

3. Remove the engine.

4. Remove the cylinder head.

5. Remove the idler pulley bracket mounting bolt.

6. Remove the water pump.

7. Remove the crankshaft pulley.

8. On 2002–03 models, remove the engine oil cooler bypass hoses.

9. Remove the oil pan.

10. Remove the oil screen and oil pump assembly.

11. Remove the timing chain.

12. Remove the crankshaft sprocket.

13. Remove the timing chain tensioning arm and chain guide.

14. Remove the Crankshaft Position Sensor (CKP) pulser plate.

To install:

15. Install the CKP sensor plate.

16. Install the timing chain guide and tighten the bolt to 105 inch lbs. (12 Nm).

17. Install the timing chain tensioning arm and tighten the bolts to 16 ft. lbs. (22 Nm).

18. Set the crankshaft sprocket so no. 1 piston is at TDC. Align the TDC mark (A) on the crankshaft pulser with the pointer (B) on the block.

19. Align the timing chain colored link (A) with the punch mark (B) on the crankshaft sprocket.

20. Clean and dry the oil pump mating surfaces.

21. Apply liquid gasket to the oil pump mating surface on the cylinder block and the bolt holes as shown.

22. Install the dowel pins (A) and align the inner rotor with the crankshaft and install the oil pump (B) using new o-rings (C).

23. Install the oil screen (D) using a new gasket (E).

24. Clean and dry the oil pan mating surfaces.

25. Apply liquid gasket to the oil pan mating surface on the cylinder block and the bolt holes as shown.

26. Using new gaskets and o-rings, install the oil pan. Tighten the bolts in sequence as shown. Tighten M8 bolts to 16 ft. lbs. (22 Nm). Tighten the M6 bolts to 105 inch lbs. (12 Nm).

27. Lubricate the crankshaft pulley bolt with clean engine oil and install the pulley. Tighten the bolt to 14 ft. lbs. (20 Nm), plus an additional 90°.

28. Install or connect the following:
- On 2000–03 models, the engine oil cooler bypass hoses

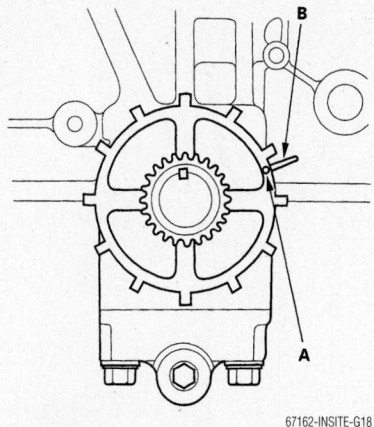

67162-INSITE-G18

Aligning the crankshaft sprocket timing marks—Insight

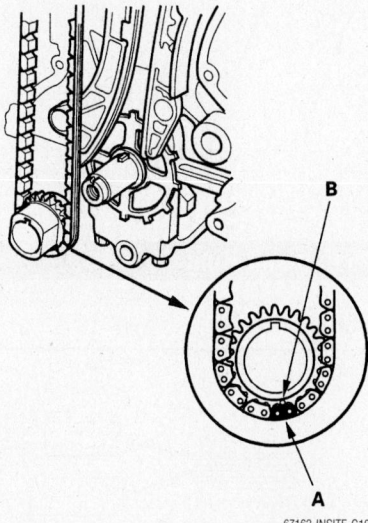

67162-INSITE-G19

Aligning the timing chain link with the crankshaft sprocket timing mark—Insight

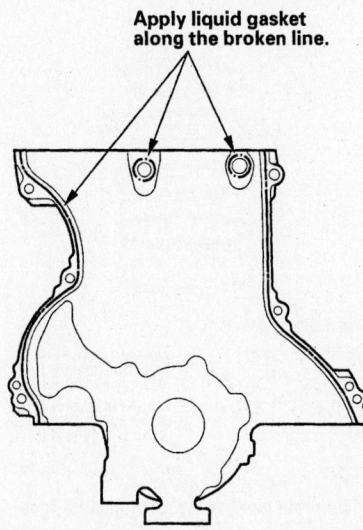

Apply liquid gasket along the broken line.

67162-INSITE-G20

Oil pump sealant application areas on cylinder block—Insight

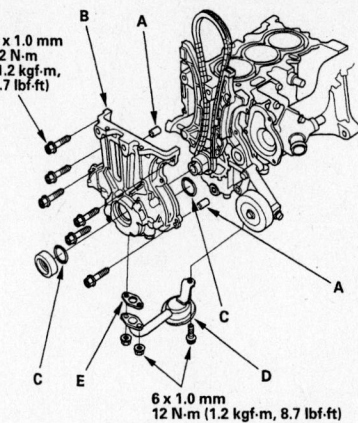

6 x 1.0 mm
12 N·m
(1.2 kgf·m,
8.7 lbf·ft)

6 x 1.0 mm
12 N·m (1.2 kgf·m, 8.7 lbf·ft)

67162-INSITE-G21

Exploded view of the oil pump installation—Insight

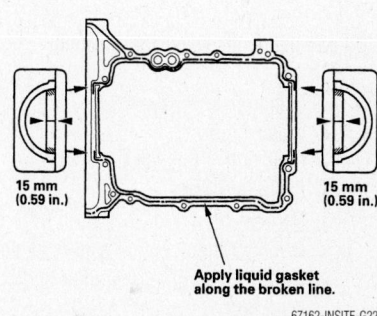

15 mm
(0.59 in.)

15 mm
(0.59 in.)

Apply liquid gasket along the broken line.

67162-INSITE-G22

Oil pan sealant application areas on cylinder block—Insight

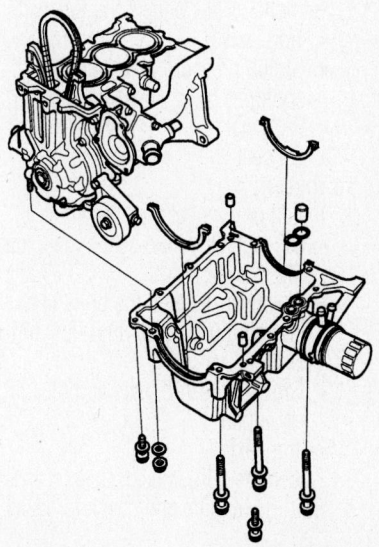

67162-INSITE-G23

Exploded view of the oil pan installation—Insight

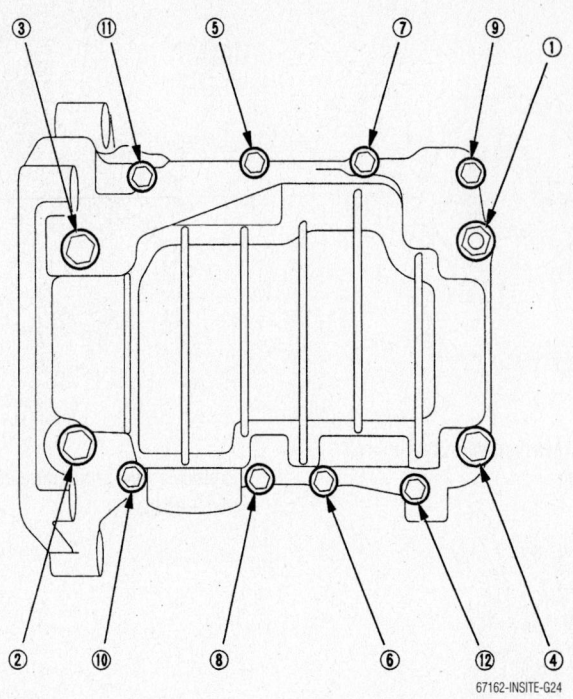

Oil pan tightening sequence—Insight

- Crankshaft pulley
- Water pump
- Idler pulley bracket bolt
- Cylinder head
- Engine
- Connect the negative battery cable.
- Check the engine operation and road test.

Rear Main Seal

REMOVAL & INSTALLATION

1. Remove or disconnect the following:
   - Transaxle
   - Clutch pressure plate and disc, if equipped
   - Flywheel
   - Oil seal

To install:
2. Install or connect the following:
   - Oil seal. Drive the seal square into the seal case.
   - Flywheel. Tighten the bolts in a crossing pattern to 33 ft. lbs. (44 Nm).
   - Clutch pressure plate and disc, if equipped
   - Transaxle
3. Check the fluid levels.
4. Start the engine and check for leaks.

Piston and Ring

POSITIONING

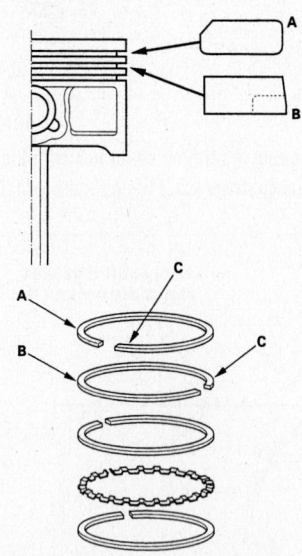

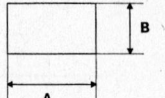

Piston Ring Dimensions:

Top Ring (Standard):
A: 2.3 mm (0.09 in.)
B: 1.0 mm (0.04 in.)

Second Ring (Standard):
A: 3.0 mm (0.12 in.)
B: 1.2 mm (0.05 in.)

67162-INSITE-G25

Piston ring positioning and top mark location

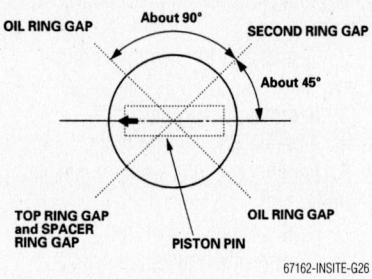

Piston ring end-gap spacing

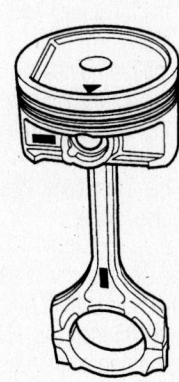

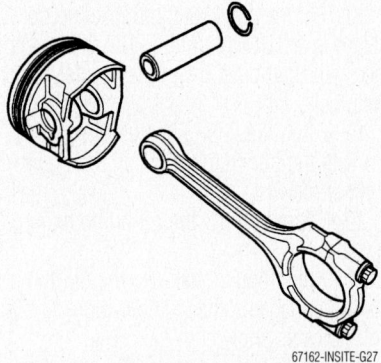

Piston and connecting rod assembly

INTEGRATED MOTOR ASSEMBLY

Integrated Motor Assembly Service Precautions

Before servicing any vehicle, please be sure to read all of the following precautions, which deal with personal safety, prevention of component damage, and important points to take into consideration when servicing the integrated motor assembly:

• The IMA system uses 144 volt circuits. Always shut off electrical circuits and isolate the IMA system and related parts before servicing the system.

• The high voltage cables and their covers are identified by an Orange coloring.

Caution labels are attached to high voltage and related parts. Be careful not to touch these cables and parts without adequate protective gear.

• If the 12 volt battery has been discharged or disconnected or the Motor Control Module (MCM) has been reset, the IMA battery level indicator will not show the

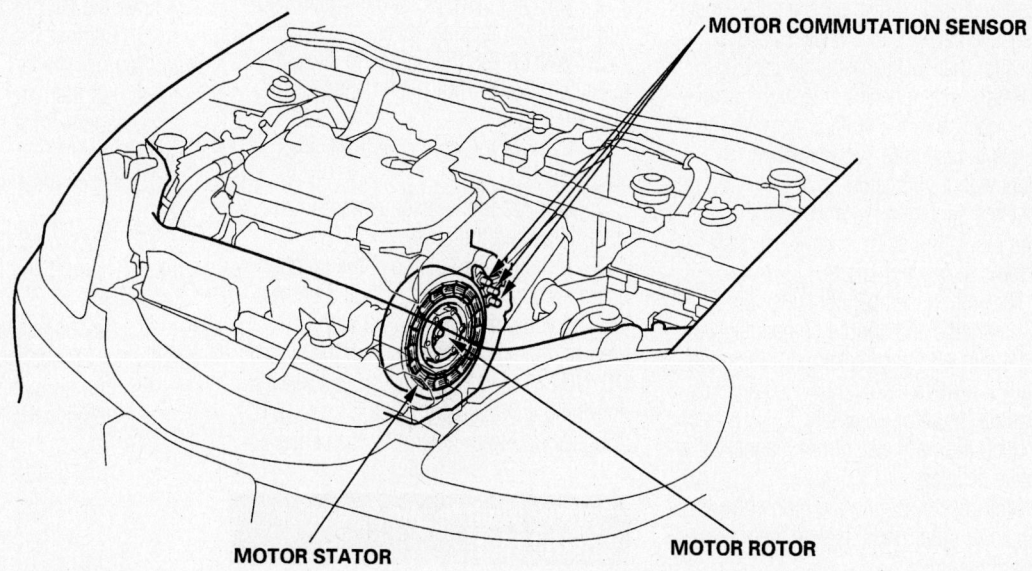

MOTOR COMMUTATION SENSOR

MOTOR STATOR

MOTOR ROTOR

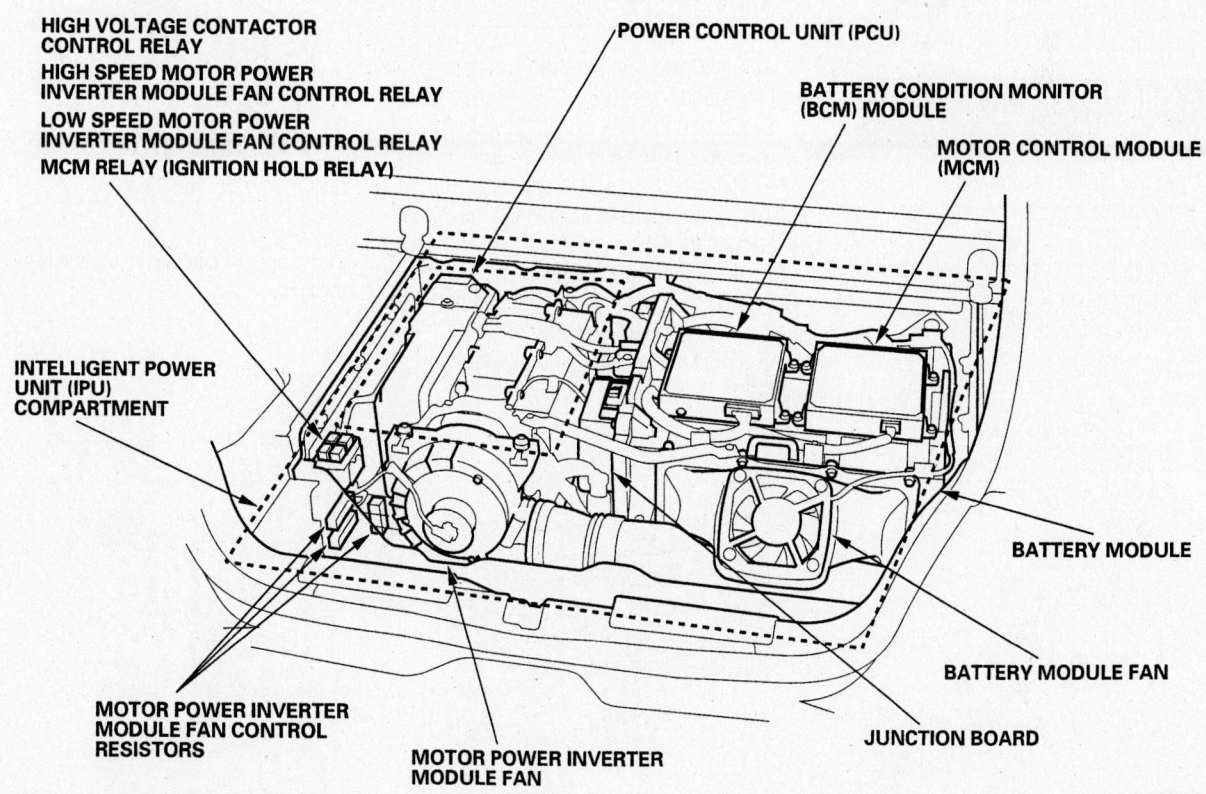

HIGH VOLTAGE CONTACTOR
CONTROL RELAY
HIGH SPEED MOTOR POWER
INVERTER MODULE FAN CONTROL RELAY
LOW SPEED MOTOR POWER
INVERTER MODULE FAN CONTROL RELAY
MCM RELAY (IGNITION HOLD RELAY)

POWER CONTROL UNIT (PCU)

BATTERY CONDITION MONITOR
(BCM) MODULE

MOTOR CONTROL MODULE
(MCM)

INTELLIGENT POWER
UNIT (IPU)
COMPARTMENT

BATTERY MODULE

BATTERY MODULE FAN

MOTOR POWER INVERTER
MODULE FAN CONTROL
RESISTORS

MOTOR POWER INVERTER
MODULE FAN

JUNCTION BOARD

67162-INSITE-G28

IMA system component locations—Insight

state of charge when the engine is started. Start the engine and hold the idle at 3500 rpm with no load until the battery level indicator reads full charge.

• Wear insulated gloves with no tears, pin holes or other damage when inspecting or servicing the IMA.

• Turn the battery module switch **OFF** and secure the switch in the **OFF** position with the locking cover before servicing the IMA system.

• Wait for 5 minutes or more after turning the battery module switch off, then disconnect the negative battery cable, It takes about 5 minutes for the PDU capacitor to discharge.

• Before disconnecting the high voltage cables, make sure the voltage between the terminals is less than 30 volts when checked with a voltmeter.

• When servicing parts without the Orange insulating color, always use insulated tools to prevent shorts.

• The rotor assembly contains very strong magnets and should be handled with care. People with pacemakers or other magnetically sensitive medical devices should not handle the rotor assembly.

• Keep the rotor away from magnetically sensitive devices.

• After disconnecting the high voltage terminals or other parts, isolate them with insulated parts.

• Attach a sign saying "WORKING ON HIGH VOLTAGE PARTS. DO NOT TOUCH," to the steering wheel as a safety measure.

Disabling The IMA System Power

1. Turn the ignition switch off.
2. Remove the luggage compartment floor mat.
3. Remove the battery module cover (A) and the locking cover (B).

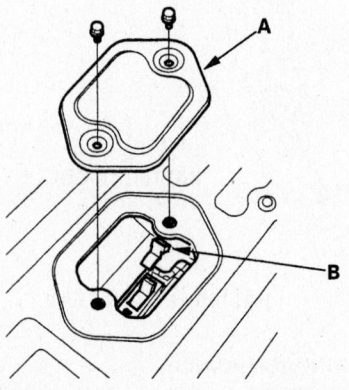

IMA battery module switch and locking cover location—Insight

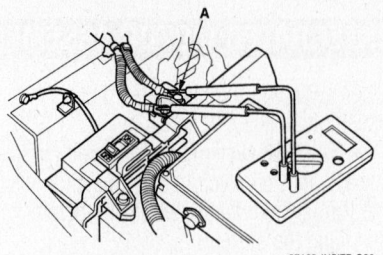

67162-INSITE-G30

Measuring IMA battery voltage—Insight

4. Turn the battery module switch **OFF** and install the locking cover.
5. Wait at least 5 minutes to allow the Intelligent Power Unit (IPU) capacitor to discharge.
6. Remove the right side trunk shelf support.
7. Remove the luggage compartment cover clips and the cover.
8. Measure the voltage at the junction board terminals (A). There should be less than 30 volts. If more than 30 volts are present, there is a problem with the system. **DO NOT** proceed with any procedure without doing a Diagnostic Trouble Code (DTC) troubleshooting to determine the cause.

Battery Module and Power Control Unit

REMOVAL & INSTALLATION

1. Before servicing the vehicle, refer to the Precautions Section.
2. Disable the IMA system power.
3. Remove the foam inserts.
4. Remove the mid frame (A), the front and rear braces (B) and disconnect the high voltage cables (C).
5. Wrap the cables with insulating tape.
6. Remove the air duct mounting bolt and push the air duct forward.
7. Remove the battery module mounting bolts.

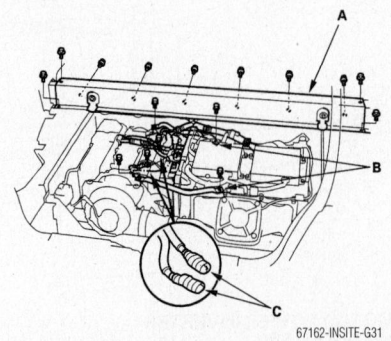

67162-INSITE-G31

Removing the battery module braces and high voltage cables—Insight

8. Disconnect the 3 connectors at the left side and secure the connectors.
9. Install battery lifting tool 07YAK-001010A on the battery module.
10. With the aid of an assistant, lift the battery module from the vehicle and place on a flat surface.
11. To remove the Power Control Unit (PCU), disconnect the connector (A) from the cooling fan assembly (B) and remove the clip from the fan shroud (C).
12. Remove the 4 fan bolts (D) and remove the fan bracket (E).
13. Remove the fan duct (F) from the fan and remove the fan.
14. Lift the relay pack (A) from its holder and disconnect the harness from the resistors (B).
15. Disconnect the Y condenser ground (C).
16. Disconnect the connector (D) from the PCU and then disconnect the battery cables (E).
17. Disconnect the high voltage DC-DC converter 2P connector (F).
18. Pull away the intake duct (G).
19. Remove the PCU terminal cover (H)

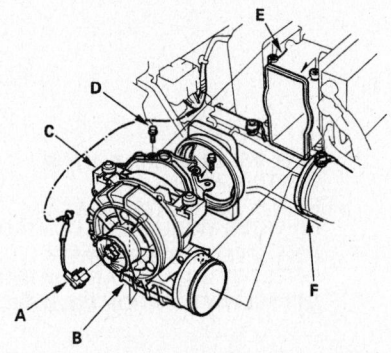

67162-INSITE-G32

Removing the cooling fan assembly—Insight

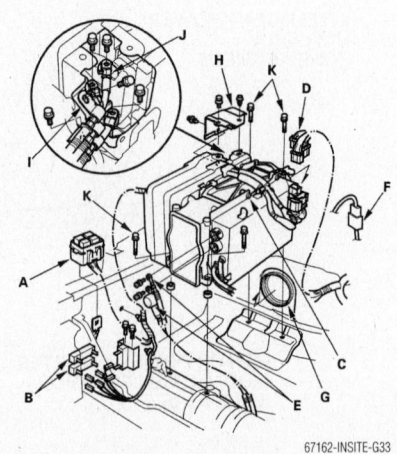

67162-INSITE-G33

Removing the power control unit—Insight

and remove the clip (J) and bolts that hold the three phase cables (J) in place.

20. Remove the 4 PCU mounting bolts (K).

21. Remove the PCU and place it on a flat surface.

To install:

22. Installation is the reverse of the removal procedure.

IMA Motor

REMOVAL & INSTALLATION

1. Before servicing the vehicle, refer to the Precautions Section.

2. Disable the IMA system power.

3. Remove the transaxle and clutch, if equipped.

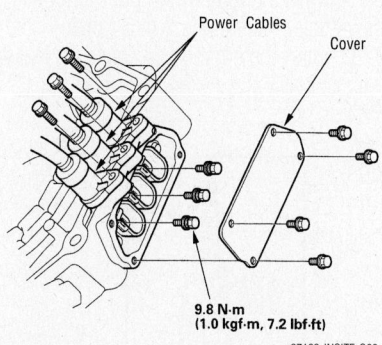

9.8 N·m
(1.0 kgf·m, 7.2 lbf·ft)

67162-INSITE-G03

IMA power cable locations—Insight

4. Remove 8 bolts and the stator cover.

5. Remove 3 of the opposing rotor attaching bolts.

6. Install guide pins in the removed bolt holes and then remove the remaining 3 bolts.

7. Install rotor puller tool 07YAC-PHM1010A over the guide pins and tighten the puller bolts.

8. Turn the tool clockwise and remove the rotor.

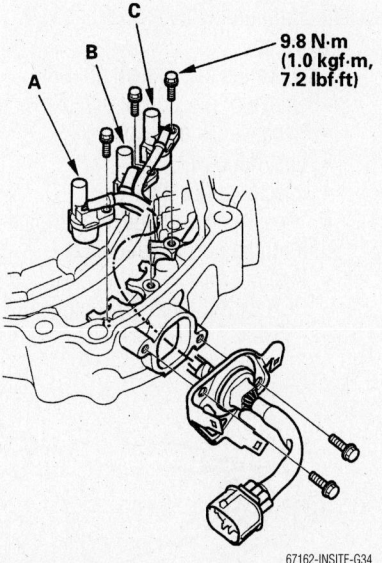

9.8 N·m
(1.0 kgf·m,
7.2 lbf·ft)

67162-INSITE-G34

Removing IMA motor commutation sensors—Insight

9. Remove the 2 stator mounting bolts.

10. Remove the stator.

11. Remove the cover (A) and the motor power cables (B).

12. Remove the motor commutation sensors (A), (B) and (C).

13. Installation is the reverse of the removal procedure, noting the following items:

a. Install the communication sensors in the correct locations.

b. Install the motor power cables in the correct locations.

c. Set the rotor on the special tool and install the rotor with the end of the tool extended as shown

d. Turn the handle of the tool slowly when inserting the rotor into the stator. The rotor is drawn into the stator by magnetic force.

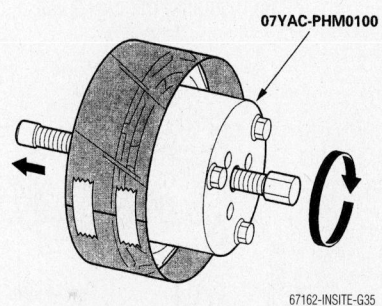

07YAC-PHM0100

67162-INSITE-G35

Installing rotor into the stator—Insight

FUEL SYSTEM

Fuel System Pressure

RELIEVING

2002–2005 Models

1. Before servicing the vehicle, refer to the Precautions Section. Also see the IMA system disabling before proceeding with any procedure.

2. Disconnect the negative battery cable.

3. Remove the fuel filler cap.

4. Hold the fuel rail inlet banjo bolt with a flare nut wrench.

5. Place a shop towel over the fitting to absorb leakage.

6. Loosen the service bolt 1 turn.

7. When repairs are complete, replace the sealing washers and tighten the service bolt to 16 ft. lbs. (22 Nm) on 2002–03

models, or 105 inch lbs. (12 Nm) on 2004–2005 models.

8. Install the fuel filler cap.

9. Connect the negative battery cable.

10. Start the engine and check for leaks.

2006 Models

1. Before servicing the vehicle, refer to the Precautions Section. Also see the IMA system disabling before proceeding with any procedure.

2. Remove the PGM-FI (Fuel Pump) main relay from the multi relay box.

3. Start the engine and let it idle until it stalls.

4. Turn off the ignition.

5. Remove the fuel fill cap.

6. Disconnect the negative battery cable.

7. Remove the quick-connect fitting cover from fuel supply line.

8. Place a shop towel over the fitting to absorb leakage and disconnect the fitting.

9. When repairs are complete, replace the fitting.

10. Reconnect the negative battery cable and replace the PGM-FI main relay. Install the fuel filler cap.

11. Turn the ignition to the **ON** position and allow the fuel pressure to rise. Repeat two or three times and check for leaks.

Fuel Filter and Pump

REMOVAL & INSTALLATION

1. Before servicing the vehicle, refer to the Precautions Section. Also see the IMA system disabling before proceeding with any procedure.

2. Relieve the fuel system pressure.

3. Drain the fuel tank into a suitable approved container.

4. Remove or disconnect the following:
- Negative battery cable
- Muffler
- Place a support under the fuel tank
- Fuel tank shield and straps
- Fuel supply and return lines
- Fuel pump connector
- Fuel tank
- Fuel pump unit locknut using Special Tool 07AAA-S0XA100
- Fuel pump
- Fuel pump filter

To install:

5. Installation is the reverse of the removal procedure noting the following items:

   a. Install a new fuel pump base gasket.

   b. Do not push on the suction filter.

   c. Ensure the wiring harness connections are secure.

   d. When installing the fuel pump align the marks on the fuel tank with the marks on the fuel pump.

   e. Tighten the fuel tank unit locknut to 69 ft. lbs. (93 Nm).

   f. Tighten the fuel tank straps bolts to 28 ft. lbs. (38 Nm).

Fuel Injector

REMOVAL & INSTALLATION

1. Before servicing the vehicle, refer to the Precautions Section. Also see the IMA system disabling before proceeding with any procedure.

2. Relieve the fuel system pressure.

3. Remove or disconnect the following:
- Negative battery cable
- Fuel rail cover
- Injector connectors
- Vacuum hose and fuel return line from the pressure regulator
- Fuel rail nuts
- PCV valve (2002–2004 models)
- PCV hose (2005–2006 models)
- Fuel rail
- Fuel injectors from the cylinder head

To install:

4. Install new cushion rings (A) on the fuel injectors (B).

5. Coat new O-rings (C) with clean engine oil.

6. Install the fuel injectors to the fuel rail (D).

7. Coat new seal rings (E) with clean engine oil and press them in.

8. Install or connect the following:
- Fuel rail. Tighten the nuts to 105 inch lbs. (12 Nm).
- Fuel line to rail with new washers
- Vacuum line and fuel return line
- PCV valve (2002–2004 models)
- PCV hose (2005–2006 models)
- Negative battery cable

9. Turn the ignition on but do not start the vehicle. Repeat this 2 or 3 times.

10. Start the engine and check for leaks.

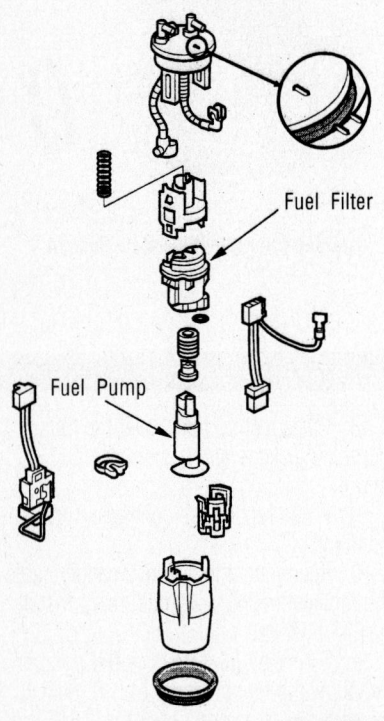

Fuel Filter

Fuel Pump

67162-INSITE-G36

Exploded view of the fuel pump module assembly—Insight; A/T shown M/T similar

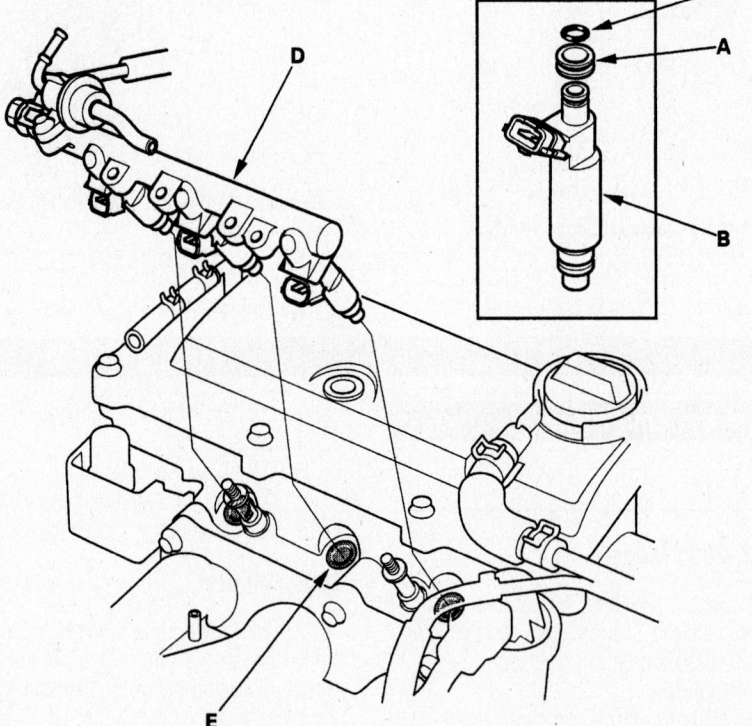

67162-INSITE-G37

Exploded view of the fuel injector assembly—Insight

DRIVE TRAIN

Transaxle Assembly

REMOVAL & INSTALLATION

Automatic Transaxle

1. Before servicing the vehicle, refer to the Precautions Section. Also see the IMA system disabling before proceeding with any procedure.
2. Drain the transaxle fluid.
3. Remove or disconnect the following:
   - Negative and then positive battery cables
   - Battery tray
   - Air intake assembly
   - Engine splash shield and bracket
   - Starter motor
   - Transaxle ground cable
   - Solenoid valve connector
   - Input shaft speed sensor connector
   - Shift cable and bracket
   - Range switch connector
   - Output shaft speed sensor connectors
   - Upper transaxle housing mounting bolts
   - Front wheels
   - Axle shaft nuts
   - Lower arms from knuckle
   - Axle shafts from differential
   - Transaxle oil cooler hoses and cap the openings
   - Rear mount and bracket
   - Drive plate bolts through the starter opening
4. Install a suitable engine lifting device and lift and support the engine.
5. Raise the transaxle enough to take the weight off the mounts.
6. Install a transmission jack.
7. Remove the transaxle ground cable from the transaxle mount bracket, then remove the mount bracket.
8. Remove the heat shield and exhaust manifold bracket.
9. Remove the lower transmission housing mounting bolts.
10. Pull the transaxle back until it clears the dowel pins and remove the transaxle.
   To install:

➡**Use new circlips, split pins and self-locking nuts for assembly.**

11. Install or connect the following:
    - Transaxle
    - Lower transaxle housing mounting bolts. Tighten the bolts to 33 ft. lbs. (44 Nm).
    - Exhaust manifold bracket and heat shield. Tighten the bolts to 33 ft. lbs. (44 Nm).
    - Transaxle mount bracket. Tighten the nuts to 76 ft. lbs. (103 Nm), then the bolt to 40 ft. lbs. (54 Nm).
    - Ground cable
    - Upper transaxle housing mounting bolts. Tighten the bolts to 47 ft. lbs. (64 Nm).
    - Rear mount bracket. Tighten the horizontal bolts to 36 ft. lbs. (49 Nm). Tighten the vertical bolts to 54 ft. lbs. (74 Nm).
    - Remove the engine hangar
12. Install the drive plate bolts and tighten them in a cross cross pattern to 52 inch lbs. (6 Nm). Tighten them again in a criss cross pattern to 105 inch lbs. (12 Nm).
13. Install or connect the following:
    - Transaxle oil cooler hoses
    - Axle shafts to differential using new circlips
    - Lower arms to knuckle. Tighten the nut to 40–47 ft. lbs. (54–64 Nm).
    - Axle shaft nut. Tighten the nut to 134 ft. lbs. (181 Nm).
    - Front wheels
    - Splash shield
    - Output shaft speed sensor connectors
    - Range switch connector
    - Shift cable and bracket
    - Input shaft speed sensor connector
    - Transaxle ground cable
    - Starter motor
    - Engine splash shield
    - Air intake assembly
    - Battery tray
    - Negative and then positive battery cables
14. Check the engine timing.
15. Fill the transaxle to the correct level.
16. Start the engine and check for leaks.
17. Check the wheel alignment and adjust as necessary.
18. Connect the Honda Diagnostic System (HDS) tool to the data link connector.
19. Follow the prompts on the tester to perform the Start Clutch Control system calibration procedure.

Manual Transaxle

1. Before servicing the vehicle, refer to the Precautions Section. Also see the IMA system disabling before proceeding with any procedure.
2. Drain the transaxle.
3. Remove or disconnect the following:
   - Negative and then positive battery cables
   - Battery tray
   - Air intake assembly
   - Engine splash shield
   - Speed sensor connector
   - Back-up light switch connector
   - Neutral safety switch connector
   - Breather tube
   - Air cleaner bracket
   - Clutch line bracket
   - Shift cables and bracket
   - Throttle drum cover
   - Clutch slave cylinder and hose bracket
   - Starter motor
4. Install a suitable engine lifting device and lift and support the engine.
5. Raise the transaxle enough to take the weight off the mounts.
6. Remove or disconnect the following:
   - Transaxle mount bracket
   - Upper 2 transaxle mounting bolts
   - Splash shield bracket
   - Front wheels
   - Axle shafts
7. Install a transmission jack.
8. Remove the rear engine mount bracket.
9. Remove the 4 lower transmission mounting bolts
10. Pull the transaxle back until it clears the dowel pins and remove the transaxle.
    To install:

➡**Use new circlips, split pins and self-locking nuts for assembly.**

11. Install or connect the following:
    - Transaxle
    - Lower transaxle housing mounting bolts. Tighten the bolts to 47 ft. lbs. (65 Nm).
    - Rear engine mount bracket. Tighten the bolts to 36 ft. lbs. (49 Nm). Tighten the through bolt to 66 ft. lbs. (89 Nm).
    - Upper transaxle housing mounting bolts. Tighten the bolts to 47 ft. lbs. (65 Nm).
    - Transaxle mount bracket. Tighten the fasteners to 40 ft. lbs. (54 Nm).
    - Remove the engine hangar
    - Starter

➡**Apply super high temperature urea grease to the end of the slave cylinder.**

    - Slave cylinder. Tighten the bolts to 16 ft. lbs. (22 Nm).

- Throttle drum cover
- Shift cables and bracket
- Clutch line bracket
- Air cleaner bracket
- Breather tube
- Neutral safety switch connector
- Back-up light switch connector
- Speed sensor connector
- Engine splash shield
- Air intake assembly
- Battery tray
- Negative and then positive battery cables

12. Test drive the vehicle and check for smooth shifting and clutch operation.

Clutch

ADJUSTMENTS

The Insight is equipped with a hydraulic clutch system. No adjustment is necessary.

REMOVAL & INSTALLATION

1. Before servicing the vehicle, refer to the Precautions Section. Also see the IMA system disabling before proceeding with any procedure.
2. Remove or disconnect the following:
   - Negative battery cable
   - Transaxle
   - Pressure plate. Loosen the bolts evenly in a crossing pattern.
   - Clutch disc

To install:

3. Install the clutch disc and pressure plate. Tighten the pressure plate bolts in a crossing pattern and in several steps to 19 ft. lbs. (25 Nm).
4. Install or connect the following:
   - Transaxle
   - Negative battery cable

Hydraulic Clutch System

BLEEDING

1. Before servicing the vehicle, refer to the Precautions Section.

✴✴ WARNING

Overtightening the bleeder screw can damage the slave cylinder.

2. Attach a hose to the bleeder screw and suspend the other end in a container of clean brake fluid.
3. Open the bleeder screw.
4. Slowly pump the clutch pedal until no

more air bubbles appear at the bleeder hose.
5. Tighten the bleeder screw to 70 inch lbs. (8 Nm).
6. Refill the clutch master cylinder as necessary.
7. Check for leaks and proper clutch operation.

Halfshaft

REMOVAL & INSTALLATION

1. Before servicing the vehicle, refer to the Precautions Section. Also see the IMA system disabling before proceeding with any procedure.
2. Drain the transaxle.
3. Remove or disconnect the following:
   - Negative battery cable
   - Front wheels
   - Spindle nut
   - Stabilizer link
   - Brake hose clamp
   - Wheel speed sensor clamp
4. Turn the front knuckle outward.
5. Tap the halfshaft inward with a plastic hammer.
6. Remove the lower ball joint.
7. Remove the inboard joint heat shield.
8. Pry the inboard joint from the transaxle or bearing support.
9. Pull the axle shaft straight out to avoid damaging the oil seal.

To install:

➡**Use new circlips, split pins and self-locking nuts for assembly.**

10. Install the new circlip into the circlip groove of the driveshaft
11. Install the inner CV-joint to the transaxle or bearing support until the circlip locks in the retaining groove.
12. Install or connect the following:
   - Inboard joint heat cover
   - Outboard joint into the front hub
   - Lower ball joint. Tighten the nut to 40–47 ft. lbs. (54–64 Nm).
   - Wheel speed sensor clamp
   - Brake hose clamp
   - Stabilizer link
   - Axle shaft nut. Tighten the nut to 134 ft. lbs. (181 Nm).
   - Front wheels
   - Battery cables
13. Fill the transaxle to the correct level and check for leaks.

CV-Joint

OVERHAUL

Outboard Joint

1. Before servicing the vehicle, refer to the Precautions Section. Also see the IMA system disabling before proceeding with any procedure.
2. Remove or disconnect the following:
   - Axle halfshaft from the vehicle and place it in a vise
   - Outboard joint boot clamps and push the boot back
   - Outboard joint by using a slide hammer to pull it off the driveshaft
   - Snap ring
   - Outboard joint boot

To install:

➡**Use new circlips and boot clamps for assembly.**

3. Install the outboard joint boot and clamps to the axle shaft.
4. Install a new snap ring on the drive-shaft groove.
5. Insert the driveshaft into the joint until the snap ring closes.
6. Seat the joint by dropping it straight down onto a hard surface from about 4 inches. DO NOT use a hammer to seat the joint.
7. Fill the outboard joint with grease.
8. Adjust the length of the driveshaft to the dimensions shown. Adjust the boots to half-way between full extension and full compression.
9. On the right driveshaft, position the dynamic damper to the length shown.
10. Fit the boot ends onto the driveshaft.
11. Crimp the clamps onto the boots.
12. Install the axle halfshaft to the vehicle.

Inboard Joint

1. Before servicing the vehicle, refer to the Precautions Section. Also see the IMA system disabling before proceeding with any procedure.
2. Remove or disconnect the following:
   - Axle halfshaft from the vehicle.
   - Circlip from the joint
   - Inboard joint boot clamps and push the boot back

➡**The inboard joint and rollers must be installed in their original positions. Matchmark the inboard joint in relation to each roller and mark each roller and its position on the spider.**

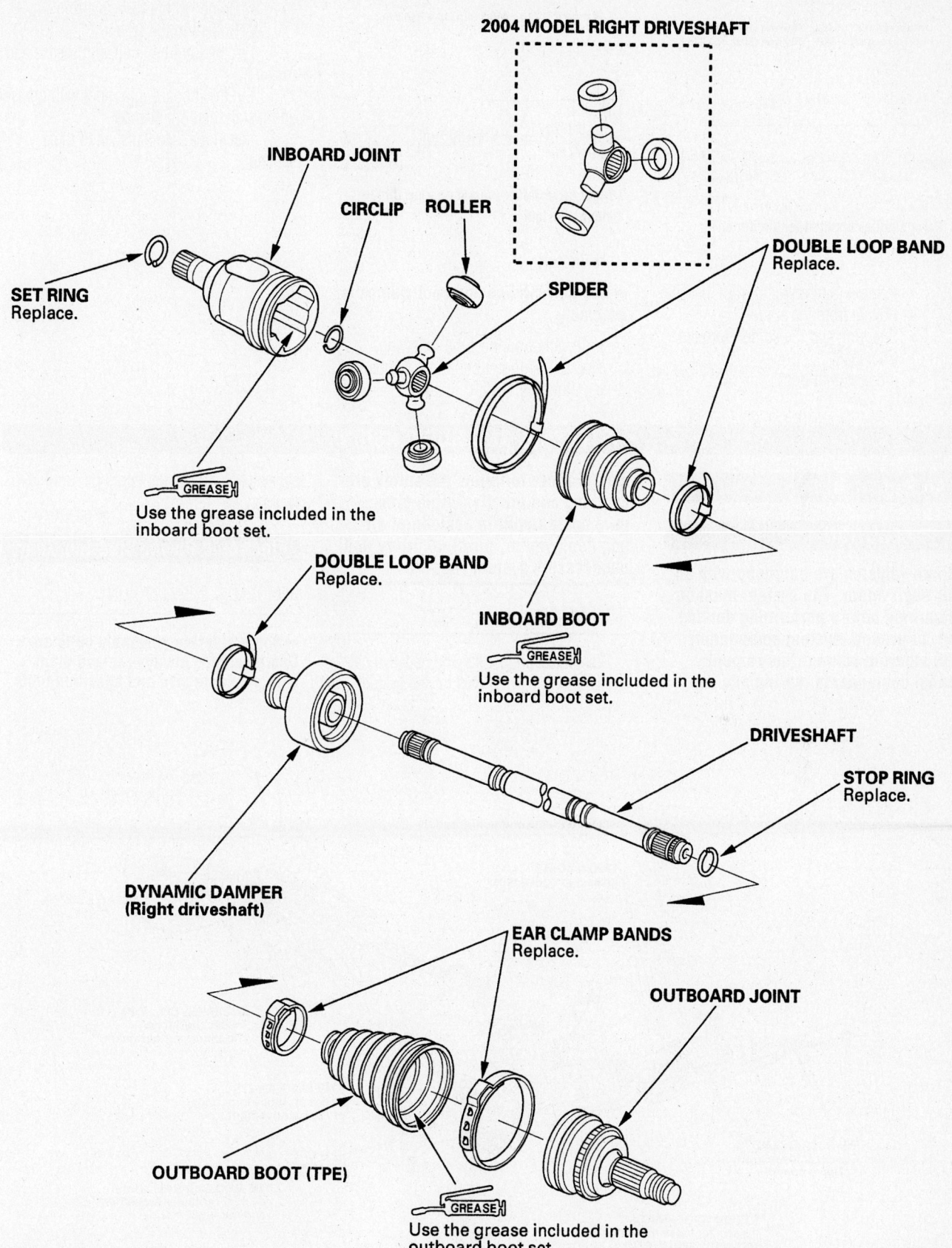

2004 MODEL RIGHT DRIVESHAFT

INBOARD JOINT

CIRCLIP   **ROLLER**

SET RING
Replace.

SPIDER

DOUBLE LOOP BAND
Replace.

GREASE
Use the grease included in the
inboard boot set.

DOUBLE LOOP BAND
Replace.

INBOARD BOOT

GREASE
Use the grease included in the
inboard boot set.

DRIVESHAFT

STOP RING
Replace.

DYNAMIC DAMPER
(Right driveshaft)

EAR CLAMP BANDS
Replace.

OUTBOARD JOINT

OUTBOARD BOOT (TPE)

GREASE
Use the grease included in the
outboard boot set.

67162-INSITE-G38

Exploded view of the driveshaft/CV-joint assembly—Insight

Left driveshaft: 481 – 486 mm (18.9 – 19.1 in.)
Right driveshaft: 761 – 766 mm (30.0 – 30.2 in.)

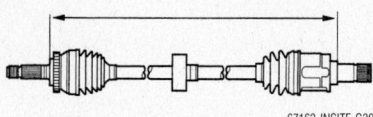

67162-INSITE-G39

Drive shaft assembled length dimensions—Insight

- Inboard joint housing from the axle
- Rollers from the spider
- Snapring and the spider from the axle shaft
- Inboard joint boot

Standard: 415.5 – 419.5 mm (16.4 – 16.5 in.)

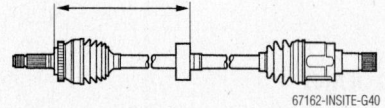

67162-INSITE-G40

Right driveshaft dynamic damper dimensions—Insight

To install:

➡ **Use new circlips and boot clamps for assembly.**

3. Install or connect the following:
   - Inboard joint boot and clamps to the axle shaft

- Spider with a new snapring
- Rollers to the spider

4. Fill the joint housing with grease and install it.

5. Fill the inboard joint boot with grease and install the boot clamps.

6. Install the axle halfshaft to the vehicle.

STEERING AND SUSPENSION

Air Bag

❊❊ CAUTION

Some vehicles are equipped with an air bag system. The system must be disarmed before performing service on, or around, system components, the steering column, instrument panel components, wiring and sensors. Failure to follow the safety precautions and the disarming procedure could result in accidental air bag deployment, possible injury and unnecessary system repairs.

DISARMING

Disconnect and isolate the negative battery cable. Wait 3 minutes for the system capacitor to discharge before performing any service.

Electronic Power Steering Gear

REMOVAL & INSTALLATION

➡ **Some steering assembly bolts are Dacro coated. Always replace them with the same size and specified bolts.**

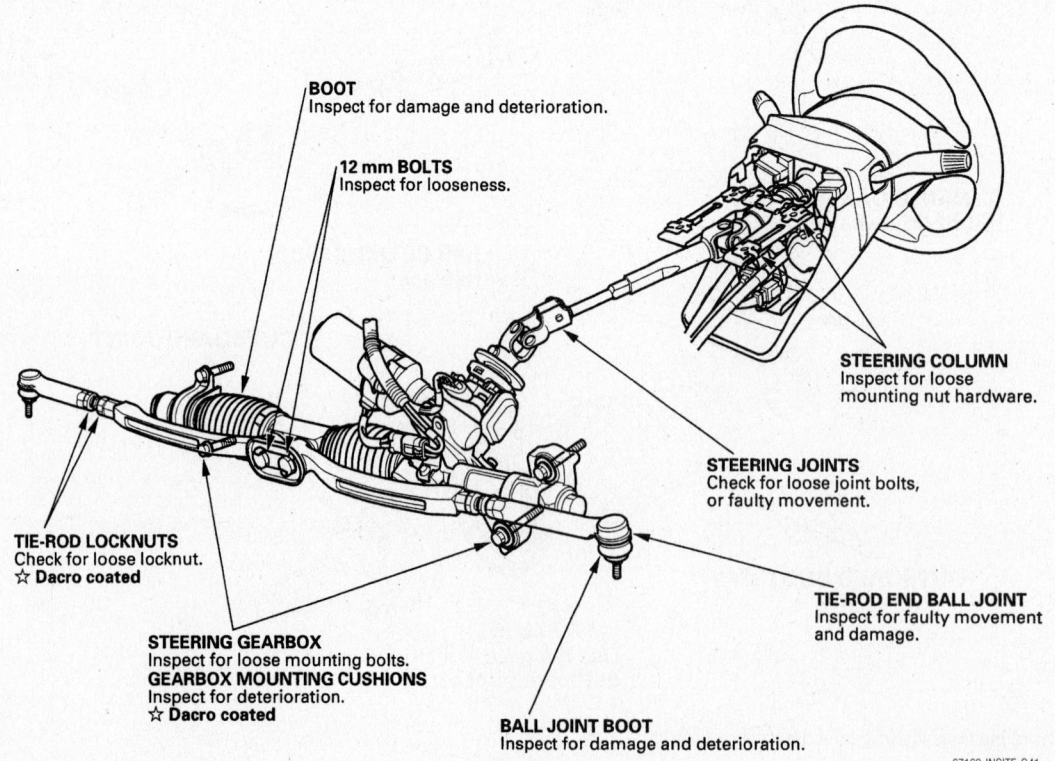

BOOT
Inspect for damage and deterioration.

12 mm BOLTS
Inspect for looseness.

STEERING COLUMN
Inspect for loose mounting nut hardware.

STEERING JOINTS
Check for loose joint bolts, or faulty movement.

TIE-ROD LOCKNUTS
Check for loose locknut.
☆ Dacro coated

STEERING GEARBOX
Inspect for loose mounting bolts.
GEARBOX MOUNTING CUSHIONS
Inspect for deterioration.
☆ Dacro coated

BALL JOINT BOOT
Inspect for damage and deterioration.

TIE-ROD END BALL JOINT
Inspect for faulty movement and damage.

67162-INSITE-G41

Exploded view of steering gear —Insight

✳✳ WARNING

Do not permit the steering wheel to turn whenever the steering gear is disconnected from the steering column. Damage to the air bag wiring can result.

1. Before servicing the vehicle, refer to the Precautions Section. Also see the IMA system disabling before proceeding with any procedure.

2. Center the steering wheel and lock it in position.

3. Raise and support the vehicle.

4. Remove or disconnect the following:
   - Negative battery cable
   - Air bag and steering wheel
   - Steering flexible joint
   - Front wheels
   - Battery box
   - Front damper base beam in the engine compartment

5. Unlock the locking tabs on the tie rod stop plate.

6. Remove the bolts, stop plate, tie rod plate and seal washers to separate the tie-rod ends.

7. Remove or disconnect the following:
   - Steering gear ground cable and electronic connectors
   - Steering gear mounting bolts

8. Slide the steering gear to the passenger side until the left end clears the subframe, then slide the gear to to the driver side and out of the vehicle.

To install:

9. Installation is the reverse of the removal procedure, noting the following:

   a. Use Dacro bolts to mount the steering gear and tighten them to 43 ft. lbs. (58 Nm).

   b. Tighten the stop plate bolts to 80 ft. lbs. (108 Nm). Ensure the bolt head flats line up with the lock tabs, then bend the lock tabs over the bolt heads.

Strut Damper and Spring

REMOVAL & INSTALLATION

Front

1. Before servicing the vehicle, refer to the Precautions Section. Also see the IMA system disabling before proceeding with any procedure.

2. Remove or disconnect the following:
   - Front wheel
   - Tie rod end nuts
   - Upper ball joint

- Stabilizer link
- Brake hose retainer
- Speed sensor bracket
- Lower damper pinch bolts
- Strut top nut
- Strut and damper

To install:

➡**Use new self-locking fasteners for assembly.**

3. Install or connect the following:
   - Strut. Loosely tighten the mounting nuts and pinch bolts
   - Stabilizer link and hand tighten the nut.

4. Place a jack under the lower arm and raise the suspension to load it.

5. Tighten the strut top nuts to 40 ft. lbs. (54 Nm).

6. Tighten the lower pinch bolts to 72 ft. lbs. (98 Nm).

7. Tighten the stabilizer link nut to 22 ft. lbs. (29 Nm).

8. Tighten the tie rod nut to 32 ft. lbs. (43 Nm).

9. Install the speed sensor bracket and brake line retainer.

10. Install the front wheels and check the wheel alignment.

Coil Spring and Shock Absorber

REMOVAL & INSTALLATION

Rear

1. Before servicing the vehicle, refer to the Precautions Section. Also see the IMA system disabling before proceeding with any procedure.

2. Raise and support the rear of the vehicle.

3. Remove the fender skirts and rear wheels.

4. Disconnect the wheel speed sensors

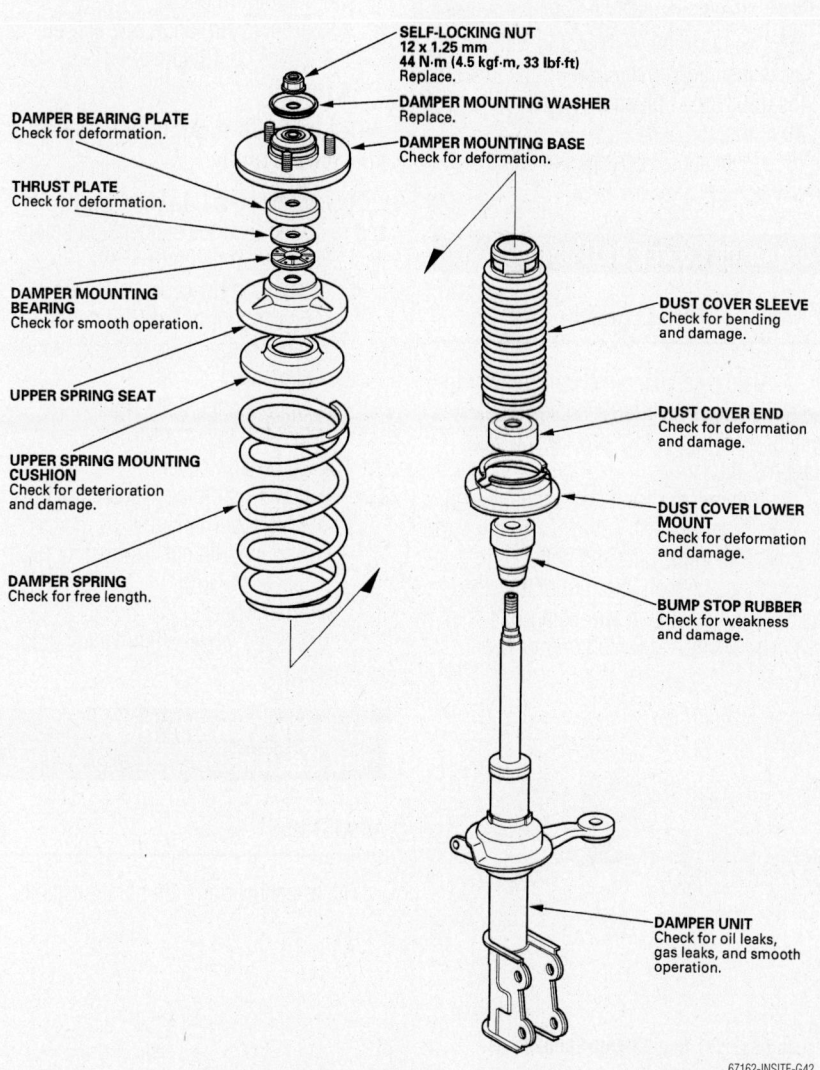

Exploded view of front strut and damper—Insight

67162-INSITE-G42

and remove the sensor and brake line brackets.

5. Place a jack under each end of the rear axle beam.

6. Remove the lower shock absorber mounting bolts.

7. Slowly lower the jacks and remove the coil springs, spring cushion and bump stop.

8. Remove the shock upper bolts and remove the shock.

To install:

➡️**Use a new self-locking nut.**

9. Installation is the reverse of the removal procedure, noting the following items:

a. Install the shock and upper mounting bolts.

b. Align the upper spring end with the stepped part of the upper spring cushion. Align the lower spring end with the stepped part of the spring seat of the rear axle beam.

c. Install the lower shock bolt loosely, raise and load the suspension with the jacks then tighten the lower bolts to 43 ft. lbs. (59 Nm). Tighten the upper bolts to 40 ft. lbs. (54 Nm).

10. Check the wheel alignment and adjust as necessary.

Upper or Lower Ball Joint

REMOVAL & INSTALLATION

1. Before servicing the vehicle, refer to the Precautions Section. Also see the IMA system disabling before proceeding with any procedure.

2. Remove or disconnect the following:
- Front wheel
- Ball joints nuts and cotter pin

3. Press the ball joint out of the steering arm or suspension arm using Ball Joint Remover 07MAC-SL00200 or equivalent.

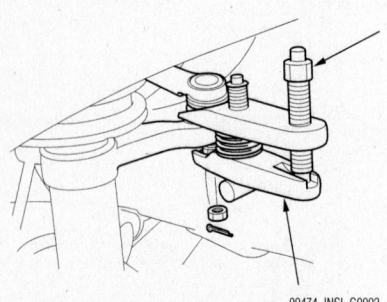

09474_INSI_G0002

Using Special Tool 07MAC-SL00200 to press out the ball joint—Insight

To install:

➡️**Use new ball joint nuts and cotter pins.**

4. Press the ball joint into the steering arm or suspension arm.

5. Install the ball joint nuts. Tighten the nuts to 47 ft. lbs. (64 Nm).

6. Install the front wheel.

7. Check the wheel alignment and adjust as necessary.

Lower Control Arm

REMOVAL & INSTALLATION

1. Before servicing the vehicle, refer to the Precautions Section. Also see the IMA system disabling before proceeding with any procedure.

2. Remove or disconnect the following:
- Front wheel
- Spindle nut
- Lower ball joint
- Lower arm cover
- Front inner flange bolt and nut
- Lower control arm

To install:

➡️**Use new self-locking nuts and split pins for assembly.**

3. Loosely install the mounting bolts and raise and load the suspension before final tightening of the bolts.

4. Install or connect the following:
- Lower control arm. Tighten the inner flange bolts to 51 ft. lbs. (69 Nm).
- Upper flange bolt. Tighten the bolt to 54 ft. lbs. (74 Nm).
- Lower arm cover
- Lower ball joint. Tighten the nuts to 47 ft. lbs. (64 Nm).
- New spindle nut. Tighten to 134 ft. lbs. (181 Nm).
- Front wheel

5. Check the wheel alignment and adjust as necessary.

Steering Knuckle, Hub and Bearing

ADJUSTMENT

The wheel bearings are not adjustable.

REMOVAL & INSTALLATION

Front

1. Before servicing the vehicle, refer to the Precautions Section. Also see the IMA system disabling before proceeding with any procedure.

2. Remove or disconnect the following:
- Front wheel
- Brake hose bracket
- Brake caliper
- Spindle nut
- Brake rotor
- Wheel speed sensor, if equipped
- Lower ball joint
- Steering knuckle

3. Press the hub out of the knuckle.

4. Press the inner bearing race out of the hub.

5. Remove the splash guard.

6. Remove the bearing unit from the knuckle.

To install:

➡️**Use new ball joint nuts, split pins, snapring and a new spindle nut for assembly.**

7. Install the hub unit and tighten the bolts to 43 ft. lbs. (59 Nm).

8. Install the splash guard.

9. Press the bearing unit into the hub.

10. Install the splash guard.

11. Press the hub into the bearing.

12. Install or connect the following:
- Steering knuckle. Tighten the knuckle-to-damper unit bolts to 72 ft. lbs. (98 Nm)
- Lower ball joint. Tighten the nut 40–47 ft. lbs. (54–64 Nm).
- Wheel speed sensor, if equipped
- Brake caliper and rotor. Tighten the caliper bracket bolts to 80 ft. lbs. (108 Nm).
- Brake hose bracket
- Spindle nut. Tighten the nut to 134 ft. lbs. (181 Nm).
- Front wheel

13. Check the wheel alignment and adjust as necessary.

Rear

1. Before servicing the vehicle, refer to the Precautions Section. Also see the IMA system disabling before proceeding with any procedure.

2. Remove or disconnect the following:
- Rear wheel
- Fender skirts

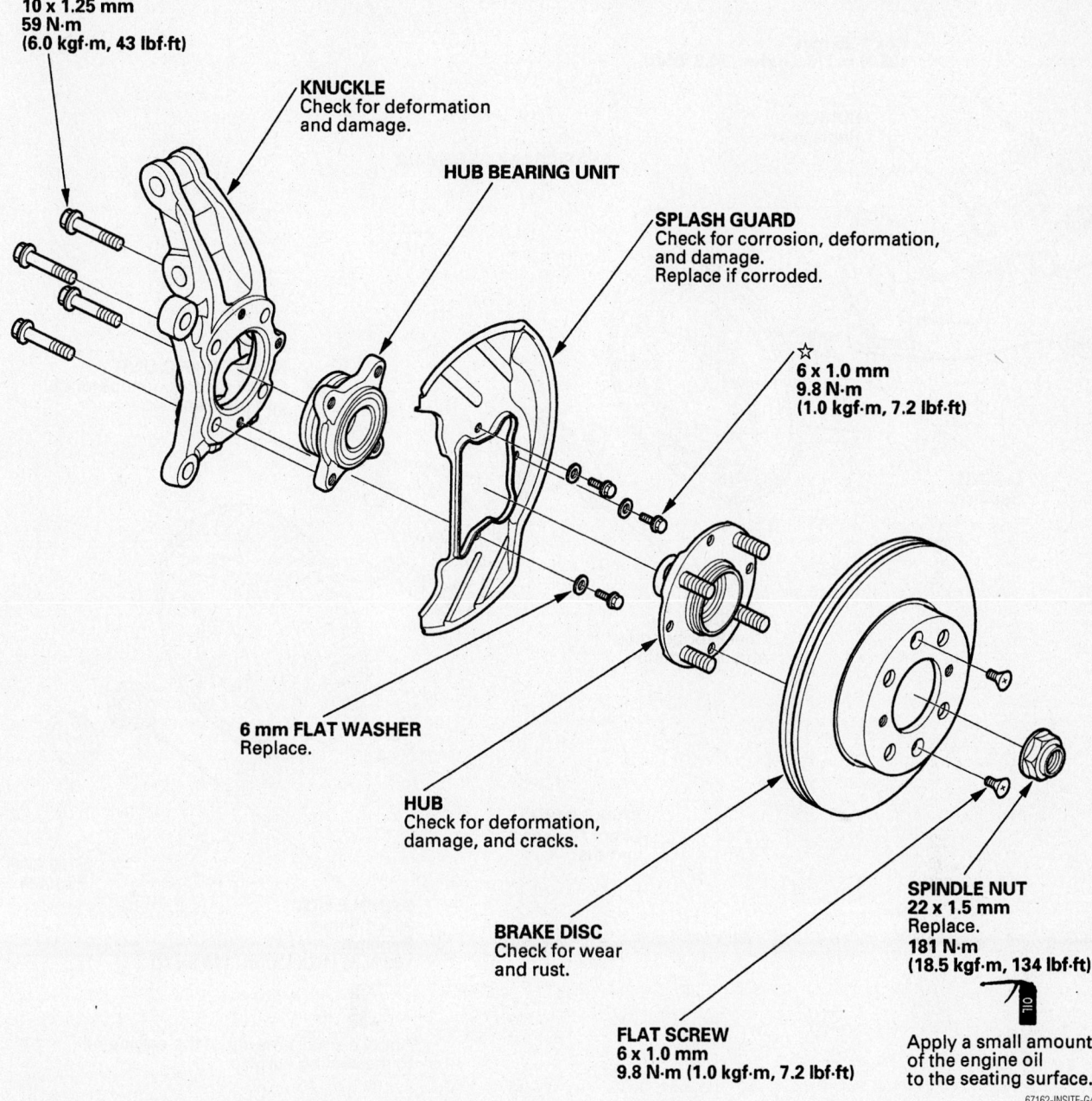

10 x 1.25 mm
59 N·m
(6.0 kgf·m, 43 lbf·ft)

KNUCKLE
Check for deformation
and damage.

HUB BEARING UNIT

SPLASH GUARD
Check for corrosion, deformation,
and damage.
Replace if corroded.

☆
6 x 1.0 mm
9.8 N·m
(1.0 kgf·m, 7.2 lbf·ft)

6 mm FLAT WASHER
Replace.

HUB
Check for deformation,
damage, and cracks.

BRAKE DISC
Check for wear
and rust.

SPINDLE NUT
22 x 1.5 mm
Replace.
181 N·m
(18.5 kgf·m, 134 lbf·ft)

Apply a small amount
of the engine oil
to the seating surface.

FLAT SCREW
6 x 1.0 mm
9.8 N·m (1.0 kgf·m, 7.2 lbf·ft)

67162-INSITE-G43

Exploded view of front steering knuckle, hub and wheel bearing—Insight

- Spindle nut
- Brake drum
- Hub and bearing assembly

To install:

➡**Use a new spindle nut for assembly.**

3. Install or connect the following:
   - Hub and bearing assembly. Tighten
     the flange bolts to 76 ft. lbs. (103
     Nm).
   - Spindle nut. Tighten the nut to 119
     ft. lbs. (162 Nm).
   - Brake drum
   - Rear wheel

Rear Axle Beam

REMOVAL & INSTALLATION

1. Before servicing the vehicle, refer to
the Precautions Section. Also see the IMA
system disabling before proceeding with
any procedure.

2. Place a jack at each end of the rear
axle beam.

3. Remove or disconnect the following:
   - Rear wheel
   - Fender skirts

- Wheel speed sensors and remove
  the sensor and brake line brackets
- Parking brake cable brackets
- Speed sensor harnesses
- Lower shock bolts
- Coil springs
- Spring bump stop
- Spindle nut
- Brake drum
- Bearing hub
- Brake assembly
- Axle beam mounting bolts
- Axle beam

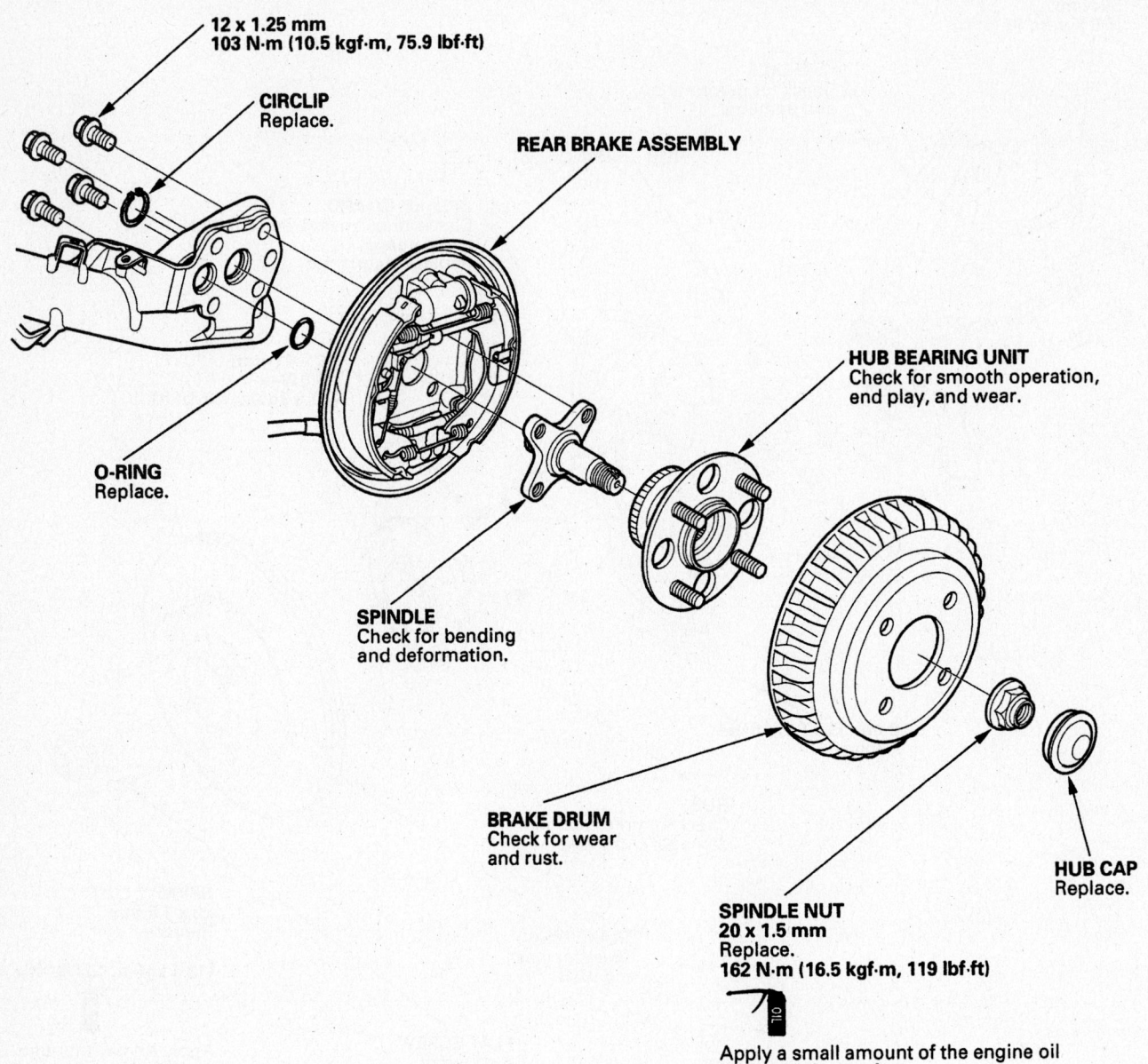

12 x 1.25 mm
103 N·m (10.5 kgf·m, 75.9 lbf·ft)

CIRCLIP
Replace.

REAR BRAKE ASSEMBLY

HUB BEARING UNIT
Check for smooth operation,
end play, and wear.

O-RING
Replace.

SPINDLE
Check for bending
and deformation.

BRAKE DRUM
Check for wear
and rust.

SPINDLE NUT
20 x 1.5 mm
Replace.
162 N·m (16.5 kgf·m, 119 lbf·ft)

Apply a small amount of the engine oil
to the seating surface.

HUB CAP
Replace.

67162-INSITE-G44

Exploded view of rear hub and bearing—Insight

To install:
4. Installation is the reverse of the removal procedure, noting the following items:

a. Tighten the axle beam bolts to 43 ft. lbs. (59 Nm).
b. Install the lower shock bolt loosely, raise and load the suspension with the jacks then tighten the lower bolts to 43 ft.

lbs. (59 Nm). Tighten the upper bolts to 40 ft. lbs. (54 Nm).
5. Check the wheel alignment and adjust as necessary.

BRAKES

Brake Caliper

REMOVAL & INSTALLATION

1. Before servicing the vehicle, refer to the Precautions Section. Also see the IMA system disabling before proceeding with any procedure.
2. Remove or disconnect the following:
   • Wheel
   • Brake hose bracket
   • Caliper mounting bolts
   • Caliper
3. Installation is the reverse of removal. Torque the lower bolt to 16 ft. lbs. (22 Nm) and upper bolt to 23 ft. lbs. (31 Nm).

Disc Brake Pads

REMOVAL & INSTALLATION

1. Remove the lower bolt caliper bolt.
2. Pivot the caliper up and hold the pads.
3. Remove the pad springs.
4. Remove the pads and shims.
5. Remove the pad retainers.
6. Installation is the reverse of removal. Coat both sides of the shims and the backs of the pads with brake grease. Torque the lower bolt to 16 ft. lbs. (22 Nm).

Brake Drums

REMOVAL & INSTALLATION

1. Raise and safely support the vehicle. Release the parking brake.

2. Remove the fender skirts.
3. Remove the rear wheels.
4. Use chalk to mark the brake drum to one of the wheel studs as an index mark for reinstallation.
5. Remove the retaining screw that holds the brake drum to the axle flange.
6. Pull the brake drum from the axle flange.
   To install:
7. Align the index mark and install the brake drum to the axle flange.
8. Install the retaining screw to secure the brake drum to the axle flange.
9. Install the rear wheels.

Rear Brake Shoes

REMOVAL & INSTALLATION

1. Raise and safely support the vehicle.
2. Remove the rear wheels.

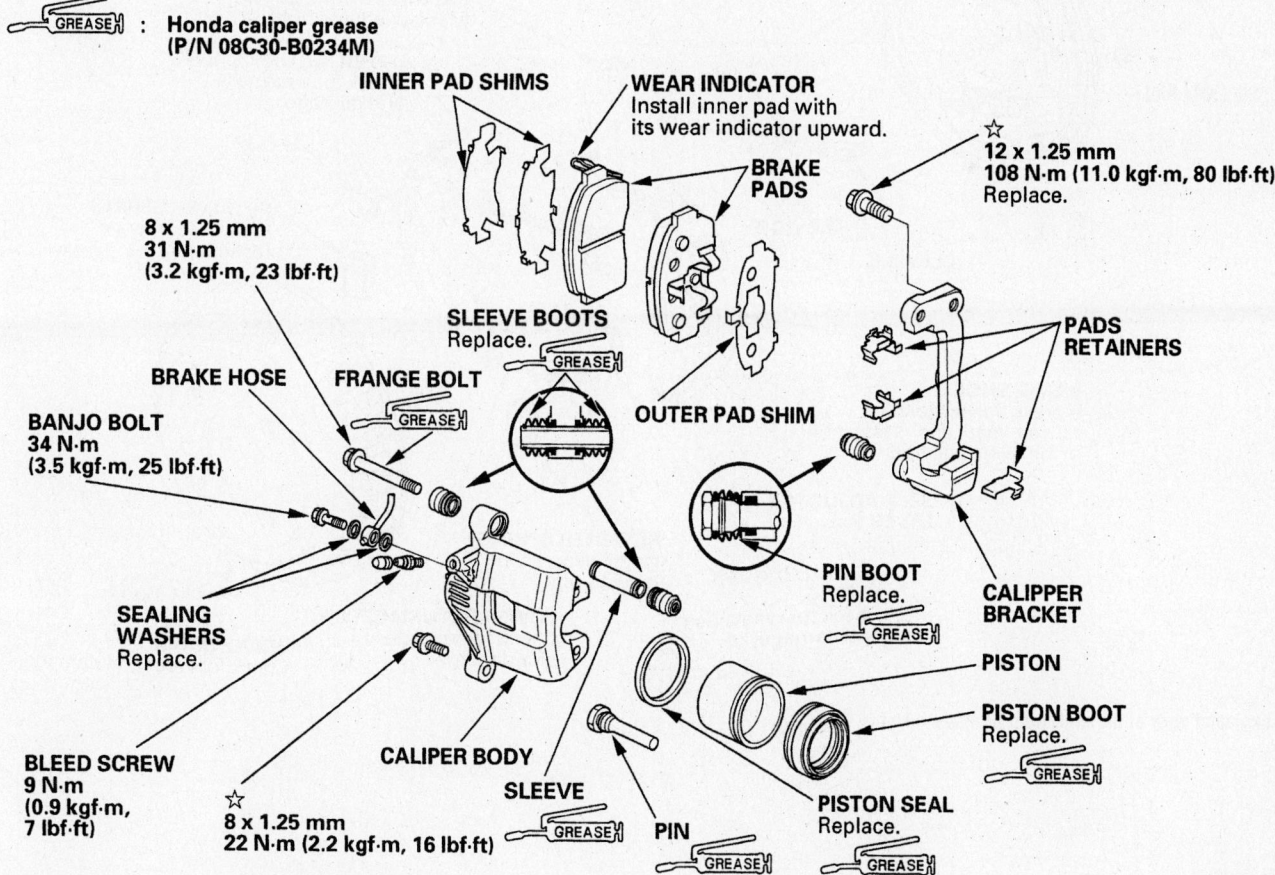

Exploded view of the front disc brakes—Insight

67162-INSITE-G45

3. Remove the brake drums.

4. Remove the brake return springs.

5. Remove the leading shoe holding pin and spring, and then the leading shoe.

6. Remove the self-adjuster and the adjuster lever.

7. Remove the trailing shoe holding pin and spring.

8. Disconnect the parking brake cable from the trailing shoe and remove the trailing shoe. Remove the parking brake lever from the trailing shoe.

To install:

9. Attach the parking brake lever to the trailing shoe.

10. Connect the parking brake cable to the parking brake lever.

11. Apply a thin coat of high temperature grease to the shoe contact points on the brake backing plate contact surface, and self-adjuster.

12. Position the trailing shoe on the backing plate and install the hold-down pin, spring, and retainer. Be careful not to stretch the return spring when fitting the shoes onto the backing plate.

13. Connect the upper return spring and the leading shoe to the trailing shoe and position the leading brake shoe on the backing plate.

14. Install the adjuster assembly and the hold-down pin, spring, and retainer.

15. Use a brake spring tool to install the lower return spring.

16. Install the self-adjuster lever and adjuster spring.

17. Check the brake drum for scoring or other wear. Machine or replace as necessary. Check the maximum brake drum diameter specification when machining.

18. Install the rear wheels. Lower the vehicle.

19. Road-test the vehicle.

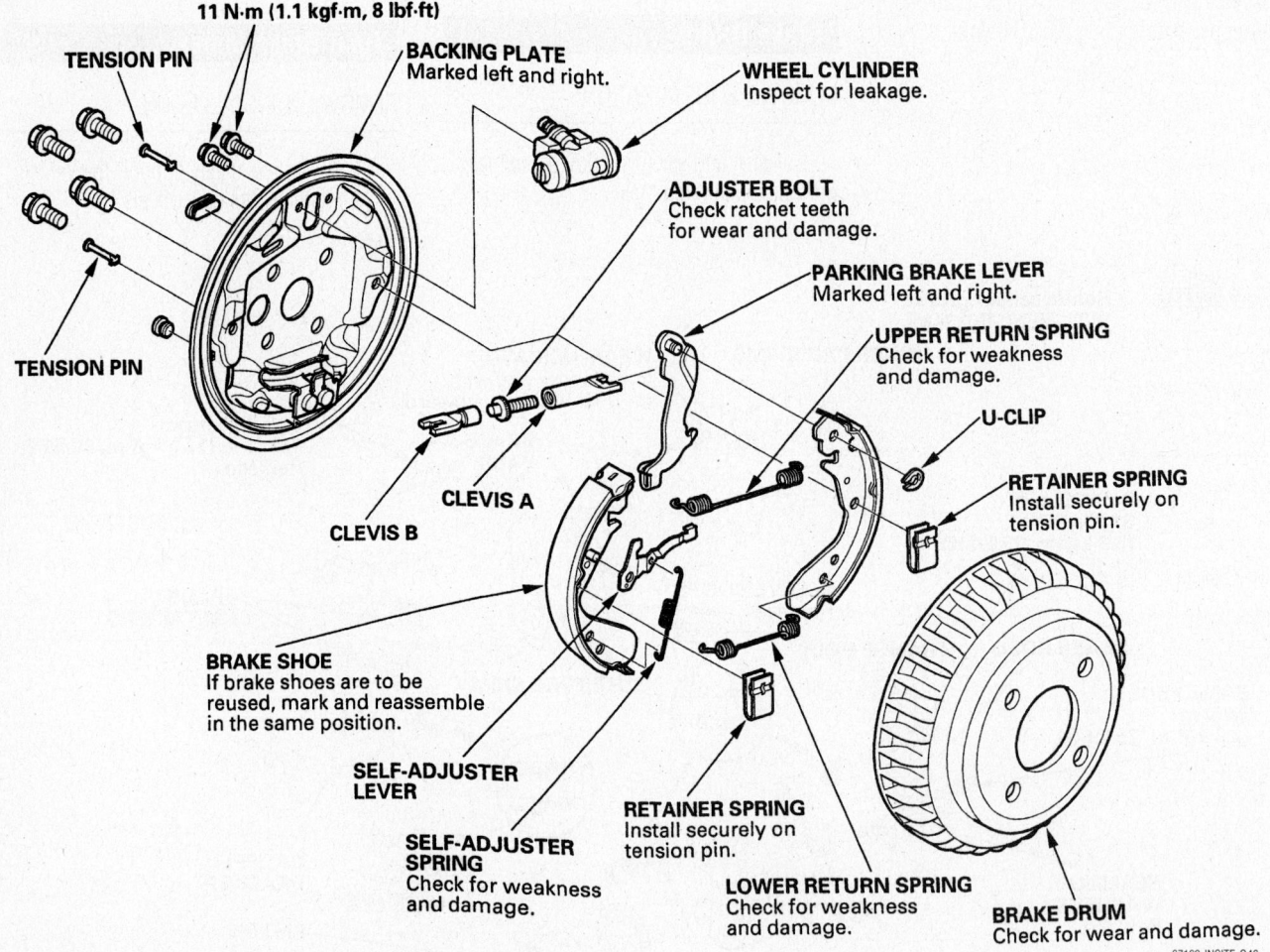

11 N·m (1.1 kgf·m, 8 lbf·ft)

TENSION PIN

BACKING PLATE
Marked left and right.

WHEEL CYLINDER
Inspect for leakage.

ADJUSTER BOLT
Check ratchet teeth for wear and damage.

PARKING BRAKE LEVER
Marked left and right.

UPPER RETURN SPRING
Check for weakness and damage.

U-CLIP

RETAINER SPRING
Install securely on tension pin.

TENSION PIN

CLEVIS A

CLEVIS B

BRAKE SHOE
If brake shoes are to be reused, mark and reassemble in the same position.

SELF-ADJUSTER LEVER

SELF-ADJUSTER SPRING
Check for weakness and damage.

RETAINER SPRING
Install securely on tension pin.

LOWER RETURN SPRING
Check for weakness and damage.

BRAKE DRUM
Check for wear and damage.

67162-INSITE-G46

Exploded view of the rear drum brakes—Insight

SPECIFICATIONS AND MAINTENANCE CHARTS

ENGINE AND VEHICLE IDENTIFICATION CHART

Engine Code						
Code	Liters (cc)	Cu. In.	Cyl.	Fuel Sys.	Engine Type	Eng. Mfg.
J35A4	3.5 (3471)	212	6	SMFI	SOHC	Honda
J35A6	3.5 (3471)	212	6	SMFI	SOHC	Honda
J35A7 ②	3.5 (3471)	212	6	SMFI	SOHC	Honda

Model Year	
Code ①	Year
2	2002
3	2003
4	2004
5	2005
6	2006

SOHC: Single Overhead Cam

SMFI: Sequential Multi-port Fuel Injection

① 10th position of VIN

② Variable cylinder management

09474_ODYS_C0001

GENERAL ENGINE SPECIFICATIONS

Year	Model	Engine Displacement Liters	Engine ID	Net Horsepower @ rpm	Net Torque @ rpm (ft. lbs.)	Bore x Stroke (in.)	Com-pression Ratio	Oil Pressure @ rpm
2002	Odyssey	3.5	J35A4	240@5200	242@3500	3.50x3.66	10.0:1	71@3000
2003	Odyssey	3.5	J35A4	250@5200	242@4500	3.50x3.66	10.0:1	71@3000
2004	Odyssey	3.5	J35A4	250@5200	242@4500	3.50x3.66	10.0:1	71@3000
2005	Odyssey ①	3.5	J35A6	244@5750	240@5000	3.50x3.66	10.0:1	71@3000
	Odyssey ②	3.5	J35A7	244@5750	240@4500	3.50x3.66	10.0:1	71@3000
2006	Odyssey ①	3.5	J35A6	244@5750	240@5000	3.50x3.66	10.0:1	71@3000
	Odyssey ②	3.5	J35A7	244@5750	240@4500	3.50x3.66	10.0:1	71@3000

① LX and EX models

② EX-L and EX-L Touring models

09474_ODYS_C0002

ENGINE TUNE-UP SPECIFICATIONS

Year	Engine Displacement Liters	Engine ID	Spark Plug Gap (in.)	Ignition Timing (deg.)		Fuel Pump (psi)	Idle Speed (rpm)		Valve Clearance (in.)	
				MT	AT		MT	AT	In.	Ex.
2002	3.5	J35A4	0.039-0.043	—	10B	48-54	—	680-780	0.008-0.009	0.011-0.013
2003	3.5	J35A4	0.039-0.040	—	10B	41-48	—	680-780	0.008-0.009	0.011-0.013
2004	3.5	J35A4	0.039-0.040	—	10B	41-48	—	680-780	0.008-0.009	0.011-0.013
2005	3.5	J35A6 J35A7	0.039-0.043	—	10B	41-48	—	680-780	0.008-0.009	0.011-0.013
2006	3.5	J35A6 J35A7	0.039-0.043	—	10B	41-48	—	680-780	0.008-0.009	0.011-0.013

NOTE: The Vehicle Emission Control Information label often reflects changes made during production and must be used if they differ from this chart.

NOTE: The fuel pressure readings are given with the vacuum hose connected to the regulator and the engine running

B: Before top dead center

09474_ODYS_C0003

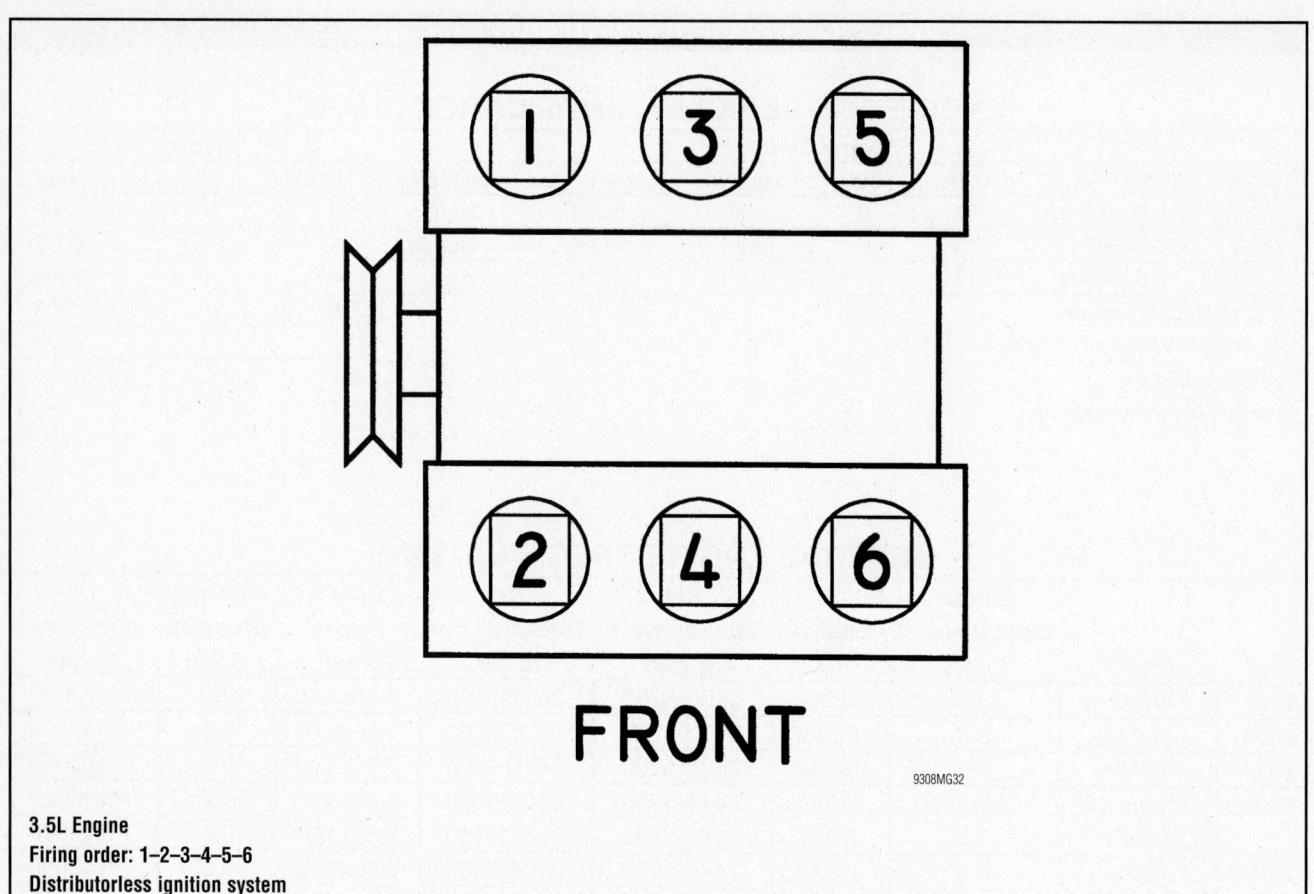

FRONT

9308MG32

3.5L Engine
Firing order: 1–2–3–4–5–6
Distributorless ignition system

CAPACITIES

Year	Model	Engine Displacement Liters	Engine ID	Engine Oil with Filter (qts.)	Transmission (qts.)		Fuel Tank (gal.)	Cooling System (qts.)
					5-Spd	Auto.		
2002	Odyssey	3.5	J35A4	4.6	—	8.3	20.0	7.0
2003	Odyssey	3.5	J35A4	4.6	—	8.3	20.0	8.0
2004	Odyssey	3.5	J35A4	4.6	—	8.3	20.0	8.0
2005	Odyssey	3.5	J35A6/J35A7	4.5	—	6.6	21.0	7.4
2006	Odyssey	3.5	J35A6/J35A7	4.5	—	6.6	21.0	7.4

NOTE: All capacities are approximate. Add fluid gradually and check to be sure a proper fluid level is obtained.

09474_ODYS_C0004

VALVE SPECIFICATIONS

Year	Engine Displacement Liters	Engine ID	Seat Angle (deg.)	Face Angle (deg.)	Spring Test Pressure (lbs. @ in.)	Spring Installed Height (in.)	Stem-to-Guide Clearance (in.)		Stem Diameter (in.)	
							Intake	Exhaust	Intake	Exhaust
2002	3.5	J35A4	45	45	NA	①	0.0008-0.0018	0.0022-0.0031	0.2159-0.2163	0.2146-0.2150
2003	3.5	J35A4	45	45	NA	①	0.0008-0.0018	0.0022-0.0031	0.2159-0.2163	0.2146-0.2150
2004	3.5	J35A4	45	45	NA	①	0.0008-0.0018	0.0022-0.0031	0.2159-0.2163	0.2146-0.2150
2005	3.5	J35A6	NA	NA	NA	②	0.0008-0.0018	0.0022-0.0031	0.2159-0.2163	0.2146-0.2150
		J35A7	NA	NA	NA	③	0.0008-0.0018	0.0022-0.0031	0.2159-0.2163	0.2146-0.2150
2006	3.5	J35A6	NA	NA	NA	②	0.0008-0.0018	0.0022-0.0031	0.2159-0.2163	0.2146-0.2150
		J35A7	NA	NA	NA	③	0.0008-0.0018	0.0022-0.0031	0.2159-0.2163	0.2146-0.2150

NA: Not Available

① Valve spring free length:
   Intake: 1.9713 in.
   Exhaust: 2.1060 in.

② Valve spring free length:
   Intake: 2.029 in.
   Exhaust: 2.010 in.

③ Valve spring free length:
   Intake: 2.029 in.
   Exhaust: 2.069 in.

09474_ODYS_C0005

CRANKSHAFT AND CONNECTING ROD SPECIFICATIONS
All measurements are given in inches

Year	Engine Displacement Liters	Engine ID	Crankshaft				Connecting Rod		
			Main Brg. Journal Dia.	Main Brg. Oil Clearance	Shaft End-play	Thrust on No.	Journal Diameter	Oil Clearance	Side Clearance
2002	3.5	J35A4	2.8337-2.8346	0.0008-0.0017	0.0040-0.0140	3	2.1644-2.1654	0.0008-0.0017	0.0060-0.0140
2003	3.5	J35A4	2.8337-2.8346	0.0008-0.0017	0.0040-0.0140	3	2.1644-2.1654	0.0008-0.0017	0.0060-0.0140
2004	3.5	J35A4	2.8337-2.8346	0.0008-0.0017	0.0040-0.0140	3	2.1644-2.1654	0.0008-0.0017	0.0060-0.0140
2005	3.5	J35A6 J35A7	2.8337-2.8346	0.0008-0.0017	0.0040-0.0140	3	2.1644-2.1654	0.0008-0.0017	0.0060-0.0140
2006	3.5	J35A6 J35A7	2.8337-2.8346	0.0008-0.0017	0.0040-0.0140	3	2.1644-2.1654	0.0008-0.0017	0.0060-0.0140

09474_ODYS_C0006

PISTON AND RING SPECIFICATIONS

All measurements are given in inches

Year	Engine Displacement Liters	Engine ID	Piston Clearance	Ring Gap			Ring Side Clearance		
				Top Compression	Bottom Compression	Oil Control	Top Compression	Bottom Compression	Oil Control
2002	3.5	J35A4	0.0006-0.0016	0.0080-0.0140	0.0160-0.0220	0.0080-0.0280	0.0022-0.0031	0.0012-0.0022	NA
2003	3.5	J35A4	0.0006-0.0016	0.0080-0.0140	0.0160-0.0220	0.0080-0.0280	0.0022-0.0031	0.0012-0.0022	NA
2004	3.5	J35A4	0.0006-0.0016	0.0080-0.0140	0.0160-0.0220	0.0080-0.0280	0.0022-0.0031	0.0012-0.0022	NA
2005	3.5	J35A6 J35A7	0.0006-0.0016	0.0080-0.0140	0.0160-0.0220	0.0080-0.0280	0.0022-0.0031	0.0012-0.0022	NA
2006	3.5	J35A6 J35A7	0.0006-0.0016	0.0080-0.0140	0.0160-0.0220	0.0080-0.0280	0.0022-0.0031	0.0012-0.0022	NA

NA: Not Applicable

09474_ODYS_C0007

TORQUE SPECIFICATIONS

All readings in ft. lbs.

Year	Engine Displacement Liters	Engine ID	Cylinder Head Bolts	Main Bearing Bolts	Rod Bearing Bolts	Crankshaft Damper Bolts	Flywheel Bolts	Manifold		Spark Plugs	Oil Pan Drain Plug
								Intake	Exhaust		
2002	3.5	J35A4	①	②	③	181	54	16	23	13	29
2003	3.5	J35A4	①	②	③	181	54	16	23	13	29
2004	3.5	J35A4	①	②	③	181	54	16	23	13	29
2005	3.5	J35A6	④	⑤	③	⑥	54	16	23	13	29
		J35A7	⑦	②	③	⑥	54	16	23	13	29
2006	3.5	J35A6	④	⑤	③	⑥	54	16	23	13	29
		J35A7	⑦	②	③	⑥	54	16	23	13	29

NOTE: Dip main bearing bolts and crankshaft damper bolt in clean engine oil prior to tightening.

① Step 1: 29 ft. lbs.
  Step 2: 51 ft. lbs.
  Step 3: 72 ft. lbs.

② 11mm bolt 56 ft. lbs.
  10mm bolt 36 ft. lbs.

③ Step 1: 14 ft. lbs.
  Step 2: 90 degrees

④ 6-point head bolts
  Step 1: 29 ft. lbs.
  Step 2: 51 ft. lbs.
  Step 3: 72 ft. lbs.
  Perform each step twice
  12-point head bolts
  Step 1: Torque all bolts to 22 ft. lbs.
  Step 2: Torque all bolts an additional 90 deg.
  Step 3: Torque all bolts an additional 90 deg.
  New Bolt Only: an additional 90 deg.

⑤ Cap bolts: 54 ft. lbs.
  Side bolts: 36 ft. lbs.

⑥ Step 1: 47 ft. lbs.
  Step 2: plus 60 degrees

⑦ Step 1: 29 ft. lbs.
  Step 2: 51 ft. lbs.
  Step 3: 72 ft. lbs.
  Perform each step twice

09474_ODYS_C0008

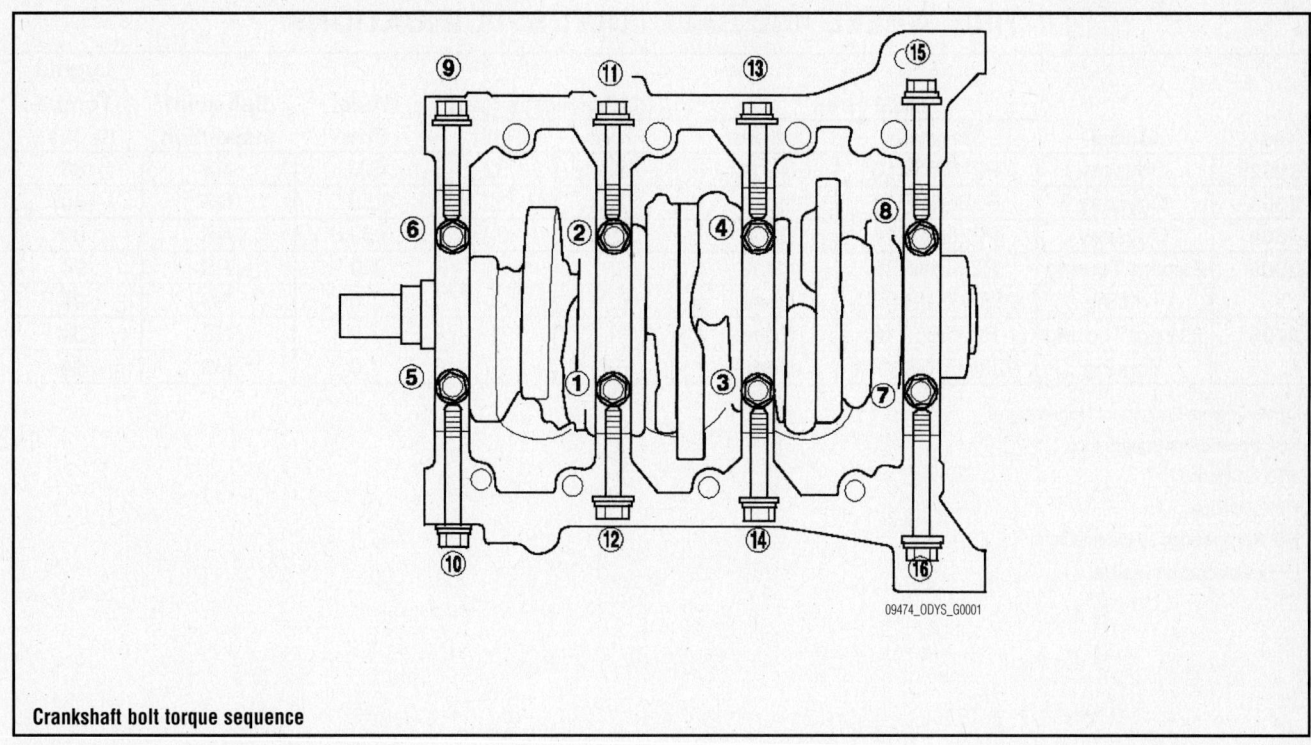

09474_ODYS_G0001

Crankshaft bolt torque sequence

WHEEL ALIGNMENT

Year	Model		Caster Range (+/-Deg.)	Caster Preferred Setting (Deg.)	Camber Range (+/-Deg.)	Camber Preferred Setting (Deg.)	Toe-in (in.)
2002	Odyssey	F	1.00	+2.07	1.00	0	0+/-0.08
		R	—	—	0.75	-0.50	0+/-0.08
2003	Odyssey	F	1.00	+2.07	1.00	0	0+/-0.08
		R	—	—	0.75	-0.50	0+/-0.08
2004	Odyssey	F	1.00	+2.07	1.00	0	0+/-0.08
		R	—	—	0.75	-0.50	0+/-0.08
2005	Odyssey	F	1.00	+2.53	0.50	0	0+/-0.08
		R	—	—	0.75	-0.50	0.08+/-0.08
2006	Odyssey	F	1.00	+2.53	0.50	0	0+/-0.08
		R	—	—	0.75	-0.50	0.08+/-0.08

09474_ODYS_C0009

TIRE, WHEEL AND BALL JOINT SPECIFICATIONS

Year	Model	OEM Tires Standard	OEM Tires Optional	Tire Pressures (psi) Front	Tire Pressures (psi) Rear	Wheel Size	Ball Joint Inspection	Lugnut Torque (ft. lbs.)
2002	Odyssey	P215/65R16	None	32	32	6JJ	NS	80
2003	Odyssey	P225/60R16	None	32	32	6.5JJ	NS	80
2004	Odyssey	P225/60R16	None	32	32	6.5JJ	NS	80
2005	Except Touring	P235/65R16	None	①	①	7.0	NS	94
	Touring	P235-710R460A	None	①	①	7.0	NS	94
2006	Except Touring	P235/65R16	None	①	①	7.0	NS	94
	Touring	P235-710R460A	None	①	①	7.0	NS	94

OEM: Original Equipment Manufacturer

PSI: Pounds Per Square Inch

STD: Standard

OPT: Optional

NS: Not specified by manufacturer

① See placard on vehicle

09474_ODYS_C0010

BRAKE SPECIFICATIONS

All measurements in inches unless noted

Year	Model		Brake Disc Original Thickness	Brake Disc Minimum Thickness	Brake Disc Maximum Runout	Parking Brake Brake Drum Diameter Original Inside Diameter	Parking Brake Brake Drum Diameter Max. Wear Limit	Parking Brake Brake Drum Diameter Maximum Machine Diameter	Minimum Lining Thickness Front	Minimum Lining Thickness Rear	Brake Caliper Bracket Bolts (ft. lbs.)	Brake Caliper Mounting Bolts (ft. lbs.)
2002	Odyssey	F	1.100	1.020	0.004	—	—	—	0.060	—	80	27
		R	0.440	0.350	0.004	8.268	8.272	—	0.060	0.040 ①	41	27
2003	Odyssey	F	1.100	1.020	0.004	—	—	—	0.060	—	80	20
		R	0.440	0.350	0.004	8.268	8.272	—	0.060	0.040 ①	41	27
2004	Odyssey	F	1.100	1.020	0.004	—	—	—	0.060	—	80	20
		R	0.440	0.350	0.004	8.268	8.272	—	0.060	0.040 ①	41	27
2005	Odyssey	F	1.100	1.020	0.004	—	—	—	0.060	—	101	37
		R	0.440	0.350	0.004	8.270	8.307	—	0.060	0.040 ①	65	16
2006	Odyssey	F	1.100	1.020	0.004	—	—	—	0.060	—	101	37
		R	0.440	0.350	0.004	8.270	8.307	—	0.060	0.040 ①	65	16

F: Front

R: Rear

① Parking brake shoe

09474_ODYS_C0011

SCHEDULED MAINTENANCE INTERVALS
2002-04 Honda Odyssey

TO BE SERVICED	TYPE OF SERVICE	VEHICLE MILEAGE INTERVAL (x1000)															
		7.5	15	22.5	30	37.5	45	52.5	60	67.5	75	82.5	90	97.5	105	112	120
Accessory drive belts	I & A				✓				✓				✓				✓
Air cleaner element	R				✓				✓				✓				✓
Air conditioning filter	R				✓				✓				✓				✓
Brake fluid	R						✓						✓				
Brake hoses & lines	I		✓		✓		✓		✓		✓		✓		✓		✓
Cooling system hoses	I		✓		✓		✓		✓		✓		✓		✓		✓
Engine coolant	R						✓						✓				
Engine oil	R	✓	✓	✓	✓	✓	✓	✓	✓	✓	✓	✓	✓	✓	✓	✓	✓
Engine oil and coolant levels	I	Inspect at each fuel stop															
Engine oil filter	R		✓		✓		✓		✓		✓		✓		✓		✓
Exhaust system	I		✓		✓		✓		✓		✓		✓		✓		✓
Fluid levels and condition	I		✓		✓		✓		✓		✓		✓		✓		✓
Front and rear brakes	I		✓		✓		✓		✓		✓		✓		✓		✓
Fuel lines & connection	I		✓		✓		✓		✓		✓		✓		✓		✓
Halfshaft boots	I		✓		✓		✓		✓		✓		✓		✓		✓
Idle speed	I & A														✓		
Parking brake system	I & A		✓		✓		✓		✓		✓		✓		✓		✓
Rear differential fluid	R												✓				
Rotate and inspect tires	I	✓	✓	✓	✓	✓	✓	✓	✓	✓	✓	✓	✓	✓	✓	✓	✓
Spark plugs	R				✓				✓				✓				✓
Supplemental Restraint system	I	Inspect the SRS 10 years after production															
Suspension components	I		✓		✓		✓		✓		✓		✓		✓		✓
Steering components	I		✓		✓		✓		✓		✓		✓		✓		✓
Timing balancer belt	R														✓		
Timing belt	R														✓		
Transmission fluid	R						✓				✓				✓		
Valve clearance	I					✓			✓				✓				✓
Water pump	S/I														✓		

R: Replace   I: Inspect   A: Adjust

FREQUENT OPERATION MAINTENANCE (SEVERE SERVICE)
If a vehicle is operated under any of the following conditions it is considered severe service:
- Towing a trailer or using a camper or car-top carrier.
- Repeated short trips of less than 5 miles in temperatures below freezing, or trips of less than 10 miles in any temperature.
- Extensive idling or low-speed driving for long distances as in heavy commercial use, such as delivery, taxi or police cars.
- Operating on rough, muddy or salt-covered roads.
- Operating on unpaved or dusty roads.
- Driving in extremely hot (over 90°) conditions.

Air cleaner element: replace every 15,000 miles

Engine oil and filter: replace every 3750 miles or 6 months, whichever occurs first.

Timing belt: replace every 60,000 miles if the vehicle is regularly driven in temperatures above 110°F or below -20°F.

Transmission fluid: replace every 30,000 miles.

Rear differential fluid: replace every 60,000 miles.

Front and rear brakes: inspect every 7500 miles or 6 months, whichever occurs first.

Locks and hinges: lubricate every 15,000 miles.

09474_ODYS_C0012

MAINTENANCE MINDER SCHEDULE
2005-06 Honda Odyssey

The Ridgeline displays engine oil life and maintenance service items in the information display to indicate when to perform maintenance service. If the engine oil life is 15% or less, based on the onboard computer's caluculations, you will see SERVICE DUE SOON in the information display every time the ignition key is turned to ON. The maintenance minder indicator will also come on and the maintenance code(s) for other scheduled maintenance items needing service will be displayed below the message.

Symbol	Item	Service
A	Engine oil ①	Change
B	Engine oil and filter	Change
	Tires	Rotate
	Brakes	Inspect
	Parking brake adjustment	Check
	Steering gear and linkage	Inspect
	Suspension components	Inspect
	Driveshaft boots	Inspect
	Brake hoses and lines	Inspect
	All fluid levels and condition	Inspect
	Exhaust system	Inspect
	Fuel lines and connections	Inspect
1	Tires	Rotate
2	Engine air filter ②	Replace
	Dust and pollen filter ③	Replace
	Accessory drive belt	Inspect
3	Transmission fluid ④	Replace
4	Spark plugs	Replace
	Timing belt ⑤	Replace
	Water pump	Inspect
	Valve clearance ⑥	Inspect
5	Engine coolant	Replace

① If the message SERVICE DUE NOW does not appear more than 12 months after the display is reset, change every year.

② If driven in dusty conditions, replace every 15,000 miles.

③ If driven in urban areas that have a high concentration of soot from industry and diesel, replace every 15,000 miles

④ If regularly driven in mountainous areas at very low speed or trailer towing, change the fluid every 30,000 miles.

⑤ If driven regularly in temperatures over 110 deg.F or below -20 deg.F, or towing a trailer, replace every 60,000 miles.

⑥ Adjust if necessary.

Additionally, replace the brake fluid every 3 years, and inspect the idle speed every 160,000 miles.

To reset the Engine Oil Life Display on LX, EX and EX-L models:

1. Turn the ignition switch to ON.

2. Press the SELECT/RESET button repeatedly until the engine oil life display or the service message is displayed.

3. Press the SELECT/RESET button for about 10 seconds. The engine oil life idicator and the codes will blink.

4. Press the SELECT/RESET knob for more than 5 seconds. The codes will disappear and the indicator will reset to 100.

To reset the Engine Oil Life Display on Touring models:

1. Turn the ignition switch to ON.

2. Press the SEL/RESET button on the steering wheel until the engine oil life is displayed.

3. Press the SEL/RESET button on the steering wheel for 10 seconds. The display will change to the CUSTOM SETUP mode

4. Press the SEL/RESET button on the steering wheel. The codes will disappear and the indicator will reset to 100.

ENGINE REPAIR

➡ Disconnecting the negative battery cable on some vehicles may interfere with the functions of the on board computer system. The computer may undergo a relearning process once the negative battery cable is reconnected.

Alternator

REMOVAL & INSTALLATION

1. Before servicing the vehicle, refer to the precautions in the beginning of this section.
2. Remove or disconnect the following:
   - Negative battery cable
   - Accessory drive belt
   - Alternator wiring harness connectors
   - Alternator mounting bolts
   - Wiring harness clamp
   - Alternator

To install:

3. Install or connect the following:
   - Alternator
   - Wiring harness clamp. Tighten the bolt to 105 inch lbs. (12 Nm).

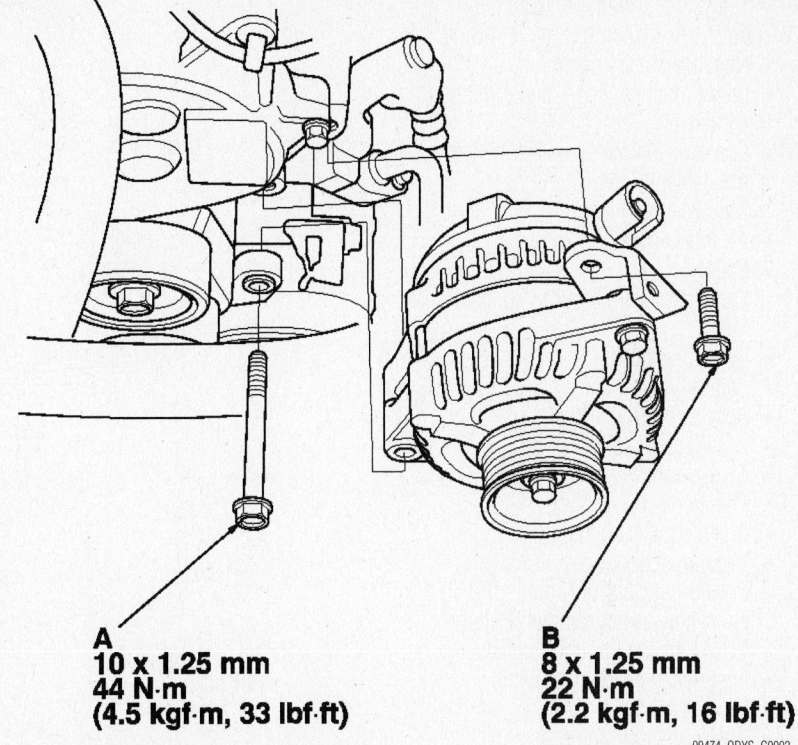

A
10 x 1.25 mm
44 N·m
(4.5 kgf·m, 33 lbf·ft)

B
8 x 1.25 mm
22 N·m
(2.2 kgf·m, 16 lbf·ft)

09474_ODYS_G0002

Alternator mounting—J35A6 and J35A7 engines

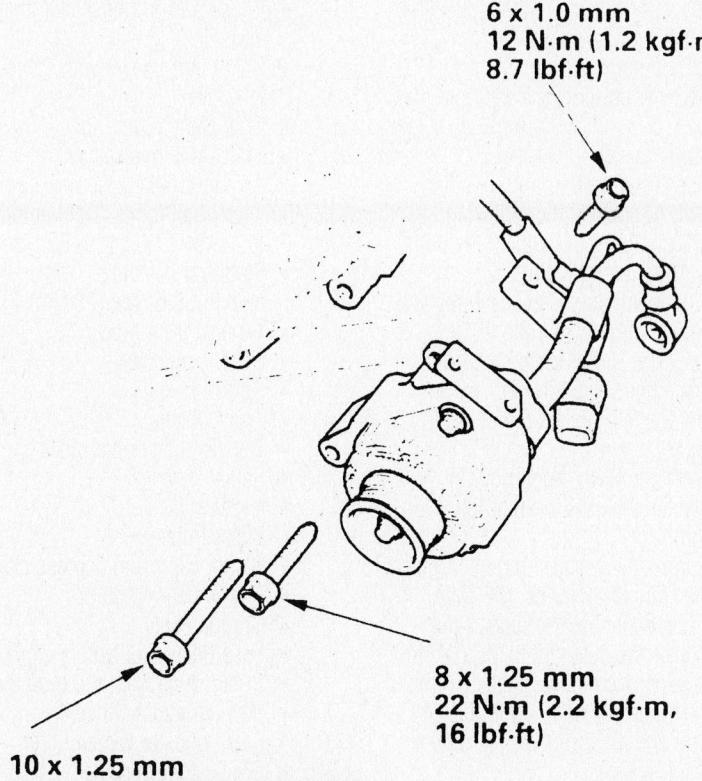

6 x 1.0 mm
12 N·m (1.2 kgf·m, 8.7 lbf·ft)

8 x 1.25 mm
22 N·m (2.2 kgf·m, 16 lbf·ft)

10 x 1.25 mm
44 N·m (4.5 kgf·m, 33 lbf·ft)

9358MG01

Alternator mounting—J35A4 engine

- Alternator mounting bolts. Tighten the 10mm bolt to 33 ft. lbs. (44 Nm) and the 8mm bolt to 16 ft. lbs. (22 Nm).
- Alternator wiring harness connectors. Tighten the battery terminal nut to 105 inch lbs. (12 Nm).
- Accessory drive belt
- Negative battery cable

Ignition Timing

ADJUSTMENT

The 3.5L engine is equipped with a Distributorless Ignition System (DIS). The ignition timing is controlled by the Powertrain Control module (PCM). No adjustment is necessary.

Engine Assembly

REMOVAL & INSTALLATION

2002–04

➡ The engine and transaxle are removed from the vehicle as a unit.

1. Before servicing the vehicle, refer to the precautions in the beginning of this section.

2. Drain the cooling system.
3. Drain the transaxle fluid.
4. Drain the engine oil.
5. Relieve fuel system pressure.
6. Remove or disconnect the following:
   - Negative battery cable
   - Intake manifold cover and ignition coil cover
   - Evaporative Emissions (EVAP) control canister or vacuum hose, whichever applies
   - Air intake duct
   - Battery
   - Left engine wire harness connectors
   - Relay bracket
   - Battery tray
   - Accessory drive belts
   - Accelerator cable
   - Cruise control cable
   - Fuel lines
   - Brake booster vacuum line
   - Vacuum supply hose
   - Powertrain Control Module (PCM) connectors and grommet. Pull the PCM harness through the firewall.
   - Fuse/Relay box battery cable
   - Ground cable
   - Power steering pump
   - Starter cable and harness clamp
   - Radiator hoses
   - Heater hoses
   - Bypass hose
   - Transaxle oil cooler lines
   - Front wheels
   - Splash shield
   - Heated Oxygen Sensor (HO$_2$S) connector
   - Exhaust front pipe
   - Stabilizer bar links
   - Lower ball joints
7. Separate the inner CV-joints from the transaxle and support the axle halfshafts out of the work area with safety wire.
8. Remove or disconnect the following:
   - Shift cable bracket
   - Shift cable cover
   - Shift control lever with cable attached
   - Power steering hose clamp and clips
   - Transaxle lower front mount
   - Transaxle lower rear mount
   - Steering rack and pinion gear. Support the steering gear with safety wire.
9. Attach a hoist to the engine lifting eyes and support the powertrain weight.
10. Remove or disconnect the following:
    - Side engine mount bracket
    - Front mount bracket support nut
11. Matchmark the front subframe to the mounting points.

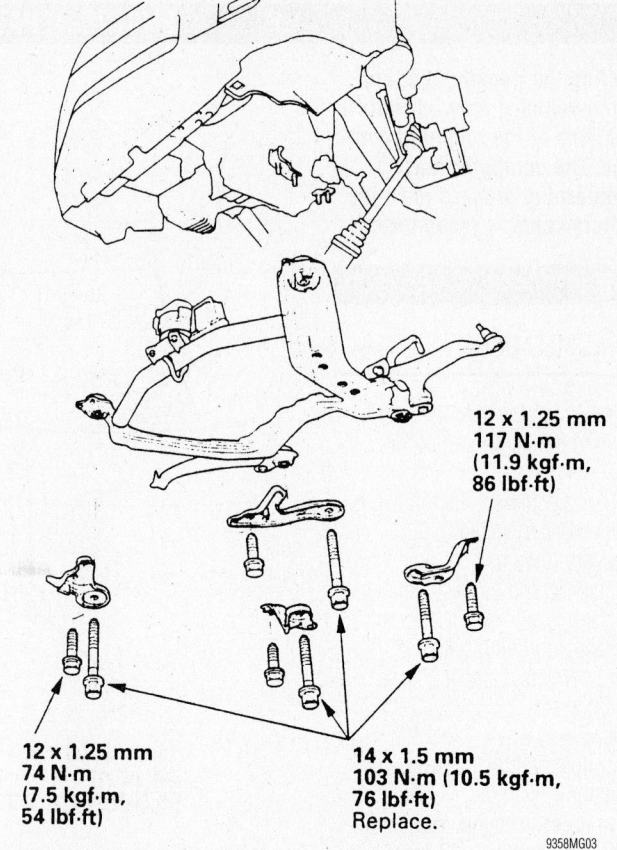

12 x 1.25 mm
117 N·m
(11.9 kgf·m,
86 lbf·ft)

12 x 1.25 mm
74 N·m
(7.5 kgf·m,
54 lbf·ft)

14 x 1.5 mm
103 N·m (10.5 kgf·m,
76 lbf·ft)
Replace.

9358MG03

Sub-frame fastener locations and tightening torque—2002–04

12. Remove or disconnect the following:
    - Front subframe
    - A/C compressor
13. Lower the powertrain away from the vehicle.

To install:
14. Raise the powertrain into position.
15. Install or connect the following:
    - A/C compressor. Tighten the bolts to 16 ft. lbs. (22 Nm).
    - Front subframe. Use new bolts and tighten the 14mm bolts to 76 ft. lbs. (103 Nm). Tighten the front brace bolts to 54 ft. lbs. (74 Nm) and the rear brace bolts to 86 ft. lbs. (117 Nm).
    - Transaxle lower front mount. Tighten the nuts to 28 ft. lbs. (38 Nm).
    - Transaxle lower rear mount. Tighten the bolts to 28 ft. lbs. (38 Nm).
    - Front mount bracket support nut. Tighten the nut to 40 ft. lbs. (54 Nm).
    - Side engine mount bracket. Tighten the bracket bolts to 33 ft. lbs. (44 Nm) and the through bolt to 40 ft. lbs. (54 Nm).
    - Steering rack and pinion gear. Tighten the bolts to 29 ft. lbs. (39 Nm).
    - Power steering hose clamp and clips
    - Shift control lever with cable attached
    - Shift cable cover
    - Shift cable bracket
    - Axle halfshafts. Use new circlips.
    - Lower ball joints. Tighten the nuts to 43–51 ft. lbs. (59–69 Nm).
    - Stabilizer bar links. Tighten the nuts to 58 ft. lbs. (78 Nm).
    - Exhaust front pipe
    - HO$_2$S connector
    - Splash shield
    - Front wheels
    - Transaxle oil cooler lines
    - Radiator hoses
    - Heater hoses
    - Bypass hose
    - Starter cable and harness clamp
    - Power steering pump
    - Ground cable
    - Fuse/Relay box battery cable
    - PCM connectors and grommet
    - Vacuum supply hose
    - Brake booster vacuum line
    - Cruise control cable
    - Accelerator cable
    - Accessory drive belts
    - Fuel lines

- Battery tray
- Relay bracket
- Left engine wire harness connectors
- Battery
- Air intake duct
- EVAP control canister hose
- Negative battery cable

16. Fill the engine crankcase to the correct level.

17. Fill the transaxle to the correct level.

18. Fill the cooling system.

19. Start the engine and check for leaks.

20. Check the wheel alignment and adjust as necessary.

2005–06

1. Make sure that you have the anti-theft codes for the radio and navigation system.

2. Drain the power steering fluid.

3. Relive the fuel system pressure.

4. Disconnect the battery.

5. Remove the grille.

6. Remove the battery.

7. Remove the intake manifold cover.

8. Remove the air cleaner housing and related tubes.

9. Remove the air intake duct.

10. Remove the battery tray.

11. Disconnect the starter cables.

12. Disconnect the fuel supply line.

13. Tag and disconnect all wires and hoses connected to the engine.

14. Remove the driver's side console lower panel and pull back the carpet.

15. Remove the steering joint cover.

16. Lock the steering wheel and match-mark the steering shaft joint and the steering gear shaft.

17. Remove the bolt and disconnect the steering joint from the shaft.

➡**Do not turn the steering wheel with the joint disconnected.**

18. Remove the PCM cover.

19. Remove the accessory drive belt.

20. Disconnect the power steering outlet hose at the pump.

21. Remove the power steering reservoir and disconnect the return hose.

22. Remove the radiator cap.

23. Raise the vehicle on a hoist.

24. Remove the front wheels.

25. Remove the splash shield.

26. Drain the cooling system.

27. Drain the transmission.

28. Drain the engine oil.

29. Remove the front exhaust pipe.

30. Disconnect the stabilizer links.

31. Disconnect the tie rod ends from the knuckles.

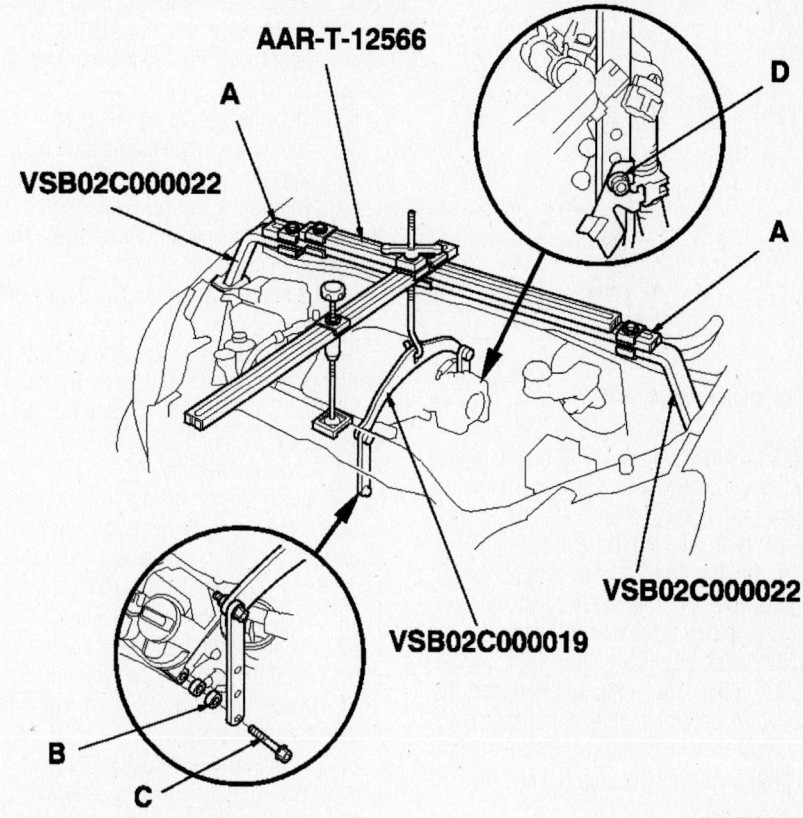

Engine support and hanger installed

32. Disconnect the lower control arms from the knuckles.

33. Remove the halfshafts.

34. Disconnect the shift cable and remove the control lever.

35. Disconnect the power steering hoses from the gear.

36. Disconnect the power steering pressure switch.

37. Remove the nuts from the transmission lower front and rear mounts.

38. Lower the vehicle.

39. Disconnect the cooling fan wiring.

40. Remove the coolant reservoir.

41. Remove the fan and shroud.

42. Remove the upper and lower radiator hoses.

43. Remove the heater hoses.

44. Remove the transmission cooler hoses.

45. Remove the radiator.

46. Remove the A/C compressor and secure it aside without disconnecting the hoses.

47. Install engine hanger adapters VSB02C000022 on the engine support beams (A). Then, install the engine support hanger AAR-T-12566 with the adapters.

48. Lift and support the engine with the support hanger and the balance bar

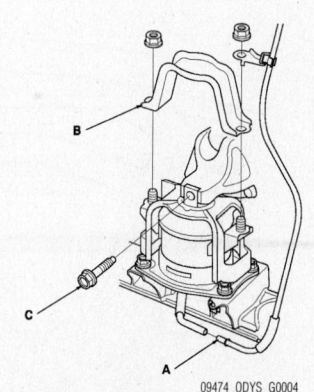

Front engine mount—J35A6 engine

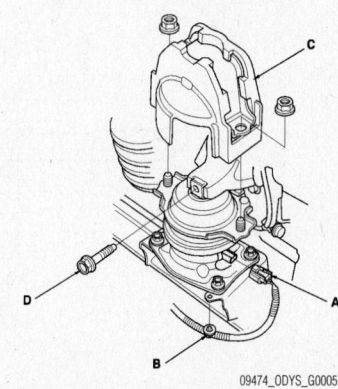

Front engine mount—J35A7 engine

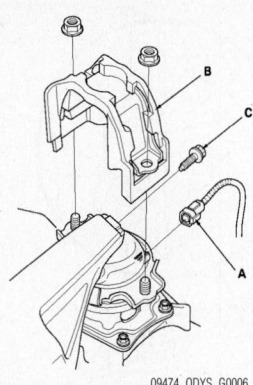

Rear engine mount—J35A7 engine

09474_ODYS_G0006

VSB02C000019. Attach the front arm to the front cylinder head with a spacer (B) and the 10x1.25mm bolt (C). Attach the rear arm to the rear head with the 8x1.25mm bolt (D).

49. On the J35A6 engine, remove the vacuum hose (A) from the front engine mount stop (B), then remove the front engine mount bolt (C).

50. On the J35A7 engine, disconnect the front active control engine mount actuator connector (A) and clamp (B), then remove the front engine mount stop (C) and the bolt (D).

51. On the J35A6 engine, remove the heat shield and the rear engine mount bolts.

52. On the J35A7 engine, disconnect the rear ACM connector (A), remove the rear engine mount stop (B) and the rear engine mount bolt (C).

53. Raise the vehicle on the hoist.

54. Matchmark the subframe and body.

55. Attach the subframe holding fixture (A) by hanging the belt (B) over the front of the subframe, then secure the belt with the retaining pin (C).

56. Raise the jack and install the 4 bolts at the corners.

57. Remove the six 12x1.25mm bolts (A) securing the subframe stiffeners (B), the 4 subframe mounting bolts (C) and the stiffeners. Lower the subframe (D).

58. Lower the vehicle.

59. Attach a chain hoist to the engine hook and the transmission hook. Lift the engine/transmission assembly until it is supported by the hoist, then, remove the engine support hanger and adapters.

60. Remove the mounting bolts from the upper half of the side engine mount bracket.

61. Check that all hoses, wires and lines are free from the engine.

62. Slowly lower the engine/transmission assembly about 6 inches. Check for anything that might hinder removal, then remove the engine/transmission assembly.

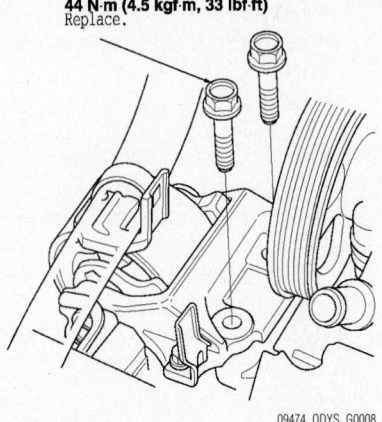

10 x 1.25 mm
44 N·m (4.5 kgf·m, 33 lbf·ft)
Replace.

09474_ODYS_G0008

Engine mount bracket torques—J35A6 and J35A7 engines

63. Raise the vehicle and remove the unit.

To install:

64. Install the engine mount brackets and tighten to the torques shown in the accompanying illustration.

65. Position the engine/transmission under the vehicle. Make sure they are aligned and lower the vehicle until it is properly positioned. Using the hoist, raise the assembly into place.

➡**Install the mounting bolts/nuts in the order given in the following steps. Otherwise, noise and vibration will result.**

66. Install the engine hanger adapters on the end of the support beams. Install the support hanger with adapters.

67. Install the support hanger and balance bar to the front and rear heads.

68. Install NEW mounting bolts into the upper half of the engine mount bracket. Torque to 33 ft. lbs. (44 Nm).

69. Remove the chain hoist.

70. Raise the subframe into position and loosely install all bolts.

71. Align all reference marks and tighten the bolts as shown in the accompanying illustration.

72. Remove the subframe support tool.

73. Install the transmission lower rear mount. Torque to 28 ft. lbs. (38 Nm).

74. On the J35A6 engine, tighten the rear engine bolts and heat shield. Torque the mount bolts to 31 ft. lbs. (42 Nm); torque the heat shield bolts to 43 ft. lbs. (58 Nm).

75. On the J35A7 engine, torque the rear engine mount bolt to 40 ft. lbs. (54 Nm), then torque the rear engine stop nuts to 54 ft. lbs. (73 Nm). Connect the ACM.

76. On the J35A6 engine, tighten the

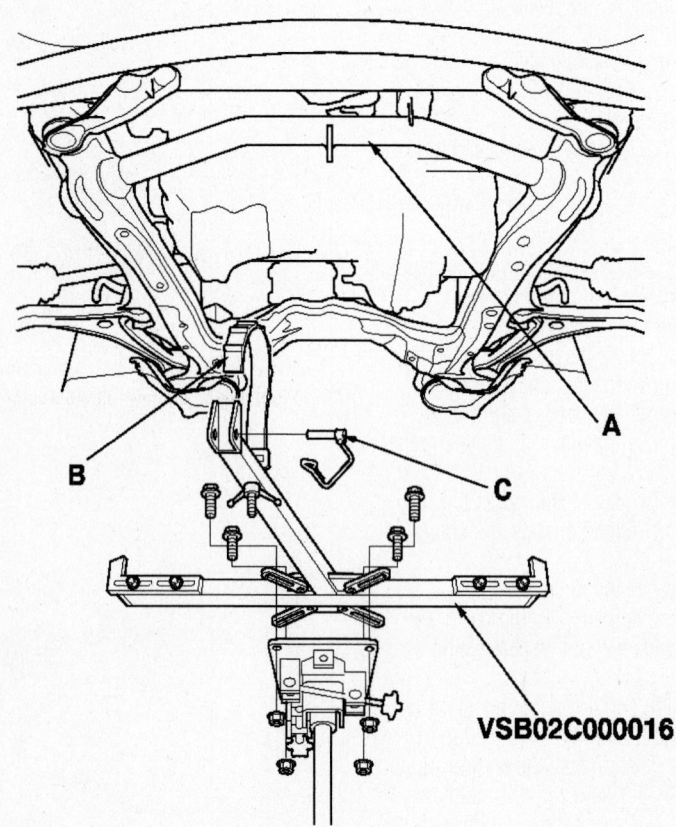

VSB02C000016

09474_ODYS_G0007

Subframe holding fixture—J35A6 and J35A7 engines

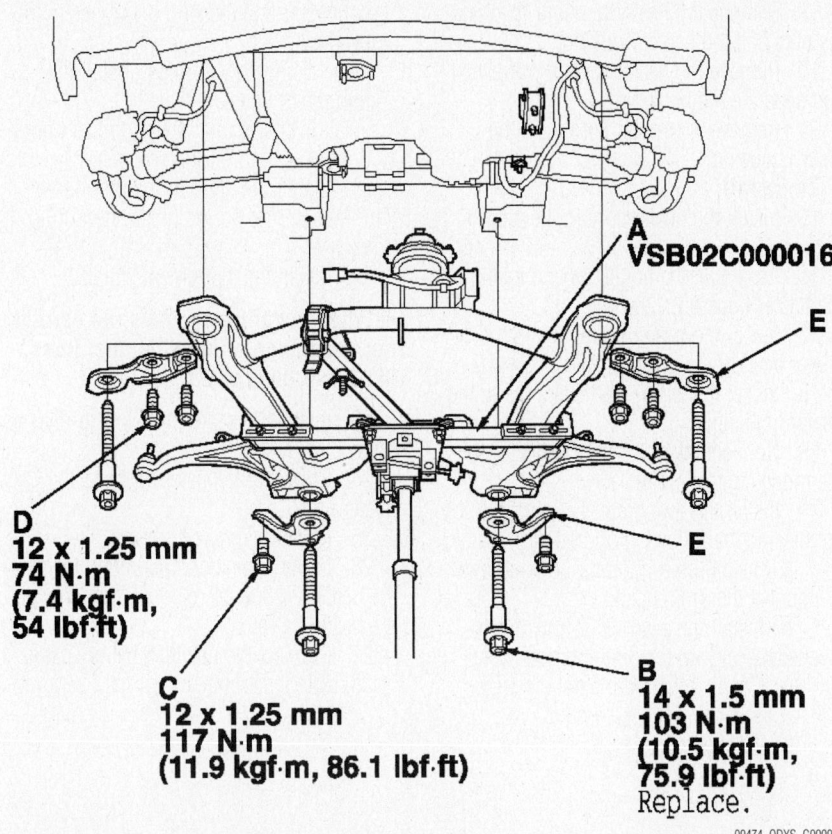

A VSB02C000016

E

E

D
12 x 1.25 mm
74 N·m
(7.4 kgf·m,
54 lbf·ft)

C
12 x 1.25 mm
117 N·m
(11.9 kgf·m, 86.1 lbf·ft)

B
14 x 1.5 mm
103 N·m
(10.5 kgf·m,
75.9 lbf·ft)
Replace.

09474_ODYS_G0009

Subframe installation torques—J35A6 and J35A7 engines

front mount bolt to 40 ft. lbs. (54 Nm), then the front mount stop nuts to 54 ft. lbs. (73 Nm). Connect the vacuum hose.

77. On the J35A7 engine, tighten the front engine mount bolt to 40 ft. lbs. (54 Nm), then tighten the front mount stop nuts to 54 ft. lbs. (73 Nm). Connect the ACM actuator and install the harness clamp.

78. Loosen the bolts at the upper half of the side mount and retighten them to 33 ft. lbs. (44 Nm).

79. Remove the engine support pieces.

80. Install the A/C compressor. Torque to 16 ft. lbs. (22 Nm).

81. Install the radiator.

82. The remainder of installation is the reverse of removal. Observe the following torques:

- Exhaust pipe-to-manifold: 40 ft. lbs. (54 Nm)
- Exhaust pipe hanger nut: 16 ft. lbs. (22 Nm)
- Exhaust pipe-to-converter: 25 ft. lbs. (33 Nm)
- Steering joint bolt: 21 ft. lbs. (28 Nm)
- Battery tray bolts: 16 ft. lbs. (22 Nm)

83. Refill all fluids and check for leaks and proper operation.

Heater Core

REMOVAL & INSTALLATION

2002–04 Models

FRONT UNIT

➡Make sure to acquire the anti-theft code for the radio and write down the frequencies for the radio's preset buttons.

1. Disconnect the negative battery cable.

✷✷ CAUTION

Wait at least 3 minutes for the air bag to deplete its energy before working on the steering wheel or instrument panel.

2. In the engine compartment, remove the heater valve cable clamp; then, disconnect the heater valve cable and rotate the heater valve to the fully open position.

3. Drain the engine coolant into a clean container for reuse.

4. Disconnect the heater hoses from the heater unit.

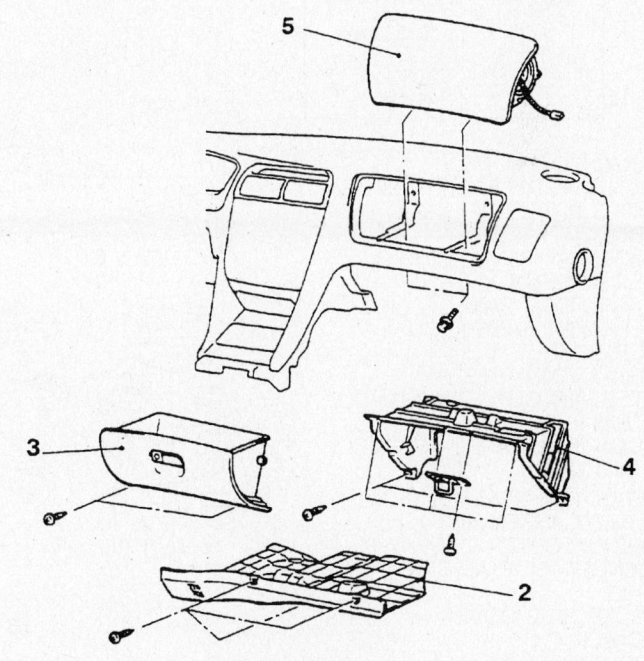

2. UNDERCOVER
3. GLOVE BOX ASSEMBLY
4. GLOVE BOX CASE
5. AIR BAG MODULE

93112GG1

View of the heater housing, evaporator housing and related components—2002–04

5. Remove the heater housing-to-chassis nuts.

6. Remove the center console.

7. Remove the instrument panel.

8. Remove the steering hanger beam mounting bolts and the steering hanger beam.

9. Remove the evaporator housing by performing the following procedure:

a. Discharge and recover the air conditioning system refrigerant.

b. Remove the refrigerant lines. Discard the O-rings. Plug the openings to prevent contamination.

c. Disconnect the thermostat electrical connector and the wiring harness from the evaporator.

d. Remove the evaporator housing-to-chassis screws, the bolt and nuts.

e. Disconnect the drain hose and remove the evaporator housing.

10. Disconnect the electrical connector from the mode control motor.

11. Remove the wiring harness clips from the heater housing.

12. Remove the heater housing-to-chassis nuts and the heater housing.

13. Remove the heater housing screws and separate the housings.

14. Remove the heater core from the heater housing.

To install:

15. Install the heater core to the heater housing.

16. Assemble the housings and install the heater housing screws.

17. Install the heater housing and the heater housing-to-chassis nuts.

18. Install the wiring harness clips to the heater housing.

19. Connect the electrical connector to the mode control motor.

20. Install the evaporator housing by performing the following procedure:

a. Install the evaporator housing and connect the drain hose.

b. Install the evaporator housing-to-chassis screws, the bolt and nuts.

c. Connect the thermostat electrical

connector and the wiring harness to the evaporator.

d. Using new O-rings, install the refrigerant lines.

e. Evacuate and charge the air conditioning system refrigerant.

21. Install the steering hanger beam and the steering hanger beam mounting bolts.

22. Install the instrument panel.

➡**When installing the nuts, be careful not to damage or bend the fuel lines, the brake lines or etc.**

23. Install the heater housing-to-chassis nuts.

24. Connect the heater hoses to the heater unit.

25. Refill the cooling system.

26. In the engine compartment, install the heater valve cable clamp; then, connect the heater valve cable.

27. Connect the negative battery cable.

28. Run the engine to normal operating

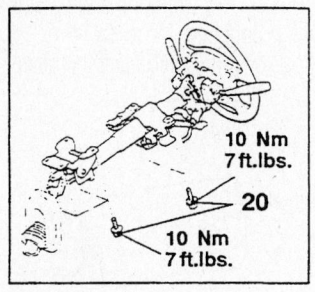

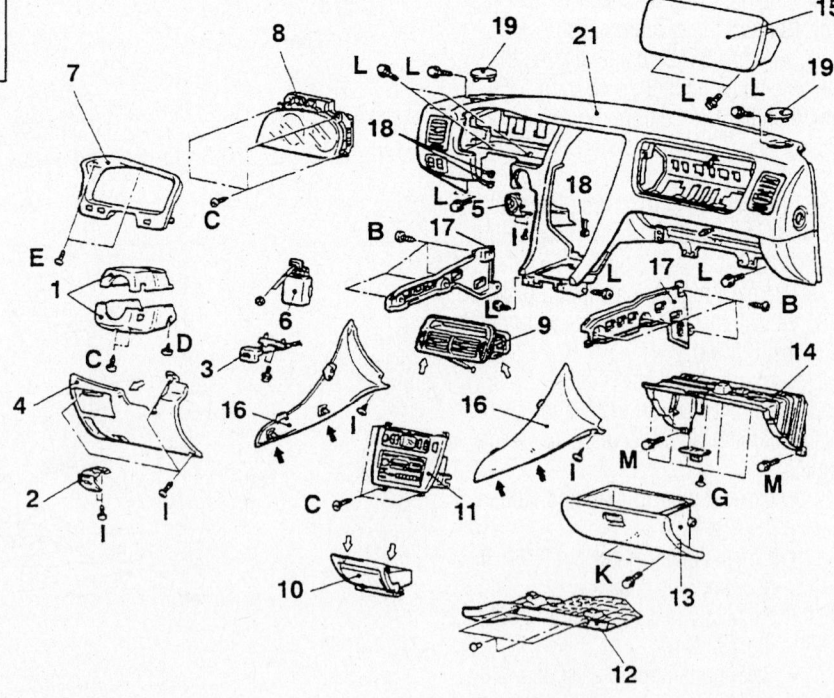

1. COLUMN COVER
2. HOOD LOCK RELEASE HANDLE
3. PARKING BRAKE RELEASE HANDLE
4. INSTRUMENT PANEL LOWER COVER ASSEMBLY (LH)
5. KEY CYLINDER PANEL
6. INSTRUMENT PANEL ECU
7. METER BEZEL
8. COMBINATION METER
9. CENTER AIR OUTLET ASSEMBLY
10. ASHTRAY
11. AIR CONTROL PANEL ASSEMBLY & AUDIO UNIT
12. UNDERCOVER ASSEMBLY
13. GLOVEBOX ASSEMBLY
14. GLOVEBOX OUTER CASE
15. PASSENGER SIDE AIRBAG MODULE
16. CONSOLE SIDE COVER ASSEMBLY
17. FLOOR CARPET REAR REINFORCEMENT
18. HARNESS CONNECTOR
19. PLUG
20. STEERING COLUMN MOUNTIN BOLT
21. INSTRUMENT PANEL

NOTE
(1) ⟵ : metal clip position
(2) ⬅ : plastic clip position

93112GG2

View of the steering hanger beam and related components—2002-04

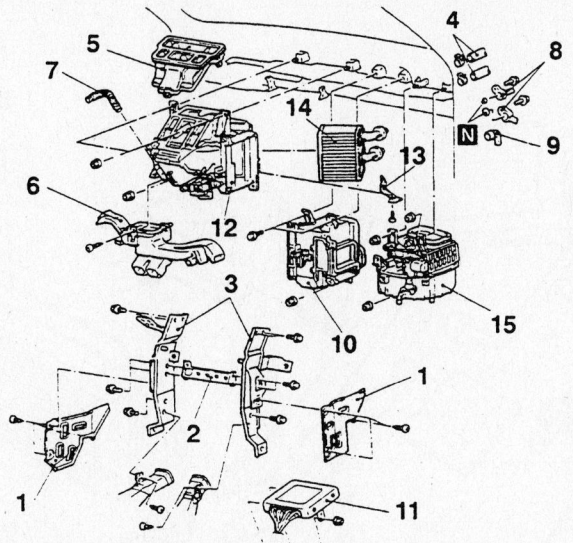

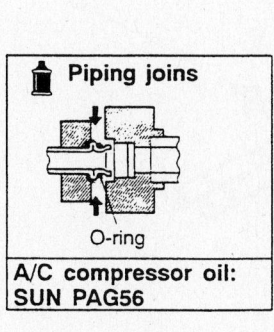

Piping joins

O-ring

A/C compressor oil:
SUN PAG56

1. FLOOR CARPET FRONT
   REINFORCEMENT
3. ECU BRACKET
4. CENTER STAY ASSEMBLY
5. HEATER HOSE CONNECTION
6. CENTER DUCT ASSEMBLY
7. FOOT DISTRIBUTION DUCT
8. BREATHER HOSE
9. SUCTION PIPE, LIQUID PIPE B
   AND COOLING UNIT
   CONNECTION
10. DRAIN HOSE
11. EVAPORATOR
12. ENGINE CONTROL MODULE
13. HEATER UNIT
14. HEATER CORE SUPPORT
15. HEATER CORE

93112GG3

Exploded view of the heater core, the heater housing and related components—2002–04

temperatures; then, check the climate control operation and check for leaks.

REAR UNIT

1. When the engine is cool, drain the engine coolant from the radiator.

2. Recover the refrigerant with a recovery/recycling/charging station.

➡**Put on gloves to protect your hands. When prying with a flat-tip screwdriver, wrap it with protective tape to prevent damage. Take care not to bend or scratch the trim and panels.**

3. Remove the trim as shown.
   - To remove the rear side trim panel, remove the speaker grille and rear speaker.
   - When removing the side trim panels, be sure to remove the mounting screws behind the rear speakers.
   - Be sure to disconnect the rear accessory socket and brake light failure sensor connectors when removing the left rear trim panel.

4. Disconnect the connectors (A) from the rear blower motor and the rear blower resistor, then remove the self-tapping screws, the mounting bolts and the rear blower unit (B).

5. Remove the clips (A) and the rear heater duct (B).

6. Remove the bolt from the bracket (A). Disconnect the suction line (B) and receiver line (C) from the rear evaporator. Slide the hose clamps (D) back, then disconnect the inlet heater hose (E) and the outlet heater hose (F) from the rear heater

unit. Engine coolant will run out when the hoses are disconnected; drain it into a clean drip pan. Be sure not to let coolant spill on the electrical parts or the painted surfaces. If any coolant spills, rinse it off immediately. Plug or cap the lines immediately after disconnecting them to avoid moisture and dust contamination.

7. Remove the clip (A), then lifting the upper duct (B) up. Disconnect the connector (C) from the rear mode control motor. Remove the mounting bolts and the rear evaporator-heater unit (D).

8. Remove the self-tapping screws and the clamp, then remove the bolts and the evaporator lines together with the expansion valve. If necessary, remove the expansion valve. Use a second wrench to hold the other fitting on the valve so the evaporator lines won't twist. Leave the first fitting loosely connected so you can use it to hold the valve while you loosen the second fitting.

9. Remove the self-tapping screws and the clamp, then be careful not to bend the inlet and outlet pipes during the heater core removal, and pull out the heater core.

10. Install the unit in the reverse order of removal, and note these items.
   - If you're installing a new evaporator, add refrigerant oil (DENSO ND-OIL 8).
   - Replace the O-rings with new ones at each fitting, and apply a thin coat of refrigerant oil before installing them. Be sure to use the correct O-rings for HFC-134a (R-134a) to avoid leakage.
   - Immediately after using the oil,

reinstall the cap on the container, and seal it to avoid moisture absorption.
   - Do not spill the refrigerant oil on the vehicle; it may damage the paint. If the refrigerant oil contacts the paint, wash it off immediately.
   - Do not interchange the inlet and outlet heater hoses. Install the hose clamps securely.
   - Refill the cooling system with engine coolant.
   - Make sure that there is no coolant leakage.
   - Make sure that there is no air leakage.
   - Charge the system.

2005–06 Models

FRONT UNIT

1. Make sure you have the anti-theft codes for the radio and the navigation system, then write down the XM radio channel presets.

2. Make sure the ignition is OFF, then disconnect the negative cable from the battery. Wait at least 3 minutes before proceeding.

3. Recover the refrigerant with a recovery/recycling/charging station.

4. Remove the intake manifold cover. See Intake Manifold Removal and Installation.

5. Remove the bolts and nut, then disconnect the suction line (A) and the receiver line (B) from the evaporator core.

6. From under the hood, open the cable clamp (A), then disconnect the heater valve

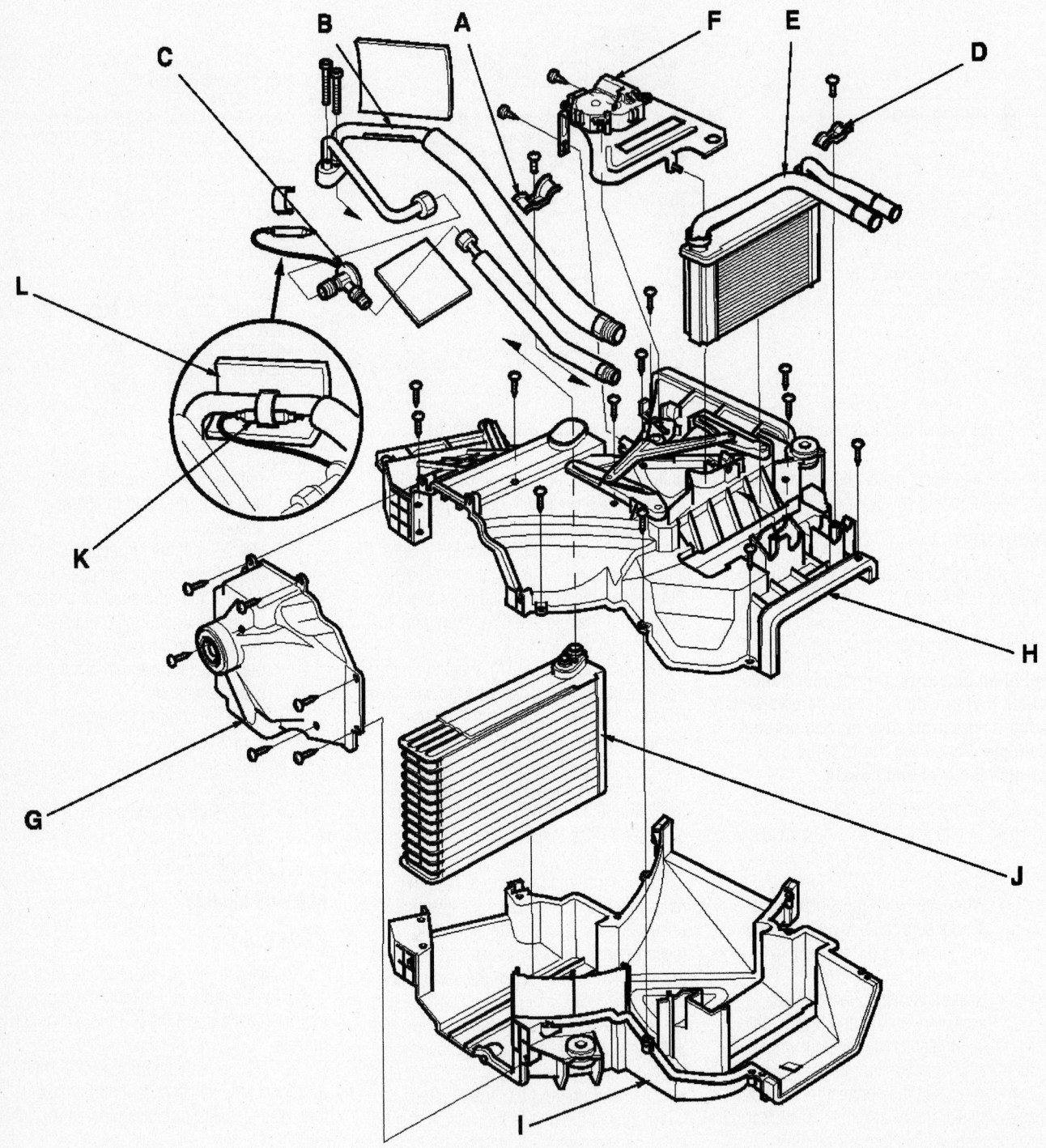

A. Clamp
B. Evaporator lines
C. Expansion valve
D. Clamp

E. Heater core
F. Rear mode control motor
G. Lower housing
H. Right housing

I. Left housing
J. Evaporator core
K. Capillary tube
L. Electrical tape

09474_ODYS_G0311

Rear HVAC unit—2002–04 models

Fastener Locations

A ▷ : Clip (Black)
Left side, 8
Right side, 9

B ▷ : Clip (White)
Except Van, 8
Van, 6

C ▷ : Clip
Except Van, 2

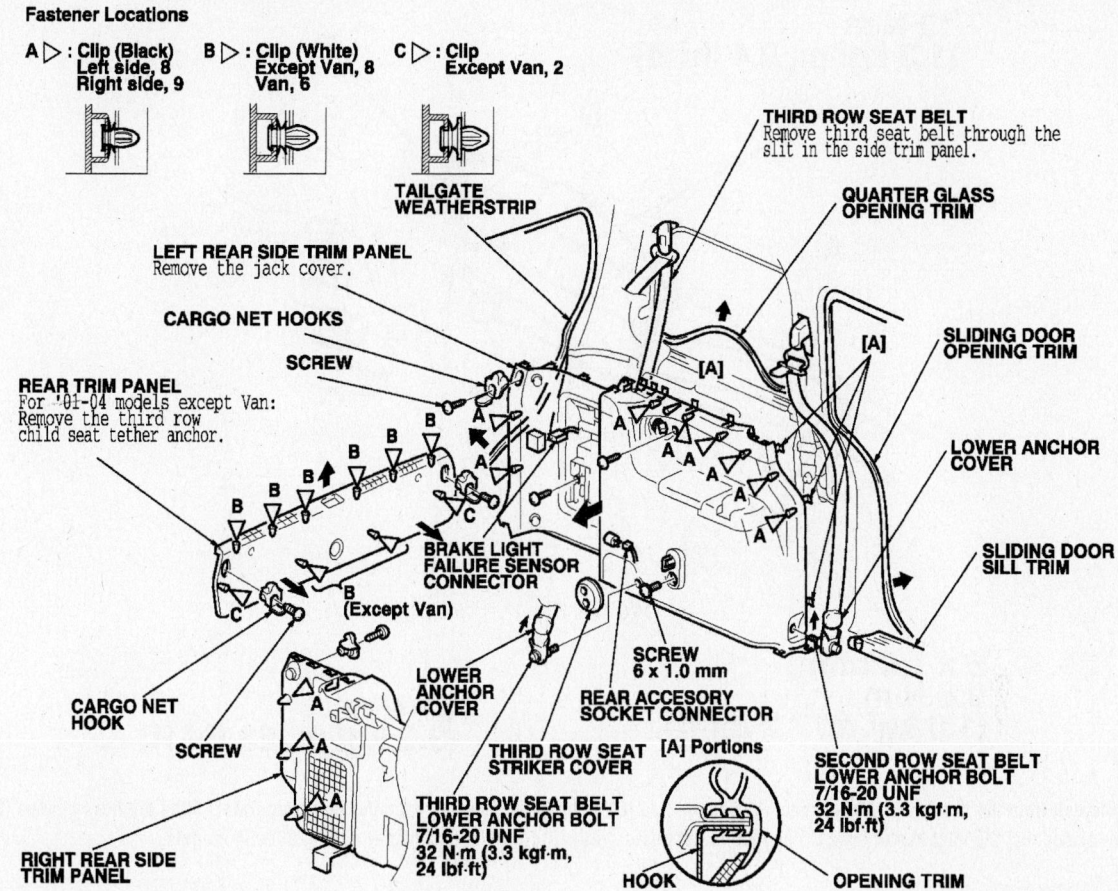

THIRD ROW SEAT BELT
Remove third seat belt through the slit in the side trim panel.

QUARTER GLASS OPENING TRIM

TAILGATE WEATHERSTRIP

SLIDING DOOR OPENING TRIM

LEFT REAR SIDE TRIM PANEL
Remove the jack cover.

CARGO NET HOOKS

LOWER ANCHOR COVER

SCREW

[A]

REAR TRIM PANEL
For '01-04 models except Van:
Remove the third row child seat tether anchor.

SLIDING DOOR SILL TRIM

BRAKE LIGHT FAILURE SENSOR CONNECTOR
(Except Van)

SCREW
6 x 1.0 mm

CARGO NET HOOK

LOWER ANCHOR COVER

REAR ACCESORY SOCKET CONNECTOR

SCREW

THIRD ROW SEAT STRIKER COVER

SECOND ROW SEAT BELT LOWER ANCHOR BOLT
7/16-20 UNF
32 N·m (3.3 kgf·m, 24 lbf·ft)

RIGHT REAR SIDE TRIM PANEL

THIRD ROW SEAT BELT LOWER ANCHOR BOLT
7/16-20 UNF
32 N·m (3.3 kgf·m, 24 lbf·ft)

[A] Portions

HOOK

OPENING TRIM

09474_ODYS_G0316

Rear trim panel removal—2002–04 models

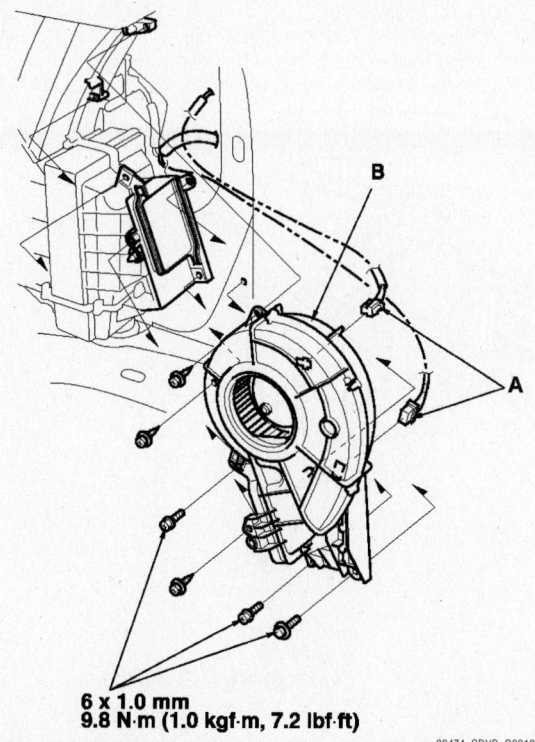

B

A

6 x 1.0 mm
9.8 N·m (1.0 kgf·m, 7.2 lbf·ft)

09474_ODYS_G0312

Rear blower motor removal—2002–04 models

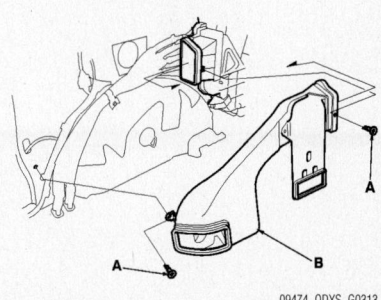

A

A

B

09474_ODYS_G0313

Rear heater duct removal—2002–04 models

cable (B) from the heater valve arm (C). Turn the heater valve arm to the fully opened position as shown.

7. When the engine is cool, drain the engine coolant from the radiator.

8. Slide the hose clamps (A) back. Remove the nut and the water valve (B), then disconnect the inlet heater hose (C) and the outlet heater hose (D) from the heater unit. Engine coolant will run out when the hoses are disconnected; drain it into a clean drip pan. Be sure not to let coolant spill on the electrical parts or the

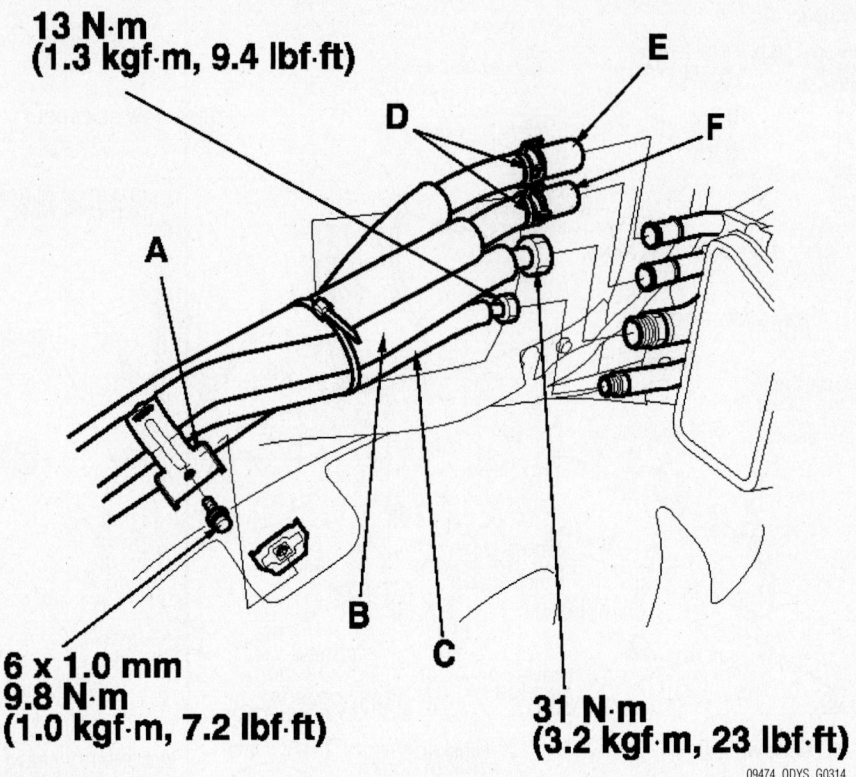

**13 N·m
(1.3 kgf·m, 9.4 lbf·ft)**

E

D

F

A

**6 x 1.0 mm
9.8 N·m
(1.0 kgf·m, 7.2 lbf·ft)**

B

C

**31 N·m
(3.2 kgf·m, 23 lbf·ft)**

09474_ODYS_G0314

Remove the bolt from the bracket (A). Disconnect the suction line (B) and receiver line (C) from the rear evaporator. Slide the hose clamps (D) back, then disconnect the inlet heater hose (E) and the outlet heater hose (F) from the rear heater unit—2002–04 models

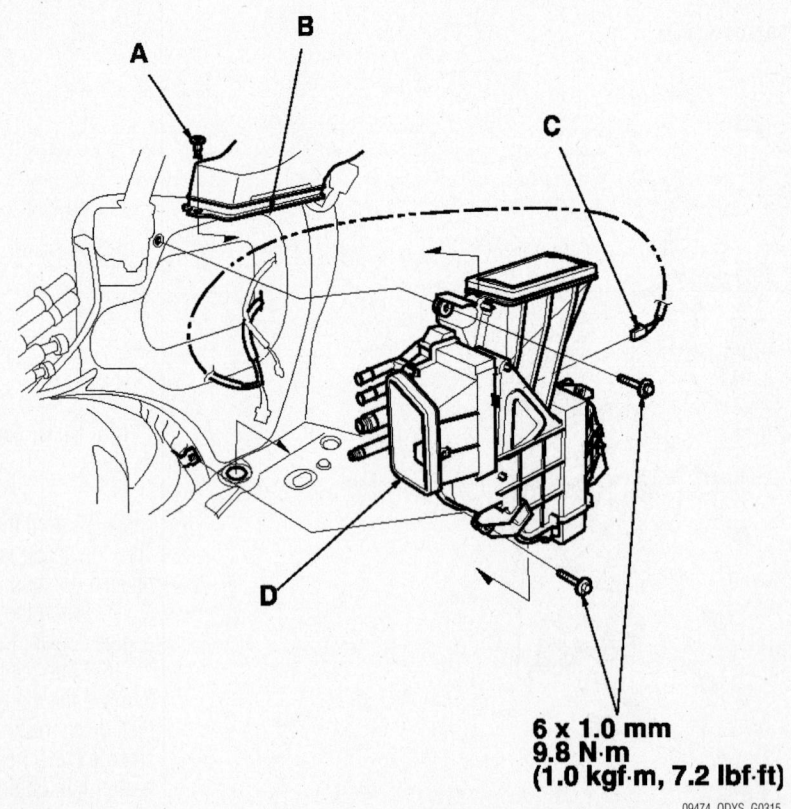

A

B

C

D

**6 x 1.0 mm
9.8 N·m
(1.0 kgf·m, 7.2 lbf·ft)**

09474_ODYS_G0315

Remove the clip (A), then lifting the upper duct (B) up. Disconnect the connector (C) from the rear mode control motor. Remove the mounting bolts and the rear evaporator-heater unit (D)—2002–04 models

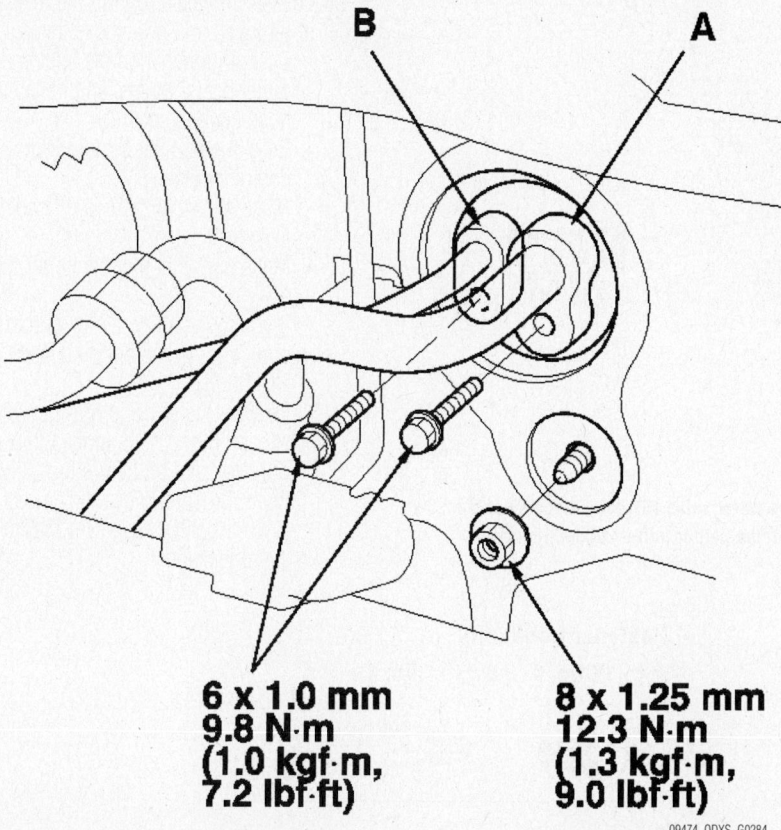

6 x 1.0 mm
9.8 N·m
(1.0 kgf·m,
7.2 lbf·ft)

8 x 1.25 mm
12.3 N·m
(1.3 kgf·m,
9.0 lbf·ft)

09474_ODYS_G0284

Remove the bolts and nut, then disconnect the suction line (A) and the receiver line (B) from the evaporator core—2005–06 models

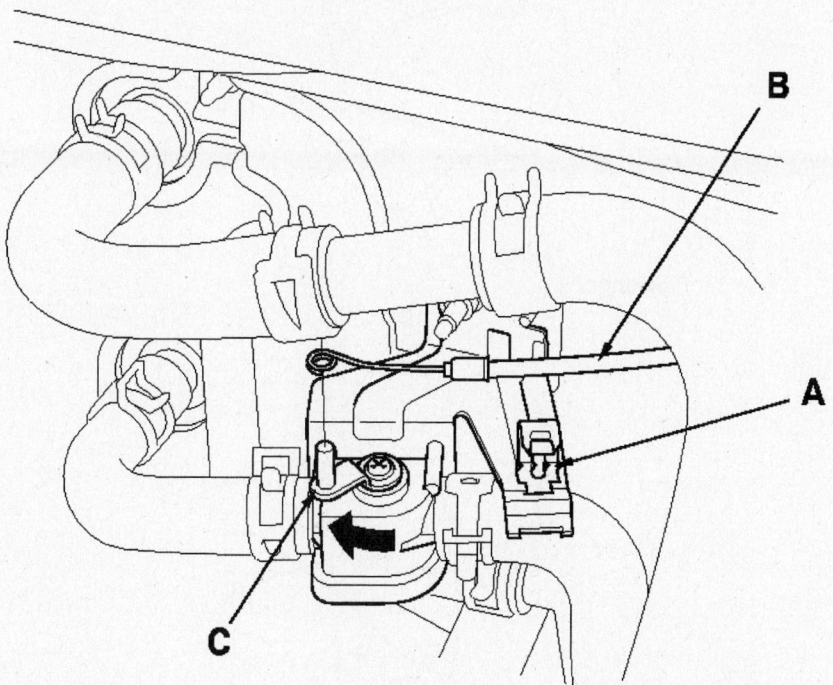

09474_ODYS_G0285

From under the hood, open the cable clamp (A), then disconnect the heater valve cable (B) from the heater valve arm (C). Turn the heater valve arm to the fully opened position as shown—2005–06 models

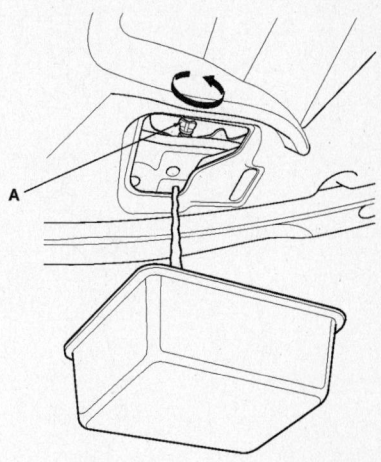

09474_ODYS_G0297

Drain the engine coolant from the radiator

painted surfaces. If any coolant spills, rinse it off immediately.

9. Remove the mounting nut from the heater unit. Take care not to damage or bend the fuel lines and the brake lines, etc.

➡**Put on gloves to protect your hands. Take care not to scratch the dashboard and related parts. Take care not to bend the brackets. Use the appropriate tool from the KTC trim tool set to avoid damage when prying components.**

10. From under the dash, remove the clip (A) and detach the clip (B), release the hook (C) and stud (D), then remove the driver's heater lower cover (E) and passenger's heater lower cover (F).

11. Open the beverage holder (A), and remove the screws.

12. Remove the dashboard center lower cover (A).

a. Pull out the cover to release the clips (B, C).

b. Disconnect the seat heater connectors (D) and accessory socket connectors (E).

13. Remove the dashboard center lower cover (A). Pull out the cover to release the clips (B, C).

14. Remove the screws, then remove the driver's A/C duct (A) from the dashboard (B).

15. Remove the screws, then remove the passenger's A/C duct (A) from the dashboard (B).

16. Remove the screws, then remove the center A/C duct from the dashboard.

17. Remove the screws, then remove the defogger duct (A) from the dashboard (B).

18. Disconnect the connectors (A) from

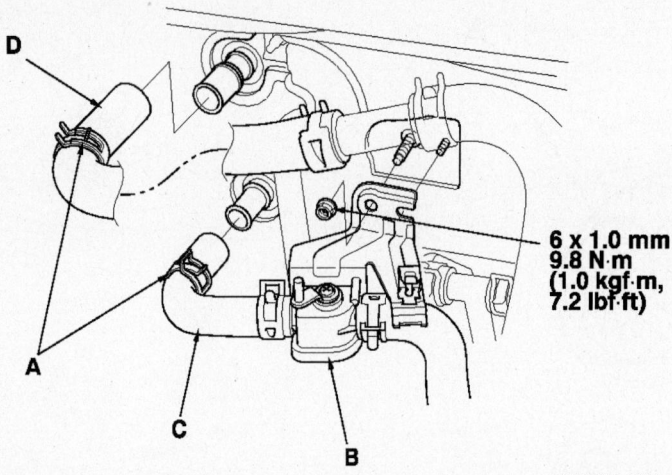

6 x 1.0 mm
9.8 N·m
(1.0 kgf·m,
7.2 lbf·ft)

09474_ODYS_G0286

Slide the hose clamps (A) back. Remove the nut and the water valve (B), then disconnect the inlet heater hose (C) and the outlet heater hose (D) from the heater unit—2005–06 models

24. Install the heater unit in the reverse order of removal, and note these items:
- Do not interchange the inlet and outlet heater hoses, and install the hose clamps securely.
- Refill the cooling system with engine coolant.
- Make sure that there is no coolant leakage.
- Make sure that there is no air leakage.
- If you're installing a new evaporator core, add refrigerant oil (DENSO ND-OIL 8).
- Replace the O-rings with new ones at each fitting, and apply a thin coat of refrigerant oil before installing them. Be sure to use the correct O-rings for HFC-134a (R-134a) to avoid leakage.

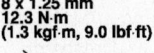
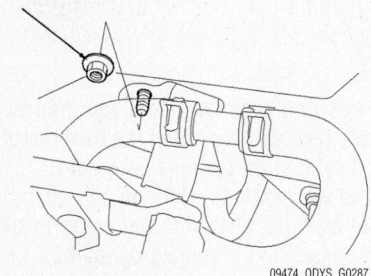

8 x 1.25 mm
12.3 N·m
(1.3 kgf·m, 9.0 lbf·ft)

09474_ODYS_G0287

Remove the mounting nut from the heater unit—2005–06 models

the blower motor and the power transistor, then remove the wire harness clips (B).

19. Disconnect the connectors (A) from the mode control motor and the recirculation control motor, then remove the wire harness clips (B) and the connector clip (C).

20. Disconnect the connectors from the evaporator sensor and the air mix control motor, then remove the wire harness clips and the wire harness.

21. Remove the drain hose (A), then remove the nuts and the blower-heater unit (B).

22. Remove the self-tapping screws, the joint duct (A), and seal (B). Remove the self-tapping screws, then remove the passenger's heater outlet (C), and the heater core cover (D). Remove the self-tapping screws, the heater pipe brackets (E), the grommets (F), and carefully pull out the heater core (G) so you don't bend the inlet and outlet pipes.

23. Install the heater core in the reverse order of removal.

Fastener Locations

A ▷ : Clip, 1   B ▷ : Clip, 1

Driver's

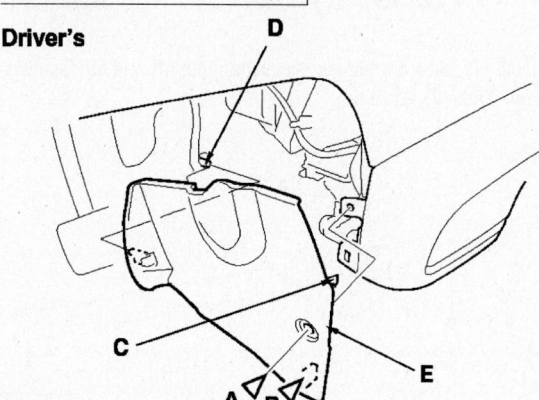

Passenger's

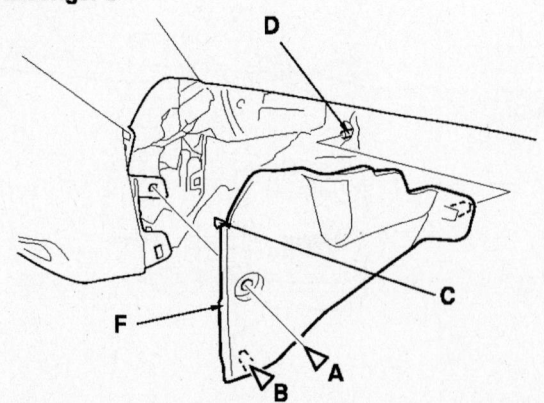

09474_ODYS_G0288

From under the dash, remove the clip (A) and detach the clip (B), release the hook (C) and stud (D), then remove the driver's heater lower cover (E) and passenger's heater lower cover (F)—2005–06 models

Fastener Locations
▶ : Screw, 4

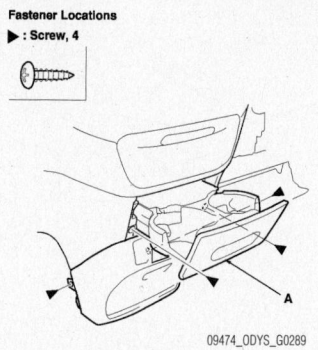

09474_ODYS_G0289

Open the beverage holder (A), and remove the screws—2005–06 models

Fastener Locations
B ▷ : Clip, 2   C ▷ : Clip, 2

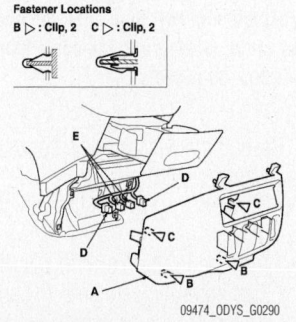

09474_ODYS_G0290

Disconnect the seat heater connectors (D) and accessory socket connectors (E)—2005–06 models

Fastener Locations
▶ : Screw, 5

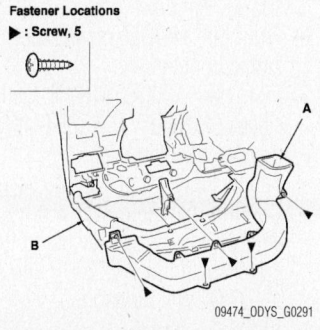

09474_ODYS_G0291

Remove the screws, then remove the driver's A/C duct (A) from the dashboard (B)—2005–06 models

Fastener Locations
▶ : Screw, 6

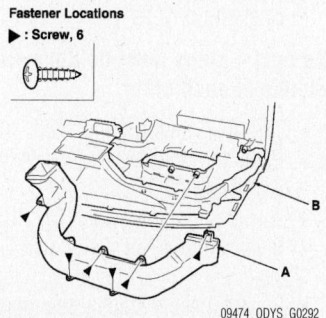

09474_ODYS_G0292

Remove the screws, then remove the passenger's A/C duct (A) from the dashboard (B)—2005–06 models

Fastener Locations
▶ : Screw, 20

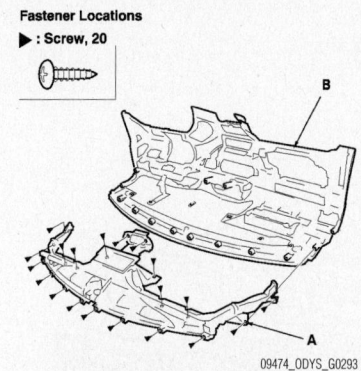

09474_ODYS_G0293

Remove the screws, then remove the defogger duct (A) from the dashboard (B)—2005–06 models

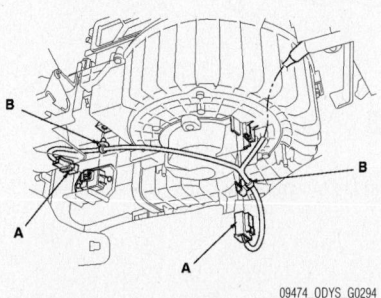

09474_ODYS_G0294

Disconnect the connectors (A) from the blower motor and the power transistor, then remove the wire harness clips (B)—2005–06 models

- Immediately after using the oil, reinstall the cap on the container, and seal it to avoid moisture absorption.
- Do not spill the refrigerant oil on the vehicle; it may damage the paint; if the refrigerant oil contacts the paint, wash it off immediately.
- Make sure that there is no air leakage.
- Charge the system.

25. Adjust the heater valve cable.

a. Set the temperature control dial to Max Cool (Lo) with the ignition switch ON (II).

b. Attach the heater valve cable (B) to the air mix control linkage (C) as shown. Hold the end of the heater valve cable housing against the stop (D), then snap the heater valve cable housing into the cable clamp (A).

➡**Make sure the ring-end of the cable is pushed all the way to the base of the pin on air mix control linkage.**

c. From under the hood, turn the heater valve arm (C) to the fully closed position as shown, and hold it. Attach the heater valve cable (B) to the heater valve arm, and gently pull on the heater valve cable housing to take up any slack, then install the heater valve cable housing into the cable clamp (A).

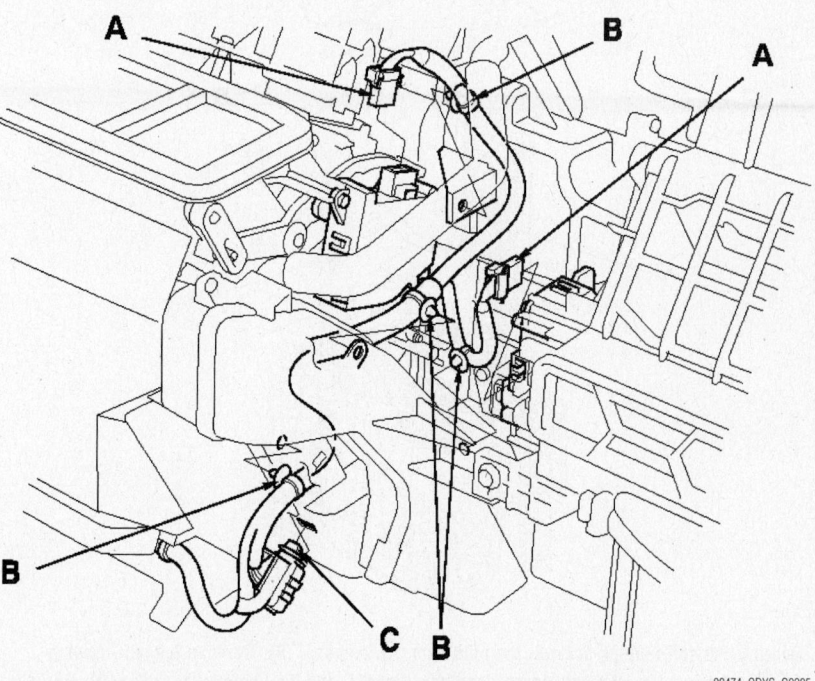

09474_ODYS_G0295

Disconnect the connectors (A) from the mode control motor and the recirculation control motor, then remove the wire harness clips (B) and the connector clip (C)—2005–06 models

6 x 1.0 mm
9.8 N·m (1.0 kgf·m, 7.2 lbf·ft)

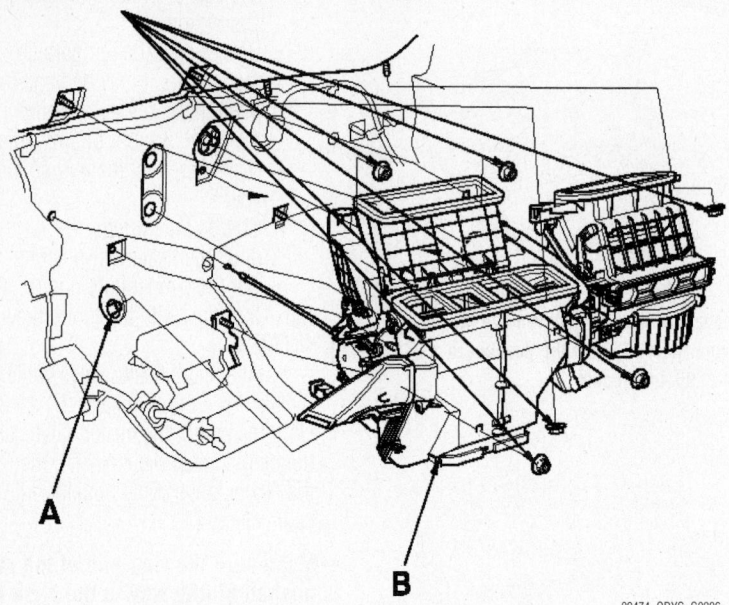

09474_ODYS_G0296

Remove the drain hose (A), then remove the nuts and the blower-heater unit (B)—2005–06 models

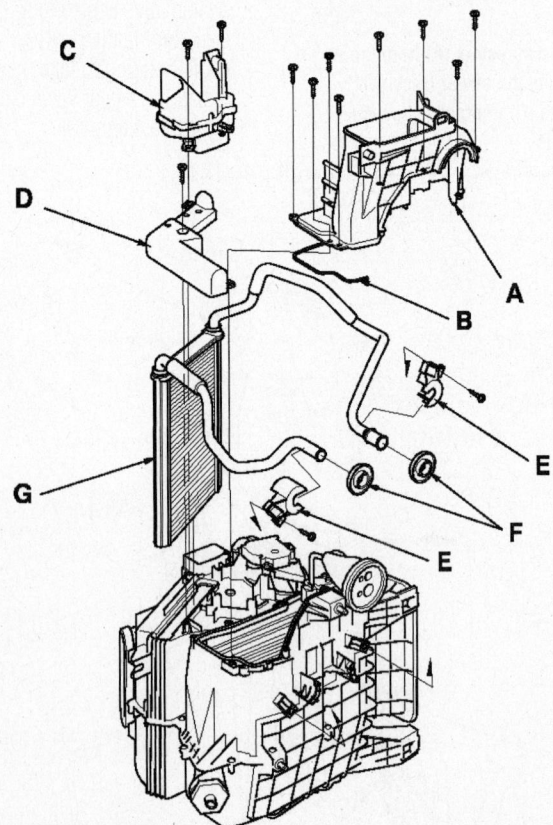

09474_ODYS_G0298

Remove the self-tapping screws, the joint duct (A), and seal (B). Remove the self-tapping screws, then remove the passenger's heater outlet (C), and the heater core cover (D). Remove the self-tapping screws, the heater pipe brackets (E), the grommets (F), and carefully pull out the heater core (G)—2005–06 models

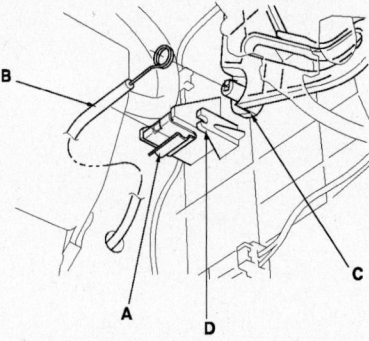

09474_ODYS_G0299

Attach the heater valve cable (B) to the air mix control linkage (C) as shown. Hold the end of the heater valve cable housing against the stop (D), then snap the heater valve cable housing into the cable clamp (A)—2005–06 models

26. Reset the power window control unit:
   a. Using the HDS:
   • Connect the HDS to the vehicle's DLC.
   • Turn the ignition switch ON (II), then enter the vehicle's VIN and mileage at the prompts.
   • Select "BODY ELECTRICAL" from the "System Selection" menu.
   • From the "BODY ELECTRICAL SYSTEM SELECT" menu, select "Power Windows".
   • From the "MODE" menu, select "Adjustments".
   • From the "ADJUSTMENT" menu, select "WINDOW P RESET" for driver's side window.
   • Follow the prompts on the screen.
   • Confirm that the power window control unit is reset by using the driver's window AUTO UP and AUTO DOWN function.
   b. Without the HDS:
   • Turn the ignition switch ON (II).
   • Move the driver's window all the way down by using the driver's window DOWN switch.
   • Open the driver's door.

➡ **The next 4 steps must be done within 5 seconds of each other.**

   • Turn the ignition switch OFF.
   • Push and hold the driver's window DOWN switch.
   • Turn the ignition switch ON (II).
   • Release the driver's window DOWN switch.
   • Repeat these 4 steps three more times.
   • Wait 1 second.
   • Confirm that AUTO UP and AUTO

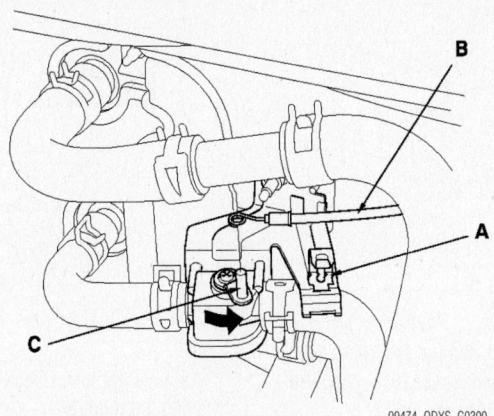

09474_ODYS_G0300

From under the hood, turn the heater valve arm (C) to the fully closed position as shown, and hold it. Attach the heater valve cable (B) to the heater valve arm, and gently pull on the heater valve cable housing to take up any slack, then install the heater valve cable housing into the cable clamp (A)—2005–06 models

DOWN do not work. If AUTO UP and DOWN work, go back to the first step.

- Move the driver's window all the way down by using the driver's window DOWN switch.
- Pull up and hold the driver's window UP switch until the window reaches the fully closed position, then continue to hold the switch for 1 second.
- Confirm that the power window control unit is reset by using the driver's window AUTO UP and AUTO DOWN function.
- If the window still does not work in AUTO, repeat the procedure several times, paying close attention to the 5 second time limit between steps.

27. Enter the anti-theft codes for the radio and the navigation system, then enter the XM radio channel presets.

REAR UNIT

1. Recover the refrigerant with a recovery/recycling/charging station.
2. Remove these items:
    - Sliding door sill trim, as needed
    - C-pillar trim
    - Sliding door opening trim, as needed
3. Pull the lower anchor cover (A) back, and remove the lower anchor bolt (B).
4. Detach the clip and release the hooks (A) then remove the pivot cover (B).
5. Remove the lower anchor bolt (A).
6. Remove the rear side trim panel (A).
7. Remove the cap (B) by releasing the two hooks (C).
8. Remove the rear side bolt (D).
9. Pull out the rear upper edge of the

trim panel by hand to release the hooks (E) from the D-pillar trim (F).

10. Pull the trim panel back by hand to detach the clips (G, H).

11. Disconnect the accessory socket connectors (I) and the rear entertainment system auxiliary jack connector (J).

12. Remove the rear side trim panel beverage holder (A).

13. Detach the clips and release the hooks (B), then remove the rear speaker grille (C).

14. Remove the screws.

15. Slide the beverage holder forward to release the rear hooks (D).

16. Remove the wire harness clip (A), the clip (B) and remove the side duct (C).

17. Remove the bolts, then hang the rear junction box (A) down.

18. Disconnect the rear evaporator

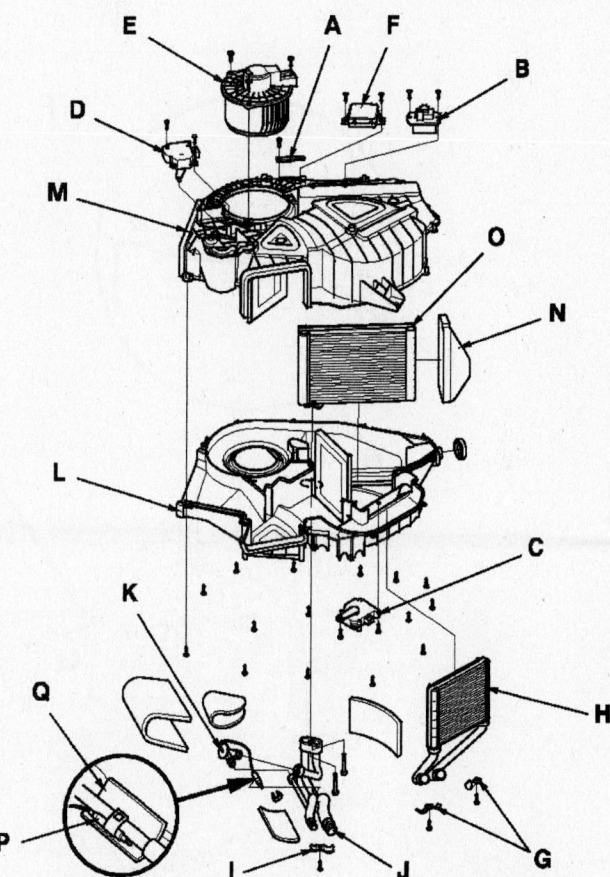

A. Rear in-car temperature sensor	J. Evaporator lines
B. Rear power transistor	K. Expansion valve
C. Rear air mix control motor	L. Right housing
D. Rear mode control motor	M. Left housing
E. Rear blower motor	N. Insulator
F. Rear climate control unit	O. Evaporator core
G. Clamps	P. Capillary tube
H. Rear heater core	Q. Electrical tape
I. Clamp	

09474_ODYS_G0301

Rear HVAC unit components—2005–06 models

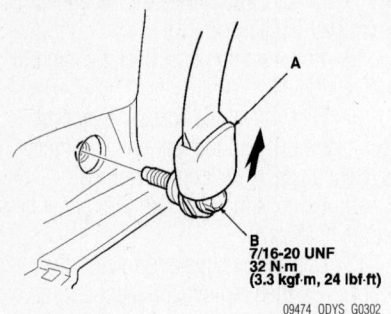

09474_ODYS_G0302

Pull the lower anchor cover (A) back, and remove the lower anchor bolt (B)—2005–06 models

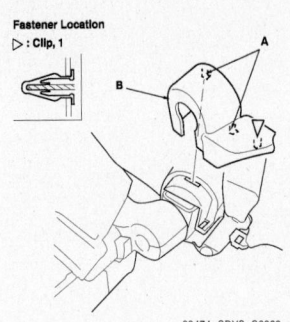

09474_ODYS_G0303

Detach the clip and release the hooks (A) then remove the pivot cover (B)—2005–06 models

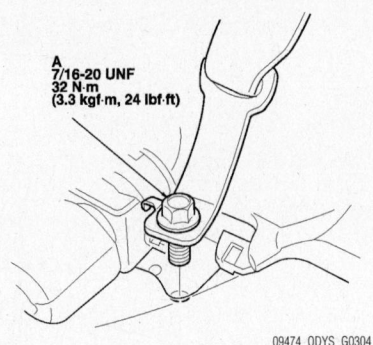

09474_ODYS_G0304

Remove the lower anchor bolt (A)—2005–06 models

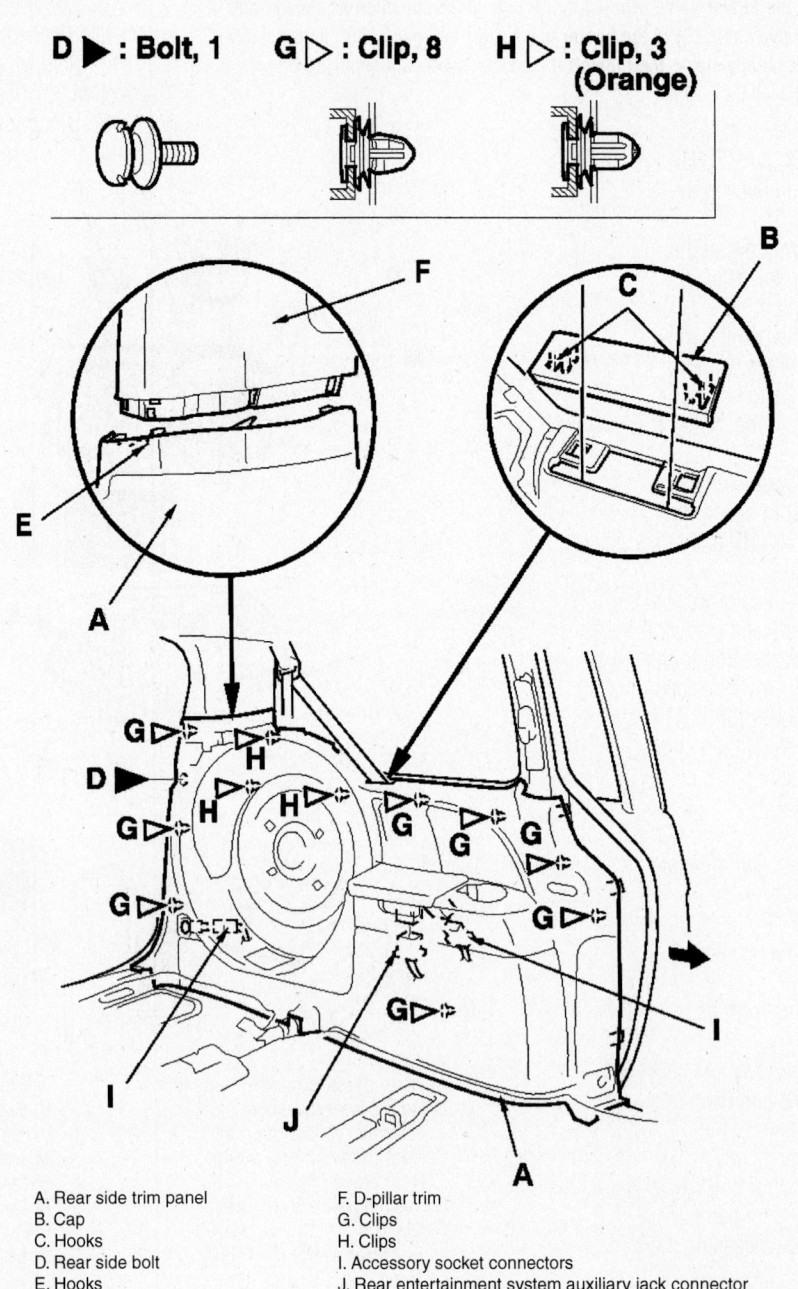

A. Rear side trim panel
B. Cap
C. Hooks
D. Rear side bolt
E. Hooks
F. D-pillar trim
G. Clips
H. Clips
I. Accessory socket connectors
J. Rear entertainment system auxiliary jack connector

09474_ODYS_G0305

Rear trim panel removal—2005–06 models

Fastener Locations

▷ : Clip, Left, 2
   Right, 3

▶ : Screw, Left, 6
   Right, 8

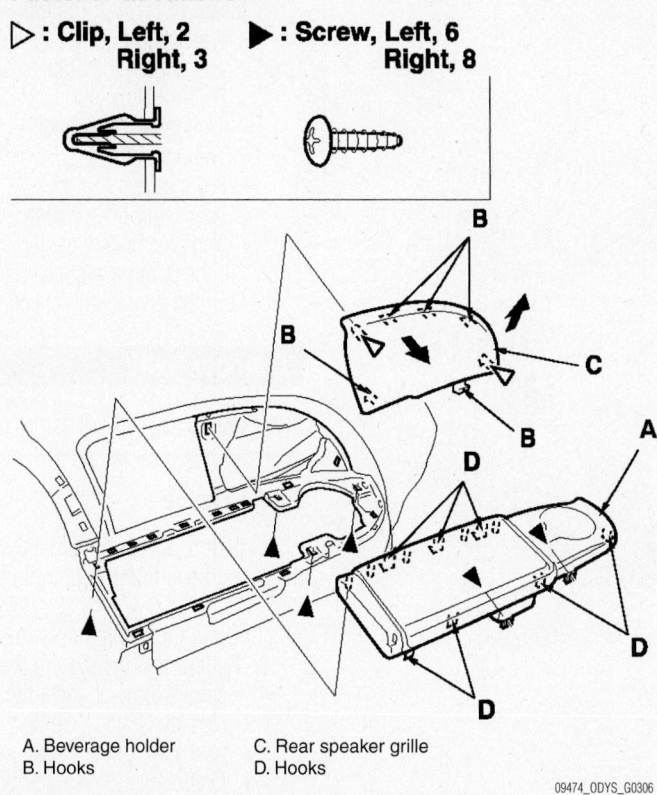

A. Beverage holder
B. Hooks
C. Rear speaker grille
D. Hooks

09474_ODYS_G0306

Beverage holder removal—2005–06 models

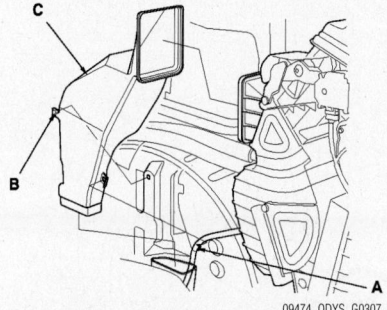

09474_ODYS_G0307

Remove the wire harness clip (A), the clip (B) and remove the side duct (C)— 2005–06 models

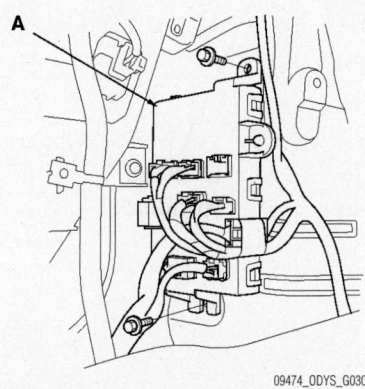

09474_ODYS_G0308

Rear junction box—2005–06 models

receiver line (A) and the suction line (B) connections. Slide the hose clamps (C) back, then disconnect the inlet heater hose (D) and the outlet heater hose (E) from the rear heater core. Engine coolant will run out when the hoses are disconnected; drain it into a clean drip pan. Be sure not to let coolant spill on the electrical parts, the carpet, or the painted surfaces. If any coolant spills, rinse it off immediately.

19. Disconnect the connectors from the rear mode control motor, the rear blower motor, rear in-car temperature sensor and rear climate control unit, then remove the wire harness clips, from the rear HVAC unit.

20. Remove the bolts, then pull out the rear HVAC unit (A). Disconnect the rear air mix control motor connector (B), then remove the rear HVAC unit.

21. Remove the self-tapping screws, the clamps and the rear heater core.

22. Installation is the reverse of removal.
- Make sure no air is leaking from the right housing, the left housing and the lower housing fitting.
- Before reassembly, make sure that the rear air mix control linkage and door move smoothly without binding.

B
24 x 1.5 mm
32 N·m
(3.3 kgf·m,
23.5 lbf·ft)

A
16 x 1.5 mm
13 N·m (1.4 kgf·m, 9.8 lbf·ft)

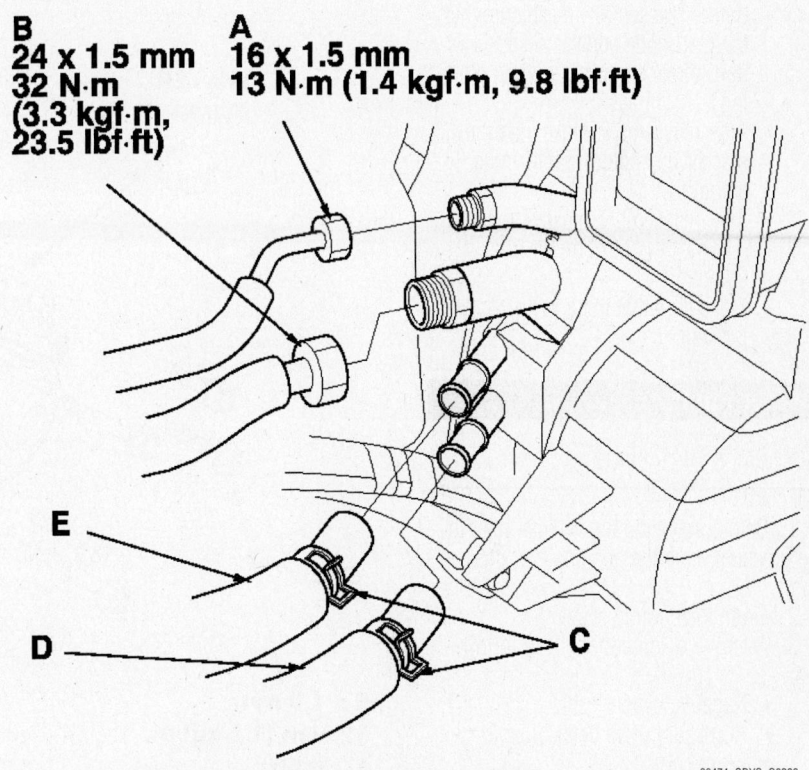

09474_ODYS_G0309

Disconnect the rear evaporator receiver line (A) and the suction line (B) connections. Slide the hose clamps (C) back, then disconnect the inlet heater hose (D) and the outlet heater hose (E) from the rear heater core—2005–06 models

6 x 1.0 mm
9.8 N·m (1.0 kgf·m, 7.2 lbf·ft)

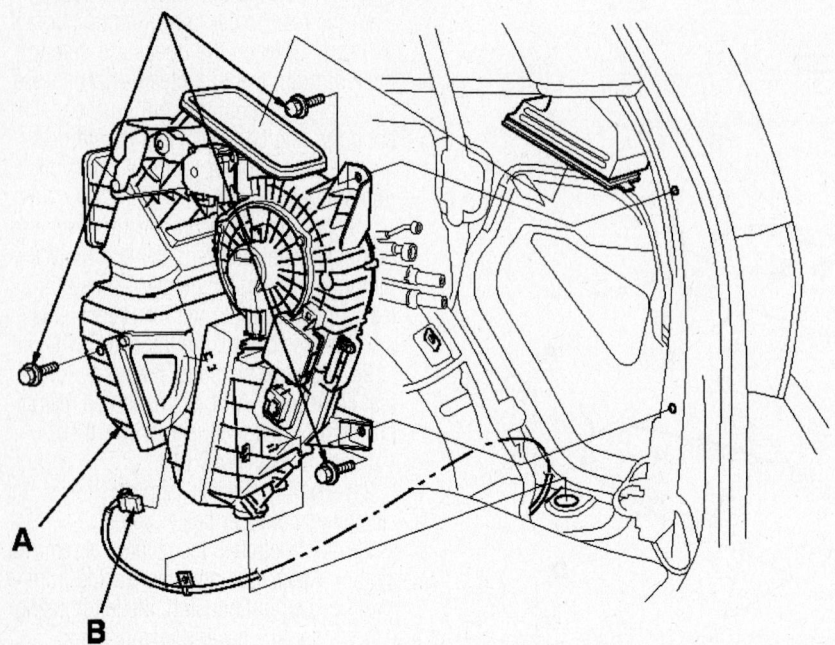

09474_ODYS_G0310

Remove the bolts, then pull out the rear HVAC unit (A). Disconnect the rear air mix control motor connector (B), then remove the rear HVAC unit—2005–06 models

- Before reassembly, make sure that the rear mode control linkage and door move smoothly without binding.
- After reassembly, make sure the rear air mix control motor runs smoothly.
- After reassembly, make sure the rear mode control motor runs smoothly.
- Make sure that there is no coolant leakage.

Water Pump

REMOVAL & INSTALLATION

1. Before servicing the vehicle, refer to the precautions in the beginning of this section.
2. Drain the cooling system.
3. Remove or disconnect the following:
- Negative battery cable
- Accessory drive belts
- Front cover
- Timing belt. Refer to the timing belt procedure.
- Timing belt tensioner
- Water pump

To install:
4. Install or connect the following:
- Water pump. Use a new O-ring seal and tighten the bolts to 105 inch lbs. (12 Nm).
- Timing belt tensioner
- Timing belt
- Front cover
- Accessory drive belts
- Negative battery cable
5. Fill the cooling system.
6. Start the engine and check for leaks.

Cylinder Head

REMOVAL & INSTALLATION

J35A4 Engine

1. Before servicing the vehicle, refer to the precautions in the beginning of this section.
2. Drain the cooling system.
3. Relieve the fuel system pressure.
4. Remove or disconnect the following:
- Negative battery cable
- Accessory drive belts
- Evaporative Emissions (EVAP) control canister hose and vacuum hose
- Air intake tube

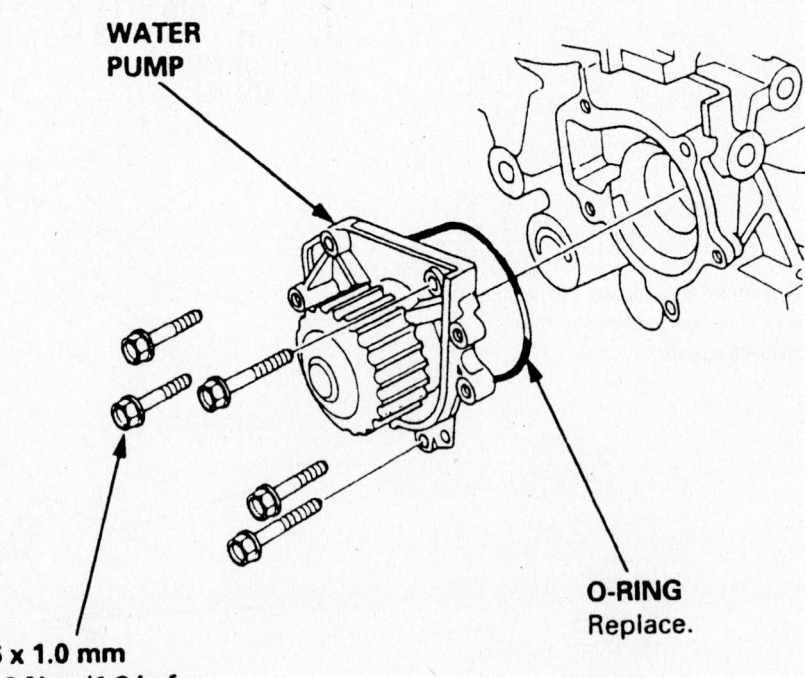

WATER PUMP

O-RING
Replace.

6 x 1.0 mm
12 N·m (1.2 kgf·m, 8.7 lbf·ft)

7924MG10

Water pump mounting—J35A4 engines

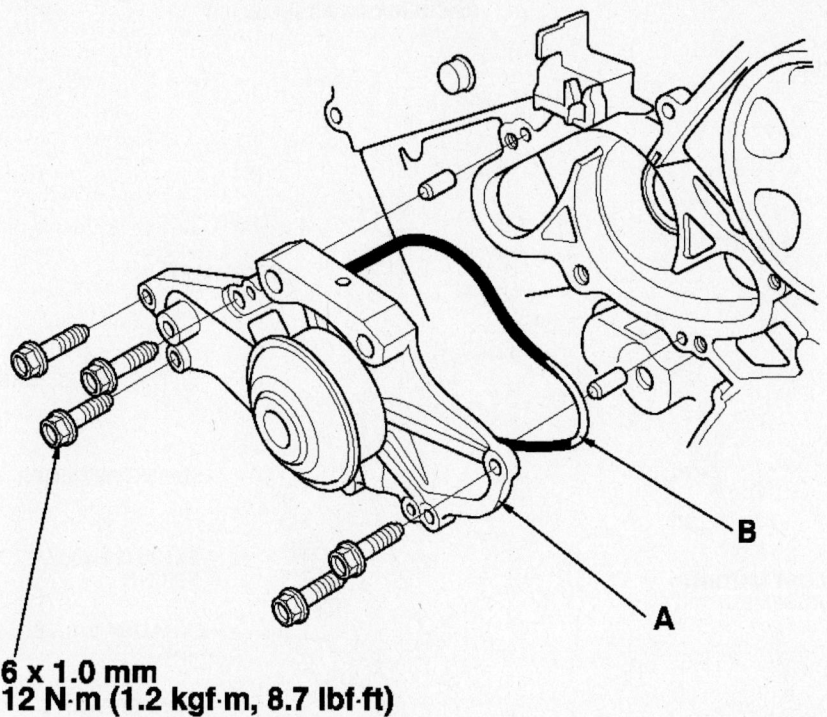

6 x 1.0 mm
12 N·m (1.2 kgf·m, 8.7 lbf·ft)

09474_ODYS_G0010

Water pump mounting—J35A6 and J35A7 engines

- Ignition coil covers
- Intake manifold cover
- Accelerator cable
- Cruise control cable
- Fuel lines
- Brake booster vacuum line
- Intake manifold vacuum line
- Positive Crankcase Ventilation (PCV) valve and hose
- Side engine mount bracket
- Power steering pump
- Power steering hose clamp
- Alternator
- Intake Air Temperature (IAT) sensor connector
- Idle Air Control (IAC) valve connector
- Throttle Position (TP) sensor connector
- Manifold Absolute Pressure (MAP) sensor connector
- Engine Coolant Temperature (ECT) sensor connector
- Radiator fan switch connectors
- ECT gauge sending unit connector
- Crankshaft Position (CKP) sensor connector
- Top Dead Center (TDC) sensor connector
- Exhaust Gas Recirculation (EGR) connector
- Variable Valve Timing and Valve

Lift Electronic Control (VTEC) solenoid valve connector
- VTEC oil pressure switch connector
- Oil pressure switch connector
- Ignition coils
- Intake manifold
- Fuel injector connectors
- Fuel supply manifold
- Fuel injection air control valve vacuum lines
- Front cover
- Timing belt. Refer to the timing belt procedure.
- Radiator hoses
- Heater hoses
- Front and rear exhaust manifolds
- Coolant cross-over pipe
- Valve covers

5. Loosen the cylinder head bolts in sequence and 1/3 turns until all bolts are loose.

6. Remove the cylinder head.

To install:

7. Align the crankshaft and camshaft sprocket TDC marks. See the Timing Belt Procedure.

8. Install the cylinder heads with new gaskets.

9. Apply clean engine oil to the cylinder head bolt threads and flanges.

10. Tighten the cylinder head bolts in sequence as follows:

    a. Step 1: 29 ft. lbs. (39 Nm)

    b. Step 2: 51 ft. lbs. (69 Nm)

    c. Step 3: 72 ft. lbs. (98 Nm)

11. Install or connect the following:
- Valve covers
- Coolant cross-over pipe
- Front and rear exhaust manifolds
- Heater hoses
- Radiator hoses
- Timing belt
- Front cover
- Fuel injection air control valve vacuum lines
- Fuel supply manifold
- Fuel injector connectors
- Intake manifold
- Ignition coils
- Oil pressure switch connector
- VTEC oil pressure switch connector
- VTEC solenoid valve connector
- EGR connector
- TDC sensor connector
- CKP sensor connector
- ECT gauge sending unit connector
- Radiator fan switch connectors
- ECT sensor connector
- MAP sensor connector
- TP sensor connector
- IAC valve connector
- IAT sensor connector
- Alternator
- Power steering hose clamp
- Power steering pump
- Side engine mount bracket
- PCV valve and hose
- Intake manifold vacuum line
- Brake booster vacuum line
- Fuel lines
- Cruise control cable
- Accelerator cable
- Intake manifold cover
- Ignition coil covers
- Accessory drive belts
- Air intake tube
- EVAP control canister hose and vacuum hose
- Negative battery cable

12. Fill the cooling system.

13. Start the engine and check for leaks.

J35A6 Engine

➡**The engine coolant temperature must be below 100°F (38°C) before removing the head bolts.**

1. Make sure that you have the anti-theft codes for the radio and navigation system.

2. Relieve the fuel system pressure.

3. Disconnect the battery ground.

4. Remove the accessory drive belt.

5. Drain the coolant.

6. Remove the power steering pump and the hose bracket.

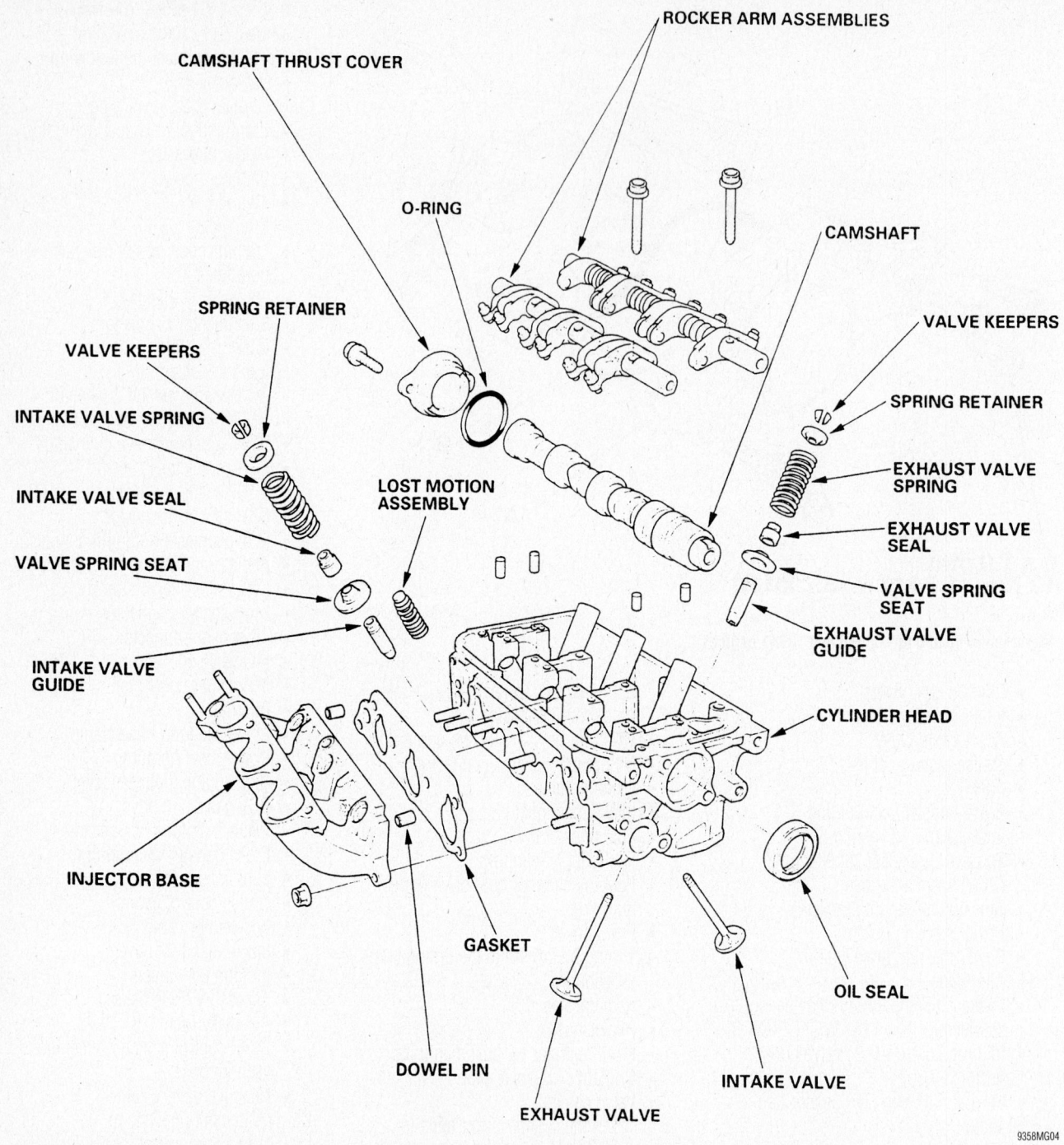

ROCKER ARM ASSEMBLIES

CAMSHAFT THRUST COVER

O-RING

CAMSHAFT

SPRING RETAINER

VALVE KEEPERS

VALVE KEEPERS

SPRING RETAINER

INTAKE VALVE SPRING

EXHAUST VALVE SPRING

INTAKE VALVE SEAL

LOST MOTION ASSEMBLY

EXHAUST VALVE SEAL

VALVE SPRING SEAT

VALVE SPRING SEAT

EXHAUST VALVE GUIDE

INTAKE VALVE GUIDE

CYLINDER HEAD

INJECTOR BASE

GASKET

OIL SEAL

DOWEL PIN

INTAKE VALVE

EXHAUST VALVE

9358MG04

Exploded view of the cylinder head assembly and related components—J35A4 engine

7. Remove the alternator.
8. Remove the timing belt.
9. Remove the intake manifold.
10. Remove the ignition coils.
11. Tag and disconnect all wiring from the head.
12. Remove the front and rear warm-up converters.
13. Disconnect the fuel feed line.
14. Remove the wiring harness bracket.
15. Remove the upper and lower radiator hoses.

16. Remove the heater and bypass hoses.
17. Remove the EVAP canister hose joint bracket.
18. Remove the harness brackets from the front and rear heads.
19. Remove the fuel rail.
20. Remove the water passage assembly.
21. Remove the camshaft pulley.
22. Remove the cylinder head cover.
23. Remove the head bolts in the sequence shown ⅓ turn at a time.
24. Remove the head.

To install:
25. Clean all mating surfaces thoroughly.
26. Clean and install the oil control orifices with new O-rings.
27. Install the dowel pins and a new head gasket.
28. Set the crankshaft pulley to TDC by aligning the TDC mark (A) with the pointer (B). See the Timing Belt Procedure.
29. Set the camshaft pulley(s) to TDC by aligning the mark (A) with the pointer (B). See the Timing Belt Procedure.

30. Coat the threads and flanges of head bolts with clean engine oil.

❋❋ WARNING

There are 2 types of head bolts in service, 6-point and 12-point. Do not mix them on the same head.

➡There are 2 different tightening methods based on which type of head bolt is used.

❋❋ WARNING

When tightening the bolts, tighten them slowly. If any bolt makes a noise while tightening, loosen the bolt and retighten from Step 1.

31. For 6-points bolts, tighten the bolts, in sequence, in 3 steps. Perform each step twice.
- Step 1: 29 ft. lbs. (39 Nm)
- Step 2: 51 ft. lbs. (69 Nm)
- Step 3: 72 ft. lbs. (98 Nm)

32. For 12-point bolts:
- Step 1: Torque all bolts in sequence to 22 ft. lbs.
- Step 2: Torque all bolts in sequence an additional 90°
- Step 3: Torque all bolts in sequence an additional 90°
- If new bolts are used, torque them in sequence an additional 90°

33. The remainder of installation is the reverse of removal. Note the following torques:
- Water passage 8mm bolts: 16 ft. lbs. (22 Nm)
- Water passage 6mm bolts: 105 inch lbs. (12 Nm)
- Front head bracket 10mm bolt: 33 ft. lbs. (44 Nm)
- Rear head bracket 8mm bolt: 16 ft. lbs. (22 Nm)
- EVAP canister joint bracket: 105 inch lbs. (12 Nm)
- Power steering pump bolts: 16 ft. lbs. (22 Nm)

J35A7 Engine

➡The engine coolant temperature must be below 100°F (38°C) before removing the head bolts.

1. Make sure that you have the anti-theft codes for the radio and navigation system.
2. Relieve the fuel system pressure.
3. Disconnect the battery ground.
4. Remove the accessory drive belt.
5. Drain the coolant.

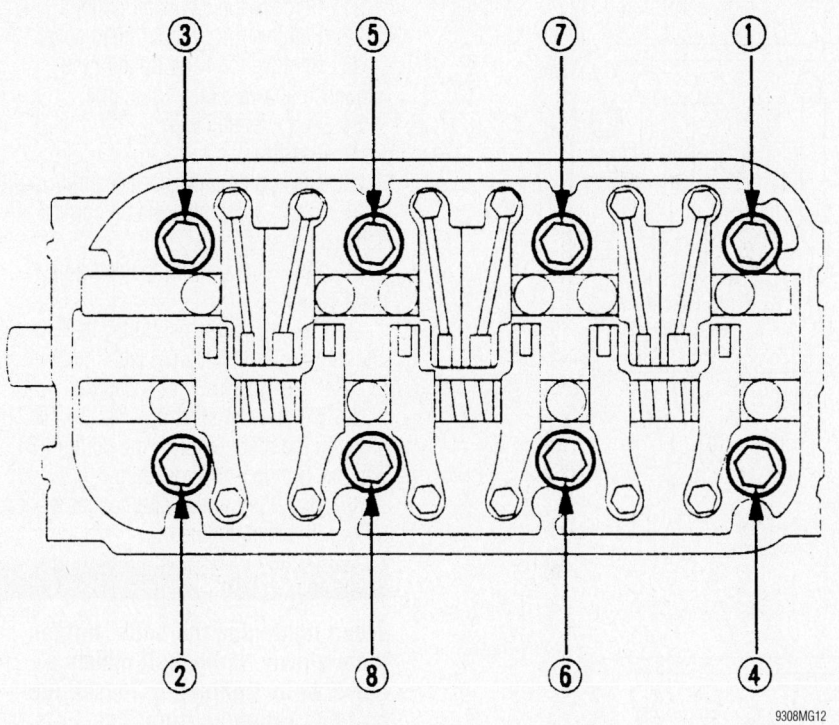

Cylinder head bolt loosening sequence—J35A4 engine

9308MG12

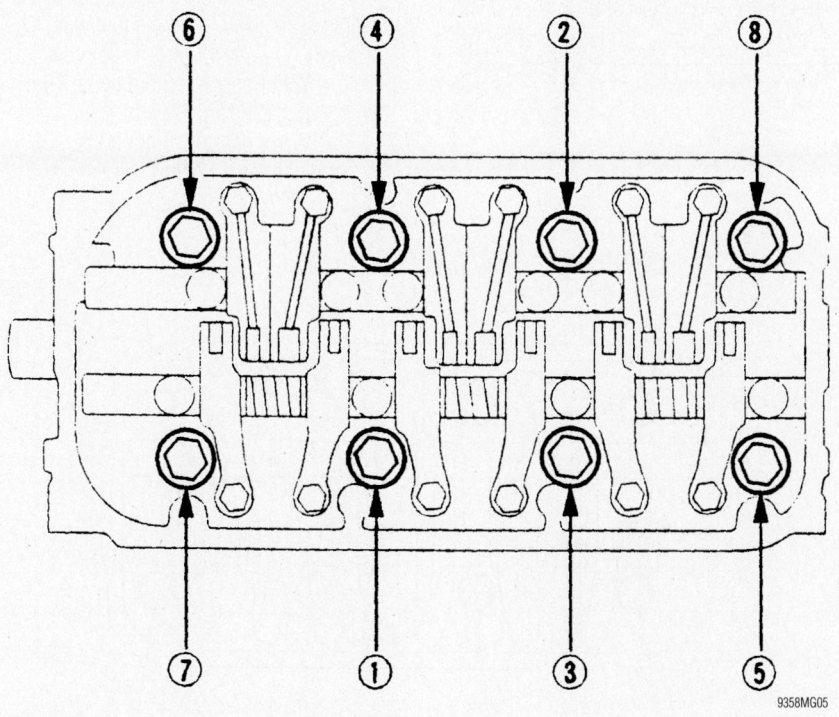

Cylinder head bolt torque sequence—J35A4

9358MG05

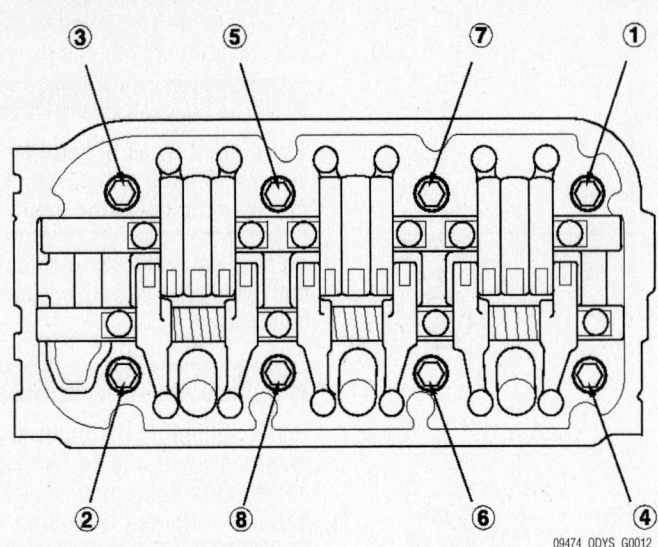

Cylinder head bolt removal sequence—J35A6 engine

09474_ODYS_G0012

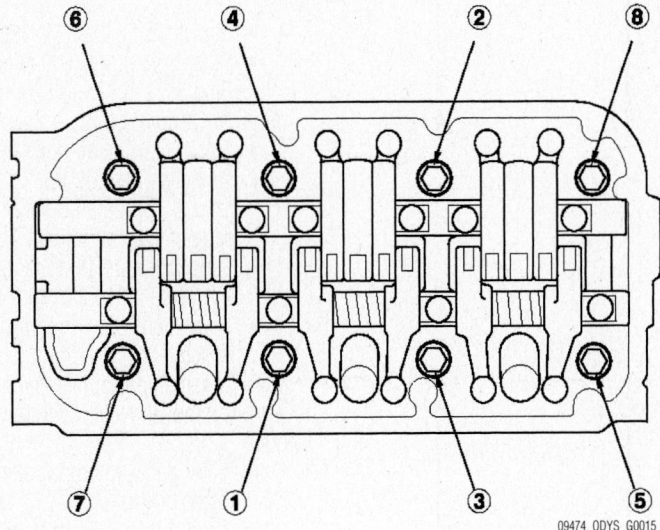

Head bolt torque sequence—J35A6 engine

09474_ODYS_G0015

6. Remove the power steering pump and the hose bracket.

7. Remove the alternator.

8. Remove the timing belt.

9. Remove the intake manifold.

10. Remove the ignition coils.

11. Tag and disconnect all wiring from the head.

12. Remove the front and rear warm-up converters.

13. Disconnect the fuel feed line.

14. Remove the wiring harness bracket.

15. Remove the upper and lower radiator hoses.

16. Remove the heater and bypass hoses.

17. Remove the EVAP canister hose joint bracket.

18. Remove the harness brackets from the front and rear heads.

19. Remove the fuel rail.

20. Remove the water passage assembly.

21. Remove the camshaft pulley.

22. Remove the cylinder head cover.

23. Remove the head bolts in the sequence shown ⅓ turn at a time.

24. Remove the head.

To install:

25. Clean all mating surfaces thoroughly.

26. Clean and install the oil control orifices with new O-rings.

27. Install the dowel pins and a new head gasket.

28. Set the crankshaft pulley to TDC by aligning the TDC mark (A) with the pointer (B). See the Timing Belt Procedure.

29. Set the camshaft pulley(s) to TDC by aligning the mark (A) with the pointer (B). See the Timing Belt Procedure.

30. Coat the threads and flanges of head bolts with clean engine oil.

✸✸ WARNING

When tightening the bolts, tighten them slowly. If any bolt makes a noise while tightening, loosen the bolt and retighten from Step 1.

31. Tighten the bolts, in sequence, in 3 steps. Perform each step twice.
   - Step 1: 29 ft. lbs. (39 Nm)
   - Step 2: 51 ft. lbs. (69 Nm)
   - Step 3: 72 ft. lbs. (98 Nm)

32. The remainder of installation is the reverse of removal. Note the following torques:
   - Water passage 8mm bolts: 16 ft. lbs. (22 Nm)
   - Water passage 6mm bolts: 105 inch lbs. (12 Nm)

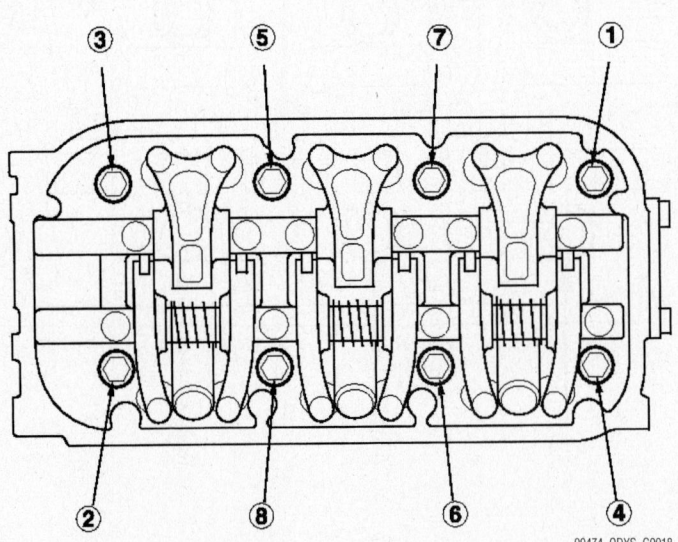

Front cylinder head bolt removal sequence—J35A7 engine

09474_ODYS_G0018

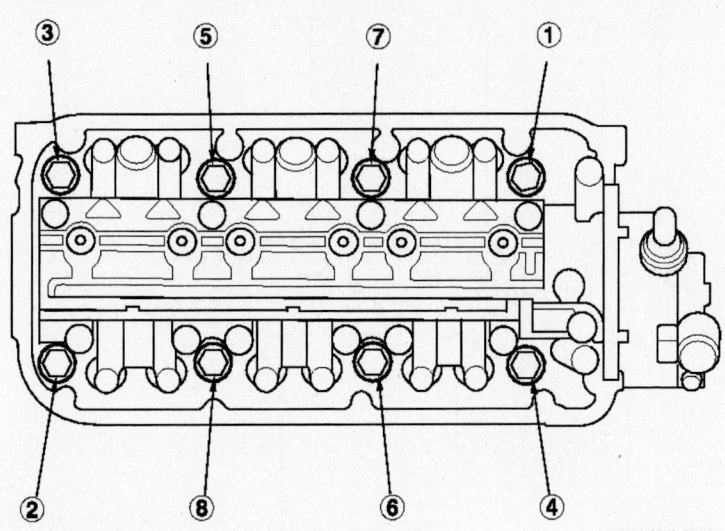

Rear cylinder head bolt removal sequence—J35A7 engine

09474_ODYS_G0018A

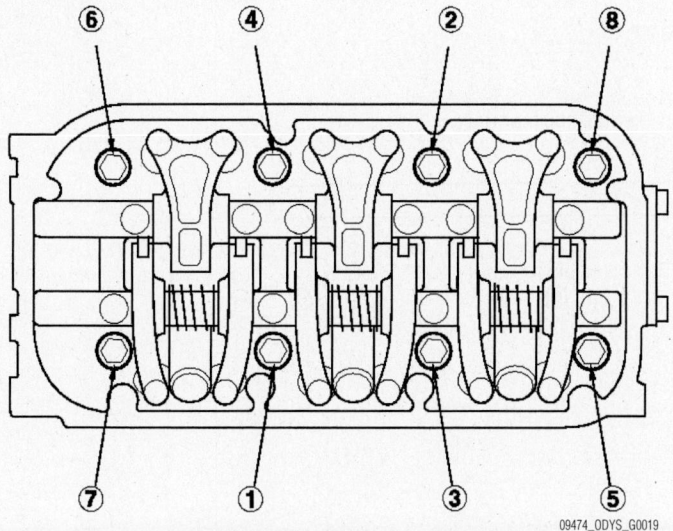

Front head bolt torque sequence—J35A7 engine

09474_ODYS_G0019

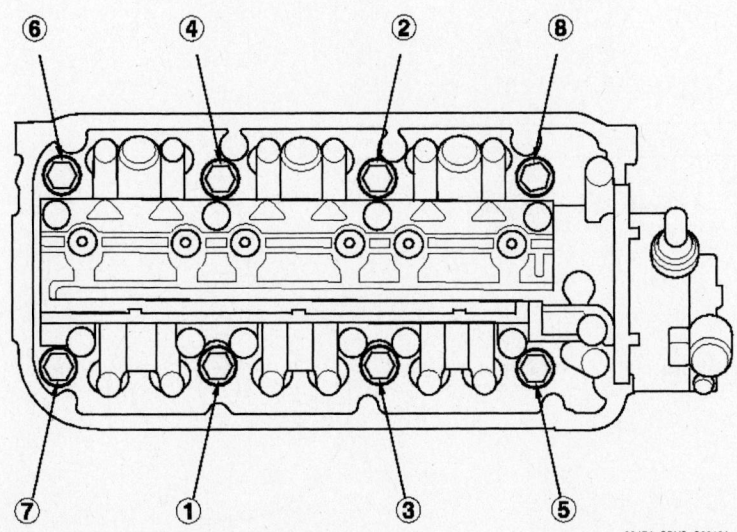

Rear head bolt torque sequence—J35A7 engine

09474_ODYS_G0019A

- Front head bracket 10mm bolt: 33 ft. lbs. (44 Nm)
- Rear head bracket 8mm bolt: 16 ft. lbs. (22 Nm)
- EVAP canister joint bracket: 105 inch lbs. (12 Nm)
- Power steering pump bolts: 16 ft. lbs. (22 Nm)

Rocker Arms/Shafts

REMOVAL & INSTALLATION

J35A4 and J35A6 Engines

1. Before servicing the vehicle, refer to the precautions in the beginning of this section.
2. Remove or disconnect the following:
   - Negative battery cable
   - Air intake tube
   - Ignition coil covers
   - Intake manifold cover
   - Intake manifold
   - Valve cover
3. Loosen the valve adjuster locknuts and screws so that all valves are closed.
4. Loosen the rocker arm shaft mounting bolts 2 turns at a time in sequence. Do not remove the mounting bolts. They will keep the springs and rocker arms on the shaft.
5. Remove the rocker arms and shafts from the vehicle as an assembly.

➡**Keep all valvetrain components in order for assembly.**

6. Remove the rocker arms and springs from the rocker arm shafts.
   To install:
7. Assemble the rocker arms and springs to the rocker arm shafts in their original positions.
8. Install the rocker arm assemblies. Tighten the bolts in sequence 2 turns at a time to 17 ft. lbs. (24 Nm).
9. Adjust the valve clearance.
10. Install or connect the following:
    - Valve covers
    - Intake manifold
    - Intake manifold cover
    - Ignition coil covers
    - Air intake tube
    - Negative battery cable
11. Start the engine and check for leaks.

J35A7 Engine

FRONT HEAD

1. Before servicing the vehicle, refer to the precautions in the beginning of this section.
2. Remove or disconnect the following:

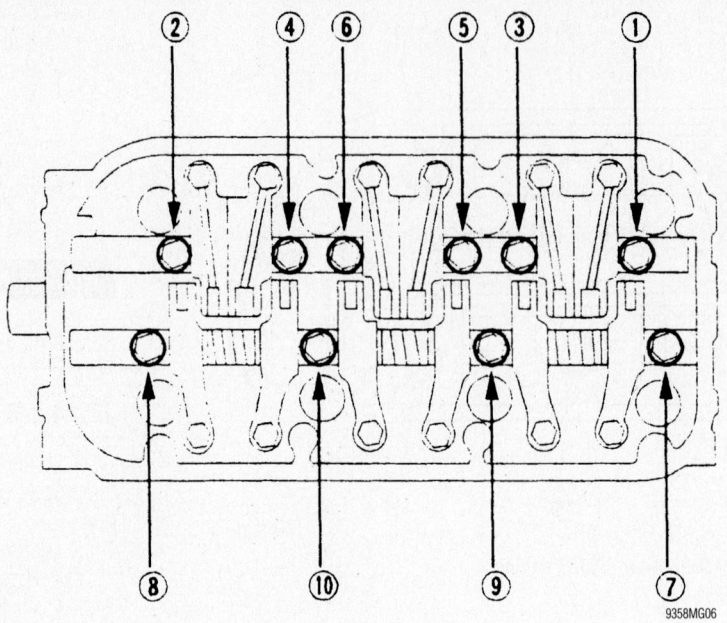

Rocker arm shaft loosening sequence—J35A4 and J35A6 engines

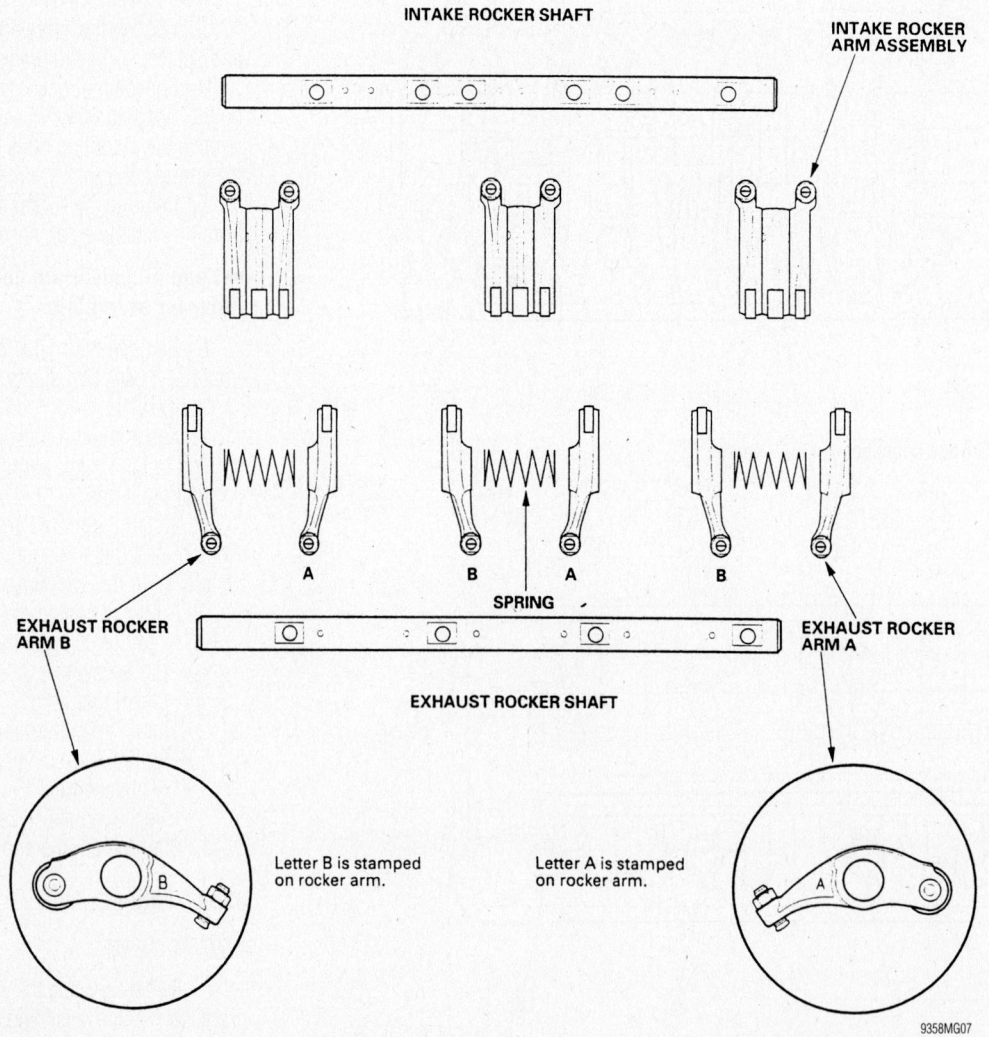

INTAKE ROCKER SHAFT

INTAKE ROCKER ARM ASSEMBLY

EXHAUST ROCKER ARM B

SPRING

EXHAUST ROCKER ARM A

EXHAUST ROCKER SHAFT

Letter B is stamped on rocker arm.

Letter A is stamped on rocker arm.

Exploded view of the rocker arms and shafts—J35A4 and J35A6 engines

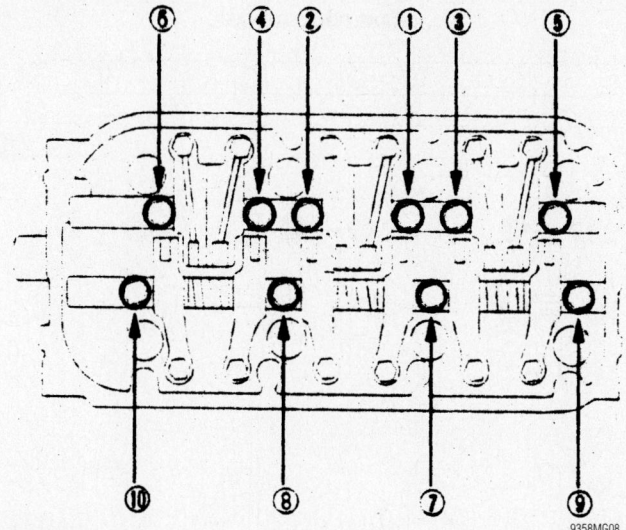

9358MG08

Rocker shaft tightening sequence—J35A4 engines

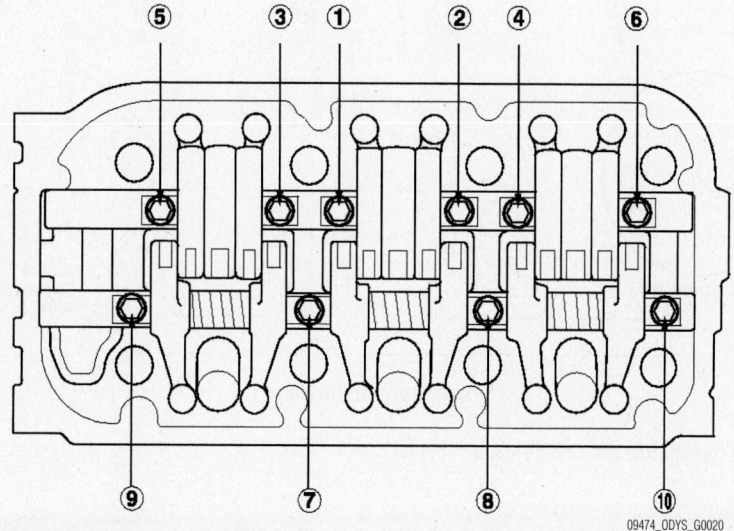

09474_ODYS_G0020

Rocker shaft tightening sequence—J35A6 engines

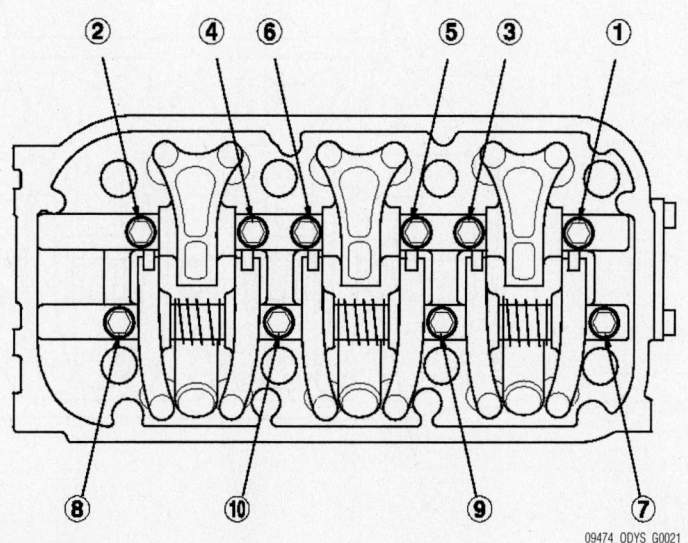

09474_ODYS_G0021

Front head rocker arm shaft loosening sequence—J35A7 engines

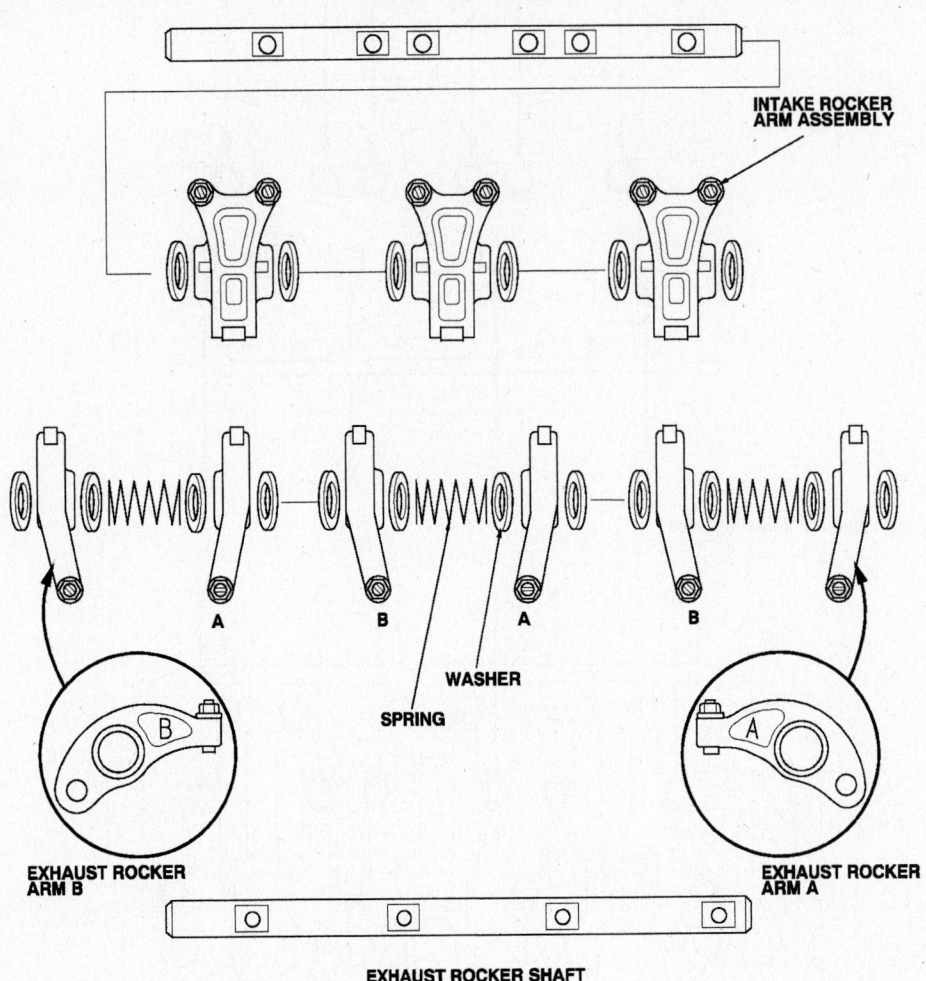

INTAKE ROCKER SHAFT

INTAKE ROCKER
ARM ASSEMBLY

A   B   A   B

WASHER

SPRING

EXHAUST ROCKER
ARM B

EXHAUST ROCKER
ARM A

EXHAUST ROCKER SHAFT

09474_ODYS_G0024

Exploded view of the front head rocker arms and shafts—J35A7 engines

- Negative battery cable
- Air intake tube
- Ignition coil covers
- Intake manifold cover
- Intake manifold
- Valve cover

3. Loosen the valve adjuster locknuts and screws so that all valves are closed.

4. Loosen the rocker arm shaft mounting bolts 2 turns at a time in sequence. Do not remove the mounting bolts. They will keep the springs and rocker arms on the shaft.

5. Remove the rocker arms and shafts from the vehicle as an assembly.

➡ **Keep all valvetrain components in order for assembly.**

6. Remove the rocker arms and springs from the rocker arm shafts.

To install:

7. Assemble the rocker arms and springs to the rocker arm shafts in their original positions.

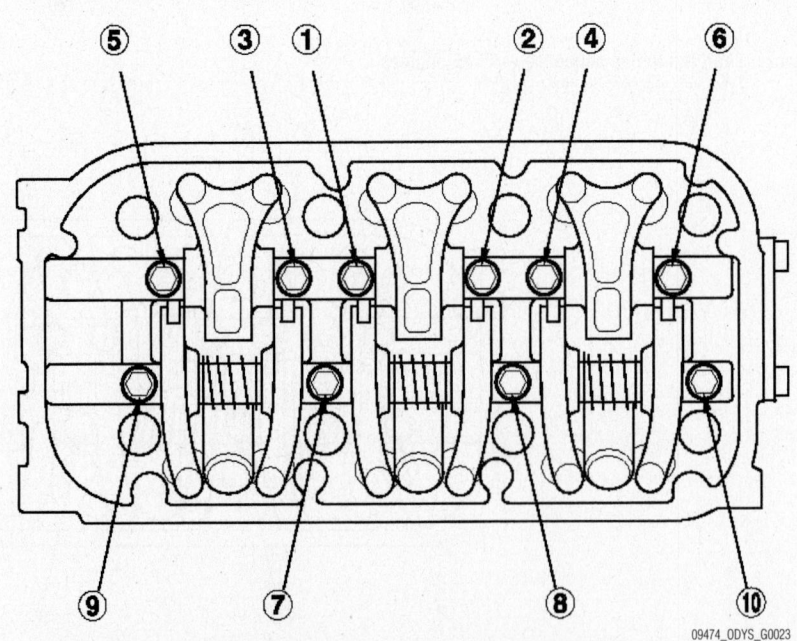

Front head rocker shaft tightening sequence—J35A7 engines

09474_ODYS_G0023

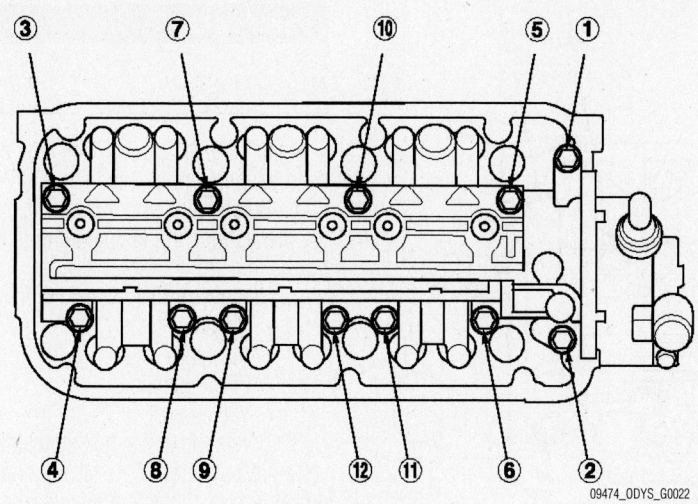

Rear head rocker arm shaft loosening sequence—J35A7 engines

8. Install the rocker arm assemblies. Tighten the bolts in sequence 2 turns at a time to 17 ft. lbs. (24 Nm).

9. Adjust the valve clearance.

10. Install or connect the following:
- Valve covers
- Intake manifold
- Intake manifold cover
- Ignition coil covers
- Air intake tube
- Negative battery cable

11. Start the engine and check for leaks.

REAR HEAD

1. Before servicing the vehicle, refer to the precautions in the beginning of this section.

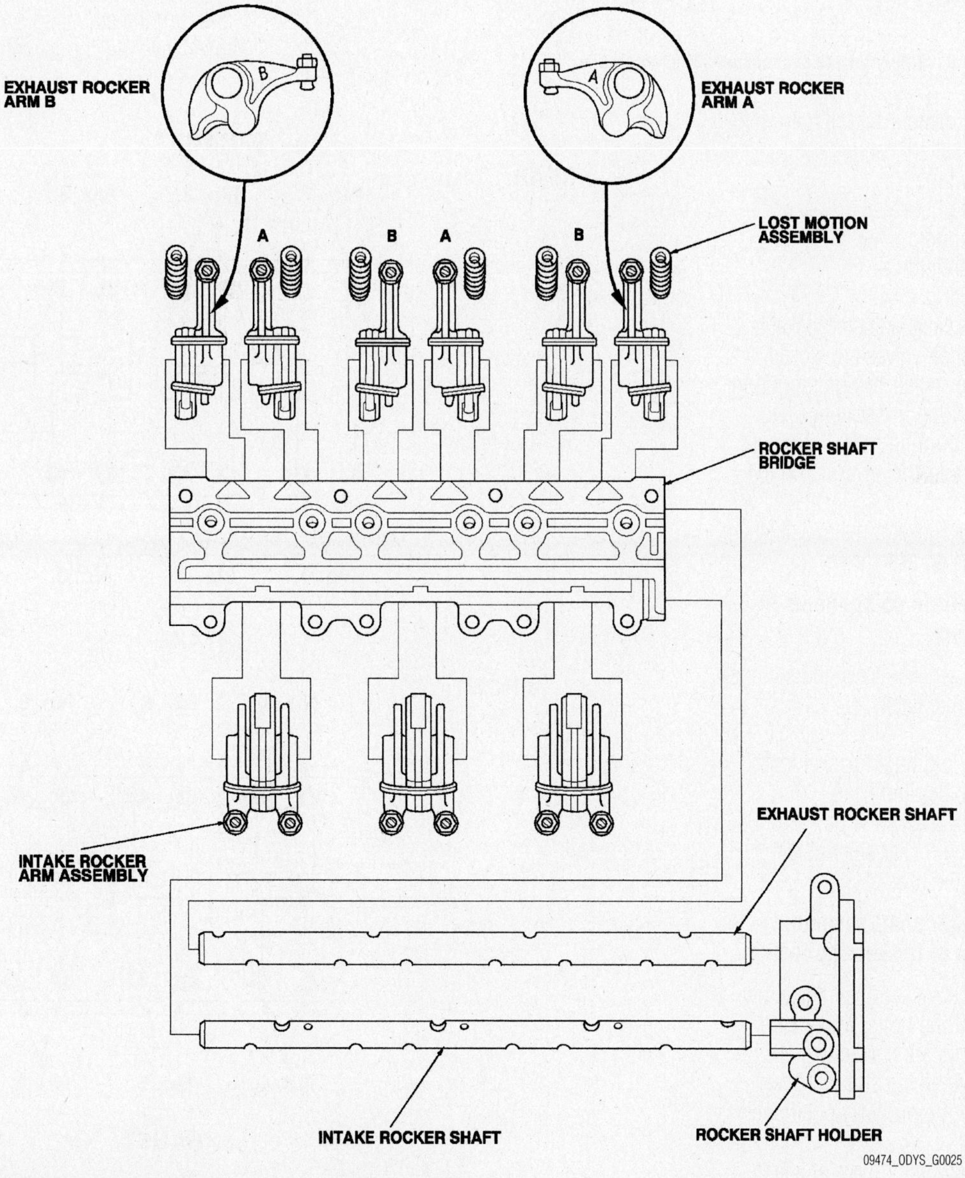

EXHAUST ROCKER ARM B

EXHAUST ROCKER ARM A

LOST MOTION ASSEMBLY

ROCKER SHAFT BRIDGE

INTAKE ROCKER ARM ASSEMBLY

EXHAUST ROCKER SHAFT

INTAKE ROCKER SHAFT

ROCKER SHAFT HOLDER

09474_ODYS_G0025

Exploded view of the rear head rocker arms and shafts—J35A7 engines

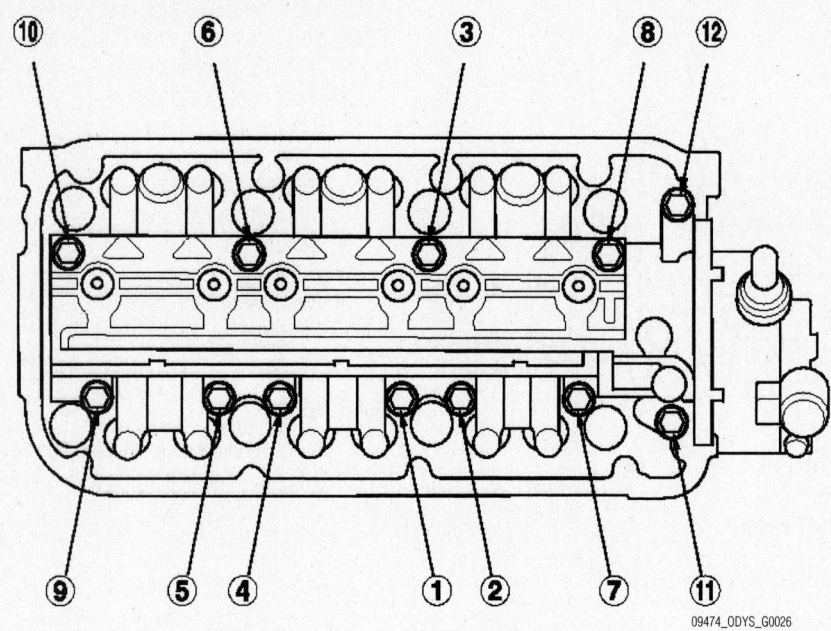

Rear head rocker assembly torque sequence—J35A7 engine

09474_ODYS_G0026

Valve Lash

ADJUSTMENT

Adjust the valves only when the cylinder head temperature is less than 100°F (38°C).

1. Before servicing the vehicle, refer to the precautions in the beginning of this section.

2. Remove or disconnect the following:
- Negative battery cable
- Air intake tube
- Intake manifold
- Valve cover

3. Rotate the crankshaft so that the valves to be adjusted are closed and the rocker arm is contacting the camshaft lobe base circle.

4. Measure the valve clearance. If adjustment is necessary, loosen the locknut and turn the adjusting screw as necessary to achieve the correct valve clearance.

2. Remove or disconnect the following:
- Negative battery cable
- Air intake tube
- Ignition coil covers
- Intake manifold cover
- Intake manifold
- Valve cover

3. Loosen the valve adjuster locknuts and screws so that all valves are closed.

4. Loosen the rocker bridge mounting bolts 2 turns at a time in sequence. Do not remove the mounting bolts. They will keep the springs and rocker arms on the shaft.

5. Remove the rocker arms and shafts from the vehicle as an assembly.

➡**Keep all valvetrain components in order for assembly.**

6. Remove the rocker arms and springs from the rocker arm shafts.

To install:

7. Remove all old liquid gasket from the rocker shaft holder and head.

8. Apply liquid gasket PN 08717-0004, or equivalent, to the rocker shaft holder mating surface of the head.

➡**Install the rocker shaft assembly within 4 minutes of gasket application.**

9. Position the rocker shaft assembly and loosely install the bolts. Make sure that the rocker arms are correctly positioned on the valve stems and the dowel pins in the bridge are properly positioned in the head.

10. Tighten each bolt 2 turns at a time, in sequence to 16 ft. lbs. (22 Nm).

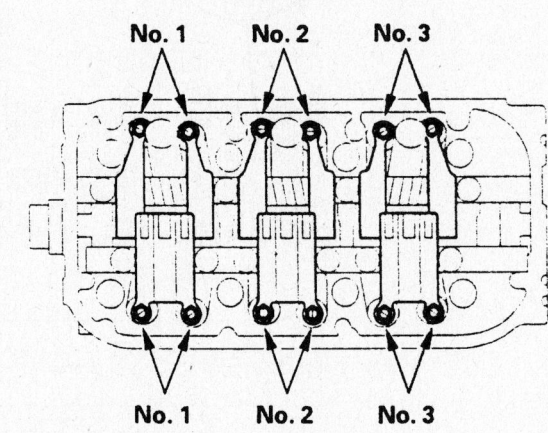

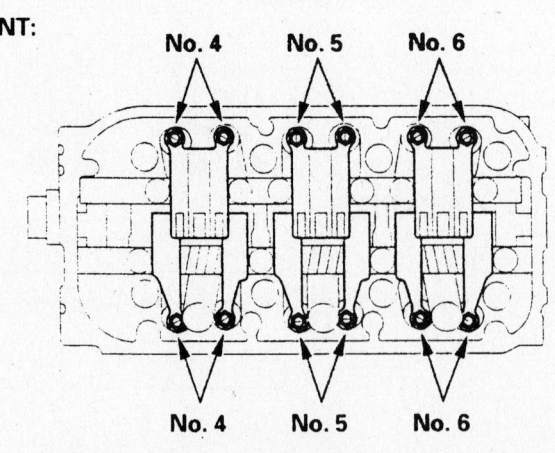

Adjusting screw locations—J35A4 and J35A6 engines

9358MG16

REAR:

EXHAUST

No. 1   No. 2   No. 3

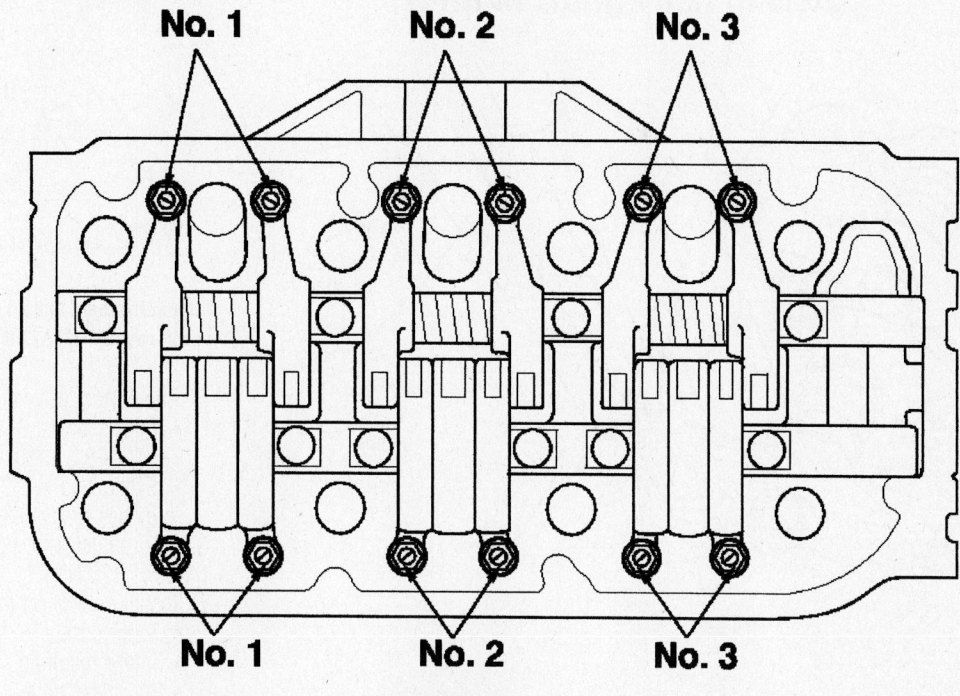

No. 1   No. 2   No. 3

INTAKE

FRONT:

INTAKE

No. 4   No. 5   No. 6

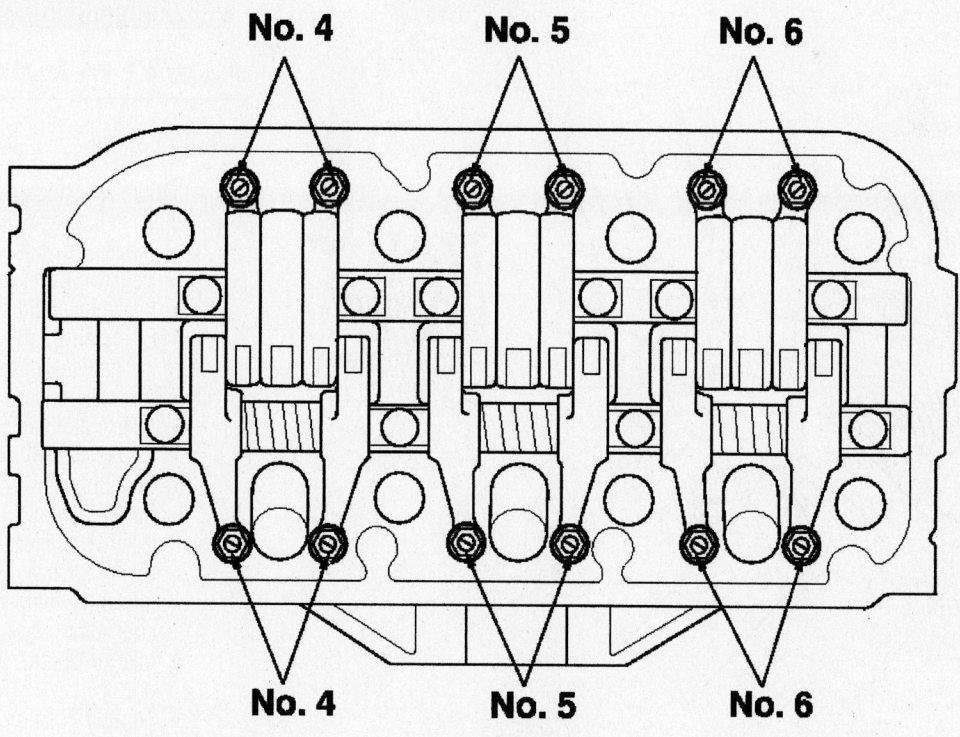

No. 4   No. 5   No. 6

EXHAUST

09474_ODYS_G0027

Adjusting screw locations—J35A7 engines

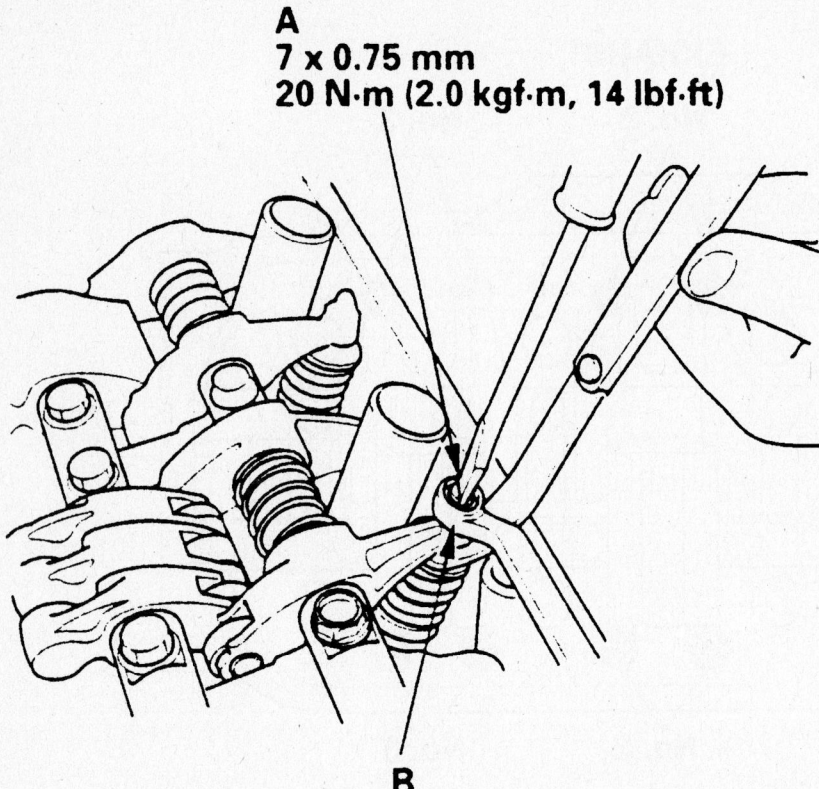

A
7 x 0.75 mm
20 N·m (2.0 kgf·m, 14 lbf·ft)

B

9358MG15

After adjustment tighten the locknut to specification—J35A4 and J35A6 engines

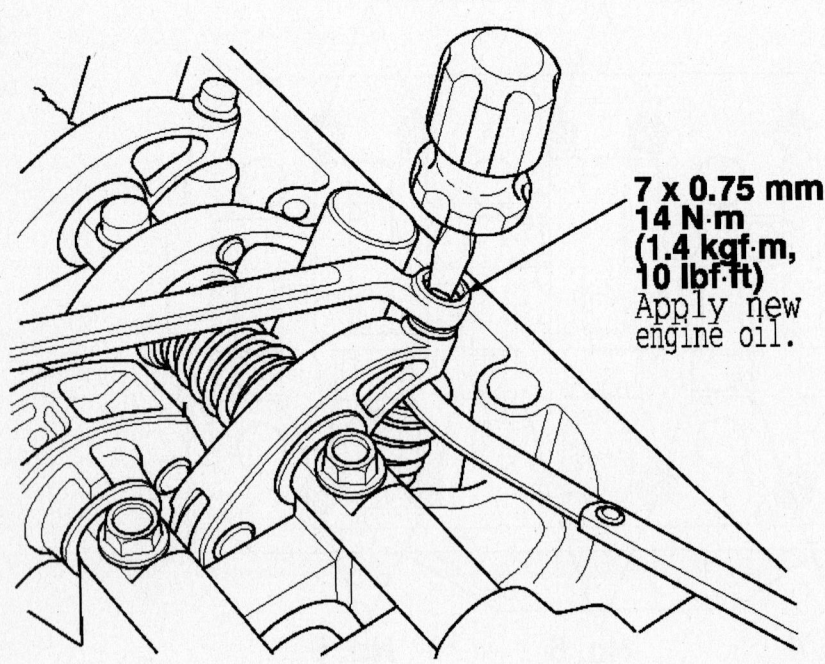

7 x 0.75 mm
14 N·m
(1.4 kgf·m,
10 lbf·ft)
Apply new
engine oil.

09474_ODYS_G0028

After adjustment tighten the locknut to specification—J35A7 engine front head

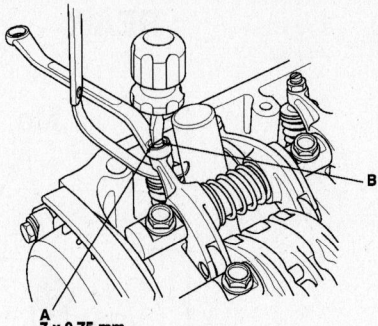

A
7 x 0.75 mm
20 N·m (2.0 kgf·m, 14 lbf·ft)

09474_ODYS_G0028A

After adjustment tighten the locknut to specification—J35A7 engine rear head

5. The correct valve clearance is:
  • Intake valves: 0.008–0.009 inches (0.20–0.24mm)
  • Exhaust valves: 0.011–0.013 inches (0.28–0.32mm)
6. After adjustment, tighten the locknuts to 14 ft. lbs. (20 Nm).
7. Install or connect the following:
  • Valve cover
  • Intake manifold
  • Air intake tube
  • Negative battery cable
8. Start the engine and check for proper operation.

Intake Manifold

REMOVAL & INSTALLATION

J35A4 Engine

1. Before servicing the vehicle, refer to the precautions in the beginning of this section.
2. Remove or disconnect the following:
  • Negative battery cable
  • Intake manifold cover
  • Evaporative Emissions (EVAP) control canister hose or vacuum hose
  • Air intake tube
  • Accelerator cable
  • Cruise control cable
  • Brake booster vacuum line
  • Intake manifold vacuum line
  • Positive Crankcase Ventilation (PCV) valve and hose
  • Intake Air Temperature (IAT) sensor connector
  • Idle Air Control (IAC) valve connector
  • Throttle Position (TP) sensor connector
  • Manifold Absolute Pressure (MAP) sensor connector
  • Intake manifold

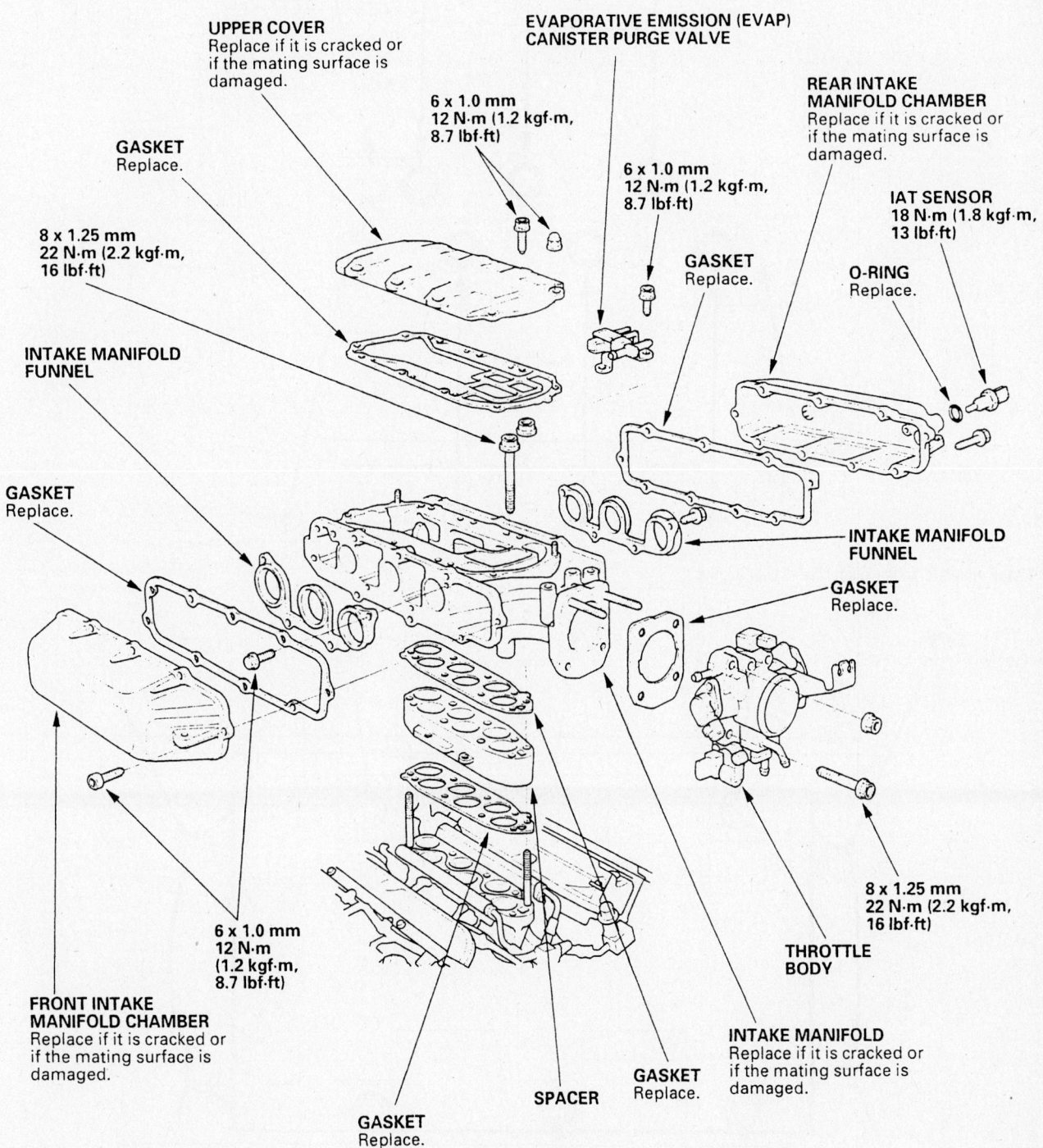

UPPER COVER
Replace if it is cracked or
if the mating surface is
damaged.

EVAPORATIVE EMISSION (EVAP)
CANISTER PURGE VALVE

REAR INTAKE
MANIFOLD CHAMBER
Replace if it is cracked or
if the mating surface is
damaged.

GASKET
Replace.

6 x 1.0 mm
12 N·m (1.2 kgf·m,
8.7 lbf·ft)

6 x 1.0 mm
12 N·m (1.2 kgf·m,
8.7 lbf·ft)

IAT SENSOR
18 N·m (1.8 kgf·m,
13 lbf·ft)

8 x 1.25 mm
22 N·m (2.2 kgf·m,
16 lbf·ft)

GASKET
Replace.

O-RING
Replace.

INTAKE MANIFOLD
FUNNEL

INTAKE MANIFOLD
FUNNEL

GASKET
Replace.

GASKET
Replace.

FRONT INTAKE
MANIFOLD CHAMBER
Replace if it is cracked or
if the mating surface is
damaged.

6 x 1.0 mm
12 N·m
(1.2 kgf·m,
8.7 lbf·ft)

8 x 1.25 mm
22 N·m (2.2 kgf·m,
16 lbf·ft)

THROTTLE
BODY

INTAKE MANIFOLD
Replace if it is cracked or
if the mating surface is
damaged.

GASKET
Replace.

SPACER

GASKET
Replace.

9358MG09

Exploded view of the intake manifold—J35A4 engine

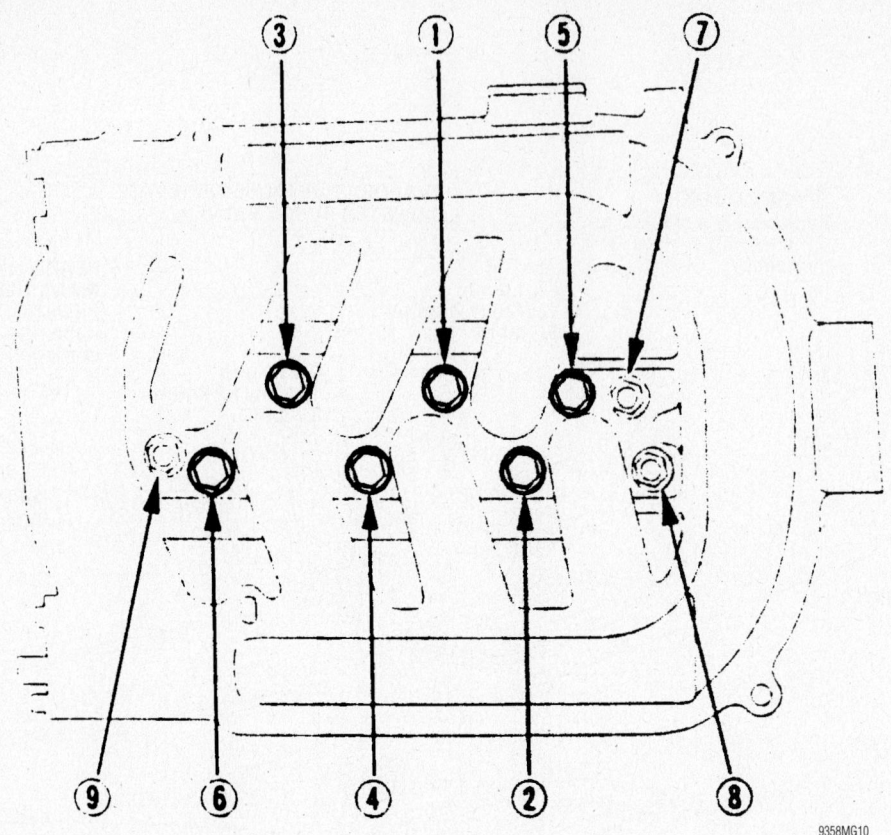

Intake manifold torque sequence—J35A4 engine

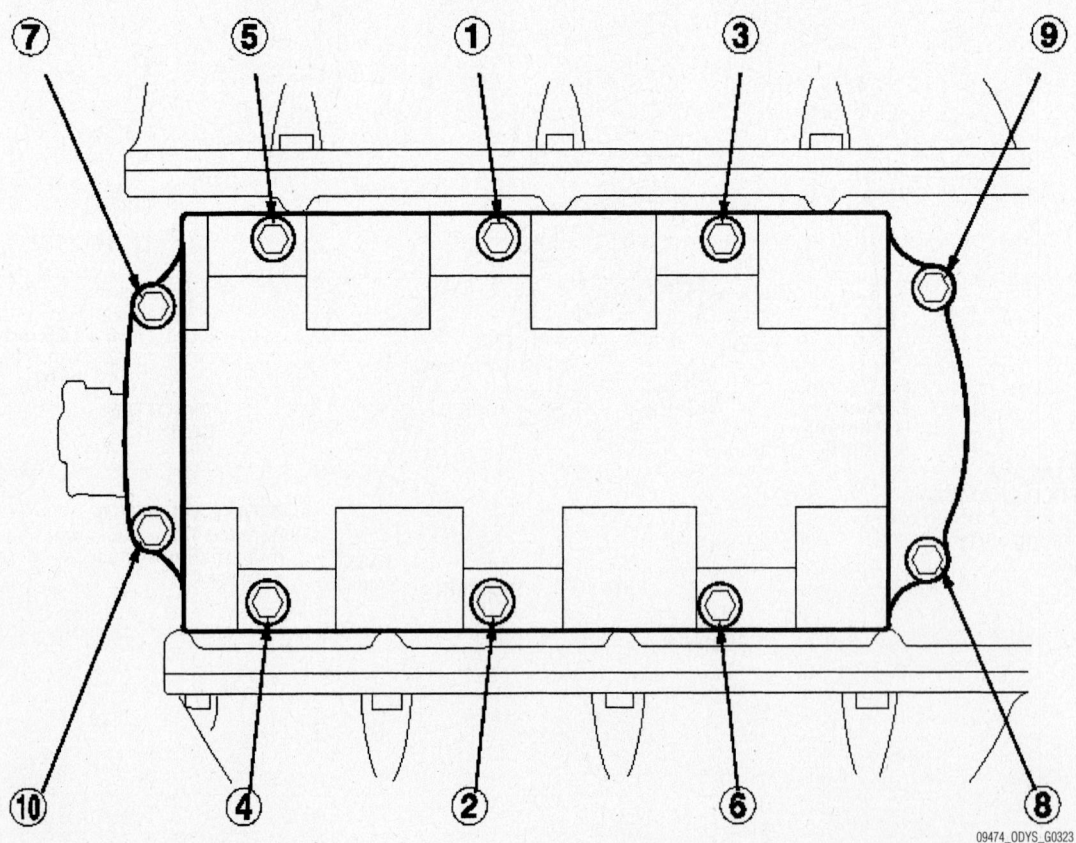

Upper cover torque sequence—2002–04 models

J35A6 engine:

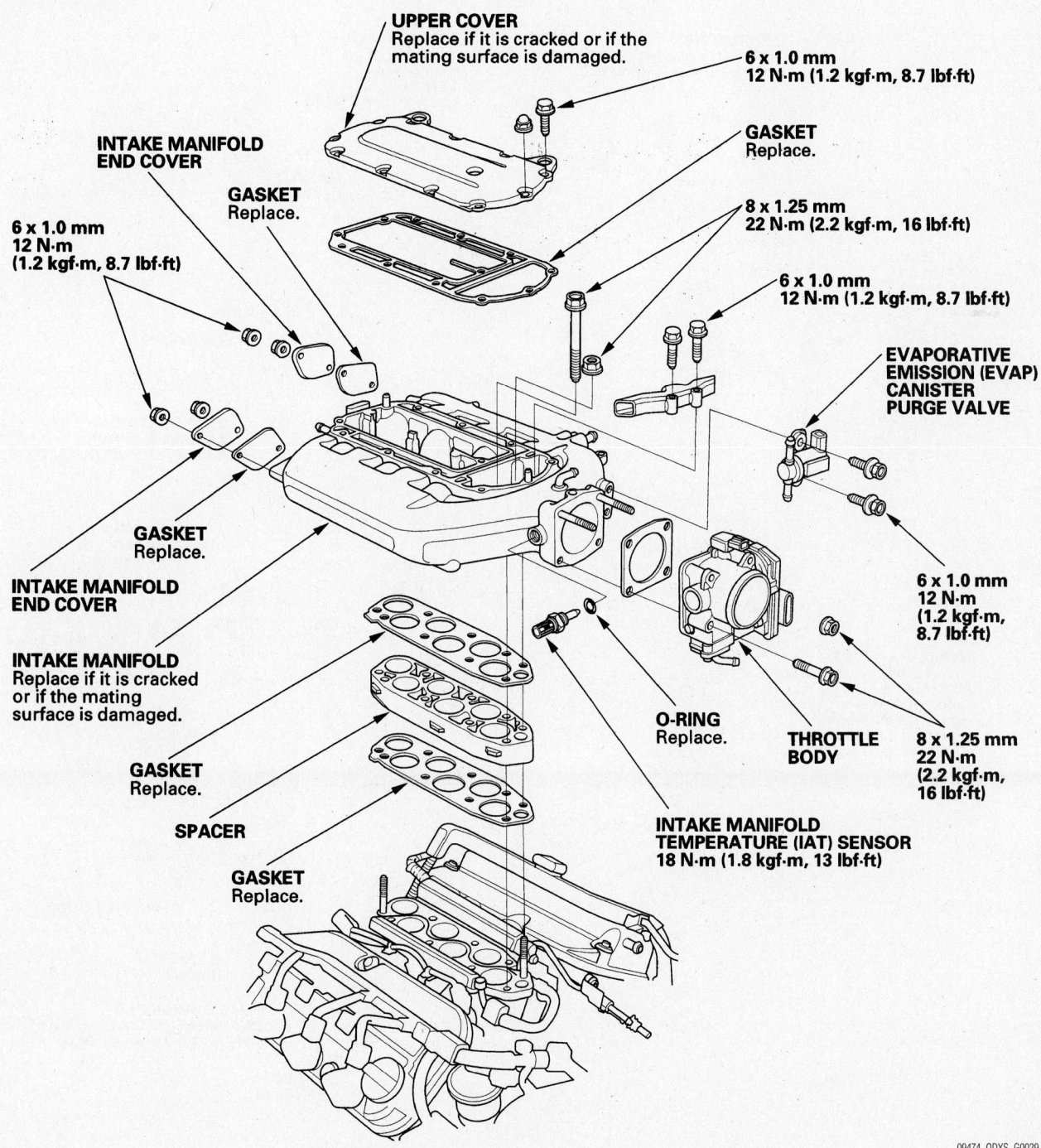

UPPER COVER
Replace if it is cracked or if the
mating surface is damaged.

6 x 1.0 mm
12 N·m (1.2 kgf·m, 8.7 lbf·ft)

GASKET
Replace.

8 x 1.25 mm
22 N·m (2.2 kgf·m, 16 lbf·ft)

6 x 1.0 mm
12 N·m (1.2 kgf·m, 8.7 lbf·ft)

**EVAPORATIVE
EMISSION (EVAP)
CANISTER
PURGE VALVE**

**INTAKE MANIFOLD
END COVER**

GASKET
Replace.

6 x 1.0 mm
12 N·m
(1.2 kgf·m, 8.7 lbf·ft)

6 x 1.0 mm
12 N·m
(1.2 kgf·m,
8.7 lbf·ft)

GASKET
Replace.

**INTAKE MANIFOLD
END COVER**

INTAKE MANIFOLD
Replace if it is cracked
or if the mating
surface is damaged.

GASKET
Replace.

SPACER

GASKET
Replace.

O-RING
Replace.

**THROTTLE
BODY**

**INTAKE MANIFOLD
TEMPERATURE (IAT) SENSOR**
18 N·m (1.8 kgf·m, 13 lbf·ft)

8 x 1.25 mm
22 N·m
(2.2 kgf·m,
16 lbf·ft)

09474_ODYS_G0029,

Intake manifold and related parts—J35A6 engine

J35A7 engine:

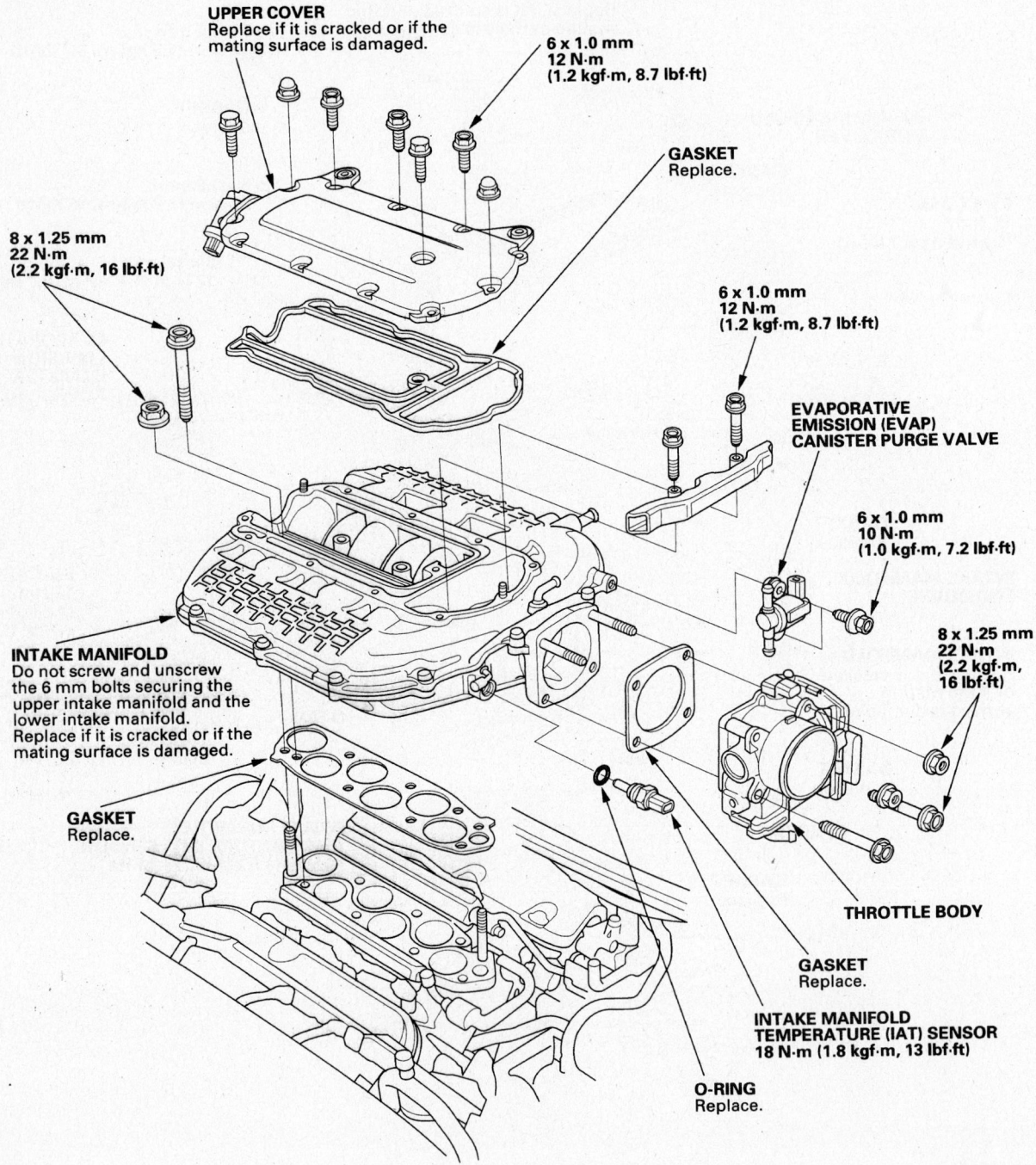

UPPER COVER
Replace if it is cracked or if the mating surface is damaged.

6 x 1.0 mm
12 N·m
(1.2 kgf·m, 8.7 lbf·ft)

GASKET
Replace.

8 x 1.25 mm
22 N·m
(2.2 kgf·m, 16 lbf·ft)

6 x 1.0 mm
12 N·m
(1.2 kgf·m, 8.7 lbf·ft)

EVAPORATIVE EMISSION (EVAP) CANISTER PURGE VALVE

6 x 1.0 mm
10 N·m
(1.0 kgf·m, 7.2 lbf·ft)

8 x 1.25 mm
22 N·m
(2.2 kgf·m, 16 lbf·ft)

INTAKE MANIFOLD
Do not screw and unscrew the 6 mm bolts securing the upper intake manifold and the lower intake manifold.
Replace if it is cracked or if the mating surface is damaged.

GASKET
Replace.

THROTTLE BODY

GASKET
Replace.

INTAKE MANIFOLD TEMPERATURE (IAT) SENSOR
18 N·m (1.8 kgf·m, 13 lbf·ft)

O-RING
Replace.

09474_ODYS_G0030

Intake manifold and related parts—J35A7 engine

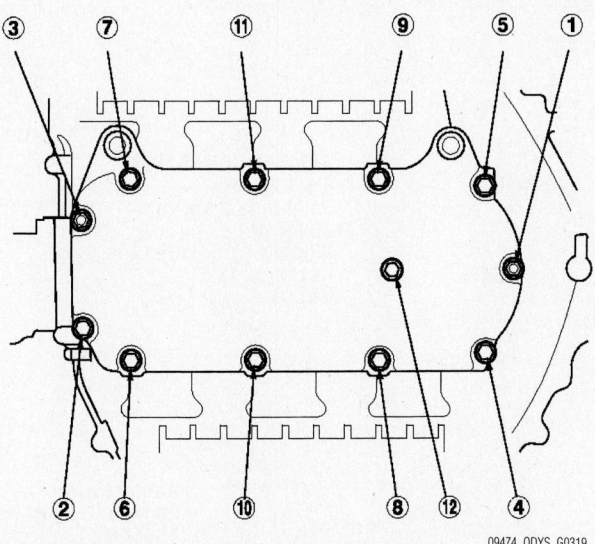

Upper cover loosening sequence—J35A6 and J35A7 engines

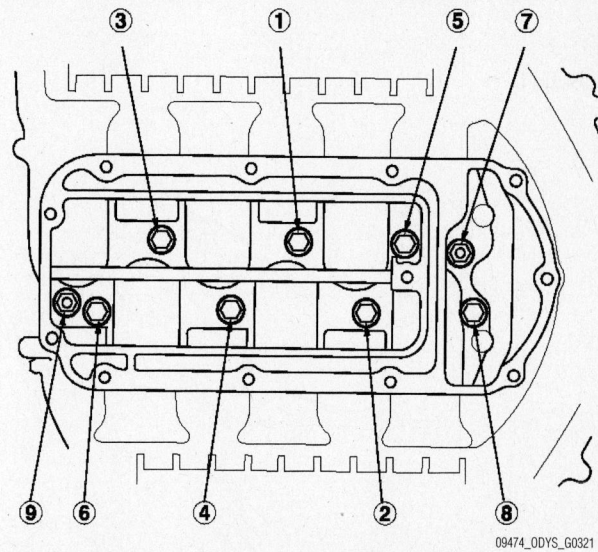

Intake manifold torque sequence—J35A6 and J35A7 engines

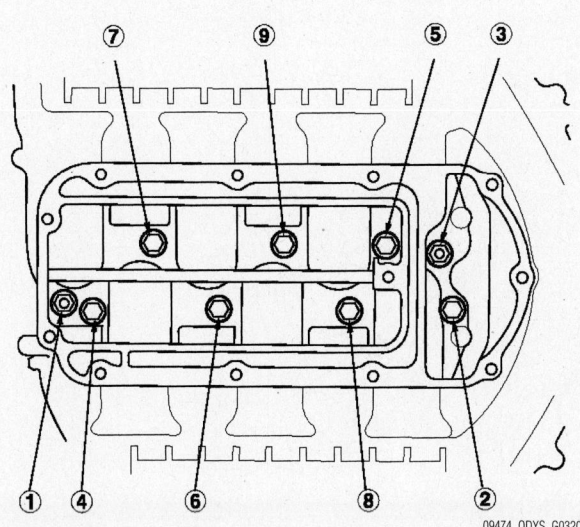

Intake manifold loosening sequence—J35A6 and J35A7 engines

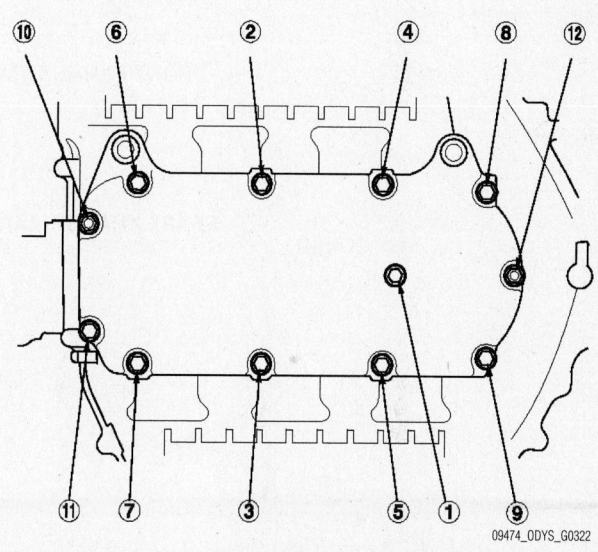

Upper cover torque sequence—J35A6 and J35A7 engines

To install:

3. Install or connect the following:
- New intake manifold gasket
- Intake manifold. Tighten the fasteners in sequence and in several passes to 16 ft. lbs. (22 Nm).
- MAP sensor connector
- TP sensor connector
- IAC valve connector
- IAT sensor connector
- PCV valve and hose
- Intake manifold vacuum line
- Brake booster vacuum line
- Cruise control cable
- Accelerator cable
- Intake manifold cover
- Air intake tube
- EVAP control canister hose and vacuum hose

- Negative battery cable

4. Start the engine and check for proper operation.

J35A6 and J35A7 Engines

1. Remove the intake manifold cover.
2. Remove the air inlet duct.
3. Remove the PCV hose and brake booster hose.
4. Remove the EVAP canister hose and water bypass hoses. Plug the bypass hose.
5. Tag and disconnect all wiring connected to the manifold.
6. Remove the upper cover mounting bolts and nuts, in 2 equal steps, in the sequence shown. Remove the cover.
7. Remove the intake manifold mounting

bolts and nuts, in 2 equal steps, in the sequence shown. Remove the manifold.

8. Installation is the reverse of removal. Tighten the manifold bolts and nuts, in the sequence shown, in 2 equal steps, to 16 ft. lbs. (22 Nm). Tighten the cover bolts and nuts, in the sequence shown, in 2 equal steps, to 105 inch lbs. (12 Nm).

Exhaust Manifold

REMOVAL & INSTALLATION

J35A4 Engine

1. Before servicing the vehicle, refer to the precautions in the beginning of this section.

FRONT:

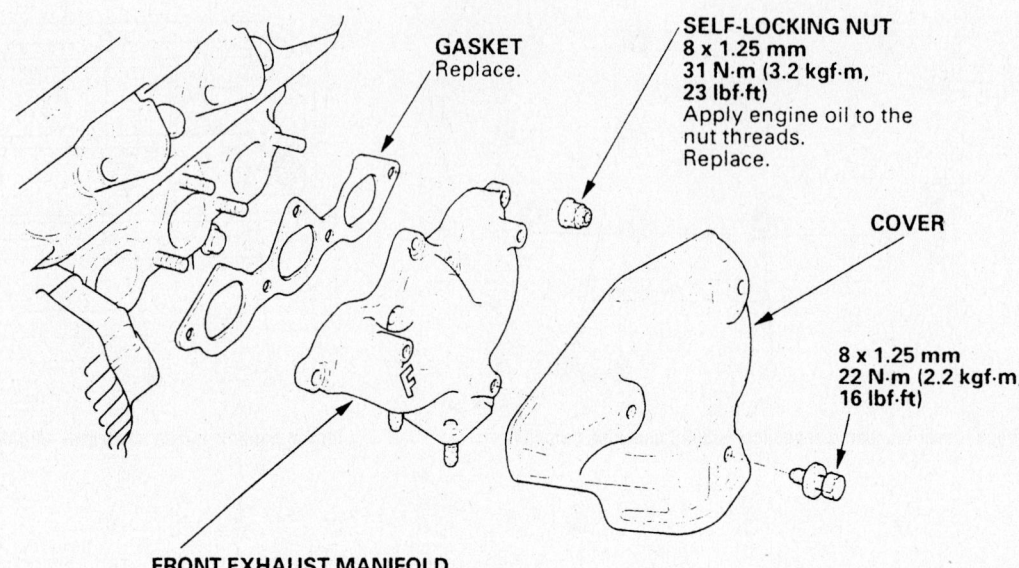

GASKET
Replace.

SELF-LOCKING NUT
8 x 1.25 mm
31 N·m (3.2 kgf·m, 23 lbf·ft)
Apply engine oil to the nut threads.
Replace.

COVER

8 x 1.25 mm
22 N·m (2.2 kgf·m, 16 lbf·ft)

FRONT EXHAUST MANIFOLD

REAR:

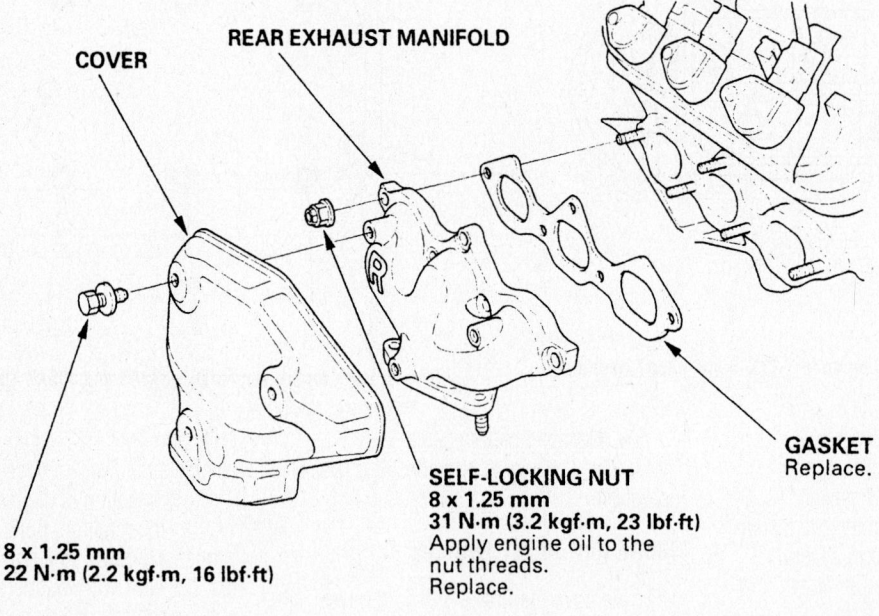

COVER

REAR EXHAUST MANIFOLD

SELF-LOCKING NUT
8 x 1.25 mm
31 N·m (3.2 kgf·m, 23 lbf·ft)
Apply engine oil to the nut threads.
Replace.

8 x 1.25 mm
22 N·m (2.2 kgf·m, 16 lbf·ft)

GASKET
Replace.

9358MG12

Exploded view of the exhaust manifolds—J35A4 engine

2. Remove or disconnect the following:
- Negative battery cable
- Exhaust manifold heat shield
- Heated Oxygen Sensor (HO2S) connector
- Exhaust front pipe
- Exhaust manifold bracket, if equipped
- Exhaust manifold

To install:
3. Install or connect the following:
- Exhaust manifold. Tighten the fasteners to 23 ft. lbs. (31 Nm).
- Exhaust manifold bracket, if equipped. Tighten the bolts to 33 ft. lbs. (44 Nm).
- Exhaust front pipe. Tighten the nuts to 40 ft. lbs. (55 Nm).
- HO2S connector
- Exhaust manifold heat shield
- Negative battery cable

Warm-Up Catalytic Converter

➡**The J35A6 and J35A7 engines don't have conventional exhaust manifolds. Instead, warm-up converters are attached directly to the heads.**

REMOVAL & INSTALLATION

Front

1. Remove the radiator.
2. Remove the condenser fan assemblies.

3. Disconnect the exhaust pipe.
4. Disconnect the air/fuel ratio sensor.
5. Remove the converter mounting nuts and carefully remove the converter.
6. Installation is the reverse of removal. Torque the nuts in a criss-cross pattern, in 2 equal steps, to 23 ft. lbs. (31 Nm). Torque the exhaust pipe nuts to 40 ft. lbs. (54 Nm).

Rear

1. Disconnect the exhaust pipe.
2. Remove the power steering heat baffle.

3. Remove the intermediate shaft shield.
4. Disconnect the rear air/fuel ratio sensor and the HO2 sensor.
5. Remove the converter bracket.
6. Remove the converter mounting nuts and carefully remove the converter.
7. Installation is the reverse of removal. Torque the nuts in a criss-cross pattern, in 2 equal steps, to 23 ft. lbs. (31 Nm). Torque the exhaust pipe nuts to 40 ft. lbs. (54 Nm).

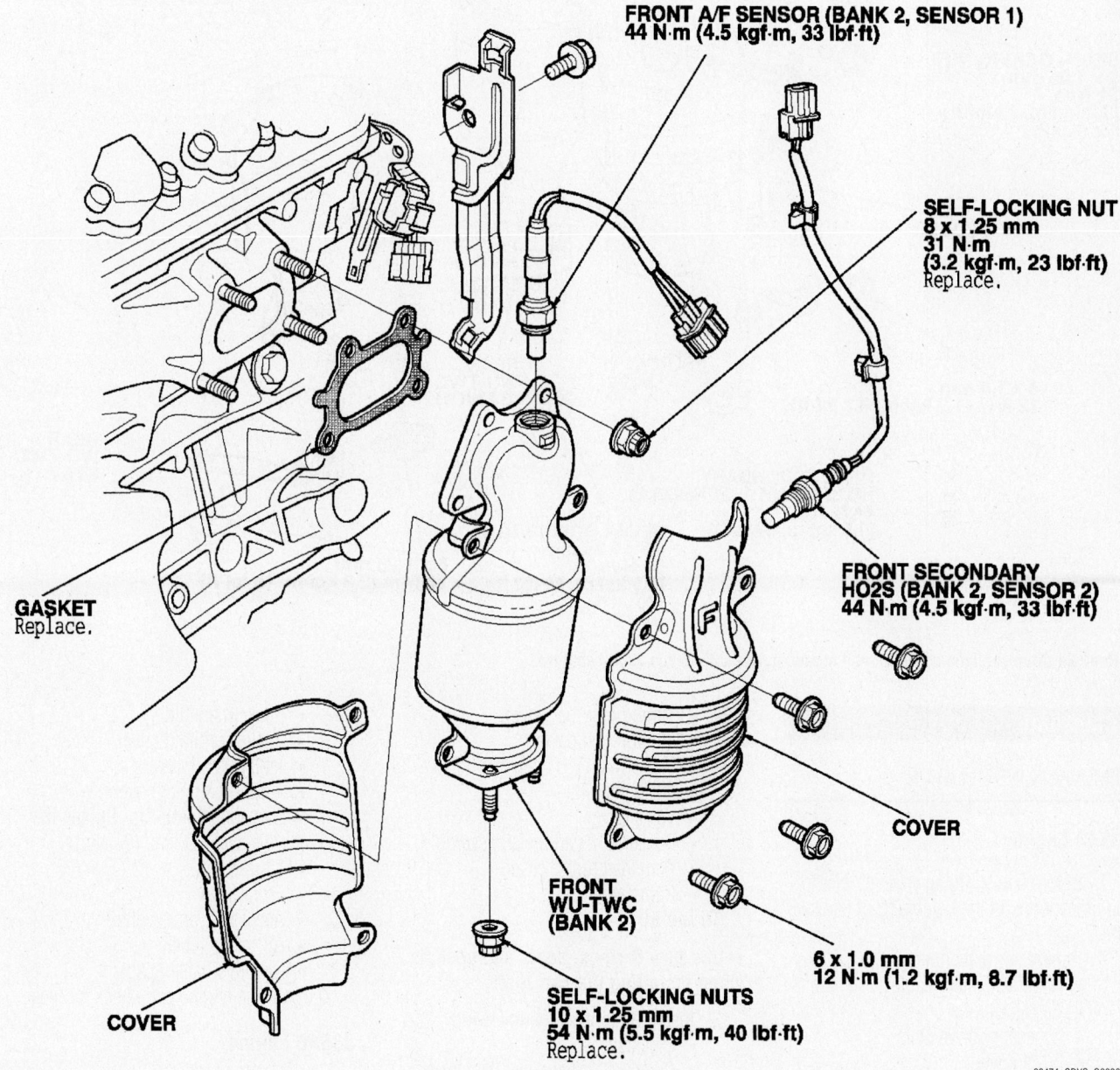

FRONT A/F SENSOR (BANK 2, SENSOR 1)
44 N·m (4.5 kgf·m, 33 lbf·ft)

SELF-LOCKING NUT
8 x 1.25 mm
31 N·m
(3.2 kgf·m, 23 lbf·ft)
Replace.

GASKET
Replace.

FRONT SECONDARY
HO2S (BANK 2, SENSOR 2)
44 N·m (4.5 kgf·m, 33 lbf·ft)

COVER

FRONT
WU-TWC
(BANK 2)

SELF-LOCKING NUTS
10 x 1.25 mm
54 N·m (5.5 kgf·m, 40 lbf·ft)
Replace.

6 x 1.0 mm
12 N·m (1.2 kgf·m, 8.7 lbf·ft)

COVER

09474_ODYS_G0036

Front warm-up catalytic converter—J35A6 and J35A7 engines

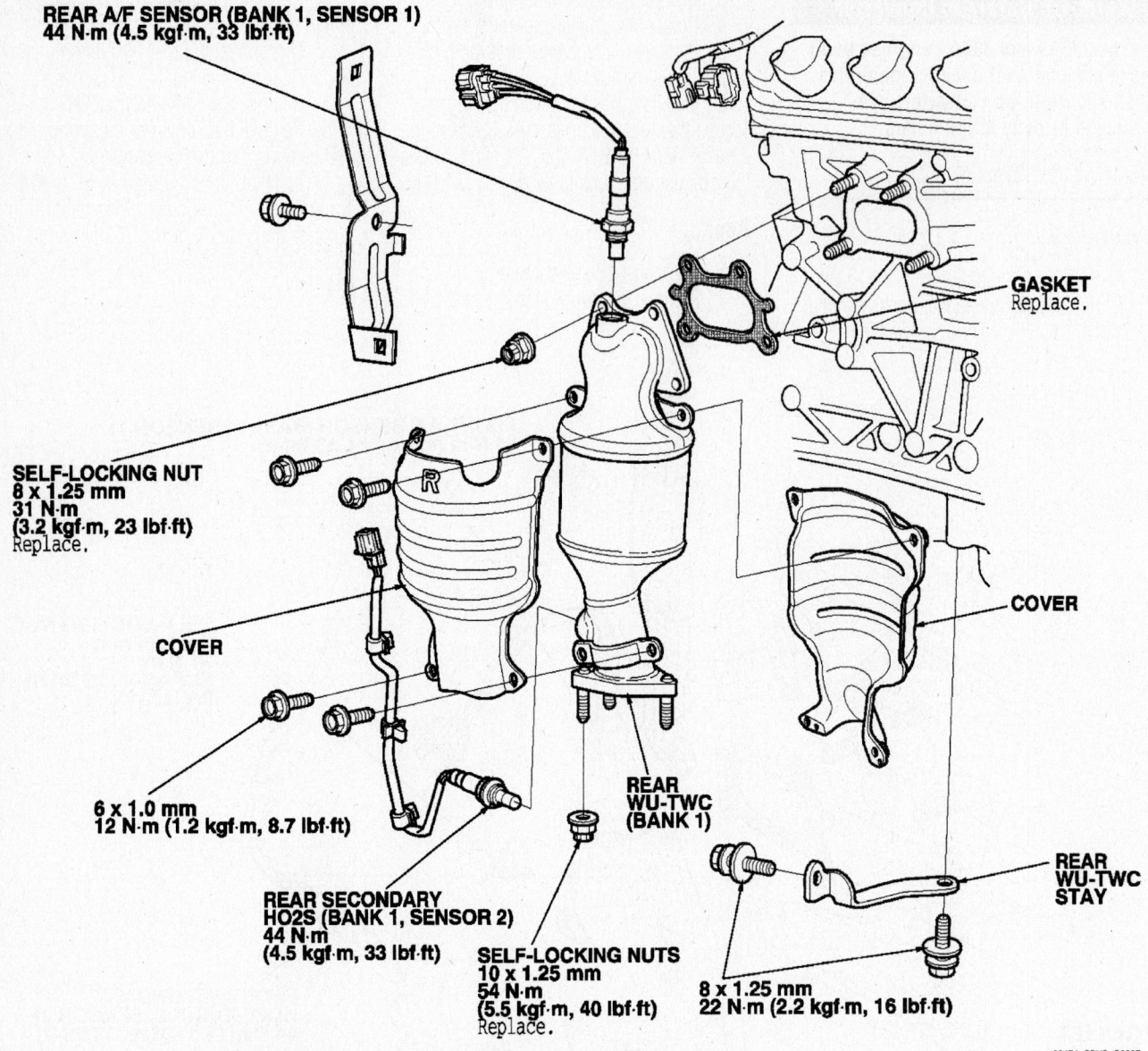

REAR A/F SENSOR (BANK 1, SENSOR 1)
44 N·m (4.5 kgf·m, 33 lbf·ft)

GASKET
Replace.

SELF-LOCKING NUT
8 x 1.25 mm
31 N·m
(3.2 kgf·m, 23 lbf·ft)
Replace.

COVER

COVER

6 x 1.0 mm
12 N·m (1.2 kgf·m, 8.7 lbf·ft)

REAR WU-TWC (BANK 1)

REAR WU-TWC STAY

REAR SECONDARY HO2S (BANK 1, SENSOR 2)
44 N·m
(4.5 kgf·m, 33 lbf·ft)

SELF-LOCKING NUTS
10 x 1.25 mm
54 N·m
(5.5 kgf·m, 40 lbf·ft)
Replace.

8 x 1.25 mm
22 N·m (2.2 kgf·m, 16 lbf·ft)

09474_ODYS_G0037

Rear warm-up catalytic converter and related parts—J35A6 and J35A7 engines

Camshaft

REMOVAL & INSTALLATION

J35A4 Engine

1. Before servicing the vehicle, refer to the precautions in the beginning of this section.

2. Remove or disconnect the following:
   - Negative battery cable
   - Air intake tube
   - Accessory drive belts
   - Front cover
   - Timing belt. Refer to the timing belt procedure.
   - Camshaft sprockets

- Timing belt rear covers
- Ignition coil covers
- Intake manifold cover
- Intake manifold
- Valve cover
- Rocker arms and shaft assembly
- Camshaft thrust cover
- Camshaft

To install:

➡**Use new O-rings, seals and gaskets when installing the camshaft.**

3. Install or connect the following:
   - Camshaft
   - Camshaft thrust cover. Tighten the bolts to 16 ft. lbs. (22 Nm).
   - Rocker arms and shaft assembly
   - Valve cover

- Intake manifold
- Intake manifold cover
- Ignition coil covers
- Timing belt rear covers
- Camshaft sprockets. Tighten the bolts to 67 ft. lbs. (90 Nm).
- Timing belt
- Front cover
- Accessory drive belts
- Air intake tube
- Negative battery cable

4. Start the engine and check for leaks.

J35A6 Engine

FRONT

1. Make sure you have the anti-theft codes for the radio and navigation systems.

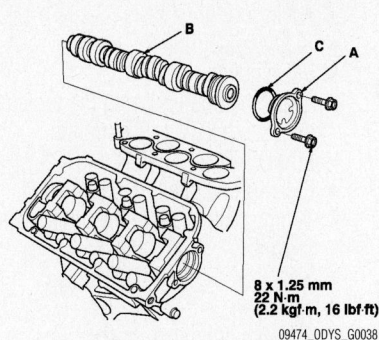

Camshaft removal—J35A6 and J35A7 engines

2. Remove the battery.
3. Drain the coolant.
4. Remove the upper radiator hose.
5. Remove the EGR valve.
6. Remove the timing belt.
7. Remove the rocker arm assembly.
8. Remove the camshaft pulley.
9. Remove the thrust cover (A).
10. Carefully remove the camshaft (B).
11. Installation is the reverse of removal. Always use new O-rings (C). Apply clean engine oil to the journals and lobes.

REAR

1. Relieve the fuel system pressure.
2. Remove the air cleaner assembly.
3. Remove the intake manifold.
4. Disconnect the fuel feed line.
5. Disconnect the brake lines at the master cylinder. Plug the openings.
6. Remove the timing belt.
7. Remove the rocker arm assembly.
8. Remove the camshaft pulley.
9. Remove the EVAP canister hose joint and bracket.
10. Remove the thrust cover (A) and camshaft (B).
11. Installation is the reverse of removal. Always use new O-rings (C). Apply clean engine oil to the journals and lobes. Bleed the brakes. Perform the CKP pattern clear/learn procedure.

J35A7 Engine

FRONT

1. Make sure you have the anti-theft codes for the radio and navigation systems.
2. Remove the battery.
3. Drain the coolant.
4. Remove the upper radiator hose.
5. Remove the EGR valve.
6. Remove the timing belt.

7. Remove the rocker arm assembly.
8. Remove the camshaft pulley.
9. Remove the thrust cover (A).
10. Carefully remove the camshaft (B).
11. Installation is the reverse of removal. Always use new O-rings (C). Apply clean engine oil to the journals and lobes.

REAR

1. Relieve the fuel system pressure.
2. Remove the air cleaner assembly.
3. Remove the intake manifold.
4. Disconnect the fuel feed line.
5. Disconnect the brake lines at the master cylinder. Plug the openings.
6. Remove the timing belt.
7. Remove the rocker arm assembly.
8. Remove the camshaft pulley.
9. Remove the thrust cover (A) and camshaft (B).
10. Installation is the reverse of removal. Always use new O-rings (C). Apply clean engine oil to the journals and lobes. Remove all traces of the old liquid gasket and apply new liquid gasket. Seat the rocker shaft assembly within 4 minutes of applying the liquid gasket. Bleed the brakes. Perform the CKP pattern clear/learn procedure.

Starter Motor

REMOVAL & INSTALLATION

J35A4 engine

1. Before servicing the vehicle, refer to the precautions in the beginning of this section.
2. Remove or disconnect the following:
   - Negative battery cable
   - Transmission fluid cooler line clamp
   - Starter wiring harness connectors
   - Starter motor

To install:
3. Install or connect the following:
   - Starter motor. Tighten the upper bolt to 33 ft. lbs. (44 Nm) and the lower bolt to 47 ft. lbs. (64 Nm).
   - Starter wiring harness connectors. Tighten the battery cable nut to 79 inch lbs. (9 Nm).
   - Transmission fluid cooler line clamp
   - Negative battery cable

J35A6 and J35A7 Engines

1. Before servicing the vehicle, refer to the precautions in the beginning of this section.

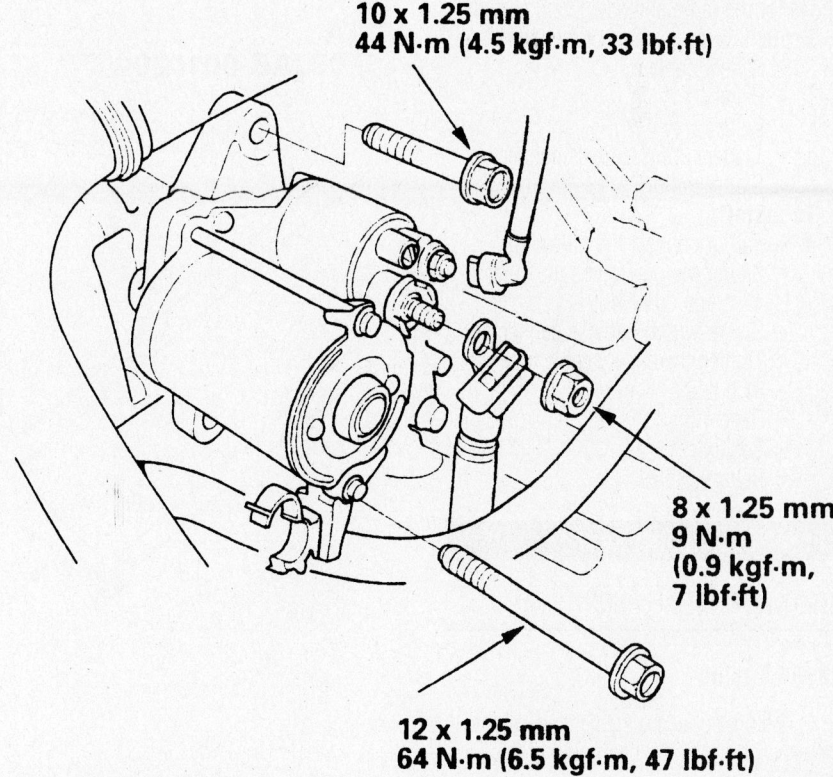

Starter mounting—J35A4 engine

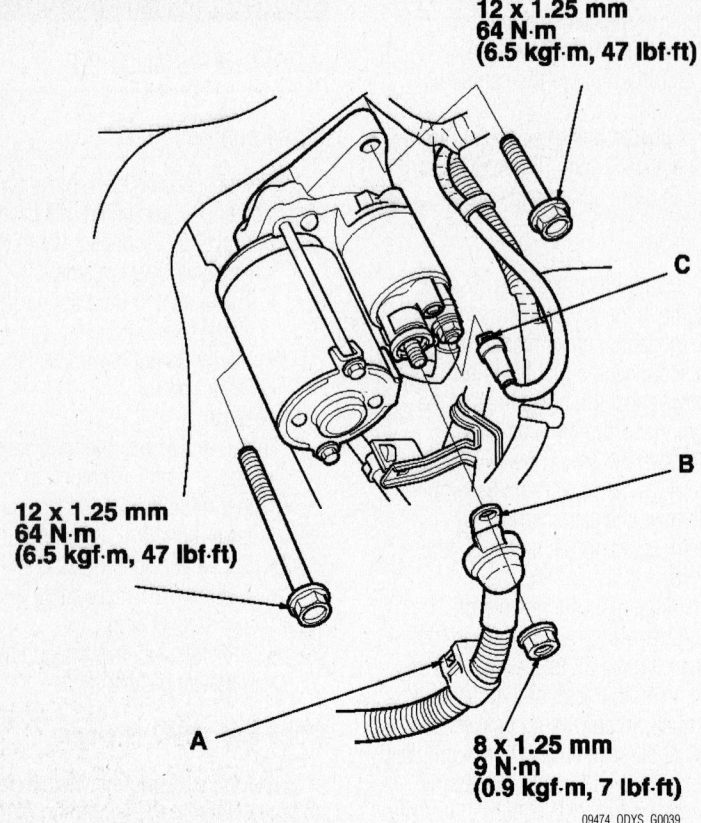

**12 x 1.25 mm
64 N·m
(6.5 kgf·m, 47 lbf·ft)**

C

B

**12 x 1.25 mm
64 N·m
(6.5 kgf·m, 47 lbf·ft)**

A

**8 x 1.25 mm
9 N·m
(0.9 kgf·m, 7 lbf·ft)**

09474_ODYS_G0039

Starter mounting—J35A6 and J35A7 engines

2. Make sure you have the anti-theft codes for the radio and navigation system.
3. Remove or disconnect the following:
- Battery cables
- Battery
- Battery tray
- Starter wiring harness connectors
- Starter motor

To install:
4. Install or connect the following:
- Starter motor. Tighten the bolts to 47 ft. lbs. (64 Nm).
- Starter wiring harness connectors. Tighten the battery cable nut to 84 inch lbs. (9 Nm).
- Battery tray
- Battery
- Battery cables

Front Crankshaft Seal

REMOVAL & INSTALLATION

J35A4 Engine

1. Before servicing the vehicle, refer to the precautions in the beginning of this section.
2. Remove or disconnect the following:
- Negative battery cable

- Accessory drive belts
- Side engine mount
- Valve cover
- Crankshaft pulley using the tools in the accompanying illustration
- Front cover
- Balance shaft belt, if equipped
- Timing belt. Refer to the timing belt procedure.
- Top Dead Center (TDC) sensor, if equipped
- Crankshaft timing sprocket
- Front crankshaft seal

To install:
3. Lubricate the crankshaft seal lip with grease prior to installation.
4. Install the front crankshaft seal so that it is flush with the surface of the oil pump housing.
5. Install or connect the following:
- Crankshaft timing sprocket
- Top Dead Center (TDC) sensor, if equipped
- Timing belt. Refer to the timing belt procedure.
- Balance shaft belt, if equipped
- Front cover
- Crankshaft pulley. Tighten the bolt to 181 ft. lbs. (245 Nm) using the tools in the accompanying illustration.

A
07JAB-001020A

B
07MAB-PY3010A

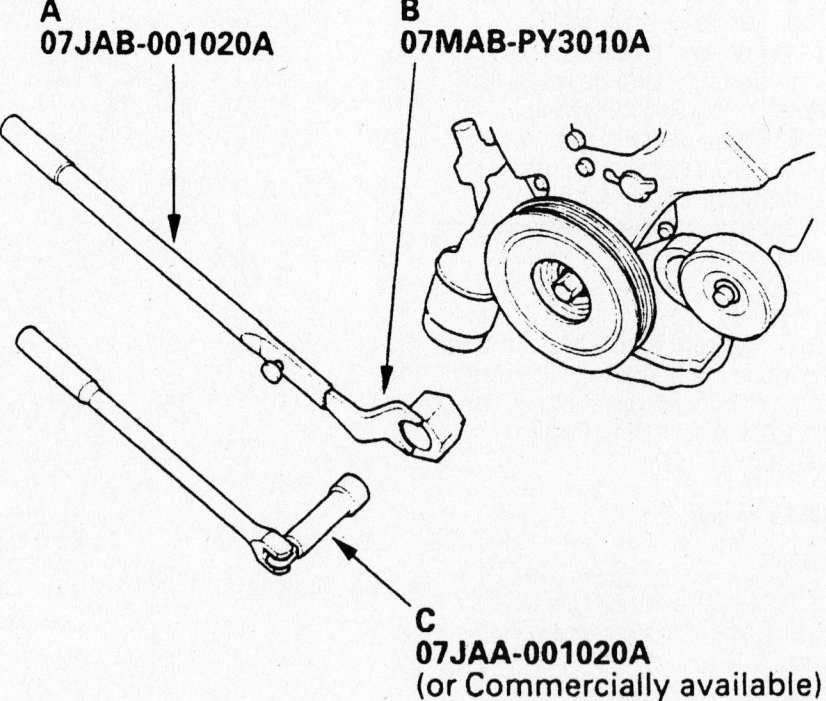

C
**07JAA-001020A
(or Commercially available)**

9358MG13

Remove the crankshaft pulley using tools (A) Holder handle 07JAB-001020A, (B) attachment 07MAB-PY3010A and (C) a breaker bar and 19mm socket—J35A4 engine

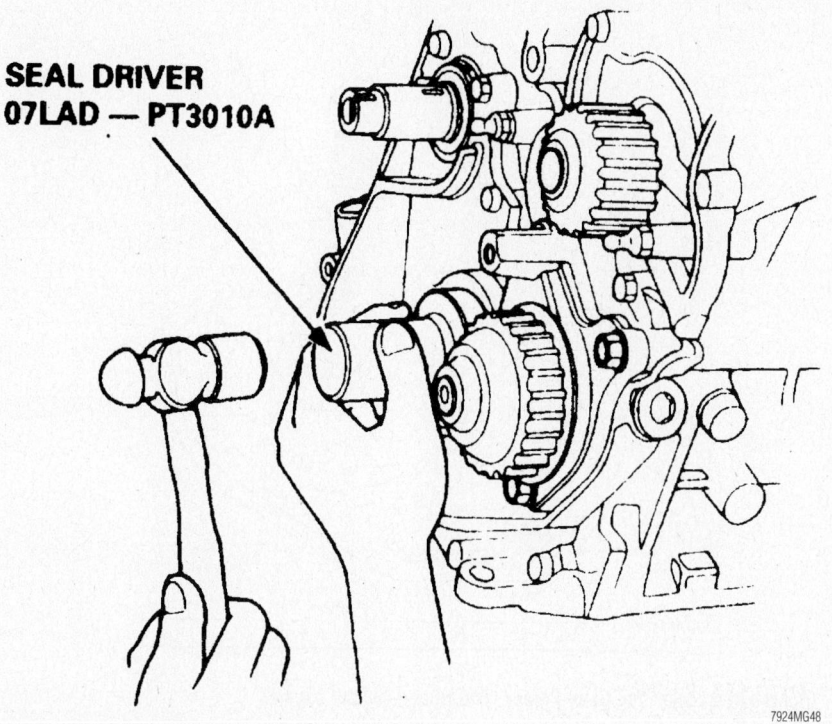

**SEAL DRIVER
07LAD — PT3010A**

7924MG48

Typical front crankshaft seal installation—J35A4 engine

- Valve cover
- Side engine mount
- Accessory drive belts
- Negative battery cable

6. Check the engine oil level and add if necessary.

7. Start the engine and check for leaks.

J35A6 and J35A7 Engines

1. Remove the CKP sensor.
2. Remove the timing belt.
3. Remove the timing belt drive sprocket.
4. Remove the oil seal.

To install:

5. Dry and clean the oil seal housing.
6. Apply a light coating of multi-purpose grease to the crankshaft and oil seal lip.
7. Using a seal driver, install the seal until it bottoms.
8. The remainder of installation is the reverse of removal.

Crankshaft Damper, Front Cover, and Timing Belt

REMOVAL & INSTALLATION

J35A4 Engine

➡The radio may contain a coded theft protection circuit. Always make note the code number before disconnecting the battery.

1. Disconnect the negative battery terminal.

2. Turn the crankshaft so the white mark on the crankshaft pulley aligns with the pointer on the oil pump housing cover.

3. Open the inspection plugs on the upper timing belt covers and check that the camshaft sprocket marks align with the upper cover marks.

✳✳ WARNING

Align the camshaft and crankshaft sprockets with their alignment marks before removing the timing belt. Failure to align the timing marks correctly may result in valve damage.

4. Raise and safely support the vehicle and remove both front tires/wheels.

5. Remove the front lower splash shield.

6. Move the alternator tensioner with a Belt Tensioner Release Arm tool YA9317, or equivalent, to release tension from the belt and remove the alternator drive belt.

7. Remove the alternator belt tensioner release arm.

FRONT CAMSHAFT PULLEY:

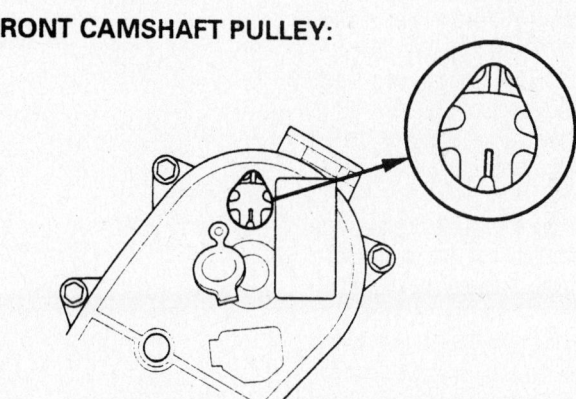

REAR CAMSHAFT PULLEY:

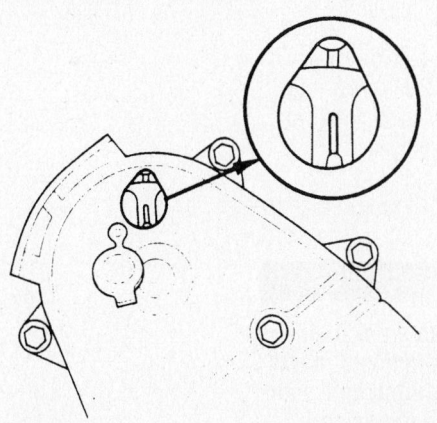

93025G04

Crankshaft and camshaft timing marks at Top Dead Center (TDC)—J35A4 engine

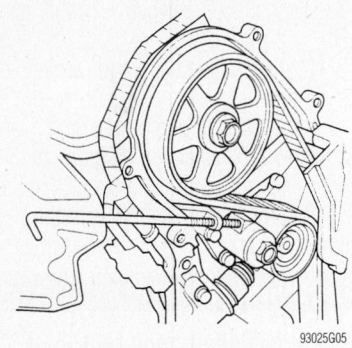

Battery hold-down bolt installed to hold auto-tensioner—J35A4 engine

8. Loosen the power steering pump adjustment nut, adjustment locknut and mounting bolt, then remove the power steering pump with the hoses attached.

9. Support the weight of the engine by placing a wood block on a floor jack and carefully lift on the oil pan.

10. Remove the bolts from the side engine mount bracket and remove the bracket.

11. Remove the dipstick, the dipstick tube and discard the O-ring.

12. Hold the crankshaft pulley with the Handle tool 07JAB-001020A and Crankshaft Holding tool 07MAB-PY3010A, or equivalent. While holding the crankshaft pulley, remove the crankshaft pulley bolt using a heavy duty ¾ in. (19mm) socket and breaker bar.

13. Remove the crankshaft pulley, the upper timing belt covers and the lower timing belt cover.

14. Remove one of the battery clamp fasteners from the battery tray and grind a 45 degree bevel on the threaded end of the battery clamp bolt.

15. Screw in the battery hold-down bolt into the threaded bracket just above the auto-tensioner (automatic timing belt adjuster) and tighten the bolt hand-tight to hold the auto-tensioner adjuster in its current position.

16. Remove the engine mount bracket bolts and the bracket.

17. Loosen the timing belt idler pulley bolt (located on the right side across from the auto-tensioner pulley) about 5–6 revolutions and remove the timing belt.

To install:

18. Clean the timing belt sprockets and the timing belt covers.

❋❋ WARNING

Align the camshaft and crankshaft sprockets with their alignment marks before installing the timing belt. Failure to align the timing marks correctly may result in valve damage.

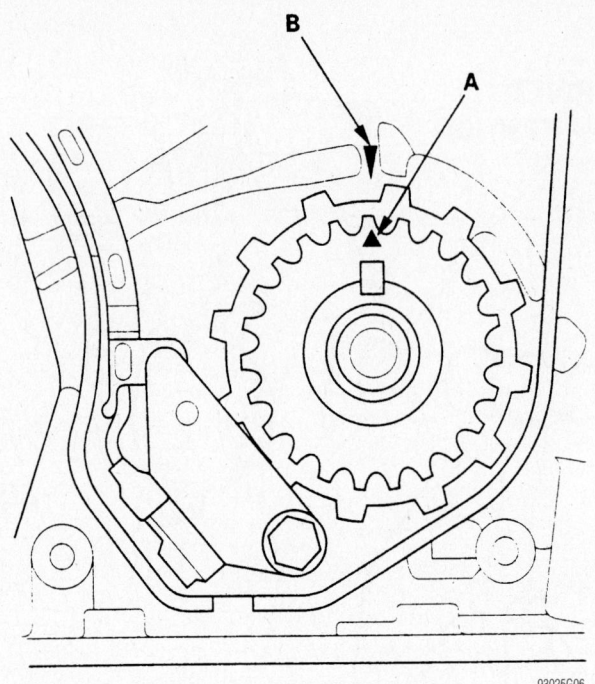

Crankshaft sprocket Top Dead Center (TDC) mark—J35A4 engine

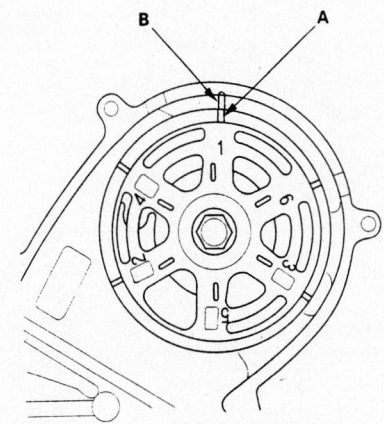

FRONT:

REAR:

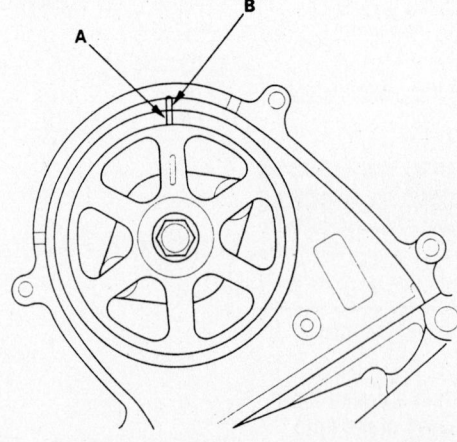

Camshaft sprocket Top Dead Center (TDC) mark—J35A4 engine

19. Align the timing mark on the crankshaft sprocket with the oil pump pointer.

20. Align the camshaft sprocket TDC timing marks with the pointers on the rear cover.

21. If installing a new belt or if the auto-tensioner has extended or if the timing belt cannot be reinstalled easily, the auto-tensioner must be collapsed before installation of the timing belt, perform the following procedures:

    a. Remove the battery hold-down bolt from the auto-tensioner bracket.

    b. Remove the timing belt auto-tensioner bolts and the auto-tensioner.

    c. Secure the auto-tensioner in a soft jawed vise, clamping onto the flat surface of one of the mounting bolt holes with the maintenance bolt facing upward.

    d. Remove the maintenance bolt and use caution not to spill oil from the tensioner assembly.

    e. Should oil spill from the tensioner, be sure the tensioner is filled with 0.22 ounces (6.5 ml) of fresh engine oil.

    f. Using care not to damage the threads or the gasket sealing surface, insert a flat-blade screwdriver through the tensioner maintenance hole and turn the screwdriver clockwise to compress the auto-tensioner bottom while the Tensioner Holder tool 14540-P8A-A01, or equivalent, is installed on the auto-tensioner assembly.

    g. Install the auto-tensioner maintenance bolt with a new gasket and tighten to a torque 72 inch lbs. (8 Nm).

    h. Install the auto-tensioner on the engine with the tensioner holder tool installed and torque the mounting bolts to 104 inch lbs. (12 Nm).

22. Install the timing belt in a counter-clockwise pattern starting with the crankshaft drive sprocket. Install the timing belt counterclockwise in the following sequence:

- Crankshaft drive sprocket.
- Idler pulley.
- Left side camshaft sprocket.
- Water pump.
- Right side camshaft sprocket.
- Auto-tensioner adjustment pulley.

23. Torque the timing belt idler pulley bolt to 33 ft. lbs. (44 Nm).

24. Remove the auto-tensioner holding tool to allow the tensioner to extend.

25. Install the engine mount bracket to the engine and torque the bolts to 33 ft. lbs. (44 Nm).

26. Install the lower timing belt cover and both upper timing belt covers.

27. Hold the crankshaft pulley with special tools 07JAB-001020A handle and

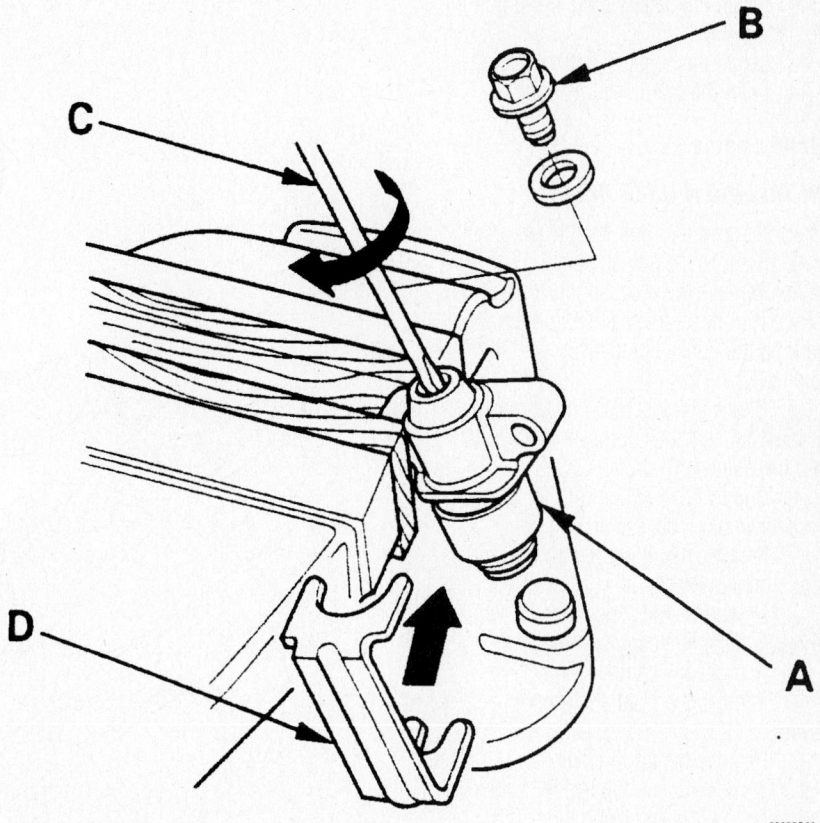

Adjusting the auto-tensioner—J35A4 engine

93025G08

07MAB-PY3010A crankshaft holding tool, or equivalent tools. While holding the crankshaft pulley, install the crankshaft pulley bolt using a heavy duty ¾ in. (19mm) socket and a torque wrench and torque the bolt to 181 ft. lbs. (245 Nm).

✳✳ WARNING

If any binding is felt while moving the crankshaft pulley, STOP turning the crankshaft pulley immediately because the pistons may be hitting the valves.

28. Rotate the crankshaft pulley clockwise 5–6 revolutions to allow the timing belt to be seated in the pulleys.

29. Move the crankshaft pulley to the white TDC mark and inspect the camshaft TDC marks to ensure proper timing of the camshafts.

✳✳ WARNING

If the timing marks do not align, the timing belt removal and installation procedure must be performed again.

30. Install the engine dipstick tube using a new O-ring.

31. Install the power steering pump, and loosely install the mounting bolt, adjustment locknut and adjustment nut.

32. Adjust the power steering belt to a tension such that a 22 lb. (98 N) pull halfway between the 2 drive pulleys will allow the belt to move 0.51–0.65 in. (13.0–16.5mm).

33. Tighten the power steering pump mounting bolt and adjustment locknut.

➡ **If a new belt is used, set the deflection to 0.33–0.43 in. (8.5–11.0mm) and after engine has run for 5 minutes, readjust the new belt to the used belt specification.**

34. Install the alternator belt tensioner arm.

35. Move the alternator tensioner with a Belt Tensioner Release Arm tool YA9317, or equivalent, to release tension from the belt and install the alternator drive belt.

36. Install both engine mount bracket bolts and torque to 33 ft. lbs. (44 Nm).

37. Install the bushing through bolt and tighten to 40 ft. lbs. (54 Nm).

38. Release and carefully remove the floor jack.

39. Install the front lower splash shield.

40. Install both front tires/wheels.

41. Carefully lower the vehicle.

42. Install the battery hold-down bolt in the battery tray.

43. Install the negative battery cable.

44. Enter the radio security code.

J35A6 Engine

INSTALLING A USED BELT

1. Remove the right front wheel.

2. Remove the splash shield.

3. Remove the accessory drive belt.

4. Turn the crankshaft so that the white mark on the pulley (A) lines up with the pointer (B).

5. Check that the TDC timing mark (A) on the front camshaft sprocket is aligned with the pointer (B) on the cover.

6. Support the engine with a jack and a block of wood under the oil pan.

7. Remove the upper half of the side engine mount bracket.

8. Using the tools shown and a 19mm socket, remove the damper bolt.

9. Remove the damper.

10. Remove the front and rear upper belt covers.

11. Remove the lower cover.

12. Remove one of the battery hold-down clamp bolts and grind the end as shown.

13. Screw the battery clamp bolt in as shown to hold the timing belt adjuster in its current position. Tighten it by hand. Do not use a wrench.

14. Remove the timing belt guide plate.

15. Remove the lower side engine mount bracket.

16. Remove the idler pulley bolt and idler pulley. Discard the bolt.

17. Remove the timing belt.

To install:

18. Clean all parts.

19. Set the crankshaft sprocket to TDC by aligning the TDC mark (A) with the pointer (B) on the oil pump.

20. Set the camshaft pulleys to TDC by

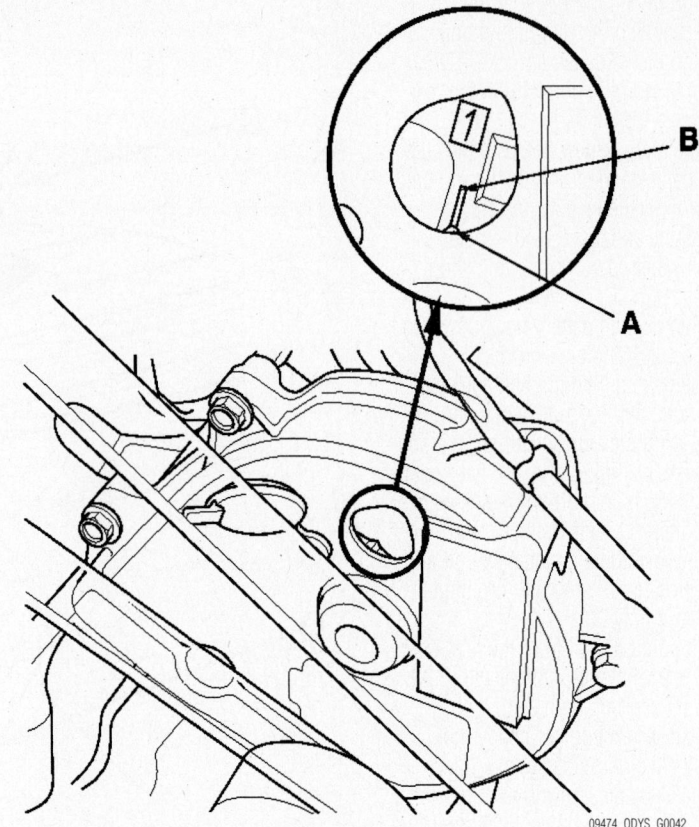

Front camshaft timing marks aligned—J35A6 engine

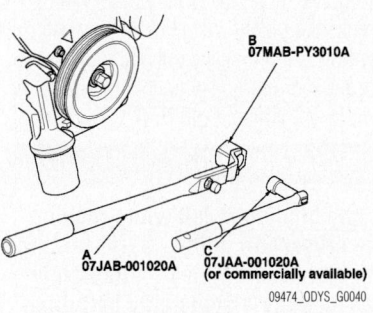

09474_ODYS_G0040

Damper bolt removal tools

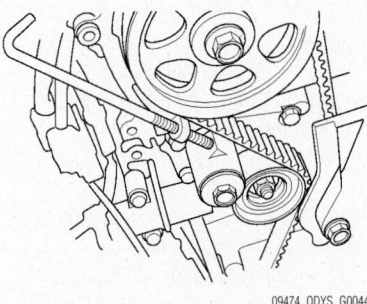

09474_ODYS_G0044

Battery clamp bolt screwed into position—J35A6 engine

aligning the TDC marks (A) on the camshaft pulleys with the pointers (B) on the covers. See the illustrations in INSTALLING A NEW BELT.

21. Loosely install the idler pulley with a new bolt so that the pulley can move but won't come off.

22. If the auto-tensioner has extended, and the timing belt can't be installed, see the procedure for INSTALLING A NEW BELT.

23. Install the timing belt in a counter-clockwise sequence, starting with the crank-shaft drive pulley.

24. Tighten the idler pulley bolt to 33 ft. lbs. (44 Nm).

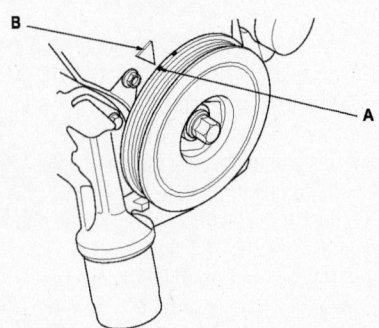

09474_ODYS_G0041

Crankshaft timing marks aligned—J35A6 engine

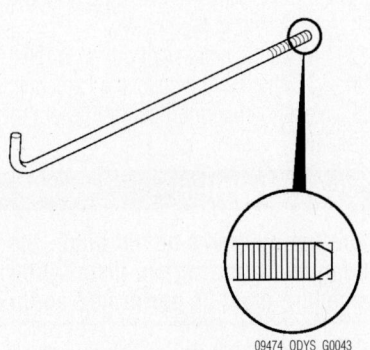

09474_ODYS_G0043

Remove one of the battery hold-down clamp bolts and grind the end as shown

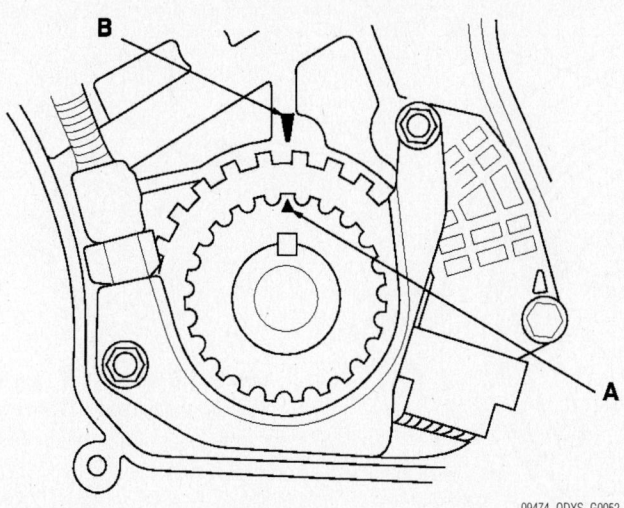

Set the crankshaft sprocket to TDC by aligning the TDC mark (A) with the pointer (B) on the oil pump—J35A6 engine

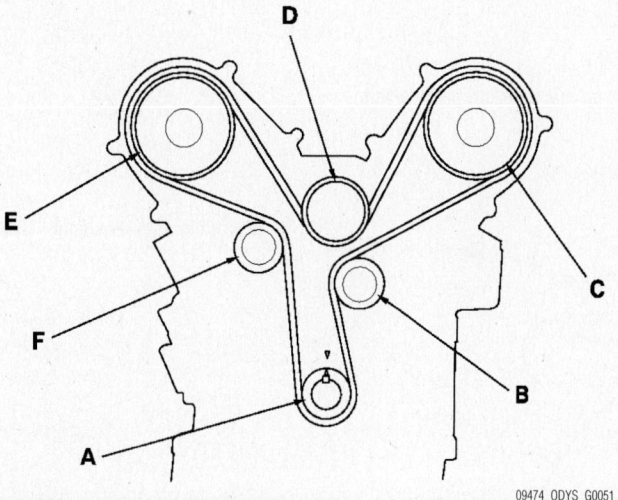

Timing belt installation sequence—J35A6 engine

25. The remainder of installation is the reverse of removal. Note the following torques:

- Lower half engine mount bracket: 33 ft. lbs. (44 Nm)
- Lower cover: 105 inch lbs. (12 Nm)
- Upper covers: 105 inch lbs. (12 Nm)
- Crankshaft damper: Tighten the bolt to 47 ft. lbs. (64 Nm). Mark the bolt head and pulley, then tighten the bolt an additional 60°.

INSTALLING A NEW BELT

1. Remove the timing belt.
2. Clean the timing belt pulleys, timing belt guide plate, and the upper and lower covers.
3. Set the timing belt drive pulley to top dead center (TDC) by aligning the TDC mark (A) on the tooth of the timing belt drive pulley with the pointer (B) on the oil pump.

4. Set the camshaft pulleys to TDC by aligning the TDC marks (A) on the camshaft pulleys with the pointers (B) on the back covers.
5. Remove the battery clamp bolt from the back cover.
6. Remove the auto-tensioner.
7. Align the holes on the rod and housing of the auto-tensioner.
8. Use a hydraulic press to slowly compress the auto-tensioner. Insert a 2.0 mm (0.08 in.) pin through the housing and the rod.

➡**The compression pressure should not exceed 9,800 N (2,200 lbs.).**

9. Align the holes on the rod and housing of the auto-tensioner.
10. Use a hydraulic press to slowly compress the auto-tensioner. Insert a 2.0 mm (0.08 in.) pin through the housing and the rod.

➡**The compression pressure should not exceed 9,800 N (2,200 lbs.).**

11. Install the auto-tensioner.

➡**Make sure the pin stays in place.**

12. Screw the battery clamp bolt in as shown to hold the timing belt adjuster. Tighten it by hand; do not use a wrench.
13. Loosely install the idler pulley with a new idler pulley bolt so the pulley can move but does not come off.
14. Install the timing belt in a counter-clockwise sequence starting with the drive pulley.

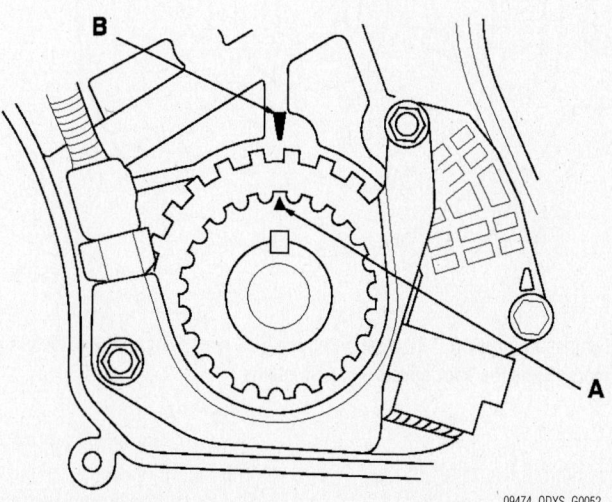

Set the timing belt drive pulley to top dead center (TDC) by aligning the TDC mark (A) on the tooth of the timing belt drive pulley with the pointer (B) on the oil pump—J35A6 engine

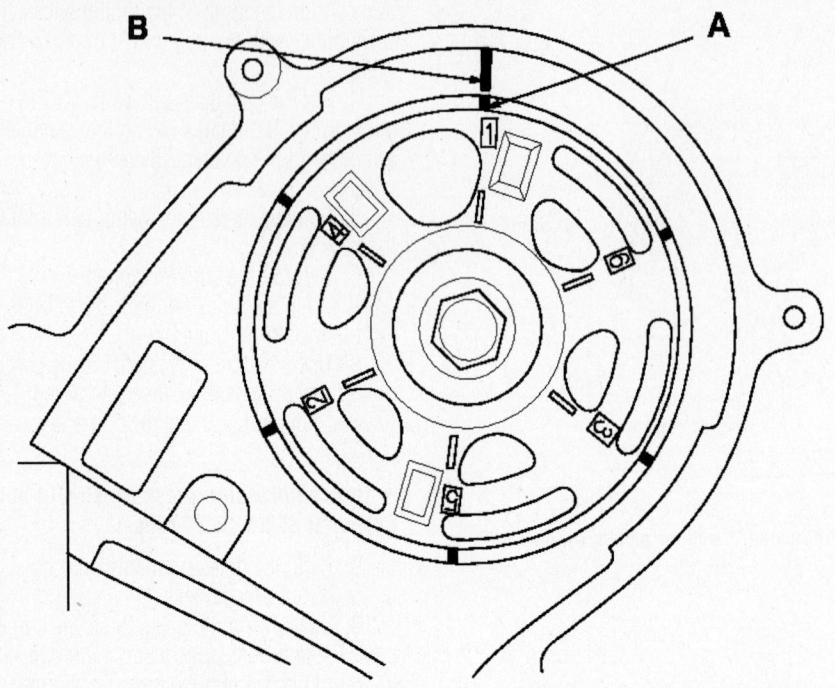

Set the left camshaft pulley to TDC by aligning the TDC marks (A) on the camshaft pulleys with the pointers (B) on the back covers—J35A6 engine

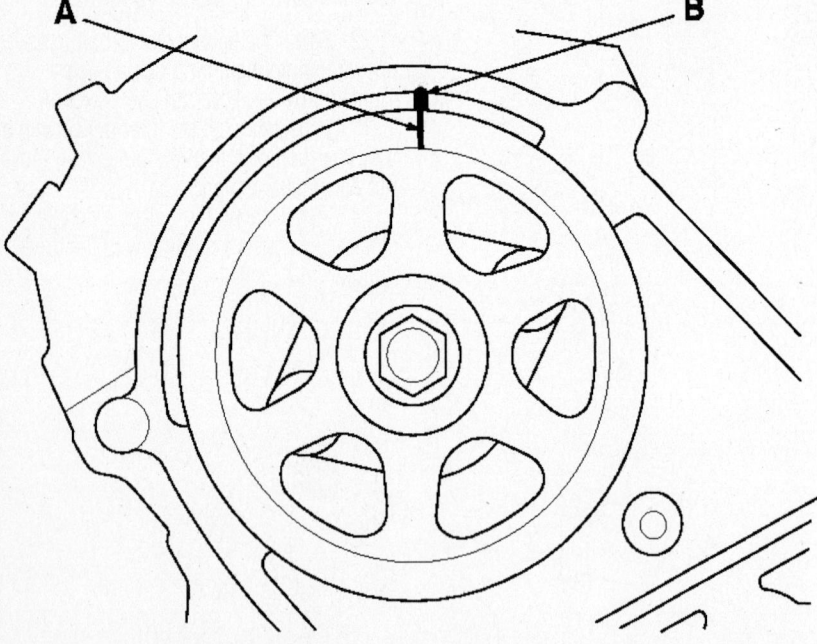

Set the right camshaft pulley to TDC by aligning the TDC marks (A) on the camshaft pulleys with the pointers (B) on the back covers—J35A6 engine

15. Tighten the idler pulley bolt.
16. Remove the pin from the auto-tensioner.
17. Remove the battery clamp bolt from the back cover.

18. Install the lower half of the side engine mount bracket.
19. Install the timing belt guide plate as shown.
20. Install the lower cover.

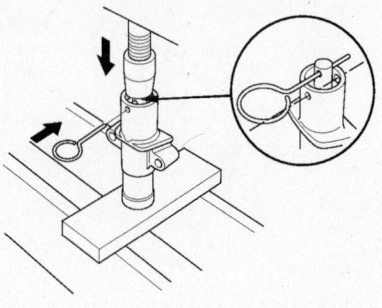

Insert a 2.0 mm (0.08 in.) pin through the housing and the rod—J35A6 engine

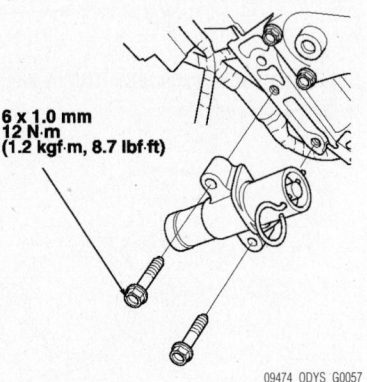

6 x 1.0 mm
12 N·m
(1.2 kgf·m, 8.7 lbf·ft)

Install the auto-tensioner—J35A6 engine

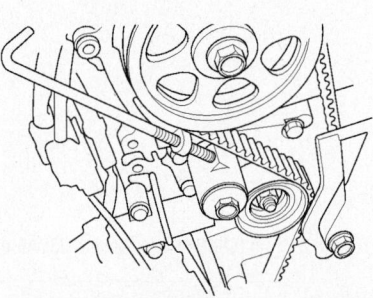

Screw the battery clamp bolt in as shown to hold the timing belt adjuster. Tighten it by hand—J35A6 engine

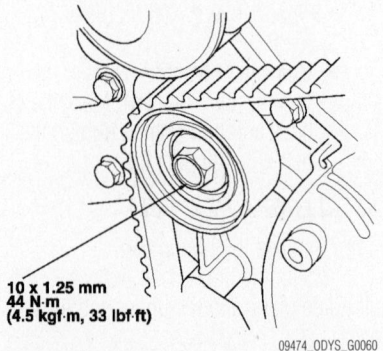

10 x 1.25 mm
44 N·m
(4.5 kgf·m, 33 lbf·ft)

Tighten the idler pulley bolt—J35A6 engine

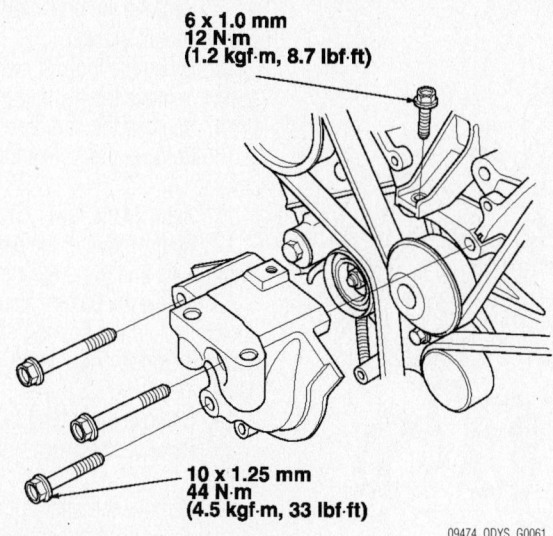

6 x 1.0 mm
12 N·m
(1.2 kgf·m, 8.7 lbf·ft)

10 x 1.25 mm
44 N·m
(4.5 kgf·m, 33 lbf·ft)

09474_ODYS_G0061

Install the lower half of the side engine mount bracket—J35A6 engine

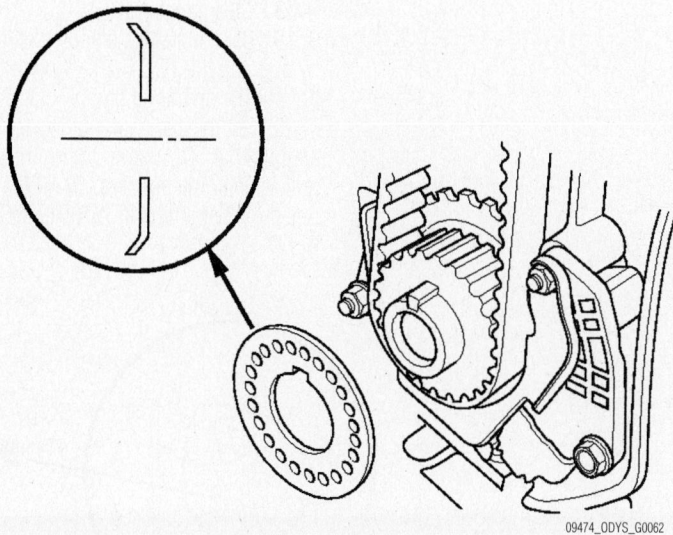

09474_ODYS_G0062

Install the timing belt guide plate as shown—J35A6 engine

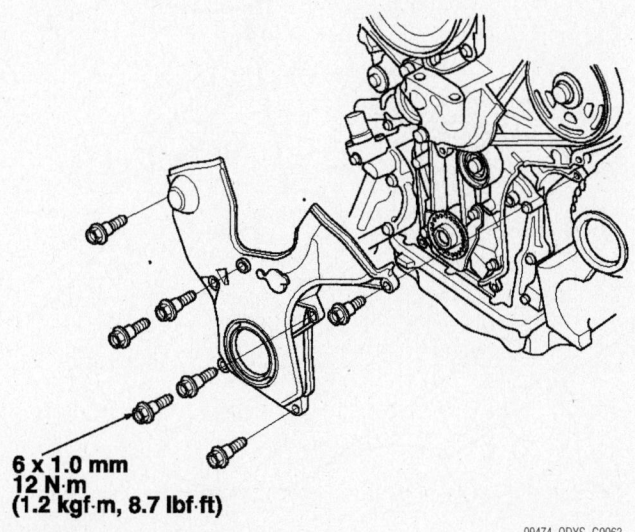

6 x 1.0 mm
12 N·m
(1.2 kgf·m, 8.7 lbf·ft)

09474_ODYS_G0063

Install the lower cover—J35A6 engine

21. Install the front upper cover (A) and rear upper cover (B).

22. Install the crankshaft pulley.

23. Rotate the crankshaft pulley six turns clockwise so the timing belt positions itself on the pulleys.

24. Turn the crankshaft pulley so its white mark (A) lines up with the pointer (B).

25. Check the camshaft pulley marks.

➡**If the marks are not aligned, rotate the crankshaft 360 degrees, and recheck the camshaft pulley mark.**

   a. If the camshaft pulley marks are at TDC, go to step 24.

   b. If the camshaft pulley marks are not at TDC, remove the timing belt and repeat step 3 through 22.

26. Install the upper half of the side engine mount bracket, and tighten the new mounting bolts (A), then the mass damper mounting bolt (B).

27. Install the drive belt.

28. Install the splash shield.

29. Install the right front wheel.

30. Do the crankshaft position (CKP) pattern clear/CKP pattern learn procedure.

➡**The ECT needs to be at 176°F (80°C) or higher.**

   a. Clear the CKP pattern while the engine is stopped.

   b. Turn the ignition switch OFF.

   c. Turn the ignition switch ON (II), and wait 30 seconds.

   d. Test-drive the vehicle on a level road: decelerate (with the throttle fully closed) from an engine speed of 2,500 rpm to 1,000 rpm with the A/T in 2 position.

   e. Stop the vehicle, but keep the engine running.

   f. Check PULSER F/B LEARN in the DATA LIST with the HDS. If it is NOT COMPLETED, go to step 4. If it is COMPLETED, go to step 7.

   g. Test-drive the vehicle on a level road: decelerate (with the throttle fully closed) from an engine speed of 5,000 rpm to 3,000 rpm with the A/T in 2 position.

   h. Stop the vehicle, but keep the engine running.

   i. Check the PULSER F/B LEARN (HIGH RPM) in the DATA LIST with the HDS. If it is NOT COMPLETED, go to step 7. If it is COMPLETED, go to step 10.

   j. Turn the ignition switch OFF.

   k. Turn the ignition switch ON (II), and wait 30 seconds. The CKP learning procedure is complete.

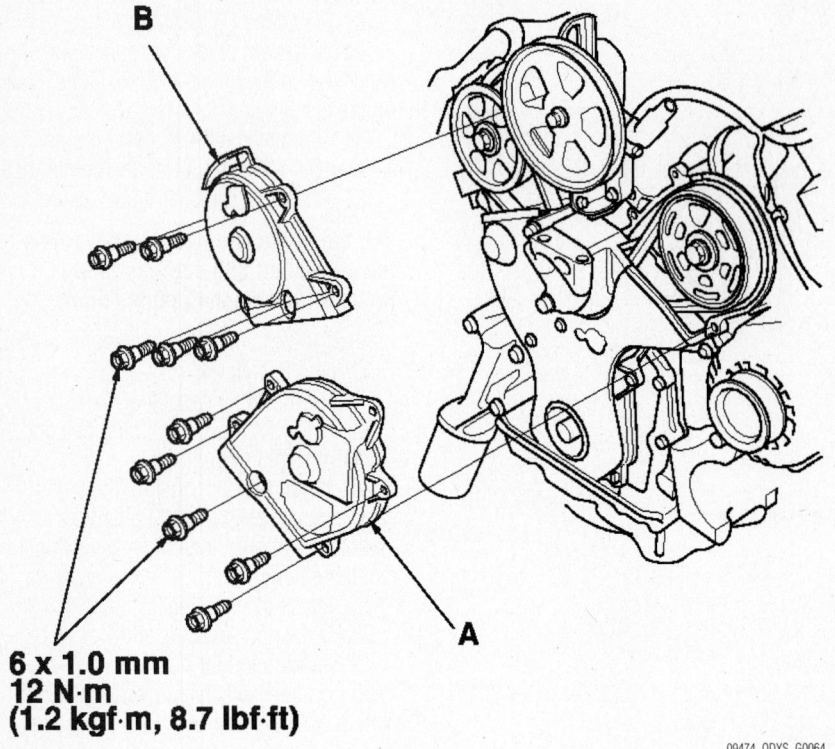

6 x 1.0 mm
12 N·m
(1.2 kgf·m, 8.7 lbf·ft)

09474_ODYS_G0064

Install the front upper cover (A) and rear upper cover (B)—J35A6 engine

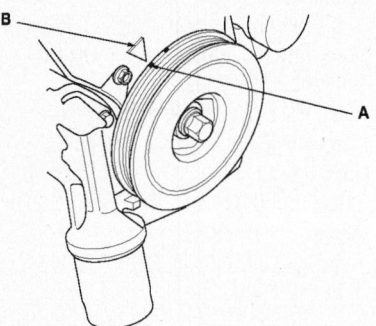

09474_ODYS_G0041

Turn the crankshaft pulley so its white mark (A) lines up with the pointer (B) — J35A6 engine

J35A7 Engine

INSTALLING A USED BELT

1. Remove the right front wheel.
2. Remove the splash shield.
3. Remove the accessory drive belt.
4. Turn the crankshaft so that the white mark on the pulley (A) lines up with the pointer (B).
5. Check that the TDC timing mark (A) on the front camshaft sprocket is aligned with the pointer (B) on the cover.
6. Support the engine with a jack and a block of wood under the oil pan.

7. Remove the upper half of the side engine mount bracket.
8. Using the tools shown and a 19mm socket, remove the damper bolt.
9. Remove the damper.
10. Remove the front and rear upper belt covers.
11. Remove the lower cover.
12. Remove one of the battery hold-down clamp bolts and grind the end as shown.
13. Screw the battery clamp bolt in as shown to hold the timing belt adjuster in its current position. Tighten it by hand. Do not use a wrench.
14. Remove the timing belt guide plate.
15. Remove the lower side engine mount bracket.
16. Remove the idler pulley bolt and idler pulley. Discard the bolt.
17. Remove the timing belt.

To install:
18. Clean all parts.
19. Set the crankshaft sprocket to TDC by aligning the TDC mark (A) with the pointer (B) on the oil pump.
20. Set the camshaft pulleys to TDC by aligning the TDC marks (A) on the camshaft pulleys with the pointers (B) on the covers. See the illustrations in INSTALLING A NEW BELT.

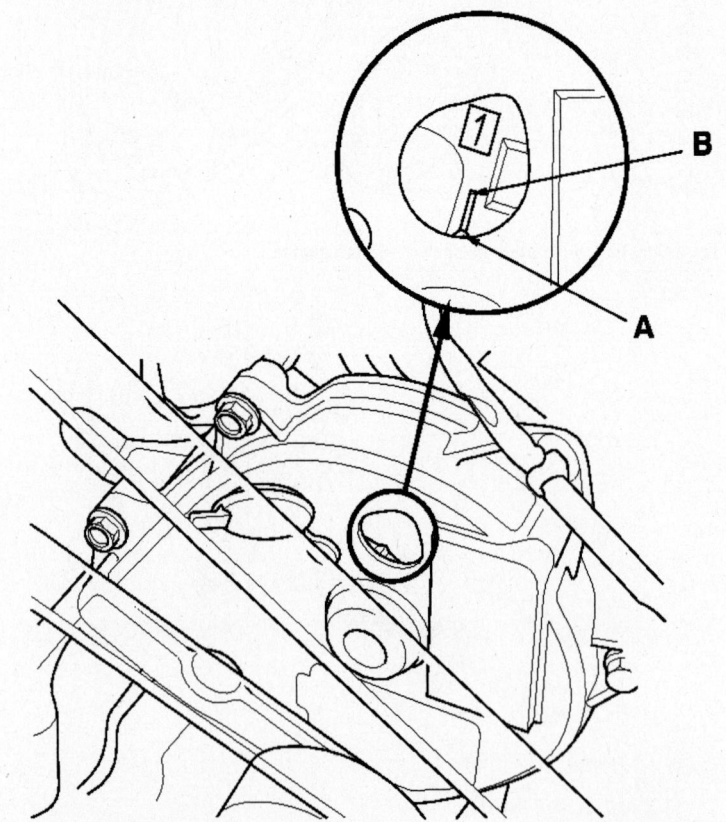

09474_ODYS_G0042

Check the camshaft pulley marks—J35A6 engine

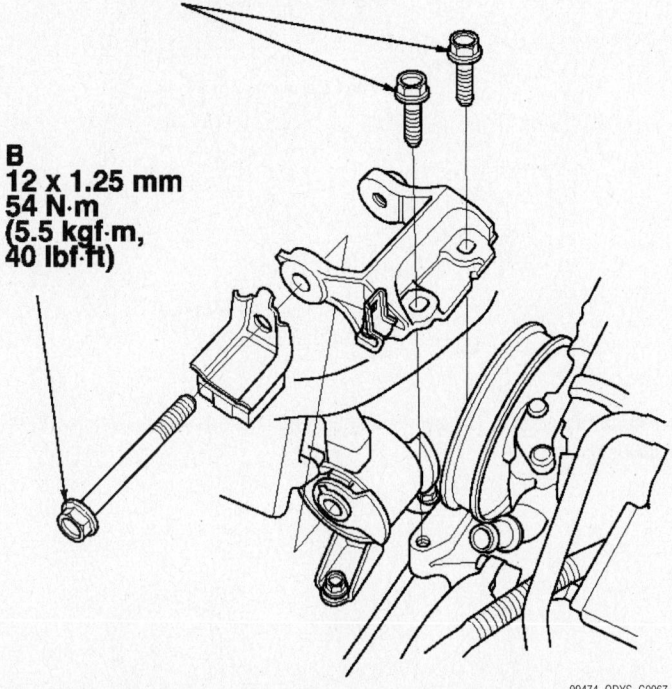

A
10 x 1.25 mm
44 N·m (4.5 kgf·m, 33 lbf·ft)

B
12 x 1.25 mm
54 N·m
(5.5 kgf·m,
40 lbf·ft)

09474_ODYS_G0067

Install the upper half of the side engine mount bracket, and tighten the new mounting bolts (A), then the mass damper mounting bolt (B) —J35A6 engine

21. Loosely install the idler pulley with a new bolt so that the pulley can move but won't come off.

22. If the auto-tensioner has extended, and the timing belt can't be installed, see the procedure for INSTALLING A NEW BELT.

23. Install the timing belt in a counterclockwise sequence, starting with the crankshaft drive pulley.

24. Tighten the idler pulley bolt to 33 ft. lbs. (44 Nm).

25. The remainder of installation is the reverse of removal. Note the following torques:

- Lower half engine mount bracket: 33 ft. lbs. (44 Nm)
- Lower cover: 105 inch lbs. (12 Nm)
- Upper covers: 105 inch lbs. (12 Nm)
- Crankshaft damper: Tighten the bolt to 47 ft. lbs. (64 Nm). Mark the bolt head and pulley, then tighten the bolt an additional 60°.

26. Do the crankshaft position (CKP) pattern clear/CKP pattern learn procedure.

➡ **The ECT needs to be at 176ºF (80ºC) or higher.**

    a. Clear the CKP pattern while the engine is stopped.

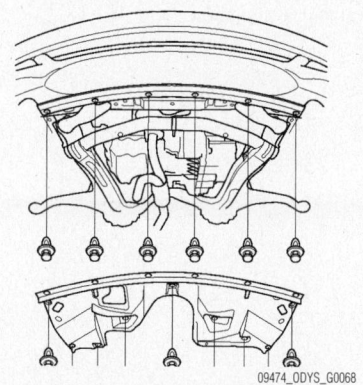

09474_ODYS_G0068

Install the splash shield—J35A6 engine

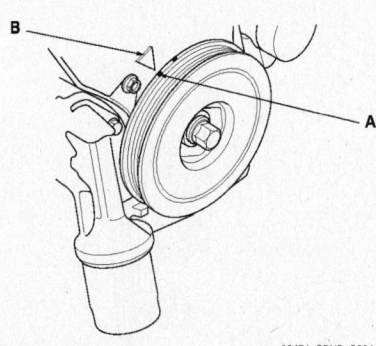

09474_ODYS_G0041

Crankshaft timing marks aligned—J35A7 engine

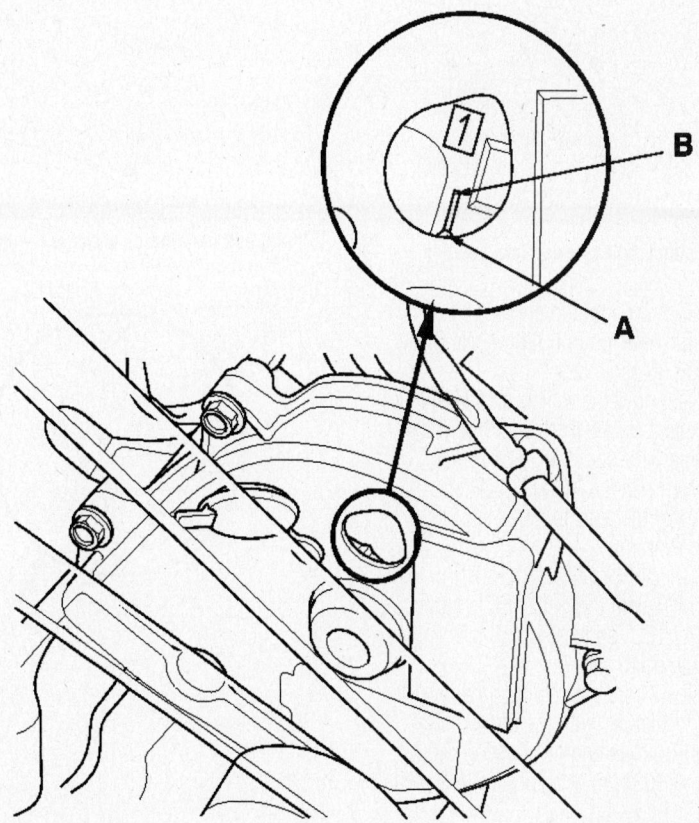

09474_ODYS_G0042

Front camshaft timing marks aligned—J35A7 engine

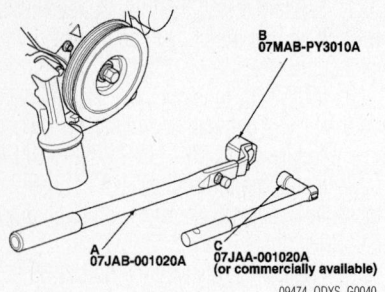

Damper bolt removal tools

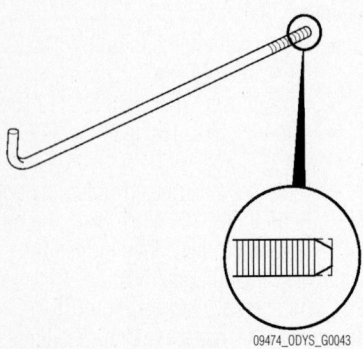

Remove one of the battery hold-down clamp bolts and grind the end as shown

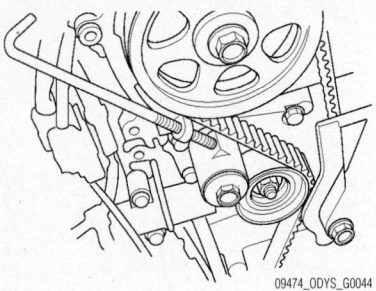

Battery clamp bolt screwed into position

b. Turn the ignition switch OFF.

c. Turn the ignition switch ON (II), and wait 30 seconds.

d. Test-drive the vehicle on a level road: decelerate (with the throttle fully closed) from an engine speed of 2,500 rpm to 1,000 rpm with the A/T in 2 position.

e. Stop the vehicle, but keep the engine running.

f. Check PULSER F/B LEARN in the DATA LIST with the HDS. If it is NOT COMPLETED, go to step 4. If it is COMPLETED, go to step 7.

g. Test-drive the vehicle on a level road: decelerate (with the throttle fully closed) from an engine speed of 5,000 rpm to 3,000 rpm with the A/T in 2 position.

h. Stop the vehicle, but keep the engine running.

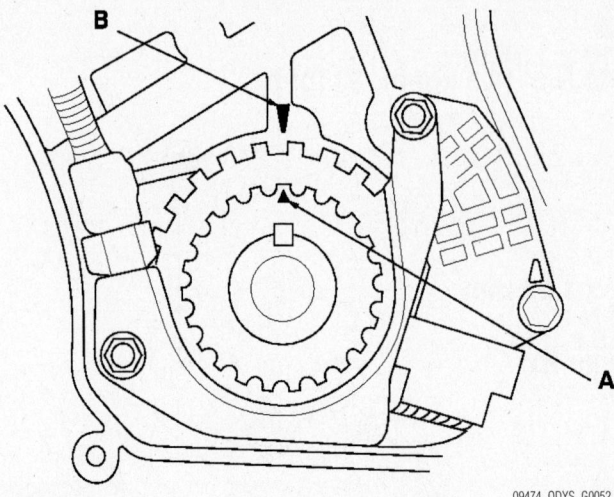

Set the crankshaft sprocket to TDC by aligning the TDC mark (A) with the pointer (B) on the oil pump—J35A7 engine

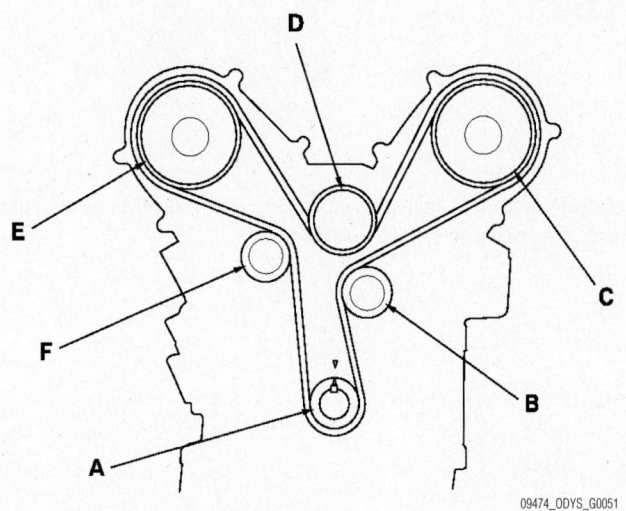

Timing belt installation sequence—J35A7 engine

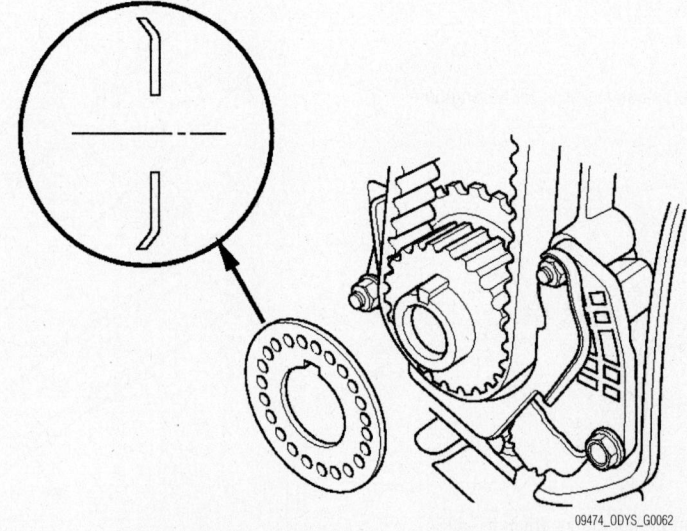

Install the timing belt guide plate as shown—J35A7 engine

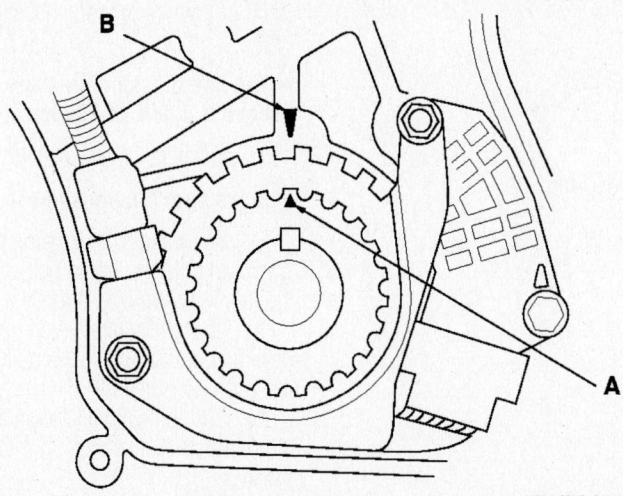

09474_ODYS_G0052

Set the timing belt drive pulley to top dead center (TDC) by aligning the TDC mark (A) on the tooth of the timing belt drive pulley with the pointer (B) on the oil pump—J35A7 engine

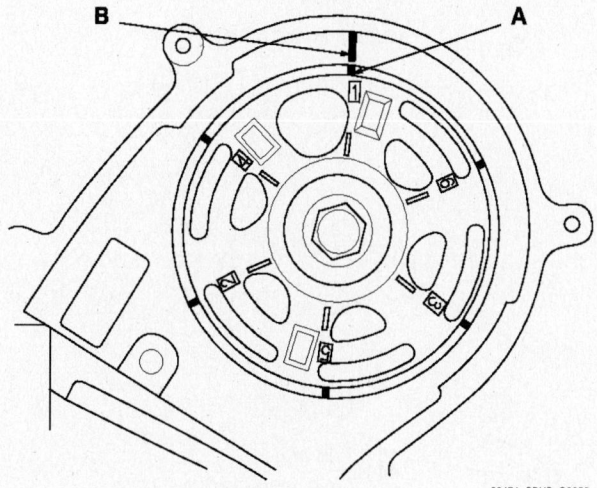

09474_ODYS_G0053

Set the left camshaft pulley to TDC by aligning the TDC marks (A) on the camshaft pulleys with the pointers (B) on the back covers—J35A7 engine

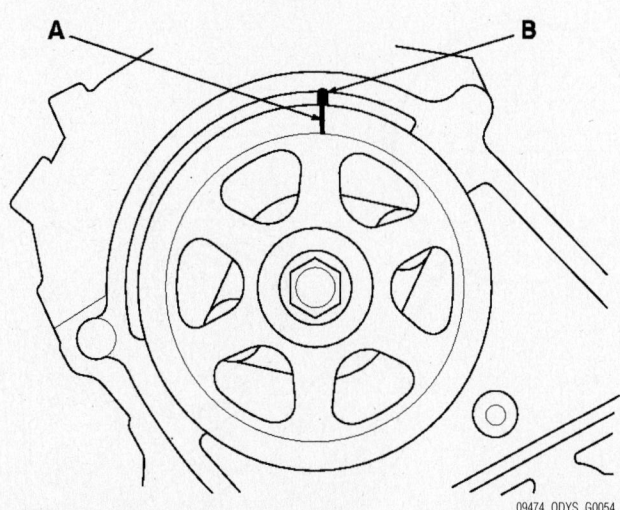

09474_ODYS_G0054

Set the right camshaft pulley to TDC by aligning the TDC marks (A) on the camshaft pulleys with the pointers (B) on the back covers—J35A7 engine

i. Check the PULSER F/B LEARN (HIGH RPM) in the DATA LIST with the HDS. If it is NOT COMPLETED, go to step 7. If it is COMPLETED, go to step 10.

j. Turn the ignition switch OFF.

k. Turn the ignition switch ON (II), and wait 30 seconds. The CKP learning procedure is complete.

INSTALLING A NEW BELT

1. Remove the timing belt.
2. Clean the timing belt pulleys, timing belt guide plate, and the upper and lower covers.

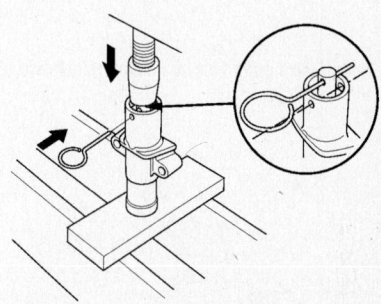

09474_ODYS_G0056

Insert a 2.0 mm (0.08 in.) pin through the housing and the rod—J35A7 engine

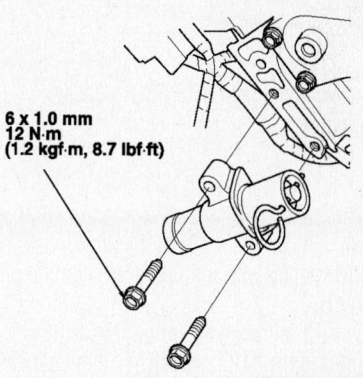

6 x 1.0 mm
12 N·m
(1.2 kgf·m, 8.7 lbf·ft)

09474_ODYS_G0057

Install the auto-tensioner—J35A7 engine

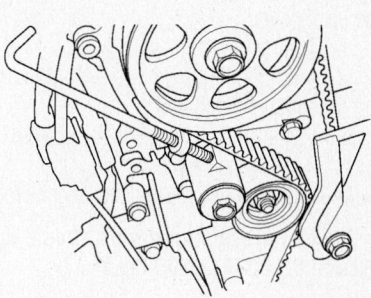

09474_ODYS_G0044

Screw the battery clamp bolt in as shown to hold the timing belt adjuster. Tighten it by hand—J35A7 engine

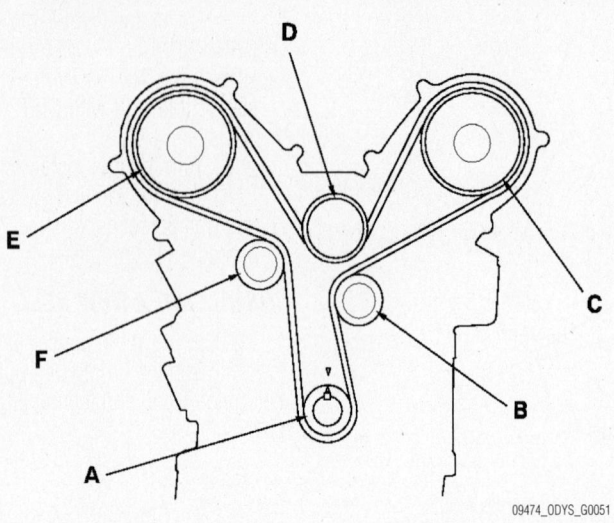

Install the timing belt in a counterclockwise sequence starting with the drive pulley—J35A7 engine

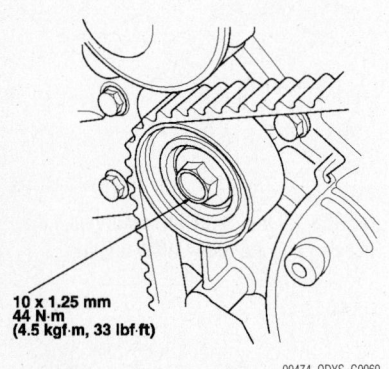

10 x 1.25 mm
44 N·m
(4.5 kgf·m, 33 lbf·ft)

Tighten the idler pulley bolt—J35A7 engine

3. Set the timing belt drive pulley to top dead center (TDC) by aligning the TDC mark (A) on the tooth of the timing belt drive pulley with the pointer (B) on the oil pump.

4. Set the camshaft pulleys to TDC by aligning the TDC marks (A) on the camshaft pulleys with the pointers (B) on the back covers.

5. Remove the battery clamp bolt from the back cover.

6. Remove the auto-tensioner.

7. Align the holes on the rod and housing of the auto-tensioner.

8. Use a hydraulic press to slowly compress the auto-tensioner. Insert a 2.0 mm (0.08 in.) pin through the housing and the rod.

➡ **The compression pressure should not exceed 9,800 N (2,200 lbs.).**

9. Align the holes on the rod and housing of the auto-tensioner.

10. Use a hydraulic press to slowly compress the auto-tensioner. Insert a 2.0 mm

(0.08 in.) pin through the housing and the rod.

➡ **The compression pressure should not exceed 9,800 N (2,200 lbs.).**

11. Install the auto-tensioner.

➡ **Make sure the pin stays in place.**

12. Screw the battery clamp bolt in as shown to hold the timing belt adjuster. Tighten it by hand; do not use a wrench.

13. Loosely install the idler pulley with a new idler pulley bolt so the pulley can move but does not come off.

14. Install the timing belt in a counter-clockwise sequence starting with the drive pulley.

- 1. Drive pulley (A)
- 2. Idler pulley (B)

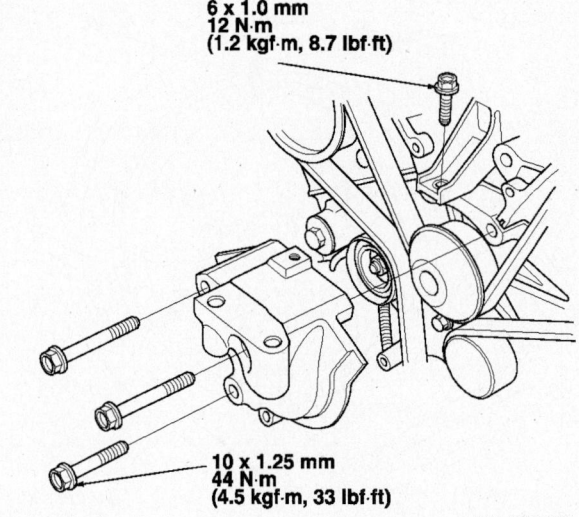

6 x 1.0 mm
12 N·m
(1.2 kgf·m, 8.7 lbf·ft)

10 x 1.25 mm
44 N·m
(4.5 kgf·m, 33 lbf·ft)

Install the lower half of the side engine mount bracket—J35A7 engine

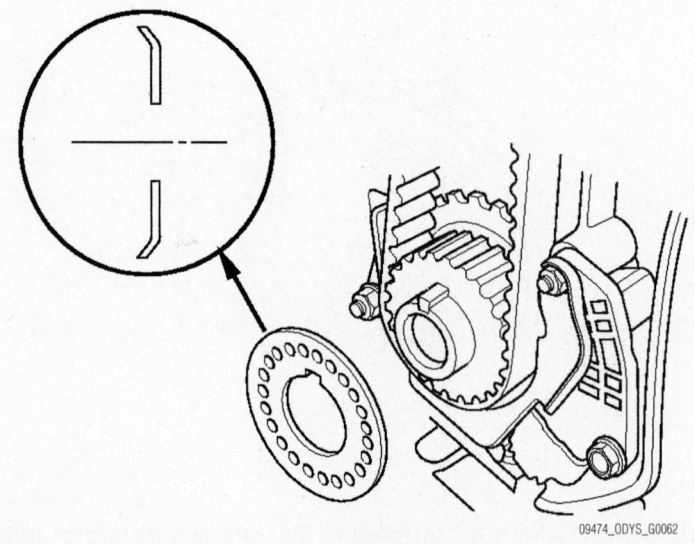

Install the timing belt guide plate as shown—J35A7 engine

- 3. Front camshaft pulley (C)
- 4. Water pump pulley (D)
- 5. Rear camshaft pulley (E)
- 6. Adjusting pulley (F)

15. Tighten the idler pulley bolt.

16. Remove the pin from the auto-tensioner.

17. Remove the battery clamp bolt from the back cover.

18. Install the lower half of the side engine mount bracket.

19. Install the timing belt guide plate as shown.

20. Install the lower cover.

21. Install the front upper cover (A) and rear upper cover (B).

22. Install the crankshaft pulley.

23. Rotate the crankshaft pulley six turns clockwise so the timing belt positions itself on the pulleys.

24. Turn the crankshaft pulley so its white mark (A) lines up with the pointer (B).

25. Check the camshaft pulley marks.

➡**If the marks are not aligned, rotate the crankshaft 360 degrees, and recheck the camshaft pulley mark.**

    a. If the camshaft pulley marks are at TDC, go to step 24.

    b. If the camshaft pulley marks are not at TDC, remove the timing belt and repeat step 3 through 22.

26. Install the upper half of the side engine mount bracket, and tighten the new mounting bolts (A), then the mass damper mounting bolt (B).

27. Install the drive belt.

28. Install the splash shield.

29. Install the right front wheel.

30. Do the crankshaft position (CKP) pattern clear/CKP pattern learn procedure.

➡**The ECT needs to be at 176°F (80°C) or higher.**

    a. Clear the CKP pattern while the engine is stopped.

    b. Turn the ignition switch OFF.

    c. Turn the ignition switch ON (II), and wait 30 seconds.

    d. Test-drive the vehicle on a level road: decelerate (with the throttle fully closed) from an engine speed of 2,500 rpm to 1,000 rpm with the A/T in 2 position.

    e. Stop the vehicle, but keep the engine running.

    f. Check PULSER F/B LEARN in the DATA LIST with the HDS. If it is NOT COMPLETED, go to step 4. If it is COMPLETED, go to step 7.

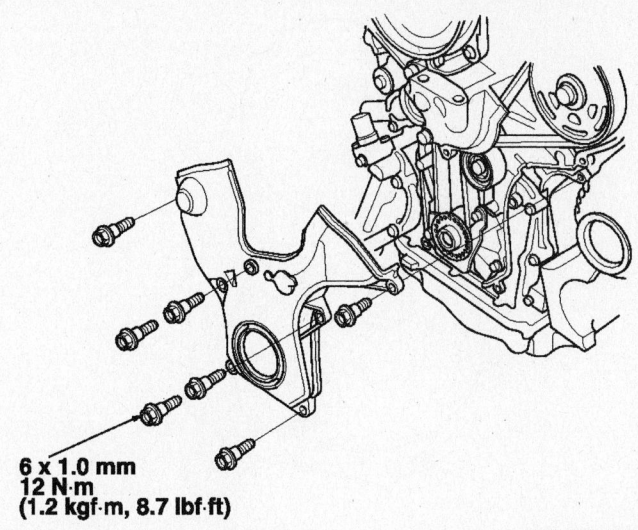

**6 x 1.0 mm
12 N·m
(1.2 kgf·m, 8.7 lbf·ft)**

09474_ODYS_G0063

Install the lower cover—J35A7 engine

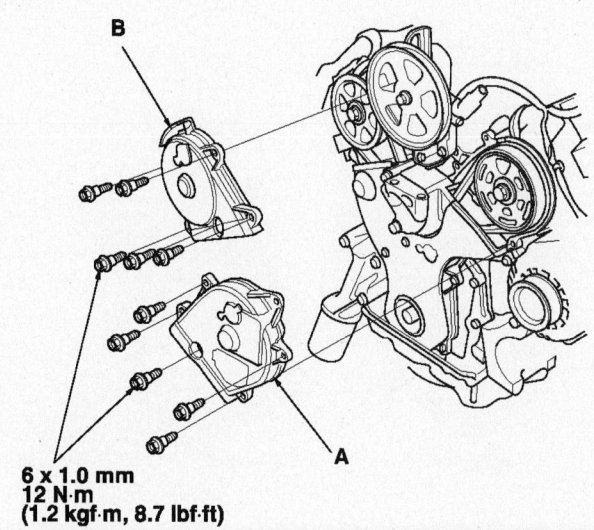

**6 x 1.0 mm
12 N·m
(1.2 kgf·m, 8.7 lbf·ft)**

09474_ODYS_G0064

Install the front upper cover (A) and rear upper cover (B)—J35A7 engine

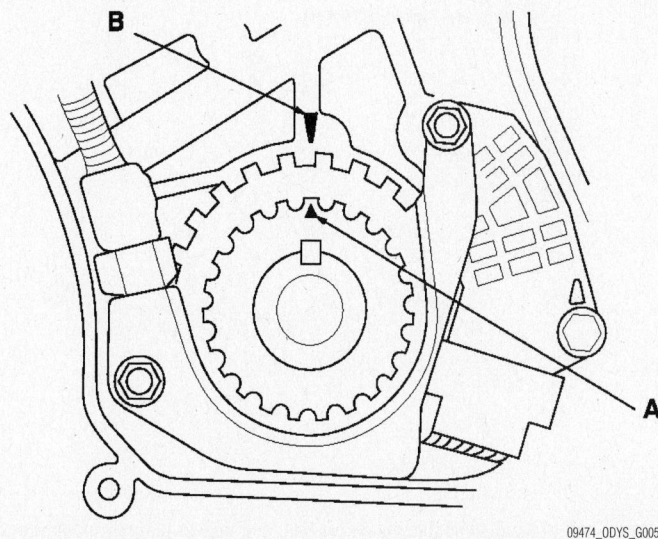

09474_ODYS_G0052

Turn the crankshaft pulley so its white mark (A) lines up with the pointer (B) —J35A7 engine

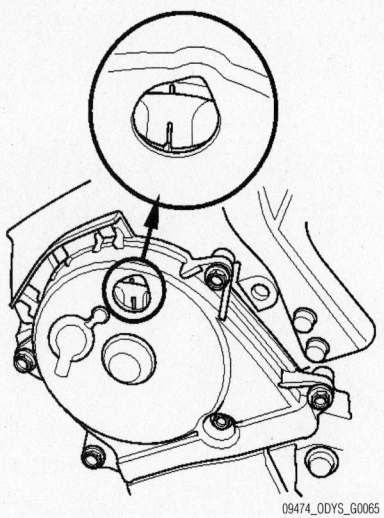

09474_ODYS_G0065

Camshaft pulley marks—J35A7 engine rear head

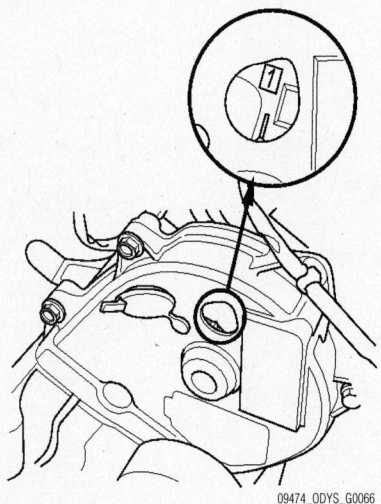

09474_ODYS_G0066

Camshaft pulley marks—J35A7 engine front head

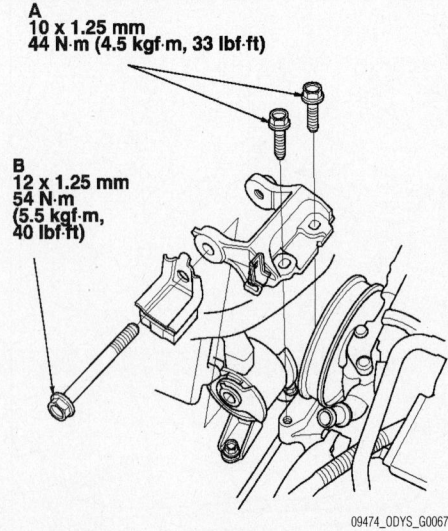

A
10 x 1.25 mm
44 N·m (4.5 kgf·m, 33 lbf·ft)

B
12 x 1.25 mm
54 N·m
(5.5 kgf·m,
40 lbf·ft)

09474_ODYS_G0067

Install the upper half of the side engine mount bracket, and tighten the new mounting bolts (A), then the mass damper mounting bolt (B) —J35A7 engine

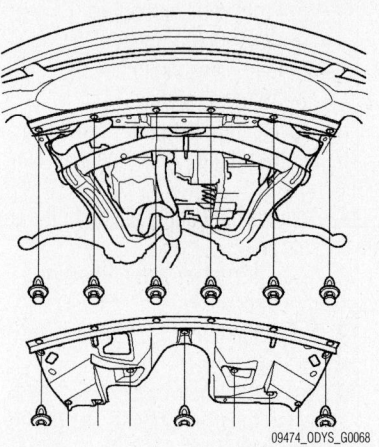

09474_ODYS_G0068

Install the splash shield—J35A7 engine

g. Test-drive the vehicle on a level road: decelerate (with the throttle fully closed) from an engine speed of 5,000 rpm to 3,000 rpm with the A/T in 2 position.

h. Stop the vehicle, but keep the engine running.

i. Check the PULSER F/B LEARN (HIGH RPM) in the DATA LIST with the HDS. If it is NOT COMPLETED, go to step 7. If it is COMPLETED, go to step 10.

j. Turn the ignition switch OFF.

k. Turn the ignition switch ON (II), and wait 30 seconds. The CKP learning procedure is complete.

Oil Pan

REMOVAL & INSTALLATION

2002–04 Models

1. Before servicing the vehicle, refer to the precautions in the beginning of this section.

2. Drain the engine oil.

3. Set the engine at Top Dead Center (TDC).

4. Remove or disconnect the following:
   - Negative battery cable
   - Timing belt
   - Idler pulley
   - VTEC solenoid valve and oil filter
   - Oil pan bolts and the pan.

To install:

5. Apply liquid gasket to the inner threads of the bolt holes and the engine block along the area indicated by the broken line in the accompanying illustration.

6. Install the oil and tighten the bolts to 105 inch lbs. (12 Nm).

7. Install the remaining components in the reverse order of removal.

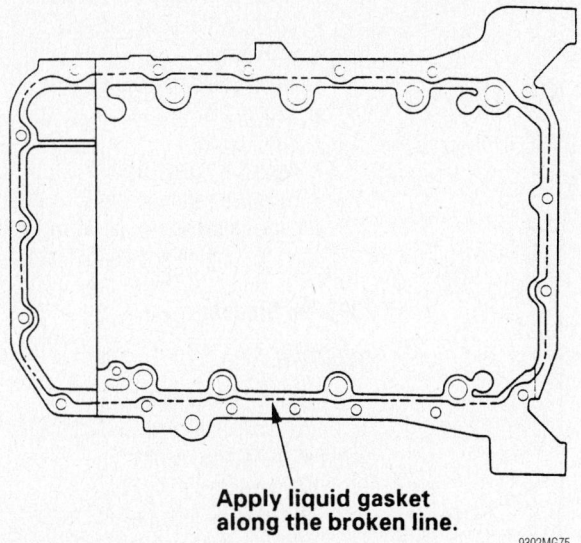

**Apply liquid gasket
along the broken line.**

9302MG75

Apply liquid gasket to the inner threads of the bolt holes and the
engine block along the area indicated by the broken line—2002–06

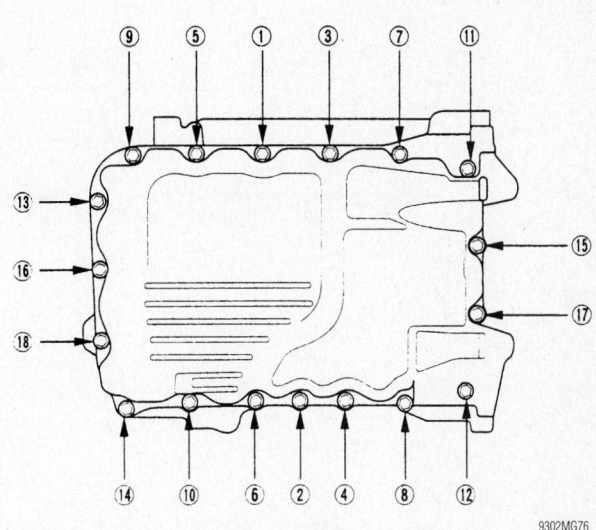

9302MG76

Oil pan tightening sequence—2002–04

2005–06 Models

1. Raise the vehicle on a hoist to full
height.

2. Drain the engine oil.

3. Remove the splash shield.

4. Remove exhaust pipe A.

5. Remove the torque converter cover
(A) and the two bolts (B) securing the transmission.

6. Remove the bolts securing the oil
pan.

7. Using a flat blade screwdriver, separate the oil pan from the block in the places
shown.

8. Remove the oil pan.

To install:

9. Remove all of the old liquid gasket
from the oil pan mating surfaces, bolts, and
bolt holes.

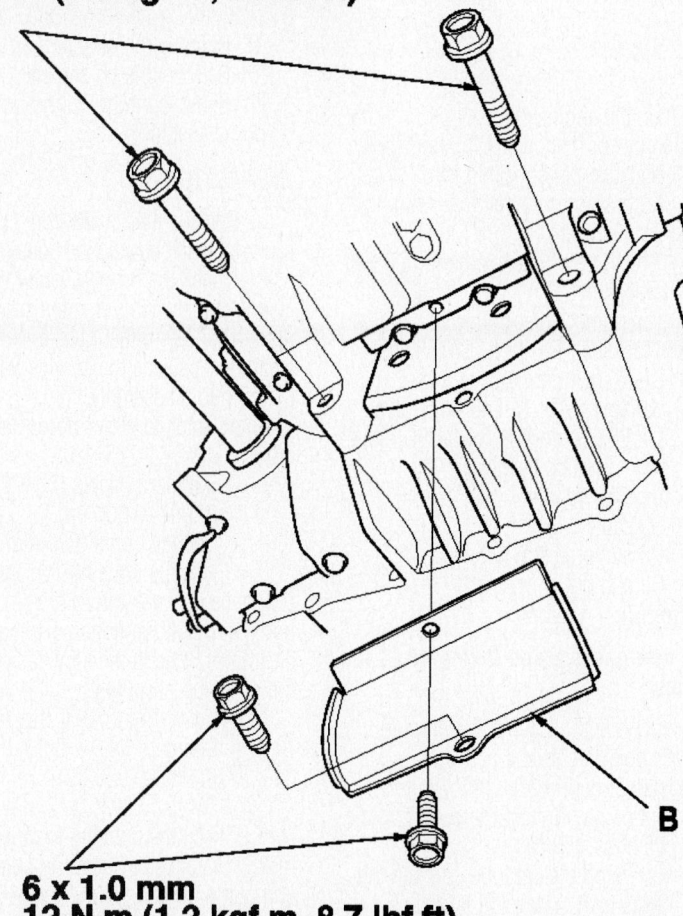

**A
12 x 1.25 mm
74 N·m (7.5 kgf·m, 54 lbf·ft)**

**6 x 1.0 mm
12 N·m (1.2 kgf·m, 8.7 lbf·ft)**

B

09474_ODYS_G0069

Remove the torque converter cover (A) and the two bolts (B) securing the transmission

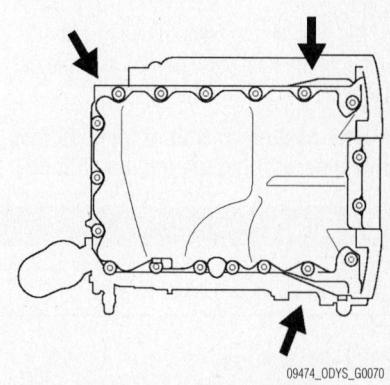

09474_ODYS_G0070

**Using a flat blade screwdriver, separate
the oil pan from the block in the places
shown**

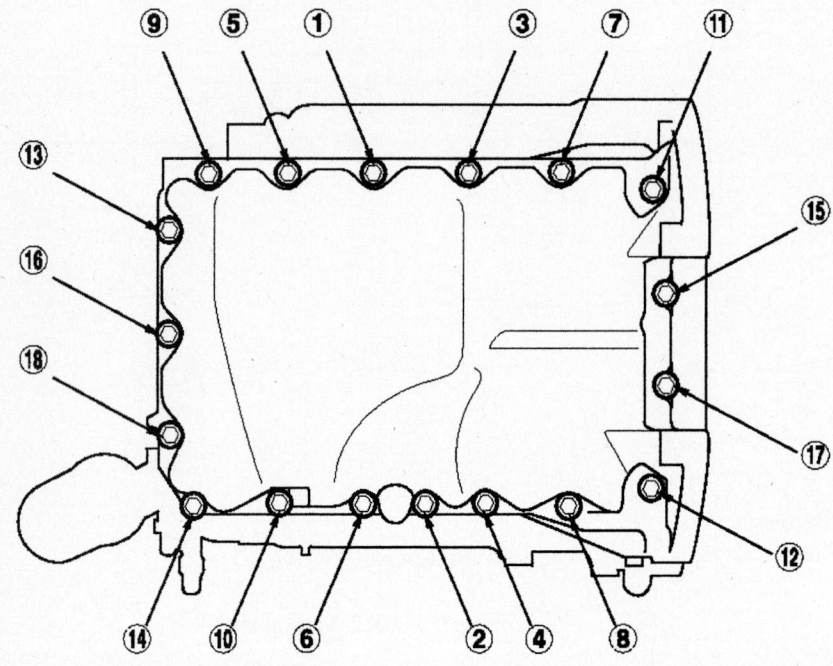

09474_ODYS_G0072

Oil pan bolt tightening sequence

10. Clean and dry the oil pan mating surfaces.

11. Apply liquid gasket, P/N 08717-0004, 08718-0001, 08718-0002, 08718-0003, or 08718-009, evenly to the oil pan mating surface of the engine block.

➡**Do not install components if too much time has passed after applying the liquid gasket (for P/N 08718-0002, no more than 4 minutes, for all others, no more than 5 minutes). Instead, remove the old residue and reapply liquid gasket.**

12. Install the oil pan on the engine block.

13. Tighten the bolts in two or three steps. In the final step, tighten all bolts, in sequence, to 12 Nm (104 inch lbs.).

➡**After assembly, wait at least 30 minutes before filling the engine with oil.**

Oil Pump

REMOVAL & INSTALLATION

2002–04 Models

1. Before servicing the vehicle, refer to the precautions in the beginning of this section.

2. Drain the engine oil.
3. Remove or disconnect the following:
- Negative battery cable
- Accessory drive belts
- Front cover
- Timing belt. Refer to the timing belt procedure.
- Timing belt idler pulley
- Crankshaft Position (CKP) sensor
- Crankshaft timing sprocket
- Variable Valve Timing and Valve Lift Electronic Control (VTEC) solenoid valve connector
- Oil filter adapter
- Oil pan
- Oil pump pickup tube
- Oil pump

To install:

➡**Use new gaskets and O-ring seals for assembly.**

4. Apply liquid gasket to the oil pump and to the bolt hole threads.
5. Install or connect the following:
- Oil pump. Tighten the bolts to 105 inch lbs. (12 Nm).
- Oil pump pickup tube. Tighten the bolts to 105 inch lbs. (12 Nm).
- Oil pan
- Oil filter adapter

- VTEC solenoid valve connector
- Crankshaft timing sprocket
- CKP sensor
- Timing belt idler pulley
- Timing belt
- Front cover
- Accessory drive belts
- Negative battery cable

6. Fill the crankcase to the correct level.
7. Start the engine and check for leaks.

2005–06 Models

REMOVAL

1. Drain the engine oil.
2. Turn the crankshaft so that the No. 1 piston is at top dead center:
3. Remove the timing belt:
4. Remove the idler pulley.
5. Remove the crankshaft position (CKP) sensor A/B.
6. Attach the chain hoist to the engine hanger on the power steering (P/S) pump bracket.
7. Remove the jack from under the oil pan.
8. Remove the rocker arm oil control solenoid (VTEC solenoid valve)/oil filter assembly (J35A6 engine) or oil filter base/oil filter assembly (J35A7 engine).
9. Remove the oil pan.
10. Remove the oil screen.
11. Remove the mounting bolts and the oil pump assembly.

INSPECTION

1. Remove the screws from the pump housing, then separate the housing and cover.
2. Check the inner-to-outer rotor radial clearance between the inner rotor (A) and outer rotor (B). If the inner-to-outer rotor clearance exceeds the service limit, replace the oil pump assembly.

Inner Rotor-to-Outer Rotor Radial Clearance:
- Standard (New): 0.04–0.16 mm (0.002–0.006 in.)
- Service Limit: 0.20 mm (0.008 in.)

3. Check the housing-to-rotor axial clearance between the rotors (A) and pump housing (B). If the housing-to-rotor axial clearance exceeds the service limit, replace the oil pump assembly.

Housing-to-Rotor Axial Clearance:
- Standard (New): 0.02–0.07 mm (0.001–0.003 in.)
- Service Limit: 0.12 mm (0.005 in.)

4. Check the housing-to-outer rotor radial clearance between the outer rotor (A) and pump housing (B). If the housing-to-outer rotor radial clearance exceeds the service limit, replace the oil pump assembly.

Housing-to-Outer Rotor Radial Clearance:

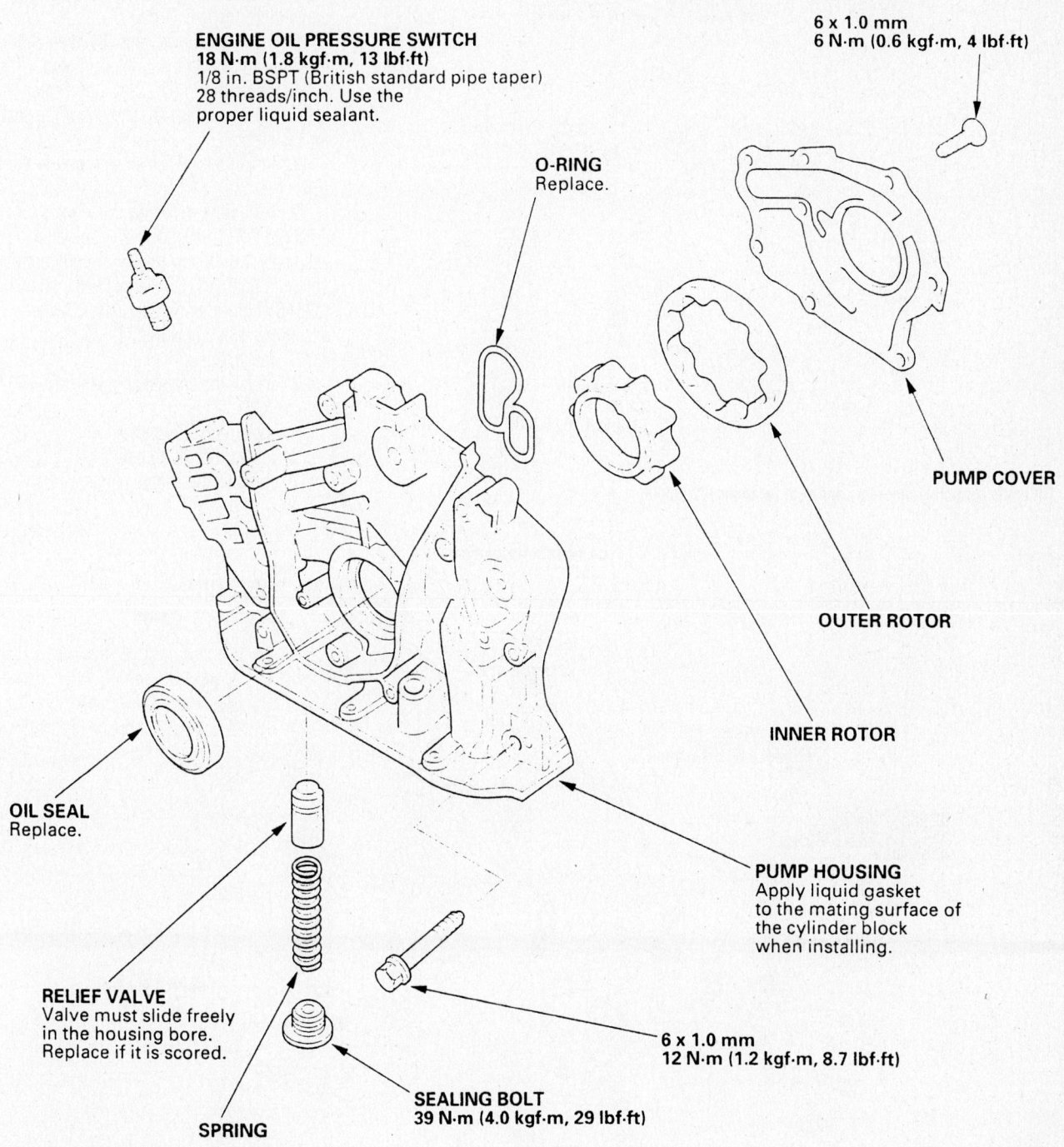

ENGINE OIL PRESSURE SWITCH
18 N·m (1.8 kgf·m, 13 lbf·ft)
1/8 in. BSPT (British standard pipe taper)
28 threads/inch. Use the
proper liquid sealant.

6 x 1.0 mm
6 N·m (0.6 kgf·m, 4 lbf·ft)

O-RING
Replace.

PUMP COVER

OUTER ROTOR

INNER ROTOR

OIL SEAL
Replace.

PUMP HOUSING
Apply liquid gasket
to the mating surface of
the cylinder block
when installing.

RELIEF VALVE
Valve must slide freely
in the housing bore.
Replace if it is scored.

6 x 1.0 mm
12 N·m (1.2 kgf·m, 8.7 lbf·ft)

SPRING

SEALING BOLT
39 N·m (4.0 kgf·m, 29 lbf·ft)

9358MG17

Exploded view of the oil pump assembly—2002–04

- Standard (New): 0.10–0.19 mm
  (0.004–0.007 in.)
- Service Limit: 0.20 mm (0.008 in.)

INSTALLATION

1. Remove the old oil seal from the oil pump.

2. Gently tap in the new oil seal until the oil seal driver bottoms on the pump.

3. Remove all of the old liquid gasket from the oil pump mating surfaces, bolts, and bolt holes.

4. Clean and dry the oil pump mating surfaces.

5. Inspect both rotors and pump housing for scoring or other damage. Replace the parts, if necessary.

6. Apply liquid thread lock to the pump housing screws, then install the oil pump cover.

7. Check that the oil pump turns freely.

8. Apply liquid gasket, P/N 08717-0004, 08718-0001, 08718-0002, 08718-0003, or 08718-0009, evenly to the engine block mating surface of the oil pump.

➡**Do not install components if too much time has passed after applying the liquid gasket (for P/N 08718-0002, no more than 4 minutes, for all others, no more than 5 minutes). Instead, remove the old residue and reapply liquid gasket.**

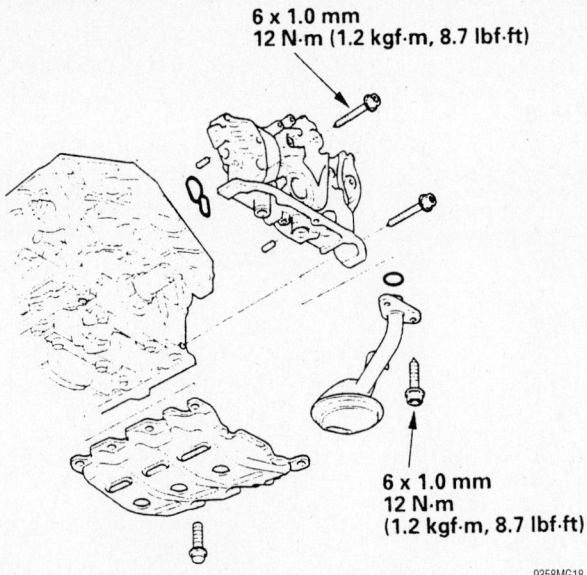

6 x 1.0 mm
12 N·m (1.2 kgf·m, 8.7 lbf·ft)

6 x 1.0 mm
12 N·m
(1.2 kgf·m, 8.7 lbf·ft)

9358MG18

Oil pump assembly mounting and seal locations—2002–06

9. Grease the lip of the oil seal, and apply oil to the new O-ring (A).

10. Install the dowel pins (B), then align the inner rotor with the crankshaft, and install the oil pump (C). Clean the excess grease off the crankshaft, and check the seal for distortion.

11. Install the oil screen with new O-ring (E).

12. Install the rocker arm oil control solenoid (VTEC solenoid valve)/oil filter assembly, with a new rocker arm oil control solenoid filter (VTEC solenoid valve filter) (J35A6 engine), or oil filter base/oil filter assembly, with a new O-ring.

13. Install the oil pan.

14. Install the crankshaft position (CKP) sensor A/B.

15. Install the idler pulley.

16. Install the timing belt:

17. Remove the engine hanger.

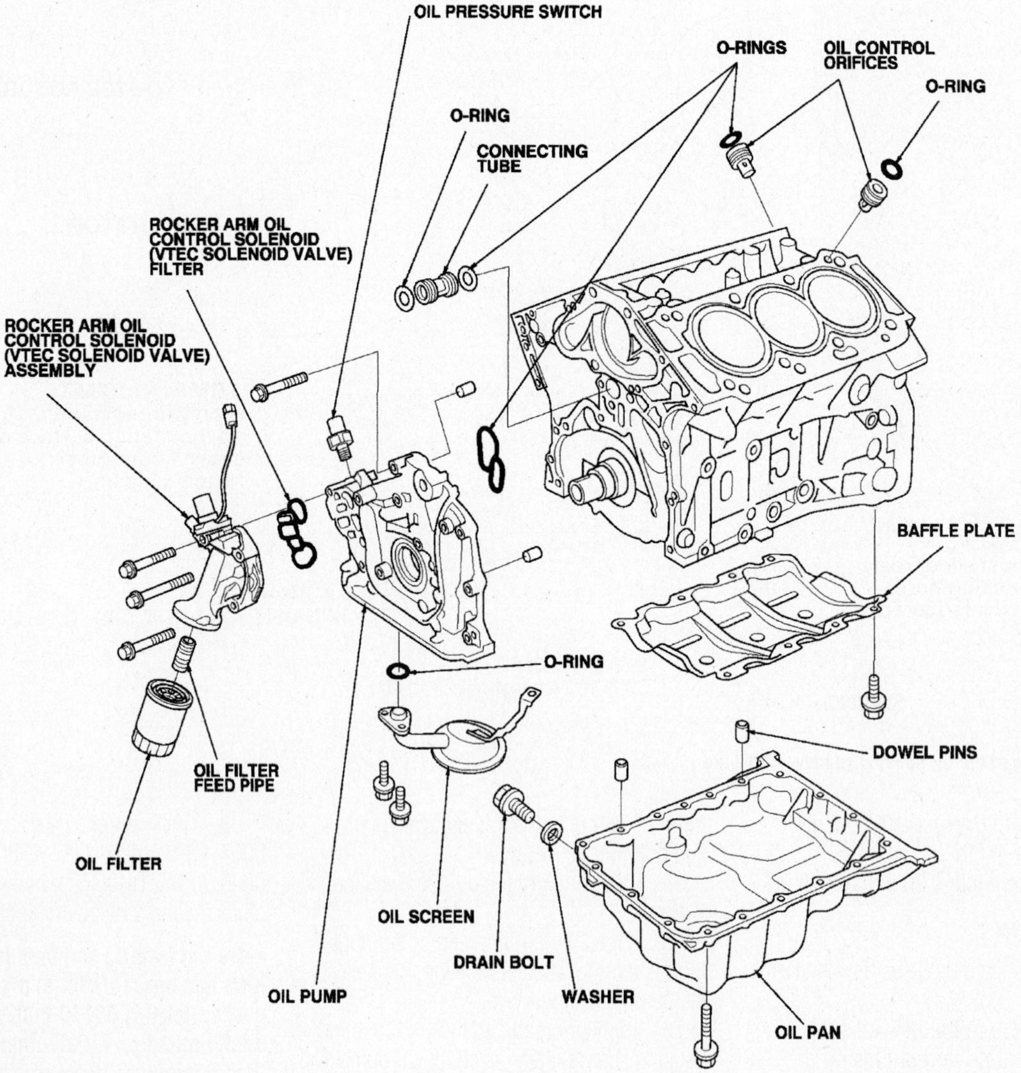

OIL PRESSURE SWITCH

O-RINGS

OIL CONTROL ORIFICES

O-RING

O-RING

CONNECTING TUBE

ROCKER ARM OIL CONTROL SOLENOID (VTEC SOLENOID VALVE) FILTER

ROCKER ARM OIL CONTROL SOLENOID (VTEC SOLENOID VALVE) ASSEMBLY

BAFFLE PLATE

O-RING

DOWEL PINS

OIL FILTER FEED PIPE

OIL FILTER

OIL PUMP

OIL SCREEN

DRAIN BOLT

WASHER

OIL PAN

09474_ODYS_G0317

Oil pump and related parts—J35A6 engine

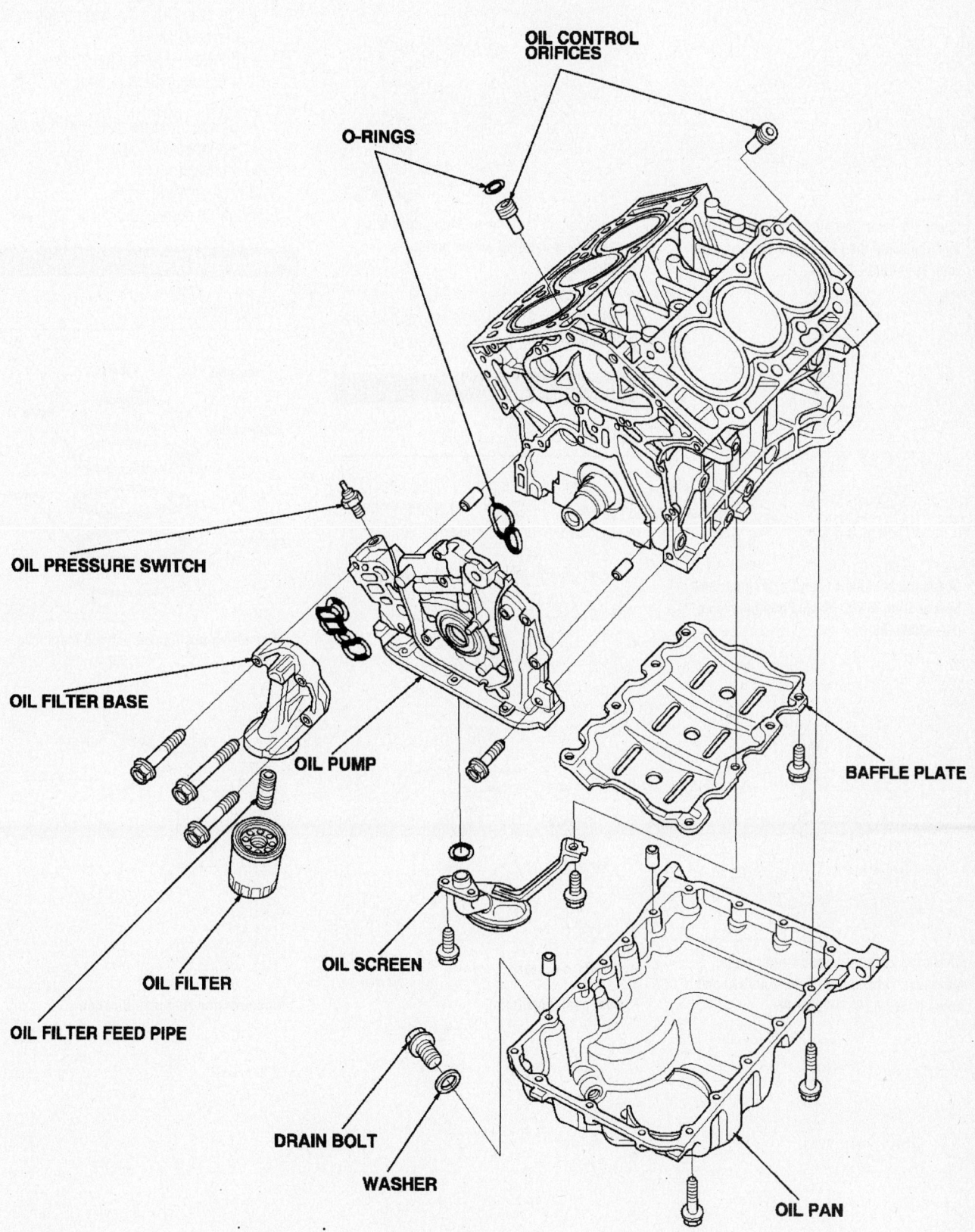

OIL CONTROL
ORIFICES

O-RINGS

OIL PRESSURE SWITCH

OIL FILTER BASE

OIL PUMP

BAFFLE PLATE

OIL FILTER

OIL SCREEN

OIL FILTER FEED PIPE

DRAIN BOLT

WASHER

OIL PAN

09474_ODYS_G0318

Oil pump and related parts—J35A7 engine

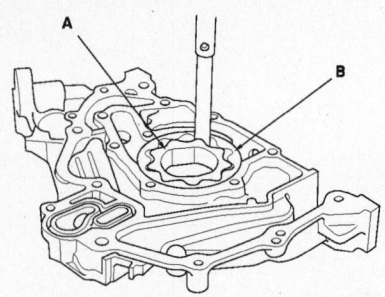

Check the inner-to-outer rotor radial clearance between the inner rotor (A) and outer rotor (B)—2005–06

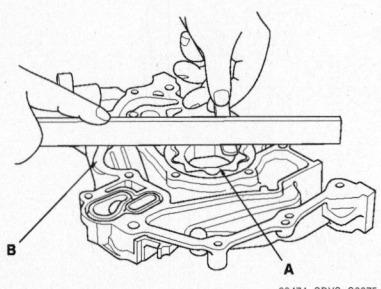

Check the housing-to-rotor axial clearance between the rotors (A) and pump housing (B)—2005–06

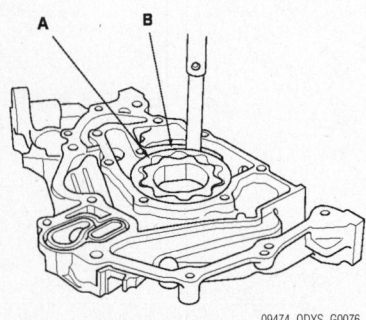

Check the housing-to-outer rotor radial clearance between the outer rotor (A) and pump housing (B) —2005–06

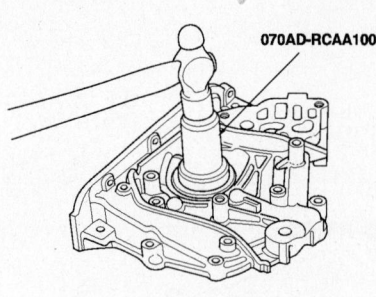

Gently tap in the new oil seal until the oil seal driver bottoms on the pump—2005–06

18. After assembly, wait at least 30 minutes before filling the engine with oil.

Rear Main Seal

REMOVAL & INSTALLATION

1. Before servicing the vehicle, refer to the precautions in the beginning of this section.
2. Remove or disconnect the following:
   - Transaxle
   - Clutch pressure plate and disc, if equipped
   - Flywheel
   - Oil seal

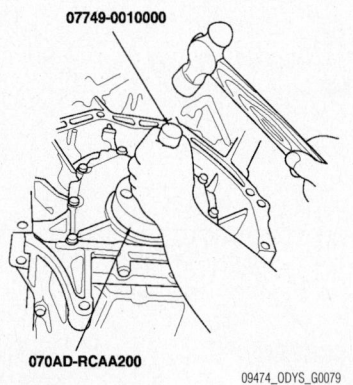

Rear seal installation

To install:
3. Install or connect the following:
   - Oil seal. Drive the seal square into the seal case.
   - Flywheel. Tighten the bolts in a crossing pattern to 54 ft. lbs. (73 Nm).
   - Clutch pressure plate and disc, if equipped
   - Transaxle
4. Check the fluid levels.
5. Start the engine and check for leaks.

Piston and Ring

POSITIONING

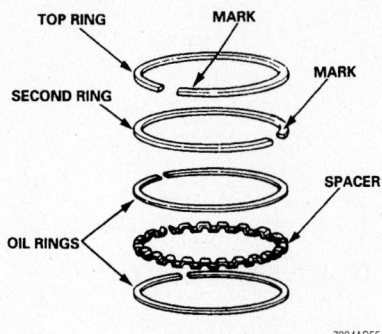

Piston ring positioning and top mark location

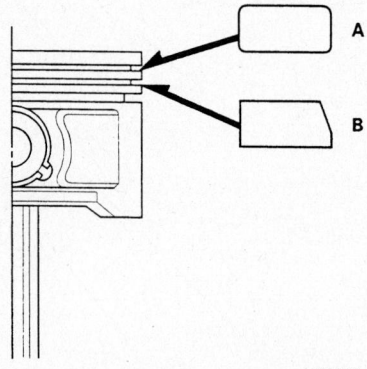

Compression ring identification

FUEL SYSTEM

Fuel System Service Precautions

Safety is the most important factor when performing not only fuel system maintenance, but any type of maintenance. Failure to conduct maintenance and repairs in a safe manner may result in serious personal injury or death. Maintenance and testing of the vehicle's fuel system components can be accomplished safely and effectively by adhering to the following rules and guidelines:

• To avoid the possibility of fire and personal injury, always disconnect the negative battery cable unless the repair or test procedure requires that battery voltage be applied.

• Always relieve the fuel system pressure prior to disconnecting any fuel system component (injector, fuel rail, pressure regulator, etc.), fitting or fuel line connection. Exercise extreme caution whenever relieving fuel system pressure to avoid exposing skin, face and eyes to fuel spray. Please be advised that fuel under pressure may penetrate the skin or any part of the body that it contacts.

• Always place a shop towel or cloth around the fitting or connection prior to loosening to absorb any excess fuel due to spillage. Ensure that all fuel spillage (should it occur) is quickly removed from engine surfaces. Ensure that all fuel soaked cloths or towels are deposited into a suitable waste container.

• Always keep a dry chemical (Class B) fire extinguisher near the work area.

• Do not allow fuel spray or fuel vapors to come into contact with a spark or open flame.

• Always use a backup wrench when loosening and tightening fuel line connection fittings. This will prevent unnecessary stress and torsion to fuel line piping. Always follow the proper torque specifications.

• Always replace worn fuel fitting O-rings with new. Do not substitute fuel hose or equivalent, where fuel pipe is installed.

Fuel System Pressure

RELIEVING

2002–04 Models

1. Before servicing the vehicle, refer to the precautions in the beginning of this section.
2. Disconnect the negative battery cable.
3. Remove the fuel filler cap.

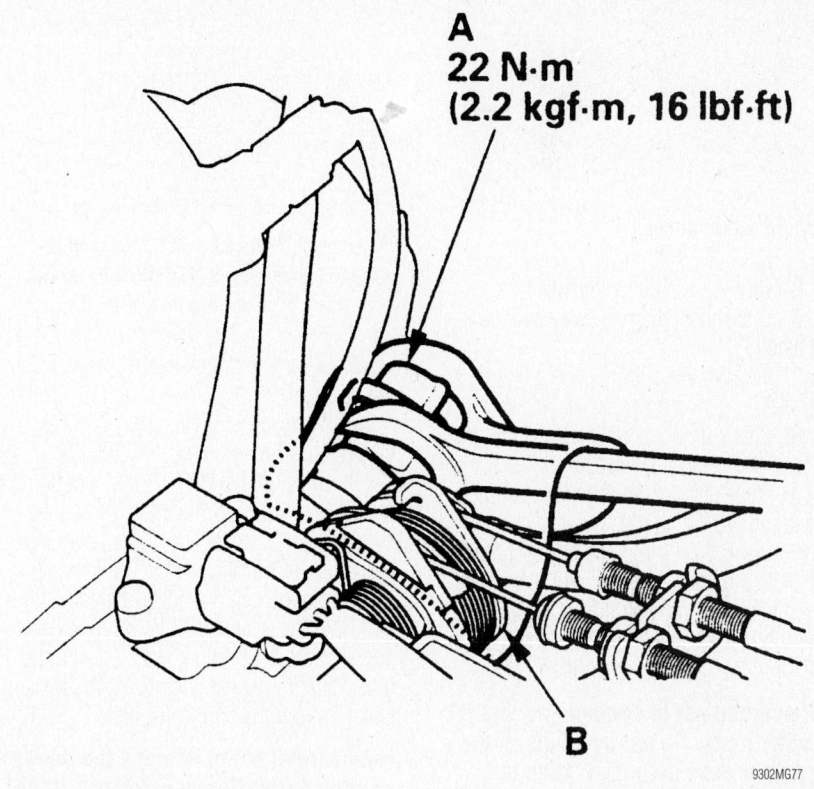

A
22 N·m
(2.2 kgf·m, 16 lbf·ft)

B

9302MG77

Use a wrench on the fuel pulsation damper (A). Place a rag over the damper (B) when relieving residual fuel pressure—2002–04

4. Place a shop towel over the fuel pulsation damper.
5. Loosen the fuel pulsation damper 1 turn.
6. When service is completed, replace the sealing washer and tighten the pulsation damper to 16 ft. lbs. (22 Nm).
7. Replace the fuel filler cap.
8. Connect the negative battery cable.
9. Start the engine and check for leaks.

2005–06 Models

Before disconnecting fuel lines or hoses, relieve pressure from the system by stopping the fuel pump and then disconnecting the fuel tube/quick connect fitting in the engine compartment.

WITH THE HDS

1. Make sure you have the anti-theft codes for the radio and the navigation system (if equipped) then write down the customer's radio station and XM radio channel presets.
2. Remove the fuel fill cap, and relieve the pressure in the fuel tank.

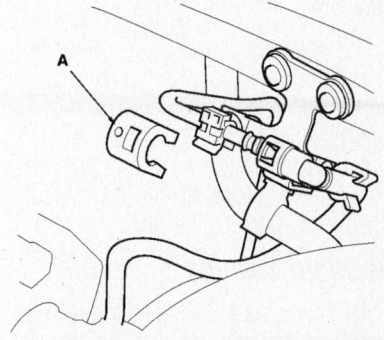

A

09474_ODYS_G0080

Remove the quick-connect fitting cover (A)

3. Turn the ignition switch ON (II).
4. From the INSPECTION MENU of the HDS, select Fuel Pump OFF, then start the engine and let it idle until it stalls.
5. Turn the ignition switch OFF.

➡Do not allow the engine to idle above 1,000 rpm or the PCM will continue to operate the fuel pump. A DTC or a Temporary DTC may be set during this procedure. Check for DTCs, and clear them as needed.

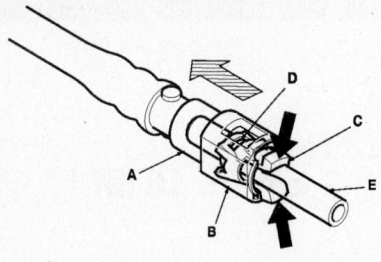

09474_ODYS_G0082

Quick-connect fitting

6. Turn the ignition switch OFF.

7. Disconnect the negative cable from the battery.

8. Remove the quick-connect fitting cover (A).

9. Check the fuel quick-connect fitting for dirt, and clean it if needed

10. Place a rag or shop towel over the quick-connect fitting (A).

11. Disconnect the quick-connect fitting (A): Hold the connector (B) with one hand, and squeeze the retainer tabs (C) with the other hand to release them from the locking tabs (D). Pull the connector off.

➡**Be careful not to damage the line (E) or other parts. Do not use tools. If the connector does not move, keep the retainer tabs pressed down, and alternately pull and push the connector until it comes off easily. Do not remove the retainer from the line; once removed, the retainer must be replaced with a new one.**

12. After disconnecting the quick-connect fitting, check it for dirt or damage.

13. Reconnect the negative cable to the battery. Enter the anti-theft codes for the radio and the navigation system, then enter the customer's radio station and XM radio channel presets. Set the clock.

WITHOUT THE HDS

1. Make sure you have the anti-theft codes for the radio and the navigation system (if equipped) then write down the customer's radio station and XM radio channel presets.

2. Remove the left kick panel, then remove PGM-FI main relay 2 (FUEL PUMP) (A) from the driver's under-dash fuse/relay box.

3. Start the engine, and let it idle until it stalls.

➡**If any DTCs are stored, clear and ignore them.**

4. Turn the ignition switch OFF.

5. Remove the fuel fill cap.

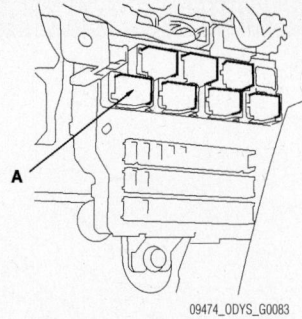

09474_ODYS_G0083

Remove the left kick panel, then remove PGM-FI main relay 2 (FUEL PUMP) (A) from the driver's under-dash fuse/relay box

6. Disconnect the negative cable from the battery.

7. Remove the quick-connect fitting cover (A).

8. Check the fuel quick-connect fitting for dirt, and clean it if needed.

9. Place a rag or shop towel over the quick-connect fitting (A).

10. Disconnect the quick-connect fitting (A): Hold the connector (B) with one hand, and squeeze the retainer tabs (C) with the other hand to release them from the locking tabs (D). Pull the connector off.

➡**Be careful not to damage the line (E) or other parts. Do not use tools. If the connector does not move, keep the retainer tabs pressed down, and alternately pull and push the connector until it comes off easily. Do not remove the retainer from the line; once removed, the retainer must be replaced with a new one.**

11. After disconnecting the quick-connect fitting, check it for dirt or damage.

12. Reconnect the negative cable to the battery. Enter the anti-theft codes for the radio and the navigation system, then enter the customer's radio station and XM radio channel presets. Set the clock.

Fuel Filter

REMOVAL & INSTALLATION

2002–04 Models

1. Before servicing the vehicle, refer to the precautions in the beginning of this section.

2. Relieve the fuel system pressure.

3. Remove or disconnect the following:
- Negative battery cable
- Rear seats and carpet
- Access panel
- Fuel lines
- Fuel pump locknut using wrench

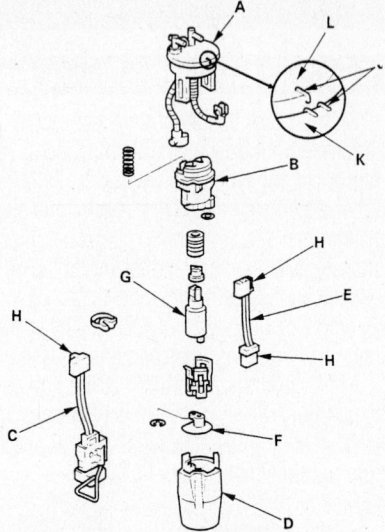

A. Bracket
B. Fuel filter
C. Fuel gauge sender
D. Case
E. Wire harness
F. Suction filter
G. Fuel pump
H. Connectors
J. Alignment marks
K. Fuel tank
L. Fuel pump module

9308MG26

Exploded view of the fuel pump module—2002–04 Models

07XAA-001010A as shown in the accompanying illustration.
- Fuel pump module

4. Disassemble the fuel pump module and remove the fuel filter.

To install:

5. Install the fuel filter and assemble the fuel pump module.

6. Install or connect the following:
- Fuel pump module
- Fuel pump locknut and tighten to 69 ft. lbs. (93 Nm) using wrench 07XAA-001010A
- Fuel lines
- Access panel
- Rear seats and carpet
- Negative battery cable

7. Start the engine and check for leaks.

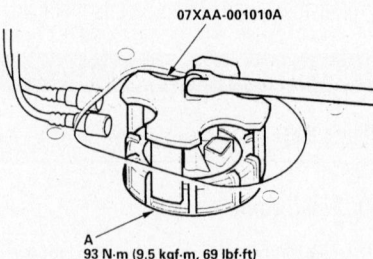

07XAA-001010A

93 N·m (9.5 kgf·m, 69 lbf·ft)

A

9358MG19

Use wrench 07XAA-001010A to remove and install the fuel pump locknut—2002–04 Models

2005–06 Models

The fuel filter should be replaced whenever the fuel pressure drops below the specified value, after making sure that the fuel pump and fuel pressure regulator are OK. The pressure should be 380–430 kPa (55–63 psi).

1. Relieve the fuel pressure.
2. Remove the fuel fill cap.
3. Remove the second row seat.
4. Remove the access panel (A) from the floor.
5. Disconnect the fuel pump 5P connector (B).
6. Disconnect the quick-connect fitting (C) from the fuel tank unit.
7. Using the special tool, loosen the locknut (A).
8. Remove the locknut (A) and the fuel tank unit.
9. Remove the fuel filter (B), fuel pressure regulator (C), pressure regulator bracket (D) and linkage (E).
10. Remove the fuel pump (F) and suction filter (G) from the reservoir (H).
11. When installing the fuel pump, make sure the connection is secure and the suc-

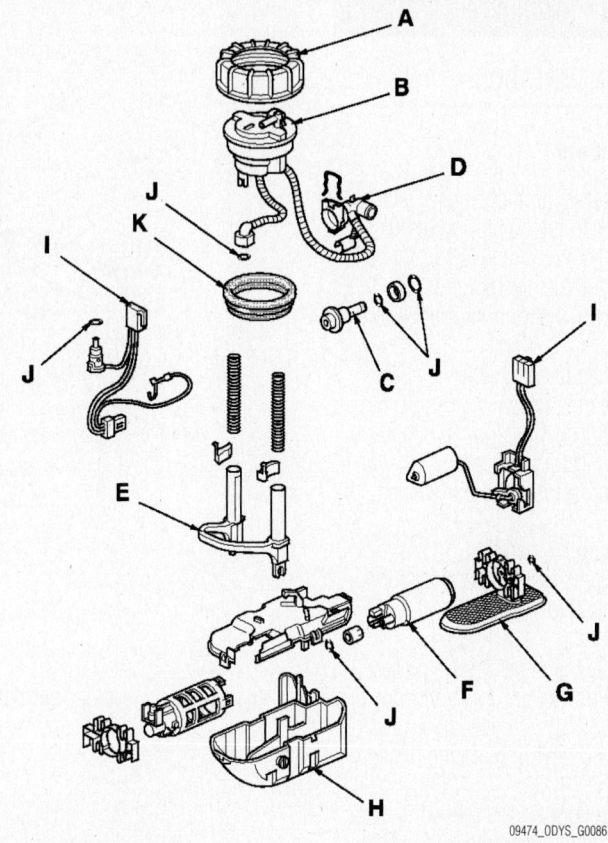

Fuel tank unit—2005–06 models

tion filter is firmly connected to the fuel pump.

12. Check these items before installing the fuel tank unit:
   - When connecting the wire harness, make sure the connection is secure and the connectors (I) are firmly locked into place.
   - When installing the fuel gauge sending unit make sure the connection is secure and the connector is firmly locked into place. Be careful not to bend or twist it excessively.

13. Install the parts in the reverse order of removal with new O-rings (J) a new base gasket (K), and a new locknut.

➡ **After installation, check the base gasket visually or by hand to be sure the gasket is not pinched. Coat the O-rings with clean engine oil.**

Fuel Pump

REMOVAL & INSTALLATION

2002–04 Models

1. Before servicing the vehicle, refer to the precautions in the beginning of this section.

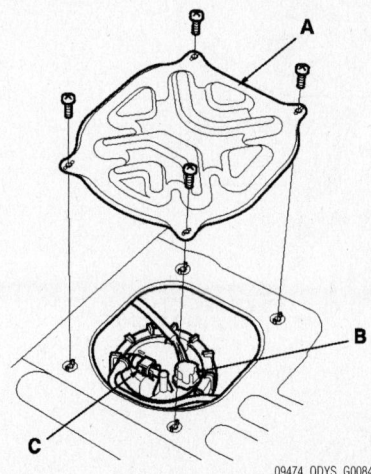

Access panel removal—2005–06 models

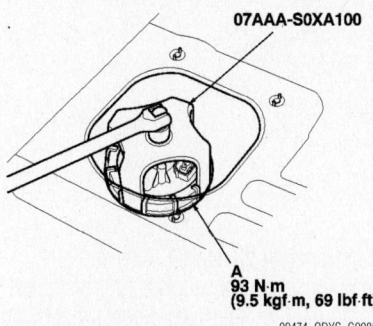

07AAA-S0XA100

A
93 N·m
(9.5 kgf·m, 69 lbf·ft)

09474_ODYS_G0085

Using the special tool, loosen the locknut (A)—2005–06 models

2. Relieve the fuel system pressure.
3. Remove or disconnect the following:
   - Negative battery cable
   - Rear seats and carpet
   - Access panel
   - Fuel pump module wiring connector
   - Fuel supply and return lines
   - Fuel pump locknut using wrench 07XAA-001010A as shown in the accompanying illustration.
   - Fuel pump module

To install:
4. Install or connect the following:
   - Fuel pump module. Use a new seal and align the matchmarks.
   - Fuel pump locknut and tighten to 69 ft. lbs. (93 Nm) using wrench 07XAA-001010A
   - Fuel supply and return lines
   - Fuel pump module wiring connector
   - Access panel
   - Rear seats and carpet
   - Negative battery cable
5. Start the engine and check for leaks.

2005–06 Models

See the procedure under Fuel Filter.

Fuel Injector

REMOVAL & INSTALLATION

2002–04 Models

1. Before servicing the vehicle, refer to the precautions in the beginning of this section.
2. Relieve the fuel system pressure.
3. Remove or disconnect the following:

- Negative battery cable
- Intake manifold
- Fuel lines
- Fuel injector connectors
- Fuel pressure regulator vacuum line
- Fuel supply manifold

4. Separate the fuel injectors from the fuel supply manifold.

To install:

5. Install the fuel injectors to the fuel supply manifold with new cushion rings and O-rings.
6. Install new seal rings to the intake manifold.
7. Install or connect the following:

- Fuel supply manifold and injector assembly. Tighten the bolts to 86 inch lbs. (10 Nm).
- Fuel pressure regulator vacuum line
- Fuel injector connectors
- Fuel lines
- Intake manifold
- Negative battery cable

8. Start the engine and check for leaks.

2005–06 Models

1. Relieve fuel pressure.
2. Remove the intake manifold.
3. Disconnect the connectors from the injectors (A).
4. Disconnect the quick-connect fitting (B).
5. Remove the fuel rail mounting bolts (C) from the fuel rail (D).
6. Remove the injector clip (E) from the fuel rail.
7. Remove the injectors from the rails.

To install:

8. Coat the new O-ring with clean engine oil, and insert the injectors (B) into the fuel rail.
9. Install the injector clip.
10. Coat the new injector O-ring with clean engine oil.
11. Install the injectors in the injector base.

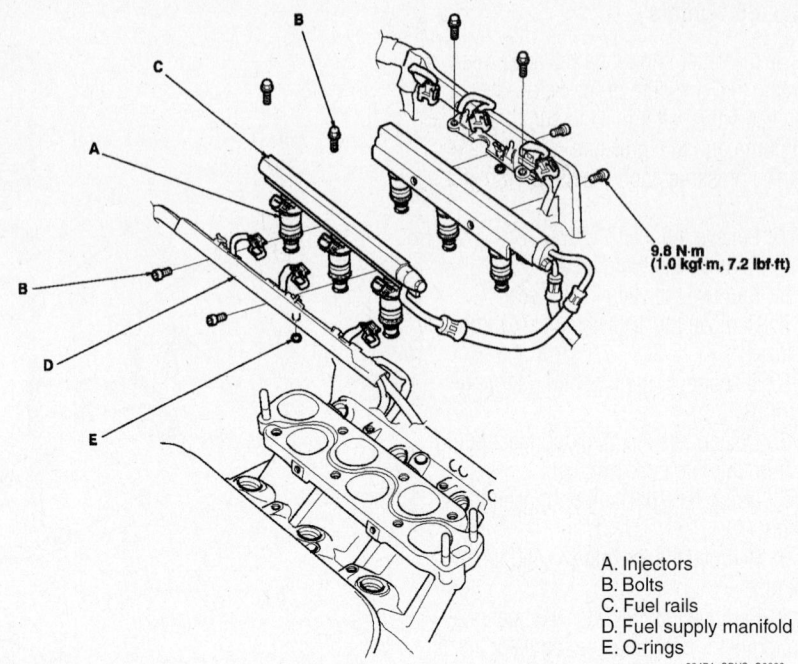

9.8 N·m (1.0 kgf·m, 7.2 lbf·ft)

A. Injectors
B. Bolts
C. Fuel rails
D. Fuel supply manifold
E. O-rings

09474_ODYS_G0088

Fuel rail and injectors—2002–04 models

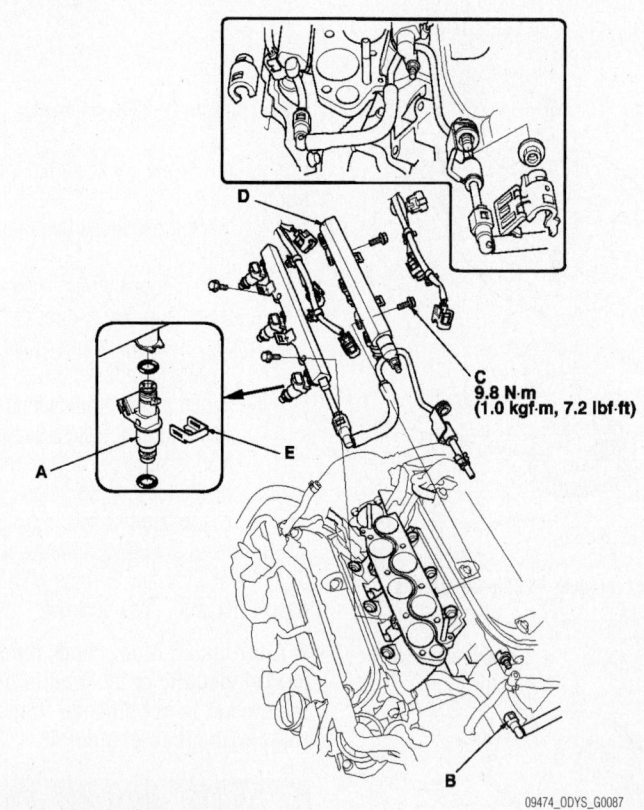

9.8 N·m (1.0 kgf·m, 7.2 lbf·ft)

09474_ODYS_G0087

Fuel rail and injectors—2005–06 models

12. Install the fuel rail mounting bolts.
13. Install the connectors on the injectors.
14. Connect the quick-connect fitting.
15. Turn the ignition switch ON (II), but do not operate the starter. After the fuel pump runs about 2 seconds, the fuel pressure in the fuel line rises. Repeat this two or three times, then check for fuel leakage.
16. Install the intake manifold.

DRIVE TRAIN

Transaxle Assembly

REMOVAL & INSTALLATION

Automatic Transaxle

2002–04 MODELS

1. Before servicing the vehicle, refer to the precautions in the beginning of this section.

2. Drain the transaxle.

3. Remove the engine appearance covers and install a support fixture to the engine lifting eyes.

4. Remove or disconnect the following:
- Air intake assembly
- Battery
- Battery tray
- Transaxle oil cooler lines
- Starter motor
- Transaxle ground cable
- Shift control solenoid valve connectors
- Clutch pressure switch connectors
- Mainshaft speed sensor connector
- Pressure control solenoid valve connectors
- Connector bracket
- Wiring harness cover
- Countershaft speed sensor connector
- Gear position switch connector
- Front motor mount
- Vacuum tube
- Splash shield
- Heated Oxygen Sensor (HO2S) connectors
- Exhaust front pipe
- Stabilizer bar links
- Lower ball joints

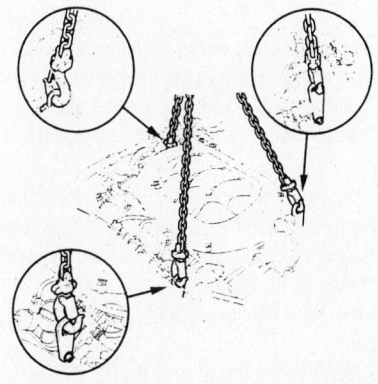

9302MG83

Support the engine while removing the transaxle—2002–04 models

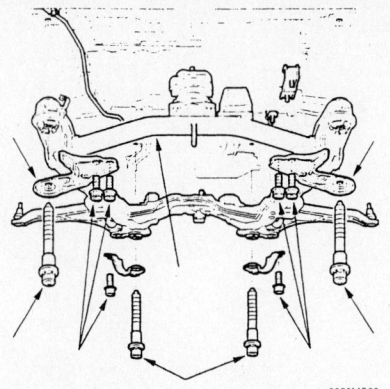

9358MG20

Remove the subframe bolts—2002–04 models

- Shift cable bracket
- Shift cable cover
- Shift control lever
- Torque converter
- Power steering hose bracket
- Power steering gear and brace
- Rear engine mount
- Transaxle lower mounts

5. Matchmark the front subframe to the vehicle body.

6. Support the subframe with a jack.

7. Support the steering gear with safety wire and remove the subframe.

8. Separate the inner CV-joints from the transaxle and intermediate shaft and support the axle halfshafts out of the work area with safety wire.

9. Remove or disconnect the following:
- Intermediate shaft
- Transaxle flange bolts
- Transaxle

To install:

➡**Use new circlips, split pins and self-locking nuts for assembly.**

10. Install or connect the following:
- Transaxle. Tighten the flange bolts to 47 ft. lbs. (64 Nm).
- Intermediate shaft. Tighten the bolts to 29 ft. lbs. (39 Nm).
- Axle halfshafts

11. Raise the subframe into position and align the matchmarks. Tighten the subframe bolts to 76 ft. lbs. (103 Nm). Tighten the front subframe bracket bolts to 54 ft. lbs. (74 Nm) and the rear bracket bolts to 86 ft. lbs. (117 Nm).

12. Install or connect the following:
- Transaxle lower mounts. Tighten the nuts to 28 ft. lbs. (38 Nm).
- Rear engine mount. Tighten the bolts to 28 ft. lbs. (38 Nm).

- Power steering gear. Tighten the bolts to 29 ft. lbs. (39 Nm).
- Power steering gear brace. Tighten the bolts to 43 ft. lbs. (58 Nm).
- Power steering hose bracket
- Torque converter. Tighten the bolts to 105 inch lbs. (12 Nm).
- Shift control lever
- Shift cable cover
- Shift cable bracket
- Lower ball joints. Tighten the nuts to 43–51 ft. lbs. (59–69 Nm).
- Stabilizer bar links. Tighten the nuts to 58 ft. lbs. (78 Nm).
- Exhaust front pipe
- HO2S connectors
- Splash shield
- Vacuum tube
- Front motor mount. Tighten the nut to 40 ft. lbs. (54 Nm).
- Gear position switch connector
- Countershaft speed sensor connector
- Wiring harness cover
- Connector bracket
- Pressure control solenoid valve connectors
- Mainshaft speed sensor connector
- Clutch pressure switch connectors
- Shift control solenoid valve connectors
- Transaxle ground cable
- Starter motor
- Transaxle oil cooler lines
- Battery tray
- Battery
- Air intake assembly

13. Fill the transaxle to the correct level.

14. Start the engine and check for leaks.

15. Check the wheel alignment and adjust as necessary.

2005–06 MODELS

The following special tools, or their equivalents, will be necessary:
- Engine support hanger, A and Reds AAR-T-12566
- Engine hanger adapter VSB02C000022
- Engine hanger balancer bar VSB02C000019
- Front subframe adapter VSB02C000016

➡**Use fender covers to avoid damaging painted surfaces. Special tool Reds engine support hanger AAR-T-12566 must be used with the side engine mount installed.**

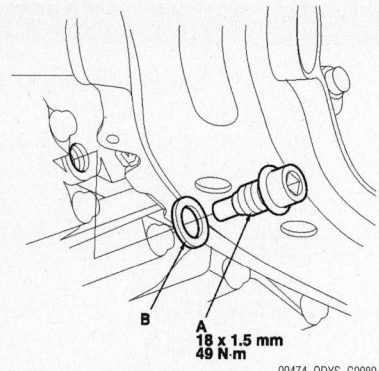

B
A
18 x 1.5 mm
49 N·m

09474_ODYS_G0089

Drain plug—2005–06 models

1. Make sure you have the anti-theft codes for the radio and navigation system, then write down the audio presets.

2. Lift the vehicle up on a hoist, and make sure it is securely supported.

3. Remove the front inner fender and splash shield.

4. Remove the drain plug (A), and drain the automatic transmission fluid (ATF).

5. Reinstall the drain plug with a new sealing washer (B).

6. Disconnect the negative cable from the battery first, then disconnect the positive cable.

7. Remove the battery hold-down bracket, battery, and battery tray.

8. Remove the intake air duct and air cleaner housing.

9. Loosen the two bolts securing the rear of the battery base, and remove the two bolts securing the front of the battery base, then remove the battery base.

10. Disconnect the ATF cooler hoses from the ATF cooler lines. Turn the ends of the ATF cooler hoses up to prevent ATF from flowing out, and plug the ATF cooler

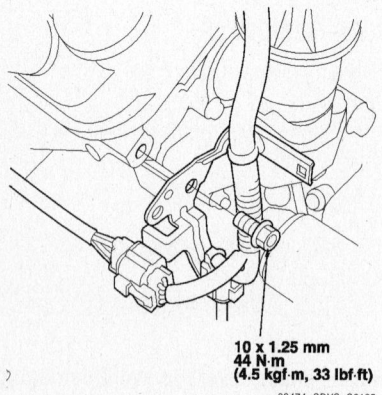

10 x 1.25 mm
44 N·m
(4.5 kgf·m, 33 lbf·ft)

09474_ODYS_G0125

Remove the connector bracket from the engine front cylinder head; use the bracket bolt hole to attach engine hanger balancer bar front arm—2005–06 models

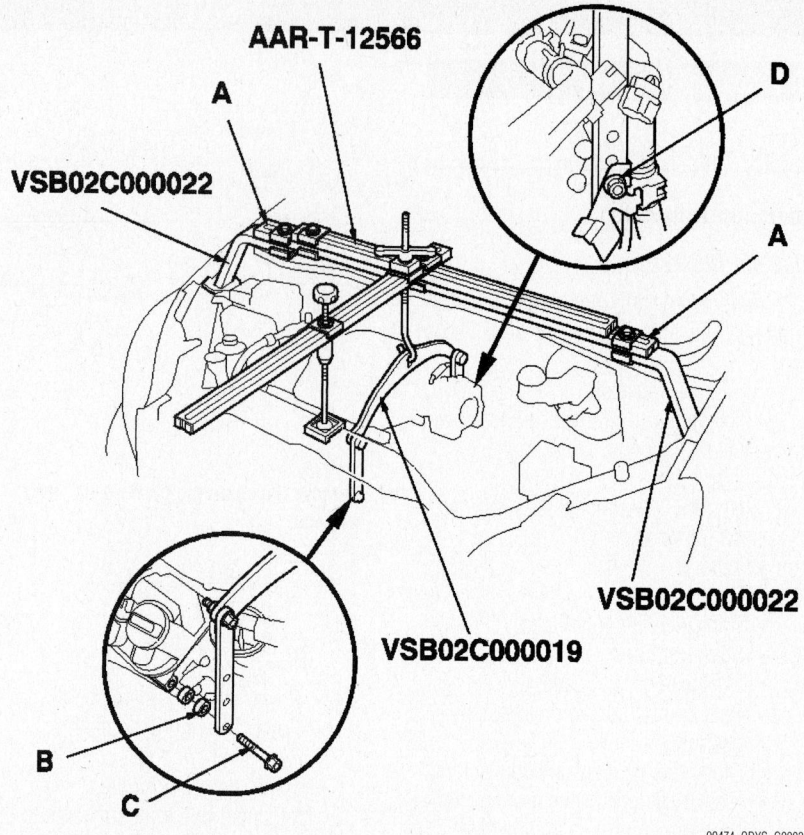

AAR-T-12566

A

VSB02C000022

D

A

B

C

VSB02C000019

VSB02C000022

09474_ODYS_G0003

Engine hangers and support beams—2005–06 models

hoses and lines. Check for any signs of leakage at the hose joints.

11. Disconnect and tag all wiring connected to the transmission or in the way of removal.

12. Remove the connector bracket from the engine front cylinder head; use the bracket bolt hole to attach engine hanger balancer bar front arm.

13. Remove the harness clamp bracket from the engine rear cylinder head; use the bracket bolt hole to attach engine hanger balancer bar rear arm.

14. Remove the front bulkhead cover.

15. Install the engine hanger adapters (VSB02C000022) on the end of the engine support beams (A), then install the engine support hanger (AAR-T-12566) with the adapters to the vehicle.

16. Lift and support the engine with the engine support hanger and engine balancer bar (VSB02C000019). Attach the front arm to the front cylinder head with a spacer (B) and the 10 x 1.25 mm bolt (C). Attach the rear arm to the rear cylinder head with the 8 x 1.25 mm bolt (D).

17. For models with active engine mount control system: Remove the front engine mount stop (A), and remove the front engine mount bolt (B).

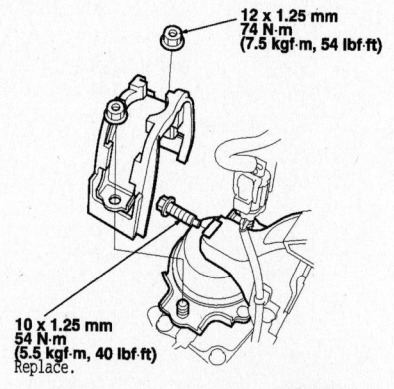

12 x 1.25 mm
74 N·m
(7.5 kgf·m, 54 lbf·ft)

10 x 1.25 mm
54 N·m
(5.5 kgf·m, 40 lbf·ft)
Replace.

09474_ODYS_G0122

For models with active engine mount control system: Remove the front engine mount stop (A), and remove the front engine mount bolt (B)—2005–06 models

18. For models with engine mount control system: Remove the front engine mount stop (A), and vacuum tube clamp bracket (B), and remove the front engine mount bolt (C).

19. Remove the exhaust pipe A and its mount.

20. Remove the lock pins and castle nuts, and separate the lower arms from the knuckles. See the procedures in the Steering and Suspension parts.

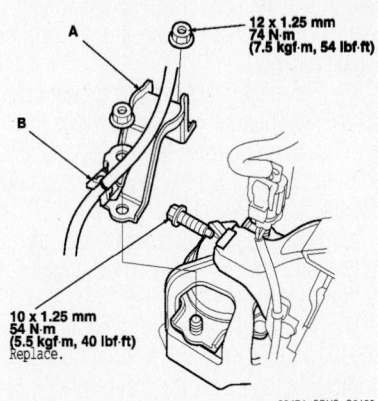

09474_ODYS_G0123

For models with engine mount control system: Remove the front engine mount stop (A), and vacuum tube clamp bracket (B), and remove the front engine mount bolt (C)—2005–06 models

21. Insert a 6 mm Allen wrench in the top of the joint pin, remove the nuts, then separate the stabilizer links.

22. Remove the bolts securing the shift cable holder, then remove the shift cable cover.

23. Remove the lock bolt securing the control lever, and remove the shift cable and the control lever. Do not bend the shift cable excessively.

24. Install a 6 x 1.0-14 mm bolt (A) and nut (B) on the shift cable cover (C), and reinstall the shift cable cover to the torque converter housing. If you don't do this, the bolt head of the cable cover may prevent you from removing the torque converter during transmission removal.

25. Remove the torque converter cover, and remove the drive plate bolts while rotating the crankshaft pulley.

26. Remove the engine-to-torque converter housing mounting bolts.

27. Remove the power steering fluid line bracket bolts, and remove the steering gearbox mounting bolt.

28. Remove the steering gearbox mounting bolt and nut, and remove the steering gearbox stiffener (A).

29. Remove the heat shield (A), and remove the steering gearbox mounting bolt (B).

30. Remove the steering gearbox mounting bolt and nut, and remove the steering gearbox stiffener (C).

31. Remove the rear mount mounting bolts.

32. Unclamp the power steering fluid line clamps on the front subframe.

33. Remove the transmission lower mount nuts.

34. For models with active engine mount

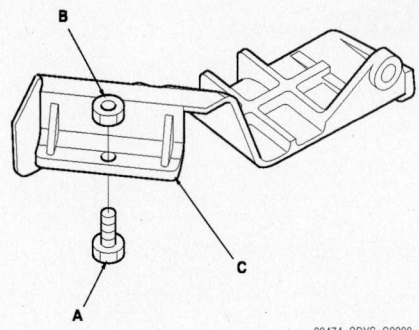

09474_ODYS_G0098

Install a 6 x 1.0-14 mm bolt (A) and nut (B) on the shift cable cover (C)—2005–06 models

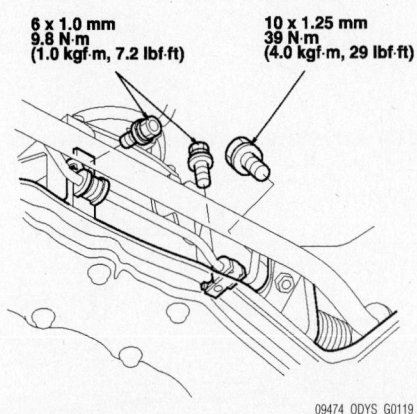

09474_ODYS_G0119

Remove the power steering fluid line bracket bolts, and remove the steering gearbox mounting bolt—2005–06 models

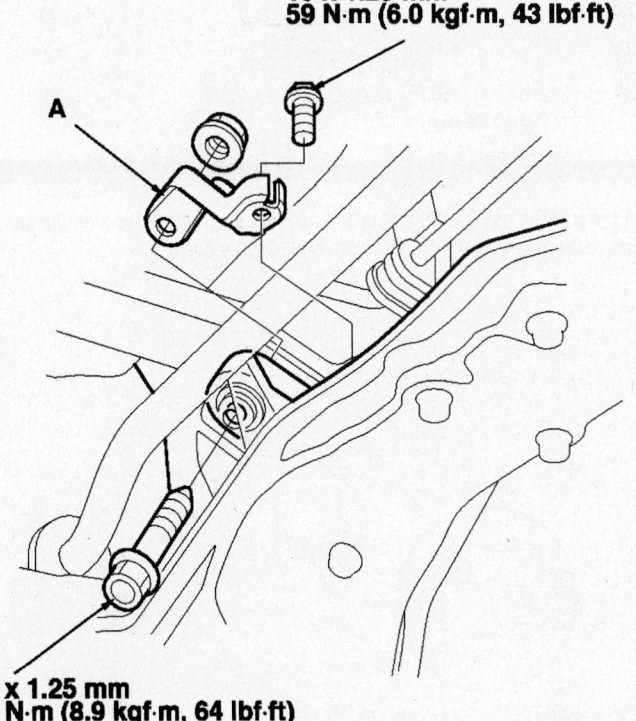

09474_ODYS_G0117

Remove the steering gearbox mounting bolt and nut, and remove the steering gearbox stiffener (A)—2005–06 models

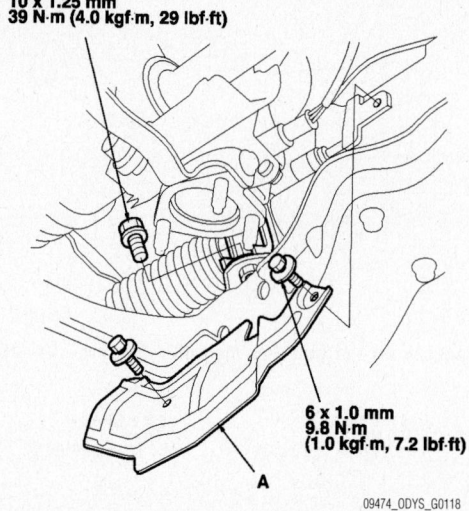

10 x 1.25 mm
39 N·m (4.0 kgf·m, 29 lbf·ft)

6 x 1.0 mm
9.8 N·m
(1.0 kgf·m, 7.2 lbf·ft)

A

09474_ODYS_G0118

Remove the heat shield (A), and remove the steering gearbox mounting bolt—2005–06 models

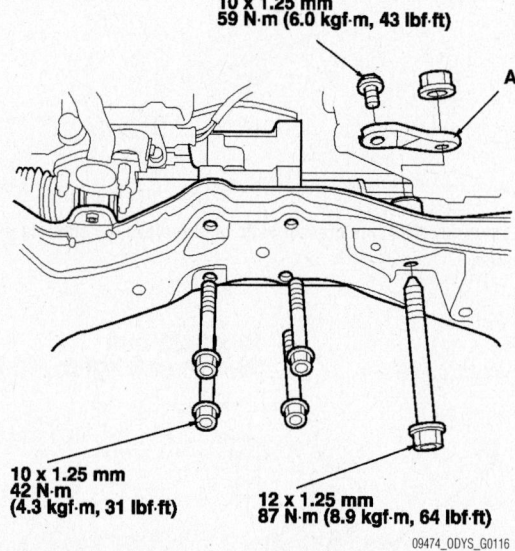

10 x 1.25 mm
59 N·m (6.0 kgf·m, 43 lbf·ft)

A

10 x 1.25 mm
42 N·m
(4.3 kgf·m, 31 lbf·ft)

12 x 1.25 mm
87 N·m (8.9 kgf·m, 64 lbf·ft)

09474_ODYS_G0116

Remove the steering gearbox mounting bolt and nut, and remove the steering gearbox stiffener (C). Remove the rear mount mounting bolts—2005–06 models

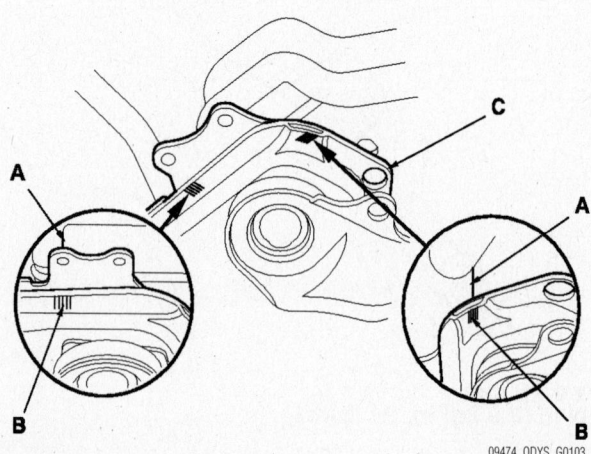

C

A

A

B

B

09474_ODYS_G0103

Make reference marks (A) on the body across the marks (B) on the edge of the front subframe (C)—2005–06 models

control system: Disconnect engine mount control solenoid valve connector at the front engine mount.

35. For models with engine mount control system: Remove the vacuum tube joint from its clamp from the front engine mount, disconnect the vacuum tube joint, then reinstall the joint in the clamp.

36. Make reference marks (A) on the body across the marks (B) on the edge of the front subframe (C).

37. Attach the front subframe adapter (VSB02C000016) to the front subframe by looping the belt over the front of the subframe, then securing the belt with its stop.

38. Raise the jack and line up the slots in the arms with the bolt holes on the corner of the jack base, then attach them with bolts.

39. Remove the four bolts securing the front stiffeners, two bolts securing the rear stiffeners, subframe mounting bolts, and remove the front and rear stiffeners.

40. Lower the front subframe with the jack.

41. Remove the transmission lower mounts.

42. Remove the left driveshaft from the differential and the right driveshaft from the intermediate shaft.

43. Move the left driveshaft to the front side. Coat all precision finished surfaces with clean engine oil, then put plastic bags over the driveshaft ends.

44. Remove the rear warm-up three-way catalytic converter (WU-TWC) stay (A) and heat shield (B).

45. Remove the intermediate shaft (C). Coat all precision finished surfaces with clean engine oil, then put plastic bags over the intermediate shaft ends.

46. Remove the upper and front transmission housing mounting bolts.

47. Remove the bolt (A) securing the harness clamp bracket, and remove the front mount bracket (B).

48. Remove the rear transmission housing mounting bolts.

49. Slide the transmission away from the engine to remove it from the vehicle, and lower it on the transmission jack. If the torque converter is stuck to the drive plate, pull it toward the transmission housing through the starter opening.

50. Remove the shift cable cover, then remove the torque converter assembly from the torque converter housing.

51. Inspect the drive plate, and replace it if it is damaged.

52. Remove the 6 x 1.0-14 mm bolt (A) and nut (B) from the shift cable cover (C).

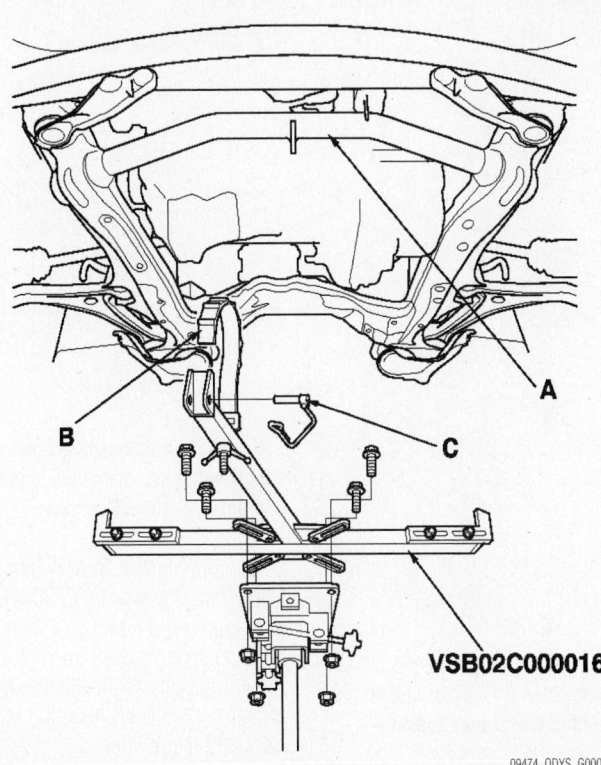

VSB02C000016

09474_ODYS_G0007

Attach the front subframe adapter (VSB02C000016) to the front subframe (A) by looping the belt (B) over the front of the subframe, then securing the belt with its pin (C)—2005–06 models

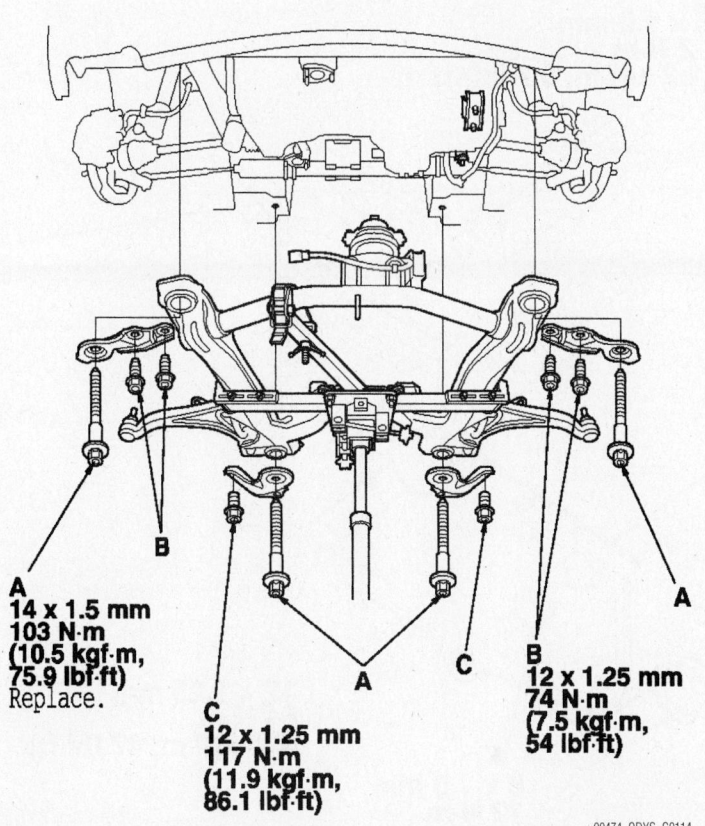

A
14 x 1.5 mm
103 N·m
(10.5 kgf·m,
75.9 lbf·ft)
Replace.

B
12 x 1.25 mm
74 N·m
(7.5 kgf·m,
54 lbf·ft)

C
12 x 1.25 mm
117 N·m
(11.9 kgf·m,
86.1 lbf·ft)

09474_ODYS_G0114

Remove the four bolts securing the front stiffeners, two bolts securing the rear stiffeners, subframe mounting bolts, and remove the front and rear stiffeners—2005–06 models

To install:

53. Clean the ATF cooler.

54. Install the torque converter assembly (A) on the mainshaft (B) with the new O-ring (C).

55. Install the 14 x 20 mm dowel pin (D) and 10 x 20 mm dowel pin (E) in the torque converter housing.

56. Place the transmission on the jack, and raise the transmission to the engine level.

57. Attach the transmission to the engine, then install the rear transmission housing mounting bolts.

58. Install the upper and front transmission housing mounting bolts.

59. Remove the jack from the transmission.

60. Install the front mount bracket with the new bolts, and install the harness clamp bracket.

61. Install the engine-to-torque converter housing mounting bolts.

62. Attach the torque converter to the drive plate (A) with eight bolts (B). Rotate the crankshaft pulley as necessary to tighten the bolt to 1/2 of the specified torque, then to the final torque, in a crisscross pattern. After tightening the last bolt, check that the crankshaft rotates freely.

63. Install the torque converter cover (C).

64. Clean the areas where the intermediate shaft and driveshaft contact the transmission (differential) with carburetor cleaner, and dry with compressed air.

65. Install the new set ring (A) on the intermediate shaft (B), and install the intermediate shaft in the differential. While installing the intermediate shaft in the differential, be sure not to allow dust or other foreign particles to enter the transmission.

66. Install the rear warm-up three-way catalytic converter (WU-TWC) stay (C) and heat shield (D).

67. Install the new set ring on the left driveshaft, then install the left driveshaft in the differential. While installing the driveshaft in the differential, be sure not to allow dust or other foreign particles to enter the transmission. Install the right driveshaft over the intermediate shaft.

➡**Turn the right and left steering knuckle fully outward, and slide the driveshaft and intermediate shaft into the differential and over the intermediate shaft until you feel the set ring engage the side gear.**

68. Install the transmission lower front mount (A), and install the lower rear mount (B) with the new bolts.

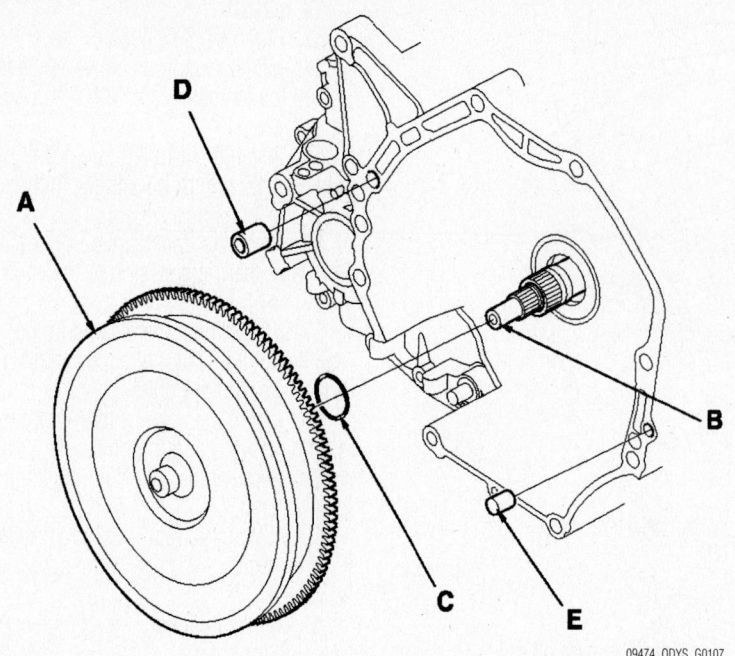

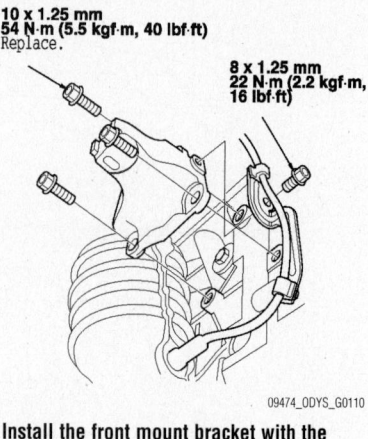

09474_ODYS_G0107

Install the torque converter assembly (A) on the mainshaft (B) with the new O-ring (C). Install the 14 x 20 mm dowel pin (D) and 10 x 20 mm dowel pin (E) in the torque converter housing—2005–06 models

10 x 1.25 mm
54 N·m (5.5 kgf·m, 40 lbf·ft)
Replace.

8 x 1.25 mm
22 N·m (2.2 kgf·m, 16 lbf·ft)

09474_ODYS_G0110

Install the front mount bracket with the new bolts, and install the harness clamp bracket—2005–06 models

69. Support the front subframe with the front subframe adapter (VSB02C000016) and a jack, and lift it up to body.

70. Loosely install the new subframe mounting bolts, front stiffener mounting bolts, rear stiffener mounting bolts, and front and rear stiffeners.

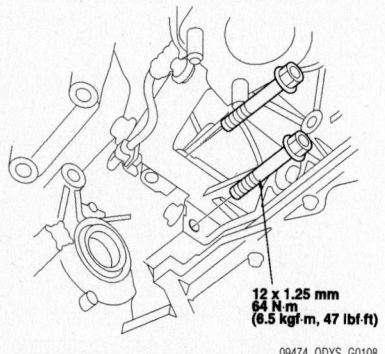

12 x 1.25 mm
64 N·m
(6.5 kgf·m, 47 lbf·ft)

09474_ODYS_G0108

Attach the transmission to the engine, then install the rear transmission housing mounting bolts—2005–06 models

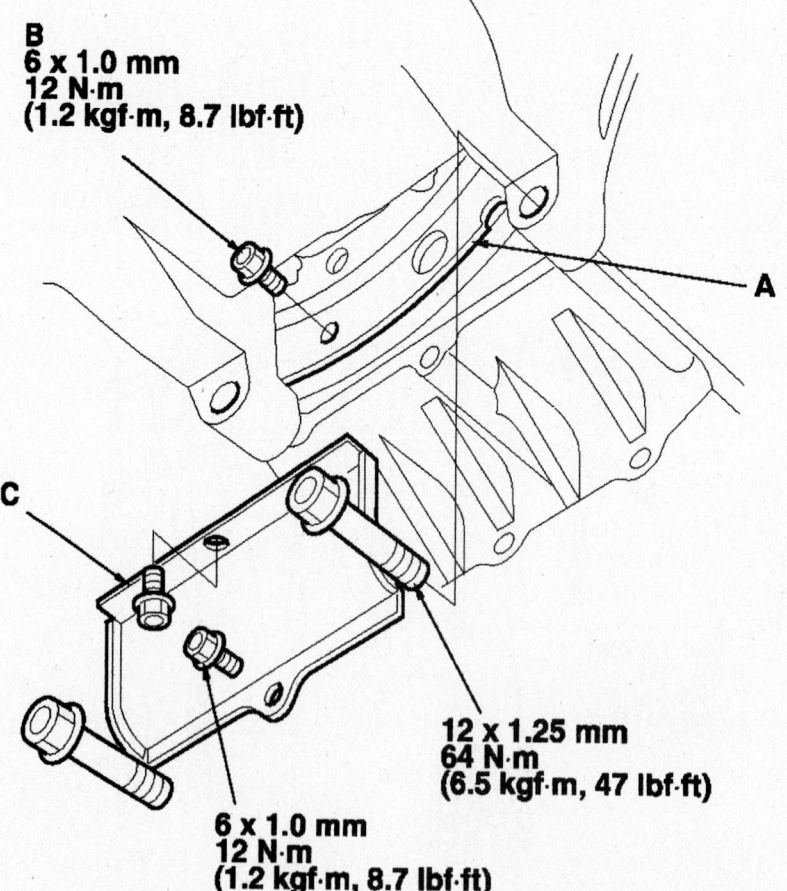

B
6 x 1.0 mm
12 N·m
(1.2 kgf·m, 8.7 lbf·ft)

A

C

12 x 1.25 mm
64 N·m
(6.5 kgf·m, 47 lbf·ft)

6 x 1.0 mm
12 N·m
(1.2 kgf·m, 8.7 lbf·ft)

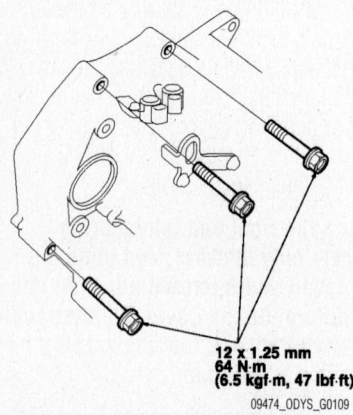

12 x 1.25 mm
64 N·m
(6.5 kgf·m, 47 lbf·ft)

09474_ODYS_G0109

Install the upper and front transmission housing mounting bolts—2005–06 models

09474_ODYS_G0111

Install the engine-to-torque converter housing mounting bolts—2005–06 models

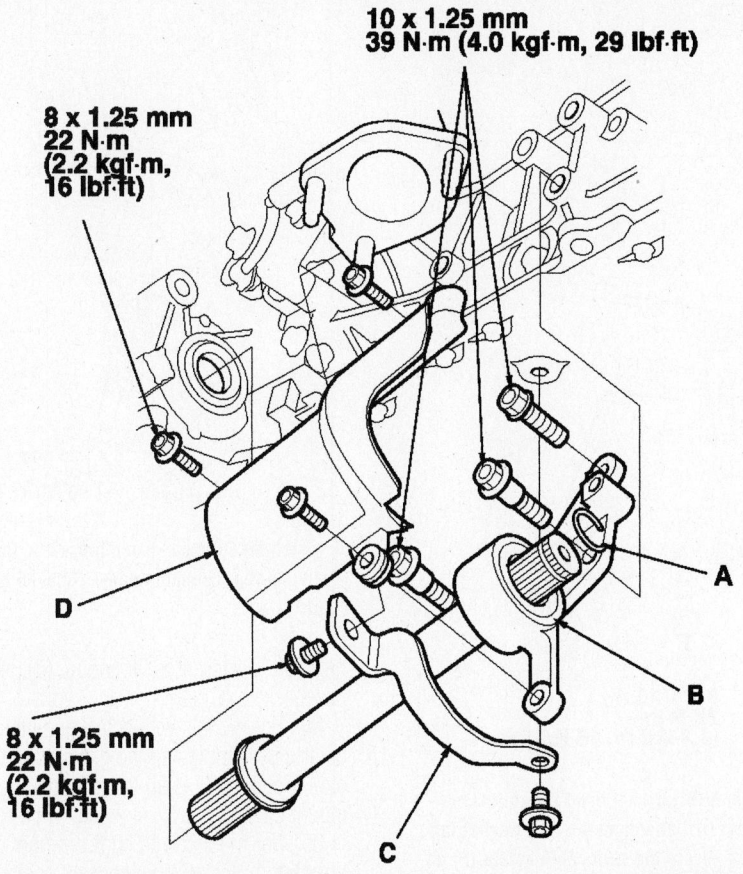

10 x 1.25 mm
39 N·m (4.0 kgf·m, 29 lbf·ft)

8 x 1.25 mm
22 N·m
(2.2 kgf·m,
16 lbf·ft)

D

A

B

8 x 1.25 mm
22 N·m
(2.2 kgf·m,
16 lbf·ft)

C

09474_ODYS_G0112

Install the new set ring (A) on the intermediate shaft (B), and install the intermediate shaft in the differential. While installing the intermediate shaft in the differential, be sure not to allow dust or other foreign particles to enter the transmission. Install the rear warm-up three-way catalytic converter (WU-TWC) stay (C) and heat shield (D)—2005–06 models

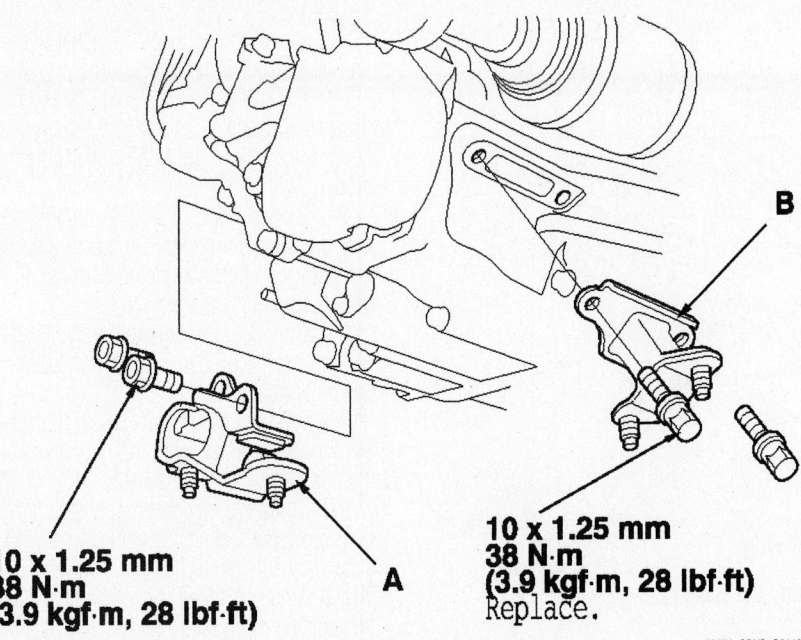

B

10 x 1.25 mm
38 N·m
(3.9 kgf·m, 28 lbf·ft)

A

10 x 1.25 mm
38 N·m
(3.9 kgf·m, 28 lbf·ft)
Replace.

09474_ODYS_G0113

Install the transmission lower front mount (A), and install the lower rear mount (B) with the new bolts—2005–06 models

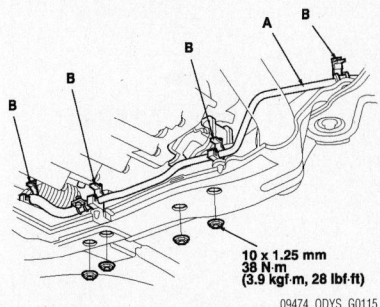

A
B
B
B
B

10 x 1.25 mm
38 N·m
(3.9 kgf·m, 28 lbf·ft)

09474_ODYS_G0115

Install the transmission lower mount nuts.
Install the power steering fluid line (A) on the clamps (B)—2005–06 models

71. Align all reference marks on the body and subframe. Tighten the rear sub-frame mounting bolts, then front bolts, and tighten the stiffener bolts to the specified torque.

72. Install the transmission lower mount nuts.

73. Install the power steering fluid line (A) on the clamps (B).

74. Install the rear mount mounting bolts.

75. Install the steering gearbox stiffener (A), and loosely install the bolts.

76. Install the steering gearbox stiffener, and loosely install the bolts.

77. Install the steering gearbox mounting bolt and heat shield (A).

78. Install the steering gearbox mounting bolt and power steering fluid line bracket bolts.

79. Tighten the steering gearbox mounting bolts and gearbox stiffener mounting bolts to the specified torque.

80. For models with active engine mount control system: Connect engine mount control solenoid valve connector.

81. For models with engine mount control system: Connect the vacuum tube joint, and install the joint in the clamp at the engine mount.

82. Install selector control lever (A) on the selector control shaft (B). Do not bend the shift cable excessively.

83. Install the lock bolt (C) and a new lock washer (D), and bend the lock washer tab against the bolt head.

84. Install the shift cable cover (E), and secure the shift cable holder (F) on the shift cable cover.

85. Install the ball joint on each lower arm with the new castle nut and lock pin.

86. Install the stabilizer links to the lower arms, and install the nuts. Insert a 6 mm Allen wrench in the ball joint pin, and tighten the nuts.

87. Install the exhaust pipe (A) with the

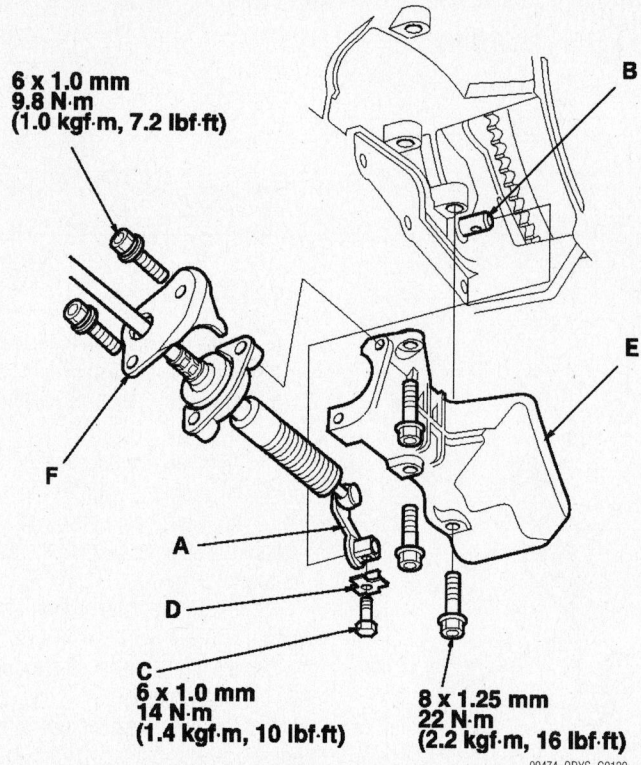

6 x 1.0 mm
9.8 N·m
(1.0 kgf·m, 7.2 lbf·ft)

B

E

F

A

D

C
6 x 1.0 mm
14 N·m
(1.4 kgf·m, 10 lbf·ft)

8 x 1.25 mm
22 N·m
(2.2 kgf·m, 16 lbf·ft)

09474_ODYS_G0120

Install selector control lever (A) on the selector control shaft (B). Do not bend the shift cable excessively. Install the lock bolt (C) and a new lock washer (D), and bend the lock washer tab against the bolt head. Install the shift cable cover (E), and secure the shift cable holder (F) on the shift cable cover—2005–06 models

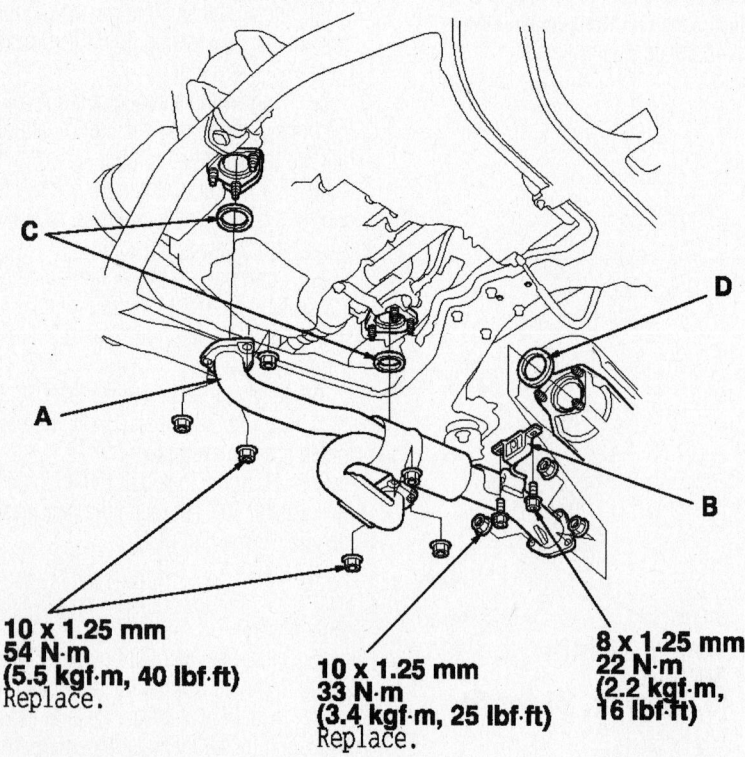

C

D

A

B

10 x 1.25 mm
54 N·m
(5.5 kgf·m, 40 lbf·ft)
Replace.

10 x 1.25 mm
33 N·m
(3.4 kgf·m, 25 lbf·ft)
Replace.

8 x 1.25 mm
22 N·m
(2.2 kgf·m,
16 lbf·ft)

09474_ODYS_G0121

Install the exhaust pipe (A) with the new self-locking nuts, its mount (B), and new gaskets (C) (D)—2005–06 models

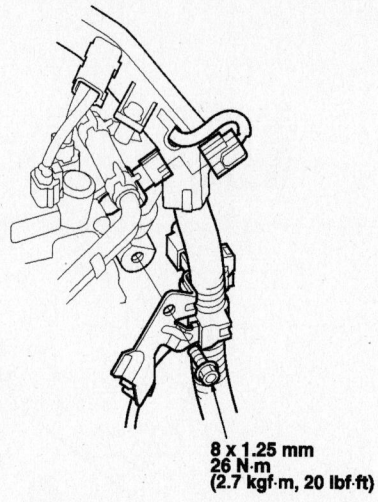

8 x 1.25 mm
26 N·m
(2.7 kgf·m, 20 lbf·ft)

09474_ODYS_G0124

Install the harness clamp bracket to the engine rear cylinder head—2005–06 models

new self-locking nuts, its mount (B), and new gaskets (C) (D).

88. For models with active engine mount control system: Install the new front engine mount bolt, and install the front engine mount stop.

89. For models with engine mount control system: Install the new front engine mount bolt, and install the front engine mount stop (A) and vacuum tube clamp bracket (B).

90. Remove the engine support hanger and engine balancer bar.

91. Install the harness clamp bracket to the engine rear cylinder head.

92. Install the connector bracket to the engine front cylinder head.

93. Install the harness cover on the harness cover bracket, and secure the harness cover bracket with the bolt on the transmission.

94. Connect output shaft (countershaft) speed sensor connector. Connect transmission range switch connector, and install it on the connector bracket.

95. Connect shift solenoid valve B connector and torque converter clutch solenoid valve connector.

96. Connect transmission sub-harness 6P connector, and install the harness clamps on the clamp brackets.

97. Connect the connectors to the shift solenoid valve A, shift solenoid valve C, A/T clutch pressure control solenoid valve C, input shaft (mainshaft) speed sensor, and 4th clutch transmission fluid pressure switch. Install the harness clamps on the clamp brackets.

98. Install the transmission ground terminal.

99. Install the starter on the torque converter housing. Connect the starter cables, and install the harness clamp on the clamp bracket.

100. Connect the ATF cooler hoses to the ATF cooler lines, and secure the hoses with the clips.

101. Install the battery base.

102. Install the air cleaner housing and intake air duct.

103. Install the battery tray and battery, and secure the battery with its hold-down bracket.

104. Clean the battery posts and cable terminals with sandpaper, then assemble them and apply grease to prevent corrosion.

105. Install the front bulkhead cover.

106. Set the parking brake. Start the engine, and shift the transmission through all positions three times.

107. Check the shift lever operation, A/T gear position indication operation, and shift cable adjustment.

108. Check and adjust the front wheel alignment.

109. Install the splash shield and front inner fender.

110. Start the engine in the P or N position, and warm it up to normal operating temperature (the radiator fan comes on).

111. Turn off the engine, and check the ATF level.

112. Road test the vehicle.

113. Enter the audio and navigation antitheft codes, then enter the customer's audio presets, and set the clock.

Halfshaft

REMOVAL & INSTALLATION

2002–04 Models

1. Before servicing the vehicle, refer to the precautions in the beginning of this section.

2. Drain the transaxle.

3. Remove or disconnect the following:
   - Negative battery cable
   - Front wheels

➡️**If removing the left halfshaft, drain the transmission fluid, it is not necessary to drain the fluid if removing the right halfshaft.**

   - Stabilizer bar link
   - Lower ball joint
   - Spindle nut

4. Pry the inboard joint from the transaxle or intermediate shaft.

5. Remove the outer CV-joint stub shaft

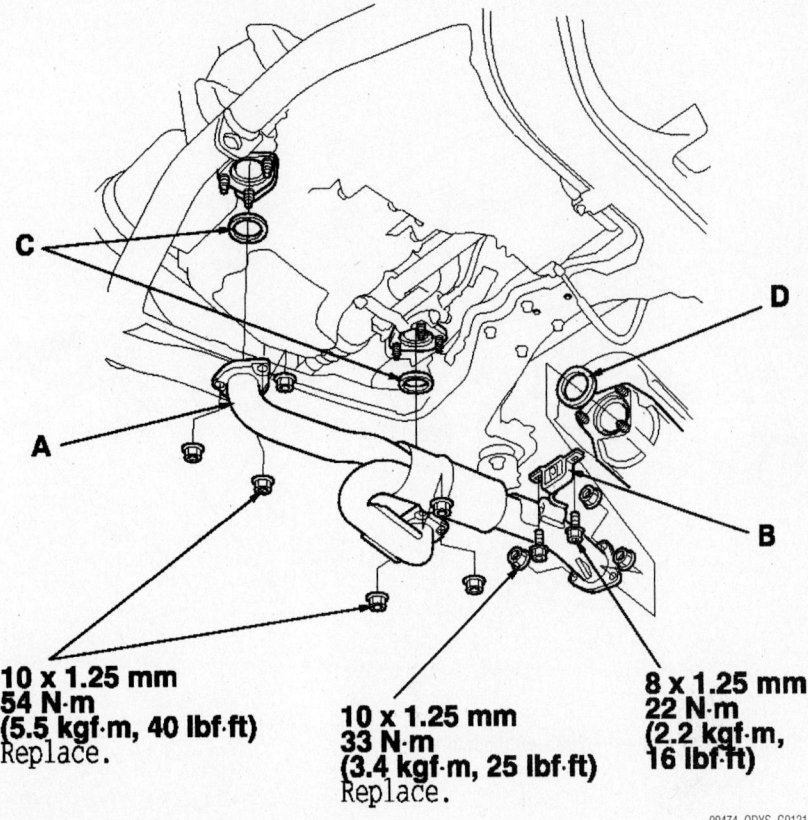

10 x 1.25 mm
54 N·m
(5.5 kgf·m, 40 lbf·ft)
Replace.

10 x 1.25 mm
33 N·m
(3.4 kgf·m, 25 lbf·ft)
Replace.

8 x 1.25 mm
22 N·m
(2.2 kgf·m, 16 lbf·ft)

09474_ODYS_G0121

Remove exhaust pipe A—2005–06 models

from the hub by tapping the stub shaft with a plastic hammer.

To install:

➡️**Use new circlips, split pins and self-locking nuts for assembly.**

6. Install the outer CV-joint stub shaft into the hub.

7. Install the inner CV-joint to the transaxle or intermediate shaft until the circlip locks in the retaining groove.

8. Install or connect the following:
   - Lower ball joint. Tighten the nut to 43–51 ft. lbs. (59–69 Nm).
   - Stabilizer bar link. Tighten the nut to 58 ft. lbs. (78 Nm).
   - Spindle nut. Tighten the nut to 181 ft. lbs. (245 Nm).
   - Front wheels
   - Negative battery cable

9. Fill the transaxle to the correct level and check for leaks.

2005–06 Models

1. Loosen the wheel nuts slightly.

2. Raise the front of the vehicle, and support it with safety stands in the proper locations.

3. Remove the wheel nuts and front wheels.

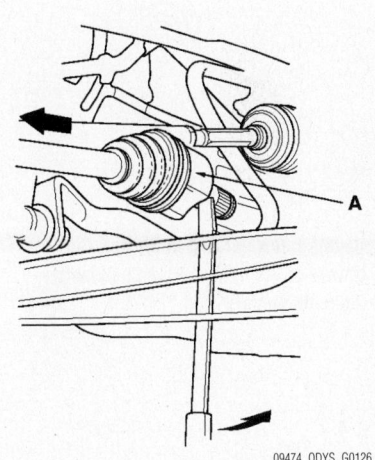

09474_ODYS_G0126

Left halfshaft removal—2005–06 models

4. Lift up the locking tab on the spindle nut, then remove the nut.

5. Drain the automatic transmission fluid Reinstall the drain plug using a new washer.

6. Separate the front stabilizer link from the damper. See the procedure under Front Suspension.

7. Remove the lock pin from the lower arm ball joint castle nut, and remove the nut, then separate the ball joint from the lower arm with the ball joint thread protector and remover.

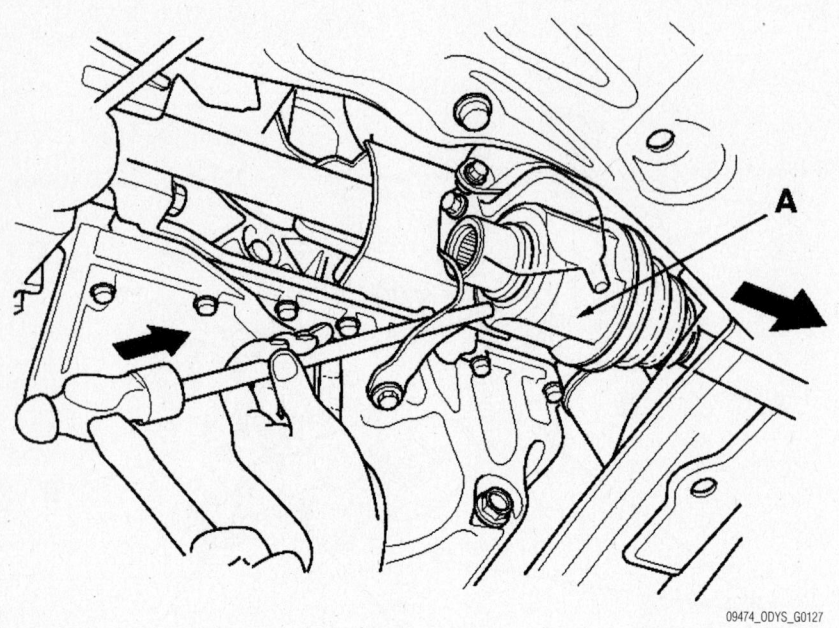

Right halfshaft removal—2005–06 models

➡ To avoid damaging the ball joint, install the ball joint thread protector onto the threads of the ball joint. Be careful not to damage the ball joint boot when installing the remover.

  8. Pull the knuckle outward, and remove the outboard joint from the front wheel hub using a plastic hammer.
  9. Remove exhaust pipe A.
  10. Left halfshaft: Pry the inboard joint (A) from the transmission housing with a prybar. Remove the halfshaft as an assembly.

➡ Do not pull on the halfshaft (B) or the inboard joint may come apart. Pull the halfshaft straight out to avoid damaging the oil seal.

  11. Right halfshaft: Drive the inboard joint (A) off of the intermediate shaft with a drift and hammer. Remove the halfshaft as an assembly.

➡ Do not pull on the halfshaft (B) or the inboard joint may come apart. Pull the halfshaft straight out to avoid damaging the oil seal.

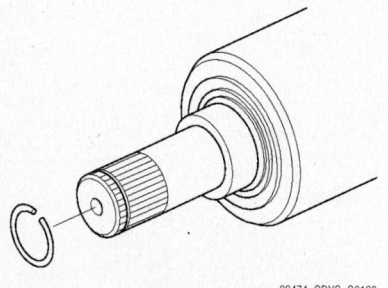

Install a new set ring in the set ring groove of the halfshaft (left halfshaft)—2005–06 models

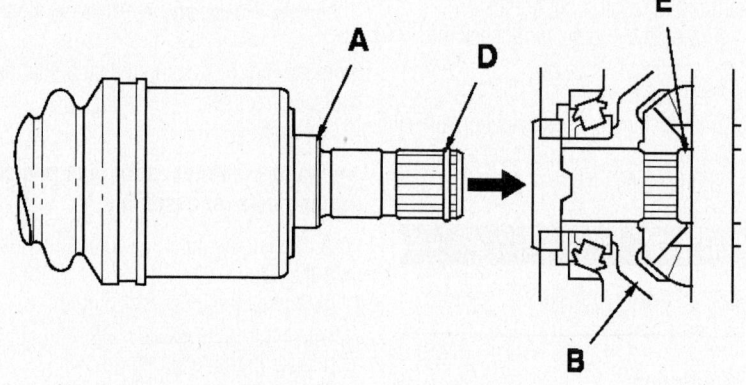

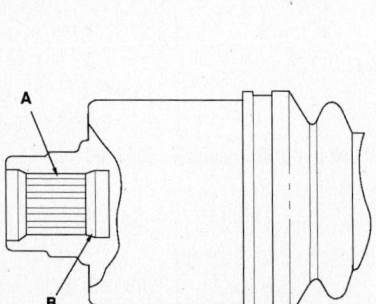

Apply 2.0-3.0 g (0.07-0.11 oz) of grease to the whole splined surface (A) of the right halfshaft. After applying grease, remove the grease from the splined grooves at intervals of 2-3 splines and from the set ring groove (B) so that air can bleed from the intermediate shaft—2005–06 models

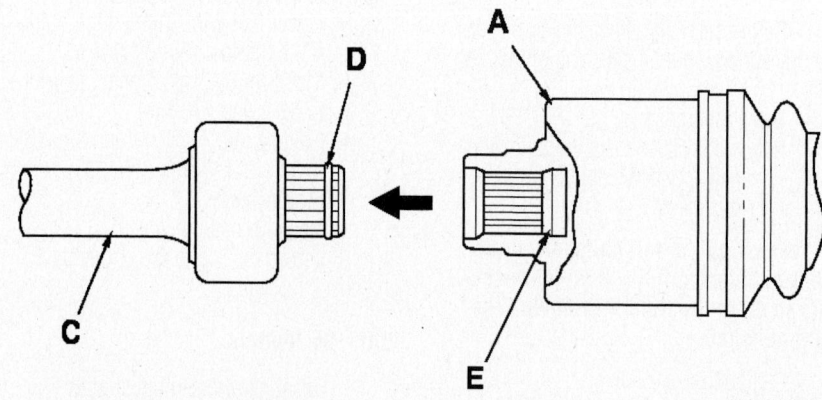

Clean the areas where the halfshaft contacts the differential thoroughly with brake cleaner, and dry with compressed air. Do not wash the rubber parts with solvent. Insert the inboard end (A) of the halfshaft into the differential (B) or intermediate shaft (C) until the new set ring (D) locks in the groove (E)—2005–06 models

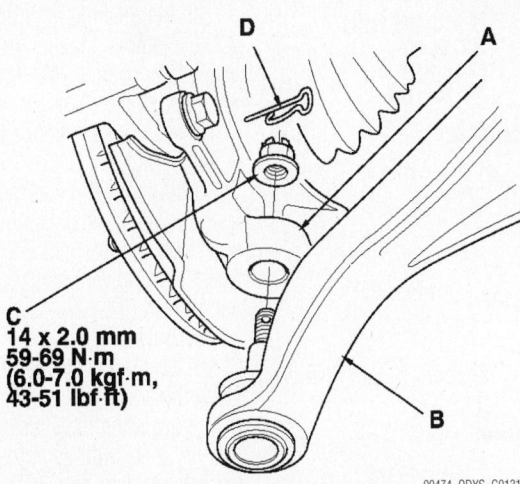

09474_ODYS_G0131

Clean off any grease contamination from the ball joint tapered section and threads, then install the knuckle (A) onto the lower arm (B). Torque the new castle nut (C) to the lower torque specification, then tighten it only far enough to align the slot with the ball joint pin hole—2005–06 models

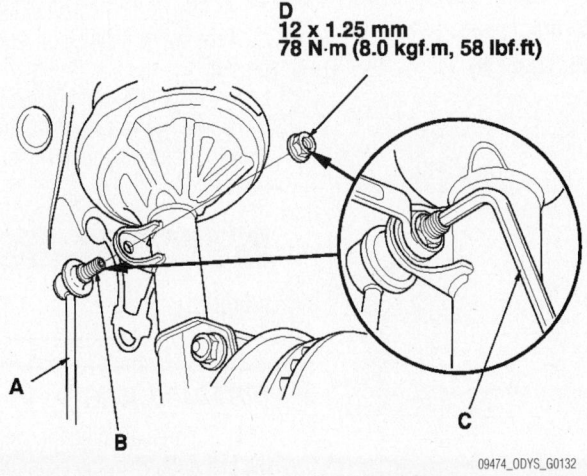

09474_ODYS_G0132

Connect the front stabilizer link (A) to the damper. Hold the stabilizer link ball joint pin (B) with a hex wrench (C), and tighten the new flange nut (D)—2005–06 models

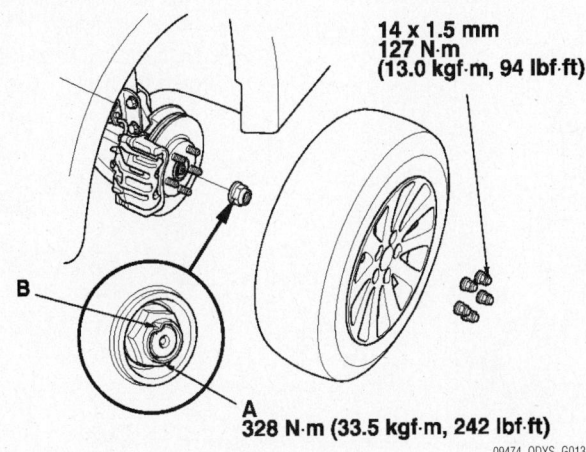

09474_ODYS_G0133

After tightening, use a drift to stake the spindle nut shoulder (B) against the halfshaft—2005–06 models

To install:

12. Install a new set ring in the set ring groove of the halfshaft (left halfshaft).

13. Apply 2.0–3.0 g (0.07–0.11 oz) of grease to the whole splined surface (A) of the right halfshaft. After applying grease, remove the grease from the splined grooves at intervals of 2–3 splines and from the set ring groove (B) so that air can bleed from the intermediate shaft.

14. Clean the areas where the halfshaft contacts the differential thoroughly with brake cleaner, and dry with compressed air. Do not wash the rubber parts with solvent. Insert the inboard end (A) of the halfshaft into the differential (B) or intermediate shaft (C) until the new set ring (D) locks in the groove (E).

15. Install the outboard joint (A) into the front hub (B).

16. Install exhaust pipe A.

17. Clean off any grease contamination from the ball joint tapered section and threads, then install the knuckle (A) onto the lower arm (B). Torque the new castle nut (C) to the lower torque specification, then tighten it only far enough to align the slot with the ball joint pin hole. Do not align the nut by loosening it.

➥**Make sure the ball joint boot is not damaged or cracked.**

18. Install the new lock pin (D) into the ball joint pin hole.

19. Connect the front stabilizer link (A) to the damper. Hold the stabilizer link ball joint pin (B) with a hex wrench (C), and tighten the new flange nut (D).

20. Install a new spindle nut (A), then tighten the nut. After tightening, use a drift to stake the spindle nut shoulder (B) against the halfshaft.

21. Clean the mating surfaces of the brake disc and the front wheel, then install the front wheel with the wheel nuts.

22. Install a new spindle nut (A), then tighten the nut.

23. Turn the front wheel by hand, and make sure there is no interference between the halfshaft and surrounding parts.

24. Refill the automatic transmission with recommended transmission fluid.

25. Check the front wheel alignment, and adjust it if necessary

Intermediate Shaft

REMOVAL & INSTALLATION

1. Drain the automatic transmission fluid Reinstall the drain plug, using a new washer.

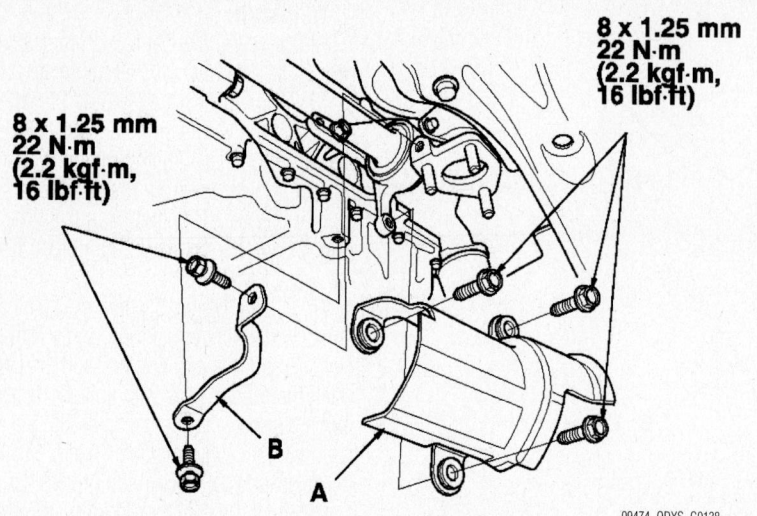

8 x 1.25 mm
22 N·m
(2.2 kgf·m,
16 lbf·ft)

8 x 1.25 mm
22 N·m
(2.2 kgf·m,
16 lbf·ft)

B

A

09474_ODYS_G0138

Remove the rear warm-up three-way catalytic converter (WU-TWC) bracket (B) and heat shield (A)

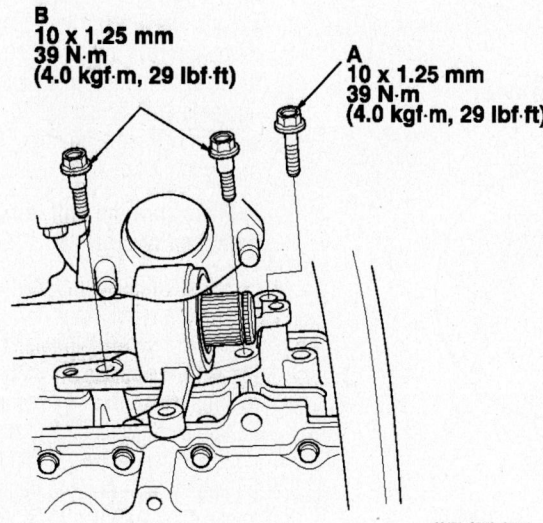

B
10 x 1.25 mm
39 N·m
(4.0 kgf·m, 29 lbf·ft)

A
10 x 1.25 mm
39 N·m
(4.0 kgf·m, 29 lbf·ft)

09474_ODYS_G0137

Remove the flange bolt (A) and two dowel bolts (B)

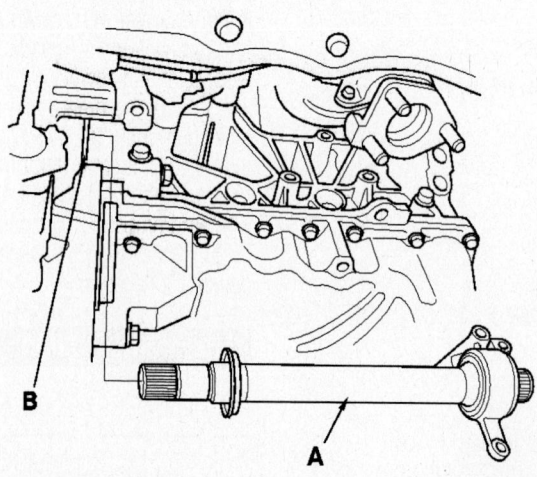

B

A

09474_ODYS_G0136

Remove the intermediate shaft (A) from the differential. Hold the intermediate shaft horizontally until it is clear of the differential to prevent damage to the differential oil seal (B)

2. Remove the right driveshaft.

3. Remove the rear warm-up three-way catalytic converter (WU-TWC) bracket (B) and heat shield (A).

4. Remove the flange bolt (A) and two dowel bolts (B).

5. Remove the intermediate shaft (A) from the differential. Hold the intermediate shaft horizontally until it is clear of the differential to prevent damage to the differential oil seal (B).

To install:

6. Use brake cleaner to thoroughly clean the areas where the intermediate shaft (A) contacts the transmission (differential), and dry them with compressed air. Do not wash the rubber parts with solvent. Insert the intermediate shaft assembly into the differential. Hold the intermediate shaft horizontally to prevent damage to the differential oil seal (B).

7. Install the flange bolt (A) and two dowel bolts (B).

8. Install the heat shield and rear warm-up three-way catalytic converter (WU-TWC) bracket.

9. Install the right driveshaft.

10. Refill the automatic transmission with the recommended transmission fluid.

CV-Joints

OVERHAUL

OUTBOARD JOINT

1. Before servicing the vehicle, refer to the precautions in the beginning of this section.

2. Remove or disconnect the following:

- Axle halfshaft from the vehicle and place it in a vise
- Outboard joint boot clamps and push the boot back

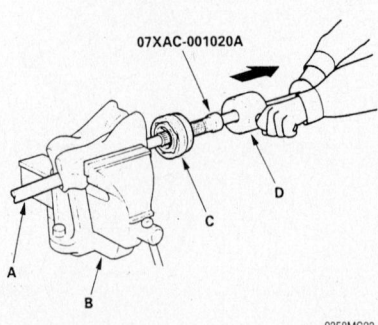

07XAC-001020A

A

B

C

D

9358MG23

Removing the outboard joint using tool 07XAC-001020A

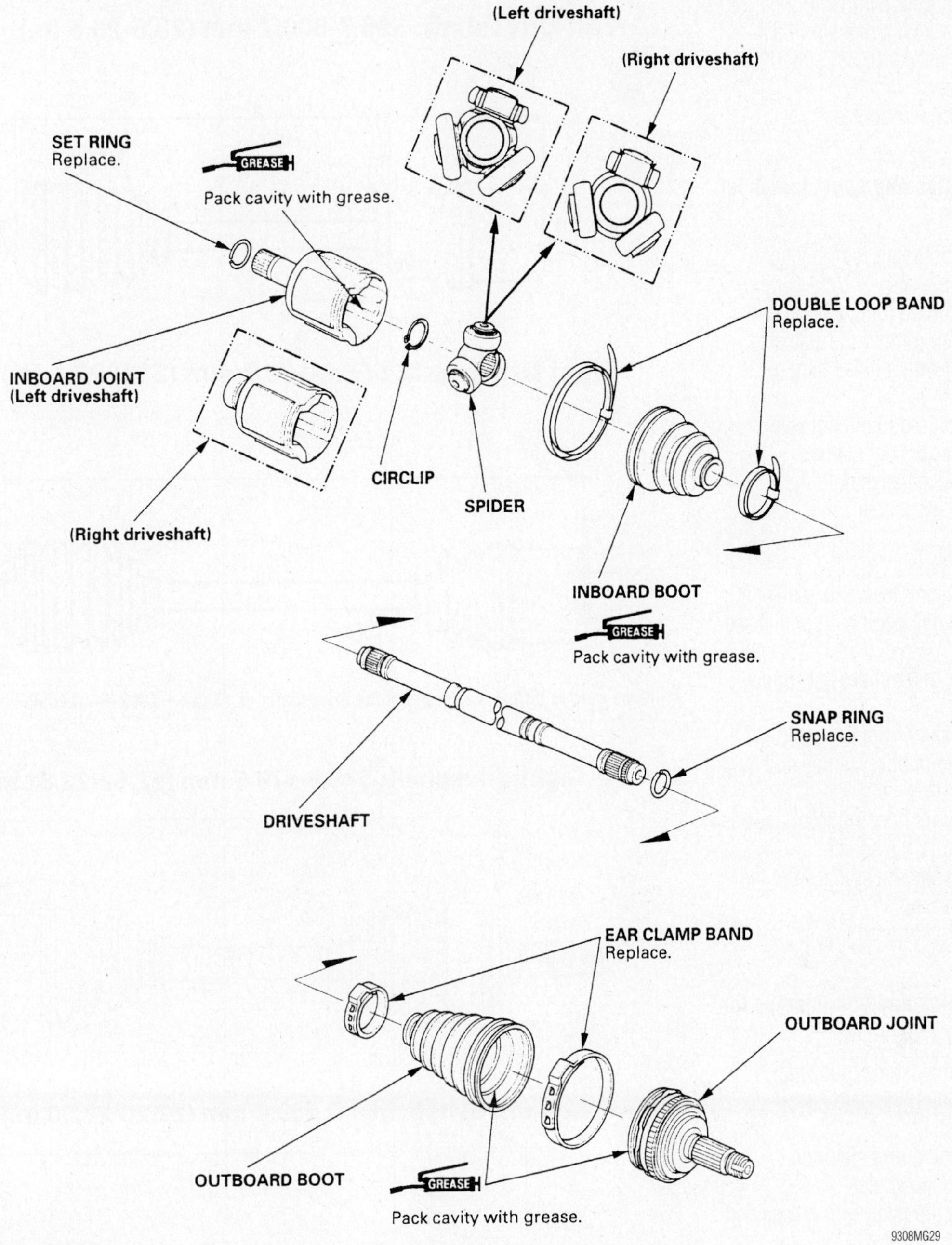

(Left driveshaft)

(Right driveshaft)

SET RING
Replace.

GREASE

Pack cavity with grease.

DOUBLE LOOP BAND
Replace.

INBOARD JOINT
(Left driveshaft)

(Right driveshaft)

CIRCLIP

SPIDER

INBOARD BOOT

GREASE

Pack cavity with grease.

SNAP RING
Replace.

DRIVESHAFT

EAR CLAMP BAND
Replace.

OUTBOARD JOINT

OUTBOARD BOOT

GREASE

Pack cavity with grease.

9308MG29

Front axle exploded view

Locking tab type

A

B

Double loop type

D

C

9358MG21

View of the locking tab type boot band

9358MG22

View of the double loop tab type boot band

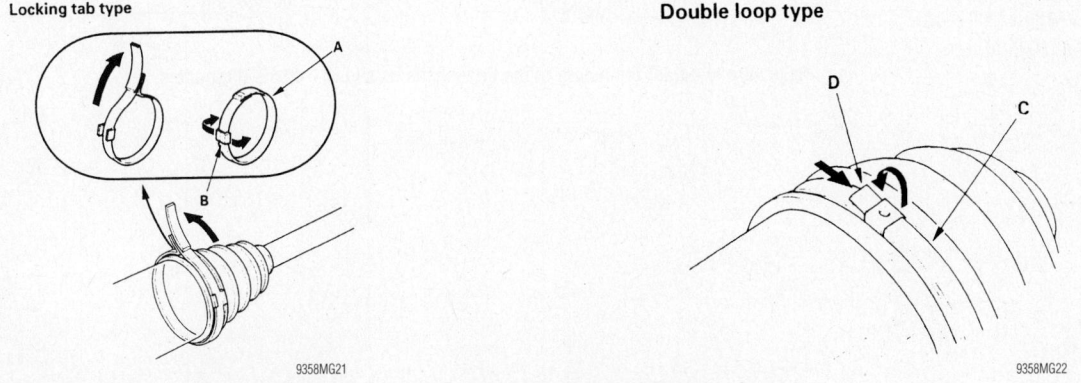

- Outboard joint by driving it off the axle shaft with a brass drift and hammer or tool 07XAC-001020A, if available
- Outboard joint boot

To install:

➡ **Use new circlips and boot clamps for assembly.**

3. Install the outboard joint boot and clamps to the axle shaft.

4. Fill the outboard joint with grease. Install the outboard joint to the axle shaft. Tap the stub shaft with a brass hammer to seat the circlip.

5. Fill the outboard joint boot with grease and install the boot clamps.

6. Install the axle halfshaft to the vehicle.

INBOARD JOINT

1. Before servicing the vehicle, refer to the precautions in the beginning of this section.

2. Remove or disconnect the following:

- Axle halfshaft from the vehicle.
- Inboard joint boot clamps and push the boot back
- Inboard joint housing from the axle
- Rollers from the spider
- Snapring and the spider from the axle shaft
- Inboard joint boot

To install:

➡ **Use new circlips and boot clamps for assembly.**

3. Install or connect the following:
- Inboard joint boot and clamps to the axle shaft
- Spider with a new snapring
- Rollers to the spider

4. Fill the joint housing with grease and install it.

5. Fill the inboard joint boot with grease and install the boot clamps. Make sure to adjust the length of the driveshafts as shown in the accompanying illustration.

6. Install the axle halfshaft to the vehicle.

Left Driveshaft: 595.7–600.7 mm (23.5–23.6 in.)

Right Driveshaft: 555.8–560.8 mm (21.9–22.1 in.)

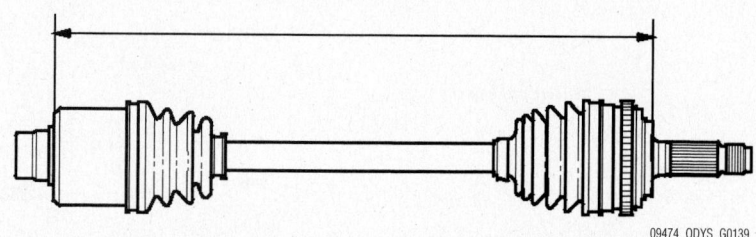

09474_ODYS_G0139

Make sure to adjust the length of the driveshafts as shown—2002–04 models

Right driveshaft: 574.5–579.5 mm (22.62–22.81 in.)

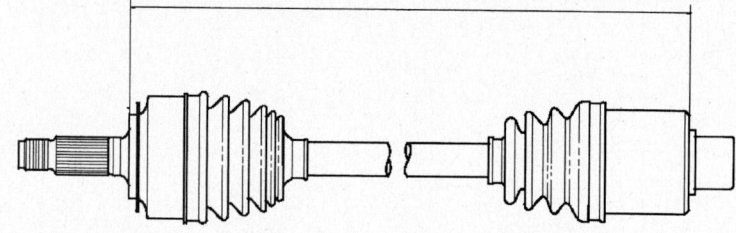

Left driveshaft: 564.0–569.0 mm (22.20–22.40 in.)

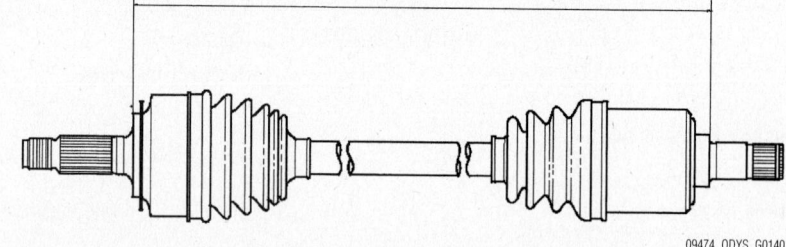

09474_ODYS_G0140

Make sure to adjust the length of the driveshafts as shown—2005–06 models

STEERING

Air Bag

✳✳ CAUTION

Some vehicles are equipped with an air bag system. The system must be disarmed before performing service on, or around, system components, the steering column, instrument panel components, wiring and sensors. Failure to follow the safety precautions and the disarming procedure could result in accidental air bag deployment, possible injury and unnecessary system repairs.

PRECAUTIONS

Several precautions must be observed when handling the inflator module to avoid accidental deployment and possible personal injury.

• Never carry the inflator module by the wires or connector on the underside of the module.

• When carrying a live inflator module, hold securely with both hands, and ensure that the bag and trim cover are pointed away.

• Place the inflator module on a bench or other surface with the bag and trim cover facing up.

• With the inflator module on the bench, never place anything on or close to the module that may be thrown in the event of an accidental deployment.

Before servicing the vehicle, also make sure to refer to the precautions in the beginning of this section as well.

DISARMING

Disconnect and isolate the negative battery cable. Wait 3 minutes for the system capacitor to discharge before performing any service.

Power Rack and Pinion Steering Gear

REMOVAL & INSTALLATION

2002–04 Models

✳✳ WARNING

Do not permit the steering wheel to turn whenever the steering gear is

disconnected from the steering column. Damage to the air bag wiring can result.

1. Before servicing the vehicle, refer to the precautions in the beginning of this section.

2. Center the steering wheel and lock it in position.

3. Attach a support fixture to the engine lifting eyes.

4. Remove or disconnect the following:
• Negative battery cable
• Drivers side air bag assembly, if equipped
• Steering joint cover
• Steering flexible joint
• Front wheels
• Outer tie rod ends
• Splash shield
• Heated Oxygen Sensor (HO2S) connectors
• Exhaust front pipe
• Fule feed line
• Power steering fluid lines
• Rear engine mount

5. Support the front subframe with a jack.

6. Loosen the 14mm subframe bolts and remove the 12mm stiffener plate bolts.

7. Lower the subframe about 1 ³⁄₁₆ inches (30mm).

8. Remove or disconnect the following:
• Right steering gear mounting bracket
• Left steering gear mounting bolts
• Steering gear

To install:

9. Position the steering gear in the vehicle.

10. Install or connect the following:
• Left steering gear mounting bolts. Tighten the bolts to 43 ft. lbs. (58 Nm).
• Right steering gear mounting bracket. Tighten the bolts to 29 ft. lbs. (39 Nm).

11. Raise the subframe into position. Tighten the 14mm bolts to 76 ft. lbs. (103 Nm) and the 12mm bolts to 54 ft. lbs. (74 Nm).

12. Install or connect the following:
• Rear engine mount
• Power steering fluid lines
• Exhaust front pipe
• HO2S connectors
• Splash shield
• Outer tie rod ends
• Front wheels. Position the wheels straight-ahead.

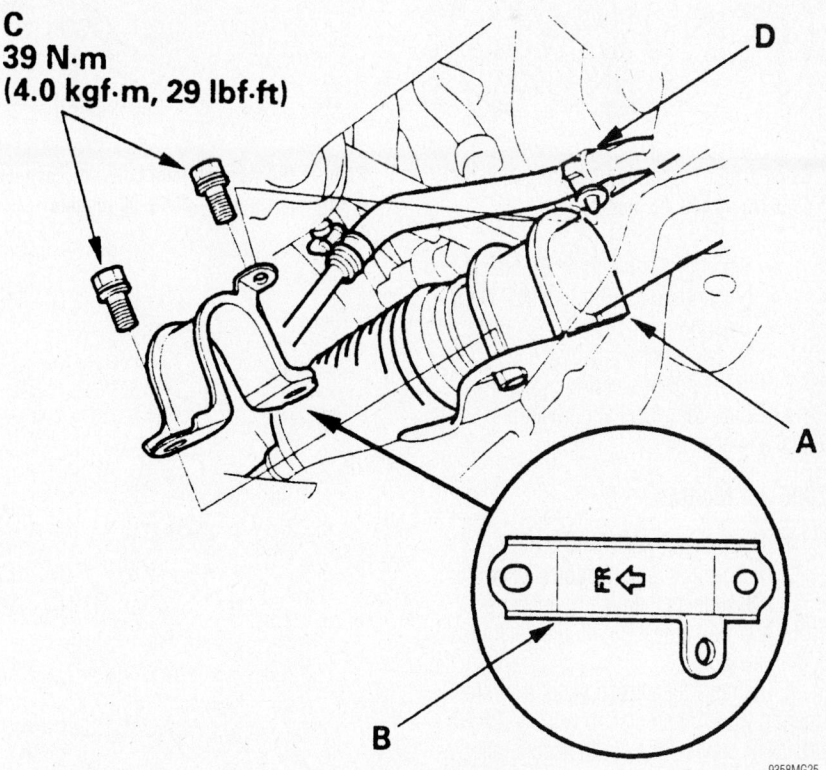

**C
39 N·m
(4.0 kgf·m, 29 lbf·ft)**

9358MG25

Make sure the right steering gear mounting bracket is oriented properly—2002–04 models

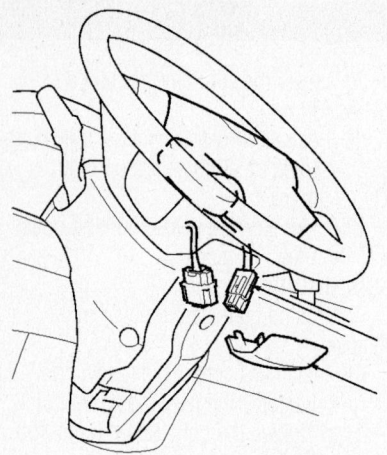

09474_ODYS_G0141

Remove the access panel from the steering wheel, then disconnect the driver's airbag 4P connector from the cable reel—2005–06 models

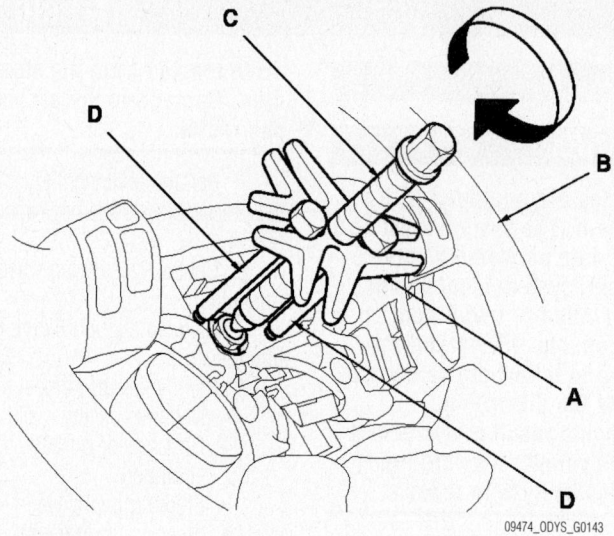

09474_ODYS_G0143

Steering wheel puller

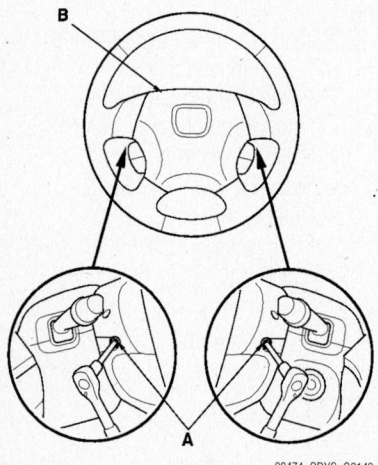

09474_ODYS_G0142

Using a Torx T30 bit, remove the two Torx bolts (A)—2005–06 models

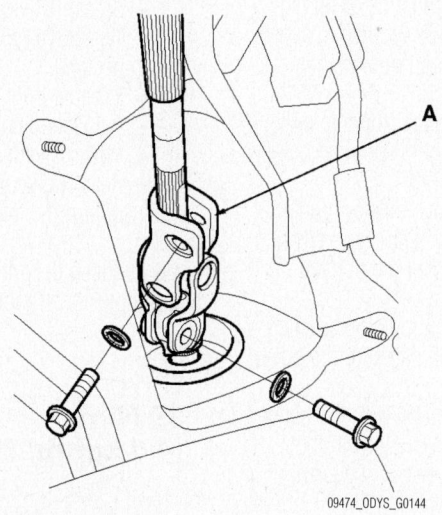

09474_ODYS_G0144

Remove the steering joint bolts, disconnect the steering joint by moving the steering joint (A) toward the column—2005–06 models

- Steering flexible joint. Tighten the pinch bolts to 16 ft. lbs. (22 Nm).
- Steering joint cover
- Negative battery cable
13. Fill the power steering system.
14. Check the wheel alignment and adjust as necessary.

2005–06 Models

1. Drain the power steering fluid.
2. Record the radio station preset.
3. Disconnect the battery negative cable, and wait at least 3 minutes before beginning work.
4. Remove the access panel from the steering wheel, then disconnect the driver's airbag 4P connector from the cable reel.
5. Using a Torx T30 bit, remove the two Torx bolts (A).

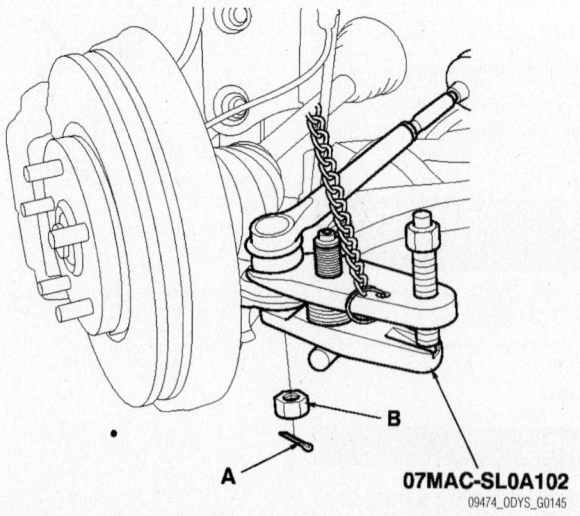

07MAC-SL0A102
09474_ODYS_G0145

Remove the cotter pin (A) from the 12 mm nut (B), and loosen the nut—2005–06 models

not

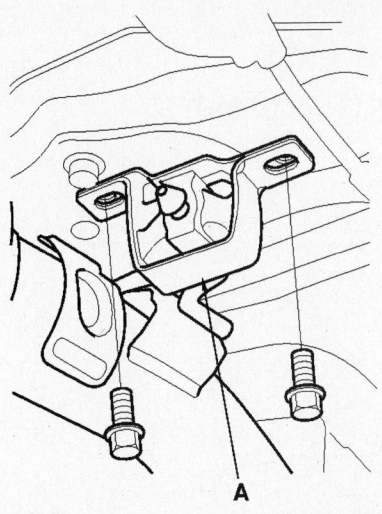

09474_ODYS_G0146

Remove the 10 mm flange bolts of the exhaust rubber mount (A)—2005–06 models

6. Disconnect the horn switch connector (1P), then remove the driver's airbag.

7. Align the front wheels straight ahead.

8. Disconnect the cable reel sub-harness connector from the cable reel.

9. Loosen the steering wheel nut three full turns.

10. Install a commercially available steering wheel puller (A) on the steering wheel (B). Free the steering wheel from the steering column shaft by turning the pressure bolt (C) of the puller.

11. Note these items when removing the steering wheel:

  a. Do not tap on the steering wheel or the steering column shaft when removing the steering wheel.

  b. If you thread the puller bolts (D) into the wheel hub more than five threads, the bolts will hit the cable reel and damage it. To prevent this, install a pair of jam nuts five threads up on each puller bolt.

12. Remove the steering wheel puller, then remove the steering wheel nut and steering wheel from the steering column.

13. Remove the steering joint cover.

14. Remove the steering joint bolts, disconnect the steering joint by moving the steering joint (A) toward the column.

15. Remove the cotter pin (A) from the 12 mm nut (B), and loosen the nut.

16. Separate the tie-rod ball joint and knuckle using the special tool.

17. Remove the 10 mm flange bolts of the exhaust rubber mount (A).

18. Remove the three self-locking nuts, and disconnect the three way catalytic converter (TWC) from the muffler.

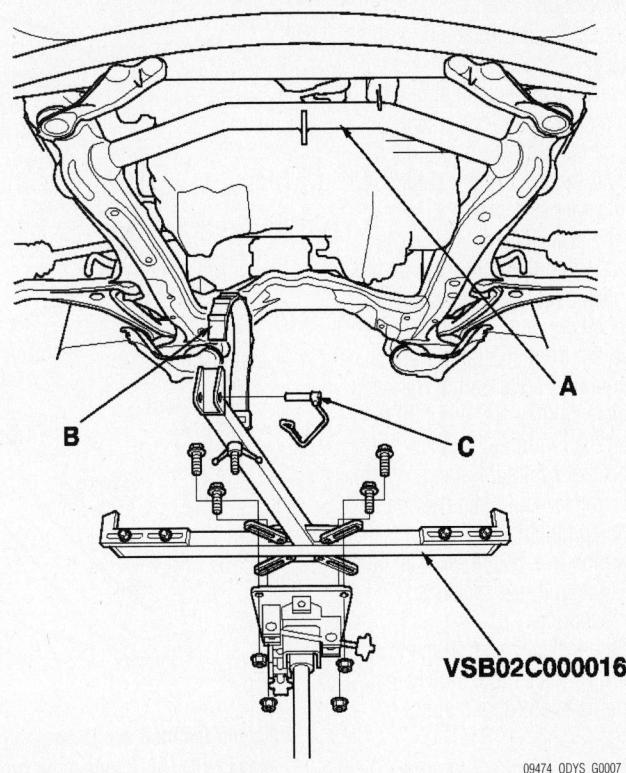

VSB02C000016

09474_ODYS_G0007

Attach the special tool to the front subframe (A) by hanging the belt (B) of the special tool over the front of the subframe, secure it with the pin (C), then tighten the special tool screw—2005–06 models

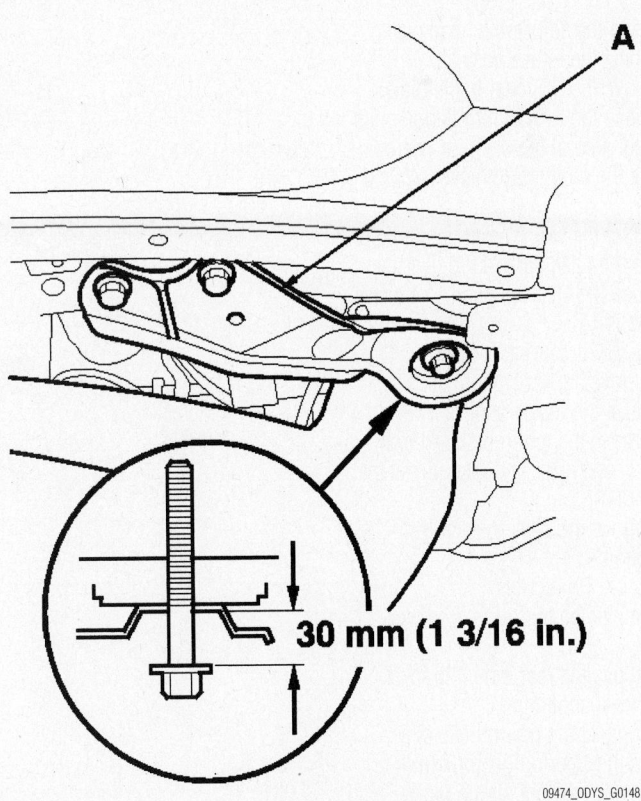

30 mm (1 3/16 in.)

09474_ODYS_G0148

Loosen the front subframe front bracket (A) mounting bolts on the right and left of the vehicle so they are about 30 mm (1³⁄₁₆ in.) from the mounting surface—2005–06 models

19. Remove exhaust pipe A.

20. Disconnect the power steering pressure (PSP) switch connector (A).

21. Remove the front splash shield (A).

22. Attach the special tool to the front subframe (A) by hanging the hook of the special tool over the front of the subframe, then tighten the special tool screw.

23. Raise the jack (B) and line up the slots in the arms with the bolt holes on the corner of the jack base, then attach them with bolts securely.

24. Loosen the front subframe front bracket (A) mounting bolts on the right and left of the vehicle so they are about 30 mm (1 3/16 in.) from the mounting surface.

25. Support the front subframe securely by raising the special tool, then remove the two 12 mm flange bolts (A).

26. Remove the two 14 mm special bolts (B) and front subframe rear brackets (C) from the front subframe.

27. Lower the jack supporting the front subframe with the special tool slowly until the front subframe has dropped about 50 mm (1 15/16 in.).

28. Remove the P/S line mounting brackets from the front subframe and gearbox mounting bracket.

29. Loosen the 16 mm flare nut, and disconnect the feed line from the gear box.

30. Loosen the adjustable hose clamp and disconnect the return hose.

31. Remove the P/S heat baffle plate.

32. Remove the two 10 mm flange bolts from the right side of the steering gearbox, then remove the mounting bracket (A) and cushion (B).

33. Remove the two 10 mm flange bolts and nuts from the left side of the gearbox.

34. Loosen the steering gearbox bracket mounting bolts (A).

35. EX-L, EX-L Touring models:

   a. Remove the rear mount stop (A), then remove the rear mount bolt (B).

   b. Remove the rear mount (A) from the base bracket (B), and disconnect the connector (C).

   c. Remove the base bracket from the front subframe.

36. VAN, LX, EX models

   a. Remove the rear mount bracket bolt (A).

   b. Remove the rear mount bracket from the front subframe.

37. Disconnect the return hose clip.

38. Move the steering gearbox to the driver's side, and rotate it so the pinion shaft points toward the front of the vehicle.

39. Carefully move the steering gearbox as an assembly toward the left side of the

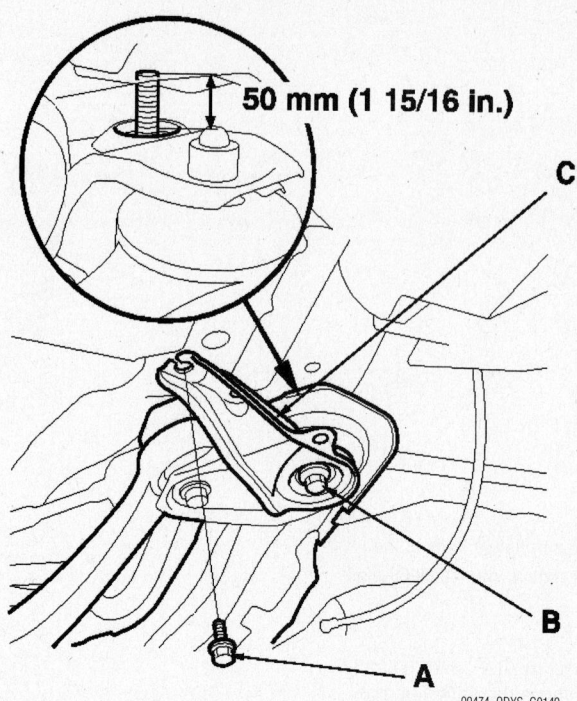

50 mm (1 15/16 in.)

09474_ODYS_G0149

Support the front subframe securely by raising the special tool, then remove the two 12 mm flange bolts (A). Remove the two 14 mm special bolts (B) and front subframe rear brackets (C) from the front subframe. Lower the jack supporting the front subframe with the special tool slowly until the front subframe has dropped about 50 mm (1 15/16 in.)—2005–06 models

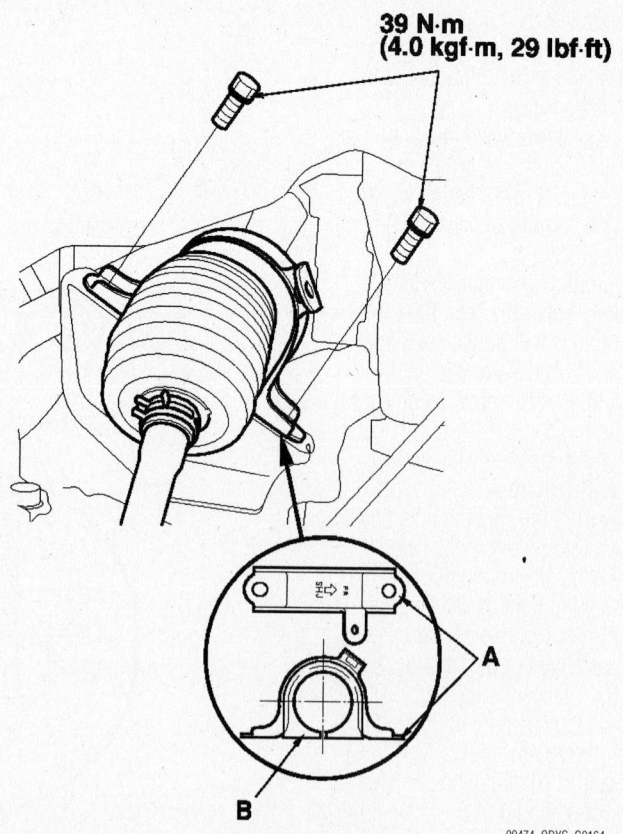

39 N·m (4.0 kgf·m, 29 lbf·ft)

09474_ODYS_G0164

Remove the two 10 mm flange bolts from the right side of the steering gearbox, then remove the mounting bracket (A) and cushion (B)—2005–06 models

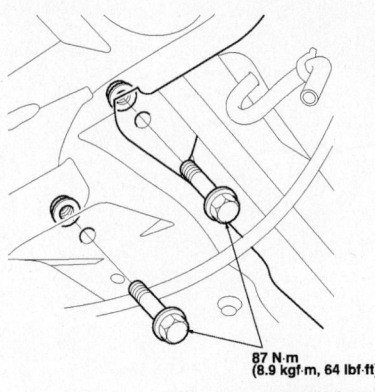

87 N·m
(8.9 kgf·m, 64 lbf·ft)

09474_ODYS_G0163

Remove the two 10 mm flange bolts and nuts from the left side of the gearbox—2005–06 models

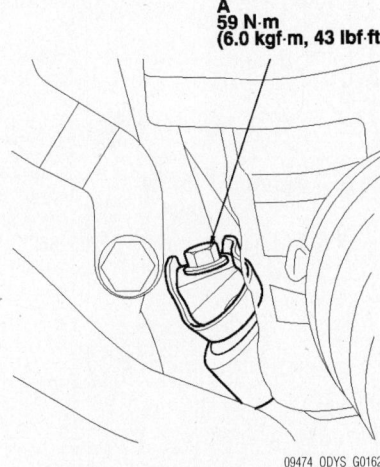

A
59 N·m
(6.0 kgf·m, 43 lbf·ft

09474_ODYS_G0162

Loosen the steering gearbox bracket mounting bolts (A)—2005–06 models

vehicle until the pinion shaft clears the wheel well opening. Be careful not to damage the brake lines with the pinion shaft.

40. Remove the steering gearbox through the wheel well opening on the driver's side.

41. Remove the pinion shaft grommet from the steering joint cover B.

42. After removing the steering gearbox, make sure that no power steering fluid gets on the gearbox mount cushions, gearbox housing, surface of the front subframe and stiffener. Wipe off any spilled fluid at once.

To install:

43. Install the pinion shaft grommet on the valve housing.

44. Slide the steering gearbox between the front subframe and body from the driver's side. Place the gearbox in position on the front suspension subframe.

45. Rotate the steering gearbox so the pinion shaft points upward.

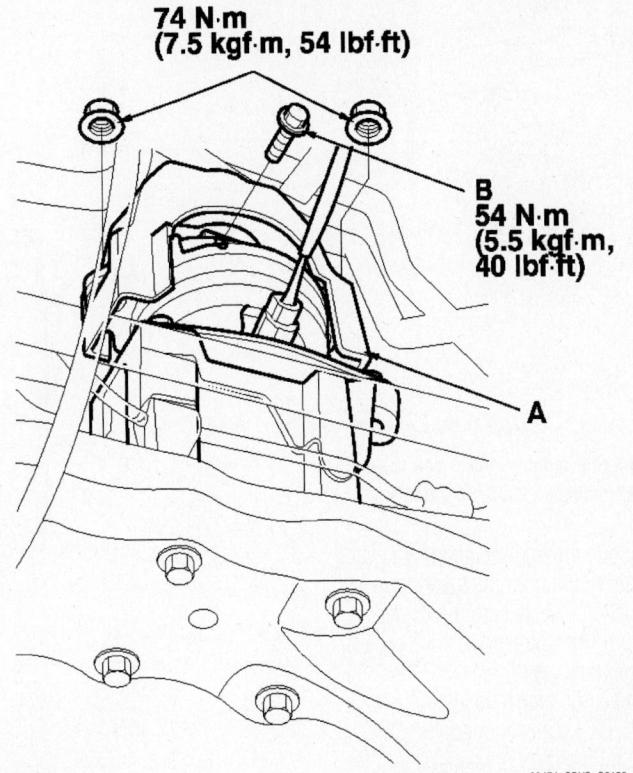

74 N·m
(7.5 kgf·m, 54 lbf·ft)

B
54 N·m
(5.5 kgf·m, 40 lbf·ft)

A

09474_ODYS_G0159

Remove the rear mount stop (A), then remove the rear mount bolt (B)—EX-L, EX-L Touring models—2005–06 models

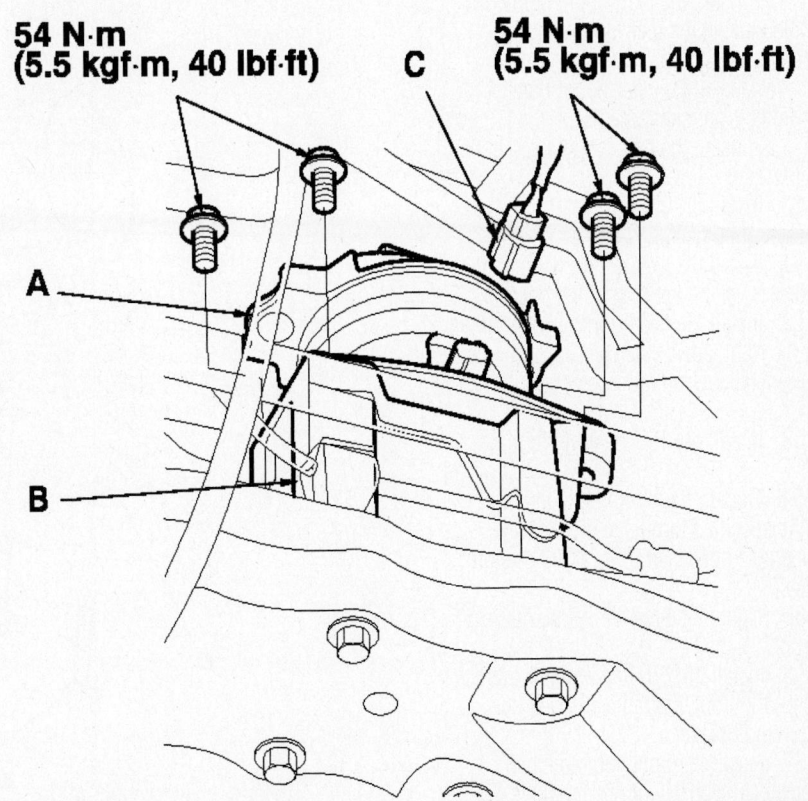

54 N·m
(5.5 kgf·m, 40 lbf·ft)

C

54 N·m
(5.5 kgf·m, 40 lbf·ft)

A

B

09474_ODYS_G0158

Remove the rear mount (A) from the base bracket (B), and disconnect the connector (C)—EX-L, EX-L Touring models—2005–06 models

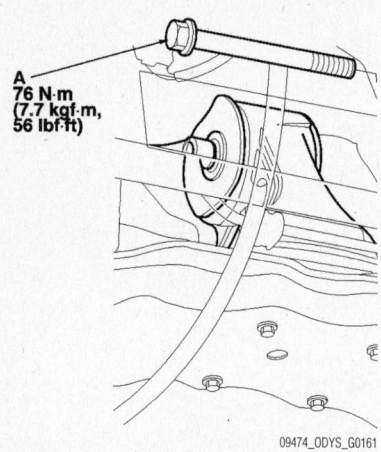

Remove the rear mount bracket bolt (A)—Van, LX, EX models—2005–06 models

46. Continue moving the gearbox toward the passenger's side until the steering gearbox is in position. Make sure the power steering return line and feed line are routed above the gearbox.

47. Connect the return hose clip.

48. EX-L, EX-L Touring models:

    a. Install the base bracket (A) on the front subframe (B).

    b. Install the rear mount (A) on the base bracket (B), and connect the connector (C).

    c. Install the rear mount stop (A), then install the rear mount bolt (B).

49. Van, LX, EX models:

    a. Install the rear mount bracket (A) on the front subframe (B).

    b. Install the rear mount bracket bolt (A).

50. Install the steering gearbox bracket mounting bolts (A).

51. Install the two 12 mm flange bolts and nuts on the left side of the gearbox.

52. Install the two 10 mm flange bolts on the right side of the steering gearbox, then install the mounting bracket and cushion.

53. Install the P/S heat baffle plate (A).

54. Tighten the return hose flare nut (A) to the specified torque, and install the adjustable hose clamp (B) and the return hose (C).

55. Tighten the feed line flare nut to 37 Nm (27 ft. lbs.).

56. Install the P/S line mounting brackets on the front subframe and gearbox mounting bracket.

57. Install the front subframe rear bracket (A) with 12 mm flange bolts (B) and 14 mm special bolts (C), and tighten to specified torque.

58. Install the front subframe front

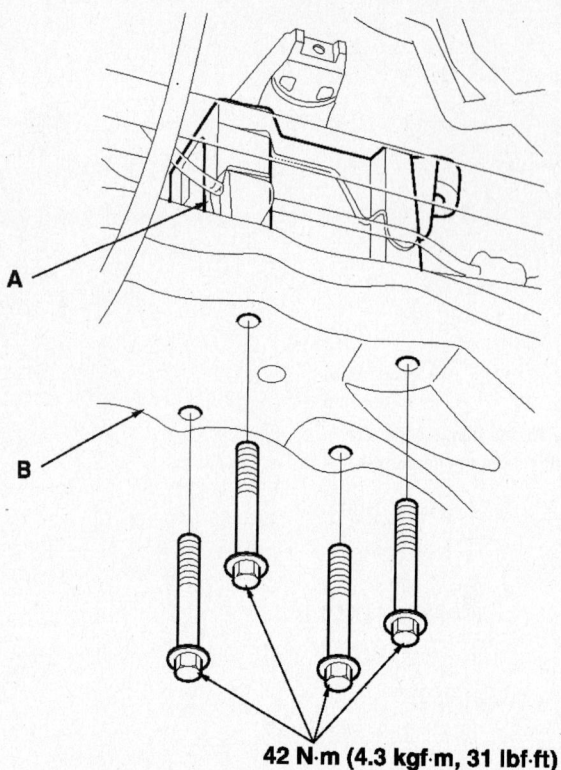

42 N·m (4.3 kgf·m, 31 lbf·ft)

Install the base bracket (A) on the front subframe (B)—EX-L, EX-L Touring models—2005–06 models

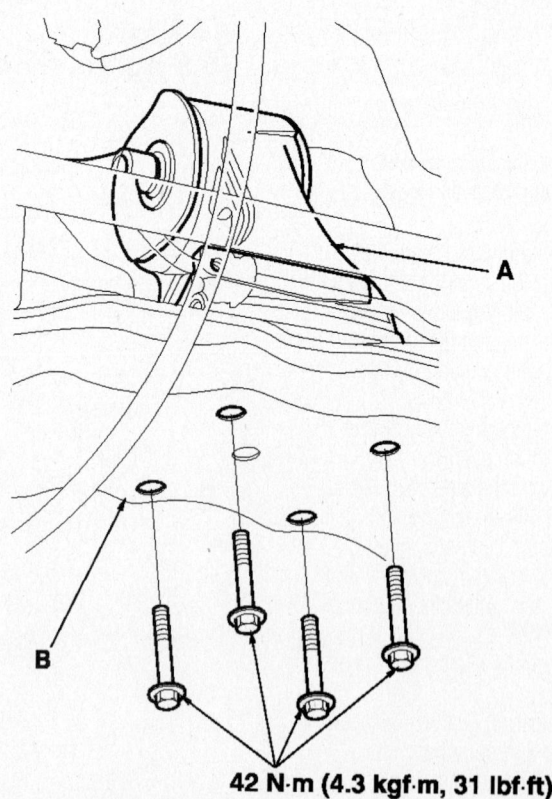

42 N·m (4.3 kgf·m, 31 lbf·ft)

Install the rear mount bracket (A) on the front subframe (B)—Van, LX, EX models—2005–06 models

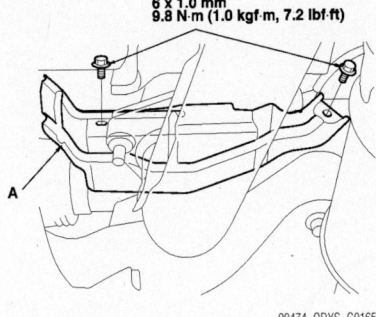

6 x 1.0 mm
9.8 N·m (1.0 kgf·m, 7.2 lbf·ft)

09474_ODYS_G0165

Install the P/S heat baffle plate (A)—
2005–06 models

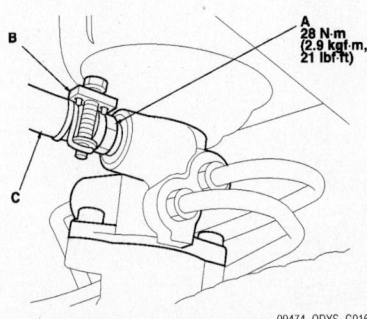

A
28 N·m
(2.9 kgf·m,
21 lbf·ft)

09474_ODYS_G0166

Tighten the return hose flare nut (A) to the
specified torque, and install the
adjustable hose clamp (B) and the return
hose (C)—2005–06 models

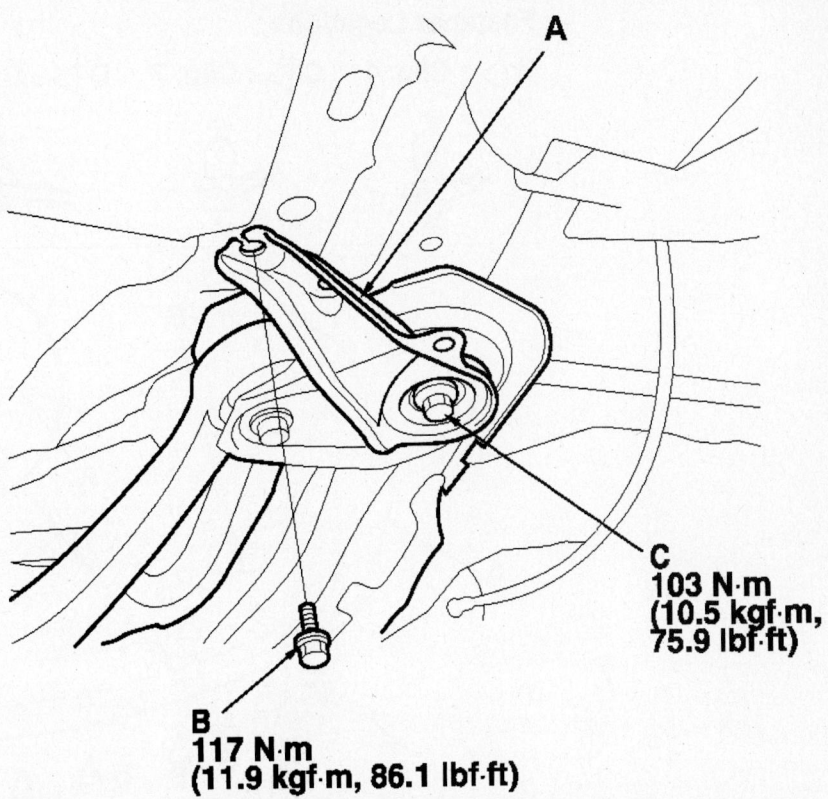

C
103 N·m
(10.5 kgf·m,
75.9 lbf·ft)

B
117 N·m
(11.9 kgf·m, 86.1 lbf·ft)

09474_ODYS_G0167

Install the front subframe rear bracket (A) with 12 mm flange bolts (B) and 14 mm special bolts
(C), and tighten to specified torque—2005–06 models

bracket (A) with 12 mm flange bolts (B) and
14 mm special bolts (C), and tighten to
specified torque.

59. Install the front splash shield.

60. Connect the power steering pressure
(PSP) switch connector.

61. Install exhaust pipe A using new
gaskets and new self-locking nuts.

62. Connect the three way catalytic converter (TWC) (A) to the muffler (B).

63. Install the new 10 mm self-locking
nuts, and tighten them to 33 Nm (25 ft. lbs.).

64. Install the exhaust rubber mount on
the frame. Torque to 22 Nm (16 ft. lbs.).

65. Wipe off any grease contamination
from the ball joint tapered section and
threads. Reconnect the tie-rod ends (A) to
the steering knuckles. Install the 12 mm nut
(B) and tighten it.

66. Install a new cotter pin (C), and bend
it as shown.

67. Center the steering rack within its
stroke.

68. Insert the upper end of the steering
joint onto the steering shaft (A) (line up the
bolt hole (B) with the flat portion (C) on the
shaft), and loosely install the upper joint
bolt.

69. Slip the lower end of the steering

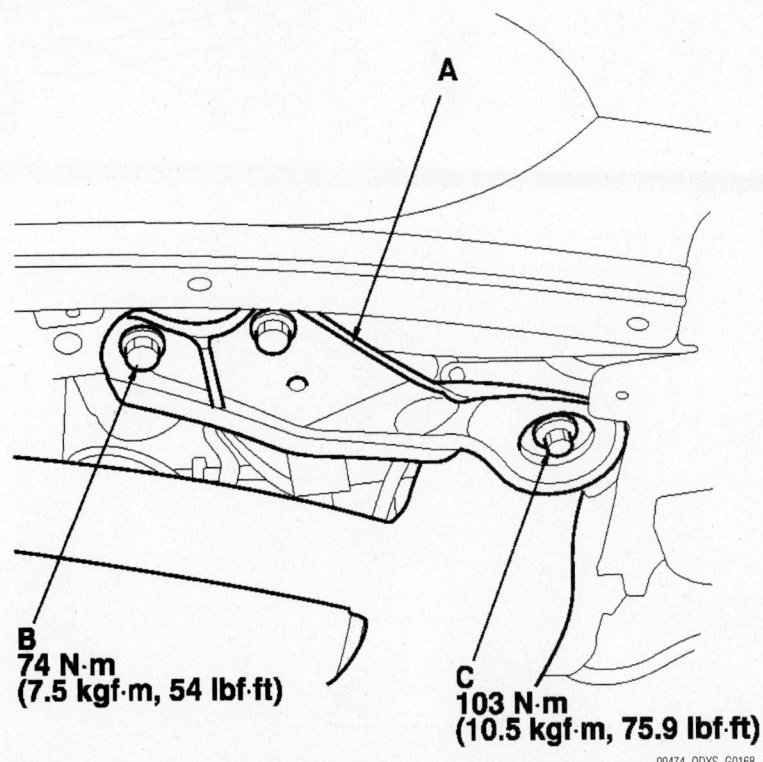

A

B
74 N·m
(7.5 kgf·m, 54 lbf·ft)

C
103 N·m
(10.5 kgf·m, 75.9 lbf·ft)

09474_ODYS_G0168

Install the front subframe front bracket (A) with 12 mm flange bolts (B) and 14 mm special
bolts (C), and tighten to specified torque—2005–06 models

Fastener Locations

B ▷ : Clip, 4   C ▷ : Clip, 7   D ▷ : Clip, 6

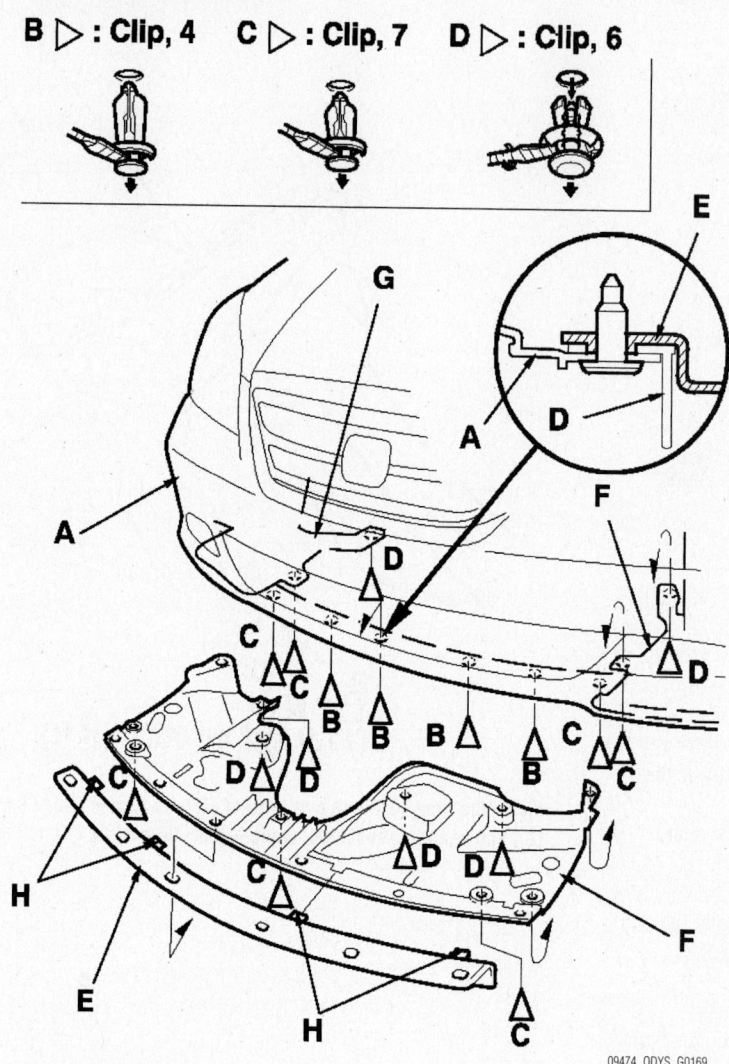

Front splash shield installation—2005–06 models

09474_ODYS_G0169

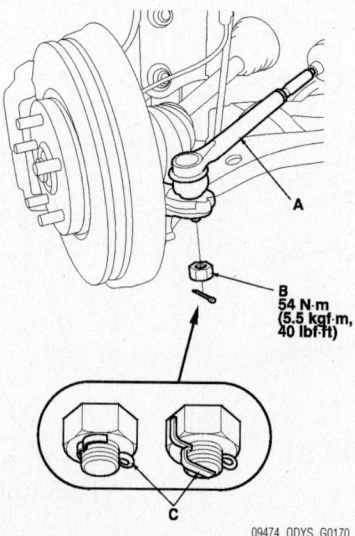

09474_ODYS_G0170

Reconnect the tie-rod ends (A) to the steering knuckles. Install the 12 mm nut (B) and tighten it. Install a new cotter pin (C), and bend it as shown—2005–06 models

joint onto the pinion shaft (D) taking care to align the gap (E) within the angle.

70. Align the bolt hole (A) on the steering joint with the groove (B) around the pinion shaft, then loosely install the lower joint bolt (C).

71. Pull on the steering joint to make sure that the steering joint is fully seated, then tighten the lower joint bolt to the specified torque.

72. Tighten the upper joint bolt (D) to the specified torque.

73. Reinstall the steering joint cover.

74. Before installing the steering wheel, make sure the front wheels are aligned straight ahead, then center the cable reel (A). Do this by first rotating the cable reel clockwise until it stops. Then rotate it counterclockwise about three full turns. The arrow mark (B) on the cable reel label point should point straight up.

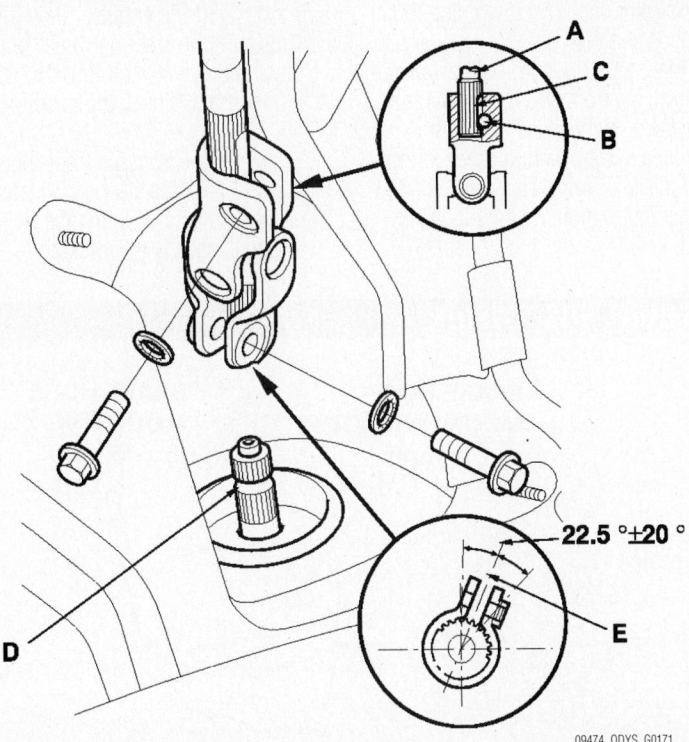

09474_ODYS_G0171

Insert the upper end of the steering joint onto the steering shaft (A) (line up the bolt hole (B) with the flat portion (C) on the shaft), and loosely install the upper joint bolt. Slip the lower end of the steering joint onto the pinion shaft (D) taking care to align the gap (E) within the angle—2005–06 models

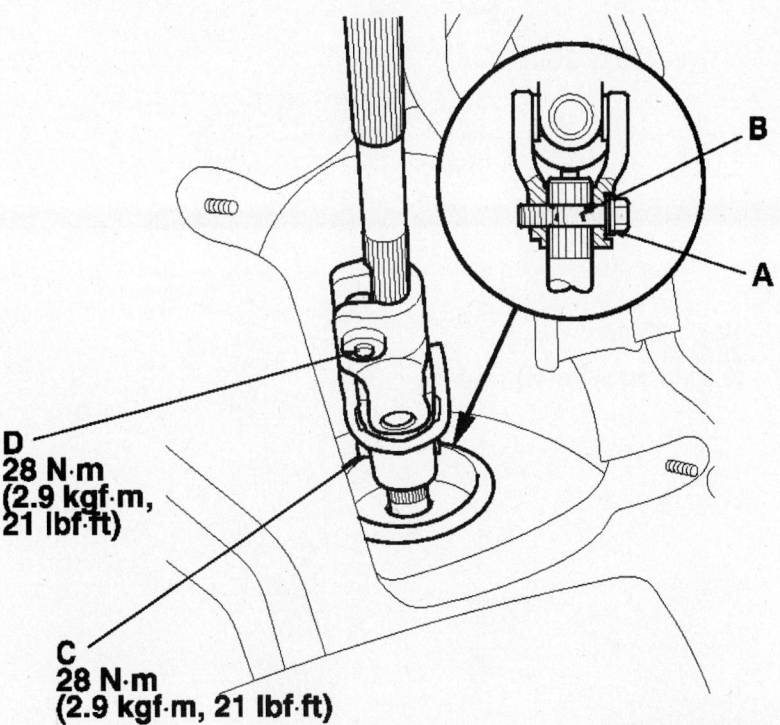

D
28 N·m
(2.9 kgf·m,
21 lbf·ft)

C
28 N·m
(2.9 kgf·m, 21 lbf·ft)

09474_ODYS_G0172

Align the bolt hole (A) on the steering joint with the groove (B) around the pinion shaft, then loosely install the lower joint bolt (C). Pull on the steering joint to make sure that the steering joint is fully seated, then tighten the lower joint bolt to the specified torque. Tighten the upper joint bolt (D) to the specified torque—2005–06 models

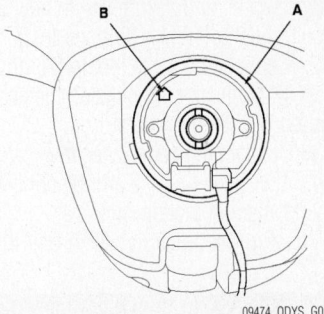

09474_ODYS_G0173

Make sure the front wheels are aligned straight ahead, then center the cable reel (A). Do this by first rotating the cable reel clockwise until it stops. Then rotate it counterclockwise about three full turns. The arrow mark (B) on the cable reel label point should point straight up—2005–06 models

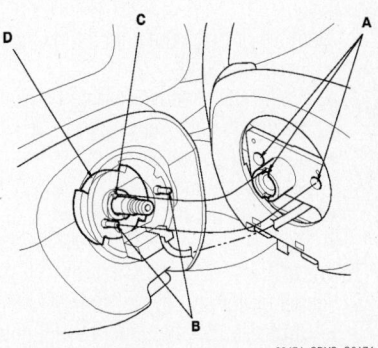

09474_ODYS_G0174

Install the steering wheel on to the steering column shaft, making sure the steering wheel hub (A) engages the pins (B) of the cable reel and tabs (C) of the turn signal canceling sleeve (D). Do not tap on the steering wheel or steering column shaft when installing the steering wheel—2005–06 models

75. Install the steering wheel on to the steering column shaft, making sure the steering wheel hub (A) engages the pins (B) of the cable reel and tabs (C) of the turn signal canceling sleeve (D). Do not tap on the steering wheel or steering column shaft when installing the steering wheel.

76. Install the steering wheel nut and tighten it to 49 Nm (36 ft. lbs.).

77. Connect the cable reel sub-harness connector.

78. Enter the anti-theft codes for the radio and the navigation system, then enter the audio presets.

79. Without navigation: Reset the clock.

80. Check the cruise control, radio remote, navigation system, and turn signal canceling for proper operation

81. Place the new driver's airbag in the

steering wheel, and secure it with new Torx bolts. Tighten to 10 Nm (84 inch lbs.).

82. Connect the cable reel to the driver's airbag 4P connector, then install the access panel on the steering wheel.

83. Connect the battery negative cable.

84. After installing the airbag, confirm proper system operation:

a. Turn the ignition switch ON (II); the SRS indicator should come on for about 6 seconds and then go off.

b. Make sure the horn button works.

85. Install the front wheel, then set the wheels in the straight ahead position.

86. Fill the system with power steering fluid, and bleed air from the system.

87. After installation, do the following checks.

a. Start the engine, allow it to idle, and turn the steering wheel from lock-to-lock several times to warm up to the fluid. Check the gearbox for leaks.

b. Do the front toe inspection.

c. Check the steering wheel spoke angle. Adjust by turning the right and left tie-rods equally if necessary.

FRONT SUSPENSION

Strut

REMOVAL & INSTALLATION

2002–04 Models

1. Before servicing the vehicle, refer to the precautions in the beginning of this section.

2. Remove or disconnect the following:
- Front wheel
- Wheel speed sensor wiring bracket
- Brake hose bracket
- Stabilizer bar link
- Strut pinch bolts
- Upper mount nuts
- Strut

To install:

3. Install or connect the following:
- Strut. Tighten the upper mount nuts to 43 ft. lbs. (59 Nm).
- Strut pinch bolts. Tighten the nuts to 116 ft. lbs. (157 Nm).
- Stabilizer bar link. Tighten the nut to 58 ft. lbs. (78 Nm).
- Brake hose bracket
- Wheel speed sensor wiring bracket
- Front wheel

4. Check the wheel alignment and adjust as necessary.

2005–06 Models

➡When compressing the damper spring, use a commercially available strut spring compressor (Branick MST-580A or Model 7200 or equivalent), according to the manufacturer's instructions.

1. Raise the vehicle, and support it with safety stands in the proper locations.

2. Remove the front wheel.

3. Disconnect the stabilizer link (A) from the damper (B).

4. Remove the wheel sensor harness (C) and the brake hose bracket (D) from the damper. Do not disconnect the wheel sensor connector.

5. Remove the flange nuts (A) and damper pinch bolts (B) from the damper.

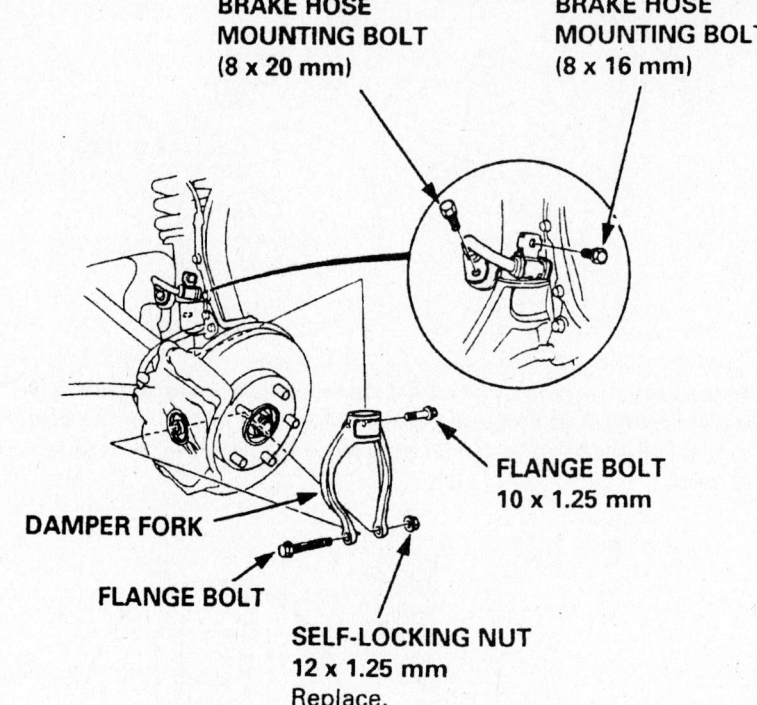

BRAKE HOSE MOUNTING BOLT (8 x 20 mm)

BRAKE HOSE MOUNTING BOLT (8 x 16 mm)

FLANGE BOLT 10 x 1.25 mm

DAMPER FORK

FLANGE BOLT

SELF-LOCKING NUT 12 x 1.25 mm Replace.

7924MG36

Identification of some of the front suspension components—2002–04 models

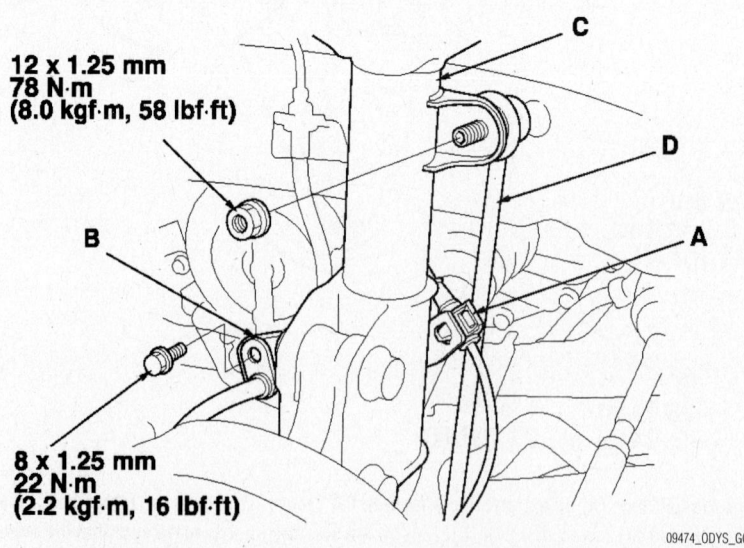

12 x 1.25 mm 78 N·m (8.0 kgf·m, 58 lbf·ft)

8 x 1.25 mm 22 N·m (2.2 kgf·m, 16 lbf·ft)

09474_ODYS_G0175

Disconnect the stabilizer link (A) from the damper (B). Remove the wheel sensor harness (C) and the brake hose bracket (D) from the damper—2005–06 models

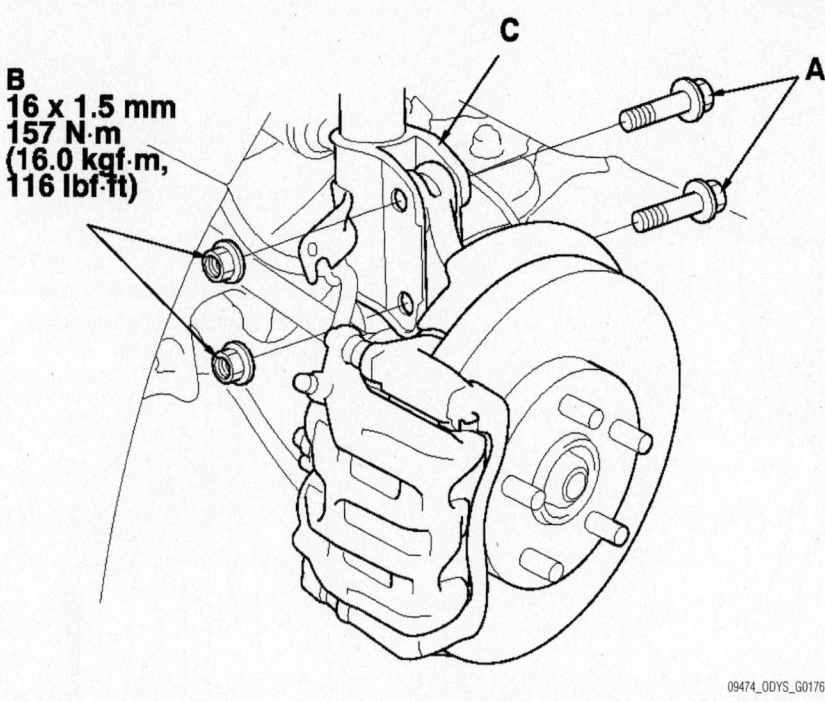

B
16 x 1.5 mm
157 N·m
(16.0 kgf·m,
116 lbf·ft)

C

A

09474_ODYS_G0176

Remove the flange nuts (A) and damper pinch bolts (B) from the damper—2005–06 models

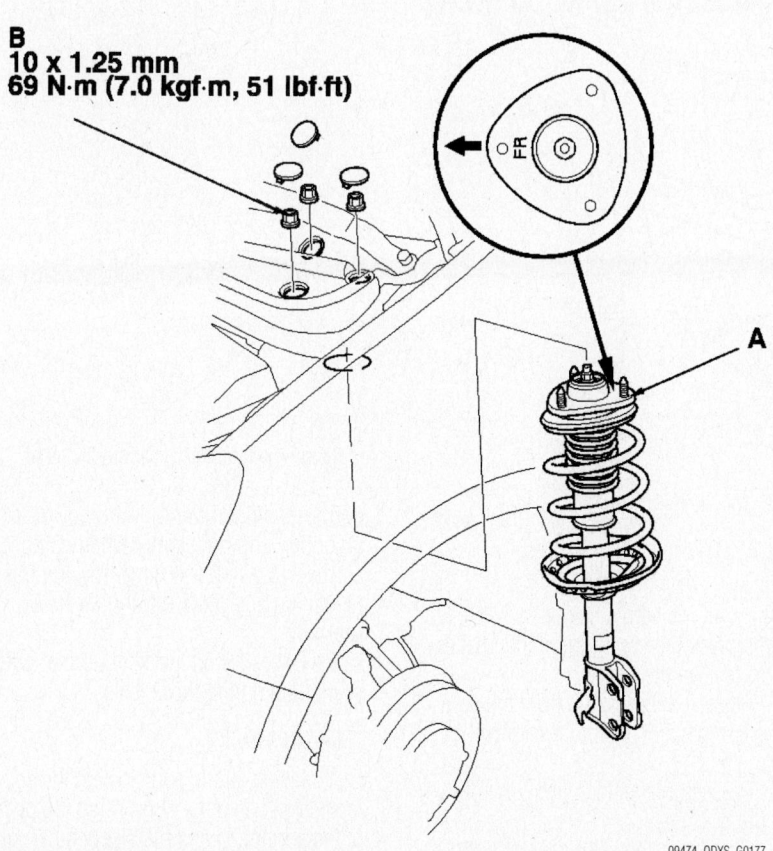

B
10 x 1.25 mm
69 N·m (7.0 kgf·m, 51 lbf·ft)

FR

A

09474_ODYS_G0177

Remove the service caps, and remove the damper by removing the three flange nuts (B)—2005–06 models

6. Remove the service caps, and remove the damper by removing the three flange nuts (B).

7. Remove the damper.

➡ **Damper springs are different, left and right. Mark the springs L and R before you continue.**

To install:

8. Install the damper onto the frame, then loosely install the three flange nuts.

9. Loosely install the damper pinch bolts and flange nuts to the damper.

10. Install the wheel sensor harness and the brake hose bracket to the damper.

11. Loosely install the stabilizer link to the damper.

12. Raise the front suspension with a floor jack to load the suspension with the vehicle's weight.

13. Tighten the damper pinch bolts and flange nuts to the specified torque value.

14. Tighten the flange nuts on top of the damper to the specified torque value.

15. Tighten the stabilizer link nuts to the specified torque value.

16. Install the service caps.

17. Install the front wheel.

18. Check the front wheel alignment, and adjust it if necessary.

Coil Spring

REMOVAL & INSTALLATION

2002–04 Models

1. Before servicing the vehicle, refer to the precautions in the beginning of this section.

2. Remove the strut from the vehicle and install in a strut spring compressor. Compress the spring until the end of the spring comes away from the spring seat.

3. Remove the upper strut mount, spring seat and related components.

4. Remove the coil spring from the strut spring compressor.

To install:

➡ **Use a new self-locking nut.**

5. Compress the spring and position the strut so that the end of the spring aligns with the notch in the spring seat.

6. Install the upper strut mounting components and tighten the nut to 22 ft. lbs. (29 Nm) for vehicles with 4 cylinder engines or to 33 ft. lbs. (44 Nm) for vehicles with 6 cylinder engines.

7. Install the strut to the vehicle.

8. Check the wheel alignment and adjust as necessary.

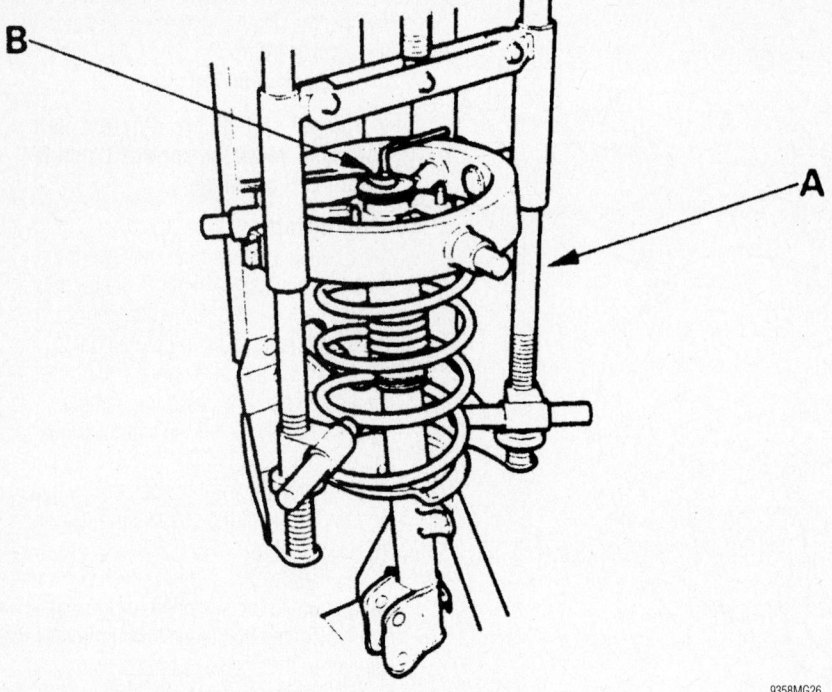

Install the strut assembly into a compressor—2002–04 models

9358MG26

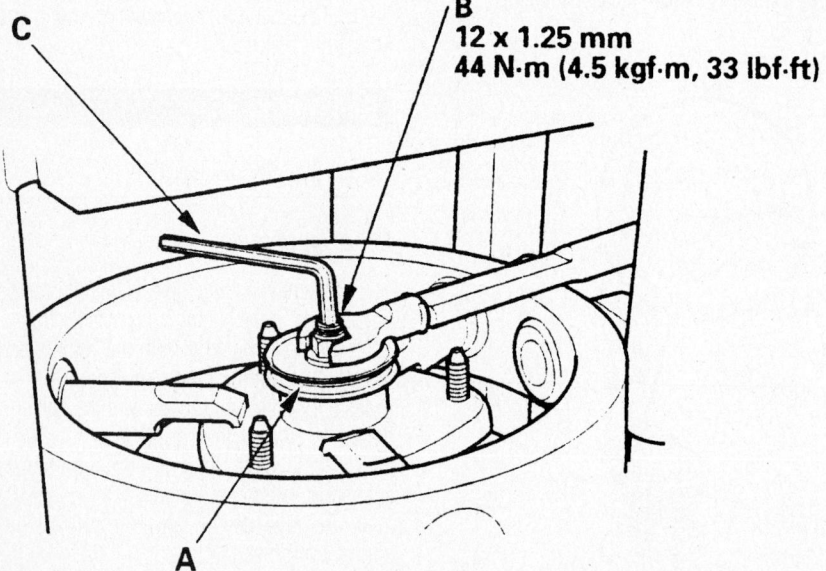

B
12 x 1.25 mm
44 N·m (4.5 kgf·m, 33 lbf·ft)

Tighten the strut nut to specification—2002–04 models

9358MG27

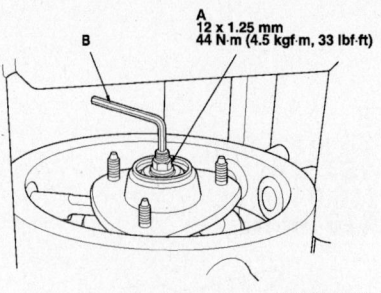

A
12 x 1.25 mm
44 N·m (4.5 kgf·m, 33 lbf·ft)

09474_ODYS_G0178

Compress the strut spring, then remove
the self-locking nut (A) while holding the
strut rod with a hex wrench (B)

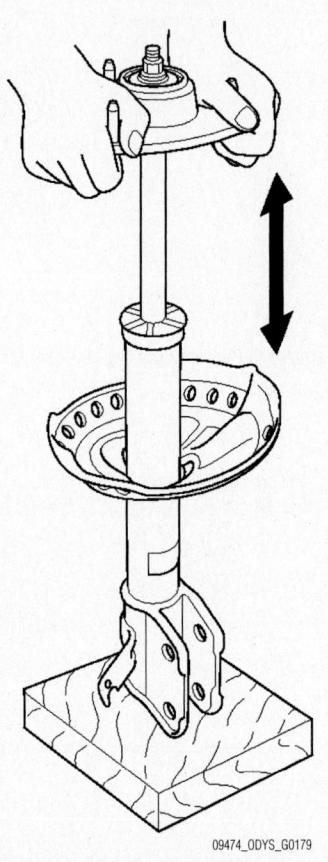

09474_ODYS_G0179

Compress the strut assembly by hand

2005–06 Models

➡ **When compressing the strut spring,
use a commercially available strut
spring compressor (Branick MST-580A
or Model 7200 or equivalent), according
to the manufacturer's instructions.**

DISASSEMBLY/INSPECTION

1. Compress the strut spring, then
remove the self-locking nut (A) while holding the strut rod with a hex wrench (B). Do
not compress the spring more than necessary to remove the nut.

2. Release the pressure from the strut
spring compressor, then disassemble the
strut.

3. Reassemble all the parts, except for
the spring.

4. Compress the strut assembly by
hand, and check for smooth operation
through a full stroke, both compression

and extension. The strut should extend
smoothly and constantly when compression is released. If it does not, the gas
is leaking and the strut should be
replaced.

5. Check for oil leaks, abnormal noises,
or binding during these tests

REASSEMBLY

1. Install all the parts except the self-
locking nut onto the strut. Align the bottom
of the spring (A) and the stepped part of the
lower spring seat (B).

2. Align the stamp of the upper spring

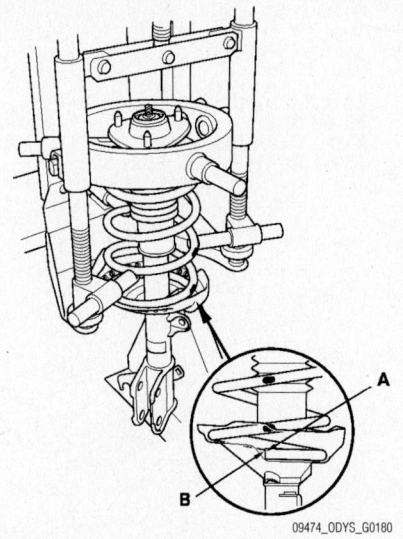

Align the bottom of the spring (A) and the stepped part of the lower spring seat (B)

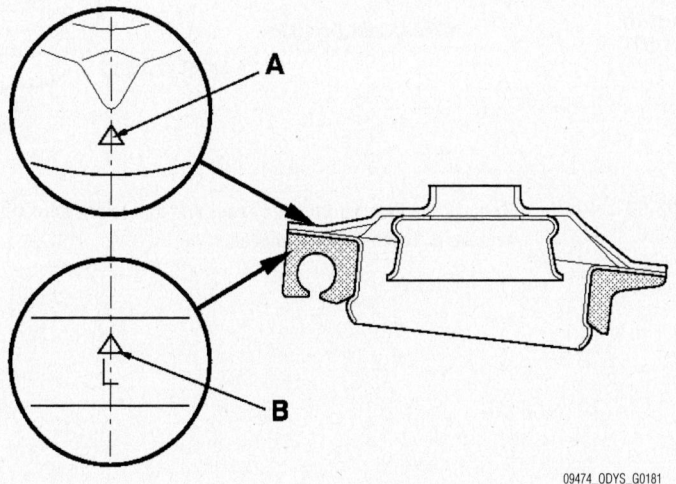

Align the stamp of the upper spring seat (A) and the spring mount rubber (B) as shown, and then set it on the spring

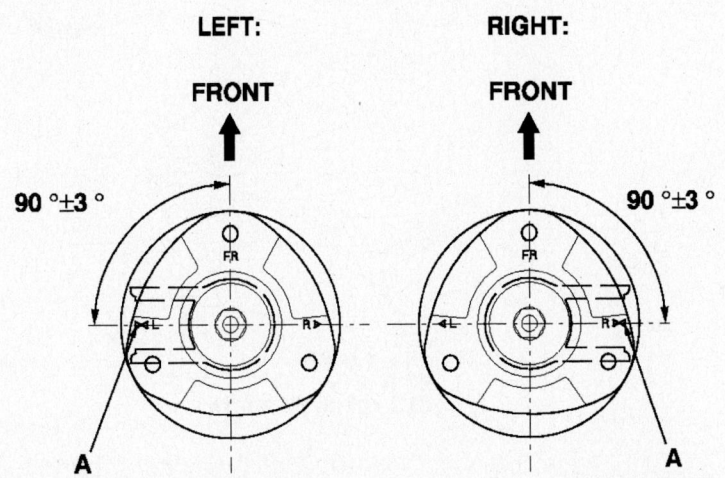

Align the stamp (A) on the upper spring seat with the spring mounting cushion as shown. Install the strut assembly on a commercially available strut spring compressor

seat (A) and the spring mount rubber (B) as shown, and then set it on the spring.

3. Align the stamp (A) on the upper spring seat with the spring mounting cushion as shown. Install the strut assembly on a commercially available strut spring compressor.

4. Compress the strut spring.

5. Install a new self-locking nut.

6. Hold the strut rod using a hex wrench, and tighten the self-locking nut.

Lower Ball Joint

REMOVAL & INSTALLATION

2002–04 Models

The lower ball joints are replaced with the lower control arms as an assembly.

Lower Control Arm

REMOVAL & INSTALLATION

2002–04 Models

1. Before servicing the vehicle, refer to the precautions in the beginning of this section.

2. Remove or disconnect the following:

- Front wheel
- Lower ball joint
- Front inner flange bolt
- Rear inner flange bolt
- Lower control arm

To install:

➡ **Use a new split pin for assembly.**

3. Install or connect the following:
- Lower control arm. Tighten the inner flange bolt (A) to 69 ft. lbs. (93 Nm) and bolt (B) to 90 ft. lbs. (122 Nm) (69 ft. lbs. for 2004 models). Refer to the accompanying illustration for the locations of bolts A and B.
- Lower ball joint. Tighten the nut to 43–51 ft. lbs. (59–69 Nm).
- Front wheel

4. Check the wheel alignment and adjust as necessary.

2005–06 models

Special Tools Required:
- Ball joint remover, 32 mm 07MAC-SL0A102
- Ball joint thread protector, 14 mm 071AF-S3VA000

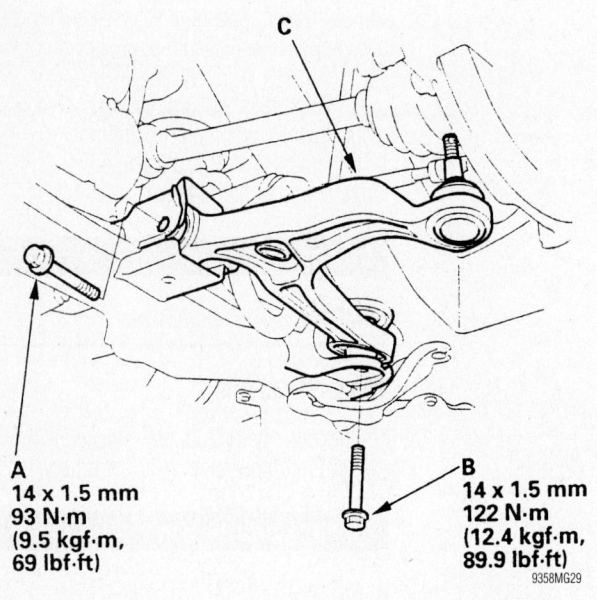

A
14 x 1.5 mm
93 N·m
(9.5 kgf·m,
69 lbf·ft)

B
14 x 1.5 mm
122 N·m
(12.4 kgf·m,
89.9 lbf·ft)

9358MG29

Tighten the lower control arm bolts to the specifications—2002–04 models shown

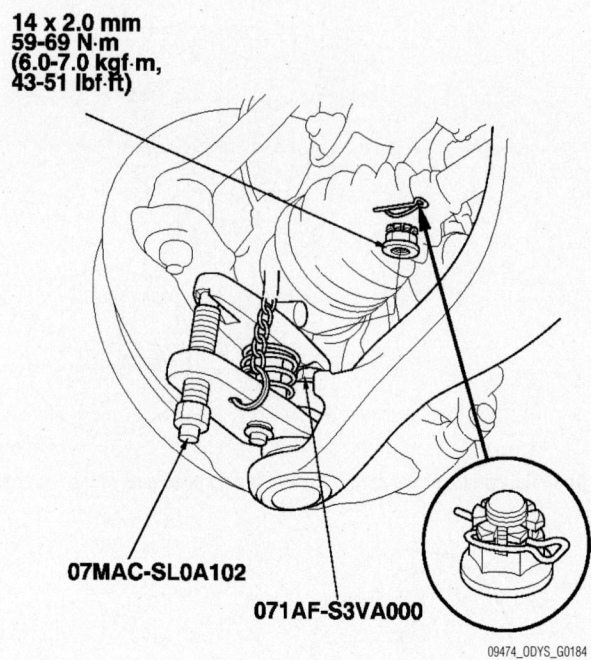

14 x 2.0 mm
59-69 N·m
(6.0-7.0 kgf·m,
43-51 lbf·ft)

07MAC-SL0A102

071AF-S3VA000

09474_ODYS_G0184

Remove the lock pin from the lower arm ball joint castle nut, and remove the nut—2005–06 models

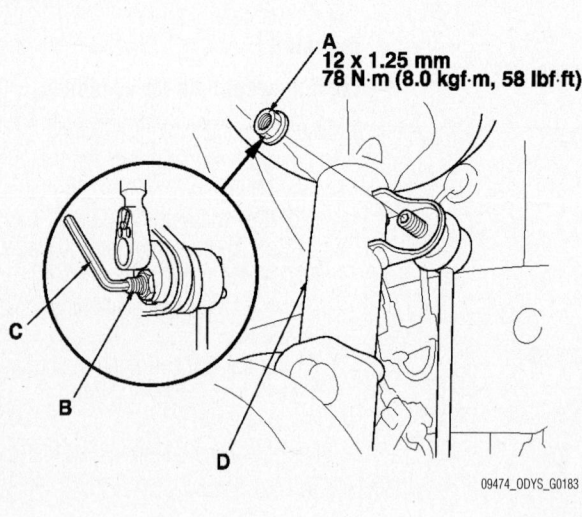

A
12 x 1.25 mm
78 N·m (8.0 kgf·m, 58 lbf·ft)

C

B

D

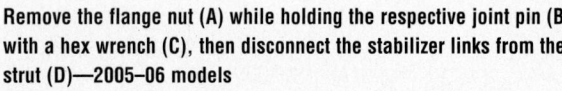

09474_ODYS_G0183

Remove the flange nut (A) while holding the respective joint pin (B) with a hex wrench (C), then disconnect the stabilizer links from the strut (D)—2005–06 models

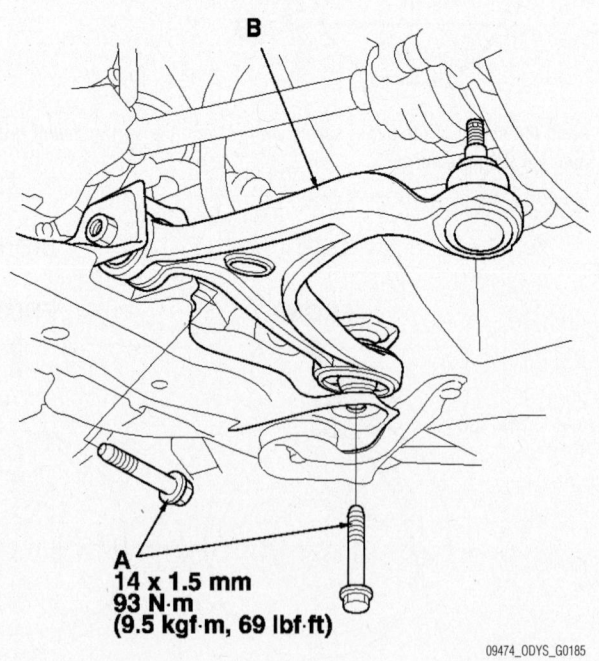

B

A
14 x 1.5 mm
93 N·m
(9.5 kgf·m, 69 lbf·ft)

09474_ODYS_G0185

Remove the flange bolts (A), then remove the lower arm (B)—2005–06 models

1. Raise the front of the vehicle, and support it with safety stands in the proper locations.

2. Remove the front wheel.

3. Remove the flange nut (A) while holding the respective joint pin (B) with a hex wrench (C), then disconnect the stabilizer links from the strut (D).

4. Turn the stabilizer bar backward to gain easier access to the front side of the lower arm mounting bolt.

5. Remove the lock pin from the lower arm ball joint castle nut, and remove the nut.

6. Remove the lower ball joint from the knuckle using the special tools.

7. Remove the flange bolts (A), then remove the lower arm (B).

8. Install the lower arm in the reverse order of removal, and note these items:

- Be careful not to damage the ball joint boot when connecting the lower arm to the knuckle.
- Place a jack under the lower arm, and raise the lower arm to load the suspension with the vehicle's weight.
- Tighten all mounting hardware to the specified torque values.
- Torque the castle nut to the lower torque specification, then tighten it only far enough to align the slot with the hole in the stud. Do not align the castle nut by loosening it.
- Use a new lock pin on the castle nut.
- Connect the struts to the links.
- Before installing the wheel, clean the mating surface on the brake disc and the inside of the wheel.
- Check the front wheel alignment, and adjust it if necessary.

CONTROL ARM BUSHING REPLACEMENT

The lower control arm bushings are serviced with the lower control arm as an assembly.

Stabilizer Bar

REMOVAL & INSTALLATION

2005–06 Models

Special tool required: front subframe adapter VSB02C000016

1. Raise the front of the vehicle, and support it with safety stands in the proper locations.

2. Remove the front wheels.

3. Disconnect the stabilizer links from the stabilizer bar on the right and left sides.

4. Remove the splash shield (A).

5. Attach the special tool to the front subframe (A) by hanging the belt (B) of the special tool over the front of the subframe, secure it with the pin (C), then tighten the special tool screw.

6. Raise the jack (B) and line up the slots in the arms with the bolt holes on the corner of the jack base, then attach them with bolts securely.

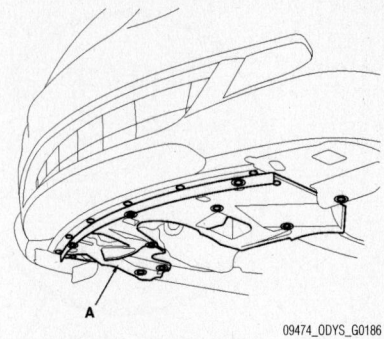

09474_ODYS_G0186

Remove the splash shield (A)

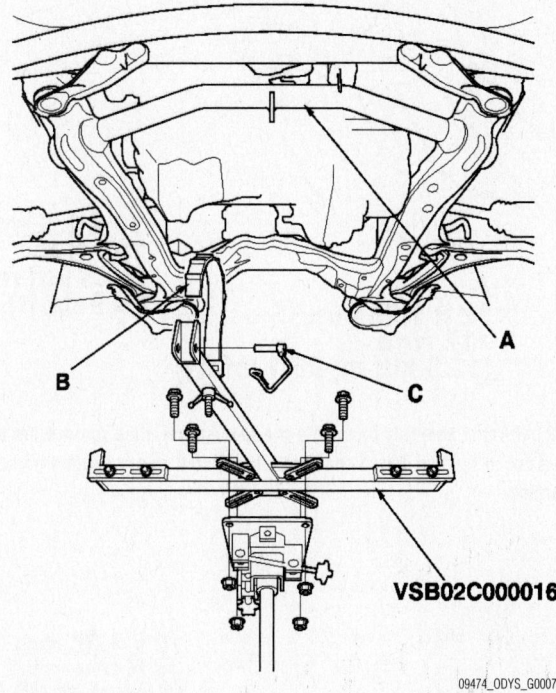

VSB02C000016

09474_ODYS_G0007

Raise the jack (B) and line up the slots in the arms with the bolt holes on the corner of the jack base, then attach them with bolts securely. Raise the jack (B) and line up the slots in the arms with the bolt holes on the corner of the jack base, then attach them with bolts securely

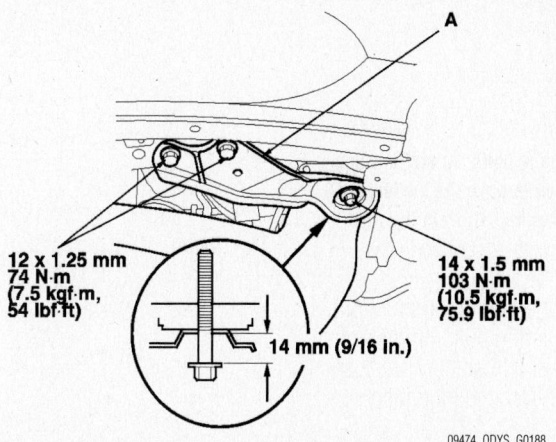

12 x 1.25 mm
74 N·m
(7.5 kgf·m,
54 lbf·ft)

14 x 1.5 mm
103 N·m
(10.5 kgf·m,
75.9 lbf·ft)

14 mm (9/16 in.)

09474_ODYS_G0188

Loosen the front subframe front bracket (A) mounting bolts on the right and left of the vehicle so they are about 14mm (9/16 in.) from the mounting surface

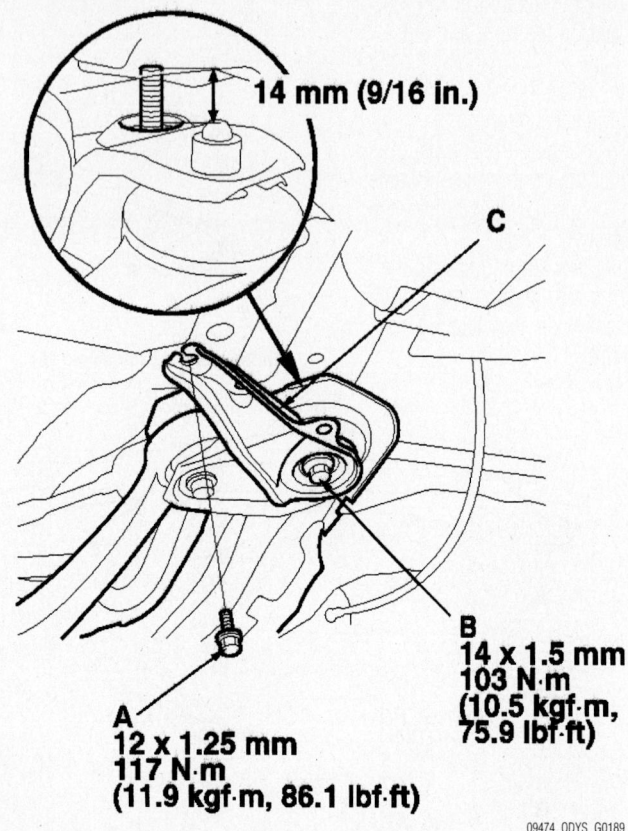

14 mm (9/16 in.)

C

B
14 x 1.5 mm
103 N·m
(10.5 kgf·m,
75.9 lbf·ft)

A
12 x 1.25 mm
117 N·m
(11.9 kgf·m, 86.1 lbf·ft)

09474_ODYS_G0189

Support the front subframe securely by raising the special tool, then remove the two 12mm flange bolts (A). Loosen the two 14mm special bolts (B) so they are about 14mm (9/16 in.) from the mounting surface

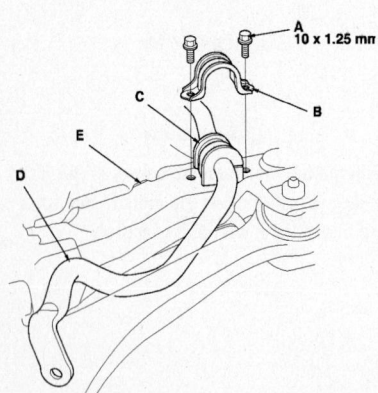

A
10 x 1.25 mm

C
B
E
D

09474_ODYS_G0190

Remove the flange bolts (A) and bushing holders (B), then remove the bushings (C) and the stabilizer bar (D) from the front subframe (E)

7. Loosen the front subframe front bracket (A) mounting bolts on the right and left of the vehicle so they are about 14mm (9/16 in.) from the mounting surface.

8. Support the front subframe securely by raising the special tool, then remove the two 12mm flange bolts (A).

9. Loosen the two 14mm special bolts (B) so they are about 14mm (9/16 in.) from the mounting surface.

10. Lower the jack supporting the front subframe with the special tool slowly until the front subframe has dropped about 14mm (9/16 in.).

11. Remove the flange bolts (A) and bushing holders (B), then remove the bushings (C) and the stabilizer bar (D) from the front subframe (E).

12. Install the stabilizer bar in the reverse order of removal, and note these items:

- Note the right and left direction of the stabilizer bar.
- Align the paint marks (A) on the stabilizer bar with the sides of the bushings (B).
- Note the fore/aft direction of the bushing holders.
- Raise the front subframe up with the jack and special tool until it contacts the body frame, then tighten the mounting bolts to the specified torque.
- Refer to Stabilizer Link Removal/Installation to connect the stabilizer bar to the links.
- Do the front toe inspection, and adjust it if necessary.

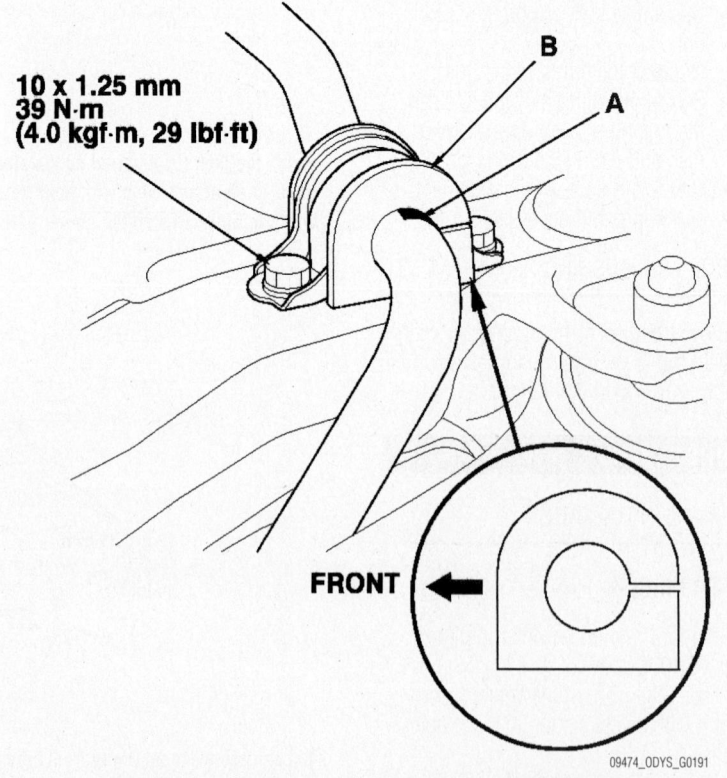

B
A
10 x 1.25 mm
39 N·m
(4.0 kgf·m, 29 lbf·ft)

FRONT

09474_ODYS_G0191

Align the paint marks (A) on the stabilizer bar with the sides of the bushings (B)

Stabilizer Link

REMOVAL & INSTALLATION

2005–06 Models

1. Raise the front of the vehicle, and support it with safety stands in the proper locations.

2. Remove the front wheel.

3. Remove the flange nuts (A) while holding the respective joint pin (B) with a hex wrench (C). Remove the stabilizer link (D).

To install:

4. Install the stabilizer link (A) on the stabilizer bar (B) and damper (C) with the joint pin (D) set at the center of each moving range.

5. Install the flange nuts, and lightly tighten them.

6. Place a jack under the lower arm, and raise the suspension to load it with the vehicle's weight.

7. Tighten the flange nuts (A) to the specified torque value while holding the respective joint pin (B) with a hex wrench (C).

8. Reinstall all removed parts and test-drive the vehicle.

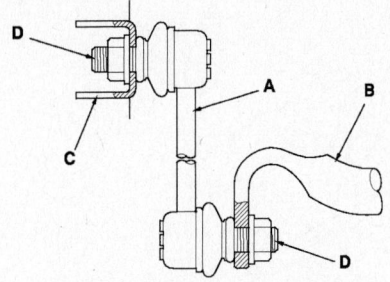

09474_ODYS_G0193

Install the stabilizer link (A) on the stabilizer bar (B) and damper (C) with the joint pin (D) set at the center of each moving range

9. After 5 minutes of driving, tighten the flange nuts again to the specified torque value.

Knuckle

2002–04 Models

1. Raise the front of the vehicle, and support it with safety stands in the proper locations.

2. Remove the wheel nuts (A) and front wheel (B).

3. Raise the locking tab on the spindle nut (C), then remove the nut.

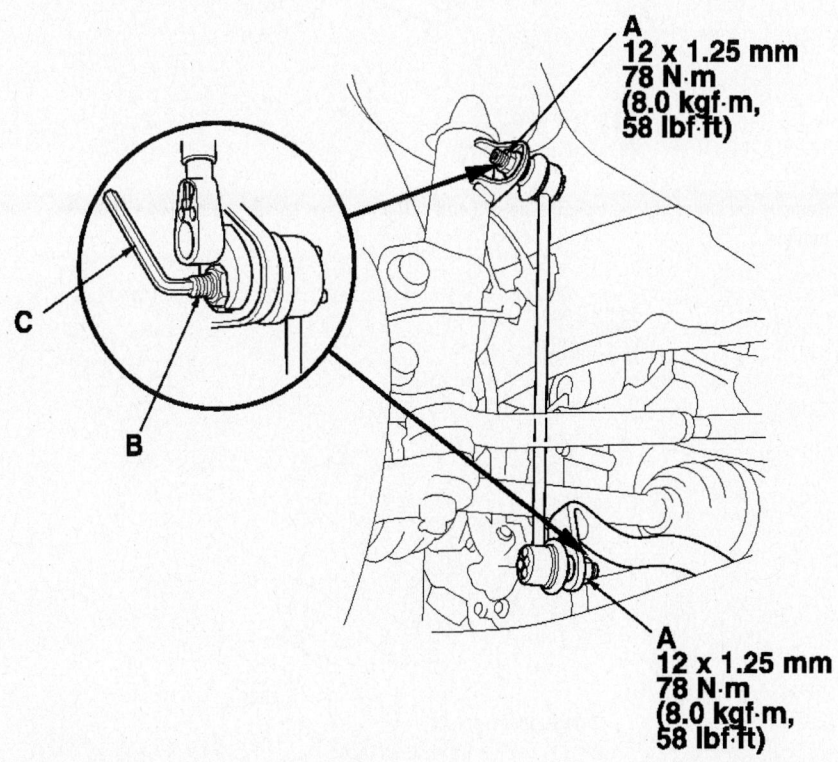

09474_ODYS_G0194

Remove the flange nuts (A) while holding the respective joint pin (B) with a hex wrench (C). Remove the stabilizer link (D)

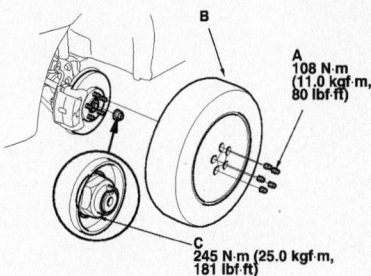

09474_ODYS_G0202

Remove the wheel nuts (A) and front wheel (B). Raise the locking tab on the spindle nut (C), then remove the nut— 2002–04 models

4. Remove the brake hose mounting bolt (A).

5. Remove the caliper mounting bolts (B), and hang the caliper assembly (C) to one side. To prevent damage to the caliper assembly or brake hose, use a short piece of wire to hang the caliper from the undercarriage.

6. Remove the wheel sensor (A) from the knuckle (B). Do not disconnect the wheel sensor connector.

7. Remove the 6 mm brake disc retaining screws (A).

8. Screw two 8 x 1.25 mm bolts (B) into the disc to push it away from the hub. Turn each bolt two turns at a time to prevent cocking the disc excessively.

9. Remove the brake disc from the knuckle.

10. Check the front hub for damage and cracks.

11. Remove the cotter pin (A) from the tie-rod end ball joint, then loosen the nut (B).

12. Remove the tie-rod ball joint from the knuckle using the special tool.

13. Remove the lock pin (A) from the lower arm ball joint castle nut (B), and remove the nut.

14. Remove the lower ball joint from the knuckle using the special tool.

15. Remove the damper pinch bolts (A) and flange nuts (B) from the damper.

16. Pull the knuckle outward, and remove the driveshaft outboard joint (C) from the knuckle (D) by tapping the driveshaft end (E) with a plastic hammer, then remove the knuckle.

To install:

17. Install the knuckle in the reverse order of removal, and pay particular attention to the following items:

• Be careful not to damage the ball joint boots when installing the knuckle.

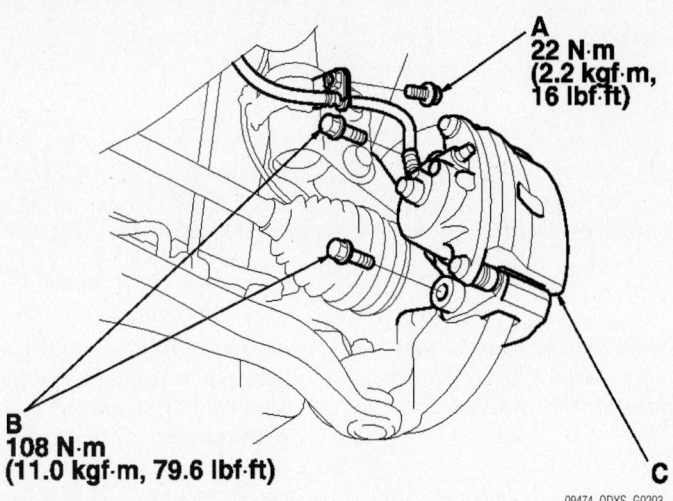

A
22 N·m
(2.2 kgf·m,
16 lbf·ft)

B
108 N·m
(11.0 kgf·m, 79.6 lbf·ft)

C

09474_ODYS_G0203

Remove the brake hose mounting bolt (A). Remove the caliper mounting bolts (B), and hang the caliper assembly (C) to one side—2002–04 models

- Install new lock pins on the castle nuts after torquing.
- Raise the locking tab on the spindle nut, then remove the nut.
- Before installing the brake disc, clean the mating surface of the front hub and the inside of the brake disc.
- Before installing the new spindle nut, apply a small amount of engine oil to the seating surface of the nut. After tightening, use a drift to stake the spindle nut shoulder against the driveshaft.
- Before installing the wheel, clean the mating surface of the brake disc and the inside of the wheel.
- Check the front wheel alignment, and adjust it if necessary.

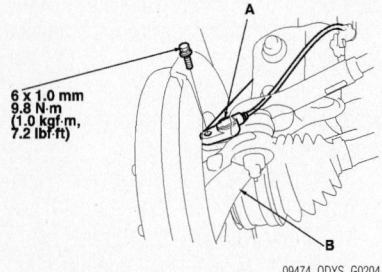

A
6 x 1.0 mm
9.8 N·m
(1.0 kgf·m,
7.2 lbf·ft)

B

09474_ODYS_G0204

Remove the wheel sensor (A) from the knuckle (B). Do not disconnect the wheel sensor connector—2002–04 models

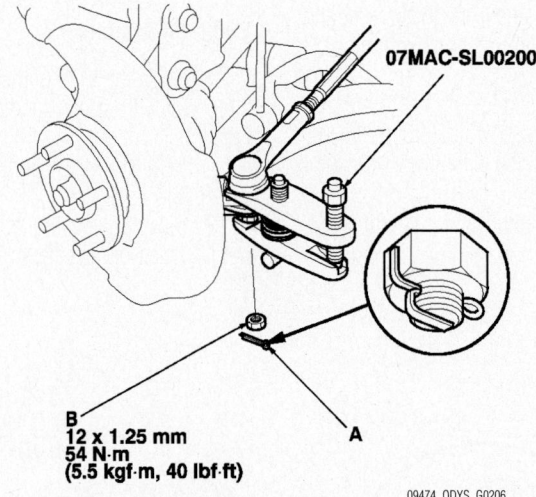

07MAC-SL00200

B
12 x 1.25 mm
54 N·m
(5.5 kgf·m, 40 lbf·ft)

A

09474_ODYS_G0206

Remove the cotter pin (A) from the tie-rod end ball joint, then loosen the nut (B)—2002–04 models

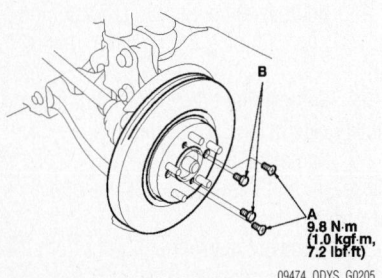

B

A
9.8 N·m
(1.0 kgf·m,
7.2 lbf·ft)

09474_ODYS_G0205

Remove the 6 mm brake disc retaining screws (A). Screw two 8 x 1.25 mm bolts (B) into the disc to push it away from the hub—2002–04 models

- Torque all mounting hardware to the specified torque values.
- Torque the castle nuts to the lower torque specifications, then tighten them only far enough to align the slot with the ball joint pin hole. Do not align the castle nut by loosening it.

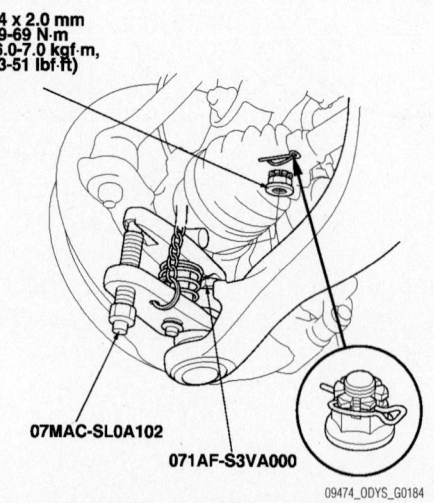

14 x 2.0 mm
59–69 N·m
(6.0–7.0 kgf·m,
43–51 lbf·ft)

07MAC-SL0A102

071AF-S3VA000

09474_ODYS_G0184

Remove the lock pin (A) from the lower arm ball joint castle nut (B), and remove the nut—2002–04 models

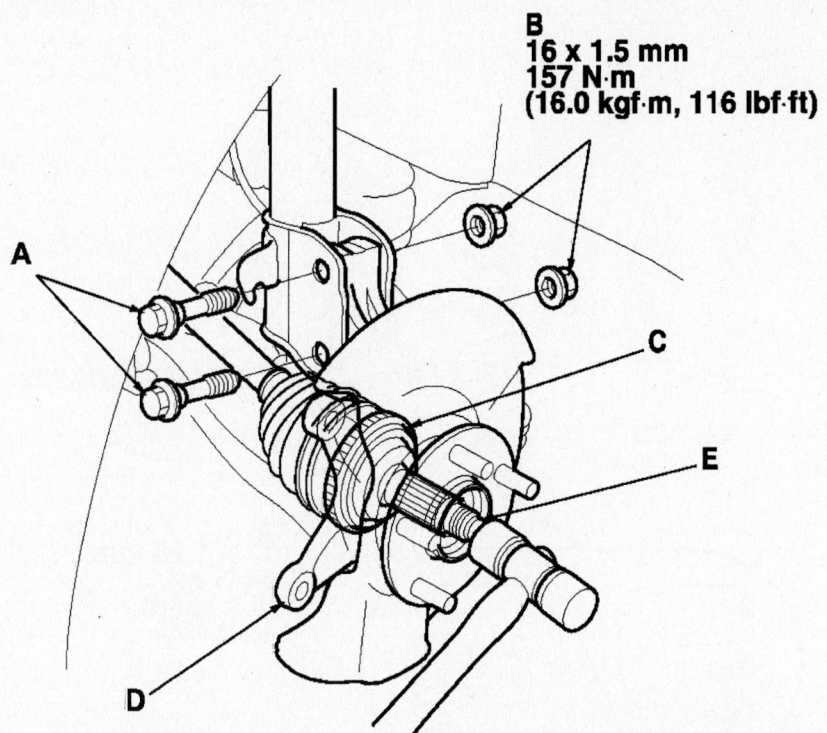

B
16 x 1.5 mm
157 N·m
(16.0 kgf·m, 116 lbf·ft)

A

C

E

D

09474_ODYS_G0208

Remove the damper pinch bolts (A) and flange nuts (B) from the damper. Pull the knuckle out-ward, and remove the driveshaft outboard joint (C) from the knuckle (D) by tapping the drive-shaft end (E) with a plastic hammer—2002–04 models

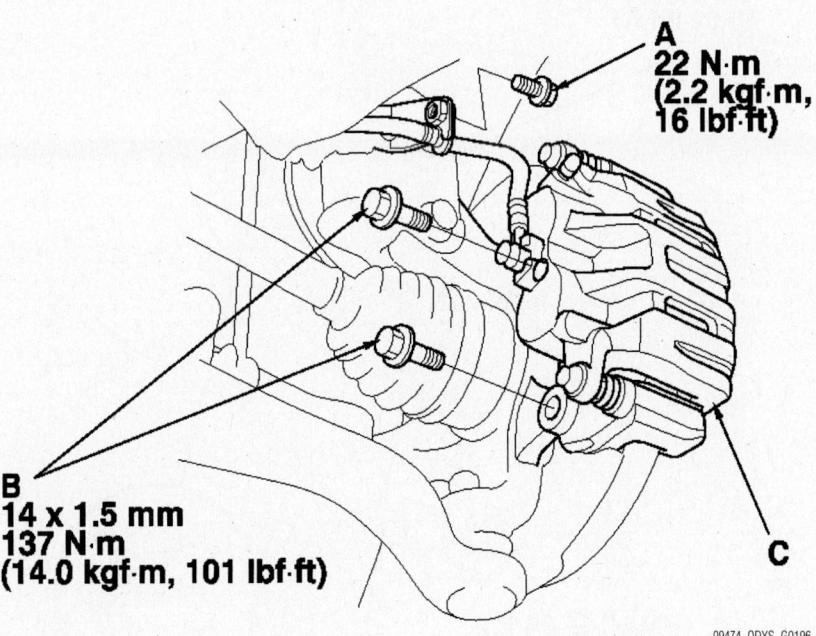

A
22 N·m
(2.2 kgf·m, 16 lbf·ft)

B
14 x 1.5 mm
137 N·m
(14.0 kgf·m, 101 lbf·ft)

C

09474_ODYS_G0196

Remove the caliper mounting bolts (B), and hang the caliper assembly (C) to one side—2005–06 Models

2005–06 Models

1. Raise the front of the vehicle, and support it with safety stands in the proper locations.

2. Remove the wheel nuts and front wheel.

3. Remove the brake hose mounting bolt (A).

4. Remove the caliper mounting bolts (B), and hang the caliper assembly (C) to one side. To prevent damage to the caliper assembly or brake hose, use a short piece of wire to hang the caliper from the under-carriage.

5. Remove the wheel sensor (A) from the knuckle (B). Do not disconnect the wheel sensor connector.

6. Raise the stake, and then remove the spindle nut.

7. Remove the 6mm brake disc retaining screws (A).

8. Screw two 8 x 1.25mm bolts (B) into the brake disc to push it away from the hub.

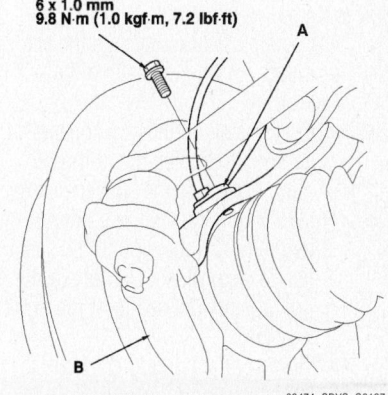

6 x 1.0 mm
9.8 N·m (1.0 kgf·m, 7.2 lbf·ft)

A

B

09474_ODYS_G0197

Remove the wheel sensor (A) from the knuckle (B). Do not disconnect the wheel sensor connector—2005–06 Models

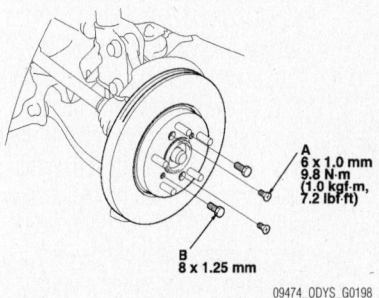

A
6 x 1.0 mm
9.8 N·m
(1.0 kgf·m, 7.2 lbf·ft)

B
8 x 1.25 mm

09474_ODYS_G0198

Remove the 6mm brake disc retaining screws (A). Screw two 8 x 1.25mm bolts (B) into the brake disc to push it away from the hub. Turn each bolt two turns at a time to prevent cocking the disc exces-sively—2005–06 Models

Turn each bolt two turns at a time to prevent cocking the disc excessively.

9. Remove the brake disc from the knuckle.

10. Check the front hub for damage and cracks.

11. Remove the cotter pin (A) from the tie-rod end ball joint, then loosen the nut (B).

12. Remove the tie-rod ball joint from the knuckle using the special tool.

13. Remove the lock pin (A) from the lower arm ball joint castle nut (B), and remove the nut.

14. Remove the lower ball joint from the knuckle using the special tools.

15. Remove the damper pinch bolts (A) and flange nuts (B) from the damper.

16. Pull the knuckle outward, and remove the driveshaft outboard joint (C) from the knuckle (D) by tapping the drive-shaft end (E) with a plastic hammer, then remove the knuckle.

To install:

17. Install the knuckle in the reverse order of removal, and pay particular attention to the following items:

- Be careful not to damage the ball joint boot when installing the knuckle.
- First, install all the components and lightly tighten the bolts and nuts, then raise the suspension to load it with the vehicle's weight before fully tightening to the specified torque values. Do not place the jack against the ball joint pin of the knuckle.
- Tighten all mounting hardware to the specified torque values.
- Before connecting the ball joint to the knuckle, degrease the threaded section and tapered portion of the ball joint pin, the connecting hole, the threaded section and mating surface of the castle nut.
- Torque the castle nut to the lower torque specification, then tighten it only far enough to align the slot with the ball joint pin hole. Do not align the castle nut by loosening it.
- Use a new lock pin on the castle nut.
- Before installing the brake disc, clean the mating surface of the front hub and the inside of the brake disc.
- Use a new spindle nut on reassembly.
- Before installing the new spindle nut, apply a small amount of engine oil to the seating surface of

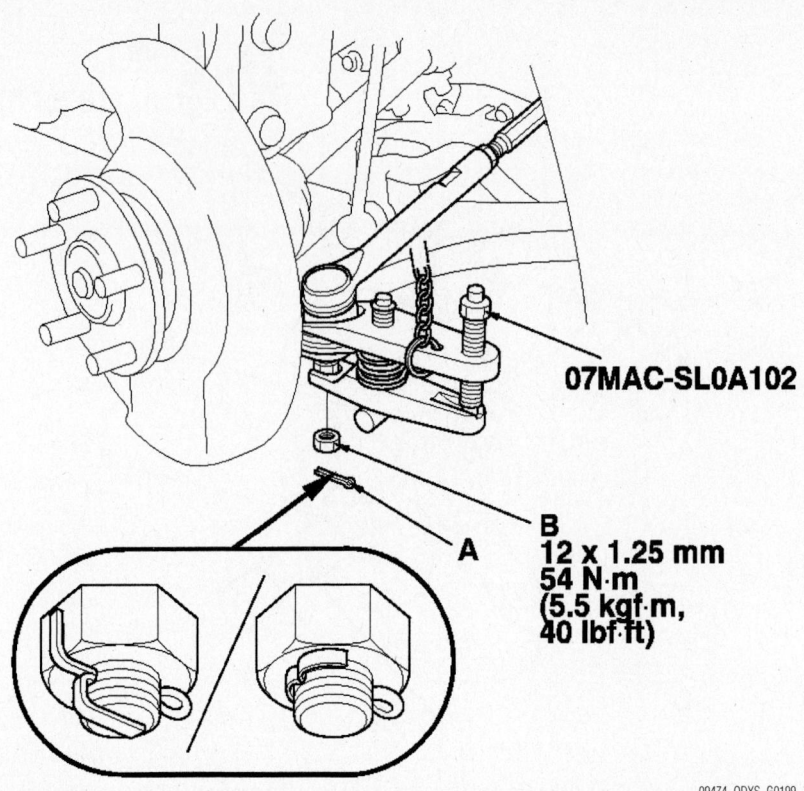

07MAC-SL0A102

B
12 x 1.25 mm
54 N·m
(5.5 kgf·m,
40 lbf·ft)

A

09474_ODYS_G0199

Remove the cotter pin (A) from the tie-rod end ball joint, then loosen the nut (B)—2005–06 Models

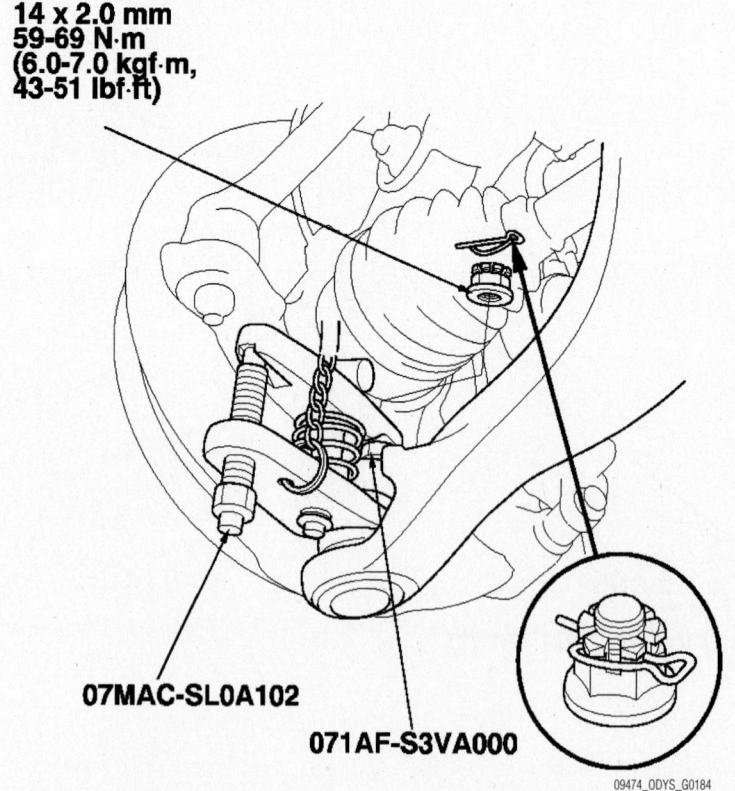

14 x 2.0 mm
59-69 N·m
(6.0-7.0 kgf·m,
43-51 lbf·ft)

07MAC-SL0A102

071AF-S3VA000

09474_ODYS_G0184

Remove the lock pin (A) from the lower arm ball joint castle nut (B), and remove the nut—2005–06 Models

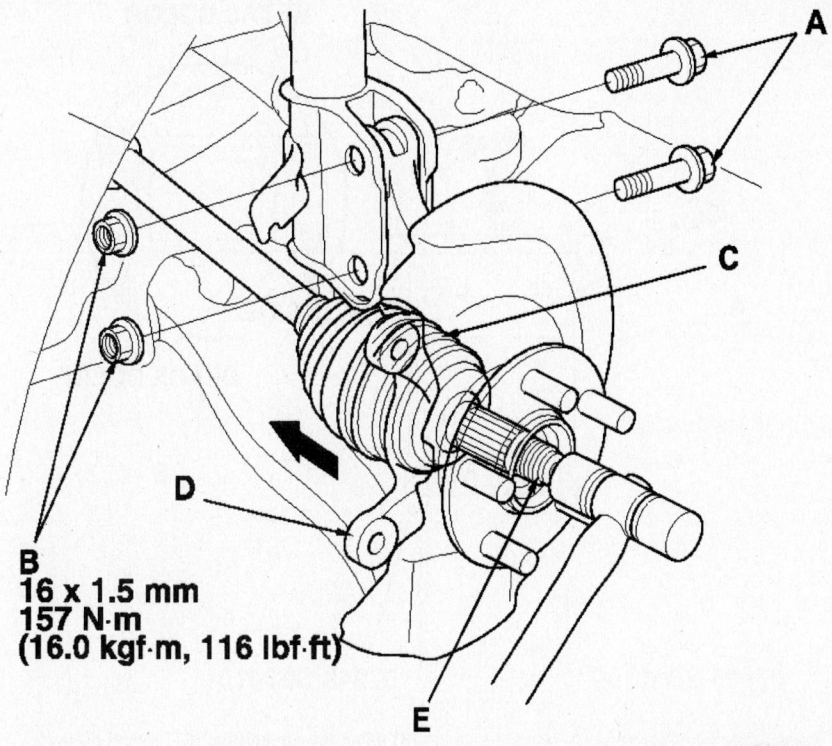

Remove the damper pinch bolts (A) and flange nuts (B) from the damper. Pull the knuckle outward, and remove the driveshaft outboard joint (C) from the knuckle (D) by tapping the driveshaft end (E) with a plastic hammer, then remove the knuckle—2005–06 Models

the nut. After tightening, use a drift to stake the spindle nut shoulder against the driveshaft.
- Before installing the wheel, clean the mating surface of the brake disc and the inside of the wheel.
- Check the front wheel alignment, and adjust it if necessary.

Wheel Bearings

INSPECTION

The wheel bearings are not adjustable.
1. Raise the vehicle, and support it with safety stands in the proper locations.
2. Remove the wheels, then reinstall the wheel nuts.
3. Attach the dial gauge. Place the dial gauge against the hub flange.
4. Measure the bearing end play by moving the disc inward or outward. If the bearing end play measurement is more than 0–0.05 mm (0–0.002 in.), replace the wheel bearing.

REMOVAL & INSTALLATION

1. Before servicing the vehicle, refer to the precautions in the beginning of this section.

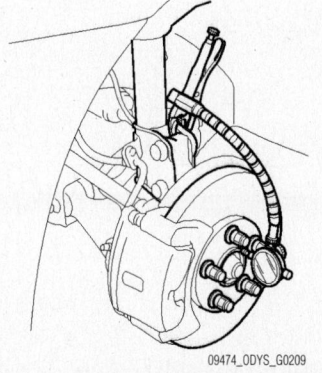

Place the dial gauge against the hub flange

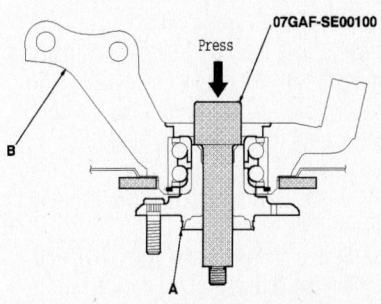

Separate the hub (A) from the knuckle (B) using the special tool and a hydraulic press—2002–04 models

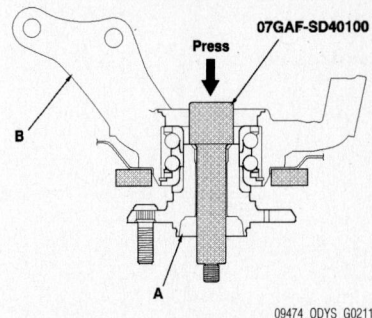

Separate the hub (A) from the knuckle (B) using the special tool and a hydraulic press—2005–06 models

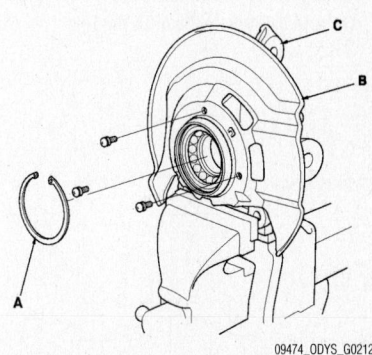

Remove the snap ring (A) and the splash guard (B) from the knuckle (C)

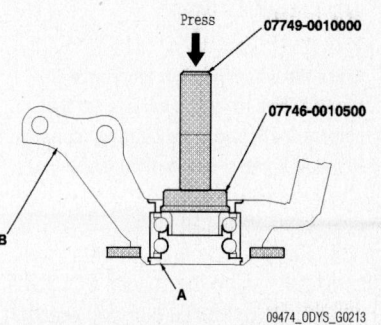

Press the wheel bearing (A) out of the knuckle (B) using the special tools and a press—2002–04 models

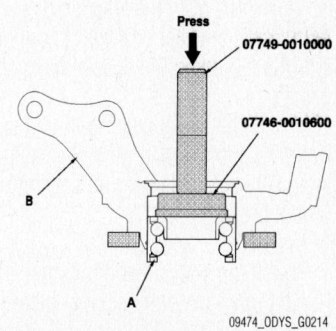

Press the wheel bearing (A) out of the knuckle (B) using the special tools and a press—2005–06 models

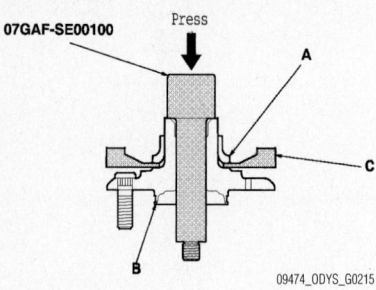

Press the wheel bearing inner race (A) from the hub (B) using the special tool, commercially available bearing separator (C), and a press—2002–04 models

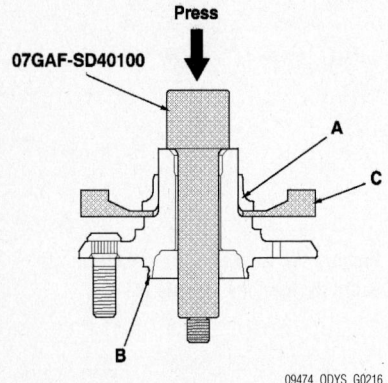

Press the wheel bearing inner race (A) from the hub (B) using the special tool, commercially available bearing separator (C), and a press—2005–06 models

2. Remove the knuckle.
3. Separate the hub (A) from the knuckle (B) using the special tool and a hydraulic press. Be careful not to distort the splash guard. Hold onto the hub to keep it from falling when pressed clear.
4. Remove the snap ring (A) and the splash guard (B) from the knuckle (C).
5. Press the wheel bearing (A) out of the knuckle (B) using the special tools and a press.
6. Press the wheel bearing inner race (A) from the hub (B) using the special tool, commercially available bearing separator (C), and a press.

To install:
7. Wash the knuckle and hub thoroughly in high flash point solvent before reassembly.
8. Press a new wheel bearing (A) into the knuckle (B) using the old bearing (C), a steel plate (D), the special tools, and a

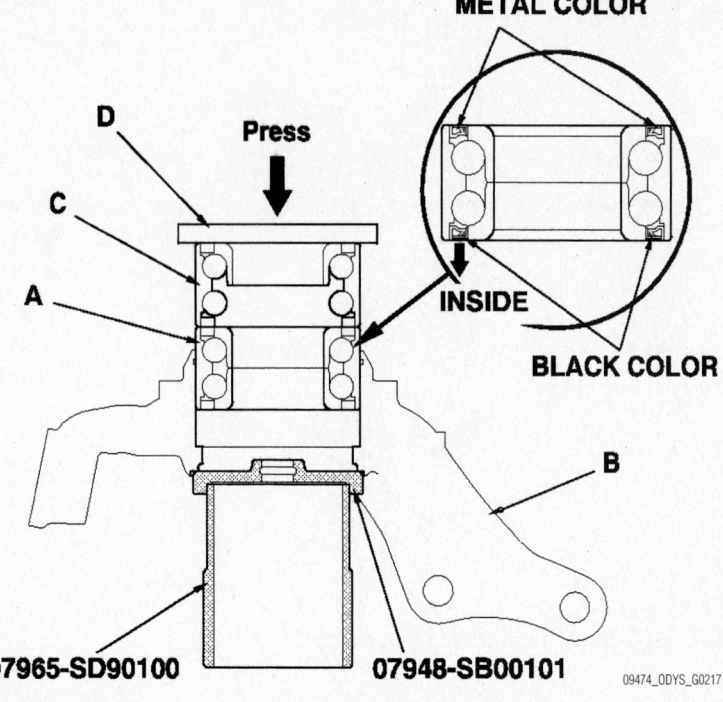

07965-SD90100   07948-SB00101

Press a new wheel bearing (A) into the knuckle (B) using the old bearing (C), a steel plate (D), the special tools, and a press. Place the wheel bearing on the knuckle with the magnetic encoder side (black color) facing toward the inside

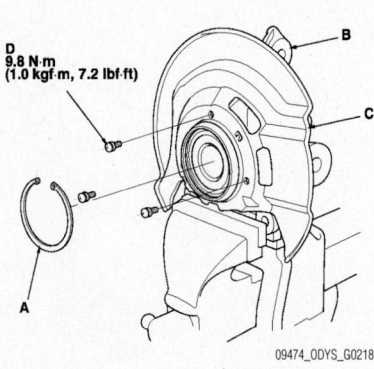

Install the snap ring (A) securely in the knuckle (B). Install the splash guard (C), and tighten the screws (D)

press. Place the wheel bearing on the knuckle with the magnetic encoder side (black color) facing toward the inside. Be careful not to damage the magnetic encoder.
9. Install the snap ring (A) securely in the knuckle (B).
10. Install the splash guard (C), and tighten the screws (D).
11. Install the hub (A) onto the knuckle (B) using the special tools shown and a hydraulic press. Be careful not to distort the splash guard (C).

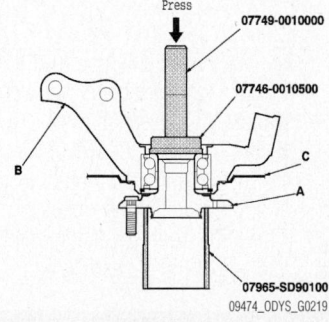

Install the hub (A) onto the knuckle (B) using the special tools shown and a hydraulic press. Be careful not to distort the splash guard (C)—2002–04 models

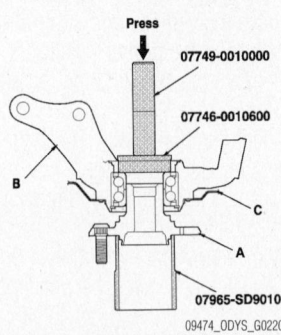

Install the hub (A) onto the knuckle (B) using the special tools shown and a hydraulic press. Be careful not to distort the splash guard (C)—2005–06 models

REAR SUSPENSION

Shock Absorber

REMOVAL & INSTALLATION

2002–04 Models

1. Before servicing the vehicle, refer to the precautions in the beginning of this section.
2. Support the vehicle under the lower control arm.
3. Remove or disconnect the following:
   - Rear wheel
   - Upper shock absorber flange bolt (B)
   - Lower shock absorber nut (C)
   - Shock absorber (A)

   To install:
4. Install or connect the following:
   - Shock absorber. Tighten the fasteners to 47 ft. lbs. (64 Nm).
   - Rear wheel

2005–06 Models

1. Raise the rear of the vehicle, and support it with safety stands in the proper locations.
2. Remove the rear wheel.
3. Position a floor jack at the connecting point of the lower control arm and knuckle (A). Raise the floor jack until the suspension begins to compress.
4. Remove the flange bolt and nut from the body.
5. Remove the self-locking nut from the knuckle.
6. Compress the damper by hand, and remove it from the vehicle.
7. Compress the damper assembly by hand, and check for smooth operation through a full stroke, both compression and extension. The damper should move smoothly and constantly when compression is released. If it does not, the gas is leaking and the damper should be replaced.
8. Check for oil leaks, abnormal noises, or binding during these tests.

   To install:
9. Lower the rear suspension. Compress the damper (A) by hand, and move it into position. Loosely install the flange nut (B), bolt (C), and new self-locking nut (D).
10. Raise the rear suspension with a floor jack until the vehicle just lifts off the safety stands.
11. Tighten the flange bolt and self-locking nut on the bottom of the damper to the specified torque value.

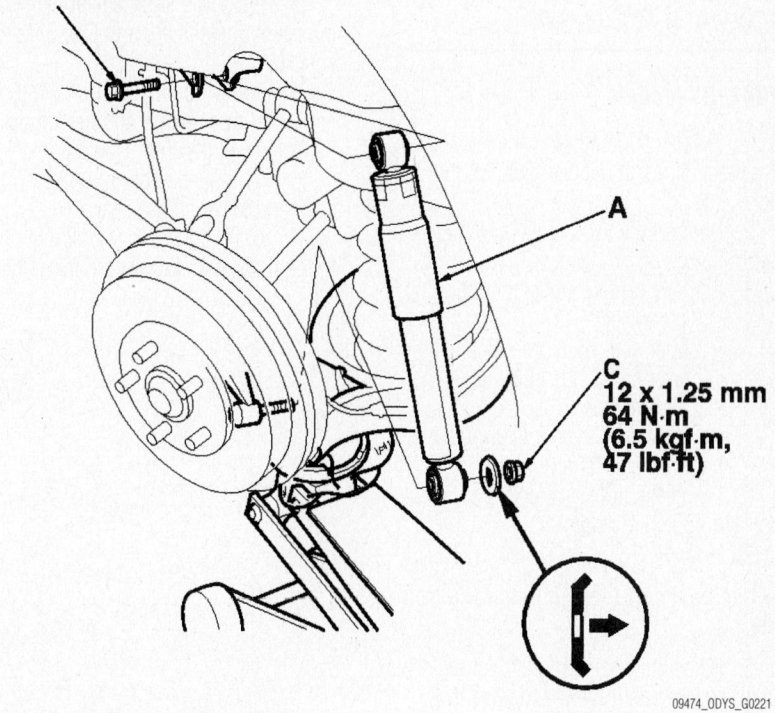

B
12 x 1.25 mm
64 N·m (6.5 kgf·m, 47 lbf·ft)

A

C
12 x 1.25 mm
64 N·m
(6.5 kgf·m,
47 lbf·ft)

09474_ODYS_G0221

Rear shock absorber—2002–04 models

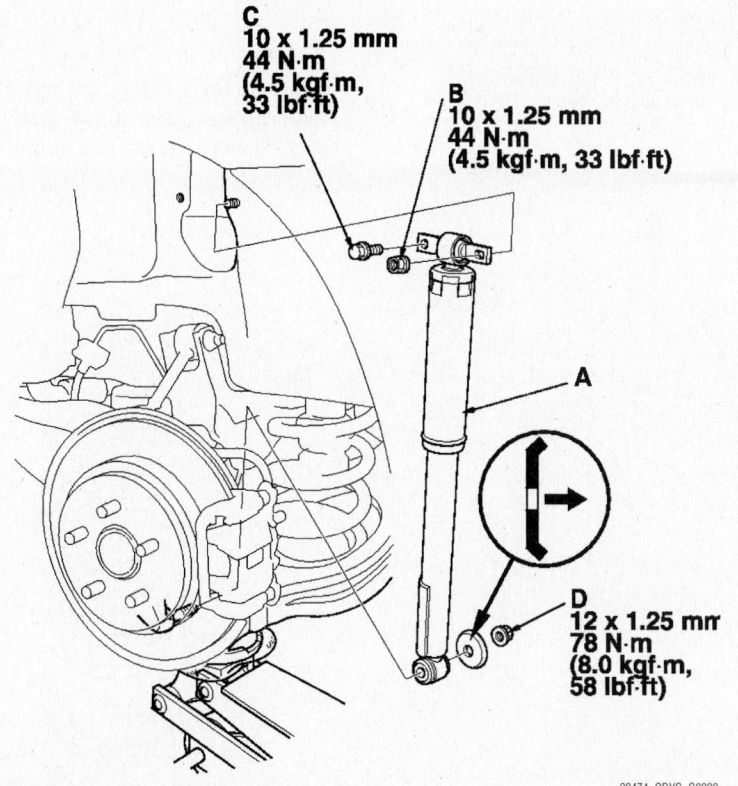

C
10 x 1.25 mm
44 N·m
(4.5 kgf·m,
33 lbf·ft)

B
10 x 1.25 mm
44 N·m
(4.5 kgf·m, 33 lbf·ft)

A

D
12 x 1.25 mm
78 N·m
(8.0 kgf·m,
58 lbf·ft)

09474_ODYS_G0222

Rear shock absorber—2005–06 models

12. Install the rear wheel.

13. Check the rear wheel alignment, and adjust it if necessary.

Coil Spring

REMOVAL & INSTALLATION

2002–04 Models

1. Before servicing the vehicle, refer to the precautions in the beginning of this section.

2. Support the vehicle under the lower control arm.

3. Remove or disconnect the following:
- Rear wheel
- Upper shock absorber flange bolt
- Lower shock absorber nut
- Shock absorber
- Wheel speed sensor wiring harness
- Lower control arm bolts

4. Lower the floor jack and remove the coil spring and spring seats.

To install:

➡ **Use new self-locking nuts for assembly.**

5. Place the coil spring and spring seats on the lower control arm and raise into position. Tighten the inboard bolt to 61 ft. lbs. and the outer bolt to 47 ft. lbs. (64 Nm).

6. Install or connect the following:
- Shock absorber. Tighten the fasteners to 47 ft. lbs. (64 Nm).
- Rear wheel

2005–06 Models

1. Raise the rear of the vehicle, and support it with safety stands in the proper locations.

2. Remove the rear wheel.

3. Position a floor jack at the connecting point of the lower arm B and the knuckle.

4. Remove the flange bolt (A) that connects the lower arm B and the knuckle.

5. Lower the floor jack gradually.

6. Remove the spring, upper spring seat, and lower spring seat.

7. Remove the flange bolt that connects to the body, and remove the bump stop.

To install:

8. Install the bump stop (A), and tighten the flange bolt (C) to the specified torque value.

9. Install the upper spring seat (D) and spring (E). Align the bottom of the spring and the lower spring seat (F) with lower arm B as shown.

10. Position a floor jack at the connecting point of the lower arm B and the knuckle.

11. Slowly raise the jack until you can align the bolt hole with the holes in the lower arm B and the knuckle and install the flange bolt (A).

12. Raise the rear suspension with a

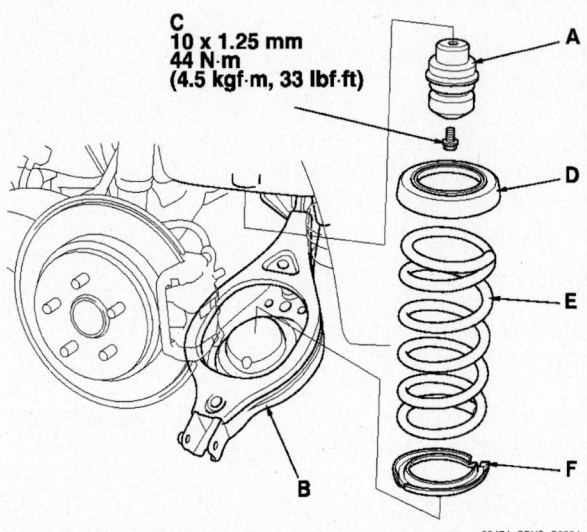

C
10 x 1.25 mm
44 N·m
(4.5 kgf·m, 33 lbf·ft)

09474_ODYS_G0224

Install the bump stop (A), and tighten the flange bolt (C) to the specified torque value. Install the upper spring seat (D) and spring (E). Align the bottom of the spring and the lower spring seat (F) with lower arm B as shown—2005–06 Models

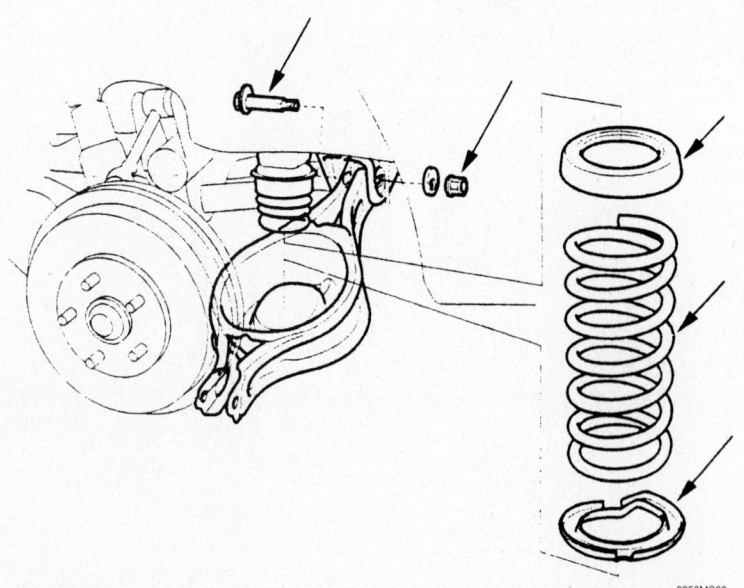

9358MG28

Exploded view of the rear spring assembly—2002–04 models

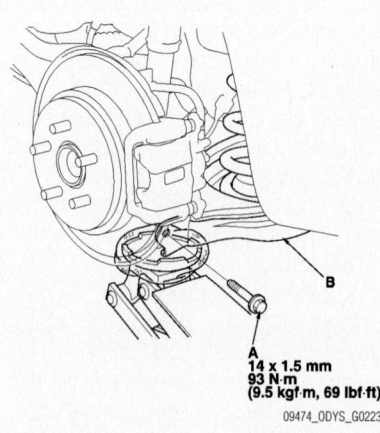

A
14 x 1.5 mm
93 N·m
(9.5 kgf·m, 69 lbf·ft)

09474_ODYS_G0223

Remove the flange bolt (A) that connects the lower arm B and the knuckle—2005–06 Models floor jack until the vehicle just lifts off the safety stands.

13. Tighten the flange bolt.

14. Install the rear wheel.

15. Check the rear wheel alignment, and adjust it if necessary.

Lower Arm A

REMOVAL & INSTALLATION

2002–04

1. Raise the rear of the vehicle, and support it with safety stands in the proper locations. Remove the rear wheel.

2. Remove the lower arm A mounting bolt (B).

3. Remove the lower arm mounting nut (C).

4. Remove the lower arm A from the vehicle.

5. Install the lower arm A.

6. Hand tighten the mounting bolt and nut.

7. Position a floor jack at the connecting point of lower arm B and the knuckle (A).

8. Raise the rear suspension with the floor jack until the vehicle just lifts off the safety stands.

9. Tighten the mounting bolt and nut to the specified torque values, and note these items:

- Before installing the wheel, clean the mating surfaces on the brake drum or the drum in brake disc and the inside of the wheel.
- Check the rear wheel alignment, and adjust it if necessary.

2005–06 Models

1. Raise the rear of the vehicle, and support it with safety stands in the proper locations.

2. Remove the rear wheel.

3. Remove the wheel sensor harness from the lower arm A.

4. Remove the lower arm A mounting bolt (B).

5. Remove and discard the washer (C) and the lower arm mounting special self-locking nut (D).

6. Remove the lower arm A from the vehicle.

To install:

7. Install the lower arm A and a new washer.

8. Lightly tighten the mounting bolt and a new special self-locking nut.

9. Position a floor jack at the connecting point of lower arm B and the knuckle.

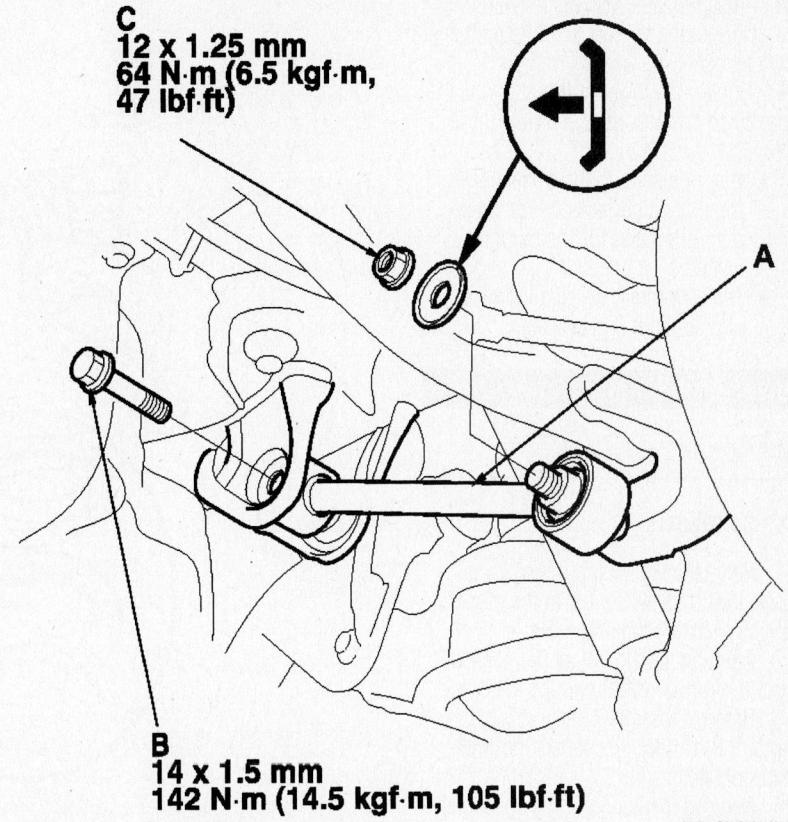

C
12 x 1.25 mm
64 N·m (6.5 kgf·m,
47 lbf·ft)

A

B
14 x 1.5 mm
142 N·m (14.5 kgf·m, 105 lbf·ft)

09474_ODYS_G0239

Lower arm A—2002–04 models

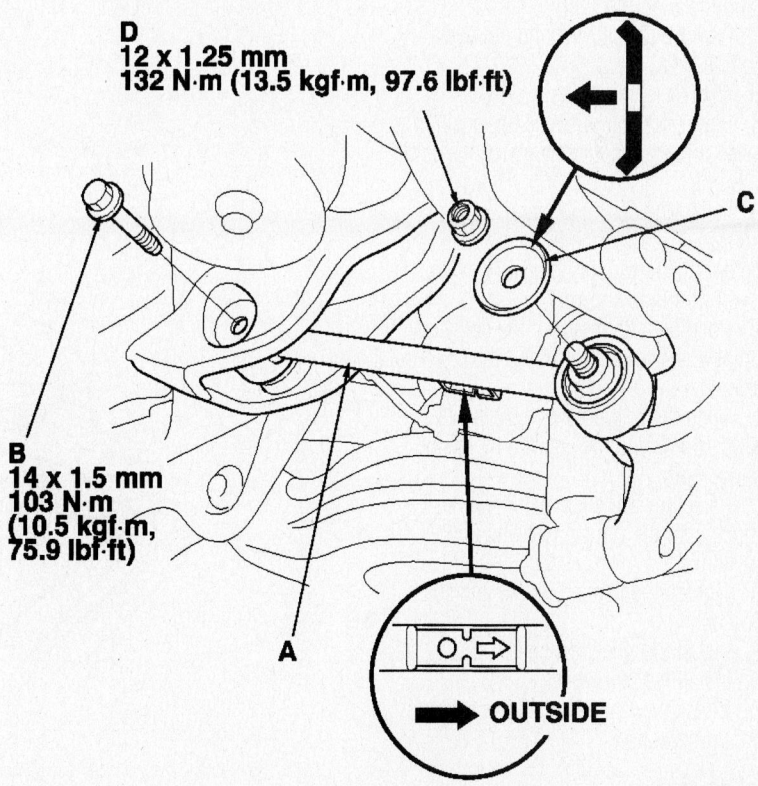

D
12 x 1.25 mm
132 N·m (13.5 kgf·m, 97.6 lbf·ft)

C

B
14 x 1.5 mm
103 N·m
(10.5 kgf·m,
75.9 lbf·ft)

A

OUTSIDE

09474_ODYS_G0225

Remove the lower arm A mounting bolt (B). Remove and discard the washer (C) and the lower arm mounting special self-locking nut (D)—2005–06 models

10. Raise the rear suspension with the floor jack until the vehicle just lifts off the safety stands.

11. Tighten the lower arm bolt and nut to the specified torque value, and note these items:

- Before installing the wheel, clean the mating surfaces on the brake disc/drum and the inside of the wheel.
- Check the rear wheel alignment, and adjust it if necessary

Lower Arm B

REMOVAL & INSTALLATION

2002–04 Models

1. Raise the rear of the vehicle, and support it with safety stands in the proper locations. Remove the rear wheel.

2. Position a floor jack at the connecting point of lower arm B and the knuckle.

3. Remove the wheel sensor from the lower arm B. Do not disconnect the wheel sensor connector.

4. Remove the flange bolt that connects the lower arm B and the knuckle.

5. Lower the floor jack gradually.

6. Remove the spring, upper spring seat and lower spring seat.

7. Remove the self-locking nut and flange bolt.

To install:

8. Lower the rear suspension, position the lower arm B, and loosely install the self-locking nut.

9. Install the spring (A), and upper spring seat (C). Align the bottom of the spring and the lower spring seat (D) with the lower arm B as shown.

10. Position a floor jack at the connecting point of the lower arm B and the knuckle.

11. Align the bolt hole with the holes in the lower arm B and the knuckle by raising the flange bolt.

12. Raise the rear suspension with a floor jack until the vehicle just lifts off the safety stands.

13. Tighten the flange bolt (A) and self-locking nut.

14. Install the wheel sensor.

15. Install the rear wheel.

16. Check the rear wheel alignment, and adjust it if necessary.

2005–06 Models

1. Raise the rear of the vehicle, and support it with safety stands in the proper locations.

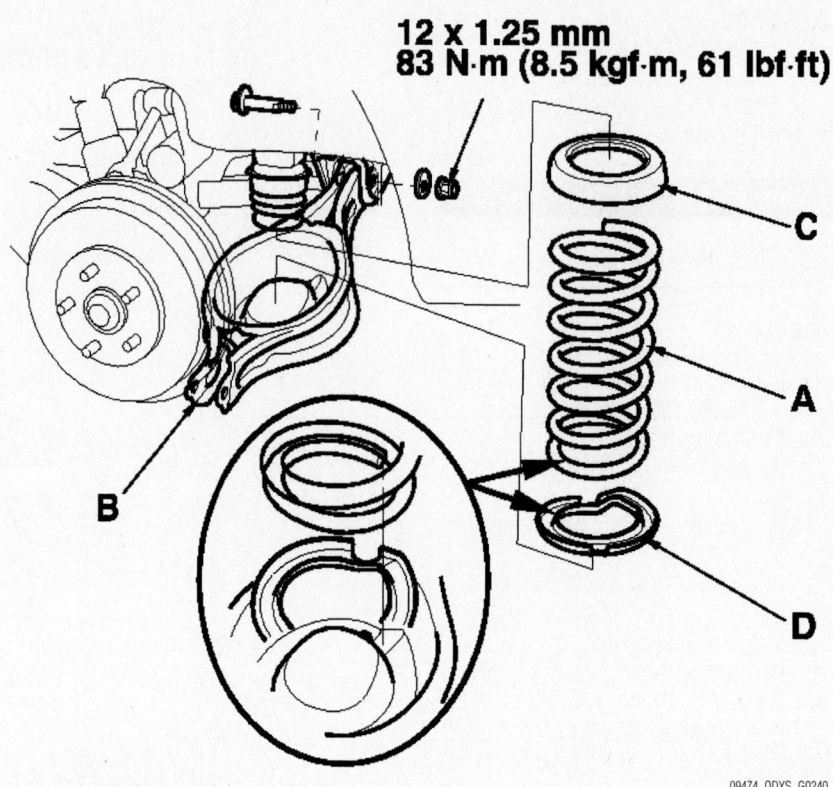

09474_ODYS_G0240

Install the spring (A), and upper spring seat (C). Align the bottom of the spring and the lower spring seat (D) with the lower arm B as shown—2002–04 models

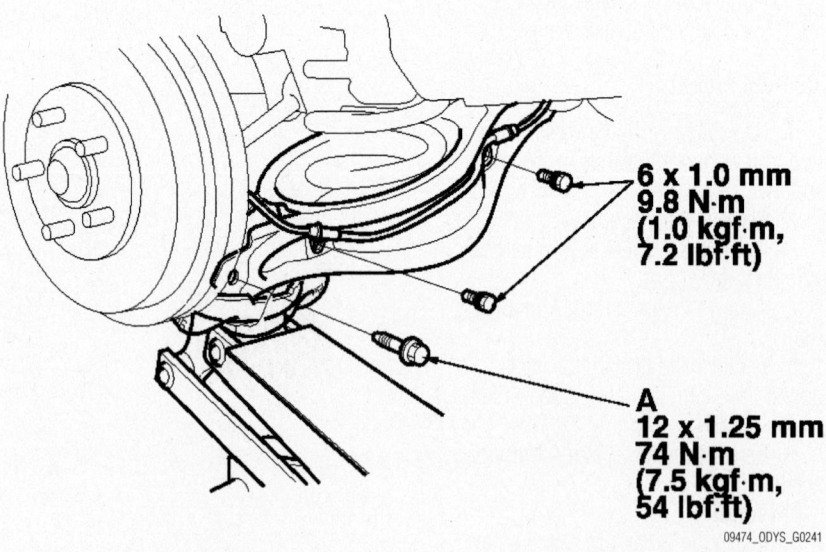

09474_ODYS_G0241

Tighten the flange bolt (A) and self-locking nut—2002–04 models

**14 x 1.5 mm
98.1 N·m
(10.0 kgf·m, 72.3 lbf·ft)**

C

A

D

09474_ODYS_G0226

Install the spring (A), and upper spring seat (C). Align the bottom of the spring and the lower spring seat (D) with the lower arm B as shown—2005–06 models

2. Remove the rear wheel.

3. Position a floor jack at the connecting point of lower arm B and the knuckle.

4. Remove the flange bolt (A) that connects the lower arm B and the knuckle.

5. Lower the floor jack gradually.

6. Remove the spring, upper spring seat and lower spring seat.

7. Remove the self-locking nut and flange bolt.

To install:

8. Position the lower arm B, and loosely install the flange bolt and the self-locking nut.

9. Install the spring (A), and upper spring seat (C). Align the bottom of the spring and the lower spring seat (D) with the lower arm B as shown.

Upper Control Arm

REMOVAL & INSTALLATION

2002–04 Models

1. Raise the rear of the vehicle, and support it with safety stands in the proper locations. Remove the rear wheel.

2. Position a floor jack at the connecting point of the lower arm B and the knuckle.

3. Remove the lock pin from the upper ball joint castle nut, and remove the nut.

4. Remove the upper ball joint from the knuckle using the special tool.

5. Remove the upper arm bolt (A). Remove the upper arm (B) from the vehicle.

To install:

6. Install the upper arm.

7. Install the upper arm bolt and castle nut. Tighten finger tight.

8. Position a floor jack at the connecting point of the lower arm B and the knuckle.

9. Raise the rear suspension with the floor jack until the vehicle just lifts off the safety stands.

10. Tighten the upper arm bolt to the specific torque value, and note these items:

a. Be careful not to damage the ball joint boot when connecting the upper arm to the knuckle.

b. Tighten the castle nut to the lower torque specifications, then tighten it only far enough to align the slot with the hole in the stud. Do not align the castle nut by loosening it.

c. Use a new lock pin in the castle nut.

d. Before installing the wheel, clean the mating surfaces on the brake drum or

07MAC-SL00200

**12 x 1.25 mm
49-59 N·m
(5.0-6.0 kgf·m,
36-43 lbf·ft)**

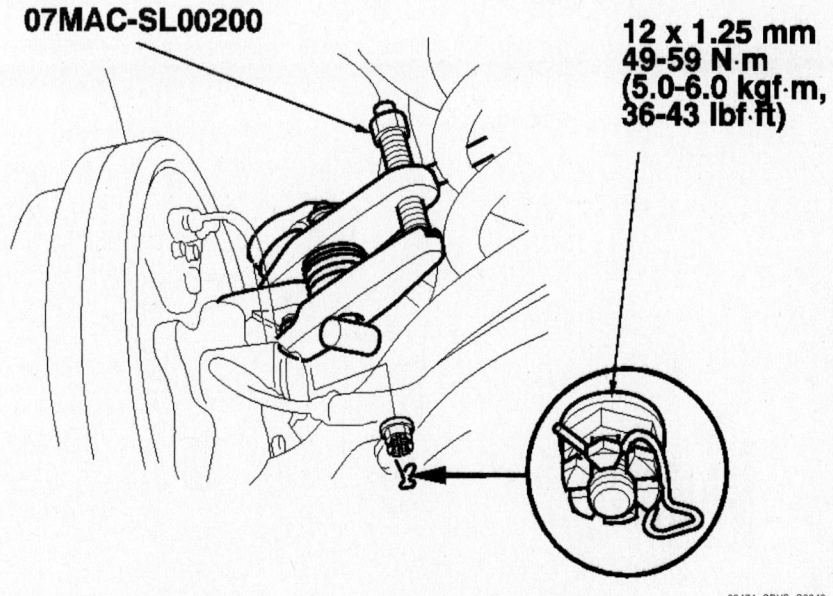

09474_ODYS_G0242

Remove the upper ball joint from the knuckle using the special tool—2002–04 models

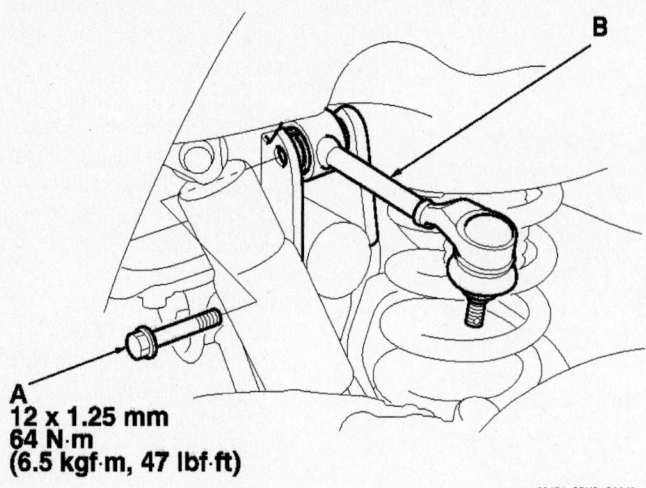

A
12 x 1.25 mm
64 N·m
(6.5 kgf·m, 47 lbf·ft)

09474_ODYS_G0243

Remove the upper arm bolt (A). Remove the upper arm (B) from the vehicle—2002–04 models

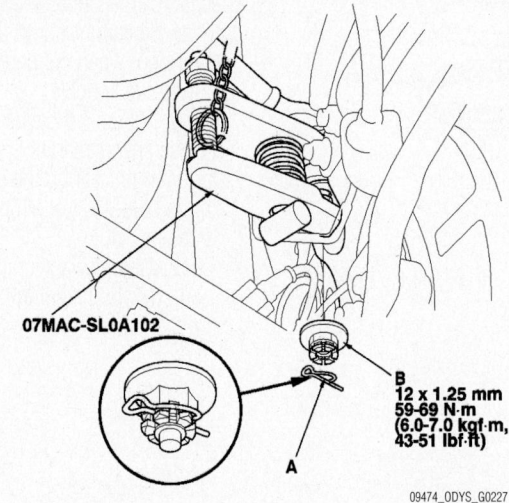

07MAC-SL0A102

B
12 x 1.25 mm
59-69 N·m
(6.0-7.0 kgf·m,
43-51 lbf·ft)

A

09474_ODYS_G0227

Remove the lock pin (A) from the upper ball joint castle nut (B), and remove the nut—2005–06 models

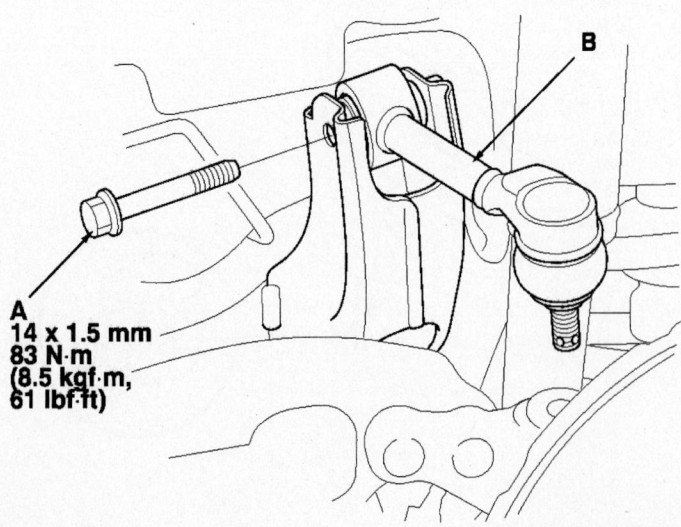

A
14 x 1.5 mm
83 N·m
(8.5 kgf·m,
61 lbf·ft)

09474_ODYS_G0228

Remove the upper arm bolt (A). Remove the upper arm (B) from the vehicle—2005–06 models

the drum in brake disc and the inside of the wheel.

   e. Check the rear wheel alignment, and adjust it if necessary.

2005–06 Models

1. Raise the rear of the vehicle, and support it with safety stands in the proper locations.

2. Remove the rear wheel.

3. Position a floor jack at the connecting point of the lower arm B and the knuckle.

4. Remove the lock pin (A) from the upper ball joint castle nut (B), and remove the nut.

5. Remove the upper ball joint from the knuckle using the special tool.

6. Remove the upper arm bolt (A). Remove the upper arm (B) from the vehicle.

To install:

7. Install the upper arm.

8. Lightly tighten the upper arm bolt and castle nut.

9. Position a floor jack at the connecting point of the lower arm B and the knuckle.

10. Raise the rear suspension with the floor jack until the vehicle just lifts off the safety stands.

11. Tighten the upper arm bolt to the specified torque value, and note these items:

- Be careful not to damage the ball joint boot when connecting the upper arm to the knuckle.
- Before connecting the ball joint to the knuckle, degrease the threaded section and tapered portion of the ball joint pin, the connecting hole, and the threaded section and mating surface of the castle nut.
- Torque the castle nut to the lower torque specification, then tighten it only far enough to align the slot with the hole in the stud. Do not align the castle nut by loosening it.
- Use a new lock pin on the castle nut.
- Before installing the wheel, clean the mating surfaces on the brake disc/drum and the inside of the wheel.
- Check the rear wheel alignment, and adjust it if necessary.

Trailing Arm

REMOVAL & INSTALLATION

2002–04 Models

1. Raise the rear of the vehicle, and support it with safety stands in the proper locations.

2. Remove the brake drum or brake disc/drum.

3. Remove the parking brake cable from the trailing arm.

4. Remove the brake hose mounting bolts from the trailing arm.

5. Position a floor jack at the connecting point of the lower arm and the knuckle.

6. Remove the flange bolts (A) and (B) from the trailing arm.

7. Install the trailing arm in the reverse order of removal, and note these items:

   a. Tighten all mounting hardware to the specified torque values.

   b. Check the brake hose and line joint for leaks, and tighten if necessary.

   c. Check the brake hose for interference and twisting.

   d. Before installing the wheel, clean the mating surfaces on the brake drum or the drum in brake disc and the inside of the wheel.

   e. Fill up the brake reservoir, and bleed the brake system.

   f. After installation, check for leaks at the line joint, and retighten if necessary. Check the rear wheel alignment, and adjust it if necessary.

2005–06 Models

1. Raise the rear of the vehicle, and support it with safety stands in the proper locations.

2. Remove the rear wheel.

3. Remove the brake disc.

4. Remove the parking brake cable from the trailing arm.

5. Remove the brake hose mounting nut from the trailing arm.

6. Position a floor jack at the connecting point of the lower arm and the knuckle.

7. Remove the self-locking nuts (A) and bolts (B) from the trailing arm.

8. Install the trailing arm in the reverse order of removal, and note these items:

   • Use a new self-locking nut on reassembly.

   • First install all the components and lightly tighten the bolts and nuts, then raise the suspension to load it with the vehicle's weight before fully tightening to the specified torque values.

   • Tighten all mounting hardware to the specified torque values.

   • Check the brake hose and line joint for leaks, and tighten if necessary.

   • Check the brake hose for interference and twisting.

   • Before installing the wheel, clean the mating surfaces on the brake

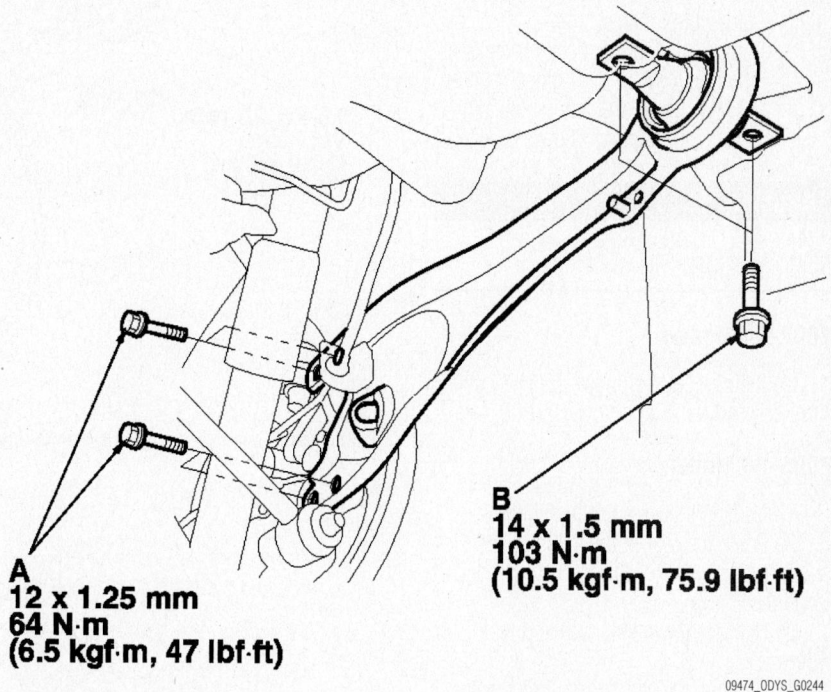

A
12 x 1.25 mm
64 N·m
(6.5 kgf·m, 47 lbf·ft)

B
14 x 1.5 mm
103 N·m
(10.5 kgf·m, 75.9 lbf·ft)

09474_ODYS_G0244

Remove the flange bolts (A) and (B) from the trailing arm—2002–04 models

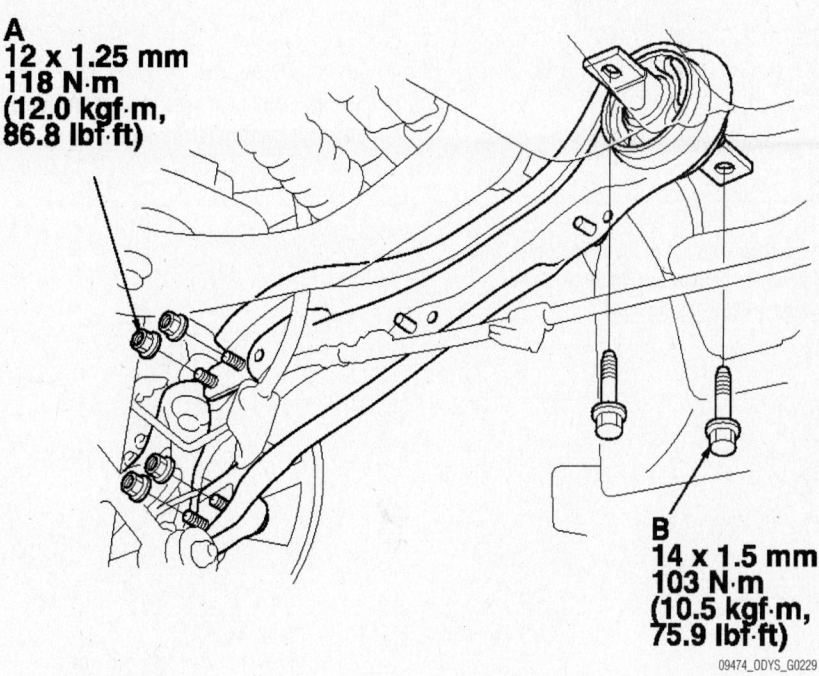

A
12 x 1.25 mm
118 N·m
(12.0 kgf·m, 86.8 lbf·ft)

B
14 x 1.5 mm
103 N·m
(10.5 kgf·m, 75.9 lbf·ft)

09474_ODYS_G0229

Remove the self-locking nuts (A) and bolts (B) from the trailing arm—2005–06 models

disc/drum and the inside of the wheel.
- Fill up the brake reservoir, and bleed the brake system.
- After installation, check for leaks at the line joint, and retighten if necessary. Check the rear wheel alignment, and adjust it if necessary.

Knuckle

REMOVAL & INSTALLATION

2002–04 Models

See the procedures under Upper Control Arm, Lower Arm A, and Lower Arm B.

2005–06 Models

1. Loosen the wheel nuts slightly.
2. Raise the rear of the vehicle, and support it with safety stands in the proper locations.
3. Remove the wheel nuts and rear wheel.
4. Remove the caliper mounting bolts, and hang the caliper assembly (A) to one side. To prevent damage to the caliper assembly or brake hose, use a short piece

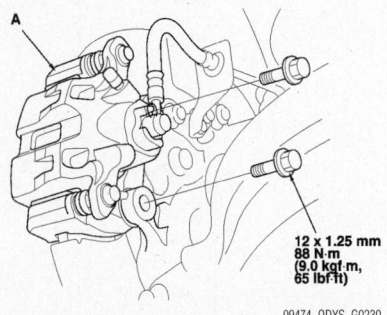

12 x 1.25 mm
88 N·m
(9.0 kgf·m,
65 lbf·ft)

09474_ODYS_G0230

Remove the caliper mounting bolts, and hang the caliper assembly (A) to one side—2005–06 models

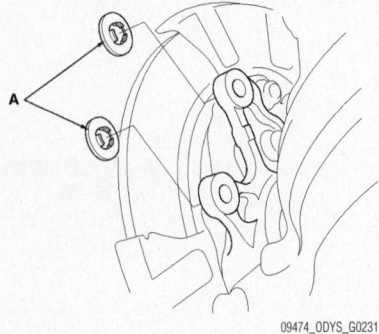

09474_ODYS_G0231

Remove the two washers (A)—2005–06 models

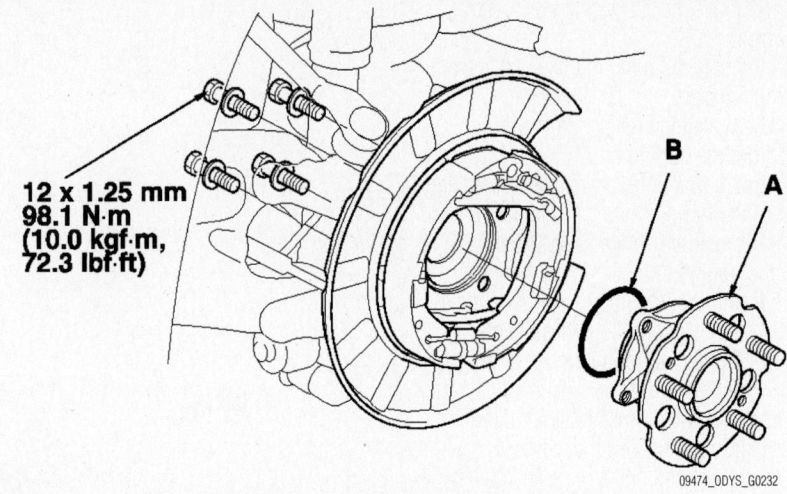

12 x 1.25 mm
98.1 N·m
(10.0 kgf·m,
72.3 lbf·ft)

09474_ODYS_G0232

Remove the hub bearing unit (A) and O-ring (B)—2005–06 models

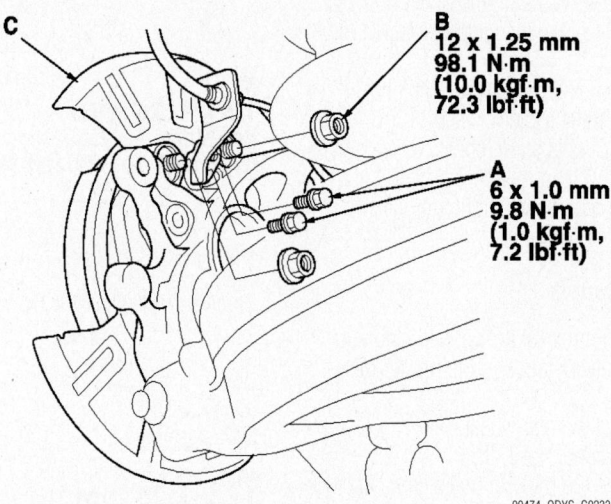

B
12 x 1.25 mm
98.1 N·m
(10.0 kgf·m,
72.3 lbf·ft)

A
6 x 1.0 mm
9.8 N·m
(1.0 kgf·m,
7.2 lbf·ft)

09474_ODYS_G0233

Remove the brake hose bracket mounting bolts (A) from the knuckle. Remove the flange nuts (B), then remove the backing plate (C)—2005–06 models

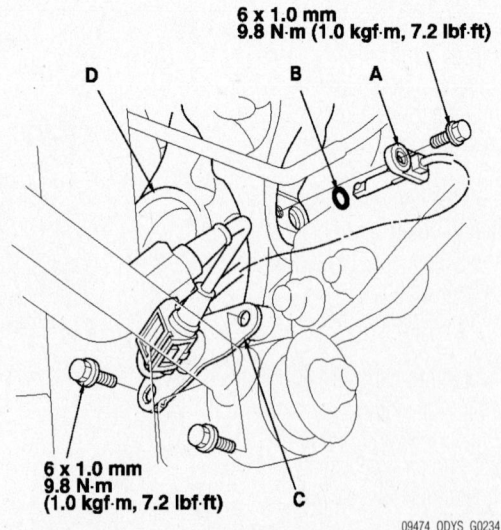

6 x 1.0 mm
9.8 N·m (1.0 kgf·m, 7.2 lbf·ft)

6 x 1.0 mm
9.8 N·m
(1.0 kgf·m, 7.2 lbf·ft)

09474_ODYS_G0234

Remove the wheel sensor (A), the O-ring (B), and the sensor harness (C) from the knuckle (D). Do not disconnect the wheel sensor connector—2005–06 models

of wire to hang the caliper from the undercarriage.

5. Remove the two washers (A).

6. Remove the brake disc retaining
screws.

7. Release the parking brake, and
remove the brake disc/drum.

8. Screw two 8 x 1.25mm bolts into the
brake disc/drum to push it away from the
hub. Turn each bolt two turns at a time to
prevent cocking the brake disc/drum excessively.

9. Remove the hub bearing unit (A) and
O-ring (B).

10. Check the rear hub for damage and
cracks.

11. Remove the brake hose bracket
mounting bolts (A) from the knuckle.

12. Remove the flange nuts (B), then
remove the backing plate (C).

13. Remove the wheel sensor (A), the O-
ring (B), and the sensor harness (C) from
the knuckle (D). Do not disconnect the
wheel sensor connector.

➡**Use a new O-ring on reassembly.**

14. Remove the lock pin (A) from the
upper arm ball joint, and loosen the nut (B).

➡**During installation, install the new
lock pin and nut after tightening the nut
as shown.**

15. Disconnect the upper arm ball joint
from the knuckle using the special tool.

16. Remove the special self-locking nut
(B), washer, and flange bolt (C), then
remove the lower arm A.

17. Remove the self-locking nuts (A),
and separate the knuckle from the trailing
arm (B).

18. Place a floor jack under lower arm B,
and remove the self-locking nut (A), washer,
and flange bolt (C).

19. Install the knuckle in the reverse
order of removal, and note these items:

a. Be careful not to damage the ball
joint boot when connecting the upper
arm to the knuckle.

b. Use a new self-locking nut when
reinstalling the bottom of the damper.

c. Use a new washer, and a new special self-locking nut when reinstalling the
lower arm A.

d. Before connecting the ball joint to
the knuckle, degrease the threaded section and tapered portion of the ball joint
pin, the connecting hole, and the
threaded section and mating surface of
the castle nut.

e. Torque the castle nut to the lower
torque specification, then tighten it only
far enough to align the slot with the hole

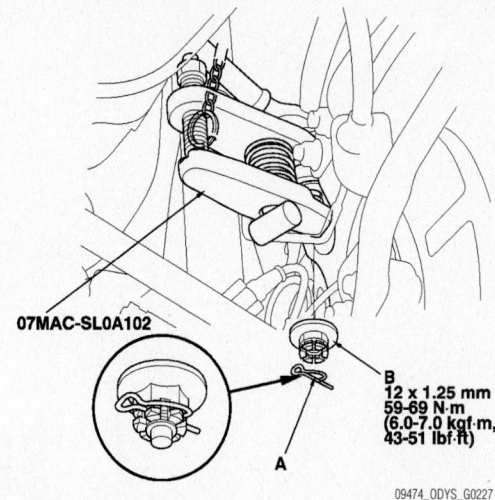

Remove the lock pin (A) from the upper arm ball joint, and loosen the nut (B)—2005–06 models

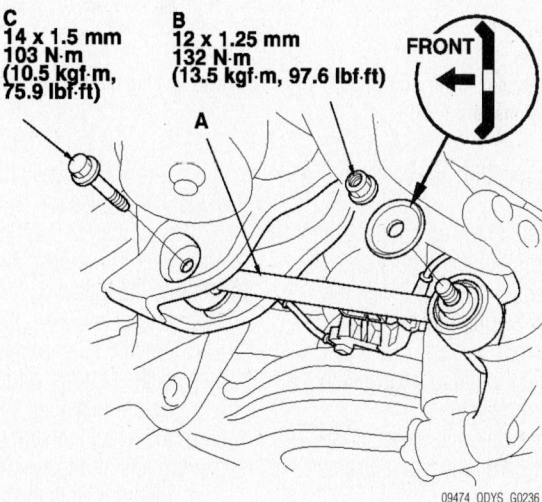

Remove the special self-locking nut (B), washer, and flange bolt (C), then remove the lower
arm A—2005–06 models

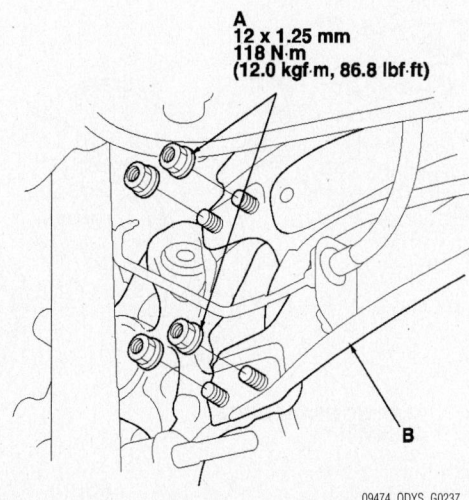

Remove the self-locking nuts (A), and separate the knuckle from the trailing arm (B)—2005–06
models

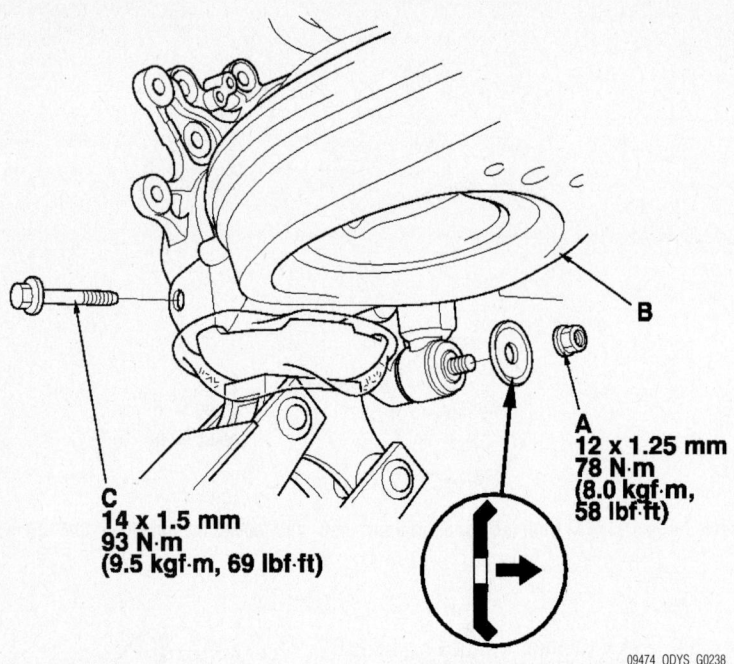

Place a floor jack under lower arm B, and remove the self-locking nut (A), washer, and flange bolt (C)—2005–06 models

in the stud. Do not align the castle nut by loosening it.

f. First install all the components and lightly tighten the bolts and nuts, then raise the suspension to load it with the vehicle's weight before fully tightening to the specified torque values.

g. Tighten all mounting hardware to the specified torque values.

h. Before installing the wheel, clean the mating surface on the brake disc and the inside of the wheel.

i. Check the wheel alignment, and adjust it if necessary.

20. Install the hub bearing unit in the reverse order of removal, and pay particular attention to the following items:

a. Use a new O-ring on reassembly.

b. Tighten all mounting hardware to the specified torque values.

c. Before installing the brake disc, clean the matching surfaces of the hub bearing unit and brake disc.

d. Before installing the wheel, clean the mating surfaces of the brake disc and the inside of the wheel.

Hub/Bearings

ADJUSTMENT

The wheel bearings are sealed units and are not adjustable.

REMOVAL & INSTALLATION

2002–04 Models

1. Before servicing the vehicle, refer to the precautions in the beginning of this section.

2. Remove or disconnect the following:
   - Rear wheel
   - Brake caliper and rotor
   - Spindle cap, nut and washer
   - Hub and bearing assembly

To install:

→**Use a new spindle nut for assembly.**

3. Install or connect the following:
   - Hub and bearing assembly. Tighten the spindle nut to 181 ft. lbs. (245 Nm).
   - Spindle cap
   - Brake caliper and rotor. Tighten the caliper bracket bolts to 28 ft. lbs. (38 Nm).
   - Rear wheel

2005–05 Models

See the procedure under Knuckle Removal and Installation.

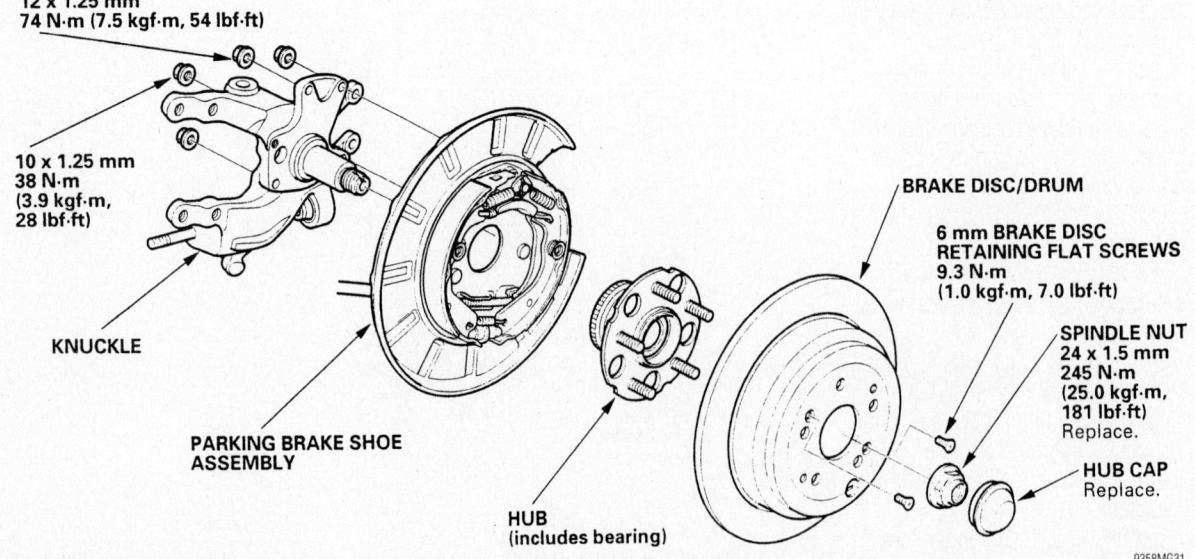

Exploded view of the rear hub/bearing assembly—2002–04 models

FRONT BRAKES

Brake Caliper

REMOVAL AND INSTALLATION

2002–04 Models

1. Remove some fluid from the reservoir with a suction pump.
2. Raise and safely support the vehicle.
3. Remove the front wheels.
4. Remove the banjo bolt and disconnect the brake hose from the caliper. Plug the hose to prevent fluid loss and contamination.
5. Remove the mounting bolts and remove the caliper from its mounting bracket.

To Install:

6. Fit the caliper over the pads and onto its mounting bracket.
7. Torque both caliper bolts to 27 ft. lbs. (36 Nm).
8. Reconnect the brake hose to the caliper using new sealing washers. Carefully torque the banjo bolt to 25 ft. lbs. (35 Nm).
9. Fill the reservoir with fluid and bleed the brakes.
10. Install the front wheels and lower the vehicle.

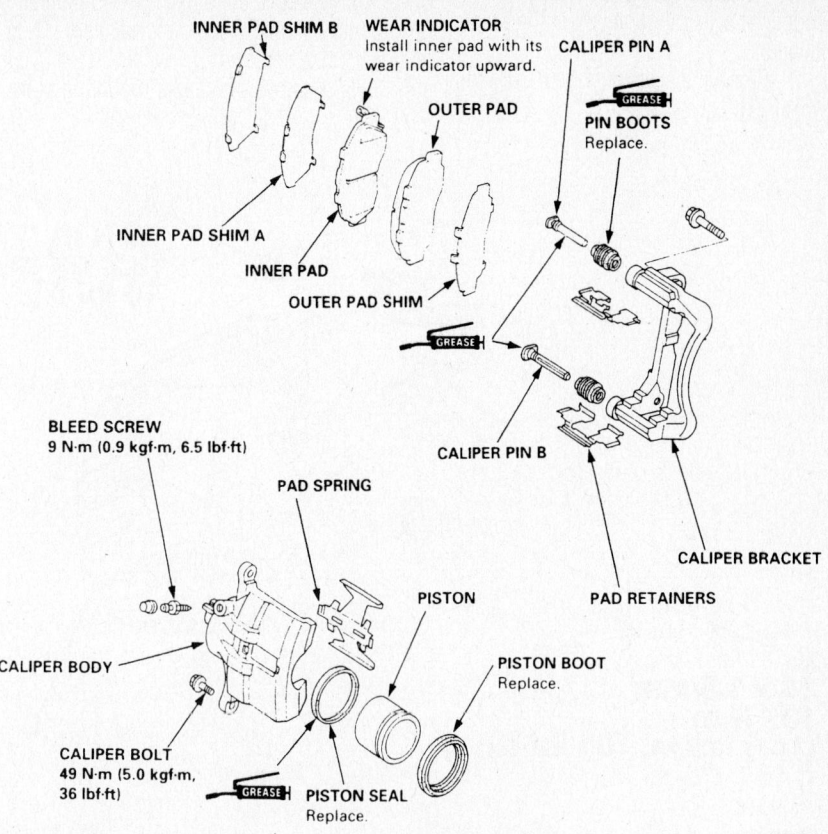

Exploded view of the front caliper components—2002–04 models

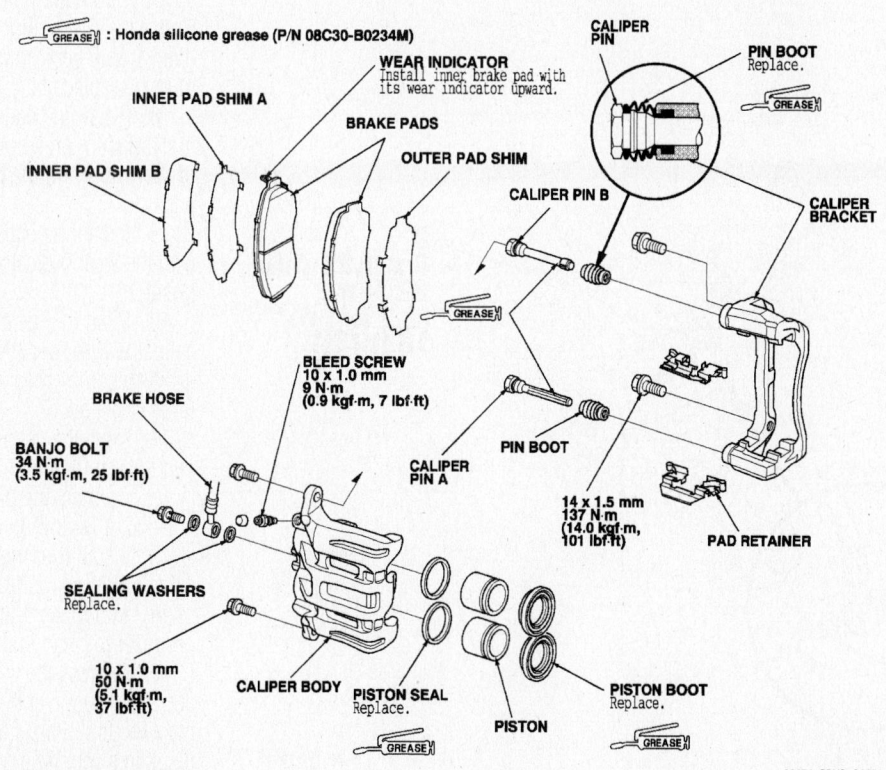

Exploded view of the front caliper—2005–06 models

2005–06 Models

1. Raise the front of the vehicle, and support it with safety stands in the proper locations.

2. Remove the wheel nuts (A) and front wheel (B).

3. Remove the brake hose mounting bolt (A).

4. Remove the brake hose-to-caliper union bolt and discard the washers. Plug the brake hose.

5. Remove the caliper mounting bolts (B).

6. Installation is the reverse of removal. Install the brake hose (A) on the damper bracket with flange bolt (B) first, then connect the brake hose to the caliper with the banjo bolt (C) and new sealing washers (D). Observe the torques given in the accompanying illustrations.

7. Bleed the brakes.

8. Press the brake pedal several times to make sure the brakes work.

Disc Brake Pads

REMOVAL AND INSTALLATION

2002–04 Models

1. Raise and support the vehicle safely.

2. Remove the front wheels.

3. Remove a small amount of brake fluid from the reservoir using a suction pump.

4. Unbolt the brake hose clamp from the knuckle by removing the retaining bolts.

5. Remove the lower caliper retaining bolt and pivot the caliper upward, off of the pads.

6. Remove the pad shim and pad retainers. Remove the disc brake pads from the caliper.

To install:

7. Clean the caliper thoroughly; remove any rust from the lip of the disc or rotor. Check the brake rotor for grooves or cracks. If any heavy scoring is present, the rotor must be replaced.

8. Install the pad retainers. Apply molybdenum brake grease to both surfaces of the shims and the back of the disc brake pads.

9. Install the pads and shims. The pad with the wear indicator goes in the inboard position.

10. Push in the caliper piston so the caliper will fit over the pads. This is most easily accomplished with a pad spreader or large C-clamp.

11. Pivot the caliper down into position and tighten the mounting bolt.

12. Connect the brake hose to the knuckle, if removed.

13. Install the wheel and lower the vehicle to the ground.

14. Add brake fluid to the master cylinder reservoir and install the cap.

15. Depress the brake pedal several times and make sure that the movement feels normal. The first brake pedal application may result in a very long pedal action due to the pistons being retracted. Always make several

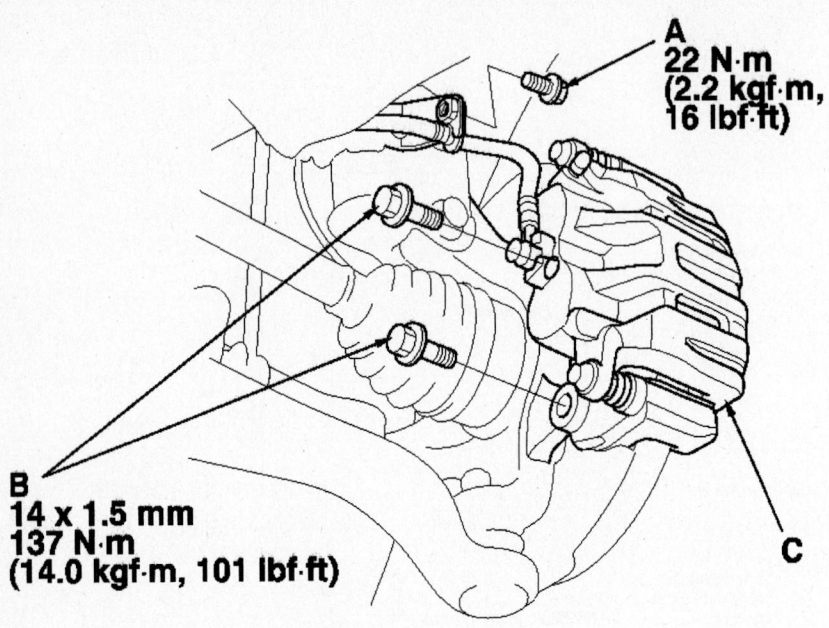

A
22 N·m
(2.2 kgf·m, 16 lbf·ft)

B
14 x 1.5 mm
137 N·m
(14.0 kgf·m, 101 lbf·ft)

C

09474_ODYS_G0196

Remove the caliper mounting bolts (B) —2005–06 Models

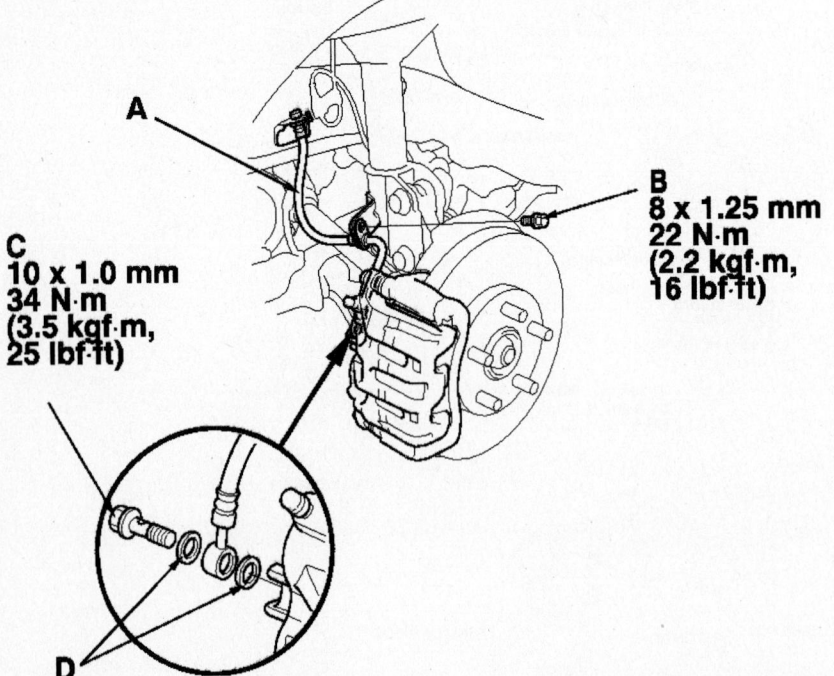

A

B
8 x 1.25 mm
22 N·m
(2.2 kgf·m, 16 lbf·ft)

C
10 x 1.0 mm
34 N·m
(3.5 kgf·m, 25 lbf·ft)

D

09474_ODYS_G0245

Brake hose attachment—2005–06 Models

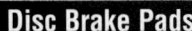

brake applications before starting the vehicle. Bleed the system if necessary.

2005–06 Models

1. Raise the front of the vehicle, and support it with safety stands in the proper locations.

2. Remove the front wheels.

3. While holding the caliper pin (A) with a wrench, remove the flange bolt (B). Be careful not to damage the pin boot. Then pivot the caliper (C) up out of the way. Check the hose and pin boots for damage and deterioration.

4. Remove the pad shims and brake pads.

5. Remove the pad retainers (A) from the caliper bracket (B).

6. Clean the caliper thoroughly; remove any rust, and check for grooves and cracks.

7. Check the brake disc for damage and cracks.

To install:

8. Apply a thin coat of M-77 assembly paste (P/N 08798-9010) to the retainers on their mating surfaces (indicated by the arrows) against the caliper bracket.

9. Install the pad retainers. Wipe excess paste off the retainers. Keep paste off the discs and pads.

10. Apply a thin coat of M-77 assembly paste (P/N 08798-9010) to the pad side of the shims, the back of the brake pads and the other areas indicated by the arrows. Wipe excess assembly paste off the shims and brake pads. Contaminated brake discs or brake pads reduce stopping ability. Keep grease and assembly paste off the brake discs and brake pads.

11. Install the brake pads and pad shims correctly. Install the brake pad with the wear

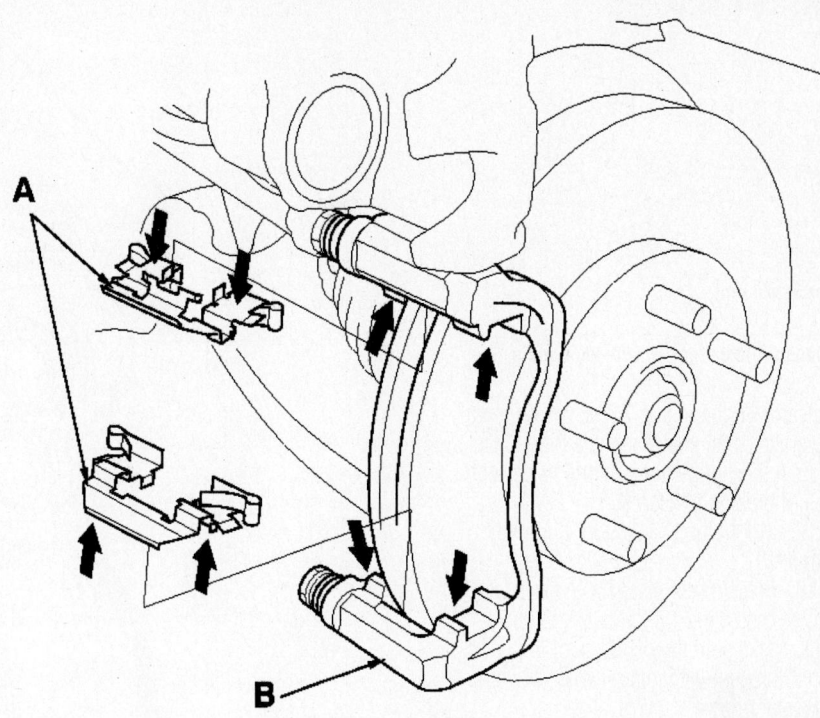

09474_ODYS_G0247

Remove the pad retainers (A) from the caliper bracket (B)—2005–06 models

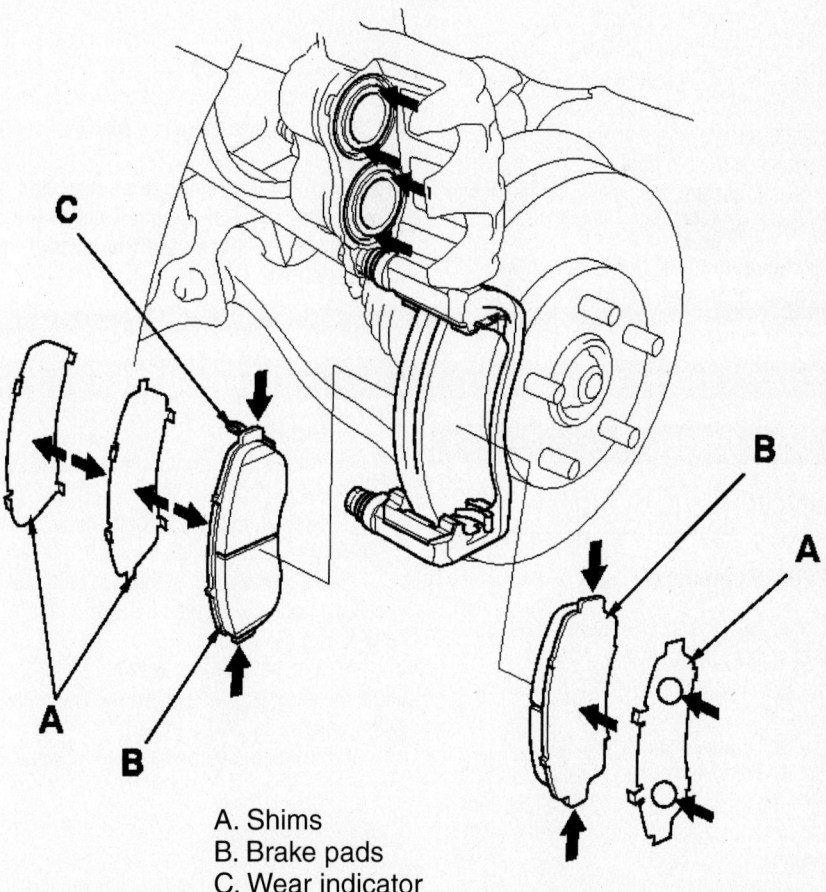

A. Shims
B. Brake pads
C. Wear indicator

09474_ODYS_G0248

Front brake pads and shims—2005–06 models

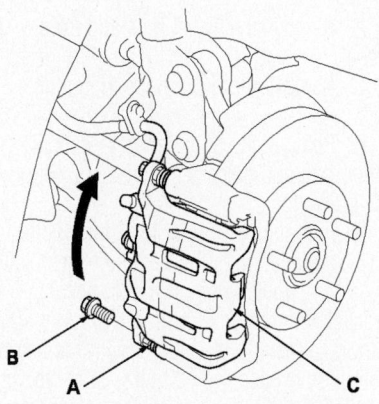

09474_ODYS_G0246

While holding the caliper pin (A) with a wrench, remove the flange bolt (B). Then pivot the caliper (C) up out of the way— 2005–06 models

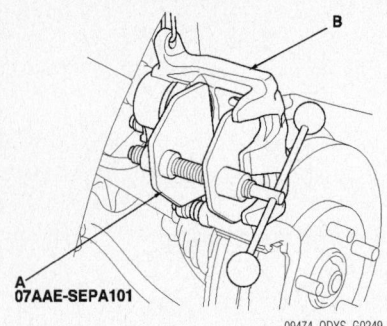

07AAE-SEPA101

09474_ODYS_G0249

Caliper piston tool—2005–06 models

indicator on the inside. If you are reusing the brake pads, always reinstall the brake pads in their original positions to prevent a loss of braking efficiency.

12. Mount the special tool (A) on the caliper (B).

13. Press in the piston with the special tool so the caliper will fit over the brake pads. Make sure the piston boot is in position to prevent damaging it when pivoting the caliper down.

➡**Be careful when pressing in the piston, brake fluid might overflow from the master cylinder's reservoir.**

14. Remove the special tool.

15. Pivot the caliper down into position. Install the flange bolt (A), and torque it to the specified torque while holding the caliper pin (B) with a wrench. Be careful not to damage the pin boot.

16. Press the brake pedal several times to make sure the brakes work.

➡**Engagement of the brakes may require a greater pedal stroke immedi-**

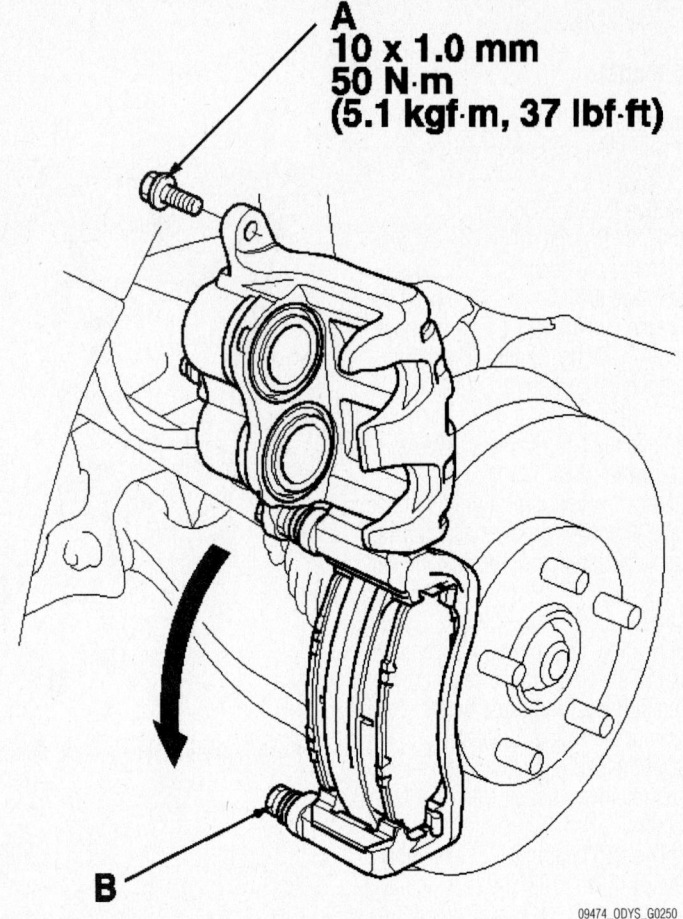

A
10 x 1.0 mm
50 N·m
(5.1 kgf·m, 37 lbf·ft)

B

09474_ODYS_G0250

Pivot the caliper down into position—2005–06 models

ately after the brake pads have been replaced. Several applications of the brake pedal will restore the normal pedal stroke.

17. After installation, check for leaks at hose and line joints or connections, and retighten if necessary.

18. Install the front wheels, then test-drive the vehicle.

REAR BRAKES

Brake Caliper

REMOVAL AND INSTALLATION

2002–04 Models

1. Remove some fluid from the reservoir with a suction pump.

2. Raise and safely support the vehicle.

3. Remove the rear wheels.

4. Remove the banjo bolt and disconnect the brake hose from the caliper. Plug the hose to prevent fluid loss and contamination.

5. Remove the 2 caliper mounting bolts. Remove the caliper from its mounting bracket.

To Install:

6. Fit the caliper over the pads and onto its mounting bracket.

7. Tighten the caliper bolts to 27 ft. lbs. (36 Nm).

8. Reconnect the brake hose with new sealing washers. Tighten the banjo bolt to 17 ft. lbs. (34 Nm).

9. Fill the reservoir with fluid and bleed the brake system. Adjust the parking brake if necessary.

10. Install the rear wheels and lower the vehicle.

2005–06 Models

1. Raise the front of the vehicle, and support it with safety stands in the proper locations.

2. Remove the wheel nuts and front wheel.

3. Remove the brake hose mounting bolt.

4. Remove the brake hose-to-caliper union bolt and discard the washers. Plug the brake hose.

5. Remove the caliper mounting bolts.

6. Installation is the reverse of removal. Install the brake hose on the damper bracket with flange bolt first, then connect the brake hose to the caliper with the banjo bolt and new sealing washers. Observe the torques given in the accompanying illustrations.

7. Bleed the brakes.

8. Press the brake pedal several times to make sure the brakes work.

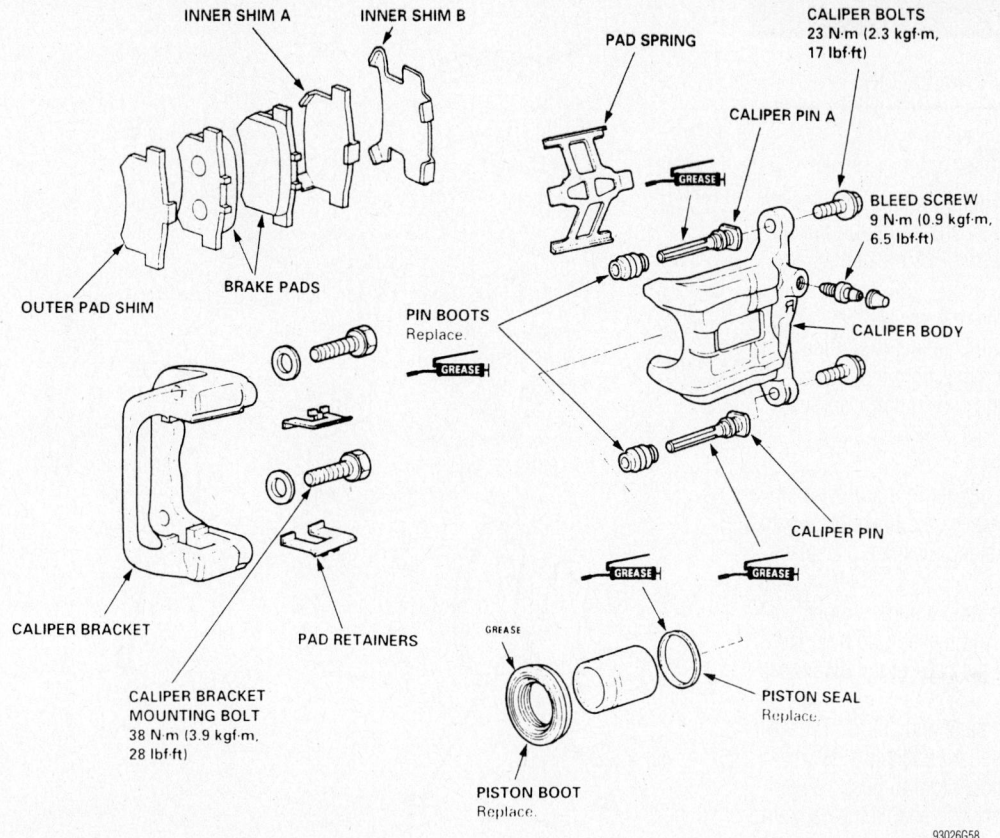

Exploded view of the rear caliper components—2002–04 models

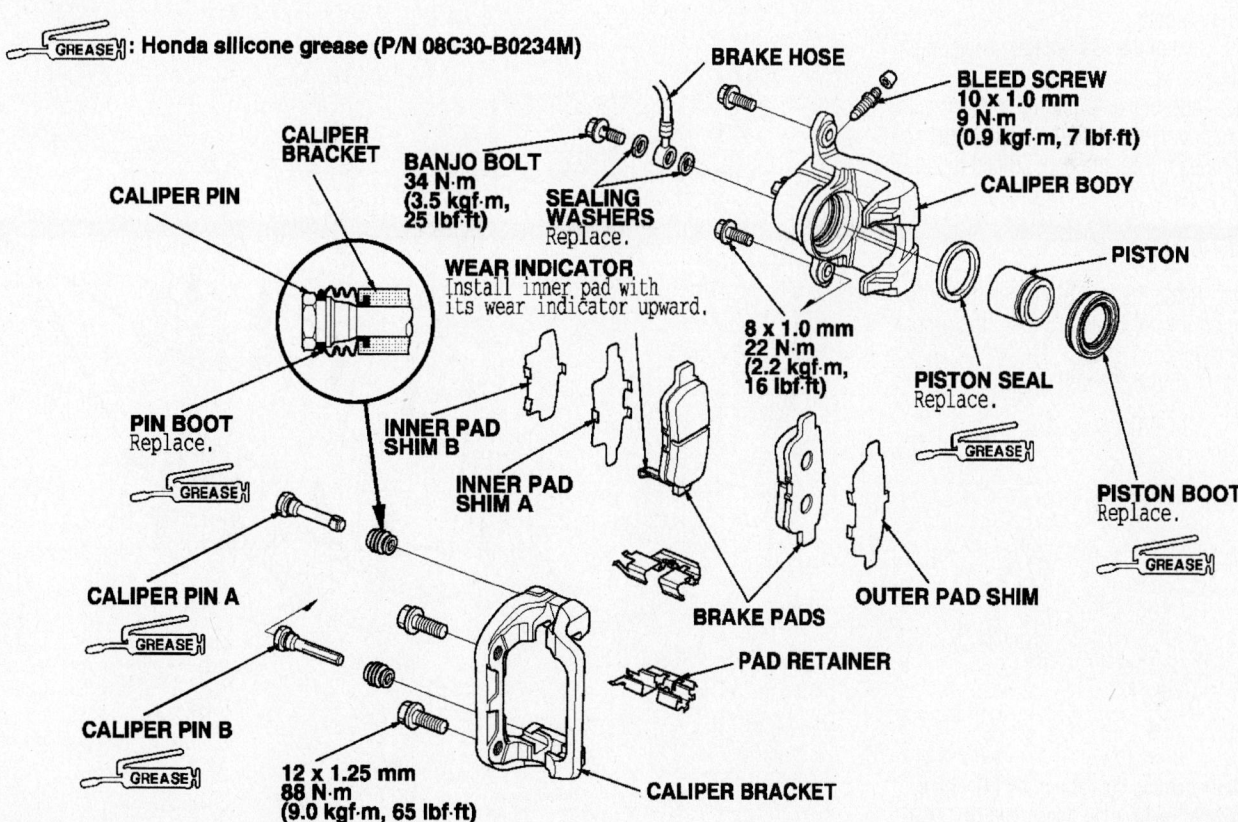

Exploded view of the rear caliper—2005–06 models

09474_ODYS_G0252

Disc Brake Pads

REMOVAL AND INSTALLATION

2002–04 Models

1. Raise and safely support the vehicle.
2. Remove a small amount of brake fluid from the reservoir using a suction pump.
3. Remove the rear wheels.
4. Remove the 2 caliper mounting bolts and remove the caliper from the bracket.
5. Remove the pads, shims, and pad retainers.

To install:

6. Clean the caliper thoroughly; remove any dirt or dust. Check the brake rotor for grooves or cracks and machine or replace, as necessary.
7. Install the pad retainers. Apply molybdenum brake grease to both surfaces of the shims and the back of the disc brake pads.
8. Install the pads and shims. The wear retainer on the inboard pad faces down.
9. Use a suitable tool to push caliper piston into its bore and enable the caliper to fit over the pads. Lubricate the piston boot with silicon grease. Avoid twisting the boot.
10. Install the brake caliper. Tighten the mounting bolts.
11. Install the rear wheels. Lower the vehicle.
12. Add brake fluid to the master cylinder reservoir. Depress the brake pedal several times to seat the pads. Bleed the brakes if necessary.

2005–06 Models

1. Raise the rear of the vehicle, and support it with safety stands in the proper locations.
2. Remove the rear wheels.

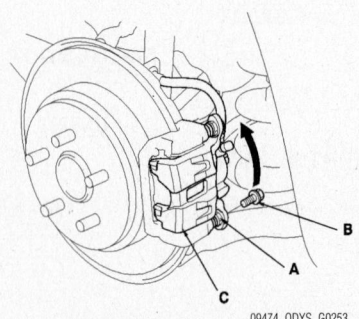

While holding the caliper pin (A) with a wrench, remove the flange bolt (B). Then pivot the caliper (C) up out of the way—2005–06 rear brakes

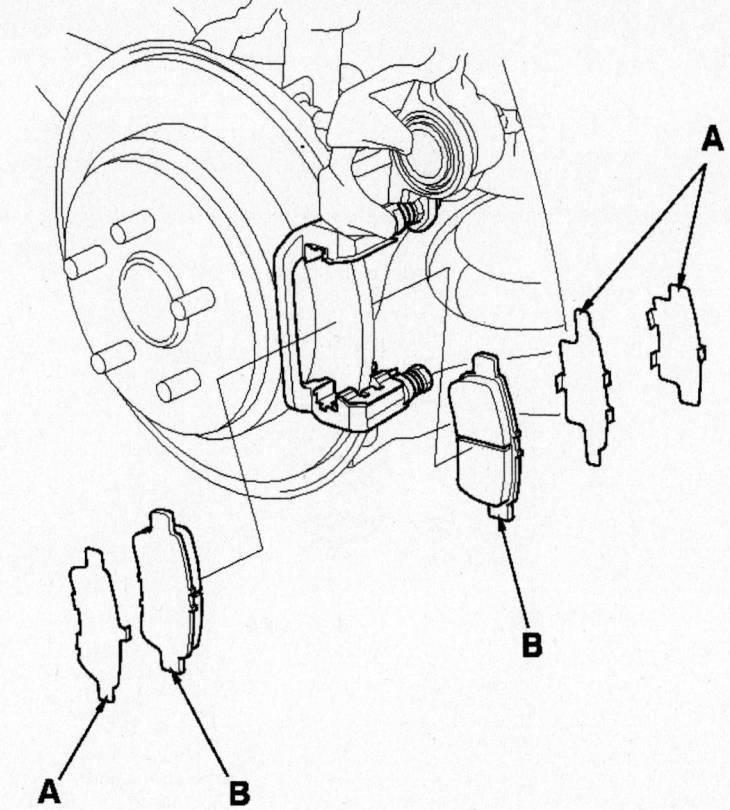

Remove the pad shims (A) and brake pads (B)—2005–06 rear brakes

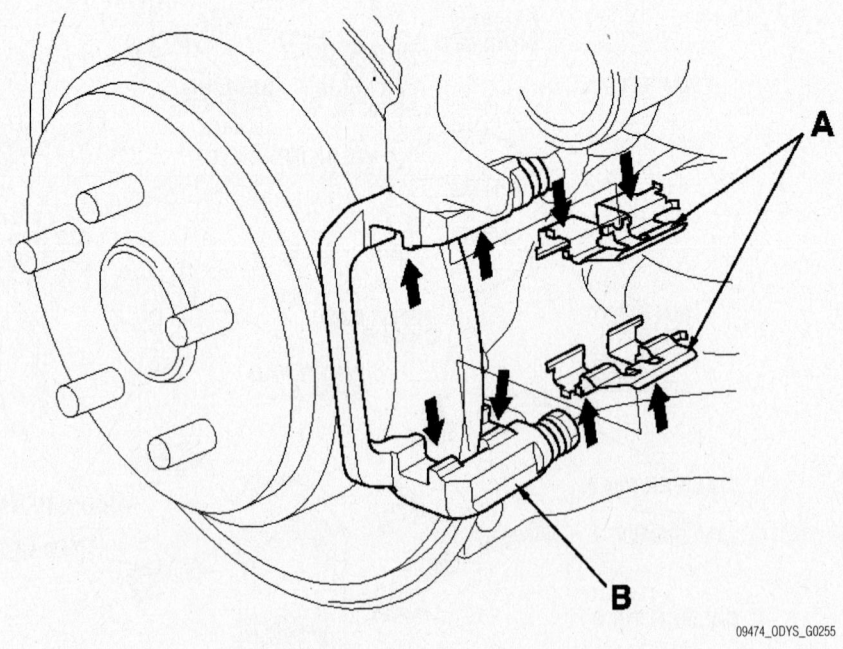

Remove the pad retainers (A) from the caliper bracket (B)—2005–06 rear brakes

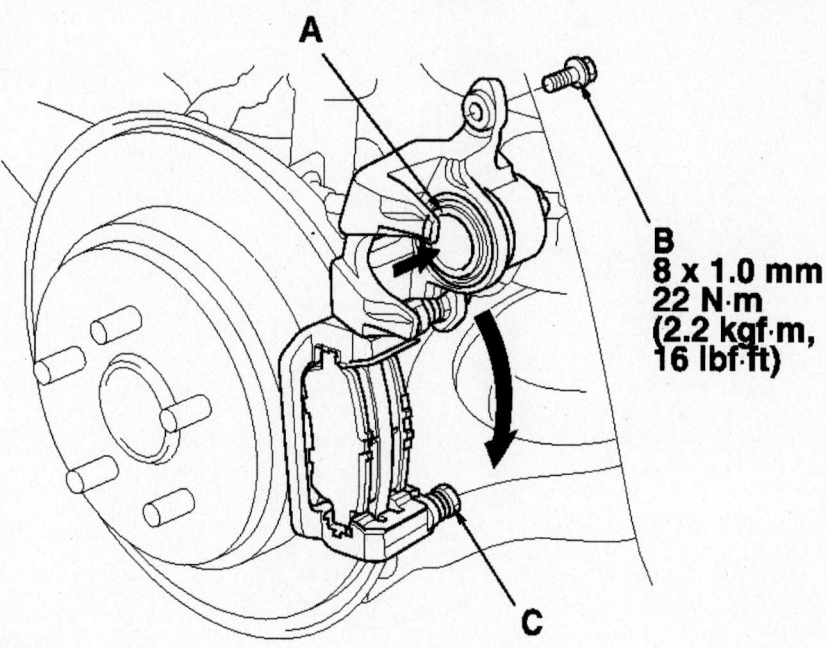

09474_ODYS_G0256

Push in the piston (A) so the caliper will fit over the brake pads. Install the flange bolt (B), and torque it to the specified torque while holding the caliper pin (C) with a wrench—2005–06 rear brakes

3. While holding the caliper pin (A) with a wrench, remove the flange bolt (B). Be careful not to damage the pin boot. Then pivot the caliper (C) up out of the way. Check the hose and pin boots for damage and deterioration.

4. Remove the pad shims (A) and brake pads (B).

5. Remove the pad retainers (A) from the caliper bracket (B).

To install:

6. Clean the caliper bracket thoroughly; remove any rust, and check for grooves and cracks.

7. Check the brake disc for damage and cracks.

8. Apply a thin coat of M-77 assembly paste (P/N 08798-9010) to the retainers on their mating surfaces (indicated by the arrow) against the caliper bracket.

9. Install the pad retainers. Wipe excess assembly paste off the retainers. Keep any assembly paste off the discs and pads.

10. Apply a thin coat of M-77 assembly paste (P/N 08798-9010) to the pad side of the shims, the back of the brake pads, and the other areas indicated by the arrows. Wipe excess assembly paste off the pad shims and brake pads. Contaminated brake discs or pads reduce stopping ability. Keep assembly paste off the brake discs and pads.

11. Install the brake pads and pad shims correctly. Install the brake pad with the wear indicator on the inside. If you are reusing the brake pads, always reinstall the brake pads in their original positions to prevent a momentary loss of braking efficiency.

12. Push in the piston (A) so the caliper will fit over the brake pads. Make sure the piston boot is in position to prevent damaging it when pivoting the caliper down.

13. Pivot the caliper down into position. Install the flange bolt (B), and torque it to the specified torque while holding the caliper pin (C) with a wrench. Be careful not to damage the pin bool.

14. Press the brake pedal several times to make sure the brakes work.

➡ **Brake engagement may require a greater pedal stroke immediately after the brake pads have been replaced. Several applications of the brake pedal will restore the normal pedal stroke.**

15. After installation, check for leaks at hose and line joints or connections, and retighten if necessary.

PARKING BRAKE

Parking Brake Shoes

REMOVAL & INSTALLATION

2002–04 Models

1. Raise the rear of the vehicle, and support it with safety stands in the proper locations. Remove the rear wheels.

2. Release the parking brake, and remove the rear brake caliper and brake disc/drum.

3. Disconnect and remove the upper return springs (A).

4. Remove the tension pins (A) by pushing and turning the retainers (B).

5. Remove the connecting rod (A).

6. Lower the parking brake shoe assembly.

7. Remove the forward brake shoe by

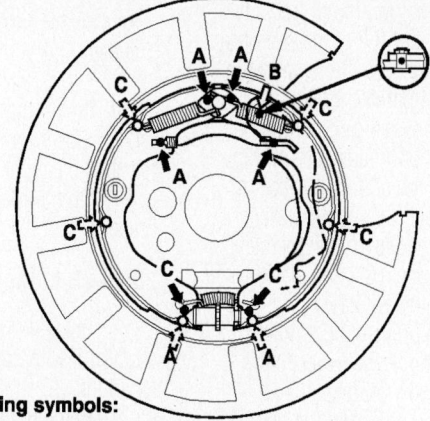

Greasing symbols:

➡● Brake shoe ends and connecting rod ends

⌐⫶○ Opposite edge of the shoe

⌐⫶● Sliding surface

09474_ODYS_G0278

Parking brake shoes—2002–04 models

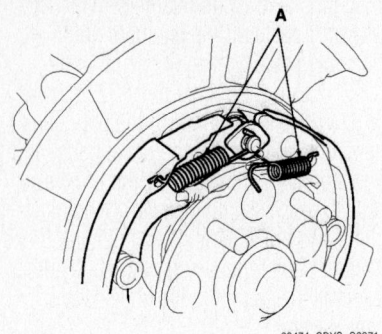

09474_ODYS_G0271

Disconnect and remove the upper return springs (A)—2002–04 models

removing the lower return spring (A) and adjuster assembly (B).

8. Remove the rearward brake shoe by disconnecting the parking brake cable (A) from the parking brake lever (B).

9. Remove the U-clip, wave washer, and parking brake lever from the brake shoe.

To Assemble

10. Apply Molykote 44 MA grease to the sliding surface of the pivot pin (A) of the rearward brake shoe (B).

11. Install the parking brake lever (C) and wave washer (D) on the pivot pin, and secure with a new U-clip (E).

  a. Install the wave washer with its convex side facing out.

  b. Pinch the U-clip securely to prevent the parking brake lever from coming out from the brake shoe.

12. Connect the parking brake cable to the parking brake lever.

13. Apply Molykote 44 MA grease to the shoe ends and connecting rod ends, sliding surfaces, and opposite edges of the parking brake shoe as shown. Wipe off any excess grease Keep grease off the brake linings.

14. Install the tension pin (A), retainer spring (B), and retainer (C) of the rearward brake shoe (D). Make sure the tension pin does not contact the parking brake lever.

15. Clean the threaded portions of the clevis A, and coat the threads with multipurpose grease. Clean the sliding surface of the clevis B, and coat the sliding surface with multipurpose grease. Install the clevis A and B on the adjuster (C), and shorten the clevis A by turning the adjuster.

16. Reinstall the brake shoe adjuster assembly (D), and hook the lower return spring (E) on the parking brake shoes.

17. Install the rod spring to the connecting rod first. Then install the connecting rod on the parking brake shoes.

18. Install the tension pin, retainer spring, and retainer of the forward brake shoe.

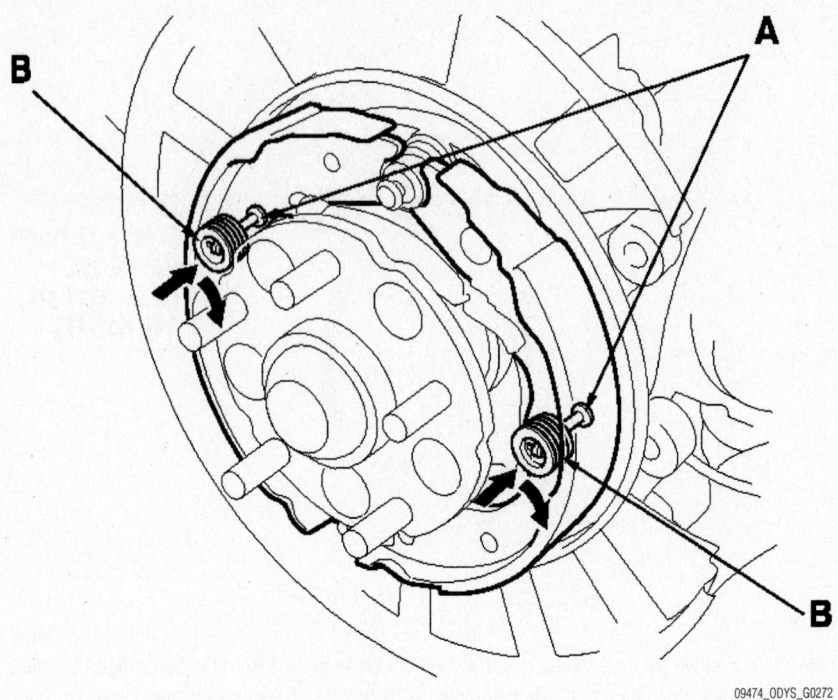

09474_ODYS_G0272

Remove the tension pins (A) by pushing and turning the retainers (B)—2002–04 models

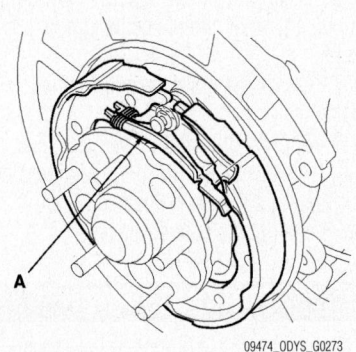

09474_ODYS_G0273

Remove the connecting rod (A)—2002–04 models

19. Install the upper return springs.

20. Install the rear brake disc/drum and rear brake caliper.

21. Do the major parking adjustment brake.

2005–06 Models

1. Raise the rear of the vehicle, and support it with safety stands in the proper locations.

2. Remove the rear wheels.

3. Release the parking brake, and remove the rear brake caliper and brake disc.

4. Disconnect and remove the upper return springs (A).

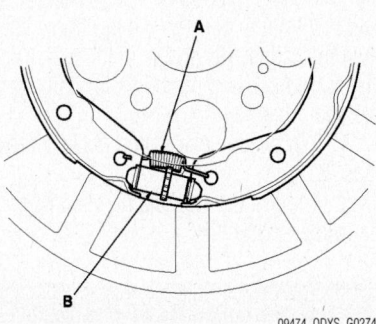

09474_ODYS_G0274

Remove the forward brake shoe by removing the lower return spring (A) and adjuster assembly (B)—2002–04 models

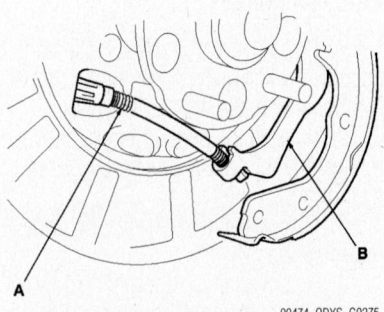

09474_ODYS_G0275

Remove the rearward brake shoe by disconnecting the parking brake cable (A) from the parking brake lever (B)—2002–04 models

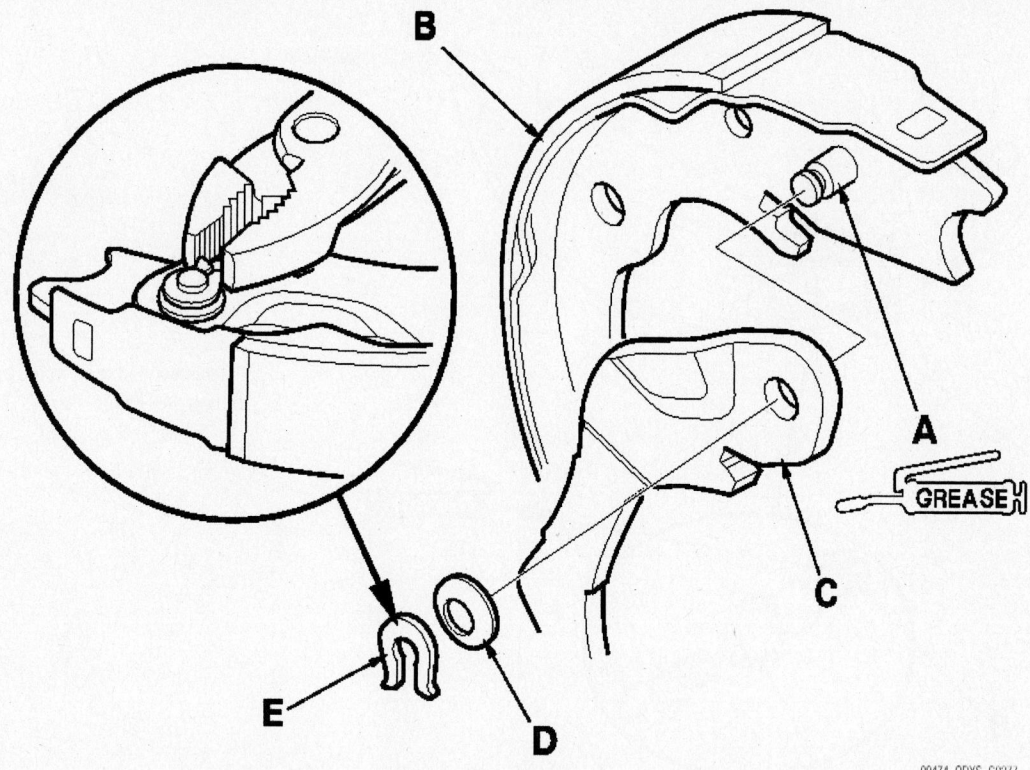

09474_ODYS_G0277

Apply Molykote 44 MA grease to the sliding surface of the pivot pin (A) of the rearward brake shoe (B). Install the parking brake lever (C) and wave washer (D) on the pivot pin, and secure with a new U-clip (E)—2002–04 models

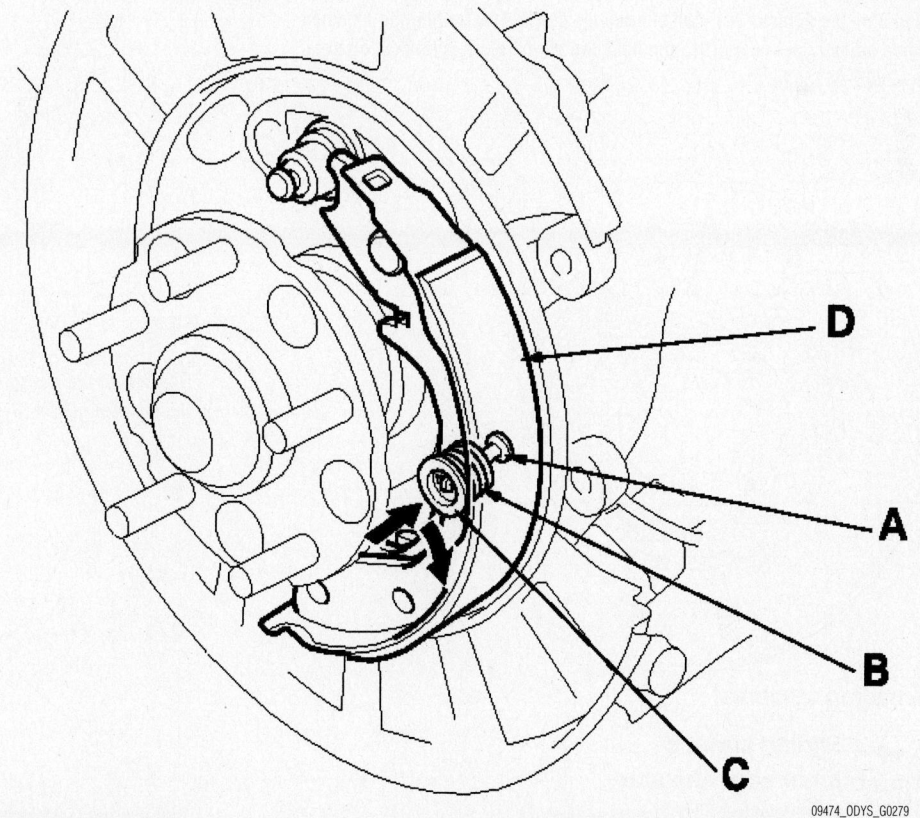

09474_ODYS_G0279

Install the tension pin (A), retainer spring (B), and retainer (C) of the rearward brake shoe (D). Make sure the tension pin does not contact the parking brake lever—2002–04 models

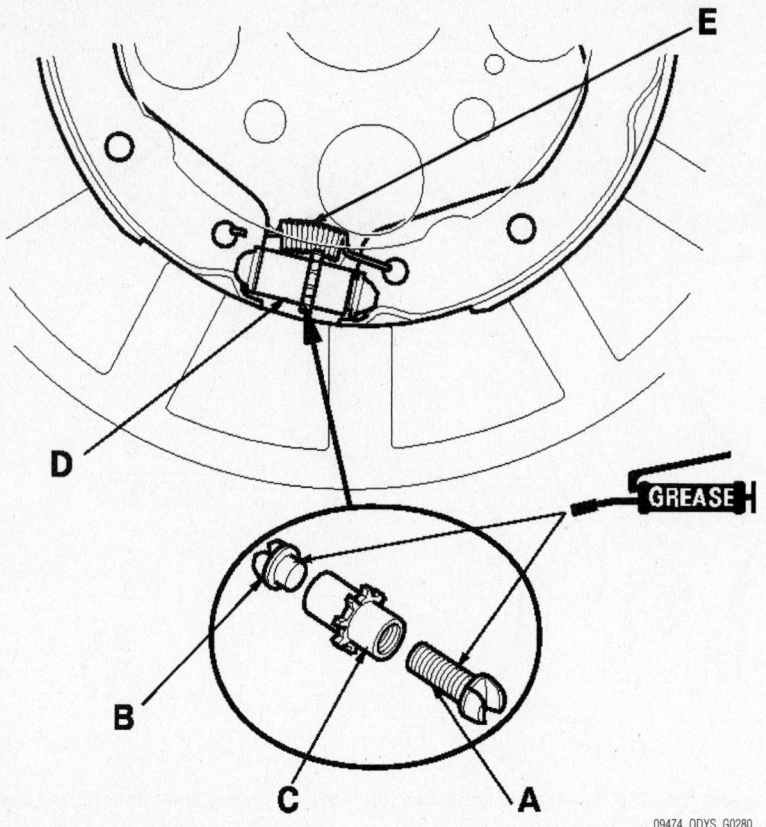

09474_ODYS_G0280

Clean the threaded portions of the clevis A, and coat the threads with multipurpose grease. Clean the sliding surface of the clevis B, and coat the sliding surface with multipurpose grease. Install the clevis A and B on the adjuster (C), and shorten the clevis A by turning the adjuster. Reinstall the brake shoe adjuster assembly (D), and hook the lower return spring (E) on the parking brake shoes—2002–04 models

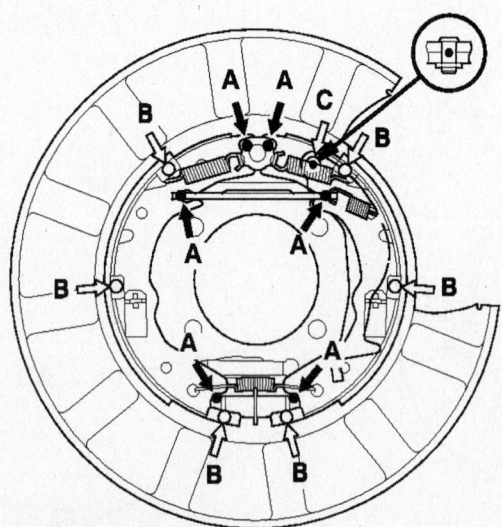

Greasing symbols:

A ➡• **Sliding surface**
B ⇨○ **Inner edge the shoe**
C ⇨• **Pivot of parking brake lever**

09474_ODYS_G0257

Parking brake components—2005–06

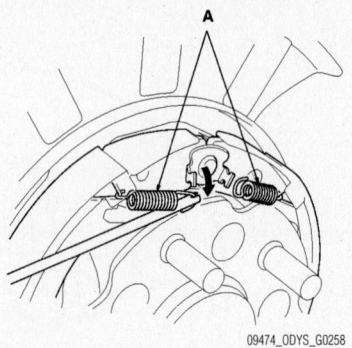

09474_ODYS_G0258

Disconnect and remove the upper return springs (A)—2005–06

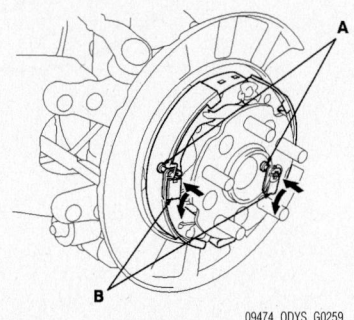

09474_ODYS_G0259

Remove the tension pins (A) by pushing and turning the retainer springs (B)—2005–06

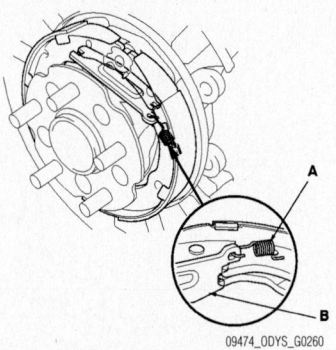

09474_ODYS_G0260

Disconnect the rod spring (A), and remove the connecting rod (B)—2005–06

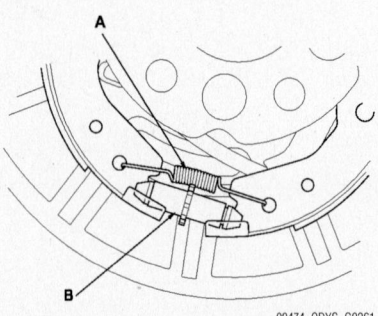

09474_ODYS_G0261

Remove the forward brake shoe by removing the lower return spring (A) and adjuster assembly (B)—2005–06

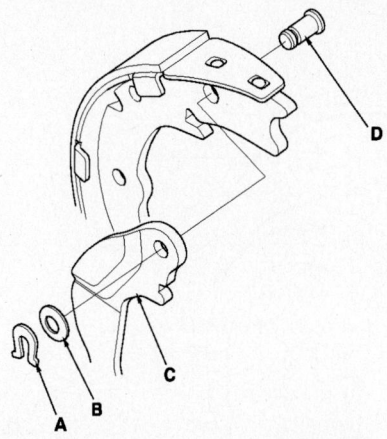

09474_ODYS_G0263

Remove the U-clip (A), wave washer (B), parking brake lever (C), and pivot pin (D) from the brake shoe—2005–06

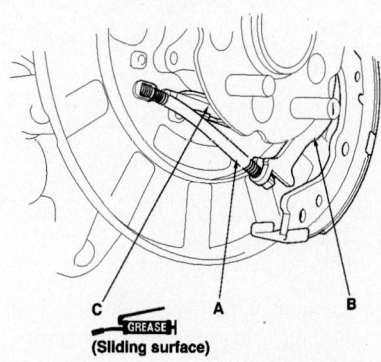

09474_ODYS_G0264

Connect the parking brake cable (A) to the parking brake lever (B). Apply silicone grease to the cable contact surface (C) on the backing plate—2005–06

5. Remove the tension pins (A) by pushing and turning the retainer springs (B).

6. Disconnect the rod spring (A), and remove the connecting rod (B).

7. Lower the parking brake shoe assembly.

8. Remove the forward brake shoe by removing the lower return spring (A) and adjuster assembly (B).

9. Remove the rearward brake shoe by disconnecting the parking brake cable from the parking brake lever.

10. Remove the U-clip (A), wave washer (B), parking brake lever (C), and pivot pin (D) from the brake shoe.

To assemble:

11. Apply Molykote 44 MA grease to the sliding surface of the pivot pin and insert the pin into the brake shoe from the rear side.

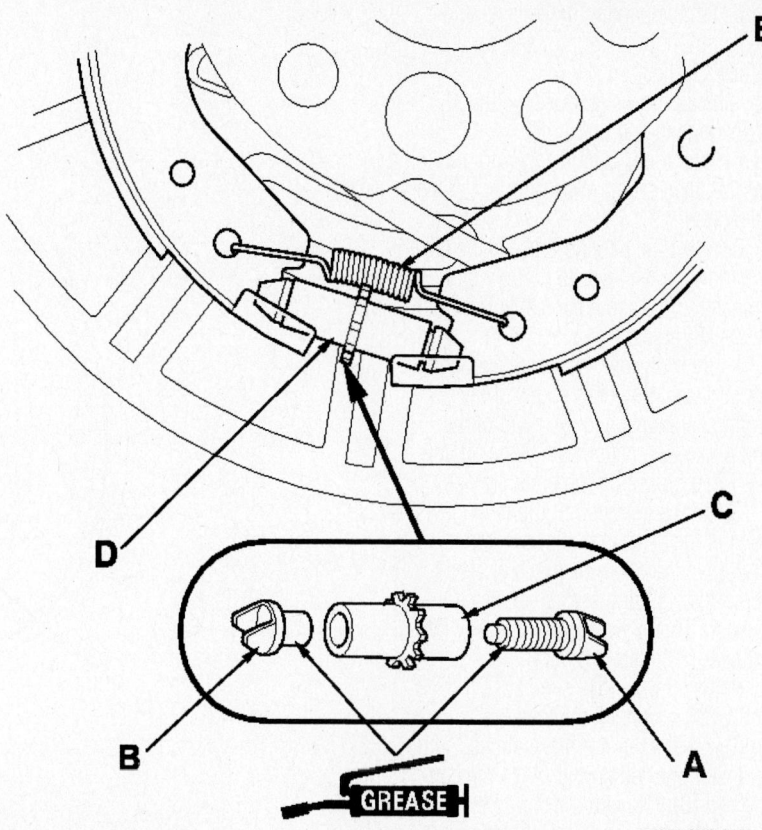

09474_ODYS_G0265

Clean the threaded portions of clevis A, and coat the threads of clevis A with grease. Clean the sliding surface of clevis B, and coat the sliding surface of clevis B with grease. Thread clevis A all the way into the adjuster (C). Install clevis B. Reinstall the brake shoe adjuster assembly (D), and hook the lower return spring (E) on the parking brake shoes—2005–06

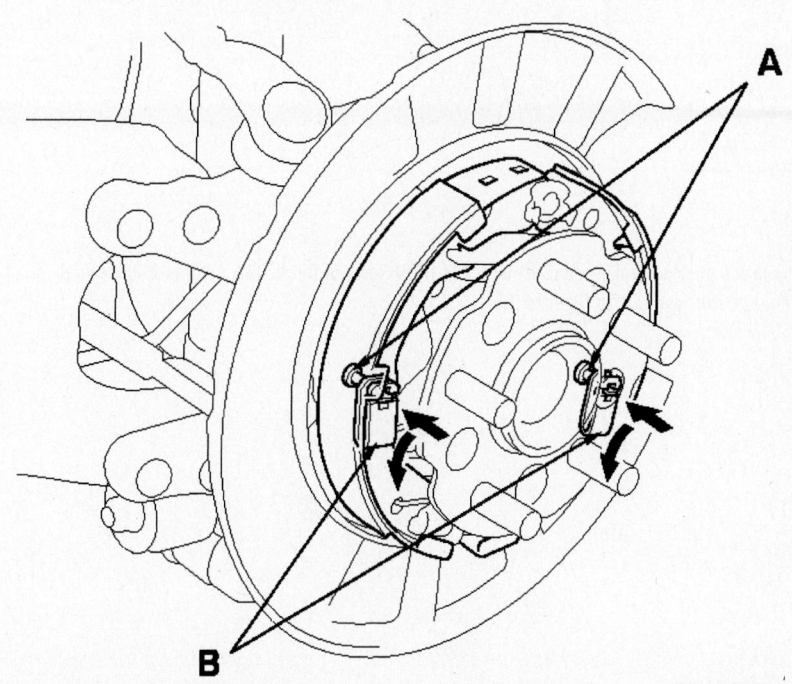

09474_ODYS_G0267

Install the tension pins (A), and retainer springs (B). Make sure the tension pin does not contact to the parking brake lever—2005–06

12. Install the parking brake lever and wave washer on the pivot pin, and secure with a new U-clip.

   a. Install the wave washer with its convex side facing out.

   b. Pinch the U-clip securely to prevent the pivot pin from coming out of the brake shoe.

13. Connect the parking brake cable (A) to the parking brake lever (B). Apply silicone grease to the cable contact surface (C) on the backing plate.

14. Apply Molykote 44 MA grease to the sliding surfaces, the inner edges of the parking brake shoes, and the pivot of the parking brake lever as shown. Wipe off any excess. Keep grease off the brake linings.

15. Clean the threaded portions of clevis A, and coat the threads of clevis A with grease. Clean the sliding surface of clevis B, and coat the sliding surface of clevis B with grease. Thread clevis A all the way into the adjuster (C). Install clevis B.

16. Reinstall the brake shoe adjuster assembly (D), and hook the lower return spring (E) on the parking brake shoes.

17. Install the rod spring (A) to the connecting rod (B) first with the spring end (C) pointing downward. Then install the connecting rod on the parking brake shoes.

18. Install the tension pins (A), and

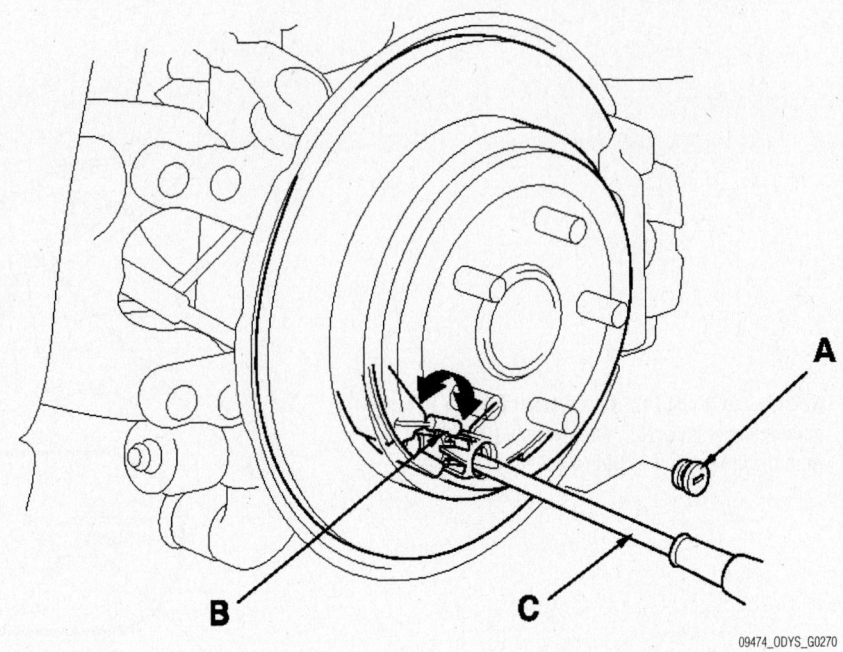

09474_ODYS_G0270

Remove the access plug (A). Turn the adjuster (B) with a flat-tip screwdriver (C) until the shoes lock against the drum. Then back off ten clicks, and install the access plug

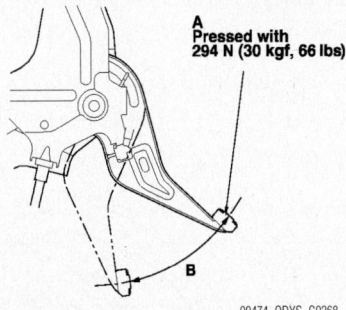

A
Pressed with
294 N (30 kgf, 66 lbs)

B

09474_ODYS_G0268

Press the parking brake pedal (A) with 294 N (66 lbs.) of force. The parking brake pedal should travel within 4 to 5 clicks (B)

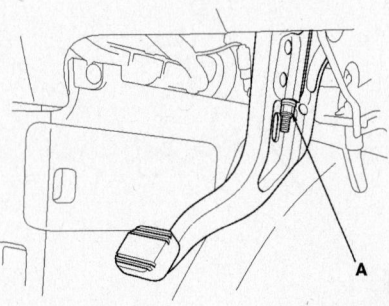

A

09474_ODYS_G0269

Tighten the adjusting nut (A) until the parking brake drags slightly when the rear wheels are turned

retainer springs (B). Make sure the tension pin does not contact to the parking brake lever.

19. Install the upper return springs.

20. Install the brake disc/drum and rear brake caliper.

21. Adjust the parking brake.

ADJUSTMENT

Inspection

1. Press the parking brake pedal (A) with 294 N (66 lbs.) of force. The parking brake pedal should travel within 3 to 5 clicks (2002–04 or 4 to 5 clicks (2005–06 models) (B).

2. If the number of pedal clicks is excessive, adjust the parking brake.

Minor Adjustment

1. Raise the rear of the vehicle, and support it with safety stands in the proper locations.

2. Tighten the adjusting nut (A) until the parking brake drags slightly when the rear wheels are turned.

3. Back off the adjusting nut in half-turn increments, and check for proper adjustment at a pedal force of 294 N (66 lbs.).

Major Adjustment

To be done when replacing brake shoes.

1. Raise the rear of the vehicle, and support it with safety stands in the proper locations.

2. Remove the rear wheels.

3. Release the parking brake, and back off the adjusting nut (A).

4. Remove the access plug (A).

5. Turn the adjuster (B) with a flat-tip screwdriver (C) until the shoes lock against the drum. Then back off ten clicks, and install the access plug.

6. Do the minor adjustment procedure.

7. Install the rear wheels.

PARKING BRAKE SHOE BREAK-IN PROCEDURE

➡ **Do the brake linings surface break-in when replacing shoes and/or the rear brake disc/drum.**

✳✳ CAUTION

Do this operation in a safe area

1. Park the vehicle on a firm, level surface.

2. Do the major parking brake adjustment.

3. Do the minor parking brake adjustment.

4. Pull the parking brake release lever, while driving the vehicle.

5. Press the parking brake pedal five clicks.

6. Drive the vehicle for ¼ mile (400m) at no more than 31 mph (50 km/h).

7. Stop the vehicle and release the parking brake for 5–10 minutes or drive the vehicle for 5–10 minutes to allow the brake disc/drum to cool.

8. Repeat step 4 through 6 three more times.

9. Check the parking brake pedal adjustment.

BRAKE HYDRAULIC SYSTEM

Brake System Bleeding

➡ **Do not reuse the drained fluid. Use only clean Honda DOT 3 Brake Fluid from an unopened container. Using a non-Honda brake fluid can cause corrosion and shorten the life of the system.**

Do not mix different brands of brake fluid; they may not be compatible.

Make sure no dirt or other foreign matter is allowed to contaminate the brake fluid.

Do not spill brake fluid on the vehicle, it may damage the paint; if brake fluid does contact the paint, wash it off immediately with water.

The reservoir on the master cylinder must be at the MAX (upper) level mark at the start of the bleeding procedure and

9.8 N·m (1.0 kgf·m, 7.2 lbf·ft)

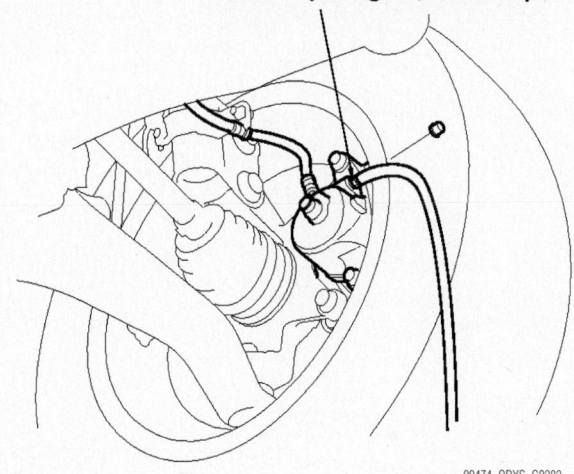

09474_ODYS_G0282

Bleed screw

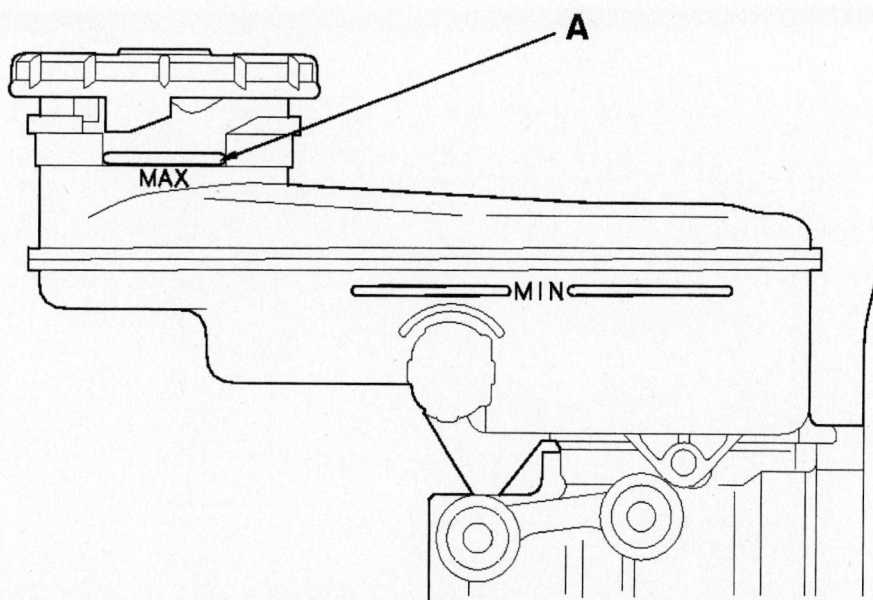

09474_ODYS_G0281

The reservoir on the master cylinder must be at the MAX (upper) level mark at the start of the bleeding procedure

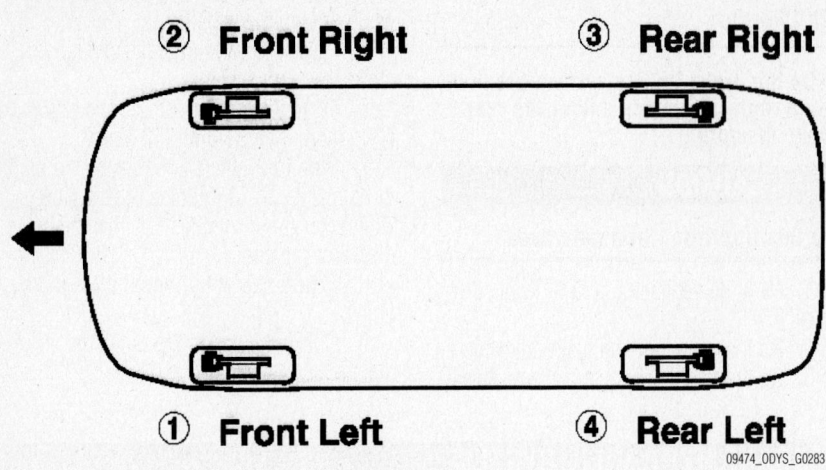

② **Front Right**   ③ **Rear Right**

① **Front Left**   ④ **Rear Left**

09474_ODYS_G0283

Bleeding sequence

checked after bleeding each brake caliper. Add fluid as required.

1. Make sure the brake fluid level in the reservoir is at the MAX (upper) level line (A).

2. Attach a length of clear drain tube to the bleed screw.

3. Have someone slowly pump the brake pedal several times, then apply steady pressure.

4. Starting at the left-front, loosen the brake bleed screw to allow air to escape from the system. Then tighten the bleed screw securely.

5. Repeat the procedure for each caliper until no air bubbles are in the fluid. Bleed the calipers in the sequence shown.

6. Refill the master cylinder reservoir to the MAX (upper) level line.

HONDA

Pilot

8

SPECIFICATION AND MAINTENANCE CHARTS

ENGINE AND VEHICLE IDENTIFICATION CHART

		Engine Code						Model Year	
Code	Liters (cc)	Cu. In.	Cyl.	Fuel Sys.	Engine Type	Eng. Mfg.	Code ①		Year
J35A4	3.5 (3471)	212	6	SMFI	SOHC	Honda	3		2003
J35A6	3.5 (3471)	212	6	SMFI	SOHC	Honda	4		2004
SOHC: Single Overhead Cam							5		2005
SMFI: Sequential Multi-port Fuel Injection							6		2006

① 10th position of VIN

09474_PILO_C0001

GENERAL ENGINE SPECIFICATIONS

Year	Model	Engine Displacement Liters (VIN)	Net Horsepower @ rpm	Net Torque @ rpm (ft. lbs.)	Bore x Stroke (in.)	Com-pression Ratio	Oil Pressure @ rpm
2003	Pilot	3.5 (J35A4)	210@5200	229@4300	3.50x3.66	9.4:1	71@3000
2004	Pilot	3.5 (J35A4)	210@5200	229@4300	3.50x3.66	9.4:1	71@3000
2005-06	Pilot	3.5 (J35A6)	255@5600	250@4500	3.50x3.66	10.0:1	71@3000

SMFI: Sequential Multi-port Fuel Injection

09474_PILO_C0002

ENGINE TUNE-UP SPECIFICATIONS

Year	Engine Displacement Liters (VIN)	Spark Plug Gap (in.)	Ignition Timing (deg.) MT	Ignition Timing (deg.) AT	Fuel Pump (psi)	Idle Speed (rpm) MT	Idle Speed (rpm) AT	Valve Clearance (in.) In.	Valve Clearance (in.) Ex.
2003	3.5 (J35A4)	0.039-0.043	—	8-12B	32-40	—	680-780	0.008-0.009	0.011-0.013
2004	3.5 (J35A4)	0.039-0.043	—	8-12B	32-40	—	680-780	0.008-0.009	0.011-0.013
2005-06	3.5 (J35A6)	0.039-0.043	—	8-12B	32-40	—	680-780	0.008-0.009	0.011-0.013

NOTE: The Vehicle Emission Control Information label often reflects changes made during production and must be used if they differ from this chart.

NOTE: The fuel pressure readings are given with the vacuum hose connected to the regulator and the engine running

B: Before top dead center

09474_PILO_C0003

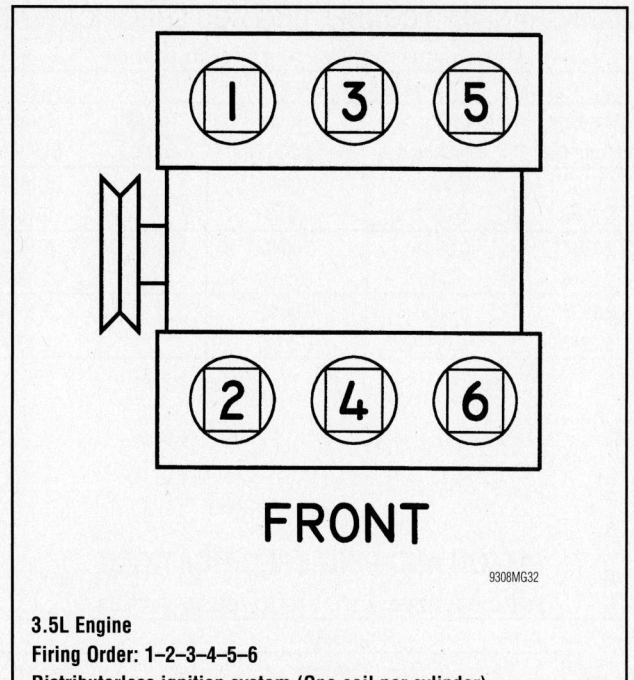

FRONT

9308MG32

3.5L Engine
Firing Order: 1–2–3–4–5–6
Distributorless ignition system (One coil per cylinder)

CAPACITIES

Year	Model	Engine Displacement Liters (VIN)	Engine Oil with Filter (qts.)	Transmission (pts.) 5-Spd	Transmission (pts.) Auto.	Transfer Case (pts.)	Drive Axle Front (pts.)	Drive Axle Rear (pts.)	Fuel Tank (gal.)	Cooling System (qts.)
2003	Pilot	3.5 (J35A4)	5.0	—	①	—	—	5.5	19.2	8.0
2004	Pilot	3.5 (J35A4)	5.0	—	①	—	—	5.5	19.2	8.0
2005-06	Pilot	3.5 (J35A6)	4.5	—	①	—	—	5.5	19.3	8.0

NOTE: All capacities are approximate. Add fluid gradually and check to be sure a proper fluid level is obtained.

① Without filter change: 6.4 pts;

  With filter Change: 16.4 pts.

09474_PILO_C0004

VALVE SPECIFICATIONS

Year	Engine Displacement Liters (VIN)	Seat Angle (deg.)	Face Angle (deg.)	Spring Test Pressure (lbs. @ in.)	Spring Installed Height (in.)	Stem-to-Guide Clearance (in.) Intake	Stem-to-Guide Clearance (in.) Exhaust	Stem Diameter (in.) Intake	Stem Diameter (in.) Exhaust
2003	3.5 (J35A4)	45	45	NA	①	0.0008-0.0018	0.0022-0.0031	0.2159-0.2163	0.2146-0.2150
2004	3.5 (J35A4)	45	45	NA	①	0.0008-0.0018	0.0022-0.0031	0.2159-0.2163	0.2146-0.2150
2005-06	3.5 (J35A6)	45	45	NA	②	0.0008-0.0018	0.0022-0.0031	0.2159-0.2163	0.2146-0.2150

NA: Not Available

① Valve spring free length:

  Intake: 1.9713 in.

  Exhaust: 2.1060 in.

② Valve spring free length:

  Intake: 2.029 in.

  Exhaust: 2.010 in.

09474_PILO_C0005

CRANKSHAFT AND CONNECTING ROD SPECIFICATIONS

All measurements are given in inches

| Year | Engine Displacement Liters (VIN) | Crankshaft | | | | Connecting Rod | | |
		Main Brg. Journal Dia.	Main Brg. Oil Clearance	Shaft End-play	Thrust on No.	Journal Diameter	Oil Clearance	Side Clearance
2003	3.5 (J35A4)	2.8337- 2.8346	0.0008- 0.0017	0.0040- 0.0140	3	2.1644- 2.1654	0.0008- 0.0017	0.0060- 0.0140
2004	3.5 (J35A4)	2.8337- 2.8346	0.0008- 0.0017	0.0040- 0.0140	3	2.1644- 2.1654	0.0008- 0.0017	0.0060- 0.0140
2005-06	3.5 (J35A6)	2.8337- 2.8346	0.0008- 0.0017	0.0040- 0.0140	3	2.1644- 2.1654	0.0008- 0.0017	0.0060- 0.0140

09474_PILO_C0006

PISTON AND RING SPECIFICATIONS

All measurements are given in inches

| Year | Engine Displacement Liters (VIN) | Piston Clearance | Ring Gap | | | Ring Side Clearance | | |
			Top Compression	Bottom Compression	Oil Control	Top Compression	Bottom Compression	Oil Control
2003	3.5 (J35A4)	0.0006- 0.0016	0.0080- 0.0140	0.0160- 0.0220	0.0080- 0.0280	0.0014- 0.0024	0.0012- 0.0022	NA
2004	3.5 (J35A4)	0.0006- 0.0016	0.0080- 0.0140	0.0160- 0.0220	0.0080- 0.0280	0.0014- 0.0024	0.0012- 0.0022	NA
2005-06	3.5 (J35A6)	0.0006- 0.0016	0.0080- 0.0140	0.0160- 0.0220	0.0080- 0.0280	0.0022- 0.0031	0.0012- 0.0022	NA

NA: Not Available

09474_PILO_C0007

TORQUE SPECIFICATIONS

All readings in ft. lbs.

| Year | Engine Displacement Liters (VIN) | Cylinder Head Bolts | Main Bearing Bolts | Rod Bearing Bolts | Crankshaft Damper Bolts | Flywheel Bolts | Manifold | | Spark Plugs | Oil Pan Drain Plug |
							Intake	Exhaust		
2003	3.5 (J35A4)	①	②	③	181	54	16	23	13	29
2004	3.5 (J35A4)	①	②	③	181	54	16	23	13	29
2005-06	3.5 (J35A6)	①	②	③	④	54	16	23	13	29

NOTE: Dip main bearing bolts and crankshaft damper bolt in clean engine oil prior to tightening.

① Step 1: 29 ft. lbs.
Step 2: 51 ft. lbs.
Step 3: 72 ft. lbs.

② 11mm bolt 54 ft. lbs.
10mm bolt 36 ft. lbs.

③ Step 1: 14 ft. lbs.
Step 2: 90 degrees

④ Step 1: 47 ft. lbs.
Step 2: plus 60 degrees

09474_PILO_C0008

WHEEL ALIGNMENT

Year	Model		Caster Range (+/-Deg.)	Caster Preferred Setting (Deg.)	Camber Range (+/-Deg.)	Camber Preferred Setting (Deg.)	Toe-in (in.)
2003	Pilot	F	1.00	1.00	1.00	30	0+/-1/16
		R	—	—	0	30	0+/-1/16
2004	Pilot	F	1.00	1.00	1.00	30	0+/-1/16
		R	—	—	0	30	0+/-1/16
2005-06	Pilot	F	1.00	1.00	1.00	30	0+/-1/16
		R	—	—	0	30	0+/-1/16

09474_PILO_C0009

TIRE, WHEEL AND BALL JOINT SPECIFICATIONS

Year	Model	OEM Tires Standard	OEM Tires Optional	Tire Pressures (psi) Front	Tire Pressures (psi) Rear	Wheel Size	Ball Joint Inspection	Lug Nut
2003	Pilot	P235/65R16	None	32	32	R16	NS	80
2004	Pilot	P235/65R16	None	32	32	R16	NS	80
2005-06	Pilot	P235/65R16	None	32	32	R16	NS	80

OEM: Original Equipment Manufacturer

PSI: Pounds Per Square Inch

NS: Not specified by manufacturer

09474_PILO_C0010

BRAKE SPECIFICATIONS
Honda Pilot
All measurements in inches unless noted

Year	Model		Brake Disc Original Thickness	Brake Disc Minimum Thickness	Brake Disc Maximum Runout	Brake Drum Diameter Original Inside Diameter	Max. Wear Limit	Brake Drum Diameter Maximum Machine Diameter	Minimum Lining Thickness Front	Minimum Lining Thickness Rear	Brake Caliper Bracket Bolts (ft. lbs.)	Brake Caliper Mounting Bolts (ft. lbs.)
2003	Pilot	F	1.100	1.020	0.004	—	—	—	0.060	—	80	27
		R	0.430	0.350	0.004	—	—	—	—	0.41	41	27
2004	Pilot	F	1.100	1.020	0.004	—	—	—	0.060	—	80	27
		R	0.430	0.350	0.004	—	—	—	—	0.41	41	27
2005-06	Pilot	F	1.100	1.020	0.004	—	—	—	0.060	—	80	27
		R	0.430	0.350	0.004	—	—	—	—	0.41	41	27

F: Front

R: Rear

09474_PILO_C0011

SCHEDULED MAINTENANCE INTERVALS
HONDA—PILOT

TO BE SERVICED	TYPE OF SERVICE	VEHICLE MILEAGE INTERVAL (x1000)															
		7.5	15	22.5	30	37.5	45	52.5	60	67.5	75	82.5	90	97.5	105	112.5	120
Accessory drive belts	I & A				✓				✓				✓				✓
Air cleaner element	R				✓				✓				✓				✓
Brake fluid	R	Every 3 years															
Brake hoses & lines (including ABS)	I		✓		✓		✓		✓		✓		✓		✓		✓
Cooling system hoses & connections	I		✓		✓		✓		✓		✓		✓		✓		✓
Engine coolant ①	R						✓						✓				
Engine oil	R	✓	✓	✓	✓	✓	✓	✓	✓	✓	✓	✓	✓	✓	✓	✓	✓
Engine oil and coolant levels	I	Inspect at each fuel stop															
Engine oil filter	R		✓		✓		✓		✓		✓		✓		✓		✓
Exhaust system	I		✓		✓		✓		✓		✓		✓		✓		✓
Fluid levels and condition	I		✓		✓		✓		✓		✓		✓		✓		✓
Front and rear brakes	I		✓		✓		✓		✓		✓		✓		✓		✓
Fuel lines & connection	I		✓		✓		✓		✓		✓		✓		✓		✓
Halfshaft boots	I		✓		✓		✓		✓		✓		✓		✓		✓
Idle speed	I & A														✓		
Parking brake system	I & A		✓		✓		✓		✓		✓		✓		✓		✓
Rear differential fluid	R	✓			✓				✓				✓				✓
Rotate and inspect tires	I	✓	✓	✓	✓	✓	✓	✓	✓	✓	✓	✓	✓	✓	✓	✓	✓
Spark plugs	R														✓		
Supplemental Restrain system (SRS)	I	Inspect the SRS 10 years after production															
Suspension components	I		✓		✓		✓		✓		✓		✓		✓		✓
Tie rod ends, steering gear box & boots	I		✓		✓		✓		✓		✓		✓		✓		✓
Timing belt	R														✓		
Transmission fluid	R						✓				✓				✓		
Valve clearance	I	Adjust if valves are noisy															
Water pump	S/I														✓		

R: Replace   I: Inspect   A: Adjust

① Every 12,000 miles or 10 years, then every 60,000 miles or 5 years

FREQUENT OPERATION MAINTENANCE (SEVERE SERVICE)

If a vehicle is operated under any of the following conditions it is considered severe service:

- Towing a trailer or using a camper or car-top carrier.
- Repeated short trips of less than 5 miles in temperatures below freezing, or trips of less than 10 miles in any temperature.
- Extensive idling or low-speed driving for long distances as in heavy commercial use, such as delivery, taxi or police cars.
- Operating on rough, muddy or salt-covered roads.
- Operating on unpaved or dusty roads.
- Driving in extremely hot (over 90°) conditions.

Air cleaner element: replace every 15,000 miles

Engine oil and filter: replace every 3750 miles or 6 months, whichever occurs first.

Timing belt: replace every 60,000 miles if the vehicle is regularly driven in temperatures above 110°F or below -20°F, or if frequently towing a trailer.

Transmission fluid: replace every 30,000 miles.

Rear differential fluid: replace every 60,000 miles.

Front and rear brakes: inspect every 7500 miles or 6 months, whichever occurs first.

Locks and hinges: lubricate every 15,000 miles.

Tie rods, steering gear box, boots: inspect every 7500 miles or 6 months, whichever occurs first.

Suspension components: inspect every 7500 miles or 6 months, whichever occurs first.

Halfshaft boots: inspect every 7500 miles or 6 months, whichever occurs first.

ENGINE REPAIR

→Disconnecting the negative battery cable on some vehicles may interfere with the functions of the on board computer system. The computer may undergo a relearning process once the negative battery cable is reconnected.

Distributor

The Pilot is equipped with a Distributorless Ignition System (DIS).

Alternator

REMOVAL

1. Before servicing the vehicle, refer to the precautions section.
2. Remove or disconnect the following:
   - Negative battery cable
   - Accessory drive belt
   - Intake manifold and ignition coil covers
   - Alternator wiring harness connectors
   - Alternator mounting bolts
   - Wiring harness clamp
   - Alternator

INSTALLATION

1. Install or connect the following:
   - Alternator
   - Wiring harness clamp. Tighten the bolt to 105 inch lbs. (12 Nm).
   - Alternator mounting bolts. Tighten the 10mm bolt to 33 ft. lbs. (44 Nm) and the 8mm bolt to 16 ft. lbs. (22 Nm).
   - Alternator wiring harness connectors. Tighten the battery terminal nut to 105 inch lbs. (12 Nm).
   - Accessory drive belt
   - Intake manifold and ignition coil covers
   - Negative battery cable

Ignition Timing

ADJUSTMENT

The Pilot is equipped with a Distributorless Ignition System (DIS). The ignition timing is controlled by the Powertrain Control module (PCM). No adjustment is necessary.

Engine Assembly

REMOVAL & INSTALLATION

→The engine and transaxle are removed from the vehicle as a unit.

1. Before servicing the vehicle, refer to the precautions section.
2. Drain the cooling system.
3. Drain the power steering system.
4. Drain the transaxle fluid.
5. Drain the engine oil.
6. Relieve fuel system pressure.
7. Remove or disconnect the following:
   - Negative battery cable
   - Battery
   - Intake and ignition coil covers
   - Air intake duct
   - Left engine wire harness connectors
   - Relay bracket
   - Battery and tray
   - Starter cable and harness clamp
   - Accelerator cable
   - Cruise control cable
   - Fuel lines
   - EVAP canister hose
8. Remove the drivers side center console lower panel and pull back the cover to access steering joint cover.
   - Steering joint bolt
   - Powertrain Control Module (PCM) connectors
   - Heated Oxygen (HO2S) sensor connector and grommet. Pull the PCM harness through the firewall.
   - Brake booster vacuum line
   - Clamps and clips from power steering hoses
   - Fuse/relay box battery cable
   - Accessory drive belts
   - Front wheels
   - Splash shield
   - Front sub-frame stiffener
   - Exhaust front pipe
   - Propeller shaft
   - Shift control cable
   - Transfer assembly
   - Ball joints

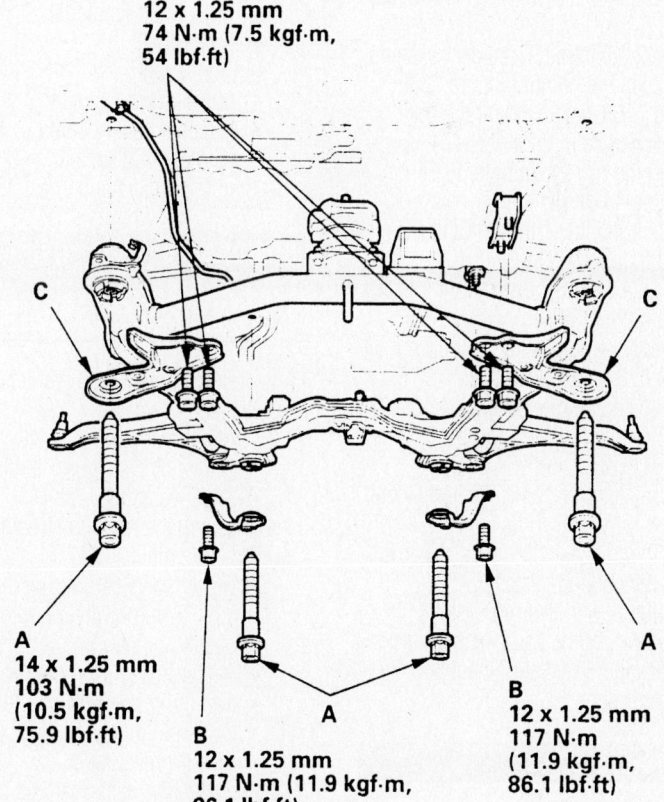

Sub-frame fastener locations and tightening torque—Pilot

9302MG69

- Stabilizer bar links
- Halfshafts
- Power steering hose and pressure switch connector
- Transaxle lower front mount
- Transaxle lower rear mount
- A/C compressor
- Heater hoses
- Radiator hoses
- Ground cable
- Transaxle oil cooler lines
- Radiator

9. Attach a hoist to the engine lifting eyes and support the powertrain weight.

10. Remove or disconnect the following:
- Side engine mount bracket
- Front mount bracket support nut

11. Matchmark the front subframe to the mounting points.

12. Remove or disconnect the following:
- Front subframe
- All remaining hoses and electrical connections

13. Lower the powertrain away from the vehicle.

To install:

14. Raise the powertrain into position.

15. Installation is the reverse of removal but please note the following steps:
- A/C compressor bolts to 16 ft. lbs. (22 Nm)
- Front subframe. Use new bolts and tighten the 14mm bolts to 76 ft. lbs. (103 Nm). Tighten the front brace bolts to 54 ft. lbs. (74 Nm) and the rear brace bolts to 86 ft. lbs. (117 Nm).
- Transaxle lower front mount nuts to 28 ft. lbs. (38 Nm)
- Transaxle lower rear mount bolts to 28 ft. lbs. (38 Nm)
- Front mount bracket support nut to 40 ft. lbs. (54 Nm)
- Side engine mount bracket bolts to 33 ft. lbs. (44 Nm) and the through bolt to 40 ft. lbs. (54 Nm)

16. Fill the engine crankcase to the correct level.

17. Fill the transaxle to the correct level.

18. Fill the cooling system.

19. Fill the power steering system.

20. Start the engine and check for leaks.

21. Check the wheel alignment and adjust as necessary.

Water Pump

REMOVAL & INSTALLATION

1. Before servicing the vehicle, refer to the precautions section.

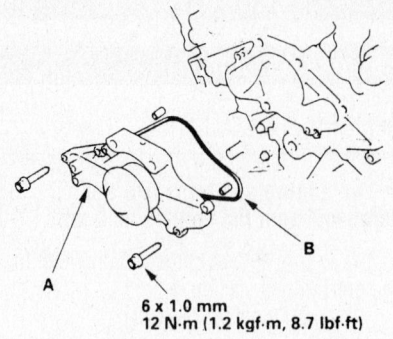

6 x 1.0 mm
12 N·m (1.2 kgf·m, 8.7 lbf·ft)

93552G01

Exploded view of the water pump mounting

2. Drain the cooling system.

3. Remove or disconnect the following:
- Negative battery cable
- Accessory drive belts
- Front cover
- Timing belt
- Timing belt tensioner
- Water pump

To install:

4. Install or connect the following:
- Water pump. Use a new O-ring seal and tighten the bolts to 105 inch lbs. (12 Nm).
- Timing belt tensioner
- Timing belt
- Front cover
- Accessory drive belts
- Negative battery cable

5. Fill the cooling system.

6. Start the engine and check for leaks.

Heater Core

REMOVAL & INSTALLATION

1. Before servicing the vehicle, refer to the precautions section.

2. Drain the cooling system.

3. Remove or disconnect the following:
- Negative battery cable

4. Recover the refrigerant using approved equipment.
- Heater valve cable from the valve arm. Turn the valve arm to the fully opened position.
- Heater hoses from the heater unit
- Mounting nut from the heater unit. Be careful not to bend or damage fuel or brake lines.

5. Remove the dashboard as follows:
   a. Remove the center console by unlatching the clips.
   b. Remove the dashboard lower cover screw, gently pull down on the cover to disengage the clips and disconnect the electrical connections.
   c. Remove the dashboard side cover by gently pulling and turning to unfasten the clips.
   d. Remove the right kick panel.
   e. While holding the glove box, remove the box stop from each side, then disconnect the lock from the damper.
   f. Remove the glove box bolts and the glove box.
   g. Remove the front door trim, kick panels and A-pillar trim from both sides.
   h. Remove the cap from the front pillar corner trim. Unfasten the screw, slide the trim upward along the pillar and remove it. Remove the remaining clips from the body.
   i. On the drivers side, remove the fuel/relay box nut and pull out the box.
   j. Remove the steering column
   k. On the passenger side remove the fuse/relay bolt and pull out the box.
   l. Disconnect all electrical connections from the dashboard.
   m. If equipped with a navigation system, remove the passenger seat, pull back the carpet, remove the harness cushions and then pull out the GPS harness.
   n. Remove all harness and connector clips.
   o. Remove all the bolts and lift up on the dashboard to release the dashboard and steering hanger beam from the guide pins.
   p. Remove the dashboard through the door.

6. Remove the evaporator as follows:
   a. Disconnect the receiver and suction lines from the evaporator.
   b. Remove the mounting nuts and plug the lines to avoid system contamination.
   c. Remove the plastic brace and glove box frame.
   d. Disconnect the wire harness and evaporator temperature sensor connector.
   e. Remove the self-tapping screws, the nuts and the evaporator.

7. Remove or disconnect the following:
- Mounting bolts and the heater unit
- Self-tapping screws and the clamp, then pull the heater core from the case being careful not to bend the pipes

To install:

8. Install or connect the following:
- Heater core in the case
- Clamp and the screws
- Heater unit and tighten the bolts to 7 ft. lbs. (10 Nm)

Fastener Locations

A ▶ : Bolt, 2   B ▶ : Bolt, 1   C ▶ : Bolt, 3

D ▶ : Bolt, 5   E ▶ : Bolt, 2   F ▶ : Bolt, 2

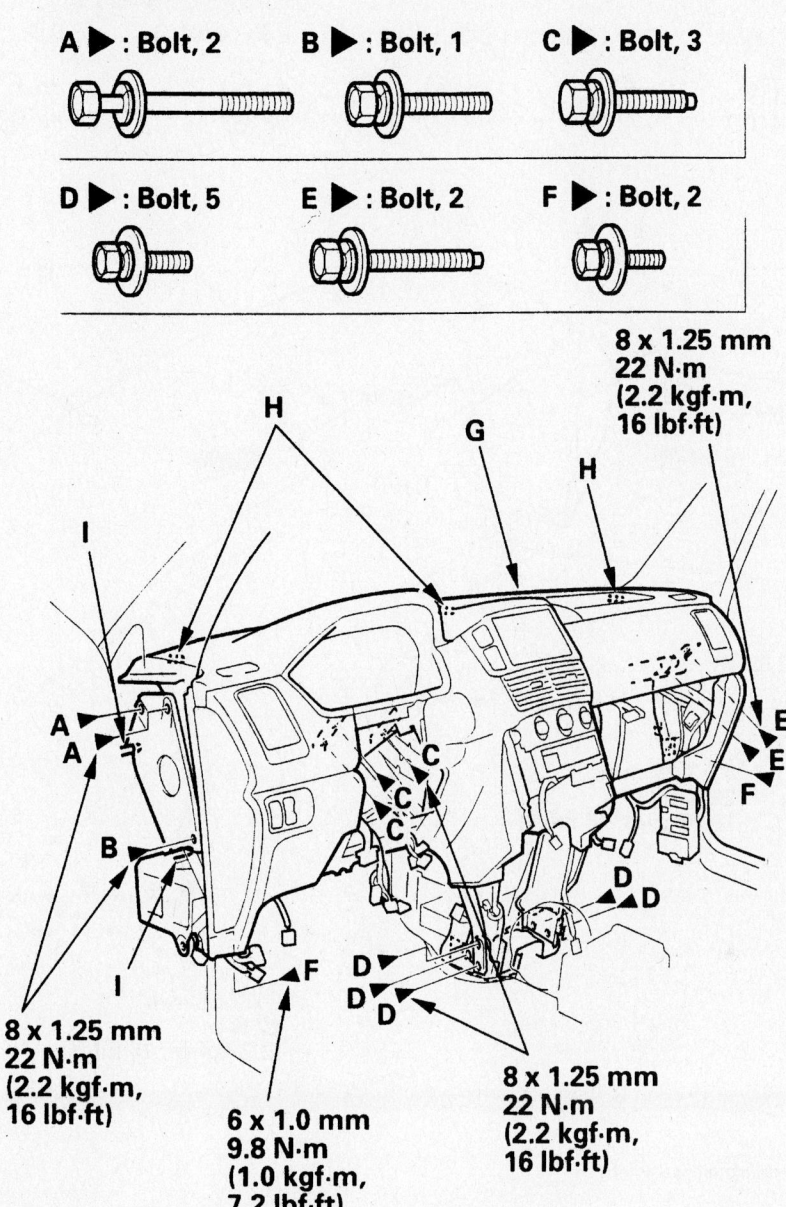

8 x 1.25 mm
22 N·m
(2.2 kgf·m,
16 lbf·ft)

8 x 1.25 mm
22 N·m
(2.2 kgf·m,
16 lbf·ft)

6 x 1.0 mm
9.8 N·m
(1.0 kgf·m,
7.2 lbf·ft)

8 x 1.25 mm
22 N·m
(2.2 kgf·m,
16 lbf·ft)

42356-HPIL-G03

Tighten the dashboard bolts as illustrated

• Evaporator in the reverse order of removal. Tighten all the retainers to 7 ft. lbs. (10 Nm) .

9. Install the dashboard in the reverse order of removal keeping in mind the following points:

a. Make sure the dashboard is seated properly and that the wiring harness and steering hanger beam wire harness are not pinched.

b. Referring to the accompanying illustration, tighten bolts **(A, B, C, D and E)** to 16 ft. lbs. (22 Nm). Tighten bolts **F** to 7 ft. lbs. (10 Nm). Apply thread lock to the **B** bolts before installation.

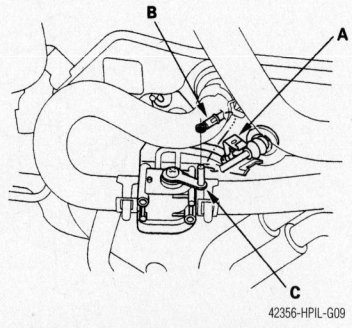

42356-HPIL-G09

In the engine compartment, open the cable clamp (A), then disconnect the heater valve cable (B) from the valve arm (C)

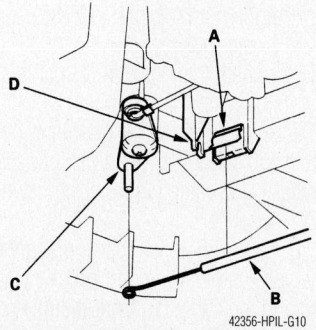

42356-HPIL-G10

Under the dashboard, disconnect the valve cable housing from the clamp (A) and the cable (B) from the mix control linkage (C)

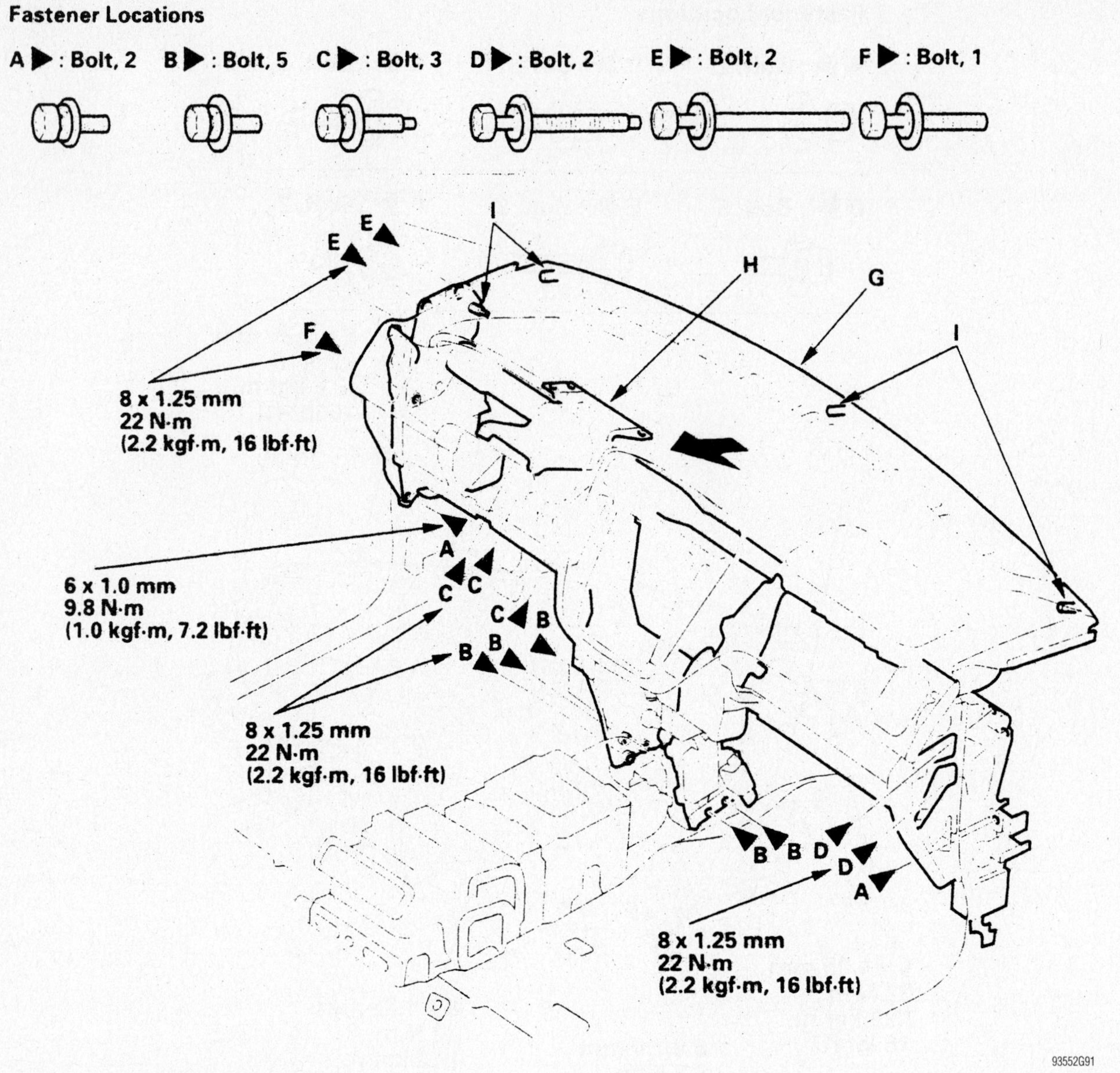

Fastener Locations
A ▶ : Bolt, 2
B ▶ : Bolt, 5
C ▶ : Bolt, 3
D ▶ : Bolt, 2
E ▶ : Bolt, 2
F ▶ : Bolt, 1
E
E
F
I
H
G
I
8 x 1.25 mm
22 N·m
(2.2 kgf·m, 16 lbf·ft)
A
C
C
6 x 1.0 mm
9.8 N·m
(1.0 kgf·m, 7.2 lbf·ft)
C
B
B
B
8 x 1.25 mm
22 N·m
(2.2 kgf·m, 16 lbf·ft)
B
B
D
D
A
8 x 1.25 mm
22 N·m
(2.2 kgf·m, 16 lbf·ft)
93552G91

Exploded view of the dashboard mounting—Pilot

c. Ensure that all electrical connectors are properly connected.

10. Install or connect the following:

- Mounting nut to the heater unit and tighten to 7 ft. lbs. (10 Nm)
- Heater hoses

11. Connect the heater valve cable and adjust as follows:

a. In the engine compartment, open the cable clamp (A), then disconnect the heater valve cable (B) from the valve arm (C).

b. Under the dashboard, disconnect the valve cable housing from the clamp (A) and the cable (B) from the mix control linkage (C).

c. Set the temperature control button to the MAX COOL position with the ignition switch in the on position.

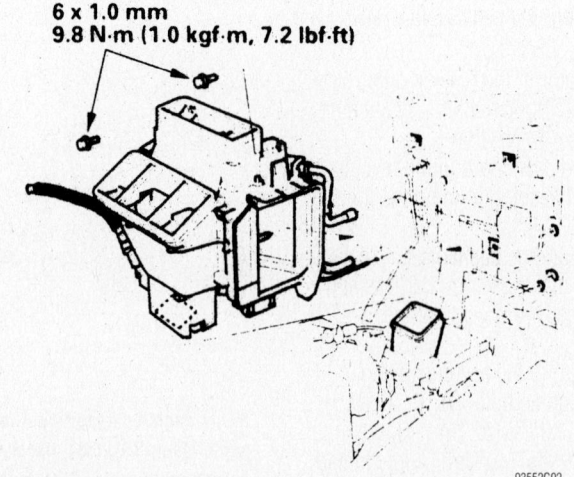

6 x 1.0 mm
9.8 N·m (1.0 kgf·m, 7.2 lbf·ft)
93552G92

Exploded view of the evaporator mounting—Pilot

d. Attach the valve cable (B) to the mix control linkage (C) as shown in the illustration, hold the end of the cable housing against the stop, then snap the cable housing into the clamp.

e. In the engine compartment, turn the valve arm (C) to the fully closed position as shown in the accompanying illustration and hold it there. Attach the cable (B) to the valve arm and pull gently on the cable housing to take up the slack, then install the cable housing into the clamp (A).

12. Fill the cooling system

13. Connect the battery cable.

Cylinder Head

REMOVAL & INSTALLATION

1. Before servicing the vehicle, refer to the precautions section.

2. Drain the cooling system.

3. Relieve the fuel system pressure.

4. Remove or disconnect the following:
- Negative battery cable
- Drive belts
- Power steering and pump
- Power steering hose clamp
- Alternator
- Intake manifold cover
- Ignition coil covers
- Ignition coils
- Timing belt
- Intake manifold
- Fuel injector connectors
- Engine Coolant Temperature (ECT) sensor connector
- Crankshaft Position (CKP) sensor connector
- Camshaft Position (CMP) sensor connector
- Exhaust Gas Recirculation (EGR) connector
- Radiator fan switch connectors
- Valve Lift Electronic Control (VTEC) solenoid valve connector and oil pressure switch connections
- Oil pressure switch connector
- Two Air/Fuel (A/F) connectors
- Two Heated Oxygen (HO2S) sensor connectors
- Radiator hoses
- Heater hose
- Water bypass hose
- Harness bracket
- EVAP canister hose
- Fuel feed and return lines
- Fuel rails
- Exhaust manifolds
- Water passage

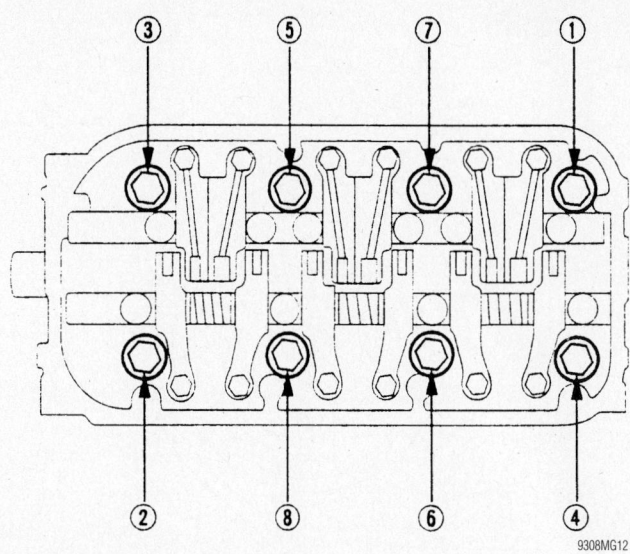

Cylinder head bolt loosening sequence—Pilot

- Vacuum hoses from the intake air bypass control valve
- Ground cable
- Camshaft pulleys and back covers
- Valve covers

5. Loosen the cylinder head bolts in sequence and $\frac{1}{3}$ turns until all bolts are loose.

6. Remove the cylinder head.

To install:

7. Align the crankshaft and camshaft sprocket TDC marks as shown.

8. Install the cylinder heads with new gaskets.

9. Apply clean engine oil to the cylinder head bolt threads and flanges.

10. Tighten the cylinder head bolts in sequence as follows:

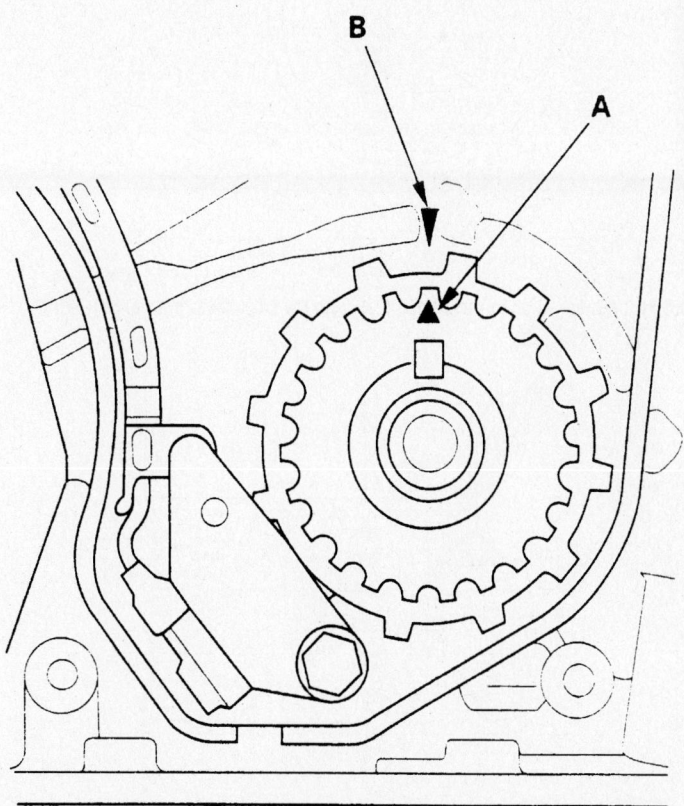

Crankshaft timing belt sprocket TDC marks. Align sprocket mark (A) with pointer (B)—Pilot

FRONT:

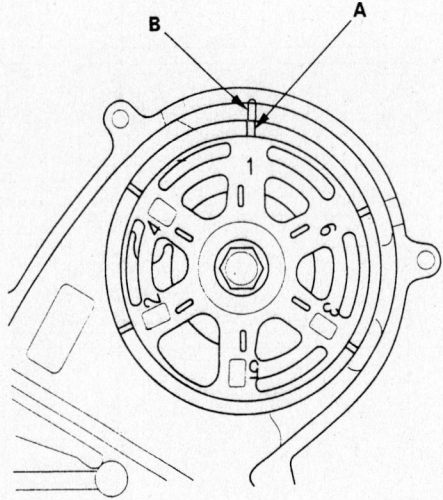

REAR:

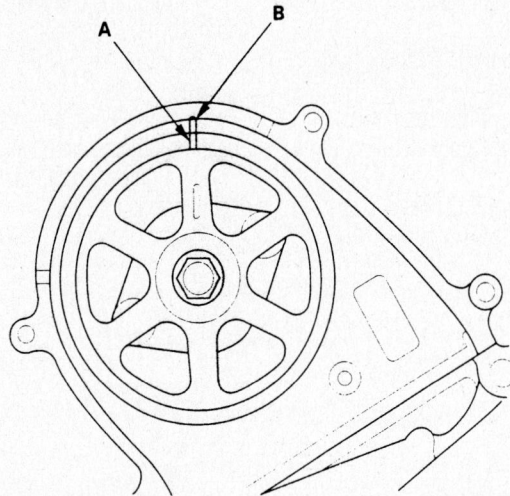

9302MG85

Camshaft TDC marks. Align sprocket mark (A) with the back cover pointer (B)—Pilot

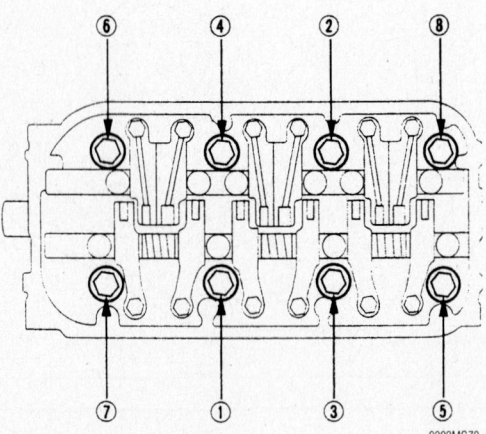

9302MG70

Cylinder head bolt tightening sequence—Pilot

a. Step 1: 29 ft. lbs. (39 Nm).
b. Step 2: 51 ft. lbs. (69 Nm).
c. Step 3: 72 ft. lbs. (98 Nm).
11. Install or connect the following:
- Timing belt and adjust the valve clearance
- Valve covers
- Exhaust manifolds
- Water passage
- Fuel rails
- EVAP canister hose
- Vacuum hoses to the intake air bypass control valve
- Fuel feed and return lines
- Heater hose
- Radiator hoses
- Intake manifold
- Two A/F connectors
- Two HO2S sensor connectors
- Oil pressure switch connector
- VTEC solenoid valve connector and oil pressure switch connections
- EGR connector
- CMP sensor connector
- CKP sensor connector
- Radiator fan switch connectors
- ECT sensor connector
- Fuel injector connectors
- Ignition coils
- Power steering pump and belt
- Power steering hose clamp
- Alternator
- Ground cable
- Ignition coil covers
- Intake manifold cover
- Alternator belt
- Negative battery cable
12. Fill the cooling system.
13. Start the engine and check for leaks.

Rocker Arms/Shafts

REMOVAL & INSTALLATION

1. Before servicing the vehicle, refer to the precautions section.
2. Remove or disconnect the following:
- Negative battery cable
- Intake manifold
- Ignition coils
- Valve cover
- Rocker arm adjusting screws. refer to the illustration for location.
3. Remove the rocker arm assembly as follows:
a. Unscrew the rocker shaft bolts 2 turns at a time in a criss-cross pattern to avoid damaging the valves or rocker assembly.
b. Do not remove the rocker shaft

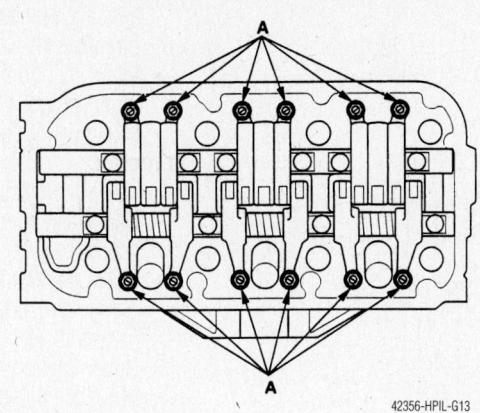

42356-HPIL-G13

Rocker arm shaft adjusting screw locations

bolts. These bolts keep the springs and rocker arms on the shafts.

4. Loosen the valve adjuster locknuts and screws so that all valves are closed.

5. Remove the rocker arms and shafts from the vehicle as an assembly.

➡**Keep all valvetrain components in order for assembly.**

6. Remove the rocker arms and springs from the rocker arm shafts.

To install:

7. Assemble the rocker arms and springs to the rocker arm shafts in their original positions.

8. Install the rocker arm assemblies.

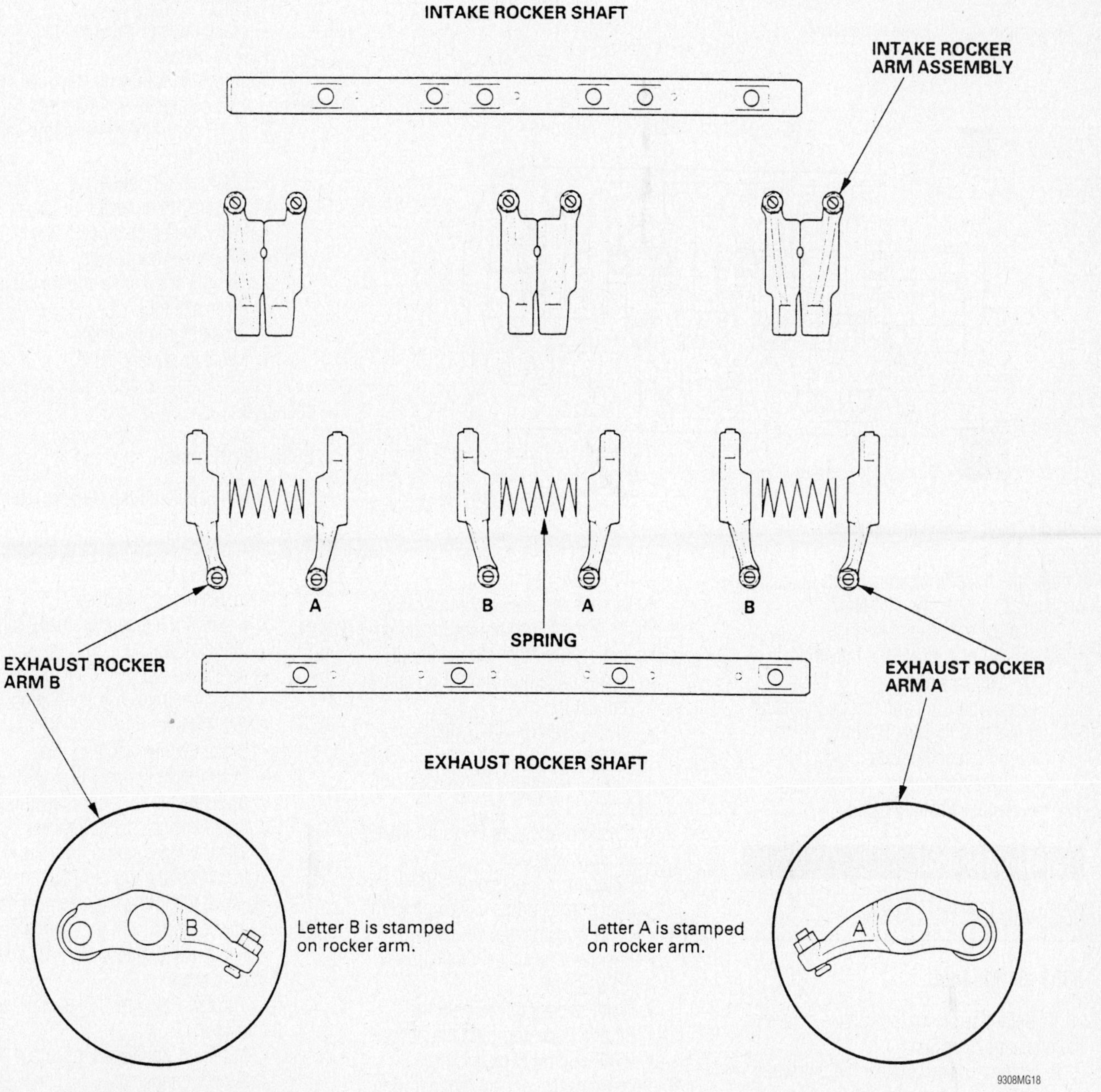

INTAKE ROCKER SHAFT

INTAKE ROCKER ARM ASSEMBLY

A   B   A   B

SPRING

EXHAUST ROCKER ARM B

EXHAUST ROCKER ARM A

EXHAUST ROCKER SHAFT

Letter B is stamped on rocker arm.

Letter A is stamped on rocker arm.

B

A

9308MG18

Exploded view of the rocker arms and shafts—Pilot

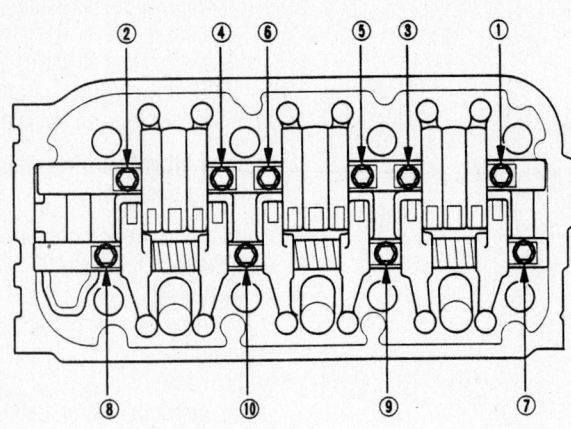

Rocker arm shaft loosening sequence

42356-HPIL-G14

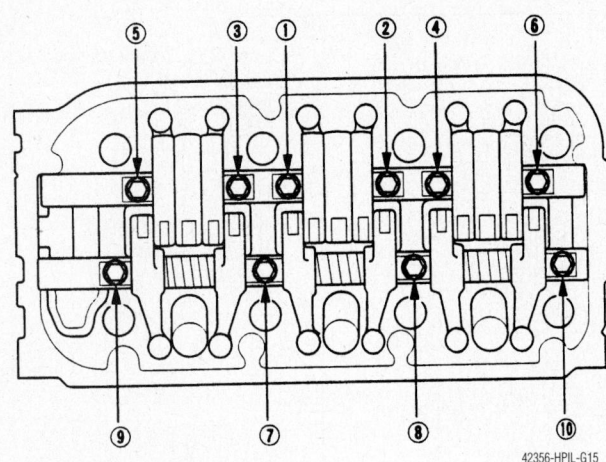

Rocker shaft tightening sequence

42356-HPIL-G15

Tighten the bolts in sequence and in multiple passes to 17 ft. lbs. (24 Nm).
9. Adjust the valve clearance.
10. Install or connect the following:
   - Valve covers
   - Ignition coils and torque the retainers to 9 ft. lbs. (12 Nm)
   - Ignition coil covers
   - Intake manifold
   - Negative battery cable

Intake Manifold

REMOVAL & INSTALLATION

2003-04 Models

1. Before servicing the vehicle, refer to the precautions section.
2. Remove or disconnect the following:
   - Negative battery cable
   - Intake manifold cover

- Air intake tube
- Throttle and cruise control cables
3. Remove the following electrical connections and clamps from the manifold:
   - Intake Air Temperature (IAT) sensor connector
   - Idle Air Control (IAC) valve connector
   - Throttle Position (TP) sensor connector
   - Manifold Absolute Pressure (MAP) sensor connector
   - Evaporative Emissions (EVAP) control canister purge valve connector
   - Brake booster vacuum line
   - Positive Crankcase Ventilation (PCV) hose
   - Water bypass hoses from the throttle body and plug the hoses
   - EVAP control canister hose
4. Remove or disconnect the following:
   - Upper cover bolts and nuts in the

sequence illustrated using two or three passes
   - Intake manifold bolts using two or three passes
   - Intake manifold and spacer

To install:
5. Install or connect the following:
   - New intake manifold gasket and spacer
   - Intake manifold. Tighten the fasteners in sequence and in several passes to 16 ft. lbs. (22 Nm).
   - Upper cover bolts and nuts in the sequence illustrated using two or three passes to 9 ft. lbs. (12 Nm)
   - EVAP control canister hose
   - Water bypass hoses toe throttle body
   - Brake booster vacuum line
   - PCV hose
6. Connect the following electrical connections and clamps to the manifold:
   - EVAP control canister purge valve connector
   - MAP sensor connector
   - TP sensor connector
   - IAC valve connector
   - IAT sensor connector
   - Throttle and cruise control cables
   - Air intake tube
   - Intake manifold cover
   - Negative battery cable
7. Start the engine and check for proper operation.

2005-06 Models

1. Before servicing the vehicle, refer to the precautions section.
2. Remove or disconnect the following:
   - Negative battery cable
   - Intake manifold cover
   - Intake Air Temperature (IAT) sensor 2 connector
   - Air intake tube
   - Positive Crankcase Ventilation (PCV) hose
   - Brake booster vacuum line
   - Evaporative Emissions (EVAP) control canister hose and transmission breather hose clamp bracket
   - Water bypass hoses from the throttle body and plug the hoses
3. Remove the following electrical connections and clamps from the manifold:
   - Intake Air Temperature (IAT) sensor connector
   - Throttle Position (TP) sensor connector
   - Manifold Absolute Pressure (MAP) sensor connector
   - EVAP canister purge valve connector

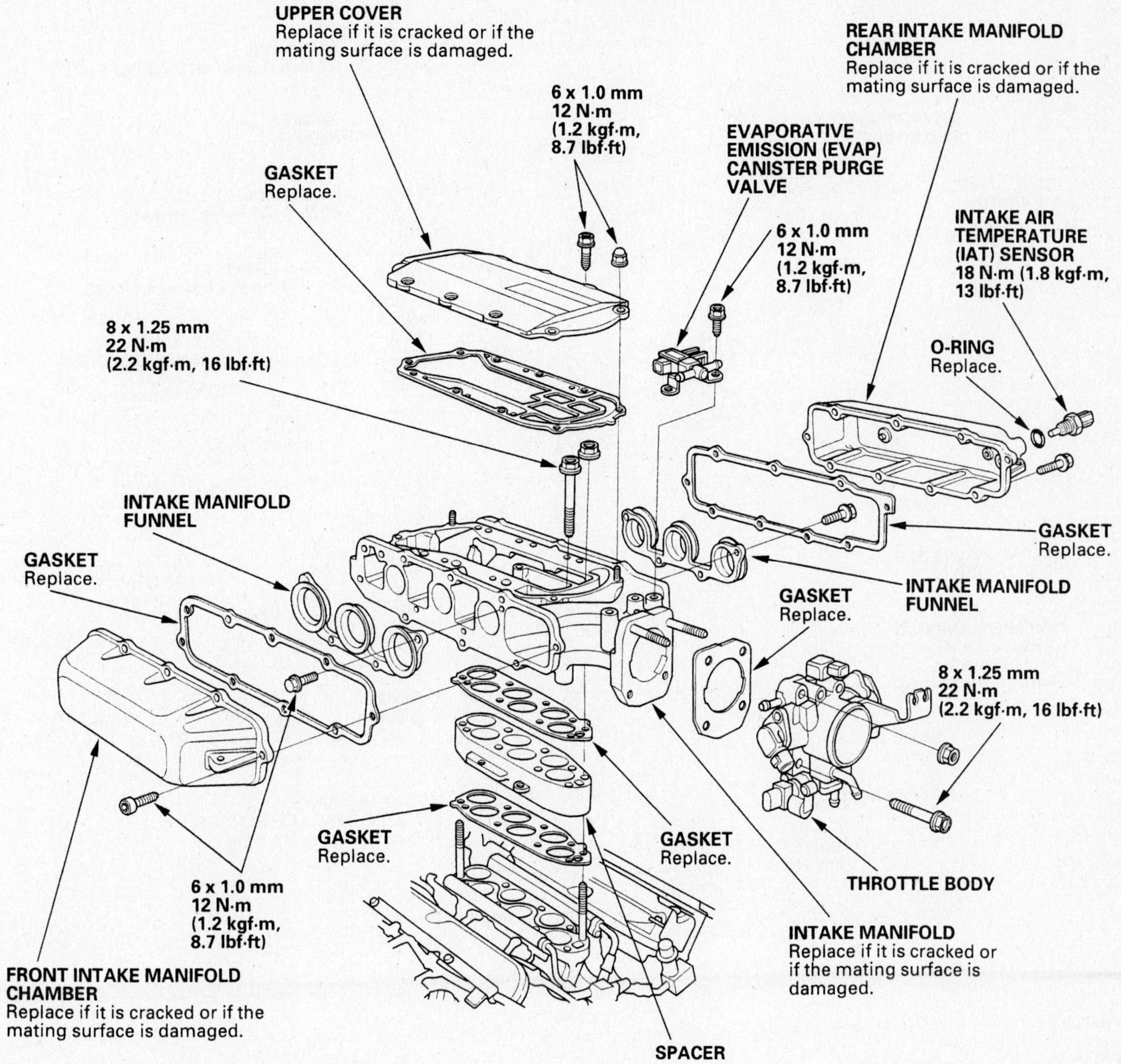

UPPER COVER
Replace if it is cracked or if the mating surface is damaged.

GASKET
Replace.

6 x 1.0 mm
12 N·m
(1.2 kgf·m,
8.7 lbf·ft)

REAR INTAKE MANIFOLD CHAMBER
Replace if it is cracked or if the mating surface is damaged.

EVAPORATIVE EMISSION (EVAP) CANISTER PURGE VALVE

6 x 1.0 mm
12 N·m
(1.2 kgf·m,
8.7 lbf·ft)

INTAKE AIR TEMPERATURE (IAT) SENSOR
18 N·m (1.8 kgf·m, 13 lbf·ft)

8 x 1.25 mm
22 N·m
(2.2 kgf·m, 16 lbf·ft)

O-RING
Replace.

INTAKE MANIFOLD FUNNEL

GASKET
Replace.

GASKET
Replace.

INTAKE MANIFOLD FUNNEL

GASKET
Replace.

8 x 1.25 mm
22 N·m
(2.2 kgf·m, 16 lbf·ft)

GASKET
Replace.

6 x 1.0 mm
12 N·m
(1.2 kgf·m,
8.7 lbf·ft)

GASKET
Replace.

THROTTLE BODY

INTAKE MANIFOLD
Replace if it is cracked or if the mating surface is damaged.

FRONT INTAKE MANIFOLD CHAMBER
Replace if it is cracked or if the mating surface is damaged.

SPACER

42356-HPIL-G04

Exploded view of the intake manifold—2003–04 models

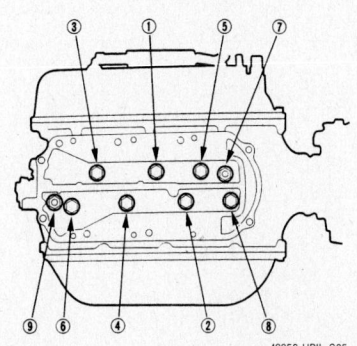

Intake manifold torque sequence—2003–04 models

42356-HPIL-G05

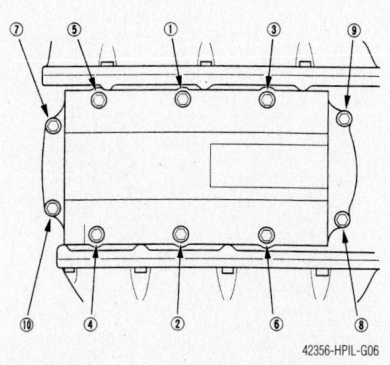

Upper cover torque sequence—2003–04 models

42356-HPIL-G06

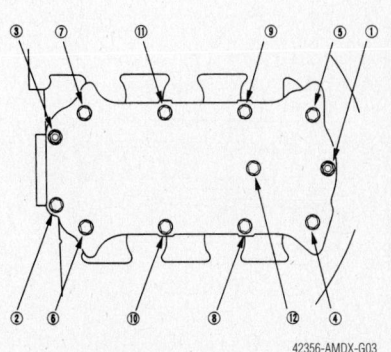

Upper cover loosening sequence—2005–06 models

42356-AMDX-G03

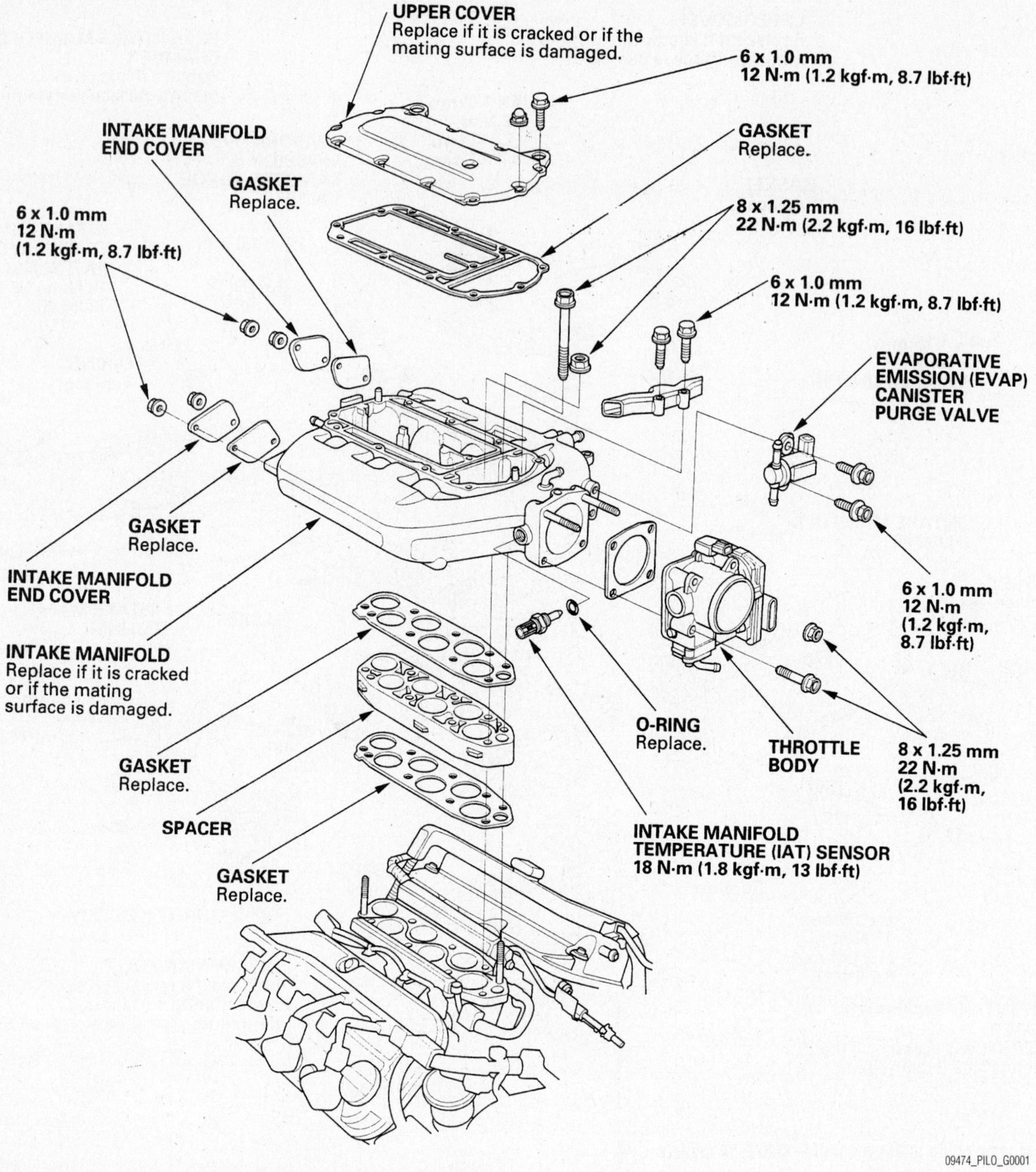

UPPER COVER
Replace if it is cracked or if the mating surface is damaged.
6 x 1.0 mm
12 N·m (1.2 kgf·m, 8.7 lbf·ft)
INTAKE MANIFOLD END COVER
GASKET
Replace.
GASKET
Replace.
6 x 1.0 mm
12 N·m
(1.2 kgf·m, 8.7 lbf·ft)
8 x 1.25 mm
22 N·m (2.2 kgf·m, 16 lbf·ft)
6 x 1.0 mm
12 N·m (1.2 kgf·m, 8.7 lbf·ft)
EVAPORATIVE EMISSION (EVAP) CANISTER PURGE VALVE
GASKET
Replace.
INTAKE MANIFOLD END COVER
6 x 1.0 mm
12 N·m
(1.2 kgf·m,
8.7 lbf·ft)
INTAKE MANIFOLD
Replace if it is cracked or if the mating surface is damaged.
O-RING
Replace.
THROTTLE BODY
8 x 1.25 mm
22 N·m
(2.2 kgf·m,
16 lbf·ft)
GASKET
Replace.
SPACER
INTAKE MANIFOLD TEMPERATURE (IAT) SENSOR
18 N·m (1.8 kgf·m, 13 lbf·ft)
GASKET
Replace.
09474_PILO_G0001

Exploded view of the intake manifold—2005–06 models

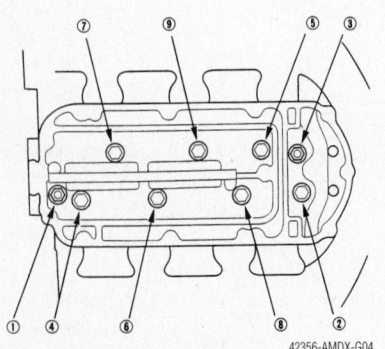

⑦ ⑨ ⑤ ③
① ④ ⑥ ⑧ ②
42356-AMDX-G04

Intake manifold loosening sequence—2005–06 models

③ ① ⑤ ⑦
⑨ ⑥ ④ ② ⑧
42356-AMDX-G34

Intake manifold torque sequence—2005–06 models

⑩ ⑥ ② ④ ⑧ ⑫
⑪ ⑦ ③ ⑤ ① ⑨
42356-AMDX-G05

Upper cover torque sequence—2005–06 models

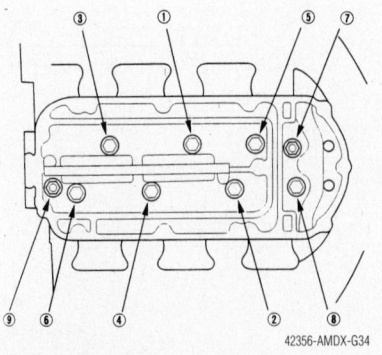

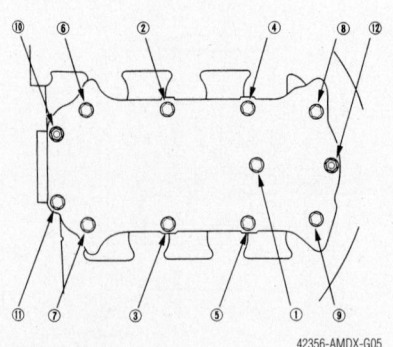

- Upper cover bolts and nuts in the sequence illustrated using two or three passes
- Intake manifold bolts in the sequence illustrated
- Intake manifold and spacer

To install:

4. Install or connect the following:
   - New intake manifold gasket and spacer
   - Intake manifold. Tighten the fasteners in sequence and in several passes to 16 ft. lbs. (22 Nm).
   - Upper cover bolts and nuts in the sequence illustrated using two or three passes to 9 ft. lbs. (12 Nm)

5. Connect the following electrical connections and clamps to the manifold:
   - EVAP canister purge valve connector
   - MAP sensor connector
   - TP sensor connector
   - IAT sensor 1 connector
   - Water bypass hoses to the throttle body
   - EVAP control canister hose and clamp bracket
   - Intake manifold vacuum line
   - Brake booster vacuum line
   - PCV hose
   - Air intake tube
   - IAT sensor 2 connector
   - Intake manifold cover
   - Negative battery cable

6. Start the engine and check for proper operation.

Exhaust Manifold

REMOVAL & INSTALLATION

1. Before servicing the vehicle, refer to the precautions section.
2. Remove or disconnect the following:
   - Negative battery cable
   - Exhaust manifold heat shield
   - Heated Oxygen (HO2S) sensor connector
   - Exhaust front pipe
   - Exhaust manifold bracket, if equipped
   - Exhaust manifold

To install:

3. Install or connect the following:
   - Exhaust manifold. Tighten the fasteners to 23 ft. lbs. (31 Nm).
   - Exhaust manifold bracket, if equipped. Tighten the bolts to 33 ft. lbs. (44 Nm).
   - Exhaust front pipe. Tighten the nuts to 40 ft. lbs. (55 Nm).

- Heated Oxygen (HO2S) sensor connector
- Exhaust manifold heat shield
- Negative battery cable

Front Crankshaft Seal

REMOVAL & INSTALLATION

1. Before servicing the vehicle, refer to the precautions section.
2. Remove or disconnect the following:
   - Negative battery cable
   - Accessory drive belts
   - Side engine mount
   - Valve cover
   - Crankshaft pulley
   - Front cover
   - Balance shaft belt, if equipped
   - Timing belt
   - Top Dead Center (TDC) sensor, if equipped
   - Crankshaft timing sprocket
   - Front crankshaft seal

To install:

3. Lubricate the crankshaft seal lip with grease prior to installation.
4. Install the front crankshaft seal so that it is flush with the surface of the oil pump housing.
5. Install or connect the following:
   - Crankshaft timing sprocket
   - Top Dead Center (TDC) sensor, if equipped
   - Timing belt
   - Balance shaft belt, if equipped
   - Front cover
   - Crankshaft pulley. Tighten the bolt to 181 ft. lbs. (245 Nm), on 2003–04 models or 47 ft. lbs. (64 Nm) plus 60 degrees on 2005–06 models.
   - Valve cover
   - Side engine mount
   - Accessory drive belts
   - Negative battery cable

07LAD-PT3010A

93552G02

Front crankshaft seal installation

6. Check the engine oil level and add if necessary.
7. Start the engine and check for leaks.

Camshaft

REMOVAL & INSTALLATION

Front

1. Before servicing the vehicle, refer to the precautions section.
2. Remove or disconnect the following:
   - Negative and positive battery cables
   - Battery
3. Drain the coolant.
   - Exhaust Gas Recirculation (EGR) valve
   - Timing belt
   - Rocker arm assembly
   - Front camshaft pulley
   - Thrust plate and camshaft

To install:

4. Install or connect the following:
   - Camshaft using a new O-ring. Tighten the thrust plate to 16 ft. lbs. (22 Nm).
   - Front camshaft pulley
   - Rocker arm assembly
   - Timing belt
   - Exhaust Gas Recirculation (EGR) valve
   - Battery
   - Positive, then negative battery cables
5. Fill the cooling system.
   - Camshaft

To install:

6. Start the engine and check for leaks.

Rear

1. Before servicing the vehicle, refer to the precautions section.
2. Drain the cooling system.
3. Relieve the fuel system pressure.

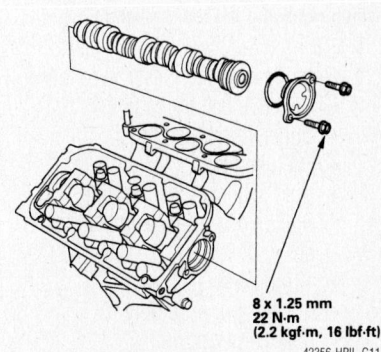

8 x 1.25 mm
22 N·m
(2.2 kgf·m, 16 lbf·ft)

42356-HPIL-G11

Front camshaft assembly

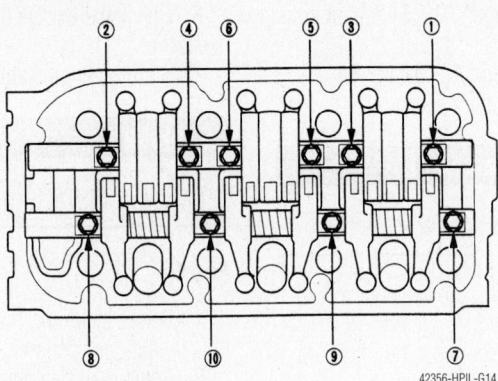

Rocker arm shaft loosening sequence

42356-HPIL-G14

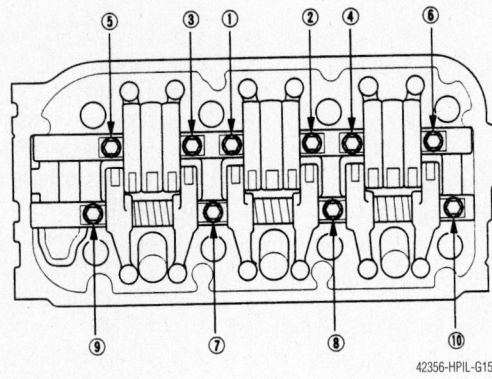

Rocker shaft tightening sequence

42356-HPIL-G15

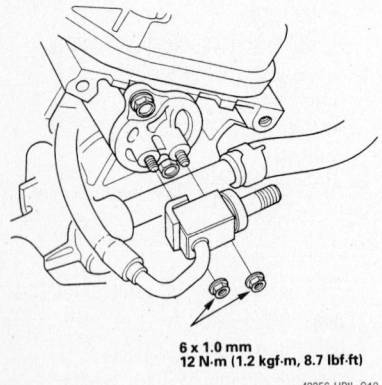

6 x 1.0 mm
12 N·m (1.2 kgf·m, 8.7 lbf·ft)

42356-HPIL-G12

Remove the nuts attaching the fuel line when removing the rear camshaft

4. Remove or disconnect the following:
   • Negative battery cable
   • Under-hood fuse box
   • Fuel feed hose
   • Nuts securing the fuel line
   • Brake lines from the master cylinder
   • Timing belt
   • Rocker arm assembly
   • Rear camshaft pulley
   • Thrust plate and camshaft

To install:

5. Install or connect the following:
   • Camshaft using a new O-ring. Tighten the thrust plate to 16 ft. lbs. (22 Nm).
   • Rear camshaft pulley
   • Rocker arm assembly
   • Timing belt
   • Brake lines to the master cylinder
   • Nuts securing the fuel line
   • Fuel feed hose
   • Under-hood fuse box
   • Negative battery cable

Valve Lash

ADJUSTMENT

Adjust the valves only when the cylinder head temperature is less than 100°F (38°C).

1. Before servicing the vehicle, refer to the precautions section.

2. Remove or disconnect the following:
   • Negative battery cable
   • Air intake tube
   • Intake manifold
   • Valve cover

3. Rotate the crankshaft so that the valves to be adjusted are closed and the rocker arm is contacting the camshaft lobe base circle.

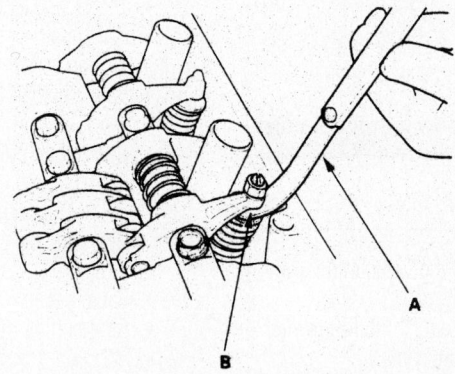

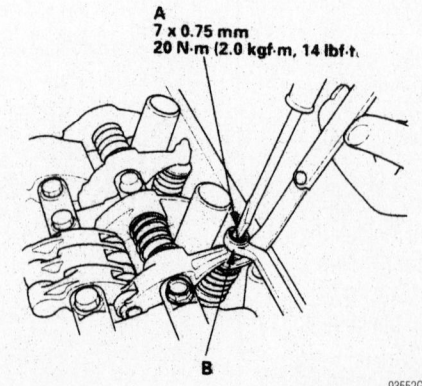

A
7 x 0.75 mm
20 N·m (2.0 kgf·m, 14 lbf·ft)

93552G04

Inspect the valve clearance, adjust to specification and tighten the retainer to specification

Adjusting screw locations:

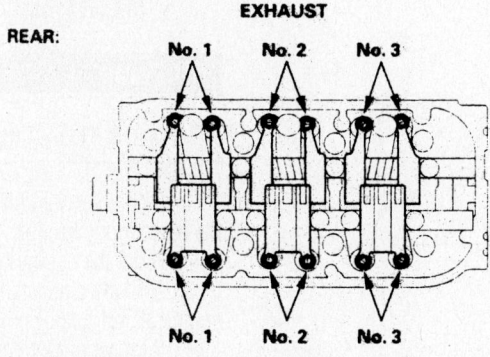

Valve adjusting retainer locations

4. Measure the valve clearance. If adjustment is necessary, loosen the locknut and turn the adjusting screw as necessary to achieve the correct valve clearance.

5. The correct valve clearance is:
- Intake valves: 0.008–0.009 inches (0.20–0.24mm)
- Exhaust valves: 0.011–0.013 inches (0.28–0.32mm)

6. After adjustment, tighten the locknuts to 14 ft. lbs. (20 Nm).

7. Install or connect the following:
- Valve cover
- Intake manifold
- Air intake tube
- Negative battery cable

8. Start the engine and check for proper operation.

Starter Motor

REMOVAL & INSTALLATION

2003–04 Models

1. Before servicing the vehicle, refer to the precautions section.
2. Remove or disconnect the following:
- Negative battery cable and wait at least 3 minutes

- Unlock the transmission fluid cooler hose clamp
- Starter wiring harness connectors
- Starter motor

To install:

3. Install or connect the following:
- Starter motor. Tighten the 10mm bolt to 33 ft. lbs. (44 Nm) and the 12mm bolt to 47 ft. lbs. (64 Nm).
- Starter wiring harness connectors. Tighten the battery cable nut to 79 inch lbs. (9 Nm).
- Lock the transmission fluid cooler hose clamp
- Negative battery cable

2005–06 Models

1. Before servicing the vehicle, refer to the precautions section.
2. Remove or disconnect the following:
- Negative battery cable, then positive battery cable and wait at least 3 minutes
- Harness clamp
- Starter wiring harness connectors
- Starter motor

To install:

3. Install or connect the following:
- Starter motor. Tighten the bolts to 54 ft. lbs. (74 Nm).

- Starter wiring harness connectors. Tighten the battery cable nut to 79 inch lbs. (9 Nm).
- Harness clamp
- Negative battery cable

Oil Pan

REMOVAL & INSTALLATION

2003–04 Models

1. Before servicing the vehicle, refer to the precautions section.
2. Drain the engine oil and power steering system.
3. Remove or disconnect the following:
- Negative battery cable
- Power steering pump outlet hose from the pump and the hose clamp
- Steering joint cover, mark the steering joint-to-gearbox pinion shaft for reference
- Steering joint from the pinion shaft
- Splash shield
- Transfer assembly
- Tie rod ends from the knuckles
- Lower arm ball joints from the knuckles
- Power steering hose
- Power steering pressure switch connector
- Nuts attaching the transmission lower front and rear mount
- Bolt attaching the rear mount
4. Support the engine with a hoist.
- Nut attaching the front mount bracket
5. Make reference marks on the body across the marks on the edge of the front subframe.
- Front subframe

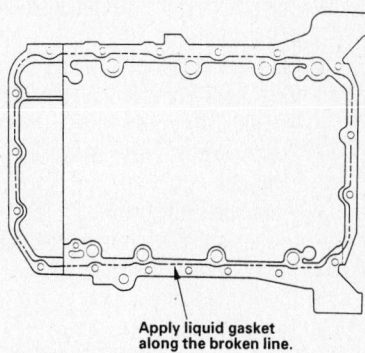

Apply liquid gasket along the broken line.

Apply liquid gasket to the inner threads of the bolt holes and the engine block along the area indicated by the broken line— Pilot

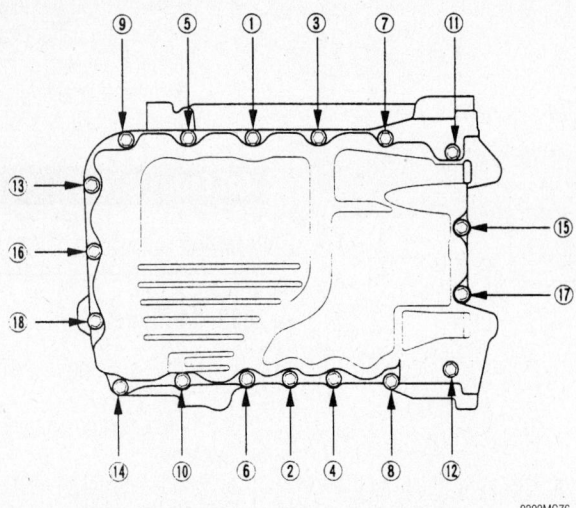

Oil pan tightening sequence—Pilot

9302MG76

- Torque converter cover and the 2 bolts retaining the transmission
- Oil pan

To install:

6. Install the oil pan. Apply liquid gasket as shown.

7. Tighten the bolts in sequence to 105 inch lbs. (12 Nm) using several passes.

8. Install or connect the following:
  - 2 bolts retaining the transmission and tighten to 28 ft. lbs. (38 Nm)
  - Torque converter cover and tighten the bolts to 9 ft. lbs. (12 Nm)

9. Align the reference marks on the body across the marks on the edge of the front subframe.
  - Front subframe. Use new bolts and tighten the 14mm bolts to 76 ft. lbs. (103 Nm). Tighten the front brace bolts to 54 ft. lbs. (74 Nm) and the rear brace bolts to 86 ft. lbs. (117 Nm).
  - Nut attaching the front mount bracket to 40 ft. lbs. (54 Nm)
  - Bolt attaching the rear mount to 28 ft. lbs. (38 Nm)
  - Nuts attaching the transmission lower front and rear mount to 28 ft. lbs. (38 Nm)
  - Power steering pressure switch connector
  - Power steering hose
  - Lower arm ball joints to the knuckles
  - Tie rod ends to the knuckles
  - Transfer assembly
  - Splash shield
  - Steering joint to the pinion shaft
  - Steering joint cover
  - Power steering pump outlet hose clamp and hose
  - Negative battery cable

2005–06 Models

1. Before servicing the vehicle, refer to the precautions section.

2. Drain the engine oil.

3. Remove or disconnect the following:
  - Negative battery cable
  - Exhaust front pipe
  - Torque converter cover
  - Oil pan

To install:

4. Install the oil pan. Apply liquid gasket as shown.

5. Tighten the bolts in sequence to 105 inch lbs. (12 Nm) using several passes.

6. Wait at least 30 minutes before adding oil to the engine.

7. Install or connect the following:

- Torque converter cover, if removed
- Exhaust front pipe
- Negative battery cable

Oil Pump

REMOVAL & INSTALLATION

1. Before servicing the vehicle, refer to the precautions section.

2. Drain the engine oil.

3. Turn the crankshaft and place the engine at Top Dead Center (TDC).

4. Remove or disconnect the following:
  - Negative battery cable
  - Accessory drive belts
  - Front cover
  - Timing belt
  - Timing belt idler pulley
  - Crankshaft Position (CKP) sensor
  - Crankshaft timing sprocket
  - Variable Valve Timing and Valve Lift Electronic Control (VTEC) solenoid valve connector
  - Oil filter adapter
  - Oil pan
  - Oil pump pickup tube
  - Oil pump

To install:

➡**Use new gaskets and O-ring seals for assembly.**

5. Apply liquid gasket to the oil pump and to the bolt hole threads.

6. Install or connect the following:
  - Oil pump. Tighten the bolts to 9 ft. lbs. (12 Nm).

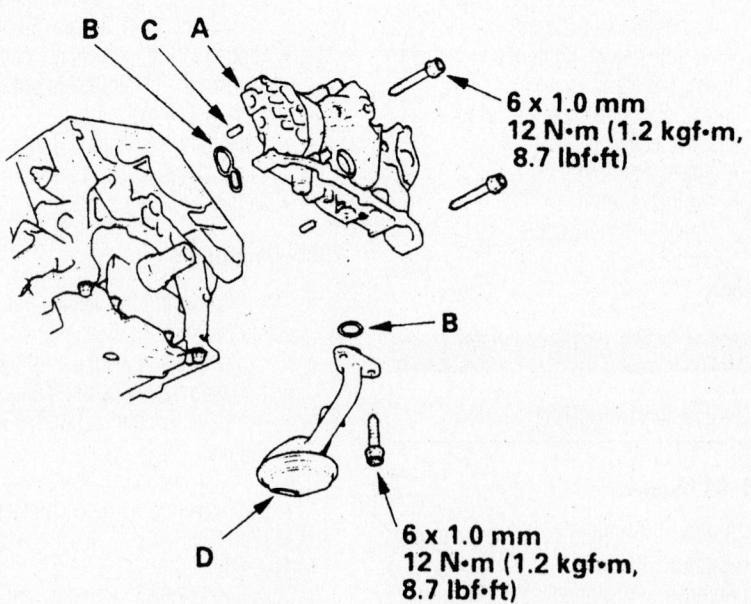

Exploded view of the oil pump assembly

6 x 1.0 mm
12 N·m (1.2 kgf·m, 8.7 lbf·ft)

6 x 1.0 mm
12 N·m (1.2 kgf·m, 8.7 lbf·ft)

93552G05

- Oil pump pickup tube. Tighten the bolts to 9 ft. lbs. (12 Nm).
- Oil pan
- Oil filter adapter
- VTEC solenoid valve connector
- Crankshaft timing sprocket
- CKP sensor
- Timing belt idler pulley
- Timing belt
- Front cover
- Accessory drive belts
- Negative battery cable

7. Fill the crankcase to the correct level.
8. Start the engine and check for leaks.

Rear Main Seal

REMOVAL & INSTALLATION

1. Before servicing the vehicle, refer to the precautions section.
2. Remove or disconnect the following:
    - Transaxle
    - Flywheel
    - Oil seal

To install:
3. Install or connect the following:
    - Oil seal. Drive the seal square into the seal case.
    - Flywheel. Tighten the bolts in a crossing pattern to 54 ft. lbs. (73 Nm).
    - Transaxle
4. Check the fluid levels.
5. Start the engine and check for leaks.

Timing Belt

REMOVAL & INSTALLATION

1. Before servicing the vehicle, refer to the precautions section.
2. Turn the crankshaft so the white mark aligns with the pointer.
3. Make sure the number 1 piston is at Top Dead center (TDC).
4. Remove or disconnect the following:
    - Negative battery cable
    - Wheels and splash shield
    - Drive belts
5. Support the engine with a block of wood and a jack under the oil pan.
    - Upper side engine mount
    - Dipstick tube
6. Remove the crankshaft pulley using holder tool shown in the accompanying illustration and a breaker and socket, loosen the 19mm bolt and remove the pulley.
    - Front upper cover, rear upper cover and the lower cover

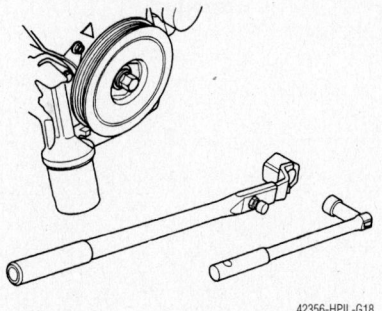

42356-HPIL-G18

Remove the crankshaft pulley using holder tool shown

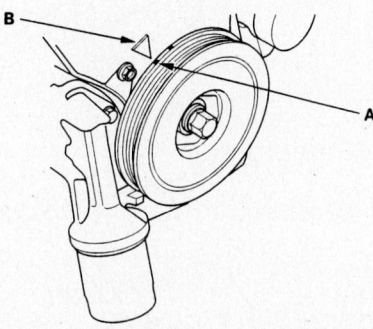

42356-HPIL-G16

Turn the crankshaft so the white mark (A) aligns with the pointer (B)

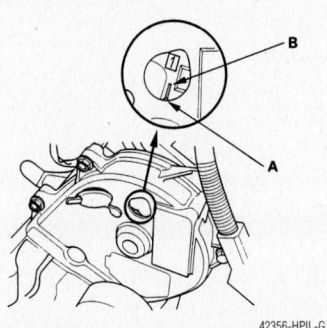

42356-HPIL-G17

Make sure the number 1 piston is at top dead center (A) on the front camshaft pulley and pointer (B)

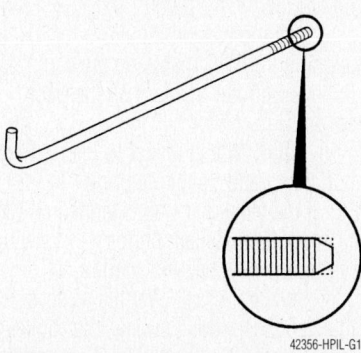

42356-HPIL-G19

Remove a battery clamp bolt and grind the end as shown

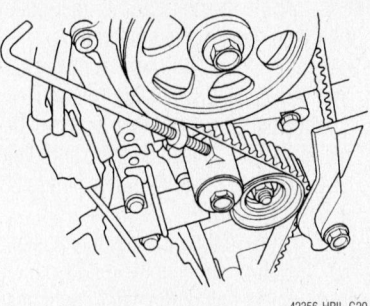

42356-HPIL-G20

Install the battery clamp bolt as shown to hold the belt adjuster in position

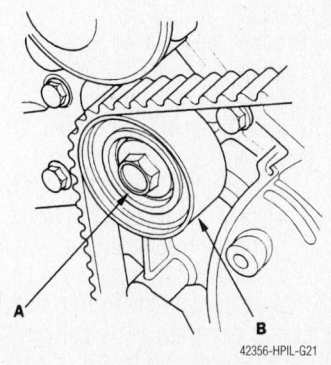

42356-HPIL-G21

Remove the idler pulley bolt (A), pulley (B) and the timing belt

- One of the battery clamp bolts and grind the end as illustrated
7. Screw the battery clamp bolt as illustrated to hold the belt adjuster in position. Do not use a wrench, hand tighten only.
    - Lower side engine mount
    - Idler pulley bolt and the pulley
    - Timing belt

To install:
8. If installing a new belt, perform the following steps:
    a. Clean the pulleys, belt guide plate and the upper and lower covers.
    b. Set the timing belt drive pulley to TDC by aligning the TDC mark on the tooth of the belt drive pulley with the pointer on the oil pump.
    c. Set the camshaft pulleys to TDC by aligning the TDC marks on the camshaft pulleys with the pointers on the back covers.
    d. Remove the battery clamp bolt.
    e. Remove the belt tensioner.
    f. Align the holes on the rod and housing of the tensioner.
    g. Using a press or other suitable device, slowly compress the tensioner and insert a 0.08 inch (2mm) pin through the housing and rod.
    h. Install the tensioner making sure the pin is still installed.

i. Apply thread locker to idler pulley bolt then hand tighten the bolt.

j. Install the belt over the pulleys in this sequence; drive pulley, idler pulley, front camshaft pulley, water pump pulley, rear camshaft pulley and adjusting pulley.

k. Tighten the idler pulley bolt to 33 ft. lbs. (44 Nm).

l. Remove the pin from the tensioner.

9. Install or connect the following:
- Lower half of the side mount and tighten the 3 long bolts to 33 ft. lbs. (44 Nm) and the one short bolt to 9 ft. lbs. (12 Nm)
- Timing belt guide plate as illustrated
- Lower timing cover and tighten the bolts to 9 ft. lbs. (12 Nm)
- Front and rear upper timing covers and tighten the bolts to 9 ft. lbs. (12 Nm)
- Crankshaft pulley and tighten the bolts to 181 ft. lbs. (245 Nm), using the holding tool to prevent the unit from turning

10. Rotate the crankshaft pulley about 5 or 6 degrees clockwise so the belt positions itself on the pulleys.

11. Turn the crankshaft pulley so the white mark aligns with the pointer.

12. Check the camshaft pulley marks are aligned. If the marks are aligned, proceed to the next step. If the marks are not aligned, remove the timing belt and reinstall using the steps outlined before this step.

13. Remove or disconnect the following:
- Drive belt
- Upper side mount and tighten the bolts in the sequence illustrated to the specifications in the illustration

14. Using a suitable scan tool, perform the Powertrain Control Module (PCM) reset and the Crankshaft position (CKP) pattern clear/learn procedures, following the scan tool manufactures instructions.

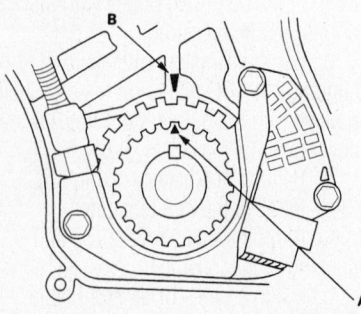

Set the timing belt pulley to TDC by aligning the TDC mark (A) on the tooth of the belt pulley with the pointer (B) on the oil pump

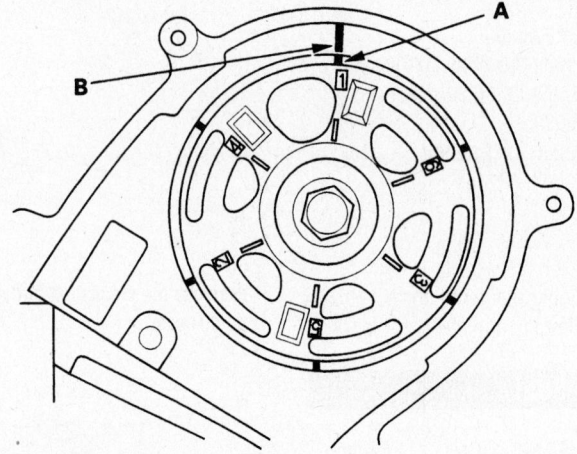

FRONT:

REAR:

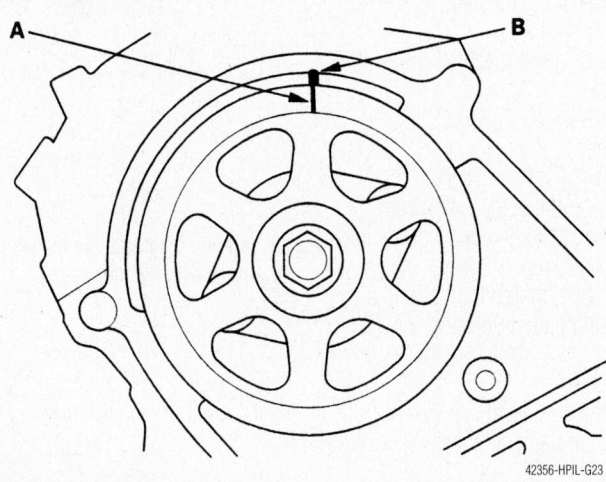

42356-HPIL-G23

Set the camshaft pulleys to TDC by aligning the TDC marks (A) on the camshaft pulleys with the pointers (B) on the back covers

15. If installing the old belt, perform the following steps:

a. Clean the pulleys, belt guide plate and the upper and lower covers.

b. Set the timing belt drive pulley to TDC by aligning the TDC mark on the tooth of the belt drive pulley with the pointer on the oil pump.

c. Set the camshaft pulleys to TDC by aligning the TDC marks on the camshaft pulleys with the pointers on the back covers.

d. Apply thread locker to idler pulley bolt then hand tighten the bolt.

16. If the tensioner was extended and the belt cannot be installed, perform the steps above for the new belt installation.

a. Install the belt over the pulleys in this sequence; drive pulley, idler pulley, front camshaft pulley, water pump pulley, rear camshaft pulley and adjusting pulley.

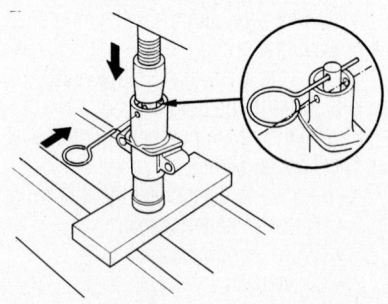

42356-HPIL-G24

Insert a 0.08 inch (2mm) pin through the tensioner housing and rod

b. Tighten the idler pulley bolt to 33 ft. lbs. (44 Nm).

c. Remove the battery clamp bolt.

17. Install or connect the following:
- Lower half of the side mount and tighten the 3 long bolts to 33 ft.

-1 Drive pulley (A).
-2 Idler pulley (B).
-3 Front camshaft pulley (C).
-4 Water pump pulley (D).
-5 Rear camshaft pulley (E).
-6 Adjusting pulley (F).

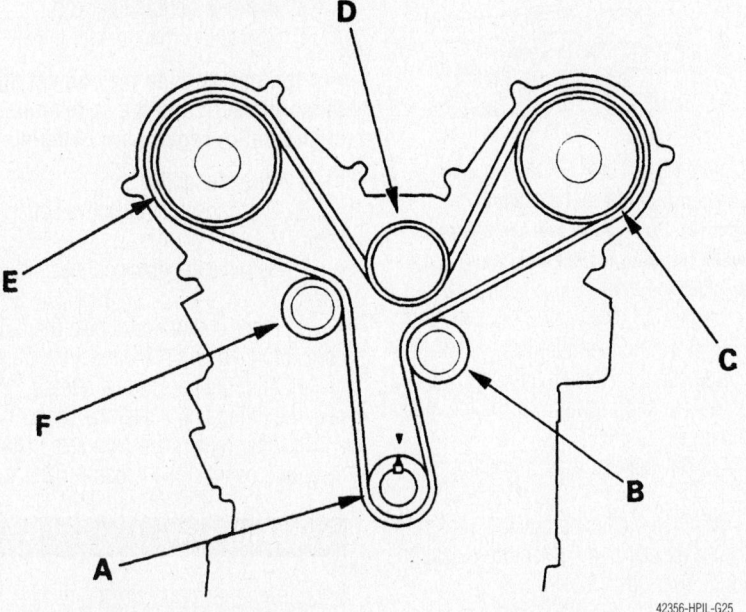

42356-HPIL-G25

Route the belt as shown in the sequence listed

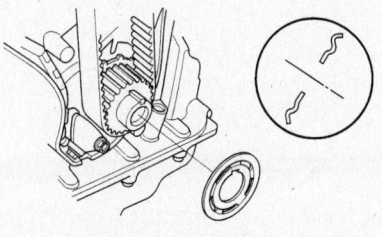

42356-HPIL-G07

Install the timing belt guide plate as shown

lbs. (44 Nm) and the one short bolt to 9 ft. lbs. (12 Nm)
- Timing belt guide plate as illustrated
- Lower timing cover and tighten the bolts to 9 ft. lbs. (12 Nm)
- Front and rear upper timing covers and tighten the bolts to 9 ft. lbs. (12 Nm)
- Crankshaft pulley and tighten the bolts to 181 ft. lbs. (245 Nm), using the holding tool to prevent the

18. Rotate the crankshaft pulley about 5 or 6 degrees clockwise so the belt positions itself on the pulleys.

FRONT CAMSHAFT PULLEY:

REAR CAMSHAFT PULLEY:

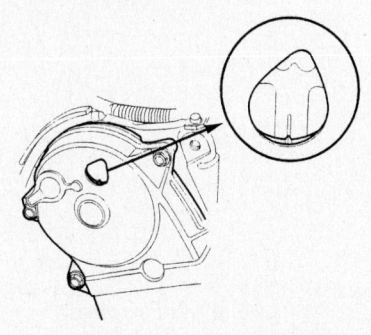

42356-HPIL-G08

Check that the camshaft pulley marks are aligned as shown

19. Turn the crankshaft pulley so the white mark aligns with the pointer.
20. Check the camshaft pulley marks are aligned. If the marks are aligned, proceed to the next step. If the marks are not aligned, remove the timing belt and reinstall using the steps outlined before this step.
21. Install or connect the following:
- Drive belt
- Upper side mount and tighten the bolts in the sequence illustrated to the specifications in the illustration
- Dipstick tube

22. Using a suitable scan tool, perform the Powertain Control Module (PCM) reset and the Crankshaft position (CKP) pattern clear/learn procedures, following the scan tool manufactures instructions.

Piston and Ring

POSITIONING

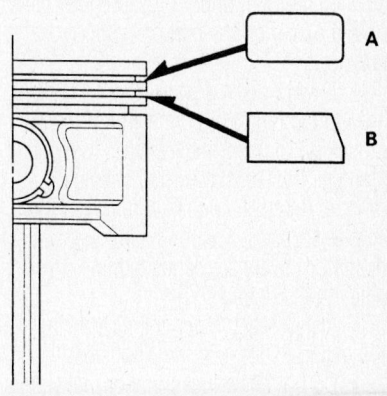

9302AG06

Compression ring identification—3.5L engine

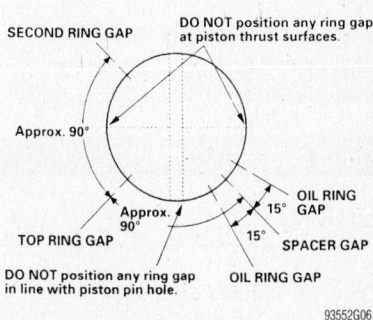

93552G06

Ring end gap positioning

FUEL SYSTEM

Fuel System Service Precautions

Safety is the most important factor when performing not only fuel system maintenance, but any type of maintenance. Failure to conduct maintenance and repairs in a safe manner may result in serious personal injury or death. Maintenance and testing of the vehicle's fuel system components can be accomplished safely and effectively by adhering to the following rules and guidelines:

• To avoid the possibility of fire and personal injury, always disconnect the negative battery cable unless the repair or test procedure requires that battery voltage be applied.

• Always relieve the fuel system pressure prior to disconnecting any fuel system component (injector, fuel rail, pressure regulator, etc.), fitting or fuel line connection. Exercise extreme caution whenever relieving fuel system pressure to avoid exposing skin, face and eyes to fuel spray. Please be advised that fuel under pressure may penetrate the skin or any part of the body that it contacts.

• Always place a shop towel or cloth around the fitting or connection prior to loosening to absorb any excess fuel due to spillage. Ensure that all fuel spillage (should it occur) is quickly removed from engine surfaces. Ensure that all fuel soaked cloths or towels are deposited into a suitable waste container.

• Always keep a dry chemical (Class B) fire extinguisher near the work area.

• Do not allow fuel spray or fuel vapors to come into contact with a spark or open flame.

• Always use a backup wrench when loosening and tightening fuel line connection fittings. This will prevent unnecessary stress and torsion to fuel line piping. Always follow the proper torque specifications.

• Always replace worn fuel fitting O-rings with new. Do not substitute fuel hose or equivalent, where fuel pipe is installed.

Fuel System Pressure

RELIEVING

2003–04 Models

1. Before servicing the vehicle, refer to the precautions section.
2. Disconnect the negative battery cable.

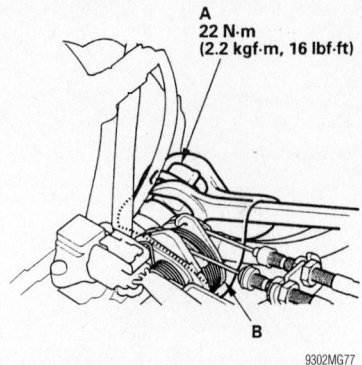

A
22 N·m
(2.2 kgf·m, 16 lbf·ft)

B

9302MG77

Use a wrench on the fuel pulsation damper (A). Place a rag over the damper (B) when relieving residual fuel pressure—2003–04 Pilot

3. Remove the fuel filler cap.
4. Remove the intake manifold cover.
5. Place a shop towel over the fuel pulsation damper.
6. Loosen the fuel pulsation damper 1 turn.
7. When service is completed, replace the sealing washer and tighten the pulsation damper to 16 ft. lbs. (22 Nm).
8. Replace the fuel filler cap.
9. Connect the negative battery cable.
10. Start the engine and check for leaks.

2005–06 Models

1. Before servicing the vehicle, refer to the precautions section.
2. Remove the drivers side dashboard lower cover and disconnect the PGM-FI main relay.
3. Remove the fuel filler cap.
4. Start the engine and let it stall.

➡**A temporary code may be set during this procedure and the codes must be cleared after repairs are completed.**

5. Turn the ignition off.
6. Disconnect the negative battery cable.
7. Remove the quick connect fitting cover.
8. Place a shop towel over the quick connect fitting and disconnect the fitting.
9. After the pressure is release, reconnect the fitting and install the cover.
10. After repairs are complete install the relay and connect the negative battery cable.

Fuel Filter

REMOVAL & INSTALLATION

1. Before servicing the vehicle, refer to the precautions section.

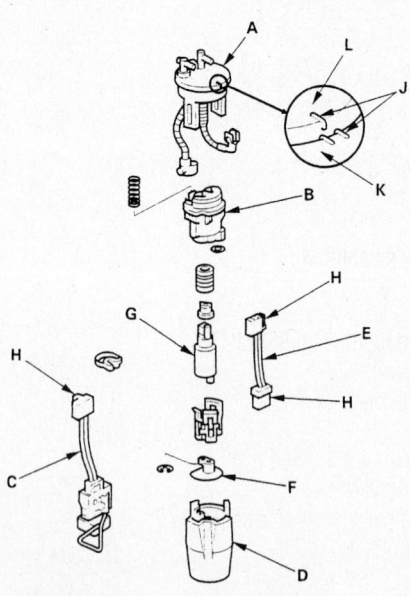

A. Bracket	F. Suction filter
B. Fuel filter	G. Fuel pump
C. Fuel gauge sender	H. Connectors
D. Case	J. Alignment marks
E. Wire harness	K. Fuel tank
	L. Fuel pump module

9308MG26

Exploded view of the fuel pump module—Pilot

2. Relieve the fuel system pressure.
3. Remove or disconnect the follow-ing:
   - Negative battery cable
   - Driver's side second row seat and cut the carpet along the dotted line. Be careful not to cut the wiring harness under the carpet.
   - Access panel
   - Fuel pump module
4. Disassemble the fuel pump module and remove the fuel filter.

To install:

5. Install the fuel filter and assemble the fuel pump module.
6. Install or connect the following:
   - Fuel pump module
   - Access panel
   - Carpet and seat
   - Negative battery cable
7. Start the engine and check for leaks.

Fuel Pump

REMOVAL & INSTALLATION

1. Before servicing the vehicle, refer to the precautions section.
2. Relieve the fuel system pressure.
3. Remove or disconnect the following:

- Negative battery cable
- Driver's side second row seat and cut the carpet along the dotted line. Be careful not to cut the wiring harness under the carpet.
- Access panel
- Fuel pump module wiring connector
- Fuel supply and return lines
- Fuel pump locknut
- Fuel pump module

To install:

4. Install or connect the following:
   - Fuel pump module. Use a new seal and align the matchmarks.
   - Fuel pump locknut
   - Fuel supply and return lines
   - Fuel pump module wiring connector
   - Access panel
   - Carpet and seat
   - Negative battery cable
5. Start the engine and check for leaks.

Fuel Injector

REMOVAL & INSTALLATION

1. Before servicing the vehicle, refer to the precautions section.

2. Relieve the fuel system pressure.
3. Remove or disconnect the following:
   - Negative battery cable
   - Intake manifold
   - Fuel lines
   - Fuel injector connectors
   - Fuel pressure regulator vacuum line
   - Fuel supply manifold
4. Separate the fuel injectors from the fuel supply manifold.

To install:

5. Install the fuel injectors to the fuel supply manifold with new cushion rings and O-rings.
6. Install new seal rings to the intake manifold.
7. Install or connect the following:
   - Fuel supply manifold and injector assembly. Tighten the bolts to 86 inch lbs. (10 Nm).
   - Fuel pressure regulator vacuum line
   - Fuel injector connectors
   - Fuel lines
   - Intake manifold
   - Negative battery cable
8. Start the engine and check for leaks.

DRIVE TRAIN

Transaxle Assembly

REMOVAL & INSTALLATION

1. Before servicing the vehicle, refer to the precautions section.
2. Drain the transaxle.
3. Drain the power steering system.
4. Remove the engine appearance covers.
5. Remove the drivers side center console lower panel and pull back the cover to access steering joint cover.
6. Remove or disconnect the following:
   - Steering joint bolt

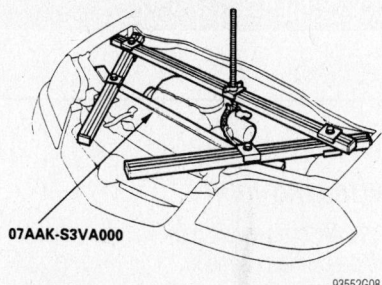

07AAK-S3VA000

93552G08

Support the engine while removing the transaxle—Pilot

- Steering joint from the steering gearbox pinion shaft
- Battery
- Battery tray
- Intake manifold cover
- Air intake assembly
- Power steering pump hose and the clamp bolt
- Splash shield
- Transmission breather tube
- Transaxle oil cooler lines
- Cooler hose from the clamp on the starter, if equipped
- Starter motor
- Shift control solenoid valve connectors
- Transaxle ground cable
- Connector from the bracket and the connector
- Harness clip from the brackets
- Clutch pressure switch connectors
- Joint connector and transmission range switch connector from the brackets
- Countershaft speed sensor connector
- Heated Oxygen (HO2S) sensor connectors
- Transmission housing mounting bolts

- Nut from the front mount and the ground cable from the engine
- Bulkhead cover, windshield wiper arms, cowl cover sealing and cover
- Install a support fixture to the engine lifting eyes.
- Front sub-frame stiffener
- Primary HO2S sensor clamp bracket from the transmission and harness from the clamp
- Exhaust front pipe
- Lower control arms from the knuckle
- Stabilizer bar links
- Tie rod ends from the knuckle
- Left driveshaft from the differential
- Right driveshaft from the intermediate shaft
- Propeller shaft from the companion flange
- Shift cable cover and holder
- Shift control cable and lever
7. Install a 6 x 1 x 14mm bolt and nut on the cable cover, then reinstall the cable cover to the torque converter housing. If this is not done, the bolt head of the cable cover may prevent torque converter removal.
   - Transfer assembly
   - Engine-to-torque converter bolts

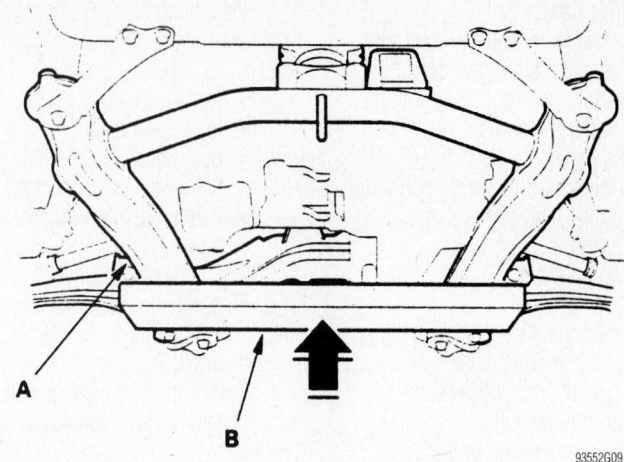

Support the sub-frame with a 4 x 4 x 50 inch piece of wood and a jack

- Power steering pressure switch connection
- Power steering hose clamp, then the hose from the pipe at the sub-frame
- Transmission lower mount nuts

8. Matchmark the front subframe to the vehicle body.

- Rear mount bracket bolts

9. Support the sub-frame with a 4 x 4 x 50 inch piece of wood and a jack.

- Sub-frame
- Transaxle lower mounts
- Driveshafts from the differential and intermediate shaft
- Intermediate shaft
- Transmission front mount bracket
- Transmission flange bolts
- Transmission

To install:

➡**Use new circlips, split pins and self-locking nuts for assembly.**

10. Installation is the reverse of removal. Please note the following specifications:

- Transmission housing bolts and harness clamp bolts to 47 ft. lbs. (64 Nm)
- Transmission housing bolts to 40 ft. lbs. (54 Nm)
- Front mount bracket bolts to 28 ft. lbs. (38 Nm)
- Intermediate shaft bolts to 29 ft. lbs. (39 Nm)
- Transfer assembly bolts to 33 ft. lbs. (44 Nm)

11. Raise the subframe into position and align the matchmarks. Tighten the subframe bolts to 76 ft. lbs. (103 Nm). Tighten the front subframe bracket bolts to 54 ft. lbs. (74 Nm) and the rear bracket bolts to 86 ft. lbs. (117 Nm).

- Rear engine mount bolts to 28 ft. lbs. (38 Nm)

- Engine-to-torque converter bolts. Tighten the 6 x 1 mm bolts to 105 inch lbs. (12 Nm), 10 x 1.25mm bolt to 28 ft. lbs. (38 Nm).
- Front motor mount nut to 40 ft. lbs. (54 Nm)

12. Fill the transaxle to the correct level.

13. Start the engine and check for leaks.

14. Check the wheel alignment and adjust as necessary.

Transfer Assembly

REMOVAL & INSTALLATION

1. Before servicing the vehicle, refer to the precautions section.

2. Drain the transmission fluid.

3. Remove or disconnect the following:

- Negative battery cable
- Heated Oxygen (HO2S) sensor connectors
- Front sub-frame stiffener
- Exhaust front pipe
- Breather tube bracket bolt, then the tube from the breather pipe
- Propeller shaft from the transfer assembly
- Transfer assembly bolts and the assembly

To install:

4. Install or connect the following:

- New O-ring on the transfer cover
- Dowel pin on the assembly
- Transfer assembly and tighten the bolts to 33 ft. lbs. (44 Nm) in a star pattern
- Propeller shaft
- Breather tube bracket, attach the tube with the dot facing outwards and tighten the bolt to 9 ft. lbs. (12 Nm)
- Exhaust front pipe

- Front sub-frame stiffener and tighten the bolts to 40 ft. lbs. (54 Nm)
- Heated Oxygen (HO2S) sensor connectors
- Negative battery cable

Halfshaft

REMOVAL & INSTALLATION

1. Before servicing the vehicle, refer to the precautions section.

2. Drain the transaxle if removing the left halfshaft. If is not necessary to drain the fluid if removing the right halfshaft.

3. Remove or disconnect the following:

- Negative battery cable
- Front wheels
- Spindle nut
- Stabilizer bar link
- Lower ball joint

4. Pry the inboard joint from the transaxle or intermediate shaft.

5. Remove the outer CV-joint stub shaft from the hub by tapping the stub shaft with a plastic hammer.

To install:

➡**Use new circlips, split pins and self-locking nuts for assembly.**

6. Install the outer CV-joint stub shaft into the hub.

7. Install the inner CV-joint to the transaxle or intermediate shaft until the circlip locks in the retaining groove.

8. Install or connect the following:

- Lower ball joint. Tighten the nut to 47 ft. lbs. (64 Nm).
- Stabilizer bar link. Tighten the nut to 58 ft. lbs. (78 Nm).
- Spindle nut. Tighten the nut to 181 ft. lbs. (245 Nm) on 2003–04 models, or 210 ft. lbs. (285 Nm) on 2005–06 models.
- Front wheels
- Negative battery cable

9. Fill the transaxle to the correct level and check for leaks.

CV-Joint

OVERHAUL

Front

OUTBOARD JOINT

1. Before servicing the vehicle, refer to the precautions section.

2. Remove or disconnect the follow-ing:

- Axle halfshaft from the vehicle and place it in a vise

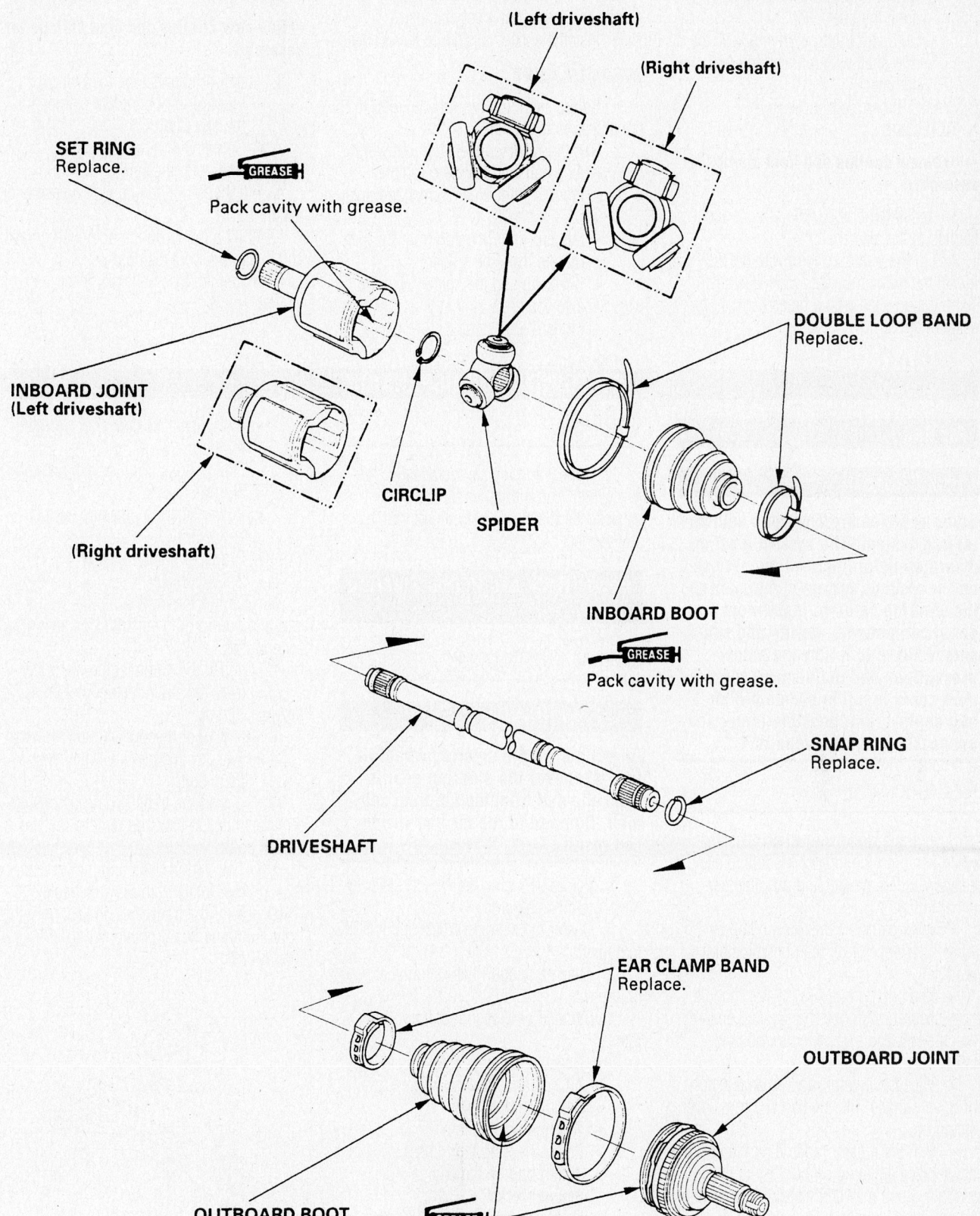

(Left driveshaft)

(Right driveshaft)

SET RING
Replace.

GREASE
Pack cavity with grease.

INBOARD JOINT
(Left driveshaft)

CIRCLIP

SPIDER

(Right driveshaft)

DOUBLE LOOP BAND
Replace.

INBOARD BOOT

GREASE
Pack cavity with grease.

SNAP RING
Replace.

DRIVESHAFT

EAR CLAMP BAND
Replace.

OUTBOARD JOINT

OUTBOARD BOOT

GREASE
Pack cavity with grease.

9308MG29

Front axle exploded view—Pilot

- Outboard joint boot clamps and push the boot back
- Outboard joint by driving it off the axle shaft with a brass drift and hammer
- Outboard joint boot

To install:

➡ **Use new circlips and boot clamps for assembly.**

3. Install the outboard joint boot and clamps to the axle shaft.

4. Fill the outboard joint with grease. Install the outboard joint to the axle shaft. Tap the stub shaft with a brass hammer to seat the circlip.

5. Fill the outboard joint boot with grease and install the boot clamps.

6. Install the axle halfshaft to the vehicle.

INBOARD JOINT

1. Before servicing the vehicle, refer to the precautions section.

2. Remove or disconnect the following:
- Axle halfshaft from the vehicle.
- Inboard joint boot clamps and push the boot back
- Inboard joint housing from the axle
- Rollers from the spider
- Snapring and the spider from the axle shaft
- Inboard joint boot

To install:

➡ **Use new circlips and boot clamps for assembly.**

3. Install or connect the following:
- Inboard joint boot and clamps to the axle shaft
- Spider with a new snapring
- Rollers to the spider

4. Fill the joint housing with grease and install it.

5. Fill the inboard joint boot with grease and install the boot clamps.

6. Install the axle halfshaft to the vehicle.

STEERING AND SUSPENSION

Air Bag

✳✳ CAUTION

Some vehicles are equipped with an air bag system. The system must be disarmed before performing service on, or around, system components, the steering column, instrument panel components, wiring and sensors. Failure to follow the safety precautions and the disarming procedure could result in accidental air bag deployment, possible injury and unnecessary system repairs.

PRECAUTIONS

Several precautions must be observed when handling the inflator module to avoid accidental deployment and possible personal injury.

- Never carry the inflator module by the wires or connector on the underside of the module.

- When carrying a live inflator module, hold securely with both hands, and ensure that the bag and trim cover are pointed away.

- Place the inflator module on a bench or other surface with the bag and trim cover facing up.

- With the inflator module on the bench, never place anything on or close to the module which may be thrown in the event of an accidental deployment.

Before servicing the vehicle, also make sure to refer to the precautions in the beginning of this section as well.

DISARMING

Disconnect and isolate the negative battery cable. Wait 3 minutes for the system capacitor to discharge before performing any service.

Power Rack and Pinion Steering Gear

REMOVAL & INSTALLATION

✳✳ WARNING

Do not permit the steering wheel to turn whenever the steering gear is disconnected from the steering column. Damage to the air bag wiring can result.

1. Before servicing the vehicle, refer to the precautions section.

2. Center the steering wheel and lock it in position.

3. Attach a support fixture to the engine lifting eyes.

4. Remove or disconnect the following:

- Negative battery cable
- Air bag and steering wheel
- Steering joint cover
- Steering flexible joint
- Power steering fluid lines
- 10mm bolt on the engine side mount bracket
- Front wheels
- Outer tie rod ends
- Sub-frame stiffener
- Heated Oxygen (HO₂S) sensor connectors

- 3 way catalytic converter from the mufflers
- Flange bolts from the exhaust rubber mount
- Power steering pressure switch connector
- Propeller shaft protector
- Splash shield

5. Support the front subframe with a jack and support the transmission with a second jack.

6. Loosen the 14mm subframe bolts.

7. Lower the subframe about 1 ³⁄₁₆ inches (30mm).

8. Remove or disconnect the following:
- Two 12mm and two 14 stiffener plate bolts

9. Support the transfer case by raising the transmission jack and remove the two 12mm bolts.
- Two 14mm bolts and the rear stiffener plats from the sub-frame

10. Lower the transmission jack until the front subframe has dropped about 1 ¹⁵⁄₁₆ inch (50mm).

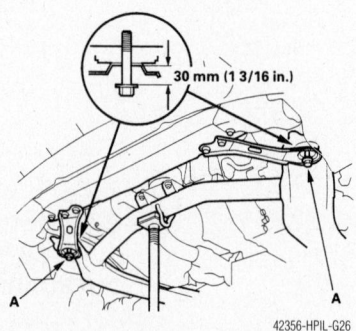

42356-HPIL-G26

Loosen the 14mm subframe bolts and lower the subframe about 1 ³⁄₁₆ inches (30mm)

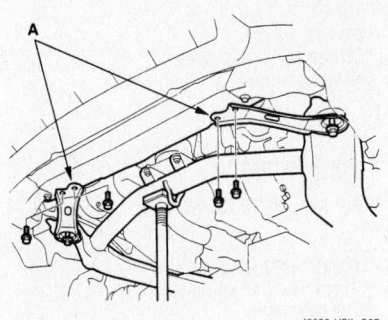

Remove the Four 12mm stiffener plate bolts

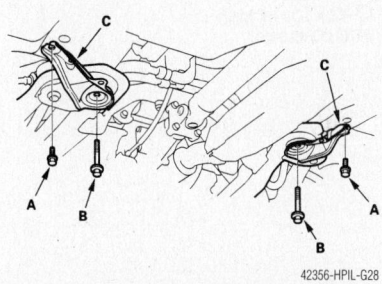

Remove the Two 12mm bolts (A), then the 14mm bolts (B) and the rear stiffener plates (C) from the sub-frame

- Power steering line brackets
- Feed line
- Return hose
- Two 10mm bolts from the right side gearbox
- Mounting bracket and cushion
- Two 10mm bolts from the left side gearbox

11. Lower the transmission jack until the front subframe has dropped about 3 $^{15}/_{16}$ inch (100mm).

- Gearbox stiffener bracket

12. Slide the gearbox between the body and front sub-frame towards the left and from the vehicle.

To install:

13. Position the steering gear in the vehicle.

14. Install or connect the following:

- Left steering gear mounting bolts. Tighten the bolts to 43 ft. lbs. (58 Nm).
- Right steering gear mounting bracket. Tighten the bolts to 29 ft. lbs. (39 Nm).
- Return hose
- Feed line
- Power steering line mounting brackets and tighten the bolts to 7 ft. lbs. (10 Nm)

15. Raise the subframe into position. Tighten the 14mm bolts to 76 ft. lbs. (103

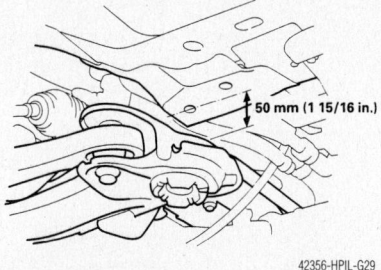

Lower the transmission jack until the front subframe has dropped about 1 $^{15}/_{16}$ inch (50mm)

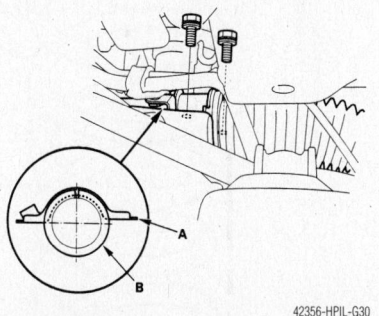

Remove the two 10mm bolts from the right side gearbox and the mounting bracket and cushion

Nm) and the 12mm bolts to 86 ft. lbs. (117 Nm).

16. Install or connect the following:

- Front stiffener plates. Tighten the 14mm bolts to 76 ft. lbs. (103 Nm)

and the 12mm bolts to 54 ft. lbs. (74 Nm).

- Splash shield
- Propeller shaft protector
- Power steering pressure switch
- 3 way catalytic converter and mufflers. Tighten the nuts to 25 ft. lbs. (33 Nm)
- Rubber exhaust mount and tighten the bolts to 28 ft. lbs. (38 Nm)
- HO$_2$S sensor connectors
- Sub-frame stiffener plate
- 10mm flange bolts on the engine side mount bracket to 33 ft. lbs. (44 Nm)
- Power steering hoses
- Outer tie rod ends
- Front wheels. Position the wheels straight-ahead.
- Steering flexible joint. Tighten the pinch bolts to 16 ft. lbs. (22 Nm)
- Steering joint cover
- Negative battery cable

17. Fill the power steering system.

18. Check the wheel alignment and adjust as necessary.

Strut

REMOVAL & INSTALLATION

Front

1. Before servicing the vehicle, refer to the precautions section.

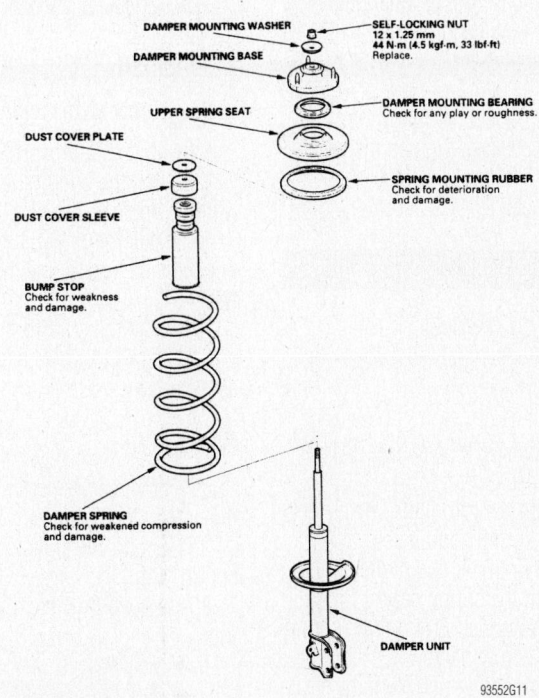

Exploded view of the front strut assembly

2. Remove or disconnect the following:
- Front wheel
- Wheel speed sensor wiring bracket
- Brake hose bracket
- Stabilizer bar link
- Strut pinch bolts
- Upper mount nuts
- Strut

To install:

3. Install or connect the following:
- Strut. Tighten the upper mount nuts to 43 ft. lbs. (59 Nm).
- Strut pinch bolts. Tighten the nuts to 116 ft. lbs. (157 Nm).
- Stabilizer bar link. Tighten the nut to 58 ft. lbs. (78 Nm).
- Brake hose bracket
- Wheel speed sensor wiring bracket
- Front wheel

4. Check the wheel alignment and adjust as necessary.

Shock Absorber

REMOVAL & INSTALLATION

Rear

1. Before servicing the vehicle, refer to the precautions section.
2. Support the vehicle under the lower control arm.
3. Remove or disconnect the following:
- Rear wheel
- Upper shock absorber flange bolt
- Lower shock absorber nut
- Shock absorber

To install:

4. Install or connect the following:
- Shock absorber. Tighten the fasteners to 47 ft. lbs. (64 Nm).
- Rear wheel

Coil Spring

REMOVAL & INSTALLATION

Front

1. Before servicing the vehicle, refer to the precautions section.
2. Remove the strut from the vehicle and install in a strut spring compressor. Compress the spring until the end of the spring comes away from the spring seat.
3. Remove the upper strut mount, spring seat and related components.
4. Remove the coil spring from the strut spring compressor.

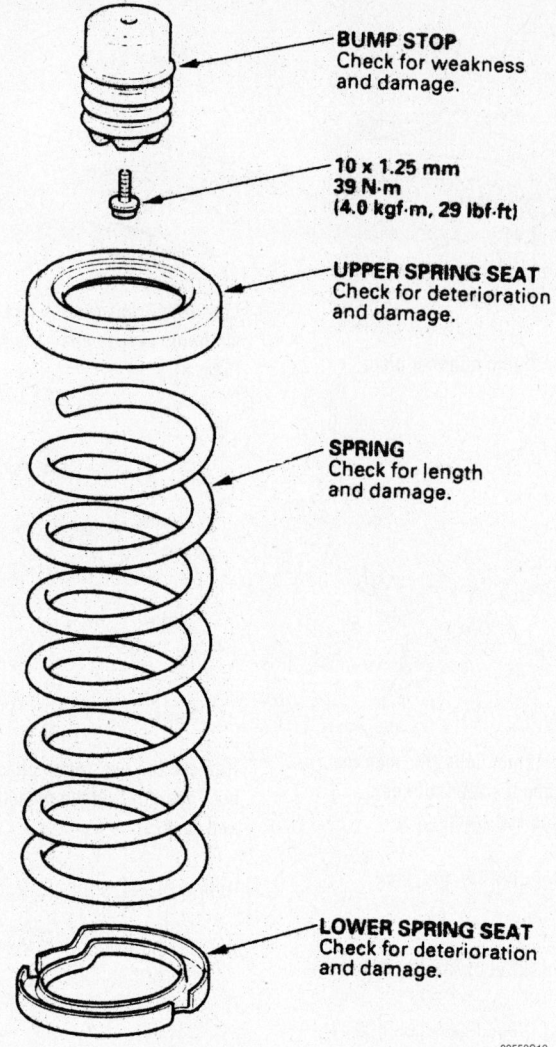

BUMP STOP
Check for weakness and damage.

**10 x 1.25 mm
39 N·m
(4.0 kgf·m, 29 lbf·ft)**

UPPER SPRING SEAT
Check for deterioration and damage.

SPRING
Check for length and damage.

LOWER SPRING SEAT
Check for deterioration and damage.

93552G12

Exploded view of the rear spring assembly

To install:

➡**Use a new self-locking nut.**

5. Compress the spring and position the strut so that the end of the spring aligns with the notch in the spring seat.
6. Install the upper strut mounting components and tighten the nut to 33 ft. lbs. (44 Nm).
7. Install the strut to the vehicle.
8. Check the wheel alignment and adjust as necessary.

Rear

1. Before servicing the vehicle, refer to the precautions section.
2. Support the vehicle under the lower control arm.
3. Remove or disconnect the following:
- Rear wheel
- Stabilizer link from the lower arm
- Wheel speed sensor wiring harness

from the lower arm. Do not disconnect the connector.
- Upper shock absorber flange bolt
- Lower control arm bolts

4. Lower the floor jack and remove the coil spring and spring seats.

To install:

➡**Use new self-locking nuts for assembly.**

5. Place the coil spring and spring seats on the lower control arm and raise into position. Tighten the inboard bolt to 61 ft. lbs. and the outer bolt to 54 ft. lbs. (74 Nm).
6. Install or connect the following:
- Rear wheel

Ball Joint

REMOVAL & INSTALLATION

The lower ball joints are replaced with the control arms as an assembly.

Upper Control Arm

REMOVAL & INSTALLATION

Rear

1. Before servicing the vehicle, refer to the precautions section.
2. Support the control arm at the knuckle.
3. Remove or disconnect the following:
   - Wheel
   - Upper ball joint from the knuckle
   - Upper arm bolt and the arm
4. Installation is the reverse of removal. Tighten the arm bolt to 47 ft. lbs. (64 Nm) on 2003–04 models or 69 ft. lbs. (93 Nm) on 2005–06 models. Tighten the ball joint nut to 36–43 ft. lbs. (49–59 Nm).

Lower Control Arm

REMOVAL & INSTALLATION

Front

1. Before servicing the vehicle, refer to the precautions section.
2. Remove or disconnect the following:
   - Front wheel
   - Lower ball joint
   - Front inner flange bolt
   - Rear inner flange bolt
   - Lower control arm

To install:

➡**Use a new split pin for assembly.**

3. Install or connect the following:
   - Lower control arm. Tighten the inner flange bolts to 69 ft. lbs. (93 Nm).
   - Lower ball joint. Tighten the nut to 43–51 ft. lbs. (59–69 Nm).
   - Front wheel
4. Check the wheel alignment and adjust as necessary.

Rear

LOWER ARM (A)

1. Before servicing the vehicle, refer to the precautions section.
2. Remove or disconnect the following:
   - Lower arm mounting bolt and nut
   - Lower arm
3. Installation is the reverse of removal. Tighten the bolt to 105 ft. lbs. (142 Nm) and the nut to 47 ft. lbs. (64 Nm).

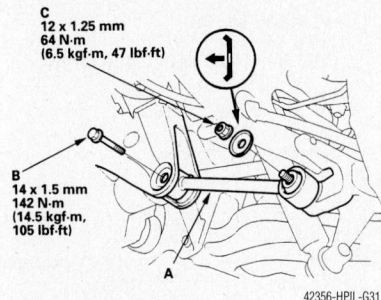

C
12 x 1.25 mm
64 N·m
(6.5 kgf·m, 47 lbf·ft)

B
14 x 1.5 mm
142 N·m
(14.5 kgf·m, 105 lbf·ft)

A

42356-HPIL-G31

Rear lower arm (A) mounting

LOWER ARM (B)

1. Before servicing the vehicle, refer to the precautions section.
2. Support the control arm with a jack.
3. Remove or disconnect the following:
   - Wheel
   - Stabilizer link from the lower arm
   - Wheel speed sensor wiring harness from the lower arm. Do not disconnect the connector.
   - Flange bolts that attaches the lower arm to the knuckle
4. Spring assembly
   - Inner nuts and bolts and the arm
5. Install or connect the following:
   - Arm, inner bolt and loosely install the nut
   - Spring assembly
6. Raise the arm into position and install the flange bolt.
7. Raise the rear suspension with a floor jack to load the vehicle weight.
   - Tighten the flange bolt to 54 ft. lbs. (74 Nm) and the inner nut and bolt to 61 ft. lbs. (83 Nm).
   - Wheel speed sensor harness
   - Wheel
8. Check the vehicle alignment.

CONTROL ARM BUSHING REPLACEMENT

The control arm bushings are serviced with the control arms as an assembly.

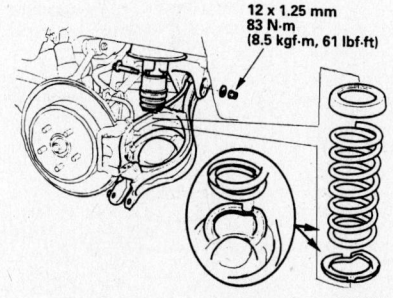

12 x 1.25 mm
83 N·m
(8.5 kgf·m, 61 lbf·ft)

93552G13

Rear lower arm (B) mounting

Wheel Bearings

ADJUSTMENT

The wheel bearings are sealed units and are not adjustable.

REMOVAL & INSTALLATION

Front

1. Before servicing the vehicle, refer to the precautions section.
2. Remove or disconnect the following:
   - Front wheel
   - Brake hose mounting bolt
   - Brake caliper
   - Wheel speed sensor
   - Spindle nut
   - Brake rotor
   - Outer tie rod end
   - Lower ball joint
   - Steering knuckle
3. Press the hub out of the wheel bearing.
   - Splash guard
   - Snapring and press the wheel bearing out of the steering knuckle
4. If necessary, press the inner bearing race off of the hub.

To install:

➡**Use a new ball joint nut, split pin, snapring and spindle nut for assembly.**

5. Press the bearing into the steering knuckle and install the snapring.
6. Install the splash guard.
7. Press the hub into the bearing.
8. Install or connect the following:
   - Steering knuckle. Tighten the ball joint nut to 43–51 ft. lbs. (59–69 Nm) and the damper flange bolts to 116 ft. lbs. (157 Nm).
   - Outer tie rod end. Tighten the nut to 40 ft. lbs. (54 Nm).
   - Wheel speed sensor, if equipped
   - Brake caliper and rotor
   - Brake hose
   - Spindle nut. Tighten the nut to 210 ft. lbs. (285 Nm).
   - Front wheel
9. Check the wheel alignment and adjust as necessary.

Rear

1. Before servicing the vehicle, refer to the precautions section.
2. Remove or disconnect the following:
   - Rear wheel
   - Brake hose bracket mounting bolts

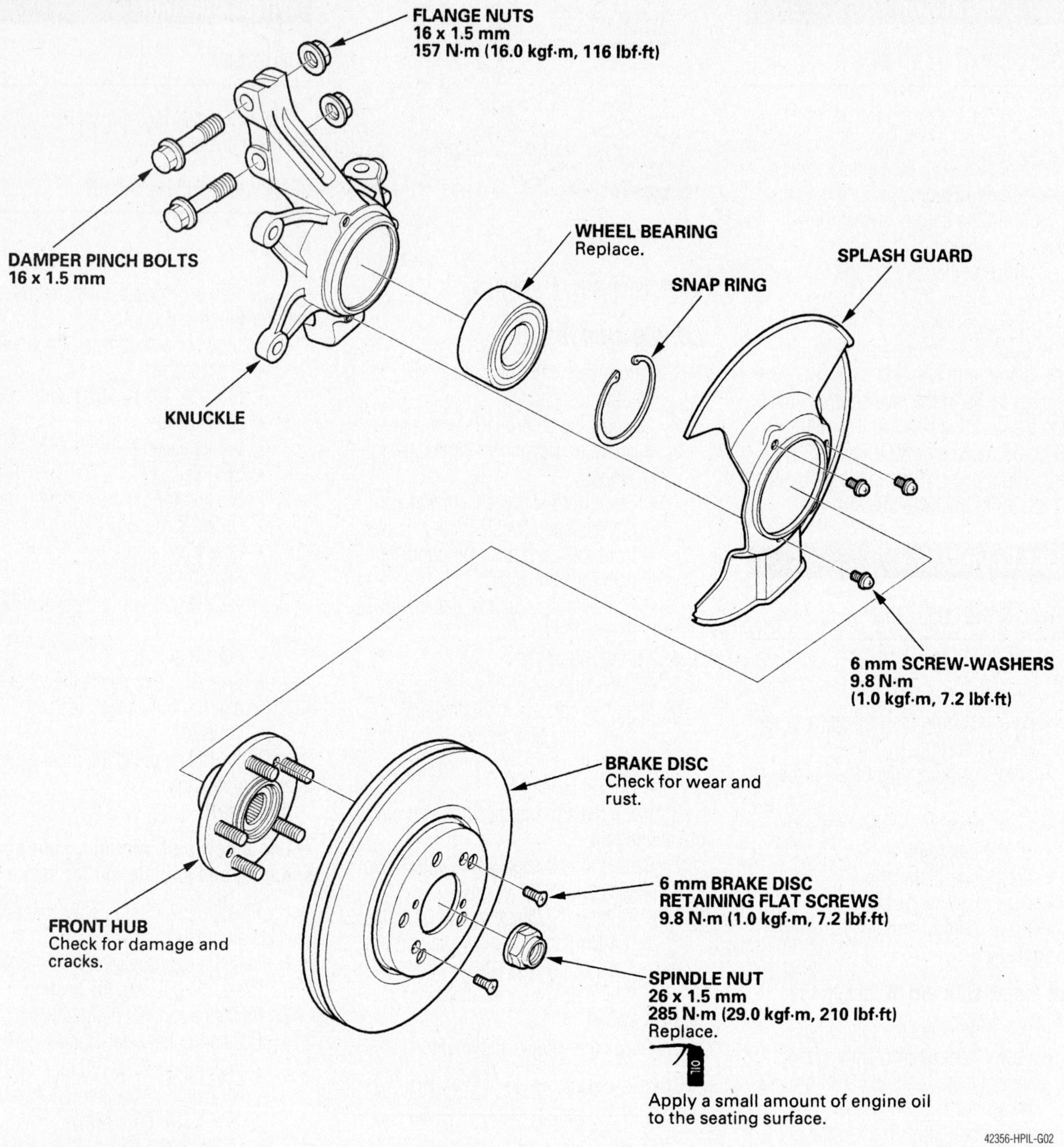

FLANGE NUTS
16 x 1.5 mm
157 N·m (16.0 kgf·m, 116 lbf·ft)
DAMPER PINCH BOLTS
16 x 1.5 mm
KNUCKLE
WHEEL BEARING
Replace.
SNAP RING
SPLASH GUARD
6 mm SCREW-WASHERS
9.8 N·m
(1.0 kgf·m, 7.2 lbf·ft)
BRAKE DISC
Check for wear and rust.
FRONT HUB
Check for damage and cracks.
6 mm BRAKE DISC
RETAINING FLAT SCREWS
9.8 N·m (1.0 kgf·m, 7.2 lbf·ft)
SPINDLE NUT
26 x 1.5 mm
285 N·m (29.0 kgf·m, 210 lbf·ft)
Replace.
OIL
Apply a small amount of engine oil to the seating surface.
42356-HPIL-G02

Front wheel bearing assembly

from the trailing arm and the knuckle

- Brake caliper
- Wheel speed sensor
- Spindle nut
- Brake rotor
- Upper ball joint
- Lower arm (A)
- Lower arm (B) from the trailing arm

3. Support the lower arm (B)
   - Steering knuckle
4. Press the hub out of the wheel bearing.
   - Splash guard

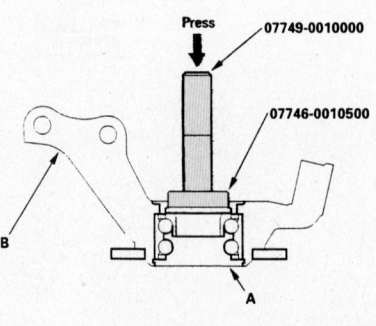

Press
07749-0010000
07746-0010500
B
A
93552G14

Press the wheel bearing out of the knuckle

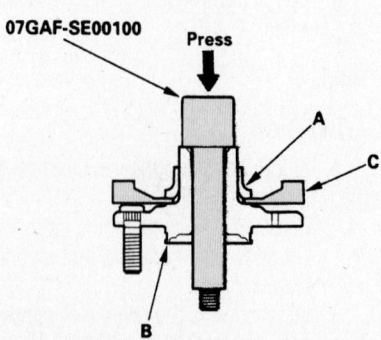

07GAF-SE00100
Press
A
C
B
93552G15

Press the wheel bearing inner race from the hub

- Snapring and press the wheel bearing out of the steering knuckle
5. If necessary, press the inner bearing race off of the hub.

To install:

➡Use a new ball joint nut, split pin, snapring and spindle nut for assembly.

6. Press the bearing into the steering knuckle and install the snapring.
7. Install the splash guard.
8. Press the hub into the bearing.
9. Install or connect the following:
- Steering knuckle. Tighten the flange bolt to 54 ft. lbs. (74 Nm) and the lower shock nut to 47 ft. lbs. (64 Nm)
- Lower arm (B) to the trailing arm and tighten the bolts to 47 ft. lbs. (64 Nm)
- Lower arm (A)
- Upper ball joint and tighten the nut to 40 ft. lbs. (54 Nm)
- Brake rotor and tighten the screws to 7 ft. lbs. (9 Nm)
- Spindle nut and tighten to 181 ft, lbs. (245 Nm)
- Wheel speed sensor
- Brake caliper and tighten the bolts to 41 ft. lbs. (55 Nm)
- Brake hose bracket mounting bolts to the knuckle and trailing arm
- Rear wheel
10. Check the wheel alignment and adjust as necessary.

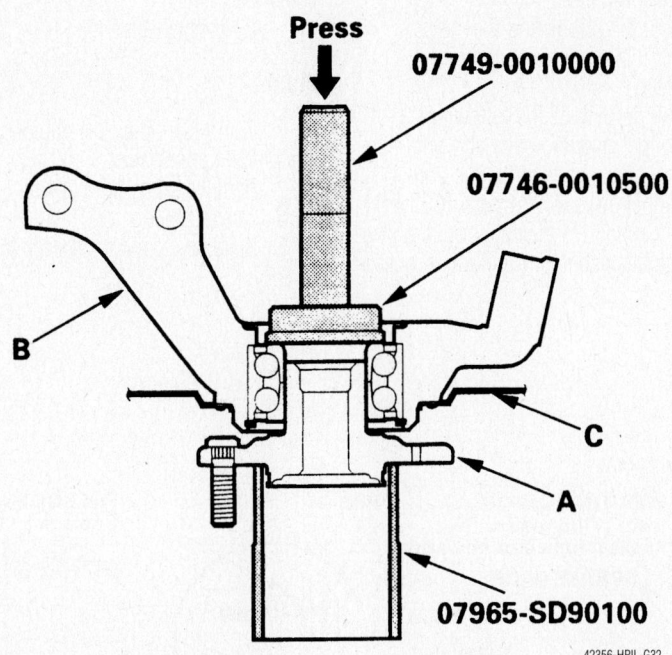

Press

07749-0010000

07746-0010500

B

C

A

07965-SD90100

42356-HPIL-G32

Press the wheel bearing into the knuckle

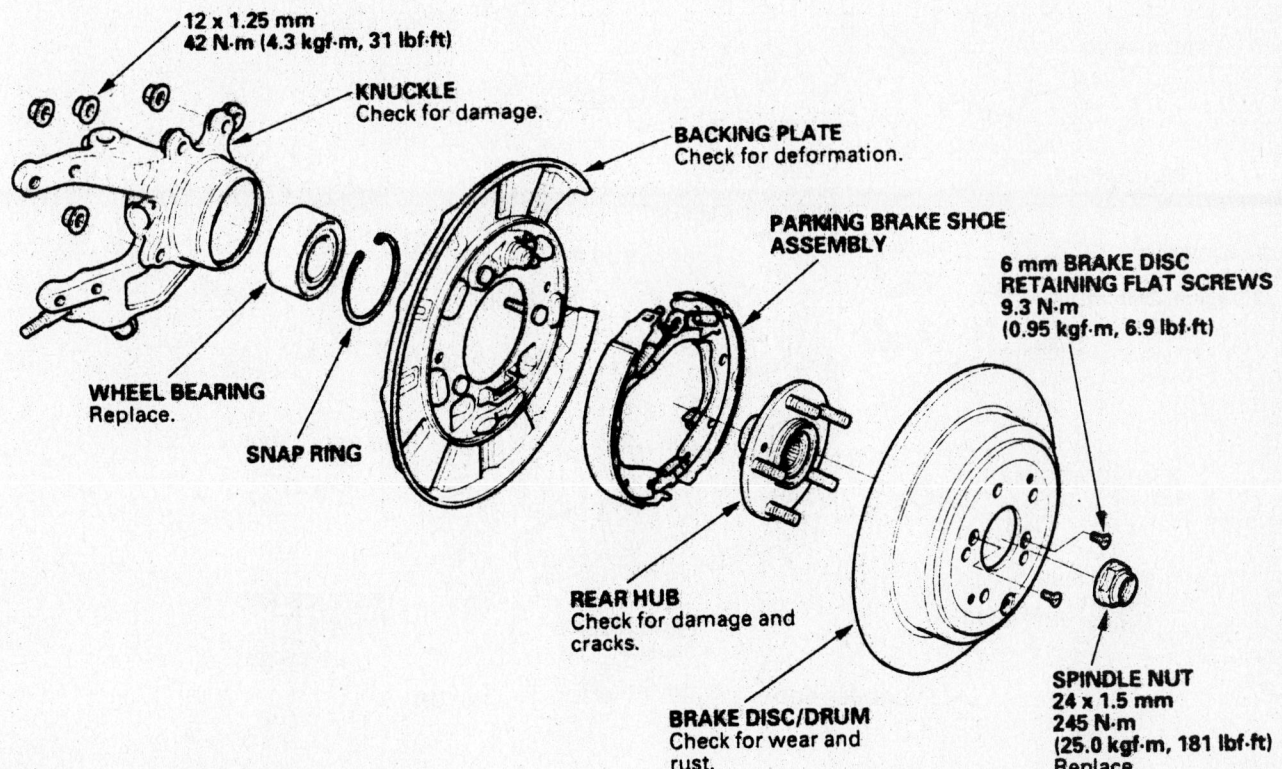

12 x 1.25 mm
42 N·m (4.3 kgf·m, 31 lbf·ft)

KNUCKLE
Check for damage.

BACKING PLATE
Check for deformation.

PARKING BRAKE SHOE ASSEMBLY

6 mm BRAKE DISC RETAINING FLAT SCREWS
9.3 N·m
(0.95 kgf·m, 6.9 lbf·ft)

WHEEL BEARING
Replace.

SNAP RING

REAR HUB
Check for damage and cracks.

BRAKE DISC/DRUM
Check for wear and rust.

SPINDLE NUT
24 x 1.5 mm
245 N·m
(25.0 kgf·m, 181 lbf·ft)
Replace.

93552G17

Exploded view of the rear wheel bearing assembly

BRAKES

Brake Caliper

REMOVAL AND INSTALLATION

Front

1. Before servicing the vehicle, refer to the precautions section.
2. Remove some fluid from the reservoir with a suction pump.
3. Remove or disconnect the following:
   • Front wheels

• Banjo bolt and disconnect the brake hose from the caliper. Plug the hose to prevent fluid loss and contamination.
• Mounting bolts and the caliper from its mounting bracket

To Install:

4. Install or connect the following:
   • Caliper over the pads and onto its mounting bracket. Torque both caliper bolts to 27 ft. lbs. (36 Nm).
   • Brake hose to the caliper using new sealing washers. Carefully torque

the banjo bolt to 25 ft. lbs. (34 Nm).
5. Fill the reservoir with fluid and bleed the brakes.
   • Front wheels

Rear

1. Before servicing the vehicle, refer to the precautions section.
2. Remove some fluid from the reservoir with a suction pump.
3. Remove or disconnect the following:

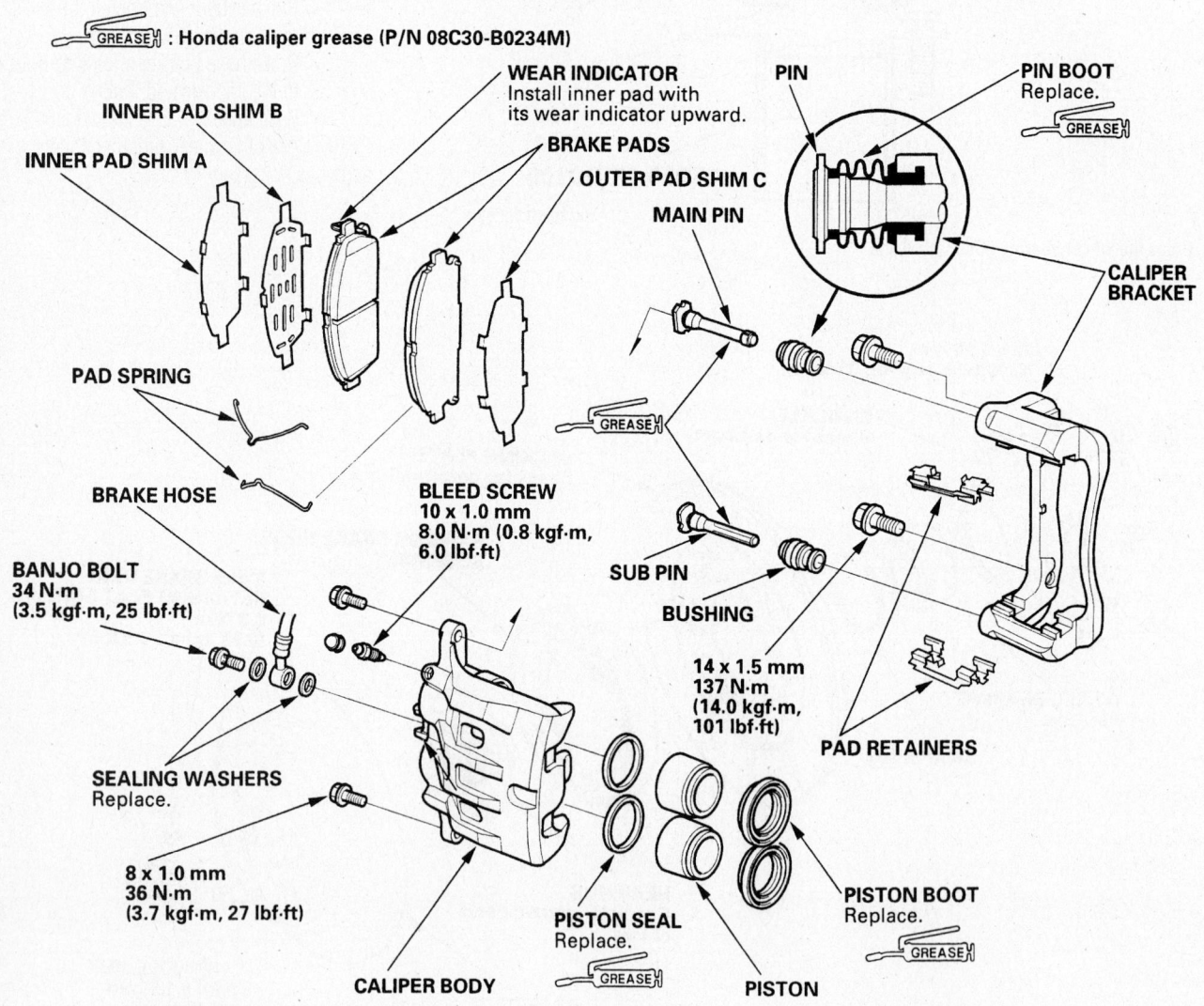

Exploded view of the front caliper components

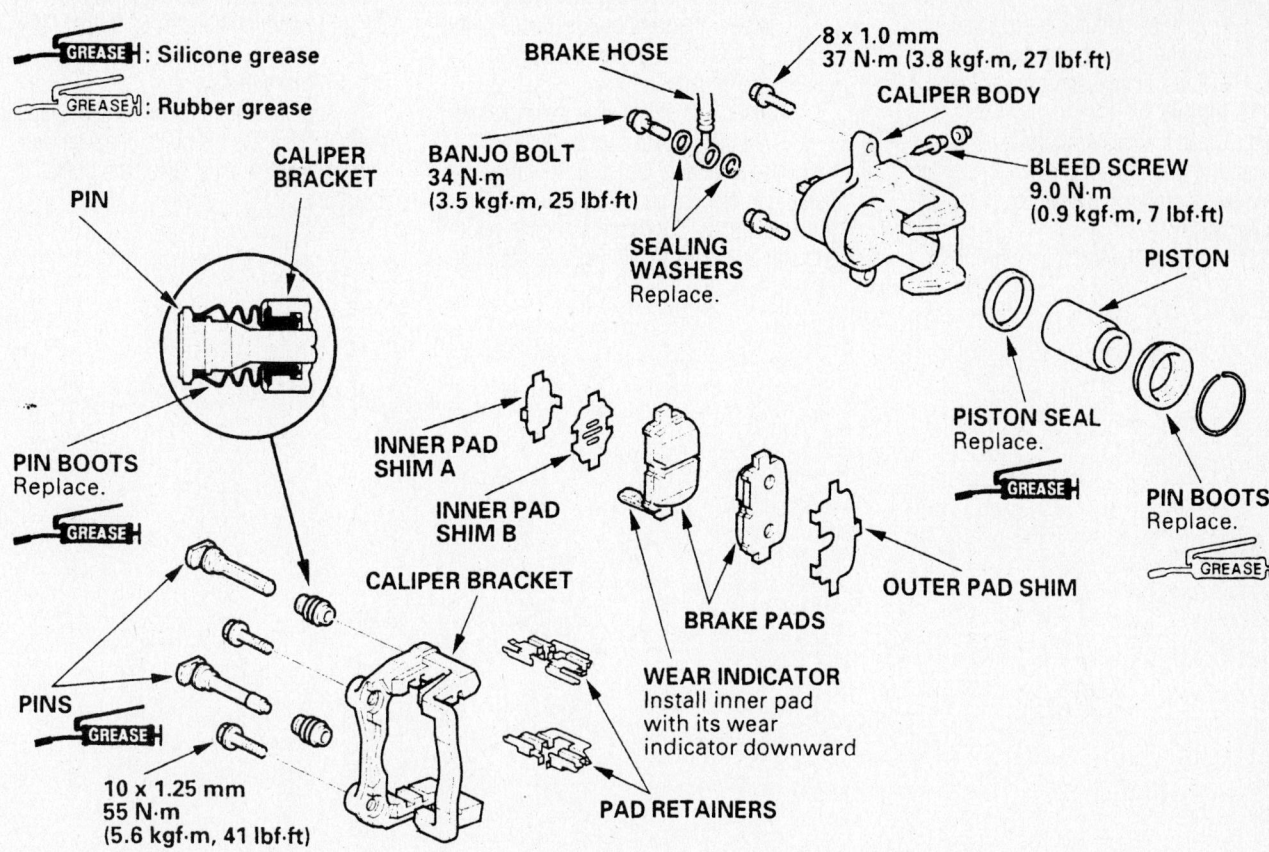

GREASE : Silicone grease

GREASE : Rubber grease

PIN

CALIPER BRACKET

BRAKE HOSE

8 x 1.0 mm
37 N·m (3.8 kgf·m, 27 lbf·ft)

CALIPER BODY

BANJO BOLT
34 N·m
(3.5 kgf·m, 25 lbf·ft)

BLEED SCREW
9.0 N·m
(0.9 kgf·m, 7 lbf·ft)

SEALING WASHERS
Replace.

PISTON

PIN BOOTS
Replace.

GREASE

INNER PAD SHIM A

INNER PAD SHIM B

CALIPER BRACKET

PISTON SEAL
Replace.

GREASE

PIN BOOTS
Replace.

GREASE

PINS

GREASE

BRAKE PADS

OUTER PAD SHIM

10 x 1.25 mm
55 N·m
(5.6 kgf·m, 41 lbf·ft)

WEAR INDICATOR
Install inner pad
with its wear
indicator downward

PAD RETAINERS

93352GZB

Exploded view of the rear caliper components—Pilot

- Rear wheels
- Banjo bolt and disconnect the brake hose from the caliper. Plug the hose to prevent fluid loss and contamination.
- 2 caliper mounting bolts and the caliper from its mounting bracket

To Install:

4. Install or connect the following:
- Caliper over the pads and onto its mounting bracket. Tighten the caliper bolts to 27 ft. lbs. (37 Nm).
- Brake hose with new sealing washers. Tighten the banjo bolt to 25 ft. lbs. (34 Nm).

5. Fill the reservoir with fluid and bleed the brake system. Adjust the parking brake if necessary.
- Rear wheels

Disc Brake Pads

REMOVAL AND INSTALLATION

Front

1. Before servicing the vehicle, refer to the precautions section.

2. Remove or disconnect the following:
- Front wheels

3. Remove a small amount of brake fluid from the reservoir using a suction pump.
- Brake hose clamp from the knuckle by unfastening the retaining bolts
- Lower caliper retaining bolt and pivot the caliper upward, off of the pads
- Pad springs while holding the pads
- Pad shim and pad retainers
- Disc brake pads from the caliper

To install:

4. Clean the caliper thoroughly; remove any rust from the lip of the disc or rotor. Check the brake rotor for grooves or cracks. If any heavy scoring is present, the rotor must be replaced.

5. Install or connect the following:
- Pad retainers. Apply molybdenum brake grease to both surfaces of the shims and the back of the disc brake pads.
- Pads and shims. The pad with the wear indicator goes in the inboard position.
- Pad springs while holding the pads

6. Push in the caliper piston so the caliper will fit over the pads. This is most easily accomplished with a pad spreader or large C-clamp.
- Caliper down into position and tighten the mounting bolt to 27 ft. lbs. (37 Nm)
- Brake hose to the knuckle, if removed
- Wheels

7. Add brake fluid to the master cylinder reservoir and install the cap.

8. Depress the brake pedal several times and make sure that the movement feels normal. The first brake pedal application may result in a very long pedal action due to the pistons being retracted. Always make several brake applications before starting the vehicle. Bleed the system if necessary.

Rear

1. Before servicing the vehicle, refer to the precautions section.

2. Remove a small amount of brake fluid from the reservoir using a suction pump.

3. Remove or disconnect the following:

- Rear wheels
- 2 caliper mounting bolts and the caliper from the bracket
- Pads, shims, and pad retainers

To install:

4. Clean the caliper thoroughly; remove any dirt or dust. Check the brake rotor for grooves or cracks and machine or replace, as necessary.

5. Install or connect the following:

- Pad retainers. Apply molybdenum brake grease to both surfaces of the shims and the back of the disc brake pads.
- Pads and shims. The wear retainer on the inboard pad faces down.

6. Use a suitable tool to push caliper piston into its bore and enable the caliper to fit over the pads. Lubricate the piston boot with silicon grease. Avoid twisting the boot.

- Brake caliper and tighten the mounting bolts to 27 ft. lbs. (37 Nm)
- Rear wheels

7. Add brake fluid to the master cylinder reservoir. Depress the brake pedal several times to seat the pads. Bleed the brakes if necessary.

HONDA

Ridgeline

9

SPECIFICATIONS AND MAINTENANCE CHARTS

ENGINE AND VEHICLE IDENTIFICATION CHART

		Engine Code					Model Year	
Code	Liters (cc)	Cu. In.	Cyl.	Fuel Sys.	Engine Type	Eng. Mfg.	Code ①	Year
J35A9	3.5 (3471)	222	6	SMFI	SOHC	Honda	6	2006

SOHC: Single Overhead Cam

SMFI: Sequential Multi-port Fuel Injection

① 10th position of VIN

09474_RIDGE_C0001

GENERAL ENGINE SPECIFICATIONS

Year	Model	Engine Displacement Liters	Engine ID	Net Horsepower @ rpm	Net Torque @ rpm (ft. lbs.)	Bore x Stroke (in.)	Com-pression Ratio	Oil Pressure @ rpm
2006	Ridgeline	3.5	J35A9	255@5750	252@4500	3.50x3.66	10:01	71@3000 ①

① At idle: 10 psi

09474_RIDGE_C0002

ENGINE TUNE-UP SPECIFICATIONS

Year	Engine Displacement Liters	Engine ID	Spark Plug Gap (in.)	Ignition Timing (deg. BTDC)	Fuel Pump (psi)	Idle Speed (rpm)	Valve Clearance (in.) In.	Valve Clearance (in.) Ex.
2006	3.5	J35A9	0.039-0.043	8-12	57-64	680-780	0.008-0.009	0.011-0.013

NOTE: The Vehicle Emission Control Information label often reflects changes made during production and must be used if they differ from this chart.

BTDC: Before top dead center

09474_RIDGE_C0003

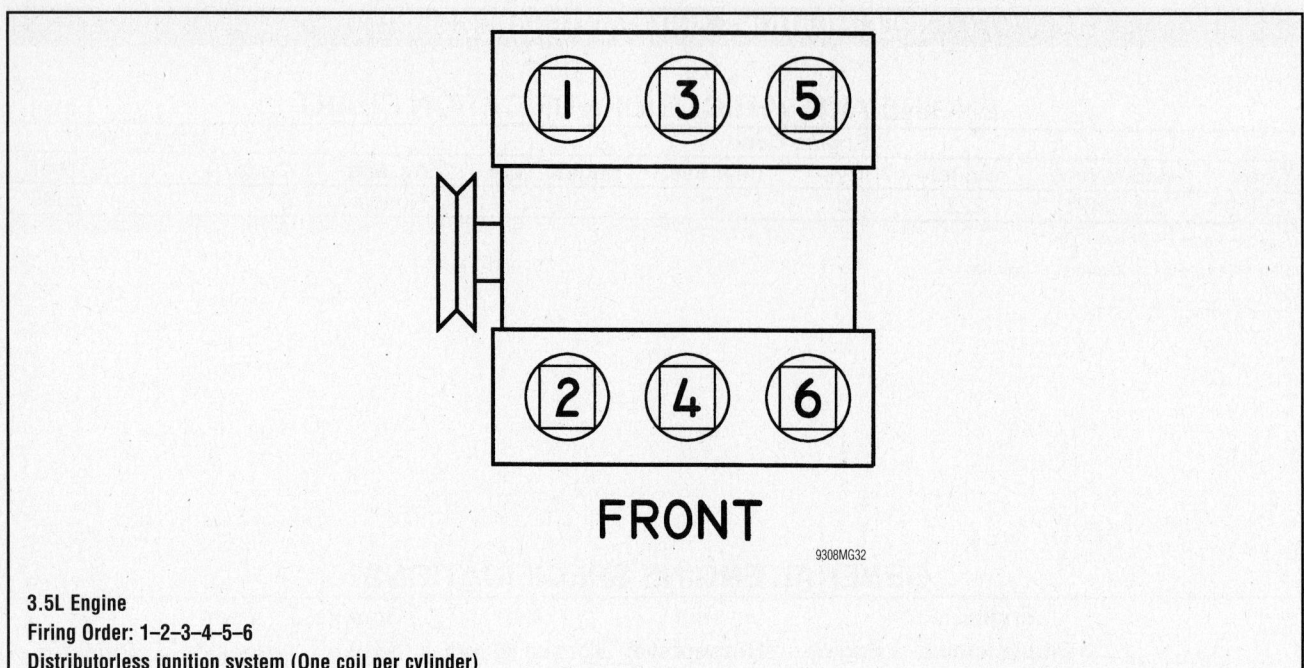

FRONT

9308MG32

3.5L Engine
Firing Order: 1–2–3–4–5–6
Distributorless ignition system (One coil per cylinder)

CAPACITIES

Year	Model	Engine Displacement Liters	Engine ID	Engine Oil with Filter (qts.)	Transmission (pts.) *	Transfer Case (pts.)	Rear Drive Axle (pts.)	Fuel Tank (gal.)	Cooling System (qts.)
2006	Ridgeline	3.5	J35A9	4.5	6.6	0.9	5.58	22.0	①

NOTE: All capacities are approximate. Add fluid gradually and check to be sure a proper fluid level is obtained.

* Fluid change

① Total capacity: 8.56 qts.

At coolant change: 6.26 qts.

09474_RIDGE_C0004

VALVE SPECIFICATIONS

Year	Engine Displacement Liters	Engine ID	Seat Angle (deg.)	Face Angle (deg.)	Spring Test Pressure (lbs. @ in.)	Spring Free Length (in.)	Stem-to-Guide Clearance (in.) Intake	Stem-to-Guide Clearance (in.) Exhaust	Stem Diameter (in.) Intake	Stem Diameter (in.) Exhaust
2006	3.5	J35A9	45	45	NA	①	0.0008-0.0018	0.0022-0.0031	0.2159-0.2163	0.2146-0.2150

NA: Information not available

① Intake: 1.9713 in.

Exhaust: 2.1060 in.

09474_RIDGE_C0005

CRANKSHAFT AND CONNECTING ROD SPECIFICATIONS

All measurements are given in inches

Year	Engine Displacement Liters	Engine ID	Crankshaft				Connecting Rod		
			Main Brg. Journal Dia.	Main Brg. Oil Clearance	Shaft End-play	Thrust on No.	Journal Diameter	Oil Clearance	Side Clearance
2006	3.5	J35A9	2.8337-2.8346	0.0008-0.0017	0.0040-0.0140	3	2.1644-2.1654	0.0008-0.0017	0.0060-0.0140

NA: Information not available

09474_RIDGE_C0006

PISTON AND RING SPECIFICATIONS

All measurements are given in inches

Year	Engine Displacement Liters	Engine ID	Piston Clearance	Ring Gap			Ring Side Clearance		
				Top Compression	Bottom Compression	Oil Control	Top Compression	Bottom Compression	Oil Control
2006	3.5	J35A9	0.0006-0.0016	0.0080-0.0140	0.0160-0.0220	0.0080-0.0280	0.0022-0.0031	0.0012-0.0022	snug

09474_RIDGE_C0007

TORQUE SPECIFICATIONS

All readings in ft. lbs.

Year	Engine Displacement Liters	Engine ID	Cylinder Head Bolts	Main Bearing Bolts	Rod Bearing Bolts	Crankshaft Damper Bolts	Flywheel Bolts	Manifold		Spark Plugs	Oil Pan Drain Plug
								Intake	Exhaust		
2006	3.5	J35A9	①	②	③	④	54	16	23	13	29

① Step 1: 29 ft. lbs.
　Step 2: 51 ft. lbs.
　Step 3: 72 ft. lbs.

② Cap bolts: 54 ft. lbs.
　Side bolts: 36 ft. lbs.

③ Step 1: 14 ft. lbs.
　Step 2: plus 90 degrees

④ Step 1: 47 ft. lbs.
　Step 2: plus 60 degrees

09474_RIDGE_C0008

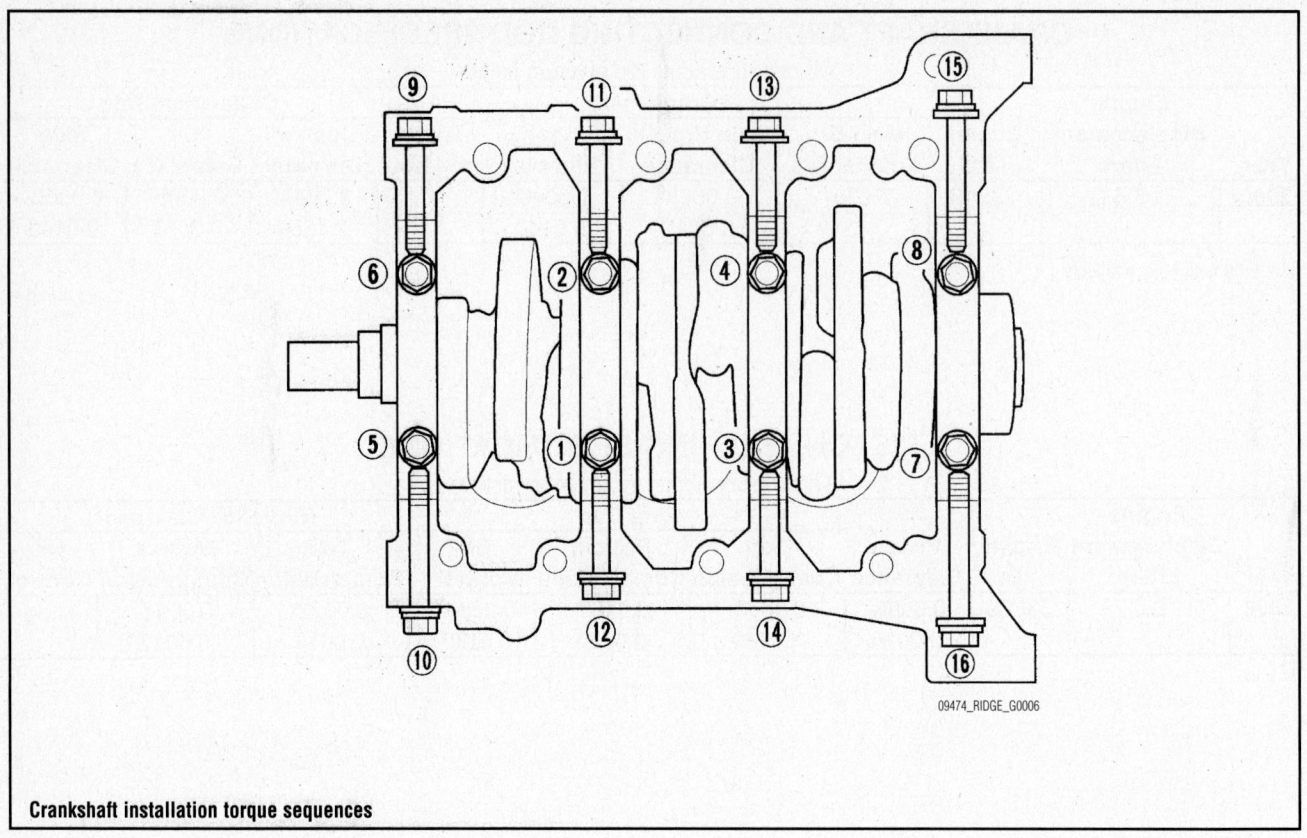

Crankshaft installation torque sequences

09474_RIDGE_G0006

WHEEL ALIGNMENT

Year	Model		Caster Range (+/-Deg.)	Caster Preferred Setting (Deg.)	Camber Range (+/-Deg.)	Camber Preferred Setting (Deg.)	Toe-in (in.)
2006	Ridgeline	F	1.00	+1.88	1.00	-0.50	0 +/- 1/16
		R	—	—	0.75	-0.50	0 +/- 1/16

09474_RIDGE_C0009

TIRE, WHEEL AND BALL JOINT SPECIFICATIONS

Year	Model	OEM Tires		Tire Pressures (psi)		Wheel Size	Ball Joint Inspection	Lugnut Torque (ft. lbs.)
		Standard	Optional	Front	Rear			
2006	Ridgeline	P245/65R17	None	32	32	7.5	NA	94

NA: Information not available

OEM: Original Equipment Manufacturer

PSI: Pounds Per Square Inch

09474_RIDGE_C0010

BRAKE SPECIFICATIONS

All measurements in inches unless noted

Year	Model		Brake Disc			Minimum Lining Thickness	Brake Caliper	
			Original Thickness	Minimum Thickness	Maximum Runout		Bracket Bolts (ft. lbs.)	Mounting Bolts (ft. lbs.)
2006	Ridgeline	F	1.105	1.020	0.0006	0.040	101	53
		R	0.435	0.350	0.0006	0.040	80	16

F: Front

R: Rear

09474_RIDGE_C0011

MAINTENANCE MINDER SCHEDULE
2006 Honda Ridgeline

The Ridgeline displays engine oil life and maintenance service items in the information display to indicate when to perform maintenance service. If the engine oil life is 15% or less, based on the onboard computer's caluculations, you will see SERVICE DUE SOON in the information display every time the ignition key is turned to ON. The maintenance minder indicator will also come on and the maintenance code(s) for other scheduled maintenance items needing service will be displayed below the message.

Symbol	Item	Service
A	Engine oil ①	Change
B	Engine oil and filter	Change
	Tires	Rotate
	Brakes	Inspect
	Parking brake adjustment	Check
	Steering gear and linkage	Inspect
	Suspension components	Inspect
	Driveshaft boots	Inspect
	Brake hoses and lines	Inspect
	Exhaust system	Inspect
	Fuel lines and connections	Inspect
1	Tires	Rotate
2	Engine air filter ②	Replace
	Dust and pollen filter ③	Replace
	Accessory drive belt	Inspect
3	Transmission fluid ④	Replace
	Transfer case fluid ④	Replace
4	Spark plugs	Replace
	Timing belt ⑤	Replace
	Water pump	Inspect
	Valve clearance ⑥	Inspect
5	Engine coolant	Replace
6	VTM-4 rear differential fluid	Replace

① If the message SERVICE DUE NOW does not appear more than 12 months after the display is reset, change every year.
② If driven in dusty conditions, replace every 15,000 miles.
③ If driven in urban areas that have a high concentration of soot from industry and diesel, replace every 15,000 miles
④ If regularly driven in mountainous areas at very low speed or trailer towing, change the fluid every 30,000 miles.
⑤ If driven regularly in temperatures over 110 deg.F or below -20 deg.F, or towing a trailer, replace every 60,000 miles.
⑥ Adjust if necessary.
Additionally, replace the brake fluid every 3 years, and inspect the idle speed every 160,000 miles.
To reset the Engine Oil Life Display:
1. Turn the ignition switch to ON.
2. Press the SELECT button repeatedly until the engine oil life display or the service message is displayed.
3. Press the RESET button for about 10 seconds. You will see a MAINT RESET message.
4. Select the appropriate answer, MAINT RESET >N (NO) or MAINT RESET > y (YES) by pressing the SELECT button repeat >N or >Y is displayed on the outside temperature >N or >Y is displayed on the outside temperature display.
5. Select the MAINT RESET > Y (YES), and press and hold the RESET button again to reset the engine oil life to 100%.

ENGINE REPAIR

➡**Disconnecting the negative battery cable on some vehicles may interfere with the functions of the on board computer system. The computer may undergo a relearning process once the negative battery cable is reconnected.**

Distributor

The Ridgeline is equipped with a Distributorless Ignition System (DIS).

Alternator

REMOVAL & INSTALLATION

1. Before servicing the vehicle, refer to the Precautions Section.
2. Make sure you have the anti-theft codes for the radio and navigation systems.
3. Remove or disconnect the following:
   - Negative battery cable
   - Positive battery cable
   - Intake manifold cover
   - Accessory drive belt
   - Alternator wiring harness connectors
   - Alternator mounting bolts
   - Wiring harness clamp
   - Alternator

To install:
4. Install or connect the following:
   - Alternator

- Wiring harness clamp. Tighten the bolt to 105 inch lbs. (12 Nm).
- Alternator mounting bolts. Tighten the 10mm bolt to 33 ft. lbs. (44 Nm) and the 8mm bolt to 16 ft. lbs. (22 Nm).
- Alternator wiring harness connectors. Tighten the battery terminal nut to 105 inch lbs. (12 Nm).
- Accessory drive belt
- Intake manifold cover
- Positive cable
- Negative battery cable
5. Enter the theft codes

Ignition Timing

ADJUSTMENT

The engine is equipped with a Distributorless Ignition System (DIS). The ignition timing is controlled by the Powertrain Control module (PCM). No adjustment is necessary.

Engine Assembly

REMOVAL & INSTALLATION

1. Before servicing the vehicle, refer to the Precautions Section.
2. Relieve the fuel system pressure.

3. Place the vehicle on a hoist.
4. Drain the power steering fluid.
5. Remove the air intake duct.
6. Remove the bulkhead cover.
7. Make sure you have the anti-theft codes for the radio and navigation systems.
8. Disconnect the negative battery cable then the positive cable.
9. Remove the battery.
10. Remove the intake manifold cover.
11. Remove the battery box.
12. Disconnect the starter wiring.
13. Disconnect the fuel line.
14. Remove the transfer case breather box.
15. Disconnect the brake booster hose.
16. Disconnect the EVAP canister hose.
17. Disconnect the battery ground cable and wait at least 3 minutes.
18. Place the wheels in the straight-ahead position.
19. Remove the steering wheel air bag access cover and disconnect the air bag connector from the cable reel.
20. Remove the 2 Torx bolts, one each side of the steering wheel hub.
21. Disconnect the horn switch connector.
22. Remove the air bag.
23. Disconnect the cable reel harness.
24. Remove the steering wheel nut.
25. Using a puller, remove the steering wheel.
26. Remove the steering coupler cover from the floor.
27. Matchmark and disconnect the joint coupler.
28. Disconnect the underhood fuse/relay box.
29. Remove the coolant reservoir.
30. Remove the PCM cover.
31. Disconnect the PCM.
32. Remove the drive belt.
33. Disconnect the power steering hoses.
34. Remove the radiator cap.
35. Raise the hoist full height.
36. Remove the front wheels.
37. Remove the splash shield.
38. Drain the coolant.
39. Drain the transaxle.
40. Drain the engine oil.
41. Remove the front subframe stiffener.
42. Remove the front exhaust pipe.
43. Disconnect the stabilizer bar links.
44. Disconnect the tie rod ends from the knuckles.
45. Disconnect the knuckles from the lower arms.

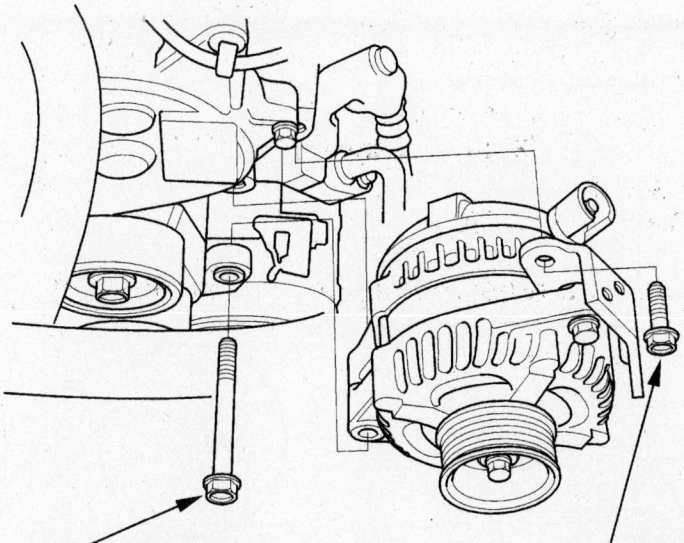

**10 x 1.25 mm
44 N·m (4.5 kgf·m,
33 lbf·ft)**

**8 x 1.25 mm
22 N·m (2.2 kgf·m,
16 lbf·ft)**

09474_RIDGE_G0001

Alternator mounting

46. Remove the halfshafts.
47. Remove the transfer case.
48. Disconnect the fluid lines from the steering gear.
49. Disconnect the power steering pressure switch.
50. Remove the transaxle lower front mount bolts.
51. Remove the rear mount bolts.
52. Remove the A/C compressor and set it aside without disconnecting the hoses.
53. Lower the vehicle.
54. Remove the heater hoses.
55. Remove the radiator hoses.
56. Disconnect the transaxle cooler lines.
57. Remove the radiator.
58. Remove the connector bracket from the front cylinder head and use that bolt hole to attach the balancer bar front arm.
59. Remove the harness clamp bracket from the rear head and use that bolt hole to attach the balancer bar rear arm.
60. Remove the caps from the front strut mounting nuts. Position the engine hanger adapters (VSB02C000024) with the FRONT mark facing forward, over the flange nuts.
61. Install the balancer bar (VSB02C000019):
    a. Attach the front arm (A) to the front head with spacer (B) and a 10x1.25mm bolt (C)
    b. Attach the rear arm (D) to the rear head with an 8x1.25mm bolt (E).
62. Install the support hanger (AAR-T-12566):
    a. Attach the hook to the slotted hole in the balancer bar.
    b. Tighten the wing nut (F) by hand to lift and support the engine/transaxle assembly.
63. Remove and discard the front engine mount nut.
64. Matchmark the body and subframe as shown.
65. Install the subframe adapter (A) as shown.
66. Remove the six 12x1.25mm bolts (A) securing the subframe stiffeners (B), the 4 subframe mounting bolts (C) and the stiffeners. Then, lower the subframe (D).
67. Lower the vehicle and attach an engine crane as shown. Lift the engine until the crane has the weight and remove the support hanger.
68. Remove the upper side engine mount bolts.
69. Slowly lower the engine about 6 inches, make sure everything is clear, then lower it all the way.
70. Remove the crane.
71. Raise the vehicle and remove the engine assembly.

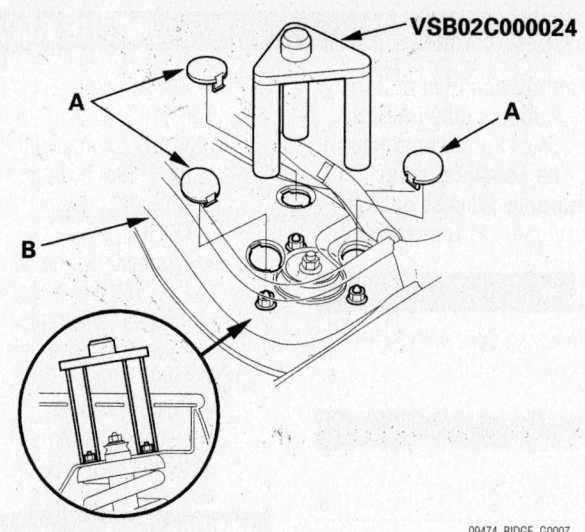

09474_RIDGE_G0007

Engine hanger adapters installed

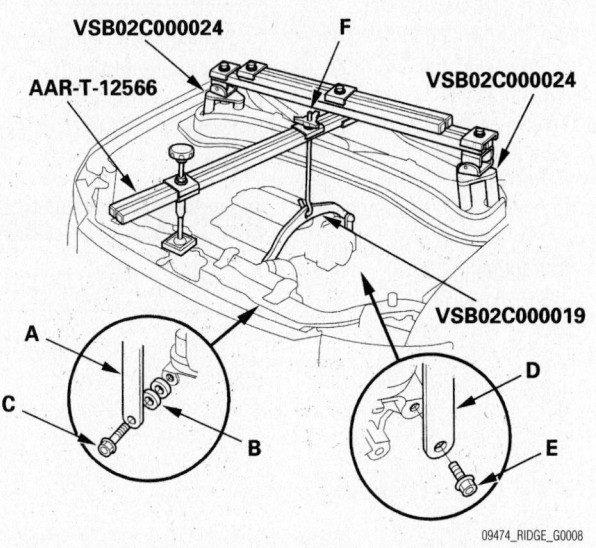

09474_RIDGE_G0008

Balancer bar installation and support hanger

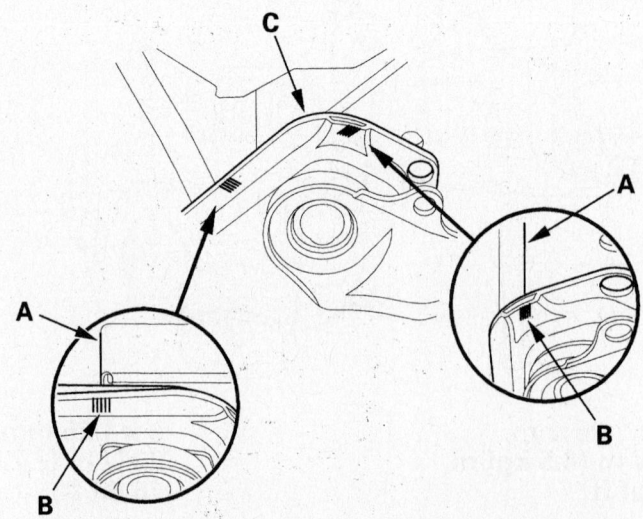

09474_RIDGE_G0009

Matchmark the body and subframe as shown

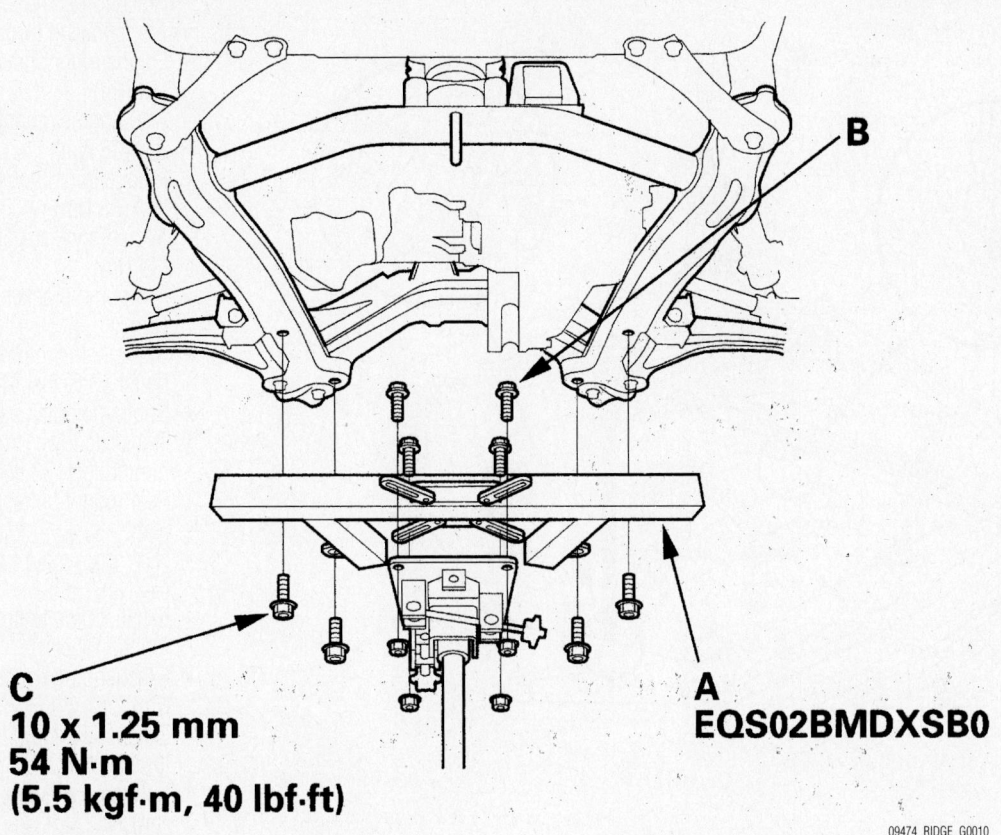

B

C
10 x 1.25 mm
54 N·m
(5.5 kgf·m, 40 lbf·ft)

A
EQS02BMDXSB0

09474_RIDGE_G0010

Install the subframe adapter as shown

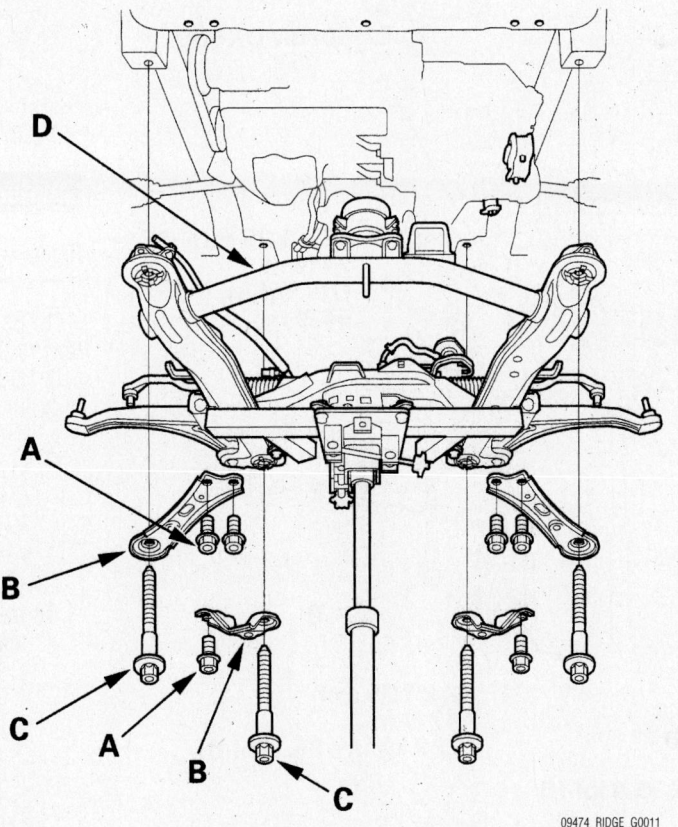

D

A

B

C

A

B

C

09474_RIDGE_G0011

Subframe removal

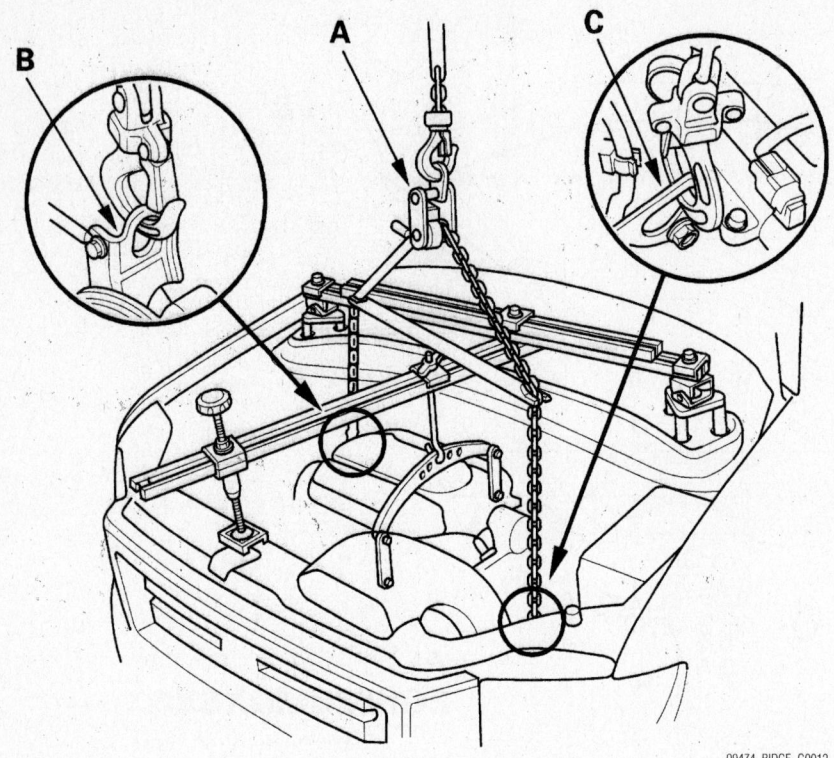

Engine crane attachment

09474_RIDGE_G0012

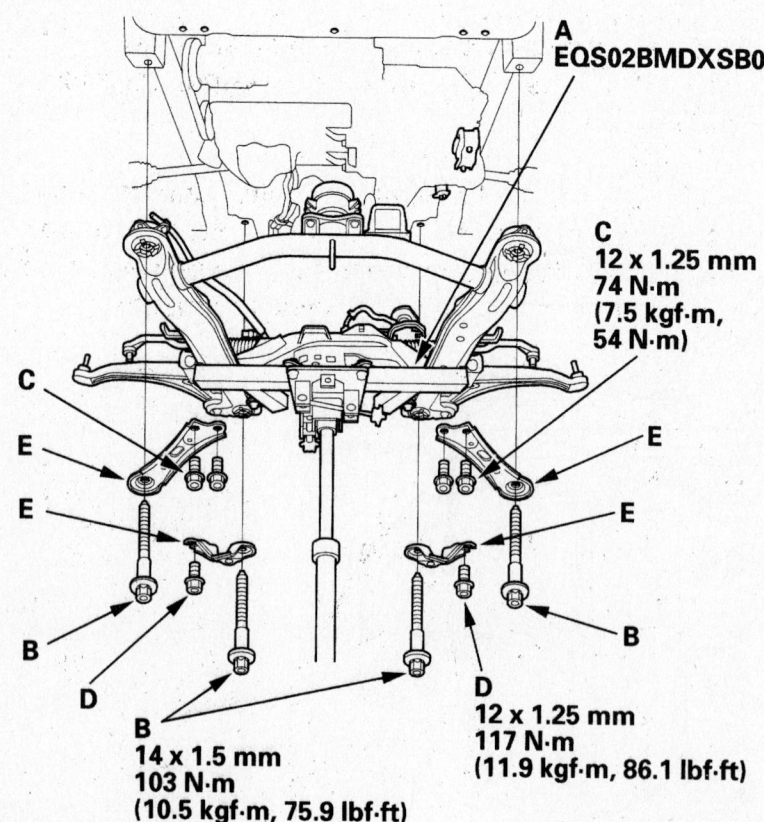

Subframe installation

09474_RIDGE_G0013

72. Installation is the reverse of removal. Observe the following torques:

- Engine mount bracket: 33 ft. lbs. (44 Nm)
- Rear engine mount bracket: 65 ft. lbs. (88 Nm)
- Front engine mount bracket: 28 ft. lbs. (38 Nm)
- A/C compressor bracket: 33 ft. lbs. (44 Nm)
- P/S pump bracket: 16 ft. lbs. (22 Nm)
- Upper half engine mount bracket bolts: 33 ft. lbs. (44 Nm)
- Subframe bolts: as shown
- Transaxle lower front and rear lower mount nuts: 33 ft. lbs. (44 Nm)
- Rear mount bolts: 28 ft. lbs. (38 Nm)
- NEW front mount nut: 40 ft. lbs. (54 Nm)
- Exhaust pipe to crossover pipe: 40 ft. lbs. (54 Nm)
- Exhaust pipe hanger: 16 ft. lbs. (22 Nm)
- Exhaust pipe to converter: 25 ft. lbs. (33 Nm)
- Subframe stiffener: 40 ft. lbs. (54 Nm)
- Connector bracket to front head: 33 ft. lbs. (44 Nm)
- Harness clamp to rear head: 16 ft. lbs. (22 Nm)
- Steering shaft coupler: 16 ft. lbs. (22 Nm)
- Steering wheel nut to 36 ft. lbs. (49 Nm)

Water Pump

REMOVAL & INSTALLATION

1. Before servicing the vehicle, refer to the Precautions Section.
2. Drain the cooling system.
3. Remove or disconnect the following:
   - Negative battery cable
   - Accessory drive belts
   - Front cover
   - Timing belt
   - Timing belt tensioner
   - Water pump

To install:

4. Install or connect the following:
   - Water pump. Use a new O-ring seal and tighten the bolts to 105 inch lbs. (12 Nm).
   - Timing belt tensioner
   - Timing belt
   - Front cover
   - Accessory drive belts

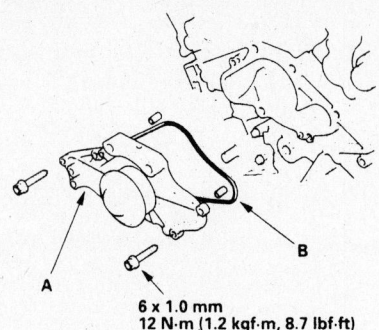

Exploded view of the water pump mounting. A is the pump; B is the seal

6 x 1.0 mm
12 N·m (1.2 kgf·m, 8.7 lbf·ft)

93352G01

- Negative battery cable
5. Fill the cooling system.
6. Start the engine and check for leaks.

Heater Core

REMOVAL & INSTALLATION

1. Make sure you have the anti-theft codes for the radio and navigation systems.
2. Disconnect the battery ground cable and wait at least 3 minutes.
3. Properly discharge and recover the refrigerant.

4. Disconnect the suction and receiver lines from the evaporator. Cap the openings.
5. Disconnect the heater coolant valve cable at the valve and fully open the valve.
6. Drain the coolant.
7. Disconnect the heater hoses at the core tubes.
8. Remove the heater unit mounting nut.
9. Remove the:
   - Driver's side dashboard lower cover
   - Dashboard center lower cover
   - Center console
   - Glove box
   - Driver's side dashboard side cover
   - Both kick panels
   - Both A-pillar trim panels
10. Place the wheels in the straight-ahead position.
11. Remove the steering wheel air bag access cover and disconnect the air bag connector from the cable reel.
12. Remove the 2 Torx bolts, one each side of the steering wheel hub.
13. Disconnect the horn switch connector.
14. Remove the air bag.
15. Disconnect the cable reel harness.
16. Remove the steering wheel nut.
17. Using a puller, remove the steering wheel.

18. Remove the steering coupler cover from the floor.
19. Set the column shaft to the neutral position by raising the column to the uppermost position, then lower it 8mm (0.31 in.). Tighten the tilt lever.
20. Remove the column covers.
21. Move the shift lever to N and remove the shift cable from the column.
22. Remove the combination switch.
23. Disconnect the ignition switch.
24. Disconnect the immobilizer receiver unit, the park pin switch and the shift lock solenoid.
25. Matchmark and disconnect the steering joint from the column shaft.
26. Remove the attaching bolts and nuts and remove the steering column.
27. Disconnect the shift cable.
28. Remove the parking brake release lever bolt.
29. Remove the passenger's side dashboard side cover.
30. Remove the right speaker grille.
31. Remove the console front bracket.
32. Remove the rear vent duct.
33. Remove the rear heater duct.
34. From under the dash on the driver's side, disconnect:
   - Cabin wiring harness

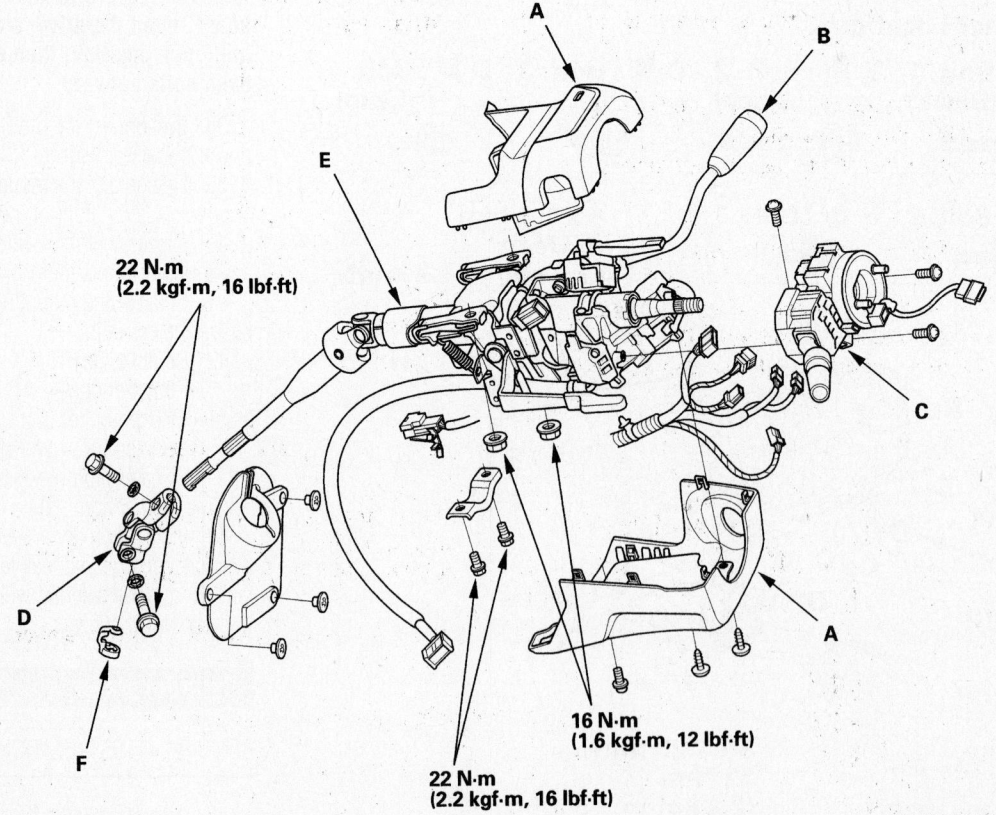

22 N·m
(2.2 kgf·m, 16 lbf·ft)

16 N·m
(1.6 kgf·m, 12 lbf·ft)

22 N·m
(2.2 kgf·m, 16 lbf·ft)

09474_RIDGE_G0017

Steering column exploded view

- Interior wiring harness
- Left side wiring harness
- Parking brake switch
- Brake switch
- Dashboard wiring harness
- Under-dash fuse/relay box

➡ **Lift the large harness connector locks before trying to remove the connectors from the fuse/relay box.**

35. From under the dash center, disconnect:
- SRS control unit
- GPS antenna
- Floor wiring harness, and remove the ground bolt

36. From under the dash on the passenger's side, disconnect:
- Door wiring harness
- Radio antenna
- Cabin wiring harness
- Right side harness
- Front glass defogger

37. Open the driver's door. See the illustration and completely loosen upper bolts A & B. Then, remove bolts C, D, E and F. Lift up on the dashboard (G) to release it from the guide pins (H&I).

➡ **Don't open the door fully with the driver's side upper dash bolts partially**

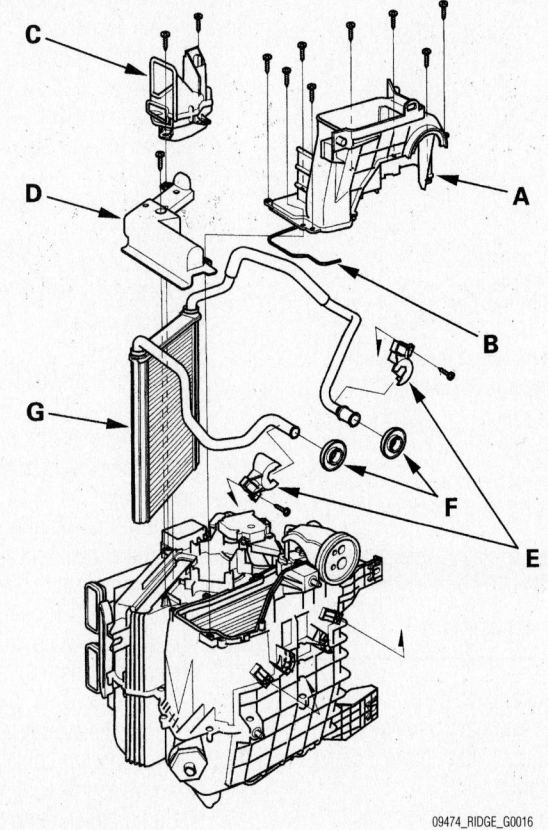

09474_RIDGE_G0016

Heater case disassembly

removed. **Before removing the dashboard, open the driver's door to the half open position, then pull the upper dash bolts outward.**

38. Disconnect the blower motor and power transistor wiring.

39. Disconnect the mode control motor and recirculation control motor.

40. Disconnect the evaporator sensor and air control mix motor.

41. Remove the mounting nuts and the blower/heater unit.

42. Remove the:
- Joint duct (A)
- Heater outlet (C)
- Heater core cover (D)
- Heater pip brackets (E)
- Heater core (G)

43. Installation is the reverse of removal. Evacuate, charge and leak test the A/C system. Torque the steering wheel nut to 36 ft. lbs. (49 Nm).

Cylinder Head

REMOVAL & INSTALLATION

1. Before servicing the vehicle, refer to the Precautions Section.

Fastener Locations

A ▶ : Bolt, 1 (Black)   B ▶ : Bolt, 2 (Black)   C ▶ : Bolt, 3   D ▶ : Bolt, 4 (Gold)

E ▶ : Bolt, 2   F ▶ : Bolt, 1   6 x 1.0 mm 9.8 N·m (1.0 kgf·m, 7.2 lbf·ft)

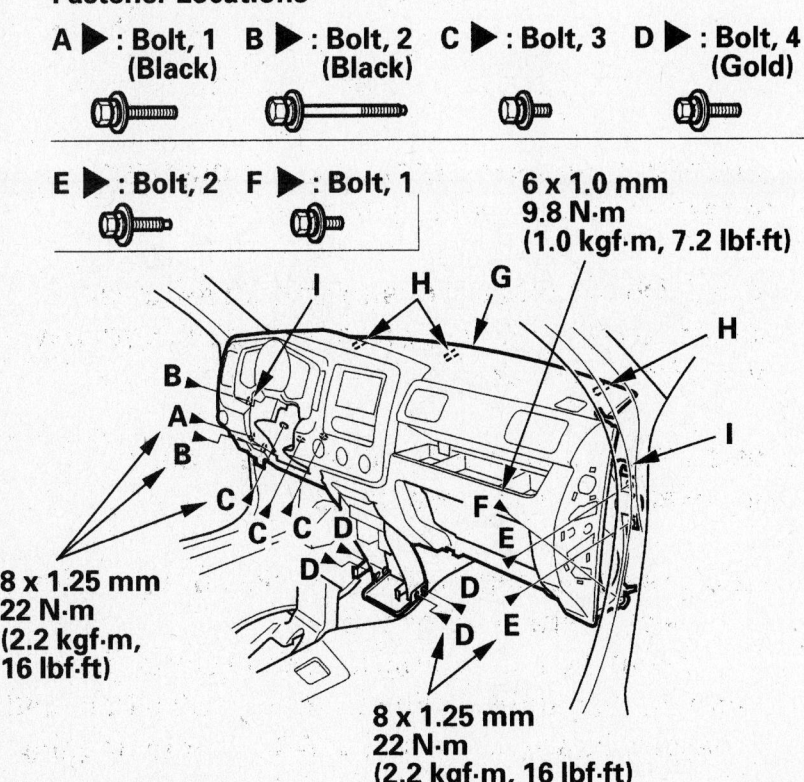

8 x 1.25 mm 22 N·m (2.2 kgf·m, 16 lbf·ft)

8 x 1.25 mm 22 N·m (2.2 kgf·m, 16 lbf·ft)

09474_RIDGE_G0015

Dashboard fastener locations

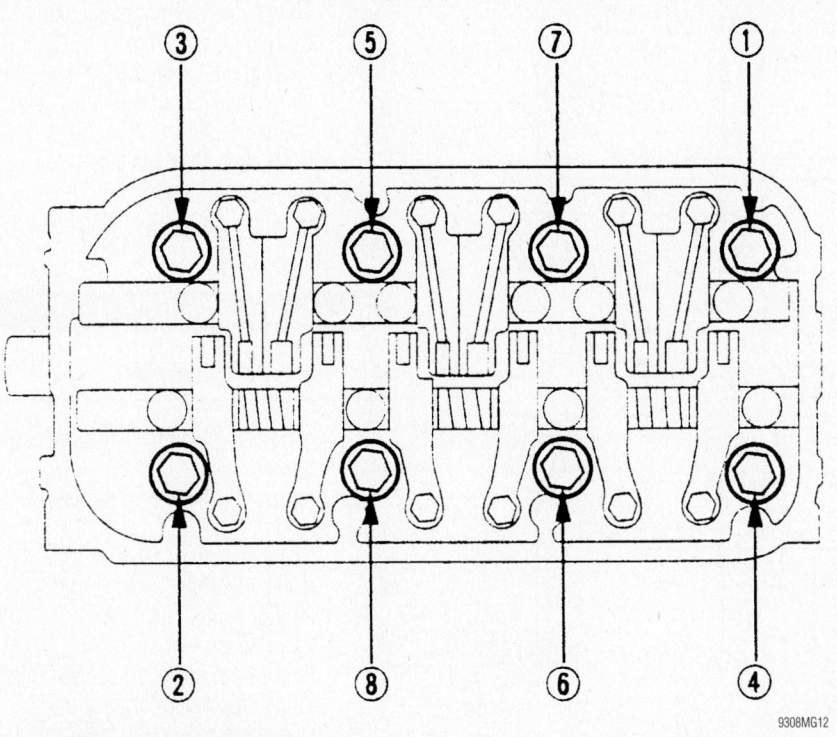

9308MG12

Cylinder head bolt loosening sequence

- Ignition coils
- Timing belt
- Fuel injector connectors
- Engine Coolant Temperature (ECT) sensor connector
- Radiator fan switch connectors
- Crankshaft Position (CKP) sensor connector
- Camshaft Position (CMP) sensor connector
- Exhaust Gas Recirculation (EGR) connector
- Valve Lift Electronic Control (VTEC) solenoid valve connector and oil pressure switch connections
- Oil pressure switch connector
- Vacuum hoses from the intake air bypass control valve
- Fuel rails
- Heater hose
- Radiator hoses
- Ground cable
- Exhaust manifolds
- Water passage
- Camshaft pulleys and back covers
- Valve covers

2. Drain the cooling system.
3. Relieve the fuel system pressure.
4. Remove or disconnect the following:
   - Negative battery cable
   - Alternator belt
   - Intake manifold cover
   - Ignition coil covers
   - Power steering belt and pump
   - Power steering hose clamp
   - Alternator
   - Fuel feed and return lines
   - EVAP canister hose
   - Intake manifold

FRONT:

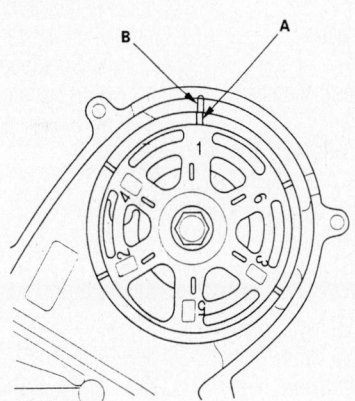

REAR:

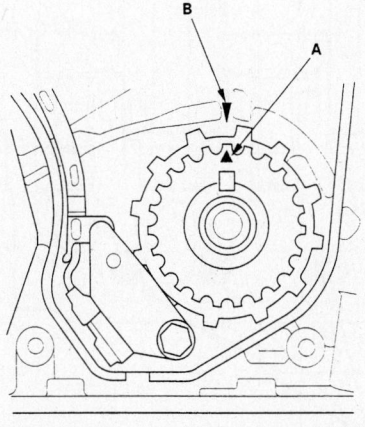

9302MG74

Crankshaft timing belt sprocket TDC marks. Align sprocket mark (A) with pointer (B)

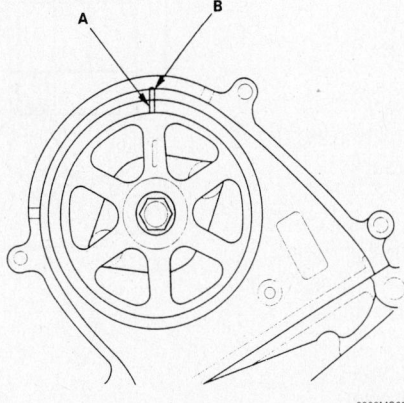

9302MG85

Camshaft TDC marks. Align sprocket mark (A) with the back cover pointer (B)

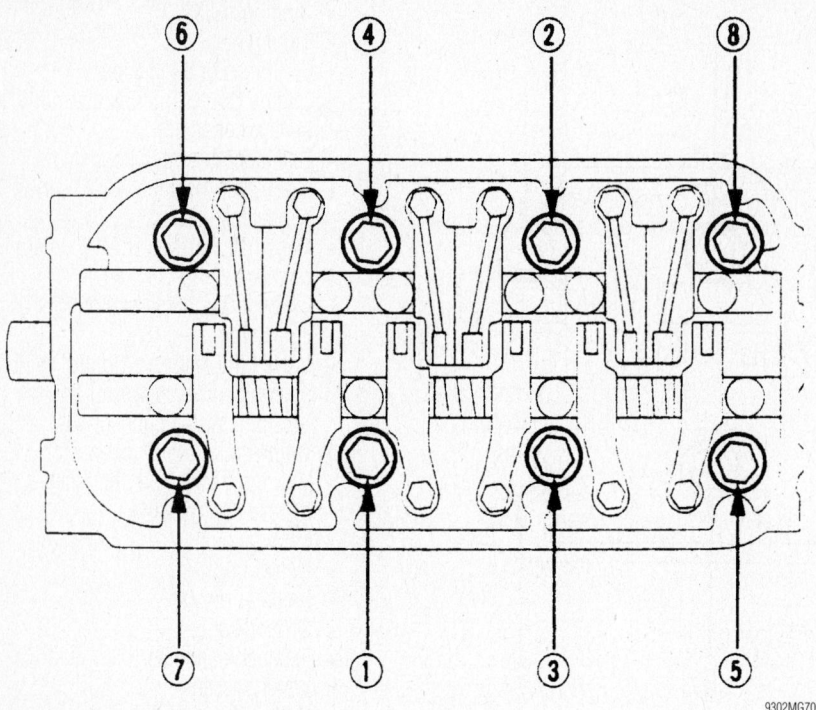

Cylinder head bolt tightening sequence

5. Loosen the cylinder head bolts in sequence and ⅓ turns until all bolts are loose.

6. Remove the cylinder head.

To install:

7. Align the crankshaft and camshaft sprocket TDC marks as shown.

8. Install the cylinder heads with new gaskets.

9. Apply clean engine oil to the cylinder head bolt threads and flanges.

10. Tighten the cylinder head bolts in sequence as follows:

   a. Step 1: 29 ft. lbs. (39 Nm).

   b. Step 2: 51 ft. lbs. (69 Nm).

   c. Step 3: 72 ft. lbs. (98 Nm).

11. Install or connect the following:

- Timing belt and adjust the valve clearance
- Valve covers
- Exhaust manifolds
- Water passage
- Fuel rails
- Vacuum hoses to the intake air bypass control valve
- Radiator hoses
- Heater hose
- Oil pressure switch connector
- VTEC solenoid valve connector and oil pressure switch connections
- EGR connector
- CMP sensor connector
- CKP sensor connector
- Radiator fan switch connectors
- ECT sensor connector

- Fuel injector connectors
- Ignition coils
- Intake manifold
- EVAP canister hose
- Fuel feed and return lines
- Alternator
- Power steering hose clamp

- Power steering pump and belt
- Ground cable
- Ignition coil covers
- Intake manifold cover
- Alternator belt
- Negative battery cable

12. Fill the cooling system.

13. Start the engine and check for leaks.

Rocker Arms/Shafts

REMOVAL & INSTALLATION

1. Before servicing the vehicle, refer to the Precautions Section.

2. Remove or disconnect the following:

- Negative battery cable
- Intake manifold
- Ignition coils
- Valve cover
- Rocker arm adjusting screws. refer to the illustration for location.

3. Remove the rocker arm assembly as follows:

   a. Unscrew the rocker shaft bolts 2 turns at a time in a criss-cross pattern to avoid damaging the valves or rocker assembly.

   b. Do not remove the rocker shaft bolts. These bolts keep the springs and rocker arms on the shafts.

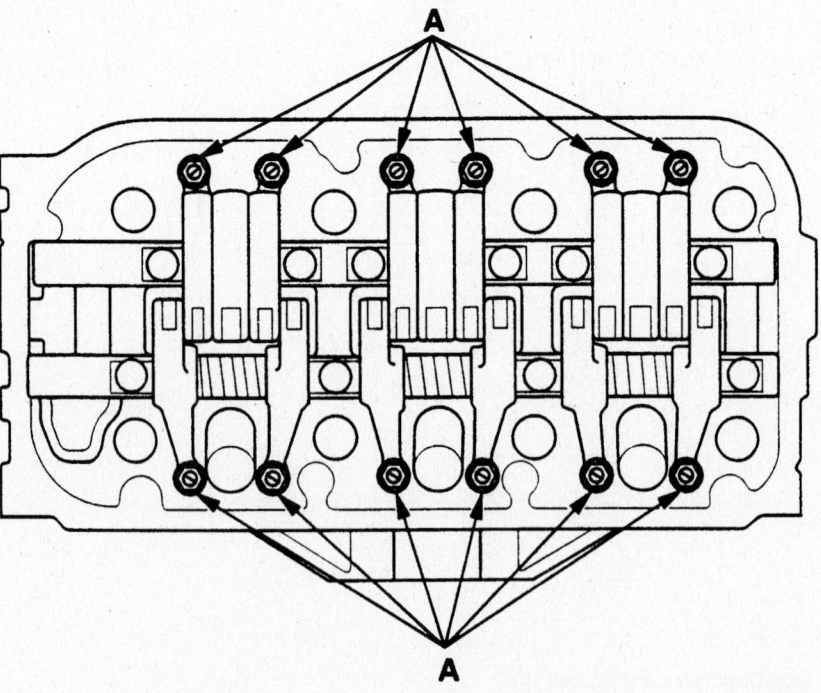

Rocker arm shaft adjusting screw locations

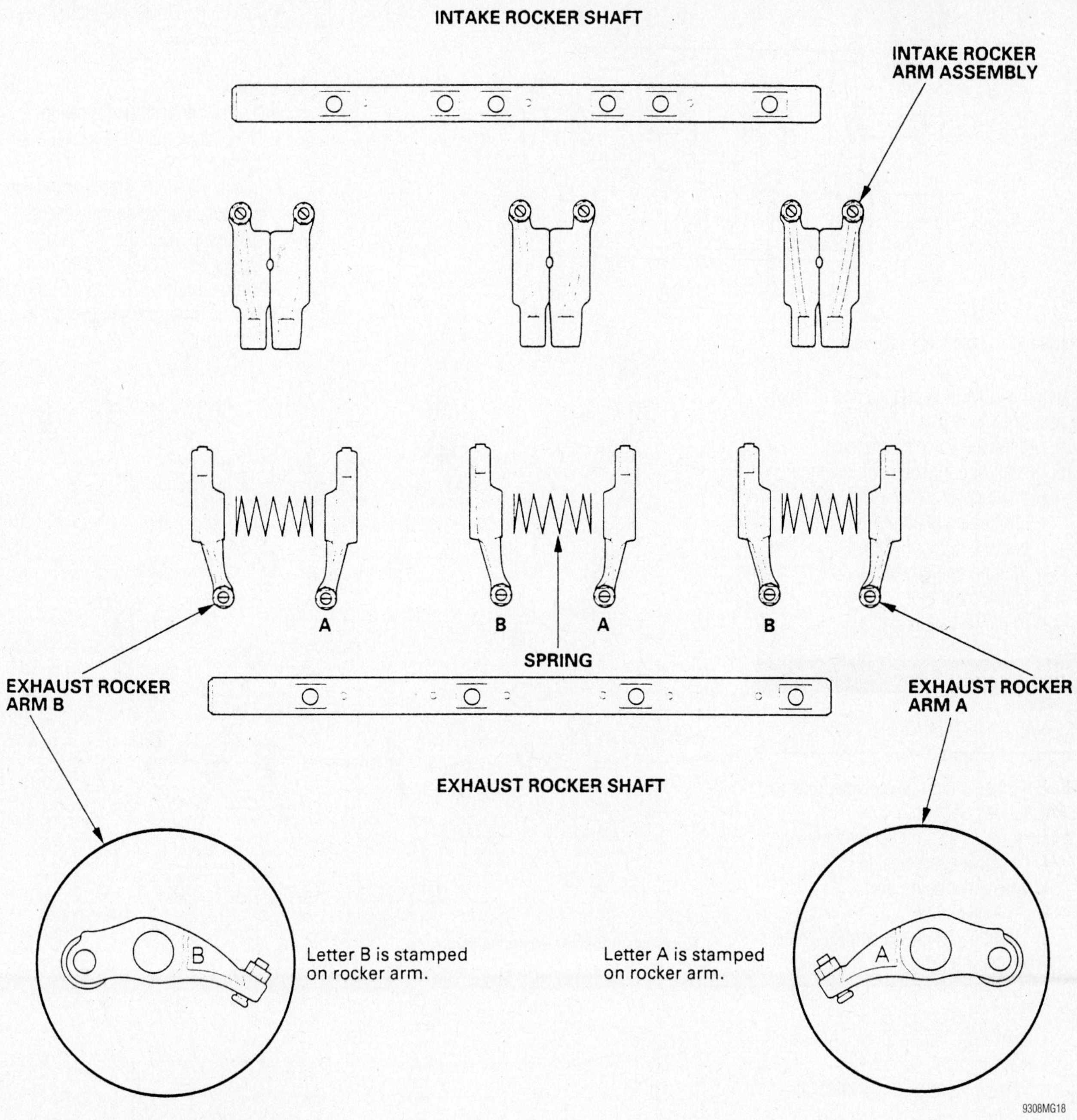

INTAKE ROCKER SHAFT

INTAKE ROCKER ARM ASSEMBLY

A   B   A   B

SPRING

EXHAUST ROCKER ARM B

EXHAUST ROCKER ARM A

EXHAUST ROCKER SHAFT

Letter B is stamped on rocker arm.

Letter A is stamped on rocker arm.

9308MG18

Exploded view of the rocker arms and shafts

4. Loosen the valve adjuster lock-nuts and screws so that all valves are closed.

5. Remove the rocker arms and shafts from the vehicle as an assembly.

➡**Keep all valvetrain components in order for assembly.**

6. Remove the rocker arms and springs from the rocker arm shafts.

To install:

7. Assemble the rocker arms and springs to the rocker arm shafts in their original positions.

8. Install the rocker arm assemblies.

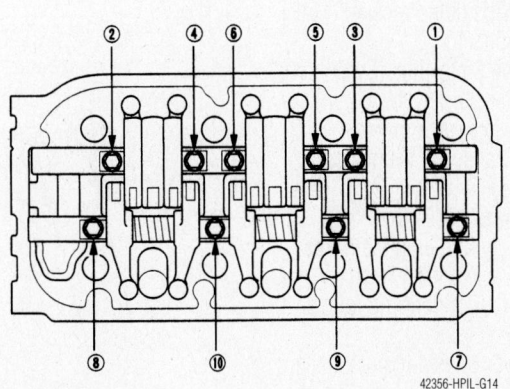

42356-HPIL-G14

Rocker arm shaft loosening sequence

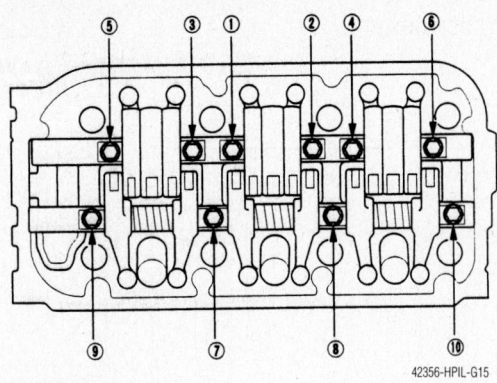

Rocker shaft tightening sequence

Tighten the bolts in sequence and in multiple passes to 17 ft. lbs. (24 Nm).
9. Adjust the valve clearance.
10. Install or connect the following:
   • Valve covers
   • Ignition coils and torque the retainers to 9 ft. lbs. (12 Nm)
   • Ignition coil covers
   • Intake manifold
   • Negative battery cable

Intake Manifold

REMOVAL & INSTALLATION

1. Before servicing the vehicle, refer to the Precautions Section.
2. Remove or disconnect the following:
   • Negative battery cable
   • Intake manifold cover
   • Air intake tube
   • Throttle and cruise control cables
3. Remove the following electrical connections and clamps from the manifold:
   • Intake Air Temperature (IAT) sensor connector
   • Idle Air Control (IAC) valve connector
   • Throttle Position (TP) sensor connector
   • Manifold Absolute Pressure (MAP) sensor connector
   • Evaporative Emissions (EVAP) control canister purge valve connector
   • Brake booster vacuum line
   • Positive Crankcase Ventilation (PCV) hose
   • Water bypass hoses from the throttle body and plug the hoses
   • EVAP control canister hose
4. Remove or disconnect the following:
   • Upper cover bolts and nuts in the sequence illustrated using two or three passes

   • Intake manifold bolts using two or three passes
   • Intake manifold and spacer
To install:
5. Install or connect the following:
   • New intake manifold gasket and spacer
   • Intake manifold. Tighten the fasteners in sequence and in several passes to 16 ft. lbs. (22 Nm).
   • Upper cover bolts and nuts in the sequence illustrated using two or three passes to 9 ft. lbs. (12 Nm)

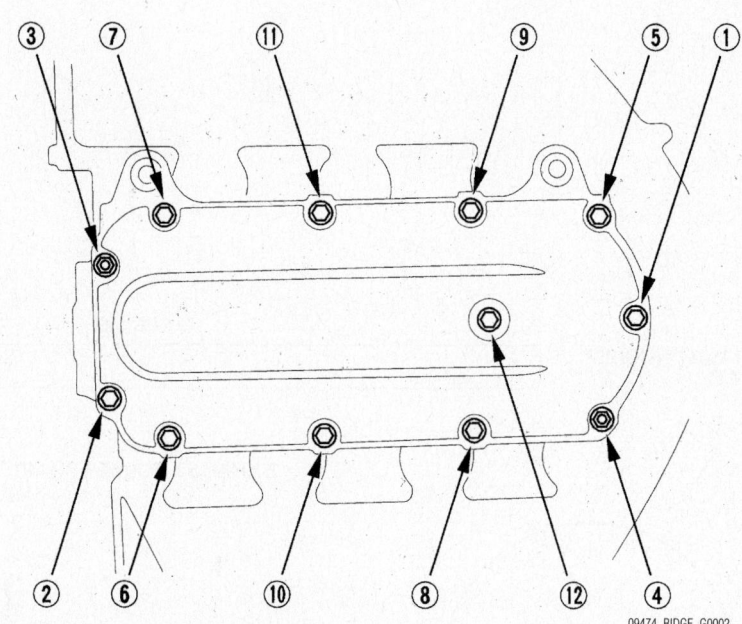

Upper cover fastener removal sequence

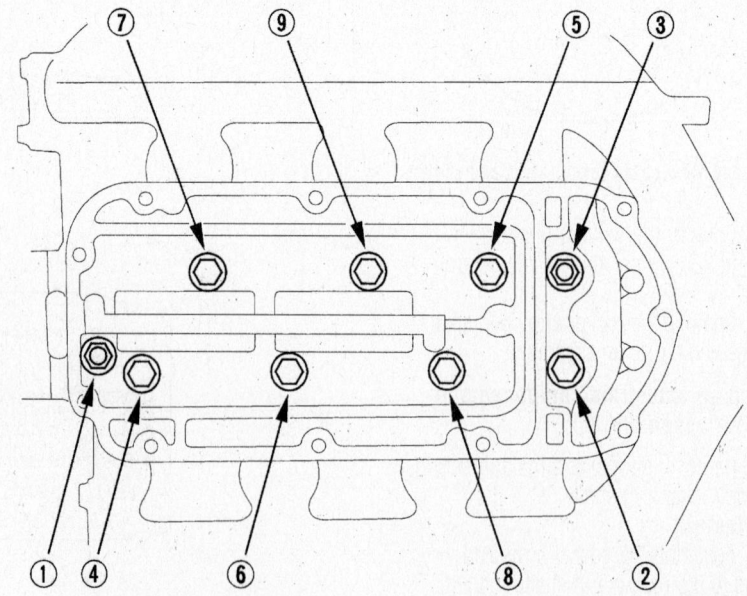

Intake manifold fastener removal sequence

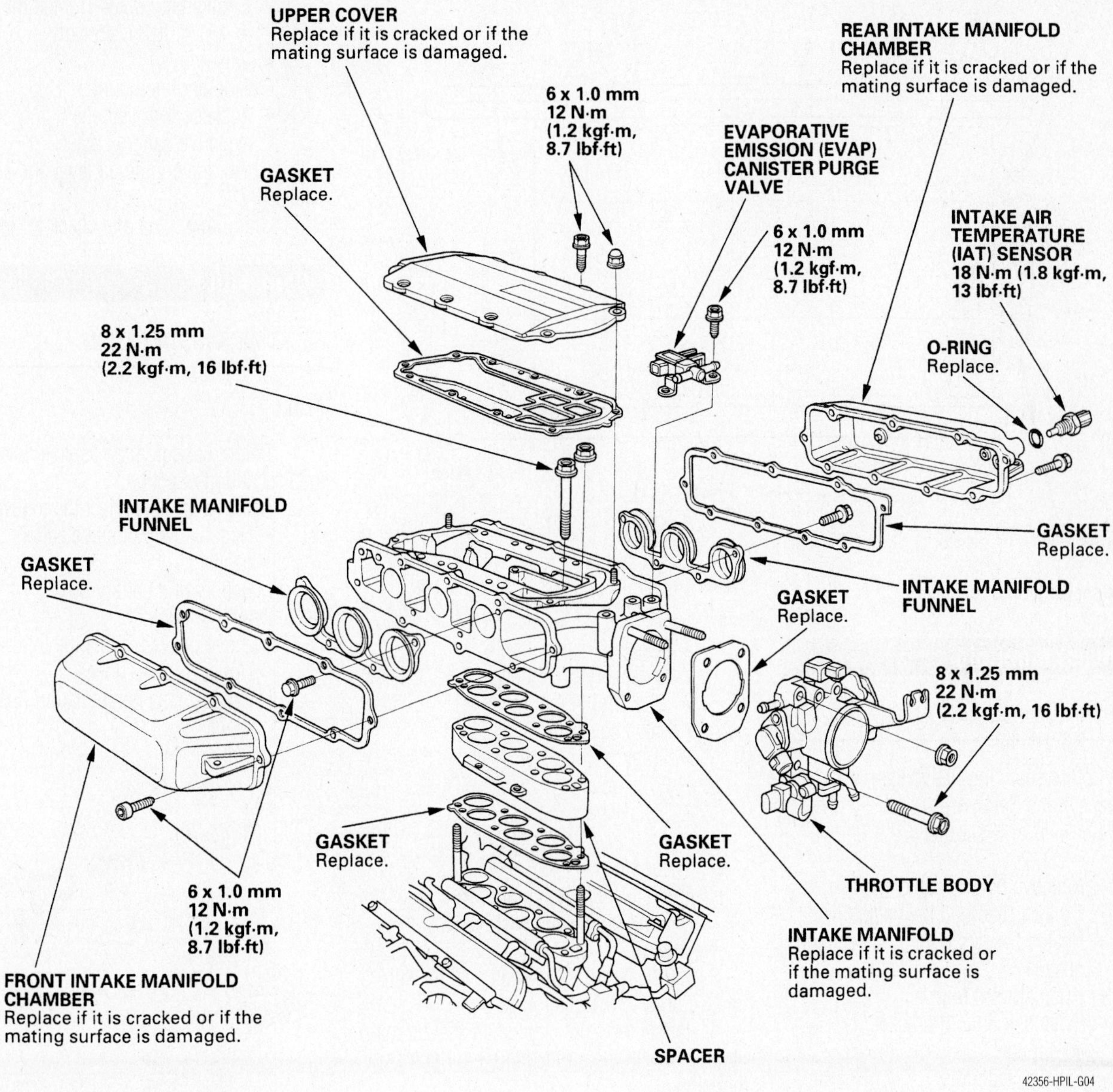

UPPER COVER
Replace if it is cracked or if the mating surface is damaged.

REAR INTAKE MANIFOLD CHAMBER
Replace if it is cracked or if the mating surface is damaged.

6 x 1.0 mm
12 N·m
(1.2 kgf·m, 8.7 lbf·ft)

EVAPORATIVE EMISSION (EVAP) CANISTER PURGE VALVE

GASKET
Replace.

6 x 1.0 mm
12 N·m
(1.2 kgf·m, 8.7 lbf·ft)

INTAKE AIR TEMPERATURE (IAT) SENSOR
18 N·m (1.8 kgf·m, 13 lbf·ft)

O-RING
Replace.

8 x 1.25 mm
22 N·m
(2.2 kgf·m, 16 lbf·ft)

INTAKE MANIFOLD FUNNEL

GASKET
Replace.

GASKET
Replace.

INTAKE MANIFOLD FUNNEL

GASKET
Replace.

8 x 1.25 mm
22 N·m
(2.2 kgf·m, 16 lbf·ft)

6 x 1.0 mm
12 N·m
(1.2 kgf·m, 8.7 lbf·ft)

GASKET
Replace.

GASKET
Replace.

THROTTLE BODY

FRONT INTAKE MANIFOLD CHAMBER
Replace if it is cracked or if the mating surface is damaged.

INTAKE MANIFOLD
Replace if it is cracked or if the mating surface is damaged.

SPACER

42356-HPIL-G04

Exploded view of the intake manifold

- EVAP control canister hose
- Water bypass hoses toe throttle body
- Brake booster vacuum line
- PCV hose

6. Connect the following electrical connections and clamps to the manifold:
- EVAP control canister purge valve connector
- MAP sensor connector
- TP sensor connector
- IAC valve connector
- IAT sensor connector
- Throttle and cruise control cables
- Air intake tube
- Intake manifold cover
- Negative battery cable

7. Start the engine and check for proper operation.

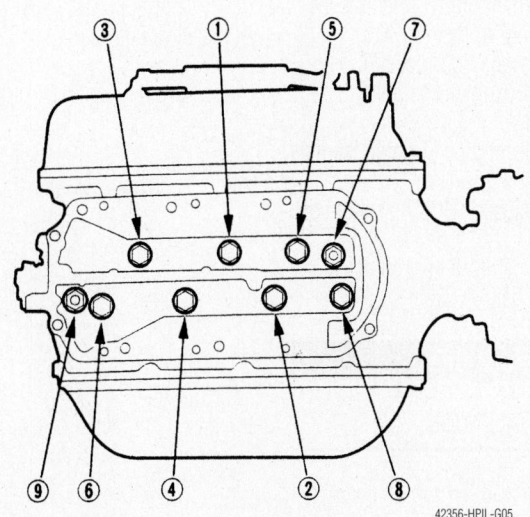

42356-HPIL-G05

Intake manifold torque sequence

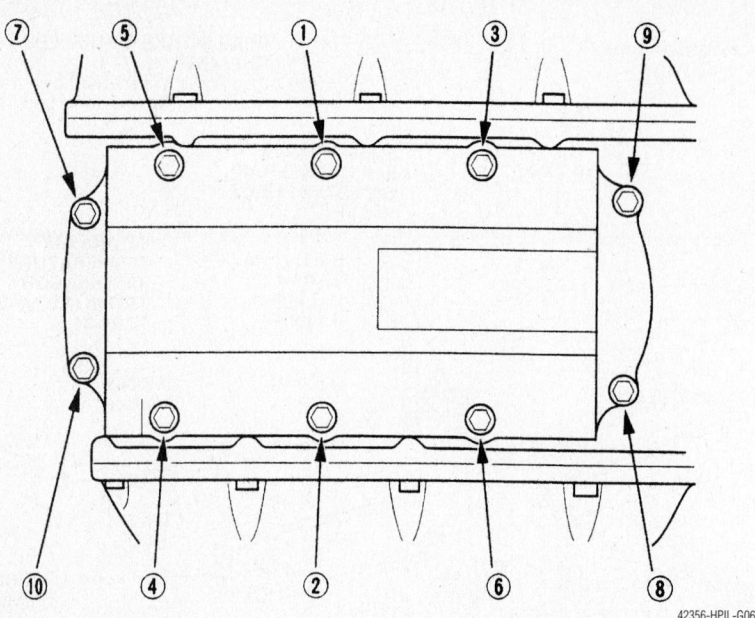

Upper cover torque sequence

42356-HPIL-G06

Exhaust Manifold

REMOVAL & INSTALLATION

1. Before servicing the vehicle, refer to the Precautions Section.
2. Remove or disconnect the following:
   - Negative battery cable
   - Exhaust manifold heat shield
   - Heated Oxygen (HO2S) sensor connector
   - Exhaust front pipe
   - Exhaust manifold bracket, if equipped
   - Exhaust manifold

To install:

3. Install or connect the following:
   - Exhaust manifold. Tighten the fasteners to 23 ft. lbs. (31 Nm).
   - Exhaust manifold bracket, if equipped. Tighten the bolts to 33 ft. lbs. (44 Nm).
   - Exhaust front pipe. Tighten the nuts to 40 ft. lbs. (55 Nm).
   - Heated Oxygen (HO2S) sensor connector
   - Exhaust manifold heat shield
   - Negative battery cable

Front Crankshaft Seal

REMOVAL & INSTALLATION

1. Before servicing the vehicle, refer to the Precautions Section.
2. Remove or disconnect the following:

   - Negative battery cable
   - Accessory drive belts
   - Side engine mount
   - Valve cover
   - Crankshaft pulley
   - Front cover
   - Timing belt
   - Top Dead Center (TDC) sensor, if equipped
   - Crankshaft timing sprocket
   - Front crankshaft seal

To install:

3. Lubricate the crankshaft seal lip with grease prior to installation.
4. Install the front crankshaft seal so that it is flush with the surface of the oil pump housing.
5. Install or connect the following:
   - Crankshaft timing sprocket
   - Top Dead Center (TDC) sensor, if equipped
   - Timing belt
   - Front cover

07LAD-PT3010A

Front crankshaft seal installation

93352G02

   - Crankshaft pulley. Tighten the bolt to 181 ft. lbs. (245 Nm).
   - Valve cover
   - Side engine mount
   - Accessory drive belts
   - Negative battery cable

6. Check the engine oil level and add if necessary.
7. Start the engine and check for leaks.

Camshaft

REMOVAL & INSTALLATION

Front

1. Before servicing the vehicle, refer to the Precautions Section.
2. Remove or disconnect the following:
   - Negative and positive battery cables
   - Battery and battery box
3. Drain the coolant.
4. Remove or disconnect the following:
   - Upper radiator hose
   - Exhaust Gas Recirculation (EGR) valve
   - Timing belt

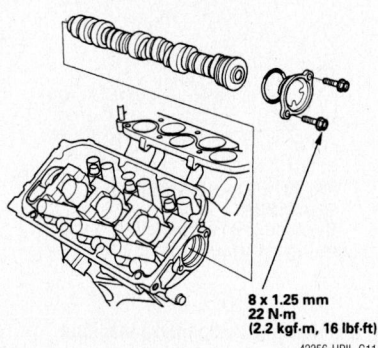

8 x 1.25 mm
22 N·m
(2.2 kgf·m, 16 lbf·ft)

42356-HPIL-G11

Front camshaft assembly

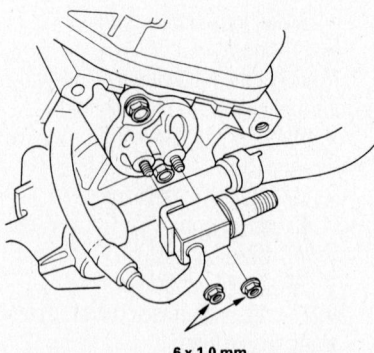

6 x 1.0 mm
12 N·m (1.2 kgf·m, 8.7 lbf·ft)

42356-HPIL-G12

Remove the nuts attaching the fuel line when removing the rear camshaft

- Rocker arm assembly
- Front camshaft pulley
- Thrust plate and camshaft

To install:

5. Install or connect the following:
  - Camshaft using a new O-ring. Tighten the thrust plate to 16 ft. lbs. (22 Nm).
  - Front camshaft pulley
  - Rocker arm assembly
  - Timing belt
  - Exhaust Gas Recirculation (EGR) valve
  - Battery
  - Positive, then negative battery cables
6. Fill the cooling system.
7. Start the engine and check for leaks.

Rear

1. Before servicing the vehicle, refer to the Precautions Section.
2. Drain the cooling system.
3. Relieve the fuel system pressure.
4. Remove or disconnect the following:
  - Negative battery cable
  - Under-hood fuse box
  - Fuel feed hose
  - Nuts securing the fuel line
  - Brake lines from the master cylinder
  - Timing belt
  - Rocker arm assembly
  - Rear camshaft pulley
  - Thrust plate and camshaft

To install:

5. Install or connect the following:
  - Camshaft using a new O-ring. Tighten the thrust plate to 16 ft. lbs. (22 Nm).
  - Rear camshaft pulley
  - Rocker arm assembly
  - Timing belt
  - Brake lines to the master cylinder
  - Nuts securing the fuel line
  - Fuel feed hose
  - Under-hood fuse box
  - Negative battery cable

Valve Lash

ADJUSTMENT

Adjust the valves only when the cylinder head temperature is less than 100°F (38°C).

1. Before servicing the vehicle, refer to the Precautions Section.
2. Remove or disconnect the following:
  - Negative battery cable
  - Air intake tube
  - Intake manifold

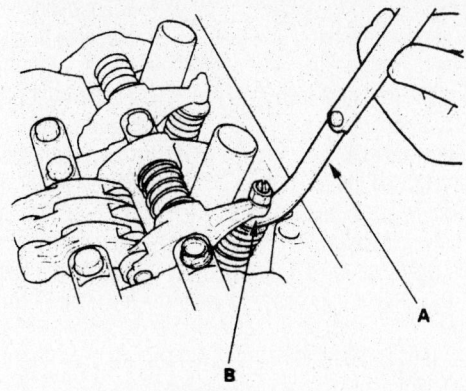

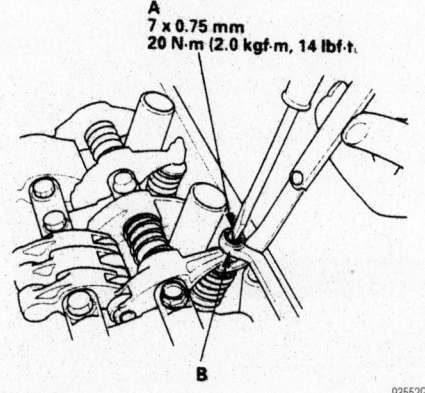

93552G04

Inspect the valve clearance, adjust to specification and tighten the retainer to specification

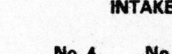

Adjusting screw locations:

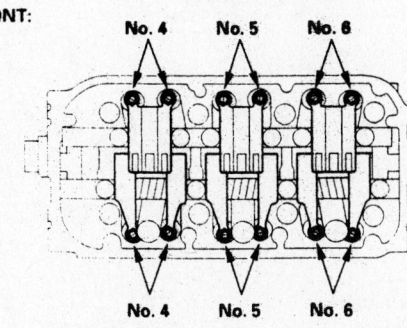

93552G03

Valve adjusting retainer locations

- Valve cover

3. Rotate the crankshaft so that the valves to be adjusted are closed and the rocker arm is contacting the camshaft lobe base circle.

4. Measure the valve clearance. If adjustment is necessary, loosen the locknut and turn the adjusting screw as necessary to achieve the correct valve clearance.

5. The correct valve clearance is:
- Intake valves: 0.008–0.009 inches (0.20–0.24mm)
- Exhaust valves: 0.011–0.013 inches (0.28–0.32mm)

6. After adjustment, tighten the locknuts to 14 ft. lbs. (20 Nm).

7. Install or connect the following:
- Valve cover
- Intake manifold
- Air intake tube
- Negative battery cable

8. Start the engine and check for proper operation.

Starter Motor

REMOVAL & INSTALLATION

1. Before servicing the vehicle, refer to the Precautions Section.

2. Make sure you have the anti-theft codes for the radio and navigation system.

3. Remove or disconnect the following:
- Negative battery cable
- Positive battery cable
- Battery
- ATF dipstick
- Starter wiring clamp
- Starter wiring harness connectors
- Starter motor bolts

To install:

4. Install or connect the following:
- Starter motor. Tighten the bolts to 33 ft. lbs. (44 Nm).
- Starter wiring harness connectors. Tighten the battery cable nut to 79 inch lbs. (9 Nm).
- ATF dipstick
- Battery
- Positive battery cable
- Negative battery cable

Oil Pan

REMOVAL & INSTALLATION

1. Before servicing the vehicle, refer to the Precautions Section.

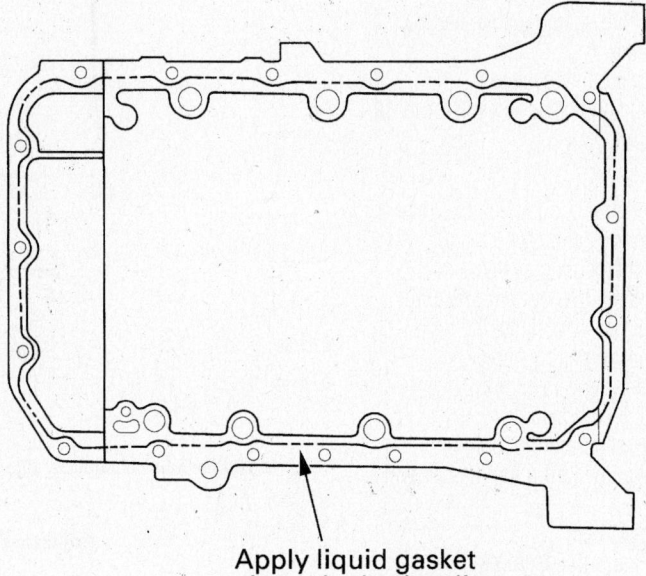

Apply liquid gasket along the broken line.

09474_RIDGE_G0018

Oil pan sealer application

2. Raise the vehicle on a hoist.
3. Drain the oil.
4. Remove the splash shield.
5. Remove the front subframe stiffener.
6. Remove the front exhaust pipe.
7. Remove the catalytic converter bracket.
8. Remove the torque converter cover.
9. Remove the 4 lower transaxle-to-engine bolts.
10. Remove the oil pan bolts.
11. Pry at the pry-points to break loose the oil pan.

To install:

12. Clean the pan and all mounting surfaces thoroughly.

13. Apply RTV gasket material to the oil pan flange as shown.

14. Position the pan on the block and install the bolts. Torque the bolts, in the sequence shown, in 2 even steps, to 104 inch lbs. (12 Nm).

➡**Wait at least 30 minutes before filling with oil.**

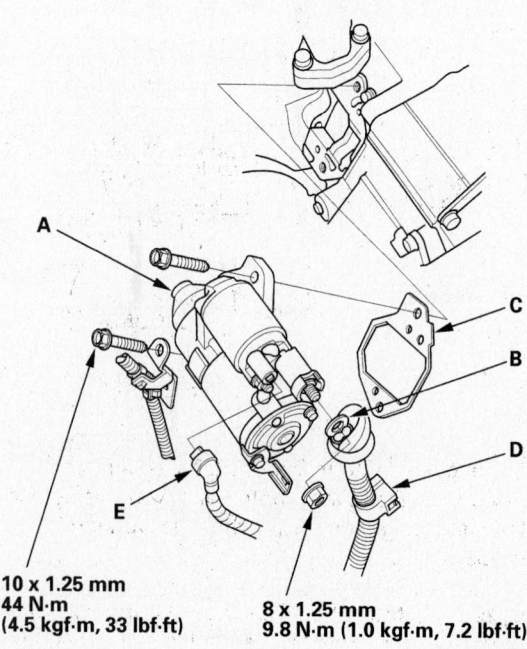

10 x 1.25 mm
44 N·m
(4.5 kgf·m, 33 lbf·ft)

8 x 1.25 mm
9.8 N·m (1.0 kgf·m, 7.2 lbf·ft)

09474_RIDGE_G0005

Starter mounting

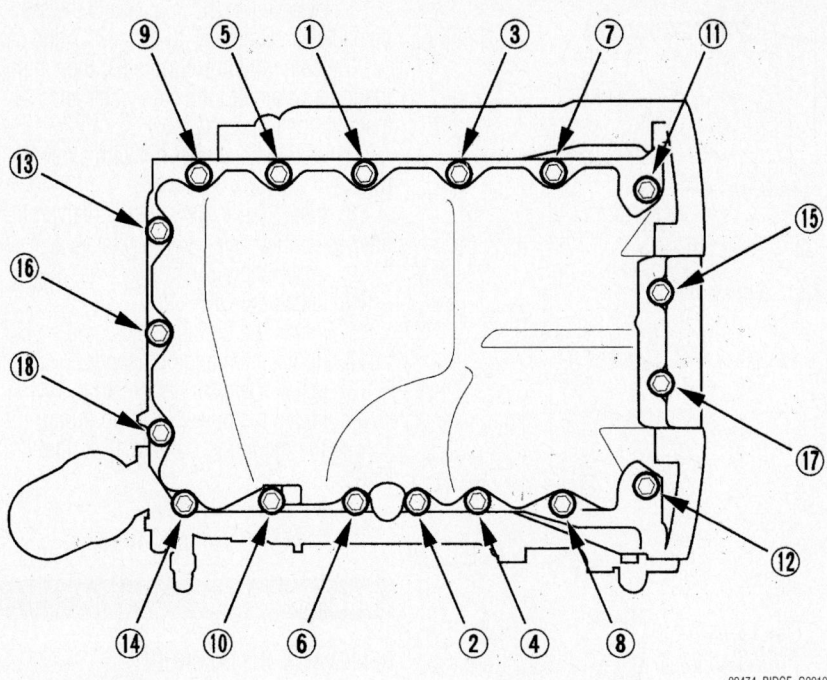

09474_RIDGE_G0019

Oil pan bolt torque sequence

15. Torque the transaxle bolts to 54 ft. lbs. (74 Nm).

16. Install the cover.

17. Install the converter bracket.

18. Install the exhaust pipe using new gaskets and new self-locking nuts. Torque the flange-to-crossover nuts to 40 ft. lbs. (54 Nm), the hanger bolts to 16 ft. lbs. (22 Nm) and the flange-to-converter nuts to 25 ft. lbs. (33 Nm).

19. Install the stiffener. Torque the bolts to 40 ft. lbs. (54 Nm).

20. Install the splash shield.

21. Refill the engine oil.

Oil Pump

REMOVAL

1. Remove the bulkhead cover.

2. Remove the intake manifold cover.

3. Remove the drive belt.

4. Remove the power steering pump and the line bracket. Set the pump aside without disconnecting the lines.

5. Remove the caps from the front strut mounting nuts. Position the engine hanger adapters (VSB02C000024) with the FRONT mark facing forward, over the flange nuts.

6. Install the balancer bar (VSB02C000019):

    a. Attach the front arm (A) to the front head with spacer (B) and a 10x1.25mm bolt (C)

    b. Attach the rear arm (D) to the rear head with an 8x1.25mm bolt (E).

7. Install the support hanger (AAR-T-12566):

    a. Attach the hook to the slotted hole in the balancer bar.

    b. Tighten the wing nut (F) by hand to

lift and support the engine/transaxle assembly.

8. Remove the timing belt.

9. Remove the crankshaft position sensor.

10. Remove the VTEC solenoid valve/oil filter assembly.

11. Remove the oil pan.

12. Remove the oil screen.

13. Remove the oil pump.

INSPECTION

Check the oil pump components for wear or damage. The following measurements are the service limits:

- Inner rotor-to-outer rotor: 0.008 in.
- Housing-to-rotor axial clearance: 0.005 in.
- Housing-to-rotor radial clearance: 0.008 in.

INSTALLATION

1. Discard the oil seal.

2. Tap a new seal into place until it bottoms.

3. Thoroughly clean and dry all gasket surfaces.

4. Apply a bead of RTV gasket material to the pump mating surface of the block.

➡**The oil pump must be installed within 4 minutes of applying the gasket material.**

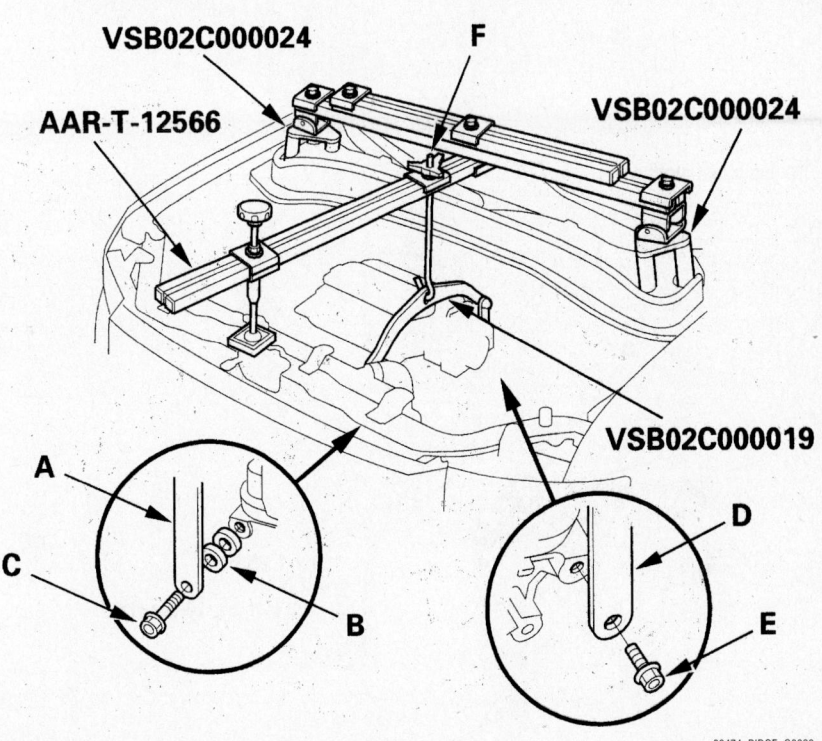

09474_RIDGE_G0008

Balancer bar installation and support hanger

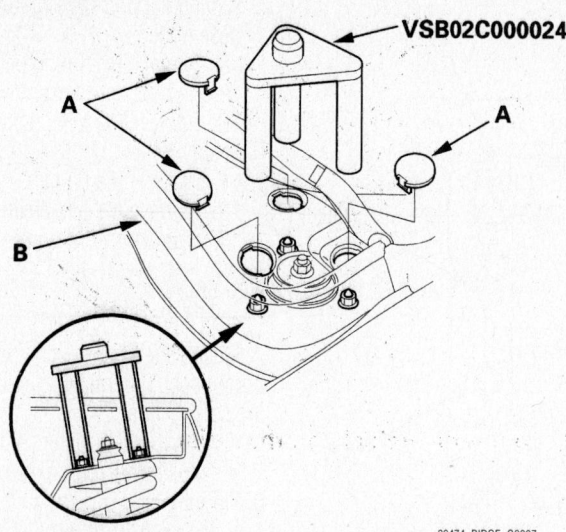

VSB02C000024

09474_RIDGE_G0007

Engine hanger adapters installed

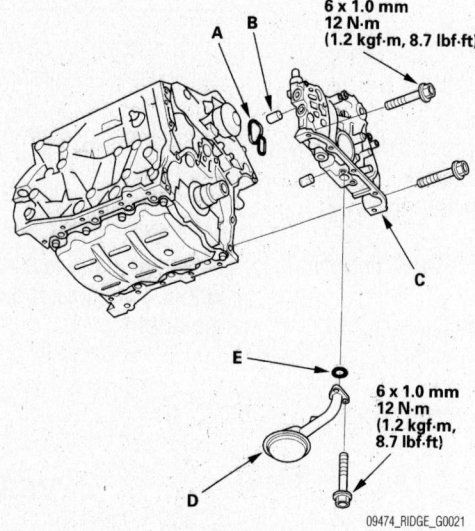

6 x 1.0 mm
12 N·m
(1.2 kgf·m, 8.7 lbf·ft)

6 x 1.0 mm
12 N·m
(1.2 kgf·m,
8.7 lbf·ft)

09474_RIDGE_G0021

Oil pump installation

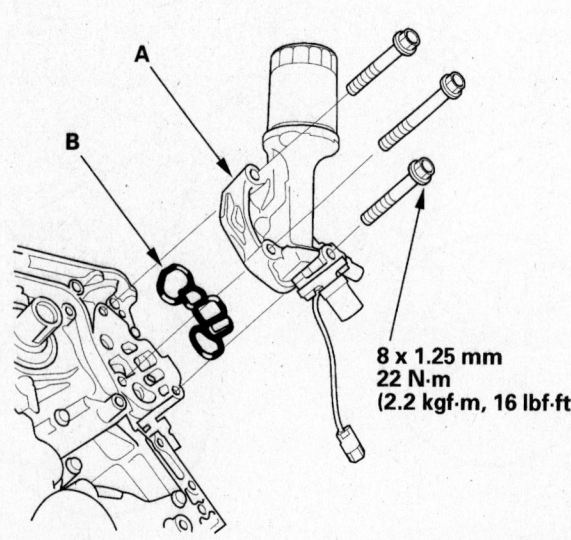

8 x 1.25 mm
22 N·m
(2.2 kgf·m, 16 lbf·ft)

09474_RIDGE_G0022

VTEC solenoid valve/oil filter assembly

5. Apply all-purpose grease to the seal lip and clean engine oil to the new O-ring (A).

6. Install the dowel pins (B) then align the inner rotor with the crankshaft and install the oil pump (C).

7. Install the screen (D) with a new O-ring (E).

8. Install the VTEC solenoid valve/oil filter assembly, using a new filter.

9. Install the oil pan.

10. Install the CKP.

11. Install the timing belt.

12. Remove the engine hanger.

13. Install the caps on the strut.

14. Install the power steering pump. Torque the mounting bolts to 16 ft. lbs. (22 Nm).

15. Install the drive belt.

16. Install the bulkhead cover.

Rear Main Seal

REMOVAL & INSTALLATION

1. Before servicing the vehicle, refer to the Precautions Section.

2. Remove or disconnect the following:
   - Transaxle
   - Flywheel
   - Oil seal

To install:

3. Install or connect the following:
   - Oil seal. Drive the seal square into the seal case.
   - Flywheel. Tighten the bolts in a crossing pattern to 54 ft. lbs. (73 Nm).
   - Transaxle

4. Check the fluid levels.

5. Start the engine and check for leaks.

Timing Belt

REMOVAL & INSTALLATION

1. Before servicing the vehicle, refer to the Precautions Section.

2. Turn the crankshaft so the white mark aligns with the pointer.

3. Make sure the number 1 piston is at Top Dead center (TDC).

4. Remove or disconnect the following:
   - Negative battery cable
   - Wheels and splash shield
   - Drive belts

5. Support the engine with a block of wood and a jack under the oil pan.
   - Ground cable bracket
   - Upper side engine mount
   - Dipstick tube

6. Remove the crankshaft pulley using holder tool shown in the accompanying illustration and a breaker and socket, loosen the 19mm bolt and remove the pulley.

- Front upper cover, rear upper cover and the lower cover
- One of the battery clamp bolts and grind the end as illustrated

7. Screw the battery clamp bolt as illustrated to hold the belt adjuster in position. Do not use a wrench, hand tighten only.

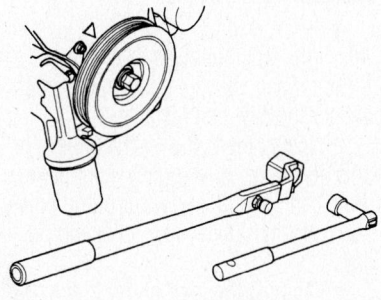

Remove the crankshaft pulley using holder tool shown

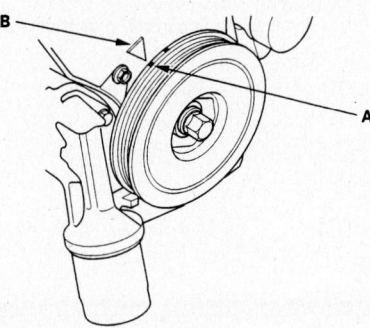

Turn the crankshaft so the white mark (A) aligns with the pointer (B)

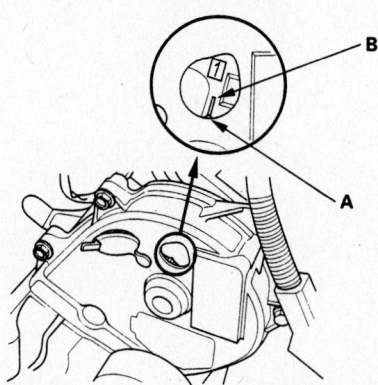

Make sure the number 1 piston is at top dead center (A) on the front camshaft pulley and pointer (B)

- Lower side engine mount
- Idler pulley bolt and the pulley
- Timing belt

To install:

8. If installing a new belt, perform the following steps:

a. Clean the pulleys, belt guide plate and the upper and lower covers.

b. Set the timing belt drive pulley to TDC by aligning the TDC mark on the tooth of the belt drive pulley with the pointer on the oil pump.

c. Set the camshaft pulleys to TDC by

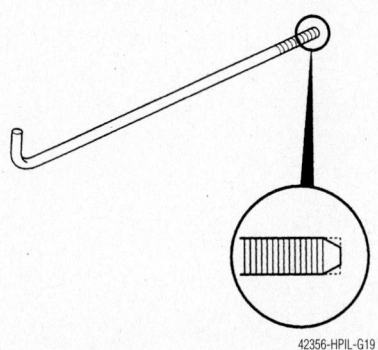

Remove a battery clamp bolt and grind the end as shown

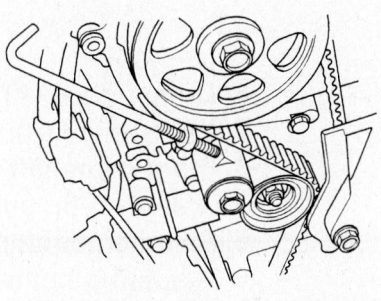

Install the battery clamp bolt as shown to hold the belt adjuster in position

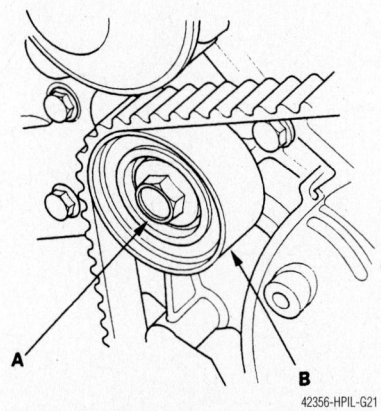

Remove the idler pulley bolt (A), pulley (B) and the timing belt

aligning the TDC marks on the camshaft pulleys with the pointers on the back covers.

d. Remove the battery clamp bolt.

e. Remove the belt tensioner.

f. Align the holes on the rod and housing of the tensioner.

g. Using a press or other suitable device, slowly compress the tensioner and insert a 0.08 inch (2mm) pin through the housing and rod.

h. Install the tensioner making sure the pin is still installed.

i. Apply thread locker to idler pulley bolt then hand tighten the bolt.

j. Install the belt over the pulleys in this sequence; drive pulley, idler pulley, front camshaft pulley, water pump pulley, rear camshaft pulley and adjusting pulley.

k. Tighten the idler pulley bolt to 33 ft. lbs. (44 Nm).

l. Remove the pin from the tensioner.

9. Install or connect the following:

- Lower half of the side mount and tighten the 3 long bolts to 33 ft. lbs. (44 Nm) and the one short bolt to 9 ft. lbs. (12 Nm)
- Timing belt guide plate as illustrated
- Lower timing cover and tighten the bolts to 9 ft. lbs. (12 Nm)
- Front and rear upper timing covers and tighten the bolts to 9 ft. lbs. (12 Nm)
- Crankshaft pulley and tighten the bolts to 181 ft. lbs. (245 Nm), using the holding tool to prevent the unit from turning

10. Rotate the crankshaft pulley about 5 or 6 degrees clockwise so the belt positions itself on the pulleys.

11. Turn the crankshaft pulley so the white mark aligns with the pointer.

12. Check the camshaft pulley marks are

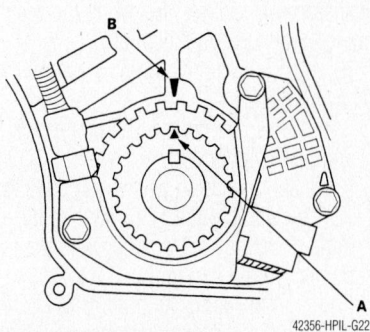

Set the timing belt pulley to TDC by aligning the TDC mark (A) on the tooth of the belt pulley with the pointer (B) on the oil pump

FRONT:

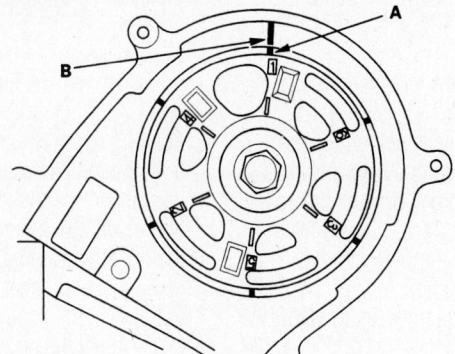

REAR:

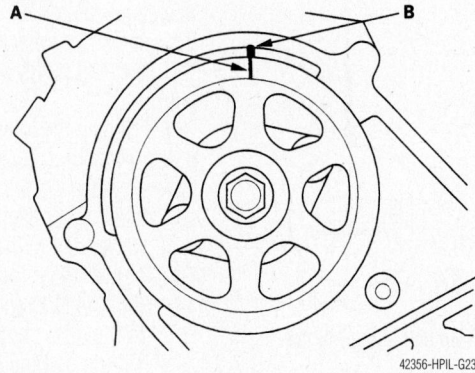

42356-HPIL-G23

Set the camshaft pulleys to TDC by aligning the TDC marks (A) on the camshaft pulleys with the pointers (B) on the back covers

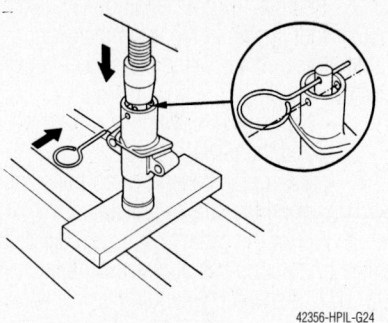

42356-HPIL-G24

Insert a 0.08 inch (2mm) pin through the tensioner housing and rod

16. If the tensioner was extended and the belt cannot be installed, perform the steps above for the new belt installation.

   a. Install the belt over the pulleys in this sequence; drive pulley, idler pulley, front camshaft pulley, water pump pulley, rear camshaft pulley and adjusting pulley.

   b. Tighten the idler pulley bolt to 33 ft. lbs. (44 Nm).

   c. Remove the battery clamp bolt.

17. Install or connect the following:
- Lower half of the side mount and tighten the 3 long bolts to 33 ft. lbs. (44 Nm) and the one short bolt to 9 ft. lbs. (12 Nm)

aligned. If the marks are aligned, proceed to the next step. If the marks are not aligned, remove the timing belt and reinstall using the steps outlined before this step.

13. Remove or disconnect the following:
- Drive belt
- Upper side mount and tighten the bolts in the sequence illustrated to the specifications in the illustration

14. Using a suitable scan tool, perform the Powertain Control Module (PCM) reset and the Crankshaft position (CKP) pattern clear/learn procedures, following the scan tool manufactures instructions.

15. If installing the old belt, perform the following steps:

   a. Clean the pulleys, belt guide plate and the upper and lower covers.

   b. Set the timing belt drive pulley to TDC by aligning the TDC mark on the tooth of the belt drive pulley with the pointer on the oil pump.

   c. Set the camshaft pulleys to TDC by aligning the TDC marks on the camshaft pulleys with the pointers on the back covers.

   d. Apply thread locker to idler pulley bolt then hand tighten the bolt.

-1 Drive pulley (A).
-2 Idler pulley (B).
-3 Front camshaft pulley (C).
-4 Water pump pulley (D).
-5 Rear camshaft pulley (E).
-6 Adjusting pulley (F).

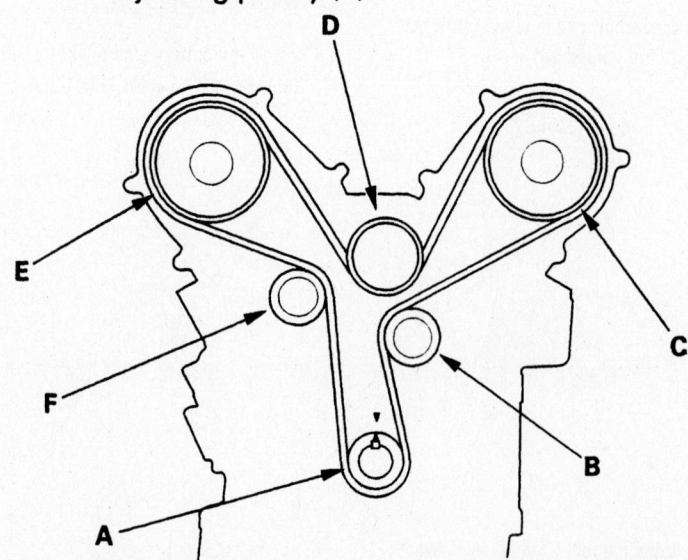

42356-HPIL-G25

Route the belt as shown in the sequence listed

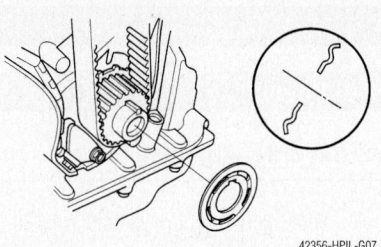

Install the timing belt guide plate as shown

- Timing belt guide plate as illustrated
- Lower timing cover and tighten the bolts to 9 ft. lbs. (12 Nm)
- Front and rear upper timing covers

and tighten the bolts to 9 ft. lbs. (12 Nm)
- Crankshaft pulley and tighten the bolts to 181 ft. lbs. (245 Nm), using the holding tool to prevent the

18. Rotate the crankshaft pulley about 5 or 6 degrees clockwise so the belt positions itself on the pulleys.

19. Turn the crankshaft pulley so the white mark aligns with the pointer.

20. Check the camshaft pulley marks are aligned. If the marks are aligned, proceed to the next step. If the marks are not aligned, remove the timing belt and reinstall using the steps outlined before this step.

FRONT CAMSHAFT PULLEY:

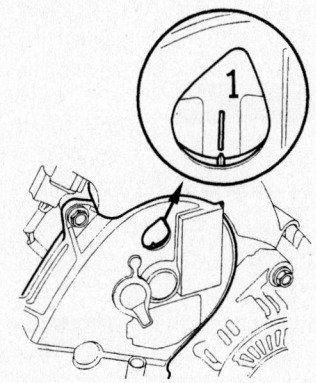

REAR CAMSHAFT PULLEY:

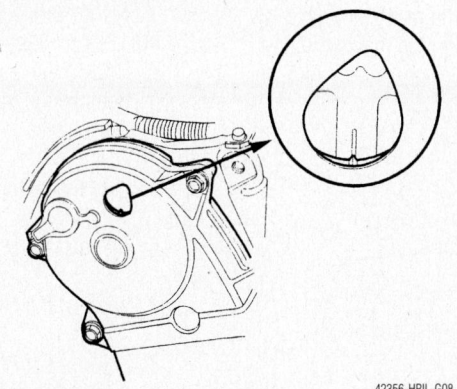

Check that the camshaft pulley marks are aligned as shown

21. Install or connect the following:
- Drive belt
- Upper side mount and tighten the bolts in the sequence illustrated to the specifications in the illustration
- Dipstick tube

22. Using a suitable scan tool, perform the Powertrain Control Module (PCM) reset and the Crankshaft position (CKP) pattern clear/learn procedures, following the scan tool manufactures instructions.

Piston and Ring

POSITIONING

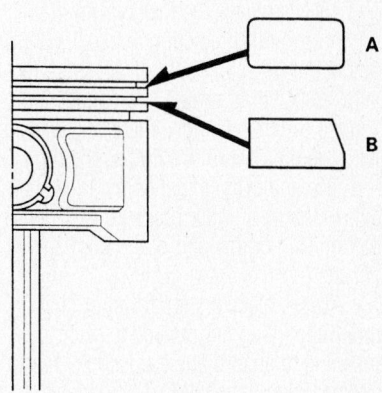

Compression ring identification—3.5L engine

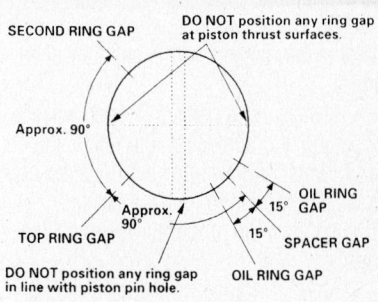

Ring end gap positioning

FUEL SYSTEM

Fuel System Service Precautions

Safety is the most important factor when performing not only fuel system maintenance, but any type of maintenance. Failure to conduct maintenance and repairs in a safe manner may result in serious personal injury or death. Maintenance and testing of the vehicle's fuel system components can be accomplished safely and effectively by adhering to the following rules and guidelines:

• To avoid the possibility of fire and personal injury, always disconnect the negative battery cable unless the repair or test procedure requires that battery voltage be applied.

• Always relieve the fuel system pressure prior to disconnecting any fuel system component (injector, fuel rail, pressure regulator, etc.), fitting or fuel line connection. Exercise extreme caution whenever relieving fuel system pressure to avoid exposing skin, face and eyes to fuel spray. Please be advised that fuel under pressure may penetrate the skin or any part of the body that it contacts.

• Always place a shop towel or cloth around the fitting or connection prior to loosening to absorb any excess fuel due to spillage. Ensure that all fuel spillage (should it occur) is quickly removed from engine surfaces. Ensure that all fuel soaked cloths or towels are deposited into a suitable waste container.

• Always keep a dry chemical (Class B) fire extinguisher near the work area.

• Do not allow fuel spray or fuel vapors to come into contact with a spark or open flame.

• Always use a backup wrench when loosening and tightening fuel line connection fittings. This will prevent unnecessary stress and torsion to fuel line piping. Always follow the proper torque specifications.

• Always replace worn fuel fitting O-rings with new. Do not substitute fuel hose or equivalent, where fuel pipe is installed.

Fuel System Pressure

RELIEVING

1. Before servicing the vehicle, refer to the Precautions Section.
2. Remove the left kick panel, then remove the PGM-FI main relay 2 (FUEL PUMP) (A).
3. Start the engine and let it idle until it stalls.
4. Turn the ignition off.
5. Remove the fuel filler cap.
6. Disconnect the battery ground.
7. Remove the quick-connect fitting cover from the fuel supply line, cover the fitting with a shop rag and disconnect the fitting.
8. After recharging, use the HDS to reset the power window control unit.

Fuel Filter

REMOVAL & INSTALLATION

➡The fuel filter should be replaced whenever the fuel pressure drops below 57 psi.

1. Before servicing the vehicle, refer to the Precautions Section.
2. Relieve the fuel system pressure.
3. Remove the fuel module.
4. Remove the fuel filter.
5. Installation is the reverse of removal. Use new O-rings. Align the marks on the module and tank

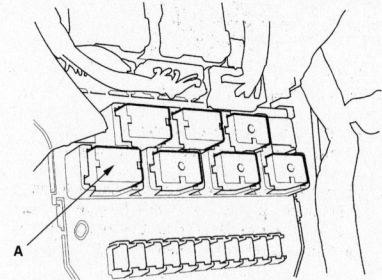

Fuel tank module

09474_RIDGE_G0023

Fuel Pump

REMOVAL & INSTALLATION

1. Before servicing the vehicle, refer to the Precautions Section.
2. Relieve the fuel system pressure.
3. Remove the fuel filler cap.
4. Remove the rear back seat.
5. Remove the access panel on the floor.
6. Disconnect the fuel pump connector.
7. Disconnect the quick-connect fitting from the module.
8. Using a lockring tool and breaker bar, loosen the lockring.
9. Remove the lockring (A) and the module.
10. Remove the filter (B), the sending unit (C), the case (D), the harness (E) and the pressure regulator (F).
11. Installation is the reverse of removal. Make sure that the suction filter (G) is firmly connected to the pump (H). Use a new gasket (J), a new lockring and new O-rings (K). Align the marks (L) on the unit (M) and the tank (N). Torque the lockring to 69 ft. lbs. (93 Nm).

Fuel Injector

REMOVAL & INSTALLATION

1. Before servicing the vehicle, refer to the Precautions Section.
2. Relieve the fuel system pressure.
3. Remove the intake manifold.
4. Disconnect the wiring from the injectors (A).
5. Disconnect the quick-connect fitting (B).
6. Remove the bolts (C) from the fuel rail (D).
7. Remove the clip (E) from the fuel rail.
8. Remove the injectors from the rail
9. Installation is the reverse of removal. Coat the new O-rings with clean engine oil. Torque the bolts to 89 inch lbs. (9.8 Nm).
10. Start the engine and check for leaks.

DRIVE TRAIN

Transaxle Assembly

REMOVAL & INSTALLATION

1. Before servicing the vehicle, refer to the Precautions Section.

2. Make sure you have the radio and navigation system anti-theft codes.

3. Set the wheels in the straight-ahead position and lock the steering.

4. Drain the power steering reservoir.

5. Make sure that the ignition switch is in the OFF position.

6. Disconnect the negative, then the positive battery cables.

7. Remove the battery and battery tray.

8. Remove the intake manifold cover.

9. Remove the intake air duct, resonator and air cleaner housing.

10. Remove the battery base and bracket.

11. Remove the front bulkhead cover.

12. Raise and support the vehicle.

13. Remove the splash shield.

14. Drain the transaxle fluid. When the fluid is drained, install the drain plug and new sealing washer. Torque the plug to 36 ft. lbs. (49 Nm).

15. Matchmark and disconnect the steering coupler.

16. Disconnect the power steering outlet line at the pump and remove the hose clamp bolt.

17. Disconnect the transaxle breather hose at the transaxle.

18. Disconnect the solenoid wiring.

19. Disconnect the starter wiring at the starter.

20. Remove the dipstick.

21. Remove the starter.

22. Remove the shift cable bracket nuts.

23. Disconnect the shift cable from the control lever.

24. Disconnect all wiring from the transaxle.

25. Disconnect the cooler hoses.

26. Disconnect the transfer case breather.

27. Remove the connector bracket from the front cylinder head and use that bolt hole to attach the balancer bar front arm.

28. Remove the harness clamp bracket from the rear head and use that bolt hole to attach the balancer bar rear arm.

29. Remove the caps from the front strut mounting nuts. Position the engine hanger adapters (VSB02C000024) with the FRONT mark facing forward, over the flange nuts.

30. Install the balancer bar (VSB02C000019):

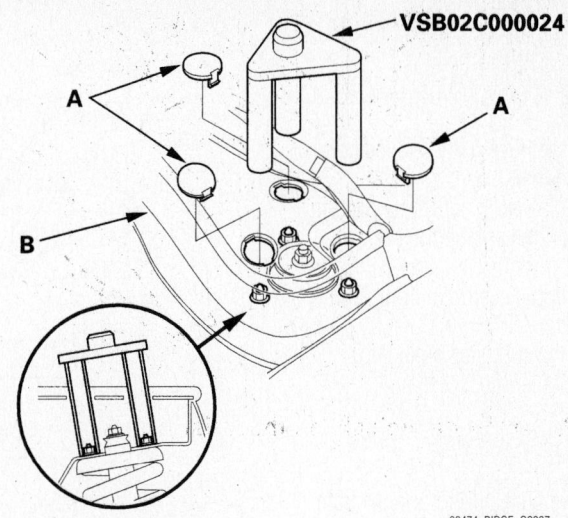

Engine hanger adapters installed

09474_RIDGE_G0007

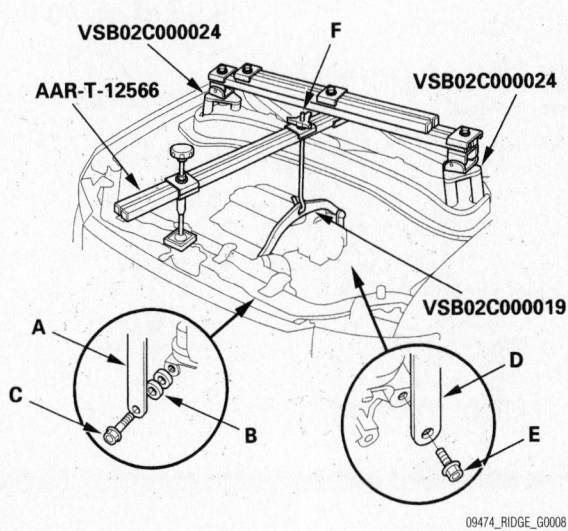

Balancer bar installation and support hanger

09474_RIDGE_G0008

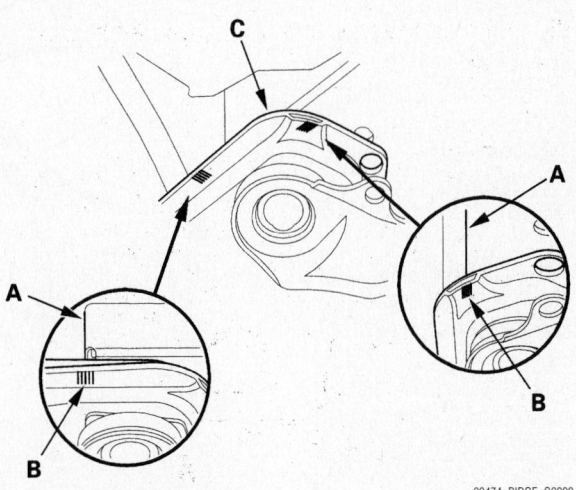

Matchmark the body and subframe as shown

09474_RIDGE_G0009

a. Attach the front arm (A) to the front head with spacer (B) and a 10x1.25mm bolt (C).

b. Attach the rear arm (D) to the rear head with an 8x1.25mm bolt (E).

31. Install the support hanger (AAR-T-12566):

a. Attach the hook to the slotted hole in the balancer bar.

b. Tighten the wing nut (F) by hand to lift and support the engine/transaxle assembly.

32. Remove and discard the front engine mount nut.

33. Remove the subframe stiffener.

34. Remove the exhaust pipe and bracket.

35. Separate the lower arms from the knuckles.

36. Separate the stabilizer bar links from the knuckles.

37. Separate the tie rod ends from the knuckles.

38. Disconnect the transfer case breather.

39. Matchmark and disconnect the driveshaft from the transfer case.

40. Remove the transfer case.

41. Remove the torque converter cover.

42. Remove the 8 torque converter-to-flexplate bolts.

43. Disconnect the power steering pressure switch.

44. Remove the power steering hose clamp from the subframe.

45. Remove the 4 transaxle lower mount nuts.

46. Matchmark the body and subframe as shown.

47. Install the subframe adapter (A) as shown.

48. Remove the six 12x1.25mm bolts (A) securing the subframe stiffeners (B), the 4 subframe mounting bolts (C) and the stiffeners. Then, lower the subframe (D).

49. Remove the 3 rear mount bracket bolts.

50. Remove the transaxle lower mounts.

51. Remove the halfshafts.

52. Remove the exhaust manifold bracket and heat shield.

53. Remove the intermediate shaft.

54. Remove the front mount bracket.

55. Remove the transaxle-to-engine bolts.

56. lower the transaxle by loosing the wingnut of the support and tile the engine just enough for the transaxle to clear the frame.

57. Slide a jack under the transaxle and roll it out.

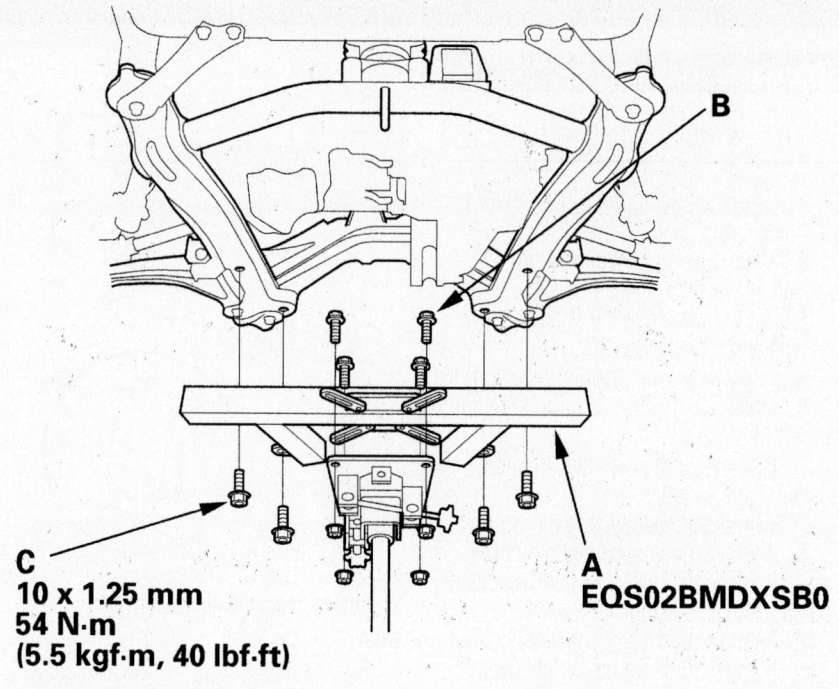

C
10 x 1.25 mm
54 N·m
(5.5 kgf·m, 40 lbf·ft)

A
EQS02BMDXSB0

09474_RIDGE_G0010

Install the subframe adapter as shown

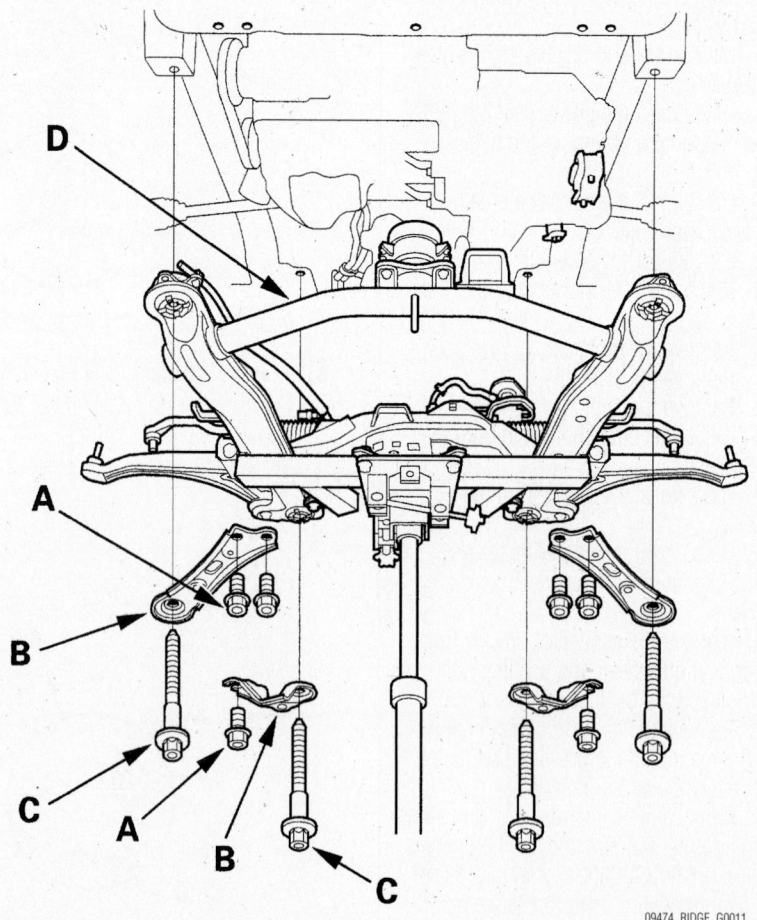

09474_RIDGE_G0011

Subframe removal

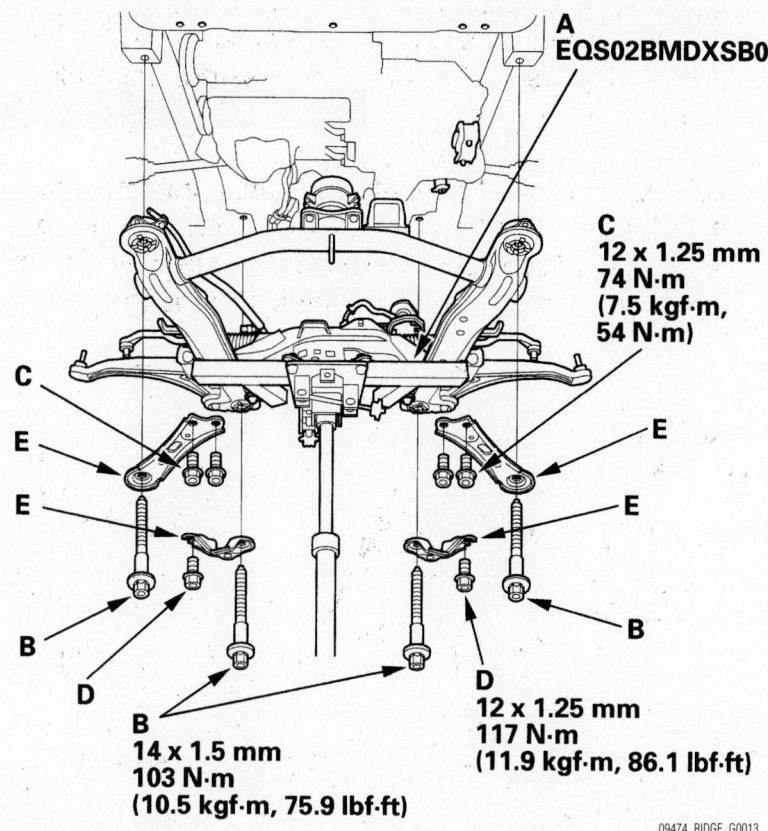

A
EQS02BMDXSB0

C
12 x 1.25 mm
74 N·m
(7.5 kgf·m, 54 N·m)

B
14 x 1.5 mm
103 N·m
(10.5 kgf·m, 75.9 lbf·ft)

D
12 x 1.25 mm
117 N·m
(11.9 kgf·m, 86.1 lbf·ft)

09474_RIDGE_G0013

Subframe installation

To install:

58. Make sure the dowel pins are installed in the converter housing.

59. Install the transmission lower mounts. Torque to 33 ft. lbs. (44 Nm).

60. Mate the transaxle to the engine. Install the bolts and torque to 47 ft. lbs. (64 Nm).

61. Install the front bracket using new bolts. Torque to 40 ft. lbs. (54 Nm).

62. Install the intermediate shaft.

63. Install the halfshafts.

64. Install the subframe, aligning the matchmarks and tightening the bolts as shown.

65. Install the rear mount bracket bolts. Torque to 28 ft. lbs. (38 Nm).

66. The remainder of installation is the reverse of removal. Observe the following torques:

- Lower mount nuts: 33 ft. lbs. (44 Nm)
- Torque converter bolts, in 2 equal steps, alternating the tightening sequence, to 104 inch lbs. (12 Nm).
- Torque converter cover: 104 inch lbs. (12 Nm)
- Transfer case: 38 ft. lbs. (51 Nm)
- Driveshaft-to-transfer case: 53 ft. lbs. (72 Nm)

- Tie rod end ball stud nut: 40 ft. lbs. (54 Nm)
- Lower ball joint stud nut: 65–72 ft. lbs. (88–98 Nm)
- Stabilizer link nut: 58 ft. lbs. (78 Nm)
- Exhaust pipe-to-converter: 25 ft. lbs. (33 Nm)
- Exhaust pipe hanger: 16 ft. lbs. (22 Nm)
- Exhaust pipe-to-manifold: 40 ft. lbs. (54 Nm)
- Stiffener (new bolts): 40 ft. lbs. (54 Nm)
- Front mount nut (new): 40 ft. lbs. (54 Nm)
- Connector bracket-to-front head: 33 ft. lbs. (44 Nm)
- Harness bracket-to-rear head: 20 ft. lbs. (26 Nm)

Transfer Case

REMOVAL & INSTALLATION

1. Before servicing the vehicle, refer to the Precautions Section.

2. Raise and support the vehicle.

3. Place the transaxle in **N**.

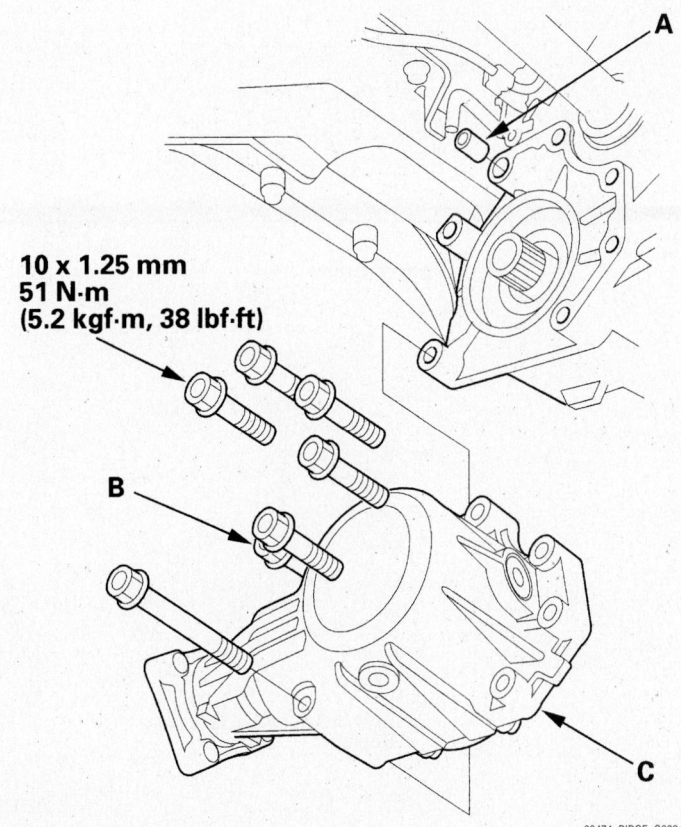

A

10 x 1.25 mm
51 N·m
(5.2 kgf·m, 38 lbf·ft)

B

C

09474_RIDGE_G0024

Transfer case installation

4. Drain the transaxle. When the fluid is drained, install the drain plug, using a new washer and torque to 36 ft. lbs. (49 Nm).

5. Remove the subframe stiffener.

6. Remove the exhaust pipe and bracket.

7. Disconnect the transfer case breather.

8. Matchmark the driveshaft and disconnect it from the transfer case.

9. Remove the bolts and remove the transfer case.

10. Installation is the reverse of removal. Observe the following torques:
- Transfer case bolts: 38 ft. lbs. (51 Nm)
- Driveshaft bolts: 53 ft. lbs. (72 Nm)
- Exhaust pipe-to-converter: 25 ft. lbs. (33 Nm)
- Exhaust pipe hanger: 16 ft. lbs. (22 Nm)
- Exhaust pipe-to-manifold: 40 ft. lbs. (54 Nm)
- Stiffener (new bolts): 40 ft. lbs. (54 Nm)

Front Halfshaft

REMOVAL & INSTALLATION

1. Before servicing the vehicle, refer to the Precautions Section.

2. Drain the transaxle if removing the

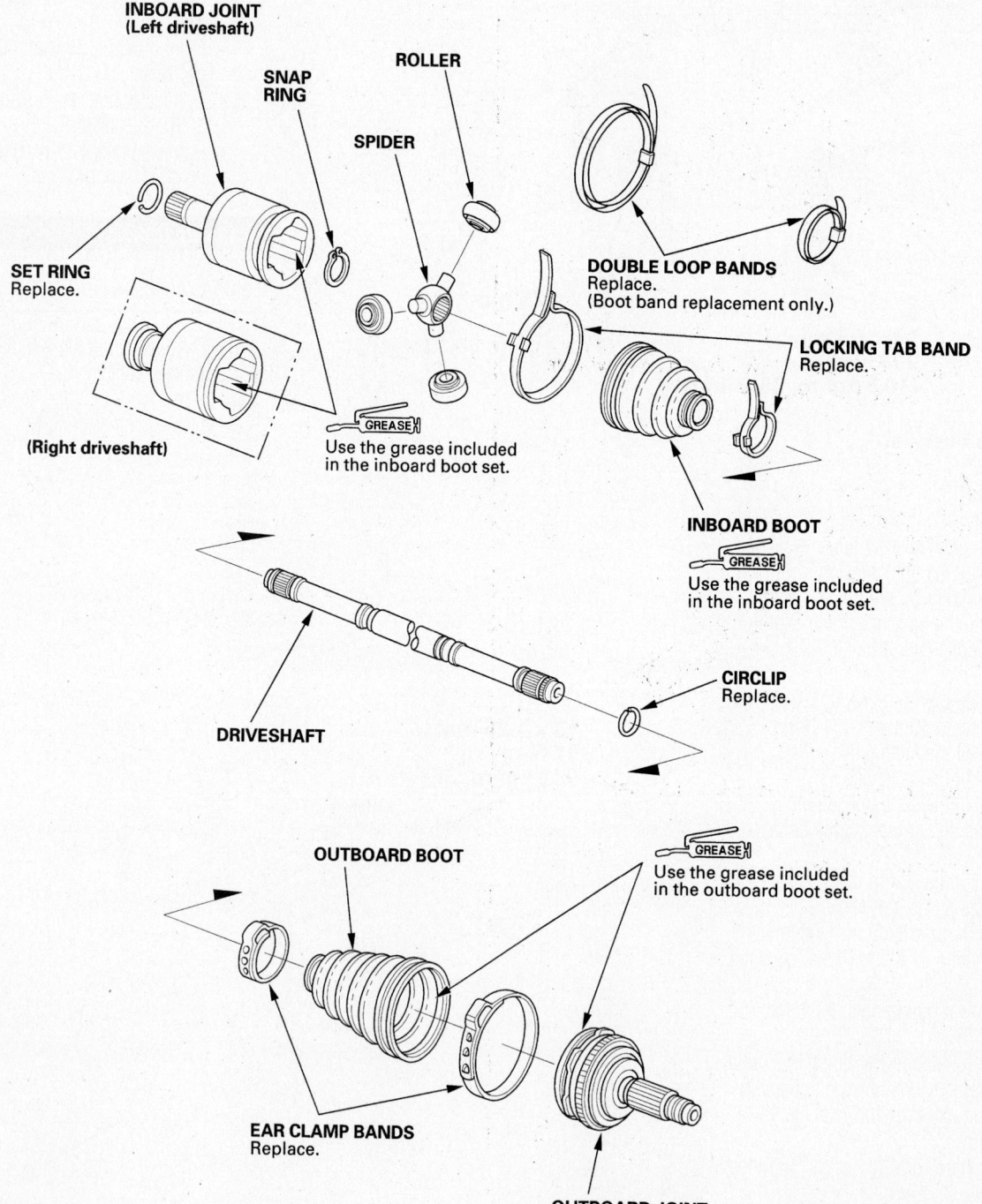

Front halfshaft exploded view

09474_RIDGE_G0027

left halfshaft. If is not necessary to drain the fluid if removing the right halfshaft.

3. Remove or disconnect the following:
- Negative battery cable
- Front wheels
- Spindle nut
- Lower ball joint

4. Pry the inboard joint from the transaxle or intermediate shaft.

5. Remove the outer CV-joint stub shaft from the hub by tapping the stub shaft with a plastic hammer.

To install:

➡**Use new circlips, split pins and self-locking nuts for assembly.**

6. Install the outer CV-joint stub shaft into the hub.

7. Install the inner CV-joint to the transaxle or intermediate shaft until the circlip locks in the retaining groove.

8. Install or connect the following:
- Lower ball joint. Tighten the nut to 69 ft. lbs. (93 Nm). Always advance castellated nuts to align cotter pin holes.
- Spindle nut (new). Tighten the nut to 242 ft. lbs. (328 Nm).
- Front wheels
- Negative battery cable

9. Fill the transaxle to the correct level and check for leaks.

Intermediate Shaft

REMOVAL & INSTALLATION

1. Before servicing the vehicle, refer to the Precautions Section.
2. Drain the transaxle.
3. Remove the right halfshaft.
4. Remove the subframe stiffener.
5. Remove the exhaust pipe and bracket.
6. Remove the heat shield.
7. Remove the intermediate shaft from the differential.

➡**Hold the intermediate shaft horizontally to avoid damage to the seal.**

8. Installation is the reverse of removal. Torque the heat shield bolts to 29 ft. lbs. (39 Nm).

Rear Halfshaft

REMOVAL & INSTALLATION

1. Before servicing the vehicle, refer to the Precautions Section.
2. Raise and support the rear of the vehicle.
3. Drain the differential.
4. Remove the wheels.
5. Remove and discard the spindle nut.
6. Remove the VSA rear wheel sensor.
7. Separate the upper ball joint from the upper arm.
8. Remove the lower track rod.
9. Separate the lower arm from the knuckle.

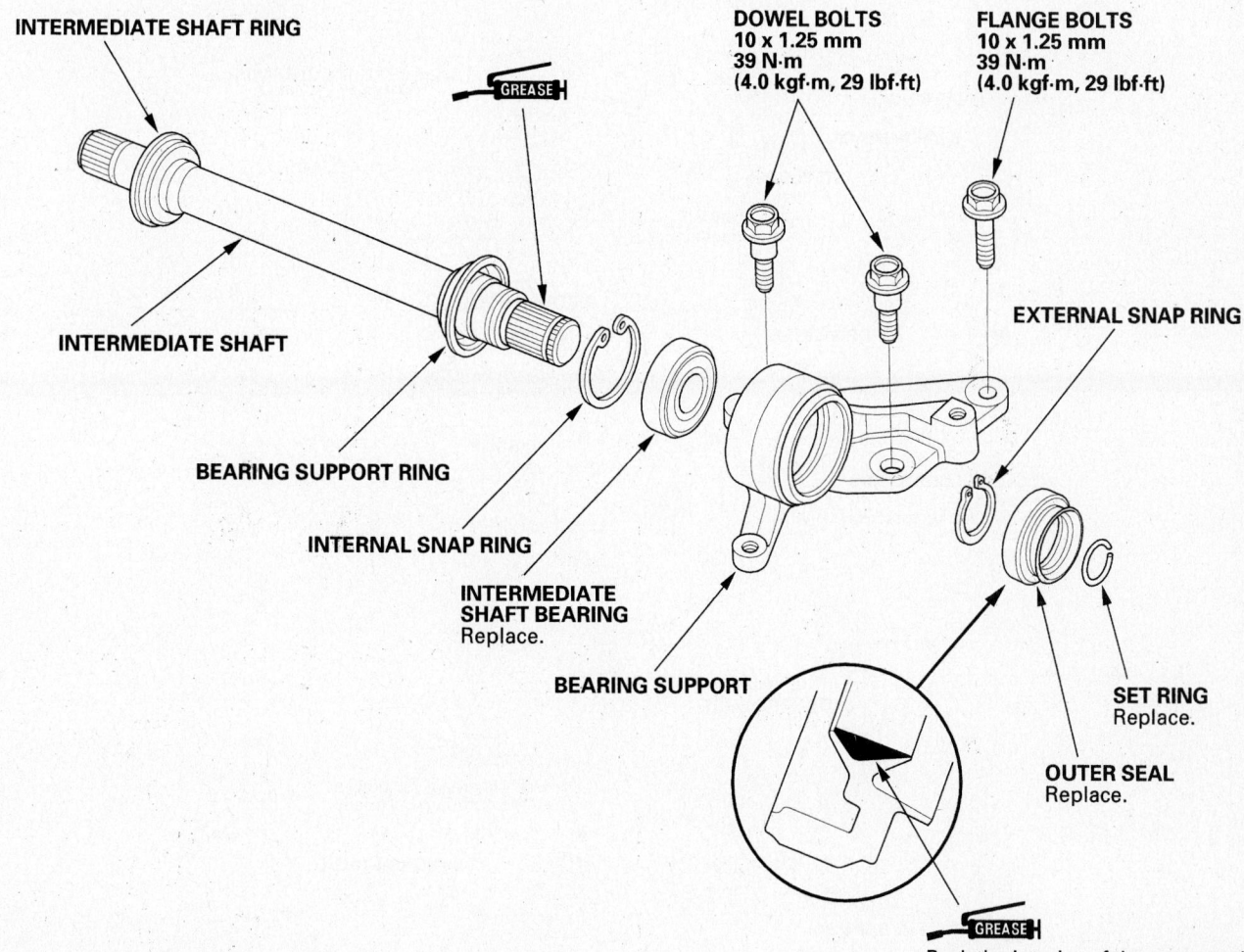

INTERMEDIATE SHAFT RING

GREASE

INTERMEDIATE SHAFT

BEARING SUPPORT RING

INTERNAL SNAP RING

INTERMEDIATE SHAFT BEARING
Replace.

BEARING SUPPORT

DOWEL BOLTS
10 x 1.25 mm
39 N·m
(4.0 kgf·m, 29 lbf·ft)

FLANGE BOLTS
10 x 1.25 mm
39 N·m
(4.0 kgf·m, 29 lbf·ft)

EXTERNAL SNAP RING

SET RING
Replace.

OUTER SEAL
Replace.

GREASE
Pack the interior of the outer seal.

09474_RIDGE_G0020

Intermediate shaft exploded view

10. Pull the knuckle outwards and separate the halfshaft from the hub.

11. Pry the halfshaft from the differential.

12. Installation is the reverse of removal. Use new snaprings and cotter pins. Always advance castellated nuts to align cotter pin holes. Observe the following torques:

- Lower arm-to-knuckle bolt: 105 ft. lbs. (142 Nm)
- Track rod inner end: 69 ft. lbs. (93 Nm)
- Track rod outer end: 74 ft. lbs. (101 Nm)
- Upper arm ball stud nut: 40 ft. lbs. (54 Nm)
- Halfshaft end nut: 181 ft. lbs. (245 Nm)

CV-Joint

OVERHAUL

Front

➡**When reassembled, check the halfshaft lengths. On the left side, the measurement from the inner face of the inner joint to the outer face of the outer joint, should be 22.7–22.97 inches (578.4–583.4mm). On the right side, the distance should be 22.70–22.90 inches (576.7–581.7mm).**

OUTBOARD JOINT

1. Before servicing the vehicle, refer to the Precautions Section.

2. Remove or disconnect the following:

- Axle halfshaft from the vehicle and place it in a vise
- Outboard joint boot clamps and push the boot back
- Outboard joint by driving it off the axle shaft with a brass drift and hammer
- Outboard joint boot

To install:

➡**Use new circlips and boot clamps for assembly.**

3. Install the outboard joint boot and clamps to the axle shaft.

4. Fill the outboard joint with grease. Install the outboard joint to the axle shaft.

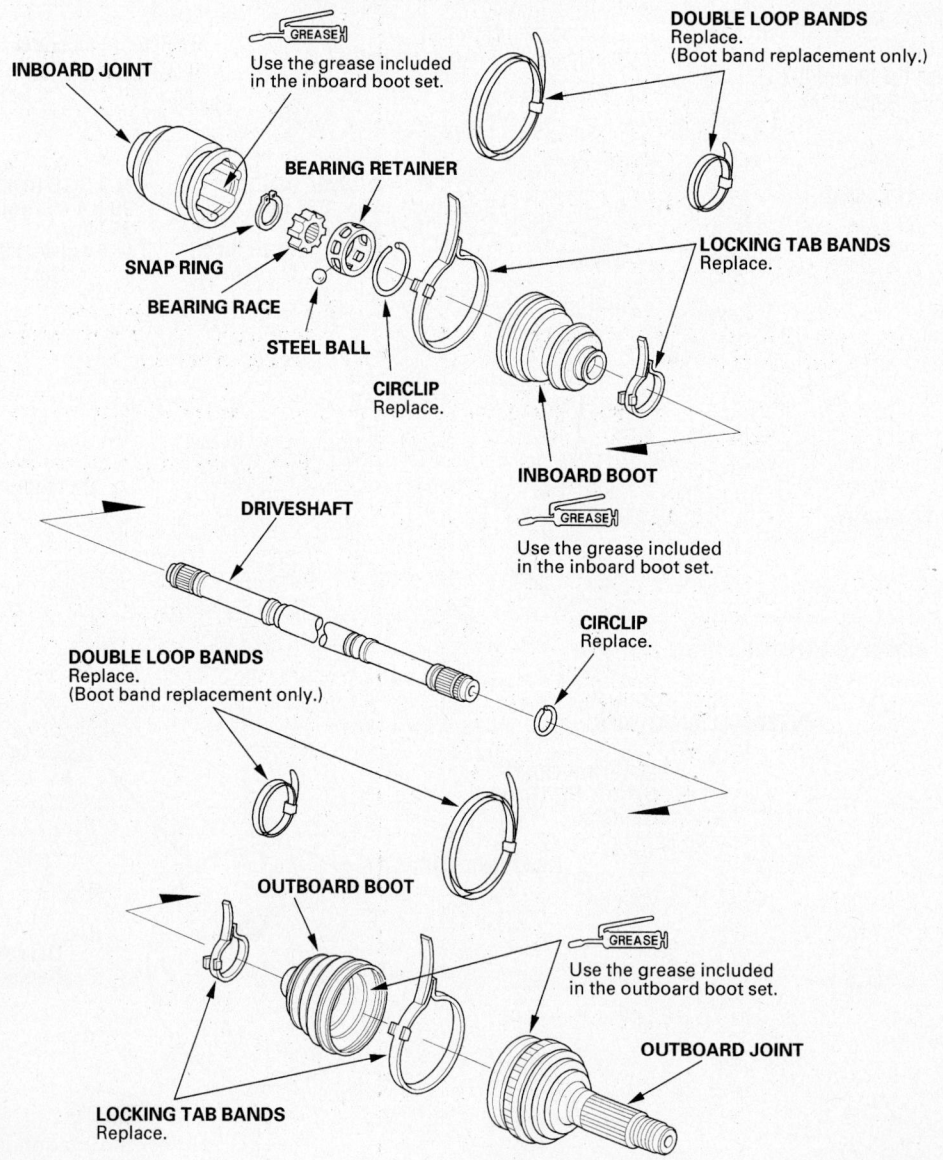

Rear halfshaft exploded view

09474_RIDGE_G0026

Tap the stub shaft with a brass hammer to seat the circlip.

5. Fill the outboard joint boot with 5–6 oz. of grease and install the boot clamps.

6. Install the axle halfshaft to the vehicle.

INBOARD JOINT

1. Before servicing the vehicle, refer to the Precautions Section.

2. Remove or disconnect the following:
- Axle halfshaft from the vehicle.
- Inboard joint boot clamps and push the boot back
- Inboard joint housing from the axle
- Rollers from the spider
- Snapring and the spider from the axle shaft
- Inboard joint boot

To install:

➡**Use new circlips and boot clamps for assembly.**

3. Install or connect the following:
- Inboard joint boot and clamps to the axle shaft
- Spider with a new snapring
- Rollers to the spider

4. Fill the joint housing with 7–8 oz. of grease and install it.

5. Fill the inboard joint boot with grease and install the boot clamps.

6. Install the axle halfshaft to the vehicle.

Rear

➡**When reassembled, check the half-shaft lengths. On the left side, the measurement from the inner end of the inner joint to the outer face of the outer joint, should be 24.19–24.38 inches (614.5–619.5mm). On the right side, the distance should be 25.33–25.53 inches (643.5–648.5mm).**

OUTBOARD JOINT

1. Before servicing the vehicle, refer to the Precautions Section.

2. Remove or disconnect the following:
- Axle halfshaft from the vehicle and place it in a vise
- Outboard joint boot clamps and push the boot back
- Outboard joint by driving it off the axle shaft with a brass drift and hammer
- Outboard joint boot

To install:

➡**Use new circlips and boot clamps for assembly.**

3. Install the outboard joint boot and clamps to the axle shaft.

4. Fill the outboard joint with grease. Install the outboard joint to the axle shaft. Tap the stub shaft with a brass hammer to seat the circlip.

5. Fill the outboard joint boot with 2–3 oz. of grease and install the boot clamps.

6. Install the axle halfshaft to the vehicle.

INBOARD JOINT

1. Before servicing the vehicle, refer to the Precautions Section.

2. Remove or disconnect the following:
- Axle halfshaft from the vehicle.
- Inboard joint boot clamps and push the boot back
- Inboard joint housing from the axle
- Rollers from the spider
- Snapring and the spider from the axle shaft
- Inboard joint boot

To install:

➡**Use new circlips and boot clamps for assembly.**

3. Install or connect the following:
- Inboard joint boot and clamps to the axle shaft
- Spider with a new snapring
- Rollers to the spider

4. Fill the joint housing with 4–5 oz. of grease and install it.

5. Fill the inboard joint boot with grease and install the boot clamps.

6. Install the axle halfshaft to the vehicle.

Driveshaft

REMOVAL & INSTALLATION

1. Before servicing the vehicle, refer to the Precautions Section.

2. Raise and support the vehicle.

3. Remove the front and rear driveshaft protector loops.

4. Matchmark the front driveshaft and transfer case flange.

5. Remove the bolts.

6. Remove the center bearing bolts.

7. Matchmark the rear driveshaft and differential flange.

8. Remove the bolts and remove the driveshaft assembly.

9. Installation is the reverse of removal. Align all matchmarks. Torque the flange bolts to 53 ft. lbs. (72 Nm). Torque the center bearing bolts to 29 ft. lbs. (39 Nm). Torque the loop bolts to 16 ft. lbs. (22 Nm).

Rear Differential

REMOVAL & INSTALLATION

1. Before servicing the vehicle, refer to the Precautions Section.

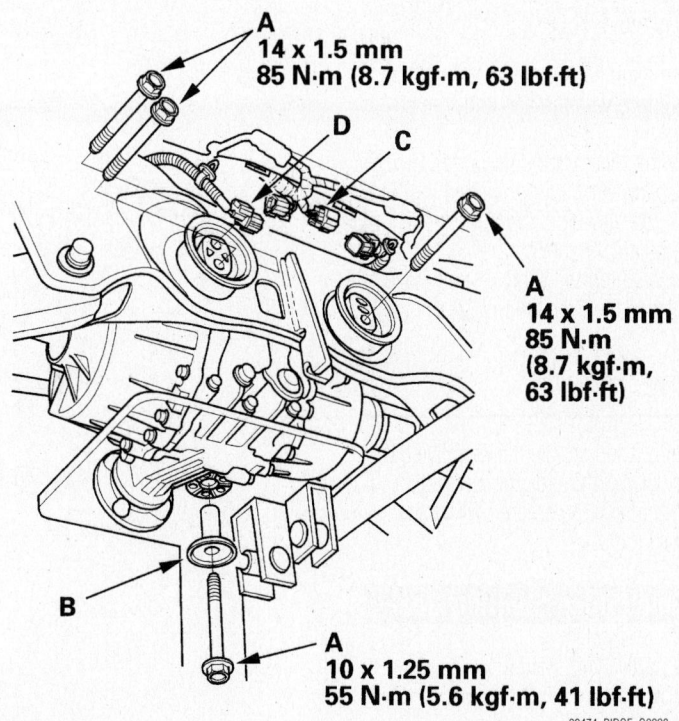

A
14 x 1.5 mm
85 N·m (8.7 kgf·m, 63 lbf·ft)

A
14 x 1.5 mm
85 N·m
(8.7 kgf·m,
63 lbf·ft)

A
10 x 1.25 mm
55 N·m (5.6 kgf·m, 41 lbf·ft)

Rear differential installation

09474_RIDGE_G0028

2. Matchmark and disconnect the drive-shaft.

3. Remove the halfshafts.

4. Support the differential with a transmission jack.

5. Disconnect the wiring at the differential.

6. Remove the mounting bolts.

7. Lower the unit slightly and disconnect the breather hose.

8. Remove the differential.

9. Installation is the reverse of removal. Torque the mounting bolts as shown in the illustration.

Air Bag

✳✳ CAUTION

Some vehicles are equipped with an air bag system. The system must be disarmed before performing service on, or around, system components, the steering column, instrument panel components, wiring and sensors. Failure to follow the safety precautions and the disarming procedure could result in accidental air bag deployment, possible injury and unnecessary system repairs.

PRECAUTIONS

Several precautions must be observed when handling the inflator module to avoid accidental deployment and possible personal injury.

• Never carry the inflator module by the wires or connector on the underside of the module.

• When carrying a live inflator module, hold securely with both hands, and ensure that the bag and trim cover are pointed away.

• Place the inflator module on a bench or other surface with the bag and trim cover facing up.

• With the inflator module on the bench, never place anything on or close to the module which may be thrown in the event of an accidental deployment.

Before servicing the vehicle, also make sure to refer to the Precautions Section as well.

DISARMING

Disconnect and isolate the negative battery cable. Wait 3 minutes for the system capacitor to discharge before performing any service.

Rack and Pinion Steering Gear

REMOVAL & INSTALLATION

1. Before servicing the vehicle, refer to the Precautions Section.

STEERING

2. Drain the power steering fluid.

3. Disconnect the battery ground cable and wait at least 3 minutes.

4. Place the wheels in the straight-ahead position.

5. Remove the steering wheel air bag access cover and disconnect the air bag connector from the cable reel.

6. Remove the 2 Torx bolts, one each side of the steering wheel hub.

7. Disconnect the horn switch connector.

8. Remove the air bag.

9. Disconnect the cable reel harness.

10. Remove the steering wheel nut.

11. Using a puller, remove the steering wheel.

12. Remove the steering coupler cover from the floor.

13. Matchmark and disconnect the joint coupler. If equipped with a center guide, discard it. It's for factory assembly only.

14. Disconnect the power steering outlet hose from the pump and remove the clamp.

15. Remove the 10mm flange bolts from the engine side mount bracket.

16. Remove the front wheels.

17. Separate the tie rod ends from the knuckle.

18. Remove the front subframe stiffener.

19. Disconnect the converter from the muffler.

20. Disconnect the power steering pressure switch.

21. Remove the driveshaft protector loop.

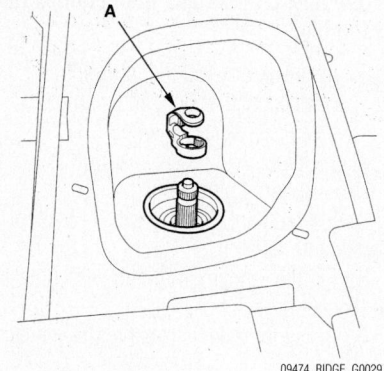

09474_RIDGE_G0029

Center guide (A)

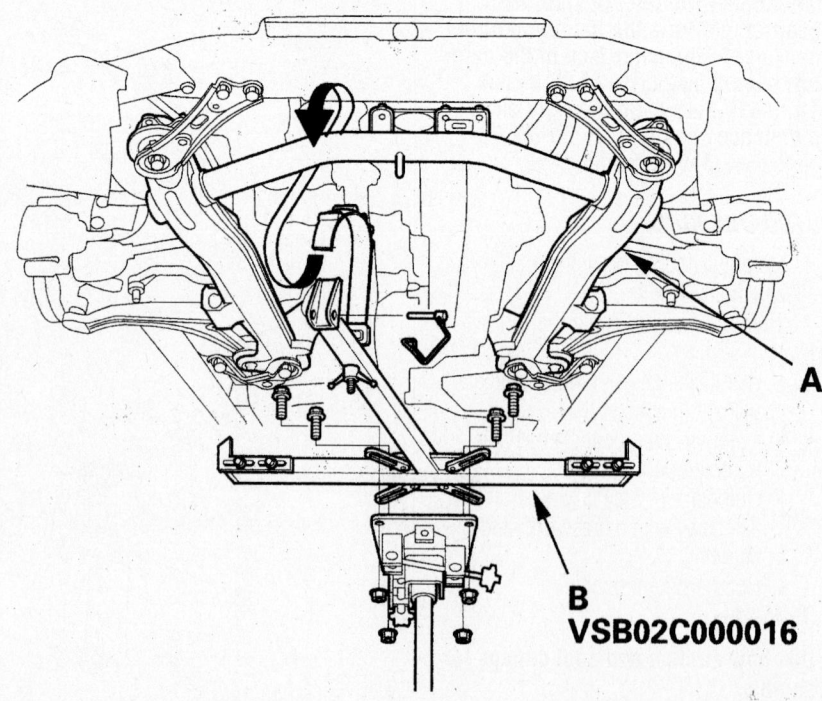

Subframe adapter installation

09474_RIDGE_G0030

22. Attach front subframe adapter VSB02C000016 to the subframe and attach the power train lift, OTC-1585, to the adapter as shown.

23. Remove the four 12mm flange bolts (A) from the subframe front brackets (B).

24. Loosen the two 14mm flange bolts (C) so they are $1\frac{3}{16}$ in. (about 30mm) from the mounting surface. Don't loosen the more than is necessary.

25. Raise the jack slightly and remove the 12mm bolts from the subframe rear brackets.

26. Remove the two 14mm bolts from the subframe rear brackets.

27. Lower the jack slowly until the subframe has dropped about $1\frac{15}{16}$ in. (about 50mm).

28. Remove the power steering line brackets.

29. Disconnect the return hose.

30. Remove the two 10mm bolts from the right side of the steering gear, then remove the mounting bracket and cushion.

31. Remove the two 10mm bolts from the left side of the steering gear.

32. Lower the jack another 2 inches.

33. Remove the steering gear stiffener from the left side of the subframe.

34. Disconnect the feed line from the gear.

35. Slide the gear towards the left side and out.

36. Installation is the reverse of removal. Observe the following torques:

- Feed line to gear: 31 ft. lbs. (42 Nm)
- Steering gear stiffener bolts: 28 ft. lbs. (38 Nm)
- Left side gear mounting bolts (loose until right side bolts are installed) then: 43 ft. lbs. (58 Nm)
- Right side gear mounting bolts: 29 ft. lbs. (39 Nm)
- Rear subframe brackets 12mm bolts: 86 ft. lbs. (117 Nm)
- Rear subframe brackets 14mm bolts: 76 ft. lbs. (103 Nm)
- Front subframe brackets 12mm bolts: 54 ft. lbs. (74 Nm)
- Front subframe brackets 14mm bolts: 76 ft. lbs. (103 Nm)
- Converter-to-muffler: 25 ft. lbs. (33 Nm)
- Subframe stiffener: 40 ft. lbs. (54 Nm)
- Engine side mount bolts: 33 ft. lbs. (44 Nm)
- Tie rod nuts: 40 ft. lbs. (54 Nm)
- Coupler bolt: 16 ft. lbs. (22 Nm)
- Steering wheel nut to 36 ft. lbs. (49 Nm)

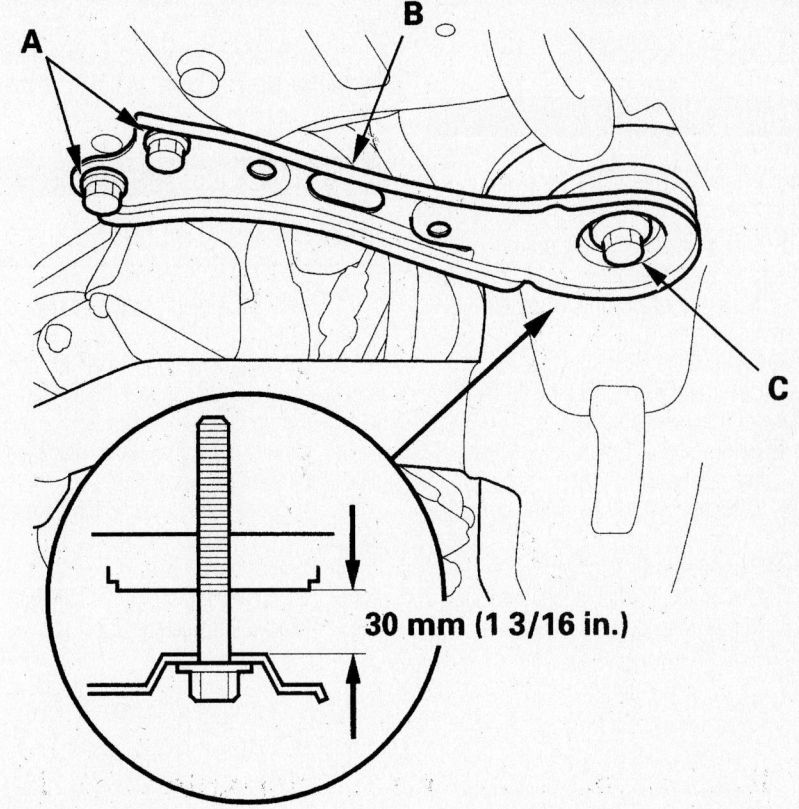

30 mm (1 3/16 in.)

09474_RIDGE_G0031

Subframe bracket bolts

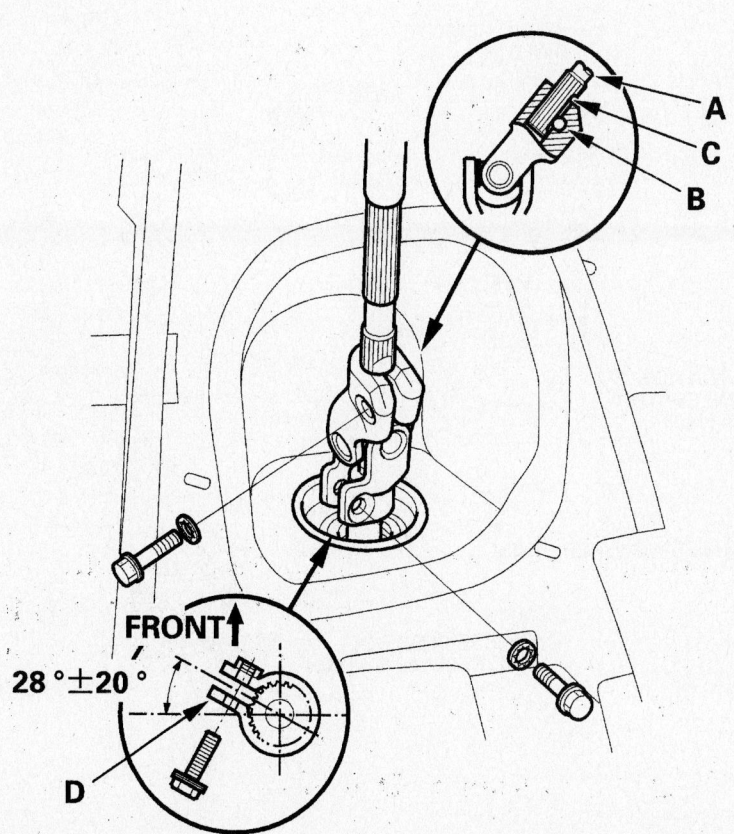

FRONT

$28°\pm20°$

09474_RIDGE_G0032

Coupler alignment. (A) Steering shaft, (B) bolt hole, (C) flat, (D) gap

37. Fill the system with fluid and bleed the air.
38. Reset the alignment.

Steering Wheel

REMOVAL & INSTALLATION

1. Disconnect the battery ground cable and wait at least 3 minutes.
2. Place the wheels in the straight-ahead position.
3. Remove the steering wheel air bag access cover and disconnect the air bag connector from the cable reel.
4. Remove the 2 Torx bolts, one each side of the steering wheel hub.
5. Disconnect the horn switch connector.
6. Remove the air bag.
7. Disconnect the cable reel harness.
8. Remove the steering wheel nut.

9. Using a puller, remove the steering wheel.
10. Installation is the reverse of removal. Torque the steering wheel nut to 36 ft. lbs. (49 Nm).

Steering Column

REMOVAL & INSTALLATION

1. Place the wheels in the straight-ahead position.
2. Remove the steering wheel air bag access cover and disconnect the air bag connector from the cable reel.
3. Remove the 2 Torx bolts, one each side of the steering wheel hub.
4. Disconnect the horn switch connector.
5. Remove the air bag.
6. Disconnect the cable reel harness.
7. Remove the steering wheel nut.

8. Using a puller, remove the steering wheel.
9. Remove the steering coupler cover from the floor.
10. Set the column shaft to the neutral position by raising the column to the uppermost position, then lower it 8mm (0.31 in.). Tighten the tilt lever.
11. Remove the column covers.
12. Move the shift lever to N and remove the shift cable from the column.
13. Remove the combination switch.
14. Disconnect the ignition switch.
15. Disconnect the immobilizer receiver unit, the park pin switch and the shift lock solenoid.
16. Matchmark and disconnect the steering joint from the column shaft.
17. Remove the attaching bolts and nuts and remove the steering column.
18. Installation is the reverse of removal. See the coupler alignment illustration. Torque the coupler bolt to 16 ft. lbs. (22 Nm).

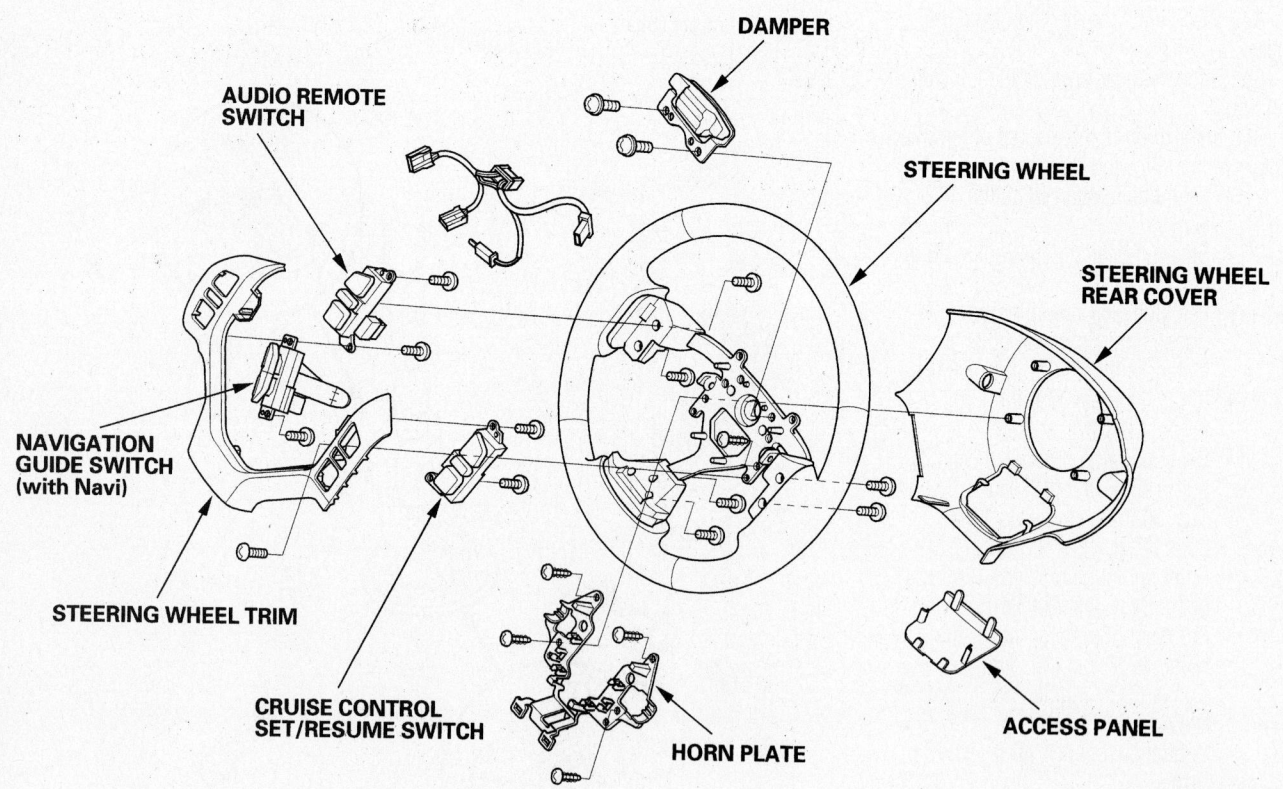

AUDIO REMOTE SWITCH

DAMPER

STEERING WHEEL

STEERING WHEEL REAR COVER

NAVIGATION GUIDE SWITCH (with Navi)

STEERING WHEEL TRIM

CRUISE CONTROL SET/RESUME SWITCH

HORN PLATE

ACCESS PANEL

09474_RIDGE_G0035

Steering wheel and related parts

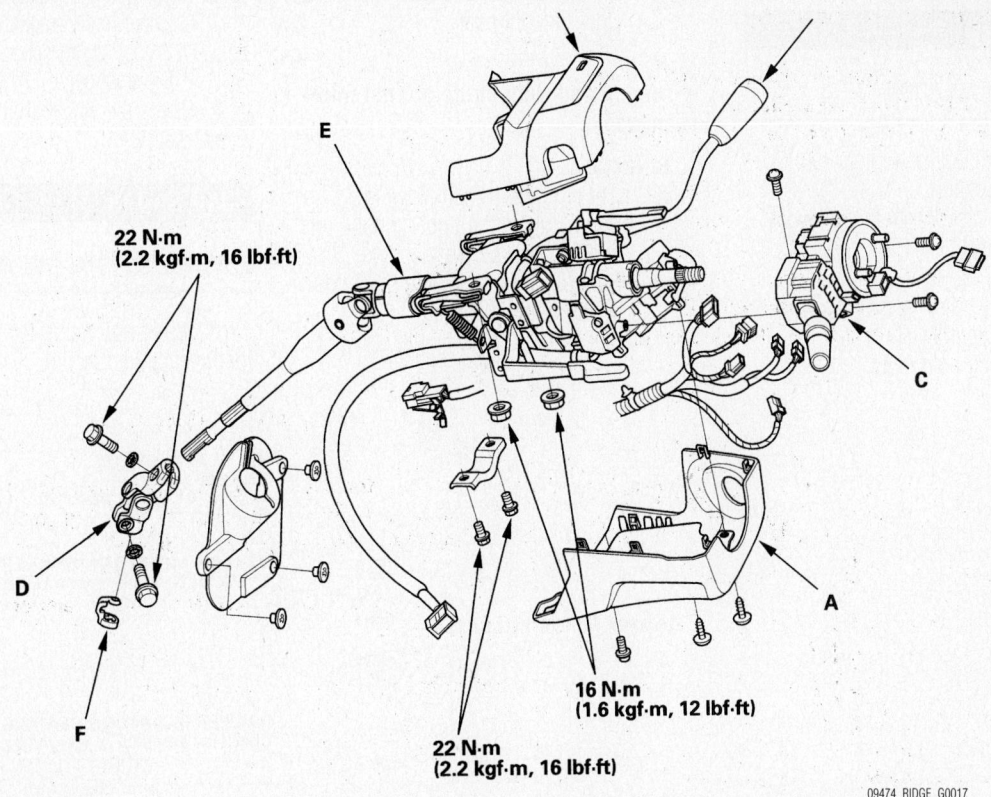

22 N·m
(2.2 kgf·m, 16 lbf·ft)

E

D

F

C

A

16 N·m
(1.6 kgf·m, 12 lbf·ft)

22 N·m
(2.2 kgf·m, 16 lbf·ft)

09474_RIDGE_G0017

Steering column exploded view

FRONT SUSPENSION

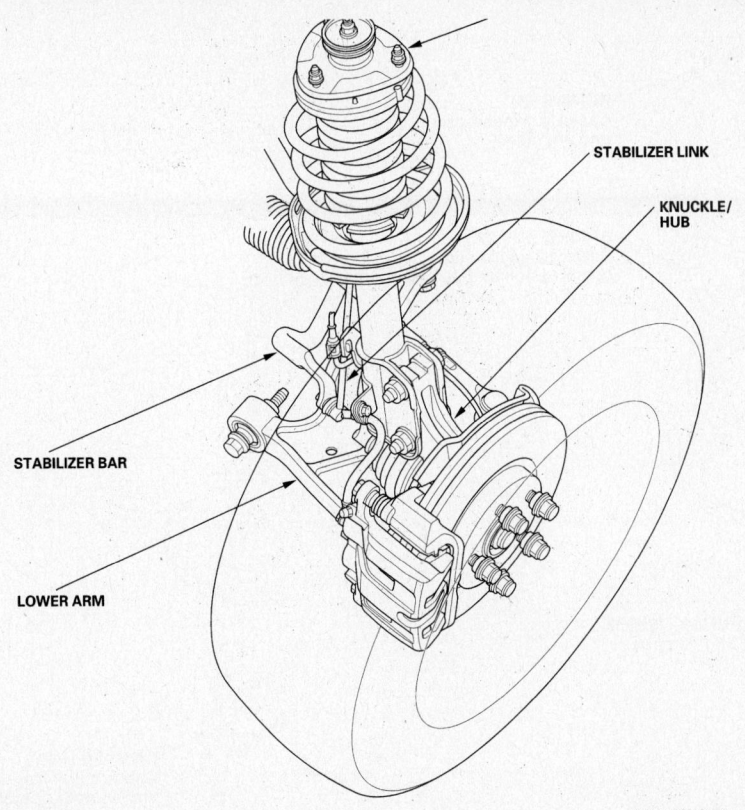

STABILIZER LINK

KNUCKLE/
HUB

STABILIZER BAR

LOWER ARM

09474_RIDGE_G0039

Front suspension components

Strut

REMOVAL & INSTALLATION

1. Before servicing the vehicle, refer to the Precautions Section.
2. Remove or disconnect the following:
   - Front wheel
   - Stabilizer bar link
   - Wheel speed sensor wiring bracket
   - Brake hose bracket
   - Strut pinch bolts/nuts

- Upper mount nuts
- Strut

➡**The left and right struts are not interchangeable.**

To install:
3. Install or connect the following:
   - Strut. Tighten the upper mount nuts to 43 ft. lbs. (59 Nm).
   - Strut pinch bolts. Tighten the nuts to 156 ft. lbs. (211 Nm).
   - Stabilizer bar link. Tighten the nut to 58 ft. lbs. (78 Nm).

- Brake hose bracket
- Wheel speed sensor wiring bracket
- Front wheel

4. Check the wheel alignment and adjust as necessary.

Coil Spring

REMOVAL & INSTALLATION

1. Before servicing the vehicle, refer to the Precautions Section.

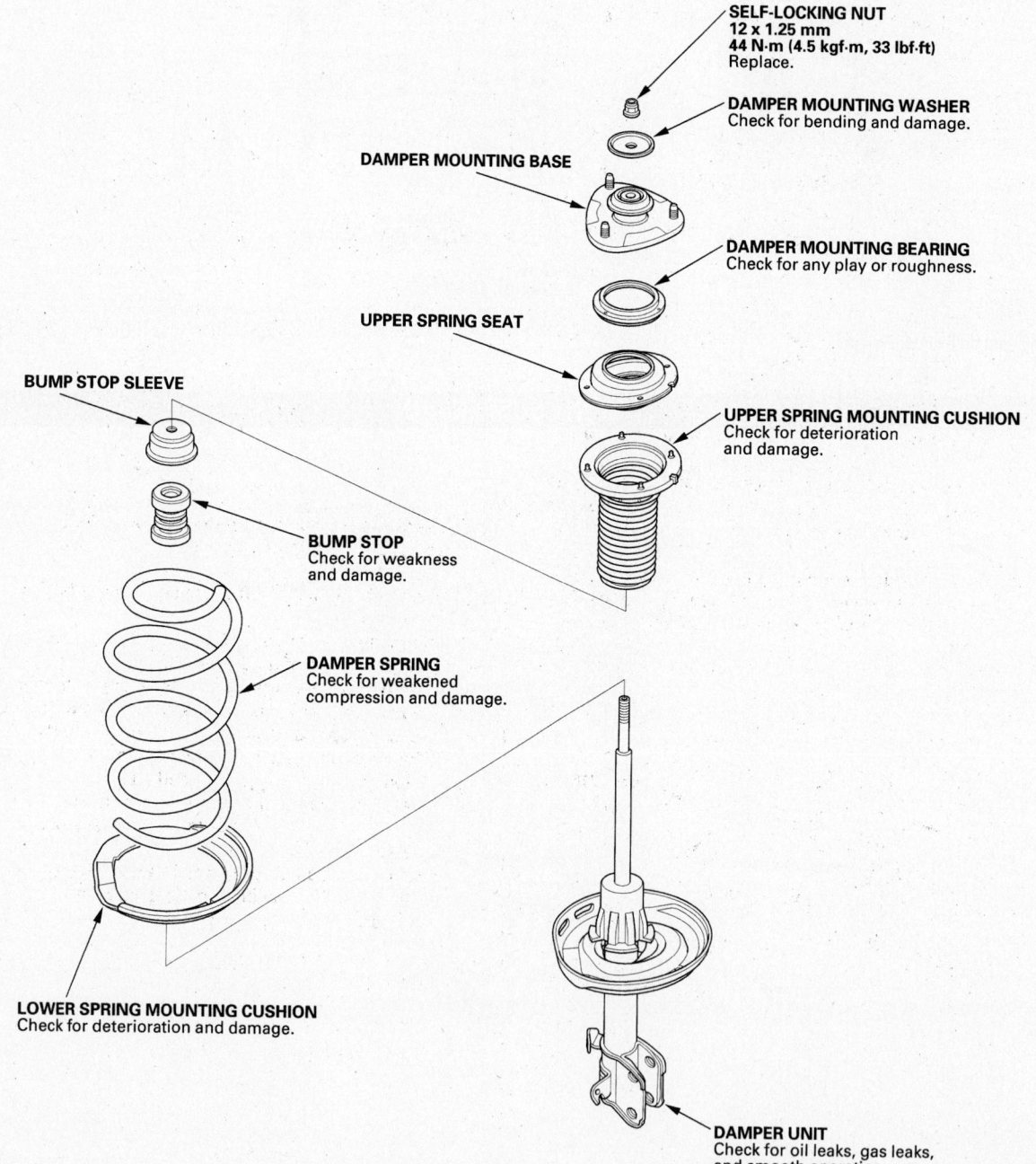

SELF-LOCKING NUT
12 x 1.25 mm
44 N·m (4.5 kgf·m, 33 lbf·ft)
Replace.

DAMPER MOUNTING WASHER
Check for bending and damage.

DAMPER MOUNTING BASE

DAMPER MOUNTING BEARING
Check for any play or roughness.

UPPER SPRING SEAT

UPPER SPRING MOUNTING CUSHION
Check for deterioration and damage.

BUMP STOP SLEEVE

BUMP STOP
Check for weakness and damage.

DAMPER SPRING
Check for weakened compression and damage.

LOWER SPRING MOUNTING CUSHION
Check for deterioration and damage.

DAMPER UNIT
Check for oil leaks, gas leaks, and smooth operation.

09474_RIDGE_G0033

Exploded view of the front strut assembly

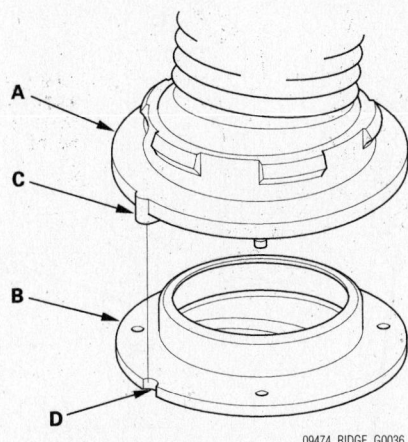

09474_RIDGE_G0036

Install the cushion (A) on the upper spring seat (B) by aligning the tab (C) with the notch (D) in the seat

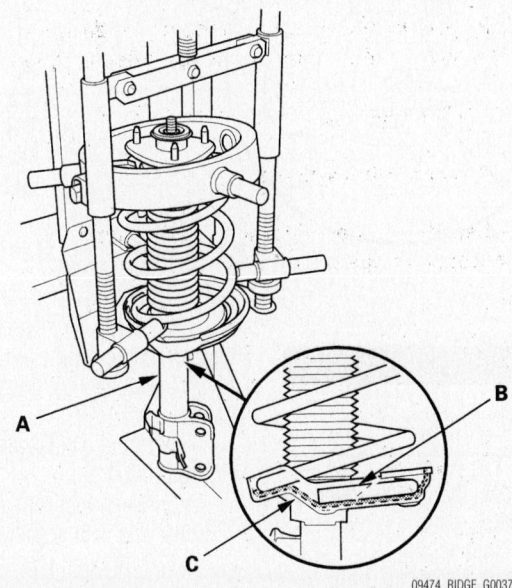

09474_RIDGE_G0037

Align the bottom of the spring (B) and the stepped portion of the lower spring seat (C)

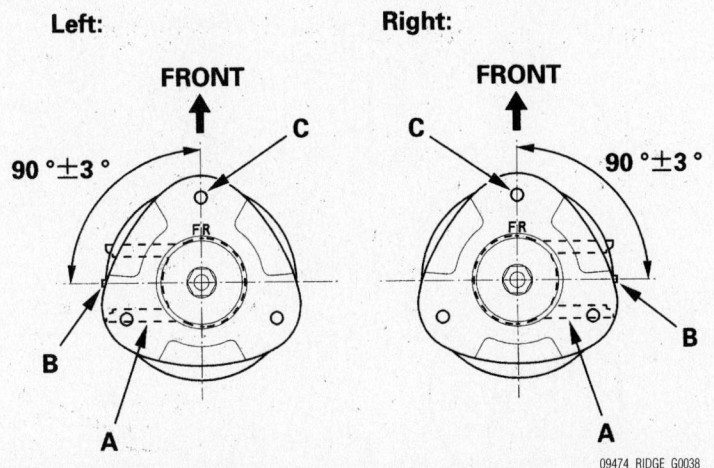

09474_RIDGE_G0038

Align an angle of the bracket (A) and the tab portion (B) on the cushion. Position the angle of the tab portion and the stud bolt (C) near the FR stamp on the upper spring seat

2. Remove the strut from the vehicle and install in a strut spring compressor. Compress the spring until the end of the spring comes away from the spring seat.

3. Remove the upper strut mount, spring seat and related components.

4. Remove the coil spring from the strut spring compressor.

To install:

➥**Use a new self-locking nut.**

5. Install the cushion (A) on the upper spring seat (B) by aligning the tab (C) with the notch (D) in the seat.

6. Compress the spring.

7. Align an angle of the bracket (A) and the tab portion (B) on the cushion. Position the angle of the tab portion and the stud bolt (C) near the **FR** stamp on the upper spring seat.

8. Install the upper strut mounting components and tighten the nut to 33 ft. lbs. (44 Nm).

9. Install the strut to the vehicle.

10. Check the wheel alignment and adjust as necessary.

Ball Joint

REMOVAL & INSTALLATION

The lower ball joints are replaced with the control arms as an assembly.

Lower Control Arm

REMOVAL & INSTALLATION

1. Before servicing the vehicle, refer to the Precautions Section.

2. Remove the front wheels.

3. Disconnect the stabilizer link from the strut.

4. Remove and discard the lower ball joint nut.

5. Separate the lower ball stud from the knuckle.

6. Remove and discard the lower arm horizontal and vertical bolts.

7. Remove the lower arm.

8. Installation is the reverse of removal. Use new bolts and nuts. Install all fasteners loosely, then, final torque all fasteners with the suspension loaded (weight of the car). Never back off the ball stud nut to align the cotter pin holes. Always tighten it to align.

- Vertical lower arm bolt: 83 ft. lbs. (113 Nm)
- Horizontal lower arm bolt: 119 ft. lbs. (162 Nm)

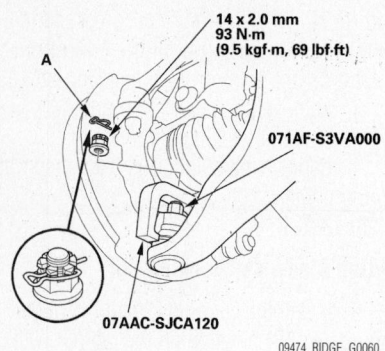

14 x 2.0 mm
93 N·m
(9.5 kgf·m, 69 lbf·ft)

071AF-S3VA000

07AAC-SJCA120

09474_RIDGE_G0060

Lower control arm ball stud

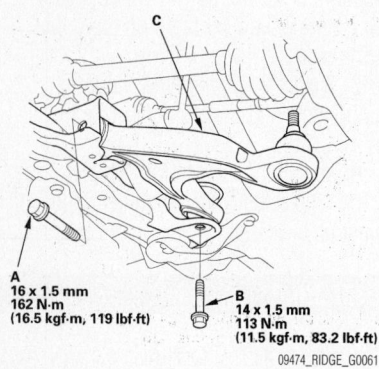

A
16 x 1.5 mm
162 N·m
(16.5 kgf·m, 119 lbf·ft)

B
14 x 1.5 mm
113 N·m
(11.5 kgf·m, 83.2 lbf·ft)

09474_RIDGE_G0061

Lower control arm mounting. (A) horizontal bolt, (B) vertical bolt, (C) control arm

- Ball stud nut: 69 ft. lbs. (93 Nm)
- Stabilizer link nut: 58 ft. lbs. (78 Nm)

CONTROL ARM BUSHING REPLACEMENT

The control arm bushings are serviced with the control arms as an assembly.

Stabilizer Link

REMOVAL & INSTALLATION

1. Before servicing the vehicle, refer to the Precautions Section.
2. Remove the front wheels.
3. Disconnect the stabilizer link from the strut.
4. Disconnect the stabilizer link from the stabilizer bar.
5. Installation is the reverse of removal. Torque the nuts to 58 ft. lbs. (78 Nm).

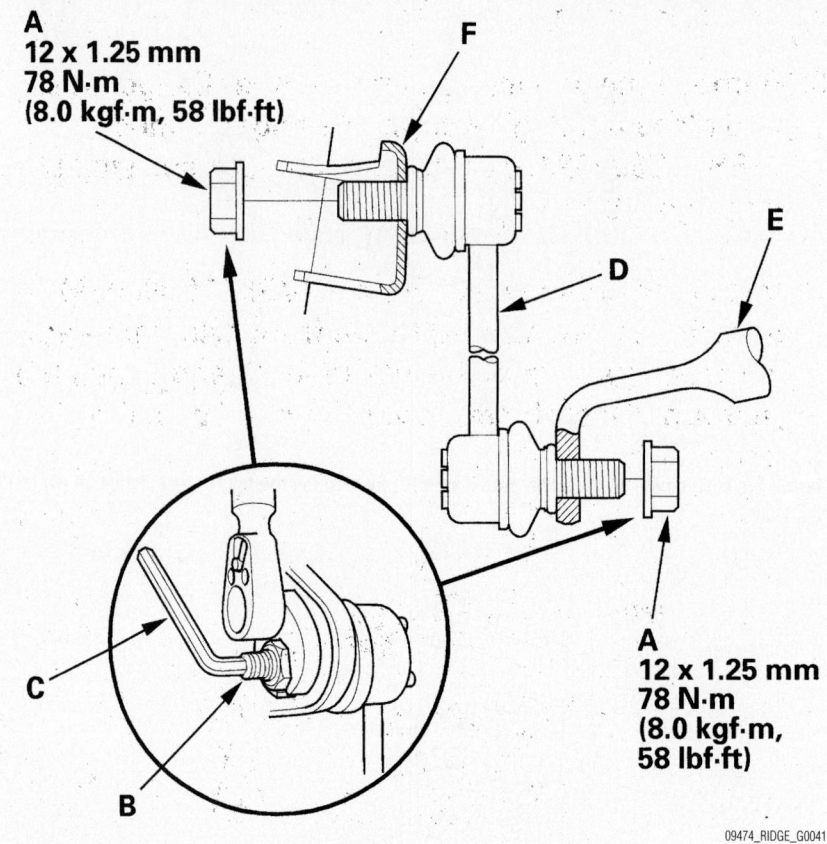

A
12 x 1.25 mm
78 N·m
(8.0 kgf·m, 58 lbf·ft)

A
12 x 1.25 mm
78 N·m
(8.0 kgf·m, 58 lbf·ft)

09474_RIDGE_G0041

Stabilizer link

Stabilizer Bar

REMOVAL & INSTALLATION

1. Before servicing the vehicle, refer to the Precautions Section.
2. Remove the front wheels.
3. Remove the caps from the front strut mounting nuts. Position the engine

hanger adapters (VSB02C000024) with the FRONT mark facing forward, over the flange nuts.
4. Install the balancer bar (VSB02C000019):
   a. Attach the front arm (A) to the front head with spacer (B) and a 10x1.25mm bolt (C)
   b. Attach the rear arm (D) to the rear head with an 8x1.25mm bolt (E).

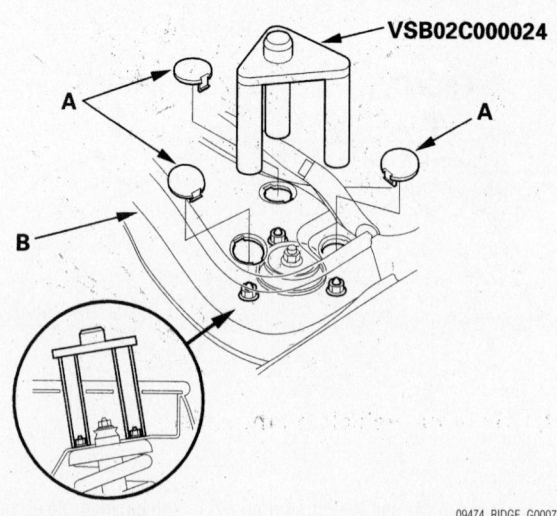

VSB02C000024

09474_RIDGE_G0007

Engine hanger adapters installed

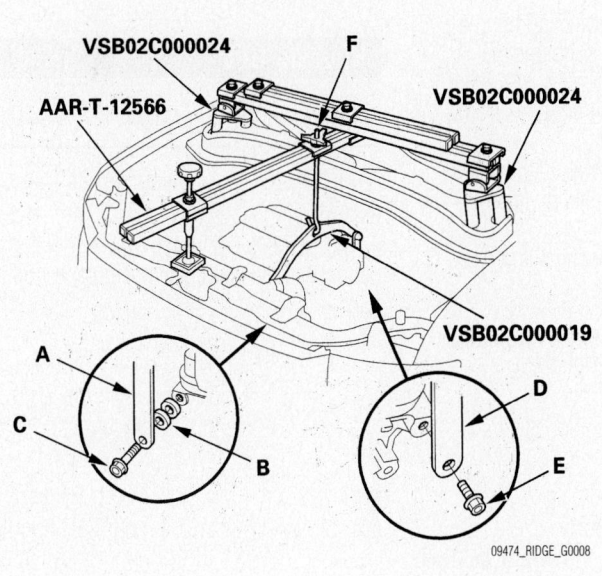

Balancer bar installation and support hanger

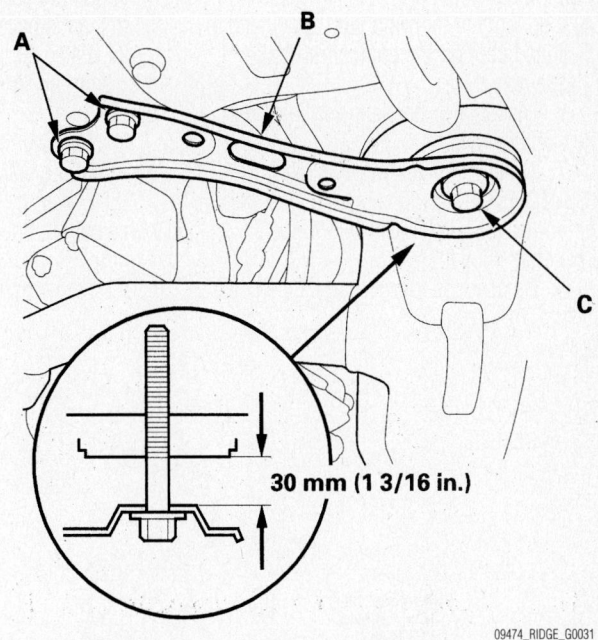

Subframe bracket bolts

30 mm (1 3/16 in.)

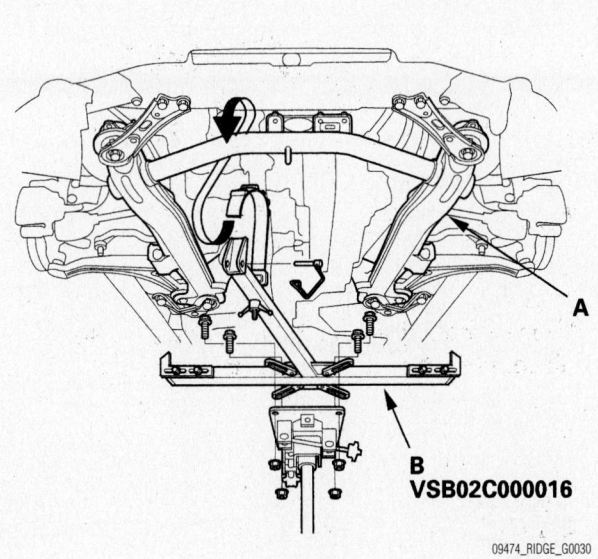

Subframe adapter installation

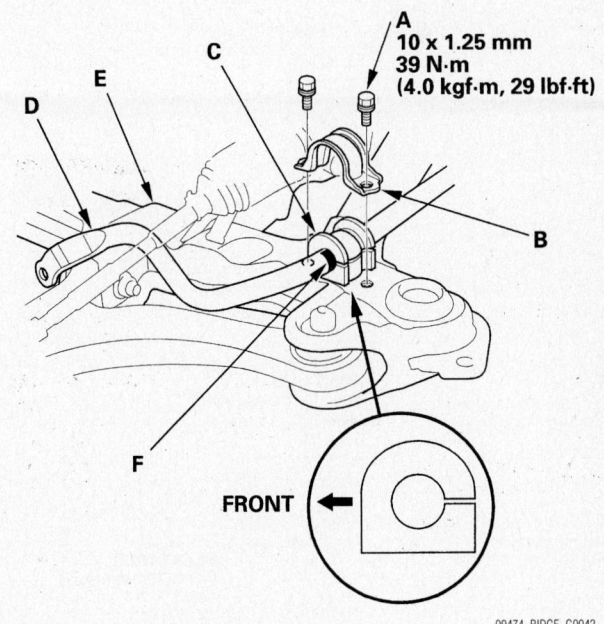

Stabilizer bar brackets

A
10 x 1.25 mm
39 N·m
(4.0 kgf·m, 29 lbf·ft)

FRONT

5. Install the support hanger (AAR-T-12566):

a. Attach the hook to the slotted hole in the balancer bar.

b. Tighten the wing nut (F) by hand to lift and support the engine/transaxle assembly.

6. Disconnect the stabilizer links from the stabilizer bar.

7. Attach front subframe adapter VSB02C000016 to the subframe and attach the power train lift, OTC-1585, to the adapter as shown.

8. Remove the four 12mm flange bolts

(A) from the subframe front brackets (B).

9. Loosen the two 14mm flange bolts (C) so they are 9/16 in. (about 14mm) from the mounting surface. Don't loosen the more than is necessary.

10. Raise the jack slightly and remove the 12mm bolts from the subframe rear brackets.

11. Remove the two 14mm bolts from the subframe rear brackets.

12. Lower the jack slowly until the subframe has dropped about 9/16 in. (about 14mm).

13. Matchmark the bushings and bar.

Remove the stabilizer bar brackets and bar.

14. Installation is the reverse of removal. Align the subframe and perform a front end alignment. Torque the bracket bolts to 29 ft. lbs. (39 Nm).

Knuckle/Hub

REMOVAL & INSTALLATION

1. Before servicing the vehicle, refer to the Precautions Section.
2. Remove the wheel.
3. Remove the brake hose bracket bolt.

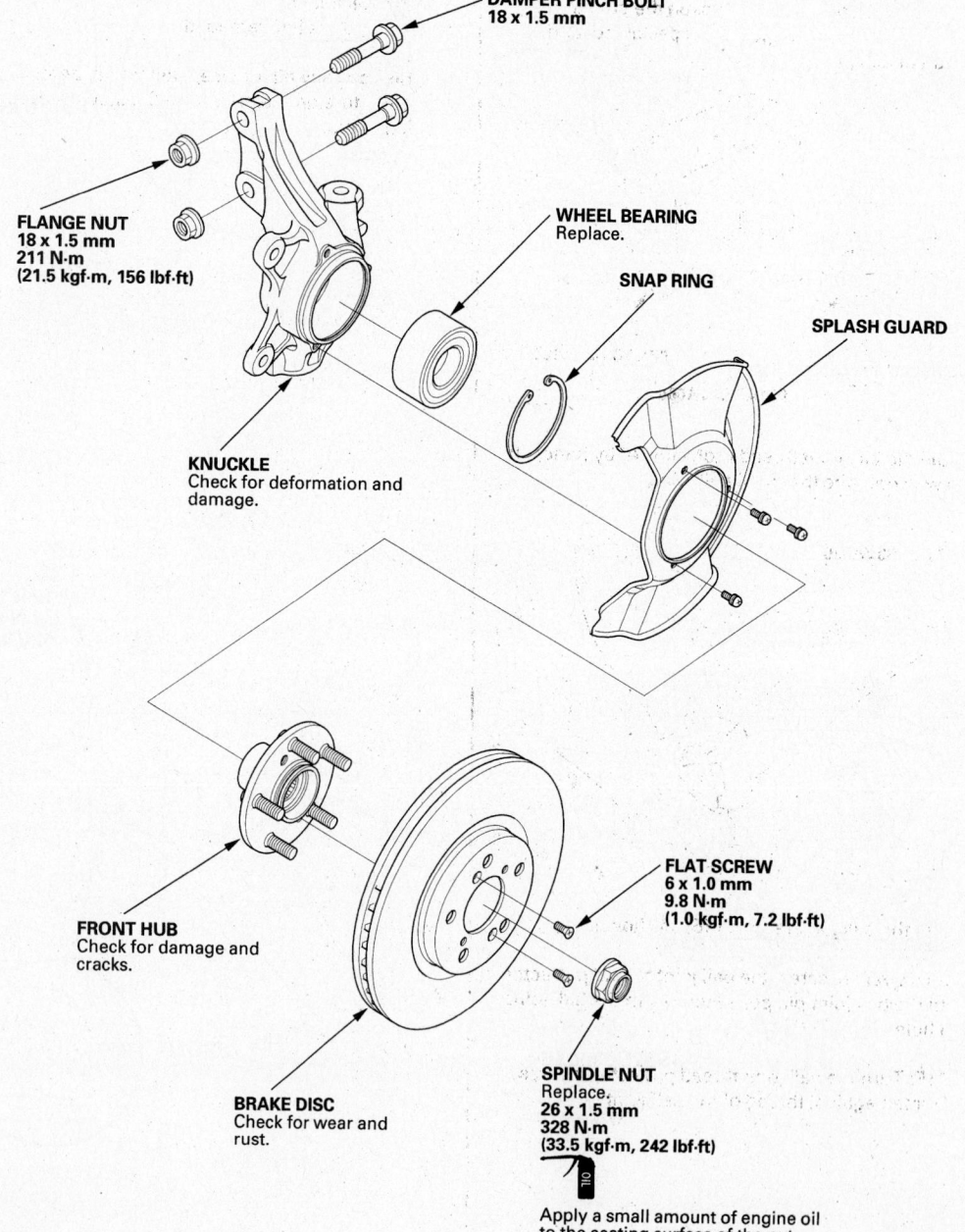

Hub/knuckle exploded view

4. Remove the caliper bracket bolts and caliper and support it out of the way.

5. Remove the sensor from the knuckle.

6. Unstake and remove the halfshaft end nut.

7. Remove the brake rotor.

➡️**It may be necessary to push off the rotor by using two 8x1.25mm bolts in the holes provided.**

8. Remove the nut and disconnect the tie rod end from the knuckle.

9. Remove the ball stud nut and separate the lower control arm from the knuckle.

10. Remove the strut-to-knuckle bolts/nuts.

11. Pull the knuckle outwards while tapping the end of the halfshaft with a plastic hammer.

12. Installation is the reverse of removal. Use new bolts and nuts. Install all fasteners loosely, then, final torque all fasteners with the suspension loaded (weight of the car). Never back off the ball stud nut to align the cotter pin holes. Always tighten it to align. Observe the following torques:

- Strut-to-knuckle: 156 ft. lbs. (211 Nm)
- Lower arm ball stud nut: 69 ft. lbs. (93 Nm)
- Tie rod end nut: 40 ft. lbs. (54 Nm)
- Halfshaft end nut: 242 ft. lbs. (328 Nm)
- Caliper bracket bolts: 101 ft. lbs. (137 Nm)

Wheel Bearings

REMOVAL & INSTALLATION

1. Before servicing the vehicle, refer to the Precautions Section.

2. Remove the knuckle.

3. Place the knuckle in a press. Separate the hub (A) from the knuckle (B). Hold the knuckle with the press attachment (C).

4. Press the inner race from the hub.

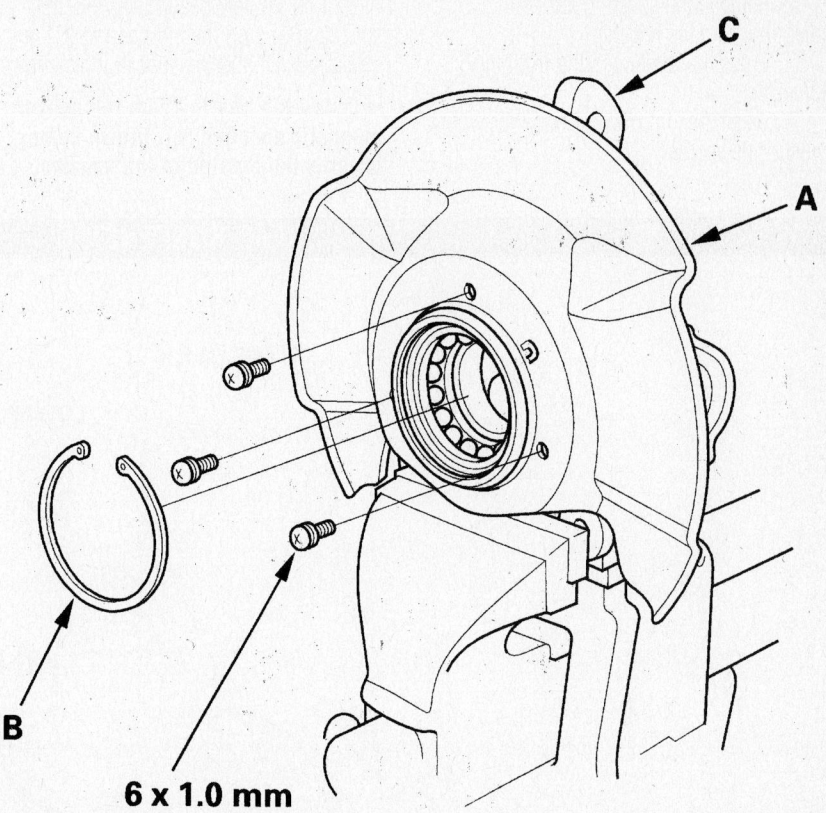

6 x 1.0 mm

Remove the splash guard and snapring from the knuckle

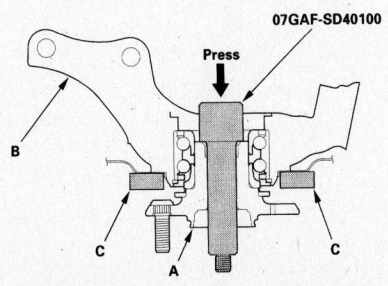

Press the hub from the knuckle

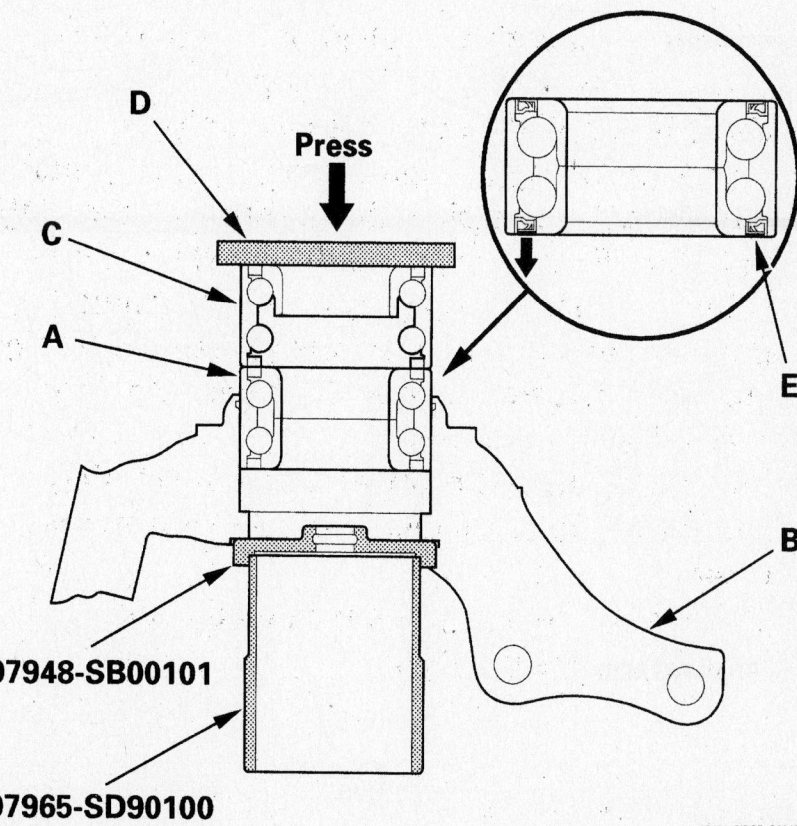

07948-SB00101

07965-SD90100

Installing the new bearing

5. Remove the splash guard and snapring from the knuckle.

6. Press the wheel bearing from the knuckle.

7. Wash the knuckle and hub thoroughly with a safe solvent.

8. Press a new bearing (A) into the knuckle (B) using the old bearing (C), a steel plate (D) and the special tools shown.

→**Install the bearing with the sensor magnetic encoded (E), brown color, towards the outside of the knuckle.**

Take care to avoid damaging the encoder surface. Keep all magnetic tools away from the encoder.

9. The remainder of installation is the reverse of removal. Torque the splash shield bolts to 86 inch lbs. (9.8 Nm).

REAR SUSPENSION

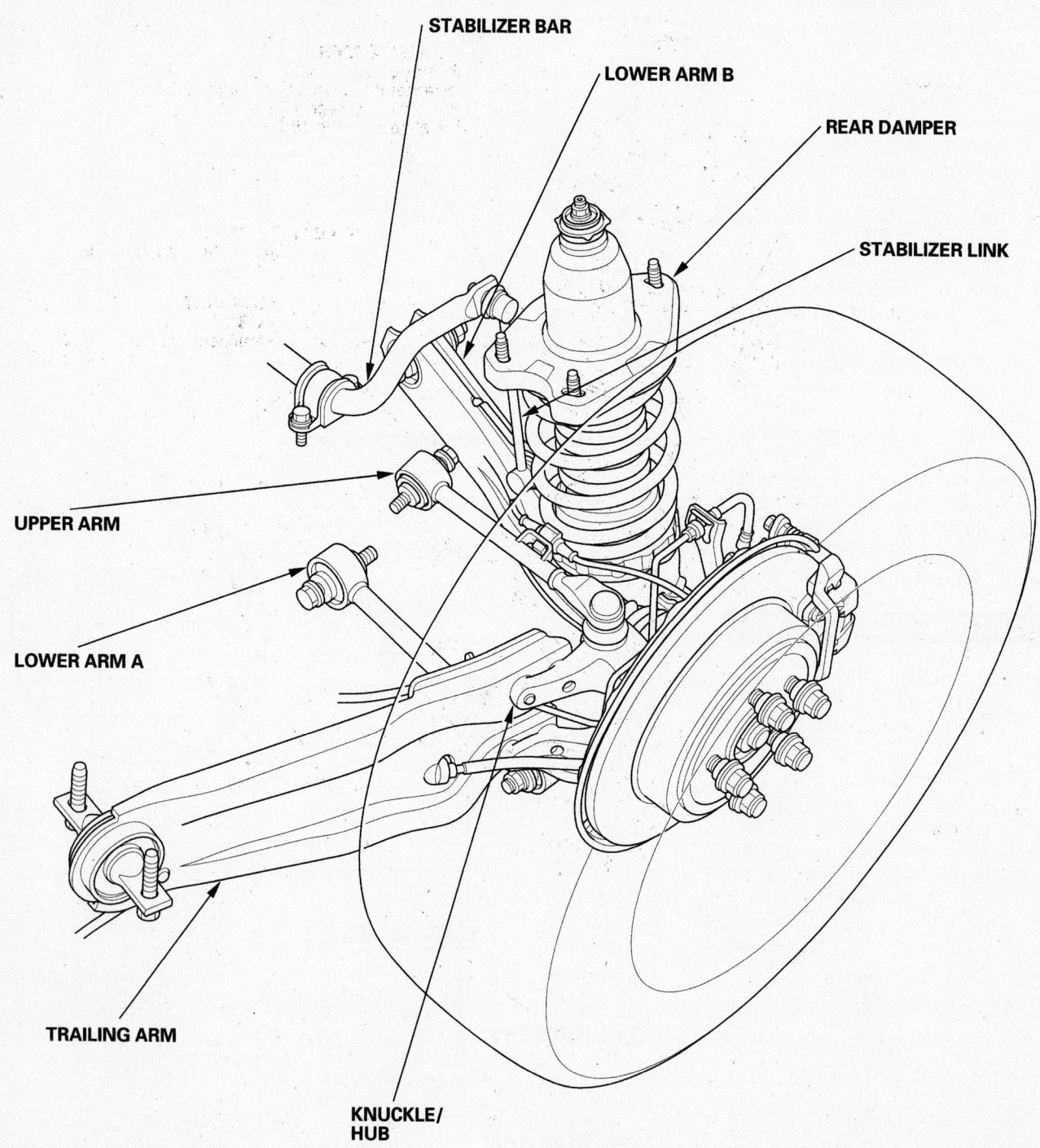

Rear suspension components

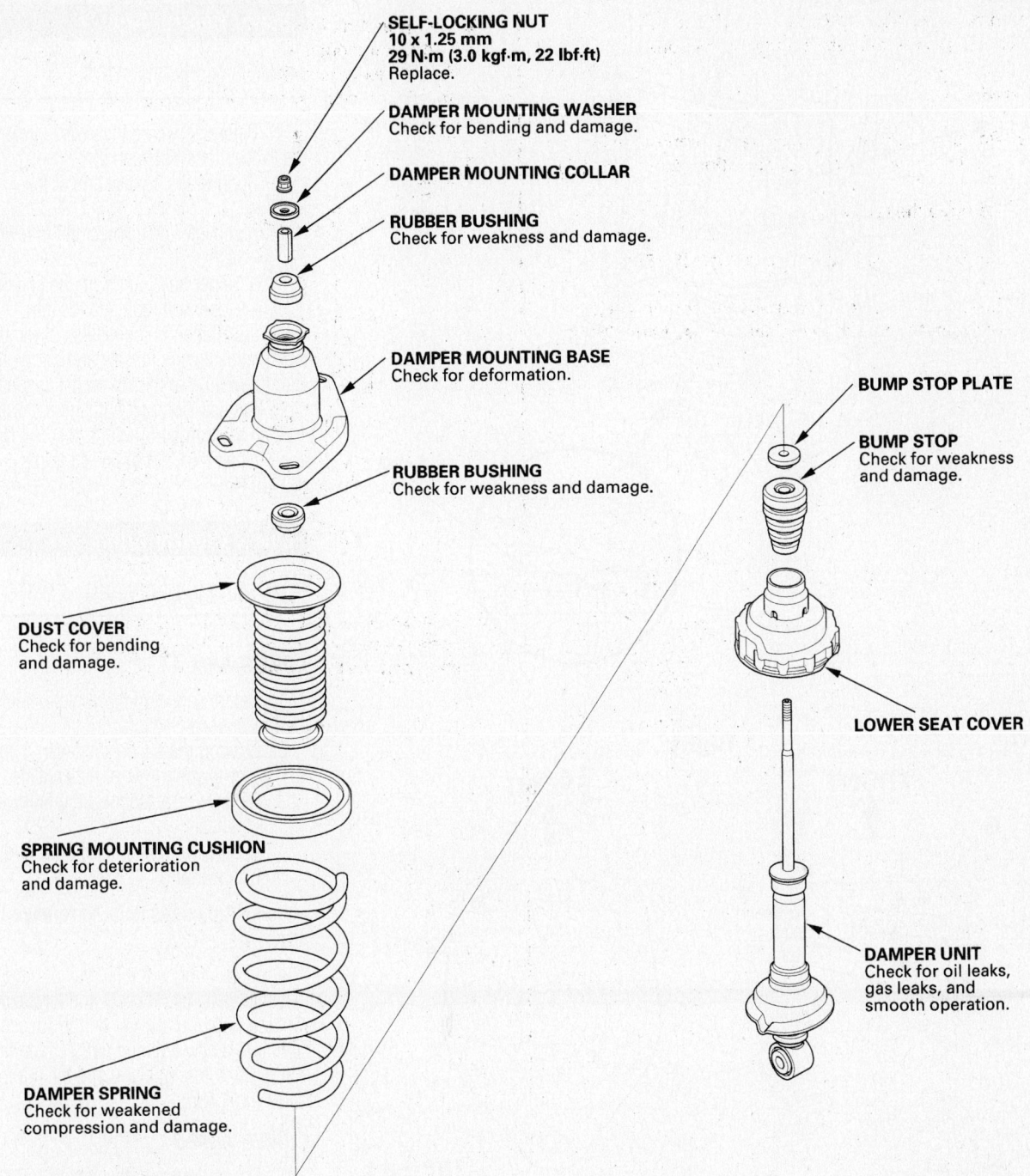

SELF-LOCKING NUT
10 x 1.25 mm
29 N·m (3.0 kgf·m, 22 lbf·ft)
Replace.

DAMPER MOUNTING WASHER
Check for bending and damage.

DAMPER MOUNTING COLLAR

RUBBER BUSHING
Check for weakness and damage.

DAMPER MOUNTING BASE
Check for deformation.

BUMP STOP PLATE

BUMP STOP
Check for weakness
and damage.

RUBBER BUSHING
Check for weakness and damage.

DUST COVER
Check for bending
and damage.

LOWER SEAT COVER

SPRING MOUNTING CUSHION
Check for deterioration
and damage.

DAMPER UNIT
Check for oil leaks,
gas leaks, and
smooth operation.

DAMPER SPRING
Check for weakened
compression and damage.

09474_RIDGE_G0034

Rear strut exploded view

Strut

REMOVAL & INSTALLATION

1. Before servicing the vehicle, refer to the Precautions Section.
2. Raise and support the rear end.
3. Remove the wheel(s).
4. Remove lower arm B.
5. Remove the 3 bolts from the top of the strut.

6. Installation is the reverse of removal. Torque the upper bolts to 25 ft. lbs. (34 Nm).

Coil Spring

REMOVAL & INSTALLATION

1. Before servicing the vehicle, refer to the Precautions Section.
2. Place the strut in a spring compressor.

3. Compress the spring and remove the shaft nut.
4. Release the spring pressure
To assemble:
5. Compress the spring.
6. Assemble all parts except the shaft washer and nut.
7. Align the spring as shown.
8. Install the washer and NEW nut. Torque to 22 ft. lbs. (29 Nm).

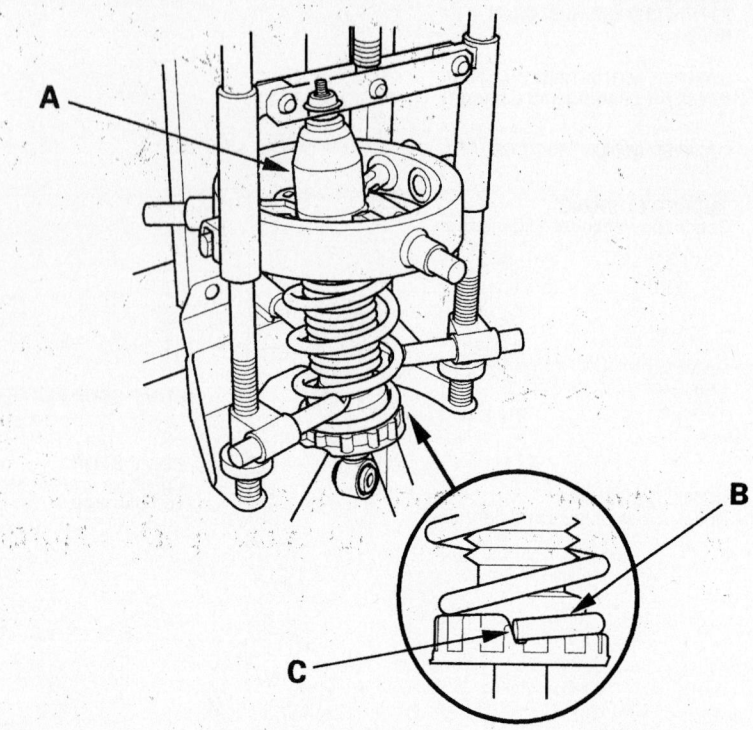

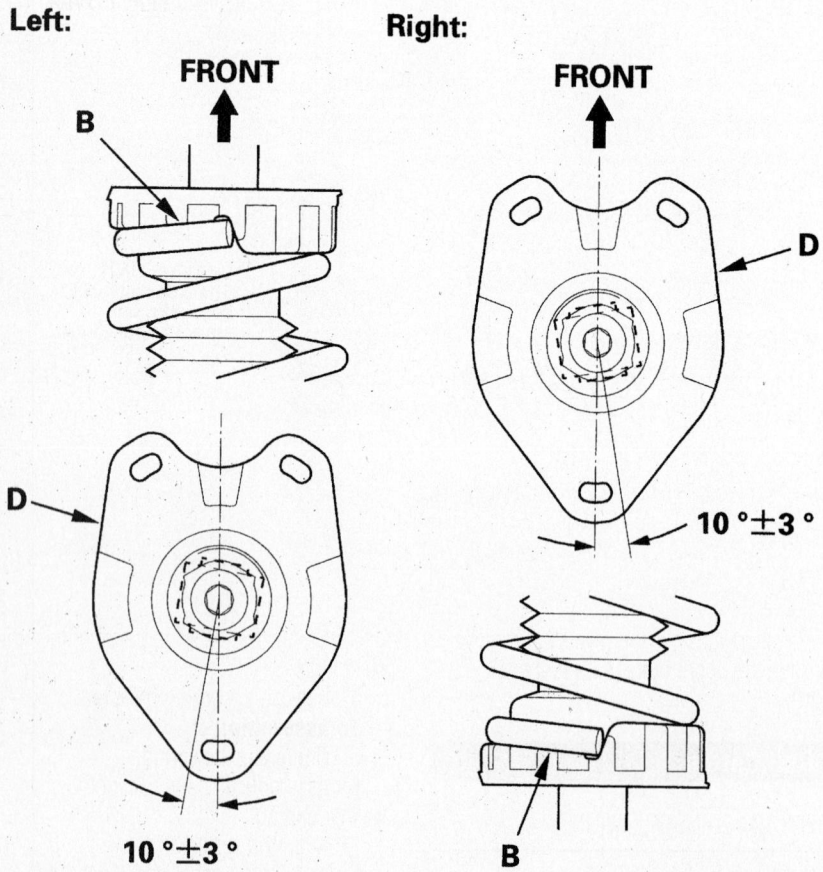

Left: **Right:**

FRONT   **FRONT**

B

D

D

10 °±3 °

10 °±3 °

B

09474_RIDGE_G0047

Spring alignment. (A) strut, (B) spring, (C) lower spring seat, (D) mounting base

Upper Control Arm

REMOVAL & INSTALLATION

   1. Before servicing the vehicle, refer to the Precautions Section.
   2. Support the control arm at the knuckle.
   3. Remove or disconnect the following:
   * Wheel
   * Upper ball joint from the knuckle
   * Upper arm bolt and the arm
   4. Installation is the reverse of removal. Install all fasteners loosely and torque them with the weight of the car on the suspension. Use new bolt and nut, and cotter pin. Tighten the arm bolt to 69 ft. lbs. (94 Nm) and the ball joint nut to 36–43 ft. lbs. (49–59 Nm).

Lower Control Arm

REMOVAL & INSTALLATION

LOWER ARM A

   1. Before servicing the vehicle, refer to the Precautions Section.
   2. Support the suspension with a floor jack under lower arm B, at the knuckle.
   3. Remove or disconnect the following:
   * Wheel
   * Lower arm mounting bolts and nuts
   * Lower arm

➡**Raise the suspension to remove the arm.**

   4. Installation is the reverse of removal. Install all fasteners loosely and torque them with the weight of the car on the suspension. Use a new bolt and nut. Tighten the bolt to 69 ft. lbs. (93 Nm) and the nut to 74 ft. lbs. (101 Nm).

LOWER ARM B

   1. Before servicing the vehicle, refer to the Precautions Section.
   2. Support the control arm with a jack at the knuckle.
   3. Remove or disconnect the following:
   * Wheel
   * Locknut (A)
   * Bolt (C)
   * Bolt (D)
   4. Gradually lower the arm.
   5. Matchmark the adjusting cam.
   6. Remove the locknut, adjusting bolt and adjusting cam at the frame.
   7. Spring assembly
   * Inner nuts and bolts and the arm

B
14 x 1.5 mm
101 N·m
(10.3 kgf·m, 74.5 lbf·ft)

C

D
14 x 1.5 mm
93 N·m
(9.5 kgf·m, 69 lbf·ft)

A

09474_RIDGE_G0048

Rear lower arm (A) mounting

8. Installation is the reverse of removal. Install all fasteners loosely and torque them with the weight of the car on the suspension. Use new bolts and nuts.

9. Observe the following torques:
   - Adjusting cam nut: 61 ft. lbs. (83 Nm)
   - Nut A: 36 ft. lbs. (49 Nm)
   - Bolt C: 156 ft. lbs. (211 Nm)
   - Bolt D: 105 ft. lbs. (142 Nm)

CONTROL ARM BUSHING REPLACEMENT

The control arm bushings are serviced with the control arms as an assembly.

Trailing Arm

REMOVAL & INSTALLATION

1. Before servicing the vehicle, refer to the Precautions Section.
2. Support control arm B with a jack at the knuckle.
3. Remove the caliper and support it out of the way.
4. Remove the rotor.
5. Disconnect the brake hose from the brake pipe and discard the clip. Plug the lines.

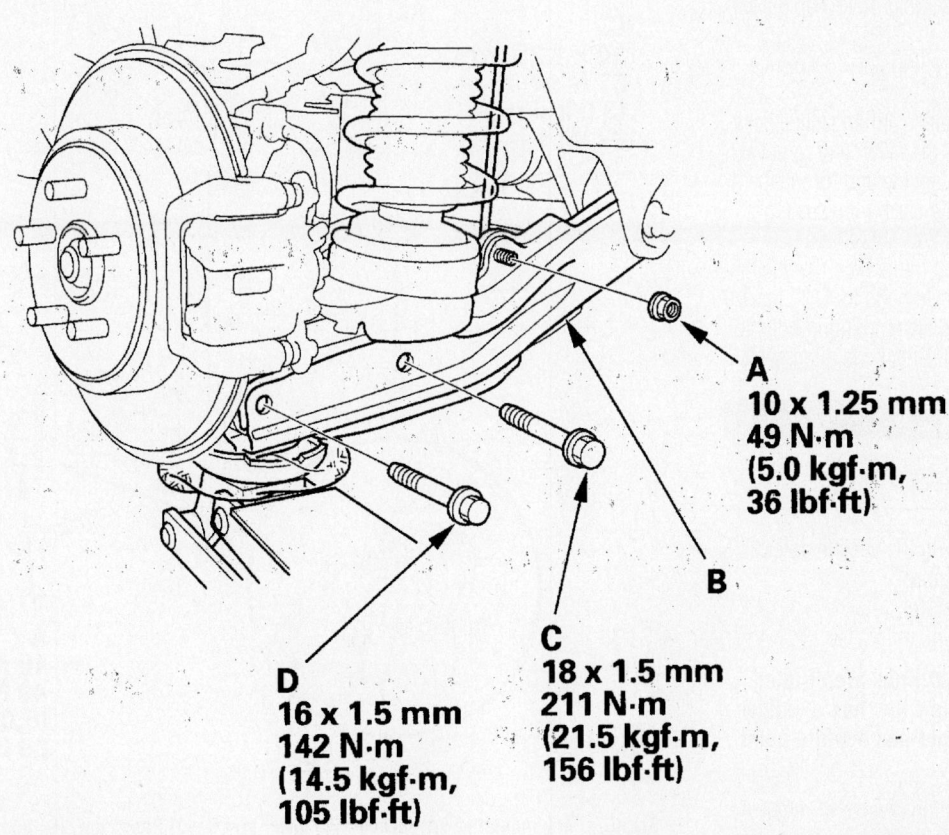

A
10 x 1.25 mm
49 N·m
(5.0 kgf·m, 36 lbf·ft)

B

C
18 x 1.5 mm
211 N·m
(21.5 kgf·m, 156 lbf·ft)

D
16 x 1.5 mm
142 N·m
(14.5 kgf·m, 105 lbf·ft)

09474_RIDGE_G0049

Rear lower arm B mounting

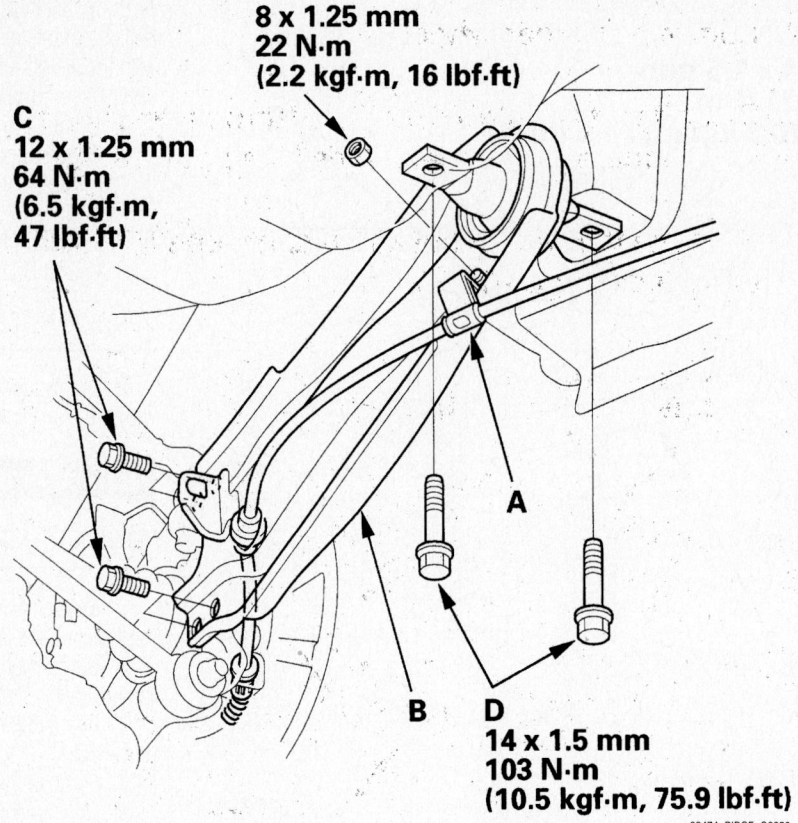

8 x 1.25 mm
22 N·m
(2.2 kgf·m, 16 lbf·ft)

C
12 x 1.25 mm
64 N·m
(6.5 kgf·m, 47 lbf·ft)

A

B

D
14 x 1.5 mm
103 N·m
(10.5 kgf·m, 75.9 lbf·ft)

09474_RIDGE_G0050

Trailing arm mounting

6. Remove the parking brake cable from the training arm.

7. Remove the trailing arm-to-knuckle bolts.

8. Remove the trailing arm mounting bolts.

9. Installation is the reverse of removal. Install all fasteners loosely and torque them with the weight of the car on the suspension. Use new bolts and nuts and a new brake line clip. Bleed the brakes. Torque the trailing arm-to-knuckle bolts to 47 ft. lbs. (64 Nm); the trailing arm mounting bolts to 76 ft. lbs. (103 Nm). Check the alignment.

Stabilizer Link

REMOVAL & INSTALLATION

1. Before servicing the vehicle, refer to the Precautions Section.

2. Remove the wheel.

3. Remove the link nuts.

➡ The left and right links aren't interchangeable. The left link has a yellow paint mark; the right has a white paint mark.

4. Installation is the reverse of removal. Install the link on the bar with the joint pins set at he center of their movement. Use new

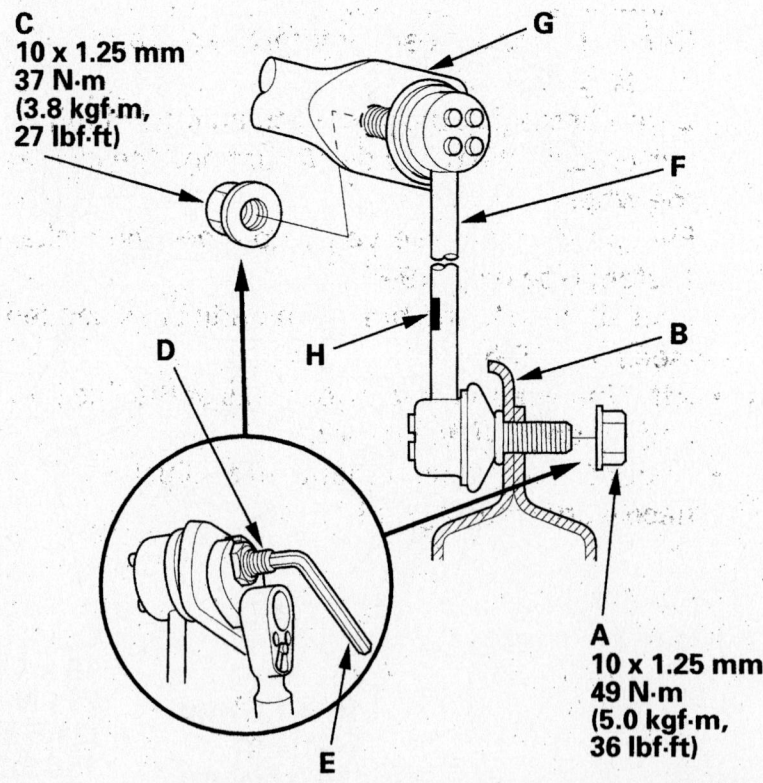

C
10 x 1.25 mm
37 N·m
(3.8 kgf·m, 27 lbf·ft)

G

F

D

H

B

E

A
10 x 1.25 mm
49 N·m
(5.0 kgf·m, 36 lbf·ft)

09474_RIDGE_G0051

Stabilizer link mounting. (A) locknut, (B) lower arm B, (C) flange nut, (D) joint pin, (E) hex wrench nuts. Torque the link-to-bar nut to 27 ft. lbs. (37 Nm); the link-to-lower arm B nut to 36 ft. lbs. (49 Nm).

Stabilizer Bar

REMOVAL & INSTALLATION

1. Before servicing the vehicle, refer to the Precautions Section.

2. Remove the wheel.
3. Remove the link-to-bar nuts.
4. Remove the rear subframe. See the illustration.
5. Remove the stabilizer bar bracket bolts. Matchmark the bushings, and remove the bar.
6. Installation is the reverse of removal. Torque the bracket bolts to 16 ft. lbs. (22 Nm).

Knuckle/Hub

REMOVAL & INSTALLATION

1. Before servicing the vehicle, refer to the Precautions Section.
2. Remove the wheel.
3. Remove and discard the brake hose clip.

Rear Subframe Torque

NOTE:
- When installing, align both installation reference holes in the subframe with the reference holes in the body using a screwdriver or tapered punch as a guide.
- After removing the subframe mounting bolts, be sure to replace them with new ones.

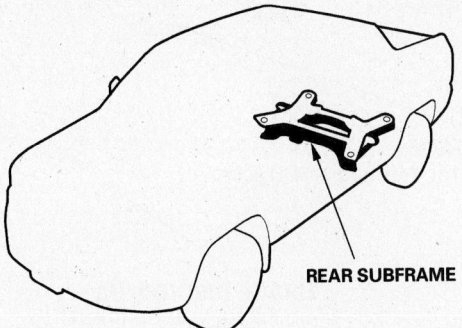

REAR SUBFRAME

Reference hole alignment

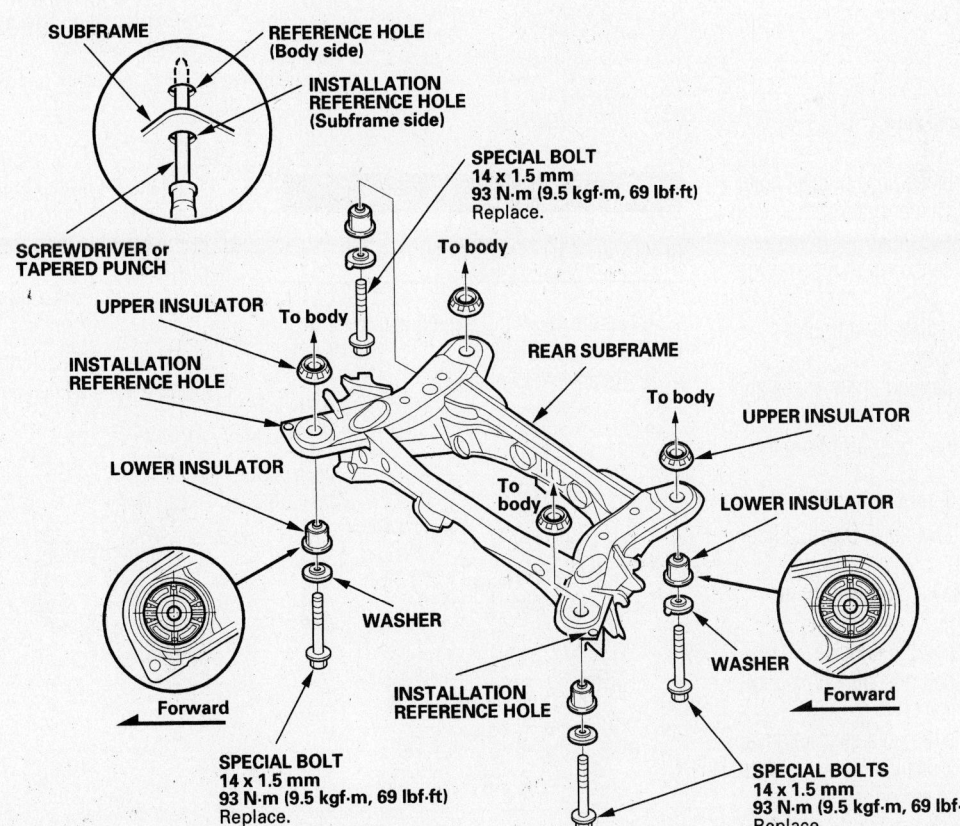

Rear subframe mounting

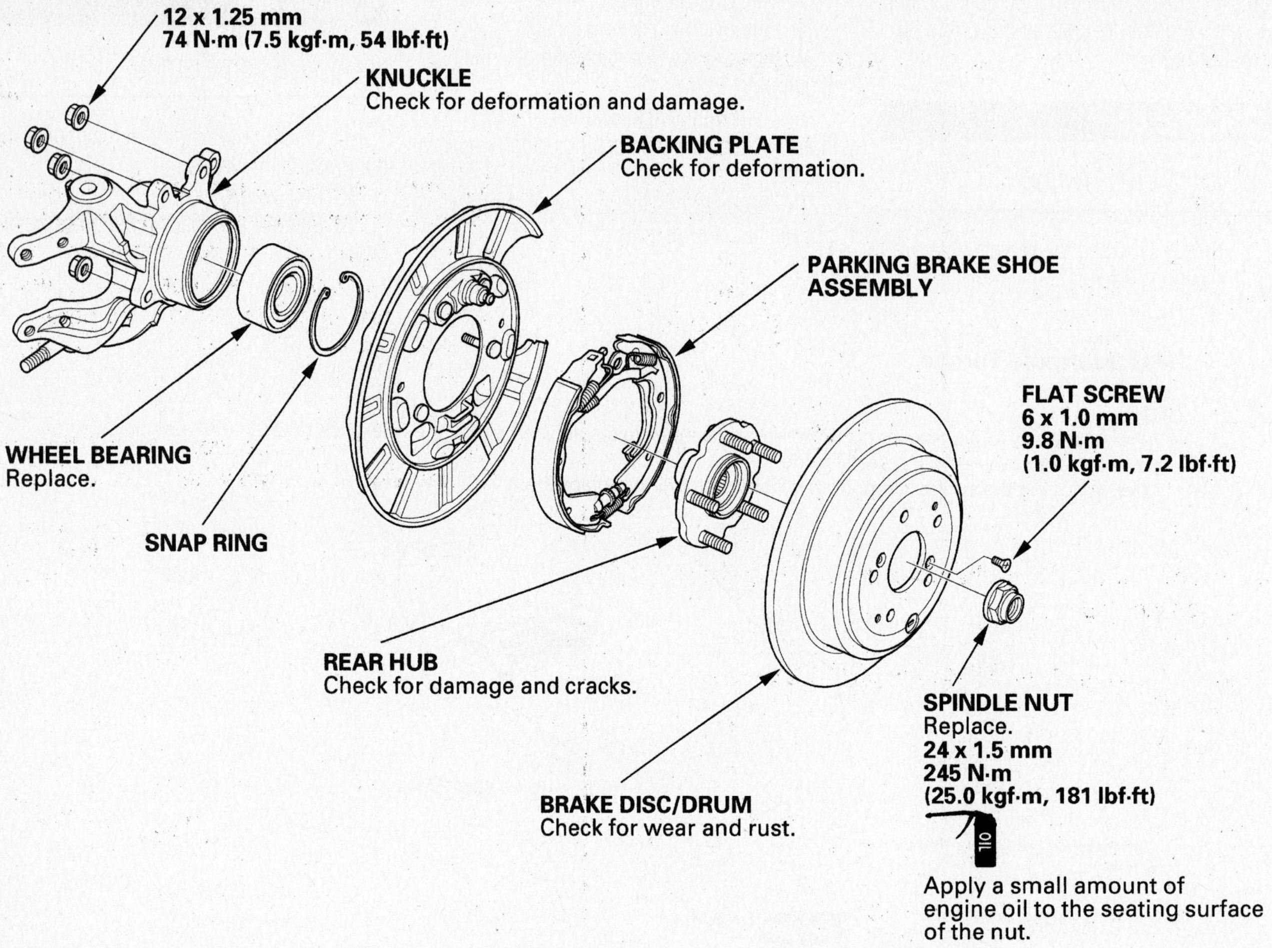

12 x 1.25 mm
74 N·m (7.5 kgf·m, 54 lbf·ft)

KNUCKLE
Check for deformation and damage.

BACKING PLATE
Check for deformation.

**PARKING BRAKE SHOE
ASSEMBLY**

WHEEL BEARING
Replace.

SNAP RING

FLAT SCREW
6 x 1.0 mm
9.8 N·m
(1.0 kgf·m, 7.2 lbf·ft)

REAR HUB
Check for damage and cracks.

BRAKE DISC/DRUM
Check for wear and rust.

SPINDLE NUT
Replace.
24 x 1.5 mm
245 N·m
(25.0 kgf·m, 181 lbf·ft)

Apply a small amount of
engine oil to the seating surface
of the nut.

09474_RIDGE_G0053

Knuckle/hub exploded view

4. Remove the brake hose bracket bolts.
5. Remove the caliper bracket mounting bolts and support the caliper out of the way.
6. Remove the sensor from the knuckle. Don't disconnect it.
7. Remove the halfshaft end nut.
8. Remove the rotor.
9. Remove the parking brake shoes and cable.
10. Disconnect the upper arm from the knuckle.
11. Remove lower arm A.
12. Support the lower arm B with a floor jack.
13. Disconnect lower arm B from the knuckle.
14. Pull outward on the knuckle while tapping the end of the halfshaft with a plastic hammer.
15. Installation is the reverse of removal. Install all fasteners loosely and torque them with the weight of the car on the suspension. Use new bolts and nuts and a new brake line clip. See the relevant procedures in this chapter for torque values.

Wheel Bearings

REMOVAL & INSTALLATION

1. Before servicing the vehicle, refer to the Precautions Section.
2. Remove the knuckle.

3. Place the knuckle in a press. Separate the hub (A) from the knuckle (B). Hold the knuckle with the press attachment (C).
4. Press the inner race from the hub.
5. Remove the backing plate and snapring from the knuckle.

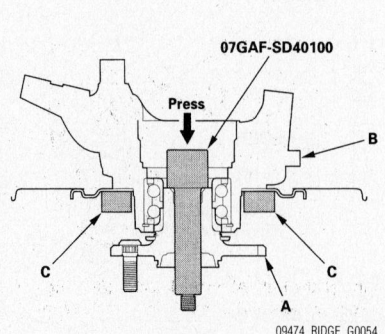

09474_RIDGE_G0054

Press the hub from the knuckle

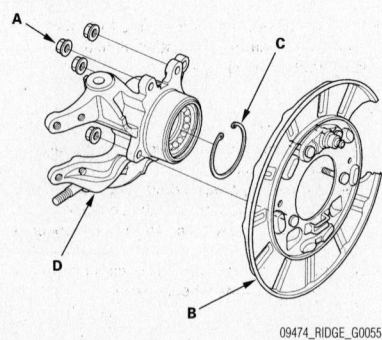

09474_RIDGE_G0055

Remove the backing plate and snapring from the knuckle

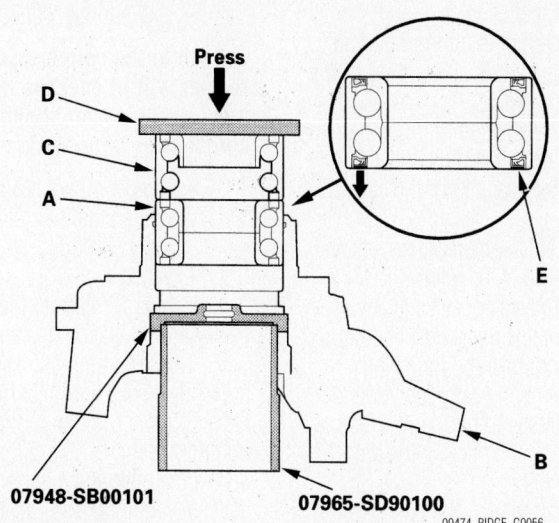

Press

D
C
A
E

B

07948-SB00101   07965-SD90100

09474_RIDGE_G0056

Installing the new bearing

6. Press the wheel bearing from the knuckle.

7. Wash the knuckle and hub thoroughly with a safe solvent.

8. Press a new bearing (A) into the knuckle (B) using the old bearing (C), a steel plate (D) and the special tools shown.

➡**Install the bearing with the sensor magnetic encoded (E), brown color, towards the outside of the knuckle. Take care to avoid damaging the encoder surface. Keep all magnetic tools away from the encoder.**

9. The remainder of installation is the reverse of removal. Torque the backing plate bolts to 54 ft. lbs. (74 Nm).

FRONT BRAKES

Brake Caliper

REMOVAL & INSTALLATION

Front

1. Before servicing the vehicle, refer to the Precautions Section.

2. Remove some fluid from the reservoir with a suction pump.

3. Remove or disconnect the following:
   • Front wheels
   • Banjo bolt and disconnect the brake hose from the caliper. Plug the hose to prevent fluid loss and contamination.

   • Caliper pin bolts and the caliper from its mounting bracket

To install:

4. Install or connect the following:
   • Caliper over the pads and onto its mounting bracket. Torque both caliper pin bolts to 53 ft. lbs. (72 Nm).

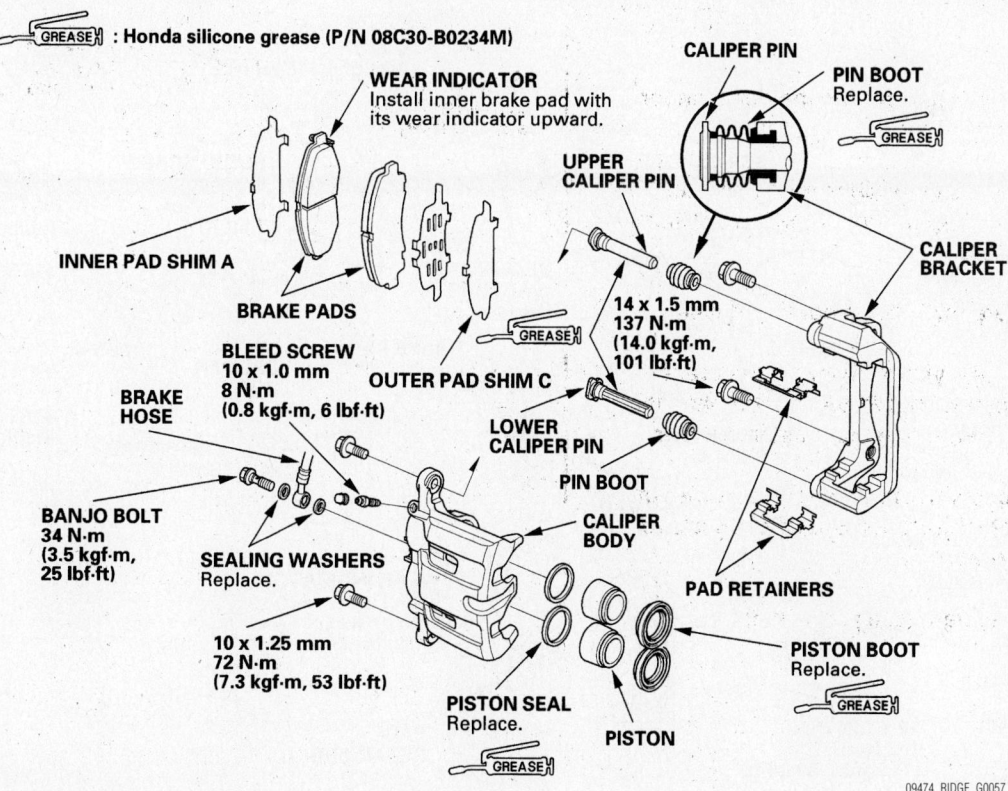

[GREASE] : Honda silicone grease (P/N 08C30-B0234M)

WEAR INDICATOR
Install inner brake pad with its wear indicator upward.

CALIPER PIN

PIN BOOT
Replace.

UPPER CALIPER PIN

CALIPER BRACKET

INNER PAD SHIM A

BRAKE PADS

14 x 1.5 mm
137 N·m
(14.0 kgf·m,
101 lbf·ft)

BLEED SCREW
10 x 1.0 mm
8 N·m
(0.8 kgf·m, 6 lbf·ft)

OUTER PAD SHIM C

BRAKE HOSE

LOWER CALIPER PIN

PIN BOOT

BANJO BOLT
34 N·m
(3.5 kgf·m,
25 lbf·ft)

SEALING WASHERS
Replace.

CALIPER BODY

PAD RETAINERS

10 x 1.25 mm
72 N·m
(7.3 kgf·m, 53 lbf·ft)

PISTON BOOT
Replace.

PISTON SEAL
Replace.

PISTON

09474_RIDGE_G0057

Exploded view of the front caliper components

- Brake hose to the caliper using new sealing washers. Carefully torque the banjo bolt to 25 ft. lbs. (34 Nm).
5. Fill the reservoir with fluid and bleed the brakes.
  - Front wheels

Disc Brake Pads

REMOVAL & INSTALLATION

Front

1. Before servicing the vehicle, refer to the Precautions Section.
2. Remove a small amount of brake fluid from the reservoir using a suction pump.
3. Remove or disconnect the following:

- Front wheels
- Lower caliper retaining bolt and pivot the caliper upward, off of the pads
- Pad springs while holding the pads
- Pad shim and pad retainers
- Disc brake pads from the caliper

To install:

4. Clean the caliper thoroughly; remove any rust from the lip of the rotor. Check the brake rotor for grooves or cracks. If any heavy scoring is present, the rotor must be replaced.
5. Install or connect the following:

- Pad retainers. Apply molybdenum brake grease to both surfaces of the shims and the back of the disc brake pads.
- Pads and shims. The pad with the wear indicator goes in the inboard position.

- Pad springs while holding the pads

➡ **Push in the caliper piston so the caliper will fit over the pads. This is most easily accomplished with a pad spreader or large C-clamp.**

- Caliper down into position and tighten the mounting bolt to 53 ft. lbs. (72 Nm)
- Wheels

6. Add brake fluid to the master cylinder reservoir and install the cap.
7. Depress the brake pedal several times and make sure that the movement feels normal. The first brake pedal application may result in a very long pedal action due to the pistons being retracted. Always make several brake applications before starting the vehicle. Bleed the system if necessary.

REAR BRAKES

Brake Caliper

REMOVAL & INSTALLATION

1. Before servicing the vehicle, refer to the Precautions Section.

2. Remove some fluid from the reservoir with a suction pump.
3. Remove or disconnect the following:

- Rear wheels
- Banjo bolt and disconnect the brake hose from the caliper. Plug

the hose to prevent fluid loss and contamination.
- 2 caliper mounting bolts and the caliper from its mounting bracket

To Install:

4. Install or connect the following:
- Caliper over the pads and onto its

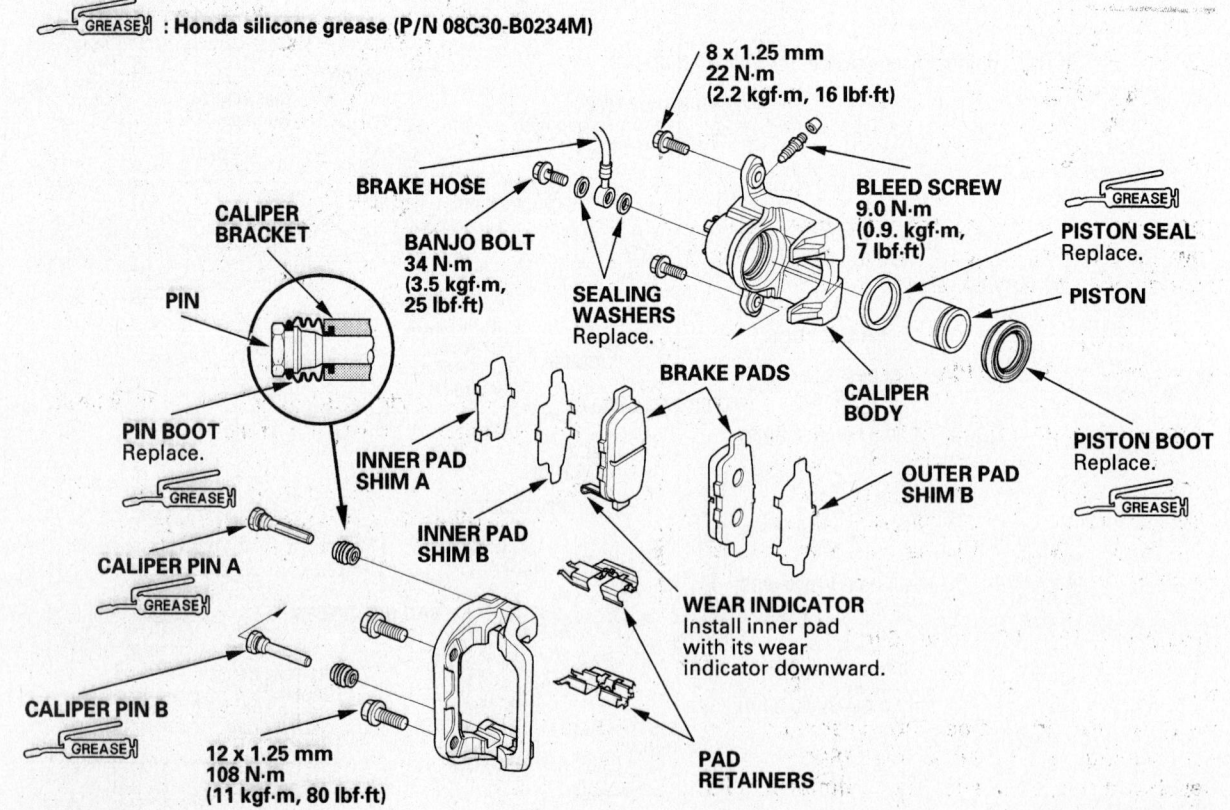

GREASE : Honda silicone grease (P/N 08C30-B0234M)

8 x 1.25 mm
22 N·m
(2.2 kgf·m, 16 lbf·ft)

BRAKE HOSE

BLEED SCREW
9.0 N·m
(0.9. kgf·m,
7 lbf·ft)

GREASE

PISTON SEAL
Replace.

CALIPER BRACKET

BANJO BOLT
34 N·m
(3.5 kgf·m,
25 lbf·ft)

SEALING WASHERS
Replace.

PISTON

PIN

BRAKE PADS

CALIPER BODY

PIN BOOT
Replace.

PISTON BOOT
Replace.

GREASE

INNER PAD SHIM A

OUTER PAD SHIM B

GREASE

CALIPER PIN A

INNER PAD SHIM B

GREASE

WEAR INDICATOR
Install inner pad with its wear indicator downward.

CALIPER PIN B

GREASE

12 x 1.25 mm
108 N·m
(11 kgf·m, 80 lbf·ft)

PAD RETAINERS

Exploded view of the rear caliper components

mounting bracket. Tighten the caliper bolts to 16 ft. lbs. (22 Nm).
- Brake hose with new sealing washers. Tighten the banjo bolt to 25 ft. lbs. (34 Nm).

5. Fill the reservoir with fluid and bleed the brake system. Adjust the parking brake if necessary.
- Rear wheels

Disc Brake Pads

REMOVAL & INSTALLATION

1. Remove a small amount of brake fluid from the reservoir using a suction pump.

2. Remove the lower caliper pin bolt and pivot the caliper upward.

3. Remove the pads, shims, and pad retainers.

To install:

4. Clean the caliper thoroughly; remove any dirt or dust. Check the brake rotor for grooves or cracks and machine or replace, as necessary.

5. Install the pad retainers. Apply molybdenum brake grease to both surfaces of the shims and the back of the disc brake pads.

6. Install the pads and shims. The wear retainer on the inboard pad faces down.

7. Use a suitable tool to push caliper piston into its bore and enable the caliper to fit over the pads. Lubricate the piston boot with silicon grease. Avoid twisting the boot.

8. Rotate the caliper down and tighten the mounting bolts to 16 ft. lbs. (22 Nm)

9. Install the rear wheels

10. Add brake fluid to the master cylinder reservoir. Depress the brake pedal several times to seat the pads. Bleed the brakes if necessary.

Parking Brake Shoes

REMOVAL & INSTALLATION

1. Before servicing the vehicle, refer to the Precautions Section.

2. Remove the rear wheels.

3. Release the parking brake tension.

4. Remove the caliper and support it out of the way.

5. Remove the rotor/drum assembly.

6. Remove the upper return springs.

7. Remove the hold-down pins and retainers.

8. Remove the connecting rod.

9. Remove the lower return spring and adjuster.

10. Remove the forward shoe.

11. Disconnect the cable and remove the rear shoe.

12. Installation is the reverse of removal. Clean the backing plate thoroughly and apply a suitable brake grease to all mounting points and the adjuster threads. Adjust the parking brake

ADJUSTMENT

1. Before servicing the vehicle, refer to the Precautions Section.

2. With the parking brake released, back off the adjusting nut at the pedal.

3. Remove the access plug from the drum.

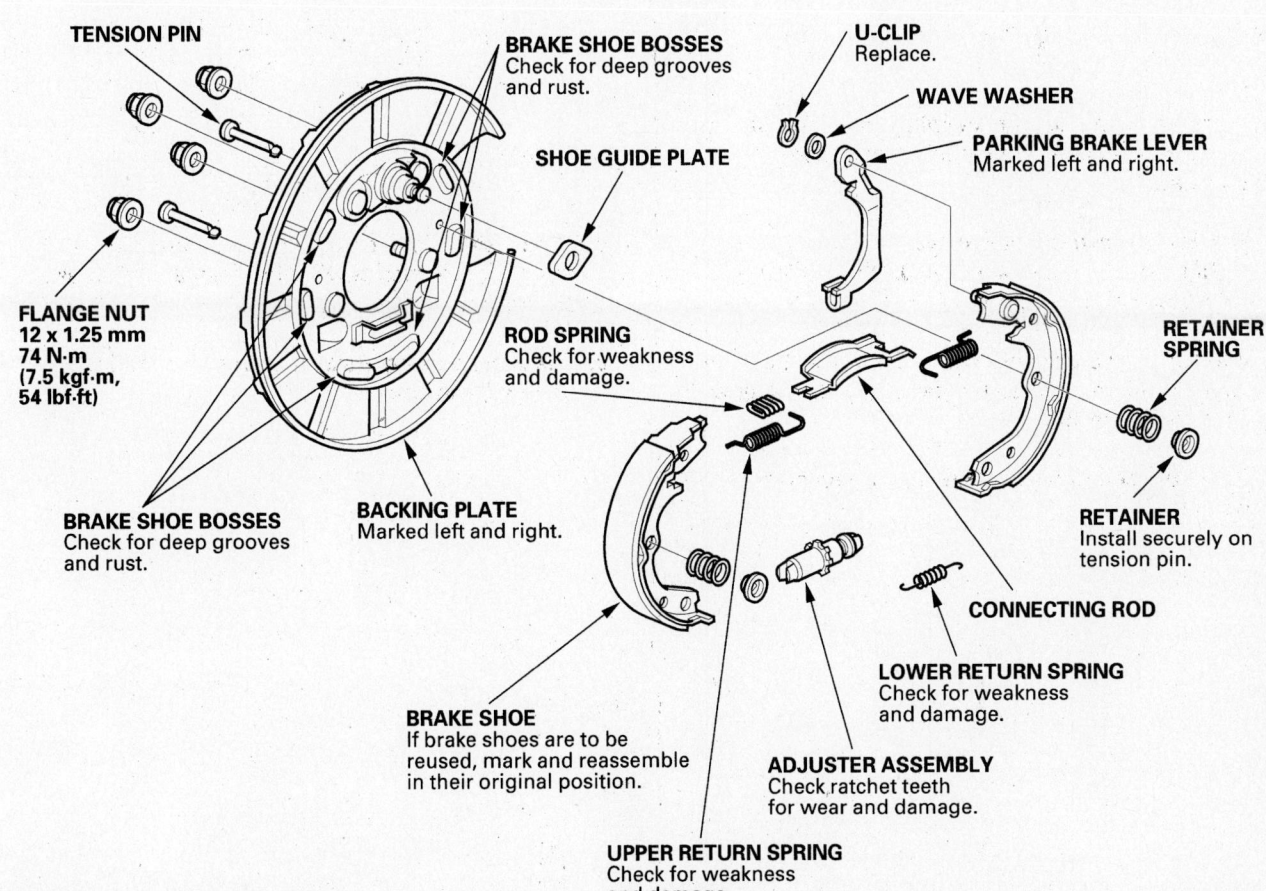

TENSION PIN

BRAKE SHOE BOSSES
Check for deep grooves and rust.

U-CLIP
Replace.

WAVE WASHER

SHOE GUIDE PLATE

PARKING BRAKE LEVER
Marked left and right.

FLANGE NUT
12 x 1.25 mm
74 N·m
(7.5 kgf·m,
54 lbf·ft)

ROD SPRING
Check for weakness and damage.

RETAINER SPRING

BRAKE SHOE BOSSES
Check for deep grooves and rust.

BACKING PLATE
Marked left and right.

RETAINER
Install securely on tension pin.

CONNECTING ROD

LOWER RETURN SPRING
Check for weakness and damage.

BRAKE SHOE
If brake shoes are to be reused, mark and reassemble in their original position.

ADJUSTER ASSEMBLY
Check ratchet teeth for wear and damage.

UPPER RETURN SPRING
Check for weakness and damage.

09474_RIDGE_G0059

Parking brake shoes and related parts

4. Using an adjusting spoon, turn up the ratchet teeth on the adjuster until the shoes lock the drum. Then, back off 10 clicks and install the plug.

5. Press the parking brake pedal with about 66 lbs. of force. The pedal should travel 10–12 clicks. If it travels more than that:

   a. Turn the adjusting nut at the pedal until the brake shoes drag slightly when the rear wheels are turned.

   b. Back off the adjusting nut in half-turn increments until the proper pressure gives the proper number of clicks.

PARKING BRAKE SHOE BREAK-IN PROCEDURE

1. Before servicing the vehicle, refer to the Precautions Section.

2. Adjust the parking brake shoes.

3. While driving the vehicle safely, pull the parking brake release lever.

4. Press the parking brake pedal 2–4 clicks.

5. Drive the vehicle for one-quarter mile at no more that 30 mph.

6. Stop the vehicle and release the parking brake for 10 minutes to allow the drums to cool.

7. Repeat the procedure 3 more times.

8. Recheck the adjustment.

HYUNDAI

Accent • Elantra • Sonata • Tiburon • XG 350

SPECIFICATIONS AND MAINTENANCE CHARTS

ENGINE AND VEHICLE IDENTIFICATION

Code	Liters (cc)	Cu. In.	Cyl.	Fuel Sys.	Engine Type	Eng. Mfg.
G	1.5 (1495)	91.17	4	MPFI	SOHC	Hyundai
C	1.6 (1599)	97.54	4	MPFI	DOHC	Hyundai
D ①	2.0 (1975)	120.52	4	MPFI	DOHC	Hyundai
S	2.4 (2351)	143.46	4	MPFI	DOHC	Hyundai
H ②	2.7 (2656)	164.30	6	MPFI	DOHC	Hyundai
F	2.7 (2656)	164.30	6	MPFI	DOHC	Hyundai
E	3.5 (3496)	211.60	6	MPFI	DOHC	Hyundai

Model Year	
Code	Year
2	2002
3	2003
4	2004
5	2005
6	2006

MPFI: Multi-Point Fuel Injection

SOHC: Single Overhead Camshaft

DOHC: Double Overhead Camshafts

① Elantra and 2003-2005 Tiburon

② Sonota

09474_HYUN_C0001

GENERAL ENGINE SPECIFICATIONS

Year	Engine Displacement Liters	Engine ID/VIN	Net Horsepower @ rpm	Net Torque @ rpm (ft. lbs.)	Bore x Stroke (in.)	Compression Ratio	Oil Pressure @ rpm
2002	1.5	G	92@5500	97@4000	2.97 x 3.29	10.0:1	21@Idle
	1.6	C	104@5800	106@3000	3.01 x 3.43	10.0:1	21@Idle
	2.0	D	140@6000	133@4800	3.23 x 3.68	10.3:1	24@Idle
	2.4	S	137@6000	129@4000	3.41 x 3.94	10.0:1	12@Idle
	2.7	H	172@6000	181@4000	3.41 x 2.95	10.0:1	12@Idle
	3.5	E	194@5500	216@3500	3.66 x 3.38	10.0:1	12@Idle
2003	1.6	C	104@5800	106@3000	3.01 x 3.43	10.0:1	21@Idle
	2.0	D	140@6000	133@4800	3.23 x 3.68	10.3:1	24@Idle
	2.4	S	137@6000	129@4000	3.41 x 3.94	10.0:1	12@Idle
	2.7	H	172@6000	181@4000	3.41 x 2.95	10.0:1	12@Idle
	2.7	F	172@6000	181@4000	3.41 x 2.95	10.0:1	12@Idle
	3.5	E	194@5500	216@3500	3.66 x 3.38	10.0:1	12@Idle
2004	1.6	C	104@5800	106@3000	3.01 x 3.43	10.0:1	21@Idle
	2.0	D	140@6000	133@4800	3.23 x 3.68	10.3:1	24@Idle
	2.4	S	137@6000	129@4000	3.41 x 3.94	10.0:1	12@Idle
	2.7	H	172@6000	181@4000	3.41 x 2.95	10.0:1	12@Idle
	2.7	F	172@6000	181@4000	3.41 x 2.95	10.0:1	12@Idle
	3.5	E	194@5500	216@3500	3.66 x 3.38	10.0:1	12@Idle
2005	1.6	C	104@5800	106@3000	3.01 x 3.43	10.0:1	21@Idle
	2.0	D	140@6000	133@4800	3.23 x 3.68	10.3:1	24@Idle
	2.4	S	137@6000	129@4000	3.41 x 3.94	10.0:1	12@Idle
	2.7	H	172@6000	181@4000	3.41 x 2.95	10.0:1	12@Idle
	2.7	F	172@6000	181@4000	3.41 x 2.95	10.0:1	12@Idle
	3.5	E	194@5500	216@3500	3.66 x 3.38	10.0:1	12@Idle

09474_HYUN_C0002

GASOLINE ENGINE TUNE-UP SPECIFICATIONS

Year	Engine Displacement Liters	Engine ID/VIN	Spark Plugs Gap (in.)	Ignition Timing (deg.) MT	AT	Fuel Pump (psi)	Idle Speed (rpm) MT	AT	Valve Clearance In.	Ex.
2002	1.5	G	0.039-0.043	6-16B	6-16B	43	700-900	700-900	HYD	HYD
	1.6	C	0.039-0.043	4-14B	4-14B	50	700-900	700-900	HYD	HYD
	2.0	D	0.039-0.043	5-15B	5-15B	43	700-900	700-900	HYD	HYD
	2.4	S	0.039-0.043	3-7B	3-7B	48	650-850	650-850	HYD	HYD
	2.7	H	0.039-0.043	7-17B	7-17B	48	600-800	600-800	HYD	HYD
	3.5	E	0.039-0.043	—	3-7B	48	600-800	600-800	HYD	HYD
2003	1.6	C	0.039-0.043	4-14B	4-14B	50	700-900	700-900	HYD	HYD
	2.0	D	0.039-0.043	5-15B	5-15B	43	700-900	700-900	HYD	HYD
	2.4	S	0.039-0.043	3-7B	3-7B	48	650-850	650-850	HYD	HYD
	2.7	H	0.039-0.043	7-17B	7-17B	48	600-800	600-800	HYD	HYD
	2.7	F	0.039-0.043	7-17B	7-17B	48	600-800	600-800	HYD	HYD
	3.5	E	0.039-0.043	—	3-7B	48	600-800	600-800	HYD	HYD
2004	1.6	C	0.039-0.043	4-14B	4-14B	50	700-900	700-900	HYD	HYD
	2.0	D	0.039-0.043	5-15B	5-15B	43	700-900	700-900	HYD	HYD
	2.4	S	0.039-0.043	3-7B	3-7B	48	650-850	650-850	HYD	HYD
	2.7	H	0.039-0.043	7-17B	7-17B	48	600-800	600-800	HYD	HYD
	2.7	F	0.039-0.043	7-17B	7-17B	48	600-800	600-800	HYD	HYD
	3.5	E	0.039-0.043	—	3-7B	48	600-800	600-800	HYD	HYD
2005	1.6	C	0.039-0.043	4-14B	4-14B	50	700-900	700-900	HYD	HYD
	2.0	D	0.039-0.043	5-15B	5-15B	43	700-900	700-900	HYD	HYD
	2.4	S	0.039-0.043	3-7B	3-7B	48	650-850	650-850	HYD	HYD
	2.7	H	0.039-0.043	7-17B	7-17B	48	600-800	600-800	HYD	HYD
	2.7	F	0.039-0.043	7-17B	7-17B	48	600-800	600-800	HYD	HYD
	3.5	E	0.039-0.043	—	3-7B	48	600-800	600-800	HYD	HYD

HYD: Hydraulic Valve Lifters

B: Before Top Dead Center

09474_HYUN_C0003

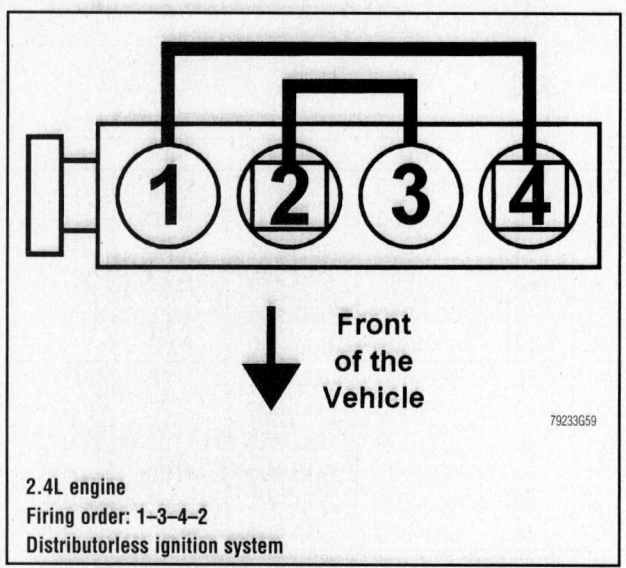

2.4L engine
Firing order: 1–3–4–2
Distributorless ignition system

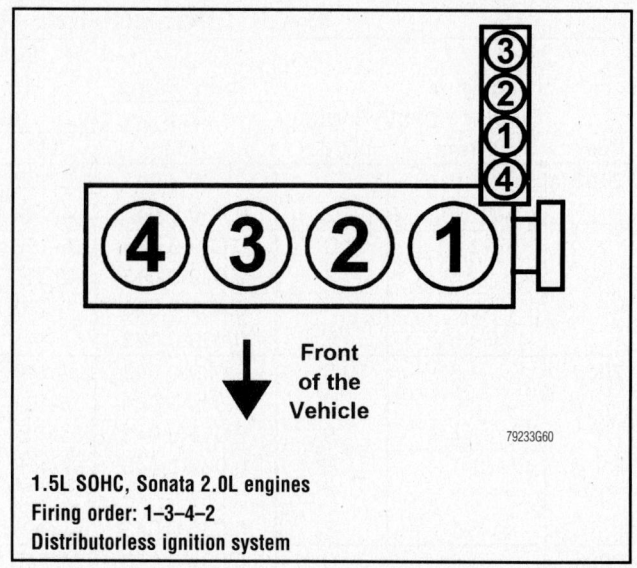

1.5L SOHC, Sonata 2.0L engines
Firing order: 1–3–4–2
Distributorless ignition system

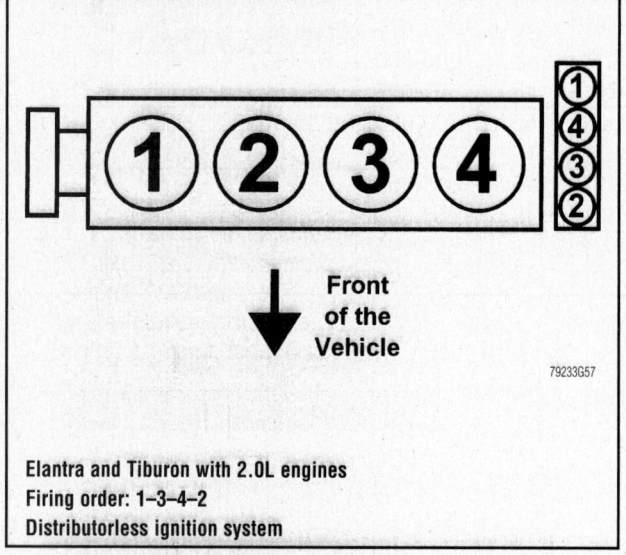

Elantra and Tiburon with 2.0L engines
Firing order: 1–3–4–2
Distributorless ignition system

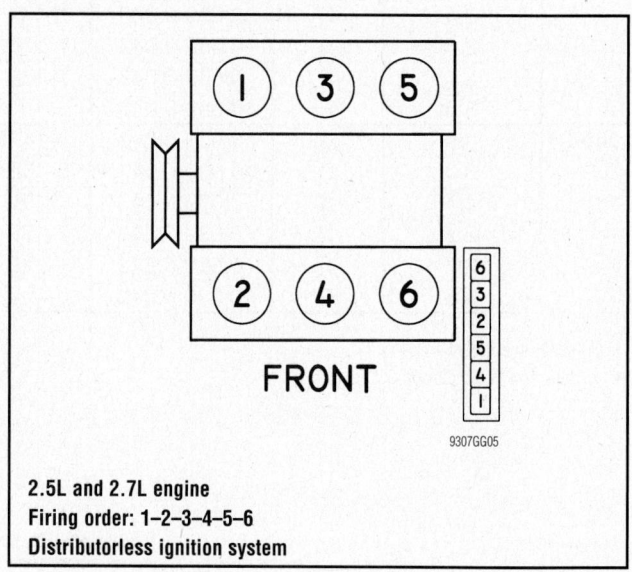

2.5L and 2.7L engine
Firing order: 1–2–3–4–5–6
Distributorless ignition system

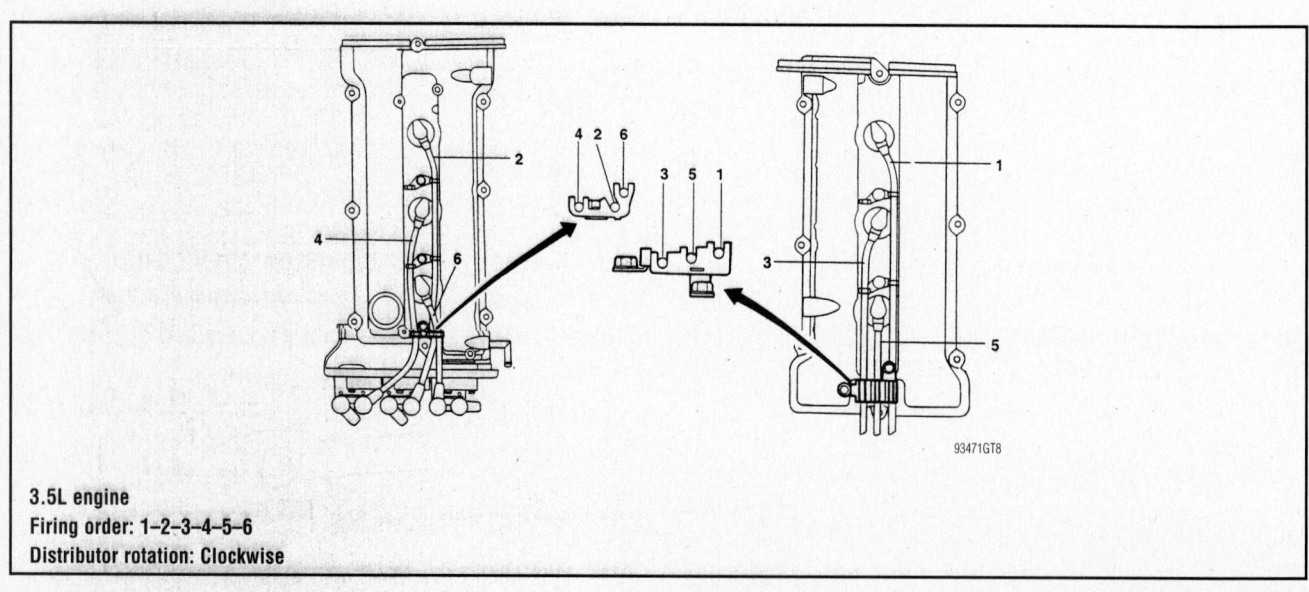

3.5L engine
Firing order: 1–2–3–4–5–6
Distributor rotation: Clockwise

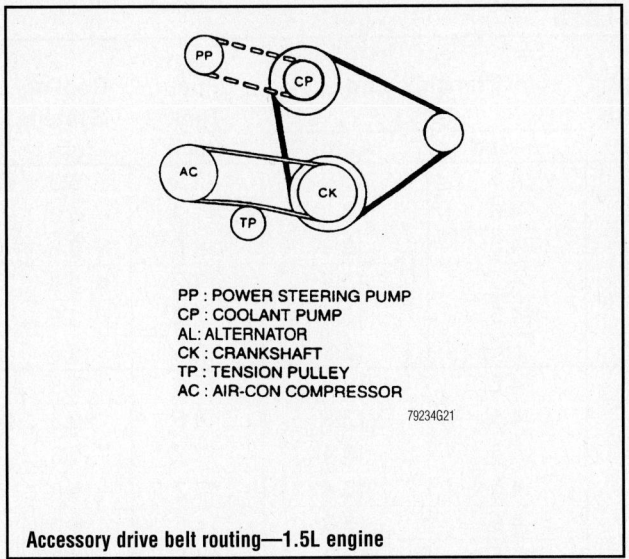

PP : POWER STEERING PUMP
CP : COOLANT PUMP
AL: ALTERNATOR
CK : CRANKSHAFT
TP : TENSION PULLEY
AC : AIR-CON COMPRESSOR

79234G21

Accessory drive belt routing—1.5L engine

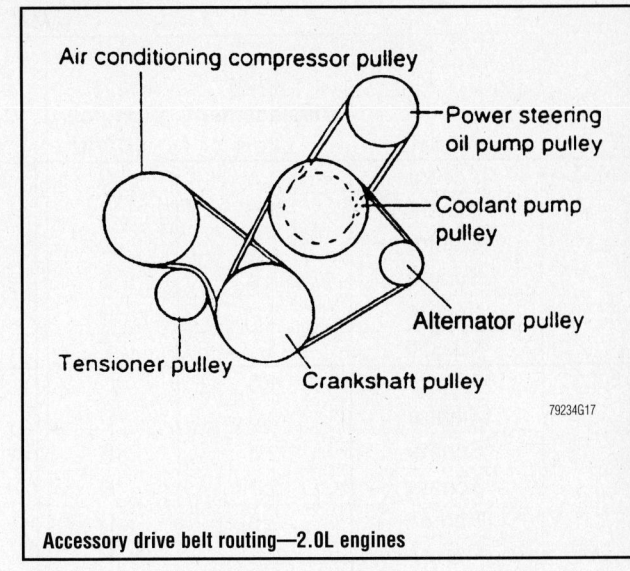

Accessory drive belt routing—2.0L engines

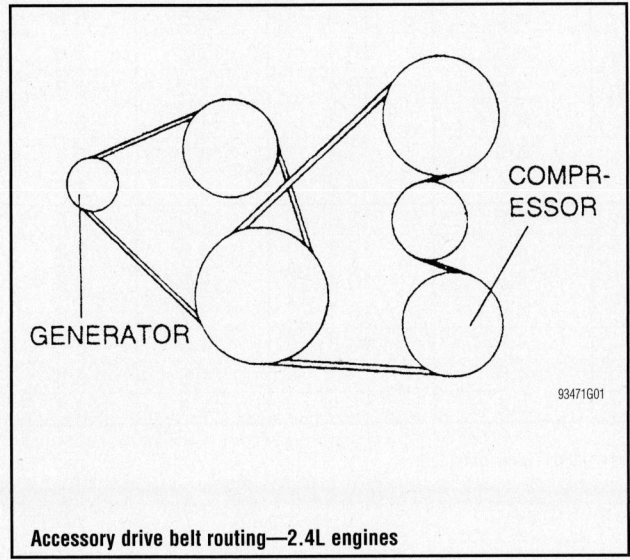

Accessory drive belt routing—2.4L engines

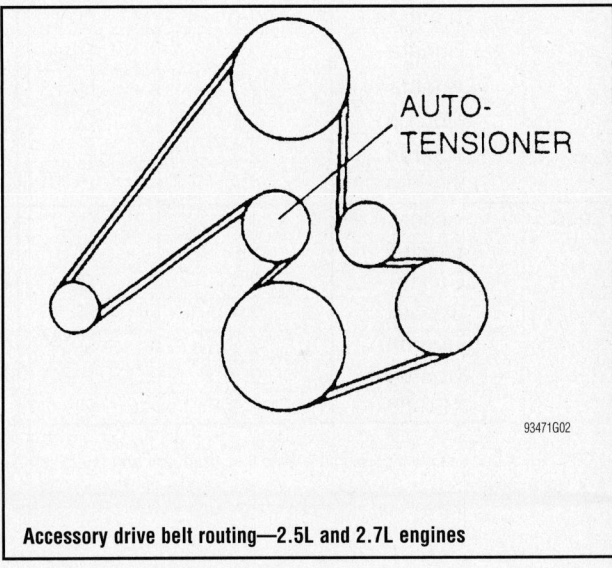

Accessory drive belt routing—2.5L and 2.7L engines

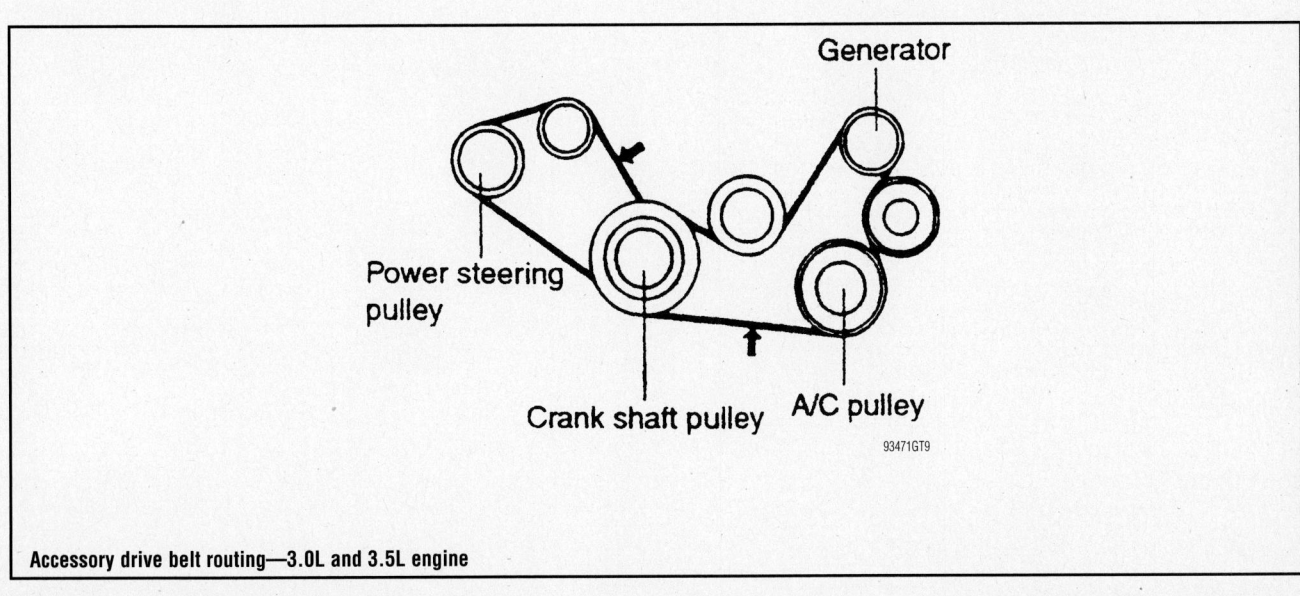

Accessory drive belt routing—3.0L and 3.5L engine

CAPACITIES

Year	Model	Engine Displacement Liters	Engine ID/VIN	Engine Oil with Filter	Transmission (pts.)		Fuel Tank (gal.)	Cooling System (qts.)
					5–Spd	Auto.		
2002	Accent	1.5	G	3.5	4.6	13.6	11.9	6.3
	Accent	1.6	C	3.5	4.6	12.6	11.9	6.8
	Elantra	2.0	D	4.2	4.5	12.8	14.5	6.3
	Sonata	2.4	S	4.2	4.5	16.4	17.2	5.8
	Sonata	2.7	H	4.5	4.5	16.4	17.2	5.8
	XG 350	3.5	E	4.3	NA	15.8	18.5	8.6
2003	Accent	1.6	C	3.5	4.6	12.6	11.9	6.8
	Elantra	2.0	D	4.2	4.5	12.8	14.5	6.3
	Sonata	2.4	S	4.2	4.5	16.4	17.2	5.8
	Sonata	2.7	H	4.5	4.5	16.4	17.2	5.8
	Tiburon	2.0	D	4.2	4.5	13.8	14.5	6.3
	Tiburon	2.7	F	4.2	4.5	13.8	14.5	6.3
	XG 350	3.5	E	4.3	NA	15.8	18.5	8.6
2004	Accent	1.6	C	3.5	4.6	12.6	11.9	6.8
	Elantra	2.0	D	4.2	4.5	12.8	14.5	6.3
	Sonata	2.4	S	4.2	4.5	16.4	17.2	5.8
	Sonata	2.7	H	4.5	4.5	16.4	17.2	5.8
	Tiburon	2.0	D	4.2	4.5	13.8	14.5	6.3
	Tiburon	2.7	F	4.2	4.5	13.8	14.5	6.3
	XG 350	3.5	E	4.3	NA	15.8	18.5	8.6
2005	Accent	1.6	C	3.5	4.6	12.6	11.9	6.8
	Elantra	2.0	D	4.2	4.5	12.8	14.5	6.3
	Sonata	2.4	S	4.2	4.5	16.4	17.2	5.8
	Sonata	2.7	H	4.5	4.5	16.4	17.2	5.8
	Tiburon	2.0	D	4.2	4.5	13.8	14.5	6.3
	Tiburon	2.7	F	4.2	4.5	13.8	14.5	6.3
	XG 350	3.5	E	4.3	NA	15.8	18.5	8.6

NOTE: All capacities are approximate. Add fluid gradually and check to be sure a proper fluid level is obtained.
NA: Not Available

09474_HYUN_C0004

VALVE SPECIFICATIONS

Year	Engine Displacement Liters	Engine ID/VIN	Seat Angle (deg.)	Face Angle (deg.)	Spring Test Pressure (lbs. @ in.)	Spring Installed Height (in.)	Stem-to-Guide Clearance (in.)		Stem Diameter (in.)	
							Intake	Exhaust	Intake	Exhaust
2002	1.5	G	45	45	54@1.358	1.358	0.0012-0.0024	0.0014-0.0026	0.3920	0.3960
	1.6	C	45	N/A	47.6@1.378	N/A	0.0012-0.0024	0.0020-0.0031	0.2344-0.2350	0.2337-0.2343
	2.0	F	45	45	56@1.457	1.358	0.0008-0.0019	0.0019-0.0033	0.2348-0.2354	0.2334-0.2342
	2.4	S	45	45-45.5	56@1.457	1.358	0.0008-0.0019	0.0020-0.0033	0.2585-0.2891	0.2571-0.2579
	2.7	H	45	45-45.5	49@1.378	1.378	0.0009-0.0020	0.0014-0.0026	0.2348-0.2354	0.2343-0.2348
	3.5	E	45	45-45.5	53@1.492	1.826	0.0009-0.0020	0.0020-0.0033	0.258-0.2590	0.257-0.258
2003	1.6	C	45	N/A	47.6@1.378	N/A	0.0012-0.0024	0.0020-0.0031	0.2344-0.2350	0.2337-0.2343
	2.0	D	45	45	56@1.457	1.358	0.0008-0.0019	0.0019-0.0033	0.2348-0.2354	0.2334-0.2342
	2.4	S	45	45-45.5	56@1.457	1.358	0.0008-0.0019	0.0020-0.0033	0.2585-0.2891	0.2571-0.2579
	2.7	H	45	45	49@1.378	1.378	0.0009-0.0020	0.0014-0.0026	0.2348-0.2354	0.2343-0.2348
	3.5	E	45	45-45.5	53@1.492	1.826	0.0009-0.0020	0.0020-0.0033	0.258-0.2590	0.257-0.258
2004	1.6	C	45	N/A	47.6@1.378	N/A	0.0012-0.0024	0.0020-0.0031	0.2344-0.2350	0.2337-0.2343
	2.0	D	45	45	56@1.457	1.358	0.0008-0.0019	0.0019-0.0033	0.2348-0.2354	0.2334-0.2342
	2.4	S	45	45-45.5	56@1.457	1.358	0.0008-0.0019	0.0020-0.0033	0.2585-0.2891	0.2571-0.2579
	2.7	H	45	45	49@1.378	1.378	0.0009-0.0020	0.0014-0.0026	0.2348-0.2354	0.2343-0.2348
	3.5	E	45	45-45.5	53@1.492	1.826	0.0009-0.0020	0.0020-0.0033	0.258-0.2590	0.257-0.258
2005	1.6	C	45	N/A	47.6@1.378	N/A	0.0012-0.0024	0.0020-0.0031	0.2344-0.2350	0.2337-0.2343
	2.0	D	45	45	56@1.457	1.358	0.0008-0.0019	0.0019-0.0033	0.2348-0.2354	0.2334-0.2342
	2.4	S	45	45-45.5	56@1.457	1.358	0.0008-0.0019	0.0020-0.0033	0.2585-0.2891	0.2571-0.2579
	2.7	H	45	45	49@1.378	1.378	0.0009-0.0020	0.0014-0.0026	0.2348-0.2354	0.2343-0.2348
	3.5	E	45	45-45.5	53@1.492	1.826	0.0009-0.0020	0.0020-0.0033	0.258-0.2590	0.257-0.258

09474_HYUN_C0006

CRANKSHAFT AND CONNECTING ROD SPECIFICATIONS

All measurements are given in inches.

| Year | Engine Displacement Liters | Engine ID/VIN | Crankshaft | | | | Connecting Rod | | |
			Main Brg. Journal Dia.	Main Brg. Oil Clearance	Shaft End-play	Thrust on No.	Journal Diameter	Oil Clearance	Side Clearance
2002	1.5	G	2.2440	0.0011-0.0018	0.0019-0.0068	3	1.7700	0.0009-0.0016	0.0039-0.0098
	1.6	C	1.9700	N/A	0.0019-0.0068	3	1.7700	0.0007-0.0014	0.0039-0.0098
	2.0	D	2.2400	0.0011-0.0018	0.0023-0.0100	3	1.7700	0.0009-0.0016	0.0039-0.0098
	2.4	S	2.2434-2.2442	0.0007-0.0014 ①	0.0020-0.0098	3	1.7709-1.7717	0.0008-0.0020	0.0040-0.0098
	2.7	H	2.4402-2.4409	0.0002-0.0009	0.0028-0.0098	3	1.8891-1.8898	0.0007-0.0014	0.0039-0.0098
	3.5	E	2.5190-2.5197	0.00086-0.00157	0.0020-0.0098	3	2.1650-2.1653	0.0010-0.0017	0.0039-0.0098
2003	1.6	C	1.9700	N/A	0.0019-0.0068	3	1.7700	0.0007-0.0014	0.0039-0.0098
	2.0	D	2.2400	0.0011-0.0018	0.0023-0.0100	3	1.7700	0.0009-0.0016	0.0039-0.0098
	2.4	S	2.2434-2.2442	0.0007-0.0014 ①	0.0020-0.0098	3	1.7709-1.7717	0.0008-0.0020	0.0040-0.0098
	2.7	H	2.4402-2.4409	0.0002-0.0009	0.0028-0.0098	3	1.8891-1.8898	0.0007-0.0014	0.0039-0.0098
	2.7	F	2.4402-2.4409	0.0002-0.0009	0.0028-0.0098	3	1.8891-1.8898	0.0007-0.0014	0.0039-0.0098
	3.5	E	2.5190-2.5197	0.00086-0.00157	0.0020-0.0098	3	2.1650-2.1653	0.0010-0.0017	0.0039-0.0098
2004	1.6	C	1.9700	N/A	0.0019-0.0068	3	1.7700	0.0007-0.0014	0.0039-0.0098
	2.0	D	2.2400	0.0011-0.0018	0.0023-0.0100	3	1.7700	0.0009-0.0016	0.0039-0.0098
	2.4	S	2.2434-2.2442	0.0007-0.0014 ①	0.0020-0.0098	3	1.7709-1.7717	0.0008-0.0020	0.0040-0.0098
	2.7	H	2.4402-2.4409	0.0002-0.0009	0.0028-0.0098	3	1.8891-1.8898	0.0007-0.0014	0.0039-0.0098
	2.7	F	2.4402-2.4409	0.0002-0.0009	0.0028-0.0098	3	1.8891-1.8898	0.0007-0.0014	0.0039-0.0098
	3.5	E	2.5190-2.5197	0.00086-0.00157	0.0020-0.0098	3	2.1650-2.1653	0.0010-0.0017	0.0039-0.0098
2005	1.6	C	1.9700	N/A	0.0019-0.0068	3	1.7700	0.0007-0.0014	0.0039-0.0098
	2.0	D	2.2400	0.0011-0.0018	0.0023-0.0100	3	1.7700	0.0009-0.0016	0.0039-0.0098
	2.4	S	2.2434-2.2442	0.0007-0.0014 ①	0.0020-0.0098	3	1.7709-1.7717	0.0008-0.0020	0.0040-0.0098
	2.7	H	2.4402-2.4409	0.0002-0.0009	0.0028-0.0098	3	1.8891-1.8898	0.0007-0.0014	0.0039-0.0098
	2.7	F	2.4402-2.4409	0.0002-0.0009	0.0028-0.0098	3	1.8891-1.8898	0.0007-0.0014	0.0039-0.0098
	3.5	E	2.5190-2.5197	0.00086-0.00157	0.0020-0.0098	3	2.1650-2.1653	0.0010-0.0017	0.0039-0.0098

① No. 3: 0.0009 - 0.0016

N/A: Not available

PISTON AND RING SPECIFICATIONS
All measurements are given in inches.

Year	Engine Displacement Liters	Engine ID/VIN	Piston Clearance	Ring Gap			Ring Side Clearance		
				Top Compression	Bottom Compression	Oil Control	Top Compression	Bottom Compression	Oil Control
2002	1.5	G	0.0008-0.0016	0.008-0.020	0.008-0.020	0.008-0.039	0.0016-0.0033	0.0016-0.0033	snug
	1.6	C	N/A	0.006-0.012	0.012-0.018	0.008-0.028	0.0015-0.0033	0.0015-0.0033	snug
	2.0	D	0.0008-0.0016	0.009-0.015	0.013-0.019	0.008-0.024	0.0015-0.0031	0.0012-0.0027	snug
	2.4	S	0.0008-0.0012	0.010-0.014	0.006-0.022	0.004-0.016	0.0008-0.0024	0.0008-0.0024	snug
	2.7	H	0.0004-0.0012	0.008-0.014	0.015-0.020	0.008-0.028	0.0016-0.0031	0.0012-0.0028	snug
	3.5	E	0.0012-0.0020	0.0078-0.012	0.0177-0.0236	0.0079-0.0276	0.0016-0.0031	0.0008-0.0024	snug
2003	1.6	C	N/A	0.006-0.012	0.012-0.018	0.008-0.028	0.0015-0.0033	0.0015-0.0033	snug
	2.0	D	0.0008-0.0016	0.009-0.015	0.013-0.019	0.008-0.024	0.0015-0.0031	0.0012-0.0027	snug
	2.4	S	0.0008-0.0012	0.010-0.014	0.006-0.022	0.004-0.016	0.0008-0.0024	0.0008-0.0024	snug
	2.7	H	0.0004-0.0012	0.008-0.014	0.015-0.020	0.008-0.028	0.0016-0.0031	0.0012-0.0028	snug
	2.7	F	0.0004-0.0012	0.008-0.014	0.015-0.020	0.008-0.028	0.0016-0.0031	0.0012-0.0028	snug
	3.5	E	0.0012-0.0020	0.0078-0.012	0.0177-0.0236	0.0079-0.0276	0.0016-0.0031	0.0008-0.0024	snug
2004	1.6	C	N/A	0.006-0.012	0.012-0.018	0.008-0.028	0.0015-0.0033	0.0015-0.0033	snug
	2.0	D	0.0008-0.0016	0.009-0.015	0.013-0.019	0.008-0.024	0.0015-0.0031	0.0012-0.0027	snug
	2.4	S	0.0008-0.0012	0.010-0.014	0.006-0.022	0.004-0.016	0.0008-0.0024	0.0008-0.0024	snug
	2.7	H	0.0004-0.0012	0.008-0.014	0.015-0.020	0.008-0.028	0.0016-0.0031	0.0012-0.0028	snug
	2.7	F	0.0004-0.0012	0.008-0.014	0.015-0.020	0.008-0.028	0.0016-0.0031	0.0012-0.0028	snug
	3.5	E	0.0012-0.0020	0.0078-0.012	0.0177-0.0236	0.0079-0.0276	0.0016-0.0031	0.0008-0.0024	snug
2005	1.6	C	N/A	0.006-0.012	0.012-0.018	0.008-0.028	0.0015-0.0033	0.0015-0.0033	snug
	2.0	D	0.0008-0.0016	0.009-0.015	0.013-0.019	0.008-0.024	0.0015-0.0031	0.0012-0.0027	snug
	2.4	S	0.0008-0.0012	0.010-0.014	0.006-0.022	0.004-0.016	0.0008-0.0024	0.0008-0.0024	snug
	2.7	H	0.0004-0.0012	0.008-0.014	0.015-0.020	0.008-0.028	0.0016-0.0031	0.0012-0.0028	snug
	2.7	F	0.0004-0.0012	0.008-0.014	0.015-0.020	0.008-0.028	0.0016-0.0031	0.0012-0.0028	snug
	3.5	E	0.0012-0.0020	0.0078-0.012	0.0177-0.0236	0.0079-0.0276	0.0016-0.0031	0.0008-0.0024	snug

TORQUE SPECIFICATIONS
All readings in ft. lbs.

Year	Engine Displacement Liters	Engine ID/VIN	Cylinder Head Bolts	Main Bearing Bolts	Rod Bearing Bolts	Crankshaft Damper Bolts	Flywheel Bolts	Manifold Intake	Manifold Exhaust	Spark Plugs	Oil Pan Drain Plug
2002	1.5	G	①	40-44	25-28	110-118	94-101	11-14	11-14	18	30-33
	1.6	C	⑪	41-44	24-26	103-111	89-96	11-14	18-22	15-22	30-33
	2.0	D	②	③	34-39	125-133	88-95	11-14	17-22	18	30-33
	2.4	S	④	⑤	⑤	80-94	94-101	⑥	⑦	15-22	25-33
	2.7	H	①	⑧	⑨	130-138	53-55	14-15	18-22	15-22	25-33
	3.5	E	75-82	65-72	⑩	NA	NA	9-10	20-40	15-22	26-32
2003	1.6	C	⑪	41-44	24-26	103-111	89-96	11-14	18-22	15-22	30-33
	2.0	D	②	③	34-39	125-133	88-95	11-14	17-22	18	30-33
	2.4	S	④	⑤	⑤	80-94	94-101	⑥	⑦	15-22	25-33
	2.7	H	①	⑧	⑨	130-138	53-55	14-15	18-22	15-22	25-33
	2.7	F	①	⑧	⑨	130-138	53-55	14-15	18-22	15-22	25-33
	3.5	E	75-82	65-72	⑩	NA	NA	9-10	20-40	15-22	26-32
2004	1.6	C	⑪	41-44	24-26	103-111	89-96	11-14	18-22	15-22	30-33
	2.0	D	②	③	34-39	125-133	88-95	11-14	17-22	18	30-33
	2.4	S	④	⑤	⑤	80-94	94-101	⑥	⑦	15-22	25-33
	2.7	H	①	⑧	⑨	130-138	53-55	14-15	18-22	15-22	25-33
	2.7	F	①	⑧	⑨	130-138	53-55	14-15	18-22	15-22	25-33
	3.5	E	75-82	65-72	⑩	NA	NA	9-10	20-40	15-22	26-32
2005	1.6	C	⑪	41-44	24-26	103-111	89-96	11-14	18-22	15-22	30-33
	2.0	D	②	③	34-39	125-133	88-95	11-14	17-22	18	30-33
	2.4	S	④	⑤	⑤	80-94	94-101	⑥	⑦	15-22	25-33
	2.7	H	①	⑧	⑨	130-138	53-55	14-15	18-22	15-22	25-33
	2.7	F	①	⑧	⑨	130-138	53-55	14-15	18-22	15-22	25-33
	3.5	E	75-82	65-72	⑩	NA	NA	9-10	20-40	15-22	26-32

① Step 1: 18 ft. lbs.
　Step 2: Plus 60-64 degrees
　Step 3: Plus 45-49 degrees

② Step 1: M10 bolts to 22 ft. lbs. and M12 bolts to 26 ft. lbs.
　Step 2: Plus 60-65 degrees
　Step 3: Plus 60-65 degrees

③ Step 1: 20-24 ft. lbs.
　Step 2: Plus 60-65 degrees

④ Step 1: 14 ft. lbs.
　Step 2: Plus 90 degrees
　Step 3: Loosen to 0 ft. lbs.
　Step 4: 14 ft. lbs.
　Step 5: Plus 90 degrees
　Step 6: Plus 90 degrees

⑤ 13-16 ft. lbs. plus 90-94 degrees

⑥ M8: 11-14 ft. lbs.
　M10: 13-18 ft. lbs.
　Nuts: 22-30 ft. lbs.

⑦ M8: 18-22 ft. lbs.
　M10: 25-40 ft. lbs.

⑧ M10: 20-24 ft. lbs. plus 90-94 degrees
　M7: 10-13 ft. lbs. plus 90-94 degrees

⑨ 12-14 ft. lbs. plus 90-94 degrees

⑩ 26 ft. lbs. + 90 degrees

⑪ Step 1: 22 ft. lbs.
　Step 2: Plus 90 degrees
　Step 3: Loosen to 0 ft. lbs.
　Step 4: 22 ft. lbs.
　Step 5: Plus 90 degrees

09474_HYUN_C0008

WHEEL ALIGNMENT

Year	Model		Caster Range (+/-Deg.)	Caster Preferred Setting (Deg.)	Camber Range (+/-Deg.)	Camber Preferred Setting (Deg.)	Toe-in (in.)
2002	Accent	F	0.50	+1.80	0.50	0	0 +/- 0.12
		R	—	—	0.50	-0.68	0.12 +/- 0.08
	Elantra	F	0.50	+2.35	2.00	+4.00	0 +/- 0.12
		R	—	—	0.50	-0.70	0.14 +/- 0.02
	Sonata	F	1.00	+3.25	0.50	0	0 +/- 0.12
		R	—	—	0.50	-0.50	0.08 +/- 0.08
	XG 350	F	1.00	+3.15	0.50	0	0 +/- 0.12
		R	—	—	0.50	-0.50	0.08 +/- 0.08
2003	Accent	F	0.50	+1.80	0.50	0	0 +/- 0.12
		R	—	—	0.50	-0.68	0.12 +/- 0.08
	Elantra	F	0.50	+2.35	2.00	+4.00	0 +/- 0.12
		R	—	—	0.50	-0.70	0.14 +/- 0.02
	Sonata	F	1.00	+3.25	0.50	0	0 +/- 0.12
		R	—	—	0.50	-0.50	0.08 +/- 0.08
	Tiburon	F	0.50	+2.45	2.00	+4.00	0 +/- 0.12
		R	—	—	0.50	-0.90	0.14 +/- 0.02
	XG 350	F	1.00	+3.15	0.50	0	0 +/- 0.12
		R	—	—	0.50	-0.50	0.08 +/- 0.08
2004	Accent	F	0.50	+1.80	0.50	0	0 +/- 0.12
		R	—	—	0.50	-0.68	0.12 +/- 0.08
	Elantra	F	0.50	+2.35	2.00	+4.00	0 +/- 0.12
		R	—	—	0.50	-0.70	0.14 +/- 0.02
	Sonata	F	1.00	+3.25	0.50	0	0 +/- 0.12
		R	—	—	0.50	-0.50	0.08 +/- 0.08
	Tiburon	F	0.50	+2.45	2.00	+4.00	0 +/- 0.12
		R	—	—	0.50	-0.90	0.14 +/- 0.02
	XG 350	F	1.00	+3.15	0.50	0	0 +/- 0.12
		R	—	—	0.50	-0.50	0.08 +/- 0.08
2005	Accent	F	0.50	+1.80	0.50	0	0 +/- 0.12
		R	—	—	0.50	-0.68	0.12 +/- 0.08
	Elantra	F	0.50	+2.81	0.50	0	0 +/- 0.08
		R	—	—	0.50	-0.91	0.10 +/- 0.04
	Sonata	F	1.00	+3.25	0.50	0	0 +/- 0.08
		R	—	—	0.50	-0.50	0.08 +/- 0.08
	Tiburon	F	0.50	+3.38	0.50	-0.20	0.08 +/- 0.08
		R	—	—	0.50	-1.18	0.16 +0.12 -0.04
	XG 350	F	1.00	+3.25	0.50	0	0 +/- 0.08
		R	—	—	0.50	-0.50	0.08 +/- 0.08

TIRE, WHEEL AND BALL JOINT SPECIFICATIONS

Year	Model	OEM Tires Standard	OEM Tires Optional	Tire Pressures (psi) Front	Tire Pressures (psi) Rear	Wheel Size	Ball Joint Inspection	Lug Nut Torque ①
2002	Accent	P155/80R13	P175/70R13 P175/65R14	30	30	5-J	②	67-81
	Elantra	P195/60R14	None	30	30	5.5-JJ	②	65-80
	Sonata	P195/70R14	P205/60HR15	30	30	Std: 5.5-JJ Opt: 6-JJ	②	67-81
	XG 350	P205/60R16	None	30		6.0J	②	67-81
2003	Accent	P155/80R13	P175/70R13 P175/65R14	30	30	5-J	②	65-80
	Elantra	P195/60R14	None	30	30	5.5-JJ	②	67-81
	Sonata	P195/70R14	P205/60HR15	30	30	Std: 5.5-JJ Opt: 6-JJ	②	67-81
	Tiburon	P195/60R14	None	30	30	5.5JJ	②	67-81
	XG 350	P205/60R16	None	30		6.0J	②	65-80
2004	Accent	P155/80R13	P175/70R13 P175/65R14	30	30	5-J	②	65-80
	Elantra	P195/60R14	None	30	30	5.5-JJ	②	67-81
	Sonata	P195/70R14	P205/60HR15	30	30	Std: 5.5-JJ Opt: 6-JJ	②	67-81
	Tiburon	P195/60R14	None	30	30	5.5JJ	②	67-81
	XG 350	P205/60R16	None	30		6.0J	②	67-81
2005	Accent	P175/70R13	P185/60HR14	③	③	N/A	②	66-81
	Elantra	P195/60R15	None	③	③	N/A	②	67-81
	Sonata	P215/60R16	P225/50R17	③	③	N/A	②	67-81
	Tiburon	P205/55R16	P215/45R17	③	③	N/A	②	67-81
	XG 350	P205/60R16	None	③	③	N/A	②	67-81

OEM: Original Equipment Manufacturer

PSI: Pounds Per Square Inch

STD: Standard

OPT: Optional

① Ft. Lbs.

② Replace if any measurable movement is found.

③ See label on driver's side of center pillar outer panel

BRAKE SPECIFICATIONS
All measurements in inches unless noted

Year	Model		Brake Disc Original Thickness	Brake Disc Minimum Thickness	Brake Disc Maximum Run-out	Brake Drum Diameter Original Inside Diameter	Brake Drum Diameter Max. Wear Limit	Brake Drum Diameter Maximum Machine Diameter	Minimum Lining Thickness Front	Minimum Lining Thickness Rear	Brake Caliper Bracket Bolts (ft. lbs.)	Brake Caliper Mounting Bolts (ft. lbs.)
2002	Accent	F	0.750	0.670	0.002	—	—	—	0.039	—	50	①
		R	—	—	—	7.100	—	7.165	—	0.039	—	—
	Elantra	F	0.750	0.670	0.002	—	—	—	0.039	—	50	①
		R	—	—	—	8.000	—	8.079	—	0.039	—	—
	Elantra w/rear disc	F	0.750	0.670	0.002	—	—	—	0.039	—	50	①
		R	0.354	NA	NA	—	—	—	—	0.031	—	23
	Sonata	F	0.945	0.787	0.002	—	—	—	0.079	—	51-63	16-24
		R	—	—	—	9.000	—	8.936	—	0.059	—	—
	Sonata w/rear disc	F	0.945	0.880	0.003	—	—	—	0.079	—	51-63	16-24
		R	0.390	0.350	0.005	—	—	—	—	0.079	—	23
	Tiburon	F	0.866	0.787	0.002	—	—	—	0.079	—	55	①
		R	—	—	—	8.000	—	8.079	—	0.059	—	—
	Tiburon w/rear disc	F	0.866	0.787	0.002	—	—	—	0.079	—	55	①
		R	0.354	NA	NA	—	—	—	—	0.031	—	23
	XG 350	F	0.413	0.096	0.002	—	—	—	0.079	—	51-63	16-24
		R	0.390	0.080	NA	—	—	—	—	—	51-63	16-24
2003	Accent	F	0.750	0.670	0.002	—	—	—	0.039	—	50	①
		R	—	—	—	7.100	—	7.165	—	0.039	—	—
	Elantra	F	0.750	0.670	0.002	—	—	—	0.039	—	50	①
		R	—	—	—	8.000	—	8.079	—	0.039	—	—
	Elantra w/rear disc	F	0.750	0.670	0.002	—	—	—	0.039	—	50	①
		R	0.354	NA	NA	—	—	—	—	0.031	—	23
	Sonata	F	0.945	0.787	0.002	—	—	—	0.079	—	51-63	16-24
		R	—	—	—	9.000	—	8.936	—	0.059	—	—
	Sonata w/rear disc	F	0.945	0.880	0.003	—	—	—	0.079	—	51-63	16-24
		R	0.390	0.350	0.005	—	—	—	—	0.079	—	23
	Tiburon	F	0.866	0.787	0.002	—	—	—	0.079	—	55	①
		R	—	—	—	8.000	—	8.079	—	0.059	—	—
	Tiburon w/rear disc	F	0.866	0.787	0.002	—	—	—	0.079	—	55	①
		R	0.354	NA	NA	—	—	—	—	0.031	—	23
	XG 350	F	0.413	0.096	0.002	—	—	—	0.079	—	51-63	16-24
		R	0.390	0.080	NA	—	—	—	—	—	51-63	16-24

NA: Not Available

F: Front

R: Rear

① Upper: 28 ft. lbs.
  Lower: 19 ft. lbs.

09474_HYUN_C0011

BRAKE SPECIFICATIONS
All measurements in inches unless noted

Year	Model		Brake Disc Original Thickness	Brake Disc Minimum Thickness	Brake Disc Maximum Run-out	Brake Drum Diameter Original Inside Diameter	Brake Drum Diameter Max. Wear Limit	Brake Drum Diameter Maximum Machine Diameter	Minimum Lining Thickness Front	Minimum Lining Thickness Rear	Brake Caliper Bracket Bolts (ft. lbs.)	Brake Caliper Mounting Bolts (ft. lbs.)
2004	Accent	F	0.750	0.670	0.002	—	—	—	0.039	—	50	①
		R	—	—	—	7.100	—	7.165	—	0.039	—	—
	Elantra	F	0.750	0.670	0.002	—	—	—	0.039	—	50	①
		R	—	—	—	8.000	—	8.079	—	0.039	—	—
	Elantra w/rear disc	F	0.750	0.670	0.002	—	—	—	0.039	—	50	①
		R	0.354	NA	NA	—	—	—	—	0.031	—	23
	Sonata	F	0.945	0.787	0.002	—	—	—	0.079	—	51-63	16-24
		R	—	—	—	9.000	—	8.936	—	0.059	—	—
	Sonata w/rear disc	F	0.945	0.880	0.003	—	—	—	0.079	—	51-63	16-24
		R	0.390	0.350	0.005	—	—	—	—	0.079	—	23
	Tiburon	F	0.866	0.787	0.002	—	—	—	0.079	—	55	①
		R	—	—	—	8.000	—	8.079	—	0.059	—	—
	Tiburon w/rear disc	F	0.866	0.787	0.002	—	—	—	0.079	—	55	①
		R	0.354	NA	NA	—	—	—	—	0.031	—	23
	XG 350	F	0.413	0.096	0.002	—	—	—	0.079	—	51-63	16-24
		R	0.390	0.080	NA	—	—	—	—	—	51-63	16-24
2005	Accent	F	0.750	0.670	0.002	—	—	—	0.079	—	51-63	16-24
		R	—	—	—	8.000	—	7.170	—	0.040	—	—
	Elantra	F	0.945	0.882	0.004	—	—	—	0.079	—	51-63	16-24
		R	—	—	—	8.000	—	8.079	—	0.039	—	—
	Elantra w/rear disc	F	0.945	0.882	0.004	—	—	—	0.079	—	51-63	16-24
		R	0.400	0.315	0.004	—	—	—	—	0.079	—	16-24
	Sonata	F	0.960	0.880	0.001	—	0.078	—	0.080	—	48-55	16-24
		R	0.390	0.330	—	—	0.078	—	—	0.080	—	23
	Tiburon	F	1.024	0.961	0.003	—	—	—	0.079	—	48-54	16-24
		R	0.394	0.315	NA	—	—	—	—	0.079	—	16-24
	XG 350	F	1.020	0.960	0.001	—	—	—	0.080	—	48-55	16-24
		R	0.390	0.330	0.001	—	—	—	—	0.080	48-55	16-24

NA: Not Available

F: Front

R: Rear

① Upper: 28 ft. lbs.
   Lower: 19 ft. lbs.

09474_HYUN_C0012

SCHEDULED MAINTENANCE INTERVALS
HYUNDAI—ACCENT, ELANTRA, SONATA, TIBURON & XG350

TO BE SERVICED	TYPE OF SERVICE	VEHICLE MILEAGE INTERVAL (x1000)												
		7.5	15	22.5	30	37.5	45	52.5	60	67.5	75	82.5	90	97.5
Engine oil & filter	R	✓	✓	✓	✓	✓	✓	✓	✓	✓	✓	✓	✓	✓
Automatic transaxle fluid	S/I		✓		✓		✓		✓		✓		✓	
Brake pads, calipers & rotors	S/I		✓		✓		✓		✓		✓		✓	
Driveshafts & boots	S/I		✓		✓		✓		✓		✓		✓	
Wheel bearing grease	S/I				✓				✓				✓	
Air cleaner filter	R				✓				✓				✓	
Brake fluid	S/I				✓				✓				✓	
Engine coolant	R								✓				✓	
Fuel hose, vapor hose & fuel filter cap	S/I				✓				✓				✓	
Spark plugs	R				✓				✓				✓	
Spark plugs (Sonata 3.0L V6)	R								✓					
Bolts & nuts on chassis & body (Accent)	S/I				✓				✓				✓	
Drive belts	S/I				✓				✓				✓	
Exhaust pipe connections, muffler & suspension bolts	S/I		✓		✓		✓		✓		✓		✓	
Manual transaxle oil	S/I				✓				✓				✓	
Power steering pump, belt, & hoses	S/I				✓				✓				✓	
Rear brake drums, linings & parking brake	S/I				✓				✓				✓	
Steering gear rack, linkage & boots	S/I				✓				✓				✓	
Suspension ball joints & dust covers (Accent)	S/I		✓		✓		✓		✓		✓		✓	
Timing belt (Accent & Elantra)	S/I				✓				✓				✓	
Timing belt (except Accent & Elantra)	R								✓					
Fuel filter	R							✓						
Fuel lines & connections	S/I				✓				✓				✓	
Vacuum & crankcase ventilation hoses	S/I							✓						

R: Replace   S/I: Service or Inspect

Automatic transaxle fluid: replace at 105,000 miles.

Air conditioner filter (For Evap and Blower unit): Replace every 12 months or 12,000 miles.

FREQUENT OPERATION MAINTENANCE (SEVERE SERVICE)

If a vehicle is operated under any of the following conditions it is considered severe service

- **Extremely dusty areas.**

- **50% or more of the vehicle operation is in 32°C (90°F) or higher temperatures, or constant operation in temperatures below 0°C (32°F).**

- **Prolonged idling (vehicle operation in stop and go traffic).**

- **Frequent short running periods (engine does not warm to normal operating temperatures).**

- **Police, taxi, delivery usage or trailer towing usage.**

Oil & oil filter: change every 3000 miles.

Brake pads, calipers & rotors: service or inspect every 7500 miles.

Driveshaft boots: service or inspect every 7500 miles

Steering gear rack, linkage & boots: service or inspect every 7500 miles.

Air cleaner filter: service or inspect every 15,000 miles.

Automatic transaxle fluid & filter: replace every 15,000 miles.

Rear brake drums & linings: service or inspect every 15,000 miles.

Spark plugs: replace every 24,000 miles.

ENGINE REPAIR

➡**Disconnecting the negative battery cable on some vehicles may interfere with the functions of the on board computer system. The computer may undergo a relearning process once the negative battery cable is reconnected.**

Distributor

All engines are equipped with a Distributorless Ignition System (DIS).

Alternator

REMOVAL

1. Before servicing the vehicle, refer to the Precautions Section.
2. Remove or disconnect the following:
   • Negative battery cable
   • Alternator drive belt
   • Alternator wiring harness connectors
   • Alternator. It may be necessary to raise the radiator

INSTALLATION

1. Before servicing the vehicle, refer to the Precautions Section.
2. Install or connect the following:
   • Alternator and reposition the radiator, if raised
   • Alternator wiring harness connectors
   • Alternator drive belt
3. Adjust the alternator belt and torque the bolts to the following specifications:
   a. Accent: Pivot bolt to 14–18 ft. lbs. (19–25 Nm) and adjustment bolt to 14–20 ft. lbs. (19–28 Nm).
   b. 2.0L, 2.5L and 2.7L Sonata, Elantra and Tiburon: Pivot bolt to 14–18 ft. lbs. (19–25 Nm) and adjustment bolt to 105–132 inch lbs. (12–15 Nm).
   c. 2.4L Sonata: Pivot bolt to 26–41 ft. lbs. (34–54 Nm) and the adjustment bolt to 14–18 ft. lbs. (19–25 Nm).
   d. XG 350: Pivot bolt to 14–22 ft. lbs. (19–25 Nm) and adjustment bolt to 11–16 ft. lbs. (15–22 Nm).

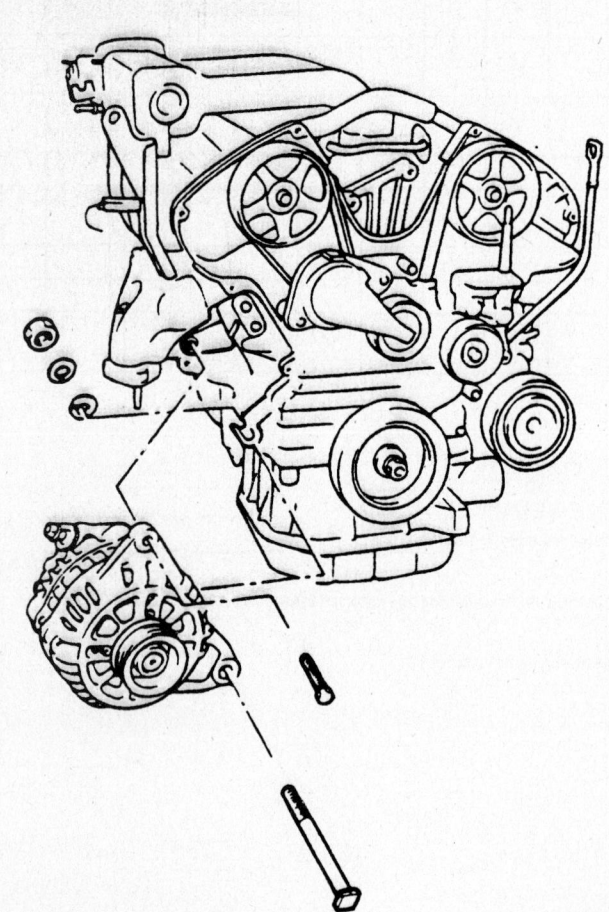

Exploded view of the alternator—XG 350

9347KG01

4. Install the radiator mounting bolts, if removed.
5. Connect the negative battery cable.

Ignition Timing

ADJUSTMENT

These engines are equipped with a Distributorless Ignition System (DIS). No adjustment is necessary.

Engine Assembly

REMOVAL & INSTALLATION

➡**Hyundai recommends that the engine and transaxle be removed as a single unit on all models.**

1. Before servicing the vehicle, refer to the Precautions Section.
2. Drain the cooling system.
3. Drain the transaxle.
4. Drain the engine oil.
5. Relieve fuel system pressure.
6. Remove or disconnect the following:
   • Battery
   • Hood
   • Air intake assembly
   • Accessory drive belts
   • Engine wiring harness connectors
   • Reverse lamp switch connector, if equipped
   • Speedometer cable
   • Alternator harness connectors
   • Oil pressure gauge sender connector
   • Radiator hoses
   • Cooling fan
   • Fuel lines
   • Control cable, if equipped
   • Brake booster vacuum line
   • Intake manifold vacuum lines
   • Heater hoses
   • Accelerator cable
   • Cruise control cable, if equipped
   • Engine ground cable
7. If equipped with a manual transaxle, disconnect or remove the following:
   • Clutch cable
   • Select control valve connector
   • Shift linkage rods
8. If equipped with an automatic transaxle, disconnect or remove the following:
   • Transaxle oil cooler lines
   • Shift cable
   • Transaxle wiring connectors

9. For all vehicles, remove or disconnect the following:
- Radiator
- Power steering pump
- A/C compressor, if equipped
- Exhaust front pipe
- Lower ball joints
- Stabilizer bar links

10. Separate the inner CV-joints from the transaxle and suspend the halfshafts out of the work area with safety wire.

11. Attach a hoist to the engine lifting eyes.

12. Remove or disconnect the following:
- Front and rear roll stoppers
- Engine mount and bracket
- Transaxle mount and bracket

13. Lift the powertrain out of the vehicle.

To install:

14. Lower the powertrain into position.

15. Install the motor mount bracket and torque the fasteners as follows:

    a. V6 engines: 43–58 ft. lbs. (60–80 Nm).

    b. All others: 37–48 ft. lbs. (45–60 Nm).

16. Install the transaxle mount bracket and torque the fasteners as follows:

    a. Sonata: 29–36 ft. lbs. (40–50 Nm).

    b. Tiburon and Elantra: 33–43 ft. lbs. (45–60 Nm).

    c. Accent: 22–30 ft. lbs. (30–40 Nm).

    d. XG 350: 65–79 ft. lbs. (90–110 Nm).

17. Install or connect the following:
- Front and rear roll stoppers
- Engine mount
- Transaxle mount

18. Remove the engine hoist.

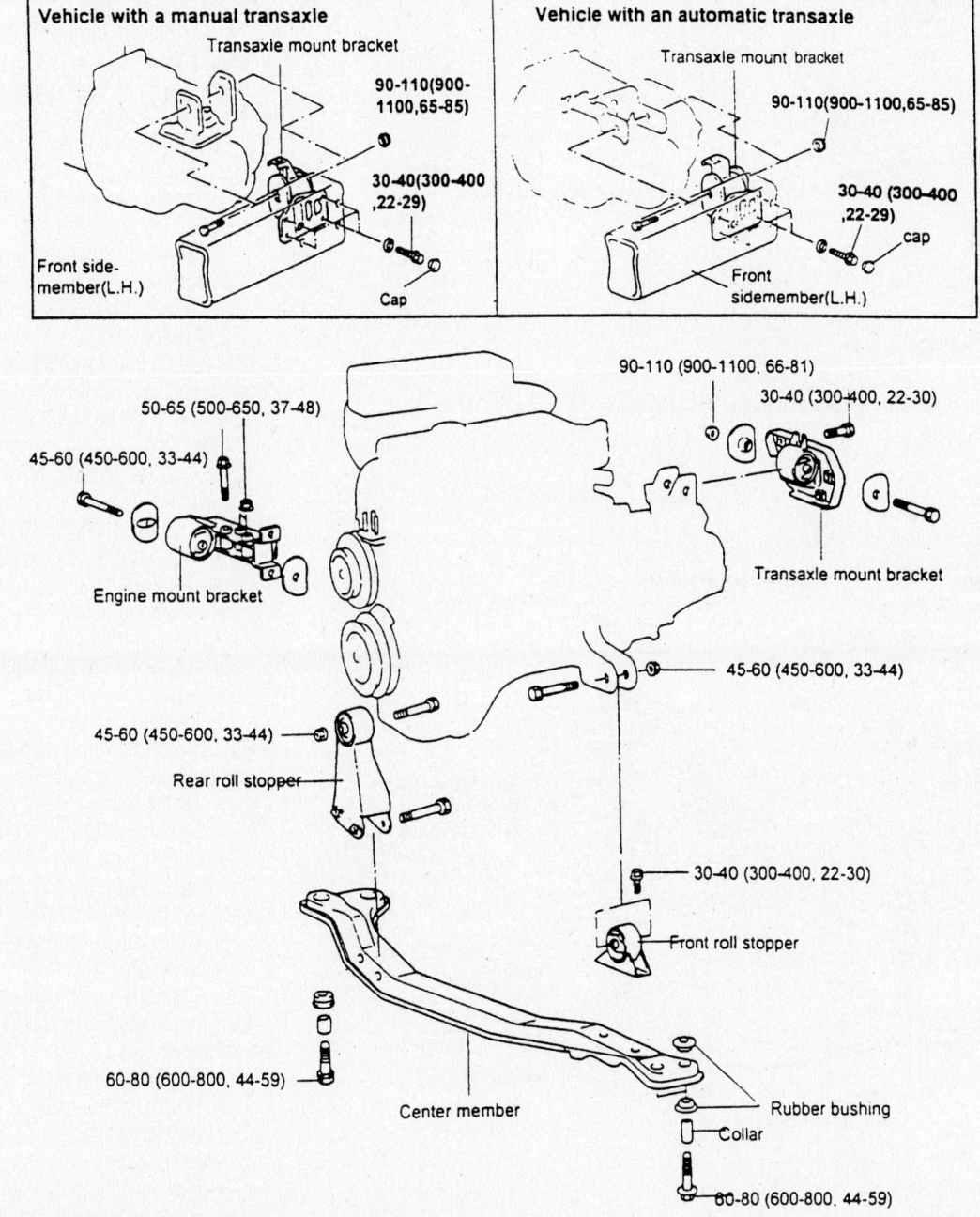

Exploded view of the engine mounts and torque specifications—Accent

7923GG09

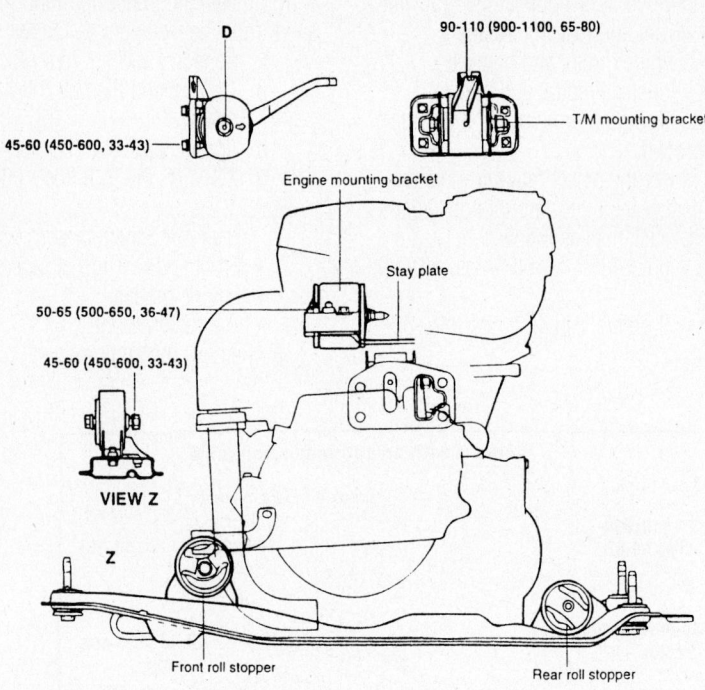

TORQUE: Nm (kg.cm, lb.ft)

7923GG13

Exploded view of the engine mounts and torque specifications—Tiburon and Elantra

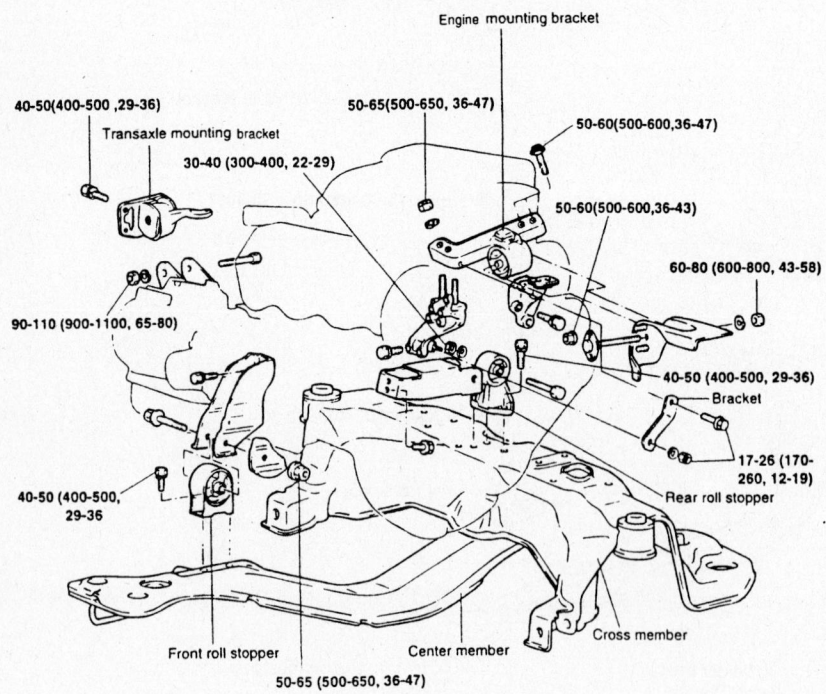

TORQUE : Nm (kg.cm, lb.ft)

7923GG15

Exploded view of the engine mounts and torque specifications—4 cylinder Sonata

19. For Accent, torque the mount through bolts as follows:
   a. Engine mount: 33–43 ft. lbs. (45–60 Nm).
   b. Transaxle mount: 65–80 ft. lbs. (90–110 Nm).
   c. Front and rear roll stoppers: 33–43 ft. lbs. (45–60 Nm).

20. For Elantra and Tiburon, torque the mount through bolts as follows:
   a. Engine mount: 36–47 ft. lbs. (50–65 Nm).
   b. Transaxle mount: 65–80 ft. lbs. (90–110 Nm).
   c. Front and rear roll stoppers: 33–43 ft. lbs. (45–60 Nm).

21. For Sonata, torque the mount through bolts as follows:
   a. 4 cylinder engine mount: 43–58 ft. lbs. (60–80 Nm).
   b. V6 engine mount: 65–80 ft. lbs. (90–110 Nm).
   c. Transaxle mount: 65–80 ft. lbs. (90–110 Nm).
   d. Front roll stopper: 36–47 ft. lbs. (50–65 Nm).
   e. Rear roll stopper: 22–29 ft. lbs. (30–40 Nm).

22. For XG 350, torque the mount through bolts as follows:
   a. Front roll stopper: 36–47 ft. lbs. (50–65 Nm).
   b. Rear roll stopper: 36–47 ft. lbs. (50–65 Nm).

23. Install or connect the following:
- Axle halfshafts using new circlips
- Stabilizer bar links
- Lower ball joints
- Exhaust front pipe
- A/C compressor, if equipped
- Power steering pump
- Radiator

24. If equipped with a manual transaxle, install or connect the following:
- Clutch cable
- Select control valve connector
- Shift linkage rods

25. If equipped with an automatic transaxle, install or connect the following:
- Transaxle oil cooler lines
- Shift cable
- Transaxle wiring connectors

26. For all vehicles, install or connect the following:
- Engine ground cable
- Cruise control cable, if equipped
- Accelerator cable
- Heater hoses
- Intake manifold vacuum lines
- Brake booster vacuum line
- Fuel lines
- Cooling fan

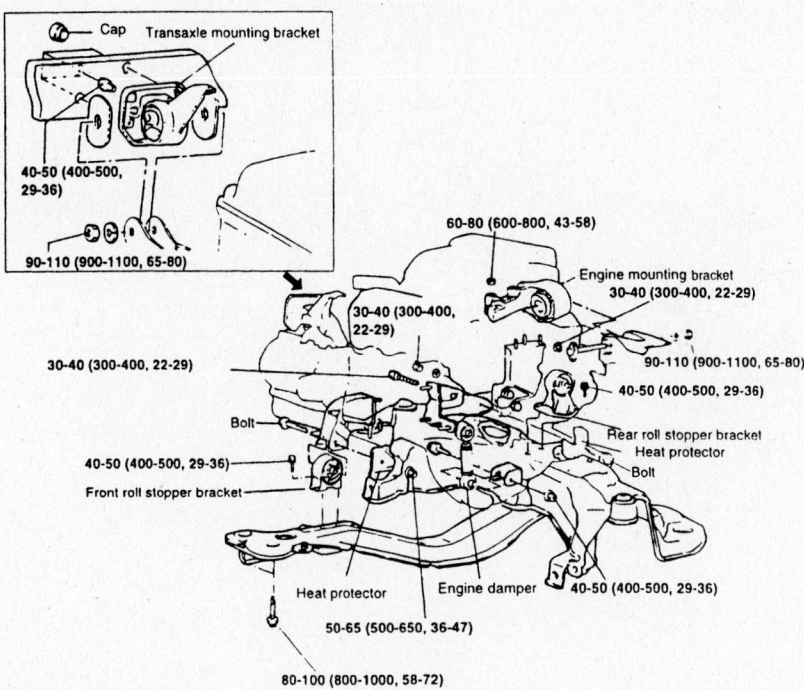

TORQUE : Nm (kg.cm, lb.ft)

7923GG16

Exploded view of the engine mounts and torque specifications—V6 Sonata

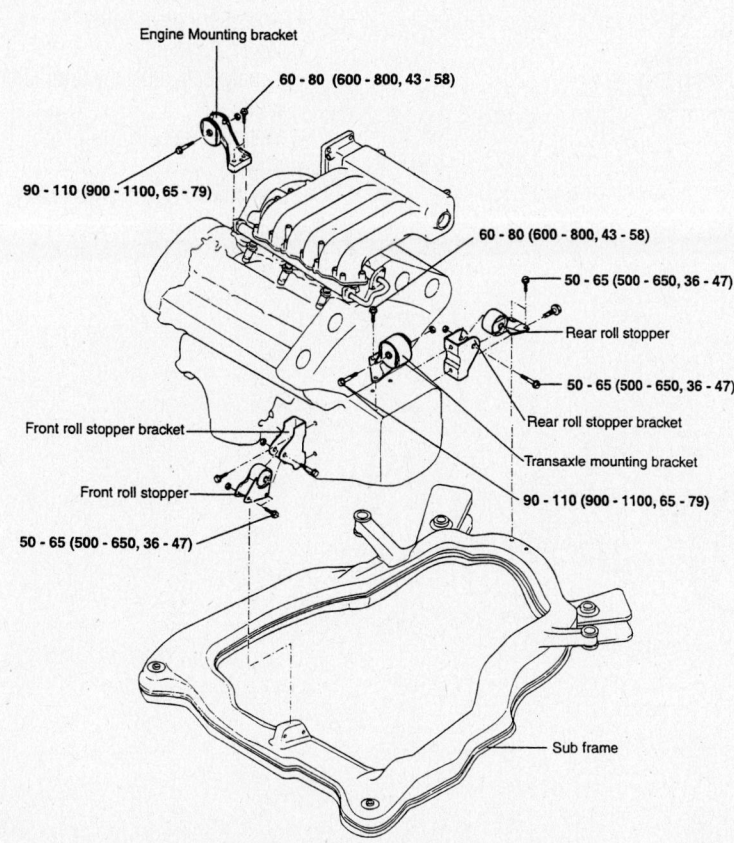

9347KG02

Exploded view of the engine mounts and torque specifications— XG 350

- Radiator hoses
- Oil pressure gauge sender connector
- Alternator harness connectors
- Speedometer cable
- Reverse lamp switch connector
- Engine wiring harness connectors
- Accessory drive belts
- Air intake assembly
- Hood
- Battery

27. Fill the engine with clean oil.
28. Fill the transaxle to the correct level.
29. Fill the cooling system to the proper level.
30. Start the engine and check for leaks.

Water Pump

REMOVAL & INSTALLATION

Except XG 350

1. Before servicing the vehicle, refer to the Precautions Section.
2. Drain the cooling system.
3. Remove or disconnect the following:
   - Negative battery cable
   - Accessory drive belts
   - Radiator hose
   - Bypass hose, if equipped
   - Water pump pulley
   - Front cover
   - Timing belt
   - Alternator bracket
   - Water pump

➡ **The water pump bolts are different lengths. Note the bolt location for assembly.**

To install:

4. Install the water pump with new gaskets and O-rings. Tighten the bolts to the following specifications:
   a. 1.5L and 1.6L engines: 105–132 inch lbs. (12–15 Nm).
   b. 2.0L, and 2.4L engines: 15–20 ft. lbs. (20–27 Nm).
   c. 2.5L and 2.7L engine: 11–16 ft. lbs. (15–22 Nm).
5. Install or connect the following:
   - Alternator bracket
   - Timing belt and front cover
   - Water pump pulley
   - Bypass hose, if equipped
   - Radiator hose
   - Accessory drive belts
   - Negative battery cable
6. Fill the cooling system to the proper level.
7. Start the engine and check for leaks.

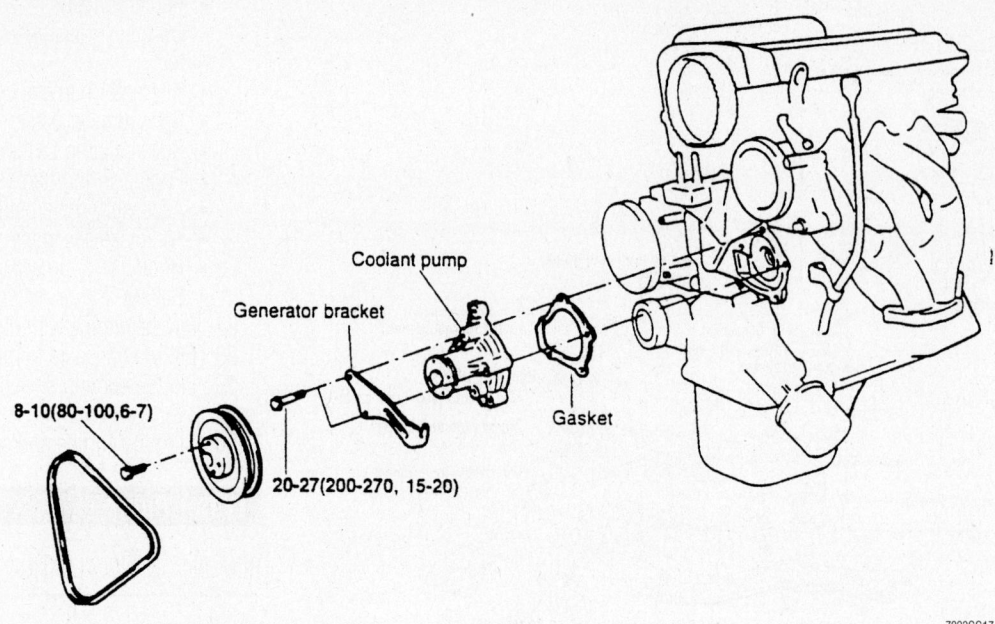

Exploded view of the water pump assembly—1.5L engines

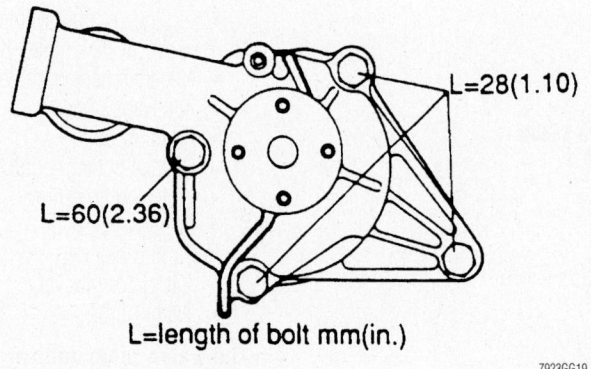

Water pump bolt lengths—1.5L engines

XG 350

1. Before servicing the vehicle, refer to the Precautions Section.
2. Drain the cooling system.
3. Remove or disconnect the following:
   - Negative battery cable
   - Accessory drive belts
   - Water pump pulley
   - Timing belt, tensioner and idler pulley
   - Water pump

To install:

4. Clean all mating surfaces of any residual gasket material.

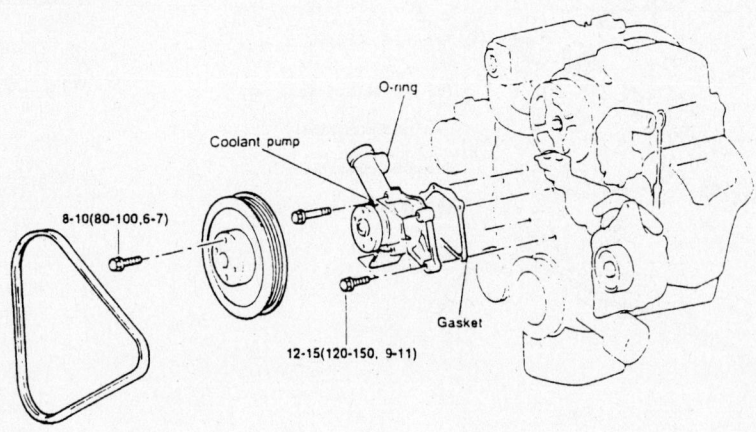

TORQUE : Nm (kg.cm, lb.ft)

Water pump assembly—2.0L engines

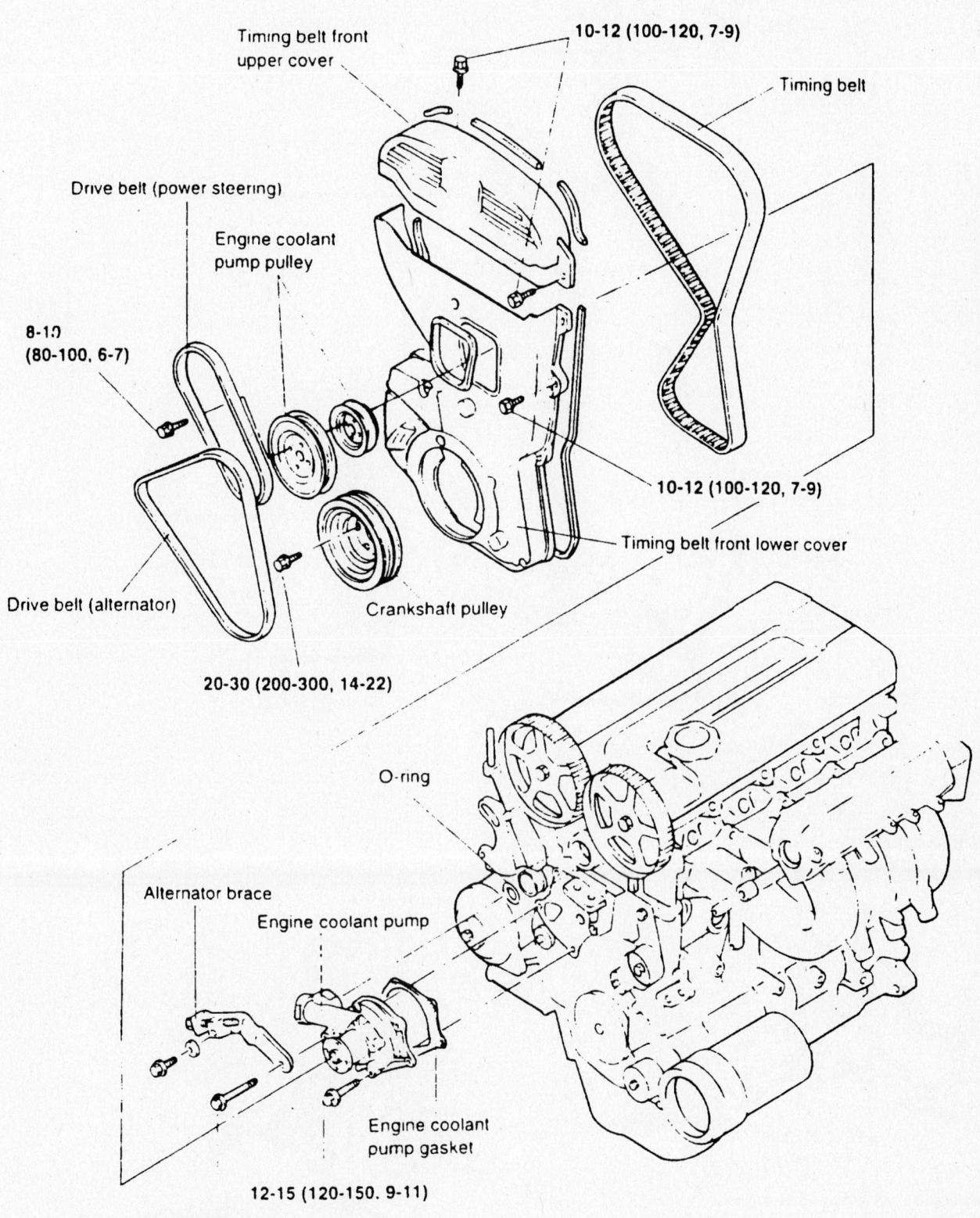

Timing belt front upper cover

10-12 (100-120, 7-9)

Timing belt

Drive belt (power steering)

Engine coolant pump pulley

8-10 (80-100, 6-7)

10-12 (100-120, 7-9)

Timing belt front lower cover

Drive belt (alternator)

Crankshaft pulley

20-30 (200-300, 14-22)

O-ring

Alternator brace

Engine coolant pump

Engine coolant pump gasket

12-15 (120-150. 9-11)

20-27 (200-270, 14-20)

TORQUE : Nm (kg.cm, lb.ft)

7923GG21

Exploded view of the water pump assembly and related components—2.0L engines

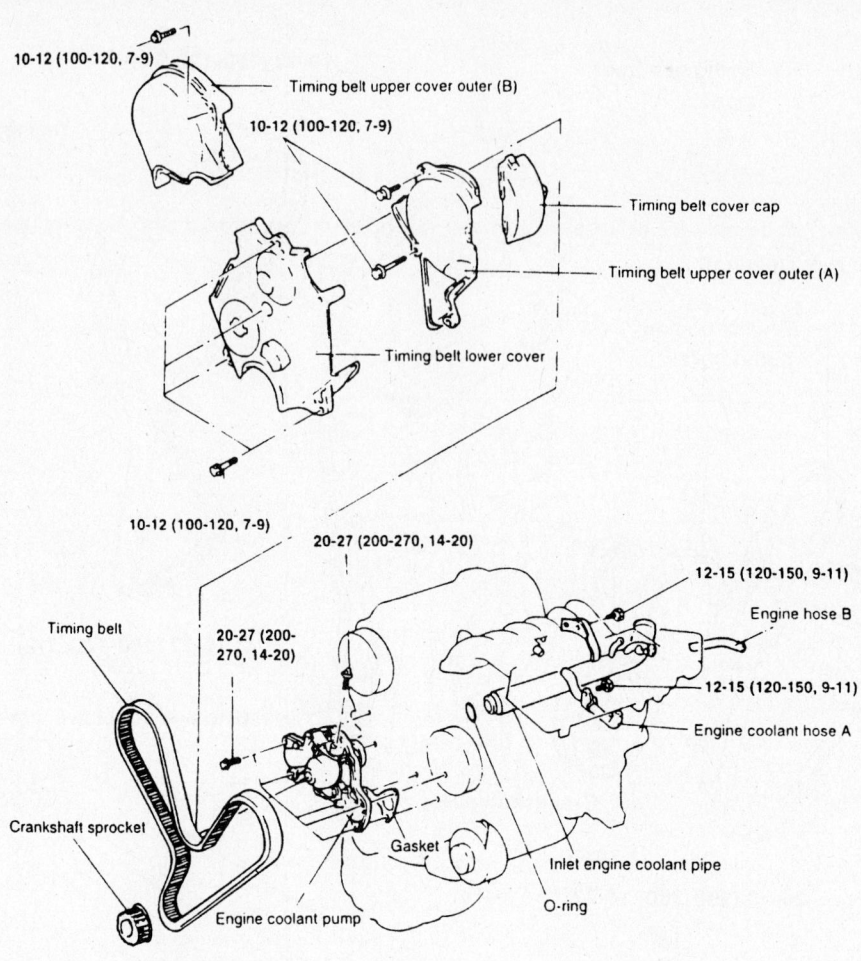

10-12 (100-120, 7-9)

Timing belt upper cover outer (B)

10-12 (100-120, 7-9)

Timing belt cover cap

Timing belt upper cover outer (A)

Timing belt lower cover

10-12 (100-120, 7-9)

20-27 (200-270, 14-20)

12-15 (120-150, 9-11)

Engine hose B

Timing belt

20-27 (200-270, 14-20)

12-15 (120-150, 9-11)

Engine coolant hose A

Crankshaft sprocket

Gasket

Inlet engine coolant pipe

O-ring

Engine coolant pump

TORQUE : Nm (kg.cm, lb.ft)

7923GG22

Water pump assembly— 3.5L engine

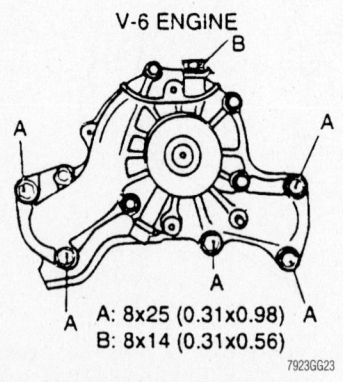

V-6 ENGINE

B

A   A

A   A

A: 8x25 (0.31x0.98)
B: 8x14 (0.31x0.56)

7923GG23

Water pump bolt lengths—3.5L engine

5. Install or connect the following:
- Water pump with a new gasket and torque the bolts to 16 ft. lbs. (22 Nm)
- Tensioner and timing belt and idler pulley
- Water pump pulley and drive belts
- Negative battery cable

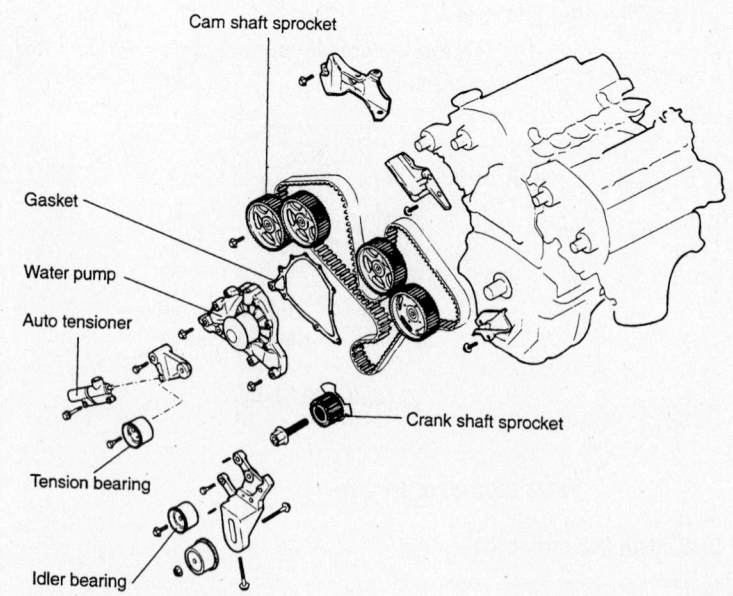

Cam shaft sprocket

Gasket

Water pump

Auto tensioner

Tension bearing

Idler bearing

Crank shaft sprocket

9347KG03

Water pump assembly—XG 350

6. Fill the cooling system to the proper level.

7. Start the vehicle, check for leaks and repair if necessary.

Heater Core

REMOVAL & INSTALLATION

Accent

1. Before servicing the vehicle, refer to the Precautions Section.

2. Disconnect the negative battery cable and wait 90 seconds for the SRS memory battery to drain.

❈❈ CAUTION

After disconnecting the negative battery cable, wait for at least 30 seconds for the SRS module to deplete its stored energy.

3. Drain the cooling system into a clean container for reuse.

4. Remove or disconnect the following:
- Heater hoses with the vacuum hose from the heater housing
- Discharge and recover the air conditioning system refrigerant
- Suction and discharge hoses from the evaporator assembly

5. Remove the SRS module and the steering wheel:
- Steering wheel-to-SRS module nuts
- SRS module from the steering wheel and disconnect the electrical connector
- Steering wheel-to-steering column nut
- Steering wheel from the steering column

6. Remove or disconnect the following:
- Multi-function switch assembly
- Front and rear console assemblies
- Lower left side crash pad
- Center fascia panel and disconnect the connectors and vacuum connector from the heater control assembly
- Heater control assembly and the audio system
- Glove box
- 4 mounting bolts from the passenger air bag mounting bracket, if equipped
- Main crash pad assembly
- Cables from the heater housing and the thermostatic switch connector from the evaporator housing

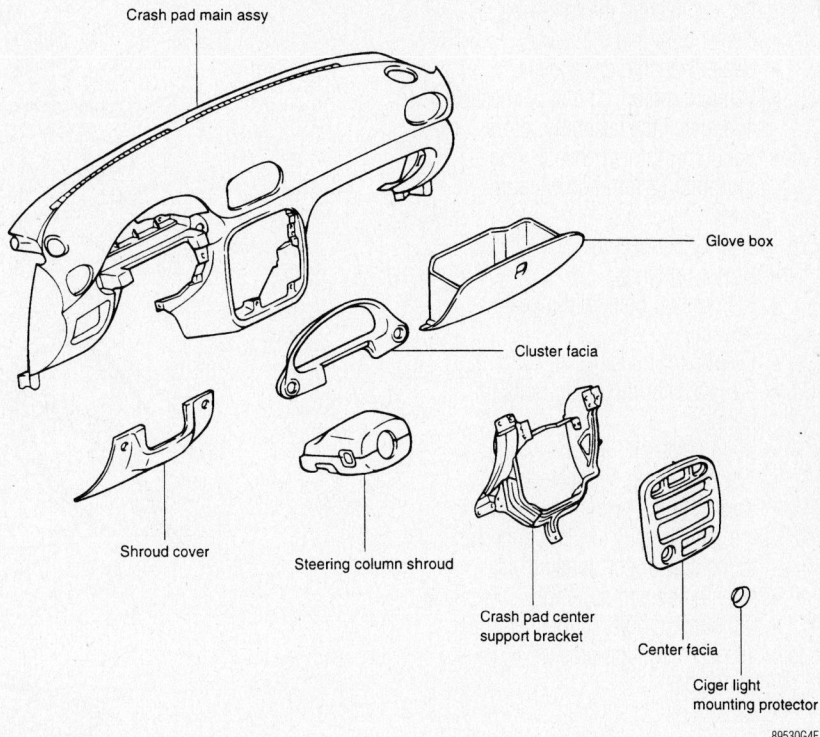

Instrument panel assembly—Accent

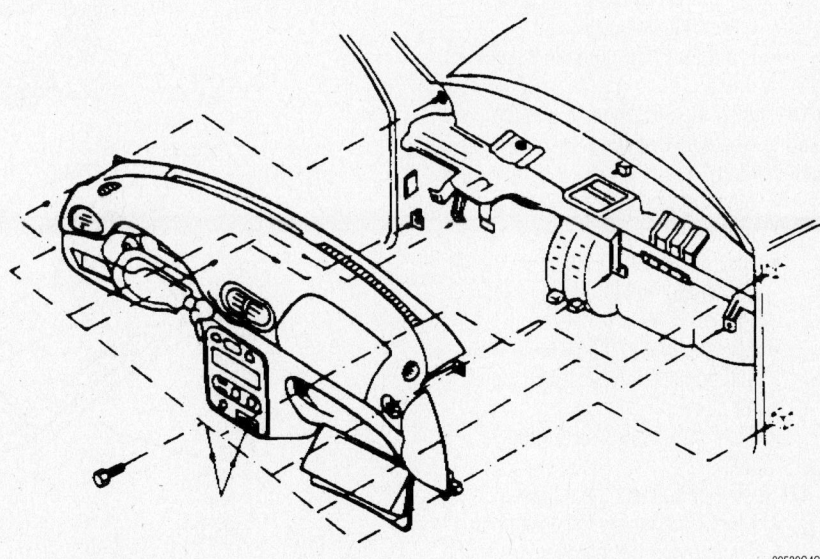

Instrument panel screw locations—Accent

- Any remaining connectors
- Main crash pad assembly
- 3 evaporator mounting bolts (or nuts)
- Evaporator housing
- 3 mounting bolts from the heater housing
- Heater housing

7. Disassemble the heater housing by removing or disconnecting the following:
- Vacuum motor-to-heater housing bolts (2 for each vacuum motor)
- Vacuum motor rod end connection and remove the vacuum motors
- Heater housing cover clips
- Cover and the heater core

To install:

8. Assemble the heater housing by installing or connecting the following:
- Heater core and the cover
- Heater housing cover clips
- Vacuum motor rod end connection and install the vacuum motors
- Vacuum motor-to-heater housing bolts (2 for each vacuum motor)

9. Install or connect the following:
- Heater housing
- 3 mounting bolts to the heater housing
- Evaporator housing
- 3 evaporator mounting bolts (or nuts)
- Main crash pad assembly
- Any remaining connectors
- Cables to the heater housing and the thermostatic switch connector to the evaporator housing
- Main crash pad assembly
- 4 mounting bolts to the passenger air bag mounting bracket, if equipped
- Glove box
- Heater control assembly and the audio system
- Connectors and vacuum connector to the heater control assembly. Install the center fascia panel.
- Lower left side crash pad
- Front and rear console assemblies

10. Install the SRS module and the steering wheel by installing or connecting the following:
- Steering wheel to the steering column
- Steering wheel-to-steering column nut and torque to 30–37 ft. lbs. (40–50 Nm)
- SRS module to the steering wheel and connect the electrical connector
- Steering wheel-to-SRS module nuts

11. Install or connect the following:
- Multi-function switch assembly
- Suction and discharge hoses to the evaporator assembly
- Heater hoses with the vacuum hose to the heater housing

12. Refill the cooling system.

13. Connect the negative battery cable.

14. Evacuate, charge and leak test the air conditioning system refrigerant.

15. Operate the engine to normal operating temperatures; then, check the climate control operation and check for leaks.

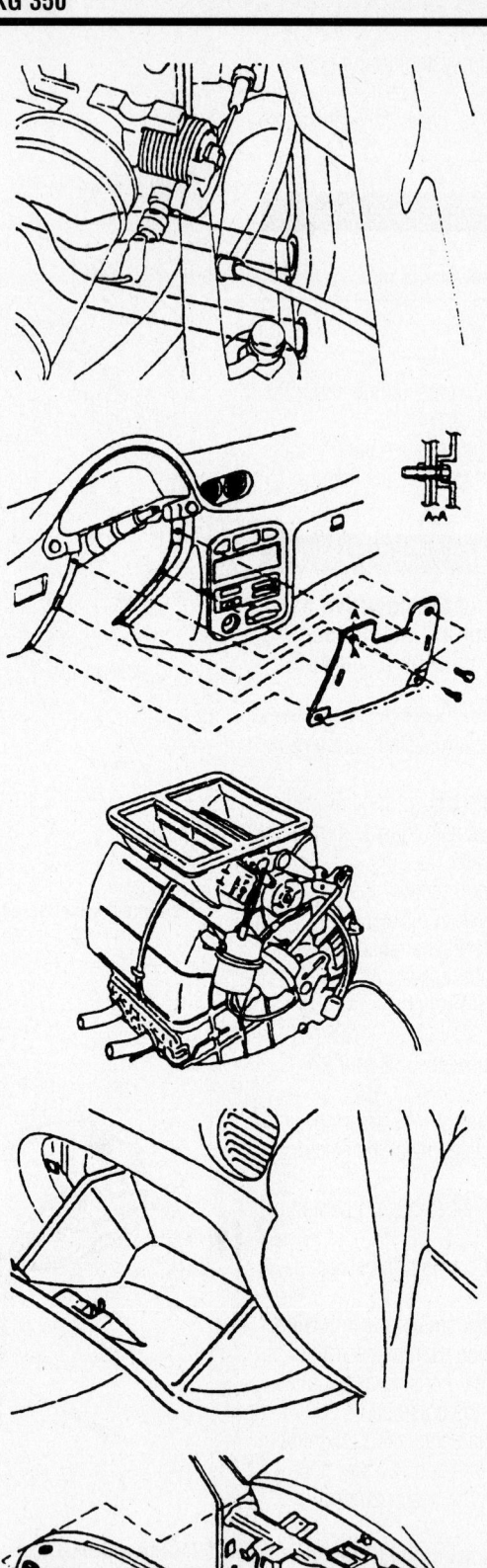

93112G30

View of the heater housing assembly and related components—Accent

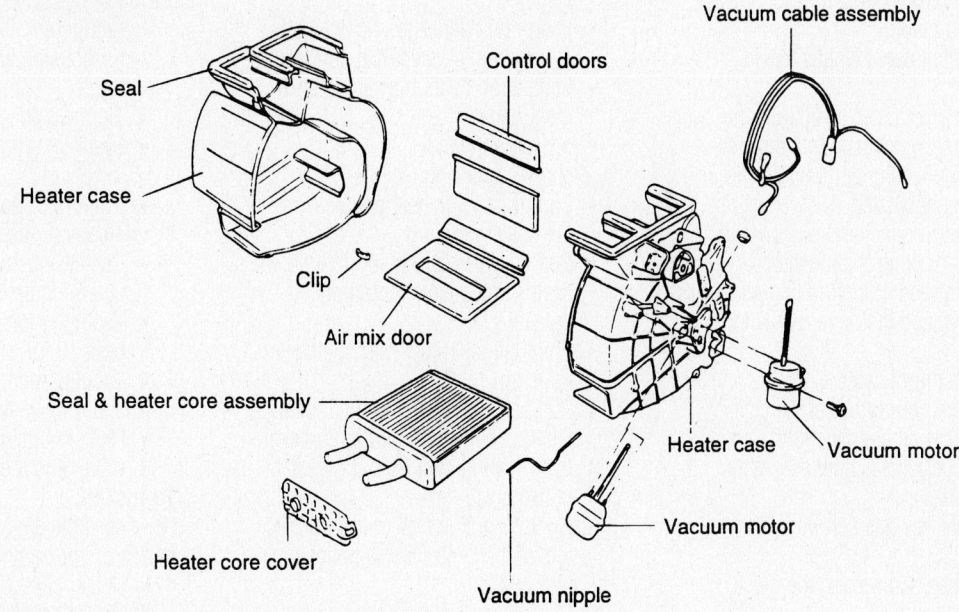

* Heater Assembly

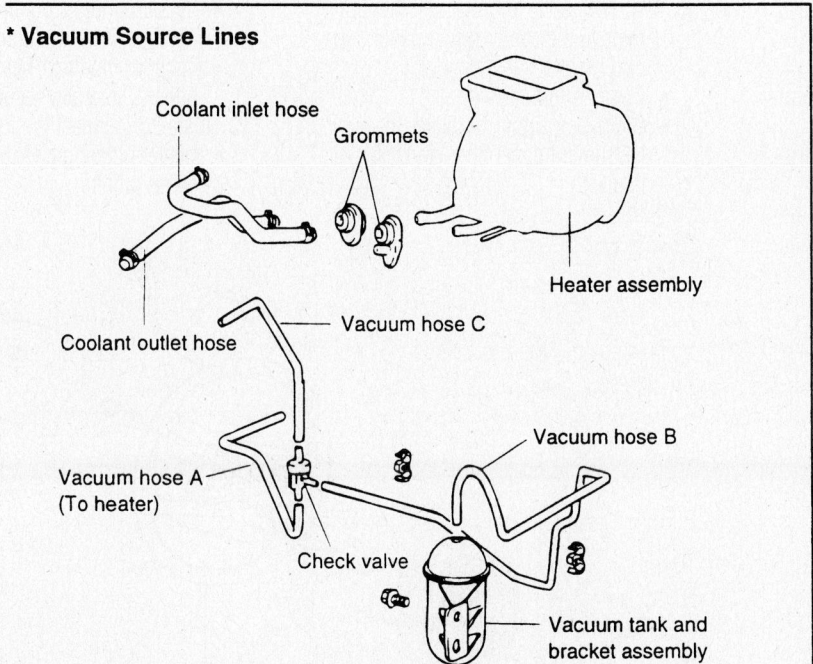

* Vacuum Source Lines

93112GP4

Exploded view of the heater core, heater housing and related components—Accent

Elantra

1. Before servicing the vehicle, refer to the Precautions Section.

2. Disconnect the negative battery cable.

❈❈ CAUTION

After disconnecting the negative battery cable, wait for at least 30 seconds for the SRS module to deplete its stored energy.

3. Discharge and recover the air conditioning system refrigerant.

4. Drain the engine coolant into a clean container for reuse.

5. Disconnect the heater hoses from the heater core. Plug the openings.

6. Disconnect the vacuum line from the heater housing vacuum nipple.

7. Remove the SRS module and the steering wheel by removing or disconnecting the following:

- Steering wheel-to-SRS module nuts

- SRS module from the steering wheel and disconnect the electrical connector
- Steering wheel-to-steering column nut
- Steering wheel from the steering column

8. Remove the instrument panel by removing or disconnecting the following:

- Steering column cover screws and the covers
- Instrument panel lower cover at the driver's side

- Multi-function switch and disconnect the electrical connector at the steering column
- Instrument panel cluster fascia panel
- Instrument cluster-to-instrument panel screws, disconnect the electrical connectors and remove the instrument cluster
- Side fascia panel and disconnect the mirror control connector
- Hood release mounting screws
- Rheostat and the upper console cover
- Heater control cable
- Electrical connectors at the center of the instrument panel and remove the center fascia panel assembly
- Console
- Radio-to-chassis screws and the radio
- Glove box screws and the glove box
- Glove box striker screws and the upper glove box cover
- Defroster nozzle
- Loosen the speedometer drive gear sleeve and disconnect the speedometer cable from the instrument panel

- Passenger's side SRS module connector
- Instrument panel-to-chassis bolts
- Any remaining electrical connectors
- Ventilation ducts from the instrument panel
- Instrument panel

9. Remove or disconnect the following:
- Front right side heating duct from the heater housing
- Pull back the carpet and remove the right side console mounting bracket
- Front left side duct from the heater housing
- Pull back the carpet and remove the left side console mounting bracket
- Rear heating duct from the heater housing
- Control modules electrical connectors at the center fascia panel support bracket
- Center fascia panel support bracket screws, bolts and/or nuts; then, remove the center fascia panel support bracket
- Center support bars
- Glove box support bracket-to-instrument panel bolts and the bracket

10. Remove the evaporator housing by removing or disconnecting the following:
- Refrigerant lines from the evaporator housing and discard the O-rings
- Thermostatic switch connector
- Evaporator housing upper and lower bolts
- Evaporator housing

11. Remove or disconnect the following:
- Heater housing-to-chassis bolts and the housing
- Vacuum motor-to-heater housing bolts (2 for each vacuum motor)
- Vacuum motor rod end connection and remove the vacuum motors
- Heater housing cover clips
- Cover and the heater core

To install:

12. Assemble the heater housing by installing or connecting the following:
- Heater core and the cover
- Heater housing cover clips
- Vacuum motor rod end connection and install the vacuum motors
- Vacuum motor-to-heater housing bolts (2 for each vacuum motor)

13. Install or connect the following:
- Heater housing and the housing-to-chassis bolts

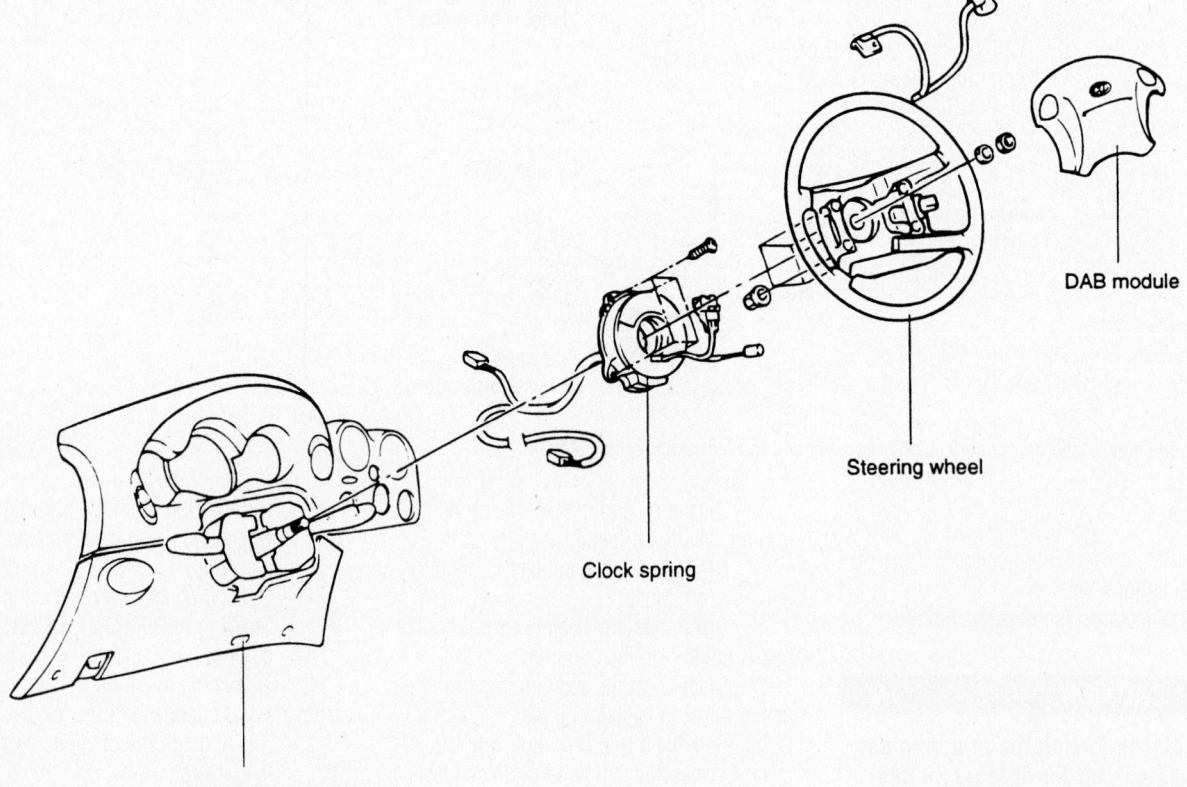

DAB module

Steering wheel

Clock spring

Data link connector

Exploded view of the SRS module, steering wheel and related components—Elantra

93112G07

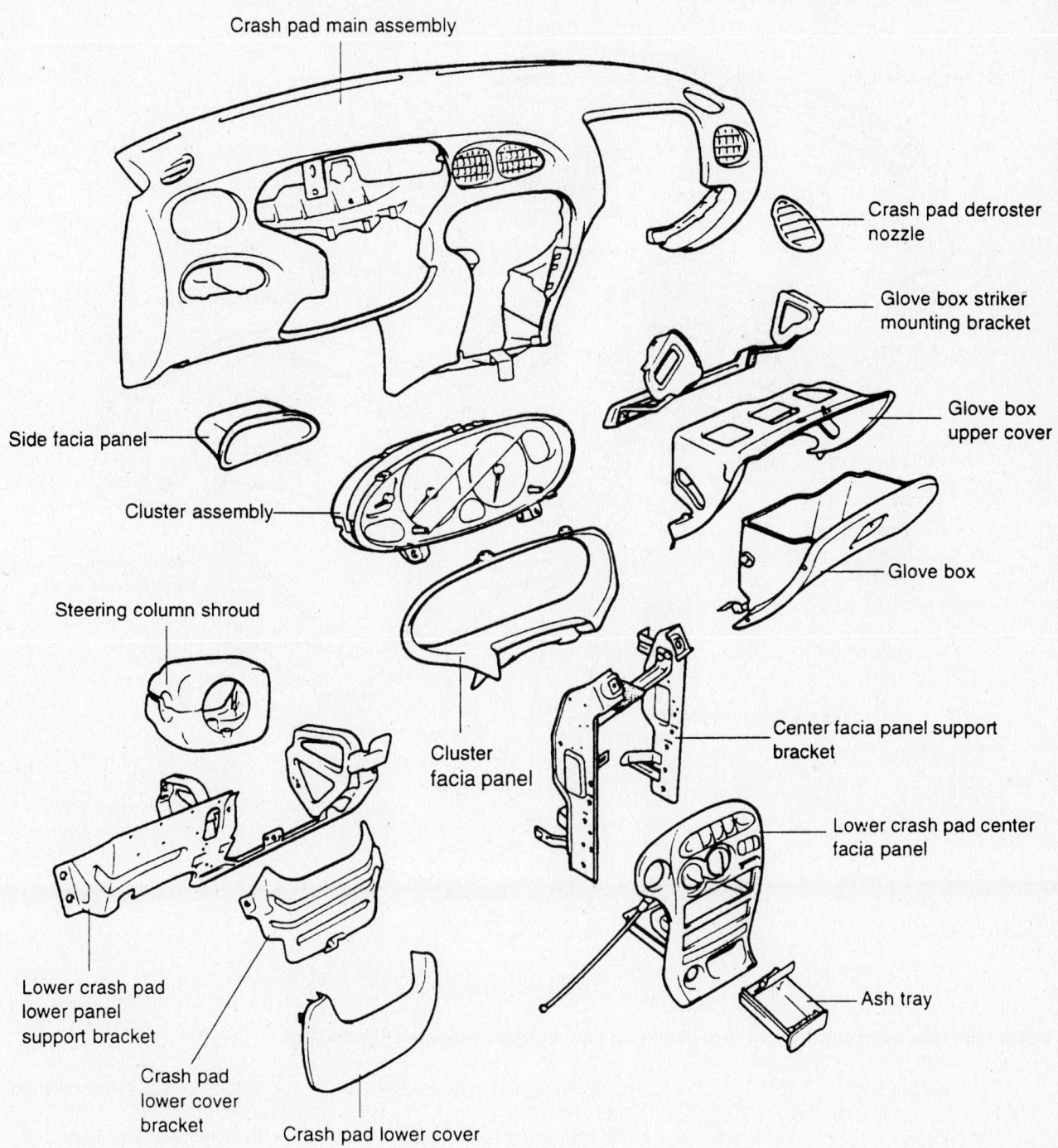

Crash pad main assembly

Crash pad defroster nozzle

Glove box striker mounting bracket

Glove box upper cover

Side facia panel

Cluster assembly

Glove box

Steering column shroud

Cluster facia panel

Center facia panel support bracket

Lower crash pad center facia panel

Lower crash pad lower panel support bracket

Crash pad lower cover bracket

Crash pad lower cover

Ash tray

TORQUE : Nm (kg·cm, lb·ft)

93112G08

Exploded view of the instrument panel and related components—Elantra

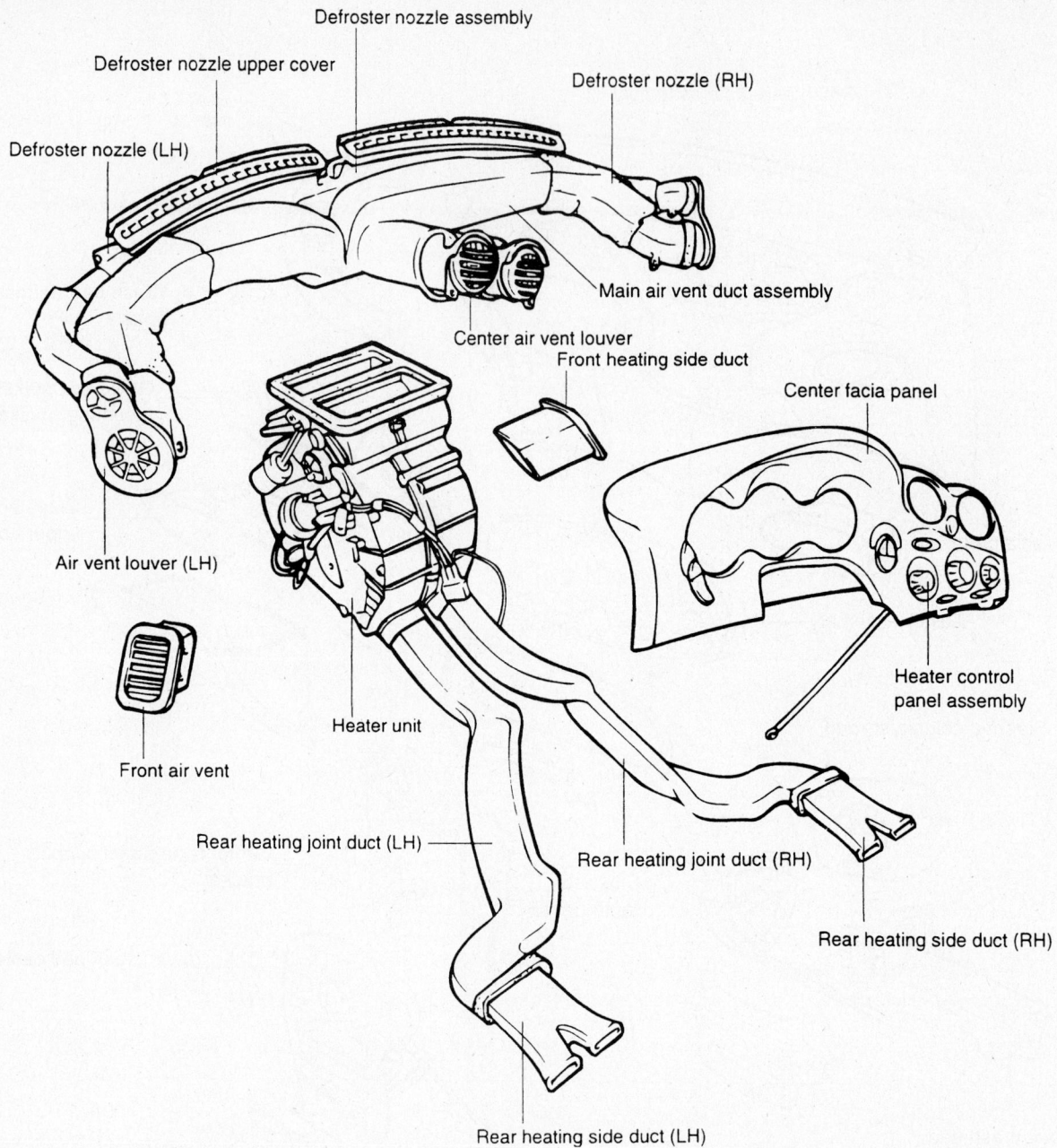

Defroster nozzle assembly

Defroster nozzle upper cover

Defroster nozzle (RH)

Defroster nozzle (LH)

Main air vent duct assembly

Center air vent louver

Front heating side duct

Center facia panel

Air vent louver (LH)

Heater control panel assembly

Front air vent

Heater unit

Rear heating joint duct (LH)

Rear heating joint duct (RH)

Rear heating side duct (RH)

Rear heating side duct (LH)

93112G00

Exploded view of the heater housing, center fascia, distribution ducts and related components—Elantra Coupe

- Evaporator housing
- Evaporator housing upper and lower bolts
- Thermostatic switch connector
- Connect the refrigerant lines to the evaporator housing
14. Install or connect the following:
  - Glove box support bracket and the bracket-to-instrument panel bolts
  - Center support bars
  - Center fascia panel support bracket; then, install the center fascia panel support bracket screws, bolts and/or nuts
  - Center fascia panel support bracket,

  connect the control modules electrical connectors
  - Rear heating duct to the heater housing
  - Left side console mounting bracket and install the carpet
  - Front left side duct to the heater housing
  - Right side console mounting bracket and install the carpet
  - Front right side heating duct to the heater housing
15. Install the instrument panel by installing or connecting the following:
  - Instrument panel

- Ventilation ducts to the instrument panel
- Instrument panel-to-chassis bolts
- Passenger's side SRS module connector
- Speedometer cable to the instrument panel and tighten the speedometer drive gear sleeve
- Defroster nozzle
- Upper glove box cover and the glove box striker screws
- Glove box and the glove box screws
- Radio and the radio-to-chassis screws

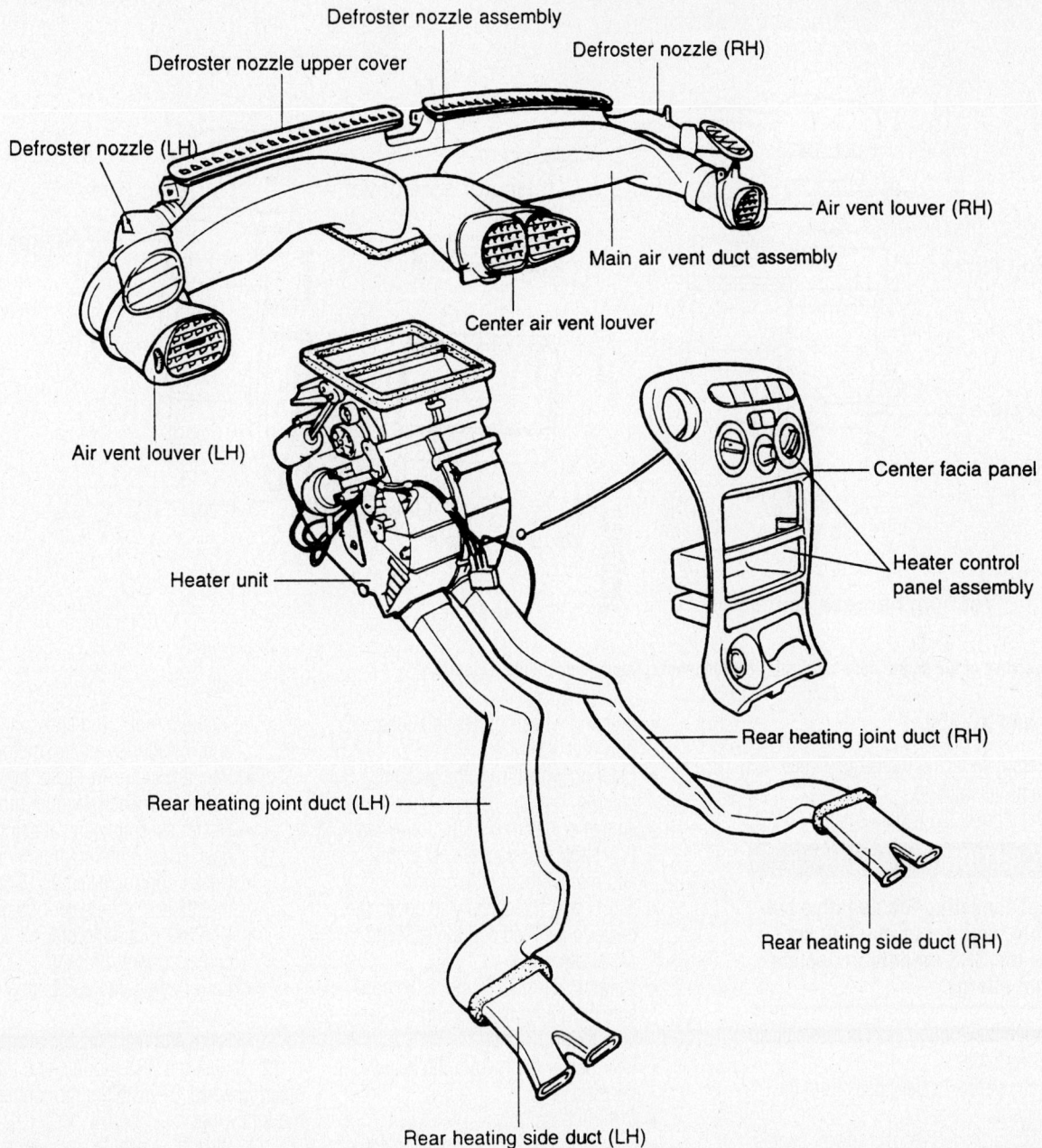

Defroster nozzle assembly

Defroster nozzle upper cover

Defroster nozzle (RH)

Defroster nozzle (LH)

Air vent louver (RH)

Main air vent duct assembly

Center air vent louver

Air vent louver (LH)

Center facia panel

Heater control panel assembly

Heater unit

Rear heating joint duct (RH)

Rear heating joint duct (LH)

Rear heating side duct (RH)

Rear heating side duct (LH)

93112GP1

Exploded view of the heater housing, center fascia, distribution ducts and related components—Elantra Sedan and Wagon

- Console
- Electrical connectors and install the center fascia panel assembly
- Heater control cable
- Rheostat and the upper console cover
- Hood release mounting screws
- Side fascia panel and connect the mirror control connector
- Instrument cluster, connect the electrical connectors and install the instrument cluster-to-instrument panel screws
- Instrument panel cluster fascia panel

- Multi-function switch and connect the electrical connector
- Instrument panel lower cover
- Steering column cover and the cover screws

16. Install the SRS module and the steering wheel by installing or connecting the following:

- Steering wheel to the steering column
- Steering wheel-to-steering column nut and torque to 30–37 ft. lbs. (40–50 Nm)
- SRS module to the steering wheel and connect the electrical connector

- Steering wheel-to-SRS module nuts

17. Connect the vacuum line to the heater housing vacuum nipple.

18. Connect the heater hoses to the heater core.

19. Refill the cooling system.

20. Connect the negative battery cable.

21. Evacuate, charge and leak test the air conditioning system.

22. Operate the engine to normal operating temperatures; then, check the climate control operation and check for leaks.

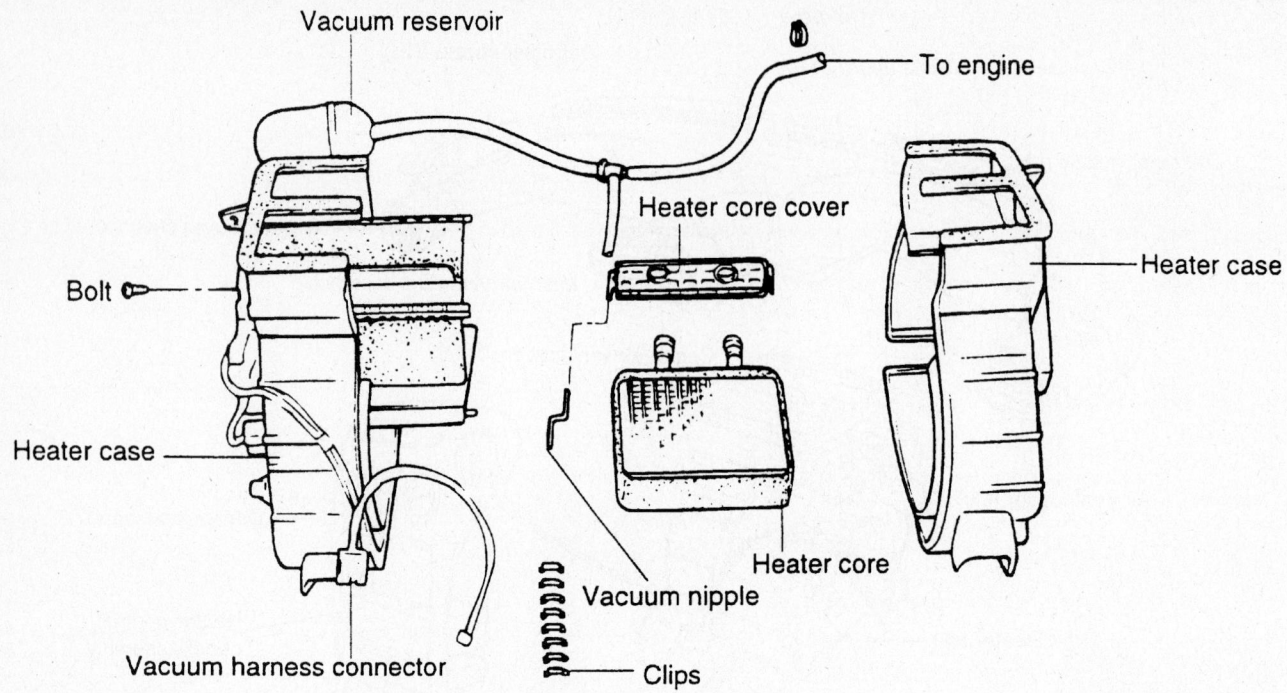

Vacuum reservoir

To engine

Heater core cover

Heater case

Bolt

Heater case

Heater core

Vacuum nipple

Vacuum harness connector

Clips

93112GP2

Exploded view of the heater core and heater housing and related components—Elantra

Sonata and XG 350

1. Before servicing the vehicle, refer to the Precautions Section.
2. Disconnect the negative battery cable.

✳✳ CAUTION

After disconnecting the negative battery cable, wait for at least 30 seconds for the SRS module to deplete its stored energy.

3. Drain the cooling system into a clean container for reuse.
4. Remove the heater hoses from the heater housing.
5. Discharge and recover the air conditioning system refrigerant.
6. Remove the suction and discharge hoses from the evaporator assembly. Cap the hoses to minimize contamination.
7. Remove the evaporator drain hose.
8. Remove the SRS module and the steering wheel by removing or disconnecting the following:
- Steering wheel-to-SRS module nuts
- SRS module from the steering wheel and disconnect the electrical connector

✳✳ CAUTION

Store the SRS module in a safe place with the front facing upward.

- Steering wheel-to-steering column nut
- Steering wheel from the steering column

9. Remove or disconnect the following:
- Front and rear console assembly and remove both side covers
- Glove box, the center pad cover, the center crash pad and the cassette assembly
- Lower crash pad. Remove the console mounting bracket and the center support bracket
- Rear heater ducts from the heater housing
- Control assembly
- Blower speed control actuator connector and the blend door actuator connector, if equipped with semi-automatic temperature control
- 4 retaining bolts and remove the heater assembly

10. Disassemble the heater housing by removing or disconnecting the following:
- Vacuum motor-to-heater housing bolts (2 for each vacuum motor)
- Vacuum motor rod end connection and remove the vacuum motors
- Heater housing cover clips
- Cover and the heater core

To install:

11. Install or connect the following:
- Heater core and the cover
- Heater housing cover clips

- Vacuum motor rod end connection and install the vacuum motors
- Vacuum motor-to-heater housing bolts (2 for each vacuum motor)
- Heater assembly and attach it to the dash panel with the mounting bolts
- Heater control assembly. Connect the ducts to the heater housing
- Console mounting bracket and the center support bracket
- Lower crash pad and both side covers
- Front and rear console assembly

12. Install the SRS module and the steering wheel by installing or connecting the following:
- Steering wheel to the steering column
- Steering wheel-to-steering column nut and torque to 30–37 ft. lbs. (40–50 Nm)
- SRS module to the steering wheel and connect the electrical connector
- Steering wheel-to-SRS module nuts
- Evaporator tubes, the heater hoses and the drain hose

13. Refill the cooling system.
14. Connect the negative battery cable.
15. Evacuate, charge and leak test the air conditioning system.
16. Operate the engine to normal operating temperatures; then, check the climate control operation and check for leaks.

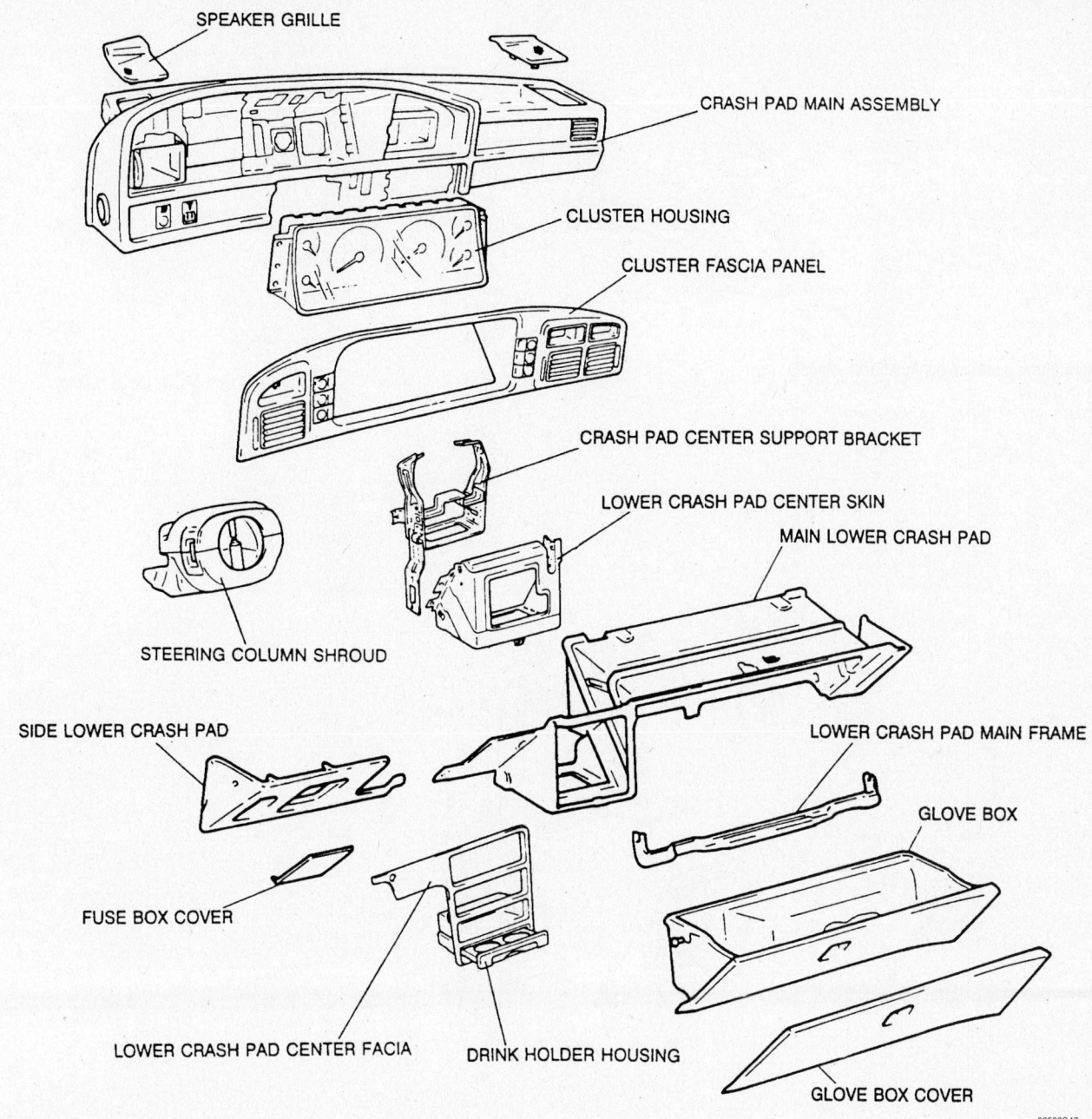

SPEAKER GRILLE

CRASH PAD MAIN ASSEMBLY

CLUSTER HOUSING

CLUSTER FASCIA PANEL

CRASH PAD CENTER SUPPORT BRACKET

LOWER CRASH PAD CENTER SKIN

MAIN LOWER CRASH PAD

STEERING COLUMN SHROUD

SIDE LOWER CRASH PAD

LOWER CRASH PAD MAIN FRAME

GLOVE BOX

FUSE BOX COVER

LOWER CRASH PAD CENTER FACIA

DRINK HOLDER HOUSING

GLOVE BOX COVER

89530G47

Instrument panel assembly—Sonata

Tiburon

1. Before servicing the vehicle, refer to the Precautions Section.

2. Disconnect the negative battery cable.

✳✳ CAUTION

After disconnecting the negative battery cable, wait for at least 30 seconds for the SRS module to deplete its stored energy.

3. Discharge and recover the air conditioning system refrigerant.

4. Drain the engine coolant into a clean container for reuse.

5. Disconnect the heater hoses from the heater core. Plug the openings.

6. Disconnect the vacuum line from the heater housing vacuum nipple.

7. Remove the SRS module and the steering wheel by removing or disconnecting the following:
- Steering wheel-to-SRS module nuts
- SRS module from the steering wheel and disconnect the electrical connector
- Steering wheel-to-steering column nut

- Steering wheel from the steering column

8. Remove the instrument panel by removing or disconnecting the following:
- Upper console cover
- Center fascia panel and disconnect the cigar lighter connector
- 3 lower instrument panel screws and the lower instrument panel
- Radio-to-bracket bolts and the radio
- Rheostat switch, the hood release handle and DLC from the lower instrument panel
- 5 cluster fascia panel-to-instrument

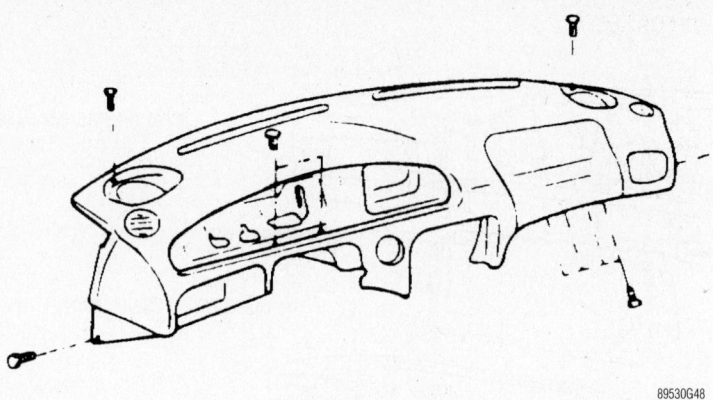

89530G48

Instrument panel screw locations—Sonata

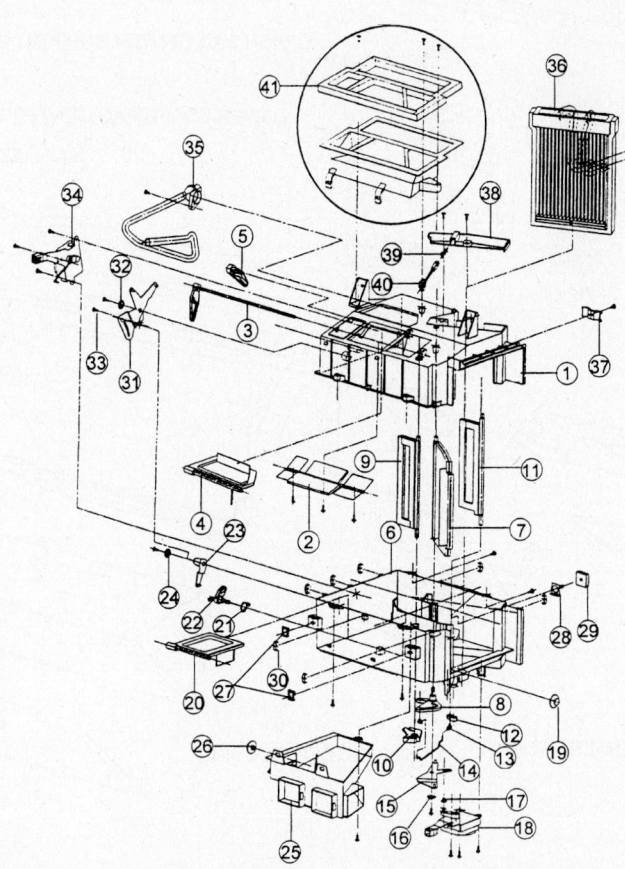

1 Case-heater upper	11 Door ass'y-by pass	21 Spring	31 Cam-mode
2 Door ass'y-vent	12 Arm-By pass door	22 Arm-floor door	32 Spring-washer
3 Shaft ass'y-vent door	13 Holder-rod link	23 Lever-floor door	33 Holder-rod link
4 Door ass'y-defrost	14 Link	24 Spring washer	34 Mode actuator
5 Arm defrost door	15 Lever-temp. door	25 Duct-floor	35 Aspirator & hose ass'y
6 Case-heater lower	16 Spring washer	26 Guide bush	36 Heater core
7 Door ass'y-temp.	17 Guide bush	27 U-nut	37 Clip
8 Arm-temp. door	18 Blend door actuator	28 Clip & Bolt ass'y	38 Cover-heater core
9 Door ass'y (A)-temp. door	(For AUTO A/C only)	29 Seal (A)-heater to D/panel	39 Stopper
10 Arm (A)-temp. door (A)	19 Guide bush	30 Clip	40 Sensor
	20 Door ass'y-floor		41 Plenum duct ass'y

93112GP3

Exploded view of the heater housing assembly—Sonata

panel screws; then, disconnect the heater control cable and the cluster electrical connectors and remove the cluster fascia panel

- 4 instrument cluster-to-instrument panel screws and the instrument cluster
- 2 glove box-to-instrument panel bolts and the glove box
- 4 upper glove box cover-to-instrument panel screws, the 2 glove box striker screws and the upper glove box cover
- Upper instrument panel speaker grille
- 2 upper speaker-to-instrument panel screws
- Instrument panel-to-chassis bolts, disconnect the electrical connectors and remove the instrument panel

9. Remove or disconnect the following:
- Front right side heating duct from the heater housing
- Pull back the carpet and remove the right side console mounting bracket
- Front left side duct from the heater housing

- Pull back the carpet and remove the left side console mounting bracket
- Rear heating duct from the heater housing
- Control modules electrical connectors at the center fascia panel support bracket
- Center fascia panel support bracket screws, bolts and/or nuts; then, remove the center fascia panel support bracket
- Center support bars
- Glove box support bracket-to-instrument panel bolts and the bracket

10. Remove the evaporator housing by removing or disconnecting the following:
- Refrigerant lines (located in the engine compartment), from the evaporator housing and discard the O-rings
- Thermostatic switch connector
- Evaporator housing upper and lower bolts
- Evaporator housing
- Heater housing-to-chassis bolts and the housing

11. Disassemble the heater housing by removing or disconnecting the following:
- Vacuum motor-to-heater housing bolts (2 for each vacuum motor)
- Vacuum motor rod end connection and remove the vacuum motors
- Heater housing cover clips
- Cover and the heater core

To install:

12. Assemble the heater housing by installing or connecting the following:
- Heater core and the cover
- Heater housing cover clips
- Vacuum motor rod end connection and install the vacuum motors
- Vacuum motor-to-heater housing bolts (2 for each vacuum motor)
- Heater housing and the housing-to-chassis bolts

13. Install the evaporator housing by installing or connecting the following:
- Evaporator housing
- Evaporator housing upper and lower bolts
- Thermostatic switch connector

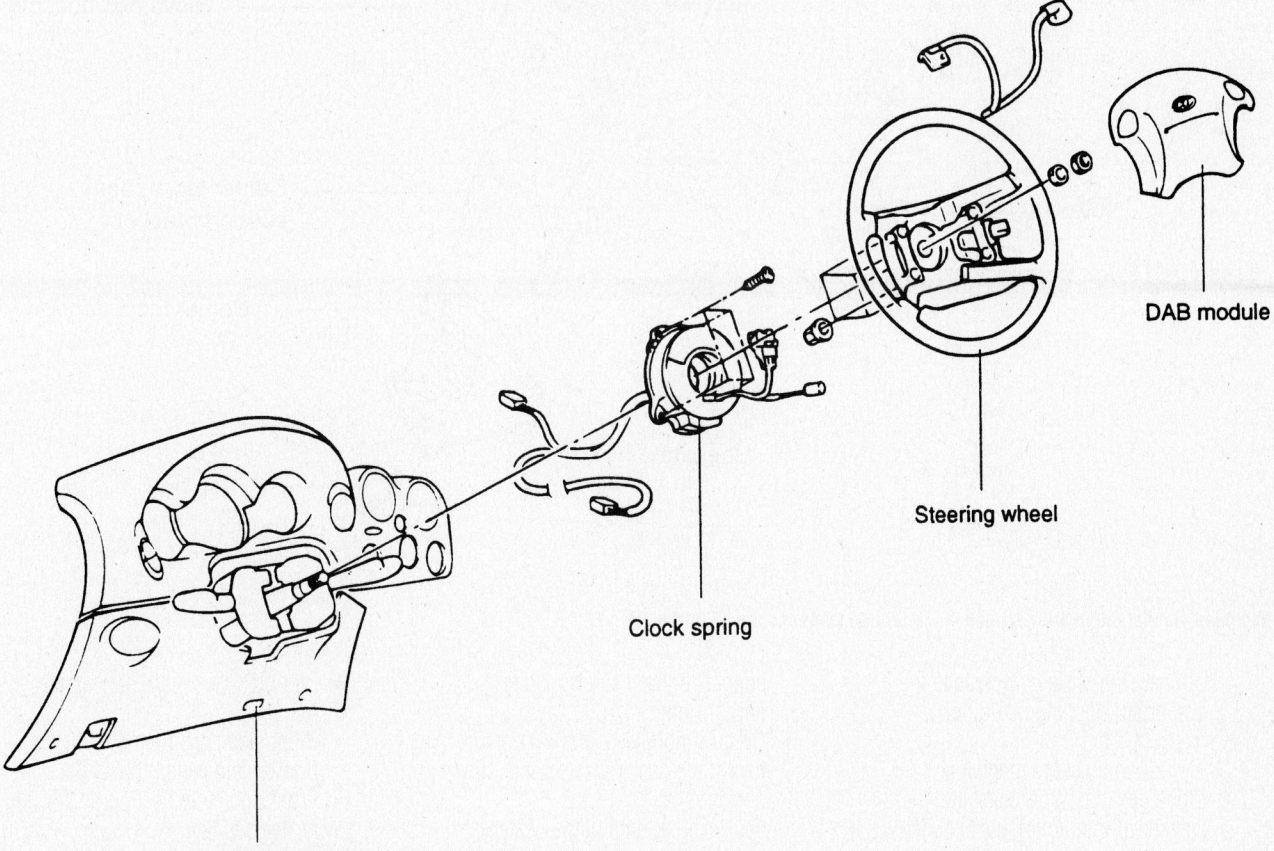

DAB module

Steering wheel

Clock spring

Data link connector

93112G07

Exploded view of the SRS module, steering wheel and related components—Tiburon

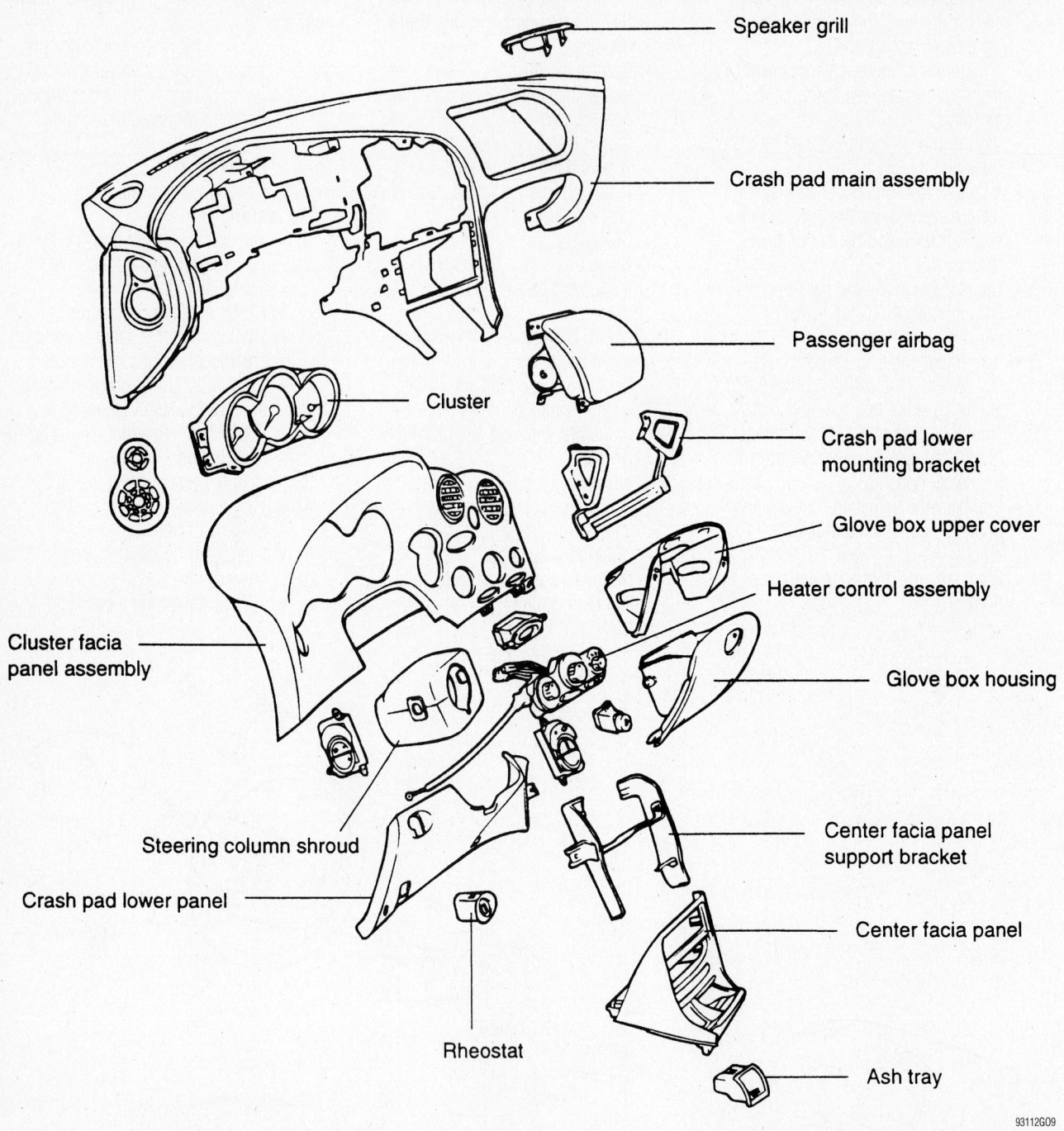

Speaker grill

Crash pad main assembly

Passenger airbag

Cluster

Crash pad lower mounting bracket

Glove box upper cover

Heater control assembly

Cluster facia panel assembly

Glove box housing

Steering column shroud

Center facia panel support bracket

Crash pad lower panel

Center facia panel

Rheostat

Ash tray

93112G09

Exploded view of the instrument panel and related components—Tiburon

- Connect the refrigerant lines to the evaporator housing using new O-rings
14. Install or connect the following:
  - Glove box support bracket and the bracket-to-instrument panel bolts
  - Center support bars
  - Center fascia panel support bracket; then, install the center fascia panel

support bracket screws, bolts and/or nuts
  - Control modules electrical connectors at the center fascia panel support bracket
  - Rear heating duct to the heater housing
  - Left side console mounting bracket and install the carpet

- Front left side duct to the heater housing
  - Right side console mounting bracket and install the carpet
  - Front right side heating duct to the heater housing
15. Install the instrument panel by installing or connecting the following:
  - Instrument panel, connect the elec-

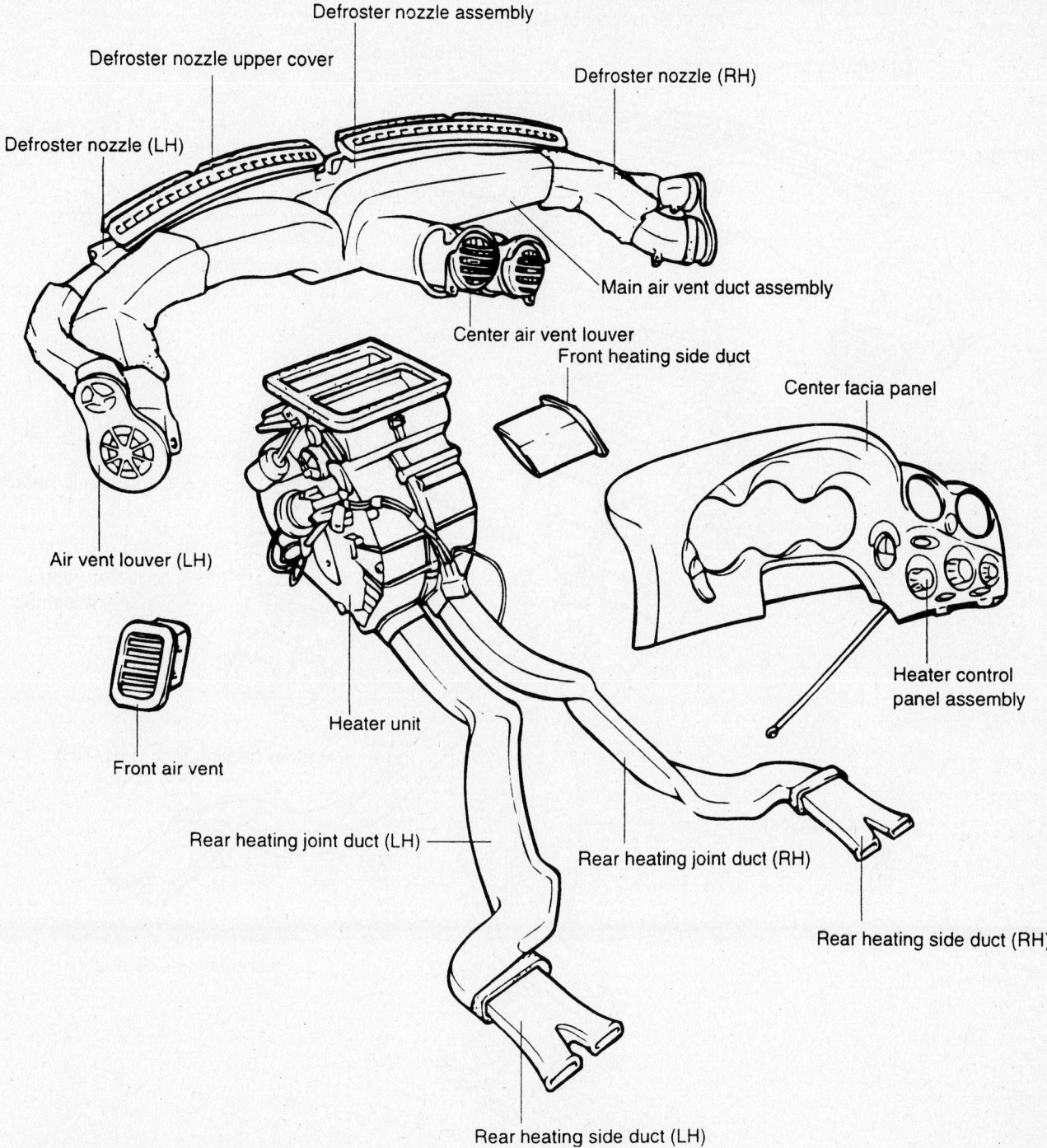

Defroster nozzle assembly

Defroster nozzle upper cover

Defroster nozzle (RH)

Defroster nozzle (LH)

Main air vent duct assembly

Center air vent louver

Front heating side duct

Center facia panel

Air vent louver (LH)

Front air vent

Heater control panel assembly

Heater unit

Rear heating joint duct (LH)

Rear heating joint duct (RH)

Rear heating side duct (RH)

Rear heating side duct (LH)

93112G00

Exploded view of the heater housing, center fascia, distribution ducts and related components—Tiburon Coupe

trical connectors and install the instrument panel-to-chassis bolts
• 2 upper speaker-to-instrument panel screws
• Upper instrument panel speaker grille
• Upper glove box cover, the 2 glove box striker screws and the 4 upper glove box cover-to-instrument panel screws

• Glove box and the 2 glove box-to-instrument panel bolts
• Instrument cluster and the 4 instrument cluster-to-instrument panel screws
• Cluster fascia panel; then, connect the heater control cable and the cluster electrical connectors and install the 5 cluster fascia panel-to-instrument panel screws

• Rheostat switch, the hood release handle and DLC to the lower instrument panel
• Radio and the radio-to-bracket bolts
• Lower instrument panel and the 3 lower instrument panel screws
• Cigar lighter connector and install the center fascia panel
• Upper console cover

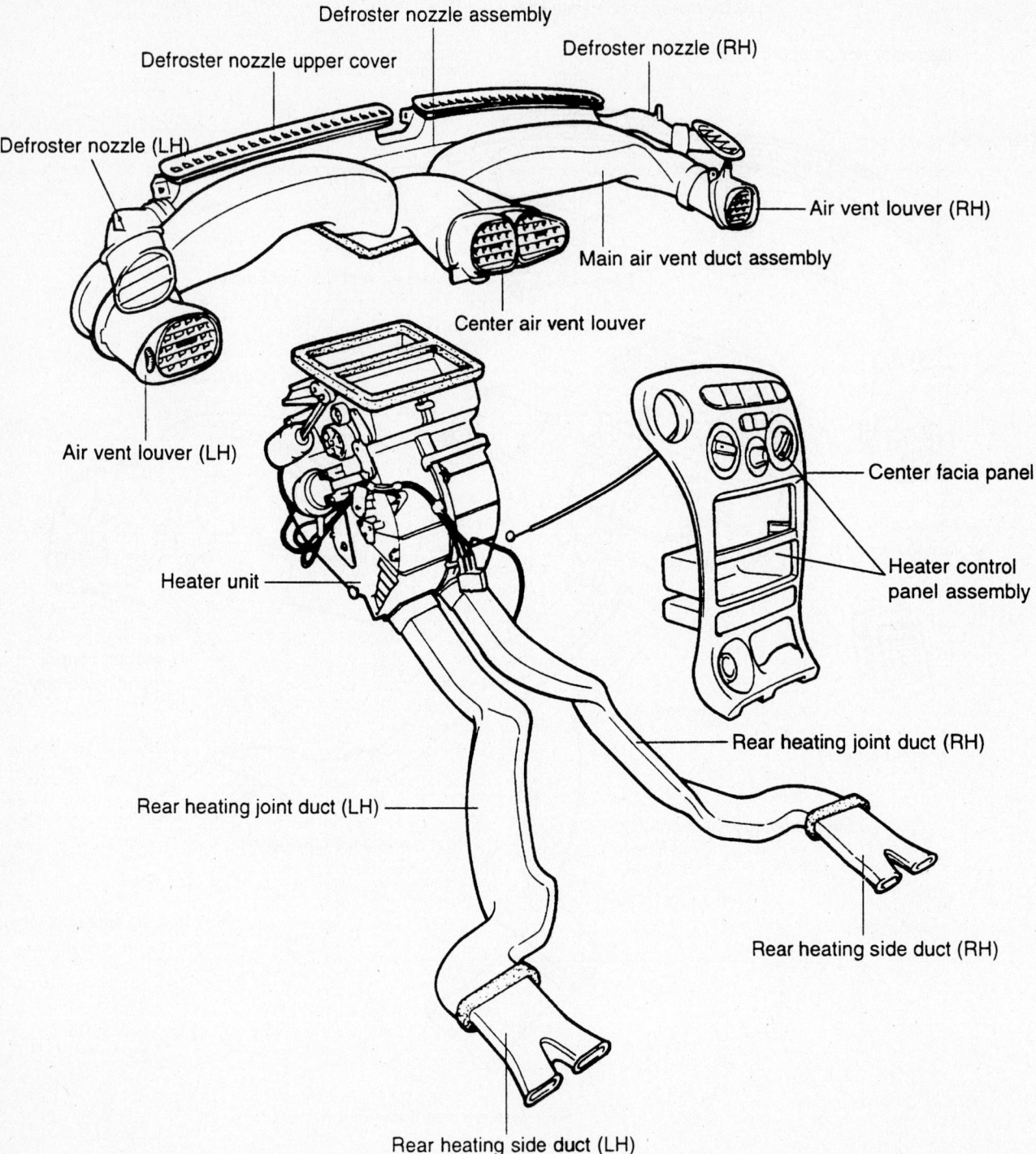

Defroster nozzle assembly

Defroster nozzle upper cover

Defroster nozzle (RH)

Defroster nozzle (LH)

Air vent louver (RH)

Main air vent duct assembly

Center air vent louver

Air vent louver (LH)

Center facia panel

Heater control panel assembly

Heater unit

Rear heating joint duct (RH)

Rear heating joint duct (LH)

Rear heating side duct (RH)

Rear heating side duct (LH)

93112GP1

Exploded view of the heater housing, center fascia, distribution ducts and related components—Tiburon Sedan and Wagon

16. Install the SRS module and the steering wheel by installing or connecting the following:
- Steering wheel to the steering column
- Steering wheel-to-steering column nut and torque to 30–37 ft. lbs. (40–50 Nm)
- SRS module to the steering wheel and connect the electrical connector
- Steering wheel-to-SRS module nuts

17. Install or connect the following:
- Vacuum line to the heater housing vacuum nipple
- Heater hoses to the heater core

18. Refill the cooling system.

19. Connect the negative battery cable.

20. Evacuate, charge and leak test the air conditioning system.

21. Operate the engine to normal operating temperatures; then, check the climate control operation and check for leaks.

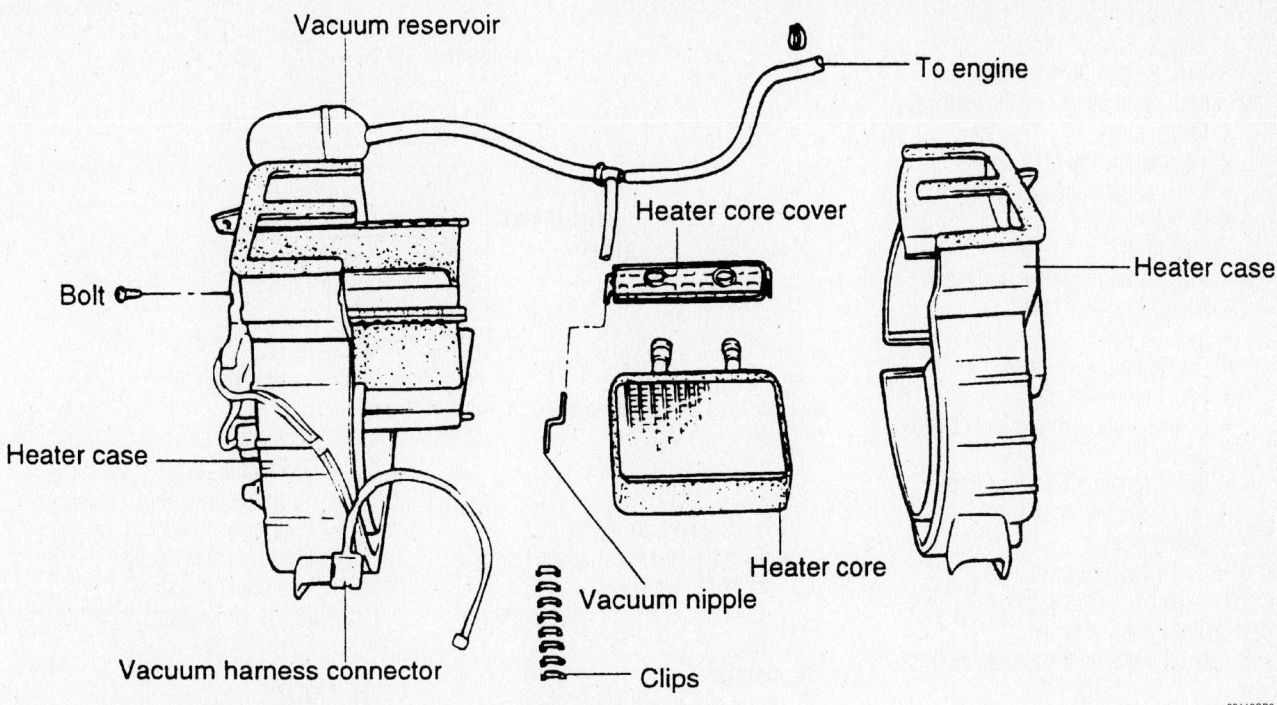

Exploded view of the heater core and heater housing and related components—Tiburon

Cylinder Head

REMOVAL & INSTALLATION

4 Cylinder Engines

1. Before servicing the vehicle, refer to the Precautions Section.
2. Drain the cooling system.
3. Relieve the fuel system pressure.
4. Remove or disconnect the following:

- Negative battery cable
- Upper radiator hose
- Heater hose
- Air cleaner assembly
- Intake manifold vacuum lines
- Engine control wiring harness connectors
- Spark plug wires
- Ignition coil
- Accessory drive belts
- Power steering pump and bracket
- Fuel lines
- Intake manifold
- Exhaust manifold
- Front cover
- Timing belt
- Valve cover bolts
- Cylinder head by loosening the bolts in sequence
- Cylinder head and discard the gasket

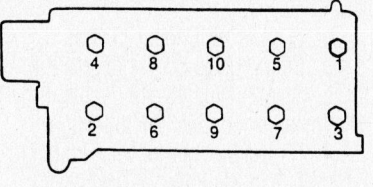

Cylinder head bolt loosening sequence—1.5L, Elantra and Tiburon 2.0L engines

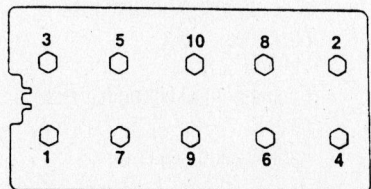

Cylinder head bolt loosening sequence—Sonata 2.0L and 2.4L engines

To install:

5. Install the cylinder head with a new gasket.

6. For 1.5L engines, tighten the bolts in sequence to 51–54 ft. lbs. (71–75 Nm).

7. For 2.0L Elantra and Tiburon engines, tighten the bolts in sequence as follows:

   a. Step 1: M10 bolts to 22 ft. lbs. (30 Nm) and M12 bolts to 26 ft. lbs. (35 Nm).

   b. Step 2: Plus 60–65 degrees.

   c. Step 3: Plus 60–65 degrees.

8. For Sonata 2.0L engines, tighten the bolts in sequence to 65–72 ft. lbs. (90–100 Nm).

9. For 2.4L engines, tighten the bolts in sequence as follows:

   a. Step 1: 14 ft. lbs. (20 Nm).

   b. Step 2: Plus 90 degrees.

   c. Step 3: Loosen all bolts in reverse of tightening order.

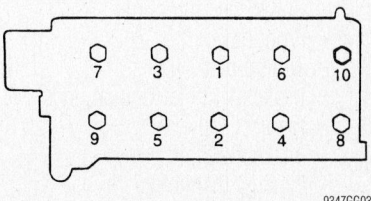

Cylinder head torque sequence—1.5L, and Elantra and Tiburon 2.0L Engines

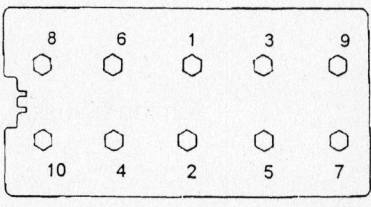

Cylinder head torque sequence—Sonata 2.0L and 2.4L Engines

d. Step 4: 14 ft. lbs. (20 Nm).

e. Step 5: Plus 90 degrees.

f. Step 6: Plus 90 degrees.

10. Install or connect the following:

- Valve cover
- Timing belt and front cover
- Exhaust manifold
- Intake manifold
- Fuel lines
- Power steering pump and bracket
- Accessory drive belts
- Ignition coil
- Distributor, if equipped
- Spark plug wires
- Engine control wiring harness connectors
- Intake manifold vacuum lines
- Air cleaner assembly
- Heater hose
- Upper radiator hose
- Negative battery cable

11. Fill the cooling system.

12. Start the engine and check for leaks.

V6 Engines

2.7L ENGINE

1. Before servicing the vehicle, refer to the Precautions Section.

2. Drain the cooling system.

3. Relieve the fuel system pressure.

4. Remove or disconnect the following:

- Negative battery cable
- Accessory drive belts
- Air intake assembly
- Radiator and heater hoses
- Engine control wiring harness connectors
- Fuel lines
- Brake booster vacuum hose
- Power steering pump
- Front covers
- Timing belt
- Intake manifold vacuum lines
- Ignition coil
- Intake manifold
- Exhaust manifolds

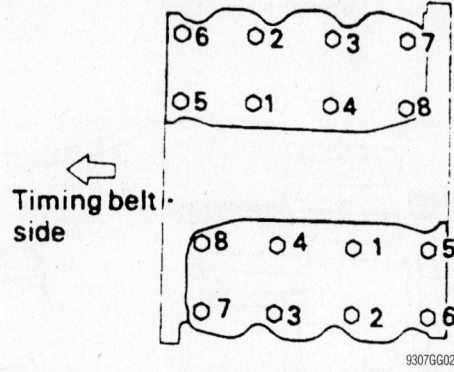

Cylinder head torque sequence—2.5L and 2.7L Sonata

- Valve covers
- Camshafts
- Timing belt rear cover
- Cylinder heads by loosening the bolts in sequence
- Cylinder head and discard the gasket

To install:

5. Install or connect the following:

- New head gaskets
- Cylinder heads

6. Tighten the bolts in sequence as follows:

a. Step 1: 18 ft. lbs. (25 Nm).

b. Step 2: Plus 60–64 degree turn.

c. Step 3: Plus 45–49 degree turn.

7. Install or connect the following:

- Timing belt rear cover. Tighten to 7–9 ft. lbs. (10–12 Nm).
- Camshafts
- Valve covers
- Timing belt
- Exhaust manifolds
- Power steering pump
- Intake manifold
- Brake booster vacuum hose
- Fuel lines
- Ignition coil
- Engine control wiring harness connectors
- Radiator and heater hoses

- Spark plug wires
- Air intake assembly
- Accessory drive belts
- Negative battery cable

8. Fill the cooling system.

9. Start the engine and check for leaks.

3.5L ENGINE

1. Before servicing the vehicle, refer to the Precautions Section.

2. Drain the cooling system.

3. Relieve the fuel system pressure.

4. Remove or disconnect the following:

- Negative battery cable
- Upper radiator hose
- Breather hose
- Air intake hose
- Vacuum hose
- Fuel hoses
- Intake manifold
- Spark plug wires
- Ignition coil
- Upper and lower timing belt covers
- Timing belt
- Camshaft sprockets
- Heat protector and exhaust manifold
- Water pump
- Rocker arm cover
- Camshafts

5. The cylinder head bolts should be loosened in sequence as shown.

6. Remove the cylinder head and discard the gasket.

7. Clean all mating surfaces of any residual gasket material.

To install:

8. Install or connect the following:

- New gasket with the identification mark facing the cylinder head
- Cylinder head and torque the bolts in sequence to 82 ft. lbs. (115 Nm)
- Camshafts
- Rocker arm cover and torque the bolts to 89 inch lbs. (10 Nm)

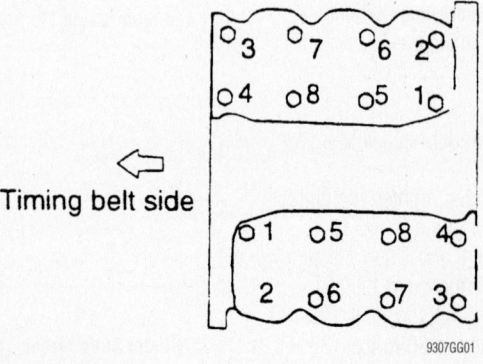

Cylinder head bolt loosening sequence— 2.7L engine

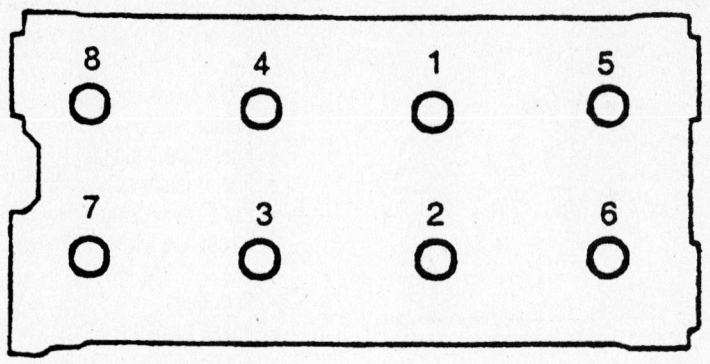

Cylinder head torque sequence—3.5L engine

- Water pump and torque the bolts to 16 ft. lbs. (22 Nm)
- Heat protector and exhaust manifold and torque the bolts to 14 ft. lbs. (19 Nm)
- Camshaft sprockets
- Timing belt
- Upper and lower timing belt covers
- Ignition coil and spark plug wires
- Intake manifold and torque the bolts to 10 ft. lbs. (15 Nm)
- Fuel hoses
- Vacuum hose
- Air intake hose
- Breather hose
- Upper radiator hose

- Negative battery cable

9. Fill the cooling system to the proper level.

10. Start the vehicle, check for leaks and repair if necessary.

Rocker Arms/Shafts

REMOVAL & INSTALLATION

Except 1.5L SOHC

These engines are not equipped with rocker arms. The camshaft acts directly on the valves through hydraulic lash adjusters.

1.5L SOHC Engine

1. Before servicing the vehicle, refer to the Precautions Section.

2. Remove or disconnect the following:
- Negative battery cable
- Valve cover
- Rocker arm shaft bolts by loosening them evenly in several steps
- Rocker arm and shaft assemblies

➡ Keep all valvetrain components in order for assembly.

- Rocker arms and springs from the shafts

To install:

3. Install or connect the following:
- Rocker arms and springs in their original positions
- Rocker arm and shaft assemblies and torque the bolts evenly to 14–20 ft. lbs. (20–26 Nm)
- Valve cover
- Negative battery cable

Intake Manifold

REMOVAL & INSTALLATION

4 Cylinder Engines

1. Before servicing the vehicle, refer to the Precautions Section.

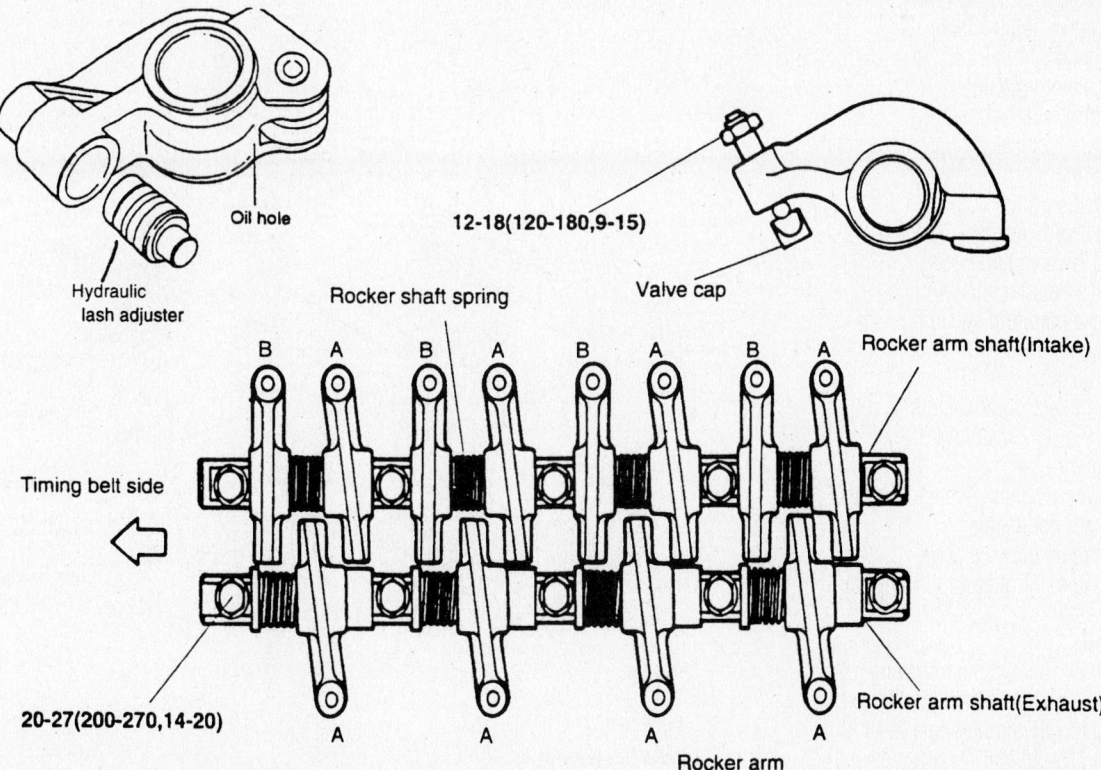

Rocker assembly components and arrangement. Rockers marked "A" and "B" must be returned to their original positions—1.5L SOHC engines

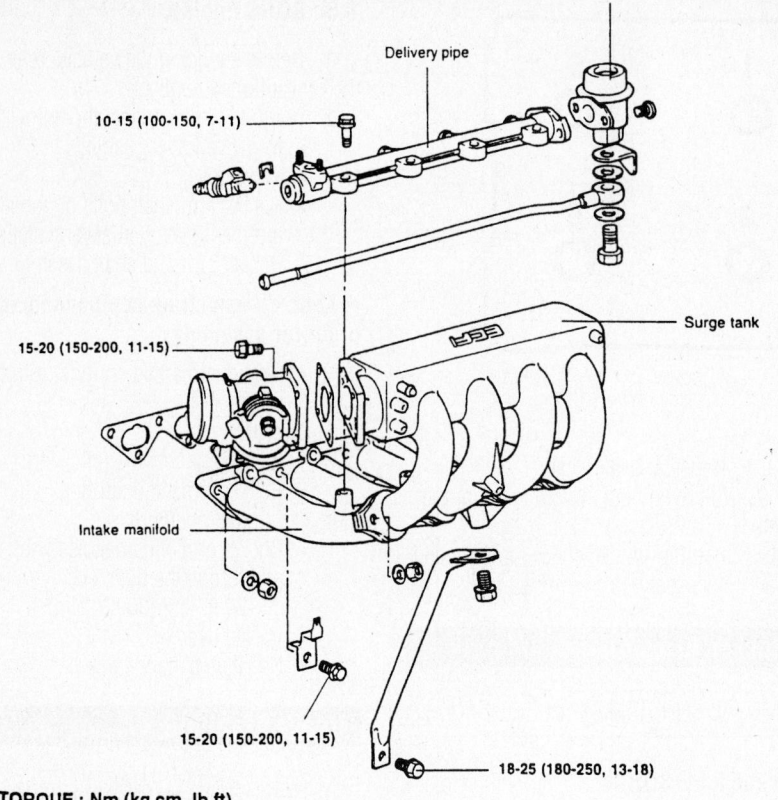

TORQUE : Nm (kg.cm, lb.ft)

Surge tank and intake manifold components—1.5L engine

2. Relieve the fuel system pressure.
3. Drain the cooling system.
4. Remove or disconnect the following:
   - Negative battery cable
   - Air intake hose
   - Accelerator cable
   - Upper radiator hose
   - Engine control wiring harness connectors
   - Throttle body
   - Positive Crankcase Ventilation (PCV) valve and hose
   - Brake booster vacuum line
   - Intake manifold vacuum hoses
   - Fuel lines
   - Surge tank
   - Fuel injector connectors
   - Fuel supply manifold
   - Heater hose
   - Engine Coolant Temperature (ECT) sensor connector
   - Intake manifold bracket
   - Intake manifold and discard the gasket

To install:

5. Install or connect the following:
   - Intake manifold using a new gasket and torque the nuts to 11–14 ft. lbs. (15–20 Nm), starting from the center and working outwards

   - Intake manifold bracket and torque the bolts to 13–18 ft. lbs. (18–25 Nm)
   - ECT sensor connector
   - Heater hose
   - Fuel supply manifold
   - Fuel injector connectors
   - Surge tank using a new gasket and torque the bolts to 11–16 ft. lbs. (15–22 Nm)
   - Fuel lines
   - Intake manifold vacuum hoses
   - Brake booster vacuum line
   - PCV valve and hose
   - Throttle body
   - Engine control wiring harness connectors
   - Upper radiator hose
   - Accelerator cable
   - Air intake hose
   - Negative battery cable
6. Fill the cooling system.
7. Start the engine and check for leaks.

V6 Engines

1. Before servicing the vehicle, refer to the Precautions Section.
2. Relieve the fuel system pressure.
3. Drain the cooling system.

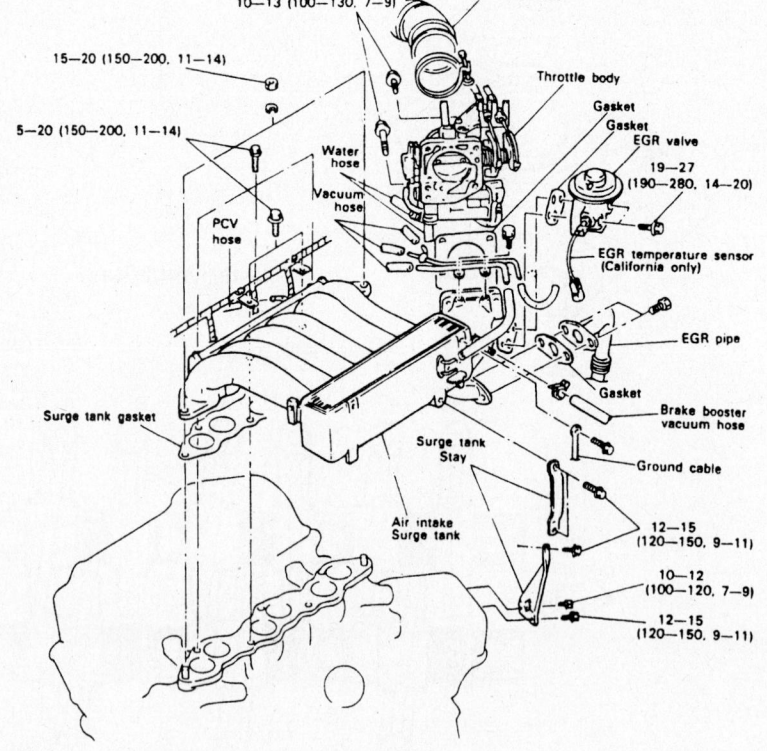

TORQUE : Nm (kg.cm, lb.ft)

Surge tank and intake manifold components—Sonata 2.0L and 2.4L engines

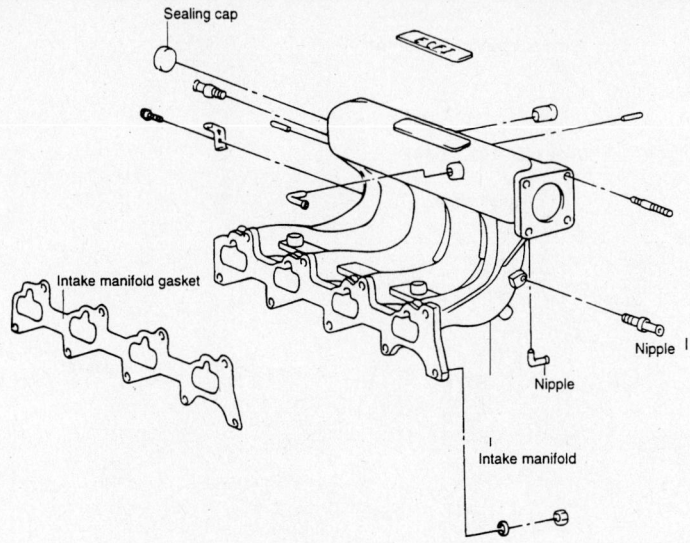

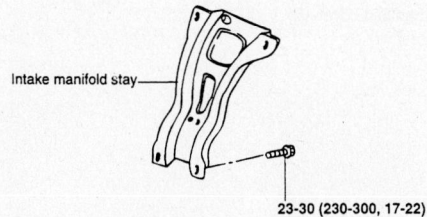

23-30 (230-300, 17-22)

TORQUE : Nm (kg.cm, lb.ft)

7923GG39

Surge tank and intake manifold components—Elantra and Tiburon 2.0L engine

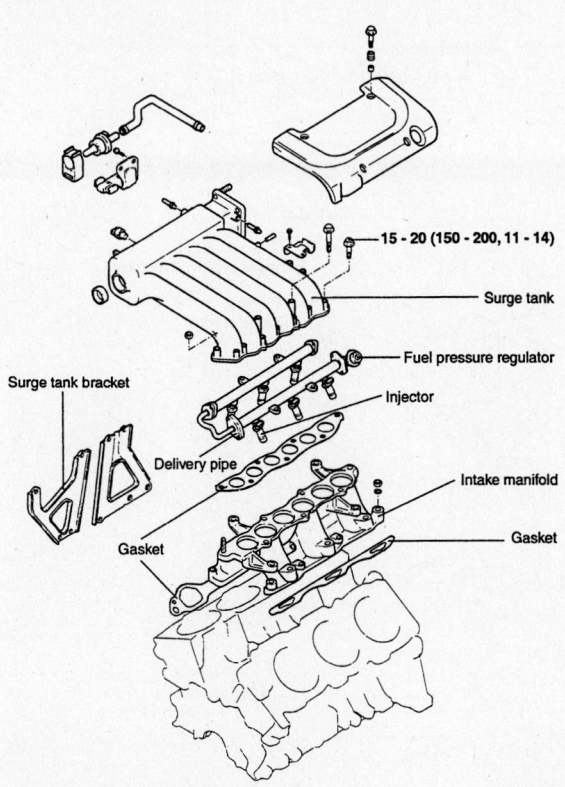

15 - 20 (150 - 200, 11 - 14)

9347KG05

Surge tank and intake manifold components—3.5L Engine

4. Remove or disconnect the following:
   - Negative battery cable
   - Air intake hose
   - Accelerator cable
   - Upper radiator hose
   - Engine control wiring harness
   - Throttle body
   - Positive Crankcase Ventilation (PCV) valve and hose
   - Intake manifold vacuum hoses
   - Exhaust Gas Recirculation (EGR) pipe
   - Surge tank
   - Fuel lines
   - Fuel injector connectors
   - Fuel supply manifold
   - Thermostat housing
   - Intake manifold and discard the gasket

To install:

5. Install or connect the following:
   - Intake manifold using a new gasket and torque the nuts to 11–14 ft. lbs. (15–20 Nm) for 3.5L engines or to 14–15 ft. lbs. (15–20 Nm) for 2.5L and 2.7L engines starting from the center and working outwards
   - Thermostat housing and torque the bolts to 12–14 ft. lbs. (17–20 Nm)
   - Fuel supply manifold and torque the bolts to 84–108 inch lbs. (10–13 Nm)
   - Fuel injector connectors
   - Fuel lines
   - Surge tank and torque the bolts to 11–14 ft. lbs. (15–20 Nm)
   - EGR pipe
   - Intake manifold vacuum hoses
   - PCV valve and hose

➡**One throttle body bolt is shorter than the rest. This bolt is installed in the upper left hole when viewed from the front of the throttle body.**

   - Throttle body and torque the bolts to 11–16 ft. lbs. (15–22 Nm)
   - Engine control wiring harness
   - Upper radiator hose
   - Accelerator cable
   - Air intake hose
   - Negative battery cable
6. Fill the cooling system.
7. Start the engine and check for leaks.

Exhaust Manifold

REMOVAL & INSTALLATION

4 Cylinder Engines

1. Before servicing the vehicle, refer to the Precautions Section.

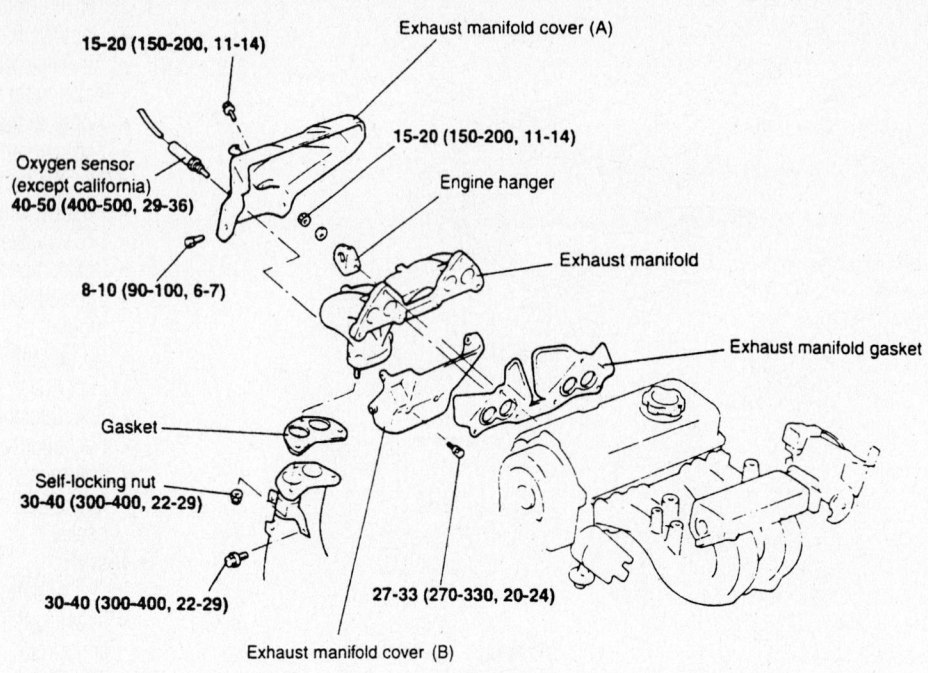

15-20 (150-200, 11-14)

Exhaust manifold cover (A)

15-20 (150-200, 11-14)

Oxygen sensor
(except california)
40-50 (400-500, 29-36)

Engine hanger

8-10 (90-100, 6-7)

Exhaust manifold

Exhaust manifold gasket

Gasket

Self-locking nut
30-40 (300-400, 22-29)

30-40 (300-400, 22-29)

27-33 (270-330, 20-24)

Exhaust manifold cover (B)

TORQUE : Nm (kg.cm, lb.ft)

7923GG41

Exploded view of the exhaust manifold and related components—1.5L engines

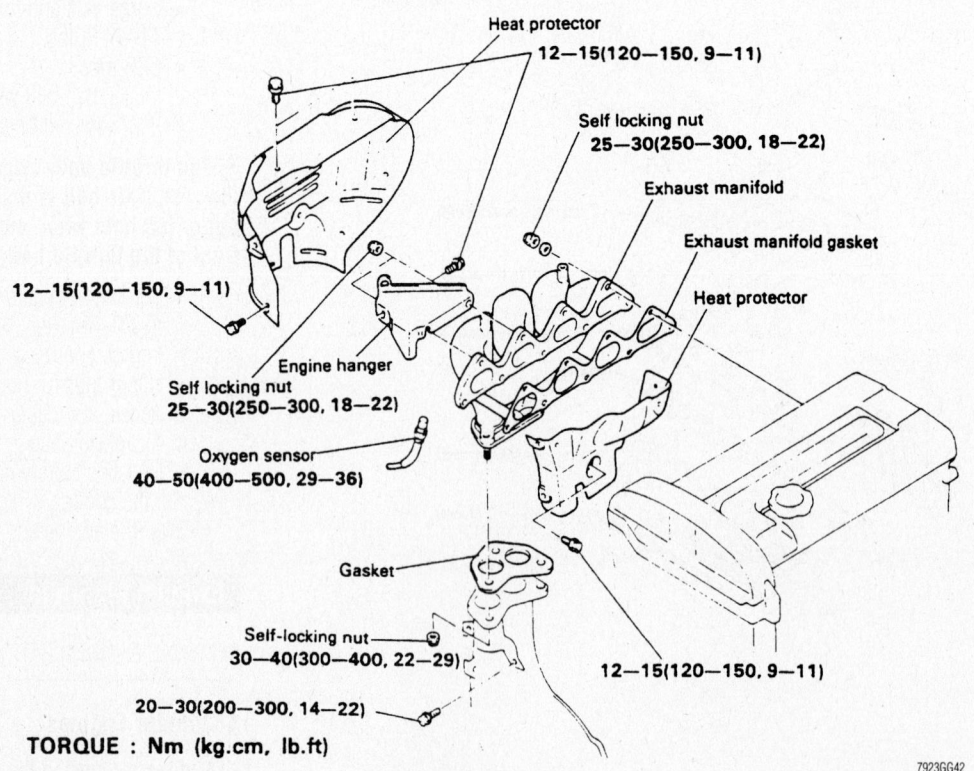

Heat protector
12—15(120—150, 9—11)

Self locking nut
25—30(250—300, 18—22)

Exhaust manifold

Exhaust manifold gasket

Heat protector

12—15(120—150, 9—11)

Engine hanger

Self locking nut
25—30(250—300, 18—22)

Oxygen sensor
40—50(400—500, 29—36)

Gasket

Self-locking nut
30—40(300—400, 22—29)

12—15(120—150, 9—11)

20—30(200—300, 14—22)

TORQUE : Nm (kg.cm, lb.ft)

7923GG42

Exploded view of the exhaust manifold components—Sonata 2.0L engines

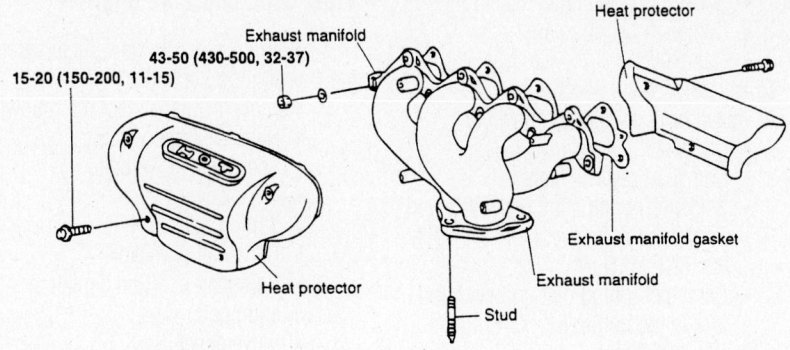

TORQUE: Nm (kg.cm, lb.ft)

7923GG43

Exploded view of the exhaust manifold components—Elantra and Tiburon with 2.0L engine

2. Remove or disconnect the following:
   - Negative battery cable
   - Heated Oxygen (HO2S) sensor
   - Exhaust manifold heat shield
   - Exhaust front pipe
   - Exhaust manifold and discard the gasket

To install:

3. Install the exhaust manifold with a new gasket and torque the nuts to the following specifications:
   a. 1.5L and 1.6L engines: 11–15 ft. lbs. (15–20 Nm).
   b. Sonata 2.0L engines: 18–22 ft. lbs. (25–30 Nm).
   c. Elantra and Tiburon 2.0L engine: 32–41 ft. lbs. (43–50 Nm).
   d. 2.4L engine: M8 fasteners to 18–22 ft. lbs. (25–30 Nm) and M10 fasteners to 25–40 ft. lbs. (34–55 Nm).

4. Install or connect the following:
   - Exhaust front pipe
   - Exhaust manifold heat shield
   - HO2S sensor
   - Negative battery cable

5. Start the engine and check for leaks.

6 Cylinder Engines

SONATA

1. Before servicing the vehicle, refer to the Precautions Section.
2. Remove or disconnect the following:
   - Negative battery cable
   - Exhaust front pipe
   - Exhaust Gas Recirculation (EGR) tube
   - Oil dipstick tube
   - Exhaust manifold heat shields
   - Exhaust manifolds and discard the gasket

To install:

3. Install or connect the following:
   - Exhaust manifolds and torque the nuts to 11–16 ft. lbs. (15–22 Nm) for 3.0L and 3.5L engines or to 18–22 ft. lbs. (25–30 Nm) for 2.5L and 2.7L engines
   - Exhaust manifold heat shields
   - Oil dipstick tube
   - EGR tube
   - Exhaust front pipe and torque the nuts to 22–29 ft. lbs. (30–40 Nm)
   - Negative battery cable

4. Start the engine and check for leaks.

XG 350

1. Before servicing the vehicle, refer to the Precautions Section.
2. Remove or disconnect the following:
   - Negative battery cable
   - Heat protector
   - Heated Oxygen (HO2S) sensor
   - Exhaust manifold and discard the gaskets

3. Clean and residual gasket material from all mating surfaces.

To install:

4. Install or connect the following:
   - Exhaust manifold with a new gasket and torque the bolts to 22 ft. lbs. (30 Nm)
   - Heat protector and torque the bolts to 11 ft. lbs. (15 Nm)
   - HO2S sensor
   - Negative battery cable

Front Crankshaft Seal

REMOVAL & INSTALLATION

1. Before servicing the vehicle, refer to the Precautions Section.
2. Remove or disconnect the following:
   - Negative battery cable
   - Accessory drive belts
   - Front cover
   - Timing belt
   - Crankshaft timing sprocket
   - Front crankshaft seal

To install:

3. Install the front crankshaft seal so that it is flush with the oil pump housing.

4. Install or connect the following:
   - Crankshaft timing sprocket
   - Timing belt
   - Front cover
   - Accessory drive belts
   - Negative battery cable

5. Start the engine and check for leaks.

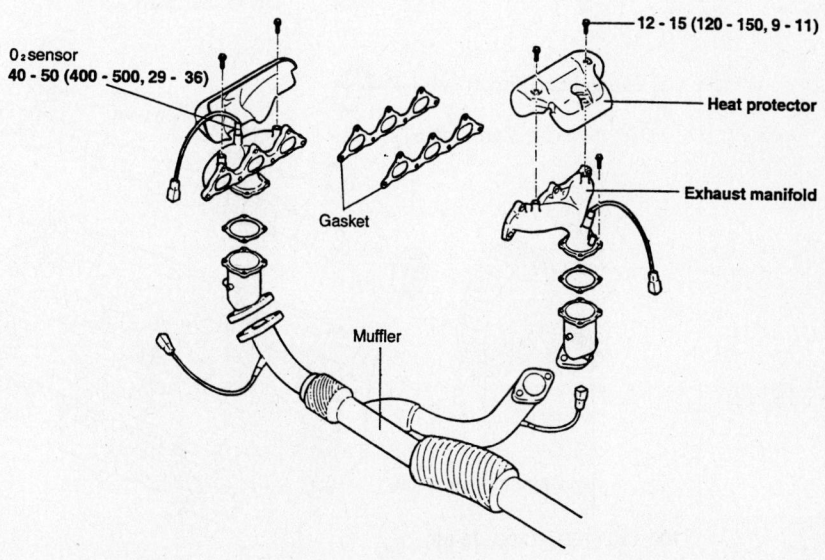

Exploded view of the exhaust manifold—XG 350

9347KG07

Camshaft and Valve Lifters

REMOVAL & INSTALLATION

1.5L SOHC Engines

➡ **The hydraulic lash adjusters are housed in the rocker arms.**

1. Before servicing the vehicle, refer to the Precautions Section.
2. Remove or disconnect the following:
   - Negative battery cable
   - Accessory drive belts
   - Ignition coil
   - Valve cover
   - Front cover
   - Timing belt
   - Camshaft timing belt sprocket
   - Rocker arm and shaft assembly
   - Camshaft bearing caps
   - Camshaft

To install:

3. Install or connect the following:
   - Camshaft
   - Camshaft bearing caps
   - Rocker arm and shaft assembly and torque the bolts evenly to 14–20 ft. lbs. (20–26 Nm)
   - Camshaft timing belt sprocket and torque the bolt to 58–72 ft. lbs. (80–100 Nm)
   - Timing belt
   - Front cover
   - Valve cover
   - Ignition coil
   - Accessory drive belts
   - Negative battery cable

1.6L, 2.0L and 2.4L Engines

1. Before servicing the vehicle, refer to the Precautions Section.
2. Remove or disconnect the following:
   - Negative battery cable
   - Accessory drive belts
   - Front cover
   - Timing belt
   - Camshaft sprocket
   - Camshaft position sensor, if equipped
   - Valve cover
   - Camshaft bearing caps and timing chain
   - Intake and exhaust camshafts
   - Hydraulic lash adjusters

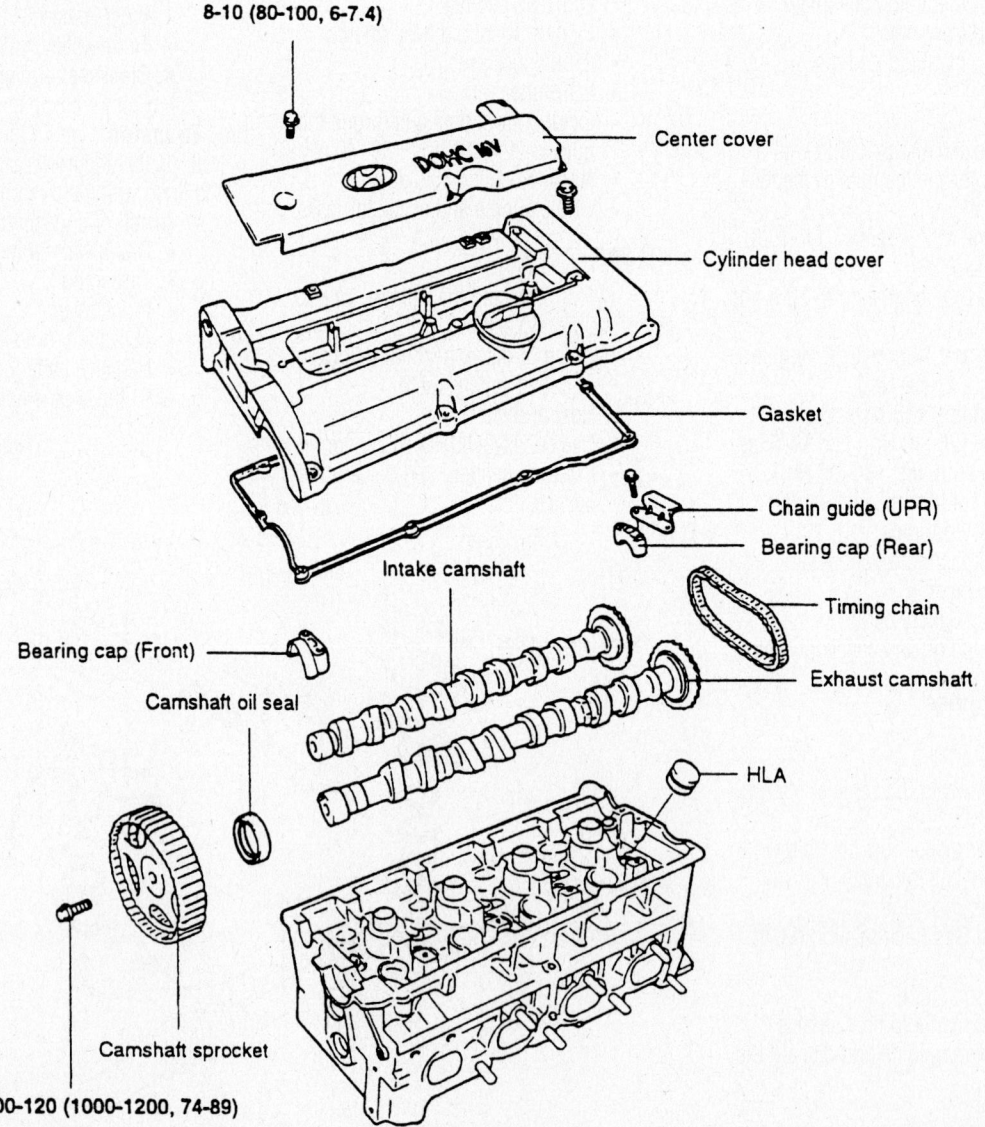

8-10 (80-100, 6-7.4)

Center cover

Cylinder head cover

Gasket

Chain guide (UPR)

Bearing cap (Rear)

Timing chain

Intake camshaft

Bearing cap (Front)

Camshaft oil seal

Exhaust camshaft

HLA

Camshaft sprocket

100-120 (1000-1200, 74-89)

Camshaft assembly components—1.6L, Elantra and Tiburon 2.0L engines

7923GG48

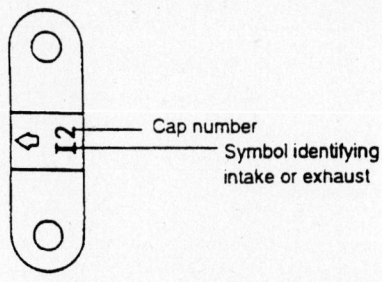

7923GG49

The camshaft bearing caps are identified with a letter and number stamp. The letter indicates either intake or exhaust and the number is sequential from the cylinder head end opposite the timing chain—1.6L, Elantra and Tiburon 2.0L engines

➡**Keep all valvetrain components in order for assembly.**

To install:

3. Install or connect the following:
- Hydraulic lash adjusters in their original positions
- Intake and exhaust camshafts with the secondary chain aligned as shown
- Camshaft bearing caps and timing chain and torque the bolts to 15 ft. lbs. (21 Nm) for 2.0L (VIN P) engines or to 10 ft. lbs. (14 Nm) for all others

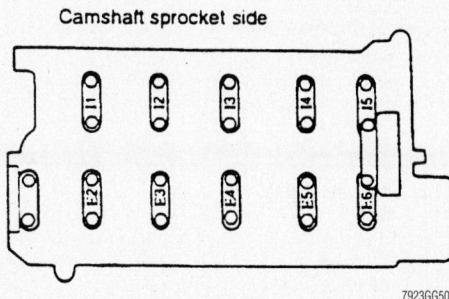

7923GG50

The camshaft bearing caps are arranged on the cylinder head as illustrated—1.6L, Elantra and Tiburon 2.0L engines

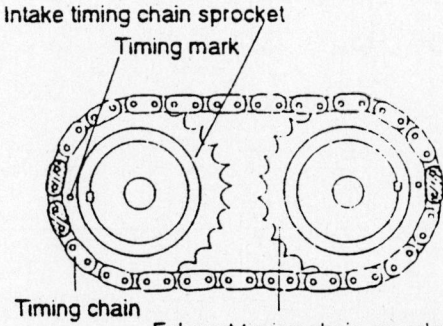

7923GG52

Align the timing chain and camshaft sprockets as illustrated—DOHC engines

- Valve cover
- Distributor, if equipped
- Camshaft position sensor, if equipped
- Camshaft sprocket and torque the bolt to 60–74 ft. lbs. (80–100 Nm)
- Timing belt
- Front cover
- Accessory drive belts
- Negative battery cable

2.5L and 2.7L Engine

1. Before servicing the vehicle, refer to the Precautions Section.
2. Remove or disconnect the following:
- Negative battery cable
- Accessory drive belts
- Front cover
- Timing belt
- Camshaft sprocket
- Camshaft Position Sensor (CMP)
- Valve cover
- Camshaft bearing caps
- Intake and exhaust camshafts
- Hydraulic lash adjusters

➡**Keep all valvetrain components in order for assembly.**

To install:

3. Install or connect the following:
- Hydraulic lash adjusters in their original positions

- Intake and exhaust camshafts with the secondary chain aligned as shown
- Camshaft bearing caps and torque the bolts to 10 ft. lbs. (14 Nm)
- Valve cover
- CMP sensor
- Camshaft sprocket and torque the bolt to 60–74 ft. lbs. (80–100 Nm)
- Timing belt
- Front cover
- Accessory drive belts
- Negative battery cable

XG 350

1. Before servicing the vehicle, refer to the Precautions Section.
2. Remove or disconnect the following:

- Negative battery cable
- Engine cover
- Intake manifold
- Breather hose and engine harness
- Power steering pulley
- A/C pulley
- Crankshaft pulley
- Idler pulley and tensioner pulley
- Timing belt cover and loosen the auto tensioner
- Timing belt
- Spark plug cables
- Rocker arm cover
- Camshaft sprockets
- Camshaft bearing caps
- Camshafts

To install:

3. Rotate the crankshaft so that the No. 1 cylinder is at the Top Dead Center (TDC) position.

4. Make certain that the rocker arm is installed properly on the lash adjuster and valve.

5. Install the camshaft dowel pin.

6. Install or connect the following:
- Hydraulic lash adjusters in their original locations
- Lash adjuster retaining clips
- Camshafts
- Rocker arm and shaft assemblies and torque the bearing cap bolts 14–15 ft. lbs. (19–21 Nm) starting from the center and working out

7. Remove the lash adjuster retaining clips.

- Camshaft sprockets and torque the bolts to 79 ft. lbs. (110 Nm)
- Rocker arm cover with a new gasket
- Spark plug cables
- Timing belt, cover and tensioner
- Idler pulley
- Tensioner pulley

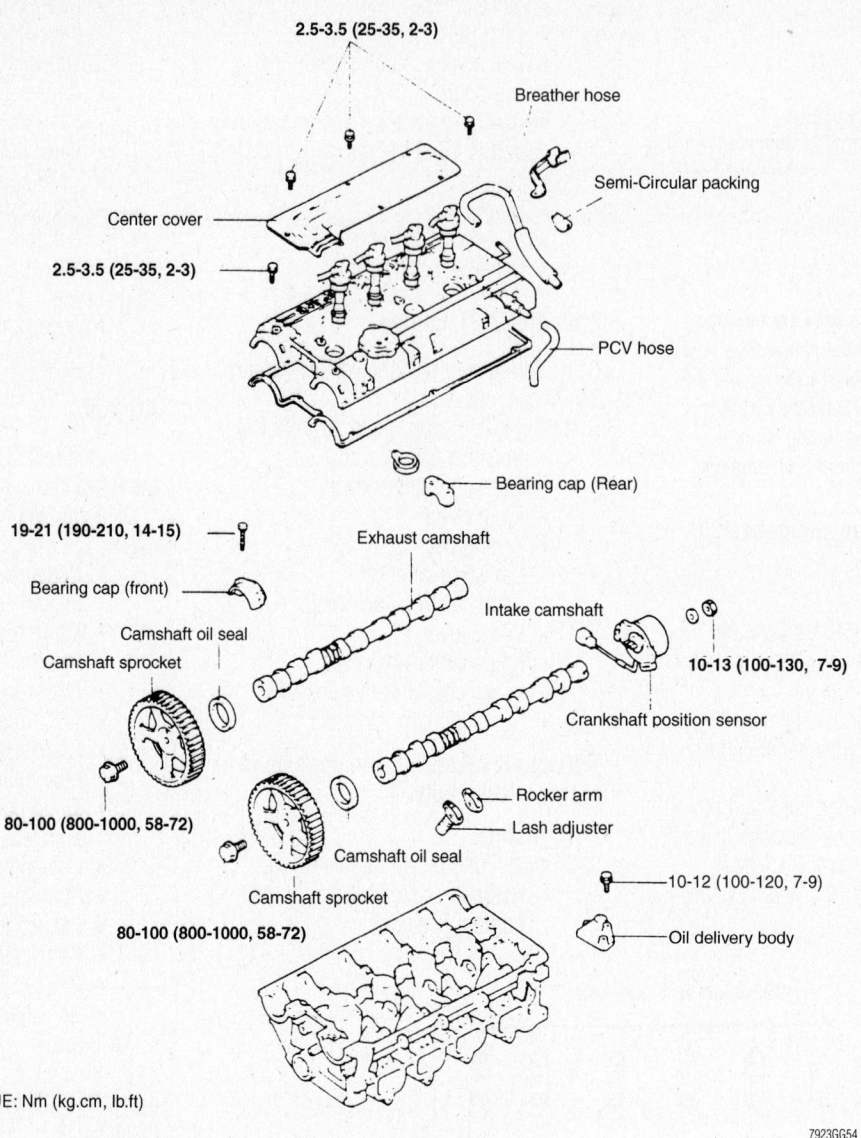

TORQUE: Nm (kg.cm, lb.ft)

7923GG54

Exploded view of the camshaft and rocker arm assembly components—Sonata with 2.0L engine

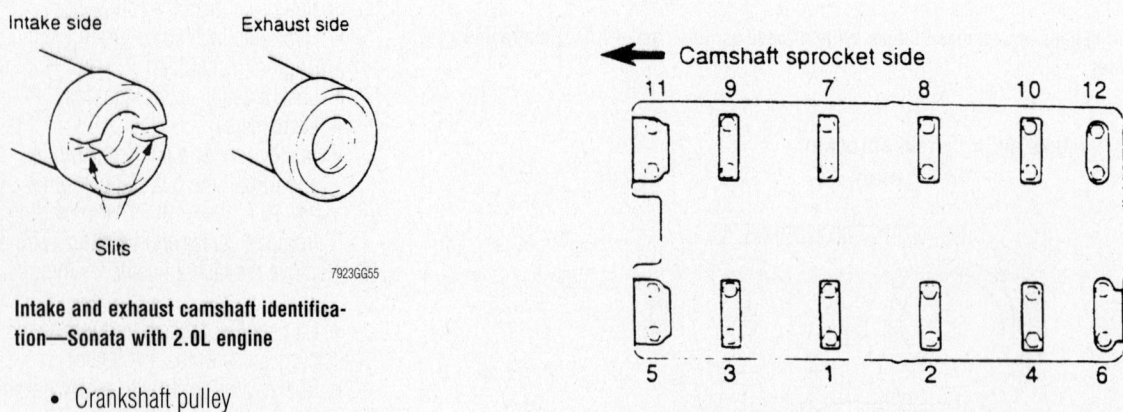

Intake and exhaust camshaft identification—Sonata with 2.0L engine

Bearing cap torque sequence—Sonata with 2.0L engine

7923GG58

- Crankshaft pulley
- A/C pulley
- Power steering pump pulley
- Breather hose and engine harness
- Intake manifold

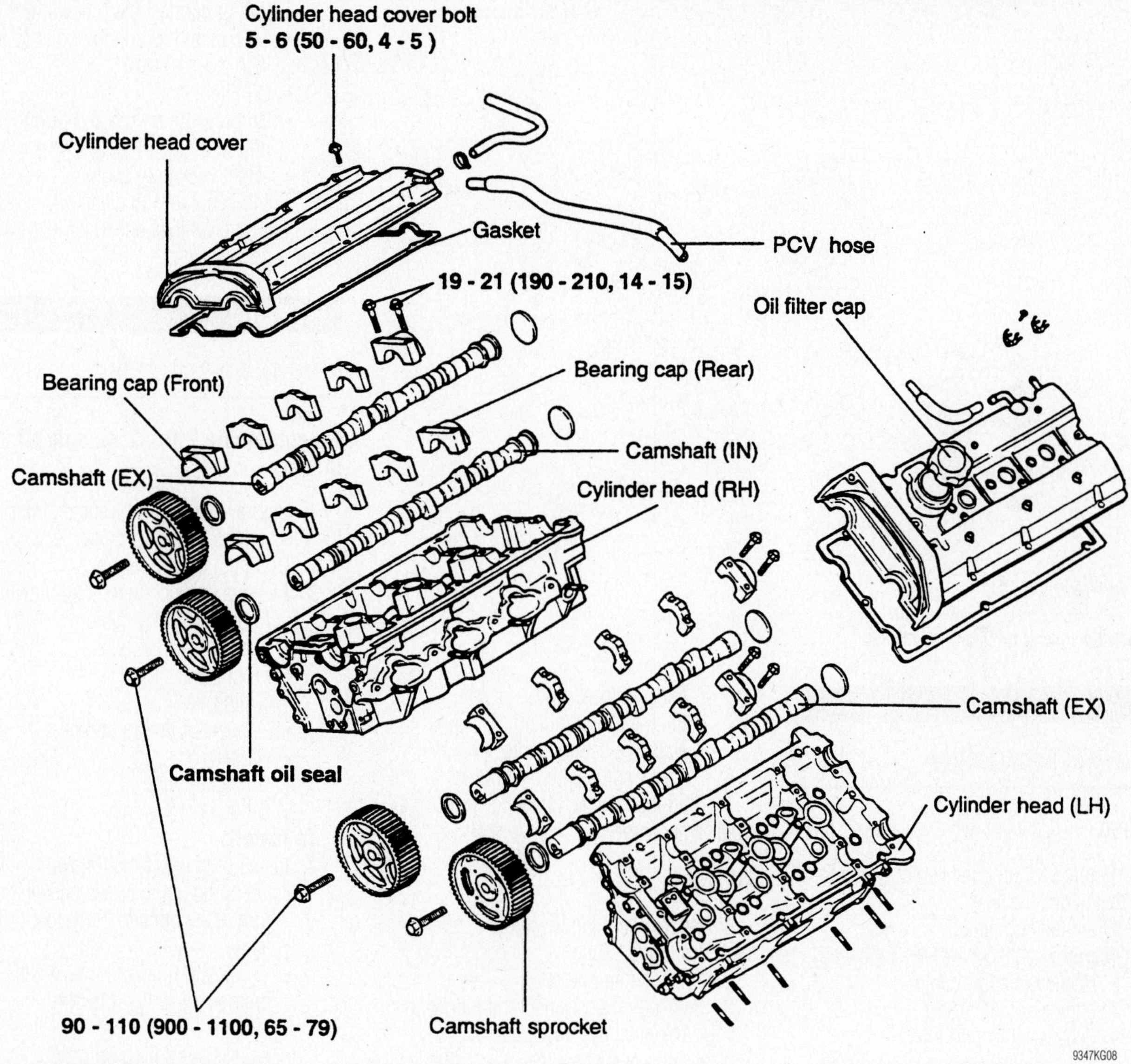

Cylinder head cover bolt
5 - 6 (50 - 60, 4 - 5)

Cylinder head cover

Gasket

PCV hose

19 - 21 (190 - 210, 14 - 15)

Oil filter cap

Bearing cap (Front)

Bearing cap (Rear)

Camshaft (IN)

Camshaft (EX)

Cylinder head (RH)

Camshaft oil seal

Camshaft (EX)

Cylinder head (LH)

90 - 110 (900 - 1100, 65 - 79)

Camshaft sprocket

9347KG08

Exploded view of the camshaft and related components—XG 350

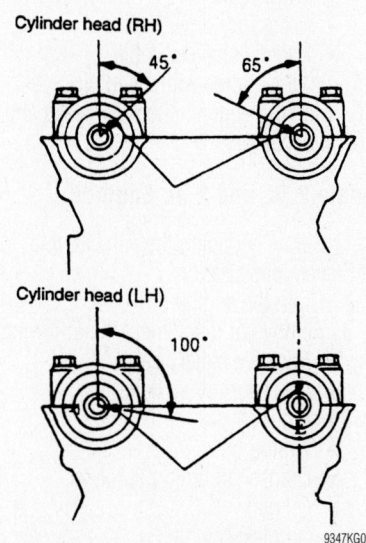

Cylinder head (RH)

45°   65°

Cylinder head (LH)

100°

9347KG09

**Install the camshaft dowel pin as shown—
XG 350**

- Engine cover
- Negative battery cable

Valve Lash

ADJUSTMENT

All engines use hydraulic valve lash adjusters. Valve lash adjustments are not necessary or possible on these engines.

Starter Motor

REMOVAL & INSTALLATION

1. Before servicing the vehicle, refer to the Precautions Section.

2. Remove or disconnect the following:
   - Negative battery cable
   - Speedometer and shift control cables
   - Starter electrical connectors
   - Starter motor

To install:

3. Install or connect the following:
   - Starter motor and torque the bolts to 25 ft. lbs. (34 Nm)
   - Starter electrical connectors
   - Speedometer and shift control cables
   - Negative battery cable

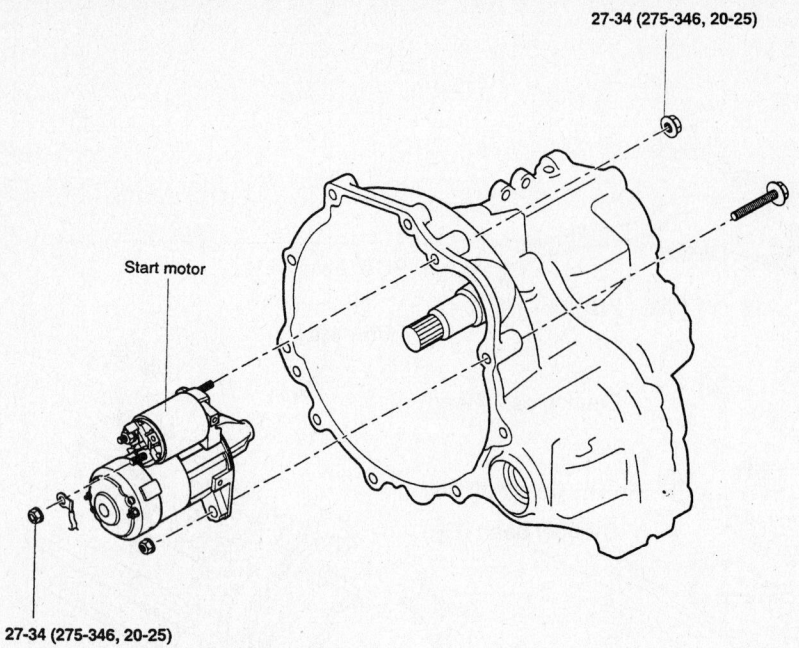

27-34 (275-346, 20-25)

27-34 (275-346, 20-25)

Start motor

Exploded view of the starter—XG 350

Oil Pan

REMOVAL & INSTALLATION

XG 350

1. Before servicing the vehicle, refer to the Precautions Section.
2. Drain the engine oil.
3. Remove or disconnect the following:
   - Negative battery cable
   - Engine splash shield
   - Exhaust front pipe
   - Timing belt
   - Oil pan

To install:

4. Apply a ⅛ in. (3mm) bead of RTV sealer along the groove in the oil pan.
5. Install or connect the following:
   - Oil pan and torque the bolts to 48–72 inch lbs. (6–8 Nm)
   - Timing belt

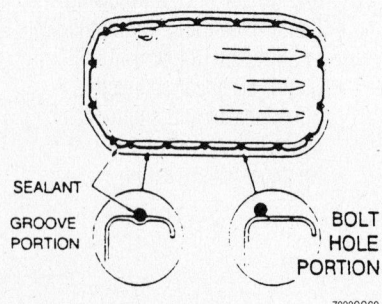

Oil pan sealant applications points— except 3.5L engine

- Exhaust front pipe
- Engine splash shield
- Negative battery cable
6. Fill the engine with clean oil.
7. Start the vehicle, check for leaks and repair if necessary.

XG 350

1. Before servicing the vehicle, refer to the Precautions Section.
2. Drain the engine oil.
3. Remove or disconnect the following:
   - Negative battery cable
   - Oil pressure switch
   - Oil filter
   - Oil pan and discard the gasket
4. Clean all mating surfaces of any residual gasket material.

To install:

5. Apply a ⅛ in. (3mm) bead of RTV sealer along the groove in the oil pan.

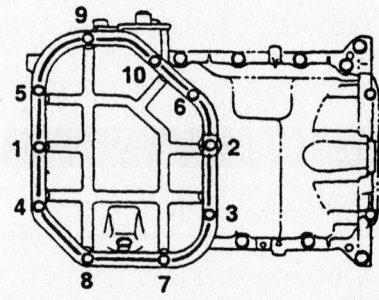

Tighten the oil pan bolts, in sequence, as shown

6. Install or connect the following:
   - Oil pan and torque the bolts to 89 inch lbs. (10 Nm)
   - Oil filter
   - Oil pressure switch and torque the bolt to 89 inch lbs. (10 Nm)
   - Negative battery cable
7. Fill the engine with clean oil.
8. Start the vehicle, check for leaks and repair if necessary.

Oil Pump

REMOVAL & INSTALLATION

Except Sonata 2.0L, 2.4L and XG 350 Models

1. Before servicing the vehicle, refer to the Precautions Section.
2. Drain the engine oil.
3. Remove or disconnect the following:
   - Negative battery cable
   - Accessory drive belts
   - Front cover
   - Timing belt
   - Crankshaft timing sprocket
   - Oil pan
   - Oil pickup tube
   - Oil pump

To install:

4. Install or connect the following:
   - Oil pump using a new gasket and torque the bolts to 11 ft. lbs. (15 Nm)
   - Oil pickup tube and torque the bolts to 11 ft. lbs. (15 Nm)
   - Oil pan
   - Crankshaft timing sprocket
   - Timing belt
   - Front cover
   - Accessory drive belts
   - Negative battery cable
5. Fill the engine with clean oil.
6. Start the engine, check for leaks and repair if necessary.

Sonata 2.0L and 2.4L Engines

1. Before servicing the vehicle, refer to the Precautions Section.
2. Drain the engine oil.
3. Remove or disconnect the following:
   - Negative battery cable
   - Accessory drive belts
   - Front cover
   - Timing belt
   - Crankshaft timing sprocket
   - Oil pan
   - Oil pickup tube
   - Pressure relief valve
   - Oil pressure switch

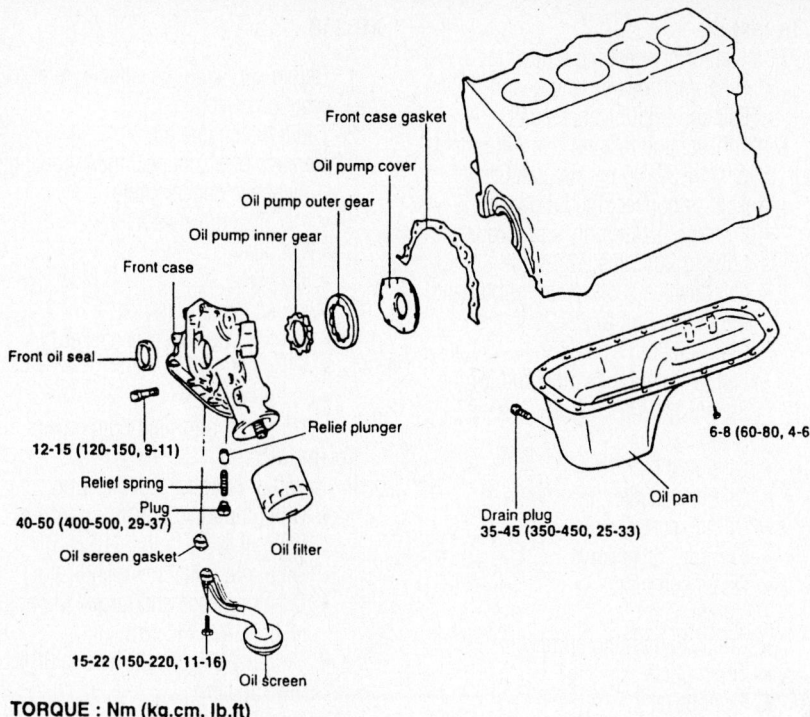

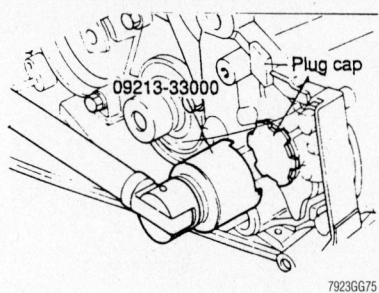

A special socket (PN 09213-33000) is available to remove the plug cap from the oil pump portion of the case—Sonata 2.0L and 2.4L engines

- Oil filter adapter
- Oil pump plug cap
- Left cylinder block plug

4. Insert a prytool with a $^{5}/_{16}$in. (8mm) diameter shaft into the plug hole to hold the shaft while removing the balance shaft bolt.

5. Remove or disconnect the following:
- Left balance shaft retaining bolt
- Front case assembly
- Left and right balance shafts
- Oil pump housing and gears

TORQUE : Nm (kg.cm, lb.ft)

Exploded view of the oil pump and pan—1.5L, and Elantra and Tiburon with 2.0L engine

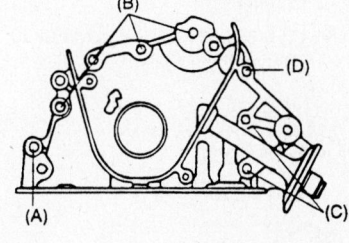

(A) ....25 mm (0.98 in)

(B)..............................20 mm (0.787 in)

(C)..............................38 mm (1.496 in)

(D)..............................45 mm (1.771 in)

Oil pump cover bolt lengths and locations— Elantra and Tiburon with 2.0L engines

(A) .... 25 mm (0.98 in.)
(B) .... 30 mm (1.18 in.)
(C) .... 45 mm (1.77 in.)
(D) .... 60 mm (2.36 in.)

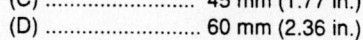
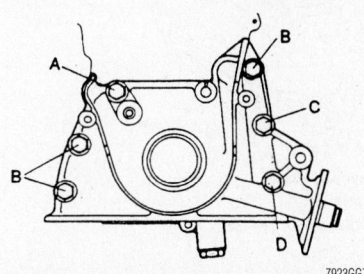

Oil pump cover bolt lengths and locations—1.5L engines

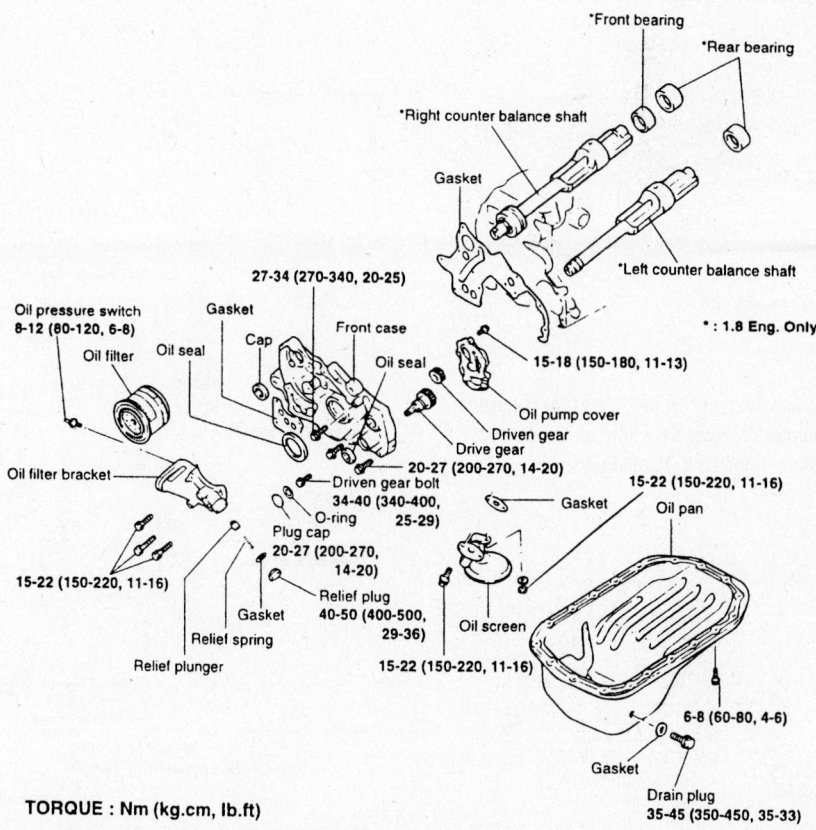

TORQUE : Nm (kg.cm, lb.ft)

Exploded view of the oil pump and pan—Sonata 2.0L engine

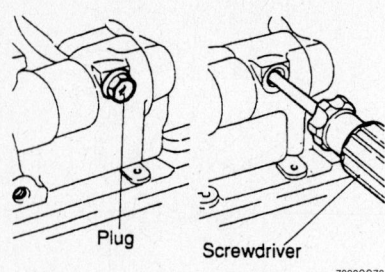

Remove the left side cylinder block plug and insert a screwdriver into the hole to hold the balance shaft from turning—Sonata 2.0L and 2.4L engines

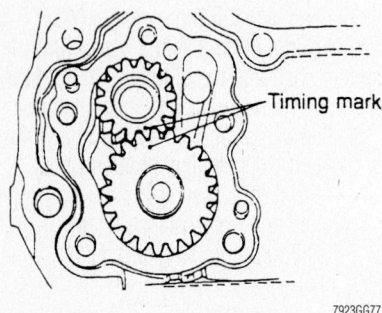

Align the timing marks on the gears during assembly—Sonata 2.0L and 2.4L engines

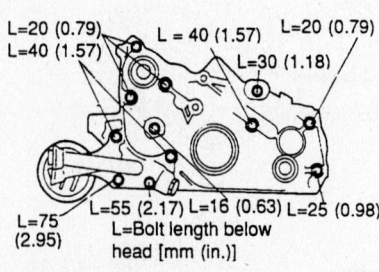

A special tool (PN 09214-32100) is used to center the front case hole on the crankshaft—Sonata 2.0L engines

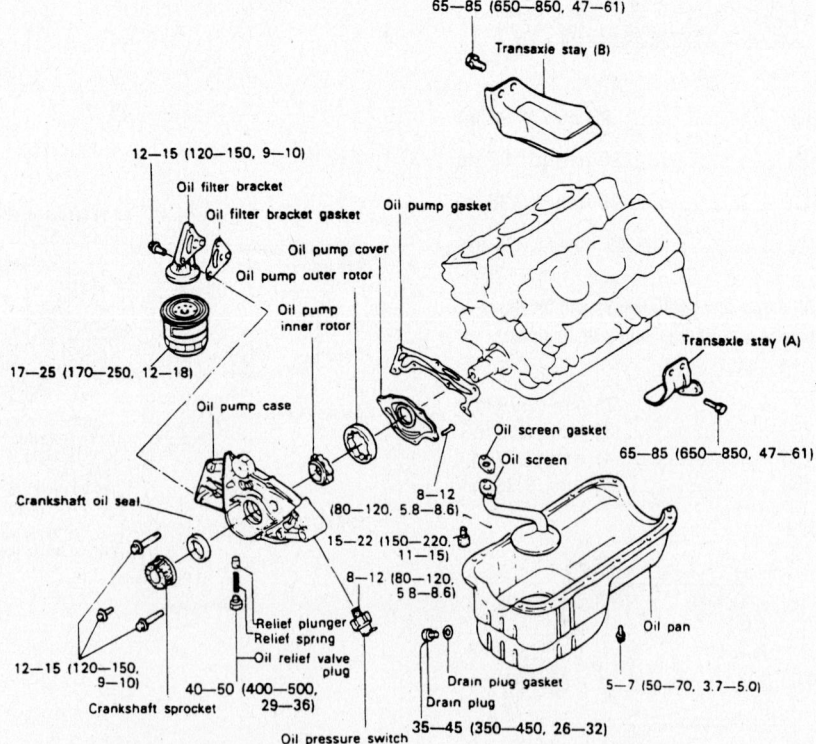

Oil pump cover bolt lengths and locations—Sonata 2.0L engines

To install:

6. Install or connect the following:
   - Oil pump housing and gears
   - Left and right balance shafts
7. Lubricate and install Case Alignment Tool, PN 09214-32100 on the crankshaft.
8. Install or connect the following:
   - Front case assembly and torque the bolts to 20 ft. lbs. (27 Nm)
   - Left balance shaft retaining bolt and torque the bolt to 25–29 ft. lbs. (34–40 Nm)
   - Left cylinder block plug and torque the plug to 14–20 ft. lbs. (20–27 Nm)
   - Oil pump plug cap
   - Oil filter adapter
   - Oil pressure switch
   - Pressure relief valve
   - Oil pickup tube
   - Oil pan
   - Crankshaft timing sprocket
   - Timing belt
   - Front cover
   - Accessory drive belts
   - Negative battery cable
9. Fill the engine with clean oil.
10. Start the engine, check for leaks and repair if necessary.

XG 350

1. Before servicing the vehicle, refer to the Precautions Section.
2. Drain the engine oil.
3. Remove or disconnect the following:
   - Negative battery cable
   - Oil pressure switch
   - Oil filter
   - Oil pan
   - Oil screen and gasket
   - Oil filter bracket and gasket
   - Oil relief plug
   - Oil pump case
   - Oil pump rotor and both covers

To install:

4. Install or connect the following:
   - Oil pump inner cover and torque the bolt to 11 ft. lbs. (15 Nm)
   - Oil pump outer cover and rotor
   - Oil pump case and torque the bolts to 11 ft. lbs. (15 Nm)
   - Oil relief plug and torque to 36 ft. lbs. (50 Nm)
   - Oil filter bracket with a new gasket
   - Oil screen and gasket and torque the bolt to 15 ft. lbs. (20 Nm)
   - Oil pan
   - Oil filter
   - Oil pressure switch and torque to 89 inch lbs. (10 Nm)

Exploded view of the oil pump and pan—3.0L and 3.5L engine

- Negative battery cable
5. Fill the engine with clean oil.
6. Start the vehicle, check for leaks and repair if necessary.

Rear Main Seal

REMOVAL & INSTALLATION

1. Before servicing the vehicle, refer to the Precautions Section.
2. Remove or disconnect the following:
   - Transaxle
   - Flywheel
   - Oil seal case
   - Oil separator, if equipped
   - Oil seal

To install:
3. Install or connect the following:
   - Oil seal
   - Oil separator, if equipped
   - Oil seal case and torque the case bolts to 84–108 inch lbs. (8–10 Nm)
   - Flywheel
   - Transaxle

Timing Belt

REMOVAL & INSTALLATION

✳✳ CAUTION

Timing belt maintenance is extremely important. All Hyundai models use interference-type non-freewheeling engines. Should the timing belt break in these engines, the valves in the cylinder head will come in contact with the pistons, causing major engine damage. The recommended replacement interval for timing belts is 60,000 miles.

1.5L Engines

1. Before servicing the vehicle, refer to the Precautions Section.
2. Remove or disconnect the following:
   - Negative battery cable
   - Engine coolant
   - Water pump pulley bolts
   - Alternator bolt, loosen only
   - Water pump pulley and drive belts
   - Crankshaft pulley
   - Timing belt cover
3. Rotate the crankshaft clockwise and align the timing marks so No. 1 piston will be at Top Dead Center (TDC) of the compression stroke.
4. Loosen the tensioning bolt and the

pivot bolt on the timing belt tensioner. Move the tensioner as far as it will go toward the water pump. Tighten the adjusting bolt.
5. Mark the timing belt with an arrow showing direction of rotation.
6. Move the timing belt tensioner pulley toward the water pump and temporarily secure it.
7. Remove or disconnect the following:
   - Timing belt
   - Crankshaft sprocket bolts, sprocket and flange
   - Timing belt tensioner, if defective

To install:
8. Install the flange and crankshaft sprocket. Tighten the crankshaft sprocket bolt to 103–111 ft. lbs. (140–150 Nm).
9. Align the timing marks of the camshaft sprocket and check that the crankshaft timing marks are still in alignment.
10. If removed, install the timing belt tensioner, spring and spacer with the bottom end of the spring free. Tighten the adjusting bolt slightly with the tensioner moved as far as possible away from the water pump.

11. Install the free end of the spring into the locating tang on the front case.
12. Position the timing belt over the crankshaft sprocket, then over the camshaft sprocket. Slip the back of the belt over the tensioner wheel.
13. Turn the camshaft sprocket in the opposite of its normal direction of rotation until the straight side of the belt is tight and be sure the timing marks align.

➡**If the timing marks are not properly aligned, shift the belt 1 tooth at a time in the appropriate direction until they are aligned.**

14. Loosen the tensioner mounting bolts so the tensioner works, without the interference of any friction, under spring pressure. Be sure the belt follows the curve of the camshaft pulley so the teeth are engaged all the way around. Correct the path of the belt, if necessary.
15. Tighten the tensioner adjusting bolt, then the tensioner pivot bolt to 15–18 ft. lbs. (20–26 Nm).

➡**Bolts must be tightened in the stated order or tension won't be correct.**

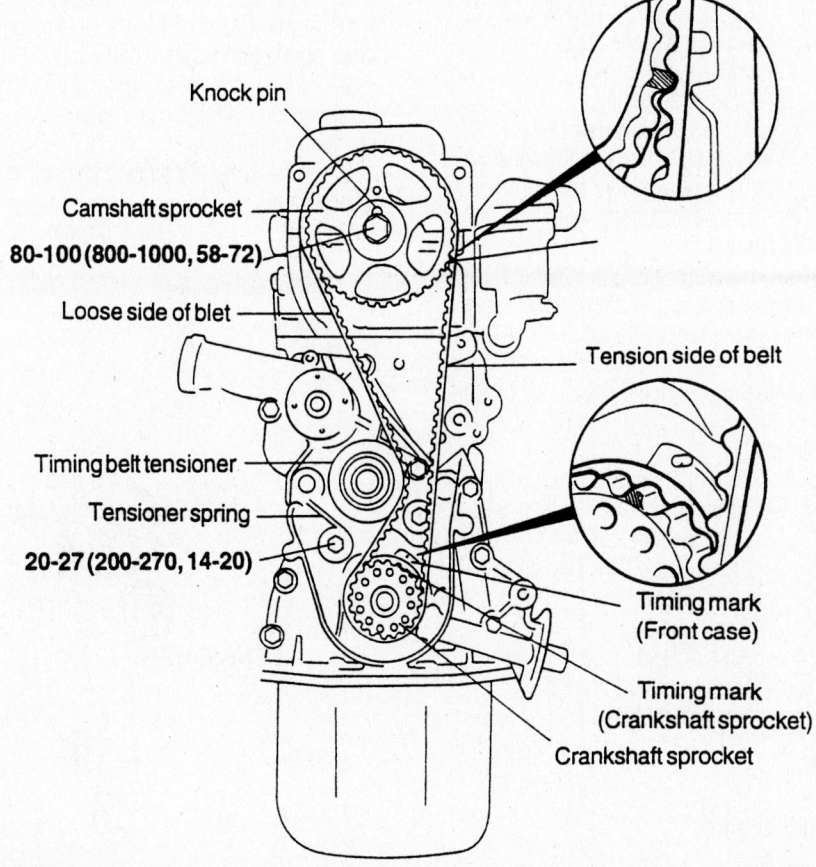

Knock pin

Camshaft sprocket
80-100 (800-1000, 58-72)

Loose side of blet

Tension side of belt

Timing belt tensioner

Tensioner spring
20-27 (200-270, 14-20)

Timing mark (Front case)

Timing mark (Crankshaft sprocket)

Crankshaft sprocket

93015G26

Proper pulley alignment for timing belt installation at TDC—Hyundai 1.5L SOHC engines, 1.6L similar

16. Turn the crankshaft 1 turn clockwise until timing marks again align to seat the belt.

17. Loosen both tensioner attaching bolts and let the tensioner position itself under spring tension. Retighten the bolts.

18. Check belt tension by putting a finger on the water pump side of the tensioner wheel and pull the belt toward the water pump. The belt should move toward the pump until the teeth are approximately ½ of the way across the head of the tensioner adjusting bolt. Re-tension the belt, if necessary.

19. Install or connect the following:
- Timing belt cover
- Crankshaft pulley
- A/C compressor belt
- Water pump pulley
- V belt
- Negative battery cable
- Engine with coolant

1.6L Engines

1. Before servicing the vehicle, refer to the Precautions Section.

2. Remove or disconnect the following:
- Negative battery cable
- Engine coolant
- Engine support bracket
- Accessory drive belts
- Water pump pulley
- Crankshaft pulley
- Timing belt cover

3. Move the timing belt tensioner pulley toward the water pump and secure it.

4. Remove the timing belt.

To install:

5. Install the timing belt and turn the crankshaft sprocket in a reverse direction and align the timing marks.

6. Turn the crankshaft two turns in its operating direction and realign the camshaft sprocket timing mark to Top Dead Center (TDC).

7. Install or connect the following:
- Timing belt cover. Tighten the bolts to 84 inch lbs. (10 Nm).
- Crankshaft pulley. Tighten to 111 ft. lbs. (150 Nm).
- Water pump pulley
- Accessory drive belts
- Engine support bracket
- Negative battery cable

8. Refill the engine with coolant to the correct level.

2.0L Engines

1. Before servicing the vehicle, refer to the Precautions Section.

2. Remove or disconnect the following:
- Negative battery cable
- Engine coolant
- Water pump pulley bolts
- Alternator bolt, loosen only
- Water pump pulley and drive belts
- Crankshaft pulley
- Timing belt cover(s)

3. Rotate the crankshaft clockwise and align the timing marks so No. 1 piston will be at Top Dead Center (TDC) of the compression stroke.

4. Remove the timing belt tensioner and idler pulley.

5. Mark the timing belt with an arrow showing direction of rotation.

6. Remove the timing belt.

To install:

7. Align the timing marks of the camshaft sprocket and check that the crankshaft timing marks are still in alignment.

8. Install the timing belt tensioner.

9. Install the idler pulley, if equipped. Tighten bolt to 32–41 ft. lbs. (43–55 Nm).

10. Position the timing belt over the camshaft sprocket, then over the crankshaft sprocket.

11. Tension the timing belt and tighten the tensioner pulley bolt to 32–41 ft. lbs. (43–55 Nm). When properly tensioned, the timing belt should deflect 0.16–0.24 in. (4–6mm) when a force of 5 lbs. (2.2kg) is placed on the longest span of the belt.

12. Turn the crankshaft sprocket one turn clockwise and realign the crankshaft sprocket timing mark.

13. Recheck the belt tension and adjust as necessary.

14. Install or connect the following:
- Timing belt cover(s)
- Crankshaft pulley
- A/C compressor belt
- Water pump pulley

- V belt
- Negative battery cable
- Engine with coolant

3.5L Engine

1. Before servicing the vehicle, refer to the Precautions Section.

2. Remove or disconnect the following:
- Negative battery cable
- Engine coolant
- Water pump pulley bolts
- Alternator bolt, loosen only
- Water pump pulley and drive belts
- Crankshaft pulley
- Timing belt cover(s)

3. Turn the crankshaft until the timing marks on the camshaft sprocket and cylinder head are aligned.

4. Loosen the timing belt tensioner bolt and turn the tensioner counterclockwise as far as it will go. Tighten the adjusting bolt.

5. Mark the timing belt with an arrow showing direction of rotation.

6. Remove the timing belt.

7. If defective, remove the timing belt tensioner.

To install:

8. If necessary, install the timing belt tensioner.

9. Attach the top of the tensioner spring on the engine coolant pump pin. Ensure the hook on the pin is facing down and the hook on the tensioner is facing away from the engine

10. Rotate the timing belt tensioner to the extreme counterclockwise position. Temporarily lock the tensioner in place.

11. Align the timing marks of the camshaft and crankshaft sprockets.

12. Install the timing belt on the crank-

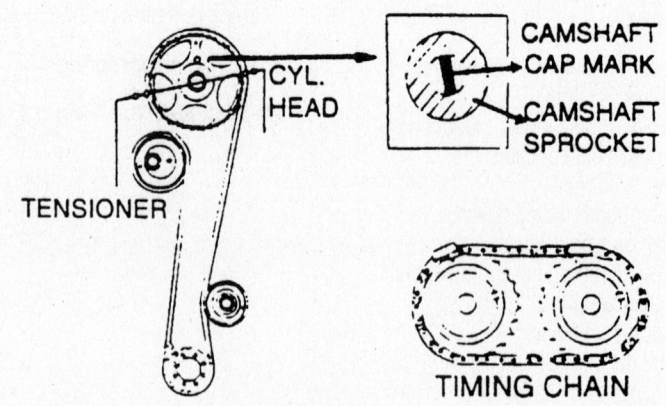

Proper alignment of the timing belt alignment marks for belt removal and installation—Hyundai 2.0L engines

79235G43

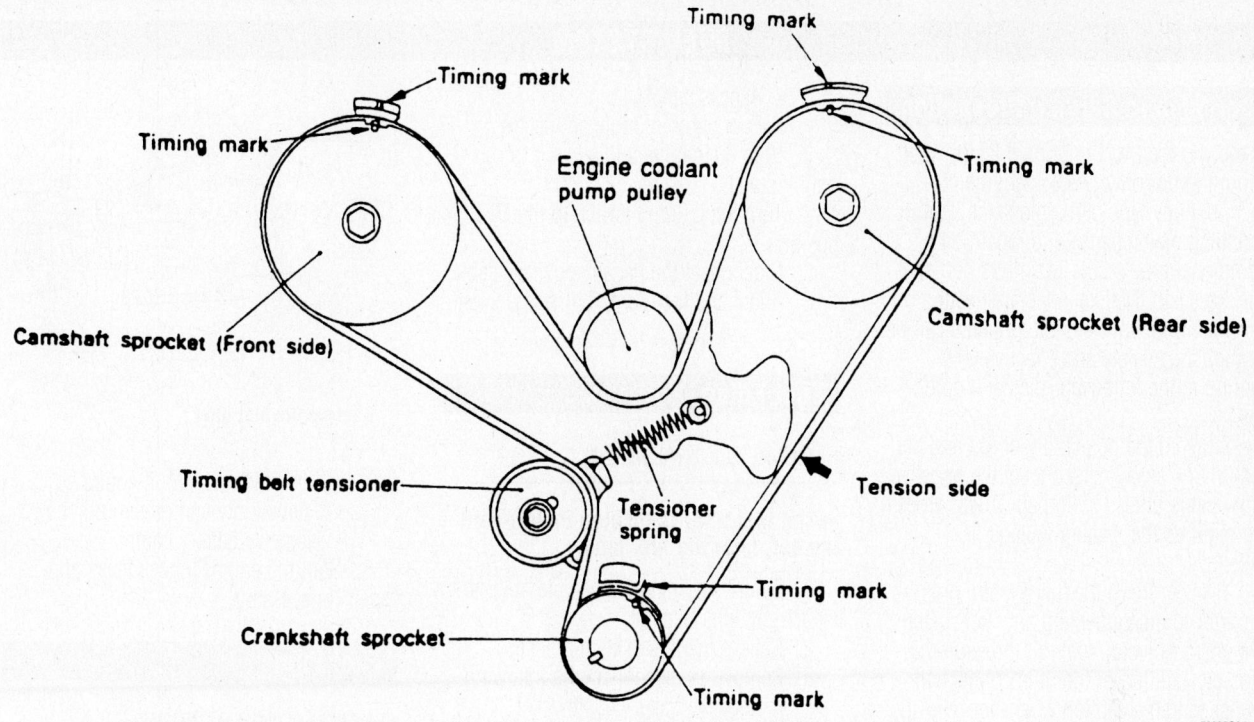

Timing mark
Timing mark
Timing mark
Timing mark
Engine coolant pump pulley
Camshaft sprocket (Front side)
Camshaft sprocket (Rear side)
Timing belt tensioner
Tension side
Tensioner spring
Timing mark
Crankshaft sprocket
Timing mark

79235G34

Timing belt sprocket alignment mark positioning for belt removal and installation—3.5L engine

shaft sprocket, then onto the rear camshaft sprocket.

13. Route the belt to the coolant pump pulley, the front camshaft sprocket and the timing belt tensioner.

14. Apply force counterclockwise to the rear camshaft sprocket with tension on the tight side of the belt and check that timing marks are aligned.

15. Loosen the tensioner bolt 1–2 turns and tighten the timing belt to a tension of 57–84 lbs. (260–380 N).

16. Turn the crankshaft two turns clockwise.

17. Readjust the sprocket timing marks and tighten the tensioner bolts.

18. Install the timing belt covers.

19. Install the crankshaft pulley and tighten to 108–116 ft. lbs. (150–160 Nm).

20. Install or connect the following:
- Timing belt cover(s)
- Crankshaft pulley
- A/C compressor belt
- Water pump pulley
- V belt
- Negative battery cable
- Engine with coolant

Piston and Ring

POSITIONING

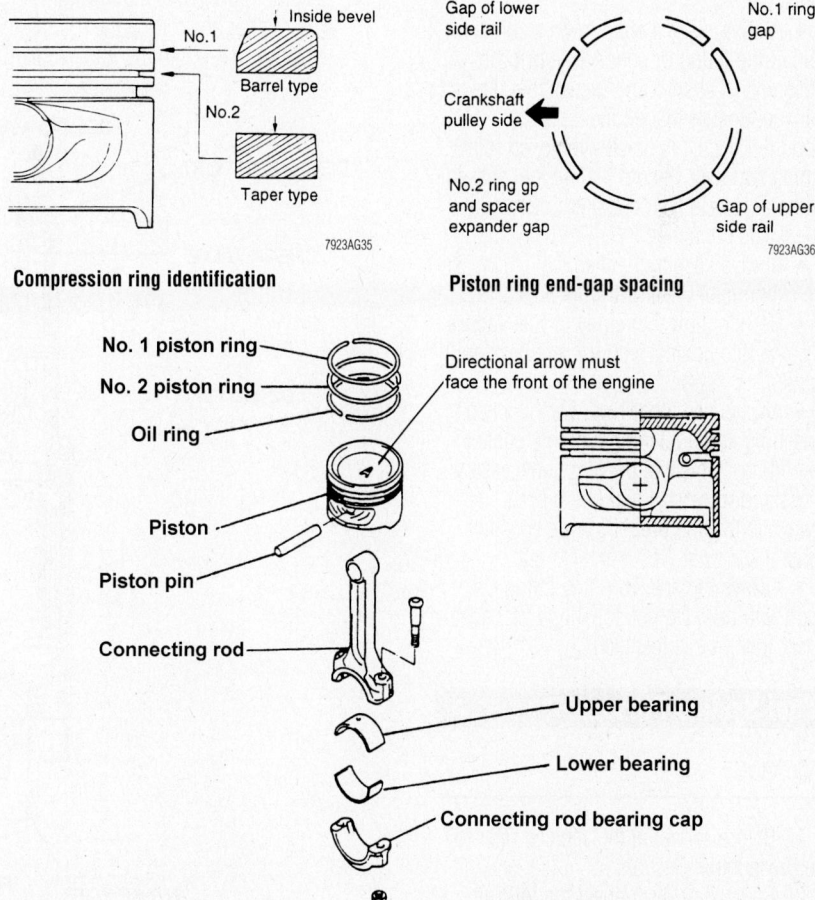

Inside bevel
No.1
Barrel type
No.2
Taper type

7923AG35

Compression ring identification

Gap of lower side rail
No.1 ring gap
Crankshaft pulley side
No.2 ring gp and spacer expander gap
Gap of upper side rail

7923AG36

Piston ring end-gap spacing

No. 1 piston ring
No. 2 piston ring
Oil ring
Directional arrow must face the front of the engine
Piston
Piston pin
Connecting rod
Upper bearing
Lower bearing
Connecting rod bearing cap

7923AG38

Piston and connecting rod assembly

FUEL SYSTEM

Fuel System Service Precautions

Safety is the most important factor when performing not only fuel system maintenance, but any type of maintenance. Failure to conduct maintenance and repairs in a safe manner may result in serious personal injury or death. Maintenance and testing of the vehicle's fuel system components can be accomplished safely and effectively by adhering to the following rules and guidelines.

• To avoid the possibility of fire and personal injury, always disconnect the negative battery cable unless the repair or test procedure requires that battery voltage be applied.

• Always relieve the fuel system pressure prior to disconnecting any fuel system component (injector, fuel rail, pressure regulator, etc.), fitting or fuel line connection. Exercise extreme caution whenever relieving the fuel system pressure, to avoid exposing skin, face and eyes to fuel spray. Please be advised that fuel under pressure may penetrate the skin or any part of the body that it contacts.

• Always place a shop towel or cloth around the fitting or connection prior to loosening to absorb any excess fuel due to spillage. Ensure that all fuel spillage (should it occur) is quickly removed from engine surfaces. Ensure that all fuel soaked cloths or towels are deposited into a suitable waste container.

• Always keep a dry chemical (Class B) fire extinguisher near the work area.

• Do not allow fuel spray or fuel vapors to come into contact with a spark or open flame.

• Always use a back-up wrench when loosening and tightening fuel line connection fittings. This will prevent unnecessary stress and torsion to fuel line piping. Always follow the proper torque specifications.

• Always replace worn fuel fitting O-rings with new. Do not substitute fuel hose where fuel pipe is installed.

Fuel System Pressure

RELIEVING

1. Before servicing the vehicle, refer to the Precautions Section.
2. Remove or disconnect the following:
   • Rear seat cushion

• Access panel
• Fuel pump module connector
3. Start the engine and allow it to run until it stalls.
4. Turn the ignition switch to the **OFF** position.
5. Disconnect the negative battery cable.
6. Attach the fuel pump harness connector.

Fuel Filter

REMOVAL & INSTALLATION

➡**The fuel filter is located underneath the car, near the fuel tank.**

1. Before servicing the vehicle, refer to the Precautions Section.
2. Relieve the fuel system pressure.
3. Remove or disconnect the following:
   • Negative battery cable
   • Fuel supply and pressure lines
   • Fuel filter bracket, if equipped
   • Fuel filter
To install:
4. Install or connect the following:
   • Fuel filter and torque the mounting bolts to 18–25 ft. lbs. (25–35 Nm)

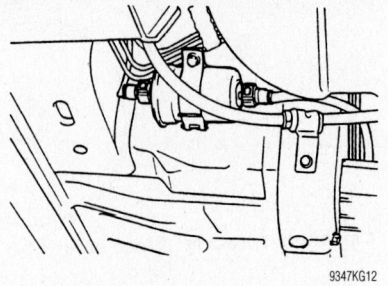
9347KG12

Remove the fuel filter

   • Fuel filter bracket, if equipped
   • Fuel supply and pressure lines
   • Negative battery cable
5. Start the engine, check leaks and repair if necessary.

Fuel Pump

REMOVAL & INSTALLATION

1. Before servicing the vehicle, refer to the Precautions Section.
2. Relieve the fuel system pressure.
3. Drain the fuel tank.

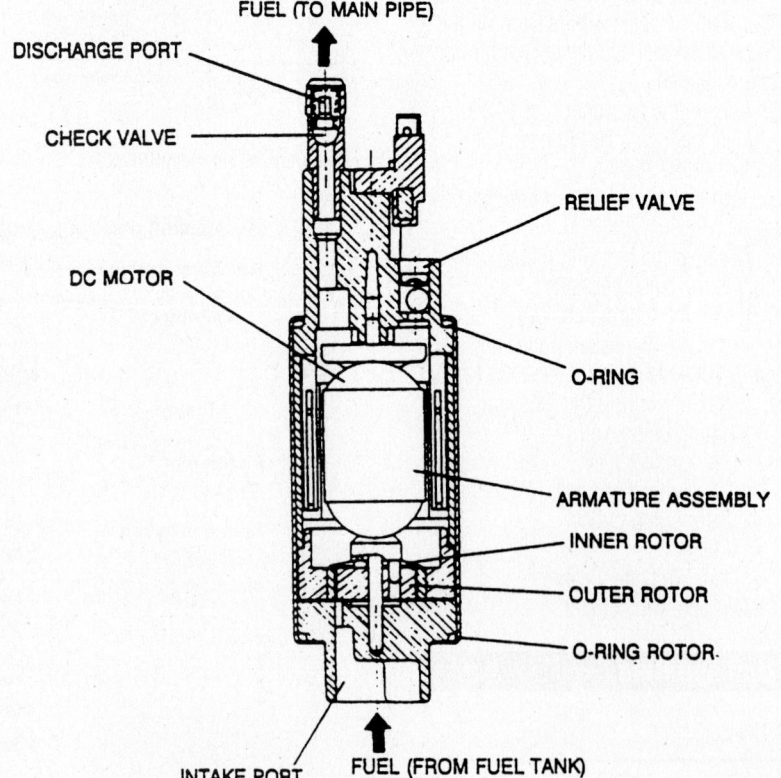
7923GG81

Cut away view of the electric fuel pump

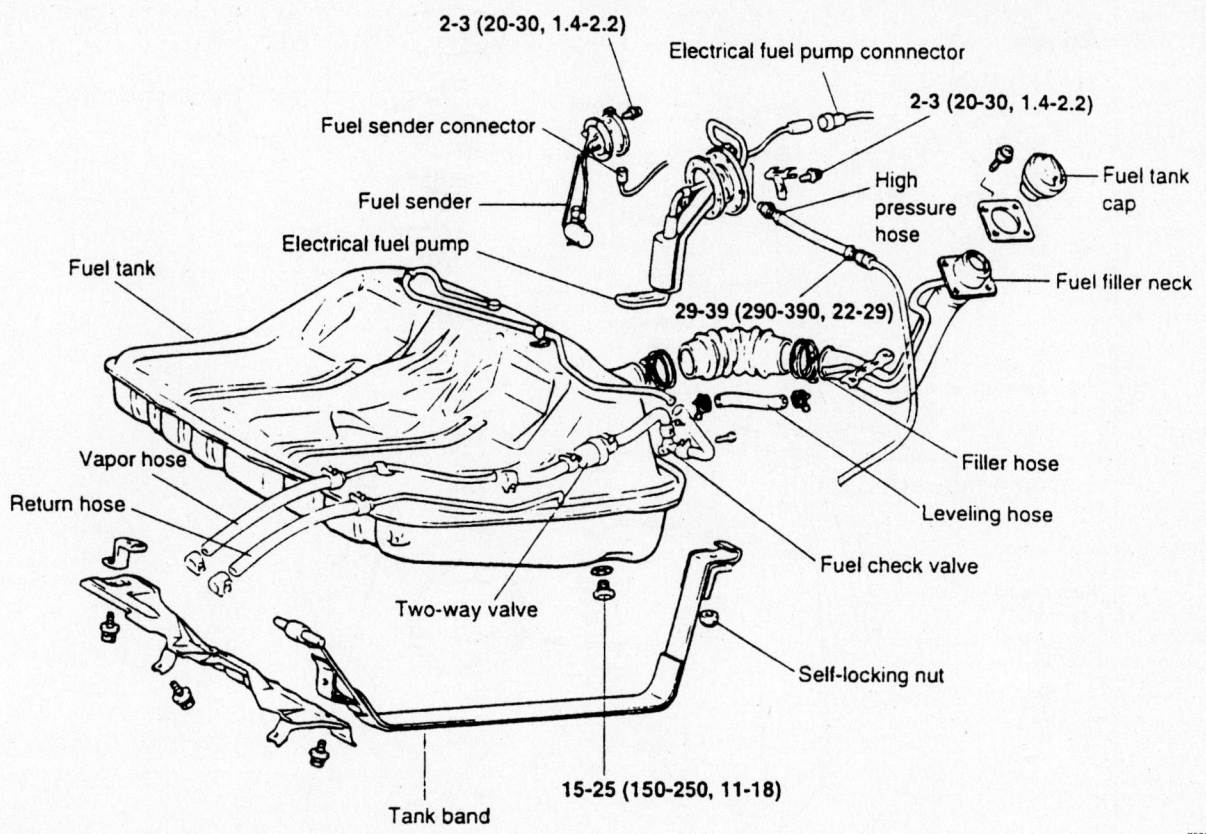

2-3 (20-30, 1.4-2.2)
Electrical fuel pump connnector
Fuel sender connector
2-3 (20-30, 1.4-2.2)
Fuel sender
High pressure hose
Fuel tank cap
Electrical fuel pump
Fuel tank
29-39 (290-390, 22-29)
Fuel filler neck
Vapor hose
Filler hose
Return hose
Leveling hose
Fuel check valve
Two-way valve
Self-locking nut
Tank band
15-25 (150-250, 11-18)

7923GG84

Exploded view of the fuel tank and fuel pump assembly—Sonata

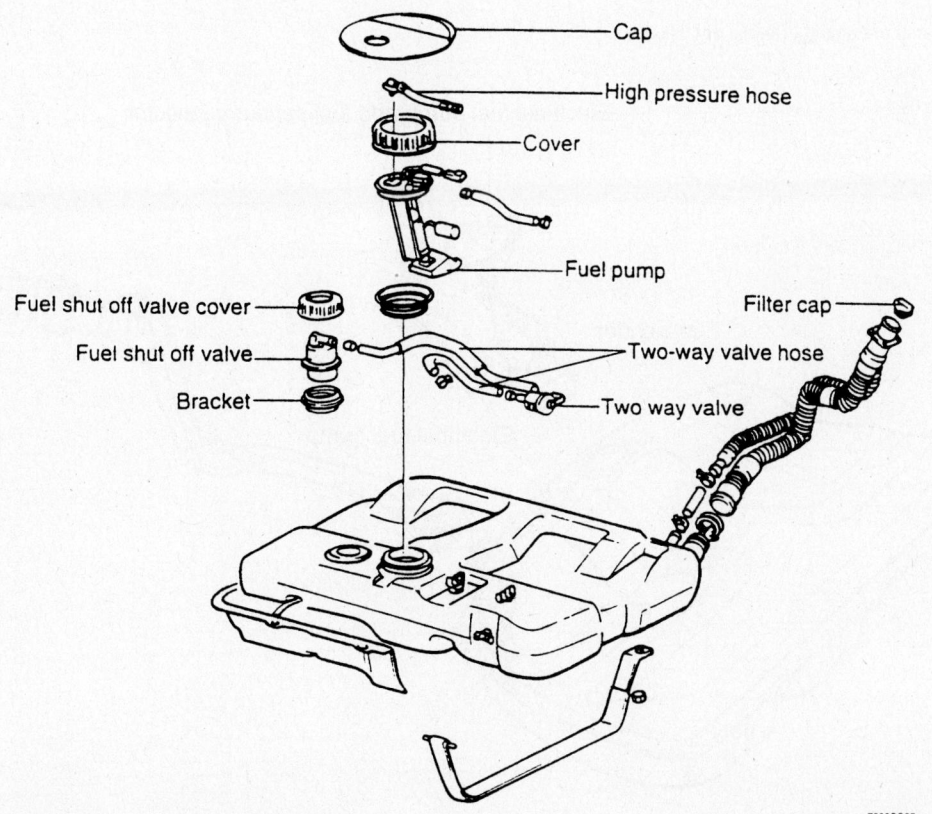

Cap
High pressure hose
Cover
Fuel pump
Fuel shut off valve cover
Filter cap
Fuel shut off valve
Two-way valve hose
Bracket
Two way valve

7923GG85

Exploded view of the fuel tank and fuel pump assembly—Accent

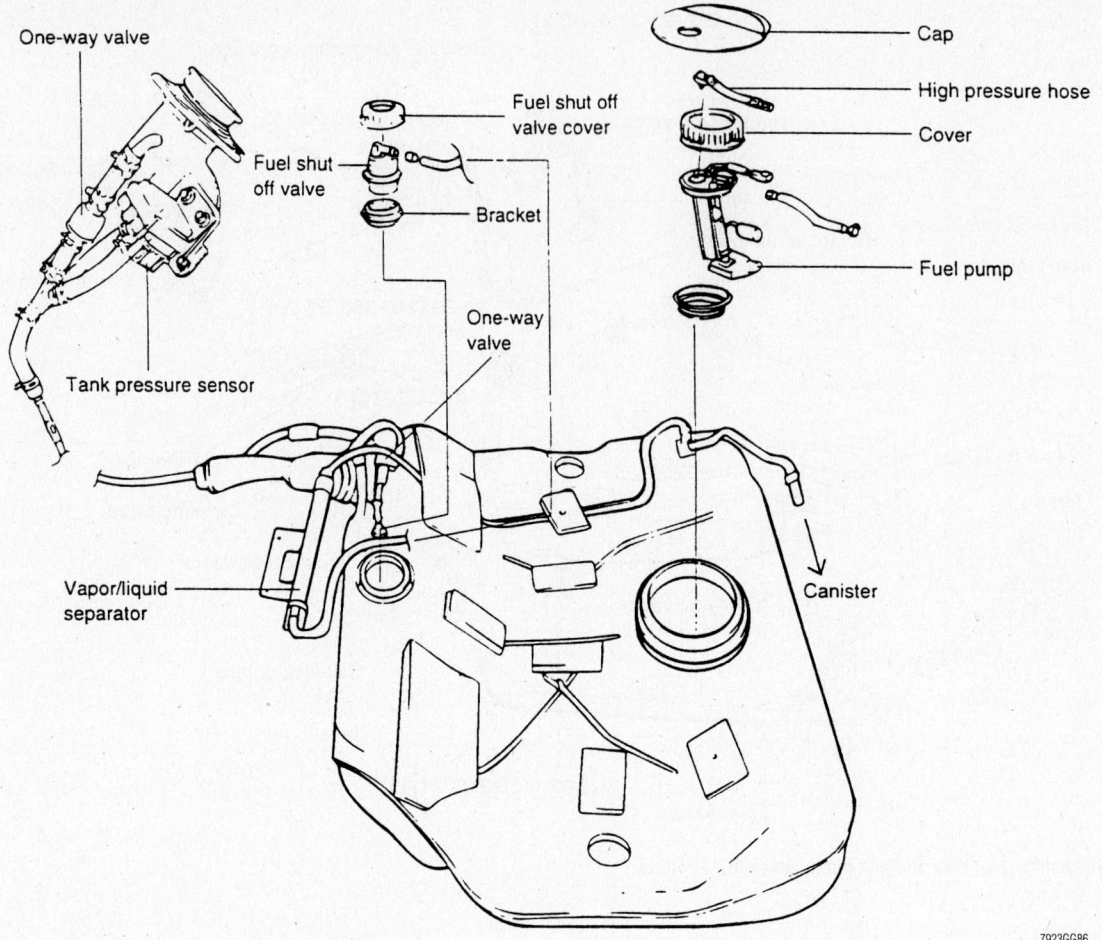

Fuel tank and fuel pump assembly—Tiburon and Elantra

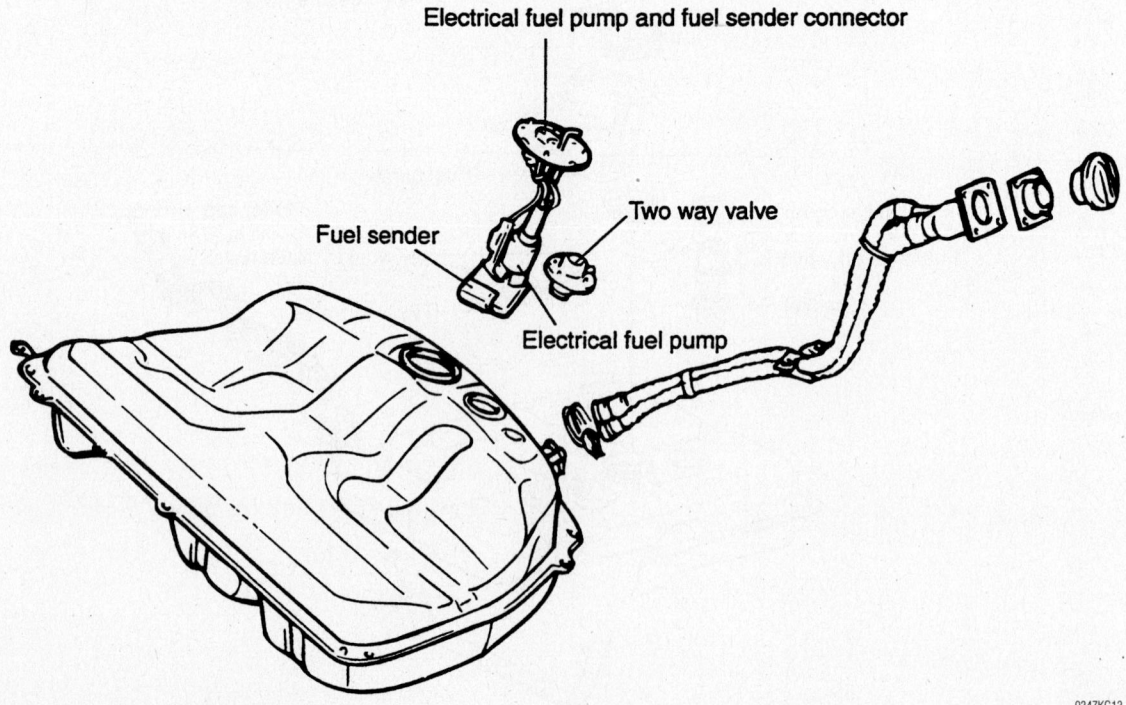

Fuel tank and fuel pump assembly—XG 350

4. Remove or disconnect the following:
- Negative battery cable
- Fuel supply, return and vapor lines
- Fuel fill and vent hoses
- Fuel pump module connector
- Fuel level sender connector
- Fuel tank straps
- Fuel tank
- Fuel pump module

To install:
5. Install or connect the following:
- Fuel pump module and torque the mounting bolts to 12–24 inch lbs. (1–3 Nm)
- Fuel tank
- Fuel tank straps
- Fuel level sender connector
- Fuel pump module connector
- Fuel fill and vent hoses

- Fuel supply, return and vapor lines
- Negative battery cable

6. Fill the tank with fuel and check for proper fuel pump operation.

Fuel Injector

REMOVAL & INSTALLATION

1. Before servicing the vehicle, refer to the Precautions Section.
2. Relieve the fuel system pressure.
3. Remove or disconnect the following:
- Negative battery cable
- Air intake surge tank, if necessary
- Fuel lines
- Fuel injector connectors
- Pressure regulator vacuum line

- Fuel supply manifold with injectors attached

4. Separate the injectors from the supply manifold.

To install:
5. Install or connect the following:
- Injectors to the fuel supply manifold using new O-rings
- Fuel supply manifold with injectors attached and torque the bolts to 84–132 inch lbs. (10–15 Nm)
- Fuel injector connectors
- Pressure regulator vacuum line
- Fuel lines
- Air intake surge tank, if removed
- Negative battery cable

6. Start the engine and check for leaks.

DRIVE TRAIN

Transaxle Assembly

REMOVAL & INSTALLATION

Manual

1. Before servicing the vehicle, refer to the Precautions Section.
2. Attach a support fixture to the engine lifting eyes.
3. Drain the transaxle.
4. Remove or disconnect the following:
- Negative battery cable
- Air intake assembly
- Back-up lamp switch
- Clutch slave cylinder
- Speedometer cable
- Shift cables
- Steering shaft joint
- Starter motor
- Axle halfshafts
- Flywheel cover
- Transaxle mount
- Transaxle flange bolts
- Transaxle

To install:

➡**Use new circlips, split pins and self-locking fasteners for assembly.**

5. Position the transaxle to the engine and tighten the flange bolts to the following specifications:
  a. M8 bolts: 72–84 inch lbs. (8–10 Nm).
  b. M10 bolts: 22–25 ft. lbs. (30–35 Nm).
  c. M12 bolts: 32–39 ft. lbs. (43–55 Nm).

6. Install or connect the following:
- Transaxle mount and torque the bolts to 65–80 ft. lbs. (90–110 Nm)
- Flywheel cover and torque the bolts to 72–84 inch lbs. (8–10 Nm)
- Axle halfshafts
- Starter motor
- Steering shaft joint
- Shift cables
- Speedometer cable
- Clutch slave cylinder
- Back-up lamp switch
- Air intake assembly
- Negative battery cable

7. Fill the transaxle.

Automatic

EXCEPT XG 350

1. Before servicing the vehicle, refer to the Precautions Section.
2. Attach a support fixture to the engine lifting eyes.
3. Drain the transaxle.
4. Remove or disconnect the following:
- Negative battery cable
- Air intake assembly
- Transaxle oil cooler lines
- Shift cable
- Speedometer cable
- Pulse generator connector
- Inhibitor connector
- Kickdown servo connector
- Solenoid valve connector
- Oil temperature sensor connector
- Starter motor
- Axle halfshafts
- Flywheel cover

- Torque converter
- Transaxle mount
- Transaxle flange bolts
- Transaxle

To install:
5. Position the transaxle to the engine and tighten the flange bolts to the following specifications:
  a. M8 bolts: 72–84 inch lbs. (8–10 Nm).
  b. M10 bolts: 22–25 ft. lbs. (30–35 Nm).
  c. M12 bolts: 32–39 ft. lbs. (43–55 Nm).

6. Install or connect the following:
- Transaxle mount and torque the bolts to 65–80 ft. lbs. (90–110 Nm)
- Torque converter and torque the bolts to 34–39 ft. lbs. (46–53 Nm)
- Flywheel cover and torque the bolts to 72–84 inch lbs. (8–10 Nm)
- Axle halfshafts
- Starter motor
- Oil temperature sensor connector
- Solenoid valve connector
- Kickdown servo connector
- Inhibitor connector
- Pulse generator connector
- Speedometer cable
- Shift cable
- Transaxle oil cooler lines
- Air intake assembly
- Negative battery cable

7. Fill the transaxle to the correct level.

XG 350

1. Before servicing the vehicle, refer to the Precautions Section.

2. Attach a support fixture to the engine lifting eyes.

3. Drain the transaxle.

4. Remove or disconnect the following:
- Negative battery cable
- Air cleaner assembly
- Control cable
- Speedometer sensor connector
- Transaxle range switch connector
- Solenoid connector
- Oil temperature sensor connector
- Oil cooler lines
- Steering gear
- Sway bar
- Tie rod end
- Ball joints and drive shafts
- Steering U-joint and return tube mounting bolts
- Subframe
- Starter
- Engine-to-transaxle bolts
- Transaxle

To install:

5. Install or connect the following:
- Engine-to-transaxle and torque the bolts to 39 ft. lbs. (54 Nm)
- Starter
- Subframe and torque the subframe to transaxle bolts to 58 ft. lbs. (80 Nm) and the roll stopper bolts to 38 ft. lbs. (55 Nm)
- Steering U-joint and return tube mounting bolts
- Ball joints and driveshafts
- Tie rod end
- Sway bar
- Gear box
- Oil cooler lines
- Oil temperature sensor connector
- Solenoid connector
- Transaxle range switch connector
- Speedometer sensor connector and torque to 19 ft. lbs. (26 Nm)
- Control cable and torque the bracket bolt to 18 ft. lbs. (25 Nm)
- Air cleaner assembly
- Negative battery cable

6. Fill the transaxle fluid to the proper level.

7. Start the vehicle, check for leaks and repair if necessary.

Clutch

ADJUSTMENTS

These vehicles are equipped with a hydraulic clutch system. No adjustment is necessary.

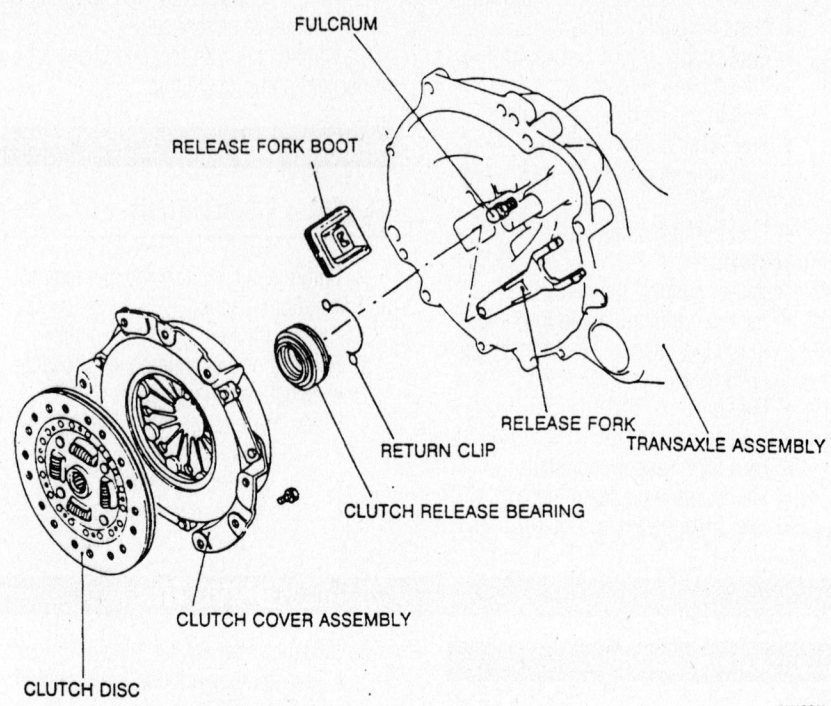

Exploded view of the clutch disc and pressure plate components

REMOVAL & INSTALLATION

1. Before servicing the vehicle, refer to the Precautions Section.

2. Remove or disconnect the following:
- Transaxle
- Pressure plate bolts by loosening them evenly in a crossing pattern
- Pressure plate and clutch disc

To install:

3. Install or connect the following:
- Clutch disc on the flywheel
- Pressure plate and torque the bolts evenly in a crossing pattern to 11–15 ft. lbs. (15–21 Nm)
- Transaxle

Hydraulic Clutch System

BLEEDING

1. Connect a hose to the bleeder screw and place the other end of hose into a container of clean brake fluid. Open the bleeder screw.

2. Have an assistant pump the clutch pedal slowly until no air bubbles are present at the bleeder screw.

3. Close the bleeder screw.

4. Fill the clutch master cylinder.

5. Check the clutch operation.

Halfshaft

REMOVAL & INSTALLATION

Except XG 350

1. Before servicing the vehicle, refer to the Precautions Section.

2. Remove or disconnect the following:
- Front wheel
- Spindle nut
- Wheel speed sensor, if equipped
- Outer tie rod end
- Stabilizer bar link
- Lower ball joint

3. Press the stub shaft out of the hub.

4. Pry the inner joint out of the transaxle or intermediate shaft.

To install:

➡️**Use new circlips, split pins and self-locking nuts for assembly.**

5. Install the inner joint so that the circlip is felt to seat in the retaining groove.

6. Guide the stub shaft into the hub.

7. Install or connect the following:
- Lower ball joint and torque the nut to 43–52 ft. lbs. (58–70 Nm)
- Stabilizer bar link
- Outer tie rod end and torque the nut to 17–25 ft. lbs. (23–34 Nm)
- Wheel speed sensor, if equipped
- Spindle nut and torque the nut to 144–187 ft. lbs. (195–253 Nm)

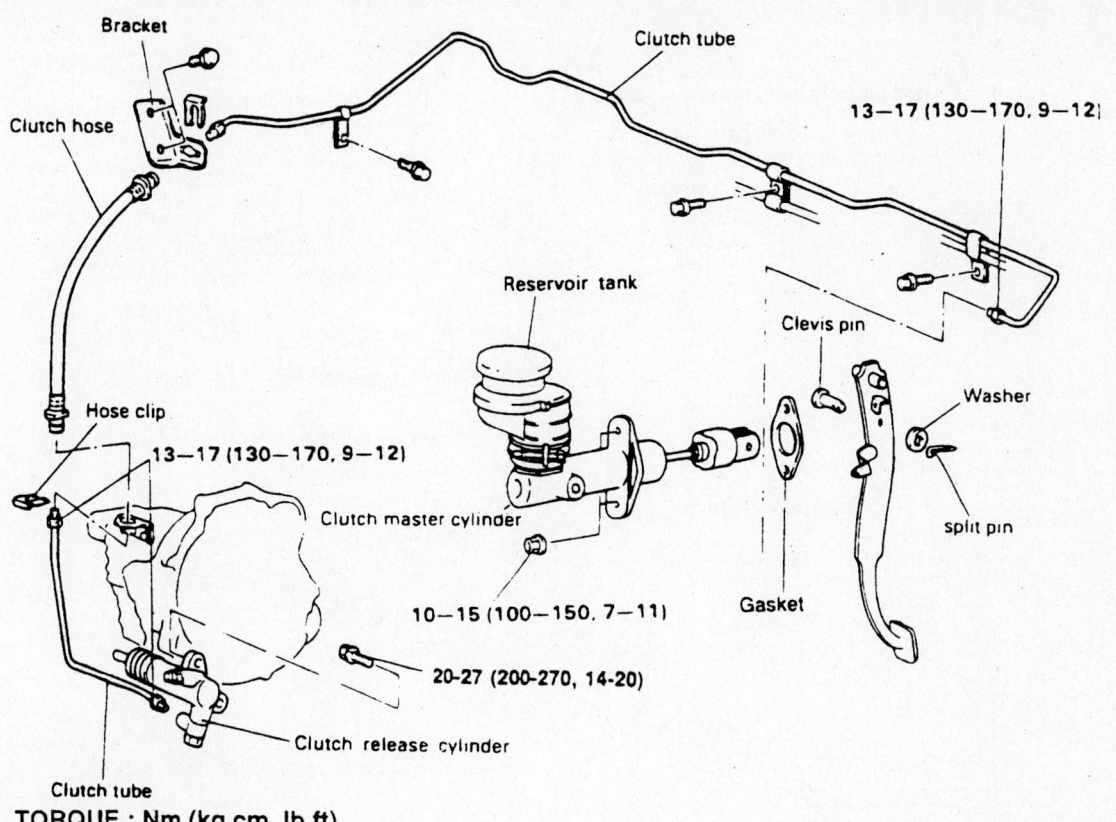

Bracket

Clutch tube

Clutch hose

13—17 (130—170, 9—12)

Reservoir tank

Clevis pin

Washer

Hose clip

13—17 (130—170, 9—12)

Clutch master cylinder

split pin

10—15 (100—150, 7—11)

Gasket

20-27 (200-270, 14-20)

Clutch release cylinder

Clutch tube

TORQUE : Nm (kg.cm, lb.ft)

7923GG90

Exploded view of the clutch hydraulic system

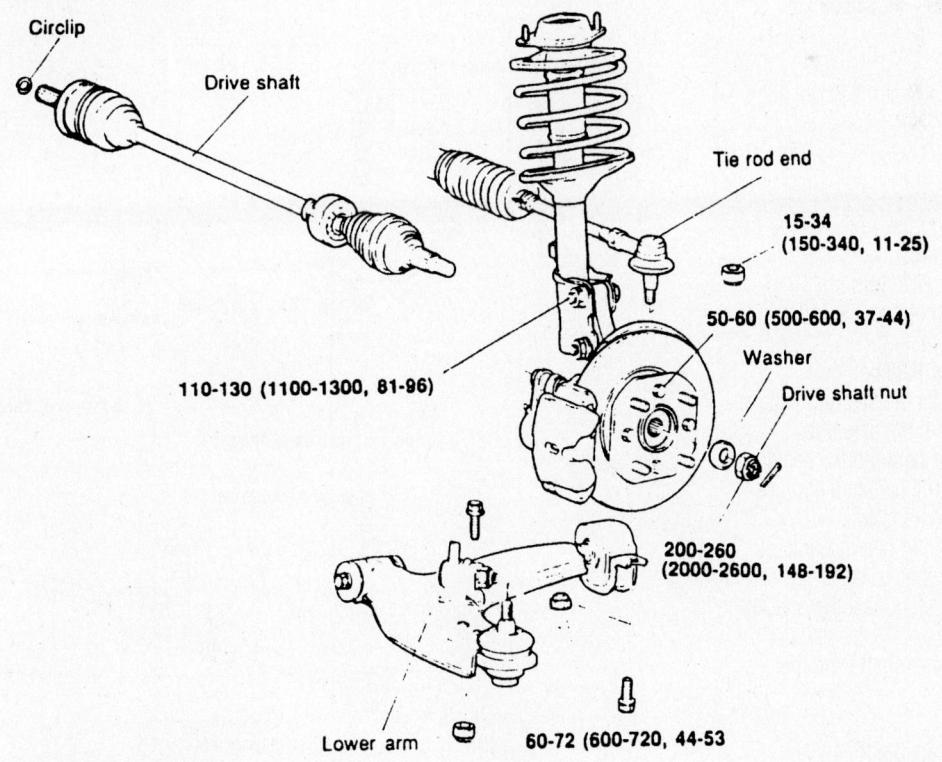

Circlip

Drive shaft

Tie rod end

15-34
(150-340, 11-25)

50-60 (500-600, 37-44)

Washer

110-130 (1100-1300, 81-96)

Drive shaft nut

200-260
(2000-2600, 148-192)

Lower arm

60-72 (600-720, 44-53)

TORQUE : Nm (kg.cm, lb.ft)

7923GG94

Halfshaft components—except V6 Sonata

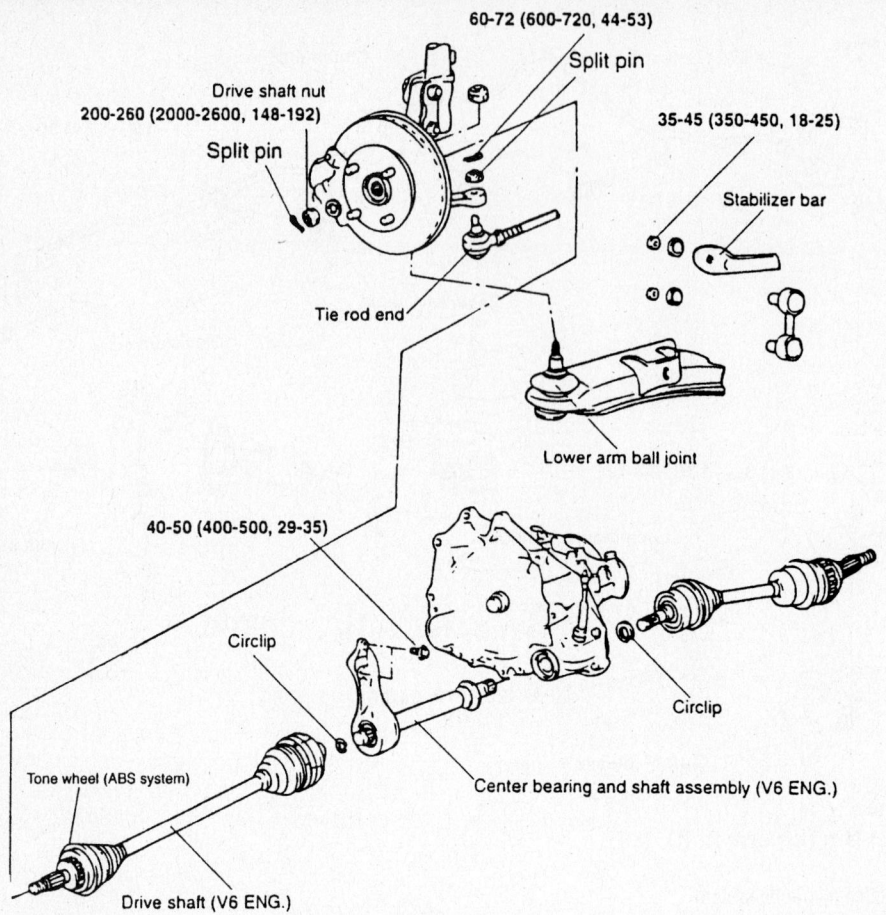

Halfshaft components—V6 Sonata

- Front wheel

8. Check and/or adjust the wheel alignment.

XG 350

1. Before servicing the vehicle, refer to the Precautions Section.
2. Drain the transaxle fluid.
3. Remove or disconnect the following:
   - Front wheel
   - Split pin and halfshaft nut
   - Ball joint from the steering knuckle
   - Halfshaft from the axle hub
4. Install a prybar between the center bearing bracket and the halfshaft and pry the halfshaft from the transaxle.
5. Remove the center bearing bracket bolts and install a prybar between bracket and the engine and disconnect the bracket from the engine.
6. Remove the inner shaft from the transaxle.

To install:
7. Install or connect the following:
   - Inner halfshaft to the transaxle
   - Center bearing bracket and torque the bolt to 36 ft. lbs. (50 Nm)

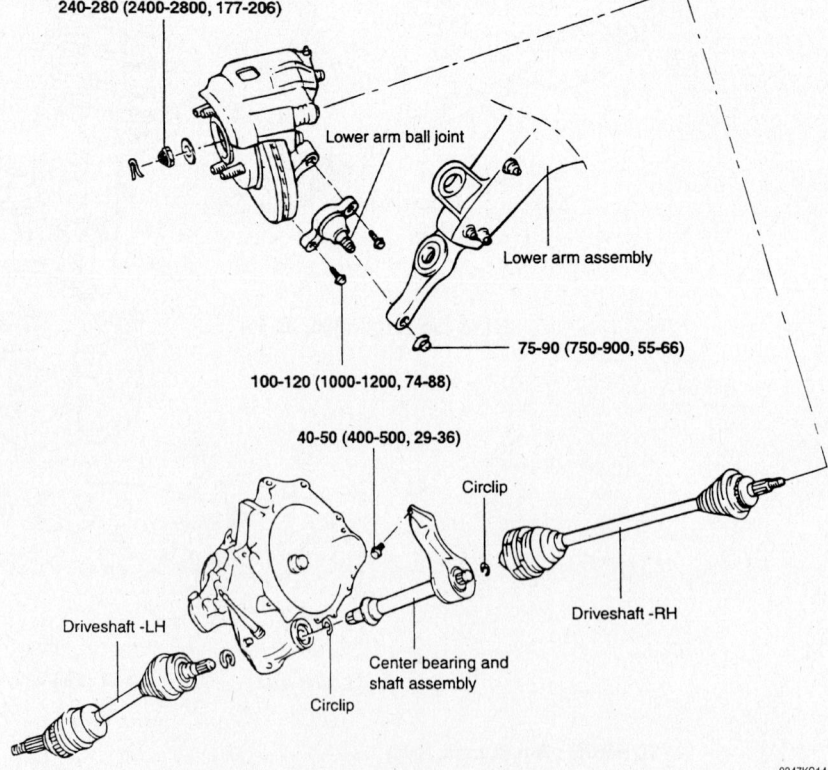

Exploded view of the halfshaft assembly— XG 350

- Halfshaft to the axle hub
- Ball joint to the steering knuckle and torque 88 ft. lbs. (110 Nm)
- Split pin and halfshaft nut and torque the nut to 206 ft. lbs. (280 Nm)
- Front wheel

8. Fill the transaxle fluid to the proper level.

9. Check and/or adjust the wheel alignment.

CV-Joints

OVERHAUL

Outer CV-Joint

The outer CV-joint is serviced with the axle halfshaft as an assembly. The outer CV-joint boot may be replaced by first removing the inner joint.

Inner CV-Joint

TRIPOD JOINT

1. Before servicing the vehicle, refer to the Precautions Section.

2. Remove or disconnect the following:
- Axle halfshaft from the vehicle
- Inner boot clamps and push the boot back
- CV-joint housing
- Snapring
- Spider and rollers
- CV-joint boot

➡**Do not disassemble the spider and rollers.**

To install:

3. Install or connect the following:
- CV-joint boot
- Spider and rollers
- Snapring

4. Apply clean grease to the CV-joint housing and the boot.
- CV-joint housing and tighten the boot clamps
- Axle to the vehicle

DOUBLE OFFSET JOINT

1. Before servicing the vehicle, refer to the Precautions Section.

2. Remove or disconnect the following:
- Axle halfshaft from the vehicle

- Inner boot clamps and push the boot back
- Circlip
- CV-joint housing
- Snapring
- Double offset joint inner race, cage and ball assembly
- CV-joint boot

➡**Do not disassemble the inner race, cage and ball assembly.**

To install:

3. Install or connect the following:
- CV-joint boot
- Double offset joint inner race, cage and ball assembly
- Snapring

4. Apply clean grease to the CV-joint housing and the boot.
- CV-joint housing
- Circlip
- Boot clamps by tightening them
- Axle to the vehicle

STEERING AND SUSPENSION

Air Bag

❈❈ CAUTION

Some vehicles are equipped with an air bag system. The system must be disarmed before performing service on, or around, system components, the steering column, instrument panel components, wiring and sensors. Failure to follow the safety precautions and the disarming procedure could result in accidental air bag deployment, possible injury and unnecessary system repairs.

PRECAUTIONS

Several precautions must be observed when handling the inflator module to avoid accidental deployment and possible personal injury.
- Never carry the inflator module by the wires or connector on the underside of the module.
- When carrying a live inflator module, hold securely with both hands, and ensure

that the bag and trim cover are pointed away.
- Place the inflator module on a bench or other surface with the bag and trim cover facing up.
- With the inflator module on the bench, never place anything on or close to the module which may be thrown in the event of an accidental deployment.

Before servicing the vehicle, also make sure to refer to the precautions in the beginning of this section as well.

DISARMING

Disconnect and isolate the negative battery cable. Wait 3 minutes for the system capacitor to discharge before performing any service.

Rack and Pinion Steering Gear

REMOVAL & INSTALLATION

1. Before servicing the vehicle, refer to the Precautions Section.

2. Remove or disconnect the following:
- Negative battery cable
- Front wheels
- Outer tie rod ends
- Steering column flexible coupler
- Power steering fluid hoses, if equipped with power steering
- Subframe center beam
- Exhaust front pipe
- Left lower control arm
- Stabilizer bar
- Steering gear

To install:

3. Install or connect the following:
- Steering gear and torque the bolts to 44–59 ft. lbs. (60–80 Nm)
- Stabilizer bar
- Left lower control arm
- Exhaust front pipe
- Subframe center beam
- Power steering fluid hoses, if equipped with power steering
- Steering column flexible coupler and torque the bolt to 11–14 ft. lbs. (15–19 Nm)
- Outer tie rod ends and torque the nuts to 17–25 ft. lbs. (23–34 Nm)

COMPONENTS

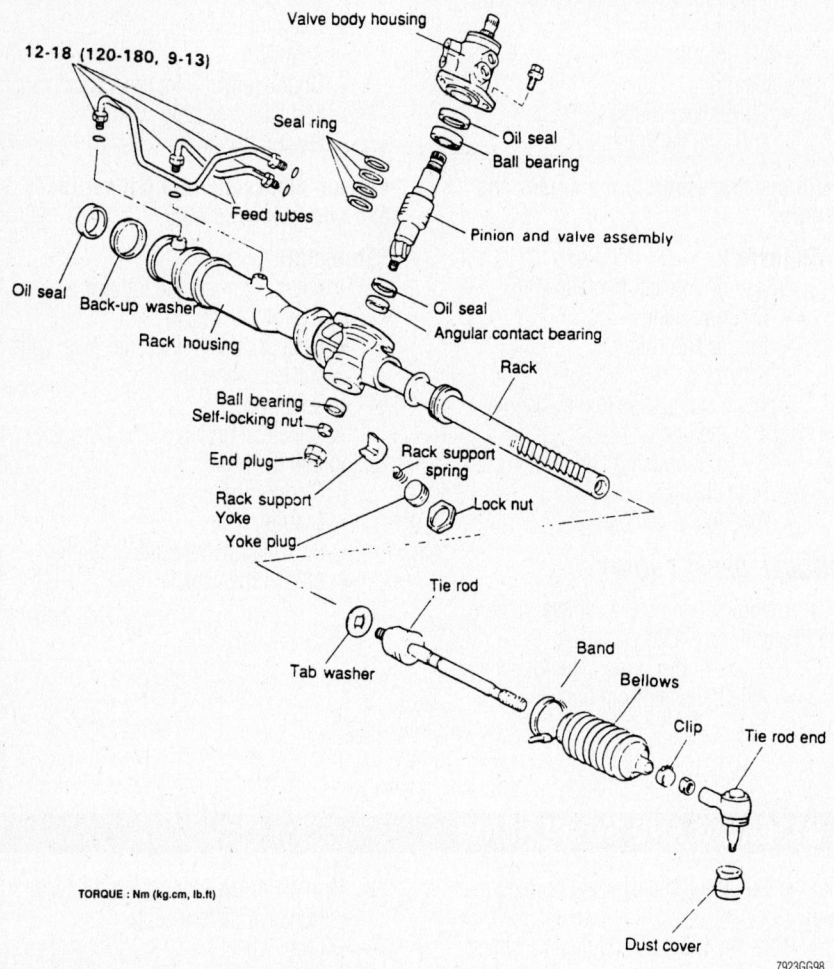

12-18 (120-180, 9-13)

Valve body housing

Seal ring

Oil seal
Ball bearing

Feed tubes

Pinion and valve assembly

Oil seal

Back-up washer

Oil seal
Angular contact bearing

Rack housing

Rack

Ball bearing
Self-locking nut

End plug

Rack support
spring

Rack support
Yoke

Lock nut

Yoke plug

Tie rod

Band

Bellows

Clip

Tie rod end

Tab washer

Dust cover

TORQUE : Nm (kg.cm, lb.ft)

7923GG98

Exploded view of the rack and pinion assembly

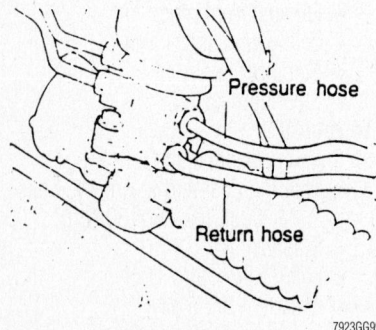

Pressure hose

Return hose

7923GG99

Pressure and return hose location on the rack

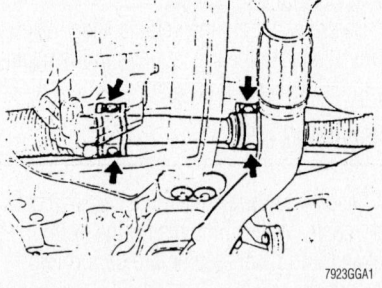

7923GGA1

Power rack and pinion mounting bolt locations

- Front wheels
- Negative battery cable
4. Fill the power steering system.
5. Start the engine and check for leaks.

Struts

REMOVAL & INSTALLATION

Front

ACCENT, ELANTRA AND TIBURON

1. Before servicing the vehicle, refer to the Precautions Section.
2. Remove or disconnect the following:
   - Front wheel
   - Brake hose bracket
   - Strut upper mounting bolts
   - Lower mounting bolts
   - Strut assembly

To install:

3. Install or connect the following:
   - Strut assembly and torque the lower bolts to 66 ft. lbs. (90 Nm)
   - Strut upper mounting bolts and torque to 22 ft. lbs. (30 Nm)
   - Brake hose bracket
   - Front wheel
4. Check and/or adjust the wheel alignment.

9347KG15

Exploded view of the strut assembly

SONATA AND XG 350

1. Before servicing the vehicle, refer to the Precautions Section.
2. Remove or disconnect the following:
   - Front wheel
   - Brake hose bracket from the mounting fork
   - Mounting fork and lower arm connecting bolt
   - Strut upper mounting nuts
   - Strut assembly

To install:

3. Install or connect the following:
   - Strut assembly
   - Fork to the strut and torque the bolts to 59 ft. lbs. (80 Nm)
   - Fork to the lower arm and torque the bolt to 88 ft. lbs. (120 Nm)
   - Upper strut mounting nuts and torque to 36 ft. lbs. (50 Nm)
   - Brake hose bracket

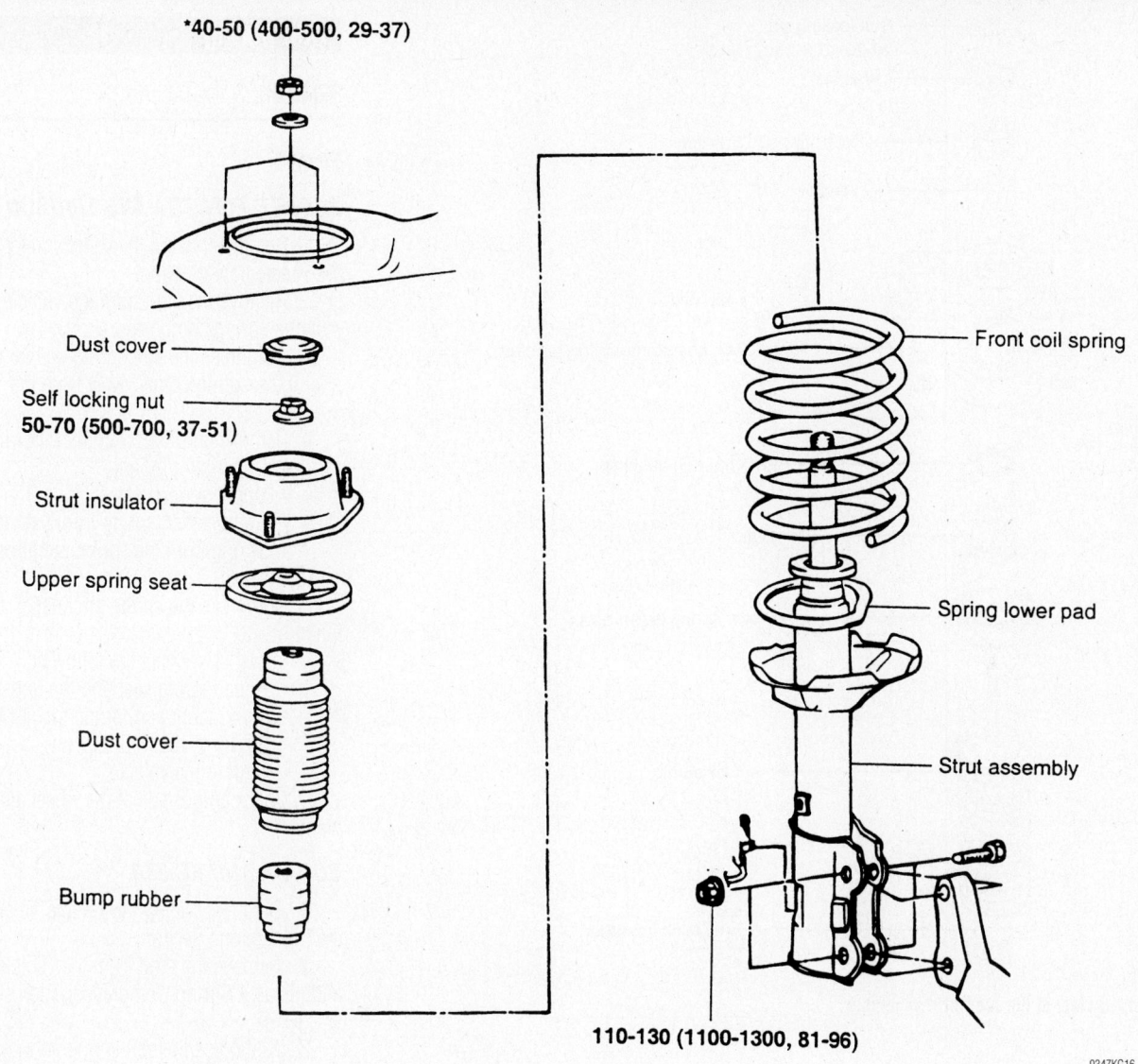

*40-50 (400-500, 29-37)

Dust cover

Self locking nut
50-70 (500-700, 37-51)

Strut insulator

Upper spring seat

Dust cover

Bump rubber

Front coil spring

Spring lower pad

Strut assembly

110-130 (1100-1300, 81-96)

9347KG16

Exploded view of the front strut assembly

- Front wheel
4. Check and/or adjust the wheel alignment.

Rear

ACCENT, ELANTRA AND TIBURON

1. Before servicing the vehicle, refer to the Precautions Section.
2. Remove or disconnect the following:
   - Rear seatback assembly and wheel house cover
   - Rear wheel
   - Upper mounting nuts
   - Brake hose and wheel speed sensor connectors
   - Carrier mounting nuts
   - Strut assembly

To install:
3. Install or connect the following:
   - Strut assembly and torque the carrier mounting nuts to 66 ft. lbs. (90 Nm)
   - Brake hose and wheel speed sensor connectors
   - Upper mounting nuts and torque the nuts to 22 ft. lbs. (30 Nm) on the Accent and to 37 ft. lbs. (50 Nm) for the Elantra
   - Wheel house cover and wheel
   - Rear seatback

SONATA AND XG 350

1. Before servicing the vehicle, refer to the Precautions Section.
2. Remove or disconnect the following:

- Rear wheel
- Lower mounting bolt
- Upper arm and rear carrier bolt
- Strut mounting bracket
- Strut assembly

To install:
3. Install or connect the following:
   - Strut assembly and mounting bracket and torque the bolt to 36 ft. lbs. (50 Nm)
   - Upper arm and rear carrier bolt and torque the bolt to 88 ft. lbs. (120 Nm)
   - Lower mounting bolt
   - Rear wheel

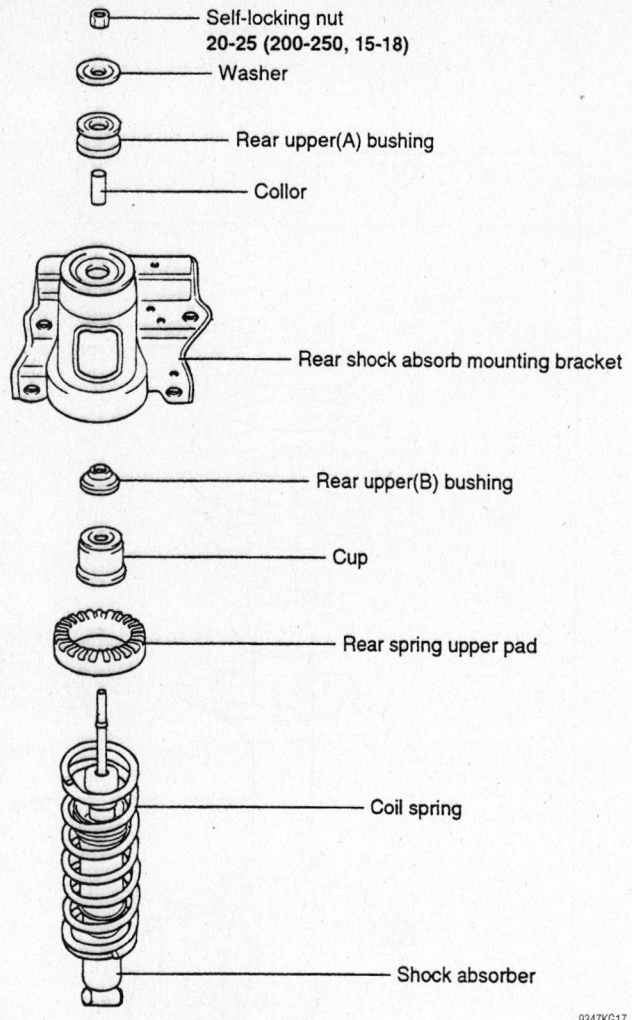

- Self-locking nut
  20-25 (200-250, 15-18)
- Washer
- Rear upper(A) bushing
- Collor
- Rear shock absorb mounting bracket
- Rear upper(B) bushing
- Cup
- Rear spring upper pad
- Coil spring
- Shock absorber

9347KG17

Exploded view of the rear strut assembly

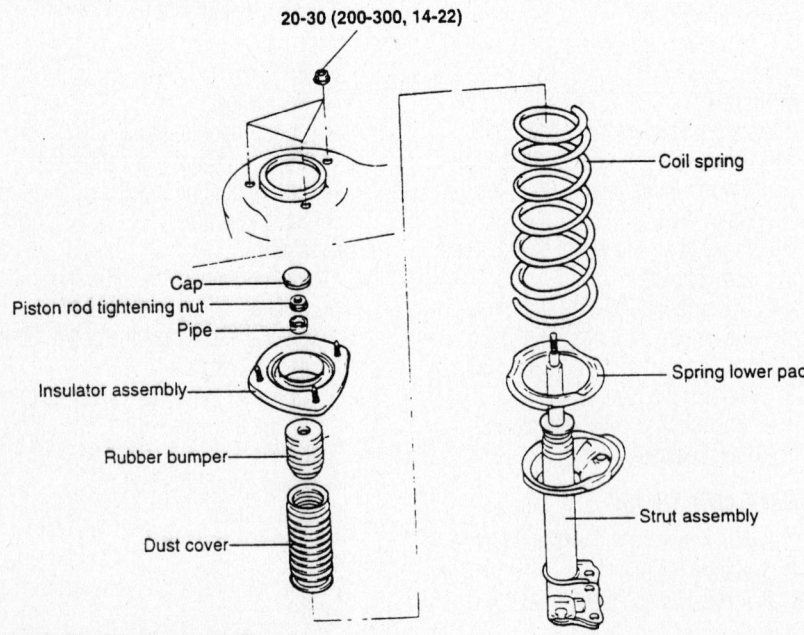

20-30 (200-300, 14-22)

- Cap
- Piston rod tightening nut
- Pipe
- Insulator assembly
- Rubber bumper
- Dust cover
- Coil spring
- Spring lower pad
- Strut assembly

TORQUE : Nm (kg·cm, lb·ft)

7923GGA4

Rear strut components

Coil Spring

REMOVAL & INSTALLATION

Front

ACCENT, ELANTRA AND TIBURON

1. Before servicing the vehicle, refer to the Precautions Section.
2. Remove the strut from the vehicle and install a spring compressor.
3. Compress the coil spring so that the end of the spring comes away from the spring seat.
4. Remove or disconnect the following:
   - Upper strut mount
   - Upper spring seat
   - Compressed spring from the strut
   - Spring from the spring compressor

To install:

5. Compress the spring and install it on the strut.
6. Install or connect the following:
   - Upper spring seat and the upper strut mount and torque the nut to 29–36 ft. lbs. (40–50 Nm)
   - Strut to the vehicle
7. Check and/or adjust the wheel alignment.

SONATA AND XG 350

1. Before servicing the vehicle, refer to the Precautions Section.
2. Remove the strut from the vehicle and install a Spring Compressor Tool, such as J38402.
3. Compress the coil spring so that the end of the spring comes away from the spring seat.
4. Remove or disconnect the following:
   - Self locking nut
5. Install the Compressor Tool, J38402.
6. Remove the bracket, spring pad and coil spring.

To install:

7. Compress the coil spring with Compressor Tool J38402.
8. Install or connect the following:
   - Coil spring to the strut
   - Dust cover, upper spring pad, bushing and hand tighten the lock nut
9. Remove the compressor tool when the coil spring is properly aligned and torque the lock nut to 18 ft. lbs. (25 Nm).
10. Install the strut assembly.

Upper Ball Joint

The upper ball joints are replaced with the upper control arms as an assembly.

Lower Ball Joint

REMOVAL & INSTALLATION

Bolt-On Type

1. Before servicing the vehicle, refer to the Precautions Section.
2. Remove or disconnect the following:
   - Front wheel
   - Ball joint stud from the knuckle
   - Ball joint from the lower control arm

To install:

➡ **Use a new split pin for assembly.**

3. Install or connect the following:
   - Ball joint and torque the mounting bolts to 69–87 ft. lbs. (95–120 Nm)
   - Stud nut and torque to 43–52 ft. lbs. (60–72 Nm)
   - Front wheel
4. Check and/or adjust the wheel alignment.

Press-In Type

1. Before servicing the vehicle, refer to the Precautions Section.
2. Remove or disconnect the following:
   - Front wheel
   - Lower control arm
   - Ball joint dust cover
3. Press the ball joint out of the lower control arm.

To install:

4. Press the ball joint into the control arm.
5. Install or connect the following:
   - Ball joint dust cover
   - Lower control arm and torque the stud nut to 43–52 ft. lbs. (60–72 Nm)
   - Front wheel
6. Check and/or adjust the wheel alignment.

Upper Control Arm

REMOVAL & INSTALLATION

Sonata and XG 350

1. Before servicing the vehicle, refer to the Precautions Section.
2. Support the lower control arm assembly with a floor jack.
3. Remove or disconnect the following:
   - Front wheel
   - Ball joint nut, loosen only
   - Upper arm ball joint from the steer-

ing knuckle with Special Tool 09568-34000
   - Wheel house panel nuts
   - Upper arm assembly
   - Upper arm shaft

To install:

4. Install or connect the following:
   - Upper control arm shaft
   - Upper control arm assembly and torque the bolts to 73 ft. lbs. (100 Nm)
   - Wheel house panel nuts and torque the nuts to 48 ft. lbs. (65 Nm)
   - Upper arm ball joint to the steering knuckle and torque the bolts to 33 ft. lbs. (45 Nm)
   - Front wheel

CONTROL ARM BUSHING REPLACEMENT

1. Before servicing the vehicle, refer to the Precautions Section.

2. Remove or disconnect the following:
   - Control arm from the vehicle
   - Control arm bushings by unbolting them

To install:

3. Install or connect the following:
   - New bushings and torque the bolts to 40–48 ft. lbs. (55–65 Nm)
   - Control arm to the vehicle

Lower Control Arm

REMOVAL & INSTALLATION

Except Sonata and XG 350

1. Before servicing the vehicle, refer to the Precautions Section.
2. Remove or disconnect the following:
   - Front wheel
   - Stabilizer bar link

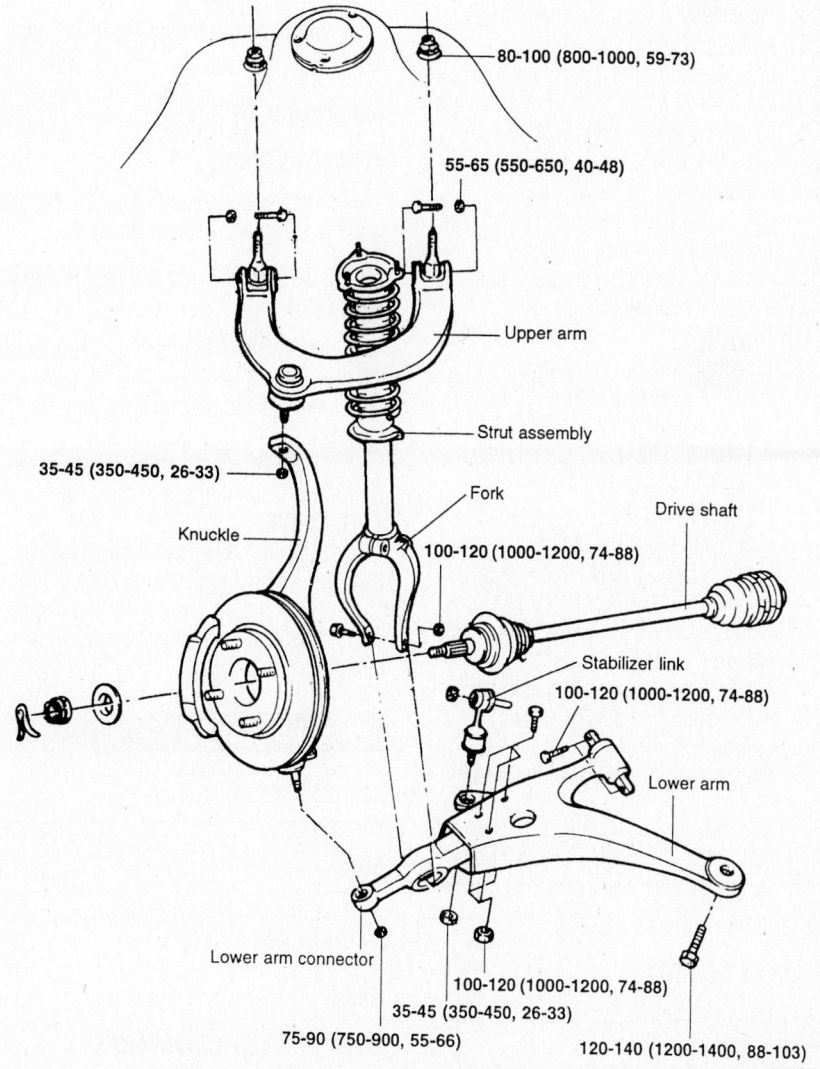

80-100 (800-1000, 59-73)

55-65 (550-650, 40-48)

Upper arm

Strut assembly

Fork

Drive shaft

35-45 (350-450, 26-33)

Knuckle

100-120 (1000-1200, 74-88)

Stabilizer link

100-120 (1000-1200, 74-88)

Lower arm

Lower arm connector

100-120 (1000-1200, 74-88)

35-45 (350-450, 26-33)

75-90 (750-900, 55-66)

120-140 (1200-1400, 88-103)

9347KG18

Exploded view of the upper control arm assembly—Sonata

- Lower ball joint
- Rear bushing bracket
- Front bolt
- Lower control arm

To install:

3. Install or connect the following:
- Lower control arm and torque the front bolt to 72–87 ft. lbs. (100–120 Nm)
- Rear bushing bracket. Tighten the bolts to 58–72 ft. lbs. (80–100 Nm)
- Lower ball joint and torque the nut to 43–52 ft. lbs. (60–72 Nm)
- Stabilizer bar link and torque the nut to 25–33 ft. lbs. (35–45 Nm)
- Front wheel

4. Check and/or adjust the wheel alignment.

Sonata and XG 350

1. Before servicing the vehicle, refer to the Precautions Section.
2. Remove or disconnect the following:
- Front wheel
- Lower ball joint nut, loosen only
- Lower arm ball joint from the lower arm connector with Special Tool 09445-21000
- Ball joint
- Fork from the lower arm connector
- Stabilizer bar link
- Control arm inner bushing bolts
- Lower control arm

To install:

3. Install or connect the following:
- Lower control arm and torque the front bushing bolts to 74–88 ft. lbs. (100–120 Nm) and the rear bushing bolt to 88–103 ft. lbs. (120–140 Nm)
- Stabilizer bar link and torque the nut to 26–33 ft. lbs. (35–45 Nm)
- Damper fork lower bolt and torque the nut to 74–88 ft. lbs. (100–120 Nm)
- Lower ball joint and torque the nut to 55–66 ft. lbs. (75–90 Nm)
- Front wheel

4. Check and/or adjust the wheel alignment.

CONTROL ARM BUSHING REPLACEMENT

Except Sonata and XG 350

FRONT BUSHING

1. Before servicing the vehicle, refer to the Precautions Section.
2. Remove the lower control arm from the vehicle.

3. Press the front bushing out of the control arm.

To install:

4. Lubricate the front bushing with soap and press into the control arm.
5. Install the control arm to the vehicle.
6. Check and/or adjust the wheel alignment.

REAR BUSHING

1. Before servicing the vehicle, refer to the Precautions Section.
2. Remove or disconnect the following:
- Front wheel
- Rear bushing bracket
- Rear bushing nut
- Rear bushing

To install:

3. Install or connect the following:
- Rear bushing and torque the nut to 25–33 ft. lbs. (35–45 Nm)
- Rear bushing bracket and torque the bolts to 58–72 ft. lbs. (80–100 Nm)
- Front wheel

4. Check and/or adjust the wheel alignment.

Sonata and XG 350

FRONT BUSHING

The front control arm bushing is serviced with the control arm as an assembly.

REAR BUSHING AND DAMPER FORK BUSHING

1. Before servicing the vehicle, refer to the Precautions Section.
2. Remove the control arm from the vehicle.
3. Press the rear bushing and the damper fork bushing out of the control arm.

To install:

4. Press the rear bushing and the damper fork bushing into the control arm.
5. Install the control arm to the vehicle.
6. Check and/or adjust the wheel alignment.

Wheel Bearings

ADJUSTMENT

Front

The front wheel bearing is a sealed unit and is not adjustable.

Rear

WITH REAR DRUM BRAKES

1. Before servicing the vehicle, refer to the Precautions Section.

2. Remove the rear wheels.
3. Loosen the spindle nut.
4. Torque the nut to 108–145 ft. lbs. (150–200 Nm). Check for correct bearing end-play by placing a dial indicator on the hub surface and moving the hub outward. Note the movement of the gauge and compare it to the desired reading of 0.008 in. (0.2mm) or less. If end-play exceeds the desired reading, retighten the rear hub bearing nut and recheck the end-play. If the reading is still excessive, replace the hub unit.
5. If end-play is correct, check the starting torque by attaching a spring balance to the hub lug bolts and pulling at a 90 degree angle while noting the required force to turn the hub. If the force required is above the desired reading of 5 lbs. (2.3 kg) or less, loosen the nut and again tighten to the desired torque. Recheck the starting torque. If the torque is still above the desired reading, replace the rear bearings.
6. Install the rear wheels.

WITH REAR DISC BRAKES

The rear wheel bearing is an integral part of the rear hub. No adjustment is possible.

REMOVAL & INSTALLATION

Front

1. Before servicing the vehicle, refer to the Precautions Section.
2. Remove or disconnect the following:
- Front wheel
- Brake caliper
- Lower ball joint
- Spindle nut
- Knuckle pinch bolts
- Steering knuckle

3. Press the hub out of the wheel bearing.
4. Press the wheel bearings out of the steering knuckle.
5. If necessary, press the inner race off the hub.

To install:

6. Press the wheel bearings into the steering knuckle.
7. Install the outer grease seal and press the hub into the wheel bearings.
8. Install or connect the following:
- Inner grease seal
- Steering knuckle and torque the knuckle pinch bolts to 65–76 ft. lbs. (95–105 Nm)
- Lower ball joint and torque the stud nut to 43–52 ft. lbs. (60–72 Nm)
- Spindle nut and torque the nut to 144–187 ft. lbs. (195–253 Nm)

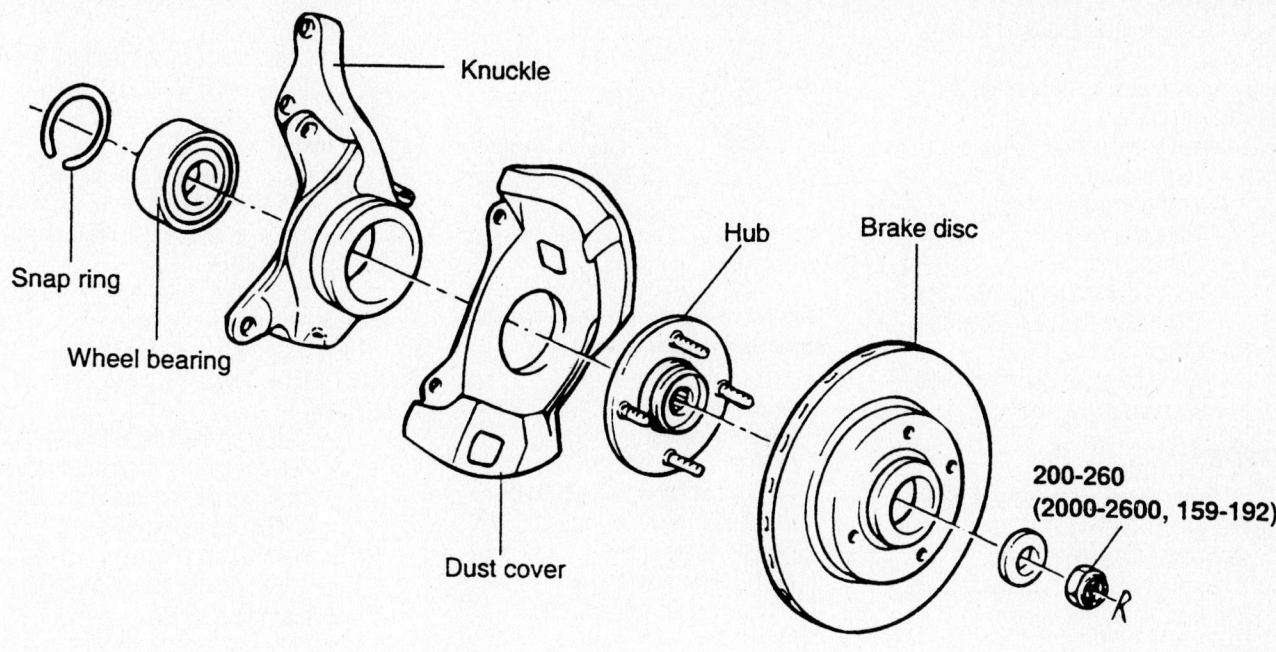

Exploded view of the front hub assembly

- Brake caliper and torque the bracket bolts to 50 ft. lbs. (68 Nm)
- Front wheel

9. Check and/or adjust the wheel alignment.

Rear

DRUM BRAKES

1. Before servicing the vehicle, refer to the Precautions Section.
2. Remove or disconnect the following:
   - Rear wheel
   - Speed sensor, if equipped
   - Grease cap
   - Flange nut
   - Outer bearing
   - Brake drum
   - Inner grease seal
   - Inner bearing

3. Drive the bearing races out of the drum hub.

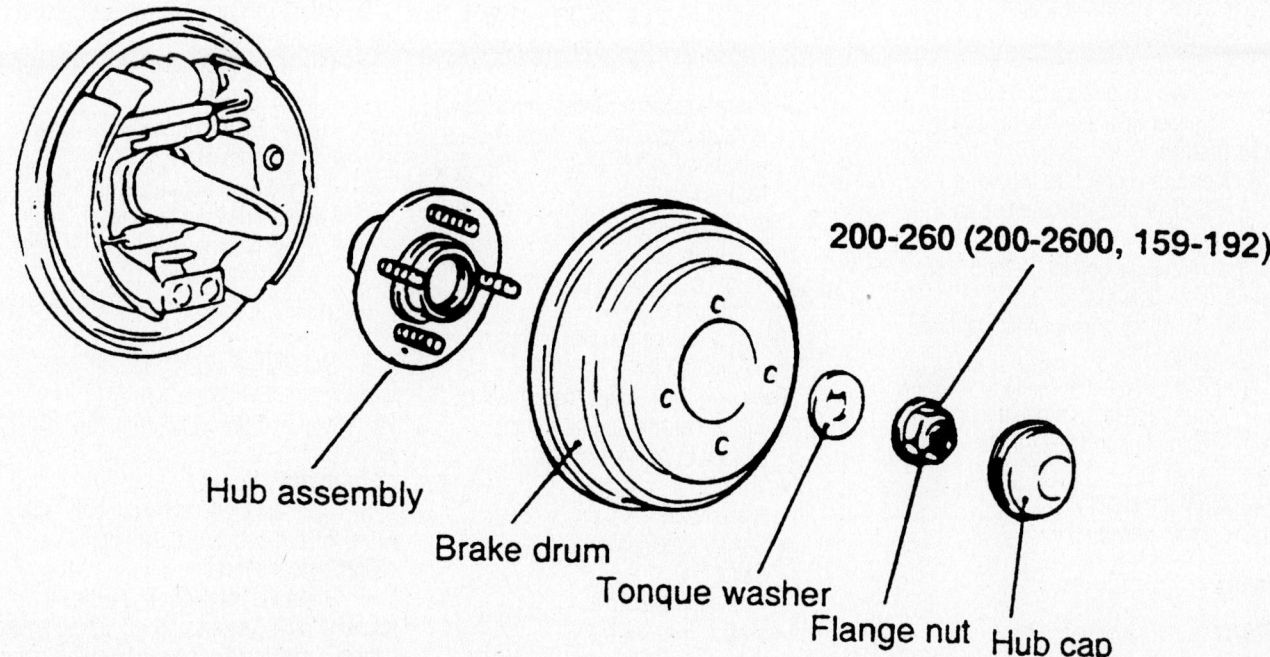

Exploded view of the rear hub assembly—with drum brakes

To install:

4. Install the inner and outer bearing races.

5. Apply grease to the bearings and to the cavity in the hub.

6. Install or connect the following:
- Inner bearing
- Inner grease seal
- Brake drum
- Outer bearing
- Flange nut and torque the nut to 159–192 ft. lbs. (200–260 Nm)
- Grease cap
- Wheel speed sensor, if equipped
- Rear wheel

DISC BRAKES

1. Before servicing the vehicle, refer to the Precautions Section.
2. Release the parking brake.
3. Remove or disconnect the following:
- Rear wheel

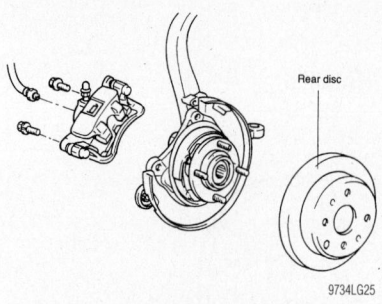

9734LG25

Exploded view of the rear wheel bearing assembly—with disc brakes

- Wheel speed sensor, if equipped
- Brake caliper and rotor
- Rear axle hub bolts
- Tone wheel with Tool 09445-21000
- Carrier assembly
- Nut after unstaking it
4. Press out the rear axle hub.

5. Remove the bearing inner race with Tool 09445-21000.

6. Remove the bushings from the carrier with Tools 09453-33000B and 09545-21100.

To install:

7. Press in the bushings to the carrier with Tools 09453-33000B and 09545-21100.

8. Press in the bearing to the hub with Tool 09221-21000.

9. Tighten the flange nut to meet the concave portion of the spindle.

10. Press in the tone wheel with Tool 09221-21000. Torque the nut to 191 ft. lbs. (260 Nm).

11. Install or connect the following:
- Hub and bearing assembly to the backing plate and torque the bolts to 88 ft. lbs. (120 Nm)
- Brake caliper and rotor
- Wheel speed sensor, if equipped
- Rear wheel

BRAKES

Brake Caliper

REMOVAL & INSTALLATION

Accent

1. Before servicing the vehicle, refer to the Precautions Section.
2. Remove or disconnect the following:
- Front wheels
- Brake line at the caliper
- Brake pads
- Pin and sleeve boots
- Lower caliper bolt and raise the caliper up and out to remove it

To install:

3. Install or connect the following:
- Caliper onto its mounting and install the lower mounting bolt. Torque the bolt to 16–24 ft. lbs. (22–32 Nm).
- Pin boots, sleeve boots and brake pads
- Brake line to the caliper with 2 new metal gaskets. Torque the brake line union bolt to 18–22 ft. lbs. (25–30 Nm).
4. Bleed the system.
- Front wheels

Elantra

FRONT

1. Before servicing the vehicle, refer to the Precautions Section.

2. Remove or disconnect the following:
- Front wheels
- Brake line at the caliper
- Brake pads
- Pin and sleeve boots
- Lower caliper bolt and raise the caliper up and out to remove it

To install:

3. Install or connect the following:
- Caliper onto its mounting
- Lower mounting bolt. Torque the bolt to 16–24 ft. lbs. (22–32 Nm).
- Pin boots, sleeve boots and brake pads
- Brake line to the caliper with 2 new metal gaskets. Torque the brake line union bolt to 18–22 ft. lbs. (24–30 Nm).
4. Bleed the system.
- Front wheels

REAR

1. Before servicing the vehicle, refer to the Precautions Section.

2. Remove or disconnect the following:
- Center console and loosen the parking brake adjustment
- Wheels
- Brake hose
- Caliper assembly mounting bolts
- Caliper
- Parking brake cable

To install:

3. Install or connect the following:
- Parking brake cable

- Caliper. Tighten the mounting bolts to 16–23 ft. lbs. (22–32 Nm).
- Brake hose
- Master cylinder with clean fluid and bleed the hydraulic system
- Wheel. Adjust the parking brake.
- Center console

Sonata and XG 350

FRONT

1. Before servicing the vehicle, refer to the Precautions Section.

2. Remove or disconnect the following:
- Front wheels
- Brake tube from the brake hose
- Brake hose clip
- Brake hose from the strut
- Brake line from the caliper
- Small retaining pin from the lower part of the caliper
3. Swing the caliper up until it clears the rotor and pads.
4. Slide the caliper inboard until the locating pin disengages from its groove in the caliper. Pull the caliper from the locating pin.

To install:

5. Lubricate the locating pin bore with white silicone compound and mount the caliper onto the locating pin.

6. Lower the caliper until the small retaining pin holes are aligned. Install a new retaining pin into the lower part of the caliper. Tighten the pin.

7. Install or connect the following:
- Brake line to the caliper and bleed the brakes
- Brake hose to the strut
- Brake hose clip
- Brake tube to the brake hose
- Front wheels

REAR

1. Before servicing the vehicle, refer to the Precautions Section.
2. Remove or disconnect the following:
- Center console and loosen the parking brake adjustment
- Wheels
- Brake hose
- Caliper assembly mounting bolts
- Caliper
- Parking brake cable

To install:
3. Install or connect the following:
- Parking brake cable
- Caliper. Tighten the mounting bolts to 16–23 ft. lbs. (22–32 Nm).
- Brake hose
- Master cylinder with clean fluid and bleed the hydraulic system
- Wheel. Adjust the parking brake.
- Center console

Tiburon

FRONT

1. Before servicing the vehicle, refer to the Precautions Section.
2. Remove or disconnect the following:
- Wheels
- Brake hose
- Caliper
- Caliper support

To install:
3. Install or connect the following:
- Caliper support and tighten the bolts to 44–63 ft. lbs. (69–85 Nm)
- Caliper. Tighten the guide rod bolts to 16–24 ft. lbs. (22–32 Nm).
- Brake hose and tighten to 18–22 ft. lbs. (25–30 Nm)
- Wheels
4. Fill the master cylinder with clean brake fluid and bleed the hydraulic system.

REAR

1. Before servicing the vehicle, refer to the Precautions Section.
2. Remove or disconnect the following:
- Wheels
- Brake hose
- Caliper
- Parking brake cable

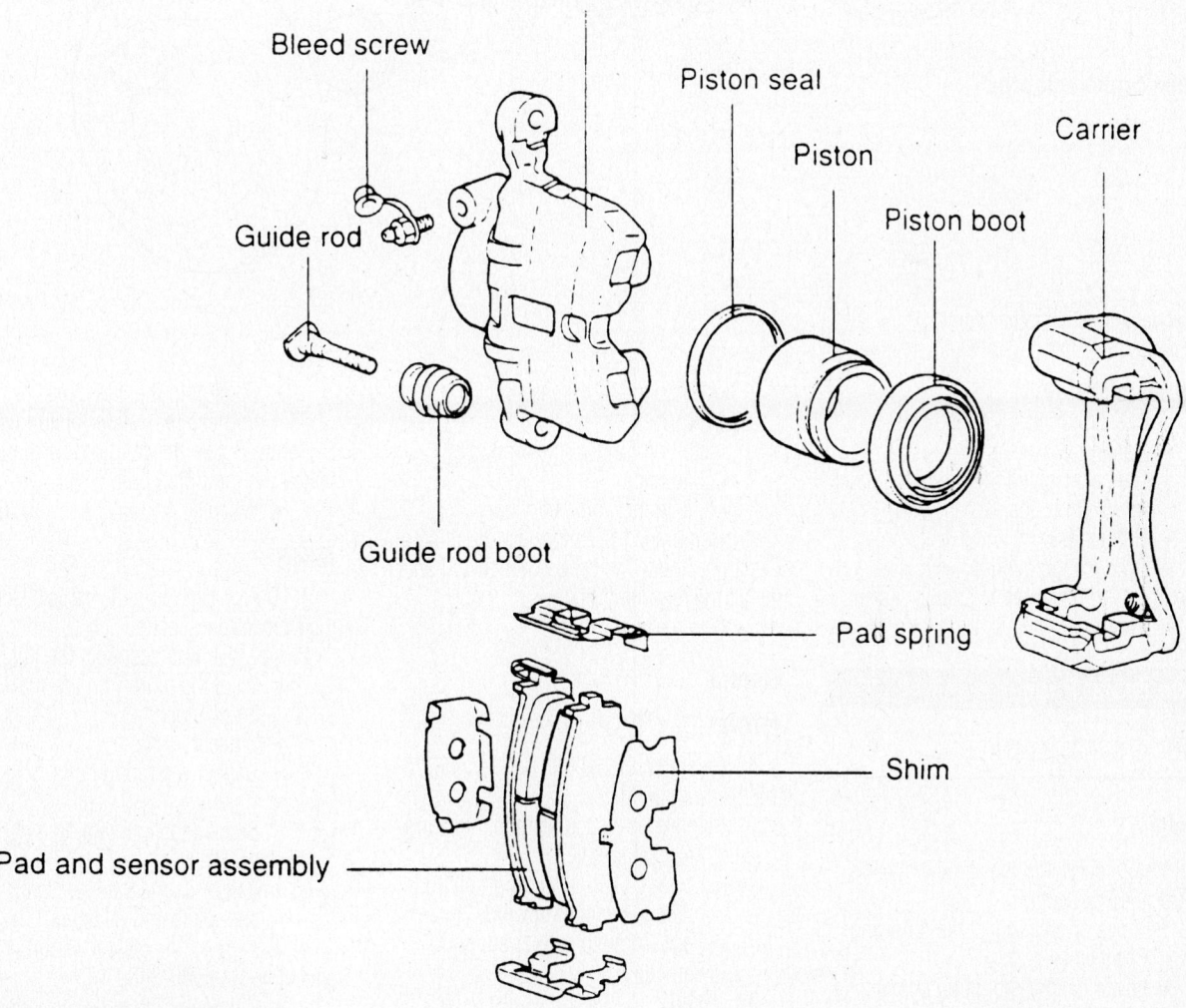

Cylinder assembly

Bleed screw

Piston seal

Piston

Carrier

Guide rod

Piston boot

Guide rod boot

Pad spring

Shim

Pad and sensor assembly

93016G10

Front caliper—Sonata shown

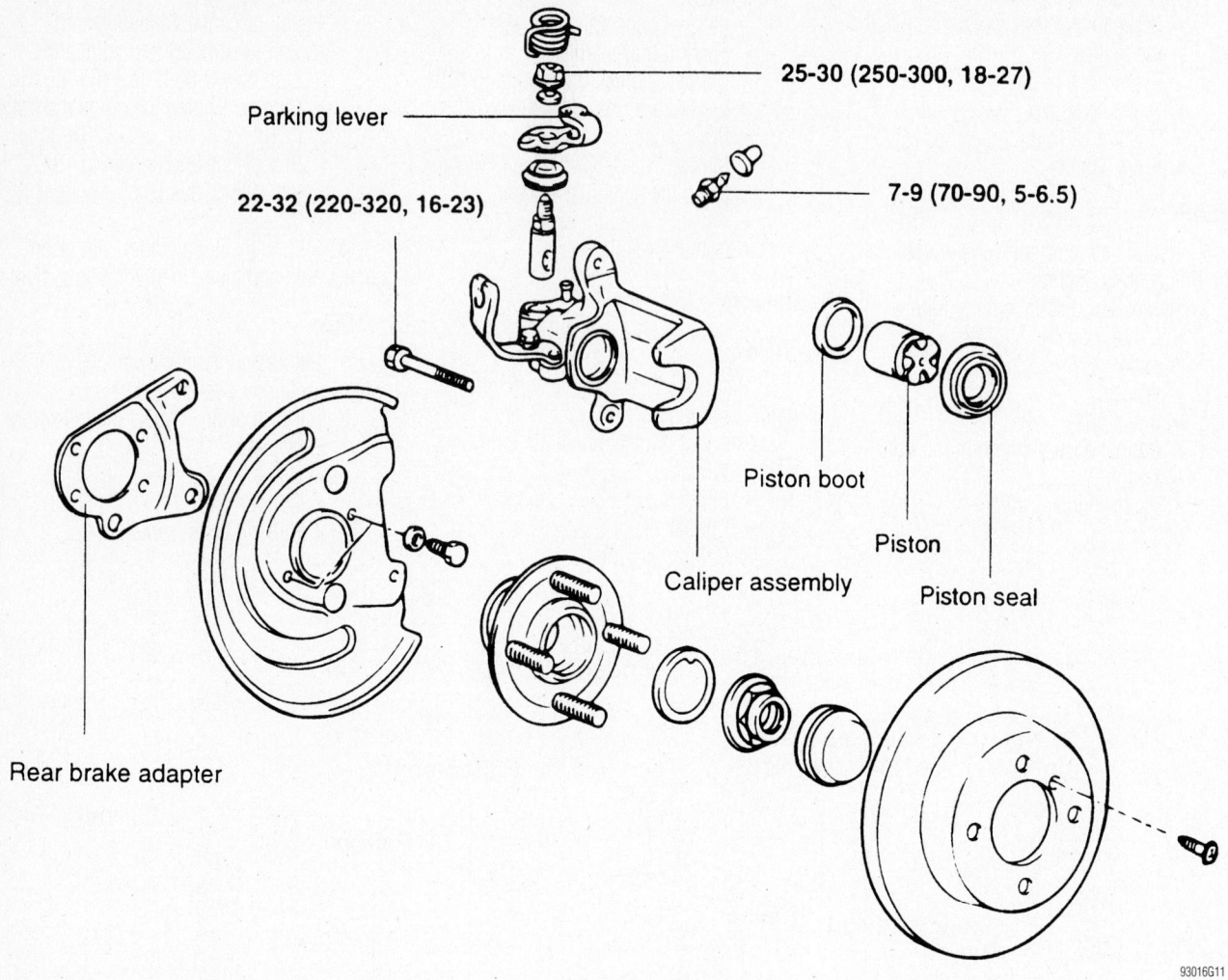

25-30 (250-300, 18-27)

Parking lever

22-32 (220-320, 16-23)

7-9 (70-90, 5-6.5)

Piston boot

Piston

Caliper assembly

Piston seal

Rear brake adapter

93016G11

Rear caliper—Elantra shown

To install:
3. Install or connect the following:
- Parking brake cable
- Caliper. Tighten the mounting bolts to 16–24 ft. lbs. (22–32 Nm).
- Brake hose
4. Fill the master cylinder with clean brake fluid and bleed the hydraulic system.
- Wheels

Disc Brake Pads

REMOVAL & INSTALLATION

Accent

1. Before servicing the vehicle, refer to the Precautions Section.
2. Remove or disconnect the following:
- Front wheels
- Lower caliper mounting bolt and rotate the caliper upward
- Pads from the caliper support

- Pad clips, if necessary

To install:
3. Install or connect the following:
- Pad clips
- Pads onto the pad clips
4. Compress the caliper piston using a C-clamp. Rotate the caliper downward and install the mounting bolt.
- Wheels

Elantra

FRONT

1. Before servicing the vehicle, refer to the Precautions Section.
2. Remove or disconnect the following:
- Front wheels
- Lower caliper mounting bolt and rotate the caliper upward
- Pads from the caliper support
- Pad clips

To install:
3. Install or connect the following:
- Pad clips

- Pads onto the pad clips
4. Compress the caliper piston using a C-clamp. Rotate the caliper downward and install the mounting bolt.
- Wheels

REAR

1. Before servicing the vehicle, refer to the Precautions Section.
2. Remove or disconnect the following:
- ½ of the fluid from the brake master cylinder
- Wheels
- Caliper mounting bolts
- Caliper
- Brake pads and retaining clips

To install:
3. Install or connect the following:
- New pads and retainers
4. Compress the caliper piston using special tool 09580-3400.
- Caliper. Tighten the mounting bolts to 16–24 ft. lbs. (22–32 Nm).

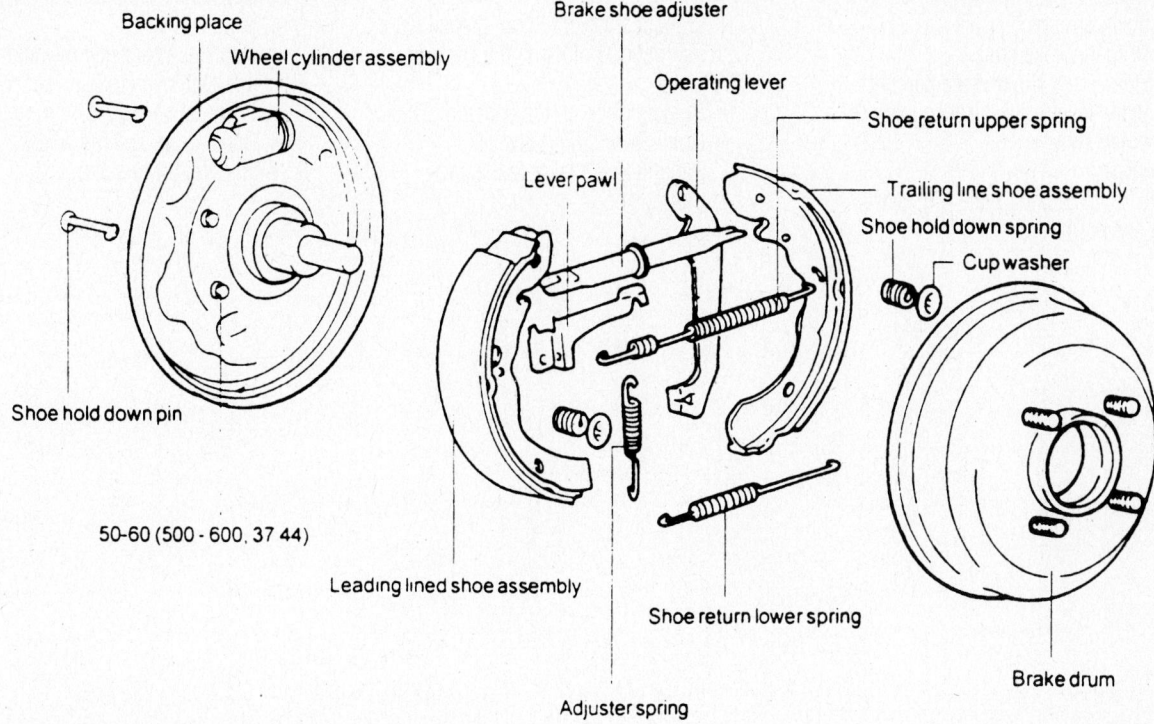

Backing place
Wheel cylinder assembly
Brake shoe adjuster
Operating lever
Shoe return upper spring
Lever pawl
Trailing line shoe assembly
Shoe hold down spring
Cup washer
Shoe hold down pin
50-60 (500 - 600, 37 44)
Leading lined shoe assembly
Shoe return lower spring
Adjuster spring
Brake drum
93016G12

Rear drum brakes—Accent shown

- Master cylinder with clean brake fluid and bleed the hydraulic system
- Wheels

Sonata, Tiburon and XG 350

FRONT

1. Before servicing the vehicle, refer to the Precautions Section.
2. Remove or disconnect the following:
   - ½ of the fluid from the brake master cylinder
   - Front wheels
   - Small retaining pin from the lower part of the caliper
3. Swing the caliper up until it clears the rotor and pads.
   - Pads and anti-squeal spring from the caliper support

To install:
4. Install or connect the following:
   - Pads and anti-rattle spring
5. Compress the caliper piston using special tool 09580-3400.
   - Caliper. Tighten the mounting bolts to 16–24 ft. lbs. (22–32 Nm).
   - Master cylinder with clean brake fluid and bleed the hydraulic system
   - Wheels

REAR

1. Before servicing the vehicle, refer to the Precautions Section.
2. Remove or disconnect the following:
   - ½ of the fluid from the brake master cylinder
   - Rear wheels
   - Caliper mounting bolts and remove the caliper
   - Brake pads and retaining clips

To install:
3. Install or connect the following:
   - New pads and retainers
4. Compress the caliper piston using special tool 09580-3400.
   - Caliper. Tighten the mounting bolts to 16–24 ft. lbs. (22–32 Nm).
   - Master cylinder with clean brake fluid and bleed the hydraulic system
   - Wheels

Brake Drums

REMOVAL & INSTALLATION

1. Before servicing the vehicle, refer to the Precautions Section.
2. Remove or disconnect the following:
   - Wheels
   - Dust cap, cotter pin, nut lock,

wheel bearing nut and washer from the spindle
   - Outer wheel bearing
   - Drum with the inner wheel bearing from the spindle

To install:
3. Install or connect the following:
   - Lubricated inner wheel bearing
   - New grease seal
   - Drum to the spindle
   - Lubricated outer wheel bearing, washer and nut. Adjust the bearing preload as required.
   - Nut lock and a new cotter pin
   - Grease cap
   - Wheels. Adjust the rear brakes as required.

Brake Shoes

REMOVAL & INSTALLATION

1. Before servicing the vehicle, refer to the Precautions Section.
2. Remove or disconnect the following:
   - Wheels
   - Brake drum
   - Self-adjuster spring and the adjuster lever
   - Spread the shoes and remove the adjuster strut

- Shoe-to-shoe spring and the hold-down springs
- Primary brake shoe
- Horseshoe clip and the parking brake lever from the secondary brake shoe

To install:

3. Clean the backing plate with brake cleaning solvent.

4. Install or connect the following:
   - Light coating of lithium grease to the friction points on the backing plate
   - Primary shoe on the backing plate
   - Hold-down spring and pin
   - Parking brake lever to the secondary shoe
   - Secondary shoe to the backing plate
   - Adjuster strut assembly and the adjuster lever and spring
   - Lower shoe-to-shoe spring
   - Brake drum and the wheel

5. Adjust the brake shoes.

HYUNDAI

Santa Fe

11

SPECIFICATION AND MAINTENANCE CHARTS

ENGINE AND VEHICLE IDENTIFICATION CHART

Engine Code							Model Year	
Code	Liters (cc)	Cu. In.	Cyl.	Fuel Sys.	Engine Type	Eng. Mfg.	Code ①	Year
B	2.4 (2351)	120	4	MFI	DOHC	Hyundai	2	2002
D	2.7 (2656)	120	6	MFI	DOHC	Hyundai	3	2003
E	3.5 (3496)	120	6	MFI	DOHC	Hyundai	4	2004
DOHC: Double Overhead Cam							5	2005
MFI: Multi-port Fuel Injection							6	2006

① 8th position of VIN

09474_SAFE_C0001

GENERAL ENGINE SPECIFICATIONS

Year	Model	Engine Displacement Liters (VIN)	Net Horsepower @ rpm	Net Torque @ rpm (ft. lbs.)	Bore x Stroke (in.)	Compression Ratio	Oil Pressure @ rpm
2002	Santa Fe	2.4 (B)	149@5500	156@3000	3.41x3.94	10:01	①
		2.7 (D)	181@6000	177@4000	3.41x2.95	10:01	②
2003	Santa Fe	2.4 (B)	149@5500	156@3000	3.41x3.94	10:01	①
		2.7 (D)	181@6000	177@4000	3.41x2.95	10:01	②
2004	Santa Fe	2.4 (B)	149@5500	156@3000	3.41x3.94	10:01	①
		2.7 (D)	181@6000	177@4000	3.41x2.95	10:01	②
		3.5 (E)	195@5500	219@3500	3.66x3.77	10:01	③
2005-06	Santa Fe	2.4 (B)	149@5500	156@3000	3.41x3.94	10:01	①
		2.7 (D)	170@6000	181@4000	3.41x2.95	10:01	②
		3.5 (E)	200@5500	219@3500	3.66x3.77	10:01	③

MFI: Multi-port Fuel Injection

① 11.6 Psi (80 kPa) @ idle.

② 7.3 Psi (50 kPa) or more @ idle.

③ 11.4 Psi (80 kPa) or more @ 700 RPM with engine oil temperature at 204 degrees F (95 degrees C)

09474_SAFE_C0002

ENGINE TUNE-UP SPECIFICATIONS

Year	Engine Displacement Liters (VIN)	Spark Plug Gap (in.)	Ignition Timing (deg.)		Fuel Pump (psi)	Idle Speed (rpm)		Valve Clearance (in.)	
			MT	AT		MT	AT	In.	Ex.
2002	2.4 (B)	0.039-0.043	2-12B	2-12B	37	625-825	625-825	NA	NA
	2.7 (D)	0.040-0.043	7-19B	7-19B	37	625-825	625-825	NA	NA
2003	2.4 (B)	0.039-0.043	2-12B	2-12B	37	625-825	625-825	NA	NA
	2.7 (D)	0.040-0.043	7-19B	7-19B	37	625-825	625-825	NA	NA
2004	2.7 (D)	0.040-0.043	7-19B	7-19B	37	625-825	625-825	NA	NA
	3.5 (E)	0.039-0.043	①	①	37	②	②	NA	NA
2005-06	2.7 (D)	0.040-0.043	7-19B	7-19B	37	625-825	625-825	NA	NA
	3.5 (E)	0.039-0.043	①	①	37	②	②	NA	NA

NOTE: The Vehicle Emission Control Information label often reflects changes made during production and must be used if they differ from this chart.

NOTE: The fuel pressure readings are given with the vacuum hose connected to the regulator and the engine running

B: Before top dead center

HYD: Hydraulic

NA: Not Availible

① The timing is controlled by the Powertrain Control Module (PCM) and is not adjustable.

② 600-700 with the A/C OFF or 750-850 with the A/C ON

09474_SAFE_C0003

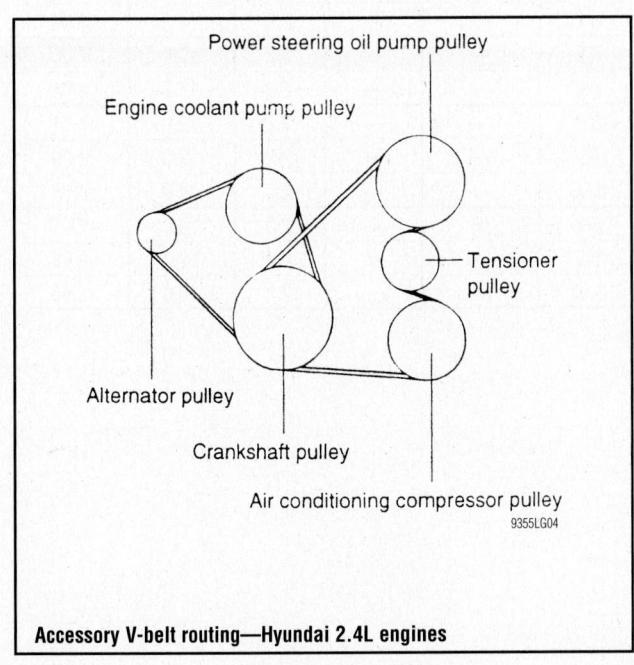

Accessory V-belt routing—Hyundai 2.4L engines

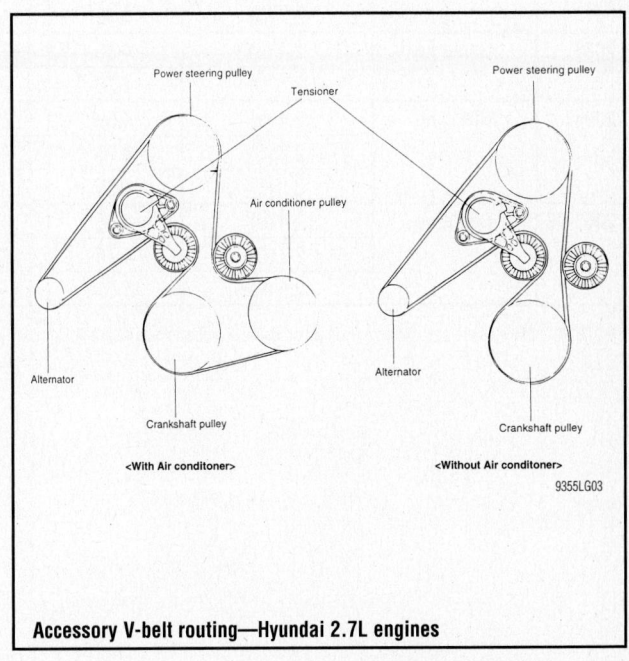

Accessory V-belt routing—Hyundai 2.7L engines

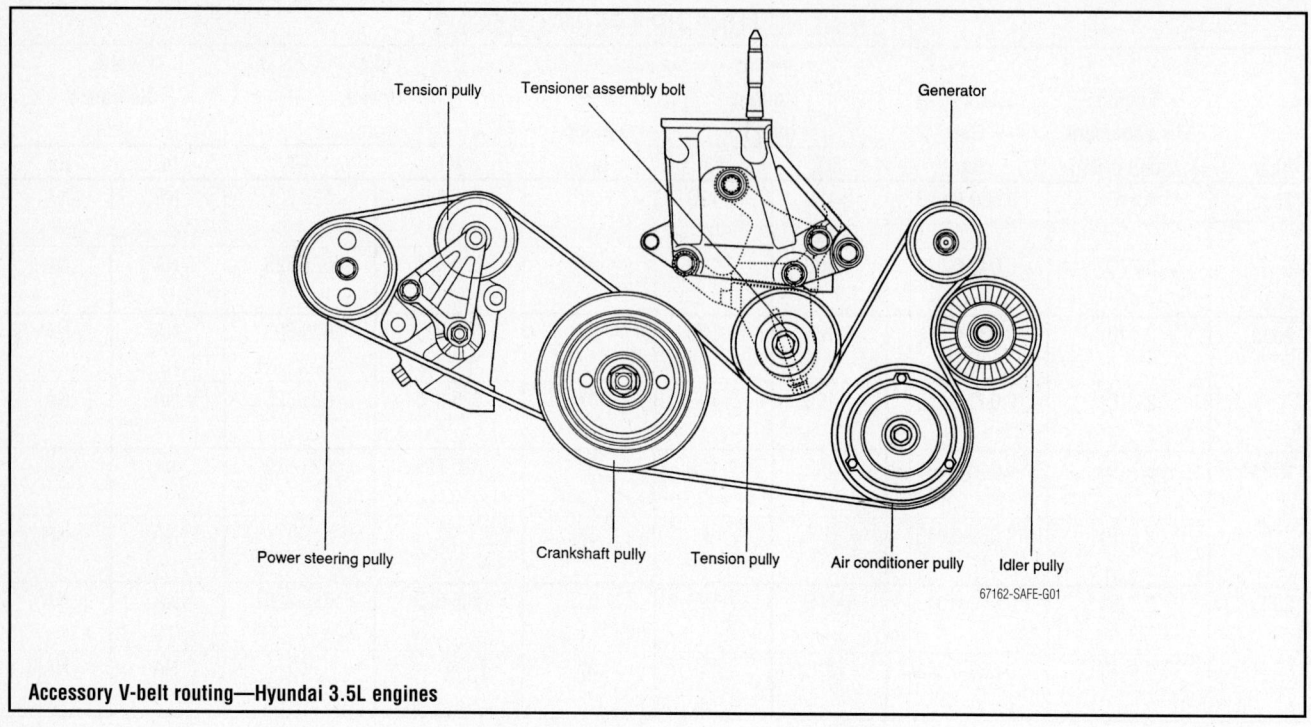

Accessory V-belt routing—Hyundai 3.5L engines

67162-SAFE-G01

CAPACITIES

Year	Model	Engine Displacement Liters (VIN)	Engine Oil with Filter (qts.)	Transmission (qts.) Man.	Transmission (qts.) Auto.	Transfer Case (pts.)	Drive Axle Front (pts.)	Drive Axle Rear (pts.)	Fuel Tank (gal.)	Cooling System (qts.)
2002	Santa Fe	2.4 (B)	4.53	2.3	8.2	—	—	2.2	17.2	7.35
		2.7 (D)	4.76	2.3	8.2	—	—	2.2	17.2	8.94
2003	Santa Fe	2.4 (B)	4.53	2.3	8.2	—	—	2.2	17.2	7.35
		2.7 (D)	4.76	2.3	8.2	—	—	2.2	17.2	8.94
2004	Santa Fe	2.4 (B)	4.53	2.3	8.2	—	—	2.2	19.0	7.71
		2.7 (D)	4.76	2.3	8.2	—	—	2.2	19.0	9.09
		3.5 (E)	4.55	—	8.98	—	—	2.2	19.0	8.66
2005-06	Santa Fe	2.4 (B)	4.55	2.3	8.2	—	—	2.2	19.0	7.71
		2.7 (D)	4.76	2.3	8.2	—	—	2.2	19.0	9.09
		3.5 (E)	4.55	—	8.98	—	—	2.2	19.0	8.66

NOTE: All capacities are approximate. Add fluid gradually and check to be sure a proper fluid level is obtained.

09474_SAFE_C0004

VALVE SPECIFICATIONS

Year	Engine Displacement Liters (VIN)	Seat Angle (deg.)	Face Angle (deg.)	Spring Test Pressure (lbs. @ in.)	Spring Free Length (in.)	Stem-to-Guide Clearance (in.)		Stem Diameter (in.)	
						Intake	Exhaust	Intake	Exhaust
2002	2.4 (B)	45	45	NA	1.804	0.0008-0.0019	0.0020-0.0030	0.2580-0.2590	0.2571-0.2579
	2.7 (D)	45	45	NA	1.670	0.008-0.0020	0.0014-0.0026	0.2350-0.2354	0.2340-0.2350
2003	2.4 (B)	45	45	NA	1.804	0.0008-0.0019	0.0020-0.0030	0.2580-0.2590	0.2571-0.2579
	2.7 (D)	45	45	NA	1.670	0.008-0.0020	0.0014-0.0026	0.2350-0.2354	0.2340-0.2350
2004	2.4 (B)	45	45	NA	1.804	0.0008-0.0019	0.0020-0.0030	0.2580-0.2590	0.2571-0.2579
	2.7 (D)	45	45	NA	1.670	0.008-0.0020	0.0014-0.0026	0.2350-0.2354	0.2340-0.2350
	3.5 (E)	43.5-44	43.5-44	NA	1.826	0.009-0.0020	0.0020-0.0033	0.2580-0.2590	0.2570-0.2580
2005-06	2.4 (B)	45	45	NA	1.804	0.0008-0.0019	0.0020-0.0030	0.2580-0.2590	0.2571-0.2579
	2.7 (D)	45	45-45.5	NA	1.673	0.008-0.0020	0.0014-0.0026	0.2350-0.2354	0.2340-0.2350
	3.5 (E)	43.5-44	45-45.5	NA	1.826	0.009-0.0020	0.0020-0.0033	0.2580-0.2590	0.2570-0.2580

NA: Not Available

09474_SAFE_C0005

CRANKSHAFT AND CONNECTING ROD SPECIFICATIONS

All measurements are given in inches

Year	Engine Displacement Liters (VIN)	Crankshaft				Connecting Rod		
		Main Brg. Journal Dia.	Main Brg. Oil Clearance	Shaft End-play	Thrust on No.	Journal Diameter	Oil Clearance	Side Clearance
2002	2.4 (B)	2.2434-2.2441	①	0.0020-0.0098	3	2.2434-2.2411	0.0007-0.0014	0.004-0.0098
	2.7 (D)	2.4402-2.4409	0.0002-0.0009	0.0024-0.0094	3	2.2434-2.2411	0.0007-0.0014	0.0039-0.0098
2003	2.4 (B)	2.2434-2.2441	①	0.0020-0.0098	3	2.2434-2.2411	0.0007-0.0014	0.004-0.0098
	2.7 (D)	2.4402-2.4409	0.0002-0.0009	0.0024-0.0094	3	2.2434-2.2411	0.0007-0.0014	0.0039-0.0098
2004	2.4 (B)	2.2434-2.2441	①	0.0020-0.0098	3	2.2434-2.2411	0.0007-0.0014	0.004-0.0098
	2.7 (D)	2.4402-2.4409	0.0002-0.0009	0.0024-0.0094	3	2.2434-2.2411	0.0007-0.0014	0.0039-0.0098
	3.5 (E)	2.5190-2.5197	0.00086-0.00157	0.0020-0.0098	3	NA	0.0001-0.0017	0.0039-0.0098
2005-06	2.4 (B)	2.2434-2.2441	①	0.0020-0.0098	3	2.2434-2.2411	0.0007-0.0014	0.004-0.0098
	2.7 (D)	2.4402-2.4409	0.0002-0.0009	0.0028-0.0098	3	2.2434-2.2411	0.0007-0.0014	0.0039-0.0098
	3.5 (E)	2.5190-2.5197	0.00086-0.00157	0.0020-0.0098	3	NA	0.0001-0.0017	0.0039-0.0098

NA: Not Available

① Nos. 1, 2, 4 and 5: 0.0007-0.0014

No. 3: 0.0009-0.0016

09474_SAFE_C0006

PISTON AND RING SPECIFICATIONS

All measurements are given in inches

Year	Engine Displacement Liters (VIN)	Piston Clearance	Ring Gap			Ring Side Clearance		
			Top Compression	Bottom Compression	Oil Control	Top Compression	Bottom Compression	Oil Control
2002	2.4 (B)	0.0008-0.0016	0.0098-0.0138	0.0157-0.0216	0.0039-0.0157	0.0012-0.0028	0.0008-0.0024	0.0024-0.0059
	2.7 (D)	0.0004-0.0012	0.0079-0.0138	0.0146-0.0205	0.0079-0.0276	0.0016-0.0031	0.0012-0.0028	NA
2003	2.4 (B)	0.0008-0.0016	0.0098-0.0138	0.0157-0.0216	0.0039-0.0157	0.0012-0.0028	0.0008-0.0024	0.0024-0.0059
	2.7 (D)	0.0004-0.0012	0.0079-0.0138	0.0146-0.0205	0.0079-0.0276	0.0016-0.0031	0.0012-0.0028	NA
2004	2.4 (B)	0.0008-0.0016	0.0098-0.0138	0.0157-0.0216	0.0039-0.0157	0.0012-0.0028	0.0008-0.0024	0.0024-0.0059
	2.7 (D)	0.0004-0.0012	0.0079-0.0138	0.0146-0.0205	0.0079-0.0276	0.0016-0.0031	0.0012-0.0028	NA
	3.5 (E)	0.0012-0.0020	0.0078-0.0118	0.0157-0.0216	0.0079-0.0276	0.0016-0.0031	0.0008-0.0024	NA
2005-06	2.4 (B)	0.0008-0.0016	0.0098-0.0138	0.0157-0.0216	0.0039-0.0157	0.0012-0.0028	0.0008-0.0024	0.0024-0.0059
	2.7 (D)	0.0004-0.0012	0.0079-0.0138	0.0146-0.0205	0.0079-0.0276	0.0016-0.0031	0.0012-0.0028	NA
	3.5 (E)	0.0012-0.0020	0.0078-0.0118	0.0157-0.0216	0.0079-0.0276	0.0016-0.0031	0.0008-0.0024	NA

NA: Not Applicable

09474_SAFE_C0007

TORQUE SPECIFICATIONS
All readings in ft. lbs.

Year	Engine Displacement Liters (VIN)	Cylinder Head Bolts	Main Bearing Bolts	Rod Bearing Bolts	Crankshaft Damper Bolts	Flywheel Bolts	Manifold Intake	Manifold Exhaust	Spark Plugs	Oil Pan Drain Plug
2002	2.4 (B)	①	②	③	58-72	94-101	④	⑤	15-22	25-33
	2.7 (D)	⑥	⑦	⑧	65-80	53-56	14-15	18-22	15-22	25-33
2003	2.4 (B)	①	②	③	58-72	94-101	④	⑤	15-22	25-33
	2.7 (D)	⑥	⑦	⑧	65-80	53-56	14-15	18-22	15-22	25-33
2004	2.4 (B)	①	②	③	58-72	94-101	④	⑤	15-22	25-33
	2.7 (D)	⑥	⑦	⑧	65-80	53-56	14-15	18-22	15-22	25-33
	3.5 (E)	75-82	52-59	⑨	130-138	NA	14-16	29-33	15-22	25-33
2005-06	2.4 (B)	①	②	③	58-72	94-101	④	⑤	15-22	25-33
	2.7 (D)	⑥	⑦	⑧	65-80	53-56	14-15	22-26	15-22	25-33
	3.5 (E)	75-82	52-59	⑨	130-138	NA	14-16	29-33	15-22	26-32

NOTE: Dip main bearing bolts and crankshaft damper bolt in clean engine oil prior to tightening.

NA: Not Available

① If using used parts:
  Step 1: 14 ft. lbs. (20 Nm).
  Step 2: plus an additional 90 degrees.
  Step 3: plus an additional 90 degrees.
  If using new parts:
  Step 1: 46 ft. lbs. (64 Nm)
  Step 2: Release the bolts.
  Step 3: 14 ft. lbs. (20 Nm)
  Step 4: plus an additional 90 degrees.
  Step 5: plus an additional 90 degrees.

② 15 ft. lbs. Plus 90 degrees

③ 14 ft. lbs. Plus 90 degrees

④ Bolt (M8): 11-14 ft. lbs.
  Nut: 22-30 ft. lbs.

⑤ Bolt (M8): 18-2 ft. lbs.
  Bolt (M10): 25-40

⑥ Step 1: 18 ft. lbs.
  Step 2: plus an additional 58<en dash>62 degrees.
  Step 3: plus an additional 43<en dash>47 degrees.

⑦ Bolt (M10): 19.5-24 ft. lbs. Plus 90 degrees
  Bolt (M7): 10-14 plus 90 degrees

⑧ 12-15 ft. lbs. Plus 90-94 degrees

⑨ 26 ft. lbs. Plus 90 degrees

09474_SAFE_C0008

WHEEL ALIGNMENT

Year	Model		Caster Range (+/-Deg.)	Caster Preferred Setting (Deg.)	Camber Range (+/-Deg.)	Camber Preferred Setting (Deg.)	Toe-in (in.)
2002	Santa Fe	F	2 + or - 30'	2.00	0 + or - 30'	0	0.008
		R	—	—	0 + or - 30'	0	0.008
2003	Santa Fe	F	2 + or - 30'	2.00	0 + or - 30'	0	0.008
		R	—	—	0 + or - 30'	0	0.008
2004	Santa Fe	F	2 + or - 30'	2.00	0 + or - 30'	0	0.008
		R	—	—	0 + or - 30'	0	0.008
2005-06	Santa Fe	F	2 + or - 30'	2.00	0 + or - 30'	0	0.008
		R	—	—	0 + or - 30'	0	0.008

F: Front

R: Rear

09474_SAFE_C0009

TIRE, WHEEL AND BALL JOINT SPECIFICATIONS

| Year | Model | OEM Tires | | Tire Pressures (psi) | | Wheel | Ball Joint | Lug |
		Standard	Optional	Front	Rear	Size	Inspection	Nut
2002	Santa Fe	P225/70R16	None	30	30	6.5Jx16	NS	66-81
2003	Santa Fe	P225/70R16	None	30	30	6.5Jx16	NS	66-81
2004	Santa Fe	P225/70R16	None	30	30	6.5Jx16	NS	66-81
2005-06	Santa Fe	P225/70R16	None	30	30	6.5Jx16	NS	66-81

OEM: Original Equipment Manufacturer

PSI: Pounds Per Square Inch

NS: Not specified by manufacturer

09474_SAFE_C0010

BRAKE SPECIFICATIONS
Hyundai Santa Fe
All measurements in inches unless noted

| Year | Model | | Brake Disc | | | Brake Drum Diameter | | | Minimum Lining Thickness | | Brake Caliper | |
			Original Thickness	Minimum Thickness	Maximum Runout	Original Inside Diameter	Max. Wear Limit	Maximum Machine Diameter	Pad	Shoe	Bracket Bolts (ft. lbs.)	Mounting Bolts (ft. lbs.)
2002	Santa Fe	F	1.0200	0.960	0.002	—	—	—	0.079	—	58-73	16-24
		R	0.390	0.330	0.040	10.00	10.08	—	0.080	0.590	37-44	16-24
2003	Santa Fe	F	1.0200	0.960	0.002	—	—	—	0.079	—	58-73	16-24
		R	0.390	0.330	0.040	10.00	10.08	—	0.080	0.590	37-44	16-24
2004	Santa Fe	F	1.0200	0.960	0.002	—	—	—	0.079	—	58-73	16-24
		R	0.390	0.330	0.040	10.00	10.08	—	0.080	0.590	37-44	16-24
2005-06	Santa Fe	F	1.0200	0.960	0.002	—	—	—	0.079	—	59-73	16-24
		R	0.390	0.330	0.040	10.00	10.08	—	0.080	0.590	37-44	16-24

F: Front

R: Rear

09474_SAFE_C0011

ENGINE REPAIR

➡️**Disconnecting the negative battery cable on some vehicles may interfere with the functions of the on board computer system. The computer may undergo a relearning process once the negative battery cable is reconnected.**

Distributor

The Santa Fe is equipped with a Distributorless Ignition System (DIS).

Alternator

REMOVAL

2.4L Engine

1. Before servicing the vehicle, refer to the precautions section.
2. Remove or disconnect the following:
   • Negative battery cable
   • Drive belt
   • Alternator electrical connections
   • Alternator mounting bolts
   • Alternator

2.7L And 3.5L Engine

1. Before servicing the vehicle, refer to the precautions section.
2. Remove or disconnect the following:
   • Negative battery cable
   • Drive belt
   • Alternator electrical connections
   • Alternator mounting bolts
   • Alternator

INSTALLATION

2.4L Engine

1. Install or connect the following:
   • Alternator
   • Alternator mounting bolts. Tighten the adjuster (upper) bolt to 15–18 ft. lbs. (20–25 Nm) and the lower bolt and nut to 26–41 ft. lbs. (34–54 Nm).
   • Alternator electrical connections
   • Drive belt
   • Negative battery cable

2.7L And 3.5L Engine

1. Install or connect the following:
   • Alternator
   • Alternator mounting bolts. Tighten the adjuster (upper) bolt to 9–11 ft. lbs. (12–15 Nm) and the lower bolt and nut to 15–18 ft. lbs. (20–25 Nm).
   • Alternator electrical connections
   • Drive belt
   • Negative battery cable

Ignition Timing

ADJUSTMENT

The Santa Fe is equipped with a Distributorless Ignition System (DIS). The ignition timing is controlled by the Powertrain Control module (PCM). No adjustment is necessary.

Engine Assembly

REMOVAL & INSTALLATION

Except 3.5L Engine

1. Before servicing the vehicle, refer to the precautions section.
2. Remove the battery and air cleaner assembly.
3. Drain the cooling system.
4. Drain the engine oil.
5. Drain the transaxle fluid.
6. Relieve the fuel system pressure.
7. Disconnect the following electrical connections:
   • Starter
   • Alternator
   • Throttle Position Sensor (TPS)
   • Power steering switch connector
   • Oil pressure gauge connector
   • Back-up lamp switch connector
   • A/T solenoid inhibitor switch connector
   • Coolant Temperature Sensor (CTS)
   • Ignition coil
   • Idle Speed Control (ISC) valve connector
   • Manifold Absolute Pressure (MAP) sensor
   • Oxygen (O_2S) sensor connector
8. If equipped with an automatic transmission, disconnect the oil cooler lines.
9. Remove or disconnect the following:
   • Radiator hoses from the engine
   • Radiator

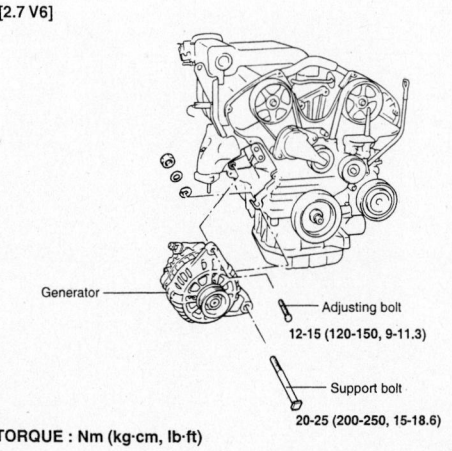

[2.4 I4]

Adjusting bolt
20-25 (200-250, 15-18.6)

34-54 (340-540, 26-41) — Generator

Support bolt

[2.7 V6]

Generator

Adjusting bolt
12-15 (120-150, 9-11.3)

Support bolt
20-25 (200-250, 15-18.6)

TORQUE : Nm (kg·cm, lb·ft)

9355LG01

Exploded view of the alternator mounting for both engines used in the Santa Fe

- Engine ground
- Brake vacuum hose
- Heater hoses at the engine
- Throttle cable at the engine
- Cruise control cable at the engine, if equipped
- Main fuel line at the supply/return pipe
- Speedometer cable at the transaxle
- Clutch or control cable from the transaxle
- Power steering hoses from the pump
- Steering dust cover in the engine compartment
- Gear box universal joint bolt
- Front wheel
- Brake caliper and support with wire
- Strut lower bolt
- Front muffler bolts
- Transaxle control rod and extension rod, if equipped with a manual transmission

10. Support the transmission with a jack using the special attachment shown in the accompanying illustration.

11. Make sure all cable, harness connectors and hoses are disconnected from the engine and transmission.

- Engine and transaxle mounting brackets
- Sub frame bolts
- Drive shaft

12. Lower the engine and transaxle assembly enough so the front and rear roll stoppers can be removed.

13. Remove the engine assembly

14. Installation is the reverse of removal but please note the following steps:

- Tighten the roll stopper bolts to 36–47 ft. lbs. (50–65 Nm)
- Tighten the transaxle mounting bracket bolts to 65–80 ft. lbs. (90–110 Nm)

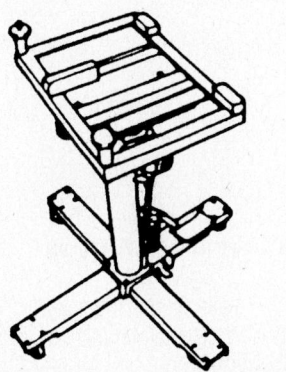

9355LG02

Attach the special tool to the transmission jack and support the transmission

- Tighten the engine mount bracket bolts to 43–58 ft. lbs. (60–80 Nm)

15. Fill the engine crankcase to the correct level.

16. Fill the transaxle to the correct level.

17. Fill the cooling system.

18. Fill the power steering system.

19. Start the engine and check for leaks.

20. Check the wheel alignment and adjust as necessary.

3.5L Engine

1. Before servicing the vehicle, refer to the precautions section.

2. Remove the battery and engine cover assembly.

3. Remove the battery stay.

4. Remove the air cleaner.

5. Drain the cooling system.

6. Drain the engine oil.

7. Drain the transaxle fluid.

8. Relieve the fuel system pressure.

9. Disconnect the following electrical connections:

- Alternator
- Starter
- Power steering switch connector
- Oil pressure gauge connector
- A/C switch
- Fuel injector connectors
- Back-up lamp switch connector
- A/T solenoid inhibitor switch connector
- Ignition coils
- Power TR selector connector
- Idle Speed Control (ISC) valve connector
- AFS and ATS connectors
- Oxygen (O$_2$S) sensor connector

10. Remove any remaining electrical connections that would interfere with engine removal.

11. If equipped with an automatic transmission, disconnect the oil cooler lines.

12. Disconnect the radiator hoses from the engine.

13. Disconnect the engine and transaxle grounds.

14. Disconnect the brake booster vacuum hose.

15. Disconnect the heater hoses from the engine.

16. Disconnect the fuel delivery and return lines.

17. Disconnect the speedometer cable from the transaxle.

18. Disconnect the control cable from the transaxle.

19. Disconnect the power steering hose from the engine mount bracket.

20. Disconnect the steering dust cover in the engine compartment.

21. Disconnect the gear box universal joint bolt. Mark the locations prior to removal to aid in installation.

22. Remove the front wheel.

23. Remove the brake caliper and support with wire.

24. Remove the strut lower bolt and disconnect the strut from the knuckle.

25. Remove the wheel speed sensor from the knuckle.

26. Remove the front muffler bolts.

27. Support the transmission with a jack using the special attachment shown in the accompanying illustration.

28. Make sure all cable, harness connectors and hoses are disconnected from the engine and transmission.

29. Remove the engine and transaxle mounting brackets.

30. Remove the sub frame bolts.

31. Lower the transaxle side down, then lift the engine and transaxle assembly from the vehicle.

32. Installation is the reverse of removal but please note the following steps:

a. Tighten the front roll stopper–to–transaxle bolts to 43–58 ft. lbs. (60–80 Nm).

b. Tighten the front roll stopper insulator bolt and nut to 36–47 ft. lbs. (50–65 Nm).

c. Tighten the front roll stopper–to–subframe bolts to 43–58 ft. lbs. (60–80 Nm).

d. Tighten the rear roll stopper–to–subframe bolts to 43–58 ft. lbs. (60–80 Nm).

e. Tighten the rear roll stopper–to–transaxle bolt and nut to 36–47 ft. lbs. (50–65 Nm).

f. Tighten the rear roll stopper insulator bolt and nut to 36–47 ft. lbs. (50–65 Nm).

g. Tighten the transaxle mounting sub–bracket bolts to 43–58 ft. lbs. (60–80 Nm).

h. Tighten the transaxle mounting bracket bolts to 43–58 ft. lbs. (60–80 Nm).

i. Tighten the transaxle mounting insulator bolt to 65–80 ft. lbs. (90–110 Nm).

j. Tighten the engine mount bracket bolts to 43–58 ft. lbs. (60–80 Nm).

k. Tighten the engine mount insulator bolt to 65–80 ft. lbs. (90–110 Nm).

33. Fill the engine crankcase to the correct level.

34. Fill the transaxle to the correct level.

35. Fill the cooling system.

36. Fill the power steering system.
37. Start the engine and check for leaks.
38. Check the wheel alignment and adjust as necessary.

Water Pump

REMOVAL & INSTALLATION

2.4L Engine

1. Before servicing the vehicle, refer to the precautions section.
2. Drain the cooling system.
3. Remove or disconnect the following:
   - Negative battery cable
   - Water pump inlet pipe
   - Drive belt and water pump pulley
   - Timing belt covers
   - Timing belt tensioner
   - Water pump bolts
   - Alternator brace
   - Water pump and gasket
4. Clean the gasket mating surfaces.

To install:

5. Install or connect the following:
   - New O-ring onto the groove on the front end of the coolant pipe and wet the O-ring with water
   - Water pump and gasket
   - Bolts and alternator bracket. Tighten the bolts to 14–20 ft. lbs. (20–27 Nm).
   - Timing belt tensioner
   - Timing belt covers
   - Water pump pulley and drive belt
   - Water pump inlet pipe
   - Negative battery cable
6. Fill the cooling system.
7. Start the engine and check for leaks.

2.7L Engine

1. Before servicing the vehicle, refer to the precautions section.
2. Drain the cooling system.
3. Remove or disconnect the following:

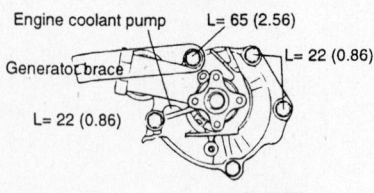

[DOHC ENGINE]

Engine coolant pump   L= 65 (2.56)

Generator brace   L= 22 (0.86)

L= 22 (0.86)

L= 22 (0.86)

L=Length of bolt mm (in.)

9355LG05

Make sure the water pump bolts are positioned in their original positions as the bolts are different lengths—2.4L engine

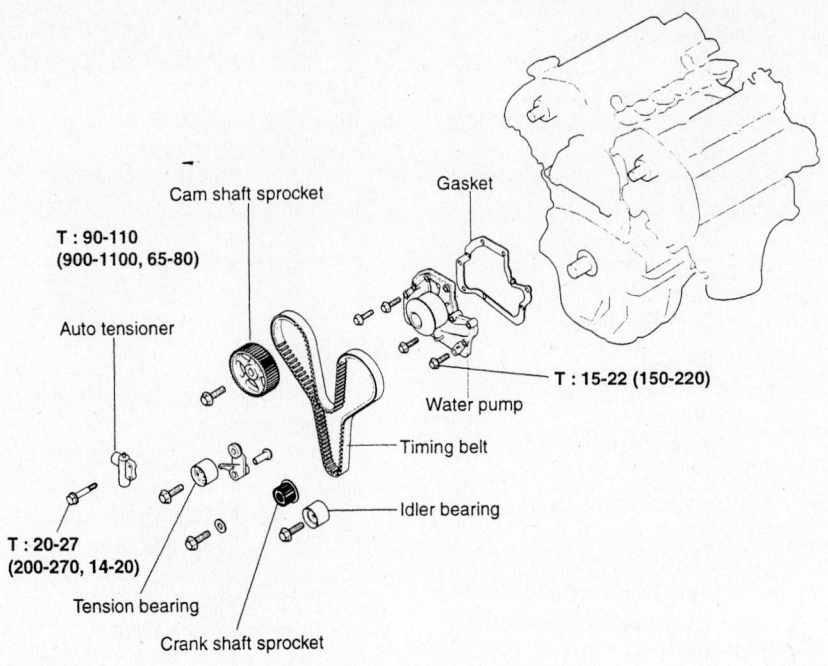

Cam shaft sprocket   Gasket

T : 90-110
(900-1100, 65-80)

Auto tensioner

T : 15-22 (150-220)

Water pump

Timing belt

Idler bearing

T : 20-27
(200-270, 14-20)

Tension bearing

Crank shaft sprocket

TORQUE : Nm (kg·cm, lb·ft)

9355LG06

Exploded view of the water pump mounting and related components—2.7L engine

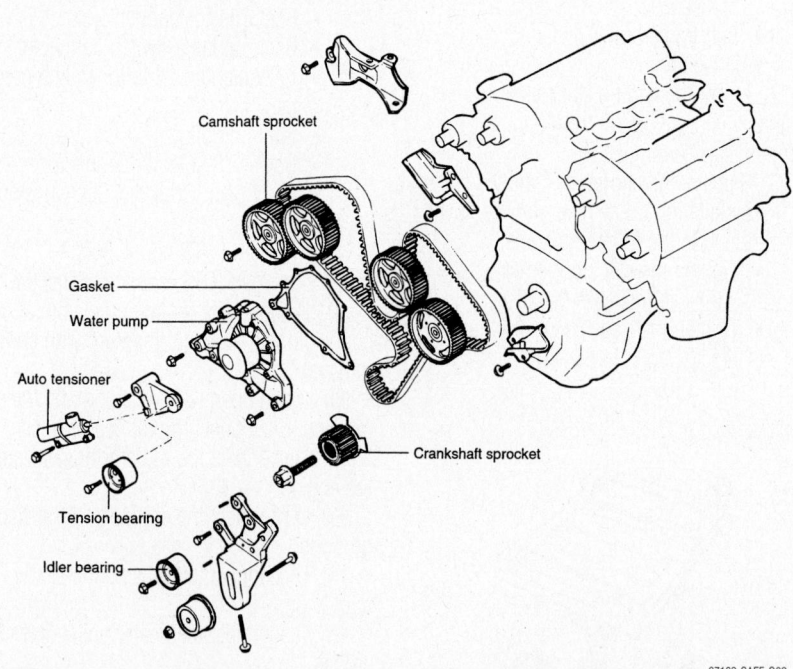

Camshaft sprocket

Gasket

Water pump

Auto tensioner

Crankshaft sprocket

Tension bearing

Idler bearing

67162-SAFE-G02

Exploded view of the water pump mounting and related components—3.5L engine

   - Negative battery cable
   - Drive belt and water pump pulley
   - Timing belt covers
   - Timing belt tensioner
   - Idler pulley
   - Water pump bolts
   - Water pump and gasket
4. Clean the gasket mating surfaces.

To install:

5. Install or connect the following:
   - Water pump and gasket
   - Bolts and alternator bracket.

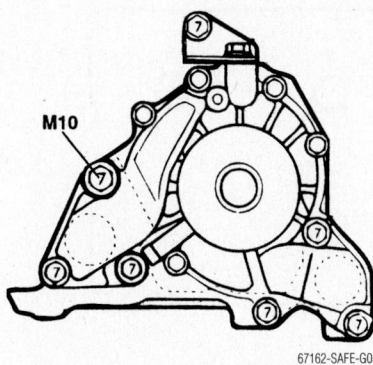

Location of the water pump mounting bolts—3.5L engine

Tighten the bolts to 11–16 ft. lbs. (15–22 Nm).
- Idler pulley
- Timing belt tensioner
- Timing belt covers
- Water pump pulley and drive belt
- Negative battery cable

6. Fill the cooling system.
7. Start the engine and check for leaks.

3.5L Engine

1. Before servicing the vehicle, refer to the precautions section.
2. Drain the cooling system.
3. Disconnect the negative battery cable.
4. Remove the drive belt.
5. Remove the timing belt covers.
6. Remove the timing belt tensioner.
7. Remove the idler pulley.
8. Remove the water pump bolts.
9. Remove the water pump and gasket.
10. Clean the gasket mating surfaces.

To install:

11. Install the water pump and gasket.
12. Install the bolts. Tighten the M8 bolts to 11–16 ft. lbs. (15–22 Nm) and the M10 bolt to 24–36 ft. lbs. (33–50 Nm).
13. Install the idler pulley.
14. Install the timing belt tensioner.
15. Install the timing belt covers.
16. Install the drive belt.
17. Connect the negative battery cable.
18. Fill the cooling system.
19. Start the engine and check for leaks.

Cylinder Head

REMOVAL & INSTALLATION

2.4L Engine

1. Before servicing the vehicle, refer to the precautions section.

2. Drain the cooling system.
3. Relieve the fuel system pressure.
4. Remove or disconnect the following:
- Negative battery cable
- All necessary electrical connections, hoses and cables
- Air cleaner
- Intake manifold
- Ignition coil
- Timing belt
- Exhaust manifold
- Rocker cover
- Camshafts

5. Loosen the cylinder head bolts in the sequence illustrated
6. Remove the cylinder head and gasket.

To install:

7. Clean the gasket mating surfaces
8. Install the cylinder head gasket so the surface with the identification mark faces towards the head.
9. Measure the head bolts. The bolt length should be 3.9 inch (99.4mm). If the bolts do not meet specification they must be replaced.
10. Install the cylinder head.
11. If using used parts (bolts, head or block), tighten the cylinder head bolts in sequence as follows:
   a. Step 1: 14 ft. lbs. (20 Nm).
   b. Step 2: plus an additional 90 degrees.
   c. Step 3: plus an additional 90 degrees.

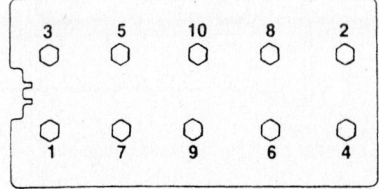

Cylinder head bolt loosening sequence—2.4L engine

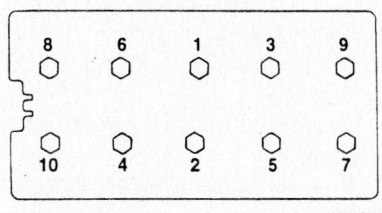

Cylinder head bolt tightening sequence—2.4L engine

12. If using new parts (even if only one thing is replaced), tighten the cylinder head bolts in sequence as follows:
   a. Step 1: 46 ft. lbs. (64 Nm).
   b. Step 2: release the bolts.
   c. Step 3: 14 ft. lbs. (20 Nm).
   d. Step 4: plus an additional 90 degrees.
   e. Step 5: plus an additional 90 degrees.
13. Install or connect the following:
- Camshafts
- Timing belt
- Rocker cover
- Ignition coil
- Exhaust manifold
- Intake manifold
- Air cleaner
- All necessary electrical connections, hoses and cables
- Negative battery cable
14. Fill the cooling system.
15. Start the engine and check for leaks.

2.7L Engine

1. Before servicing the vehicle, refer to the precautions section.
2. Drain the cooling system.
3. Relieve the fuel system pressure.
4. Remove or disconnect the following:
- Negative battery cable
- All necessary electrical connections, hoses and cables
- Air cleaner
- Intake manifold
- Ignition coil
- Timing belt
- Exhaust manifold
- Water pump pulley and rocker cover
- Camshafts
5. Loosen the cylinder head bolts in the reverse order of the tightening sequence.
6. Remove the cylinder head and gasket.

To install:

7. Clean the gasket mating surfaces
8. Install the cylinder head gasket so

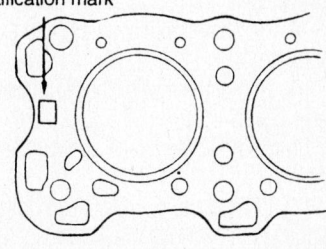

Location of the cylinder head gasket identification mark—2.7L engine

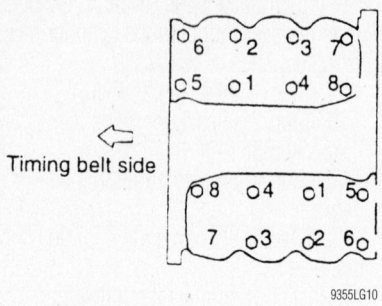

Cylinder head bolt tightening sequence—2.7L engine

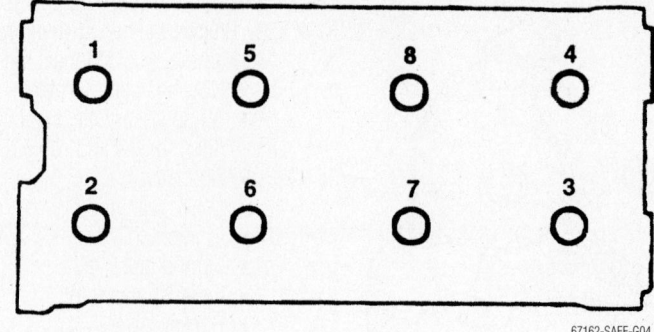

Cylinder head bolt loosening sequence—3.5L engine

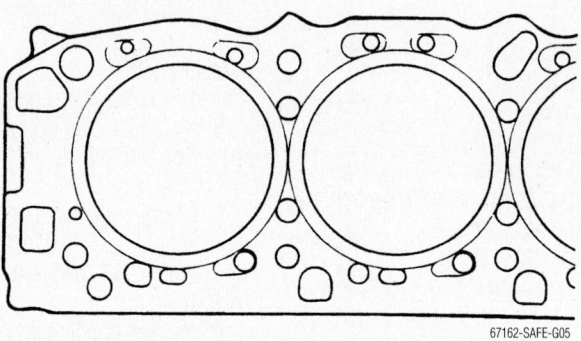

Location of the cylinder head gasket identification mark—3.5L engine

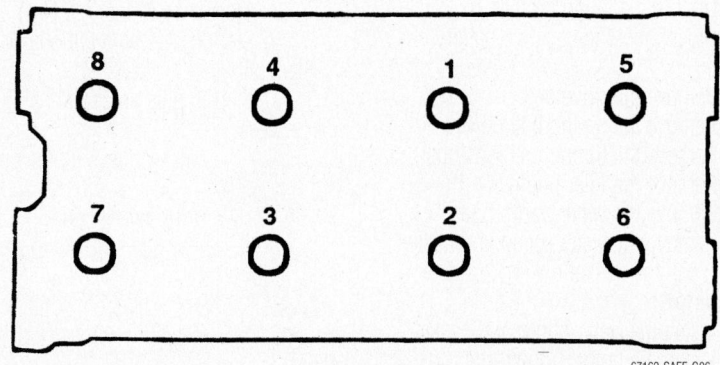

Cylinder head bolt tightening sequence—3.5L engine

the surface with the identification mark faces towards the head.

9. Install the cylinder head.

10. Tighten the cylinder head bolts in sequence as follows:

　a. Step 1: 18 ft. lbs. (25 Nm).

　b. Step 2: plus an additional 58–62 degrees.

　c. Step 3: plus an additional 43–47 degrees.

11. Install or connect the following:

- Camshafts
- Timing belt
- Rocker cover and water pump pulley
- Ignition coil
- Exhaust manifold
- Intake manifold
- Air cleaner
- All necessary electrical connections, hoses and cables
- Negative battery cable

12. Fill the cooling system.

13. Start the engine and check for leaks.

3.5L Engine

1. Before servicing the vehicle, refer to the precautions section.

2. Drain the cooling system.

3. Relieve the fuel system pressure.

4. Remove the engine from the vehicle.

5. Disconnect the spark plug wires.

6. Remove the ignition coil.

7. Remove the timing belt covers.

8. Remove the timing belt and camshaft sprockets.

9. Remove the heat shield and exhaust manifold.

10. Remove the water pump pulley and valve cover.

11. Remove the intake and exhaust camshafts.

12. Loosen the cylinder head bolts in the sequence shown using a 12mm socket in 2–3 steps.

13. Remove the cylinder head and gasket.

To install:

14. Clean the gasket mating surfaces

15. Install the cylinder head gasket so the surface with the identification mark faces towards the head.

16. Install the cylinder head.

17. Tighten the cylinder head bolts to 75–82 ft. lbs. (105–115 Nm).

18. Install the intake and exhaust camshafts.

19. Install the water pump pulley and valve cover.

20. Install the heat shield and exhaust manifold.

21. Install the timing belt and camshaft sprockets.

22. Install the timing belt covers.

23. Install the ignition coil.

24. Connect the spark plug wires.

25. Install the engine in the vehicle.

26. Fill the cooling system.

27. Change the oil and filter.

28. Start the engine and check for leaks.

Rocker Arms/Shafts

REMOVAL & INSTALLATION

Refer to the camshaft removal and installation procedure.

Intake Manifold

REMOVAL & INSTALLATION

2.4L Engine

1. Before servicing the vehicle, refer to the precautions section.
2. Remove or disconnect the following:
   - Negative battery cable
   - Air breather hose from the throttle body
   - Throttle cable
   - Engine coolant hose and throttle body
   - Positive Crankcase Ventilation (PCV) valve and brake booster vacuum hose
   - Vacuum hose connector
   - Injector cover
   - High pressure fuel hose
   - Fuel injector harness connector
   - Delivery pipe with the injectors and the pressure regulator as an assembly
   - Intake manifold stay
   - Intake manifold

To install:

3. Install or connect the following:
   - New intake manifold gasket
   - Intake manifold and bolts and nuts. Tighten the bolts to 11–14 ft. lbs. (15–20 Nm) and the nuts to 22–30 ft. lbs. (30–42 Nm).
   - Delivery pipe and injector assembly
   - Intake manifold stay and tighten the bolts to 13–18 ft. lbs. (18–25 Nm)
   - Fuel injector harness connector
   - High pressure fuel hose
   - Injector cover
   - Vacuum hose connector
   - PCV valve and brake booster vacuum hose
   - Throttle body and engine coolant hose
   - Throttle cable
   - Air breather hose from the throttle body
   - Negative battery cable

4. Start the engine and check for proper operation.

2.7L Engine

1. Before servicing the vehicle, refer to the precautions section.
2. Remove or disconnect the following:
   - Negative battery cable
   - Air breather hose from the throttle body
   - Throttle and cruise control cables
   - Engine coolant hose and throttle body
   - Positive Crankcase Ventilation (PCV) valve and brake booster vacuum hose
   - Vacuum hose connector
   - Surge tank stay

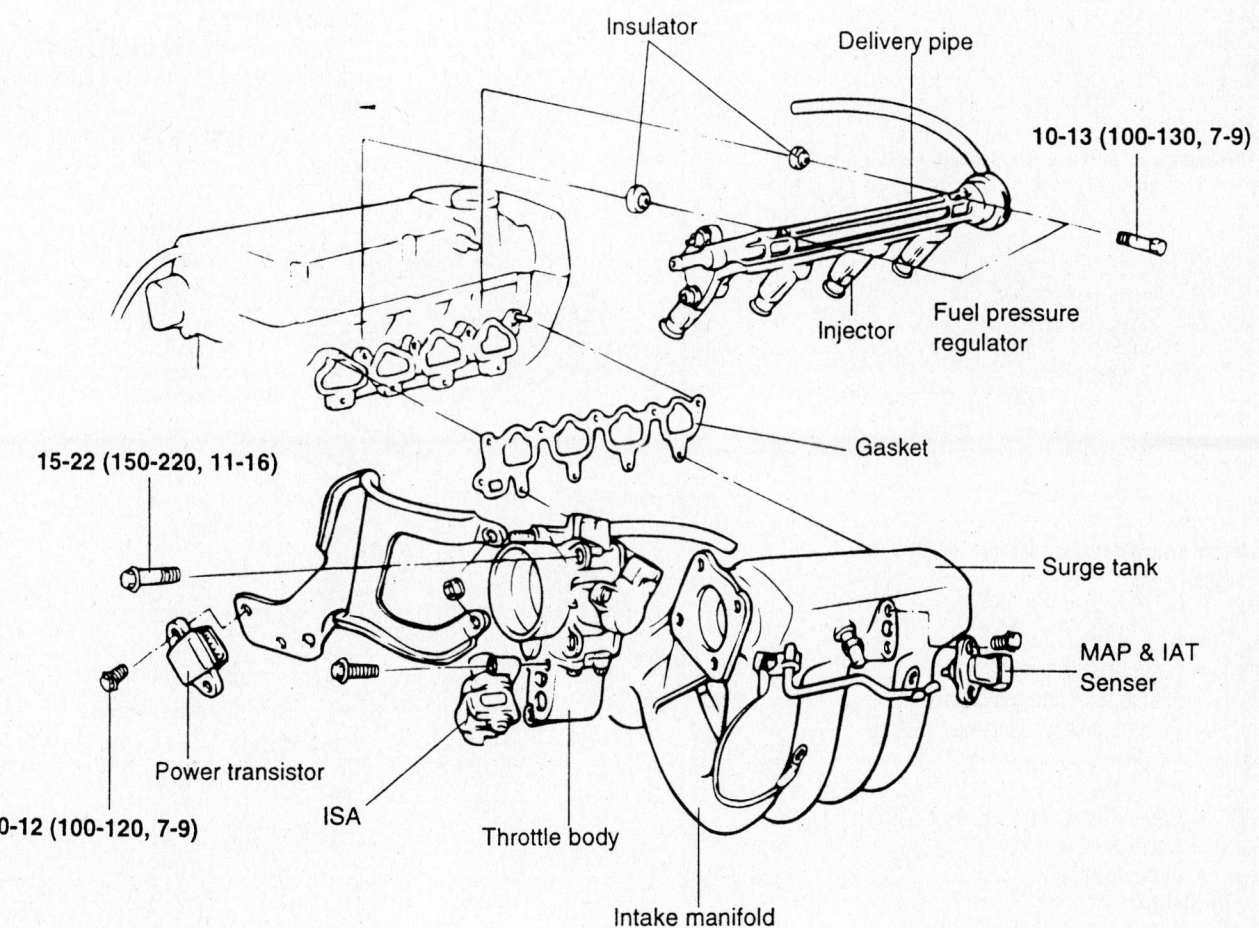

TORQUE : Nm (kg·cm, lb·ft)

9355LG12A

Exploded view of the intake manifold—2.4L engine

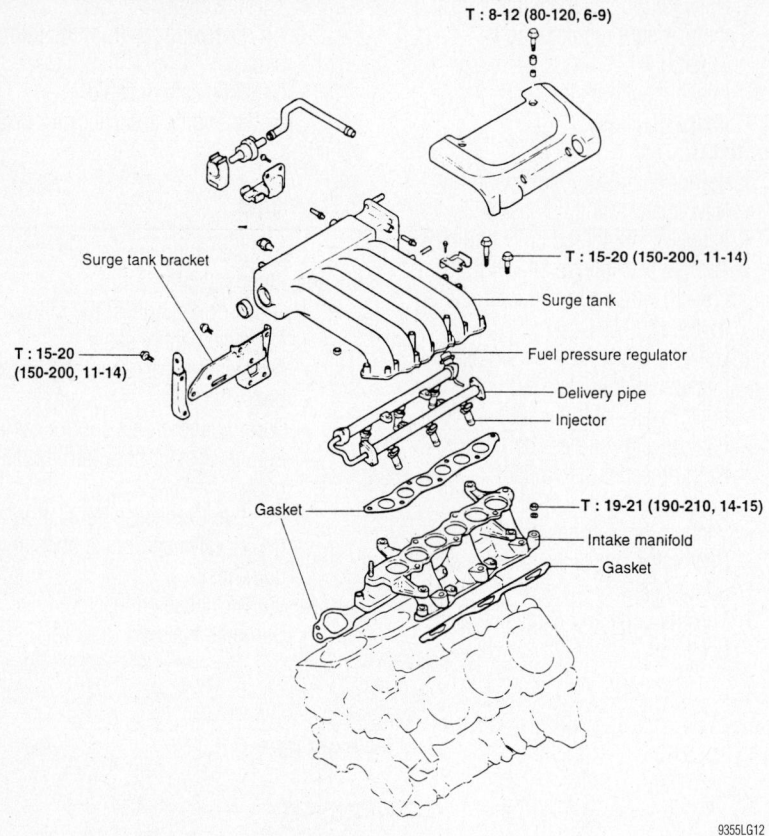

T : 8-12 (80-120, 6-9)

Surge tank bracket

T : 15-20
(150-200, 11-14)

T : 15-20 (150-200, 11-14)

Surge tank

Fuel pressure regulator

Delivery pipe

Injector

Gasket

T : 19-21 (190-210, 14-15)

Intake manifold

Gasket

9355LG12

Exploded view of the intake manifold—2.7L engine

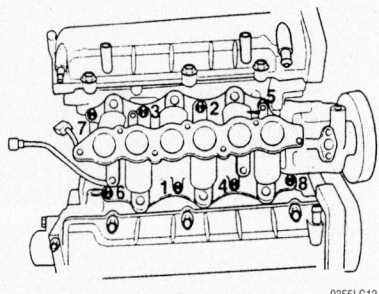

9355LG13

Intake manifold torque sequence—2.7L engine

- High pressure fuel hose
- Surge tank and gasket
- Fuel injector harness connector
- Delivery pipe with the injectors and the pressure regulator as an assembly
- Coolant Temperature Sensor (CTS) electrical connector
- Intake manifold

To install:

3. Install or connect the following:
- New intake manifold gasket
- Intake manifold and tighten the bolts to 14–15 ft. lbs. (19–21 Nm)
- CTS electrical connector
- Delivery pipe with the injectors and

the pressure regulator as an assembly
- Fuel injector harness connector
- Surge tank and gasket, tighten the bolts to 11–14 ft. lbs. (15–20 Nm
- High pressure fuel hose
- Surge tank stay
- Vacuum hose connector
- PCV valve and brake booster vacuum hose
- Engine coolant hose and throttle body
- Throttle and cruise control cables
- Air breather hose from the throttle body
- Negative battery cable

4. Start the engine and check for proper operation.

3.5L Engine

1. Before servicing the vehicle, refer to the precautions section.
2. Disconnect the negative battery cable.
3. Remove the air breather hose from the throttle body.
4. Remove the Positive Crankcase Ventilation (PCV) valve and brake booster vacuum hoses.
5. Disconnect the vacuum hose connections.

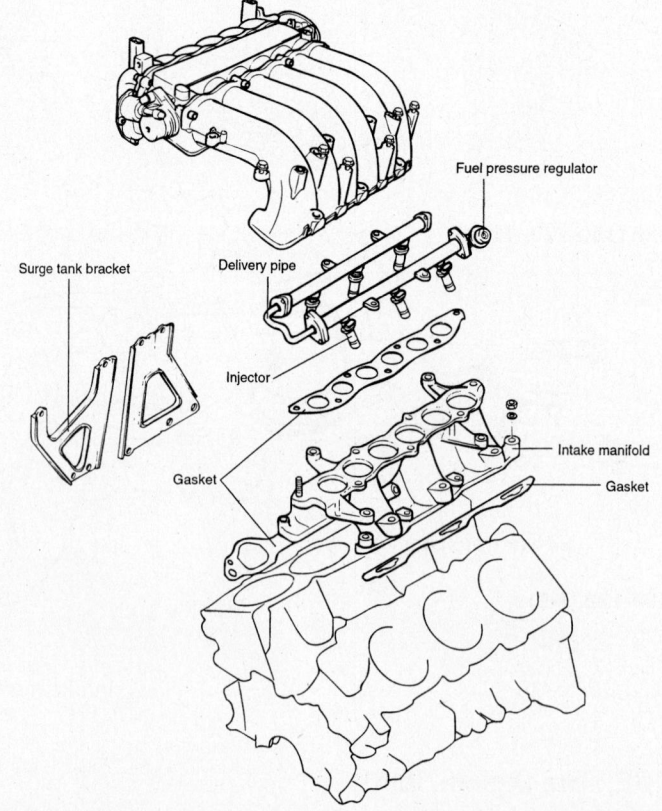

Fuel pressure regulator

Surge tank bracket

Delivery pipe

Injector

Gasket

Intake manifold

Gasket

67162-SAFE-G07

Exploded view of the intake manifold—3.5L engine

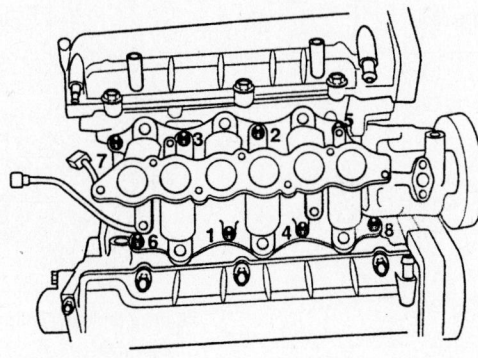

67162-SAFE-G08

Intake manifold torque sequence—3.5L engine

6. Remove the surge tank stay.

7. Remove the surge tank and gasket.

8. Disconnect the fuel injector harness connector.

9. Remove the delivery pipe with the injectors and the pressure regulator as an assembly.

10. Disconnect the Coolant Temperature Sensor (CTS) electrical connector.

11. Remove the intake manifold bolts, manifold and gasket.

To install:

12. Install a new intake manifold gasket.

13. Install the intake manifold and tighten the bolts in the sequence illustrated to 14–16 ft. lbs. (20–23 Nm)

14. Attach the CTS electrical connector.

15. Install the delivery pipe with the injectors and the pressure regulator as an assembly.

16. Attach the fuel injector harness connector.

17. Install the surge tank and gasket.

- Surge tank stay and tighten the retainers to 11–14 ft. lbs. (15–20 Nm).

18. Attach the vacuum hose connectors.

19. Connect the PCV valve and brake booster vacuum hoses.

20. Connect the engine coolant hose to the throttle body.

21. Connect the air breather hose to the throttle body.

22. Connect the negative battery cable.

23. Start the engine and check for proper operation.

Exhaust Manifold

REMOVAL & INSTALLATION

2.4L Engine

1. Before servicing the vehicle, refer to the precautions section.

2. Remove or disconnect the following:

- Negative battery cable
- Heat shield
- Exhaust manifold retainers
- Manifold and gasket

To install:

3. Install or connect the following:

- New gasket and the manifold. Tighten the manifold M8 bolts to 18–22 ft. lbs. (25–30 Nm) and the M10 bolts to 25–40 ft. lbs. (35–55 Nm).
- Heat shield
- Negative battery cable

2.7L Engine

1. Before servicing the vehicle, refer to the precautions section.

2. Remove or disconnect the following:

- Negative battery cable
- Heat shield
- Exhaust manifold retainers
- Manifold and gasket

To install:

3. Install or connect the following:

- New gasket and the manifold. Tighten the manifold bolts to 22–26 ft. lbs. (30–35 Nm).
- Heat shield
- Negative battery cable

3.5L Engine

1. Before servicing the vehicle, refer to the precautions section.

2. In necessary, disconnect the Oxygen ($O_2$2) sensor connector and remove the sensor.

3. Remove the front muffler.

4. Remove the heat shield.

5. Remove the exhaust manifold retainers.

6. Remove the manifold and gasket.

To install:

7. Install a new gasket and the manifold. Tighten the manifold bolts to 29–33 ft. lbs. (40–45 Nm).

8. Install the heat shield.

9. If removed, install the O_2 sensor, tighten to 29–36 ft. lbs. (40–50 Nm) and attach the electrical connector.

10. Install the front muffler.

Camshaft

REMOVAL & INSTALLATION

2.4L Engine

1. Before servicing the vehicle, refer to the precautions section.

2. Drain the cooling system.

3. Remove or disconnect the following:

- Negative battery cable
- Breather hose from the air cleaner and rocker cover
- Air cleaner
- Timing belt cover
- Rocker cover and the Crankshaft Position (CKP) sensor
- Camshaft sprocket bolts and the sprockets
- Timing belt
- Camshaft bearing cap bolts using several passes
- Camshaft bearing caps, camshafts, rocker arms and lash adjusters

To install:

4. Inspect all parts for wear and damage.

5. Install or connect the following:

- Camshafts and apply engine oil to the journals. Do not install the rocker arms yet.

➡**The exhaust camshaft has a slit on the rear end for the CKP sensor.**

- Bearing caps. The caps are marked **I** for intake and **E** for exhaust, they also contain the cap number. For example **I2** would be intake cap number 2.

6. Make sure the camshafts can turn freely, then remove the caps and the camshafts.

7. Make sure the dowel pins on the ends of camshaft sprockets are facing up.

- Rocker arms
- Camshafts and bearing caps. Tighten the bearing cap bolts uniformly to 14–15 ft. lbs. (19–21 Nm).

8. Using special tools, camshaft oil seal installer and guide 09221-21000, 09221-21100 install the oil seal. Coat the outside of the seal with oil prior to installation, then slide the seal along the front end of the camshaft and using the driver and a hammer install the seal until it is full seated.

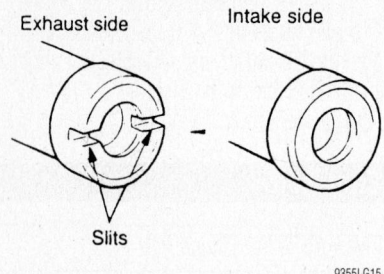

The exhaust camshaft has a slit on the rear end for the CKP sensor—2.4L engine

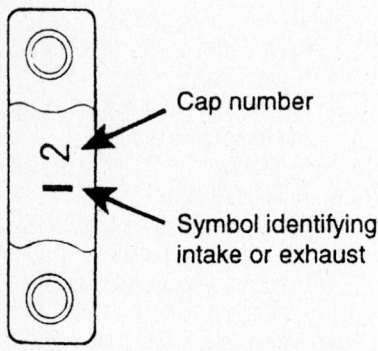

Cap number

Symbol identifying intake or exhaust

The caps are marked I for intake and E for exhaust, they also contain the cap number. For example I2 would be intake cap number 2—2.4L engine

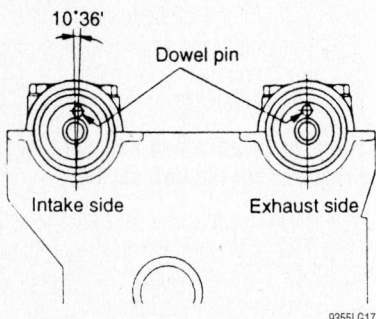

10°36'

Dowel pin

Intake side   Exhaust side

Make sure the dowel pins on the ends of camshaft sprockets are facing up—2.4L engine

- Camshaft sprockets and bolts. Tighten the bolts to 58–72 ft. lbs. (80–100 Nm).
- CKP sensor and rocker cover
- Breather hose to the air cleaner and rocker cover
- Air cleaner
- Negative battery cable

9. Start the engine and check for leaks.

2.7L Engine

1. Before servicing the vehicle, refer to the precautions section.

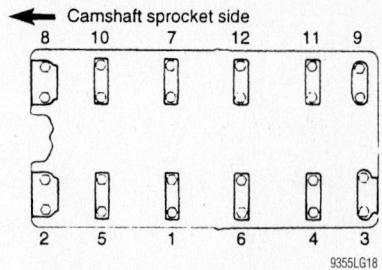

Camshaft sprocket side

Camshaft bearing cap locations—2.4L engines

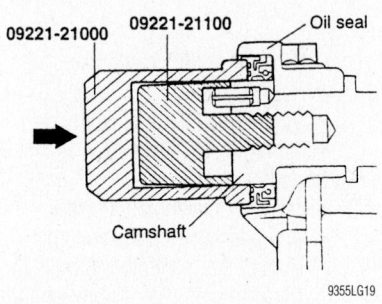

09221-21000   09221-21100   Oil seal

Camshaft

Installing the camshaft oil seal—2.4L engine

2. Remove or disconnect the following:
- Negative battery cable
- Engine cover
- Intake manifold
- Breather hose and engine harness
- Power steering pulley
- A/C pulley
- Crankshaft pulley
- Idler pulley
- Tensioner pulley
- Timing belt cover
- Timing belt from the camshaft sprocket(s)
- Spark plug wires
- Rocker arm cover
- Camshaft sprockets
- Camshaft bearing caps
- Camshafts

To install:

3. Align the camshaft timing chain with the intake timing chain sprocket and exhaust sprocket as shown in the accompanying illustration.

4. Lubricate the camshaft journals with oil and install them.

➡**To check the press fit, the camshaft (IN) and timing chain sprocket should be separable by a force greater than 1000kg minimum at room temperature.**

5. Install the bearing caps. The caps are marked **I** for intake and **E** for exhaust, they also contain the cap number. For example **I2** would be intake cap number 2.

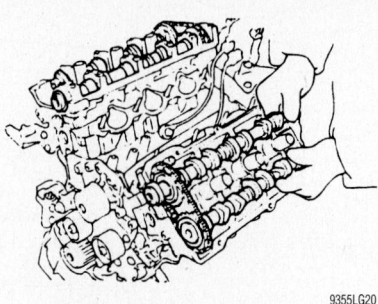

Remove the camshafts from the head—2.7L engine

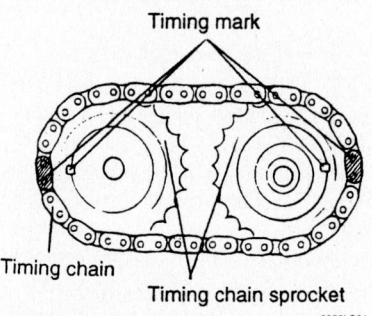

Timing mark

Timing chain

Timing chain sprocket

Align the camshaft timing chain with the intake timing chain sprocket and exhaust sprocket—2.7L engine

6. Tighten the bearing caps M6 (38mm) bolts to 7–9 ft. lbs. (10–12 Nm) and the M6 (50mm) bolts to 10–12 ft. lbs. (14–16 Nm) using several passes.

7. Using special tool, camshaft oil seal installer 09221-21000 install the oil seal. Coat the outside of the seal with oil prior to installation, then insert the seal along the front end of the camshaft and using the driver and a hammer install the seal until it is full seated.

8. Install or connect the following:
- Camshaft sprocket and tighten the bolt to 65–80 ft. lbs. (90–110 Nm).
- Rocker cover in two steps to 5.8–7.2 ft. lbs. (8–10 Nm)
- Spark plug wires
- Timing belt
- Timing belt cover
- Power steering pulley
- A/C pulley
- Crankshaft pulley
- Idler pulley
- Tensioner pulley
- Breather hose and engine harness
- Intake manifold
- Engine cover
- Negative battery cable

3.5L Engine

1. Before servicing the vehicle, refer to the precautions section.

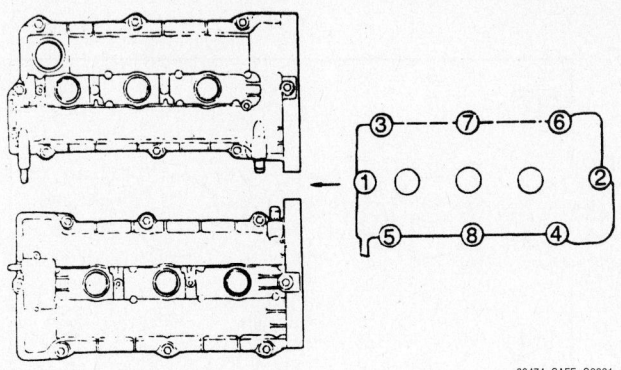

Rocker cover torque sequence—2.7L engine

09474_SAFE_G0001

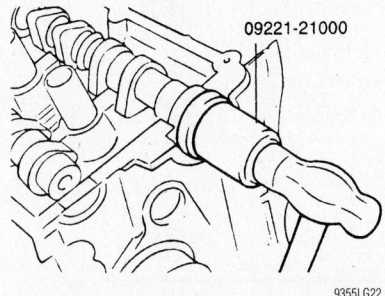

09221-21000

9355LG22

Install the camshaft oil seal—2.7L engine

2. Disconnect the negative battery cable.

3. Remove the engine cover.

4. Remove the intake manifold.

5. Disconnect the breather hose and engine harness.

6. Remove the timing belt.

7. Disconnect the spark plug wires.

8. Remove the rocker arm cover.

9. Remove the camshaft sprockets.

10. Remove the camshaft bearing caps. Make sure to note the cap location prior to removal.

11. Remove the camshafts.

12. If necessary, remove the rocker arms and lifters.

To install:

13. Rotate the crankshaft until the number one cylinder is at Top Dead Center (TDC).

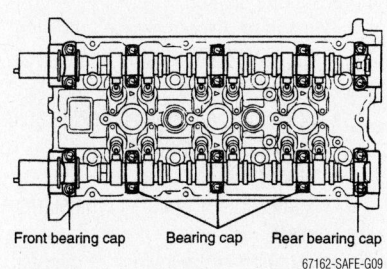

Front bearing cap   Bearing cap   Rear bearing cap

67162-SAFE-G09

Remove the camshaft bearing caps—3.5L engine

14. Check the position of the rocker arm to make sure it is proper installed on the lash adjuster and valve.

15. Install the camshaft dowel pin as illustrated.

16. When installing the camshafts be careful to place them in their correct positions. The are marked as follows:

- Left bank intake is marked with an (I)
- Left bank exhaust is marked with an (E)
- Right bank intake is marked with an (J)

- Right bank exhaust is marked with an (H)

17. Make sure the camshaft caps are installed in their original locations. Bearing caps 3, 4 and 5 have the front mark on them. All caps are marked with and I for intake or E for exhaust.

18. Install the caps, check once more they are installed in the original locations and tighten the caps using 2–to–3 passes. Tighten the outer cap bolts (identified by an * mark in the illustration) to 14 ft. lbs. (19 Nm) and the inner cap bolts to 7.4 ft. lbs. (10 Nm).

19. Install a new rocker cover gasket.

20. Clean the camshaft cap sealing sur-

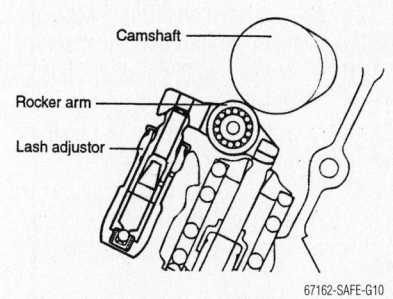

Camshaft

Rocker arm

Lash adjustor

67162-SAFE-G10

Camshaft and related component positioning—3.5L engine

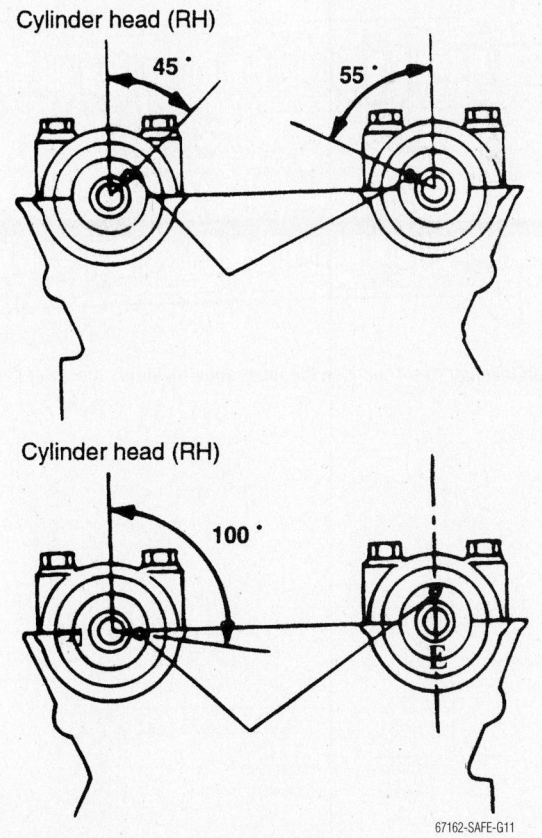

Cylinder head (RH)

45°   55°

Cylinder head (RH)

100°

67162-SAFE-G11

Install the camshaft dowel pin as shown—3.5L engine

Identification mark
Intake : I
Exhaust : E

67162-SAFE-G12

Camshaft cap identification—3.5L engine

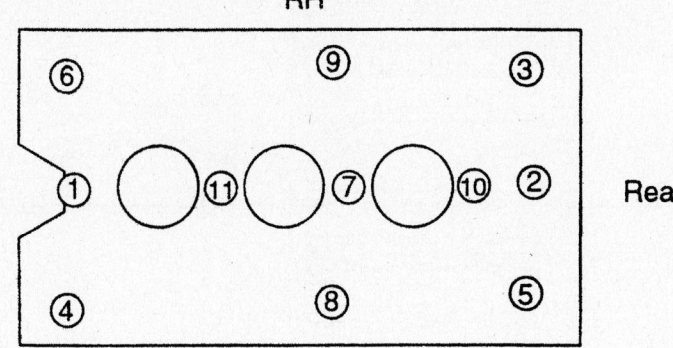

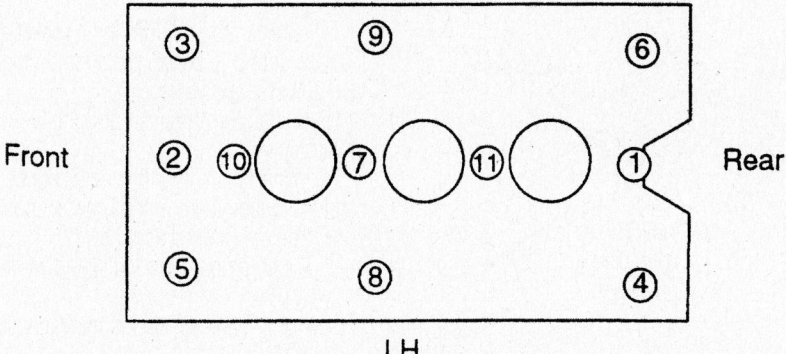

67162-SAFE-G15

Valve cover tightening sequence—3.5L engine

face using a plastic scraper to avoid damage.

21. Apply a 0.4 inch (10mm) bead of LT 5900 sealant to the locations illustrated.

22. Be careful that the gasket is positioned properly when installing the rocker arm cover and that it stays in its positioned when the cover bolts are tightened.

23. Install the cover, making sure to use the washers when installing the bolts and tighten in the sequence illustrated to 6 ft. lbs. (8 Nm).

24. Connect the spark plug wires.

25. Install the timing belt.

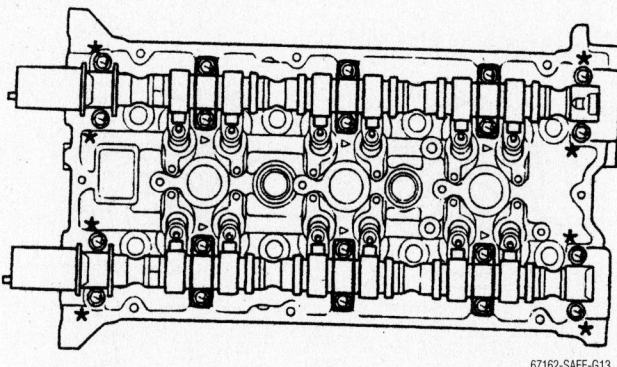

67162-SAFE-G13

Camshaft cap bolt location (the * mark on the illustration identifies the outer cap bolts)—3.5L engine

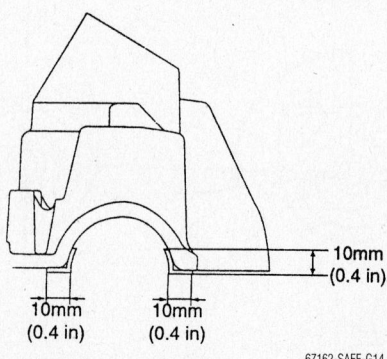

67162-SAFE-G14

Apply a 0.4 inch (10mm) bead of sealant to the locations shown—3.5L engine

26. Connect the breather hose and engine harness.

27. Install the intake manifold.

28. Install the engine cover.

29. Connect the negative battery cable.

Valve Lash

ADJUSTMENT

The valve lash is controlled by hydraulic adjusters and no adjustment is possible.

Starter Motor

REMOVAL & INSTALLATION

1. Before servicing the vehicle, refer to the precautions section.

2. Remove or disconnect the following:
- Negative battery cable and wait at least 3 minutes
- Speedometer cable and shift cable from the transaxle
- Starter motor wiring
- Starter motor bolts and the starter

To install:

3. Installation is the reverse of removal. Tighten motor retainers to 20–25 ft. lbs. (27–34 Nm).

4. Remove the oil pan bolts, note the bolts length and location.

5. Tap the oil pan with a rubber mallet and remove the upper and lower pan components.

6. Clean the gasket mating surfaces.

To install:

7. Apply 0.16 inch (4mm) of sealant to the oil pan groove as illustrated. Install the within 15 minutes of sealant installation.

8. Install the upper and lower oil pans and the bolts making sure the proper length is installed in its original position. Tighten the bolt in sequence to 7–9 ft. lbs. (10–12 Nm).

9. Connect the negative battery cable.

10. Refill the crankcase with oil.

2.7L Engine

1. Before servicing the vehicle, refer to the precautions section.

2. Drain the engine oil.

3. Remove or disconnect the following:
- Negative battery cable
- Lower oil pan bolts and the pan
- Upper oil pan bolts and the pan

4. Clean the gasket mating surfaces

To install:

5. Apply 0.16 inch (4mm) of sealant to the lower oil pan groove. Install the pan within 15 minutes of sealant installation.

6. Install the upper oil pan and the tighten the bolts in sequence as follows:

a. 0.937 x 1.4961 inch (10 x 38mm) bolt to 22–30 ft. lbs. (30–42 Nm).

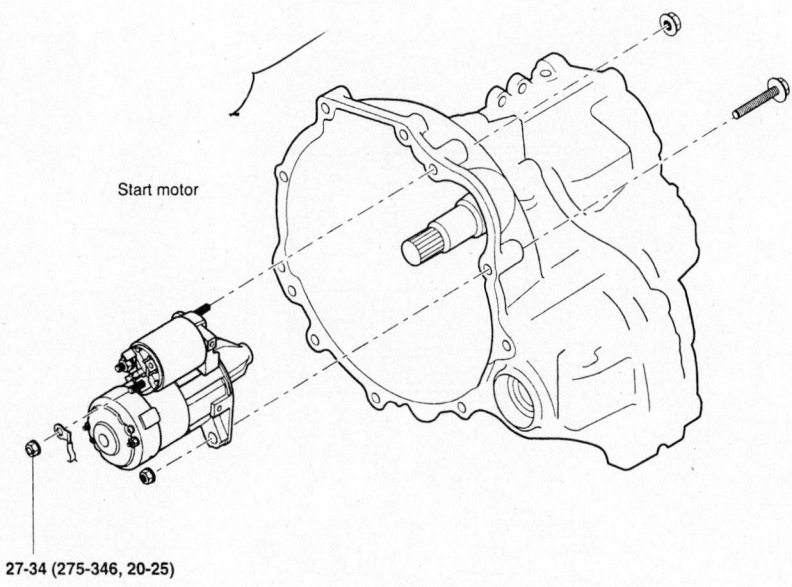

27-34 (275-346, 20-25)

Start motor

27-34 (275-346, 20-25)

TORQUE : Nm (kg.cm, lb.ft)

Starter motor mounting

9355LG23

Oil Pan

REMOVAL & INSTALLATION

2.4L Engine

1. Before servicing the vehicle, refer to the precautions section.

2. Drain the engine oil.

3. Disconnect the negative battery cable.

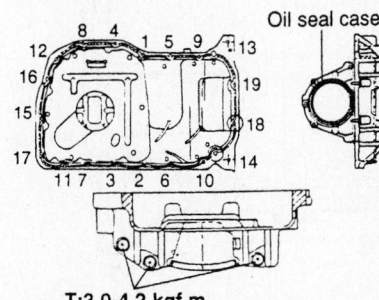

Oil seal case

T:3.0-4.2 kgf.m

9355LG31

Oil pan torque sequence and sealant application points—2.4L engine

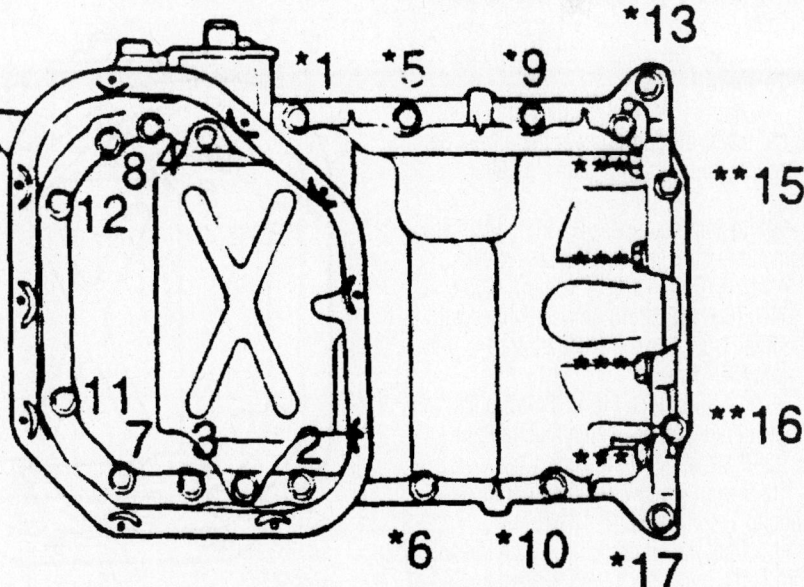

9355LG32

Upper oil pan torque sequence (tighten bolts indicated with * to 14–20 inch (19–28 Nm), bolts indicated with a ** to 4–5 ft. lbs. (5–7 Nm) and bolts indicated with a * to 22–30 ft. lbs. (30–42 Nm—2.7L engine**

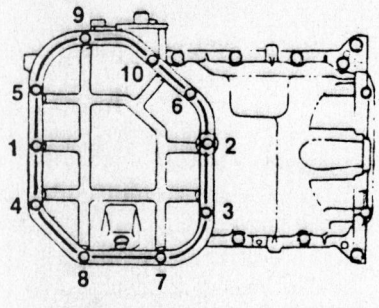

9355LG33

Lower oil pan bolt torque sequence—2.7L engine

    b. 0.3150 x 0.866a inch (8 x 22mm) bolt 14–20 inch (19–28 Nm).

    c. 6.7519 inch (171.5mm) bolt to 4–5 ft. lbs. (5–7 Nm).

    d. 6.7520 inch (152.5mm) bolt to 4–5 ft. lbs. (5–7 Nm).

    7. Install the lower pan and tighten the bolts to 7–9 ft. lbs. (10–12 Nm).

    8. Connect the negative battery cable.

    9. Refill the crankcase with oil.

3.5L Engine

    1. Before servicing the vehicle, refer to the precautions section.

    2. Drain the engine oil.

    3. Remove the oil pressure switch.

    4. Disconnect the negative battery cable.

    5. Remove the lower oil pan bolts and the pan.

    6. Remove the upper oil pan bolts and the pan.

    7. Clean the gasket mating surfaces

To install:

    8. Apply 0.16 inch (4mm) of sealant to the lower oil pan groove. Install the pan within 15 minutes of sealant installation.

    9. Install the upper oil pan and the tighten the bolts in sequence as follows:

    a. 0.7087 inch (6 x 18mm), 6.004 inch (6 x 152.5mm) bolts identified with either * or ** in the illustration to 4–5 ft. lbs. (5–7 Nm).

    b. 1.4961 inch (10 x 38mm) bolts identified with *** in the illustration to 22–30 ft. lbs. (30–42 Nm).

    10. Install the lower pan and tighten the bolts to 7–9 ft. lbs. (10–12 Nm).

    11. Coat the oil pressure switch threads with Three bond No. 1104E sealant and tighten to 6 ft. lbs. (8 Nm).

    12. Connect the negative battery cable.

    13. Refill the crankcase with oil.

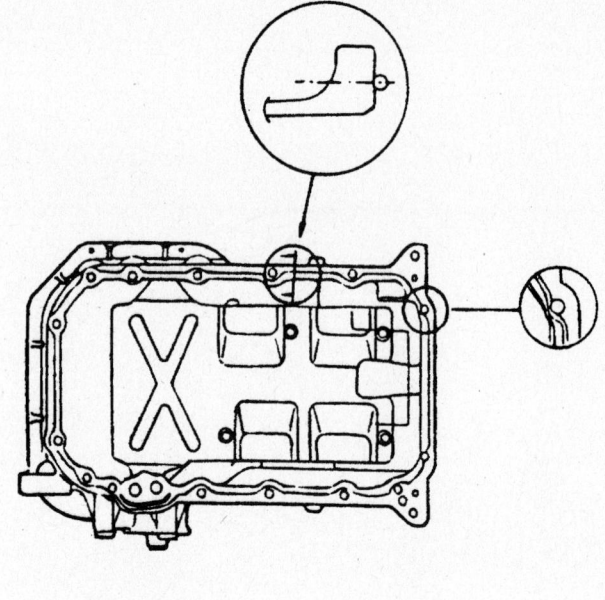

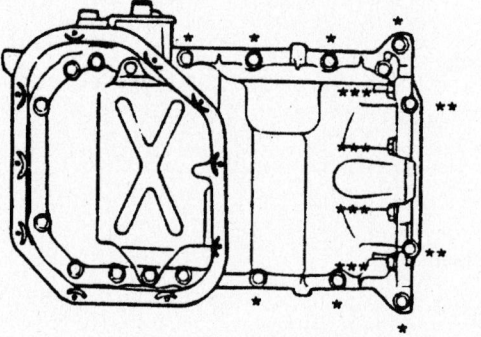

67162-SAFE-G16

Apply 0.16 inch (4mm) of sealant to the lower oil pan groove—3.5L engine

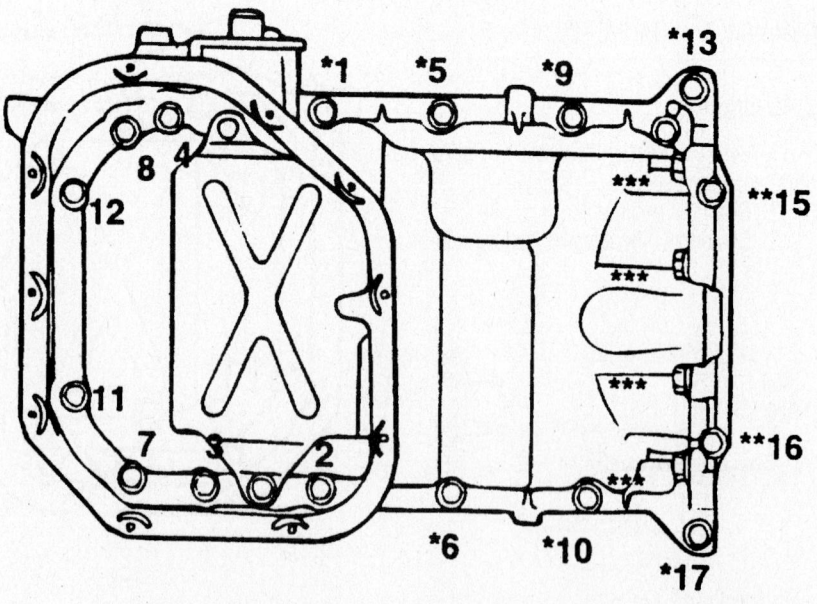

67162-SAFE-G17

Upper oil pan torque sequence (tighten bolts indicated with * or ** in the illustration to 4–5 lbs. (5–7 Nm), and bolts indicated with a * to 22–30 ft. lbs. (30–42 Nm—3.5L engine**

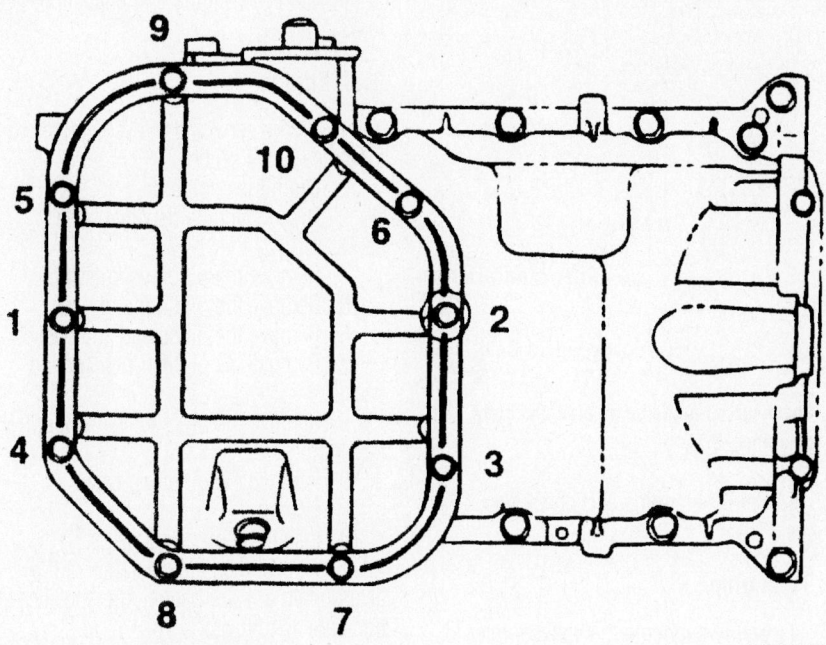

Lower oil pan bolt torque sequence—3.5L engine

67162-SAFE-G18

Oil Pump

REMOVAL & INSTALLATION

2.4L Engine

1. Before servicing the vehicle, refer to the precautions section.
2. Drain the engine oil.

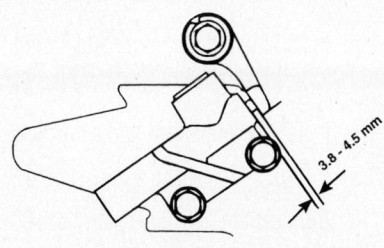

9355LG24

Use Tool 09213-33000, remove the plug cap from the oil pump portion of the case—2.4L engine

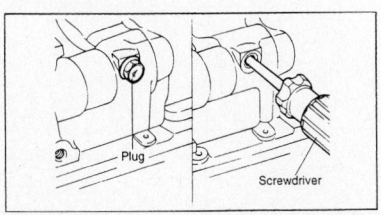

9355LG25

Insert an 0.32 inch (8mm) screwdriver at least 2.4 inch (60mm) into the plug hole—2.4L engine

3. Remove or disconnect the following:
   - Negative battery cable
   - Timing belt
   - Oil pan
   - Oil screen and gasket
   - Oil pressure switch
   - Oil filter bracket and gasket
4. Using Tool 09213-33000, remove

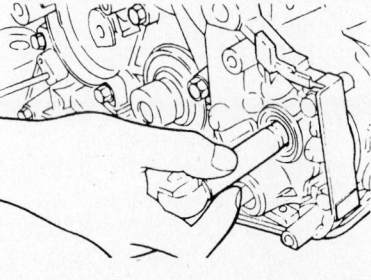

9355LG26

Remove the pump driven gear the left counter balance shaft bolt—2.4L engine

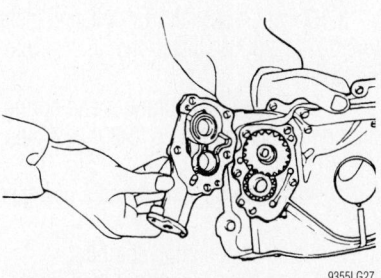

9355LG27

Remove the oil pump cover from the case—2.4L engine

the plug cap from the oil pump portion of the case.
   - Plug from the left side of the block and insert a 0.32 inch (8mm) screwdriver into the plug hole. The screwdriver must be inserted at least 2.4 inch (60mm).
   - Pump driven gear the left counter balance shaft bolt
   - Front case bolts (noting the bolt length and location), the case and gasket.
   - Two counter balance shafts from the block
   - Oil pump cover from the case
   - Oil pump gears from the case
   - Screwdriver from the plug hole

To install:

5. Install the oil pump gears.
6. Inspect the tip clearance of the gears using a feeler gauge. The specifications are as follows:
   a. Standard value drive gear: 0.0063–0.0083 inch (0.16–0.21mm).
   b. Standard value driven gear: 0.0071–0.0083 inch (0.18–0.21mm).
   c. Limit drive gear: 0.0098 inch (0.25mm).
   d. Limit driven gear: 0.0098 inch (0.25mm).
7. Inspect the side clearance of the gears using a feeler gauge. The specifications are as follows:
   a. Standard value drive gear: 0.0031–0.0055 inch (0.08–0.14mm).
   b. Standard value driven gear: 0.0024–0.0047 inch (0.06–0.12mm).
   c. Limit drive gear: 0.0098 inch (0.25mm).
   d. Limit driven gear: 0.0098 inch (0.25mm).
8. Apply engine oil to both gears and align the gear timing marks.
9. Install the oil pump case.
10. Using crankshaft front oil seal install Tool 09214-32000, install the oil seal into the case.

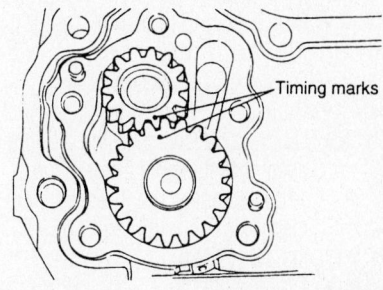

9355LG28

Apply engine oil to both gears and align the gear timing marks—2.4L engine

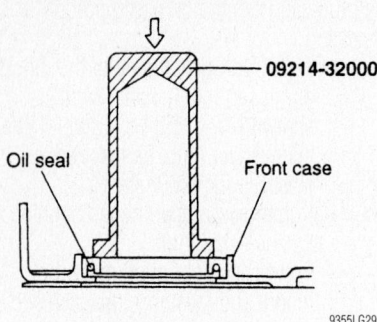

Using crankshaft front oil seal install Tool 09214-32000, install the oil seal into the case—2.4L engine

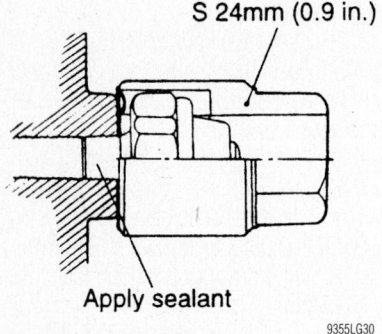

Front case bolt length and location—2.4L engine

11. Place special tool 09214-32100 on the front of the crankshaft and apply a coat of oil to the outside of the tool to aid in case installation.

12. Install or connect the following:
- New front case gasket and temporarily tighten the flange bolts
- Front case and tighten the bolts to 14–20 ft. lbs. (20–27 Nm), making sure the correct length bolt is installed in the correct location.

13. Insert an 0.32 inch (8mm) screwdriver into the plug hole. The screwdriver must be inserted at least 2.4 inch (60mm). Verify the shaft is in place and install the bolt.
- New O-ring on the groove on the front case
- Plug case and tighten to 14–20 ft. lbs. (20–27 Nm)
- Oil screen and gasket
- Oil pan
- Oil pressure switch using a 24mm deep socket. Apply Three bond 1104 sealant to the threads before installation and tighten to 6–9 ft. lbs. (8–12 Nm).
- Timing belt
- Negative battery cable

14. Fill the crankcase to the correct level.

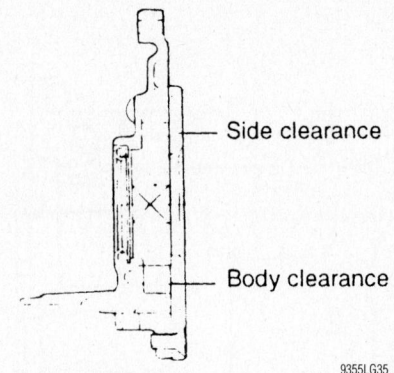

Check the oil pump gears side and body clearance—2.7L

15. Start the engine and check for leaks.

2.7L Engine

1. Before servicing the vehicle, refer to the precautions section.

2. Drain the engine oil.

3. Remove or disconnect the following:
- Negative battery cable
- Oil pressure switch
- Oil filter and pans
- Oil screen and gasket
- Oil filter bracket and gasket
- Oil relief valve plug from the pump case
- Oil pump case

To install:

4. Install the oil pump gears.

5. Inspect the side clearance of the gears using a feeler gauge. The specifications are as follows:
  a. Standard body clearance: 0.0039–0.0071 inch (0.100–0.181mm).
  b. Standard side clearance: 0.0016–0.0037 inch (0.040–0.095mm).

6. Install the oil pump case with a new gasket. Tighten the bolt to 9–11 ft. lbs. (12–15 Nm) and the screw to 6–9 ft. lbs. (8–12 Nm).

7. Install a new oil seal into the pump as tightly as possible.

8. Using crankshaft front oil seal install Tool 09214-33000, install the oil seal into the case.

9. Install the relief plunger and spring and tighten the valve plug to 29–36 ft. lbs. (40–50 Nm).

10. Install the oil screen and a new gasket.

11. Install the oil pans and filter.

12. Connect the negative battery cable.

13. Fill the crankcase to the correct level.

14. Start the engine and check for leaks.

3.5L Engine

1. Before servicing the vehicle, refer to the precautions section.

2. Drain the engine oil.

3. Disconnect the negative battery cable.

4. Remove the oil pressure switch.

5. Remove the oil filter and pans.

6. Remove the oil screen and gasket.

7. Remove the oil filter bracket and gasket.

8. Remove the oil relief valve plug from the pump case.

9. Remove the oil pump case.

To install:

10. Install the oil pump gears.

11. Inspect the side clearance of the gears using a feeler gauge. The specifications are as follows:
  a. Standard body clearance: 0.0039–0.0071 inch (0.100–0.181mm).
  b. Standard side clearance: 0.0016–0.0037 inch (0.040–0.095mm).
  c. Oil tip clearance: 0.0024–0.0071 inch (0.06–0.18mm).

12. Install the oil pump case with a new gasket. Tighten the bolt to 11–14 ft. lbs. (15–20 Nm) and the screw to 6–9 ft. lbs. (8–12 Nm).

13. Install a new oil seal into the pump as tightly as possible.

14. Using crankshaft front oil seal install Tool 09214-33000, install the oil seal into the case.

15. Install the relief plunger and spring and tighten the valve plug to 29–36 ft. lbs. (40–50 Nm).

16. Install the oil screen and a new gasket.

17. Install the oil pans and filter.

18. Connect the negative battery cable.

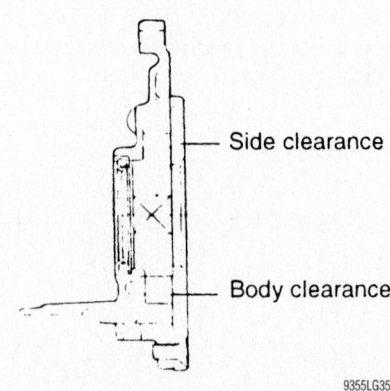

Check the oil pump gears side and body clearance—3.5L

19. Fill the crankcase to the correct level.
20. Start the engine and check for leaks.

Rear Main Seal

REMOVAL & INSTALLATION

2.4L Engine

1. Before servicing the vehicle, refer to the precautions section.
2. Remove or disconnect the following:
- Transaxle
- Clutch pressure plate and disc, if equipped
- Flywheel
- Oil seal case bolts and the case
- Oil seal

To install:
3. Install or connect the following:
- Oil seal. Drive the seal square into the seal case.
- Oil seal case so that the oil hole in the separator may be directed downwards. Tighten the bolts to 7–9 ft. lbs. (10–12 Nm).
- Flywheel
- Clutch pressure plate and disc, if equipped
- Transaxle
4. Check the fluid levels.
5. Start the engine and check for leaks.

2.7L And 3.5L Engines

1. Before servicing the vehicle, refer to the precautions section.
2. Remove or disconnect the following:
- Transaxle
- Clutch pressure plate and disc, if equipped
- Flywheel
- Drive plate and adapter plate
- Oil seal case bolts and the case
- Oil seal

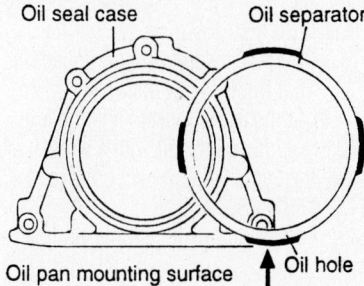

Position the oil seal case so that the oil hole in the separator may be directed downwards—2.4L engine

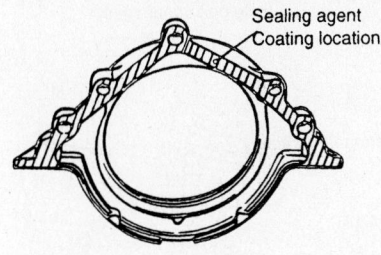

Apply sealant to the areas shown—2.7L and 3.5L engines

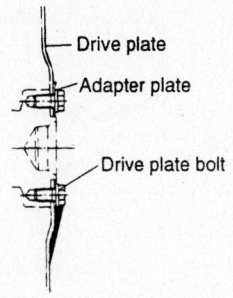

Install the drive plate and adapter plate—2.7L and 3.5L engines

To install:
3. Install or connect the following:
- Oil seal. Drive the seal square into the seal case.
- Oil seal case so that the oil hole in the separator may be directed downwards. Tighten the bolts to 7–9 ft. lbs. (10–12 Nm).
- Drive plate and adapter plate. Tighten the bolt to 53–56 ft. lbs. (73–77 Nm).
- Flywheel
- Clutch pressure plate and disc, if equipped
- Transaxle
4. Check the fluid levels.
5. Start the engine and check for leaks.

Timing Belt

REMOVAL & INSTALLATION

2.4L Engine

1. Before servicing the vehicle, refer to the precautions in the beginning of this section.
2. Align the timing marks to set the No. 1 piston to Top Dead Center (TDC) by rotating the crankshaft clockwise. The timing marks of the camshaft sprocket and the cylinder head cover should be aligned and

the dowel pin of the camshaft sprocket should be at the upper side.
3. Remove or disconnect the following:
- Crankshaft pulley, water pump pulley and drive belt
- Timing belt cover
- Auto tensioner
4. Mark the timing belt is being reused, mark an arrow on the belt noting the direction of rotation or the front of the engine to make sure the belt is reinstalled in its original position.
- Timing belt
5. Hold the camshaft with a wrench and loosen the camshaft sprocket bolts.
- Sprockets
6. When removing the oil pump socket nut, first remove the plug at the side of the block and insert a 0.3 inch (8mm) diameter screwdriver to keep the left counterbalance shaft in position. Insert the screwdriver at least 2.36 inch (60 mm).
- Oil pump sprocket nut and the sprocket
- Loosen the right counterbalance shaft sprocket bolt until you can loosen it by hand
- Tensioner **B** and timing belt **B**. Refer to the accompanying illustration for tensioner and belt identification.

✳✳ CAUTION

Do not attempt to loosen bolts while holding the sprocket with pliers or any tool after removing timing belt B.

- Crankshaft sprocket **B** from the crankshaft

To install:
7. Install or connect the following:
- Crankshaft sprocket **B** to the crankshaft

✳✳ CAUTION

Pay attention to the direction of the flange. If it is installed in the wrong direction, the belt will break.

8. Apply engine oil to the outer surface of the spacer lightly and install the spacer to the right counterbalance shaft. Be sure to install spacer correctly.
- Counterbalance shaft sprocket onto the right counterbalance shaft and then tighten the flange bolt by hand until it is tight
9. Align the timing mark on each sprocket with its corresponding timing mark on the front case.
- Timing belt **B** and make sure there is no slack

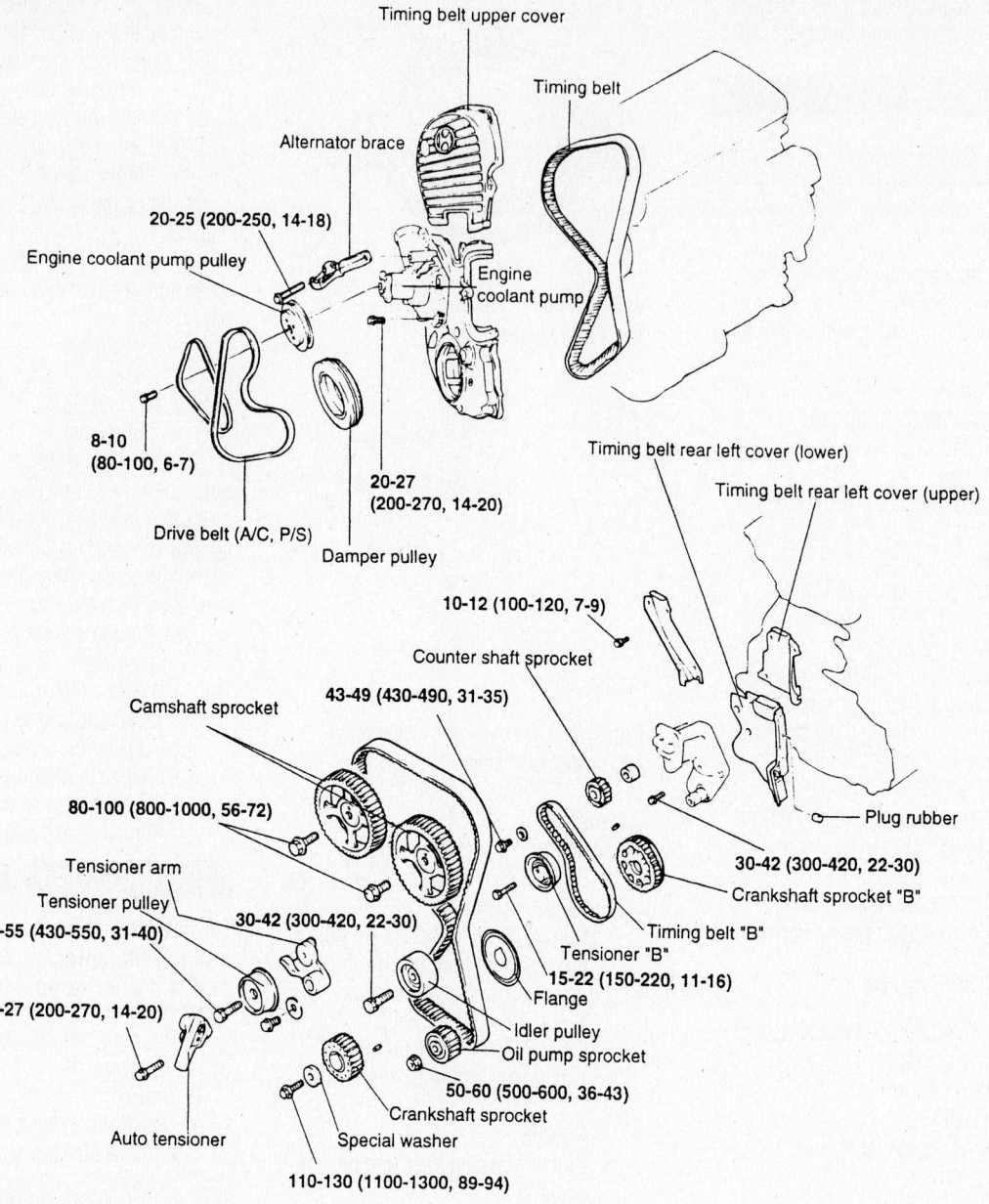

Timing belt upper cover

Timing belt

Alternator brace

20-25 (200-250, 14-18)

Engine coolant pump pulley

Engine coolant pump

8-10 (80-100, 6-7)

20-27 (200-270, 14-20)

Drive belt (A/C, P/S)

Damper pulley

Timing belt rear left cover (lower)

Timing belt rear left cover (upper)

10-12 (100-120, 7-9)

Counter shaft sprocket

43-49 (430-490, 31-35)

Camshaft sprocket

80-100 (800-1000, 56-72)

Tensioner arm

Tensioner pulley

43-55 (430-550, 31-40)

30-42 (300-420, 22-30)

20-27 (200-270, 14-20)

Auto tensioner

Special washer

Crankshaft sprocket

110-130 (1100-1300, 89-94)

50-60 (500-600, 36-43)

Oil pump sprocket

Idler pulley

Flange

15-22 (150-220, 11-16)

Tensioner "B"

Timing belt "B"

Crankshaft sprocket "B"

30-42 (300-420, 22-30)

Plug rubber

TORQUE : Nm (kg·cm, lb·ft)

9355LG73

Exploded view of the timing belt assembly and related components—2.4L engine

- Tensioner **B** so that the center of the pulley is located on the left side of the mounting bolt and the pulley flange faces the front of the engine

10. Align the timing mark on the right counterbalance shaft sprocket with the timing mark on the front case.

11. Lift the tensioner **B** to tighten tensioner **B** so that its tension side is pulled tight. Tighten the bolt on tensioner **B**. As the bolt is being tightened, make sure the shaft does not turn. If the shaft turns, the belt will be over tightened.

12. Make sure the timing marks are aligned.

13. Check the belt tension by depressing the center of the belt span with an index finger. The deflection should be 0.20–0.28 inch (5–7mm).

14. Install or connect the following:
- Flange and crankshaft sprocket making sure it is installed properly. Installing the flange incorrectly will cause the belt to break.
- Crankshaft washer and bolt. Tighten the bolt to 80–94 ft. lbs. (110–130 Nm).

15. Insert a 0.3 inch (8mm) diameter screwdriver through the plug hole on the left side of the block to keep the left counterbalance shaft in position. Insert the screwdriver at least 2.36 inch (60 mm).
- Oil pump sprocket and tighten the nut to 36–43 ft. lbs. (50–60 Nm)
- Camshaft sprockets and the bolts

16. Hold the camshaft with a wrench and tighten the camshaft sprocket bolts to 56–72 ft. lbs. (80–100 Nm).

17. Reset the auto tensioner as follows:

a. Place the tensioner in a soft jawed vice in a level position. If there is a plug at the bottom of the tensioner use a plain washer.

b. Compress the rod slowly using

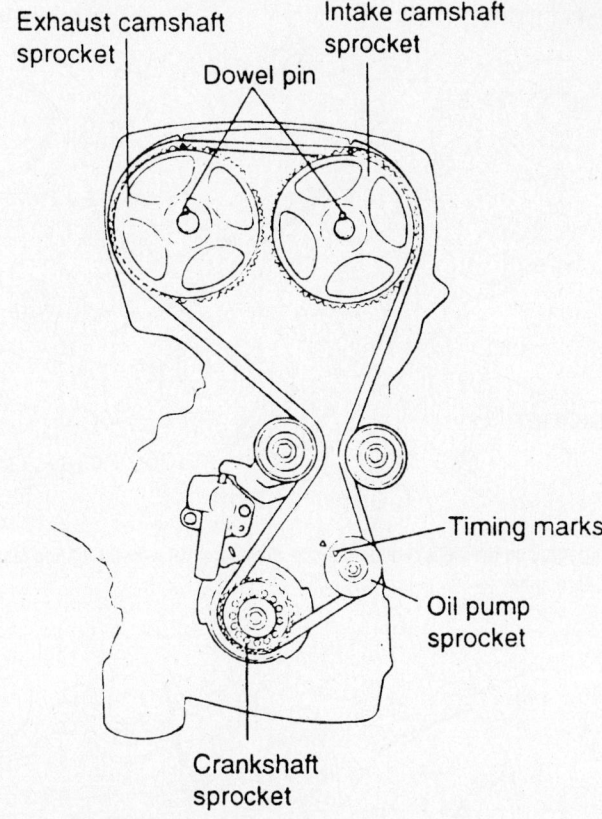

Correct sprocket alignment when the belt is installed—2.4L engine

20. Align the crankshaft sprocket timing marks.

21. Align the pump sprocket timing marks.

22. Install the timing belt counterclockwise around the tensioner pulley and crankshaft sprocket. Hold the belt onto the tensioner pulley using your hand.

23. Pull the belt around the oil pump sprocket using your other hand.

24. Install the belt around the right-hand idler pulley, then the intake camshaft sprocket.

25. Turn the exhaust camshaft sprocket one tooth clockwise to align its timing mark with the cylinder top surface, then pull the belt around the exhaust camshaft sprocket.

26. Raise the tensioner pulley gently so that the belt does not sag and temporarily tighten the pulley center bolt.

27. Recheck that all timing marks are correct.

28. Remove the set pin from the auto tensioner.

29. Rotate the crankshaft two turns clockwise and let it sit for around 15 minutes. After 15 minutes, measure the auto tensioner protrusion **A** (the distance between the tensioner arm and tensioner) as shown in the accompanying illustration. The

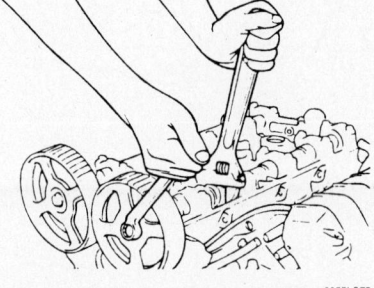

Hold the camshaft with a wrench and loosen the camshaft sprocket bolts—2.4L engine

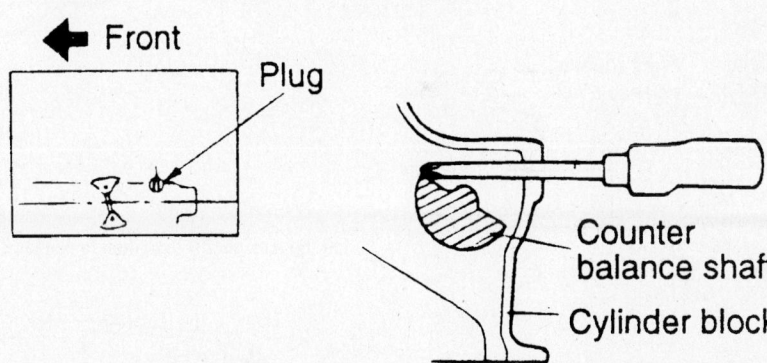

Insert a 0.3 inch (8mm) diameter screwdriver to keep the left counterbalance shaft in position—2.4L engine

the vice until the set hole in the rod is aligned with the set hole on the cylinder.

c. Insert a set pin through the body and rod and leave the pin installed.

18. Install or connect the following:
- Tensioner and tighten the bolts to 14–20 ft. lbs. (20–27 Nm). Leave the set pin in place.
- Tensioner pulley and tighten the bolt to 31–40 ft. lbs. (43–55 Nm)

19. Rotate the camshaft sprockets so that the dowel pin is at the upper side. Set the timing mark of the sprocket correctly.

➡Before installing the belt, the timing mark of the camshaft sprocket doers not coincide with that of the rocker cover, do not rotate the cam sprocket more than 2 teeth of the sprocket in any direction. Rotating the sprocket more that 2 teeth may make the valve and piston contact each other. If it is necessary to rotate the sprocket more that 2 teeth, rotate the crankshaft sprocket counterclockwise first based on the timing mark. After the camshaft sprocket is properly timed, return the crankshaft to TDC.

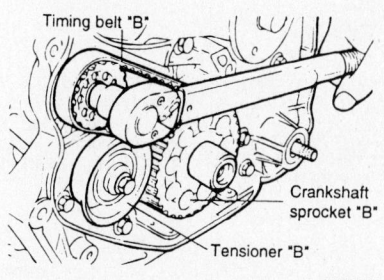

Remove tensioner B and timing belt B—2.4L engine

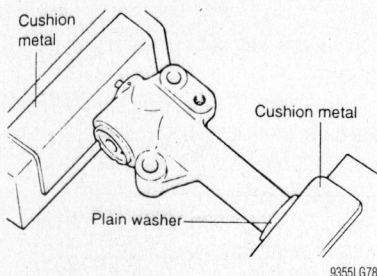

Place the tensioner in a soft jawed vice in a level position. If there is a plug at the bottom of the tensioner use a plain washer—2.4L engine

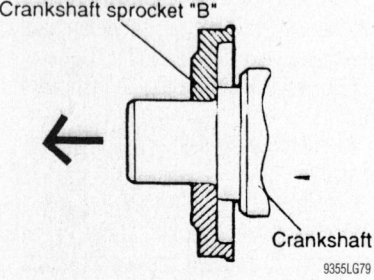

Pay attention to the direction of the flange, if it is installed in the wrong direction, the belt will break—2.4L engine

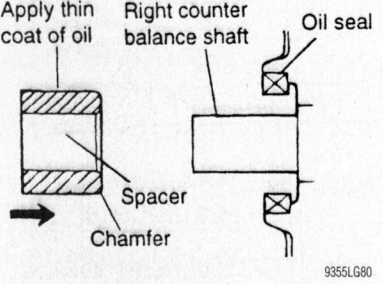

Apply engine oil to the outer surface of the spacer lightly and install the spacer to the right counterbalance shaft. Be sure to install spacer correctly—2.4L engine

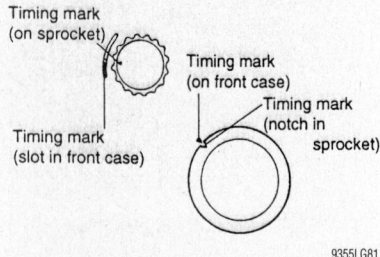

Align the timing mark on each sprocket with its corresponding timing mark on the front case—2.4L engine

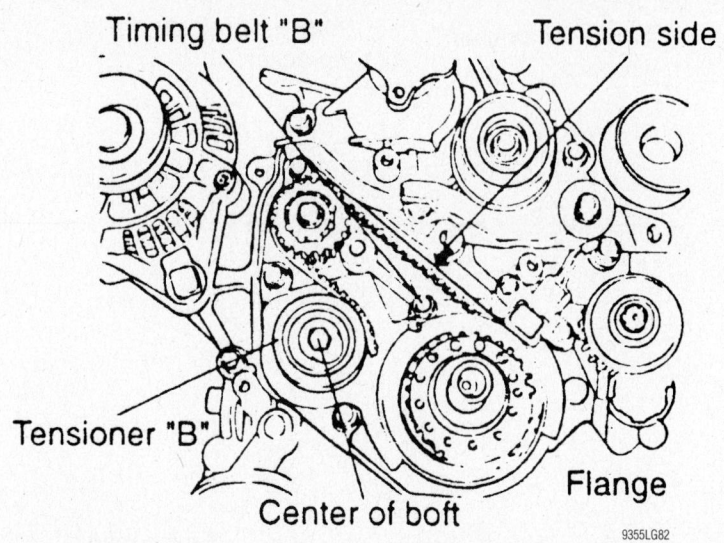

Align the timing mark on the right counterbalance shaft sprocket with the timing mark on the front case—2.4L engine

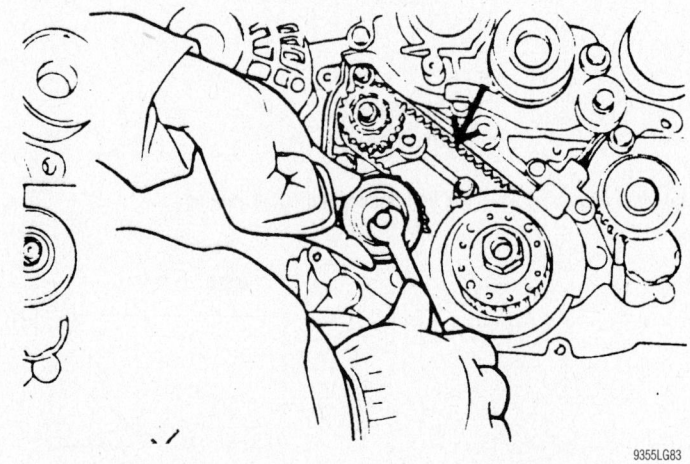

Lift the tensioner B to tighten tensioner B so that its tension side is pulled tight—2.4L engine

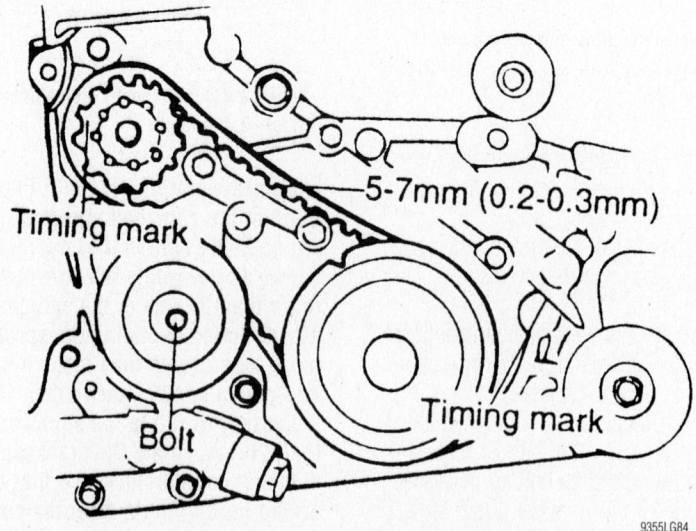

Make sure the timing belt B marks are aligned and the belt tensioner is correct—2.4L engine

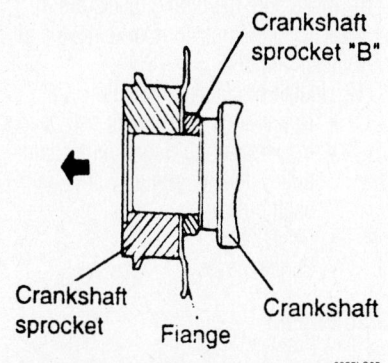

Install the flange and crankshaft sprocket making sure it is installed properly—2.4L engine

specification should be 0.24–0.35 inch (6–9mm).

30. Install the timing covers. Tighten the bolts as shown in the accompanying illustration to specification as shown in the accompanying illustration.

2.7L Engine

1. Before servicing the vehicle, refer to the precautions section.

2. Remove or disconnect the following:
   • Engine cover

3. Using a 16mm wrench, rotate the tensioner arm clockwise about 14 degrees and remove the drive belt.
   • Power steering pump pulley, idler

pulley, tensioner pulley and crankshaft pulley
   • Upper and lower timing covers
   • Timing belt tensioner

4. Rotate the crankshaft clockwise and align the timing mark to set the No. 1 cylinder to Top Dead Center (TDC). Make sure the timing marks of the camshaft sprocket and cylinder head cover should align with each other.

➡**If reusing the belt, mark the direction of rotation on the belt to ensure proper belt installation.**

5. Unbolt the tensioner and remove the belt.

6. Hold the flange of the camshaft with a wrench, unfasten the sprocket bolts and remove the sprockets.

To install:

7. Install or connect the following:
   • Idler pulley on the water pump boss
   • Idler pulley to the roll pin that is pressed in the water pump boss
   • Tensioner arm and plain washer to the block
   • Tensioner pulley to the tensioner arm
   • Camshaft sprockets and align the timing marks. Tighten the bolts to 65–80 ft. lbs. (90–110 Nm).

8. Compress the tensioner and install a set pin to keep the plunger in position.
   • Tensioner and tighten the bolt to 14–20 ft. lbs. (20–27 Nm)

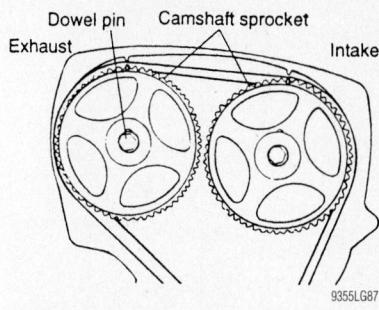

Rotate the camshaft sprockets so that the dowel pin is at the upper side, set the timing mark of the sprocket correctly—2.4L engine

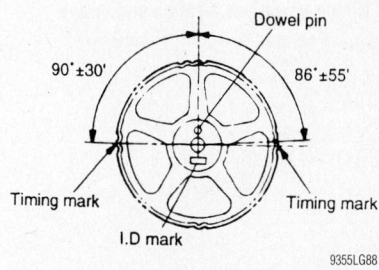

Camshaft sprocket alignment—2.4L engine

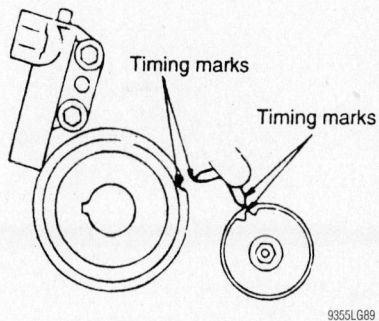

Align the oil pump sprocket timing marks—2.4L engine

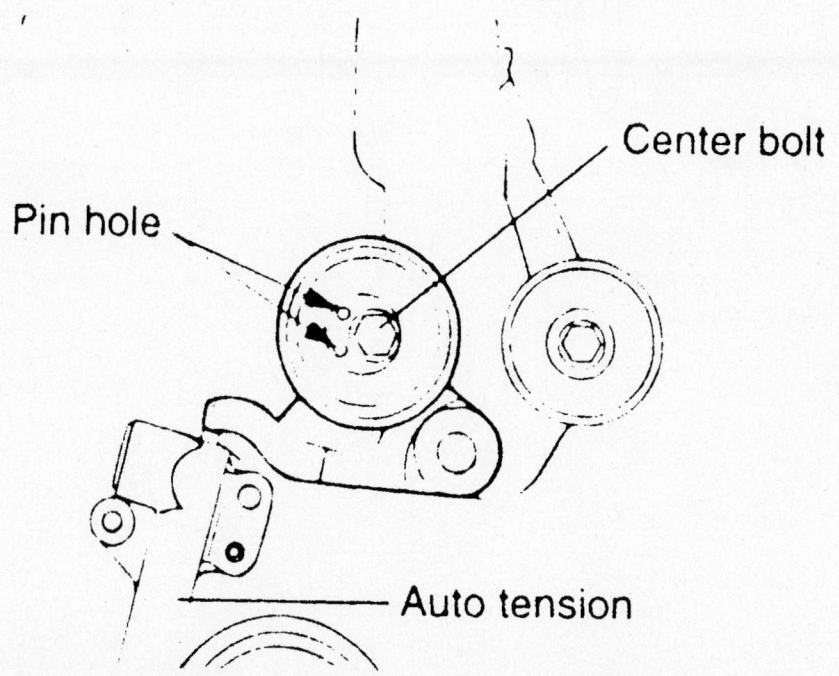

Install the tensioner pulley—2.4L engine

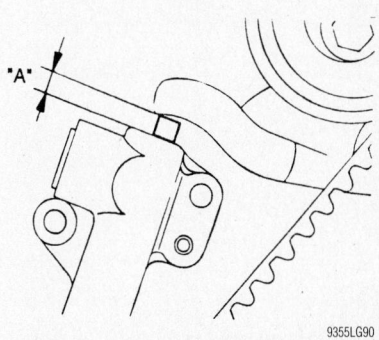

Measure the auto tensioner protrusion A—2.4L engine

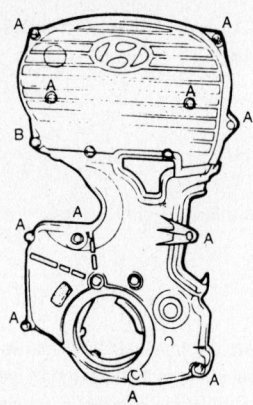

A:8-10 N.m (80-100 kg.cm, 6-7 lb.ft)
B:10-12 N.m (100-120 kg.cm, 7-9 lb.ft)

9355LG91

Timing cover bolt location and torque specifications A —2.4L engine

9. Align the sprocket timing marks and install the belt in the following order:
- Crankshaft sprocket
- Idler pulley
- Camshaft sprocket on the left hand side
- Water pump pulley
- Camshaft sprocket on the right hand side
- Tensioner pulley

10. Remove the tensioner set pin.

11. Adjust the timing belt tension as follows:

a. Rotate the crankshaft 2 turns clockwise and measure the projected length of the auto tensioner at TDC after being installed for 5 minutes.

b. The projected length should be 7–9mm.

Make sure the timing marks are in their specified positions or remove and reinstall the belt.

12. Install or connect the following:
- Upper and lower timing belt covers
- Power steering pump pulley, idler pulley, tensioner pulley and crankshaft pulley
- Drive belt
- Engine cover

3.5L Engine

1. Before servicing the vehicle, refer to the precautions section.

2. Remove the passenger side wheel.

3. Remove the side cover.

4. Remove the engine cover.

5. Disconnect the power steering hose from the engine mount bracket.

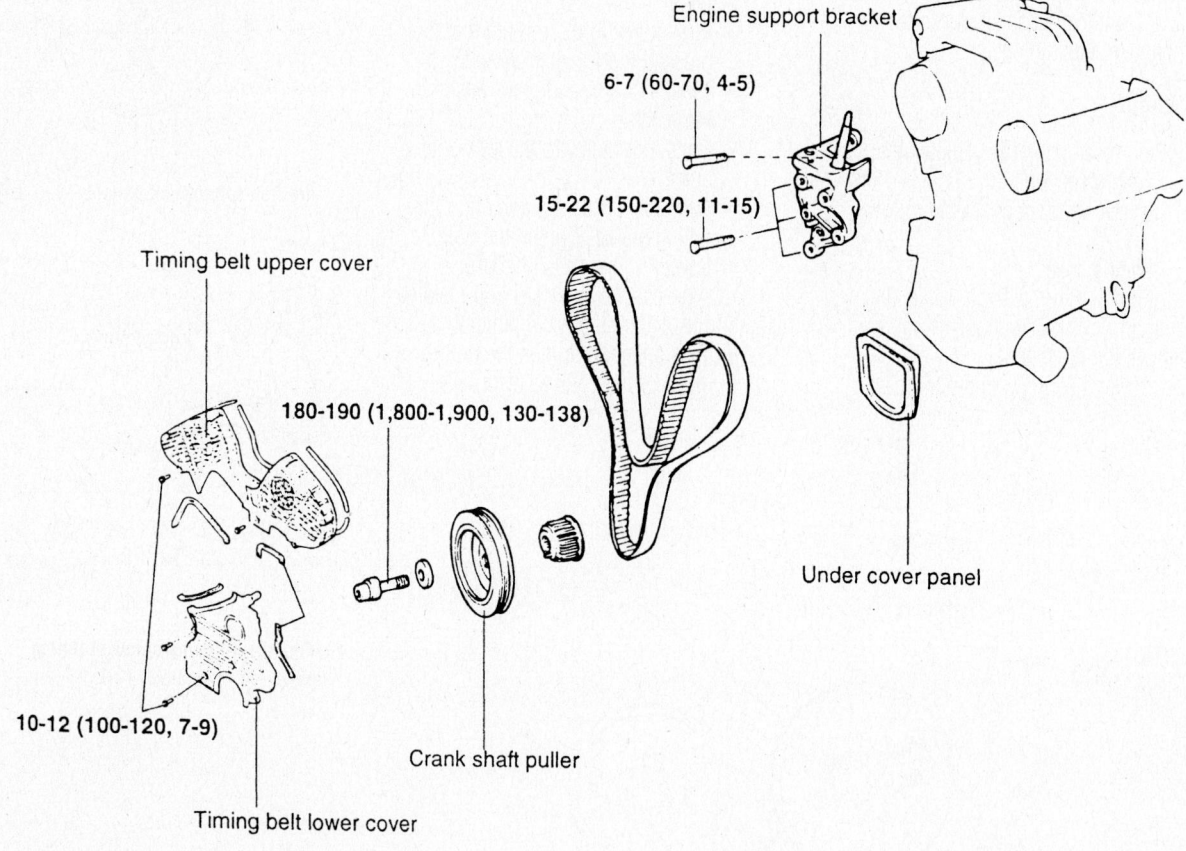

TORQUE : Nm (kg.cm, lb.ft)

9355LG92

Exploded view of the timing belt assembly—2.7L engine

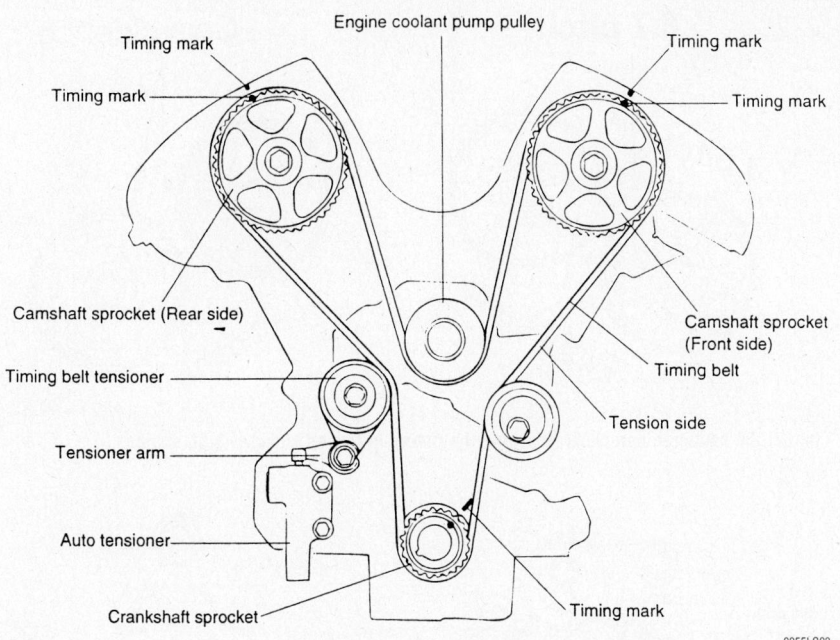

Correct timing marks alignment with the timing belt installed—2.7L engine

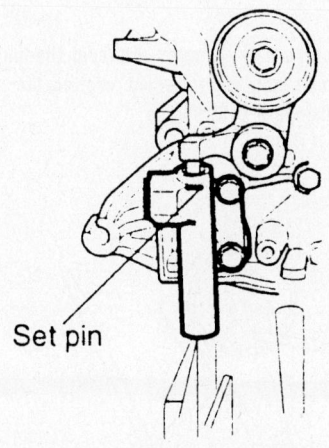

Set pin

9355LG94

Remove the set pin from the auto tensioner—2.7L engine

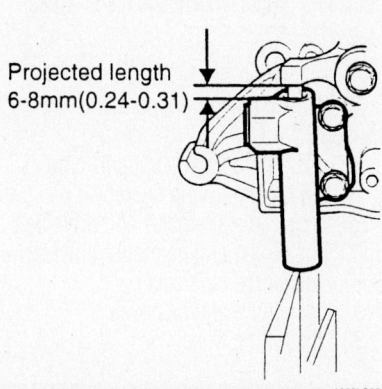

Projected length
6-8mm(0.24-0.31)

9355LG95

The projected length of the auto tensioner at Top Dead Center (TDC) should be 7–9mm—2.7L engine

by moving the engine up and down slightly and remove the bracket upwards.

14. Remove the auto tensioner.

15. Rotate the crankshaft clockwise and align the timing mark to set the No. 1 cylinder to Top Dead Center (TDC). Make sure the timing marks of the camshaft sprocket and cylinder head cover should align with each other.

➡**If reusing the belt, mark the direction of rotation on the belt to ensure proper belt installation.**

16. Unbolt the tensioner and remove the belt.

17. Hold the flange of the camshaft with a wrench, unfasten the sprocket bolts and remove the sprockets.

To install:

18. Install the idler pulley to the engine support bracket.

19. Install the tensioner arm, shaft and washer to the block. Tighten to 25–40 ft. lbs. (35–55 Nm).

20. Install the crankshaft sprocket and align the timing mark as illustrated. Tighten the bolts to 60–74 ft. lbs. (80–100 Nm). Be careful not to bend the crankshaft sensing blade.

21. Install the auto tensioner to the oil pump case. If the auto tensioner is in its fully extended position, perform the following:

a. Place the tensioner in a soft jawed vise in a level position. Use a plain washer if there is a plug at the bottom of the tensioner.

b. Slowly compress the rod until the set hole in the rod is aligned with the set hole in the tensioner.

6. Support the engine with a jack and a wooden block and remove the engine mount bracket.

7. Disconnect the connectors from the upper timing belt cover.

8. Remove the upper timing belt cover.

9. Remove the drive belt.

10. Remove the 2 alternator mounting nuts and 2 bolts attaching the engine support bracket.

11. Loosen the 5 engine support bracket bolts.

12. Remove the alternator.

13. Remove the engine support bracket

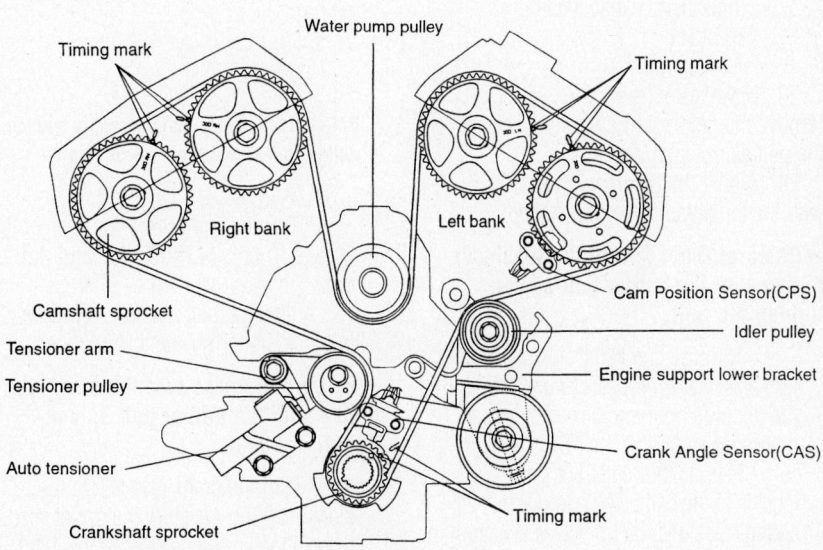

Exploded view of the timing belt assembly—3.5L engine

67162-SAFE-G19

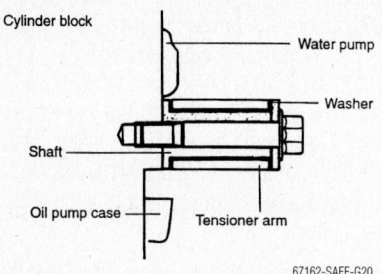

Install the tensioner arm, shaft and washer to the block—3.5L engine

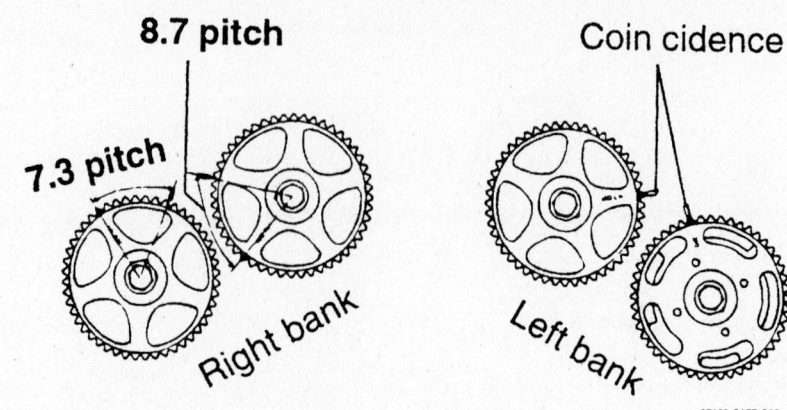

Install the camshaft sprockets and align the timing marks as shown—3.5L engine

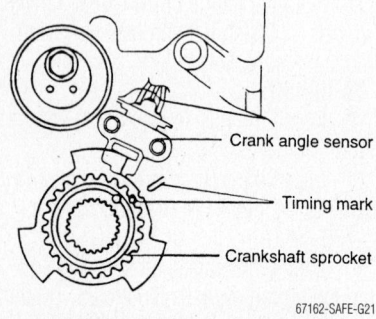

After installing the crankshaft sprocket, align the timing mark as shown—3.5L engine

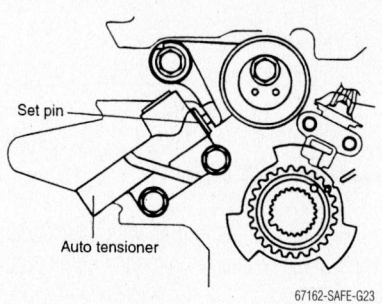

Install the auto tensioner—3.5L engine

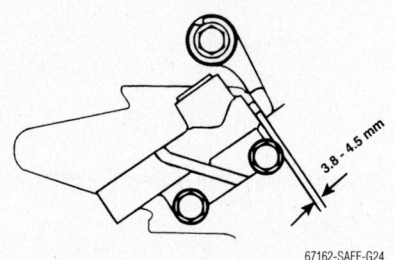

Insert a 3.8—4.5mm pin through the tensioner body and rod when resetting the tensioner—3.5L engine

c. Insert a 3.8—4.5mm pin through the tensioner body and rod.

22. Align the timing marks of each sprocket and install the timing belt in this order:

- Crankshaft sprocket
- Idler pulley
- Left hand exhaust camshaft sprocket
- Right hand intake camshaft sprocket
- Right hand exhaust camshaft sprocket
- Tensioner pulley

23. Make sure the engine is at TDC or remove the belt reset to TDC and reinstall the belt as outlined above.

24. Adjust the belt tension as follows with the tensioner pin still installed:

➡**Be careful not to turn the tensioner pulley with the center bolt as you tighten the bolt.**

a. Turn the crankshaft ¼ turn counterclockwise, turn it clockwise to fit it in the position counterclockwise, turn it clockwise to fit it in the TDC position.

b. Release the center bolt and apply tension to the belt using a torque wrench and the tensioner pulley socket as shown in the accompanying illustration and tighten the center bolt to 3.71 ft. lbs. (5 Nm).

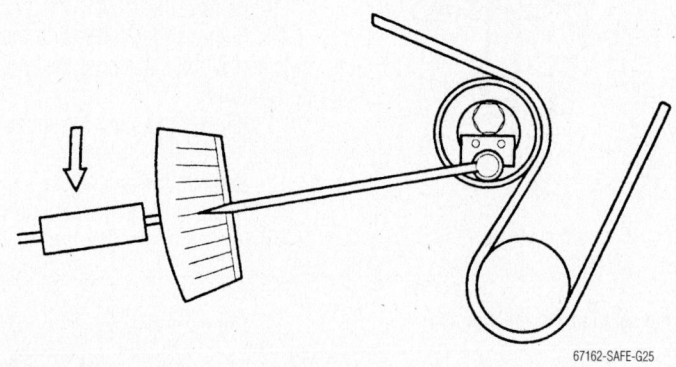

Release the center bolt and apply tension to the belt using a torque wrench and the tensioner pulley socket as shown—3.5L engine

c. Tighten the auto tensioner bolt to 31–39 ft. lbs. (43–55 Nm) and pull out

25. Adjust the belt tension as follows with the tensioner pin removed:

➡**Be careful not to turn the tensioner pulley with the center bolt as you tighten the bolt.**

a. Turn the crankshaft 2 rotations clockwise and measure the projected load of the auto tensioner in the TDC position after 5 minutes. The projected length should be 3.8–4.5mm.

26. Check that all timing marks are aligned.

27. Install the engine support bracket.
28. Install the timing belt cover.
29. Install the alternator and drive belt.
30. Install the engine mount bracket.
31. Install the side cover.
32. Install the engine cover.
33. Install the wheel.

Piston and Ring

POSITIONING

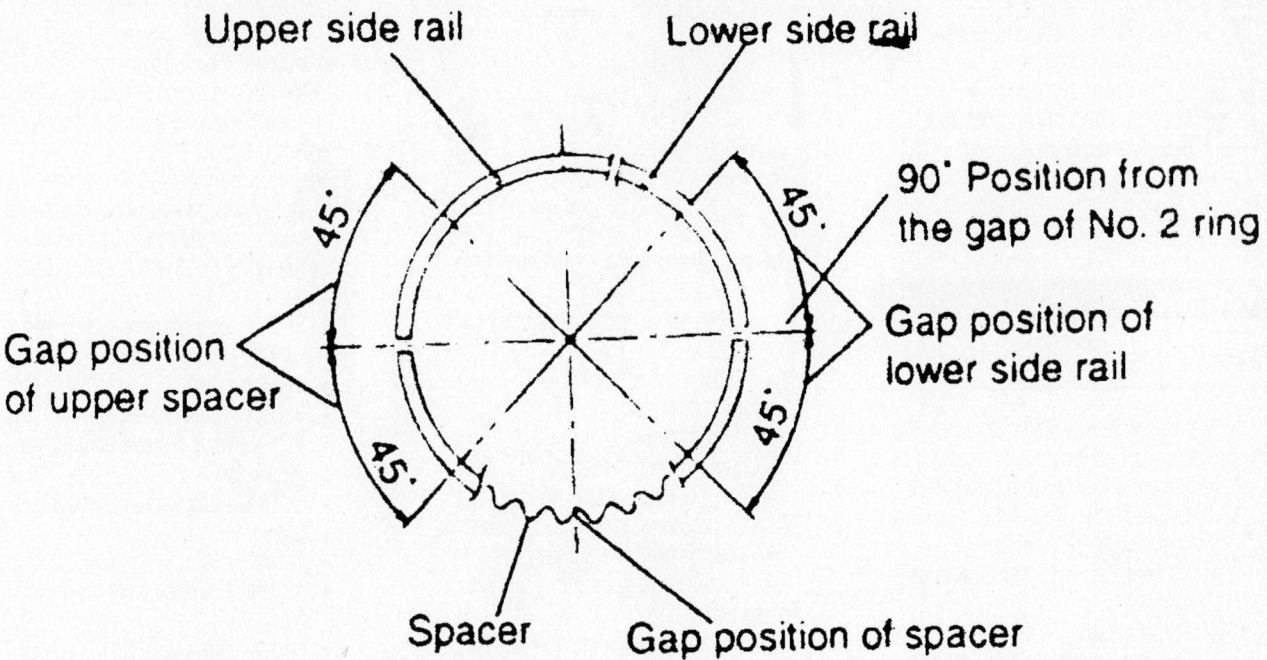

Ring end gap positioning

FUEL SYSTEM

Fuel System Service Precautions

Safety is the most important factor when performing not only fuel system maintenance, but any type of maintenance. Failure to conduct maintenance and repairs in a safe manner may result in serious personal injury or death. Maintenance and testing of the vehicle's fuel system components can be accomplished safely and effectively by adhering to the following rules and guidelines:

• To avoid the possibility of fire and personal injury, always disconnect the negative battery cable unless the repair or test procedure requires that battery voltage be applied.

• Always relieve the fuel system pressure prior to disconnecting any fuel system component (injector, fuel rail, pressure regulator, etc.), fitting or fuel line connection. Exercise extreme caution whenever relieving fuel system pressure to avoid exposing skin, face and eyes to fuel spray. Please be advised that fuel under pressure may penetrate the skin or any part of the body that it contacts.

• Always place a shop towel or cloth around the fitting or connection prior to loosening to absorb any excess fuel due to spillage. Ensure that all fuel spillage (should it occur) is quickly removed from engine surfaces. Ensure that all fuel soaked cloths or towels are deposited into a suitable waste container.

• Always keep a dry chemical (Class B) fire extinguisher near the work area.

• Do not allow fuel spray or fuel vapors to come into contact with a spark or open flame.

• Always use a backup wrench when loosening and tightening fuel line connection fittings. This will prevent unnecessary stress and torsion to fuel line piping. Always follow the proper torque specifications.

• Always replace worn fuel fitting O-rings with new. Do not substitute fuel hose or equivalent, where fuel pipe is installed.

Fuel System Pressure

RELIEVING

1. Before servicing the vehicle, refer to the precautions section.
2. Remove the fuel filler cap.
3. Remove the fuel pump fuse and crank the engine until it stalls.
4. Disconnect the negative battery cable.
5. Replace the fuse.

Fuel Filter

The fuel filter is part of the fuel pump assembly located in the tank.

REMOVAL & INSTALLATION

1. Before servicing the vehicle, refer to the precautions section.
2. Relieve the fuel system pressure.
3. Disconnect the negative battery cable.
4. Remove the two fitting nuts while holding the filter.
5. Remove the filter mounting bolts and the filter from the clamp.
6. Installation is the reverse of removal. Tighten the fitting nuts to 18–25 ft. lbs. (25–35 Nm).
7. Connect the negative battery cable.
8. Start the engine and check for leaks.

Fuel Pump

REMOVAL & INSTALLATION

1. Before servicing the vehicle, refer to the precautions section.
2. Relieve the fuel system pressure.
3. Remove or disconnect the following:

• Negative battery cable
• Fuel pump connector
• Fuel feed and return lines from the pump assembly
• Pump cover screws
• Pump assembly from the tank

To install:

4. Install or connect the following:
- Pump into the tank
- Pump cover screws
- Fuel feed and return lines to the pump assembly
- Fuel pump connector
- Negative battery cable

5. Start the engine and check for leaks.

Fuel Injector

REMOVAL & INSTALLATION

1. Before servicing the vehicle, refer to the precautions section.
2. Relieve the fuel system pressure.
3. Remove or disconnect the following:
- Negative battery cable
- Air breather hose from the throttle body
- Throttle cable
- Engine coolant hose and throttle body
- Positive Crankcase Ventilation

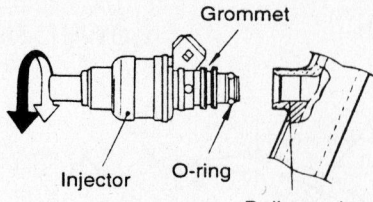

Install the injector using a twisting motion

(PCV) valve and brake booster vacuum hose
- Vacuum hose connector
- Injector cover
- High pressure fuel hose
- Fuel injector harness connector
- Delivery pipe with the injectors and the pressure regulator as an assembly
- Injector from the delivery pipe
- Injector O-ring, grommet and discard

To install:

4. Install a new grommet and O-ring
5. Apply a coating of spindle oil or gasoline to the injector O-ring

6. Install the injector into the delivery pipe while turning the injector left and right making sure the injector turns smoothly. If the injector does not turn smoothly check for a jammed O-ring, remove the injector and reinsert it again.

7. Install or connect the following:
- Delivery pipe and injector assembly
- Intake manifold stay and tighten the bolts to 13–18 ft. lbs. (18–25 Nm)
- Fuel injector harness connector
- High pressure fuel hose
- Injector cover
- Vacuum hose connector
- PCV valve and brake booster vacuum hose
- Throttle body and engine coolant hose
- Throttle cable
- Air breather hose from the throttle body
- Negative battery cable

8. Start the engine and check for proper operation.

DRIVE TRAIN

Transaxle Assembly

REMOVAL & INSTALLATION

Manual

2002–03 Models

1. Before servicing the vehicle, refer to the precautions section.
2. Drain the transaxle.
3. Remove or disconnect the following:
- Negative battery cable
- Air cleaner duct
- Air cleaner and air flow hose
- Back-up light connector
- Clutch line and clip
- Clutch release cylinder
- Speedometer cable
- Gear select or shift cables
- Starter motor mounting bolts
- Transaxle upper bolts

4. Attach an engine support fixture to the engine.
- Transaxle mounting bracket and insulator
- Front wheels
- Tie rod end, lower ball joint and drive shaft
- Gear box u-joint bolt and the return tube mounting bolts
- Front muffler

- Sub-frame mounting bolts and the frame
- Transaxle front and rear mounting brackets
- Transaxle side mounting bolts

5. Using a suitable jack, remove the transaxle from the vehicle.

To install:

6. Installation is the reverse of removal keeping in mind the following torques:
- Transaxle case bolts to 15–20 ft. lbs. (20–27 Nm)
- Transaxle mounting sub-bracket nut 43–58 ft. lbs. (60–80 Nm)
- Transaxle mounting bracket bolts 29–40 ft. lbs. (40–55 Nm)
- Transaxle mounting insulator bolt 65–80 ft. lbs. (90–110 Nm)
- Clutch release cylinder retainers to 11–16 ft. lbs. (15–22 Nm)

7. Fill the transaxle to the correct level.
8. Start the engine and check for leaks.
9. Check the wheel alignment and adjust as necessary.

2004–06 Models

1. Before servicing the vehicle, refer to the precautions section.
2. Drain the transaxle.
3. Remove the battery and tray.
4. Remove the air cleaner duct.
5. Remove the air cleaner and air flow hose.

6. Disconnect the back-up light connector, Crankshaft Position (CKP) sensor, Oxygen (O2s) sensor and oil pressure switch connectors.
7. Disconnect the speedometer cable.
8. Remove the clutch release cylinder bolt.
9. Remove the cotter pin from the shift cable at the transaxle.
10. Remove the clip from the shift cable on the transaxle side.
11. Remove the clip from the select cable at the transaxle side.
12. Separate the steering column shaft joint.
13. Separate the power steering pump hose.
14. Separate the hose after removing the clip from the power steering return hose.
15. Remove the upper transaxle bolt.
16. Remove the starter motor.
17. Install an engine support fixture.
18. Remove the front wheels and calipers.
19. Disconnect the tie rod end, wheel speed sensor, and knuckle mounting bolt.
20. Remove the transaxle mounting bracket and insulator.
21. Remove the front roll stopper insulator bolt, upper and lower stopper bolts and the front roll stopper.
22. Remove the rear roll stopper insulator bolt, and stopper bolt and the roll stopper.

23. Remove the drive shaft.

24. Remove the subframe bolts and sub-frame.

25. Install a transaxle jack.

26. Remove the transaxle lower bolts.

27. Remove the transaxle–to–engine bolts and the transaxle.

To install:

28. Installation is the reverse of removal keeping in mind the following steps and torques:

   a. Transaxle case bolts to 15–20 ft. lbs. (20–27 Nm)

   b. Tighten the transaxle mounting bracket bolts to 43–58 ft. lbs. (60–80 Nm).

   c. Transaxle mounting bracket bolts 29–40 ft. lbs. (40–55 Nm)

   d. Transaxle mounting insulator bolt 65–80 ft. lbs. (90–110 Nm)

   e. Tighten the front roll stopper–to–transaxle bolts to 43–58 ft. lbs. (60–80 Nm).

   f. Tighten the front roll stopper insulator bolt and nut to 36–47 ft. lbs. (50–65 Nm).

   g. Tighten the front roll stopper–to–subframe bolts to 43–58 ft. lbs. (60–80 Nm).

   h. Tighten the rear roll stopper–to–subframe bolts to 43–58 ft. lbs. (60–80 Nm).

   i. Tighten the rear roll stopper–to–transaxle bolt and nut to 36–47 ft. lbs. (50–65 Nm).

   j. Tighten the rear roll stopper insulator bolt and nut to 36–47 ft. lbs. (50–65 Nm).

   k. Clutch release cylinder retainers to 11–16 ft. lbs. (15–22 Nm)

29. Fill the transaxle to the correct level.

30. Start the engine and check for leaks.

31. Check the wheel alignment and adjust as necessary.

Automatic

2002–03 MODELS

1. Before servicing the vehicle, refer to the precautions section.

2. Drain the transaxle.

3. Remove or disconnect the following:
   - Negative battery cable
   - Air cleaner assembly
   - Control cable
   - Speedometer sensor connector
   - Transaxle range switch, solenoid, and oil temperature sensor connectors
   - Oil cooler hose

4. Attach an engine support fixture to the engine.

   - Gear box, stabilizer bar, tie rod end, lower ball joint and drive shaft
   - Gear box u-joint bolt and the return tube mounting bolts
   - Sub-frame mounting bolts and the frame
   - Starter motor
   - Transaxle mounting bolts

5. Using a suitable jack, remove the transaxle from the vehicle.

To install:

6. Installation is the reverse of removal keeping in mind the following torques:
   - Transaxle case bolts to 15–20 ft. lbs. (20–27 Nm)
   - Transaxle mounting sub-bracket nut 43–58 ft. lbs. (60–80 Nm)
   - Transaxle mounting bracket bolts 29–40 ft. lbs. (40–55 Nm)
   - Transaxle mounting insulator bolt 65–80 ft. lbs. (90–110 Nm)

7. Adjust the control cable as follows:

   a. Move the shift lever and the transaxle range switch to the **N** position and install the cable.

   b. When attaching the control cable to the bracket, make sure the clip is installs so it contacts the cable.

   c. Adjust the nut to remove any free-play in the cable and make sure the lever moves freely.

8. Fill the transaxle to the correct level.

9. Start the engine and check for leaks.

10. Check the wheel alignment and adjust as necessary.

2004–06 Models

1. Before servicing the vehicle, refer to the precautions section.

2. Drain the transaxle.

3. Remove the battery and tray.

4. Remove the air cleaner duct.

5. Remove the air cleaner and air flow hose.

6. Disconnect the back-up light connector, Crankshaft Position (CKP) sensor and oil pressure switch connectors.

7. Disconnect the speedometer cable.

8. Remove the cotter pin from the shift cable at the transaxle.

9. Remove the clip from the shift cable on the transaxle side.

10. Remove the clip from the select cable at the transaxle side.

11. Separate the steering column shaft joint.

12. Separate the power steering pump hose.

13. Separate the hose after removing the clip from the power steering return hose.

14. Remove the upper transaxle bolt.

15. Remove the starter motor.

16. Install an engine support fixture.

17. Remove the front wheels and calipers.

18. Disconnect the tie rod end, wheel speed sensor, and knuckle mounting bolt.

19. Remove the transaxle mounting bracket and insulator.

20. Remove the front roll stopper insulator bolt, upper and lower stopper bolts and the front roll stopper.

21. Remove the rear roll stopper insulator bolt, and stopper bolt and the roll stopper.

22. Remove the drive shaft.

23. Remove the subframe bolts and sub-frame.

24. Install a transaxle jack.

25. Remove the transaxle lower bolts.

26. Remove the transaxle–to–engine bolts and the transaxle.

To install:

27. Installation is the reverse of removal keeping in mind the following steps and torques:

   a. Transaxle case bolts to 15–20 ft. lbs. (20–27 Nm)

   b. Tighten the transaxle mounting bracket bolts to 43–58 ft. lbs. (60–80 Nm).

   c. Transaxle mounting bracket bolts 29–40 ft. lbs. (40–55 Nm)

   d. Transaxle mounting insulator bolt 65–80 ft. lbs. (90–110 Nm)

   e. Tighten the front roll stopper–to–transaxle bolts to 43–58 ft. lbs. (60–80 Nm).

   f. Tighten the front roll stopper insulator bolt and nut to 36–47 ft. lbs. (50–65 Nm).

   g. Tighten the front roll stopper–to–subframe bolts to 43–58 ft. lbs. (60–80 Nm).

   h. Tighten the rear roll stopper–to–subframe bolts to 43–58 ft. lbs. (60–80 Nm).

   i. Tighten the rear roll stopper–to–transaxle bolt and nut to 36–47 ft. lbs. (50–65 Nm).

   j. Tighten the rear roll stopper insulator bolt and nut to 36–47 ft. lbs. (50–65 Nm).

   k. Move the shift lever and the transaxle range switch to the **N** position and install the cable.

   l. When attaching the control cable to the bracket, make sure the clip is installs so it contacts the cable.

   m. Adjust the nut to remove any free-play in the cable and make sure the lever moves freely.

28. Fill the transaxle to the correct level.

29. Start the engine and check for leaks.
30. Check the wheel alignment and adjust as necessary.

Transfer Case

REMOVAL & INSTALLATION

1. Before servicing the vehicle, refer to the precautions section.
2. Attach an engine support fixture to the engine.
3. Drain the transfer case fluid.
4. Disconnect the negative battery cable.

5. Disconnect the Oxygen (O_2S) sensor connector and remove the sensor connector from the mounting bracket.
6. Remove the O_2S sensor wire bracket.
7. Remove the alternator wire terminal mounting nut.
8. Remove the right hand wheel assembly.
9. Remove the right hand side engine cover.
10. Remove the lower ball joint mounting bolt and disconnect the ball joint from the knuckle.

11. Remove the drive shaft from the transfer case and support with a piece of wire.
12. Remove the exhaust heat shield.
13. Remove the exhaust manifold.
14. Remove the propeller shaft.
15. Remove the transfer case mounting bracket.
16. Remove the transfer case mounting bolts.
17. Using a flat bladed prytool, remove the transfer case from its mounting by prying it from side-to-side.
18. Remove the transfer case from the vehicle.

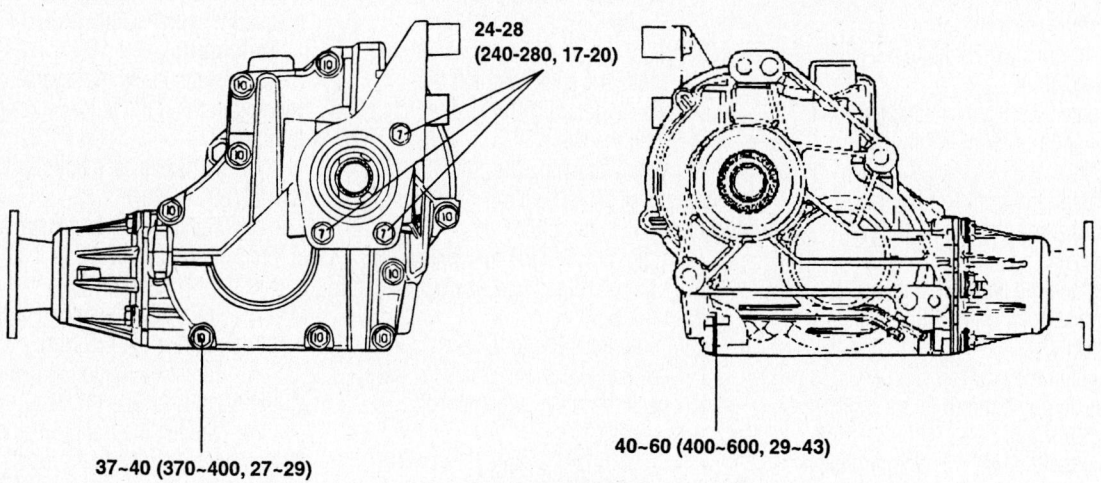

<Right view>

<Left view>

24-28
(240-280, 17-20)

37~40 (370~400, 27~29)

40~60 (400~600, 29~43)

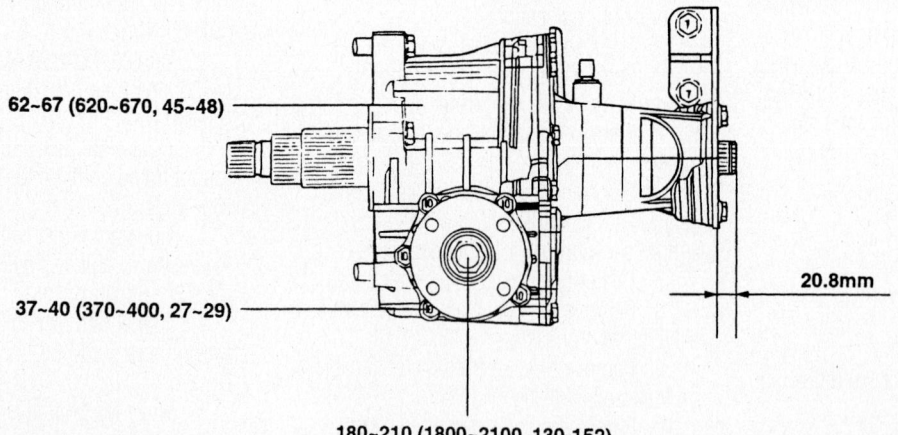

<Rear view>

62~67 (620~670, 45~48)

37~40 (370~400, 27~29)

20.8mm

180~210 (1800~2100, 130-152)

TORQUE : N·m (kg·cm, lb·ft)

Transfer case bolt torque spoliations

67162-SAFE-G26

19. Installation is the reverse of removal. Refer to the illustration for the transfer case bolt torque specifications.

Clutch

ADJUSTMENTS

1. Before servicing the vehicle, refer to the precautions section.

2. Measure the clutch pedal height from the face of the pedal pad to the floorboard. The proper measurement (8.597 inch (218.9mm).

3. Measure the clutch pedal clevis pin play. The measurement should be 0.04–0.11 inch (1–3mm).

4. Adjust the height and clevis pin if necessary as follows:

   a. Turn and adjust the bolt, then secure it by tightening the lock nut.

➡**After adjustment, tighten the bolt until it reaches the pedal stopper and tighten the lock nut.**

   b. Turn the push rod until the proper specification is reached and tighten the lock nut.

➡**When adjusting the clevis pin play or the pedal height, make sure not to push the push rod towards the master cylinder.**

   c. After the adjustments are made, check that the clutch pedal free play is 0.2–0.5 inch (6–13mm). The free play is measured at the face of the pedal pad

5. If the adjustments do not bring the pedal into specifications, check the hydraulic system for air or a faulty component.

REMOVAL & INSTALLATION

1. Before servicing the vehicle, refer to the precautions section.

2. Disconnect the negative battery cable.

3. Remove the transmission assembly.

4. If the pressure plate is attached to the flywheel, remove the release bearing using snap-ring pliers as follows:

   a. Insert the snap-ring pliers under the wave washer and place it in the center of the snap-ring

   b. Spread the snap-ring by pushing down on the bearing assembly.

   c. Once the snap-ring is fully expanded, remove the release bearing

5. Install a clutch disc alignment tool.

6. Remove pressure plate retaining bolts using a star pattern in several passes.

7. Remove the pressure plate with clutch disc.

To install:

8. Apply multi-purpose grease to clutch splines.

9. Install or connect the following:
   - Clutch plate to the flywheel
   - Pressure plate and cover. Do not torque the bolts at this time.

10. Align the bearing to the release fork and install it until it is fully engaged.

11. Torque the pressure plate bolts using a star pattern torque sequence to 11–16 ft. lbs. (15–22 Nm) on models with a single mass flywheel or 14–19 ft. lbs. (20–27 Nm) on models with a dual mass flywheel.

12. Install transmission.

13. Connect the negative battery cable.

Hydraulic Clutch System

BLEEDING

Bleeding air from the hydraulic clutch system is necessary whenever any part of the system has been disconnected or the fluid level (in the reservoir) has been allowed to fall so low that air has been drawn into the master cylinder.

❄ WARNING

NEVER use fluid that has been bled from a clutch system to fill the master cylinder reservoir, as it may be aerated, contain excessive moisture and/or be contaminated in some other way.

1. Before servicing the vehicle, refer to the precautions section.

2. Fill the clutch master cylinder reservoir with new hydraulic clutch fluid.

3. Attach a hose to the bleeder on the clutch actuator and submerge the other end of the hose in a container of hydraulic clutch fluid.

4. Have an assistant slowly depress and hold the clutch pedal.

5. Loosen the bleeder to purge air.

6. Tighten the bleeder.

7. Repeat the above 3 steps until all air is completely purged from the system.

8. Refill the clutch master cylinder reservoir.

Halfshaft

REMOVAL & INSTALLATION

Front

1. Before servicing the vehicle, refer to the precautions section.

2. Drain the transaxle.

3. Remove or disconnect the following:
   - Negative battery cable
   - Aluminum wheel cover
   - Split pin and halfshaft nut
   - Wheel Speed Sensor (WSS) from the bracket, if equipped
   - Brake hose from the bracket
   - Knuckle from the strut by removing the flange bolts
   - Halfshaft from the hub by tapping the end with a plastic mallet
   - Halfshaft from the transaxle using a pry bar.

4. Insert a plug in the transaxle opening.

To install:

➡**Use new circlips, split pins and self-locking nuts for assembly.**

5. Remove the plug from the transaxle opening.

6. Coat the halfshaft splines and sliding surfaces with gear oil.

7. Make sure the gap of the circlip is facing downwards.

8. Install the inner CV-joint to the transaxle until the circlip locks in the retaining groove. Pull on the shaft by hand to make sure it is properly engaged.

9. Install or connect the following:
   - Halfshaft into the hub
   - Knuckle to the strut
   - Flange bolts and tighten to 74–88 ft. lbs. (100–120 Nm)
   - Halfshaft nut and tighten to 146–190 ft. lbs. (200–260 Nm)
   - New split pin
   - Brake hose to the bracket
   - WSS to the bracket, if equipped
   - Aluminum wheel cover
   - Negative battery cable

10. Fill the transaxle to the correct level and check for leaks.

Rear

1. Before servicing the vehicle, refer to the precautions section.

2. Remove or disconnect the following:
   - Rear wheel
   - Split pin and nut
   - Spare tire and support the hanger of the main muffler to avoid interfering with the carrier during right hand shaft removal

3. Matchmark the propeller rubber coupling and differential flange, then remove the bolts and nuts.

4. Support the differential carrier with a jack and remove the differential carrier mounting nuts and bolts.

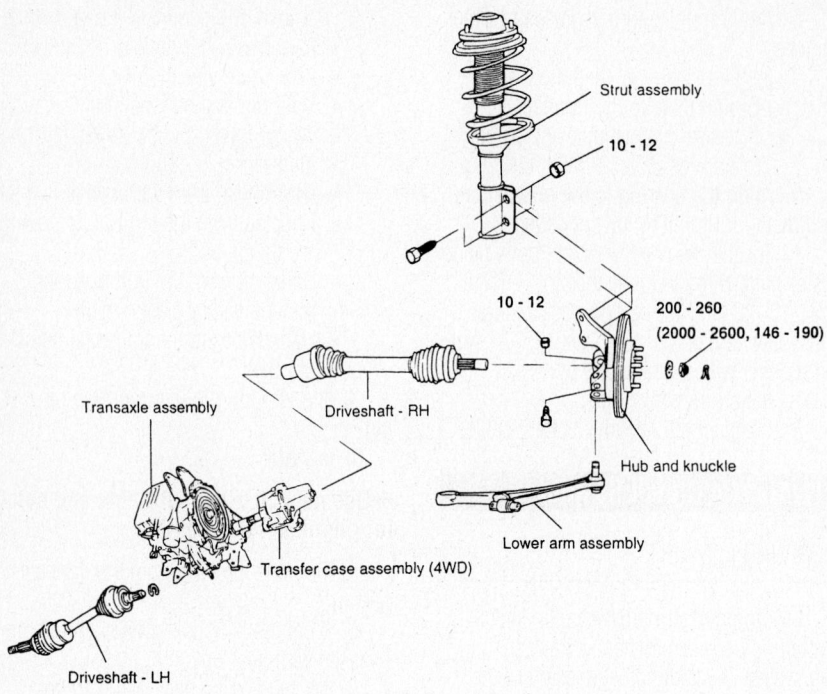

Strut assembly

10 - 12

10 - 12

200 - 260
(2000 - 2600, 146 - 190)

Transaxle assembly

Driveshaft - RH

Hub and knuckle

Lower arm assembly

Transfer case assembly (4WD)

Driveshaft - LH

TORQUE : Nm (kg·cm, lb·ft)

Exploded view of the halfshaft mounting and related components

- Shaft from the carrier by inserting a prybar between the carrier and the shaft
- Differential carrier to the rear after lowering the jack
- Shaft from the axle hub using a plastic mallet
- Shaft from the vehicle

To install:
5. Install or connect the following:
- Shaft into the axle hub
- Differential carrier to the rear using a jack
- Shaft to the carrier
- Differential carrier mounting nuts and bolts. Tighten the carrier nuts and bolts to 51–58 ft. lbs. (70–80 Nm) and the carrier rear bracket bolt to 58–73 ft. lbs. (80–100 Nm).
- Propeller shaft
- Spare tire and remove the support from the main muffler hanger
- New shaft nut and tighten to 148–192 ft. lbs. (200–260 Nm)
- New split pin
- Rear wheel

CV-Joint

OVERHAUL

DOJ-BJ Type

The DOJ-BJ type joint is used on both the front and rear of the vehicle and the following overhaul procedure is used.

1. Before servicing the vehicle, refer to the precautions section.

➡**Do not disassemble the BJ assembly.**

2. Remove or disconnect the following:
- Axle halfshaft from the vehicle and place it in a vise
- DOJ boot clamps and push the boot back from the outer race
- Circlip with a flat bladed prytool
- Driveshaft from the DOJ outer race
- Snap-ring and take out the inner race, cage and balls as an assembly

3. Clean the outer race, cage and balls without disassembling them.
- BJ boot clamps
- DOJ and BJ boots

To install:

➡**Use new circlips and boot clamps for assembly.**

4. Apply some of the grease supplied with the kit to the halfshaft.
5. Install the boots.

9355LG40

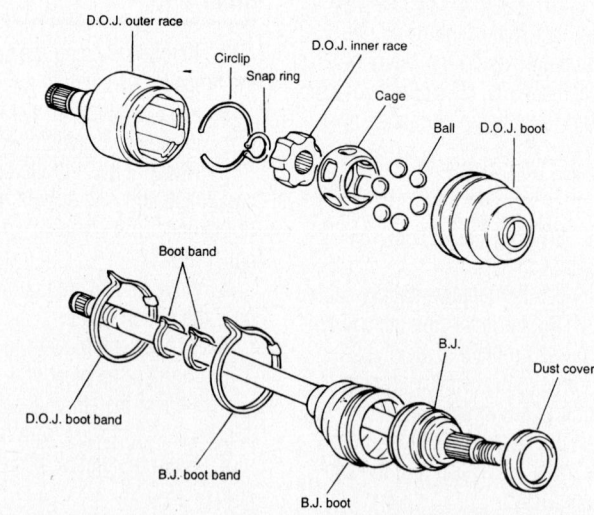

D.O.J. outer race

Circlip

Snap ring

D.O.J. inner race

Cage

Ball   D.O.J. boot

Boot band

D.O.J. boot band

B.J.

Dust cover

B.J. boot band

B.J. boot

TORQUE : Nm (kg·cm, lb·ft)

9355LG41

Exploded view of the DOJ-BJ type CV-joint assembly

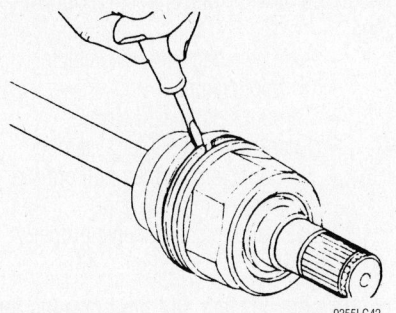

Remove the boot clamps from the DOJ

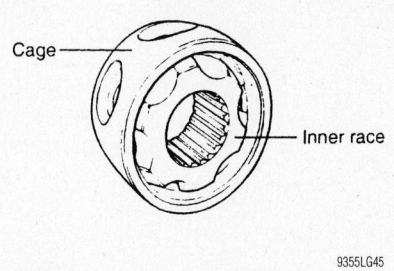

Install the cage so that it is offset on race

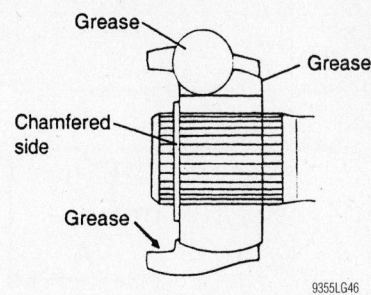

Install the chamfered side of the cage as shown

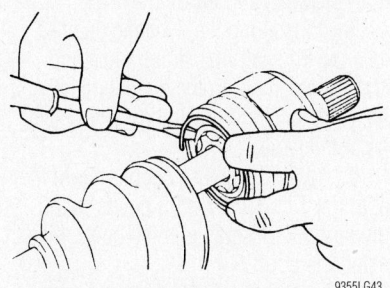

Remove the circlip with a suitable tool

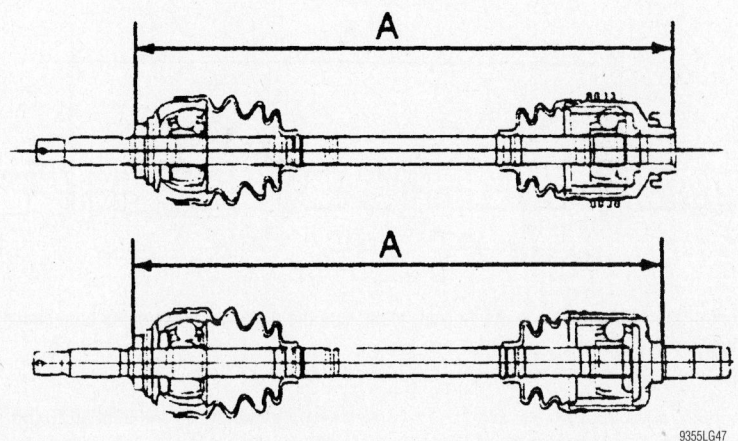

To control the volume of air in the DOJ boot, make sure the distance between the boot bands is as shown

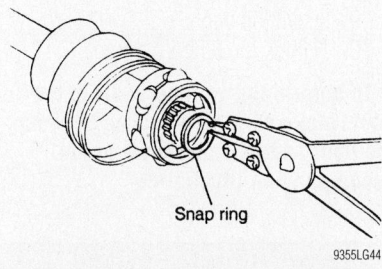

Remove snap-ring and take out the inner race, cage and balls as an assembly

6. Apply the grease supplied with the kit to the inner race and cage.

7. Install the cage so that it is offset on the race.

8. Apply the grease supplied with the kit to the cage and install the balls

9. Install the chamfered side of the cage as shown in the accompanying illustration, then insert the inner race onto the shaft and install the snap-ring.

10. Apply the grease supplied with the kit to the BJ outer race (67–73 grams of grease in the joint and 62–68 grams of grease in the boot); and install the outer race onto the shaft.

11. Apply the grease supplied with the kit to the DOJ outer race (62–68 grams in the joint and 37–43 grams in the boot); and install the circlip.

12. Tighten the DOJ boot clamps.

13. Apply the grease supplied with the kit to the BJ.

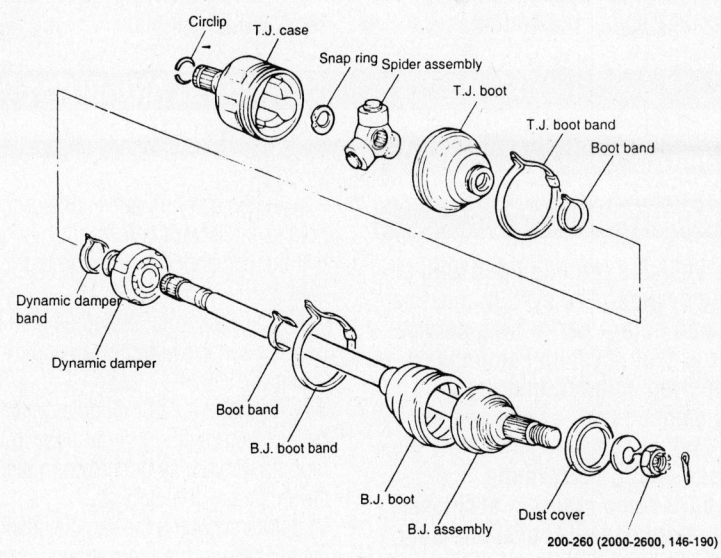

TORQUE : Nm (kg·cm, lb·ft)

Exploded view of the TJ-BJ type CV assembly

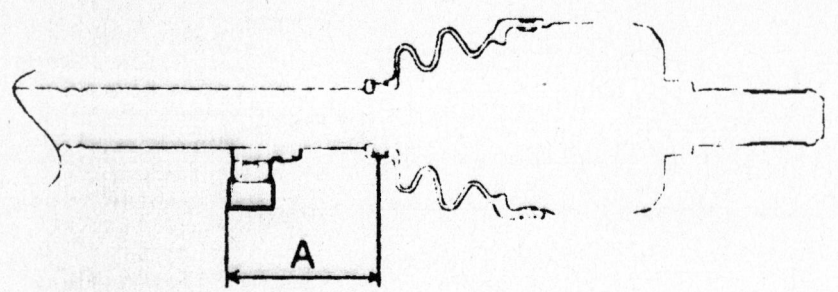

If installing the dynamic damper, keep the BJ shaft in a straight line and attach the damper in the direction shown

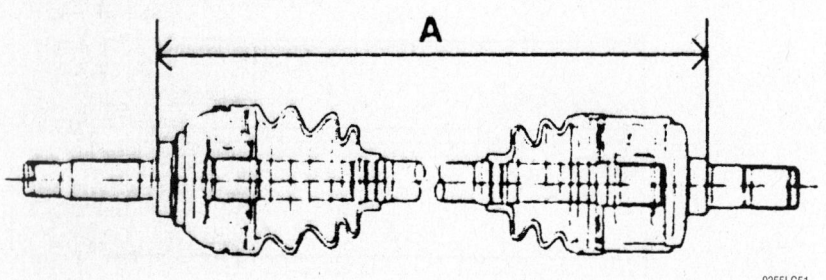

To control the volume of air in the TJ boot, make sure the distance between the boot bands is as shown

14. Install the boots.
15. Tighten the BJ boot clamps

➡ **To control the volume of air in the DOJ boot, make sure the distance between the boot bands is as shown in the accompanying illustration.**

16. Install the axle halfshaft to the vehicle.

TJ-BJ Type

1. Before servicing the vehicle, refer to the precautions section.

2. Remove or disconnect the following:

- Axle halfshaft from the vehicle.
- TJ boot clamps
- TJ boot from the TJ case
- Snap-ring and spider assembly
- BJ boot clamps, then pull out the TJ boot and the BJ boot

3. Clean all the components properly

To install:

➡ **Use new circlips and boot clamps for assembly.**

4. Apply the grease supplied with the kit to the shaft and install the boots.
5. If installing the dynamic damper, keep the BJ shaft in a straight line and attach the damper in the direction shown in the accompanying illustration, then install the boot clamp.
6. Apply the grease supplied with the kit to the TJ boot (97–103 grams in the joint and 42–48 grams in the boot), then install the boot.
7. Tighten the TJ boot clamps.
8. Add the grease supplied with the kit to the BJ as much as was wiped out during cleaning and inspection.
9. Install the boots.
10. Tighten the BJ boot clamps.

➡ **To control the volume of air in the TJ boot, make sure the distance between the boot bands is as shown in the accompanying illustration**

STEERING AND SUSPENSION

Air Bag

✳✳ CAUTION

Some vehicles are equipped with an air bag system. The system must be disarmed before performing service on, or around, system components, the steering column, instrument panel components, wiring and sensors. Failure to follow the safety precautions and the disarming procedure could result in accidental air bag deployment, possible injury and unnecessary system repairs.

PRECAUTIONS

Several precautions must be observed when handling the inflator module to avoid accidental deployment and possible personal injury.

- Never carry the inflator module by the

wires or connector on the underside of the module.

- When carrying a live inflator module, hold securely with both hands, and ensure that the bag and trim cover are pointed away.
- Place the inflator module on a bench or other surface with the bag and trim cover facing up.
- With the inflator module on the bench, never place anything on or close to the module which may be thrown in the event of an accidental deployment.

Before servicing the vehicle, also make sure to refer to the precautions in the beginning of this section as well.

DISARMING

1. Turn the wheel to the straight ahead position.
2. Disconnect the negative battery cable and wait at least 30 seconds for the air bag energy to deplete.

3. After work has been completed, connect the battery cable.

Power Rack and Pinion Steering Gear

REMOVAL & INSTALLATION

1. Before servicing the vehicle, refer to the precautions section.
2. Center the steering wheel and lock it in position.
3. Drain the power steering system.
4. Remove or disconnect the following:

- Negative battery cable
- Pressure and return hoses
- Joint assembly connecting bolt
- Tie rod end from the knuckle
- Feed tube
- Gear box mounting bolts
- Gear box assembly with the rubber mounts

To install:

5. Install or connect the following:
    *   Gear box assembly with the rubber mounts
    *   Gear box mounting bolts and tighten to 66–81 ft. lbs. (90–110 Nm)
    *   Feed tube and tighten to 7–11 ft. lbs. (10–16 Nm)
    *   Tie rod end from the knuckle and tighten to 18–25 ft. lbs. (24–34 Nm)
    *   Joint assembly connecting bolt and tighten to 11–14 ft. lbs. (15–20 Nm)
    *   Pressure and return hoses. Tighten the fittings to 9–13 ft. lbs. (12–18 Nm).
    *   Negative battery cable
6. Fill the power steering system.
7. Check the wheel alignment and adjust as necessary.

Strut

REMOVAL & INSTALLATION

Front

1. Before servicing the vehicle, refer to the precautions section.
2. Remove or disconnect the following:
    *   Front wheel
    *   Brake hose clip from the strut mounting bracket
    *   Wheel Speed Sensor (WSS) from the knuckle
    *   Stabilizer bar link
    *   Strut-to-knuckle bolts
    *   3 upper mounting nuts
    *   Strut

To install:

3. Install or connect the following:
    *   Strut. Tighten the upper mount nuts to 30–37 ft. lbs. (40–50 Nm) on 2002–03 models or 33–44 ft. Lbs. (45–60 Nm) on 2004–06 models.

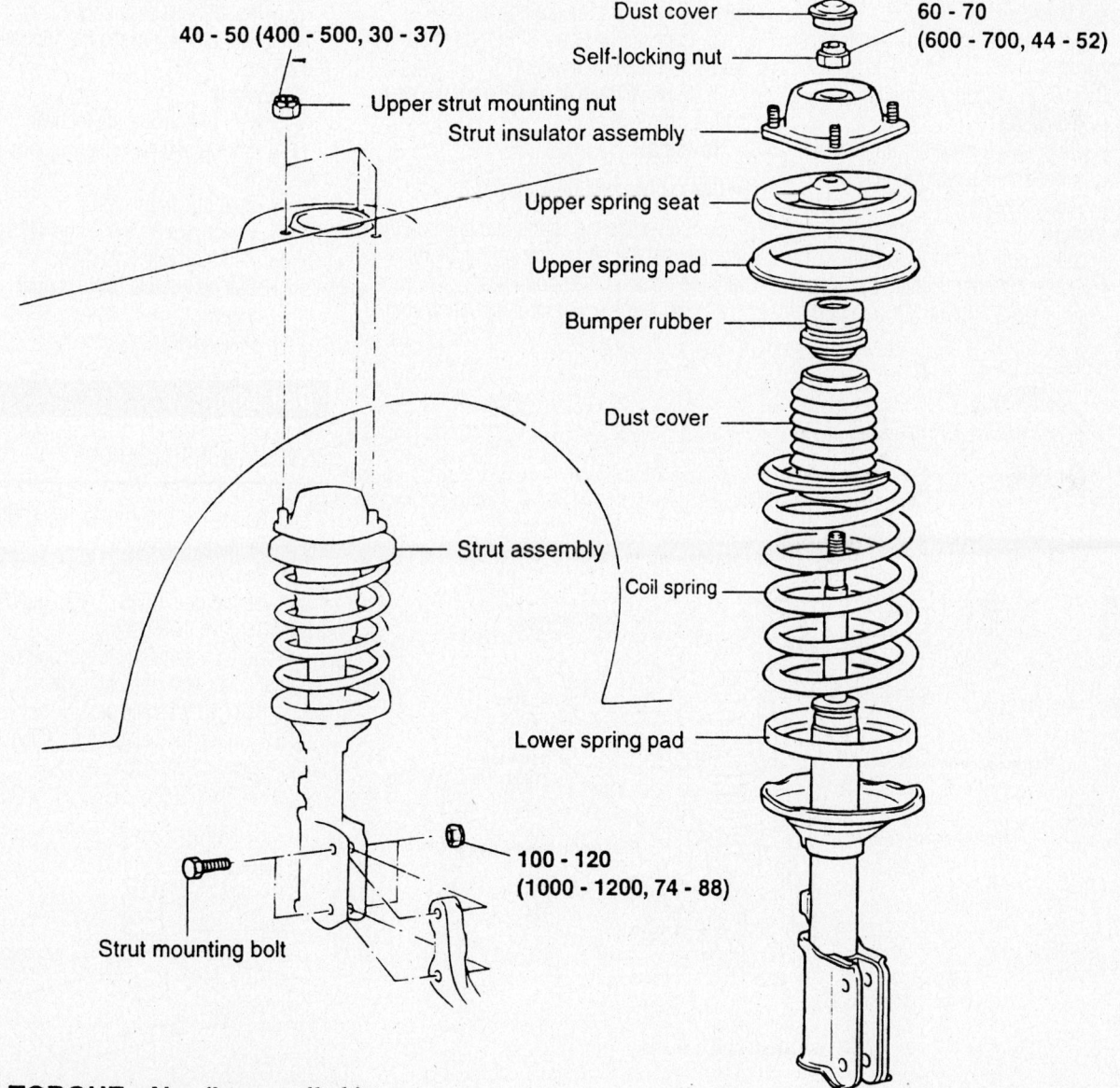

40 - 50 (400 - 500, 30 - 37)

Upper strut mounting nut

Strut assembly

Strut mounting bolt

100 - 120
(1000 - 1200, 74 - 88)

Dust cover

Self-locking nut

60 - 70
(600 - 700, 44 - 52)

Strut insulator assembly

Upper spring seat

Upper spring pad

Bumper rubber

Dust cover

Coil spring

Lower spring pad

TORQUE : Nm (kg·cm, lb·ft)

Exploded view of the front strut assembly and mounting

9355LG52

- Strut-to-knuckle bolts and tighten to 74–88 ft. lbs. (100–120 Nm)
- Stabilizer bar link and tighten to 30–37 ft. lbs. (30–50 Nm)
- WSS from the knuckle
- Brake hose clip to the strut mounting bracket
- Front wheel

4. Check the wheel alignment and adjust as necessary.

Shock Absorber

REMOVAL & INSTALLATION

Rear

1. Before servicing the vehicle, refer to the precautions section.

2. Support the vehicle under the lower control arm.

3. Remove or disconnect the following:
- Rear wheel
- Upper shock absorber upper nut
- Lower shock absorber nut
- Shock absorber

To install:

4. Install or connect the following:
- Shock absorber. Tighten the upper nut to 15–22 ft. lbs. (20–30 Nm)

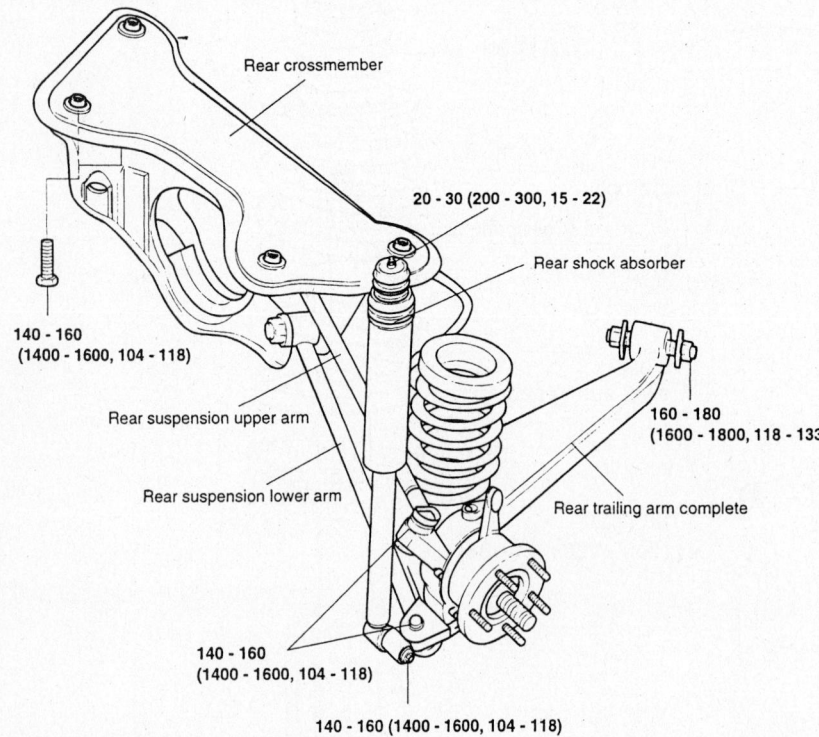

TORQUE : Nm (kg·cm, lb·ft)

Exploded view of the rear suspension assembly

9355LG53

- Tighten the lower nut to 104–118 ft. lbs. (140–160 Nm) on 2002–03 models or 88–103 ft. lbs. (120–140 Nm) on 2004–06 models.
- Rear wheel

Coil Spring

REMOVAL & INSTALLATION

Front

1. Before servicing the vehicle, refer to the precautions section.

2. Remove the strut from the vehicle and install in a strut spring compressor. Compress the spring until the end of the spring comes away from the spring seat.

3. Remove the upper strut mount, spring seat and related components.

4. Remove the coil spring from the strut spring compressor.

To install:

➡**Use a new self-locking nut.**

5. Compress the spring and position the strut so that the end of the spring aligns with the notch in the spring seat.

6. Install the upper strut mounting com-

ponents and tighten the nut to 44–52 ft. lbs. (60–70 Nm).

7. Install the strut to the vehicle.

8. Check the wheel alignment and adjust as necessary.

Rear

1. Before servicing the vehicle, refer to the precautions section.

2. Support the vehicle under the lower control arm.

3. Remove or disconnect the following:
- Rear wheel
- Flange nut and brake caliper assembly
- Parking brake assembly
- Wheel Speed Sensor (WSS) and the parking brake cable
- Rear shock assembly

4. Lower the jack assembly and remove the spring.

To install:

5. Install or connect the following:
- Spring and raise the jack into position
- Rear shock assembly
- Parking brake cable and WSS
- Parking brake assembly
- Flange nut and brake caliper assembly
- Rear wheel

Ball Joint

REMOVAL & INSTALLATION

1. Before servicing the vehicle, refer to the precautions section.

2. Remove the control arm.

3. Using tools 09551-3100 and 09216-21100, remove the bushing.

4. Using a suitable prytool and remove the dust cover from the ball joint.

5. Remove the snap-ring.

6. Remove the ball joint from the arm using a plastic hammer.

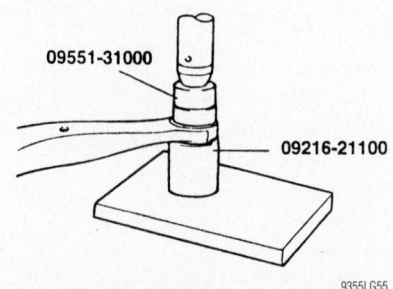

9355LG55

Removing the bushing from the control arm using the appropriate removal and installation tools

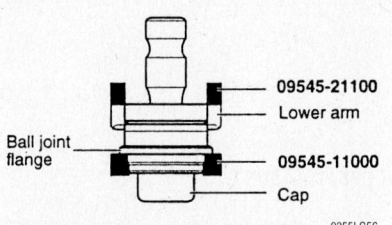

09545-21100
Lower arm
Ball joint flange
09545-11000
Cap

9355LG56

Installing the ball joint dust cover

To install:

7. Install the ball joint.

8. Install the bushings into the arm using the appropriate tools. Make sure that the ball joint flange is supported while pressing down on the bushing until the flange touches the arm surface.

9. Install the ball joint snap-ring. Be careful to keep the snap-ring expansion as small as possible during installation.

10. Apply multi-purpose grease to the dust cover lip and inside of the cover.

11. Using tool 09545-11000, install the dust cover until it is completely seated on the snap ring.

12. Install the control arm.

Lower Control Arm

REMOVAL & INSTALLATION

Front

1. Before servicing the vehicle, refer to the precautions section.

2. Remove or disconnect the following:
   - Front wheel
   - Ball joint-to-knuckle bolt
   - Sub frame bolts and the frame
   - Lower arm bolts and the arm

3. Installation is the reverse of removal. Tighten the fasteners as follows:

   a. Tighten the arm bolt **A** to 74–88 ft. lbs. (100–120 Nm).

   b. Tighten and bolt **B** to 66–81 ft. lbs. (90–110 Nm) on 2002–03 models, or 74–88 ft. lbs. (100–120 Nm) on 2004–06 models.

   c. Tighten the sub frame bolts to 118–148 ft. lbs. (160–200 Nm).

   d. Tighten the ball joint-to-knuckle bolt to 74–88 ft. lbs. (100–120 Nm)

4. Check the wheel alignment and adjust as necessary.

Trailing Arm

REMOVAL & INSTALLATION

Rear

1. Before servicing the vehicle, refer to the precautions section.

2. Support the vehicle under the lower control arm.

3. Remove or disconnect the following:
   - Rear wheel
   - Flange nut and brake caliper assembly

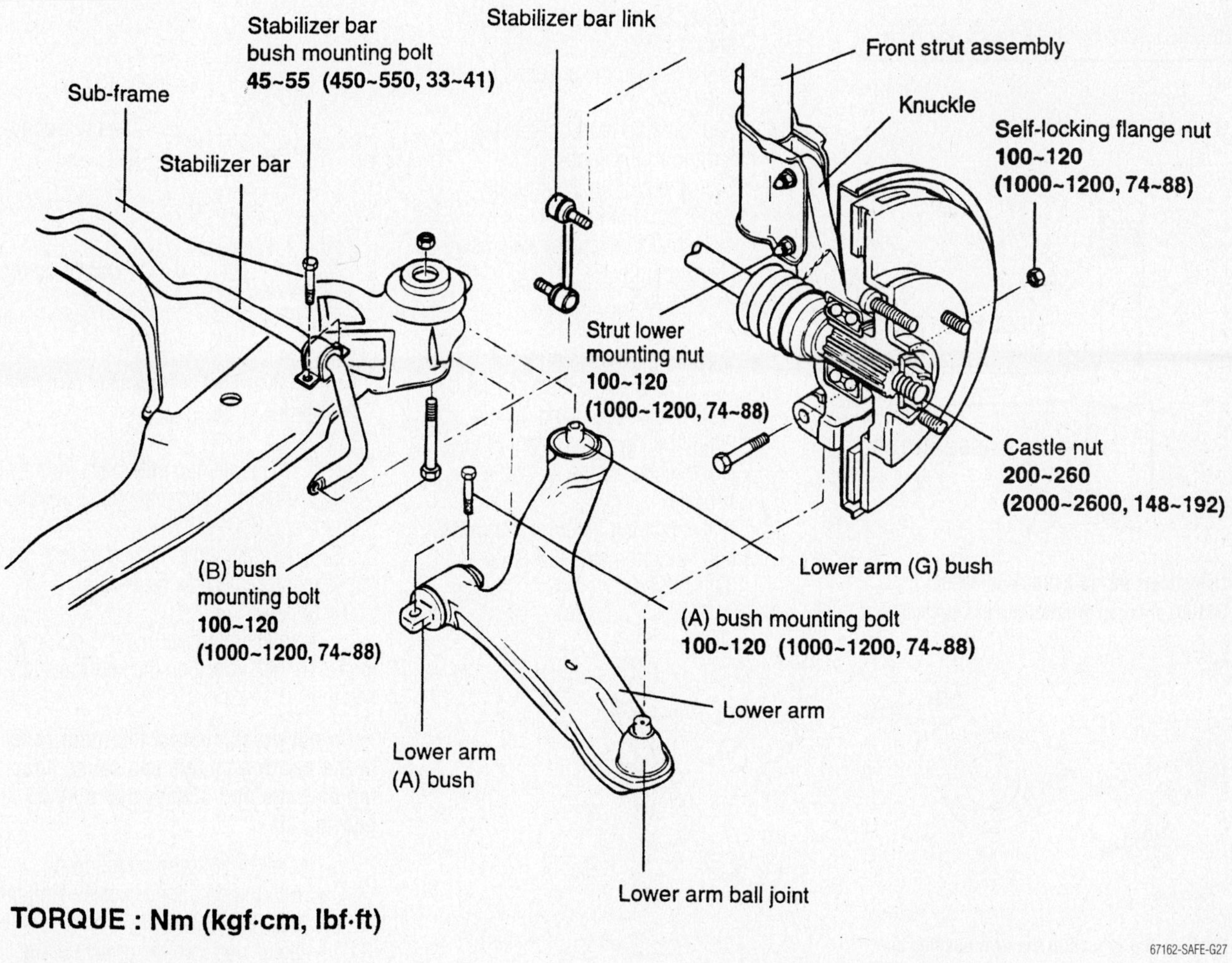

Sub-frame

Stabilizer bar bush mounting bolt
45~55 (450~550, 33~41)

Stabilizer bar

Stabilizer bar link

Front strut assembly

Knuckle

Self-locking flange nut
**100~120
(1000~1200, 74~88)**

Strut lower mounting nut
**100~120
(1000~1200, 74~88)**

Castle nut
**200~260
(2000~2600, 148~192)**

(B) bush mounting bolt
**100~120
(1000~1200, 74~88)**

Lower arm (A) bush

Lower arm (G) bush

(A) bush mounting bolt
100~120 (1000~1200, 74~88)

Lower arm

Lower arm ball joint

TORQUE : Nm (kgf·cm, lbf·ft)

67162-SAFE-G27

Lower control arm assembly

- Parking brake assembly
- Wheel Speed Sensor (WSS) and the parking brake cable
- Rear shock assembly
- Spring
- Rear driveshaft from the rear axle
- Upper and lower arm using tool 09517-43001.
- Trailing arm bolt and the arm

To install:

4. Install or connect the following:
- Trailing arm and tighten the trailing arm bolt to 118–133 (160–180 Nm)
- Upper and lower arm. Tighten the ball joint nut to 103–118 ft. lbs. (140–160 Nm).
- Spring and raise the jack into position
- Rear shock assembly
- Parking brake cable and WSS
- Parking brake assembly
- Flange nut and brake caliper assembly
- Rear wheel

TRAILING ARM BUSHING REPLACEMENT

1. Before servicing the vehicle, refer to the precautions section.

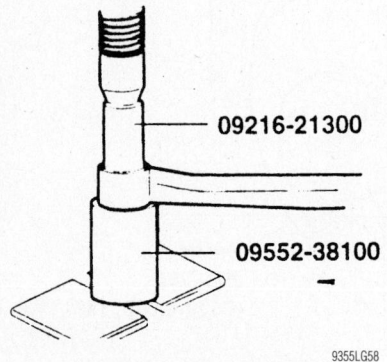

09216-21300

09552-38100

9355LG58

Using tools 09216-21300 and 09552-38100, press fit the trailing arm bushing

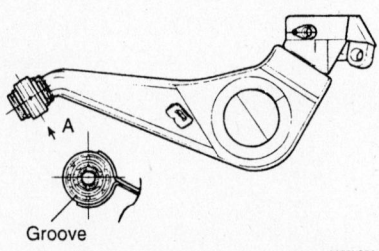

Groove

9355LG59

Position the groove in the arm bushing so that it is aligned as shown, before pressing the bushing into position

2. Remove the trailing arm.
3. Press the bushing from the trailing arm.

➡**Position the groove in the arm bushing so that it is aligned as shown in the accompanying illustration, then press fit the bushing.**

4. Using tools 09216-21300 and 09552-38100, press fit the bushing.

Wheel Bearings

ADJUSTMENT

The wheel bearings are sealed units and are not adjustable.

REMOVAL & INSTALLATION

Front

1. Before servicing the vehicle, refer to the precautions section.
2. Remove or disconnect the following:
- Front wheel
- Wheel Speed Sensor (WSS) from the knuckle
- Brake caliper and suspend it aside using wire
- Split pin and nut from the axle
- Strut from the knuckle
- Tie rod end from the knuckle
- Lower ball joint bolt
- Axle shaft from the knuckle using a plastic hammer
- Brake disc
- Knuckle assembly
- Snap-ring from the hub
- Hub from the knuckle by installing tools 09517-3A00 and 09517-2900, then tighten the nut of the tool to separate the tool from the knuckle
- Wheel bearing inner race from the hub using tools 09455-2100 and 09545-34100

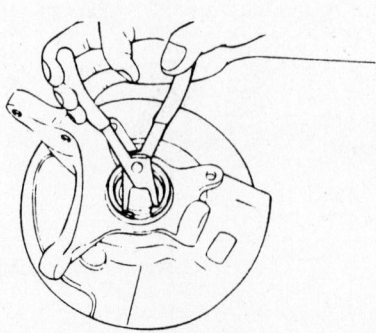

9355LG60

Remove the snap-ring from the hub

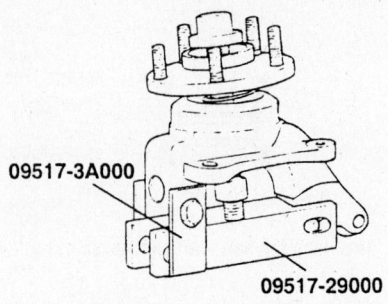

09517-3A000

09517-29000

9355LG61

Remove the hub from the knuckle

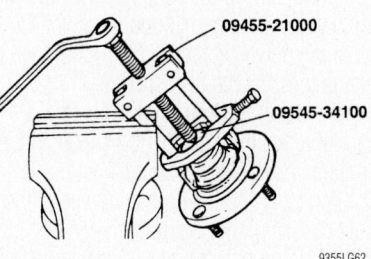

09455-21000

09545-34100

9355LG62

Remove the wheel bearing inner race from the hub

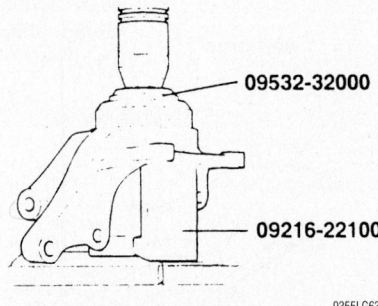

09532-32000

09216-22100

9355LG63

Remove the wheel bearing outer race from the knuckle

- Wheel bearing outer race from the knuckle using tools 09532-3200 and 09216-22100

3. Check all components for wear or damage and replace as necessary.

To install:

4. Apply a thin coat of multi-purpose grease to the surface on the knuckle and bearing.

➡**Do not press against the outer race of the bearing as this can cause bearing damage and always use a new bearing kit.**

5. Install or connect the following:
- Bearing onto the knuckle using tool 09216-21100.
- Snap-ring into the groove of the knuckle
- Backing plate onto the knuckle

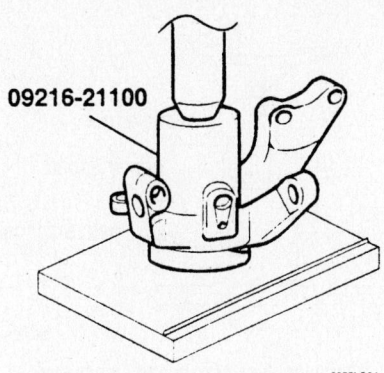

09216-21100

Install the bearing onto the knuckle

9355LG64

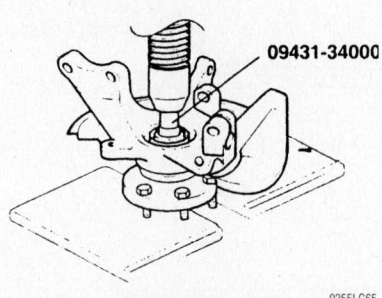

09431-34000

Press the hub onto the knuckle

9355LG65

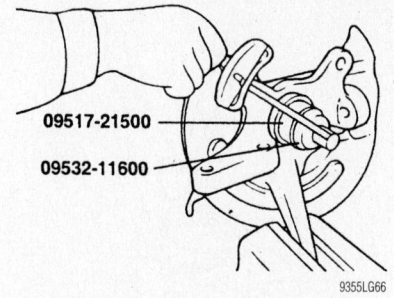

09517-21500

09532-11600

Check the wheel bearing starting torque

9355LG66

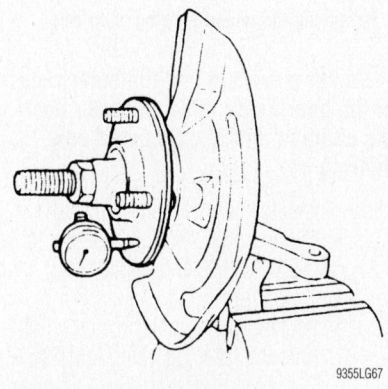

Check the hub end end-play

9355LG67

➥**Do not press against the outer race of the bearing as this can cause bearing damage and always use a new bearing kit.**

• Hub onto the knuckle by pressing it into position using tool 09431-3400

6. Rotate the bearing several times to seat the bearing.

7. Measure the wheel bearing torque using an inch lb. torque wrench. The measurement is 16.64 inch lbs. (1.88 Nm).

8. Measure the end-play of the hub using a dial gauge. The specification is 0.003-0.008mm on 2002–03 models or 0.0025–0.0035 inch (0.064–0.088mm) on 2004–06 models.

9. Install the remaining components in the reverse order of removal.

10. Check the wheel alignment and adjust as necessary.

Rear

1. Before servicing the vehicle, refer to the precautions section.

2. Remove or disconnect the following:

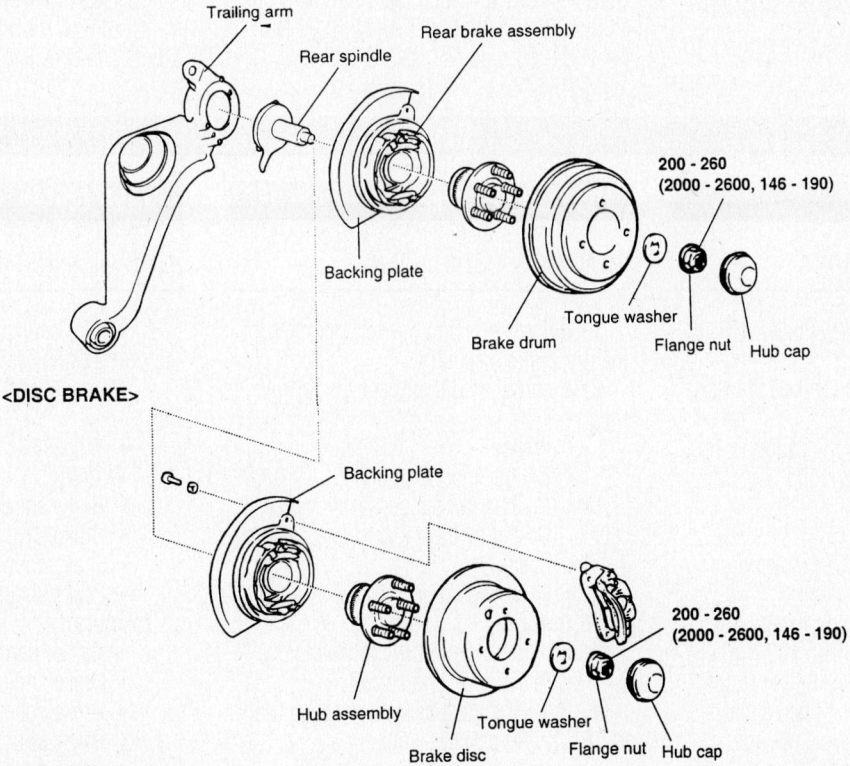

<DRUM BRAKE>

Trailing arm

Rear spindle

Rear brake assembly

Backing plate

200 - 260
(2000 - 2600, 146 - 190)

Tongue washer

Brake drum

Flange nut

Hub cap

<DISC BRAKE>

Backing plate

Hub assembly

Brake disc

Tongue washer

Flange nut

Hub cap

200 - 260
(2000 - 2600, 146 - 190)

TORQUE : Nm (kg·cm, lb·ft)

9355LG68

Exploded view of the rear hub assembly

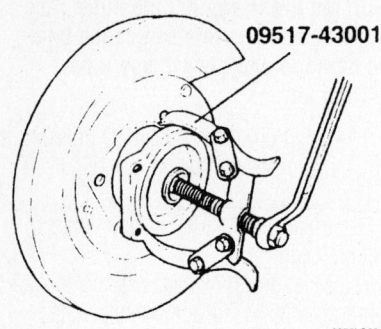

09517-43001

9355LG69

Removing the rear hub from housing

- Rear wheel
- Flange nut and washer
- Drum or rotor
- Brake line
- Parking brake assembly
- Parking brake cable
- Spindle bolts and the spindle
- Rear hub from the housing using tool 09517-43001
- Wheel bearing snap-ring
- Wheel bearing inner race from the housing using tools 09500-2100, 09527-33000 and 09216-22100

3. Inspect the components for damage and replace as necessary.

To install:

4. Apply a thin coat of multi-purpose grease to the surface on the housing and bearing.

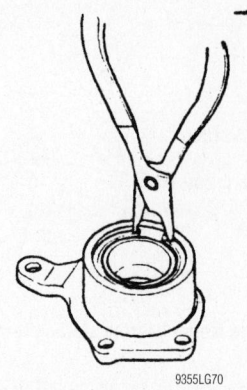

9355LG70

Removing the wheel bearing snap-ring

➡ **Do not press against the outer race of the bearing as this can cause bearing damage and always use a new bearing kit.**

5. Install or connect the following:
- Bearing onto the spindle using tools 09216-21100 and 09532-3200
- Snap-ring
- Backing plate, then press the hub onto the housing using tool 09517-21500

6. Rotate the bearing several times to seat the bearing.

7. Measure the wheel bearing torque using an inch lb. torque wrench. The measurement is 16.64 inch lbs. (1.88 Nm) on 2002–03 models, or 0.73 inch (0.99 Nm) on 2004–06 models.

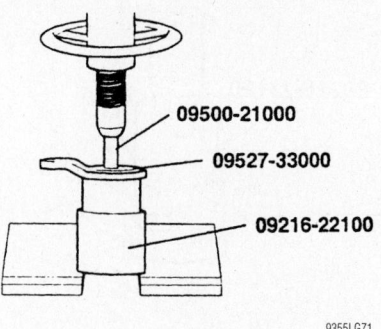

09500-21000
09527-33000
09216-22100

9355LG71

Removing the wheel bearing inner race from housing

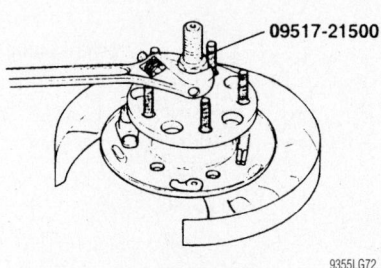

09517-21500

9355LG72

Press the hub onto the housing

8. Measure the end-play of the hub using a dial gauge. The specification is 0.003-0.008mm.

9. Install the remaining components in the reverse order of removal.

10. Check the wheel alignment and adjust as necessary.

BRAKES

Brake Caliper

REMOVAL & INSTALLATION

1. Before servicing the vehicle, refer to the precautions section.
2. Remove or disconnect the following:
- Wheel
- Brake hose from the caliper
- Caliper mounting bolts
- Caliper

To install:

3. Install or connect the following:
- Caliper
- Caliper mounting bolts and tighten to 58–73 ft. lbs. (80–100 Nm) on the front caliper, or 37–44 ft. lbs. 50–60 Nm) on the rear caliper.
- Brake hose to the caliper and tighten the fitting too 18–22 ft. lbs. (25–30 Nm)
- Wheel

4. Bleed the brake system.

Disc Brake Pads

REMOVAL & INSTALLATION

1. Before servicing the vehicle, refer to the precautions section.
2. Remove or disconnect the following:
- Wheel
- Caliper mounting bolts
- Caliper and support aside with wire. Do not let the caliper hang by the hose.
- Pads and shims

To install:

3. Install or connect the following:
- Pads, clips and shims.

4. Bottom the caliper piston using tool 09581-11000 or a C-clamp
- Caliper mounting bolts and tighten to 16–24 ft. lbs. (22–32Nm)
- Wheel

Brake Shoes

REMOVAL & INSTALLATION

1. Before servicing the vehicle, refer to the precautions section.
2. Remove or disconnect the following:
- Wheel
- Drum
- Brake shoe hold-down spring
- Brake strut
- Brake shoe return spring
- Brake shoes

To install:

3. Install or connect the following:
- Brake shoes
- Brake shoe return spring
- Brake strut
- Brake shoe hold-down spring
- Brake drum
- Wheel

4. Adjust the brake shoes.

HYUNDAI

12

Tucson

SPECIFICATIONS AND MAINTENANCE CHARTS

ENGINE AND VEHICLE IDENTIFICATION

		Engine						Model Year	
Code	Liters (cc)	Cu. In.	Cyl.	Fuel Sys.	Engine Type	Eng. Mfg.		Code	Year
B	2.0 (1975)	120.52	4	MPFI	DOHC	Hyundai		5	2005
D	2.7 (2656)	164.30	6	MPFI	DOHC	Hyundai		6	2006

MPFI: Multi-Point Fuel Injection

DOHC: Double Overhead Camshafts

09474_TUCS_C0001

GENERAL ENGINE SPECIFICATIONS

Year	Engine Displacement Liters	Engine ID/VIN	Net Horsepower @ rpm	Net Torque @ rpm (ft. lbs.)	Bore x Stroke (in.)	Compression Ratio	Oil Pressure @ rpm
2005	2.0	B	140@6000	136@4500	3.23 x 3.68	10.1:1	36@1500
	2.7	D	173@6000	178@4000	3.41 x 2.95	10.0:1	NA

NA: Not Available

09474_TUCS_C0002

GASOLINE ENGINE TUNE-UP SPECIFICATIONS

			Spark Plugs Gap	Ignition Timing (deg.)		Fuel Pump	Idle Speed (rpm)		Valve Clearance	
Year	Engine Displacement Liters	Engine ID/VIN	(in.)	MT	AT	(psi)	MT	AT	In.	Ex.
2005	2.0	B	0.039-0.043	8B	8B	50	700	700	0.0079	0.011
	2.7	D	0.039-0.043	12B	12B	50	650	650	HYD	HYD

HYD: Hydraulic Valve Lifters

B: Before Top Dead Center

09474_TUCS_C0003

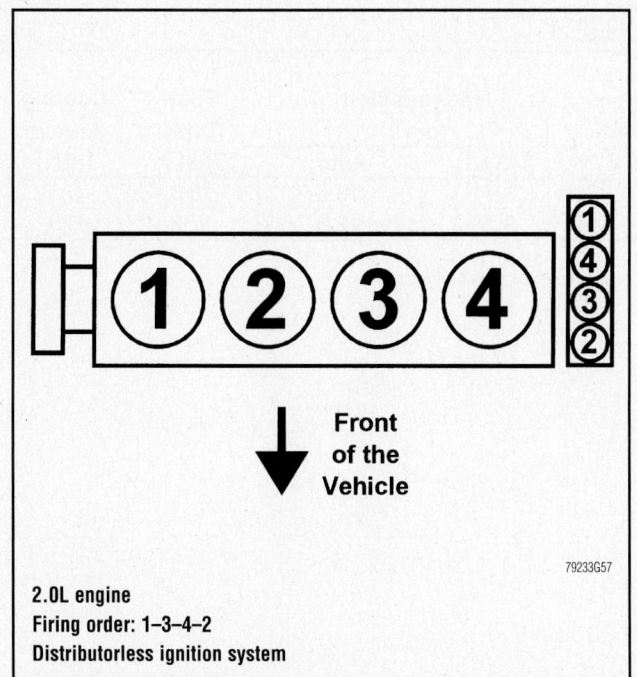

Front of the Vehicle

2.0L engine
Firing order: 1–3–4–2
Distributorless ignition system

79233G57

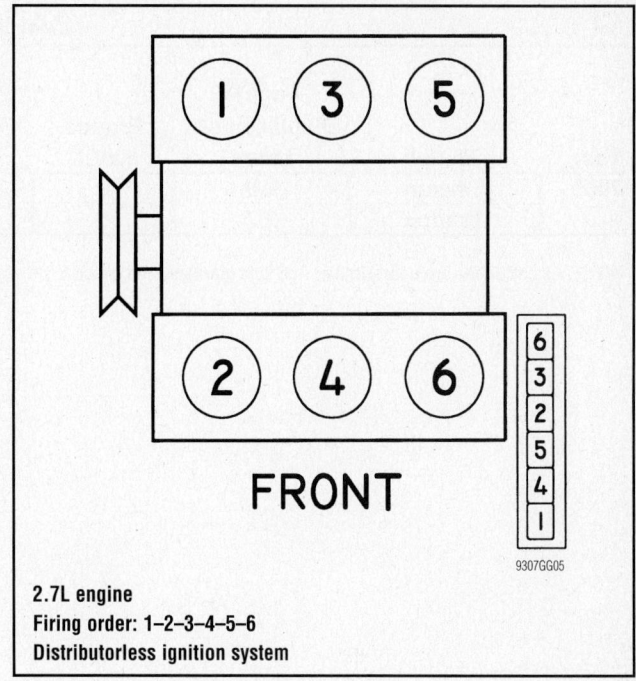

FRONT

2.7L engine
Firing order: 1–2–3–4–5–6
Distributorless ignition system

9307GG05

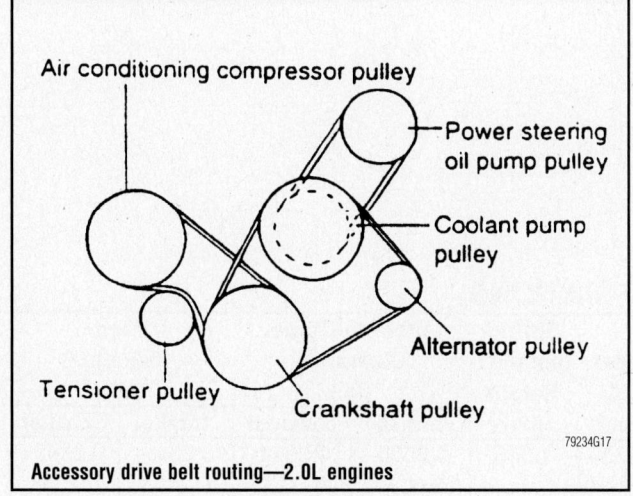

Air conditioning compressor pulley

Power steering oil pump pulley

Coolant pump pulley

Alternator pulley

Tensioner pulley

Crankshaft pulley

79234G17

Accessory drive belt routing—2.0L engines

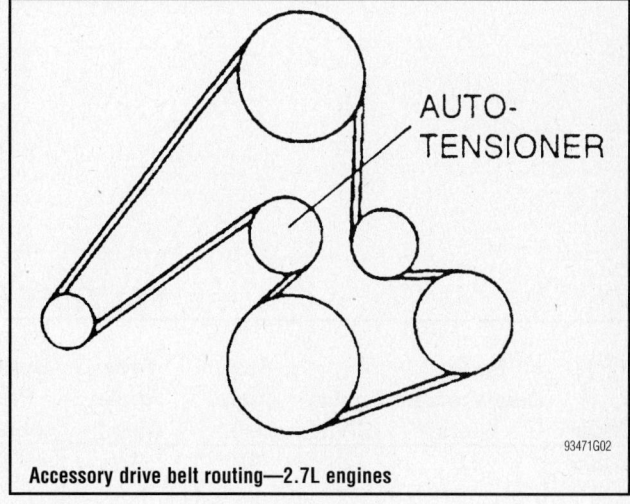

AUTO-TENSIONER

93471G02

Accessory drive belt routing—2.7L engines

CAPACITIES

Year	Model	Engine Displacement Liters	Engine ID/VIN	Engine Oil with Filter	Transmission (pts.) 5–Spd	Transmission (pts.) Auto.	Fuel Tank (gal.)	Cooling System (qts.)
2005	Tiburon	2.0	B	4.0	3.7	13.6	15.3	5.3
	Tiburon	2.7	D	4.5	3.7	13.6	17.2	6.8

NOTE: All capacities are approximate. Add fluid gradually and check to be sure a proper fluid level is obtained.

09474_TUCS_C0004

CRANKSHAFT AND CONNECTING ROD SPECIFICATIONS
All measurements are given in inches.

Year	Engine Displacement Liters	Engine ID/VIN	Crankshaft Main Brg. Journal Dia.	Crankshaft Main Brg. Oil Clearance	Crankshaft Shaft End-play	Crankshaft Thrust on No.	Connecting Rod Journal Diameter	Connecting Rod Oil Clearance	Connecting Rod Side Clearance
2005	2.0	B	2.2440	0.0011-0.0018	0.0023-0.0100	3	1.7700	0.0009-0.0016	0.0039-0.0100
	2.7	D	2.4402-2.4409	0.0002-0.0009	0.0028-0.0098	3	1.8891-1.8898	0.0007-0.0014	0.0039-0.0098

09474_TUCS_C0005

VALVE SPECIFICATIONS

Year	Engine Displacement Liters	Engine ID/VIN	Seat Angle (deg.)	Face Angle (deg.)	Spring Test Pressure (lbs. @ in.)	Spring Installed Height (in.)	Stem-to-Guide Clearance (in.) Intake	Stem-to-Guide Clearance (in.) Exhaust	Stem Diameter (in.) Intake	Stem Diameter (in.) Exhaust
2005	2.0	B	45	45	41@1.535	1.535	0.0008-0.0019	0.0014-0.0026	0.2348-0.2354	0.2343-0.2348
	2.7	D	45	45	49@1.378	N/A	0.0008-0.0020	0.0012-0.0026	0.2350-0.2354	0.2340-0.2350

NA: Not Available

09474_TUCS_C0006

PISTON AND RING SPECIFICATIONS
All measurements are given in inches.

Year	Engine Displacement Liters	Engine ID/VIN	Piston Clearance	Ring Gap			Ring Side Clearance		
				Top Compression	Bottom Compression	Oil Control	Top Compression	Bottom Compression	Oil Control
2005	2.0	B	0.0008-0.0016	0.009-0.015	0.013-0.019	0.008-0.024	0.0015-0.0031	0.0012-0.0027	NA
	2.7	D	0.0004-0.0012	0.008-0.014	0.015-0.020	0.008-0.028	0.0016-0.0031	0.0012-0.0028	NA

NA: Not Available

09474_TUCS_C0007

TORQUE SPECIFICATIONS
All readings in ft. lbs.

Year	Engine Displacement Liters	Engine ID/VIN	Cylinder Head Bolts	Main Bearing Bolts	Rod Bearing Bolts	Crankshaft Damper Bolts	Flywheel Bolts	Manifold		Spark Plugs	Oil Pan Drain Plug
								Intake	Exhaust		
2005	2.0	B	②	③	37-39	125-133	88-95	12-17	32-40	15-22	30-33
	2.7	D	①	④	⑤	130-138	53-56	14-15	22-26	15-22	25-33

① Step 1: 18 ft. lbs.
  Step 2: Plus 58-62 degrees
  Step 3: Plus 43-47 degrees
② Step 1: M10 bolts to 18 ft. lbs. and M12 bolts to 22 ft. lbs.
  Step 2: Plus 60-65 degrees
  Step 3: Plus 60-65 degrees
③ Step 1: 20-24 ft. lbs.
  Step 2: Plus 60-65 degrees

④ M10: 20-24 ft. lbs. plus 90-94 degrees
  M7: 10-14 ft. lbs. plus 90-94 degrees
⑤ 12-15 ft. lbs. plus 90-94 degrees

09474_TUCS_C0008

WHEEL ALIGNMENT

Year	Model		Caster		Camber		Toe-in (in.)
			Range (+/-Deg.)	Preferred Setting (Deg.)	Range (+/-Deg.)	Preferred Setting (Deg.)	
2005	Tucson	F	0.50	3.86	0.50	0	0 +/- 0.08
		R	—	—	0.50	-0.91	0.18 +/- 0.12

09474_TUCS_C0009

TIRE, WHEEL AND BALL JOINT SPECIFICATIONS

| Year | Model | OEM Tires | | Tire Pressures (psi) | | Wheel Size | Ball Joint Inspection | Lug Nut Torque ① |
		Standard	Optional	Front	Rear			
2005	Tuscon	P215/65R16	P235/60R16	30	30	6.5Jx16	N/A	67-81

OEM: Original Equipment Manufacturer

PSI: Pounds Per Square Inch

① Ft. Lbs.

09474_TUCS_C0010

BRAKE SPECIFICATIONS

All measurements in inches unless noted

| Year | Model | | Brake Disc | | | Brake Drum Diameter | | | Minimum Lining Thickness | | Brake Caliper | |
			Original Thickness	Minimum Thickness	Maximum Run-out	Original Inside Diameter	Max. Wear Limit	Maximum Machine Diameter	Front	Rear	Bracket Bolts (ft. lbs.)	Mounting Bolts (ft. lbs.)
2005	Tucson	F	1.014	0.961	0.001	—	—	—	0.079	—	59-74	16-24
		R	0.394	0.315	0.001	—	—	—	—	0.079	59-74	16-24

F: Front

R: Rear

09474_TUCS_C0011

SCHEDULED MAINTENANCE INTERVALS
HYUNDAI—TUCSON

TO BE SERVICED	TYPE OF SERVICE	VEHICLE MILEAGE INTERVAL (x1000)												
		7.5	15	22.5	30	37.5	45	52.5	60	67.5	75	82.5	90	97.5
Engine oil & filter	R	✔	✔	✔	✔	✔	✔	✔	✔	✔	✔	✔	✔	✔
Automatic transaxle fluid	S/I		✔		✔		✔		✔		✔		✔	
Brake pads, calipers & rotors	S/I		✔		✔		✔		✔		✔		✔	
Driveshafts & boots	S/I		✔		✔		✔		✔		✔		✔	
Wheel bearing grease	S/I				✔				✔				✔	
Air cleaner filter	R				✔				✔				✔	
Brake fluid	S/I				✔				✔				✔	
Engine coolant	R								✔				✔	
Fuel hose, vapor hose & fuel filter cap	S/I				✔				✔				✔	
Spark plugs	R				✔				✔				✔	
Drive belts	S/I				✔				✔				✔	
Exhaust pipe connections, muffler & suspension bolts	S/I		✔		✔		✔		✔		✔		✔	
Manual transaxle oil	S/I				✔				✔				✔	
Power steering pump, belt, & hoses	S/I				✔				✔				✔	
Rear brake drums, linings & parking brake	S/I				✔				✔				✔	
Steering gear rack, linkage & boots	S/I				✔				✔				✔	
Timing belt	R								✔					
Fuel filter	R							✔						
Fuel lines & connections	S/I				✔				✔				✔	
Vacuum & crankcase ventilation hoses	S/I							✔						

R: Replace   S/I: Service or Inspect

Automatic transaxle fluid: replace at 105,000 miles.

Air conditioner filter (For Evap and Blower unit): Replace every 12 months or 12,000 miles

FREQUENT OPERATION MAINTENANCE (SEVERE SERVICE)
If a vehicle is operated under any of the following conditions it is considered severe service

- **Extremely dusty areas.**
- **50% or more of the vehicle operation is in 32°C (90°F) or higher temperatures, or constant operation in temperatures below 0°C (32°F).**
- **Prolonged idling (vehicle operation in stop and go traffic).**
- **Frequent short running periods (engine does not warm to normal operating temperatures).**
- **Police, taxi, delivery usage or trailer towing usage.**

Oil & oil filter: change every 3000 miles.

Brake pads, calipers & rotors: service or inspect every 7500 miles.

Driveshaft boots: service or inspect every 7500 miles

Steering gear rack, linkage & boots: service or inspect every 7500 miles.

Air cleaner filter: service or inspect every 15,000 miles.

Automatic transaxle fluid & filter: replace every 15,000 miles.

Rear brake drums & linings: service or inspect every 15,000 miles.

Spark plugs: replace every 24,000 miles.

ENGINE REPAIR

➡Disconnecting the negative battery cable on some vehicles may interfere with the functions of the on board computer system. The computer may undergo a relearning process once the negative battery cable is reconnected.

Distributor

All engines are equipped with a Distributorless Ignition System (DIS).

Alternator

REMOVAL

1. Before servicing the vehicle, refer to the Precautions Section.
2. Remove or disconnect the following:
   • Negative battery cable
   • Alternator wiring harness connectors
   • Alternator drive belt
   • Alternator

INSTALLATION

1. Before servicing the vehicle, refer to the Precautions Section.
2. Install or connect the following:
   • Alternator
   • Alternator drive belt
3. Adjust the alternator belt and torque the bolts to the following specifications:
   a. Lock bolt to 14–18 ft. lbs. (19–25 Nm)
   b. Adjustment bolt to 105–132 inch lbs. (12–15 Nm)
4. Connect the alternator wiring harness connectors.
5. Connect the negative battery cable.

Ignition Timing

ADJUSTMENT

These engines are equipped with a Distributorless Ignition System (DIS). No adjustment is necessary.

Engine Assembly

REMOVAL & INSTALLATION

➡Hyundai recommends that the engine and transaxle be removed as a single unit on all models.

1. Before servicing the vehicle, refer to the Precautions Section.
2. Drain the cooling system.
3. Drain the transaxle.
4. Drain the engine oil.
5. Relieve fuel system pressure.
6. Remove or disconnect the following:
   • Negative battery cable
   • Air intake assembly
   • Engine cover
   • Upper and lower radiator hoses
   • Heater hoses
7. Remove the following engine wiring harnesses:
   a. Oil control valve (OCV) connector
   b. Oil temperature sensor connector
   c. Engine coolant temperature (ECT) sensor
   d. Ignition coil connector
   e. Throttle position sensor (TPS) connector
   f. Idle Speed Actuator (ISA) connector
   g. Camshaft position sensor (CMP) connector
   h. Fuel injector connectors
   i. Ground cable from intake manifold
   j. Compressor switch
   k. Front heated oxygen sensors
   l. Crankshaft position (CKP) connector
   m. Oil pressure switch connector
   n. Purge control solenoid valve (PCSV) connector
8. Remove or disconnect the following:
   • Fuel supply hose
   • PCSV hose
   • Brake booster vacuum hose
   • Throttle cable
   • Power steering pump
   • Battery and battery bracket
   • Transmission wiring harnesses
   • Transmission cooler hose
   • Engine splash guard
   • Front exhaust pipe
   • ABS wheel speed sensor from front knuckles
   • Front strut lower mounting bolts
   • Front calipers
   • Steering U-joint mounting bolt
9. Using a suitable jack, support the engine and transmission assembly.
10. Remove or disconnect the following:
    • Engine mounting bracket
    • Transmission mounting bracket
    • Sub-frame mounting bolts
11. Jack up the vehicle and remove the engine/transmission assembly.

To install:
12. Move the powertrain assembly into position and lower the vehicle into place.

13. Install or connect the following:
    • Sub-frame mounting bolts and nuts. Tighten to nuts to 118–133 ft. lbs. (160–180 Nm) and bolts to 52–66 ft. lbs. (70–90 Nm).
    • Transmission mounting bracket. Tighten to 118–133 ft. lbs. (160–180 Nm).
    • Engine mounting bracket. Tighten to 118–133 ft. lbs. (160–180 Nm).
    • Steering U-joint mounting bolt
    • Front calipers
    • Front strut lower mounting bolts
    • ABS wheel speed sensors
    • Front exhaust pipe
    • Engine splash guard
    • Transmission cooler hose
    • Transmission wiring harness
    • Battery and battery bracket
    • Power steering pump
    • Throttle cable
    • Brake booster vacuum hose
    • PCSV hose
    • Fuel supply hose
    • Engine wiring harnesses
    • Heater hoses
    • Upper and lower radiator hoses
    • Air intake assembly
    • Engine cover
    • Negative battery cable
14. Fill the engine with clean oil.
15. Fill the transaxle to the correct level.
16. Fill the cooling system to the proper level.
17. Start the engine and check for leaks.

Water Pump

REMOVAL & INSTALLATION

1. Before servicing the vehicle, refer to the Precautions Section.
2. Drain the cooling system.
3. Remove or disconnect the following:
   • Negative battery cable
   • Accessory drive belts
   • Timing belt
   • Timing belt idler
   • Water pump pulley
   • Alternator bracket, 2.0L engine only
   • Water pump and gasket

➡The water pump bolts are different lengths. Note the bolt locations for assembly.

To install:
4. Install the water pump with new gasket. Tighten the bolts to the following specifications:

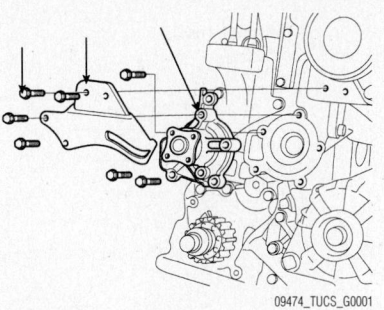

Exploded view of water pump and alternator bracket—2.0L engine

   a. 2.0L engine: 15–20 ft. lbs. (20–27 Nm).

   b. 2.7L engine: 11–16 ft. lbs. (15–22 Nm).

5. Install or connect the following:
- Alternator bracket, 2.0L engine only
- Water pump pulley
- Timing belt idler
- Timing belt
- Accessory drive belts
- Negative battery cable

6. Fill the cooling system to the proper level.

7. Start the engine and check for leaks.

Heater Core

REMOVAL AND INSTALLATION

1. Before servicing the vehicle, refer to the Precautions Section.

2. Discharge and recover the air conditioning system refrigerant.

3. Drain the engine coolant into a clean container for reuse.

4. Disconnect the negative battery cable.

✳✳ CAUTION

After disconnecting the negative battery cable, wait for at least 30 seconds for the SRS module to deplete its stored energy.

5. Remove the bolts and expansion valve from evaporate core and plug the lines.

6. Disconnect the heater hoses from the heater unit.

7. Remove or disconnect the following:
- Front seats
- Center console end cover
- Center console mounting bolts
- Center console
- Crash pad (dashboard) side trim
- Front pillar trim

8. Tilt the steering column down to the lowest position.

9. Remove the mounting screws for cluster trim panel.

10. Disconnect the trip sensor connector.

11. Remove the gauge cluster fascia panel.

12. Disconnect the hood release cable from hood release handle.

13. Remove the screws, bolts and clips from lower crash pad panels.

14. Disconnect the self-diagnosis connector from lower crash pad panel.

15. Remove the lower crash pad panel.

16. Remove the front console side trim.

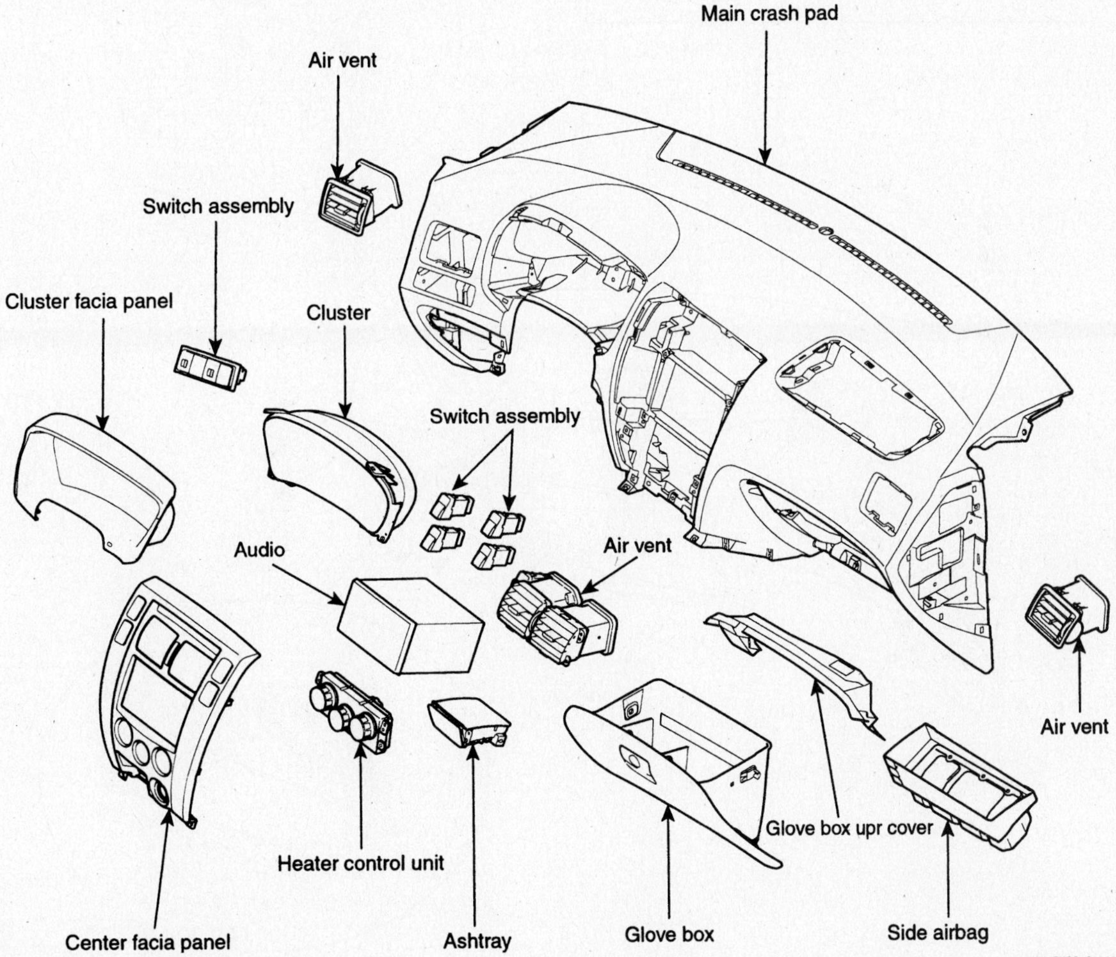

Exploded view of the crash pad components

17. Remove the front console upper cover.

18. Remove front console mounting screws.

19. Remove the front console.

20. Remove the mounting screws and clips for the center fascia panel.

21. Disconnect the electrical connectors to the center fascia panel.

22. Remove the center fascia panel.

23. Remove the radio and disconnect the electrical connectors.

24. Remove the heater control unit.

25. Remove the gauge cluster mounting screws.

26. Disconnect the cluster connectors and remove the gauge cluster.

27. Disconnect the air damper wire and guide from the glove box.

28. Disengage the hinge pins and remove the glove box.

29. Remove the driver airbag module by removing the two mounting bolts.

30. Remove the steering wheel center lock nut.

31. Align the marks on the steering shaft and wheel.

32. Install Special Tool 09561-11002 to remove the steering wheel.

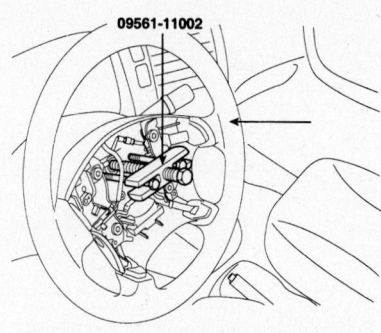

09474_TUCS_G0002

Use steering wheel removal tool 09561-11002 to remove the wheel from the column.

✳✳ WARNING

Do not hammer on the wheel to remove. Damage to the steering column may occur.

33. Remove the upper and lower steering column shrouds.

34. Disconnect the two lower tightening bolts and remove the lower crash pad.

35. Disconnect the passenger's air bag connector.

36. Remove the main crash pad.

37. Remove the cross member.

38. Remove the heater and evaporator unit.

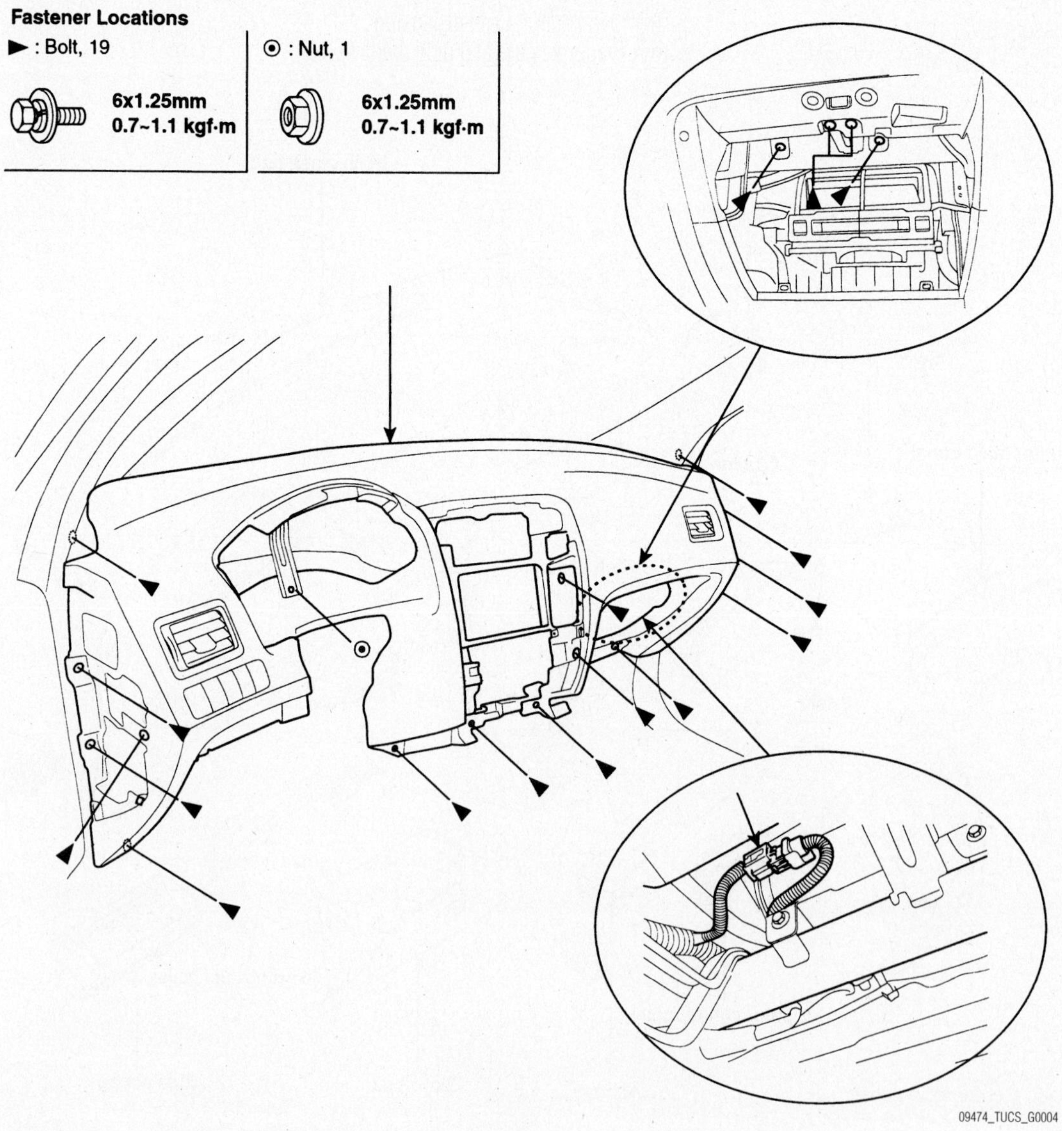

Fastener Locations

▶ : Bolt, 19

6x1.25mm
0.7~1.1 kgf·m

⊙ : Nut, 1

6x1.25mm
0.7~1.1 kgf·m

09474_TUCS_G0004

Location of the crash pad mounting screws and passenger air bag connector.

39. Remove the side bracket and lower cover for the unit to access the heater core.

40. Installation is the reverse of removal. Observe the following:
- Ensure the crash pad fits on the guide pins correctly and no wiring harnesses are pinched.
- Any damaged trim clips must be replaced.

Cylinder Head

REMOVAL & INSTALLATION

2.0L Engine

1. Before servicing the vehicle, refer to the Precautions Section.
2. Drain the cooling system.
3. Relieve the fuel system pressure.
4. Remove or disconnect the following:
- Negative battery cable
- Engine cover
- Air intake assembly
- Upper and lower radiator hose
- Heater hoses
- Engine wiring harnesses connectors and clamps from the cylinder head and intake manifold
- Fuel supply hose
- Purge control solenoid valve (PCSV) hose
- Brake booster vacuum hose
- Throttle cable
- Power steering pump and mounting bracket
- Spark plug cables
- PCV hose
- Cylinder head cover
- Timing belt
- Exhaust manifold
- Intake manifold
- Camshaft sprocket
- Timing chain auto tensioner
- Camshaft bearing caps
- Camshafts
- Oil control valve (OCV)
- OCV filter
- Coolant hose from coolant pipe

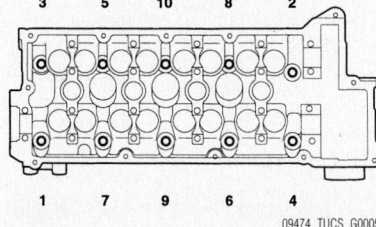

Cylinder head removal sequence—2.0L engine

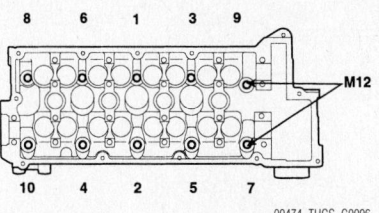

09474_TUCS_G0006

Cylinder head bolt tightening sequence—2.0L engine

5. Remove the cylinder head mounting bolts in sequence as shown.
6. Remove the cylinder head.

To install:

7. Install the cylinder head using a new gasket.
8. Apply a light coat of engine oil to the mounting bolts and tighten in sequence as shown:
   a. Step 1: M10 bolts to 18 ft. lbs. (25 Nm). M12 bolts to 22 ft. lbs. (30 Nm).
   b. Step 2: Plus 60–65 degrees.
   c. Step 3: Plus an additional 60–65 degrees.
9. Install or connect the following:
- OCV filter and tighten to 30–38 ft. lbs. (41–51 Nm).
- OCV and tighten to 7–9 ft. lbs. (10–12 Nm).
- Camshafts and bearing caps
- Timing chain auto tensioner
- Timing belt
10. Install the cylinder head cover and new gasket.
   a. Apply liquid gasket to the head cover gasket at the corners of the recesses.
   b. Tighten the cylinder head cover mounting bolts uniformly in several passes to 6–7.4 ft. lbs. (8–10 Nm).
11. Install or connect the following:
- Intake manifold
- Exhaust manifold
- PCV
- Spark plug wires
- Power steering pump mounting bracket. Tighten to 26–37 ft. lbs. (35–50 Nm).
- Power steering pump
- Throttle cable
- Brake booster hose
- PCSV hose
- Fuel supply hose
- Engine wiring harnesses to the cylinder head and intake manifold
- Heater hoses
- Upper and lower radiator hoses
- Air intake assembly
- Engine cover
- Negative battery cable

12. Fill the engine with coolant to the correct level.
13. Start the engine and check for leaks.

2.7L Engine

1. Before servicing the vehicle, refer to the Precautions Section.
2. Drain the cooling system.
3. Relieve the fuel system pressure.
4. Remove or disconnect the following:
- Negative battery cable
- Engine cover
- Air intake assembly
- Upper and lower radiator hoses
- Heater hoses
- Engine wiring harnesses from the cylinder head and intake manifold
- Fuel supply hoses
- Purge control solenoid valve (PCSV) hose
- Brake booster vacuum hose
- Throttle cable
- PCV hose
- Intake manifold
- Power steering pump
- Exhaust manifold
- Timing belt
- Spark plug cable
- Cylinder head covers
- Camshaft bearing caps
- Camshafts
- Timing belt rear cover
- Water temperature control assembly and water pipe
5. Loosen the cylinder head bolts in several passes in the sequence shown.

✳✳ WARNING

Head warpage or cracking could result from removing bolts in an incorrect order.

6. Remove the cylinder bolts and plate washer.
7. Remove the cylinder heads.

To install:

8. Install the cylinder heads with new gaskets.
9. Apply a light coat of engine oil the cylinder head bolts and tighten in sequence as follows:
   a. Step 1: Tighten to 18 ft. lbs. (25 Nm).
   b. Step 2: Plus 60 degrees.
   c. Step 3: Plus 45 degrees.
10. Install or connect the following:
- Water pipe and water temperature control assembly. Tighten to 11–14 ft. lbs. (15–20 Nm).
- Timing belt rear cover. Tighten to 7–9 ft. lbs. (10–12 Nm).

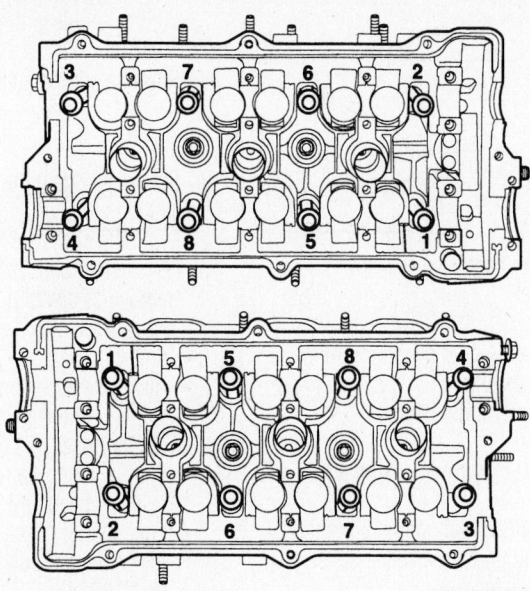

Cylinder head bolt removal sequence—2.7L

09474_TUCS_G0007

- Camshafts and camshaft bearing caps
- Cylinder head cover gaskets. Apply liquid gasket material to the gasket at the corners of the recess.
- Cylinder head covers. Tighten bolts in sequence to 6–7.4 ft. lbs. (8–10 Nm).
- Spark plug cable
- Timing belt
- Exhaust manifold
- Power steering pump
- Intake manifold
- PCV hose
- Throttle cable

- Brake booster vacuum hose
- PCSV hose
- Fuel supply hose
- Engine wiring harnesses to the cylinder head and intake manifold
- Heater hoses
- Upper and lower radiator hoses
- Air intake assembly
- Engine cover
- Negative battery cable

11. Refill the engine with coolant to the correct level.

12. Start the engine and check for leaks.

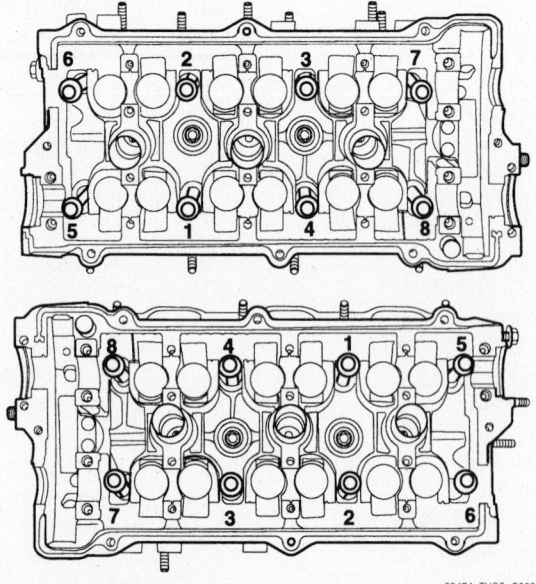

Cylinder head tightening sequence-2.7L

09474_TUCS_G0008

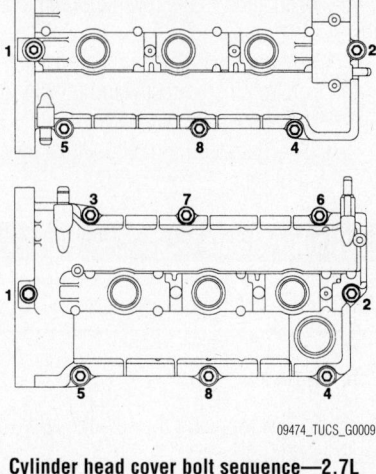

09474_TUCS_G0009

Cylinder head cover bolt sequence—2.7L

Rocker Arms/Shafts

REMOVAL & INSTALLATION

These engines are not equipped with rocker arms. The camshaft acts directly on the valves through hydraulic/mechanical lash adjusters.

Intake Manifold

REMOVAL & INSTALLATION

2.0L Engine

1. Before servicing the vehicle, refer to the Precautions Section.
2. Relieve the fuel system pressure.
3. Drain the cooling system.
4. Remove or disconnect the following:
   - Negative battery cable
   - Engine cover
   - Air intake assembly
   - TPS and ISA connectors
   - PCV and breather hoses
   - Throttle cable
   - Fuel rail
   - Heater hoses
   - PCSV hose
   - Brake booster vacuum hose
   - Intake manifold braces
   - Intake manifold and gasket.
5. Installation is the reverse of removal. Tighten the intake manifold nuts to 12–17 ft. lbs. (16–23 Nm).

2.7L Engine

1. Before servicing the vehicle, refer to the Precautions Section.
2. Relieve the fuel system pressure.

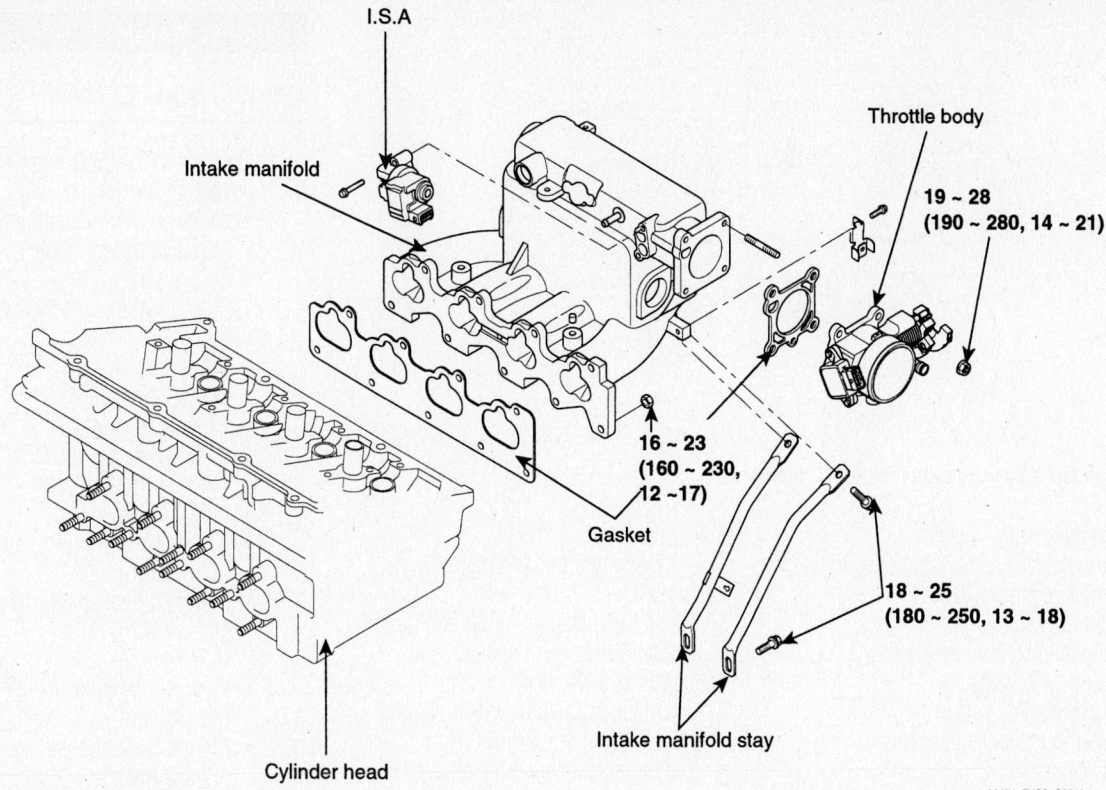

I.S.A

Intake manifold

Throttle body

19 ~ 28
(190 ~ 280, 14 ~ 21)

16 ~ 23
(160 ~ 230,
12 ~17)

Gasket

18 ~ 25
(180 ~ 250, 13 ~ 18)

Intake manifold stay

Cylinder head

09474_TUCS_G0010

Exploded view of intake manifold—2.0L engine

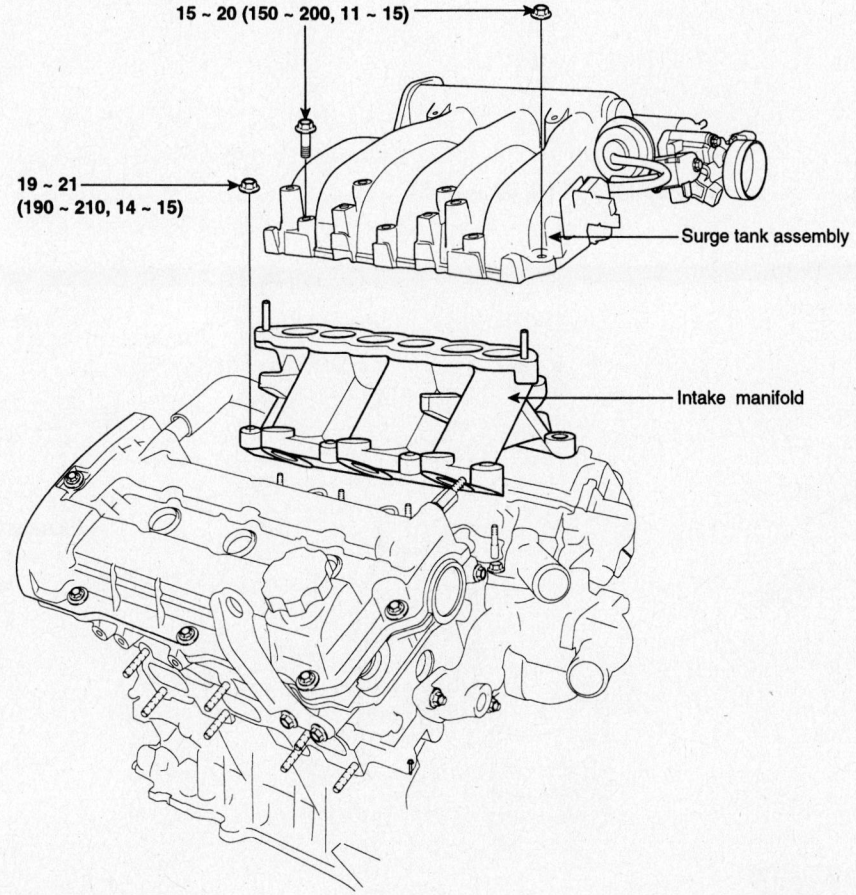

15 ~ 20 (150 ~ 200, 11 ~ 15)

19 ~ 21
(190 ~ 210, 14 ~ 15)

Surge tank assembly

Intake manifold

09474_TUCS_G0011

Exploded view of intake manifold—2.7L engine

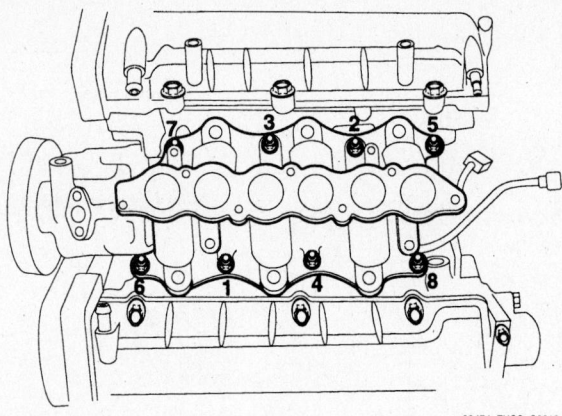

09474_TUCS_G0012

Intake manifold tightening sequence—2.7L engine

3. Drain the cooling system.
4. Remove or disconnect the following:
   - Negative battery cable
   - Throttle cable
   - All electrical connectors to the surge tank assembly
   - PCSV hose
   - Brake booster vacuum hose
   - PCV hose
   - Surge tank brace
   - Surge tank assembly

   - Fuel rail
   - Intake manifold and gasket

To install:

5. Install the intake manifold and gasket.
6. Tighten the bolts in sequence to 14–15 ft. lbs. (19–21 Nm).
7. Install the surge tank assembly and tighten to 11–15 ft. lbs. (15–20 Nm).
8. The remainder of the installation is the reverse of removal.

Exhaust Manifold

REMOVAL & INSTALLATION

1. Before servicing the vehicle, refer to the Precautions Section.
2. Remove or disconnect the following:
   - Negative battery cable
   - Engine cover
   - Front oxygen sensor (O_2S) connector
   - Front muffler
   - Manifold heat shield
   - Exhaust manifold
3. Installation is the reverse of removal. Observe the following torques:
   a. 2.0L Engine
   - Exhaust manifold: 32–40 ft. lbs. (43–55 Nm).
   - Manifold heat shield: 12–16 ft. lbs. (17–22 Nm).
   b. 2.7L Engine
   - Exhaust manifold: 22–26 ft. lbs. (30–35 Nm).
   - Manifold heat shield: 12–16 ft. lbs. (17–22 Nm).
   - Front exhaust pipe: 22–30 ft. lbs. (30–40 Nm).

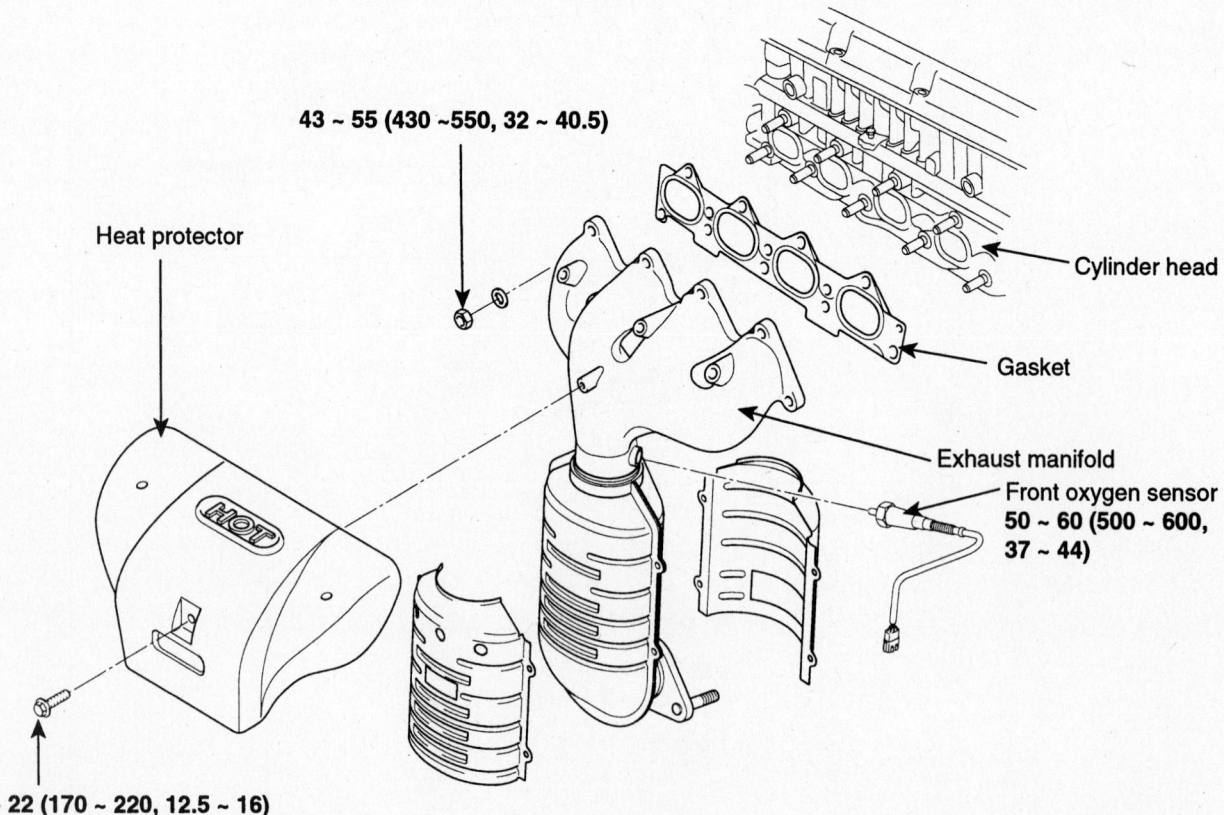

43 ~ 55 (430 ~550, 32 ~ 40.5)

Heat protector

Cylinder head

Gasket

Exhaust manifold

Front oxygen sensor
50 ~ 60 (500 ~ 600, 37 ~ 44)

17 ~ 22 (170 ~ 220, 12.5 ~ 16)

Exhaust manifold exploded view—2.0L engine

09474_TUCS_G0013

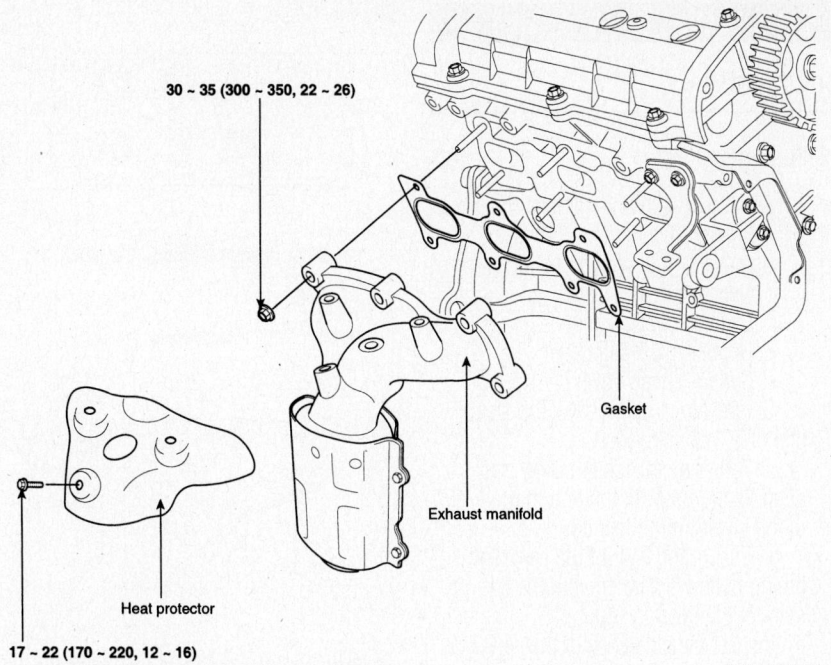

30 ~ 35 (300 ~ 350, 22 ~ 26)

Gasket

Exhaust manifold

Heat protector

17 ~ 22 (170 ~ 220, 12 ~ 16)

09474_TUCS_G0014

Exhaust manifold exploded view—2.7L engine

Front Crankshaft Seal

REMOVAL & INSTALLATION

1. Before servicing the vehicle, refer to the Precautions Section.
2. Remove or disconnect the following:
   - Negative battery cable
   - Accessory drive belts
   - Front covers
   - Timing belt
   - Crankshaft timing sprocket
   - Front crankshaft seal

To install:

3. Install the front crankshaft seal using Special Tool 09214-33000 seal installer
4. Install or connect the following:
   - Crankshaft timing sprocket

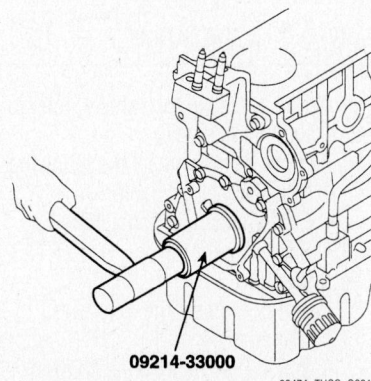

09214-33000

09474_TUCS_G0015

Installing the front crankshaft seal—2.0L engine, 2.7L engine similar

- Timing belt
- Front covers
- Accessory drive belts
- Negative battery cable
5. Start the engine and check for leaks.

Camshaft and Valve Lifters

REMOVAL & INSTALLATION

2.0L Engine

1. Before servicing the vehicle, refer to the Precautions Section.
2. Remove or disconnect the following:
   - Negative battery cable
   - Accessory drive belts
   - Cylinder head cover
   - Timing belt
   - Camshaft sprocket
   - Timing chain auto tensioner
   - Camshaft bearing caps and camshaft timing chain
   - Intake and exhaust camshafts
   - Mechanical lash adjusters

➡**Keep all valvetrain components in order for assembly.**

To install:

3. Install or connect the following:
   - Mechanical lash adjusters in their original positions
   - Intake and exhaust camshafts with the timing chain aligned as shown
   - Camshaft bearing caps. Tighten to 10 ft. lbs. (14 Nm).

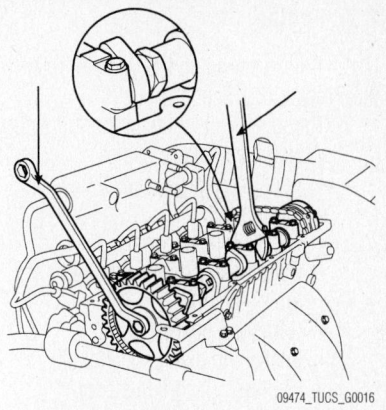

09474_TUCS_G0016

Removing the camshaft sprocket—2.0L, 2.7L similar

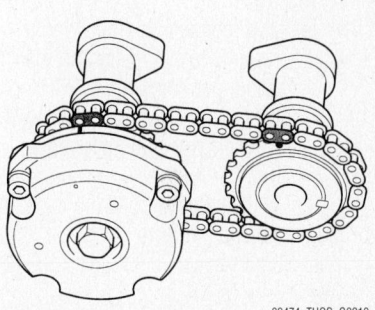

09474_TUCS_G0019

Camshaft timing chain alignment marks—2.0L engine

- Timing chain auto tensioner. Tighten to 6–7 ft. lbs. (8–10 Nm).
4. Using Special Tool 09221-21000 seal installer, install the camshaft bearing oil seal.
5. Install or connect the following:
   - Camshaft sprocket and torque the bolt to 74–89 ft. lbs. (100–120 Nm)
   - Timing belt
   - Cylinder head cover
   - Accessory drive belts
   - Negative battery cable
6. Refill the cooling system to the correct level.
7. Refill the engine oil to the correct level.
8. Start the engine and check for leaks.

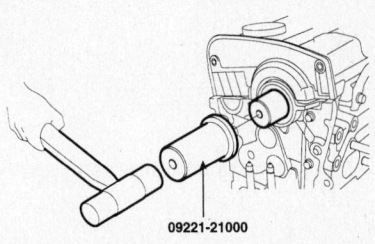

09221-21000

09474_TUCS_G0020

Camshaft bearing oil seal install tool—2.0L, 2.7L similar

2.7L Engine

1. Before servicing the vehicle, refer to the Precautions Section.

2. Remove or disconnect the following:
   - Negative battery cable
   - Accessory drive belts
   - Timing belt
   - Cylinder head cover
   - Camshaft sprocket and camshaft timing chain
   - Camshaft bearing caps
   - Intake and exhaust camshafts
   - Hydraulic lash adjusters

➡**Keep all valvetrain components in order for assembly.**

To install:

3. Install or connect the following:
   - Hydraulic lash adjusters in their original positions
   - Intake and exhaust camshafts with the secondary chain aligned as shown

4. Install the camshaft bearing caps and torque the bolts as follows:

   a. 38 mm bolts: 7–9 ft. lbs. (10–12 Nm).

   b. 50 mm bolts: 10–12 ft. lbs. (14–16 Nm).

5. Using Special Tool 09214-21000, install the camshaft bearing oil seal.

6. Install or connect the following:
   - Camshaft sprocket and torque the bolt to 65–80 ft. lbs. (90–110 Nm)
   - Cylinder head cover
   - Timing belt
   - Accessory drive belts
   - Negative battery cable

7. Refill the cooling system to the correct level.

8. Fill the engine with oil to the correct level.

9. Start the engine and check for leaks.

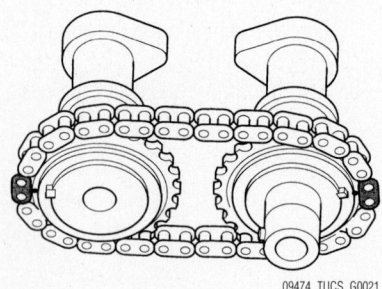

Camshaft timing chain alignment marks—2.7L

Valve Lash

ADJUSTMENT

2.0L Engine

1. Before servicing the vehicle, refer to the Precautions Section.

2. Remove or disconnect the following:
   - Negative battery cable
   - Engine cover
   - Upper timing belt cover
   - Cylinder head cover

3. Set the No. 1 cylinder to Top Dead Center (TDC) as follows:

   a. Turn the crankshaft pulley and align its groove with the timing mark "T" of the lower timing belt cover.

   b. Ensure the hole of the camshaft timing pulley is aligned with the timing mark of the bearing cap.

4. Inspect the clearance of the valves as shown.

5. Turn the crankshaft one revolution and align the timing mark "T" of the lower timing belt cover to set the No. 4 cylinder and TDC.

6. Inspect the clearance of the valves as shown.

7. If adjustment is necessary:

   a. Turn the crankshaft so that the cam lobe on the adjusting valve is upward.

   b. Using Special Tool 09220-2D000, press down on the valve lifter and place a stopper between the camshaft and valve lifter and remove the special tool.

   c. Remove the adjusting shim with a small screwdriver and magnet.

   d. Measure the thickness of the removed shim using a micrometer.

   e. Calculate the thickness of a new shim so that the valve clearance comes within the specified valve.

   f. Place a new adjusting shim on the valve lifter.

   g. Using Special Tool 09220-2D000, press down on the valve lifter and remove the stopper.

   h. Recheck the valve clearance.

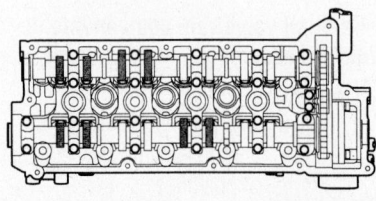

Valves to inspect with no. 1 cylinder at TDC

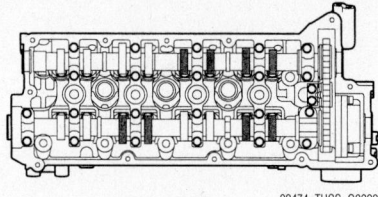

Valves to inspect with no. 4 at TDC

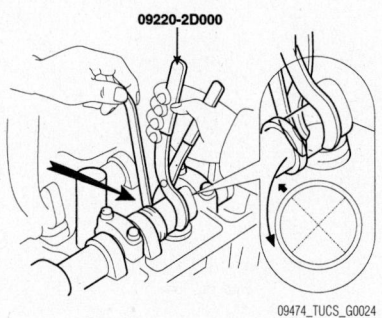

Press down on the valve lifter with special tool and place a stopper between the camshaft and valve lifter.

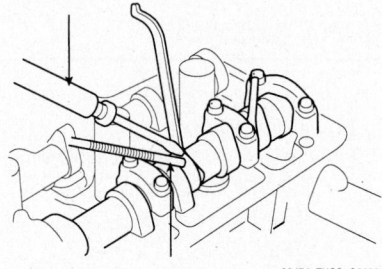

Remove adjusting shim with screwdriver and magnet

2.7L engines use hydraulic valve lash adjusters. Valve lash adjustments are not necessary or possible on these engines.

Starter Motor

REMOVAL & INSTALLATION

1. Before servicing the vehicle, refer to the Precautions Section.

2. Remove or disconnect the following:
   - Negative battery cable
   - Starter electrical connectors
   - Starter motor

To install:

3. Install or connect the following:
   - Starter motor
   - Starter electrical connectors
   - Negative battery cable

Oil Pan

REMOVAL & INSTALLATION

2.0L Engine

1. Before servicing the vehicle, refer to the Precautions Section.
2. Remove the engine from the vehicle and separate from the transaxle.
3. Install the engine to an engine stand.
4. Remove the oil pan mounting bolts.
5. Remove the oil pan.

To install:

6. Using liquid gasket MS 721-40A or equivalent, apply an even bead along the groove in the oil pan.
7. Install the oil pan with the 19 bolts. Uniformly tighten the bolts in several passes to 7.2–8.8 ft. lbs. (10–12 Nm).
8. The remainder of installation is the reverse of removal.
9. Fill the engine with clean oil.
10. Start the vehicle and check for leaks.

2.7L Engine

1. Before servicing the vehicle, refer to the Precautions Section.
2. Remove the engine from the vehicle and separate from the transaxle.
3. Install the engine to an engine stand.
4. Remove the lower oil pan.
5. Remove the oil screen.
6. Remove the upper oil pan.

To install:

7. Using liquid gasket MS 721-40A or equivalent, apply an even bead along the groove in the upper oil pan.
8. Install the upper oil pan. Tighten the bolts uniformly in several passes as follows:
   a. Tighten bolts 1-15 to 14–20 ft. lbs. (19–28 Nm).
   b. Tighten bolts 16 and 17 to 4–5 ft. lbs. (5–7 Nm).
9. Install the oil screen. Tighten the mounting bolts to 11–16 ft. lbs. (15–22 Nm).

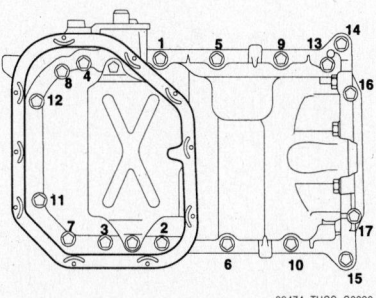

Upper oil pan bolts—2.7L engine

10. Using liquid gasket MS 721-40A or equivalent, apply an even bead along the groove in the lower oil pan.
11. Install the lower oil pan and tighten the bolts uniformly in several passes to 7.3–8.8 ft. lbs. (10–12 Nm).
12. The remainder of installation is the reverse of removal.
13. Fill the engine with clean oil.
14. Start the vehicle and check for leaks.

Oil Pump

REMOVAL & INSTALLATION

2.0L Engine

1. Before servicing the vehicle, refer to the Precautions Section.
2. Drain the engine oil.
3. Remove or disconnect the following:
- Negative battery cable
- Accessory drive belts
- Timing belt
- Oil pan
- Oil screen
- Front case
- Cover from the front case
- Inner and outer rotors from the front case

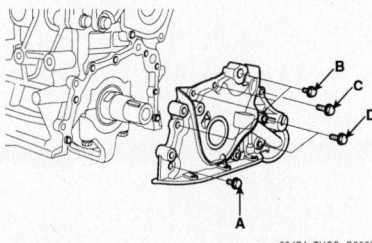

Front case mounting—2.0L engine

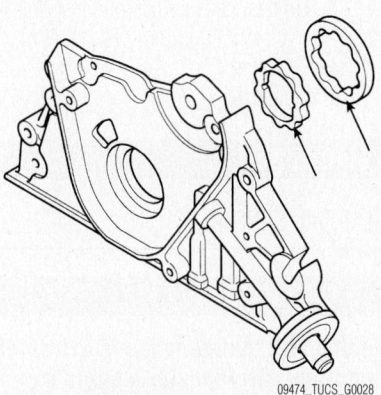

Inner and outer rotors removed from case—2.0L engine

To install:

4. Place the inner and outer rotors into the front case with the marks facing the oil pump cover side.
5. Install the oil pump cover to the front case. Tighten the bolts to 4.4–6.6 ft. lbs. (6–9 Nm).
6. Ensure that the oil pump moves freely.
7. Install the oil pump to the engine block. Note the bolts lengths and installation locations as shown. Tighten the bolts to 14.5–20 ft. lbs. (20–27 Nm).
8. Install or connect the following:
- Front crankshaft seal
- Oil screen
- Oil pan
- Timing belt
- Accessory drive belts
- Negative battery cable
9. Fill the engine with oil to the correct level.
10. Start the engine and check for leaks.

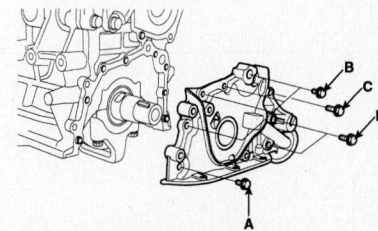

Bolt length
(A) : 25mm (0.98in)
(B) : 20mm (0.787in)
(C) : 38mm (1.496in)
(D) : 45mm (1.771in)

Oil pump bolt locations—2.0L engine

2.7L Engine

1. Before servicing the vehicle, refer to the Precautions Section.
2. Drain the engine oil.
3. Remove or disconnect the following:
- Negative battery cable
- Right-hand front wheel
- Right-hand side cover
- Front exhaust pipe
- Alternator
- Accessory drive belts
- Timing belt
- Lower oil pan
- Oil screen
- Oil pump case
- Cover from oil pump housing
- Inner and outer rotors

To install:

4. Place the inner and outer rotors into front case with the marks facing the oil pump cover side.

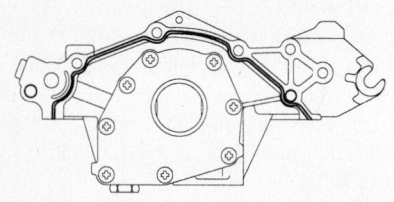

Liquid gasket installation—2.7L engine

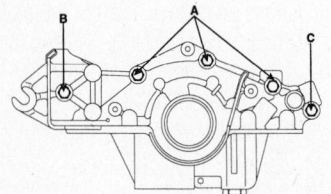

Bolt	Size	Number
A	8 × 25	3
B	8 × 35	1
C	8 × 45	1

Oil pump bolt locations—2.7L engine

5. Install the oil pump cover to the front case. Tighten the bolts to 6–9 ft. lbs. (8–12 Nm).

6. Ensure the oil pump turns freely.

7. Apply liquid gasket to the oil pump housing as shown.

8. Place a new O-ring on the cylinder block.

9. Engage the spline teeth of the oil pump drive gear with the large teeth of the crankshaft and slide the oil pump into place.

10. Tighten the mounting bolts to 9–11 ft. lbs. (12–15 Nm). Note the bolt lengths and mounting locations as shown.

11. Install or connect the following:
- Front crankshaft seal
- Oil screen
- Lower oil pan
- Accessory drive belts
- Alternator
- Front exhaust pipe
- Right-hand side cover
- Right-hand wheel
- Negative battery cable

12. Fill the engine with oil to the correct level.

13. Start the engine and check for leaks.

Rear Main Seal

REMOVAL & INSTALLATION

1. Before servicing the vehicle, refer to the Precautions Section.

2. Remove or disconnect the following:
- Transaxle

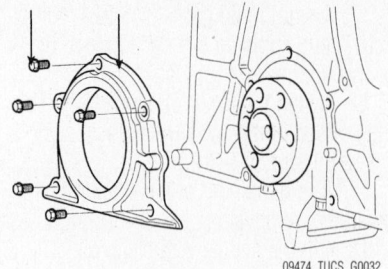

Oil seal case—2.0L engine

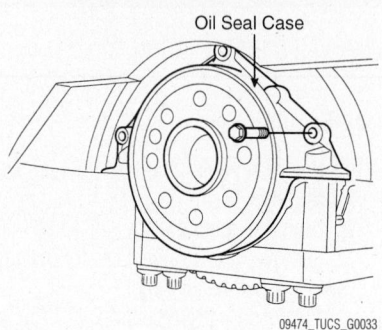

Oil seal case—2.7L engine

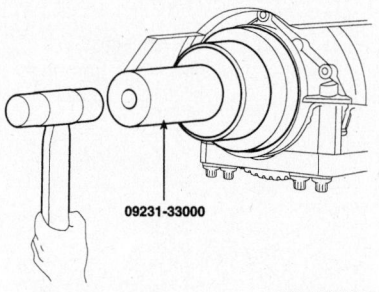

Seal installer tool—2.7L, 2.0L similar

- Flywheel
- Oil seal case
- Oil seal

To install:

3. Install or connect the following:
- Oil seal, using Seal Installer tool
- Oil seal case and torque the case bolts to 7–9 ft. lbs. (10–12 Nm)
- Flywheel
- Transaxle

Timing Belt

REMOVAL & INSTALLATION

❊❊ CAUTION

Timing belt maintenance is extremely important. All Hyundai models use interference-type non-freewheeling engines. Should the timing belt break in these engines, the valves in the cylinder head will come in contact with the pistons, causing major engine damage. The recommended replacement interval for timing belts is 60,000 miles.

2.0L Engine

1. Before servicing the vehicle, refer to the Precautions Section.

2. Remove or disconnect the following:
- Negative battery cable
- Engine cover
- Right-hand front wheel
- Engine splash guard

3. Using a block of wood under the oil pan, support the engine with a suitable jack.

4. Remove the engine mount bracket.

5. Loosen the water pump pulley bolts.

6. Remove or disconnect the following:
- Accessory drive belts
- Water pump pulley
- Upper timing belt cover

7. Rotate the crankshaft clockwise and align the groove with the timing mark "T" of the timing belt cover.

8. Remove or disconnect the following:
- Lower timing belt cover
- Timing belt tensioner
- Timing belt

➡ **If timing belt is being reused, mark the belt indicating its turning direction to ensure the belt is reinstalled in the same direction.**

- Timing belt idler
- Crankshaft sprocket

To install:

9. Install the crankshaft sprocket.

10. Align the timing marks of the

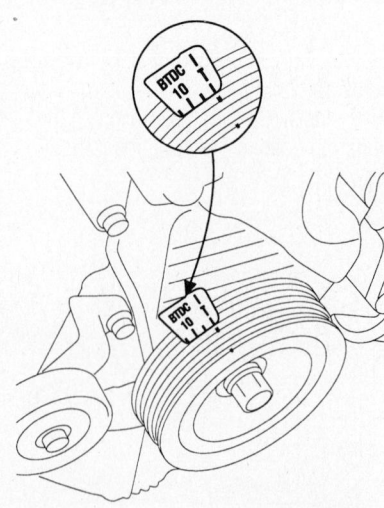

Crankshaft pulley timing marks—2.0L engine

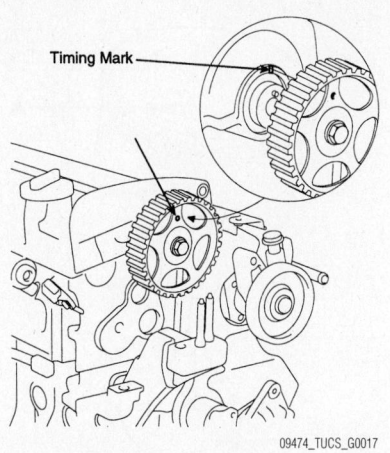

Camshaft sprocket alignment marks—2.0L engine

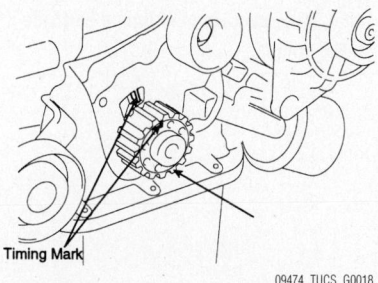

Crankshaft sprocket timing mark—2.0L engine

camshaft sprocket and crankshaft sprocket with the No. 1 piston placed at top dead center of its compression stroke.

11. Install the idler pulley and tighten the bolt to 32–40 ft. lbs. (43–55 Nm).

12. Temporarily install the timing belt tensioner with a plain washer.

13. Position the timing belt over the crankshaft sprocket, inside the idler pulley, over the camshaft sprocket and then inside the timing belt tensioner.

14. Rotate the crankshaft clockwise the angle equivalent of two teeth of the camshaft sprocket.

15. Give tension to the timing belt by rotating the tensioner in the direction of the arrow. Tighten the tensioner bolt to 32–40 ft. lbs. (43–55 Nm).

16. Check the belt tension by depressing the center of the belt span with an index finger. The deflection should be 0.16–0.24 inches (4–6 mm). Adjust if necessary.

17. Turn the crankshaft clockwise two turns and realign the crankshaft and camshaft sprocket timing marks.

18. Install or connect the following:
- Lower timing belt cover. Tighten to 6–7 ft. lbs. (8–10 Nm).
- Flange and crankshaft pulley.

Tighten to 125–133 ft. lbs. (170–180 Nm).
- Upper timing belt cover. Tighten to 6–7 ft. lbs. (8–10 Nm).
- Water pump pulley
- Accessory drive belts
- Engine mount bracket. Tighten engine side bolts and nuts to 37–48 ft. lbs. (50–60 Nm). Tighten body side nut to 44–59 ft. lbs. (60–80 Nm).
- Splash guard
- Right hand wheel
- Engine cover
- Negative battery cable

2.7L Engine

1. Before servicing the vehicle, refer to the Precautions Section.

2. Remove or disconnect the following:
- Negative battery cable
- Engine cover
- Right-hand front wheel
- Right-hand side cover

3. Rotate the crankshaft clockwise and align the groove with the timing mark "T" of the timing belt cover.

4. Remove the accessory drive belts and belt tensioner.

5. Using a block of wood under the oil pan, support the engine with a suitable jack.

6. Remove or disconnect the following:
- Engine mount bracket
- Power steering pump

- Upper timing belt cover
- Crankshaft pulley
- Drive belt idler pulley
- Lower timing belt cover
- Front engine support bracket
- Timing belt tensioner
- Timing belt

➡ **If timing belt is being reused, mark the belt indicating its turning direction to ensure the belt is reinstalled in the same direction.**

- Tensioner pulley
- Timing belt idler pulley
- Crankshaft sprocket

To install:

7. Align the pulley set key with the key groove on the crankshaft sprocket and slide on the crankshaft sprocket. Tighten to 65–80 ft. lbs. (90–110 Nm).

8. Install the timing idler pulley and tighten to 36–43 ft. lbs. (50–60 Nm).

9. Install the timing belt tensioner pulley and tighten the arm fixed bolt to 25–40 ft. lbs. (35–55 Nm).

10. Align the timing marks of the camshaft sprocket and crankshaft sprocket with the No. 1 piston placed at top dead center of its compression stroke.

11. Using a suitable press, slowly press in the push rod of the timing belt tensioner. Align the holes of the push rod and housing. Pass a set pin through the holes to keep the position of the push rod.

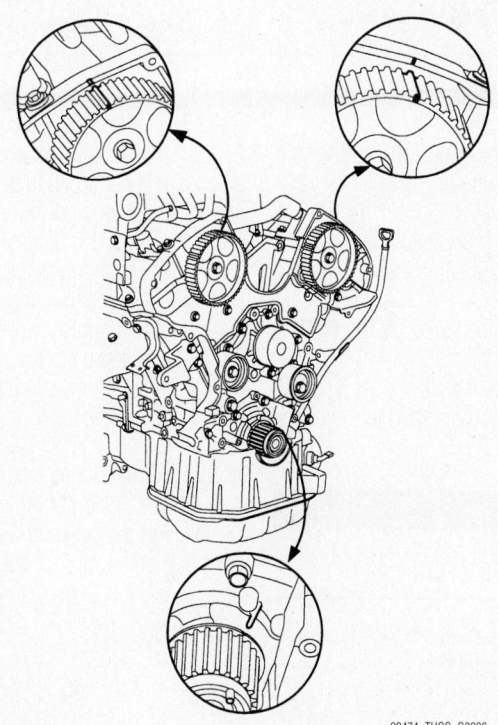

Timing marks—2.7L engine

12. Install the tensioner and alternately tighten the bolts to 14–20 ft. lbs. (20–27 Nm).

13. To install the timing belt, position it over the crankshaft sprocket, then the idler pulley, over the left side camshaft sprocket, under the water pump pulley, over the right side camshaft sprocket and then side the tensioner pulley.

14. Remove the set pin from the tensioner.

15. Turn the crankshaft clockwise two turns and realign the crankshaft and camshaft sprocket timing marks.

16. Check the belt tension by depressing the center of the belt span with an index finger. The deflection should be 0.27–0.31 inches (7–9 mm). Adjust if necessary.

17. Install or connect the following:
- Front engine support bracket. Tighten large bolts to 43–51 ft. lbs. (60–70 Nm). Tighten the small bolt to 11–16 ft. lbs. (15–22 Nm).
- Lower timing belt cover. Tighten to 7–9 ft. lbs. (10–12 Nm).
- Idler pulley. Tighten to 25–40 ft. lbs. (35–55 Nm).
- Crankshaft pulley. Tighten to 130–138 ft. lbs. (180–190 Nm).

- Timing belt upper cover
- Power steering pump
- Drive belt tensioner
- Accessory drive belt
- Engine mount bracket. Tighten nuts and bolts to 44–59 ft. lbs. (60–80 Nm).
- Right-hand side cover
- Right-hand front wheel
- Engine cover
- Negative battery cable

Piston and Ring

POSITIONING

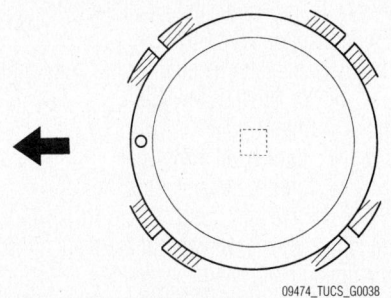

09474_TUCS_G0038

Piston ring end-gap spacing

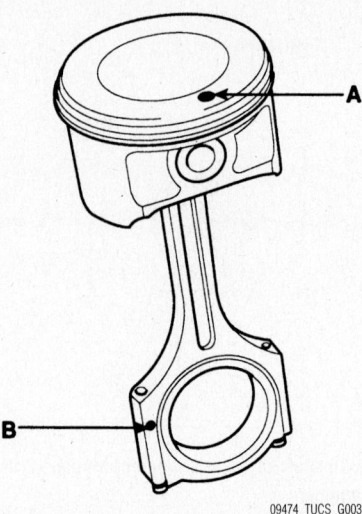

09474_TUCS_G0037

The piston front mark and connecting rod front mark must face the timing belt side of the engine.

FUEL SYSTEM

Fuel System Pressure

RELIEVING

1. Before servicing the vehicle, refer to the Precautions Section.
2. Remove or disconnect the following:
- Rear seat cushion
- Access panel
- Fuel pump module connector
3. Start the engine and allow it to run until it stalls.
4. Turn the ignition switch to the **OFF** position.
5. Disconnect the negative battery cable.
6. Attach the fuel pump harness connector.

Fuel Filter

REMOVAL & INSTALLATION

The fuel filter is part of the fuel pump assembly located in the tank.

Fuel Pump

REMOVAL & INSTALLATION

1. Before servicing the vehicle, refer to the Precautions Section.
2. Relieve the fuel system pressure.
3. Drain the fuel tank.
4. Remove or disconnect the following:
- Negative battery cable
- Second seat
- Carpet covering service cover for fuel pump
- Service cover
- Fuel pump wiring connector
- Fuel hoses
- Fuel pump plate mounting bolts
- Fuel pump module
5. Installation is the reverse of removal.
6. Fill the tank with fuel and check for proper fuel pump operation.

Fuel Injector

REMOVAL & INSTALLATION

2.0L Engine

1. Before servicing the vehicle, refer to the Precautions Section.
2. Relieve the fuel system pressure.
3. Remove or disconnect the following:
- Negative battery cable
- Engine cover
- Air intake assembly
- TPS and ISA connectors
- PCV and breather hoses
- Throttle cable
- Fuel rail
- Fuel injector
4. Installation is the reverse order of removal.

2.7L Engine

1. Before servicing the vehicle, refer to the Precautions Section.
2. Relieve the fuel system pressure.

3. Drain the cooling system.
4. Remove or disconnect the following:
   - Negative battery cable
   - Throttle cable

- All electrical connectors to the surge tank assembly
- PCSV hose
- Brake booster vacuum hose
- PCV hose

- Surge tank brace
- Surge tank assembly
- Fuel rail

5. Installation is the reverse order of removal.

DRIVE TRAIN

Transaxle Assembly

REMOVAL & INSTALLATION

Manual

1. Before servicing the vehicle, refer to the Precautions Section.
2. Drain the transaxle.
3. Remove or disconnect the following:
   - Negative battery cable
   - Air intake assembly
   - Battery and battery tray
   - Back-up lamp connector
   - Vehicle speed sensor
   - Clutch release cylinder and lever
   - Shift cable from transaxle assembly
   - Steering column from the universal joint in the gear box
   - Clutch housing upper mounting bolts
   - Front, rear and left transaxle mounting brackets
4. Using a suitable engine support fixture, support the engine assembly.
5. Remove or disconnect the following:
   - Power steering pressure hose from the pump
   - Front wheel
   - Strut assembly
   - Tie rod and sway bar link from the knuckle
   - Wheel speed sensor
   - Brake caliper
   - Engine splash guard
   - Front muffler
   - Power steering hose on the front cross member
6. Using a suitable jack, support the sub-frame cross member.
7. Remove the cross member mounting bolts, and lower the cross member assembly with the steering gear and stabilizer bar attached.
8. Using a suitable jack, support the transaxle assembly.
9. Remove the front and rear roll stoppers.
10. Remove the engine and transaxle mounting bolts.
11. Slowly lower the transaxle from the vehicle.

12. Installation is the reverse order of removal.

Automatic

1. Before servicing the vehicle, refer to the Precautions Section.
2. Drain the transaxle.
3. Remove or disconnect the following:
   - Negative battery cable
   - Air intake assembly
   - Battery and battery tray
   - Intercooler inlet pipe
   - Transaxle wiring harnesses
   - Bolt which mounts the clutch release cylinder to the inhibitor switch
   - Clutch release cylinder clip
   - Oil cooler hose clamps
4. Using Special Tool 09200-38001 to support the engine.
5. Remove or disconnect the following:
   - Transaxle mounting bracket bolts
   - Transaxle upper mounting bolts
   - Bolts which mounts the transaxle to the sub-frame
6. Support the transaxle with a suitable jack.
7. Remove or disconnect the following:
   - Steering column bolt
   - Halfshafts
   - Bolt which mounts the transaxle to the rear sub-frame
   - Drive shaft, if equipped with 4WD
   - Lower transaxle mounting bolts
   - Transaxle assembly
8. Installation is the reverse of removal.

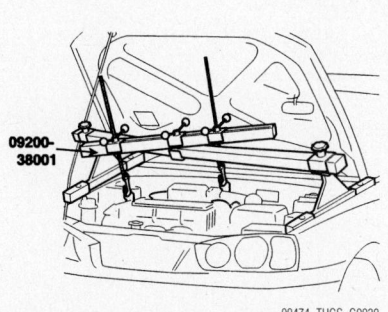

09474_TUCS_G0039

Engine support fixture

Transfer Case

1. Before servicing the vehicle, refer to the Precautions Section.
2. Drain the transfer case.
3. Remove or disconnect the following:
   - Negative battery cable
   - Drive shaft
   - Front muffler
   - Right-hand halfshaft
4. Support the transfer case assembly with a suitable jack.
5. Remove the transfer case mounting bolts and slowly lower the transfer case assembly from the vehicle.
6. Installation is the reverse of removal.

Clutch

ADJUSTMENTS

These vehicles are equipped with a hydraulic clutch system. No adjustment is necessary.

REMOVAL & INSTALLATION

1. Before servicing the vehicle, refer to the Precautions Section.
2. Remove the transaxle assembly.
3. Insert Special Tool 09411-11000 in the clutch disc to prevent the disc from falling.

09411-11000

09474_TUCS_G0040

Installing special tool 09411-11000

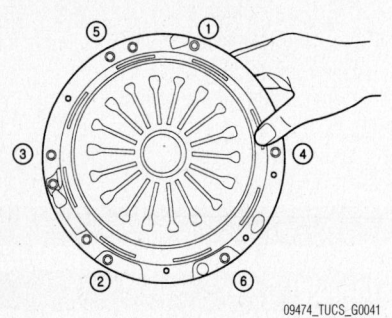

Pressure plate tightening sequence

4. Remove or disconnect the following:
   - Pressure plate bolts by loosening them evenly in a star pattern.
   - Release fork shaft and bushing
   - Pressure plate and clutch disc

To install:

5. Install or connect the following:
   - Clutch disc on the flywheel
   - Pressure plate and torque the bolts evenly in a crossing pattern to 11–16 ft. lbs. (15–22 Nm)
6. Align the bearing to the release fork

and then install it to the sleeve of the housing.

7. Install the release lever to the release fork.
8. Install the transaxle assembly.

Hydraulic Clutch System

BLEEDING

1. Connect a hose to the bleeder screw and place the other end of hose into a container of clean brake fluid. Open the bleeder screw.
2. Have an assistant pump the clutch pedal slowly until no air bubbles are present at the bleeder screw.
3. Close the bleeder screw.
4. Fill the clutch master cylinder.
5. Check the clutch operation.

Halfshaft

REMOVAL & INSTALLATION

Front

1. Before servicing the vehicle, refer to the Precautions Section.
2. Remove or disconnect the following:
   - Front wheel
   - Spindle nut
   - Wheel speed sensor, if equipped
   - Lower ball joint
   - Lower arm ball joint mounting nuts
3. Press the stub shaft out of the hub.
4. Using a suitable pry bar, pry the inner joint out of the transaxle.

To install:

➡**Use new circlips, split pins and self-locking nuts for assembly.**

5. Install the inner joint so that the circlip is felt to seat in the retaining groove.
6. Guide the stub shaft into the hub.
7. Install or connect the following:
   - Lower ball joint. Tighten the mounting bolts to 74–89 ft. lbs. (100–120 Nm)
   - Wheel speed sensor, if equipped
   - Spindle nut and torque the nut to 148–207 ft. lbs. (200–280 Nm)
   - Front wheel
8. Check and/or adjust the wheel alignment.

Rear

1. Before servicing the vehicle, refer to the Precautions Section.

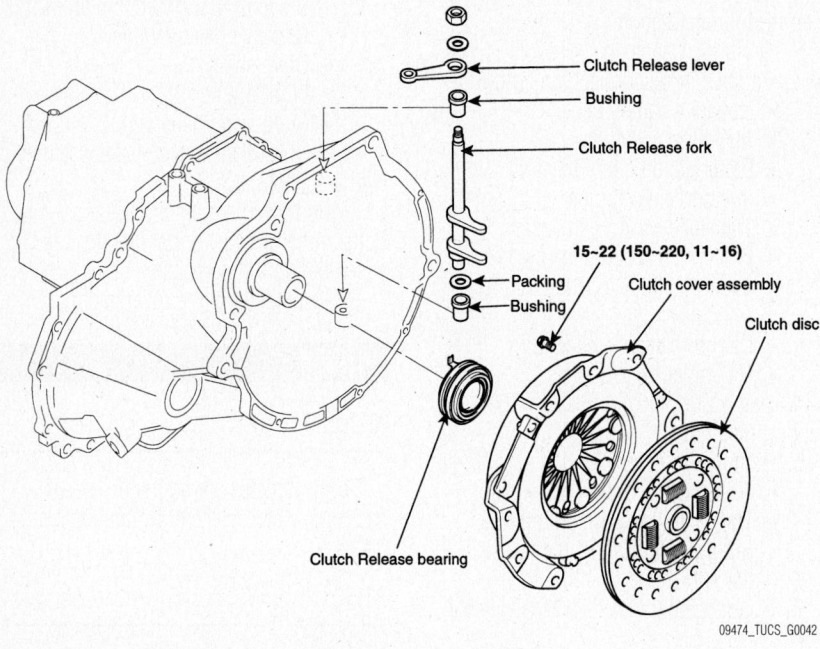

Exploded view of the clutch disc and pressure plate components

- Clutch Release lever
- Bushing
- Clutch Release fork
- 15~22 (150~220, 11~16)
- Packing
- Bushing
- Clutch cover assembly
- Clutch disc
- Clutch Release bearing

1. Driveshaft (LH)
2. Circlip
3. Transaxle
4. Driveshaft (RH)

Exploded view of the halfshaft mounting and related components—Front

2. Remove or disconnect the following:
- Rear wheel
- Wheel speed sensor, if equipped
- Spindle nut
- Trailing arm mounting bolt
- Lower arm mounting nuts

3. Press the stub shaft out of the hub.

4. Using a suitable pry bar, pry the inner joint out of the differential.

To install:

➡**Use new circlips, split pins and self-locking nuts for assembly.**

5. Install the inner joint to the differential so that the circlip is felt to seat in the retaining groove and the halfshaft cannot be removed by hand.

6. Install or connect the following:
- Lower arm mounting nuts. Tighten to 104–118 ft. lbs. (140–160 Nm).
- Trailing arm mounting bolt. Tighten to 74–89 ft. lbs. (100–120 Nm).
- Spindle nut. Tighten to 148–207 ft. lbs. (200–280 Nm).
- Wheel speed sensor, if equipped
- Rear wheel

Driveshaft

1. Before servicing the vehicle, refer to the Precautions Section.

2. Matchmark the rubber coupling and rear differential companion flange.

3. Remove the rear differential driveshaft mounting bolts.

4. Remove the center bearing bracket.

5. Matchmark the flange yoke and transaxle companion flange.

6. Remove the mounting bolts and remove the driveshaft.

7. Installation is the reverse of removal, aligning the matchmarks made during removal. Note the following torques:

a. Front driveshaft mounting bolts: 37–44 ft. lbs. (50–60 Nm).

b. Center bearing bracket: 30–37 ft. lbs. (40–50 Nm).

c. Rear driveshaft mounting bolts: 74–89 ft. lbs. (100–120 Nm).

CV-Joints

OVERHAUL

Double Offset Joint-Birfield Joint Type

1. Before servicing the vehicle, refer to the Precautions Section.

2. Remove the halfshaft and place into a vise.

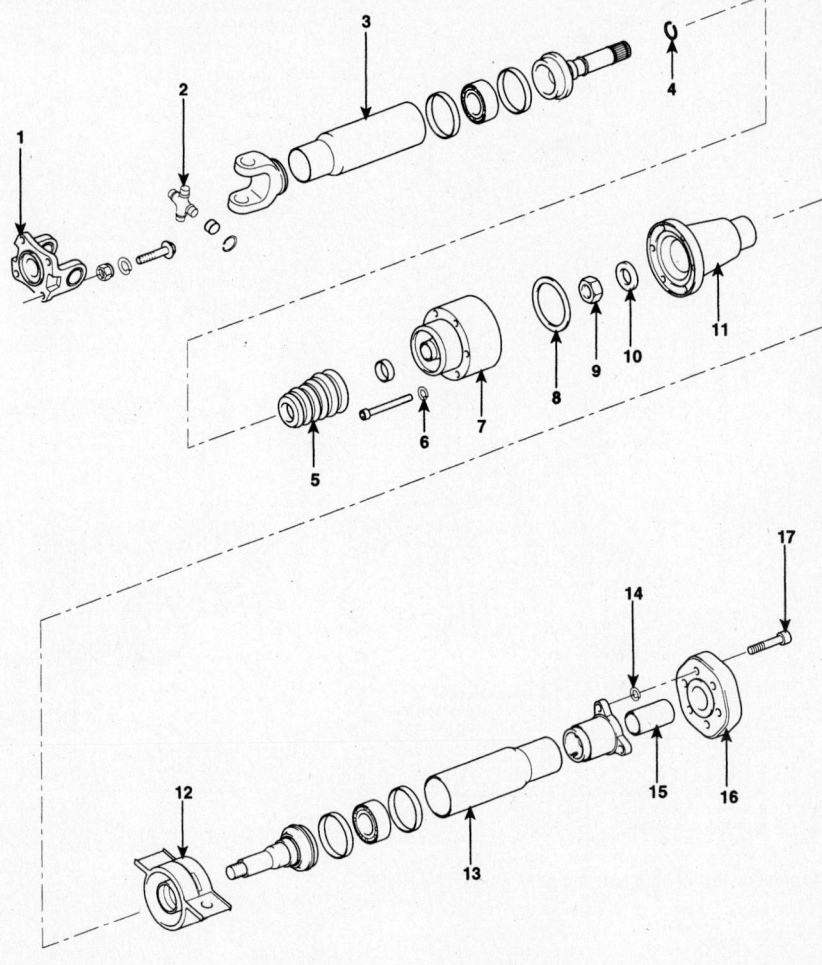

1. Flange yoke	6. Spring washer	11. Companion flange	16. Rubber coupling
2. Universal joint assembly	7. VL joint	12. Center bearing	17. Bolt
3. Front tube	8. Sealing	13. Rear tube	
4. Snap ring(VL)	9. Nut	14. Plain washer	
5. LJ boot	10. Spring washer	15. Center device	

09474_TUCS_G0044

Exploded view of the drive shaft

3. Remove the Double Offset Joint (DOJ) boot bands.

4. Pull the DOJ boot from the outer race.

5. Remove the circlip from the DOJ and pull the halfshaft from the outer race.

6. Remove the snap ring and take out the inner race assembly.

7. Remove the Birfield Joint (BJ) boot bands.

8. Pull out the DOJ boot and BJ boot.

➡**If the boots are to be reused, wrap tape around the halfshaft splines.**

To install:

➡**Use new circlips and boot clamps for assembly.**

9. Apply some boot grease to the halfshaft and install the boots.

10. Apply the specified kit grease to the inner race and cage.

11. Install the cage so that it is offset on the inner race as shown.

12. Apply the specified grease to the cage and fit the balls into the cage.

13. Position the chamfered side as shown. Install the inner race on the halfshaft and install the snap ring.

14. Apply the specified grease to the BJ outer race and install the outer race onto the halfshaft.

15. Apply the specified grease into the DOJ boot and install the boot onto the halfshaft.

16. Tighten the DOJ boot bands.

17. Add the specified grease to the BJ to replace what was wiped away.

18. Install the boots and tighten the BJ boot bands.

19. Install the halfshaft.

[2WD]

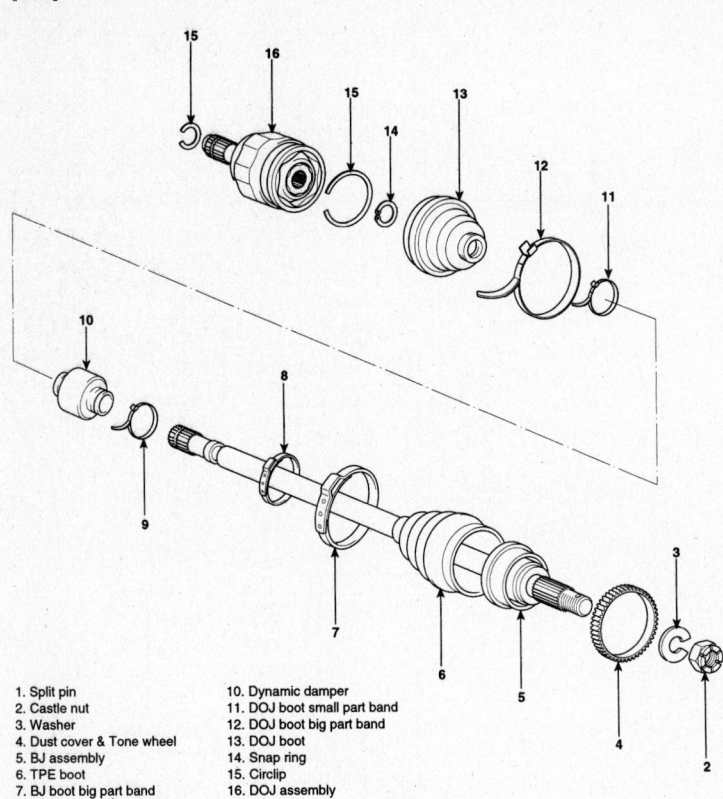

1. Split pin
2. Castle nut
3. Washer
4. Dust cover & Tone wheel
5. BJ assembly
6. TPE boot
7. BJ boot big part band
8. BJ boot small part band
9. Dynamic damper band
10. Dynamic damper
11. DOJ boot small part band
12. DOJ boot big part band
13. DOJ boot
14. Snap ring
15. Circlip
16. DOJ assembly

09474_TUCS_G0045

Exploded view of halfshaft components—DOJ-BJ Type

[2WD]

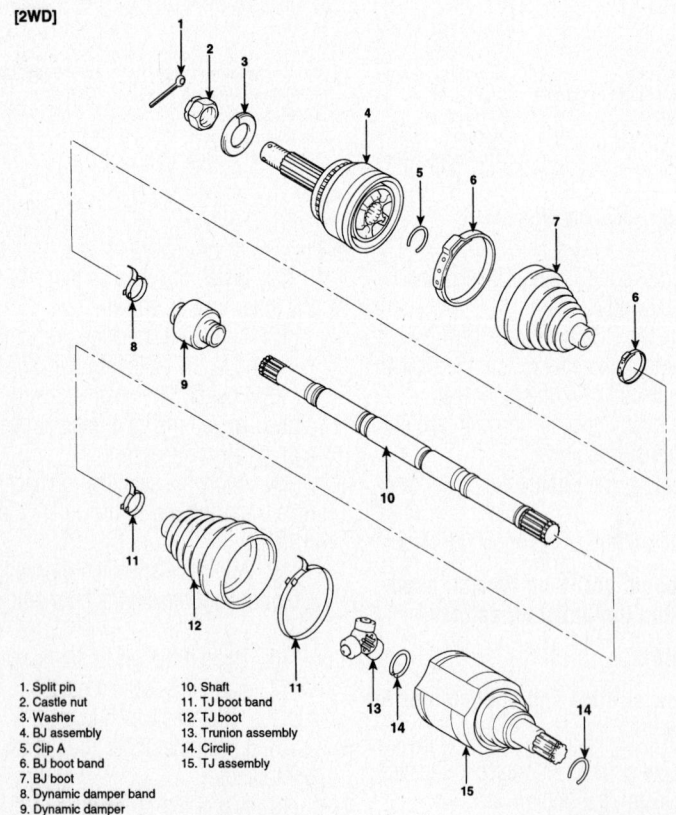

1. Split pin
2. Castle nut
3. Washer
4. BJ assembly
5. Clip A
6. BJ boot band
7. BJ boot
8. Dynamic damper band
9. Dynamic damper
10. Shaft
11. TJ boot band
12. TJ boot
13. Trunion assembly
14. Circlip
15. TJ assembly

09474_TUCS_G0046

Exploded view of halfshaft components—TJ-BJ Type

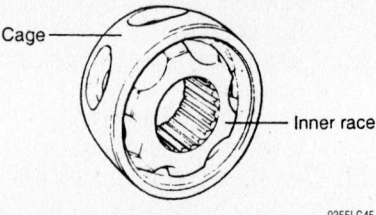

9355LG45

Install the cage so that it is offset on the race

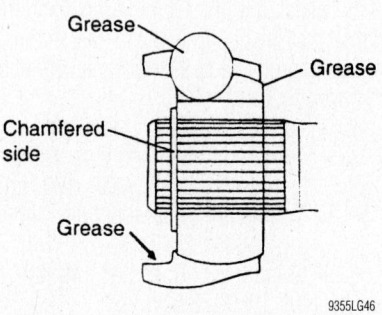

9355LG46

Install the chamfered side of the cage as shown

Tripot Joint-Birfield Joint Type

1. Before servicing the vehicle, refer to the Precautions Section.

2. Remove the halfshaft and place into a vise.

3. Remove the circlip from the splines of the transaxle side Tripot Joint (TJ) case.

4. Remove both boot clamps from the transaxle side TJ case.

5. Pull off the outer TJ boot.

6. Matchmark the trunion assembly to the TJ case and halfshaft spline.

7. Remove the circlip from the splines of the trunion assembly side.

8. Using Special Tool 09495-33000, remove the trunion assembly from the half-shaft.

9. Remove the inner TJ boot.

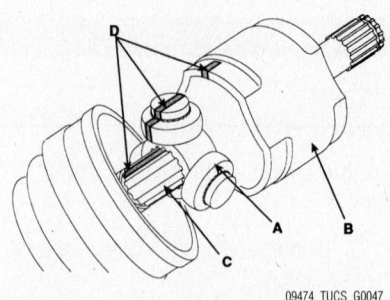

09474_TUCS_G0047

Mark the joint assembly as shown—TJ-BJ type

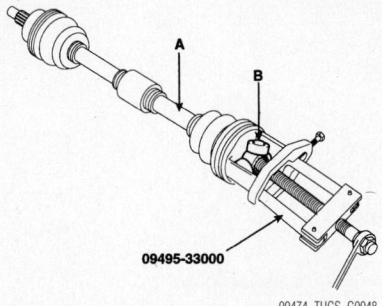

09495-33000

09474_TUCS_G0048

Using special tool 09495-33000 to remove the trunion assembly

10. Remove the dynamic damper clamps.

11. Separate the dynamic damper from the halfshaft, being careful not to damage the shaft splines.

12. Remove the Birfield Joint (BJ) boot clamp.

13. Remove the BJ being careful not to damage the boot.

To install:

➡**Wrap tape around the axle shaft splines to prevent damage to the boots during assembly.**

➡**Use new circlips and boot clamps for assembly.**

14. Apply grease to the halfshaft and install the BJ, boots and clamps.

15. Install the dynamic damper, keep the halfshaft straight and tighten the dynamic band.

16. Install the inner TJ boot and band.

17. Install the trunion assembly and circlip, aligning the matchmarks with each other.

18. Add the specified grease to the TJ, replaced all that was wiped away.

19. Install the outer TJ boot and clamp.

20. Install the halfshaft.

STEERING AND SUSPENSION

Air Bag

DISARMING

Disconnect and isolate the negative battery cable. Wait 3 minutes for the system capacitor to discharge before performing any service.

Rack and Pinion Steering Gear

REMOVAL & INSTALLATION

1. Before servicing the vehicle, refer to the Precautions Section.

2. Drain the power steering fluid.

3. Remove or disconnect the following:
   - Negative battery cable.
   - Universal joint from the gear box
   - Front tires
   - Tie rod from the knuckle using Special Tool 09568-34000 or equivalent
   - Engine splash guard
   - Front muffler
   - Pressure and return hoses from the gear box
   - Pressure and return hose brackets
   - Gear box mounting clamp and bolts

4. Pull the gear box assembly toward the right side of the vehicle and remove.

To install:

5. Install or connect the following:
   - Gear box assembly in from the right side of the vehicle
   - Gear box mounting clamp and bolts
   - Pressure and return hose clamps

- Pressure and return hoses into the gear box
- Front muffler
- Engine splash guard
- Tie rod to the knuckle
- Front tires
- Steering box assembly to the universal joint assembly
- Negative battery cable

6. Fill and bleed the power steering system.

7. Check the alignment and adjust as necessary.

Struts

REMOVAL & INSTALLATION

1. Before servicing the vehicle, refer to the Precautions Section.

2. Remove or disconnect the following:
   - Front wheel
   - Brake hose bracket
   - Speed sensor
   - Stabilizer bar link mounting nut
   - Upper strut mounting nuts
   - Lower strut mounting bolts
   - Strut assembly

To install:

3. Install or connect the following:
   - Strut assembly and tighten the lower mounting bolts to 103–118 ft. lbs. (140–160 Nm).
   - Upper strut mounting bolts. Front struts: Tighten to 33–44 ft. lbs. (45–60 Nm). Rear struts: Tighten to 22–30 ft. lbs. (30–40 Nm).
   - Stabilizer bar link mounting nut. Tighten to 74–89 ft. lbs. (100–120 Nm).

- Speed sensor
- Brake hose bracket
- Front wheel

4. Check the alignment and adjust as necessary.

Coil Spring

REMOVAL & INSTALLATION

Front

1. Before servicing the vehicle, refer to the Precautions Section.

2. Remove the strut from the vehicle and install a spring compressor.

3. Remove the dust cover.

4. Compress the coil spring so that the end of the spring comes away from the spring seat.

5. Remove or disconnect the following:
   - Upper strut mounting nut
   - Insulator
   - Upper spring seat
   - Compressed spring from the strut
   - Spring from the spring compressor

To install:

6. Compress the spring and install it on the strut.

7. Install or connect the following:
   - Upper spring seat
   - Insulator
   - Upper strut mount and torque the nut as follows: Front strut assemblies: 44–52 ft. lbs. (60–70 Nm). Rear strut assemblies : 30–40 ft. lbs. (40–55 Nm).
   - Strut to the vehicle

8. Check and/or adjust the wheel alignment.

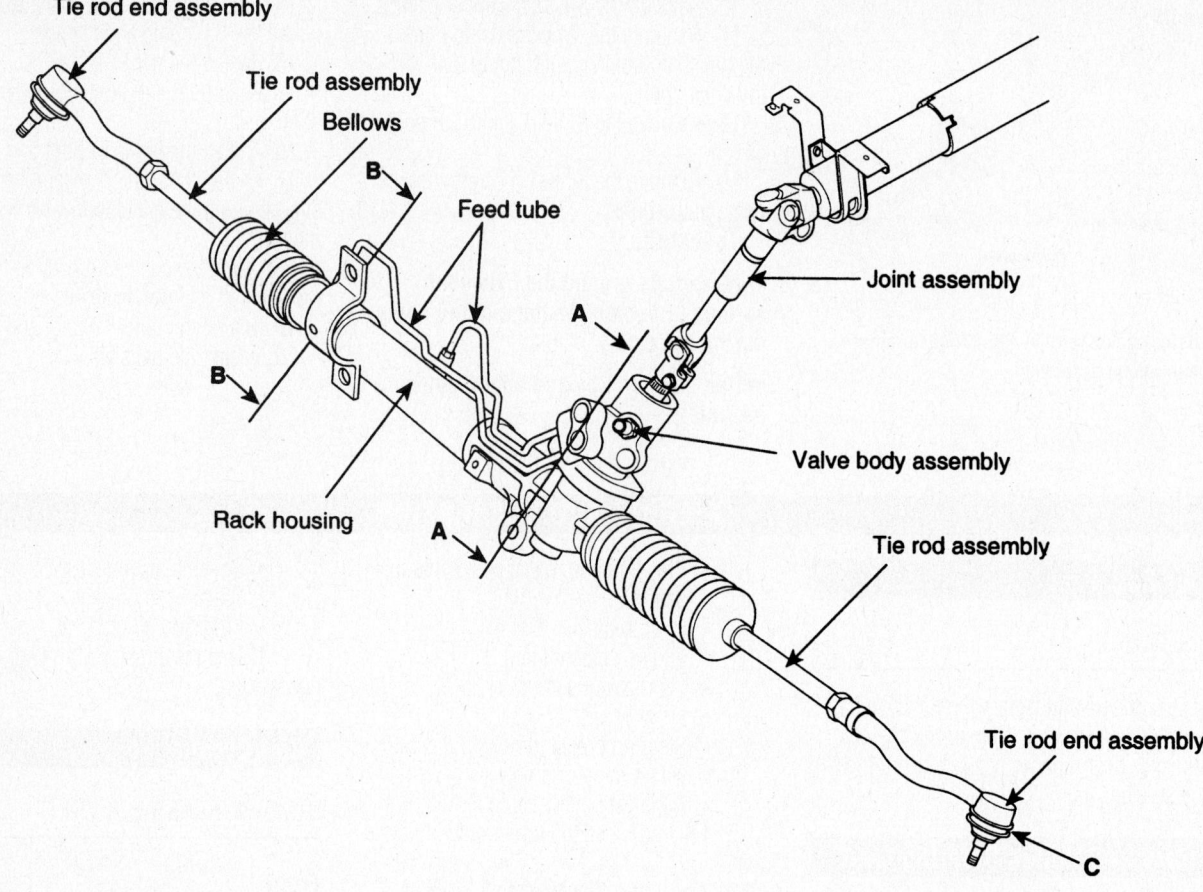

Tie rod end assembly

Tie rod assembly

Bellows

B

Feed tube

A

Joint assembly

B

Valve body assembly

Rack housing

A

Tie rod assembly

Tie rod end assembly

C

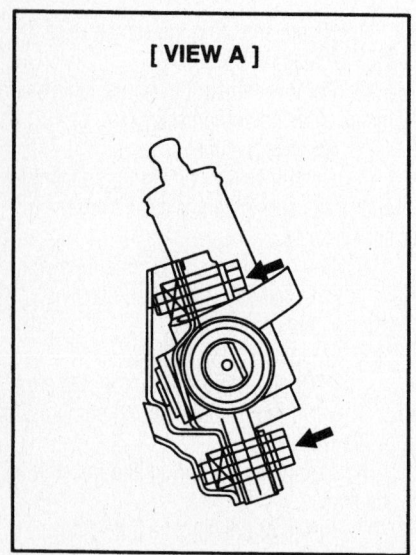

[VIEW A]

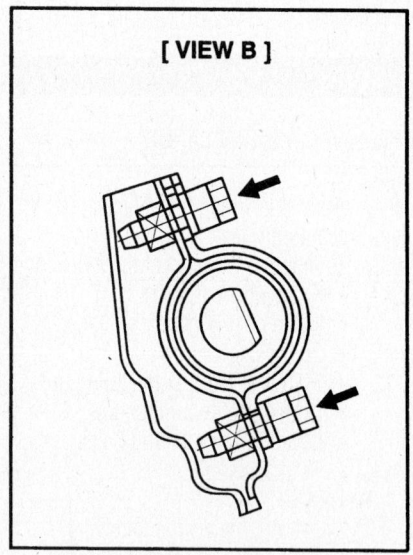

[VIEW B]

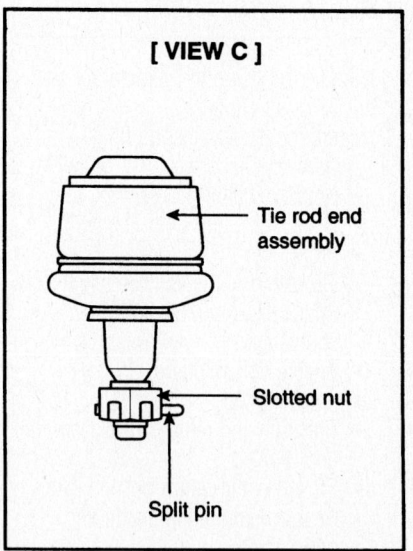

[VIEW C]

Tie rod end assembly

Slotted nut

Split pin

09474_TUCS_G0049

Exploded view of the rack and pinion assembly

Lower Control Arm

REMOVAL & INSTALLATION

1. Before servicing the vehicle, refer to the Precautions Section.
2. Remove or disconnect the following:
   • Front wheel

   • Lower ball joint
   • Lower control arm mounting bolts
   • Lower control arm
To install:
3. Install or connect the following:
   • Lower control arm and torque the front bolt to 74–89 ft. lbs. (100–120 Nm). Tighten the rear

bolt to 103–118 ft. lbs. (140–160 Nm).
   • Lower ball joint and torque the nut to 74–89 ft. lbs. (100–120 Nm)
   • Stabilizer bar link and torque the nut to 25–33 ft. lbs. (35–45 Nm)
   • Front wheel

4. Check and/or adjust the wheel alignment.

CONTROL ARM BUSHING REPLACEMENT

1. Before servicing the vehicle, refer to the Precautions Section.
2. Remove the lower control arm from the vehicle.
3. Press the front bushing out of the control arm.

To install:

4. Lubricate the front bushing with soap and press into the control arm.
5. Install the control arm to the vehicle.
6. Check and/or adjust the wheel alignment.

Trailing Arm

1. Before servicing the vehicle, refer to the Precautions Section.

2. Remove or disconnect the following:
   - Trailing arm mounting bolts
   - Trailing arm

To install:

3. Install the trailing arm. Tighten the mounting bolts and nuts to 74–89 ft. lbs. (100–120 Nm).

TRAILING ARM BUSHING REPLACEMENT

1. Before servicing the vehicle, refer to the Precautions Section.
2. Remove the trailing arm from the vehicle.
3. Press the bushing from the trailing arm using Special Tools 09529-21000 and 09216-21100.

To install:

4. Using Special Tools 09529-21000 and 09216-21100, press fit the rear trailing arm bushing into the trailing arm.

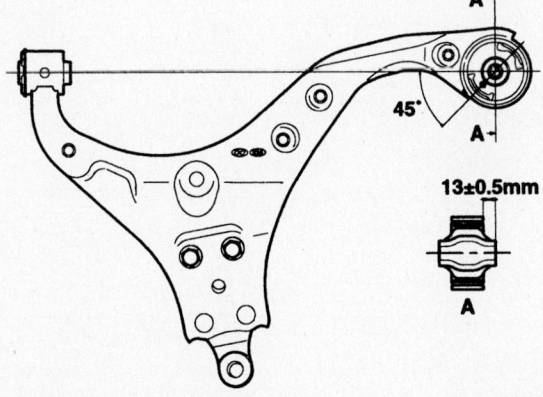

09474_TUCS_G0050

Install the bushings in the orientation shown

Wheel Bearings

ADJUSTMENT

The wheel bearings are sealed units and not adjustable.

REMOVAL & INSTALLATION

Front

The front hub and knuckle are replaced as an assembly.

Rear

1. Before servicing the vehicle, refer to the Precautions Section.
2. Remove or disconnect the following:
   - Rear wheel
   - Brake caliper and rotor
   - Wheel speed sensor, if equipped
   - Hub cap
   - Hub bearing nut and washer
   - Hub assembly

To install:

3. Install or connect the following:
   - Hub and bearing assembly to the backing plate. Torque the hub bearing nut to 148–192 ft. lbs. (200–260 Nm).
   - Hub cap
   - Wheel speed sensor, if equipped
   - Brake caliper and rotor
   - Rear wheel

BRAKES

Brake Caliper

REMOVAL & INSTALLATION

Front

1. Before servicing the vehicle, refer to the Precautions Section.
2. Remove or disconnect the following:
   - Wheel
   - Brake hose at the caliper
   - Caliper mounting bolts
   - Caliper

To install:

3. Install or connect the following:
   - Caliper. Tighten the mounting bolts to 48–55 ft. lbs. (65–75 Nm).
   - Brake line to the caliper with 2 new

metal gaskets. Torque the brake line union bolt to 18–22 ft. lbs. (25–30 Nm).
4. Bleed the system.
5. Install the wheel.

Rear

1. Before servicing the vehicle, refer to the Precautions Section.
2. Release the parking brake.
3. Remove or disconnect the following:
   - Wheel
   - Brake line at the caliper
   - Brake pads
   - Caliper mounting bolts
   - Caliper

To install:

4. Install or connect the following:

   - Caliper onto its mounting
   - Mounting bolts. Torque the bolts to 59–74 ft. lbs. (80–100 Nm).
   - Brake pads
   - Brake line to the caliper with 2 new metal gaskets. Torque the brake line union bolt to 18–22 ft. lbs. (24–30 Nm).
5. Bleed the system.
6. Install the wheel.

Disc Brake Pads

REMOVAL & INSTALLATION

1. Before servicing the vehicle, refer to the Precautions Section.

2. Remove or disconnect the following:
    *   Front wheel
    *   Lower caliper mounting bolt and rotate the caliper upward
    *   Pads from the caliper support
    *   Pad retainers, if necessary

To install:

3. Install or connect the following:
    *   Pad retainers, if removed
    *   Pads onto the pad retainers

4. Compress the caliper piston using a C-clamp. Rotate the caliper downward and install the mounting bolt. Tighten to 16–24 ft. lbs. (22–32 Nm).

5. Install the wheel.

ISUZU

Ascender

13

SPECIFICATION AND MAINTENANCE CHARTS

ENGINE AND VEHICLE IDENTIFICATION

Code ①	Liters (cc)	Cu. In.	Cyl.	Fuel Sys.	Engine Type	Eng. Mfg.
S	4.2 (4200)	256	6	MFI	DOHC	CPC
P	5.3 (5326)	325	8	SFI	OHV	CPC

Code ②	Year
1	2002
2	2003
3	2004
4	2005
5	2006

CPC: Chevrolet/Pontiac/Canada

MFI: Multi-port Fuel Injection

① 8th position of VIN

② 10th position of VIN

09474_ASCE_C0001

GENERAL ENGINE SPECIFICATIONS
All measurements are given in inches.

Year	Model	Engine Displacement Liters (VIN)	Net Horsepower @ rpm	Net Torque B rpm (ft. lbs.)	Bore x Stroke (in.)	Compression Ratio	Oil Pressure @ rpm
2003	Ascender	4.2 (S)	275@6000	275@3600	3.66x4.02	10.0:1	12@1200
		5.3 (P)	290@5300	325@4000	3.78x3.62	9.45:1	6@1000
2004	Ascender	4.2 (S)	275@6000	275@3600	3.66x4.02	10.0:1	12@1200
		5.3 (P)	290@5300	325@4000	3.78x3.62	9.45:1	6@1000
2005	Ascender	4.2 (S)	275@6000	275@3600	3.66x4.02	10.0:1	12@1200
		5.3 (P)	300@5200	330@4000	3.78x3.62	9.95:1	6@1000

09474_ASCE_C0002

GASOLINE ENGINE TUNE-UP SPECIFICATIONS

Year	Engine Displacement Liters (VIN)	Spark Plugs Gap (in.)	Ignition Timing (deg.) MT	Ignition Timing (deg.) AT	Fuel Pump (psi)	Idle Speed (rpm) MT	Idle Speed (rpm) AT	Valve Clearance In.	Valve Clearance Ex.
2003	4.2 (S)	0.050	—	①	50-57 ②	—	③	HYD	HYD
	5.3 (P)	0.060	④	④	55-62 ②	—	625	HYD	HYD
2004	4.2 (S)	0.050	—	①	50-57 ②	—	③	HYD	HYD
	5.3 (P)	0.060	④	④	55-62 ②	—	625	HYD	HYD
2005	4.2 (S)	0.050	—	①	50-57 ②	—	③	HYD	HYD
	5.3 (P)	0.060	④	④	55-62 ②	—	625	HYD	HYD

NOTE: The Vehicle Emission Control Information label often reflects specification changes made during production. The label figures must be used if they differ from those in this chart.

HYD: Hydraulic

① Distributorless ignition, cannot be adjusted

② With key ON and engine OFF

③ Idle speed is maintained by the PCM

④ Distributorless ignition, cannot be adjusted

09474_ASCE_C0003

Accessory serpentine belt routing—4.2L engines

42372-BLAZ-G01

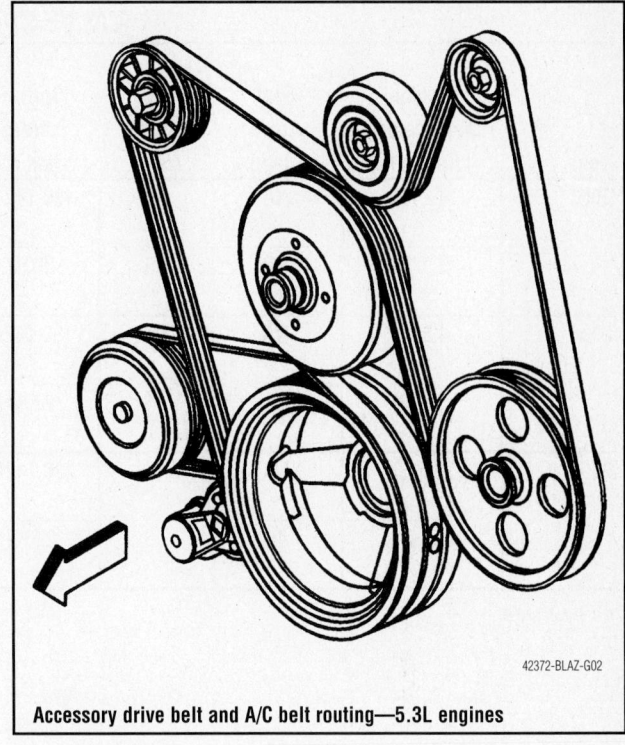

Accessory drive belt and A/C belt routing—5.3L engines

42372-BLAZ-G02

CAPACITIES

Year	Model	Engine Displacement Liters (VIN)	Engine Oil with Filter (qts.)	Transmission (pts.)		Transfer Case (pts.)	Drive Axle		Fuel Tank (gal.)	Cooling System (qts.)
				5-Spd	Auto.		Front (pts.)	Rear (pts.)		
2003	Ascnder	4.2 (S)	7.0	—	10.0	2.6	2.6	4.0	25.0	13.9
		5.3 (P)	6.0	—	10.0	4.0	2.6	4.0	25.0	14.9
2004	Ascnder	4.2 (S)	7.0	—	10.0	2.6	2.6	4.0	25.0	13.9
		5.3 (P)	6.0	—	10.0	4.0	2.6	4.0	25.0	14.9
2005	Ascnder	4.2 (S)	7.0	—	10.0	2.6	2.6	4.0	25.0	13.9
		5.3 (P)	6.0	—	10.0	4.0	2.6	4.0	25.0	14.9

NOTE: All capacities are approximate. Add fluid gradually and check to be sure a proper fluid level is obtained.

09474_ASCE_C0004

VALVE SPECIFICATIONS

Year	Engine Displacement Liters (VIN)	Seat Angle (deg.)	Face Angle (deg.)	Spring Test Pressure (lbs. @ in.)	Spring Installed Height (in.)	Stem-to-Guide Clearance (in.)		Stem Diameter (in.)	
						Intake	Exhaust	Intake	Exhaust
2003	4.2 (S)	NA	NA	130-142@1.26	NA	0.0011-0.0025	0.0015-0.0030	NA	NA
	5.3 (P)	46	45	220@1.32	1.80	0.0010-0.0026	0.0010-0.0026	0.313-0.314	0.313-0.314
2004	4.2 (S)	NA	NA	130-142@1.26	NA	0.0011-0.0025	0.0015-0.0030	NA	NA
	5.3 (P)	46	45	220@1.32	1.80	0.0010-0.0026	0.0010-0.0026	0.313-0.314	0.313-0.314
2005	4.2 (S)	NA	NA	130-142@1.26	NA	0.0011-0.0025	0.0015-0.0030	NA	NA
	5.3 (P)	46	45	220@1.32	1.80	0.0010-0.0026	0.0010-0.0026	0.313-0.314	0.313-0.314

NA: Not Available

09474_ASCE_C0005

CRANKSHAFT AND CONNECTING ROD SPECIFICATIONS
All measurements are given in inches.

Year	Engine Displacement Liters (VIN)	Crankshaft				Connecting Rod		
		Main Brg. Journal Dia.	Main Brg. Oil Clearance	Shaft End-play	Thrust on No.	Journal Diameter	Oil Clearance	Side Clearance
2003	4.2 (S)	2.7567-2.7574	0.0004-0.0025	0.0044-0.0153	4	2.2337-2.2342	0.0008-0.0025	0.0019-0.0137
	5.3 (P)	2.558	0.0008-0.0021	0.0015-0.0078	4	2.0991-2.0999	0.0009-0.0003	0.0043-0.0200
2004	4.2 (S)	2.7567-2.7574	0.0004-0.0025	0.0044-0.0153	4	2.2337-2.2342	0.0008-0.0025	0.0019-0.0137
	5.3 (P)	2.558	0.0008-0.0021	0.0015-0.0078	4	2.0991-2.0999	0.0009-0.0003	0.0043-0.0200
2005	4.2 (S)	2.7567-2.7574	0.0004-0.0025	0.0044-0.0153	4	2.2337-2.2342	0.0008-0.0025	0.0019-0.0137
	5.3 (P)	2.558	0.0008-0.0021	0.0015-0.0078	4	2.0991-2.0999	0.0009-0.0003	0.0043-0.0200

09474_ASCE_C0006

PISTON AND RING SPECIFICATIONS
All measurements are given in inches.

Year	Engine Displacement Liters (VIN)	Piston Clearance	Ring Gap			Ring Side Clearance		
			Top Compression	Bottom Compression	Oil Control	Top Compression	Bottom Compression	Oil Control
2003	4.2 (S)	0.0004-0.0017	0.0079-0.0157	0.0118-0.0197	0.0098-0.0299	0.0017-0.0037	0.0017-0.0037	0.0023-0.0085
	5.3 (P)	-0.0014 0.0006	0.009-0.017	0.017-0.027	0.007-0.029	0.0015-0.0033	0.0015-0.0031	0.0005-0.0078
2004	4.2 (S)	0.0004-0.0017	0.0079-0.0157	0.0118-0.0197	0.0098-0.0299	0.0017-0.0037	0.0017-0.0037	0.0023-0.0085
	5.3 (P)	-0.0014 0.0006	0.009-0.017	0.017-0.027	0.007-0.029	0.0015-0.0033	0.0015-0.0031	0.0005-0.0078
2005	4.2 (S)	0.0004-0.0017	0.0079-0.0157	0.0118-0.0197	0.0098-0.0299	0.0017-0.0037	0.0017-0.0037	0.0023-0.0085
	5.3 (P)	-0.0014 0.0006	0.009-0.017	0.017-0.027	0.007-0.029	0.0015-0.0033	0.0015-0.0031	0.0005-0.0078

09474_ASCE_C0007

TORQUE SPECIFICATIONS
All readings in ft. lbs.

Year	Engine Displacement Liters (VIN)	Cylinder Head Bolts	Main Bearing Bolts	Rod Bearing Bolts	Crankshaft Damper Bolts	Flywheel Bolts	Manifold		Spark Plugs	Oil Pan Drain Plug
							Intake *	Exhaust		
2003	4.2 (S)	①	②	③	④	⑤	12	⑥	13	19
	5.3 (P)	⑦	⑧	⑨	⑩	⑪	⑫	⑬	11	19
2004	4.2 (S)	①	②	③	④	⑤	12	⑥	13	19
	5.3 (P)	⑦	⑧	⑨	⑩	⑪	⑫	⑬	11	19
2005	4.2 (S)	①	②	③	④	⑤	12	⑥	13	19
	5.3 (P)	⑦	⑧	⑨	⑩	⑪	⑫	⑬	11	19

* NOTE: Applies to Lower Manifold only.

① Cylinder head bolts (14)
  1st pass: 22 ft. lbs.
  2nd pass: Plus 155 degrees
  2 short end bolts: 62 inch. lbs.
  2 short end bolts: plus 60 degrees
  1 long end bolt: 62 inch lbs.
  1 long end bolt: plus 120 degrees

② 18 ft. lbs., plus 155 depress

③ 18 ft. lbs., plus 110 degrees

④ 111 ft. lbs., plus 180 degrees

⑤ 18 ft. lbs., plus 50 degrees

⑥ 1st pass: 18 ft. lbs.
  2nd pass: 18 ft. lbs.
  3rd pass: 18 ft. lbs.

⑦ 1st pass: 22 ft. lbs.
  2nd pass: Plus 90 degrees
  Final pass:
    Except medium bolts at front and rear of head: Plus 90 degrees
    Medium bolts at front and rear of head: Plus 50 degrees

⑧ Inner bolts:
  1st pass: 15 ft. lbs.
  Final pass: Plus 80 degrees
  Outer bolts:
  1st pass: 15 ft. lbs.
  Final pass: Plus 51 degrees

⑨ 15 ft. lbs. plus 75 degrees

⑩ Installation pass: 240 ft. lbs. (discard bolt)
  First pass: 37 ft. lbs. (new bolt)
  Final pass: Plus 140 degrees

⑪ 1st pass: 15 ft. lbs.
  2nd pass: 37 ft. lbs.

⑫ 1st pass: 44 inch lbs.
  2nd pass: 89 inch lbs.

⑬ 1st pass: 11 ft. lbs.
  2nd pass: 18 ft. lbs.

09474_ASCE_C0008

WHEEL ALIGNMENT

Year	Model		Caster Range (+/-Deg.)	Caster Preferred Setting (Deg.)	Camber Range (+/-Deg.)	Camber Preferred Setting (Deg.)	Toe-in (in.)
2003	Ascender	2WD	1.00	+3.50	1.00	-0.5	-0.5+/-0.5
		4WD	1.00	+3.50	1.00	-0.5	-0.5+/-0.5
2004	Ascender	2WD	1.00	+3.50	1.00	-0.5	-0.5+/-0.5
		4WD	1.00	+3.50	1.00	-0.5	-0.5+/-0.5
2005	Ascender	2WD	1.00	+3.50	1.00	-0.5	-0.5+/-0.5
		4WD	1.00	+3.50	1.00	-0.5	-0.5+/-0.5

09474_ASCE_C0009

TIRE, WHEEL AND BALL JOINT SPECIFICATIONS

Year	Model	OEM Tires Standard	OEM Tires Optional	Tire Pressures (psi) Front	Tire Pressures (psi) Rear	Wheel Size	Ball Joint Inspection	Lug Nut
2003	Ascender	P245/65SR17	NA	36	36	17x7	L ①	103
2004	Ascender	P245/65SR17	NA	36	36	17x7	L ①	103
2005	Ascender	P245/65SR17	NA	36	36	17x7	L ①	103

NA: Not Available

OEM: Original Equipment Manufacturer

PSI: Pounds Per Square Inch

STD: Standard

OPT: Optional

L: Lower

U: Upper

① Do not lift truck. Inspect the boss into which the grease fitting is threaded. Replace if the boss is flush or receded below the surface of the ball joint

09474_ASCE_C0010

BRAKE SPECIFICATIONS
All measurements in inches unless noted

Year	Model		Brake Disc Original Thickness	Brake Disc Minimum Thickness	Brake Disc Maximum Runout	Brake Drum Diameter Original Inside Diameter	Max. Wear Limit	Maximum Machine Diameter	Minimum Lining Thickness	Brake Caliper Bracket Bolts (ft. lbs.)	Brake Caliper Mounting Bolts (ft. lbs.)
2003	Ascender	F	1.030	0.965	0.003	—	—	—	0.030	52	①
		R	0.787	0.728	0.004	9.50	9.59	9.56	0.030	NA	—
2004	Ascender	F	1.030	0.965	0.003	—	—	—	0.030	52	①
		R	0.787	0.728	0.004	9.50	9.59	9.56	0.030	NA	—
2005	Ascender	F	1.030	0.965	0.003	—	—	—	0.030	52	①
		R	0.787	0.728	0.004	9.50	9.59	9.56	0.030	NA	—

NA: Not Available

① 2WD: 38 ft. lbs.

4WD: 77 ft. lbs.

09474_ASCE_C0011

SCHEDULED MAINTENANCE INTERVALS

ISUZU—ASCENDER

TO BE SERVICED	TYPE OF SERVICE	VEHICLE MILEAGE INTERVAL (x1000)															
		7.5	15	22.5	30	37.5	45	52.5	60	67.5	75	82.5	90	97.5	105	112.5	120
Accessory drive belt	S/I								✓								✓
Air cleaner filter	R				✓				✓				✓				✓
Automatic transmission fluid	R	Every 50,000 miles															
Brake system ①	S/I	✓	✓	✓	✓	✓	✓	✓	✓	✓	✓	✓	✓	✓	✓	✓	✓
Chassis & suspension grease points	L	✓	✓	✓	✓	✓	✓	✓	✓	✓	✓	✓	✓	✓	✓	✓	✓
CV-joint boots & axle seals	S/I	✓	✓	✓	✓	✓	✓	✓	✓	✓	✓	✓	✓	✓	✓	✓	✓
Engine coolant system ②	S/I	Every 150,000 miles															
Engine oil & filter	R	✓	✓	✓	✓	✓	✓	✓	✓	✓	✓	✓	✓	✓	✓	✓	✓
Front wheel bearings	S/I & L				✓				✓				✓				✓
Fuel filter	R				✓				✓				✓				✓
Fuel tank, cap & lines	S/I								✓								✓
PCV valve	S/I	Every 100,000 miles															
Rear/front axle fluid level	S/I	✓	✓	✓	✓	✓	✓	✓	✓	✓	✓	✓	✓	✓	✓	✓	✓
Rotate tires	S/I	✓	✓	✓	✓	✓	✓	✓	✓	✓	✓	✓	✓	✓	✓	✓	✓
Spark plug wires	S/I	Every 100,000 miles															
Spark plugs	R	Every 100,000 miles															

R: Replace   S/I: Inspect and service, if necessary   L: Lubricate

① This should be performed when the tires are removed for rotation.

② Drain, flush and refill the cooling system, inspect the system hoses, and clean the radiator and condenser.

③ 2-wheel drive models only.

FREQUENT OPERATION MAINTENANCE (SEVERE SERVICE)

If a vehicle is operated under any of the following conditions it is considered severe service:

- Towing a trailer or using a camper or car-top carrier.

- Repeated short trips of less than 5 miles in temperatures below freezing, or trips of less than 10 miles in any temperature.

- Extensive idling or low-speed driving for long distances as in heavy commercial use, such as delivery, taxi or police cars.

- Operating on rough, muddy or salt-covered roads.

- Operating on unpaved or dusty roads.

- Driving in extremely hot (over 90°) conditions.

Engine oil & filter: replace every 3000 miles or 3 months, whichever occurs first.

Chassis and suspension grease points: lubricate every 3000 miles.

Rear/front axle fluid level: inspect every 3000 miles.

Rotate the tires ever 6000 miles.

Brake system components: inspect ever 6000 miles.

Front wheel bearings (2-wheel drive only): clean, inspect and repack every 15,000 miles.

Air cleaner filter: inspect every 15,000 miles.

Automatic transmission fluid & filter: replace every 15,000 miles.

09474_ASCE_C0012

ENGINE REPAIR

Alternator

REMOVAL

4.2L Engine

1. Before servicing the vehicle, refer to the precautions section.
2. Remove or disconnect the following:
   - Negative battery cable
   - Accessory belt
   - Positive battery cable nut from the generator
   - A/C line mounting bracket bolt at the engine lift hook
   - Right engine lift hook bolts
   - Engine lift hook
   - Mounting bolts
   - Alternator

5.3L Engine

1. Before servicing the vehicle, refer to the precautions section.
2. Remove or disconnect the following:

   - Negative battery cable
   - Accessory belt
   - Electrical connector
   - Terminal stud nut, after sliding boot down
   - Alternator cable
   - Mounting bolts
   - Alternator

INSTALLATION

4.2L Engine

1. Install or connect the following:
   - Alternator and loosely install the mounting blots
   - Tighten the alternator mounting bolts to 37 ft. lbs. (50 Nm)
   - Positive battery cable and secure with the nut; tighten the nut to 80 inch lbs. (9 Nm)
   - Engine lift hook and bolts; tighten the bolts to 37 ft. lbs. (50 Nm)
   - A/C line bracket to the lift hook, then tighten the retaining bolt to 89 inch lbs. (10 Nm)
   - Accessory belt
   - Negative battery cable

5.3L Engine

1. Install or connect the following:
   - Alternator and loosely install the mounting bolts

   - Tighten the bolts to 37 ft. lbs. (50 Nm)
   - Alternator cable
   - Terminal stud nut and tighten to 80 inch lbs. (9 Nm)
   - Boot back over terminal stud
   - Electrical connector
   - Accessory belt
   - Negative battery cable

Ignition Timing

ADJUSTMENT

The ignition timing is preset and cannot be adjusted.

Engine Assembly

REMOVAL & INSTALLATION

4.2L Engine

1. Before servicing the vehicle, refer to the precautions section.
2. Drain the engine cooling system

➡ **Keep the oil drain plug removed during the engine removal and installation.**

3. Drain the engine oil. Install a suitable plug into the oil pan to prevent oil leakage during the remainder of the procedure.
4. Using the proper equipment, discharge and recover the refrigerant from the A/C system, if equipped.
5. Remove or disconnect the following:
   - Hood
   - Negative battery cable
   - Fuel system pressure
   - Air cleaner assembly
   - Throttle body
   - Manifold Absolute Pressure (MAP) sensor
   - Windshield washer solvent container
   - Air intake baffle
   - Grille
   - Headlight housing
   - Radiator support brace
   - Hood latch
   - A/C lines from the condenser
   - Transmission cooler lines from the engine, not the radiator
6. Remove the cooling fan and shroud, tilting the radiator forward, and the cooling fan and shroud rearward for clearance.
   - Accessory belt

   - Power steering pump bolts; position the pump aside
   - Heater hoses from the heater core
   - Transmission filler tube bracket nut from the Air Injector Reactor (AIR) adapter
   - AIR adapter
7. Install a suitable lift hook to the AIR adapter
8. Remove or disconnect the following:
   - Oxygen (O$_2$) sensor connector
   - A/C line from the accumulator
   - Front axle actuator electrical connector
   - Camshaft phaser actuator valve electrical connector
   - Transmission cooler lines from the clips on the right side of the engine block
   - Ignition coil harness connectors
   - Harness retainer from the clips
   - Power brake hose from the booster
   - Powertrain Control Module (PCM)
   - Fuel lines from the fuel pressure regulator. Cap the lines to avoid excessive fuel leakage.
   - All harnesses from the engine harness bracket
   - Engine harness bracket bolt and bracket
   - Starter electrical connections
   - A/C pressure sensor and clutch electrical connector
   - Alternator electrical connector and battery lead
   - Knock Sensor (KS), Crankshaft Position (CKP) and Camshaft Position (CMP) sensor electrical connectors
   - 4 ground on the left side of the block
9. Raise and safely support the vehicle.
   - Left and right side driveshafts
   - Propeller shaft from the front axle pinion yoke
   - Engine protection shield
   - Exhaust pipe from the exhaust manifold. Slide the exhaust pipe backward slightly.
   - Fuel tank shield, if equipped
   - Torque converter access cover and bolts
10. Place a jack on the transmission fluid pan for support.
11. Remove the transmission support.
12. Lower the transmission enough to reach the top bell housing bolts.
13. Remove the top 4 bell housing bolts, there may be 2 harness clips that will need

to be removed in order to have access to 2 of the top bolts.

14. Raise the transmission.

15. Reinstall the transmission support using only 2 through bolts.

16. Remove or disconnect the following:
- Remaining bell housing bolts (11 total)
- Left and right engine lower mount nuts
- Oil level sensor electrical connector
- Oil pressure switch electrical connector

17. Carefully lower the vehicle.

18. Remove the left, then the right upper engine mount nut.

19. Install a suitable engine hoist.

20. Raise the engine out of the compartment slowly, keeping the transmission supported.

21. Remove both engine mounts for clearance.

22. Continue raising the engine out of the vehicle.

23. Place the engine on a suitable engine stand.

To install:

24. Remove the engine from the engine stand.

25. Slowly install the engine into the engine compartment, aligning the engine mounts with the brackets.

26. When the engine mounts are aligned, install the engine mounts, putting the mount up through the engine mount brackets before inserting into the chassis mount brackets.

27. Lower the engine onto the mounts and install the upper engine mounting nuts. Tighten the nuts to 51 ft. lbs. (71 Nm).

28. Remove the engine hoist.

29. Lay the radiator into the radiator support, but do not install the radiator completely.

30. Raise and safely support the vehicle.

31. Install the lower bell housing bolts, except the top four.

32. Remove the 2 through bolts secure the transmission support, then lower the transmission.

33. Install the top 4 bell housing bolts and tighten all 11 bolts to 37 ft. lbs. (50 Nm).

34. Raise the transmission.

35. Install or connect the following:
- Transmission support
- 3 torque converter bolts and tighten to 44 ft. lbs. (60 Nm)
- Torque converter bolt cover
- Fuel tank shield, if equipped
- Engine protection shield
- Propeller shaft to the front axle pinion yoke

- Exhaust pipe to the manifold and tighten the bolts to 37 ft. lbs. (50 Nm)
- Oil level switch and oil pressure sender electrical connectors
- Oil pan drain plug and tighten to 19 ft. lbs. (26 Nm)
- Lower radiator hose
- Left and right wheel driveshafts

36. Lower the vehicle.
- 4 grounds on the left side of the block
- CMP, CKP and knock sensor electrical connectors
- Alternator and starter electrical connectors and battery leads. Torque the nuts to 80 inch lbs. (9 Nm).
- Fuel lines at the fuel pressure regulator
- Engine harness bracket and bolt. Torque the bolt to 37 ft. lbs. (50 Nm).
- Front differential vent hose, to the engine harness bracket
- PCM
- Power brake hose to the booster
- Harness retainer to its original location
- Ignition coil harness connectors
- Transmission cooler lines to clips on right side of engine block
- Camshaft phase actuator valve electrical connector
- Front axle actuator electrical connector
- A/C line at the accumulator

37. Remove the lift hook.

38. Install or connect the following:
- AIR adapter and secure with the studs. Tighten to 18 ft. lbs. (25 Nm).
- Transmission filler tube bracket to AIR adapter stud and secure the bracket with the nut. Torque the nut to 89 inch lbs. (10 Nm).
- Heater hoses to the heater core
- Power steering pump and tighten the bolts to 18 ft. lbs. (25 Nm).

39. The remainder of installation is the reverse of removal, but please note the following important steps:

40. Connect the negative battery cable

41. Check all powertrain fluid levels and add, as necessary.

42. Refill the engine crankcase.

43. Refill the engine cooling system.

44. Perform the CKP System Variation Learn Procedure, as follows:

a. Install a suitable scan tool and check for Diagnostic Trouble Codes (DTCs). If any DTCs, other than P1336 are set, resolve those codes first, before proceeding with this procedure.

b. With the scan tool, select the crankshaft position variation learn procedure.

c. Observe the fuel cut-off for the 4.2L engine.

d. The scan tool will instruct you to perform certain steps, make sure you follow all directions given by the scan tool exactly.

e. Enable the crankshaft position system variation learn procedure.

➡**While the learn procedure is in progress, release the throttle immediately when the engine started to decelerate. The engine control is returned to the operator and the engine responds to throttle position after the learn procedure is complete.**

f. Slowly increase the engine speed to the RPM that you observed.

g. Immediately release the throttle when fuel cut-out is reached.

h. The scan tool displays: Learn Status: Learned this ignition. If the scan tool does NOT display this message and not other DTCs set, you must perform further troubleshooting.

i. Turn the ignition **OFF** for 30 seconds after the learn procedure has been completed successfully.

45. Start and run the engine, then check for leaks.

5.3L Engine

1. Before servicing the vehicle, refer to the precautions section.

2. Drain the engine cooling system

3. Drain the engine oil.

4. Remove and recover the refrigerant, if equipped with A/C.

5. Remove or disconnect the following:
- Negative battery cable
- Hood
- Radiator
- Radiator support brace
- Front axle, if 4WD
- Drive shafts
- Intake manifold
- Oil pressure sensor connector
- Oxygen (O_2) sensor connector
- Camshaft Position (CMP) sensor connector
- A/C compressor hose
- Rear auxiliary A/C compressor pipe fitting
- Rear auxiliary A/C compressor pipe nut and bolt. Tie the pipe out of the way.
- Engine Coolant Temperature (ECT) sensor

- Ground terminal bolt
- Retaining clips from the brackets
- A/C pressure switch electrical connector
- Retaining clip from the cylinder head
- Ground terminal bolts
- Starter
- Battery cable channel bolt
- Battery cable channel from the oil pan
- A/C compressor electrical connector

6. Collect all branches of the engine wiring harness, then position the harness out of the way.

- Alternator cable from the alternator
- Alternator bracket bolts, then position the bracket and alternator assembly aside
- Inlet and outlet hoses from the water outlet, using J 38185 to move the hose clamps
- Auxiliary heater inlet and outlet hose/pipe assembly from the heater water shutoff valve pipes
- Auxiliary heater inlet and outlet hoses/pipes from the water pump, using Hose Clamp Pliers J 38185
- Remove ignition coils, if necessary,

to install Engine Lifting Brackets J 41798 to the cylinder heads

7. Install Engine Lifting Brackets J 41798 to the cylinder heads. Tighten the M8 bolts to 18 ft. lbs. (25 Nm) and the M10 bolts to 37 ft. lbs. (50 Nm).

- Catalytic converter
- 3 frame engine mount bracket bolts from the right and left sides
- Torque converter bolts
- Transmission oil level dipstick tube nut and tube
- Transmission bolt and stud on the right side
- Lower transmission bolt/studs
- 3 upper transmission bolts/studs

8. Install a suitable engine hoist to the engine lifting brackets.

9. Place a floor jack under the transmission for support.

10. Separate the engine from the transmission.

11. Remove the engine from the vehicle and place on a suitable engine stand.

12. Install Converter Holding Strap J 21366 to the transmission to hold the torque converter.

To install:

13. Remove Converter Holding Strap J 21366 from the transmission.

14. Attach the engine to a hoist and remove it from the engine stand

15. Install or connect the following:

- Engine into the vehicle. Match the transmission up to the engine, then remove the floor jack.
- 3 upper transmission bolts/studs and tighten to 37 ft. lbs. (50 Nm)
- Lower transmission bolts/studs and tighten to 37 ft. lbs. (50 Nm)
- Transmission bolt and stud on the right side and tighten to 37 ft. lbs. (50 Nm)
- Transmission oil level dipstick tube and nut. Torque to 89 inch lbs. (10 Nm).
- Torque converter bolts and tighten to 44 ft. lbs. (60 Nm)
- 3 frame engine mount bracket bolts to both the right and left sides. Torque the bolts to 37 ft. lbs. (50 Nm).
- Catalytic converter

16. Remove the engine lifting brackets from the cylinder heads

- Ignition coils, if removed, and tighten the bolts to 71 inch lbs. (8 Nm)
- Auxiliary heater inlet and outlet hoses
- Auxiliary heater inlet and outlet hose/pipe assembly to the heater water shutoff valve pipes
- Outlet and inlet hoses to the water outlet
- Bracket and alternator assembly. Tighten the bolts to 37 ft. lbs. (50 Nm).
- Cable to the alternator
- Position the engine wiring harness back over the engine
- A/C compressor electrical connector
- Battery cable channel to the oil pan and secure with the bolt. Torque to 106 inch lbs. (12 Nm).
- Starter
- Ground terminal bolts and tighten to 18 ft. lbs. (25 Nm)
- Retaining clip to the cylinder head
- A/C pressure switch electrical connector
- Retaining clips to the brackets
- Ground terminal bolt and tighten to 18 ft. lbs. (25 Nm)
- ECT sensor connector
- Rear auxiliary A/C compressor pipe nut and bolt. Torque to 15 ft. lbs. (20 Nm).
- A/C compressor hose
- Oil pressure sensor connector
- O₂ sensor connector

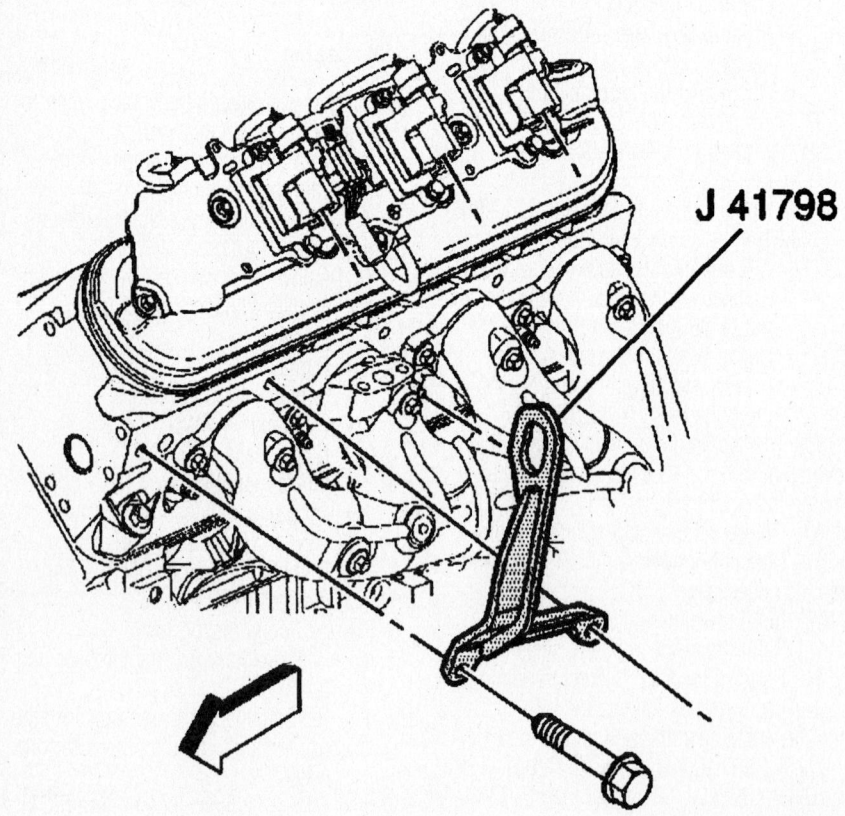

J 41798

If necessary, remove ignition coil(s) to install the engine lifting brackets—5.3L engine

42372-BLAZ-G03

- CMP sensor connector
- Intake manifold
- Drive shafts
- Front axle, if removed
- Radiator support brace
- Radiator

17. Recharge the A/C system
- Negative battery cable
- Hood

18. Check all powertrain fluid levels and add, as necessary.

19. Refill the engine crankcase.

20. Refill the engine cooling system.

21. Start and run the engine, then check for leaks.

Water Pump

REMOVAL & INSTALLATION

1. Before servicing the vehicle, refer to the precautions section.

2. Disconnect the negative battery cable.

3. Drain the engine cooling system.

4. For 5.3L engines, loosen the air cleaner outlet duct clamps at the throttle body and Mass Airflow/Intake Air Temperature (MAP/IAT) sensor. Remove the bolt and air cleaner outlet duct.

5. Relieve the belt tension and remove the accessory drive belts or the serpentine drive belt, as applicable.

6. Remove or disconnect the following:
- Upper fan shroud
- Fan or fan and clutch assembly, as applicable

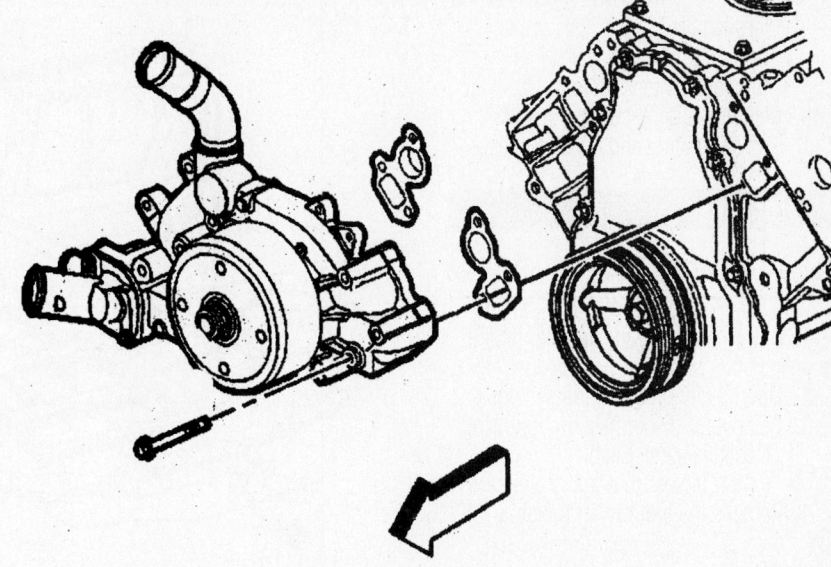

42372-BLAZ-G04

Exploded view of the water pump assembly mounting—5.3L engine

- Water pump pulley; use a suitable tool to hold the pulley while removing the bolts
- Coolant hose(s) from the water pump

➡For the hoses on some engines, removal may be easier if the hose is left attached until the pump is free from the block. Once the pump is removed from the engine, the pump may be pulled (giving a better grip and greater leverage) from the tight hose connection.

- Water pump retainers
- Water pump from the engine

❄❄ WARNING

Note the positions of all retainers as some engines will utilize different length fasteners in different locations and/or bolts and studs in different locations.

To install:

7. Clean the gasket mounting surfaces.

➡The water pumps on some of the engines covered may have been installed using sealer only, no gasket, at the factory. If a gasket is supplied with the replacement part, it should be used. Otherwise, a 1/8 in. (3mm) bead of RTV sealer should be used around the sealing surface of the pump.

8. Apply sealant to the water pump retainer threads.

9. Install or connect the following:
- Water pump using a new gasket. Tighten the water pump retainers to 89 inch lbs. (10 Nm) for 4.2L engines. For 5.3L engines, tighten the bolts to 11 ft. lbs. (15 Nm), then to 22 ft. lbs. (30 Nm).
- Coolant hose(s)
- Water pump pulley. Tighten the pulley bolts to 18 ft. lbs. (25 Nm).
- Fan or fan and clutch assembly
- Serpentine drive belt (if equipped) by positioning the belt over the pulleys and carefully allow the tensioner back into contact with the belt.

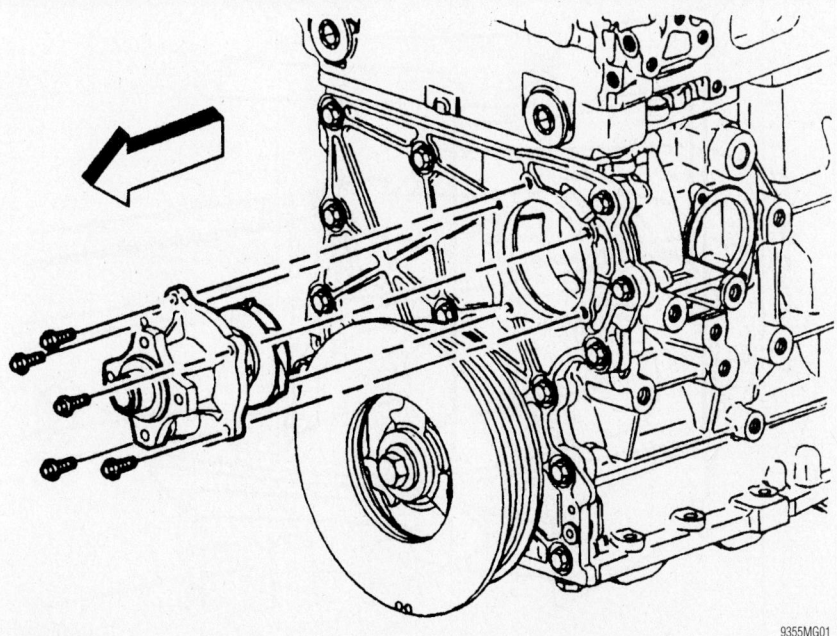

9355MG01

Exploded view of the water pump assembly mounting—4.2L engine

- V-belts (if equipped) and adjust the tension
- Upper fan shroud
- Negative battery cable

10. Refill the engine cooling system.
11. Run the engine and check for leaks.

Heater Core

REMOVAL & INSTALLATION

1. Before servicing the vehicle, refer to the precautions section.

2. Drain the engine cooling system.

3. Remove or disconnect the following:
- Negative battery cable
- Heater hoses from the heater core

4. Remove the instrument panel as follows:

a. Disable the air bag system.

b. Set the parking brake and block the wheels.

c. Disconnect the parking brake release cable from the parking brake lever.

d. Unfasten the screws that retain the DLC instrument panel left side sound insulator. Feed the DLC through the hole in the sound insulator.

e. Unfasten the right side sound insulator panel screws and remove the panel.

f. Unfasten the screws that attach the instrument panel left side sound insulator to the knee bolster and cowl panel.

g. Unfasten the nut that attaches the left side sound insulator to the accelerator pedal bracket.

h. Unplug the remote control door lock receiver module electrical connector.

i. Remove the door lock receiver module from the left side sound insulator. Remove the left side sound insulator.

j. Unfasten the screws that attach the instrument panel center sound insulator to the knee bolster, instrument panel, heater assembly and floor duct.

k. Remove the center sound insulator.

l. Unfasten the screws that attach the courtesy lamp to the knee bolster.

m. Unfasten the screws that attach the knee bolster to the instrument panel.

n. Disconnect the lap cooler duct from the knee bolster.

o. Unplug the lighter electrical connection and remove the knee bolster.

p. Unfasten the steering column-to-instrument panel nuts and lower the column.

q. Unfasten the screws that attach the instrument panel accessory trim plate to the instrument panel.

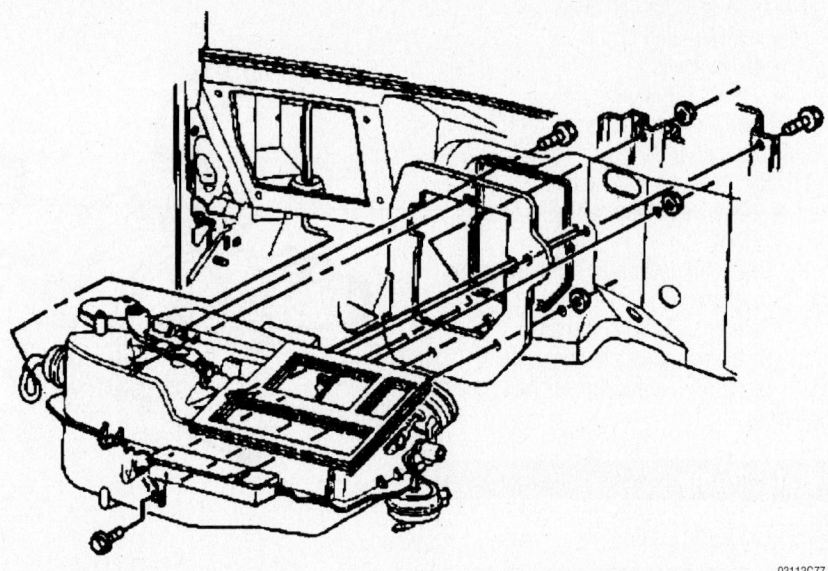

View of the heater case assembly

93113G77

r. Remove the trim plate and unplug all necessary electrical connection.

s. Remove the heater and/or air conditioning control assembly.

t. Remove the radio and the storage compartment assembly (if equipped).

u. If necessary, remove the instrument cluster.

v. Unfasten the left and right instrument panel pivot bolts and the panel lower support bolt.

w. Unfasten the speaker grilles retaining screws and remove the speaker grilles.

x. Remove the windshield defroster grille using a flat-bladed prytool. Start at one end of the grille and work your way down the grille.

y. Unfasten the 4 instrument panel upper support screws.

z. Tag and unplug all necessary electrical connections.

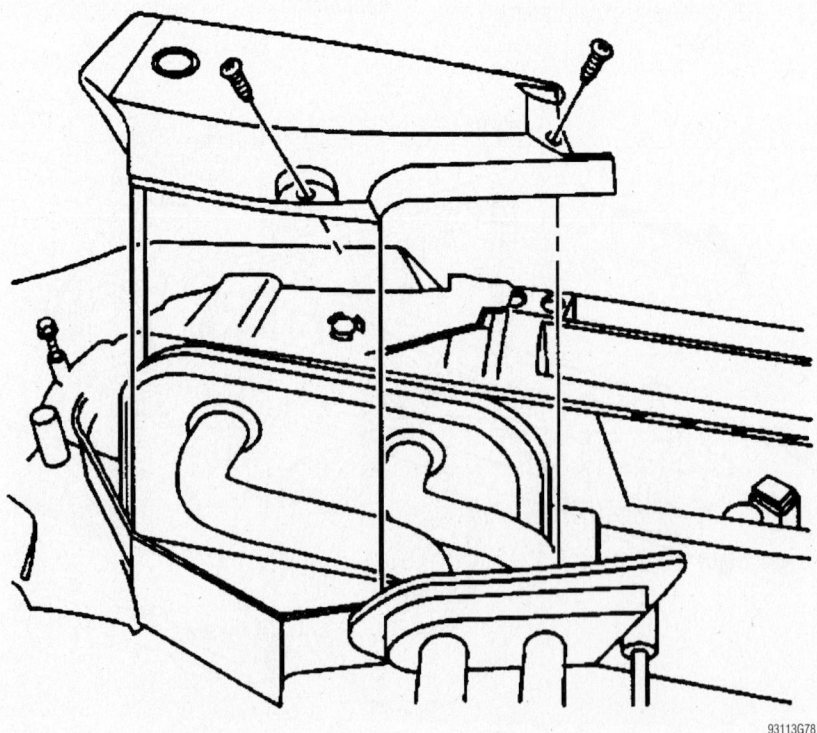

View of the heater case cover

93113G78

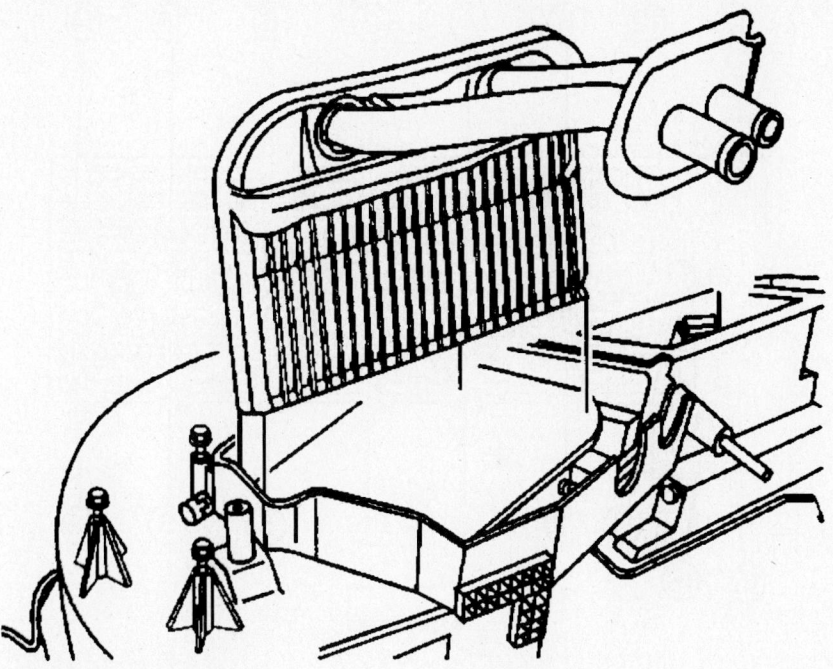

93113G79

View of the heater core

aa. Remove the instrument panel from the vehicle.

5. Remove or disconnect the following:
- Air inlet assembly, if equipped
- Vacuum hoses
- Heater assembly studs, from inside the engine compartment
- Blower motor resistor
- Stud from inside the heater case assembly; the stud is located behind the blower motor resistor
- Heater assembly-to-chassis screws
- Heater assembly from the vehicle
- Access cover screws and cover from the heater assembly
- Heater core from the heater case assembly

To install:

6. Install or connect the following:
- Heater core to the heater case assembly
- Access cover to the heater assembly and the cover screws
- Heater assembly to the vehicle
- Heater assembly-to-chassis screws and torque them to 40 inch lbs. (4.5 Nm)
- Stud, working from inside the heater case assembly; the stud is located behind the blower motor resistor
- Blower motor resistor
- Heater assembly studs, working inside the engine compartment, torque them to 17 inch lbs. (1.9 Nm)

- Vacuum hoses
- Air inlet assembly, if equipped

7. Install the instrument panel as follows:

a. Rest the instrument panel on the lower pivot studs.

b. Attach the electrical connections.

c. Install but do not tighten the 4 upper instrument panel support screws.

d. Install the left and right panel pivot bolts. Tighten the bolts to 102 inch lbs. (11.5 Nm).

e. Install the panel lower support bolt. Tighten the bolt to 102 inch lbs. (11.5 Nm).

f. Tighten the upper support screws to 17 inch lbs. (1.9 Nm).

g. Install the windshield defroster grille and the speaker grilles.

h. Install the radio and storage compartment assembly (if equipped).

i. If removed, install the instrument cluster.

j. Install the heater and/or air conditioning control assembly.

k. attach the electrical connections to the instrument panel accessory trim plate.

l. Place the trim plate in position and install its retaining screws. Tighten the screws to 17 inch lbs. (1.9 Nm).

m. Place the steering column into position and install its retaining nuts. Tighten the nuts to 22 ft. lbs. (30 Nm).

n. Attach the lighter electrical connec-

tion and the lap cooler duct to the knee bolster.

o. Place the knee bolster into position and install its retaining screws. Tighten the Torx® head screws to 80 inch lbs. (9 Nm) and the hex head screws to 17 inch lbs. (1.9 Nm).

p. Place the courtesy lamp in position and install its screws. Tighten the screws to 17 inch lbs. (1.9 Nm).

q. Place the instrument panel center sound insulator in position. Install the screws that attach the center sound insulator to the knee bolster, instrument panel and the floor duct. Tighten the screws to 17 inch lbs. (1.9 Nm).

r. Install the screw that attaches the center sound insulator to the heater assembly. Tighten the screw to 13 inch lbs. (1.5 Nm).

s. Install the remote control door lock receiver module to the instrument panel left side sound insulator.

t. Attach the door lock receiver electrical connection.

u. Install the nut that attaches the left side sound insulator to the accelerator pedal bracket. Tighten the nut to 35 inch lbs. (4 Nm).

v. Install the screw that attaches the left side sound insulator to cowl panel. Tighten the screw to 13 inch lbs. (1.5 Nm).

w. Install the screws that attach the left side sound insulator to knee bolster. Tighten the screw to 17 inch lbs. (1.9 Nm).

x. Feed the DLC through the hole in the sound insulator, place the DLC in position and install its retaining screws. Tighten the screws to 21 inch lbs. (2.4 Nm).

y. Install the right side sound insulator and tighten the screws

z. Connect the parking brake release cable to the lever.

aa. Enable the air bag system.

8. Install the heater hoses to the heater core.

9. Refill the cooling system.

10. Connect the negative battery cable.

11. Run the engine to normal operating temperatures; then, check the climate control operation and check for leaks.

Cylinder Head

REMOVAL & INSTALLATION

4.2L Engine

1. Before servicing the vehicle, refer to the precautions section.

2. Disconnect the negative battery cable.

3. Drain the engine cooling system.

4. Remove or disconnect the following:
- Camshaft cover
- Exhaust manifold
- Front cover
- Cylinder head access hole plugs
- Timing chain tensioner shoe bolt and shoe
- Timing chain tensioner guide bolts and guide
- Timing chain and sprockets

5. Unfasten the cylinder head bolts by loosening them in the reverse of the torque sequence, then carefully remove the cylinder head.

6. Remove the cylinder head gasket.

To install:

7. Carefully clean and inspect the cylinder head and the gasket mounting surfaces.

➡ **The gasket surfaces on both the head and block must be clean of any foreign matter and free of nicks or heavy scratches. The cylinder bolt threads in the block and thread on the bolts must be cleaned (dirt will affect the bolt torque).**

➡ **DO NOT apply sealer to composition steel-asbestos gaskets.**

❊❊ WARNING

Make sure the number 1 cylinder is at Top Dead Center (TDC).

8. If using a steel only gasket, apply a thin and even coat of sealer to both sides of the gaskets.

9. Place a new gasket over the dowel pins with the bead or the words "This Side Up" facing upwards (as applicable), then carefully lower the cylinder head into position over the gasket and dowels.

10. Apply a coating of 12345493 or equivalent sealer to the threads of the cylinder head bolts, then thread the bolts into position until finger-tight.

11. Tighten the cylinder head bolts in sequence as follows:

   a. Tighten the long bolts (1-14), in sequence, to 22 ft. lbs. (30 Nm).

   b. Tighten the long bolts, in sequence, an additional 155 degrees.

   c. Tighten the 2 short end bolt to 62 inch lbs. (7 Nm).

   d. Tighten the 2 short end bolts an additional 60 degrees.

   e. Tighten the 1 long end bolts to 62 inch lbs. (7 Nm).

   f. Tighten the 1 long end bolt an additional 120 degrees.

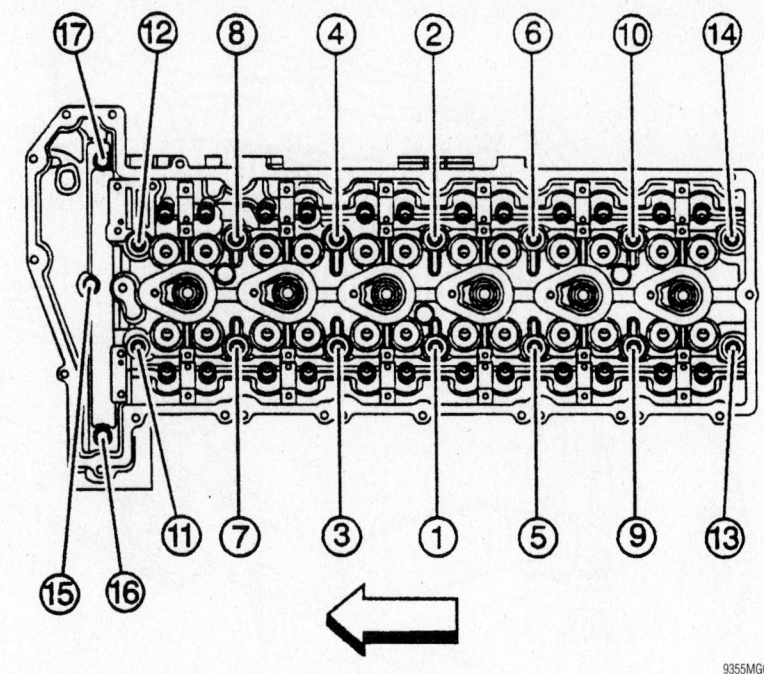

Cylinder head bolt tightening sequence—4.2L engine

12. Install or connect the following:
- Cylinder head access hole plugs and tighten to 44 inch lbs. (5 Nm)
- Timing chain and sprockets
- Front cover
- Camshaft cover
- Exhaust manifold
- Negative battery cable

13. Properly refill the engine cooling system.

14. Run the engine to check for leaks.

5.3L Engine

LEFT SIDE

1. Before servicing the vehicle, refer to the precautions section.

2. Drain the engine cooling system.

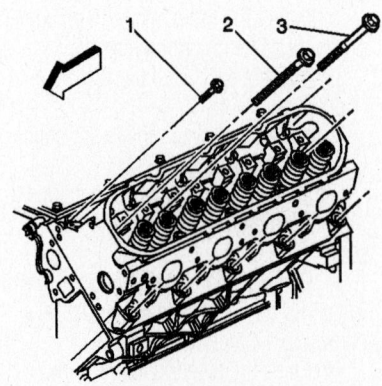

Remove and discard cylinder head bolts 1, 2 and 3—5.3L engine

3. Remove or disconnect the following:
- Negative battery cable
- Alternator bracket
- Coolant air bleed pipe
- Left exhaust manifold
- Pushrods
- Auxiliary A/C bracket bolt, if equipped
- Cylinder head bolts 1, 2 and 3. Discard the bolts
- Cylinder head
- Cylinder head gasket and discard

To install:

4. Carefully clean and inspect the cylinder head and the gasket mounting surfaces.

➡ **The gasket surfaces on both the head and block must be clean of any foreign matter and free of nicks or heavy scratches. The cylinder bolt threads in the block and thread on the bolts must be cleaned (dirt will affect the bolt torque).**

➡ **DO NOT apply any type sealer to cylinder head gasket, unless otherwise specified.**

5. Check the cylinder head locating pins for proper installation, location (a) 0.327 in. (8.3mm), as shown.

6. Place a new gasket over the dowel pins. When installed properly, the word "FRONT" on the left side, the tab on the gasket should be left of center or closer to the front of the engine.

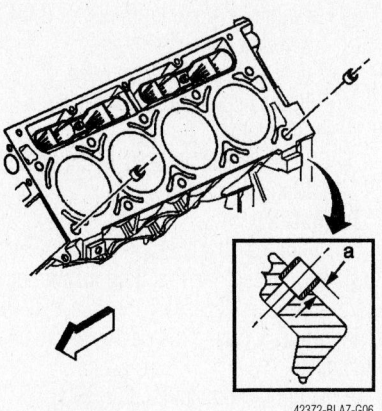

42372-BLAZ-G06

Make sure the cylinder head locating pins are properly installed, see dimension (a)

7. Install or connect the following:
- Cylinder head

➡️**You must use new cylinder head bolts during reassembly. Do NOT reuse the old head bolts.**

- NEW cylinder head bolts 1, 2 and 3.

8. Tighten the cylinder head bolts in sequence as follows:

a. Tighten the M11 bolts to 22 ft. lbs. (30 Nm).

b. Tighten the M11 an additional 90 degrees.

c. Tighten M11 bolts 1–8, an additional 90 degrees and M11 bolts 9 and 10 an additional 50 degrees.

d. Tighten the M8 bolts (11–15) to 22 ft. lbs. (30 Nm). Tighten all the bolts beginning with the center bolt and working outward, alternating sides

9. Install or connect the following:

- Auxiliary A/C bracket, if equipped. Torque the bolt to 15 ft. lbs. (20 Nm).
- Pushrods
- Left exhaust manifold
- Coolant air bleed pipe
- Alternator bracket

10. Properly refill the engine cooling system.

11. Run the engine to check for leaks.

RIGHT SIDE

1. Before servicing the vehicle, refer to the precautions section.

2. Drain the engine cooling system.

3. Remove or disconnect the following:
- Negative battery cable
- Oil level dipstick
- Coolant air bleed pipe
- Right exhaust manifold
- Pushrods
- Auxiliary A/C bracket nut, if equipped
- Cylinder head bolts 1, 2 and 3. Discard the bolts
- Cylinder head
- Cylinder head gasket and discard

To install:

4. Carefully clean and inspect the cylinder head and the gasket mounting surfaces.

➡️**The gasket surfaces on both the head and block must be clean of any foreign matter and free of nicks or heavy scratches. The cylinder bolt threads in the block and thread on the bolts must be cleaned (dirt will affect the bolt torque).**

➡️**DO NOT apply any type sealer to the cylinder head gasket, unless otherwise specified.**

5. Check the cylinder head locating pins for proper installation, location (a) 0.327 in. (8.3mm), as shown.

6. Place a new gasket over the dowel pins. When installed properly, the word "FRONT" on the right side, the tab on the gasket should be right of center or closer.

7. Install or connect the following:
- Cylinder head

➡️**You must use new cylinder head bolts during reassembly. Do NOT reuse the old head bolts.**

- NEW cylinder head bolts 1, 2 and 3.

8. Tighten the cylinder head bolts in sequence as follows:

a. Tighten the M11 bolts to 22 ft. lbs. (30 Nm).

b. Tighten the M11 an additional 90 degrees.

c. Tighten M11 bolts 1–8, an additional 90 degrees and M11 bolts 9 and 10 an additional 50 degrees.

d. Tighten the M8 bolts (11–15) to 22 ft. lbs. (30 Nm). Tighten all the bolts beginning with the center bolt and working outward, alternating sides

9. Install or connect the following:
- Auxiliary A/C bracket, if equipped. Torque the nut to 15 ft. lbs. (20 Nm).
- Pushrods
- Right exhaust manifold
- Coolant air bleed pipe
- Oil level dipstick

10. Properly refill the engine cooling system.

11. Run the engine to check for leaks.

Rocker Arms/Shafts

REMOVAL & INSTALLATION

4.2L Engine

1. Before servicing the vehicle, refer to the precautions section.

2. Remove or disconnect the following:
- Camshaft cover

➡️**Make sure to place the camshaft caps in a rack to keep them in order, so they may be installed in their original locations.**

- Camshaft cap bolts and caps
- Camshafts

➡️**If valve train components, such as the rocker arms or lash adjusters, are to be reused, they must be tagged or arranged to insure installation in their original locations.**

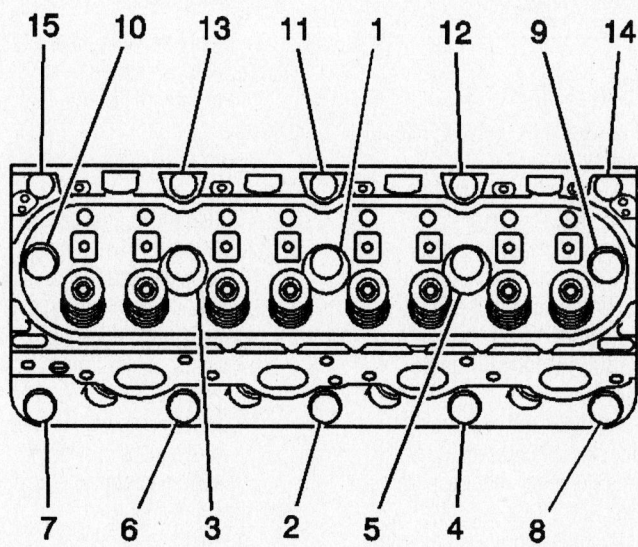

67162-ISU-ASC-G01

Cylinder head bolt torque sequence—5.3L engine

- Rocker arms
- Valve lash adjusters

To install:

3. Lubricate and fill the valve lash adjusters and the rocker arm roller with engine oil.

4. Install or connect the following

- Valve lash adjusters, in their original locations
- Rocker arm rollers in their original positions
- Camshafts
- Camshaft cap bolts
- Camshaft cover

5.3L Engine

1. Before servicing the vehicle, refer to the precautions section.

2. Remove or disconnect the following:

- Rocker arm cover(s)

➡ **If valve train components, such as the rocker arms, pushrods or pivot supports, are to be reused, they must be tagged or arranged to insure installation in their original locations.**

- Rocker arm bolts
- Rocker arms
- Rocker arm pivot support
- Pushrods

To install:

3. Inspect and replace components if worn or damaged.

4. Coat the bearing surfaces of the rocker arms, pushrods and the flange of the rocker arm bolts with clean engine oil.

5. Install or connect the following:

- Rocker arm pivot support
- Pushrods making sure they seat properly in the lifter sockets

➡ **Make sure the pushrods are seated properly to the ends of the rocker arms, but do not tighten the bolts yet.**

- Rocker arms and bolts

6. Rotate the crankshaft until the No. 1 piston is at Top Dead Center (TDC) of the compressor stroke. In this position, the cylinder No. 1 rocker arms will be off lobe lift, and the crankshaft sprocket key will be at the 1:30 position.

➡ **The engine firing order is: 1–8–7–2–6–5–4–3. Cylinders 1, 3, 5 and 7 are the left bank. Cylinders 2, 4, 6 and 8 are the right bank.**

7. With the engine in the No. 1 firing position, tighten the following rocker arm bolts:

a. Tighten cylinders 1, 2, 7 and 8 exhaust valve rocker arm bolts to 22 ft. lbs. (30 Nm).

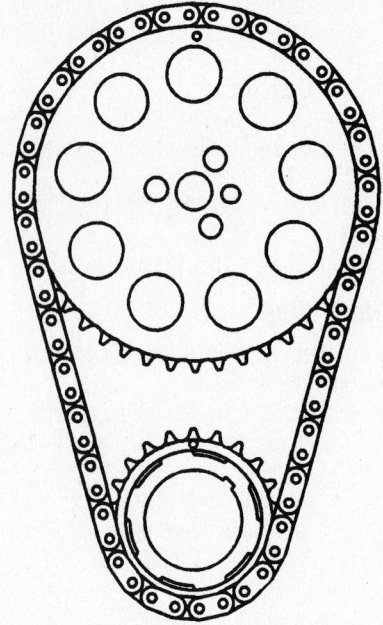

42372-BLAZ-G07

View of the crankshaft key with the No. 1 piston at TDC—5.3L engine

b. Tighten cylinder 1, 3, 4 and 5 intake valve rocker arm bolts to 22 ft. lbs. (30 Nm).

8. Rotate the crankshaft 360 degrees, then tighten the following rocker arm bolts:

a. Tighten cylinders 3, 4, 5 and 6 exhaust valve rocker arm bolts to 22 ft. lbs. (30 Nm).

b. Tighten cylinder 2, 6, 7 and 8 intake valve rocker arm bolts to 22 ft. lbs. (30 Nm).

9. Install the rocker arm cover(s).

Intake Manifold

REMOVAL & INSTALLATION

4.2L Engine

1. Before servicing the vehicle, refer to the precautions section.

2. Properly relieve the fuel system pressure.

3. Disconnect the negative battery cable.

4. Drain the engine cooling system.

5. Remove or disconnect the following:

- Throttle body
- Powertrain Control Module (PCM)
- All electrical harnesses from the engine harness bracket
- Front differential vent hose from the bracket clip
- Engine harness bracket bolt and bracket

- Manifold Absolute Pressure (MAP) sensor connector
- Crankcase ventilation hose
- Brake hose from the booster
- Alternator
- Intake manifold bolts and manifold.
- Manifold gasket

To install:

6. Clean the gasket mounting surfaces. Be sure to inspect the manifold for warpage and/or cracks. If necessary, replace it.

7. Properly position a new intake manifold gasket.

8. Install or connect the following:

- Intake manifold and bolts. Torque the bolts to 16 ft. lbs. (22 Nm).
- Alternator
- Brake hose to the booster
- Crankcase ventilation hose, lubricating the inner diameter first with 12345884, or equivalent lubricant
- MAP sensor electrical connector
- Engine harness bracket. Tighten the retaining bolt to 37 ft. lbs. (50 Nm).
- Front differential vent hose to the engine harness bracket clip
- All harnesses to their original locations onto the engine harness bracket
- PCM
- Throttle body
- Negative battery cable

9. Refill the engine cooling system.

5.3L Engine

➡ **The intake manifold, throttle body, fuel rail and injectors can be removed as an assembly. If you are not servicing these components individually, remove the intake manifold as a complete assembly.**

1. Before servicing the vehicle, refer to the precautions section.

2. Properly relieve the fuel system pressure.

3. Disconnect the negative battery cable.

4. Drain the engine cooling system.

5. Remove or disconnect the following:

- Air cleaner outlet duct
- A/C compressor pressure switch electrical connector
- Harness clip from the cylinder head and fuel raid
- Mass Airflow/Intake Air Temperature sensor connector

6. Disconnect the electrical connectors from the following:

a. Main coil

b. Electronic Throttle Control (ETC)

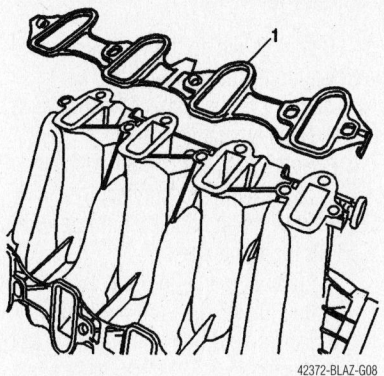

42372-BLAZ-G08

Make sure to use NEW intake manifold gaskets (1)—5.3L engine

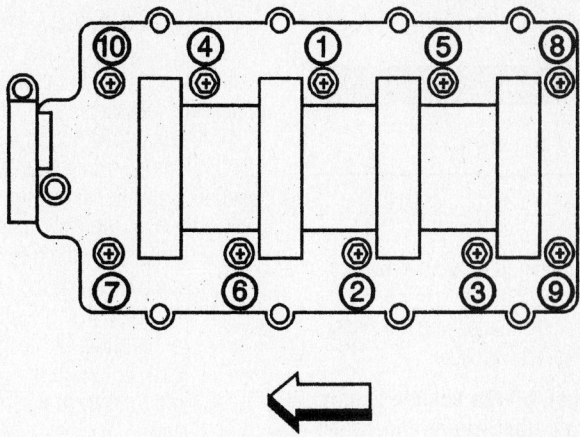

42372-BLAZ-G10

Intake manifold bolt tightening sequence—5.3L engine

c. Fuel injectors. Matchmark the connectors, pull the Connector Position Assurance (CPA) retainer up 1 click. Push the tab on the connector in, then detach the injector connector.
- Alternator connector
- Evaporative emission (EVAP) purge solenoid electrical connector
- Knock Sensor (KS) electrical connector
- Main coil
- Fuel injector electrical connector
- Electrical harness clips from the fuel rail

- KS harness electrical connector from the intake manifold
- Positive Crankcase Ventilation (PCV) valve hose and valve
- Heater water shutoff valve actuator inlet hose from the intake manifold
- EVAP purge solenoid vent tube
- Vacuum brake booster hose from the rear of the intake manifold
- Upper engine wire harness retainer nut. Position the wire harness aside.

- Intake manifold bolts
- Intake manifold and gaskets. Discard the gaskets.

To install:

7. Clean the gasket mounting surfaces. Be sure to inspect the manifold for warpage and/or cracks. If necessary, replace it.

8. Properly position a new intake manifold gasket.

9. Apply a 0.20 in. (5mm) band of a suitable threadlocking material to the intake manifold bolt threads.

10. Install or connect the following:
- Intake manifold and bolts. Torque the bolts, in sequence to 44 inch lbs. (5 Nm), then to 89 inch lbs. (10 Nm).
- Route the electrical harness into position over the engine.
- Engine harness bracket nut and tighten to 89 inch lbs. (10 Nm)
- Vacuum brake booster hose to the rear of the intake manifold
- EVAP purge solenoid valve
- Heater water shutoff valve actuator inlet hose to the intake manifold
- PCV valve and hose
- EVAP purge solenoid, KS, MAP sensor, main coil & fuel injector electrical connectors
- Harness clips to the fuel rail
- Alternator electrical connector
- Main coil, ETC, fuel injector electrical connectors
- Electrical harness clips to the fuel rail
- A/C compressor pressure switch electrical connector
- Harness clip to the cylinder head
- Mass Airflow/Intake Air Temperature sensor connector
- Air cleaner outlet duct
- Fuel fill cap

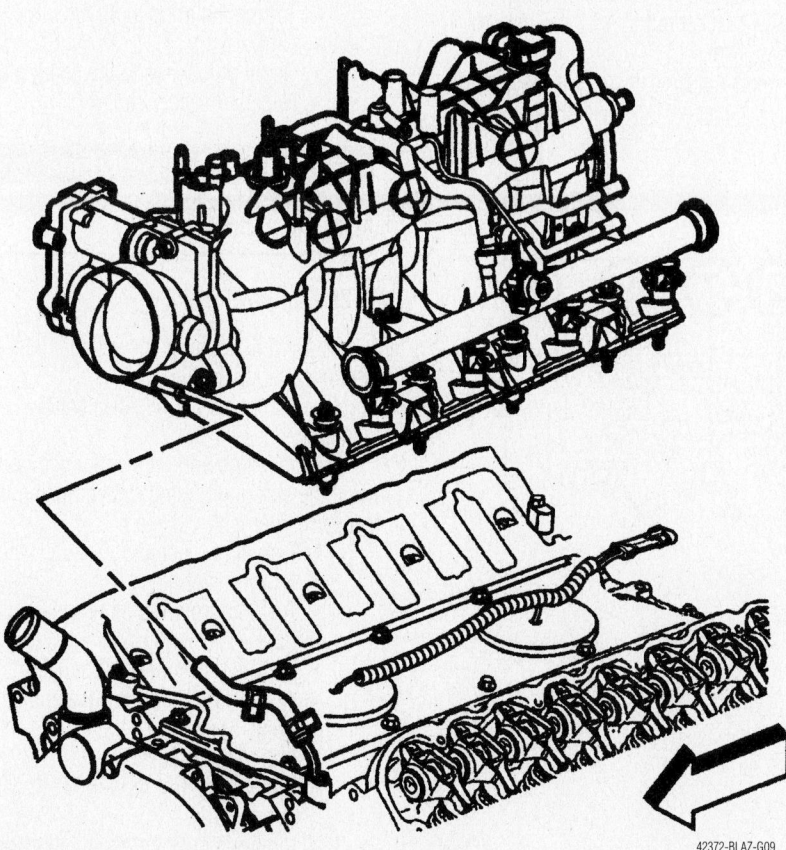

42372-BLAZ-G09

Exploded view of the intake manifold—5.3L engine

- Negative battery cable
11. Refill the engine cooling system.

Exhaust Manifold

REMOVAL & INSTALLATION

4.2L Engine

1. Before servicing the vehicle, refer to the precautions section.
2. Remove or disconnect the following:
   - Negative battery cable

➡It will be easier if the vehicle is only supported to a height where underhood access is still possible, the vehicle may be left in position for the entire procedure. If the vehicle is raised too high for underhood access, it will have to lowered, raised and lowered again during the procedure.

- Air cleaner resonator outlet duct
- Transmission filler tube stud nut from the Air Injector Reactor (AIR) adapter and move the tube aside
- Oil level indicator tube
- Oxygen (O_2) sensor from the exhaust manifold
- 4 manifold heat shield nuts and shield
- Exhaust pipe bolts from the exhaust manifold
- Exhaust manifold bolts, and manifold
- Old gaskets and discard

To install:

3. Using a putty knife, clean the gasket mounting surfaces. Inspect the exhaust manifold for distortion, cracks or damage; replace if necessary.
4. Apply a threadlock such as GM 12345493 to the threads of the manifold retainers prior to installation.
5. Install or connect the following:
   - Exhaust manifold to the cylinder using a new gasket, then tighten the bolts, in 3 passes, in sequence, to 18 ft. lbs. (25 Nm)
   - Heat shield studs, if necessary, and tighten to 89 inch lbs. (10 Nm)
   - O_2 sensor
   - Exhaust manifold heat shield

➡Apply a suitable anti-seize compound to the exhaust manifold heat shield nuts prior to installation.

- Heat shield nuts and tighten to 44 inch lbs. (5 Nm)
- Exhaust pipe to the manifold with seal and retaining nuts. Tighten the nuts to 37 ft. lbs. (50 Nm).
- Oil level indicator tube
- Transmission filler tube back onto the AIR adapter block stud and secure with the nut. Tighten the bracket nut to 89 inch lbs. (10 Nm).
- Air cleaner resonator outlet duct
- Negative battery cable.

5.3L Engine

1. Before servicing the vehicle, refer to the precautions section.
2. Remove or disconnect the following:
   - Negative battery cable
   - Spark plug wires from the spark plugs. Don't disconnect the wires from the ignition coil unless necessary for clearance.
   - Exhaust manifold bolts, manifold and gasket. Discard the gasket.
   - Heat shield bolt and shield, if necessary

To install:

3. Apply a 0.2 inch (5mm) bead of threadlock GM P/N 12345493, or equivalent to the threads of the exhaust manifold bolts. Do NOT apply sealer to the first 3 threads of the bolts.
4. Install or connect the following:
   - New exhaust manifold gasket
   - Exhaust manifold
   - Exhaust manifold bolts. Tighten in two passes. First to 11 ft. lbs. (15 Nm), then to 18 ft. lbs. (25 Nm) starting with the center bolts and working outward.
5. Bend over the exposed edge of the gasket at the rear of the cylinder head using a flat punch or equivalent tool.
   - Heat shield and bolts, if removed. Torque the bolts to 80 inch lbs. (9 Nm).
   - Spark plug wires to the spark plugs
   - Negative battery cable

Camshaft and Valve Lifters

REMOVAL & INSTALLATION

4.2L Engine

1. Before servicing the vehicle, refer to the precautions section.
2. Disconnect the negative battery cable.
3. Discharge and recover the refrigerant from the air conditioning system, using the proper equipment.
4. Remove or disconnect the following:
   - Intake manifold
   - A/C line from the oil level indicator tube
   - A/C line from the accumulator
   - A/C bracket bolt from the engine lift hook
   - Engine lift bracket
   - Ignition control module electrical connectors
   - Ignition control module bolts and module

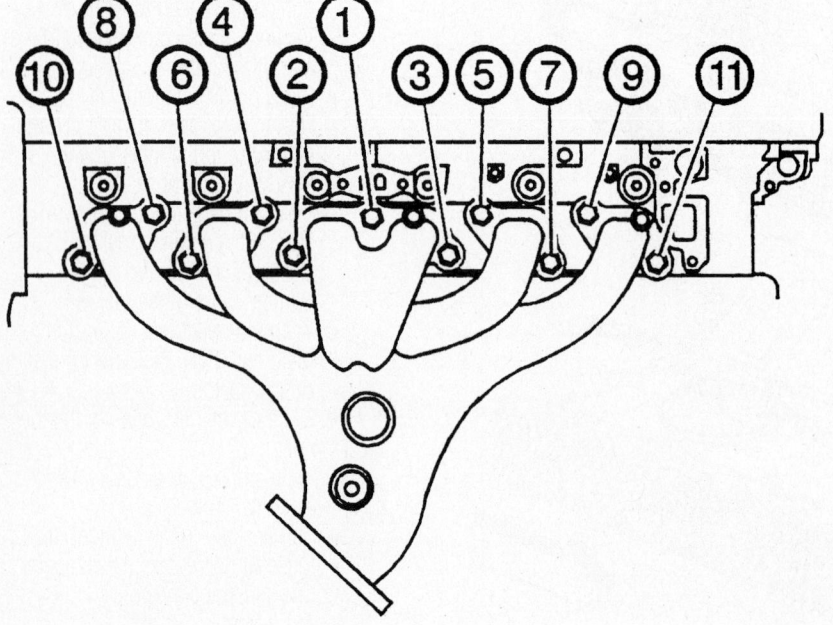

9355MG03

Exhaust manifold bolt tightening sequence—4.2L engine

Be careful not to damage the clips that hold the harness housing in place.

- Engine electrical harness housing from the camshaft cover
- Fuel injection harness electrical connector
- Camshaft cover bolts and cover
- Exhaust and intake sprocket bolts

5. Install a suitable sprocket holding tool onto the cylinder head and adjust the horizontal bolts into the camshaft sprockets to maintain timing chain tension and avoid disturbing the timing chain components.

6. Carefully move the sprockets with the timing chain off of the camshafts.

➡Make sure to place the camshaft caps in a rack to keep them in order, so they may be installed in their original locations.

7. Remove or disconnect the following:
- Camshaft cap bolts and caps
- Camshafts

To install:

8. Coat the camshaft journals with engine oil.
- Camshafts, in their original position
- Camshaft caps, in their original locations. Tighten the bolts to 106 inch lbs. (12 Nm).

9. Carefully place the camshaft sprockets back onto the camshafts and remove the holding tool.

10. Install or connect the following:
- Intake camshaft sprocket washer and bolt and the exhaust camshaft actuator bolt. Tighten the intake camshaft sprocket bolt to 22 ft. lbs. (30 Nm), plus an additional 135 degrees and the exhaust camshaft actuator bolt to 18 ft. lbs. (25 Nm), plus an additional 135 degrees.
- New camshaft cover seal
- New rubber ignition control module seals
- Camshaft cover and bolts. Tighten the bolts to 89 inch lbs. (10 Nm).
- Ignition control module. Tighten the bolts to 89 inch lbs. (10 Nm).
- Ignition control module electrical connectors
- Fuel injector electrical connectors
- Engine electrical harness housing
- A/C line bracket to the oil level indicator tube stud and secure with the nut. Tighten the nut to 62 inch lbs. (7 Nm).

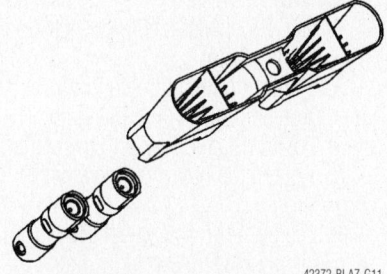

42372-BLAZ-G11

Remove the lifters from the guides, making sure to keep them in order—5.3L engine

- Engine lift bracket and secure the lift hook with the bolts. Tighten the bolts to 37 ft. lbs. (50 Nm).
- A/C line bracket to the engine lift bracket. Tighten the bolt to 89 inch lbs. (10 Nm).
- Intake manifold

11. Using the proper equipment, recharge the A/C system.

5.3L Engine

1. Before servicing the vehicle, refer to the precautions section.

2. Disconnect the negative battery cable.

3. Discharge and recover the refrigerant from the air conditioning system, using the proper equipment.

4. Remove or disconnect the following:
- Condenser
- Cylinder head and gasket
- Valve lifter guide bolts
- Valve lifters and guide

➡If the lifters are stuck in the bores due to built up deposits, use Valve Lifter Remover tool No. J 3049-A or equivalent to remove the lifters

- Valve lifters from the guide

➡Make sure to keep the lifters in order as you are removing them. They must be installed in their original locations.

5. Clean and inspect the lifters for damage.
- Camshaft sensor bolt and sensor

6. Rotate the crankshaft until the timing marks on the crankshaft and camshaft sprockets are aligned.
- Camshaft sprocket bolts

Do NOT turn the crankshaft after the timing chain has been removed to avoid damaging the pistons or valves!

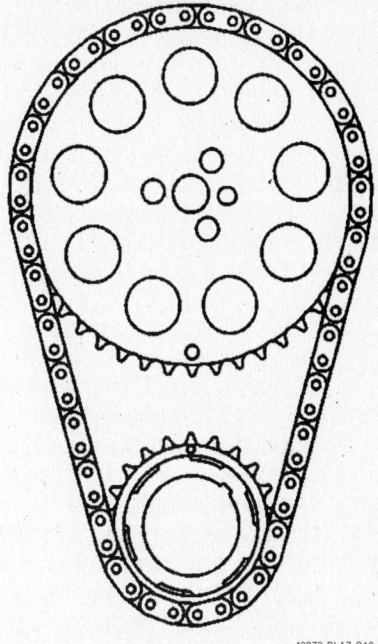

42372-BLAZ-G12

Make sure the crankshaft and camshaft timing marks are aligned

- Camshaft sprocket and reposition the timing chain
- Camshaft retaining bolts and retainer
- Camshaft by installing three M8-1.25 x 4.0 in. (M8-1.25 x 1.00mm) bolts in the front of the camshaft to act as a handle; then, remove the camshaft while turning slightly from side to side, as necessary. Remove the bolts from the camshaft.

➡Take care not to damage the camshaft bearings when removing the camshaft.

7. Clean and inspect the camshaft and bearings.

To install:

➡If the camshaft must be replaced, you must also replace the lifters.

8. Lubricate the camshaft journals with clean engine oil.

9. Install or connect the following:
- Three bolts used during removal into the bolt hold in the front of the camshaft
- Camshaft carefully into the engine block, using the bolts as a handle. Remove the bolts.
- Camshaft retainer and bolts. Make sure the retaining plate is installed with the sealing gasket facing the engine block. Tighten the bolts to 18 ft. lbs. (25 Nm).

10. Properly locate the camshaft sprocket locating pin with the cam sprocket alignment hole. The sprocket teeth and timing chain must mesh. The camshaft and crankshaft sprocket alignment marks MUST be aligned properly. Locate the camshaft sprocket alignment mark in the 6 o'clock position. It may be necessary to rotate the camshaft or crankshaft to align the marks.

- Camshaft sprocket and timing chain
- Camshaft sprocket bolts and tighten to 26 ft. lbs. (35 Nm)
- Camshaft sensor O-ring, after making sure it is not damaged and lubricating it with clean engine oil
- Camshaft sensor and bolt. Torque the bolt to 18 ft. lbs. (25 Nm).

11. Lubricate the valve lifters and engine block lifter bores with clean engine oil.

12. Install or connect the following:
- Lifters into the lifter guides. Align the area on top of the lifter with the flat area in the lifter guide bore. Push the lifter completely into the guide bore.
- Valve lifters and guide to the engine block
- Valve lifter guide bolt and tighten to 106 inch lbs. (12 Nm)
- Cylinder head and gasket
- Condenser

13. Using the proper equipment, recharge the A/C system.

Valve Lash

ADJUSTMENT

The engines covered in this section do not require a periodic valve lash adjustment.

Starter Motor

REMOVAL & INSTALLATION

4.2L Engine

1. Before servicing the vehicle, refer to the precautions section.
2. Disconnect the negative battery cable
3. Raise and safely support the vehicle.
4. Remove the left front tire and wheel assembly.
5. Working in the left fender area, disconnect the positive battery lead from the solenoid.
6. Remove or disconnect the following:
- Starter mount bolt and nut
- Starter motor

To install:
7. Install or connect the following:
- Starter motor
- Starter mounting bolt and nut. Tighten to 37 ft. lbs. (50 Nm).
- Positive battery cable to the starter. Tighten the nut to 80 inch lbs. (9 Nm).
- Left front tire and wheel assembly
8. Carefully lower the vehicle, then connect the negative battery cable.

5.3L Engine

1. Before servicing the vehicle, refer to the precautions section.
2. Remove or disconnect the following:
- Negative battery cable
- Catalytic converter
- Engine shield bolts and shield
- Right transmission cover bolt
- Starter bolts
- Transmission cover and shield, after repositioning the starter
3. Position the starter down, with the terminals facing toward the front of the vehicle.
- Starter solenoid nut
- Starter lead from the solenoid stud
- Starter lead nut
- Positive cable from the starter stud
- Starter

To install:
4. Install or connect the following:
- Starter in the vehicle. Position the starter down , with the terminals facing toward the front of the vehicle.
- Positive cable to the starter stud.
- Starter lead nut and tighten to 80 inch lbs. (9 Nm)
- Starter solenoid lead to the stud
- Starter solenoid nut and tighten to 30 inch lbs. (3.4 Nm)

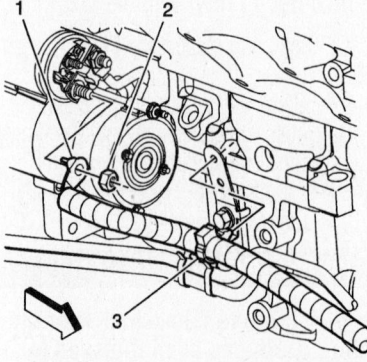

View of the starter, positive cable (1) and starter lead nut —5.3L engine

42372-BLAZ-G13

- Install the shield and transmission cover, after repositioning the starter
5. Slide the starter rearward.
- Starter bolts and tighten to 37 ft. lbs. (50 Nm)
- Right transmission cover bolt and tighten to 80 inch lbs. (9 Nm)
- Catalytic converter
- Negative battery cable
6. Start the vehicle to check for proper operation.

Oil Pan

REMOVAL & INSTALLATION

4.2L Engine

1. Before servicing the vehicle, refer to the precautions section.
2. Disconnect the negative battery cable.
3. Remove or disconnect the following:
- A/C compressor bottom bolts and loosen the top bolts
- Oil dipstick and tube
4. Raise and safely support the vehicle.
5. Drain the engine crankcase oil.
6. Remove or disconnect the following:
- Left and right front tire and wheel assemblies
- Engine protection shield mounting bolts and shield
- Front steering gear crossmember
- Left and right driveshafts
- Front drive axle clutch fork assembly
- Prop shaft from the front axle pinion yoke
- Unclip the transmission cooler lines from the engine block
- Front differential bolts and position the differential aside
- 4 transmission bell housing-to-oil pan bolts
- Remaining oil pan bolts
- Oil pan, by placing 2 oil pan bolts in the jack screws on the oil pan and tighten evenly to release the oil pan from the engine

To install:
7. Clean the gasket mounting surfaces.

➡**The alignment between the rear of the oil pan and the rear of the block is critical. When the oil pan is installed it could be inadvertently shifted front or back a small amount which could cause a transmission alignment problem. The back to the oil pan needs to be flush with the engine block.**

8. Apply a 0.12 in. (3mm) bead of sealant to engine block, rather than the oil pan.

➡**The oil pan MUST be installed within 10 minutes of applying the sealant to the engine block.**

9. Install or connect the following:
- Oil pan, maneuvering it to clear the oil pump and screen assembly

➡**After the bolts are installed, before tightening them to specifications, check the oil pan alignment. Use a straight edge on the back to the block and the oil pan transmission mounting surface.**

- Oil pan bolts; tighten the side bolts to 18 ft. lbs. (25 Nm) and the end bolts to 89 inch lbs. (10 Nm)
- Transmission bell housing-to-oil pan bolts and tighten to 35 ft. lbs. (47 Nm)
- A/C compressor bottom bolts. Tighten to 37 ft. lbs. (50 Nm)
- Front differential bolts and tighten to 63 ft. lbs. (85 Nm)
- Front drive axle and clutch fork assembly
- Transmission cooler lines to block
- Prop shaft to front differential
- Steering gear crossmember
- Left and right driveshaft
- Oil pan drain plug. Tighten to 19 ft. lbs. (26 Nm)
- Engine protection shield. Tighten the bolts to 18 ft. lbs. (25 Nm)
- Left and right front wheel and tire assemblies

10. Carefully lower the vehicle.
11. Refill the crankcase with fresh oil. Start the engine, establish normal operating temperatures and check for leaks.

5.3L Engine

1. Before servicing the vehicle, refer to the precautions section.
2. Disconnect the negative battery cable.
3. Drain the engine crankcase oil and differential oil.
4. Remove or disconnect the following:
- Oil level dipstick
- Front shock upper retaining nuts
- Tires and wheels
- Engine shield bolts and shield
- Power steering gear
- Left and right Antilock Brake System (ABS) wiring harnesses from the retainers

- Wheel Speed Sensor (WSS) electrical connectors
- Brake hose retaining bolts from the frame
- Sway bar link pins from the lower control arm on both sides

5. Place an adjustable jackstand under the lower control arm.
- Upper ball joint pinch bolt and nut
- Upper control arm from the upper ball joint

6. Lower and remove the jackstand, letting the suspension hang.
- Left driveshaft
- Right driveshaft from the intermediate shaft bearing only. Do not remove the driveshaft from the steering knuckle. Position the driveshaft aside.

7. Using wire or hooks, secure the front shock modules to the frame. Do NOT let the shocks and steering knuckle hang without being supported.
8. Matchmark the position of the propeller shaft to the front axle pinion yoke.
9. Remove or disconnect the following:
- Yoke retainer bolt and yoke retainers from the front axle pinion yoke. Wrap the bearing caps with tap to avoid losing the bearing rollers. Secure the propeller shaft to the frame.
- Transmission oil cooler lines from the retainer
- Transmission oil cooler line retaining bracket bolt and bracket
- Inner axle shaft
- Starter
- Flywheel inspection cover from the left side of the transmission
- Battery cable channel bolt from the front of the oil pan
- Battery cable channel from the oil pan
- Loosen the 2 upper A/C compressor bracket bolts
- 2 lower A/C compressor bracket bolts
- Front differential attachment bolts. Secure the front differential to the frame.
- 2 lower bellhousing bolts
- Oil pan bolts
- Oil pan by tilting the rear of the oil pan down to clear the transmission, pull the oil pan rearward past the front wire harness, then lower the oil pan clear of the vehicle

➡**The oil pan gasket is reusable if it is not damaged.**

10. Drill out the oil pan gasket retaining rivets, if necessary. Remove the gaskets.

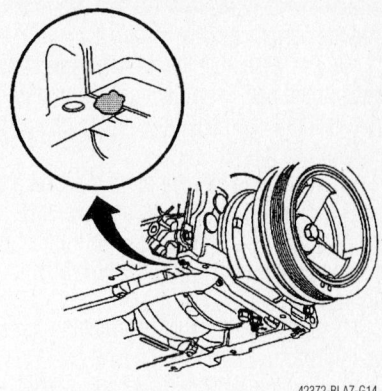

42372-BLAZ-G14

Proper sealant application to the front cover gasket

Discard the rivets. Inspect the gasket, if it is damaged, it must be replaced.

To install:

➡**The proper alignment of the oil pan is very important. The rear bolt hold location of the oil pan provide mounting points for the transmission bellhousing. To ensure the rigidity of the powertrain and correct transmission alignment, make sure that the rear of the block and rear of the oil pan NEVER protrude beyond the engine block and transmission bellhousing plane.**

➡**If replacing the oil pan gasket, it is not necessary to rivet the NEW gasket to the pan.**

11. Apply a 0.20 in. (5mm) bead of sealant 0.80 in. (20mm) long to the engine block. Apply the sealant directly onto the tabs of the front cover gasket that protrudes into the oil pan surface.

12. Apply a 0.20 in. (5mm) bead of sealant 0.80 in. (20mm) long to the engine block. Apply the sealant directly onto the

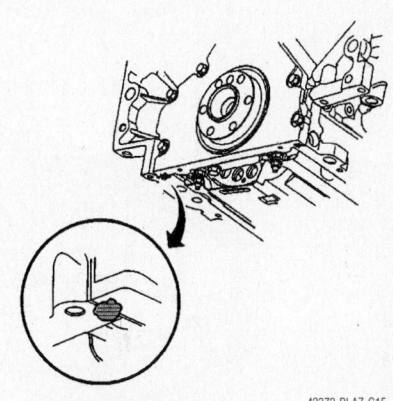

42372-BLAZ-G15

Proper sealant application to the rear cover gasket

tabs of the rear cover gasket that protrudes into the oil pan surface.

13. Pre-assemble the oil pan gasket and bolts to the pan. Install the gasket onto the pan. Install the oil pan bolts to the pan and through the gasket.

14. Install the oil pan, oil pan gasket and bolts to the engine block as an assembly.

15. Hand-start the bolts into the engine block snug-tight. Do not fully tighten yet.

16. Install the 2 lower bellhousing bolts and tighten to 37 ft. lbs. (50 Nm).

17. Tighten the 2 rear oil pan-to-rear cover bolts to 106 inch lbs. (12 Nm) and the remaining oil pan bolts to 18 ft. lbs. (25 Nm).

18. Release the differential from the frame and install to the oil pan. Install and tighten the bolts to 63 ft. lbs. (85 Nm).

19. Install or connect the following:
- 2 lower A/C compressor bracket bolts. Tighten the lower and upper compressor bolts to 37 ft. lbs. (50 Nm).
- Battery cable channel to the oil pan
- Battery cable channel bolt and tighten to 106 inch lbs. (12 Nm)
- Flywheel inspection cover to the left side of the transmission
- Starter

- Inner axle shaft
- Transmission oil cooler line retaining bracket and bolt. Torque the bolt to 80 inch lbs. (9 Nm).
- Transmission oil cooler lines to the retainer

20. Unhook the right driveshaft from the frame.
- Left and right driveshafts

21. Unsecure the shocks from the frame. Put adjustable jackstand under the lower control arm. Using the jackstand, raise the lower control arm and knuckle assembly in order to connect the upper ball joint to the upper control arm.
- Upper ball joint pinch nut and bolt and tighten to 30 ft. lbs. (40 Nm). Remove the jackstand.
- Sway bar link pins to the lower control arm on both sides
- Steering gear

22. Unsecure the prop shaft from the frame. Align the matchmarks on the prop shaft to the marks on the front axle pinion yoke.
- Propeller shaft to the front axle pinion yoke
- Yoke retainers and yoke retainer bolts to the front axle pinion yoke. Torque the bolts to 15 ft. lbs. (20 Nm).

- Brake hose retaining bolts to the frame and tighten to 18 ft. lbs. (25 Nm).
- WSS electrical connectors
- Left and right ABS wiring harnesses to the retainers
- Differential with oil
- Engine shield and bolts. Tighten the bolts to 18 ft. lbs. (25 Nm).
- Tires and wheels

23. Fill the engine with oil. Fill the power steering system with fluid.
- Upper shock nuts and tighten to 74 ft. lbs. (100 Nm).
- Oil dipstick
- Negative battery cable

Oil Pump

REMOVAL & INSTALLATION

4.2L Engine

1. Before servicing the vehicle, refer to the precautions section.

2. Remove or disconnect the following:
- Engine front cover
- Oil pump cover bolts
- Oil pump cover. Mark the inner and outer gears in relation to the pump housing.
- Inner and outer pump gears
- Oil pump pressure relief valve plug
- Oil pump pressure relief valve and spring

To install:

3. Install or connect the following:
- Oil pump pressure relief valve and spring
- Oil pump pressure relief valve plug. Tighten to 10 ft. lbs. (14 Nm).
- Oil pump outer and inner gears, as marked during removal
- Oil pump cover and bolts. Tighten the bolts to 89 inch lbs. (10 Nm).
- Front cover

5.3L Engine

1. Before servicing the vehicle, refer to the precautions section.

2. Remove or disconnect the following:
- Oil pan
- Engine front cover
- Oil pump screen bolt and nuts
- Oil pump screen with O-ring seal
- O-ring seal from the pump screen. Discard the O-ring seal.
- Remaining crankshaft oil deflector nuts
- Crankshaft oil deflector

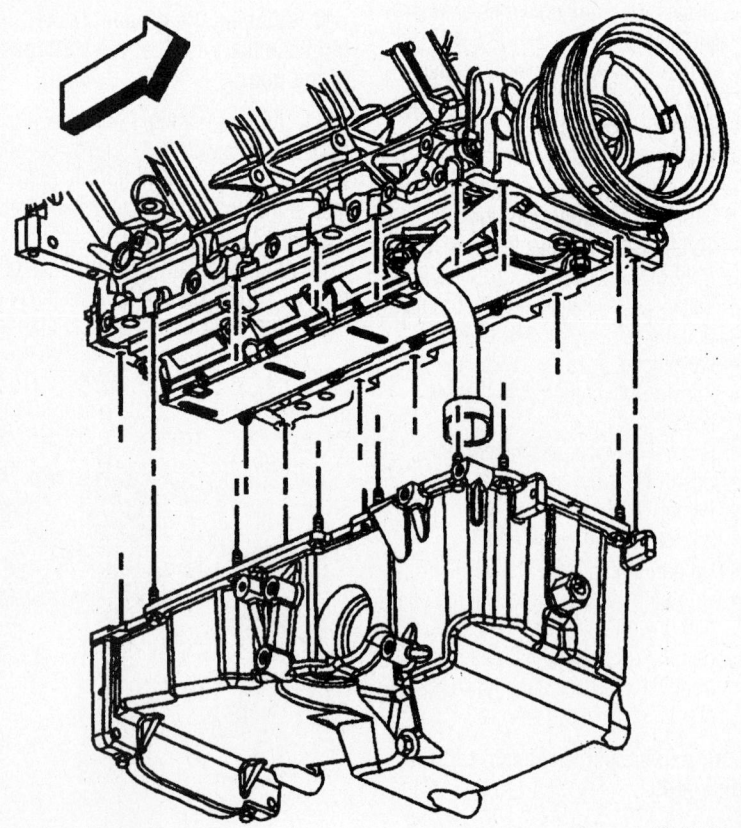

Oil pan mounting—5.3 engine

42372-BLAZ-G16

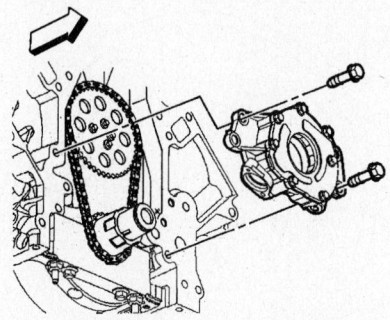

Exploded view of the oil pump mounting—
5.3L engine

- Oil pump bolts
- Oil pump

➡**Do not let any dirt or debris into the oil pump or cap end.**

- Clean and inspect the oil pump.

To install:

3. Align the splined surfaces of the crankshaft sprocket and the oil pump drive gear and install the oil pump.

4. Install or connect the following:
- Oil pump onto the crankshaft sprocket until the pump housing contacts the face of the engine block
- Oil pump bolts and tighten to 18 ft. lbs. (25 Nm)
- Crankshaft oil deflector and nuts until snug
- New oil pump screen O-ring seal into the oil pump screen, after lubricating with clean engine oil

➡**Push the oil pump screen tube completely into the oil pump prior to tightening the bolt. Do not let the bolt pull the tube into the pump.**

5. Align the oil pump screen mounting brackets with the correct crankshaft bearing cap studs.
- Oil pump screen
- Oil pump screen bolts and nuts. Tighten the bolts to 106 inch lbs. (12 Nm) and the nuts to 18 ft. lbs. (25 Nm).
- Engine front cover
- Oil pan

Rear Main Seal

REMOVAL & INSTALLATION

4.2L Engine

Please note that the transmission assembly must be removed to perform this procedure.

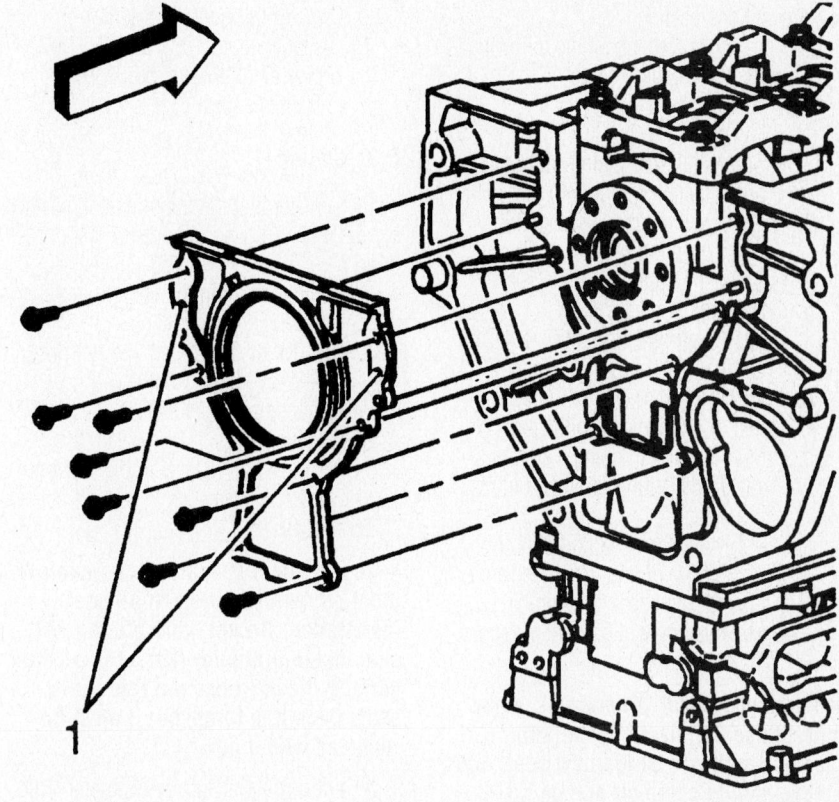

Install 2 bolts into the jackscrew holes (1) to push the cover off of the block—4.2L engine

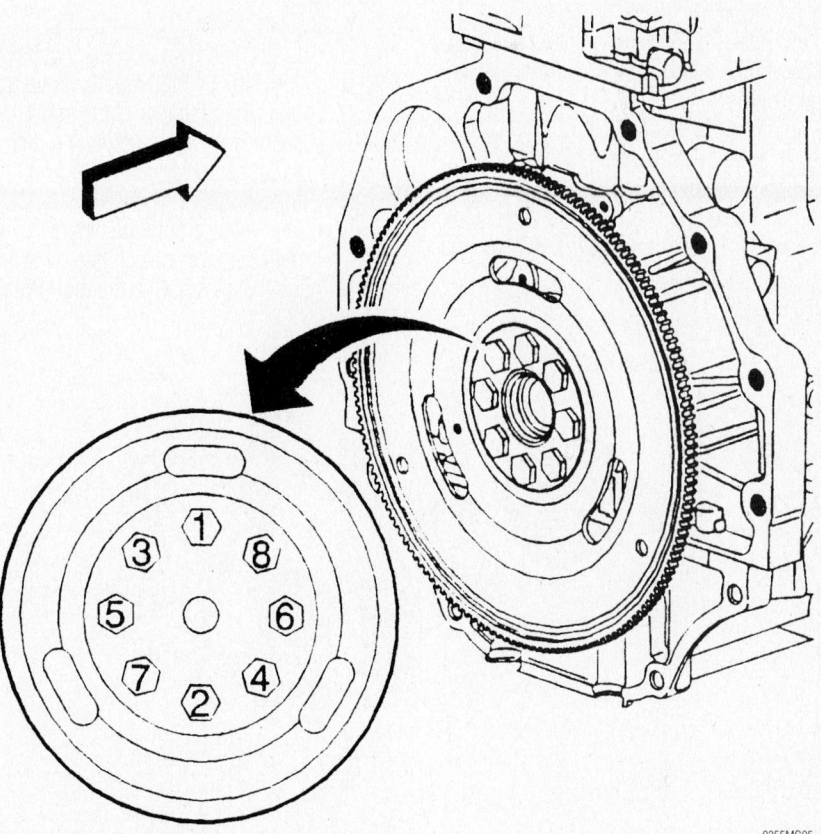

Flywheel bolt tightening sequence—4.2L engine

1. Before servicing the vehicle, refer to the precautions section.
2. Remove or disconnect the following:
   - Negative battery cable
   - Transmission
   - Flywheel
   - Crankshaft rear main seal housing bolts. Install 2 bolts into the jackscrew holes to release the cover from the block
   - Crankshaft and rear main seal housing
   - Rear main seal from the crankshaft snout

To install:

3. Install or connect the following:
   - Rear main seal, using a suitable seal installation tool, then remove the tool
   - Apply a 0.12 in. (3mm) bead of 12378521, or equivalent sealant to the rear mail seal housing
   - Suitable cover alignment pins into the block

➡**When you install a new seal, make sure to use the plastic installation sleeve supplies with the new seal. The sleeve should come off and be discarded after the seal is installed.**

4. Slide the crankshaft rear main seal housing over the alignment pins and crankshaft.
5. Install the crankshaft rear main seal housing bolts, except the 2 in place of the guide pins.
6. Remove the guide pins.
7. Install or connect the following:
   - Remaining 2 crankshaft rear main seal housing bolts and tighten to 89 inch lbs. (10 Nm). Wipe off any excess sealant.

- Flywheel and secure with the mounting bolts. Tighten, in sequence, to 18 ft. lbs. (25 Nm), plus an additional 50 degrees.
- Transmission

5.3L Engine

Please note that the transmission assembly must be removed to perform this procedure.

1. Before servicing the vehicle, refer to the precautions section.
2. Remove or disconnect the following:
   - Negative battery cable
   - Transmission
   - Flywheel
   - Crankshaft rear main oil seal from the rear cover

To install:

➡**The flywheel spacer (if applicable) must be removed prior to oil seal installation. Do not lubricate the oil seal Inside Diameter (ID) or crankshaft surface. Never reuse the rear main seal. Once it is removed, it must be replaced with a new seal.**

3. Lubricate the Outside Diameter (OD) of the rear main seal and the rear cover oil seal bore with clean engine oil. Do NOT let oil contact the seal surface or the crankshaft surface.
4. Install or connect the following:
   - Crankshaft Rear Oil Seal Installer Tool No. J 41479 tapered cone and bolts onto the rear of the crankshaft. Tighten the bolts until just snug, being careful not to overtighten.
   - Rear oil seal onto the tapered cone until the tool contacts the oil seal
5. Align the oil seal into the tool, Rotate

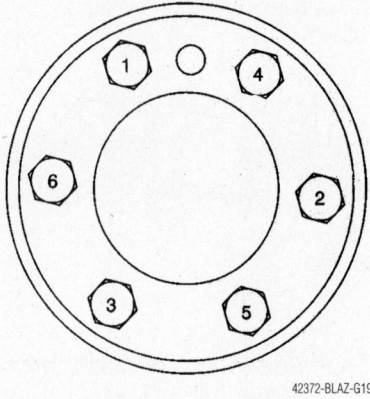

42372-BLAZ-G19

Flywheel bolt tightening sequence—5.3L engine

the handle of the tool clockwise until the seal enters the rear cover and bottoms into the cover bore. Remove the tool.
   - Flywheel and secure with the mounting bolts.
6. Tighten the flywheel mounting bolts, in sequence, as follows:
   a. 1st pass: 15 ft. lbs. (20 Nm)
   b. 2nd pass: 37 ft. lbs. (50 Nm)
   c. Final pass: 74 ft. lbs. (100 Nm)
   - Transmission
   - Negative battery cable
7. Start the engine and verify no oil leaks.

Timing Chain, Sprockets, Front Cover and Seal

REMOVAL & INSTALLATION

Front Cover and Seal

4.2L ENGINE

1. Before servicing the vehicle, refer to the precautions section.
2. Remove or disconnect the following:
   - Negative battery cable
   - Drain the engine cooling system.
   - Cooling fan and shroud
   - Accessory belt
   - Water pump
   - Crankshaft balancer

✳✳ WARNING

When removing the seal, be careful not to damage the front cover or crankshaft.

   - Seal from the front cover, using a suitable prytool in the slots provided
   - Power steering pump
3. Raise and safely support the vehicle.

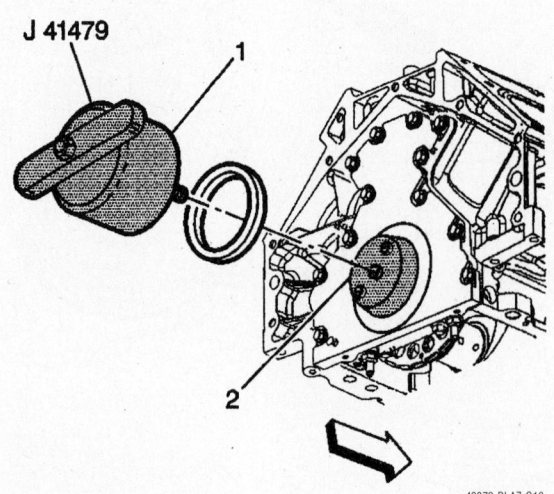

J 41479

42372-BLAZ-G18

View of the rear main seal installation—5.3L engine

- Oil pan, then carefully lower the vehicle
- 7mm center bolt
- Remaining front cover bolts. Place two of the front cover bolts in the jackscrew holes on the front cover and tighten the bolts evenly to release the front cover from the engine.
- 2 bolts from the front cover
- Oil pump

To install:

4. Clean the gasket mating surfaces of the engine and cover of all remaining gasket or sealer material. Be careful not to score or damage the surfaces.

5. Install or connect the following:
- Suitable cover alignment pins, onto the engine

➡**The front cover MUST be installed within 10 minutes of applying the sealant.**

- Apply a 0.12 in. (3mm) beat of 12378521 or equivalent sealant to the trace grooves on the back side of the engine front cover. Apply sealant on the inside 3 bolt hole bosses on the cover also.
- Oil pump to the crankshaft splines
- Front cover and bolts, tighten the center bolt last. Tighten to 89 inch lbs. (10 Nm).

6. Remove the alignment pins and raise and safely support the vehicle. Install the oil pan, then lower the vehicle.
- Power steering pump
- Crankshaft balancer
- Water pump
- Accessory belt
- Cooling fan and shroud
- Negative battery cable

7. Properly refill the engine cooling system.

8. Run the engine until normal operating temperature has been reached, then check for leaks.

5.3L ENGINE

1. Before servicing the vehicle, refer to the precautions section.
2. Properly discharge the A/C system.
3. Drain the engine cooling system.
4. Remove or disconnect the following:
- Negative battery cable
- A/C compressor and bracket
- Water pump
- Crankshaft balancer
- Oil pan-to-front cover bolts
- Front cover bolts
- Front cover and gasket. Discard the gasket.

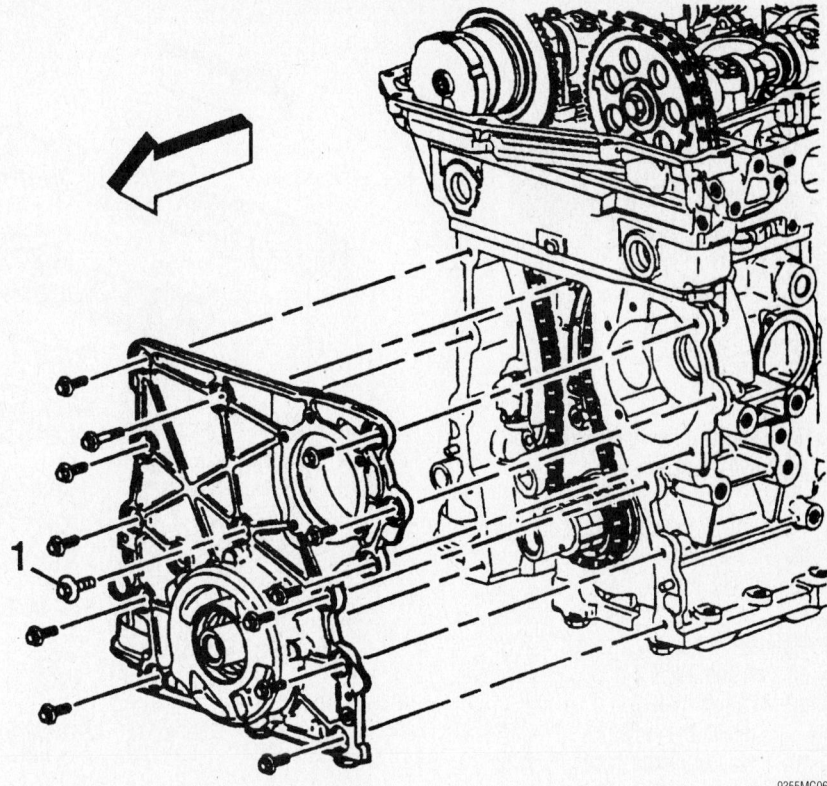

Place 2 front cover bolts in the jackscrew holes on the cover and tighten to push the cover off of the engine—4.2L engine

5. Clean and inspect the front cover.

To install:

6. Apply a 0.20 in. (5mm) bead of sealant 0.80 in. (20mm) long to the oil pan-to-engine block junction.

7. Install or connect the following:
- New front cover gasket and cover
- Front cover bolts, finger-tight
- Oil pan-to-front cover bolts, finger-tight
- Front and Rear Cover Alignment

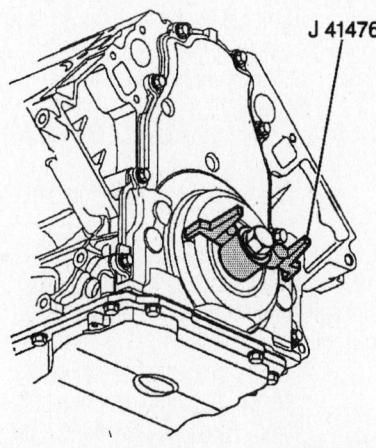

Align the tapered legs of the tool with the machined alignment surfaces on the front cover—5.3L engine

Tool No. J 41476 to the front cover. Align the tapered legs of the tool with the machined alignment surfaces on the front cover
- Crankshaft balancer bolt, finger-tight
- Oil pan-to-front cover bolts to 18 ft. lbs. (25 Nm)
- Front cover bolts to 18 ft. lbs. (25 Nm)

8. Remove the tool.

9. Install a NEW crankshaft front oil seal as follows:

a. Remove the radiator for access.

b. Remove the crankshaft balancer.

c. Remove the crankshaft oil seal.

d. Lubricate the outer edge ONLY of the NEW crankshaft oil seal with clean engine oil.

e. Install the crankshaft front oil seal into the Crankshaft Front Seal Installation Tool No. J 41478 guide.

f. Install the J 41478 threaded rod (with nut, washer, guide and oil seal) into the end of the crankshaft.

g. Use J 41478 to install the oil seal into the cover bore. Use a wrench and hold the hex on the installer bolt. Use a second wrench to rotate the installer nut clockwise until the seal bottoms in the cover bore. Remove the tool.

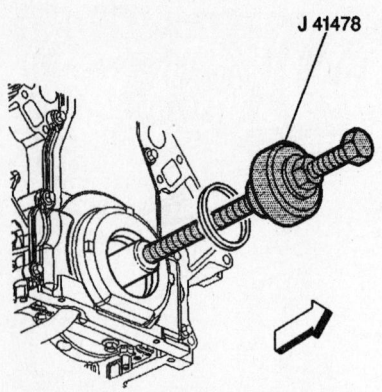

Front cover seal installation using the proper tool—5.3L engine

h. Check the seal for proper installation. It should be installed evenly and completely into the front cover bore.

i. Install the crankshaft balancer. Tighten the bolt to 37 ft. lbs. (50 Nm), plus an additional 140 degrees using a torque angle meter.

j. Install the radiator.

10. Install or connect the following:
- Water pump
- A/C compressor and bracket
- Cooling system with coolant
- Negative battery cable

11. Properly recharge the A/C system

Timing Chain and Sprockets

4.2L ENGINE

➡The following procedure requires the use of the Crankshaft Holding tool No. J-44221 and a suitable torque angle meter.

1. Before servicing the vehicle, refer to the precautions section.

2. Remove or disconnect the following:
- Camshaft cover
- Timing chain (front) cover
- Tension on the timing chain by moving the tensioner shoe in. Place a tee into the tension to hold the shoe in place.
- Top chain guide bolts and guide
- Exhaust camshaft position actuator bolt and actuator
- Intake camshaft sprocket bolt and sprocket
- Timing chain
- Crankshaft sprocket
- Cylinder head access hole plugs
- Timing chain tensioner shoe bolt and shoe
- Timing chain tensioner guide bolts and guide
- Timing chain tensioner bolts and tensioner

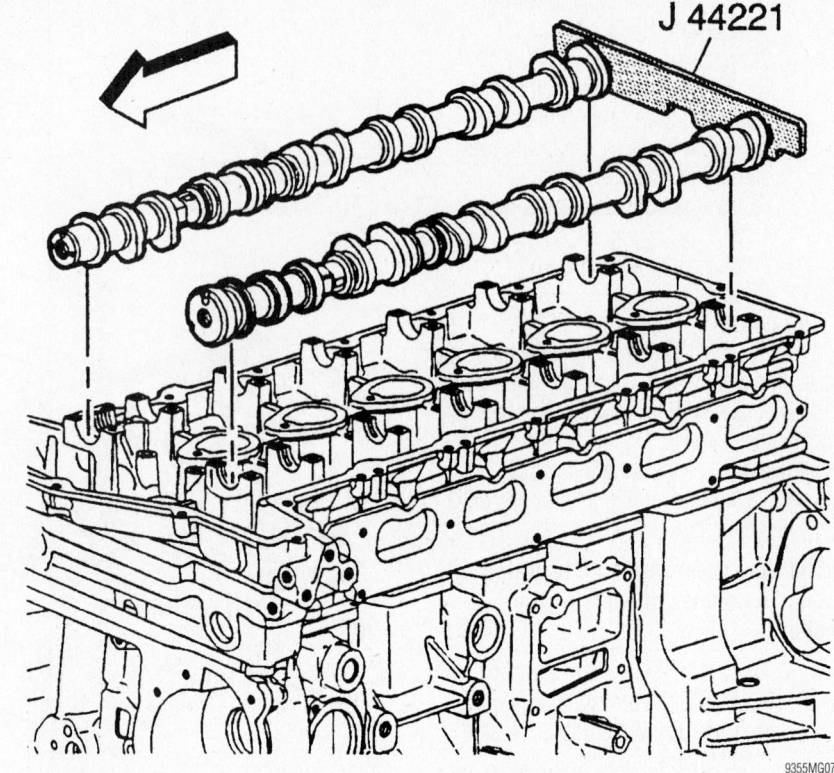

Proper installation of the crankshaft holding tool with the No. 1 cylinder at TDC—4.2L engine

To install:

➡Every seventh link of the timing chain is darkened to help in aligning the timing marks.

3. Install or connect the following:
- Timing chain tensioner and bolts. Tighten to 18 ft. lbs. (25 Nm).
- Timing chain guide and bolts. Tighten to 89 inch lbs. (10 Nm).
- Timing chain tensioner shoe and bolt. Tighten to 19 ft. lbs. (26 Nm).
- Cylinder head access hole plugs and tighten to 44 inch lbs. (5 Nm)
- Crankshaft Holding tool No. J-44221, or equivalent with the camshaft flats up and the No. 1 cylinder at Top Dead Center (TDC)
- Crankshaft sprocket
- Intake camshaft sprocket into the timing chain

4. Align the dark link of the timing chain with the timing mark on the intake camshaft sprocket. Feed the timing chain down through the opening in the head.
- Timing chain onto the crankshaft sprocket. Align the dark link of the timing chain with the timing mark on the crankshaft sprocket.

➡It may be necessary to remove the crankshaft holding tool to rotate and hold the camshaft hex to align the pin to the camshaft sprocket

- Intake camshaft sprocket onto the intake camshaft
- Intake camshaft washer and bolt
- Exhaust camshaft actuator into the timing chain. Align the dark link of the timing chain with the timing mark on the exhaust camshaft actuator.

➡It may be necessary to remove the crankshaft holding tool to rotate and hold the camshaft hex to align the pin to the camshaft sprocket

- Exhaust camshaft actuator onto the exhaust camshaft

➡Rotate the camshaft actuator clockwise relative to the camshaft prior to tightening the bolt.

5. Rotate the camshaft actuator clockwise (as seen from the front of the vehicle).

✳✳ WARNING

The camshaft actuator must be fully advanced during installation. Engine damage may occur if the camshaft actuator is not fully advanced.

6. Install the exhaust camshaft actuator bolt and tighten to 18 ft. lbs. (25 Nm), plus

Rotate the camshaft actuator clockwise—
4.2L engine

an additional 135 degrees, using a torque
angle meter.

7. Tighten the intake camshaft sprocket
bolt to 22 ft. lbs. (30 Nm), plus an additional 135 degrees, using a torque angle
meter.

8. Remove the tee from the timing
chain tensioner to regain tension on the
timing chain.

9. Remove the crankshaft holding tool.
The dark lines on the timing chain should
be aligned with the marks on the sprockets.

10. Install or connect the following:
- Top chain guide
- Suitable threadlock to the top chain
  guide bolt threads, then install and
  tighten to 89 inch lbs. (10 Nm)
- Engine front cover
- Camshaft cover

5.3L ENGINE

1. Before servicing the vehicle, refer to
the precautions section.

2. Remove the oil pump.

3. Rotate the crankshaft until the timing
marks on the crankshaft and the camshaft
sprockets are aligned.

✳✳ WARNING

**Do NOT turn the crankshaft after the
timing chain has been removed to
prevent damage to the pistons and
valves.**

4. Remove or disconnect the following:
- Camshaft sprocket bolts
- Camshaft sprocket and timing
  chain
- Crankshaft sprocket using Pulley
  Puller No. J 8433, Crankshaft End
  Protector Tool No. J 41816-2 and
  Crankshaft Sprocket Removal Tool
  No. J 41558

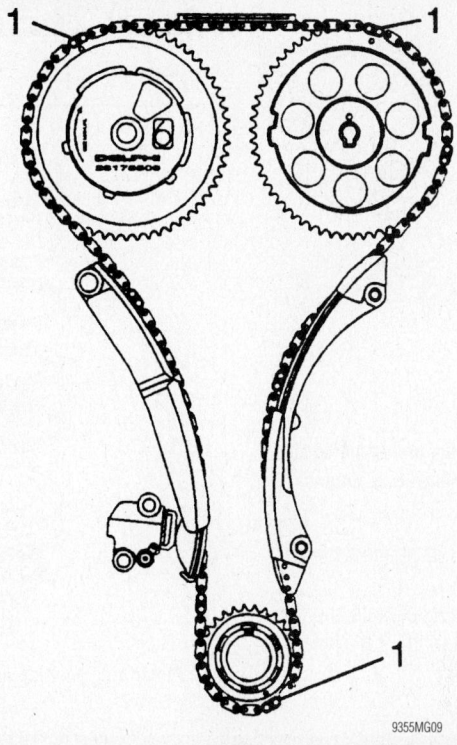

The dark lines on the timing chain should be aligned with the marks on the sprockets—4.2L
engine

- Crankshaft sprocket key, if necessary

5. Clean and inspect the timing chain
and sprockets.

To install:

6. Install or connect the following:
- Key into the crankshaft keyway, if
  removed. Tap the key into the keyway until both ends of the key bottom into the crankshaft.
- Crankshaft sprocket onto the front
  of the crankshaft. Align the crankshaft key with the sprocket keyway.
- Crankshaft sprocket using Sprocket
  Installation Tool No. J 41665.
  Install the sprocket onto the crank-

shaft until fully seated against the
crankshaft flange. Rotate the crankshaft sprocket until the alignment
mark is in the 12 o'clock position.

➡**Properly locate the camshaft
sprocket locating pin with the cam
sprocket alignment hole. The sprocket
teeth and timing chain must mesh. The
camshaft and crankshaft sprocket
alignment marks MUST be aligned
properly. Locate the camshaft sprocket
alignment mark in the 6 o'clock position. It may be necessary to rotate the
camshaft or crankshaft to align the
marks.**

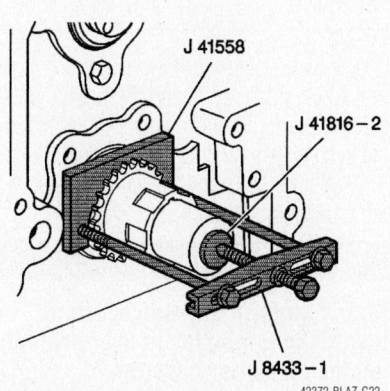

Use the proper tools to remove the crankshaft sprocket—5.3L engine

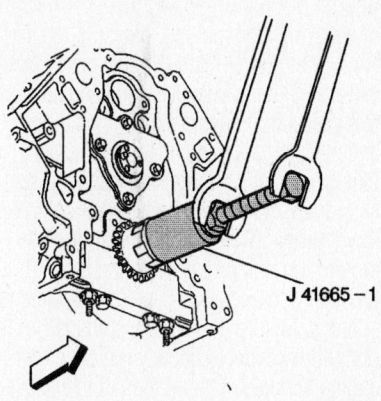

Crankshaft sprocket installation—5.3L
engine

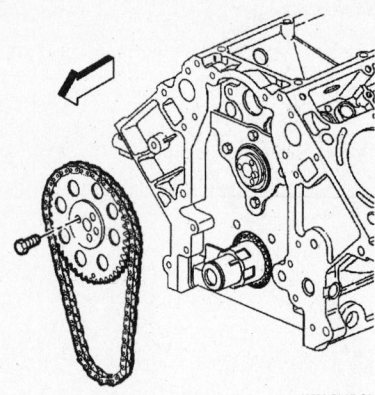

42372-BLAZ-G24

Proper alignment of the timing marks for timing chain installation—5.3L engine

- Camshaft sprocket and timing chain
- Camshaft sprocket bolts and tighten to 26 ft. lbs. (35 Nm)
- Oil pump

Piston and Ring

POSITIONING

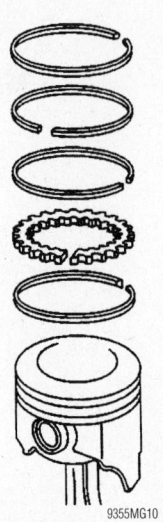

9355MG10

Piston ring positioning—4.2L engine

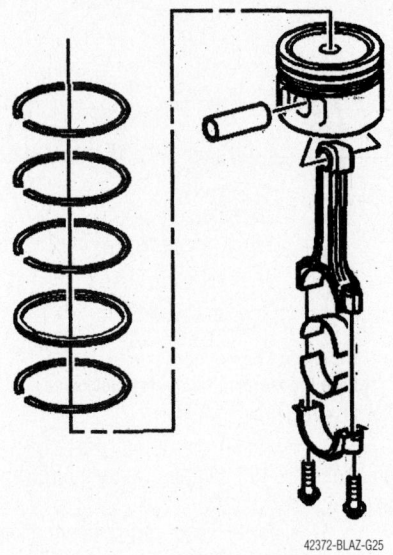

42372-BLAZ-G25

Piston ring positioning—5.3L engine

FUEL SYSTEM

Fuel System Service Precautions

Safety is the most important factor when performing not only fuel system maintenance but also any type of maintenance. Failure to conduct maintenance and repairs in a safe manner may result in serious personal injury or death. Maintenance and testing of the vehicle's fuel system components can be accomplished safely and effectively by adhering to the following rules and guidelines.

- To avoid the possibility of fire and personal injury, always disconnect the negative battery cable unless the repair or test procedure requires that battery voltage be applied.
- Always relieve the fuel system pressure prior to disconnecting any fuel system component (injector, fuel rail, pressure regulator, etc.), fitting or fuel line connection. Exercise extreme caution whenever relieving fuel system pressure, to avoid exposing skin, face and eyes to fuel spray. Please be advised that fuel under pressure may penetrate the skin or any part of the body that it contacts.
- Always place a shop towel or cloth around the fitting or connection prior to loosening to absorb any excess fuel due to spillage. Ensure that all fuel spillage (should it occur) is quickly removed from engine surfaces. Ensure that all fuel soaked cloths or towels are deposited into a suitable waste container.

- Always keep a dry chemical (Class B) fire extinguisher near the work area.
- Do not allow fuel spray or fuel vapors to come into contact with a spark or open flame.
- Always use a back-up wrench when loosening and tightening fuel line connection fittings. This will prevent unnecessary stress and torsion to fuel line piping. Always follow the proper torque specifications.
- Always replace worn fuel fitting O-rings with new. Do not substitute fuel hose or equivalent where fuel pipe is installed.

Fuel System Pressure

RELIEVING

The fuel systems operate under high fuel pressures. It is very important that the pressure be properly relieved prior to servicing the system or any of its components.

1. Before servicing the vehicle, refer to the precautions section.

✳✳ WARNING

Do not perform this procedure for more than 2 minutes to avoid damaging the catalytic converter.

2. Loosen the fuel filler cap to release the fuel tank pressure.

3. Remove the fuel pump relay from the junction block.
4. Crank the engine, allowing it to start and stall.
5. Crank the engine for an additional 3 seconds to relieve any remaining fuel pressure.
6. Disconnect the negative battery cable to avoid repressurizing the fuel system.
7. Install the fuel pump relay in the junction block.
8. Tighten the fuel filler cap.
9. After you are finished working on the fuel system, connect the negative battery cable.

Fuel Filter

REMOVAL & INSTALLATION

1. Before servicing the vehicle, refer to the precautions section.
2. Properly relieve the fuel system pressure.
3. Remove or disconnect the following:
   - Negative battery cable and fuel filler cap, if not already done
4. Raise and support the vehicle.
   - Fuel tank shield, if equipped
   - Quick connect fittings from the filter
   - Filter feed nut and the clamp bolt
   - Filter and the clamp from the vehicle

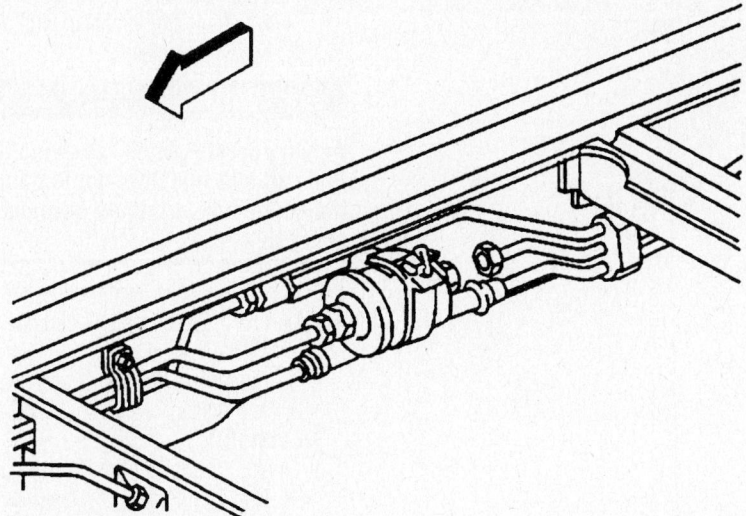

7924JG43

Typical fuel filter location along frame rail

To install:

5. Install or connect the following:
- Filter and clamp with the directional arrow facing away from the fuel tank, towards the throttle body

➡**The filter has an arrow (fuel flow direction) on the side of the case, be sure to install it correctly in the system, the with arrow facing away from the fuel tank.**

- Tighten the fuel feed nut
- Tighten the filter clamp assembly bolt
- Fuel quick disconnect fittings to the filter
- Fuel tank shield, if equipped
- Fuel filler cap
- Negative battery cable

6. Start the engine and check for leaks.

Fuel Pump

REMOVAL & INSTALLATION

1. Before servicing the vehicle, refer to the precautions section.
2. Properly relieve the fuel system pressure.
3. Drain the fuel tank.
4. Support the fuel tank.
5. Remove or disconnect the following:
- Negative battery cable
- Filler neck from the tank
- Frame brace
- Shield from tank and tank straps
- Fuel lines and vapor hose from pump
- Electrical connection from fuel pump

- Fuel tank
- Fuel pump/sending unit assembly by turning the locking ring (located on top of the fuel tank) counterclockwise using a spanner wrench
- Fuel pump from the fuel lever sending device

To install:

6. Install or connect the following:
- Fuel pump in tank with new seal around opening

➡**The fuel pump strainer must be in a horizontal position when the fuel sender is installed in the tank. When installing the sender assembly, make sure that the fuel pump strainer does not block full travel of the float arm.**

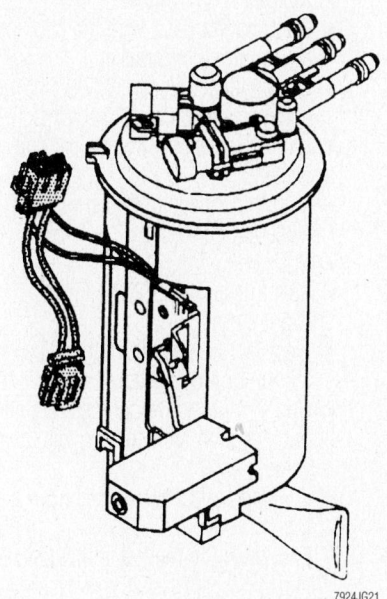

7924JG21

View of the in-tank fuel pump assembly

- Tank and connect fuel lines and vapor hose
- Tank and torque the strap fasteners to 24 ft. lbs. (32 nm).
- Shield
- Fuel filler neck and clamp
- Frame brace and tighten to 33 ft. lbs. (45 Nm)
- Negative battery cable

7. Refill the tank.
8. Run the engine and check for leaks.

Fuel Injector

REMOVAL & INSTALLATION

4.2L Engine

1. Before servicing the vehicle, refer to the precautions section.
2. Relieve the fuel system pressure. Refer to the fuel system relief procedure in this section.
3. Remove or disconnect the following:
- Negative battery cable, if not done already
- Intake manifold

➡**Clean the fuel rail assembly with a suitable spray cleaner before proceeding. Never soak the fuel rail in a cleaning solvent.**

- Fuel pressure regulator vacuum line
- Fuel feed and return pipes
- Fuel injector in-line electrical connector
- Fuel rail attaching bolts and fuel rail
- Fuel injector harness connector from the fuel injectors
- Injector retaining clip
- Injector from the fuel rail
- Retainer clip and O-ring seals from each end of the injector and discard

To install:

➡**Each injector is calibrated. When replacing the fuel injectors, be sure to replace it with the correct injector.**

4. Lubricate the new injector O-ring seats with engine oil.
5. Install or connect the following:
- O-rings on the injector
- New retainer clip on the injector

6. Push the fuel injector into the fuel rail socket, making sure the connector faces outward. The retainer clip locks to a flange on the fuel rail injector socket.
- Fuel rail assembly. Tighten the bolts to 89 inch lbs. (10 Nm).

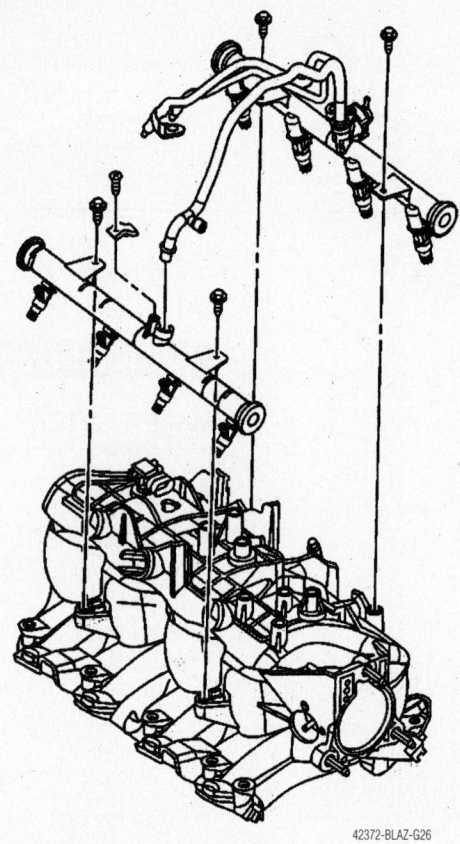

42372-BLAZ-G26

Exploded view of the fuel rail mounting—5.3L engine

- Fuel feed and return lines to the rail
- Fuel injector electrical connectors
- Fuel pressure regulator vacuum line
- Intake manifold
- Negative battery cable

7. Turn the ignition **ON** for 2 seconds and then turn it **OFF** for 10 seconds. Again turn the ignition **ON** and check for leaks.

5.3L Engine

1. Before servicing the vehicle, refer to the precautions section.

2. Relieve the fuel system pressure. Refer to the fuel system relief procedure in this section.

3. Remove or disconnect the following:

- Negative battery cable, if not done already
- A/C compressor pressure switch electrical connector
- Wire harness from the clip on the cylinder head
- Mass Airflow/Intake Air Temperature (MAF/IAT) sensor connector
- Alternator electrical connector
- Right side electrical connectors

from the coil main electrical harness, Electronic Throttle Control (ETC) and fuel injectors.

4. To detach the injector connector: Matchmark the connectors, pull the Connector Position Assurance (CPA) retainer up 1 click. Push the tab on the connector in, then detach the injector connector.

- Alternator connector
- Electrical harness from the clips on the ignition coil bracket
- Evaporative emission (EVAP) purge solenoid electrical connector
- Knock Sensor (KS) electrical connector
- Manifold Absolute Pressure (MAP) electrical connector
- Main coil
- Fuel injector electrical connector (right side)
- Electrical harness from the clips on the ignition coil bracket
- Upper engine wire harness retainer nut. Position the wire harness aside.
- Fuel feed and return pipes from the rail
- Fuel pressure regulator vacuum line

- Fuel rail bolts
- Fuel rail, after cleaning with a spray-type cleaner

- Fuel injector from the fuel rail
- Fuel injector retainer clip and discard
- Fuel injector lower O-ring seals and discard

To install:

5. Install or connect the following:

- New O-ring seals on the injectors, after lubricating with clean engine oil
- New retainer clip on the injector
- Fuel injector by pushing it into the fuel rail socket
- Fuel rail
- Apply 0.20 inch (5mm) band of threadlock to the threads of the fuel rail bolts
- Fuel rail bolts and tighten to 89 inch lbs. (10 Nm)
- Fuel pressure regulator vacuum line
- Fuel feel and return pipes

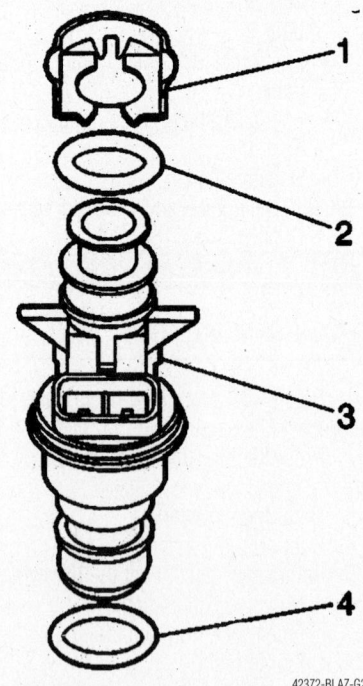

42372-BLAZ-G27

Exploded view of the fuel injector (3), retainer (1) and O-ring seals (2, 4)—5.3L engine

- Route the upper electrical harness into position over the engine.
- Engine harness bracket nut and tighten to 89 inch lbs. (10 Nm)
- PCV valve and hose
- EVAP purge solenoid, KS, MAP sensor, main coil & fuel injector electrical connectors

- Harness to the clips on the ignition coil bracket
- Main coil, ETC, fuel injector electrical connectors
- Harness to the clips on the ignition coil bracket
- Alternator electrical connector
- MAF/IAT sensor connector

- Wire harness to the clip on the cylinder head
- A/C compressor switch electrical connector
- Air cleaner outlet duct
- Fuel fill cap
- Negative battery cable
6. Refill the engine cooling system.

DRIVE TRAIN

Automatic Transmission Assembly

REMOVAL & INSTALLATION

LL8 Transmission

➡**This procedure requires the use of a Converter Holding Strap tool No. J 21366 to secure the torque converter to the transmission during removal and installation.**

1. Before servicing the vehicle, refer to the precautions section.
2. Remove or disconnect the following:
   - Negative, then the positive battery cables
   - Fill tube nut, located on the right side of the engine
3. Drain the transmission fluid.
   - Rear propeller shaft
4. Support the transmission with a transmission jack.
   - Nuts securing the transmission mount to the transmission support
   - Evaporative emission (EVAP) canister from its mounting bracket on the left inside of the frame to get access to the transmission support bolts. Do not disconnect the canister lines.
   - Fuel tank shield
   - Transmission support
   - Transmission mount bolts and mount
   - Front exhaust pipe assembly
5. Lower the transmission for access to the top and sides of the transmission.
   - Transfer case, if equipped
   - Range selector cable end from the transmission range selector lever ball stud and bracket
   - Transmission heat shield, transmission vent hose park/neutral position switch connector, and main connector from the transmission
   - Bolt that secures the fuel line bracket to the left side of the transmission
   - Torque converter access plug

6. Matchmark the flywheel and torque converter orientation for reassembly.
   - Flywheel-to-torque converter bolts. Be careful not to drop the bolts into the bell housing!
   - Disconnect the transmission oil cooler lines from the transmission. Plug the transmission oil cooler lines connectors in the transmission case.
7. Install a safety chain around the transmission.
   - Bolt that secures the fuel line bracket to the bell housing
   - Bolts that secure the coolant pipe to the bell housing
   - Remaining nuts, studs and/or bolts that secure the transmission to the engine
8. Install Converter Holding Strap tool No. J 21366 onto the transmission bell housing to hold the torque converter.
9. Pull the transmission straight back and remove it from the vehicle.

To install:

Installation is the reverse of removal, but please note the following important steps.

10. Make sure the torque converter is fully seated in the pump drive. If not, the transmission will not fit tightly to the rear of the engine block.

11. Raise the transmission into position and remove the torque converter holding strap. Carefully slide the transmission forward until the dowel pins are engaged while lining up the marks on the flywheel made during removal.

12. The torque converter should be flush with the flywheel and turn freely by hand.

13. Install the transmission-to-engine nuts, studs and or bolts. Tighten the studs and/or bolts to 37 ft. lbs. (50 Nm).

14. Tighten the bolts securing the heat shield to the transmission to 13 ft. lbs. (17 Nm).

15. Tighten the bolts and washers securing the transmission mount to 18 ft. lbs. (25 Nm).

16. Tighten the nut and washer securing the transmission mount to the transmission support to 35 ft. lbs. (46 Nm).

17. Refill the transmission with the proper amount and type of fluid.

18. Connect the negative battery cable. Start the vehicle and allow to warm while checking for leaks. Road test the vehicle to check for shift quality.

LM4 Transmission

➡**This procedure requires the use of a Converter Holding Strap tool No. J 21366 to secure the torque converter to the transmission during removal and installation.**

1. Before servicing the vehicle, refer to the precautions section.
2. Remove or disconnect the following:
   - Negative, then the positive battery cables
3. Drain the transmission fluid.
   - Rear propeller shaft
4. Support the transmission with a transmission jack.
   - Nuts securing the transmission mount to the transmission support and remove the support
   - Transmission mount bolts and mount
   - Front exhaust pipe assembly
5. Lower the transmission for access to the top and sides of the transmission.
   - Transfer case, if equipped
   - Range selector cable end from the transmission range selector lever ball stud and bracket
   - Transmission heat shield, transmission vent hose park/neutral position switch connector, and main connector from the transmission
   - Transmission harness from its retainers
   - Bolt that secures the fuel line bracket to the left side of the transmission
   - Torque converter access plug
6. Matchmark the flywheel and torque converter orientation for reassembly.
   - Flywheel-to-torque converter bolts. Be careful not to drop the bolts into the bell housing!
   - Disconnect the transmission oil

cooler lines from the transmission. Plug the transmission oil cooler lines connectors in the transmission case.

7. Install a safety chain around the transmission.

- Bolt that attaches the fuel tube to the bell housing and remove the tube
- Remaining nuts, studs and/or bolts that secure the transmission to the engine

8. Install Converter Holding Strap tool No. J 21366 onto the transmission bell housing to hold the torque converter.

9. Pull the transmission straight back and remove it from the vehicle.

To install:

Installation is the reverse of removal, but please note the following important steps.

10. Make sure the torque converter is fully seated in the pump drive. If not, the transmission will not fit tightly to the rear of the engine block.

11. Raise the transmission into position and remove the torque converter holding strap. Carefully slide the transmission forward until the dowel pins are engaged while lining up the marks on the flywheel made during removal.

12. The torque converter should be flush with the flywheel and turn freely by hand.

13. Install the transmission-to-engine nuts, studs and or bolts. Tighten the studs and/or bolts to 37 ft. lbs. (50 Nm).

14. Tighten the bolts securing the heat shield to the transmission to 13 ft. lbs. (17 Nm).

15. Tighten the bolts and washers securing the transmission mount to 18 ft. lbs. (25 Nm).

16. Tighten the nut and washer securing the transmission mount to the transmission support to 35 ft. lbs. (46 Nm).

17. Refill the transmission with the proper amount and type of fluid.

18. Connect the negative battery cable. Start the vehicle and allow to warm while checking for leaks. Road test the vehicle to check for shift quality.

Transfer Case Assembly

REMOVAL & INSTALLATION

1. Before servicing the vehicle, refer to the precautions section.

2. Disconnect the negative battery cable.

3. Raise and support the vehicle. Drain the transfer case.

4. Remove or disconnect the following:

- Fuel tank shield mounting bolts and shield
- Front and rear propeller shaft. Matchmark the shafts prior to removal.
- Fuel lines from the retainer
- Electrical harness from the retainers on the right and left sides
- Speed sensor electrical connectors
- Motor/encoder electrical connector
- Transfer case wiring harness
- Vent hose

5. Install a transmission jack to support the transfer case.

- Transfer case mounting bolts
- Transfer case from the vehicle
- Transfer case gasket and discard if damaged

To install:

6. Install or connect the following:

➡**You must replace the transfer case gasket if it is damaged. Never use silicone sealant in place of, or with the transfer case gasket.**

- Transfer case, using a new gasket if necessary
- Transfer case mounting bolts and tighten to 35 ft. lbs. (47 Nm)

7. Remove the transmission jack.

8. The remainder of installation is the reverse of removal.

9. Refill the transfer case.

Halfshaft

REMOVAL & INSTALLATION

1. Before servicing the vehicle, refer to the precautions section.

2. Remove or disconnect the following:

- Front wheel

➡**Place a drift through the caliper into the edge of the rotor to keep the rotor from turning when the nut is removed**

- Wheel center cap, if equipped
- Halfshaft nut and discard. A new nut must be used for installation.
- Drift from the rotor
- Brake caliper and support it with a piece of wire to avoid damaging the brake hose
- Brake rotor

3. To remove the steering knuckle, remove or disconnect the following:

- Wheel hub and bearing
- Outer tie rod retaining nut
- Outer tie rod end from the steering knuckle using a puller

- Brake hose bracket retaining bolts
- Brake hose bracket
- Anti-lock Brake System (ABS) wheel speed sensor wiring harness bracket, if necessary
- Upper control arm-to-steering knuckle pinch bolt and nut
- Upper control arm from the steering knuckle
- Lower ball joint retaining nut
- Steering knuckle from the control arm using a puller
- Steering knuckle

4. Remove the left side halfshaft from differential carrier, or right halfshaft from the clutch fork housing as follows:

a. Place a brass drift against the tripot housing.

b. Use a hammer to strike the drift outward from the case, striking hard enough to overcome the snaping tension holding the halfshaft.

5. Pull the halfshaft straight out of the differential carrier or clutch fork housing.

To install:

6. Install the halfshaft as follows:

a. With both hands on the tripot housing, align the splines on the shaft with the differential carrier assembly (left) or clutch fork housing (right).

b. Center the halfshaft into the differential carrier or clutch fork housing assembly seal.

c. Firmly push the shaft straight into the differential carrier or clutch fork housing assembly until the snapring is properly seated.

7. To install the steering knuckle, install or connect the following:

- Steering knuckle to the lower control arm
- Lower ball joint retaining nut and tighten to 81 ft. lbs. (110 Nm)
- Upper control arm to the steering knuckle
- Upper control arm pinch bolt and nut and tighten to 30 ft. lbs. (40 Nm)
- ABS wheel speed sensor harness bracket
- Brake hose bracket. Tighten the bolts to 7 ft. lbs. (10 Nm).
- Outer tie rod to the steering knuckle and tighten the nut to 33 ft. lbs. (45 Nm)
- Hub and bearing

8. Install or connect the following:

- New halfshaft nut and tighten to 103 ft. lbs. (140 Nm)
- Wheel

9. Lower the vehicle. Adjust the front toe.

CV-Joints

OVERHAUL

Outer CV-Joint

1. Before servicing the vehicle, refer to the precautions section.

2. Remove or disconnect the following:
- Front wheel
- Halfshaft and position it in a vise
- Large CV-joint boot clamp and discard it
- Small CV-joint boot clamp and discard it
- CV-joint boot and slide it back on the shaft

- Outer race from the halfshaft, by spreading the outer race-to-halfshaft retaining ring, using Snapring Pliers J-8059
- Retaining ring from the halfshaft and discard it
- CV-joint boot from the halfshaft and discard it, if damaged

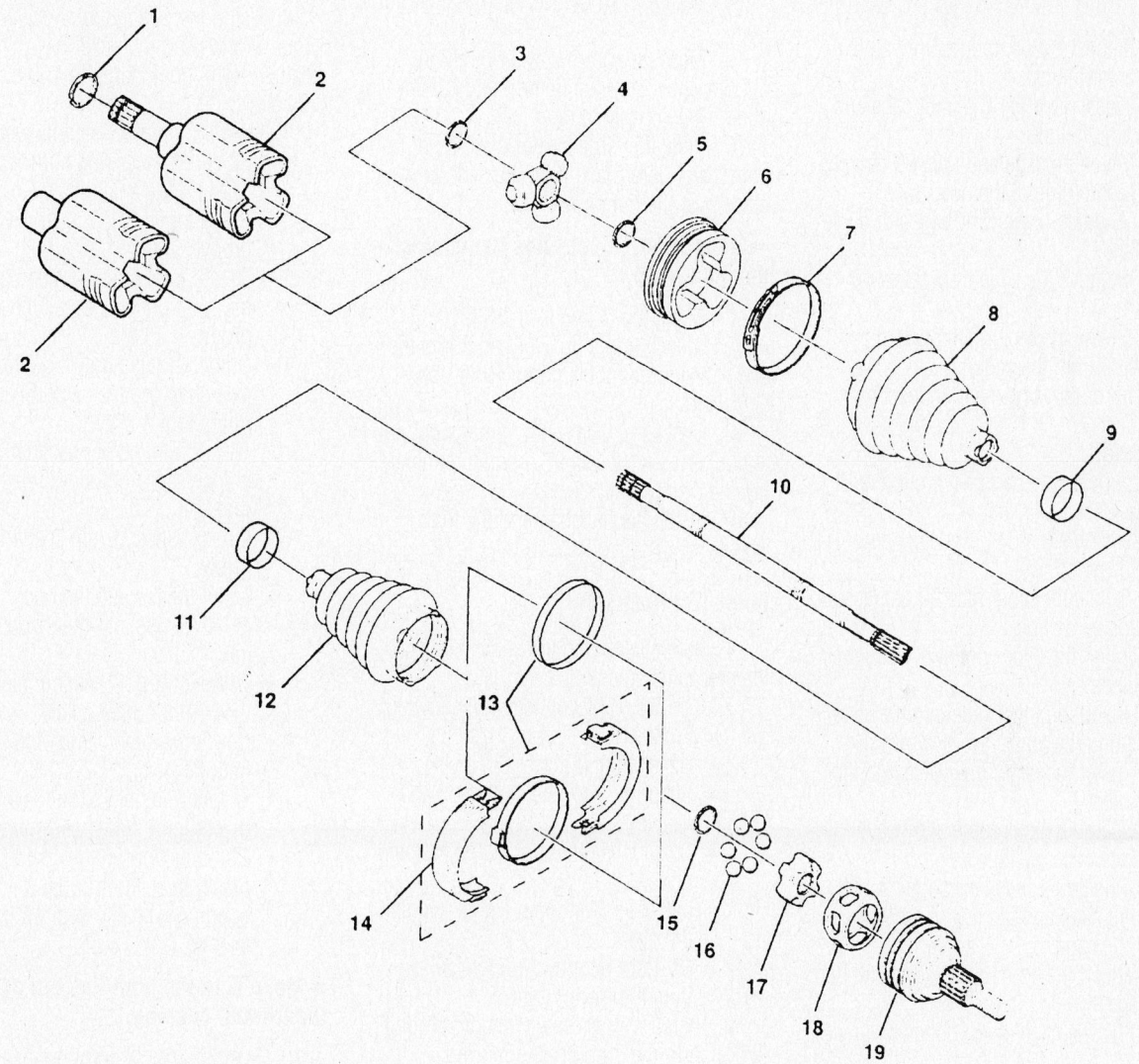

(1) Differential Shaft Ring	(11) Halfshaft Swage Ring
(2) Tripot Housing Assembly	(12) CV Joint Boot
(3) Spacer Ring	(13) Swage Ring
(4) Tripot Joint Spider Assembly	(14) Clamp Protector
(5) Spacer Ring	(15) Race Retaining Ring
(6) Tripot Bushing	(16) Ball
(7) Boot Retaining Clamp	(17) CV Joint Inner Race
(8) Tripot Joint Boot	(18) CV Joint Cage
(9) Halfshaft Swage Ring	(19) CV Joint Outer Race
(10) Halfshaft Bar	

Exploded view of the CV-Joint Assembly

9308JG09

3. Disassemble the chrome alloy balls from the CV-joint cage as follows:

a. Position a brass drift against the CV-joint cage and tap it with a hammer to tilt the cage.

b. Remove the 1st chrome alloy ball from the cage.

c. Tilt the cage in the opposite direction.

d. Remove the opposite chrome alloy ball.

e. Repeat the procedure until all 6 balls are removed.

4. Disassemble the CV-joint cage and inner race as follows:

a. Pivot the cage and race 90 degrees to the center line of the outer race.

b. Align the cage windows with outer race lands.

c. Remove the cage from the outer race.

d. Rotate the inner race upward and remove it from the cage.

5. Thoroughly clean and inspect all parts.

To install:

6. Lubricate the parts with a light coat of grease.

7. Assemble the CV-joint cage and inner race, as follows:

a. Rotate the inner race 90 degrees to the cage centerline.

b. Align the cage windows with inner race lands.

c. Insert the inner race into the cage by rotating the inner race downward.

d. Insert the cage/inner race into the outer race.

8. Assemble the chrome alloy balls into the CV-joint cage, as follows:

a. Position a brass drift against the CV-joint cage and tap it with a hammer to tilt the cage.

b. Insert the 1st chrome alloy ball into the cage.

c. Tilt the cage in the opposite direction.

d. Insert the opposite chrome alloy ball.

e. Repeat the procedure until all 6 balls are inserted.

9. Install ½ kit grease into the CV-joint.

10. Install or connect the following:
- Small ring clamp on the CV boot
- New retaining ring on the halfshaft
- Large ring clamp on the CV boot
- Outer race assembly onto the halfshaft until the ring engages the halfshaft groove

11. Slide the small end of the CV-joint boot/clamp into place, with the seal lip in the halfshaft groove

➡**Make sure the boot lies flat against the halfshaft.**

12. Using the Crimp tool J-35910, a torque wrench and a breaker bar, crimp the small CV-joint boot clamp to 100 ft. lbs. (136 Nm).

13. Check the clamp gap dimension; if it is not 0.085 in. (2.15mm), continue tightening the clamp until it is.

14. Install ½ kit grease into the CV-joint boot.

15. Measure approximately 0.687 in. (17.5mm) up from the bottom edge of the outer CV-joint assembly.

16. Slide the large end of the CV boot/clamp into place, with the seal lip in place over the outer race.

➡**Make sure the boot lies flat against the outer race.**

17. Using the Crimp tool J-35910, a torque wrench and a breaker bar, crimp the large CV-joint boot clamp to 130 ft. lbs. (176 Nm).

18. Check the clamp gap dimension; if it is not 0.102 in. (2.60mm), continue tightening the clamp until it is.

19. Install the halfshaft and the front wheel.

Inner (Tri-Pot) Joint

1. Before servicing the vehicle, refer to the precautions section.

2. Remove or disconnect the following:
- Front wheel
- Halfshaft and place it in a vise
- Snapring from the stub shaft and discard it
- Small CV-joint boot clamp, cut and discard it
- Large CV-joint boot clamp, cut and discard it
- CV-joint boot by sliding it away from the tri-pot joint

3. Install a Stub Shaft Removal tool J-38868-A to the stub shaft snapring groove.

4. Using a slide hammer puller, press the stub shaft from the tri-pot housing.

5. Remove or disconnect the following:
- Tri-pot housing from the tri-pot spider
- Inboard spacer ring slide it rearward on the shaft using Snapring Pliers tool J-8059
- Outboard retaining ring using Snapring Pliers tool J-8059 and discard it
- Tri-pot joint spider assembly
- Inboard spacer ring and discard it

- CV-joint boot
- Trilobal tri-pot bushing from the housing

6. Thoroughly clean and inspect all parts.

To install:

7. Install or connect the following:
- New snapring onto the stub shaft
- Small boot clamp
- CV-joint boot

8. Using the Crimp tool J-35910, a torque wrench and a breaker bar, crimp the small CV-joint boot clamp to 100 ft. lbs. (136 Nm).

9. Install or connect the following:
- Inboard spacer ring slide it rearward on the shaft using Snapring Pliers tool J-8059, past the 2nd groove
- Tri-pot joint spider assembly onto the shaft until it passes the 2nd groove
- Outboard retaining ring into the axle shaft groove using Snapring Pliers tool J-8059
- Tri-pot joint spider assembly, slide it against the outboard retaining ring
- Inboard spacer ring, seat it in the groove
- ½ kit grease into the boot
- ½ kit grease into the tri-pot housing
- Trilobal tip-pot bushing flush with the tri-pot housing face
- New large seal clamp onto the CV-joint boot
- Tri-pot housing, slide it over the tri-pot joint spider assembly
- CV-joint boot/clamp, slide it into place, over the trilobal tri-pot bushing with the seal lip in the groove

➡**Make sure the boot lies flat against the trilobal bushing.**

10. Position the CV-joint boot so it measures 4.9 in. (125mm).

11. Using the Crimp tool J-35566, latch the large CV-joint boot clamp.

12. Install the halfshaft and the front wheel.

Axle Shaft, Bearing and Seal

REMOVAL & INSTALLATION

For the Axle Shaft, Bearing and Seal, Removal and Installation, please refer to Wheel Bearing procedure located in the section.

Pinion Seal

REMOVAL & INSTALLATION

1. Before servicing the vehicle, refer to the precautions section.
2. Remove the wheels.
3. Remove the rear calipers and rotors.
4. remove the driveshaft from the pinion flange. Matchmark the driveshaft prior to removal.
5. Using an inch lb. torque wrench, measure the amount of torque required to rotate the pinion. Record the measurement for assembly as this will give the combined preload for the following components:
   - Pinion bearings
   - Pinion oil seal
   - Differential case bearings
   - Axle bearings
   - Axle seals
6. Place an alignment mark between the pinion and the yoke.
7. Install holding tool J 8614-01 as shown.
8. Remove the pinion nut while holding the holding tool.
9. Remove the washer.
10. Install tool J 8614-3 (2), J 8614-2 (3) into the holding tool (1) as illustrated.
11. Remove the pinion yoke by turning tool J 8614-3 clockwise while holding tool

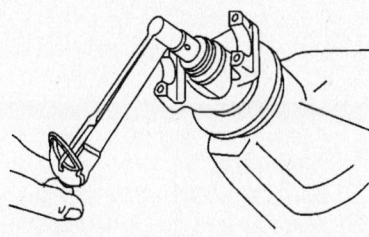

67162-ISU-ASC-G02

Using an inch lb. torque wrench, measure the amount of torque required to rotate the pinion

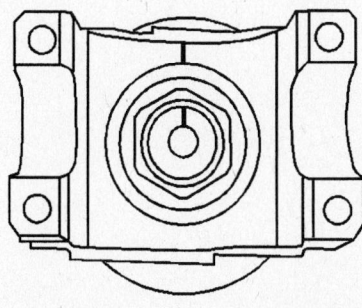

67162-ISU-ASC-G03

Place an alignment mark between the pinion and the yoke

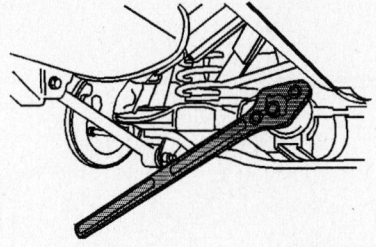

67162-ISU-ASC-G04

Install holding tool J 8614-01 as shown

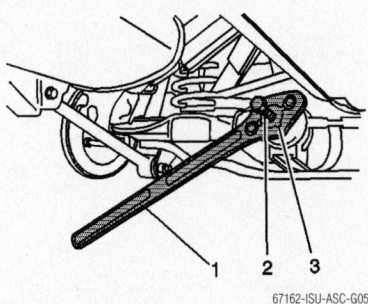

1 2 3

67162-ISU-ASC-G05

Install tool J 8614-3 (2), J 8614-2 (3) into the holding tool (1) as shown

J 8614-01. Use a pan to catch any fluid that leaks.

12. Using a suitable tool, remove the seal being careful not to damage the housing.

To install:

13. Examine the seal surface of pinion flange for tool marks, nicks or damage, such as a groove worn by the seal. If damaged, replace flange.
14. Examine the carrier bore and remove any burrs that might cause leaks around the O.D. of the seal.
15. Apply GM seal lubricant 1050169 to the outside diameter of the pinion flange and sealing lip of new seal.
16. Install or connect the following:
   - New pinion oil seal using a seal installer tool such as J 33782 for the 8 inch axle or tool J 38694 for the 8.6 inch axle.
   - Pinion flange and tighten nut to the same position as marked earlier. Tighten the nut a little at a time and turn the pinion flange several times after each tightening in order to set the rollers.
17. Measure the torque necessary to turn the pinion and compare this to the reading taken during removal. Tighten the nut additionally, as necessary to achieve the same preload as measured earlier. The rotating torque should be 3–5 inch lbs. (0.40–0.57 nm).
18. Install the driveshaft assembly to the pinion flange.

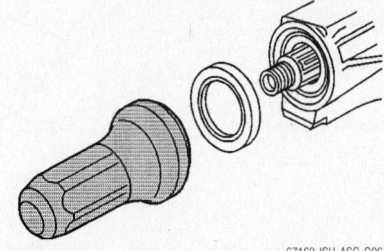

67162-ISU-ASC-G06

Use the appropriately sized installation tool to drive the new seal into position.

➡**The original matchmarks MUST be aligned to assure proper shaft balance and prevent vibration.**

19. Install the rotors and calipers.
20. Install the wheels

➡**If fluid was lost from the differential housing during this procedure, be sure to check and add additional fluid, as necessary.**

Axle Housing

REMOVAL & INSTALLATION

1. Before servicing the vehicle, refer to the precautions section.
2. Support the rear axle housing. If a floor jack is being used, take care when removing the U-bolts to keep the axle from suddenly dislodging.
3. Remove or disconnect the following:
   - Rear wheels and drums for clearance and to remove some weight from the axle housing
   - Axle vibration dampener, if equipped
   - Rear driveshaft from the pinion flange. Either remove the shaft completely from the vehicle or support it aside from the undercarriage using safety wire, but DO NOT allow the shaft to hang from the slip joint.
   - Shock absorber-to-axle housing retainers, then swing the shock absorbers away from the axle housing
   - Brake lines from the axle housing clips and the backing plates (wheel cylinders)

➡**When disconnecting the brake lines from the wheel cylinders, immediately plug or cap the lines to prevent system contamination or excessive fluid loss.**

   - Speed sensor connectors at the junction block, if applicable

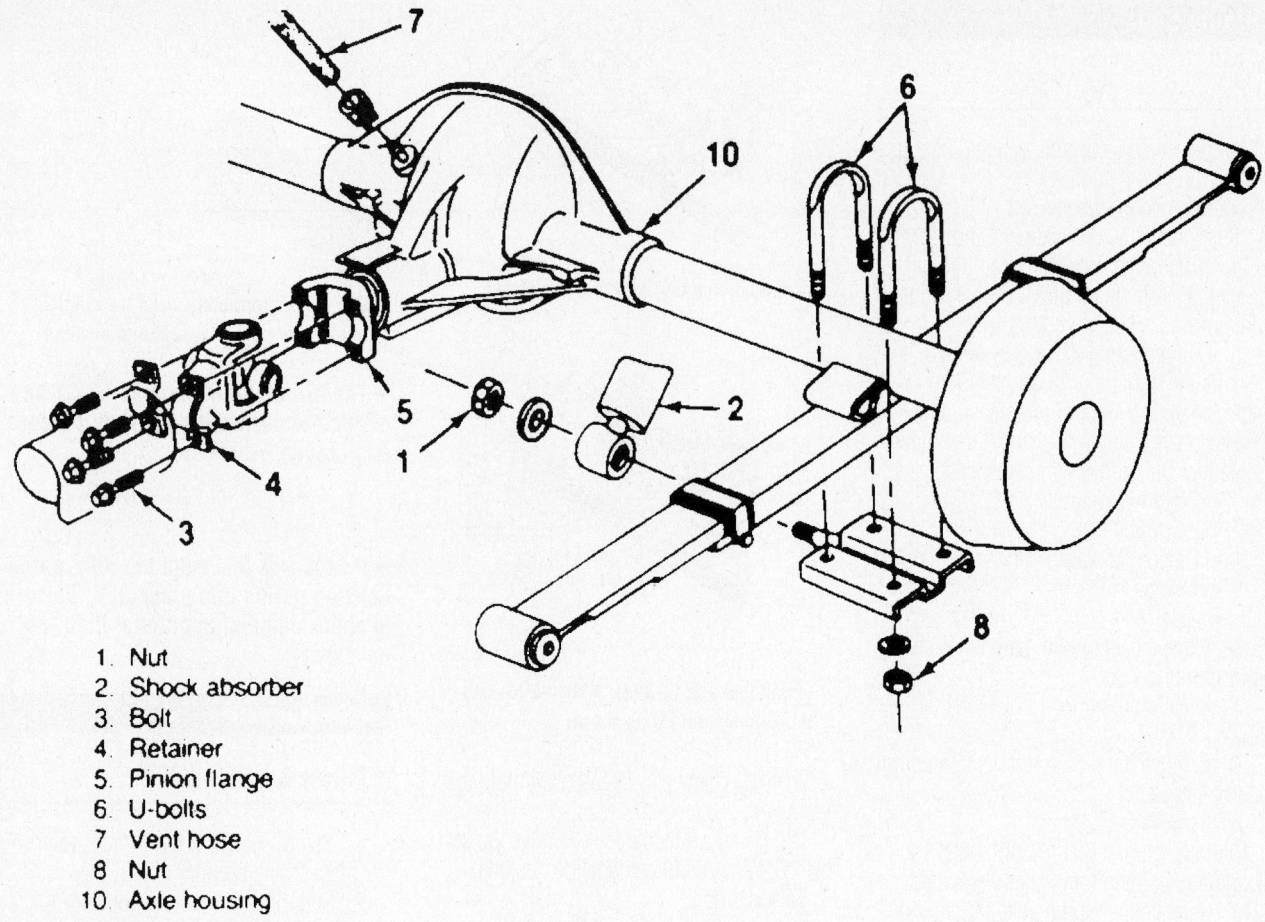

1. Nut
2. Shock absorber
3. Bolt
4. Retainer
5. Pinion flange
6. U-bolts
7. Vent hose
8. Nut
10. Axle housing

88457G85

Exploded view of the rear axle mounting

- Parking brake cable(s)
- Axle housing-to-spring U-bolt nuts, washers, U-bolts and the anchor plates
- Vent hose from the top of the axle housing
- Axle with the help of an assistant by moving it to clear the leaf spring

To install:

4. With the help of an assistant, carefully position the rear axle into the vehicle.

5. Install or connect the following:
   - Vent hose to the axle housing

6. Be sure the housing is properly positioned on the leaf spring, then loosely install the U-bolts, anchor plates, washers and nuts.
   - Tighten the U-bolt nuts in a cross pattern to 18 ft. lbs. (25 Nm) to made sure everything is evenly seated. Then tighten the nuts in steps to 74 ft. lbs. (100 Nm).
   - Brakes lines secure them to the axle housing
   - Parking brake cable(s), if removed
   - Speed sensor connectors to the junction block, if equipped
   - Driveshaft assembly

- Shock absorbers to the lower mounts, then tighten the mount nuts
- Axle vibration dampener, if equipped
- Brake drums and the tire/wheel assemblies

7. Bleed the hydraulic brake system.

8. Check the fluid level in the rear axle assembly and add, as necessary. Make sure the vehicle is level when checking and adding fluid.

STEERING AND SUSPENSION

Air Bag

✳✳ CAUTION

Some vehicles are equipped with an air bag system, also known as the Supplemental Inflatable Restraint (SIR) system. The system must be disabled before performing service on or around system components, steering column, instrument panel components, wiring and sensors. Failure to follow safety and disabling procedures could result in accidental air bag deployment, possible personal injury and unnecessary system repairs.

PRECAUTIONS

Several precautions must be observed when handling the inflator module to avoid accidental deployment and possible personal injury.

- Never carry the inflator module by the wires or connector on the underside of the module.
- When carrying a live inflator module, hold securely with both hands, and ensure that the bag and trim cover are pointed away.
- Place the inflator module on a bench or other surface with the bag and trim cover facing up.
- With the inflator module on the bench, never place anything on or close to the module, that may be thrown in the event of an accidental deployment.

DISARMING

1. Turn the steering wheel so that the vehicle's wheels are pointing straight ahead.
2. Turn the ignition switch to **LOCK**, remove the key, then disconnect the negative battery cable.
3. Remove the AIR BAG or SIR fuse from the fuse block.
4. Remove the steering column filler panel or knee bolster.
5. Unplug the Connector Position Assurance (CPA) and yellow two way connector at the base of the steering column.
6. Open the glove compartment door, lift the stop and let the door fully open.
7. Remove the Connector Position Assurance (CPA) from the passenger yellow two way connector located behind the glove box.

8. Unplug the yellow two way connector located behind the glove box.
9. Connect the negative battery cable.

➡ **With the AIR BAG fuse removed, the battery cable connected and the ignition in the ON position, the AIR BAG warning lamp will be ON. This is normal and does not indicate a system malfunction.**

ARMING

1. Disconnect the negative battery cable.
2. Attach the yellow two way connector located behind the glove box.
3. Install the Connector Position Assurance (CPA) to the passenger yellow two way connector located behind the glove box.
4. Close the glove compartment door.
5. Turn the ignition switch to **LOCK**, then remove the key.
6. Attach the two way connector at the base of the steering column and the Connector Position Assurance (CPA).
7. Install the steering column filler panel or knee bolster.
8. Install the AIR BAG fuse to the fuse block.
9. Connect the negative battery cable.
10. From the passenger seat, turn the ignition switch to **RUN** and make sure that the AIR BAG warning lamp flashes seven times and then shuts off. If the warning lamp does not shut off, make sure that the wiring is properly connected. If the light remains on, take the vehicle to a reputable repair facility for service.

Power Steering Gear

REMOVAL & INSTALLATION

1. Before servicing the vehicle, refer to the precautions section.
2. Raise and support the vehicle.
3. Position a fluid catch pan under the power steering gear.
4. Remove or disconnect the following:
   - Front tire and wheel assemblies
   - Outer tie rod retaining nuts

✳✳ WARNING

Do not try to separate a steering linkage joint by driving a wedge between the joint and the attached part. Doing this can cause seal damage and premature failure of the part.

- Outer tie rods from the steering knuckles using a suitable steering linkage and tie rod puller
- Lower intermediate shaft retaining bolt and shaft from the power steering gear
- Steering gear crossmember
- Feed and return fluid hoses from the steering gear. Immediately cap or plug all openings to prevent system contamination or excessive fluid loss.
5. Support the power steering gear.
   - Power steering gear mounting bolts, then remove the gear from the vehicle
6. Loosen the outer tie rod jam nuts, then remove the outer tie rods from the inner tie rods and discard the jam nut.
   To install:
7. Lubricate the inner tie rod threads with a suitable lubricant before installing the outer tie rod.
8. Install or connect the following:
   - New jam nuts to the outer tie rods
   - Outer tie rods to the inner tie rods
   - Power steering gear to the vehicle. Tighten the retaining bolts to 81 ft. lbs. (110 Nm).
9. Remove the support from the power steering gear.
   - Power steering hose(s) to the gear. Tighten the retaining bolt to 9 ft. lbs. (12 Nm).
   - Steering gear crossmember
   - Lower intermediate shaft to the power steering gear. Tighten the retaining bolt to 30 ft. lbs. (40 Nm).
   - Outer tie rod ends to the steering knuckles. Tighten the retaining nuts to 33 ft. lbs. (45 Nm).
   - Front tire and wheel assemblies
10. Remove the drain pan, then lower the vehicle.
11. Bleed the power steering system and adjust the front toe as necessary.

Strut/Shock Module

REMOVAL & INSTALLATION

Front

➡ **In these models a "shock module", similar to a strut was used on these vehicles. This procedure requires the use of a suitable steering linkage and tie rod puller.**

1. Before servicing the vehicle, refer to the precautions section.
2. Remove or disconnect the following:
   - Shock module upper retaining nuts
   - Tire and wheel
   - Shock module-to-lower control arm retaining nut
   - Shock module yoke from the lower control arm using a suitable puller such as J 24319-B
   - Shock module from the shock tower and lower control arm

To install:
3. Install or connect the following:
   - Shock module to the shock tower and lower control arm
   - Shock module yoke to the lower control arm
   - Shock module upper retaining nuts and tighten to 33 ft. lbs. (45 Nm)
   - Shock module-to-lower control arm retaining nut and tighten to 81 ft. lbs. (110 Nm)
   - Tire and wheel

Shock Absorbers

REMOVAL & INSTALLATION

Rear

1. Before servicing the vehicle, refer to the precautions section.
2. Properly support the rear axle assembly.
3. Remove or disconnect the following:
   - Automatic level control air lines from the shock absorber, if equipped
   - Shock absorber-to-frame retainer(s) at the top of the shock
   - Shock-to-axle retainer(s) at the bottom of the shock
   - Shock absorber

To install:
4. Install the shock in the vehicle and loosely install the upper mounting fasteners to retain it

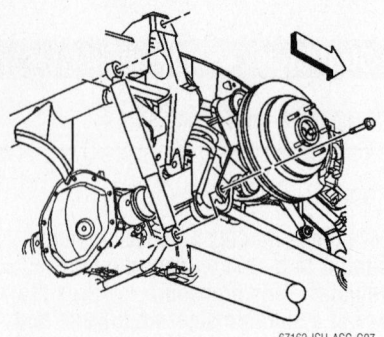

67162-ISU-ASC-G07

Rear shock absorber mounting

5. Align the lower-end of the shock absorber with the axle mounting, then loosely install the retainers.
6. Tighten the upper and lower shock retainers to 59 ft. lbs. (80 Nm).
7. If equipped, attach the automatic level control air lines to the shock absorber.

Coil Springs

REMOVAL & INSTALLATION

Front

➡**This procedure requires the use of a suitable spring compressor.**

1. Before servicing the vehicle, refer to the precautions section.
2. Remove or disconnect the following:
   - Wheel
   - Shock module
   - Shock module yoke-to-shock absorber pinch bolt and nut
3. Spread the shock module yoke at the pinch bolt using a suitable flat-bladed tool.
   - Shock module yoke from the shock absorber
4. Install pieces of heater hose or equivalent material to the shock module spring where the spring compressor contacts the lower part of the spring.
5. Install the shock module into the spring compressor.

➡**The spring is compressed when the shock absorber moves freely.**

6. Turn the spring compressor forcing screw until the coil spring is compressed.
7. Remove or disconnect the following:
   - Shock absorber upper retaining nut
   - Shock absorber from the shock module
8. Loosen the compressor forcing screw until the upper mounting plate and coil spring can be removed.
   - Upper mounting plate and coil spring from the spring compressor

To install:
9. Install or connect the following:
   - Coil spring and upper mounting plate to the spring compressor
10. Turn the compressor forcing screw until the coil spring is compressed.
    - Shock absorber to the shock module. Tighten the retaining nut to 33 ft. lbs. (45 Nm)
11. Remove the shock module from the spring compressor. Remove the pieces of heater hose from the spring.
    - Shock module yoke to the shock absorber

- Shock module yoke-to-shock pinch bolt and nut and tighten to 52 ft. lbs. (70 Nm)
- Shock module to the vehicle
- Tire and wheel
12. Lower the vehicle

Rear

1. Before servicing the vehicle, refer to the precautions section.
2. Raise and support the vehicle.
3. Support the rear axle.
4. Remove the shock absorber lower mounting bolts.
5. Lower the rear axle, then remove the coil springs.

To install:
6. Install the coil springs, then raise the rear axle.
7. Install the shock absorber lower mounting bolts and tighten to 59 ft. lbs. (80 Nm).
8. Remove the rear axle support.
9. Lower the vehicle.

Torsion Bar

REMOVAL & INSTALLATION

1. Before servicing the vehicle, refer to the precautions section.

➡**The following procedure requires the use of the Torsion Bar Unloader tool J-36202.**

2. Remove or disconnect the following:
   - Transmission shield, if equipped
   - Torsion bar unloader tool to relax the tension on the torsion bar adjusting arm screw; record the number of turns necessary to properly install the tool. Remove the adjusting screw and the unloader tool.
   - Lower link mount nut from one side
   - Torsion bars by disengaging them

➡**Note the direction of the forward end and side of the torsion bar being removed**

   - Lower link nut from the opposite side
   - Lower link mount, upper link mount nut
   - Upper link mount
   - Torsion bar from the frame

To install:
3. Install or connect the following:
   - Torsion bar and support
   - Upper link mount. Torque the nut to 48 ft. lbs. (68 Nm).

4. Place a jack under the torsion bar to release tension.
5. Install or connect the following:
- Lower link mount bushing and nut. Torque the nut to 37 ft. lbs. (50 Nm).
- Torsion bar unloader tool. Tighten the tool against the adjusting arm the same number turns recorded earlier and remove the tool. This loads the torsion bars.
- Transmission shield, if removed

Upper Ball Joints

REMOVAL & INSTALLATION

➡This procedure requires the use of the following special tools: J 9519-E Lower Ball Joint Remover and Installer, J 21474-01 Control Arm Bushing Set and J 45117 Ball Joint Installation Spacer.

1. On 4WD vehicles, remove the wheel center cap and drive axle nut.
2. Raise and support the vehicle.
3. Remove or disconnect the following:
- Tire and wheel
- Wheel hub and bearing, if necessary
- Outer tie rod retaining nut
- Out tie rod from the steering knuckle using a suitable puller
- Brake hose bracket retaining bolts and bracket
- Upper control arm-to-steering knuckle pinch bolt and nut
- Upper control arm from the steering knuckle
- Lower ball joint retaining nut
- Steering knuckle from the lower control arm using a suitable ball joint removal tool

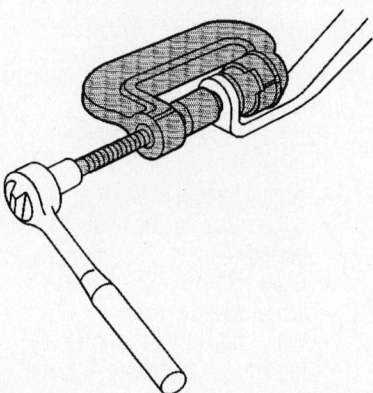

Remove the upper ball joint from the steering knuckle using tool No. J 9519-E

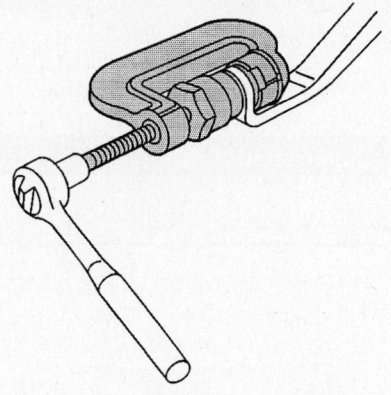

Install the upper ball joint to the steering knuckle using J 9519-E, J 21474-01 and J 45117

- Steering knuckle from the vehicle
- Upper ball joint retaining clip
- Upper ball joint from the steering knuckle using Lower Ball Joint Removal and Installer tool No. J 9519-E

To install:
4. Install or connect the following:
- Upper ball joint to the steering knuckle using J 9519-E, J 21474-01 and J 45117
- Upper ball joint retaining clip
- Steering knuckle to the lower control arm
- Lower ball joint retaining nut and tighten to 81 ft. lbs. (110 Nm)
- Upper control arm to the steering knuckle
- Upper control arm pinch bolt and nut and tighten to 30 ft. lbs. (41 Nm)
- Brake hose bracket to the steering knuckle
- Brake hose bracket retaining nuts and tighten to 7 ft. lbs. (10 Nm)
- Outer tie rod to the steering knuckle
- Outer tie rod retaining nut and tighten to 33 ft. lbs. (45 Nm)
- Wheel hub and bearing, if removed
- Tire and wheel
5. Lower the vehicle
- Drive axle nut, if 4WD, and tighten to 103 ft. lbs. (140 Nm)
- Wheel enter cap, if removed
6. Check the front wheel alignment.

Lower Ball Joints

REMOVAL & INSTALLATION

➡This procedure requires the use of the following special tools: J 9519-E Lower Ball Joint Remover and Installer, J

34874 Booster Seal Remover/Installer, J 41435 Ball Joint Installer, J 45105-1 Ball Joint Flaring Adapter and J 45105-2 Receiver.

1. On 4WD vehicles, remove the wheel center cap and drive axle nut.
2. Raise and support the vehicle.
3. Remove or disconnect the following:
- Tire and wheel
- Wheel hub and bearing, if necessary
- Outer tie rod retaining nut
- Out tie rod from the steering knuckle using a suitable puller
- Brake hose bracket retaining bolts and bracket
- Upper control arm-to-steering knuckle pinch bolt and nut
- Upper control arm from the steering knuckle
- Lower ball joint retaining nut
- Steering knuckle from the lower control arm using a suitable ball joint removal tool

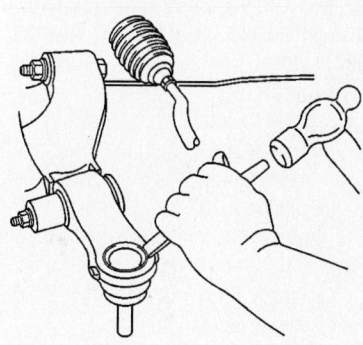

Remove the lower ball joint flange with a chisel

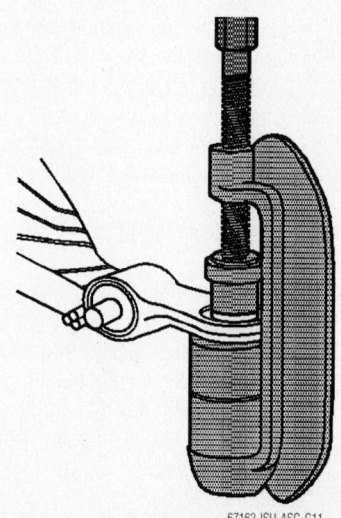

Use tools J 9519-E and J 34874 to remove the lower ball joint

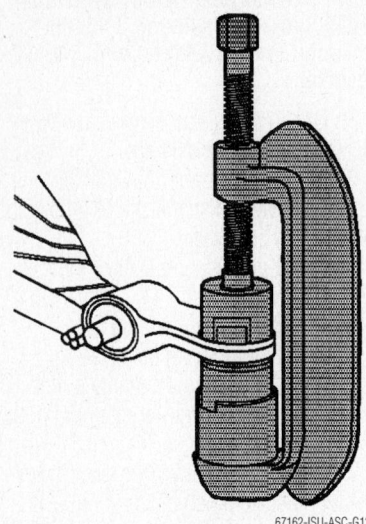

67162-ISU-ASC-G12

Use tools J 9519-E and J 45105-1 to install the lower ball joint

- Steering knuckle from the vehicle
- Lower ball joint flange with a chisel

4. Install tools J 9519-E and J 34874 to the lower ball joint, then use those tools to remove the lower ball joint from the lower control arm.

To install:

5. Install or connect the following:
- Lower ball joint to the lower control arm, using tools J 9519-E, J 41435 and J 45105-2

6. Remove the tools from the lower control arm.
- Tools J 9519-E and J 45105-1 to the lower ball joint

7. Flare the lower ball joint flange with J 9519-E and J 45105-1, then remove the tools from the lower ball joint.
- Steering knuckle to the lower control arm
- Lower ball joint retaining nut and tighten to 81 ft. lbs. (110 Nm)
- Upper control arm to the steering knuckle
- Upper control arm pinch bolt and nut and tighten to 30 ft. lbs. (41 Nm)
- Brake hose bracket to the steering knuckle
- Brake hose bracket retaining nuts and tighten to 7 ft. lbs. (10 Nm)
- Outer tie rod to the steering knuckle
- Outer tie rod retaining nut and tighten to 33 ft. lbs. (45 Nm)
- Wheel hub and bearing, if removed
- Tire and wheel

8. Lower the vehicle

- Drive axle nut, if 4WD, and tighten to 103 ft. lbs. (140 Nm)
- Wheel enter cap, if removed

9. Check the front wheel alignment.

Upper Control Arm

REMOVAL & INSTALLATION

1. Before servicing the vehicle, refer to the precautions section.
2. Remove or disconnect the following:
- Tire and wheel assembly
- Upper ball joint-to-upper control arm pinch bolt and nut
- Upper control arm from the knuckle
- Anti-lock Brake System (ABS) wheel speed sensor wiring harness
- Upper control arm mounting bolts
- Upper control arm

To install:

3. Install or connect the following:
- Upper control arm and tighten the bolts to 111 ft. lbs. (150 Nm)
- ABS wheel speed sensor wiring harness
- Upper control arm to the steering knuckle
- Upper ball joint-to-upper control arm pinch bolt and nut and tighten to 30 ft. lbs. (40 Nm)
- Tire and wheel

4. Check the front wheel alignment.

Lower Control Arm

REMOVAL & INSTALLATION

1. Before servicing the vehicle, refer to the precautions section.
2. Raise the vehicle.
3. Remove or disconnect the following:
- Tire and wheel
- Tie rod from the steering knuckle
- Stabilizer shaft link lower nut and remove the shaft and washer from the control arm
- Shock module lower mounting bolt and disconnect the module from the lower control arm using puller J 24319-B
- Lower control arm–to–bracket bolts. Note the direction of the bolts prior to removal.
- Lower ball joint from the steering knuckle.

➡**On 4WD models, do not disengage the axle shaft from the transmission.**

- Pivot the lower control arm out and down to disconnect it from the lower bracket and remove the control arm.

To install:

4. Install or connect the following:
- Lower control arm to the knuckle. Pivot the arm out and up to engage the arm to the lower bracket.
- Lower control arm–to–lower bracket bolts. Make sure the arm is parallel to the control arm bracket when installing and tightening of the bolts and nuts to ensure correct alignment of the bushings. Tighten the nuts to 81 ft. lbs. (110 Nm).
- Shock module yoke to the lower control arm and install the mounting nut.

➡**There is a washer between the stabilizer shaft and the lower control arm which is made of hardened steel with a felt inner liner. Make sure if replacing this washer it is replaced with the identical washer only.**

- Stabilizer link and washer. Tighten the nut to 74 ft. lbs. (100 Nm).
- Outer tie rod to the knuckle and tighten the nut to 33 ft. lbs. (40 Nm).
- Wheel

5. Check the wheel alignment.

Wheel Bearings

ADJUSTMENT

The wheel bearings on these vehicles are not adjustable. If the bearings become loose or make noise, they must be replaced.

REMOVAL & INSTALLATION

Front

REAR WHEEL DRIVE

1. Before servicing the vehicle, refer to the precautions section.
2. Raise and support the vehicle.
3. Remove or disconnect the following:
- Tire and wheel
- Caliper, leaving the fluid lines connected
- Brake rotor
- Wheel speed sensor
- Wheel hub and bearing-to-steering knuckle bolts and hub and bearing

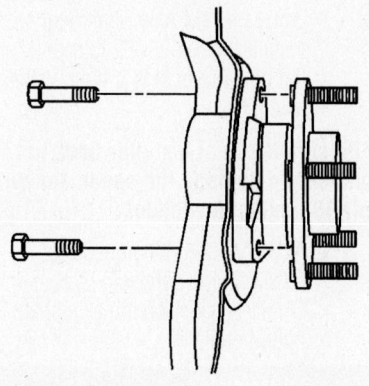

67162-ISU-ASC-G13

Front wheel bearing and hub mounting—rear wheel drive models

➡Lay the hub and bearing on the wheel studs on the outboard side. This will avoid damaging the bearing seal.

- Splash shield from the steering knuckle

To install:

4. Install or connect the following:
- Splash shield to the steering knuckle, making sure it's properly aligned
- Hub and bearing to the steering knuckle, aligning the threaded holes
- Hub and bearing bolts and tighten to 77 ft. lbs. (105 Nm)
- Wheel speed sensor. Tighten the bolt to 13 ft. lbs. (18 Nm).
- Rotor and brake caliper
- Tire and wheel
5. Lower the vehicle

FOUR WHEEL DRIVE

1. Before servicing the vehicle, refer to the precautions section.
2. Remove the wheel center cap, if equipped, and the drive axle nut and washer
3. Raise and support the vehicle.
4. Remove or disconnect the following:
- Tire and wheel
- Caliper, leaving the fluid lines connected
- Brake rotor
- Halfshaft from the hub and bearing. Place a brass drift against the outer edge of the halfshaft to protect the shaft threads. Use a hammer to sharply strike the brass drift, but to do not remove the halfshaft at this time.
- Wheel speed sensor

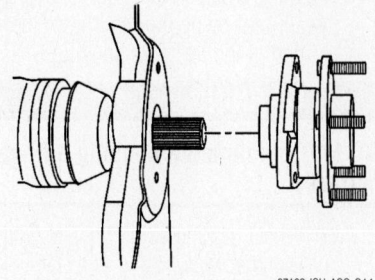

67162-ISU-ASC-G14

Front wheel bearing and hub mounting—four wheel drive models

- Wheel hub and bearing-to-steering knuckle bolts and hub and bearing

➡Lay the hub and bearing on the wheel studs on the outboard side. This will avoid damaging the bearing seal.

- Splash shield from the steering knuckle
- Seal from the hub and bearing

To install:

5. Install or connect the following:
- Wheel hub and bearing seal
- Splash shield to the steering knuckle, making sure it's properly aligned
- Hub and bearing to the steering knuckle, aligning the threaded holes
- Hub and bearing bolts and tighten to 77 ft. lbs. (105 Nm)

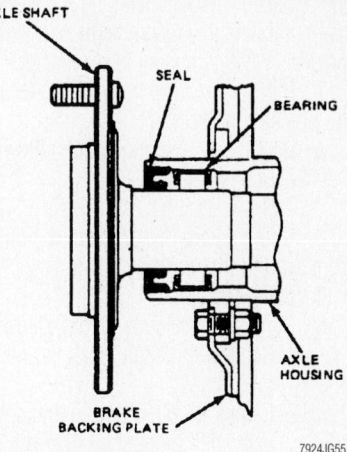

7924JG55

Cross-sectional view of the rear axle, bearing and seal assembly

- Wheel speed sensor. Tighten the bolt to 13 ft. lbs. (18 Nm).
- Rotor and brake caliper
- Tire and wheel
6. Lower the vehicle
7. Install the drive axle nut and tighten to 103 ft. lbs. (140 Nm), then install the center cap.

Rear

A new pinion shaft lockbolt should be installed whenever either of the axle shafts is removed.

The axle shaft and seal may be removed

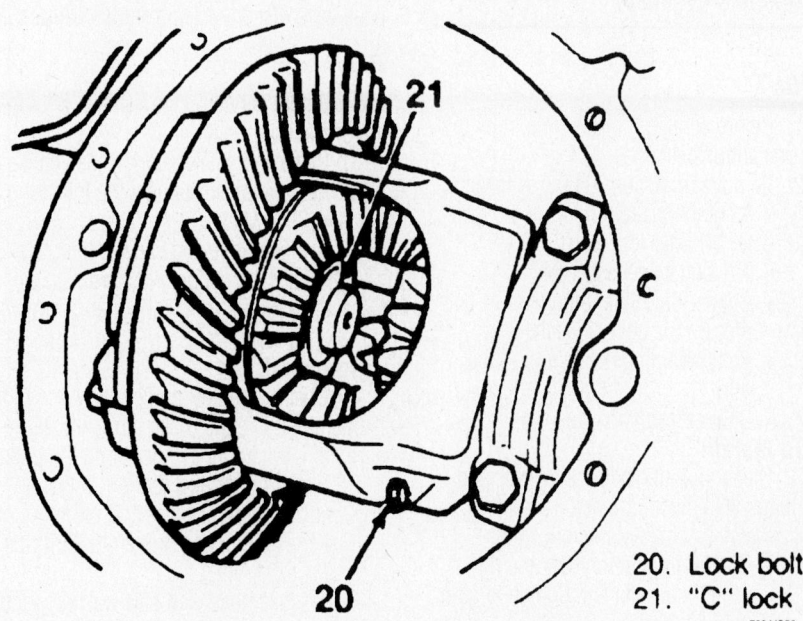

20. Lock bolt
21. "C" lock

7924JG56

Pinion shaft lockbolt and axle C-lock locations, inside the differential

and replaced without disturbing the bearing or seal but it is highly recommended to replace the seals when removing the axle shaft.

1. Before servicing the vehicle, refer to the precautions section.
2. Remove or disconnect the following:
   • Rear wheels
   • Brake drums
3. Using a wire brush, clean the dirt/rust from around the rear axle cover.
4. Drain the fluid.
5. Remove or disconnect the following:
   • Rear pinion shaft lockbolt and the pinion shaft
   • C-lock from the button end of the axle shaft by pushing the axle shaft inward
   • Axle shaft from the axle housing

➡**Be careful not to damage the oil seal.**

❊❊ WARNING

If equipped with an Anti-Lock Brake System (ABS), be careful not to damage the reflector ring on the axle shaft or the speed sensor bolted to the backing plate, immediately adjacent to the shaft.

6. Remove or disconnect the following:
   • Oil seal by prying the it from the end of the rear axle housing

❊❊ WARNING

DO NOT damage the housing oil seal surface.

   • Wheel bearing using the GM Slide Hammer tool J-2619, the GM Adapter tool J-2619-4 and the GM Axle Bearing Puller tool J-22813-01

To install:
7. Clean and inspect the components for excessive wear or damage and replace them, if necessary.
8. Install or connect the following:
   • New or reused bearing, coated with gear lubricant, using the Axle Shaft Bearing Installer tool J-34974 to drive the bearing in until it bottoms against the seat

❊❊ WARNING

Be sure the bearing installer does not contact and damage the speed sensor on ABS equipped vehicles.

   • New seal lubricated with gear oil using the GM Axle Shaft Seal Installer tool J-33782 to seat it in the housing until it is flush with the axle tube

➡**Be sure the seal installer does not contact and damage the speed sensor on ABS equipped vehicles.**

   • Axle shaft into the housing by engaging the splines
   • C-lock retainer on the axle shaft button end

❊❊ WARNING

BE CAREFUL not to damage the wheel bearing seal.

   • Axle shaft by pulling it outward to seat the C-lock retainer in the counterbore of the side gears
   • Pinion shaft through the case and the pinions. Tighten the new lock-bolt to 27 ft. lbs. (36 Nm).
   • New rear axle cover gasket
   • Housing cover
   • Brake drums
   • Wheels
9. Refill the housing.

BRAKES

Brake Caliper

REMOVAL & INSTALLATION

FRONT

1. Before servicing the vehicle, refer to the precautions section.
2. Remove or disconnect the following:
   • ⅔ of the brake fluid from the master cylinder reservoir
   • Tire and wheel assembly
   • Caliper fluid line, plug the line and discard the copper washers
   • Bolts retaining the caliper to the rotor
   • Caliper from the rotor

To install:
3. Clean and lubricate the sleeves and bushings with silicon grease.
4. Install or connect the following:
   • Caliper in position over the rotor
   • Mounting bolts and tighten to 38 ft. lbs. (51 Nm)
   • Fluid lines to the caliper using new washers and tighten to 33 ft. lbs. (45 Nm)

   • Wheel and tire assembly
5. Refill the master cylinder to the correct level. Bleed the brake system if the fluid lines were disconnected from the caliper.

REAR

1. Before servicing the vehicle, refer to the precautions section.
2. Raise and safely support the vehicle.
3. Remove or disconnect the following:
   • Rear wheels
   • Brake hose and cap the line. Discard the copper washers.
   • Retainers from caliper and remove caliper

To install:
4. Coat the caliper guide pin with a high temperature silicone brake lubricant. Make sure the lubricant does not get on the pads.
5. Install or connect the following:
   • Caliper over rotor, and onto mounts
   • Retainers, and tighten to 23 ft. lbs. or (31 Nm)
   • Brake hose with new washers, and tighten to 33 ft. lbs. (44 Nm)
6. Bleed brake system.
7. Install tires.
8. Refill the master cylinder and pump

pedal to attain full brake pedal before Road-testing the vehicle.

Disc Brake Pads

REMOVAL & INSTALLATION

FRONT

1. Before servicing the vehicle, refer to the precautions section.
2. Remove or disconnect the following:
   • ⅔ of the brake fluid from the master cylinder
3. Place a C-clamp around the outer pad and caliper; tighten the C-clamp until the piston is fully compressed in the caliper.
   • Brake pads
   • Inboard pad and retaining spring from the caliper
   • Outboard pad from the caliper
   • Sleeves and bushings

To install:
4. Clean and lubricate the sleeves and bushing with silicone lubricant and install them in the caliper.

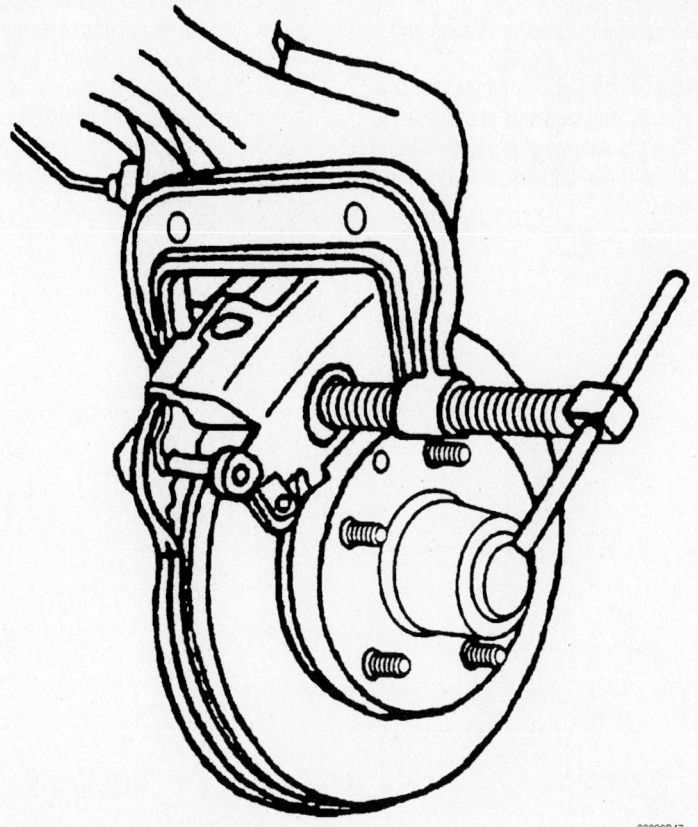

Compressing the caliper piston with a C-clamp

93026G47

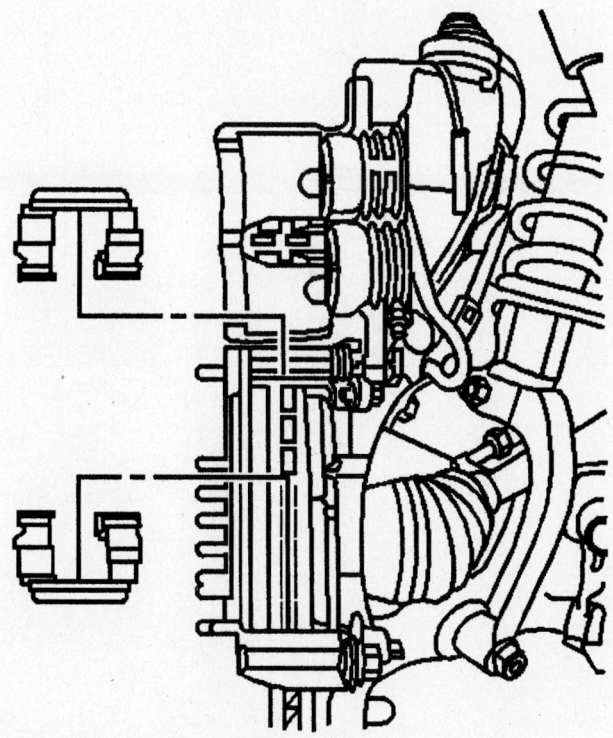

67162-ISU-ASC-G15

Install the pad retaining clips onto the caliper mounting bracket

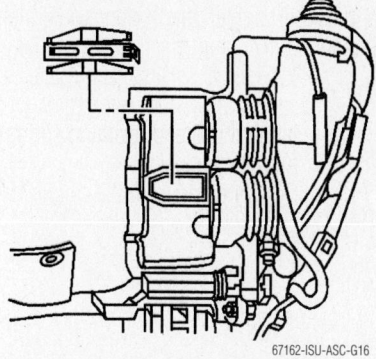

67162-ISU-ASC-G16

Install the brake pad clip to the caliper

5. Install the pad retaining clips onto the caliper mounting bracket.

6. Install the brake pad clip to the caliper.

7. Make sure the clip in the caliper are seated correctly.

8. Install the inboard pad in the caliper.

9. Install or connect the following:
- Outboard pad into the caliper
- Caliper in position over the rotor and install the mounting bolts. Bend the tabs, on the outboard brake pad, over the caliper.
- Wheel and tire assemblies

10. Refill the master cylinder and pump pedal to attain full brake pedal before Road-testing the vehicle.

REAR

1. Before servicing the vehicle, refer to the precautions section.

2. Remove or disconnect the following:
- ⅔ of the brake fluid from the master cylinder
- Wheels

67162-ISU-ASC-G17

Make sure the clip in the caliper are seated correctly

3. Place a C-clamp around the outer pad and caliper; tighten the C-clamp until the piston is fully compressed in the caliper.
- Top caliper retainer, and rotate caliper away from rotor
- Inboard pad and retaining spring from the caliper
- Outboard pad from the caliper

To install:

4. Clean and lubricate the sleeves and bushing with silicone lubricant

5. Install or connect the following:
- Sleeves and bushings into the caliper
- Clip the retaining spring onto the inboard pad and install the pad in the caliper
- Outboard pad into the caliper
- Caliper in position over the rotor and install the mounting bolts and tighten to 23 ft. lbs. (31 Nm).
- Wheel and tire assemblies

6. Refill the master cylinder and pump pedal to attain full brake pedal before Road-testing the vehicle.

ISUZU

Axiom

14

SPECIFICATIONS AND MAINTENANCE CHARTS

ENGINE AND VEHICLE IDENTIFICATION

		Engine							Model Year	
Code	Liters (cc)	Cu. In.	Cyl.	Fuel Sys.	Engine Type	Eng. Mfg.		Code ①		Year
X	3.5 (3494)	213	6	MFI	DOHC	Isuzu		2		2002

Code ①	Year
2	2002
3	2003
4	2004
5	2005

NA: Not available

MFI: Multi-port Fuel Injection

DOHC: Double Overhead Camshaft

① 10th position of VIN

09474_AXIO_C0001

GENERAL ENGINE SPECIFICATIONS

Year	Model	Engine Displacement Liters	Engine Series VIN	Net Horsepower @ rpm	Net Torque @ rpm (ft. lbs.)	Bore x Stroke (in.)	Com- pression Ratio	Oil Pressure @ rpm
2002	Axiom	3.5	X	230@5400	230@3000	3.68x3.35	9.1:1	60-80@3000
2003	Axiom	3.5	X	230@5400	230@3000	3.68x3.35	9.1:1	60-80@3000

MFI: Multiport fuel injection

09474_AXIO_C0002

ENGINE TUNE-UP SPECIFICATIONS

Year	Engine Displacement Liters	Engine VIN	Spark Plug Gap (in.)	Ignition Timing (deg.) MT	Ignition Timing (deg.) AT	Fuel Pump (psi)	Idle Speed (rpm) MT	Idle Speed (rpm) AT	Valve Clearance In.	Valve Clearance Ex.
2002	3.5	X	0.040	①	①	48-55	750	750	0.009-0.013	0.010-0.014
2003	3.5	X	0.040	①	①	48-55	750	750	0.009-0.013	0.010-0.014

NOTE: The Vehicle Emission Control Information label often reflects specification changes made during production.

The label figures must be used if they differ from those in this chart.

B: Before top dead center

HYD: Hydraulic

① Controlled by the PCM

09474_AXIO_C0003

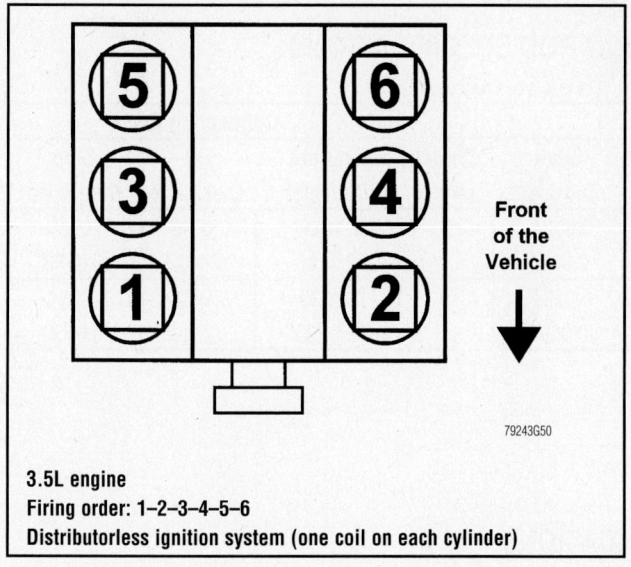

3.5L engine
Firing order: 1–2–3–4–5–6
Distributorless ignition system (one coil on each cylinder)

79243G50

Front
of the
Vehicle

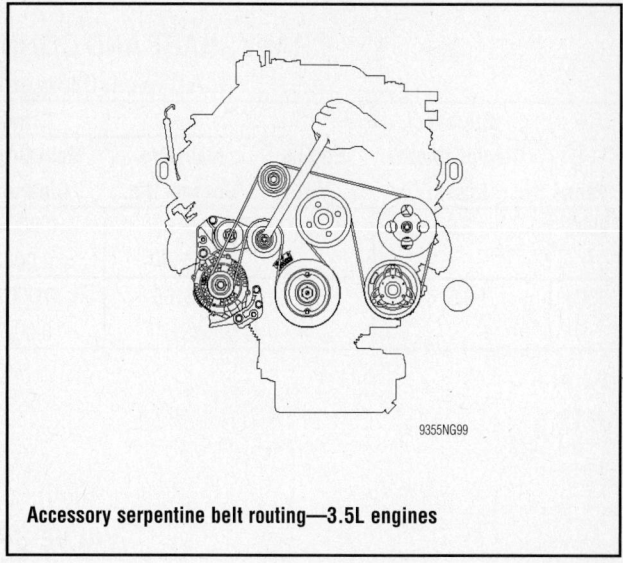

9355NG99

Accessory serpentine belt routing—3.5L engines

CAPACITIES

Year	Model	Engine Displacement Liters	Engine VIN	Oil with Filter (qts.)	Engine Transmission (pts.) Man.	Engine Transmission (pts.) Auto.	Transfer Case (pts.)	Drive Axle Front (pts.)	Drive Axle Rear (pts.)	Fuel Tank (gal.)	Cooling System (qts.)
2002	Axiom	3.5	X	5.0	—	18.2	2.8	2.66	3.64	19.5	11.7
2003	Axiom	3.5	X	5.0	—	18.2	2.8	2.66	3.64	19.5	11.7

NOTE: All capacities are approximate. Add fluid gradually and check to ensure a proper level has been reached.

09474_AXIO_C0004

CRANKSHAFT AND CONNECTING ROD SPECIFICATIONS

All measurements are given in inches.

Year	Engine Displacement Liters	Engine VIN	Crankshaft				Connecting Rod		
			Main Brg. Journal Dia.	Main Brg. Oil Clearance	Shaft End-play	Thrust on No.	Journal Diameter	Oil Clearance	Side Clearance
2002	3.5	X	2.5165-2.5170	0.0007-0.0017	0.0024-0.0094	3	2.1229-2.1235	0.0010-0.0023	0.0050-0.0150
2003	3.5	X	2.5165-2.5170	0.0007-0.0017	0.0024-0.0094	3	2.1229-2.1235	0.0010-0.0023	0.0050-0.0150

09474_AXIO_C0005

VALVE SPECIFICATIONS

Year	Engine Displacement Liters	Engine VIN	Seat Angle (deg.)	Face Angle (deg.)	Spring Test Pressure (lbs. @ in.)	Spring Installed Height (in.)	Stem-to-Guide Clearance (in.)		Stem Diameter (in.)	
							Intake	Exhaust	Intake	Exhaust
2002	3.5	X	45	45	41-44@ 1.38	1.38	0.0002-0.0009	0.0012-0.0025	0.2346-0.2353	0.2343-0.2350
2003	3.5	X	45	45	41-44@ 1.38	1.38	0.0002-0.0009	0.0012-0.0025	0.2346-0.2353	0.2343-0.2350

NA: Not Available

09474_AXIO_C0006

PISTON AND RING SPECIFICATIONS

All measurements are given in inches.

Year	Engine Displacement Liters	Engine VIN	Piston Clearance	Ring Gap			Ring Side Clearance		
				Top Compression	Bottom Compression	Oil Control	Top Compression	Bottom Compression	Oil Control
2002	3.5	X	NA	0.0118-0.0157	0.0177-0.0236	0.006-0.018	0.0006-0.0015	0.0006-0.0015	NA
2003	3.5	X	NA	0.0118-0.0157	0.0177-0.0236	0.006-0.018	0.0006-0.0015	0.0006-0.0015	NA

NA: Not Available

09474_AXIO_C0007

TORQUE SPECIFICATIONS
All readings in ft. lbs.

Year	Engine Displacement Liters	Engine VIN	Cylinder Head Bolts	Main Bearing Bolts	Rod Bearing Bolts	Crankshaft Damper Bolts	Flywheel Bolts	Manifold Intake	Manifold Exhaust	Spark Plugs	Oil Drain Plug
2002	3.5	X	①	②	40	123	40	18	38	13	58
2003	3.5	X	①	②	40	123	40	18	38	13	58

① Step 1: 21 ft. lbs.

   Step 2: 47 ft. lbs.

② Step 1: 22 ft. lbs.

   Step 2: Plus 55-65 degrees

   Step 3: Crankcase side bolts to 29 ft. lbs.

09474_AXIO_C0008

WHEEL ALIGNMENT

Year	Model	Caster Range (+/-Deg.)	Caster Preferred Setting (Deg.)	Camber Range (+/-Deg.)	Camber Preferred Setting (Deg.)	Toe-in (in.)
2002	Axiom	1.00	+2.50	0.50	0	0+/-0.08
2003	Axiom	1.00	+2.50	0.50	0	0+/-0.08

09474_AXIO_C0009

TIRE, WHEEL AND BALL JOINT SPECIFICATIONS

Year	Model	OEM Tires Standard	OEM Tires Optional	Tire Pressures (psi) Front	Tire Pressures (psi) Rear	Wheel Size	Ball Joint Inspection	Wheel Lug Torque (ft.lbs.)
2002	Axiom	P235/65R17	none	26	26	7JJ	U: 4-28 ① L: 4-55	87
2003	Axiom	P235/65R17	none	26	26	7JJ	U: 4-28 ① L: 4-55	87

L: Lower

U: Upper

NA: Not available

NS: Not specified by manufacturer

① Torque required in inch lbs. to rotate ball joint when removed from the knuckle

09474_AXIO_C0010

BRAKE SPECIFICATIONS

All measurements in inches unless noted

Year	Model		Brake Disc				Brake Drum Diameter			Minimum Lining Thickness	Brake Caliper	
			Original Thickness	Machine Thickness	Minimum Thickness	Maximum Runout	Original Inside Diameter	Max. Wear Limit	Maximum Machine Diameter		Bracket Bolts (ft. lbs.)	Mounting Bolts (ft. lbs.)
2002	Axiom	F	1.020	0.983	0.969	0.005	—	—	—	0.039	115	54
		R	0.710	0.668	0.654	0.005	11.6	11.67	NA	0.039	76	32
2003	Axiom	F	1.020	0.983	0.969	0.005	—	—	—	0.039	115	54
		R	0.710	0.668	0.654	0.005	11.6	11.67	NA	0.039	76	32

NA: Not Available

09474_AXIO_C0011

SCHEDULED MAINTENANCE INTERVALS
Isuzu—Axiom

TO BE SERVICED	TYPE OF SERVICE	VEHICLE MILEAGE INTERVAL (x1000)															
		7.5	15	22.5	30	37.5	45	52.5	60	67.5	75	82.5	90	97.5	105	112.5	120
Accelerator linkage ①	L	✓	✓	✓	✓	✓	✓	✓	✓	✓	✓	✓	✓	✓	✓	✓	✓
Accessory drive belts ②	S/I				✓				✓				✓				✓
Air cleaner filter	R				✓				✓				✓				✓
Auto cruise control linkage & hose ③	S/I		✓		✓		✓		✓		✓		✓		✓		✓
Automatic transmission fluid level ③	S/I	✓		✓		✓		✓		✓		✓		✓		✓	
Battery fluid level ③	S/I	✓	✓	✓	✓	✓	✓	✓	✓	✓	✓	✓	✓	✓	✓	✓	✓
Body and chassis ①	L	✓	✓	✓	✓	✓	✓	✓	✓	✓	✓	✓	✓	✓	✓	✓	✓
Brake fluid level ③	S/I	✓	✓	✓	✓	✓	✓	✓	✓	✓	✓	✓	✓	✓	✓	✓	✓
Brake lines & hoses ③	S/I	✓	✓	✓	✓	✓	✓	✓	✓	✓	✓	✓	✓	✓	✓	✓	✓
Brake pedal play ③	S/I		✓		✓		✓		✓		✓		✓		✓		✓
Clutch fluid level ③	S/I	✓	✓	✓	✓	✓	✓	✓	✓	✓	✓	✓	✓	✓	✓	✓	✓
Clutch lines & hose ③	S/I				✓				✓				✓				✓
Clutch pedal free-play ③	S/I		✓		✓		✓		✓		✓		✓		✓		✓
Clutch pedal spring, bushing and clevis pin ①	S/I		✓		✓		✓		✓		✓		✓		✓		✓
Cooling and heating system hoses ③	S/I		✓		✓		✓		✓		✓		✓		✓		✓
Driveshaft flange torque ③	S/I	✓		✓		✓		✓		✓		✓		✓		✓	
Drum and disc brakes ③	S/I		✓		✓		✓		✓		✓		✓		✓		✓
Engine coolant	R				✓				✓				✓				✓
Engine coolant level ③	S/I	✓	✓	✓	✓	✓	✓	✓	✓	✓	✓	✓	✓	✓	✓	✓	✓
Engine oil & filter ③	R	✓	✓	✓	✓	✓	✓	✓	✓	✓	✓	✓	✓	✓	✓	✓	✓
Exhaust system ③	S/I	✓	✓	✓	✓	✓	✓	✓	✓	✓	✓	✓	✓	✓	✓	✓	✓
Front and rear axle lubricant	R		✓		✓				✓				✓				✓
Front and rear driveshafts ①	S/I	✓	✓	✓	✓	✓	✓	✓	✓	✓	✓	✓	✓	✓	✓	✓	✓
Front wheel bearings	S/I & L				✓				✓				✓				✓
Fuel lines & tank cap ③	S/I								✓								✓
Inspect for fluid leaks ③	S/I	✓	✓	✓	✓	✓	✓	✓	✓	✓	✓	✓	✓	✓	✓	✓	✓
Key lock cylinder ③	L		✓		✓		✓		✓		✓		✓		✓		✓
Manual transmission and transfer case fluid ④	R		✓		✓				✓				✓				✓
Parking brake system ③	S/I		✓		✓		✓		✓		✓		✓		✓		✓
Power steering fluid	R				✓				✓				✓				✓
Radiator core and A/C condenser	S/I & C								✓								✓
Rotate tires	S/I	✓	✓	✓	✓	✓	✓	✓	✓	✓	✓	✓	✓	✓	✓	✓	✓
Shift-on-the-fly system gear fluid ③	S/I		✓		✓		✓		✓		✓		✓		✓		✓
Spark plug wires ⑤	S/I								✓								✓

09474_AXIO_C0012

SCHEDULED MAINTENANCE INTERVALS
Isuzu—Axiom

TO BE SERVICED	TYPE OF SERVICE	VEHICLE MILEAGE INTERVAL (x1000)														
		7.5	15	22.5	30	37.5	45	52.5	60	67.5	75	82.5	90	97.5	105	112.5	120
Spark plugs	R	Every 100,000 miles														
Starter safety switch ③	S/I	✓	✓	✓	✓	✓	✓	✓	✓	✓	✓	✓	✓	✓	✓	✓	✓
Steering operation ③	S/I	✓	✓	✓	✓	✓	✓	✓	✓	✓	✓	✓	✓	✓	✓	✓	✓
Suspension & steering ③	S/I	✓	✓	✓	✓	✓	✓	✓	✓	✓	✓	✓	✓	✓	✓	✓	✓
Throttle linkage ③	S/I		✓		✓		✓		✓		✓		✓		✓		✓
Timing belt	R										✓						
Tires and wheels ③	S/I	✓	✓	✓	✓	✓	✓	✓	✓	✓	✓	✓	✓	✓	✓	✓	✓
Valve clearance ④	A								✓								✓

R: Replace   S/I: Service or Inspect   L: Lubricate   A: Adjust   C: Clean

① Perform this at the mileage indicated or every 6 months, whichever occurs first.

② Perform this at the mileage indicated or every 24 months, whichever occurs first.

③ Perform this at the mileage indicated or every 12 months, whichever occurs first.

④ 3.2L V6 engine.

⑤ 2.2L 4 cyl. engine.

FREQUENT OPERATION MAINTENANCE (SEVERE SERVICE)

If a vehicle is operated under any of the following conditions it is considered severe service:

- Towing a trailer or using a camper or car-top carrier.

- Repeated short trips of less than 5 miles in temperatures below freezing.

- Extensive idling or low-speed driving for long distances as in heavy commercial use, such as delivery, taxi or police cars.

- Operating on rough, muddy or salt-covered roads.

- Operating on unpaved or dusty roads.

Air cleaner element: replace every 15,000 miles

Engine oil and filter: replace every 3000 miles or 3 months, whichever occurs first.

Automatic transmission fluid: replace every 20,000 miles.

Rear axle lubricant: replace every 15,000 miles.

09474_AXIO_C0013

ENGINE REPAIR

➡**Disconnecting the negative battery cable on some vehicles may interfere with the functions of the on board computer system. The computer may undergo a relearning process once the negative battery cable is reconnected.**

Distributor

REMOVAL

These engines are equipped with a Distributorless Ignition System (DIS).

Alternator

REMOVAL

1. Before servicing the vehicle, refer to the precautions section.
2. Remove or disconnect the following:
   • Negative battery cable
   • Accessory drive belt
   • Alternator wiring connectors
   • Alternator

INSTALLATION

1. Before servicing the vehicle, refer to the precautions section.
2. Install or connect the following:
   • Alternator. Tighten the 10mm bolts to 30 ft. lbs. (41 Nm) and the 8mm bolts to 15 ft. lbs. (21 Nm).
   • Alternator wiring connectors
   • Accessory drive belt
   • Negative battery cable

Ignition Timing

ADJUSTMENT

These engines are equipped with a Distributorless Ignition System (DIS). No adjustment is possible.

Engine Assembly

REMOVAL & INSTALLATION

1. Before servicing the vehicle, refer to the precautions section.
2. Drain the cooling system.
3. Remove or disconnect the following:
   • Battery
   • Hood
   • Air cleaner assembly

• Canister vacuum line
• Brake booster vacuum line
• Engine wiring harness connectors
• Transmission harness connectors and bracket
• Engine ground cable
• Bonding cable connectors
• Starter harness connector
• Alternator harness connector
• Coolant reservoir tank hose
• Radiator hoses
• Upper fan shroud
• Cooling fan
• Accessory drive belt
• Power steering pump
• A/C compressor
• Heated Oxygen (HO$_2$S) sensor connectors
• Left and right exhaust front pipes
• Flywheel dust cover
• Heater hoses
• Fuel lines
• Transmission. Refer to the transmission procedure in this section.
• Accelerator cable
• Cruise control cable
• Left and right engine mounts
• Engine

To install:
4. Install or connect the following:
   • Engine
   • Left and right engine mounts. Tighten the bolts to 30 ft. lbs. (41 Nm).
   • Transmission
   • Flywheel dust cover
   • Fuel lines
   • Left and right exhaust front pipes
   • HO$_2$S sensor connectors
   • A/C compressor

• Power steering pump
• Accessory drive belt
• Cooling fan. Tighten the nuts to 16 ft. lbs. (22 Nm).
• Radiator
• Upper fan shroud
• Heater hoses
• Radiator hoses
• Coolant reservoir tank hose
• Alternator harness connector
• Starter harness connector
• Engine ground cable
• Transmission harness connectors and bracket
• Engine wiring harness connectors
• Brake booster vacuum line
• Canister vacuum line
• Cruise control cable
• Accelerator cable
• Air cleaner assembly
• Hood
• Battery

5. Fill the cooling system. Check all fluid levels and adjust as necessary.
6. Start the engine and check for leaks.

Water Pump

REMOVAL & INSTALLATION

1. Before servicing the vehicle, refer to the precautions section.
2. Drain the cooling system.
3. Remove or disconnect the following:
   • Negative battery cable
   • Upper radiator hose
   • Timing belt. Refer to the Timing Belt Unit Repair Section.
   • Idler pulley
   • Water pump

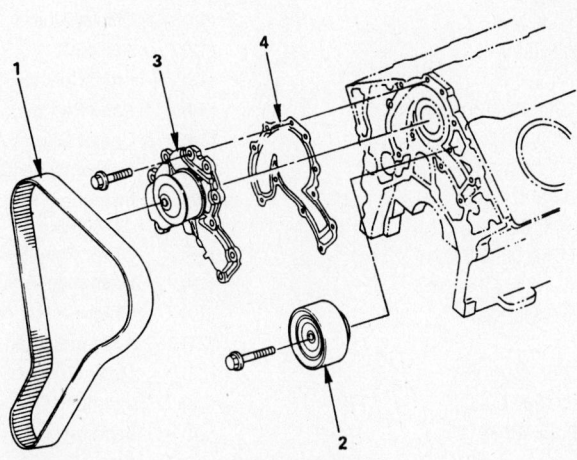

1. Timing belt
2. Idle pulley
3. Water pump assembly
4. Gasket

Exploded view of the water pump mounting

7924BG41

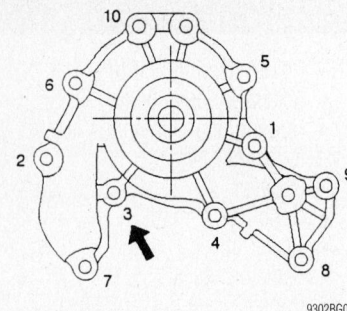

Water pump torque sequence. Apply LOC-TITE® 262 to bolt number 3 (arrow)—3.5L engine

To install:

➡ **Apply Loctite® 262 to bolt number 3 prior to installation.**

4. Install or connect the following:
   - Water pump. Tighten the bolts in two passes, in sequence, to 13 ft. lbs. (18 Nm) for 3.2L engines or to 18 ft. lbs. (25 Nm) for 3.5L engines.
   - Idler pulley
   - Timing belt
   - Upper radiator hose
   - Negative battery cable
5. Fill the cooling system.
6. Start the engine and check for leaks.

Heater Core

REMOVAL & INSTALLATION

1. Before servicing the vehicle, refer to the precautions section.
2. Disconnect the battery ground.
3. Drain the coolant.
4. Discharge and recover the refrigerant.
5. Remove the instrument panel knee pads.
6. Remove the upper cluster and connections.
7. Remove the center cluster assembly.
8. Disconnect the cigarette lighter, ash tray light, and hazard switch connectors.
9. Remove the display unit and radio.
10. Remove the front and rear consoles.
11. Remove the dash side trim panels.
12. Remove the glove box.
13. Remove the driver side lower trim panel.
14. Remove the meter cluster assembly.
15. Remove the driver side knee bolster.
16. Remove the instrument panel as follows:
    a. Disconnect the six driver's side harness connectors.

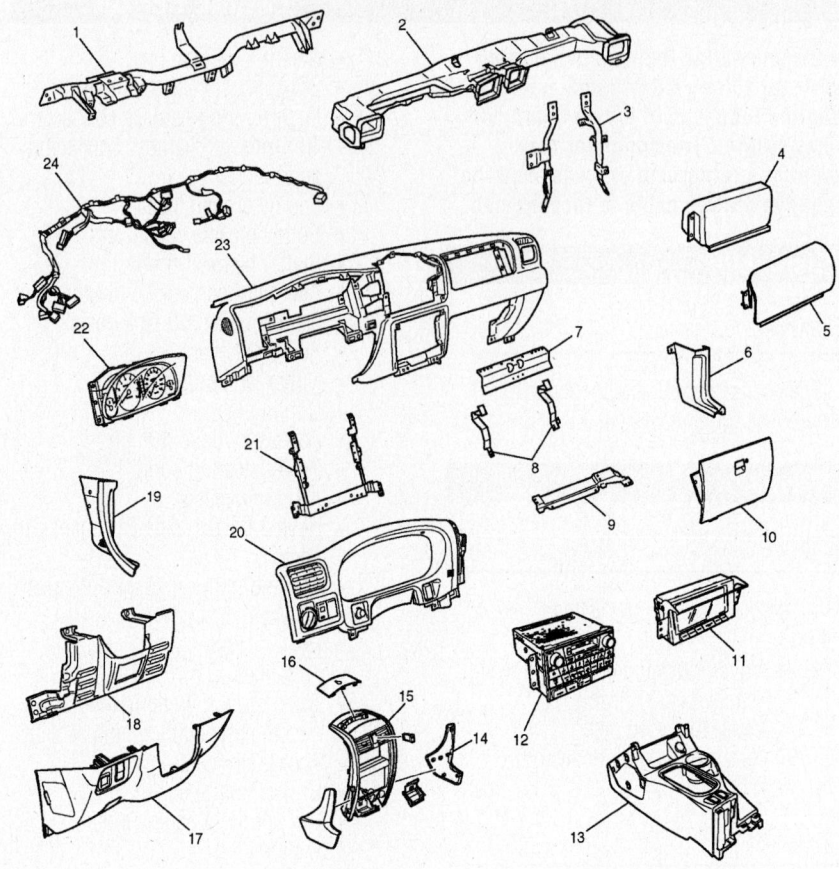

(1)	Cross Beam
(2)	Vent Duct Assembly
(3)	Instrument Panel Stay
(4)	Passenger Air Bag
(5)	Passenger Air Bag Cover
(6)	Dash Side Trim Panel (RH)
(7)	Instrument Upper Reinforcement
(8)	Glove Box Side Reinforcement
(9)	Passenger Lower Bracket
(10)	Glove Box
(11)	Display Unit
(12)	Audio Kit
(13)	Front Console Assembly
(14)	Knee Pad
(15)	Center Cluster Assembly
(16)	Cluster Upper Cover
(17)	Instrument Panel Driver Lower Cover Assembly
(18)	Driver Knee Bolster Assembly
(19)	Dash Side Trim Panel (LH)
(20)	Meter Cluster Assembly
(21)	Instrument Panel Center Reinforcement
(22)	Meter Assembly
(23)	Instrument Panel Assembly
(24)	Instrument Harness Assembly

67162-AXIO-G01

Exploded view of the instrument panel mounting

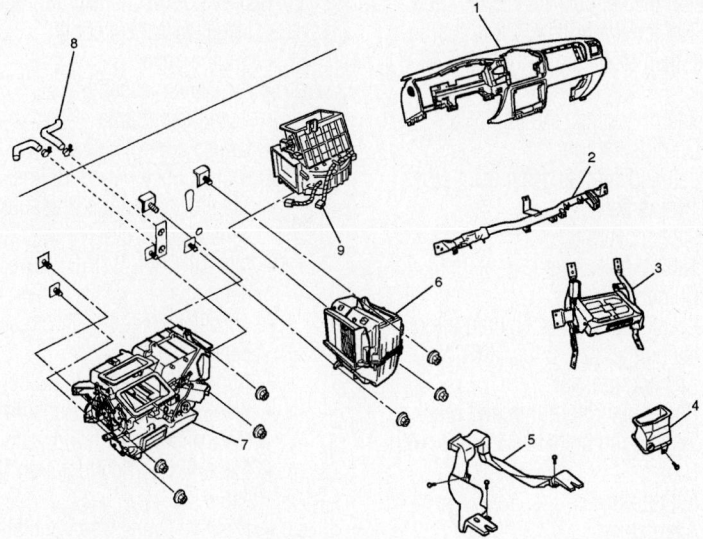

(1)   Instrument Panel Assembly
(2)   Cross Beam Assembly
(3)   Instrument Panel Bracket w/Suspension Control Unit
(4)   Ventilation Lower Duct
(5)   Rear Heater Duct
(6)   Evaporator Assembly
(7)   Heater Unit Assembly
(8)   Heater Hose
(9)   Power Transistor Connector

67162-AXIO-G02

Exploded view of the HVAC mounting

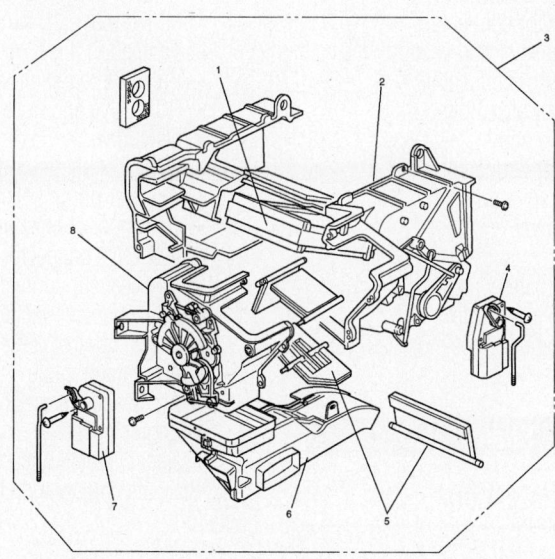

(1)   Heater Core
(2)   Case (Temperature Control)
(3)   Heater Unit
(4)   Mix Actuator
(5)   Mode Door
(6)   Duct
(7)   Mode Actuator
(8)   Case (Mode Control)

67162-AXIO-G03

Exploded view of the heater case

b. Disconnect the three passenger's side harness connectors.

c. Disconnect the two center harness connectors.

d. Disconnect the antenna cable.

e. Disconnect the ground cable bolts.

f. Disconnect the passenger side air bag.

g. Unbolt the fuse box.

h. Remove the three nuts at the base of the windshield.

i. Remove the six retaining bolts and one nut.

j. Remove the passenger air bag.

k. Remove the vent duct assembly.

l. Remove the passenger side lower bracket.

m. Remove the glove box side reinforcements.

n. Remove the instrument panel upper reinforcement.

o. Remove the instrument panel center reinforcement.

p. Remove the instrument panel harness assembly.

q. Remove the instrument panel stays.

r. Remove the instrument panel.

17. Remove the instrument panel with the suspension control unit.

18. Remove the cross beam.

19. Disconnect the power transistor connector.

20. Remove the evaporator assembly.

21. Remove the lower duct.

22. Remove the rear heater duct.

23. Remove the heater unit assembly.

24. Remove the duct, mix actuator, and mode actuator.

25. Remove the mode control case.

26. Separate the heater case and remove the heater core.

To install:

27. Installation is the reverse of the removal procedure.

Cylinder Head

REMOVAL & INSTALLATION

1. Before servicing the vehicle, refer to the precautions section.

2. Drain the cooling system.

3. Remove or disconnect the following:

- Negative battery cable
- Hood
- Engine cover
- Mass Air Flow (MAF) sensor connector
- Intake Air Temperature (IAT) sensor connector

- Positive Crankcase Ventilation (PCV) valve and hose
- Air cleaner assembly
- Manifold Absolute Pressure (MAP) sensor connector
- Vacuum Switching Valve (VSV) connector and vacuum line
- Fuel injector connectors
- Throttle Position (TP) sensor connector
- Idle Air Control (IAC) valve connector
- Ignition coils
- Brake booster vacuum line
- Canister purge vacuum line
- Duty solenoid valve
- Fuel lines
- Intake manifold
- Radiator hoses
- Engine coolant manifold
- Upper fan shroud
- Accessory drive belt and tensioner
- Cooling fan and pulley
- Alternator
- Idler pulley
- Power steering pump and bracket
- A/C compressor
- Crankshaft pulley
- Oil cooler hoses
- Timing belt cover
- Valve covers
- Timing belt. Refer to the Timing Belt Unit Repair Section.
- Left and right exhaust front pipes
- Oil dipstick tube
- Cylinder heads

To install:

➡ **Use new head bolts when installing the cylinder head. Do not apply oil to the head bolt threads.**

➡ **The left and right cylinder head gaskets are not interchangeable.**

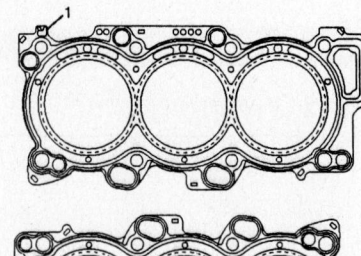
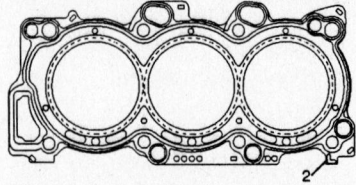

7924BG04

Right (1) and left (2) head gasket identification mark locations—3.5L engine

4. Install the cylinder heads with new gaskets. Tighten the bolts to 22 ft. lbs. in sequence, then to 47 ft. lbs. (64 Nm) in sequence.

5. Install or connect the following:
- Oil dipstick tube
- Left and right exhaust front pipes
- Timing belt
- Valve covers
- Timing belt cover
- Oil cooler hoses
- Crankshaft pulley. Tighten the pulley bolt to 123 ft. lbs. (167 Nm).
- A/C compressor
- Power steering pump and bracket. Tighten the bolts to 34 ft. lbs. (46 Nm).
- Idler pulley
- Alternator
- Cooling fan and pulley
- Accessory drive belt and tensioner
- Upper fan shroud
- Engine coolant manifold
- Radiator hoses
- Intake manifold
- Fuel lines
- Duty solenoid valve
- Canister purge vacuum line
- Brake booster vacuum line
- Ignition coils
- IAC valve connector
- TP sensor connector
- Fuel injector connectors
- VSV connector and vacuum line
- MAP sensor connector
- Air cleaner assembly
- PCV valve and hose
- IAT sensor connector
- MAF sensor connector
- Engine cover
- Hood
- Negative battery cable

6. Fill the cooling system.

7. Start the engine. Check for leaks and proper operation.

Rocker Arms/Shafts

REMOVAL & INSTALLATION

➡ **These engines are not equipped with rocker arms. The camshaft lobes act directly on the valve shims.**

Intake Manifold

REMOVAL & INSTALLATION

1. Before servicing the vehicle, refer to the precautions section.

2. Remove or disconnect the following:
- Negative battery cable
- Engine cover
- Air cleaner assembly
- Accelerator cable
- Cruise control cable
- Brake booster vacuum line
- Manifold Absolute Pressure (MAP) sensor connector
- Idle Air Control (IAC) valve connector
- Throttle Position (TP) sensor connector
- Canister purge solenoid connector
- Electronic Vacuum Sensing Valve (EVSV) connector and vacuum line
- Exhaust Gas Recirculation (EGR) valve
- Positive Crankcase Ventilation (PCV) valve and hose
- Pressure regulator vacuum line
- Ventilation hose
- Throttle body
- Fuel lines
- Fuel injector connectors
- Intake manifold

To install:

3. Install or connect the following:
- Intake manifold. Tighten the fasteners to 18 ft. lbs. (25 Nm).
- Fuel injector connectors
- Fuel lines
- Throttle body. Tighten the bolts to 88 inch lbs. (10 Nm).
- Ventilation hose
- Pressure regulator vacuum line
- PCV valve and hose
- EGR valve
- EVSV connector and vacuum line
- Canister purge solenoid connector
- TP sensor connector
- IAC valve connector
- MAP sensor connector
- Brake booster vacuum line
- Cruise control cable
- Accelerator cable
- Air cleaner assembly
- Engine cover
- Negative battery cable

4. Start the engine and check for proper operation.

Exhaust Manifold

REMOVAL & INSTALLATION

1. Before servicing the vehicle, refer to the precautions section.

2. Remove or disconnect the following:
- Negative battery cable
- Air cleaner assembly

- Heated Oxygen (HO2S) sensor connectors
- Right torsion bar
- Exhaust Gas Recirculation (EGR) pipe and bracket
- Left and right exhaust front pipes
- Heat shields
- Accessory drive belt
- A/C compressor and bracket
- Exhaust manifolds

To install:

3. Install or connect the following:
- Exhaust manifolds. Tighten the bolts to 38 ft. lbs. (52 Nm).
- A/C compressor and bracket
- Accessory drive belt
- Heat shields
- Left and right exhaust front pipes
- EGR pipe and bracket
- Right torsion bar
- HO2S sensor connectors
- Air cleaner assembly
- Negative battery cable

4. Start the engine and check for leaks.

Front Crankshaft Seal

REMOVAL & INSTALLATION

1. Before servicing the vehicle, refer to the precautions section.
2. Remove or disconnect the following:

- Negative battery cable
- Air cleaner assembly
- Upper fan shroud
- Accessory drive belt and tensioner
- Cooling fan and pulley
- Idler pulley
- Power steering pump and move it aside
- Crankshaft pulley
- Timing belt cover
- Timing belt. Refer to the Timing Belt Unit Repair Section.
- Crankshaft timing sprocket
- Oil seal

To install:

3. Install or connect the following:
- Oil seal so that it is flush with the oil pump housing
- Crankshaft timing sprocket
- Timing belt
- Timing belt cover
- Crankshaft pulley. Tighten the bolt to 123 ft. lbs. (167 Nm).
- Power steering pump
- Idler pulley
- Cooling fan and pulley
- Accessory drive belt and tensioner
- Upper fan shroud
- Air cleaner assembly
- Negative battery cable

4. Start the engine and check for leaks.

Camshaft and Valve Lifters

REMOVAL & INSTALLATION

1. Before servicing the vehicle, refer to the precautions section.
2. Remove or disconnect the following:
- Negative battery cable
- Air cleaner assembly
- Upper fan shroud
- Accessory drive belt and tensioner
- Cooling fan and pulley
- Idler pulley
- Power steering pump and move it aside
- Crankshaft pulley
- Timing belt cover
- Timing belt. Refer to the Timing Belt Unit Repair Section.
- Ignition coils
- Valve covers
- Camshafts
- Valve shims and tappets

➡**Keep the valve shims and tappets in order for installation.**

To install:

3. Install the valve tappets and shims in their original locations.
4. Using Gear Spring Lever J-42686, turn the sub gear clockwise to align the

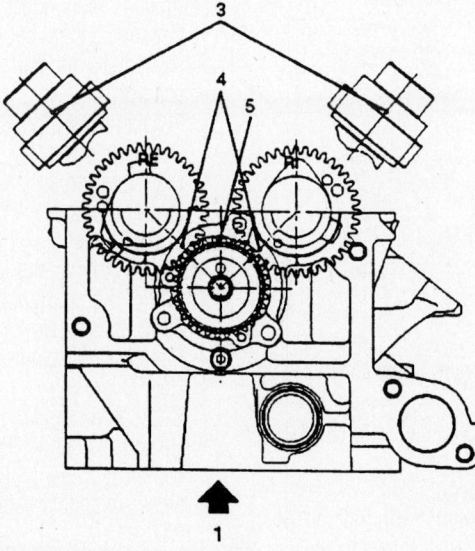

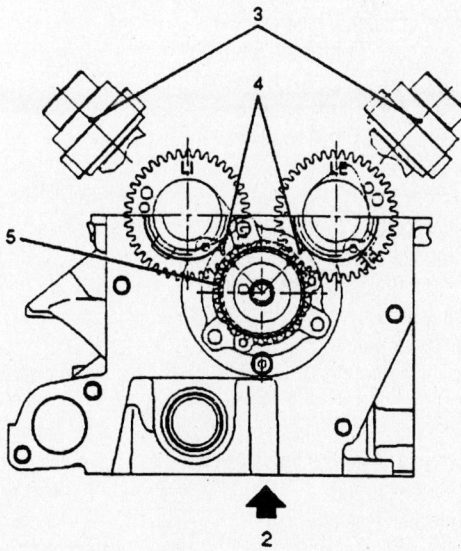

1 Right Bank	3 Alignment Mark on Camshaft Drive Gear
2 Left Bank	4 Alignment Mark on Camshaft
	5 Alignment Mark on Retainer

Camshaft alignment marks for the left and right cylinder heads—3.5L engine

7924BG11

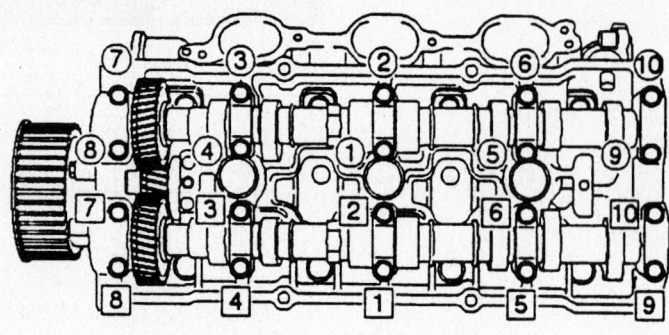

: Intake   : Exhaust

7924BG12

Camshaft retaining bracket tightening sequence—3.5L engine

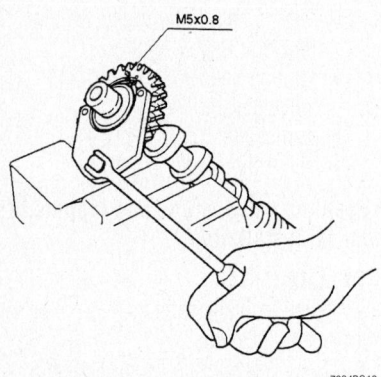

7924BG13

Aligning the sub gear with the Gear Spring Lever J-42686—3.5L engine

5mm bolt holes in the sub gear and the camshaft driven gear. Tighten the 5mm bolt.

5. Install the camshafts. Align the timing marks as shown. Tighten the bolts in sequence to 89 inch lbs. (10 Nm).

6. Install or connect the following:
- Valve covers
- Ignition coils
- Timing belt. Refer to the Timing Belt Unit Repair Section.
- Timing belt cover
- Crankshaft pulley
- Power steering pump
- Idler pulley
- Cooling fan and pulley
- Accessory drive belt and tensioner
- Upper fan shroud
- Air cleaner assembly
- Negative battery cable

Valve Lash

ADJUSTMENT

➥**Measure valve clearance with the engine cold.**

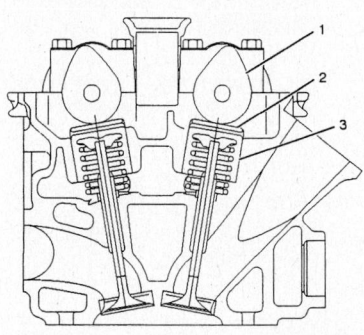

9302BG02

Cross section of the 3.5L cylinder head. Note the position of the camshaft lobe (1), adjustment shim (2) and the tappet (3)

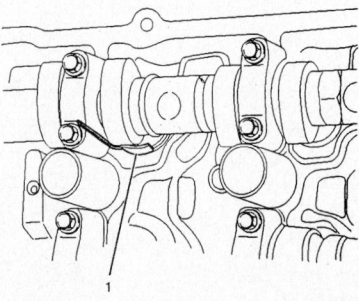

9302BG03

Valve clearance adjusting tool J–42689 (1)

1. Before servicing the vehicle, refer to the precautions section.

2. Remove the valve covers.

3. Check the valve clearance with the camshafts positioned as shown. Intake valve clearance should be 0.0091–0.0130 inches. Exhaust valve clearance should be 0.0098–0.0138 inches.

4. If adjustment is required, replace the shims as follows:

a. Step 1: Position special tool J-42689 on the edge of the tappet.

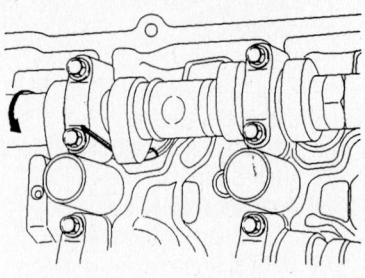

9302BG04

Using the valve clearance adjusting tool to hold the tappet for shim replacement

b. Step 2: Rotate the crankshaft until the maximum lift portion of the camshaft lobe contacts the upper edge of the special tool and presses the tappet down to create enough clearance between the adjustment shim and the camshaft for the shim to be removed.

c. Step 3: Replace shims as necessary to achieve correct valve clearance.

d. Step 4: Repeat for each valve to be adjusted.

5. Replace the valve covers. Tighten the bolts to 80 inch lbs. (9 Nm).

Starter Motor

REMOVAL & INSTALLATION

1. Before servicing the vehicle, refer to the precautions section.

2. Remove or disconnect the following:
- Negative battery cable
- Heated Oxygen (HO$_2$S) sensor connectors
- Exhaust front pipe
- Heat shield
- Starter wiring connectors
- Starter motor

To install:

3. Install or connect the following:
- Starter motor. Tighten the bolts to 30 ft. lbs. (40 Nm).
- Starter wiring connectors
- Heat shield
- Exhaust front pipe
- Heated Oxygen (HO$_2$S) sensor connectors
- Negative battery cable

Oil Pan

REMOVAL & INSTALLATION

1. Before servicing the vehicle, refer to the precautions section.

2. Drain the engine oil.

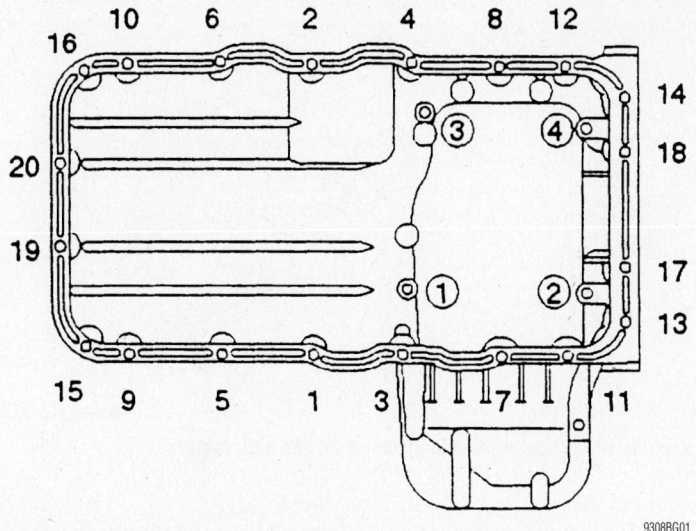

Lower crankcase torque sequence—3.5L engine

3. Remove or disconnect the following:
- Negative battery cable
- Front wheels
- Oil level dipstick
- Stone guard
- Radiator under fan shroud
- Shift-on-the-fly from the axle
- Suspension crossmember

4. If equipped with 4 wheel drive, unbolt and lower the front axle housing assembly for clearance.

5. Remove or disconnect the following:
- Steering gear
- Starter
- Oil pan
- Lower crankcase

To install:

6. Apply a bead of silicone sealant to the crankcase flange and install the crankcase. Tighten the fasteners in sequence to 89 inch lbs. (10 Nm).

7. Apply a bead of silicone sealant to the oil pan flange and install the oil pan. Tighten the fasteners to 89 inch lbs. (10 Nm).

8. Install or connect the following:
- Starter. Torque to 30 ft. lbs. (40 Nm)

9. If equipped, raise the axle housing assembly into position. Tighten the axle case bolts to 61 ft. lbs. (82 Nm) and the mounting bolts to 112 ft. lbs. (152 Nm).

10. Install or connect the following:
- Steering gear
- Suspension crossmember. Tighten the bolts to 58 ft. lbs. (78 Nm).
- Radiator under fan shroud
- Stone guard
- Oil level dipstick
- Front wheels
- Negative battery cable

11. Fill the crankcase with engine oil.
12. Start the engine and check for leaks.

Oil Pump

REMOVAL & INSTALLATION

1. Before servicing the vehicle, refer to the precautions section.
2. Remove or disconnect the following:
- Timing belt
- Oil pan
- Oil pick-up tube
- Oil filter adapter
- Oil pump

To install:
3. Apply silicone sealant to the oil pump mounting surface and install the oil pump. Tighten the bolts in sequence to 18 ft. lbs. (25 Nm).

4. Install or connect the following:
- Oil filter adapter
- Oil pickup tube
- Oil pan
- Timing belt

Rear Main Seal

REMOVAL & INSTALLATION

1. Before servicing the vehicle, refer to the precautions section.
2. Remove or disconnect the following:
- Negative battery cable
- Transmission
- Clutch assembly, if equipped with a manual transmission
- Flywheel by loosening the flywheel bolts in a 2-step crisscross sequence
- Rear main seal, using a seal puller

➡**Do not damage the crankshaft sealing surface.**

To install:
3. Install or connect the following:
- New rear main seal, by lubricating it with engine oil

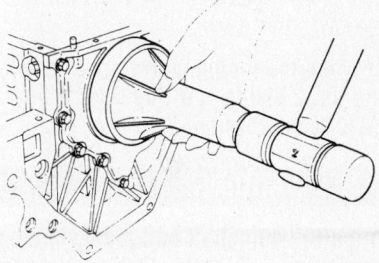

Installing a one-piece rear crankshaft oil seal

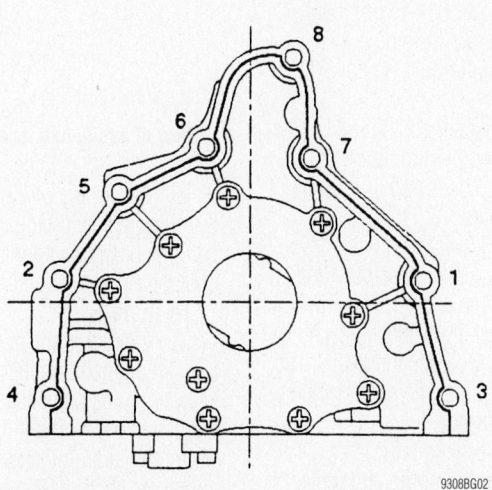

Oil pump torque sequence

- Flywheel, using new flywheel bolts. Tighten the bolts, in a 2-step criss-cross pattern, to 40 ft. lbs. (54 Nm).
- Clutch assembly, if removed
- Transmission
- Negative battery cable

4. Check the oil and refill as necessary.

Timing Belt and Cover

REMOVAL & INSTALLATION

1. Disconnect the negative battery cable.

2. Remove the air cleaner assembly and intake air duct.

3. Remove the upper fan shroud from the radiator.

4. Remove the 4 nuts retaining the cooling fan assembly. Remove the cooling fan from the fan pulley.

5. Loosen and remove the drive belts.

6. Remove the upper timing belt covers.

7. Remove the fan pulley assembly.

8. Rotate the crankshaft to align the camshaft timing marks with the pointer dots on the back covers. Verify that the pointer on the crankshaft aligns with the mark on the lower timing cover.

➡ **When the timing marks are aligned, the No. 2 piston is at Top Dead Center (TDC) compression.**

❋ WARNING

Align the camshaft and crankshaft sprockets with their alignment marks before removing the timing belt. Failure to align the belt and sprocket marks may result in valve damage.

9. Use tool No. J-8614-01, or a suitable pulley holding tool to remove the crankshaft pulley center bolt. Remove the crankshaft pulley.

10. If present, disconnect the 2 oil cooler hose bracket bolts on the timing cover. Move the oil cooler hoses and bracket off of the lower timing cover.

11. Remove the lower timing belt cover.

12. Remove the pusher assembly (tensioner) from below the belt tensioner pulley. The pusher rod must always face upward to prevent oil leakage. Depress the pusher rod, and insert a wire pin into the hole to keep the pusher rod retracted.

13. Remove the timing belt.

14. Inspect the water pump and replace it if there is any doubt about its condition.

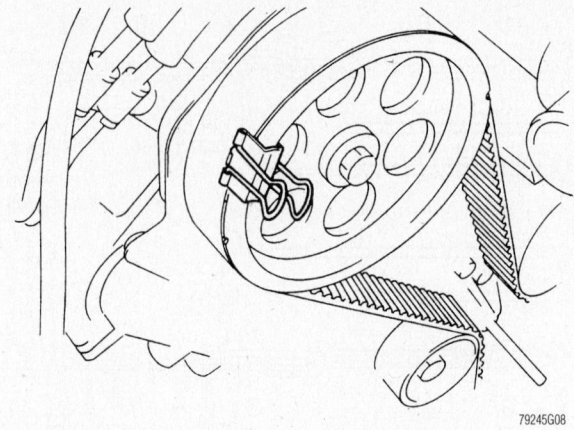

Using a double clip to hold the belt in place—3.2L and 3.5L engines

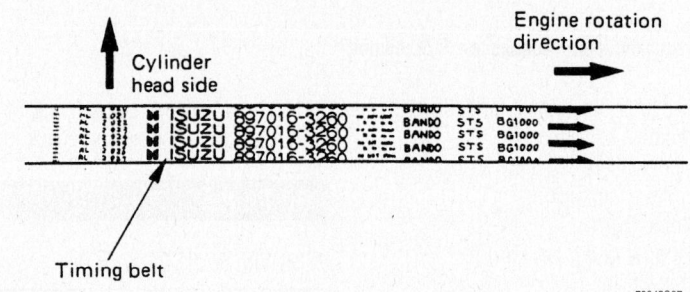

For maximum timing belt life, install the belt as shown—3.2L and 3.5L engines

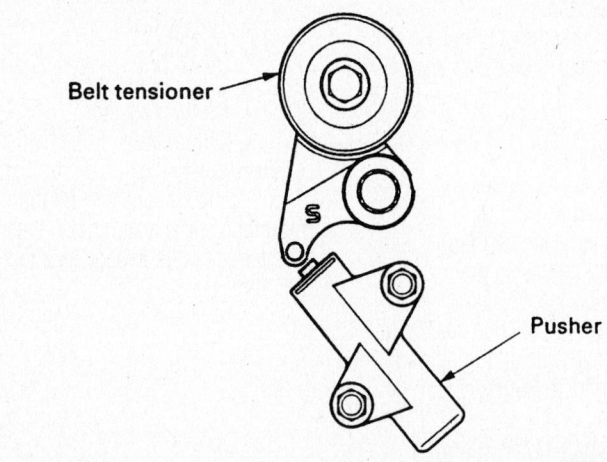

View of timing belt tensioner and pusher—3.2L and 3.5L engines

15. Repair any oil or coolant leaks before installing a new timing belt. If the timing belt has been contaminated with oil or coolant, or is damaged, it must be replaced.

To install:

16. Verify that the sprocket timing marks are still aligned and that the groove and the keyway on the crankshaft timing sprocket align with the mark on the oil pump. The white pointers on the camshaft timing sprockets should align with the dots on the front plate.

17. Install the timing belt. Use clips to secure the belt onto each sprocket until the installation is complete. Align the dotted marks on the timing belt with the timing mark opposite the groove on the crankshaft sprocket.

➡ **The arrows on the timing belt must follow the belt's direction of rotation. The manufacturer's trademark on the belt's spine should be readable left-to-right when the belt is installed.**

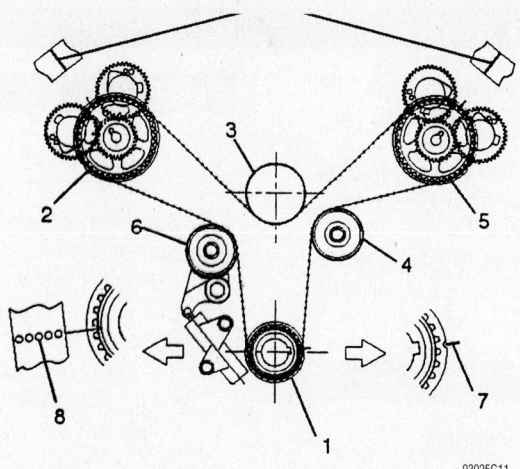

Timing mark alignment and timing belt routing—3.2L (VIN W) and 3.5L DOHC engines

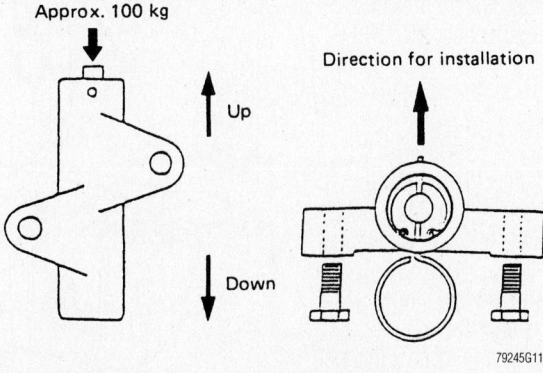

View of timing belt tensioner pusher—3.2L and 3.5L engines

18. Align the white line on the timing belt with the alignment mark on the right bank camshaft timing pulley. Secure the belt with a clip.

✳✳ WARNING

If any binding is felt when adjusting the timing belt tension by turning the crankshaft, STOP turning the engine, because the pistons may be hitting the valves.

19. Rotate the crankshaft counterclockwise to remove the slack between the crankshaft sprocket and the right camshaft timing belt sprocket.

20. Install the belt around the water pump pulley.

21. Install the belt on the idler pulley.

22. Align the white alignment mark on the timing belt with the alignment mark on the left bank camshaft timing belt sprocket.

23. Install the crankshaft pulley and tighten the center bolt by hand. Rotate the crankshaft pulley clockwise to give slack between the crankshaft timing belt pulley and the right bank camshaft timing belt pulley.

24. Insert a 1.4mm piece of wire through the hole in the pusher to hold the rod in. Install the pusher assembly while pushing the tension pulley toward the belt.

25. Pull the pin out from the pusher to release the rod.

26. Remove the clamps from the sprockets. Rotate the crankshaft pulley clockwise 2 turns. Measure the rod protrusion to ensure it is between 0.16 – 0.24 in. (4 – 6mm).

27. If the tensioner pulley bracket pivot bolt was removed, tighten it to 31 ft. lbs. (42 Nm).

28. Tighten the pusher bolts to 14 ft. lbs. (19 Nm).

29. Remove the crankshaft pulley. Install the lower and upper timing belt covers and tighten their bolts to 12 ft. lbs. (17 Nm).

30. Fit the oil cooler hose onto the timing cover and tighten its mounting bracket bolts to 16 ft. lbs. (22 Nm).

31. Install the crankshaft pulley and tighten the pulley bolt to 123 ft. lbs. (167 Nm).

32. Install fan pulley assembly and tighten the bolts to 16 ft. lbs. (22 Nm).

33. Install and adjust the accessory drive belts.

34. Install the cooling fan assembly and tighten the bolts to 72 inch lbs. (8 Nm).

35. Install the upper fan shroud.

36. Install the air cleaner assembly and intake air duct.

37. Connect the negative battery cable.

Piston and Ring

POSITIONING

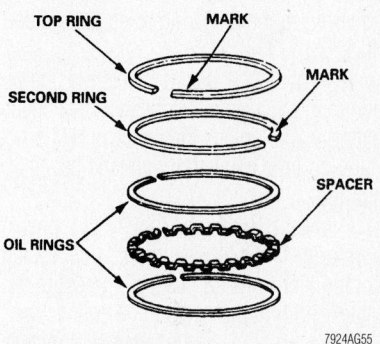

Piston ring positioning and top mark locations—3.5L engine

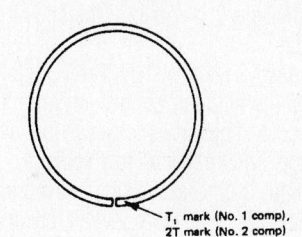

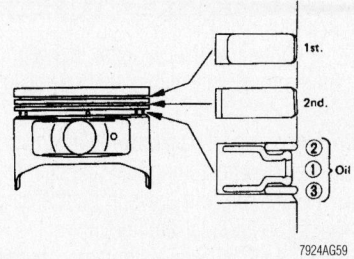

Piston ring positioning—3.2L and 3.5L engines

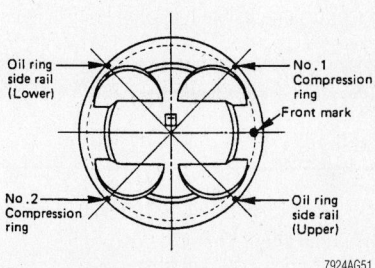

Piston ring end-gap spacing—3.2L and 3.5L engines

FUEL SYSTEM

Fuel System Service Precautions

Safety is the most important factor when performing not only fuel system maintenance but any type of maintenance. Failure to conduct maintenance and repairs in a safe manner may result in serious personal injury or death. Maintenance and testing of the vehicle's fuel system components can be accomplished safely and effectively by adhering to the following rules and guidelines:

• To avoid the possibility of fire and personal injury, always disconnect the negative battery cable unless the repair or test procedure requires that battery voltage be applied.

• Always relieve the fuel system pressure prior to disconnecting any fuel system component (injector, fuel rail, pressure regulator, etc.), fitting or fuel line connection. Exercise extreme caution whenever relieving fuel system pressure, to avoid exposing skin, face and eyes to fuel spray. Please be advised that fuel under pressure may penetrate the skin or any part of the body that it contacts.

• Always place a shop towel or cloth around the fitting or connection prior to loosening to absorb any excess fuel due to spillage. Ensure that all fuel spillage (should it occur) is quickly removed from engine surfaces. Ensure that all fuel soaked cloths or towels are deposited into a suitable waste container.

• Always keep a dry chemical (Class B) fire extinguisher near the work area.

• Do not allow fuel spray or fuel vapors to come into contact with a spark or open flame.

• Always use a backup wrench when loosening and tightening fuel line connection fittings. This will prevent unnecessary stress and torsion to fuel line piping. Always follow the proper tightening specifications.

• Always replace worn fuel fitting O-rings with new. Do not substitute fuel hose or equivalent, where fuel pipe is installed.

Fuel System Pressure

RELIEVING

1. Before servicing the vehicle, refer to the precautions section.
2. Remove the fuel filler cap.
3. Remove the fuel pump relay from the underhood relay box.

4. Start the engine and let it run until it stalls, then crank the engine for an additional 30 seconds.
5. Turn the ignition switch to the **OFF** position and remove the key. Disconnect the negative battery cable.
6. When service is completed, install the fuel pump relay and connect the negative battery cable.

Fuel Filter

REMOVAL & INSTALLATION

1. Before servicing the vehicle, refer to the precautions section.
2. Relieve the fuel system pressure.
3. Remove or disconnect the following:
• Fuel lines from the fuel filter
• Fuel filter

To install:
4. Install or connect the following:
• Fuel filter and tighten the bracket bolt. Note the fuel flow directional arrow.
• Fuel lines to the fuel filter
• Negative battery cable
5. Start the engine and inspect the fuel filter connections for leaks.

Fuel Pump

REMOVAL & INSTALLATION

1. Before servicing the vehicle, refer to the precautions section.
2. Relieve fuel system pressure.
3. Drain the fuel tank.
4. Remove or disconnect the following:
• Negative battery cable
• Fuel filler and vent hoses
• Fuel tank skid plate

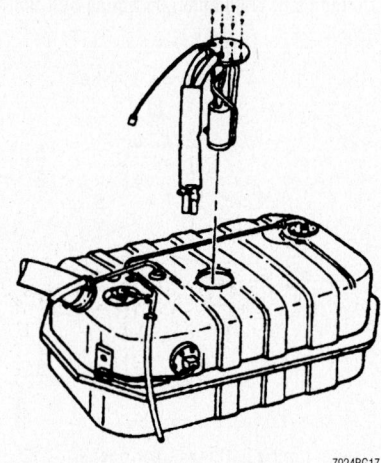

7924BG17

Fuel pump assembly mounting

Fuel Hose

Fuel Filter

Filter Mounting Bolt

7924NG25

Fuel filter mounting location under the vehicle

- Fuel tank wiring connectors
- Fuel supply and return lines
- Fuel tank
- Fuel pump assembly

To install:

5. Install or connect the following:
   - Fuel pump assembly
   - Fuel tank. Tighten the bolts to 27 ft. lbs. (36 Nm).
   - Fuel supply and return lines
   - Fuel tank wiring connectors
   - Fuel tank skid plate
   - Fuel filler and vent hoses
   - Negative battery cable
6. Start the engine and check for leaks.

Fuel Injector

REMOVAL & INSTALLATION

1. Before servicing the vehicle, refer to the precautions section.
2. Relieve fuel system pressure.
3. Remove or disconnect the following:
   - Negative battery cable
   - Engine cover
   - Fuel injector wiring connectors
   - Fuel lines
   - Fuel supply manifold with injectors attached
   - Clips
   - Injectors from the supply manifold

To install:

4. Install or connect the following:
   - New O-rings on the fuel injectors
   - Fuel injectors
   - Fuel supply manifold with injectors attached. Tighten the bolts to 60 inch lbs. (6.5 Nm).
   - Fuel lines
   - Fuel injector wiring connectors
   - Engine cover
   - Negative battery cable
5. Start the engine and check for leaks.

DRIVE TRAIN

Automatic Transmission Assembly

REMOVAL & INSTALLATION

2-Wheel Drive

1. Before servicing the vehicle, refer to the precautions section.
2. Remove or disconnect the following:
   - Negative battery cable
   - Driveshaft
   - Fuel line bracket from the transmission
   - Wiring harness heat shield
   - Transmission harness connectors
3. Support the transmission with a jack.
4. Remove or disconnect the following:
   - Transmission mount and crossmember
   - Transmission oil cooler lines
   - Selector cable
   - Starter motor
   - Undercovers
   - Torque converter bolts
   - Transmission flange bolts
   - Transmission

To install:

➡**Use new torque converter bolts.**

5. Install or connect the following:
   - Transmission. Tighten the large bolts to 56 ft. lbs. (76 Nm) and the small bolts to 69 inch lbs. (8 Nm).
   - Torque converter. Tighten the bolts to 40 ft. lbs. (54 Nm).
   - Under covers
   - Starter motor. Tighten the bolts to 30 ft. lbs. (40 Nm).
   - Selector cable
   - Transmission oil cooler lines
   - Crossmember. Tighten the bolts to 85 ft. lbs. (116 Nm).
   - Transmission mount. Tighten the bolts to 37 ft. lbs. (50 Nm).
   - Transmission harness connectors
   - Wiring harness heat shield
   - Fuel line bracket
   - Driveshaft. Tighten the flange bolts to 46 ft. lbs. (63 Nm).
   - Negative battery cable

4-Wheel Drive

1. Before servicing the vehicle, refer to the precautions section.
2. Remove or disconnect the following:
   - Negative battery cable
   - Skid plate
   - Driveshafts
   - Center exhaust pipe
   - Fuel line bracket from the transmission
   - Wiring harness heat shield
   - Transmission harness connectors
3. Support the transmission with a jack.
4. Remove or disconnect the following:
   - Transmission mount and crossmember
   - Transmission oil cooler lines
   - Selector cable
   - Starter motor
   - Undercovers
   - Torque converter bolts
   - Transmission flange bolts
   - Transmission

To install:

➡**Use new torque converter bolts.**

5. Install or connect the following:
   - Transmission. Tighten the large bolts to 56 ft. lbs. (76 Nm) and the small bolts to 69 inch lbs. (8 Nm).
   - Torque converter. Tighten the bolts to 40 ft. lbs. (54 Nm).
   - Under covers

   - Starter motor. Tighten the bolts to 30 ft. lbs. (40 Nm).
   - Selector cable
   - Transmission oil cooler lines
   - Crossmember. Tighten the bolts to 85 ft. lbs. (116 Nm).
   - Transmission mount. Tighten the bolts to 37 ft. lbs. (50 Nm).
   - Transmission harness connectors
   - Wiring harness heat shield
   - Fuel line bracket
   - Center exhaust pipe
   - Driveshafts. Tighten the flange bolts to 46 ft. lbs. (63 Nm).
   - Skid plate
   - Negative battery cable

Transfer Case Assembly

REMOVAL & INSTALLATION

1. Before servicing the vehicle, refer to the precautions section.
2. Remove or disconnect the following:
   - Negative battery cable
   - Transfer case skid plate
   - Front and rear driveshafts
   - Breather hose
   - Center exhaust pipe
   - Vehicle Speed (VSS) sensor connector
   - Harness connector
3. Support the transfer case
4. Remove or disconnect the following:
   - Transfer case flange fasteners
   - Transfer case

To install:

5. Install or connect the following:
   - Transfer case. Tighten the flange fasteners to 34 ft. lbs. (46 Nm).
   - Harness connector
   - Breather hose
   - VSS sensor connector

- Center exhaust pipe
- Front and rear driveshafts. Tighten the bolts to 46 ft. lbs. (63 Nm).
- Transfer case skid plate. Tighten the bolts to 27 ft. lbs. (37 Nm).
- Negative battery cable

Halfshaft

REMOVAL & INSTALLATION

1. Before servicing the vehicle, refer to the precautions section.

2. Remove or disconnect the following:
   - Negative battery cable
   - Front wheel
   - Radiator skid plate
   - Transfer case skid plate
   - Brake calipers and mounting bracket
   - Brake rotor
   - Wheel speed sensor
   - Steering knuckle
3. Support the axle housing with a jack. Unbolt the axle mounting bracket and remove the halfshaft/bracket assembly.

To install:
4. Install or connect the following:
   - Axle/bracket assembly. Tighten the bracket flange bolts to 85 ft. lbs. (116 Nm) and the bracket mounting bolts to 112 ft. lbs. (152 Nm).
   - Steering knuckle
   - Wheel speed sensor
   - Brake rotor
   - Brake caliper and mounting bracket. Tighten the bracket bolts to 115 ft. lbs. (155 Nm).
   - Transfer case skid plate. Tighten the bolts to 27 ft. lbs. (37 Nm).

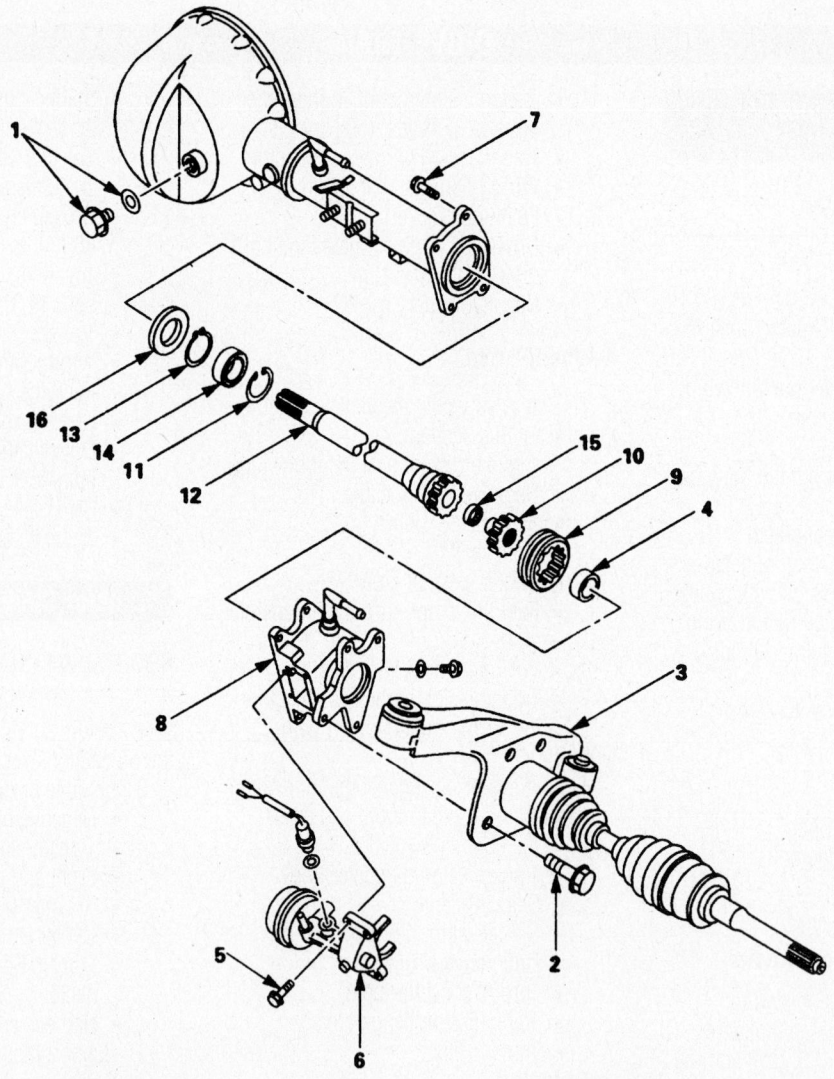

1. Filler plug	9. Sleeve
2. Bolt	10. Clutch gear
3. Front axle drive shaft (LH side)	11. Snap ring
4. Spacer	12. Inner shaft
5. Bolt	13. Snap ring
6. Actuator assembly	14. Inner shaft bearing
7. Bolt	15. Needle bearing
8. Housing	16. Oil seal

7924BG26

Exploded view of the left halfshaft, axle shaft and axle disconnect

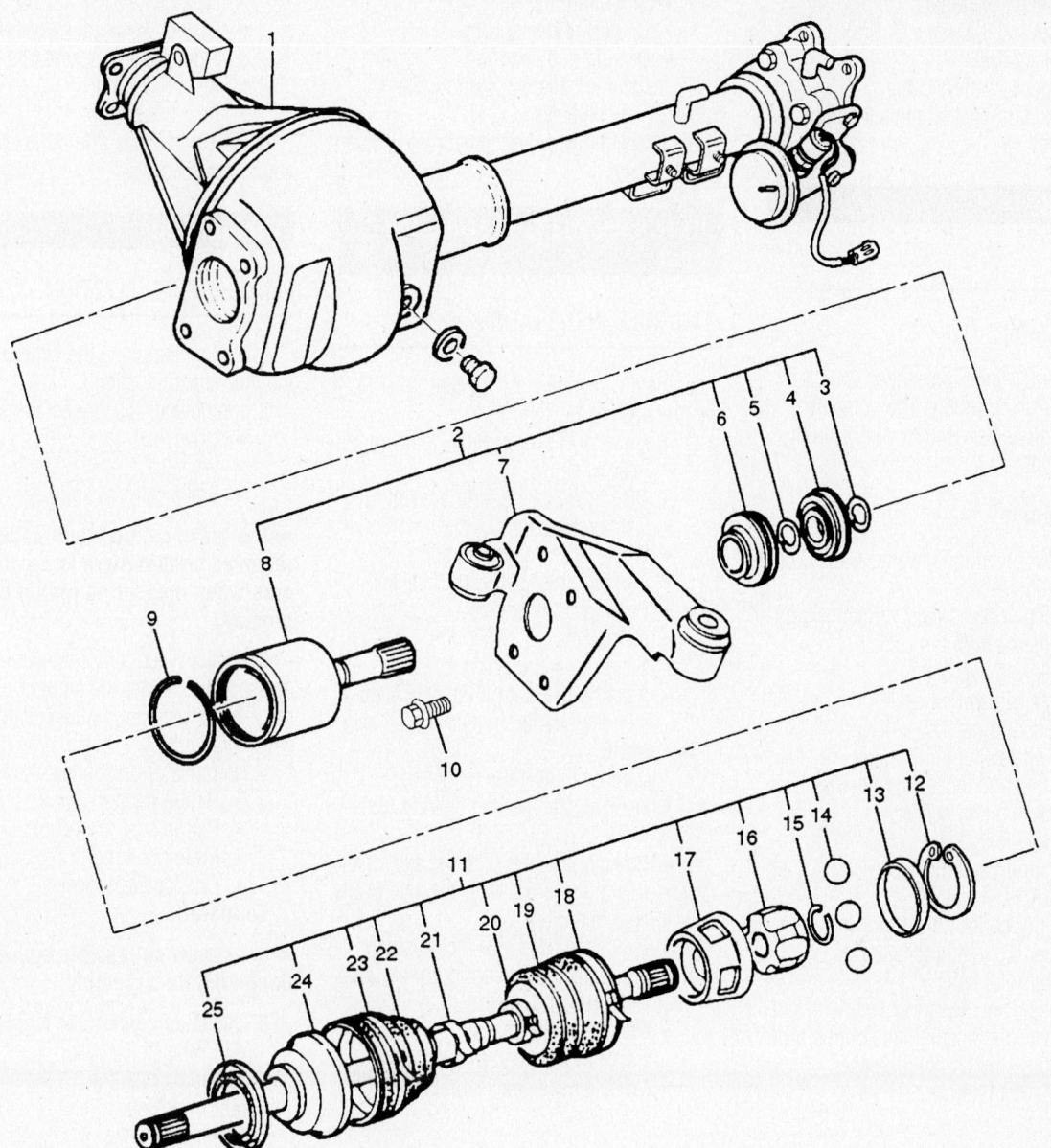

1 Axle Case and Differential
2 DOJ Case Assembly
3 Snap Ring
4 Bearing
5 Snap Ring
6 Oil Seal
7 Bracket
8 DOJ Case
9 Circlip
10 Bolt
11 Drive Shaft Joint Assembly
12 Snap Ring

13 Spacer
14 Ball
15 Snap Ring
16 Ball Retainer
17 Ball Guide
18 Band
19 Bellows
20 Band
21 Band
22 Bellows
23 Band
24 BJ Shaft
25 Dust Seal

9308BG03

Exploded view of the right halfshaft and mounting bracket

- Radiator skid plate. Tighten the bolts to 58 ft. lbs. (78 Nm).
- Front wheel
- Negative battery cable

5. Check the wheel alignment and adjust as necessary.

CV-Joints

OVERHAUL

Outer CV-Joint

The outer CV-joint is serviced with the axle shaft as an assembly. The outer CV-joint boot can be serviced by removing the inner CV-joint.

Inner CV-Joint

1. Before servicing the vehicle, refer to the precautions section.
2. Remove or disconnect the following:
   - Halfshaft from the vehicle
   - Snapring and bearing
   - Snapring and oil seal
   - Mounting bracket
   - CV-joint boot
   - Circlip and inner joint housing
   - Snapring and spacer
   - Inner joint balls
   - Snapring and inner CV-joint

To install:
3. Install or connect the following:
   - Inner CV-joint and snapring
   - Inner joint balls
   - Spacer and snapring
   - Inner joint housing and circlip. Add 150 grams CV-joint grease.
   - CV-joint boot

- Mounting bracket
- Oil seal and snapring
- Bearing and snapring

4. Install the halfshaft and mounting bracket to the vehicle.
5. Check the wheel alignment and adjust as necessary.

Rear Axle Shaft, Bearing and Seal

REMOVAL & INSTALLATION

1. Before servicing the vehicle, refer to the precautions section.
2. Remove or disconnect the following:
   - Rear wheel
   - Disc brake caliper and bracket
   - Disc brake rotor
   - Wheel speed sensor bracket
   - Parking brake cable and bracket
   - Parking brake shoes
   - Axle shaft
   - Snapring and discard it
   - Bearing, press it off the axle shaft with the bearing holder and oil seal

To install:
3. Install or connect the following:
   - New oil seal into the bearing housing
   - Bearing housing onto the axle shaft
   - Bearing, press it onto the axle shaft
   - New snapring
   - Axle shaft. Use new lockwashers and tighten the bearing holder nuts to 54 ft. lbs. (74 Nm).
   - Parking brake shoes
   - Parking brake cable and bracket
   - Wheel speed sensor bracket

- Disc brake rotor
- Disc brake caliper and bracket. Tighten the bracket bolts to 76 ft. lbs. (103 Nm).
- Rear wheel

4. Check the rear axle oil level and adjust as necessary.

Pinion Seal

REMOVAL & INSTALLATION

1. Before servicing the vehicle, refer to the precautions section.
2. Remove or disconnect the following:
   - Driveshaft
   - Wheels
   - Brake calipers and pads

➡ **The brake calipers and pads must be removed so that there is no additional drag when measuring pinion bearing preload.**

3. Use an inch lb. torque wrench and measure and record the amount of torque required to maintain pinion rotation through several revolutions.
4. Remove or disconnect the following:
   - Pinion flange
   - Pinion seal
   - Pinion bearing
   - Collapsible spacer

To install:

➡ **Use a new collapsible spacer and flange nut for assembly.**

5. Install or connect the following:
   - Collapsible spacer
   - Pinion bearing
   - Pinion seal
   - Pinion flange

6. Rotate the pinion flange occasionally while tightening the flange nut to make sure the pinion bearings seat correctly.
7. Take frequent bearing preload torque readings. Tighten the flange nut to achieve the preload torque readings originally recorded.

✳✳ CAUTION

Never loosen the pinion nut to reduce bearing preload. If it is necessary to reduce bearing preload, install a new collapsible spacer and pinion nut.

8. Install or connect the following:
   - Driveshaft
   - Brake calipers and pads
   - Wheels

9. Fill the differential with gear lubricant and check for leaks.

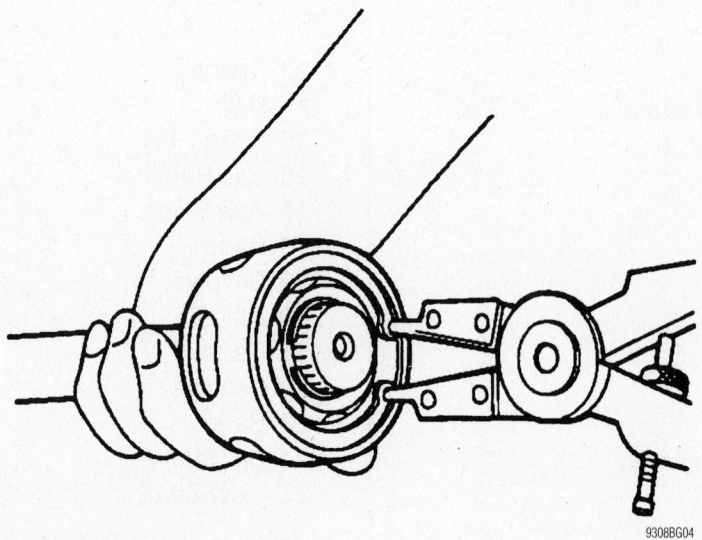

9308BG04

CV-joint spacer snapring—Inner CV-Joint

STEERING AND SUSPENSION

Air Bag

✳✳ CAUTION

Some vehicles are equipped with an air bag system. The system must be disarmed before performing service on, or around, system components, the steering column, instrument panel components, wiring and sensors. Failure to follow the safety precautions and the disarming procedure could result in accidental air bag deployment, possible injury and unnecessary system repairs.

PRECAUTIONS

Several precautions must be observed when handling the inflator module to avoid accidental deployment and possible personal injury.

• Never carry the inflator module by the wires or connector on the underside of the module.

• When carrying a live inflator module, hold securely with both hands, and ensure that the bag and trim cover are pointed away from you.

• Place the inflator module on a bench or other surface with the bag and trim cover facing up.

• With the inflator module on the bench, never place anything on or close to the module which may be thrown in the event of an accidental deployment.

DISARMING

1. Before servicing the vehicle, refer to the precautions section.
2. Turn the ignition switch to the **LOCK** position. Remove the key.
3. Disconnect the negative battery cable. Wait 1 minute before working around the air bags.
4. Disconnect the yellow 2-pin connector at the base of the steering column.
5. Disconnect the yellow 2-pin connector behind the glove box assembly.
6. When repairs are completed, connect the yellow 2-pin connectors.
7. Connect the negative battery cable.
8. Turn the ignition to the **ON** position, but don't start the engine. The AIR BAG warning light should turn on and flash on and off 7 times, and then turn off. This light sequence indicates that the SRS sys-

tem is functioning normally. If the AIR BAG light doesn't come on, or stays on longer than 7 seconds, the system must be diagnosed.

Rack and Pinion Steering Gear

REMOVAL & INSTALLATION

2-Wheel Drive

1. Before servicing the vehicle, refer to the precautions section.
2. Disable the air bag system.
3. Remove or disconnect the following:
   • Stone guard
   • Tie rod from knuckle
   • Power steering pressure and return lines
   • Steering gear

To install:
4. Install or connect the following:
   • Steering gear. Torque the mounting bolts to 85 ft. lbs. (116 Nm)
   • Fluid lines. Torque the fitting to 18 ft. lbs. (25 Nm)
   • Tie rod ends. 87 ft. lbs. (118 Nm)
   • Stone guard
5. Fill and bleed the system.

4-Wheel Drive

1. Before servicing the vehicle, refer to the precautions section.
2. Disable the air bag system.
3. Remove or disconnect the following:
   • Stone guard
   • Tie rod from knuckle
   • Power steering pressure and return lines
   • Torsion bar
   • Lower control arm frame side bolt
   • Crossmember bolts
   • Steering gear with crossmember
   • Steering gear

To install:
4. Install or connect the following:
   • Steering gear. Torque the mounting bolts to 85 ft. lbs. (116 Nm)
   • Crossmember Torque the bolts to 128 ft. lbs. (173 Nm)
   • Lower control arm bolt
   • Torsion bar
   • Fluid lines. Torque the fitting to 18 ft. lbs. (25 Nm)
   • Tie rod ends. 87 ft. lbs. (118 Nm)
   • Stone guard
5. Fill and bleed the system.

Shock Absorber

REMOVAL & INSTALLATION

Front

1. Before servicing the vehicle, refer to the precautions section.
2. Support the lower control arm with a jackstand.
3. Remove or disconnect the following:
   • Front wheels
   • Upper shock retaining nut and rubber bushing
   • ISC actuator and bracket
   • Suspension bump stops
   • Shock absorber

To install:
4. Install or connect the following:
   • Shock absorber. Tighten the lower bolt to 69 ft. lbs. (93 Nm).
   • ISC actuator and bracket
   • Bump stop. Tighten the bolts to 30 ft. lbs. (41 Nm).
   • Upper shock retaining nut and rubber bushing. Tighten the nut to 14 ft. lbs. (20 Nm).
   • Front wheels

Rear

1. Before servicing the vehicle, refer to the precautions section.
2. Support the rear axle with jackstands.
3. Remove the rear shock absorbers.

To install:
4. Install the rear shock absorbers. Tighten the upper bolt to 14 ft. lbs. (20 Nm). Tighten the lower bolt to 58 ft. lbs. (78 Nm).

➡**With ISC, do not use any grease on or near the bushings.**

5. Remove the jackstands.

Coil Spring

REMOVAL & INSTALLATION

1. Before servicing the vehicle, refer to the precautions section.
2. Support the vehicle under the frame.
3. Support the rear axle with a jack.
4. Remove or disconnect the following:
   • Rear wheels
   • Breather hose
   • Upper link fixing bolt, left side only
   • Stabilizer bar links
   • Shock absorbers

5. Lower the rear axle with the jack to release the coil spring tension. Remove the coil springs and insulators.

To install:

6. Place the coil springs on the axle assembly and the insulators on top of the springs.

7. Raise the axle assembly into position.

8. Install or connect the following:
- Shock absorbers. Torque to 58 ft. lbs. (78 Nm)
- Stabilizer bar links. Tighten the nuts to 23 ft. lbs. (31 Nm).
- Upper link. Torque to 101 ft. lbs. (137 Nm)
- Rear wheels

Torsion Bar

REMOVAL & INSTALLATION

1. Before servicing the vehicle, refer to the precautions section.

2. Matchmark the adjusting bolt and end piece, then remove the bolt, end piece, and seat.

3. Matchmark the height control arm to the torsion bar, then remove the height control arm.

4. Matchmark the torsion bar to the lower control arm, then remove the torsion bar.

To install:

5. Apply grease to the torsion bar splines.

6. Apply grease to the contact points of the height control arm, adjusting bolt end piece and seat.

7. Align the matchmarks and install the torsion bar.

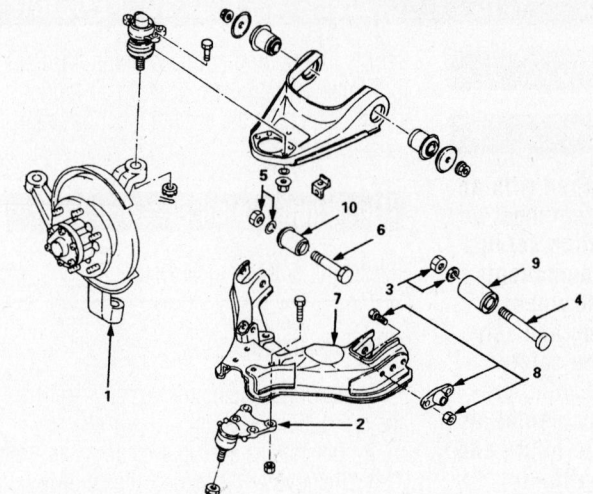

1. Knuckle
2. Lower end
3. Nut and washer, rear
4. Bolt, rear
5. Nut and washer, front
6. Bolt, front
7. Lower control arm assembly
8. Torsion bar arm bracket
9. Bushing, rear
10. Bushing, front

7924BG34

Exploded view of the control arm and ball joint components

8. Align the matchmarks and install the height control arm.

9. Install the adjusting bolt, seat and end piece.

10. Tighten the adjusting bolt to align the matchmarks.

Upper Ball Joint

REMOVAL & INSTALLATION

1. Before servicing the vehicle, refer to the precautions section.

2. Support the lower control arm with a floor jack.

3. Remove or disconnect the following:
- Front wheel
- Wheel speed sensor
- Upper ball joint

To install:

➡ **Use new nuts, bolts and split pins for assembly.**

4. Install or connect the following:
- Upper ball joint. Tighten the mounting bolts to 42 ft. lbs. (57 Nm) and the nut to 72 ft. lbs. (96 Nm).
- Wheel speed sensor
- Front wheel

Lower Ball Joint

REMOVAL & INSTALLATION

1. Before servicing the vehicle, refer to the precautions section.

2. Support the lower control arm with a jackstand.

3. Remove or disconnect the following:
- Front wheel
- Disc brake caliper and support
- Brake rotor and backing plate
- Wheel speed sensor
- Outer tie rod end
- Upper ball joint
- Steering knuckle
- Lower ball joint

To install:

➡ **Use new nuts, bolts and split pins for assembly.**

4. Install or connect the following:
- Lower ball joint. Tighten the mounting bolts to 85 ft. lbs. (116 Nm).
- Steering knuckle. Tighten the lower ball joint nut to 108 ft. lbs. (147 Nm).

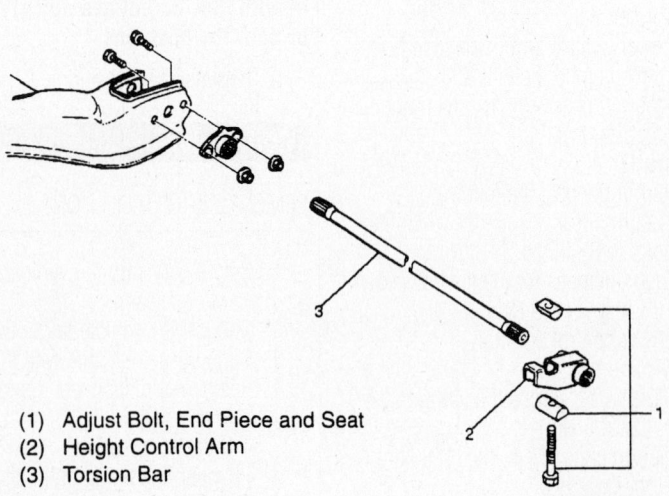

(1) Adjust Bolt, End Piece and Seat
(2) Height Control Arm
(3) Torsion Bar

9308BG05

Exploded view of the torsion bar assembly

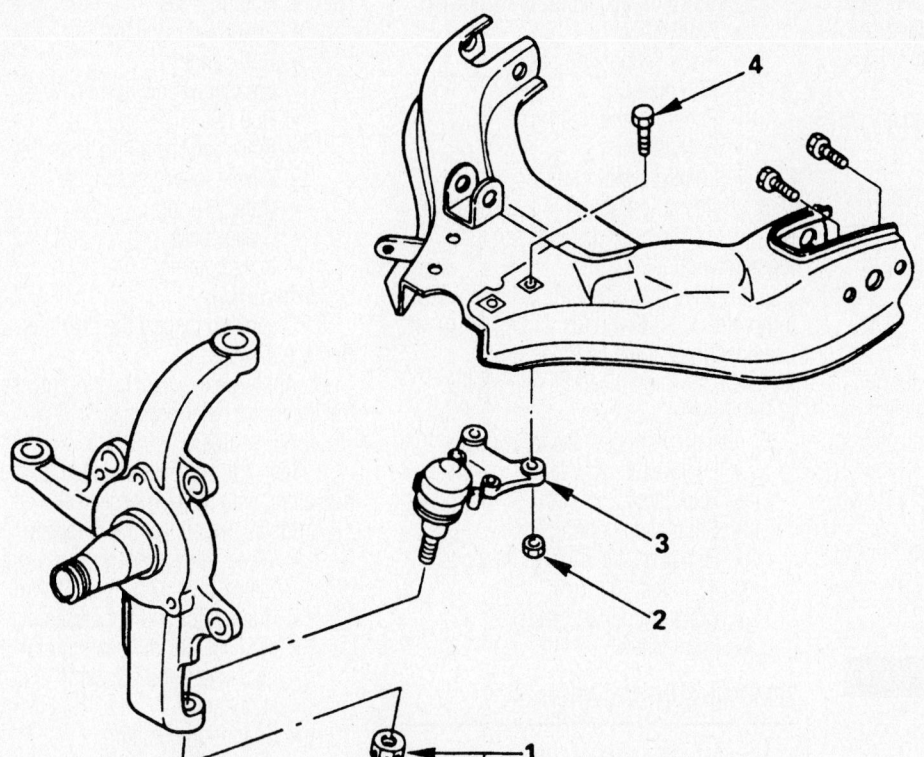

1. Nut and cotter pin
2. Nut
3. Lower ball joint
4. Bolt

7924BG35

Exploded view of the lower ball joint mounting and related components

- Upper ball joint. Tighten the nut to 72 ft. lbs. (96 Nm).
- Outer tie rod end. Tighten the nut to 72 ft. lbs. (98 Nm).
- Wheel speed sensor
- Brake rotor and backing plate
- Disc brake caliper and support. Tighten the support bolts to 115 ft. lbs. (155 Nm).
- Front wheel

5. Check the wheel alignment and adjust as necessary.

Upper Control Arm

REMOVAL & INSTALLATION

1. Before servicing the vehicle, refer to the precautions section.
2. Support the lower control arm with a jackstand.
3. Remove or disconnect the following:
   - Front wheel
   - Wheel speed sensor
   - Brake caliper
   - Upper ball joint
   - Upper control arm

➡**Note the alignment shim location for assembly.**

To install:
4. Install or connect the following:
   - Upper control arm
   - Alignment shims in their original locations. Tighten the bolts to 112 ft. lbs. (152 Nm).
   - Upper ball joint. Tighten the nut to 72 ft. lbs. (98 Nm).
   - Brake caliper
   - Wheel speed sensor
   - Front wheel

5. Check the wheel alignment and adjust as necessary.

CONTROL ARM BUSHING REPLACEMENT

1. Before servicing the vehicle, refer to the precautions section.
2. Remove the upper control arm.
3. Remove the nuts and washers from the fulcrum pin.
4. Press the bushings out of the control arm.

To install:
5. Press the bushings into the control arm.
6. Install the fulcrum pin washers and nuts.
7. Install the upper control arm.
8. Raise the suspension so that there

is 0.79 inches (20mm) between the bump stop and the lower control arm. Tighten the fulcrum pin nuts to 80 ft. lbs. (108 Nm).

9. Check the wheel alignment and adjust as necessary.

Lower Control Arm

REMOVAL & INSTALLATION

1. Before servicing the vehicle, refer to the precautions section.
2. Remove or disconnect the following:
   - Front wheel
   - Torsion bar
   - Lower ball joint
   - Stabilizer bar link
   - Shock absorber
   - Lower control arm

To install:
3. Install or connect the following:
   - Lower control arm
   - Shock absorber
   - Stabilizer bar link
   - Lower ball joint
   - Torsion bar
   - Front wheel
4. Raise the suspension so that there is

0.79 inches (20mm) between the bump stop and the lower control arm. Tighten the rear nut to 174 ft. lbs. (235 Nm); the rear nut to 137 ft. lbs. (186 Nm).

5. Check the wheel alignment and adjust as necessary.

CONTROL ARM BUSHING REPLACEMENT

1. Before servicing the vehicle, refer to the precautions section.
2. Remove or disconnect the following:
   - Lower control arm
   - Bushings, press them from the control arm, using Remover/Installer J-36833 for the front bushing and Remover/Installer J-36834 for the rear bushing.

To install:

3. Install or connect the following:
   - New bushings
   - Lower control arm

Wheel Bearings

ADJUSTMENT

2- and 4-Wheel Drive

1. Before servicing the vehicle, refer to the precautions section.

2. Remove or disconnect the following:
   - Front wheel
   - Brake caliper and pads
   - Hub dust cap
   - Snapring and shim
   - Hub flange
   - Lockscrew and washer

3. Tighten the hub nut to 22 ft. lbs. (29 Nm) to seat the bearings and then fully loosen the nut.

4. Tighten the hub nut to achieve a bearing preload of 2.6–4.0 lbs. (1.2–1.8 Kg) for used bearings. If the bearings were replaced, set the preload to 4.4–5.5 lbs. (2.0–2.5 Kg).

5. Install or connect the following:
   - Lockwasher and screw
   - Hub flange
   - Shim and snapring
   - Hub dust cap. Tighten the bolts to 43 ft. lbs. (59 Nm).
   - Brake caliper and pads
   - Front wheel

REMOVAL & INSTALLATION

1. Before servicing the vehicle, refer to the precautions section.
2. Remove or disconnect the following:
   - Front wheel
   - Brake caliper and support

- Hub dust cap
- Snapring and shim
- Hub flange
- Lockscrew and washer
- Hub nut
- Brake rotor and hub assembly
- Wheel speed sensor ring
- Outer bearing
- Grease seal
- Inner bearing

To install:

3. Clean and inspect the bearings. Replace if necessary.

4. Apply clean wheel bearing grease to the inner and outer bearings.

5. Apply grease in the hub.

6. Install the wheel bearings into the hub along with a new grease seal.

7. Install or connect the following:
   - Wheel speed sensor ring. Tighten the bolts to 13 ft. lbs. (18 Nm).
   - Brake rotor and hub assembly
   - Hub nut. Set the bearing preload.
   - Lockscrew and washer
   - Hub flange
   - Snapring and shim
   - Hub dust cap
   - Brake caliper and support. Tighten the support bolts to 115 ft. lbs. (155 Nm).
   - Front wheel

BRAKES

Brake Caliper

REMOVAL & INSTALLATION

Front

1. Raise and safely support the vehicle. Remove the front wheels.
2. Remove some brake fluid from the master cylinder reservoir.
3. Disconnect and plug the brake fluid line from the caliper.
4. Remove the brake caliper mounting bolt and guide bolt and remove the caliper from the mount. The brackets can be remove for additional work space.
5. Remove the brake pads and clips from the caliper. Inspect the brake pads for wear; replace them, if necessary.

To install:

6. Install the brake pads and clips onto the caliper.
7. If the caliper bracket was removed, tighten the bolts to 103–126 ft. lbs. (139–171 Nm).
8. Install the caliper on the mounting

bracket. Torque the caliper-to-mounting bracket bolts to:
   - 4-cylinder models: 20–27 ft. lbs. (27–37 Nm)
   - 6-cylinder models: 54 ft. lbs. (74 Nm)

9. Connect the fluid line to the caliper using new washers. Torque the brake line banjo fitting to 26 ft. lbs. (35 Nm).

✵✵ WARNING

Be sure the hook end of the flexible brake line is positioned in the anti-rotation cavity.

10. Refill the master cylinder reservoir and bleed the brake system.
11. Install the front wheels and lower the vehicle.

Rear

1. Raise and safely support the vehicle. Remove the rear wheels.
2. Remove some fluid from the master cylinder reservoir.

3. Disconnect and plug the brake fluid line from the caliper.

➡**Discard the parking brake cable mounting pin after removal.**

4. If equipped with caliper actuated parking brakes; remove the mounting pin from the parking brake cable and disconnect the parking cable from the disc caliper.

5. Remove the brake caliper mounting bolt and guide bolt and remove the caliper from the mount.

6. Remove the brake pads and clips from the caliper. Inspect the brake pads for wear; replace them, if necessary.

To install:

7. Install the brake pads and clips onto the caliper.

8. If the mounting bracket was removed, tighten the bolts to 69–84 ft. lbs. (93–114 Nm).

9. Install the caliper on the mounting bracket. Torque the caliper-to-mounting bracket bolts to 12–17 ft. lbs. (16–24 Nm), or 32 ft. lbs. (43 Nm) on vehicles with shoe-type parking brakes.

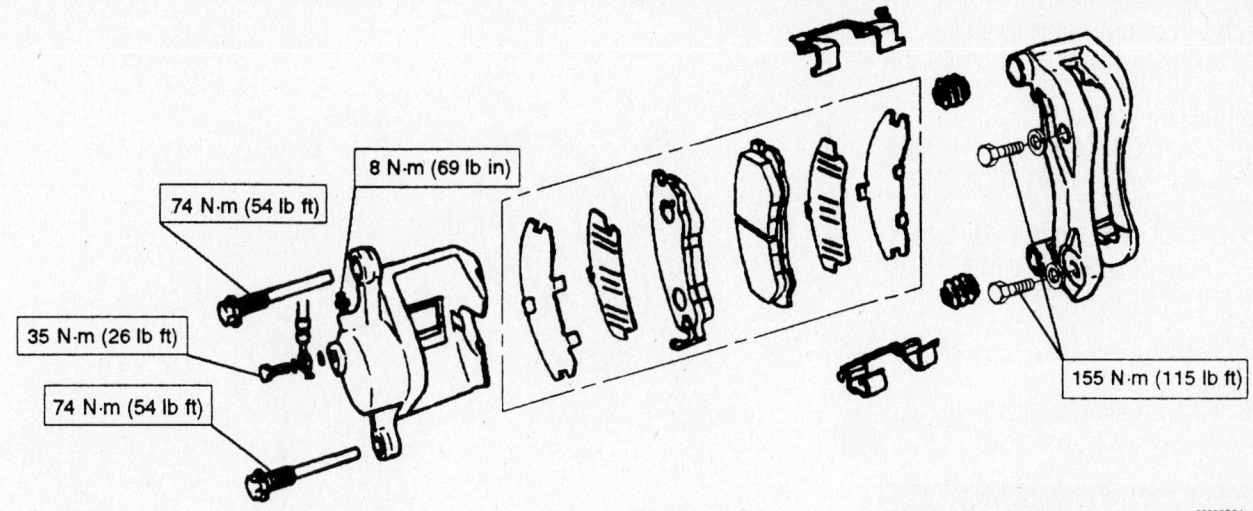

Exploded view of the front caliper components

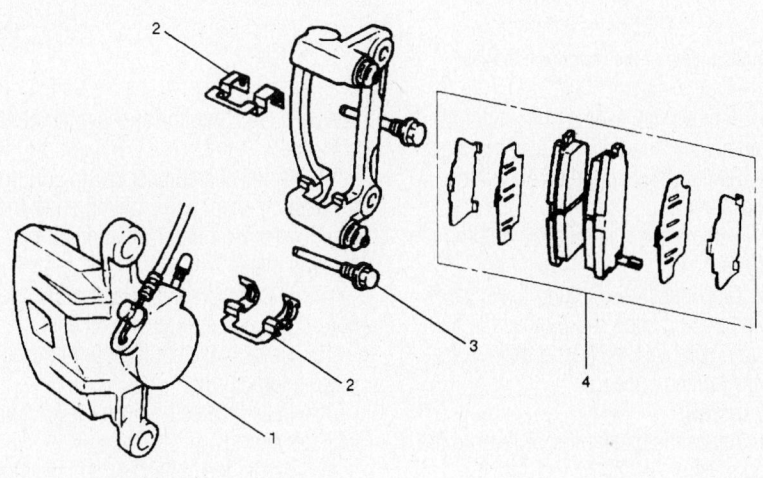

| (1) Caliper Assembly | (3) Lock Bolt |
| (2) Clip | (4) Pad Assembly |

93026G55

Exploded view of the rear caliper components

10. Connect the parking brake cable to the caliper and install a new mounting pin.

11. Connect the fluid line to the caliper using new washers. Torque the brake line banjo fitting to 26 ft. lbs. (35 Nm).

12. Refill the master cylinder reservoir and bleed the brake system.

13. Install the rear wheels and lower the vehicle.

Disc Brake Pads

REMOVAL & INSTALLATION

Front

Most disc brake pads are equipped with wear indicators. If a squealing noise occurs from the brakes while driving, check the pad wear indicator plate. If there is evidence of the indicator plate contacting the brake disc, the brake pad should be replaced.

1. Remove ½ of the volume of brake fluid from the master cylinder to prevent overflow when the caliper piston is compressed.

2. Raise and safely support the vehicle.

3. Remove the wheel and tire assemblies.

4. Remove the brake caliper without disconnecting the brake line. Support the caliper with a length of wire. Do not let the caliper hang from the brake hose.

➡On some disc brake systems it is not necessary to remove the caliper when installing new brake pads. Remove the lower slide bolt and rotate the caliper upward to remove the pads.

5. Remove the brake pads and shims. Inspect the brake rotor and machine or replace as necessary. Check the minimum thickness (specification is cast into the rotor) before machining.

To install:

6. Use a suitable tool to push the caliper piston into its bore.

7. Apply a thin coat of grease to the rear face of the brake pad and install the shim. Install the brake pads.

8. Install the calipers. Lubricate the caliper bolts and boots. If equipped with a 4-cylinder engine, tighten the caliper mounting bolts to 24 ft. lbs. (33 Nm). If equipped with a 6-cylinder engine, tighten the caliper mounting bolts to 54 ft. lbs. (74 Nm).

9. Install the wheel and tire assemblies and lower the vehicle.

10. Apply the brakes several times to seat the pads before moving the vehicle. Check the fluid in the master cylinder and add as necessary.

Rear

1. Raise and safely support the vehicle. Remove the rear wheels.

2. Remove the brake caliper mounting bolts and remove the caliper without disconnecting the brake fluid line. Support the caliper so it does not hang on the brake line.

3. Remove the brake pads and retaining clips from the caliper.

4. If equipped with caliper-activated parking brakes; use tool J-37617 or equivalent to rotate the piston clockwise until it

retracts into the bore. Align the notches of the piston face so the centerline of the notches is perpendicular to the centerline of the mounting bosses.

To install:

5. Install the new brake pads and clips in the caliper and install the caliper in the mounting bracket.

6. Tighten the caliper mounting bolts to 12–17 ft. lbs. (16–24 Nm), or 32 ft. lbs. (43 Nm) on vehicles with shoe-type parking brakes.

7. Install the rear wheels. Check the brake fluid level.

8. Pump the brake pedal until pressure is felt before moving the vehicle.

Brake Drums

REMOVAL & INSTALLATION

1. Raise and safely support the vehicle. Release the parking brake.

2. Remove the rear wheels.

3. Use chalk to mark the brake drum to one of the wheel studs as an index mark for reinstallation.

4. Remove the retaining screw that holds the brake drum to the axle flange.

5. Pull the brake drum from the axle flange.

To install:

6. Align the index mark and install the brake drum to the axle flange.

7. Install the retaining screw to secure the brake drum to the axle flange.

8. Install the rear wheels.

Rear Brake Shoes

REMOVAL & INSTALLATION

1. Raise and safely support the vehicle.
2. Remove the rear wheels.
3. Remove the brake drums.
4. Remove the brake return springs.

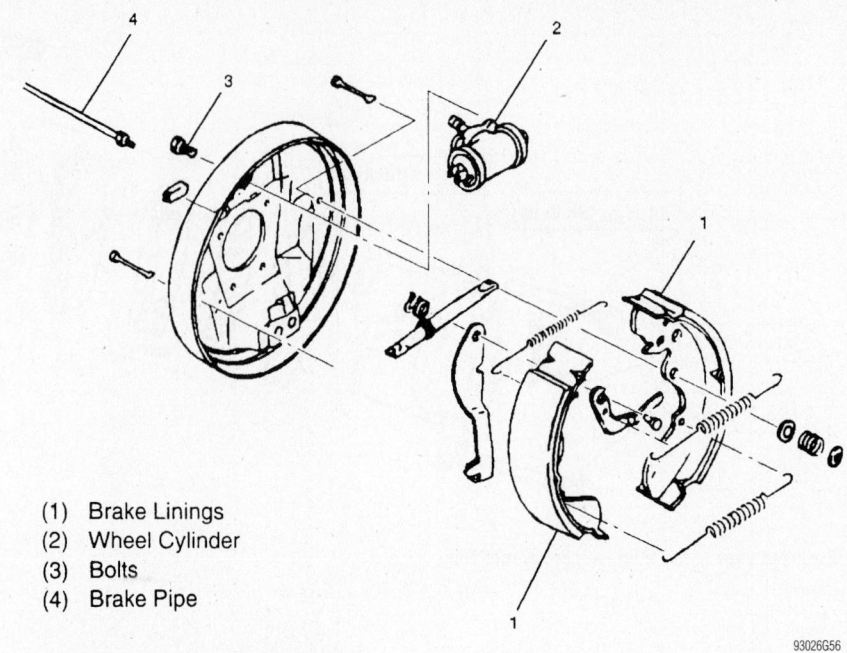

(1) Brake Linings
(2) Wheel Cylinder
(3) Bolts
(4) Brake Pipe

93026G56

Exploded view of the rear drum brakes

5. Remove the leading shoe holding pin and spring, and then the leading shoe.

6. Remove the self-adjuster and the adjuster lever.

7. Remove the trailing shoe holding pin and spring.

8. Disconnect the parking brake cable from the trailing shoe and remove the trailing shoe. Remove the parking brake lever from the trailing shoe.

To install:

9. Attach the parking brake lever to the trailing shoe.

10. Connect the parking brake cable to the parking brake lever.

11. Apply a thin coat of high temperature grease to the shoe contact points on the brake backing plate (locations A and C in the accompanying illustration), piston contact surface (B), and self-adjuster (D).

12. Position the trailing shoe on the backing plate and install the hold-down pin, spring, and retainer. Don't stretch the return

spring when fitting the shoes onto the backing plate.

13. Connect the upper return spring and the leading shoe to the trailing shoe and position the leading brake shoe on the backing plate.

14. Install the adjuster assembly and the hold-down pin, spring, and retainer.

15. Use a brake spring tool to install the lower return spring.

16. Install the self-adjuster lever and adjuster spring.

17. Adjust the shoe-to-drum clearance to 0.0098–0.0157 in. (0.25–0.40mm) and install the brake drum.

18. Check the brake drum for scoring or other wear. Machine or replace as necessary. Check the maximum brake drum diameter specification when machining.

19. Install the rear wheels. Lower the vehicle.

20. Road-test the vehicle.

ISUZU

I-280 • I-350

15

SPECIFICATIONS AND MAINTENANCE CHARTS

ENGINE AND VEHICLE IDENTIFICATION

Code ①	Liters (cc)	Cu. In.	Cyl.	Fuel Sys.	Engine Type	Eng. Mfg.	Model Year Code ②
LK5	2.8 (2786)	170	4	SFI	DOHC	GM	6
L52	3.5 (3474)	212	5	SFI	DOHC	GM	

GM: General Motors

SFI: Sequential Fuel Injection

① 8th position of VIN

② 10th position of VIN

09474_ISER_C0001

GENERAL ENGINE SPECIFICATIONS

All measurements are given in inches.

Year	Model	Engine Displacement Liters	Engine Series ID	Net Horsepower @ rpm	Net Torque @ rpm (ft. lbs.)	Bore x Stroke (in.)	Compression Ratio	Oil Pressure @ rpm
2006	i-280	2.8	LK5	175@5600	185@2800	3.66x4.02	10.0:1	12@1200
	i-350	3.5	L52	220@5600	225@2800	3.66x4.02	10.0:1	12@1200

09474_ISER_C0002

GASOLINE ENGINE TUNE-UP SPECIFICATIONS

Year	Engine Displacement Liters	Engine ID	Spark Plug Gap (in.)	Ignition Timing (deg.) MT	AT	Fuel Pump (psi)	Idle Speed (rpm) MT	AT	Valve Clearance In.	Ex.
2006	2.8	LK5	0.042	①	①	50-57	②	②	HYD	HYD
	3.5	L52	0.042	①	①	50-57	②	②	HYD	HYD

NOTE: The Vehicle Emission Control Information label often reflects specification changes made during production.

The label figures must be used if they differ from those in this chart.

HYD: Hydraulic

① Ignition timing is preset and cannot be adjusted

② Idle speed is maintained by the PCM

09474_ISER_C0003

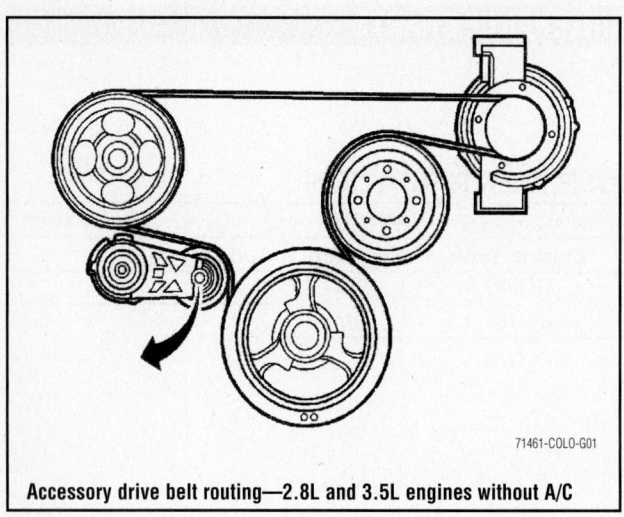

Accessory drive belt routing—2.8L and 3.5L engines without A/C

71461-COLO-G01

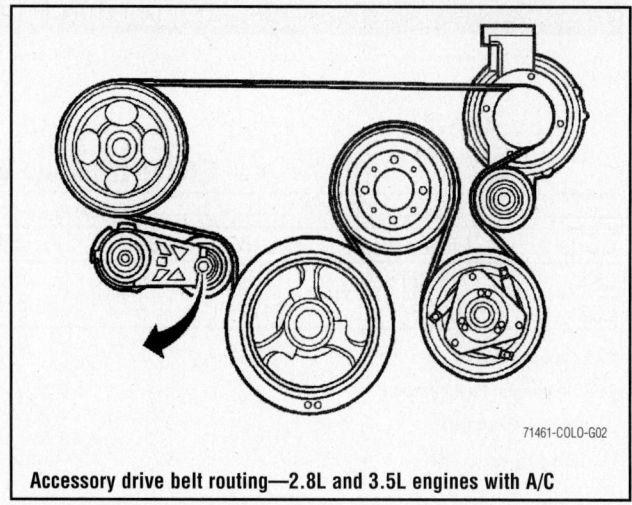

Accessory drive belt routing—2.8L and 3.5L engines with A/C

71461-COLO-G02

CAPACITIES

Year	Model	Engine Displacement Liters	Engine ID	Engine Oil with Filter (qts.)	Transmission (pts.)* 5-Spd	Transmission (pts.)* Auto.	Transfer Case (pts.)	Drive Axle Front (pts.)	Drive Axle Rear (pts.)	Fuel Tank (gal.)	Cooling System (qts.)
2006	i-280	2.8	LK5	5	①	10.0	2.7	3.2	3.4-3.8	19.6	10.4
	i-350	3.5	L52	6	①	10.0	2.7	3.2	3.4-3.8	19.6	10.6

NOTE: All capacities are approximate. Add fluid gradually and check to be sure a proper fluid level is obtained.

* Drain and refill

① RWD: 4.6
  4WD: 4.8

09474_ISER_C0004

VALVE SPECIFICATIONS

Year	Engine Displacement Liters	Engine ID	Seat Angle (deg.)	Face Angle (deg.)	Spring Test Pressure (lbs. @ in.)	Spring Installed Height (in.)	Stem-to-Guide Clearance (in.) Intake	Stem-to-Guide Clearance (in.) Exhaust	Stem Diameter (in.) Intake	Stem Diameter (in.) Exhaust
2006	2.8	LK5	NA	NA	130-142 @0.965	1.379	0.0011-0.0025	0.0015-0.0030	NA	NA
	3.5	L52	NA	NA	130-142 @0.965	1.701	0.0011-0.0025	0.0015-0.0030	NA	NA

NA: Information not Available

09474_ISER_C0005

CRANKSHAFT AND CONNECTING ROD SPECIFICATIONS

All measurements are given in inches.

Year	Engine Displacement Liters	Engine ID	Crankshaft				Connecting Rod		
			Main Brg. Journal Dia.	Main Brg. Oil Clearance	Shaft End-play	Thrust on No.	Journal Diameter	Oil Clearance	Side Clearance
2006	2.8	LK5	2.7567-2.7574	0.0004-0.0025	0.0044-0.0153	3	2.3749-2.3755	0.0008-0.0025	0.0019-0.0137
	3.5	L52	2.7567-2.7574	0.0004-0.0025	0.0044-0.0153	4	2.3749-2.3755	0.0008-0.0025	0.0019-0.0137

09474_ISER_C0006

PISTON AND RING SPECIFICATIONS

All measurements are given in inches.

Year	Engine Displ. Liters	Engine ID	Piston Clearance	Ring Gap			Ring Side Clearance		
				Top Compression	Bottom Compression	Oil Control	Top Compression	Bottom Compression	Oil Control
2006	2.8	LK5	0.0006-0.0014	0.00787-0.0157	0.0142-0.0201	0.0098-0.0299	0.0017-0.0037	0.0021-0.0037	0.0023-0.0085
	3.5	L52	0.0004-0.0017	0.0079-0.0157	0.0142-0.0201	0.0098-0.0299	0.0017-0.0037	0.0021-0.0037	0.0023-0.0085

09474_ISER_C0007

TORQUE SPECIFICATIONS

All readings in ft. lbs.

Year	Engine Displacement Liters	Engine ID	Cylinder Head Bolts	Main Bearing Bolts	Rod Bearing Bolts	Crankshaft Damper Bolts	Flywheel Bolts	Manifold		Spark Plugs	Oil Pan Drain Plug
								Intake	Exhaust		
2006	2.8	LK5	①	②	③	④	⑤	⑥	⑦	13	19
	3.5	L52	①	②	③	④	⑤	⑥	⑦	13	19

① 1st pass: 22 ft. lbs.
2nd pass: Plus 155 degrees
Short end bolt
1st pass: 62 inch lbs.
2nd pass: plus 60 degrees
Long end bolt
1st pass: 62 inch lbs.
2nd pass: plus 120 degrees

② 1st pass: 18 ft. lbs.
2nd pass: plus 180 degrees

③ 1st pass: 18 ft. lbs.
2nd pass: plus 110 degrees

④ 1st pass: 110 ft. lbs.
2nd pass: plus 180 degrees

⑤ 1st pass: 30 ft. lbs.
2nd pass: plus 45 degrees

⑥ 89 inch lbs.

⑦ 1st pass: 15 ft. lbs.
Repeat twice more in sequence

09474_ISER_C0008

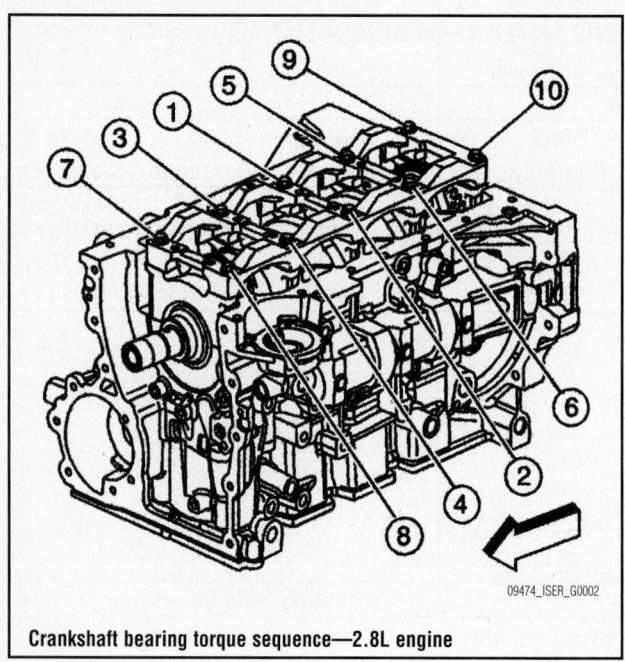

Crankshaft bearing torque sequence—2.8L engine

09474_ISER_G0002

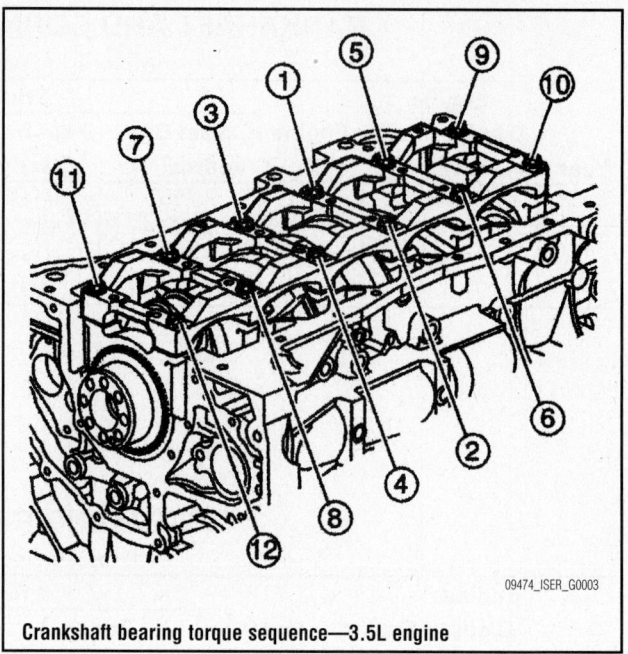

Crankshaft bearing torque sequence—3.5L engine

09474_ISER_G0003

WHEEL ALIGNMENT

Year	Model		Caster		Camber		Toe-in (Deg.)
			Range (+/-Deg.)	Preferred Setting (Deg.)	Range (+/-Deg.)	Preferred Setting (Deg.)	
2006	2WD	Left	1.0	+5.1	0.60	0	0+/-0.20
	wo/RPO Z71 & ZQ8	Right	1.0	+5.3	0.60	0	0+/-0.20
	2WD	Left	1.0	+5.5	0.60	0	0+/-0.20
	w/RPO ZQ8	Right	1.0	+5.5	0.60	0	0+/-0.20
	2WD	Left	1.0	+4.4	0.60	0	0+/-0.20
	w/RPO Z71	Right	1.0	+4.8	0.60	0	0+/-0.20
	4WD	Left	1.0	+4.6	0.60	0	0+/-0.20
		Right	1.0	+4.8	0.60	0	0+/-0.20

09474_ISER_C0009

TIRE, WHEEL AND BALL JOINT SPECIFICATIONS

Year	Model	OEM Tires		Tire Pressures (psi)		Wheel Size	Ball Joint Inspection	Lug Nut Torque (Ft. Lbs.)
		Standard	Optional	Front	Rear			
2006	i-280	P225/75R15	None	①	①	15x6.5	NA	103
	i-350	P265/75R15	None	①	①	15x7	NA	103

NA: Information not available

OEM: Original Equipment Manufacturer

OPT: Optional

PSI: Pounds Per Square Inch

STD: Standard

① Refer to placard on vehicle for proper inflation pressure

09474_ISER_C0010

BRAKE SPECIFICATIONS
All measurements in inches unless noted

Year	Model		Brake Disc			Brake Drum Diameter			Minimum Lining Thickness	Brake Caliper	
			Original Thickness	Minimum Thickness	Maximum Runout	Original Inside Diameter	Max. Wear Limit	Maximum Machine Diameter		Bracket Bolts (ft. lbs.)	Mounting Bolts (ft. lbs.)
2006	i-Series	F	1.060	1.000	0.002	—	—	—	0.070	129	29
		R	—	—	—	—	—	11.673	0.030	—	—

09474_ISER_C0011

SCHEDULED MAINTENANCE INTERVALS
Isuzu i-280 & i-350

When the CHANGE ENGINE OIL light appears, certain services and inspections are required.
Required services are described as Maintenance I and Maintenance II.
The first service on a vehicle should be Maintenance I, and the second service should be Maintenance II.
Alternate between the 2 thereafter. However, in some cases, Maintenance II may be required more often.
Maintenance I: Use Maintenance I if the CHANGE ENGINE OIL light comes on within 10 months since vehicle was purchased or, if Maintenance II was performed.
Maintenance II: Use Maintenance II if the previous service performed was Maintenance I.
Always use Maintenance II whenever the CHANGE ENGINE OIL light comes on 10 months or more since the last service, or, if the CHANGE ENGINEOIL light has not come on at all for one year.

Service	Maintenance I	Maintenance II
Change oil and oil filter, then Reset Oil Life Monitor ①	✓	✓
Visually inspect vehicle for any leaks or damage	✓	✓
Inspect engine air filter and replace as necessary		✓
Rotate tires, check for unusual wear and reset tire pressures ②	✓	✓
Inspect brake system	✓	✓
Check engine coolant and add fluid as needed	✓	✓
See additional required services	✓	✓
Inspect suspension and steering components		✓
Inspect engine cooling system		✓
Inspect wiper blades and windshield washer fluid level		✓
Inspect restraint system components		✓
Lubricate body components		✓
Check transmission and transfer case fluid level and add fluid as required		✓

① Mileage interval varies based on your driving habits.

② See Placard on Vehicle for Proper Inflation Pressure

Engine Oil Life System Reset Instructions

1). Begin with the engine off and the key in the "Lock" position.

2). Turn key to the "On" position

3). Press and release the stem located at the lower center of the instrument cluster until the "Oil Life" message is displayed.

4). Wait for the "Oil Life" and "Reset" message appear, then press and hold the stem down until several beeps are heard.

5). Turn the key to the"Lock" position.

09474_ISER_C0012

ADDITIONAL SERVICE REQUIREMENTS
Isuzu i-280 & i-350

TO BE SERVICED	TYPE OF SERVICE	VEHICLE MILEAGE INTERVAL (x1000)					
		25	50	75	100	125	150
Fuel system	I	✓	✓	✓	✓	✓	✓
Exhaust system	I	✓	✓	✓	✓	✓	✓
Air filter	R		✓		✓		✓
Automatic transmission fluid and filter (Severe Service)	S		✓		✓		✓
Automatic transmission fluid and filter (Normal Service)	S				✓		
Spark plugs	R				✓		
Cooling system	S						✓
Accessory drive belt	I						✓

09474_ISER_C0013

ENGINE REPAIR

Alternator

REMOVAL & INSTALLATION

1. Before servicing the vehicle, refer to the Precautions Section.
2. Remove or disconnect the following:

- Negative battery cable
- Accessory drive belt
- Left front wheel
- Left front inner fender liner
- A/C compressor electrical connector
- Lower A/C compressor mounting bolts

➡**The lower mounting bolts are removed to allow the engine lift bracket to be removed. Do not remove the upper A/C compressor mounting bolt.**

- Alternator wiring
- Alternator mounting bolts
- A/C compressor hose bracket from the engine lift bracket
- Engine lift bracket
- Alternator

To install:

3. Install the engine lift bracket. Tighten the bolts in sequence as follows:
   a. Step 1: 44 inch lbs. (5 Nm)
   b. Step 2: 37 ft. lbs. (50 Nm)
4. Install or connect the following:

- Alternator. Torque the bolts in sequence to 37 ft. lbs. (50 Nm).
- A/C compressor hose bracket to the engine lift bracket. Tighten the bolt to 80 inch lbs. (9 Nm).
- Alternator electrical connectors. Torque the battery feed wire nut to 15 ft. lbs. (20 Nm).
- A/C compress lower mounting bolts. Tighten the bolts to 37 ft. lbs. (50 Nm).

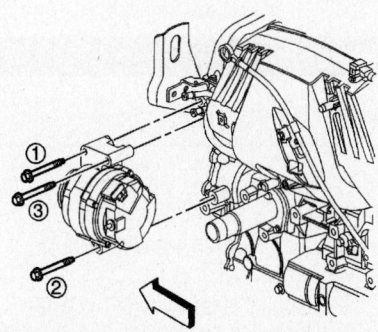

71461-COLO-G03

Alternator mounting bolt tightening sequence—2.8L and 3.5L engines

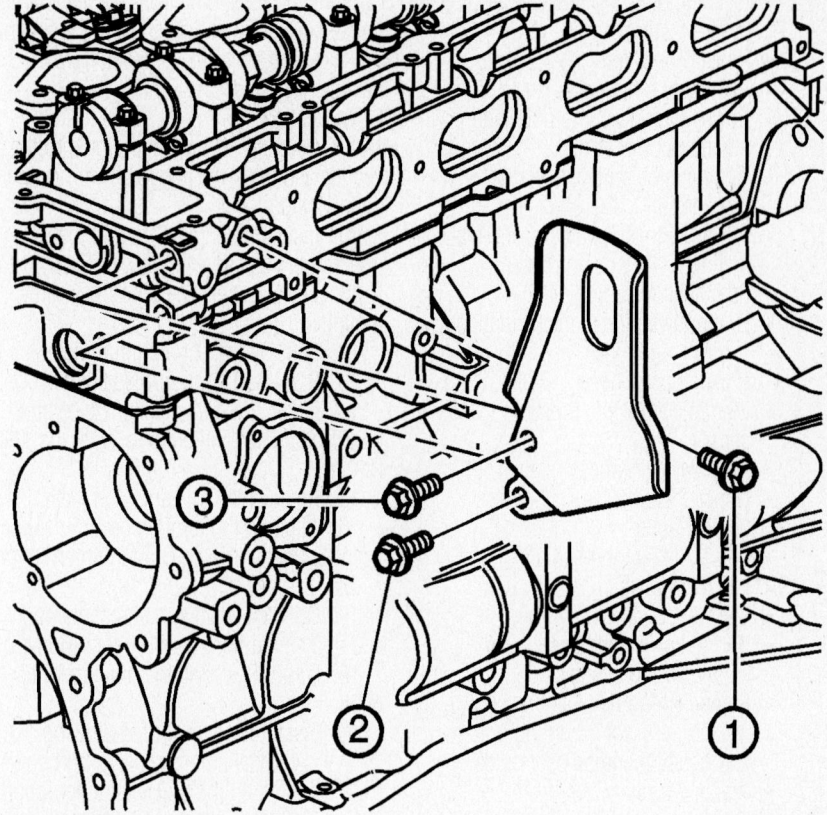

06025-COLO-G01

Engine lift bracket torque sequence

- Left front inner fender liner
- Left front wheel
- Accessory drive belt
- Negative battery cable

Ignition Timing

ADJUSTMENT

The ignition timing is preset and cannot be adjusted.

Engine Assembly

REMOVAL & INSTALLATION

1. Disconnect the battery cables and properly relieve the fuel system pressure.
2. Drain the engine cooling system.
3. Drain the engine oil.
4. Remove or disconnect the following:

- Hood
- Battery box
- Radiator hoses
- Cooling fan
- Air cleaner assembly

- Engine lifting bracket
- Alternator

➡**Reinstall the engine lift bracket after removing the alternator. Tighten the bolts to 44 inch lbs. (5 Nm) and then retighten to 37 ft. lbs. (50 Nm).**

- Washer reservoir bolts
- Engine wiring harness connector at PCM
- Wiring harness retainers at fender, power steering pump, throttle body and camshaft cover
- Oil pressure switch connector
- 4WD motor connector, if equipped
- Camshaft Position (CMP) sensor connector
- Exhaust camshaft actuator connector
- Coolant temperature sensor connector
- Injector harness connector
- Ignition coil connectors
- Oxygen Sensor (O_2S) connector
- Wiring harness conduit at camshaft cover

- Automatic transmission filler tube, if equipped
- Air injection pipe cover
- Install engine lifting eye in air pipe cover location
- Heater hoses from heater core
- A/C suction hose bracket
- Power steering pump bolts and position pump aside
- Right engine mount-to-frame bracket bolt
- Wiring harness retainer from intake manifold and position aside
- Fuel lines from fuel rail
- Evaporative pipe at intake manifold
- Dipstick and tube
- Brake booster hose
- Manifold Absolute Pressure (MAP) sensor
- MAP wiring harness retainer
- Upper 2 engine wiring harness bracket bolts
- Raise and support vehicle
- Left front wheel
- Left front fender inner liner
- Wiring harness retainers from engine wiring bracket
- Engine wiring harness bracket
- A/C compressor mounting bolts and position compressor aside
- Starter wiring
- Negative battery cable from block
- EVAP canister connector
- Knock sensor connectors
- Heater outlet hose
- Crankcase Position (CKP) sensor connector
- Engine wiring ground leads from block
- Engine wiring harness retainer at oil pan rail and position harness aside
- Catalytic converter
- Exhaust donut gasket
- Automatic transmission oil cooler and fuel line brackets, if equipped
- Left engine mount-to-frame bracket bolt
- On 2WD drive models, front cross-member
- On 4WD models, differential carrier
- On automatic transmission, torque converter bolts after marking torque converter-to-flexplate location
- Leave 2 upper transmission-to-engine bolts, but remove all other bolts

5. Lower the vehicle and place a transmission jack under the transmission

6. Remove the remaining transmission mounting bolts

7. Install a suitable lifting device to the engine.

8. Remove the engine mount bolts and carefully lift the engine from the vehicle. Pause several times while lifting the engine to make sure no wires or hoses have become snagged.

To install:

9. Carefully lower the engine into the vehicle and align the engine dowels with the transmission.

10. Install the engine mount bolts and tighten the bolts to 37 ft. lbs. (50 Nm).

11. Lower the engine onto the engine mounts.

12. Remove the engine lifting device.

13. Raise and support the vehicle.

14. Install the transmission-to-engine bolts and tighten the bolts to 37 ft. lbs. (50 Nm).

15. Remove the transmission jack.

16. Aligning the torque converter to the flexplate, install the bolts and tighten to 44 ft. lbs. (60 Nm).

17. On 2WD, install the crossmember and tighten the bolts to 44 ft. lbs. (60 Nm).

18. On 4WD, install the differential carrier.

19. Install or connect the following:

- Automatic transmission oil cooler and fuel line brackets, if equipped
- Left engine mount-to-frame bracket bolt and tighten to 63 ft. lbs. (85 Nm).
- New exhaust donut gasket
- Catalytic converter and tighten the bolts to 37 ft. lbs. (50 Nm).
- Engine wiring harness retainers at oil pan rail
- Engine wiring ground leads to block
- CKP sensor connector
- Heater outlet hose
- Knock sensor connectors
- EVAP canister connector
- Negative battery cable to block
- Starter wiring
- A/C compressor mounting bolts tighten the bolts to 37 ft. lbs. (50 Nm)
- Engine wiring harness bracket
- Wiring harness retainers from engine wiring bracket
- Left front fender inner liner
- Left front wheel
- Lower the vehicle
- Upper 2 engine wiring harness bracket bolts
- MAP wiring harness retainer
- MAP sensor
- Brake booster hose
- Dipstick and tube

- Wiring harness retainer to intake manifold
- Evaporative pipe at intake manifold
- Fuel lines to fuel rail
- Right engine mount-to-frame bracket bolt and tighten bolt to 63 ft. lbs. (85 Nm)
- Power steering pump
- A/C suction hose bracket
- Heater hoses from heater core
- Remove engine lifting eye in air pipe cover location
- Air injection pipe cover using new gasket
- Automatic transmission filler tube
- Wiring harness conduit at camshaft cover
- Wiring harness retainers at fender, power steering pump, throttle body and camshaft cover
- Coolant temperature sensor connector
- Injector harness connector
- Ignition coil connectors
- Oxygen Sensor (O$_2$S) connector
- Exhaust camshaft actuator connector
- CMP sensor connector
- 4WD motor connector, if equipped
- Oil pressure switch connector
- Engine wiring harness connector at PCM
- Washer reservoir bolts
- Alternator
- Engine lifting bracket
- Air cleaner
- Cooling fan
- Radiator hoses
- Battery box
- Hood

20. Check all powertrain fluid levels and add, as necessary. Be sure to properly fill the engine crankcase with clean engine oil.

21. Connect the battery cables and properly fill the engine cooling system.

22. Start and run the engine, then check for leaks.

Water Pump

REMOVAL & INSTALLATION

1. Before servicing the vehicle, refer to the Precautions Section.

2. Disconnect the negative battery cable.

3. Drain the engine cooling system.

4. Relieve the belt tension and remove the accessory drive belt.

5. Using Special Tool J-46406, secure

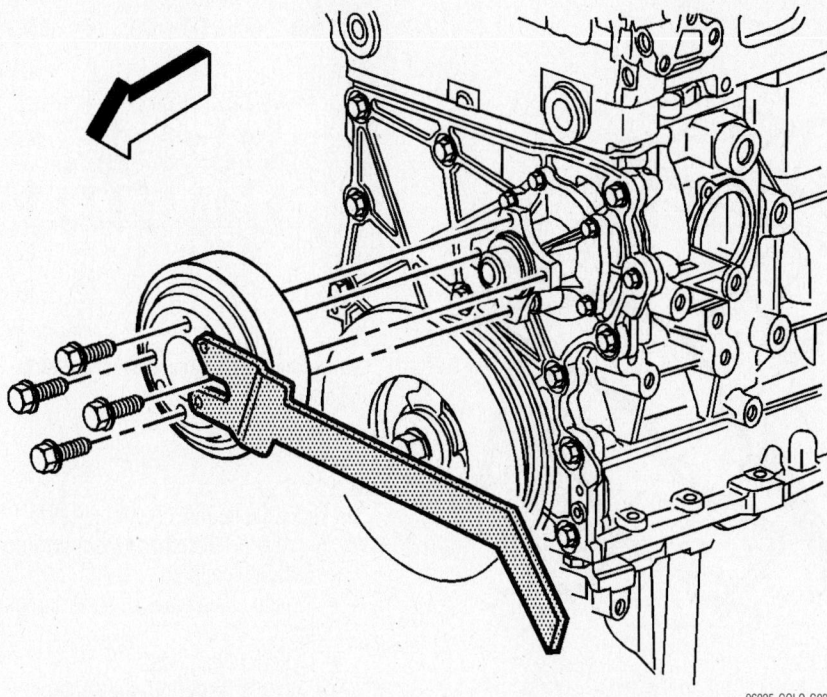

06025-COLO-G02

Use Special Tool J-46406 to secure the pump pulley—2.8L and 3.5L engines.

the water pump pulley and remove the water pump pulley bolts.

6. Remove or disconnect the following:
  • Water pump pulley
  • Coolant hose(s) from the water pump
  • Water pump bolts
  • Water pump

To install:

7. Clean the gasket mounting surfaces.

8. Install or connect the following:
  • Water pump using a new gasket. Tighten the water pump bolts to 89 inch lbs. (10 Nm).
  • Coolant hose(s)
  • Water pump pulley and tighten the bolts to 18 ft. lbs. (25 Nm).
  • Drive belt and adjust the tension
  • Negative battery cable

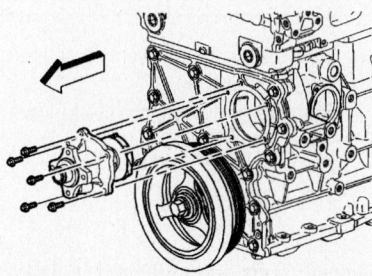

71461-COLO-G04

Exploded view of the water pump mounting

9. Refill the engine cooling system.
10. Run the engine and check for leaks.

Heater Core

REMOVAL & INSTALLATION

1. Before servicing the vehicle, refer to the Precautions Section.
2. Disable the air bag system.
3. Disconnect the negative battery cable.
4. Drain the engine cooling system.
5. Discharge and recovery the A/C refrigerant using approved recycling equipment.
6. Remove or disconnect the following:
  • Glove box door
  • A pillar trim panels
  • Door sill plates
  • Hinge pillar trim panels
  • Lower center instrument panel extension
  • Accessory trim plate panel
  • Radio
  • A/C-heater control panel
  • Center air outlets
  • Left and right air outlets
  • Knee bolster trim panel and bolster
  • Instrument luster bezel
  • Instrument cluster
  • Headlight switch
  • Daytime running light sensor

  • Upper 3 instrument panel nuts
  • Upper HVAC module screws
  • Instrument panel screws at instrument cluster and glove box openings
  • Hazard warning light connector
  • Six passenger air bag fasteners and passenger air bag module
  • Two screws in passenger air bag opening
  • Open the left side compartment and remove the screw
  • Screw at the back of the center storage compartment
  • Grasp the lower edge of the center storage compartment and the right lower edge of the instrument panel and pull out to disengage the clips
  • Partially pull out the instrument panel from the carrier
  • Release all the wiring harness clips from the instrument panel

7. With the aid of an assistant, carefully remove the instrument panel.
8. Disconnect the heater hoses from the HVAC module.
9. Disconnect the A/C lines from the thermal expansion valve.
10. Disconnect all HVAC module electrical connectors.
11. Remove the HVAC module retaining screws from the firewall
12. Remove the HVAC module.
13. Remove the core pipe clamp screws and clamp.
14. Remove the heater core from the HVAC module.

To install:

15. Install or connect the following:
  • Heater core to the HVAC module
  • Heater core pipe clamp screws and clamp
  • HVAC module
  • HVAC module retaining screws at the firewall
  • HVAC module electrical connectors
  • A/C lines at the thermal expansion valve
  • Heater hoses to HVAC module
  • With the aid of an assistant, carefully install the instrument panel
  • Wiring harness clips to instrument panel
  • Push the lower edge of the center storage compartment and the right lower edge of the instrument panel in to engage the clips
  • Screw at the back of the center storage compartment
  • Screw inside the left side compartment

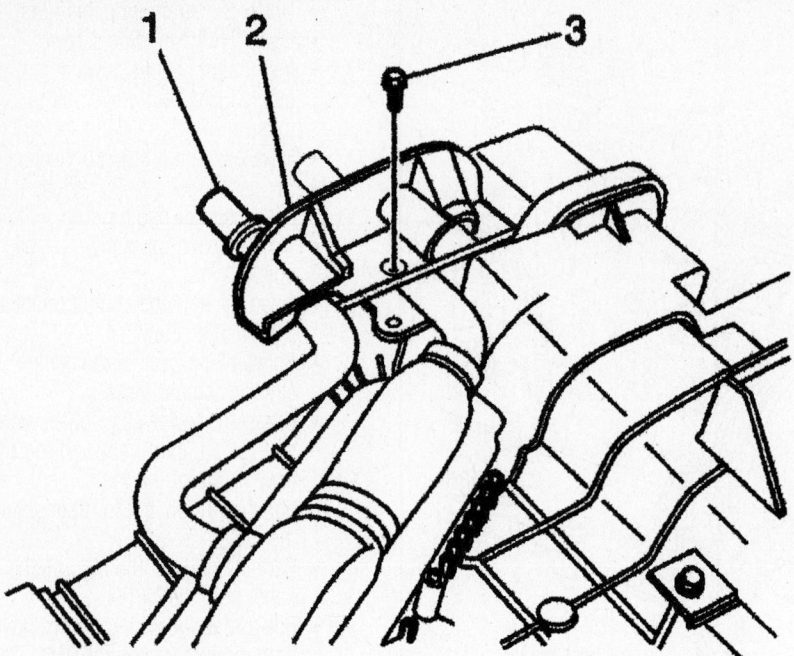

1. Heater core pipes
2. Clamp
3. Screw

09474_ISER_G0001

Heater core tubes

- Two screws in passenger air bag opening
- Hazard warning light connector
- Passenger air bag fasteners and passenger air bag module. Tighten the fasteners to 80 inch lbs. (9 Nm).
- Instrument panel screws at instrument cluster and glove box openings
- Upper HVAC module screws
- Upper 3 instrument panel nuts
- Daytime running light sensor
- Headlight switch
- Instrument cluster
- Instrument luster bezel
- Knee bolster trim panel and bolster
- Left and right air outlets
- Center air outlets
- A/C-heater control panel
- Radio
- Accessory trim plate panel
- Lower center instrument panel extension
- Hinge pillar trim panels
- Door sill plates
- A pillar trim panels
- Glove box door

16. Charge the A/C refrigerant using approved equipment.
17. Refill the cooling system.

a. Enable the air bag system.
18. Connect the negative battery cable.
19. Run the engine to normal operating temperatures; then, check the climate control operation and check for leaks.

Cylinder Head

REMOVAL & INSTALLATION

1. Before servicing the vehicle, refer to the Precautions Section.
2. Relieve the fuel system pressure.

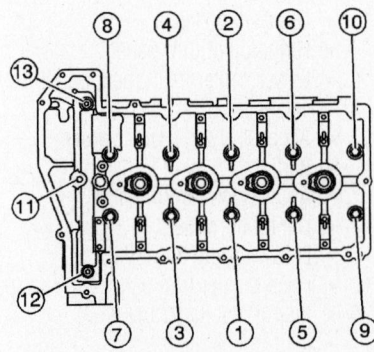

Cylinder head bolt torque sequence—2.8L engine

71461-COLO-G05

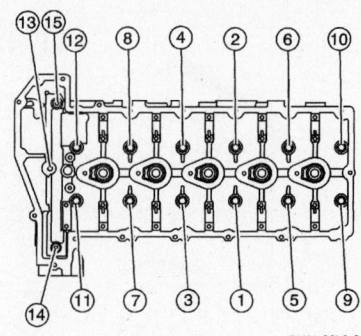

Cylinder head bolt torque sequence—3.5L engine

71461-COLO-G06

3. Disconnect the negative battery cable.
4. Drain the engine cooling system.
5. Remove or disconnect the following:
- Exhaust manifold
- Timing chain and camshaft sprockets
- Cylinder head bolts
- Cylinder head from the engine

To install:

6. Clean and inspect the gasket mounting surfaces.
7. Install the cylinder head using a new gasket. Tighten the new bolts as follows:
  a. Standard bolts in sequence to 22 ft. lbs. Plus an additional 155°.
  b. Tighten the 2 short end bolts to 62 inch lbs. (7 Nm), plus an additional 60°.
  c. Tighten the 2 long end bolts to 62 inch lbs. (7 Nm), plus an additional 120°.
8. Install or connect the following:
- Timing chain and camshaft sprockets
- Exhaust manifold
- Negative battery cable
9. Refill the engine cooling system.
10. Start the engine to check for leaks.

Rocker Arms

REMOVAL & INSTALLATION

1. Before servicing the vehicle, refer to the Precautions Section.
2. Remove or disconnect the following:

- Camshafts
- Rocker arms
- Lash adjusters

➡**Rocker arms and adjusters, being reused, must be installed in their original positions. Be sure to tag or arrange all rocker arms.**

To install:

3. Inspect the rocker arms and lash adjusters for damage or wear and replace as necessary.

4. Fill the lash adjusters with clean engine oil.

5. Install or connect the following:
- Lash adjusters in their original locations
- Rocker arms after lubricating them
- Camshafts

6. Start and run the engine to check for leaks.

Intake Manifold

REMOVAL & INSTALLATION

1. Before servicing the vehicle, refer to the Precautions Section.

2. Remove or disconnect the following:
- Negative battery cable
- Air intake assembly
- Throttle body
- Battery and battery box
- Dipstick and tube
- Brake booster hose
- Manifold Absolute Pressure (MAP) sensor connector and harness retainer
- PCV tube
- Alternator
- Engine wiring harness retainer
- Upper 2 engine wiring harness bracket bolts
- Raise and support vehicle
- Left front wheel
- Left front fender inner liner
- Wiring harness retainers for battery cable, engine and MAP sensor
- Engine wiring harness bracket from the intake manifold
- Intake manifold bolts
- Lower the vehicle
- Intake manifold and gasket

To install:

3. Install or connect the following:
- Intake manifold with new gasket

- Raise and support the vehicle
- Tighten the intake manifold bolts, working from the center outward, to 89 inch lbs. (10 Nm)
- Engine wiring harness bracket
- Wiring harness retainers for the battery cable, engine and MAP sensor
- Left front fender inner liner
- Left front wheel
- Lower the vehicle
- Upper 2 engine wiring harness bracket bolts
- Engine wiring harness retainer
- Alternator
- PCV tube
- MAP sensor connector and harness retainer
- Brake booster hose
- Dipstick and tube
- Throttle body
- Negative battery cable

4. Start the engine and check for leaks.

Exhaust Manifold

REMOVAL & INSTALLATION

1. Before servicing the vehicle, refer to the Precautions Section.

2. Remove or disconnect the following:
- Negative battery cable
- Catalytic converter from the exhaust manifold
- Exhaust seal
- Air intake assembly
- Transmission fill tube bracket nut, if equipped with automatic transmission
- Heated Oxygen Sensor (HO2S) from exhaust manifold
- Exhaust manifold heat shield
- Exhaust manifold bolts
- Exhaust manifold and gasket

To install:

3. Clean the exhaust manifold retainer threads and the gasket mating surfaces.

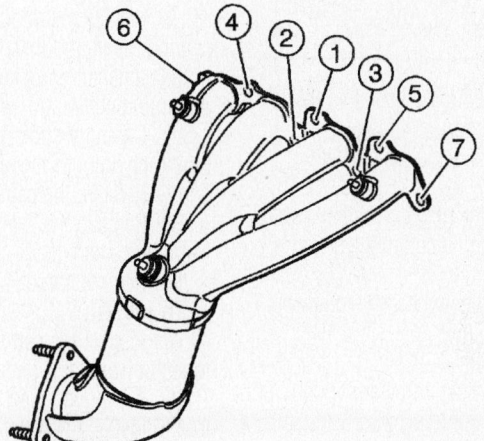

71461-COLO-G08

Exhaust manifold tightening sequence—2.8L engine

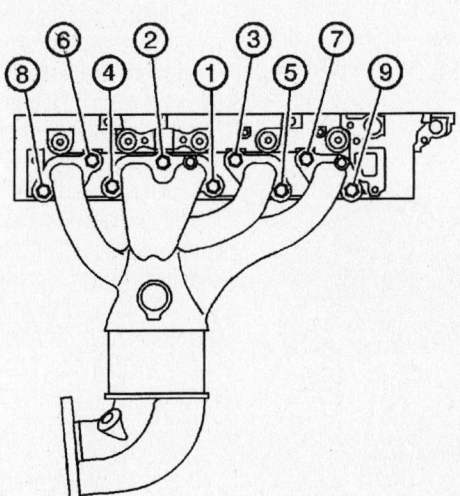

71461-COLO-G09

Exhaust manifold tightening sequence—3.5L engine

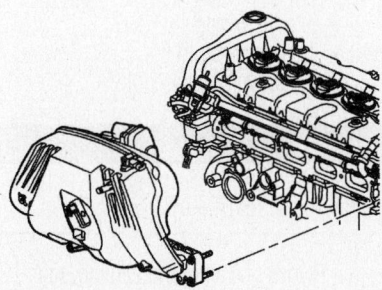

71461-COLO-G07

Intake manifold mounting

4. Coat the bolt threads with a suitable threadlock.

5. Install the exhaust manifold and tighten the bolts in sequence to 15 ft. lbs. (20 Nm), tighten the bolts again to 15 ft. lbs. (20 Nm), then tighten the bolts again to 15 ft. lbs. (20 Nm).

6. Install or connect the following:
- Heat shield and tighten the nuts to 89 inch lbs. (10 Nm)
- HO2S and tighten to 31 ft. lbs. (42 Nm)
- Transmission fill tube bracket nut, if equipped with automatic transmission.
- Air intake assembly
- New exhaust seal to the exhaust manifold flange
- Catalytic converter to the manifold. Tighten the nuts to 37 ft. lbs. (50 Nm).
- Negative battery cable

Camshafts

REMOVAL & INSTALLATION

1. Before servicing the vehicle, refer to the Precautions Section.
2. Properly relieve the fuel system pressure.
3. Disconnect the negative battery cable.
4. Drain the engine cooling system and the engine oil.
5. Remove or disconnect the following:
- Intake manifold
- Ignition coils
- Coolant temperature sensor connector
- Injector harness connector
- Ignition coil connectors
- Oxygen Sensor (O2S) connector
- Fuel pressure regulator screw
- Camshaft cover
- Both Camshaft Position (CMP) sensors

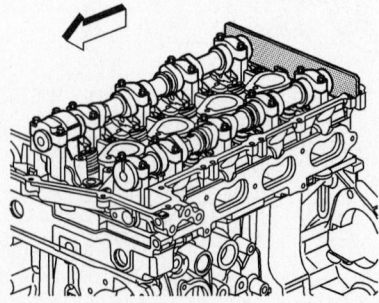

Install the camshaft locking tool J-44221 to the rear of the camshafts.

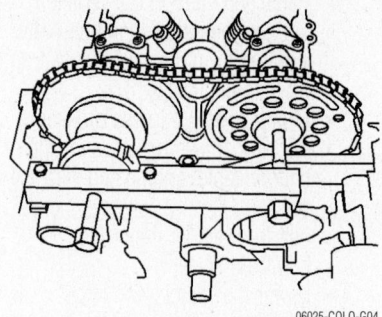

Use Special Tool J-44222 camshaft sprocket holding tool to prevent the timing chain and sprockets from turning.

6. Rotate the crankshaft clockwise until the no. 1 cylinder is at TDC on the compression stroke. The word DELPHI on the camshaft position actuator will be parallel with the cylinder head surface.
7. Install camshaft locking tool J-44221 to the rear of the camshafts.
8. Remove the camshaft sprocket bolts and discard them.
9. Install tension tool J-44222 on the cylinder head and install the holding bolts in the camshaft sprocket bolt holes to lock the timing chain and sprockets in position.
10. Carefully slide the sprockets and timing chain onto the tension tool.
11. Remove the camshaft bearing caps.
12. Remove the Camshaft Locking Tool from the camshafts.
13. Remove the camshafts.

To install:
14. Inspect the camshaft, journals and lobes for wear and replace, if necessary.
15. If removed, use the camshaft bearing tool to install a new set of bearings.
16. Coat the camshaft lobes, journals and thrust face with clean engine oil.
17. Install camshaft locking tool J-44221 to the rear of the camshafts.
18. Install the camshafts with the flats up and with cylinder no. 1 at TDC.
19. Install the bearing caps in their original position and tighten the bolts to 106 inch lbs. (12 Nm).
20. Carefully slide the sprockets and timing chain onto the camshafts, ensuring the alignment pins are engaged between the camshafts and sprockets.
21. Install new camshaft sprocket bolts and washers. Tighten the intake sprocket bolt to 15 ft. lbs. (20 Nm), plus it additional 100°. Tighten the exhaust sprocket bolt to 18 ft. lbs. (25 Nm), plus an additional 135°.
22. Remove the camshaft locking plate tool.
23. Install or connect the following:

- Both Camshaft Position (CMP) sensors
- Camshaft cover
- Fuel pressure regulator screw
- Oxygen Sensor (O2S) connector
- Ignition coil connectors
- Injector harness connector
- Coolant temperature sensor connector
- Ignition coils
- Intake manifold
- Negative battery cable

24. Refill the engine cooling system and engine oil.

Starter Motor

REMOVAL & INSTALLATION

1. Before servicing the vehicle, refer to the Precautions Section.
2. Remove or disconnect the following:
- Negative battery cable

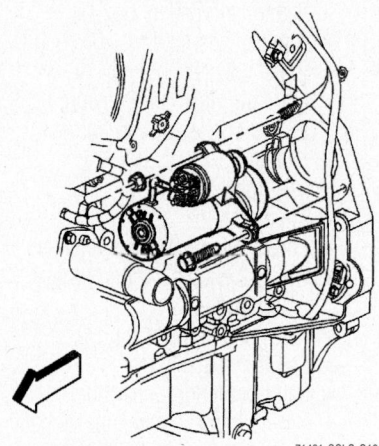

Starter motor mounting

- Intake manifold
- Starter wiring
- Starter

To install:
3. Install or connect the following:
- Starter and tighten the fasteners to 37 ft. lbs. (50 Nm)
- Starter wiring
- Intake manifold
- Negative battery cable

Oil Pan

REMOVAL & INSTALLATION

1. Before servicing the vehicle, refer to the Precautions Section.

2. Drain the engine oil.

3. Remove or disconnect the following:
- Oil dipstick tube
- Engine splash guard
- Right front halfshaft, if equipped with 3.5L Engine
- Power steering gear, if equipped with RWD

4. If equipped with 4WD:

a. Remove the front driveshaft.

b. Remove the differential carrier assembly bushing to frame bolts only.

c. Pull the differential carrier assembly downward.

d. Secure the pinion yoke to prevent the differential carrier from rotating.

5. Remove or disconnect the following:
- Service slot plug
- Fuel pipe bracket at transmission and position aside
- Four lower transmission mounting bolts attached to oil pan

6. If equipped with 4WD, remove the power steering gear mounting bolts only and pull gear down far enough to access oil pan.

7. Disconnect the engine wiring harness retainers from oil pan

8. Remove the oil pan mounting bolts.

9. Install 2 bolts in the threaded holes at the rear of the oil pan to act as jack screws. Tighten evenly to release the oil pan from the engine block.

10. Remove the oil pan and 2 bolts from the jack screw holes.

To install:

11. Apply a bead of sealant around the oil pan as shown.

➡**Install the oil pan within 10 minutes of applying the sealant.**

12. Install the oil pan, making sure that the pan if positioned fully rearward against the transmission mounting surface.

13. Install the oil pan bolts and tighten the side bolts to 18 ft. lbs. (25 Nm). Tighten the end bolts to 89 inch lbs. (10 Nm).

14. Connect the engine wiring harness retainers to oil pan.

15. If equipped with 4WD, position the steering gear upward to the frame assembly and install the steering gear mounting bolts.

16. Install the four lower transmission mounting bolts and tighten to 37 ft. lbs. (50 Nm).

17. Install the nuts securing the fuel pipe bracket at transmission and tighten to 15 ft. lbs. (20 Nm).

18. Install the service slot plug.

19. Right front halfshaft, if equipped with 3.5L Engine

20. If equipped with RWD, install the power steering gear.

21. If equipped with 4WD:

a. Position the differential carrier assembly to the frame.

b. Install the differential carrier assembly bushing to frame bolts. Tighten to 112 ft. lbs. (152 Nm).

c. Install the front driveshaft.

22. Install the engine splash shield.

23. Install the oil dipstick tube. Tighten the bolt to 89 inch lbs. (10 Nm).

24. Fill the engine with oil to the correct level.

25. Start the engine and check for leaks.

Oil Pump

REMOVAL & INSTALLATION

1. Before servicing the vehicle, refer to the Precautions Section.

2. Remove or disconnect the following:
- Engine front cover

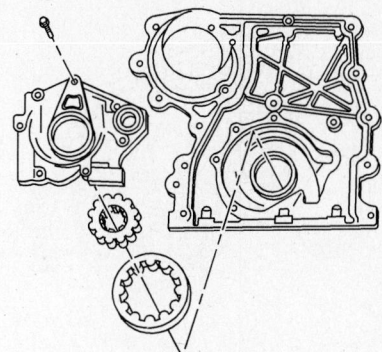

06025-COLO-G05

Exploded view of the oil pump and related components—2.8L and 3.5L engines

- Oil pump cover bolts
- Oil pump cover

3. Matchmark the inner and outer gears in relation to the oil pump housing.

4. Remove or disconnect the following:
- Inner and outer oil pump gears
- Oil pump pressure relief valve plug
- Oil pump pressure relief valve and spring

To install:

5. Install or connect the following:
- Oil pump pressure relief valve and spring
- Oil pump pressure relief valve plug and tighten to 124 inch lbs. (14 Nm)
- Oil pump outer and inner gears
- Oil pump cover and tighten the bolts to 89 inch lbs. (10 Nm).
- Engine front cover

6. Start the engine and check for leaks.

Rear Main Seal

REMOVAL & INSTALLATION

1. Before servicing the vehicle, refer to the Precautions Section.

2. Remove or disconnect the following:
- Negative battery cable
- Transmission assembly and transfer case, if equipped
- Clutch assembly, if equipped
- Flywheel
- Crankshaft seal by prying it from out oil seal housing

➡**Be careful not to damage the crankshaft seal surface with the prying tool.**

To install:

3. Install the new rear seal by lubricating it with engine oil and using a seal installer Special Tool J-44215. The spring side goes toward the engine and the seal will bottom out when installed fully.

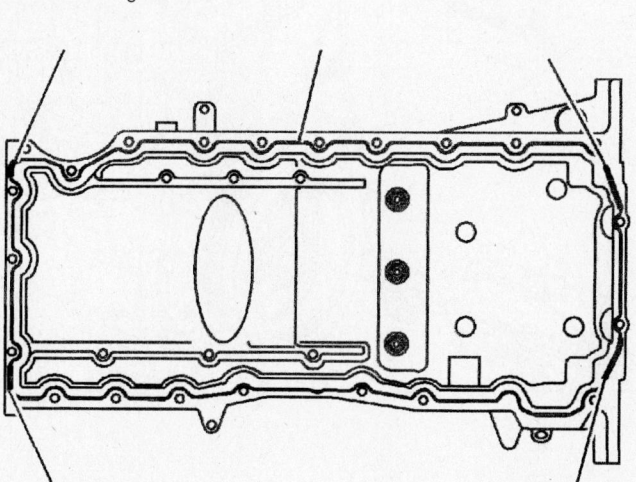

71461-COLO-G11

Oil pan sealant application areas

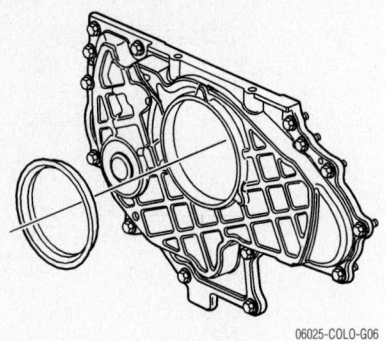

06025-COLO-G06

Remove the seal from the oil seal housing.

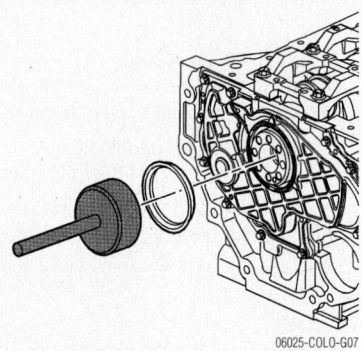

06025-COLO-G07

Use Special Tool J-44215 to install the rear oil seal–2.8L and 3.5L engines.

4. Install or connect the following:
- Flywheel/clutch assembly or flex plate
- Transmission assembly and transfer case, if equipped
- Negative battery cable
5. Start the engine and check for leaks.

Timing Chain, Sprockets, Front Cover and Seal

REMOVAL & INSTALLATION

1. Before servicing the vehicle, refer to the Precautions Section.
2. Drain the cooling system.
3. Drain the engine oil.
4. Remove or disconnect the following:
- Negative battery cable
- Number 1 cylinder spark plug
- Intake manifold
- Ignition coils
- Engine coolant temperature (ECT) sensor electrical connector from camshaft cover
- Fuel injector connector from camshaft cover
- Heated oxygen sensor (HO$_2$S) connector from camshaft cover

- Fuel pressure regulator screw
- Camshaft cover
- Camshaft position (CMP) sensor
- Water pump

5. Remove the service slot plug and install Flywheel Holding Tool EN-46547 into the flywheel teeth.
6. Remove the crankshaft balancer bolt and discard.
7. Install Crankshaft end protector J-41816-2 into the end of the crankshaft and remove the crankshaft balancer using a 3-jaw puller.
8. Remove or disconnect the following:
- Drive belt tensioner
- Power steering pump
- Oil pan
- Oil pump pipe and screen assembly
- 7mm center front cover bolt
- Remaining engine front cover bolts
9. Install 2 bolts into the threaded holes to act as jack screws and tighten evenly to release the front cover.
10. Rotate the crankshaft clockwise until the no. 1 cylinder is at TDC on the compression stroke. The word DELPHI on the camshaft position actuator will be parallel with the cylinder head surface.
11. Install camshaft locking tool J-44221 to the rear of the camshafts.
12. Release the tension on the timing chain by moving the tensioner shoe in by hand.
13. Place a tee in the tensioner to hold the shoe in place.
14. Remove the top timing chain guide.
15. Remove the exhaust camshaft position actuator bolt and actuator.

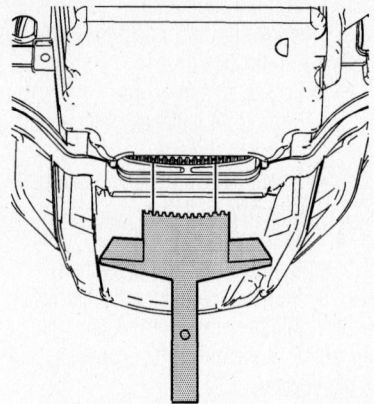

06025-COLO-G08

Use Special tool EN-46547 to prevent the flywheel from turning.

16. Remove the intake camshaft sprocket bolt and sprocket.
17. Remove the timing chain.
18. Remove the crankshaft sprocket.
To install:
19. Install the crankshaft sprocket.
20. Install the intake camshaft sprocket on the timing chain and align the dark link on the timing chain with the timing mark on the intake sprocket as shown.
21. Feed the timing chain through the opening in the cylinder head.
22. Install the timing chain on the crankshaft sprocket and align the dark link of the timing chain with the timing mark on the crankshaft sprocket.
23. Install a new intake camshaft sprocket bolt and washer and tighten the bolt to 15 ft. lbs. (20 Nm), plus an additional 100°.

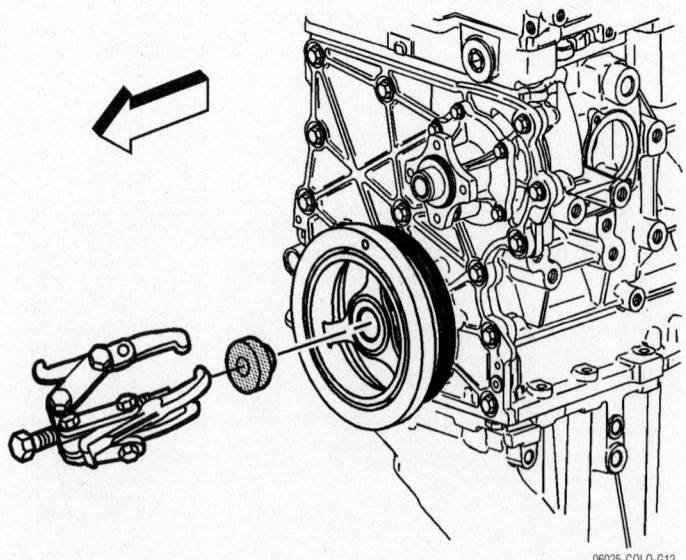

06025-COLO-G12

Remove the crankshaft balancer using a suitable puller after installing End Protector J-41816-2.

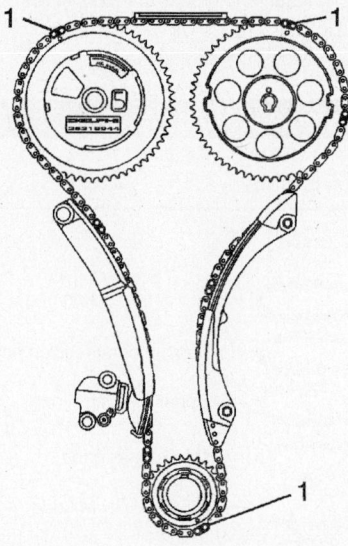

1. Timing chain dark link locations

71461-COLO-G12

Aligning timing chain dark links with camshaft and crankshaft sprockets

24. Install the exhaust camshaft actuator on the timing chain with the word DELPHI facing horizontal to the cylinder head surface and the dark link of the timing chain aligned with the timing mark on the camshaft actuator sprocket.

➡**Ensure the alignment pin is engaged between the camshaft and exhaust camshaft actuator sprocket.**

25. Install the exhaust camshaft actuator onto the exhaust camshaft.

➡**Rotate the camshaft actuator clockwise until it stops. This will fully advance the actuator. Engine damage may occur if the actuator is not fully advanced.**

26. Install a new exhaust camshaft sprocket bolt and washer and tighten the bolt to 18 ft. lbs. (25 Nm), plus an additional 135°.

27. Remove the tee in the timing chain tensioner to tension the timing chain.

28. Remove the camshaft locking tool from the camshafts.

29. The dark links should be aligned with the camshaft and crankshaft sprockets as shown.

30. Thread alignment pins into the engine block to aid front cover installation.

31. Apply sealant to the front cover surfaces as shown.

32. Align the oil pump with the crankshaft sprocket splines.

33. Install the front cover over the alignment pins, and loosely install the front cover bolts.

34. Remove the alignment pins and install the remaining 2 bolts.

35. Tighten the front cover bolts to 89 inch lbs. (10 Nm).

36. Tighten the small center cover bolt to 71 inch lbs. (8 Nm).

37. Install clean engine oil to the outside diameter of the new front crankshaft oil seal.

38. Using a seal installer, install the front oil seal.

39. Install a new o-ring to the oil pump pipe screen assembly and install the oil pump pipe screen.

40. Apply sealant to the oil pump pipe bolt threads and tighten the bolts to 89 inch lbs. (10 Nm).

41. Install or connect the following:
- Oil pan
- Power steering pump
- Drive belt tensioner
- Crankshaft damper and tighten new bolt and tighten to 111 ft. lbs. (150 Nm), plus an additional 180°.

42. Remove the flywheel locking tool.

43. Install or connect the following:
- Flywheel access service plug
- Water pump
- Accessory drive belt
- Cooling fan
- Negative battery cable

44. Fill the engine with coolant and oil.

45. Start the engine and check for leaks.

Balance Shafts

REMOVAL & INSTALLATION

1. Before servicing the vehicle, refer to the Precautions Section.

2. Remove or disconnect the following:
- Engine
- Crankshaft rear oil seal housing

3. On 2.8L engines the left hand balance shaft sprocket timing mark is at the 1:00 o'clock position, the right side sprocket is at the 1:30 position and the crankshaft sprocket is at the 5:30 position. The timing marks should line up with the dark links on the timing chain as shown.

4. On 3.5L engines the left hand balance shaft sprocket timing mark is at the 12:00 o'clock position, the right side sprocket is at the 2:30 position and the crankshaft sprocket is at the 4:30 position. The timing marks should line up with the dark links on the timing chain as shown.

5. Remove the balance shaft chain tensioner.

6. Remove the balance shaft timing chain.

➡**It may be necessary to remove the right balance shaft bolts and rotate the retainer plate counterclockwise to relieve the tension on the chain.**

7. Rotate the balance shafts to check for free rotation. If the shafts do not turn freely inspect the balance shaft bearings and bearing surface for damage.

8. Remove the balance shaft chain guide.

9. Remove the balance shaft bolts and remove the balance shafts.

To install:

10. Lubricate the balance shaft bearing journals with clean engine oil.

11. Install the balance shafts with the counterweight down to prevent damage to the shaft bearings.

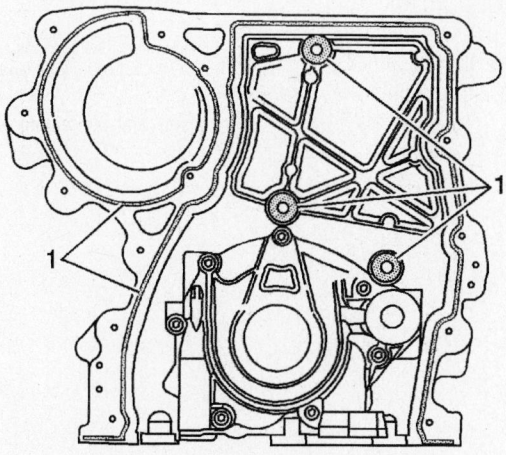

1. Sealant application areas

71461-COLO-G13

Front cover sealant application

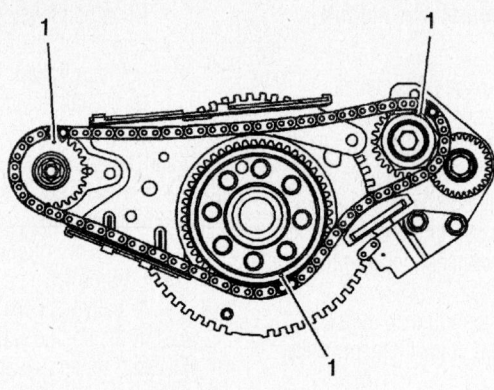

1. Sprocket timing marks

71461-COLO-G14

Balance shaft sprocket timing mark locations

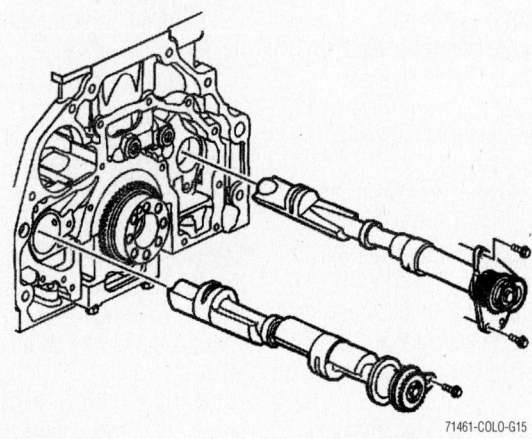

71461-COLO-G15

Removing the balance shafts

12. Install new bolts and tighten the bolts to 106 inch lbs. (12 Nm).

13. Install the chain guide and tighten the bolts to 89 inch lbs. (10 Nm).

14. Install the balance shaft timing chain and ensure the dark links of the chain are aligned with the sprocket timing marks.

15. Using both hands, rotate the timing chain tensioner ratchet release lever clockwise and hold, compress the tensioner shoe and hold, then release the ratchet lever.

16. Slowly release the pressure on the shoe until the ratchet lever moves to the first detent and a click is heard.

17. Insert a pin into the hole of the release lever to lock the tensioner shoe in the collapsed position.

18. With the release lever facing outward, install the chain tensioner and tighten the bolts to 89 inch lbs. (10 Nm).

19. Remove the pin from the tensioner.

20. Double check that the sprocket marks and timing chain dark links are aligned.

21. Install the crankshaft rear oil seal housing.

22. Install the engine.

Piston and Ring

POSITIONING

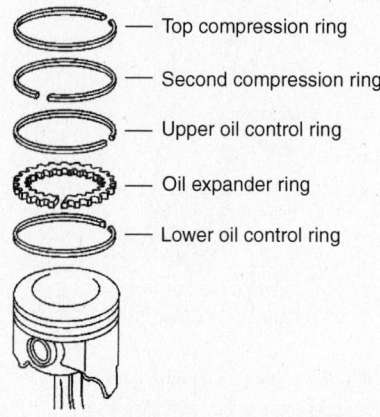

— Top compression ring

— Second compression ring

— Upper oil control ring

— Oil expander ring

— Lower oil control ring

71461-COLO-G16

Piston ring positioning

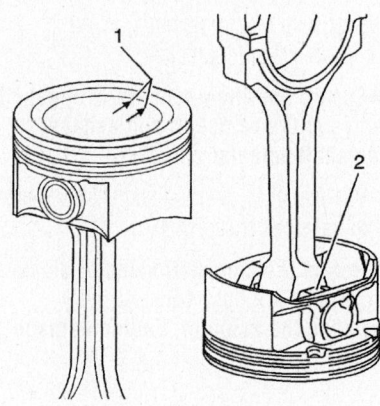

1 - Piston arrow face forward in block
2 - Flat casting surface faces forward

71461-COLO-G17

Piston and connecting rod orientation

FUEL SYSTEM

Fuel System Pressure

RELIEVING

The fuel systems operate under high fuel pressures. It is very important that the pressure be properly relieved prior to servicing the system or any of its components.

A Schrader valve is provided on these fuel systems to conveniently test or release the system pressure. A fuel pressure gauge and adapter will be necessary to connect the gauge to the fitting. This system utilizes a service valve on one end of the fuel rail assembly.

1. Before servicing the vehicle, refer to the Precautions Section.
2. Disconnect the negative battery cable to assure the prevention of fuel spillage if the ignition switch is accidentally turned **ON** while a fitting is still detached.
3. Loosen the fuel filler cap to release the fuel tank pressure.
4. Be sure the release valve on the fuel gauge is closed, then connect the fuel gauge to the pressure fitting located on the inlet fuel pipe fitting.

✳ CAUTION

When connecting the gauge to the fitting, be sure to wrap a rag around the fitting to avoid spillage. After repairs, place the rag in an approved container.

5. Install the bleed hose portion of the fuel gauge assembly into an approved container, then open the gauge release valve and bleed the fuel pressure from the system.
6. When the gauge is removed, be sure to open the bleed valve and drain all fuel from the gauge assembly.
7. When fuel service is finished, tighten the fuel filler cap and connect the negative battery cable.

Fuel Filter

REMOVAL & INSTALLATION

1. Before servicing the vehicle, refer to the Precautions Section.
2. Properly relieve the fuel system pressure.
3. Remove or disconnect the following:
   • Negative battery cable
   • Fuel filler cap

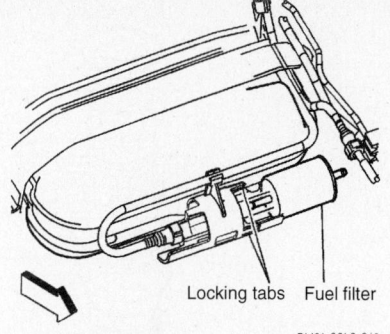
Locking tabs   Fuel filter

71461-COLO-G18

Fuel filter mounting at the fuel tank

   • Quick connect fittings from the filter
4. Pry open the locking tabs of the mounting bracket enough to remove the fuel filter and remove the filter.

To install:
5. Slide the filter into the mounting bracket until the lacking tabs are fully engaged.
6. Install or connect the following:
   • Fuel quick disconnect fittings to the filter
   • Fuel filler cap
   • Negative battery cable
7. Turn the ignition **ON** for 10 seconds and then turn it **OFF** for 10 seconds. Again turn the ignition **ON** and check for leaks.

Fuel Pump

REMOVAL & INSTALLATION

1. Before servicing the vehicle, refer to the Precautions Section.
2. Properly relieve the fuel system pressure.

3. Disconnect the negative battery cable.
4. Drain the fuel tank into an approved container.
5. Raise and support the rear of the vehicle.
6. Remove or disconnect the following:
   • Left rear tire
   • Left rear inner fender liner
   • Fuel filler tube from the fuel tank
   • EVAP hose from the filler vent tube
   • Electrical connectors and wiring harness retainers from the fuel tank
   • Fuel return line and fuel filter
   • Upper fuel tank retaining strap
7. Support the fuel tank.
8. Remove the lower tank retaining strap and lower the fuel tank.
9. Disconnect the fuel lines from the fuel pump module.
10. Using tool J-39765, rotate the fuel pump cam locking ring counterclockwise and remove the ring.
11. Raise the fuel pump and tilt it back to allow the fuel level sensor and float to clear the opening.
12. Remove and discard the fuel pump seal.

To install:
13. Install a new seal on the fuel pump.
14. Tilt the fuel pump until the fuel level sensor and float can enter the opening.
15. Lower the fuel pump and align the tang on the pump with the notch in the opening.
16. Using tool J-39765, rotate the fuel pump cam locking ring clockwise until fully seated.
17. Reconnect the fuel lines to the fuel pump module.

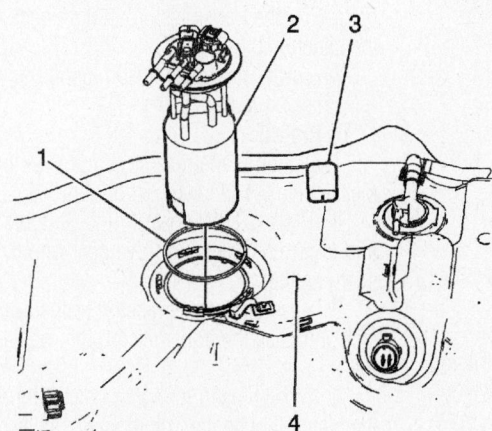

1. Seal
2. Fuel pump
3. Fuel level sensor and float
4. Fuel tank

71461-COLO-G19

Fuel pump mounting in fuel tank

18. Raise the fuel tank and install the lower tank retaining strap.

19. Install or connect the following:
- Upper fuel tank retaining strap and tighten both strap bolts to 24 ft. lbs. (32 Nm)
- Fuel return line and fuel filter
- Electrical connectors and wiring harness retainers to the fuel tank
- EVAP hose to the vent tube
- Fuel filler tube to the fuel tank
- Left rear inner fender liner
- Left rear tire

20. Lower the vehicle.

21. Fill the tank with gasoline.

22. Turn the ignition **ON** for 10 seconds and then turn it **OFF** for 10 seconds. Again turn the ignition **ON** and check for leaks.

Fuel Rail and Injectors

REMOVAL & INSTALLATION

1. Before servicing the vehicle, refer to the Precautions Section.

2. Relieve the fuel system pressure.

3. Remove or disconnect the following:
- Negative battery cable
- Fuel feed and return lines from fuel rail.
- Vent hoses from the air cleaner resonator and fuel pressure regulator
- EVAP purge hose
- Intake manifold

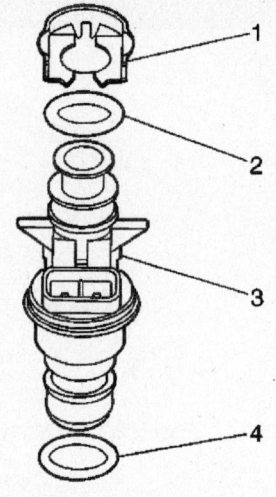

1. Retainer clip
2. O-ring
3. Injector
4. O-ring

71461-COLO-G20

Exploded view of fuel injector

- Fuel injector harness from the engine wiring harness
- Fuel rail mounting bolts
- Fuel rail
- Fuel injector connector from the injector
- Injector retaining clip
- Fuel injector from the fuel rail

4. Discard the retainer clip and remove and discard the 2 O-rings from the injector.

To install:

5. Lubricate the new injector O-ring seats with engine oil.

6. Install or connect the following:

- O-rings and retainer clip on the injector
- Fuel injector into the fuel rail socket
- Fuel rail. Torque the bolts to 89 inch lbs. (10 Nm).
- Fuel injector harness
- Intake manifold
- EVAP purge hose
- Vent hoses to air cleaner resonator and fuel pressure regulator
- Fuel feed and return lines
- Negative battery cable

7. Turn the ignition **ON** for 10 seconds and then turn it **OFF** for 10 seconds. Again turn the ignition **ON** and check for leaks.

DRIVE TRAIN

Manual Transmission Assembly

REMOVAL & INSTALLATION

1. Before servicing the vehicle, refer to the Precautions Section.

2. Remove or disconnect the following:
- Negative battery cable
- Shift lever housing and boot
- Rear driveshaft
- Front driveshaft, if equipped with 4WD
- Transfer case and shift lever, if equipped with 4WD
- All wiring harness that would interfere with transmission removal

3. Disconnect the hydraulic clutch quick-connect from the slave cylinder using special tool J–42371 to depress the white plastic sleeve on the quick connect to separate the clutch line end from the slave cylinder quick connect.

- Engine wiring harness and fuel line retainers from transmission

4. Support the transmission with a transmission jack or equivalent.

5. Remove the transmission crossmember.

6. Remove the transmission mounting bolts. Pull the transmission straight back on the clutch hub splines.

7. Lower the transmission using the transmission jack.

To install:

Installation is the reverse of removal, but please note the following important steps.

8. Place a THIN coat of high-temperature grease on the main drive gear (input shaft) splines.

9. Secure the transmission to the floor jack and raise the transmission into position.

10. Slowly insert the input shaft through the clutch. Rotate the output shaft slowly to engage the splines of the input shaft into the clutch while pushing the transmission forward into place. Do not force the transmission into position, the transmission should easily fall into place once everything is properly aligned.

11. Tighten the transmission mounting bolts to 37 ft. lbs. (50 Nm).

12. Tighten the transmission crossmember horizontal nuts to 37 ft. lbs. (50 Nm).

13. Tighten the transmission crossmember vertical bolts to 74 ft. lbs. (100 Nm).

14. Do not remove the transmission jack until the crossmember has been installed.

15. Check the transmission fluid level and replenish as necessary.

Automatic Transmission Assembly

REMOVAL & INSTALLATION

1. Before servicing the vehicle, refer to the Precautions Section.

2. Drain the transmission fluid.

3. Remove or disconnect the following:
- Negative battery cable
- Dipstick and filler tube
- Rear driveshaft
- Front driveshaft, if equipped with 4WD
- Transfer case, if equipped with 4WD
- Range selector cable from selector lever
- Transmission main harness connector
- Engine wiring harness retainers
- Park/neutral switch connector
- Vent hose retainer
- Fuel line bracket retainers and position fuel line aside
- Transmission service access plug
- 3 bolts securing the torque converter to the flywheel
- Transmission cooler lines from the transmission. Plug the lines and the ports in the transmission.

4. Place a transmission jack under the transmission.

5. Remove the transmission crossmember.

6. Inspect for any other wiring, brackets etc. which may interfere with the removal of the transmission.

7. Remove the transmission from the engine by pulling the transmission rearward to disengage it from the locator dowel pins on the back of the block. Carefully lower the transmission from the vehicle. Use care that the torque converter does not fall out of the front of the transmission.

➡**Use converter holding strap tool No. J-21366, to secure the torque converter**

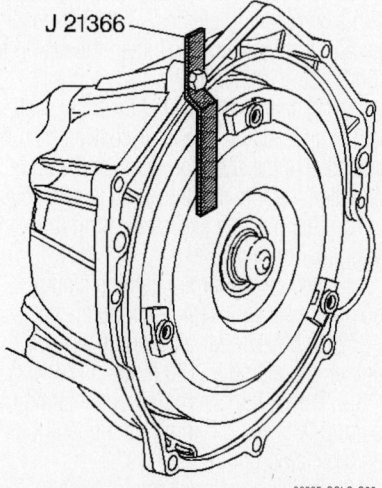

06025-COLO-G09

Install Special tool 21366 to secure the torque converter to the transmission—Automatic transmission

to the transmission during removal and installation procedures.

To install:

Installation is the reverse of removal, but please note the following important steps.

8. Make sure the torque converter is fully seated in the pump drive. If not, the transmission will not fit tightly to the rear of the engine block.

9. Raise the transmission into position and remove the torque converter holding strap and carefully. Slide the transmission forward until the dowel pins are engaged.

10. The torque converter should be flush with the flywheel and turn freely by hand.

11. Install the transmission–to–engine bolts. Tighten the bolts to 37 ft. lbs. (50 Nm).

12. Tighten the torque converter-to-flywheel bolts to 44 ft. lbs. (60 Nm).

13. Refill the transmission with the proper amount and type of fluid.

14. Connect the negative battery cable. Start the vehicle and allow to warm while checking for leaks. Road test the vehicle to check for shift quality.

Clutch

REMOVAL & INSTALLATION

1. Before servicing the vehicle, refer to the Precautions Section.

2. Remove or disconnect the following:
- Negative battery cable
- Transmission

3. Install a clutch alignment tool or a used transmission input shaft to support the clutch.

4. If the clutch assembly is going to be reused, mark the flywheel, clutch cover and a pressure plate lug for alignment when installing.

5. Remove or disconnect the following:

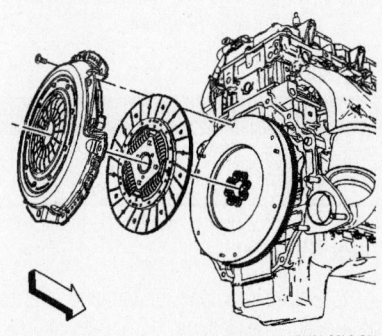

71461-COLO-G21

Exploded view of the clutch disc and related components

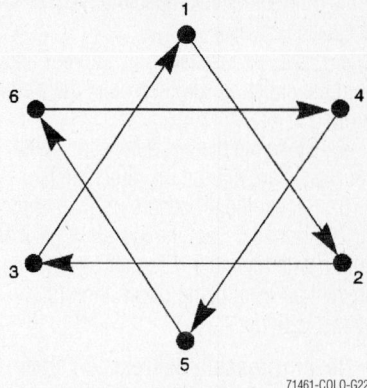

71461-COLO-G22

Clutch pressure plate bolt tightening sequence

- Clutch cover bolts and washers
- Clutch cover assembly and the clutch plate
- Clutch alignment tool

6. Clean all parts and inspect for damage.

To install:

7. Install or connect the following:
- Clutch alignment tool, to support the clutch
- Clutch plate/clutch cover assembly to the flywheel. Tighten the bolts in sequence shown to 15 ft. lbs. (20 Nm).

8. Remove the clutch alignment tool.

9. Install or connect the following:
- Transmission
- Negative battery cable

Hydraulic Clutch System

Bleeding air from the hydraulic clutch system is necessary whenever any part of the system has been disconnected or the fluid level (in the reservoir) has been allowed to fall so low, that air has been drawn into the master cylinder.

BLEEDING

1. Before servicing the vehicle, refer to the Precautions Section.

2. Fill master cylinder reservoir with new brake fluid conforming to DOT 3 specifications.

✳✳ CAUTION

Always use new fluid from a sealed container. Never, under any circumstances, use fluid that has been bled from a system to fill the reservoir as it may be aerated, have too much moisture content and possibly be contaminated.

3. Pump the clutch pedal up and down at least 15 times.

4. Have an assistant fully depress and hold the clutch pedal, then open the bleeder screw.

5. Close the bleeder screw and have your assistant release the clutch pedal.

6. Repeat the procedure until all of the air is evacuated from the system. Check and refill master cylinder reservoir as required to prevent air from being drawn through the master cylinder.

➡ **Never release a depressed clutch pedal with the bleeder screw open or air will be drawn into the system.**

7. Test the clutch for proper operation.

Transfer Case Assembly

REMOVAL & INSTALLATION

1. Before servicing the vehicle, refer to the Precautions Section.
2. Disconnect the negative battery cable.
3. Drain the transfer case fluid.
4. Support the transfer case.
5. Remove or disconnect the following:
   - Front and rear driveshafts from the transfer case. Matchmark the shafts prior to removal.
   - Vacuum lines and/or the electrical connectors, as equipped
   - Transfer case encoder motor
   - Engine wiring harness and position aside
   - Transfer case
6. Remove all traces of old gasket material from the mating surfaces.
   To install:
7. Install or connect the following:
   - New gasket using sealer to hold it in position
   - Transfer case. Torque the bolts to 37 ft. lbs. (50 Nm).
   - Engine wiring harness
   - Encoder motor

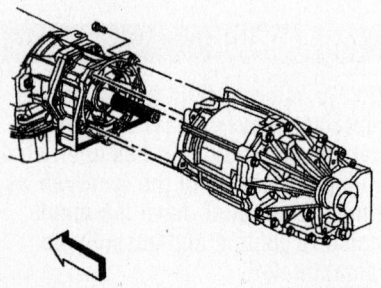

09474_ISER_G0004

Transfer case mounting

- Vacuum lines and/or electrical connections, as necessary
- Front and rear driveshafts by aligning the matchmarks
8. Refill the transfer case.
- Negative battery cable

Halfshaft

REMOVAL & INSTALLATION

1. Before servicing the vehicle, refer to the Precautions Section.
2. Unlock the steering column so the steering linkage is free to move.
3. Remove or disconnect the following:
   - Negative battery cable
   - Front wheel
   - Halfshaft nut and washer
   - Steering knuckle assembly
4. Place a drift against the tripot housing and hammer the drift to release the shaft.
5. Remove the halfshaft from the vehicle.

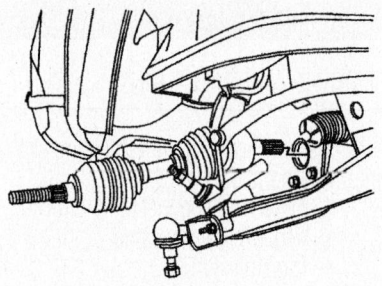

09474_ISER_G0005

Halfshaft installation

To install:
6. Install the halfshaft. A snap or pop should be heard and felt when the shaft properly seats in the differential housing.
7. Install the steering knuckle assembly.
8. Install a new halfshaft nut and tighten the nut to 191 ft. lbs. (260 Nm).
9. Install the front wheel.
10. Lower the vehicle.

CV-Joints

OVERHAUL

Outer CV-Joint

1. Before servicing the vehicle, refer to the Precautions Section.

➡ **When replacing the outer CV-joint, you must replace the inner CV-joint.**

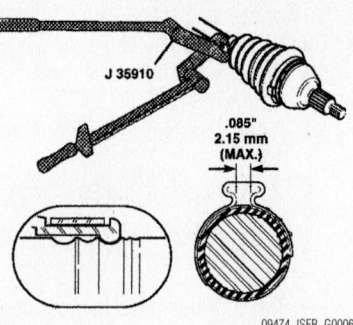

09474_ISER_G0006

Secure the small to the outer seal with seal retaining clamp J-35910, breaker bar and torque wrench

2. Remove or disconnect the following:
   - Front wheel
   - Halfshaft and position it in a vise
   - Inner CV-joint
   - Large CV-joint boot clamp and discard it
   - Small CV-joint boot clamp and discard it
3. Slide the outer seal away from the CV-joint, remove from the halfshaft and discard.

➡ **Do not remove the CV-joint from the halfshaft or you must replace the entire halfshaft assembly.**

4. Clean the old grease from the CV-joint and allow to dry.
   To install:
5. Pack the CV-joint assembly with the kit supplied grease.
6. Place new retaining clamps onto the new outer seal.
7. Position the small end of the outer seal into the CV-joint outer seal groove on the halfshaft.
8. Secure the small to the outer seal with seal retaining clamp J-35910, breaker bar and torque wrench. Tighten the clamp to 100 ft. lbs. (136 Nm).
9. Slide the large end of the outer seal with large retaining ring in place over the outside edge of the CV-joint. Position the lip of the outer seal into the groove on the CV-joint.
10. Remove any excess air from the outer seal.
11. Secure the large retaining clamp to the outer seal with seal retaining clamp J-35910, breaker bar and torque wrench. Tighten the clamp to 130 ft. lbs. (176 Nm).
12. Check the clamp gap dimension; if it is not 0.085 in. (2.15mm), continue tightening the clamp until it is.
13. Install or connect the following:
   - Inner CV-joint
   - Halfshaft
   - Front wheel

Inner (Tri-Pot) Joint

1. Before servicing the vehicle, refer to the Precautions Section.
2. Remove or disconnect the following:
   - Front wheel
   - Halfshaft and place it in a vise
   - Large and small boot clamp, cut and discard
   - Ball retaining ring using a screwdriver or equivalent
   - Tri-pot housing and bushing from the shaft
3. Using Snapring Pliers tool J-8059, spread the small retaining ring located in the cage and inner race assembly.
4. Remove the cage and inner race assembly from the halfshaft
5. Remove the balls from the cage.
6. Remove the inner race from the cage.
7. Remove the seal and discard.
8. Thoroughly clean and inspect all parts.

To install:

9. Install a new small retaining clamp onto the neck of the seal.
10. Slide the inner seal with the retaining clamp to the proper position on the halfshaft.
11. Secure the small retaining clamp retaining clamp J-35910, breaker bar and torque wrench. Tighten the clamp to 100 ft. lbs. (136 Nm).
12. Place the inner race with the retaining ring side up into the cage.
13. Place the six balls in the cage windows.
14. Slide the cage and inner race assembly, small diameter first, onto the halfshaft.
15. Using Snapring pliers J-8059, install the small retaining ring into the groove of the halfshaft.
16. Install convolute retainer 7848076 over the boot engaging 4 boot convolutions.
17. Pack the joint assembly with the kit-supplied grease.
18. Install a new large retaining clamp on the seal.
19. Slide the tri-pot joint housing over the cage and race assembly.

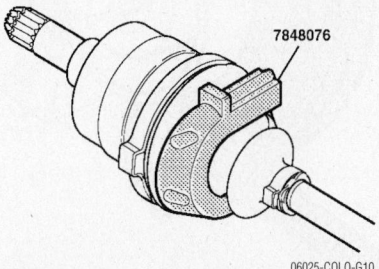

7848076

06025-COLO-G10

Install retainer 7848076 over the boot—Inner Joint

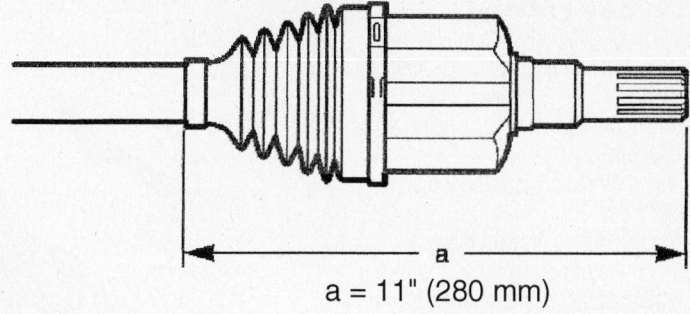

a = 11" (280 mm)

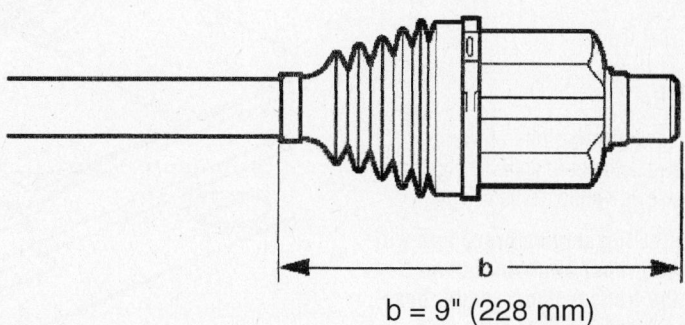

b = 9" (228 mm)

71461-COLO-G23

Inner (tri pot) joint boot dimensions

20. Insert the ball retaining ring into the groove at the top of the joint housing.
21. Install the seal onto the joint housing.
22. Ensure the boots are placed at the proper dimensions as shown.
23. Secure the large retaining clamp to the inner seal with seal retaining clamp J-35910, breaker bar and torque wrench. Tighten the clamp to 130 ft. lbs. (176 Nm).
24. Remove the convolute retainer tool.

Front Driveshaft

REMOVAL & INSTALLATION

1. Raise and support the vehicle.
2. Mark the front U-joint to the pinion yoke on the front differential.
3. Mark the relationship of the rear U-joint to the transfer case drive flange.
4. Mark the relationship of the rear U-joint to the transfer case drive flange.
5. Remove the retaining nuts and bolts from the drive flange.
6. Remove the clamps and retaining bolts from the front drive axle pinion flange.
7. Remove the front driveshaft from the vehicle.
8. Using tape and or a rubber band, wrap the front U-joint bearing caps to ensure the caps do not separate from the U-joint.

To install:

9. Remove the tape or rubber band from the front U-joint

10. Install the front driveshaft.
11. Align the reference marks on the front U-joint.

➡ **When installing the retaining bolts, only tighten them finger tight to hold the front driveshaft in position.**

12. Install the front U-joint bolts and clamps.
13. Align the reference marks on the rear U-joint.
14. Install the driveshaft to pinion flange bolts. Tighten the bolts from the driveshaft to the pinion flange 20 Nm (15 ft. lbs.). Tighten the bolts for the front propeller U-joint clamps 20 Nm (15 ft. lbs.).
15. Lower the vehicle.

Rear Driveshaft

REMOVAL & INSTALLATION

One Piece

1. Place the transmission in neutral.
2. Release the park brake, if applied.
3. Raise and support the vehicle.
4. Mark the rear universal joint to the drive shaft flange.
5. Remove the retaining bolts and clamps.
6. Remove the driveshaft from the pinion drive flange.
7. Remove the driveshaft from the transmission/transfer case.

8. Using tape or a rubber band, wrap the U-Joint bearing caps to ensure the bearing caps do not separate from the U-Joint.

To install:

9. Remove the tape or rubber band from the U-Joint.

10. Install the driveshaft in the transmission/transfer case.

11. Align the driveshaft with the reference marks on the pinion flange.

12. Install the clamps and retaining bolts. Tighten the retaining bolts to 20 Nm (15 ft. lbs.).

13. Lower the vehicle.

Two Piece

1. Place transmission in neutral.
2. Release the park brake, if applied.
3. Raise and support the vehicle.

➡ **The following service procedure will ensure the proper alignment of all U-Joint to the transmission, center bearing support and the rear drive axle pinion flange.**

4. Rotate the driveshafts so that all the external clips are aligned.

5. Mark the rear driveshaft to the pinion drive flange, center bearing support, and transmission.

6. Remove the clamps and retaining bolts for the rear driveshaft at the pinion drive flange and the center bearing.

7. Using tape or a rubber band, wrap the rear U-Joint bearing caps to ensure that they not separate from the U-Joint.

8. Remove the rear driveshaft from the center bearing.

9. Remove the mounting bolts for the center bearing.

10. Remove the center bearing from the crossmember.

11. Remove the driveshaft from the transmission.

To install:

12. Align the reference marks on the driveshaft to the transmission.

13. Install the driveshaft in the transmission.

14. Install the center bearing to the crossmember..

15. Install the center bearing mounting bolts. Tighten the center bearing mounting bolts to 80 Nm (59 ft. lbs.).

16. Rotate the driveshaft so that the alignment marks and the pinion yoke are in alignment.

17. Remove the tape and or rubber band.

18. Position the U-Joint in the pinion yoke.

19. Install the clamps and retaining

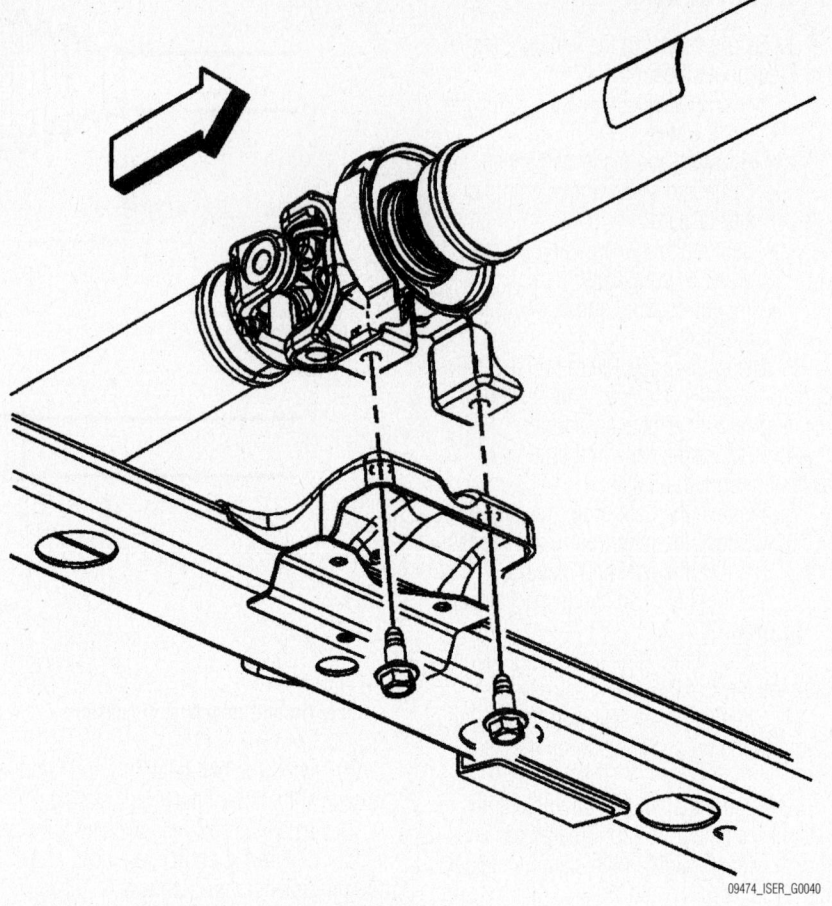

2-piece rear driveshaft center bearing attachment

09474_ISER_G0040

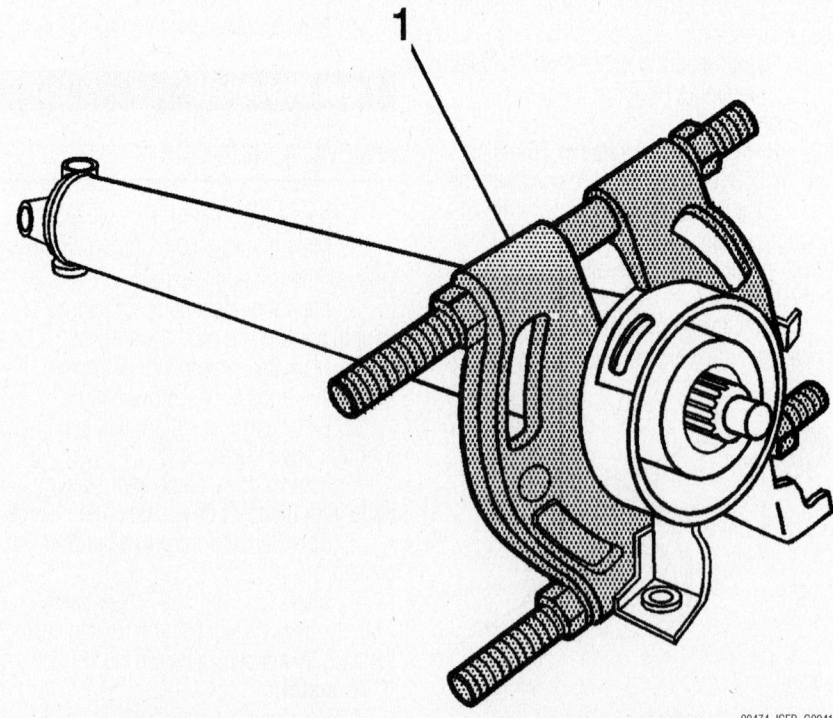

Tool J44749 installed

09474_ISER_G0041

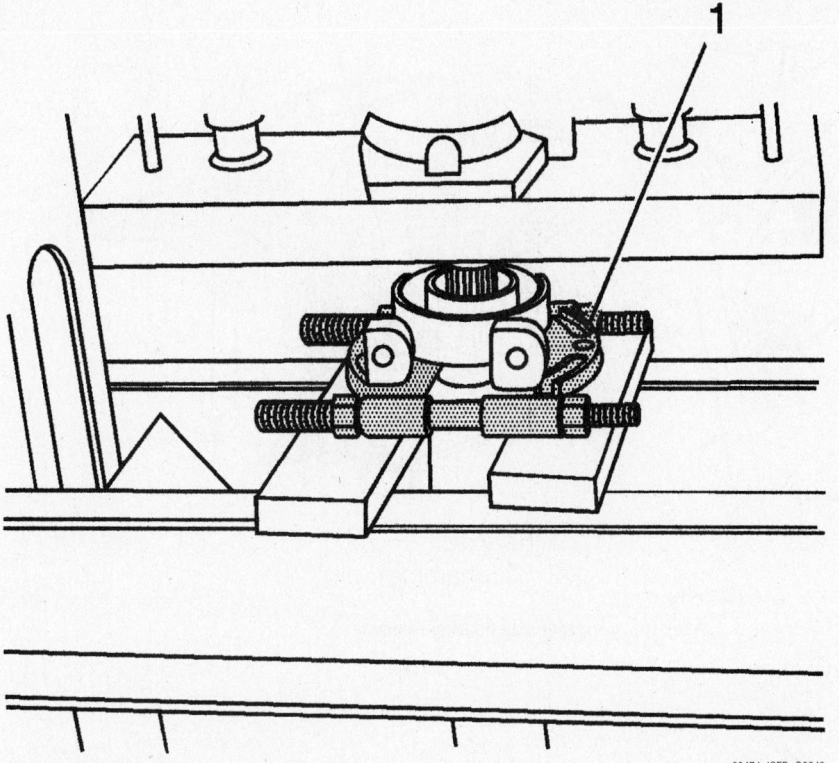

09474_ISER_G0042

Driveshaft installed in the press

bolts. Tighten the retaining clamp bolts to 30 Nm (22 ft. lbs.).

20. Double check the alignment marks to ensure that they are aligned properly

21. Lower the vehicle.

CENTER BEARING REPLACEMENT

1. Remove the driveshaft.
2. Mark the relationship of the yoke to the driveshaft.
3. Mark the relationship of the U-Joint to the yoke.
4. Remove the U-Joint from the yoke.

➡**The following service is to protect the machined portion of the slip yoke.**

5. Using 2 or 3 clean shop towels, wrap the slip yoke with the towels and secure with electrical ties or electrical tape.
6. Remove the retaining nut for the yoke.
7. Remove the washer from the yoke.
8. Remove the yoke from the driveshaft.

➡**When installing the special tool, the aid of an assistant will be required to position the tool between the center bearing and driveshaft shoulder. Ensure that the flat part of the tool is facing center bearing.**

9. Install tool J 44749 (1) to the driveshaft.

a. With the aid of an assistant, pull the center bearing toward the spline part of the driveshaft.

b. Position the tool (1) between the center bearing and the shield for the center bearing.

c. Adjust the tool so that it just comes in contact with the shoulder.

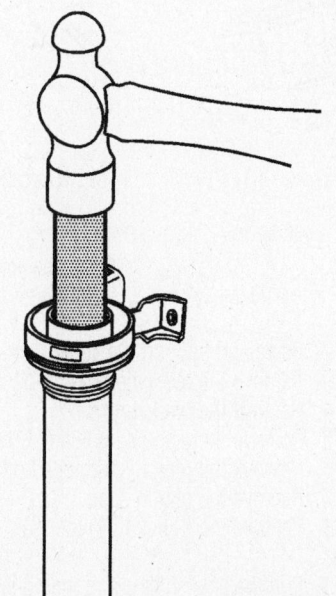

09474_ISER_G0043

Tool J37372 installed on the driveshaft

➡**It will be necessary to adjust the hydraulic press so that the slip yoke does not touch the ground.**

10. Install the driveshaft in a hydraulic press.
11. Position a piece of wood between the floor and the slip yoke.
12. Remove the center bearing from the driveshaft.
13. Remove the tool from the driveshaft.

To install:

14. Position the center bearing on the driveshaft.
15. Install tool J 37372 on the driveshaft.
16. Using the J 37372, install the center bearing on the driveshaft.

➡**Failure to align the reference marks properly will cause a vibration the driveshaft.**

17. Position the yoke on the driveshaft so that the reference marks are aligned properly.
18. Using a brass or hard rubber hammer, lightly tap the yoke on the driveshaft.
19. Install the washer.

➡**Tighten the retaining nut by hand before tightening the retaining nut to the proper specifications.**

20. Install the retaining nut. Tighten the retaining nut to 167 Nm (123 ft. lbs.).
21. Align the U-Joint to the reference marks on the driveshaft and yoke.
22. Install the U-Joint to the yoke.
23. Remove the shop towels from the driveshaft.
24. Install the driveshaft in the vehicle.

Front Axle Bearing and Seal

REMOVAL & INSTALLATION

1. Before servicing the vehicle, refer to the Precautions Section.
2. Raise and support the vehicle.
3. Remove the front wheel.
4. Remove the halfshaft.
5. Remove the sway bar link from the lower control arm.
6. Using an inside puller, remove the axle seal and then the axle bearing from the intermediate housing.

To install:

7. Position the new bearing in the housing and use a bearing installer to press it in.
8. Install a new axle seal using a seal installer.
9. Install the halfshaft.
10. Connect the sway bar link to the control arm.

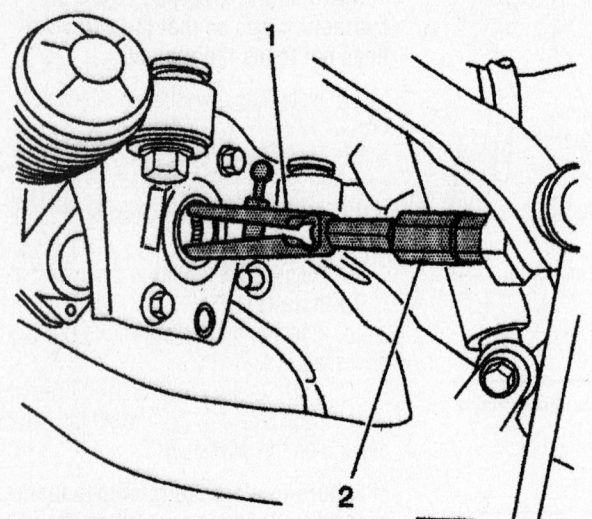

1. Adapter
2. Slidehammer

09474_ISER_G0007

Front axle seal removal

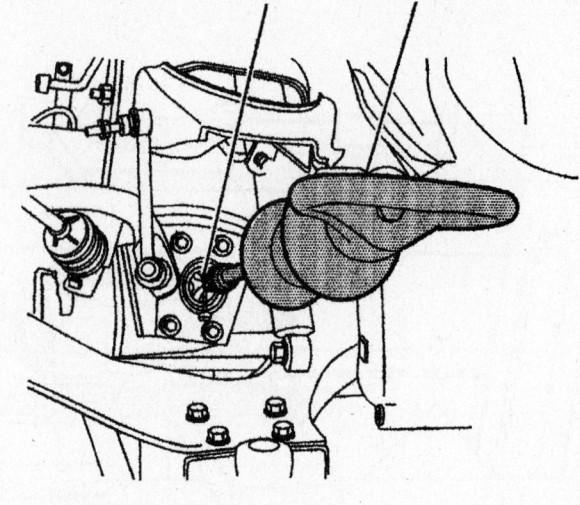

1. Bearing
2. Slidehammer

09474_ISER_G0008

Front axle bearing removal

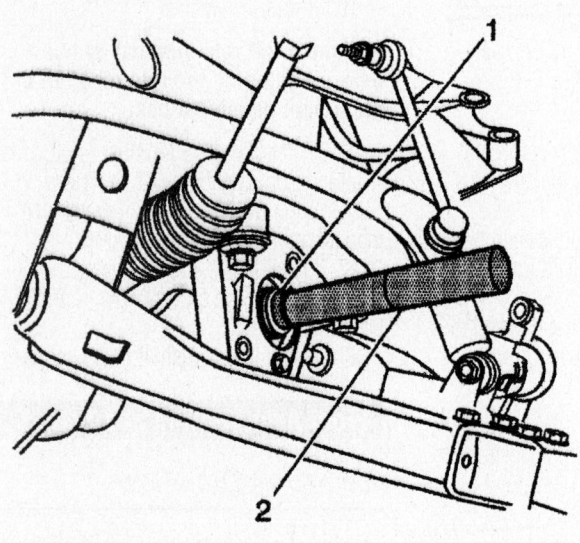

1. Bearing
2. Driver

09474_ISER_G0009

Front axle bearing installation

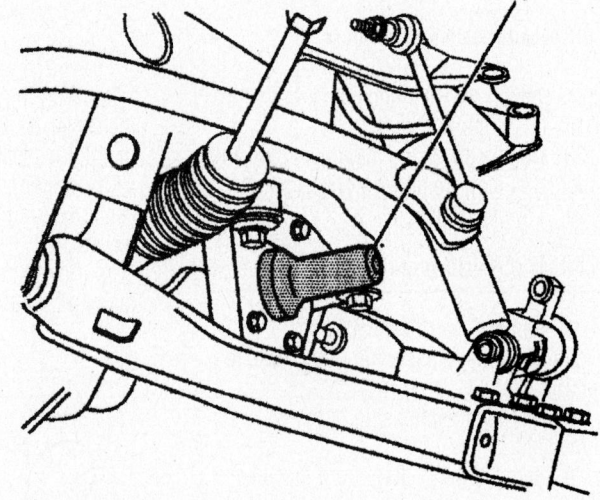

1. Installer

09474_ISER_G0010

Front axle seal installation

11. Install the wheel and tighten the lug nuts to 103 ft. lbs. (140 Nm).

Rear Axle Shaft, Bearing and Seal

REMOVAL & INSTALLATION

1. Before servicing the vehicle, refer to the Precautions Section.

2. Raise and support the vehicle.
3. Remove the rear wheel.
4. Remove the brake drum.
5. Remove the rear axle housing cover.
6. Remove the pinion shaft lock bolt.
7. Remove the pinion shaft.
8. Remove the c-lock from the rear axle.
9. Pull the axle shaft out of the housing.

10. Remove the axle shaft seal and bearing.
To install:
11. Press on a new bearing and install the seal.
12. Install the axle shaft.
13. Install the c-lock to the rear axle.
14. Install the pinion shaft.
15. Install the pinion shaft lock bolt and tighten to 18 ft. lbs. (25 Nm).

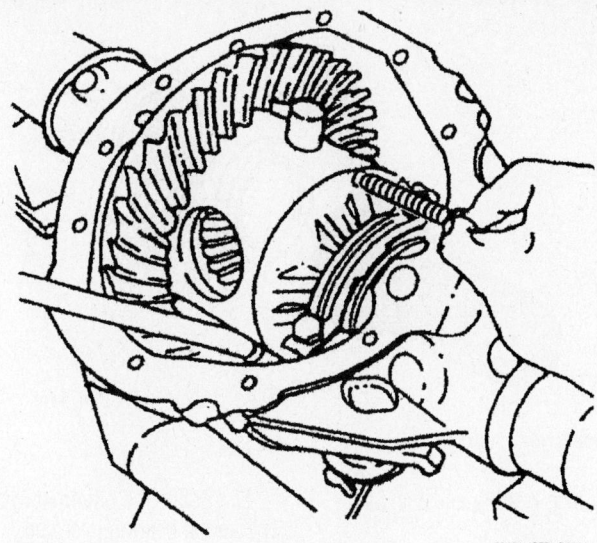

09474_ISER_G0011

Lock bolt removal

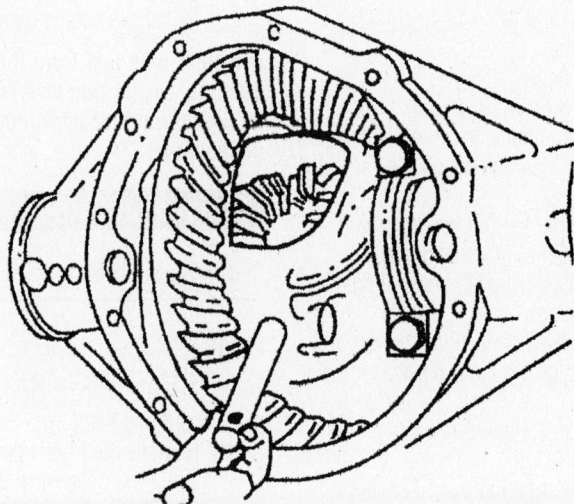

09474_ISER_G0012

Pinion shaft removal

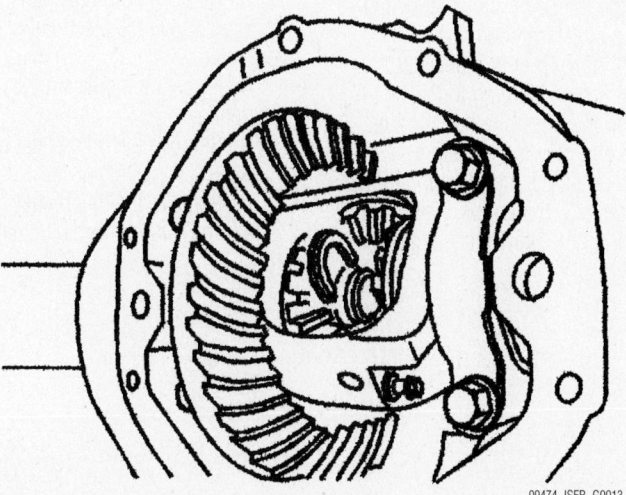

09474_ISER_G0013

C-lock removal

16. Install a new gasket and the rear axle housing cover.

17. Install the brake drum.

18. Fill the rear axle with the correct lubricant.

19. Install the wheel and tighten the lug nuts to 103 ft. lbs. (140 Nm).

20. Lower the vehicle.

Pinion Seal

REMOVAL & INSTALLATION

Front Drive Axle

1. Before servicing the vehicle, refer to the Precautions Section.

➡**The following procedure requires the use of the Pinion Holding tool J-8614-01 and the Pinion Seal Installation tool J-33782.**

2. Raise and support the vehicle.

3. Remove the front wheels.

4. Remove or disconnect the following:
- Engine undercover
- Driveshaft from the pinion flange. Matchmark the driveshaft prior to removal.
- Brake calipers and wire out of the way.

5. Mark the position of the pinion stem, flange and nut for reference.

6. Use an inch lbs. torque wrench to measure the amount of torque necessary to turn the pinion, then note this measurement as it is the combined pinion bearing, seal, carrier bearing, axle bearing and seal pre-load.

7. Remove or disconnect the following:
- Pinion flange nut and washer, using a Pinion Holding tool J-8614-01 and a Pinion Flange Removal tool J-8614-1, J-8614-2, J-8614-3, as applicable
- Pinion flange
- Pinion oil seal by driving it out of the differential with a blunt chisel; DO NOT damage the carrier

To install:

8. Examine the seal surface of pinion flange for tool marks, nicks or damage, such as a groove worn by the seal. If damaged, replace flange.

9. Examine the carrier bore and remove any burrs that might cause leaks around the O.D. of the seal.

10. Install a new pinion oil seal using a seal installer tool

11. Apply GM seal lubricant 12346004 to the splines of the pinion yoke.

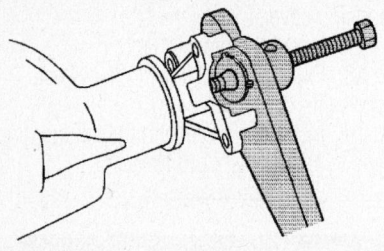

71461-COLO-G24

Removing the front axle pinion nut using a pinion holding fixture tool

12. Install or connect the following:
- Pinion flange aligning the reference marks made earlier.

13. Seat the flange on the shaft by tapping in with a soft-faced hammer.

14. Install the washer and new nut.

15. Install Pinion Holding tool J-8614-01 and tighten the nut while holding the tool until the end play is removed.

16. Continue tightening the nut until the preload torque is 3–5 inch lbs. (0.40–0.57 Nm) greater than the previously measured torque.

17. Rotate the pinion several times to ensure the bearing is seated.

18. Install the brake calipers.

19. Install the driveshaft to the pinion flange using the matchmarks made earlier.

20. Install the engine undercover.

➡**If fluid was lost from the differential housing during this procedure, be sure to check and add additional fluid, as necessary.**

Rear Axle

1. Before servicing the vehicle, refer to the Precautions Section.

2. Raise and support the vehicle.

3. Remove the rear wheels.

4. Remove or disconnect the following:
- Driveshaft
- Brake drums

5. Mark the position of the pinion stem, flange and nut for reference.

6. Use an inch lbs. torque wrench to measure the amount of torque necessary to turn the pinion, then note this measurement as it is the combined pinion bearing, seal,

09474_ISER_G0014

Pinion seal and driver

carrier bearing, axle bearing and seal pre-load.

7. Remove or disconnect the following:
- Pinion flange nut and washer, using a Pinion Holding tool J-8614-01 and a Pinion Flange Removal tool J-8614-1, J-8614-2, J-8614-3, as applicable
- Pinion flange
- Pinion oil seal by driving it out of the differential with a blunt chisel; DO NOT damage the carrier

To install:

8. Examine the seal surface of pinion flange for tool marks, nicks or damage, such as a groove worn by the seal. If damaged, replace flange.

9. Examine the carrier bore and remove any burrs that might cause leaks around the O.D. of the seal.

10. Install a new pinion oil seal using a seal installer tool

11. Apply GM seal lubricant 12346004 to the splines of the pinion yoke.

12. Install or connect the following:
- Pinion flange aligning the reference marks made earlier.

13. Seat the flange on the shaft by tapping in with a soft-faced hammer.

14. Install the washer and new nut.

15. Install Pinion Holding tool J-8614-01 and tighten the nut while holding the tool until the end play is removed.

16. Continue tightening the nut until the preload torque is 3–5 inch lbs. (0.40–0.57 Nm) greater than the previously measured torque.

17. Rotate the pinion several times to ensure the bearing is seated.

18. Install the brake calipers.

19. Install the driveshaft to the pinion flange using the matchmarks made earlier.

20. Install the engine undercover.

➡**If fluid was lost from the differential housing during this procedure, be sure to check and add additional fluid, as necessary.**

Differential Carrier

REMOVAL & INSTALLATION

Front

1. Before servicing the vehicle, refer to the Precautions Section.

2. Remove the front wheel.

3. Drain the differential fluid.

4. Remove the front driveshaft. Matchmark the driveshaft prior to removal.

5. Remove the left sway bar link from the lower control arm.

6. Remove both halfshafts.

7. Remove the electrical connectors and vent hose.

8. Place a transmission jack under the differential.

9. Remove the differential mounting bracket.

10. Remove the differential.

11. To install, reverse the removal procedure.

STEERING AND SUSPENSION

Air Bag

✳✳ CAUTION

Some vehicles are equipped with an air bag system, also known as the Supplemental Inflatable Restraint (SIR) system. The system must be disabled before performing service on or around system components, steering column, instrument panel components, wiring and sensors. Failure to follow safety and disabling procedures could result in accidental air bag deployment, possible personal injury and unnecessary system repairs.

DISABLING AND ENABLING

Zone 1

1. Turn the steering wheel so that the vehicle's wheels are pointing straight ahead.
2. Turn OFF the ignition.
3. Remove the key from the ignition.

➡With the SIR fuse removed and the ignition ON, the AIR BAG indicator illuminates. This is normal operation and does not indicate an SIR system malfunction.

4. Remove the SIR fuse from the fuse block (1) located in the underhood fuse block.

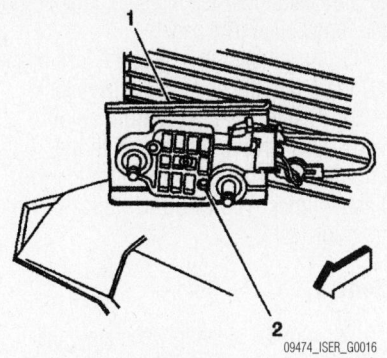

Front end sensor connector

➡This vehicle is equipped with 2 inflatable restraint front end sensors. When performing this procedure be sure to include both sensors.

5. Remove the connector position assurance (CPA) from both (2) front end sensor connectors.
6. Disconnect both sensor connectors.

Enabling Procedure

7. Remove the key from the ignition.
8. Connect the sensor connectors to both (2) front end sensors.
9. Install the CPAs to both sensor connectors.
10. Remove the SIR fuse into the fuse block (1) located in the underhood fuse block.
11. Staying well away from all air bags, turn ON the ignition, with the engine OFF. The AIR BAG indicator will flash 7 times. The AIR BAG indicator will then turn OFF.

Zone 2

1. Turn the steering wheel until the vehicles wheels are pointing straight ahead.
2. Turn OFF the ignition.
3. Remove the key from the ignition.

➡With the SIR fuse removed and the ignition ON, the AIR BAG indicator illuminates. This is normal operation and does not indicate an SIR system malfunction.

4. Remove the SIR fuse from the fuse block (1) located in the underhood fuse block.
5. Remove the left door trim panel.
6. Remove the connector position assurance (CPA) (3) from the side impact sensor yellow 2-way connector (5) located near the middle of the door.
7. Disconnect the side impact sensor yellow 2-way connector®(5) located near the middle of the door.
8. Remove the trim panel. For extended cab remove the rear access door wiring harness grommet to access the connector.
9. Remove the connector positive assurance (CPA) from the seat belt pretensioner—left connector.
10. Disconnect the seat belt pretensioner left connector from the vehicle harness connector.
11. Remove the garnish molding.
12. Disconnect the roof rail module connector (1) from the vehicle harness.

Enabling Procedure

13. Remove the key from the ignition.
14. Connect the roof rail module connector (1) and install the garnish molding.
15. Connect the seat belt pretensioner, left, and install the CPA.
16. Install the trim panel. For extended cab install the rear access door wiring harness grommet.

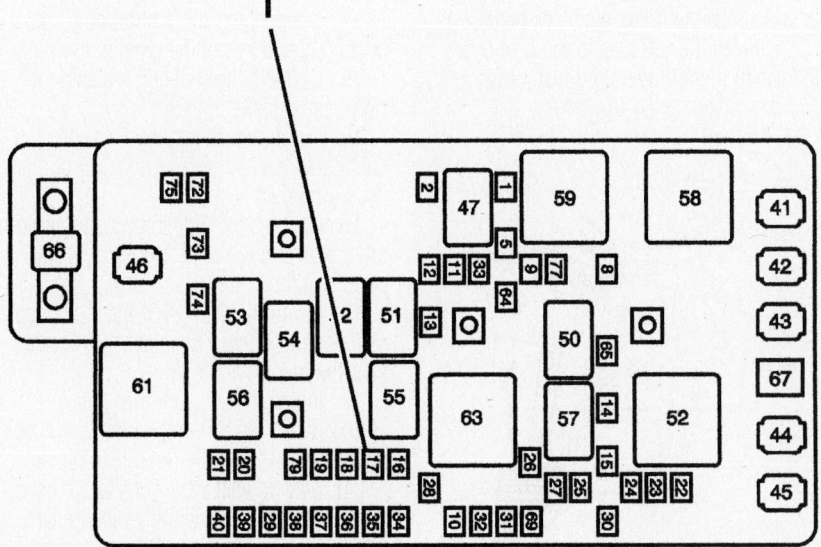

SIR fuse location

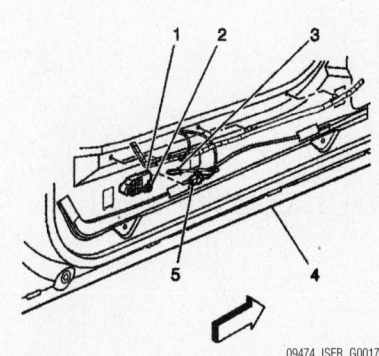

Side impact sensor yellow 2-way connector

17. Connect the side impact sensor yellow 2-way connector (5) located near the middle of the door.

18. Install the CPA (3) to the side impact sensor yellow 2-way connector (5) located near the middle of the door.

19. Install the left door trim panel.

20. Install the SIR fuse to the fuse block (1) located in the underhood fuse block.

21. Staying well away from all air bags, turn ON the ignition, with the engine OFF. The AIR BAG indicator will flash 7 times. The AIR BAG indicator will then turn OFF.

Zone 3

1. Turn the steering wheel until the vehicle's wheels are pointing straight ahead.

2. Turn OFF the ignition.

3. Remove the key from the ignition.

4. Remove the trim panel.

➡ **With the SIR fuse removed and the ignition ON, the AIR BAG indicator illuminates. This is normal operation and does not indicate an SIR system malfunction.**

5. Remove the SIR fuse from the fuse block (1) located in the underhood fuse block.

6. Remove the connector position assurance (CPA) (1) from the steering wheel module yellow 4-way connector (2) located left of the steering column above the left hinge pillar trim panel.

7. Disconnect the steering wheel module yellow 4-way connector (2) located left of the steering column above the left hinge pillar trim panel.

Enabling Procedure

8. Remove the key from the ignition.

9. Connect the steering wheel module yellow 4-way connector (2) located left of the steering column above the left hinge pillar trim panel.

10. Install the CPA (1) to the steering wheel module yellow 4-way connector (2) located left of the steering column above the left hinge pillar trim panel.

11. Install the SIR fuse to the fuse block (1) located in the underhood fuse block.

12. Staying well away from all air bags, turn ON the ignition, with the engine OFF. The AIR BAG indicator will flash 7 times. The AIR BAG indicator will then turn OFF.

Zone 5

1. Turn the steering wheel until the vehicles wheels are pointing straight ahead.

2. Turn OFF the ignition.

3. Remove the key from the ignition.

4. Access the inflatable restraint I/P module inline connector C208 above the driver side kick panel.

➡ **With the SIR fuse removed and the ignition ON, the AIR BAG indicator illuminates. This is normal operation and does not indicate an SIR system malfunction.**

5. Remove the SIR fuse from the fuse block (1) located in the underhood fuse block.

6. Remove the CPA from the I/P module yellow 4-way inline connector C208 (2) located above the driver side kick panel.

7. Disconnect the I/P module yellow 4-way inline connector C208 (2) located above the driver side kick panel.

Enabling Procedure

8. Remove the key from the ignition.

9. Connect the I/P module yellow 4-way inline connector C208 (2) located above the driver side kick panel.

10. Install the CPA to the I/P module yellow 4-way inline connector C208 (2) located above the driver side kick panel.

11. Install the SIR fuse to the fuse block (1) located in the underhood fuse block.

12. Staying well away from all air bags, turn ON the ignition, with the engine OFF. The AIR BAG indicator will flash 7 times. The AIR BAG indicator will then turn OFF.

Zone 6

1. Turn the steering wheel until the vehicles wheels are pointing straight ahead.

2. Turn OFF the ignition.

3. Remove the key from the ignition.

➡ **With the SIR fuse removed and the ignition ON, the AIR BAG indicator illuminates. This is normal operation and does not indicate an SIR system malfunction.**

4. Remove the SIR fuse from the fuse block (1) located in the underhood fuse block.

5. Remove the left door trim panel.

6. Remove the connector position assurance (CPA) (3) from the side impact sensor yellow 2-way connector (5) located near the middle of the door.

7. Disconnect the side impact sensor yellow 2-way connector (5) located near the middle of the door.

8. Remove the trim panel. For extended cab remove the rear access door wiring harness grommet to access the connector.

9. Remove the connector positive assurance (CPA) from the seat belt pretensioner—right connector.

10. Disconnect the seat belt pretensioner—right connector from the vehicle harness connector.

11. Remove the garnish molding.

12. Disconnect the roof rail module connector (1) from the vehicle harness.

Enabling Procedure

13. Remove the key from the ignition.

14. Connect the roof rail module connector (1) and install the garnish molding.

15. Connect the seat belt pretensioner, right, and install the CPA.

16. Install the trim panel. For extended cab install the rear access door wiring harness grommet.

17. Connect the side impact sensor yellow 2-way connector (5) located near the middle of the door.

18. Install the CPA (3) to the side impact sensor yellow 2-way connector (5) located near the middle of the door.

19. Install the left door trim panel.

20. Install the SIR fuse to the fuse block (1) located in the underhood fuse block.

21. Staying well away from all air bags, turn ON the ignition, with the engine OFF. The AIR BAG indicator will flash 7 times. The AIR BAG indicator will then turn OFF.

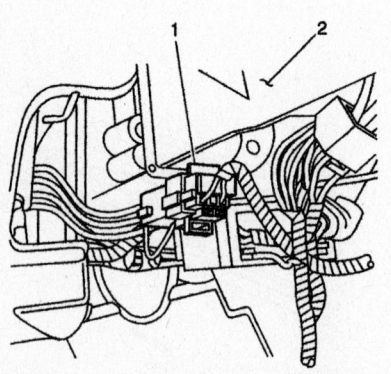

09474_ISER_G0018

Steering wheel module yellow 4-way connector

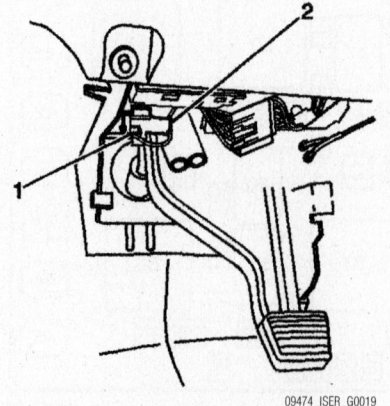

09474_ISER_G0019

I/P module yellow 4-way inline connector

Zone 8

1. Turn the steering wheel until the vehicle's wheels are pointing straight ahead.

2. Turn OFF the ignition.

3. Remove the key from the ignition.

4. Remove the trim panel.

5. Access the inflatable restraint I/P module inline connector C208 above the driver side kick panel.

➡**With the SIR fuse removed and the ignition ON, the AIR BAG indicator illuminates. This is normal operation and does not indicate an SIR system malfunction.**

6. Remove the SIR fuse from the fuse block (1) located in the underhood fuse block.

7. Remove the connector position assurance (CPA) (1) from the steering wheel module yellow 4-way connector (1) located left of the steering column (2) above the left hinge pillar trim panel.

8. Disconnect the steering wheel module yellow 4-way connector (1) located left of the steering column (2) above the left hinge pillar trim panel.

9. Remove the left door trim panel.

10. Remove the connector position assurance (CPA) (3) from the side impact sensor yellow 2-way connector (5) located near the middle of the door.

11. Disconnect the side impact sensor yellow 2-way connector (5) located near the middle of the door.

12. Remove the trim panel. For extended cab remove the rear access door wiring harness grommet to access the connector.

13. Remove the CPA from the seat belt pretensioner—left connector.

14. Disconnect the seat belt pretensioner—left connector from the vehicle harness connector.

15. Remove the garnish molding.

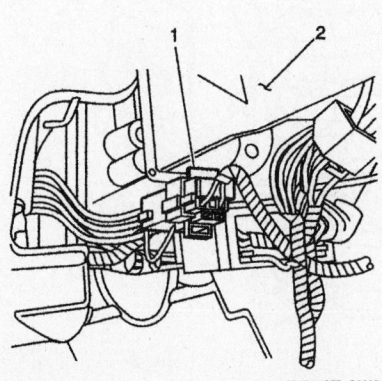

09474_ISER_G0020

Steering wheel module yellow®4-way connector

16. Disconnect the roof rail module connector (1) from the vehicle harness.

17. Remove the CPA from the I/P module yellow 4-way inline connector C208 (2) located above the driver side kick panel.

18. Disconnect the I/P module yellow 4-way inline connector C208 (2) located above the driver side kick panel.

19. Remove the right door trim panel.

20. Remove the CPA (3) from the side impact sensor yellow 2-way connector (5) located near the middle of the door.

21. Disconnect the side impact sensor yellow 2-way connector (5) located near the middle of the door.

22. Remove the trim panel. For extended cab remove the rear access door wiring harness grommet to access the connector.

23. Remove the CPA from the seat belt pretensioner—right connector.

24. Disconnect the seat belt pretensioner—right connector from the vehicle harness connector.

25. Remove the garnish molding.

26. Disconnect the roof rail module connector (1) from the vehicle harness.

Enabling Procedure

27. Remove the key from the ignition.

28. Connect the steering wheel module yellow 4-way connector (1) located left of the steering column (2) above the left hinge pillar trim panel.

29. Install the CPA (1) to the steering wheel module yellow 4-way connector (1) located left of the steering column (2) above the left hinge pillar trim panel.

30. Connect the roof rail module connector (1) and install the garnish molding.

31. Connect the seat belt pretensioner, left, and install the CPA.

32. Install the trim panel. For extended cab install the rear access door wiring harness grommet.

33. Connect the side impact sensor yellow 2-way connector (5) located near the middle of the door.

34. Install the CPA (3) to the side impact sensor yellow 2-way connector (5) located near the middle of the door.

35. Install the left door trim panel.

36. Connect the I/P module yellow 4-way inline connector C208 (2) located above the driver door kick panel.

37. Install the CPA to the I/P module yellow 4-way inline connector C208 (2) located above the driver side kick panel.

38. Connect the roof rail module connector (1) and install the garnish molding.

39. Connect the seat belt pretensioner-right and install the CPA.

40. Install the trim panel. For extended cab install the rear access door wiring harness grommet.

41. Connect the side impact sensor yellow 2-way connector (5) located near the middle of the door.

42. Install the CPA (3) to the side impact sensor yellow 2-way connector (5) located near the middle of the door.

43. Install the right door trim panel.

44. Install the SIR fuse to the fuse block (1) located in the underhood fuse block.

45. Staying well away from all air bags, turn ON the ignition, with the engine OFF. The AIR BAG indicator will flash 7 times. The AIR BAG indicator will then turn OFF.

Power Steering Gear

REMOVAL & INSTALLATION

1. Before servicing the vehicle, refer to the Precautions Section.

2. Position a fluid catch pan under the power steering gear.

3. Place the front wheels in the straight ahead position.

4. Turn the ignition key to the LOCK position and remove the key.

5. Insert Steering Column Locking Pin J-42640 into the access hole in the lower steering column trim cover.

6. Raise and support the vehicle.

7. Remove or disconnect the following:
- Front wheels
- Engine skid plate, if equipped
- Front axle housing, if equipped
- Outer tie rod ends from steering knuckle
- Feed and return fluid hoses from the steering gear. Immediately cap or plug all openings to prevent system contamination or excessive fluid loss.
- Intermediate shaft-to-lower steering column shaft pinch bolt
- Lower shaft-to-steering gear pinch bolt
- Power steering gear-to-frame bolts and washers
- Front crossmember, if equipped with RWD only
- Power steering gear

To install:

8. Install or connect the following:
- Steering gear to the vehicle. Loosely install the mounting nuts, washers and bolts.
- Crossmember, if equipped with RWD only. Tighten the mounting bolts to 44 ft. lbs. (60 Nm).
- Steering gear mounting bolts.

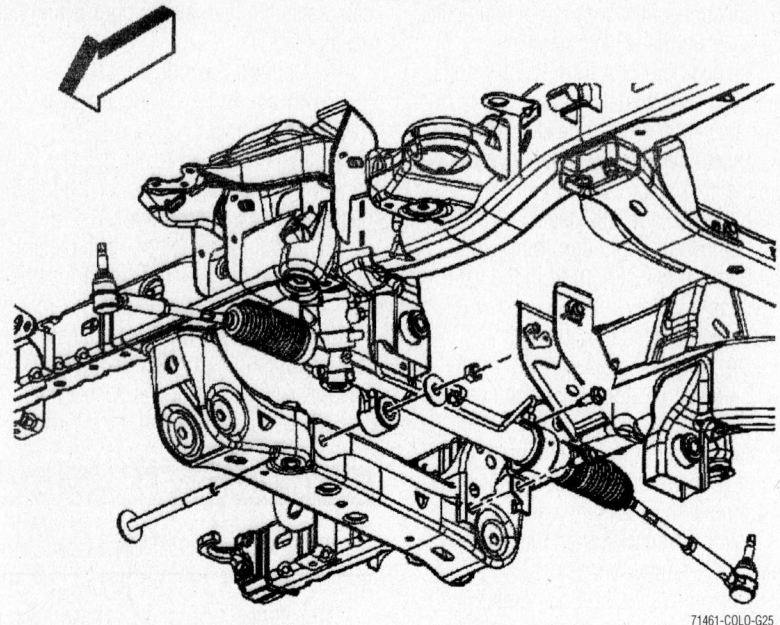

71461-COLO-G25

Power steering gear mounting

Tighten the vertical bolts to 96 ft. lbs. (130 Nm) and the horizontal bolts to 74 ft. lbs. (100 Nm).

- Lower shaft-to-steering gear pinch bolt and tighten the bolt to 33 ft. lbs. (45 Nm).
- Intermediate shaft-to-lower shaft pinch bolt, making sure the match-marks line up. Tighten the bolt to 17 ft. lbs. (23 Nm).
- Pressure and return hoses to the power steering gear. Tighten the hoses to 106 inch lbs. (12 Nm).
- Outer tie rod ends
- Front axle housing, if equipped
- Engine undercover
- Front wheels

9. Remove the steering column lock pin
10. Bleed the power steering system

Shock Absorbers

REMOVAL & INSTALLATION

Front

2WD MODELS

1. Before servicing the vehicle, refer to the Precautions Section.
2. Remove or disconnect the following:
   - Upper shock mounting nuts
   - Wheel
   - Lower mounting bolt and nut
   - Shock absorber/coil spring assembly

3. Place the shock absorber/coil spring assembly into a coil spring compressor and compress the spring.
4. Remove the upper shock retaining nut, bushings and washer.
5. Remove the shock absorber from the spring compressor.
6. If the coil spring is being replaced, remove the spring and mounting plate from the compressor, noting the mounting plate-to-spring orientation.

To install:

7. If the coil spring was replaced, align the spring to the mounting plate.
8. Align the centerline of the upper mounting studs with the centerline of the lower shock mount. Install the mounting plate and spring into the compressor and compress the spring.
9. Install the shock absorber into the coil spring and compressor.
10. Install the washers, bushings and upper retaining nut. Tighten the retaining nut to 15 ft. lbs. (20 Nm).
11. Relax the tension on the spring compressor and remove the compressor.
12. Install the shock absorber/coil spring assembly to the lower mount. Tighten the lower mounting bolt and nut to 81 ft. lbs. (110 Nm).
13. Install the wheel and lower the vehicle.
14. Install the upper mounting nuts and tighten to 20 ft. lbs. (27 Nm).

4WD MODELS

1. Before servicing the vehicle, refer to the Precautions Section.
2. Place a jack or stand under the lower control arm.
3. Remove or disconnect the following:
   - Shock absorber upper nut and isolator
   - Lower nut/bolt
   - Shock absorber through the control arm

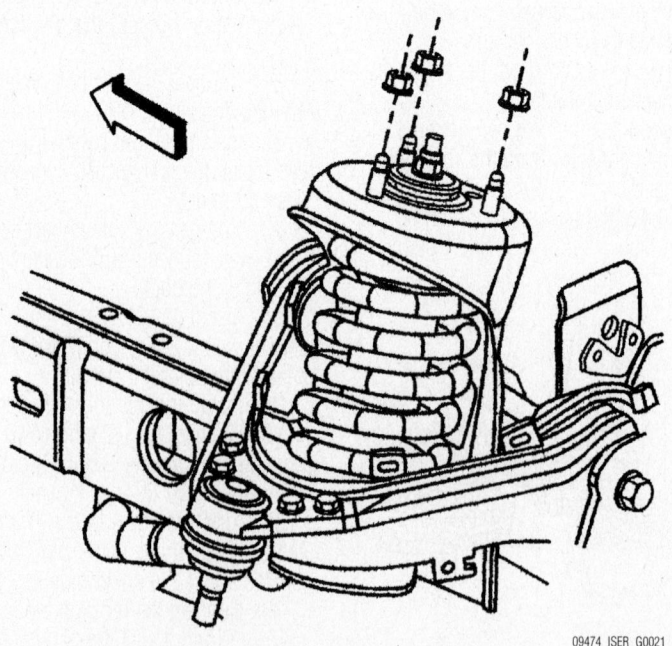

09474_ISER_G0021

Front upper shock absorber mounting nuts—2WD

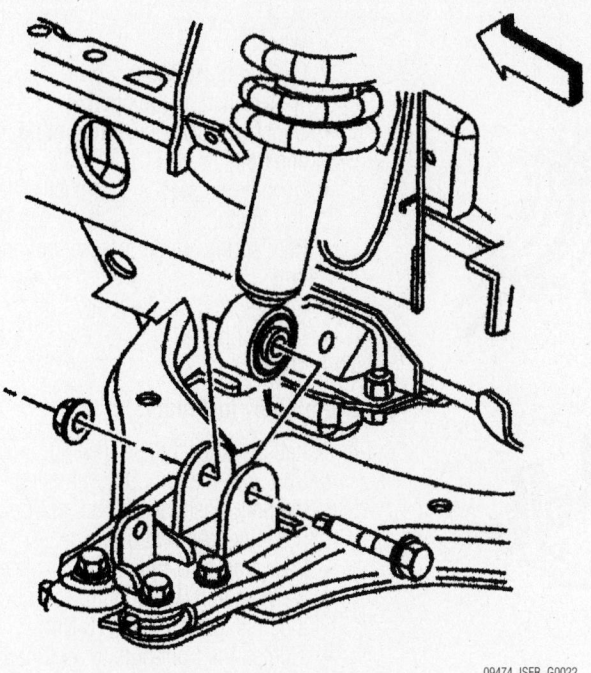

09474_ISER_G0022

Front lower shock absorber mounting nuts—2WD

09474_ISER_G0023

Front upper shock absorber mounting nut—4WD

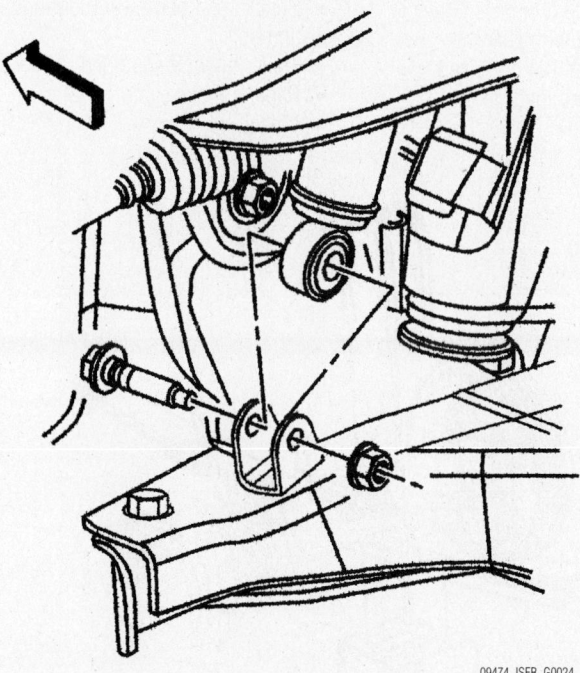

09474_ISER_G0024

Front lower shock absorber mounting nut—4WD

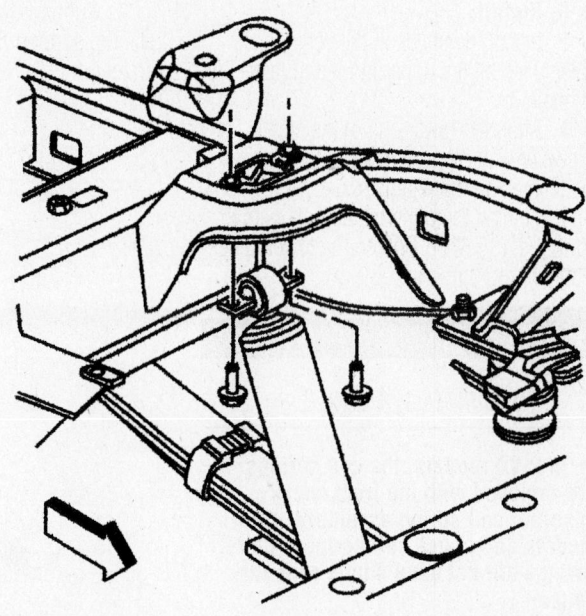

09474_ISER_G0025

Rear upper shock absorber mounting

To install:

4. Install the shock through the lower control arm and insert it through the mounting hole in the upper spring pocket.

5. Install or connect the following:
- Shock absorber to the lower control arm. Tighten the nuts/bolts to 52 ft. lbs. (70 Nm).
- Upper isolator to the shock

6. Tighten the upper mounting nut to 18 ft. lbs. (25 Nm).

Rear

1. Before servicing the vehicle, refer to the Precautions Section.

2. Support the rear axle assembly at ride height.

3. Remove or disconnect the following:
- Shock absorber-to-frame retainers at the top of the shock
- Shock-to-axle retainers at the bottom of the shock
- Shock absorber

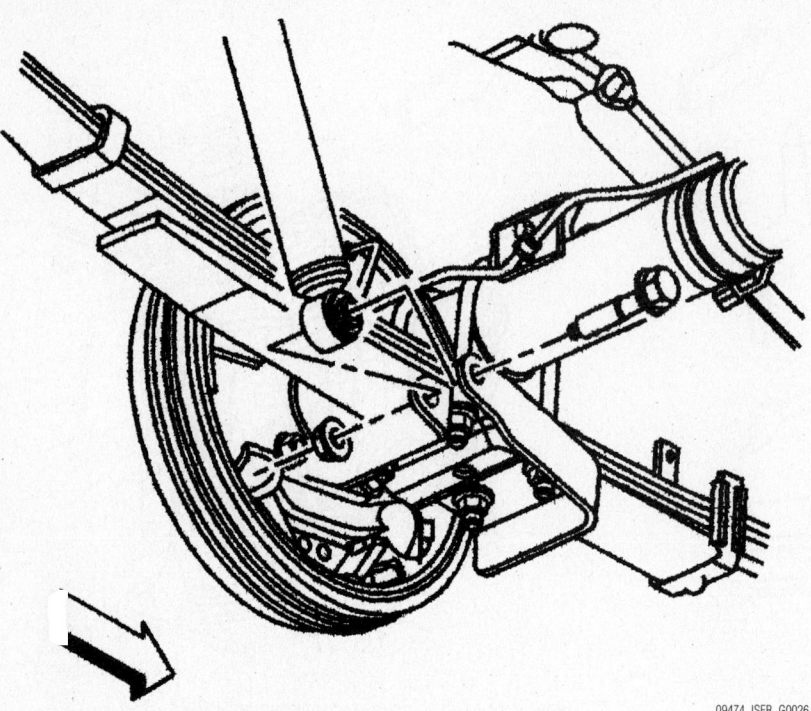

Rear lower shock absorber mounting

To install:

4. Install the shock in the vehicle and loosely install the upper mounting fasteners to retain it.

5. Align the lower-end of the shock absorber with the axle mounting, then loosely install the retainers.

6. Tighten the upper shock retainers to 26 ft. lbs. (35 Nm). Tighten the lower shock retainers to 70 ft. lbs. (95 Nm).

Coil Springs

REMOVAL & INSTALLATION

➡**On 2WD models, the coil springs are removed with the front shock absorber/coil spring assembly. 4WD models do not use coil springs. Coil springs are not used on the rear suspension.**

Leaf Springs

REMOVAL & INSTALLATION

1. Before servicing the vehicle, refer to the Precautions Section.

➡**The following procedure requires the use of two sets of jackstands.**

2. Support the rear axle with jackstands, support the axle and the body separately in order to relieve the load on the rear spring.

09474_ISER_G0026

3. Remove or disconnect the following:
    *   Wheel
    *   Parking brake cable
    *   Trailer hitch, if equipped
    *   Shock absorber lower mounting nut and bolt
    *   U-bolt nuts, washers, anchor plate and bolts
    *   Rear spring hanger bracket nut and bolt
    *   Front spring bracket bolt
    *   Leaf spring

To install:

➡**Use all new fasteners.**

4. Install or connect the following:
    *   Spring to the front bracket using the bolt, washers and nut, but do not fully tighten at this time.
    *   Rear bracket U-bolts, anchor plate, washers and U-bolt nuts. Torque the nuts to 56 ft. lbs. (76 Nm).
    *   Tighten the front nuts to 59 ft. lbs. (80 Nm), plus an additional 80°.
    *   Tighten the rear shackle nut to 63 ft. lbs. (85 Nm).
    *   Shock absorber lower mounting nut and bolt
    *   Trailer hitch, if equipped
    *   Parking brake cable
    *   Wheel

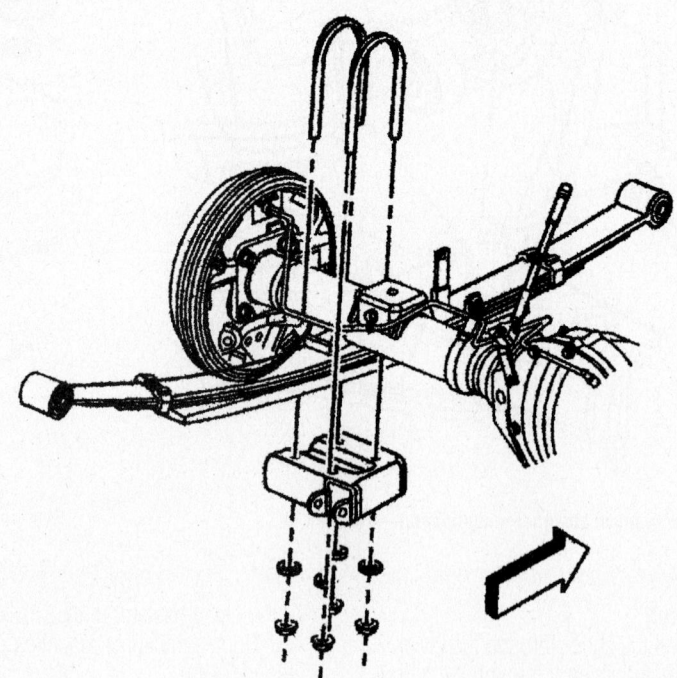

09474_ISER_G0027

Leaf spring U-bolts

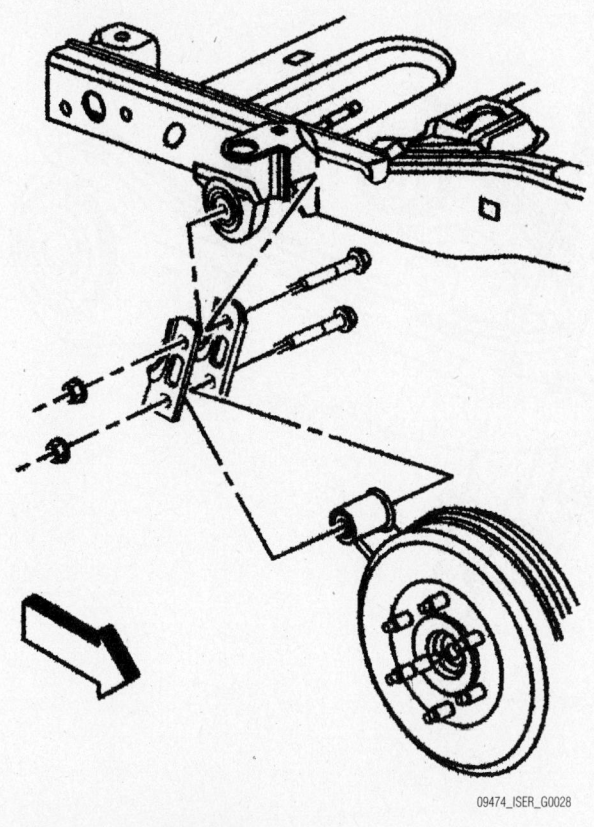

09474_ISER_G0028

Leaf spring rear shackle bolts

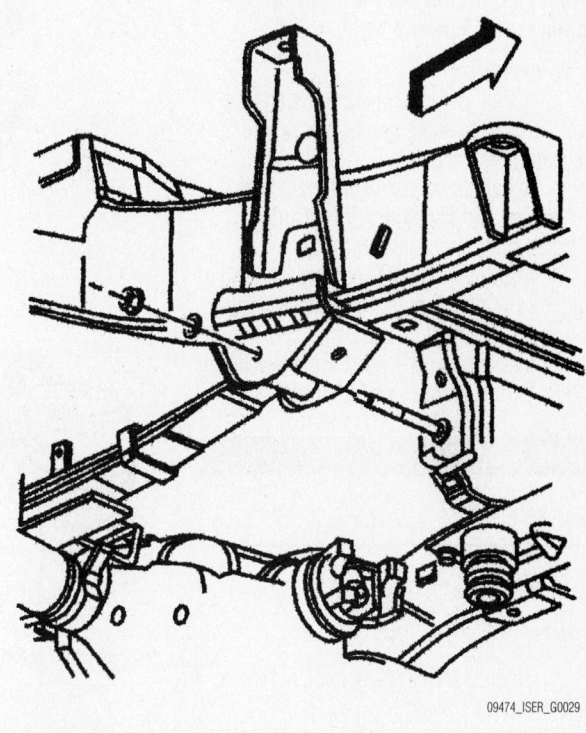

09474_ISER_G0029

Leaf spring front mounting bolt

Torsion Bar

Instead of the coil spring used on the front suspension of 2WD vehicles, the 4WD vehicles are equipped with a torsion bar.

REMOVAL & INSTALLATION

1. Before servicing the vehicle, refer to the Precautions Section.

✳✳ WARNING

Avoid damage to the torsion bar surface. Nicks and scratches will cause premature failure.

2. Raise and support the vehicle allowing the front suspension to hang in the rebound position.
3. Remove the front wheels.
4. Mark the rear torsion bar adjuster bolt location and count the number of turns required to remove the bolt.
5. Remove the adjuster bolt spacer and adjuster nut.
6. Remove the torsion bar and adjustment arms as a unit, moving it rearward to disengage it from the lower control arm.

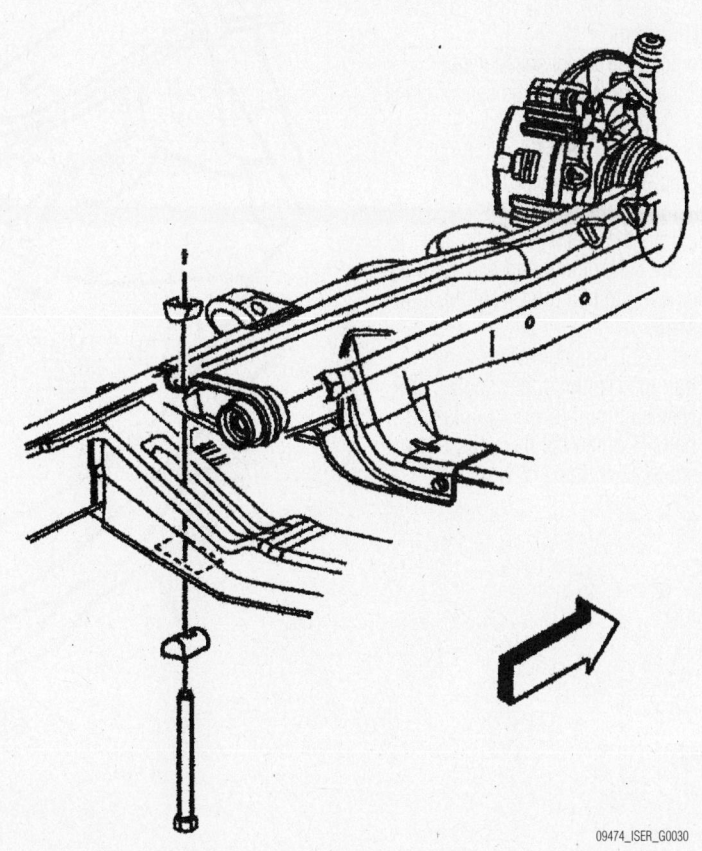

09474_ISER_G0030

Torsion bar mounting

➡Note the direction of the forward end and side of the torsion bar being removed. Torsion bars are not interchangeable, left and right.

To install:

7. Install the adjustment arms and torsion bars in the same position as where they were removed.

8. Install the adjustment arms to the torsion bar and slide the bar forward until it engages the lower control arm fully.

9. Install the adjuster bolt, spacer and adjuster nut and tighten it the same number of turns as counted on removal.

10. Install the wheels and lower the vehicle.

11. Perform a front end alignment.

Ball Joints

REMOVAL & INSTALLATION

2WD Vehicles

UPPER

1. Before servicing the vehicle, refer to the Precautions Section.

2. Raise and support the vehicle.

3. Remove the front wheels.

4. Support the lower control arm with a suitable jack.

5. Disconnect the brake lines from the upper control arm.

6. Remove the wheel speed sensor bracket bolt and disconnect the sensor brackets.

7. Remove the upper ball joint nut.

8. Separate the ball joint from the knuckle using a separator.

9. Remove the 4 ball joint nuts and bolts from the control arm and discard them.

10. Remove the ball joint from the arm.

To install:

11. Install or connect the following:

- Ball joint in the upper control arm
- New ball joint retaining nuts and bolts. Tighten the ball joint retainers to 12 ft. lbs. (16 Nm).

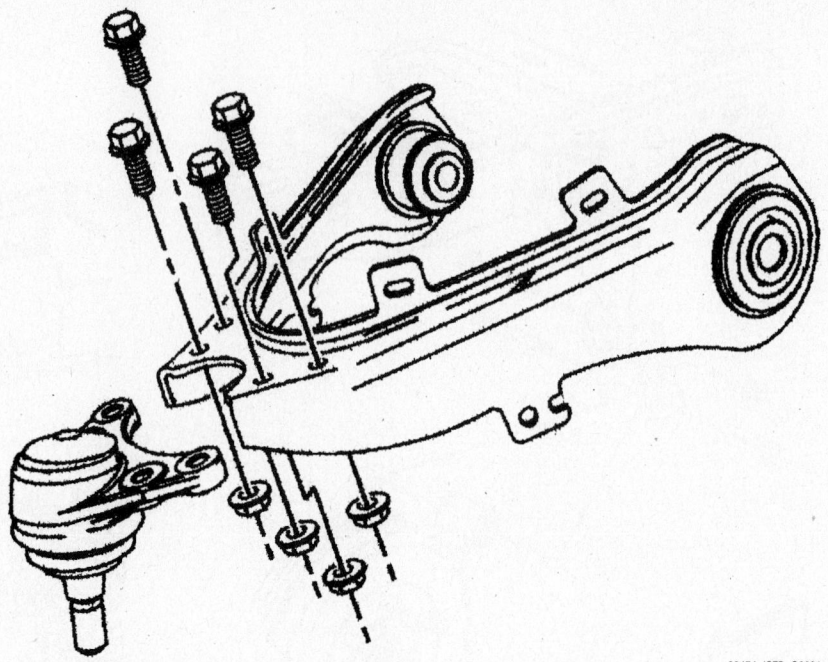

09474_ISER_G0031

Lower ball joint mounting—2WD

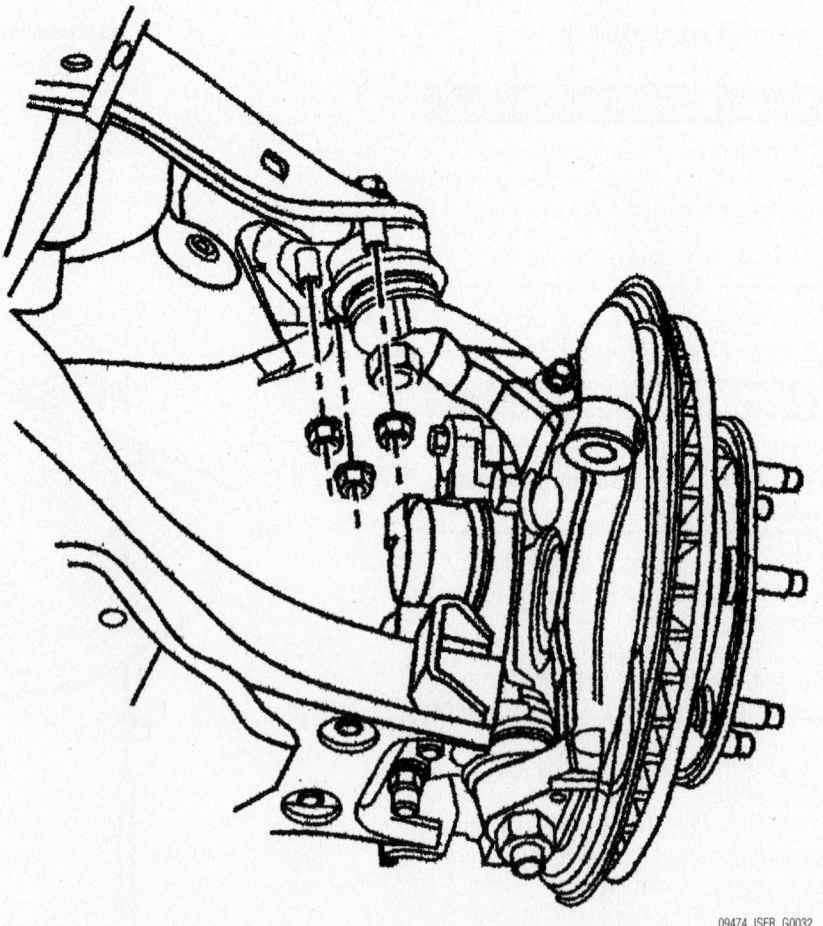

09474_ISER_G0032

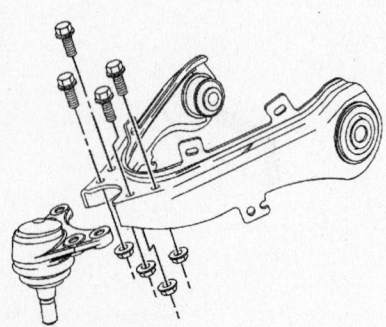

71461-COLO-G26

Upper ball joint mounting—2WD

Upper ball joint mounting—4WD

- Ball joint to steering knuckle.
  Tighten the new nut to 55 ft. lbs.
  (75 Nm).
- Speed sensor bracket and tighten
  the nuts to 15 ft. lbs. (20 Nm).
- Brake hose to the control arm
- Remove the lower control arm jack
- Tire and wheel assembly

12. Check and adjust the front end alignment, as necessary.

LOWER

1. Before servicing the vehicle, refer to the Precautions Section.
2. Remove the steering knuckle.
3. Remove the ball joint nuts and bolts.
4. Remove the ball joint.

To install:

5. Install or connect the following:
   - Ball joint nuts and bolts and tighten to 44 ft. lbs. (60 Nm)
   - Steering knuckle
6. Check and adjust the front end alignment, as necessary.

4WD Vehicles

UPPER

1. Before servicing the vehicle, refer to the Precautions Section.
2. Raise and support the vehicle.
3. Remove the front wheels.
4. Support the lower control arm with a suitable jack.
5. Disconnect the brake lines from the upper control arm.
6. Remove the wheel speed sensor bracket bolt and disconnect the sensor brackets.
7. Remove the upper ball joint nut.
8. Disconnect the upper control arm from the ball stud by removing the retaining nuts.
9. Remove and discard the upper ball joint retaining nut.
10. Separate the ball joint from the knuckle using a separator.
11. Remove the ball joint from the arm.

To install:

12. Install or connect the following:
   - Ball joint in the steering knuckle
   - Ball joint retaining nut and tighten the nut to 55 ft. lbs. (75 Nm)
   - Upper control arm to the ball stud and tighten the nut to 35 ft. lbs. (47 Nm).
   - Wheel speed sensor bracket bolt and tighten to 15 ft. lbs. (20 Nm).
   - Brake line to the upper control arm
   - Front tires
13. Check and adjust the front end alignment, as necessary.

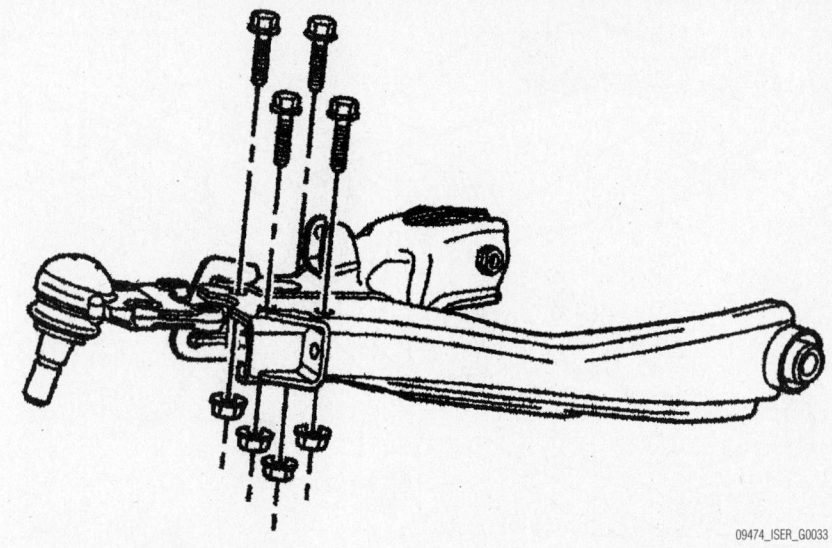

09474_ISER_G0033

Lower ball joint mounting—4WD

LOWER

1. Before servicing the vehicle, refer to the Precautions Section.
2. Remove the steering knuckle.
3. Remove the ball joint nuts and bolts.
4. Remove the ball joint.

To install:

5. Install or connect the following:
   - Ball joint nuts and bolts and tighten to 47 ft. lbs. (64 Nm)
   - Steering knuckle
6. Check and adjust the front end alignment, as necessary.

Upper Control Arm

REMOVAL & INSTALLATION

2WD Vehicles

1. Before servicing the vehicle, refer to the Precautions Section.
2. Disconnect the negative battery cable.
3. Raise and support the vehicle.
4. Place a jack stand under the lower control arm to support it at ride height.
5. Remove or disconnect the following:
   - Wheel

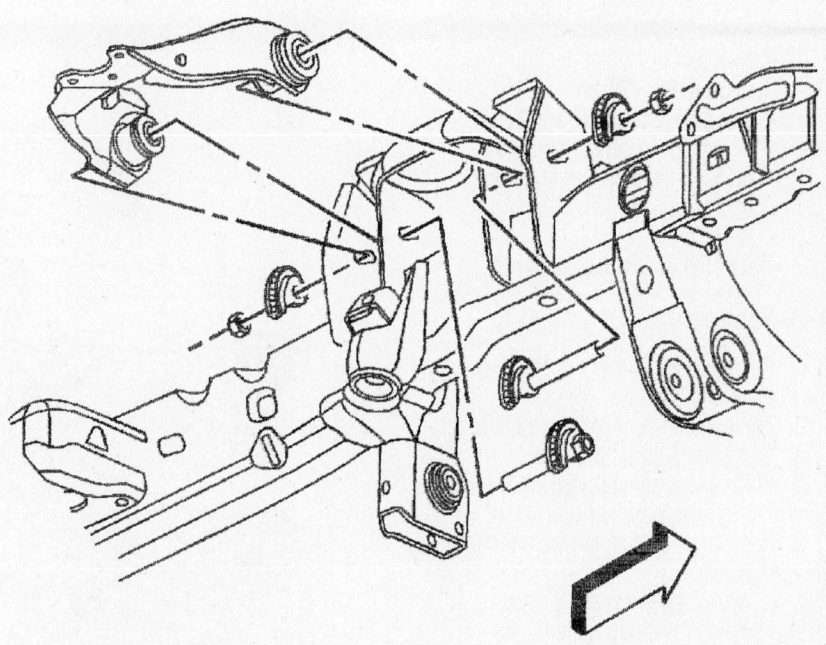

71461-COLO-G27

Upper control arm mounting—2WD

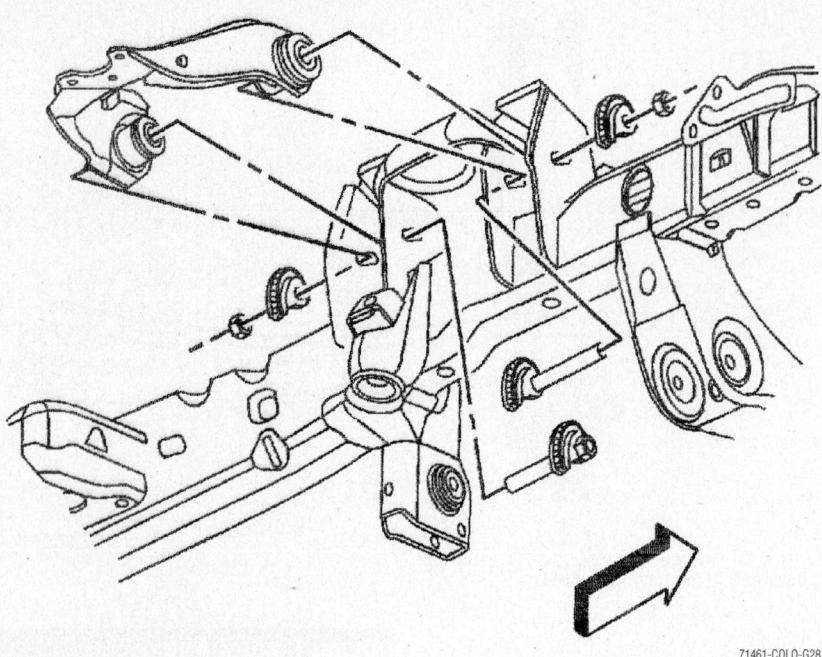

71461-COLO-G28

Upper control arm mounting—4WD

- Wheel speed sensor harness bracket retaining bolt and nut, if equipped
- Brake hose from upper arm
- Upper ball joint nut
- Upper arm from steering knuckle
- Mounting bolts
- Upper control arm

To install:

6. Install or connect the following:
- Upper control arm and tighten the bolts to 118 ft. lbs. (160 Nm).
- Steering knuckle to upper control arm ball joint
- New ball joint stud nut and tighten to 55 ft. lbs. (75 Nm)
- Brake hose to upper arm
- Wheel speed sensor harness bracket retaining bolt and nut, if equipped
- Wheel

7. Check the front wheel alignment.

4WD Vehicles

1. Before servicing the vehicle, refer to the Precautions Section.
2. Disconnect the negative battery cable.
3. Raise and support the vehicle.
4. Place a jack stand under the lower control arm to support it at ride height.
5. Remove or disconnect the following:
- Wheel
- Wheel speed sensor brackets
- Brake hose from upper arm
- Upper arm from ball joint stud nuts

➡The 4WD vehicle upper control arm bolts are equipped with cams, which are rotated to achieve caster and camber adjustments. In order to preserve adjustment and ease installation, matchmark the cams to the control arm before removal. If the control arm is being replaced, transfer the alignment marks to the new component before installation.

- Nuts and adjustments cams
- Upper arm bolts and upper arm

To install:

6. Install or connect the following:

- Upper arm and mounting bolts. Tighten the bolts to 114 ft. lbs. (155 Nm).

7. Align the cams to the reference marks made earlier, then tighten the cam nuts to 114 ft. lbs. (155 Nm).
- Ball joint stud nuts and tighten to 47 ft. lbs. (64 Nm).
- Brake hose to upper arm
- Wheel speed sensor brackets
- Wheel

Lower Control Arm

REMOVAL & INSTALLATION

2WD Vehicles

1. Before servicing the vehicle, refer to the Precautions Section.
2. Raise and support the vehicle.
Remove or disconnect the following:
- Front wheel
- Stabilizer bar links from control arm
- Lower shock nut and through bolt
- Lower ball joint stud from steering knuckle

➡The 2WD vehicle lower control arm bolts are equipped with cams, which are rotated to achieve caster and camber adjustments. In order to preserve adjustment and ease installation, matchmark the cams to the control arm before removal. If the control arm is being replaced, transfer the alignment marks to the new component before installation.

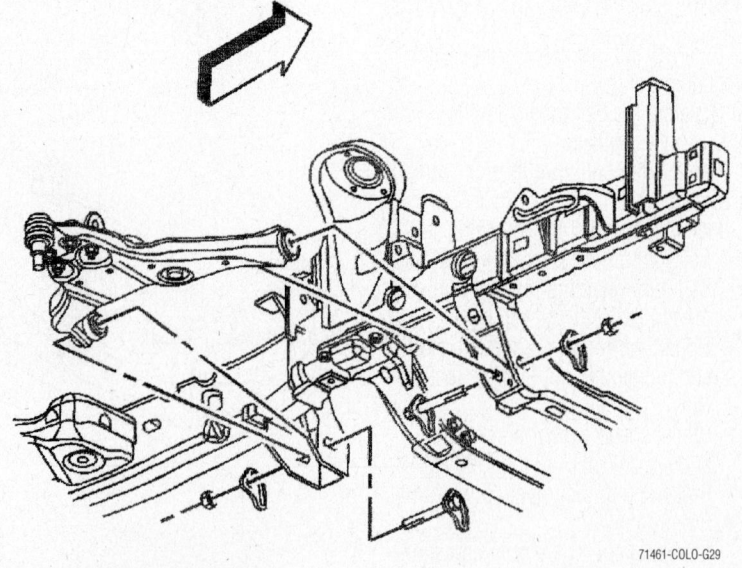

71461-COLO-G29

Lower control arm mounting—2WD

- Adjustment cam nuts and cams
- Lower control arm bolts and control arm

To install:

3. Install or connect the following:

➡**The nuts must be tightened in sequence.**

- Lower control arm. Tighten the rear nuts first, then the front nuts to 114 ft. lbs. (155 Nm).
- Lower ball joint stud into the steering knuckle
- Ball joint-to-steering knuckle nut and tighten to 107 ft. lbs. (145 Nm).
- Lower shock nut and through bolt and tighten
- Stabilizer bar links
- Front wheel

4. Check the front wheel alignment.

4WD Vehicles

1. Before servicing the vehicle, refer to the Precautions Section.
2. Raise and support the vehicle.
3. Remove or disconnect the following:

- Front wheels
- Steering knuckle
- Stabilizer bar links from the control arm
- Lower shock bolt and nut
- Torsion bar, if necessary
- Lower control arm

To install:

4. Install the control arm, bolts and nuts. Install the washer on the front bolt with the shoulder facing the control arm.

➡**The nuts must be tightened in sequence.**

5. Tighten the rear nut first to 107 ft. lbs. (145 Nm), then tighten the front nut to 122 ft. lbs. (165 Nm).
6. Install or connect the following:
   - Steering knuckle
   - Torsion bar, if removed
   - Lower shock bolt and nut
   - Stabilizer bar links
   - Front wheels

7. Check and adjust the front end alignment, as necessary.

CONTROL ARM BUSHING REPLACEMENT

1. Before servicing the vehicle, refer to the Precautions Section.
2. Remove the lower control arm.
3. Measure and record the distance from the bushing flange to the bracket.

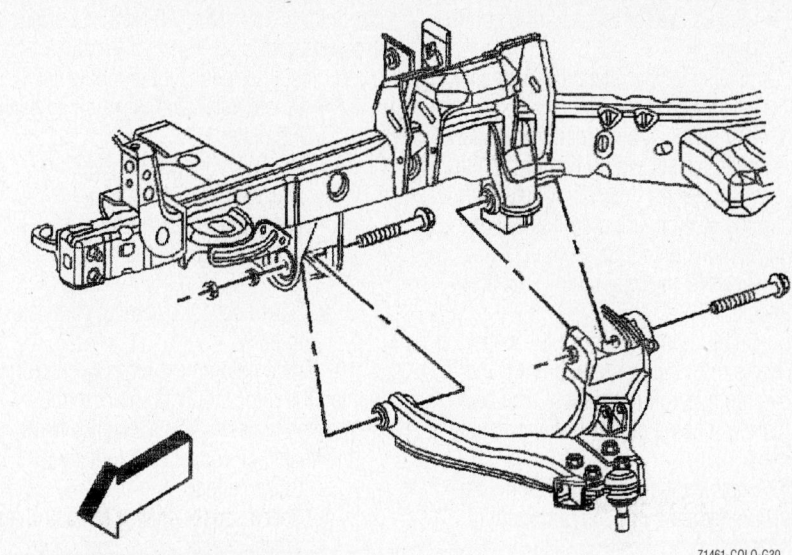

Lower control arm mounting—4WD

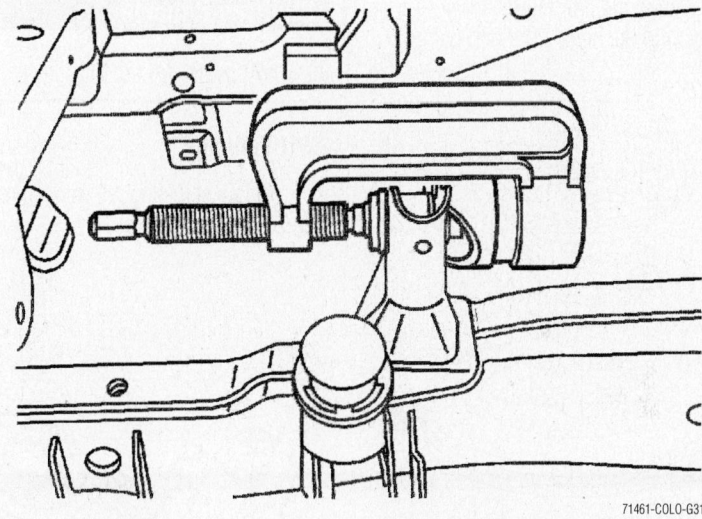

Removing and installing the lower control arm rear bushing using tool J 41805

4. Install tool J 41805 on the rear bushing and tighten until the bushing is removed.

To install:

5. Install the rear bushing into the control arm
6. Install tools J 41805. Tighten until the bushing is fully seated.
7. Install the lower control arm.

Wheel Bearings

ADJUSTMENT

➡**All models use sealed wheel bearings that are pre-adjusted. If the bearing needs replacing, replace the front wheel hub/bearing assembly.**

Front Wheel Hub/Bearing Assembly

REMOVAL & INSTALLATION

➡**To replace the rear bearing see the Axle Shaft, Bearing and Seal procedure in DRIVETRAIN.**

Front

1. Before servicing the vehicle, refer to the Precautions Section.
2. Remove or disconnect the following:
   - Wheel
   - Brake caliper with the pads without disconnecting the brake line
   - Wheel speed sensor harness from upper control arm

- Speed sensor bracket from control arm
- Speed sensor electrical connector from body

3. Using white paint, mark the location of the speed sensor wiring harness to the steering knuckle for installation reference. Coil up the sensor wiring so it is out of the way of suspension components.

4. On 4WD models, remove the steering knuckle.

5. On all models, remove the wheel hub/brake rotor rear mounting bolts and remove the hub/rotor assembly. The backing plate will come off when the assembly is removed.

6. Separate the brake rotor from the wheel hub by removing the 6 mounting bolts.

To install:

7. Clean the contact surface between the wheel hub and brake rotor.

8. Position the new hub/bearing assembly onto the brake rotor and tighten

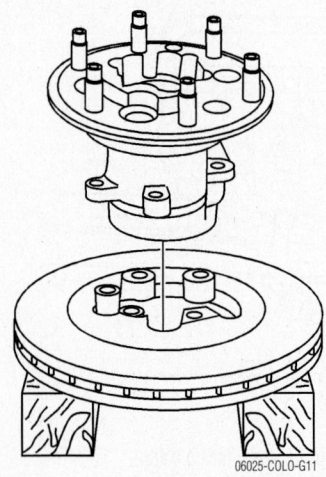

Separating the wheel hub from the brake rotor.

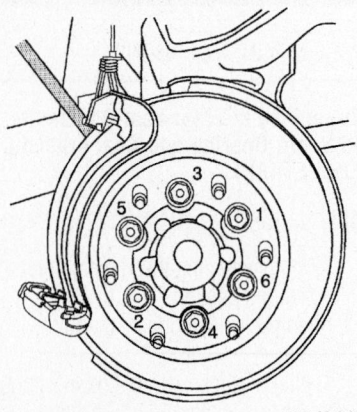

Brake rotor/wheel hub mounting bolt tightening sequence

the bolts to 15 ft. lbs. (20 Nm) in a criss-cross pattern.

9. Install or connect the following:
- Wheel hub/rotor assembly to backing plate
- On 4WD models, the steering knuckle
- Wheel hub/rotor to steering knuckle. Tighten the bolts to 92 ft. lbs. (125 Nm).
- Speed sensor electrical connector to body

10. While holding the rotor from turning, tighten the rotor mounting bolts in the sequence shown to 88 ft. lbs. (120 Nm).

11. Install or connect the following:
- Brake caliper
- Speed sensor bracket to control arm
- Wheel speed sensor harness to upper control arm
- Wheel

Steering Knuckle

REMOVAL & INSTALLATION

2WD Vehicles

1. Before servicing the vehicle, refer to the Precautions Section.

2. Raise and support the vehicle.

3. Place a jack stand under the lower control arm.

4. Remove or disconnect the following:
- Wheel
- Wheel hub/bearing assembly
- Outer tie rod end
- Upper and lower ball joint nuts and discard
- Separate the ball joints from the steering knuckle
- Steering knuckle

To install:

5. Clean the ball joints and tie rod ends. Inspect the tapered holes and mounting surfaces of the steering knuckle for damage or being out of round. Replace the knuckle if the holes are damaged.

6. Install the steering knuckle and connect the lower ball joint. Tighten the nut to 107 ft. lbs. (145 Nm).

7. Connect the upper ball joint to the knuckle and tighten the nut to 55 ft. lbs. (75 Nm).

8. Install or connect the following:
- Outer tie rod end
- Wheel hub/bearing assembly
- Wheel

9. Check the front wheel alignment.

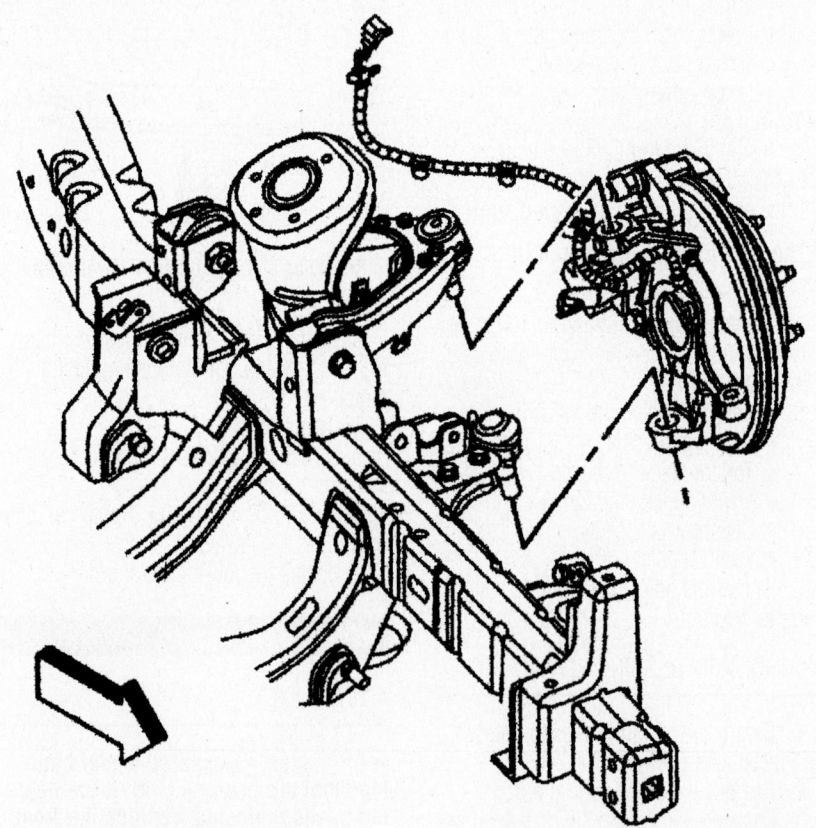

Steering knuckle mounting—2WD

10. Install or connect the following:
- Outer tie rod end
- Speed sensor harness to chassis harness and fender panel
- Brake caliper bracket
- New wheel hub nut and tighten to 191 ft. lbs. (260 Nm).
- Wheel

11. Check the front wheel alignment.

Stabilizer Bar

REMOVAL & INSTALLATION

Front

1. Before servicing the vehicle, refer to the Precautions Section.
2. Raise and support the vehicle.
3. Remove or disconnect the following:
- Front wheels
- Stabilizer link nuts from control arm
- Stabilizer links
- Stabilizer bar insulator clamps
- Stabilizer bar
- Insulators

To install:

4. Install new insulators on the stabilizer bar.
5. Install or connect the following:
- Stabilizer bar and insulator clamps. Tighten the bolts to 37 ft. lbs. (50 Nm).

6. Support the lower control arms at ride height.

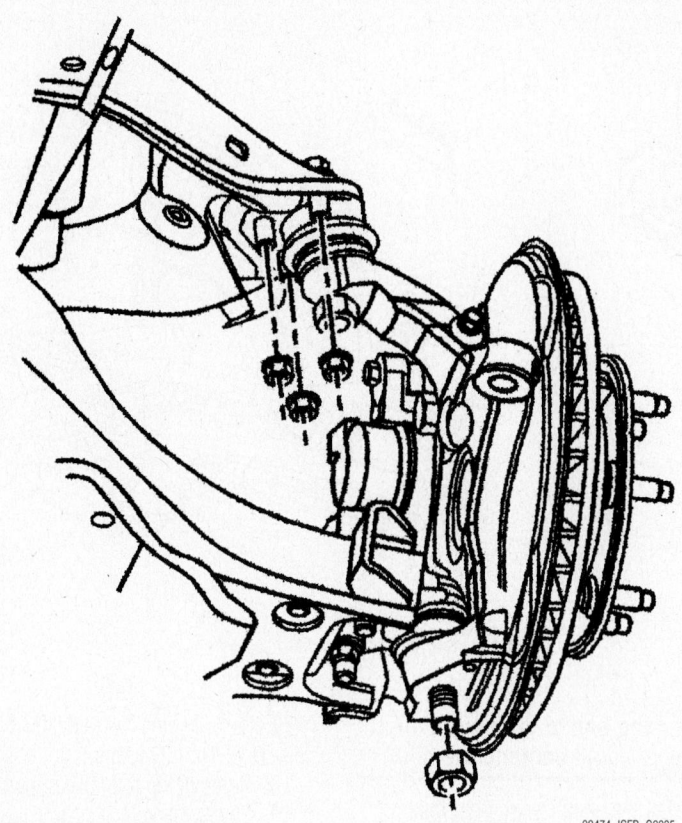

09474_ISER_G0035

Steering knuckle mounting—4WD

4WD Vehicles

1. Before servicing the vehicle, refer to the Precautions Section.
2. Raise and support the vehicle.
3. Place a jack stand under the lower control arm.
4. Remove or disconnect the following:
- Wheel
- Wheel hub nut and discard
- Brake caliper bracket
- Speed sensor harness from chassis harness and fender panel
- Outer tie rod end from steering knuckle
- Upper ball joint stud nuts
- Lower ball joint nut
- Separate ball joint from knuckle
- Steering knuckle

5. If the knuckle is being replaced, remove the wheel hub/bearing assembly.

To install:

6. Clean the ball joints and tie rod ends. Inspect the tapered holes and mounting surfaces of the steering knuckle for damage or being out of round. Replace the knuckle if the holes are damaged.
7. If removed, install the wheel hub/bearing assembly.
8. Install the steering knuckle and con-

nect the lower ball joint. Tighten the nut to 107 ft. lbs. (145 Nm).
9. Connect the upper ball joint to the knuckle and tighten the nut to 55 ft. lbs. (75 Nm).

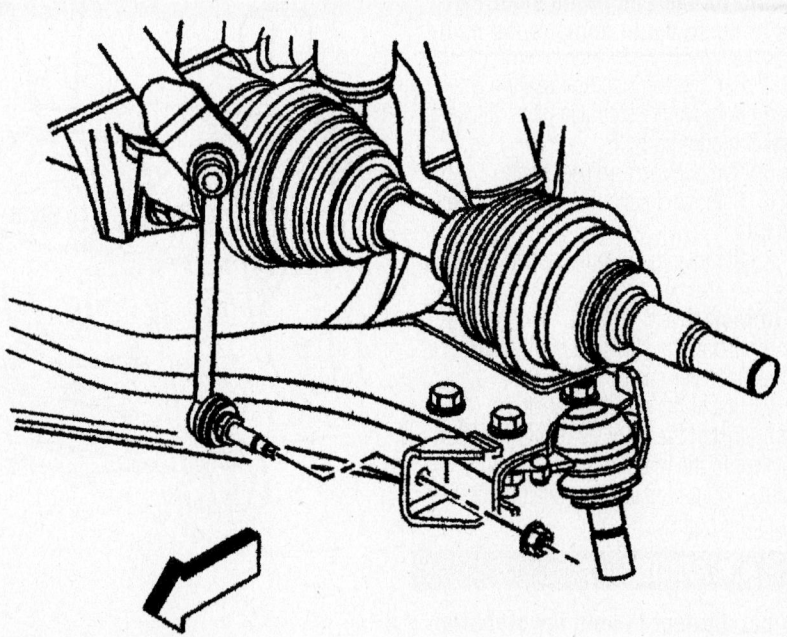

09474_ISER_G0036

Stabilizer bar end link attachment

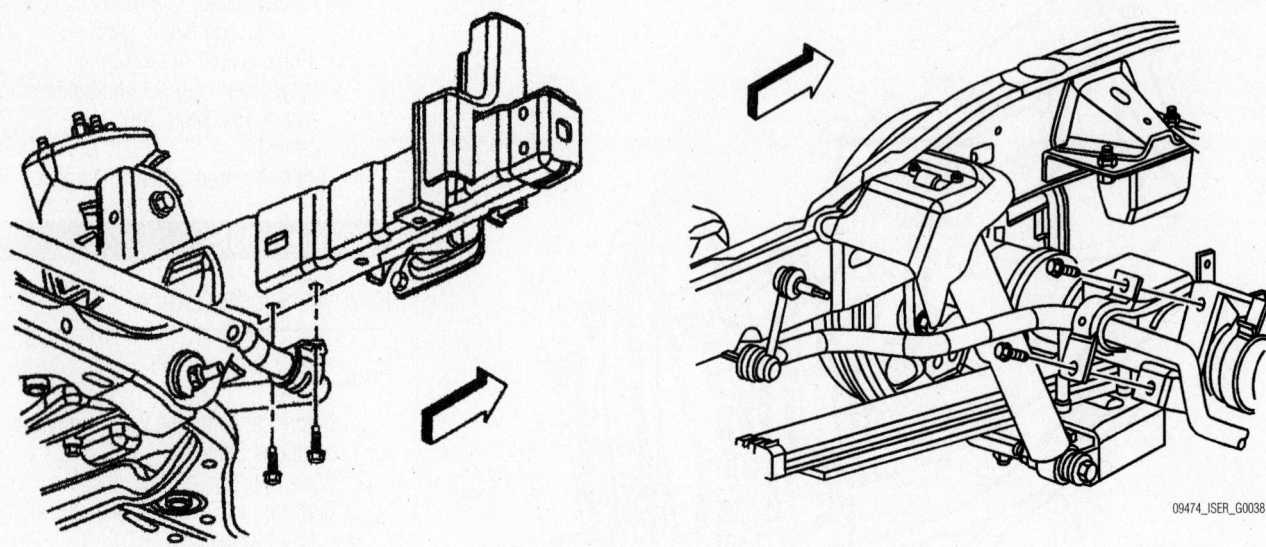

09474_ISER_G0037

Stabilizer bar mounting bracket attachment

Rear stabilizer bar bracket

09474_ISER_G0038

7. Install the stabilizer links and tighten the nuts to 32 ft. lbs. (44 Nm).

8. Install the front wheels and lower the vehicle.

Rear

1. Raise and support the vehicle.
2. Support the rear axle at curb height.

✳✳ WARNING

Do not attempt to hold the stabilizer shaft link near the boot. Use the hex feature on the end of the stud. Failure to do so could damage the boot.

3. Remove the stabilizer shaft to link nut.
4. Remove the stabilizer shaft insulator bracket mounting bolts.
5. Remove the stabilizer shaft.
6. Remove the stabilizer shaft insulator brackets.
7. Remove the stabilizer shaft insulators.

To install:

8. Install the stabilizer shaft insulators to the stabilizer shaft.
9. Install the stabilizer shaft.
10. Install the stabilizer shaft insulator brackets to the rear axle.
11. Install the stabilizer shaft bracket mounting bolts.

✳✳ WARNING

Do not attempt to hold the stabilizer shaft link near the boot. Use the hex

feature on the end of the stud. Failure to do so could damage the boot.

12. Install the stabilizer shaft to link nut. Tighten the stabilizer bar link nut to 44 Nm

(32 ft. lbs.). Tighten the insulator bracket bolts to 50 Nm (37 ft. lbs.).

13. Remove the rear axle support.
14. Lower the vehicle.

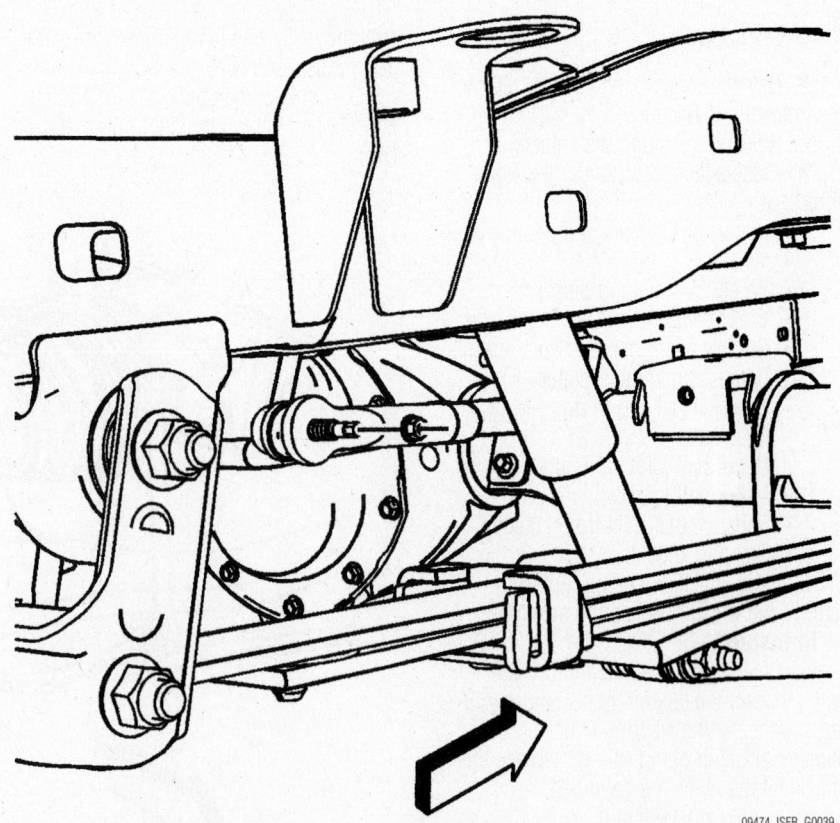

Rear stabilizer bar end link attachment

09474_ISER_G0039

DISC BRAKES

Brake Rotor

REMOVAL & INSTALLATION

1. Inspect the brake fluid level in the brake master cylinder reservoir.

2. If the brake fluid is midway between the maximum-full point and the minimum allowable level, no brake fluid needs to be removed from the reservoir before proceeding.

3. If the brake fluid level is higher than midway between the maximum-full point and the minimum allowable level, using an appropriate tool, remove the brake fluid to the midway point before proceeding.

➡**The following service procedure is for 4WD vehicles ONLY.**

4. Remove the steering knuckle from the vehicle.

5. Remove the wheel hub/speed sensor assembly.

➡**In the following service procedure, support the wheel hub/brake rotor assembly with wood when working on the bench.**

6. Using two pieces of wood, support the wheel hub/brake rotor assembly.

7. Remove the wheel hub/speed sensor to brake rotor bolts.

8. Separate the brake rotor from the wheel bearing/hub assembly.

9. Clean the surface between the wheel bearing/hub assembly and the brake rotor.

To install:

10. Position the wheel bearing/hub on the brake rotor.

➡**The following service procedure is to be performed on a flat surface and to ensure the brake rotor is securely attached to the hub assembly prior to the final torquing procedure. DO NOT use air tools of any type for this procedure.**

11. Install the wheel bearing/hub mounting bolts. Tighten the mounting bolts in a criss-cross pattern to 20 Nm (15 ft. lbs.).

➡**The following service procedure is for 4WD vehicles only.**

12. Install the steering knuckle assembly.

13. Install the wheel bearing/hub, brake rotor assembly in the steering knuckle.

14. Lower the vehicle.

15. Inspect the fluid level.

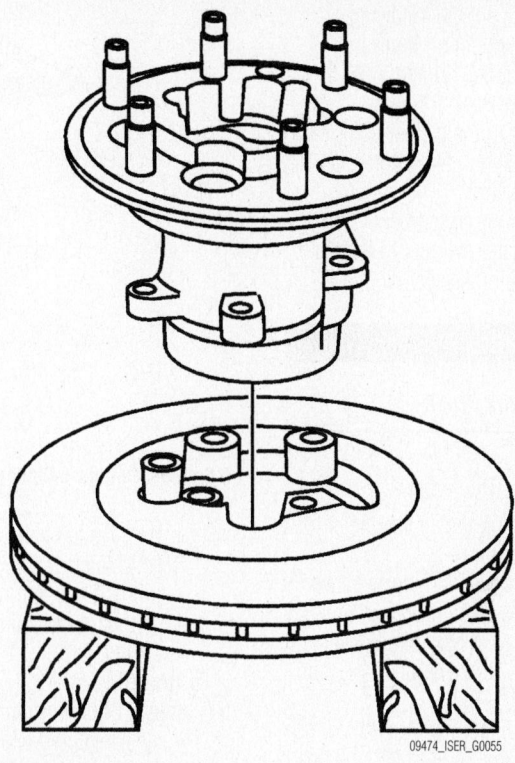

09474_ISER_G0055

Hub/rotor separation

Brake Caliper

REMOVAL & INSTALLATION

1. Before servicing the vehicle, refer to the Precautions Section.

2. If brake fluid level is midway between MAX and MIN level in the reservoir, no fluid needs to be removed. If brake fluid level is higher than midway, remove the brake fluid to the midway point.

3. Remove or disconnect the following:

- Tire and wheel assembly
- Brake caliper fluid line, then plug it
- Caliper slide pin bolts
- Caliper from the mounting bracket

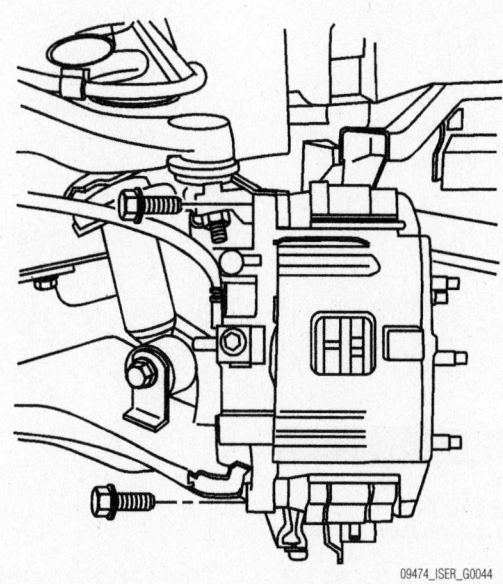

09474_ISER_G0044

Brake caliper pins

To install:

4. Clean and lubricate the sleeves and bushings with silicon grease.

5. Install or connect the following:
- Caliper in mounting bracket
- Slide pin bolts. Tighten to 29 ft. lbs. (40 Nm).
- Fluid lines to the caliper using new copper washers, and tighten to 29 ft. lbs. (40 Nm)
- Wheel and tire assembly

6. Refill the master cylinder to the correct level. Bleed the brake system.

Disc Brake Pads

REMOVAL & INSTALLATION

Front

1. Before servicing the vehicle, refer to the Precautions Section.

2. If brake fluid level is midway between MAX and MIN level in the reservoir, no fluid needs to be removed. If brake fluid level is higher than midway, remove the brake fluid to the midway point.

3. Remove or disconnect the following:

4. Place a C-clamp around the outer pad and caliper; tighten the C-clamp until the piston is fully compressed in the caliper.

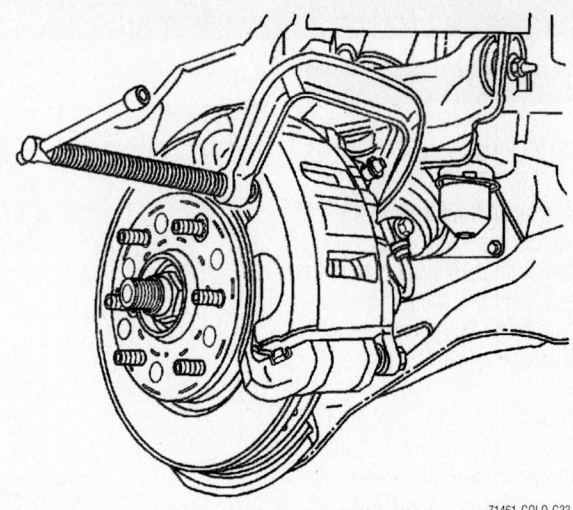

71461-COLO-G33

Compressing the caliper piston with a C-clamp

- Remove top caliper retainer, and rotate caliper away from rotor
- Inboard pad and retaining clips
- Outboard pad from the caliper
- Shims

To install:

5. Install or connect the following:
- New retaining clips onto the inboard pad
- Shims

- Outboard pad into the caliper
- Inboard pad in the caliper
- Caliper in position over the rotor
- Caliper bolts and tighten to 29 ft. lbs. (40 Nm).
- Wheel and tire

6. Refill the master cylinder and pump pedal to attain full brake pedal before road-testing the vehicle.

DRUM BRAKES

Brake Drums

REMOVAL & INSTALLATION

1. Before servicing the vehicle, refer to the Precautions Section.

2. Remove or disconnect the following:
- Wheel and tire assembly
- Retaining clip
- Brake drum. If the drum will not pull of the axle, use a rubber mallet and tap it around the edge.

To install:

3. Install or connect the following:
- Drum on the axle
- Retaining clip
- Wheel and tire assembly

4. Refill the master cylinder and pump pedal to attain full brake pedal before road-testing the vehicle.

Brake Shoes

REMOVAL & INSTALLATION

1. Before servicing the vehicle, refer to the Precautions Section.

2. Remove or disconnect the following:
- Wheel and tire assembly
- Brake drum
- Adjuster assembly
- Retractor spring from secondary shoe
- Secondary shoe
- Retractor spring from primary shoe
- Primary shoe
- Return spring
- Depress lock tab on parking brake cable

- Hold lock tab and push parking brake cable forward
- Parking brake cable from lever

To install:

3. Lubricate the contact points on the backing plate with high temperature silicone grease.

4. Install or connect the following:
- Parking brake cable in the lever
- Primary shoe
- Retractor spring on primary shoe
- Secondary shoe

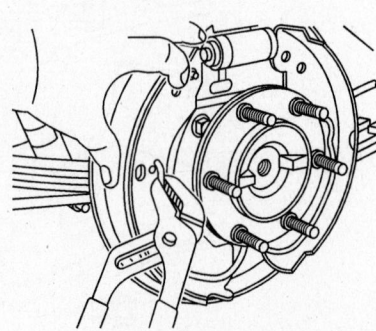

09474_ISER_G0045

Remove the retractor spring from secondary shoe

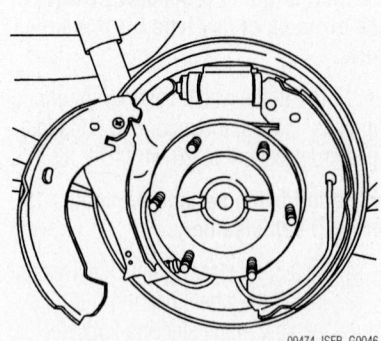

09474_ISER_G0046

Remove the secondary shoe

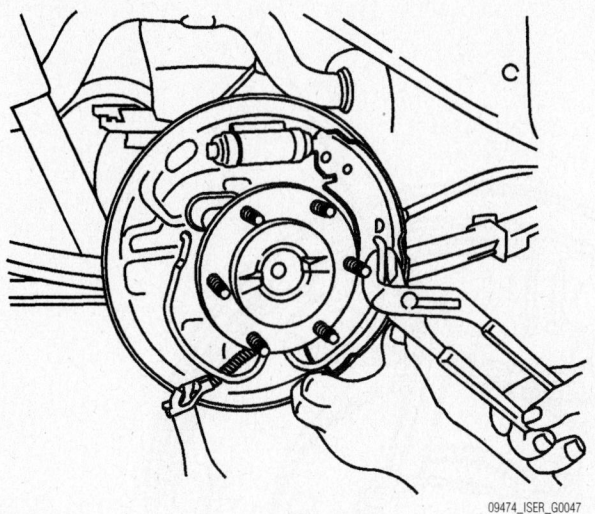

09474_ISER_G0047

Remove the retractor spring from primary shoe

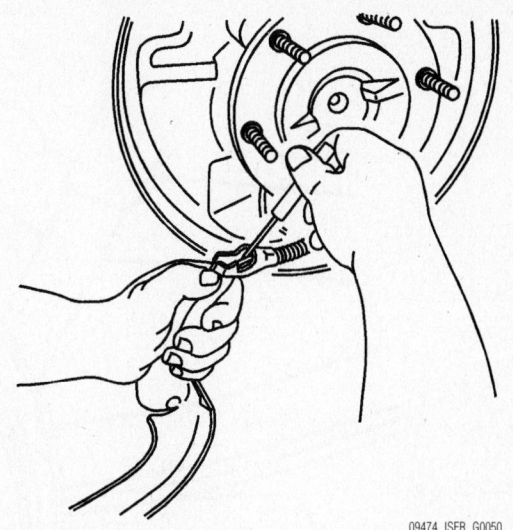

09474_ISER_G0050

Depress lock tab on parking brake cable

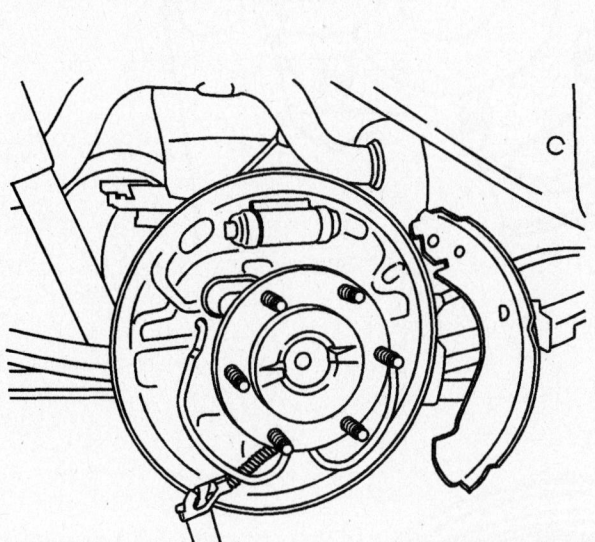

09474_ISER_G0048

Remove the primary shoe

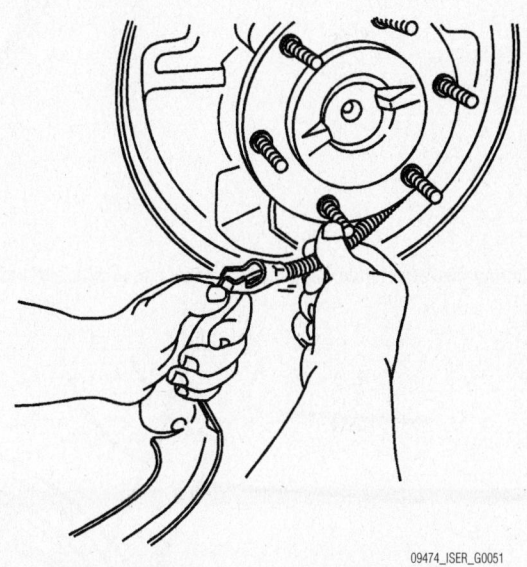

09474_ISER_G0051

Hold lock tab and push parking brake cable forward

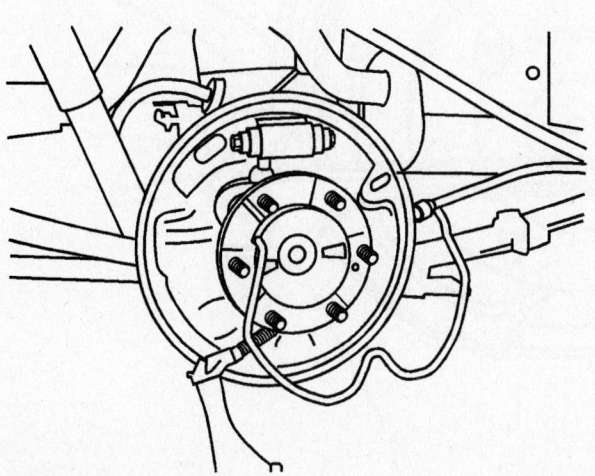

09474_ISER_G0049

Remove the return spring

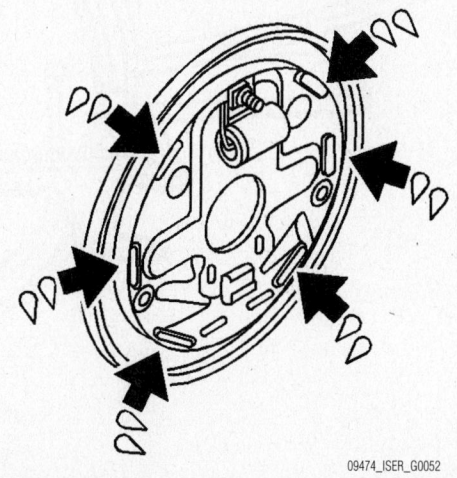

09474_ISER_G0052

Lubricate the contact points on the backing plate with high temperature silicone grease

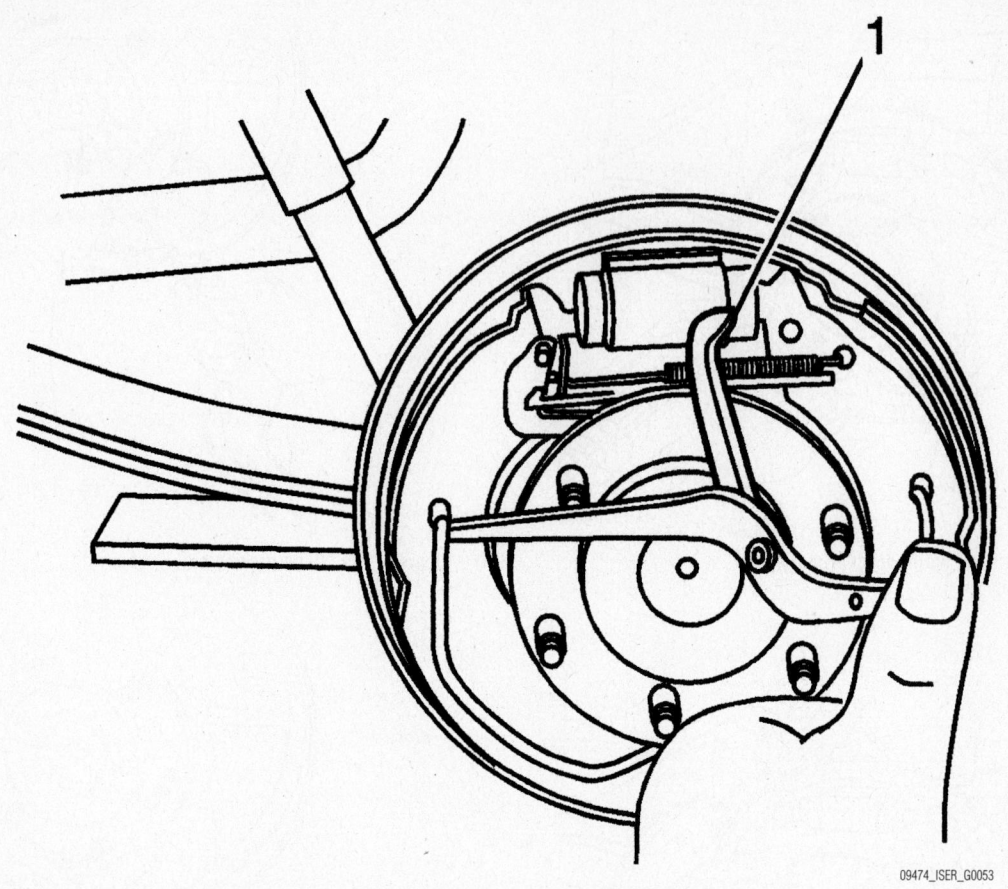

Installing the retractor spring on the secondary shoe with a brake spring tool (1)

09474_ISER_G0053

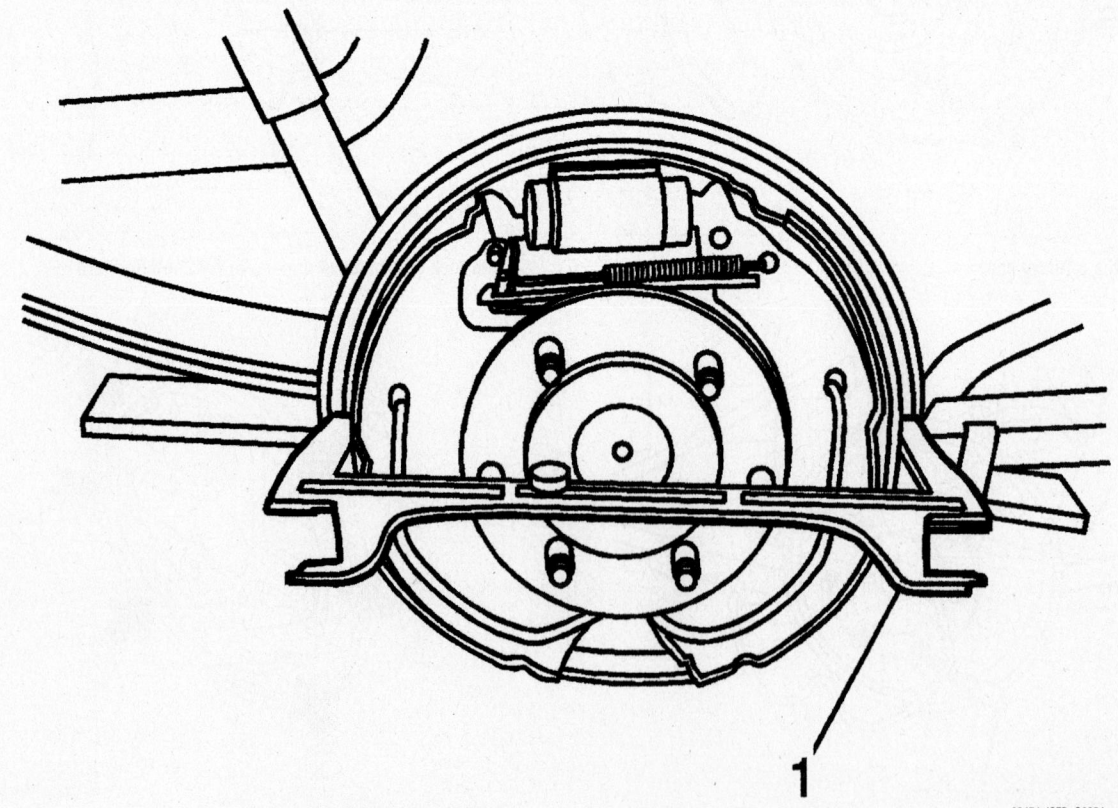

Measuring the outside diameter of the shoes with a measuring gauge

09474_ISER_G0054

- Retractor spring on secondary shoe
- Adjuster assembly

5. Adjust the brake shoes so there is 0.030 inch (0.76mm) clearance between the lining and the drum.

6. Install the brake drum.

7. Adjust the parking brake cable as necessary.

8. Install the wheel and tire assemblies.

9. Refill the master cylinder and pump pedal to attain full brake pedal before road-testing the vehicle.

BRAKE SHOE ADJUSTMENT

➡**The rear brakes must be adjusted manually after replacing the rear brake shoes.**

1. Raise the vehicle.

2. Loosen the adjuster nut for the park brakes.

3. Remove the brake drum.

4. Using a measuring gauge, measure the inside diameter of the brake drum.

➡**When performing the following service procedure, ensure that the measuring gauge is level, before taking the measurement.**

5. Adjust the brakes shoes until there is approximately 0.76mm (.030 in.) clearance between the rear brake shoes and the rear brake drum.

6. Install the brake drum.

7. With heavy force, apply the brake pedal 2 times in order to center and adjust the brake shoes.

8. Inspect the drums again for light drag. If necessary, repeat the previous step for adjusting the brakes.

9. Lower the vehicle.

10. Adjust the park brake.

PARKING BRAKE ADJUSTMENT

1. Release the park brake pedal.

2. Raise the vehicle.

3. Clean the threads on the front park brake cable.

4. Adjust the park brake until the right rear brake is locked.

5. Apply and release the park brake 5 times.

6. With the park brake in the release position, adjust the park brake until the right rear brake develops a slight drag.

7. Back off the adjusting nut 2 complete turns.

Wheel Cylinder

REMOVAL & INSTALLATION

1. Raise the vehicle.

2. Remove the rear brake shoe assembly.

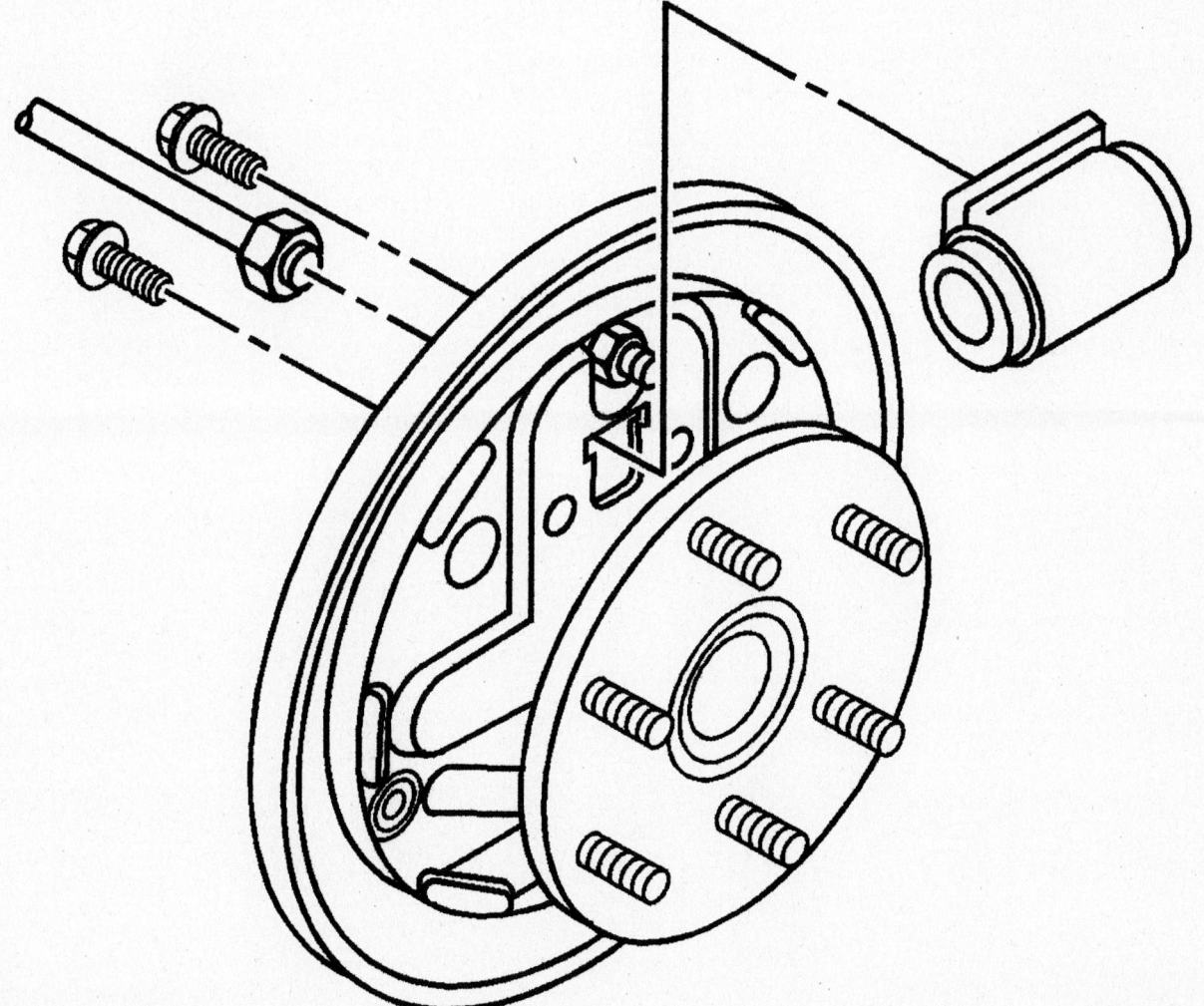

Wheel cylinder mounting

09474_ISER_G0056

3. Remove the brake line from the wheel cylinder.

4. Using a plug, plug the brake line as not to allow dirt or the brake fluid to be contaminated.

5. Remove the mounting bolts for the wheel cylinder.

6. Remove the wheel cylinder from the backing plate.

To install:

7. Install the wheel cylinder to the backing plate.

8. Install the wheel cylinder mounting bolts. Tighten the wheel cylinder mounting bolts to 18 Nm (13 ft. lbs.).

9. Remove the rubber plug from the brake line.

10. Install the brake line to the wheel cylinder.

11. Install the brake line fitting. Tighten the brake line fitting to 19 Nm (14 ft. lbs.).

12. Install the rear brake shoes.

13. Bleed the brake system.

14. Lower the vehicle.

15. Check the fluid lever in the master cylinder.

ISUZU/HONDA

Passport • Rodeo • Rodeo Sport • Trooper

SPECIFICATION AND MAINTENANCE CHARTS

ENGINE AND VEHICLE IDENTIFICATION

		Engine						Model Year	
Code	Liters (cc)	Cu. In.	Cyl.	Fuel Sys.	Engine Type	Eng. Mfg.		Code ①	Year
D	2.2 (2198)	134	4	MFI	DOHC	Isuzu		2	2002
W	3.2 (3165)	193	6	MFI	DOHC	Isuzu		3	2003
X	3.5 (3494)	213	6	MFI	DOHC	Isuzu		4	2004
Y	3.5 (3494)	213	6	MFI	DOHC	Isuzu			

MFI: Multi-port Fuel Injection

DOHC: Double Overhead Camshaft

① 10th position of VIN

09474_RODE_C0001

GENERAL ENGINE SPECIFICATIONS

Year	Model	Engine Displacement Liter	Engine Series VIN	Net Horsepower @ rpm	Net Torque @ rpm (ft. lbs.)	Bore x Stroke (in.)	Compression Ratio	Oil Pressure @ rpm
2002	Rodeo	2.2	D	130@5200	144@4000	3.39x3.72	10.0:1	22@800
		3.2	W	205@5400	214@3000	3.68x3.03	9.1:1	60-80@3000
	Rodeo Sport	2.2	D	130@5200	144@4000	3.39x3.72	10.0:1	22@800
		3.2	W	205@5400	214@3000	3.68x3.03	9.1:1	60-80@3000
	Passport	3.2	W	205@5400	214@3000	3.68x3.03	9.1:1	57-80@3000
	Trooper	3.5	X	215@5400	230@3000	3.68x3.35	9.1:1	60-80@3000
2003	Rodeo	2.2	D	130@5200	144@4000	3.39x3.72	10.0:1	22@800
		3.2	W	205@5400	214@3000	3.68x3.03	9.1:1	60-80@3000
	Rodeo Sport	2.2	D	130@5200	144@4000	3.39x3.72	10.0:1	22@800
		3.2	W	205@5400	214@3000	3.68x3.03	9.1:1	60-80@3000
2004	Rodeo	3.2	W	205@5400	214@3000	3.68x3.03	9.1:1	60-80@3000
		3.5	Y	250@NA	246@NA	3.68x3.35	10.3:1	NA

NA: Information not available

09474_RODE_C0002

ENGINE TUNE-UP SPECIFICATIONS

Year	Engine Displacement Liters	Engine VIN	Spark Plug Gap (in.)	Ignition Timing (deg.) MT	Ignition Timing (deg.) AT	Fuel Pump (psi)	Idle Speed (rpm) MT	Idle Speed (rpm) AT	Valve Clearance In.	Valve Clearance Ex.
2002	2.2	D	0.040	①	①	41-55	800	800	HYD	HYD
	3.2	W	0.040	①	①	48-55	750	750	0.009-0.013	0.010-0.014
	3.5	X	0.043	①	①	48-55	750	750	0.009-0.013	0.010-0.014
2003	2.2	D	0.040	①	①	41-55	800	800	HYD	HYD
	3.2	W	0.040	①	①	48-55	750	750	0.009-0.013	0.010-0.014
2004	3.2	W	0.040	①	①	48-55	750	750	0.009-0.013	0.010-0.014
	3.5	Y	0.032-0.035	①	①	NA	750	750	0.009-0.013	0.010-0.014

NA: Information not available

NOTE: The Vehicle Emission Control Information label figures must be used if they differ from those in this chart.

HYD: Hydraulic

① Controlled by the PCM

09474_RODE_C0003

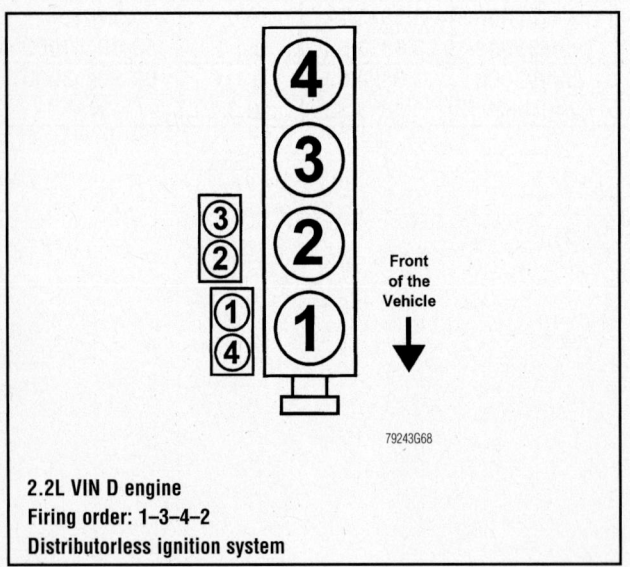

2.2L VIN D engine
Firing order: 1–3–4–2
Distributorless ignition system

79243G68

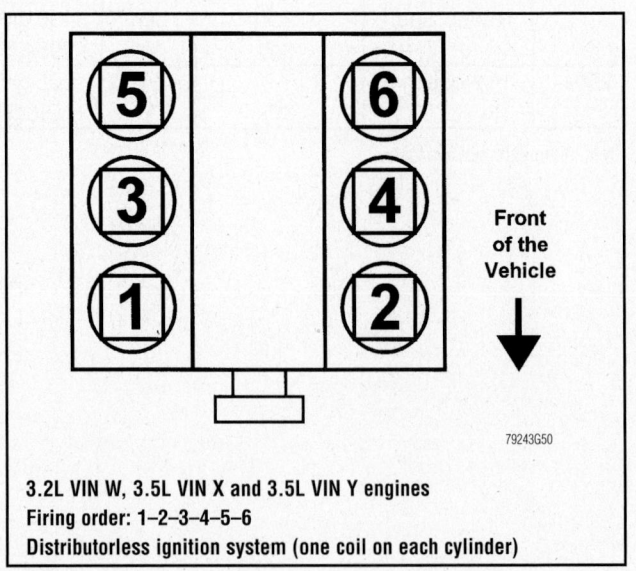

3.2L VIN W, 3.5L VIN X and 3.5L VIN Y engines
Firing order: 1–2–3–4–5–6
Distributorless ignition system (one coil on each cylinder)

79243G50

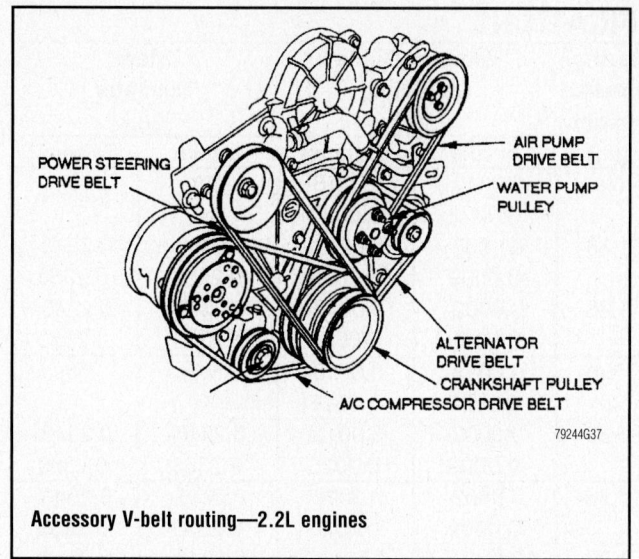

Accessory V-belt routing—2.2L engines

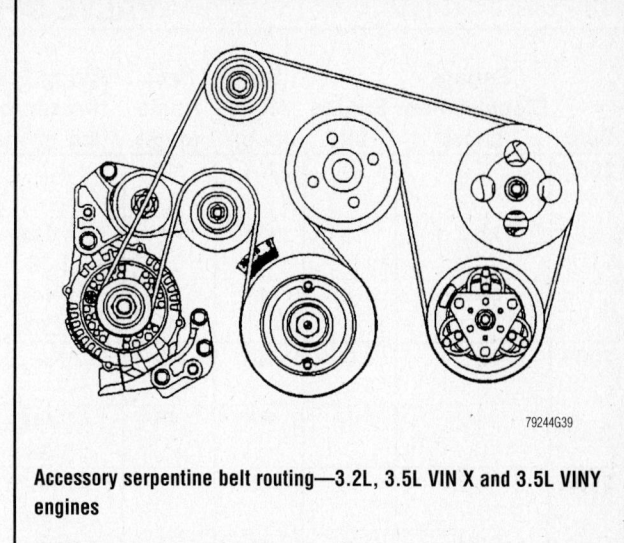

Accessory serpentine belt routing—3.2L, 3.5L VIN X and 3.5L VINY engines

CAPACITIES

Year	Model	Engine Displacement Liters	Engine VIN	Oil with Filter (qts.)	Transmission (pts.) Man.	Transmission (pts.) Auto.	Transfer Case (pts.)	Drive Axle Front (pts.)	Drive Axle Rear (pts.)	Fuel Tank (gal.)	Cooling System (qts.)
2002	Rodeo Sport	2.2	D	4.8	6.2	—	3.0	2.2	3.7	17.7	7.3
		3.2	W	5.0	6.2	18.2	3.0	2.6	3.7	17.7	11.2
	Rodeo	2.2	D	4.8	6.2	—	3.0	2.2	3.7	19.5	7.3
		3.2	W	5.0	6.2	18.2	3.0	2.6	3.7	19.5	11.6
	Passport	3.2	W	5.0	6.2	18.2	3.0	2.6	3.74	19.5	11.6
	Trooper	3.5	X	5.0	5.8	18.2	3.0 ①	3.0	6.4	22.5	②
2003	Rodeo Sport	2.2	D	4.8	6.2	—	2.7	2.2	3.7	15.6	7.3
		3.2	W	5.0	6.2	18.2	2.7	2.6	3.7	15.6	11.2
	Rodeo	2.2	D	4.8	6.2	—	2.7	2.2	3.7	19.5	7.3
		3.2	W	5.0	6.2	18.2	2.7	2.6	3.7	19.5	11.2
2004	Rodeo	3.2	W	5.0	6.2	18.2	2.7	2.6	3.7	19.5	11.2
		3.5	Y	5.0	6.2	18.4	2.7	2.6	3.7	19.5	11.2

NOTE: All capacities are approximate. Add fluid gradually and check to ensure a proper level has been reached.

① 4.0 pts. if equipped with Torque On Demand

② A/T: 7.4 qts.

  M/T: 7.0 qts.

VALVE SPECIFICATIONS

Year	Engine Displacement Liters	Engine VIN	Seat Angle (deg.)	Face Angle (deg.)	Spring Test Pressure (lbs. @ in.)	Spring Installed Height (in.)	Stem-to-Guide Clearance (in.)		Stem Diameter (in.)	
							Intake	Exhaust	Intake	Exhaust
2002	2.2	D	NA	NA	NA	NA	0.0012-0.0022	0.0016-0.0026	NA	NA
	3.2	W	45	45	41-44@1.38	1.38	0.0002-0.0009	0.0012-0.0025	0.2346-0.2353	0.2343-0.2350
	3.5	X	45	45	41-44@1.38	1.38	0.0002-0.0009	0.0012-0.0025	0.2346-0.2353	0.2343-0.2350
2003	2.2	D	NA	NA	NA	NA	0.0012-0.0022	0.0016-0.0026	NA	NA
	3.2	W	45	45	41-44@1.38	1.38	0.0002-0.0009	0.0012-0.0025	0.2346-0.2353	0.2343-0.2350
2004	3.2	W	45	45	41-44@1.38	1.38	0.0002-0.0009	0.0012-0.0025	0.2346-0.2353	0.2343-0.2350
	3.5	Y	45	45	44@1.26	1.26	0.0002-0.0009	0.0012-0.0025	0.2346-0.2353	0.2343-0.2350

NA: Not Available

09474_RODE_C0005

CRANKSHAFT AND CONNECTING ROD SPECIFICATIONS

All measurements are given in inches.

Year	Engine Displacement Liters	Engine VIN	Crankshaft				Connecting Rod		
			Main Brg. Journal Dia.	Main Brg. Oil Clearance	Shaft End-play	Thrust on No.	Journal Diameter	Oil Clearance	Side Clearance
2002	2.2	D	2.2590-2.2610	0.0007-0.0016	0.0004-0.0008	2	1.9090-1.9100	0.0002-0.0012	0.0050-0.0150
	3.2	W	2.5165-2.5170	0.0007-0.0017	0.0024-0.0094	3	2.1229-2.1235	0.0010-0.0023	0.0050-0.0150
	3.5	X	2.5165-2.5170	0.0007-0.0017	0.0024-0.0094	3	2.1229-2.1235	0.0010-0.0023	0.0050-0.0150
2003	2.2	D	2.2590-2.2610	0.0007-0.0016	0.0004-0.0008	2	1.9090-1.9100	0.0002-0.0012	0.0050-0.0150
	3.2	W	2.5165-2.5170	0.0007-0.0017	0.0024-0.0094	3	2.1229-2.1235	0.0010-0.0023	0.0050-0.0150
2004	3.2	W	2.5165-2.5170	0.0007-0.0017	0.0024-0.0094	3	2.1229-2.1235	0.0010-0.0023	0.0050-0.0150
	3.5	Y	2.5165-2.5170	0.0007-0.0017	0.0024-0.0094	3	2.1229-2.1235	①	0.0050-0.0150

① Size Mark:

A: 0.0010-0.0021 in.

B: 0.0011-0.0022 in.

C: 0.0011-0.0023 in.

09474_RODE_C0006

PISTON AND RING SPECIFICATIONS

All measurements are given in inches.

Year	Engine Displacement Liters	Engine VIN	Piston Clearance	Ring Gap			Ring Side Clearance		
				Top Compression	Bottom Compression	Oil Control	Top Compression	Bottom Compression	Oil Control
2002	2.2	D	NA	0.0118-0.0195	0.0118-0.0195	0.0159-0.0546	0.0008-0.0546	0.0008-0.0546	NA
	3.2	W	NA	0.0118-0.0157	0.0177-0.0236	0.0060-0.0177	0.0006-0.0015	0.0006-0.0015	NA
	3.5	X	NA	0.0118-0.0157	0.0177-0.0236	0.0059-0.0180	0.0006-0.0015	0.0006-0.0015	NA
2003	2.2	D	NA	0.0118-0.0195	0.0118-0.0195	0.0159-0.0546	0.0008-0.0546	0.0008-0.0546	NA
	3.2	W	NA	0.0118-0.0157	0.0177-0.0236	0.0059-0.0177	0.0006-0.0015	0.0006-0.0015	NA
2004	3.2	W	NA	0.0118-0.0157	0.0177-0.0236	0.0059-0.0177	0.0006-0.0015	0.0006-0.0015	NA
	3.5	Y	NA	0.0079-0.0118	0.0138-0.0197	0.0059-0.0138	0.0010-0.0026	0.0010-0.0026	0.0006-0.0022

NA: Not Available

09474_RODE_C0007

TORQUE SPECIFICATIONS

All readings in ft. lbs.

Year	Engine Displacement Liters	Engine VIN	Cylinder Head Bolts	Main Bearing Bolts	Rod Bearing Bolts	Crankshaft Damper Bolts	Flywheel Bolts	Manifold		Spark Plugs	Oil Pan Drain Plug
								Intake	Exhaust		
2002	2.2	D	①	②	③	④	⑤	16	⑥	18	NA
	3.2	W	⑦	29	40	123	40	18	42	13	NA
	3.5	X	⑦	⑧	40	123	40	18	38	13	NA
2003	2.2	D	①	②	⑨	④	⑤	16	⑥	14	NA
	3.2	W	⑦	29	40	123	40	18	42	13	NA
2004	3.2	W	⑦	29	40	123	40	18	42	13	NA
	3.5	Y	⑩	29	40	123	40	13	42	13	NA

NA: Information not available

① Step 1: 18 ft. lbs.
   Step 2: Plus 90 degrees
   Step 3: Plus 90 degrees
   Step 4: Plus 90 degrees
② Step 1: 37 ft. lbs.
   Step 2: Plus 45 degrees
   Step 3: Plus 15 degrees
③ Step 1: 25 ft. lbs.
   Step 2: Plus 45 degrees
   Step 3: Plus 15 degrees

④ Crankshaft sprocket:
   Step 1: 94 ft. lbs.
   Step 2: Plus 45 degrees
   Crankshaft balancer:
   Step 1: 14 ft. lbs.
   Step 2: Plus 45 degrees
⑤ Step 1: 48 ft. lbs.
   Step 2: Plus 30 degrees
   Step 3: Plus 15 degrees
⑥ Step 1: 112 inch lbs.
   Step 2: 14 ft. lbs.
   Step 3: 14 ft. lbs.

⑦ Step 1: 21 ft. lbs.
   Step 2: 47 ft. lbs.
⑧ Step 1: 22 ft. lbs.
   Step 2: Plus 55-65 degrees
   Step 3: Crankcase side bolts to 29 ft. lbs.
⑨ 26 ft. lbs. +45 degrees
⑩ M11 bolts
   Step 1: 21 ft. lbs.
   Step 2: 47 ft. lbs.
   M8 bolts: 15 ft. lbs.

09474_RODE_C0008

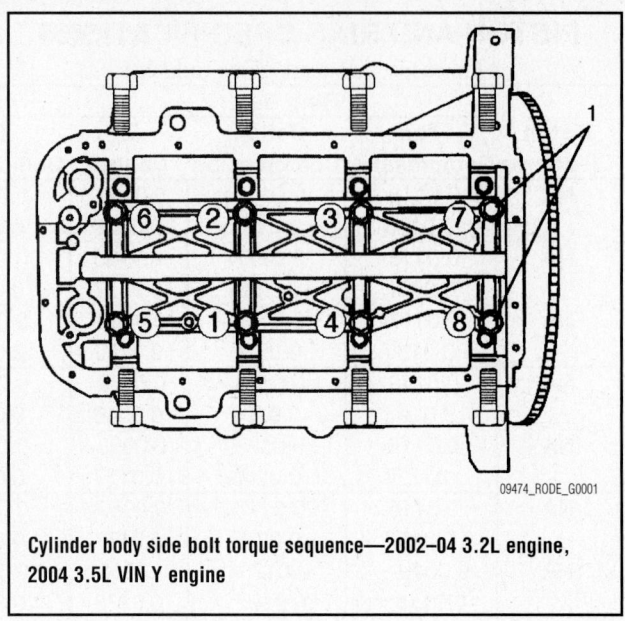

09474_RODE_G0001

Cylinder body side bolt torque sequence—2002–04 3.2L engine,
2004 3.5L VIN Y engine

WHEEL ALIGNMENT

Year	Model		Caster Range (+/-Deg.)	Caster Preferred Setting (Deg.)	Camber Range (+/-Deg.)	Camber Preferred Setting (Deg.)	Toe-in (in.)
2002	Rodeo Sport	F	1.00	+2.50	0.50	0	0+/-0.08
		R	—	—	1.00	0	0+/-0.20
	Rodeo	F	0.75	+2.50	0.50	0	0.04+/-0.04
		R	—	—	1.00	0	0+/-0.20
	Passport	F	1.00	+2.50	0.50	0	0+/-0.08
		R	—	—	1.00	0	0+/-0.20
	Trooper		0.75	+2.10	0.50	0	0+/-0.08
2003	Rodeo Sport	F	1.00	+2.50	0.50	0	0+/-0.08
		R	—	—	1.00	0	0+/-0.20
	Rodeo	F	0.75	+2.50	0.50	0	0.04+/-0.04
		R	—	—	1.00	0	0+/-0.20
2004	Rodeo	F	1.00	+2.50	0.50	0	0+0.08
		R	—	—	NA	NA	NA

NA: information not available

09474_RODE_C0009

TIRE, WHEEL AND BALL JOINT SPECIFICATIONS

Year	Model	OEM Tires Standard	OEM Tires Optional	Tire Pressures (psi) Front	Tire Pressures (psi) Rear	Wheel Size	Ball Joint Inspection	Lugnut Torque (ft. lbs.)
2002	Rodeo Sport	P245/70R16	none	26	26	7JJ	U: 4-28 ① L: 4-55	87
	Rodeo 2wd/4wd	P225/75R16	P245/70R16 P245/60R18	26	26	7JJ	U: 4-28 ① L: 4-55	87
	Passport LX 2wd	P225/75R16	None	29	29	6.5-JJ	NS	87
	Passport LX 4wd	P245/70R16	None	29	29	7J	NS	87
	Passport EX 2wd	P245/70R16	None	29	29	7J	NS	87
	Passport EX 4wd	P245/70R16	None	29	29	7-JJ	NS	87
	Trooper	P245/70R15	none	26	26	7JJ	U: 4-28 ① L: 4-55	87
2003	Rodeo Sport	P245/70R16	none	26	26	7JJ	U: 4-28 ① L: 4-55	87
	Rodeo 2wd/4wd	P225/75R16	P245/70R16	std. 29 opt. 26	std. 29 opt. 26	7JJ	U: 4-28 ①	87
2004	Rodeo	P225/75R16	P245/70R16	std. 29 opt. 26	std. 29 opt. 26	7JJ	U: 4-28 ①	87

L: Lower

U: Upper

NA: Not available

NS: Not specified by manufacturer

① Torque required in inch lbs. to rotate ball joint when removed from the knuckle

09474_RODE_C0010

BRAKE SPECIFICATIONS
All measurements in inches unless noted

Year	Model		Brake Disc Original Thickness	Brake Disc Machine Thickness	Brake Disc Minimum Thickness	Brake Disc Maximum Runout	Brake Drum Diameter Original Inside Diameter	Brake Drum Diameter Max. Wear Limit	Brake Drum Diameter Maximum Machine Diameter	Minimum Lining Thickness	Brake Caliper Bracket Bolts (ft. lbs.)	Brake Caliper Mounting Bolts (ft. lbs.)
2002	Rodeo Sport	F	1.020	0.983	0.969	0.005	—	—	—	0.039	115	33
		R	0.710	0.668	0.654	0.005	11.6	11.67	NA	0.039	76	32
	Rodeo	F	1.020	0.983	0.969	0.005	—	—	—	0.039	115	33
		R	0.710	0.668	0.654	0.005	11.6	11.67	NA	0.039	76	32
	Passport	F	1.020	0.983	0.969	0.005	—	—	—	0.039	115	54
		R	0.710	0.668	0.654	0.005	—	—	—	0.039	76	32
	Trooper	F	1.024	0.983	0.969	0.005	—	—	—	0.039	115	54
		R	0.710	0.668	0.654	0.005	—	—	—	0.039	76	32
2003	Rodeo Sport	F	1.020	0.983	0.969	0.005	—	—	—	0.039	115	33
		R	0.710	0.668	0.654	0.005	11.6	11.67	NA	0.039	76	32
	Rodeo	F	1.020	0.983	0.969	0.005	—	—	—	0.039	115	33
		R	0.710	0.668	0.654	0.005	11.6	11.67	NA	0.039	76	32
2004	Rodeo	F	1.020	0.983	0.969	0.005	—	—	—	0.039	115	33
		R	0.710	0.668	0.654	0.005	11.6	11.67	NA	0.039	76	32

NA: Not Available

09474_RODE_C0011

SCHEDULED MAINTENANCE INTERVALS
2002-03 Isuzu Rodeo, Rodeo Sport, Trooper; Honda Passport

TO BE SERVICED	TYPE OF SERVICE	7.5	15	22.5	30	37.5	45	52.5	60	67.5	75	82.5	90	97.5	105	112.5	120
Accelerator linkage ①	L	✓	✓	✓	✓	✓	✓	✓	✓	✓	✓	✓	✓	✓	✓	✓	✓
Accessory drive belts ②	S/I				✓				✓				✓				✓
Air cleaner filter	R				✓				✓				✓				✓
Auto cruise control linkage & hose ③	S/I		✓		✓		✓		✓		✓		✓		✓		✓
Automatic transmission fluid level ③	S/I	✓		✓		✓		✓		✓		✓		✓		✓	
Battery fluid level ③	S/I	✓	✓	✓	✓	✓	✓	✓	✓	✓	✓	✓	✓	✓	✓	✓	✓
Body and chassis ①	L	✓	✓	✓	✓	✓	✓	✓	✓	✓	✓	✓	✓	✓	✓	✓	✓
Brake fluid level ③	S/I	✓	✓	✓	✓	✓	✓	✓	✓	✓	✓	✓	✓	✓	✓	✓	✓
Brake lines & hoses ③	S/I	✓	✓	✓	✓	✓	✓	✓	✓	✓	✓	✓	✓	✓	✓	✓	✓
Brake pedal play ③	S/I		✓		✓		✓		✓		✓		✓		✓		✓
Clutch fluid level ③	S/I	✓	✓	✓	✓	✓	✓	✓	✓	✓	✓	✓	✓	✓	✓	✓	✓
Clutch lines & hose ③	S/I				✓				✓				✓				✓
Clutch pedal free-play ③	S/I		✓		✓		✓		✓		✓		✓		✓		✓
Clutch pedal spring, bushing and clevis pin ①	S/I		✓		✓		✓		✓		✓		✓		✓		✓
Cooling and heating system hoses ③	S/I		✓		✓		✓		✓		✓		✓		✓		✓
Driveshaft flange torque ③	S/I	✓		✓		✓		✓		✓		✓		✓		✓	
Drum and disc brakes ③	S/I		✓		✓		✓		✓		✓		✓		✓		✓
Engine coolant	R				✓				✓				✓				✓
Engine coolant level ③	S/I	✓	✓	✓	✓	✓	✓	✓	✓	✓	✓	✓	✓	✓	✓	✓	✓
Engine oil & filter ③	R	✓	✓	✓	✓	✓	✓	✓	✓	✓	✓	✓	✓	✓	✓	✓	✓
Exhaust system ③	S/I	✓	✓	✓	✓	✓	✓	✓	✓	✓	✓	✓	✓	✓	✓	✓	✓
Front and rear axle lubricant	R		✓		✓				✓				✓				✓
Front and rear driveshafts ①	S/I	✓	✓	✓	✓	✓	✓	✓	✓	✓	✓	✓	✓	✓	✓	✓	✓
Front wheel bearings	S/I & L				✓				✓				✓				✓
Fuel lines & tank cap ③	S/I								✓								✓
Inspect for fluid leaks ③	S/I	✓	✓	✓	✓	✓	✓	✓	✓	✓	✓	✓	✓	✓	✓	✓	✓
Key lock cylinder ③	L		✓		✓		✓		✓		✓		✓		✓		✓
Manual transmission and transfer case fluid ④	R		✓		✓				✓				✓				✓
Parking brake system ③	S/I		✓		✓		✓		✓		✓		✓		✓		✓
Power steering fluid	R				✓				✓				✓				✓
Radiator core and A/C condenser	S/I & C								✓								✓
Rotate tires	S/I	✓	✓	✓	✓	✓	✓	✓	✓	✓	✓	✓	✓	✓	✓	✓	✓
Shift-on-the-fly system gear fluid ③	S/I		✓		✓		✓		✓		✓		✓		✓		✓
Spark plug wires ⑤	S/I								✓								✓
Spark plugs	R	Every 100,000 miles															
Starter safety switch ③	S/I	✓	✓	✓	✓	✓	✓	✓	✓	✓	✓	✓	✓	✓	✓	✓	✓
Steering operation ③	S/I	✓	✓	✓	✓	✓	✓	✓	✓	✓	✓	✓	✓	✓	✓	✓	✓

SCHEDULED MAINTENANCE INTERVALS
2002-03 Isuzu Rodeo, Rodeo Sport, Trooper; Honda Passport

TO BE SERVICED	TYPE OF SERVICE	VEHICLE MILEAGE INTERVAL (x1000)															
		7.5	15	22.5	30	37.5	45	52.5	60	67.5	75	82.5	90	97.5	105	112.5	120
Suspension & steering ③	S/I	✓	✓	✓	✓	✓	✓	✓	✓	✓	✓	✓	✓	✓	✓	✓	✓
Throttle linkage ③	S/I		✓		✓		✓		✓		✓		✓		✓		✓
Timing belt	R										✓						
Tires and wheels ③	S/I	✓	✓	✓	✓	✓	✓	✓	✓	✓	✓	✓	✓	✓	✓	✓	✓
Valve clearance ④	A								✓								✓

R: Replace   S/I: Service or Inspect   L: Lubricate   A: Adjust   C: Clean

① Perform this at the mileage indicated or every 6 months, whichever occurs first.

② Perform this at the mileage indicated or every 24 months, whichever occurs first.

③ Perform this at the mileage indicated or every 12 months, whichever occurs first.

④ 3.2L V6 engine.

⑤ 2.2L 4 cyl. engine.

FREQUENT OPERATION MAINTENANCE (SEVERE SERVICE)

 If a vehicle is operated under any of the following conditions it is considered severe service:

- Towing a trailer or using a camper or car-top carrier.

- Repeated short trips of less than 5 miles in temperatures below freezing.

- Extensive idling or low-speed driving for long distances as in heavy commercial use, such as delivery, taxi or police cars.

- Operating on rough, muddy or salt-covered roads.

- Operating on unpaved or dusty roads.

Air cleaner element: replace every 15,000 miles

Engine oil and filter: replace every 3000 miles or 3 months, whichever occurs first.

Automatic transmission fluid: replace every 20,000 miles.

Rear axle lubricant: replace every 15,000 miles.

09474_RODE_C0013

SCHEDULED MAINTENANCE INTERVALS
2004 Isuzu Rodeo

TO BE SERVICED	TYPE OF SERVICE	VEHICLE MILEAGE INTERVAL (x1000)															
		7.5	15	22.5	30	37.5	45	52.5	60	67.5	75	82.5	90	97.5	105	112.5	120
Accelerator linkage ①	L	✓	✓	✓	✓	✓	✓	✓	✓	✓	✓	✓	✓	✓	✓	✓	✓
Accessory drive belts ②	S/I				✓				✓				✓				✓
Air cleaner filter	R				✓				✓				✓				✓
Auto cruise control linkage & hose ③	S/I		✓		✓		✓		✓		✓		✓		✓		✓
Automatic transmission fluid level	S/I				✓				✓				✓				✓
Body and chassis ①	L	✓	✓	✓	✓	✓	✓	✓	✓	✓	✓	✓	✓	✓	✓	✓	✓
Brake fluid level ③	S/I	✓	✓	✓	✓	✓	✓	✓	✓	✓	✓	✓	✓	✓	✓	✓	✓
Brake pedal play ③	S/I/A		✓		✓		✓		✓		✓		✓		✓		✓
Clutch fluid level ③	S/I	✓	✓	✓	✓	✓	✓	✓	✓	✓	✓	✓	✓	✓	✓	✓	✓
Clutch lines & hose ③	S/I				✓				✓				✓				✓
Clutch pedal free-play ③	S/I		✓		✓		✓		✓		✓		✓		✓		✓
Clutch pedal spring, bushing and clevis pin ①	S/I		✓		✓		✓		✓		✓		✓		✓		✓
Cooling and heating system hoses ③	S/I		✓		✓		✓		✓		✓		✓		✓		✓
Driveshaft flange torque ③	S/I	✓		✓		✓		✓		✓		✓		✓		✓	
Drum and disc brakes ③	S/I		✓		✓		✓		✓		✓		✓		✓		✓
Engine coolant	R				✓				✓				✓				✓
Engine coolant level ③	S/I	✓	✓	✓	✓	✓	✓	✓	✓	✓	✓	✓	✓	✓	✓	✓	✓
Engine oil & filter ③	R	✓	✓	✓	✓	✓	✓	✓	✓	✓	✓	✓	✓	✓	✓	✓	✓
Exhaust system ③	S/I	✓	✓	✓	✓	✓	✓	✓	✓	✓	✓	✓	✓	✓	✓	✓	✓
Front and rear axle lubricant	S/I		✓		✓				✓				✓				✓
Front and rear axle lubricant	R												✓				
Front and rear driveshafts ①	S/I	✓	✓	✓	✓	✓	✓	✓	✓	✓	✓	✓	✓	✓	✓	✓	✓
Front wheel bearings	S/I & L								✓								✓
Fuel lines & tank cap ③	S/I								✓								✓
Inspect for fluid leaks ③	S/I	✓	✓	✓	✓	✓	✓	✓	✓	✓	✓	✓	✓	✓	✓	✓	✓
Key lock cylinder ③	L		✓		✓		✓		✓		✓		✓		✓		✓
Manual transmission fluid	S/I		✓		✓				✓				✓				✓
Manual transmission fluid	R												✓				
Parking brake system ③	S/I		✓		✓		✓		✓		✓		✓		✓		✓
Power steering fluid Level	I	✓	✓	✓	✓	✓	✓	✓	✓	✓	✓	✓	✓	✓	✓	✓	✓
Radiator core and A/C condenser	S/I & C								✓								✓
Rotate tires	S/I	✓	✓	✓	✓	✓	✓	✓	✓	✓	✓	✓	✓	✓	✓	✓	✓
Shift-on-the-fly system gear fluid ③	S/I		✓		✓		✓		✓		✓		✓		✓		✓
Spark plugs 3.2L engine	R	Every 100,000 miles															
Spark plugs 3.5L engine	R	Every 120,000 miles															
Starter safety switch ③	S/I	✓	✓	✓	✓	✓	✓	✓	✓	✓	✓	✓	✓	✓	✓	✓	✓
Steering operation ③	S/I	✓	✓	✓	✓	✓	✓	✓	✓	✓	✓	✓	✓	✓	✓	✓	✓
Suspension & steering ③	S/I	✓	✓	✓	✓	✓	✓	✓	✓	✓	✓	✓	✓	✓	✓	✓	✓

SCHEDULED MAINTENANCE INTERVALS
2004 Isuzu Rodeo

TO BE SERVICED	TYPE OF SERVICE	VEHICLE MILEAGE INTERVAL (x1000)															
		7.5	15	22.5	30	37.5	45	52.5	60	67.5	75	82.5	90	97.5	105	112.5	120
Throttle linkage ③	S/I		✓		✓		✓		✓		✓		✓		✓		✓
Timing belt	R	Every 120,000 miles															
Tires and wheels ③	S/I	✓	✓	✓	✓	✓	✓	✓	✓	✓	✓	✓	✓	✓	✓	✓	✓
Transfer case fluid	S/I	✓		✓		✓		✓		✓		✓		✓		✓	
Transfer case fluid	R								✓								✓
Valve clearance	A	Only if noisy or every 120,000 miles															

R: Replace   S/I: Service or Inspect   L: Lubricate   A: Adjust   C: Clean

① Perform this at the mileage indicated or every 6 months, whichever occurs first.

② Perform this at the mileage indicated or every 24 months, whichever occurs first.

③ Perform this at the mileage indicated or every 12 months, whichever occurs first.

FREQUENT OPERATION MAINTENANCE (SEVERE SERVICE)

If a vehicle is operated under any of the following conditions it is considered severe service:

- Towing a trailer or using a camper or car-top carrier.

- Repeated short trips of less than 5 miles in temperatures below freezing.

- Extensive idling or low-speed driving for long distances as in heavy commercial use, such as delivery, taxi or police cars.

- Operating on rough, muddy or salt-covered roads.

- Operating on unpaved or dusty roads.

Air cleaner element: replace as needed

Engine oil and filter: replace every 3000 miles or 3 months, whichever occurs first.

Automatic transmission fluid: replace every 30,000 miles (3.2L engine) or, every 60,000 miles (3.5L engine)

Rear axle lubricant: replace every 15,000 miles.

Power steering fluid: replace every 30,000 miles

Transfer case fluid: replace every 20,000 miles

09474_RODE_C0015

ENGINE REPAIR

→Disconnecting the negative battery cable on some vehicles may interfere with the functions of the on-board computer system. The computer may undergo a relearning process once the negative battery cable is reconnected.

Distributor

REMOVAL

These engines are equipped with a Distributorless Ignition System (DIS).

Alternator

REMOVAL

2.2L Engine

1. Before servicing the vehicle, refer to the Precautions Section.
2. Remove or disconnect the following:
   • Negative battery cable
   • Accessory drive belt
   • Alternator harness connectors
   • Alternator

3.2L, 3.5L VIN X and 3.5L VINY Engines

1. Before servicing the vehicle, refer to the Precautions Section.
2. Remove or disconnect the following:
   • Negative battery cable

• Accessory drive belt (1)
• Alternator harness connectors (4)
• Bolts (3)
• Alternator (2)

INSTALLATION

2.2L Engine

Install or connect the following:
• Alternator. Tighten the long bolt to 26 ft. lbs. (35 Nm) and the short bolt to 15 ft. lbs. (20 Nm).
• Alternator harness connectors
• Accessory drive belt
• Negative battery cable

3.2L, 3.5L VIN X and 3.5L VIN Y Engines

Install or connect the following:
• Alternator. Tighten the 10mm bolt to 30 ft. lbs. (41 Nm) and the 8mm bolt to 15 ft. lbs. 21 Nm).
• Alternator harness connectors
• Accessory drive belt
• Negative battery cable

Ignition Timing

ADJUSTMENT

These engines are equipped with a Distributorless Ignition System (DIS). No adjustment is possible.

Engine Assembly

REMOVAL & INSTALLATION

2.2L Engines

1. Before servicing the vehicle, refer to the Precautions Section.
2. Drain the cooling system.
3. Relieve the fuel system pressure.
4. Remove or disconnect the following:
   • Battery
   • Hood
   • Accessory drive belt
   • Accelerator cable
   • Air intake assembly
   • Engine wiring harness connectors at left rear of the engine compartment
   • Brake booster vacuum line
   • Engine ground cables
   • Clutch fluid line bracket and slave cylinder
   • Fuel lines and bracket
   • Exhaust front pipe
   • Transmission
   • A/C compressor
   • Power steering pump
   • Chassis harness connectors at right rear of the engine compartment
   • Frame ground cable
   • Radiator hoses
   • Heater hoses
   • Cooling fan connector
   • Cooling fan and shroud
   • Radiator
   • Left and right engine mounts
5. Lift the engine from the vehicle.

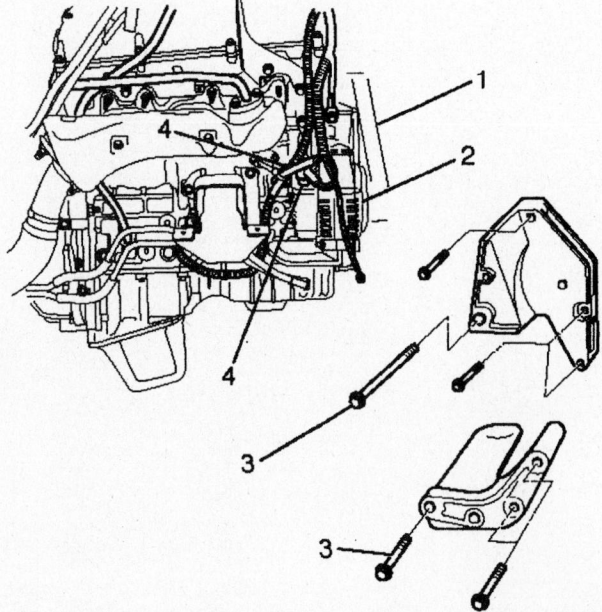

09474_RODE_G0002

Alternator removal—3.2L, 3.5L VIN X and 3.5L VINY Engines

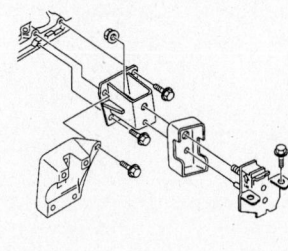

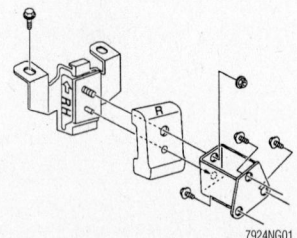

7924NG01

Left and right motor mounts—2.2L engine

To install:

6. Position the engine in the engine compartment.

7. Install or connect the following:
  - Left and right engine mounts. Tighten the fasteners to 30 ft. lbs. (41 Nm).
  - Radiator
  - Cooling fan and shroud
  - Cooling fan connector
  - Heater hoses
  - Radiator hoses
  - Frame ground cable
  - Chassis harness connectors at right rear of the engine compartment
  - Power steering pump
  - A/C compressor
  - Transmission
  - Exhaust front pipe
  - Fuel lines and bracket
  - Clutch fluid line bracket and slave cylinder
  - Engine ground cables
  - Brake booster vacuum line
  - Engine wiring harness connectors at left rear of the engine compartment
  - Air intake assembly
  - Accelerator cable
  - Accessory drive belts
  - Hood
  - Battery

8. Fill the cooling system.
9. Start the engine and check for leaks.

3.2L Engines

1. Before servicing the vehicle, refer to the Precautions Section.
2. Drain the cooling system.
3. Relieve the fuel system pressure.
4. Remove or disconnect the following:
  - Battery
  - Hood
  - Accelerator cable
  - Cruise control cable
  - Air intake assembly
  - Canister vacuum hose
  - Brake booster vacuum hose
  - Engine wiring harness connectors
  - Front axle harness connector, if equipped
  - Transmission harness connector and bracket
  - Frame ground cable
  - Firewall ground cable
  - Starter harness connectors
  - Alternator harness connectors
  - Coolant overflow reservoir hose
  - Radiator hoses
  - Cooling fan and shroud
  - Accessory drive belt
  - Power steering pump
  - A/C compressor
  - Heated Oxygen (HO$_2$S) sensor connectors
  - Exhaust front pipes
  - Heater hoses
  - Fuel lines
  - Transmission
  - Left and right engine mounts

5. Lift the engine from the vehicle.

To install:

6. Lower the engine into the vehicle.
7. Install or connect the following:
  - Left and right engine mounts. Tighten the bolts to 30 ft. lbs. (41 Nm) and the nuts to 37 ft. lbs. (50 Nm).
  - Transmission
  - Fuel lines
  - Heater hoses
  - Exhaust front pipes
  - Heated Oxygen (HO$_2$S) sensor connectors
  - A/C compressor
  - Power steering pump
  - Accessory drive belt
  - Cooling fan and shroud
  - Radiator hoses
  - Coolant overflow reservoir hose
  - Alternator harness connectors
  - Starter harness connectors
  - Firewall ground cable
  - Frame ground cable
  - Transmission harness connector and bracket
  - Front axle harness connector, if equipped
  - Engine wiring harness connectors
  - Brake booster vacuum hose
  - Canister vacuum hose
  - Air intake assembly
  - Cruise control cable
  - Accelerator cable
  - Hood
  - Battery

8. Fill the cooling system.
9. Start the engine and check for leaks.

3.5L VIN X and 3.5L VIN Y Engines

1. Before servicing the vehicle, refer to the Precautions Section.
2. Drain the cooling system.
3. Remove or disconnect the following:
  - Battery
  - Hood
  - Air cleaner assembly
  - Accelerator cable
  - Cruise control cable
  - Canister vacuum line
  - Brake booster vacuum line
  - Engine wiring harness connectors
  - Transmission harness connectors and bracket
  - Engine ground cable
  - Starter harness connector
  - Alternator harness connector
  - Coolant reservoir tank hose
  - Radiator hoses
  - Heater hoses
  - Upper fan shroud
  - Radiator
  - Cooling fan
  - Accessory drive belt
  - Power steering pump
  - A/C compressor
  - Heated Oxygen (HO$_2$S) sensor connectors
  - Left and right exhaust front pipes
  - Fuel lines
  - Flywheel dust cover
  - Transmission. Refer to the transmission procedure in this section.
  - Left and right engine mounts
  - Engine

To install:

4. Install or connect the following:
  - Engine
  - Left and right engine mounts. Tighten the bolts to 30 ft. lbs. (41 Nm).
  - Transmission
  - Flywheel dust cover
  - Fuel lines

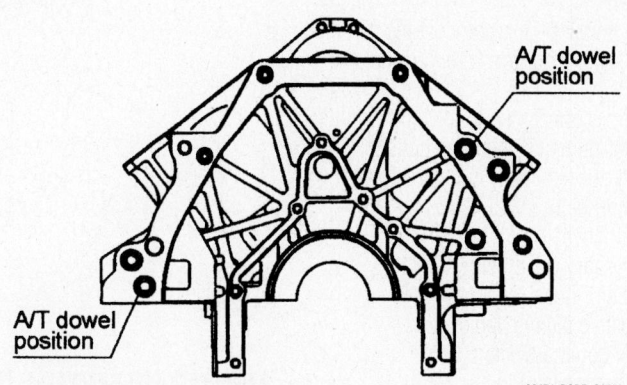

Automatic transmission dowel positions

09474_RODE_G0003

- Left and right exhaust front pipes
- HO2S sensor connectors
- A/C compressor
- Power steering pump
- Accessory drive belt
- Cooling fan. Tighten the nuts to 16 ft. lbs. (22 Nm).
- Radiator
- Upper fan shroud
- Heater hoses
- Radiator hoses
- Coolant reservoir tank hose
- Alternator harness connector
- Starter harness connector
- Engine ground cable
- Transmission harness connectors and bracket
- Engine wiring harness connectors
- Brake booster vacuum line
- Canister vacuum line
- Cruise control cable
- Accelerator cable
- Air cleaner assembly
- Hood
- Battery

5. Fill the cooling system. Check all fluid levels and adjust as necessary.
6. Start the engine and check for leaks.

Water Pump

REMOVAL & INSTALLATION

2.2L Engine

1. Before servicing the vehicle, refer to the Precautions Section.
2. Drain the cooling system.
3. Remove or disconnect the following:
- Negative battery cable
- Radiator hose
- Accessory drive belt
- Front cover
- Timing belt. Refer to the Timing Belt procedure.
- Water pump

To install:

4. Install a new O-ring and coat the water pump sealing surface with silicone grease.
5. Install or connect the following:
- Water pump. Tighten the bolts to 18 ft. lbs. (25 Nm).
- Timing belt
- Front cover
- Accessory drive belt
- Radiator hose
- Negative battery cable

6. Fill the cooling system.
7. Start the engine and check for leaks.

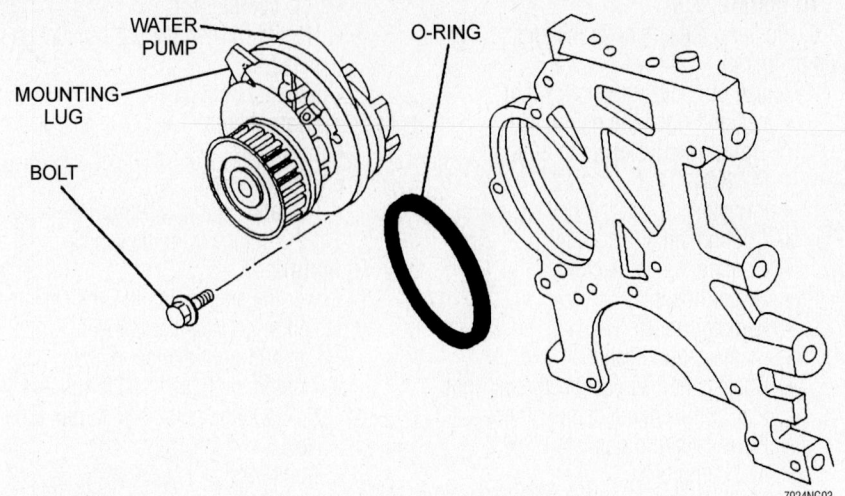

Exploded view of the water pump mounting, showing the location of the mounting lug—2.2L engine

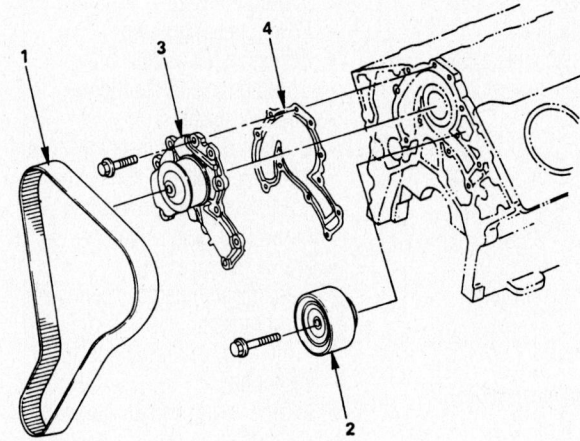

1. Timing belt
2. Idle pulley
3. Water pump assembly
4. Gasket

Exploded view of the water pump mounting

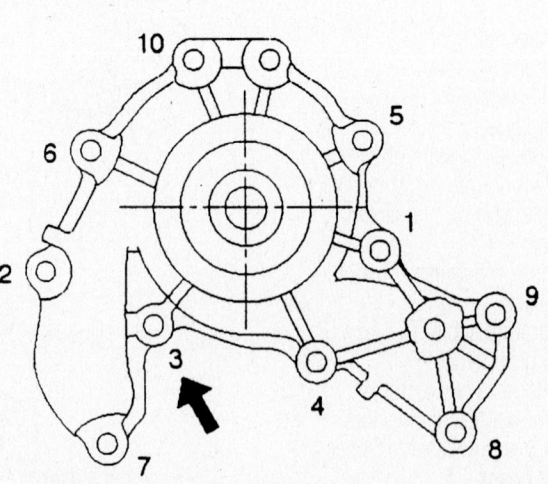

Water pump torque sequence. Apply LOCTITE® 262 to bolt number 3 (arrow)—3.2L and 3.5L engines

3.2L, 3.5L VIN X and VIN Y Engines

1. Before servicing the vehicle, refer to the Precautions Section.
2. Drain the cooling system.
3. Remove or disconnect the following:
   - Negative battery cable
   - Upper radiator hose
   - Timing belt. Refer to the Timing Belt procedure.
   - Idler pulley
   - Water pump

To install:

➡**Apply Loctite® 262 to bolt number 3 prior to installation.**

4. Install or connect the following:
   - Water pump. Tighten the bolts in two passes, in sequence, to 18 ft. lbs. (25 Nm).
   - Idler pulley. Torque to 38 ft. lbs. (52 Nm).
   - Timing belt
   - Upper radiator hose
   - Negative battery cable
5. Fill the cooling system.
6. Start the engine and check for leaks.

Heater Core

REMOVAL & INSTALLATION

Rodeo

1. Before servicing the vehicle, refer to the Precautions Section.
2. If equipped with an air bag, perform the following procedure:
   a. Turn the ignition to the LOCK position and remove the key.
   b. From the lower left dash side fuse block, remove the SRS-1 fuse.
   c. Disconnect the 2-pin yellow connector located at the base of the steering column.
   d. Remove the glove box assembly.
   e. Disconnect the 2-pin yellow connector located behind the glove box.
3. Disconnect the negative battery cable.
4. If equipped, discharge and recover the air conditioning system refrigerant.
5. Remove the evaporator lines at the firewall. Plug the air conditioning lines to minimize contamination.
6. Disconnect the cooling system hoses and drain the coolant into a clean container for reuse. Plug the cooling system hoses.
7. Remove the instrument panel by performing the following procedure:
   a. Remove the lower center cover screw and pull it out at the clip positions;

then, disconnect the cigarette lighter connector.
   b. Remove both the rear and front console.
   c. Remove the dash side trim panel sill plates and the panels.
   d. Remove the 2 glove box screws and the glove box.
   e. Remove the 2 hood release screws, the 6 instrument panel driver's lower cover assembly screws and the cover assembly.
   f. Remove 5 instrument cluster screws and the 2 clips. Then, disconnect the 8 switch connectors and remove the instrument cluster assembly.
   g. Remove the 6 driver's knee bolster assembly bolts and screws and the knee bolster assembly.
   h. Remove the 4 control lever assembly bolts; then, disconnect the 3 control cables (unit side) and the 3 harness connectors.
   i. Remove the 4 radio/audio sub box assembly screws and the radio/audio sub box assembly.
   j. Disconnect or remove the following instrument panel harness connectors or items:
   - The 6 driver's side connectors
   - The 3 passenger's side connectors
   - The 2 center connectors
   - Passenger's inflator module connector
   - Radio antenna cable plug
   - Ground cable bolt on the left dash side panel
   - The 8 instrument panel-to-chassis bolts and the 3 nuts.
   k. Remove the instrument panel assembly.
8. Remove the instrument panel bracket by performing the following procedure:
   a. Remove the 2 passenger's inflator module bolts and 4 nuts.
   b. Remove the 4 meter assembly screws. Then, disconnect the meter wiring harness connectors and remove the meter assembly.
   c. Remove the 5 vent duct assembly screws and the assembly.
   d. Remove the 3 lower passenger bracket screws and the bracket.
   e. Remove the 9 passenger knee bolster reinforcement screws and the reinforcement.
   f. Remove the 6 instrument panel center reinforcement screws and the reinforcement.
   g. Remove the instrument panel wiring harness assembly clips and the wiring harness.

   h. Remove the 2 instrument panel bracket nuts and 2 bolts for each bracket; then, remove the bracket(s).
9. Remove the 5 cross beam assembly nuts, 2 bolts and the 6 lower bolts; then, remove the crossbeam.
10. Disconnect the resistor wiring connector.
11. Remove the duct from the heater assembly.
12. If equipped with air conditioning, remove the evaporator assembly.
13. Remove the driver's lap vent.
14. Remove the lower ventilation duct.
15. Remove the footrest, the carpet, the 3 clips and the rear heater duct.
16. Remove the heater assembly.
17. Remove the mode control case-to-temperature control case screws and remove the mode control case; do not remove the link unit.
18. Remove the temperature control case screws and separate the cases.
19. Remove the heater core from the case.

To install:

20. Install the heater core to the case.
21. Assemble the temperature control cases and install the case screws.
22. Install the mode control case and the mode control case-to-temperature control case screws.
23. Install the heater assembly.
24. Install the rear heater duct, the footrest, the carpet, and the 3 clips.
25. Install the lower ventilation duct.
26. Install the driver's lap vent.
27. If equipped with air conditioning, install the evaporator assembly.
28. Install the duct to the heater assembly.
29. Connect the resistor wiring connector.
30. Install the crossbeam, the 5 crossbeam assembly nuts, 2 bolts and the 6 lower bolts.
31. Install the instrument panel bracket by performing the following procedure:
   a. Install the instrument panel bracket and the 2 nuts and 2 bolts for each bracket.
   b. Install the instrument panel wiring harness assembly and the wiring harness clips.
   c. Install the instrument panel center reinforcement and the 6 reinforcement screws.
   d. Install the passenger knee bolster reinforcement and the 9 reinforcement screws.
   e. Install the lower passenger bracket and the 3 bracket screws.

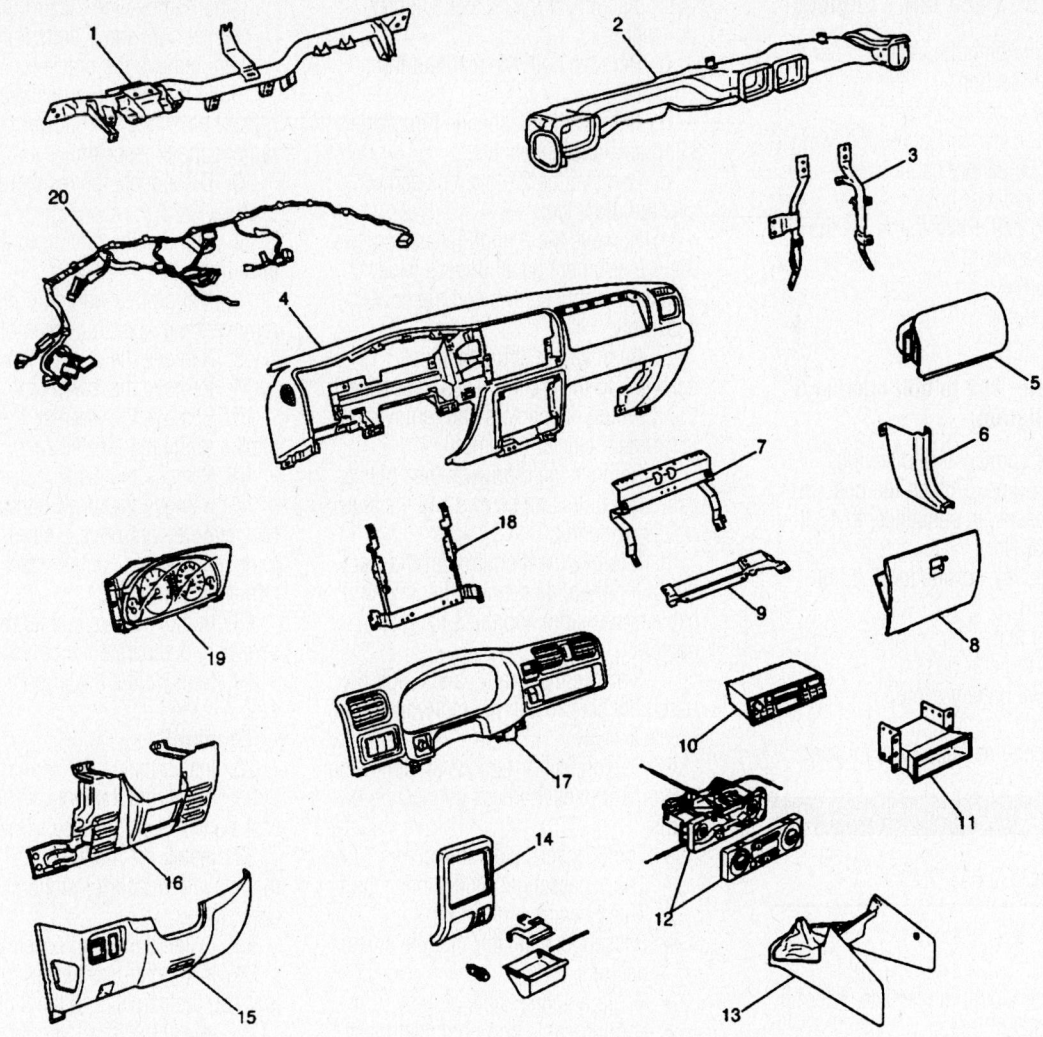

1 Cross Beam
2 Vent Duct Assembly
3 Instrument Panel Bracket
4 Instrument Panel Assembly
5 Passenger Inflator Module
6 Dash Side Trim Panel
7 Passenger Knee Bolster Reinforcement Assembly
8 Glove Box
9 Passenger Lower Bracket
10 Radio Assembly
11 Audio Sub Box
12 Control Lever Assembly
13 Front Console Assembly
14 Lower Center Cover
15 Instrument Panel Driver Lower Cover Assembly
16 Driver Knee Bolster Assembly
17 Meter Cluster Assembly
18 Instrument Panel Center Reinforcement
19 Meter Assembly
20 Instrument Harness Assembly

93113GB8

Exploded view of the instrument panel—Isuzu Rodeo

f. Install the vent duct assembly and the 5 vent duct assembly screws.

g. Install the meter assembly and the 4 meter assembly screws. Then, connect the meter wiring harness connectors.

h. Install the 2 passenger's inflator module bolts and 4 nuts.

32. Install the instrument panel by performing the following procedure:

a. Install the instrument panel assembly.

b. Connect or install the following instrument panel harness connectors or items:
- The 6 driver's side connectors
- The 3 passenger's side connectors
- The 2 center connectors
- Passenger's inflator module connector
- Radio antenna cable plug
- Ground cable bolt on the left dash side panel

- The 8 instrument panel-to-chassis bolts and the 3 nuts.

c. Install the radio/audio sub box assembly and the 4 radio/audio sub box assembly screws.

d. Connect the 3 control cables (unit side) and the 3 harness connectors. Install the 4 control lever assembly bolts.

e. Install the knee bolster assembly and the 6 driver's knee bolster assembly bolts and screws.

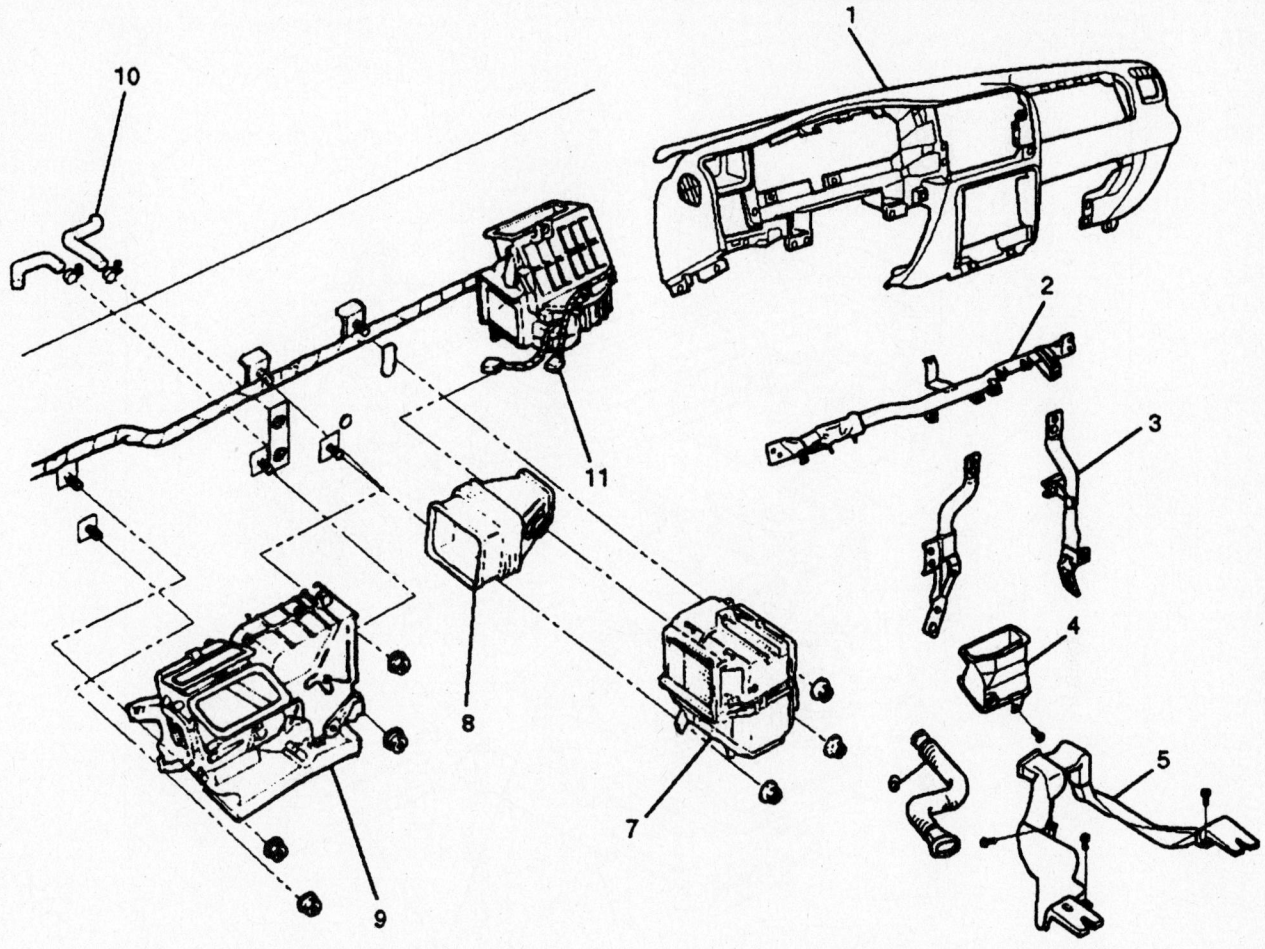

1   Instrument Panel Assembly
2   Cross Beam Assembly
3   Instrument Panel Bracket
4   Ventilation Lower Duct
5   Rear Heater Duct
6   Driver Lap Vent Duct
7   Evaporator Assembly (A/C only)
8   Duct
9   Heater Unit Assembly
10 Heater Hose
11 Resistor Connector

93113GB9

View of the heater and air conditioning housing assemblies and related components—Isuzu Rodeo

f. Install the instrument cluster assembly. Connect the 8 switch connectors. Install 5 instrument cluster screws and the 2 clips.

g. Install the 2 hood release screws, the 6 instrument panel driver's lower cover assembly screws and the cover assembly.

h. Install the glove box and the 2 glove box screws.

i. Install the dash side trim panel sill plates and the panels.

j. Install both the rear and front console.

k. Connect the cigarette lighter connector and install the lower center cover screw.

33. Connect the cooling system hoses.

34. Refill the cooling system.

35. Install the evaporator lines at the firewall.

36. If equipped, evacuate and charge the air conditioning system.

37. Connect the negative battery cable.

38. If equipped with an air bag, perform the following procedure:

a. Turn the ignition to the LOCK position and remove the key.

b. Connect the 2-pin yellow connector located behind the glove box.

c. Install the glove box assembly.

d. Connect the 2-pin yellow connector located at the base of the steering column.

e. At the lower left dash side fuse block, install the SRS-1 fuse.

f. Turn the ignition switch to ON and verify that the AIR BAG warning light flashes 7 times and turns OFF.

39. Run the engine to normal operating temperatures; then, check the climate control operation and check for leaks.

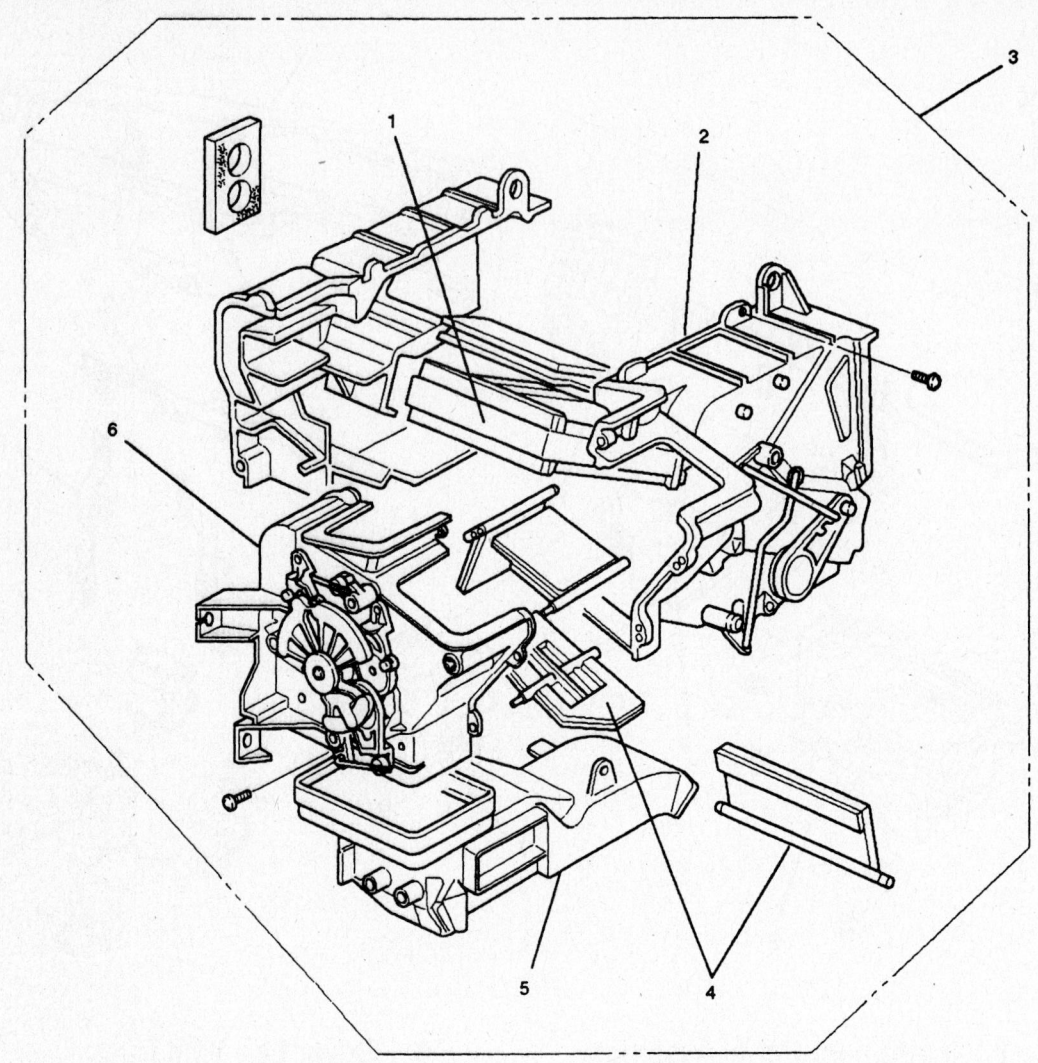

1 Heater Core
2 Case (Temperature Control)
3 Heater Unit
4 Mode Door
5 Duct
6 Case (Mode Control)

93113GB0

Exploded view of the heater housing assembly—Isuzu Rodeo

Trooper

✳✳ CAUTION

The vehicle is equipped with a driver's side and a passenger's side air bag. Before starting service procedures on components, especially under the instrument panel and/or near the steering column, disable the air bag systems. There is sufficient voltage in the system to cause a deployment for up to 15 seconds after the battery has been disconnected, the ignition turned OFF or fuse C-21 is removed from the fuse panel.

1. Before servicing the vehicle, refer to the Precautions Section.

2. If equipped with an air bag, perform the following procedures:

a. Disconnect the negative battery cable, then disconnect the positive battery cable.

b. Disconnect the yellow 2-pin connector located at the base of the steering column.

c. Remove the glove box and disconnect the yellow 2-pin connector located behind the glove box.

3. Disconnect the negative battery cable.

4. Drain the cooling system.

5. If equipped with air conditioning, discharge and recover the refrigerant.

6. Remove the instrument panel assembly by performing the following procedure:

a. At the front console assembly, disconnect the switch connectors; then, remove the console-to-chassis screws and the console.

b. At the lower cluster assembly, remove the cluster-to-instrument panel screws, disconnect the cigarette lighter and light connectors and remove the lower cluster.

c. Remove the glove box and the instrument panel lower cover and the passenger knee bolster reinforcement.

d. At the left side, remove the instrument panel lower cover and the knee bolster assembly.

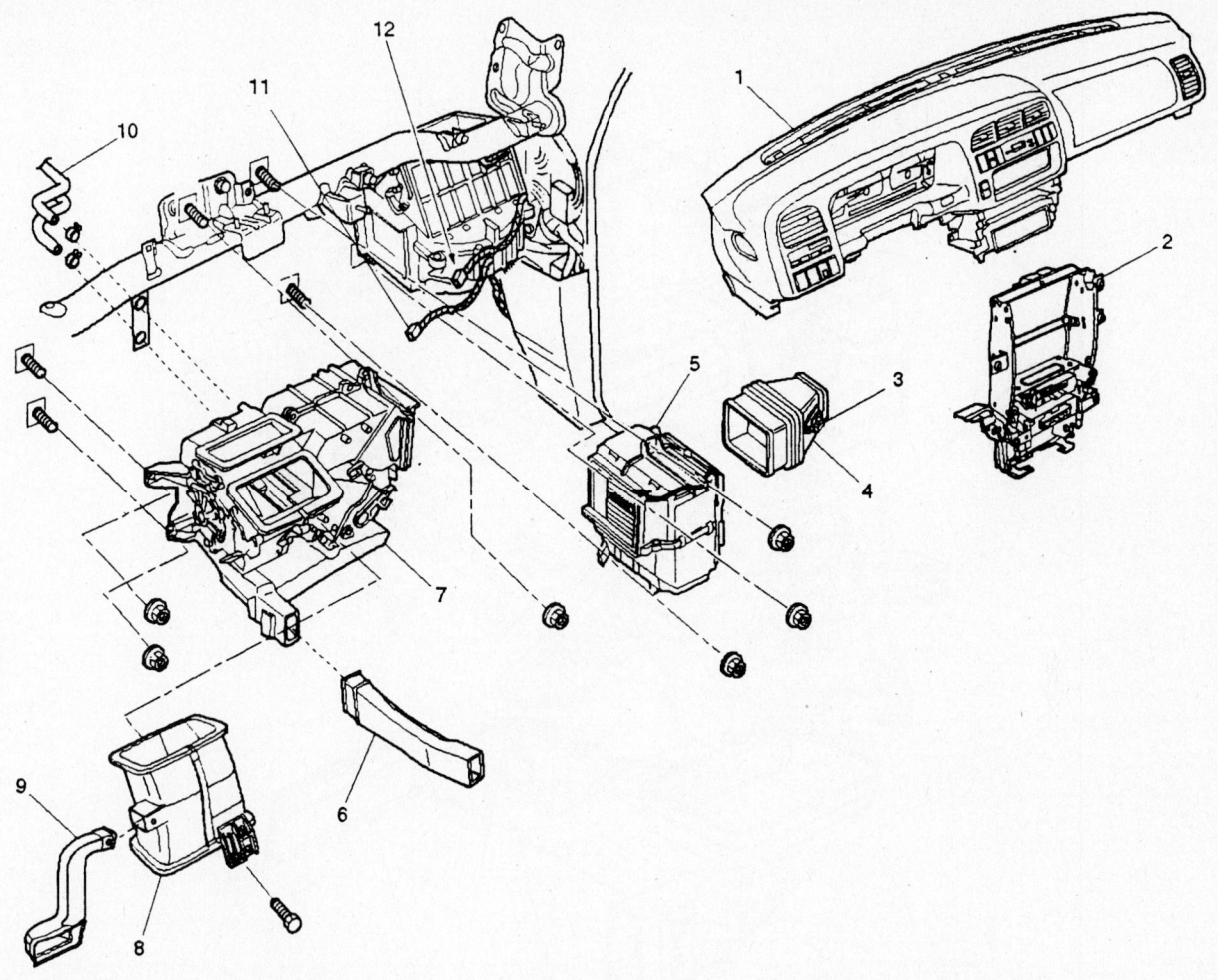

(1) Instrument Panel Assembly
(2) Instrument Panel Center Bracket
(3) Resistor
(4) Duct
(5) Evaporator Assembly (A/C only)
(6) Rear Heater Duct
(7) Heater Unit Assembly
(8) Center Ventilation Lower Duct
(9) Driver Lap Vent Nozzle
(10) Water Hose
(11) Electro Thermo Connector (With A/C)
(12) Resistor Connector

93113G01

Exploded view of the heater unit and related components—Isuzu Trooper

e. At the top of the instrument panel, pry the 8 claws on the front side toward you, raise the defroster grille and remove it.

f. At the SRS adjust bracket and cross beam under the passenger air bag module, remove the 2 fixing bolts and remove the instrument panel assembly.

g. Disconnect the air conditioning control cables from the unit.

h. Remove the instrument panel harness connectors (5 on the driver's side and 3 on the passenger's side), the passenger air bag module connector, the radio antenna plug and the center bracket ground cable bolt.

i. Remove the passenger's air bag module nuts, disconnect the connectors and remove the module.

j. Remove the instrument panel cluster assembly screws, disconnect the switch connectors and the instrument panel assembly.

7. Disconnect the heater hoses from the heater unit.

8. Disconnect the heater resistor connector and the electro thermo connector (if equipped with air conditioning).

9. Remove the heater duct.

10. If equipped with air conditioning, remove the evaporator assembly by performing the following procedure:

a. Disconnect the drain hose.

b. Using a backup wrench, disconnect the refrigerant lines from the evaporator.

c. Plug or cap the refrigerant lines.

d. Remove the evaporator assembly.

11. Remove the instrument panel center bracket (crossbeam assembly) by performing the following procedure:

a. Remove the side support bracket bolts and brackets from both sides of the vehicle.

b. Remove the crossbeam center bracket nuts, disconnect the electrical connectors and the center bracket.

12. Remove the rear heater duct and heater assembly.

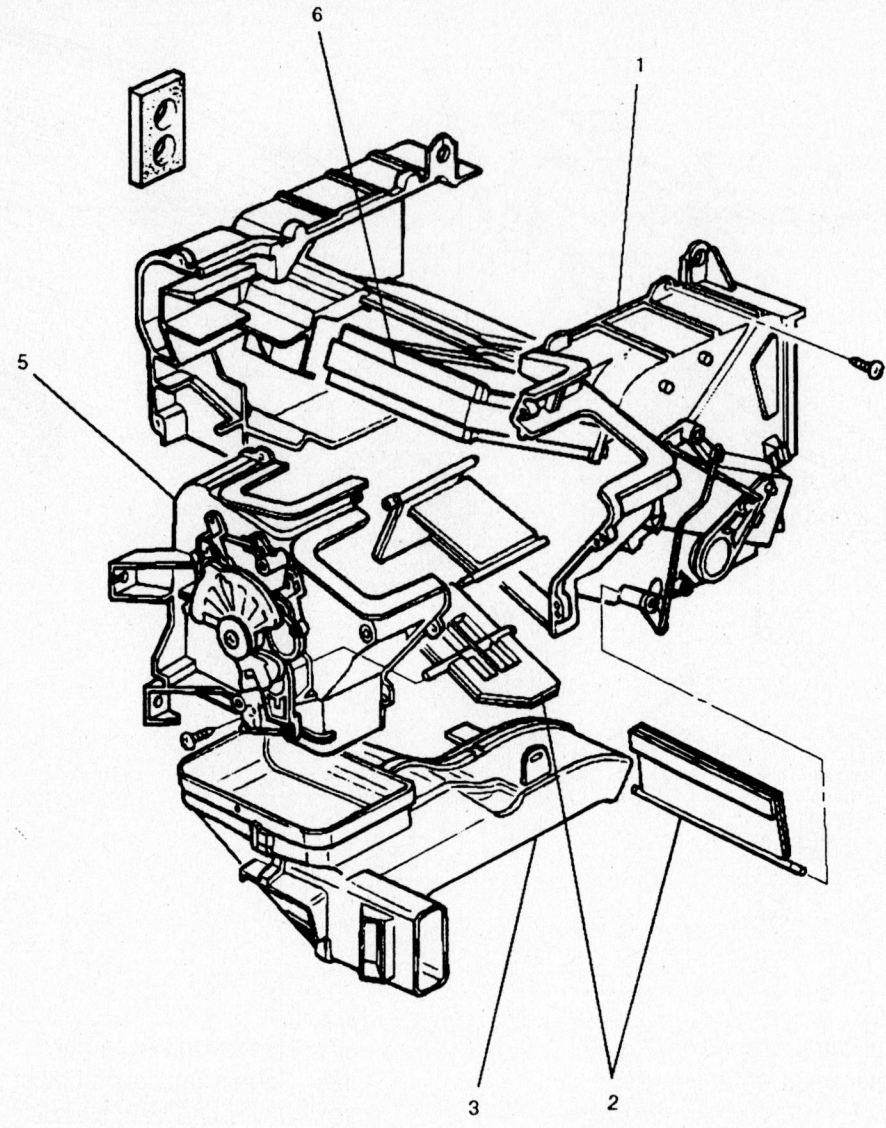

(1) Case (Temperature Control)
(2) Mode Door
(3) Duct
(5) Case (Mode Control)
(6) Heater Core

93113G02

Exploded view of the heater unit—Isuzu Trooper

13. Disassemble the heater unit assembly by performing the following procedure:
  a. Remove the lower air duct; do not remove the link unit.
  b. Remove the temperature control case screws and lift the case from the heater unit.
  c. Remove the heater core.
To install:
14. Assemble the heater unit assembly by performing the following procedure:
  a. Install the heater core into the heater unit.
  b. Install the temperature control case onto the unit and secure with screws.

  c. Install the lower air duct.
15. Install the heater unit assembly into the vehicle.
16. Install the rear heater duct.
17. Install the instrument panel cross beam assembly by reversing the removal procedures.
18. If equipped with air conditioning, install the evaporator assembly by performing the following procedures:
  a. If installing a new evaporator assembly, add 1.7 fl. oz. (50mL) of refrigerant oil to the evaporator.
  b. Using new O-rings and a backup wrench, install the refrigerant lines and

torque the outlet line to 18 ft. lbs. (25 Nm) and the inlet line to 11 ft. lbs. (15 Nm).
19. Install the heater duct.
20. Connect the heater resistor connector and the electro-thermo connector (if equipped with air conditioning).
21. Connect the heater hoses to the heater unit.
22. Install the instrument panel assembly by reversing the removal procedures.
23. If equipped with air conditioning, evacuate and charge and leak-test the system.
24. Refill the cooling system.
25. Connect the negative battery cable.

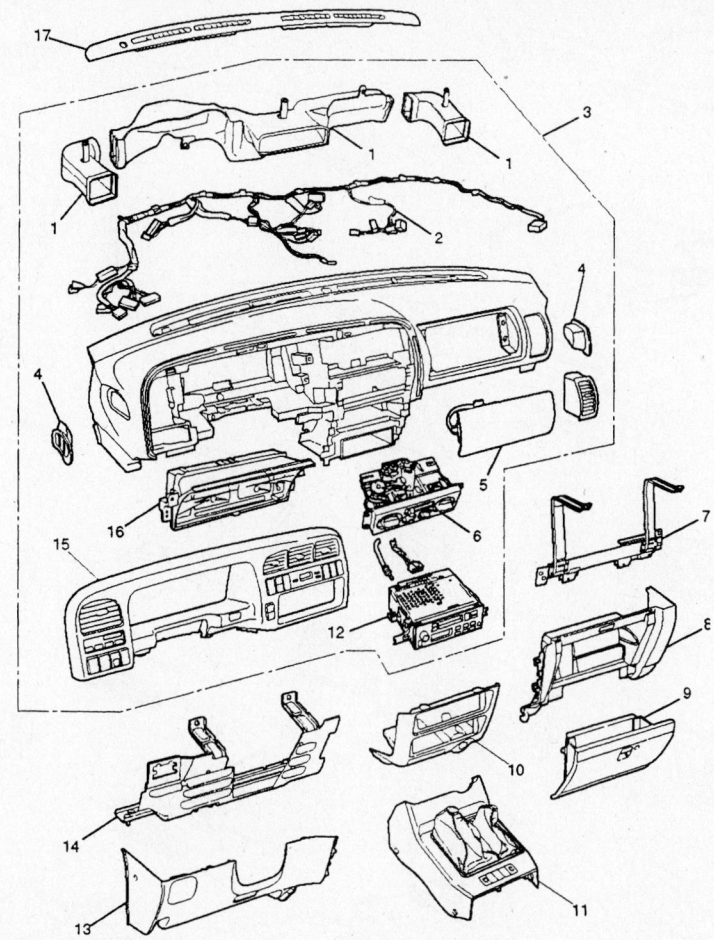

(1) Vent Duct Assembly	(9) Glove Box
(2) Instrument Harness Assembly	(10) Lower Cluster Assembly
(3) Instrument Panel Assembly	(11) Front Console Assembly
(4) Side Defroster Grille	(12) Radio Assembly
(5) Passenger Inflator Module	(13) Instrument Panel Driver Lower Cover Assembly
(6) Control Lever Assembly	(14) Driver Knee Bolster Assembly
(7) Passenger Knee Bolster Reinforcement Assembly	(15) Instrument Panel Cluster Assembly
(8) Instrument Panel Passenger Lower Cover Assembly	(16) Meter Assembly
	(17) Front Defroster Grille

93113G03

Exploded view of the instrument panel and accessories—Isuzu Trooper

❈❈ CAUTION

Never use an air bag assembly from another vehicle and/or different model year. Starting in 1999, the air bag assemblies are equipped with identification colors on the bar code label as follows: YELLOW for the driver's air bag assembly, WHITE for the passenger's air bag assembly.

26. Enable the air bags by performing the following procedure:

a. Connect the passenger's side air bag yellow 2-pin connector.

b. Install the glove box.

c. At the base of the steering column, connect the yellow 2-pin connector.

d. Install the air bag fuse C-21 (if removed) or connect the negative battery cable.

e. Turn the ignition switch ON and verify that the AIR BAG warning light flashes 7 times and then turns OFF.

27. Run the engine to normal operating temperatures and check for leaks. Check the systems for correct operation.

Passport

1. Before servicing the vehicle, refer to the Precautions Section.

2. If equipped with an air bag, perform the following procedure:

a. Turn the ignition to the LOCK position and remove the key.

b. From the lower left dash side fuse block, remove the SRS-1 fuse.

c. Disconnect the 2-pin yellow connector located at the base of the steering column.

d. Remove the glove box assembly.

e. Disconnect the 2-pin yellow connector located behind the glove box.

3. Disconnect the negative battery cable.

4. If equipped, discharge and recover the air conditioning system refrigerant.

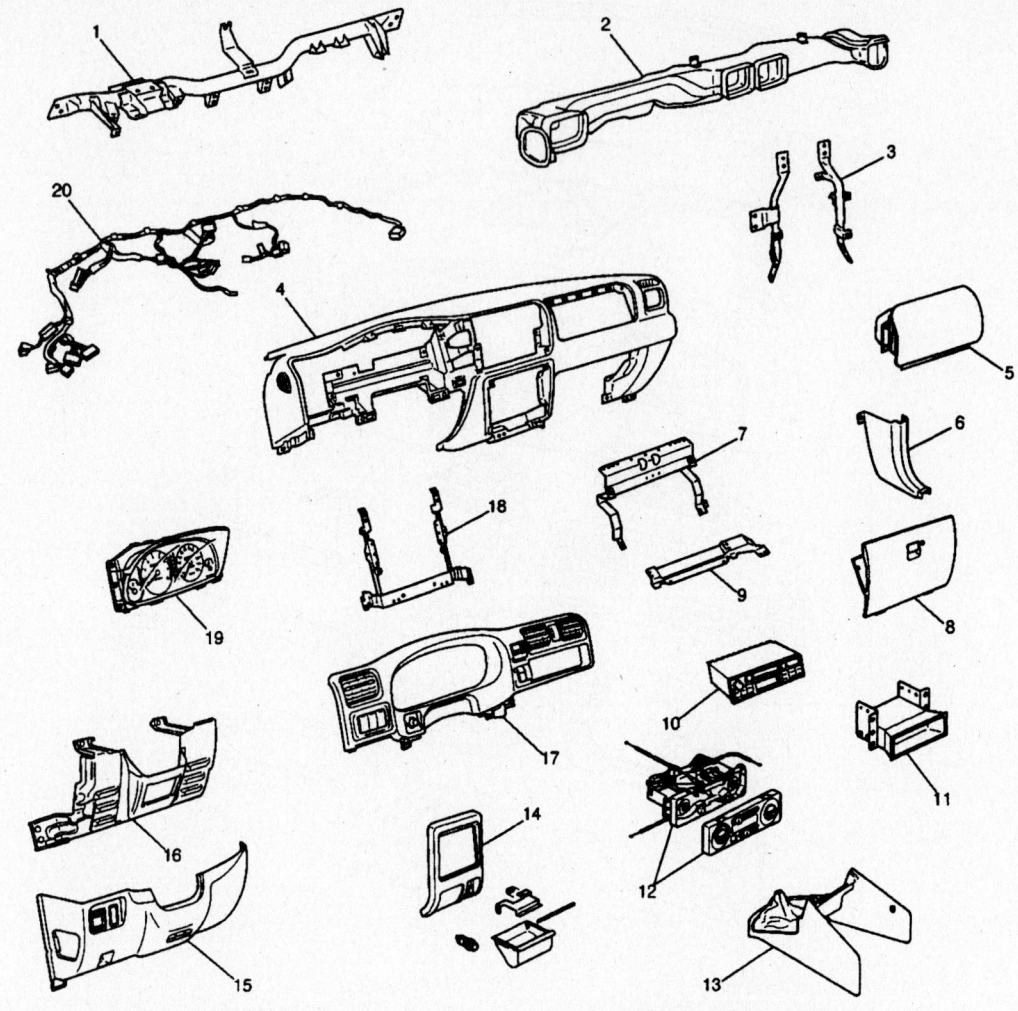

1	Cross Beam	11	Audio Sub Box
2	Vent Duct Assembly	12	Control Lever Assembly
3	Instrument Panel Bracket	13	Front Console Assembly
4	Instrument Panel Assembly	14	Lower Center Cover
5	Passenger Inflator Module	15	Instrument Panel Driver Lower Cover Assembly
6	Dash Side Trim Panel	16	Driver Knee Bolster Assembly
7	Passenger Knee Bolster Reinforcement Assembly	17	Meter Cluster Assembly
8	Glove Box	18	Instrument Panel Center Reinforcement
9	Passenger Lower Bracket	19	Meter Assembly
10	Radio Assembly	20	Instrument Harness Assembly

93113GB8

Exploded view of the instrument panel—Honda Passport

5. Remove the evaporator lines at the firewall. Plug the air conditioning lines to minimize contamination.

6. Disconnect the cooling system hoses and drain the coolant into a clean container for reuse. Plug the cooling system hoses.

7. Remove the instrument panel by performing the following procedure:

    a. Remove the lower center cover screw and pull it out at the clip positions; then, disconnect the cigarette lighter connector.

    b. Remove both the rear and front console.

    c. Remove the dash side trim panel sill plates and the panels.

    d. Remove the 2 glove box screws and the glove box.

    e. Remove the 2 hood release screws, the 6 instrument panel driver's lower cover assembly screws and the cover assembly.

    f. Remove the 5 instrument cluster screws and the 2 clips. Disconnect the 8

switch connectors and remove the instrument cluster assembly.

    g. Remove the 6 driver's knee bolster assembly bolts and screws and the knee bolster assembly.

    h. Remove the 4 control lever assembly bolts; then, disconnect the 3 control cables (unit side) and the 3 harness connectors.

    i. Remove the 4 radio/audio sub box assembly screws and the radio/audio sub box assembly.

    j. Disconnect or remove the following

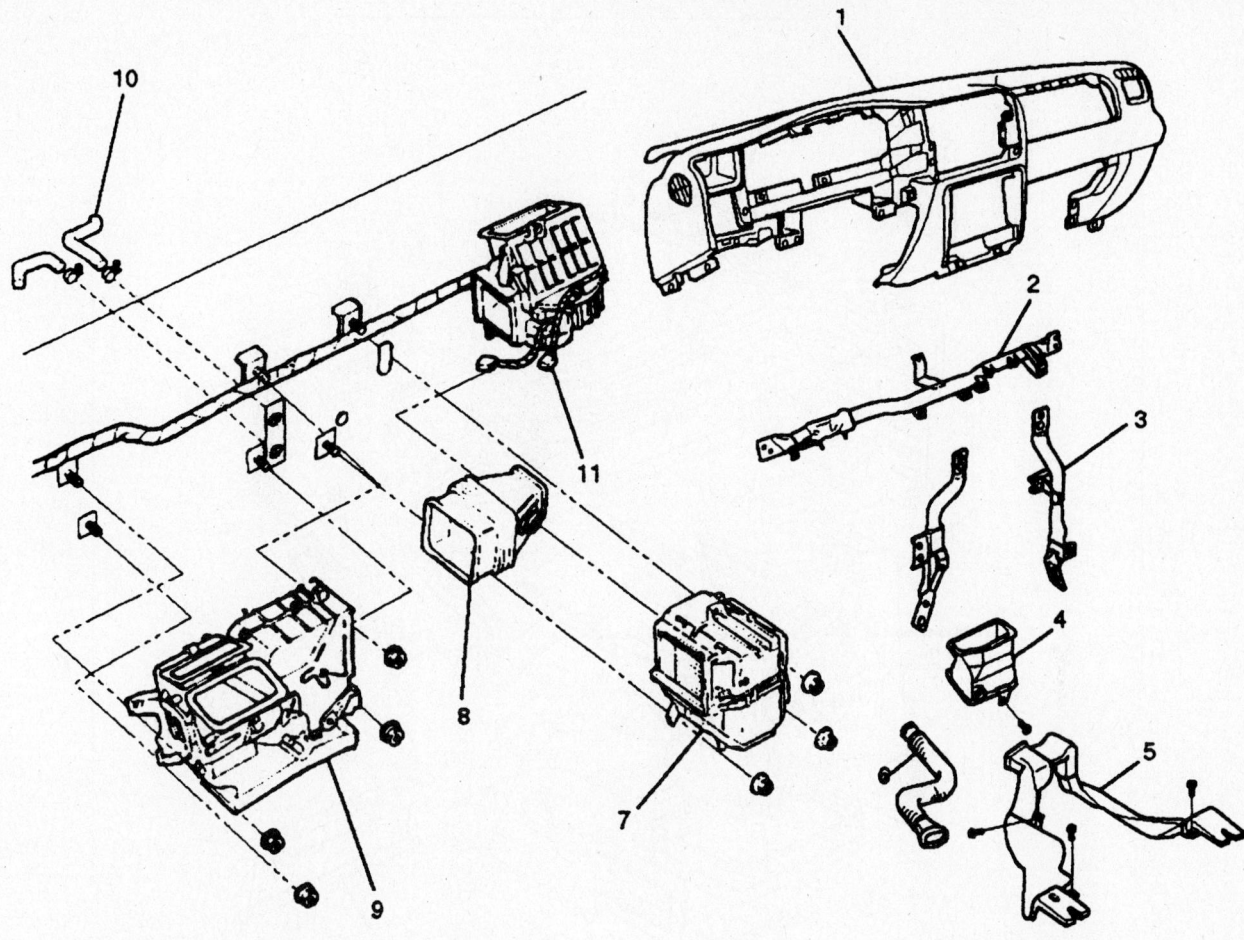

1	Instrument Panel Assembly	6	Driver Lap Vent Duct
2	Cross Beam Assembly	7	Evaporator Assembly (A/C only)
3	Instrument Panel Bracket	8	Duct
4	Ventilation Lower Duct	9	Heater Unit Assembly
5	Rear Heater Duct	10	Heater Hose
		11	Resistor Connector

93113GB9

View of the heater and air conditioning housing assemblies and related components—Honda Passport

instrument panel harness connectors or items:

- The 6 driver's side connectors
- The 3 passenger's side connectors
- The 2 center connectors
- Passenger's inflator module connector
- Radio antenna cable plug
- Ground cable bolt on the left dash side panel
- The 8 instrument panel-to-chassis bolts and the 3 nuts.

k. Remove the instrument panel assembly.

8. Remove the instrument panel bracket by performing the following procedure:

a. Remove the 2 passenger's inflator module bolts and 4 nuts.

b. Remove the 4 meter assembly screws. Then, disconnect the meter wiring harness connectors and remove the meter assembly.

c. Remove the 5 vent duct assembly screws and the assembly.

d. Remove the 3 lower passenger's bracket screws and the bracket.

e. Remove the 9 passenger's knee bolster reinforcement screws and the reinforcement.

f. Remove the 6 instrument panel center reinforcement screws and the reinforcement.

g. Remove the instrument panel wiring harness assembly clips and the wiring harness.

h. Remove the 2 instrument panel bracket nuts and 2 bolts for each bracket; then, remove the bracket(s).

9. Remove the 5 cross beam assembly nuts, 2 bolts and the 6 lower bolts; then, remove the crossbeam.

10. Disconnect the resistor wiring connector.

11. Remove the duct from the heater assembly.

12. If equipped with air conditioning, remove the evaporator assembly.

13. Remove the driver's lap vent.

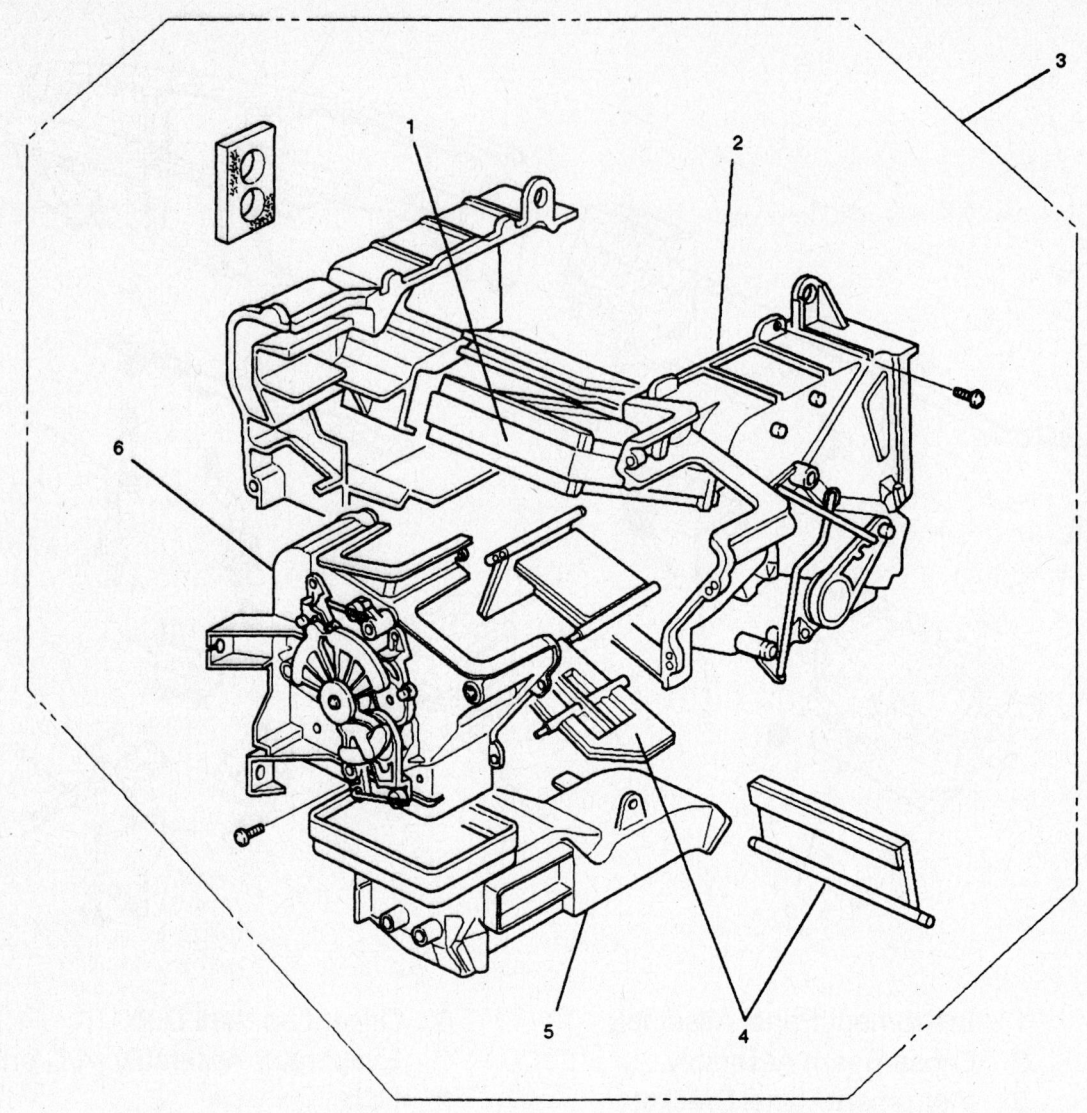

1 Heater Core
2 Case (Temperature Control)
3 Heater Unit
4 Mode Door
5 Duct
6 Case (Mode Control)

93113GB0

Exploded view of the heater housing assembly—Honda Passport

14. Remove the lower ventilation duct.

15. Remove the footrest, the carpet, the 3 clips and the rear heater duct.

16. Remove the heater assembly.

17. Remove the mode control case-to-temperature control case screws and remove the mode control case; do not remove the link unit.

18. Remove the temperature control case screws and separate the cases.

19. Remove the heater core from the case.

To install:

20. Install the heater core to the case.

21. Assemble the temperature control cases and install the case screws.

22. Install the mode control case and the mode control case-to-temperature control case screws.

23. Install the heater assembly.

24. Install the rear heater duct, the footrest, the carpet, and the 3 clips.

25. Install the lower ventilation duct.

26. Install the driver's lap vent.

27. If equipped with air conditioning, install the evaporator assembly.

28. Install the duct to the heater assembly.

29. Connect the resistor wiring connector.

30. Install the crossbeam, the 5 cross beam assembly nuts, 2 bolts and the 6 lower bolts.

31. Install the instrument panel bracket by performing the following procedure:

a. Install the instrument panel bracket and the 2 nuts and 2 bolts for each bracket.

b. Install the instrument panel wiring harness assembly and the wiring harness clips.

c. Install the instrument panel center reinforcement and the 6 reinforcement screws.

d. Install the passenger knee bolster

reinforcement and the 9 reinforcement screws.

e. Install the lower passenger bracket and the 3 bracket screws.

f. Install the vent duct assembly and the 5 vent duct assembly screws.

g. Install the meter assembly and the 4 meter assembly screws; then, connect the meter wiring harness connectors.

h. Install the 2 passenger's inflator module bolts and 4 nuts.

32. Install the instrument panel by performing the following procedure:

a. Install the instrument panel assembly.

b. Connect or install the following instrument panel harness connectors or items:

- The 6 driver's side connectors
- The 3 passenger's side connectors
- The 2 center connectors
- Passenger's inflator module connector
- Radio antenna cable plug
- Ground cable bolt on the left dash side panel
- The 8 instrument panel-to-chassis bolts and the 3 nuts.

c. Install the radio/audio sub box assembly and the 4 radio/audio sub box assembly screws.

d. Connect the 3 control cables (unit side) and the 3 harness connectors. Install the 4 control lever assembly bolts.

e. Install the knee bolster assembly and the 6 driver's knee bolster assembly bolts and screws.

f. Install the instrument cluster assembly. Connect the 8 switch connectors. Install 5 instrument cluster screws and the 2 clips.

g. Install the 2 hood release screws, the 6 instrument panel driver's lower cover assembly screws and the cover assembly.

h. Install the glove box and the 2 glove box screws.

i. Install the dash side trim panel sill plates and the panels.

j. Install both the rear and front console.

k. Connect the cigarette lighter connector and install the lower center cover screw.

33. Connect the cooling system hoses.

34. Refill the cooling system.

35. Install the evaporator lines at the firewall.

36. If equipped, evacuate and charge the air conditioning system.

37. Connect the negative battery cable.

38. If equipped with an air bag, perform the following procedure:

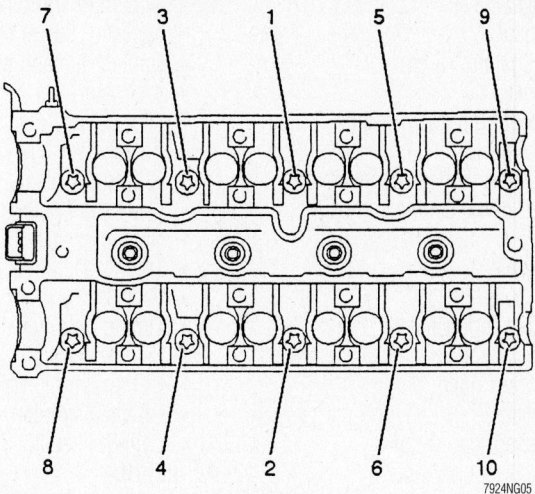

Cylinder head torque sequence—2.2L (VIN D) engine

a. Turn the ignition to the LOCK position and remove the key.

b. Connect the 2-pin yellow connector located behind the glove box.

c. Install the glove box assembly.

d. Connect the 2-pin yellow connector located at the base of the steering column.

e. At the lower left dash side fuse block, install the SRS-1 fuse.

f. Turn the ignition switch to ON and verify that the AIR BAG warning light flashes 7 times and turns OFF.

39. Run the engine to normal operating temperatures; then, check the climate control operation and check for leaks.

Cylinder Head

REMOVAL & INSTALLATION

2.2L Engine

1. Before servicing the vehicle, refer to the Precautions Section.
2. Drain the cooling system.
3. Relieve the fuel system pressure.
4. Remove or disconnect the following:
- Negative battery cable
- Intake Air Temperature (IAT) sensor connector
- Positive Crankcase Ventilation (PCV) valve and hose
- Air intake assembly
- Upper radiator hose
- Accessory drive belt
- Exhaust front pipe
- Alternator and brackets
- Crankshaft Position (CKP) sensor connector
- Knock sensor connector

- Heater hoses
- Water bypass hose
- Fuel lines
- Evaporative Emissions (EVAP) valve connector
- Canister hose
- Intake manifold
- Engine wiring harness connectors at left rear of the engine compartment
- Power steering pump pressure switch connector
- Front cover
- Spark plugs and wires
- Camshaft Position (CMP) sensor
- Valve cover
- Timing belt. Refer to the Timing Belt procedure.
- Timing belt idler pulleys
- Timing belt rear cover
- Oil pressure switch connector
- Camshafts
- Cylinder head. Remove the bolts in reverse of the tightening sequence.

To install:

➡**Use new cylinder head bolts for assembly.**

5. Install the cylinder head with a new gasket. Tighten the bolts in sequence as follows:

a. Step 1: 18 ft. lbs. (25 Nm)
b. Step 2: Plus 90 degrees
c. Step 3: Plus 90 degrees
d. Step 4: Plus 90 degrees

6. Install or connect the following:
- Camshafts
- Oil pressure switch connector
- Timing belt rear cover
- Timing belt idler pulleys. Tighten the bolts to 18 ft. lbs. (25 Nm).
- Timing belt

- Valve cover
- CMP sensor
- Spark plugs and wires
- Front cover
- Power steering pump pressure switch connector
- Engine wiring harness connectors at left rear of the engine compartment
- Intake manifold
- Canister hose
- EVAP valve connector
- Fuel lines
- Water bypass hose
- Heater hoses
- Knock sensor connector
- CKP sensor connector
- Alternator and brackets
- Exhaust front pipe
- Accessory drive belt
- Upper radiator hose
- Air intake assembly
- PCV valve and hose
- IAT sensor connector
- Negative battery cable

7. Fill the cooling system.
8. Start the engine and check for leaks.

3.2L Engines

1. Before servicing the vehicle, refer to the Precautions Section.
2. Drain the cooling system.
3. Relieve the fuel system pressure.
4. Remove or disconnect the following:
   - Negative battery cable
   - Hood
   - Engine cover
   - Mass Air Flow (MAF) sensor connector
   - Intake Air Temperature (IAT) sensor connector
   - Positive Crankcase Ventilation (PCV) valve and hose
   - Air cleaner assembly
   - Manifold Absolute Pressure (MAP) sensor connector
   - Vacuum Switching Valve (VSV) connector and vacuum line
   - Fuel injector connectors
   - Throttle Position (TP) sensor connector
   - Idle Air Control (IAC) valve connector
   - Ignition coils
   - Brake booster vacuum line
   - Canister purge vacuum line
   - Duty solenoid valve
   - Fuel lines
   - Intake manifold
   - Radiator hoses
   - Engine coolant manifold

- Upper fan shroud
- Accessory drive belt and tensioner
- Cooling fan and pulley
- Alternator
- Idler pulley
- Power steering pump and bracket
- A/C compressor
- Crankshaft pulley
- Oil cooler hoses
- Timing belt cover
- Valve covers
- Timing belt. Refer to the Timing Belt Procedure.
- Left and right exhaust front pipes
- Oil dipstick tube
- Cylinder heads

To install:

➡**Use new head bolts when installing the cylinder head.**

➡**The left and right cylinder head gaskets are not interchangeable.**

5. Install the cylinder heads with new gaskets. Tighten the bolts in sequence as follows:
   a. Step 1: 21 ft. lbs. (29 Nm)
   b. Step 2: 47 ft. lbs. (64 Nm)
6. Install or connect the following:
   - Oil dipstick tube
   - Left and right exhaust front pipes
   - Timing belt
   - Valve covers
   - Timing belt cover
   - Oil cooler hoses
   - Crankshaft pulley. Tighten the pulley bolt to 123 ft. lbs. (167 Nm).
   - A/C compressor
   - Power steering pump and bracket. Tighten the bolts to 34 ft. lbs. (46 Nm).
   - Idler pulley
   - Alternator
   - Cooling fan and pulley
   - Accessory drive belt and tensioner

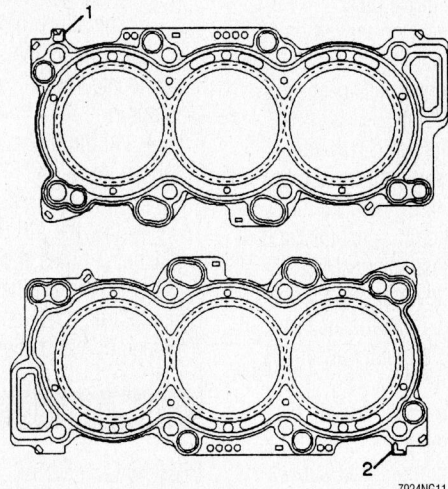

Right (1) and left (2) head gasket identification mark locations—3.2L and 3.5L VIN X engines

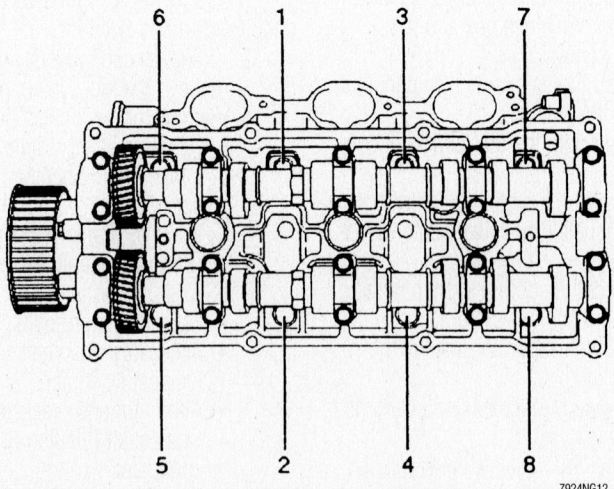

Cylinder head torque sequence—3.2L and 3.5L VIN X engines

- Upper fan shroud
- Engine coolant manifold
- Radiator hoses
- Intake manifold
- Fuel lines
- Duty solenoid valve
- Canister purge vacuum line
- Brake booster vacuum line
- Ignition coils
- IAC valve connector
- TP sensor connector
- Fuel injector connectors
- VSV connector and vacuum line
- MAP sensor connector
- Air cleaner assembly
- PCV valve and hose
- IAT sensor connector
- MAF sensor connector
- Engine cover
- Hood
- Negative battery cable
7. Fill the cooling system.
8. Start the engine and check for leaks.

3.5L VIN X Engine

1. Before servicing the vehicle, refer to the Precautions Section.
2. Drain the cooling system.
3. Remove or disconnect the following:
   - Negative battery cable
   - Hood
   - Engine cover
   - Mass Air Flow (MAF) sensor connector
   - Intake Air Temperature (IAT) sensor connector
   - Positive Crankcase Ventilation (PCV) valve and hose
   - Air cleaner assembly
   - Manifold Absolute Pressure (MAP) sensor connector
   - Vacuum Switching Valve (VSV) connector and vacuum line
   - Fuel injector connectors
   - Throttle Position (TP) sensor connector
   - Idle Air Control (IAC) valve connector
   - Ignition coils
   - Brake booster vacuum line
   - Canister purge vacuum line
   - Duty solenoid valve
   - Fuel lines
   - Intake manifold
   - Radiator hoses
   - Engine coolant manifold
   - Upper fan shroud
   - Accessory drive belt and tensioner
   - Cooling fan and pulley
   - Alternator
   - Idler pulley

- Power steering pump and bracket
- A/C compressor
- Crankshaft pulley
- Oil cooler hoses
- Timing belt cover
- Valve covers
- Timing belt. Refer to the Timing Belt Procedure.
- Left and right exhaust front pipes
- Oil dipstick tube
- Cylinder heads

To install:

➡**Use new head bolts when installing the cylinder head. Do not apply oil to the head bolt threads.**

➡**The left and right cylinder head gaskets are not interchangeable.**

4. Install the cylinder heads with new gaskets. Tighten the bolts, in sequence, to:
   a. Step 1: 21 ft. lbs. (29 Nm)
   b. Step 2: 47 ft. lbs. (64 Nm)
5. Install or connect the following:
   - Oil dipstick tube
   - Left and right exhaust front pipes
   - Timing belt
   - Valve covers
   - Timing belt cover
   - Oil cooler hoses
   - Crankshaft pulley. Tighten the pulley bolt to 123 ft. lbs. (167 Nm).
   - A/C compressor
   - Power steering pump and bracket. Tighten the bolts to 34 ft. lbs. (46 Nm).
   - Idler pulley
   - Alternator
   - Cooling fan and pulley
   - Accessory drive belt and tensioner
   - Upper fan shroud
   - Engine coolant manifold

- Radiator hoses
- Intake manifold
- Fuel lines
- Duty solenoid valve
- Canister purge vacuum line
- Brake booster vacuum line
- Ignition coils
- IAC valve connector
- TP sensor connector
- Fuel injector connectors
- VSV connector and vacuum line
- MAP sensor connector
- Air cleaner assembly
- PCV valve and hose
- IAT sensor connector
- MAF sensor connector
- Engine cover
- Hood
- Negative battery cable
6. Fill the cooling system.
7. Start the engine. Check for leaks and proper operation.

3.5L VIN Y Engine

1. Before servicing the vehicle, refer to the Precautions Section.
2. Drain the cooling system.
3. Drain the engine oil.
4. Remove or disconnect the following:
   - Negative battery cable
   - Hood
   - Exhaust manifold
   - Intake manifold
   - Camshafts
   - Oil dipstick tube
   - Engine coolant manifold
   - Fuel injector connectors
   - Fuel rail joint
   - Fuel lines
   - Radiator hoses
   - Water manifold guide tube

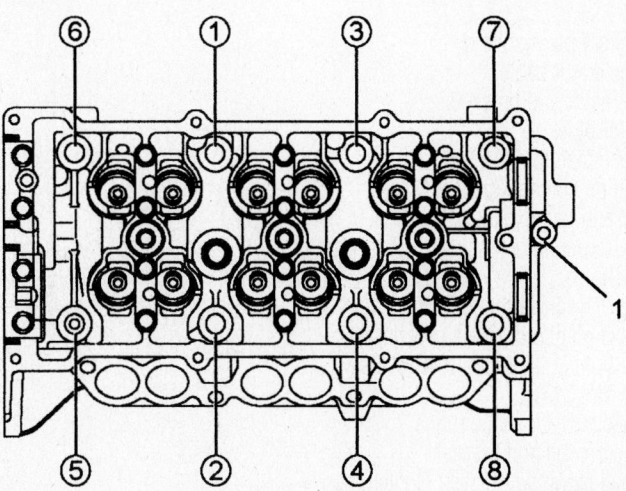

09474_RODE_G0004

Head bolt torque sequence—3.5L VIN Y engine. Note the small bolt at the right, marked 1, is the M8 bolt

5. Loosen the head bolts in the reverse of the tightening sequence. Remove the cylinder heads.

➡**Do not rotate the crankshaft while the head(s) is removed. Once removed, place the head on wood blocks.**

6. Installation is the reverse of removal. The head gaskets are marked for left (L) and right (R). Use new head gaskets. Use new head bolts. Do not apply any lubricant to the head bolts. Tighten the M11 bolts, in sequence, to:

    a. Step 1: 21 ft. lbs. (29 Nm)
    b. Step 2: 47 ft. lbs. (64 Nm)

7. Tighten the M8 bolt to 15 ft. lbs. (21 Nm).

Intake Manifold

REMOVAL & INSTALLATION

2.2L Engine

1. Before servicing the vehicle, refer to the Precautions Section.
2. Drain the cooling system.
3. Relieve the fuel system pressure.
4. Remove or disconnect the following:
- Negative battery cable
- Accessory drive belt
- Positive Crankcase Ventilation (PCV) valve and hose
- Air intake duct
- Throttle body water hoses
- Throttle Position (TP) sensor connector
- Idle Air Control (IAC) valve connector
- Fuel lines
- Fuel injector connectors
- Fuel pressure regulator vacuum line
- Fuel supply manifold
- Accelerator cable
- Alternator and brackets
- Water pipe
- Intake manifold bracket
- Ignition coil and bracket
- Brake booster vacuum line
- Intake manifold

To install:
5. Install or connect the following:
- Intake manifold. Use a new gasket and tighten the bolts to 16 ft. lbs. (22 Nm).
- Brake booster vacuum line
- Ignition coil and bracket
- Intake manifold bracket. Tighten the bolts to 16 ft. lbs. (22 Nm).
- Water pipe

- Alternator and brackets. Tighten the short bolts to 14 ft. lbs. (20 Nm) and the long bolts to 25 ft. lbs. (35 Nm).
- Accelerator cable
- Fuel supply manifold
- Fuel pressure regulator vacuum line
- Fuel injector connectors
- Fuel lines
- IAC valve connector
- TP sensor connector
- Throttle body water hoses
- Air intake duct
- PCV valve and hose
- Accessory drive belt
- Negative battery cable
6. Fill the cooling system.
7. Start the engine and check for leaks.

3.2L Engine

1. Before servicing the vehicle, refer to the Precautions Section.
2. Remove or disconnect the following:
- Negative battery cable
- Engine cover
- Air cleaner assembly
- Accelerator cable
- Cruise control cable
- Brake booster vacuum line
- Manifold Absolute Pressure (MAP) sensor connector
- Idle Air Control (IAC) valve connector
- Throttle Position (TP) sensor connector
- Canister purge solenoid connector
- Electronic Vacuum Sensing Valve (EVSV) connector and vacuum line

- Exhaust Gas Recirculation (EGR) valve
- Positive Crankcase Ventilation (PCV) valve and hose
- Pressure regulator vacuum line
- Ventilation hose
- Throttle body
- Fuel lines
- Fuel injector connectors
- Intake manifold

To install:
3. Install or connect the following:
- Intake manifold. Tighten the fasteners to 18 ft. lbs. (25 Nm).
- Fuel injector connectors
- Fuel lines
- Throttle body. Tighten the bolts to 88 inch lbs. (10 Nm).
- Ventilation hose
- Pressure regulator vacuum line
- PCV valve and hose
- EGR valve
- EVSV connector and vacuum line
- Canister purge solenoid connector
- TP sensor connector
- IAC valve connector
- MAP sensor connector
- Brake booster vacuum line
- Cruise control cable
- Accelerator cable
- Air cleaner assembly
- Engine cover
- Negative battery cable
4. Start the engine and check for proper operation.

3.5L VIN X Engine

1. Before servicing the vehicle, refer to the Precautions Section.

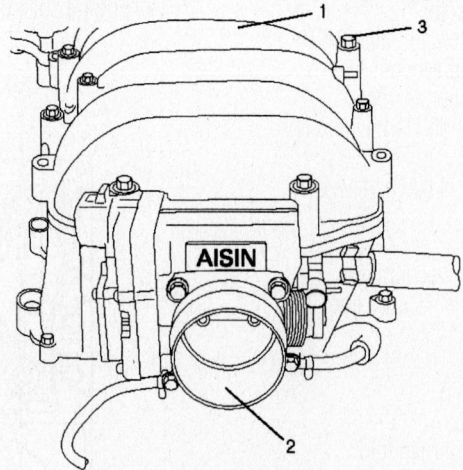

(1)	Common Chamber
(2)	Throttle Valve Assembly
(3)	Bolt

09474_RODE_G0047

Intake manifold (common chamber)—3.2L and 3.5L VIN X engines

2. Remove or disconnect the following:
- Negative battery cable
- Engine cover
- Air cleaner assembly
- Accelerator cable
- Cruise control cable
- Brake booster vacuum line
- Manifold Absolute Pressure (MAP) sensor connector
- Idle Air Control (IAC) valve connector
- Throttle Position (TP) sensor connector
- Canister purge solenoid connector
- Electronic Vacuum Sensing Valve (EVSV) connector and vacuum line
- Exhaust Gas Recirculation (EGR) valve
- Positive Crankcase Ventilation (PCV) valve and hose
- Pressure regulator vacuum line
- Ventilation hose
- Throttle body
- Fuel lines
- Fuel injector connectors
- Intake manifold

To install:

3. Install or connect the following:
- Intake manifold. Tighten the fasteners to 18 ft. lbs. (25 Nm).
- Fuel injector connectors
- Fuel lines
- Throttle body. Tighten the bolts to 88 inch lbs. (10 Nm).
- Ventilation hose
- Pressure regulator vacuum line
- PCV valve and hose
- EGR valve
- EVSV connector and vacuum line
- Canister purge solenoid connector
- TP sensor connector
- IAC valve connector
- MAP sensor connector
- Brake booster vacuum line
- Cruise control cable
- Accelerator cable
- Air cleaner assembly
- Engine cover
- Negative battery cable

4. Start the engine and check for proper operation.

3.5L VIN Y Engines

1. Before servicing the vehicle, refer to the Precautions Section.
2. Disconnect the battery ground.
3. Remove the air cleaner.
4. Tag and disconnect all wiring and hoses from the intake manifold.

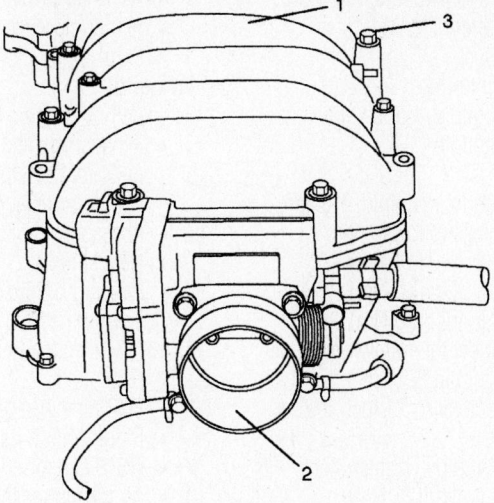

(1)   Common Chamber
(2)   Throttle Valve Assembly
(3)   Bolt

09474_RODE_G0048

Intake manifold—3.5L VIN Y engine

5. Remove the throttle body.
6. Remove the bolts and nuts and lift off the manifold.
7. Installation is the reverse of removal. Use new gaskets. Torque the intake manifold nuts and bolts to 13 ft. lbs. (18 Nm). Torque the throttle body bolts to 87 inch lbs. (10 Nm).

Exhaust Manifold

REMOVAL & INSTALLATION

2.2L Engine

1. Before servicing the vehicle, refer to the Precautions Section.

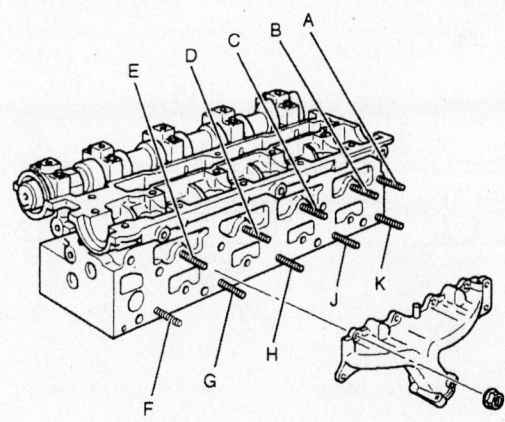

- **Tightening sequence:**
  Step1: J G H B D C J G B D
  Step2: A B C D E F G H J K
  Step3: A B C D E F G H J K
- **Tightening torque:**
  Step1: 14 N·m (10 lb ft)
  Step2: 20 N·m (14 lb ft)
  Step3: 20 N·m (14 lb ft)

7924NG17

Exhaust manifold torque sequence—2.2L engine

2. Remove or disconnect the following:
- Negative battery cable
- Air intake duct
- Exhaust front pipe
- Exhaust manifold heat shield
- Exhaust manifold

To install:

3. Install the exhaust manifold. Tighten the nuts in sequence as follows:
  a. Step 1: 10 ft. lbs. (14 Nm)
  b. Step 2: 14 ft. lbs. (20 Nm)
  c. Step 3: 14 ft. lbs. (20 Nm)
4. Install or connect the following:
- Exhaust manifold heat shield. Tighten the bolts to 71 inch lbs. (8 Nm).
- Exhaust front pipe. Tighten the bolts to 18 ft. lbs. (25 Nm).
- Air intake duct
- Negative battery cable
5. Start the engine and check for leaks.

3.2L and 3.5L VIN Y Engines

1. Before servicing the vehicle, refer to the Precautions Section.
2. Remove or disconnect the following:
- Negative battery cable
- Air cleaner assembly
- Heated Oxygen (HO2S) sensor connectors
- Right torsion bar
- Exhaust Gas Recirculation (EGR) pipe and bracket
- Left and right exhaust front pipes
- Heat shields

- Accessory drive belt
- A/C compressor and bracket
- Exhaust manifolds

To install:

3. Install or connect the following:
- Exhaust manifolds. Tighten the bolts to 42 ft. lbs. (57 Nm).
- A/C compressor and bracket
- Accessory drive belt
- Heat shields
- Left and right exhaust front pipes. Torque the stud nuts to 49 ft. lbs. (67 Nm); the nuts to 32 ft. lbs. (43 Nm).
- EGR pipe and bracket
- Right torsion bar
- HO2S sensor connectors
- Air cleaner assembly
- Negative battery cable
4. Start the engine and check for leaks.

3.5L VIN X Engine

1. Before servicing the vehicle, refer to the Precautions Section.
2. Remove or disconnect the following:
- Negative battery cable
- Air cleaner assembly
- Heated Oxygen (HO2S) sensor connectors
- Right torsion bar
- Exhaust Gas Recirculation (EGR) pipe and bracket
- Left and right exhaust front pipes
- Heat shields
- Accessory drive belt

- A/C compressor and bracket
- Exhaust manifolds

To install:

3. Install or connect the following:
- Exhaust manifolds. Tighten the bolts to 38 ft. lbs. (52 Nm).
- A/C compressor and bracket
- Accessory drive belt
- Heat shields
- Left and right exhaust front pipes. Torque the stud nuts to 49 ft. lbs. (67 Nm); the front pipe nuts to 42 ft. lbs. (57 Nm); the silencer nuts to 32 ft. lbs. (43 Nm).
- EGR pipe and bracket
- Right torsion bar
- HO2S sensor connectors
- Air cleaner assembly
- Negative battery cable
4. Start the engine and check for leaks.

Front Crankshaft Seal

REMOVAL & INSTALLATION

2.2L Engines

1. Before servicing the vehicle, refer to the Precautions Section.
2. Remove or disconnect the following:
- Negative battery cable
- Accessory drive belts
- Cooling fan
- A/C belt tensioner, if equipped
- Water pump pulley
- Power steering pump
- Crankshaft pulley
- Front cover
- Timing belt. Refer to the Timing Belt procedure.
- Crankshaft timing sprocket
- Rear timing cover
- Crankshaft oil seal

To install:

3. Install or connect the following:
- Crankshaft oil seal
- Rear timing cover
- Crankshaft timing sprocket. Tighten the bolt to 94 ft. lbs. (130 Nm) plus 45 degrees.
- Timing belt. Refer to the Timing Belt procedure.
- Front cover
- Crankshaft pulley
- Power steering pump
- Water pump pulley
- A/C belt tensioner, if equipped
- Cooling fan
- Accessory drive belts
- Negative battery cable
4. Start the engine and check for leaks.

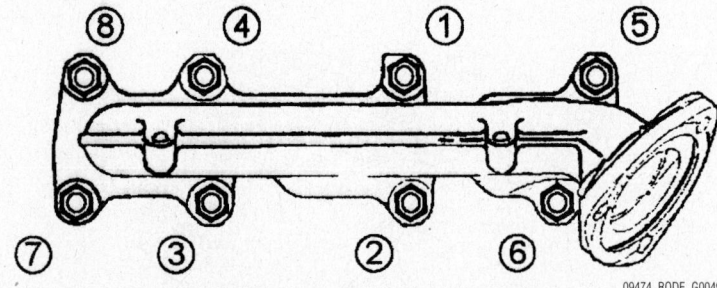

Left side exhaust manifold torque sequence—3.2L and 3.5L VIN Y Engines

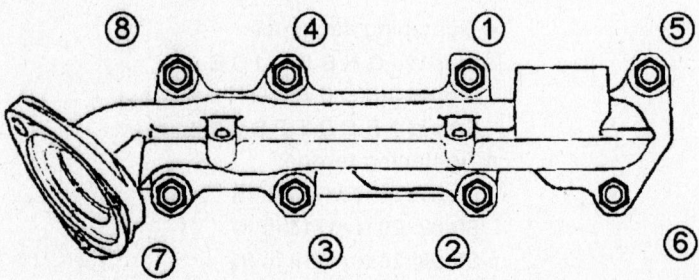

Right side exhaust manifold torque sequence—3.2L and 3.5L VIN Y Engines

3.2L Engines

1. Before servicing the vehicle, refer to the Precautions Section.
2. Remove or disconnect the following:
   - Negative battery cable
   - Air cleaner assembly
   - Upper fan shroud
   - Accessory drive belt and tensioner
   - Cooling fan and pulley
   - Idler pulley
   - Power steering pump
   - Crankshaft pulley
   - Timing belt cover
   - Timing belt. Refer to the Timing Belt Procedure.
   - Crankshaft timing sprocket
   - Oil seal

To install:

3. Install or connect the following:
   - Oil seal so that it is flush with the oil pump housing
   - Crankshaft timing sprocket
   - Timing belt
   - Timing belt cover
   - Crankshaft pulley. Tighten the bolt to 123 ft. lbs. (167 Nm).

   - Power steering pump
   - Idler pulley
   - Cooling fan and pulley
   - Accessory drive belt and tensioner
   - Upper fan shroud
   - Air cleaner assembly
   - Negative battery cable
4. Start the engine and check for leaks.

3.5L Engine

1. Before servicing the vehicle, refer to the Precautions Section.
2. Remove or disconnect the following:
   - Negative battery cable
   - Air cleaner assembly
   - Upper fan shroud
   - Accessory drive belt and tensioner
   - Cooling fan and pulley
   - Idler pulley
   - Power steering pump and move it aside
   - Crankshaft pulley
   - Timing belt cover
   - Timing belt. Refer to the Timing Belt Procedure.
   - Crankshaft timing sprocket
   - Oil seal

To install:

3. Install or connect the following:
   - Oil seal so that it is flush with the oil pump housing
   - Crankshaft timing sprocket
   - Timing belt
   - Timing belt cover
   - Crankshaft pulley. Tighten the bolt to 123 ft. lbs. (167 Nm).
   - Power steering pump
   - Idler pulley
   - Cooling fan and pulley
   - Accessory drive belt and tensioner
   - Upper fan shroud
   - Air cleaner assembly
   - Negative battery cable
4. Start the engine and check for leaks.

Camshaft and Valve Lifters

REMOVAL & INSTALLATION

2.2L Engine

1. Before servicing the vehicle, refer to the Precautions Section.
2. Remove or disconnect the following:

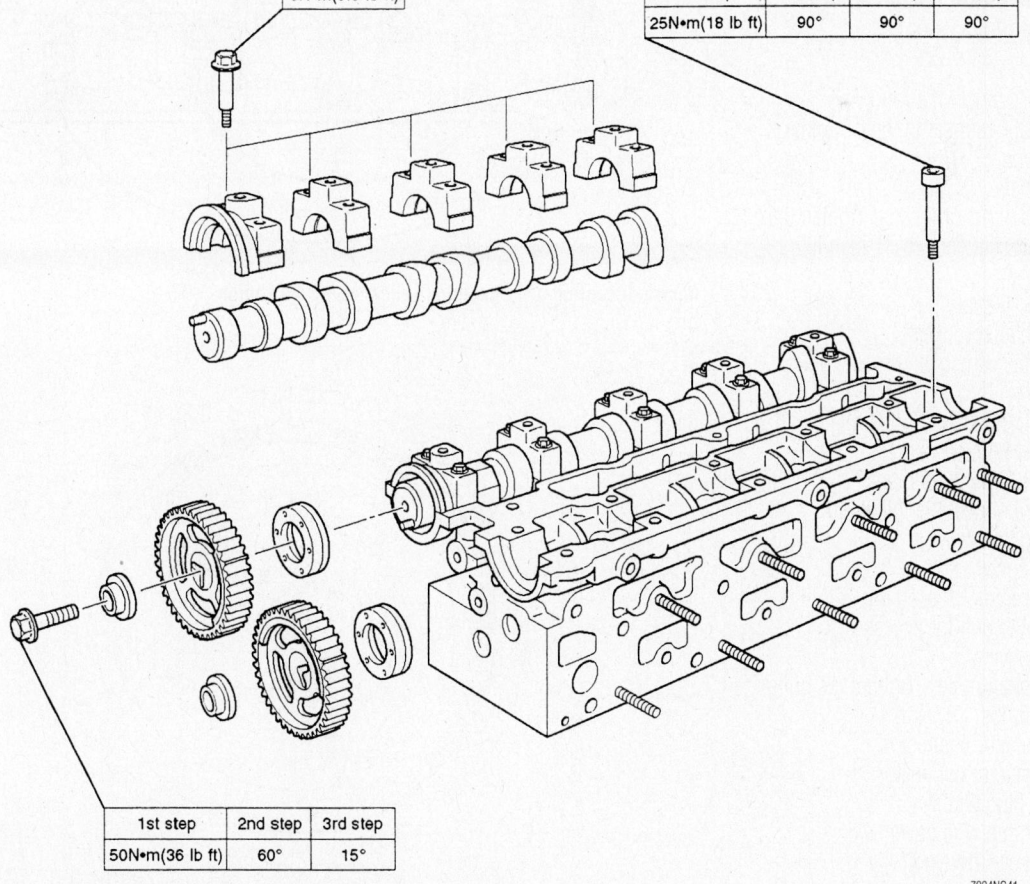

1st step	2nd step	3rd step	4th step
25N•m(18 lb ft)	90°	90°	90°

8N•m(5.9 lb ft)

1st step	2nd step	3rd step
50N•m(36 lb ft)	60°	15°

7924NG41

Exploded view of the cylinder head and camshaft components—2.2L engine

- Negative battery cable
- Positive Crankcase Ventilation (PCV) valve and hose
- Air intake duct and bracket
- Ground cables
- Engine wiring harness connectors at left rear of the engine compartment
- Cooling fan harness connector
- Accessory drive belt
- Spark plug wire cover
- Spark plug wires
- Camshaft Position (CMP) sensor connector
- Crankshaft Position (CKP) sensor connector
- Crankshaft pulley
- Front cover
- Camshaft Position (CMP) sensor. Loosen the rear timing cover bolt for access.
- Valve cover
- Timing belt. Refer to the Timing Belt procedure.
- Camshaft sprockets
- Camshaft bearing caps
- Camshaft seals
- Camshafts
- Hydraulic tappets

➡**Keep all valvetrain components in order for assembly.**

To install:
3. Install or connect the following:
   - Hydraulic tappets in their original locations
   - Camshafts
   - Camshaft bearing caps. Tighten the bolts in sequence to 71 inch lbs. (8 Nm).
   - Camshaft seals
   - Camshaft sprockets
4. Tighten the camshaft sprocket bolts as follows:
   a. Step 1: 36 ft. lbs. (50 Nm)
   b. Step 2: Plus 60 degrees
   c. Step 3: Plus 15 degrees
5. Install or connect the following:
   - Timing belt
   - Valve cover
   - CMP sensor. Loosen the rear timing cover bolt for access.
   - Front cover
   - Crankshaft pulley. Tighten the bolts to 14 ft. lbs. (20 Nm).
   - CKP sensor connector
   - CMP sensor connector
   - Spark plug wires
   - Spark plug wire cover
   - Accessory drive belt
   - Cooling fan harness connector
   - Engine wiring harness connectors

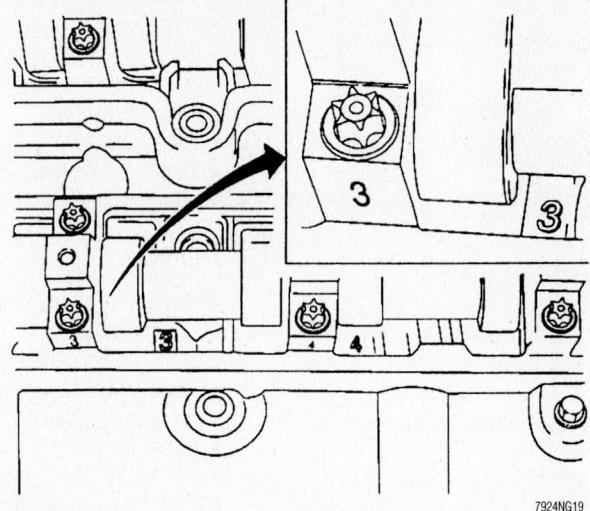

Camshaft bearing cap identification locations—2.2L engine

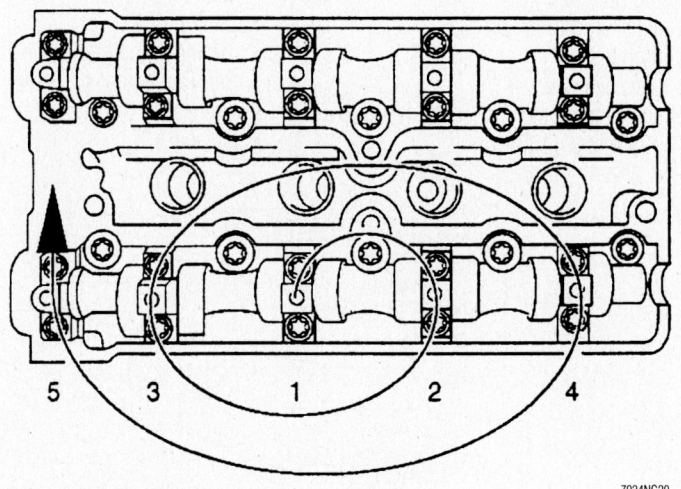

Camshaft bearing cap tightening sequence—2.2L engine

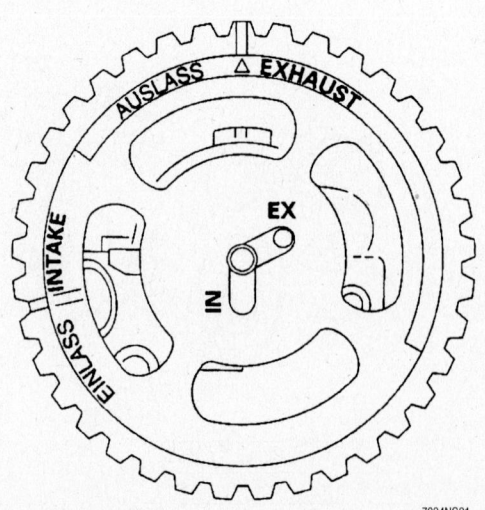

Guide pin location for the exhaust cam gear—2.2L engine

at left rear of the engine compartment
- Ground cables
- Air intake duct and bracket
- PCV valve and hose
- Negative battery cable

3.2L Engine

1. Before servicing the vehicle, refer to the Precautions Section.
2. Remove or disconnect the following:
   - Negative battery cable
   - Air cleaner assembly
   - Upper fan shroud
   - Accessory drive belt and tensioner
   - Cooling fan and pulley
   - Idler pulley
   - Power steering pump and move it aside
   - Crankshaft pulley
   - Timing belt cover
   - Timing belt. Refer to the Timing Belt Procedure.
   - Ignition coils
   - Valve covers
   - Camshafts
   - Valve shims and tappets

➡Keep the valve shims and tappets in order for installation.

To install:

3. Install the valve tappets and shims in their original locations.
4. Using Gear Spring Lever J-42686, turn the sub gear clockwise to align the 5mm bolt holes in the sub gear and the camshaft driven gear. Tighten the 5mm bolt.
5. Install or connect the following:
   - Camshafts by aligning the timing marks as shown. Tighten the bolts in sequence to 89 inch lbs. (10 Nm).
   - Valve covers
   - Ignition coils
   - Timing belt
   - Timing belt cover
   - Crankshaft pulley. Tighten the bolt to 123 ft. lbs. (167 Nm).
   - Power steering pump
   - Idler pulley
   - Cooling fan and pulley
   - Accessory drive belt and tensioner
   - Upper fan shroud
   - Air cleaner assembly
   - Negative battery cable

3.5L VIN X Engine

1. Before servicing the vehicle, refer to the Precautions Section.
2. Remove or disconnect the following:
   - Negative battery cable

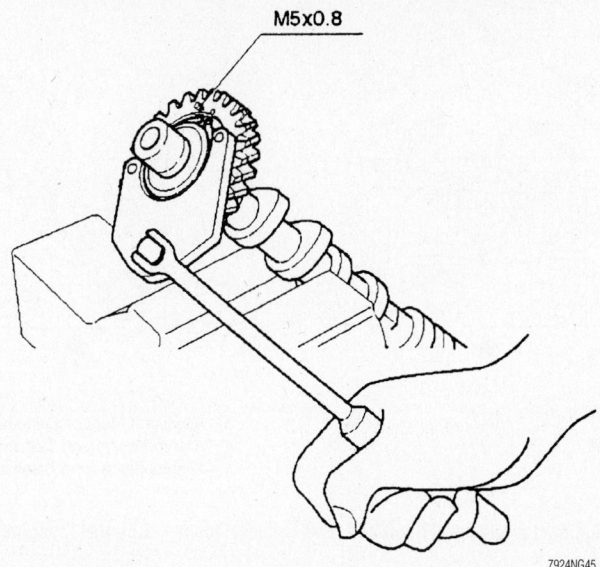

M5x0.8

Aligning the sub gear with the Gear Spring Lever J-42686—3.2L engine

7924NG45

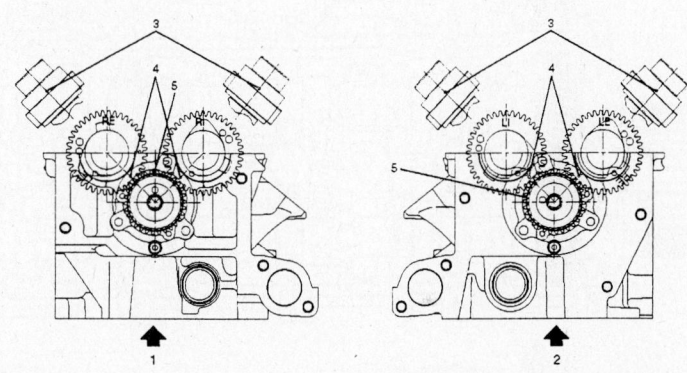

Legend
(1) Right Bank
(2) Left Bank

(3) Alignment Mark on Camshaft Drive Gear
(4) Alignment Mark on Camshaft
(5) Alignment Mark on Retainer

7924NG46

Camshaft alignment marks for the left and right cylinder heads—3.2L engine

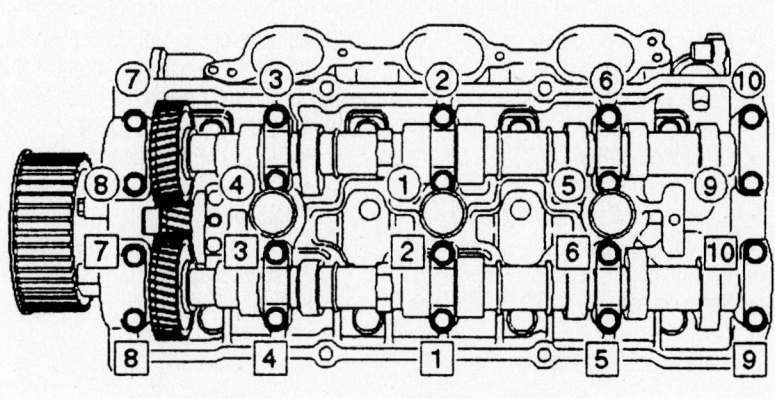

◯ : **Intake**   ▢ : **Exhaust**

7924NG47

Camshaft retaining bracket tightening sequence—3.2L engine

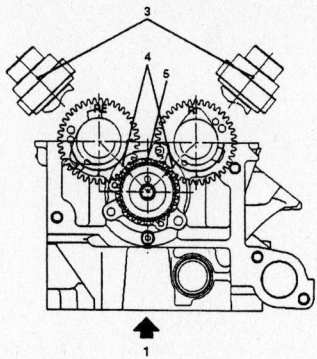 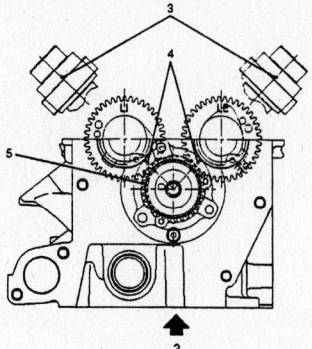

1	Right Bank	3	Alignment Mark on Camshaft Drive Gear
2	Left Bank	4	Alignment Mark on Camshaft
		5	Alignment Mark on Retainer

7924BG11

Camshaft alignment marks for the left and right cylinder heads—3.5L VIN X engine

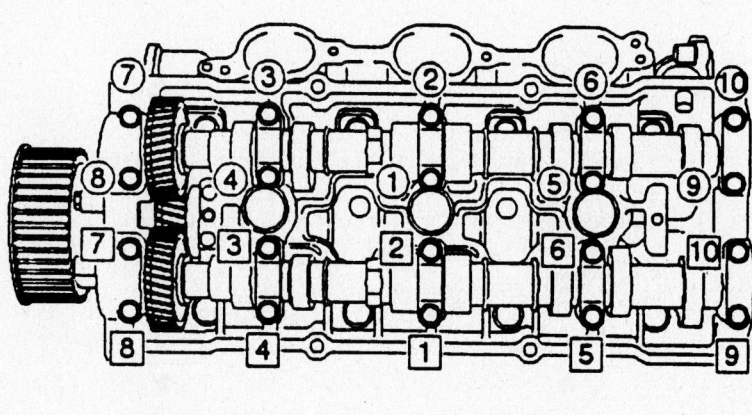

○ : Intake   □ : Exhaust

7924BG12

Camshaft retaining bracket tightening sequence—3.5L VIN X engine

- Air cleaner assembly
- Upper fan shroud
- Accessory drive belt and tensioner
- Cooling fan and pulley
- Idler pulley
- Power steering pump and move it aside
- Crankshaft pulley
- Timing belt cover
- Timing belt. Refer to the Timing Belt Procedure.
- Ignition coils
- Valve covers
- Camshafts
- Valve shims and tappets

➡**Keep the valve shims and tappets in order for installation.**

To install:

3. Install the valve tappets and shims in their original locations.

4. Using Gear Spring Lever J-42686, turn the sub gear clockwise to align the 5mm bolt holes in the sub gear and the camshaft driven gear. Tighten the 5mm bolt.

5. Install the camshafts. Align the timing marks as shown. Tighten the bolts in sequence to 89 inch lbs. (10 Nm).

6. Install or connect the following:
- Valve covers
- Ignition coils
- Timing belt. Refer to the Timing Belt Procedure.
- Timing belt cover
- Crankshaft pulley
- Power steering pump
- Idler pulley
- Cooling fan and pulley
- Accessory drive belt and tensioner
- Upper fan shroud
- Air cleaner assembly
- Negative battery cable

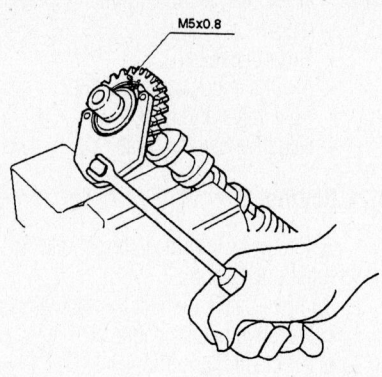

7924BG13

Aligning the sub gear with the Gear Spring Lever J-42686—3.5L VIN X engine

3.5L VIN Y Engine

1. Before servicing the vehicle, refer to the Precautions Section.
2. Disconnect the battery ground.
3. Remove the accessory drive belt.
4. Remove the timing belt.
5. Remove the camshaft covers.
6. Remove the camshaft drive sprocket.
7. Remove the retainer.

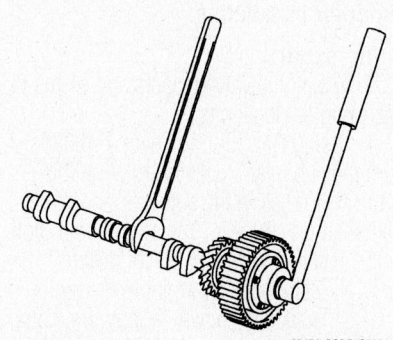

09474_RODE_G0051

Camshaft sprocket bolt removal—3.5L VIN Y engine

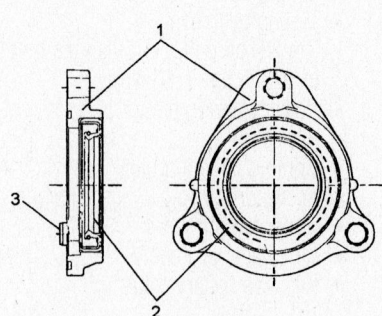

(1)	Retainer
(2)	Oil Seal
(3)	Dowel

09474_RODE_G0052

Retainer—3.5L VIN Y engine

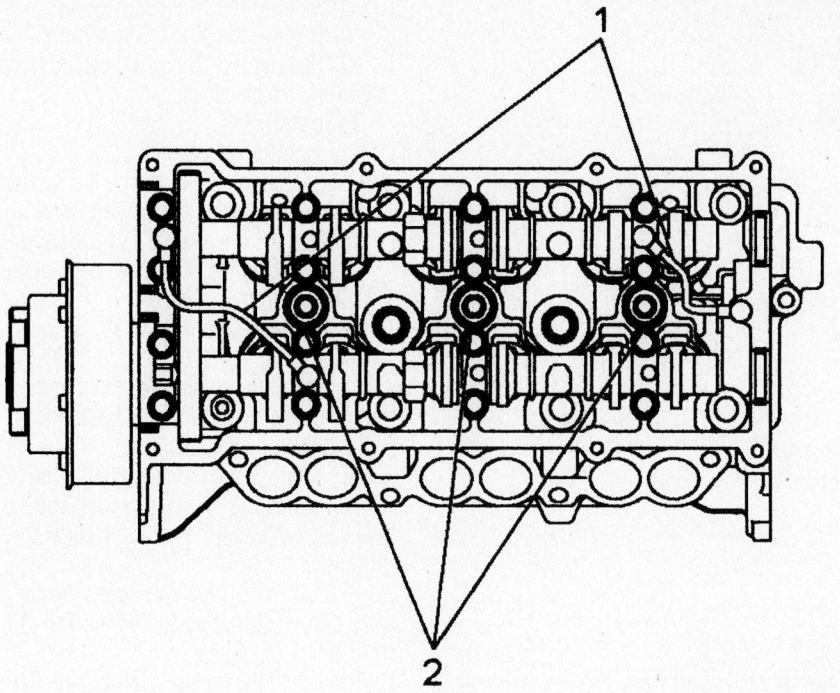

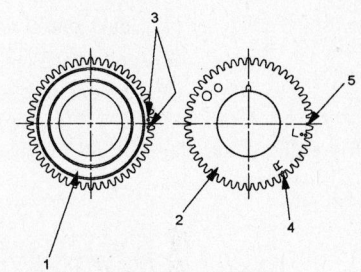

Sub gear installation—3.5L VIN Y engine

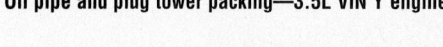

(1)   Intake Camshaft Timing Gear
(2)   Exhaust Camshaft Timing Gear
(3)   Timing Mark
(4)   Timing Mark R: Right Bank (One dot)
(5)   Timing Mark L: Left Bank (Two dot)

(1)   Oil Pipe

(2)   Plug Tower Packing

Camshaft timing marks—3.5L VIN Y engine

Oil pipe and plug tower packing—3.5L VIN Y engine

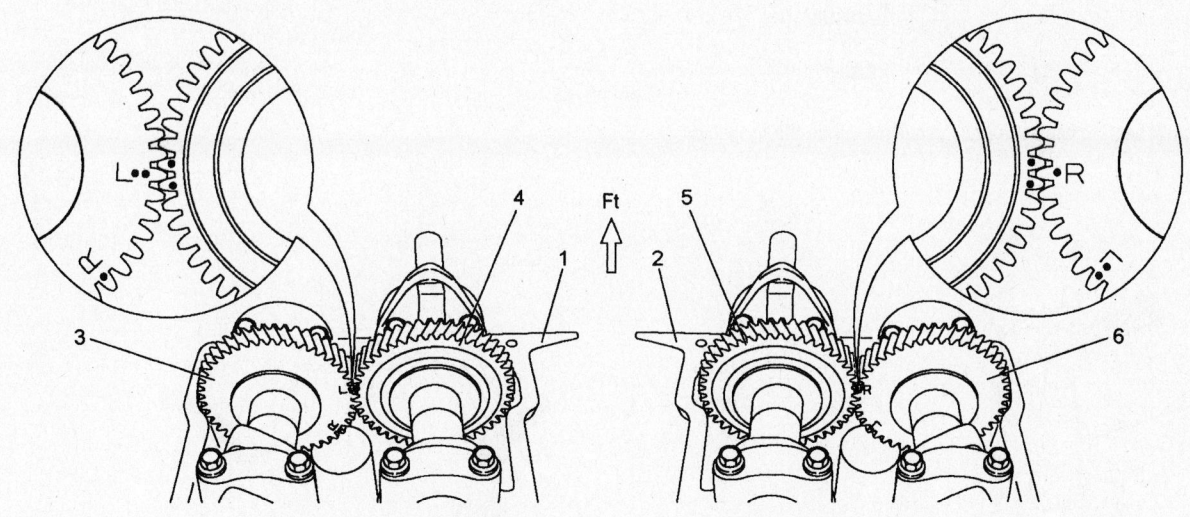

(1)   Left Bank
(2)   Right Bank
(3)   Exhaust Camshaft Timing Gear (Left Bank)
(4)   Intake Camshaft Timing Gear (Left Bank)
(5)   Intake Camshaft Timing Gear (Right Bank)
(6)   Exhaust Camshaft Timing Gear (Right Bank)

Camshaft installation—3.5L VIN Y engine

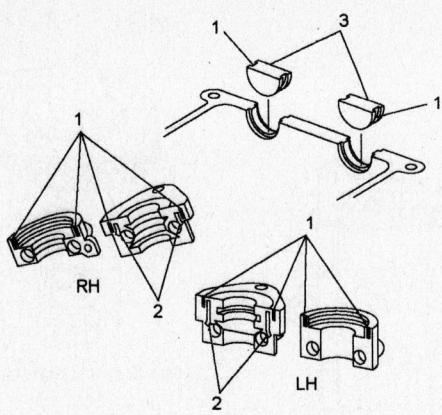

(1) Liquid gasket application area

(2) Area where liquid gasket must not be applied

(3) Cam end plug rear

09474_RODE_G0057

RTV application points—3.5L VIN Y engine

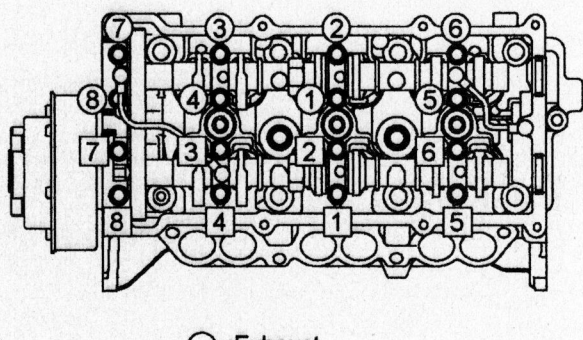

○ :Exhaust

□ :Intake

09474_RODE_G0058

Camshaft bearing cap bolt torque sequence—3.5L VIN Y engine

8. Remove the oil pipe.

9. Remove the plug tower packing.

10. Remove the 16 bearing cap bolts, the caps, and the camshafts.

To install:

11. Tighten the sub gear bolt.

   a. Use tool J7624 to turn the sub gear until it aligns with the M5 bolt hole between the driven gear and sub gear.

   b. Tighten the M5 bolt to prevent the sub gear from moving.

12. Apply clean engine oil to the journals and caps.

13. Install the camshafts, aligning the timing marks as shown: one dot on the right; two on the left.

14. Apply RTV gasket material to the mating surfaces of the cam bracket and head. Do not allow the gasket material to protrude.

15. Install the caps and bolts. Torque the bolts, in sequence, to 89 inch lbs. (10 Nm).

16. Install the oil pipe. Torque to 9 ft. lbs. (12 Nm).

17. Install the plug tower packing.

18. Install the oil seal.

19. Install the O-ring in the retainer. Apply RTV gasket material to the areas indicated. Install the retainer. Torque the bolts to 89 inch lbs. (10 Nm).

20. Torque the camshaft sprocket bolt to 58 ft. lbs. (78 Nm).

21. Install the center cap. Torque to 11 ft. lbs. (15 Nm).

22. The remainder of installation is the reverse of removal.

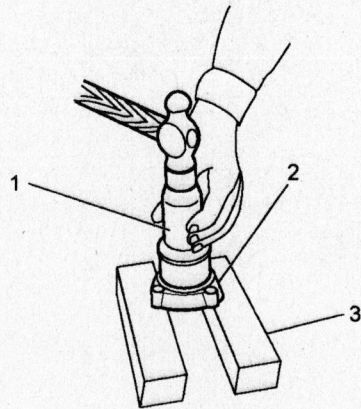

(1)   Tool J-26508

(2)   Retainer

(3)   Block

09474_RODE_G0059

Oil seal installation—3.5L VIN Y engine

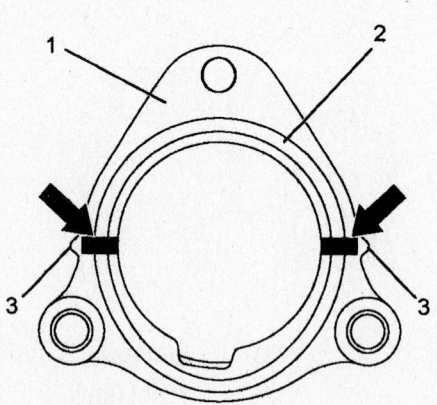

(1)   Retainer

(2)   O-Ring

(3)   Liquid gasket application area mark

09474_RODE_G0060

RTV application to the o-ring area—3.5L VIN Y engine

Valve Lash

ADJUSTMENT

2.2L Engine

The 2.2L DOHC engine is equipped with hydraulic lash adjusters. No valve adjustment is necessary.

3.2L Engines

➡**Measure valve clearance with the engine cold.**

1. Before servicing the vehicle, refer to the Precautions Section.
2. Remove the valve covers.
3. Check the valve clearance with the camshafts positioned as shown. Intake valve clearance should be 0.0091–0.0130 in. (0.2311–0.3302mm). Exhaust valve clearance should be 0.0098–0.0138 in. (0.2489–0.3505mm).
4. If adjustment is required, replace the shims as follows:
    a. Step 1: Position special tool J-42689 on the edge of the tappet.
    b. Step 2: Rotate the crankshaft until the maximum lift portion of the camshaft lobe contacts the upper edge of the special tool and presses the tappet down to create enough clearance between the adjustment shim and the camshaft for the shim to be removed.
    c. Step 3: Replace shims as necessary to achieve correct valve clearance.
    d. Step 4: Repeat for each valve to be adjusted.
5. Replace the valve covers. Tighten the bolts to 80 inch lbs. (9 Nm).

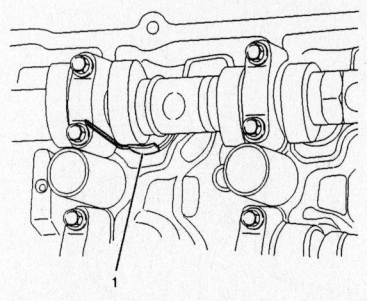

Insert special tool J-42689 (1) and use the camshaft to press the tappet down—3.2L DOHC engine

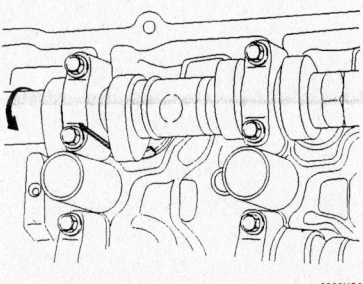

Rotate the camshaft to depress the valve with special tool J-42689—3.2L DOHC engine

3.5L Engine

➡**Measure valve clearance with the engine cold.**

1. Before servicing the vehicle, refer to the Precautions Section.
2. Remove the valve covers.
3. Check the valve clearance with the camshafts positioned as shown. Intake valve clearance should be 0.0091–0.0130 inch.

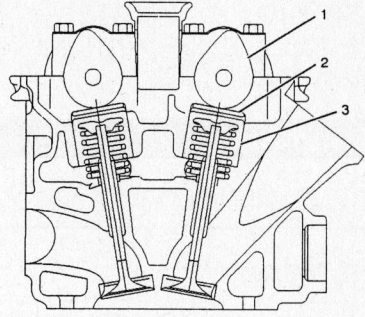

Cross section of the 3.5L cylinder head. Note the position of the camshaft lobe (1), adjustment shim (2) and the tappet (3)

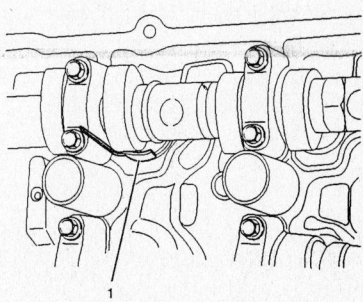

Valve clearance adjusting tool J–42689 (1)

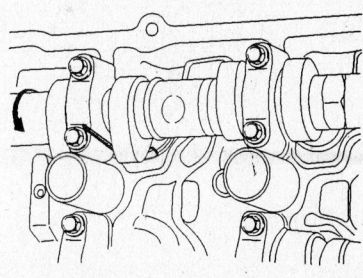

Using the valve clearance adjusting tool to hold the tappet for shim replacement

Exhaust valve clearance should be 0.0098–0.0138 inch.
4. If adjustment is required, replace the shims as follows:
    a. Step 1: Position special tool J–42689 on the edge of the tappet.
    b. Step 2: Rotate the crankshaft until the maximum lift portion of the camshaft lobe contacts the upper edge of the special tool and presses the tappet down to create enough clearance between the adjustment shim and the camshaft for the shim to be removed.
    c. Step 3: Replace shims as necessary to achieve correct valve clearance.

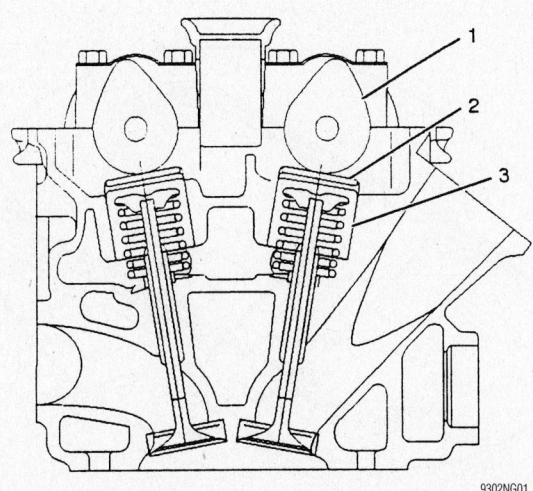

Cross section of the 3.2L cylinder head. Note the position of the camshaft lobe (1), adjustment shim (2) and the tappet (3)

d. Step 4: Repeat for each valve to be adjusted.

5. Replace the valve covers. Tighten the bolts to 80 inch lbs. (9 Nm).

Starter Motor

REMOVAL & INSTALLATION

2.2L Engines

1. Before servicing the vehicle, refer to the Precautions Section.

2. Remove or disconnect the following:
   - Negative battery cable
   - Starter harness connections
   - Starter motor

To install:

3. Install or connect the following:
   - Starter motor. Tighten the fasteners to 18 ft. lbs. (25 Nm).
   - Starter harness connections
   - Negative battery cable

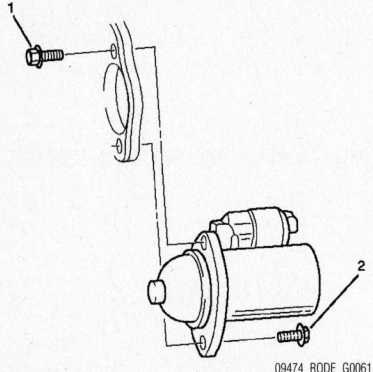

Starter mounting—2.2L engine

3.2L Engines

1. Before servicing the vehicle, refer to the Precautions Section.

2. Remove or disconnect the following:
   - Negative battery cable
   - Heated Oxygen (HO$_2$S) sensor connectors
   - Exhaust front pipe
   - Heat shield
   - Starter wiring connectors
   - Starter motor

To install:

3. Install or connect the following:
   - Starter motor. Tighten the bolts to 30 ft. lbs. (40 Nm).
   - Starter wiring connectors
   - Heat shield
   - Exhaust front pipe
   - Heated Oxygen (HO$_2$S) sensor connectors
   - Negative battery cable

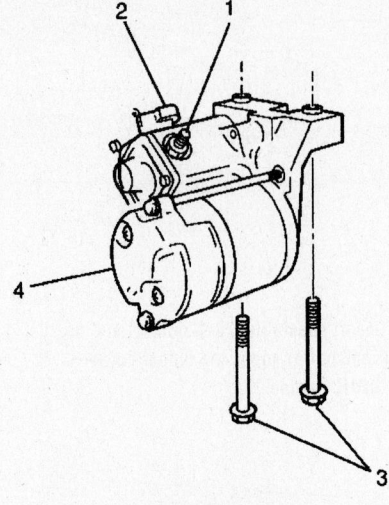

(1)	Terminal "30"
(2)	Terminal "50"
(3)	Fixing Bolts
(4)	Starter Assembly

09474_RODE_G0062

Starter mounting—3.2L and 3.5L engines

3.5L Engine

1. Before servicing the vehicle, refer to the Precautions Section.

2. Remove or disconnect the following:
   - Negative battery cable
   - Heated Oxygen (HO$_2$S) sensor connectors
   - Exhaust front pipe
   - Heat shield
   - Starter wiring connectors
   - Starter motor

To install:

3. Install or connect the following:
   - Starter motor. Tighten the bolts to 30 ft. lbs. (40 Nm).
   - Starter wiring connectors
   - Heat shield
   - Exhaust front pipe
   - Heated Oxygen (HO$_2$S) sensor connectors
   - Negative battery cable

Oil Pan

REMOVAL & INSTALLATION

2.2L Engines

1. Before servicing the vehicle, refer to the Precautions Section.

2. Drain the engine oil.

3. Remove or disconnect the following:
   - Flywheel dust cover
   - Left and right engine mounts. Raise the engine for access.
   - Oil pan
   - Oil pan support, for 2.2L engine

To install:

4. Perform the following:

   a. Install the oil pan support and tighten the bolts to 14 ft. lbs. (20 Nm).

   b. Install the oil pan and tighten the bolts to 70 inch lbs. (8 Nm) plus 30 degrees.

5. Install or connect the following:
   - Left and right engine mounts. Tighten the nuts to 41 ft. lbs. (55 Nm).
   - Flywheel dust cover

6. Fill the crankcase to the correct level.

7. Start the engine and check for leaks.

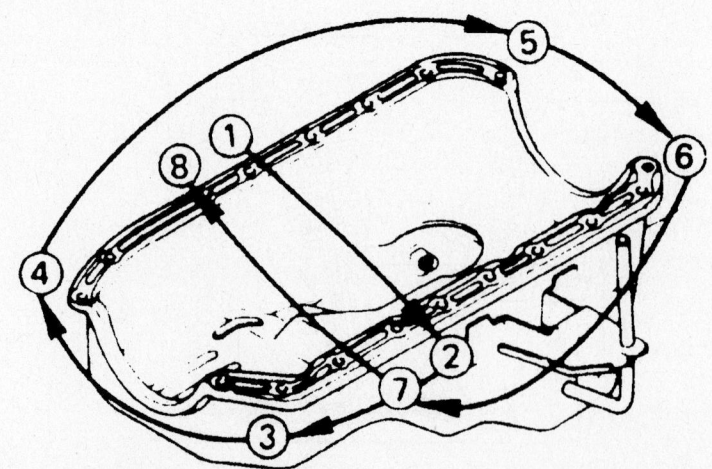

7924NG24

Oil pan bolt tightening sequence—2.2L engines

3.2L Engines

1. Before servicing the vehicle, refer to the Precautions Section.
2. Drain the engine oil.
3. Remove or disconnect the following:
   - Negative battery cable
   - Front wheels
   - Oil level dipstick
   - Stone guard
   - Radiator under fan shroud
   - Suspension crossmember
   - Flywheel dust cover
   - Pitman arm
   - Idler arm
4. If equipped with 4 wheel drive, unbolt and lower the front axle housing assembly for clearance.
5. Remove or disconnect the following:
   - Oil pan
   - Lower crankcase

To install:

6. Apply a bead of silicone sealant to the crankcase flange and install the crankcase. Tighten the fasteners in sequence to 89 inch lbs. (10 Nm).
7. Apply a bead of silicone sealant to the oil pan flange and install the oil pan. Tighten the fasteners to 89 inch lbs. (10 Nm).
8. If equipped, raise the axle housing assembly into position. Tighten the axle case bolts to 61 ft. lbs. (82 Nm) and the mounting bolts to 112 ft. lbs. (152 Nm).
9. Install or connect the following:
   - Pitman arm. Tighten the nut to 159 ft. lbs. (216 Nm).
   - Idler arm. Tighten the bolt to 33 ft. lbs. (44 Nm).
   - Flywheel dust cover
   - Suspension crossmember. Tighten the bolts to 58 ft. lbs. (78 Nm).
   - Radiator under fan shroud
   - Stone guard
   - Oil level dipstick
   - Front wheels
   - Negative battery cable
10. Fill the crankcase with engine oil.
11. Start the engine and check for leaks.

3.5L VIN X Engine

1. Before servicing the vehicle, refer to the Precautions Section.
2. Drain the engine oil.
3. Remove or disconnect the following:
   - Negative battery cable
   - Front wheels
   - Oil level dipstick
   - Stone guard
   - Radiator under fan shroud
   - Suspension crossmember

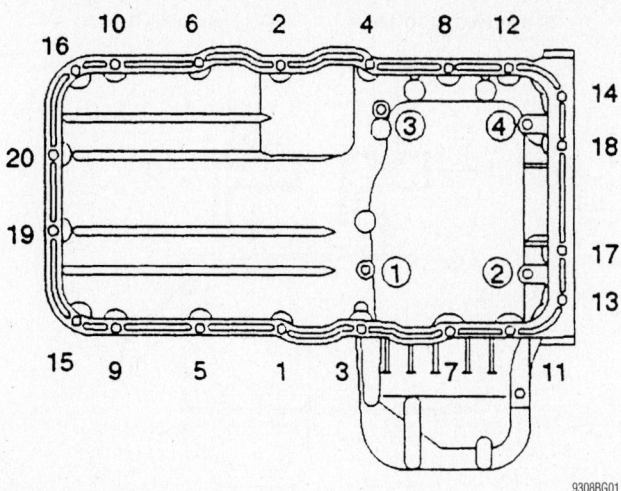

Lower crankcase torque sequence—3.2L and 3.5L engines

   - Flywheel dust cover
   - Pitman arm
   - Idler arm
4. If equipped with 4 wheel drive, unbolt and lower the front axle housing assembly for clearance.
5. Remove or disconnect the following:
   - Oil pan
   - Lower crankcase

To install:

6. Apply a bead of silicone sealant to the crankcase flange and install the crankcase. Tighten the fasteners in sequence to 89 inch lbs. (10 Nm).
7. Apply a bead of silicone sealant to the oil pan flange and install the oil pan. Tighten the fasteners to 89 inch lbs. (10 Nm).
8. If equipped, raise the axle housing assembly into position. Tighten the axle case bolts to 61 ft. lbs. (82 Nm) and the mounting bolts to 112 ft. lbs. (152 Nm).
9. Install or connect the following:
   - Pitman arm. Tighten the nut to 159 ft. lbs. (216 Nm).
   - Idler arm. Tighten the bolt to 33 ft. lbs. (44 Nm).
   - Flywheel dust cover
   - Suspension crossmember. Tighten the bolts to 58 ft. lbs. (78 Nm).
   - Radiator under fan shroud
   - Stone guard
   - Oil level dipstick
   - Front wheels
   - Negative battery cable
10. Fill the crankcase with engine oil.
11. Start the engine and check for leaks.

3.5L VIN Y Engine

1. Before servicing the vehicle, refer to the Precautions Section.
2. Drain the engine oil.

3. Remove or disconnect the following:
   - Negative battery cable
   - Front wheels
   - Oil level dipstick
   - Stone guard
   - Radiator under fan shroud
   - Right side halfshaft
   - Suspension crossmember
4. If equipped with 4 wheel drive, unbolt and support it with a jack.
5. Remove or disconnect the following:
   - Steering gear
   - Starter
6. Lower the front axle for clearance.
7. Remove or disconnect the following:
   - Oil pan
   - Lower crankcase

➡️**It may be necessary to use a gasket cutter.**

To install:

8. Apply a bead of silicone sealant to the crankcase flange and install the crankcase. Tighten the fasteners in sequence to 89 inch lbs. (10 Nm).
9. Apply a bead of silicone sealant to the oil pan flange and install the oil pan. Tighten the fasteners to 89 inch lbs. (10 Nm).
10. If equipped, raise the axle housing assembly into position. Tighten the axle case bolts to 61 ft. lbs. (82 Nm) and the mounting bolts to 112 ft. lbs. (152 Nm).
11. Install or connect the following:
    - Suspension crossmember. Tighten the bolts to 58 ft. lbs. (78 Nm).
    - Starter
    - Radiator under fan shroud
    - Stone guard
    - Oil level dipstick
    - Front wheels
    - Negative battery cable

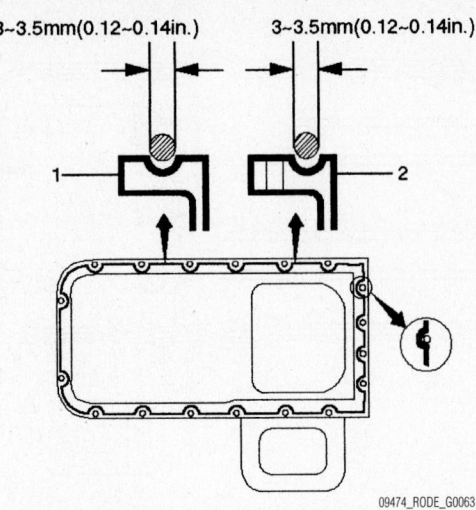

3~3.5mm(0.12~0.14in.) 3~3.5mm(0.12~0.14in.)

09474_RODE_G0063

Lower crankcase sealer application—3.2L and 3.5L engines

12. Fill the crankcase with engine oil.
13. Start the engine and check for leaks.

Oil Pump

REMOVAL & INSTALLATION

2.2L Engine

1. Before servicing the vehicle, refer to the Precautions Section.
2. Drain the engine oil.
3. Remove or disconnect the following:
   • Negative battery cable

(1)   Oil Pump Assembly
(2)   Oil Pressure Switch
(3)   Gasket
(4)   Sleeve

   • Accessory drive belts
   • Cooling fan
   • A/C belt tensioner, if equipped
   • Water pump pulley
   • Power steering pump
   • Crankshaft pulley
   • Front cover
   • Timing belt. Refer to the Timing Belt procedure.
   • Crankshaft timing sprocket
   • Rear timing cover
   • Crankshaft oil seal
   • Oil pan
   • Oil pump pickup tube
   • Oil pump

09474_RODE_G0064

Oil pump and related parts—2.2L engine

To install:
4. Install or connect the following:
   • Oil pump. Use a new gasket and tighten the bolts to 53 inch lbs. (6 Nm).
   • Oil pump pickup tube. Tighten the bolts to 70 inch lbs. (8 Nm).
   • Oil pan
   • Crankshaft oil seal
   • Rear timing cover
   • Crankshaft timing sprocket. Tighten the bolt to 94 ft. lbs. (130 Nm) plus 45 degrees.
   • Timing belt. Refer to the Timing Belt procedure.
   • Front cover
   • Crankshaft pulley. Tighten the bolts to 14 ft. lbs. (20 Nm).
   • Power steering pump
   • Water pump pulley
   • A/C belt tensioner, if equipped
   • Cooling fan
   • Accessory drive belts
   • Negative battery cable
5. Fill the crankcase to the correct level.
6. Start the engine and check for leaks.

3.2L Engine

1. Before servicing the vehicle, refer to the Precautions Section.
2. Remove or disconnect the following:
   • Timing belt
   • Oil pan
   • Oil pickup tube
   • Oil filter adapter
   • Oil pump

To install:
3. Apply silicone sealant to the oil pump mounting surface and install the oil pump. Tighten the bolts in sequence to 18 ft. lbs. (25 Nm).
4. Install or connect the following:
   • Oil filter adapter
   • Oil pickup tube
   • Oil pan
   • Timing belt
5. Fill the crankcase to the correct level.
6. Start the engine and check for leaks.

3.5L Engine

1. Before servicing the vehicle, refer to the Precautions Section.
2. Remove or disconnect the following:
   • Timing belt
   • Oil pan
   • Oil pick-up tube
   • Oil filter adapter
   • Oil pump

To install:
3. Apply silicone sealant to the oil pump mounting surface and install the oil pump.

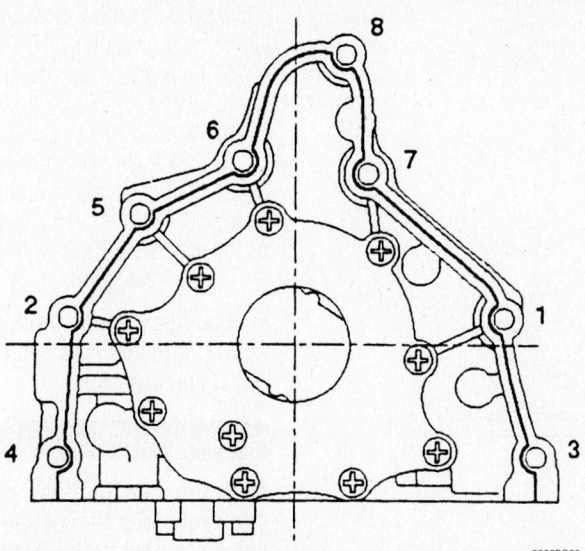

Oil pump torque sequence–3.2L and 3.5L engines

Tighten the bolts in sequence to 18 ft. lbs. (25 Nm).

  4. Install or connect the following:
  • Oil filter adapter
  • Oil pickup tube
  • Oil pan
  • Timing belt

Rear Main Seal

REMOVAL & INSTALLATION

  1. Before servicing the vehicle, refer to the Precautions Section.
  2. Remove or disconnect the following:
  • Negative battery cable
  • Transmission
  • Clutch assembly, if equipped with a manual transmission
  • Flywheel by loosening the flywheel bolts in a 2-step crisscross sequence
  • Rear main seal, using a seal puller

➡**Do not damage the crankshaft sealing surface.**

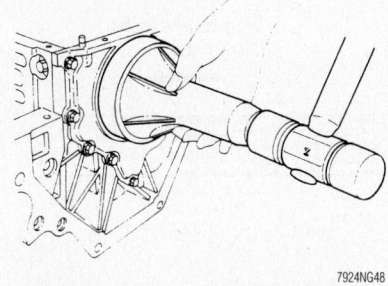

Installing a one-piece rear crankshaft oil seal

To install:
  3. Install or connect the following:
  • New rear main seal, by lubricating it with engine oil
  • Flywheel, using new flywheel bolts. Tighten the bolts, in a 2-step crisscross pattern, to 40 ft. lbs. (54 Nm).
  • Clutch assembly, if removed

  • Transmission
  • Negative battery cable
  4. Check the oil and refill as necessary.

Timing Belt and Cover

REMOVAL & INSTALLATION

2.2L Engine

  1. Before servicing the vehicle, refer to the Precautions Section.
  2. Disconnect the negative battery cable.
  3. Using a box-end wrench on the drive belt adjuster, turn the adjuster clockwise and remove the drive belt.
  4. From the left rear of the engine compartment, disconnect the 3 electrical connectors from the chassis harness.
  5. Remove the crankshaft pulley-to-crankshaft bolts and remove the pulley.
  6. From the front of the engine, remove the nut and the engine harness cover.
  7. Remove the timing belt cover.
  8. Rotate the crankshaft to position the timing marks at Top Dead Center (TDC) of the No. 1 cylinder's compression stroke.

➡**Mark the rotational direction of the timing belt for reinstallation purposes.**

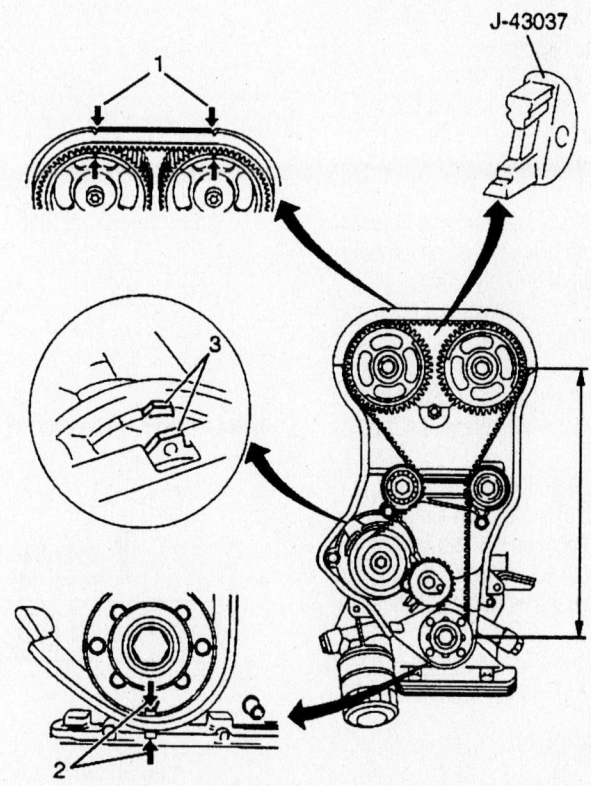

Aligning the timing marks and installing the timing belt —2.2L engine

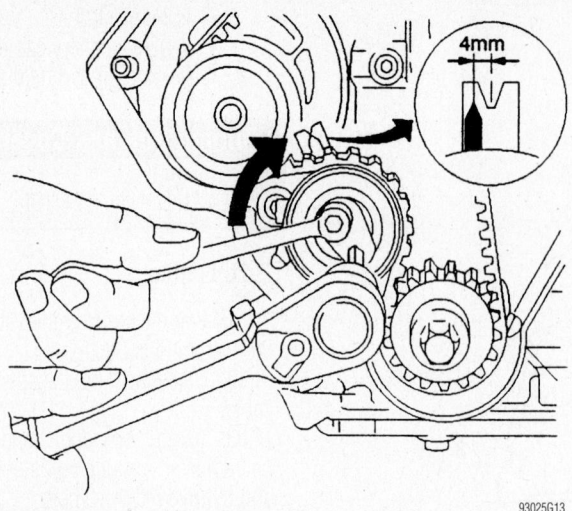

Tensioning the timing belt for a used timing belt —2.2L engine

9. Remove the timing belt tensioner adjusting bolt and the tensioner from the engine.

10. Remove the timing belt.

To install:

11. Install the timing belt tensioner and finger-tighten the tensioner bolt.

12. Inspect the timing marks to be sure that the engine is positioned at TDC of the No. 1 cylinder's compression stroke.

13. Using tool J-43037, or equivalent, place it between the intake and exhaust sprockets to prevent the camshaft gear from moving during the timing belt installation.

14. Install the timing belt.

15. Position the timing belt to ensure that the tension side of the belt is taut and move the timing belt tension adjusting lever clockwise until the tensioner pointer is flowing.

16. If installing a used timing belt (used over 60 min. from new), the pointer should be positioned approximately 0.16 in. (4mm) to the left of the "V" notch when viewed from the front of the engine.

17. If installing a new timing belt, the pointer should be positioned at the center of the "V" notch when viewed from the front of the engine.

18. Torque the timing belt tensioner adjusting bolt to 18 ft. lbs. (25 Nm).

19. Install the timing belt front cover and torque the bolts to 53 inch lbs. (6 Nm).

20. Install the engine harness connectors.

21. Install the crankshaft pulley and toque the pulley-to-crankshaft bolts to 14 ft. lbs. (20 Nm).

22. Move the drive belt tensioner to the loose side and install the drive belt to its normal position.

23. Connect the negative battery cable.

3.2L and 3.5L Engines

1. Before servicing the vehicle, refer to the Precautions Section.

2. Disconnect the negative battery cable.

3. Remove the air cleaner assembly and intake air duct.

4. Remove the upper fan shroud from the radiator.

5. Remove the 4 nuts retaining the cooling fan assembly. Remove the cooling fan from the fan pulley.

6. Loosen and remove the drive belts.

7. Remove the upper timing belt covers.

8. Remove the fan pulley assembly.

9. Rotate the crankshaft to align the camshaft timing marks with the pointer dots on the back covers. Verify that the pointer on the crankshaft aligns with the mark on the lower timing cover.

➡ **When the timing marks are aligned, the No. 2 piston is at Top Dead Center (TDC) compression.**

❋❋ WARNING

Align the camshaft and crankshaft sprockets with their alignment marks before removing the timing belt. Failure to align the belt and sprocket marks may result in valve damage.

10. Use tool No. J-8614-01, or a suitable pulley holding tool to remove the crankshaft pulley center bolt. Remove the crankshaft pulley.

11. If present, disconnect the 2 oil cooler hose bracket bolts on the timing cover. Move the oil cooler hoses and bracket off of the lower timing cover.

12. Remove the lower timing belt cover.

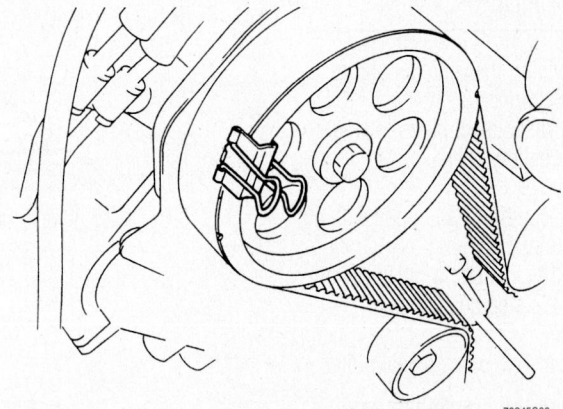

Using a double clip to hold the belt in place—3.2L and 3.5L engines

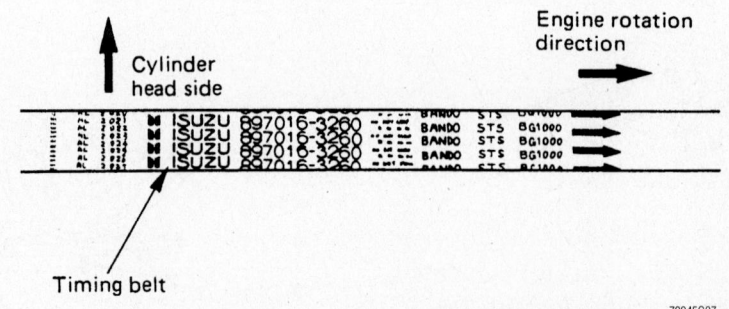

For maximum timing belt life, install the belt as shown—3.2L and 3.5L engines

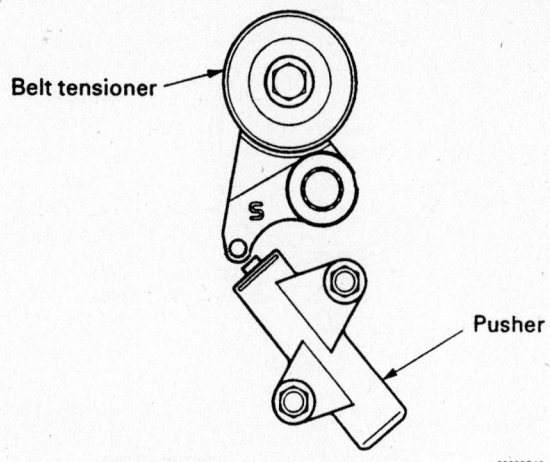

Belt tensioner

Pusher

S

93025G10

View of timing belt tensioner and pusher—3.2L and 3.5L engines

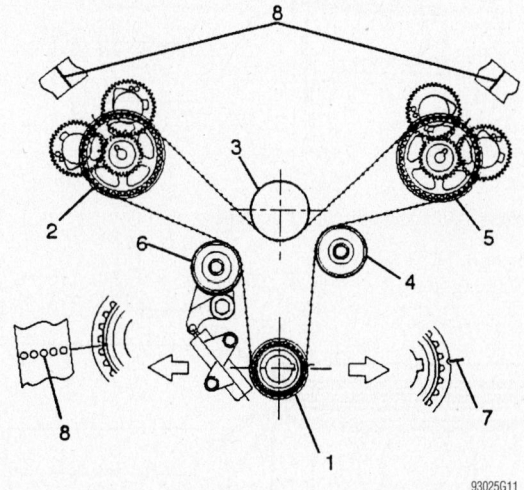

93025G11

Timing mark alignment and timing belt routing—3.2L and 3.5L engines

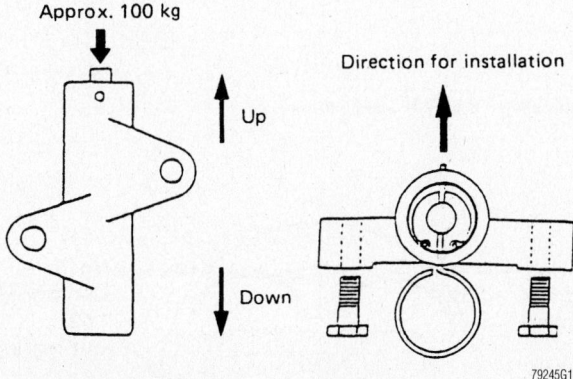

Approx. 100 kg

Up

Down

Direction for installation

79245G11

View of timing belt tensioner pusher—3.2L and 3.5L engines

13. Remove the pusher assembly (tensioner) from below the belt tensioner pulley. The pusher rod must always face upward to prevent oil leakage. Depress the pusher rod, and insert a wire pin into the hole to keep the pusher rod retracted.

14. Remove the timing belt.

15. Inspect the water pump and replace it if there is any doubt about its condition.

16. Repair any oil or coolant leaks before installing a new timing belt. If the timing belt has been contaminated with oil or coolant, or is damaged, it must be replaced.

To install:

17. Verify that the sprocket timing marks are still aligned and that the groove and the keyway on the crankshaft timing sprocket align with the mark on the oil pump. The white pointers on the camshaft timing sprockets should align with the dots on the front plate.

18. Install the timing belt. Use clips to secure the belt onto each sprocket until the installation is complete. Align the dotted marks on the timing belt with the timing mark opposite the groove on the crankshaft sprocket.

➡The arrows on the timing belt must follow the belt's direction of rotation. The manufacturer's trademark on the belt's spine should be readable left-to-right when the belt is installed.

19. Align the white line on the timing belt with the alignment mark on the right bank camshaft timing pulley. Secure the belt with a clip.

✲✲ WARNING

If any binding is felt when adjusting the timing belt tension by turning the crankshaft, STOP turning the engine, because the pistons may be hitting the valves.

20. Rotate the crankshaft counterclockwise to remove the slack between the crankshaft sprocket and the right camshaft timing belt sprocket.

21. Install the belt around the water pump pulley.

22. Install the belt on the idler pulley.

23. Align the white alignment mark on the timing belt with the alignment mark on the left bank camshaft timing belt sprocket.

24. Install the crankshaft pulley and tighten the center bolt by hand. Rotate the crankshaft pulley clockwise to give slack between the crankshaft timing belt pulley and the right bank camshaft timing belt pulley.

25. Insert a 1.4mm piece of wire through the hole in the pusher to hold the rod in. Install the pusher assembly while pushing the tension pulley toward the belt.

26. Pull the pin out from the pusher to release the rod.

27. Remove the clamps from the sprockets. Rotate the crankshaft pulley clockwise 2 turns. Measure the rod protrusion to ensure it is between 0.16 – 0.24 in. (4 – 6mm).

28. If the tensioner pulley bracket pivot bolt was removed, tighten it to 31 ft. lbs. (42 Nm).

29. Tighten the pusher bolts to 14 ft. lbs. (19 Nm).

30. Remove the crankshaft pulley. Install the lower and upper timing belt covers and tighten their bolts to 12 ft. lbs. (17 Nm).

31. Fit the oil cooler hose onto the timing cover and tighten its mounting bracket bolts to 16 ft. lbs. (22 Nm).

32. Install the crankshaft pulley and tighten the pulley bolt to 123 ft. lbs. (167 Nm).

33. Install fan pulley assembly and tighten the bolts to 16 ft. lbs. (22 Nm).

34. Install and adjust the accessory drive belts.

35. Install the cooling fan assembly and tighten the bolts to 72 inch lbs. (8 Nm).

36. Install the upper fan shroud.

37. Install the air cleaner assembly and intake air duct.

38. Connect the negative battery cable.

Piston and Ring

POSITIONING

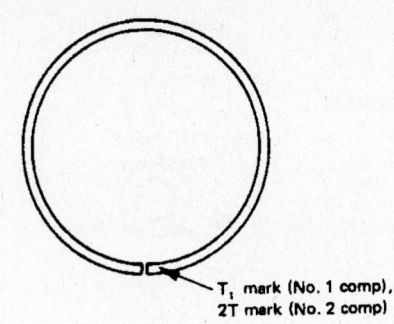

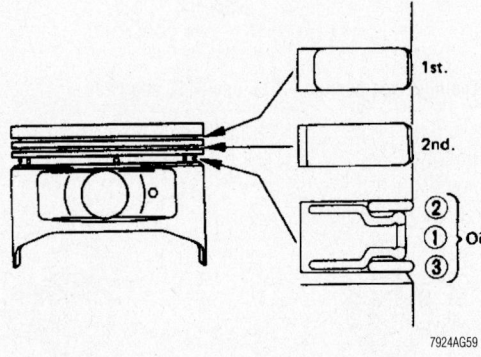

Piston ring positioning—3.2L and 3.5L VIN X engines

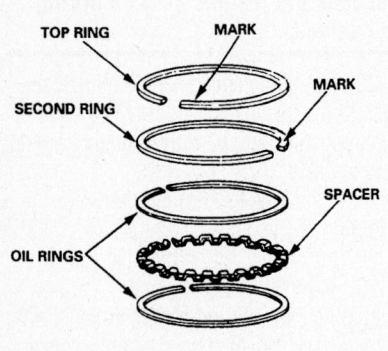

Piston ring positioning and top mark locations—all engines

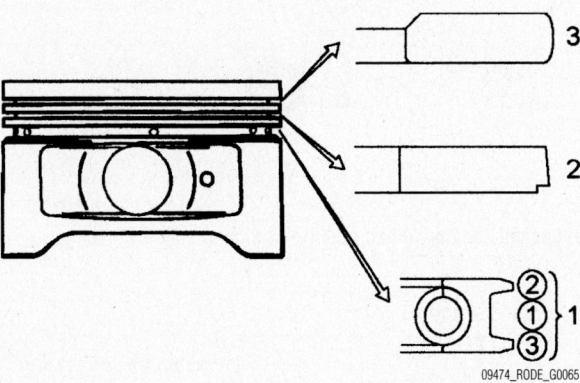

Piston ring positioning— 3.5L VIN Y engines

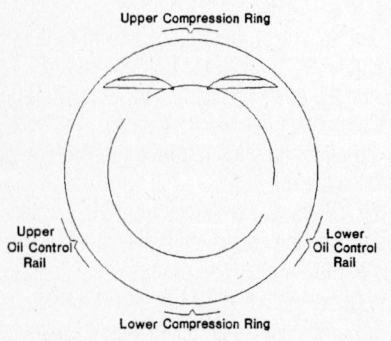

Piston ring end-gap spacing—2.2L engine

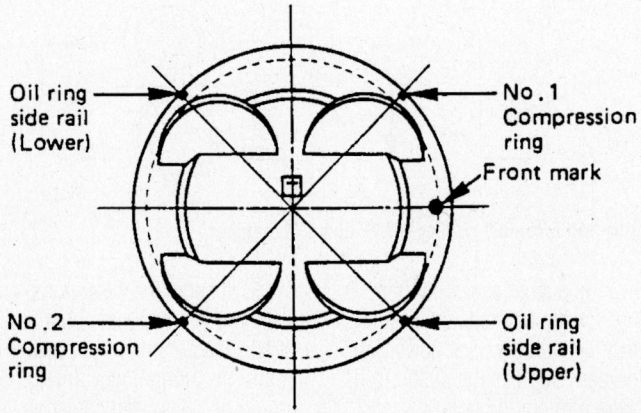

Piston ring end-gap spacing—3.2L and 3.5L engines

FUEL SYSTEM

Fuel System Service Precautions

Safety is the most important factor when performing not only fuel system maintenance but any type of maintenance. Failure to conduct maintenance and repairs in a safe manner may result in serious personal injury or death. Maintenance and testing of the vehicle's fuel system components can be accomplished safely and effectively by adhering to the following rules and guidelines:

• To avoid the possibility of fire and personal injury, always disconnect the negative battery cable unless the repair or test procedure requires that battery voltage be applied.

• Always relieve the fuel system pressure prior to disconnecting any fuel system component (injector, fuel rail, pressure regulator, etc.), fitting or fuel line connection. Exercise extreme caution whenever relieving fuel system pressure, to avoid exposing skin, face and eyes to fuel spray. Please be advised that fuel under pressure may penetrate the skin or any part of the body that it contacts.

• Always place a shop towel or cloth around the fitting or connection prior to loosening to absorb any excess fuel due to spillage. Ensure that all fuel spillage (should it occur) is quickly removed from engine surfaces. Ensure that all fuel soaked cloths or towels are deposited into a suitable waste container.

• Always keep a dry chemical (Class B) fire extinguisher near the work area.

• Do not allow fuel spray or fuel vapors to come into contact with a spark or open flame.

• Always use a backup wrench when loosening and tightening fuel line connection fittings. This will prevent unnecessary stress and torsion to fuel line piping. Always follow the proper tightening specifications.

• Always replace worn fuel fitting O-rings with new. Do not substitute fuel hose or equivalent, where fuel pipe is installed.

Fuel System Pressure

RELIEVING

1. Before servicing the vehicle, refer to the Precautions Section.
2. Remove the fuel filler cap.
3. Remove the fuel pump relay from the underhood relay box.

4. Start the engine and let it run until it stalls, then crank the engine for an additional 30 seconds.
5. Turn the ignition switch to the **OFF** position and remove the key. Disconnect the negative battery cable.
6. When service is completed, install the fuel pump relay and connect the negative battery cable.

Fuel Filter

REMOVAL & INSTALLATION

1. Before servicing the vehicle, refer to the Precautions Section.
2. Relieve the fuel system pressure.
3. Remove or disconnect the following:
   • Negative battery cable
   • Fuel filler cap

➡**To disconnect the quick connector, pull out on the connector while depressing the tab, as shown. Do not**

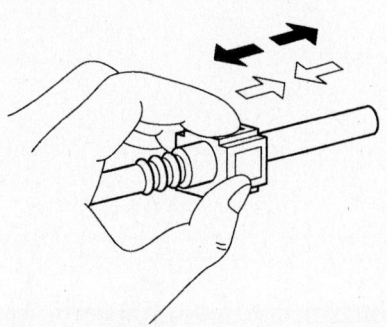

09474_RODE_G0005

Quick connector separation

use any tools! When connecting, use light engine oil.

   • Fuel lines from the fuel filter
   • Fuel filter
To install:
4. Install or connect the following:
   • Fuel filter and tighten the bracket bolt. Note the fuel flow directional arrow.
   • Fuel lines to the fuel filter
   • Negative battery cable
5. Start the engine and inspect the fuel filter connections for leaks.

Fuel Pump

REMOVAL & INSTALLATION

2002–03 Models

1. Before servicing the vehicle, refer to the Precautions Section.
2. Relieve fuel system pressure.

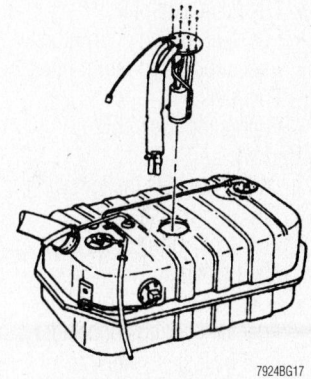

7924BG17

Fuel pump assembly mounting

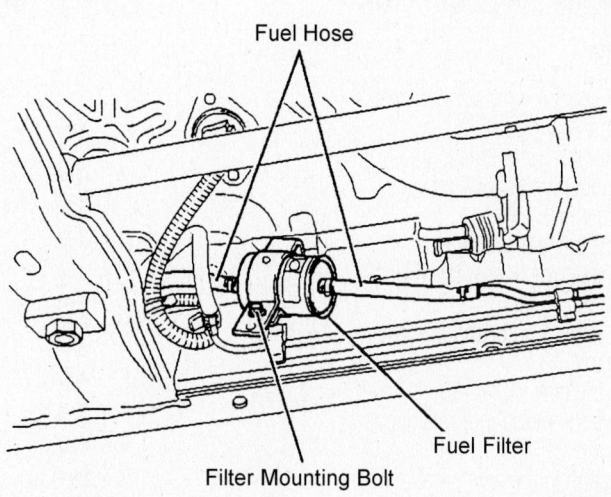

Fuel Hose

Fuel Filter

Filter Mounting Bolt

7924NG25

Fuel filter mounting location under the vehicle

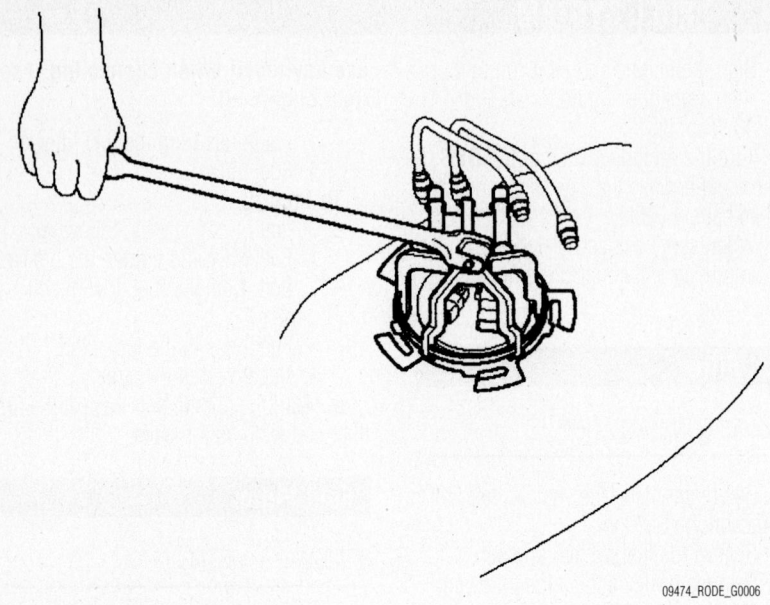

Removing the pump unit retainer

3. Drain the fuel tank.
4. Remove or disconnect the following:
   - Negative battery cable
   - Fuel filler and vent hoses
   - Fuel tank skid plate
   - Fuel tank wiring connectors
   - Fuel supply and return lines
   - Fuel tank
   - Fuel pump assembly

To install:

5. Install or connect the following:
   - Fuel pump assembly
   - Fuel tank. Tighten the bolts to 27 ft. lbs. (36 Nm).
   - Fuel supply and return lines
   - Fuel tank wiring connectors
   - Fuel tank skid plate
   - Fuel filler and vent hoses
   - Negative battery cable
6. Start the engine and check for leaks.

2004 Models

1. Before servicing the vehicle, refer to the Precautions Section.
2. Relieve fuel system pressure.
3. Disconnect the battery ground.
4. Drain the fuel tank.
5. Support the tank with a jack.
6. Disconnect the hoses and wiring at the tank module.
7. Remove the 5 bolts and one nut securing the tank.
8. Lower the tank.
9. Remove the pump unit retainer as shown.
10. Installation is the reverse of removal. Use a new seal. Torque the tank bolts and nut to 50 ft. lbs. (68 Nm).

High Pressure Fuel Pump

REMOVAL & INSTALLATION

3.5L VIN Y Engine

The high pressure pump is located at the rear of the right cylinder head.
1. Before servicing the vehicle, refer to the Precautions Section.
2. Relieve fuel system pressure.
3. Disconnect the battery ground.
4. Remove the fuel filler cap.
5. Drain the coolant.
6. Remove the intake manifold.
7. Remove the fuel rail joint pipe.
8. Disconnect the fuel pressure sensor, injectors and knock sensor.
9. Disconnect the harness clip.
10. Remove the heater hoses.
11. Remove the high pressure pump cover plate.

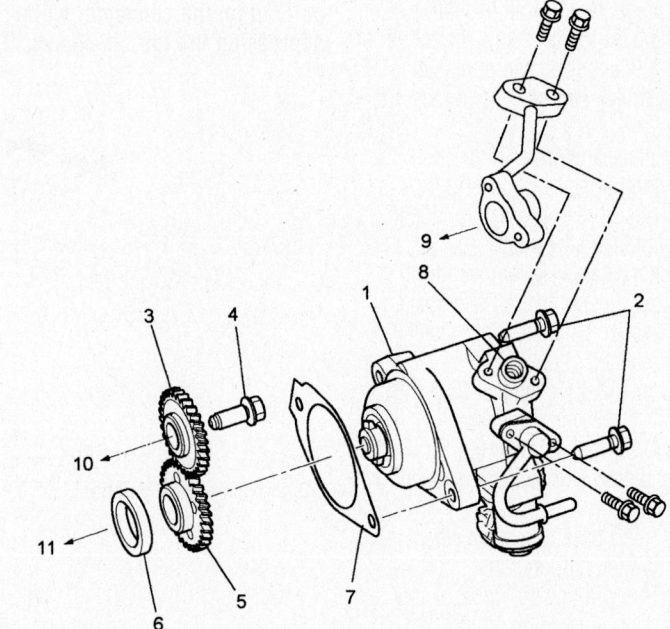

(1)	High Pressure Fuel Pump Assembly
(2)	Fixing Bolt
(3)	Fuel Pump Drive gear
(4)	Drive gear Fixing Bolt
(5)	Fuel Pump Driven gear
(6)	Driven gear Bearing
(7)	Packing
(8)	O-Ring
(9)	Fuel Rail
(10)	Camshaft
(11)	Cylinder Head

High pressure fuel pump—3.5L VIN Y engine

09474_RODE_G0007

12. Disconnect the fuel lines at the high pressure pump.

13. Remove the 2 bolts and the pump.

14. Installation is the reverse of removal. Torque the pump bolts to 18 ft. lbs. (25 Nm) and the fuel line connectors to 84 inch lbs. (10 Nm).

Fuel Rail and Injectors

REMOVAL & INSTALLATION

2.2L Engine

1. Before servicing the vehicle, refer to the Precautions Section.
2. Relieve the fuel system pressure.
3. Remove or disconnect the following:
   - Negative battery cable
   - Fuel injector harness connectors
   - Pressure regulator vacuum line
   - Fuel lines
   - Fuel supply manifold with injectors attached
   - Fuel injector retaining clips
   - Fuel injectors

To install:

4. Install or connect the following:
   - Fuel injectors. Use new O-ring seals.
   - Fuel injector retaining clips
   - Fuel supply manifold with injectors attached. Tighten the fasteners to 14 ft. lbs. (19 Nm).
   - Fuel lines
   - Pressure regulator vacuum line
   - Fuel injector harness connectors
   - Negative battery cable
5. Start the engine and check for leaks.

3.2L Engines

1. Before servicing the vehicle, refer to the Precautions Section.
2. Relieve fuel system pressure.
3. Remove or disconnect the following:
   - Negative battery cable
   - Engine cover
   - Fuel injector wiring connectors
   - Fuel lines
   - Fuel supply manifold with injectors attached
   - Fuel injector retaining clips
   - Fuel injectors

To install:

4. Install or connect the following:
   - Fuel injectors. Use new O-ring seals.
   - Fuel injector retaining clips
   - Fuel supply manifold with injectors attached. Tighten the bolts to 60 inch lbs. (6.5 Nm).

   - Fuel lines
   - Fuel injector wiring connectors
   - Engine cover
   - Negative battery cable
5. Start the engine and check for leaks.

3.5L VIN X Engine

1. Before servicing the vehicle, refer to the Precautions Section.
2. Relieve fuel system pressure.
3. Remove or disconnect the following:
   - Negative battery cable
   - Engine cover
   - Fuel injector wiring connectors
   - Fuel lines
   - Fuel supply manifold with injectors attached
   - Clips
   - Injectors from the supply manifold

To install:

4. Install or connect the following:
   - New O-rings on the fuel injectors
   - Fuel injectors
   - Fuel supply manifold with injectors attached. Tighten the bolts to 60 inch lbs. (6.5 Nm).
   - Fuel lines
   - Fuel injector wiring connectors
   - Engine cover
   - Negative battery cable
5. Start the engine and check for leaks.

3.5L VIN Y Engine

1. Before servicing the vehicle, refer to the Precautions Section.
2. Relieve fuel system pressure.
3. Disconnect the battery ground.
4. Remove the intake manifold.

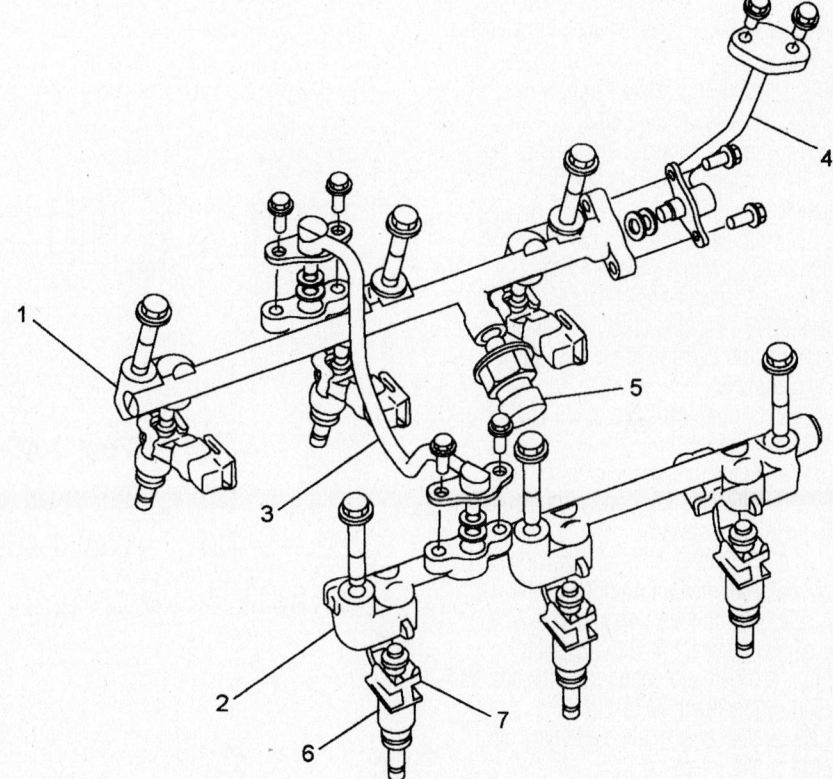

(1)	Fuel Rail RH
(2)	Fuel Rail LH
(3)	Joint Pipe
(4)	Fuel Pipe (Rail to Pump)
(5)	High Pressure Fuel Sensor
(6)	Injector ASM
(7)	Spring Clip

Fuel rail and injectors—3.5L VIN Y engine

09474_RODE_G0008

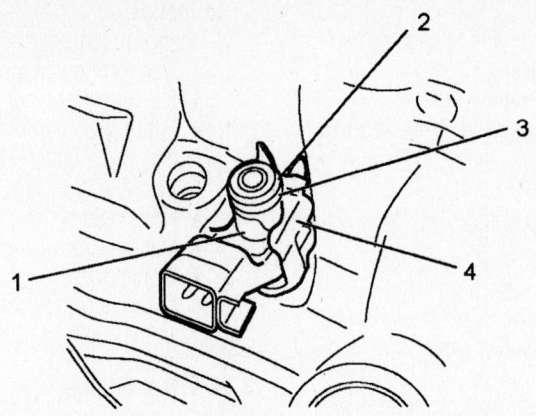

(1) Injector
(2) O-ring
(3) Back Up Ring
(4) Spring Clip

09474_RODE_G0009

Injector installed in head—3.5L VIN Y engine

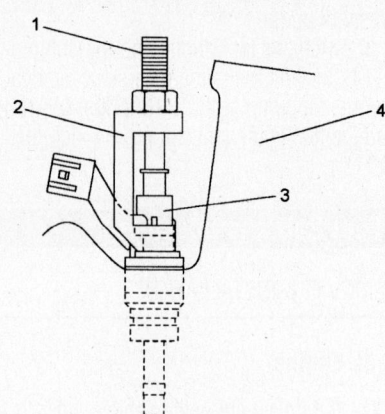

(1) Sliding hammer
(2) Tool J-46910
(3) Injector
(4) Cylinder Head

09474_RODE_G0010

Injector removal—3.5L VIN Y engine

5. Disconnect the fuel supply line.
6. Remove the joint pipe.
7. Remove the bolts and the fuel rail.
8. Remove the O-ring and back-up ring from the injectors.
9. Remove the spring clip and remove each injector with a slidehammer.
10. Use needle-nose pliers to pull and cut the TIP seal from each injector. Take care to avoid damage to the injector.

To install:

11. Install a new TIP seal with tool J46911:
   a. Place the new seal on the tool.
   b. Align the thick end of the tool with the injector nozzle.
   c. Use the pusher to install the seal.
   d. Pull the tool from the injector.
   e. Use the resizing tool to twist and shape the seal to the proper taper.
   f. Clean and debris from the seal and apply clean engine oil.
12. Install new O-rings and back-up rings on the injectors.
13. Clean the injector bores thoroughly.
14. Install new spring clips and install the injectors.
15. The remainder of installation is the reverse of removal. Torque the fuel rail bolts to 18 ft. lbs. (25 Nm). Torque the fuel lines connections to 84 inch lbs. (10 Nm).

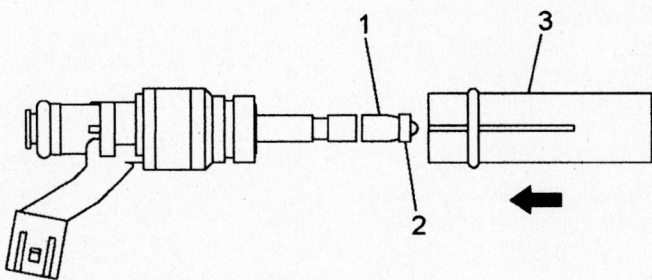

(1) Tool J-46911 (Protector Tool)
(2) TIP seal
(3) Tool J-46911 (Pusher Tool)

09474_RODE_G0011

Install a new TIP seal—3.5L VIN Y engine

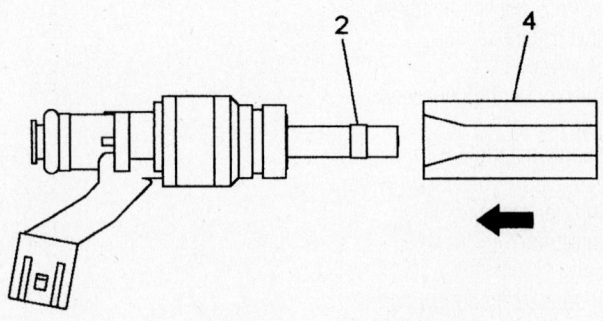

(2) TIP seal
(4) Tool J-46911 (Resizing Tool)

09474_RODE_G0012

Size the new seal—3.5L VIN Y engine

DRIVE TRAIN

Transmission Assembly

REMOVAL & INSTALLATION

2002–03 Manual Transmissions

2 WHEEL DRIVE

➡**The transmission flange bolts vary in length. Note their locations for assembly.**

1. Before servicing the vehicle, refer to the Precautions Section.
2. Remove the hood.
3. Install a support fixture to the engine lifting eyes.
4. Remove or disconnect the following:
   - Negative battery cable
   - Shift lever knob
   - Rear console assembly
   - Grommet assembly
   - Shift lever
   - Clutch slave cylinder and hose bracket
   - Driveshaft
   - Fuel line heat shield
   - Vehicle Speed (VSS) sensor connector
   - Reverse light switch connector
   - Flywheel under cover
   - Transmission mount and crossmember
   - Transmission flange bolts
   - Transmission

To install:

5. Install or connect the following:
   - Transmission. Tighten the large flange bolts to 52 ft. lbs. (71 Nm) and the small bolts to 30 ft. lbs. (41 Nm).
   - Crossmember. Tighten the bolts to 56 ft. lbs. (76 Nm).
   - Transmission mount. Tighten the fasteners to 30 ft. lbs. (41 Nm).
   - Flywheel under cover. Tighten the bolts to 69 inch lbs. (8 Nm).
   - Reverse light switch connector
   - Vehicle Speed (VSS) sensor connector
   - Fuel line heat shield
   - Driveshaft. Tighten the bolts to 37 ft. lbs. (50 Nm).

   - Clutch slave cylinder and hose bracket
   - Shift lever
   - Grommet assembly
   - Rear console assembly
   - Shift lever knob
   - Negative battery cable
6. Remove the engine support fixture and install the hood.

4 WHEEL DRIVE

➡**The transmission flange bolts vary in length. Note their locations for assembly.**

1. Before servicing the vehicle, refer to the Precautions Section.
2. Remove the hood.
3. Install a support fixture to the engine lifting eyes.
4. Remove or disconnect the following:
   - Negative battery cable
   - Shift lever knob
   - Console assembly
   - Grommet assembly
   - Shift lever

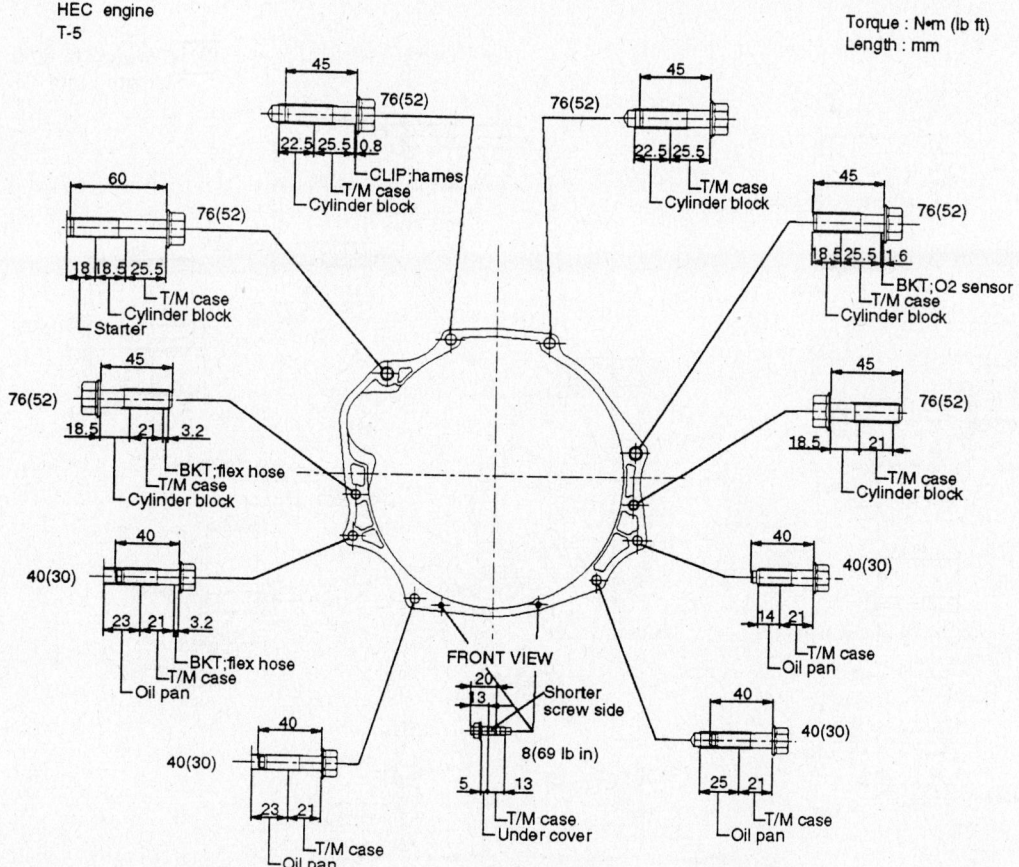

Transmission flange bolt identification and torque—2002–03 2 wheel drive transmission

9308NG02

- Transfer case control lever
- Transfer case skid plate
- Front and rear driveshafts
- Reverse lamp switch connector
- Indicator switch connectors
- Vehicle Speed (VSS) sensor connector
- 4WD actuator connector

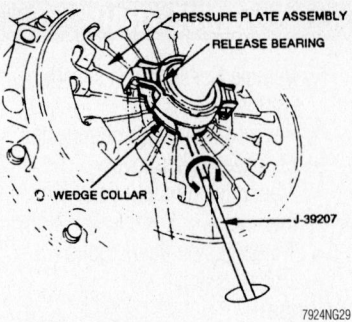

Turn the remover to separate the release bearing—2002–03

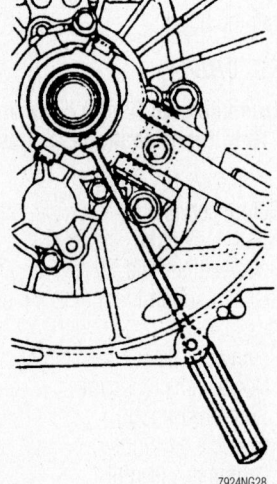

Insert the tool between the wedge collar and the release bearing—2002–03

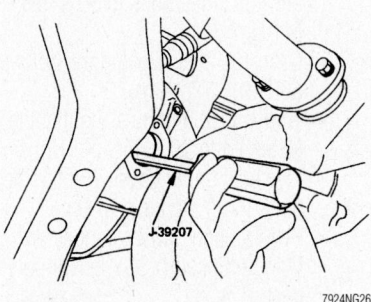

Insert the Release Bearing Remover tool J-39207 through the bell housing—4 wheel drive manual transmission—2002–03

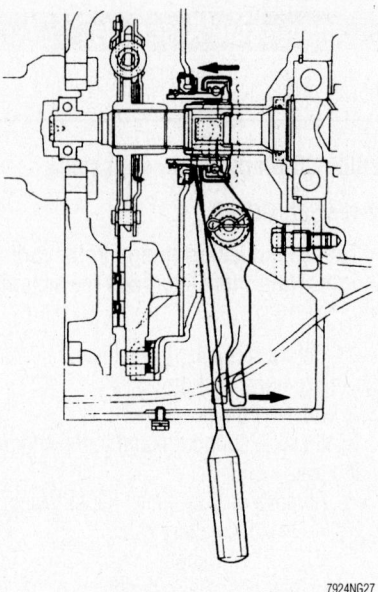

Push the release bearing fork toward the transmission to release the bearing from the pressure plate—4 wheel drive manual transmission—2002–03

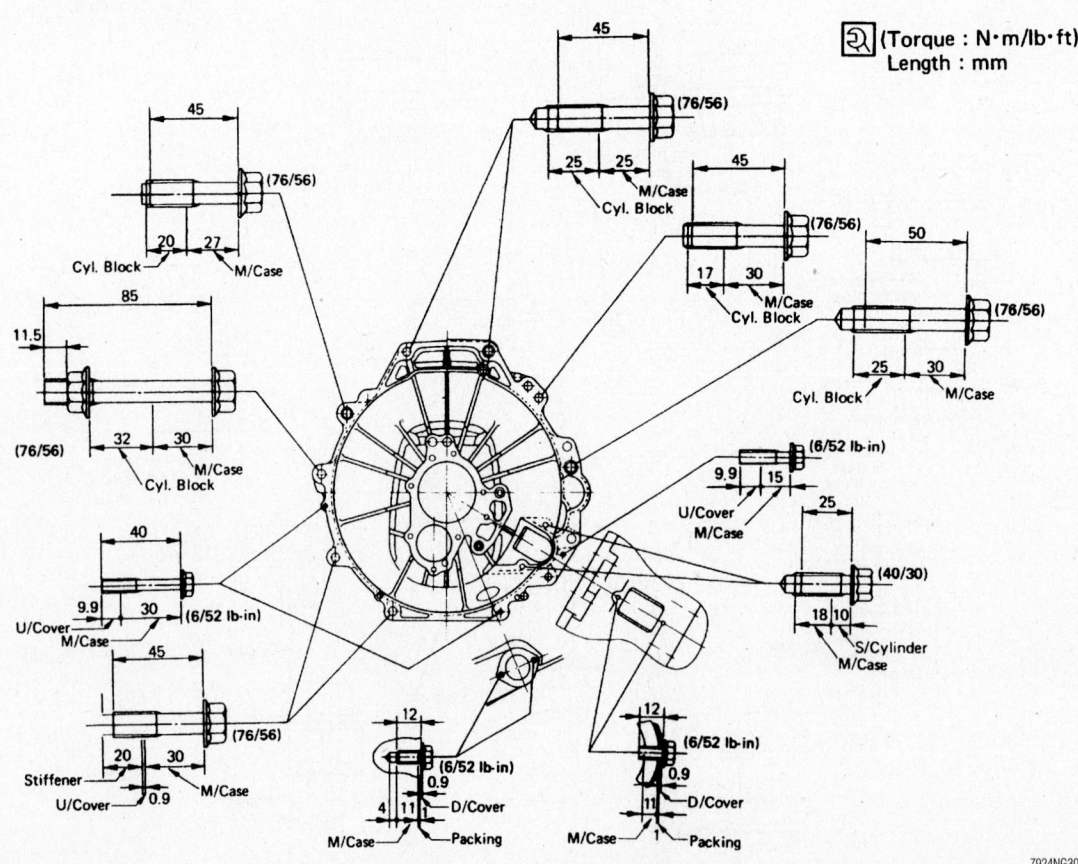

Transmission mounting bolt identification and torque specifications—V6 engine with 4 wheel drive transmission shown—2002–03

- Transmission harness clamps
- Fuel pipe bracket
- Clutch slave cylinder and heat shield
- Transmission mount and cross-member
- Heated Oxygen (HO$_2$S) sensor connectors
- Right exhaust front pipe
- Wiring harness heat shield
- Flywheel under cover

5. Release the throw out bearing from the pressure plate as shown.

6. Remove the transmission flange bolts and remove the transmission.

To install:

7. Install the transmission. Tighten the large bolts to 56 ft. lbs. (76 Nm) and the small bolts to 52 inch lbs. (6 Nm).

8. Apply 13–18 lbs. (59–78 N) of force to the clutch fork to engage the throw out bearing to the pressure plate.

9. Install or connect the following:
- Flywheel under cover
- Wiring harness heat shield
- Right exhaust front pipe. Tighten the manifold flange fasteners to 49 ft. lbs. (67 Nm) and the exhaust flange bolts to 32 ft. lbs. (43 Nm).
- HO$_2$S sensor connectors
- Crossmember. Tighten the bolts to 37 ft. lbs. (50 Nm).
- Transmission mount. Tighten the bolts to 30 ft. lbs. (41 Nm).
- Clutch slave cylinder and heat shield
- Fuel pipe bracket
- Transmission harness clamps
- 4WD actuator connector
- VSS sensor connector
- Indicator switch connectors
- Reverse lamp switch connector
- Front and rear driveshafts. Tighten

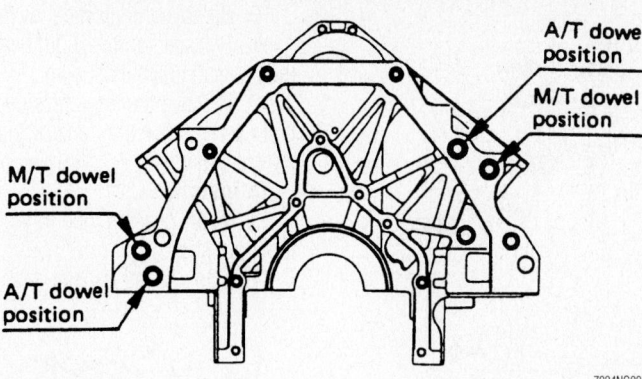

Dowel pin locations for automatic and manual transmissions—3.2L engine—2002–03

the flange bolts to 46 ft. lbs. (63 Nm).
- Transfer case skid plate. Tighten the bolts to 27 ft. lbs. (37 Nm).
- Transfer case control lever
- Shift lever
- Grommet assembly
- Console assembly
- Shift lever knob
- Negative battery cable

10. Remove the engine support fixture and install the hood.

2004 Manual Transmissions

➡The transmission flange bolts vary in length. Note their locations for assembly.

1. For 4x4 models, turn the transfer case selector knob to 2WD.

2. Before servicing the vehicle, refer to the Precautions Section.

3. Remove or disconnect the following:
- Negative battery cable
- Shift lever knob
- Lower cluster
- Rear console

- Center console assembly
- Grommet assembly
- Shift lever

4. Raise and safely support the vehicle.

5. Remove or disconnect the following:
- Transfer case skid plate
- Front and rear driveshafts. Match-mark the flanges for installation.
- Center exhaust pipe
- Transfer case control lever
- Reverse lamp switch connector
- Indicator switch connectors
- Vehicle Speed (VSS) sensor connector
- 4WD actuator connector
- Transmission harness clamps
- Fuel pipe heat shield
- Wiring harness heat shield
- Clutch slave cylinder and heat shield
- Dust covers

6. Support the transmission with a jack.

7. Remove the engine rear mount nuts.

8. Remove the 6 bolts and the 3rd crossmember.

9. Remove the starter.

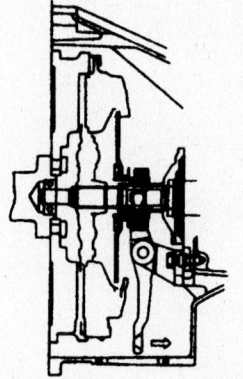

Push the release bearing fork toward the transmission to engage the release bearing with the pressure plate—2002–03

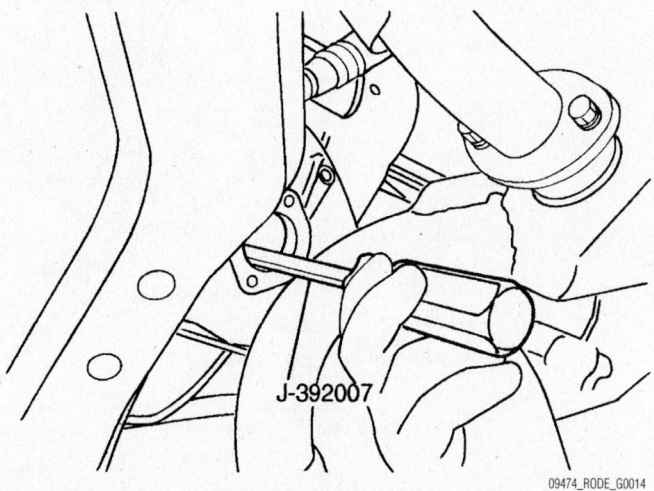

Use tool J-392007 to disconnect the release bearing from the pressure plate

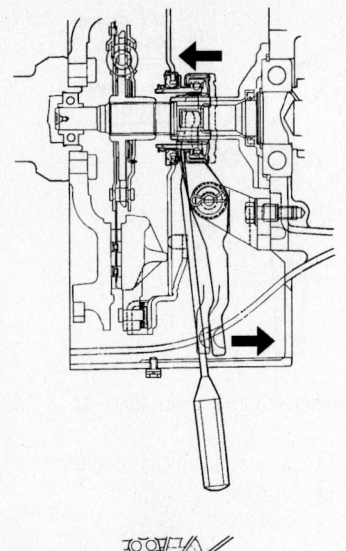

10. Remove the flywheel covers.

11. Release the throw out bearing from the pressure plate as shown.

12. Remove the transmission flange bolts and remove the transmission and transfer case.

To install:

13. Apply chassis lube to the splines of the input shaft.

14. Connect the transfer case to the transmission. Torque the bolts to 30 ft. lbs. (41 Nm).

15. Slowly raise the assembly into position. Install the transmission. Tighten the bolts as shown in the accompanying illustration.

16. Apply 13–18 lbs. (59–78 N) of force to the clutch fork to engage the throw out bearing to the pressure plate. A click will be heard when they engage.

09474_RODE_G0015

Pull the shift fork towards the transmission. Install the tool between the release bearing and the wedge collar

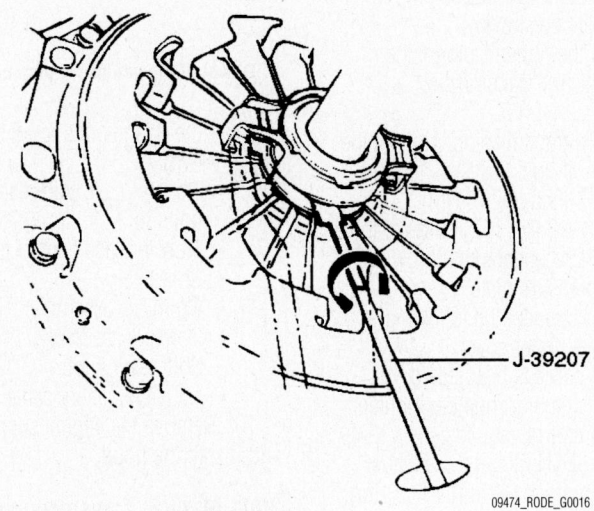

J-39207

09474_RODE_G0016

Turn the tool to separate the release bearing

Torque : N·m (lb ft)
Length : mm

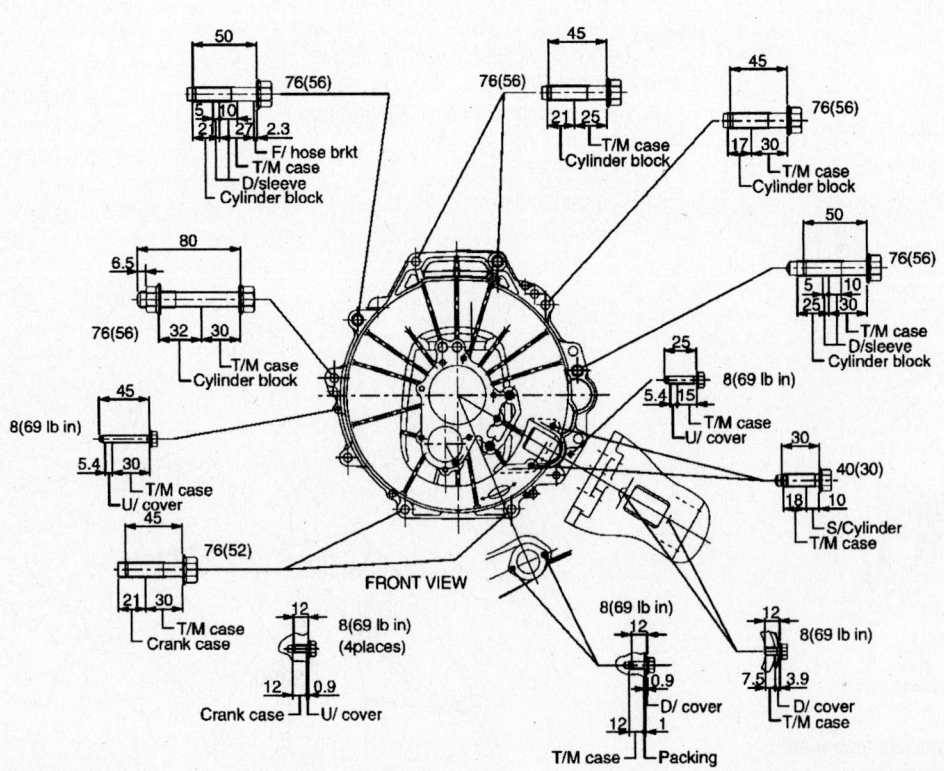

09474_RODE_G0013

Transmission installation nut and bolt torques—2004 manual transmission

17. Install or connect the following:
- Flywheel under cover
- Starter
- Crossmember. Torque to 37 ft. lbs. (50 Nm)
- Transmission mount. Tighten the bolts to 30 ft. lbs. (41 Nm).
- Clutch slave cylinder and heat shield. Torque to 32 ft. lbs. (43 Nm).
- Dust covers
- Transmission harness clamps
- Wiring harness heat shield
- HO$_2$S sensor connectors
- Crossmember. Tighten the bolts to 37 ft. lbs. (50 Nm).
- Fuel pipe bracket
- 4WD actuator connector
- VSS sensor connector
- Indicator switch connectors
- Reverse lamp switch connector
- Fuel pipe
- Center exhaust. Tighten the exhaust flange bolts to 32 ft. lbs. (43 Nm).
- Front and rear driveshafts. Tighten the flange bolts to 43 ft. lbs. (59 Nm).
- Transfer case skid plate. Tighten the bolts to 27 ft. lbs. (37 Nm).
- Transfer case control lever

- Shift lever
- Grommet assembly
- Consoles assembly
- Shift lever knob
- Negative battery cable

2002–03 Automatic Transmissions with 2-Wheel Drive

1. Before servicing the vehicle, refer to the Precautions Section.
2. Remove the hood.
3. Install a support fixture to the engine lifting eyes.
4. Remove or disconnect the following:
- Negative battery cable
- Front console assembly and wiring connectors
- Shift lock cable
- Shift control rod
- Selector lever assembly
- Driveshaft
- Wiring harness heat shield
- Transmission mount and crossmember
- Heated Oxygen (HO$_2$S) sensor connectors
- Left and right exhaust front pipes
- Transmission oil cooler lines
- Starter motor

- Fuel line bracket
- Transmission harness connectors
- Flywheel under covers
- Torque converter
- Transmission flange bolts
- Transmission

To install:

➡**Use new torque converter bolts.**

5. Install or connect the following:
- Transmission. Tighten the large bolts to 56 ft. lbs. (76 Nm) and the small bolts to 69 inch lbs. (8 Nm).
- Torque converter. Tighten the bolts to 40 ft. lbs. (54 Nm).
- Flywheel under covers
- Transmission harness connectors
- Fuel line bracket
- Starter motor. Tighten the bolts to 30 ft. lbs. (40 Nm).
- Transmission oil cooler lines
- Left and right exhaust front pipes. Tighten the manifold flange fasteners to 49 ft. lbs. (67 Nm) and the exhaust flange bolts to 32 ft. lbs. (43 Nm).
- HO$_2$S sensor connectors
- Crossmember. Tighten the bolts to 37 ft. lbs. (50 Nm).
- Transmission mount. Tighten the bolts to 30 ft. lbs. (41 Nm).

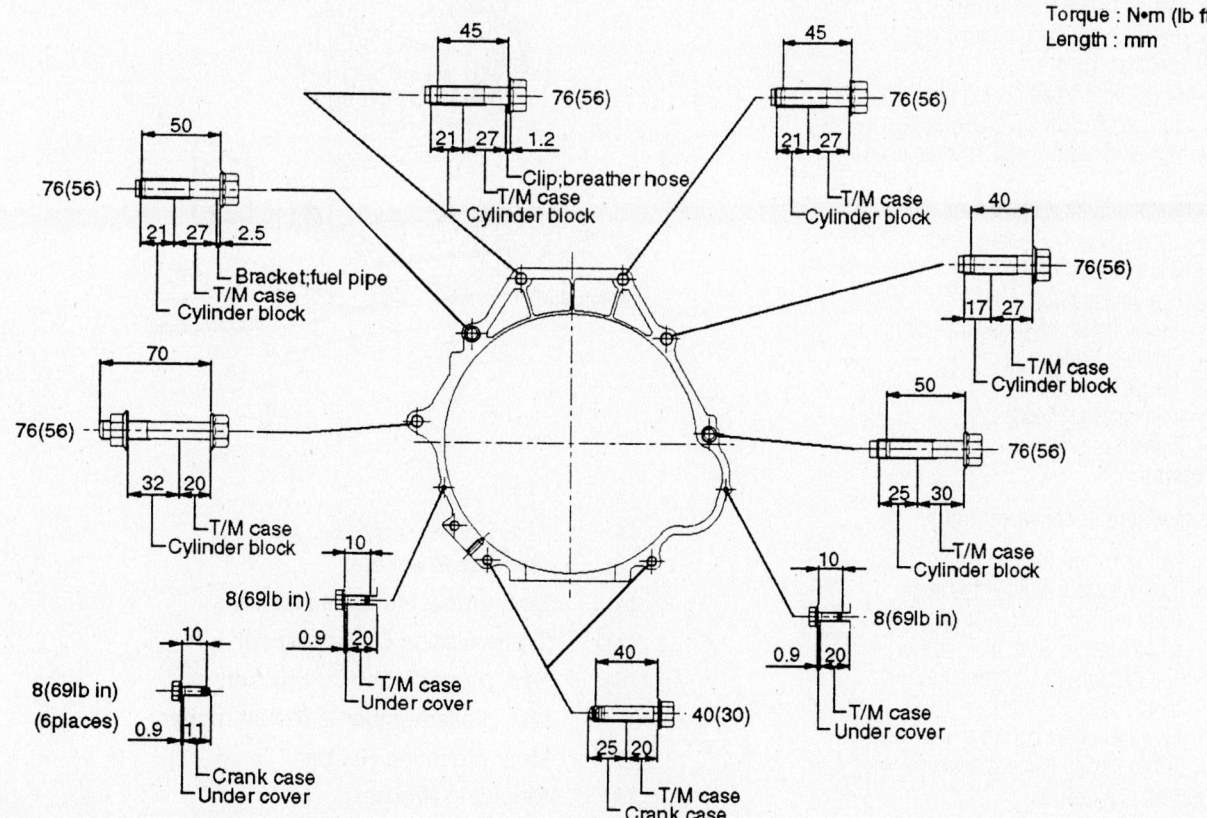

Automatic transmission mounting bolt locations and torque specifications—2002–03 models

9308NG03

- Wiring harness heat shield
- Driveshaft. Tighten the flange bolts to 46 ft. lbs. (63 Nm).
- Selector lever assembly
- Shift control rod
- Shift lock cable
- Front console assembly and wiring connectors
- Negative battery cable

6. Remove the engine support fixture and install the hood.

2002–03 Automatic Transmissions with 4-Wheel Drive

1. Before servicing the vehicle, refer to the Precautions Section.
2. Remove the hood.
3. Install a support fixture to the engine lifting eyes.
4. Remove or disconnect the following:

- Negative battery cable
- Transfer case shift lever knob
- Front console assembly and wiring connectors
- Shift lock cable
- Shift control rod
- Selector lever assembly
- Transfer case shift lever
- Transfer case skid plate
- Front and rear driveshafts
- Wiring harness heat shield
- Transmission mount and cross-member
- Right torsion bar, if equipped with Torque On Demand (TOD) system
- Heated Oxygen (HO2S) sensor connectors
- Left and right exhaust front pipes
- Transmission oil cooler lines
- Starter motor
- Fuel line bracket
- Transmission harness connectors
- Flywheel under covers
- Torque converter
- Transmission flange bolts
- Transmission

To install:

➡ Use new torque converter bolts.

5. Install or connect the following:

- Transmission. Tighten the large bolts to 56 ft. lbs. (76 Nm) and the small bolts to 30 ft. lbs. (40 Nm).
- Torque converter. Tighten the bolts to 40 ft. lbs. (54 Nm).
- Flywheel under covers
- Transmission harness connectors
- Fuel line bracket
- Starter motor. Tighten the bolts to 30 ft. lbs. (40 Nm).

- Transmission oil cooler lines
- Left and right exhaust front pipes. Tighten the manifold flange fasteners to 49 ft. lbs. (67 Nm) and the exhaust flange bolts to 32 ft. lbs. (43 Nm).
- HO2S sensor connectors
- Right torsion bar, if removed
- Crossmember. Tighten the bolts to 37 ft. lbs. (50 Nm).
- Transmission mount. Tighten the bolts to 30 ft. lbs. (41 Nm).
- Wiring harness heat shield
- Front and rear driveshafts. Tighten the flange bolts to 46 ft. lbs. (63 Nm).
- Transfer case skid plate. Tighten the bolts to 27 ft. lbs. (37 Nm).
- Transfer case shift lever
- Selector lever assembly
- Shift control rod
- Shift lock cable
- Front console assembly and wiring connectors
- Transfer case shift lever knob
- Negative battery cable

6. Remove the engine support fixture and install the hood.

2004 Automatic Transmissions

1. Before servicing the vehicle, refer to the Precautions Section.
2. Remove or disconnect the following:

- Negative battery cable
- Front and rear driveshafts. Match-mark the flanges.

3. Loosen, but don't remove, the nuts securing the front exhaust pipe to the manifold.
4. Remove or disconnect the following:

- Fuel line heat shield and bracket
- Fuel line quick connector. See Fuel Filter.
- All wiring at the transmission and transfer case
- Wiring harness heat shield
- Transfer case skid plate
- Transfer case

5. Support the transmission with a jack.
6. Remove the rear mount nuts.
7. Remove the crossmember.
8. Remove or disconnect the following:

- Transmission oil cooler lines

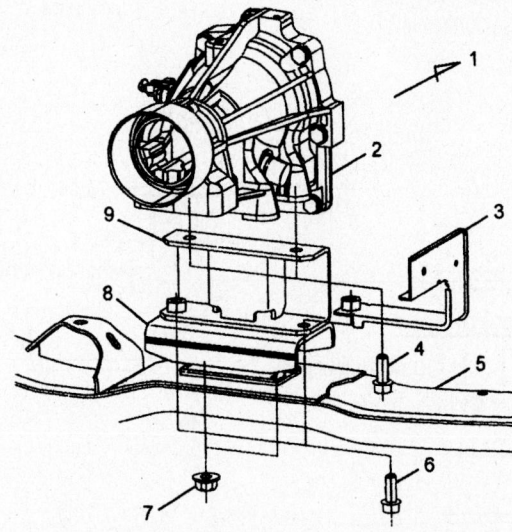

(1)	Front
(2)	Transmission (4x2)
(3)	Harness Bracket
(4)	Bolt, bracket to transmission
(5)	Transmission Crossmember
(6)	Bolt, mount rubber to bracket
(7)	Nut, crossmember to mount rubber
(8)	Rear Mounting Rubber
(9)	Mounting Bracket

09474_RODE_G0017

2wd automatic transmission rear mount—2004 models

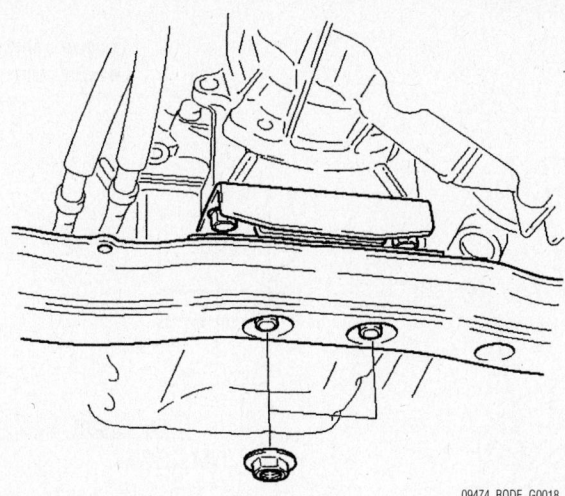

09474_RODE_G0018

4wd automatic transmission rear mount—2004 models

- Oil pipe clamp at the converter housing
- Oil pipe clamp bolt at the engine mount
- Selector lever assembly
- Oil dipstick tube
- Starter motor
- Engine and transmission undercovers

- Torque converter-to-flexplate bolts
- Transmission flange bolts
- Transmission
- Rear mount cushion

To install:

➡ **Use new torque converter bolts.**

9. Install or connect the following:
- Rear mount cushion
- Transmission. Tighten the bolts as

shown in the accompanying illustration.
- Torque converter. Tighten the bolts to 40 ft. lbs. (54 Nm).
- Engine and transmission undercovers
- Starter motor. Tighten the bolts to 30 ft. lbs. (40 Nm).
- Oil filler tube
- Selector lever assembly
- Shift control rod
- Shift lock cable
- Transmission oil cooler lines
- Crossmember. Tighten the bolts to 56 ft. lbs. (76 Nm).
- Transmission mount. Tighten the bolts to 37 ft. lbs. (57 Nm).
- Transfer case. Tighten the bolts/nuts as shown in the accompanying illustration.
- Transmission harness connectors
- Wiring harness heat shield
- Fuel line bracket
- Front exhaust pipe. Tighten to 49 ft. lbs. (67 Nm).
- Front and rear driveshafts. Tighten the flange bolts to 46 ft. lbs. (63 Nm).

10. The remainder of installation is the reverse of removal.

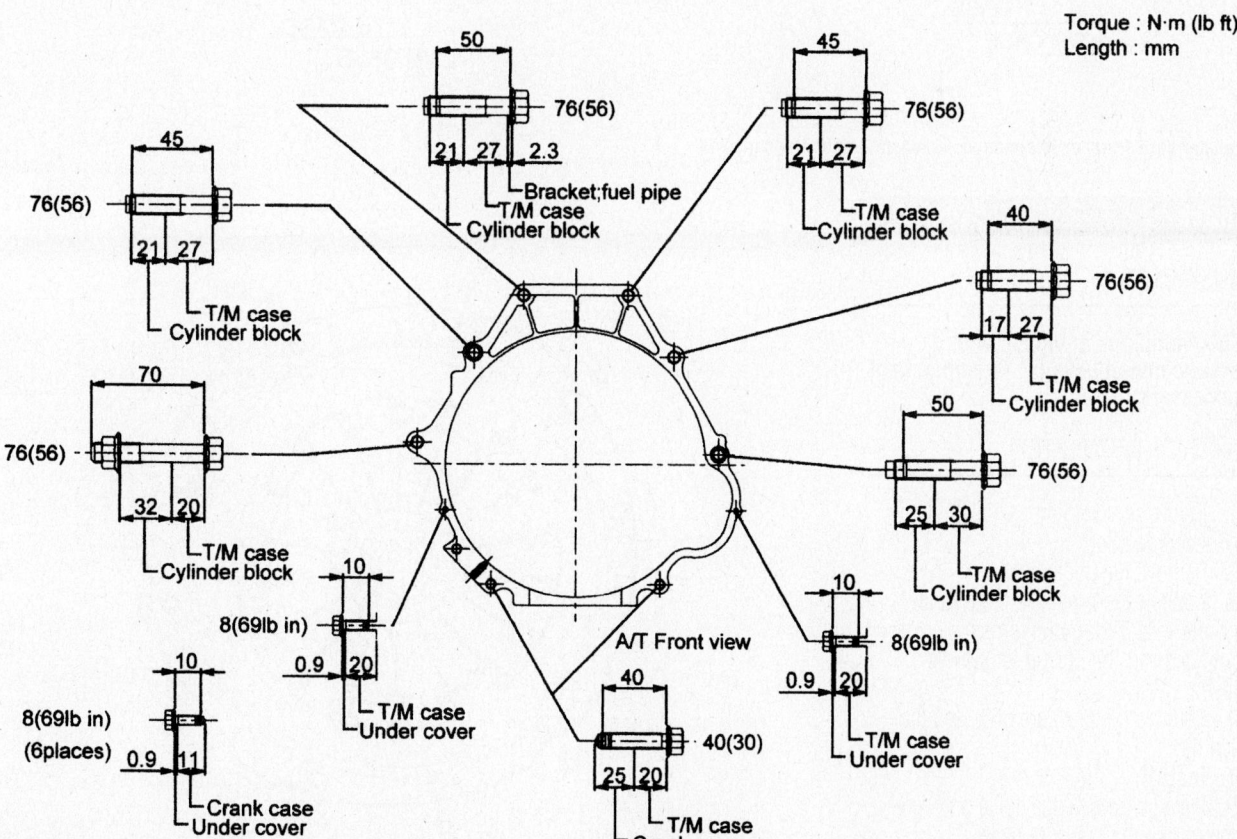

Transmission-to-engine fastener information—2004 automatic transmission

09474_RODE_G0019

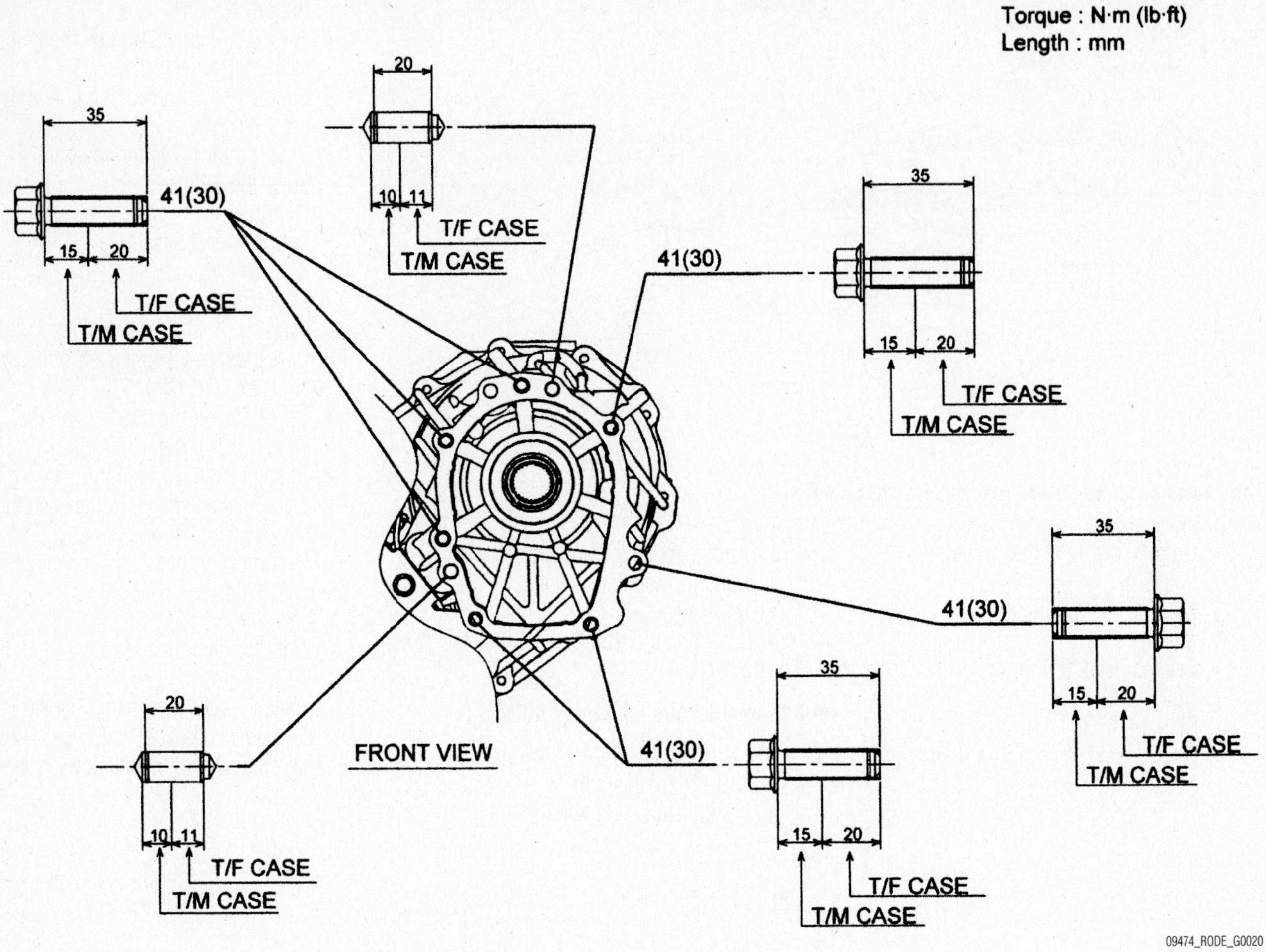

Torque : N·m (lb·ft)
Length : mm

35

20

41(30)

15 20

T/F CASE

T/M CASE

20

10 11

T/F CASE

T/M CASE

35

41(30)

15 20

T/F CASE

T/M CASE

FRONT VIEW

20

10 11

T/F CASE

T/M CASE

35

41(30)

15 20

T/F CASE

T/M CASE

35

41(30)

15 20

T/F CASE

T/M CASE

09474_RODE_G0020

Transfer case fastener information—2004 automatic transmission

Clutch

ADJUSTMENTS

➡**This vehicle is equipped with a hydraulic clutch linkage. No adjustment is necessary.**

REMOVAL & INSTALLATION

1. Before servicing the vehicle, refer to the Precautions Section.
2. Remove the transmission.
3. Loosen the pressure plate mounting bolts in a 2-step crisscross sequence until the spring tension is relieved.
4. Remove the pressure plate and the clutch disc.

To install:

5. Install a new wedge collar and wire snapring into the pressure plate.
6. Using a clutch alignment tool, assem-

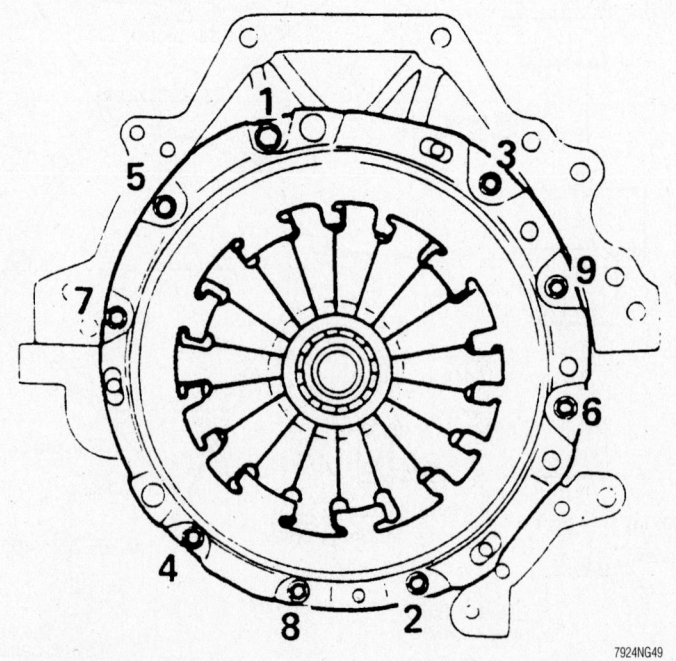

Pressure plate tightening sequence

7924NG49

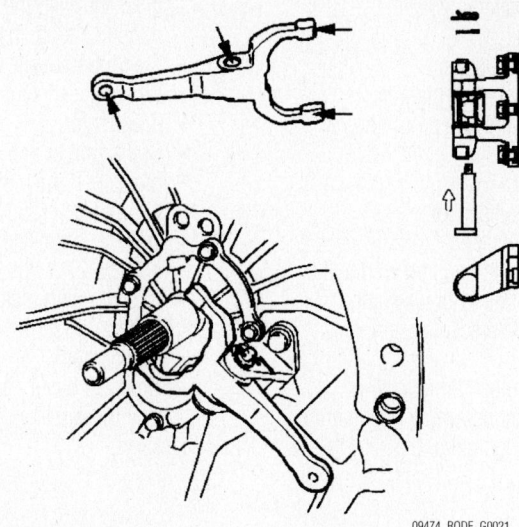

09474_RODE_G0021

Apply molybdenum-disulfide grease at the points noted by arrows

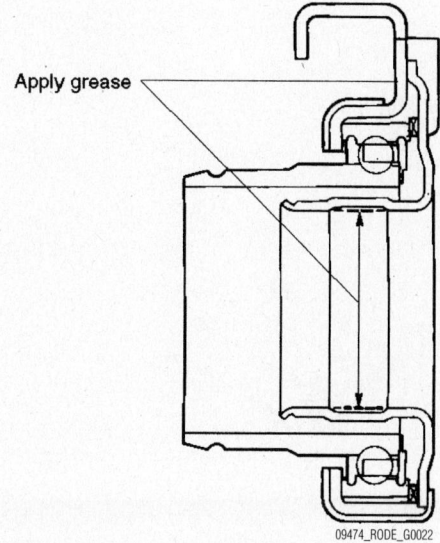

Apply grease

09474_RODE_G0022

Apply molybdenum-disulfide grease at the points noted

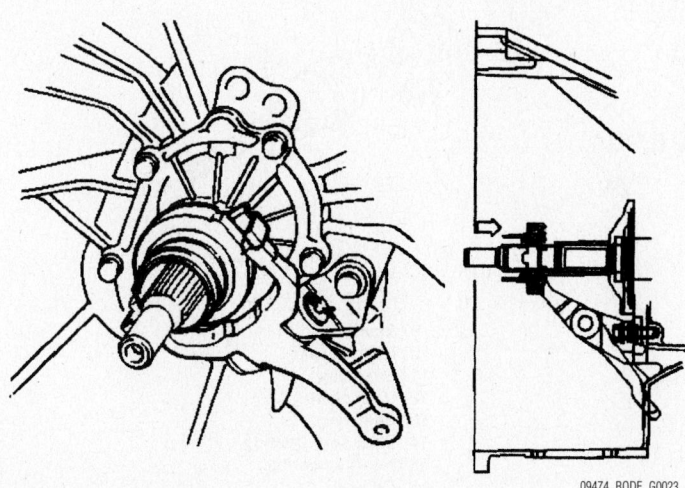

09474_RODE_G0023

Proper release bearing installation

ble the clutch disc and pressure plate onto the flywheel.

7. Tighten the pressure plate bolts in sequence and in two passes to 13 ft. lbs. (8 Nm).

8. Install the transmission.

9. Road test the vehicle and check for proper clutch operation.

Hydraulic Clutch System

BLEEDING

1. Before servicing the vehicle, refer to the Precautions Section.

2. Have an assistant pump the clutch pedal slowly several times and hold it depressed.

3. Open the slave cylinder bleeder screw and allow air to escape.

4. Close the bleeder screw before releasing the clutch pedal.

5. Repeat until all air is purged from the clutch hydraulic system.

6. Refill the reservoir to the full mark.

Transfer Case Assembly

REMOVAL & INSTALLATION

1. Before servicing the vehicle, refer to the Precautions Section.

2. Remove or disconnect the following:
   - Negative battery cable
   - Transfer case skid plate
   - Front and rear driveshafts
   - Heated Oxygen (HO_2S) sensor connectors
   - Left and right exhaust front pipes
   - Transfer case control lever knob
   - Selector lever assembly
   - Transfer case control lever
   - Vehicle Speed (VSS) sensor connector
   - 4 wheel drive switch connector
   - 4 wheel drive actuator connector
   - Transfer case flange fasteners
   - Transfer case

To install:

3. Install or connect the following:
   - Transfer case. Tighten the flange fasteners to 34 ft. lbs. (46 Nm) for 2002–03 models; 30 ft. lbs. (41 Nm) for 2004 models.
   - 4 wheel drive actuator connector
   - 4 wheel drive switch connector
   - VSS sensor connector
   - Transfer case control lever
   - Selector lever assembly
   - Transfer case control lever knob

- Left and right exhaust front pipes
- HO2S sensor connectors
- Front and rear driveshafts. Tighten the bolts to 46 ft. lbs. (63 Nm).
- Transfer case skid plate. Tighten the bolts to 27 ft. lbs. (37 Nm).
- Negative battery cable

Halfshaft

REMOVAL & INSTALLATION

1. Before servicing the vehicle, refer to the Precautions Section.
2. Remove or disconnect the following:

- Negative battery cable
- Front wheel
- Radiator skid plate
- Transfer case skid plate
- Brake calipers and mounting bracket
- Brake rotor
- Wheel speed sensor
- Steering knuckle

3. Support the axle housing with a jack. Unbolt the axle mounting bracket and remove the halfshaft/bracket assembly.

To install:

4. Install or connect the following:
- Axle/bracket assembly. Tighten the bracket flange bolts to 85 ft. lbs.

(116 Nm) and the bracket mounting bolts to 112 ft. lbs. (152 Nm).
- Steering knuckle
- Wheel speed sensor
- Brake rotor
- Brake caliper and mounting bracket. Tighten the bracket bolts to 115 ft. lbs. (155 Nm).
- Transfer case skid plate. Tighten the bolts to 27 ft. lbs. (37 Nm).
- Radiator skid plate. Tighten the bolts to 58 ft. lbs. (78 Nm).
- Front wheel
- Negative battery cable

5. Check the wheel alignment and adjust as necessary.

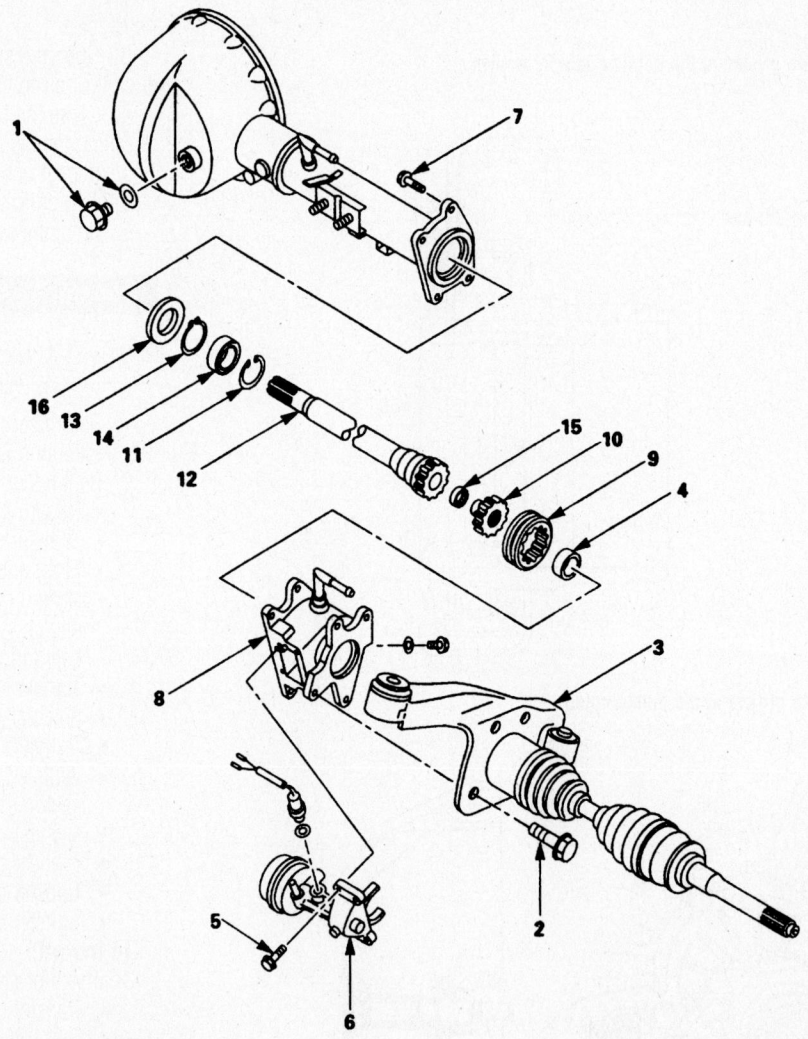

1. Filler plug
2. Bolt
3. Front axle drive shaft (LH side)
4. Spacer
5. Bolt
6. Actuator assembly
7. Bolt
8. Housing
9. Sleeve
10. Clutch gear
11. Snap ring
12. Inner shaft
13. Snap ring
14. Inner shaft bearing
15. Needle bearing
16. Oil seal

7924BG26

Exploded view of the left halfshaft, axle shaft and axle disconnect (shift on the fly system)—2002–03

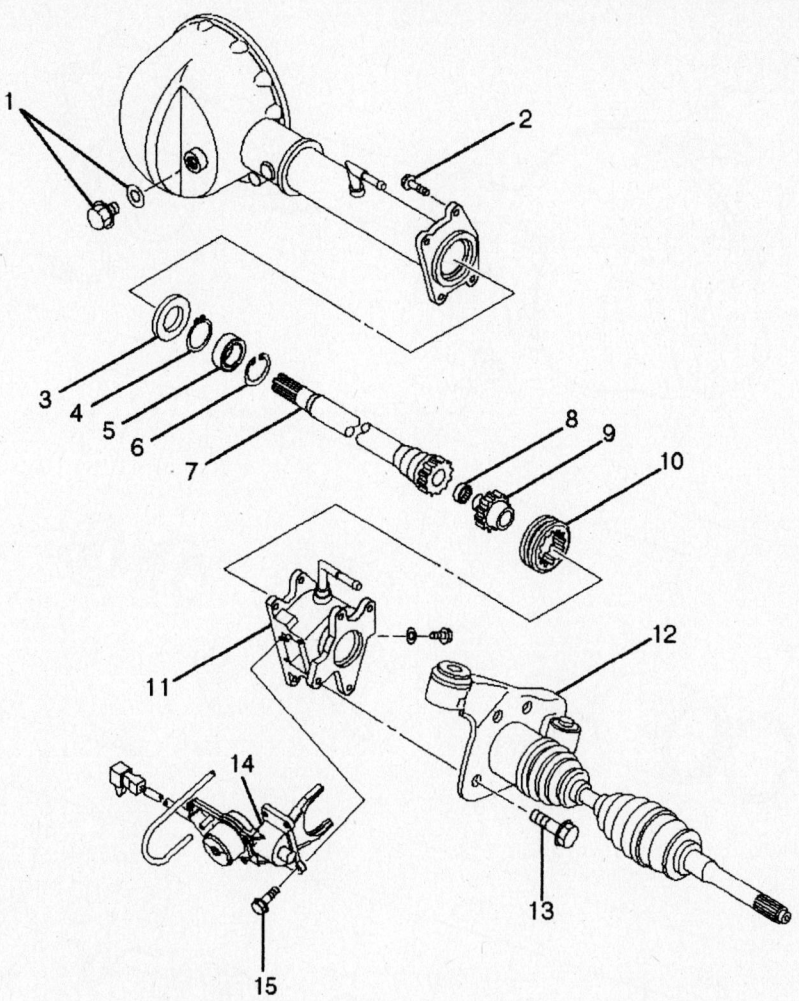

(1)	Filler Plug
(2)	Bolt
(3)	Oil Seal
(4)	Snap Ring (External)
(5)	Inner Shaft Bearing
(6)	Snap Ring (Internal)
(7)	Inner Shaft
(8)	Needle Bearing
(9)	Clutch Gear
(10)	Sleeve
(11)	Housing
(12)	Front Axle Drive Shaft (LH side) with Bracket
(13)	Bolt
(14)	Actuator Assembly
(15)	Bolt

09474_RODE_G0067

Exploded view of the left halfshaft, axle shaft and axle disconnect (shift on the fly system)—2004

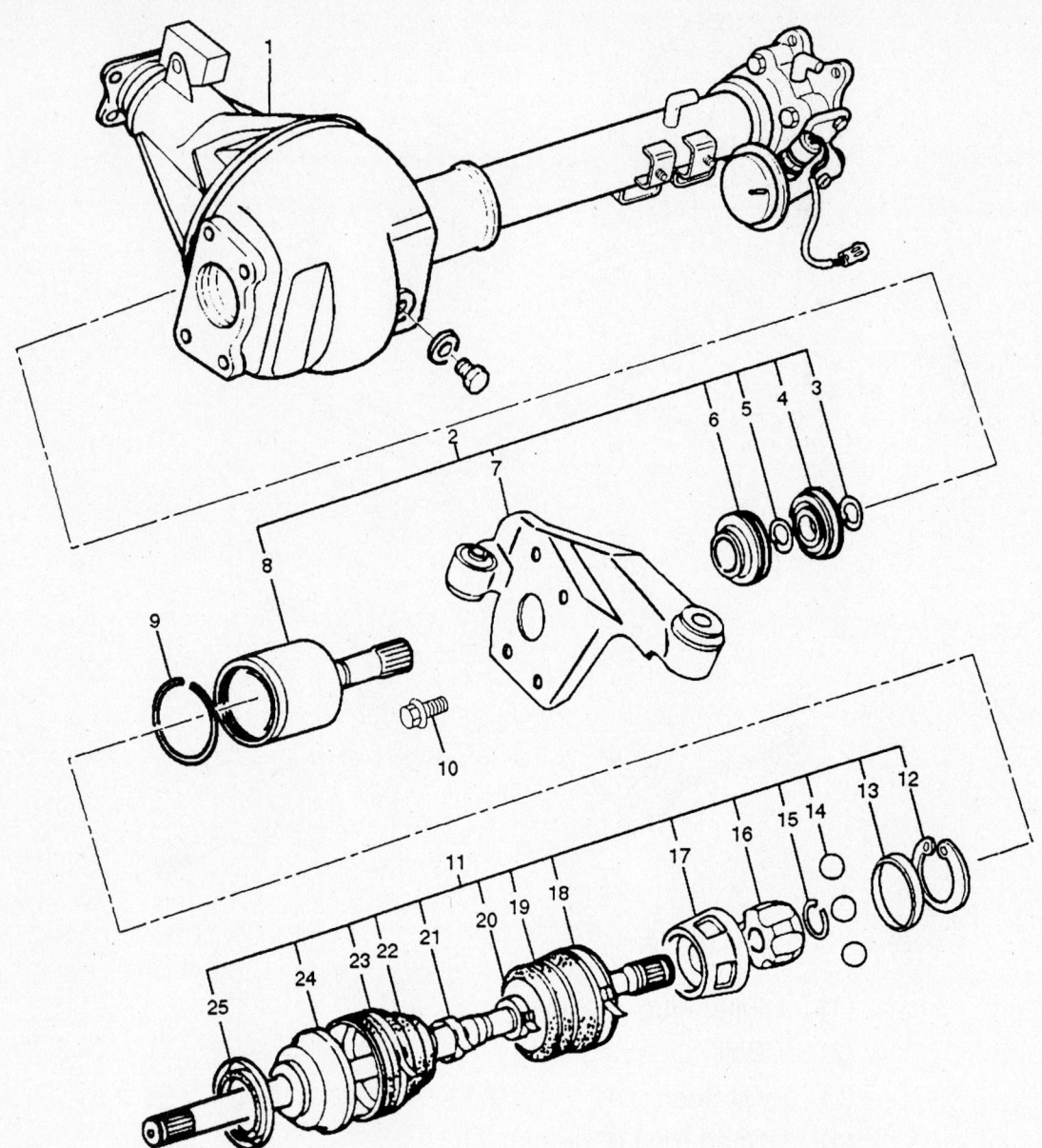

1 Axle Case and Differential
2 DOJ Case Assembly
3 Snap Ring
4 Bearing
5 Snap Ring
6 Oil Seal
7 Bracket
8 DOJ Case
9 Circlip
10 Bolt
11 Drive Shaft Joint Assembly
12 Snap Ring

13 Spacer
14 Ball
15 Snap Ring
16 Ball Retainer
17 Ball Guide
18 Band
19 Bellows
20 Band
21 Band
22 Bellows
23 Band
24 BJ Shaft
25 Dust Seal

9308BG03

Exploded view of the right halfshaft and mounting bracket—Trooper models

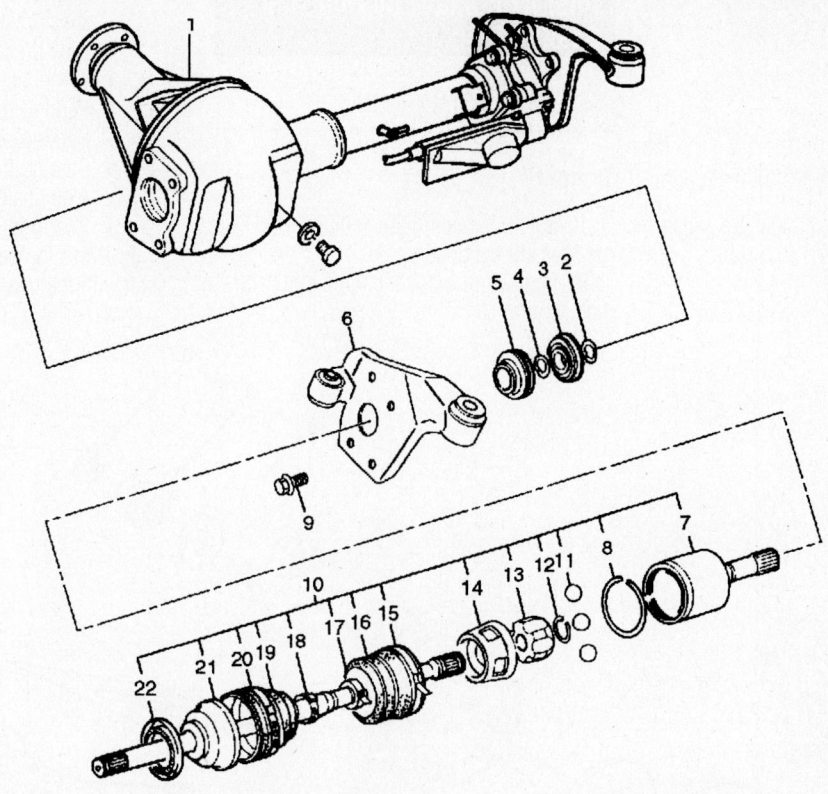

(1)	Axle Case and Differential	(12)	Snap Ring
(2)	Snap Ring	(13)	Ball Retainer
(3)	Bearing	(14)	Ball Guide
(4)	Snap Ring	(15)	Band
(5)	Oil Seal	(16)	Bellows
(6)	Bracket	(17)	Band
(7)	DOJ Case	(18)	Band
(8)	Circlip	(19)	Bellows
(9)	Bolt	(20)	Band
(10)	Drive Shaft Joint Assembly	(21)	BJ Shaft
(11)	Ball	(22)	Dust Seal

09474_RODE_G0066

Exploded view of the right halfshaft and mounting bracket—Passport, Rodeo and Rodeo Sport models

CV-Joints

OVERHAUL

Outer CV-Joint

The outer CV-joint is serviced with the axle shaft as an assembly. The outer CV-joint boot can be serviced by removing the inner CV-joint.

Inner CV-Joint

1. Before servicing the vehicle, refer to the Precautions Section.

2. Remove or disconnect the following:
   - Halfshaft from the vehicle
   - Snapring and bearing
   - Snapring and oil seal
   - Mounting bracket
   - CV-joint boot
   - Circlip and inner joint housing
   - Snapring and spacer
   - Inner joint balls
   - Snapring and inner CV-joint

To install:

3. Install or connect the following:
   - Inner CV-joint and snapring
   - Inner joint balls
   - Spacer and snapring

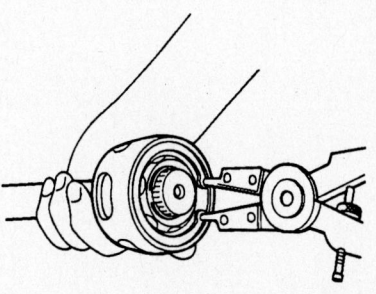

9308BG04
CV-joint spacer snapring—Inner CV-Joint

- Inner joint housing and circlip. Add 150 grams CV-joint grease.
- CV-joint boot
- Mounting bracket
- Oil seal and snapring
- Bearing and snapring

4. Install the halfshaft and mounting bracket to the vehicle.

5. Check the wheel alignment and adjust as necessary.

Rear Axle Shaft, Bearing and Seal

REMOVAL & INSTALLATION

Trooper

1. Before servicing the vehicle, refer to the Precautions Section.

2. Remove or disconnect the following:

- Rear wheel
- Disc brake caliper and bracket
- Disc brake rotor
- Wheel speed sensor bracket
- Parking brake cable and bracket
- Parking brake shoes
- Bearing holder
- Axle shaft
- Snapring and discard it

3. Press the bearing retainer off the axle shaft with the bearing holder and oil seal.

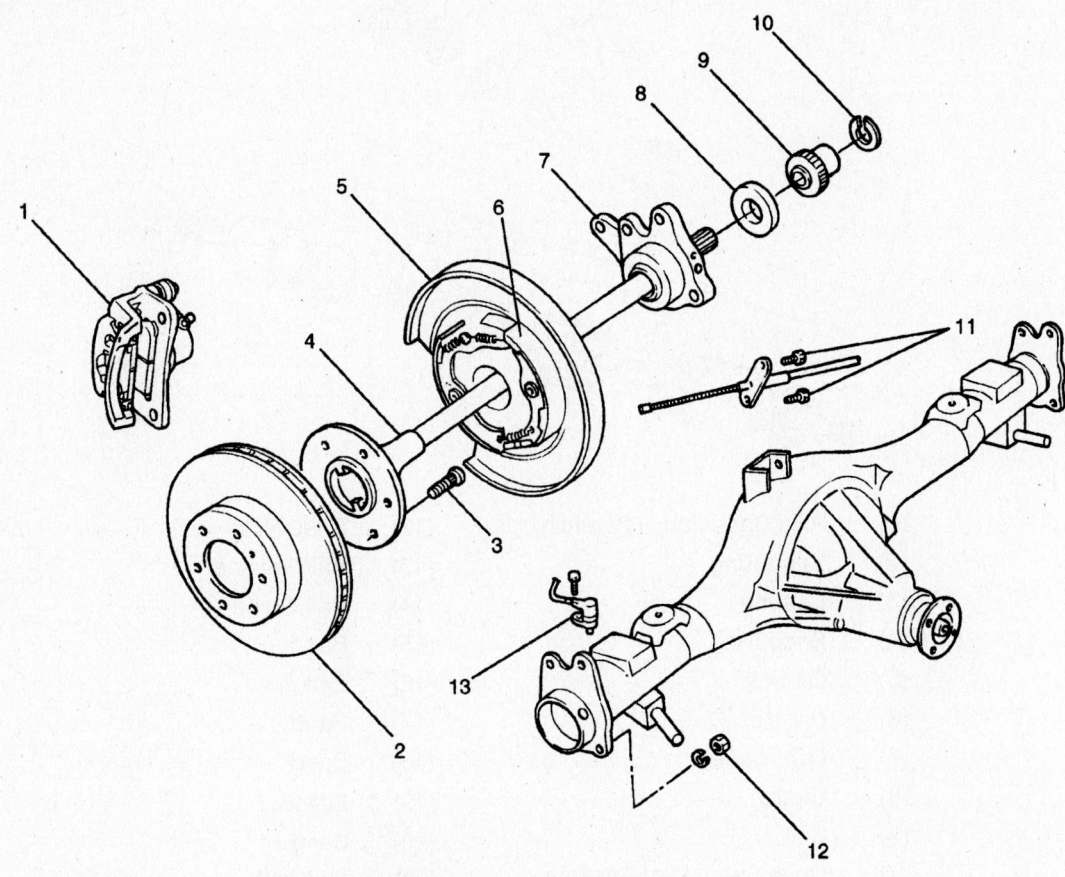

(1)	Brake Caliper
(2)	Brake Disc
(3)	Wheel Pin
(4)	Axle Shaft Assembly
(5)	Back Plate
(6)	Parking Brake Assembly
(7)	Bearing Holder
(8)	Bearing
(9)	Retainer
(10)	Snap Ring
(11)	Bolt
(12)	Nut
(13)	Antilock Brake System (ABS) Speed Sensor

09474_RODE_G0068

Rear axle shaft and related parts—Trooper

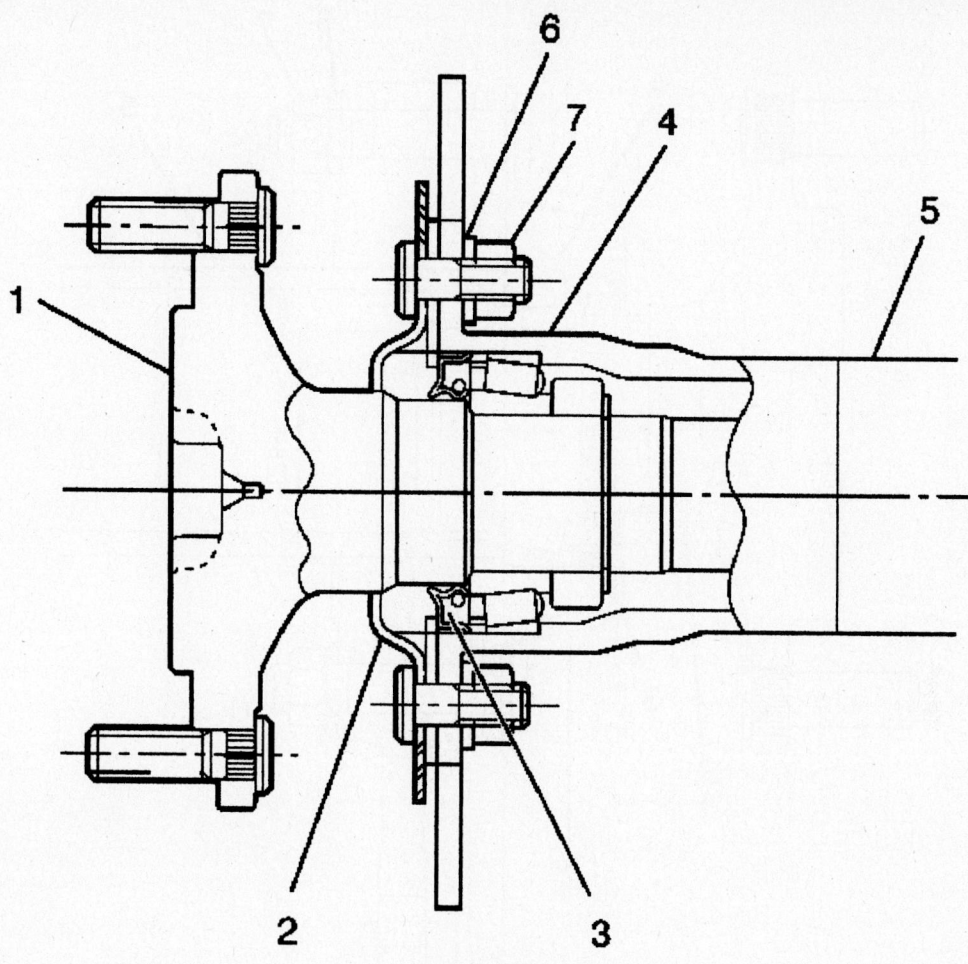

(1)   Axle Shaft

(2)   Backing Plate

(3)   Oil Seal

(4)   Bearing

(5)   Axle Housing

(6)   Lock Washer

(7)   Nut

09474_RODE_G0069

Rear axle bearing and related parts—Passport, Rodeo and Rodeo Sport w/drum brakes

4. Remove the bearing from the bearing holder and backing plate.

To install:

5. Install or connect the following:
 • New oil seal into the bearing housing
 • Bearing housing onto the axle shaft
 • Bearing, press it onto the axle shaft
 • New snapring
 • Axle shaft. Use new lockwashers and tighten the bearing holder nuts to 54 ft. lbs. (74 Nm).
 • Parking brake shoes
 • Parking brake cable and bracket
 • Wheel speed sensor bracket

 • Disc brake rotor
 • Disc brake caliper and bracket. Tighten the bracket bolts to 76 ft. lbs. (103 Nm).
 • Rear wheel

6. Check the rear axle oil level and adjust as necessary.

Passport, Rodeo and Rodeo Sport

1. Before servicing the vehicle, refer to the Precautions Section.

2. Remove or disconnect the following:
 • Rear wheel
 • Disc brake caliper and bracket

 • Disc brake rotor
 • Wheel speed sensor bracket
 • Parking brake cable and bracket
 • Parking brake shoes
 • Axle shaft
 • Snapring and discard it

3. Break the retainer ring with a chisel.

4. Break the bearing cage with a chisel.

5. Press the bearing race off the axle shaft with the bearing holder and oil seal

To install:

6. Install or connect the following:
 • New oil seal into the bearing housing
 • Bearing housing onto the axle shaft

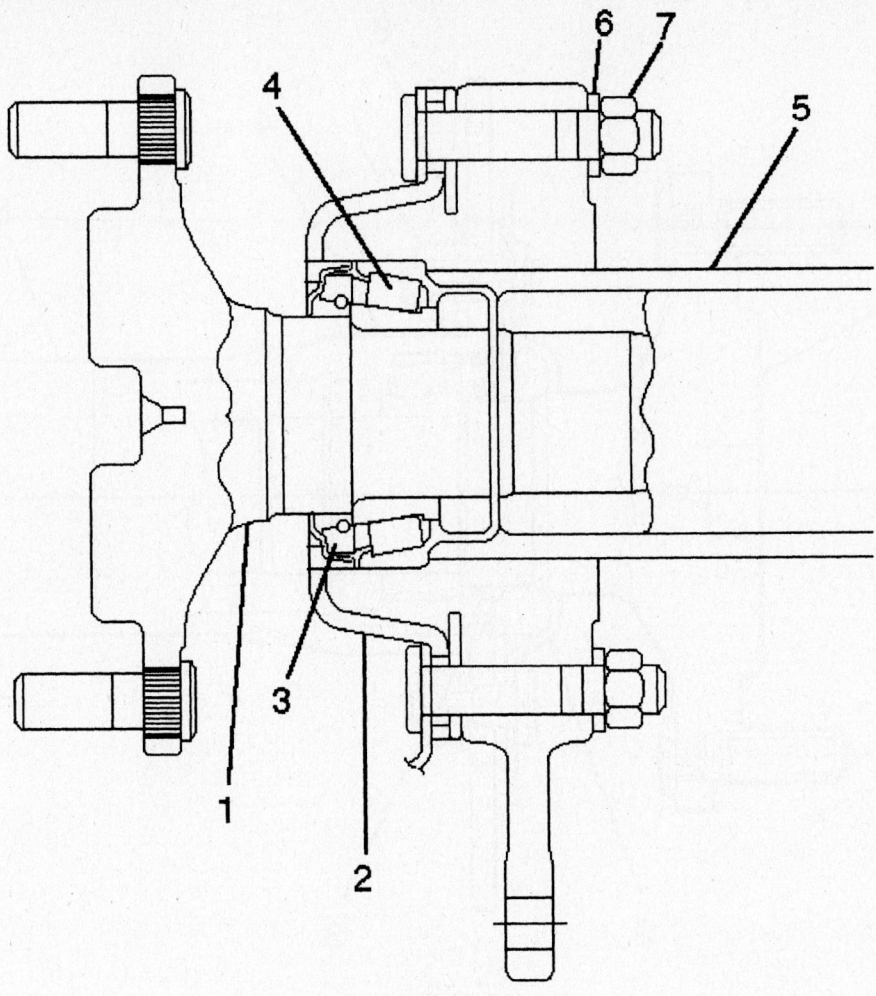

(1) Axle Shaft
(2) Backing Plate
(3) Oil Seal
(4) Bearing
(5) Axle Housing
(6) Lock Washer
(7) Nut

09474_RODE_G0070

Rear axle bearing and related parts—Passport, Rodeo and Rodeo Sport w/disc brakes

- Bearing, press it onto the axle shaft
- New snapring
- Axle shaft. Use new lockwashers and tighten the bearing holder nuts to 54 ft. lbs. (74 Nm).
- Parking brake shoes
- Parking brake cable and bracket
- Wheel speed sensor bracket
- Disc brake rotor
- Disc brake caliper and bracket. Tighten the bracket bolts to 76 ft. lbs. (103 Nm).
- Rear wheel
7. Check the rear axle oil level and adjust as necessary.

Pinion Seal

REMOVAL & INSTALLATION

Front Axle

1. Before servicing the vehicle, refer to the Precautions Section.
2. Drain the axle fluid.
3. Support the axle with a jack.
4. Remove or disconnect the following:
   - Driveshaft
   - Wheels
   - Calipers and pads

5. Remove or disconnect the following:
   - Flange nut (discard)
   - Pinion flange
   - Pinion seal
   - Pinion bearing
   - Collapsible spacer

To install:

➡**Use a new collapsible spacer and flange nut for assembly.**

6. Install or connect the following:
   - Collapsible spacer
   - Pinion bearing
   - Pinion seal
   - Pinion flange

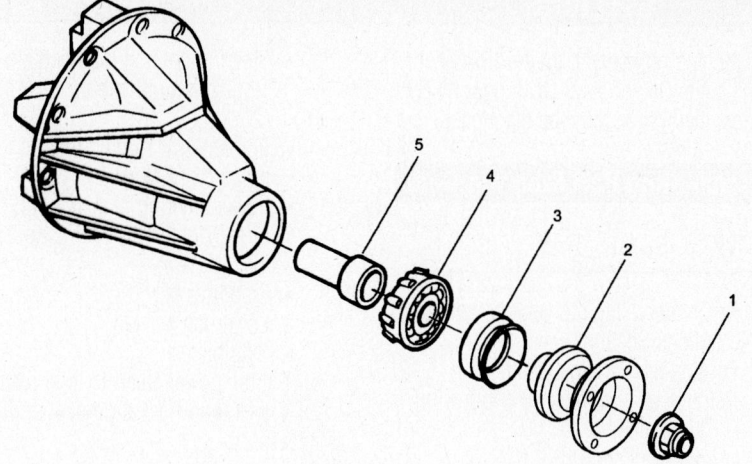

(1) Flange Nut
(2) Flange
(3) Oil Seal
(4) Outer Bearing
(5) Collapsible Spacer

09474_RODE_G0024

Front pinion seal and related parts

7. Install the new flange nut and tighten to 160 ft. lbs. (217 Nm).

8. Adjust the bearing preload:

a. Attach an inch lb. torque wrench to the nut and make several rotations. Note the torque needed to start each rotation.

b. Tighten the nut until a starting torque of 5.6–10 inch lbs. for a new bearing, or 2.8–4.9 inch lbs. for a used bearing is recorded. Do not exceed 500 ft. lbs. (678 Nm) tightening torque.

9. Using a punch, stake the nut at 2 points.

10. Install or connect the following:
- Driveshaft. Torque to 46 ft. lbs. (63 Nm).
- Brake calipers and pads
- Wheels

11. Fill the differential with gear lubricant and check for leaks.

Rear Axle

1. Before servicing the vehicle, refer to the Precautions Section.

2. Drain the rear axle fluid.

3. Remove or disconnect the following:
- Driveshaft
- Wheels
- Brake calipers and pads or drums

➡ **The brake calipers and pads or drums must be removed so that there is no additional drag when measuring pinion bearing preload.**

4. Use an inch lb. torque wrench and measure and record the amount of torque required to maintain pinion rotation through several revolutions.

5. Remove or disconnect the following:
- Flange nut
- Pinion flange
- Pinion seal
- Pinion bearing
- Collapsible spacer

To install:

➡ **Use a new collapsible spacer and flange nut for assembly.**

6. Install or connect the following:
- Collapsible spacer
- Pinion bearing
- Pinion seal
- Pinion flange

7. Rotate the pinion flange occasionally while tightening the flange nut to make sure the pinion bearings seat correctly.

8. Take frequent bearing preload torque readings. Tighten the flange nut to achieve the preload torque readings originally recorded.

✳✳ CAUTION

Never loosen the pinion nut to reduce bearing preload. If it is necessary to reduce bearing preload, install a new collapsible spacer and pinion nut. Maximum pinion nut torque should be 220–280 ft. lbs. (298–380 Nm).

9. Install or connect the following:
- Driveshaft. Torque to 46 ft. lbs. (63 Nm).
- Brake calipers and pads or drums
- Wheels

10. Fill the differential with gear lubricant and check for leaks.

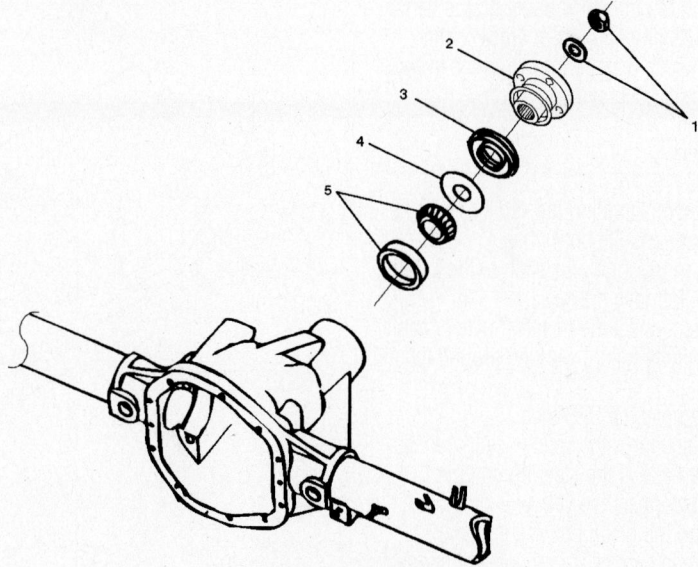

(1) Flange Nut and Washer
(2) Flange
(3) Oil Seal
(4) Outer Oil Seal Slinger
(5) Outer Pinion Bearing (Cup and Cone)

09474_RODE_G0025

Rear pinion seal and related parts

STEERING

Air Bag

❊❊ CAUTION

Some vehicles are equipped with an air bag system. The system must be disarmed before performing service on, or around, system components, the steering column, instrument panel components, wiring and sensors. Failure to follow the safety precautions and the disarming procedure could result in accidental air bag deployment, possible injury and unnecessary system repairs.

PRECAUTIONS

Several precautions must be observed when handling the inflator module to avoid accidental deployment and possible personal injury.

• Never carry the inflator module by the wires or connector on the underside of the module.

• When carrying a live inflator module, hold securely with both hands, and ensure that the bag and trim cover are pointed away from you.

• Place the inflator module on a bench or other surface with the bag and trim cover facing up.

• With the inflator module on the bench, never place anything on or close to the module which may be thrown in the event of an accidental deployment.

DISARMING

1. Before servicing the vehicle, refer to the Precautions Section.

2. Turn the ignition switch to the **LOCK** position. Remove the key.

3. Disconnect the negative battery cable. Wait 1 minute before working around the air bags.

4. Remove the SRS fuse.

5. Disconnect the yellow 2-pin connector at the base of the steering column.

6. Disconnect the yellow 2-pin connector behind the glove box assembly.

7. When repairs are completed, connect the yellow 2-pin connectors.

8. Connect the negative battery cable.

9. Turn the ignition to the **ON** position, but don't start the engine. The AIR BAG warning light should turn ON and flash ON and OFF 7 times, and then turn OFF. This light sequence indicates that the SRS system is functioning normally. If the AIR BAG light doesn't come ON, or stays ON longer than 7 seconds, the system must be diagnosed.

Recirculating Ball Steering Gear

REMOVAL & INSTALLATION

1. Before servicing the vehicle, refer to the Precautions Section.

2. Disable the air bag system.

3. Remove or disconnect the following:

• Skid plates
• Lower fan shroud
• Stabilizer bar
• Power steering pressure and return lines
• Pitman arm
• Steering column intermediate shaft
• Steering gear

To install:

4. Install or connect the following:

• Steering gear. Tighten the bolts to 33 ft. lbs. (44 Nm).

• Steering column intermediate shaft. Tighten the pinch bolt to 18 ft. lbs. (25 Nm).

• Pitman arm. Tighten the nut to 159 ft. lbs. (216 Nm).

• Power steering pressure and return lines. Tighten the fittings to 33 ft. lbs. (44 Nm).

• Stabilizer bar

• Lower fan shroud

• Skid plates

5. Fill the power steering fluid reservoir.

6. Check the wheel alignment and adjust as necessary.

Rack and Pinion Steering Gear

REMOVAL & INSTALLATION

2-Wheel Drive

1. Before servicing the vehicle, refer to the Precautions Section.

2. Disable the air bag system.

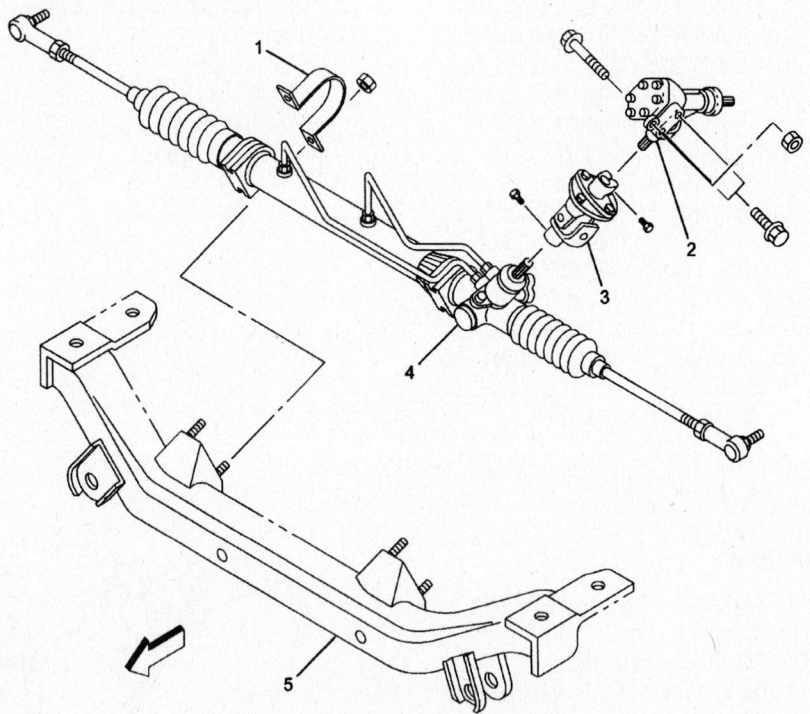

(1)	Bracket
(2)	Transfer Gear Assembly
(3)	Joint Assembly
(4)	Power Steering Unit Assembly
(5)	Crossmember

Rack and pinion steering gear

09474_RODE_G0026

3. Remove or disconnect the following:
- Stone guard
- Tie rod from knuckle
- Power steering pressure and return lines
- Steering gear

To install:

4. Install or connect the following:
- Steering gear. Torque the mounting bolts to 85 ft. lbs. (116 Nm)
- Fluid lines. Torque the fitting to 18 ft. lbs. (25 Nm)
- Tie rod ends. 87 ft. lbs. (118 Nm)
- Stone guard

5. Fill and bleed the system.

4-Wheel Drive

1. Before servicing the vehicle, refer to the Precautions Section.
2. Disable the air bag system.
3. Remove or disconnect the following:
- Stone guard
- Tie rod from knuckle
- Power steering pressure and return lines
- Torsion bar
- Lower control arm frame side bolt
- Crossmember bolts
- Steering gear with crossmember
- Steering gear

To install:

4. Install or connect the following:
- Steering gear. Torque the mounting bolts to 85 ft. lbs. (116 Nm)
- Crossmember Torque the bolts to 128 ft. lbs. (173 Nm)
- Lower control arm bolt
- Torsion bar
- Fluid lines. Torque the fitting to 18 ft. lbs. (25 Nm)
- Tie rod ends. 87 ft. lbs. (118 Nm)
- Stone guard

5. Fill and bleed the system.

FRONT SUSPENSION

Stabilizer Bar and Links

REMOVAL & INSTALLATION

Trooper

1. Before servicing the vehicle, refer to the Precautions Section.

2. Remove the wheels.
3. Remove the stone guard.
4. Disconnect the links.
5. Remove the brackets and bushings.
6. Remove the bar.
7. Installation is the reverse of removal. Torque the brackets to 16 ft. lbs. (22 Nm); torque the links to 37 ft. lbs. (50 Nm).

Passport, Rodeo and Rodeo Sport

1. Before servicing the vehicle, refer to the Precautions Section.
2. Remove the wheels.
3. Remove the stone guard.
4. Disconnect the links.
5. Remove the brackets and bushings.
6. Remove the bar.

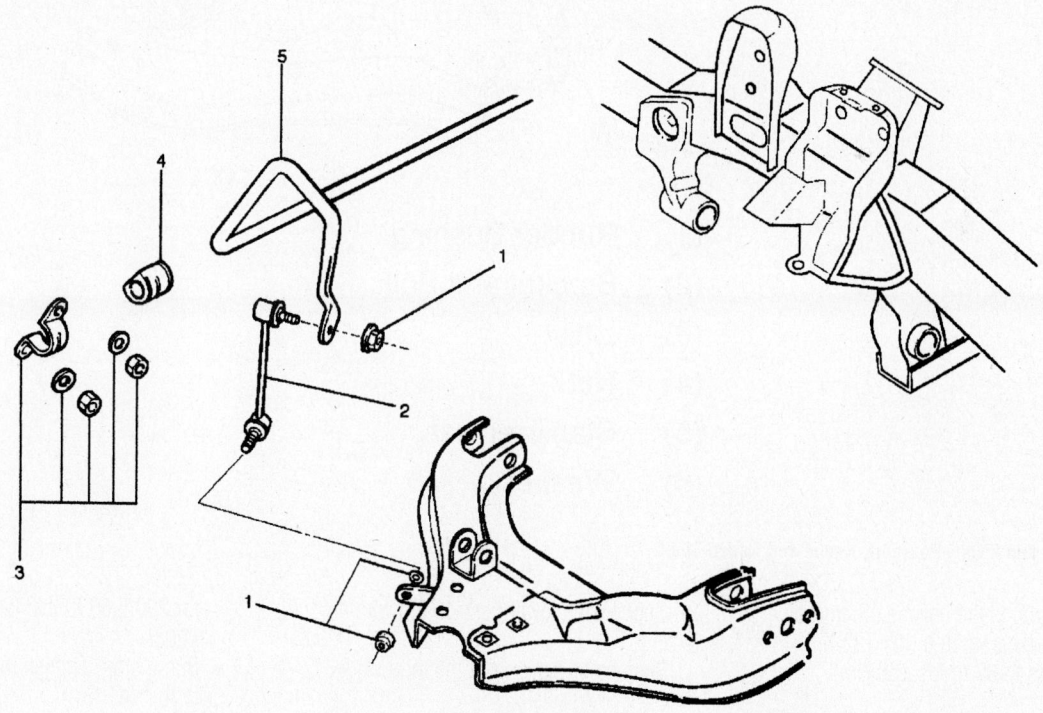

(1)	Nut and Washer
(2)	Link
(3)	Bracket
(4)	Rubber Bushing
(5)	Stabilizer Bar

09474_RODE_G0079

Stabilizer bar and links—Trooper

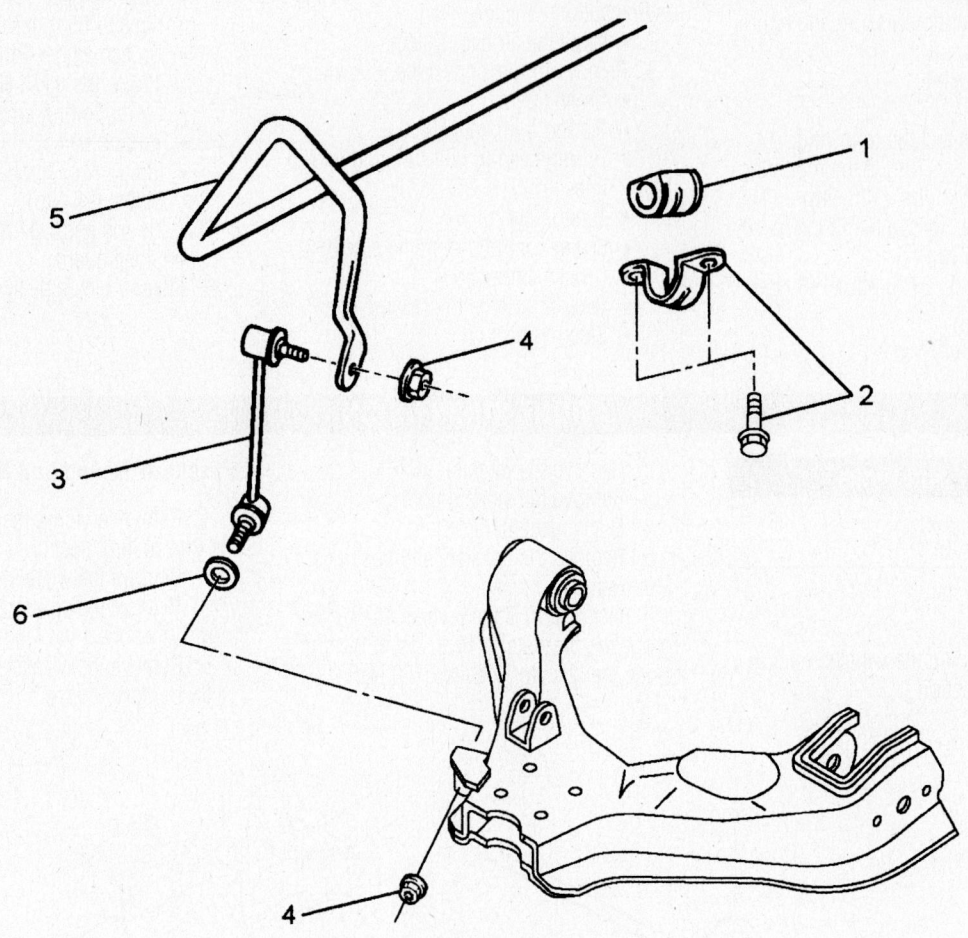

(1)   **Rubber Bushing**

(2)   **Bracket and Bolt**

(3)   **Link**

(4)   **Nut**

(5)   **Stabilizer Bar**

(6)   **Washer**

09474_RODE_G0080

Stabilizer bar and links—Passport, Rodeo and Rodeo Sport

7. Installation is the reverse of removal. Torque the brackets to 18 ft. lbs. (22 Nm); torque the links to 37 ft. lbs. (50 Nm).

Shock Absorber

REMOVAL & INSTALLATION

2002–03

1. Before servicing the vehicle, refer to the Precautions Section.

2. Support the lower control arm with a jackstand.

3. Remove or disconnect the following:
   - Front wheels
   - Upper shock retaining nut and rubber bushing
   - Suspension bump stops
   - Lower nut
   - Shock absorber

To install:

4. Install or connect the following:
   - Shock absorber. Tighten the lower bolt to 60–61 ft. lbs. (82–84 Nm)

for 2002; 69 ft. lbs. (93 Nm) for 2003.
   - Bump stop. Tighten the bolts to 30 ft. lbs. (41 Nm).
   - Upper shock retaining nut and rubber bushing. Tighten the nut to 14–15 ft. lbs. (19–20 Nm).
   - Front wheels

2004

1. Before servicing the vehicle, refer to the Precautions Section.

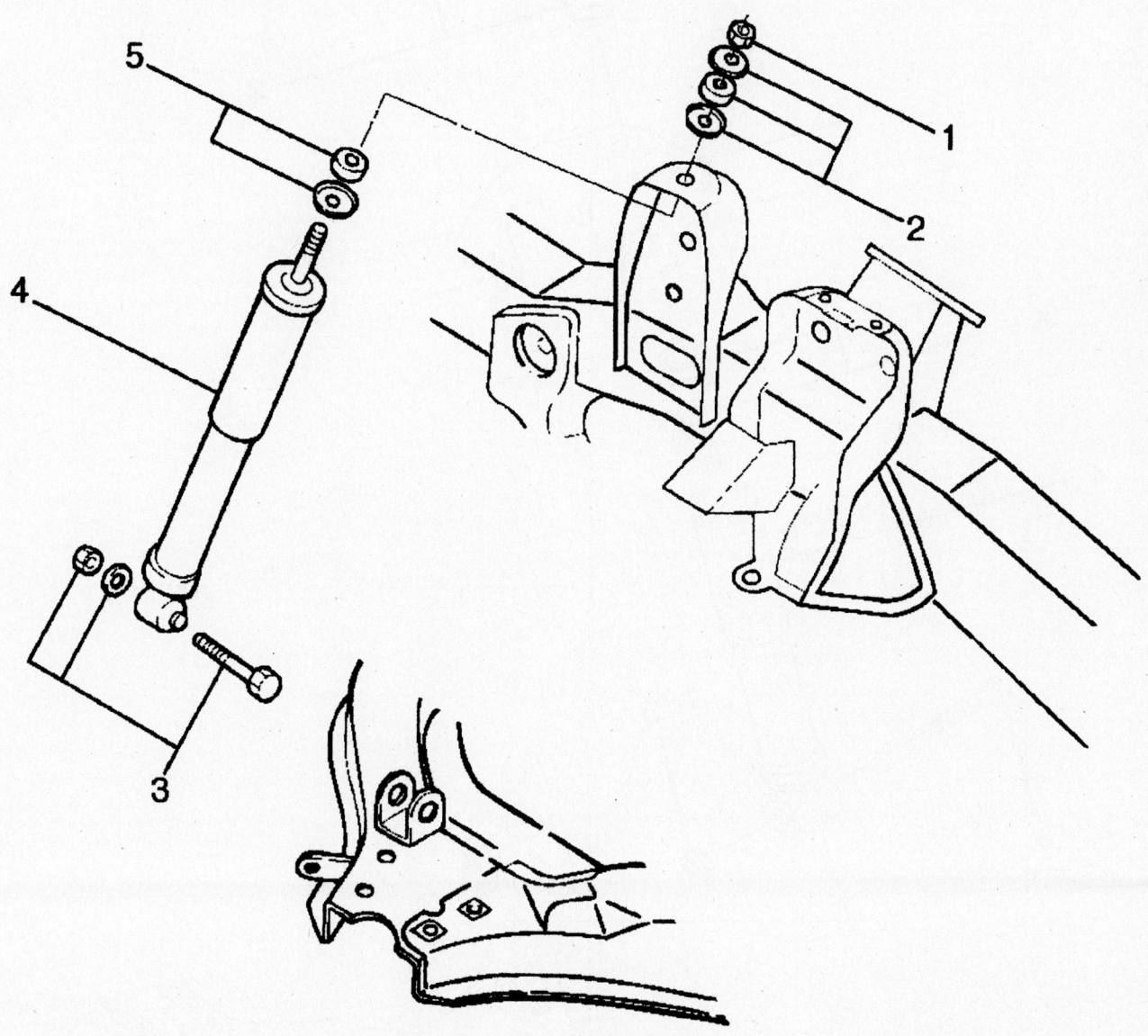

(1) Nut

(2) Rubber Bushing and Washer

(3) Bolt, Nut and Washer

(4) Shock Absorber

(5) Rubber Bushing and Washer

09474_RODE_G0071

Front shock absorber mounting—Trooper

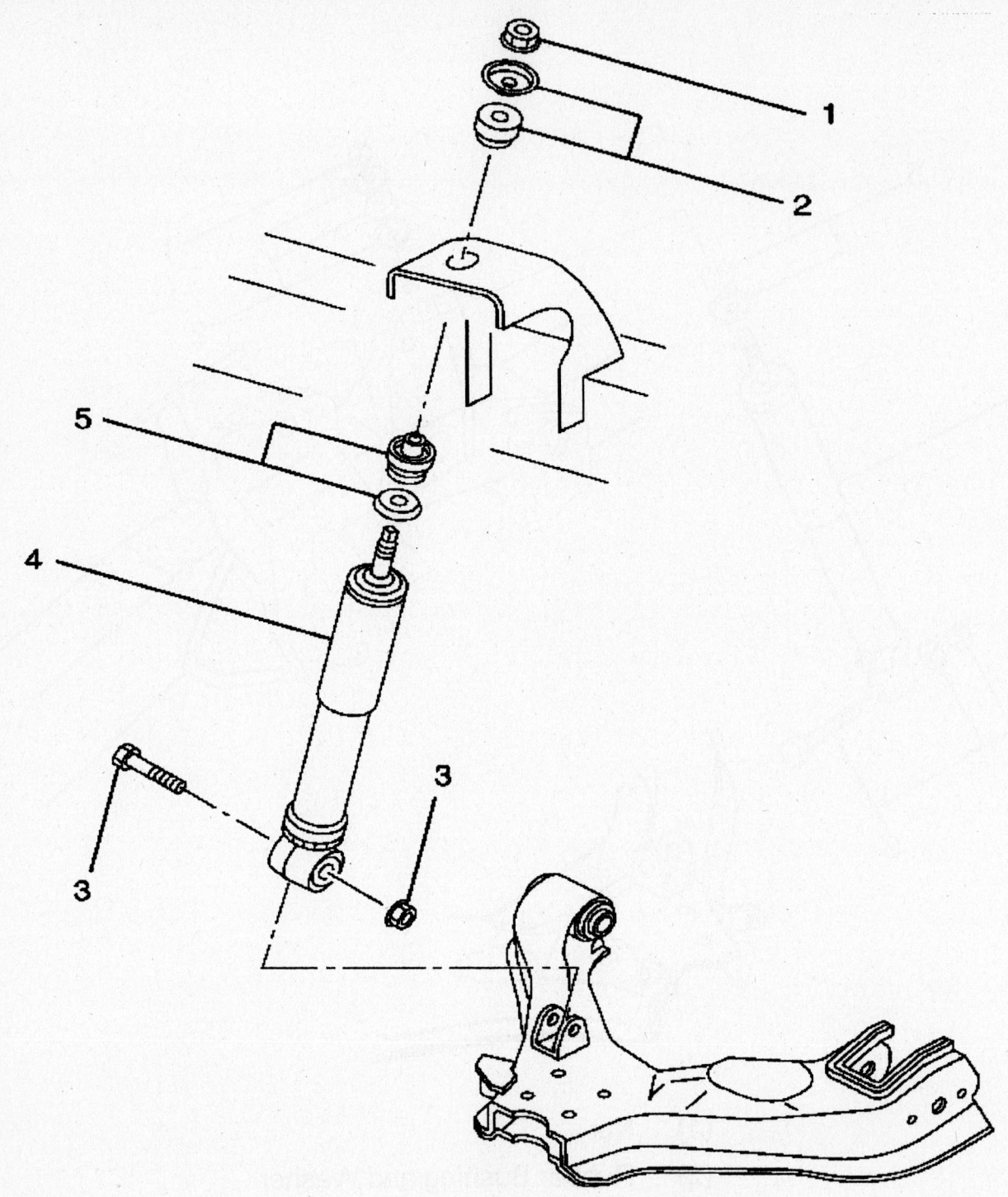

(1)   Nut

(2)   Rubber Bushing and Washer

(3)   Bolt and Nut

(4)   Shock Absorber

(5)   Rubber Bushing and Washer

09474_RODE_G0072

Front shock absorber mounting—2002-03 Passport, Rodeo and Rodeo Sport

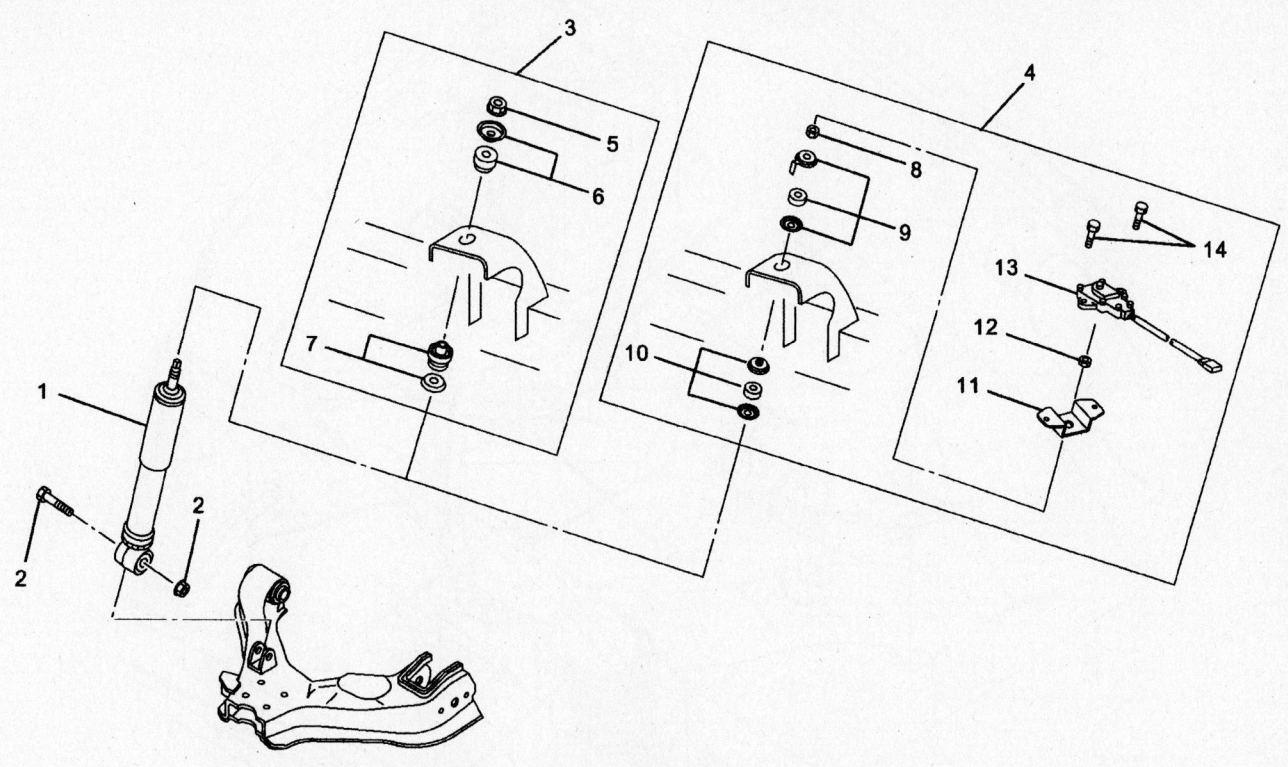

(1)   Shock Absorber

(2)   Bolt and Nut

(3)   Without Intelligent Suspension

(4)   With Intelligent Suspension

(5)   Nut

(6)   Rubber Bushing and Washer

(7)   Rubber Bushing and Washer

(8)   Nut

(9)   Rubber Bushing and Washer

(10)   Rubber Bushing and Washer

(11)   Actuator Bracket

(12)   Nut

(13)   Actuator

(14)   Screw

09474_RODE_G0029

Front shock absorber and related parts—2004 shown

2. Support the lower control arm with a jackstand.

3. Remove the front wheels.

4. Remove the upper shock retaining nut and rubber bushing.

5. With Intelligent Suspension:
   a. Disconnect the battery ground cable.

b. Remove the connector from the bracket.

c. Remove the 2 screws.

d. Disconnect the actuator.

e. Remove the nuts.

f. Remove the bracket.

6. Remove the lower bolt/nut.

7. Remove the shock absorber.

8. Installation is the reverse of removal. Torque the upper shock nut to 14 ft. lbs. (20 Nm); torque the lower bolt/nut to 69 ft. lbs. (93 Nm). Torque the actuator bracket to 29 ft. lbs. (39 Nm).

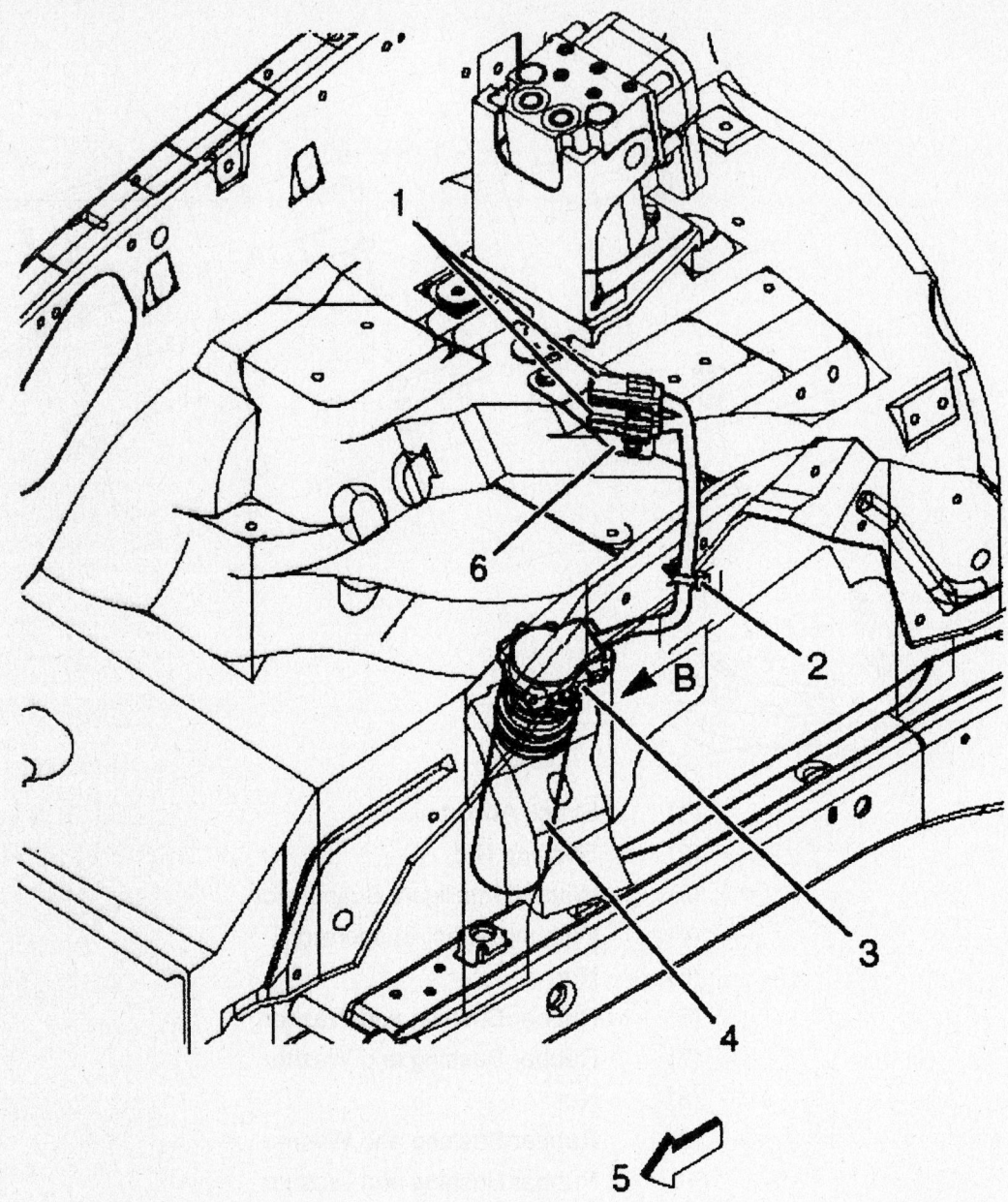

(1)	Connector
(2)	Clip
(3)	Actuator
(4)	Shock Absorber
(5)	Front
(6)	Bracket

09474_RODE_G0027

Right side front actuator and related parts

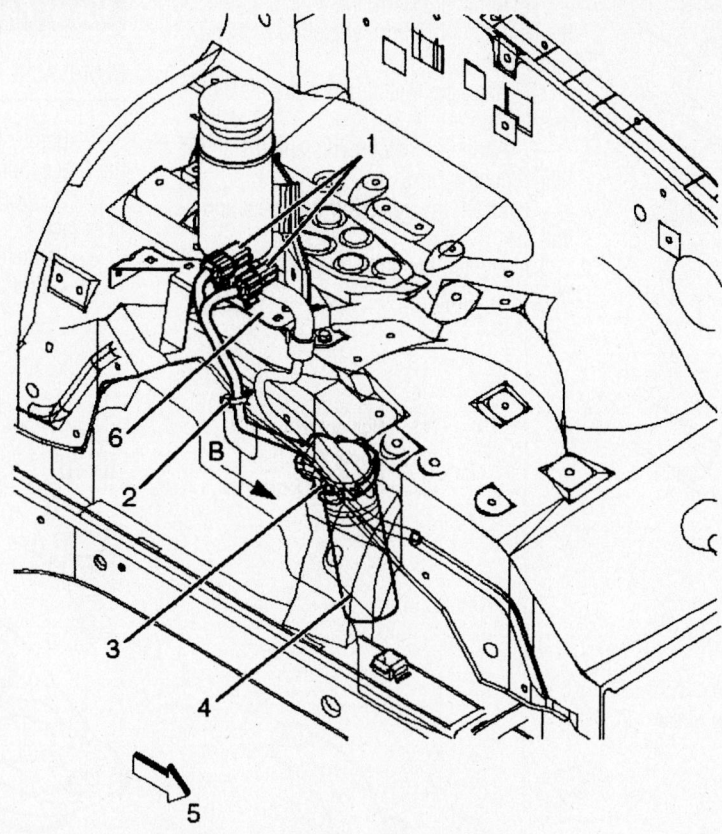

(1)	Connector
(2)	Clip
(3)	Actuator
(4)	Shock Absorber
(5)	Front
(6)	Bracket

09474_RODE_G0028

Left side front actuator and related parts

Torsion Bar

REMOVAL & INSTALLATION

1. Before servicing the vehicle, refer to the Precautions Section.

2. Matchmark the adjusting bolt and end piece, then remove the bolt, end piece, and seat.

3. Matchmark the height control arm to the torsion bar, then remove the height control arm.

4. Matchmark the torsion bar to the lower control arm, then remove the torsion bar.

To install:

5. Apply grease to the torsion bar splines.

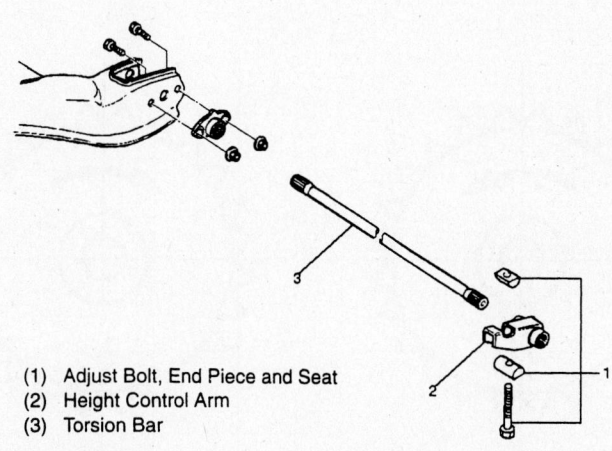

(1) Adjust Bolt, End Piece and Seat
(2) Height Control Arm
(3) Torsion Bar

9308BG05

Exploded view of the torsion bar assembly

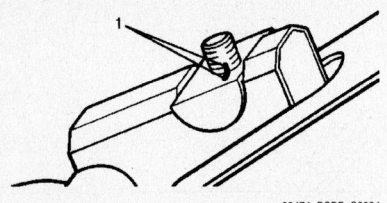

Matchmark the height control arm to the torsion bar

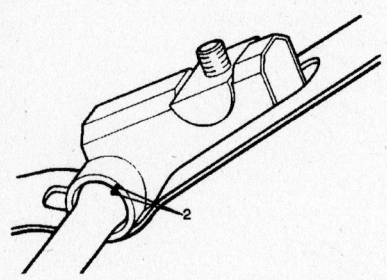

Matchmark the height control arm to the torsion bar

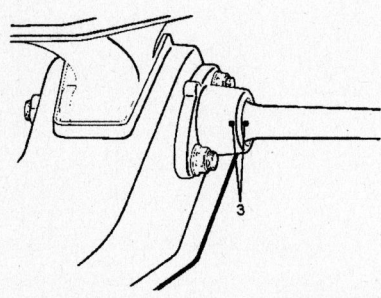

Matchmark the torsion bar to the lower control arm

6. Apply grease to the contact points of the height control arm, adjusting bolt end piece and seat.

7. Align the matchmarks and install the torsion bar.

8. Align the matchmarks and install the height control arm.

9. Install the adjusting bolt, seat and end piece.

10. Tighten the adjusting bolt to align the matchmarks.

(1) Bolt and Nut
(2) Upper Ball Joint
(3) Nut and Cotter Pin

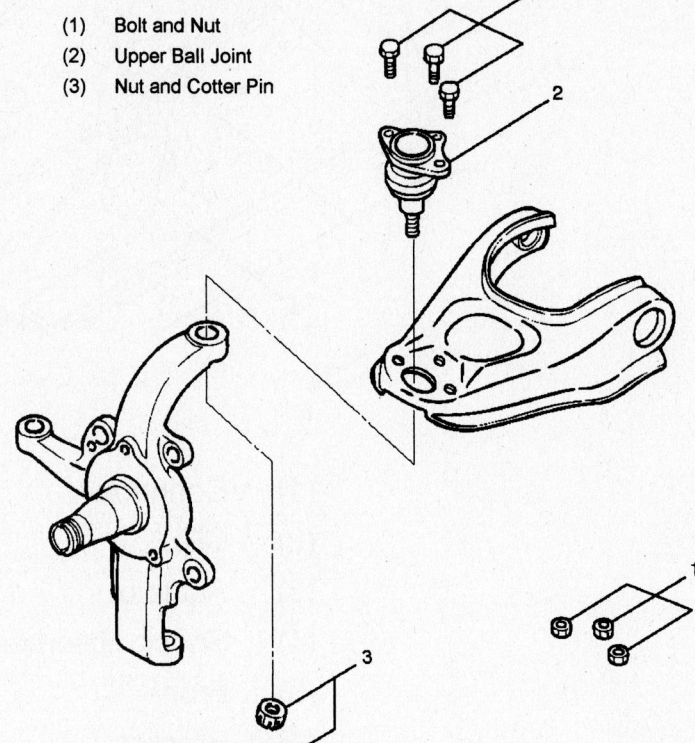

Upper ball joint installation—Passport, Rodeo and Rodeo Sport

Upper Ball Joint

REMOVAL & INSTALLATION

1. Before servicing the vehicle, refer to the Precautions Section.

2. Support the lower control arm with a floor jack.

3. Remove or disconnect the following:
   • Front wheel
   • Wheel speed sensor
   • Upper ball joint

To install:

➡**Use new nuts, bolts and split pins for assembly.**

4. Install or connect the following:
   • Upper ball joint. Tighten the mounting bolts to 42 ft. lbs. (57 Nm) and the castellated nut to 72 ft. lbs. (96 Nm). Use a new cotter pin. Never back off the nut to align the cotter pin holes.
   • Wheel speed sensor
   • Front wheel

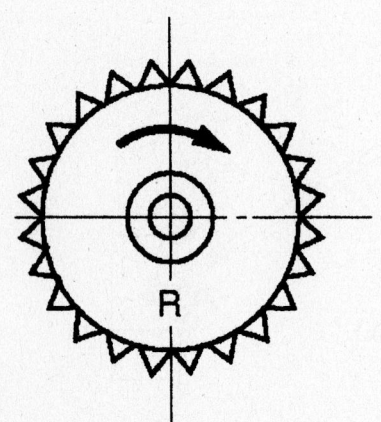

Torsion bar identification

(1) Bolt, Nut and Washer
(2) Upper Ball Joint
(3) Nut and Cotter Pin

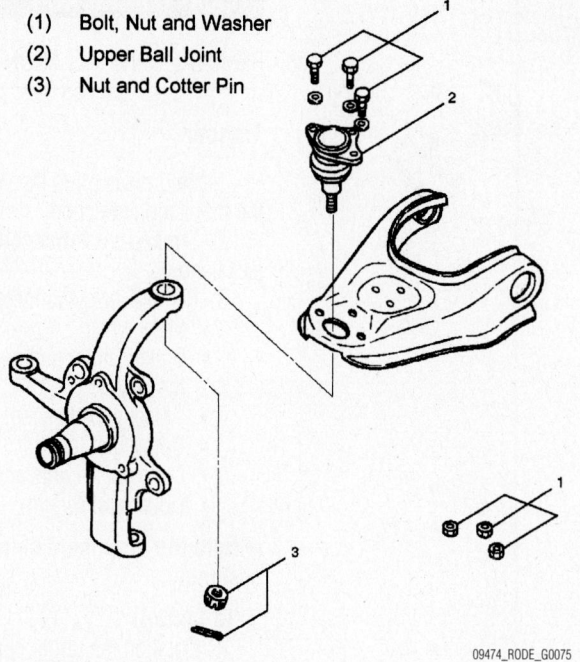

09474_RODE_G0075

Upper ball joint installation—Trooper

Lower Ball Joint

REMOVAL & INSTALLATION

1. Before servicing the vehicle, refer to the Precautions Section.

2. Remove the wheel.

3. Disconnect the tie rod end at the knuckle.

4. Remove the halfshaft from the knuckle. See Halfshaft R&I.

5. Support the lower control arm with a jack.

6. Remove the cotter pin and castellated nut then, using a screw-type forcing tool, separate the ball joint from the knuckle. Discard the cotter pin.

7. Remove the nuts and bolts and remove the ball joint from the control arm.

8. Installation is the reverse of removal. Torque the ball joint-to-control arm nuts/bolts to 85 ft. lbs. (115 Nm). Torque the castellated nut to 108 ft. lbs. (147 Nm). Never back off the nut to align the cotter pin holes.

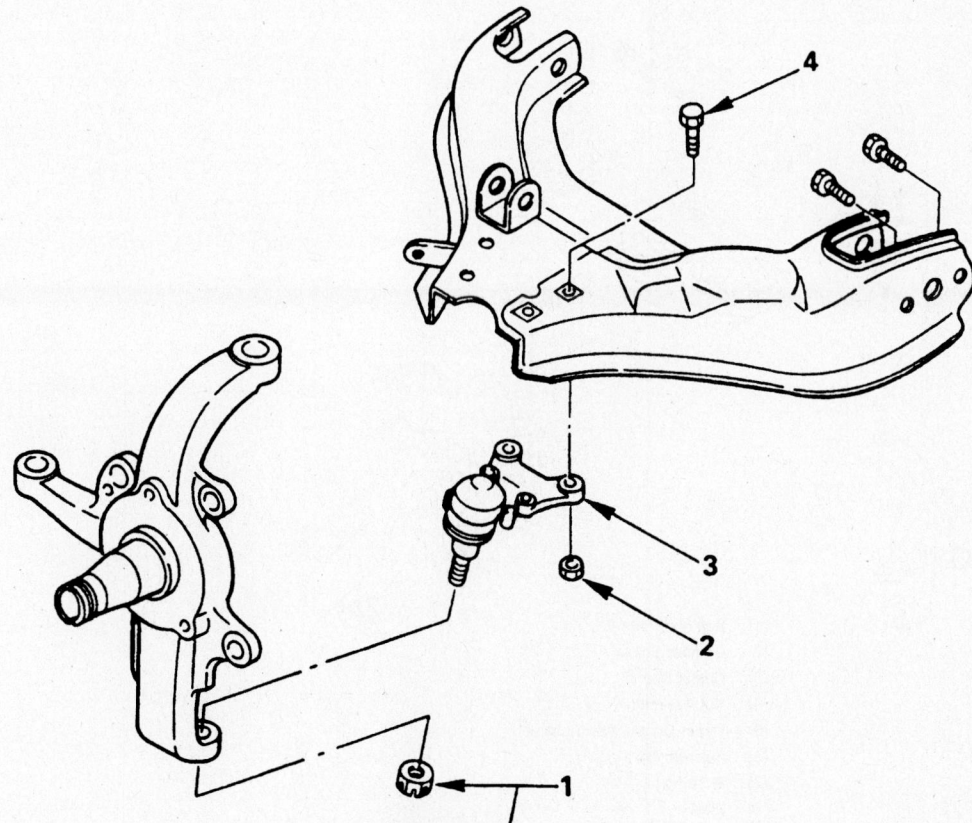

1. Nut and cotter pin
2. Nut
3. Lower ball joint
4. Bolt

7924BG35

Exploded view of the lower ball joint mounting and related components—Trooper

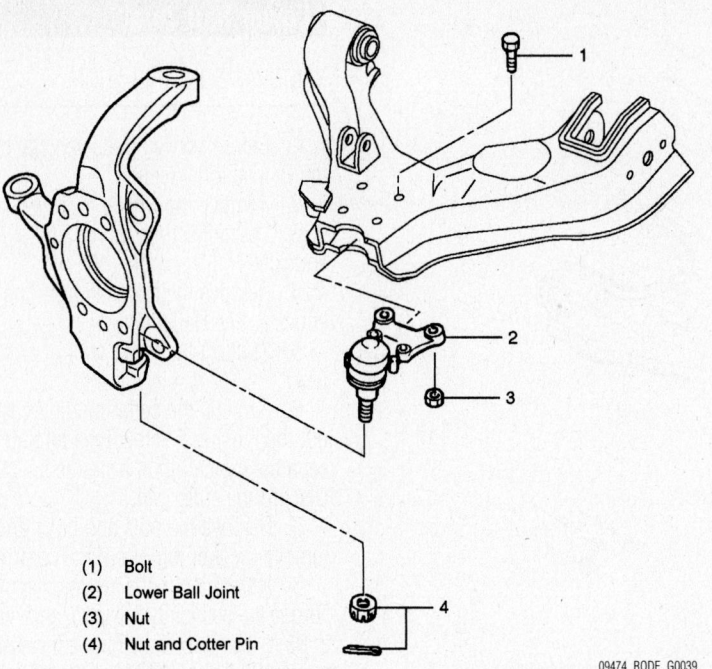

(1) Bolt
(2) Lower Ball Joint
(3) Nut
(4) Nut and Cotter Pin

09474_RODE_G0039

Exploded view of the lower ball joint mounting and related components—Passport, Rodeo and Rodeo Sport

Upper Control Arm

REMOVAL & INSTALLATION

Trooper

1. Before servicing the vehicle, refer to the Precautions Section.
2. Support the lower control arm with a jackstand.
3. Remove or disconnect the following:
   - Front wheel
   - Brake caliper and suspend it out of the way
   - Wheel speed sensor
   - Upper ball joint
   - Control arm nuts and bolts
   - Upper control arm

➡**Note the alignment shim location for assembly.**

To install:

4. Position the upper control arm and tighten the fulcrum pin nuts finger tight.
5. Adjust the bump stop clearance to

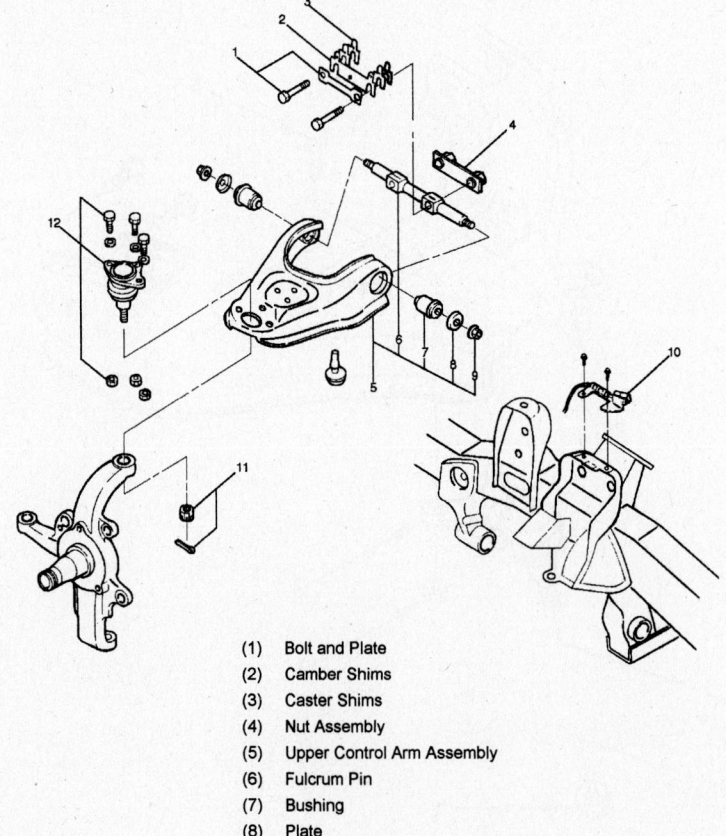

(1) Bolt and Plate
(2) Camber Shims
(3) Caster Shims
(4) Nut Assembly
(5) Upper Control Arm Assembly
(6) Fulcrum Pin
(7) Bushing
(8) Plate
(9) Nut
(10) Speed Sensor Cable
(11) Nut and Cotter Pin
(12) Upper Ball Joint

09474_RODE_G0076

Upper control arm and related parts—Trooper

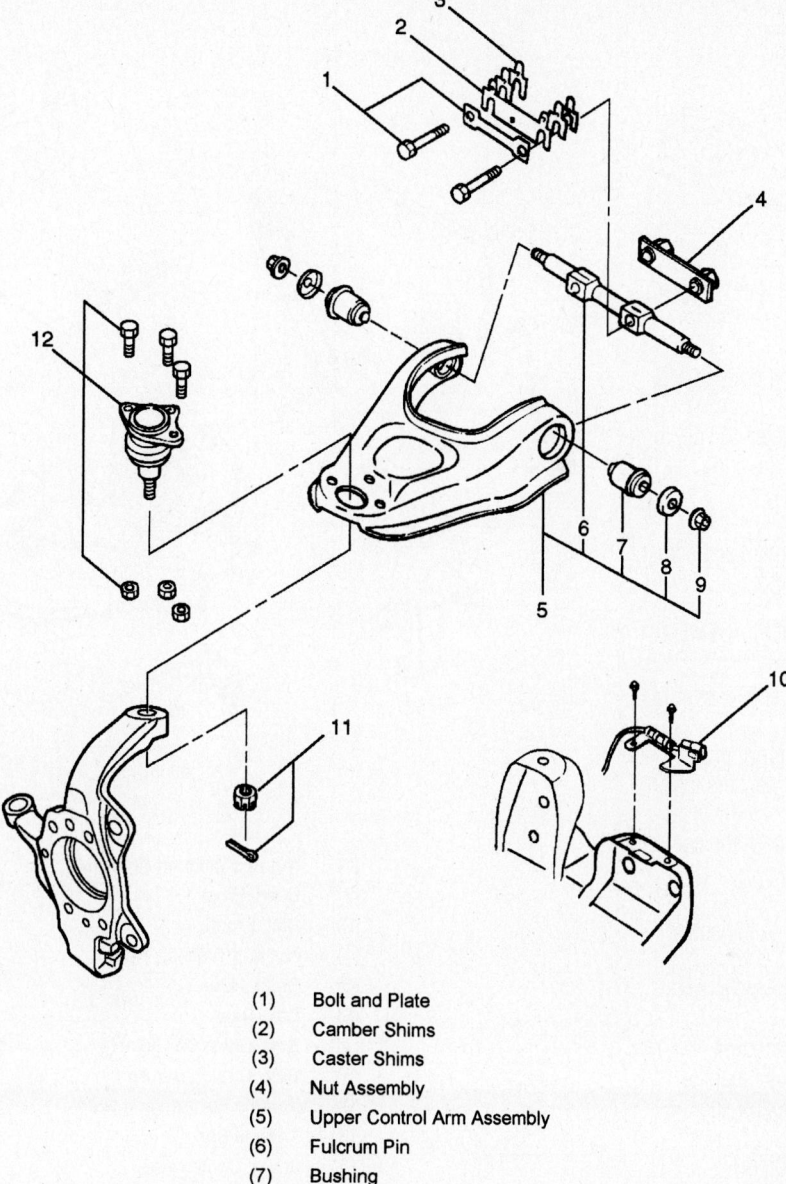

(1)	Bolt and Plate
(2)	Camber Shims
(3)	Caster Shims
(4)	Nut Assembly
(5)	Upper Control Arm Assembly
(6)	Fulcrum Pin
(7)	Bushing
(8)	Plate
(9)	Nut
(10)	Speed Sensor Cable
(11)	Nut and Cotter Pin
(12)	Upper Ball Joint, Bolt and Nut

09474_RODE_G0040

Upper control arm and related parts— Rodeo, Rodeo Sport and Passport

0.79 in. (20mm), then torque the nuts to 80 ft. lbs. (225 Nm).

6. Position the fulcrum pin with the projections inward.

7. Install the alignment shims in their original locations. Tighten the bolts to 112 ft. lbs. (152 Nm).

8. Install the upper ball joint. Tighten the nut to 72 ft. lbs. (98 Nm). Never back off the nut to align the cotter pin holes. Always use a new cotter pin.

9. Install or connect the following:

- Brake caliper
- Wheel speed sensor
- Front wheel

10. Check the wheel alignment and adjust as necessary.

Rodeo, Rodeo Sport and Passport

1. Before servicing the vehicle, refer to the Precautions Section.

2. Support the lower control arm with a jackstand.

3. Remove or disconnect the following:
- Front wheel
- Wheel speed sensor
- Upper ball joint
- Upper control arm

➡**Note the alignment shim location for assembly.**

To install:

4. Install or connect the following:
- Upper control arm
- Alignment shims in their original

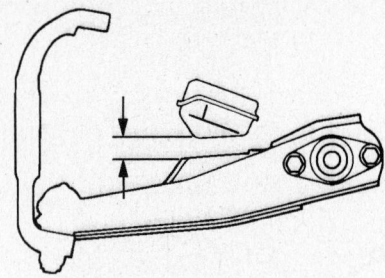

Bump stop clearance

locations. Tighten the bolts to 112 ft. lbs. (152 Nm).
- Upper ball joint. Tighten the nut to 72 ft. lbs. (98 Nm).
- Wheel speed sensor
- Front wheel

5. Check the wheel alignment and adjust as necessary.

CONTROL ARM BUSHING REPLACEMENT

1. Before servicing the vehicle, refer to the Precautions Section.
2. Remove the upper control arm.
3. Remove the nuts and washers from the fulcrum pin.
4. Press the bushings out of the control arm.

To install:

5. Press the bushings into the control arm.
6. Install the fulcrum pin washers and nuts.
7. Install the upper control arm.
8. Raise the suspension so that the clearance between the bump stop and the lower control arm is:
- Trooper: 0.79 in. (20mm)
- 2002–03 Passport, Rodeo and Rodeo Sport: 0.87 in. (22mm)
- 2004 without Intelligent Suspension: 0.59 in. (15mm)
- 2004 with Intelligent Suspension: 0.83 in. (21mm)

9. Tighten the fulcrum pin nuts to 80 ft. lbs. (108 Nm).
10. Check the wheel alignment and adjust as necessary.

Lower Control Arm

REMOVAL & INSTALLATION

Trooper

1. Before servicing the vehicle, refer to the Precautions Section.
2. Remove the wheel.

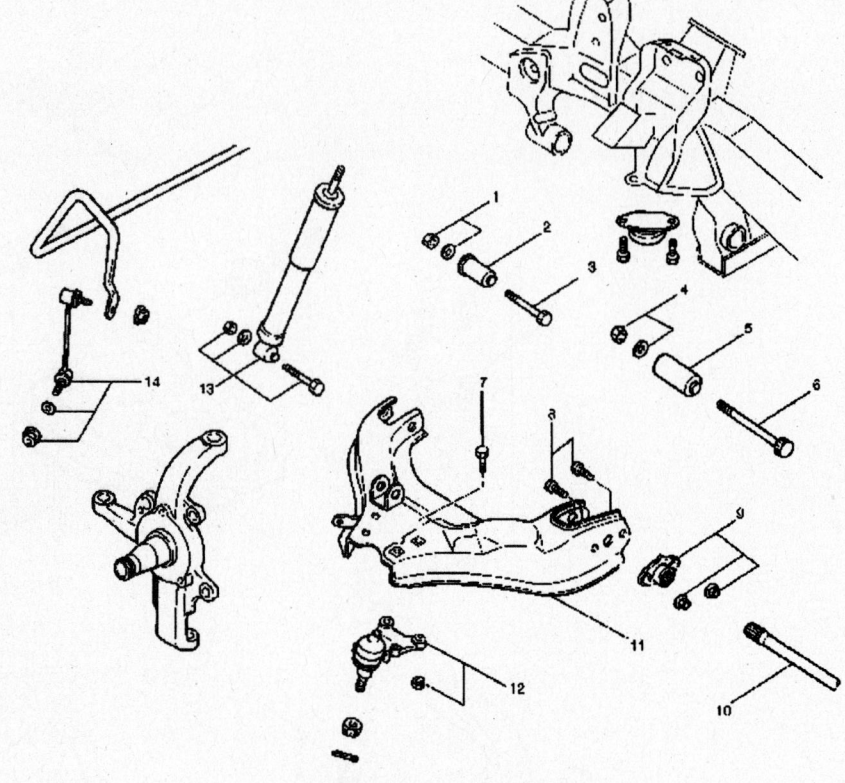

(1)	Nut and Washer, Front
(2)	Bush, Front
(3)	Bolt, Front
(4)	Nut and Washer, Rear
(5)	Bush, Rear
(6)	Bolt, Rear
(7)	Bolt, Lower Ball Joint
(8)	Bolt, Torsion Bar Arm
(9)	Torsion Bar Arm and Nut
(10)	Torsion Bar
(11)	Lower Control Arm
(12)	Lower Ball Joint and Nut
(13)	Shock Absorber, Bolt Washer and Nut
(14)	Stabilizer Link, Washer and Nut

Lower control arm and related parts—Trooper

3. Disconnect the tie rod at the knuckle.
4. Remove the halfshaft from the knuckle. See Halfshaft R&I.
5. Support the lower control arm with a jack.
6. Remove the control arm-to-frame nuts.
7. Remove the torsion bar.
8. Remove the torsion bar arm and bracket.
9. Disconnect the stabilizer link at the lower arm.
10. Disconnect the shock absorber from the arm.

11. Remove the cotter pin and castellated nut then, using a screw-type forcing tool, separate the ball joint from the knuckle. Discard the cotter pin.
12. Remove the control arm-to-frame bolts.
13. Remove the arm.

To install:

14. Install the arm.
15. Install the control arm-to-frame bolts.
16. Install the ball joint in the knuckle. Torque to 76 ft. lbs. (103 Nm). Use a new cotter pin.

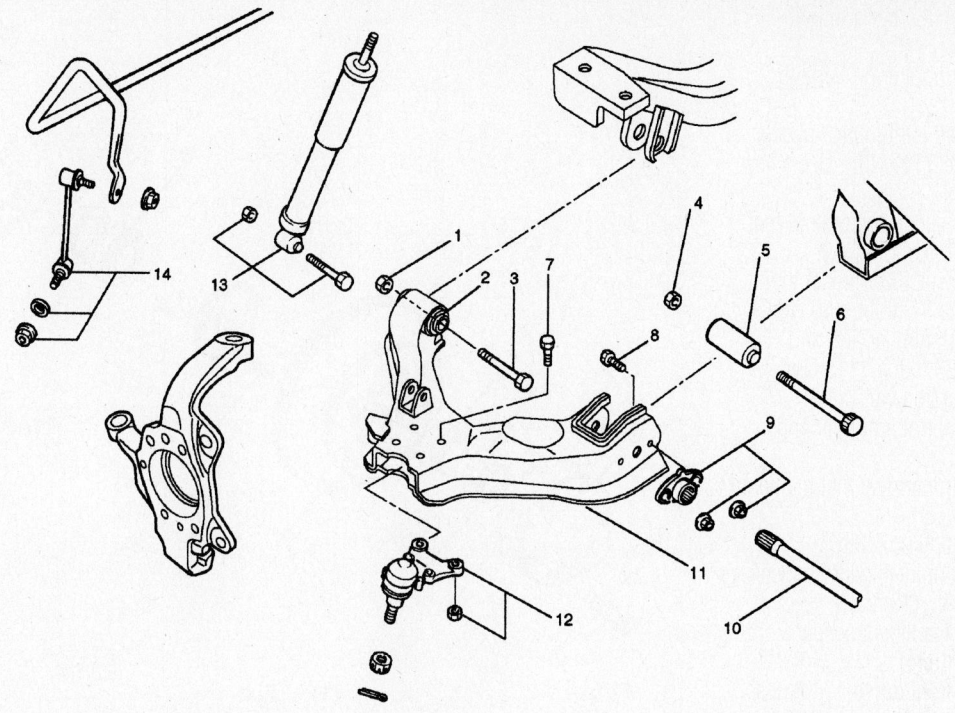

(1)	Nut, Front
(2)	Bush, Front
(3)	Bolt, Front
(4)	Nut, Rear
(5)	Bush, Rear
(6)	Bolt, Rear
(7)	Bolt, Lower Ball Joint
(8)	Bolt, Torsion Bar Arm
(9)	Torsion Bar Arm Bracket
(10)	Torsion Bar
(11)	Lower Control Arm
(12)	Lower Ball Joint and Nut
(13)	Shock Absorber
(14)	Stabilizer Link, Washer and Nut

09474_RODE_G0042

Lower control arm and related parts—Passport, Rodeo, and Rodeo Sport

17. Connect the shock absorber at the arm. Torque to 61 ft. lbs. (82 Nm).

18. Connect the stabilizer link at the lower arm. Torque to 37 ft. lbs. (50 Nm).

19. Install the torsion bar arm and bracket. Torque to 85 ft. lbs. (115 Nm).

20. Install the torsion bar.

21. Install the control arm nuts finger tight.

22. Set the control arm so that the bump stop clearance is 0.79 in. (20mm).

23. Tighten the front nut to 116 ft. lbs. (157 Nm). Tighten the rear nut to 145 ft. lbs. (196 Nm).

24. Adjust the alignment.

Passport, Rodeo, and Rodeo Sport

1. Before servicing the vehicle, refer to the Precautions Section.

2. Remove the wheel.

3. Disconnect the tie rod at the knuckle.

4. Remove the halfshaft from the knuckle. See Halfshaft R&I.

5. Support the lower control arm with a jack.

6. Remove the cotter pin and castellated nut then, using a screw-type forcing tool, separate the ball joint from the knuckle. Discard the cotter pin.

7. Remove the control arm-to-frame nuts.

8. Remove the torsion bar.

9. Remove the torsion bar arm and bracket.

10. Disconnect the stabilizer link at the lower arm.

11. Disconnect the shock absorber from the arm.

12. Remove the lower ball joint from the arm.

13. Remove the control arm-to-frame bolts.

14. Remove the arm.

To install:

15. Install the lower ball joint in the arm. Torque to 85 ft. lbs. (115 Nm).

16. Install the torsion bar arm bolt.

17. Install the arm.

18. Install the control arm-to-frame bolts.

19. Install the ball joint in the knuckle. Torque to 108 ft. lbs. (147 Nm). Use a new cotter pin.

20. Connect the shock absorber at the arm. Torque to 69 ft. lbs. (93 Nm).

21. Connect the stabilizer link at the lower arm. Torque to 37 ft. lbs. (50 Nm).

22. Install the torsion bar arm and bracket. Torque to 85 ft. lbs. (115 Nm).

23. Install the torsion bar.

24. Install the control arm nuts finger tight.

25. Set the control arm so that the bump stop clearance is:
- 2002–03: 0.87 in. (22 mm)
- 2004 without Intelligent Suspension: 0.59 in. (15mm)
- 2004 with intelligent suspension: 0.83 in. (21mm)

26. Tighten the front nut to 174 ft. lbs. (235 Nm). Tighten the rear nut to 137 ft. lbs. (186 Nm).

27. Adjust the alignment.

CONTROL ARM BUSHING REPLACEMENT

1. Before servicing the vehicle, refer to the Precautions Section.

2. Remove or disconnect the following:
- Lower control arm
- Bushings, press them from the control arm, using Remover/Installer J-36833 for the front bushing and Remover/Installer J-36834 for the rear bushing.

To install:

3. Install or connect the following:
- New bushings
- Lower control arm

Front Wheel Bearings

ADJUSTMENT

2WD Trooper

1. Before servicing the vehicle, refer to the Precautions Section.

2. Remove or disconnect the following:
- Front wheel
- Brake caliper and pads
- Hub dust cap
- Snapring and shim
- Hub flange
- Lockscrew and washer

3. Tighten the hub nut to 22 ft. lbs. (29

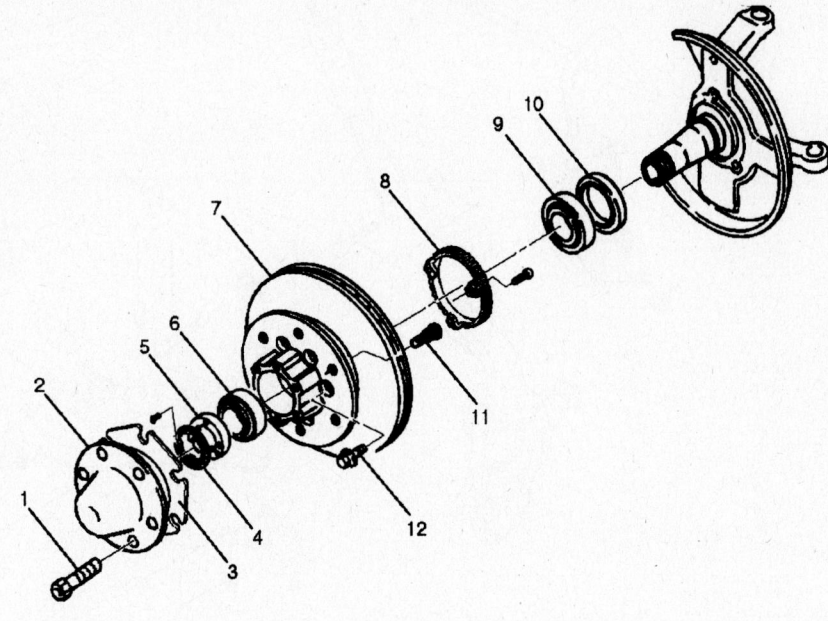

(1)	Bolt
(2)	Cover and Gasket
(3)	Packing
(4)	Lock Washer
(5)	Hub Nut
(6)	Outer Bearing
(7)	Hub and Disc Assembly
(8)	ABS Sensor Ring
(9)	Inner Bearing
(10)	Oil Seal
(11)	Bolt
(12)	Wheel Pin

09474_RODE_G0078

Front wheel hub and bearings—2wd Trooper

Nm) to seat the bearings and then fully loosen the nut.

4. Tighten the hub nut to achieve a bearing preload of 2.6–4.0 lbs. (1.2–1.8 Kg) for used bearings. If the bearings were replaced, set the preload to 4.4–5.5 lbs. (2.0–2.5 Kg).

5. Install or connect the following:
- Lockwasher and screw
- Hub flange
- Shim and snapring
- Hub dust cap. Tighten the bolts to 43 ft. lbs. (59 Nm).
- Brake caliper and pads
- Front wheel

Passport, Rodeo, Rodeo Sport

The front wheel bearings are not adjustable.

REMOVAL & INSTALLATION

Trooper

1. Before servicing the vehicle, refer to the Precautions Section.

2. Remove or disconnect the following:
- Front wheel
- Brake caliper and support
- Hub dust cap
- Snapring and shim
- Hub flange
- Lockscrew and washer
- Hub nut
- Brake rotor and hub assembly
- Wheel speed sensor ring
- Outer bearing
- Grease seal
- Inner bearing

To install:

3. Clean and inspect the bearings. Replace if necessary.

4. Apply clean wheel bearing grease to the inner and outer bearings.

5. Apply grease in the hub.

6. Install the wheel bearings into the hub along with a new grease seal.

7. Install or connect the following:
- Wheel speed sensor ring. Tighten the bolts to 13 ft. lbs. (18 Nm).
- Brake rotor and hub assembly
- Hub nut. Set the bearing preload.
- Lockscrew and washer
- Hub flange
- Snapring and shim
- Hub dust cap
- Brake caliper and support. Tighten the support bolts to 115 ft. lbs. (155 Nm).
- Front wheel

Passport, Rodeo and Rodeo Sport

2-WHEEL DRIVE

1. Before servicing the vehicle, refer to the Precautions Section.

2. Raise and support the front end.

3. Remove the caliper from the anchor plate and suspend it out of the way.

4. Disconnect the ABS sensor.

5. Remove the rotor.

6. Remove the bolts attaching the hub to the knuckle.

7. Installation is the reverse of removal. Torque the hub bolts to 76 ft. lbs. (103 Nm).

4-WHEEL DRIVE

1. Before servicing the vehicle, refer to the Precautions Section.

2. Place the transfer case in 2WD.

3. Raise and support the front end.

4. Remove the caliper from the anchor plate and suspend it out of the way.

5. Remove the rotor.

6. Loosen the caulking around the front driveshaft nuts. Remove the nuts and discard them.

7. Remove the 4 hub retaining bolts.

➡**If the hub is difficult to remove, install 2 long bolt in the bolt holes and strike them to loosen the hub.**

8. Installation is the reverse of removal. Torque the hub bolts to 76 ft. lbs. (103 Nm). Use new driveshaft nuts, torquing them to 181 ft. lbs. (245 Nm). Apply new caulking. The caulk should be free of cracks when applied.

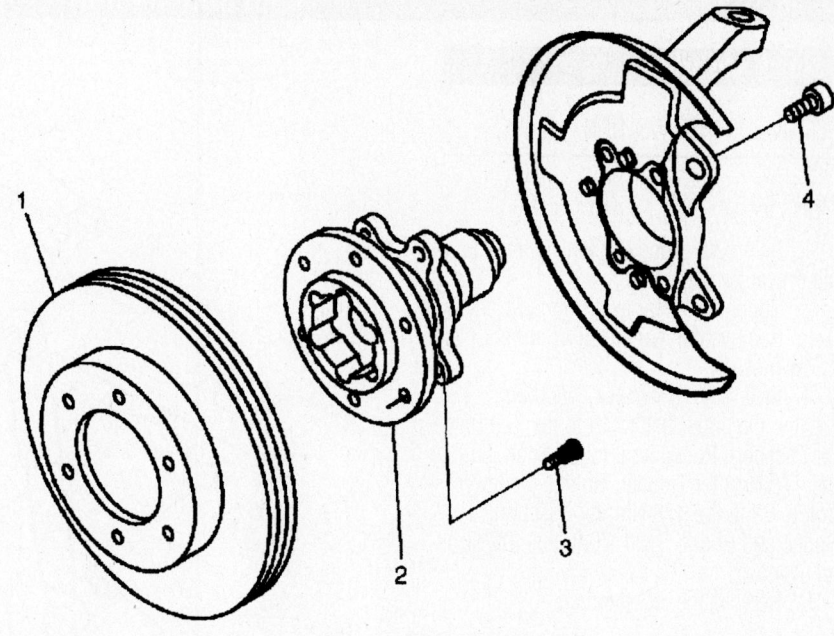

(1)	Disc Rotor
(2)	Hub unit, Bearing
(3)	Wheel Pin
(4)	Bolt

09474_RODE_G0043

Front hub—2wd Rodeo, Rodeo Sport and Passport

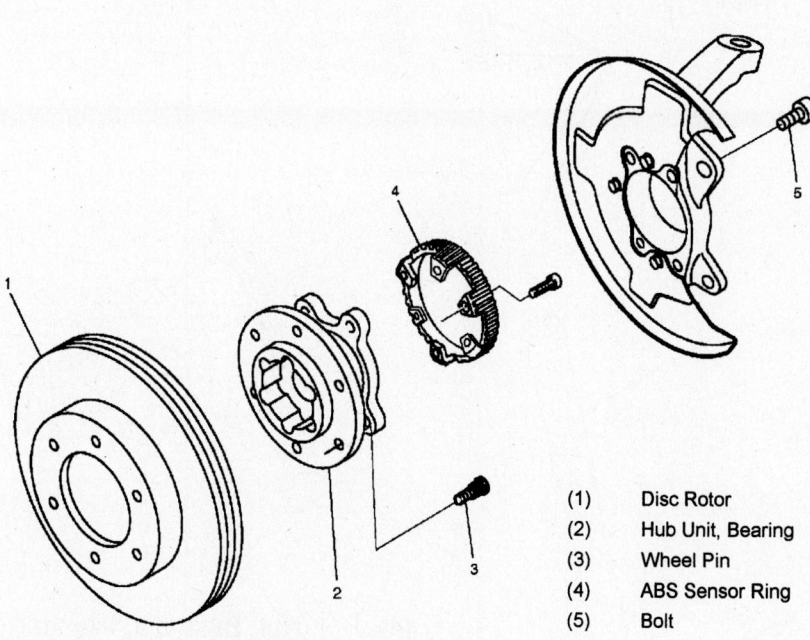

(1)	Disc Rotor
(2)	Hub Unit, Bearing
(3)	Wheel Pin
(4)	ABS Sensor Ring
(5)	Bolt

09474_RODE_G0044

Front hub—4wd Rodeo, Rodeo Sport and Passport

REAR SUSPENSION

Shock Absorber

REMOVAL & INSTALLATION

2002-03

1. Before servicing the vehicle, refer to the Precautions Section.
2. Support the rear axle with jackstands.
3. Remove the rear shock absorbers.

To install:

4. Install the rear shock absorbers. Tighten the upper bolt to 14 ft. lbs. (20 Nm) for Passport, Rodeo and Rodeo Sport; 58 ft. lbs. (78 Nm) for Trooper. Tighten the lower bolt to 58 ft. lbs. (78 Nm) for Passport, Rodeo and Rodeo Sport; 70 ft. lbs. (95 Nm) for Trooper.
5. Remove the jackstands.

2004 Without Intelligent Suspension

1. Before servicing the vehicle, refer to the Precautions Section.
2. Support the rear axle with jackstands.
3. Remove the rear shock absorbers.

To install:

4. Install the rear shock absorbers. Tighten the upper bolt to 14 ft. lbs. (20

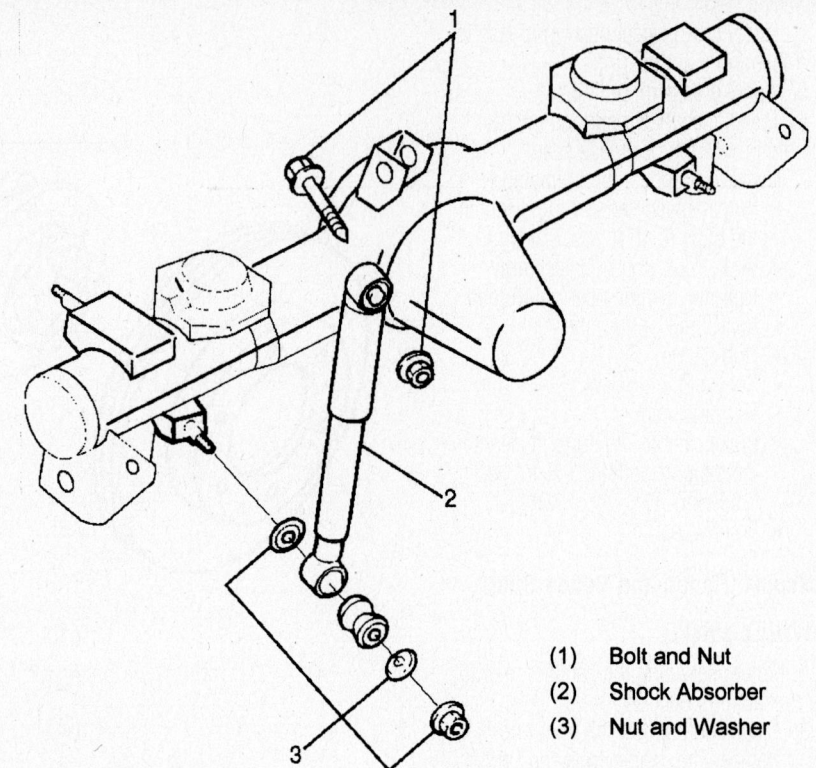

(1) Bolt and Nut
(2) Shock Absorber
(3) Nut and Washer

09474_RODE_G0073

Trooper rear shock absorber

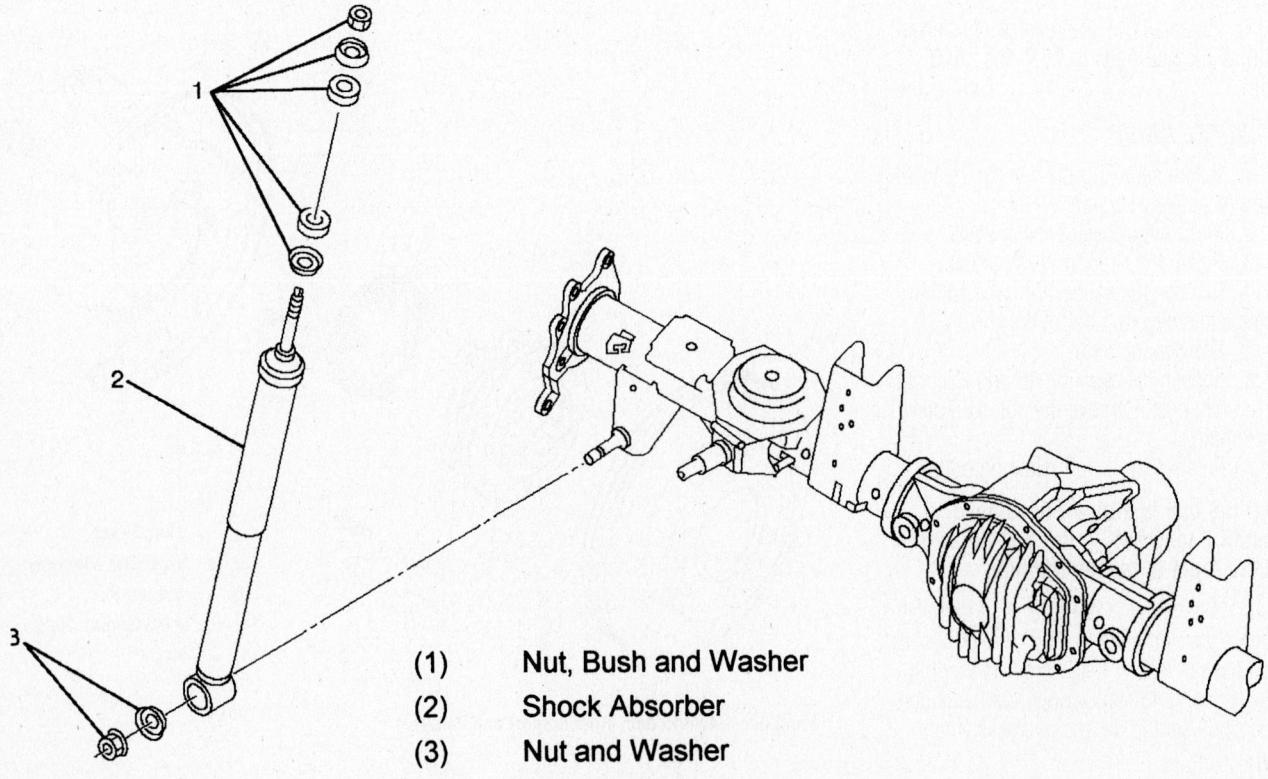

(1) Nut, Bush and Washer
(2) Shock Absorber
(3) Nut and Washer

09474_RODE_G0030

Passport, Rodeo and Rodeo Sport rear shock absorber, without Intelligent Suspension

LH

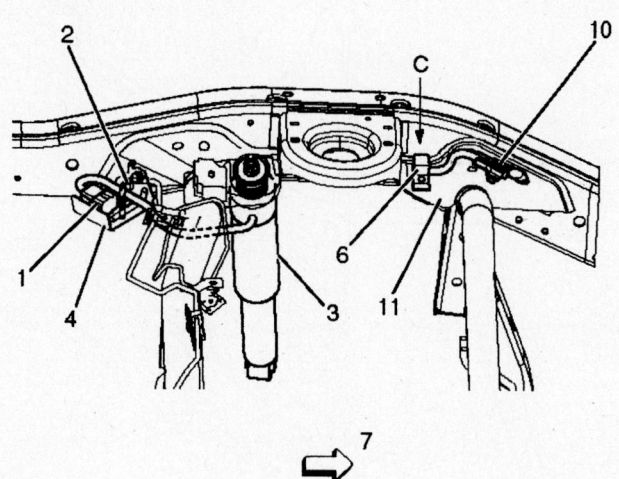

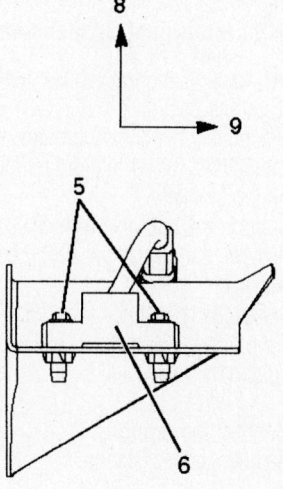

Detail C

RH

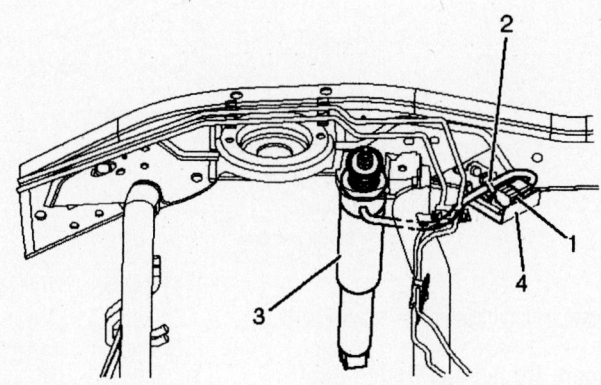

(1)	Connector
(2)	Clip
(3)	Rear Shock Absorber
(4)	Bracket
(5)	Bolt
(6)	G-Sensor
(7)	Front
(8)	Upper
(9)	Right
(10)	Connector
(11)	Gusset

09474_RODE_G0031

Rear shock absorber with Intelligent Suspension

Nm). Tighten the lower bolt to 58 ft. lbs. (78 Nm).

5. Remove the jackstands.

2004 With Intelligent Suspension

1. Before servicing the vehicle, refer to the Precautions Section.
2. Disconnect the battery ground.
3. Disconnect the wiring and remove the clip from the bracket.
4. Support the rear axle with jackstands.
5. Remove the rear shock absorbers.

To install:

6. Install the rear shock absorbers. Tighten the upper bolt to 14 ft. lbs. (20 Nm). Tighten the lower bolt to 58 ft. lbs. (78 Nm).
7. Connect the wiring.
8. Remove the jackstands.

Coil Spring

REMOVAL & INSTALLATION

Trooper

1. Before servicing the vehicle, refer to the Precautions Section.
2. Support the vehicle under the frame.
3. Support the rear axle with a jack.
4. Remove or disconnect the following:
   - Rear wheels
   - Parking brake cable at the trailing link
   - Stabilizer bar at the links
   - Shock absorbers
5. Lower the rear axle with the jack to release the coil spring tension. Remove the coil springs and insulators.

> ⚠ **WARNING**
>
> **Avoid stretching the brake hose, parking brake cable and breather hose.**

To install:

6. Place the coil springs on the axle assembly and the insulators on top of the springs. Note the position of the paint mark in the accompanying illustration.
7. Raise the axle assembly into position.
8. Install the shock absorber and just snug the nut. The nut should be tightened to 58 ft. lbs. (78 Nm) when the weight of the vehicle is on the suspension.
9. Connect the stabilizer bar at the links. Torque to 37 ft. lbs. (50 Nm).
10. The remainder of installation is the reverse of removal.

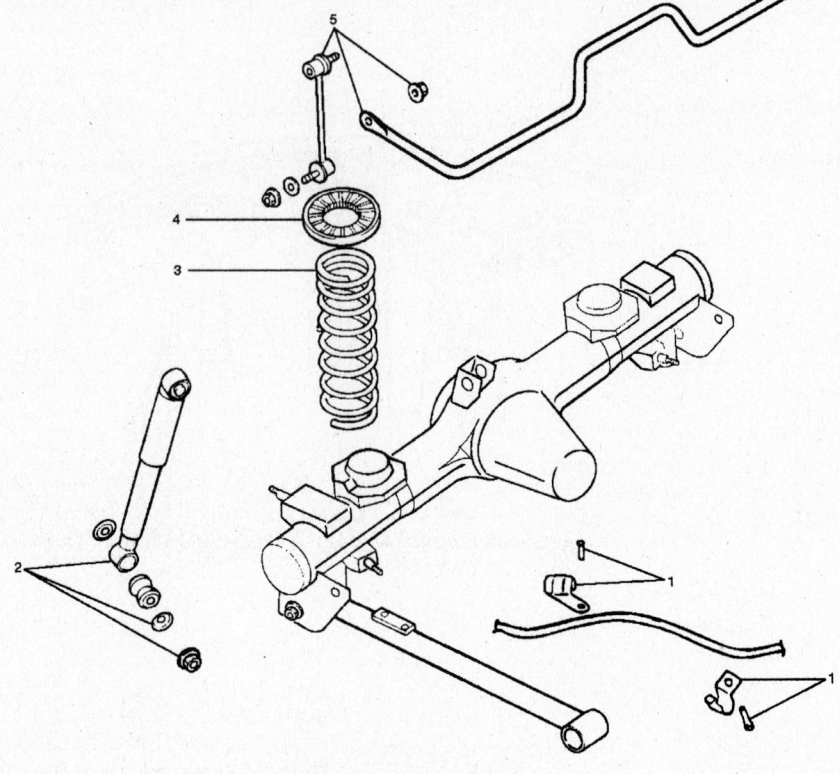

(1)	Parking Brake Cable Bracket
(2)	Shock Absorber
(3)	Coil Spring
(4)	Insulator
(5)	Stabilizer Bar

09474_RODE_G0074

Trooper rear coil spring and related parts

Passport, Rodeo and Rodeo Sport

2002–03

1. Before servicing the vehicle, refer to the Precautions Section.
2. Support the vehicle under the frame.
3. Support the rear axle with a jack.
4. Remove or disconnect the following:
   - Rear wheels
   - Brake hose at the crossmember
   - Upper link at the axle on the left side
   - Stabilizer bar at the links
   - Shock absorbers
5. Lower the rear axle with the jack to release the coil spring tension. Remove the coil springs and insulators.

To install:

6. Place the coil springs on the axle assembly and the insulators on top of the springs. Note the position of the paint mark in the accompanying illustration.
7. Raise the axle assembly into position.
8. Install the shock absorber and just snug the nut. The nut should be tightened to 58 ft. lbs. (78 Nm) when the weight of the vehicle is on the suspension.
9. Connect the stabilizer bar at the links. Torque to 23 ft. lbs. (31 Nm).
10. Connect the upper link. Torque to 101 ft. lbs. (137 Nm).
11. Connect the brake hose. Bleed the brakes.

2004

1. Before servicing the vehicle, refer to the Precautions Section.
2. Support the vehicle under the frame.
3. Support the rear axle with a jack.
4. Remove or disconnect the following:
   - Rear wheels
   - Brake hose at the crossmember
   - Upper link at the axle on the left side
   - Stabilizer bar at the links
   - Shock absorbers
5. Lower the rear axle with the jack to release the coil spring tension. Remove the coil springs and insulators.

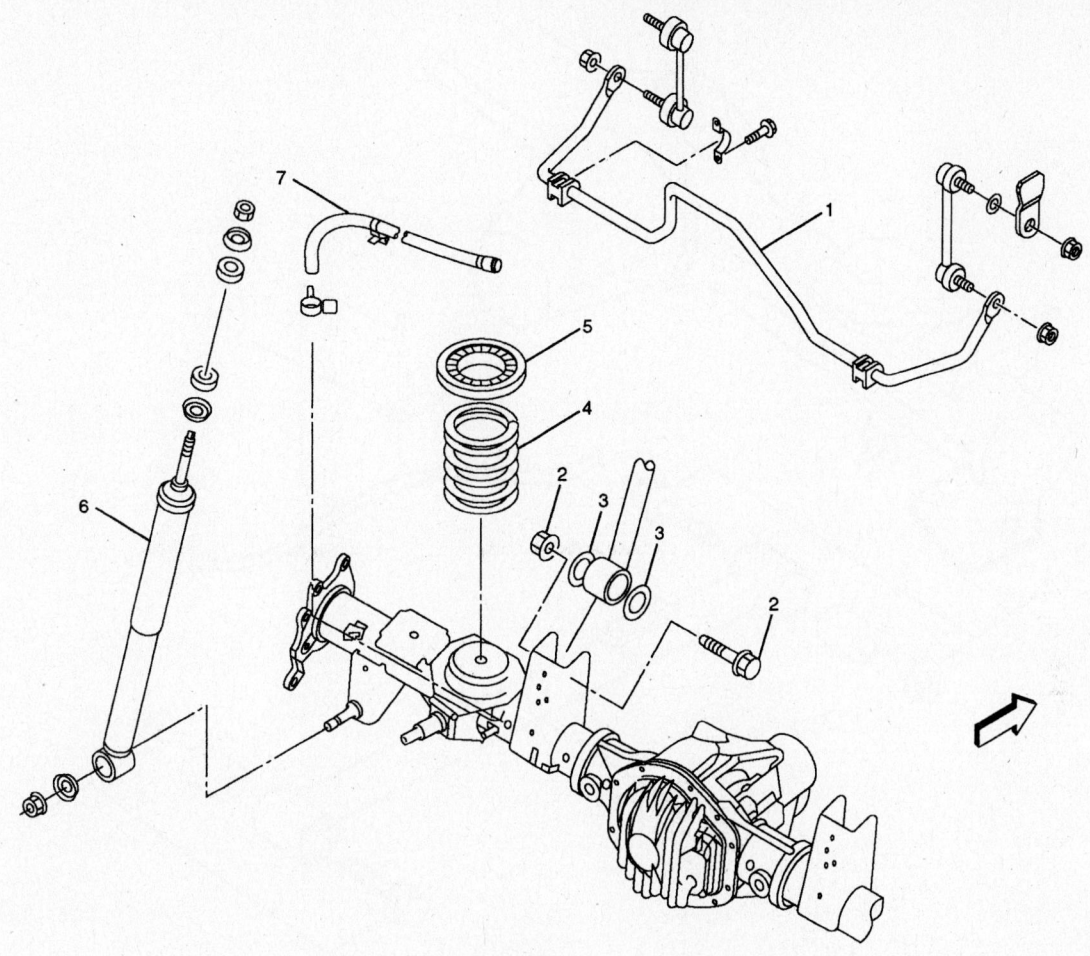

(1)   Stabilizer Bar

(2)   Upper Link Fixing Bolt and Nut

(3)   Rubber Plate

(4)   Coil Spring

(5)   Insulator

(6)   Shock Absorber

(7)   Breather Hose

09474_RODE_G0032

Passport, Rodeo and Rodeo Sport rear coil spring and related parts

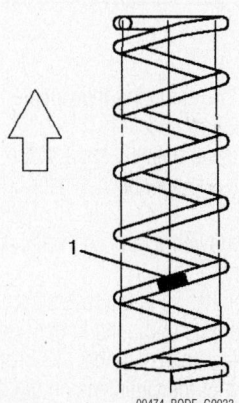

09474_RODE_G0033

Correct coil spring installation. 1 is the paint mark

To install:

6. Place the coil springs on the axle assembly and the insulators on top of the springs. Note the position of the paint mark in the accompanying illustration.

7. Raise the axle assembly into position.

8. Install the shock absorber and just snug the nut. The nut should be tightened to 58 ft. lbs. (78 Nm) when the weight of the vehicle is on the suspension.

9. Connect the stabilizer bar at the links. Torque to 37 ft. lbs. (50 Nm).

10. Connect the upper link. Torque to 101 ft. lbs. (137 Nm).

11. Connect the brake hose. Bleed the brakes.

Trailing Link

REMOVAL & INSTALLATION

1. Before servicing the vehicle, refer to the Precautions Section.

2. Raise and safely support the vehicle.

3. Remove the parking brake cable from the link.

4. Remove the nuts and bolts and remove the link.

5. Installation is the reverse of removal. Install the nuts and bolts finger tight, then tighten them to 101 ft. lbs. (137 Nm) when the weight of the vehicle is on the suspension.

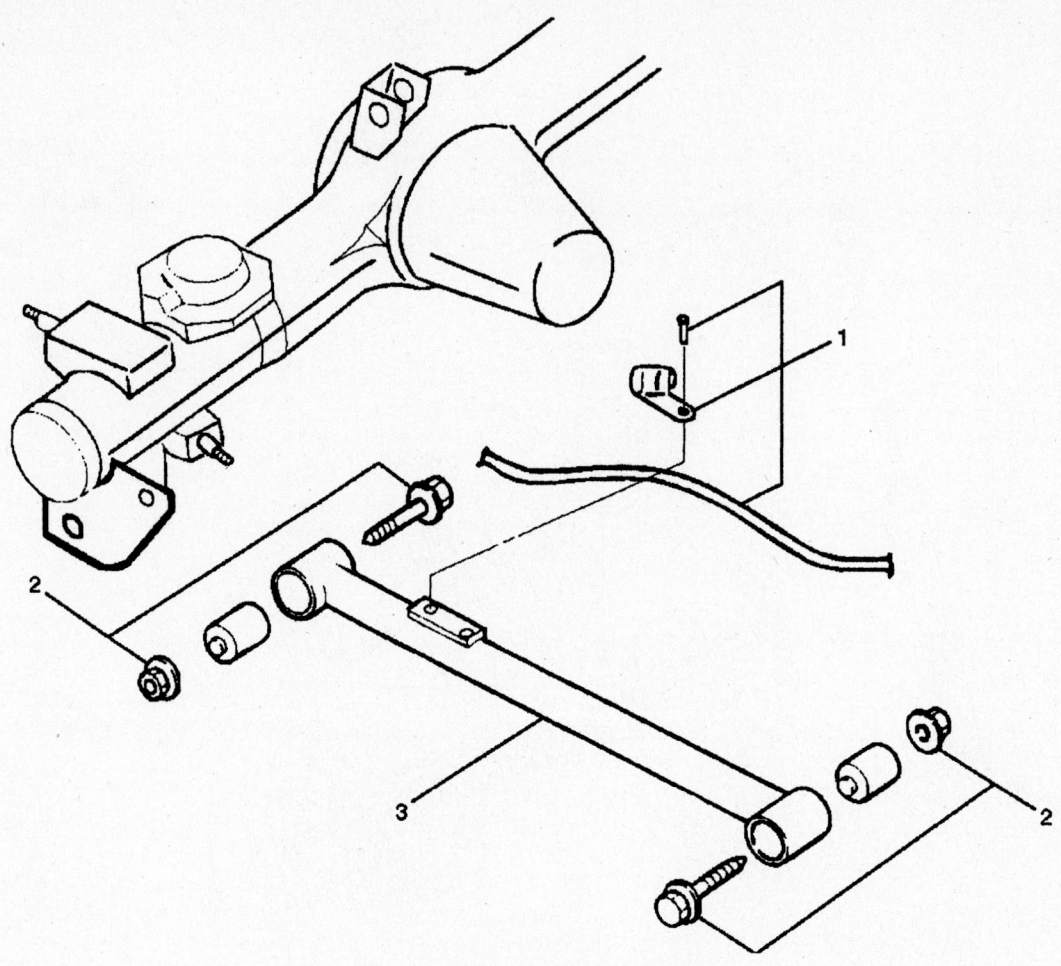

(1) Parking Brake Cable

(2) Bolt and Nut

(3) Trailing Link

09474_RODE_G0081

Trailing link—Trooper

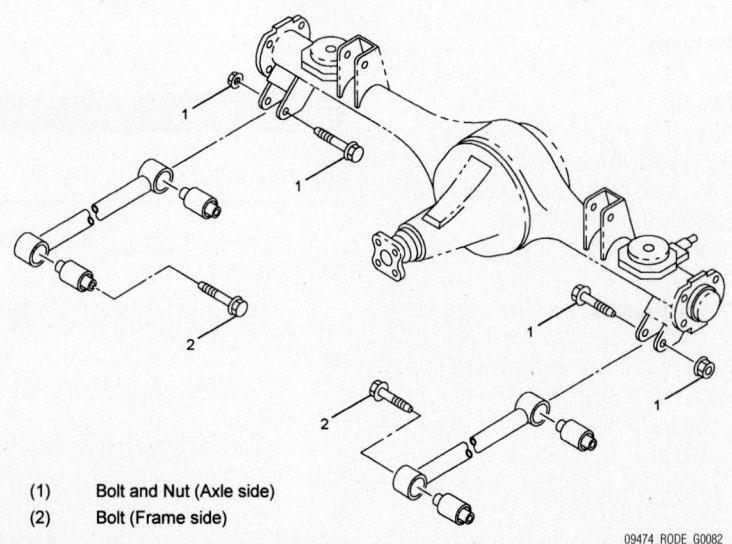

(1) Bolt and Nut (Axle side)

(2) Bolt (Frame side)

09474_RODE_G0082

Trailing link—Passport, Rodeo and Rode Sport

Center Link

REMOVAL & INSTALLATION

Trooper

1. Before servicing the vehicle, refer to the Precautions Section.

2. Raise and safely support the vehicle.

3. Remove the speed sensor cable from the link.

4. Remove the nuts and bolts and remove the link.

5. Installation is the reverse of removal. Install the nuts and bolts finger tight, then tighten them to 101 ft. lbs. (137 Nm) when the weight of the vehicle is on the suspension.

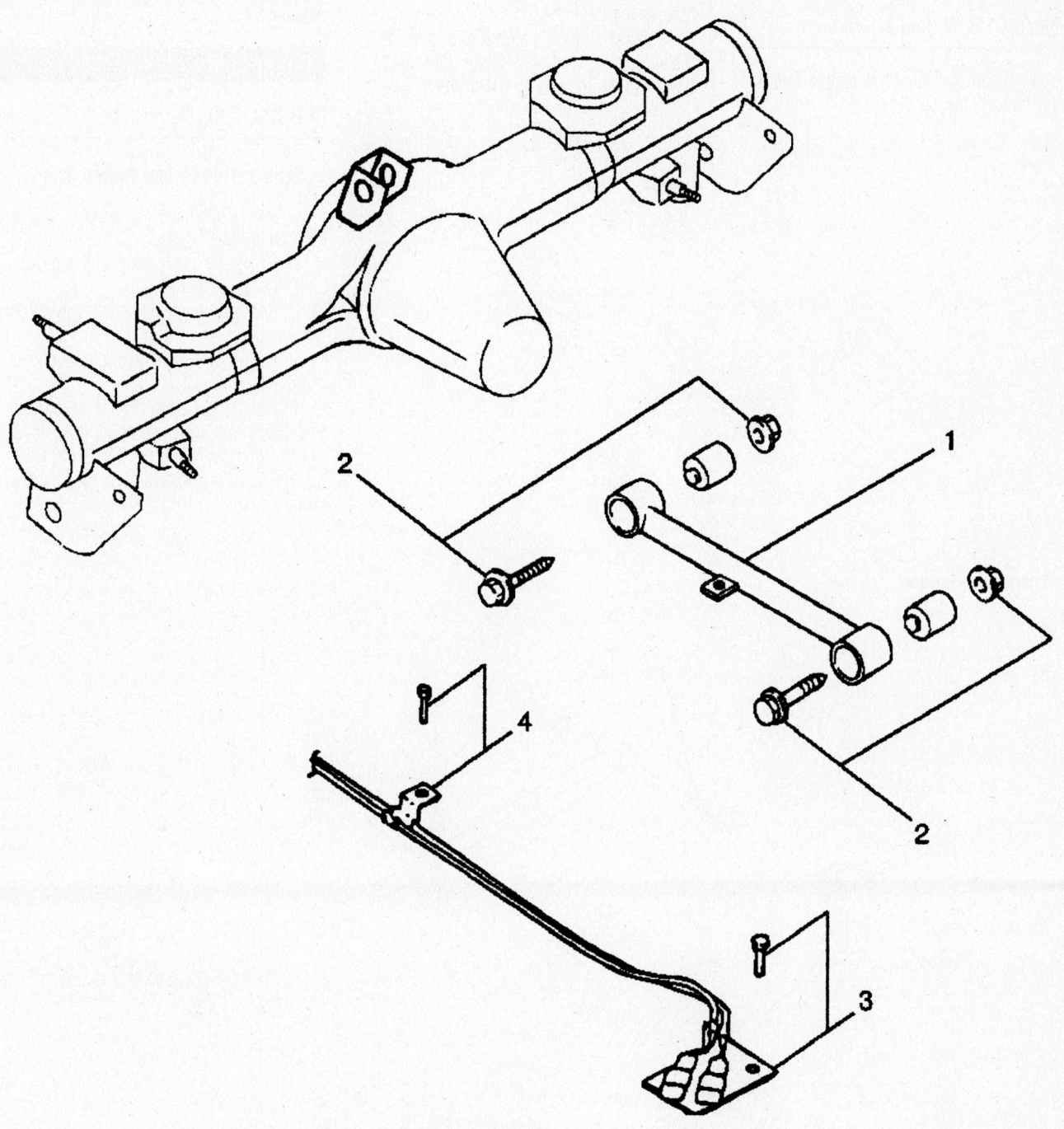

(1)   Center Link
(2)   Bolt and Nut
(3)   Speed Sensor Cable Bracket
(4)   Speed Sensor Cable

09474_RODE_G0083

Center link—Trooper

Lateral Rod

REMOVAL & INSTALLATION

1. Before servicing the vehicle, refer to the Precautions Section.

2. Raise and safely support the vehicle.

3. Remove the nuts and bolts and remove the rod.

4. Installation is the reverse of removal. Install the nuts and bolts finger tight, then tighten the frame end to 101 ft. lbs. (137 Nm) and the axle end to 58 ft. lbs. (78 Nm) when the weight of the vehicle is on the suspension.

Upper Link

REMOVAL & INSTALLATION

Passport, Rodeo and Rodeo Sport

1. Before servicing the vehicle, refer to the Precautions Section.

2. Raise and safely support the vehicle.

3. Remove the fuel tank.

4. Remove the speed sensor cable from the link.

5. Remove the nuts and bolts and remove the link.

6. Installation is the reverse of removal. Install the nuts and bolts finger tight, then tighten them to 101 ft. lbs. (137 Nm) when the weight of the vehicle is on the suspension.

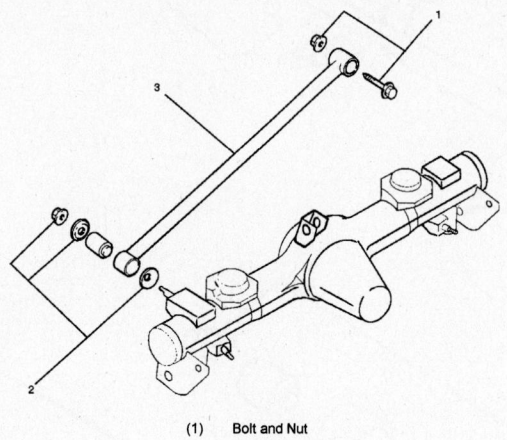

(1)   Bolt and Nut
(2)   Nut and Washer
(3)   Lateral Rod

09474_RODE_G0084

Lateral rod—Trooper

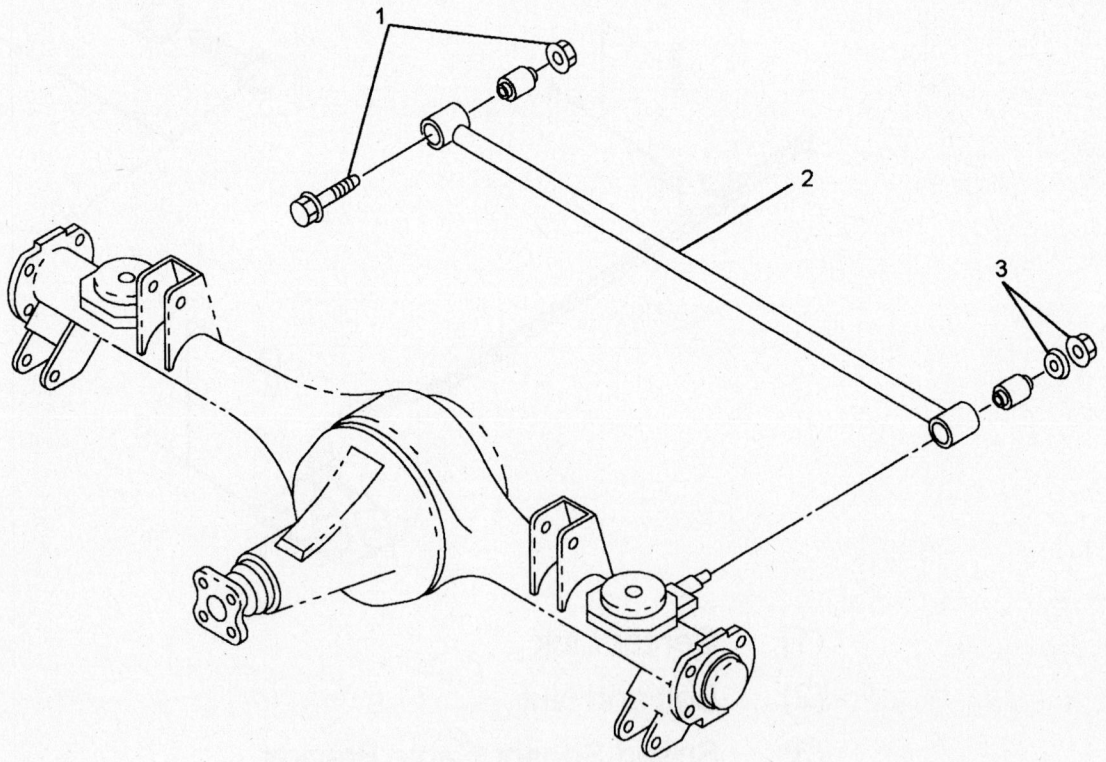

(1)   **Bolt and Nut (Frame side)**

(2)   **Lateral Rod**

(3)   **Nut and Washer (Axle side)**

09474_RODE_G0085

Lateral rod—Passport, Rodeo, and Rodeo Sport

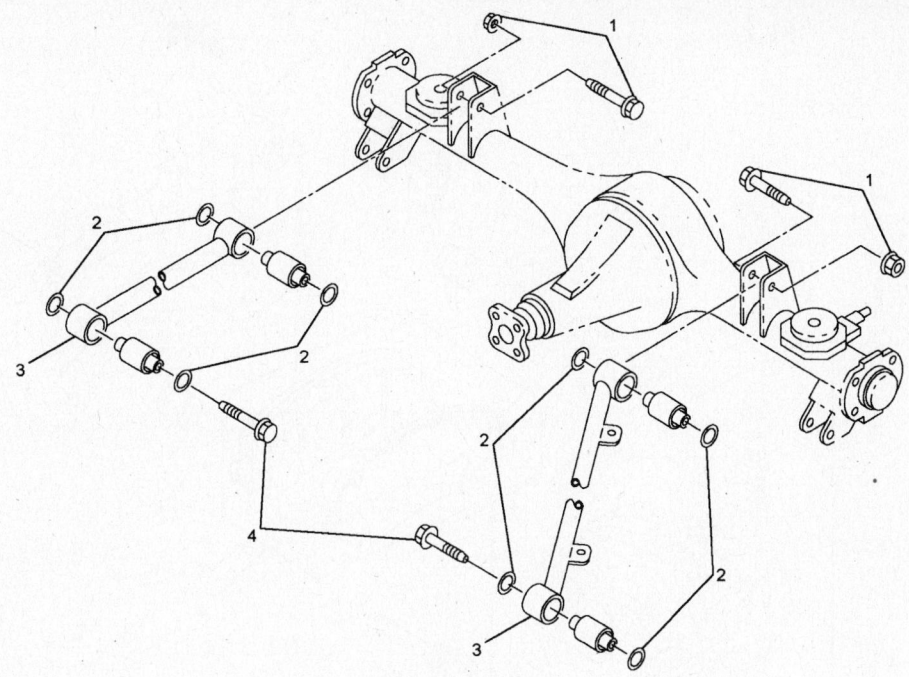

(1) Bolt and Nut (Axle side)
(2) Rubber Plate
(3) Upper Link
(4) Bolt (Frame side)

09474_RODE_G0086

Upper link—Passport, Rodeo, and Rodeo Sport

FRONT DISC BRAKES

Brake Caliper

REMOVAL & INSTALLATION

Trooper

1. Before servicing the vehicle, refer to the Precautions Section.
2. Raise and safely support the vehicle.
3. Remove some brake fluid from the reservoir.
4. Remove the front wheels.
5. Disconnect the brake fluid line from the caliper. Plug the line to prevent fluid loss.
6. Loosen the brake caliper pin bolts. Remove the caliper from the mount.
7. Remove the brake pads and clips from the caliper. Inspect the brake pads for wear and replace them if necessary.
To install:
8. Fill the brake caliper with clean brake fluid and connect the fluid line to the caliper using new washers. Tighten the brake line banjo fitting to 26 ft. lbs. (35 Nm). Install the brake pads and clips onto the caliper.
9. Install the caliper onto the mounting

bracket. Lubricate the caliper bolts and their boots. Then, install the caliper mounting bolts and tighten them to 54 ft. lbs. (74 Nm).
10. Refill and bleed the brake system.
11. Install the front wheels and lower the vehicle.

Passport, Rodeo and Rodeo Sport

1. Before servicing the vehicle, refer to the Precautions Section.
2. Raise and safely support the vehicle. Remove the front wheels.
3. Remove some brake fluid from the master cylinder reservoir.

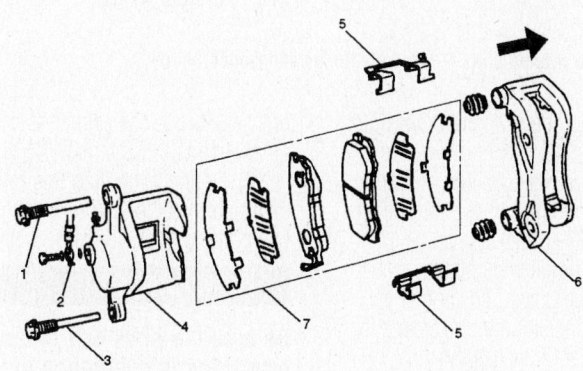

(1) Guide Bolt
(2) Brake Flexible Hose
(3) Lock Bolt
(4) Caliper Assembly
(5) Clip
(6) Support Bracket with Pad Assembly
(7) Pad Assembly

93026G02

Front caliper assembly—Trooper

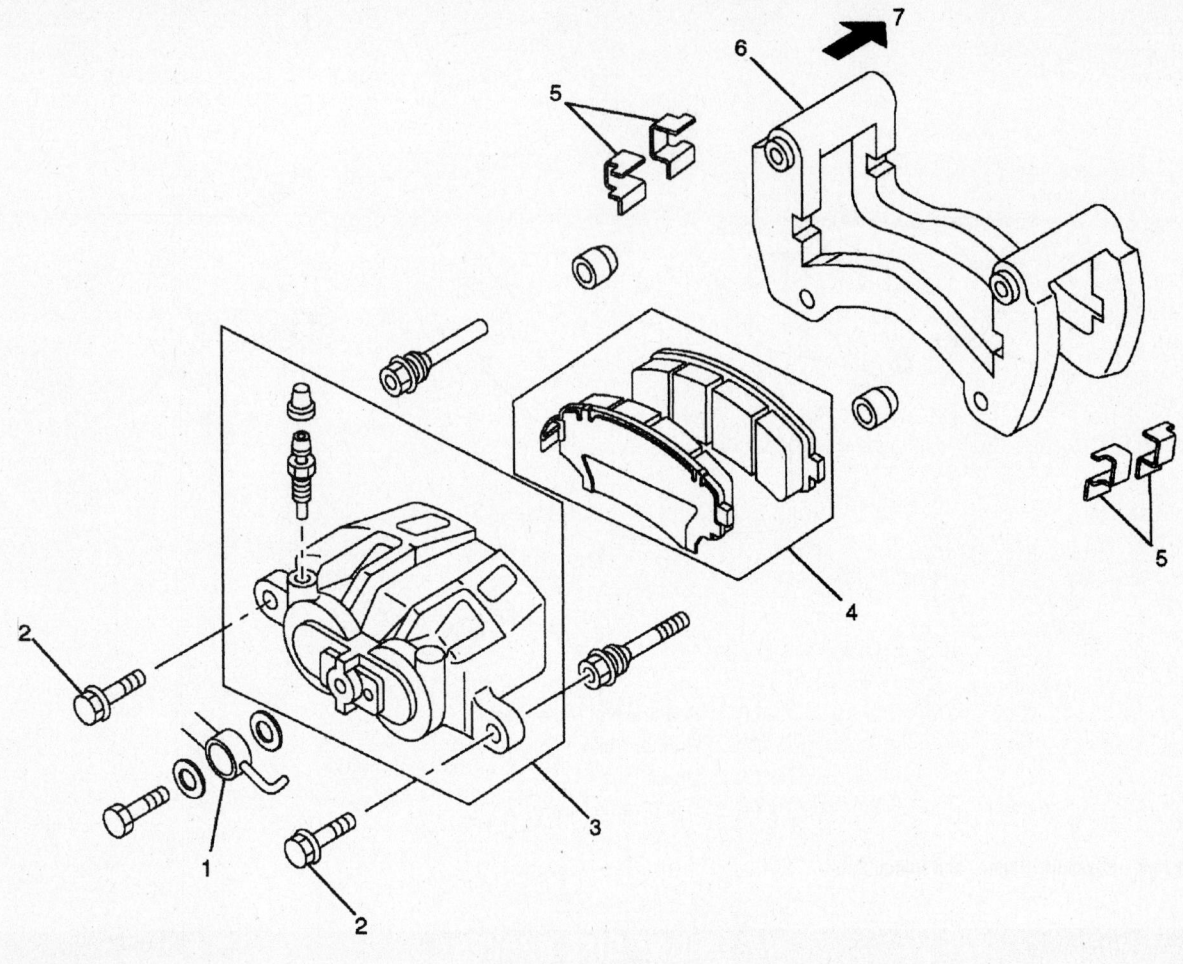

(1)   Brake Flexible Hose

(2)   Pin Bolt

(3)   Caliper Assembly

(4)   Pad Assembly

(5)   Clip

(6)   Support Bracket with Pad Assembly

(7)   Outer Side

09474_RODE_G0045

Front caliper and related parts—Passport, Rodeo and Rodeo Sport

4. Disconnect and plug the brake fluid line from the caliper.

5. Remove the brake caliper mounting bolt and guide bolt and remove the caliper from the mount. The brackets can be removed for additional work space.

6. Remove the brake pads and clips from the caliper. Inspect the brake pads for wear; replace them, if necessary.

To install:

7. Install the brake pads and clips onto the caliper.

8. If the caliper bracket was removed, tighten the bolts to 115 ft. lbs. (155 Nm).

9. Install the caliper on the mounting bracket. Torque the caliper-to-mounting bracket bolts to 33 ft. lbs. (45 Nm).

10. Connect the fluid line to the caliper using new washers. Torque the brake line banjo fitting to 26 ft. lbs. (35 Nm).

❋❋ WARNING

Be sure the hook end of the flexible brake line is positioned in the anti-rotation cavity.

11. Refill the master cylinder reservoir and bleed the brake system.

12. Install the front wheels and lower the vehicle.

Disc Brake Pads

REMOVAL & INSTALLATION

Passport, Rodeo and Rodeo Sport

Most disc brake pads are equipped with wear indicators. If a squealing noise occurs from the brakes while driving, check the pad wear indicator plate. If there is evidence of the indicator plate contacting the brake disc, the brake pad should be replaced.

1. Before servicing the vehicle, refer to the Precautions Section.

2. Remove ½ of the volume of brake fluid from the master cylinder to prevent overflow when the caliper piston is compressed.

3. Raise and safely support the vehicle.

4. Remove the wheel and tire assemblies.

5. Remove the brake caliper without disconnecting the brake line. Support the caliper with a length of wire. Do not let the caliper hang from the brake hose.

➡ **On some disc brake systems it is not necessary to remove the caliper when installing new brake pads. Remove the lower slide bolt and rotate the caliper upward to remove the pads.**

6. Remove the brake pads and shims. Inspect the brake rotor and machine or replace as necessary. Check the minimum thickness (specification is cast into the rotor) before machining.

To install:

7. Use a suitable tool to push the caliper piston into its bore.

8. Apply a thin coat of grease to the rear face of the brake pad and install the shim. Install the brake pads.

9. Install the calipers. Lubricate the caliper bolts and boots.

10. Install the wheel and tire assemblies and lower the vehicle.

11. Apply the brakes several times to seat the pads before moving the vehicle. Check the fluid in the master cylinder and add as necessary.

Trooper

1. Before servicing the vehicle, refer to the Precautions Section.

2. Remove about ½ of the brake fluid from the master cylinder reservoir to prevent overflow when the caliper piston is compressed.

3. Raise and safely support the vehicle.

4. Remove the front wheels.

5. Remove the brake caliper from the caliper bracket without disconnecting the brake line. Support the caliper with a length of wire. Do not let the caliper hang from the brake hose.

6. Remove the brake pads and shims. Inspect the brake rotor and machine or replace as necessary. Check the minimum thickness (specification is cast into the rotor) before machining.

To install:

7. Use a large C-clamp or brake piston tool to push the caliper piston into its bore.

8. Apply a thin coat of brake grease to both sides of both inner shims. Assemble the pads and shims, then install them into the caliper. The wear indicator on the inner pad must face down.

9. Install the calipers. Clean and lubricate the caliper mounting bolts and lubricate the mounting bolt boots. Install the mounting bolts and tighten them to 54 ft. lbs. (74 Nm).

10. Install the front wheels and lower the vehicle.

11. Apply the brakes several times to seat the pads before moving the vehicle. Check the fluid level in the master cylinder reservoir and add as necessary.

REAR DISC BRAKES

Brake Caliper

REMOVAL & INSTALLATION

Trooper

1. Before servicing the vehicle, refer to the Precautions Section.

2. Raise and safely support the vehicle.

3. Remove some brake fluid from the reservoir.

4. Remove the rear wheels.

5. Disconnect the brake fluid line from the caliper. Plug the line to prevent fluid loss.

6. Loosen the brake caliper mounting bolt and guide bolt. Remove the caliper from the mount bracket.

7. Remove the brake pads and clips from the caliper. Inspect the brake pads for wear; replace them if necessary.

8. If necessary for servicing, unbolt the caliper mounting bracket from the backing plate.

To install:

9. If removed, install the caliper mounting bracket and tighten its bolts to 76 ft. lbs. (103 Nm).

10. Fill the brake caliper with clean brake fluid and connect the fluid line to the caliper using new washers. Tighten the brake line banjo fitting to 26 ft. lbs. (35 Nm). Install the brake pads and clips onto the caliper.

11. Install the caliper on the mounting bracket. Lubricate the caliper bolts and their boots. Then, install the caliper mounting bolts. Tighten them to 32 ft. lbs. (44 Nm).

12. Refill and bleed the brake system.

13. Install the rear wheels and lower the vehicle.

Passport, Rodeo and Rodeo Sport

1. Before servicing the vehicle, refer to the Precautions Section.

2. Raise and safely support the vehicle. Remove the rear wheels.

3. Remove some fluid from the master cylinder reservoir.

4. Disconnect and plug the brake fluid line from the caliper.

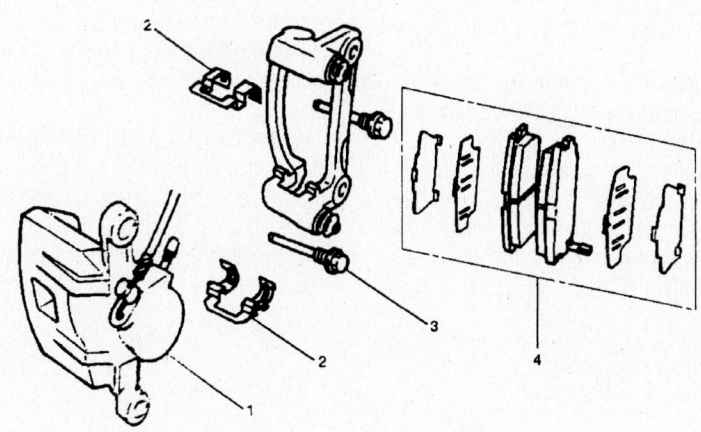

(1) Caliper Assembly
(2) Clip
(3) Lock Bolt
(4) Pad Assembly

Rear caliper assembly—Trooper

93026G01

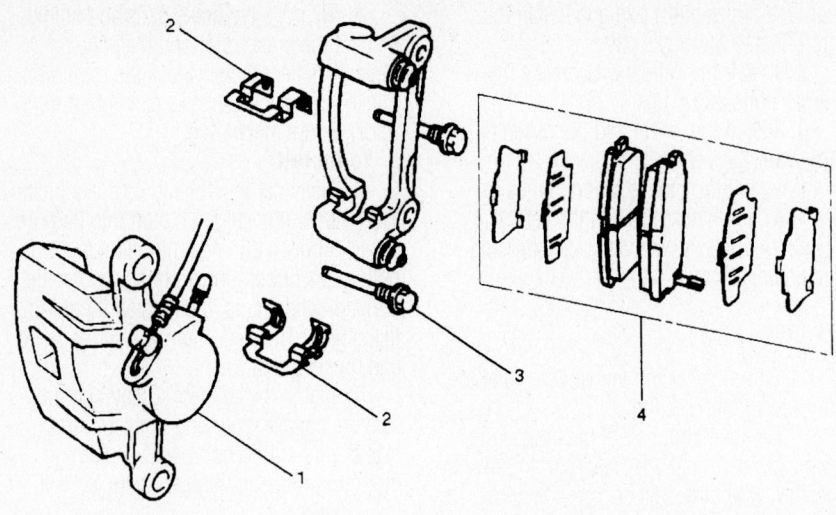

(1) Caliper Assembly
(2) Clip
(3) Lock Bolt
(4) Pad Assembly

93026G55

Exploded view of the rear caliper components—Rodeo, Rodeo Sport and Passport

➡**Discard the parking brake cable mounting pin after removal.**

5. If equipped with caliper actuated parking brakes; remove the mounting pin from the parking brake cable and disconnect the parking cable from the disc caliper.

6. Remove the brake caliper mounting bolt and guide bolt and remove the caliper from the mount.

7. Remove the brake pads and clips from the caliper. Inspect the brake pads for wear; replace them, if necessary.

To install:

8. Install the brake pads and clips onto the caliper.

9. If the mounting bracket was removed, tighten the bolts to 76 ft. lbs. (103 Nm).

10. Install the caliper on the mounting bracket. Torque the caliper-to-mounting bracket bolts to 32 ft. lbs. (43 Nm).

11. Connect the parking brake cable to the caliper and install a new mounting pin.

12. Connect the fluid line to the caliper using new washers. Torque the brake line banjo fitting to 26 ft. lbs. (35 Nm).

13. Refill the master cylinder reservoir and bleed the brake system.

14. Install the rear wheels and lower the vehicle.

Disc Brake Pads

REMOVAL & INSTALLATION

Passport, Rodeo and Rodeo Sport

1. Before servicing the vehicle, refer to the Precautions Section.

2. Raise and safely support the vehicle. Remove the rear wheels.

3. Remove the brake caliper mounting bolts and remove the caliper without disconnecting the brake fluid line. Support the caliper so it does not hang on the brake line.

4. Remove the brake pads and retaining clips from the caliper.

5. If equipped with caliper-activated parking brakes; use tool J-37617 or equivalent to rotate the piston clockwise until it retracts into the bore. Align the notches of the piston face so the centerline of the notches is perpendicular to the centerline of the mounting bosses.

To install:

6. Install the new brake pads and clips in the caliper and install the caliper in the mounting bracket.

7. Install the rear wheels. Check the brake fluid level.

8. Pump the brake pedal until pressure is felt before moving the vehicle.

Trooper

1. Before servicing the vehicle, refer to the Precautions Section.

2. Use a vacuum pump to remove some brake fluid from the master cylinder reservoir to prevent overflow when the caliper piston is compressed.

3. Raise and safely support the vehicle.

4. Remove the rear wheels.

5. Remove the brake caliper from the caliper bracket without disconnecting the brake line. Support the caliper with a length of wire. Do not let the caliper hang from the brake hose.

6. Remove the brake pads and shims. Inspect the brake rotor and machine or replace as necessary. Check the minimum thickness (specification is cast into the rotor) before machining.

To install:

7. Use a large C-clamp or brake piston tool to push the caliper piston into its bore.

8. Apply a thin coat of brake grease to both sides of both inner shims. Assemble the pads and shims, then install them into the caliper. The wear indicator on the inner pad must face down.

9. Install the calipers. Clean and lubricate the caliper mounting bolts and lubricate the mounting bolt boots. Install the mounting bolts and tighten them to 32 ft. lbs. (44 Nm).

10. Install the rear wheels and lower the vehicle.

11. Apply the brakes several times to seat the pads before moving the vehicle. Check the fluid level in the master cylinder reservoir and add as necessary.

DRUM BRAKES

Brake Drums

REMOVAL & INSTALLATION

Passport, Rodeo and Rodeo Sport

1. Before servicing the vehicle, refer to the Precautions Section.
2. Raise and safely support the vehicle. Release the parking brake.
3. Remove the rear wheels.
4. Use chalk to mark the brake drum to one of the wheel studs as an index mark for reinstallation.
5. Remove the retaining screw that holds the brake drum to the axle flange.
6. Pull the brake drum from the axle flange.

To install:

7. Align the index mark and install the brake drum to the axle flange.
8. Install the retaining screw to secure the brake drum to the axle flange.
9. Install the rear wheels.

Rear Brake Shoes

REMOVAL & INSTALLATION

Passport, Rodeo and Rodeo Sport

1. Before servicing the vehicle, refer to the Precautions Section.
2. Raise and safely support the vehicle.
3. Remove the rear wheels.
4. Remove the brake drums.
5. Remove the brake return springs.
6. Remove the leading shoe holding pin and spring, and then the leading shoe.
7. Remove the self-adjuster and the adjuster lever.
8. Remove the trailing shoe holding pin and spring.
9. Disconnect the parking brake cable from the trailing shoe and remove the trailing shoe. Remove the parking brake lever from the trailing shoe.

To install:

10. Attach the parking brake lever to the trailing shoe.
11. Connect the parking brake cable to the parking brake lever.
12. Apply a thin coat of high temperature grease to the shoe contact points on the brake backing plate (locations A and C in the accompanying illustration), piston contact surface (B), and self-adjuster (D).
13. Position the trailing shoe on the backing plate and install the hold-down pin, spring, and retainer. Don't stretch the return spring when fitting the shoes onto the backing plate.
14. Connect the upper return spring and the leading shoe to the trailing shoe and position the leading brake shoe on the backing plate.
15. Install the adjuster assembly and the hold-down pin, spring, and retainer.
16. Use a brake spring tool to install the lower return spring.
17. Install the self-adjuster lever and adjuster spring.
18. Adjust the shoe-to-drum clearance to 0.0098–0.0157 in. (0.25–0.40mm) and install the brake drum.
19. Check the brake drum for scoring or other wear. Machine or replace as necessary. Check the maximum brake drum diameter specification when machining.
20. Install the rear wheels. Lower the vehicle.
21. Road-test the vehicle.

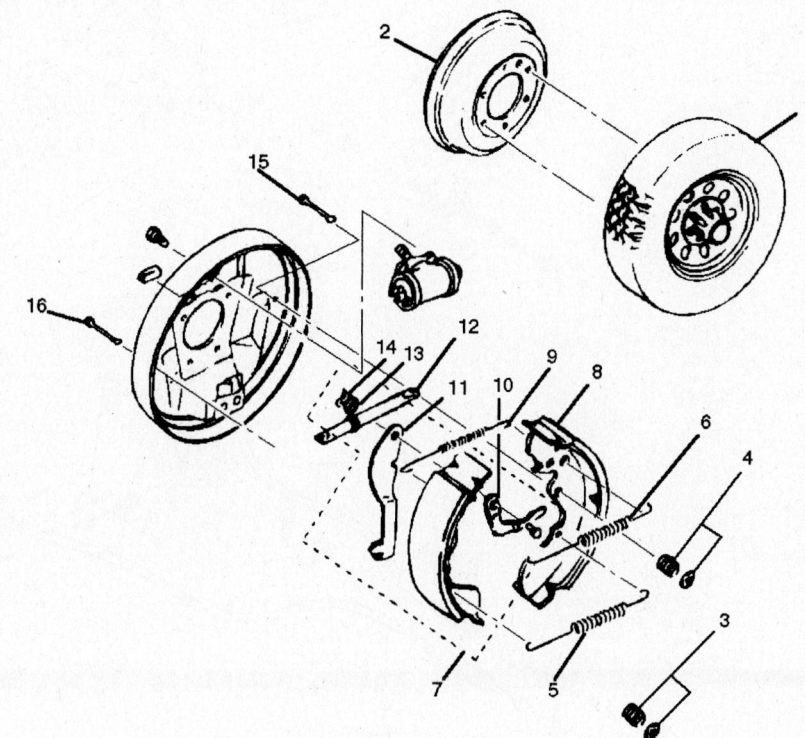

(1)	Wheel and Tire Assembly
(2)	Drum
(3)	Hold-down Spring and Cup
(4)	Hold-down Spring and Cups
(5)	Lower Return Spring
(6)	Upper (other) Return Spring
(7)	Trailing Shoe Assembly with Parking Brake Lever
(8)	Leading Shoe Assembly with Upper (inner) Return Spring
(9)	Upper (inner) Return Spring
(10)	Auto Adjuster Lever
(11)	Parking Brake Lever
(12)	Adjuster Assembly
(13)	Wave Washer
(14)	Retainer
(15)	Hold-down Pin
(16)	Hold-down Pin

09474_RODE_G0046

Exploded view of the rear drum brakes—Rodeo, Rodeo Sport, Passport

PARKING BRAKE W/REAR DISC BRAKES

Shoes

REMOVAL & INSTALLATION

1. Before servicing the vehicle, refer to the Precautions Section.
2. Raise and safely support the vehicle.

3. Remove the adjusting nut.
4. Remove the wheels.
5. Remove the caliper and support it out of the way.
6. Remove the rotor.
7. Remove the holding springs.
8. Remove the upper and lower return springs.

9. Remove the brake shoes.
10. Disconnect the cable from the lever.
11. Installation is the reverse of removal. Apply silicone brake grease to the points shown.
12. Adjust the parking brake.

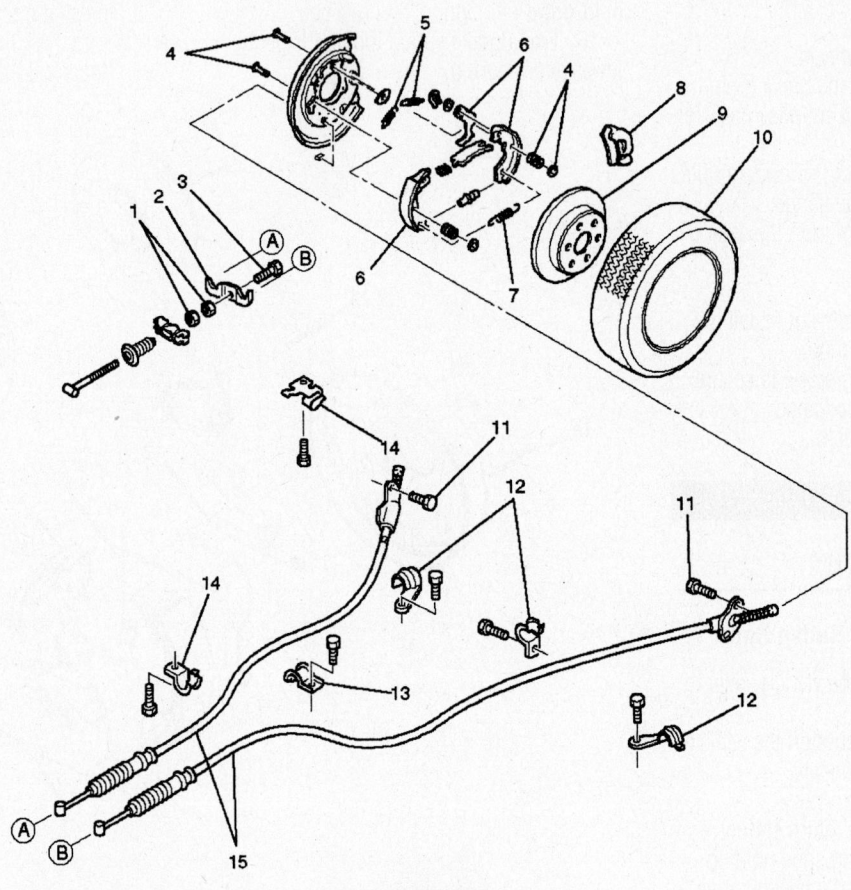

(1)	Adjust Nut
(2)	Back Plate
(3)	Bolt
(4)	Holding Spring
(5)	Return Spring; Upper
(6)	Shoe Assembly
(7)	Return Spring; Lower
(8)	Caliper Assembly
(9)	Rotor (Drum)
(10)	Rear Wheels
(11)	Cable Fixing Bolt
(12)	Clip
(13)	Clip
(14)	Clip
(15)	Rear Cable

09474_RODE_G0087

Parking brake system—Trooper

SPECIFICATIONS AND MAINTENANCE CHARTS

ENGINE AND VEHICLE IDENTIFICATION

Engine							Model Year	
Code ①	Liters (cc)	Cu. In.	Cyl.	Fuel Sys.	Engine Type	Eng. Mfg.	Code ②	Year
4	3.5 (3497)	213.32	6	MPFI	DOHC	KIA	4	2004
							5	2005
							6	2006

MPFI: Multi-Point Fuel Injection

DOHC: Double Overhead Camshafts

① 8th Digit of VIN

② 10th Digit of VIN

09474_AMAN_C0001

GENERAL ENGINE SPECIFICATIONS

Year	Engine Displacement Liters	Engine ID/VIN	Net Horsepower @ rpm	Net Torque @ rpm (ft. lbs.)	Bore x Stroke (in.)	Compression Ratio	Oil Pressure @ rpm
2004	3.5	4	200@5500	220@3500	3.66 x 3.38	10.1:1	11.4@700
2005	3.5	4	200@5500	220@3500	3.66 x 3.38	10.1:1	11.4@700
2006	3.5	4	200@5500	220@3500	3.66 x 3.38	10.1:1	11.4@700

09474_AMAN_C0002

GASOLINE ENGINE TUNE-UP SPECIFICATIONS

Year	Engine Displacement Liters	Engine ID/VIN	Spark Plugs Gap (in.)	Ignition Timing (deg.) MT	AT	Fuel Pump (psi)	Idle Speed (rpm) MT	AT	Valve Clearance In.	Ex.
2004	3.5	4	0.039-0.043	-	10B	47-50	-	700	HYD	HYD
2005	3.5	4	0.039-0.043	-	10B	47-50	-	700	HYD	HYD
2006	3.5	4	0.039-0.043	-	10B	47-50	-	700	HYD	HYD

HYD: Hydraulic Valve Lifters

B: Before Top Dead Center

09474_AMAN_C0003

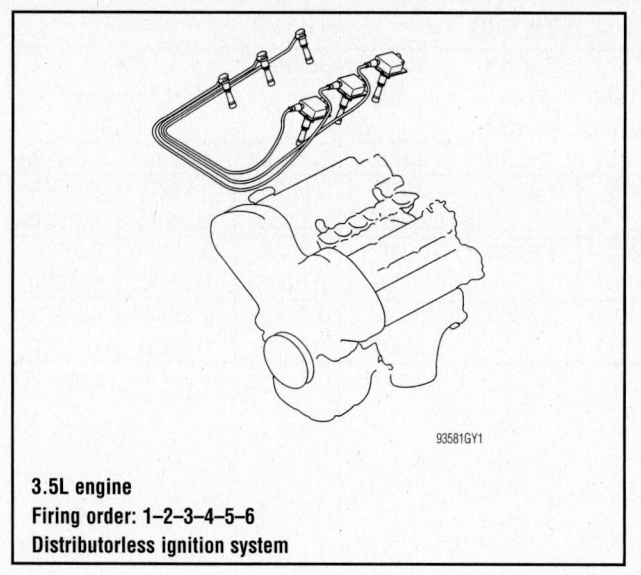

3.5L engine
Firing order: 1–2–3–4–5–6
Distributorless ignition system

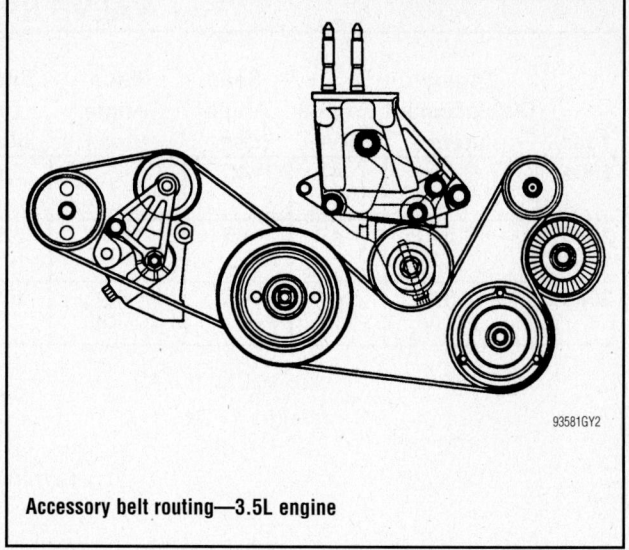

Accessory belt routing—3.5L engine

CAPACITIES

Year	Model	Engine Displacement Liters	Engine ID/VIN	Engine Oil with Filter	Transmission (pts.) 5–Spd	Transmission (pts.) Auto.	Fuel Tank (gal.)	Cooling System (qts.)
2004	Amanti	3.5	4	4.5	-	17.8	18.5	10.8
2005	Amanti	3.5	4	4.5	-	17.8	18.5	10.8
2006	Amanti	3.5	4	4.5	-	17.8	18.5	10.8

NOTE: All capacities are approximate. Add fluid gradually and check to be sure a proper fluid level is obtained.

09474_AMAN_C0004

CRANKSHAFT AND CONNECTING ROD SPECIFICATIONS
All measurements are given in inches.

Year	Engine Displacement Liters	Engine ID/VIN	Main Brg. Journal Dia.	Main Brg. Oil Clearance	Shaft End-play	Thrust on No.	Journal Diameter	Oil Clearance	Side Clearance
2004	3.5	4	2.5190-2.5197	0.0007-0.0014	0.0020-0.0098	3	2.2835-2.2842	0.0010-0.0017	0.0039-0.0098
2005	3.5	4	2.5190-2.5197	0.0007-0.0014	0.0020-0.0098	3	2.2835-2.2842	0.0010-0.0017	0.0039-0.0098
2006	3.5	4	2.5190-2.5197	0.0007-0.0014	0.0020-0.0098	3	2.2835-2.2842	0.0010-0.0017	0.0039-0.0098

09474_AMAN_C0005

VALVE SPECIFICATIONS

Year	Engine Displacement Liters	Engine ID/VIN	Seat Angle (deg.)	Face Angle (deg.)	Spring Test Pressure (lbs. @ in.)	Spring Free Height (in.)	Stem-to-Guide Clearance (in.)		Stem Diameter (in.)	
							Intake	Exhaust	Intake	Exhaust
2004	3.5	4	45	45	53@1.492	1.8268	0.0009-0.0020	0.0020-0.0033	0.2580-0.2590	0.2570-0.2580
2005	3.5	4	45	45	53@1.492	1.8268	0.0009-0.0020	0.0020-0.0033	0.2580-0.2590	0.2570-0.2580
2006	3.5	4	45	45	53@1.492	1.8268	0.0009-0.0020	0.0020-0.0033	0.2580-0.2590	0.2570-0.2580

09474_AMAN_C0006

PISTON AND RING SPECIFICATIONS

All measurements are given in inches.

Year	Engine Displacement Liters	Engine ID/VIN	Piston Clearance	Ring Gap			Ring Side Clearance		
				Top Compression	Bottom Compression	Oil Control	Top Compression	Bottom Compression	Oil Control
2004	3.5	4	0.0012-0.0020	0.0079-0.0118	0.0117-0.0236	0.0079-0.0276	0.0016-0.0031	0.0008-0.0024	NA
2005	3.5	4	0.0012-0.0020	0.0079-0.0118	0.0117-0.0236	0.0079-0.0276	0.0016-0.0031	0.0008-0.0024	NA
2006	3.5	4	0.0012-0.0020	0.0079-0.0118	0.0117-0.0236	0.0079-0.0276	0.0016-0.0031	0.0008-0.0024	NA

09474_AMAN_C0007

TORQUE SPECIFICATIONS

All readings in ft. lbs.

Year	Engine Displacement Liters	Engine ID/VIN	Cylinder Head Bolts	Main Bearing Bolts	Rod Bearing Bolts	Crankshaft Damper Bolts	Flywheel Bolts	Manifold		Spark Plugs	Oil Pan Drain Plug
								Intake	Exhaust		
2004	3.5	4	76-83	51-58	①	130-137	53-56	15-17	29-36	15-22	25-33
2005	3.5	4	76-83	51-58	①	130-137	53-56	15-17	29-36	15-22	25-33
2006	3.5	4	76-83	51-58	①	130-137	53-56	15-17	29-36	15-22	25-33

① Step 1: 25 ft. lbs.
Step 2: Plus 92 degrees

09474_AMAN_C0008

WHEEL ALIGNMENT

Year	Model		Caster Range (+/-Deg.)	Caster Preferred Setting (Deg.)	Camber Range (+/-Deg.)	Camber Preferred Setting (Deg.)	Toe-in (in.)
2004	Amanti	F	1.0	3.25	0.50	0	0 +/- 0.08
		R	—	—	0.50	-0.50	0.08 +/- 0.08
2005	Amanti	F	1.0	3.25	0.50	0	0 +/- 0.08
		R	—	—	0.50	-0.50	0.08 +/- 0.08
2006	Amanti	F	1.0	3.25	0.50	0	0 +/- 0.08
		R	—	—	0.50	-0.50	0.08 +/- 0.08

09474_AMAN_C0009

TIRE, WHEEL AND BALL JOINT SPECIFICATIONS

Year	Model	OEM Tires Standard	OEM Tires Optional	Tire Pressures (psi) Front	Tire Pressures (psi) Rear	Wheel Size	Ball Joint Inspection	Lug Nut Torque ①
2004	Amanti	P225/60R16	P235/60R16	31	30	6.5Jx16	②	67-82
2005	Amanti	P215/65R16	P235/60R16	30	30	6.5Jx16	②	67-82
2006	Amanti	P215/65R16	P235/60R16	30	30	6.5Jx16	②	67-82

OEM: Original Equipment Manufacturer

PSI: Pounds Per Square Inch

① Ft. Lbs.

② If Rotating Torque exceeds 1.1 ft. lbs., replace the ball joint assembly

09474_AMAN_C0010

BRAKE SPECIFICATIONS

All measurements in inches unless noted

Year	Model		Brake Disc Original Thickness	Brake Disc Minimum Thickness	Brake Disc Maximum Run-out	Brake Drum Diameter Original Inside Diameter	Brake Drum Diameter Max. Wear Limit	Brake Drum Diameter Maximum Machine Diameter	Minimum Lining Thickness Front	Minimum Lining Thickness Rear	Brake Caliper Bracket Bolts (ft. lbs.)	Brake Caliper Mounting Bolts (ft. lbs.)
2004	Amanti	F	1.100	1.040	0.001	—	—	—	0.079	—	49-61	16-24
		R	0.390	0.331	0.001	—	—	—	—	0.079	36-43	16-24
2005	Amanti	F	1.100	1.040	0.001	—	—	—	0.079	—	49-61	16-24
		R	0.390	0.331	0.001	—	—	—	—	0.079	36-43	16-24
2006	Amanti	F	1.100	1.040	0.001	—	—	—	0.079	—	49-61	16-24
		R	0.390	0.331	0.001	—	—	—	—	0.079	36-43	16-24

F: Front

R: Rear

09474_AMAN_C0011

SCHEDULED MAINTENANCE INTERVALS
Kia Amanti

TO BE SERVICED	TYPE OF SERVICE	VEHICLE MILEAGE INTERVAL (x1000)												
		7.5	15	22.5	30	37.5	45	52.5	60	67.5	75	82.5	90	97.5
Accessory drive belts	S/I			✓			✓			✓			✓	
Air cleaner filter	R		✓		✓		✓		✓		✓		✓	
Air conditioner system	S/I													
Brake lines, hoses and connections	S/I		✓		✓		✓		✓		✓		✓	
Chassis and body fasteners	T	✓	✓	✓	✓	✓	✓	✓	✓	✓	✓	✓	✓	✓
Cooling system hoses and coolant level	S/I		✓		✓		✓		✓		✓		✓	
CV-joint boots	S/I				✓				✓				✓	
Engine coolant	R				✓				✓				✓	
Engine oil and filter	R	✓	✓	✓	✓	✓	✓	✓	✓	✓	✓	✓	✓	✓
Exhaust system heat shields	S/I				✓				✓				✓	
Front and rear brakes	S/I				✓				✓				✓	
Front ball joints	S/I				✓				✓				✓	
Fuel filter	R								✓					
Fuel lines and hoses	S/I				✓				✓				✓	
Locks and hinges	L	✓	✓	✓	✓	✓	✓	✓	✓	✓	✓	✓	✓	✓
Spark plugs	R				✓				✓				✓	
Steering operation and linkage	S/I				✓				✓				✓	
Timing belt	R								✓					

R: Replace   S/I: Service or Inspect   L: Lubricate   T: Tighten

FREQUENT OPERATION MAINTENANCE (SEVERE SERVICE)

If a vehicle is operated under any of the following conditions it is considered severe service

- Towing a trailer or using a camper or car-top carrier

- Repeated short trips of less than 5 miles in temperatures below freezing, or trips of less than 10 miles in any temperature

- Prolonged idling (vehicle operation in stop and go traffic).

- Operating on rough, muddy, unpaved, dusty or salt-covered roads.

- Police, taxi, delivery usage or trailer towing usage.

- Driving in extremely hot (over 90°F) conditions

Oil & oil filter: change every 5000 miles or 5 months, whichever occurs first.

Air cleaner filter: inspect every 15,000 miles or 15 months and replace everything 30,000 miles or 30 months, whichever occurs firs

Fuel system hoses (California models only): replace every 105,000 miles

Emission system hoses (non-CA models): inspect every 55,000 or 55 months, whichever occurs first

Emission system hoses (CA models): inspect every 60,000 miles or 60 months, which occurs first

Front and rear brakes: inspect every 15,000 miles or 15 months, whichever occurs first

Chassis and body fasteners: tighten every 15,000 miles or 15 months, whichever occurs first

Locks and hinges: lubricate every 5000 miles or 5 months, whichever occurs first

09474_AMAN_C0012

ENGINE REPAIR

➡Disconnecting the negative battery cable on some vehicles may interfere with the functions of the on board computer system. The computer may undergo a relearning process once the negative battery cable is reconnected.

Distributor

All engines are equipped with a Distributorless Ignition System (DIS).

Alternator

REMOVAL & INSTALLATION

1. Before servicing the vehicle, refer to the Precautions Section.
2. Remove or disconnect the following:
   - Negative battery cable
   - Alternator wiring harness connectors
   - Accessory drive belt
   - Alternator support bracket
   - Alternator

To install:

3. Install the alternator. Tighten the mounting bolts as follows:
   a. Mounting bracket bolts: Tighten to 15–18 ft. lbs. (20–25 Nm)
   b. Support bracket bolt: Tighten to 13–16 ft. lbs. (18–22 Nm)
4. Install or connect the following:
   - Accessory drive belt
   - Alternator wiring harness connectors
   - Negative battery cable

Ignition Timing

ADJUSTMENT

These engines are equipped with a Distributorless Ignition System (DIS). No adjustment is necessary.

Engine Assembly

REMOVAL & INSTALLATION

1. Before servicing the vehicle, refer to the Precautions Section.
2. Drain the cooling system.
3. Drain the transaxle.
4. Drain the engine oil.
5. Relieve fuel system pressure.
6. Remove or disconnect the following:
   - Negative battery cable
   - Engine cover

- Battery and battery tray
- Air intake assembly
- Alternator wiring connector and oil pressure switch connector

7. Remove the following engine wiring harnesses:
   a. Crank angle sensor connector
   b. Fuel injector connector
   c. Power steering switch connector
   d. Variable intake motor connector
   e. Accelerator position sensor connector
   f. EGR solenoid connector
   g. Ignition coil harness connector
   h. Oxygen sensor connector
   i. Knock sensor connector
   j. Throttle position sensor
   k. Electronic throttle system (ETS) motor connector
   l. Limp-home connector
   m. Purge solenoid valve connector
   n. Water temperature sensor connector
8. Remove or disconnect the following:
   - Upper and lower radiator hoses
   - Ground wire from body frame and dash board panel
   - Transmission wiring harnesses
   - Ground wire from the transmission assembly
   - Heater hoses
   - Front and center muffler assembly
   - Starter motor wiring harness
   - Power steering gear connector
   - Steering column intermediate shaft
   - Halfshaft from the transmission
   - Right side engine under cover
   - Accessory drive belt
   - A/C compressor and secure to the frame with a wire
   - Hood release cable
   - Upper radiator support member
   - Cooling fan
   - Electric fan
   - Radiator
   - Fuel hoses
9. Support the sub-frame with a suitable jack.

■ Front

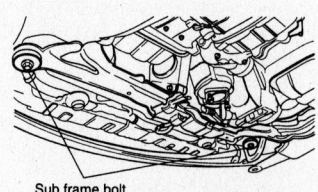

Sub frame bolt

09474_AMAN_G0001

Location of the front sub-frame mounting bolts.

■ Rear

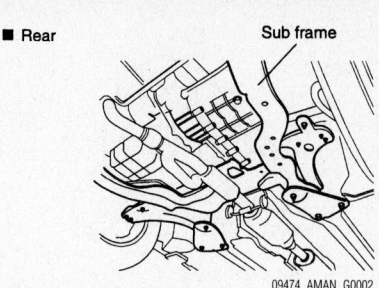

09474_AMAN_G0002

Rear sub-frame assembly.

■ Rear

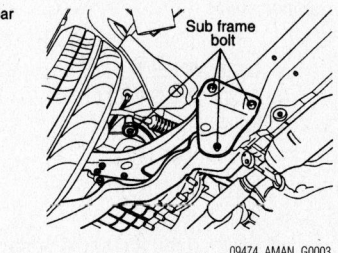

09474_AMAN_G0003

Location of rear sub-frame mounting bolts.

10. Remove the sub-frame mounting bolts.
11. Remove the engine mounting bolt.
12. Remove the transmission mounting bolt.
13. Lift the vehicle up and remove the engine, transmission and sub-frame assembly.

To install:

14. Position the engine and transmission assembly into the vehicle.
15. Install the transmission mounting bolt and tighten to 43–58 ft. lbs. (59–79 Nm).
16. Install the engine mounting bolt and tighten to 43–58 ft. lbs. (59–79 Nm).
17. Install the sub-frame bolts and tighten to 72–87 ft. lbs. (98–118 Nm).
18. The remainder of the installation is the reverse order of removal.
19. Fill the engine with oil to the correct level.
20. Fill the transaxle to the correct level.
21. Fill the cooling system to the proper level.
22. Start the engine and check for leaks.

Water Pump

REMOVAL & INSTALLATION

1. Before servicing the vehicle, refer to the Precautions Section.
2. Drain the cooling system.
3. Remove or disconnect the following:

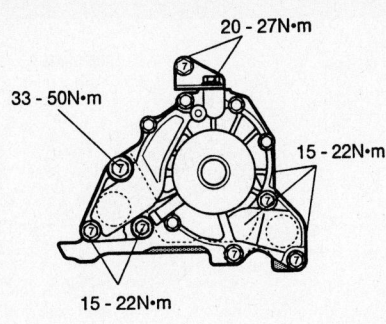

9358HG04

Water pump bolt locations and torque specifications

- Negative battery cable
- Accessory drive belt
- Water pump pulley
- Timing belt cover
- Drive belt auto tensioner
- Idler pulley
- Water pump mounting bolts
- Water pump

To install:

4. Clean the gasket surfaces of the water pump and cylinder block.

5. Install the water pump with a new gasket. Tighten the mounting bolts as shown in illustration.

6. Install or connect the following:
- Idler pulley
- Auto tensioner
- Timing belt cover
- Water pump pulley
- Accessory drive belt and adjust the auto tensioner
- Negative battery cable

7. Refill the cooling system to the correct level.

8. Start the engine and check for leaks.

Heater Core

REMOVAL AND INSTALLATION

1. Before servicing the vehicle, refer to the Precautions Section.

2. Drain the engine coolant into a clean container for reuse.

3. Disconnect the negative battery cable.

4. Disconnect the heater hoses from the heater unit.

5. Remove or disconnect the following:
- Front door scuff trim
- Hood latch release lever
- Cowl side trim
- A-pillar trim
- Left side mounting trim cover

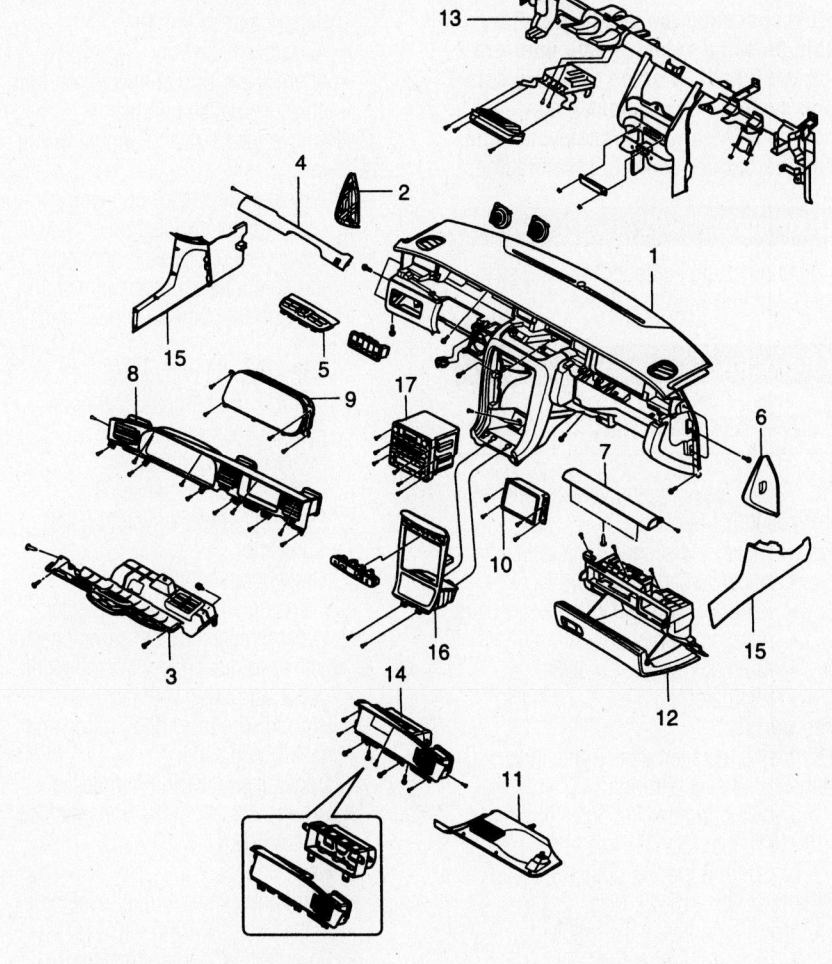

1. Instrument main panel assembly
2. Side mounting cover(LH)
3. Crash pad lower panel(LH)
4. Crash pad plate(LH)
5. Audio keyboard
6. Side mounting cover(RH)
7. Crash pad plate(RH)
8. Cluster facia panel
9. Cluster assembly
10. Audio monitor assembly
11. Under cover(RH)
12. Glove box assembly
13. Cowl cross bar assembly
14. Passenger airbag door & Airbag assembly
15. Side cover
16. Center facia panel
17. Audio & Heater controller assembly

09474_AMAN_G0004

Exploded view of crash pad components

- Parking brake release lever
- Crash pad lower panel and lower panel lamp
- Steering wheel
- Steering column shroud
- Left side crash pad plate
- Audio keyboard from main instrument panel assembly
- Right side mounting trim cover
- Right side crash pad plate
- Cluster facia panel
- Instrument cluster assembly
- Audio monitor assembly
- Under cover and under cover lamp

- Glove box
- Glove box housing
- Passenger airbag assembly
- Floor console side covers
- Floor console
- Center facia panel
- Radio and heater controller

6. Disconnect the main instrument panel connect and remove the main instrument panel.

7. Remove the cowl crossbar mounting bolts and remove the crossbar assembly.

8. Remove the heater unit.

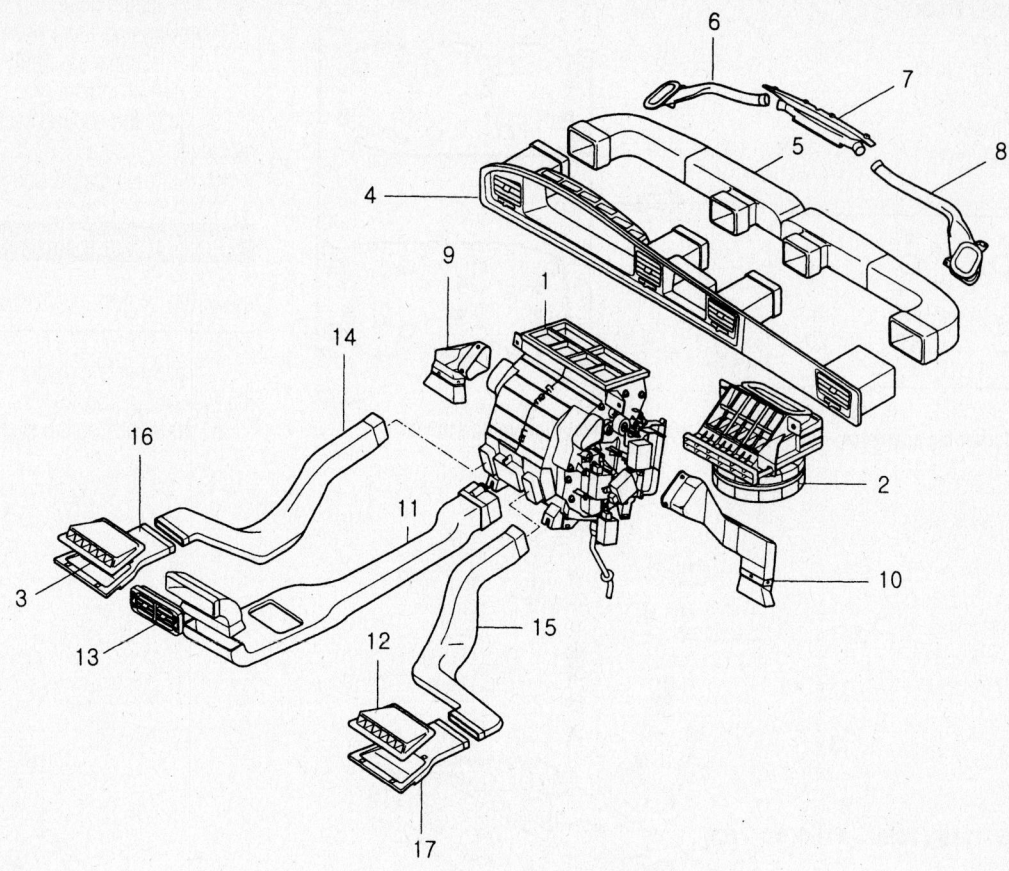

1. Heater unit
2. Blower unit
3. Rear floor duct grill-LH
4. Vent grill
5. Vent duct
6. Dip duct-LH
7. Dip duct center
8. Dip duct-RH
9. Driver side floor duct

10. Passenger side floor duct
11. Rear vent duct
12. Rear floor duct grill-RH
13. Rear vent grill
14. Rear floor duct LH (A)
15. Rear floor duct RH (A)
16. Rear floor duct LH (B)
17. Rear floor duct RH (B)

09474_AMAN_G0005

Exploded view of heater unit components

9. Installation is the reverse order of removal. Note the following torques:
   a. Upper cowl crossbar mounting bolts: 24–40 ft. lbs. (33–55 Nm).
   b. Lower cowl crossbar mounting bolts: 12–19 ft. lbs. (17–26 Nm).
10. Fill the cooling system to proper level.
11. Start the engine and check for leaks.

Cylinder Head

REMOVAL & INSTALLATION

1. Before servicing the vehicle, refer to the Precautions Section.
2. Drain the cooling system.

3. Relieve the fuel system pressure.
4. Remove or disconnect the following:
   - Negative battery cable
   - Air intake assembly
   - Upper radiator hose
   - Vacuum hose between intake manifold and cylinder head cover
   - Fuel hose
   - Heater hose
   - Intake manifold
   - Spark plug cables
   - Ignition coil
   - Upper and lower timing belt covers
   - Timing belt
   - Camshaft sprockets
   - Exhaust manifold heat shield
   - Exhaust manifold assembly
   - Water pump pulley

   - Cylinder head cover
   - Camshafts
5. Loosening the bolts in 2 or 3 steps in the order shown, remove the cylinder head assembly.
   To install:
6. Clean the cylinder head and block mating surfaces.
7. Install the cylinder head and tighten the bolts in sequence to 76–83 ft. lbs. (103–113 Nm).
8. Install or connect the following:
   - Camshafts
   - Cylinder head cover
   - Water pump pulley
   - Exhaust manifold and heat shield
   - Camshaft sprockets
   - Timing belt

- Timing belt covers
- Ignition coil
- Spark plug cables
- Intake manifold
- Heater hose
- Fuel hose

1
5
8
4
2
6
7
3
09474_AMAN_G0006

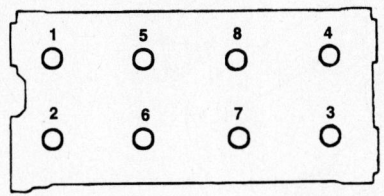

Cylinder head loosening sequence

6
2
3
7
5
1
4
8
8
4
1
5
7
3
2
6
9358HG06

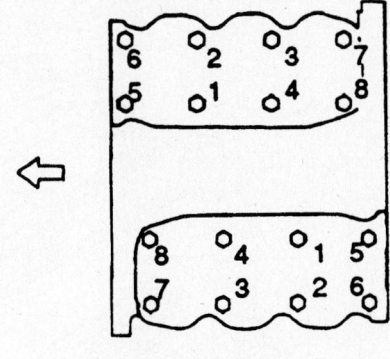

Cylinder head tightening sequence

- Vacuum hose
- Upper radiator hose
- Air intake assembly
- Negative battery

9. Refill the cooling system to the correct level.

10. Start the engine and check for leaks.

Rocker Arms/Shafts

REMOVAL & INSTALLATION

1. Before servicing the vehicle, refer to the Precautions Section.
2. Drain the cooling system.

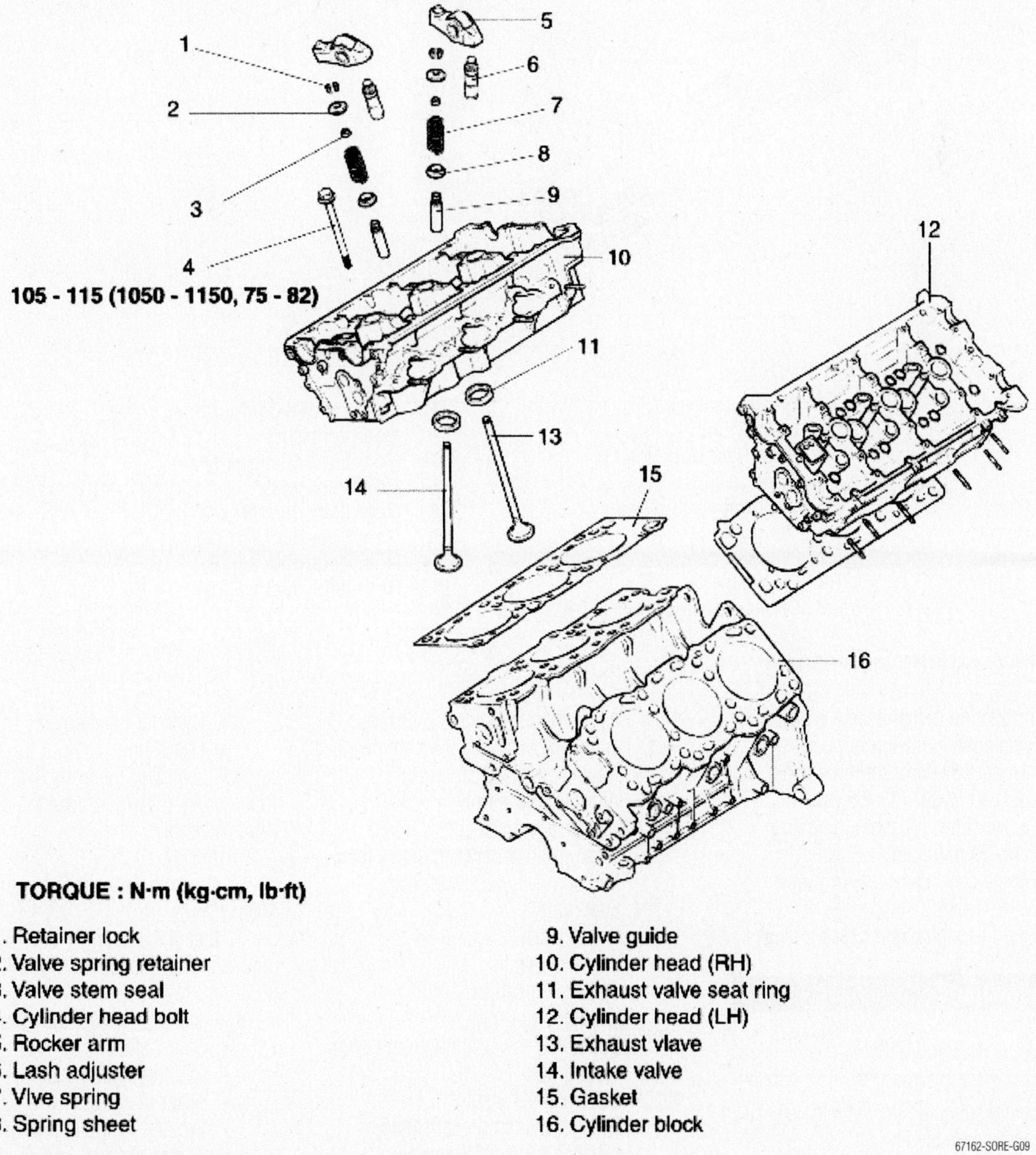

1
2
3
4
5
6
7
8
9
10
11
12
13
14
15
16
105 - 115 (1050 - 1150, 75 - 82)
TORQUE : N·m (kg·cm, lb·ft)
1. Retainer lock
2. Valve spring retainer
3. Valve stem seal
4. Cylinder head bolt
5. Rocker arm
6. Lash adjuster
7. Vlve spring
8. Spring sheet
9. Valve guide
10. Cylinder head (RH)
11. Exhaust valve seat ring
12. Cylinder head (LH)
13. Exhaust vlave
14. Intake valve
15. Gasket
16. Cylinder block
67162-SORE-G09

Exploded view of cylinder head components

3. Relieve the fuel system pressure.
4. Remove or disconnect the following:
   - Negative battery cable
   - Air intake assembly
   - Upper radiator hose
   - Vacuum hose between intake manifold and cylinder head cover
   - Fuel hose
   - Heater hose
   - Intake manifold
   - Spark plug cables
   - Ignition coil
   - Upper and lower timing belt covers
   - Timing belt
   - Camshaft sprockets
   - Exhaust manifold heat shield
   - Exhaust manifold assembly
   - Water pump pulley
   - Cylinder head cover
   - Camshafts
   - Rocker arms and lash adjusters
5. Installation is the reverse of removal.

Intake Manifold

REMOVAL & INSTALLATION

1. Before servicing the vehicle, refer to the Precautions Section.
2. Relieve the fuel system pressure.
3. Drain the cooling system.
4. Remove or disconnect the following:
   - Negative battery cable
   - Engine cover
   - Air intake assembly
5. Remove the following engine wiring harnesses:
   a. Crank angle sensor
   b. Cam angle sensor
   c. Fuel injector harness

➡**When disconnecting the injector, lift the fuel supply hose and injector assembly upward. Then unscrew the mounting bolt to lift the fuel supply hose. Reinstall the fuel hose and injector assembly after disconnecting the injector harness.**

   d. Power steering switch
   e. Variable intake motor connector
   f. Accelerator position sensor
   g. EGR solenoid
   h. Throttle position sensor
   i. Electronic throttle system (ETS) motor
   j. Limp-home connector
   k. Purge solenoid valve connector
6. Remove or disconnect the following:
   - Vacuum hoses and heater hoses between the intake manifold and cylinder head cover

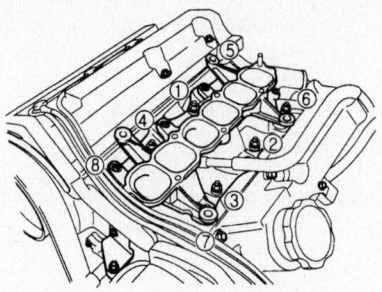

Intake manifold tightening sequence

   - Engine wiring harness bracket
   - EGR valve hose and bracket
   - Surge tank assembly
   - Fuel supply hose
   - Fuel injector assembly
   - Timing belt cover
   - Water pump bracket mounting bolts
   - Intake manifold mounting nuts
   - Intake manifold

To install:

7. Install the intake manifold with a new gasket and tighten the mounting nuts in sequence to 15–17 ft. lbs. (20–23 Nm).
8. Install the water pump bracket and tighten the bolts to 15–20 ft. lbs. (20–27 Nm).
9. Install the timing belt covers.
10. Connect the fuel supply hose and

install the fuel injector assembly. Tighten to 7–10 ft. lbs. (10–13 Nm).
11. Install the surge tank assembly. Tighten to 11–15 ft. lbs. (15–20 Nm).
12. Install the EGR valve hose and tighten the mounting bolts to 12–19 ft. lbs. (17–26 Nm).
13. Install the EGR pipe fixing bracket and tighten the bolts to 11–15 ft. lbs. (15–20 Nm).
14. Install surge tank stay mounting bolts and tighten the bolts to 11–15 ft. lbs. (15–20 Nm).
15. Install the engine wiring harness bracket.
16. Reconnect the vacuum and heater hoses between the intake manifold and cylinder head cover.
17. Reconnect all wiring harnesses that were disconnected during removal.
18. Install the air intake assembly.
19. Install the engine cover.
20. Connect the negative battery cable.
21. Start the engine and check for leaks.

Exhaust Manifold

REMOVAL & INSTALLATION

1. Before servicing the vehicle, refer to the Precautions Section.

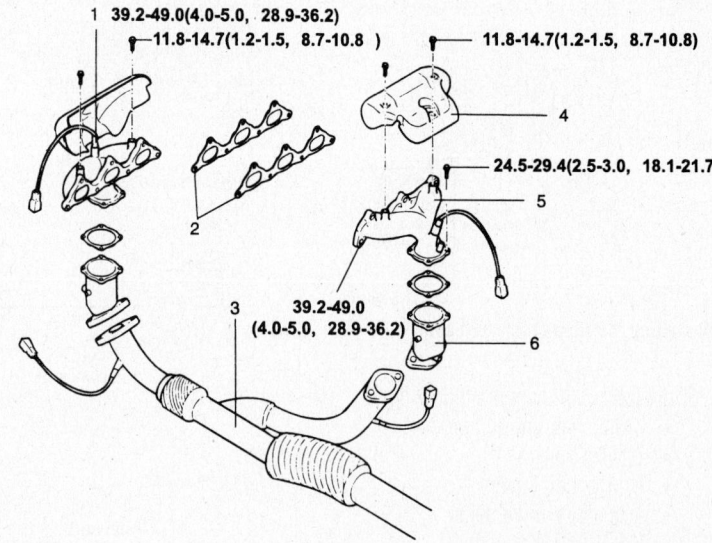

TORQUE : Nm (kg. m, lb.ft)

1. O₂ sensor
2. Gasket
3. Front muffler assembly
   (Front pipe and underbody
   catalyst converter)

4. Heat protector
5. Exhaust manifold
6. MCC (Manifold Catalyst converter)

Exhaust manifold exploded view

09474_AMAN_G0008

2. Remove or disconnect the following:
- Negative battery cable
- Manifold heat shield
- EGR pipe, if removing right-hand manifold
- Exhaust manifold and gasket

To install:

3. Install the manifold with a new gasket. Tighten the mounting bolts to 29–36 ft. lbs. (39–49 Nm).

4. Install the EGR pipe, if removed. Tighten the mounting bolts to 36–50 ft. lbs. (49–69 Nm).

5. Install the exhaust manifold heat shield. Tighten to 9–11 ft. lbs. (12–15 Nm).

Front Crankshaft Seal

REMOVAL & INSTALLATION

1. Before servicing the vehicle, refer to the Precautions Section.

2. Remove or disconnect the following:
- Negative battery cable
- Accessory drive belts
- Timing belt covers
- Timing belt
- Crankshaft timing sprocket

3. Pry the oil seal from the oil pump case.

To install:

4. Install the front crankshaft seal using Special Tool 09214-33000 seal installer.

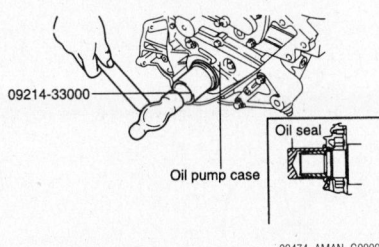

09474_AMAN_G0009

Installing the front crankshaft seal

5. Install or connect the following:
- Crankshaft timing sprocket
- Timing belt
- Timing belt covers
- Accessory drive belts
- Negative battery cable

6. Start the engine and check for leaks.

Camshaft and Valve Lifters

REMOVAL & INSTALLATION

1. Before servicing the vehicle, refer to the Precautions Section.

2. Remove or disconnect the following:
- Negative battery cable
- Engine cover
- Intake manifold
- Power steering pulley
- A/C pulley
- Crankshaft pulley
- Idler pulley
- Tensioner pulley
- Timing belt
- Spark plug cables
- Cylinder head cover
- Camshaft sprockets
- Camshaft bearing caps
- Camshafts

➡ **Keep all valvetrain components in order for assembly.**

To install:

3. Rotate the crankshaft to set the No. 1 at Top Dead Center (TDC) of the compression stroke.

4. Ensure the rocker arm is correctly positioned on the lash adjuster and valve.

5. Install the camshaft dowel pins as shown.

6. Install the camshafts and tighten the bearing caps in 2–3 steps as follows:
- a. Outer (front and rear) bearing caps to 14–15 ft. lbs. (19–21 Nm).
- b. Inner (center) bearing caps to 88–106 inch lbs. (10–12 Nm).

7. Install the camshaft sprockets and tighten to 58–72 ft. lbs. (79–98 Nm).

8. Install or connect the following:
- Cylinder head cover
- Spark plug cables

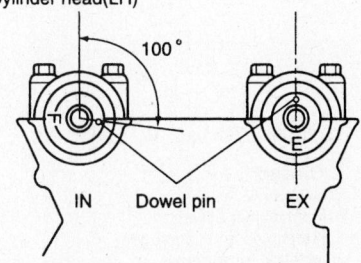

Cylinder head(RH)

45° 65°

EX   Dowel pin   IN

Cylinder head(LH)

100°

IN   Dowel pin   EX

09474_AMAN_G0010

Install the camshaft dowel pin as shown

- Timing belt
- Tensioner pulley
- Idler pulley
- Crankshaft pulley
- A/C pulley
- Power steering pulley
- Intake manifold
- Engine cover
- Negative battery

9. Start the engine and check for leaks.

Valve Lash

ADJUSTMENT

This vehicle uses hydraulic valve lash adjusters. Valve lash adjustments are not necessary or possible on these engines.

Starter Motor

REMOVAL & INSTALLATION

1. Before servicing the vehicle, refer to the Precautions Section.

2. Remove or disconnect the following:
- Negative battery cable
- Speedometer and shift cables
- Starter electrical connectors
- Starter motor

To install:

3. Install or connect the following:
- Starter motor. Tighten mounting bolt and nut to 33–40 ft. lbs. (45–55 Nm).
- Starter electrical connectors. Tighten nut to 20–25 ft. lbs. (27–34 Nm).
- Speedometer and shift cables
- Negative battery cable

Oil Pan

REMOVAL & INSTALLATION

1. Before servicing the vehicle, refer to the Precautions Section.

2. Drain the engine oil.

3. Remove or disconnect the following:
- Negative battery cable
- Oil pressure switch
- Oil filter
- Lower oil pan
- Upper oil pan

To install:

4. Clean all gasket surfaces of the oil pans and cylinder block.

5. Apply silicone sealant to the groove of the upper oil pan flange.

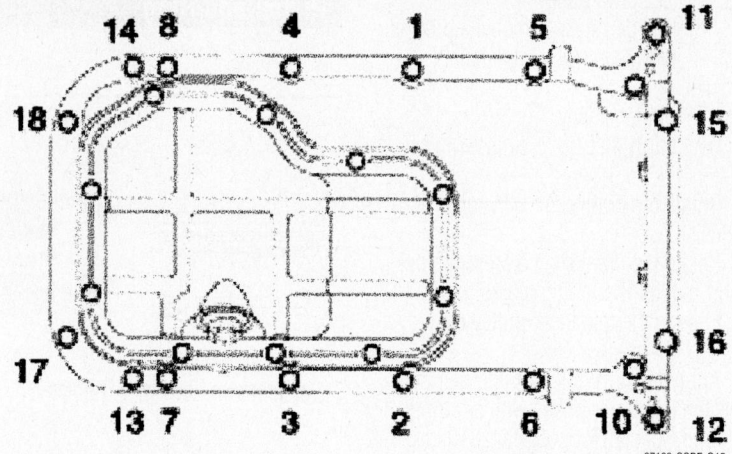

Upper oil pan torque sequence

67162-SORE-G15

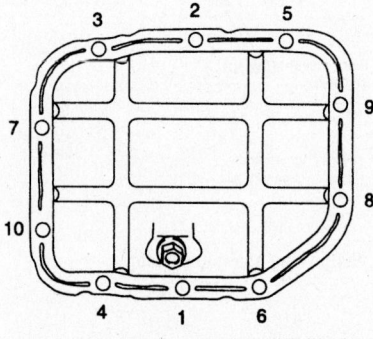

09474_AMAN_G0011

Lower oil pan torque sequence

6. Install the upper oil pan and tighten the bolts as follows:

a. Bolts 1–14: 88–106 inch lbs. (10–12 Nm).

b. Bolts 15–16: 44–62 inch lbs. (5–7 Nm).

c. Oil pan-to-transaxle bolts: 22–30 ft. lbs. (30–41 Nm).

7. Apply Loctite® 5900 or equivalent to the threaded area of the oil pressure switch. Install the switch and tighten to 70–106 inch lbs. (8–12 Nm).

8. Install the lower oil pan and tighten the bolts in sequence to 88–106 inch lbs. (10–12 Nm).

9. Install or connect the following:
- Oil filter
- Negative battery cable

10. Fill the engine with clean oil.

11. Start the vehicle and check for leaks.

Oil Pump

REMOVAL & INSTALLATION

1. Before servicing the vehicle, refer to the Precautions Section.

2. Drain the engine oil.

3. Remove or disconnect the following:
- Negative battery cable
- Accessory drive belts
- Timing belt covers
- Timing belt
- Crankshaft timing sprocket
- Oil filter
- Oil pans
- Oil filter bracket and gasket
- Oil relief valve plug from oil pump case
- Oil pump case

To install:

4. Install the oil pump case with the gasket and tighten as follows:

a. Oil pump case bolts to 106–133 inch lbs. (12–15 Nm).

b. Oil pump cover screws to 70–106 inch lbs. (8–12 Nm).

5. Install the oil relief valve plug and tighten to 29–36 ft. lbs. (39–49 Nm).

6. Install or connect the following:
- Oil filter bracket with gasket
- Oil pans
- Oil filter
- Crankshaft timing sprocket
- Timing belt
- Timing belt covers
- Accessory drive belts
- Negative battery cable

7. Fill the engine with oil to the correct level.

8. Start the engine and check for leaks.

Rear Main Seal

REMOVAL & INSTALLATION

1. Before servicing the vehicle, refer to the Precautions Section.

2. Remove or disconnect the following:

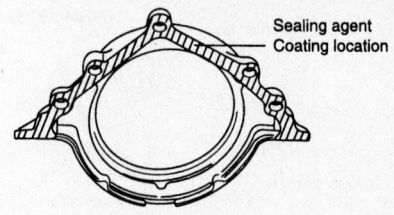

09474_AMAN_G0012

Rear case sealant application location

- Transaxle
- Flywheel
- Rear cover plate
- Oil seal case
- Oil seal

To install:

3. Install the oil seal to the oil seal case, using Seal Installer tool 09231-33000.

4. Apply a silicone sealant to the oil seal case as shown and torque the case bolts to 88–106 inch lbs. (10–12 Nm).

5. Install or connect the following :
- Rear plate. Tighten to 88–106 inch lbs. (10–12 Nm).
- Flywheel
- Transaxle

Timing Belt

REMOVAL & INSTALLATION

1. Before servicing the vehicle, refer to the Precautions Section.

2. Remove or disconnect the following:
- Negative battery cable
- Engine cover

3. Rotate the tensioner arm clockwise and remove the accessory drive belt from the pulleys.

4. Remove the upper and lower timing belt covers.

5. Remove the auto tensioner.

6. Rotate the crankshaft clockwise and align the timing marks of the camshaft sprocket and cylinder head to set the No. 1 cylinder at TDC.

7. Unbolt the tensioner pulley to remove the timing belt.

➡**If the timing belt is to be reused, mark the rotation direction so it is reinstalled correctly.**

To install:

8. Prepare the auto tensioner for installation by compressing it in a vise and installing a set pin.

9. Install the auto tensioner.

10. Tighten the tensioner pulley to 31–40 ft. lbs. (42–54 Nm).

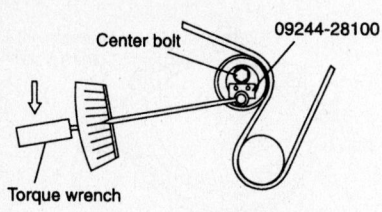

Use Special Tool 09244-28100 to apply tension to the timing belt.

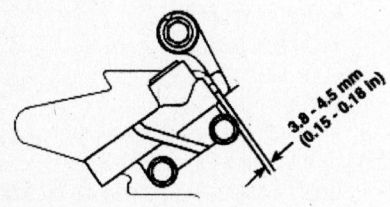

Measure the auto tensioner rod

11. Ensure the No. 1 cylinder is set to TDC.
12. Install the timing belt in the following order:
   a. Crankshaft sprocket
   b. Idler pulley
   c. Left side exhaust camshaft sprocket
   d. Left side intake camshaft sprocket

e. Water pump pulley
f. Right side intake camshaft sprocket
g. Right side exhaust camshaft sprocket
h. Tensioner pulley

13. Reconfirm that No. 1 cylinder is set at TDC.
14. Rotate the crankshaft a ¼ turn counterclockwise.
15. Rotate the crankshaft a ¼ turn clockwise and realign the timing marks.
16. Loosen the tensioner pulley center bolt.
17. Apply tension to the timing belt by rotating the tensioner pulley with a torque wrench and Special Tool 09244-28100 to 43 inch lbs.
18. Tighten the center bolt of the tensioner pulley to 31–40 ft. lbs. (42–54 Nm).
19. Release the set pin of the auto tensioner.
20. Rotate the crankshaft 2 revolutions clockwise then wait 5 minutes for the auto tensioner to adjust.
21. Measure the projected load of the auto tensioner. Measurement should be between 0.15–0.18 inches (3.8–4.5 mm).
22. Repeat the belt tightening procedure if measurement is not with specification.
23. The remainder of the installation is the reverse of removal.

Piston and Ring

POSITIONING

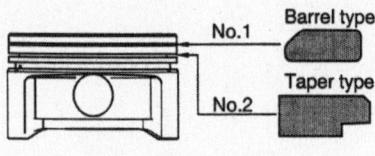

Compression ring identification

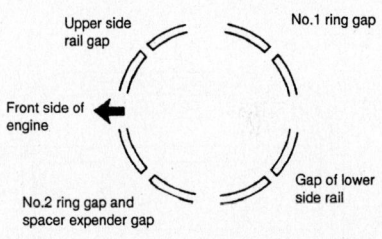

Piston ring end-gap spacing

FUEL SYSTEM

Fuel System Pressure

RELIEVING

1. Before servicing the vehicle, refer to the Precautions Section.
2. Open the service cover in the trunk area.
3. Disconnect the fuel pump wiring harness.
4. Start the engine and allow it to stall.
5. Turn the ignition switch to the **OFF** position.

6. Reconnect the electrical connections after fuel system repairs are completed.

Fuel Filter

REMOVAL & INSTALLATION

The fuel filter is part of the fuel pump assembly located in the tank.

Fuel Pump

REMOVAL & INSTALLATION

1. Before servicing the vehicle, refer to the Precautions Section.
2. Relieve the fuel system pressure.
3. Disconnect the fuel supply and return hoses from the top of the fuel pump assembly.
4. Remove the mounting bolts and remove the fuel pump assembly from the fuel tank.
5. Remove the fuel filter holder and disconnect the connecting hose and cable assembly
6. Remove the fuel filter and pump assembly.

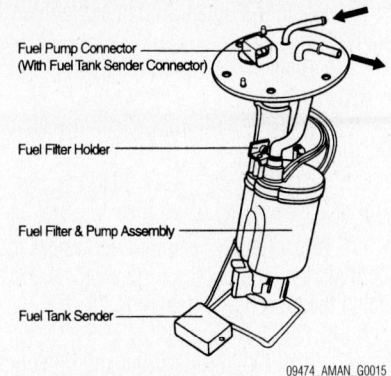

Fuel pump and filter assembly

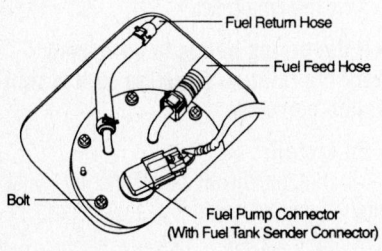

Fuel hose locations beneath the service cover

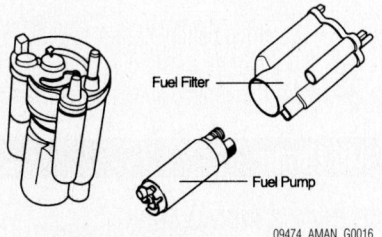

Fuel pump and filter removed from the assembly

7. Installation is the reverse of removal.

8. Fill the tank with fuel and check for proper fuel pump operation.

Fuel Injector

REMOVAL & INSTALLATION

1. Before servicing the vehicle, refer to the Precautions Section.

2. Relieve the fuel system pressure.

3. Drain the cooling system.

4. Remove or disconnect the following:

- Negative battery cable
- Engine cover
- Air intake assembly

5. Remove the following engine wiring harnesses:

   a. Crank angle sensor

   b. Cam angle sensor

   c. Fuel injector harness

➡ **When disconnecting the injector, lift the fuel supply hose and injector assembly upward. Then unscrew the mounting bolt to lift the fuel supply hose. Reinstall the fuel hose and injector assembly after disconnecting the injector harness.**

   d. Power steering switch

   e. Variable intake motor connector

   f. Accelerator position sensor

   g. EGR solenoid

   h. Throttle position sensor

   i. Electronic throttle system (ETS) motor

   j. Limp-home connector

   k. Purge solenoid valve connector

6. Remove or disconnect the following:

- Vacuum hoses and heater hoses between the intake manifold and cylinder head cover
- Engine wiring harness bracket
- EGR valve hose and bracket
- Surge tank assembly
- Fuel supply hose
- Fuel injector assembly
- Fuel injector

7. Installation is the reverse order of removal. Note the following torques:

   a. Fuel injector assembly: 7–10 ft. lbs. (10–13 Nm).

   b. Surge tank assembly: 11–15 ft. lbs. (15–20 Nm).

   c. EGR valve hose mounting bolts: 12–19 ft. lbs. (17–26 Nm).

   d. EGR pipe fixing bracket bolts: 11–15 ft. lbs. (15–20 Nm).

   e. Surge tank stay mounting bolts: 11–15 ft. lbs. (15–20 Nm).

DRIVE TRAIN

Transaxle Assembly

REMOVAL & INSTALLATION

1. Before servicing the vehicle, refer to the Precautions Section.

2. Drain the transaxle.

3. Drain the engine oil.

4. Drain the cooling system.

5. Remove the engine and transaxle as an assembly.

6. Support the transaxle with a suitable jack.

7. Separate the transaxle from the engine by removing the transaxle flange bolts.

8. Installation is the reverse of removal.

Halfshaft

REMOVAL & INSTALLATION

1. Before servicing the vehicle, refer to the Precautions Section.

2. Remove or disconnect the following:

- Front wheel
- Spindle nut
- Lower ball joint from the knuckle

3. Press the stub shaft out of the hub, use a plaster hammer if necessary.

4. If removing the left side halfshaft:

   a. Using a suitable pry bar, pry the inner joint out of the transaxle.

5. If removing the right side halfshaft:

   a. Using a suitable pry bar, pry the inner joint out of the center bearing bracket.

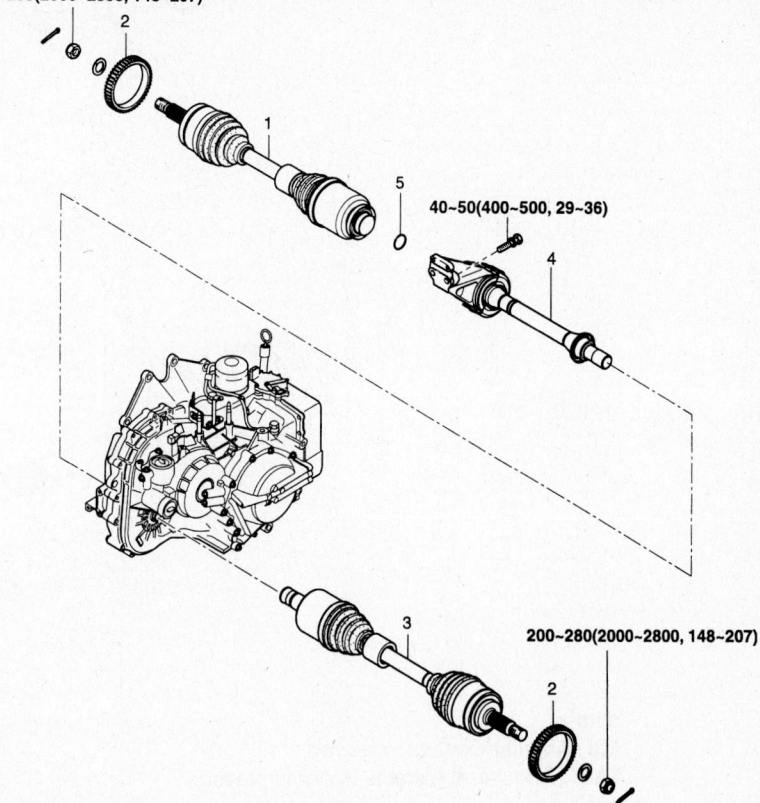

200~280(2000~2800, 148~207)

40~50(400~500, 29~36)

200~280(2000~2800, 148~207)

TORQUE : N·m (kg·cm, lb·ft)

1. Driveshaft-RH
2. Tone wheel (ABS System)
3. Driveshaft-LH

4. Center bearing and inner shaft assembly
5. Circlip

09474_AMAN_G0017

Exploded view of the halfshaft mounting and related components

b. Remove the center bearing bracket mounting bolts.

c. Using a suitable pry bar, pry the center bearing bracket from the cylinder block.

d. Remove the inner shaft from the transaxle.

To install:

➡️**Use new circlips, split pins and self-locking nuts for assembly.**

6. Apply gear oil on the halfshaft splines and differential case contact surfaces.

7. When installing the halfshaft, set the opening side of the circlip so it faces downward.

8. Install the inner joint so that the circlip is felt to seat in the retaining groove.

9. Install or connect the following:
- Lower ball joint. Tighten the mounting bolts to 74–88 ft. lbs. (100–120 Nm)
- Spindle nut and torque the nut to 148–207 ft. lbs. (200–280 Nm)
- Front wheel

10. Check and/or adjust the wheel alignment.

CV-Joints

OVERHAUL

Angular Adjustable Roller Improved-Angular Contact Improved (AARi-ACi) Type

1. Before servicing the vehicle, refer to the Precautions Section.

2. Remove the halfshaft and place into a vise.

3. Cut to remove the AARi and ACi big clamps.

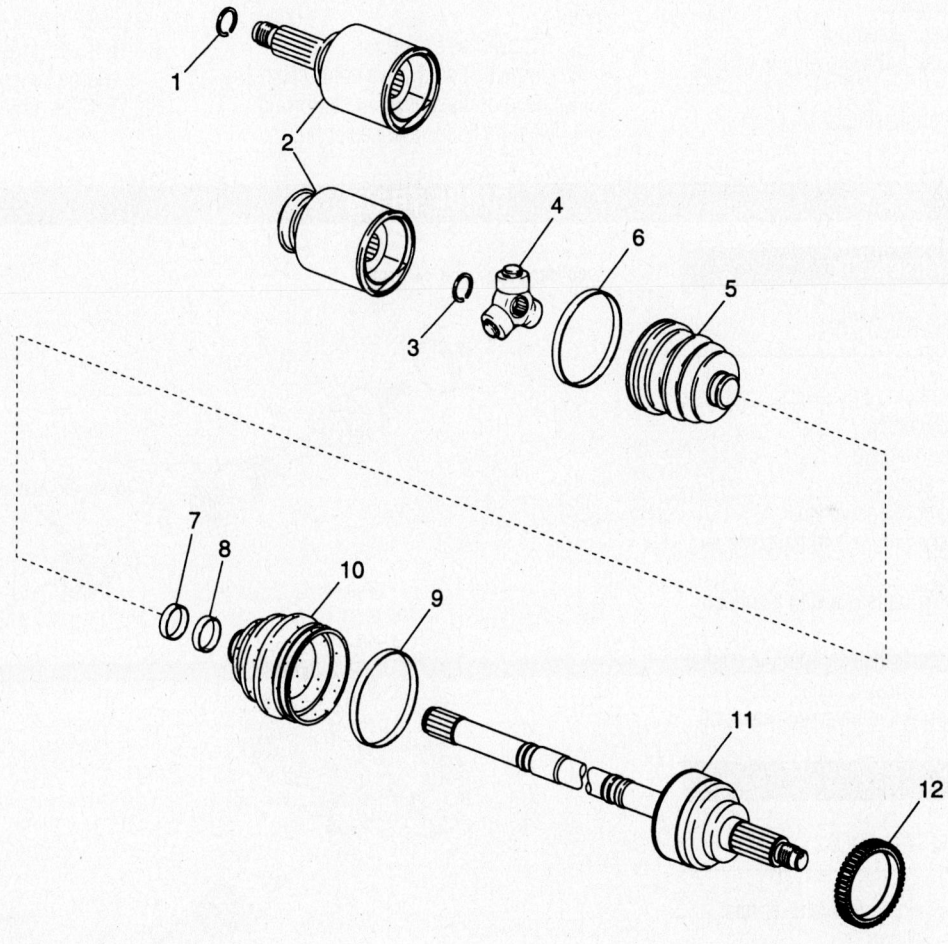

Note
ACi : Angular Contact improved
AARi : Angular Adjustable Roller improved

1. Clip
2. Tulip
3. Clip
4. Tripod assembly
5. AARi boot
6. AARi big clamp

7. AARi small clamp
8. ACi small clamp
9. ACi big clamp
10. ACi boot
11. Fixed joint
12. ABS Tone wheel

09474_AMAN_G0018

Exploded view of halfshaft components—AARi-ACi Type

Matchmarks on the tripod and tulip assemblies

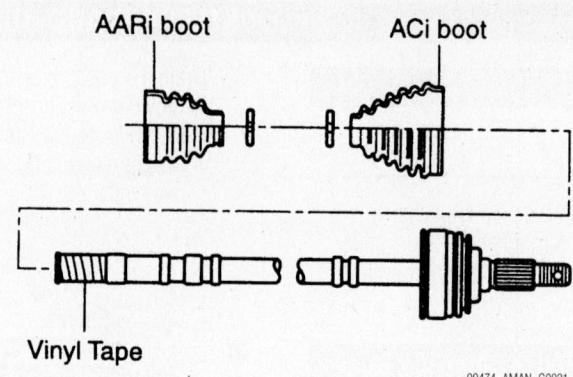

Boot identification and placement

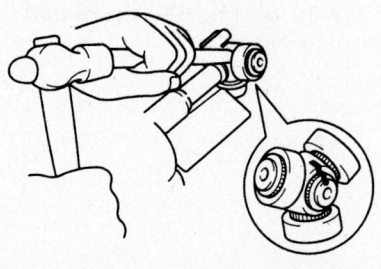

Matchmark the tripod and joint shaft

4. Using a brass drift and hammer, remove the AARi and ACi small clamps.

5. Matchmark the tripod and tulip assemblies to the halfshaft.

6. Remove the tulip assembly from the halfshaft.

7. Using a snap ring expander, remove the circlip from the tripod.

8. Matchmark the tripod and joint shaft.

9. Using a brass drift and hammer, remove the tripod from the joint shaft.

10. Place tape over the splines of the shaft (AARi side) and remove the ACi boot clamps.

11. Pull out the AARi and ACi boots.

To install:

➡**To the boots and boot clamps should be replaced when removed.**

12. Ensure there is tape over the splines of the shaft to prevent damage to the boots when installing.

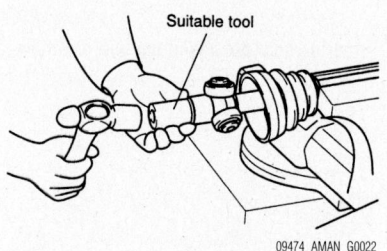

Tap the tripod onto the halfshaft

13. Using new boots, place new clamps to the small boot ends and install them onto the halfshaft.

14. Align the matchmarks and place the beveled side of the tripod axial splines toward the ACi joint. Using a suitable tool and hammer, tap the tripod onto the halfshaft.

15. Install a new circlip.

16. Fill the ACi and AARi boot with the specified grease to replace any that was removed during disassembly.

17. Align the matchmarks and install the AARi joint to the shaft.

18. Install the boot clamps, ensuring that the boots are properly seated on the halfshaft grooves.

19. Secure the large boot clamps to the specified distance of .079 inches (2 mm) as shown.

20. Secure the small clamps to the specified distance of 0.063 inches (1.6 mm) as shown.

21. Install the halfshaft into the vehicle.

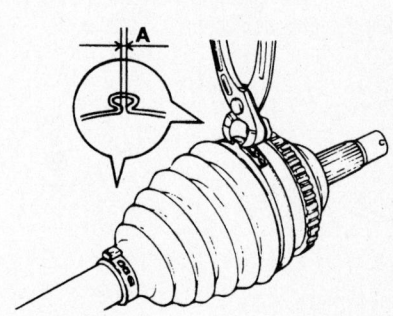

Closing the large boot clamps to the specified distance

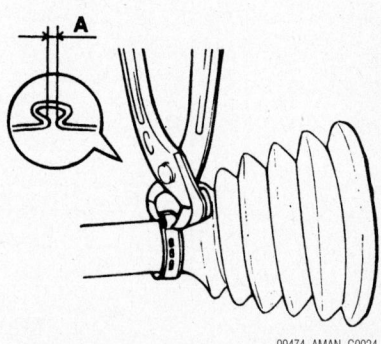

Closing the small boot clamps to the specified distance

STEERING AND SUSPENSION

Air Bag

DISARMING

Disconnect and isolate the negative battery cable. Wait 3 minutes for the system capacitor to discharge before performing any service.

Rack and Pinion Steering Gear

REMOVAL & INSTALLATION

1. Before servicing the vehicle, refer to the Precautions Section.

2. Drain the power steering fluid.
3. Remove or disconnect the following:
   - Negative battery cable.
   - Pressure hose and return tube from reservoir

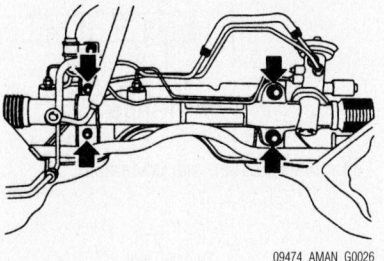

Steering gear box mounting bolt locations

09474_AMAN_G0026

Mounting rubber

Mounting clamp

09474_AMAN_G0027

Align the mounting rubber with the cross-member indentation

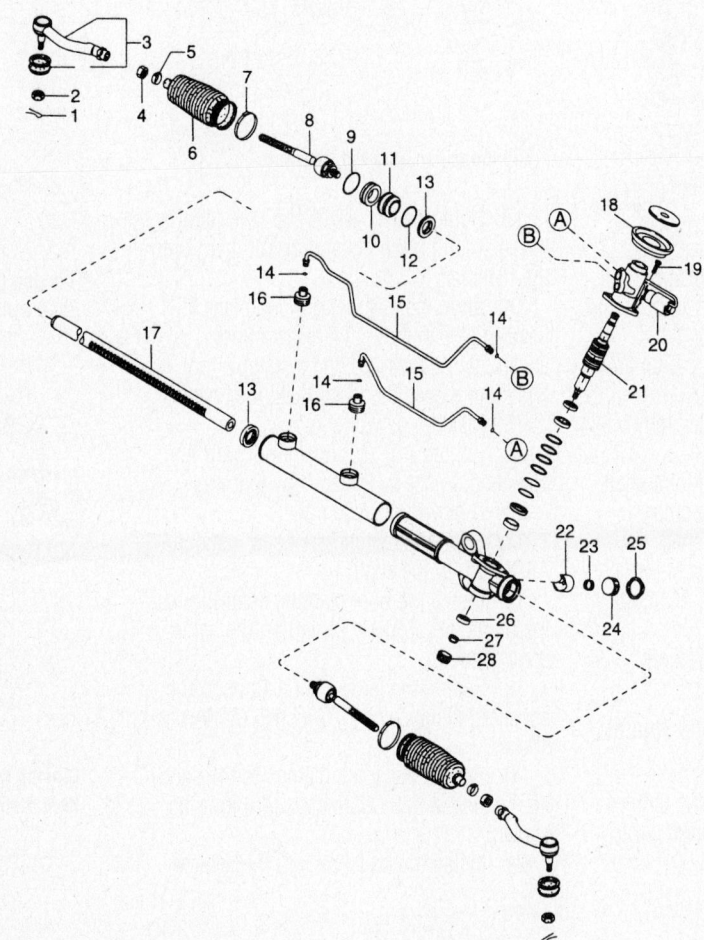

1. Cotter pin	8. Tie rod	15. Pressure pipe	22. Yoke assembly
2. Nut	9. Rack bushing	16. Damper valve assembly	23. Yoke spring
3. Tie rod end assembly	10. Insert bushing	17. Rack bar	24. Yoke cover
4. Lock nut	11. O-ring housing	18. Protector	25. Yoke nut
5. Clip	12. Oil seal	19. Cut off plug	26. Ball bearing
6. Bellows	13. Rack stopper	20. Solenoid valve assembly	27. Self locking nut
7. Steel band	14. Packing	21. Valve assembly	28. Pinion plug

09474_AMAN_G0025

Exploded view of the rack and pinion assembly

KIA

Amanti

17

ADJUSTMENT

Shoes

1. Before servicing the vehicle, refer to the Precautions Section.
2. Loosen the equalizer adjusting nut.
3. Remove the rubber plug from the backing plate.
4. Turn the adjuster until the rotor can no longer be turned by hand.
5. Back off the adjuster until the rotor turns freely.

Cable

1. Before servicing the vehicle, refer to the Precautions Section.
2. Turn the equalizer nut so that the brake lever travels 6–7 notches with an applied force of 66 lbs. (33 kg).
3. Check that the brakes are not dragging and tighten the cable locknut.

Shoe Break-in

1. Before servicing the vehicle, refer to the Precautions Section.

✳✳ CAUTION

Perform this procedure safely and in accordance with traffic laws!

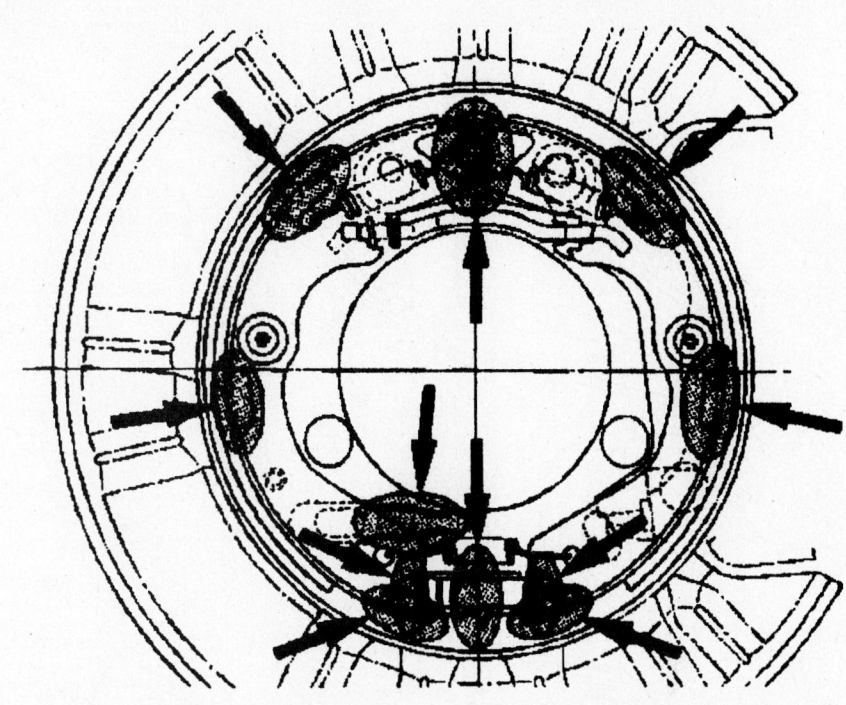

Brake grease application points

09474_RODE_G0089

2. With the parking brake lever pulled about half way, drive at 30 mph for about 30 seconds, then, drive in reverse at 6 mph for 18 seconds.

3. Check the parking brake lever stroke and adjust if necessary.

BRAKE HYDRAULIC SYSTEM

Brake Bleeding

1. Before servicing the vehicle, refer to the Precautions Section.

➡ **Do not reuse the drained fluid. Use only clean 3 Brake Fluid from an unopened container. Using a non-Honda brake fluid can cause corrosion and shorten the life of the system.**

Do not mix different brands of brake fluid; they may not be compatible.

Make sure no dirt or other foreign matter is allowed to contaminate the brake fluid.

Do not spill brake fluid on the vehicle, it may damage the paint; if brake fluid does contact the paint, wash it off immediately with water.

The reservoir on the master cylinder must be at the MAX (upper) level mark at the start of the bleeding procedure and checked after bleeding each brake caliper. Add fluid as required.

2. Make sure the brake fluid level in the reservoir is at the MAX (upper) level line.
3. Before proceeding, on vehicles with 4-wheel ABS, remove the ABS main fuse.
4. Attach a length of clear drain tube to the bleed screw.

5. Have someone slowly pump the brake pedal several times, then apply steady pressure.
6. Starting at the right rear, loosen the brake bleed screw to allow air to escape from the system. Then tighten the bleed screw securely.
7. Repeat the procedure for each caliper until no air bubbles are in the fluid. Bleed the calipers in this sequence: right rear, left rear, right front, left front.
8. Refill the master cylinder reservoir to the MAX (upper) level line.

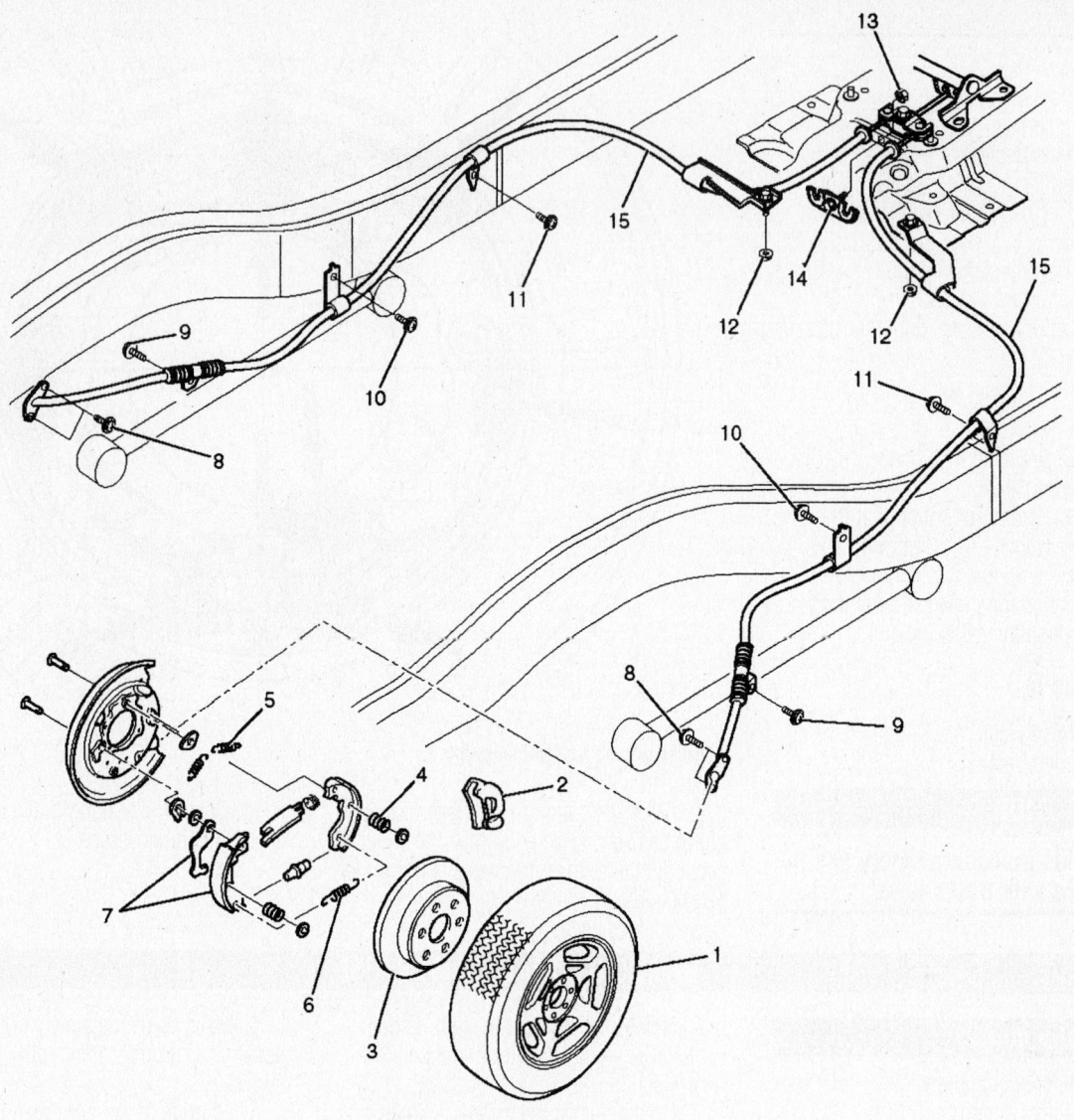

(1)　Rear Wheels

(2)　Caliper Assembly

(3)　Rotor (Drum)

(4)　Holding Spring

(5)　Upper Return Spring

(6)　Lower Return Spring

(7)　Shoe Assembly

(8)　Cable Fixing Bolt

(9)　Bolt

(10)　Bolt

(11)　Bolt (Only Long Wheel Base Model)

(12)　Nut

(13)　Nut

(14)　Retainer

(15)　Rear Cable

09474_RODE_G0088

Parking brake system— Passport, Rodeo and Rodeo Sport

- Steering joint assembly connecting bolt
- Tie rod end from steering knuckle using Special Tool 09568-34000 or equivalent.
- Front muffler
- Connecting bolts of the front and rear roll stopper
- Cross-member assembly mounting bolts
- Pressure hose and return tube from rack and pinion gear

- Steering gear box mounting bolts
- Steering gear box assembly

To install:

4. When installing the mounting rubber, align the projection of the mounting rubber with the indentation in the cross-member.

5. Installation is the reverse of removal.

6. Check the wheel alignment and adjust as necessary.

Struts

REMOVAL & INSTALLATION

1. Before servicing the vehicle, refer to the Precautions Section.

2. Remove or disconnect the following:
- Front wheel
- Brake hose bracket from shock absorber mounting fork
- Lower shock absorber mounting

(Non ECS)

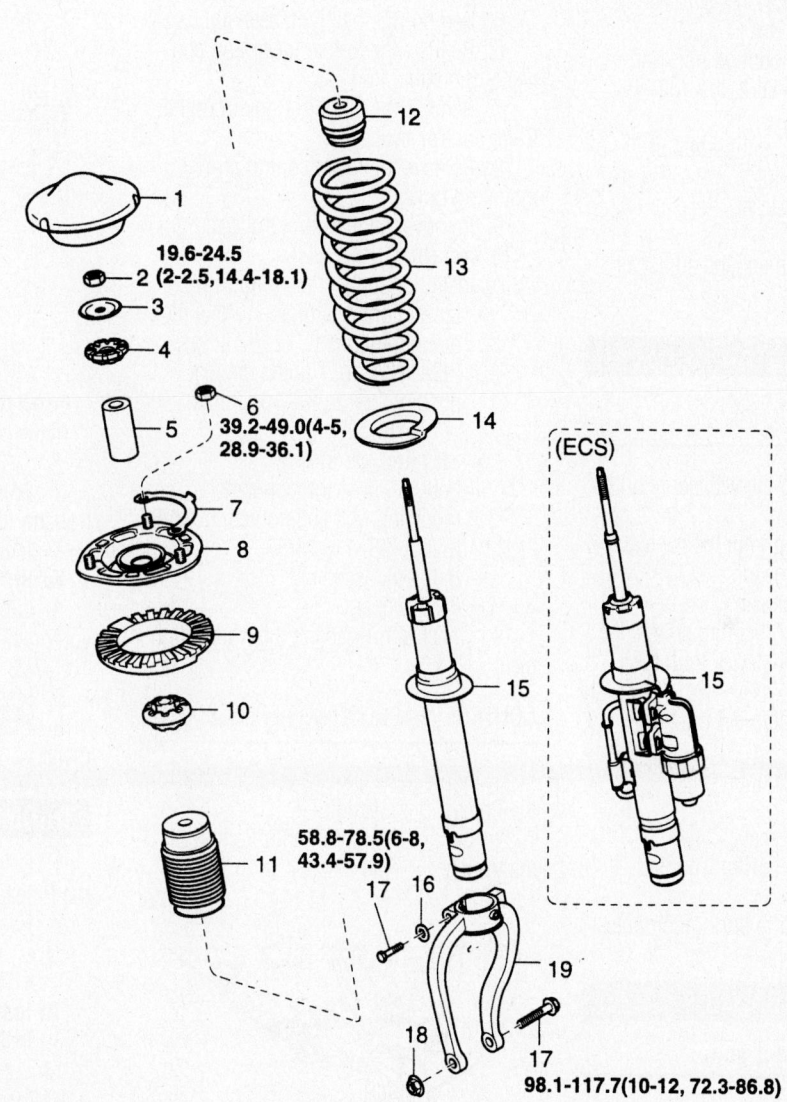

19.6-24.5 (2-2.5, 14.4-18.1)

39.2-49.0 (4-5, 28.9-36.1)

(ECS)

58.8-78.5 (6-8, 43.4-57.9)

98.1-117.7 (10-12, 72.3-86.8)

Tightening torque : N·m(kg·m, lb·ft)

1. Front cap
2. Self locking nut
3. Washer
4. Upper bushing (A)
5. Collar
6. Flange nut
7. Ring top mounting

8. Front bracket assembly
9. Front spring upper pad
10. Upper bushing (B)
11. Dust cover assembly
12. Rubber bumper
13. Front spring
14. Front spring lower pad

15. Front shock absorber assembly
16. Spring washer
17. Bolt
18. Nut
19. Front fork

Strut assembly exploded view

09474_AMAN_G0028

fork/lower arm connector mounting bolt
- Mounting fork from the shock absorber
- ECS wiring mounting bolt, if equipped
- Upper strut mounting nuts

3. Push the axle assembly upward and remove the strut assembly.

To install:

4. Install or connect the following:
- Strut assembly and tighten the lower mounting bolt to 72–87 ft. lbs. (98–118 Nm).
- Upper strut mounting nuts and tighten to 29–36 ft. lbs. (39–49 Nm).
- ECS wiring mounting bolt, if equipped
- Brake hose bracket
- Front wheel

5. Check the alignment and adjust as necessary.

Coil Spring

REMOVAL & INSTALLATION

1. Before servicing the vehicle, refer to the Precautions Section.
2. Remove the strut from the vehicle and install a spring compressor.
3. Remove the front cap.
4. Compress the coil spring so that the end of the spring comes away from the spring seat.
5. Remove or disconnect the following:
- Upper strut mounting nut
- Insulator
- Upper spring seat
- Compressed spring from the strut
- Spring from the spring compressor

✳✳ CAUTION

Do not use an impact gun.

To install:

6. Compress the spring and install it on the strut.
7. Install or connect the following:
- Upper spring seat
- Insulator
- Upper spring mount and torque the nut to 29–36 ft. lbs. (39–49 Nm).
- Strut to the vehicle

8. Check and/or adjust the wheel alignment.

Lower Control Arm

REMOVAL & INSTALLATION

1. Before servicing the vehicle, refer to the Precautions Section.
2. Remove the front wheel.
3. Loosen the ball joint nut, but do not remove.
4. Using Special Tool 09455-21000 or equivalent, disconnect the lower arm ball joint from the lower arm connector.
5. Remove the ball joint assembly.
6. Remove the fork to lower arm connector mounting bolt.
7. Remove the stabilizer link from the lower control arm.
8. Remove the lower control arm mounting bolts.
9. Remove the lower control arm.

To install:

10. Install or connect the following:
- Lower control arm and torque the mounting bolts to 72–87 ft. lbs. (98–118 Nm). Tighten the rear bushing bolt to 87–101 ft. lbs. (118–137 Nm).
- Stabilizer bar link
- Fork to lower arm connector mounting bolt and tighten to 72–87 ft. lbs. (98–118 Nm).
- Ball joint assembly
- Front wheel

11. Check and/or adjust the wheel alignment.

CONTROL ARM BUSHING REPLACEMENT

1. Before servicing the vehicle, refer to the Precautions Section.
2. Remove the lower control arm from the vehicle.

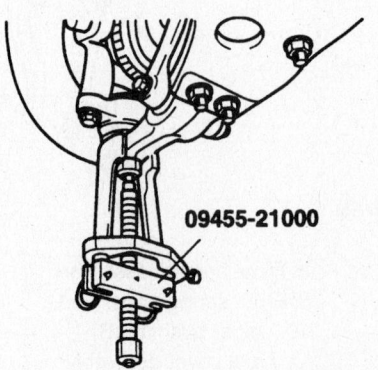

Disconnect the lower ball joint using Special Tool 094455-21000

09474_AMAN_G0029

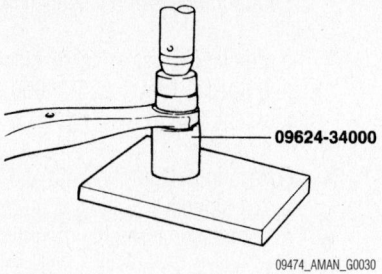

09624-34000

09474_AMAN_G0030

Removing the bushing from the control arm

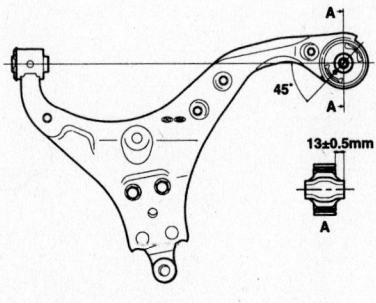

09474_TUCS_G0050

Install the bushings in the orientation shown

3. Press the rear bushing out of the control arm using Special Tool 09624-34000 or equivalent.

To install:

4. Lubricate the front bushing with soap and press into the control arm at an angle of 18° +/- 3°.
5. Install the control arm to the vehicle.
6. Check and/or adjust the wheel alignment.

Trailing Arm

1. Before servicing the vehicle, refer to the Precautions Section.
2. Remove or disconnect the following:
- Trailing arm mounting bolts
- Trailing arm

To install:

3. Install the trailing arm. Tighten the rear mounting bolt to 72–87 ft. lbs. (98–118 Nm). Tighten the front side mounting bolt to 101–116 ft. lbs. (137–157 Nm).

TRAILING ARM BUSHING REPLACEMENT

1. Before servicing the vehicle, refer to the Precautions Section.
2. Remove the trailing arm from the vehicle.
3. Press the bushing from the trailing arm using Special Tools 09529-21000 and 09216-21100.

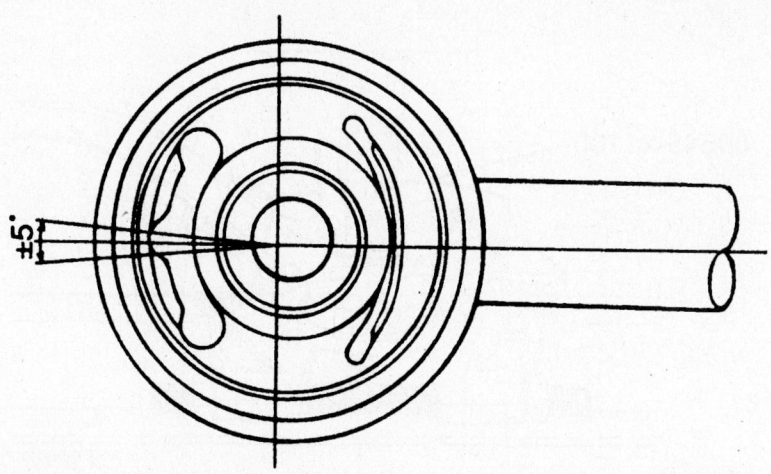

09474_AMAN_G0031

Trailing arm bushing orientation

To install:

4. Using Special Tools 09529-21000 and 09216-21100, press fit the rear trailing arm bushing into the trailing arm.

Upper Control Arm

REMOVAL & INSTALLATION

Front

1. Before servicing the vehicle, refer to the Precautions Section.
2. Remove the front wheel.
3. Loosen the ball joint nut, but do not remove.
4. Using Special Tool 09568-34000 or equivalent, disconnect the upper arm ball joint from the knuckle.
5. Remove the two nuts on the wheel house panel and remove the upper control arm assembly.
6. Installation is the reverse of removal.
7. Tighten the wheel house panel nuts to 65–80 ft. lbs. (88–108 Nm).

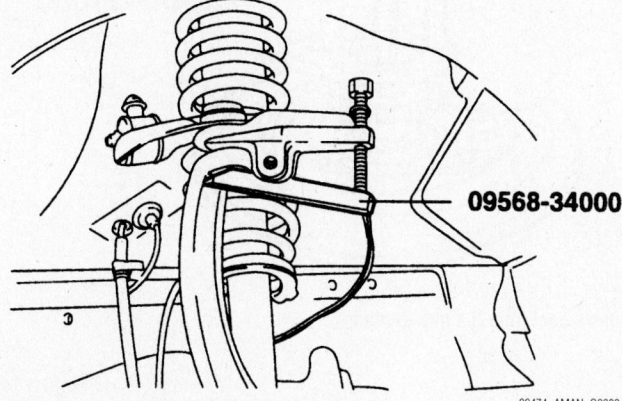

09474_AMAN_G0032

Disconnecting the upper ball joint from the knuckle

Rear

1. Before servicing the vehicle, refer to the Precautions Section.
2. Remove the rear wheel.
3. Remove upper control arm mounting bolts.
4. Remove the upper arm assembly.
5. Installation is the reverse of removal. Tighten the mounting bolts to 72–87 ft. lbs. (98–118 Nm).

Wheel Bearings

REMOVAL & INSTALLATION

Front

1. Before servicing the vehicle, refer to the Precautions Section.
2. Remove or disconnect the following:
- Rear wheel
- Brake caliper
- Wheel speed sensor, if equipped
- Tie rod end from the knuckle

- Lower ball joint from the knuckle
- Brake disc
- Spindle nut
- Halfshaft from the axle hub, using a plastic hammer if necessary.
- Upper control arm from the knuckle

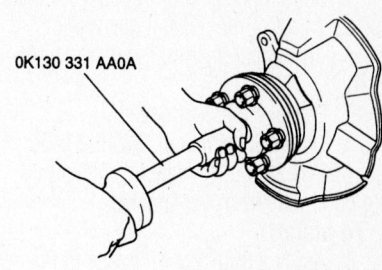

09474_AMAN_G0033

Disconnect the hub assembly from the knuckle

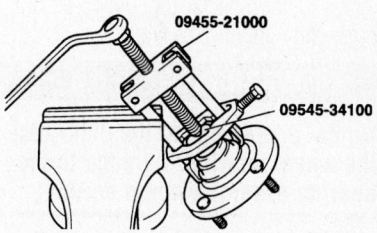

09474_AMAN_G0034

Remove the wheel bearing inner race from the hub

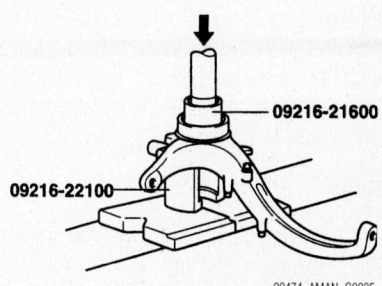

09474_AMAN_G0035

Press the outer race from the knuckle

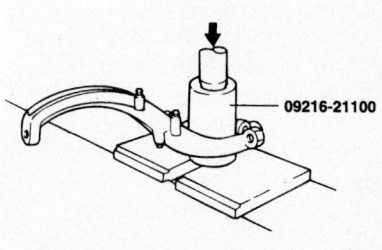

09474_AMAN_G0036

Press the bearing into the knuckle

• Front axle and knuckle as an assembly

3. Remove the snap ring from the knuckle assembly.

4. Using Special Tool 0K-130-331-AA0A or equivalent slide hammer, disconnect the hub from the knuckle.

5. Using special Tool 09455-21000 and 09545-34100 or equivalent gear puller, remove the wheel bearing inner race from the hub.

6. Remove the dust cover.

7. Using Special Tool 09216-21600 and 09216-22100, remove the wheel bearing outer race from the knuckle.

To install:

8. Apply a thin coat of grease to the knuckle and bearing contact surfaces.

9. Using Special Tool 09216-21100, press the bearing into the knuckle.

10. Install the snap ring into the groove of the knuckle.

11. Install the dust cover.

12. Using Special Tool 09545-21100, press the hub onto the knuckle.

❊❊ WARNING

Do not press against the outer race of the wheel bearing. Damage to the bearing assembly could occur.

13. Tighten the hub to the knuckle using Special Tool 09517-21500 to 148 ft. lbs. (200 Nm).

14. Rotate the hub to the seat the wheel bearing assembly.

15. Measure the wheel bearing turning torque using Special Tools 0951-21500 and 09532-11600. Torque value should equal 9 inch lbs. (1 Nm) or less.

16. The remainder of the installation is the reverse of removal.

Rear

1. Before servicing the vehicle, refer to the Precautions Section.

2. Release the parking brake.

3. Remove or disconnect the following:
• Rear wheel
• Wheel speed sensor, if equipped
• Caliper assembly
• Parking brake assembly
• Rear axle hub mounting bolts
• Hub assembly

4. Using Special Tool 09455-21000, remove the tone wheel from the hub assembly.

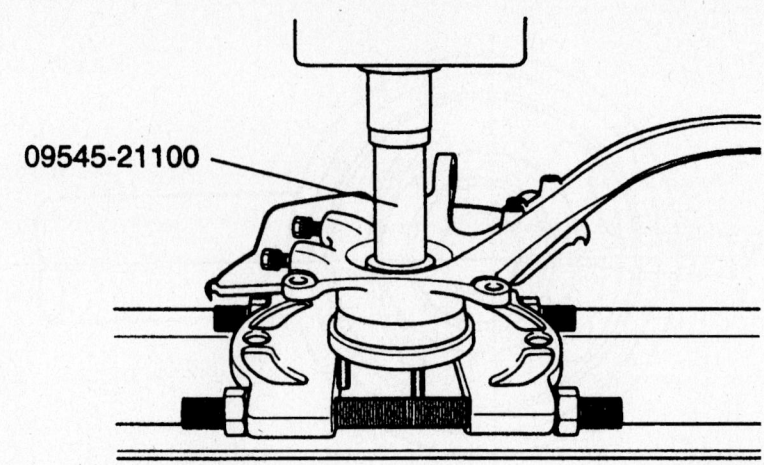

09474_AMAN_G0037

Pressing the hub into the knuckle

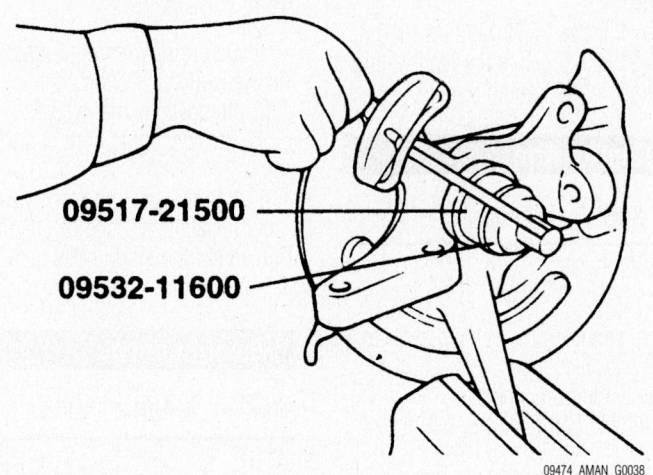

09474_AMAN_G0038

Measuring the turning torque of the wheel bearing

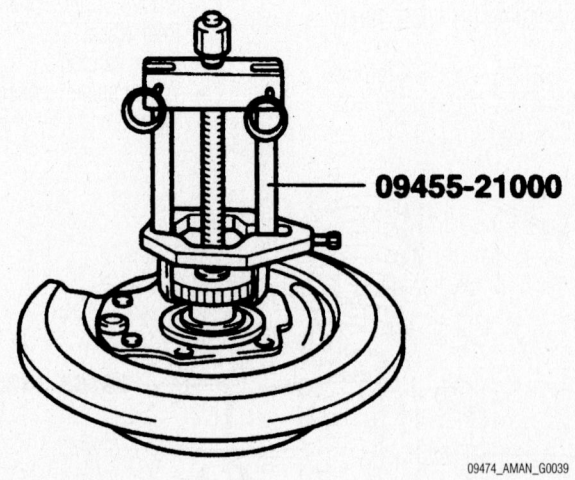

09474_AMAN_G0039

Removing tone wheel from the hub assembly

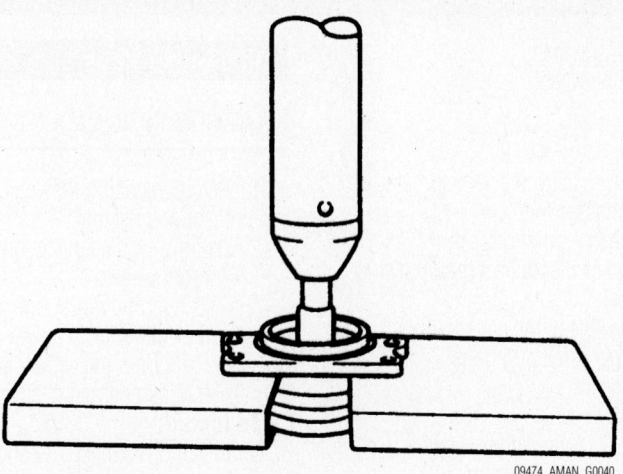

09474_AMAN_G0040

Press out the rear axle hub

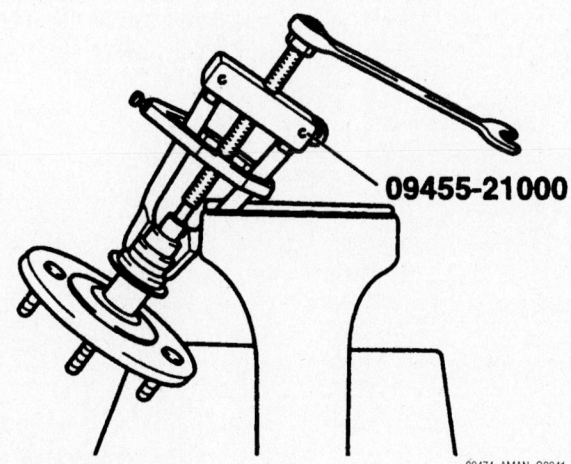

09455-21000

09474_AMAN_G0041

Remove bearing inner race from the hub

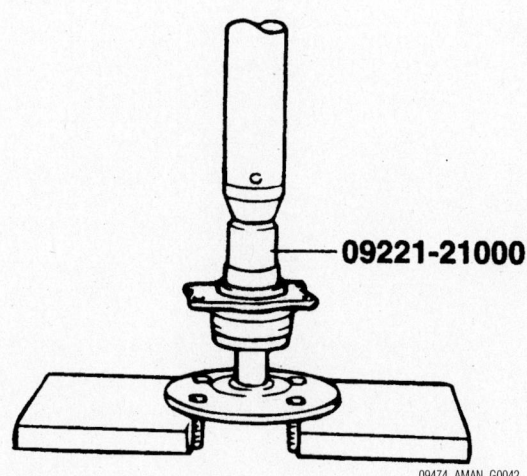

09221-21000

09474_AMAN_G0042

Press the bearing into the hub

5. Remove the flange nut from the hub assembly.

6. While supporting the flange area of the bearing outer race, press out the axle hub.

7. Using Special Tool 09544-21000, remove the bearing inner race from the axle hub.

To install:

8. Apply a thin coat of grease to the knuckle and bearing contact surfaces.

9. Using Special Tool 09221-21000, press the bearing into the hub.

❊❊ WARNING

Do not press against the outer race of the wheel bearing. Damage to the bearing assembly could occur.

10. After tightening the flange nut, stake the nut to meet the concave portion of the spindle.

11. Using Special Tool 09221-21000, press the tone wheel onto the hub assembly.

12. Fix the hub assembly to the brake backing plate so the rounded area of the bearing outer race is placed facing upward.

13. Install the hub mounting bolts and tighten to 59–74 ft. lbs. (100–120 Nm).

14. Rotate the hub to seat the bearing.

15. Using a spring balance, measure the wheel bearing turning torque. Torque should be 9 inch lbs. (1 Nm) or less.

16. The remainder of the installation is the reverse of removal.

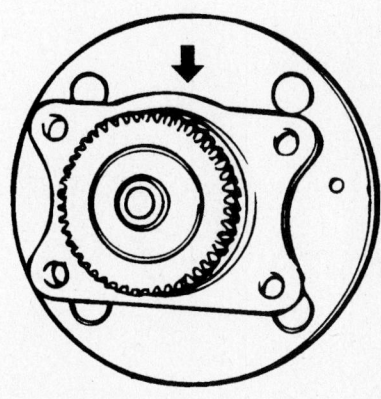

09474_AMAN_G0043

The rounded area of outer race should face upward

BRAKES

Brake Caliper

REMOVAL & INSTALLATION

Front

1. Before servicing the vehicle, refer to the Precautions Section.
2. Remove or disconnect the following:
   - Wheel
   - Brake hose at the caliper
   - Caliper mounting bolts
   - Caliper

To install:

3. Install or connect the following:
   - Caliper. Tighten the mounting bolts to 49–61 ft. lbs. (69–85 Nm).
   - Brake line to the caliper. Torque the brake line bolt to 12–14 ft. lbs. (17–20 Nm).

4. Bleed the system.
5. Install the wheel.

Rear

1. Before servicing the vehicle, refer to the Precautions Section.
2. Release the parking brake.
3. Remove or disconnect the following:
   - Wheel
   - Brake line at the caliper
   - Caliper mounting bolts
   - Caliper

To install:

4. Install or connect the following:
   - Caliper onto its mounting
   - Mounting bolts. Torque the bolts to 16–24 ft. lbs. (22–32 Nm).
   - Brake line to the caliper. Torque the brake line bolt to 12–14 ft. lbs. (17–20 Nm).

5. Bleed the system.
6. Install the wheel.

Disc Brake Pads

REMOVAL & INSTALLATION

1. Before servicing the vehicle, refer to the Precautions Section.
2. Remove or disconnect the following:
   - Front wheels
   - Caliper mounting bolt and suspend the caliper from a wire
   - Pads from the caliper support
   - Pad retainers, if necessary

To install:

3. Install or connect the following:
   - Pad retainers, if removed
   - Pads onto the pad retainers

4. Compress the caliper piston using a C-clamp. Rotate the caliper downward and install the mounting bolt. Tighten to 16–24 ft. lbs. (22–32 Nm).

5. Install the wheel.

KIA

18

Optima • Rio • Spectra

SPECIFICATIONS AND MAINTENANCE CHARTS

ENGINE AND VEHICLE IDENTIFICATION

Code ①	Liters (cc)	Cu. In.	Cyl.	Fuel Sys.	Engine Type	Eng. Mfg.
1	1.8 (1793)	109	4	EGI	DOHC	KIA
3	1.5 (1493)	91	4	EGI	DOHC	KIA
5	1.6 (1594)	97	4	EGI	DOHC	KIA
3	1.6 (1599)	97.6	4	EGI	DOHC	KIA
2	2.0 (1975)	120.5	4	EGI	DOHC	KIA
S	2.4 (2351)	144	4	EGI	DOHC	KIA
8	2.7 (2656)	163	6	EGI	DOHC	KIA

Model Year	
Code ②	Year
2	2002
3	2003
4	2004
5	2005
5	2006

EGI: Electronic Gasoline Injection

DOHC: Double Overhead Camshafts

① 8th digit of VIN

② 10th digit of VIN

09474_KIAC_C0001

GENERAL ENGINE SPECIFICATIONS

Year	Model	Engine Displacement Liters	Engine VIN	Net Horsepower @ rpm	Net Torque @ rpm (ft. lbs.)	Bore x Stroke (in.)	Compression Ratio	Oil Pressure @ rpm
2002	Rio	1.5	3	96@5800	98@4500	2.97x3.28	9.5:1	43-57@3000
	Spectra	1.8	1	125@6000	120@4500	3.19x3.43	9.5:1	43-57@3000
	Optima	2.4	S	149@6000	159@4500	3.41x3.94	10.0:1	43-57@3000
		2.7	8	178@6000	181@4000	3.41x2.95	10.0:1	43-57@3000
2003	Rio	1.6	5	104@5800	104@4700	3.07x3.28	10.0:1	43-57@3000
	Spectra	1.8	1	125@6000	120@4500	3.19x3.43	9.5:1	43-57@3000
	Optima	2.4	S	149@6000	159@4500	3.41x3.94	10.0:1	43-57@3000
		2.7	8	178@6000	181@4000	3.41x2.95	10.0:1	43-57@3000
2004	Rio	1.6	5	104@5800	104@4700	3.07x3.28	10.0:1	43-57@3000
	Spectra	1.8	1	125@6000	120@4500	3.19x3.43	9.5:1	43-57@3000
		2.0	2	138@6000	136@4500	3.23x8.68	10.1:1	43-57@3000
	Optima	2.4	S	149@6000	159@4500	3.41x3.94	10.0:1	43-57@3000
		2.7	8	178@6000	181@4000	3.41x2.95	10.0:1	43-57@3000
2005	Rio	1.6	5	104@5800	104@4700	3.07x3.28	10.0:1	43-57@3000
	Spectra	2.0	2	138@6000	136@4500	3.23x8.68	10.1:1	43-57@3000
	Optima	2.4	S	149@6000	159@4500	3.41x3.94	10.0:1	43-57@3000
		2.7	8	178@6000	181@4000	3.41x2.95	10.0:1	43-57@3000
2006	Rio	1.6	3	110@6000	107@4500	3.01x3.43	10.0:1	15.6@710-730
	Spectra	2.0	2	138@6000	136@4500	3.23x8.68	10.1:1	43-57@3000
	Optima	2.4	S	149@6000	159@4500	3.41x3.94	10.0:1	43-57@3000
		2.7	8	178@6000	181@4000	3.41x2.95	10.0:1	43-57@3000

09474_KIAC_C0002

ENGINE TUNE-UP SPECIFICATIONS

Year	Engine Displacement Liters	Engine VIN	Spark Plug Gap (in.)	Ignition Timing (deg.)		Fuel Pump (psi)	Idle Speed (rpm)		Valve Clearance	
				MT	AT		MT	AT	Intake	Exhaust
2002	1.5	3	0.039-0.043	1-11B	1-11B	65-94	700-800	700-800	HYD	HYD
	1.8	1	0.039-0.043	3-13B	3-13B	47-48	750-850	750-850	HYD	HYD
	2.4	S	0.039-0.043	3-7B	3-7B	46-49	700-900	700-900	HYD	HYD
	2.7	8	0.039-0.043	—	7-17B	46-49	—	600-800	HYD	HYD
2003	1.6	5	0.039-0.043	1-11B	1-11B	65-94	700-800	700-800	HYD	HYD
	1.8	1	0.039-0.043	3-13B	3-13B	47-48	750-850	750-850	HYD	HYD
	2.4	S	0.039-0.043	3-7B	3-7B	46-49	700-900	700-900	HYD	HYD
	2.7	8	0.039-0.043	—	7-17B	46-49	—	600-800	HYD	HYD
2004	1.6	5	0.039-0.043	1-11B	1-11B	65-94	700-800	700-800	HYD	HYD
	1.8	1	0.039-0.043	3-13B	3-13B	47-48	750-850	750-850	HYD	HYD
	2.0	2	0.039-0.043	3-7B	3-7B	49.8	750-850	750-850	HYD	HYD
	2.4	S	0.039-0.043	3-7B	3-7B	46-49	700-900	700-900	HYD	HYD
	2.7	8	0.039-0.043	—	7-17B	46-49	—	600-800	HYD	HYD
2005	1.6	5	0.039-0.043	1-11B	1-11B	65-94	700-800	700-800	HYD	HYD
	2.0	2	0.039-0.043	3-7B	3-7B	49.8	750-850	750-850	HYD	HYD
	2.4	S	0.039-0.043	3-7B	3-7B	46-49	700-900	700-900	HYD	HYD
	2.7	8	0.039-0.043	—	7-17B	46-49	—	600-800	HYD	HYD
2006	1.6	3	0.039-0.043	1-10B	1-10B	49.8	710-730	710-730	HYD	HYD
	2.0	2	0.039-0.043	3-7B	3-7B	49.8	750-850	750-850	HYD	HYD
	2.4	S	0.039-0.043	3-7B	3-7B	46-49	700-900	700-900	HYD	HYD
	2.7	8	0.039-0.043	—	7-17B	46-49	—	600-800	HYD	HYD

NOTE: The Vehicle Emission Control Information label often reflects specification changes made during production.
The label figures must be used if they differ from those in this chart.

B: Before top dead center

HYD: Hydraulic

09474_KIAC_C0003

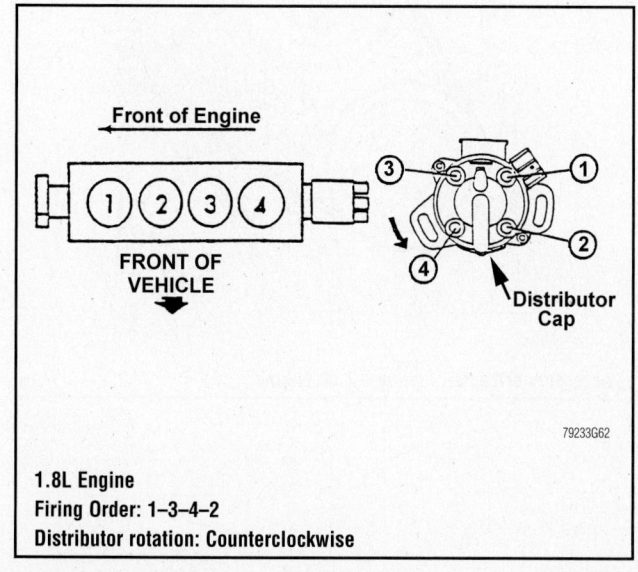

1.8L Engine
Firing Order: 1–3–4–2
Distributor rotation: Counterclockwise

79233G62

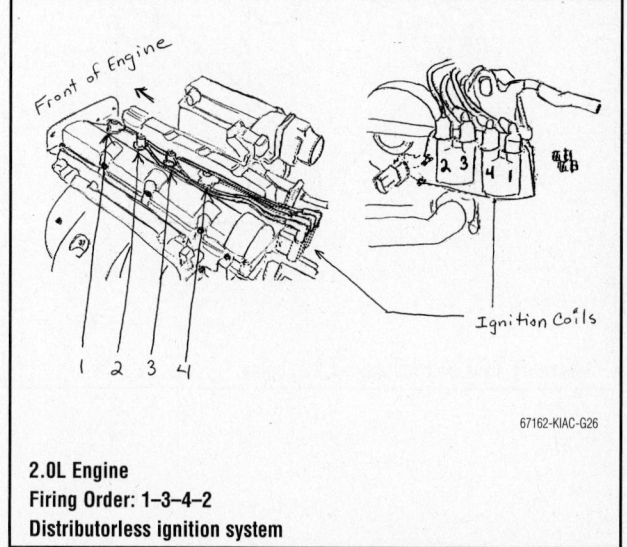

2.0L Engine
Firing Order: 1–3–4–2
Distributorless ignition system

67162-KIAC-G26

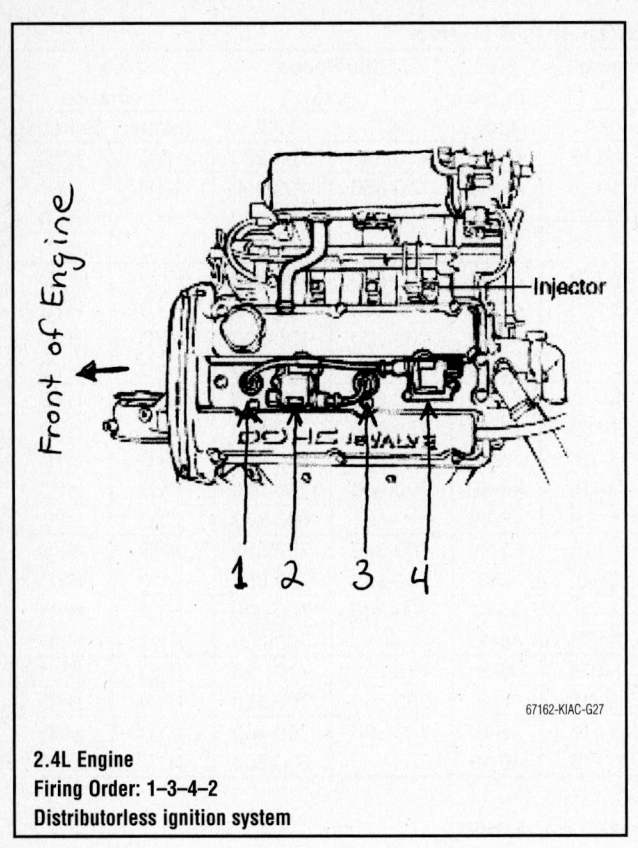

2.4L Engine
Firing Order: 1–3–4–2
Distributorless ignition system

67162-KIAC-G27

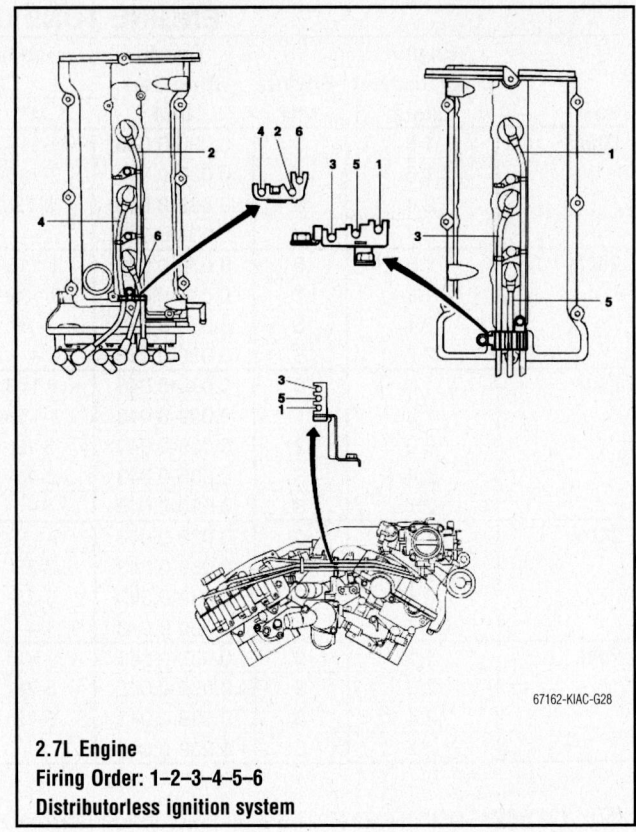

2.7L Engine
Firing Order: 1–2–3–4–5–6
Distributorless ignition system

67162-KIAC-G28

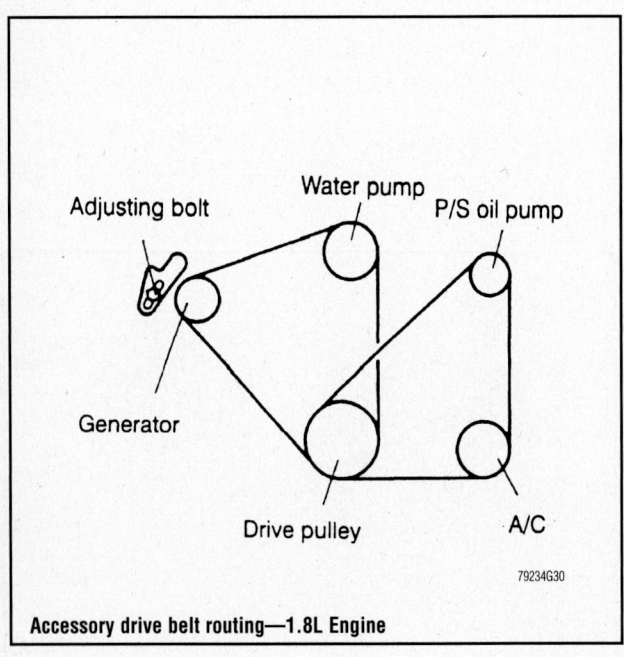

Accessory drive belt routing—1.8L Engine

79234G30

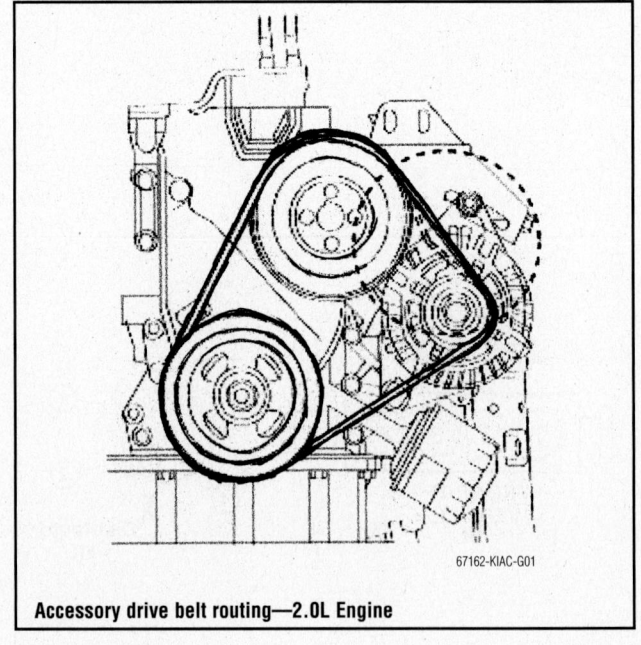

Accessory drive belt routing—2.0L Engine

67162-KIAC-G01

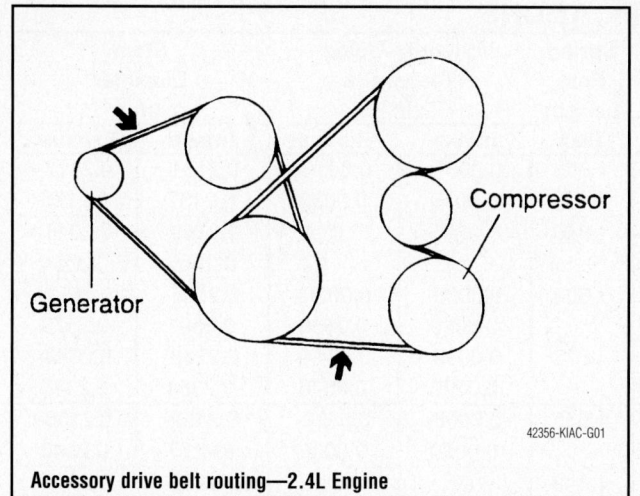

Accessory drive belt routing—2.4L Engine

42356-KIAC-G01

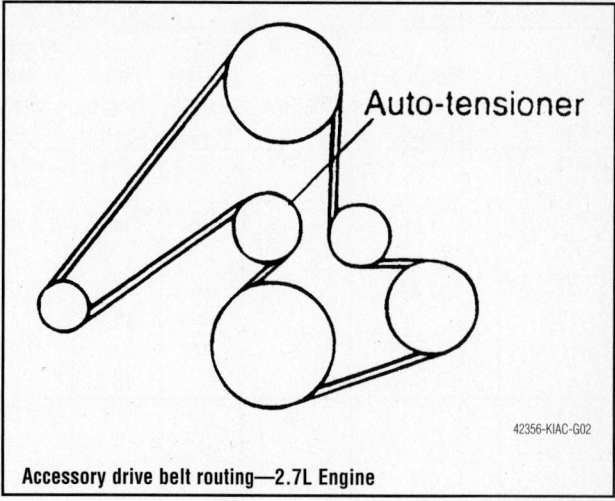

Accessory drive belt routing—2.7L Engine

42356-KIAC-G02

CAPACITIES

Year	Model	Engine Displacement Liters	Engine VIN	Engine Oil with Filter	Transaxle (pts.) Manual	Transaxle (pts.) Auto.	Fuel Tank (gal.)	Cooling System (qts.)
2002	Rio	1.5	3	3.4	5.8	12.4	11.9	6.3
	Spectra	1.8	1	4.0	5.6	11.4	13.2	6.3
	Optima	2.4	S	4.5	4.4	16.4	17.2	7.2
		2.7	8	4.75	—	16.4	17.2	9.1
2003	Rio	1.6	5	3.2	5.9	12.4	11.9	6.3
	Spectra	1.8	1	4.0	5.6	11.4	13.2	6.3
	Optima	2.4	S	4.5	4.4	16.4	17.2	7.2
		2.7	8	4.75	—	16.4	17.2	9.1
2004	Rio	1.6	5	3.2	5.9	12.4	11.9	6.3
	Spectra	1.8	1	4.0	5.6	11.4	13.2	6.3
		2.0	2	4.23	4.5	16.4	13.2	8.6
	Optima	2.4	S	4.5	4.4	16.4	17.2	7.2
		2.7	8	4.75	—	16.4	17.2	9.1
2005	Rio	1.6	5	3.2	5.9	12.4	11.9	6.3
	Spectra	2.0	2	4.23	4.5	16.4	13.2	8.6
	Optima	2.4	S	4.5	4.4	16.4	17.2	7.2
		2.7	8	4.75	—	16.4	17.2	9.1
2006	Rio	1.6	3	3.5	4.2	12.9	11.9	6.1
	Spectra	2.0	2	4.23	4.5	16.4	13.2	8.6
	Optima	2.4	S	4.5	4.4	16.4	17.2	7.2
		2.7	8	4.75	—	16.4	17.2	9.1

NOTE: All capacities are approximate. Add fluid gradually and ensure a proper level is obtained.

VALVE SPECIFICATIONS

Year	Engine Displacement Liters	Engine VIN	Seat Angle (deg.)	Face Angle (deg.)	Maximum out of Square (in.)	Spring Free Length (in.)	Stem-to-Guide Clearance (in.)		Stem Diameter (in.)	
							Intake	Exhaust	Intake	Exhaust
2002	1.5	3	45	45	0.433	1.653	0.0007-0.0019	0.0019-0.0033	0.2131-0.2137	0.2117-0.2125
	1.8	1	45	45	0.0638	1.840	①	②	0.2350-0.2356	0.2348-0.2354
	2.4	S	44-44.5	45-45.5	0.0638	1.804	0.0008-0.0019	0.0020-0.0033	0.2585-0.2591	0.2571-0.2579
	2.7	8	45	45-45.5	0.0638	1.673	0.0009-0.0020	0.0014-0.0026	0.2348-0.2354	0.2343-0.2348
2003	1.6	5	45	45	0.0402	1.535	0.0008-0.0020	0.0020-0.0033	0.2152-0.2157	0.2138-0.2146
	1.8	1	45	45	0.0638	1.840	①	②	0.2350-0.2356	0.2348-0.2354
	2.4	S	44-44.5	45-45.5	0.0638	1.804	0.0008-0.0019	0.0020-0.0033	0.2585-0.2591	0.2571-0.2579
	2.7	8	45	45-45.5	0.0638	1.673	0.0009-0.0020	0.0014-0.0026	0.2348-0.2354	0.2343-0.2348
2004	1.6	5	45	45	0.0402	1.535	0.0008-0.0020	0.0020-0.0033	0.2152-0.2157	0.2138-0.2146
	1.8	1	45	45	0.0638	1.840	①	②	0.2350-0.2356	0.2348-0.2354
	2.0	2	45-45.5	45-45.5	0.0638	1.923	0.0008-0.0020	0.0014-0.0026	0.2348-0.2354	0.2343-0.2348
	2.4	S	44-44.5	45-45.5	0.0638	1.804	0.0008-0.0019	0.0020-0.0033	0.2585-0.2591	0.2571-0.2579
	2.7	8	45	45-45.5	0.0638	1.673	0.0009-0.0020	0.0014-0.0026	0.2348-0.2354	0.2343-0.2348
2005	1.6	5	45	45	0.0402	1.535	0.0008-0.0020	0.0020-0.0033	0.2152-0.2157	0.2138-0.2146
	2.0	2	45-45.5	45-45.5	0.0638	1.923	0.0008-0.0020	0.0014-0.0026	0.2348-0.2354	0.2343-0.2348
	2.4	S	44-44.5	45-45.5	0.0638	1.804	0.0008-0.0019	0.0020-0.0033	0.2585-0.2591	0.2571-0.2579
	2.7	8	45	45-45.5	0.0638	1.673	0.0009-0.0020	0.0014-0.0026	0.2348-0.2354	0.2343-0.2348
2006	1.6	3	45-45.5	45-45.5	0.0402	1.732	0.0008-0.0020	0.0014-0.0026	0.2348-0.2354	0.2343-0.2348
	2.0	2	45-45.5	45-45.5	0.0638	1.923	0.0008-0.0020	0.0014-0.0026	0.2348-0.2354	0.2343-0.2348
	2.4	S	44-44.5	45-45.5	0.0638	1.804	0.0008-0.0019	0.0020-0.0033	0.2585-0.2591	0.2571-0.2579
	2.7	8	45	45-45.5	0.0638	1.673	0.0009-0.0020	0.0014-0.0026	0.2348-0.2354	0.2343-0.2348

NA: Not Available

① Standard range: 0.0010-0.0023 in.
  Maximum value: 0.0080 in.

② Standard range: 0.0012-0.0025 in.
  Maximum value: 0.0080 in.

CRANKSHAFT AND CONNECTING ROD SPECIFICATIONS

All measurements are given in inches.

Year	Engine Displacement Liters	Engine VIN	Crankshaft				Connecting Rod		
			Main Brg. Journal Dia.	Main Brg. Oil Clearance	Shaft End-play	Thrust on No.	Journal Diameter	Oil Clearance	Side Clearance
2002	1.5	3	1.9365-1.9562	0.0007-0.0014	0.0032-0.0111	3	1.7693-1.7700	0.0008-0.0019	0.0044-0.0103
	1.8	1	2.1629-2.1636	0.0010-0.0017	0.0032-0.0111	3	1.7693-1.7700	0.0008-0.0019	0.0044-0.0019
	2.4	S	2.2434-2.2442	①	0.0020-0.0098	3	1.7709-1.7717	0.0008-0.0020	0.0040-0.0098
	2.7	8	2.4402-2.4409	0.0002-0.0009	0.0028-0.0098	3	1.8891-1.8898	0.0007-0.0014	0.0039-0.0098
2003	1.6	5	1.9365-1.9562	0.0007-0.0014	0.0032-0.0111	3	1.7693-1.7700	0.0008-0.0019	0.0044-0.0103
	1.8	1	2.1629-2.1636	0.0010-0.0017	0.0032-0.0111	3	1.7693-1.7700	0.0008-0.0019	0.0044-0.0019
	2.4	S	2.2434-2.2442	①	0.0020-0.0098	3	1.7709-1.7717	0.0008-0.0020	0.0040-0.0098
	2.7	8	2.4402-2.4409	0.0002-0.0009	0.0028-0.0098	3	1.8891-1.8898	0.0007-0.0014	0.0039-0.0098
2004	1.6	5	1.9365-1.9562	0.0007-0.0014	0.0032-0.0111	3	1.7693-1.7700	0.0008-0.0019	0.0044-0.0103
	1.8	1	2.1629-2.1636	0.0010-0.0017	0.0032-0.0111	3	1.7693-1.7700	0.0008-0.0019	0.0044-0.0019
	2.0	2	2.2418-2.2426	0.0011-0.0019	0.0024-0.0102	3	1.8898-1.8905	0.0009-0.0017	0.0039-0.0098
	2.4	S	2.2434-2.2442	①	0.0020-0.0098	3	1.7709-1.7717	0.0008-0.0020	0.0040-0.0098
	2.7	8	2.4402-2.4409	0.0002-0.0009	0.0028-0.0098	3	1.8891-1.8898	0.0007-0.0014	0.0039-0.0098
2005	1.6	5	1.9365-1.9562	0.0007-0.0014	0.0032-0.0111	3	1.7693-1.7700	0.0008-0.0019	0.0044-0.0103
	1.8	1	2.1629-2.1636	0.0010-0.0017	0.0032-0.0111	3	1.7693-1.7700	0.0008-0.0019	0.0044-0.0019
	2.0	2	2.2418-2.2426	0.0011-0.0019	0.0024-0.0102	3	1.8898-1.8905	0.0009-0.0017	0.0039-0.0098
	2.4	S	2.2434-2.2442	①	0.0020-0.0098	3	1.7709-1.7717	0.0008-0.0020	0.0040-0.0098
	2.7	8	2.4402-2.4409	0.0002-0.0009	0.0028-0.0098	3	1.8891-1.8898	0.0007-0.0014	0.0039-0.0098
2006	1.6	3	1.9665-1.9672	②	0.0020-0.0069	3	1.8898-1.8905	0.0007-0.0014	0.0039-0.0098
	1.8	1	2.1629-2.1636	0.0010-0.0017	0.0032-0.0111	3	1.7693-1.7700	0.0008-0.0019	0.0044-0.0019
	2.0	2	2.2418-2.2426	0.0011-0.0019	0.0024-0.0102	3	1.8898-1.8905	0.0009-0.0017	0.0039-0.0098
	2.4	S	2.2434-2.2442	①	0.0020-0.0098	3	1.7709-1.7717	0.0008-0.0020	0.0040-0.0098
	2.7	8	2.4402-2.4409	0.0002-0.0009	0.0028-0.0098	3	1.8891-1.8898	0.0007-0.0014	0.0039-0.0098

① Journal Nos. 1, 2, 4 & 5: 0.0007-0.0014 in.
   Journal Nos. 3: 0.0009-0.0016 in.
② Journal Nos. 1, 2, 4 & 5: 0.0009-0.0016 in.
   Journal Nos. 3: 0.0011-0.0018 in.

PISTON AND RING SPECIFICATIONS

All measurements are given in inches.

Year	Engine Displacement Liters	Engine VIN	Piston Clearance	Ring Gap			Ring Side Clearance		
				Top Compression	Bottom Compression	Oil Control	Top Compression	Bottom Compression	Oil Control
2002	1.5	3	0.0015-0.0021	0.006-0.011	0.0016-0.0021	0.008-0.027	0.0020-0.0030	0.0010-0.0030	SNUG
	1.8	1	0.0015-0.0021	0.006-0.011	0.012-0.018	0.008-0.027	0.0020-0.0030	0.0010-0.0030	SNUG
	2.4	S	0.0008-0.0120	0.0098-0.0138	0.0157-0.0216	0.0039-0.0157	0.0012-0.0028	0.0008-0.0024	SNUG
	2.7	8	0.0004-0.0012	0.0079-0.0138	0.0146-0.0205	0.0079-0.0276	0.0016-0.0031	0.0012-0.0028	SNUG
2003	1.6	5	0.0015-0.0021	0.006-0.011	0.0016-0.0021	0.008-0.027	0.0012-0.0028	0.0012-0.0028	SNUG
	1.8	1	0.0015-0.0021	0.006-0.011	0.012-0.018	0.008-0.027	0.0020-0.0030	0.0010-0.0030	SNUG
	2.4	S	0.0008-0.0120	0.0098-0.0138	0.0157-0.0216	0.0039-0.0157	0.0012-0.0028	0.0008-0.0024	SNUG
	2.7	8	0.0004-0.0012	0.0079-0.0138	0.0146-0.0205	0.0079-0.0276	0.0016-0.0031	0.0012-0.0028	SNUG
2004	1.6	5	0.0015-0.0021	0.006-0.011	0.0016-0.0021	0.008-0.027	0.0012-0.0028	0.0012-0.0028	SNUG
	1.8	1	0.0015-0.0021	0.006-0.011	0.012-0.018	0.008-0.027	0.0020-0.0030	0.0010-0.0030	SNUG
	2.0	2	0.0008-0.0016	0.0091-0.0150	0.0130-0.0189	0.0079-0.0236	0.0016-0.0031	0.0012-0.0028	0.0024-0.0059
	2.4	S	0.0008-0.0120	0.0098-0.0138	0.0157-0.0216	0.0039-0.0157	0.0012-0.0028	0.0008-0.0024	SNUG
	2.7	8	0.0004-0.0012	0.0079-0.0138	0.0146-0.0205	0.0079-0.0276	0.0016-0.0031	0.0012-0.0028	SNUG
2005	1.6	5	0.0015-0.0021	0.006-0.011	0.0016-0.0021	0.008-0.027	0.0012-0.0028	0.0012-0.0028	SNUG
	2.0	2	0.0008-0.0016	0.0091-0.0150	0.0130-0.0189	0.0079-0.0236	0.0016-0.0031	0.0012-0.0028	0.0024-0.0059
	2.4	S	0.0008-0.0120	0.0098-0.0138	0.0157-0.0216	0.0039-0.0157	0.0012-0.0028	0.0008-0.0024	SNUG
	2.7	8	0.0004-0.0012	0.0079-0.0138	0.0146-0.0205	0.0079-0.0276	0.0016-0.0031	0.0012-0.0028	SNUG
2006	1.6	3	0.0008-0.0016	0.0059-0.0118	0.0138-0.0197	0.0079-0.0276	0.0016-0.0033	0.0016-0.0033	0.0031-0.0069
	2.0	2	0.0008-0.0016	0.0091-0.0150	0.0130-0.0189	0.0079-0.0236	0.0016-0.0031	0.0012-0.0028	0.0024-0.0059
	2.4	S	0.0008-0.0120	0.0098-0.0138	0.0157-0.0216	0.0039-0.0157	0.0012-0.0028	0.0008-0.0024	SNUG
	2.7	8	0.0004-0.0012	0.0079-0.0138	0.0146-0.0205	0.0079-0.0276	0.0016-0.0031	0.0012-0.0028	SNUG

09474_KIAC_C0007

TORQUE SPECIFICATIONS
All readings in ft. lbs.

Year	Engine Displacement Liters	Engine VIN	Cylinder Head Bolts	Main Bearing Bolts	Rod Bearing Bolts	Crankshaft Damper Bolts	Flywheel Bolts	Manifold		Spark Plugs	Oil Pan Drain Plug
								Intake	Exhaust		
2002	1.5	3	①	40-43	22-24	28-38	71-76	11-14	11-14	11-17	22-30
	1.8	1	②	③	35-37	9-13 ④	71-76	14-19	28-34	11-17	29-32.5
	2.4	S	⑤	⑥	⑦	80-94	94-101	⑧	⑨	15-21	26-32
	2.7	8	⑩	⑪	⑫	130-138	53-55	14-15	12-22	15-21	26-32
2003	1.6	5	①	40-43	22-24	28-38	71-76	11-14	11-14	11-17	22-30
	1.8	1	②	③	35-37	9-13 ④	71-76	14-19	28-34	11-17	29-32.5
	2.4	S	⑤	⑥	⑦	80-94	94-101	⑧	⑨	15-21	26-32
	2.7	8	⑩	⑪	⑫	130-138	53-55	14-15	12-22	15-21	26-32
2004	1.6	5	①	40-43	22-24	28-38	71-76	11-14	11-14	11-17	22-30
	1.8	1	②	③	35-37	9-13 ④	71-76	14-19	28-34	11-17	29-32.5
	2.0	2	⑬	⑭	36.2-38.3	123-130.2	86.8-94	13-16.5	31-40	11-17	29-32.5
	2.4	S	⑤	⑥	⑦	80-94	94-101	⑧	⑨	15-21	26-32
	2.7	8	⑩	⑪	⑫	130-138	53-55	14-15	12-22	15-21	26-32
2005	1.6	5	①	40-43	22-24	28-38	71-76	11-14	11-14	11-17	22-30
	2.0	2	⑬	⑭	36.2-38.3	123-130.2	86.8-94	13-16.5	31-40	11-17	29-32.5
	2.4	S	⑤	⑥	⑦	80-94	94-101	⑧	⑨	15-21	26-32
	2.7	8	⑩	⑪	⑫	130-138	53-55	14-15	12-22	15-21	26-32
2006	1.6	3	⑮	40-43	23-25	101-109	87-94	11-15	13-18	11-17	29-33
	2.0	2	⑬	⑭	36.2-38.3	123-130.2	87-94	13-16.5	31-40	11-17	29-33
	2.4	S	⑤	⑥	⑦	80-94	94-101	⑧	⑨	15-21	26-32
	2.7	8	⑩	⑪	⑫	130-138	53-55	14-15	12-22	15-21	26-32

① Step 1: 36 inch lbs.
   Step 2: Loosen fully
   Step 3: 18 ft. lbs.
   Step 4: Tighten 90 degrees
   Step 5: Tighten 60 degrees

② Step 1: 36 ft. lbs.
   Step 2: Loosen fully
   Step 3: 29 ft. lbs.
   Step 4: Tighten 90 degrees
   Step 5: Additional 90 degrees

③ Step 1: 29 ft. lbs.
   Step 2: Loosen fully
   Step 3: 14.5 ft. lbs.
   Step 4: Tighten 90 degrees
   Step 5: Tighten 60 degrees

④ Crankshaft pulley

⑤ Step 1: 14 ft. lbs.
   Step 2: Loosen fully
   Step 3: 14 ft. lbs.
   Step 4: Tighten 90 degrees
   Step 5: Tighten 90 degrees

⑥ 18 ft. lbs., plus 90-94 degrees

⑦ 13-16 ft. lbs., plus 90-94 degrees

⑧ M10 bolts: 13-18 ft. lbs.
   M8 bolts: 11-14 ft. lbs.
   Nut: 22-30 ft. lbs.

⑨ M10 bolts: 25-40 ft. lbs.
   M8 bolts: 18-22 ft. lbs.

⑩ Step 1: 18 ft. lbs., plus 60-65 degrees
   Step 2: Tighten 45-49 degrees
   Step 3: 26-32 ft. lbs.

⑪ M10 bolts: 19.5-23.9 ft. lbs., plus 90-95 degrees
   M8 bolts: 9.6-14 ft. lbs., plus 90-95 degrees

⑫ 11.6-14 ft. lbs., plus 90-94 degrees

⑬ M10 bolts: 17-19.5 ft. lbs., plus 60-65 degrees, plus additional 60-65 degrees
   M12 bolts: 20.5-23 ft. lbs., plus 60-65 degrees, plus additional 60-65 degrees

⑭ 20.3-23.1 ft. lbs., plus 60-64 degrees

⑮ Step 1: 22 ft. lbs.
   Step 2: Plus 90 degrees
   Step 3: Loosen fully
   Step 4: 22 ft. lbs.
   Step 5: Tighten 90 degrees

09474_KIAC_C0006

WHEEL ALIGNMENT

Year	Model		Caster Range (+/-Deg.)	Caster Preferred Setting (Deg.)	Camber Range (+/-Deg.)	Camber Preferred Setting (Deg.)	Toe-in (in.)
2002	Rio	F	0.45	+1.41	0.45	+0.6	0.12 +/- 0.12
		R	—	—	0.18	-0.50	0.20 +/- 0.24
	Spectra	F	0.75	+2.45	0.50	0	0.11 +/- 0.12
		R	—	—	0.50	-0.52	0.07 +/- 0.12
	Optima	F	1.00	+3.15	0.30	0	0.11 +/- 0.12
		R	—	—	0.30	-0.30	0.07 +/- 0.12
2003	Rio	F	0.45	+1.41	0.45	+0.6	0.12 +/- 0.12
		R	—	—	0.18	-0.50	0.20 +/- 0.24
	Spectra	F	0.75	+2.45	0.50	0	0.11 +/- 0.12
		R	—	—	0.50	-0.52	0.07 +/- 0.12
	Optima	F	1.00	+3.15	0.30	0	0.11 +/- 0.12
		R	—	—	0.30	-0.30	0.07 +/- 0.12
2004	Rio	F	0.45	+1.41	0.45	+0.6	0.12 +/- 0.12
		R	—	—	0.18	-0.50	0.20 +/- 0.24
	Spectra	F	0.75	+2.45	0.50	0	0.11 +/- 0.12
		R	—	—	0.50	-0.52	0.07 +/- 0.12
	Optima	F	1.00	+3.15	0.30	0	0.11 +/- 0.12
		R	—	—	0.30	-0.30	0.07 +/- 0.12
2005	Rio	F	0.45	+1.41	0.45	+0.6	0.12 +/- 0.12
		R	—	—	0.18	-0.50	0.20 +/- 0.24
	Spectra	F	0.50	+2.60	0.50	0	0 +/- 0.08
		R	—	—	0.50	-0.92	0.16 +/- 0.08
	Optima	F	1.00	+3.15	0.30	0	0.11 +/- 0.12
		R	—	—	0.30	-0.30	0.07 +/- 0.12
2006	Rio	F	0.50	0	4.00	+0.5	0.08 +/- 0.08
		R	—	—	1.00	-0.50	0.08 +/- 0.24
	Spectra	F	0.50	+2.60	0.50	0	0 +/- 0.08
		R	—	—	0.50	-0.92	0.16 +/- 0.08
	Optima	F	1.00	+3.15	0.30	0	0.11 +/- 0.12
		R	—	—	0.30	-0.30	0.07 +/- 0.12

09474_KIAC_C0009

TIRE, WHEEL AND BALL JOINT SPECIFICATIONS

| Year | Model | OEM Tires | | Tire Pressures (psi) | | Wheel Size | Ball Joint Inspection | Wheel Lug Torque (ft. lbs.) |
		Standard	Optional	Front	Rear			
2002	Rio	P155/80SR13	P175/70R13	①	①	Std: 5-JJ Opt: 5-JJ	②	65-87
	Spectra	P175/70SR13	P185/60HR14	29	29	Std: 5-JJ Opt: 6-JJ	②	65-87
	Optima	P175/70R14	P205/60R15	30	30	Std: 5.5-JJ Opt: 6-JJ	②	65-87
2003	Rio	P155/80SR13	P175/70R13	①	①	Std: 5-JJ Opt: 5-JJ	②	65-87
	Spectra	P175/70SR13	P185/60HR14	29	29	Std: 5-JJ Opt: 6-JJ	②	65-87
	Optima	P175/70R14	P205/60R15	30	30	Std: 5.5-JJ Opt: 6-JJ	②	65-87
2004	Rio	P155/80SR13	P175/70R13	①	①	Std: 5-JJ Opt: 5-JJ	②	65-87
	Spectra	P175/70SR13	P185/60HR14	29	29	Std: 5-JJ Opt: 6-JJ	②	65-87
	Optima	P175/70R14	P205/60R15	30	30	Std: 5.5-JJ Opt: 6-JJ	②	65-87
2005	Rio	P175/70R13	P175/65R14	29	29	Std: 5-JJ Opt: 5-JJ	②	65-87
	Spectra	P195/60R15	P205/50R16	30	30	6.0-J	②	65-87
	Optima	P175/70R14	P205/60R15	30	30	Std: 5.5-JJ Opt: 6-JJ	②	65-87
2006	Rio	P155/80SR13	P175/70R13	29	29	Std: 5.0-J Opt: 5.5-J	②	65-87
	Spectra	P195/60R15	P205/50R16	30	30	6.0-J	②	65-87
	Optima	P205/60R15	P205/55R16	30	30	6.0-J	②	65-87

OEM: Original Equipment Manufacturer

PSI: Pounds Per Square Inch

STD: Standard

OPT: Optional

① Standard front and rear tire pressure: 32 psi

   Optional front and rear tire pressure: 29 psi

② Replace if any measurable movement is found.

09474_KIAC_C0010

BRAKE SPECIFICATIONS
All measurements in inches unless noted

Year	Model		Brake Disc – Original Thickness	Brake Disc – Minimum Thickness	Brake Disc – Maximum Run-out	Brake Drum – Original Inside Diameter	Brake Drum – Max. Wear Limit	Brake Drum – Maximum Machine Diameter	Minimum Lining Thickness	Brake Caliper – Bracket Bolts (ft. lbs.)	Brake Caliper – Mounting Bolts (ft. lbs.)
2002	Rio	F	0.870	0.710	0.0040	—	—	—	0.080	33-49	19-21
		R	—	—	—	7.87	7.91	7.91	0.079	33-49	22-29
	Spectra	F	0.940	0.710	0.0040	—	—	—	0.080	33-49	19-21
		R	0.400	0.320	0.0039	7.87	7.91	7.91	0.079	33-49	22-29
	Optima	F	0.965	0.880	0.0012	—	—	—	0.079	51-63	16-24
		R	0.390	0.320	0.0039	9.00	9.08	9.08	0.079	51-63	16-24
2003	Rio	F	0.870	0.710	0.0040	—	—	—	0.080	33-49	19-21
		R	—	—	—	7.87	7.91	7.91	0.079	33-49	22-29
	Spectra	F	0.940	0.710	0.0040	—	—	—	0.080	33-49	19-21
		R	0.400	0.320	0.0039	7.87	7.91	7.91	0.079	33-49	22-29
	Optima	F	0.965	0.880	0.0012	—	—	—	0.079	51-63	16-24
		R	0.390	0.320	0.0039	9.00	9.08	9.08	0.079	51-63	16-24
2004	Rio	F	0.870	0.710	0.0040	—	—	—	0.080	33-49	19-21
		R	—	—	—	7.87	7.91	7.91	0.079	33-49	22-29
	Spectra	F	0.940	0.710	0.0040	—	—	—	0.080	33-49	19-21
		R	0.400	0.320	0.0039	7.87	7.91	7.91	0.079	33-49	22-29
	Optima	F	0.965	0.880	0.0012	—	—	—	0.079	51-63	16-24
		R	0.390	0.320	0.0039	9.00	9.08	9.08	0.079	51-63	16-24
2005	Rio	F	0.870	0.790	0.0020	—	—	—	0.080	33-49	19-21
		R	—	—	—	7.87	7.91	7.91	0.079	33-49	22-29
	Spectra	F	0.940	0.710	0.0040	—	—	—	0.080	33-49	19-21
		R	0.400	0.320	0.0039	7.87	7.91	7.91	0.079	33-49	22-29
	Optima	F	0.965	0.880	0.0012	—	—	—	0.079	51-63	16-24
		R	0.390	0.320	0.0039	9.00	9.08	9.08	0.079	51-63	16-24
2006	Rio	F	0.870	0.790	0.0012	—	—	—	0.079	58-72	16-23
		R	0.390	0.315	0.0012	7.87	7.91	7.91	0.079	62-69	16-23
	Spectra	F	0.940	0.710	0.0040	—	—	—	0.080	33-49	19-21
		R	0.400	0.320	0.0039	7.87	7.91	7.91	0.079	33-49	22-29
	Optima	F	0.965	0.880	0.0012	—	—	—	0.079	51-63	16-24
		R	0.390	0.320	0.0039	9.00	9.08	9.08	0.079	51-63	16-24

F: Front

R: Rear

09474_KIAC_C0011

SCHEDULED MAINTENANCE INTERVALS
Kia - Spectra, Rio and Optima

TO BE SERVICED	TYPE OF SERVICE	VEHICLE MILEAGE INTERVAL (x1000)												
		7.5	15	22.5	30	37.5	45	52.5	60	67.5	75	82.5	90	97.5
Accessory drive belts	S/I			✓			✓			✓			✓	
Air cleaner filter	R		✓		✓		✓		✓		✓		✓	
Air conditioner system	S/I													
Brake lines, hoses and connections	S/I		✓		✓		✓		✓		✓		✓	
Chassis and body fasteners	T	✓	✓	✓	✓	✓	✓	✓	✓	✓	✓	✓	✓	✓
Cooling system hoses and coolant level	S/I		✓		✓		✓		✓		✓		✓	
CV-joint boots	S/I				✓				✓				✓	
Engine coolant	R				✓				✓				✓	
Engine oil and filter	R	✓	✓	✓	✓	✓	✓	✓	✓	✓	✓	✓	✓	✓
Exhaust system heat shields	S/I				✓				✓				✓	
Front and rear brakes	S/I				✓				✓				✓	
Front ball joints	S/I				✓				✓				✓	
Fuel filter	R								✓					
Fuel lines and hoses	S/I				✓				✓				✓	
Locks and hinges	L	✓	✓	✓	✓	✓	✓	✓	✓	✓	✓	✓	✓	✓
Spark plugs	R				✓				✓				✓	
Steering operation and linkage	S/I				✓				✓				✓	
Timing belt	R								✓					

R: Replace   S/I: Service or Inspect   L: Lubricate   T: Tighten

FREQUENT OPERATION MAINTENANCE (SEVERE SERVICE)

If a vehicle is operated under any of the following conditions it is considered severe service

- Towing a trailer or using a camper or car-top carrier

- Repeated short trips of less than 5 miles in temperatures below freezing, or trips of less than 10 miles in any temperature

- Prolonged idling (vehicle operation in stop and go traffic).

- Operating on rough, muddy, unpaved, dusty or salt-covered roads.

- Police, taxi, delivery usage or trailer towing usage.

- Driving in extremely hot (over 90°F) conditions

Oil & oil filter: change every 5000 miles or 5 months, whichever occurs first.

Air cleaner filter: inspect every 15,000 miles or 15 months and replace everything 30,000 miles or 30 months, whichever occurs first

Fuel system hoses (California models only): replace every 105,000 miles

Emission system hoses (non-CA models): inspect every 55,000 or 55 months, whichever occurs first

Emission system hoses (CA models): inspect every 60,000 miles or 60 months, which occurs first

Front and rear brakes: inspect every 15,000 miles or 15 months, whichever occurs first

Chassis and body fasteners: tighten every 15,000 miles or 15 months, whichever occurs first

Locks and hinges: lubricate every 5000 miles or 5 months, whichever occurs first

ENGINE REPAIR

Alternator

REMOVAL

1. Before servicing the vehicle, refer to the Precautions Section.

2. Disconnect the negative battery cable.

3. Temporarily loosen the water pump pulley bolts, 1.6L (VIN 3) engine only.

4. Remove the front air intake inlet pipe bolts, 1.8L engine only.

5. Remove the top hose from the air intake inlet pipe, 1.8L engine only.

6. Remove the air intake inlet pipe clamp and the pipe, 1.8L engine only.

7. Remove the power steering pump, 2.0L engine only.

8. Remove the power steering pump bracket, 2.0L engine only.

9. Remove the alternator **B** terminal cover cap.

10. Disconnect the alternator electrical connections.

11. Loosen, but do not remove the alternator pivot bolt and the tensioner mounting bolt.

12. Remove the drive belt(s); relieve tension on the belt by rotating the adjustment bolt.

13. Remove the water pump pulley, 1.6L (VIN 3) engine only.

14. Remove the power steering pump, 1.6L (VIN 3) engine only.

15. Remove the power steering pump bracket, 1.6L (VIN 3) engine only.

16. Remove the alternator tensioner mounting bolt and the belt tensioner.

17. Remove the alternator pivot bolt.

18. Loosen the bolt at the base of the

2.4L DOHC ENGINE

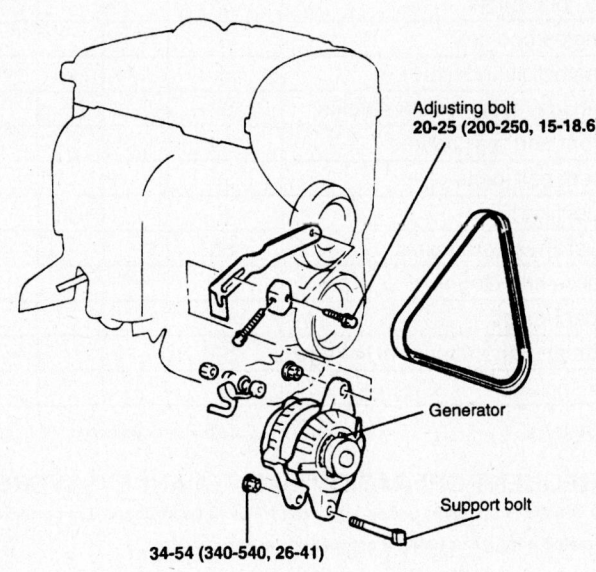

V6 ENGINE

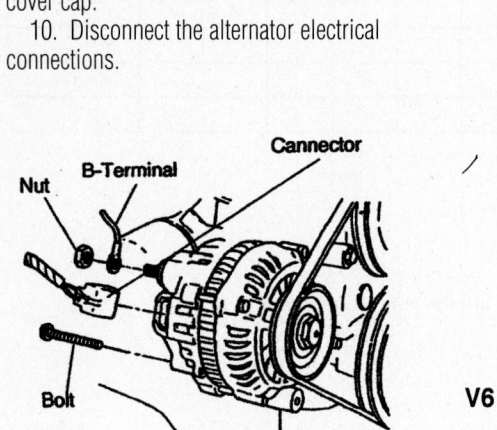

Alternator mounting—1.5L and 1.6L (VIN 5) engine shown

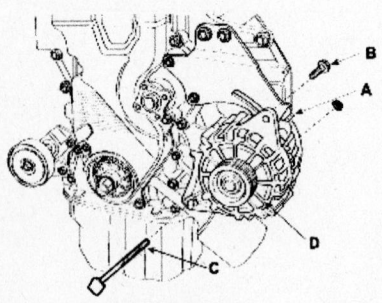

A - adjustment bolt
B - tension mounting bolt
C - pivot bolt
D - alternator

67162-KIAC-G03

Alternator mounting—2.0L and 1.6L (VIN 3) engines shown

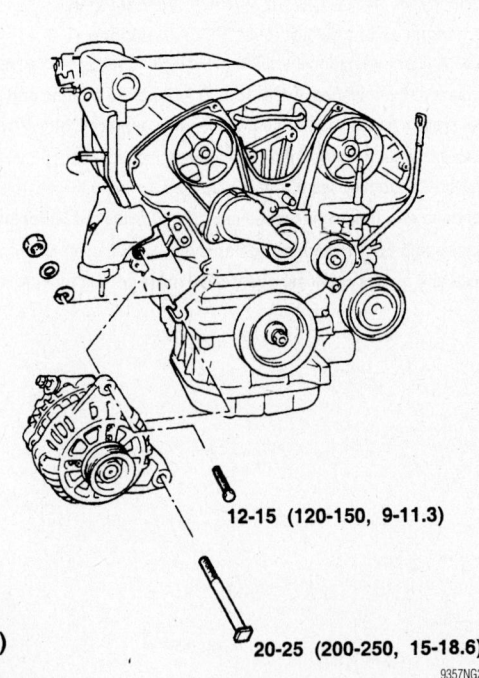

TORQUE : N·m (kg·cm, lb·ft)

Alternator mounting—2.4L and 2.7L engine

adjusting bracket and rotate the bracket up.

19. Remove the alternator.

INSTALLATION

1. Before servicing the vehicle, refer to the Precautions Section.

2. Install the alternator.

3. Install the alternator pivot bolt and hand-tighten at this time.

4. Rotate the adjusting bracket into position, place the belt tensioner into position. Hand-tighten the mounting bolt.

5. Install the drive belt.

6. Adjust the belt tension by rotating the adjustment bolt.

7. The belt deflection for 1.5L, 1.6L (VIN 5), 2.4L and 2.7L engines, is as follows:
   a. New belt: 0.22–0.28 in. (5.5–7mm).
   b. Used belt: 0.24–0.28 in. (6–7mm).

8. The belt deflection for 1.6L (VIN 3) engine is as follows :
   a. New belt: 0.13–0.15 in. (3.3–3.7mm).
   b. Used belt: 0.17–0.19 in. (4.2–4.7mm).

9. The belt deflection for the 2.0L engine, is as follows:
   a. New belt: 0.22–0.31 in. (5.5–8mm).
   b. Used belt: 0.33–0.45 in. (8.5–11.5mm).

10. The belt deflection for the 1.8L engine, is as follows:
   a. New belt: 0.23–0.31 in. (6–8mm).
   b. Used belt: 0.28–0.35 in. (7–9mm).

11. Tighten the tensioner bolt to 14–19 ft. lbs. (19–26 Nm) and the pivot bolt to 28–38 ft. lbs. (38–51 Nm).

12. Connect the alternator electrical connections.

13. Install the alternator **B** terminal cover cap.

14. Install the power steering pump and adjust the power steering belt tension, 2.0L and 1.6L (VIN 3) engines only.

15. Install the water pump pulley, 2006 1.6L engine only.

16. Install the air intake inlet pipe and fasten the clamp, 1.8L engine only.

17. Install the top hose to the air intake inlet pipe, 1.8L engine only.

18. Install the front air intake inlet pipe bolts, 1.8L engine only.

19. Connect the negative battery cable.

Ignition Timing

This vehicle is equipped with a Distributorless Ignition System (DIS). No adjustment is necessary or possible.

Engine Assembly

REMOVAL & INSTALLATION

1.5L and 1.6L (VIN 5) Engines

1. Before servicing the vehicle, refer to the Precautions Section.

2. Properly relieve the fuel system pressure.

3. Drain and recycle the engine coolant.

4. Disconnect the battery cables.

5. Remove the battery and tray.

6. Remove the fresh air intake duct.

7. Remove the upper and lower radiator hoses.

8. Disconnect the accelerator cable.

9. Disconnect the fuel hose from the fuel rail.

10. Remove the heater hose.

11. Disconnect the brake vacuum hose and purge control hose from the dynamic chamber.

12. Disconnect the injector connectors.

13. Disconnect the electrical connectors, tag before removing.

14. Remove the transaxle linkage, if equipped with an automatic transaxle.

15. Remove the manual transaxle linkage and extension bar, if equipped.

16. Remove the clutch release cylinder and pipe, if equipped.

17. Disconnect the transaxle range switch connector, solenoid valve connector and fluid cooler hose, if equipped with an automatic transaxle.

18. Disconnect the power steering pump hose.

19. Disconnect the B and S-terminal connectors from starter.

20. Disconnect the alternator B-terminal connector.

21. Remove the 4 A/C compressor mounting bolts, but do not disconnect the fluid line. Position the compressor aside.

22. Remove the front wheels.

23. Disconnect the front exhaust pipe from the manifold.

24. Remove the right and left tie rod ends from the steering knuckle.

25. Remove the bolt and nut, then separate the right and left lower control arm from the steering knuckle.

26. Remove the 2 bolts and nuts from the damper, then separate the damper from the knuckle.

27. Remove the halfshafts from the transaxle, by carefully prying them.

28. Support the engine with a suitable hoist.

29. Remove the 4 nuts and bolts from the engine mounting member.

30. Remove the 2 bolts from the No. 1 engine mounting bracket.

31. Remove the 4 bolts from the No. 2 engine mounting bracket.

32. Remove the 2 nuts from the No. 3 engine mounting rubber.

33. Remove the engine and transaxle assembly by lifting it out of the engine compartment as a unit.

To install:

34. Installation is the reverse of the removal procedure. Note the following important steps.

35. When possible, leave the engine mounting nuts/bolts loose (hand tight) until all mounts are aligned and bolted. This may help in aligning the engine/transaxle assembly in the vehicle.

36. Tighten the engine mount bolts/nuts as follows:
   - No. 3 engine mounting rubber (insulator) bolts: 49–68 ft. lbs. (68–93 Nm)
   - No. 2 engine mounting nuts: 49–68 ft. lbs. (68–93 Nm)
   - No. 2 engine mounting bolts: 28–38 ft. lbs. (38–51 Nm)
   - Engine mounting member nuts: 28–38 ft. lbs. (39–52 Nm)
   - Engine mounting member bolts: 47–66 ft. lbs. (65–91 Nm)

37. Install new circlips on the inner CV-joint stub shafts, if equipped, intermediate shaft. Grease the shaft splines before installing the halfshaft/intermediate shaft into the transaxle.

38. Always install new gaskets and/or O-rings. Use new self-locking nuts, especially on the exhaust.

39. Fill the engine and the transaxle with the proper types and quantities of oil. Fill the cooling system.

40. Connect the negative battery cable, start the engine and check for leaks. Check all fluid levels.

1.6L (VIN 3) Engine

1. Before servicing the vehicle, refer to the Precautions Section.

2. Properly relieve the fuel system pressure.

3. Drain the cooling system.

4. Drain the engine oil.

5. Drain the transaxle.

6. Recover the A/C refrigerant.

7. Remove or disconnect the following:
   - Battery cables
   - Battery
   - Engine cover

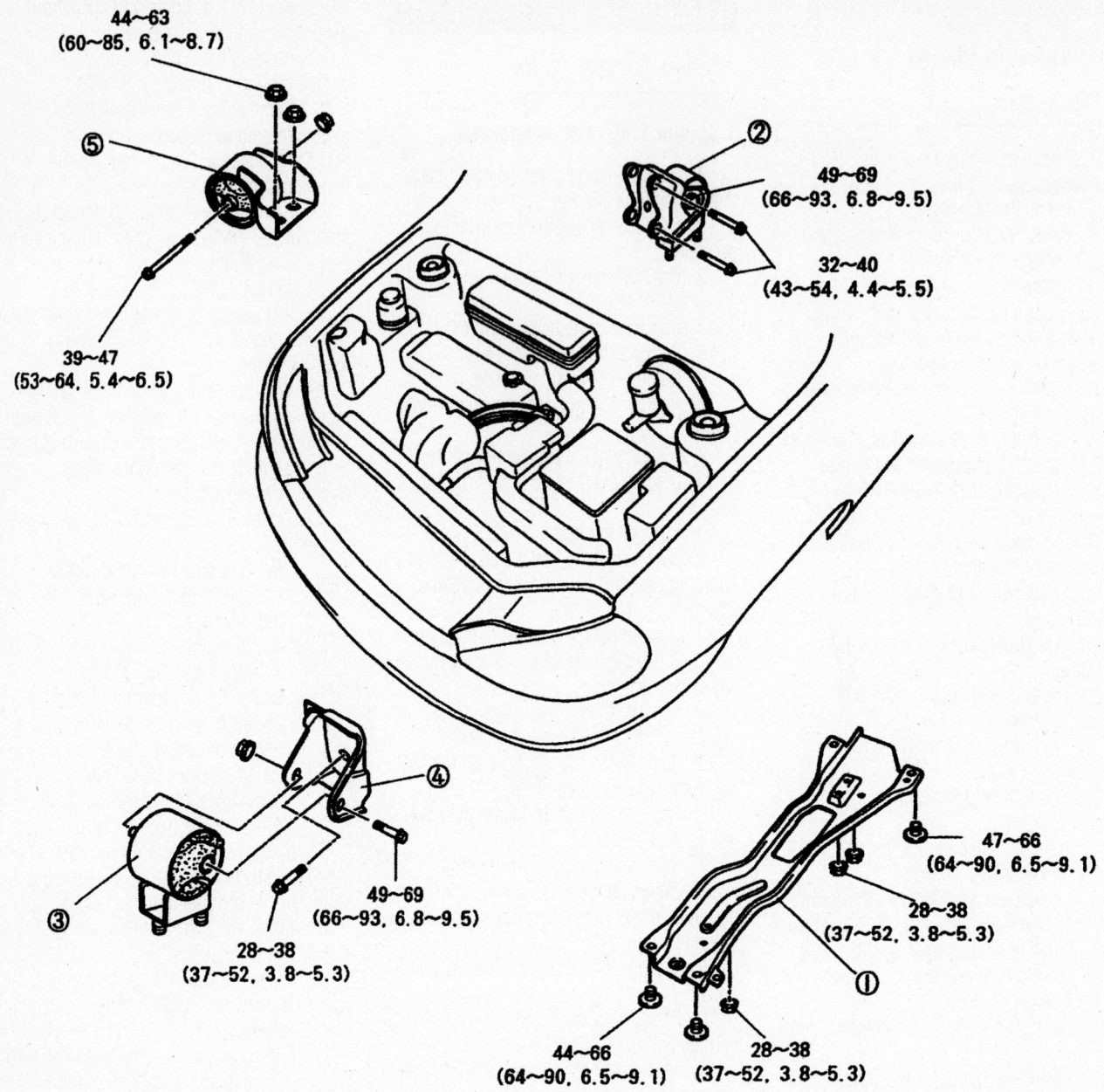

44~63
(60~85, 6.1~8.7)

⑤

49~69
(66~93, 6.8~9.5)

②

32~40
(43~54, 4.4~5.5)

39~47
(53~64, 5.4~6.5)

49~69
(66~93, 6.8~9.5)

④

28~38
(37~52, 3.8~5.3)

③

47~66
(64~90, 6.5~9.1)

28~38
(37~52, 3.8~5.3)

①

44~66
(64~90, 6.5~9.1)

28~38
(37~52, 3.8~5.3)

lb-ft(N·m, kg-m)

(1) **Engine Mounting Member**
(2) **No.1 Engine Mounting Bracket**
(3) **No.2 Engine Mounting Rubber**
(4) **No.2 Engine Mounting Bracket**
(5) **No.3 Engine Mounting Bracket**

67162-KIAC-G04

Exploded view of the engine mount and brackets—1.5L and 1.6L (VIN 5) engine

- Engine splash guard
- Air intake assembly
- Battery tray
- Radiator hoses
- Transmission cooler hoses
- Heater hoses
- Fuel hose
- Throttle cable

- Throttle Position Sensor (TPS) connector

8. Disconnect the following wiring harnesses and clamps from the cylinder head and intake manifold:
- Rear O2 sensor connector
- A/C compressor switch connector
- Knock sensor

- Fuel injectors
- Top wiring harnesses bracket
- Idle Speed Actuator (ISA) connector
- Front O2 sensor connector
- Crankshaft Position Sensor (CPS) connector
- Oil Control Valve (OCV) connector
- Ignition coil connectors

- Camshaft Position Sensor (CMP) connector
- Ground cables

9. Remove the following wiring harnesses from vehicles equipped with automatic transmissions:

- Transaxle range switch connector
- Solenoid valve connector
- ATF oil temperature sensor
- Vehicle Speed Sensor (VSS) connector
- Band server switch connector
- Pulse generator connectors

10. Remove the following wiring harnesses from vehicles equipped with manual transmissions:

- VSS connector
- Neutral switch connector
- Back-up lamp switch

11. Remove or disconnect the following:

- Transmission control cable
- Purge Control Solenoid Valve (PCSV) hose
- Brake booster vacuum hose
- Power steering fluid hoses
- High and low A/C pressure hoses
- Engine mounting support bracket
- Transaxle mounting bracket
- Alternator connectors
- Front tires
- ABS wheel speed sensors

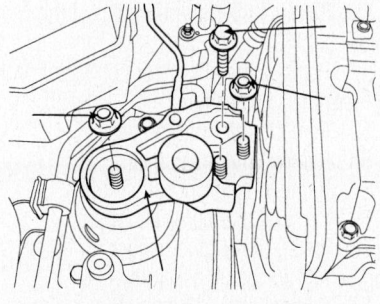

09474_KIAC_G0001

Engine mounting support bracket—1.6L (VIN 3)

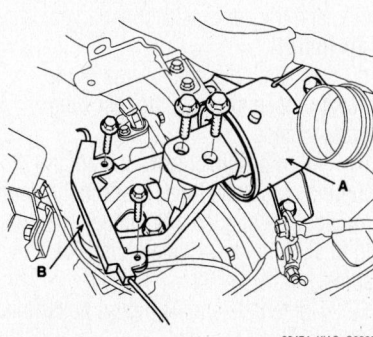

09474_KIAC_G0002

Transaxle mounting bracket-A/T—1.6L (VIN 3)

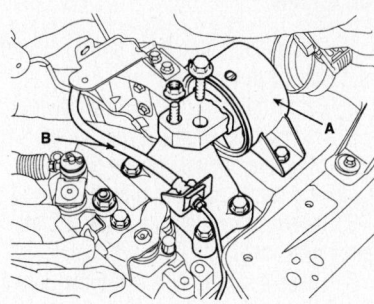

09474_KIAC_G0003

Transaxle mounting bracket-M/T—1.6L (VIN 3)

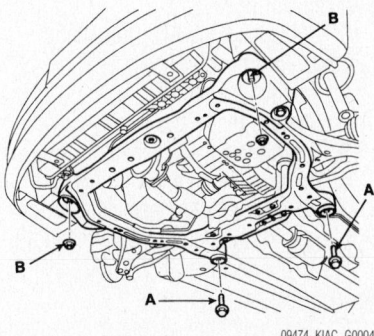

09474_KIAC_G0004

Sub frame mounting bolts and nuts—1.6L (VIN 3)

- Caliper assembly and securely hang the assembly from the vehicle
- Wheel knuckle mounting bolts
- Steering U-joint mounting bolts
- Front muffler

12. Using a suitable jack, secure the engine and transaxle assembly.

13. Remove the sub frame mounting bolts and nuts.

14. Remove the engine and transaxle assembly by lifting the vehicle.

To install:

15. Installation is the reverse of removal.

16. Tighten the mounting bolts as follows:

  a. Sub frame mounting bolts and nuts: 69–87 ft. lbs. (93–118 Nm).

  b. Transaxle mounting bracket: 51–69 ft. lbs. (69–93 Nm).

  c. Engine mounting support bracket: Body side nut to 51–69 ft. lbs. (69–93 Nm). Engine side nuts to 36–47 ft. lbs. (49–64 Nm).

17. Refill the engine with oil to the correct level.

18. Refill the cooling system to the correct level.

19. Refill the transaxle assembly to the correct level.

20. Adjust the throttle and shift cables as required.

21. Start the engine and check for leaks.

1.8L Engine

1. Before servicing the vehicle, refer to the Precautions Section.

2. Properly relieve the fuel system pressure.

3. Drain and recycle the engine coolant.

4. Disconnect the battery cables.

5. Remove the battery and tray.

6. Remove the Data Link Connector (DLC) from the Mass Air Flow (MAF) sensor bracket.

7. Disconnect the Intake Air Temperature (IAT) and Mass Air Flow (MAF) sensor connectors.

8. Remove the air intake hose from the throttle body.

9. Remove the ventilation hose and fresh air duct.

10. Disconnect the accelerator and, if equipped, cruise control cables.

11. Remove the air cleaner assembly.

12. Disconnect the brake booster and purge control vacuum hoses from the intake manifold.

13. Remove the upper and lower radiator hoses.

14. Disconnect the fuel hose from the fuel injector rail.

15. Remove the heater hoses.

16. Disconnect the Idle Air Control (IAC) and Throttle Position (TP) sensor connectors.

17. Disconnect the fuel injector electrical connectors.

18. Disconnect the starter and alternator electrical connectors.

19. Disconnect the engine ground strap.

20. Remove the left and right splash shields.

21. Remove the exhaust manifold heat shield and 3 power steering pump bracket bolts.

22. Remove the air conditioning compressor mounting bolts and position it aside leaving the hoses attached.

23. Disconnect the ground strap from the top of the transaxle.

24. Remove the No. 4 engine mount.

25. Disconnect the input/turbine speed sensor connector, if equipped with an automatic transaxle.

26. Remove the back-up light switch, if equipped with a manual transaxle.

27. Remove the Vehicle Speed Sensor (VSS).

28. Remove the U-clip from the selector cable and the nut and washer from the transaxle linkage, if equipped with an automatic transaxle.

29. Remove the linkage and extension bar, then the clutch release cylinder and hydraulic hose, if equipped with a manual transaxle.

30. Disconnect the transaxle range switch and solenoid valve connectors, then the 2 transaxle oil cooler hoses, if equipped with an automatic transaxle.

31. Remove the front wheels.

32. Disconnect the Oxygen (O₂) sensor electrical connectors.

33. Disconnect the front exhaust pipe.

34. Remove the halfshafts.

35. Properly support the engine/transaxle assembly.

36. Remove the two No. 2 engine mount-to-mounting member bolts.

37. Remove the one bolt from the No. 3 engine mount.

38. Remove the three No. 1 engine mounting bolts, then carefully lift the engine/transaxle assembly from the vehicle.

To install:

39. Installation is the reverse of the removal procedure. Note the following important steps.

40. When possible, leave the engine mounting nuts/bolts loose (hand tight) until all mounts are aligned and bolted. This may help in aligning the engine/transaxle assembly in the vehicle.

41. Tighten the engine mount bolts/nuts as follows:
- No. 1 mounting bolts: 50–70 ft. lbs. (67–93 Nm)
- No. 3 mounting bolts: 63–86 ft. lbs. (85–116 Nm)
- No. 2 mounting nuts: 28–38 ft. lbs. (38–51 Nm)
- No. 4 mounting bolts: 47–66 ft. lbs. (64–89 Nm)
- No. 4 mounting nuts: 49–68 ft. lbs. (68–93 Nm)

42. Install new circlips on the inner CV-joint stub shafts, if equipped, intermediate shaft. Grease the shaft splines before installing the halfshaft/intermediate shaft into the transaxle.

43. Always install new gaskets and/or O-rings. Use new self-locking nuts, especially on the exhaust.

44. Fill the engine and the transaxle with the proper types and quantities of oil. Fill the cooling system.

45. Connect the negative battery cable, start the engine and check for leaks. Check all fluid levels.

2.0L Engine

1. Disconnect the battery terminals and remove the heat shield.

2. Remove the battery and battery tray.

3. Remove the engine cover.

4. Drain the engine coolant. Remove the radiator cap to speed draining.

5. Remove the intake air hose and air cleaner assembly.

6. Disconnect the AFS (Air Flow Sensor) connector.

7. Disconnect the breather hose from intake air hose.

8. Remove the intake air hose and air cleaner upper cover.

9. Disconnect the upper radiator hose and lower radiator hose.

10. Disconnect the ATF (Automatic Transaxle Fluid) cooler hose.

11. Remove the radiator.

12. Remove the heater hose.

13. Disconnect the accelerator cable.

14. Disconnect the engine wire harness connectors and wire harness clamps from the cylinder head and the intake manifold.

15. Disconnect the OCV (Oil Control Valve) connector.

16. Disconnect the oil temperature sensor connector.

17. Disconnect the ECT (Engine Coolant Temperature) sensor connector.

18. Disconnect the ignition coil connector.

19. Disconnect the TPS (Throttle Position Sensor) connector.

20. Disconnect the ISA (Idle Speed Actuator) connector.

21. Disconnect the air conditioner compressor switch.

22. Disconnect the CMP (Camshaft Position Sensor) connector.

23. Disconnect the four fuel injector connectors.

24. Disconnect the knock sensor connector.

25. Disconnect the ground cables from the intake manifold and vehicle's body.

26. Disconnect the front heated oxygen sensor connector.

27. Disconnect the CKP (Crankshaft Position Sensor) connector.

28. Disconnect the oil pressure switch connector.

29. Disconnect the PCSV (Purge Control Solenoid Valve) connector.

30. Disconnect the transaxle wire harness connectors and control cable from transaxle (A/T).

31. Disconnect the transaxle range switch connector.

32. Disconnect the solenoid valve connector.

33. Disconnect the input shaft speed sensor connector.

34. Disconnect the output shaft speed sensor connector.

35. Disconnect the Vehicle Speed Sensor (VSS) connector.

36. Disconnect the transaxle ground cable.

37. Disconnect the control cable nut from transaxle range switch.

38. Disconnect the control cable.

39. Disconnect the fuel inlet hose of the delivery pipe side.

40. Disconnect the hose of the Purge Control Solenoid Valve (PCSV) side.

41. Remove the brake booster vacuum hose.

42. Remove the power steering pump drive belt.

43. Remove the power steering pump and use a wire to secure the pump to the vehicle so that it is out of the way.

44. Remove the front wheel (RH, LH).

45. Remove the bolts and RH side cover.

46. Remove the air conditioner compressor drive belt.

47. Remove the air conditioner compressor and fix the compressor to vehicle with a wire.

48. Disconnect the rear oxygen sensor connector and remove the clamp.

49. Disconnect the starter motor connector and "B" terminal connection.

50. Remove the mounting clip of the starter cable.

51. Remove the alternator connector and "B" terminal connection.

52. Remove the front muffler.

53. Remove the drive shaft.

54. Remove the front roll stopper.

55. Remove the rear roll stopper.

56. Install the engine hanger lift to the engine and transaxle assembly.

57. Remove the bolt, nuts and engine mounting support bracket.

58. Remove the transaxle mounting bracket.

59. Using engine hanger lift, remove the engine and transaxle assembly on vehicle.

To install:

60. Installation is the reverse of the removal procedure. Note the following important steps.

61. When possible, leave the engine mounting nuts/bolts loose (hand tight) until all mounts are aligned and bolted. This may help in aligning the engine/transaxle assembly in the vehicle.

62. Tighten the engine mount bolts/nuts as follows:
- Front roll stopper bolt and nut: 36–47 ft. lbs. (49–63 Nm)

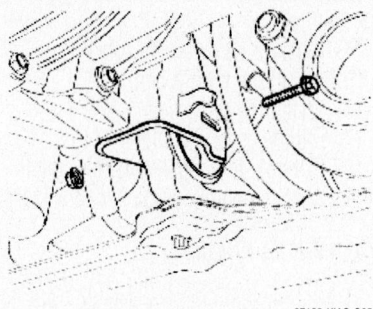

Front roll stopper—2.0L engine

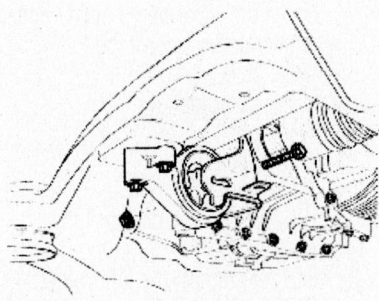

Rear roll stopper—2.0L engine

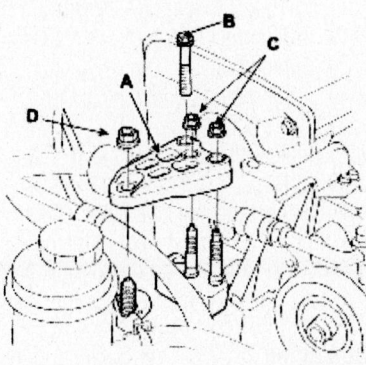

Engine mount support bracket—2.0L engine

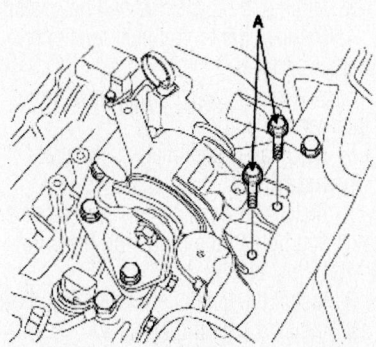

Transaxle mounting bracket bolts—2.0L engine

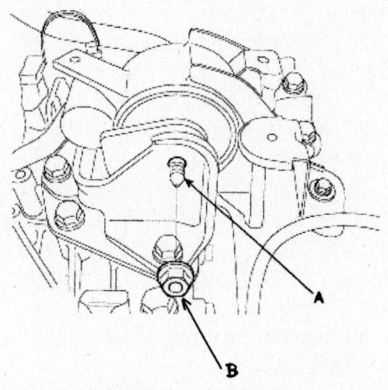

Transaxle mounting bracket bolt and nut—2.0L engine

- Rear roll stopper bolt and nut: 36–47 ft. lbs. (49–63 Nm)
- Engine mount support bracket bolt (B) and nut (C): 36–47 ft. lbs. (49–63 Nm)
- Engine mount support bracket nut (D): 51–68 ft. lbs. (69–93 Nm)
- Transaxle mounting bracket bolts (A): 29–36 ft. lbs. (39–49 Nm)
- Transaxle mounting bracket nut (B): 65–79 ft. lbs. (88–108 Nm)

63. Always install new gaskets and/or O-rings. Use new self-locking nuts, especially on the exhaust.

64. Fill the engine and the transaxle with the proper types and quantities of oil. Fill the cooling system.

65. Adjust the shift cable.

66. Adjust the throttle cable.

67. Bleed the cooling system.

68. After, assemble the fuel line, turn on the ignition switch (do not operate the starter) so that the fuel pump runs for approximately two seconds and fuel line pressurizes.

69. Repeat this operation two or three times, then check for fuel leakage at any point in the fuel line.

70. Connect the negative battery cable, start the engine and check for leaks.

71. Bleed the cooling system.

72. Check all fluid levels.

2.4L Engine

1. Before servicing the vehicle, refer to the Precautions Section.

2. Properly relieve the fuel system pressure.

3. Drain and recycle the engine coolant, transaxle fluid and engine oil.

4. Remove the battery.

5. Remove the air cleaner.

6. Disconnect the back-up lamp and engine harness connectors.

7. Disconnect the select control valve connector, if equipped with a manual transaxle.

8. Disconnect the alternator harness and oil pressure gauge wiring.

9. Disconnect the transaxle oil cooler hoses, if equipped with an automatic transaxle.

10. Remove the upper and lower radiator hoses from the engine.

11. Disconnect the engine ground.

12. Disconnect the brake booster vacuum hose.

13. Disconnect the main fuel line and return and vapor hoses from the engine side.

14. Disconnect the inlet and outlet heater hoses from the engine.

15. Disconnect the accelerator cable from the engine.

16. Disconnect the clutch cable, shift control rod and extension rod if equipped with a manual transaxle.

17. Disconnect the control cable from the transaxle, if equipped with an automatic transaxle.

18. Disconnect the Vehicle Speed Sensor (VSS) connector from the transaxle.

19. Disconnect the column shaft from the steering gear box.

20. Remove the engine and transaxle mounting insulators.

21. Remove the wheel.

22. Remove the caliper. Unbolt and support with a piece of wire. Do not disconnect the brake fluid line.

23. Remove the lower arm and fork and separate the upper arm and knuckle.

24. Disconnect the front exhaust pipe from the manifold. Use wire to hang the exhaust pipe from the bottom of the vehicle.

25. Place a suitable supporter under the sub-frame. Make sure all hoses, vacuum lines and connectors are detached from the engine.

26. Loosen the sub-frame bolts, then slowly lift the vehicle. The engine/transaxle, sub-frame, steering gear and halfshafts are removed as an assembly.

27. Remove the halfshafts, raise the engine/transaxle assembly with a hoist, and separate the engine and transaxle from the sub-frame.

28. To separate the engine/transaxle from the sub-frame, remove the engine mounting bracket and transaxle mounting bracket, then lift up the engine hoist.

To install:

29. To place the engine and transaxle assembly on the sub-frame, install the front

roll stopper and rear roll stopper among the sub-frame and engine/transaxle assembly.

30. Installation is the reverse of the removal procedure, noting the following steps:

   a. Fill the engine and the transaxle with the proper types and quantities of oil.

   b. Fill the cooling system.

   c. Start the engine and check for leaks. Check all fluid levels.

2.7L Engine

1. Before servicing the vehicle, refer to the Precautions Section.
2. Properly relieve the fuel system pressure.
3. Drain and recycle the engine coolant and engine oil.
4. Remove the battery and engine cover.
5. Remove the air cleaner.
6. Disconnect the engine harness connectors.
7. Remove the alternator harness, oil pressure switch and oil pressure sender connectors.
8. Disconnect the transaxle oil cooler hoses.
9. Remove the upper and lower radiator hoses from the engine.
10. Remove the radiator.
11. Disconnect the spark plug wires.
12. Disconnect the engine ground.
13. Disconnect the brake booster vacuum hose.
14. Disconnect the main fuel line and return and vapor hoses from the engine side.
15. Remove the inlet and outlet heater hoses from the engine.
16. Disconnect the accelerator and cruise control cables from the engine.
17. Disconnect the control cable from the transaxle.
18. Remove the speedometer cable from the transaxle.
19. Remove the power steering pump hose.
20. Remove the oil pan shield.
21. Disconnect the front exhaust pipe from the manifold. Use a wire to support the exhaust pipe from the bottom of the vehicle.
22. Remove the lower control arm ball joint bolts and upper arm bolt from the steering knuckle.
23. Remove the halfshafts from the transaxle. Plug the openings in the transaxle case and discard the halfshaft circlips.
24. Attach a cable or chain to the engine and use a chain hoist to lift the engine enough to pull the cable tight.

25. Remove the front and rear roll stoppers.
26. Disconnect the connector from the starter motor harness.
27. Remove the engine mount bolts.
28. Remove the bolts and nuts that fasten the engine mount bracket to the body.
29. Slowly raise the engine and transaxle assembly and temporarily hold in the raised position. Make sure that all hoses, cables, vacuum lines and connectors are detached from the engine.
30. Remove the transaxle mounting bracket bolts.
31. Remove the left mount insulator bolt.
32. Remove the engine and transaxle assembly. While directing the transaxle side down, lift the engine and transaxle up and out of the vehicle.

To install:

33. Installation is the reverse of the removal procedure, noting the following steps:

   a. Fill the engine and the transaxle with the proper types and quantities of oil.

   b. Fill the cooling system.

   c. Start the engine and check for leaks. Check all fluid levels.

Water Pump

REMOVAL & INSTALLATION

1.5L and 1.6L (VIN 5) Engine

1. Before servicing the vehicle, refer to the Precautions Section.
2. Disconnect the negative battery cable.
3. Drain and recycle the engine coolant.
4. Remove the drive belt.
5. Remove the timing belt.

➡ **The power steering pump must be removed to access the water inlet pipe. Do not disconnect the pump fluid lines.**

6. Remove the power steering pump and position aside.
7. Remove the water inlet pipe and gasket.
8. Remove the water bypass pipe and O-ring.
9. Remove the water pump bolts, pump and gasket.
10. Clean all gasket mating surfaces.

To install:

11. Install the water pump, using a new gasket and tighten the mounting bolts to 14–19 ft. lbs. (19–26 Nm)
12. Install the water pump bypass pump, with a new O-ring.

13. Install the water inlet pipe, with a new O-ring.
14. Install the power steering pump.
15. Install the timing belt.
16. Install the drive belt.
17. Fill the engine coolant.
18. Connect the negative battery cable.
19. Start the engine and check for leaks.

1.6L (VIN 3) Engine

1. Before servicing the vehicle, refer to the Precautions Section.
2. Drain the engine coolant.
3. Disconnect the negative battery cable.
4. Loosen the water pump pulley bolts.
5. Remove or disconnect the following:
   - Drive belts
   - Water pump pulley
   - Timing belt
   - Timing belt idler
   - Alternator brace
   - Water pump and gasket

To install:

6. Install or connect the following:
   - Water pump with a new gasket. Tighten the mounting bolts to 9–11 ft. lbs. (12–15 Nm).
   - Alternator brace. Tighten the bolts to 15–20 ft. lbs. (20–27 Nm).
   - Timing belt idler
   - Timing belt
   - Water pump pulley.
   - Drive belts

7. Tighten the pulley bolts to 6–7 ft. lbs. (8–10 Nm).
8. Connect the negative battery cable.
9. Fill the cooling system to the correct level.
10. Start the engine and check for leaks.

1.8L Engine

1. Before servicing the vehicle, refer to the Precautions Section.
2. Disconnect the negative battery cable.
3. Drain and recycle the engine coolant.
4. Remove the timing belt, tensioner and idler pulleys.
5. Remove the water pump mounting bolts, then the pump.
6. Clean all gasket mating surfaces.

To install:

7. Install the water pump, using a new gasket and tighten the mounting bolts to 14–19 ft. lbs. (19–26 Nm)
8. Install the tensioner and idler pulleys
9. Install the timing belt
10. Fill the engine coolant.
11. Connect the negative battery cable.
12. Start the engine and check for leaks.

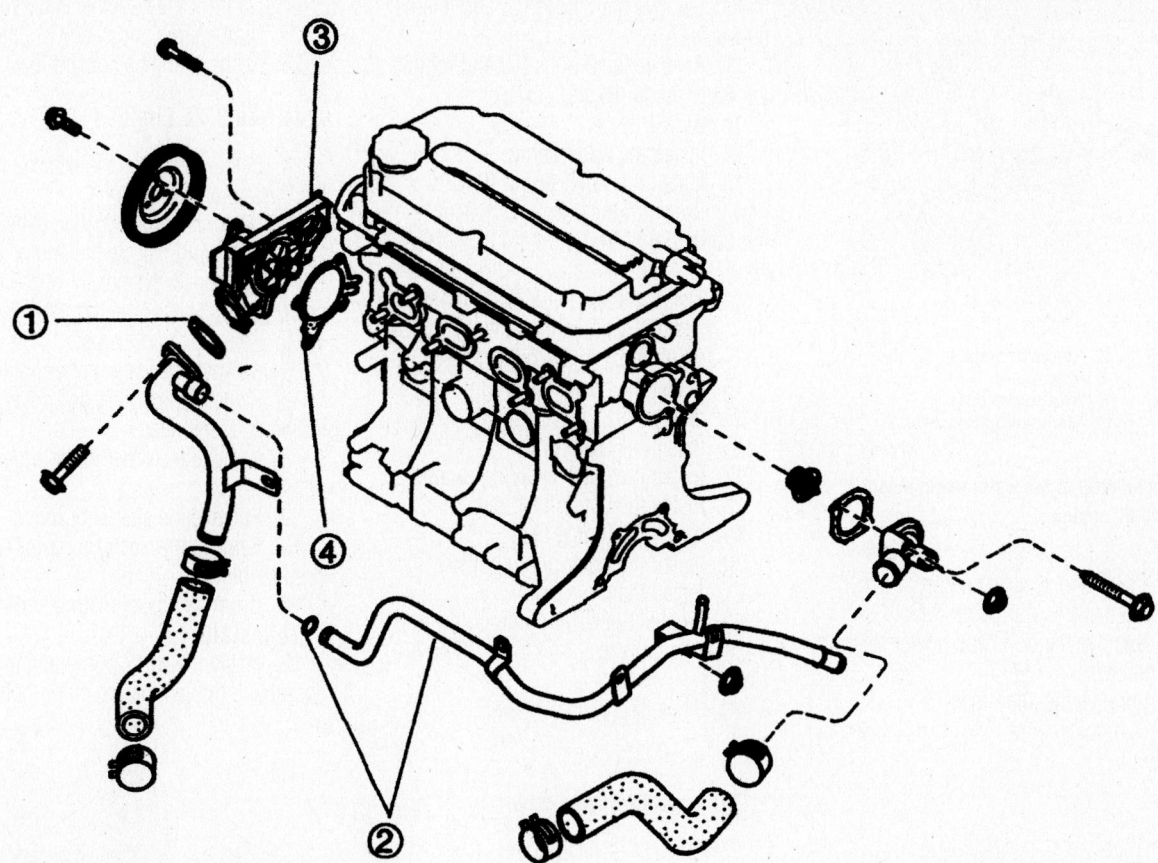

(1) Water inlet pipe and gasket
(2) Water bypass pipe and O-ring

(3) Water pump assembly
(4) Water pump gasket

67162-KIAC-G10

Water pump mounting—1.5L and 1.6L (VIN 5) engine

2.0L Engine

1. Before servicing the vehicle, refer to the Precautions Section.
2. Disconnect the negative battery cable.
3. Drain the engine coolant.
4. Remove the drive belts.
5. Remove the timing belt.
6. Remove the timing belt idler.
7. Remove the power steering pump

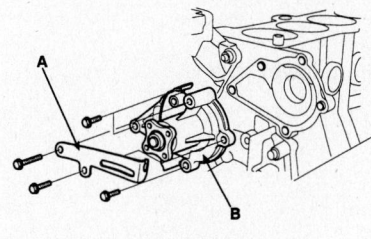

A. Alternator Brace
B. Alternator

09474_KIAC_G0005

Water pump mounting—1.6L (VIN 3) Engine

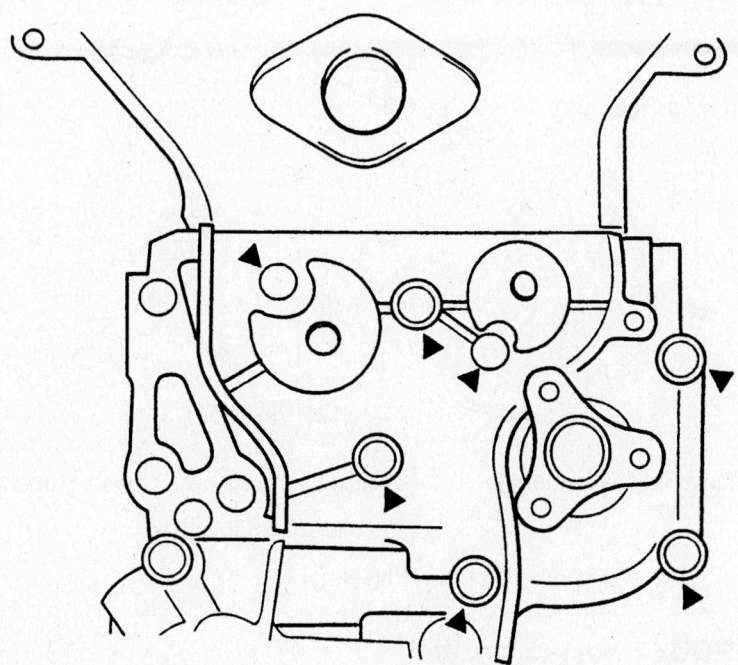

9301KG01

Water pump mounting bolt locations (arrows)—1.8L engine

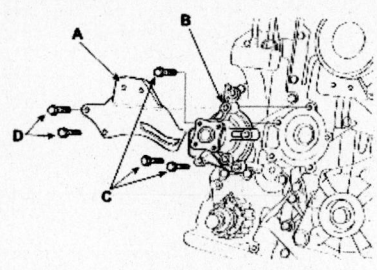

A. Alternator brace
B. Water pump
C. Water pump bolts
D. Alternator bracket bolts

67162-KIAC-G11

Exploded view of the water pump mounting—2.0L engine

and use a wire to secure the pump to the vehicle so that it is out of the way.

8. Remove the bolts and power steering pump bracket.

9. Remove the alternator.

10. Remove the 2 bolts (D) and alternator brace (A).

11. Remove the 3 bolts (C) and remove the water pump (B) and gasket.

To install:

12. Install the water pump.

13. Install the water pump (B) and a new gasket with the 3 bolts (C). Tighten the bolts to 8.5–10.5 ft. lbs. (11.5–14.5 Nm)

14. Install the alternator brace (A) with the 2 bolts (D). Tighten the bolts to 14.5–19.5 ft. lbs. (19.5–26.5 Nm)

15. Install the power steering pump bracket and bolts.

16. Install the alternator.

17. Install the power steering pump.

18. Install the timing belt idler.

19. Install the timing belt.

20. Install the water pump pulley.

21. Install the drive belts.

22. Tighten the water pump pulley bolts. Tighten the bolts to 5.5–7 ft. lbs. (7.5–10 Nm)

23. Fill with engine coolant.

24. Connect the negative battery cable.

25. Start engine and check for leaks.

26. Recheck engine coolant level.

2.4L and 2.7L Engines

1. Before servicing the vehicle, refer to the Precautions Section.

2. Disconnect the negative battery cable.

3. Drain and recycle the engine coolant.

4. Disconnect the radiator outlet hose and coolant bypass hose from the water pump.

5. Remove the drive belt.

6. Remove the water pump pulley.

7. Remove the timing belt covers and timing belt tensioner.

8. Remove the water pump mounting bolts.

9. Remove the alternator brace.

10. Remove the water pump from the cylinder block.

11. Clean all gasket mating surfaces.

To install:

12. Install the water pump, using a new gasket and tighten the mounting bolts to

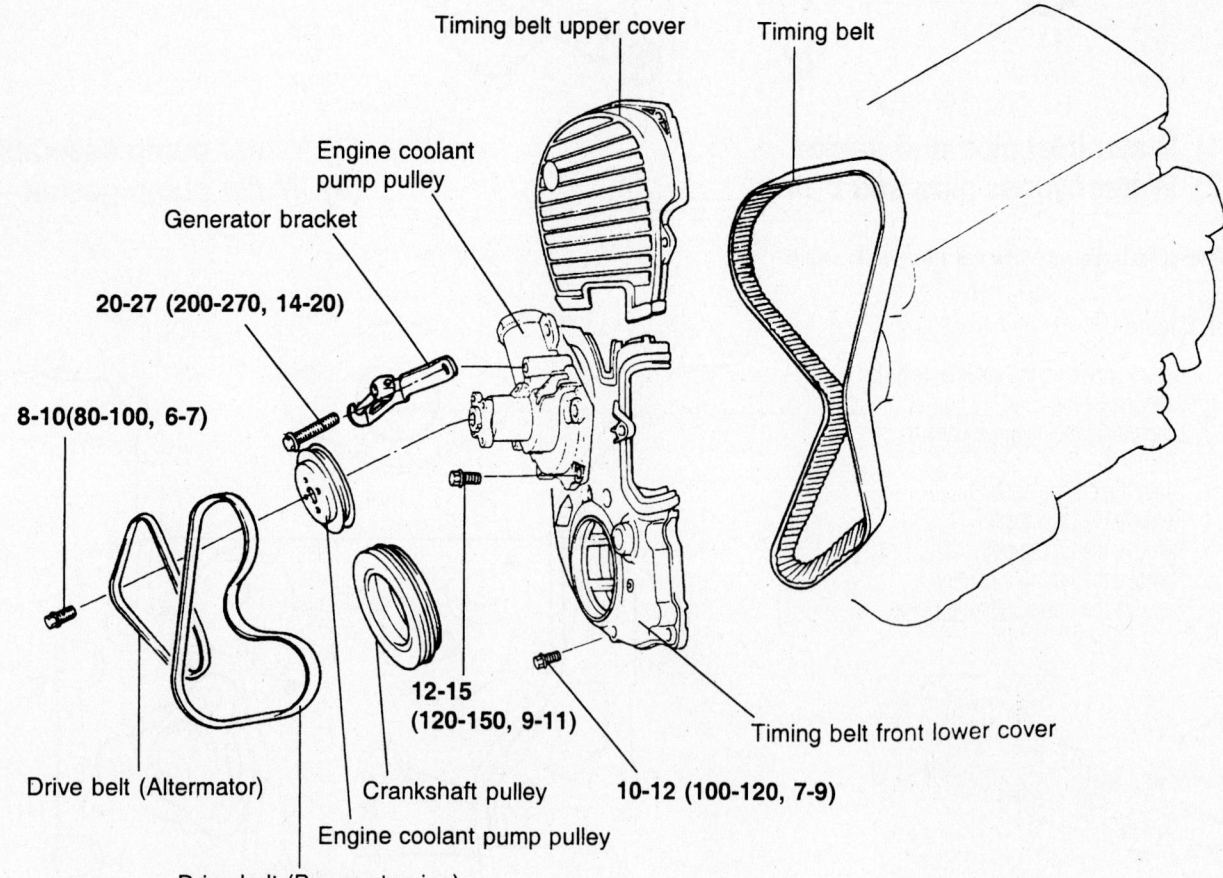

Timing belt upper cover

Timing belt

Engine coolant pump pulley

Generator bracket

20-27 (200-270, 14-20)

8-10(80-100, 6-7)

12-15 (120-150, 9-11)

Crankshaft pulley

Drive belt (Altermator)

Engine coolant pump pulley

Drive belt (Power steering)

10-12 (100-120, 7-9)

Timing belt front lower cover

TORQUE : Nm (kg.cm, lb.ft)

9357NG27

Exploded view of the water pump mounting—2.4L engine

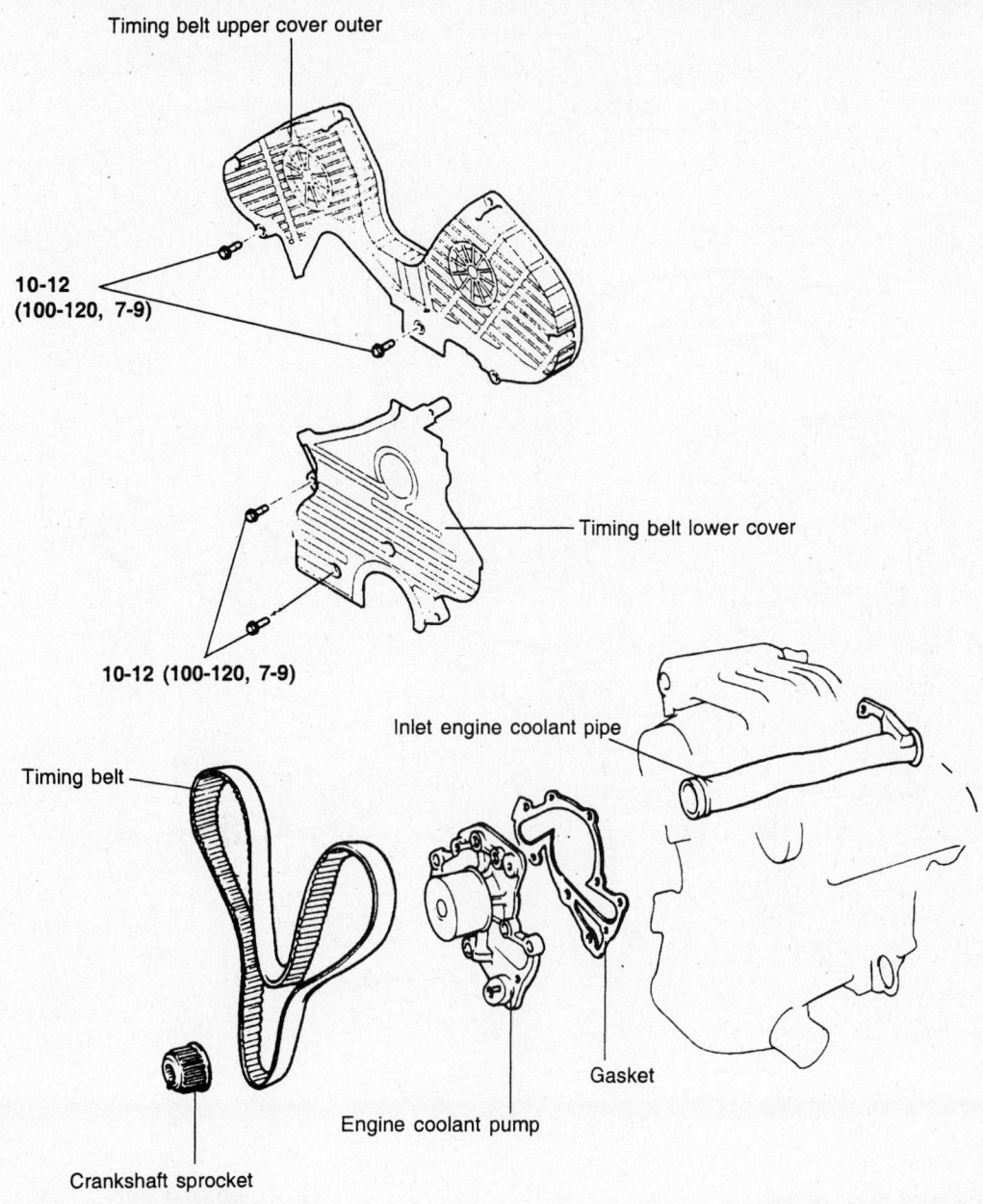

Timing belt upper cover outer

10-12 (100-120, 7-9)

Timing belt lower cover

10-12 (100-120, 7-9)

Inlet engine coolant pipe

Timing belt

Gasket

Engine coolant pump

Crankshaft sprocket

TORQUE : Nm (kg.cm, lb.ft)

9357NG28

Exploded view of the water pump mounting—2.7L engine

14–20 ft. lbs. (19–27 Nm) for 2.4L engines and 11–16 ft. lbs. (15–22 Nm) for 2.7L engines.

13. Install the timing belt tensioner and timing belt.

14. Install the timing belt covers.

15. Install the coolant pump pulley and drive belt.

16. Connect the radiator outlet hose and coolant bypass hose to the water pump.

17. Fill the engine coolant.

18. Connect the negative battery cable.

19. Start the engine and check for leaks.

Heater Core

REMOVAL & INSTALLATION

Optima

1. Before servicing the vehicle, refer to the Precautions Section.

2. Disconnect the negative battery cable.

✸✸ CAUTION

After disconnecting the negative battery cable, wait for at least 10 minutes for the SRS module to deplete its stored energy.

3. Recover the refrigerant.

4. Drain the cooling system.

5. Remove the expansion valve from the evaporator core and plug the lines.

6. Disconnect the heater hoses from the heater unit.

7. Remove the crash pad as follows:

a. Remove the front seat.

b. Tilt the steering column down and remove the center facia panel.

c. Disconnect the trip sensor connector and remove the gauge cluster.

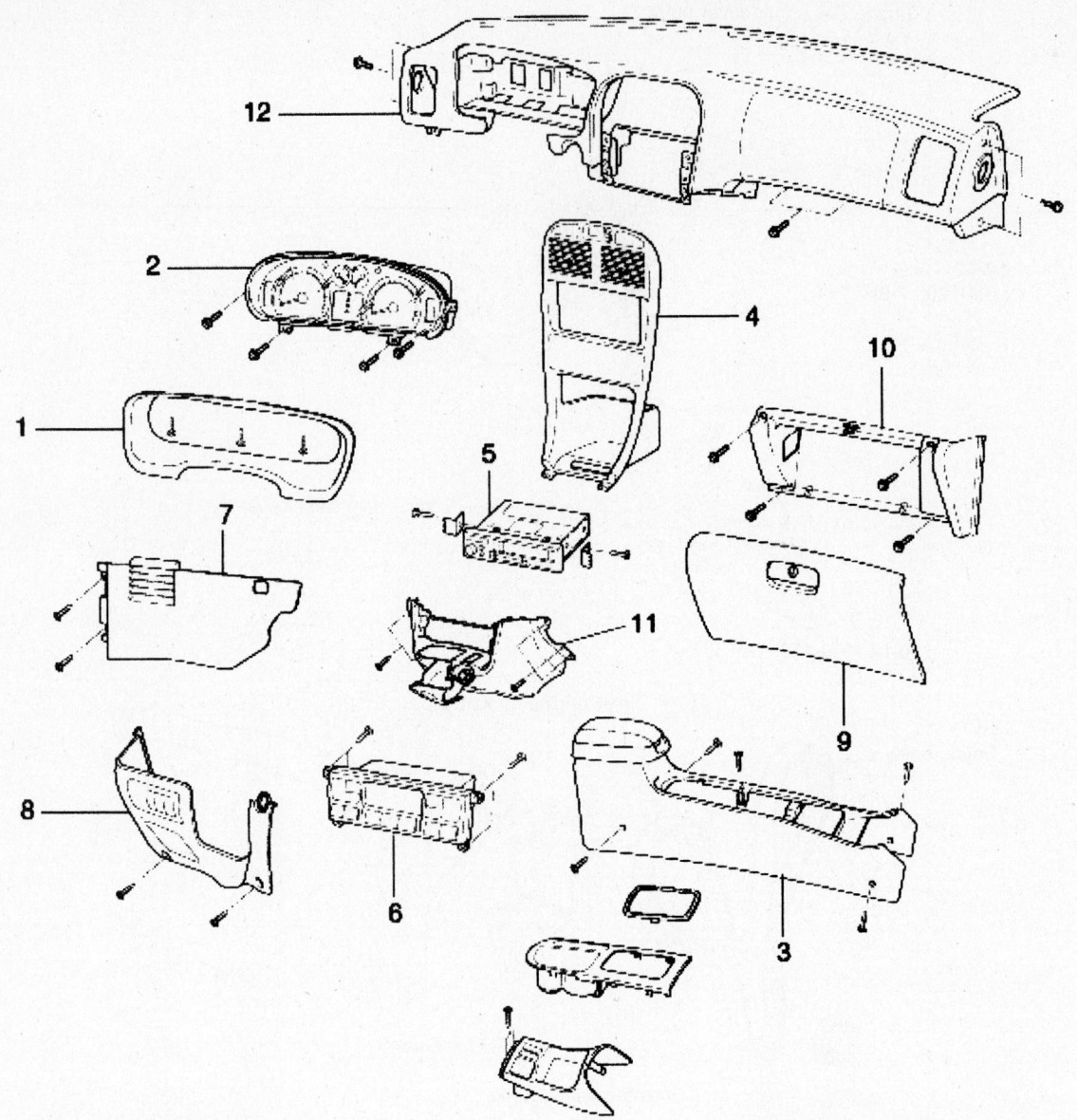

1. Cluster facia panel
2. Instrument cluster
3. Floor console
4. Center facia panel
5. Radio
6. Heater control panel

7. Cowl side trim
8. Lower panel (LH)
9. Glove box
10. Lower panel (RH)
11. Center panel
12. Instrument panel assembly

09474_KIAC_G0006

Crash pad exploded view—Optima

d. Remove the heater control unit.
e. Remove the radio unit.
f. Disconnect the damper from the glove box.
g. Remove the glove box from the lift, disconnect the pin and remove the glove box.
h. Remove the crash pad side cover.
i. Remove the center under cover.

j. Remove the console side cover.
k. Crash pad lower panel mounting.
l. Remove the A-pillar trim
m. Remove the photo sensor.
n. Disconnect the passenger air bag connector.
o. Remove the crash pad.
8. Remove the cowl cross bar assembly.

9. Remove the heater unit.
10. Disconnect the heater from the blower unit.
11. Remove the heater core cover and remove the heater core.
12. Installation is the reverse of removal.
13. Refill the cooling system to the correct level.
14. Start the engine and check for leaks.

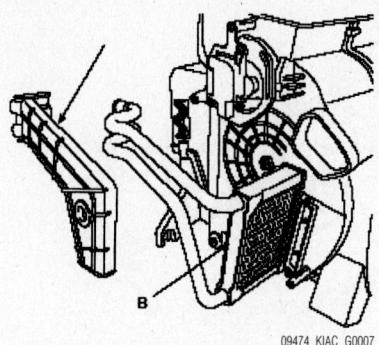

09474_KIAC_G0007

Heater core from the heater unit—Optima

Rio

1. Before servicing the vehicle, refer to the Precautions Section.
2. Disconnect the negative battery cable.

> ✳✳ **CAUTION**
>
> **After disconnecting the negative battery cable, wait for at least 10 minutes for the SRS module to deplete its stored energy.**

3. Recover the refrigerant.
4. Drain the cooling system.

5. Remove the expansion valve from the evaporator core and plug the lines.
6. Disconnect the heater hoses from the heater unit.
7. Remove the crash pad as follows:
   a. Remove the front seat.
   b. Tilt the steering column down and disconnect the trip sensor connector.
   c. Remove the gauge cluster facia panel.
   d. Disconnect the trim clips and remove the center facial panel.
   e. Remove the fuse box cover.

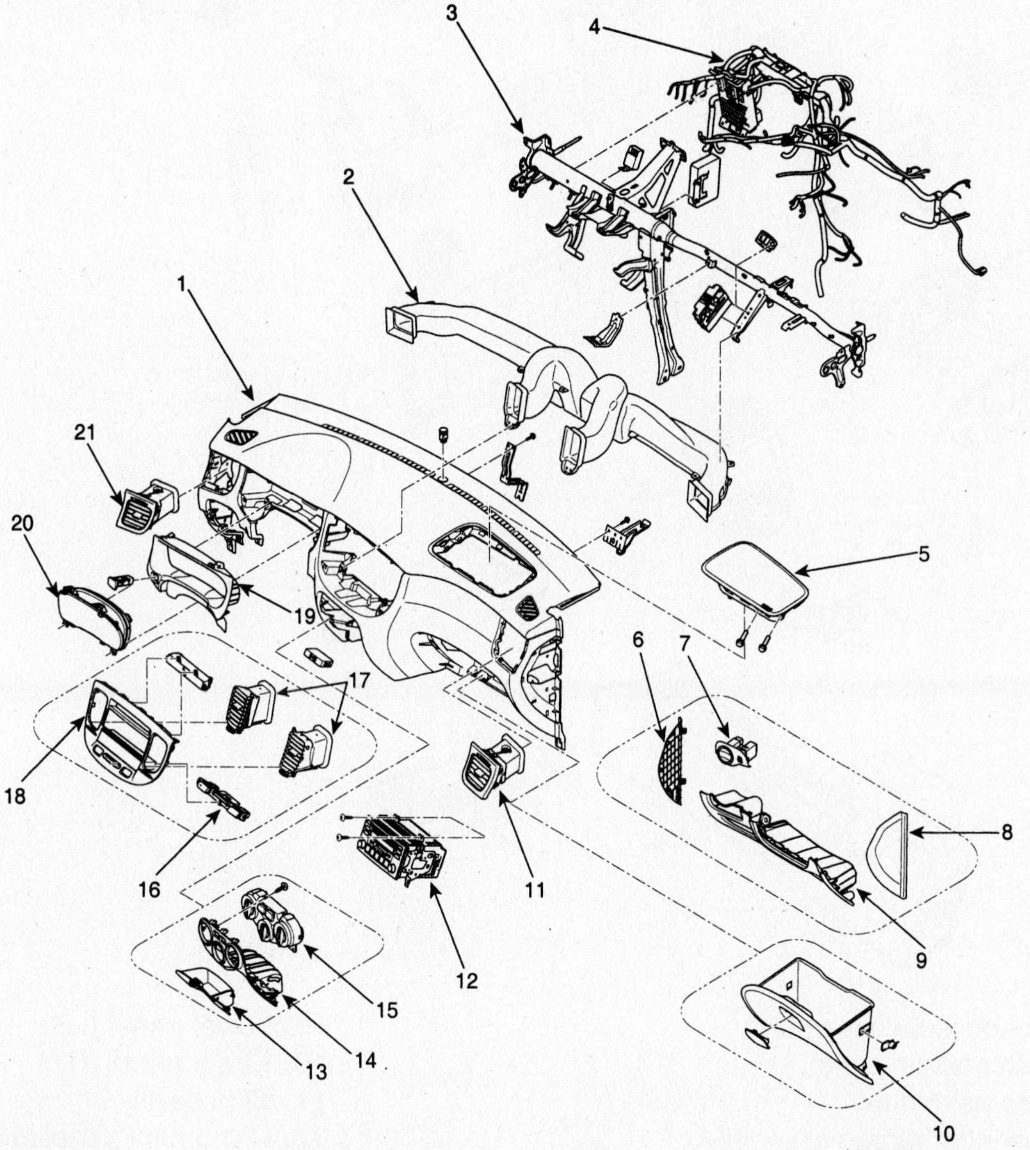

1. Crash pad main	7. Crash pad side switch
2. Air vent duct	8. Crash pad side cover(Right)
3. Cowl cross bar	9. Lower crash pad
4. Main wire harness	10. Glove box
5. Air bag moduel	11. Side air vent(Right)
6. Crash pad side cover(Left)	12. Audio
12. Audio	17. Center air vent
13. Center facia lower tray	18. Center facia upper panel
14. Center facia lower panel	19. Cluster facia panel
15. Heater control switch	20. Cluster gauge
16. Center facia switch	21. side air vent(Left)

09474_KIAC_G0008

Crash pad exploded view-Rio

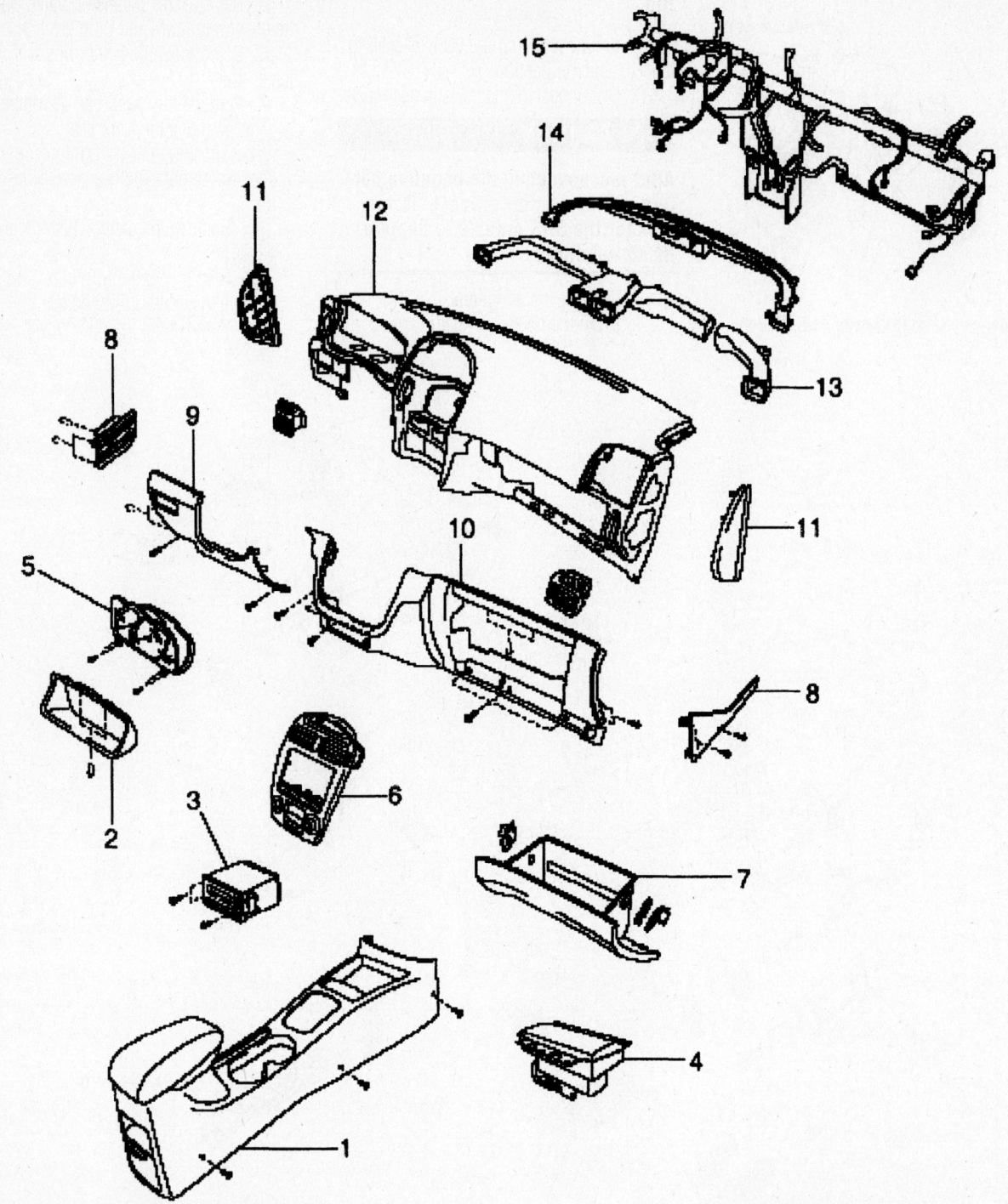

1. Floor console assembly
2. Cluster fascia panel
3. Audio assembly
4. Passenger airbag assembly
5. Instrument cluster assembly
6. Heater control & center fascia panel assembly
7. Glove box
8. Center crash pad side cover
9. Lower panel (LH)
10. Lower panel (RH)
11. Side cover
12. Instrument panel assembly
13. Air vent assembly
14. Defroster nozzle
15. Cowl crossbar

Crash pad exploded view—Spectra

09474_KIAC_G0009

f. Remove the lower crash pad panel mounting screws and remove the panel.

g. Remove the center lower tray.

h. Remove the lower center facia panel.

i. Disconnect the heater control cables and remove the heater control unit.

j. Disconnect the gauge cluster connector and remove the cluster.

k. Remove the radio assembly mounting screws and disconnect the audio connectors.

l. Remove the radio assembly.

m. Disconnect the damper from the glove box and remove the glove box.

n. Remove the left and right side crash pad side cover.

o. Remove air vents using a suitable pry tool.

p. Remove the A-pillar trim.

q. Remove the cowl side trim.

r. Remove the main crash pad assembly.

8. Remove the cowl cross bar assembly.

9. Disconnect the heater unit electrical connectors.

10. Remove the heater unit.

11. Remove the heater core cover.

12. Remove the heater core.

➡**Take care not to bend the inlet and outlet pipes of the heater core during removal.**

13. Installation is the reverse order of removal.

14. Fill the cooling system to the correct level.

15. Start the engine and check for leaks.

Spectra

1. Before servicing the vehicle, refer to the Precautions Section.

2. Disconnect the negative battery cable.

☀☀ CAUTION

After disconnecting the negative battery cable, wait for at least 10 minutes for the SRS module to deplete its stored energy.

3. Recover the refrigerant.

4. Drain the cooling system.

5. Disconnect the heater hoses from the heater unit.

6. Remove the expansion valve from the evaporator core and plug the lines.

7. Remove the crash pad as follows:

a. Remove the floor console assembly.

b. Remove the A-pillar trim.

c. Remove the cowl side trim.

d. Remove the left and right side crash pad side cover.

e. Remove the hood release handle.

f. Remove the left lower panel from the instrument panel.

g. Remove steering wheel and column cover.

h. Remove steering shaft from the cowl cross bar.

i. Remove the damper from the glove box.

j. Remove the hinge panels and remove the glove box.

k. Remove the instrument panel assembly main mounting bolts.

l. Disconnect the fuse box and radio antenna cable.

m. Remove the cluster facia panel.

n. Remove the instrument cluster.

o. Remove the center fascia panel.

p. Remove the heater control assembly.

q. Remove the radio assembly.

r. Remove the right lower panel from the instrument panel.

s. Remove the mounting bolts and remove the crash pad assembly.

8. Disconnect the heater unit electrical connectors.

9. Remove the upper pipe bracket and side cover from the heater unit.

10. Pull the heater core from the heater unit.

➡**Take care not to bend the inlet and outlet pipes of the heater core during removal.**

11. Installation is the reverse order of removal.

12. Fill the cooling system to the correct level.

13. Start the engine and check for leaks.

Cylinder Head

REMOVAL & INSTALLATION

1.5L and 1.6L (VIN 5) Engines

1. Before servicing the vehicle, refer to the Precautions Section.

2. Disconnect the negative battery cable.

3. Properly relieve the fuel system pressure.

4. Drain and recycle the engine coolant.

5. Drain and recycle the engine oil.

6. Remove the upper radiator hose.

7. Remove the breather hose from the between the air cleaner and rocker arm (valve) cover.

8. Remove the air intake hose.

9. Disconnect the vacuum hose, fuel hose and coolant hose.

10. Disconnect the spark plug wires, tag before disconnecting.

11. Remove the ignition coil.

12. Remove the power steering pump and bracket and position aside. Do not disconnect the fluid lines.

13. Remove the intake manifold.

14. Remove the heat shield from the exhaust manifold.

15. Remove the exhaust manifold.

16. Remove the surge tank.

17. Remove the water pump and crankshaft pulleys.

18. Remove the timing belt cover, belt tensioner and timing belt.

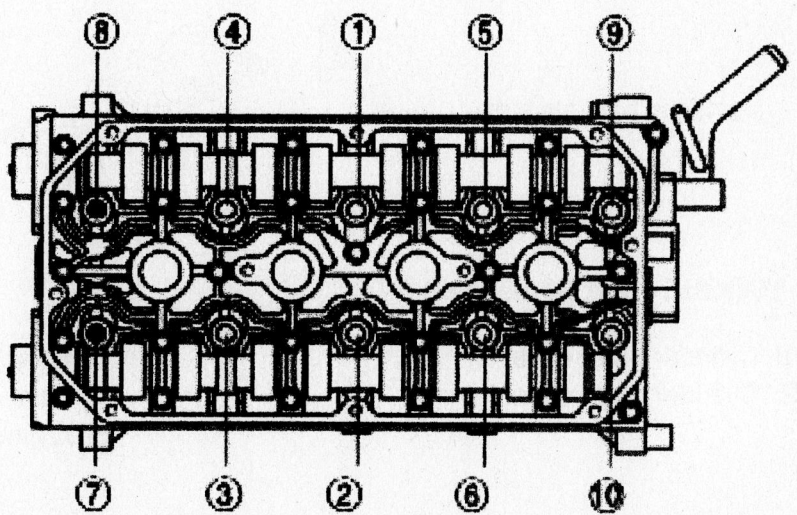

67162-KIAC-G12

Cylinder head bolt tightening (loosen the reverse order)—1.5L and 1.6L (VIN 5) engines

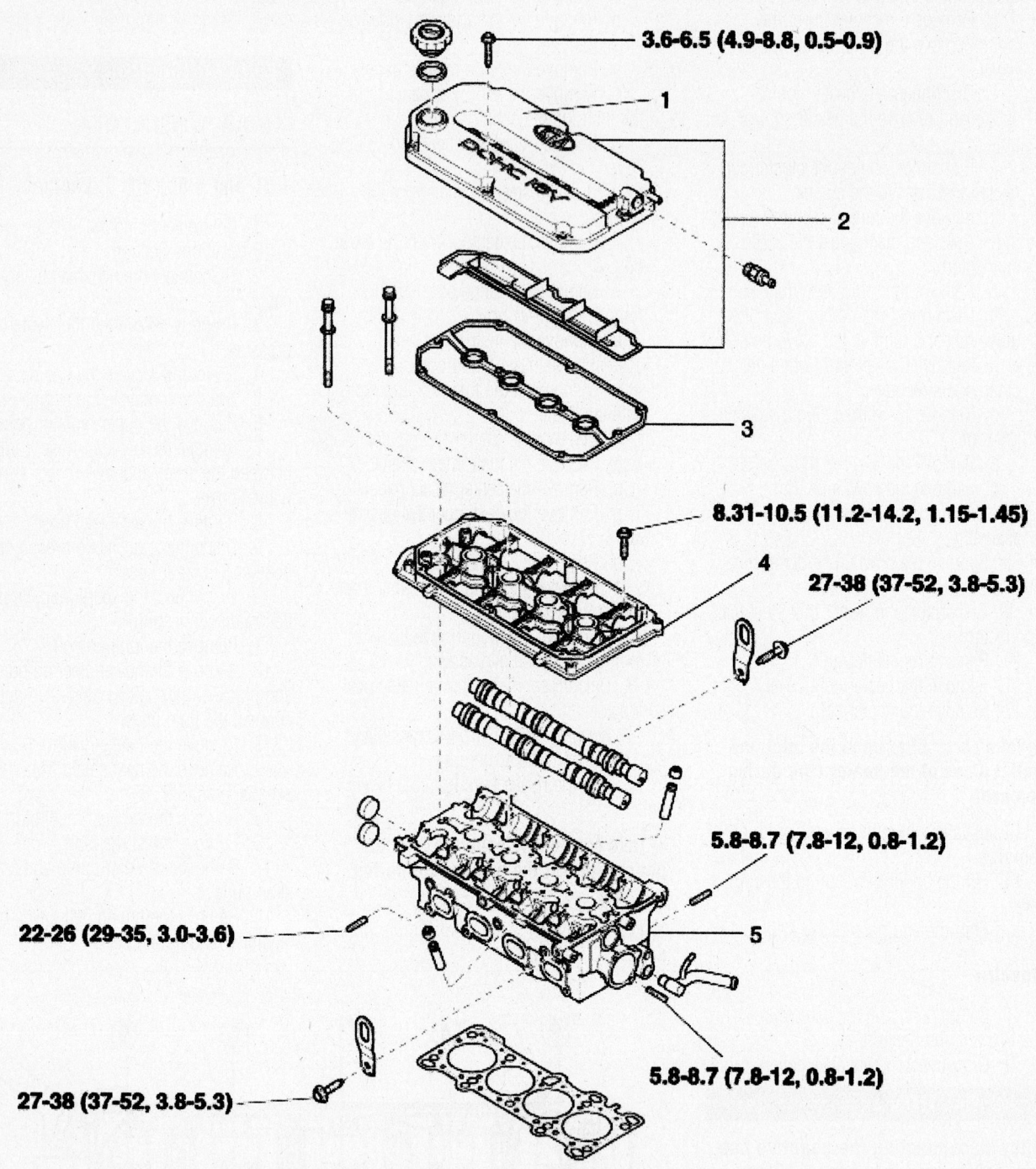

3.6-6.5 (4.9-8.8, 0.5-0.9)

8.31-10.5 (11.2-14.2, 1.15-1.45)

27-38 (37-52, 3.8-5.3)

5.8-8.7 (7.8-12, 0.8-1.2)

22-26 (29-35, 3.0-3.6)

5.8-8.7 (7.8-12, 0.8-1.2)

27-38 (37-52, 3.8-5.3)

TORQUE : lb·ft (N·m, kg·m)

1. Cylinder head cover
2. Cylinder head cover
3. Cylinder head cover gasket
4. Cam carrier assembly
5. Cylinder head

67162-KIAC-G13

Exploded view of the cylinder head and related components—1.5L and 1.6L (VIN 5) engines

19. Remove the cylinder head cover and cam carrier assembly.

20. Remove the cylinder head bolts, in several steps in the proper sequence.

21. Remove the cylinder head and gasket. Discard the gasket.

To install:

22. Thoroughly, clean the cylinder head and the block contact surfaces. Examine the head gasket and check the cylinder head for cracks. Check the cylinder head for warpage using a feeler gauge and straightedge. The maximum allowable distortion is 0.006 in. (0.15mm).

23. Clean the cylinder head bolts and the threads in the block. Be sure the bolts turn freely in the block.

24. Install a new head gasket on the engine block.

25. Install the cylinder head.

26. Install the cylinder head bolts.

27. Tighten the head bolts in the following step using the proper sequence:

    a. Step 1: Tighten to 36 ft. lbs. (49 Nm).

    b. Step 2: Loosen the bolts in the reverse order shown.

    c. Step 3: Tighten to 18 ft. lbs. (25 Nm).

    d. Step 4: Tighten 90° (¼ turn).

    e. Step 5: Tighten 90° (¼ turn).

28. Install the timing belt tensioner pulley and timing belt. Make sure all timing marks are aligned.

29. Install the cylinder head cover. Tighten the bolts to 3.6–6.5 ft. lbs. (5–9 Nm).

30. Install the timing belt cover.

31. Install the intake manifold, with a new gasket. Torque the bolts to 11–14 ft. lbs. (15–20 Nm).

32. Install the exhaust manifold with a new gasket. Torque the bolts to 11–14 ft. lbs. (15–20 Nm).

33. Install the exhaust manifold heat shield.

34. Install the surge tank and tighten the bolts to 11–14 ft. lbs. (15–20 Nm).

35. Install the power steering pump and bracket.

36. Install the ignition coil and connect the spark plug wires.

37. Install the air intake hose.

38. Connect the vacuum hose, fuel hose and water hose.

39. Install the breather hose.

40. Fill the engine oil and coolant.

41. Start the vehicle and check for leaks.

2.0L and 1.6L (VIN 3) Engines

1. Before servicing the vehicle, refer to the Precautions Section.

2. Disconnect the negative battery cable.

3. Turn the crankshaft pulley so that the No. 1 piston is at top dead center.

4. Disconnect the terminals from battery and remove the heat shield.

5. Remove the engine cover.

6. Drain the engine coolant.

7. Remove the radiator cap to speed draining.

8. Remove the intake air hose and air cleaner assembly.

9. Disconnect the AFS (Air Flow Sensor) connector.

10. Disconnect the breather hose from intake air hose.

11. Remove the intake air hose and air cleaner upper cover.

12. Remove the air cleaner element.

13. Remove the bolts and air cleaner lower cover.

14. Remove the upper radiator hose and lower radiator hose.

15. Remove the heater hoses.

16. Disconnect the accelerator cable by loosening the lock-nut, then slip the cable end out of the throttle linkage.

17. Remove the engine wire harness connectors and wire harness clamps from cylinder head and the intake manifold.

18. Disconnect the OCV (Oil Control Valve) connector.

19. Disconnect the oil temperature sensor connector.

20. Disconnect the ECT (Engine Coolant Temperature) sensor connector.

21. Disconnect the ignition coil connector.

22. Disconnect the TPS (Throttle Position Sensor) connector.

23. Disconnect the ISA (Idle Speed Actuator) connector.

24. Disconnect the CMP (Camshaft Position Sensor) connector.

25. Disconnect the four injector connectors.

26. Disconnect the knock sensor connector.

27. Disconnect the ground cables from the intake manifold and vehicle's body.

28. Disconnect the air conditioner compressor switch connector.

29. Disconnect the front heated oxygen sensor connector.

30. Disconnect the CKP (Crankshaft Position Sensor) connector.

31. Disconnect the oil pressure switch connector.

32. Disconnect the PCSV (Purge Control Solenoid Valve) connector.

33. Disconnect the fuel inlet hose of the delivery pipe side.

34. Disconnect the hose of the PCSV (Purge Control Solenoid Valve) side.

35. Remove the brake booster vacuum hose.

36. Remove the power steering pump drive belt.

37. Remove the power steering pump and use a wire to secure the pump to the vehicle so that it is out of the way.

38. Remove the bolts and power steering pump bracket.

39. Remove the spark plug cables.

40. Remove the exhaust manifold.

41. Remove the intake manifold.

42. Remove the timing belt.

43. Remove the PCV (Positive Crankcase Ventilation) hose.

44. Remove the cylinder head cover.

45. Remove the camshaft sprocket.

46. Insert a stopper pin or other device into timing chain auto tensioner and remove the auto tensioner.

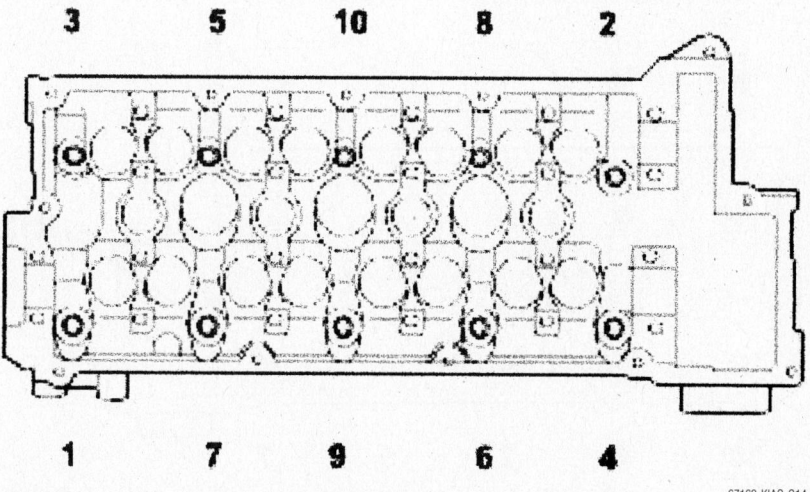

Cylinder head bolt loosening sequence—2.0L and 1.6L (VIN 3) engines

67162-KIAC-G14

47. Remove the camshaft bearing caps and camshafts.

48. Remove the OCV(Oil Control Valve) and filter.

49. Disconnect the water hose from water pipe.

50. Using 8mm and 10mm hexagon wrench, uniformly loosen and remove the 10 cylinder head bolts, in several passes, in the sequence shown.

51. Remove the 10 cylinder head bolts and plate washers.

52. Lift the cylinder head from the dowels on the cylinder block and place the cylinder head on wooden blocks on a bench.

To install:

53. Install the cylinder head gasket onto the cylinder block. Be careful of the installation direction.

54. Place the cylinder head onto the block carefully in order to prevent damaging the gasket. If the gasket is damaged, fluid leakage could occur.

55. Apply a light coat if engine oil on the threads and under the heads of the cylinder head bolts. Using an 8mm and 10mm hexagon wrench, install and tighten the 10 cylinder head bolts and plate washers, in several passes, in the proper sequence.

56. Tighten 2.0L engine as follows:
   a. Step 1: Tighten the M10 bolts to: 17–19.5 ft. lbs. (23–26.5 Nm)
   b. Step 2: Tighten 60–65°.
   c. Step 3: Tighten an additional 60–65°.
   d. Step 4: Tighten the M12 bolts to: 20.5–23 ft. lbs. (27.5–31 Nm)
   e. Step 5: Tighten 60–65°.
   f. Step 6: Tighten an additional 60–65°.

57. Tighten 1.6L (VIN 3) engine

58. Install the OCV filter along with a new filter gasket. Tighten the filter to 30–37 ft. lbs. (40–50 Nm).

59. Install the OCV. Tighten the bolt to 7–8.5 ft. lbs. (10–11.5 Nm).

60. Install the camshafts.

61. Align the camshaft timing chain with the intake timing chain sprocket and exhaust timing chain sprocket as shown.

62. Install the camshaft and bearing caps. Tighten the bearing cap bolts to 10.5 ft. lbs. (14 Nm).

63. Install the timing chain auto tensioner. Tighten the bolts to 6 ft. lbs. (8.5 Nm).

64. Remove the auto tensioner stopper pin.

65. Check and adjust valve clearance.

66. Using the SST (09221-21000), install the camshaft bearing oil seal.

67. Install the camshaft sprocket.

68. Install the cylinder head cover.

69. Install the cylinder head cover gasket in the groove of the cylinder head cover. Before installing the cylinder head cover gasket, thoroughly clean the cylinder head cover and the groove. When installing, make sure the cylinder head cover gasket is seated securely in the corners of the recesses with no gap. Apply liquid gasket to the head cover gasket at the corners of the recess.

70. Install the cylinder head cover with the 12 bolts. Tighten all cylinder head cover bolts temporarily to half of the standard torque, the re-tighten completely using the order specified in the illustration.

71. Install the PCV(Positive Crankcase Ventilation) hose.

72. Install the timing belt.

73. Install the intake manifold.

74. Install the exhaust manifold.

75. Install the spark plug cables.

76. Install the power steering pump bracket and bolts.

77. Install the power steering pump.

78. Connect the accelerator cable.

79. Install the brake booster hose.

80. Connect the hose of the PCSV side.

81. Connect the fuel inlet hose of the delivery pipe side.

82. Install the engine wire harness connectors and wire harness clamps to the cylinder head and the intake manifold.

83. Connect the PCSV connector.

84. Connect the front heated oxygen sensor connector.

85. Connect the CKP connector.

86. Connect the oil pressure switch connector.

87. Connect the air conditioner compressor switch connector.

88. Connect the ground cables to intake manifold and vehicle's body.

89. Connect the knock sensor connector.

90. Connect the fuel injector connectors.

91. Connect the CMP connector.

92. Connect the ISA connector.

93. Connect the TPS connector.

94. Connect the ignition coil connector.

95. Connect the ECT connector.

96. Connect the oil temperature sensor connector.

97. Connect the OCV connector.

98. Install the heater hose.

99. Install the upper radiator hose and lower radiator hose.

100. Install the intake air hose and air cleaner assembly.

101. Install the air cleaner element.

102. Install the intake air hose and air cleaner upper cover.

103. Install the breather hose to intake air hose.

104. Connect the AFS connector.

105. Install the engine cover.

106. Install the heat shield and connect the battery terminals to the battery.

107. Fill with engine coolant.

108. Start the engine and check for leaks.

109. Recheck engine coolant level and oil level.

1.8L and 2.4L Engines

1. Before servicing the vehicle, refer to the Precautions Section.

2. Disconnect the negative battery cable.

3. Properly relieve the fuel system pressure.

4. Drain and recycle the engine coolant.

5. Drain and recycle the engine oil.

6. Remove the Positive Crankcase Ventilation (PCV) and crankcase ventilation hoses.

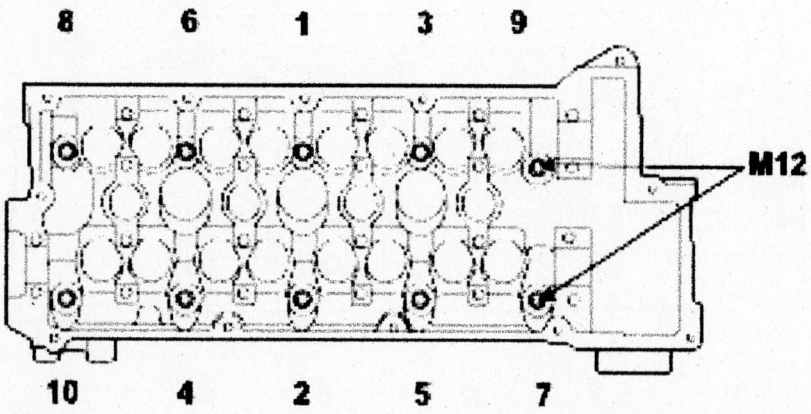

67162-KIAC-G15

Cylinder head bolt tightening sequence—2.0L and 1.6L (VIN 3) engines

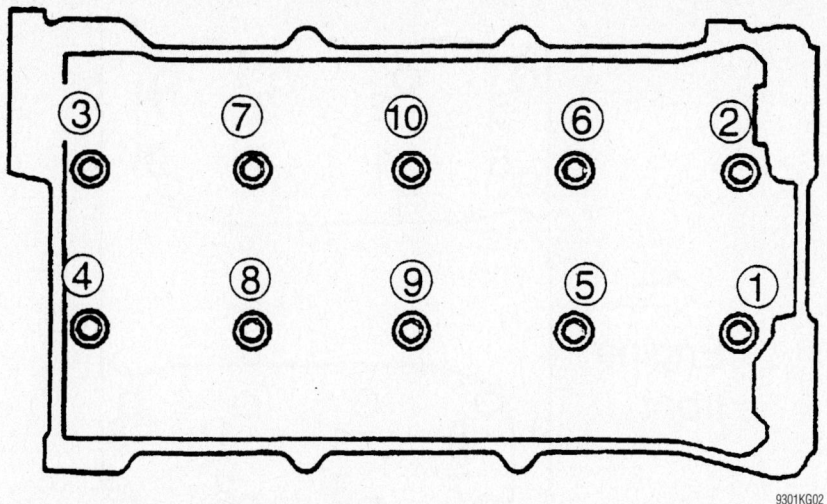

Cylinder head bolt removal sequence—1.8L engine

9301KG02

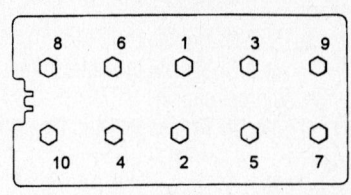
Camshaft sprocket side

Cylinder head torque sequence—2.4L engine

9357NG30

7. Disconnect the accelerator and, if equipped, the cruise control cables.

8. Disconnect the air intake hose from the throttle body.

9. Disconnect the brake vacuum hose and the purge control vacuum hose.

10. Remove the ventilation hose and fresh air duct.

11. Remove the upper radiator hose.

12. Disconnect the fuel hose from the fuel injector rail.

13. Remove the heater hoses.

14. Disconnect the Idle Air Control (IAC) and Throttle Position (TP) sensor connectors.

15. Disconnect the fuel injector electrical connectors.

16. Remove the engine ground strap.

17. Remove the exhaust manifold heat shield.

18. Remove the front exhaust pipe from the manifold.

19. Remove the exhaust manifold.

20. Remove the coolant bypass pipe from the cylinder head.

21. Remove the intake manifold support bracket.

22. Remove the camshaft and Hydraulic Lash Adjusters (HLA's).

23. Remove the cylinder head bolts in several steps, in the order illustrated.

24. Remove the cylinder head bolts, then the cylinder head.

To install:

25. Thoroughly, clean the cylinder head and the block contact surfaces. Examine the head gasket and check the cylinder head for cracks. Check the cylinder head for warpage using a feeler gauge and straightedge. The maximum allowable distortion is 0.006 in. (0.15mm).

26. Clean the cylinder head bolts and the threads in the block. Be sure the bolts turn freely in the block.

27. Install a new head gasket on the engine block.

28. Install the cylinder head.

29. Install the cylinder head bolts.

30. For 1.8L engines, tighten the head bolts in the following step using the proper sequence:

   a. Step 1: Tighten to 36 ft. lbs. (49 Nm).

   b. Step 2: Loosen the bolts in the reverse order shown.

   c. Step 3: Tighten to 29 ft. lbs. (39 Nm).

   d. Step 4: Tighten 90° (¼ turn).

   e. Step 5: Tighten 90° (¼ turn).

31. For 2.4L engines, tighten the head bolts in the following step using the proper sequence:

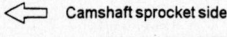
Camshaft sprocket side

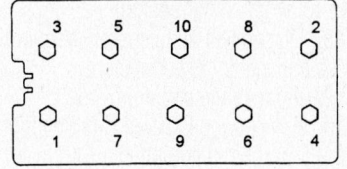

9357NG29

Cylinder head bolt removal sequence—2.4L engine

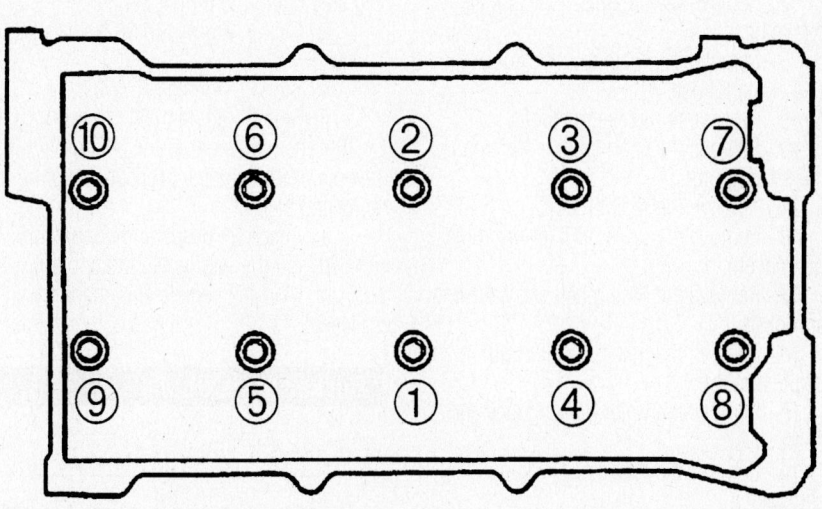

Cylinder head torque sequence—1.8L engine

7923KG15

a. Step 1: Tighten to 14 ft. lbs. (20 Nm).

b. Step 2: Tighten 90° (¼ turn).

c. Step 3: Loosen the bolts in the reverse order shown.

d. Step 4: Tighten to 14 ft. lbs. (20 Nm).

e. Step 5: Tighten 90° (¼ turn).

f. Step 6: Tighten 90° (¼ turn).

32. Install the camshaft and HLA's.

33. Install the intake manifold support bracket and tighten the mounting bolts to 28–38 ft. lbs. (37–52 Nm).

34. Install the coolant bypass pipe and tighten the mounting bolt to 66–86 ft. lbs. (89–117 Nm).

35. Install the exhaust manifold, tighten the manifold-to-cylinder head mounting nuts to 28–34 ft. lbs. (38–46 Nm) and the manifold-to-exhaust pipe mounting nuts to 16–21 ft. lbs. (22–28 Nm).

36. Install the exhaust manifold heat shield and tighten the mounting bolts to 13–22 ft. lbs. (19–30 Nm).

37. The remaining steps of the installation procedure is the reverse of the removal, while keeping in mind the following:

a. Attach all hoses and connectors.

b. Fill the engine oil and coolant.

c. Start the vehicle and check for leaks.

2.7L Engine

1. Before servicing the vehicle, refer to the Precautions Section.

2. Disconnect the negative battery cable.

3. Properly relieve the fuel system pressure.

4. Drain and recycle the engine coolant.

5. Drain and recycle the engine oil.

6. Remove the upper radiator hose.

7. Remove the breather hose and air intake hose.

8. Remove the vacuum hose, fuel hose and coolant hose.

9. Remove the intake manifold.

10. Remove the spark plug wires from the spark plugs.

11. Remove the ignition coil.

12. Remove the upper and lower timing belt covers.

13. Remove the timing belt and camshaft sprockets.

14. Remove the heat protector and exhaust manifold.

15. Remove the water pump pulley and head cover.

16. Remove the intake and exhaust camshafts.

17. Remove the cylinder head bolts,

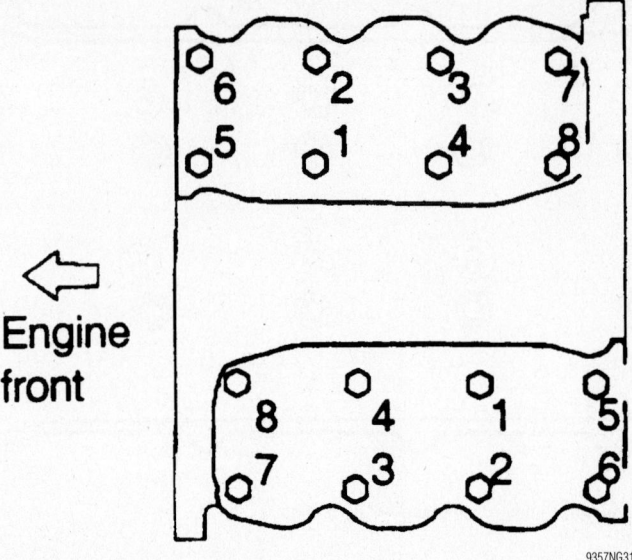

Engine front

Cylinder head tightening sequence (use the reverse for removal)—2.7L engine

9357NG31

loosening in the reverse of the torque sequence, in several steps.

18. Remove the cylinder head.

19. Remove the gaskets and discard.

To install:

20. Thoroughly, clean the cylinder head and the block contact surfaces. Examine the head gasket and check the cylinder head for cracks. Check the cylinder head for warpage using a feeler gauge and straightedge. The maximum allowable distortion is 0.006 in. (0.15mm).

21. Clean the cylinder head bolts and the threads in the block. Be sure the bolts turn freely in the block.

22. Install the new head gasket on the engine block.

23. Install the cylinder head.

24. Install the cylinder head bolts.

25. Tighten the head bolts in the following step using the proper sequence:

a. Step 1: Tighten to 18 ft. lbs. (25 Nm).

b. Step 2: Tighten 60° (⅙ turn).

c. Step 3: Tighten 45° (⅛ turn).

26. The remainder of installation is the reverse or the removal procedure, noting the following steps:

a. Attach all hoses and connectors.

b. Fill the engine oil and coolant.

c. Start the vehicle and check for leaks.

Rocker Arms/Shafts

REMOVAL & INSTALLATION

The engines covered in this section are not equipped with rocker arms/shafts. The camshafts directly actuate the valve through a bucket type follower.

Intake Manifold

REMOVAL & INSTALLATION

1. Before servicing the vehicle, refer to the Precautions Section.

2. Properly relieve the fuel system pressure. Disconnect the negative battery cable and drain the cooling system.

3. Remove the air intake hose from the throttle body.

4. Remove the air intake hose and air cleaner assembly, if necessary.

5. Remove the air intake surge tank, if necessary.

6. Disconnect the accelerator cable.

7. Disconnect the fuel lines. Plug the lines to avoid contamination.

8. Disconnect all necessary vacuum hoses and electrical connectors.

9. Remove the coolant hoses.

10. Remove the Exhaust Gas Recirculation (EGR) tube, if equipped.

11. Remove the air valve, if equipped.

12. Remove the fuel rail attaching bolts.

13. Remove the fuel rail and injectors as an assembly.

14. Remove the intake manifold support bracket.

15. Remove the bolt retaining the dipstick tube bracket to the intake manifold, if necessary.

16. Remove the intake manifold-to-cylinder bolts/nuts and the intake manifold assembly.

17. Remove the throttle body, if neces-

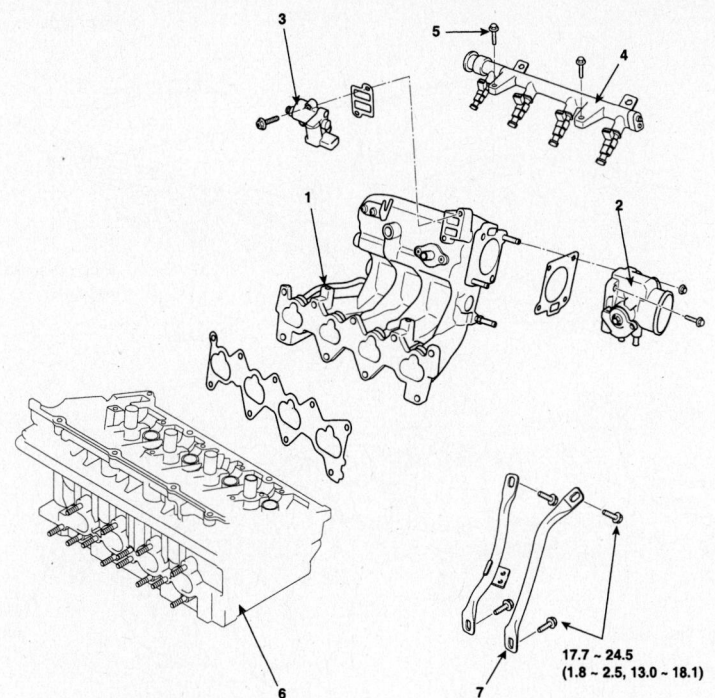

TORQUE : N.m(kgf.m, lb-ft)

1. Intake manifold
2. Throttle body
3. ISA(Idle Speed Actuator)
4. Delivery pipe

5. Gasket
6. Cylinder head
7. Intake manifold stay

09474_KIAC_G0010

Exploded view of intake manifold assembly—1.6L (VIN 3)

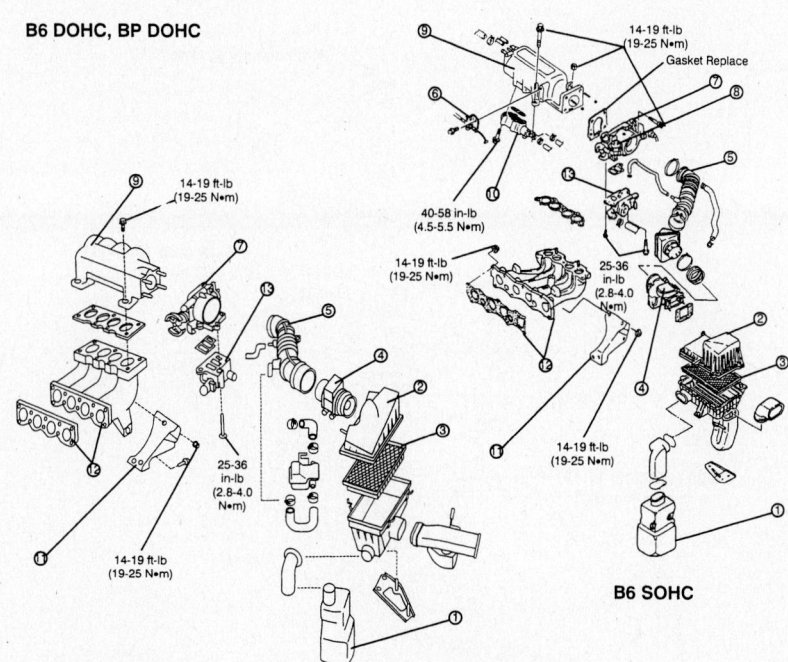

1. Resonance chamber
2. Upper air filter housing
3. Air filter (B6 SOHC)
4. Mass air flow (MAF) sensor (B6 DOHC, BP DOHC)/
   Volume air flow (VAF) sensor (B6 SOHC)
5. Intake air hose
6. Throttle cable

7. Throttle body
8. Dashpot (B6 SOHC)
9. Dynamic chamber
10. Air valve (B6 SOHC)
11. Intake manifold support bracket
12. Intake manifold and gasket (Replace)
13. Idle air control valve (B6 SOHC)/
    Bypass air control (BAC) valve (B6 DOHC, BP DOHC)

7923KG17

Exploded view of the intake manifold assembly—1.8L engine shown

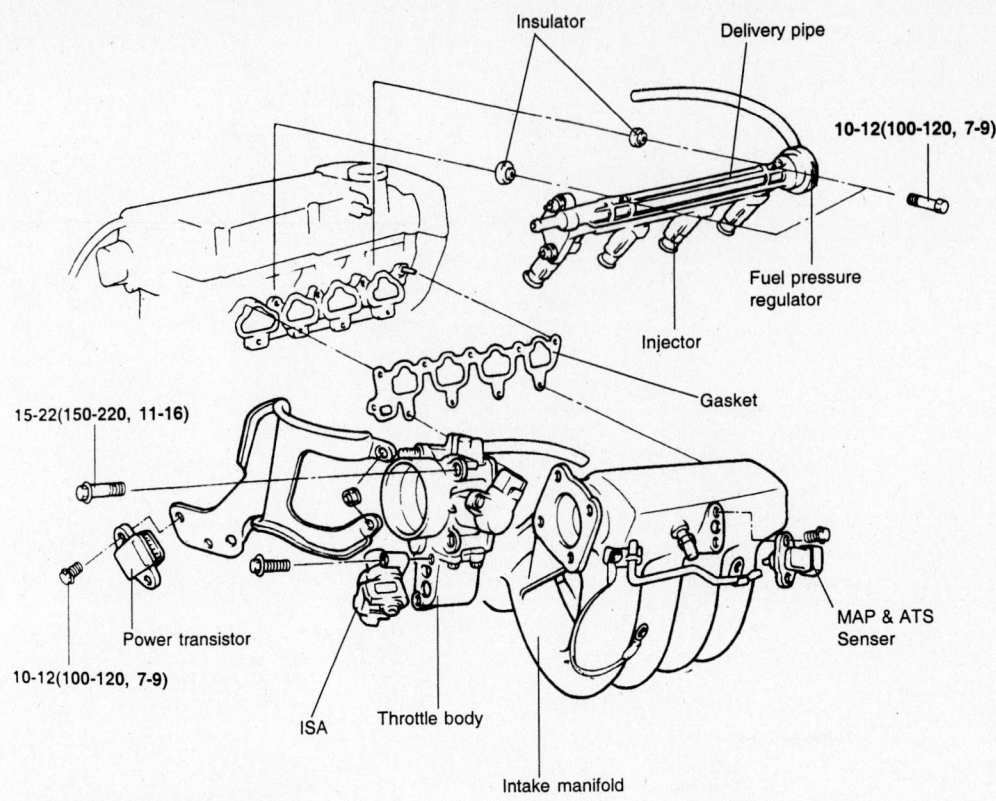

Insulator

Delivery pipe

10-12(100-120, 7-9)

Fuel pressure regulator

Injector

Gasket

15-22(150-220, 11-16)

Power transistor

10-12(100-120, 7-9)

ISA

Throttle body

Intake manifold

MAP & ATS Senser

TORQUE : Nm (kg.cm, lb.ft)

9357NG32

Exploded view of the intake manifold—2.4L engine

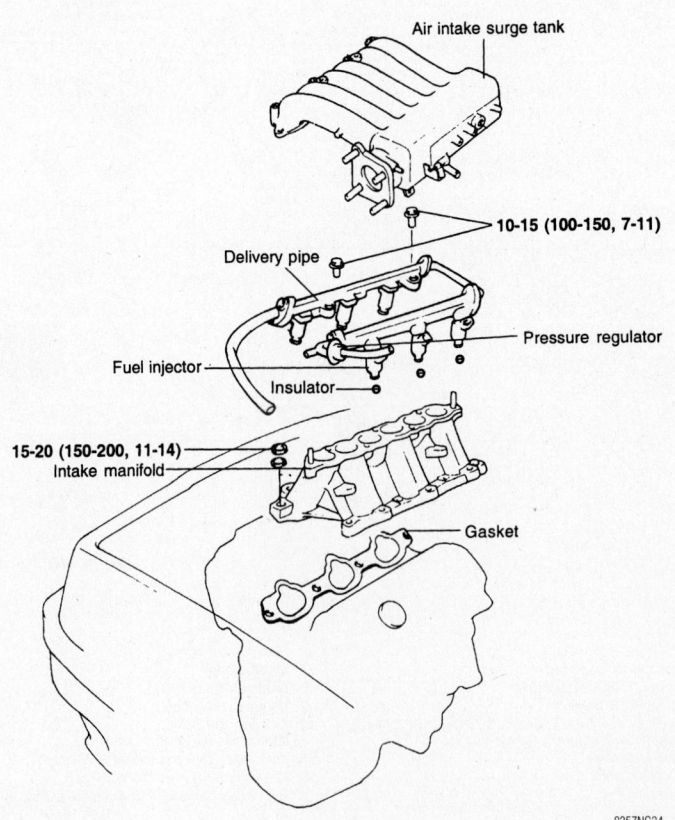

Air intake surge tank

10-15 (100-150, 7-11)

Delivery pipe

Pressure regulator

Fuel injector

Insulator

15-20 (150-200, 11-14)
Intake manifold

Gasket

9357NG34

Exploded view of the intake manifold—2.7L engine

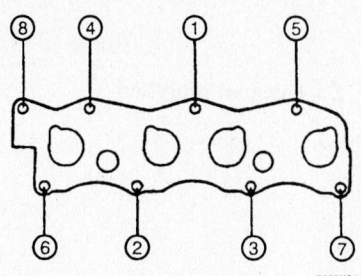

Intake manifold torque sequence—1.8L engine

7923KG18

sary and separate the intake manifold upper and lower halves.

To install:

18. Clean all gasket mating surfaces.

19. Install the upper and lower intake manifolds using a new gasket, if separated. Tighten the nuts/bolts to 19 ft. lbs. (25 Nm).

20. Install the throttle body using a new gasket, if removed. Tighten the retaining nuts/bolts to 19 ft. lbs. (25 Nm).

21. Install the intake manifold assembly to the cylinder head using a new gasket. Tighten the nuts/bolts to 11–14 ft. lbs. (15–20 Nm) for 1.5L, 1.6L and 2.7L engines, or 13–16.5 ft. lbs. (18–22.5 Nm) for 2.0L engine, or to 19 ft. lbs. (25 Nm) for 1.8L and 2.4L engines.

➡**Tighten the bolts in the center of the manifold first and work outward toward the ends.**

22. Install the bolt retaining the dipstick tube to the intake manifold, if equipped.

23. Install the intake manifold bracket. Tighten the attaching nuts/bolts to 19 ft. lbs. (25 Nm).

24. Install the EGR tube, if equipped.

25. Connect the coolant and vacuum hoses, electrical connectors and fuel lines.

26. Connect the accelerator cable.

27. Install the air intake surge tank, if removed.

28. Install the air cleaner assembly, if removed.

29. Install the air intake tube to the throttle body.

30. Connect the negative battery cable.

31. Fill the cooling system.

32. Start the engine and bring to normal operating temperature. Check for leaks. Check the idle speed.

Exhaust Manifold

REMOVAL & INSTALLATION

1. Before servicing the vehicle, refer to the Precautions Section.

2. Disconnect the negative battery cable.

16.7 ~ 21.6
(1.7 ~ 2.2, 12.3 ~ 15.9)

29.4 ~ 34.3
(3.0 ~ 3.5, 21.7 ~ 25.3)

TORQUE : N.m (kgf.m, lb-ft)

1. Cylinder head
2. Heat protector
3. Gasket
4. Exhaust manifold

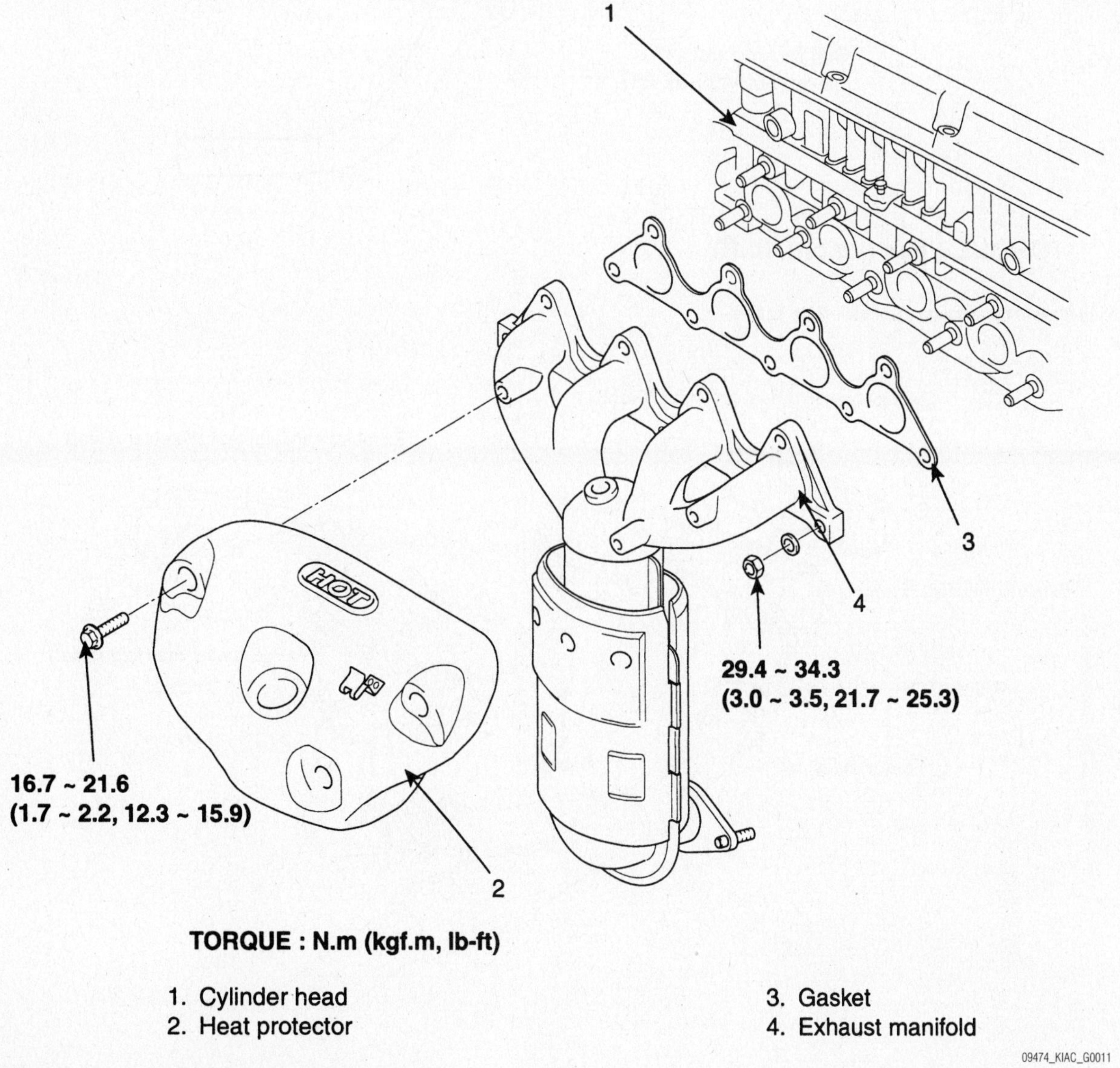

Exploded view of exhaust manifold assembly—1.6L (VIN 3)

09474_KIAC_G0011

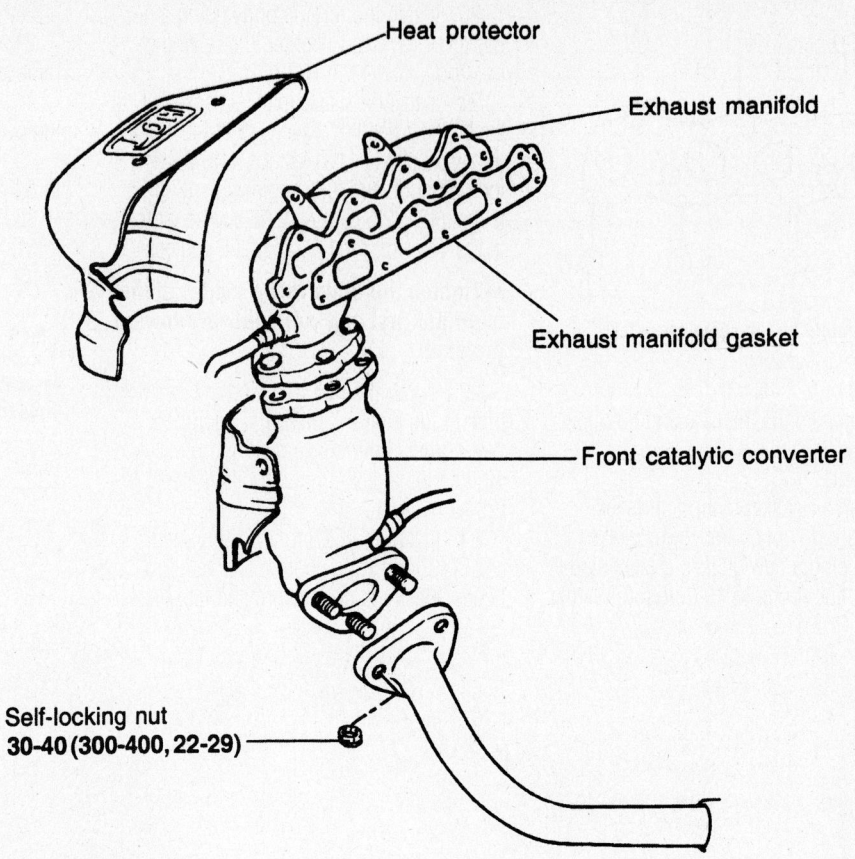

TORQUE : Nm (kg.cm, lb.ft)

9357NG33

Exploded view of the exhaust manifold—2.4L engine

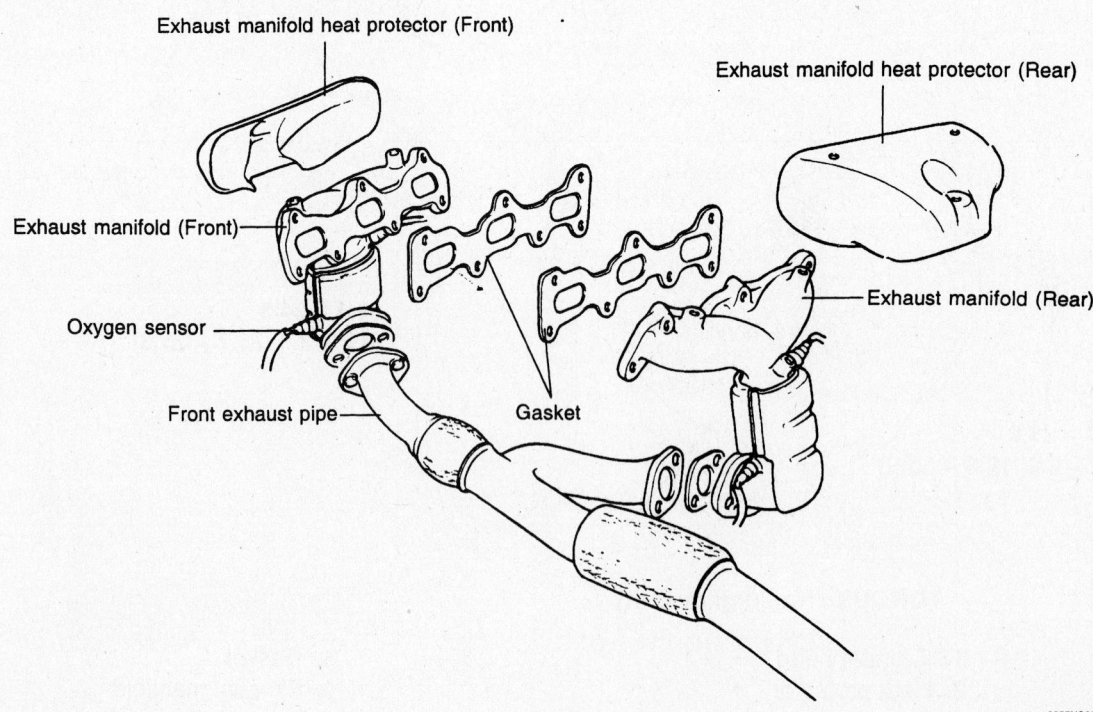

9357NG35

Exploded view of the exhaust manifolds—2.7L engine

3. Remove or disconnect the following:
- Engine cover
- Front oxygen sensors
- Front muffler heat protector
- Front muffler
- Heater protector
- Exhaust manifold assembly

To install:

4. Clean all gasket mating surfaces.

5. Install a new exhaust manifold gasket and the exhaust manifold. Tighten the mounting nuts and bolts as follows:

   a. 1.5L and 1.6L (VIN 5) engines to 11–14 ft. lbs. (15–20 Nm)

   b. 1.6L (VIN 3) engine to 13–18 ft. lbs. (18–25 Nm)

   c. 1.8L and 2.4L engines to 29–33 ft. lbs. (39–44 Nm)

   d. 2.0L engines to 31–40 ft. lbs. (42–54 Nm)

   e. 2.7L engine to 22–25 ft. lbs. (29–34 Nm)

6. The reminder of installation is the reverse order of removal.

Front Crankshaft Seal

REMOVAL & INSTALLATION

1. Before servicing the vehicle, refer to the Precautions Section.

2. Disconnect the negative battery cable.

3. Remove the timing belt covers and belt.

4. Remove the timing belt pulley using a puller.

5. Remove the oil pump bolts and the pump.

6. Wrap a suitable prytool with a rag and work the old seal from the oil pump housing.

To install:

7. Lubricate the seal lip with clean engine oil and push the seal slightly in by hand.

8. Install the seal using a seal installer. Install the seal until it is flush with the oil pump body.

9. Install the timing belt.

10. Connect the negative battery cable.

11. Start the engine and check for leaks.

Camshaft

REMOVAL & INSTALLATION

1.5L and 1.6L (VIN 5) Engines

1. Before servicing the vehicle, refer to the Precautions Section.

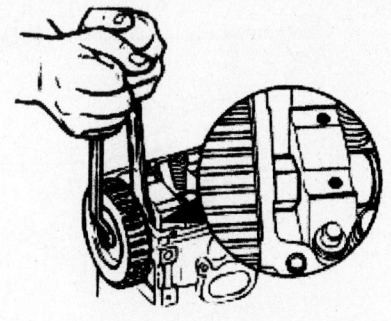

67162-KIAC-G16

Remove the camshaft pulley (sprocket)—1.5L and 1.6L engines

2. Disconnect the negative battery cable.

3. Remove the breather hose and Positive Crankcase Ventilation (PCV) hose.

4. Remove the water pump and crankshaft pulleys.

5. Remove the timing belt cover.

6. Loosen the timing bolt tensioner pulley and temporarily secure it.

7. Remove the timing belt from the camshaft sprocket.

8. Remove the center cover bolts and cover.

9. Remove the ignition coil.

10. Remove the cylinder head cover.

11. Remove the camshaft pulley (sprocket).

12. Remove the cam carrier assembly and timing belt.

13. Remove the camshafts.

14. Remove the camshaft oil seal using a seal removal tool.

➡ **The Hydraulic Lash Adjusters (HLA's) must be installed in the location from which they were removed.**

15. Mark the HLA's to identify their original positions.

16. Remove the HLA's from the cylinder head using a magnet, and store upside-down in a oil-filled container

To install:

17. Apply a coat of the clean engine oil to the sides of the HLA's.

18. Install the HLA's into the cylinder head bore. Check that the HLA moves freely in its bore.

19. Lubricate the camshaft journals and lobes with clean engine oil.

20. Install the camshafts in the cylinder head making sure that the camshaft dowel pins point straight up.

21. Install the cam carrier assembly. Tighten the bolts, in several steps, to 8–11 ft. lbs. (11–15 Nm).

22. Install the camshaft oil seal, using a suitable seal installer.

23. Install the camshaft sprockets onto the camshaft. Be sure to align the **I** mark with the intake camshaft dowel pin and the **E** mark with the exhaust camshaft dowel pin, then tighten the retaining bolt to 36–44 ft. lbs. (49–61 Nm).

24. The remainder of installation is the reverse of the removal procedure.

25. Check the engine fluid levels, start the vehicle and check for leaks.

1.8L Engine

1. Before servicing the vehicle, refer to the Precautions Section.

2. Disconnect the negative battery cable.

3. Remove the ignition coils and high-tension cords.

4. Remove the Positive Crankcase Ventilation (PCV) valve and ventilation hoses.

5. Position the engine so that No. 1 cylinder is at Top Dead Center (TDC).

6. Remove the timing belt.

7. Remove the cylinder head cover.

8. Remove the Camshaft Position (CMP) sensor.

9. Remove the camshaft sprockets.

10. Loosen the camshaft bearing cap retaining bolts in several steps, following the proper sequence.

11. Remove the camshafts.

12. Remove the camshaft oil seal using a seal removal tool.

➡ **The Hydraulic Lash Adjusters (HLA's) must be installed in the location from which they were removed.**

13. Mark the HLA's to identify their original positions.

14. Remove the HLA's from the cylinder head using a magnet, and store upside-down in a oil-filled container

To install:

15. Apply a coat of the clean engine oil to the sides of the HLA's.

16. Install the HLA's into the cylinder head bore.

17. Check that the HLA moves freely in its bore.

18. Lubricate the camshaft journals and lobes with clean engine oil.

19. Install the camshafts in the cylinder head making sure that the camshaft dowel pins point straight up.

20. Install the camshaft caps in their original positions. Loosely install the cap bolts.

21. Install the camshaft cap bolts in 5–6 steps to 13–20 ft. lbs. (18–27 Nm) in the proper sequence.

22. Oil the lip of the new camshaft oil

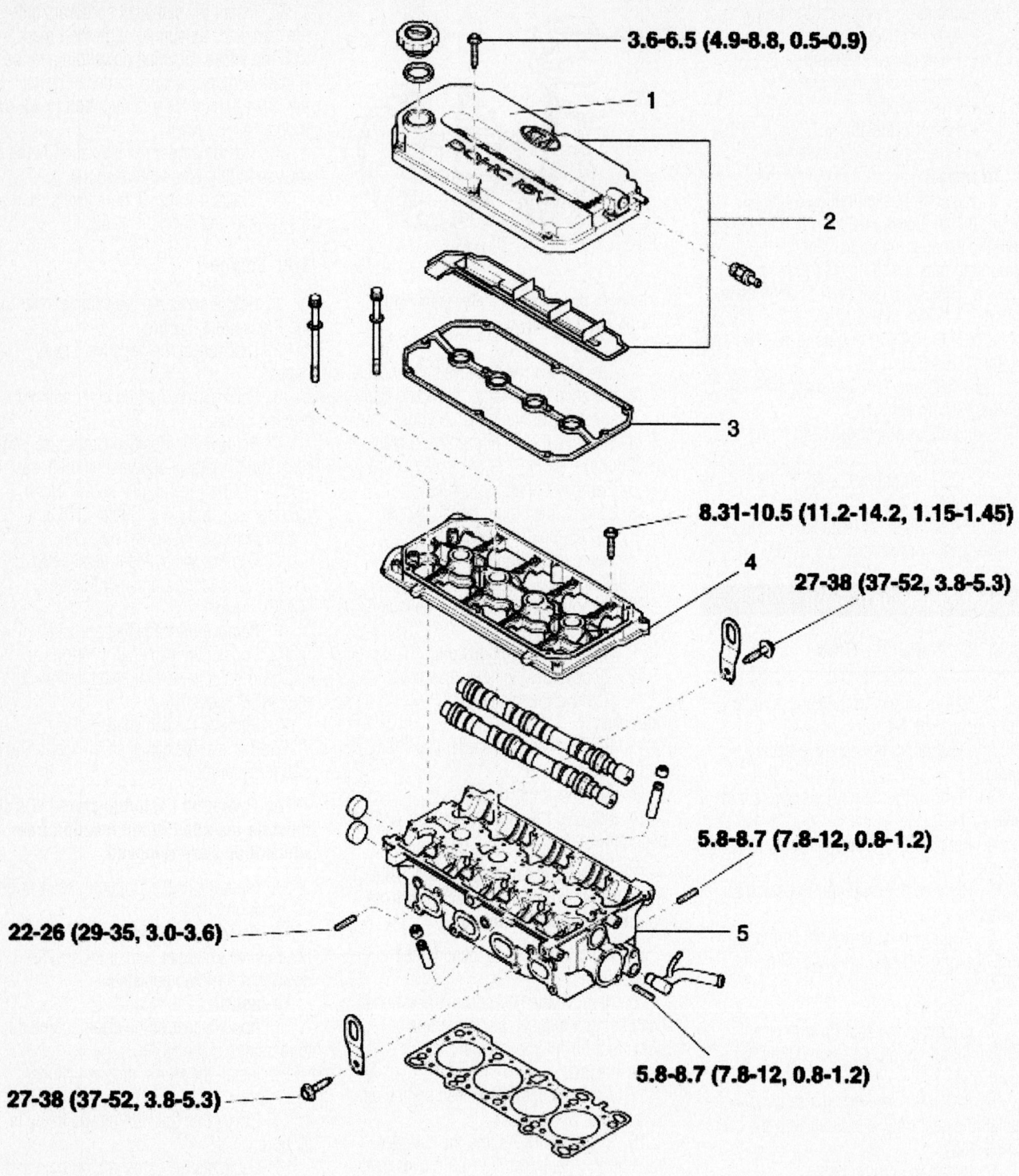

3.6-6.5 (4.9-8.8, 0.5-0.9)

1

2

3

8.31-10.5 (11.2-14.2, 1.15-1.45)

4

27-38 (37-52, 3.8-5.3)

5.8-8.7 (7.8-12, 0.8-1.2)

22-26 (29-35, 3.0-3.6)

5

5.8-8.7 (7.8-12, 0.8-1.2)

27-38 (37-52, 3.8-5.3)

TORQUE : lb·ft (N·m, kg·m)

1. Cylinder head cover
2. Cylinder head cover
3. Cylinder head cover gasket
4. Cam carrier assembly
5. Cylinder head

67162-KIAC-G17

Exploded view of the camshafts and related components—1.5L and 1.6L engines

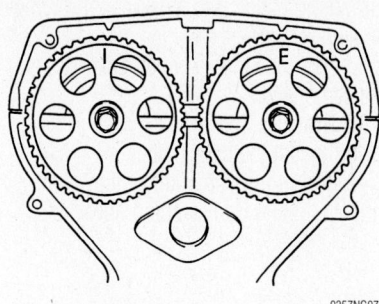

Proper alignment, prior to camshaft installation—1.8L engine

seal and, using a seal installer. Drive the seal into the cylinder head until it is flush with the edge of the camshaft bearing cap.

23. Install the camshaft sprockets onto the camshaft. Be sure to align the **I** mark with the intake camshaft dowel pin and the

E mark with the exhaust camshaft dowel pin, then tighten the retaining bolt to 36–44 ft. lbs. (49–61 Nm).

24. Install the CMP sensor.
25. Install the cylinder head cover.
26. Install the timing belt.
27. Install the PCV and ventilation hoses.
28. Install the ignition coils and high-tension cords.
29. Connect the negative battery cable.
30. Check the engine fluid levels, start the vehicle and check for leaks.

2.0L and 1.6L (VIN 3) Engines

1. Before servicing the vehicle, refer to the Precautions Section.
2. Disconnect the negative battery cable.
3. Turn the crankshaft pulley so that the No. 1 piston is at top dead center.

4. Disconnect the terminals from battery and remove the heat shield.
5. Remove the engine cover.
6. Drain the engine coolant.
7. Remove the radiator cap to speed draining.
8. Remove the intake air hose and air cleaner assembly.
9. Disconnect the AFS (Air Flow Sensor) connector.
10. Disconnect the breather hose from intake air hose.
11. Remove the intake air hose and air cleaner upper cover.
12. Remove the air cleaner element.
13. Remove the bolts and air cleaner lower cover.
14. Remove the upper radiator hose and lower radiator hose.
15. Remove the heater hoses.
16. Disconnect the accelerator cable by loosening the lock-nut, then slip the cable end out of the throttle linkage.
17. Remove the engine wire harness connectors and wire harness clamps from cylinder head and the intake manifold.
18. Disconnect the OCV (Oil Control Valve) connector.
19. Disconnect the oil temperature sensor connector.
20. Disconnect the ECT (Engine Coolant Temperature) sensor connector.
21. Disconnect the ignition coil connector.
22. Disconnect the TPS (Throttle Position Sensor) connector.
23. Disconnect the ISA (Idle Speed Actuator) connector.
24. Disconnect the CMP (Camshaft Position Sensor) connector.
25. Disconnect the four injector connectors.
26. Disconnect the knock sensor connector.
27. Disconnect the ground cables from the intake manifold and vehicle's body.
28. Disconnect the air conditioner compressor switch connector.
29. Disconnect the front heated oxygen sensor connector.
30. Disconnect the CKP (Crankshaft Position Sensor) connector.
31. Disconnect the oil pressure switch connector.
32. Disconnect the PCSV (Purge Control Solenoid Valve) connector.
33. Disconnect the fuel inlet hose of the delivery pipe side.
34. Disconnect the hose of the PCSV (Purge Control Solenoid Valve) side.
35. Remove the brake booster vacuum hose.

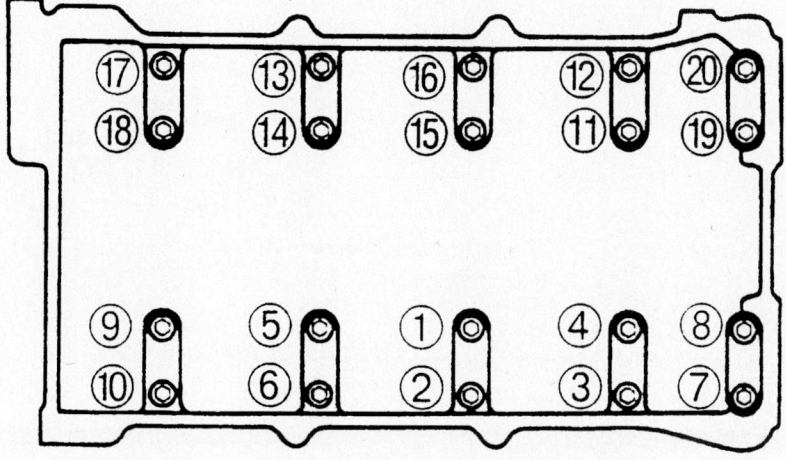

Camshaft bearing cap mounting bolt removal sequence—1.8L engine

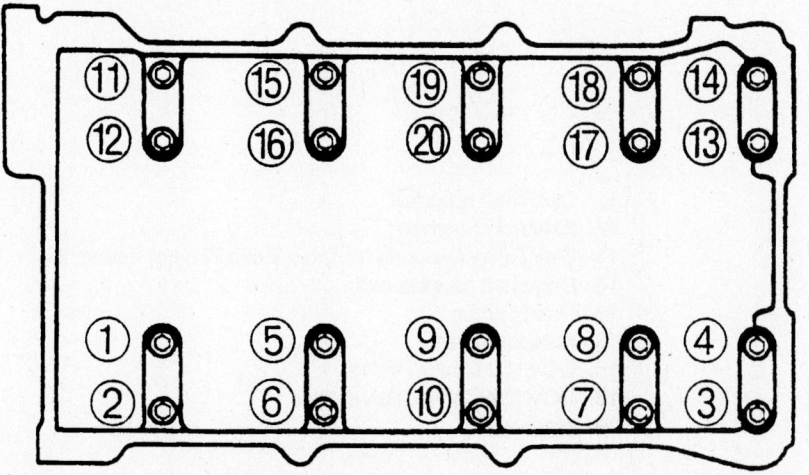

Camshaft bearing cap mounting bolt tightening sequence—1.8L engine

36. Remove the power steering pump drive belt.

37. Remove the power steering pump and use a wire to secure the pump to the vehicle so that it is out of the way.

38. Remove the bolts and power steering pump bracket.

39. Remove the spark plug cables.

40. Remove the exhaust manifold.

41. Remove the intake manifold.

42. Remove the timing belt.

43. Remove the PCV (Positive Crankcase Ventilation) hose.

44. Remove the cylinder head cover.

45. Remove the camshaft sprocket.

46. Insert a stopper pin or other device into timing chain auto tensioner and remove the auto tensioner.

47. Remove the camshaft bearing caps and camshafts.

To install:

48. Install the camshafts.

49. Align the camshaft timing chain with the intake timing chain sprocket

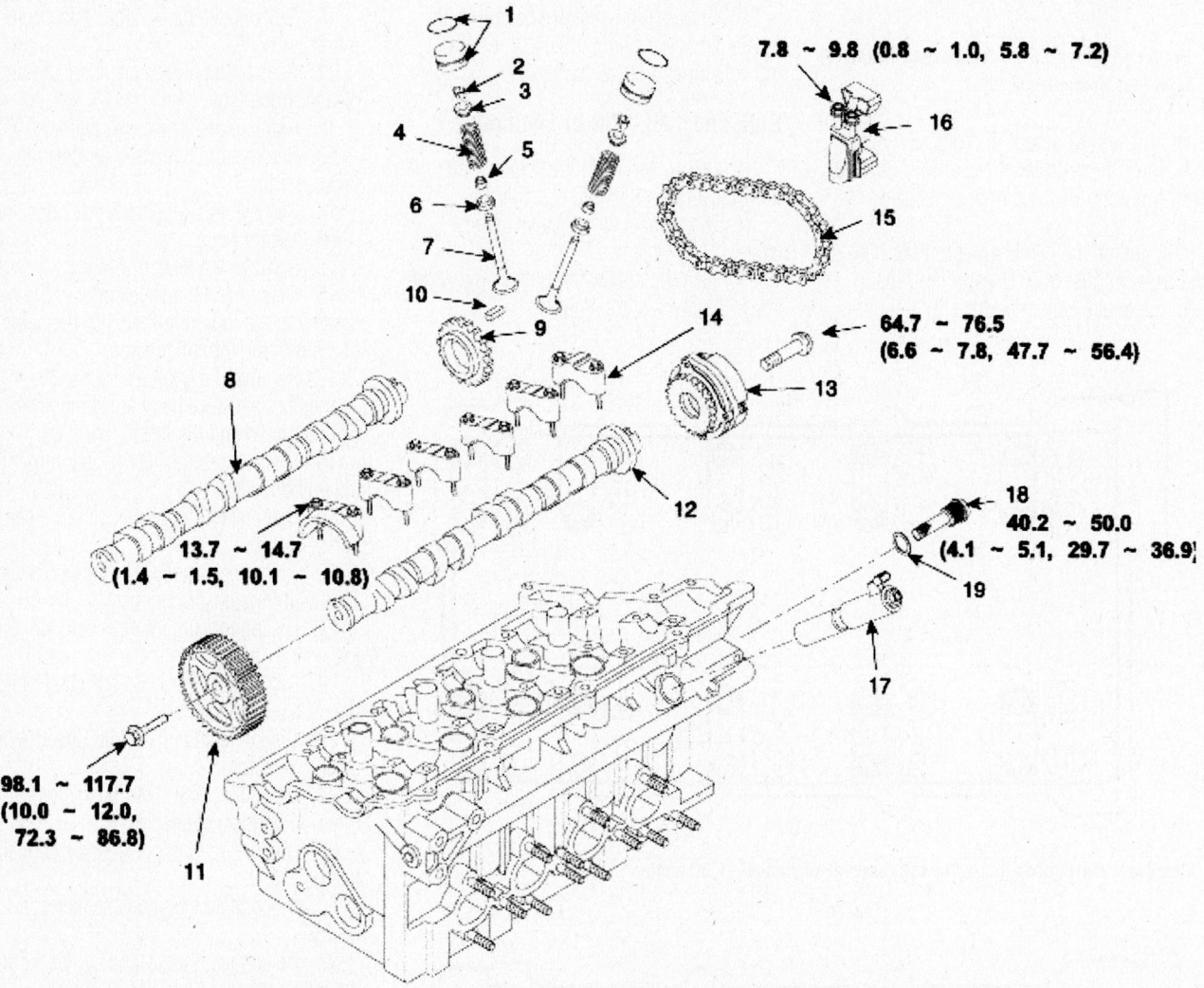

TORQUE : N-m (kg-m, lb-ft)

1. MLA (Mechanical Lash Adjuster)
2. Retainer lock
3. Retainer
4. Valve spring
5. Stem seal
6. Spring seat
7. Valve
8. Intake camshaft
9. Chain sprocket
10. Key
11. Camshaft sprocket
12. Exhaust camshaft
13. CVVT (Continuously Variable Valve Timing) assembly
14. Camshaft bearing cap
15. Timing chain
16. Auto tentioner
17. OCV (Oil Control Valve)
18. OCV (Oil Control Valve) filter
19. Washer

67162-KIAC-G18

Exploded view of the camshafts and related components—2.0L and 1.6L (VIN 3) engines

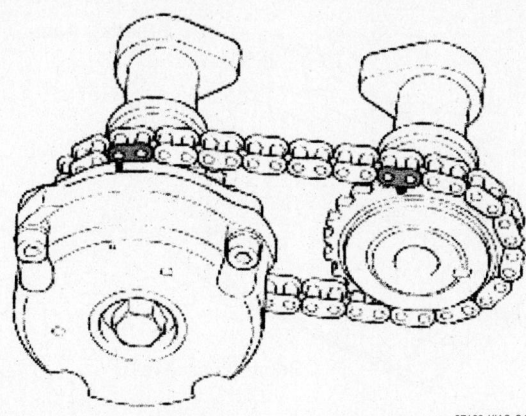

Camshaft timing chain with the intake timing chain sprocket and exhaust timing chain sprocket alignment—2.0L and 1.6L (VIN 3) engines

67162-KIAC-G19

and exhaust timing chain sprocket as shown.

50. Install the camshaft and bearing caps. Tighten the bearing cap bolts to 10.5 ft. lbs. (14 Nm).

51. Install the timing chain auto tensioner. Tighten the bolts to 6 ft. lbs. (8.5 Nm).

52. Remove the auto tensioner stopper pin.

53. Check and adjust valve clearance.

54. Using the SST (09221-21000), install the camshaft bearing oil seal.

55. Install the camshaft sprocket.

56. Install the cylinder head cover.

57. Install the cylinder head cover gasket in the groove of the cylinder head cover. Before installing the cylinder head cover gasket, thoroughly clean the cylinder head cover and the groove. When installing, make sure the cylinder head cover gasket is seated securely in the corners of the recesses with no gap. Apply liquid gasket to the head cover gasket at the corners of the recess.

58. Install the cylinder head cover with the 12 bolts. Tighten all cylinder head cover bolts temporarily to half of the standard torque, the re-tighten completely using the order specified in the illustration.

59. Install the PCV(Positive Crankcase Ventilation) hose.

60. Install the timing belt.

61. Install the intake manifold.

62. Install the exhaust manifold.

63. Install the spark plug cables.

64. Install the power steering pump bracket and bolts.

65. Install the power steering pump.

66. Connect the accelerator cable.

67. Install the brake booster hose.

68. Connect the hose of the PCSV side.

69. Connect the fuel inlet hose of the delivery pipe side.

70. Install the engine wire harness connectors and wire harness clamps to the cylinder head and the intake manifold.

71. Connect the PCSV connector.

72. Connect the front heated oxygen sensor connector.

73. Connect the CKP connector.

74. Connect the oil pressure switch connector.

75. Connect the air conditioner compressor switch connector.

76. Connect the ground cables to intake manifold and vehicle's body.

77. Connect the knock sensor connector.

78. Connect the fuel injector connectors.

79. Connect the CMP connector.

80. Connect the ISA connector.

81. Connect the TPS connector.

82. Connect the ignition coil connector.

83. Connect the ECT connector.

84. Connect the oil temperature sensor connector.

85. Connect the OCV connector.

86. Install the heater hose.

87. Install the upper radiator hose and lower radiator hose.

88. Install the intake air hose and air cleaner assembly.

89. Install the air cleaner element.

90. Install the intake air hose and air cleaner upper cover.

91. Install the breather hose to intake air hose.

92. Connect the AFS connector.

93. Install the engine cover.

94. Install the heat shield and connect the battery terminals to the battery.

95. Fill with engine coolant.

96. Start the engine and check for leaks.

97. Recheck engine coolant level and oil level.

2.4L Engine

1. Before servicing the vehicle, refer to the Precautions Section.

2. Drain the engine coolant.

3. Disconnect the negative battery cable.

4. Remove the breather hose from between the air cleaner and rocker arm cover.

5. Remove the air cleaner.

6. Remove the timing belt cover.

7. Remove the rocker arm cover and crank angle sensor.

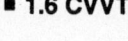

■ **1.6 CVVT**

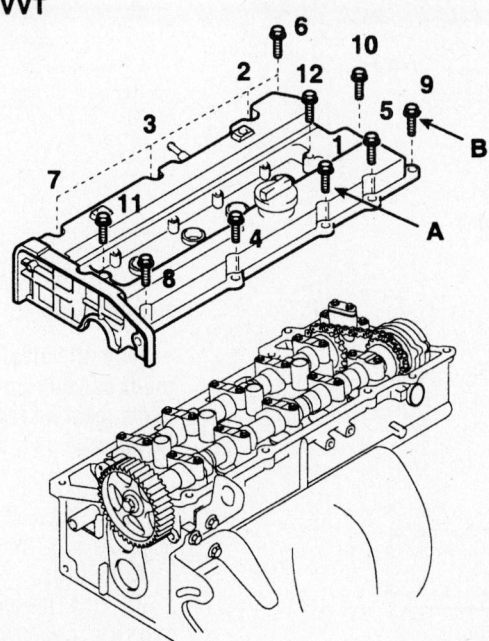

09474_KIAC_G0012

Cylinder head cover sequence—1.6L (VIN 3)

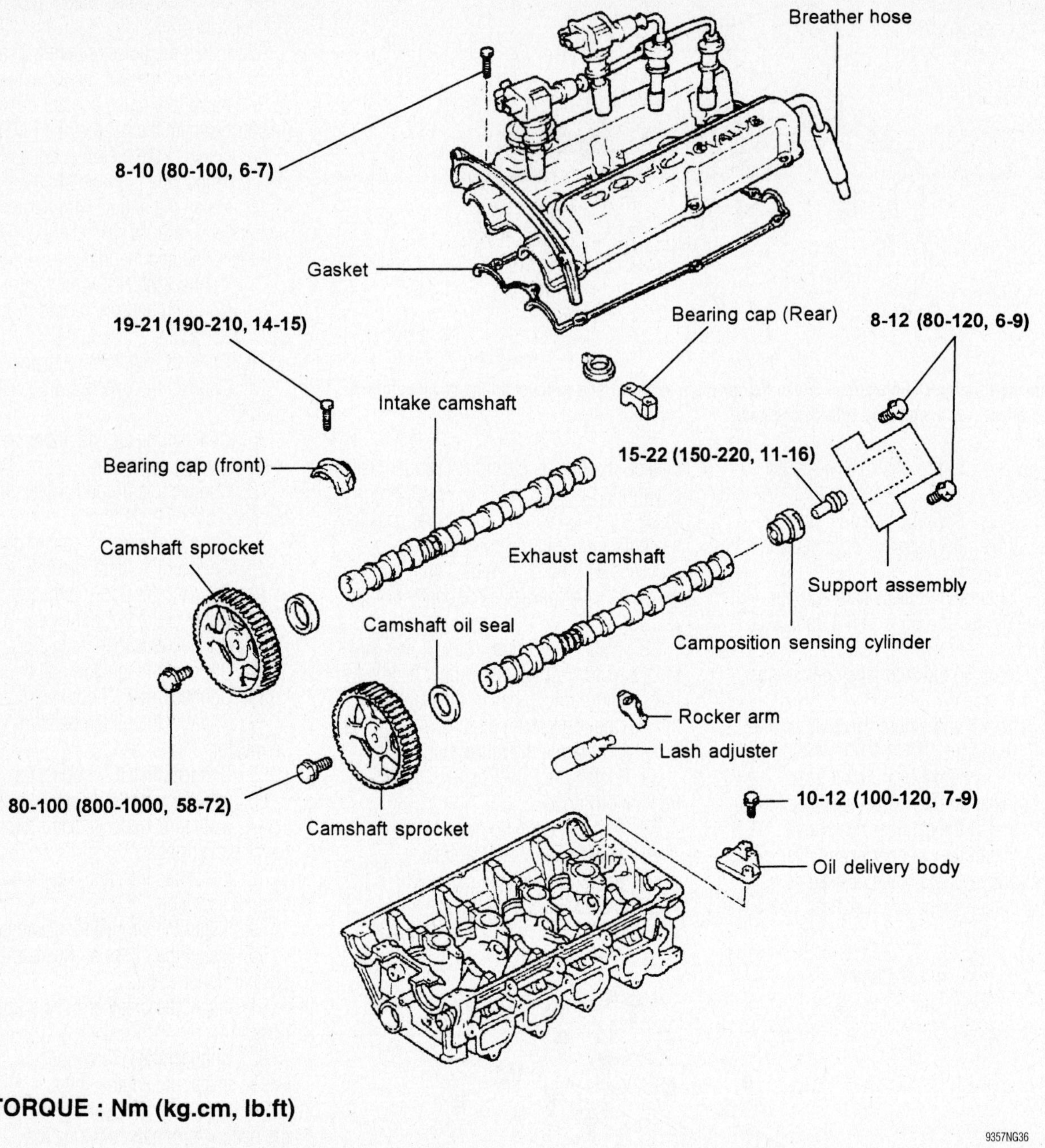

Breather hose

8-10 (80-100, 6-7)

Gasket

19-21 (190-210, 14-15)

Bearing cap (Rear)

8-12 (80-120, 6-9)

Intake camshaft

Bearing cap (front)

15-22 (150-220, 11-16)

Camshaft sprocket

Exhaust camshaft

Camshaft oil seal

Camposition sensing cylinder

Support assembly

Rocker arm

Lash adjuster

80-100 (800-1000, 58-72)

Camshaft sprocket

10-12 (100-120, 7-9)

Oil delivery body

TORQUE : Nm (kg.cm, lb.ft)

9357NG36

Exploded view of the camshafts and related components—2.4L engine

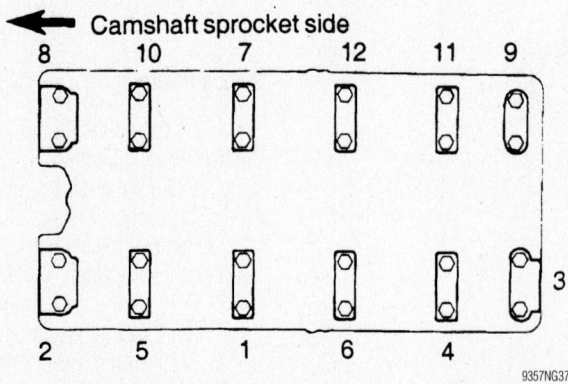

Camshaft sprocket side

| 8 | 10 | 7 | 12 | 11 | 9 |

3

| 2 | 5 | 1 | 6 | 4 |

9357NG37

Camshaft bearing cap tightening sequence–2.4L engine

8. Remove the camshaft sprocket bolts and sprockets.

➡**Keep all valvetrain components in order as you remove them. The components must be installed in their original locations.**

9. Remove the bearing cap bolts, bearing caps, camshafts, rocker arms and valve adjusters.

To install:

10. Lubricate the camshaft journals and lobes with clean engine oil.

11. Install the camshafts on the cylinder

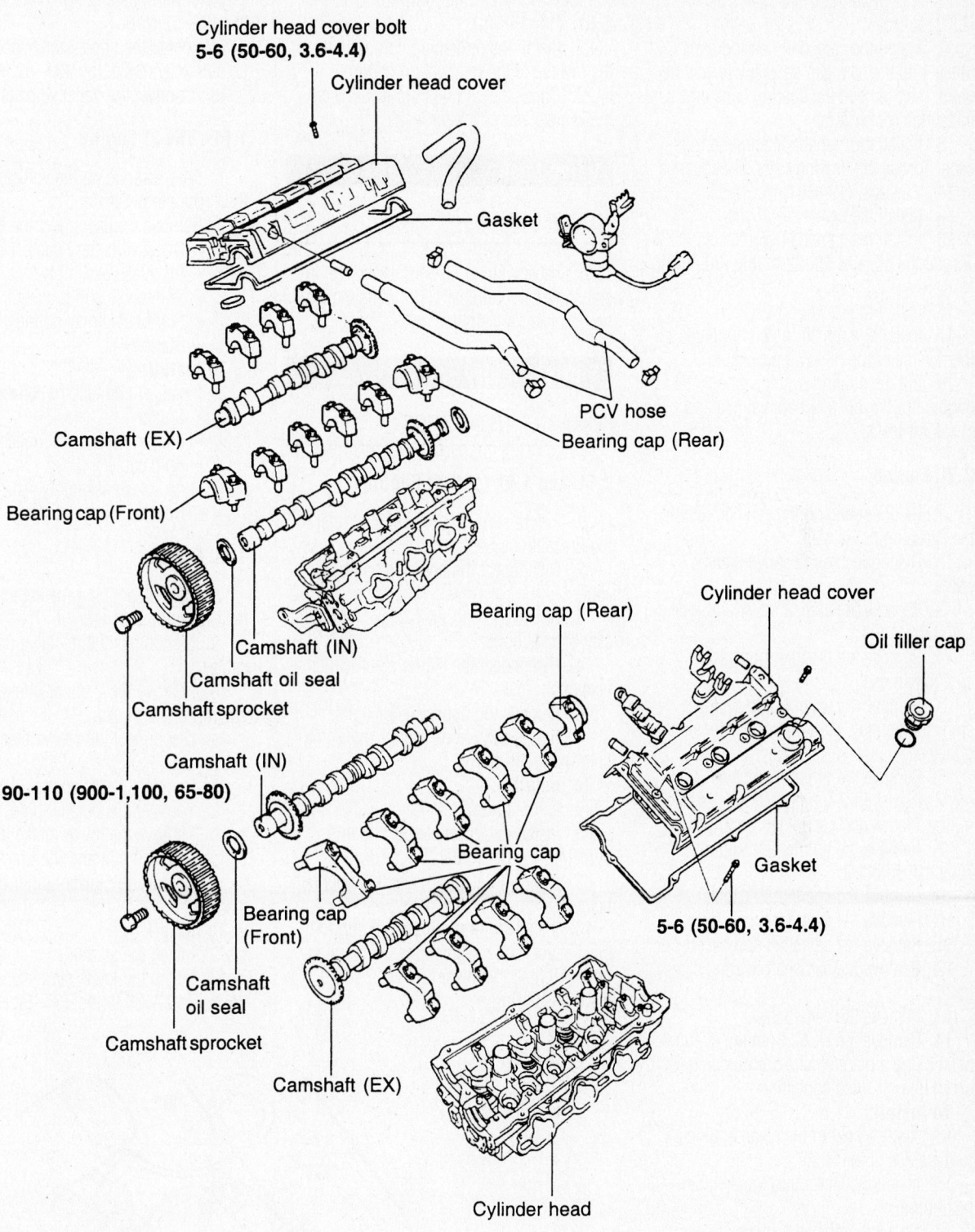

Cylinder head cover bolt
5-6 (50-60, 3.6-4.4)

Cylinder head cover

Gasket

Camshaft (EX)

Bearing cap (Front)

PCV hose

Bearing cap (Rear)

Camshaft (IN)

Camshaft oil seal

Camshaft sprocket

90-110 (900-1,100, 65-80)

Camshaft (IN)

Bearing cap
(Front)

Bearing cap (Rear)

Cylinder head cover

Oil filler cap

Bearing cap

Camshaft
oil seal

Camshaft sprocket

Camshaft (EX)

Gasket

5-6 (50-60, 3.6-4.4)

Cylinder head

TORQUE : Nm (kg.cm, lb.ft)

9357NG38

Exploded view of the camshafts and related components—2.7L engine

head. The intake camshaft has a slit on its rear side to drive the Crankshaft Position (CKP) sensor.

12. Install the camshaft bearing caps. Make sure the cam can be easily turned by hand, then remove the bearing caps and install the rocker arms.

13. Install the camshafts and bearings caps. Torque the bearing caps, in sequence, to 14–15 ft. lbs. (19–21 Nm).

14. Install the camshaft oil seal.

15. Install the camshaft sprockets. Torque the bolts to 58–72 ft. lbs. (80–100 Nm).

16. Install the rocker cover.

17. Install the remaining components in the reverse of the removal procedure.

18. Fill the cooling system. Check the engine fluid levels, start the vehicle and check for leaks.

2.7L Engine

1. Before servicing the vehicle, refer to the Precautions Section.

2. Disconnect the negative battery cable.

3. Remove the engine cover and intake manifold.

4. Remove the breather hose and engine harness.

5. Remove the water pump pulley, crankshaft pulley, idler pulley and tensioner pulley.

6. Remove the timing belt cover.

7. Remove the timing belt tensioner pulley, loosen and secure temporarily.

8. Remove the timing belt from the camshaft sprocket.

9. Remove the spark plug cables.

10. Remove the cylinder head cover.

11. Remove the camshaft sprocket.

12. Remove the camshaft bearing cap bolts.

13. Remove the camshafts.

14. Remove the HLA's from the cylinder head using a magnet, and store upside-down in a oil-filled container

To install:

15. Apply a coat of the clean engine oil to the sides of the HLA's.

16. Install the HLA's into the cylinder head bore.

17. Check that the HLA moves freely in its bore.

18. Lubricate the camshaft journals and lobes with clean engine oil.

19. Install the camshafts in the cylinder head making sure that the camshaft dowel pins point straight up.

20. Install the bearing caps. Check the marks on the caps for the Intake (I) and

Exhaust (E) markings. Torque the bolts, working from the center outward to 10–11 ft. lbs. (14–15 Nm).

21. Install the remaining components in the reverse of the removal procedure.

22. Check the engine fluid levels, start the vehicle and check for leaks.

Valve Lash

ADJUSTMENT

The valve lash on all engines is kept in adjustment hydraulically. No adjustment is necessary or possible.

Starter Motor

REMOVAL & INSTALLATION

1.5L and 1.6L (VIN 5) Engines

1. Before servicing the vehicle, refer to the Precautions Section.

2. Disconnect the negative battery cable.

3. Remove the 4 upper intake manifold stay bracket bolts.

4. Disconnect the starter electrical connections.

5. Remove the upper starter bolts.

6. Remove the lower starter mounting bolt(s) and the starter.

To install:

7. Install the starter.

8. Install the lower starter bolt and tighten to 27–38 ft. lbs. (37–52 Nm).

9. Connect the starter electrical connections.

10. Install the intake manifold support bracket and finger-tighten the 3 mounting bolts. Tighten the lower bolt to 27–38 ft. lbs. (37–52 Nm).

11. Install the 4 upper intake manifold stay bracket bolts and tighten to 27–38 ft. lbs. (37–52 Nm).

12. Install the upper starter bolts and tighten to 27–38 ft. lbs. (37–52 Nm).

13. Connect the negative battery cable.

1.6L (VIN 3) Engine

1. Before servicing the vehicle, refer to the Precautions Section.

2. Remove or disconnect the following:
- Negative battery cable
- Air intake assembly
- Shift cable and bracket
- Starter electrical connections
- Starter

To install:

3. Install or connect the following:
- Starter
- Starter electrical connections
- Shift cable
- Air intake assembly
- Negative battery cable

1.8L Engine

1. Before servicing the vehicle, refer to the Precautions Section.

2. Disconnect the negative battery cable.

3. Remove the 2 upper intake manifold support bracket bolts.

4. Disconnect the starter electrical connections.

5. Remove the 2 upper starter bolts.

6. Remove the exhaust pipe.

7. Remove the lower intake manifold support bracket bolt and the bracket.

8. Remove the lower starter bolt and the starter.

To install:

9. Install the starter.

10. Install the lower starter bolt and tighten to 27–38 ft. lbs. (37–52 Nm).

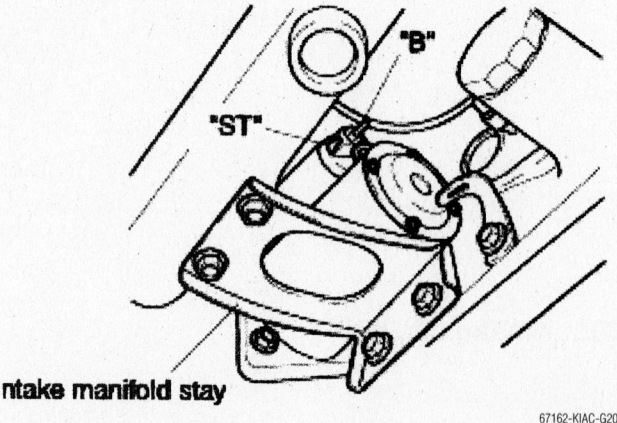

Intake manifold stay

67162-KIAC-G20

Starter mounting—1.5L and 1.6L engines

11. Connect the starter electrical connections.

12. Install the intake manifold support bracket and finger-tighten the 3 mounting bolts. Tighten the lower bolt to 27–38 ft. lbs. (37–52 Nm).

13. Install the exhaust pipe.

14. Install the 2 upper intake manifold support bracket bolts and tighten to 27–38 ft. lbs. (37–52 Nm).

15. Install the 2 upper starter bolts and tighten to 27–38 ft. lbs. (37–52 Nm).

16. Connect the negative battery cable.

2.0L Engine

1. Before servicing the vehicle, refer to the Precautions Section.

2. Disconnect the negative battery cable.

3. Disconnect the starter cable from the B terminal on the solenoid, then disconnect the connector from the S terminal.

4. Remove the 2 bolts holding the starter, then remove the starter.

To install:

5. Install the starter.

6. Install the starter bolts and tighten to 27–38 ft. lbs. (37–52 Nm).

7. Connect the starter electrical connections.

8. Connect the negative battery cable.

2.4L and 2.7L Engines

1. Before servicing the vehicle, refer to the Precautions Section.

2. Disconnect the negative battery cable.

3. Disconnect the speed meter and shift cable.

4. Disconnect the starter connector and terminal connection.

5. Remove the starter mounting bolt(s).

6. Remove the starter.

To install:

7. Install the starter and mounting bolt(s) and tighten to 20–25 ft. lbs. (27–34 Nm).

8. Install the terminal connection and electrical connection. Torque the terminal nut to 7.3–11.7 ft. lbs. (9.8–16 Nm).

9. Connect the shift cable and speed meter.

10. Connect the negative battery cable.

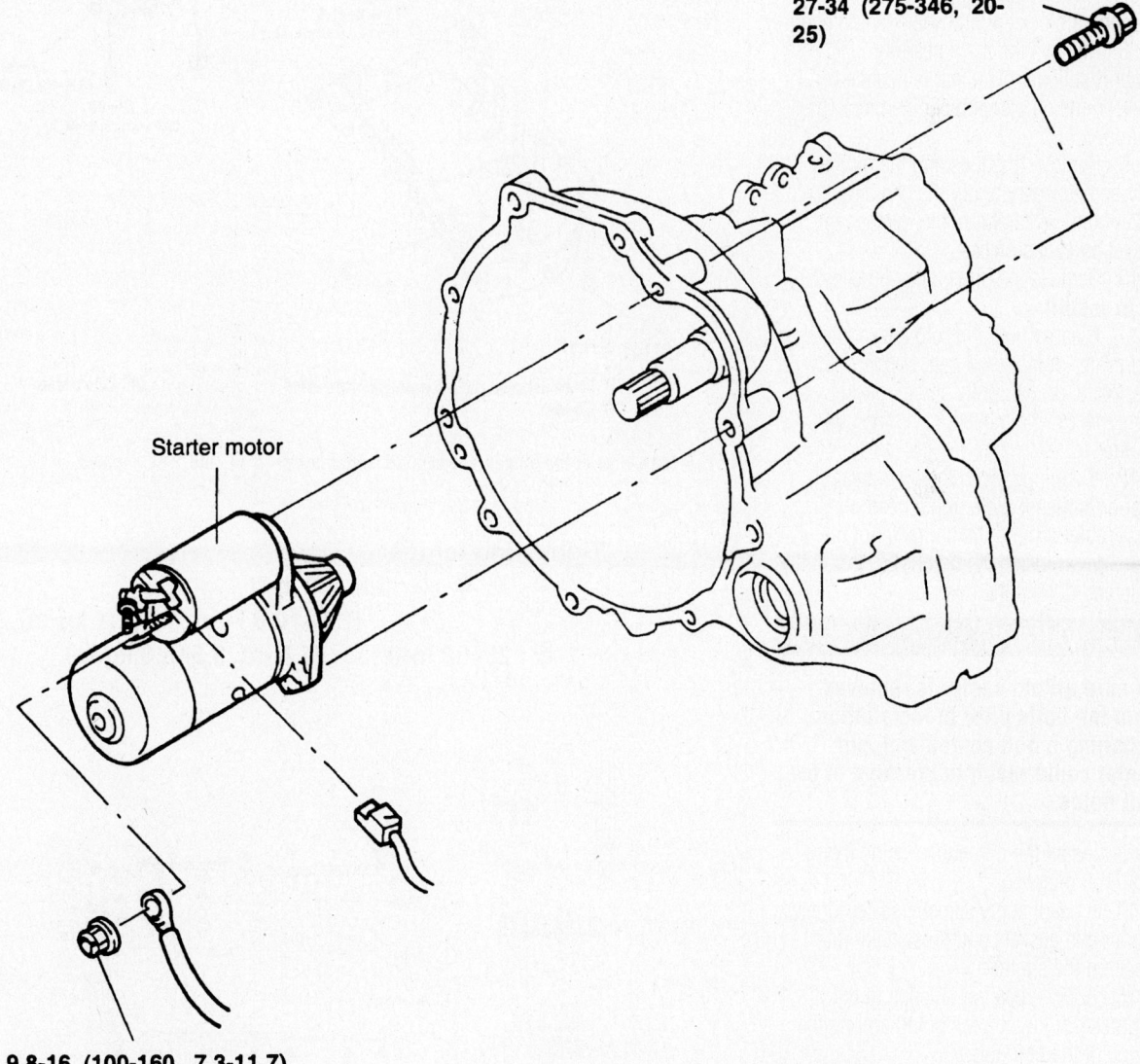

27-34 (275-346, 20-25)

Starter motor

9.8-16 (100-160, 7.3-11.7)

TORQUE : N·m (kg·cm, lb·ft)

Exploded view of the starter mounting—2.4L and 2.7L engines

9357NG39

Oil Pan

REMOVAL & INSTALLATION

1.5L, 1.6L (VIN 5) and 1.8L Engines

1. Before servicing the vehicle, refer to the Precautions Section.
2. Disconnect the negative battery cable.
3. Remove the engine undercover, if equipped.
4. Drain the engine oil.
5. Remove the exhaust pipe from the exhaust manifold and from the catalytic converter.
6. Remove the exhaust pipe bracket from the engine block, if necessary.
7. Remove the integrated stiffener from the engine block and transaxle, if equipped.
8. Remove the main bearing support/stiffener plate that is installed between the oil pan and engine block, if equipped.
9. Remove the bolts and the oil pan. It may be necessary to pry the pan away from the engine; be careful not to damage the gasket contact surfaces.
10. Remove the oil strainer, if necessary.

To install:

11. Clean all oil, dirt, old gasket material and sealer from the oil pan, support/stiffener plate, oil pan bolts and all gasket mating surfaces. If removed, clean the oil strainer.
12. If equipped with the main bearing support/stiffener plate, run a bead of silicone sealer around the perimeter of the plate, going inside the bolt holes. Install the plate and tighten the bolts.

✷✷ WARNING

Be sure all old sealer is removed from the bolts prior to installation. Installing a bolt coated with old sealer could result in cracking of the bolt holes.

13. Install the oil strainer using a new gasket, if removed.
14. If used, apply silicone sealer to new rubber end gaskets and press them into place on the engine.
15. Apply a bead of silicone to the perimeter of the oil pan, going around the inside of the bolt holes.
16. Install the pan to the engine and the oil pan bolts finger-tight.
17. Tighten the oil pan bolts to the specifications shown in the accompanying illustrations.

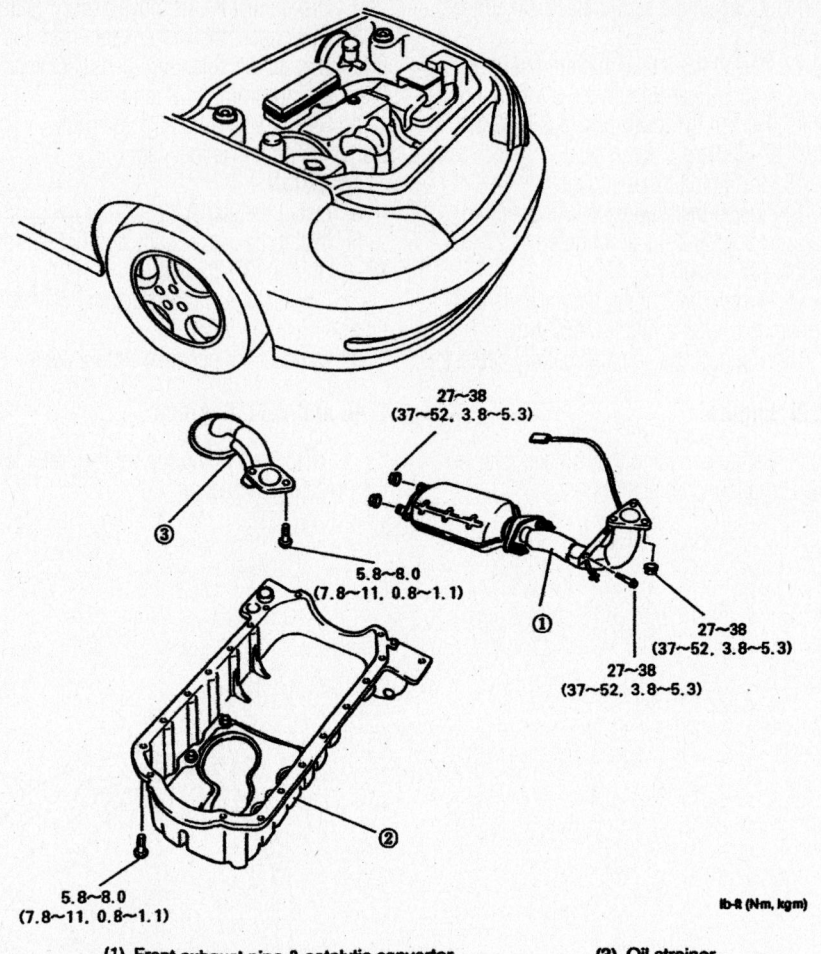

(1) Front exhaust pipe & catalytic converter
(2) Oil pan
(3) Oil strainer

67162-KIAC-G21

Exploded view of the oil pan and related components—1.5L and 1.6L engines

Tightening torque:
Ⓐ Ⓑ Ⓒ Ⓓ : 5.8~8.0 lb-ft
(7.8~10.8 N•m, 0.8~1.1 kg-m)
Ⓔ : 28~38 lb-ft (38~51 N•m, 3.8~5.3 kg-m)

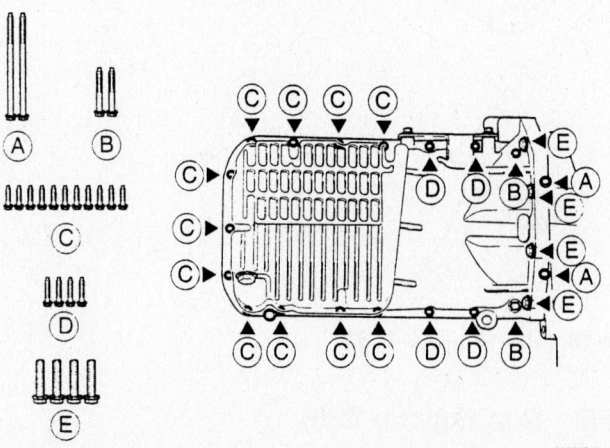

9307KG14

Oil pan mounting bolt locations and torque specifications—1.8L engine

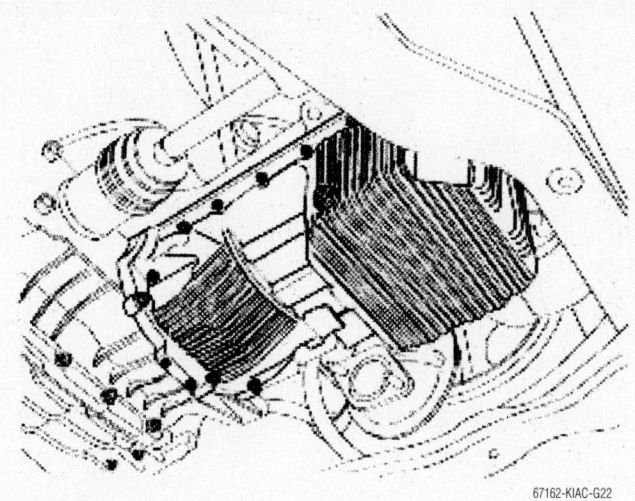

Oil pan—2.0L engine

67162-KIAC-G22

18. Install the integrated stiffener to the engine block and transaxle, if removed. Tighten the bolts to 38 ft. lbs. (52 Nm).

19. Install the transverse member, if removed. Tighten the bolts to 93 ft. lbs. (126 Nm).

20. Install the front exhaust pipe bracket, if equipped.

21. Install the front exhaust pipe, using new gaskets. Tighten the exhaust manifold flange nuts to 27–38 ft. lbs. (37–52 Nm) for

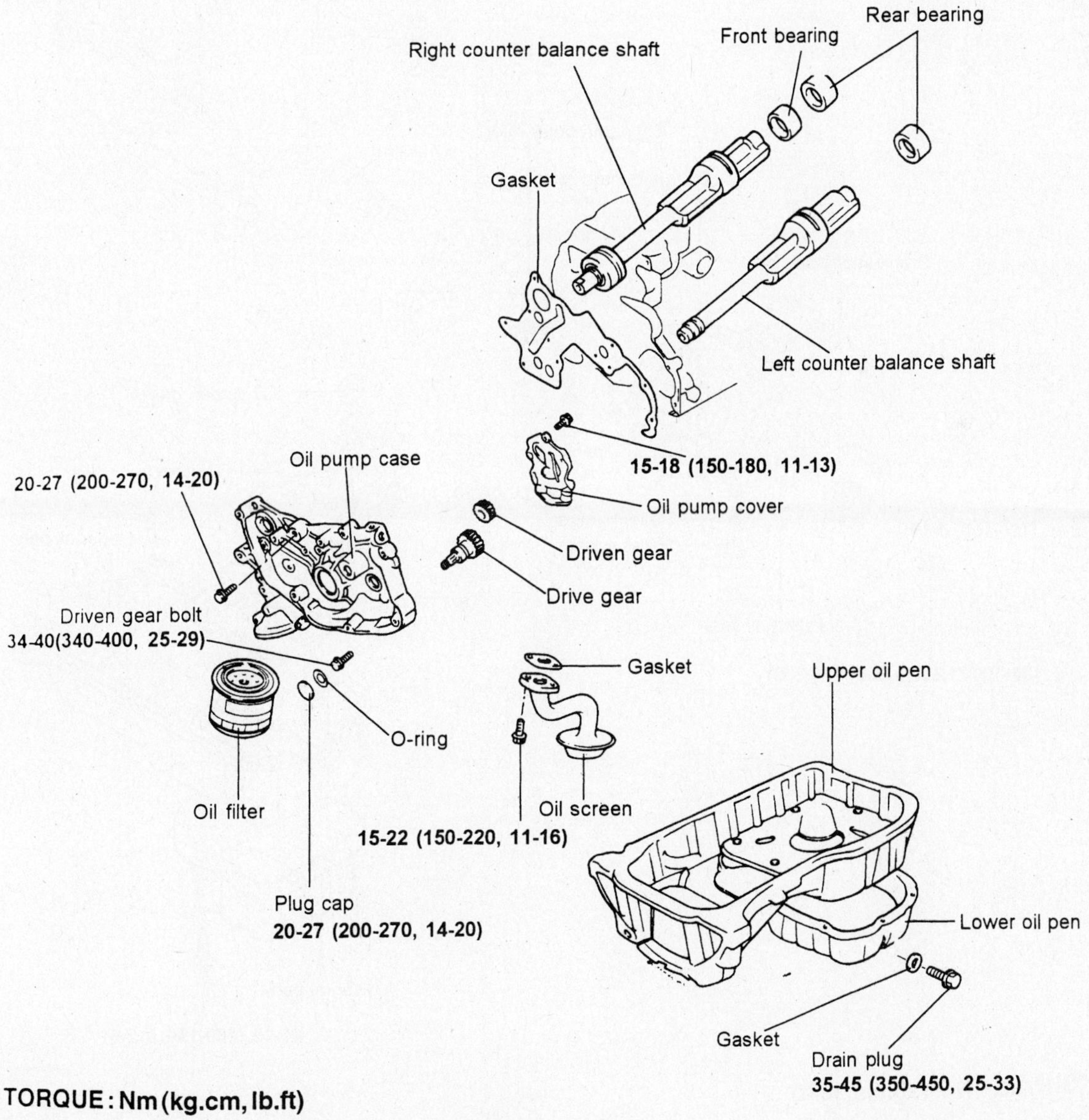

Right counter balance shaft

Front bearing

Rear bearing

Gasket

Left counter balance shaft

15-18 (150-180, 11-13)

Oil pump cover

Oil pump case

20-27 (200-270, 14-20)

Driven gear

Drive gear

Driven gear bolt
34-40(340-400, 25-29)

O-ring

Oil filter

Gasket

Oil screen

15-22 (150-220, 11-16)

Upper oil pen

Lower oil pen

Plug cap
20-27 (200-270, 14-20)

Gasket

Drain plug
35-45 (350-450, 25-33)

TORQUE: Nm(kg.cm, lb.ft)

9357NG40

Exploded view of the oil pan, pump and related components—2.4L engine

1.5L and 1.6L engines, or 34 ft. lbs. (46 Nm) for 1.8L engines.

22. Install the oil pan drain plug using a new gasket. Tighten the drain plug to 30 ft. lbs. (41 Nm).

23. Install the engine undercover.

24. Fill the engine with the proper type and quantity of oil.

25. Connect the negative battery cable. Start the engine and bring to normal operating temperature. Check for leaks.

1.6L (VIN 3) and 2.0L Engines

1. Before servicing the vehicle, refer to the Precautions Section.

2. Disconnect the negative battery cable.

3. Drain the engine oil.

4. Disconnect the rear oxygen sensor connector.

5. Remove the front muffler.

6. Remove the exhaust manifold.

7. Remove the front muffler bracket.

8. Remove the oil pan.

To install:

9. Install the oil pan.

10. Using a razor blade and gasket scraper, remove all the old packing material from the gasket surfaces.

11. Apply liquid gasket as an even bead, centered between the edges of the mating surface.

12. Install the oil pan with the bolts. Uniformly tighten the bolts in several passes to 7–8.5 ft. lbs. (10–12 Nm).

13. Install the front muffler bracket.

14. Install the exhaust manifold.

15. Install the front muffler.

16. Connect the rear oxygen sensor connector.

17. Fill with engine oil

18. Connect the negative battery cable. Start the engine and bring to normal operating temperature. Check for leaks.

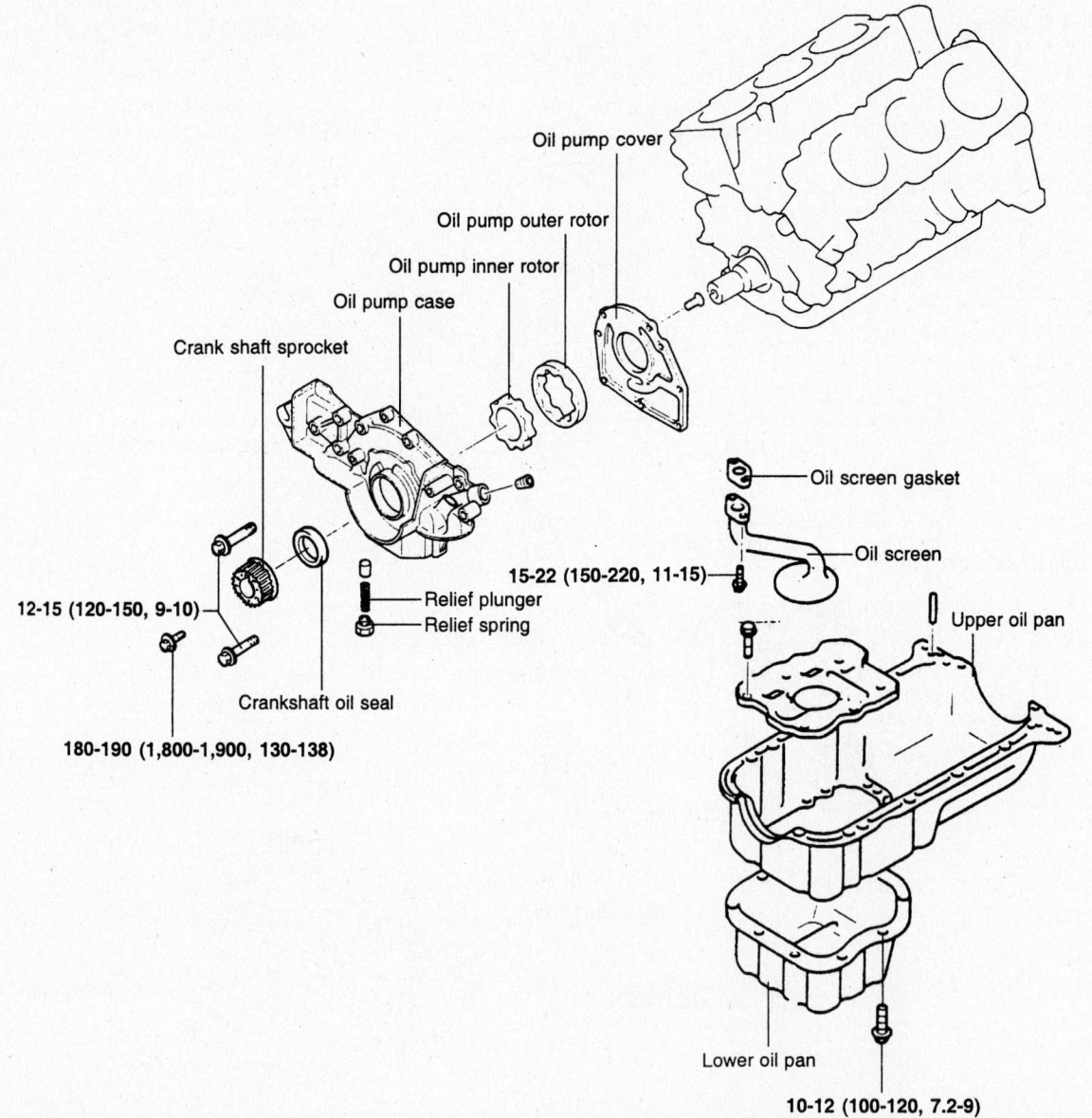

Oil pump cover

Oil pump outer rotor

Oil pump inner rotor

Oil pump case

Crank shaft sprocket

Oil screen gasket

Oil screen

15-22 (150-220, 11-15)

Upper oil pan

12-15 (120-150, 9-10)

Relief plunger

Relief spring

Crankshaft oil seal

180-190 (1,800-1,900, 130-138)

Lower oil pan

10-12 (100-120, 7.2-9)

TORQUE : Nm (kg.cm, lb.ft)

9357NG41

Exploded view of the oil pan, pump and related components—2.7L engine

2.4L and 2.7L Engines

1. Before servicing the vehicle, refer to the Precautions Section.
2. Drain the engine oil.
3. Disconnect the negative battery cable.
4. Remove the timing belt.
5. Remove the oil pan bolts.
6. Remove the oil pan.
7. Remove the oil pan screen and gasket, if necessary.

To install:

8. Install the oil pump screen. Torque the bolts to 11–16 ft. lbs. (15–22 Nm).
9. Install a 0.16 in. (44mm) bead of sealant into the groove of the oil pan flange. Install the oil pan within 15 minutes of applying the sealant.

10. Install the oil pan and tighten the bolts to 7–9 ft. lbs. (10–12 Nm).
11. Install the timing belt.
12. Fill the engine with the proper type and quantity of oil.
13. Connect the negative battery cable. Start the engine and bring to normal operating temperature. Check for leaks.

Oil Pump

REMOVAL & INSTALLATION

1.5L and 1.6L (VIN 5) Engines

1. Before servicing the vehicle, refer to the Precautions Section.

2. Drain the engine oil.
3. Disconnect the negative battery cable.
4. Remove the drive belt.
5. Remove the timing belt.
6. Remove the alternator.
7. Remove the A/C compressor and bracket and position aside. Do not disconnect the refrigerant lines.
8. Remove the timing belt pulley lockbolt.
9. Remove the timing belt pulley.
10. Remove the oil strainer.
11. Remove the oil pump.
12. Installation is the reverse of the removal procedure. Tighten the retainers to the specifications shown in the illustration.

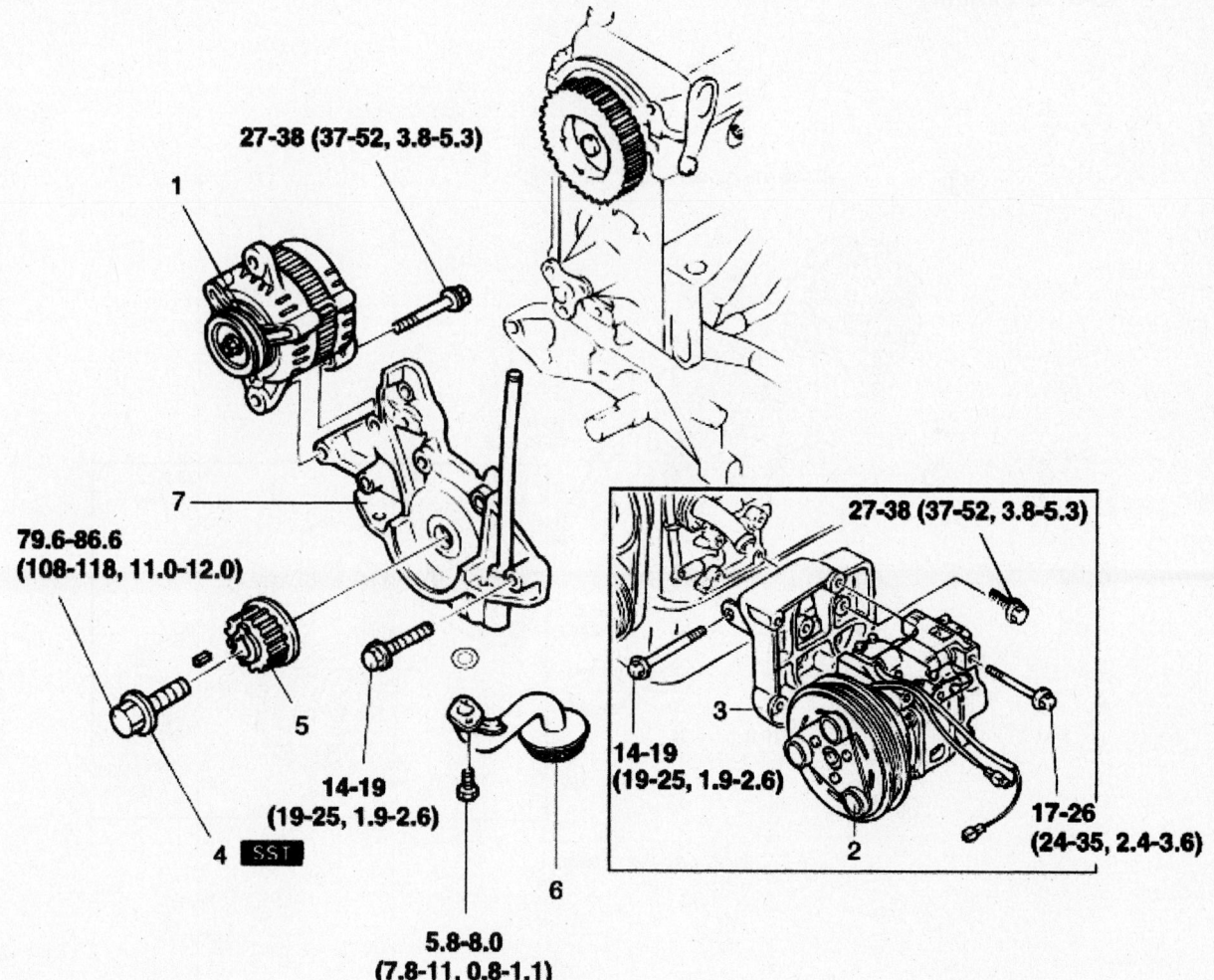

1. Alternator
2. Compressor assembly
3. Compressor bracket
4. Timing belt pulley lock bolt
5. Timing belt pulley
6. Oil strainer
7. Oil pump assembly

67162-KIAC-G23

Exploded view of the oil pump mounting—1.5L and 1.6L (VIN 5) engines

1.8L Engine

1. Before servicing the vehicle, refer to the Precautions Section.
2. Drain the engine oil.
3. Disconnect the negative battery cable.
4. Remove the crankshaft pulley, timing belt cover, belt and the crankshaft sprocket.
5. Remove the oil pan.
6. Remove the oil pick-up tube and discard the gasket.
7. Remove the oil pump attaching bolts.
8. Remove the oil pump.
9. Remove the front crankshaft seal from the oil pump if the pump is being replaced.

To install:
10. Clean the oil, dirt and old sealant from all contact surfaces.
11. Install new O-rings on the oil pump.
12. If the oil seal was removed from the oil pump, apply clean engine oil to the lip of the seal. Push the seal in lightly by hand. Press the seal, with a protrusion of 0.02–0.04 in. (0.5–1.0mm), into the oil pump with a suitable tool (49 B014 401).
13. Apply a bead of silicone to the oil pump at the cylinder block contact surface, going inside the bolt holes.
14. Install the oil pump and tighten the bolts to 14–18 ft. lbs. (19–25 Nm).
15. Install the new gasket and the oil pump pick-up tube. Tighten the mounting bolts to 70–95 inch lbs. (8–11 Nm).
16. Install the oil pan.
17. Install the crankshaft sprocket, timing belt and cover.
18. Install the crankshaft pulley.
19. Connect the negative battery cable.
20. Fill the engine with the proper type and quantity of oil. Run the engine and check for leaks.

SOHC shown (DOHC similar)

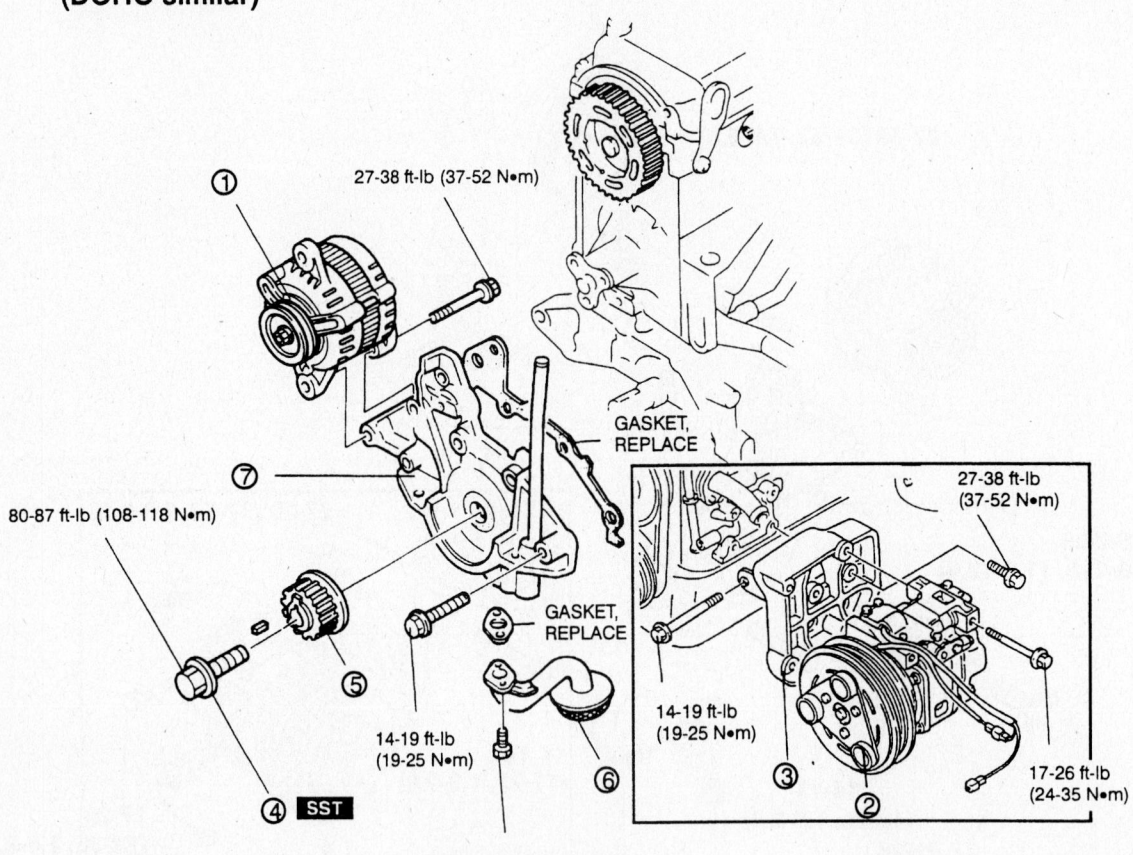

1. Generator
2. A/C compressor (if equipped)
3. A/C compressor bracket (if equipped)
4. Crankshaft pulley lock bolt
5. Timing belt pulley
6. Oil strainer
7. Oil pump

Exploded view of the oil pump—1.8L engine

7923KG26

1.6L (VIN 3) and 2.0L Engines

1. Before servicing the vehicle, refer to the Precautions Section.

2. Drain the engine oil.

3. Remove the drive belts.

4. Turn the crankshaft pulley, and align its groove with timing mark "T" of the timing belt cover.

5. Remove the timing belt.

6. Remove the bolt and timing belt idler.

7. Remove the oil pan and oil screen.

8. Remove the alternator.

9. Remove the air conditioner compressor tensioner bracket.

10. Remove the bolts and front case.

➡**Keep the bolts in order when removing. Each bolt is a different length.**

11. Remove the screw from the pump housing, then separate the housing and cover.

12. Remove the inner rotor and outer rotor.

To install:

13. Install the oil pump.

14. Place the inner and outer rotors into front case with the marks facing the oil pump cover side.

15. Install the oil pump cover to front case with the 7 screws. Tighten the screws to 4–6.5 ft. lbs. (6–9 Nm).

16. Check that the oil pump turns freely.

17. Install the oil pump on the cylinder block.

18. Place a new front case gasket on the cylinder block.

19. Apply engine oil to the lip of the oil pump seal. Then, install the oil pump onto the crankshaft.

20. When the pump is in place, clean any excess grease off the crankshaft and check that the oil seal lip is not distorted. Tighten the mounting bolts as follows:

    a. 1.6L (VIN 3) engine: 14–17 ft. lbs. (19–24 Nm).

    b. 2.0L engine: 14.5–19.5 ft. lbs. (19.5–26.5 Nm).

21. Apply a light coat of oil to the front case oil seal lip.

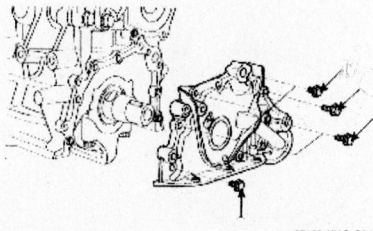

View of the oil pump—1.6L (VIN 3) and 2.0L engines

67162-KIAC-G24

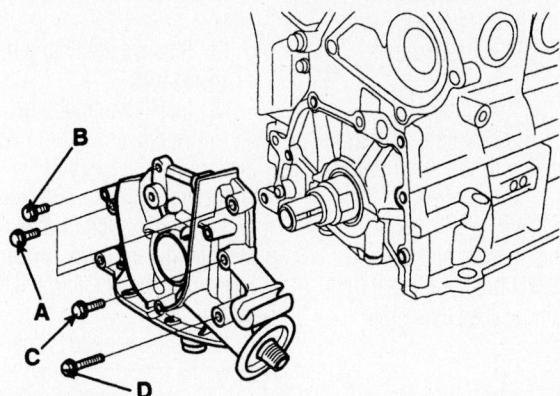

Bolt length
(A) : 30mm (1.181in), (B) : 22mm (0.866in),
(C) : 45mm (1.772in), (D) : 60mm (2.362in)

09474_KIAC_G0013

Oil pump bolt length and locations—1.6L (VIN 3)

22. Using Special Tool 09214-32000, install the front case oil seal.

23. Install or connect the following:

- Air conditioner compressor tensioner bracket
- Alternator
- Oil screen
- Oil pan
- Timing belt tensioner
- Timing belt
- Drive belts

24. Fill with engine oil to the connect level.

25. Start the engine and check for leaks.

2.4L and 2.7L Engines

1. Before servicing the vehicle, refer to the Precautions Section.

2. Drain the engine oil.

3. Disconnect the negative battery cable.

4. Remove the oil pan, screen and gasket.

5. Remove the relief plunger and gasket.

6. Remove the relief spring and relief valve from the oil filter bracket.

7. Remove the oil pressure switch.

8. Remove the oil filter bracket and gasket.

9. Plug the cap from the oil pump portion of the front case.

10. Remove the left side of the cylinder block plug and insert a screwdriver with a 0.32 in. (8mm) diameter shaft in to the plug hole. The screwdriver must be inserted more than 2.4 in. (60mm).

11. Remove the oil pump driven gear and left counter balance shaft retaining bolt.

12. Remove the front case and gasket.

13. Remove the 2 counter balance shafts from the cylinder block.

14. Remove the oil pump cover and gears from the front case.

To install:

15. Apply engine oil to the oil pump gears.

16. Install the oil pump gears, aligning the 2 timing marks.

17. Install the crankshaft front oil seal, using a proper tool.

18. Install the special tool 09214-32100 on the front end of the crankshaft and apply a thin coat of oil to the outer circumference of the special tool to install the front case.

19. Install the front case with a new gasket and temporarily tighten the flange bolts, except the bolts for the filter bracket. Torque the front case bolts to 14–20 ft. lbs. (20–27 Nm).

20. Insert a screwdriver through the plug hole in the left side of the cylinder block. After checking that the shaft is in position, tighten.

21. Install the remaining components in the reverse of the removal procedure.

22. Fill the engine with the proper type and quantity of oil. Run the engine and check for leaks.

Rear Main Seal

REMOVAL & INSTALLATION

1. Before servicing the vehicle, refer to the Precautions Section.

2. Disconnect the negative battery cable.

3. Remove the transaxle assembly.

4. Remove the clutch and flywheel assembly, if equipped with a manual transaxle.

5. Remove the flexplate-to-crankshaft bolts, the flexplate and shim plates, if equipped with an automatic transaxle.

6. Cut the oil seal lip with a knife. Install a rag to the housing and using a screwdriver, carefully pry the oil seal from the oil seal housing. Clean the gasket mounting surfaces.

To install:

7. Clean the oil seal housing. Coat the oil seal and the housing with clean engine oil.

8. Install the oil seal into the housing and tap it evenly into place with a hammer and a large diameter piece of pipe. The seal must be flush with the edge of the rear cover.

9. Install the flywheel assembly or the flexplate, as applicable, and tighten the mounting bolts to 71–76 ft. lbs. (97–102 Nm).

10. Install the clutch assembly, if applicable.

11. Install the transaxle.

12. Connect the negative battery cable.

Timing Belt

REMOVAL & INSTALLATION

1.5L, 1.6L (VIN 5) and 1.8L Engines

1. Before servicing the vehicle, refer to the Precautions Section.

2. Disconnect the negative battery cable.

3. Remove the accessory drive belts.

4. Remove the alternator.

5. Remove the water pump and crankshaft pulleys.

6. Remove the timing belt guide plate.

7. Remove the upper and lower timing belt covers.

8. Position the crankshaft so that the timing mark is aligned with the timing mark on the engine.

9. Verify that the "I" and "E" mark on the camshaft pulley align with the mark on the cylinder head.

➡**Do not move the crankshaft or camshaft once the timing marks have been correctly positioned.**

10. If the timing belt is to be reused, mark the direction of rotation on the timing belt.

11. Remove the timing belt tensioner pulley.

12. Remove the timing belt.

To install:

13. Install the timing belt tensioner pulley, move the tensioner to its furthest point and tighten the lockbolt.

14. Install the timing belt onto the pulleys, as follows, crankshaft pulley first, then the idler pulley, exhaust camshaft pulley, intake camshaft pulley, and the tensioner pulley.

15. Loosen the tensioner pulley and allow the tensioner spring to apply tension on the belt, then tighten the lockbolt to 28–38 ft. lbs. (38–51 Nm).

16. Rotate the crankshaft clockwise 2 turns and be sure all marks are still correctly aligned.

17. Install the remaining components in the revere order of the removal noting the following torque specifications:

- Crankshaft pulley: 9–13 ft. lbs. (12–17 Nm)
- Water pump pulley: 9–13 ft. lbs. (12–17 Nm)

18. Connect the negative battery cable.

1.6L (VIN 3) and 2.0L Engines

1. Before servicing the vehicle, refer to the Precautions Section.

2. Remove or disconnect the following:

- Negative battery cable
- Engine cover
- Right-hand front wheel
- Right-hand wheel well side cover
- Accessory drive belts
- Water pump pulley
- Timing belt upper cover

3. Turn the crankshaft pulley so cylinder No. 1 is at Top Dead Center (TDC). Align the pulley groove with the timing mark "T" of the timing belt cover.

4. Remove or disconnect the following:

- Crankshaft pulley
- Crankshaft flange
- Lower timing belt cover
- Timing belt tensioner
- Timing belt

➡**If the timing belt is to be reused, mark the belt to indicate its turning direction.**

To install:

5. Align the timing marks of the camshaft and crankshaft sprockets to ensure the No. 1 cylinder is at TDC.

6. Temporarily install the timing belt tensioner.

7. Install the timing belt over the crank-

shaft sprocket, then inside the idler pulley, over the camshaft sprocket and then inside the timing belt tensioner.

8. Apply tension to the timing belt by turning the tensioner in the clockwise direction so there is no slack in the belt on the tension side.

9. Tighten the tensioner mounting bolts to 20 ft. lbs. (27 Nm).

10. Check the timing belt tension as follows:

   a. 1.6L (VIN 3) engine: Push the tension side of the belt in horizontally with a moderate force, the timing belt cog end is approximately ½ of the tensioner mounting bolt head radius away from the bolt head center.

   b. 2.0L engine: Push the tension side of the belt in horizontally with a moderate force, the timing belt cog end sags approximately 0.16–0.24 inches (4–6 mm).

11. Turn the crankshaft two turns in the clockwise direction and realign the crankshaft sprocket and camshaft sprocket timing marks.

12. Install or connect the following:

- Lower timing belt cover. Tighten bolts to 6–7 ft. lbs. (8–10 Nm).
- Crankshaft flange
- Crankshaft pulley. Tighten bolt to 101–109 ft. lbs. (138–147 Nm) for 1.6L (VIN 3) engine and 123–130 ft. lbs. (168–177 Nm) for 2.0L engine.
- Upper timing belt cover. Tighten bolts to 6–7 ft. lbs. (8–10 Nm).
- Water pump pulley
- Accessory drive belts
- Right-hand wheel well side cover
- Right-hand front wheel
- Engine cover
- Negative battery cable

Piston and Ring

POSITIONING

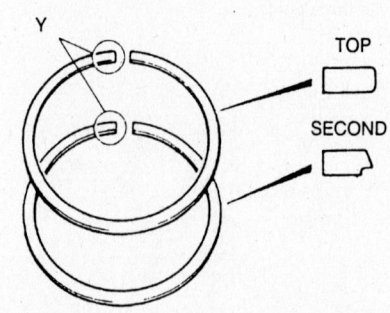

Compression ring positioning engines

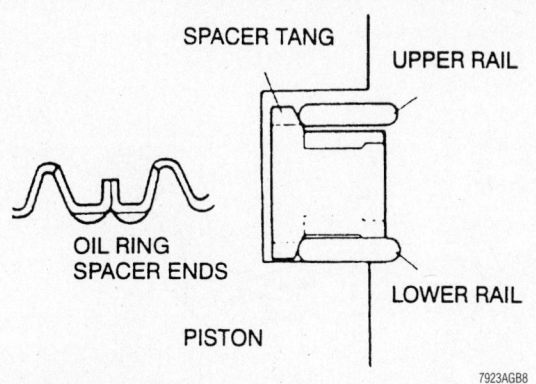

Oil control ring positioning engines

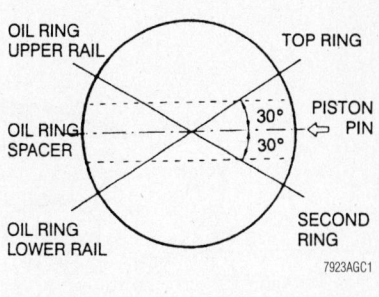

Piston ring end-gap spacing engines

FUEL SYSTEM

Fuel System Pressure

RELIEVING

1. Before servicing the vehicle, refer to the Precautions Section.

2. Release the rear seat retainers (clips or catches) and remove rear seat cushion.

3. Remove the fuel pump cover.

4. Detach the fuel pump electrical connector.

5. Start the engine, allowing it to idle until it runs out of fuel.

6. After the engine stalls, reattach the fuel pump connector and turn the ignition switch **OFF**.

Fuel Filter

REMOVAL & INSTALLATION

1.5L, 1.6L (VIN 5), 1.8L and 2.4L Engines

1. Before servicing the vehicle, refer to the Precautions Section.

2. Relieve the fuel system pressure.

3. Disconnect the negative battery cable.

4. Disconnect the fuel lines from both ends of the fuel filter. Plug the lines to prevent leakage.

5. Remove the bracket retaining nuts and bolts, as necessary.

6. Remove the filter and the mounting bracket.

To install:

7. Install the filter in the mounting bracket.

8. Connect the fuel lines to the filter.

9. Install the bracket nuts and tighten to 9.5 ft. lbs. (13 Nm) for 1.5L, 1.6L, 1.8L and 2.0L engines. For 2.4L engines, torque the nuts to 18–25 ft. lbs. (25–35 Nm).

10. Connect the negative battery cable.

11. Run the engine and check for any fuel leaks.

1.6L (VIN 3), 2.0L and 2.7L Engine

1. Before servicing the vehicle, refer to the Precautions Section.

2. Relieve the fuel system pressure.

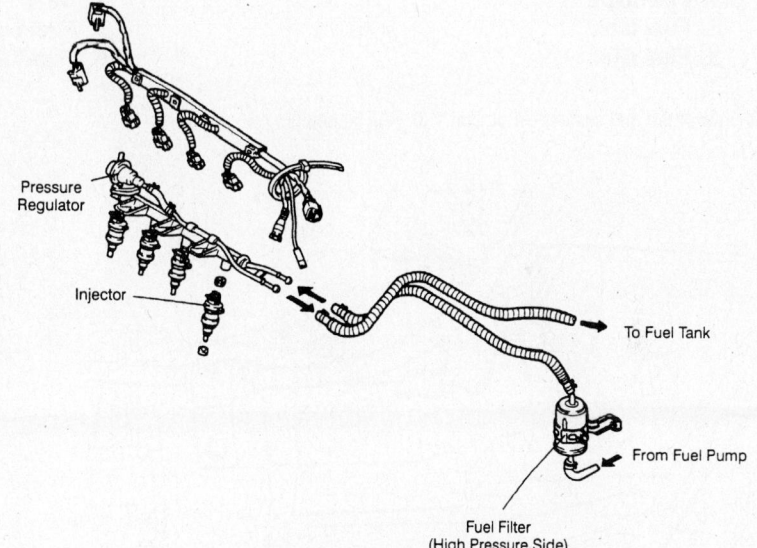

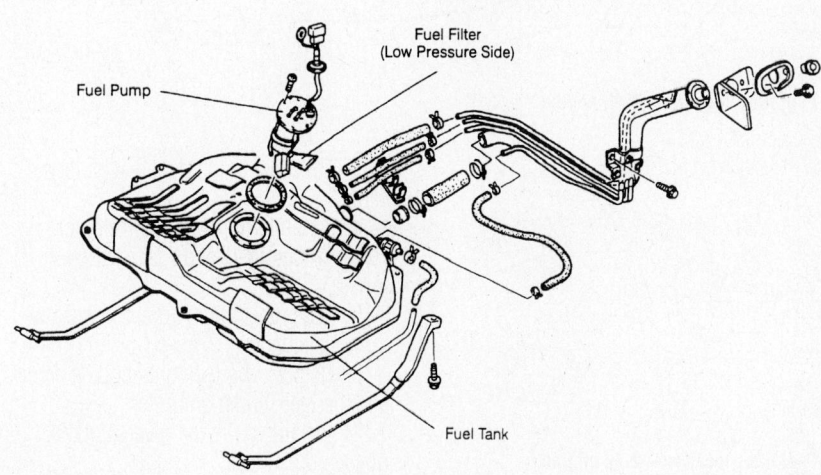

Exploded view of the fuel system–1.8L engine

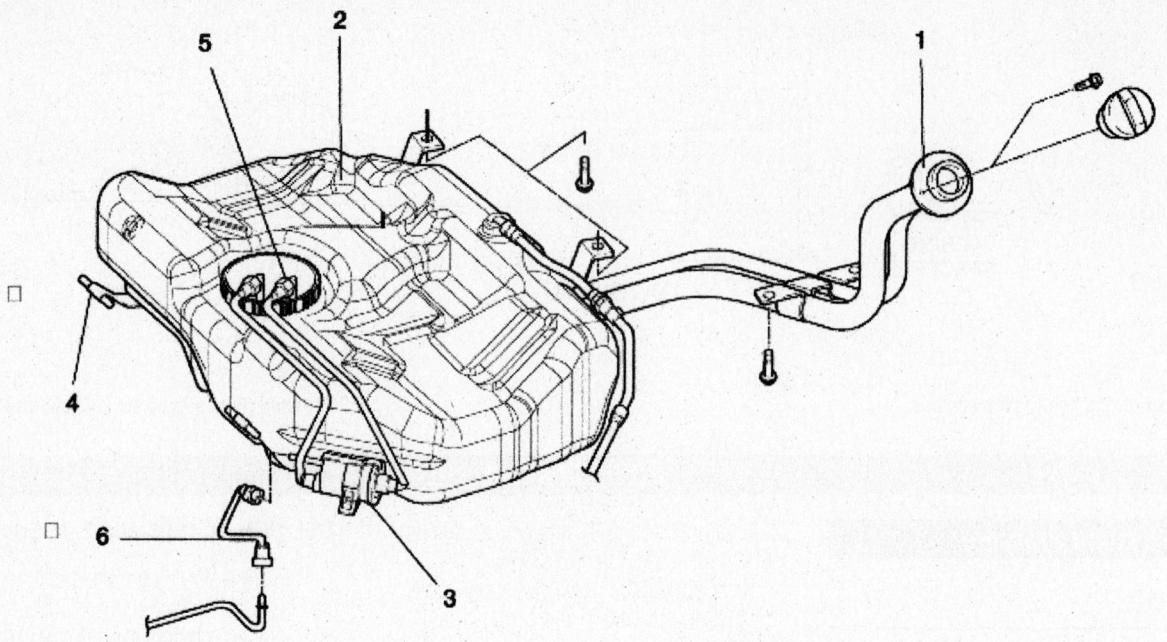

1. Filler pipe
2. Fuel tank
3. Fuel filter
4. Strap
5. Fuel pump
6. Fuel tube

67162-KIAC-G25

Exploded view of the fuel system—1.5L and 1.6L (VIN 5) engine

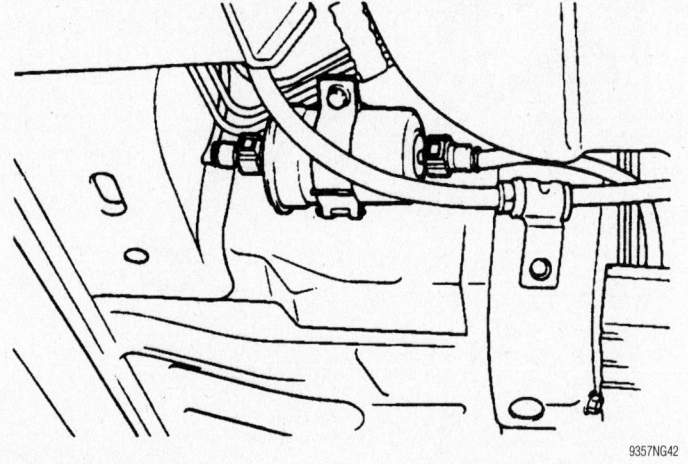

Fuel filter location—2.4L engine

9357NG42

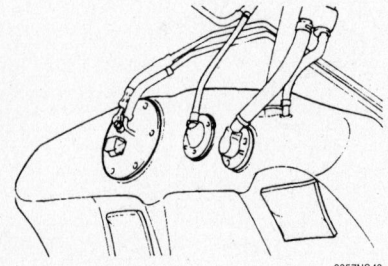

Fuel filter location—2.7L engine

9357NG43

3. Disconnect the negative battery cable.

4. Remove the rear seat retainers (clips or catches) and rear seat cushion.

5. Disconnect the harness connector from fuel sender.

6. Remove the fuel tank module screws and the fuel sender and filter.

7. Installation is the reverse of the remove procedure.

Fuel Pump

REMOVAL & INSTALLATION

1. Before servicing the vehicle, refer to the Precautions Section.

2. Relieve the fuel system pressure.

3. Disconnect the negative battery cable.

4. Remove the rear seat cushion from the vehicle.

5. Clean any dirt that has accumulated around the fuel pump cover so it will not enter the tank during pump removal and installation.

6. Remove the fuel pump cover.

7. Disconnect the fuel gauge connector, hoses, and the gauge.

8. Disconnect the fuel pump electrical connector.

9. Remove the fuel pump from the bracket assembly.

10. Remove the seal ring and discard.

To install:

11. Clean the fuel pump mounting flange, fuel tank mounting surface and seal ring groove.

12. Apply a light coating of grease on a new seal ring to hold it in place during assembly.

13. Install the seal ring.

14. Install the fuel pump to the bracket assembly carefully to ensure the filter is not damaged. Be sure the seal ring remains in the groove.

15. Hold the pump assembly in place, and pull the fuel pump down so that it is tight against the bracket.

16. Connect the fuel pump electrical connector.

17. Connect the fuel gauge, hoses, and gauge connector.

18. Install the fuel pump cover.

19. Install the rear seat cushion.

20. Connect the negative battery cable.

21. Start the engine and check for proper system operation and for fuel leaks.

Fuel Injector

REMOVAL & INSTALLATION

1. Before servicing the vehicle, refer to the Precautions Section.

2. Relieve the fuel system pressure.

3. Disconnect the negative battery cable.

4. Disconnect the injector electrical connectors.

5. Disconnect the fuel line from the fuel rail.

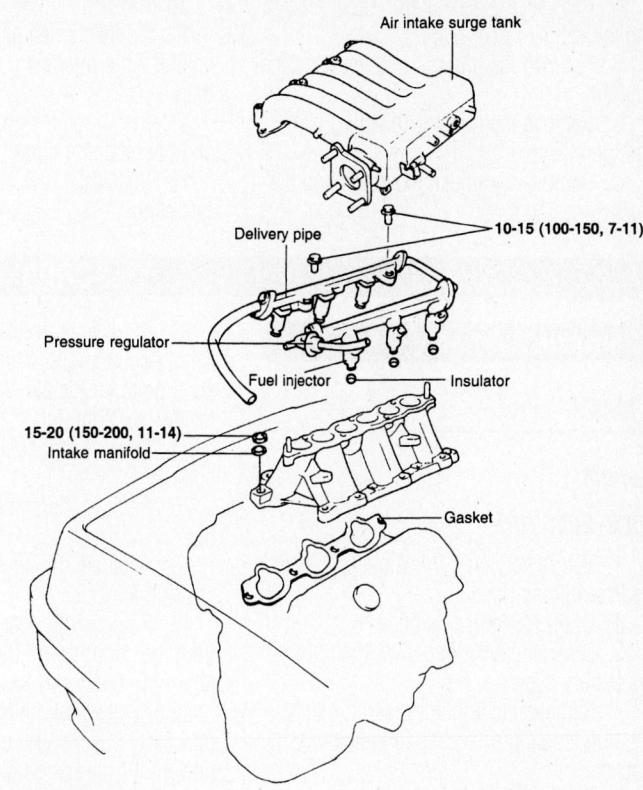

TORQUE : N·m (kg·cm, lb·ft)

9357NG45

Exploded view of the fuel rail and related components—2.7L engine

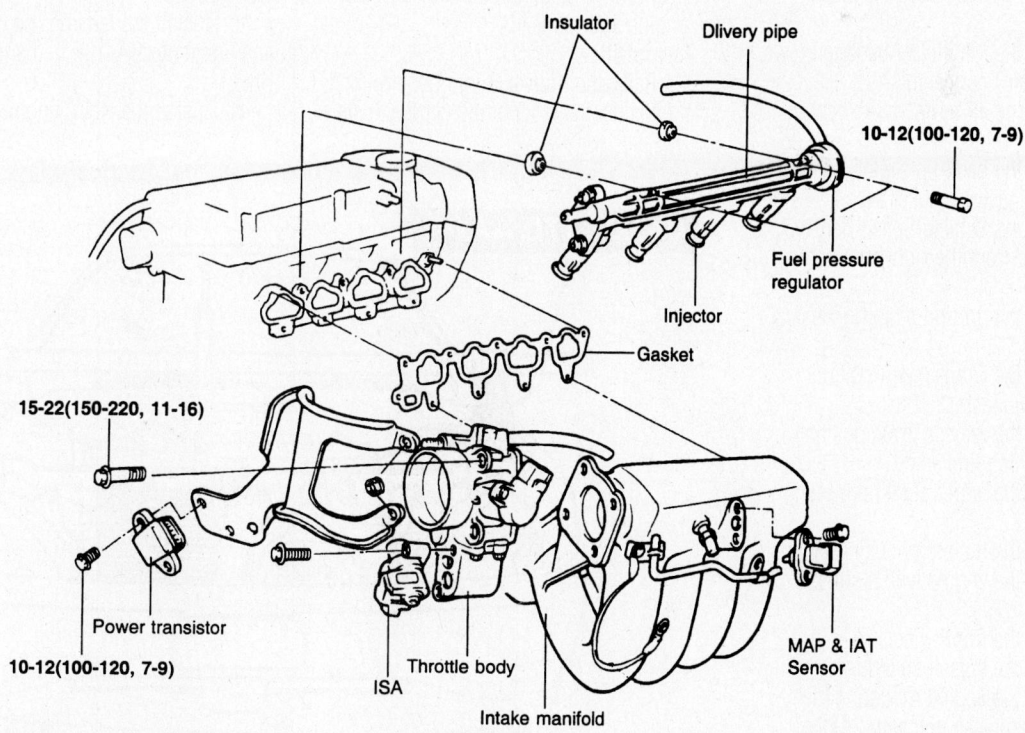

TORQUE : N·m (kg·cm, lb·ft)

9357NG44

Exploded view of the fuel rail and related components—2.4L engine

6. Remove the accelerator cable bracket and the cable, if necessary.

7. Remove the fuel rail retainers and the fuel rail.

8. Remove the injector retaining clips and the injectors.

9. Remove the injector O-rings and discard.

To install:

10. Apply a small amount of clean engine oil to the new O-rings and install them.

11. Install the injectors to the fuel rail and the retaining clips.

12. Install the fuel rail and the fuel rail retainers.

13. Install the accelerator cable and bracket, if removed.

14. Connect the fuel line to the fuel rail.

15. Connect the injector electrical connectors.

16. Connect the negative battery cable.

DRIVE TRAIN

Transaxle Assembly

REMOVAL & INSTALLATION

Manual

2002–2005 RIO

1. Before servicing the vehicle, refer to the Precautions Section.

2. Drain the transaxle oil.

3. Disconnect the negative, then positive battery cables.

4. Remove the coolant reservoir tank. Position it aside for access to the battery bracket.

5. Remove the battery bracket.

6. Remove the battery, battery tray and fixing holder.

7. Remove the fresh air duct and air cleaner assembly.

8. Remove the hose from the intake manifold.

9. Remove the Manifold Absolute Pressure (MAP) sensor connector.

10. Remove the air temperature sensor connector.

11. Remove the back-up switch connector from the connector bracket case.

12. Remove the Vehicle Speed Sensor (VSS) connector from the right side of the transaxle.

13. Remove the ground strap bolt and strap.

14. Remove the Crankshaft Position (CKP) sensor connector.

15. Remove the release lever, position the lever and clutch line aside.

16. Remove the upper starter mounting bolt.

17. Remove the 3 clutch housing bolts.

18. Support the engine with a suitable support bar.

19. Remove the front wheels.

20. Remove the splash shields.

21. Remove the engine mounting member. The member is secured with 2 bolts and 4 nuts.

22. Remove the extension bar and shift control rod.

23. Remove the bolts from the No. 1 and No. 2 engine mounts and the mounts. There are 2 bolts for the No. 1 mount and 3 bolts for the No. 2 mount.

24. Remove the tension rod mounting nuts.

25. Remove the stabilizer bar and control link.

26. Separate the tie rod ends from the steering knuckle.

27. Remove the ball joint bolt and nut from the control arm, then separate the ball joint from the control arm.

28. Remove the halfshaft from the transaxle. Support the halfshaft with wire to prevent it from hanging unsupported.

29. Remove the starter lower mounting bolt and starter.

30. Remove the 3 lower clutch housing bolts, support the transaxle with a jack, then remove the 2 remaining bolts.

31. Carefully separate the transaxle from the engine and lower it.

To install:

32. Raise the transaxle into position and seat it against the back of the engine. Install the transaxle-to-engine bolts and tighten as follows:

a. Upper bolts (1–4): 47–66 ft. lbs. (64–89 Nm).

b. Lower bolts (5–8): 28–38 ft. lbs. (37–51 Nm).

33. Remove the jack from the transaxle.

34. Install the starter and lower mounting bolt.

35. Install a new clip on the driveshaft.

36. Install the driveshaft into the transaxle with the opening of the clip pointing upward.

37. Install the lower ball joint into the steering knuckle. Torque the nut to 43–50 ft. lbs. (54–68 Nm).

38. Install the tie rod end to the steering knuckle. Torque the nut to 22–33 ft. lbs. (30–44 Nm). Install a new cotter pin.

39. Install the stabilizer bar and control link. Torque the retainers to 32–45 ft. lbs. (43–61 Nm).

40. Install the tension rod mounting nut and tighten to 87–108 ft. lbs. (118–147 Nm).

41. Install the No. 2 engine mount.

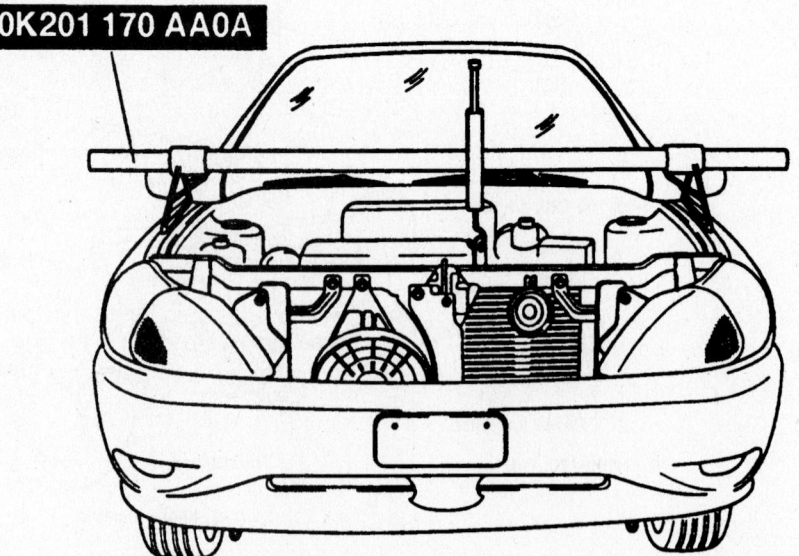

Use the proper tool to support the engine—Rio

9357NG13

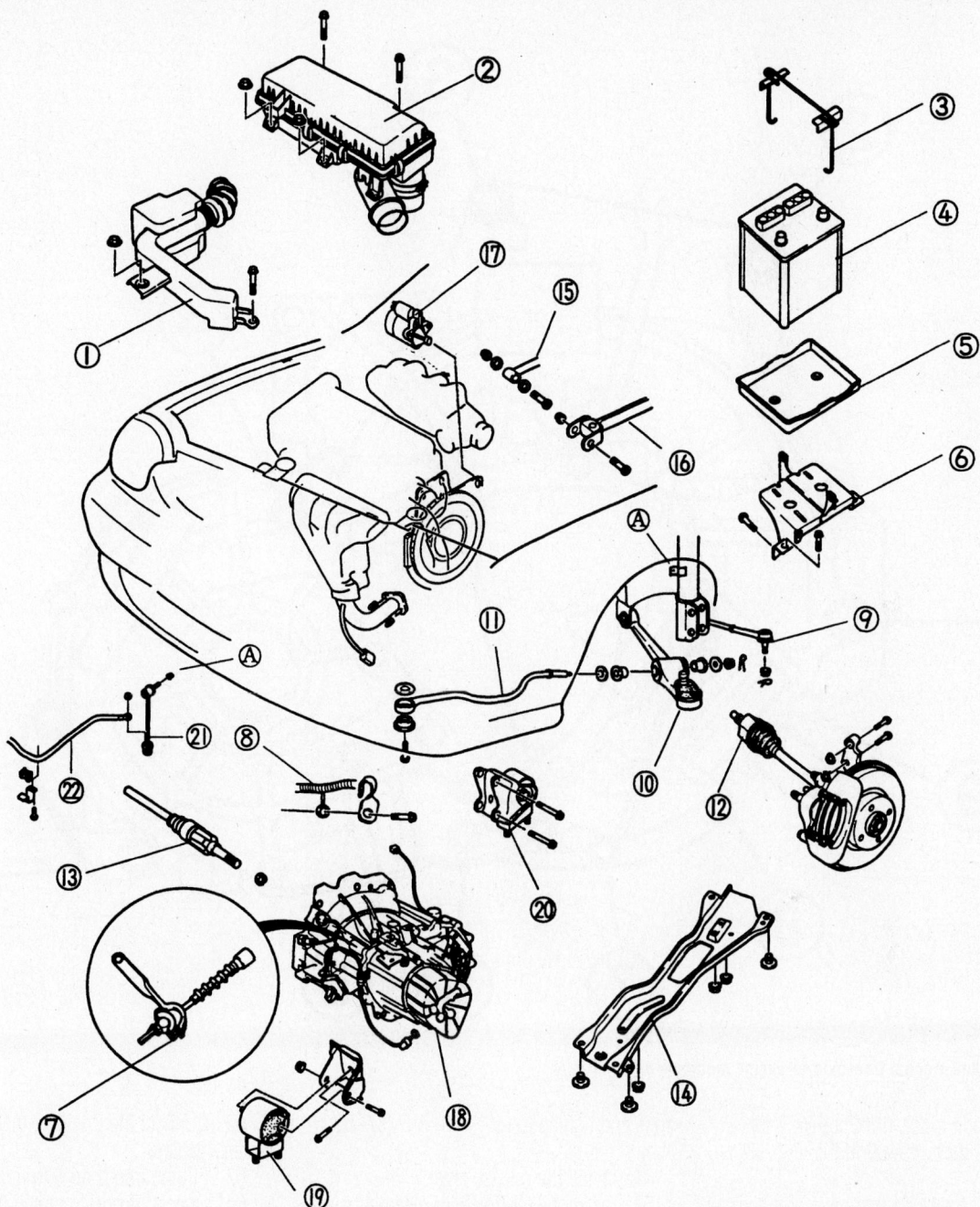

(1) Fresh air duct
(2) Air cleaner assembly
(3) Battery fixing holder
(4) Battery
(5) Battery tray
(6) Battery bracket
(7) Clutch cable
(8) Ground
(9) Tire-rod end
(10) Lower arm
(11) Torsion bar

(12) Driveshaft (Right side)
(13) Driverhaft (Left side)
(14) Engine mounting member
(15) Shift control rod
(16) Extension bar
(17) Starter
(18) Manual transaxle
(19) Engine mounting NO.2
(20) Engine mounting NO.1
(21) Shabilizer control link
(22) Stabilizer bar

9357NG12

Exploded view of the manual transaxle mounting and related components—Rio

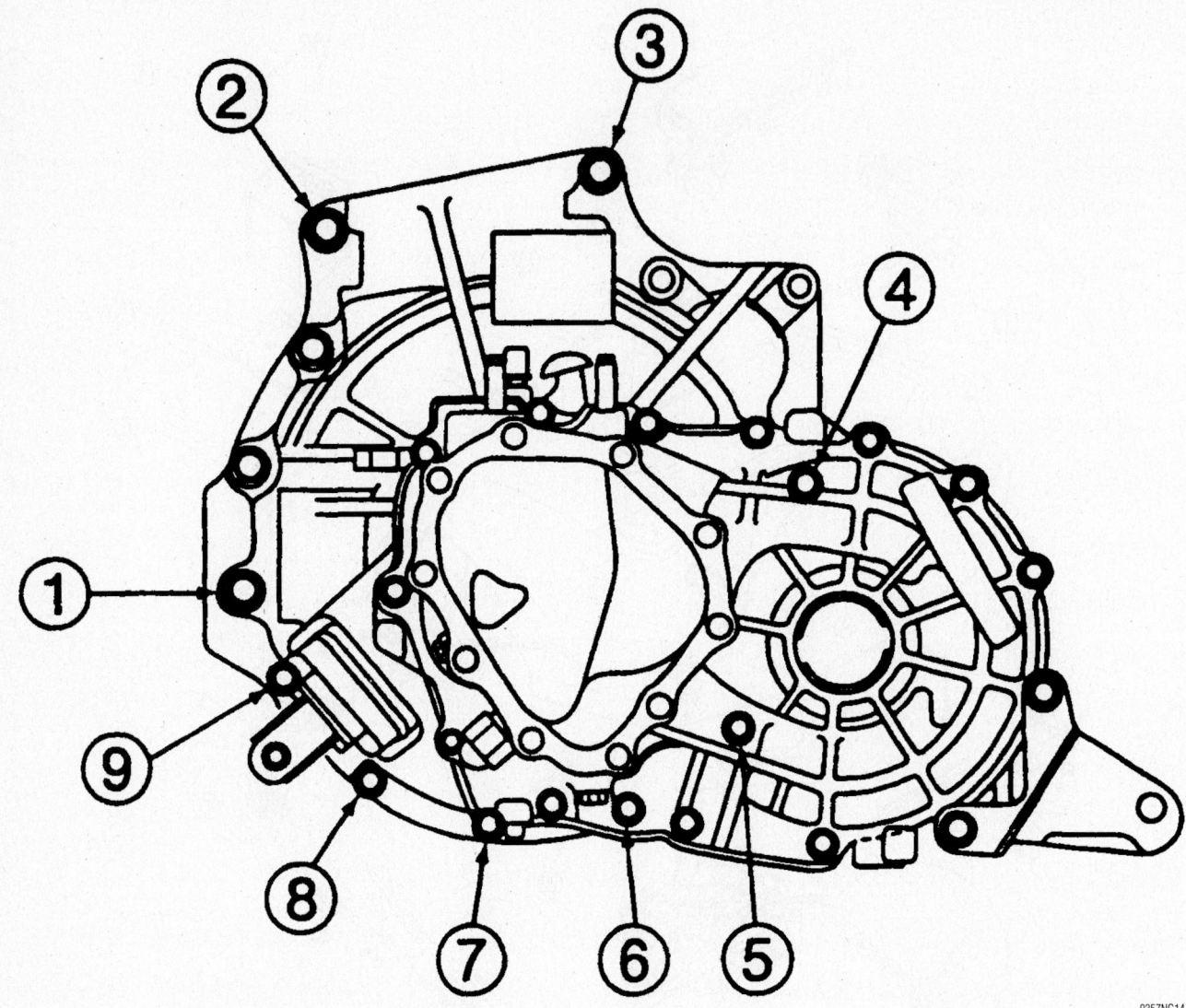

Location of the manual transaxle-to-engine mounting bolts—Rio

9357NG14

Tighten the bolt to 32–40 ft. lbs. (43–54 Nm) and the nut to 49–69 ft. lbs. (67–93 Nm).

42. Install the No. 1 engine mount and tighten to 32–40 ft. lbs. (43–54 ft. lbs.).

43. Install the extension bar and shift control rod.

44. Install the transaxle pan drain plug.

45. Install the engine mounting member-to-chassis bolts and nuts. There are 2 bolt and 4 nuts. Torque the engine mounting member bolts and nuts to 47–66 ft. lbs. (64–89 Nm) and the No. 1 and 2 engine mount nuts to 27–38 ft. lbs. (37–52 Nm).

46. Install the splash shields.

47. Install the front wheels.

48. Remove the engine support bar.

49. Install the 3 clutch housing bolts and torque to 66–86 ft. lbs. (89–116 Nm).

50. Install the upper starter mounting

bolts and tighten to 27–38 ft. lbs. (37–52 Nm).

51. Install the release lever.

52. Install the CKP sensor electrical connector.

53. Install the ground strap and bolt.

54. Install the VSS and back-up switch connectors.

55. Install the air cleaner assembly.

56. Install the air temperature and MAF sensor connectors.

57. Install the fresh air duct.

58. Install the hose from the intake manifold.

59. Install the coolant reservoir tank. Position it aside for access to the battery bracket.

60. Install the battery fixing holder, battery tray and battery.

61. Install the battery bracket.

62. Connect the Positive, then negative battery cables.

63. Fill the transaxle with fluid.

64. Check for proper clutch operation.

2006 RIO

1. Before servicing the vehicle, refer to the Precautions Section.

2. Drain the transaxle oil.

3. Remove or disconnect the following:

- Battery
- Air intake assembly
- ECM connectors
- Battery tray
- Ground cable from transaxle
- Transaxle wiring harnesses
- Transaxle shift cables
- Vehicle Speed Sensor (VSS)
- Back-up lamp switch connector
- Power steering hoses

4. Support the engine using Special Tool 09200-38001 support fixture.
5. Remove or disconnect the following:
   • Front and rear roll stoppers
   • Engine undercover
   • Side cover
   • Front tires
   • Tie rod end
   • Wheel speed sensor
   • Caliper assembly
   • Wheel knuckle mounting bolts
   • Halfshaft from the transaxle
   • Steering U-joint mounting bolt
6. Support the transaxle with a suitable jack.
7. Remove or disconnect the following:
   • Sub-frame
   • Starter mounting bolts
   • Transaxle upper mounting bolts from the engine
   • Transaxle lower mounting bolts
8. Remove the transaxle by lifting the vehicle.
9. Installation is the reverse of removal.
10. Fill the transaxle with fluid to the correct level.

1.8L SPECTRA

1. Before servicing the vehicle, refer to the Precautions Section.
2. Drain the transaxle oil.
3. Remove the battery box and disconnect the battery.
4. Remove the air cleaner assembly.
5. Remove the battery carrier.
6. Remove the back-up light switch and the bracket.
7. Disconnect the ground strap from the top of the transaxle.
8. Disconnect the neutral switch connector and the Vehicle Speed Sensor (VSS) connector.
9. Remove the wire harness bracket and ground cable.
10. Remove the Crankshaft Position (CKP) sensor.
11. Remove the wheels.
12. Remove the splash shield.
13. Remove the transverse member.
14. Remove the extension bar and the change control rod.
15. Remove the tie-rod ends.
16. Remove the stabilizer control link.
17. Remove the halfshaft and the joint shaft.
18. Remove the intake manifold bracket.
19. Remove the starter.
20. Remove the front exhaust pipe.
21. Support the engine and remove the engine mounting member.
22. Remove the rear engine/transaxle mount.

23. Remove the front engine/transaxle mount.
24. Remove the clutch release cylinder.
25. Remove the side engine/transaxle mount.
26. Support the transaxle on a jack.
27. Remove the transaxle mounting bolts.
28. Remove the transaxle.

To install:
29. Install the transaxle into position and install the mounting bolts. Tighten to 28–38 ft. lbs. (37–52 Nm).
30. Remove the support jack.
31. Install the side mount. Tighten the body side nuts and bolts to 32–44 ft. lbs. (44–60 Nm). Tighten the transaxle side nuts to 50–68 ft. lbs. (67–93 Nm).
32. Install the clutch release cylinder.
33. Install the front mount, loosely tighten the mount nut and bolt.
34. Install the rear mount, align and set all bolts, then tighten to 50–68 ft. lbs. (67–93 Nm).
35. Install the engine mounting member. Tighten the 4 outer nuts and bolts to 50–65 ft. lbs. (67–89 Nm) and the 2 remaining nuts to 28–38 ft. lbs. (38–51 Nm).
36. Tighten the front mount nut and bolt to 50–68 ft. lbs. (67–93 Nm).
37. Install the starter.
38. Install the manifold bracket and front exhaust pipe.
39. Install the joint shaft and the halfshaft.
40. Install the stabilizer control link.
41. Install the tie-rod ends.
42. Install the control rod and extension bar.
43. Install the transverse member.
44. Install the splash shield.
45. Install the wheels.
46. Install the harness bracket.
47. Install the CKP sensor.
48. Install the VSS and the neutral switch connectors.
49. Install the back-up light switch connector bracket and the switch.
50. Install the No. 4 engine mount.
51. Install the battery carrier.
52. Install the air cleaner assembly.
53. Install the battery and battery box.
54. Fill the transaxle with fluid.
55. Check for proper clutch operation.

2.0L SPECTRA

1. Before servicing the vehicle, refer to the Precautions Section.
2. Drain the transaxle oil.
3. Remove or disconnect the following:

   • Negative battery cable
   • Battery heat shield
   • Battery and battery tray
   • Air intake assembly
   • Back-up lamp switch connector
   • Clutch release cylinder
   • Shift cables
   • Vehicle speed sensor
   • Starter mounting bolts
   • Transaxle upper mounting bolts to the engine
4. Support the engine assembly with Special Tool 09200-38001 support fixture.
5. Remove or disconnect the following:
   • Transaxle mounting bracket
   • Front tire
   • Tie rod end
   • Wheel speed sensor
   • Wheel knuckle mounting bolts
   • Caliper assembly
   • Halfshaft
   • Steering U-joint mounting bolt
   • Oxygen sensor
   • Front muffler
   • Power steering hoses
   • Front and rear roll stoppers
6. Using a suitable jack, support the sub-frame assembly.
7. Remove the sub-frame.
8. Using a suitable jack, support the transaxle assembly.
9. Remove the transaxle lower mounting bolts.
10. Lower the transaxle assembly.
11. Installation is the reverse order of removal.
12. Fill the transaxle to the correct level.

OPTIMA

1. Before servicing the vehicle, refer to the Precautions Section.
2. Drain the transaxle oil.
3. Disconnect the negative battery cable.
4. Remove the air duct.
5. Remove the air cleaner and air flow assembly.
6. Remove the back-up light switch connector.
7. Remove the clutch tube and clip.
8. Remove the clutch release cylinder.
9. Disconnect the speedometer cable.
10. Disconnect the select cable and shift cable.
11. Remove the starter mounting bolts.
12. Remove the upper transaxle mounting bolts.
13. Install engine hooks and support the engine with a suitable support bar.
14. Remove the transaxle mounting bracket and insulator.

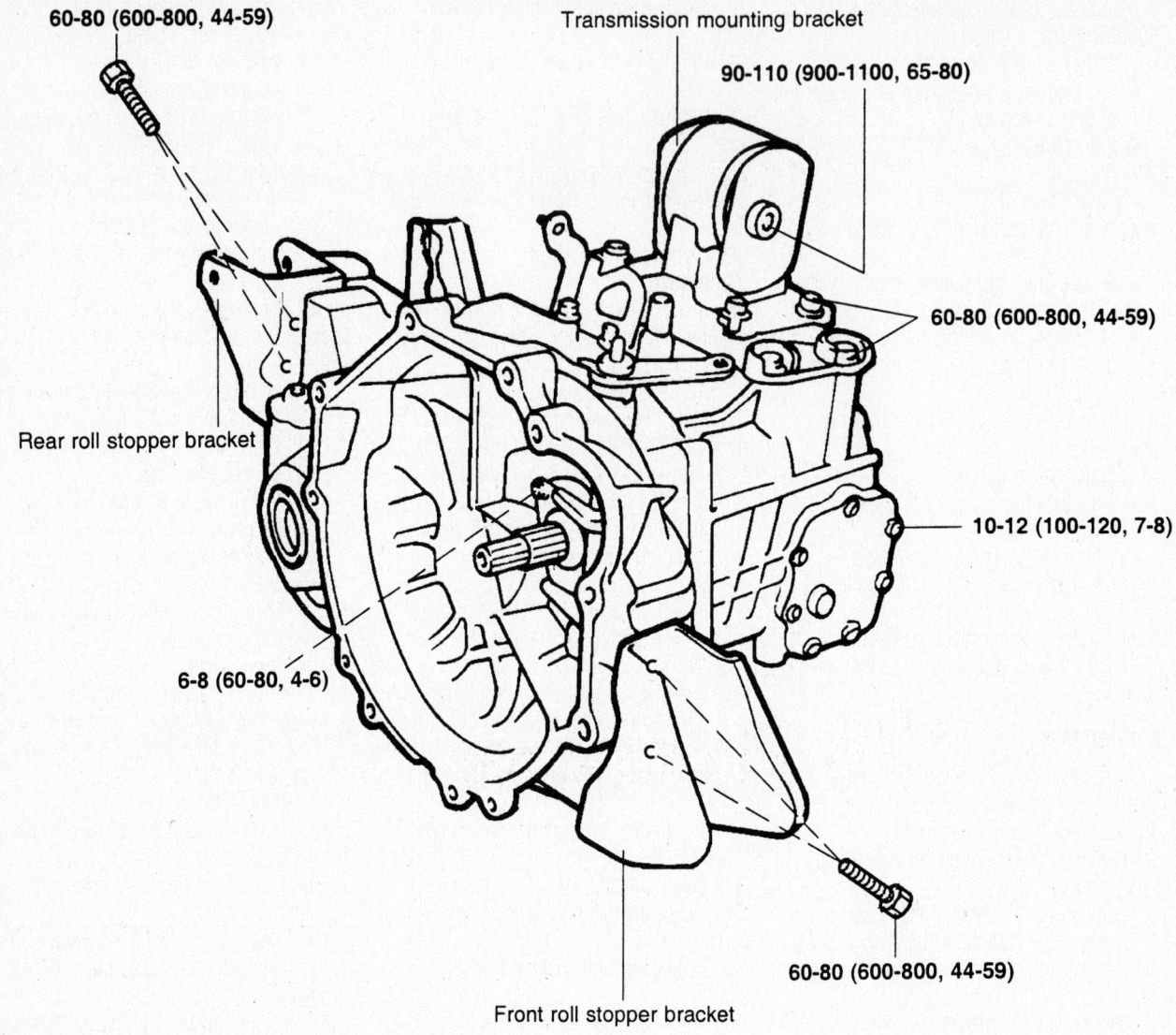

60-80 (600-800, 44-59)

Transmission mounting bracket

90-110 (900-1100, 65-80)

60-80 (600-800, 44-59)

Rear roll stopper bracket

10-12 (100-120, 7-8)

6-8 (60-80, 4-6)

60-80 (600-800, 44-59)

Front roll stopper bracket

9357NG46

TORQUE : Nm (kg·cm, lb·ft)

Manual transaxle mounting and tightening specifications—Optima

15. Remove the front wheels.
16. Remove the steering gear box U-joint bolt and return tube mounting bolts.
17. Remove the muffler.
18. Remove the sub-frame mounting bolts and sub-frame.
19. Remove the transaxle front and rear mounting bracket.
20. Remove the transaxle side mounting bolts. Support the transaxle with a jack.
21. Remove the transaxle from the vehicle.
22. Installation is the reverse of the removal procedure. Refer to the illustration for tightening specifications.
23. Fill the transaxle with fluid.
24. Check for proper clutch operation.

Automatic

2002–2005 RIO

1. Before servicing the vehicle, refer to the Precautions Section.
2. Drain the transaxle fluid.
3. Disconnect the negative, then positive battery cables.
4. Remove the fresh air duct.
5. Disconnect the input/turbine speed sensor, solenoid and transaxle range connector.
6. Remove the ground strap bolt and ground strap from the top of the transaxle.
7. Disconnect the Vehicle Speed Sensor (VSS) connector from the right side of the transaxle.

8. Remove the U-clip connecting the selector cable to the linkage.
9. Remove the nut and washer from the transaxle linkage.
10. Disconnect the Crankshaft Position (CKP) and Oxygen (O_2) sensor connectors.
11. Disconnect the 2 transaxle fluid cooler hoses.
12. Remove the top 2 upper converter housing bolts.
13. Support the engine with a suitable support bar.
14. Remove the front wheels.
15. Remove the splash shield.
16. Remove the 2 nuts from the U-bolt.
17. Remove the catalytic converter.
18. Remove the converter housing access cover and 4 drive plate-to-torque

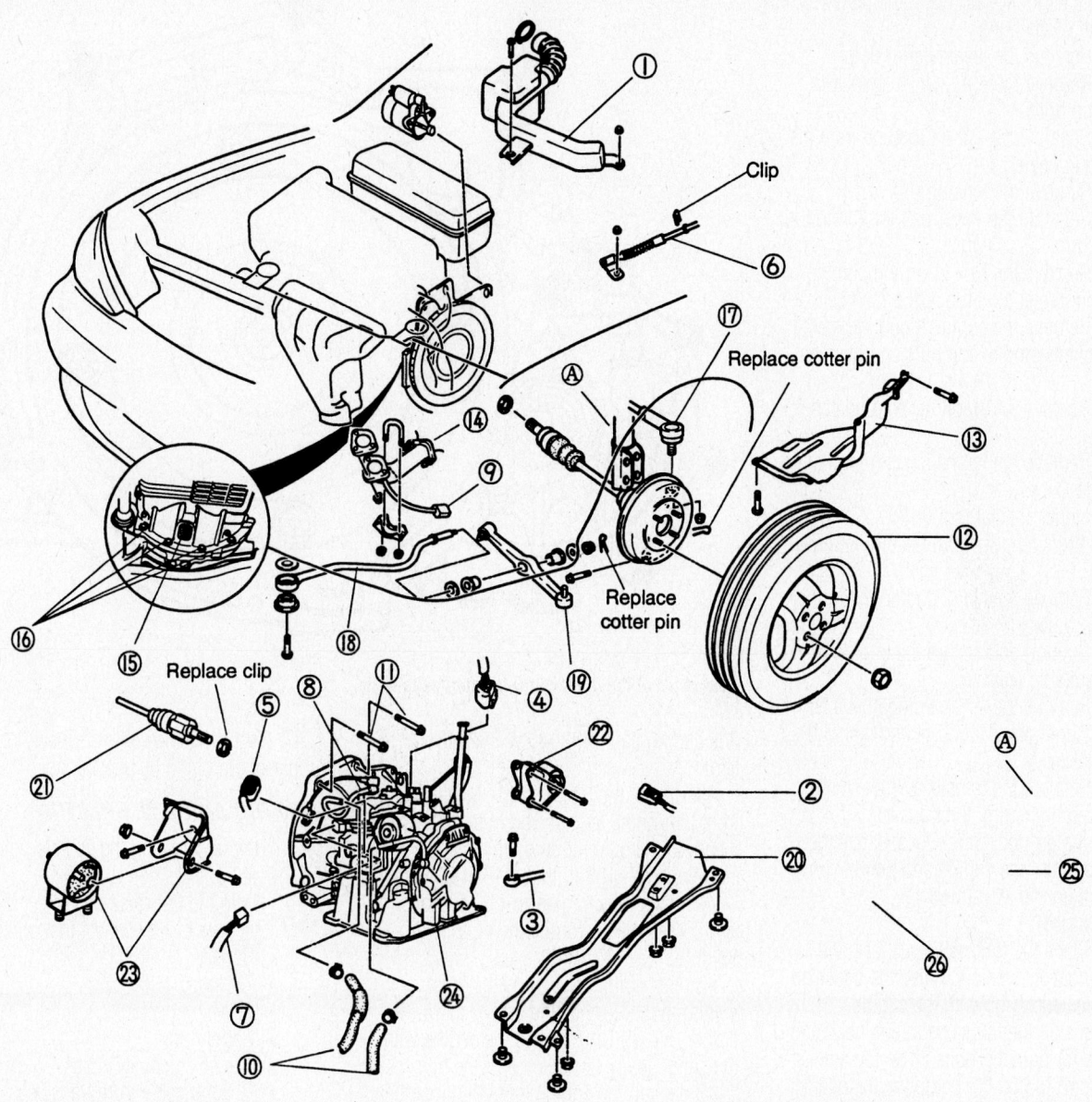

Clip

Replace cotter pin

Replace cotter pin

Replace clip

(1) Air cleaner assembly
(2) Input/turbine speed sensor connector
(3) Ground strap bolt
(4) Vehicle speed sensor connector
(5) Transaxle range switch connector
(6) Selector cable
(7) Solenoid valve connector
(8) Crankshaft position connector
(9) Oxygen sensor connector
(10) ATF cooler hose
(11) Upper converter housing bolts
(12) Wheel and tire
(13) Splash shield

(14) Catalytic converter
(15) Access cover
(16) Engine oil pan-to-transaxle mounting bolt
(17) Tie-rod end
(18) Tension rod
(19) Ball joint
(20) Engine mounting member
(21) Driveshaft
(22) No.1 engine mounting
(23) No.2 engine mounting
(24) Auto transaxle
(25) Stabilizer control link
(24) Stabilizer bar

9357NG15

Exploded view of the automatic transaxle mounting—Rio

converter mounting nuts. You will need to rotate the engine using the crank pulley to get to all of the bolts.

19. Remove the lower starter bolt.

20. Remove the 4 oil pan-to-transaxle mounting bolts.

21. Separate the left tie rod end from the steering knuckle.

22. Remove the tension rod.

23. Separate the lower ball joint from the control arm.

24. Remove the No. 1 and 2 mounting nuts from the engine mounting member.

25. Remove the engine mounting member-to-chassis bolts and nuts and the member.

26. Remove the stabilizer bar and control link.

27. Remove the left halfshaft from the transaxle.

28. Install special tool OK201 270 014 to prevent the side gear from becoming misaligned.

29. Remove the No. 2 engine mount from the transaxle.

30. Support and secure the transaxle with a suitable jack.

31. Remove the 2 remaining converter housing bolts from the front and rear sides of the transaxle

32. Remove the transaxle. Slowly lower the drivetrain, letting the transaxle tilt toward the ground. Carefully separate the transaxle from the engine and pull the unit out through the wheel well.

To install:

33. Place the transaxle on a jack and place under the vehicle. Raise the transaxle into place and align with the engine.

34. Install the transaxle to engine, using 4 converter housing bolts (2 on top and 2 on bottom) to pull the components together. Torque the bolts to 47–66 ft. lbs. (64–89 Nm).

35. Remove the jack.

36. Install the No. 2 engine mounting to the transaxle. Tighten the nut to 49–69 ft. lbs. (67–93 Nm) and the bolts to 32–40 ft. lbs. (43–54 Nm).

37. Install the starter and ground strap.

38. Install the 3 No. 1 engine mounting-to-chassis bolts and tighten to 32–46 ft. lbs. (43–52 Nm).

39. Install a new clip on the left driveshaft.

40. Install the driveshaft into transaxle with the opening of the clip facing up.

41. Install the left lower ball joint to the spindle. Torque the pinch bolt to 40–50 ft. lbs. (54–68 Nm).

42. Install the stabilizer bar and control

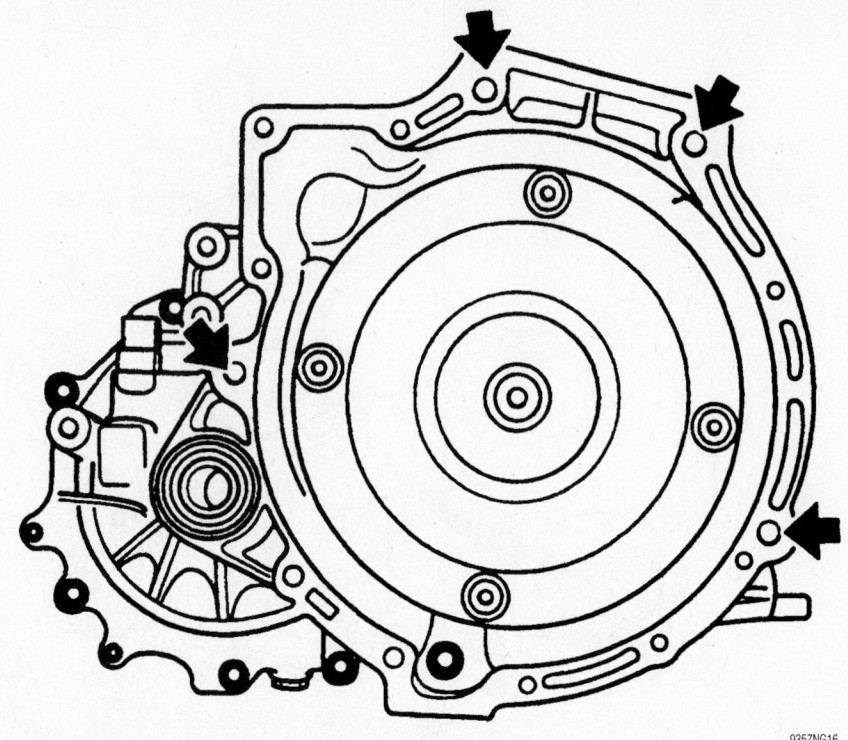

Location of the 4 converter housing bolts—Rio

9357NG16

link and tighten to 32–45 ft. lbs. (43–61 Nm).

43. Install the tension rod. Install the front part of the tension rod to the body after inserting the bushing into the tensioner rod. Tighten to 87–108 ft. lbs. (118–147 Nm).

44. Install the engine-to-oil pan bolts and access cover. Torque the bolts to 27–38 ft. lbs. (37–52 Nm).

45. Install the drive plate-to-torque converter mounting nuts and torque to 25–36 ft. lbs. (34–49 Nm). Rotate the engine with the crank pulley to get to all of the nuts.

46. Install the engine mounting member-to-chassis nuts and bolts and tighten to 48–65 ft. lbs. (64–89 Nm).

47. Install the 2 nuts to the No. 1 and 2 engine mount-to-mounting member and tighten to 28–38 ft. lbs. (38–51 Nm).

48. Install the catalytic converter and U-clip. Torque to 27–38 ft. lbs. (37–52 Nm).

49. Install the splash shield.

50. Install the front wheels.

51. Connect the transaxle fluid hoses onto the cooler.

52. Remove the engine support bar.

53. Install the CKP and O$_2$ sensor connectors.

54. Install the solenoid valve connector.

55. Install the remaining components in the reverse of the removal procedure.

56. Fill the transaxle with the proper type and amount of fluid.

57. Test drive the vehicle. Check for proper operation in all gear ranges.

2006 RIO AND 2.0L SPECTRA

1. Before servicing the vehicle, refer to the Precautions Section.

2. Drain the transaxle oil.

3. Remove or disconnect the following:
   • Battery
   • Air intake assembly
   • ECM connectors
   • Battery tray
   • Transaxle cooler hoses
   • Ground cable from transaxle
   • Transaxle wiring harnesses
   • Transaxle shift cable
   • Power steering hoses

4. Support the engine using Special Tool 09200-38001 support fixture.

5. Remove or disconnect the following:
   • Front and rear roll stoppers
   • Engine undercover
   • Side cover
   • Front tires
   • Tie rod end
   • Wheel speed sensor
   • Caliper assembly
   • Wheel knuckle mounting bolts
   • Halfshaft from the transaxle
   • Steering U-joint mounting bolt

6. Support the transaxle with a suitable jack.

7. Remove or disconnect the following:

- Sub-frame
- Starter mounting bolts
- Engine-to-torque converter mounting bolts
- Engine-to-transaxle mounting bolts

8. Remove the transaxle by lifting the vehicle.

9. Installation is the reverse of removal.

10. Note the following torques:
   a. Engine to transaxle mounting bolts—12mm: 43–58 ft. lbs. (60–80 Nm). 10mm: 32–41 ft. lbs. (43–55 Nm).
   b. Engine-to-torque converter mounting bolts: 33–38 ft. lbs. (46–53 Nm).

11. Fill the transaxle with fluid to the correct level.

1.8L SPECTRA

1. Before servicing the vehicle, refer to the Precautions Section.
2. Drain the transaxle fluid.
3. Disconnect the battery and battery cover.
4. Remove the air cleaner assembly.

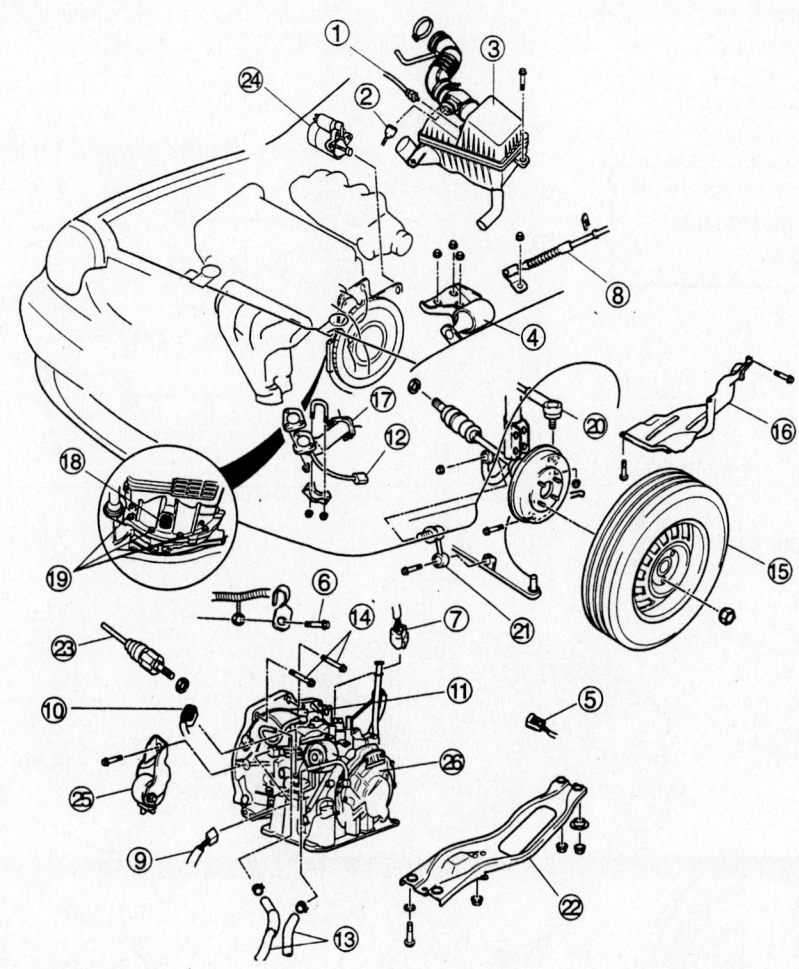

(1) Air temperature sensor connector
(2) MAF sensor connector
(3) Air cleaner assembly
(4) No. 4 mounting
(5) Input/turbine speed sensor connector
(6) Ground strap bolt
(7) Vehicle speed sensor connector
(8) Selector cable
(9) Transaxle range switch connector
(10) Solenoid valve connector
(11) Crankshaft position connector
(12) Oxygen sensor connector
(13) ATF cooler hose
(14) Upper converter housing bolts
(15) Wheel and tire

(16) Gravel shield
(17) Catalytic converter
(18) Converter housing
(19) Engine oil pan-to-transaxle mounting bolt
(20) Tie-rod end
(21) Stabilizer control link
(22) Engine mounting member
(23) Driveshaft
(24) Starter
(25) No. 2 engine mounting
(26) Auto transaxle

Exploded view of the automatic transaxle assembly mounting and related components—Spectra

9301KG08

5. Remove the battery carrier.
6. Disconnect the solenoid and the Transaxle Range Switch (TRS) connectors.
7. Disconnect the selector cable.
8. Remove the Vehicle Speed Sensor (VSS) connector.
9. Remove the harness bracket.
10. Disconnect the throttle cable.
11. Remove the front wheels.
12. Remove the splash shields.
13. Remove the front exhaust pipe.

14. Remove the transverse member, if equipped.
15. Remove the tie-rod ends.
16. Remove the stabilizer control links.
17. Remove the lower arm by removing the cinch bolt from the lower arm ball joints. Pry the lower arm out of the knuckle.
18. Support the engine.
19. Remove the engine mounting member.
20. Remove the left and right halfshaft.

Install Differential Side Gear holder K49A-4208-AT to hold the side gears.
21. Remove the joint shaft.
22. Remove the intake manifold bracket.
23. Remove the starter.
24. Remove the front engine/transaxle mount.
25. Remove the rear engine/transaxle mount.
26. Remove the inner and outer oil hoses.

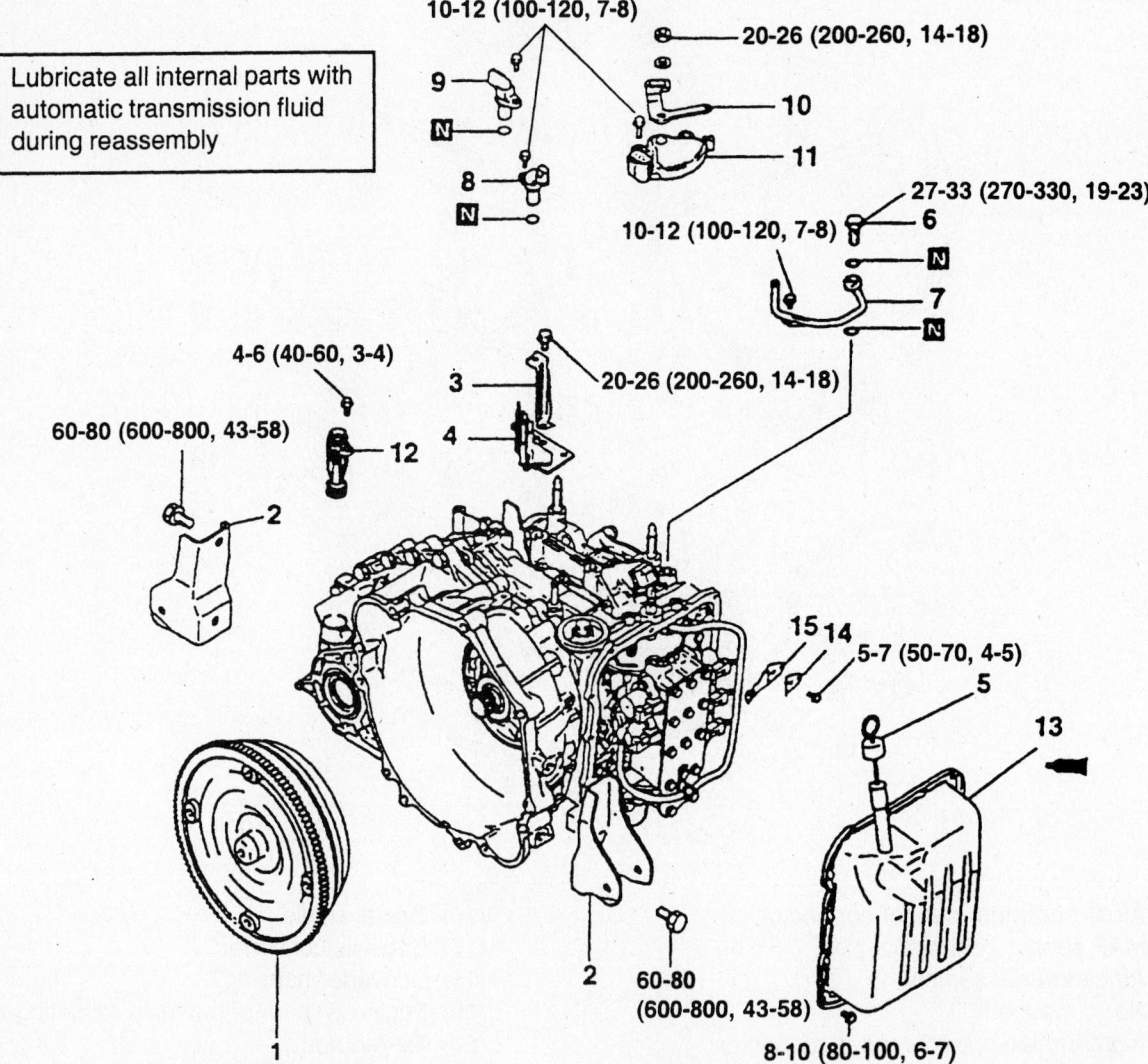

Lubricate all internal parts with automatic transmission fluid during reassembly

1. Torque converter
2. Roll stopper bracket
3. Harness bracket
4. Control cable support bracket
5. Oil level gauge
6. Eye bolt
7. Oil cooler feed tube
8. Input shaft speed sensor
9. Output shaft speed sensor
10. Manual control lever
11. Transaxle range switch
12. Vehicle speed sensor
13. Valve body cover
14. Manual control shaft detent spring
15. Manual control shaft detent

TORQUE : Nm (kg·cm, lb·ft)

9357NG47

Automatic transaxle mounting and tightening specifications—Optima

27. Remove the side engine/transaxle mount.

28. Hold the drive plate and remove the converter nuts.

29. Support the transaxle on a jack.

30. Remove the transaxle mounting bolts.

31. Remove the transaxle.

To install:

32. Support the transaxle on a jack and lift it into place. Align the transaxle with the engine and install the mounting bolts. Tighten to 65–86 ft. lbs. (89–116 Nm).

33. Hold the driveplate and install the torque converter mount nuts. 26–36 ft. lbs. (35–49 Nm).

34. Install the side engine/transaxle mount. Loosely tighten the nuts of the transaxle side. Tighten the nuts and bolts of the body side to 32–44 ft. lbs. (44–60 Nm). Tighten the nuts of the transaxle side to 50–68 ft. lbs. (67–93 Nm).

35. Install the inner and outer oil hoses.

36. Install the rear engine/transaxle mount. Tighten the bolts to 50–68 ft. lbs. (67–93 Nm).

37. Install the front engine/transaxle mount. Tighten the mount bracket to the transaxle to 28–38 ft. lbs. (38–51 Nm). Loosely tighten the nuts and bolts of the engine mount rubber, then tighten to 50–68 ft. lbs. (67–98 Nm).

38. Install the starter.

39. Install the manifold bracket.

40. Install the joint shaft into the transaxle.

41. Install the joint shaft to the cylinder block and tighten the bolts in sequence (counterclockwise). Tighten to 32–46 ft. lbs. (42–62 Nm).

42. Install the halfshafts, be sure that the shafts are properly installed and do not pull out.

43. Install the engine mounting member, the mounting nuts/bolts and tighten the nuts/bolts at the far corners to 48–65 ft. lbs. (64–89 Nm), tighten the remaining 2 nuts to 28–38 ft. lbs. (38–51 Nm).

44. Install the front exhaust pipe.

45. Install the lower arm to the knuckle.

46. Install the stabilizer control link.

47. Install the tie-rod ends.

48. Install the transverse member, if removed.

49. Install the splash shields.

50. Install the wheels.

51. Connect the throttle cable.

52. Install the harness bracket.

53. Connect the VSS connector.

54. Connect the selector cable.

55. Connect the TRS connector.

56. Connect the solenoid connector.

57. Install the battery carrier.

58. Install the air cleaner assembly.

59. Install the battery and battery cover.

60. Fill the transaxle with the proper type and amount of fluid.

61. Test drive the vehicle. Check for proper operation in all gear ranges.

OPTIMA

1. Before servicing the vehicle, refer to the Precautions Section.

2. Drain the transaxle fluid.

3. Remove the air cleaner assembly.

4. Disconnect the control cable.

5. Disconnect the Vehicle Speed Sensor (VSS) connector.

6. Disconnect the transaxle range switch connector, solenoid connector and oil temperature sensor connector.

7. Disconnect the oil cooler hose.

8. Install a suitable engine support to the engine.

9. Remove the stabilizer bar, tie rod end, lower control arm ball joint and halfshafts.

10. Remove the steering gear box U-joint bolt and return tube mounting bolts.

11. Remove the sub-frame mounting bolts and sub-frame.

12. Remove the starter.

13. Remove the automatic transaxle mounting bolts.

14. Remove the engine-to-transaxle bolts. Place a jack under the transaxle.

15. Remove the transaxle from the vehicle.

16. Installation is the reverse of the removal procedure. Refer to the illustration for tightening specifications.

17. Fill the transaxle with fluid.

18. Check for proper clutch operation.

Clutch

ADJUSTMENTS

Pedal Height

1. Before servicing the vehicle, refer to the Precautions Section.

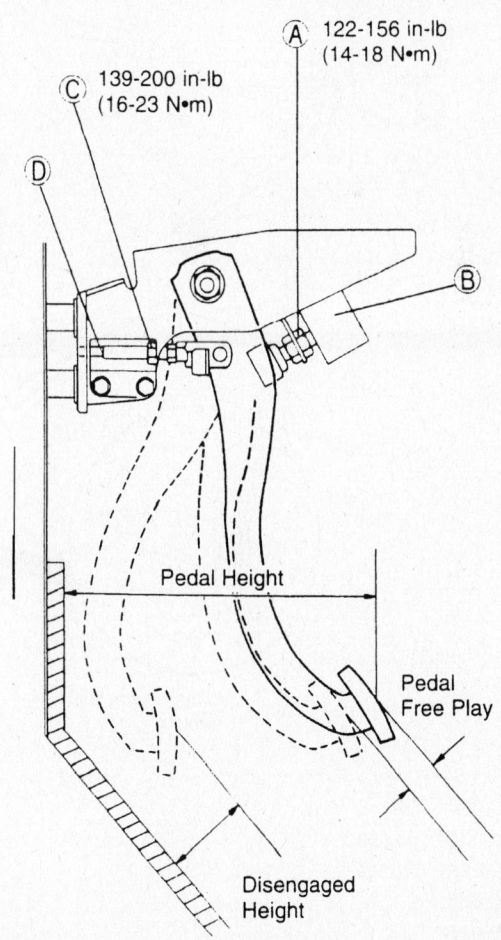

Clutch pedal measurement and adjustment points. (A) and (B) are for adjusting the pedal height, while (C) and (D) are for the free-play adjustment

2. Measure the distance from the upper surface of the pedal pad to the carpet.

3. The distance should be as follows:
   a. Spectra and Optima: 7.83–8.15 in. (199–207mm).
   b. Rio: 7.67 in. (195mm).

4. If the distance is not as specified, loosen the locknut on the stopper bolt or switch.

5. Turn the switch or bolt until the distance is correct, then tighten the locknut.

Free-Play

1. Before servicing the vehicle, refer to the Precautions Section.

2. Depress the clutch pedal by hand until resistance is felt. The free-play should be as follows:

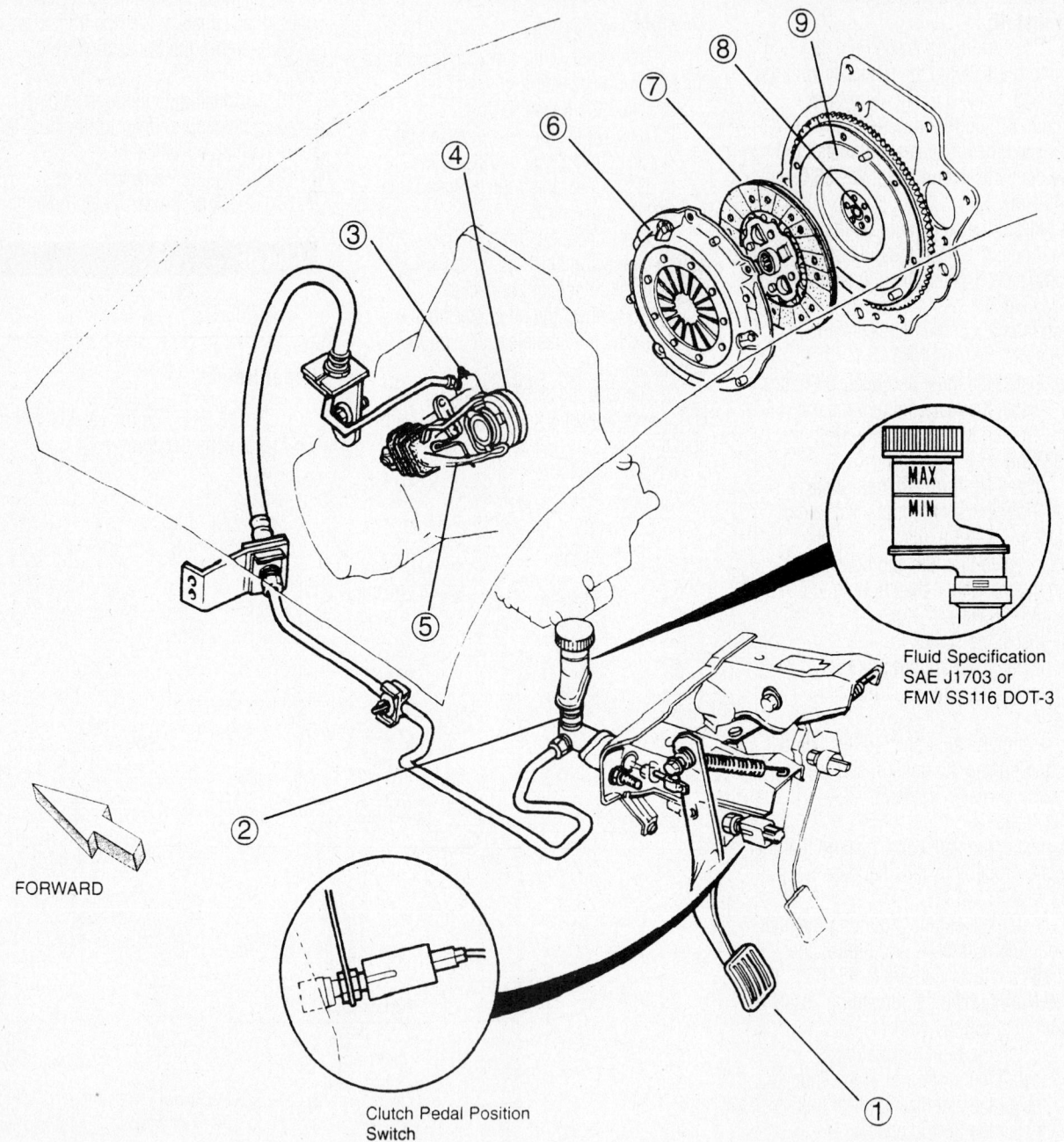

Fluid Specification
SAE J1703 or
FMV SS116 DOT-3

FORWARD

Clutch Pedal Position Switch

1. Clutch pedal
2. Clutch master cylinder
3. Clutch release cylinder
4. Release bearing
5. Clutch release fork
6. Clutch cover
7. Clutch disc
8. Pilot bearing
9. Flywheel

Structural view of the hydraulic clutch system

7923KG32

a. Spectra: 0.12–0.20 in. (3.0–5.0mm).

b. Rio: 0.35–0.59 in. (9–15mm).

c. Optima: 0.2–0.5 in. (6–13mm).

3. If the free-play is not correct, loosen the clutch master cylinder pushrod locknut and turn the pushrod to adjust.

REMOVAL & INSTALLATION

1. Before servicing the vehicle, refer to the Precautions Section.

2. Disconnect the negative battery cable.

3. Remove the transaxle.

4. Gradually loosen the clutch pressure plate bolts, in a crisscross pattern. Support the pressure plate and remove the bolts. Remove the pressure plate and clutch disc.

5. Inspect the pilot bearing. If it is worn or damaged and does not turn easily by hand, remove it using a puller/slide hammer.

6. Check the flywheel surface for scoring, cracks or burning and machine or replace, as necessary.

7. Install a flywheel holder to keep the flywheel from turning. Loosen the flywheel bolts evenly and gradually in a crisscross pattern. Remove the flywheel.

8. Inspect the clutch release bearing for wear. Replace it if it sticks or does not turn easily.

9. Inspect the release fork for wear or damage and replace as necessary.

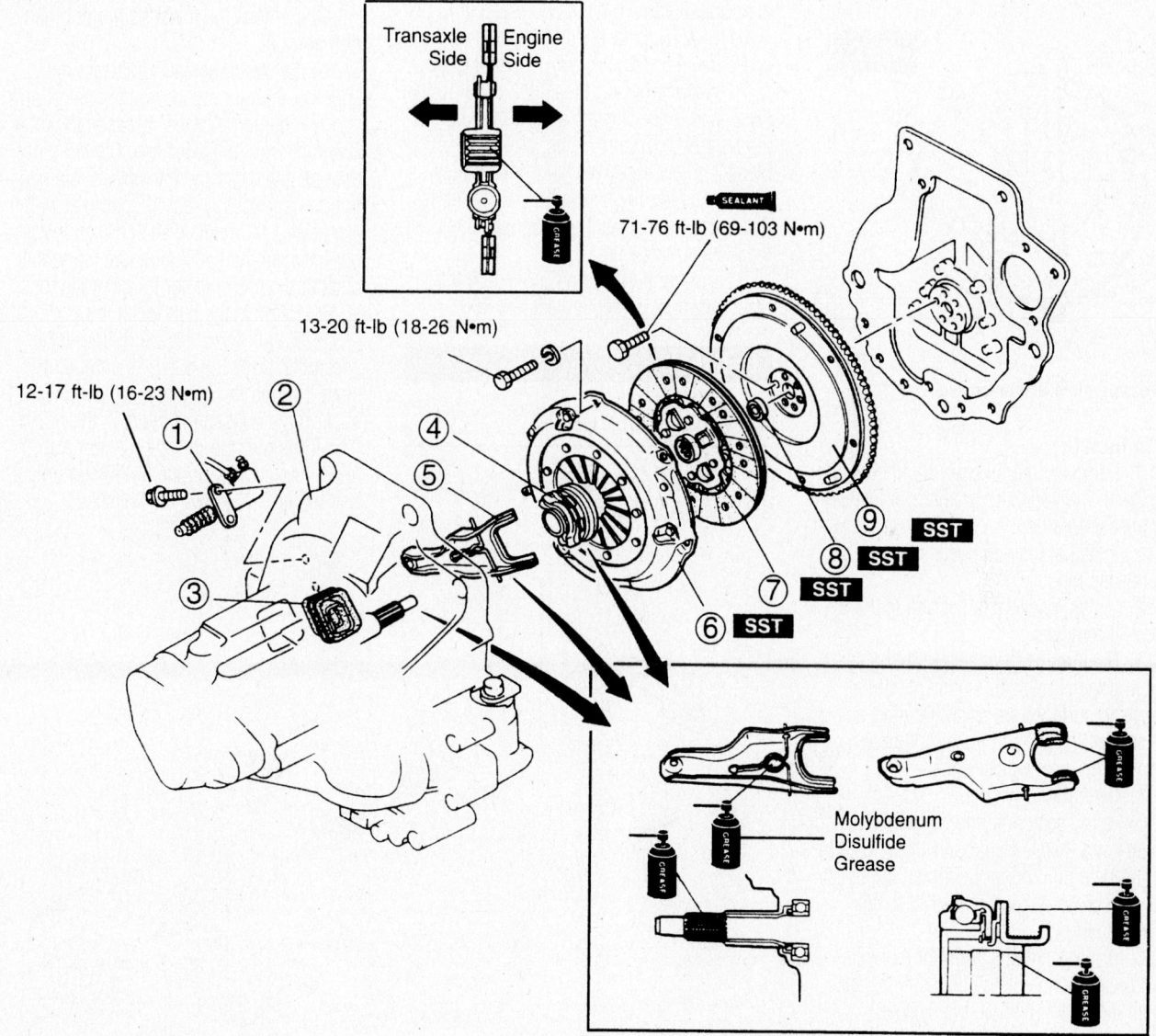

1. Clutch release cylinder
2. Transaxle housing
3. Boot
4. Release bearing
5. Clutch release fork
6. Clutch cover
7. Clutch disc
8. Pilot bearing
9. Flywheel

Exploded view of the clutch assembly

7923KG33

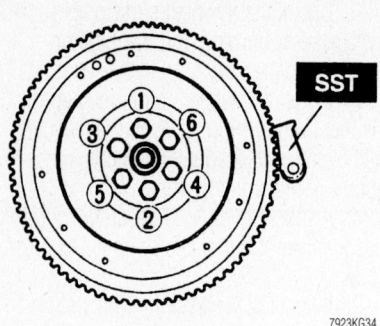

Flywheel tightening sequence

7923KG34

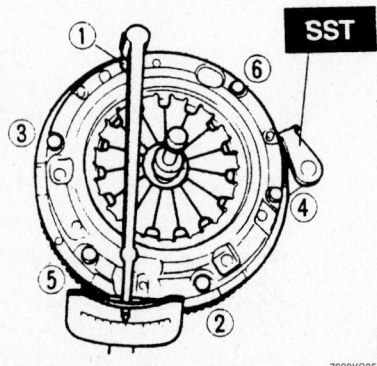

Pressure plate tightening sequence

7923KG35

To install:

10. Lubricate the release fork fingers and pivot with molybdenum grease and install in the release fork boot.

11. Install the clutch release bearing on the release fork.

12. Install a new pilot bearing in the flywheel, if removed.

13. Be sure the flywheel mounting surface and the crankshaft or eccentric shaft mounting surfaces are clean. Remove any old sealant from the flywheel bolt hole threads and the flywheel bolts.

14. Install the flywheel.

15. Apply sealant to the flywheel bolt threads and install them hand-tight. Install the flywheel holding tool. Tighten the bolts, in a crisscross pattern, to 71–76 ft. lbs. (96–103 Nm).

16. Apply a small amount of molybdenum grease to the clutch disc splines and install the clutch disc on the flywheel, spring side toward the transaxle. Install a suitable alignment tool in the pilot bearing to position the clutch disc.

17. Install the clutch pressure plate, aligning the dowel holes with the flywheel dowels.

18. Install the pressure plate bolts and gradually tighten, in a crisscross pattern to 20 ft. lbs. (26 Nm). Remove the alignment tool.

19. Install the transaxle.

Hydraulic Clutch System

BLEEDING

1. Before servicing the vehicle, refer to the Precautions Section.

2. If necessary, remove the gravel shield from the drivers side.

3. Remove the rubber cap from the bleeder screw on the release cylinder.

4. Place a bleeder tube over the end of the bleeder screw.

5. Submerge the other end of the tube in a jar half filled with hydraulic brake fluid.

6. Slowly pump the clutch pedal fully and allow it to return slowly, several times.

7. While pressing the clutch pedal to the floor, loosen the bleeder screw until the fluid starts to run out. Then, close the bleeder screw. Keep repeating this Step, while watching the hydraulic fluid in the jar. As soon as the air bubbles disappear, close the bleeder screw.

8. During the bleeding procedure the reservoir must be kept at least ¾ full.

Halfshaft

REMOVAL & INSTALLATION

1. Before servicing the vehicle, refer to the Precautions Section.

2. Remove the wheel and tire assemblies.

3. Remove the splash shield, if equipped.

4. Drain the transaxle.

5. Raise the staked portion of the hub locknut with a hammer and chisel. Lock the hub by applying the brakes and loosen the nut.

6. Separate the stabilizer bar from the lower control arm.

7. Remove the cotter pin and nut from the tie rod end ball stud.

8. Remove the tie rod end from the knuckle, using a suitable tool.

9. Remove the lower ball joint pinch bolt and nut.

10. Separate the ball joint from the knuckle, using a prybar on the control arm.

11. Position a prybar between the inner CV-joint and transaxle case. Carefully pry the halfshaft from the transaxle being careful not damage the oil seal. If equipped with a right side intermediate shaft, insert the pry-bar between the halfshaft and intermediate shaft and tap on the bar to uncouple them.

12. Remove the hub locknut and discard.

13. Pull outward on the hub/knuckle assembly, push the outer CV-joint stub shaft through the hub, and remove the half-shaft. If the halfshaft is stuck in the hub, install the old hub nut to protect the stub shaft threads. Tap on the nut, using only a soft mallet, to remove the halfshaft.

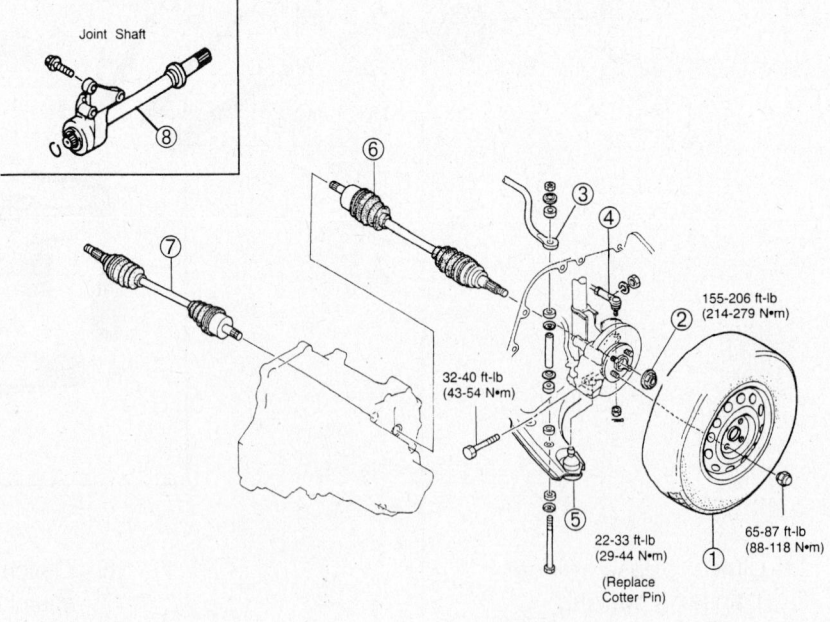

1. Wheel and tire
2. Locknut
3. Stabilizer bar
4. Tie-rod end
5. Ball joint
6. Left driveshaft
7. Right driveshaft
8. Joint shaft

7923KG36

Exploded view of the halfshafts and related components

➡**Install Differential Side Gear holder K49A-4208-AT, into the transaxle after removing the halfshaft, to keep the differential side gear in position. If the gear becomes out of position the differential may have to be removed to realign the gear.**

14. Remove the intermediate shaft, if necessary, by removing the support bearing bolts and pulling the shaft from the transaxle.

To install:

15. Install a new circlip on the end of the intermediate shaft,

    a. 1.5L, 1.6L (VIN 5), 1.8L install the circlip with the end gap facing upward.

    b. 1.6L (VIN 3), 2.0L, 2.4L and 2.7L install the circlip with end gap facing downward.

16. Install the intermediate shaft in the transaxle, being careful not to damage the oil seals.

17. Install the support bearing bolts and tighten, in sequence, to 45 ft. lbs. (61 Nm).

18. Install a new circlip on the end of the halfshaft, with the end gap facing upward.

19. Install the halfshaft into the transaxle, being careful not to damage the oil seal.

20. Install the halfshaft into the intermediate shaft, if equipped.

21. Install the other end of the halfshaft through the hub. Loosely install a new locknut.

22. Install the lower ball joint into the knuckle.

23. Install the pinch bolt and nut and tighten to 40 ft. lbs. (54 Nm).

24. Install the tie rod end to the steering knuckle and tighten the nut to 42 ft. lbs. (57 Nm). Install a new cotter pin. Tighten the nut, if necessary, to align the ball stud hole with the nut castellation.

25. Install the stabilizer bar to the lower control arm.

26. Install the splash shield and the wheel and tire assemblies.

27. Lock the hub with the brakes. Tighten the new hub nut to 155–206 ft. lbs. (214–279 Nm). After tightening, stake the locknut using a hammer and dull bladed chisel.

28. Fill the transaxle with the proper type and quantity of fluid.

CV-Joints

OVERHAUL

1.5, 1.6L (VIN 5) and 1.8L Models

1. Before servicing the vehicle, refer to the Precautions Section.

2. Remove the halfshaft.

3. Pry up the locking clip of the transaxle side boot retaining band with a suitable tool.

4. Using a pair of pliers, remove the retaining band.

5. Slide the boot away to access the CV-joint.

6. Matchmark the CV-joint housing and the shaft to ensure proper positioning during assembly.

7. Remove the outer ring.

8. Matchmark the shaft and tripod assembly to ensure proper positioning during assembly.

9. Remove the snapring.

➡**Be careful not to damage the needle bearings.**

10. Using a brass drift and a hammer, drive the tripod joint from the shaft.

➡**Cover the halfshaft serration's with tape, so as not to damage the boot.**

11. Slide the boot off the shaft.

➡**It is not necessary to remove the dynamic damper from the shaft unless replacement or repair is required.**

12. Pry up the locking clip of the dynamic damper retaining band with a suitable tool.

13. Using a pair of pliers, remove the retaining band.

14. Remove the dynamic damper.

➡**Do not remove the outer boot from the shaft unless it is necessary.**

15. Pry up the locking clip of the outer boots small and large retaining bands with a suitable tool.

16. Using a pair of pliers, remove the retaining bands.

17. Cover the halfshaft serration's with tape and remove the boot.

To install:

➡**The boot on the wheel side of the driveshaft is larger than the boot on the differential side.**

18. Cover the halfshaft serration's with tape and install the inner boot.

➡**The bands should be installed so that their pointed ends initially point in the forward direction of rotation.**

19. Install the dynamic damper and band. Fold the band back by pulling on the end of it with a pair of pliers, then lock the end of the band by bending the locking clip.

20. Install the outer boot.

21. Align the marks made during removal and install the tripod joint using a brass drift and a hammer.

22. Install the snapring.

23. Apply the grease supplied with the joint rebuild kit to the tripod joint, outer ring and the boot.

24. Install the outer ring.

25. If the outer boot was removed, fill it with the correct amount of grease as follows:

    a. Transaxle side: 4.94 oz. (140g).

    b. Wheel side: 4.58 oz. (130g).

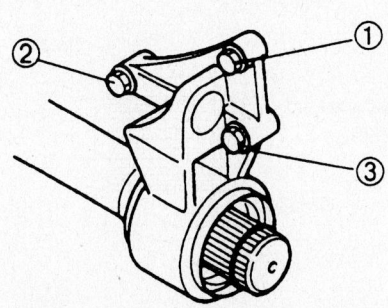

7923KG37

Support bearing bolt tightening sequence

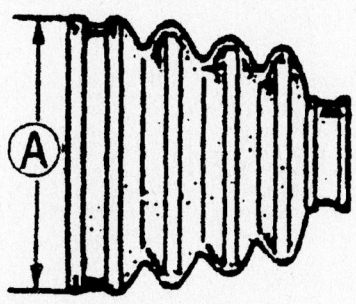

Transaxle End

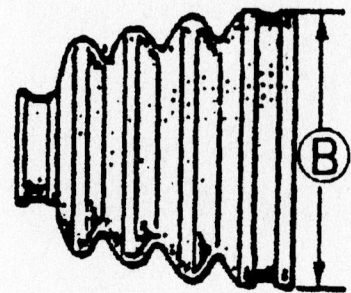

Wheel End

9307KG12

Measure the boots as shown to ensure the larger boot is placed on the wheel side of the halfshaft.

Length of driveshaft   in (mm)

Item	Right side		Left side	
	MTX	ATX	MTX	ATX
TED	24.93 (633.3)	25.54 (648.7)	25.15 (638.9)	25.16 (639)

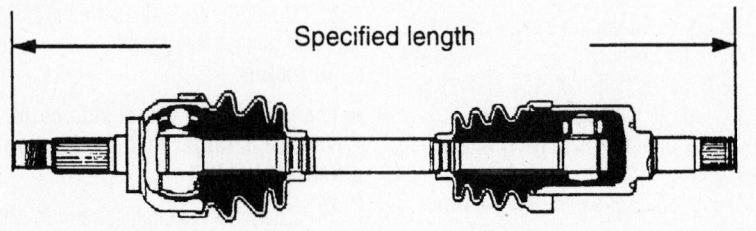

Specified length

9307KG13

Make sure the driveshaft's specified length is correct

26. Make sure the boots are not damaged, then carefully up the small end of the boots to release any trapped air.

27. Measure the length of the driveshaft to ensure it is the correct length. Refer to the accompanying illustration for the driveshaft measuring points and the correct specifications.

28. Install the boot retaining bands. Fold the bands back by pulling on the ends with a pair of pliers, then lock the end of the bands by bending the locking clips.

29. Install the halfshaft.

1.6L (VIN 3), 2.0L and 2.4L Models

1. Before servicing the vehicle, refer to the Precautions Section.

2. Remove the halfshaft and place into a vise.

3. Remove the circlip from the splines of the transaxle side Tripot Joint (TJ) case.

4. Remove both boot clamps from the transaxle side TJ case.

5. Pull off the outer TJ boot.

6. Matchmark the trunion assembly to the TJ case and halfshaft spline.

7. Remove the circlip from the splines of the spider assembly side.

8. Using Special Tool 09495-33000, remove the spider assembly from the halfshaft.

9. Remove the inner TJ boot.

10. Remove the dynamic damper clamps.

11. Separate the dynamic damper from the halfshaft, being careful not to damage the shaft splines.

12. Remove the Birfield Joint (BJ) boot clamp.

13. Remove the BJ being careful not to damage the boot.

To install:

➡**Wrap tape around the axle shaft splines to prevent damage to the boots during assembly.**

➡**Use new circlips and boot clamps for assembly.**

14. Apply grease to the halfshaft and install the BJ, boots and clamps.

15. Install the dynamic damper, keep the halfshaft straight and tighten the dynamic band.

16. Install the inner TJ boot and band.

17. Install the trunion assembly and circlip, aligning the matchmarks with each other.

18. Add the specified grease to the TJ, replaced all that was wiped away.

19. Install the outer TJ boot and clamp.

20. Install the halfshaft.

2.7L Manual Transmission Model

1. Before servicing the vehicle, refer to the Precautions Section.

2. Remove the halfshaft and place into a vise.

3. Remove the Double Offset Joint (DOJ) boot bands.

4. Pull the DOJ boot from the outer race.

5. Remove the circlip from the DOJ and pull the halfshaft from the outer race.

6. Remove the snap ring and take out the inner race assembly.

7. Remove the Birfield Joint (BJ) boot bands.

8. Pull out the DOJ boot and BJ boot.

➡**If the boots are to be reused, wrap tape around the halfshaft splines.**

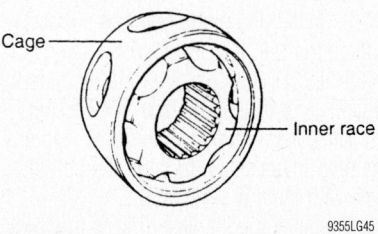

Cage — Inner race

9355LG45

Install the cage so that it is offset on the race

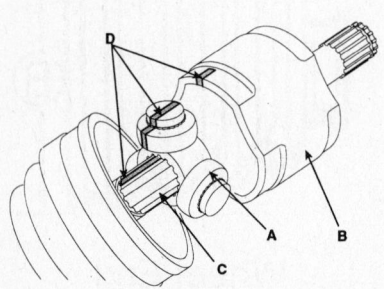

09474_TUCS_G47

Mark the joint assembly as shown—TJ-BJ type

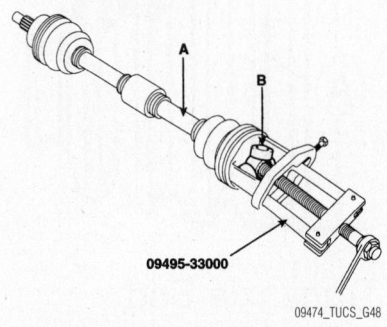

09495-33000

09474_TUCS_G48

Using special tool 09495-33000 to remove the spider assembly

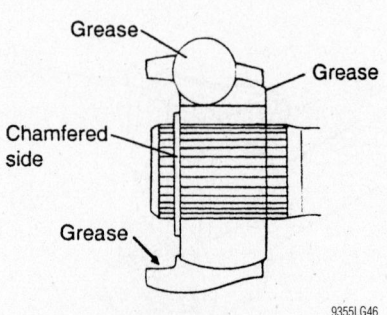

Grease — Grease — Chamfered side — Grease

9355LG46

Install the chamfered side of the cage as shown

To install:

➡**Use new circlips and boot clamps for assembly.**

9. Apply some boot grease to the half-shaft and install the boots.

10. Apply the specified kit grease to the inner race and cage.

11. Install the cage so that it is offset on the inner race as shown.

12. Apply the specified grease to the cage and fit the balls into the cage.

13. Position the chamfered side as shown. Install the inner race on the halfshaft and install the snap ring.

14. Apply the specified grease to the BJ outer race and install the outer race onto the halfshaft.

15. Apply the specified grease into the DOJ boot and install the boot onto the halfshaft.

16. Tighten the DOJ boot bands.

17. Add the specified grease to the BJ to replace what was wiped away.

18. Install the boots and tighten the BJ boot bands.

19. Install the halfshaft.

2.7L Automatic Transmission Model

1. Before servicing the vehicle, refer to the Precautions Section.

2. Remove the halfshaft and place into a vise.

3. Cut to remove the Glaenzer Interior (GI) and Angular Contact (AC) big clamps.

4. Using a brass drift and hammer, remove the GI and AC small clamps.

5. Matchmark the tripod and tulip assemblies to the halfshaft.

6. Remove the tulip assembly from the halfshaft.

7. Using a snap ring expander, remove the circlip from the tripod.

8. Matchmark the tripod and joint shaft.

9. Using a brass drift and hammer, remove the tripod from the joint shaft.

10. Place tape over the splines of the shaft (GI side) and remove the AC boot clamps.

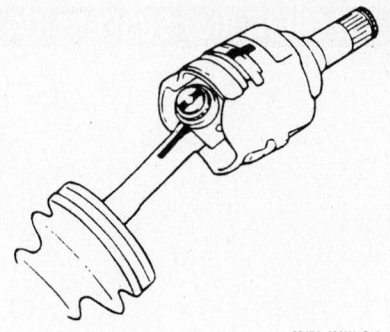

Matchmarks on the tripod and tulip assemblies

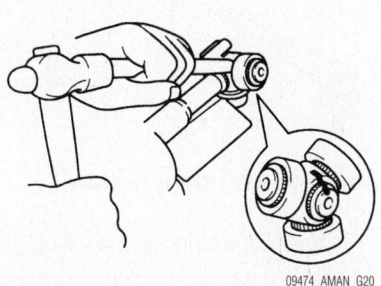

Matchmark the tripod and joint shaft

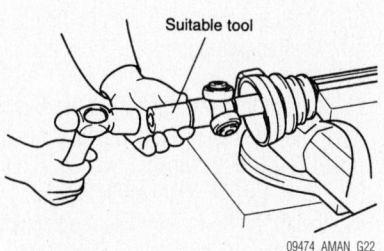

Tap the tripod onto the halfshaft

11. Pull out the GI and AC boots.

To install:

12. Ensure there is tape over the splines of the shaft to prevent damage to the boots when installing.

13. Using new boots, place new clamps to the small boot ends and install them onto the halfshaft.

14. Align the matchmarks and place the beveled side of the tripod axial splines

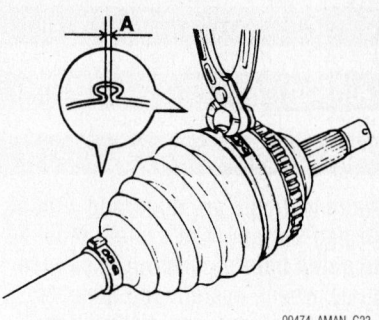

Closing the large boot clamps to the specified distance

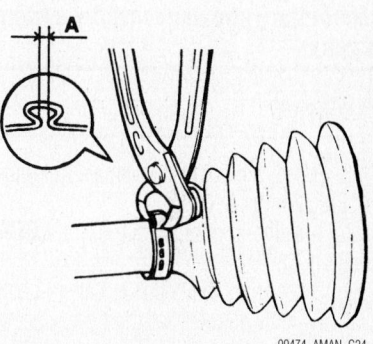

Closing the small boot clamps to the specified distance

toward the AC joint. Using a suitable tool and hammer, tap the tripod onto the half-shaft.

15. Install a new circlip.

16. Fill the AC and GI boot with the specified grease to replace any that was removed during disassembly.

17. Align the matchmarks and install the GI joint to the shaft.

18. Install the boot clamps, ensuring that the boots are properly seated on the half-shaft grooves.

19. Secure the large boot clamps to the specified distance of .079 inches (2 mm) as shown.

20. Secure the small clamps to the specified distance of 0.071 inches (1.8 mm) as shown.

21. Install the halfshaft into the vehicle.

STEERING AND SUSPENSION

Air Bag

✳✳ CAUTION

Some vehicles are equipped with an air bag system. The system must be disabled before performing service on or around system components, steering column, instrument panel components, wiring and sensors. Failure to follow safety and disabling procedures could result in accidental air bag deployment, possible personal injury and unnecessary system repairs.

DISARMING

1. Before servicing the vehicle, refer to the Precautions Section.
2. Turn the ignition switch to the **LOCK** position.
3. Disconnect the negative battery cable.
4. Wait 10 minutes for the battery back-up power to discharge.

ARMING

1. Before servicing the vehicle, refer to the Precautions Section.
2. Connect the negative battery cable.
3. Turn the ignition switch **ON**.
4. Verify that the air bag indicator illuminates for 4–8 seconds, then goes off.

Manual Rack and Pinion Steering Gear

REMOVAL & INSTALLATION

Rio

1. Before servicing the vehicle, refer to the Precautions Section.
2. Disconnect the negative battery cable.
3. Remove the front wheels.
4. Remove the cotter pins from both steering tie rod ends and the nuts.
5. Remove the tie rod out of the knuckle arm, using a suitable tool.
6. Remove the set plate from the firewall.
7. Remove the fixing bolt from the steering shaft to steering gear pinion shaft.
8. Remove the steering shaft from the steering gear.
9. Remove the steering gear mounting nuts.

7923KG39

Rack mounting tightening sequence

10. Move the steering gear to the right of the vehicle.
    To install:
11. Install the steering gear to the vehicle.
12. Install the mounting nuts in the order shown. Tighten the nuts to 23–34 ft. lbs. (31–46 Nm).
13. Install the steering shaft to the steering gear pinion shaft. Tighten the bolt/nut to 13–20 ft. lbs. (18–27Nm).
14. Install the set plate to the firewall.
15. Install the tie rod ends to the knuckle arm and tighten the nuts to 25–29 ft. lbs. (34–39 Nm)
16. Install new cotter pins.
17. Install the wheels.
18. Connect the negative battery cable.
19. Check the front end alignment.

Power Rack and Pinion Steering Gear

REMOVAL & INSTALLATION

1.5L and 1.6L (VIN 5) Models

1. Before servicing the vehicle, refer to the Precautions Section.
2. Drain the power steering fluid.
3. Remove or disconnect the following:
   - Negative battery cable
   - Wheel
   - Tie rod from the wheel knuckle
   - Catalytic converter
   - Power steering gear hoses
   - Steering intermediate shaft

   - Steering gear mounting bolts
4. Slide the steering gear to the left and pull the right tie rod through the fender opening.
5. Remove the steering gear by sliding it to the right.
   To install:
6. Position the steering gear into its mounting location.
7. Attach the steering intermediate shaft to the pinion and tighten the mounting bolt to 13–20 ft. lbs. (18–26 Nm).
8. Connect the power steering gear hoses. Tighten the bolts to 29–43 ft. lbs. (39–59 Nm).
9. Tighten the steering gear mounting bolts to 27–38 ft. lbs. (37–52 Nm).
10. Install or connect the following:
    - Catalytic converter
    - Tie rod end to the wheel knuckle
    - Wheel
    - Negative battery cable
11. Fill the power steering fluid reservoir to the correct level.

1.6L (VIN 3) Models

1. Before servicing the vehicle, refer to the Precautions Section.
2. Disconnect the negative battery cable.
3. Drain the power steering reservoir by disconnecting the return hose.
4. Remove the pressure hose from the power steering pump.
5. Disconnect the steering shaft by removing the universal joint mounting bolt.
6. Remove or disconnect the following:
   - Front wheels

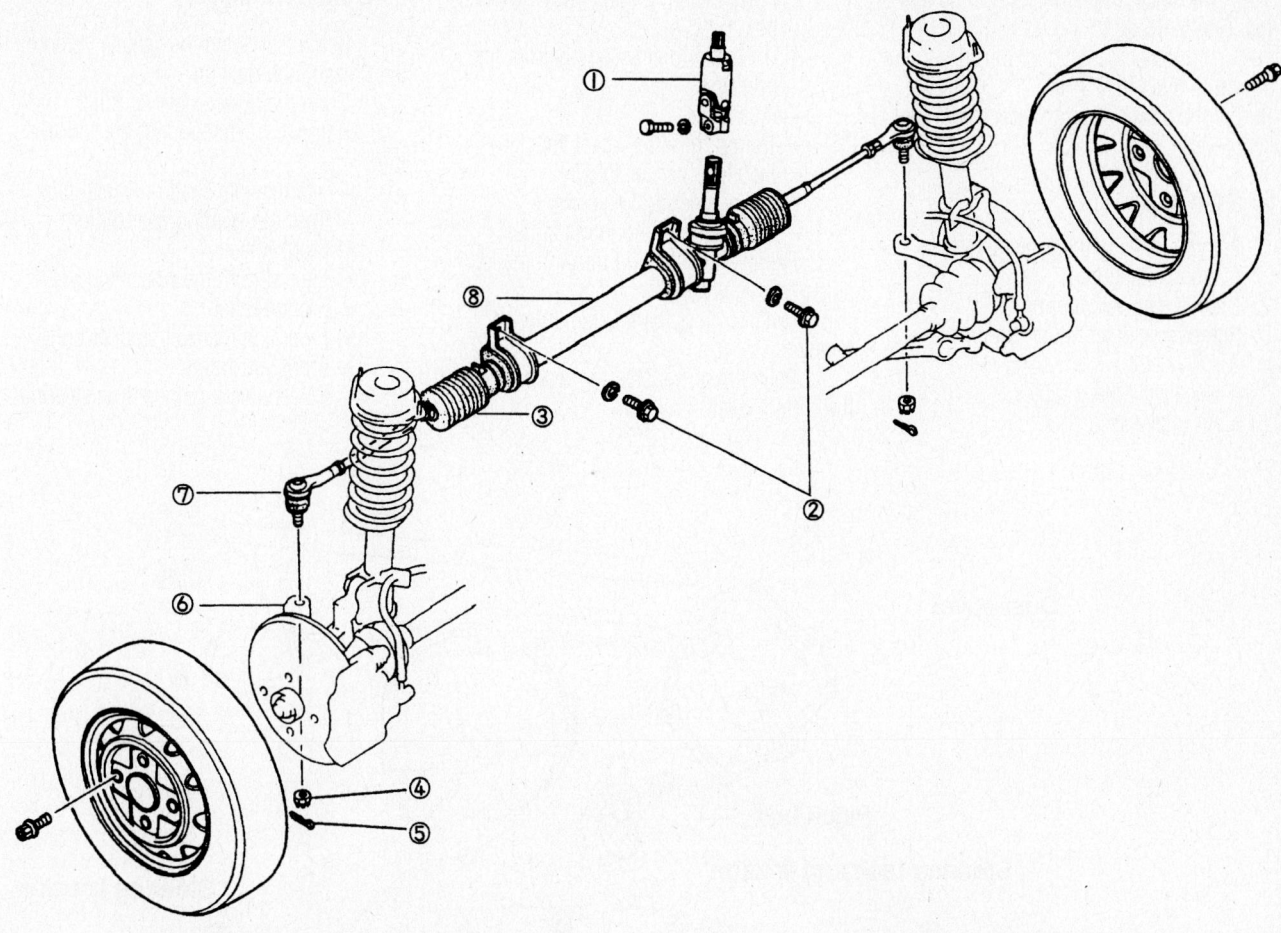

(1) Steering column intermediate shaft
(2) Steering gear mounting bolt
(3) Tie-rod boot
(4) Tie-rod nut

(5) Cotter pin
(6) Steering knuckle
(7) Tie-rod end
(8) Steering gear & Linkage

9357NG52

Power steering gear mounting—1.5L and 1.6L (VIN 5)

- Lower arm mounting bolts from the wheel knuckle assembly
- Tie rod end from the knuckle
- Intermediate shaft universal joint from the steering gear pinion shaft
- Stabilizer link from the strut assembly
- Engine mounting bolts from the sub-frame.
- Front sub-frame
- Steering gear heat cover
- Engine mounting bracket bolts
- Power steering lines from the steering gear
- Steering gear mounting bolts

7. Installation is the reverse order of removal. Note the following torques:

    a. Steering gear mounting bolts: 65–79 ft. lbs. (90–110 Nm).

    b. Heat cover: 6–9 ft. lbs. (8–12 Nm).

    c. Power steering pressure hose to power steering pump: 40–47 ft. lbs. (55–65 Nm).

8. Fill the power steering reservoir to the correct level.

1.8L Models

1. Before servicing the vehicle, refer to the Precautions Section.
2. Drain the power steering fluid.
3. Remove or disconnect the following:

- Negative battery cable
- Front wheels
- Tie rod from the wheel knuckle
- Power steering line brackets
- Power steering lines from steering gear
- Intermediate shaft from the steering gear
- Steering rack mounting nuts

- Change control rod and extension bar from transaxle
- Catalytic converter
- Cross-member
- No. 1 engine mounting bolts

4. Support the sub-frame with a suitable jack.

5. Remove the sub-frame mounting bolts.

6. Remove the steering gear and rack assembly from the right side of the vehicle.

7. Installation is the reverse order of removal. Note the following torques:

    a. Steering rack mounting nuts: 27–38 ft. lbs. (37–52 Nm).

    b. Sub-frame mounting bolts: 69–86 ft. lbs. (93–117 Nm).

    c. No. 1 engine mounting bolts: 49–69 ft. lbs. (67–93 Nm).

    d. Cross-member mounting bolts: 48–65 ft. lbs. (64–89 Nm).

e. Transaxle extension bar: 28–38 ft. lbs. (38–51 Nm).

f. Transaxle change control rod: 12–16 ft. lbs. (16–22 Nm).

8. Fill the power steering reservoir to the correct level.

2.0L Models

1. Before servicing the vehicle, refer to the Precautions Section.

2. Drain the power steering fluid.

3. Remove or disconnect the following:

- Negative battery cable
- Air intake assembly

- Power steering lines from the gear box
- Universal joint assembly from the gear box
- Front tires
- Tie rod from the wheel knuckle
- Stabilizer bar dust cover
- Power steering line clamps
- Steering gear mounting bolts.

4. Remove the steering through the left side of the vehicle.

5. Installation is the reverse order of removal.

6. Refill the power steering reservoir to the correct level.

2.4L and 2.7L Models

1. Before servicing the vehicle, refer to the Precautions Section.

2. Drain the power steering fluid.

3. Remove or disconnect the following:

- Intermediate shaft universal joint from the steering gear pinion
- Front wheels
- Tie rod from the wheel knuckle
- Front muffler
- Front and rear roll stopper bolts
- Front sub-frame
- Power steering lines from steering gear

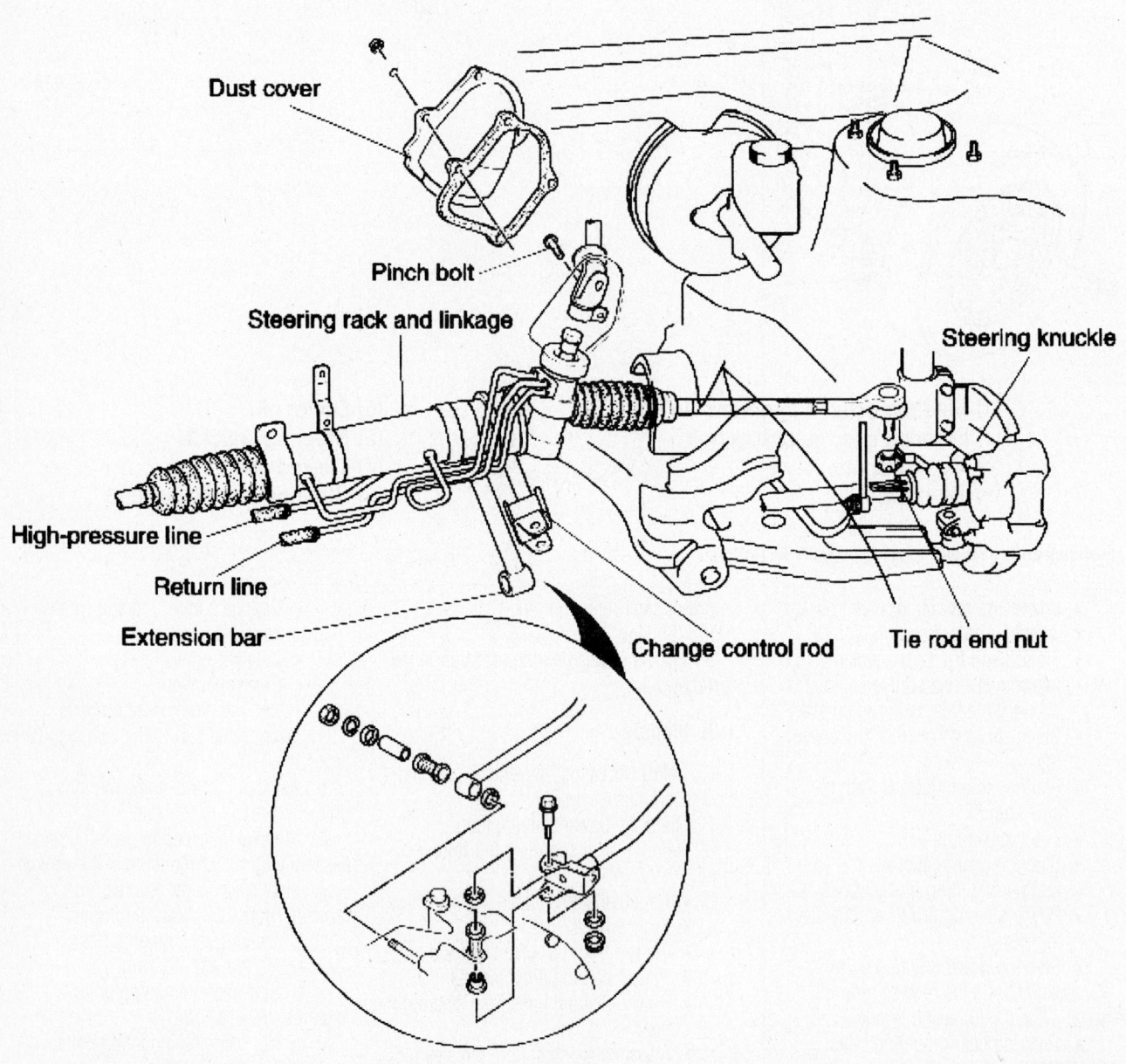

Exploded view of the steering rack assembly—1.8L models

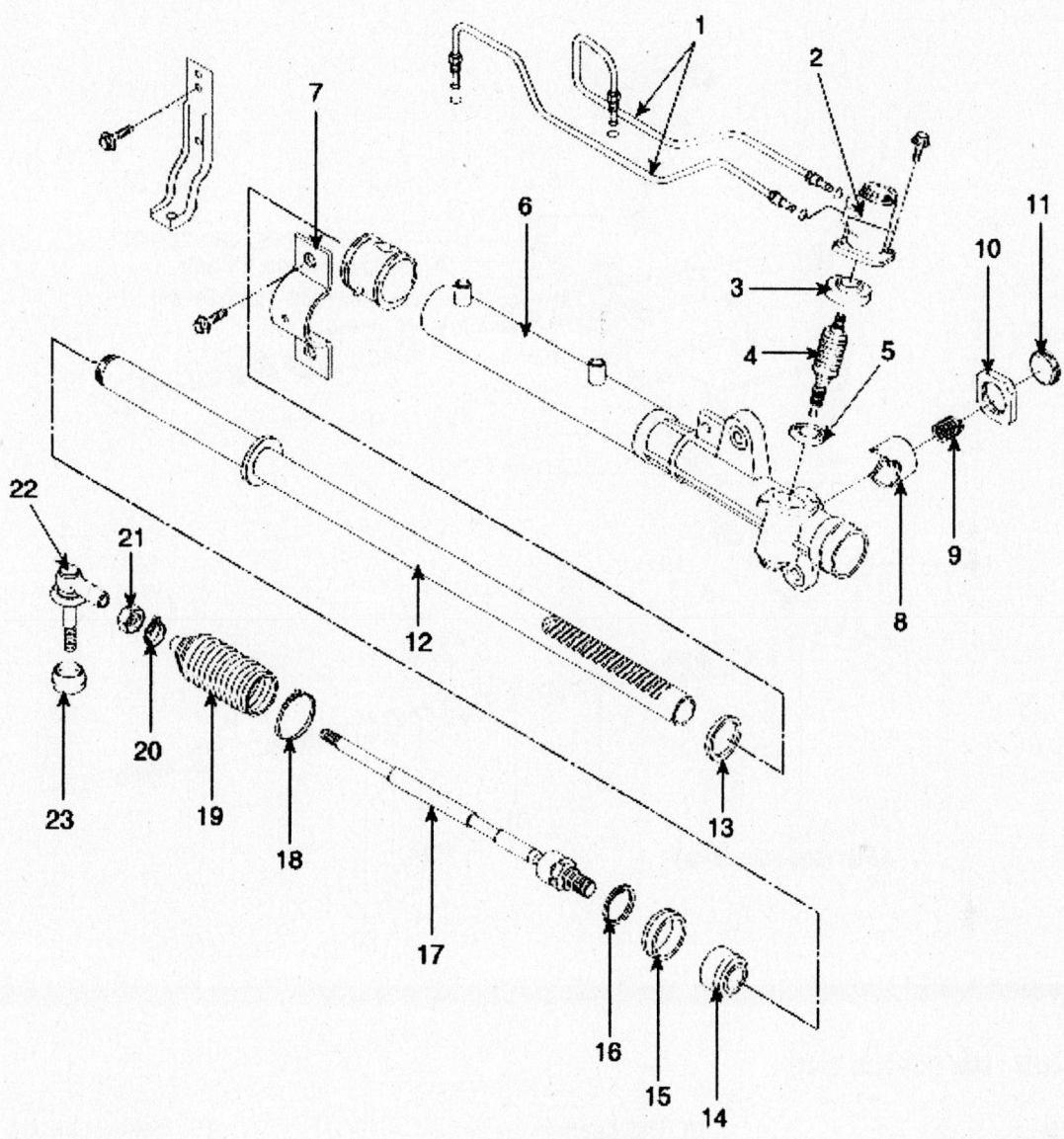

1. Feed tube
2. Valve body
3. Oil seal
4. Pinion valve assembly
5. Oil seal
6. Rack housing
7. Power steering gear box mounting clamp
8. Rack support yoke
9. Rack support spring
10. Lock nut
11. Yoke plug
12. Rack
13. Oil seal
14. Rack stopper
15. Oil seal
16. Clip
17. Tie rod
18. Bellows band
19. Bellows
20. Bellows clip
21. Lock nut
22. Tie rod end
23. Dust cover

09474_KIAC_G0015

Exploded view of steering rack assembly—2.0L models

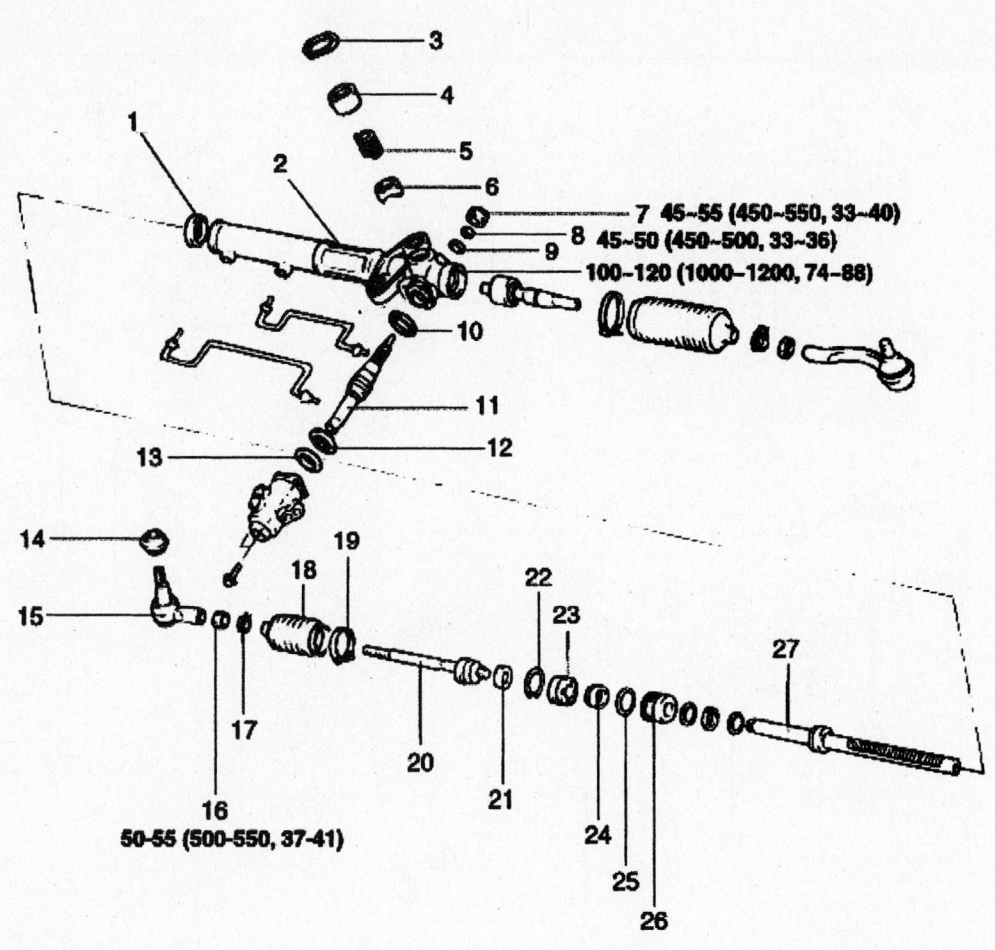

TORQUE : Nm (kg-cm, lb-ft)

1. Oil seal
2. Gear box
3. Lock nut
4. Rack support plug
5. Rack support spring
6. Rack support
7. End plug
8. Self-locking nut
9. Ball bearing

10. Ball bearing
11. Pinion and valve assembly
12. Roller bearing
13. Oil seal
14. Dust covers
15. Tie rod ends
16. Tie rod end lock nuts
17. Bellows clips
18. Bellows

19. Bellows bands
20. Tie rods
21. Tab washer
22. Circlip
23. Rack stopper
24. Oil seal
25. O-ring
26. Rack bushing assembly
27. Rack

09474_KIAC_G0016

Exploded view of the steering assembly—2.4L and 2.7L models

• Steering gear mounting brackets

4. Move the rack completely to the right and then remove the gear box to the left from the sub-frame.

5. Installation is the reverse order of removal.

Strut

REMOVAL & INSTALLATION

Front

1. Before servicing the vehicle, refer to the Precautions Section.

2. Remove the wheel and tire assembly.

3. Support the lower control arm with a jack.

4. Remove the bolts or clips attaching the brake hose and/or Anti-lock Brake System (ABS) sensor harness to the strut.

5. Remove the stabilizer control link from the bracket mounted to the strut, Rio only.

6. Paint alignment marks on the upper strut mounting block and strut tower, and on the lower strut mount-to-steering knuckle so the strut can be reinstalled in the same position.

7. Remove the lower strut-to-knuckle bolts.

8. Remove the upper strut mounting block nuts. There may be 2 or 3 nuts, depending on model.

9. Remove the strut assembly.

To install:

10. Install the strut into the strut tower, aligning the paint marks made during removal.

11. Install the strut mounting nuts. Tighten to 17–22 ft. lbs. (23–29 Nm) for Spectra, to 34–46 ft. lbs. (43–61 Nm) for Rio or to 30–36 ft. lbs. (40–50 Nm) for Optima.

12. Install the strut-to-knuckle bolts and tighten to 69–86 ft. lbs. (93–116 Nm) for

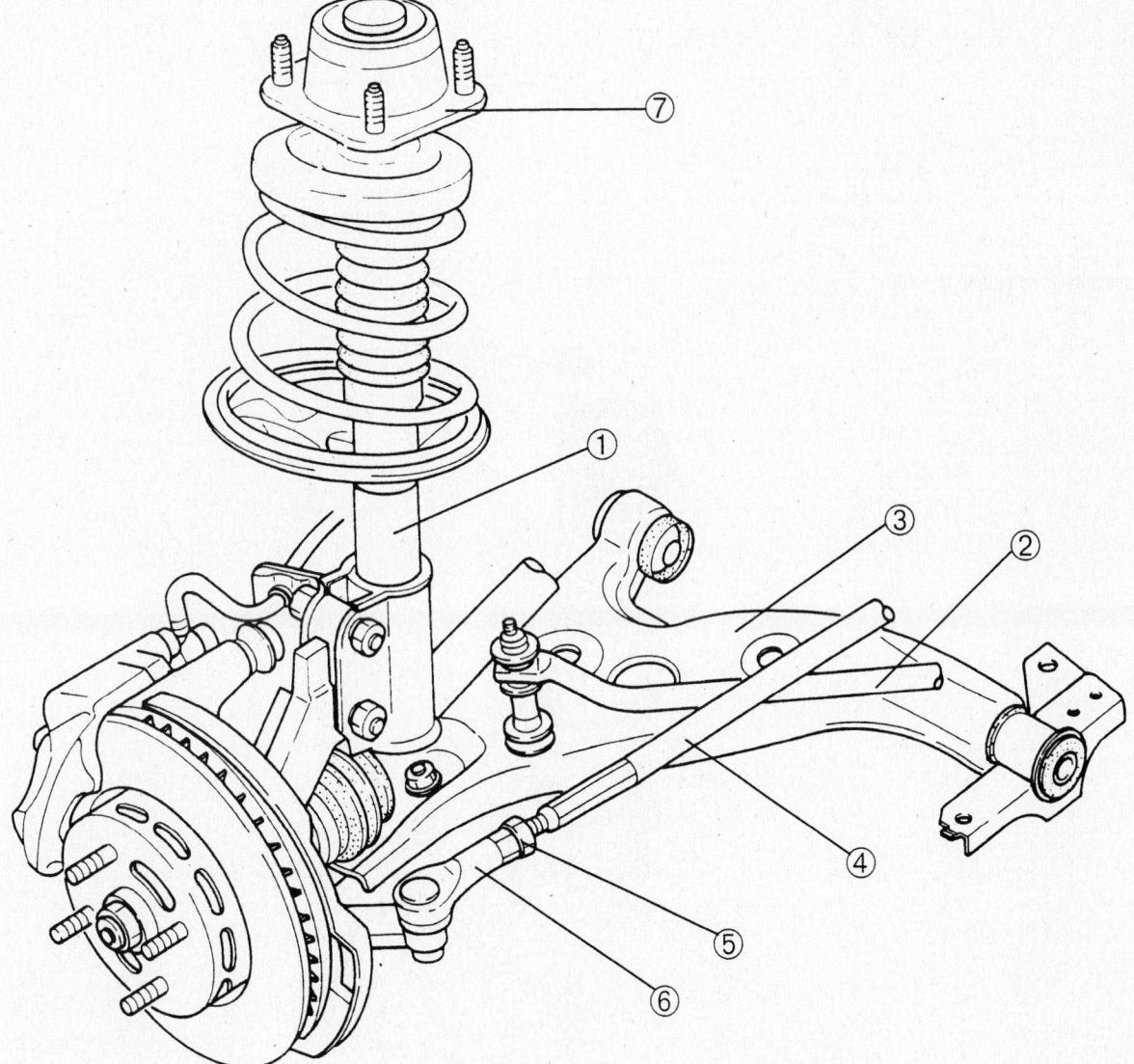

1. Front strut
2. Front stabilizer
3. Lower control arm
4. Tie-rod
5. Jam nut
6. Tie-rod end
7. Mounting block

7923KG41

Front suspension component identification

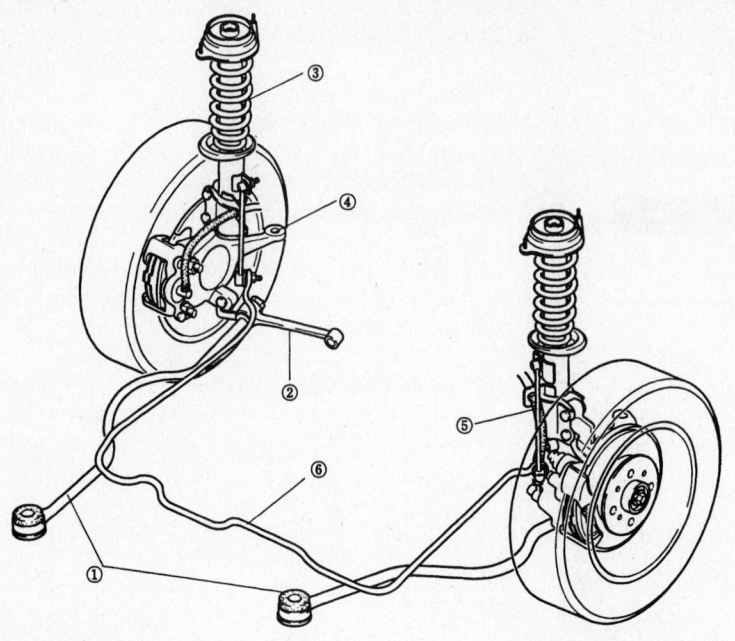

(1) Tension rod
(2) Control arm (Lower arm)
(3) Front strut

(4) Steering knuckle
(5) Stabilizer control link
(6) Stabilizer bar

9357NG17

Front suspension components—Rio

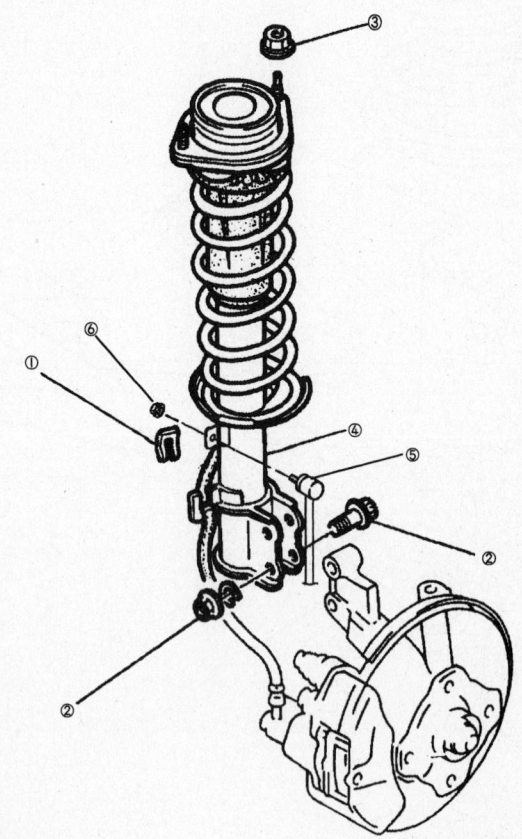

(1) Hose clip
(2) Bolt and nut
(3) Mounting block nut

(4) MacPherson strut
(5) Stabilizer control link
(6) Stabilizer control link nut

9357NG18

Exploded view of the front strut mounting—Rio

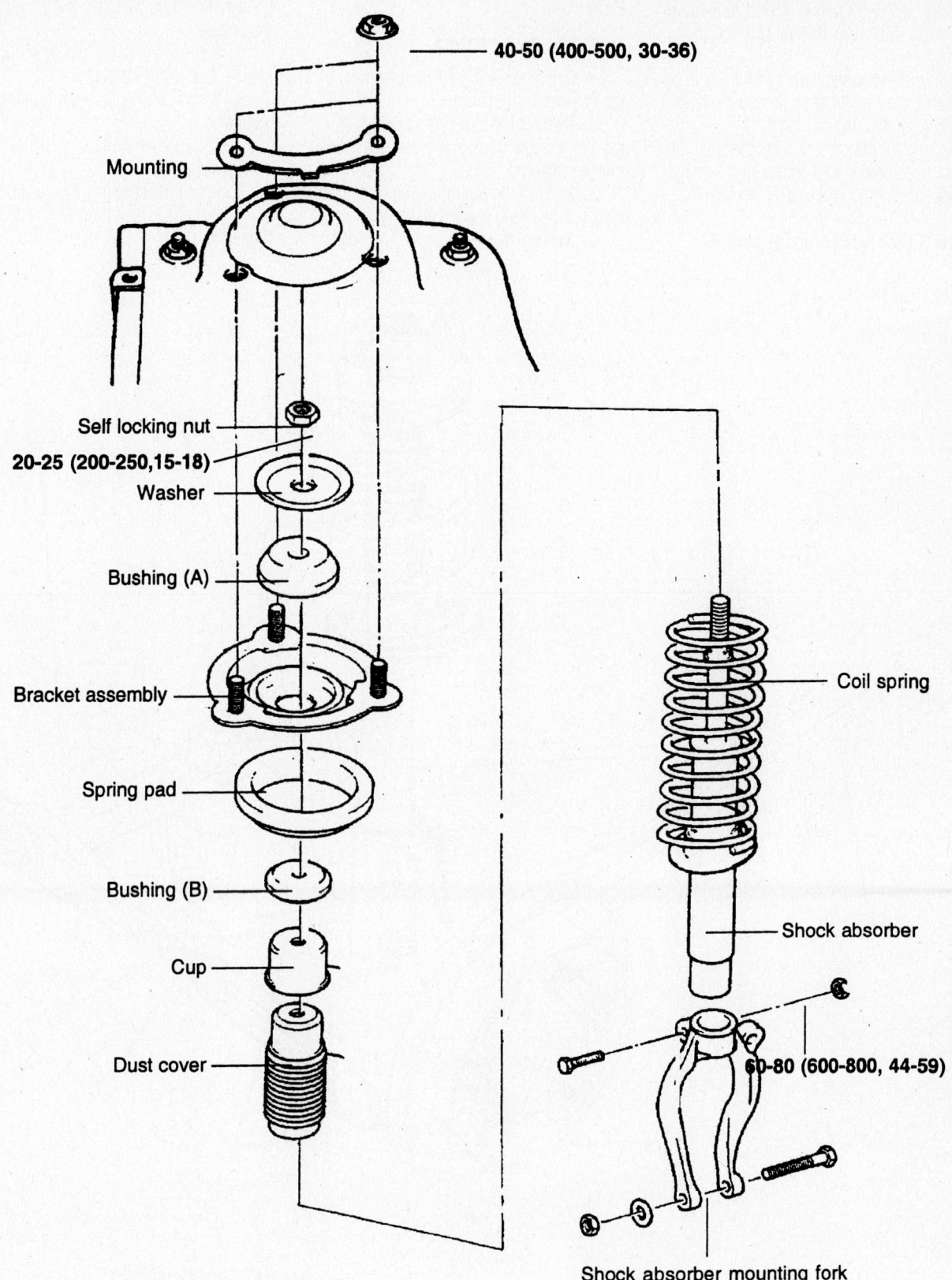

40-50 (400-500, 30-36)

Mounting

Self locking nut

20-25 (200-250,15-18)

Washer

Bushing (A)

Bracket assembly

Spring pad

Bushing (B)

Cup

Dust cover

Coil spring

Shock absorber

60-80 (600-800, 44-59)

Shock absorber mounting fork

TORQUE : Nm (kg·cm, lb·ft)

9357NG49

Exploded view of the front strut mounting—Optima

Spectra, to 76–90 ft. lbs. (103–123 Nm) for Rio or 44–59 ft. lbs. (60–80 Nm) for Optima.

13. Install the stabilizer control link to the strut mounted bracket. Torque the bolts to 32–45 ft. lbs. (43–61 Nm).

14. Install the clips or bolts attaching the brake hose and/or ABS sensor harness.

15. Install the wheel and tire assembly.

16. Check the front end alignment.

Rear

SPECTRA AND OPTIMA

1. Before servicing the vehicle, refer to the Precautions Section.

2. Remove the side trim panels from the inside of the trunk or the rear seat and trim, as required.

3. Remove the top mounting nuts from the strut mounting block assembly.

4. Remove the rear wheels. The suspension will drop when the weight lifts off the wheels.

5. Remove the brake line or wiring retainers as required.

6. Remove the bottom strut mount retainers.

7. Remove the strut.

To install:

8. Install the strut into the strut tower.

9. Install the strut mounting nuts and tighten to 17–22 ft. lbs. (23–29 Nm) for

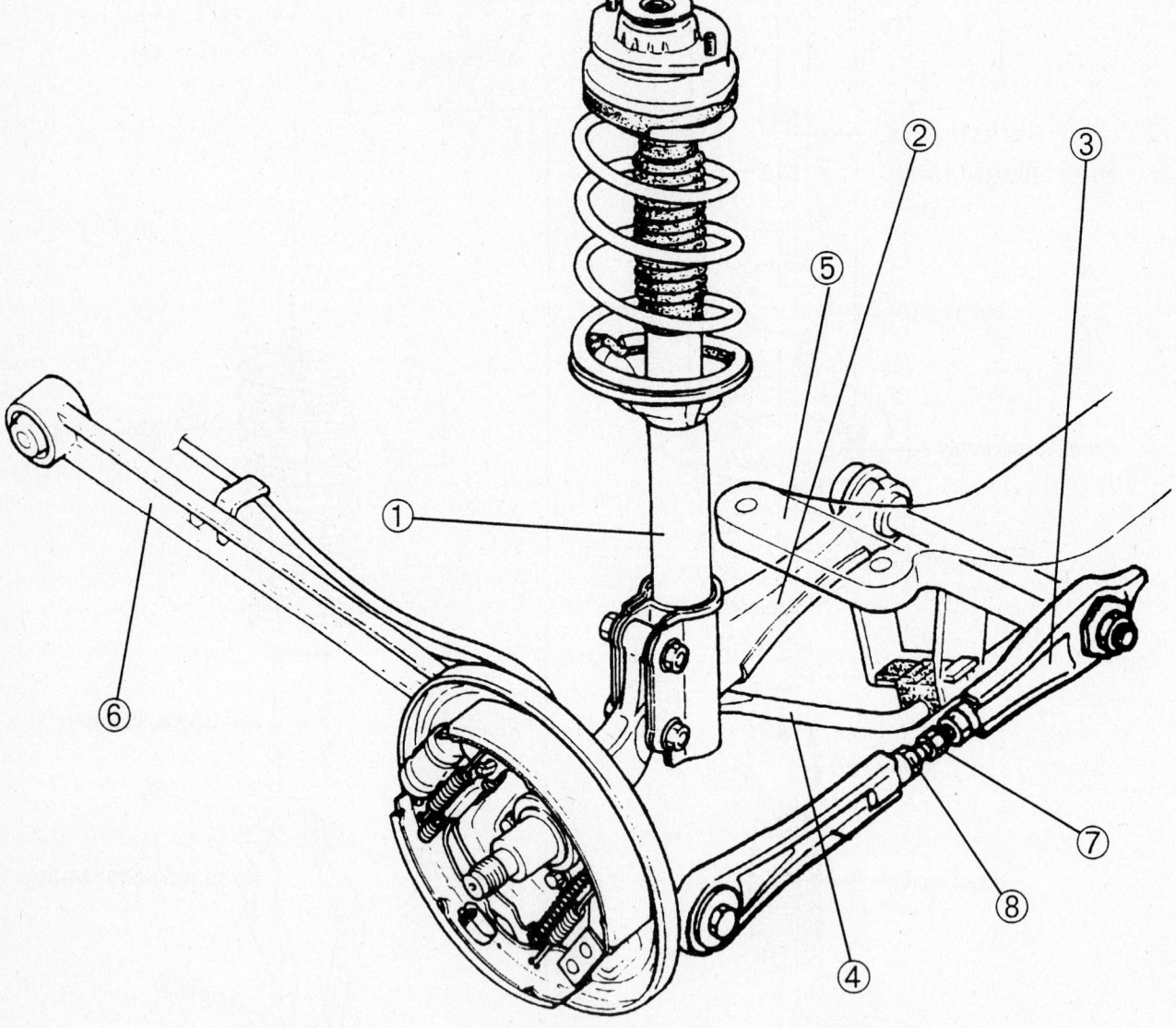

1. Rear strut
2. Front lateral link
3. Rear lateral link
4. Rear stabilizer
5. Rear crossmember
6. Trailing link
7. Adjuster
8. Jam nuts

7923KG42

Rear suspension component identification—Spectra

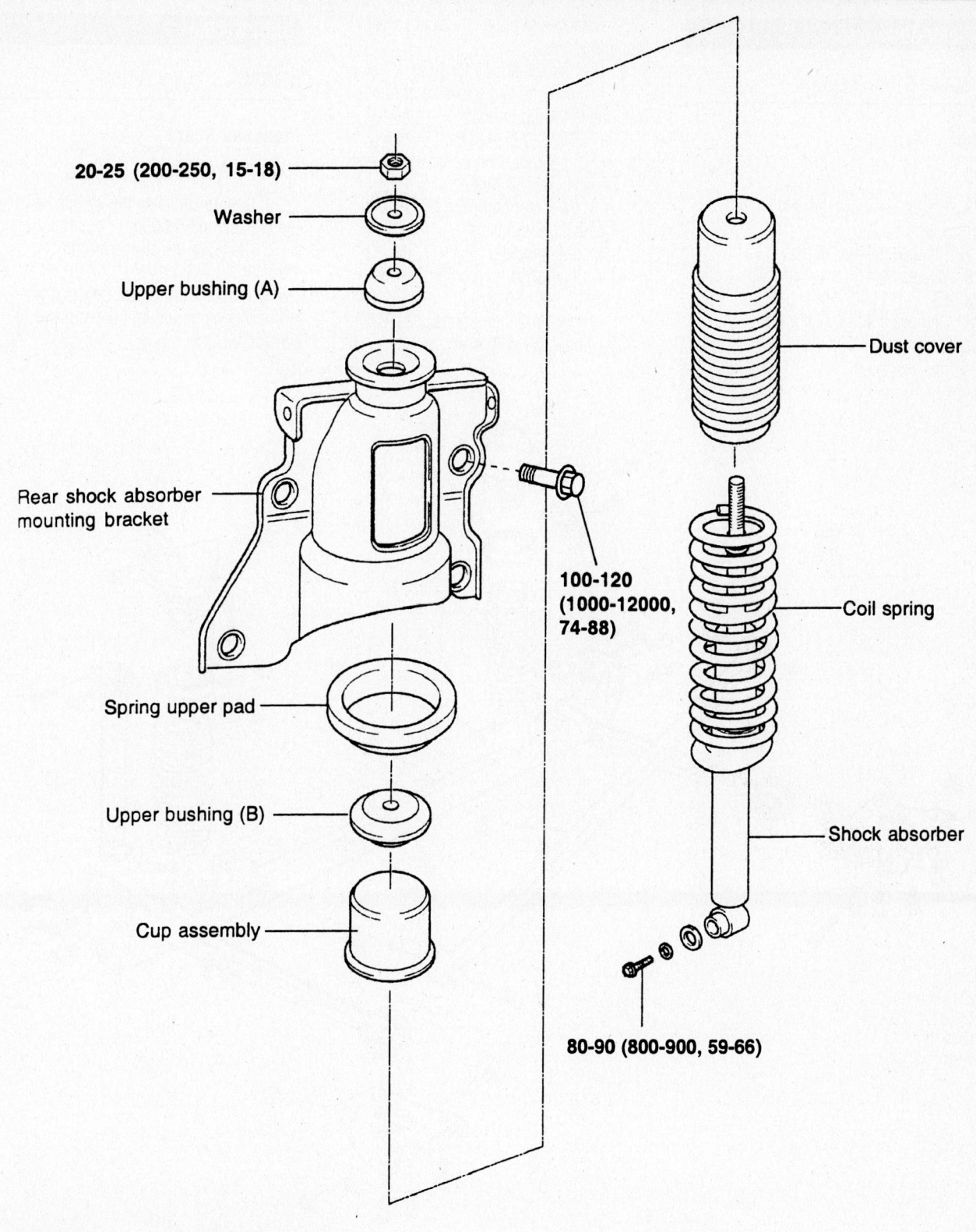

20-25 (200-250, 15-18)

Washer

Upper bushing (A)

Rear shock absorber
mounting bracket

100-120
(1000-12000,
74-88)

Spring upper pad

Upper bushing (B)

Cup assembly

Dust cover

Coil spring

Shock absorber

80-90 (800-900, 59-66)

TORQUE : Nm (kg·cm, lb·ft)

9357NG50

Exploded view of the rear strut—Optima

Spectra, or to 15–18 ft. lbs. (20–25 Nm) for Optima.

10. Install the strut-to-knuckle bolts and tighten to 69–86 ft. lbs. (93–116 Nm) for

Spectra or to 59–66 ft. lbs. (80–90 Nm) for Optima.

11. Install the clips or bolts attaching the brake hose and/or ABS sensor harness.

12. Install the wheel and tire assembly.
13. Install the side trim panels from the inside of the trunk or the rear seat and trim, if removed.

Shock Absorber

REMOVAL & INSTALLATION

Rear

RIO

1. Before servicing the vehicle, refer to the Precautions Section.
2. Remove the rear wheel.
3. Support the torsion beam with a suitable jack.
4. Remove the lower shock absorber mounting bolts.

5. Remove the upper shock absorber mounting bolts.
6. Remove the shock absorber.
7. Installations is the reverse of removal. Note the following torques:
   a. 2002–2005 models:
   - Upper mounting bolts: 34–42 ft. lbs. (46–57 Nm)
   - Lower mounting bolts: 58–72 ft. lbs. (78–98 Nm)
   b. 2006 models:
   - Upper mounting bolts: 28–43 ft. lbs. (40–60 Nm).
   - Lower mounting bolts: 72–86 ft. lbs. (100–120 Nm).

Coil Spring

REMOVAL & INSTALLATION

Front and Rear

EXCEPT RIO—REAR SPRING

1. Before servicing the vehicle, refer to the Precautions Section.
2. Remove the strut from the vehicle.
3. Install the strut securely in a vise with either aluminum or copper plates to protect the strut.

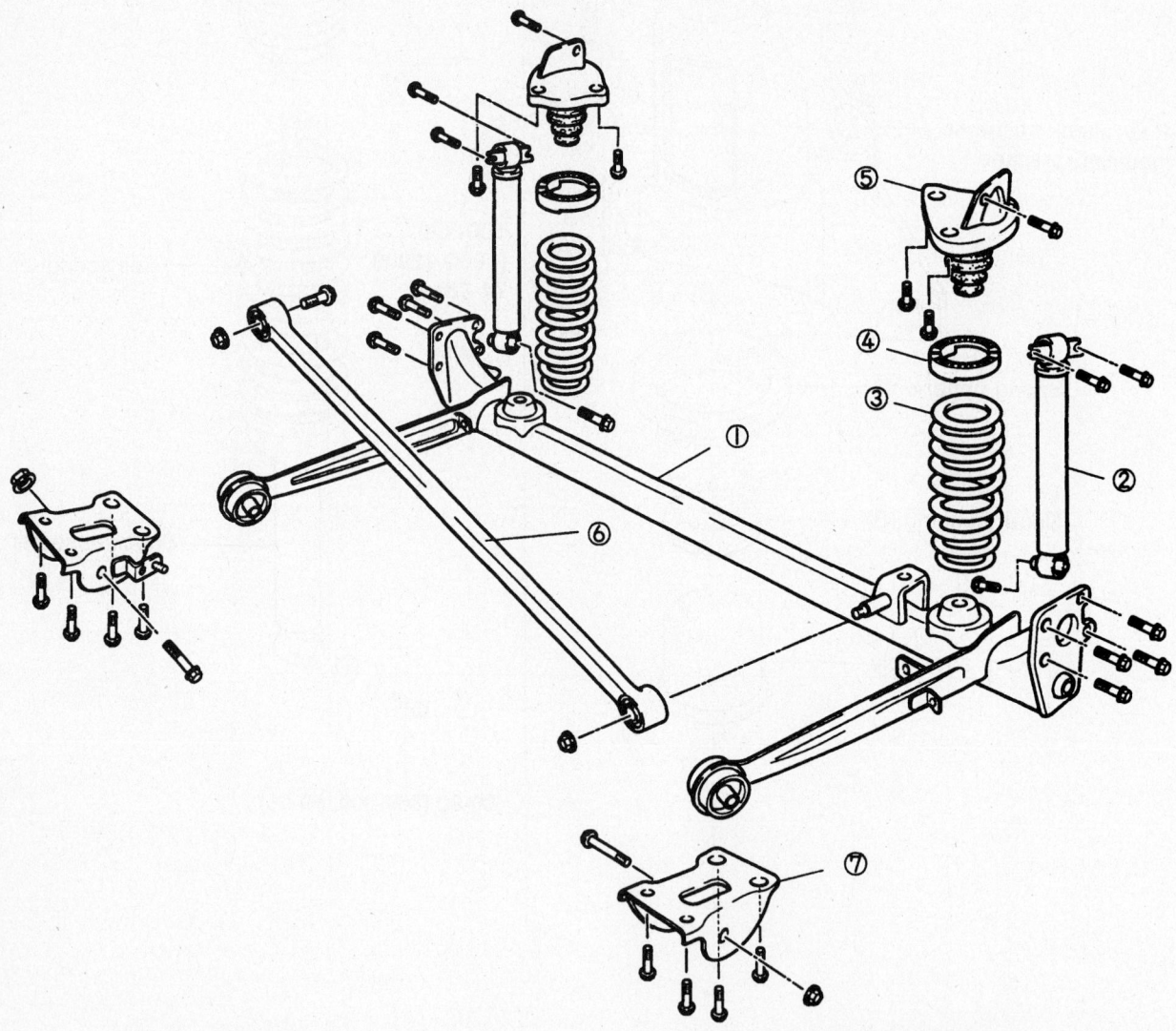

(1) Torsion Beam
(2) Shock absorber
(3) Coil spring
(4) Rubber seat
(5) Stopper assembly
(6) Lateral link
(7) Trailing bracket

9357NG19

Exploded view of the rear suspension components—2002–2005 Rio

4. Loosen the piston rod upper nut several turns but DO NOT REMOVE IT.

5. Install the lower end of the strut in the vise and install a coil spring compressor. Compress the coil spring and remove the upper nut.

✳✳ CAUTION

Failure to fully compress the spring and hold it securely can be extremely dangerous.

6. Slowly release the coil spring tension.

7. Remove the suspension support, dust seal, spring seat, spring insulators, coil spring and bumper.

8. While pushing on the piston rod, be sure that the pull stroke is even and that there is no unusual noise or resistance. Also inspect for any oil leakage around the piston rod.

9. Push the piston rod in, then release it. Be sure that the return rate is constant.

10. If the shock absorber does not operate as described, replace it.

To assemble:

11. Install the strut assembly into a vise.

12. Install the bound stopper and dust boot onto the piston rod.

13. Install the coil spring and compress the coil spring with the spring compressor.

14. Install the rubber seat, the spring upper seat, the bearing and the mounting block. Be sure that the spring upper seat notched portion is facing inward and tighten the piston rod upper nut.

15. Remove the spring compressor from the strut. Secure the upper mounting block in the vise. Tighten the nut to 41–50 ft. lbs. (55–68 Nm) for the front strut and 47–59 ft. lbs. (64–80 Nm) for the rear strut.

16. Be sure that the spring is well seated in the upper seats.

17. Install the strut to the vehicle.

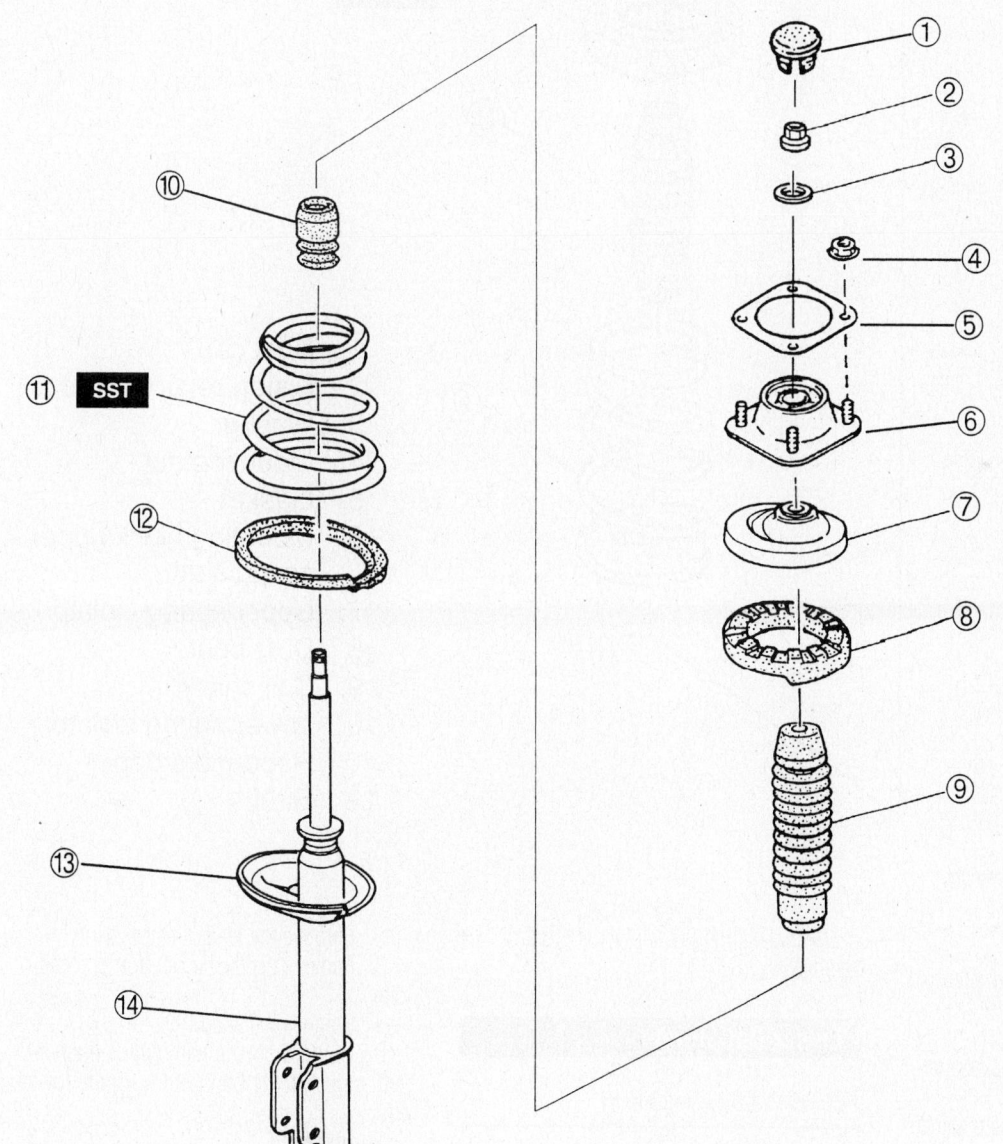

1. Dust cap
2. Piston retaining nut
3. Washer
4. Mounting nut
5. Gasket
6. Mounting block
7. Upper spring seat
8. Upper spring isolator
9. Dust boot
10. Rebound stopper
11. Coil spring
12. Lower spring isolator
13. Lower spring seat
14. Shock absorber

Exploded view of the front strut assembly

7923KG43

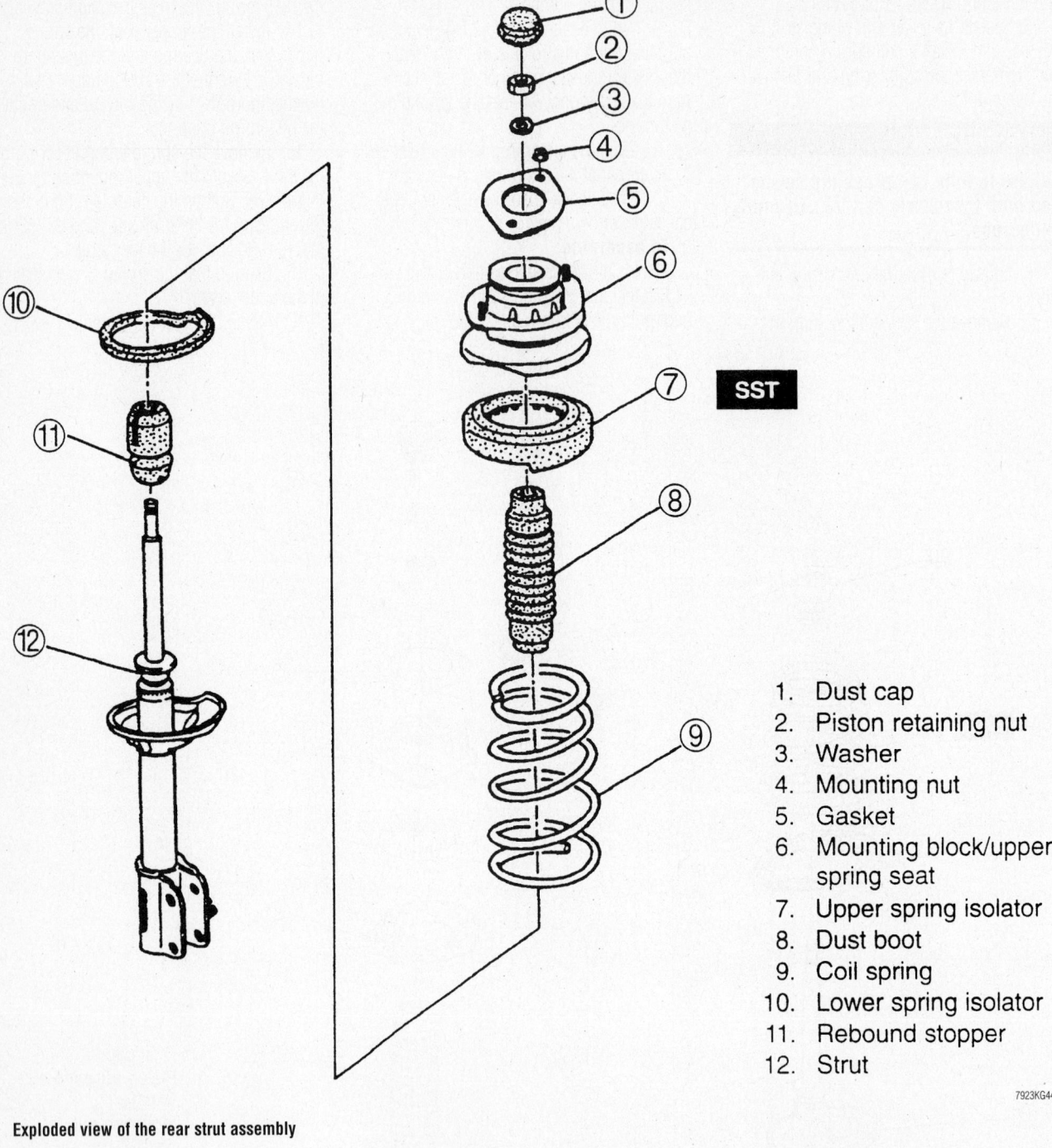

1. Dust cap
2. Piston retaining nut
3. Washer
4. Mounting nut
5. Gasket
6. Mounting block/upper spring seat
7. Upper spring isolator
8. Dust boot
9. Coil spring
10. Lower spring isolator
11. Rebound stopper
12. Strut

7923KG44

Exploded view of the rear strut assembly

RIO—REAR SPRING

1. Before servicing the vehicle, refer to the Precautions Section.
2. Remove the rear wheel.
3. Remove the brake hose bracket and wheel speed sensor wire.
4. Support the torsion beam with a suitable jack.
5. Remove the lower shock absorber mounting bolt.
6. Lower the jack and remove the spring.

7. Installation is the reverse order of removal.

Lower Ball Joint

REMOVAL & INSTALLATION

2.0L Models

1. Before servicing the vehicle, refer to the Precautions Section.

2. Remove the lower control arm.
3. Remove the dust cover from the ball joint.
4. Remove the snap ring.
5. Using a plastic hammer or equivalent, separate the ball joint from the lower arm.

To install:
6. Press fit the ball joint into the lower control arm using Special Tool 09545-11000A/B.
7. Install the snap ring.

8. Install the dust cover using Special Tool 09545-2110.

9. Install the lower control arm.

Lower Control Arm

REMOVAL & INSTALLATION

Rio

2002–2005 MODELS

1. Before servicing the vehicle, refer to the Precautions Section.

2. Remove the front wheel.

3. Remove the tension rod attaching bolt from the frame bracket.

4. Remove the tensioner rod bushing cotter pin and nut from the rear of the lower arm.

5. Remove the washer and bushing.

6. Remove the tension rod from the lower arm.

7. Remove the front bushing and washer.

8. Remove the lower arm, by lowering it and prying the ball joint from the steering knuckle after loosening the bolt and nut.

9. Remove the lower arm after loosening the lower arm attaching bolt from the frame bracket.

To install:

10. Raise the inner end of the lower arm into the pivot bracket on the frame, and loosely tighten the arm frame bracket pivot to hold in place.

11. Install the lower control arm ball joint in the clamp bore of the knuckle. Tighten the bolt and nut to 40–50 ft. lbs. (54–68 Nm).

12. Torque the lower arm frame bracket pivot bolt to 40–50 ft. lbs. (54–68 Nm).

13. Install the tension rod bolt into the lower arm, tighten washer and bushing and with a new cotter pin.

14. Install the bushing and insulator on the tensioner rod.

15. Tighten the tension rod bolt to 87–108 ft. lbs. (118–147 Nm).

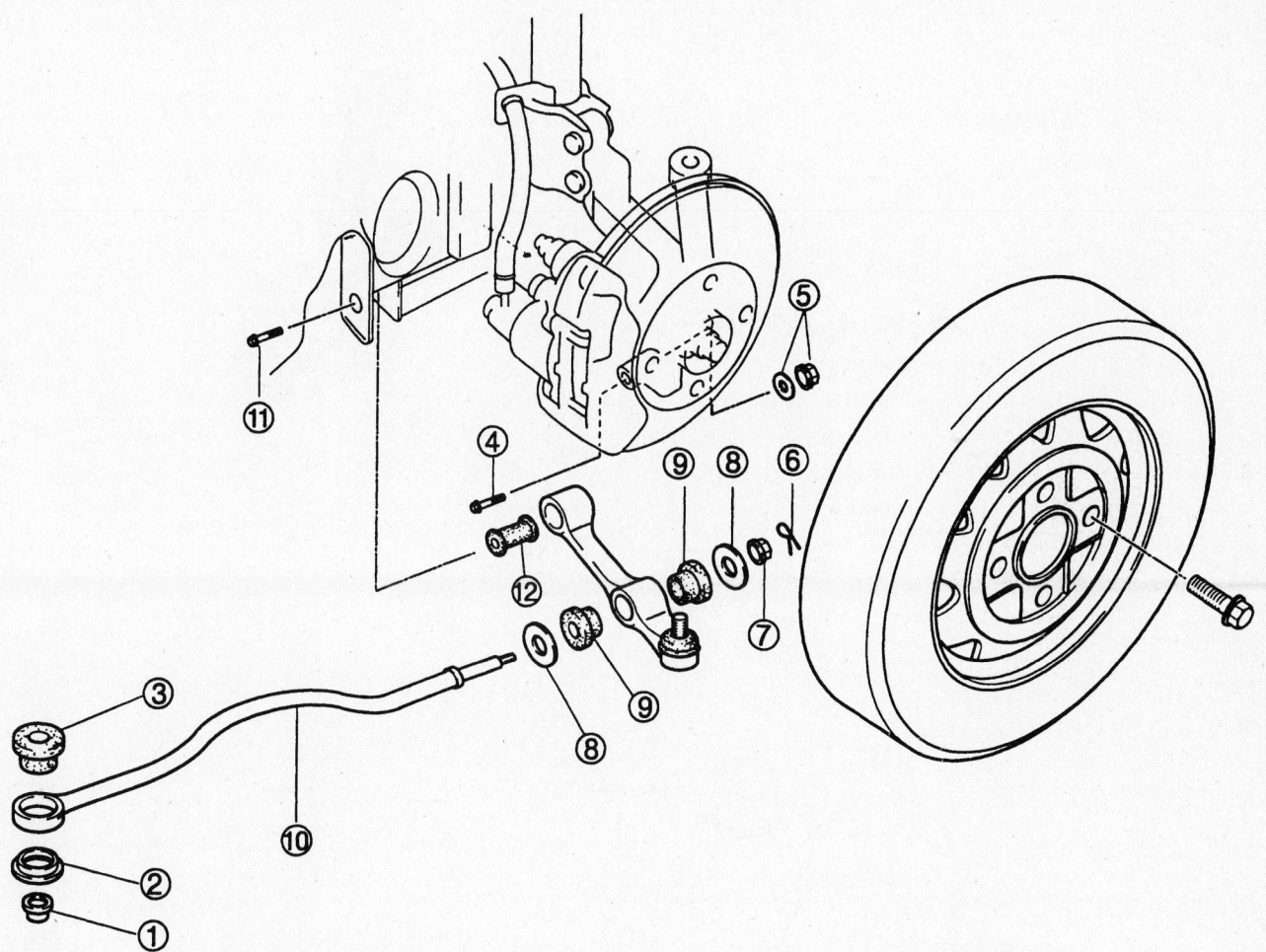

(1) Nut
(2) Stopper bushing
(3) Rubber
(4) Bolt
(5) Nut & Washer
(6) Cotter pin

(7) Nut
(8) Washer
(9) Bushing
(10) Tension rod
(11) Bolt
(12) Bushing

Exploded view of the lower control arm and tension rod—Rio

9357NG20

2006 MODEL

1. Before servicing the vehicle, refer to the Precautions Section.
2. Remove the front wheel.
3. Remove the lower control arm ball joint mounting bolts.
4. Remove the remaining lower control arm mounting bolts.

To install:

5. Install the control arm and tighten the mounting bolts as follows:

    a. G bushing mounting bolt: 86–101 ft. lbs. (120–140 Nm).

    b. A bushing mounting bolt: 72–86 ft. lbs. (100–120 Nm).

    c. Ball joint mounting bolts: 72–86 ft. lbs. (100–120 Nm).

6. Install the front wheel.

Spectra

1.8L MODEL

1. Before servicing the vehicle, refer to the Precautions Section.
2. Remove the front wheel.
3. Remove the stabilizer control link nut from the bracket on the lower control arm.
4. Remove the pivot bolt.
5. Remove the lower control arm ball joint bolt and nut from the steering knuckle.
6. Remove the 3 mounting bolts.
7. Remove the retaining nut, washer and the rear control arm bushing.
8. Remove the 2 ball joint mounting nuts and bolts from the lower control arm.
9. Remove the control arm.

To install:

10. Apply a suitable general purpose grease to the new tire dust boot and press the boot onto the ball joint using a suitable tool.
11. Install the 3 mounting bolts. Tighten the long mount bolt 86 ft. lbs. (117 Nm) and the short mounting bolts to 50 ft. lbs. (68 Nm).
12. Install the ball joint into the knuckle and tighten the pinch bolt and nut to 40 ft. lbs. (54 Nm).
13. Install the front control arm bushing.
14. Install the pivot bolt and tighten to 86 ft. lbs. (117 Nm).
15. Install the stabilizer control link nut to the bracket on the control arm and tighten the nut to 45 ft. lbs. (61 Nm).
16. Install the wheels.

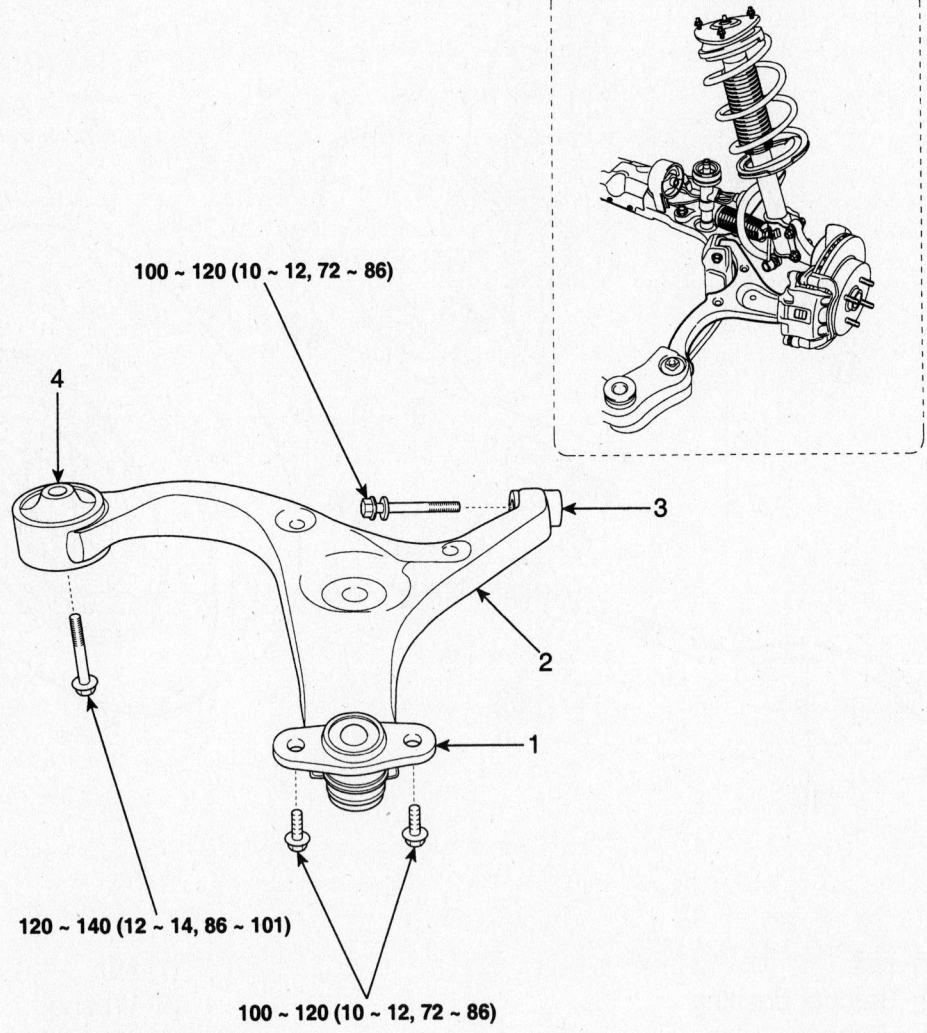

100 ~ 120 (10 ~ 12, 72 ~ 86)

120 ~ 140 (12 ~ 14, 86 ~ 101)

100 ~ 120 (10 ~ 12, 72 ~ 86)

TORQUE : Nm (kgf·m, lb-ft)

1. Front lower arm ball joint
2. Lower arm
3. A bushing
4. G bushing

09474_KIAC_G0017

Lower control arm exploded view—2006 Rio

2.0L MODEL

1. Before servicing the vehicle, refer to the Precautions Section.
2. Remove the front wheel
3. Remove the lower ball joint nut cotter pin, castle nut and washer.
4. Loosen the lower ball joint nut, but do not remove.
5. Remove the strut lower mounting bolts.
6. Using Special Tool 09568-34000 or equivalent, disconnect the lower arm ball joint from the lower arm.
7. Remove the stabilizer link nut.

8. Temporarily install the strut lower mounting bolt.
9. Remove the passenger seat side cover, if necessary.
10. Remove the lower control arm mounting bolts.
11. Remove the lower ball joint nut and remove the control arm assembly.
12. Installation is the reverse order of removal.

Optima

1. Before servicing the vehicle, refer to the Precautions Section.

2. Remove the front wheel.
3. Remove the ball joint nut. Loosen, but do not remove it.
4. Separate the ball joint from the lower control arm.
5. Remove the ball joint.
6. Remove the bolt connecting the fork to the lower control arm.
7. Remove the stabilizer link from the lower control arm.
8. Remove the 2 bolts from the lower control arm bushing.
9. Remove the lower control arm bushing bolt.

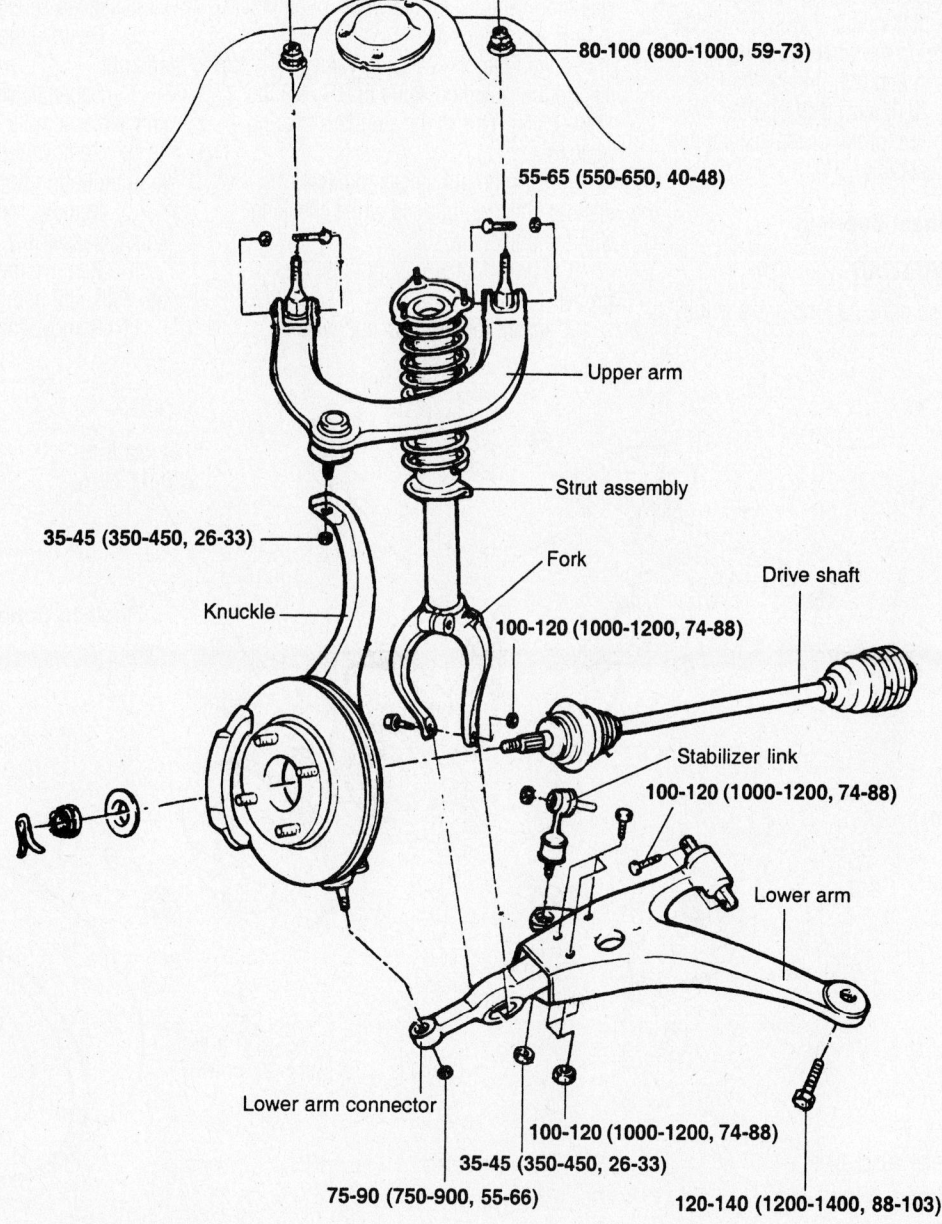

80-100 (800-1000, 59-73)

55-65 (550-650, 40-48)

Upper arm

Strut assembly

Fork

Drive shaft

35-45 (350-450, 26-33)

Knuckle

100-120 (1000-1200, 74-88)

Stabilizer link

100-120 (1000-1200, 74-88)

Lower arm

Lower arm connector

100-120 (1000-1200, 74-88)

35-45 (350-450, 26-33)

75-90 (750-900, 55-66)

120-140 (1200-1400, 88-103)

TORQUE : Nm (kg·cm, lb·ft)

9357NG51

Exploded view of the lower control arm—Optima

10. Remove the steering gear box.

11. Remove the stabilizer bar.

12. Installation is the reverse of the removal procedure. Refer to the illustration for tightening specifications.

Wheel Bearings

ADJUSTMENT

Except Rio Rear Wheel Bearings

The wheel bearings on these vehicles are not adjustable. To check if the bearing requires service, remove the wheel and tire assembly, brake caliper and disc brake rotor. Install a dial indicator with the indicator foot resting on the wheel hub. Try to move the hub in and out. If there is more than 0.002 in. (0.05mm) bearing play, check the wheel hub nut torque or replace the hub and bearing assembly.

Rio Rear Wheel Bearings

BEARING PRELOAD

1. Make sure the parking brake is fully released.

2. Remove the wheel and hub (grease) cap.

3. Rotate the brake drum to be sure there is no drag.

4. Seat the bearings by tightening the nut after raising the nut tap. Tighten to 18–22 ft. lbs. (25–29 Nm).

5. Loosen the nut slightly until it can be turned by hand.

➡ **Before the bearing preload can be set, the amount of seal drag must be measured and added to the required preload. Use a pull scale to measure the oil seal drag.**

6. Pull the scale squarely. Take the oil seal drag value when the wheel hub starters to turn and record it.

7. Add the oil seal drag value in the last step to the specified value to 0.6–1.9 lbs. (2.6–8.5 N). This is the standard bearing preload.

8. Turn the nut slowly to adjust the standard bearing preload while checking with the pull scale.

9. Firmly fix the lock nut into the groove.

10. Install the hub cap and wheel.

REMOVAL & INSTALLATION

Front

1. Before servicing the vehicle, refer to the Precautions Section.

2. Remove the front wheels.

3. Remove the center locknut. Discard the old locknut.

4. Remove the caliper assembly from the knuckle. Do not disconnect the brake lines. Support the caliper with a piece of wire. Do not allow the caliper to hang by the hose at any time. Remove the brake disc.

5. Remove the Anti-lock Brake System (ABS) speed sensor, if equipped.

6. Remove the tie rod end cotter pin and nut.

7. Separate the tie rod end from the knuckle assembly.

8. Remove the outer lower arm-to-ball joint mounting bolt and nut.

9. Remove the lower arm from the knuckle assembly.

10. Remove the knuckle assembly free of the halfshaft, using a plastic mallet.

11. Remove the knuckle assembly.

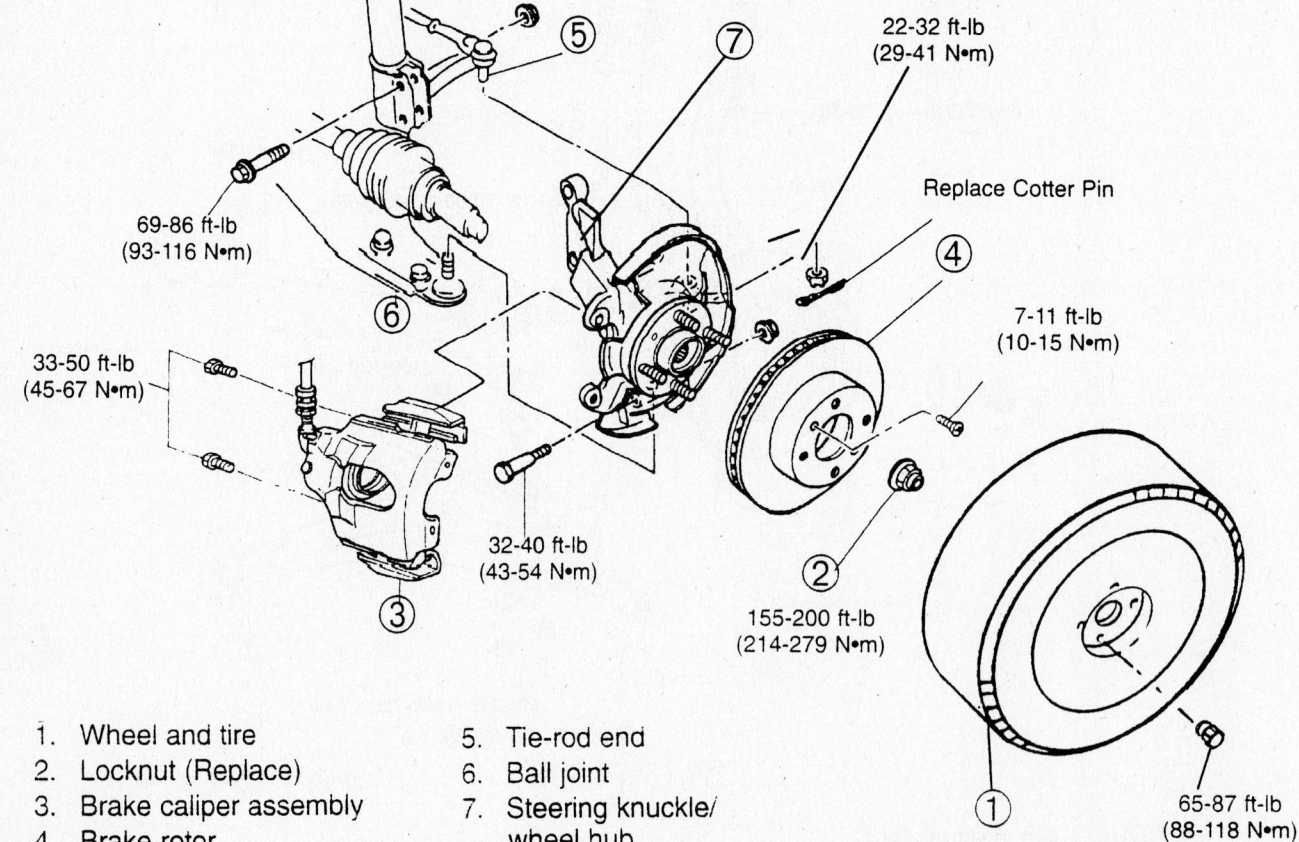

1. Wheel and tire
2. Locknut (Replace)
3. Brake caliper assembly
4. Brake rotor
5. Tie-rod end
6. Ball joint
7. Steering knuckle/ wheel hub

Exploded view of the front steering knuckle and related components—Spectra 1.8L model

7923KG46

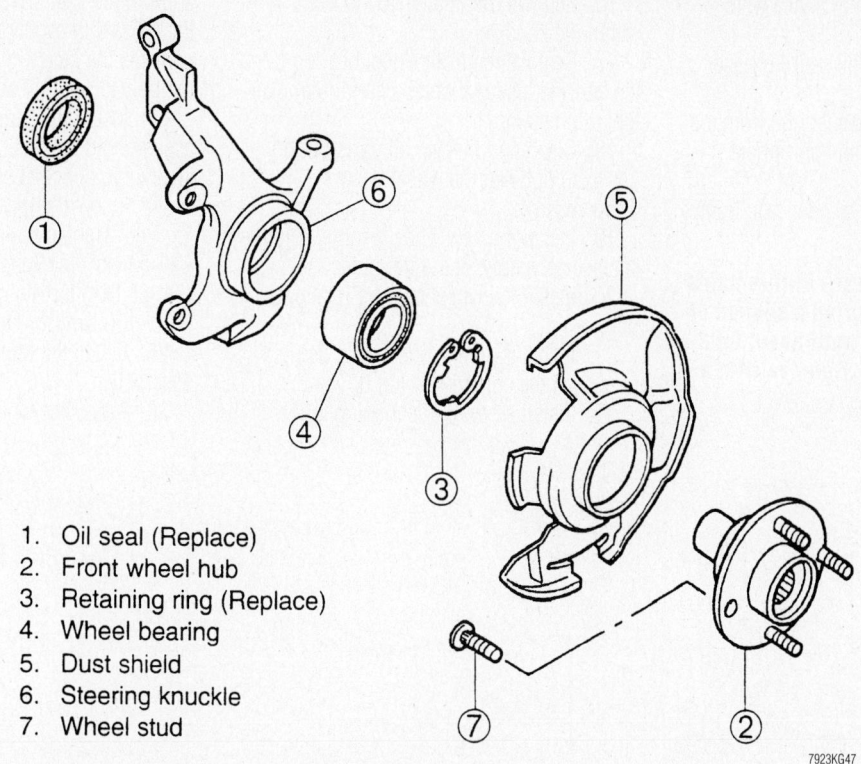

1. Oil seal (Replace)
2. Front wheel hub
3. Retaining ring (Replace)
4. Wheel bearing
5. Dust shield
6. Steering knuckle
7. Wheel stud

7923KG47

Exploded view of the front hub and bearing assembly—Spectra 1.8L model

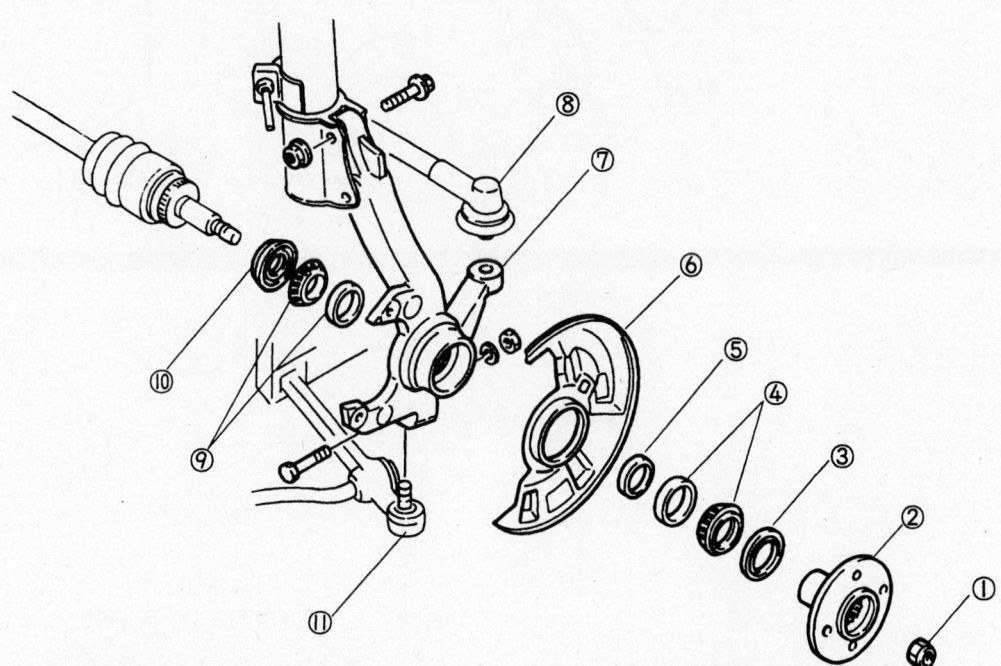

(1) Lock nut
(2) Wheel hub
(3) Outer oil seal
(4) Outer wheel bearing
(5) Spacer
(6) Dust cover

(7) Knuckle
(8) Tie-rod end
(9) Inner wheel bearing
(10) Inner oil seal
(11) Lower arm ball-joint

9357NG21

Exploded view of the front hub and bearing assembly—2002–2005 Rio

12. Clamp the knuckle in a vise with protected jaws.

13. Remove the inner oil seal from the knuckle, if equipped.

14. Remove the front wheel hub from the knuckle assembly, using an appropriate hub-puller.

15. Remove the bearing inner race from the front wheel hub.

➡ **If the bearing inner race still remains on the hub assembly, grind a section of the bearing inner race until about 0.02 in. (0.50mm) remains. Remove with a chisel.**

16. Remove the retaining ring from within the knuckle.

17. Remove the front wheel bearing from the knuckle, using a wheel bearing removal tool to press it out.

18. Clean and inspect all parts but do not wash or clean the wheel bearing.

To install:

19. Install the new wheel bearing into the knuckle assembly, using the press tools.

20. Install the wheel bearing retaining ring.

21. Install the front wheel hub, using a press and the correct bearing driver.

22. Install a new oil seal using the appropriate seal driver and a hammer. Tap the oil seal in evenly until the special tool contacts the steering knuckle. Coat the lip of the oil seal with grease.

23. Install the bearing/hub and knuckle assembly in place. Loosely tighten the knuckle to shock absorber bolt.

24. Install the lower arm ball joint to the knuckle. Tighten the nut to 32–40 ft. lbs. (43–54 Nm) for Spectra, to 40–50 ft. lbs. (54–68 Nm) for Rio or to 74–88 ft. lbs. (100–120 Nm) for Optima.

25. Install the halfshaft to the knuckle assembly.

26. Install the wheel speed sensor, if

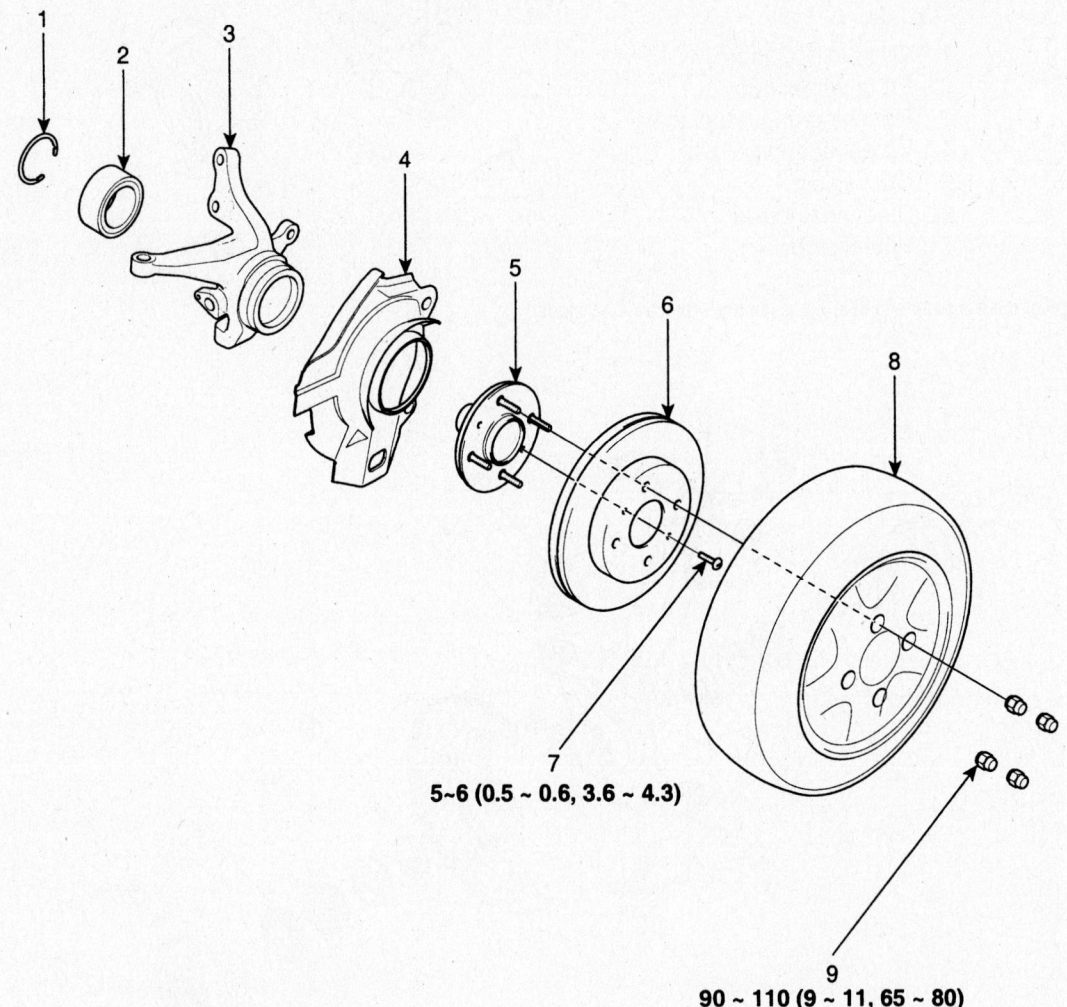

5~6 (0.5 ~ 0.6, 3.6 ~ 4.3)

90 ~ 110 (9 ~ 11, 65 ~ 80)

TORQUE : Nm (kgf·m, lb-ft)

1. Snap ring
2. Front wheel hub bearing
3. Front axle assembly
4. Front brake disc dust cover
5. Front wheel hub assembly
6. Front wheel brake disc
7. Front brake disc fixing screw
8. Front wheel/tire
9. Front wheel nut

09474_KIAC_G0018

Exploded view of the front hub and bearing assembly—2006 Rio and Spectra 2.0L

equipped with ABS and tighten the bolts to 12–17 ft. lbs. (16–23 Nm)

27. Install the tie rod end to the knuckle and tighten the nut to 22–32 ft. lbs. (29–41 Nm) for Rio, and Spectra or to 18–25 ft. lbs. (24–34 Nm) for Optima.

28. Install a New cotter pin.

29. Install a new wheel hub locknut. Tighten the nut to 155–200 ft. lbs. (214–279 Nm) for 1.8L Spectra, to 116–174 ft. lbs. (157–235 Nm) for 2002–2005 Rio, 144–188 (200–260 Nm) for 2006 Rio and 2.0L Spectra or to 148–192 ft. lbs. (200–260 Nm) for Optima.

30. Check the end-play of the wheel bearing by installing a dial indicator against the wheel hub and tire to move the brake disc back and forth. There should be no more than 0.002 in. (0.05mm) of free-play present.

31. Stake the locknut into place by bending it into the groove.

32. Install the brake caliper(s) and tighten the bolts to 33–50 ft. lbs. (45–67 Nm)

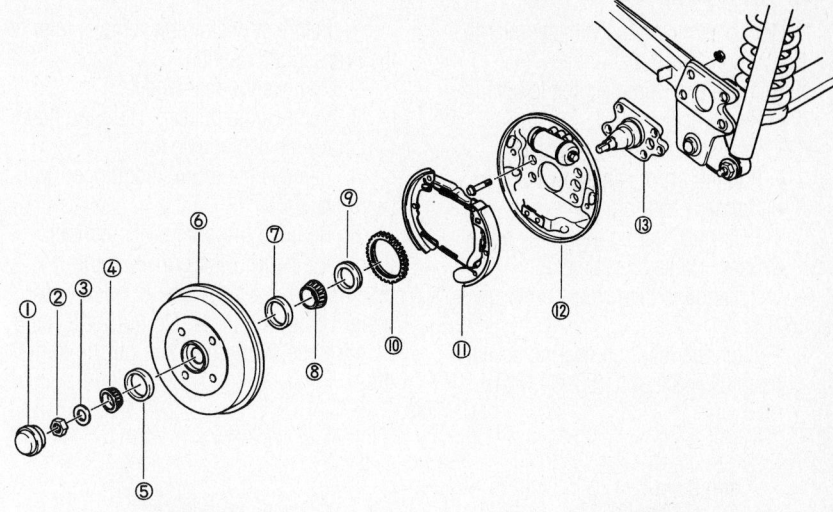

(1)	Hub cap	(8)	Inner bearing
(2)	Lock nut	(9)	Oil seal
(3)	Washer	(10)	Sensor rotor(for ABS)
(4)	Outer bearing	(11)	Brake assembly
(5)	Outer bearing outer race	(12)	Back plate
(6)	Brake drum	(13)	Spindle
(7)	Inner bearing outer race		

9357NG22

Exploded view of the rear wheel bearings and related components—2002–2005 Rio

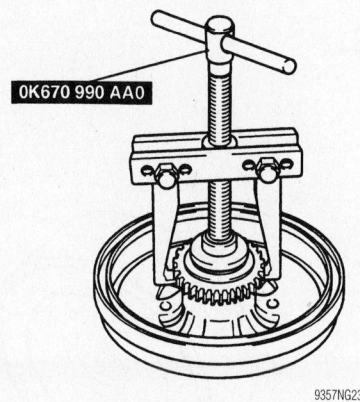

0K670 990 AA0

9357NG23

Removing the ABS rotor, using a puller— Rio

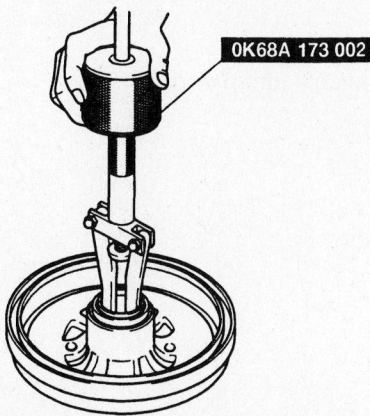

0K68A 173 002

9357NG24

View of the special tool need to remove the inner and outer bearing outer races— 2002–2005 Rio

33. Install the front wheels and lower the vehicle.

34. With the vehicle lowered check all of the bolts and retighten as necessary.

35. Inspect the front end alignment and adjust as is necessary.

Rear

2002–2005 RIO

1. Before servicing the vehicle, refer to the Precautions Section.

2. Remove the rear wheels.

3. Remove the hub cap.

4. Use a small chisel to raise the staked edge of the hub lock nut.

5. Remove the drum, washer and bearings as an assembly.

6. Remove the Anti-lock Brake System (ABS) sensor rotor using a suitable puller.

7. Remove the bearing oil seal and discard.

➡**If the bearings will be reused, they must be matchmarked so they can be installed in their original positions.**

8. Remove the inner bearings from the bearing hub.

9. Remove the inner and outer bearing outer races using special tool 0K68A 173 002.

To install:

10. Install the outer bearing race using a hammer. Tap the race in until it is fully seated in the hub.

11. Install the inner bearing in the bearing hub.

12. Install the oil seal after lubricating it with lithium grease, using a hammer and seal installation tool.

13. Install the ABS sensor rotor using a flat plate to press it into place.

14. Refer to the accompanying illustration and completely fill in the shaded area with lithium grease.

15. Install the brake drum bearings and hub on the spindle. Keep the drum centered on the spindle to prevent damage to the oil seal and spindle threads.

16. Install the outer bearing and washer.

17. Install a new hub lock nut.

18. Adjust the bearing preload to 5.6–8.6 ft. lbs. (0.63–0.98 Nm).

19. Install the hub lock nut to the groove firmly.

20. Install the hub cap.

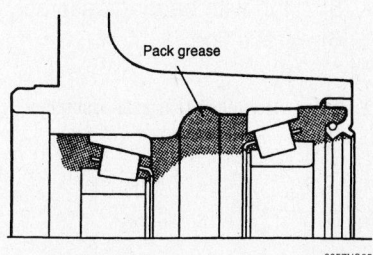

Pack grease

9357NG25

Fill in the shaded areas with lithium grease—2002–2005 Rio

2006 RIO AND 2.0L SPECTRA

1. Before servicing the vehicle, refer to the Precautions Section.

2. Remove or disconnect the following:
   - Rear wheel
   - Wheel speed sensor bracket
   - Parking brake cable bracket
   - Caliper assembly
   - Rear hub bearing assembly bolts
   - Rear hub bearing assembly

3. Installation is the reverse order of removal.

4. Tighten the rear hub bearing assembly bolts to 36–43 ft. lbs. (50–60 Nm).

1.8L SPECTRA AND OPTIMA

1. Before servicing the vehicle, refer to the Precautions Section.

2. Remove the rear wheels.

3. Remove the hubcap. Hold the brake to remove the center axle nut.

4. Remove the drum, if equipped with drum brakes.

5. Remove the disc brake caliper, if equipped with disc brakes without disconnecting the hydraulic hose and hang it from the body. Do not let it hang by the hose. Slide the disc off the spindle.

6. Remove the hub and bearing assembly off the spindle. The hub and bearing cannot be separated and must be replaced as 1 piece.

To install:

7. Install the hub and drum or rotor.

8. Install the brake caliper, if equipped.

9. Install a new spindle nut and tighten to 131–173 ft. lbs. (177–235 Nm) for disk brakes and 155–200 ft. lbs. (209–279 Nm) for drum brakes. Stake the nut into place. Replace the hubcap.

10. Install the wheel and tire assembly.

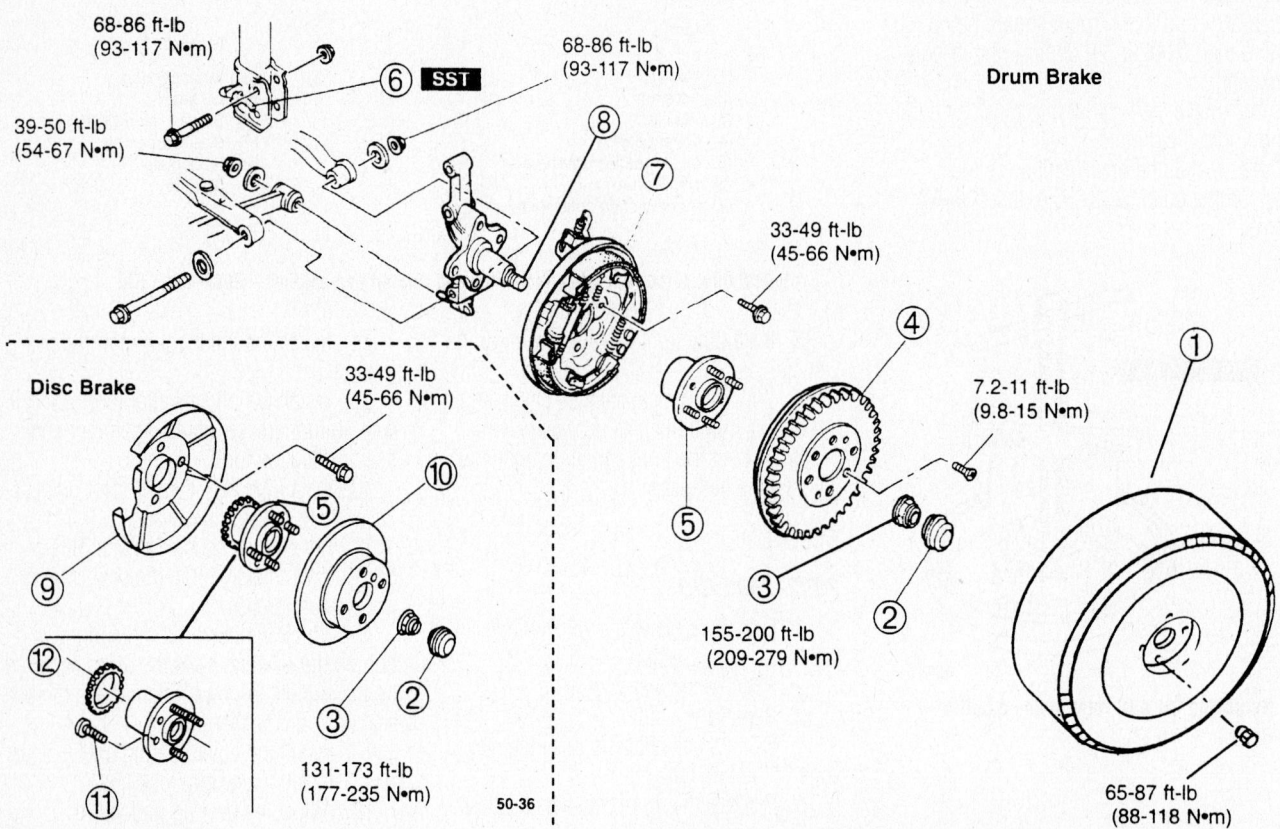

1. Wheel and tire
2. Dust cap
3. Locknut (Replace)
4. Brake drum (or disc)
5. Hub with bearing assembly
6. Brake line
7. Rear brake assembly (drum or disc)
8. Spindle
9. Dust cover
10. Brake rotor
11. Hub bolt
12. ABS sensor rotor

Exploded view of the rear axle assembly—1.8L Spectra

BRAKES

Brake Caliper

REMOVAL & INSTALLATION

Front

1. Before servicing the vehicle, refer to the Precautions Section.

2. Remove the wheels.
3. Remove the brake hose from the caliper.
4. Remove the caliper guide pin bolts and caliper.

To install:
5. Install the caliper and tighten the guide pin bolts to 19–21 ft. lbs. (26–28 Nm)

6. Install the brake hose and tighten to 9–13 ft. lbs. (13–18 Nm)
7. Bleed the brakes and fill the master cylinder with clean brake fluid.
8. Install the wheels.

Rear

1. Before servicing the vehicle, refer to the Precautions Section.

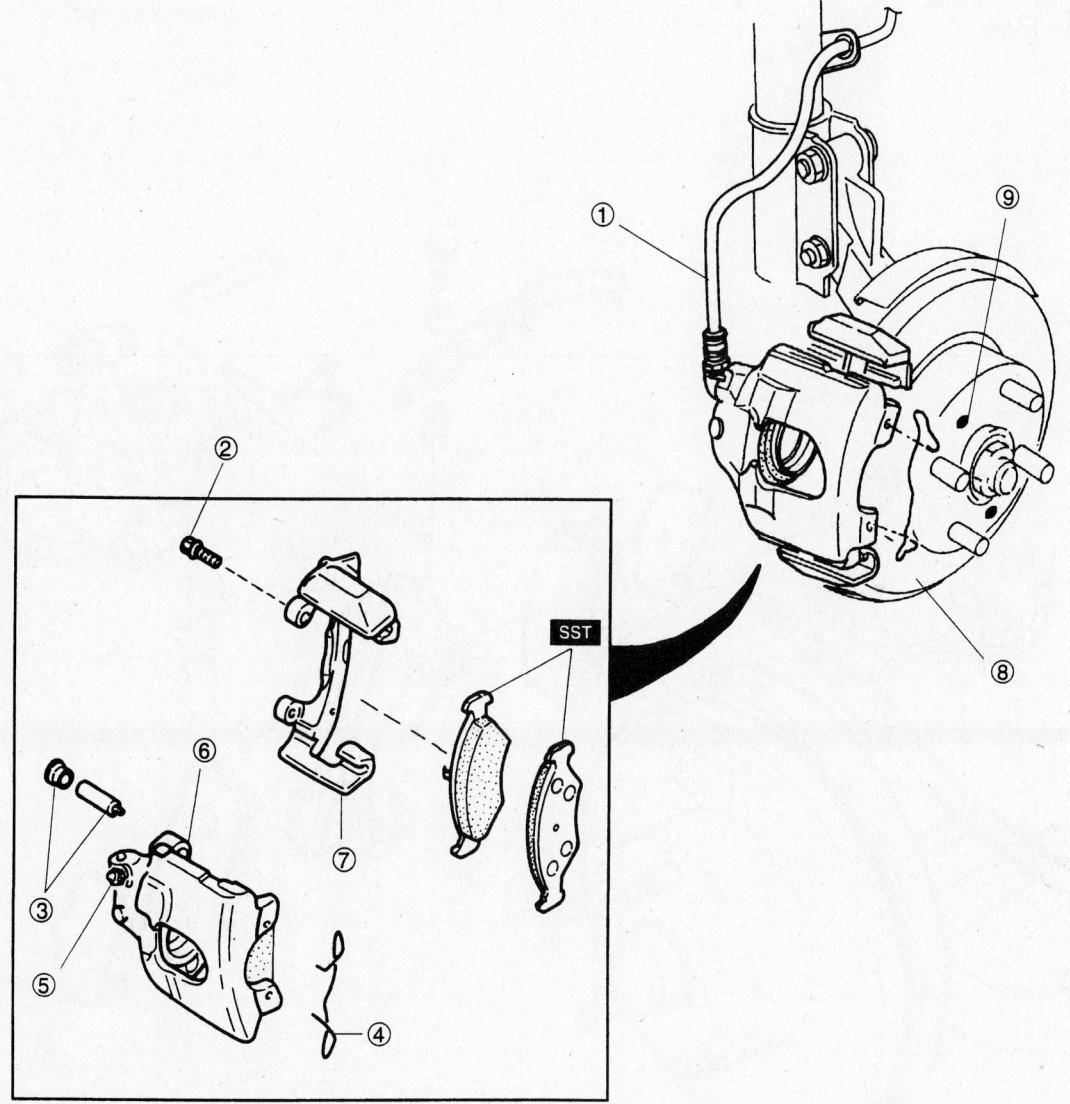

(1) Flexible hose
(2) Bolt
(3) Cap, bolt (square head) and bushing
(4) Spring
(5) Cap and bleeder screw
(6) Caliper
(7) Supporting plate
(8) Brake rotor (Disc)
(9) Mounting screws

93016G15

Exploded view of the front brakes

2. Remove the wheels.

3. Remove the parking brake cable and clip.

4. Remove the brake hose banjo bolt.

5. Remove the caliper lock bolts and the caliper.

To install:

6. Install the caliper and tighten the lock bolts to 22–29 ft. lbs. (29–39 Nm)

7. Install the brake hose and tighten the banjo bolt to 16–22 ft. lbs. (22–29 Nm)

8. Install the park brake cable and clip.

9. Bleed the brakes and fill the master cylinder with clean brake fluid.

10. Install the wheels.

Disc Brake Pads

REMOVAL & INSTALLATION

Front

1. Before servicing the vehicle, refer to the Precautions Section.

2. Remove the wheels.

3. Remove the caliper guide pin bolts and lift the caliper away from the rotor.

4. Remove the brake pads and retainer spring from the caliper.

To install:

5. Compress the caliper piston into the bore.

6. Install the brake pads and retainer spring to the caliper.

7. Install the caliper in position on the caliper support bracket.

8. Install the guide pin bolts.

9. Install the wheels.

Rear

1. Before servicing the vehicle, refer to the Precautions Section.

2. Remove the wheels.

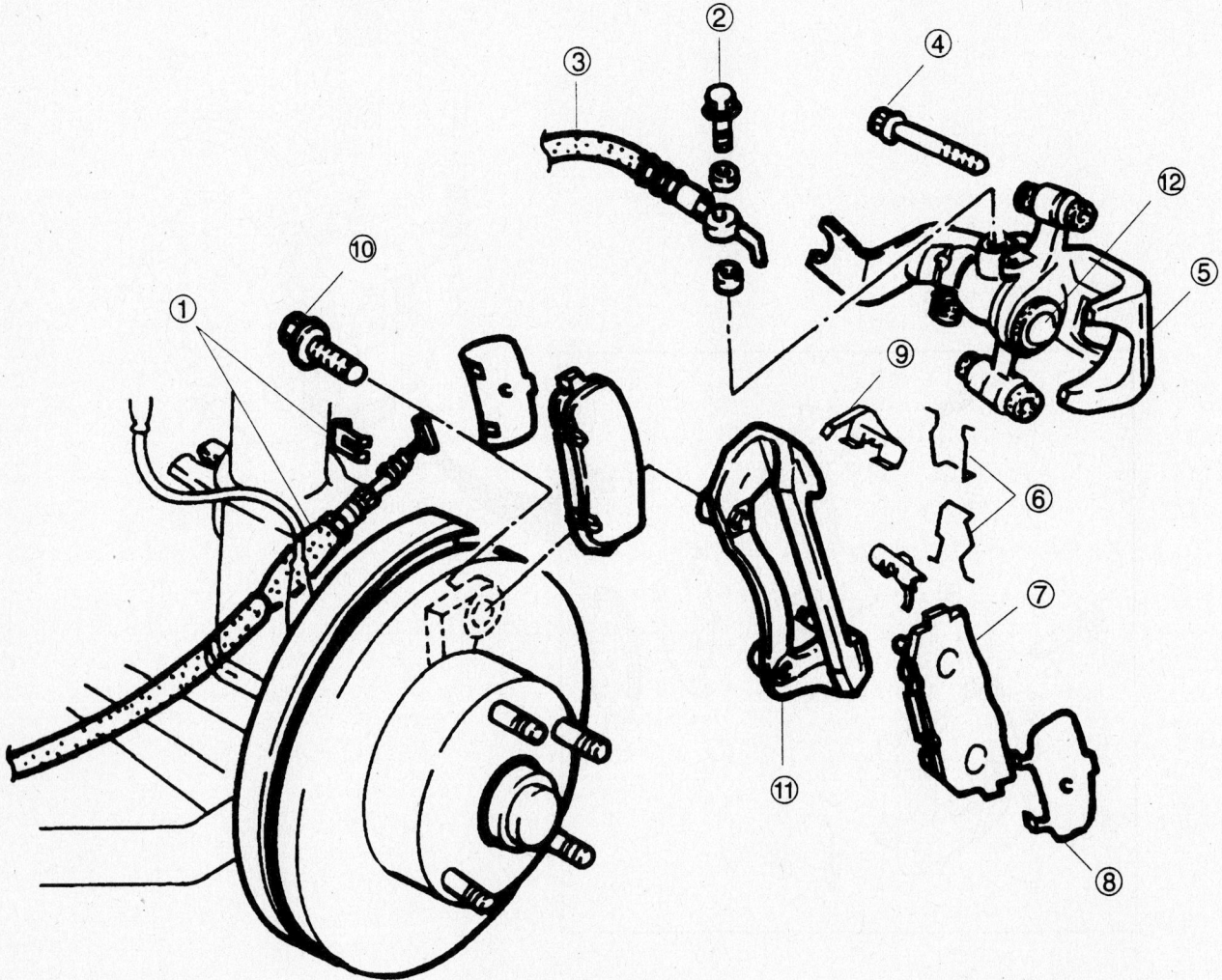

(1) Parking cable, clip	(5) Caliper
(2) Connecting bolt	(6) V-spring
(3) Brake hose	(7) Disc pad
(4) Lock bolt	(8) Shim

(9) Guide plate	
(10) Bolt	
(11) Mounting support	
(12) Caliper piston	

93016G16

Exploded view of the rear disc brakes

3. Remove the parking brake cable and clip.

4. Remove the caliper lock bolts and remove the caliper.

5. Remove the V-springs from the brake pads.

6. Remove the brake pads and shims.

To install:

7. Compress the caliper piston into the bore by rotating the piston with special tool OK9A4 263 001.

8. Install the new pads and shims.

9. Install the V-springs.

10. Install the caliper and tighten the lock bolts to 22–29 ft. lbs. (29–39 Nm)

11. Install the park brake cable and clip.

12. Install the wheels.

Brake Drums

REMOVAL & INSTALLATION

1. Before servicing the vehicle, refer to the Precautions Section.

2. Remove the wheels.

3. Remove the retaining screws and brake drum. Two 8mm x 1.25 bolts can be used to press the drum from the hub.

To install:

4. Install the brake drum to the hub.

5. Install the wheels.

Brake Shoes

REMOVAL & INSTALLATION

1. Before servicing the vehicle, refer to the Precautions Section.

2. Remove the wheels.

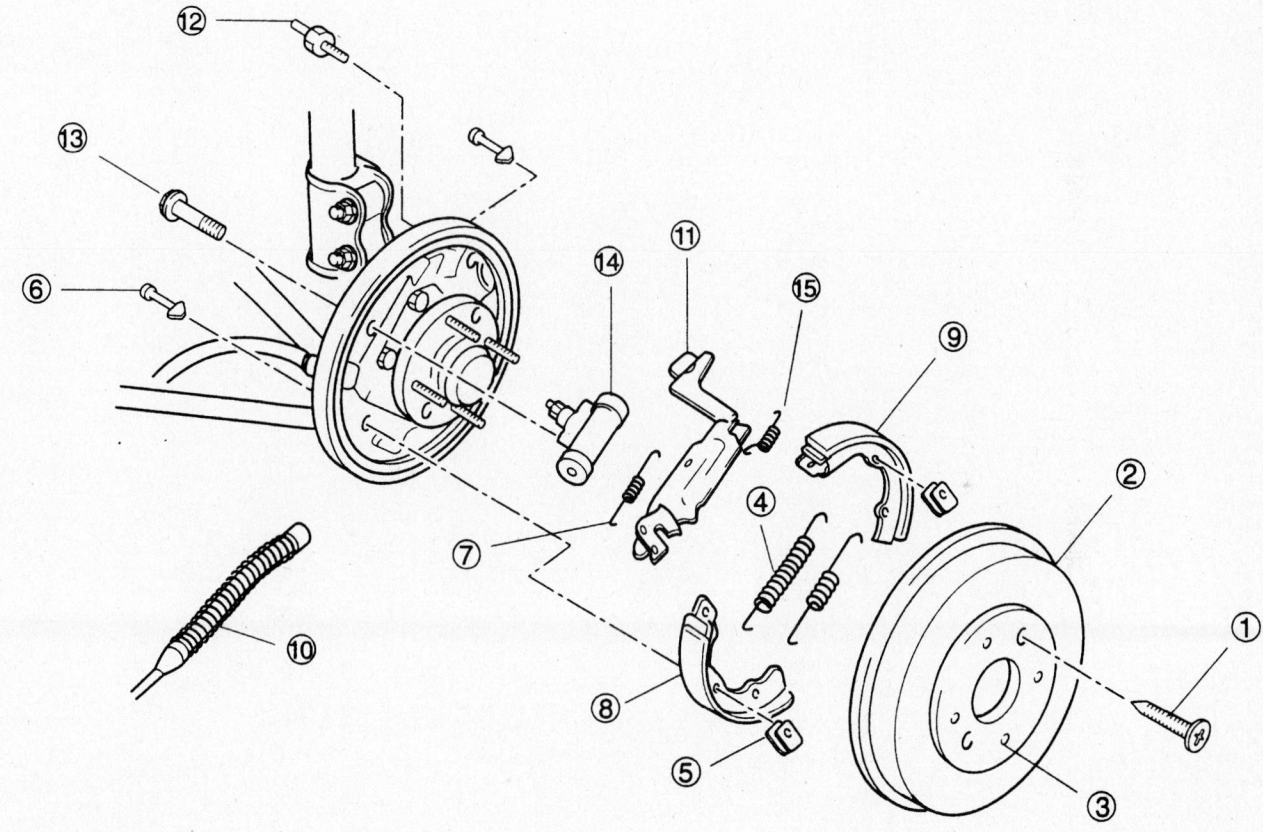

(1) Mounting screws
(2) Brake drum
(3) Drum pulling threads
(4) Return springs
(5) Spring clips
(6) Hold down pins
(7) Adjuster spring
(8) Brake shoe-leading

(9) Brake shoe-trailing
(10) Parking brake cable
(11) Operating lever assembly
(12) Brake line
(13) Bolts
(14) Wheel cylinder assembly
(15) Anti-rattle spring

93016G17

Exploded view of the rear drum brakes

3. Remove the retaining screws and brake drum. Two 8mm x 1.25 bolts can be used to press the drum from the hub.

4. Remove the top return spring.

5. Remove the shoe retainer springs and pins.

6. Remove the adjuster spring and anti-rattle spring from the operating lever assembly.

7. Remove the bottom return spring.

8. Remove the parking brake cable from the rear shoe and remove the brake shoes.

To install:

9. Position the operating lever assembly above the hub.

10. Install the parking brake cable to the rear shoe.

11. Fit the operating lever assembly between the front and rear shoes and install the bottom return spring.

12. Install the shoe retainer pins and springs.

13. Install the top return spring.

14. Install the adjuster spring and the anti-rattle spring to the operating lever assembly.

15. Install the brake drum.

16. Install the wheels and adjust the brakes.

KIA

Sedona

19

SPECIFICATION AND MAINTENANCE CHARTS

ENGINE AND VEHICLE IDENTIFICATION

		Engine							Model Year	
Code ①	Liters (cc)	Cu. In.	Cyl.	Fuel Sys.	Engine Type	Eng. Mfg.		Code ②	Year	
1	3.5 (3497)	213	6	EGI	DOHC	KIA		2	2002	
3	3.8 (3778)	231	6	EGI	DOHC	KIA		3	2003	

EGI: Electronic Gasoline Injection

DOHC: Double Overhead Camshafts

① 8th digit of VIN

② 10th digit of VIN

Code ②	Year
4	2004
5	2005
6	2006

09474_SEDO_C0001

GENERAL ENGINE SPECIFICATIONS

Year	Model	Engine Displacement Liters	Engine VIN	Net Horsepower @ rpm	Net Torque @ rpm (ft. lbs.)	Bore x Stroke (in.)	Com-pression Ratio	Oil Pressure @ rpm
2002	Sedona	3.5	1	195@5500	218@3500	3.66x3.88	10:01	43-57@3000
2003	Sedona	3.5	1	195@5500	218@3500	3.66x3.88	10:01	43-57@3000
2004	Sedona	3.5	1	195@5500	218@3500	3.66x3.88	10:01	43-57@3000
2005	Sedona	3.5	1	195@5500	218@3500	3.66x3.88	10:01	①
2006	Sedona	3.8	3	244@6000	253@3500	3.78x3.43	10:01	18.77@1000

① 11.6 at idle

09474_SEDO_C0002

ENGINE TUNE-UP SPECIFICATIONS

Year	Engine Displacement Liters	Engine VIN	Spark Plug Gap (in.)	Ignition Timing (deg.) MT	Ignition Timing (deg.) AT	Fuel Pump (psi)	Idle Speed (rpm) MT	Idle Speed (rpm) AT	Valve Clearance Intake	Valve Clearance Exhaust
2002	3.5	1	0.039-0.043	NA	①	46-49	NA	600-800	HYD	HYD
2003	3.5	1	0.039-0.043	NA	①	46-49	NA	600-800	HYD	HYD
2004	3.5	1	0.039-0.043	NA	①	46-49	NA	600-800	HYD	HYD
2005	3.5	1	0.039-0.043	NA	②	46-49	NA	600-800	HYD	HYD
2006	3.8	3	NA	NA	②	③	NA	550-750	0.0067-0.0090	0.0106-0.0129

NOTE: The Vehicle Emission Control Information label often reflects specification changes made during production.

The label figures must be used if they differ from those in this chart

NA: Not Available

B: Before top dead center

HYD: Hydraulic

① Computer controled, no adjustment possible

② 5-15 degrees

③ 54.3-55.8 at idle

09474_SEDO_C0003

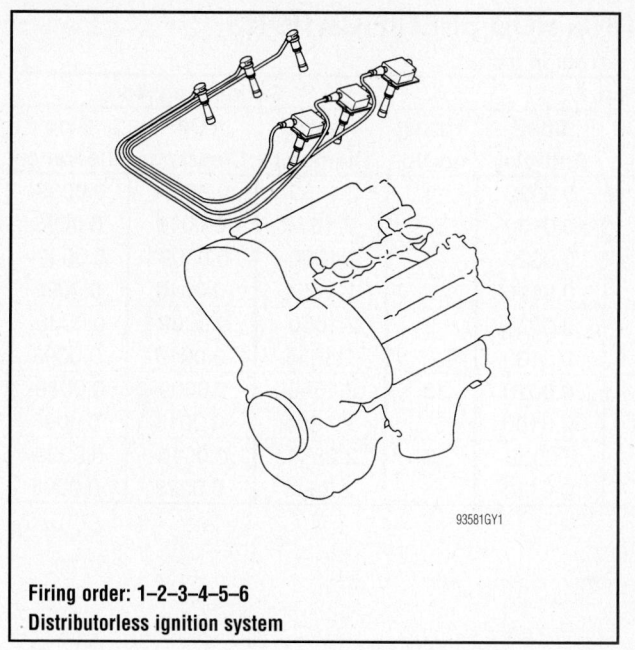

93581GY1

Firing order: 1–2–3–4–5–6
Distributorless ignition system

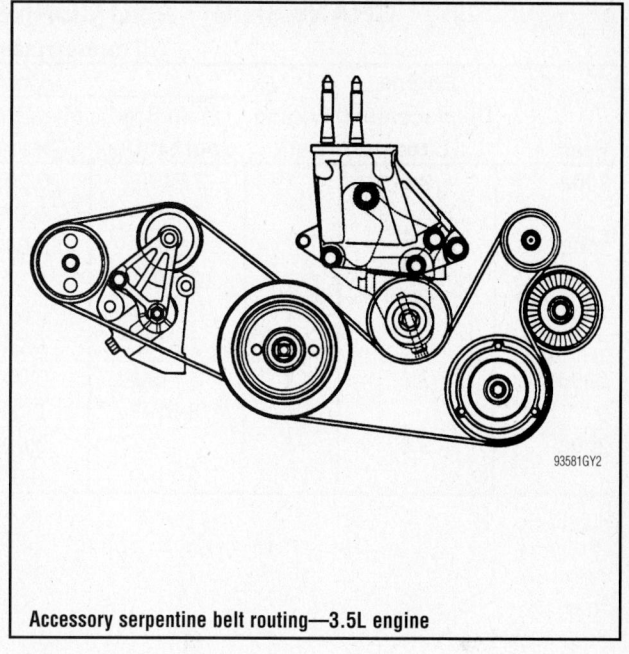

93581GY2

Accessory serpentine belt routing—3.5L engine

CAPACITIES

Year	Model	Engine Displacement Liters	Engine VIN	Engine Oil with Filter	Transaxle (pts.) Manual	Transaxle (pts.) Auto.	Fuel Tank (gal.)	Cooling System (qts.)
2002	Sedona	3.5	1	4.3	—	18.0	19.8	8.2
2003	Sedona	3.5	1	4.3	—	18.0	19.8	8.2
2004	Sedona	3.5	1	4.3	—	18.0	19.8	8.2
2005	Sedona	3.5	1	4.5	—	18.0	19.8	8.6
2006	Sedona	3.8	3	5.5	—	22.0	21.1	9.1

NOTE: All capacities are approximate. Add fluid gradually and ensure a proper level is obtained.

09474_SEDO_C0004

VALVE SPECIFICATIONS

Year	Engine Displacement Liters	Engine VIN	Seat Angle (deg.)	Face Angle (deg.)	Maximum out of Square (degrees)	Spring Free Length (in.)	Stem-to-Guide Clearance (in.) Intake	Exhaust	Stem Diameter (in.) Intake	Exhaust
2002	3.5	1	45	45	2	1.8268	0.0009-0.0020	0.0020-0.0039	0.258-0.259	0.257-0.258
2003	3.5	1	45	45	2	1.8268	0.0009-0.0020	0.0020-0.0039	0.258-0.259	0.257-0.258
2004	3.5	1	45	45	2	1.8268	0.0009-0.0020	0.0020-0.0039	0.258-0.259	0.257-0.258
2005	3.5	1	45	45	2	1.8268	0.0009-0.0020	0.0020-0.0033	0.258-0.259	0.257-0.258
2006	3.8	3	45.25-45.75	45	NA	1.7267	0.0008-0.0019	0.0012-0.0021	0.215-0.216	0.215-0.216

NA: Not Available

09474_SEDO_C0005

CRANKSHAFT AND CONNECTING ROD SPECIFICATIONS

All measurements are given in inches.

Year	Engine Displacement Liters	Engine VIN	Crankshaft				Connecting Rod		
			Main Brg. Journal Dia.	Main Brg. Oil Clearance	Shaft End-play	Thrust on No.	Journal Diameter	Oil Clearance	Side Clearance
2002	3.5	1	2.5190-2.5197	0.0071-0.0140	0.0020 0.0100	3	2.1650-2.1654	0.0009-0.0016	0.0039-0.0098
2003	3.5	1	2.5190-2.5197	0.0071-0.0140	0.0020 0.0100	3	2.1650-2.1654	0.0009-0.0016	0.0039-0.0098
2004	3.5	1	2.5190-2.5197	0.0071-0.0140	0.0020 0.0100	3	2.1650-2.1654	0.0009-0.0016	0.0039-0.0098
2005	3.5	1	2.5190-2.5197	0.0071-0.0140	0.0020 0.0100	3	2.1650-2.1654	0.0009-0.0016	0.0039-0.0098
2006	3.8	3	2.7142-2.7149	0.0008-0.0016	0.0039-0.0110	3	2.2834-2.2842	0.0015-0.0022	0.0039-0.0098

09474_SEDO_C0006

PISTON AND RING SPECIFICATIONS

All measurements are given in inches.

Year	Engine Displacement Liters	Engine VIN	Piston Clearance	Ring Gap			Ring Side Clearance		
				Top Compression	Bottom Compression	Oil Control	Top Compression	Bottom Compression	Oil Control
2002	3.5	1	0.0012-0.0020	0.0079-0.0118	0.0177-0.0236	0.0079-0.0276	0.0016-0.0315	0.0008-0.0024	SNUG
2003	3.5	1	0.0012-0.0020	0.0079-0.0118	0.0177-0.0236	0.0079-0.0276	0.0016-0.0315	0.0008-0.0024	SNUG
2004	3.5	1	0.0012-0.0020	0.0079-0.0118	0.0177-0.0236	0.0079-0.0276	0.0016-0.0315	0.0008-0.0024	SNUG
2005	3.5	1	0.0012-0.0020	0.0079-0.0118	0.0177-0.0236	0.0079-0.0276	0.0016-0.0315	0.0008-0.0024	0.0016-0.0024
2006	3.8	3	0.0012-0.0020	0.0067-0.0126	0.0126-0.0185	0.0078-0.0275	0.0012-0.0027	0.0012-0.0027	0.0024-0.0059

09474_SEDO_C0007

TORQUE SPECIFICATIONS
All readings in ft. lbs.

Year	Engine Displacement Liters	Engine VIN	Cylinder Head Bolts	Main Bearing Bolts	Rod Bearing Bolts	Crankshaft Damper Bolts	Flywheel Bolts	Manifold Intake	Manifold Exhaust	Spark Plugs	Oil Pan Drain Plug
2002	3.5	1	77-85	52-59	①	130-138	54-57	②	30-37	15-22	26-30
2003	3.5	1	77-85	52-59	①	130-138	54-57	②	30-37	15-22	26-30
2004	3.5	1	77-85	52-59	①	130-138	54-57	②	30-37	15-22	26-30
2005	3.5	1	77-85	51-58	①	130-138	54-57	14-17	29-32	15-22	26-32
2006	3.8	3	③	④	⑤	210-224	53-57	⑥	29-33	15-22	25-32

① Step 1: 24-27 ft. lbs.
   Step 2: Plus 90-94 degrees

② Upper: 15-17 ft. lbs.
   Lower: 15-22 ft. lbs.

③ Step 1: 27.5-30.4 ft. lbs.
   Step 2: Plus 120 degrees
   Step 3: Plus 90 degrees

④ Bolts 1 thru 8 (inside cap bolts): 36.16 ft. lbs., 90 degrees
   Bolts 9 thru 16 (outside cap bolts) 14.46 ft. lbs., 120 degrees
   Bolts 17thru 22 (side cap bolts) 21.7-23.2 ft. lbs.
   See illustration in text for location information

⑤ Step 1: 14.46 ft. lbs.
   Step 2: Plus 90 degrees

⑥ Step 1: 3-4 ft. lbs.
   Step 2: 14-17 ft. lbs.
   Step 3: Repeat Step 2, two more times

09474_SEDO_C0008

WHEEL ALIGNMENT

Year	Model		Caster Range (+/-Deg.)	Caster Preferred Setting (Deg.)	Camber Range (+/-Deg.)	Camber Preferred Setting (Deg.)	Toe-in (in.)
2002	Sedona	F	0.50	+1.88	0.50	0.51	-0.04
		R	—	—	—	—	—
2003	Sedona	F	0.50	+1.88	0.50	0.51	-0.04
		R	—	—	—	—	—
2004	Sedona	F	0.50	+1.88	0.50	0.51	-0.04
		R	—	—	—	—	—
2005	Sedona	F	0.50	+1.88	0.50	0.51	-0.04
		R	—	—	—	—	—
2006	Sedona	F	—	4 05' +/- 30'	—	0 +/- 30'	0 +/- .07870
		R	—	—	—	-20 +/- 30'	.1378 +/- .7870

09474_SEDO_C0009

TIRE, WHEEL AND BALL JOINT SPECIFICATIONS

| Year | Model | OEM Tires | | Tire Pressure (psi) | | Wheel Size | Ball Joint Inspection | Wheel Lug Torque (ft. lbs.) |
		Standard	Optional	Front	Rear			
2002	Sedona	P215/70R15	—	35	35	6JJ	①	65-79
2003	Sedona	P215/70R15	—	35	35	6JJ	①	65-79
2004	Sedona	P215/70R15	—	35	35	6JJ	①	65-79
2005	Sedona	P215/70R15	—	35	35	6JJ	①	65-79
2006	Sedona	P225/70R16	P235/60R17	35	35	②	①	65-79

OEM: Original Equipment Manufacturer

PSI: Pounds Per Square Inch

① Replace if any measurable movement is found.

② STD: 6.5Jx16

OPT: 6.5Jx17

09474_SEDO_C0010

BRAKE SPECIFICATIONS

All measurements in inches unless noted

| Year | Model | | Brake Disc | | | Brake Drum | | | Minimum Lining Thickness | Brake Caliper | |
			Original Thickness	Minimum Thickness	Maximum Run-out	Original Inside Diameter	Max. Wear Limit	Maximum Machine Diameter		Bracket Bolts (ft. lbs.)	Mounting Bolts (ft. lbs.)
2002	Sedona	F	1.020	0.940	0.0020	—	—	—	0.100	NA	18-26
		R	—	—	—	10.00	10.05	10.05	0.040	—	—
2003	Sedona	F	1.020	0.940	0.0020	—	—	—	0.100	NA	18-26
		R	—	—	—	10.00	10.05	10.05	0.040	—	—
2004	Sedona	F	1.020	0.940	0.0020	—	—	—	0.100	NA	18-26
		R	—	—	—	10.00	10.05	10.05	0.040	—	—
2005	Sedona	F	1.020	0.940	0.0020	—	—	—	0.100	NA	18-26
		R	—	—	—	10.00	10.05	10.05	0.040	—	—
2006	Sedona	F	1.180	1.100	0.0012	—	—	—	—	58-72	16-23
		R	0.470	NA	0.0020	—	—	—	—	36-43	16-23

NA: Not Available

F: Front

R: Rear

09474_SEDO_C0011

SCHEDULED MAINTENANCE INTERVALS
Kia—Sedona

TO BE SERVICED	TYPE OF SERVIC	VEHICLE MILEAGE INTERVAL (x1000)																
		7.5	15	22.5	30	37.5	45	52.5	60	67.5	75	82.5	90	97.5	100	105	112.5	120
Accessory drive belts	S/I	✓	✓	✓	✓	✓	✓	✓	✓	✓	✓	✓	✓	✓	✓	✓	✓	✓
Air cleaner element	I/R		✓		✓		✓		✓		✓		✓			✓		✓
Air conditioner system	S/I	Inspect the system operation and refrigerant amount annually.																
Brake lines, hoses and connections	S/I		✓		✓		✓		✓		✓		✓			✓		✓
Chassis and body fasteners	T	✓	✓	✓	✓	✓	✓	✓	✓	✓	✓	✓	✓	✓	✓	✓	✓	✓
Cooling system hoses and coolant level	S/I		✓		✓		✓		✓		✓		✓			✓		✓
CV-joint boots	S/I				✓				✓				✓					✓
Engine coolant (2002-05)	R				✓				✓				✓					✓
Engine coolant (2006)	R								✓				✓					✓
Engine oil and filter	R	✓	✓	✓	✓	✓	✓	✓	✓	✓	✓	✓	✓	✓	✓	✓	✓	✓
Exhaust system heat shields	S/I				✓				✓				✓					✓
Front and rear brakes	S/I				✓				✓				✓					✓
Front ball joints	S/I				✓				✓				✓					✓
Fuel filter (2002-05)	R								✓									
Fuel filter (2006)	R					✓					✓						✓	
Fuel tank air filter (2006)	R				✓				✓				✓					
Fuel lines and hoses	S/I				✓				✓				✓					✓
Idle speed	A				✓				✓				✓					✓
Locks and hinges	L	✓	✓	✓	✓	✓	✓	✓	✓	✓	✓	✓	✓	✓	✓	✓	✓	✓
Spark plugs (2002-05)	R				✓				✓				✓					✓
Spark plugs (2006)	R															✓		
Steering operation and linkage	S/I				✓				✓				✓					✓
Timing belt	R								✓									✓
Valve Clearance (2006)	A								✓									✓

R: Replace   S/I: Inspect and service, if needed   L: Lubricate   A: Adjust   T: Tighten

FREQUENT OPERATION MAINTENANCE (SEVERE SERVICE)

If a vehicle is operated under any of the following conditions it is considered severe service

- **Towing a trailer or using a camper or car-top carrier.**
- **Repeated short trips of less than 5 miles in temperatures below freezing, or trips of less than 10 miles in any temperature.**
- **Extensive idling or low-speed driving for long distances as in heavy commercial use, such as delivery, taxi or police cars.**
- **Operating on rough, muddy or salt-covered roads.**
- **Operating on unpaved or dusty roads.**
- **Driving in extremely hot (over 90°F) conditions.**

Engine oil and filter: replace every 5000 miles or 5 months, whichever occurs first.

Air cleaner element: inspect ever 15,000 miles or 15 months and replace every 30,000 miles or 30 months, whichever occurs first.

Fuel system hoses (California models only): replace every 105,000 miles.

Emission system hoses (non-California models): inspect every 55,000 miles or 55 months, whichever occurs first.

Emission system hoses (California models): inspect every 60,000 miles or 60 months, whichever occurs first.

Front and rear disc brakes: inspect every 15,000 miles or 15 months, whichever occurs first.

Chassis and body fasteners: tighten every 15,000 miles or 15 months, whichever occurs first.

Locks and hinges: lubricate every 5000 miles or 5 months, whichever occurs first.

ENGINE REPAIR

Alternator

REMOVAL & INSTALLATION

1. Before servicing the vehicle, refer to the precautions at the beginning of this section.
2. Disconnect the negative battery cable.
3. Disconnect the electrical connectors from the alternator.
4. Remove the accessory drive belt.
5. Remove the alternator mounting bolts.
6. Remove the alternator.

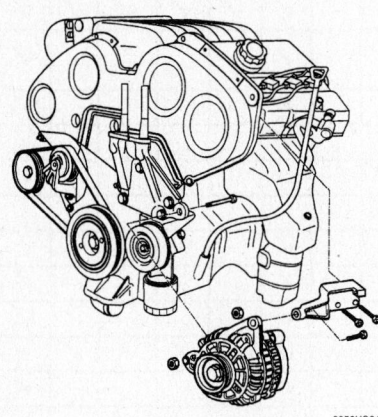

Alternator mounting exploded view

9358HG01

To install:
7. Install the alternator.
8. Connect the alternator electrical connectors. Tighten the battery terminal connector nut to 60 inch lbs. (7 Nm).
9. Install the accessory drive belt.
10. Connect the negative battery cable.

Ignition Timing

This vehicle is equipped with a Distributorless Ignition System (DIS). No adjustment is necessary or possible.

Engine Assembly

REMOVAL & INSTALLATION

3.5L Engine

1. Before servicing the vehicle, refer to the precautions at the beginning of this section.
2. Drain the cooling system.
3. Drain the transaxle fluid.

4. Relieve the fuel system pressure.
5. Remove the battery.
6. Remove the engine cover.
7. Remove the air cleaner assembly.
8. Disconnect the engine wiring harness connectors.
9. Disconnect the alternator wiring harness connectors.
10. Disconnect the oil pressure switch connector.
11. Disconnect the oil pressure sensor connector.
12. Disconnect the starter motor wiring harness connectors.
13. Disconnect the transaxle oil cooler hose.
14. Remove the upper and lower radiator hoses.
15. Remove the radiator.
16. Remove the engine ground cable.
17. Disconnect the brake booster vacuum hose.
18. Disconnect the EVAP canister hose.
19. Disconnect the fuel lines.
20. Remove the heater hoses.
21. Disconnect the throttle and cruise control cables.
22. Disconnect the transaxle shift cable.
23. Disconnect the power steering pump hose.
24. Remove the oil pan shield.
25. Remove the exhaust front pipe.
26. Remove the outer tie rod ends.
27. Remove the sway bar links.
28. Remove the lower ball joints.
29. Remove the axle halfshafts.
30. Remove the intermediate shaft bolt.
31. Remove the No. 3 and 4 engine mounts.
32. Support the front subframe with a suitable powertrain jack.
33. Remove the subframe bolts.
34. Remove the impact bar bolts.
35. Lower the powertrain and subframe away from the vehicle.

To install:
36. Installation is the reverse of the removal procedure, while using the following torque values:
- Subframe bolts: 88-101 ft. lbs. (120-137 Nm)
- Impact bar bolts: 69-85 ft. lbs. (93-115 Nm)
- Engine mount bolts: 49-69 ft. lbs. (67-93 Nm)
- Engine mount through-bolts: 63-86 ft. lbs. (85-117 Nm)
- Tie rod ends: 51-58 ft. lbs. (69-79 Nm)

3.8L Engine

1. Before servicing the vehicle, refer to the precautions at the beginning of this section.
2. Remove the engine cover. Discharge the air conditioning system. Record the radio anti theft code data. Relieve the fuel system fuel pressure.
3. Disconnect the negative battery terminal. Disconnect the positive battery terminal. Remove the battery from the vehicle.
4. Remove the air intake hose and air cleaner assembly.
5. Disconnect the MAF electrical connector. Disconnect the breather hose from the air cleaner hose. Remove the intake air hose and air cleaner assembly. Remove the battery tray.
6. Remove the radiator grille upper cover. Disconnect the left and right oxygen sensor connectors.
7. Drain the cooling system. Remove the upper and lower radiator hoses. Disconnect and plug the automatic transaxle fluid lines.
8. Disconnect the high and low pressure pipes from the radiator and compressor.
9. Disconnect the engine wiring harness. Remove the engine underhood fuse and relay box. Disconnect the FAM connectors.
10. Unscrew the FAM mounting bolts and remove the FAM from the splash shield.
11. Disconnect the transaxle wire harness connector and remove the transaxle control cable. Disconnect and remove the heater hoses.
12. Remove the brake vacuum hose. Remove the upper power steering hose.
13. Raise and support the vehicle safely. Remove the front tires and wheels.
14. Remove the radiator grille upper cover, for convenience.
15. Disconnect the power steering return hose. Remove the radiator support upper member, for convenience.

➡ **The bottom side bolt, which can be seen after removing the under cover, should be loosened for removal of the radiator support upper member assembly.**

16. Remove the under cover. Drain the engine oil. Remove the front exhaust pipe.
17. Disconnect the wheel speed sensor. Disconnect the stabilizer bar link.
18. Remove the brake caliper assemblies. Wire them to the side to prevent damage to the brake hoses.

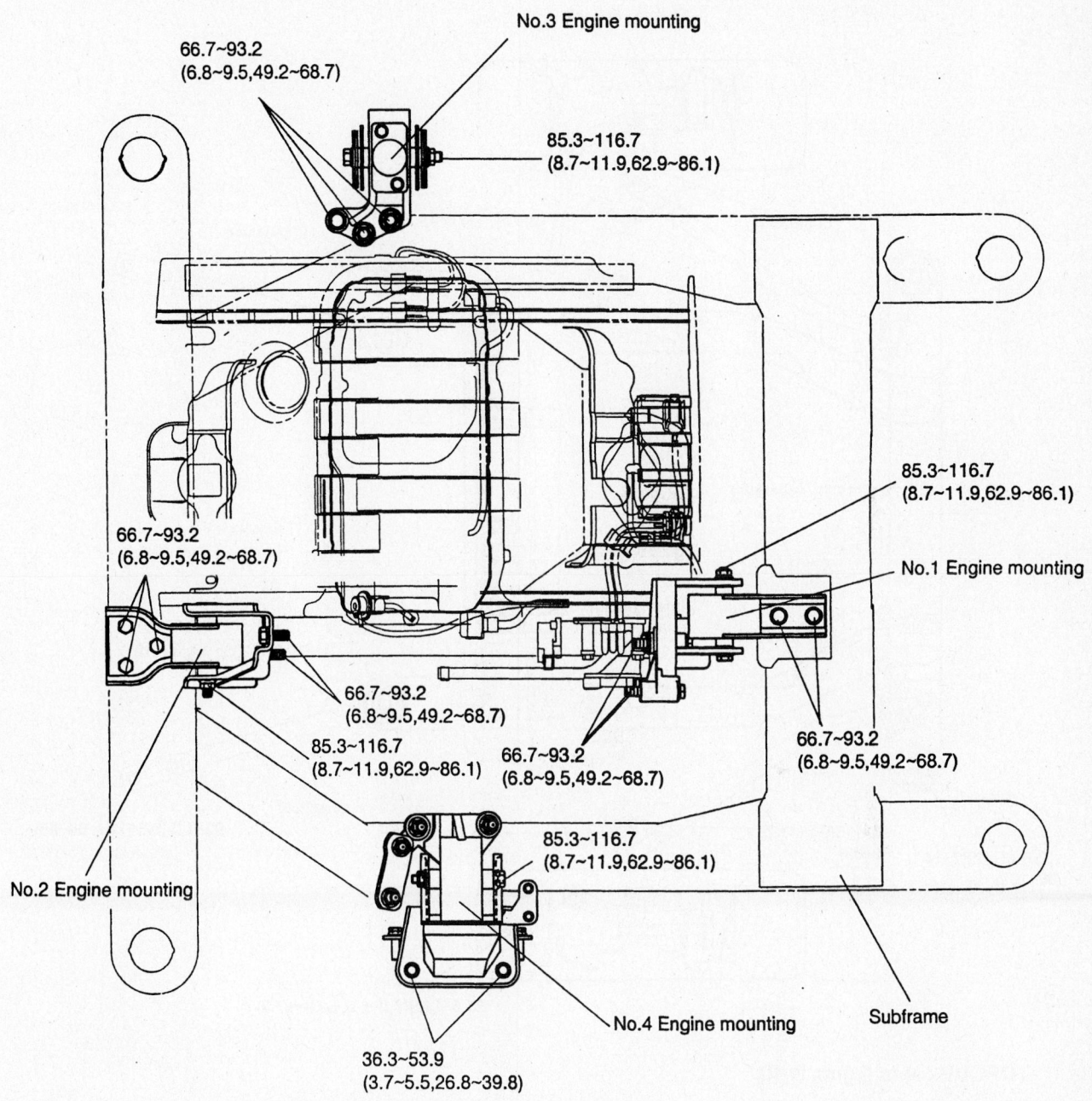

No.3 Engine mounting

66.7~93.2
(6.8~9.5,49.2~68.7)

85.3~116.7
(8.7~11.9,62.9~86.1)

85.3~116.7
(8.7~11.9,62.9~86.1)

No.1 Engine mounting

66.7~93.2
(6.8~9.5,49.2~68.7)

66.7~93.2
(6.8~9.5,49.2~68.7)

66.7~93.2
(6.8~9.5,49.2~68.7)

85.3~116.7
(8.7~11.9,62.9~86.1)

66.7~93.2
(6.8~9.5,49.2~68.7)

85.3~116.7
(8.7~11.9,62.9~86.1)

No.2 Engine mounting

No.4 Engine mounting

Subframe

36.3~53.9
(3.7~5.5,26.8~39.8)

TORQUE : N•m (kg•m, lb•ft)

9358HG02

Engine mount locations and torque specifications—3.5L Engine

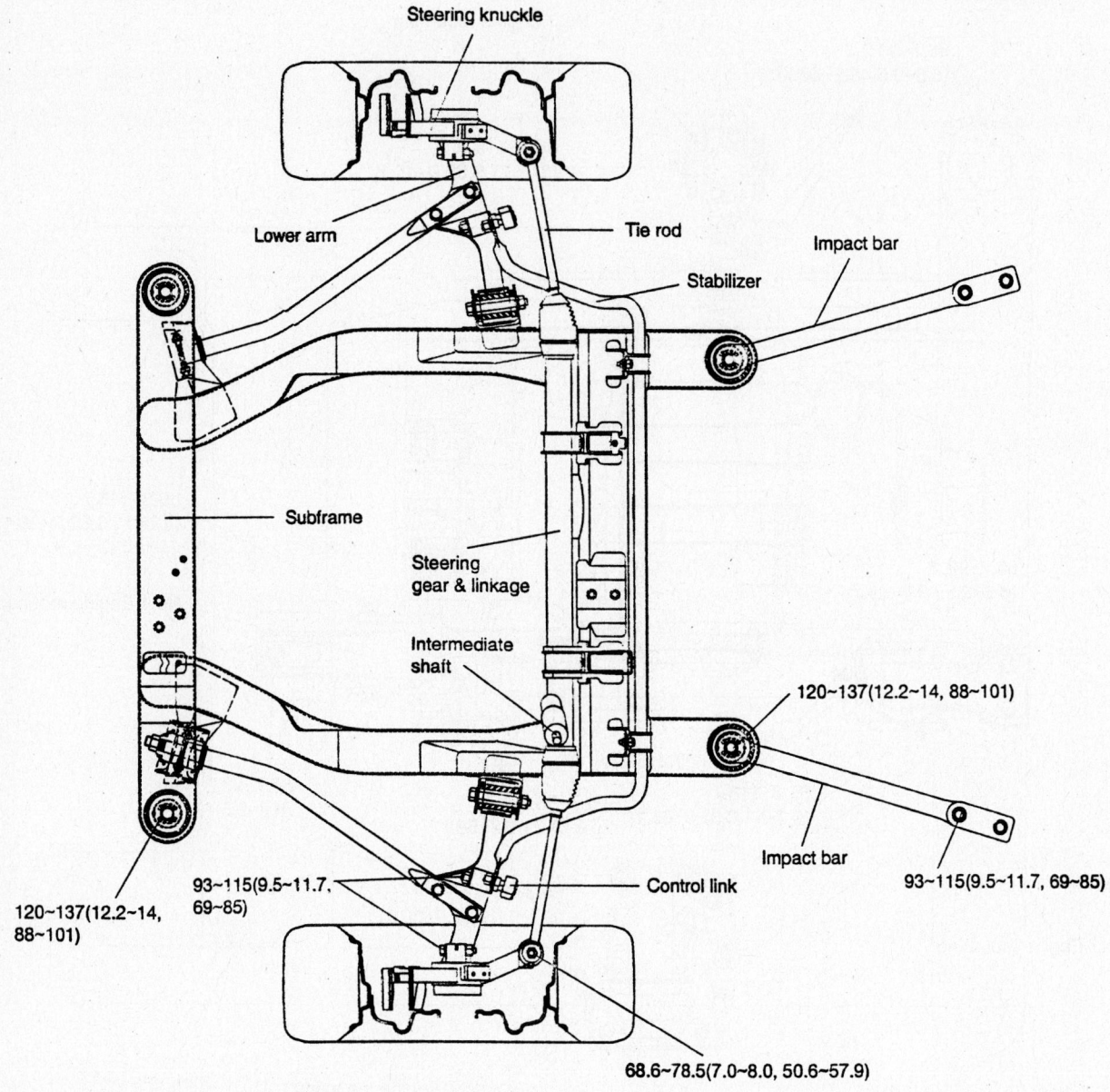

Steering knuckle

Lower arm

Tie rod

Impact bar

Stabilizer

Subframe

Steering
gear & linkage

Intermediate
shaft

120~137(12.2~14, 88~101)

Impact bar

93~115(9.5~11.7, 69~85)

Control link

120~137(12.2~14, 88~101)

93~115(9.5~11.7, 69~85)

68.6~78.5(7.0~8.0, 50.6~57.9)

TORQUE : N•m (kg•m, lb•ft)

9358HG03

Front subframe bolt locations and torque specifications—3.5L Engine

19. Properly support the engine and transaxle assembly using the proper engine/transaxle support tool.

20. Remove the engine mounting bolt. Remove the transaxle insulator mounting bolt.

21. Support the engine and transaxle assembly with a jack. Remove the sub frame with the engine and transaxle assembly.

To install:

22. Installation is the reverse of the removal procedure.

23. Adjust the shift cable. Fill the engine and transaxle to specification with the proper fluids.

24. Before starting the engine turn the ignition switch to the RUN position so that the fuel pump runs for approximately two seconds and the fuel line pressurizes.

25. Reprogram the radio anti theft code data.

26. Repeat the operation two or three more times, than check for fuel leakage.

Water Pump

REMOVAL & INSTALLATION

3.5L Engine

1. Before servicing the vehicle, refer to the precautions at the beginning of this section.

2. Drain the cooling system.

3. Disconnect the negative battery cable.

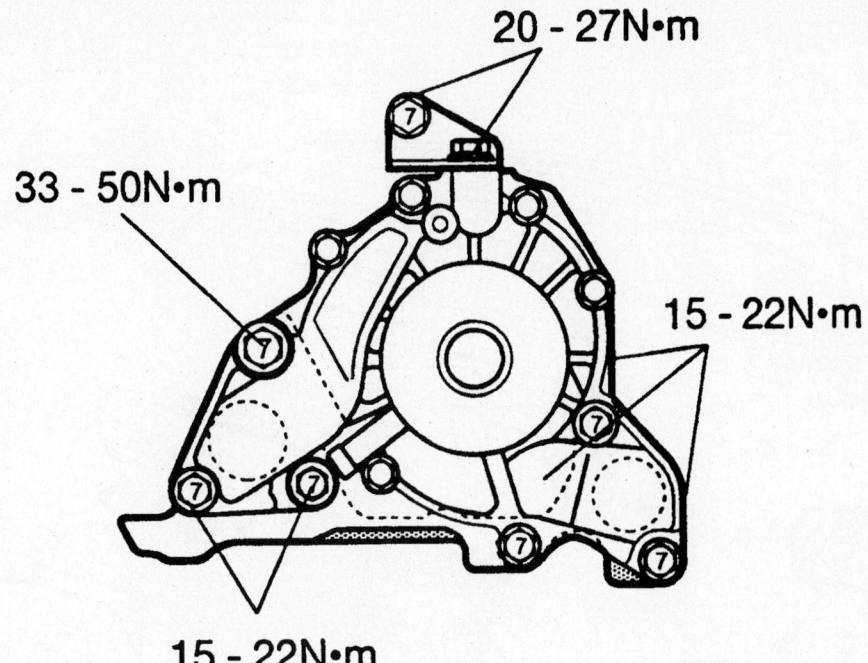

20 - 27N•m

33 - 50N•m

15 - 22N•m

15 - 22N•m

9358HG04

Water pump bolt locations and torque specifications—3.5L Engine

4. Remove the engine cover.
5. Remove the accessory drive belt.
6. Remove the idler pulley.
7. Remove the crankshaft pulley.
8. Remove the power steering pump pulley.
9. Remove the tensioner pulley.
10. Remove the upper and lower timing belt covers.
11. Remove the No. 3 engine mount.
12. Remove the timing belt.
13. Remove the timing belt tensioner.
14. Remove the water pump.

To install:
15. Installation is the reverse of the removal procedure, while using the following torque values:
- Water pump bolts: refer to the illustration
- Crankshaft pulley: 130-138 ft. lbs. (180-190 Nm)

3.8L Engine

1. Before servicing the vehicle, refer to the precautions at the beginning of this section.
2. Record the radio anti theft code data.
3. Drain the cooling system.
4. Disconnect the negative battery cable.
5. Remove the accessory drive belt. Remove the water pump pulley.
6. Remove the water pump retaining bolts. Remove the water pump from the vehicle.

To install:
7. Installation is the reverse of the removal procedure.
8. Reprogram the radio anti theft code data.
9. Be sure to use a new water pump gasket. Torque the retaining bolts to 16 ft. lbs.
10. Torque the water pump pulley bolts to 6 ft. lbs.
11. Be sure to refill the cooling system with the proper grade and type coolant. Start the engine and check for leaks.

Heater Core

REMOVAL & INSTALLATION

3.5L Engine

1. Before servicing the vehicle, refer to the precautions at the beginning of this section.
2. Drain the cooling system.
3. Disconnect the negative battery cable.
4. Remove the heater hoses.
5. Remove the driver air bag module.
6. Remove the steering wheel.
7. Remove the turn signal assembly.
8. Remove the upper and lower steering column covers.
9. Remove the A-pillar upper trim.
10. Remove the instrument cluster trim.

11. Remove the instrument cluster.
12. Remove the A-pillar lower trim.
13. Remove the hood release handle.
14. Remove the left lower trim panel.
15. Remove the multi box.
16. Remove the audio panel.
17. Remove the ventilation control panel.
18. Remove the radio.
19. Remove the glove box.
20. Remove the center console.
21. Remove the audio system speakers and mounting bolts from the top of the instrument panel.
22. Remove the side covers.
23. Remove the T-bar mounting bolts.
24. Remove the bottom mounting bolts.
25. Remove the instrument panel wiring harness connectors.
26. Remove the instrument panel.
27. Remove the heater unit.

To install:
28. Install the heater unit.
29. Install the instrument panel.
30. Connect the instrument panel wiring harness connectors.
31. Install the bottom mounting bolts.
32. Install the T-bar mounting bolts.
33. Install the side covers.
34. Install the upper mounting bolts and audio system speakers.
35. Install the center console.
36. Install the glove box.
37. Install the radio.
38. Install the ventilation control panel.
39. Install the audio panel.

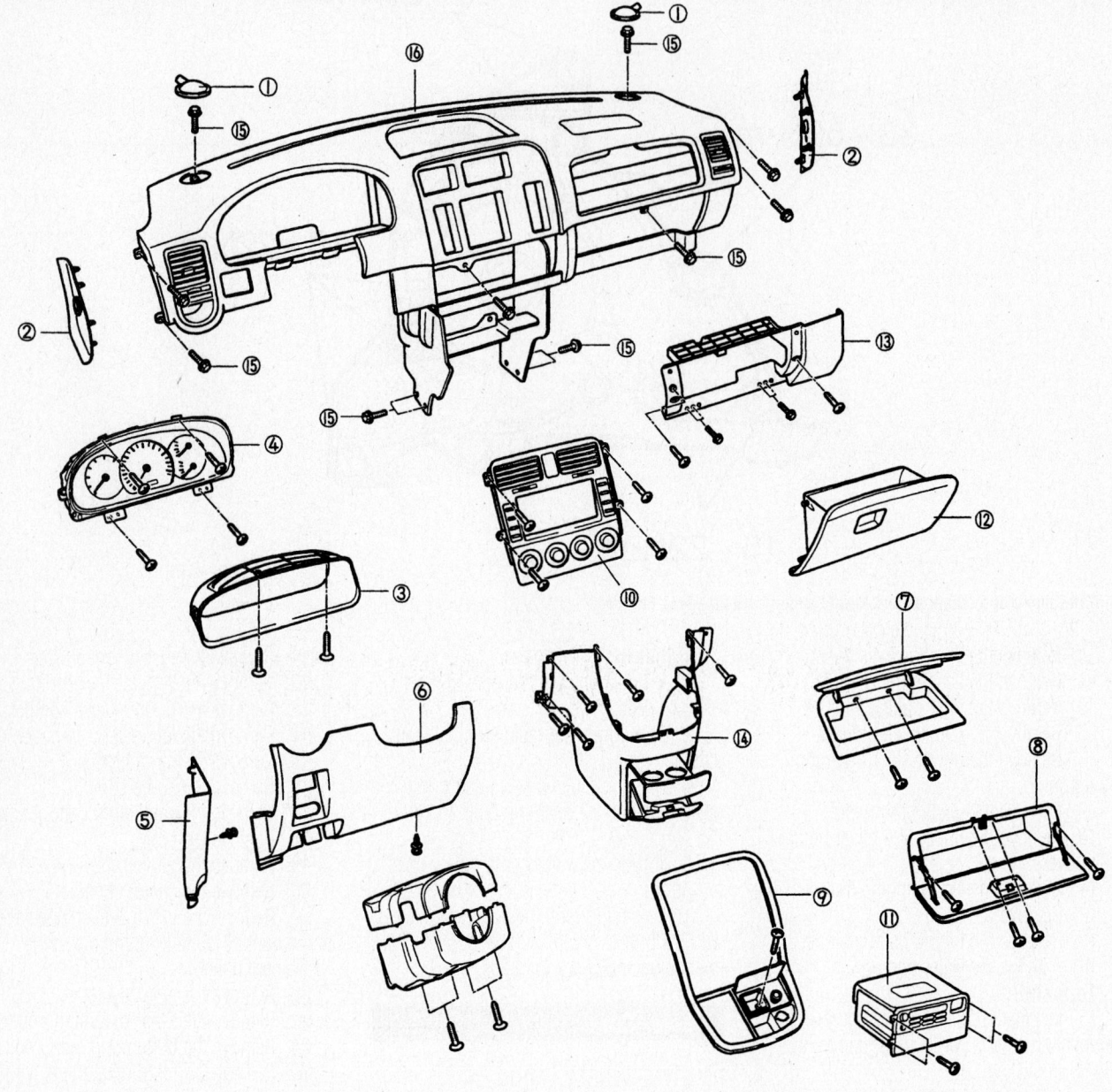

(1) Speaker assembly

(2) Side cover

(3) Instrument cluster trim

(4) Instrument cluster

(5) A-pillar lower trim

(6) Lower LH panel

(7) Center upper tray

(8) Multi box

(9) Audio panel

(10) Heater control panel

(11) Audio

(12) Glove box

(13) Lower RH panel

(14) Center console

(15) Mounting bolt

(16) Instrument panel

9358HG29

Instrument panel exploded view—3.5L Engine

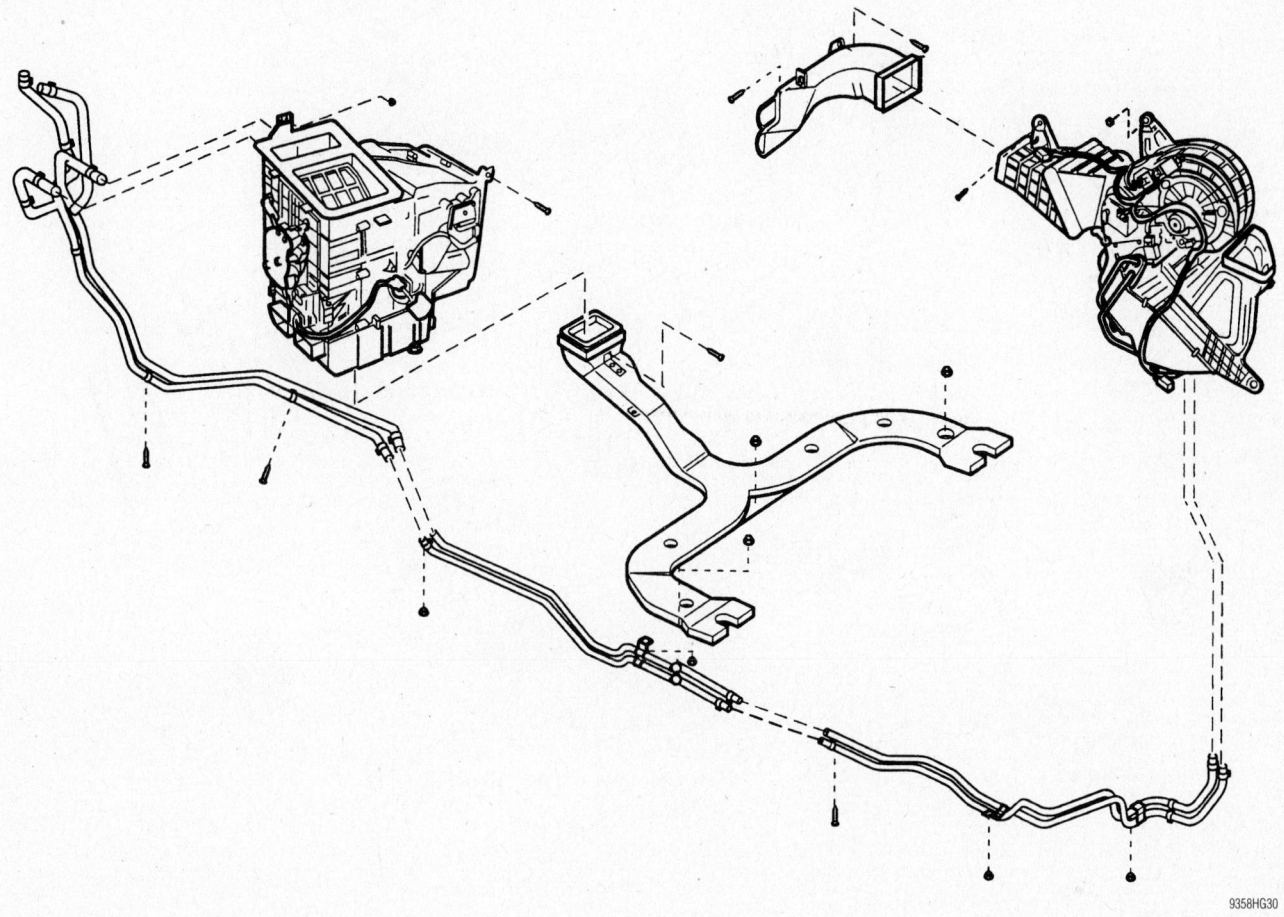

9358HG30

Front and rear heater unit mounting exploded view—2002—05 vehicles

40. Install the multi box.
41. Install the left lower trim panel.
42. Install the hood release handle.
43. Install the A-pillar lower trim.
44. Install the instrument cluster.
45. Install the instrument cluster trim.
46. Install the A-pillar upper trim.
47. Install the upper and lower steering column covers.
48. Install the turn signal assembly.
49. Install the steering wheel.
50. Install the driver air bag module.
51. Install the heater hoses.
52. Connect the negative battery cable.
53. Fill the cooling system.
54. Start the engine and check for leaks and proper heater operation.

3.8L Engine

1. Before servicing the vehicle, refer to the precautions at the beginning of this section.

2. Record the radio anti theft code data.
3. Disconnect the negative battery cable. Drain the cooling system. Discharge the air conditioning system.
4. Disconnect and plug the heater hoses.
5. Remove the front seat.
6. Tilt the steering column down. Remove the screws and detach the clips from the cluster facia panel. Disconnect the connector and remove the cluster facia panel.
7. Remove the screws and detach the clips from the center facia panel. Disconnect the connector and remove the center facia panel. Remove the radio retaining screws. Disconnect the connectors and remove the radio.
8. Disconnect the damper from the glovebox lid. Remove the glovebox lid from the lift. Disconnect the retaining pins and remove the glovebox assembly.

9. Remove the dash pad side cover, center cover and under cover. Remove the front A-pillar trim.
10. Remove the photo sensor. Remove the speaker connector.
11. Disconnect the passenger's air bag connector.
12. Loosen the bolt and nut. Remove the dash pad.
13. Disconnect the electrical connectors from the cross bar assembly. Loosen the bolts and nuts that retain the assembly in place. Remove the cross bar assembly.
14. Disconnect the connectors from the temperature control actuator, the mode control actuator and the evaporator temperature sensor.
15. Remove the heater/blower unit from the vehicle, after removing the three retaining screws.
16. Separate the blower unit from the

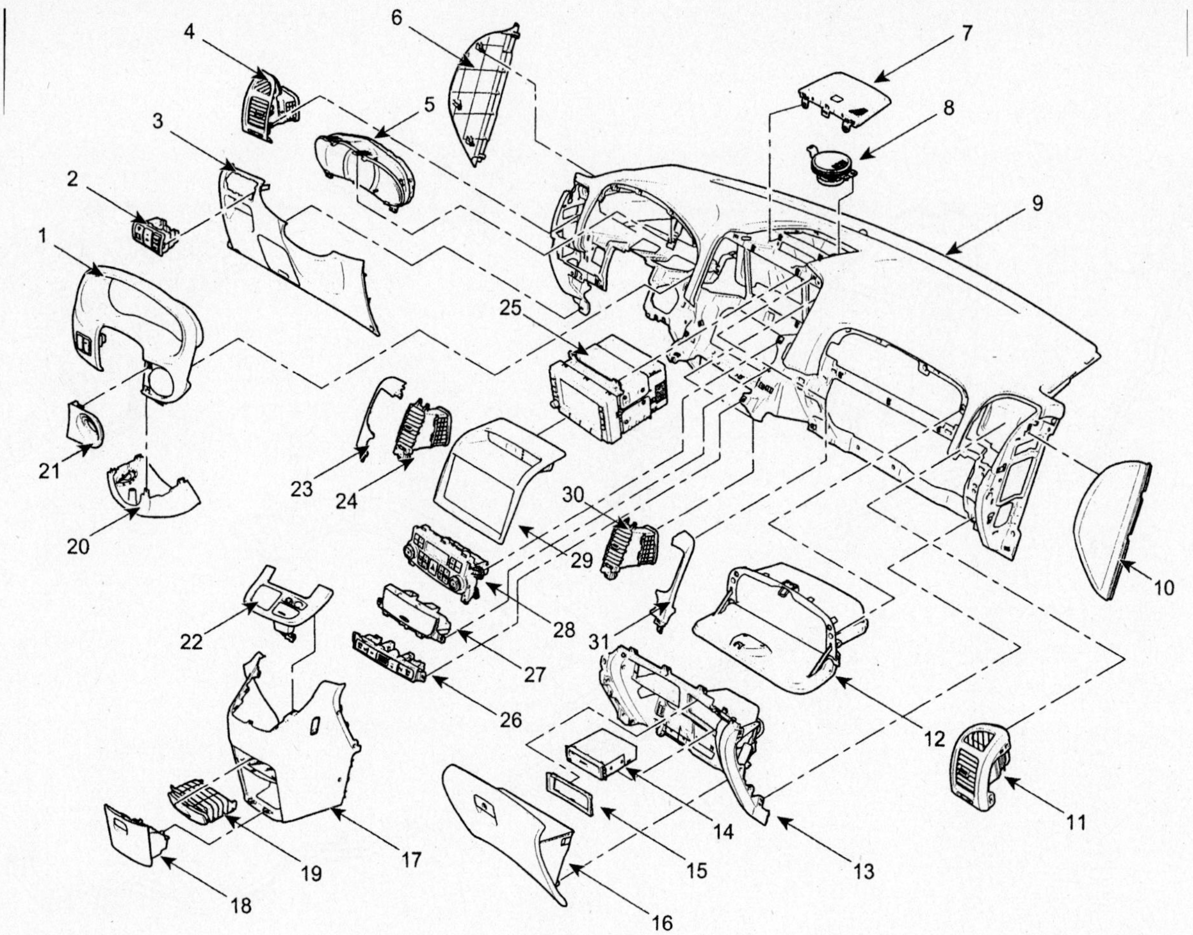

Instrument panel exploded view—2006 vehicles

1. Cluster facia panel	17. Front console
2. Switch assembly	18. Consol tray
3. Lower panel	19. Cup holders
4. Side air vent	20. Shroud
5. Cluster	21. Key box cover
6. Side cover	22. Console upper cover
7. Center speaker cover	23. Center garnish
8. Center speaker	24. Center air vent
9. Main crash pad	25. Audio assembly
10. Side cover	26. Switch assembly
11. Side air vent	27. Center tray
12. Multi box	28. Heater control unit
13. Lower crash pad panel	29. Center facia panel
14. DVD	30. Center air vent
15. DVD cover	31. Center garnish
16. Glove box	

09474_SEDO_G0001

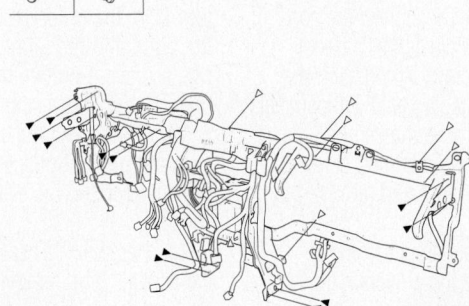

Fastener Locations
► : Bolt   ▷ : Nut

Cross bar assembly—2006 vehicles

09474_SEDO_G0002

heater unit, after removing the two retaining screws.

17. Remove the heater core cover. Remove the heater core from its mounting.

To install:

18. Install the heater core in the heater unit. Attach the heater unit to the blower housing.

19. Position the assembly in the vehicle. Attach the retaining screws.

20. Continue the installation in the reverse order of the removal procedure.

21. Make sure that you reprogram the anti theft code for the radio.

Cylinder Head

REMOVAL & INSTALLATION

3.5L Engine

1. Before servicing the vehicle, refer to the precautions at the beginning of this section.

➡ **Do not remove the cylinder head until the engine drops below normal operating temperature.**

2. Drain the cooling system.
3. Relieve the fuel system pressure.
4. Disconnect the negative battery cable.
5. Remove the engine cover.
6. Remove the timing belt.
7. Remove the intake manifold.
8. Remove the exhaust manifolds.
9. Remove the cylinder head covers.
10. Remove the camshafts.
11. Remove the rocker arms and lash adjusters.

➡ **Keep all valve train components in order for assembly.**

12. Remove the cylinder head bolts.
13. Remove the cylinder heads.

To install:

14. Install the cylinder heads with new gaskets. Tighten the bolts in sequence to 78-85 ft. lbs. (105-115 Nm).

15. Installation is the reverse of the removal procedure.

3.8L Engine

1. Before servicing the vehicle, refer to the precautions at the beginning of this section.

➡ **Do not remove the cylinder head until the engine drops below normal operating temperature.**

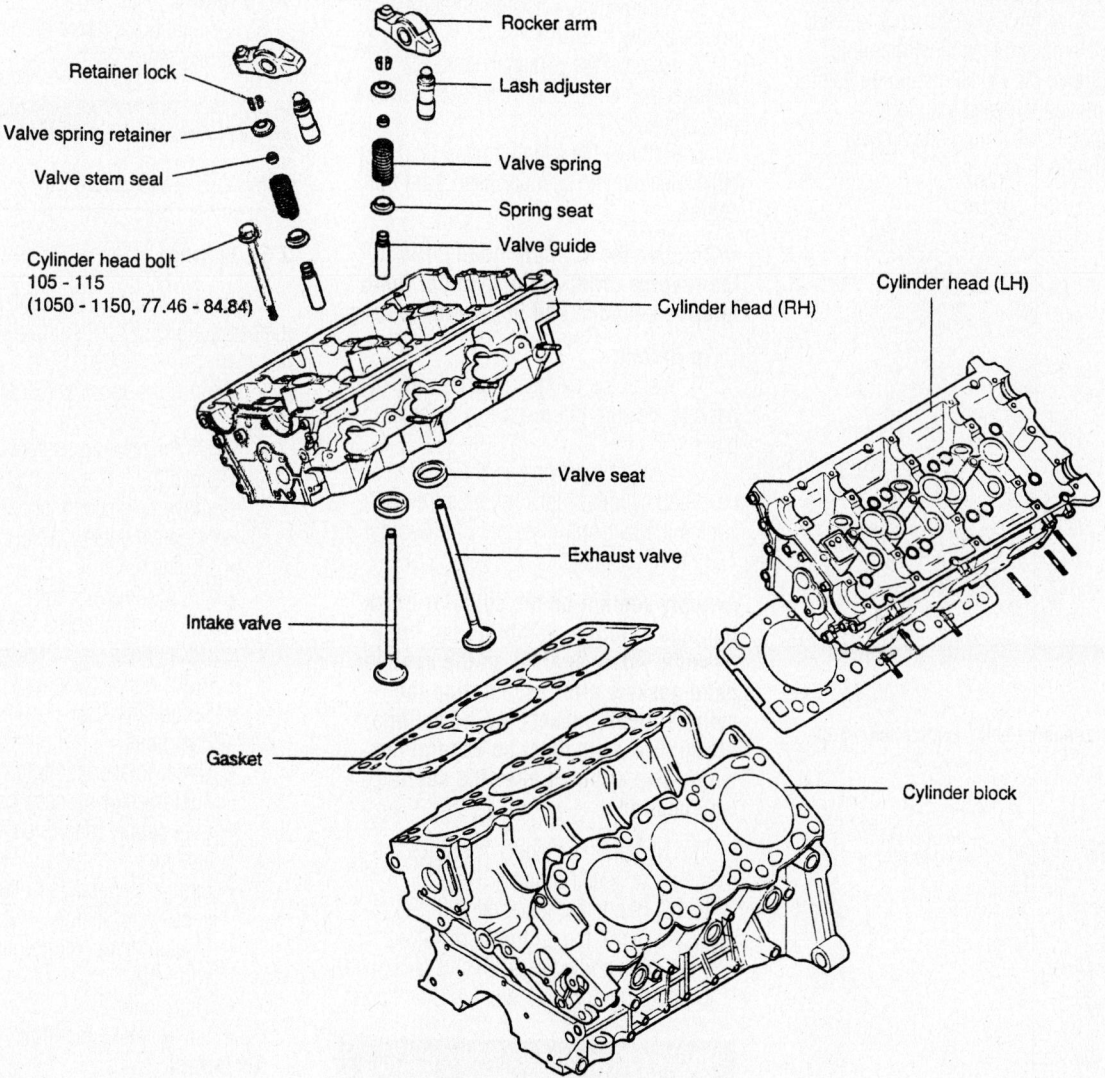

Retainer lock
Valve spring retainer
Valve stem seal
Cylinder head bolt
105 - 115
(1050 - 1150, 77.46 - 84.84)
Rocker arm
Lash adjuster
Valve spring
Spring seat
Valve guide
Cylinder head (RH)
Cylinder head (LH)
Valve seat
Exhaust valve
Intake valve
Gasket
Cylinder block

TORQUE : N•m (kg•cm, lb•ft)

9358HG05

Cylinder head exploded view—3.5L Engine

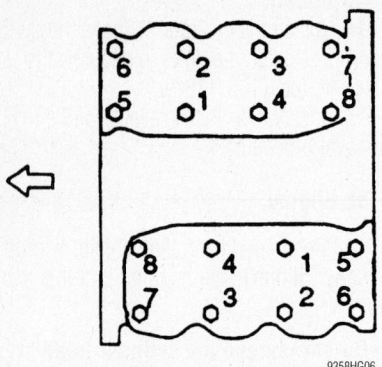

Cylinder head bolt torque sequence

2. Remove the engine from the vehicle.

3. Remove the exhaust manifold retaining bolts. Remove the exhaust manifold.

4. Remove the intake manifold retaining bolts. Remove the intake manifold.

5. Remove the timing chain cover. Remove the timing chain.

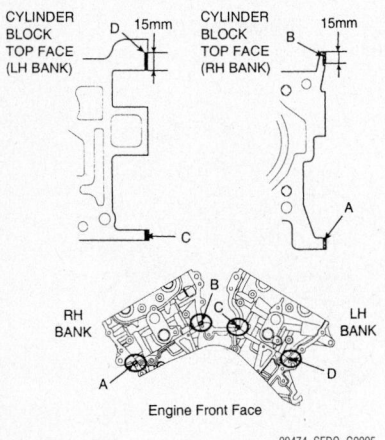

Sealant to engine block application—3.8L Engine

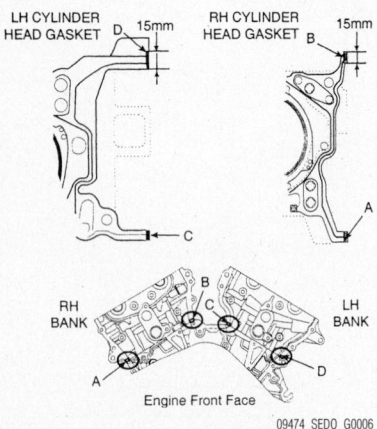

Sealant to cylinder head application—3.8L Engine

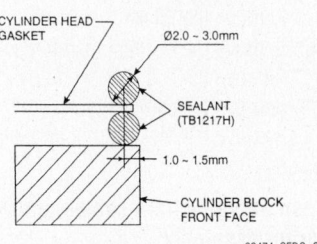

Bead width : 2.0~3.0 mm
Sealant locations : 1.0~1.5mm from block surface
Recommended sealant :Liquid sealant TB1217H

Sealant to engine block and cylinder head application—3.8L Engine

6. Remove the water temperature control assembly.

7. Remove the camshaft covers. Remove the camshaft bearing cap. Remove the camshafts.

8. Remove the cylinder head retaining bolts. Remove the cylinder head from the engine.

➡**Remove the cylinder head bolts in the reverse order of the cylinder head torque sequence and in several passes.**

To install:

9. Install the cylinder heads using new gaskets. Be sure to use new cylinder head bolts.

10. Sealant locations on the cylinder head and cylinder block must be free from contamination. Apply sealant part number TB1217H, as indicated.

➡**Apply sealant on the cylinder block top face before assembling the head gaskets. Apply sealant on the cylinder head gaskets after assembling the cylinder head gaskets on the cylinder block. The parts must be assembled within five minutes after the sealant was applied.**

11. Tighten the cylinder head bolts in sequence to specification.

12. Continue the installation in the reverse order of the removal procedure.

13. Check and adjust the valve clearance.

14. Reprogram the radio anti theft code.

Rocker Arms

REMOVAL & INSTALLATION

3.5L Engine

1. Before servicing the vehicle, refer to the precautions at the beginning of this section.

2. Remove or disconnect the following:
   - Negative battery cable
   - Engine cover
   - Timing belt
   - Cylinder head covers
   - Camshafts
   - Rocker arms and lash adjusters

➡**Keep all valvetrain components in order for installation.**

3. Inspect the roller visually. If any damage is found, replace it.

4. Check that the roller operates smoothly. If there is excessive clearance, replace it.

To install:

5. Installation is the reverse of the removal procedure.

Intake Manifold

REMOVAL & INSTALLATION

3.5L Engine

1. Before servicing the vehicle, refer to the precautions at the beginning of this section.

2. Record the radio anti theft code data.

3. Drain the cooling system.

4. Relieve the fuel system pressure.

5. Remove or disconnect the following:
   - Negative battery cable
   - Engine cover
   - Upper radiator hose
   - Positive Crankcase Ventilation (PCV) hose
   - Brake booster vacuum hose
   - Surge tank stays
   - Fuel lines
   - Upper intake manifold (surge tank)
   - Fuel injector harness connectors
   - Fuel supply manifold with injectors attached
   - Engine Coolant Temperature (ECT) sensor connector
   - Coolant temperature gauge sensor connector
   - Thermostat
   - Lower intake manifold

To install:

6. Installation is the reverse of the removal procedure, while using the following torque values:
   - Lower intake manifold nuts: 15-22 ft. lbs. (20-30 Nm)
   - Upper intake manifold bolts: 11-15 ft. lbs. (15-20 Nm)

7. Reprogram the radio anti theft codes.

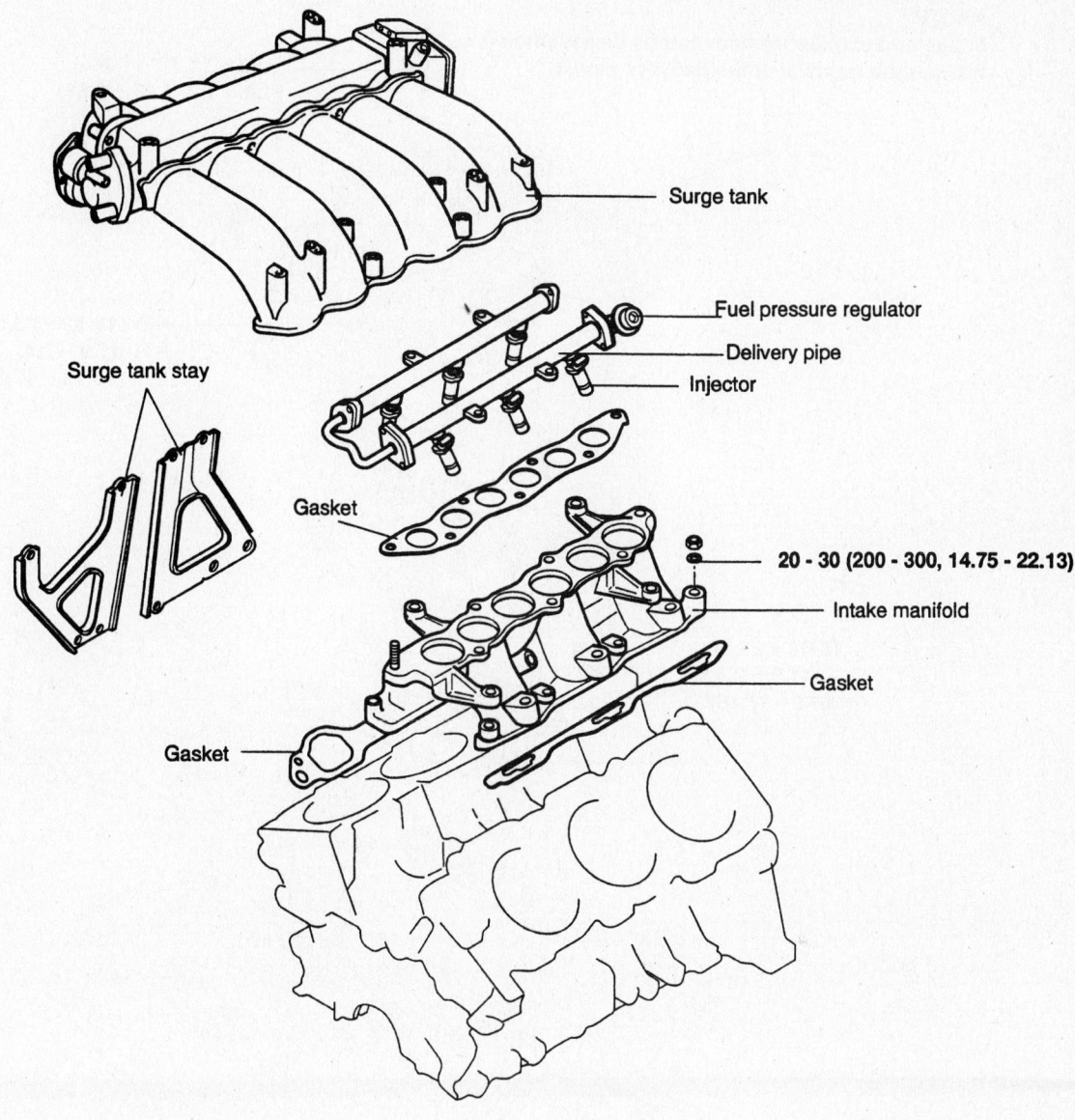

Surge tank

Fuel pressure regulator

Delivery pipe

Injector

Surge tank stay

Gasket

20 - 30 (200 - 300, 14.75 - 22.13)

Intake manifold

Gasket

Gasket

TORQUE : N•m (kg•cm, lb•ft)

9358HG07

Intake manifold and related components—3.5L Engine

3.8L Engine

1. Before servicing the vehicle, refer to the precautions at the beginning of this section.

2. Record the radio anti theft code data.

3. Remove the engine cover. Disconnect the negative battery cable. Drain the cooling system.

4. Remove the intake air hose and air cleaner assembly. Disconnect the MAF connector.

5. Disconnect the breather hose from the air cleaner. Remove the intake air hose and air cleaner assembly.

6. Disconnect the right side oxygen sensor connector. Disconnect the right side injector connector and ignition coil connector.

7. Disconnect the PCSV connector, MAP sensor connector and PCSV hose.

8. Disconnect the ETC connector and knock sensor connector. Remove the ETC bracket.

9. Disconnect the coolant hoses. Disconnect the PCV valve. Disconnect the brake vacuum hose.

10. Remove the surge tank stay. Remove the connector bracket from the surge tank. Remove the surge tank.

11. Disconnect the breather pipe assembly. Disconnect the left side injector connector.

12. Remove the intake manifold retaining bolts.

13. Remove the delivery pipe and intake manifold together, as an assembly.

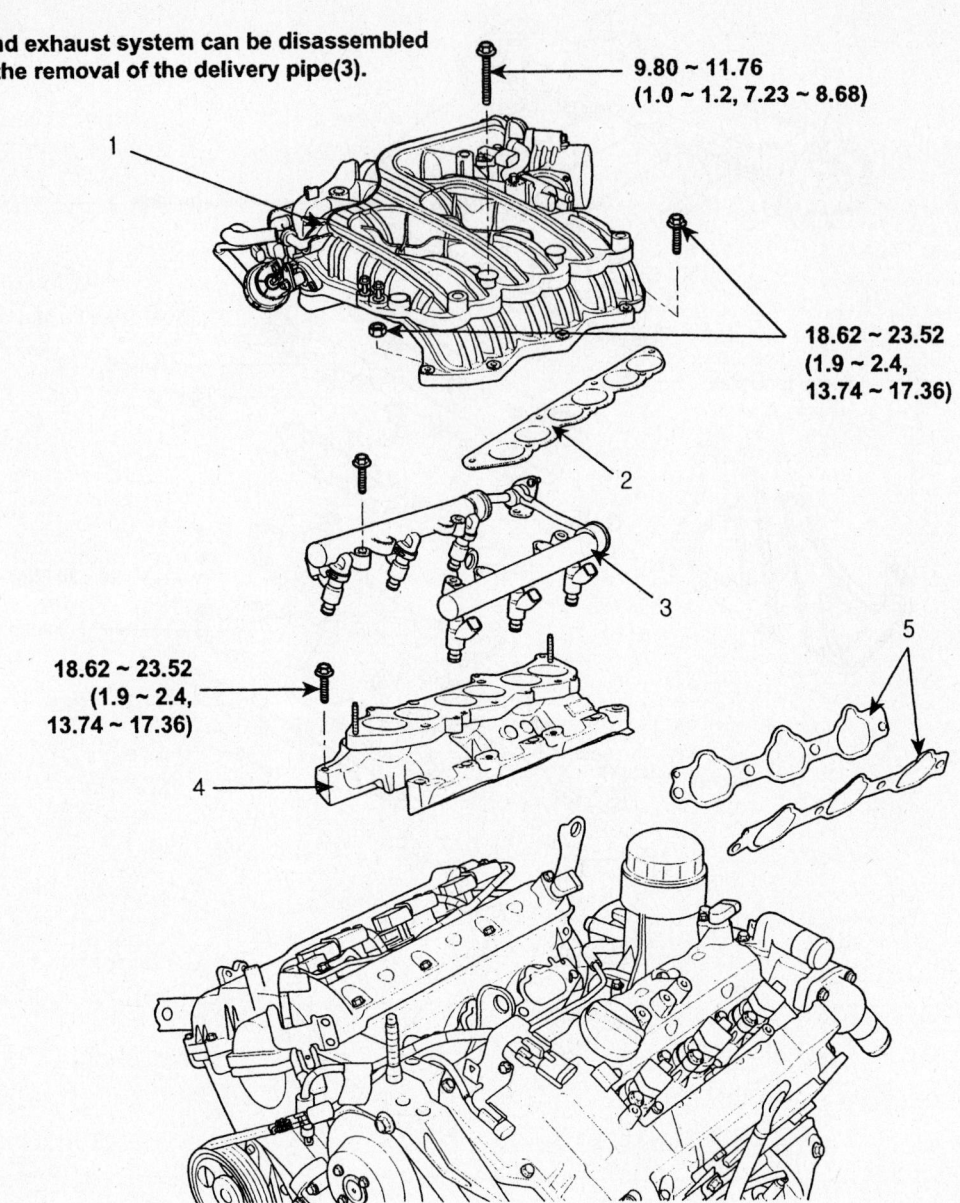

\<NOTE\>
**Intake and exhaust system can be disassembled
without the removal of the delivery pipe(3).**

9.80 ~ 11.76
(1.0 ~ 1.2, 7.23 ~ 8.68)

1

18.62 ~ 23.52
(1.9 ~ 2.4,
13.74 ~ 17.36)

2

3

5

18.62 ~ 23.52
(1.9 ~ 2.4,
13.74 ~ 17.36)

4

TORQUE : N.m (kgf.m, lb-ft)

1. Surge tank
2. Surge tank gasket
3. Delivery pipe

4. Intake manifold
5. Intake manifold gasket

09474_SEDO_G0003

Intake manifold and related components —3.8L Engine

To install:

14. Installation is the reverse of the removal procedure.

15. Be sure to use a new intake manifold gasket.

16. Torque the intake manifold bolts specification

17. Torque the surge tank bolts: 7-9 ft. lbs. (10-12 Nm).

18. Reprogram the radio anti theft codes.

Exhaust Manifold

REMOVAL & INSTALLATION

3.5L Engine

1. Before servicing the vehicle, refer to the precautions at the beginning of this section.

2. Record the radio anti theft code data.
3. Remove or disconnect the following:
   - Heated Oxygen (HO2S) sensor connectors
   - Exhaust Y pipe
   - Exhaust manifold heat shields
   - Exhaust manifolds

To install:

➡Use only new gaskets and nuts for assembly.

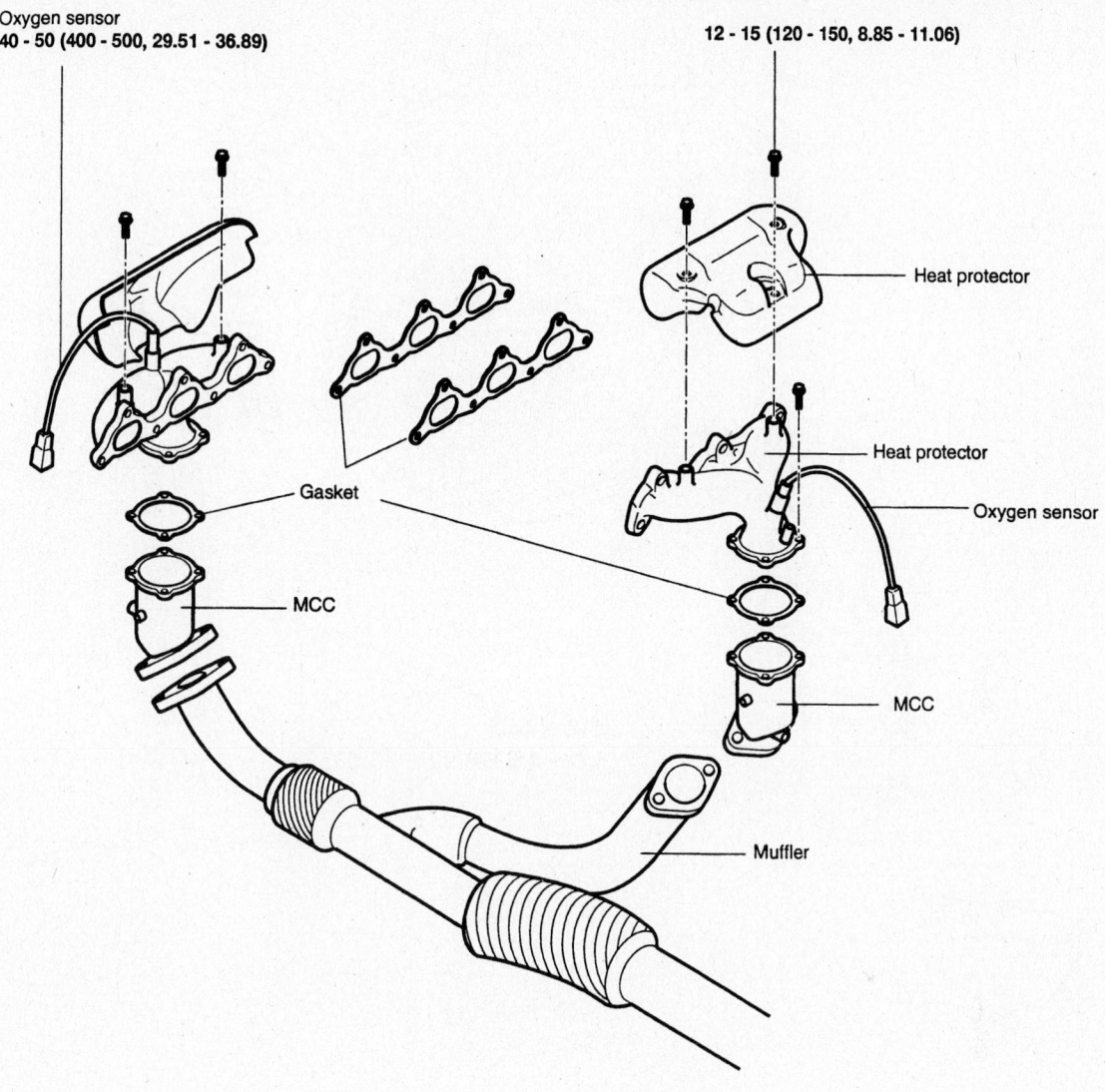

Oxygen sensor
40 - 50 (400 - 500, 29.51 - 36.89)

12 - 15 (120 - 150, 8.85 - 11.06)

Heat protector

Heat protector

Oxygen sensor

Gasket

MCC

MCC

Muffler

TORQUE : N•m (kg•cm, lb•ft)

9358HG08

Exhaust manifold and related components —3.5L Engine

4. Install or connect the following:
- Exhaust manifolds with new gaskets. Tighten the fasteners to 30-37 ft. lbs. (40-50 Nm)
- Exhaust manifold heat shields. Tighten the bolts to 106-132 inch lbs. (12-15 Nm)
- Exhaust Y pipe
- Heated Oxygen (HO$_2$S) sensor connectors

5. Reprogram the radio anti theft codes.

3.8L Engine

1. Before servicing the vehicle, refer to the precautions at the beginning of this section.

2. Record the radio anti theft code data.

3. Raise and support the vehicle safely. Remove the under cover splash shield.

4. Disconnect the left and right side oxygen sensor connector from the bracket.

5. Remove the front muffler. Remove the oil level stick.

6. Disconnect the left side oxygen sensor connector from the bracket. Remove the left side heat protector.

7. Remove the left exhaust manifold retaining bolts. Remove the left exhaust manifold.

8. Disconnect the right side oxygen sensor connector from the bracket. Remove the right side heat protector.

9. Remove the right exhaust manifold retaining bolts. Remove the right exhaust manifold.

To install:

10. Installation is the reverse of the removal procedure.

11. Be sure to use new exhaust manifold gaskets.

12. Torque the exhaust manifold bolts specification

13. Torque the heat shield bolts to 5 ft. lbs.

14. Reprogram the radio anti theft codes.

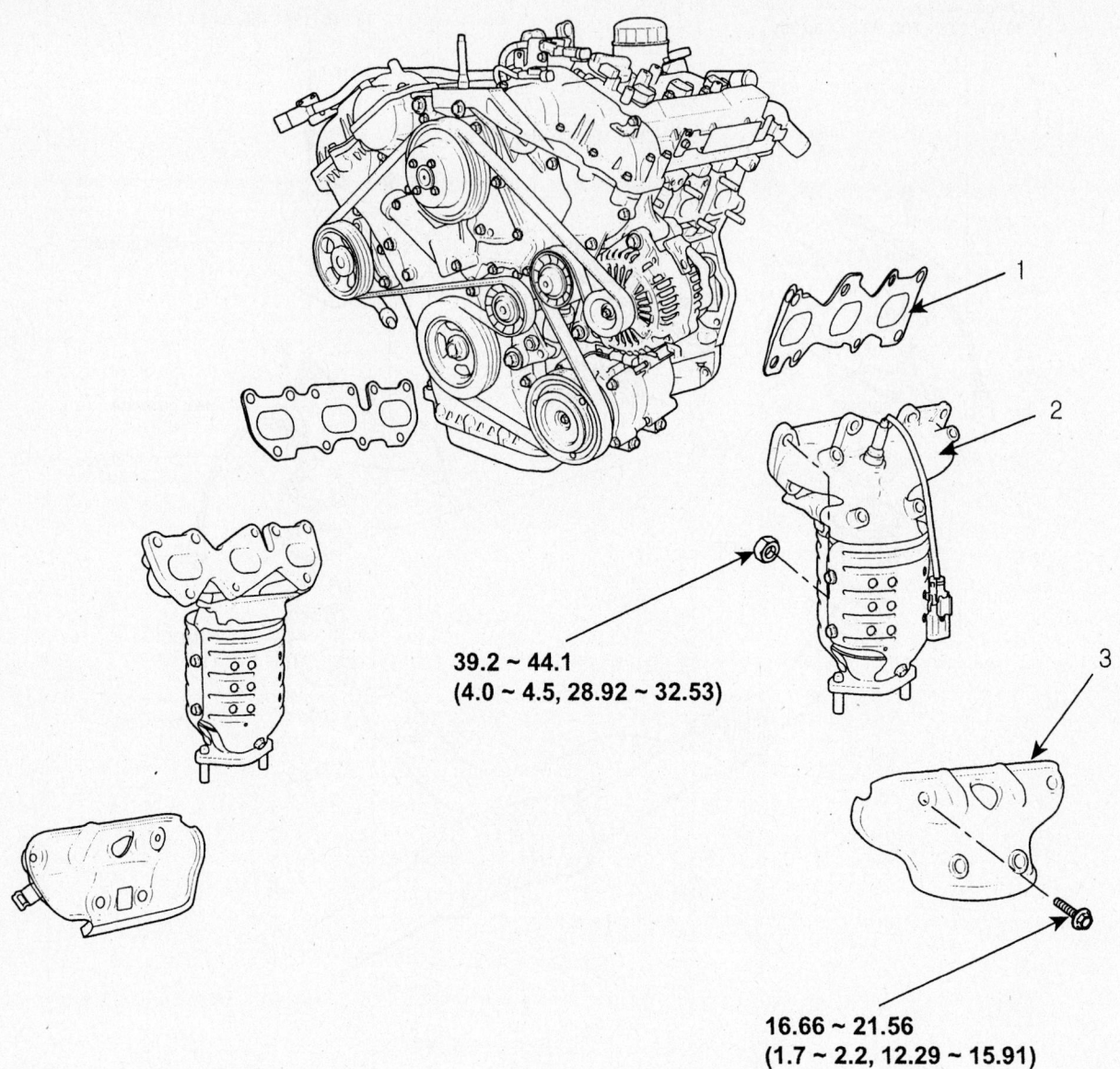

39.2 ~ 44.1
(4.0 ~ 4.5, 28.92 ~ 32.53)

16.66 ~ 21.56
(1.7 ~ 2.2, 12.29 ~ 15.91)

TORQUE : N.m (kgf.m, lb-ft)

1. Gasket
2. Exhaust manifold

3. Heat protector

09474_SEDO_G0004

Exhaust manifold and related components—3.8L Engine

Camshaft and Valve Lifters

REMOVAL & INSTALLATION

3.5L Engine

1. Before servicing the vehicle, refer to the precautions at the beginning of this section.

2. Record the radio anti theft code data.

3. Remove or disconnect the following:
   - Negative battery cable
   - Engine cover
   - Valve covers
   - Accessory drive belts
   - Idler pulley
   - Crankshaft pulley
   - Power steering pump pulley
   - Belt tensioner pulley
   - Upper and lower timing belt covers
   - No. 3 engine mount
   - Timing belt
   - Camshaft sprockets

➡**Keep all valvetrain components in order for assembly.**

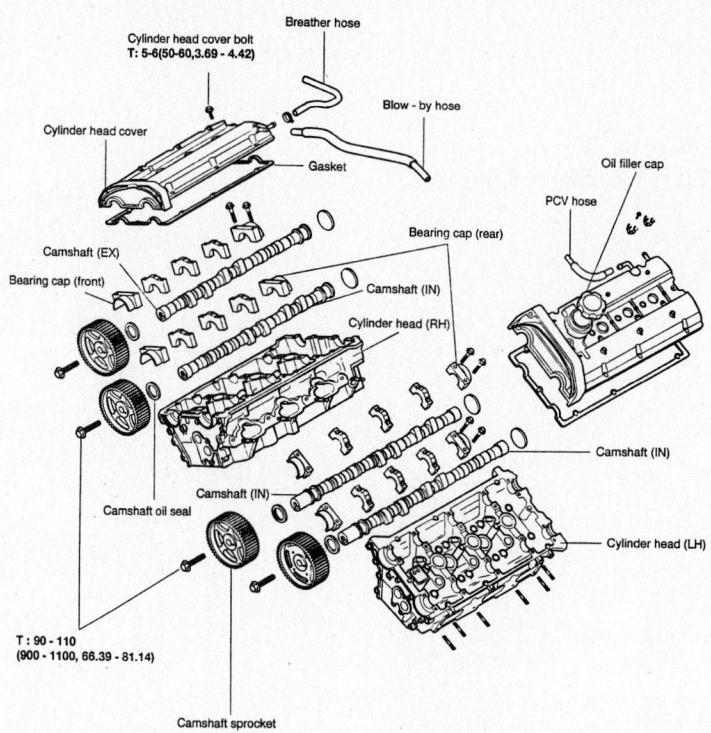

Cylinder head cover bolt
T: 5-6(50-60,3.69 - 4.42)

Breather hose

Cylinder head cover

Blow - by hose

Gasket

Oil filler cap

PCV hose

Camshaft (EX)

Bearing cap (front)

Bearing cap (rear)

Camshaft (IN)

Cylinder head (RH)

Camshaft (IN)

Camshaft oil seal

Camshaft (IN)

Cylinder head (LH)

T : 90 - 110
(900 - 1100, 66.39 - 81.14)

Camshaft sprocket

TORQUE : N•m (kg•cm, lb•ft)

9358HG09

Camshafts and related components—3.5L Engine

Cylinder head (RH)

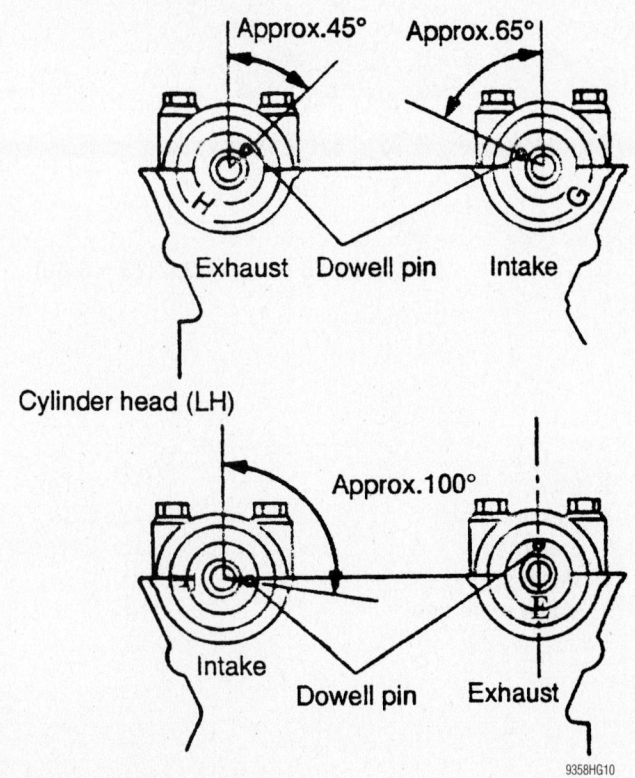

Approx.45° Approx.65°

Exhaust   Dowell pin   Intake

Cylinder head (LH)

Approx.100°

Intake

Dowell pin   Exhaust

9358HG10

Camshaft installation alignment—3.5L Engine

- Front bearing caps
- Rear bearing caps
- Center bearing caps
- Camshafts
- Rocker arms
- Hydraulic lifters

To install:

4. Set the No. 1 cylinder to Top Dead Center of the compression stroke.

5. Install the lifters and rocker arms in their original positions.

6. Install the camshafts aligned according to the illustration.

7. Install the bearing caps. Tighten the bolts evenly in several passes to the following torque specifications:

- Front and rear bearing caps: 19-21 ft. Lbs. (14-15 Nm)
- Center bearing caps: 88-106 inch lbs. (10-12 Nm)

8. Install or connect the following:

- Camshaft sprockets. Tighten the bolts to 67-81 ft. Lbs. (90-110 Nm).
- Timing belt
- No. 3 engine mount
- Upper and lower timing belt covers
- Belt tensioner pulley
- Power steering pump pulley
- Crankshaft pulley
- Idler pulley
- Accessory drive belts
- Valve covers. Tighten the bolts to 44-53 inch lbs. (5-6 Nm).
- Engine cover
- Negative battery cable

9. Reprogram the anti theft codes.

3.8L Engine

1. Before servicing the vehicle, refer to the precautions at the beginning of this section.

2. Record the radio anti theft code data.

3. Remove the engine from the vehicle.

4. Remove the cam covers.

5. Remove the timing chains.

6. Remove the camshaft retaining bolts. Remove the camshafts.

To install:

7. Apply a light coat of clean engine oil to the camshafts, prior to installation. Assemble the key groove of the camshaft rear side to the same level of head top surface.

8. Ensure that the camshaft components are installed in the correct location and direction.

9. Torque the camshaft bearing caps to specification and in the proper sequence.

10. Continue the installation is the reverse of the removal procedure.

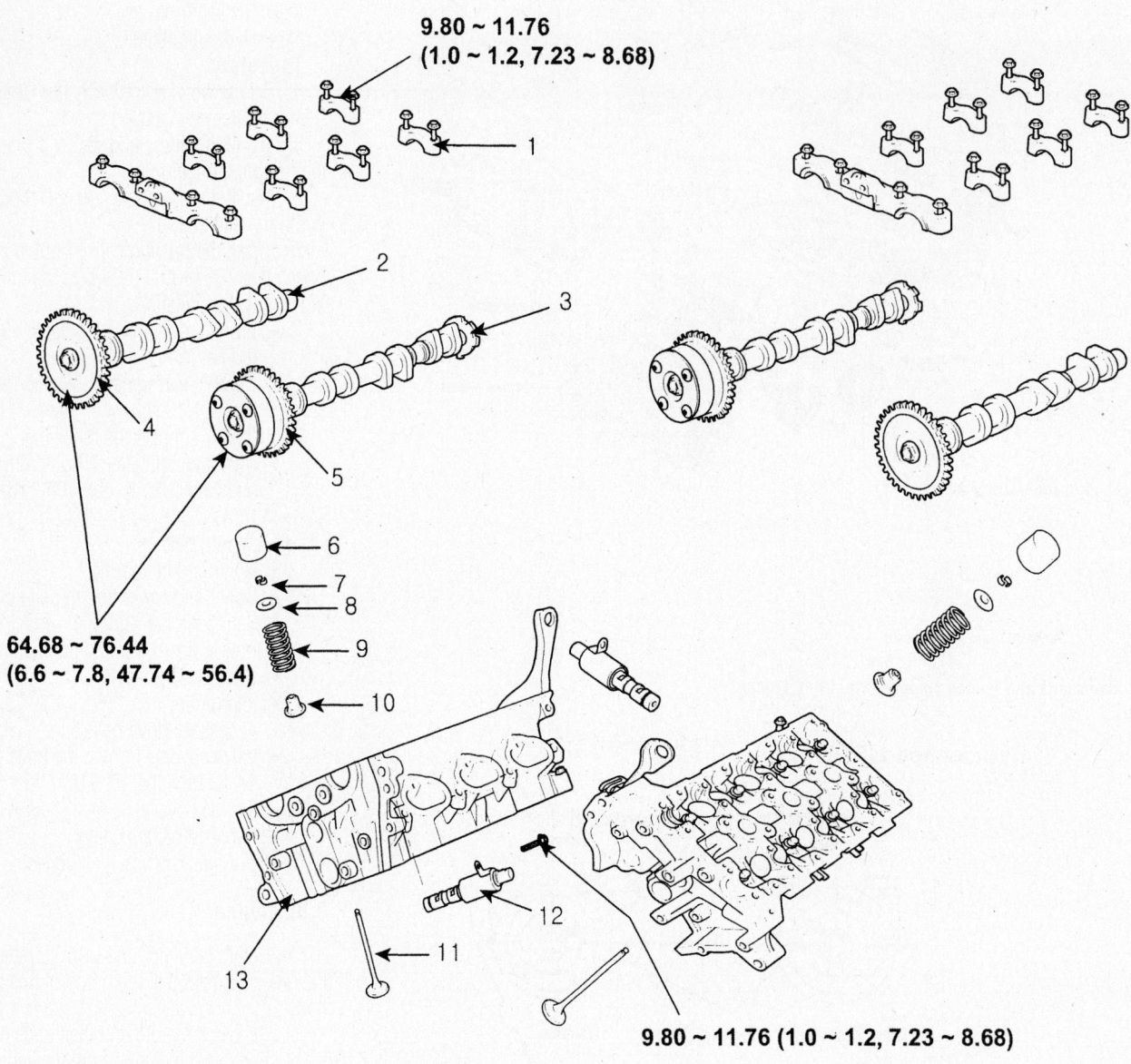

9.80 ~ 11.76
(1.0 ~ 1.2, 7.23 ~ 8.68)

64.68 ~ 76.44
(6.6 ~ 7.8, 47.74 ~ 56.4)

9.80 ~ 11.76 (1.0 ~ 1.2, 7.23 ~ 8.68)

TORQUE : N.m (kɑf.m. lb-ft)

1. Camshaft bearing cap
2. Exhaust camshaft
3. Intake camshaft
4. Exhaust camshaft sprocket
5. CVVT assembly

6. MLA
7. Retainer lock
8. Retainer
9. Valve spring
10. Valve stem seal

11. Valve
12. OCV
13. Cylinder head

09474_SEDO_G0008

Camshafts and related components—3.8L Engine

Tightening torque

5.9Nm(0.6kgf.m, 4.3lb-ft) - 1st step
9.80 ~ 11.76Nm(1.0 ~ 1.2kgf.m, 7.23 ~
8.68lb-ft) - 2nd step

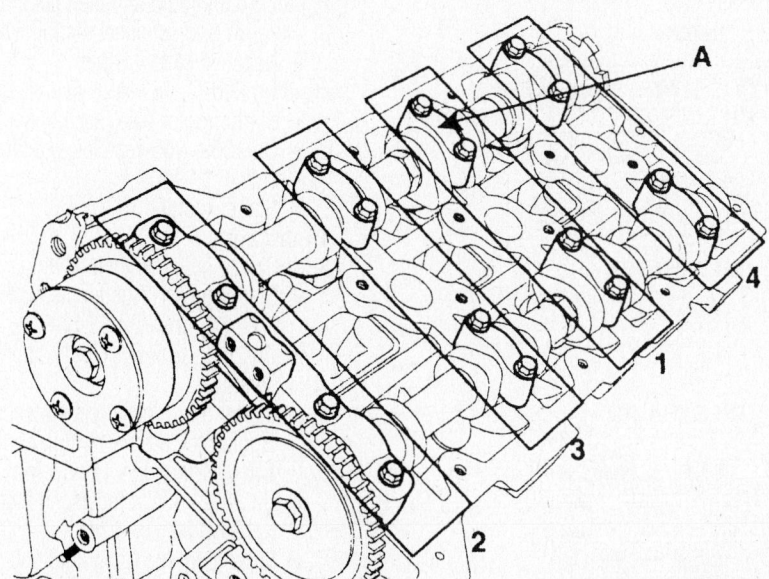

09474_SEDO_G0009

Camshaft bearing cap torque sequence—3.8L Engine

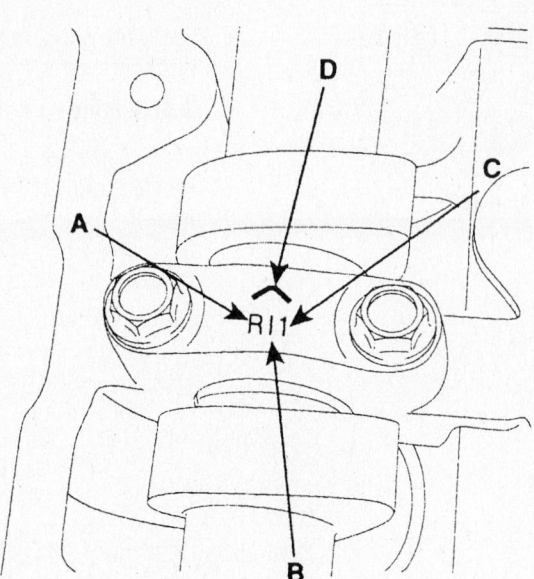

A : L(LH),R(RH)
B : I(Intake),None(Exhaust)
C : Journal number
D : Front mark

09474_SEDO_G0010

Camshaft bearing cap markings—3.8L Engine

ADJUSTMENT

3.5L Engine

This vehicle is equipped with hydraulic valve lifters. No adjustment is necessary.

3.8L Engine

1. Before servicing the vehicle, refer to the precautions at the beginning of this section.
2. Record the radio anti theft code data.
3. Disconnect the negative battery cable. Remove the engine cover. Remove the air cleaner assembly.
4. Remove the surge tank. Disconnect the ignition coil connector and remove the ignition coil.
5. Disconnect the breather pipe assembly from the cylinder head cover.
6. Remove the cylinder head cover retaining bolts. Remove the cylinder head covers from the engine.
7. Set the No. 1 cylinder to TDC on the compression stroke. Turn the crankshaft pulley and align its groove with the timing mark "T" of the lower timing chain cover.
8. Check that the mark of the camshaft timing sprockets are in straight line positioning on the cylinder head surface. If not rotate the crankshaft 360 degrees. Do not rotate the engine counterclockwise.
9. Check the valve clearance on No. 1 cylinder by measuring the clearance between the tappet and the base circle of the camshaft. Record the specification.
10. Turn the crankshaft pulley one revolution and align the groove with the timing mark "T" on the lower timing chain cover.
11. Check the valve clearance on No. 4 cylinder by measuring the clearance between the tappet and the base circle of the camshaft. Record the specification.
12. To adjust the intake and exhaust valve clearance, set the No. 1 cylinder to TDC on the compression stroke. Remove the timing chain.

➡**Before removing the timing chain mark the RH and LH timing chain with an identification based on the location of the sprocket. You must do this because the identification mark on the chain for TDC can be erased.**

13. Remove the camshaft bearing caps. Remove the camshafts. Remove the MLA's.
14. Measure the thickness of the removed tappet using a micrometer. Calculate the thickness of the new tappet so that the valve

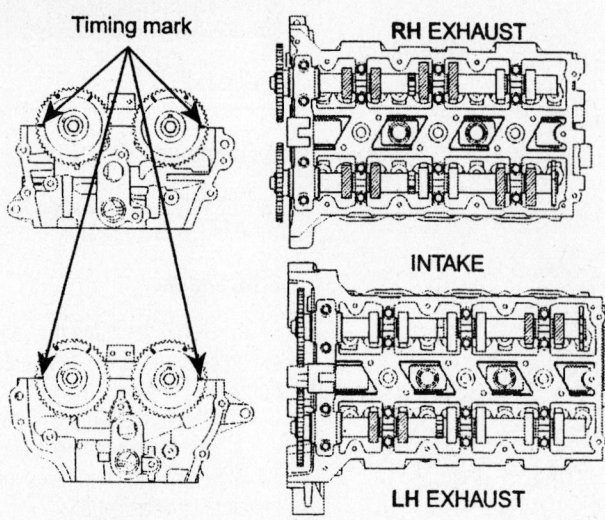

Valve adjustment No. 1 cylinder—3.8L Engine

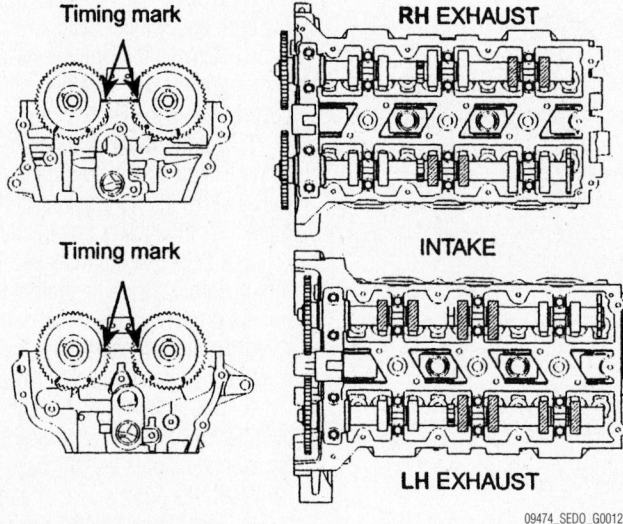

Valve adjustment No. 4 cylinder—3.8L Engine

clearance comes within the specified value. Select a new tappet with a thickness as close as possible to the calculated value.

➡Shims are available in 41 size increments ranging from 0.0006 inch to 0.118 inch.

15. Place a new tappet on the cylinder head.

➡Apply clean engine oil at the selected tappet on the periphery and top surface.

16. Install the camshafts. Install the bearing caps. Install the timing chain.

17. Turn the crankshaft two turns in the clockwise direction and realign the crankshaft sprocket and camshaft sprocket timing marks.

18. Recheck the valve clearance.

Starter Motor

REMOVAL & INSTALLATION

1. Before servicing the vehicle, refer to the precautions at the beginning of this section.

2. Record the radio anti theft code data.

3. Remove or disconnect the following:
   - Negative battery cable
   - On 2002—05 vehicles, shift cable
   - Starter motor electrical connectors
   - On 2002—05 vehicles, starter heat shield
   - Starter motor

To install:

4. Install or connect the following:
   - Starter motor. Tighten the bolts to 20-24 ft. Lbs. (27-33 Nm).
   - On 2002—05 vehicles, starter heat shield
   - Starter motor electrical connectors. Tighten the battery terminal nut to 106-141 inch lbs. (12-16 Nm).
   - On 2002—05 vehicles, shift cable
   - Negative battery cable
   - Reprogram the radio anti theft codes

Oil Pan

REMOVAL & INSTALLATION

3.5L Engine

1. Before servicing the vehicle, refer to the precautions at the beginning of this section.

2. Record the radio anti theft code data.

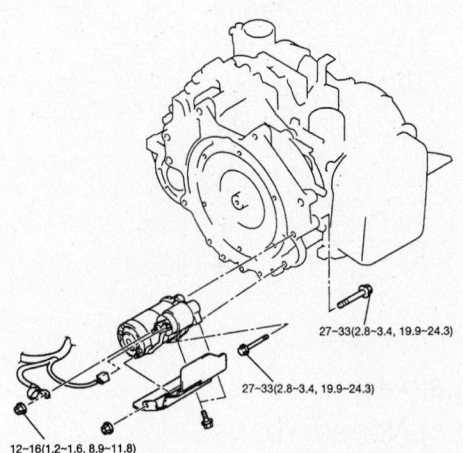

27~33(2.8~3.4, 19.9~24.3)

27~33(2.8~3.4, 19.9~24.3)

12~16(1.2~1.6, 8.9~11.8)

N•m(kg•m, lb•ft)
9358HG11

Starter motor mounting—3.5L Engine

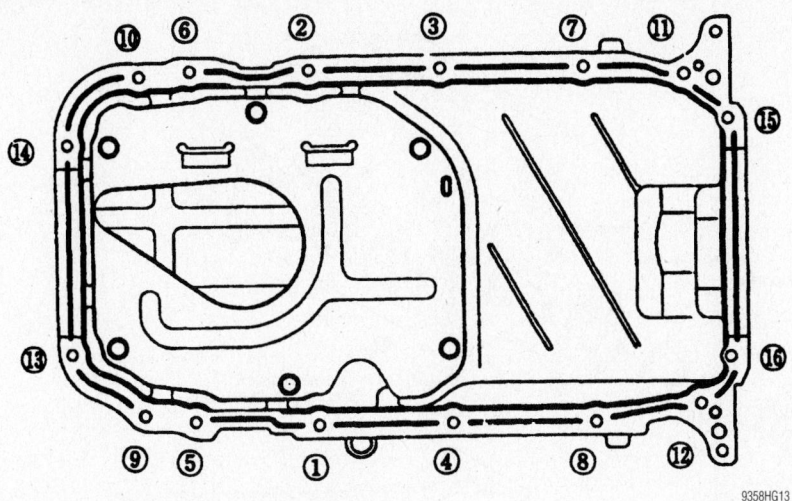

Upper oil pan torque sequence—3.5L Engine

9358HG13

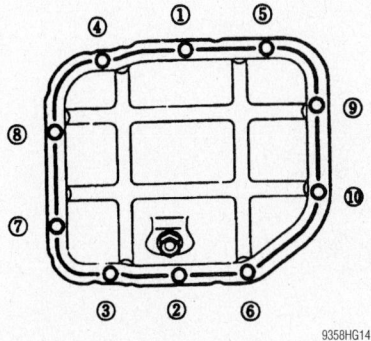

Lower oil pan torque sequence—3.5L Engine

9358HG14

3. Drain the engine oil.
4. Remove or disconnect the following:
  • Negative battery cable
  • Starter motor
  • Oil filter
  • Lower oil pan
  • Upper oil pan

To install:
5. Apply silicone sealant to the grove of the oil pan flange.
6. Install the upper oil pan and tighten the bolts in sequence as follows:
  a. Bolts 1-14: 14-20 ft. lbs. (19-28 Nm)
  b. Bolts 15 and 16: 44-62 inch lbs. (5-7 Nm)
  c. Upper oil pan-to-transaxle mounting bolts: 22-33 ft. lbs. (30-42 Nm)
7. Install the lower oil pan and tighten the bolts in sequence to 86-104 inch lbs. (10-12 Nm).
8. Install or connect the following:
  • Oil filter
  • Starter motor
  • Negative battery cable
9. Fill the crankcase to the correct level with engine oil.

Bead width : 2.5mm(0.1in.)
But marked area(*) to be 5.0mm(0.2in.)

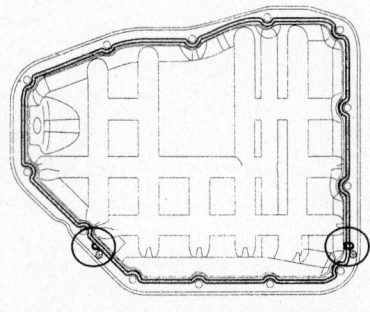

09474_SEDO_G0013

Oil pan sealant location points—3.8L Engine

10. Start the engine and check for leaks.
11. Reprogram the radio anti theft codes.

3.8L Engine
1. Before servicing the vehicle, refer to the precautions at the beginning of this section.
2. Record the radio anti theft code data.
3. Remove the engine from the vehicle.
4. Remove the oil pan retaining bolts.
5. Remove the oil pan from the engine.

To install:
6. Using a gasket scrapper, remove all old packing material from the gasket mating surfaces.
7. Before assembling the oil pan, liquid sealant TB1217H should be applied to the oil pan.

➡**After applying the sealant the part must be assembled within five minutes.**

Tightening torque
9.80 ~ 11.76Nm (1.0 ~ 1.2kgf.m, 7.23 ~ 8.68lb-ft)

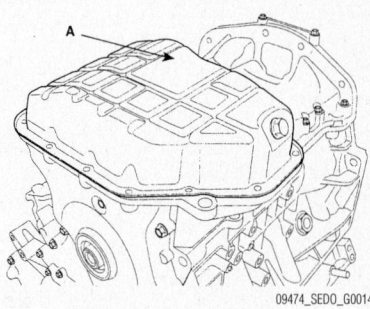

09474_SEDO_G0014

Oil pan mounting—3.8L Engine

8. Position the oil pan to the engine block. Torque the oil pan bolts to specification using an alternating sequence pattern.

➡**After assembly wait at least thirty minutes before filling the engine with clean engine oil.**

9. Continue the installation in the reverse order of the removal procedure.

Oil Pump

3.5L Engine
1. Before servicing the vehicle, refer to the precautions at the beginning of this section.
2. Record the radio anti theft code data.
3. Drain the engine oil.
4. Remove or disconnect the following:
  • Negative battery cable
  • Engine cover
  • Valve covers
  • Accessory drive belts
  • Idler pulley
  • Crankshaft pulley
  • Power steering pump pulley
  • Belt tensioner pulley
  • Upper and lower timing belt covers
  • No. 3 engine mount
  • Timing belt
  • Crankshaft sprocket
  • Crankshaft Position (CKP) sensor tone ring
  • Oil filter
  • Starter motor
  • Lower oil pan
  • Upper oil pan
  • Oil pick-up tube
  • Oil filter bracket
  • Oil relief valve plug
  • Oil pump

To install:
5. Install the oil pump with a new gasket. Tighten the oil pump case bolts to 106-133 inch lbs. (12-15 Nm). Tighten the oil

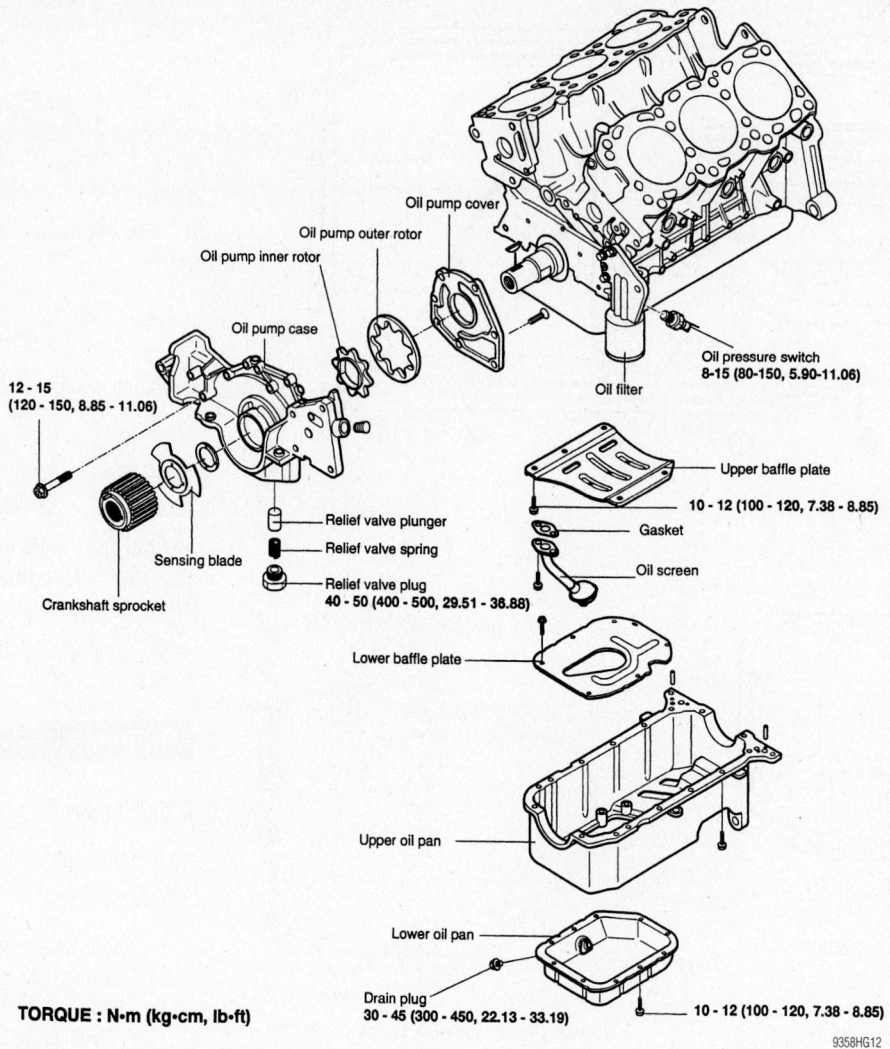

12 - 15
(120 - 150, 8.85 - 11.06)

Oil pump cover

Oil pump outer rotor

Oil pump inner rotor

Oil pump case

Oil pressure switch
8-15 (80-150, 5.90-11.06)

Oil filter

Relief valve plunger

Relief valve spring

Relief valve plug
40 - 50 (400 - 500, 29.51 - 36.88)

Sensing blade

Crankshaft sprocket

Upper baffle plate

10 - 12 (100 - 120, 7.38 - 8.85)

Gasket

Oil screen

Lower baffle plate

Upper oil pan

Lower oil pan

Drain plug
30 - 45 (300 - 450, 22.13 - 33.19)

10 - 12 (100 - 120, 7.38 - 8.85)

TORQUE : N•m (kg•cm, lb•ft)

9358HG12

Oil pump and related components—3.5L Engine

pump cover screws to 71-106 inch lbs. (8-12 Nm).

6. Install or connect the following:
- Oil relief valve plug. Tighten the plug to 30-37 ft. lbs. (40-50 Nm)
- Oil filter bracket
- Oil pick-up tube. Tighten the bolts to 11-16 ft. lbs. (15-22 Nm)
- Upper oil pan
- Lower oil pan
- Starter motor
- Oil filter
- CKP sensor tone ring
- Crankshaft sprocket
- Timing belt
- No. 3 engine mount
- Upper and lower timing belt covers
- Belt tensioner pulley
- Power steering pump pulley
- Crankshaft pulley
- Idler pulley
- Accessory drive belts

- Valve covers
- Engine cover
- Negative battery cable

7. Fill the crankcase to the correct level with engine oil.

8. Start the engine and check for leaks.

9. Reprogram the radio anti theft codes.

3.8L Engine

1. Before servicing the vehicle, refer to the precautions at the beginning of this section.

2. Record the radio anti theft code data.

3. Remove the engine from the vehicle.

4. Remove the oil pan retaining bolts.

5. Remove the oil pan from the engine.

6. Remove the oil pump chain cover. Remove the oil pump chain sprocket.

7. Remove the oil pump mounting bolts. Remove the oil pump from the engine.

To install:

8. Install the oil pump with a new gasket. Tighten the oil pump bolts to 14.5-17.5 inch lbs. (19.6-23.5 Nm).

9. Using a gasket scrapper, remove all old packing material from the gasket mating surfaces of the oil pan.

10. Before assembling the oil pan, liquid sealant TB1217H should be applied to the oil pan.

➡**After applying the sealant the part must be assembled within five minutes.**

11. Position the oil pan to the engine block. Torque the oil pan bolts to specification using an alternating sequence pattern.

➡**After assembly wait at least thirty minutes before filling the engine with clean engine oil.**

12. Continue the installation in the reverse order of the removal procedure.

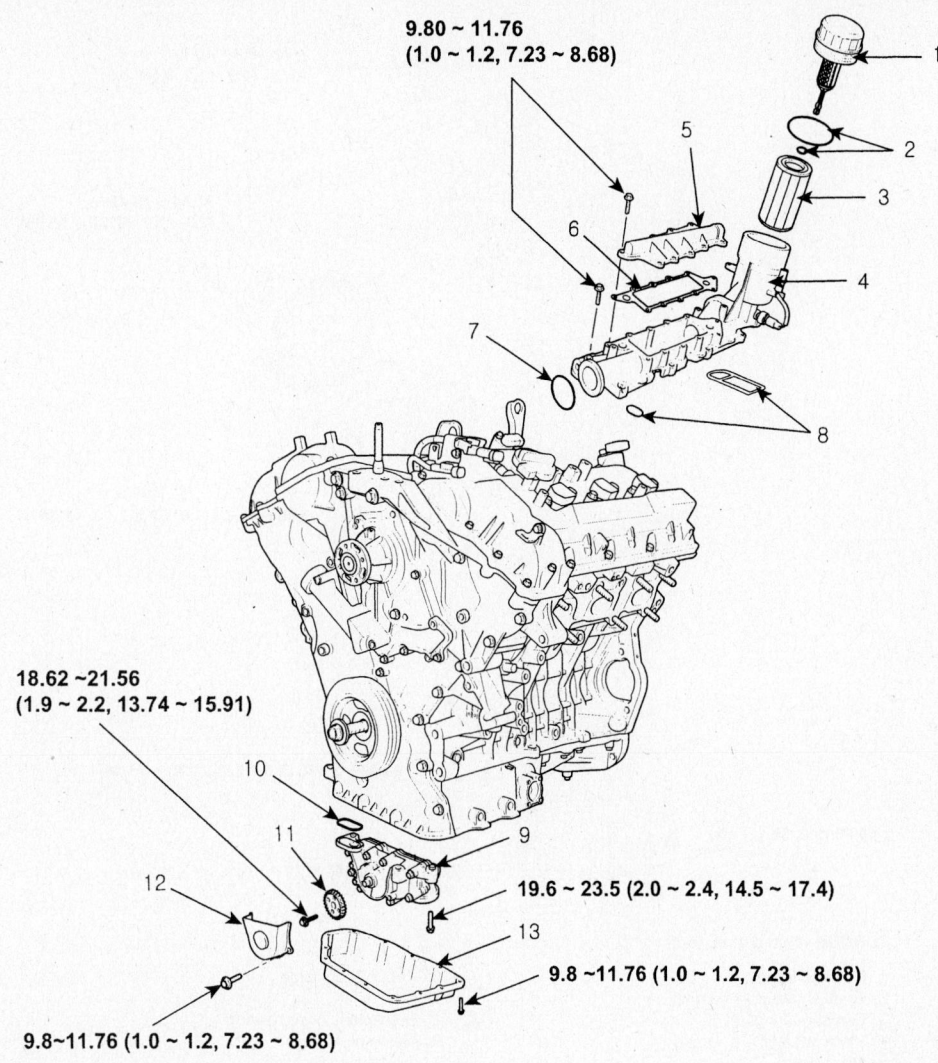

9.80 ~ 11.76
(1.0 ~ 1.2, 7.23 ~ 8.68)

18.62 ~21.56
(1.9 ~ 2.2, 13.74 ~ 15.91)

19.6 ~ 23.5 (2.0 ~ 2.4, 14.5 ~ 17.4)

9.8 ~11.76 (1.0 ~ 1.2, 7.23 ~ 8.68)

9.8~11.76 (1.0 ~ 1.2, 7.23 ~ 8.68)

TORQUE : N.m (kgf.m, lb-ft)

1. Oil filter cap
2. O - ring
3. Oil filter element
4. Oil filter body
5. Oil filter body cover
6. Gasket
7. O - ring
8. Gasket
9. Oil pump
10. Gasket
11. Oil pump sprocket
12. Oil pump chain cover
13. Lower oil paon

09474_SEDO_G0015

Oil pump and related components—3.8L Engine

Rear Main Seal

REMOVAL & INSTALLATION

3.5L Engine

1. Before servicing the vehicle, refer to the precautions at the beginning of this section.
2. Record the radio anti theft code data.
3. Remove or disconnect the following:
   - Negative battery cable
   - Front wheels
   - Starter motor
   - Axle halfshafts
   - Transaxle
   - Flexplate
   - Oil seal housing
   - Oil seal

To install:
4. Install the oil seal to the seal housing using special tool 09231-33000 or similar seal driver.
5. Apply silicone sealant to the oil seal housing flange.
6. Install the seal housing and tighten the bolts to 94-106 inch lbs. (10-12 Nm).
7. Install or connect the following:
   - Flexplate. Tighten the bolts to 54-57 ft. lbs. (73-77 Nm)
   - Transaxle
   - Axle halfshafts
   - Starter motor
   - Front wheels
   - Negative battery cable
8. Fill the crankcase to the correct level with engine oil.
9. Start the engine and check for leaks.
10. Reprogram the radio anti theft codes.

3.8L Engine

1. Before servicing the vehicle, refer to the precautions at the beginning of this section.

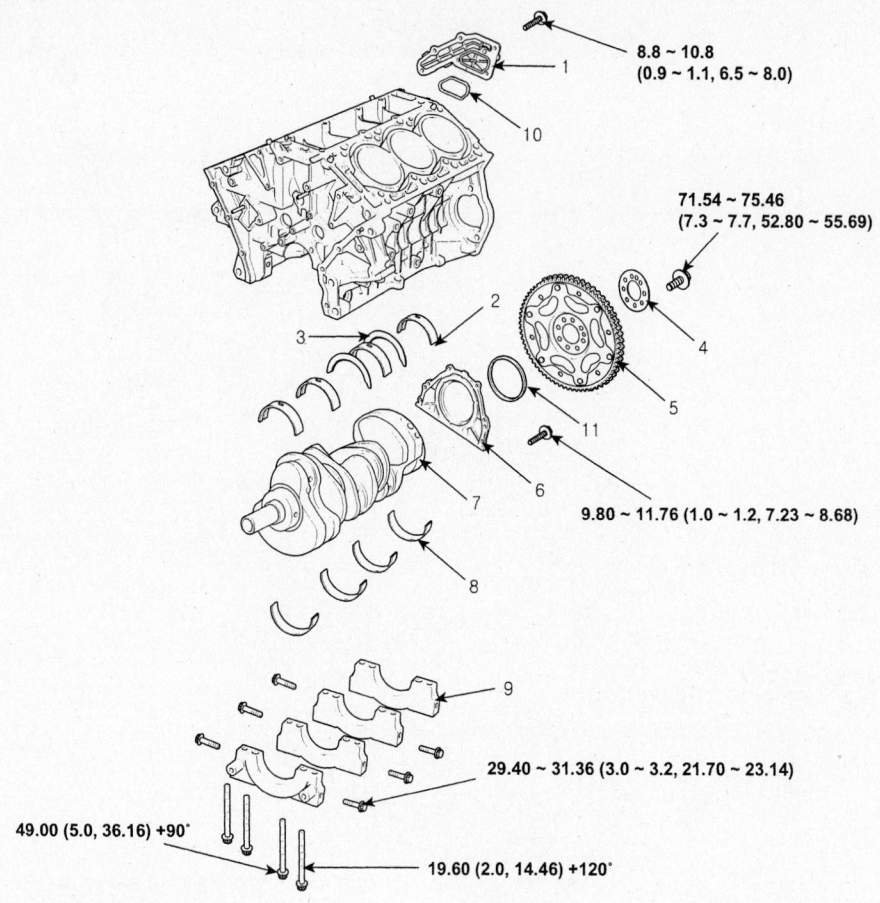

8.8 ~ 10.8
(0.9 ~ 1.1, 6.5 ~ 8.0)

71.54 ~ 75.46
(7.3 ~ 7.7, 52.80 ~ 55.69)

9.80 ~ 11.76 (1.0 ~ 1.2, 7.23 ~ 8.68)

29.40 ~ 31.36 (3.0 ~ 3.2, 21.70 ~ 23.14)

49.00 (5.0, 36.16) +90°

19.60 (2.0, 14.46) +120°

TORQUE : N.m (kgf.m, lb-ft)

1. Oil drain cover
2. Crankshaft upper bearing
3. Thrust bearing
4. Plate adapter
5. Drive plate
6. Rear oil seal case
7. Crankshaft
8. Crankshaft lower bearing
9. Main bearing cap
10. Oil drain cover gasket
11. Rear oil seal

09474_SEDO_G0016

Engine block and related components—3.8L Engine

2. Record the radio anti theft code data.

3. Remove the engine from the vehicle.

4. Remove the flex plate from the engine.

5. Remove the rear main oil seal from its mounting.

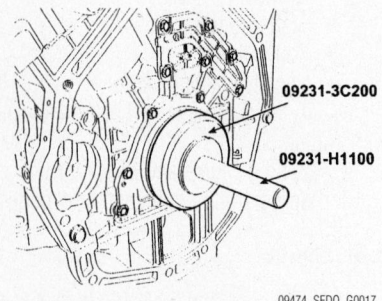

09231-3C200

09231-H1100

09474_SEDO_G0017

Rear main oil seal installation—3.8L Engine

Tightening torque
Main bearing cap bolt
49.00Nm(5.0 kgf.m, 36.16lb-ft) + 90° (1 ~ 8)
19.60 Nm(2.0 kgf.m, 14.46lb-ft)+ 120° (9 ~ 16)
29.40 ~ 31.36Nm(3.0 ~ 3.2 kgf.m, 21.70 ~
23.14lb-ft) (17 ~ 22)

🔰 NOTE
- Always use new main bearing cap bolt.
- If any of the bearing cap bolts in broken or deformed, replace it.

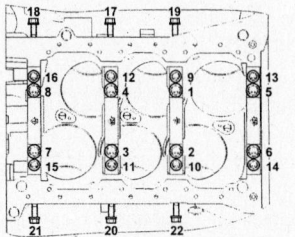

09474_SEDO_G0018

Main bearing cap torque sequence and specification—3.8L Engine

To install:

6. Install the oil seal to the oil seal housing using tool 09231-3C200 and tool 09231-H1100.

7. Continue the installation in the reverse order of the removal procedure.

Timing Belt, Cover and Crankshaft Seal

REMOVAL & INSTALLATION

3.5L Engine

1. Before servicing the vehicle, refer to the precautions at the beginning of this section.

2. Record the radio anti theft code data.

3. Remove or disconnect the following:
- Negative battery cable
- Engine cover

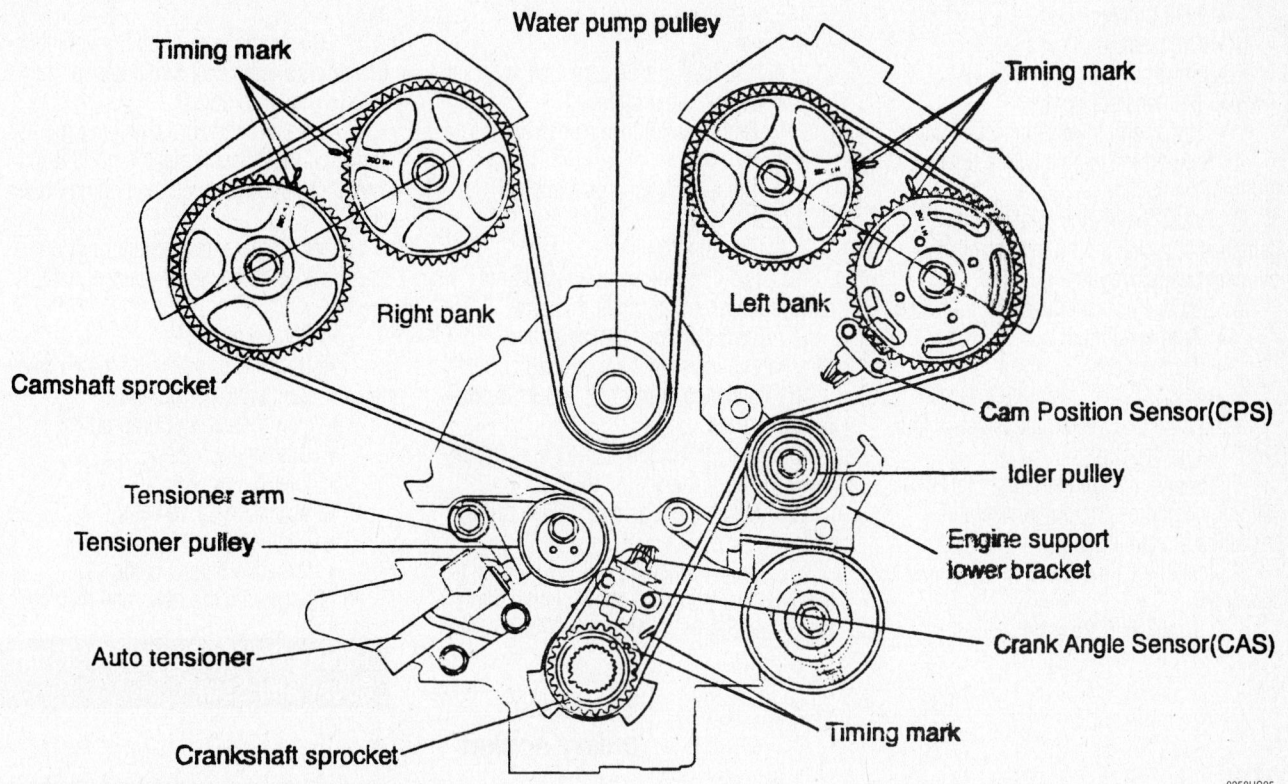

Timing belt routing and timing marks—3.5L Engine

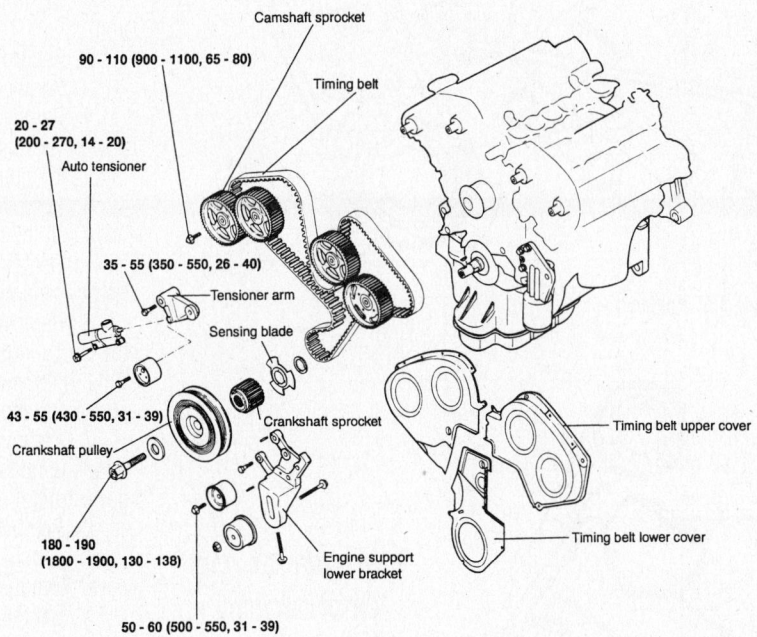

TORQUE : N•m (kg•cm, lb•ft)

Timing belt and related components—3.5L Engine

- Accessory drive belts
- Idler pulley
- Crankshaft pulley
- Power steering pump pulley
- Belt tensioner pulley
- Upper and lower timing belt covers

4. Support the engine with a floor jack and remove the engine mount.

5. Rotate the engine to align the camshaft sprocket timing marks with the cylinder head cover timing marks.

6. Remove or disconnect the following:
- Auto tensioner
- Timing belt

To install:

7. Ensure that the engine is set to Top Dead Center (TDC).

8. Prepare the auto tensioner for installation by compressing it in a vise and installing a retaining pin.

9. Install the timing belt in the following order:

a. Crankshaft sprocket

b. Idler pulley

c. Left bank exhaust camshaft sprocket

d. Left bank intake camshaft sprocket

e. Water pump pulley

f. Right bank intake camshaft sprocket

g. Right bank exhaust camshaft sprocket

h. Tensioner pulley

10. Install the auto tensioner. Do not remove the retaining pin at this time.

11. Check that the crankshaft and camshaft timing marks are aligned correctly.

12. Rotate the crankshaft ¼ turn **Counterclockwise**.

13. Rotate the crankshaft ¼ turn **Clockwise** to return the engine to TDC.

14. Loosen the tensioner pulley center bolt.

15. Apply 44 inch lbs. (5 Nm) torque to the tensioner pulley as shown and tighten the center bolt to 32-41 ft. lbs. (43-55 Nm).

16. Remove the auto tensioner retaining pin.

17. Rotate the crankshaft 2 revolutions **Clockwise**, then wait 5 minutes for the auto tensioner to adjust.

18. Measure the auto tensioner rod as shown. If the measurement is not 3.8-4.5 mm, then repeat the belt tensioning procedure.

19. When the auto tensioner measurement is correct, install or connect the following:
- Engine mount
- Upper and lower timing belt covers
- Belt tensioner pulley
- Power steering pump pulley
- Crankshaft pulley
- Idler pulley
- Accessory drive belts
- Engine cover
- Negative battery cable

20. Reprogram the radio anti theft codes.

Timing Chain, Sprockets, Front Cover and Seal

REMOVAL & INSTALLATION

3.8L Engine

1. Before servicing the vehicle, refer to the precautions at the beginning of this section.

2. Record the radio anti theft code data.

3. Remove the engine from the vehicle.

4. Remove the drive belt. Remove the power steering pump. Remove the air conditioning compressor.

5. Remove the alternator. Remove the drive belt idler. Remove the drive belt tensioner. Remove the water pump pulley. Remove the intake manifold.

6. Remove the connector bracket from the left cylinder head cover. Disconnect the right ignition coil connector, condenser connector and remove the bracket.

7. Remove the left cylinder head cover retaining bolts. Remove the cylinder head cover from the engine.

8. Remove the right cylinder head cover retaining bolts. Remove the cylinder head cover from the engine.

9. Set the No. 1 cylinder to TDC on the compression stroke. Turn the crankshaft pulley and align its groove with the timing mark "T" of the lower timing chain cover.

10. Check that the mark of the camshaft timing sprockets are in straight line positioning on the cylinder head surface. If not rotate the crankshaft 360 degrees. Do not rotate the engine counterclockwise.

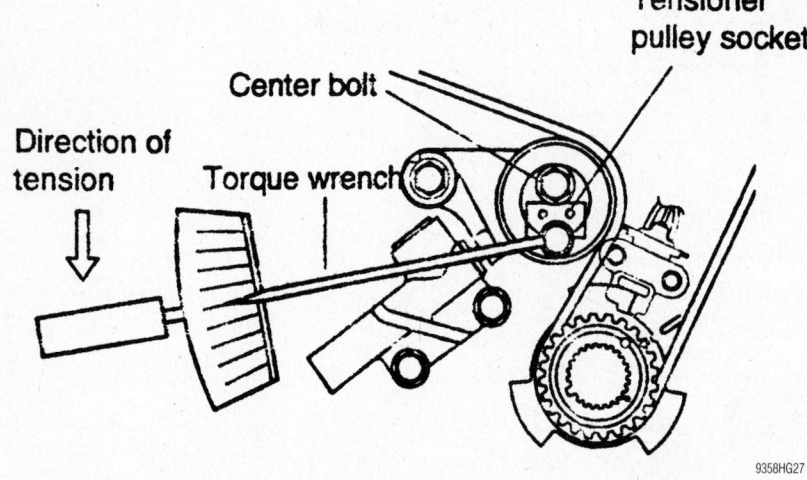

Adjusting the tensioner pulley—3.5L Engine

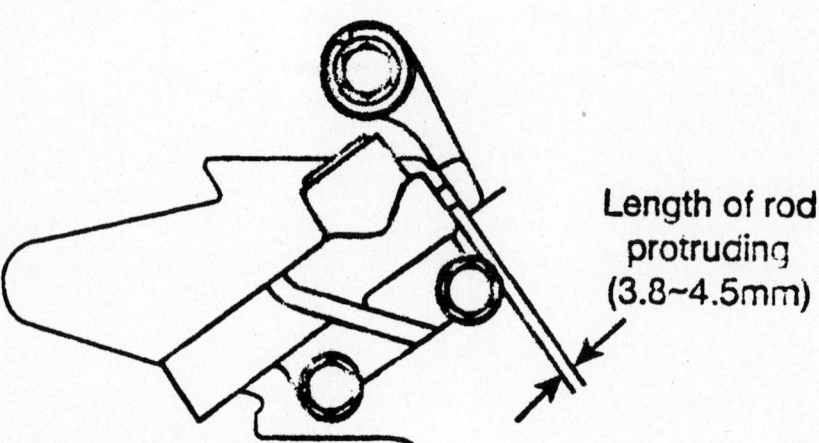

Measuring the auto tensioner rod—3.5L Engine

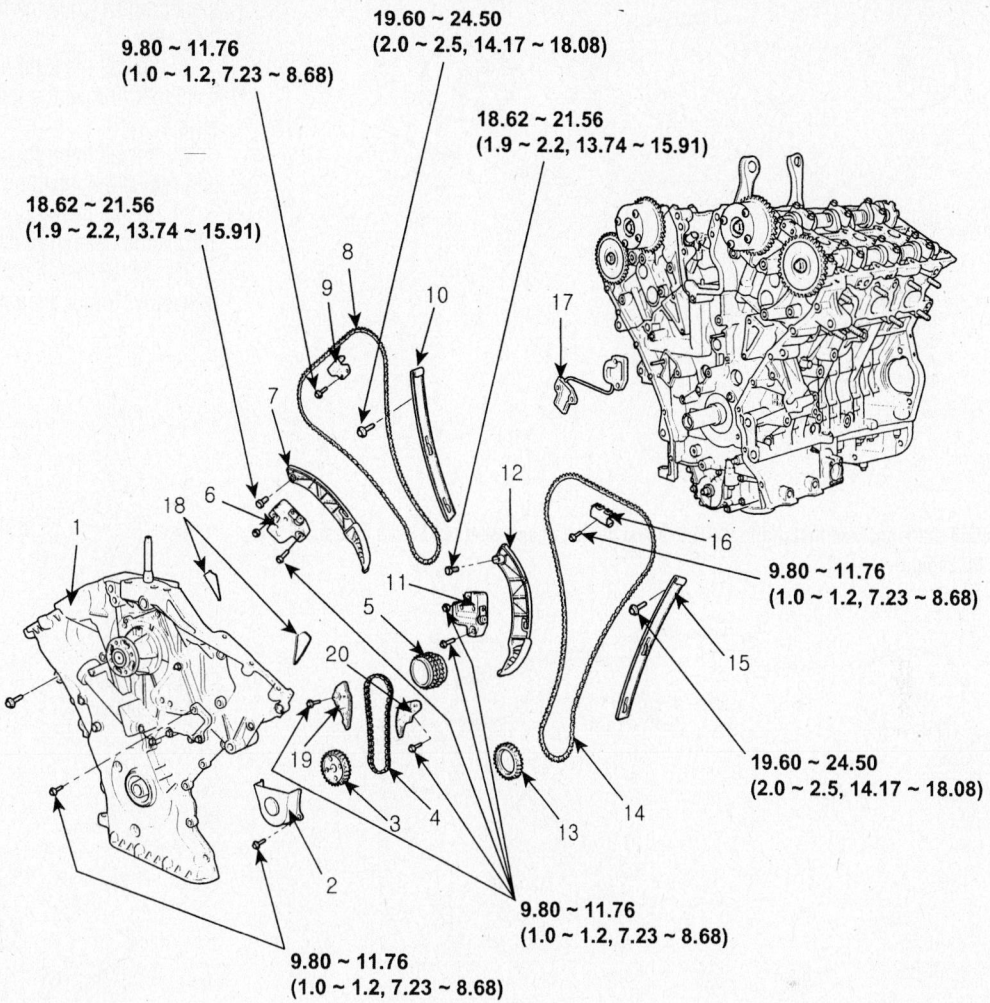

9.80 ~ 11.76
(1.0 ~ 1.2, 7.23 ~ 8.68)

19.60 ~ 24.50
(2.0 ~ 2.5, 14.17 ~ 18.08)

18.62 ~ 21.56
(1.9 ~ 2.2, 13.74 ~ 15.91)

18.62 ~ 21.56
(1.9 ~ 2.2, 13.74 ~ 15.91)

9.80 ~ 11.76
(1.0 ~ 1.2, 7.23 ~ 8.68)

19.60 ~ 24.50
(2.0 ~ 2.5, 14.17 ~ 18.08)

9.80 ~ 11.76
(1.0 ~ 1.2, 7.23 ~ 8.68)

9.80 ~ 11.76
(1.0 ~ 1.2, 7.23 ~ 8.68)

TORQUE : N.m (kgf.m, lb-ft)

1. Timing chain cover
2. Oil pump chain cover
3. Oil pump sprocket
4. Oil pump chain
5. Crankshaft sprocket
6. Timing chain auto tensioner
7. Timing chain tensioner arm
8. Timing chain
9. Cam to cam guide
10. Timing chain guide
11. Timing chain auto tensioner
12. Timing chain tensioner arm
13. Crankshaft sprocket
14. Timing chain
15. Timing chain guide
16. Cam to cam guide
17. Tensioner adapter
18. Gasket
19. Oil pump chain guide
20. Oil pump temsioner assembly

09474_SEDO_G0042

Timing chain and related components—3.8L Engine

11. Remove the oil pan. Remove the crankshaft damper pulley. Remove the timing chain cover.

➡**Be careful not to damage the contact surfaces of the cylinder block, cylinder head and timing chain cover. Before removing the timing chain, mark the right and left timing chains with an identification mark based on the location of the sprocket because the identification mark on the chain for TDC can be erased.**

12. After compressing the timing chain tensioner, install a set pin. Remove the right cam to cam guide. Remove the right timing chain auto tensioner and timing chain tensioner arm.

13. Remove the oil pump chain cover. Remove the oil pump chain tensioner assembly. Remove the oil pump chain guide. Remove the right timing chain.

14. Remove the right timing chain guide. Remove the oil pump chain sprocket and oil pump chain. Remove the crankshaft sprocket, oil pump and camshaft drive gear.

15. Remove the left cam to cam guide. Remove the left timing chain auto tensioner and timing chain tensioner arm. Remove the left timing chain.

16. Remove the left timing chain guide. Remove the crankshaft sprocket and left camshaft drive. Remove the tensioner adapter assembly.

To install:

17. Check the camshaft and crankshaft sprockets for wear and damage, replace as required.

18. Inspect the tensioner arm and chain guide for wear and damage, replace as required.

19. Check that the tensioner pin moves smoothly when the ratchet pawl is released with a thin rod.

20. The key of the crankshaft should be

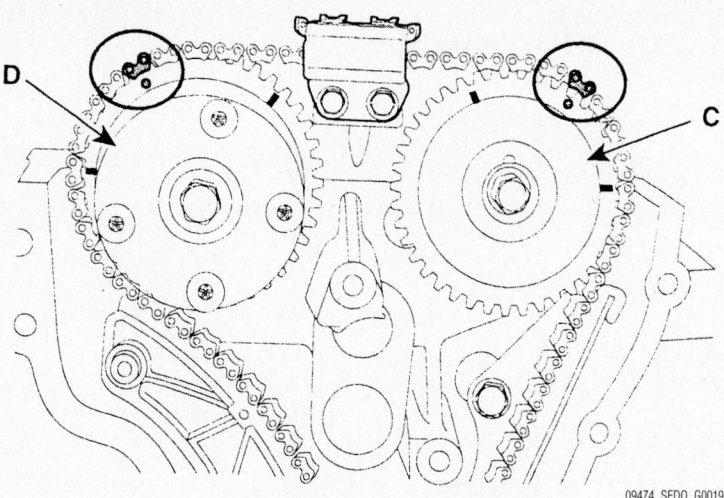

09474_SEDO_G0019

Left timing chain markings and installation: (D) exhaust camshaft sprocket (C) intake camshaft sprocket—3.8L Engine

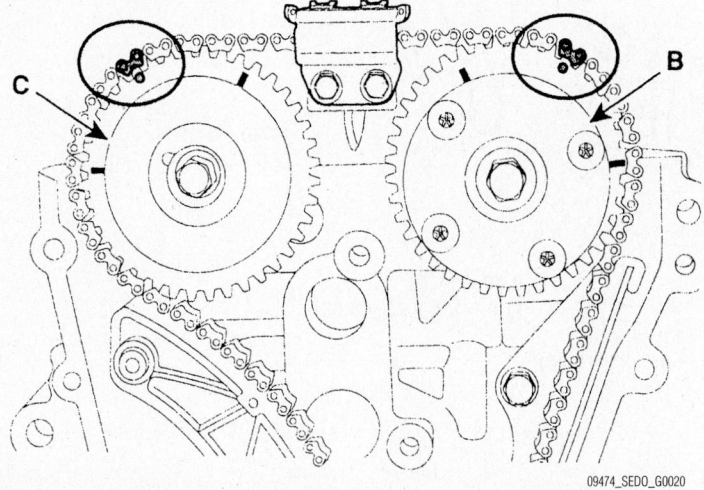

09474_SEDO_G0020

Right timing chain markings and installation: (C) exhaust camshaft sprocket (B) intake camshaft sprocket—3.8L Engine

sioner arm. Torque the retaining bolts to 14-17 ft. lbs.

32. Install the right timing chain auto tensioner. Torque the retaining bolts to 7-9 ft. lbs.

33. Install the right cam to cam guide and torque the retaining bolts to 7-9 ft. lbs.

34. Install the oil pump chain guide. Torque the retaining bolts to 7-9 ft. lbs.

35. Install the oil pump tensioner assembly. Torque the retaining bolts to 7-9 ft. lbs.

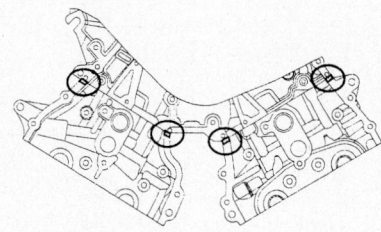

09474_SEDO_G0021

Sealant application (1 inch bead width)—3.8L Engine

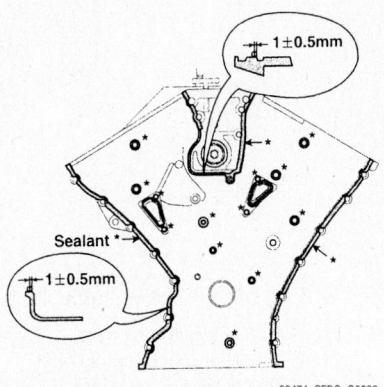

09474_SEDO_G0022

Timing cover sealant application—3.8L Engine

aligned with the timing mark of the timing chain cover. This indicates that the piston is at TDC on the compression stroke.

21. Install the tensioner adapter assembly. Install the crankshaft sprocket and left camshaft drive.

22. Install the left timing chain guide and torque the retaining bolts 14-18 ft. lbs.

23. Install the left timing chain. Be sure to install the chain with no slack between the camshaft and the crankshaft (crankshaft sprocket, timing chain guide, exhaust camshaft sprocket, intake camshaft sprocket). The timing mark of each sprocket should be matched with the timing mark (color ink) of the timing chain at installation of the timing chain.

24. Install the left timing chain tensioner arm and torque the retaining bolts 14-16 ft. lbs.

25. Install the left timing chain tensioner and torque the retaining bolts 7-9 ft. lbs.

26. Install the left cam to cam guide and torque the retaining bolts 7-9 ft. lbs.

27. Install the crankshaft sprocket, oil pump and right camshaft drive.

28. Install the oil pump chain and oil pump sprocket. Torque the retaining bolt to 14-16 ft. lbs.

29. Install the right timing chain guide. Torque the retaining bolts to 14-18 ft. lbs.

30. Install the right timing chain. Be sure to install the chain with no slack between the camshaft and the crankshaft (crankshaft sprocket, intake camshaft sprocket, exhaust camshaft sprocket). The timing mark of each sprocket should be matched with the timing mark (color ink) of the timing chain at installation of the timing chain.

31. Install the right timing chain ten-

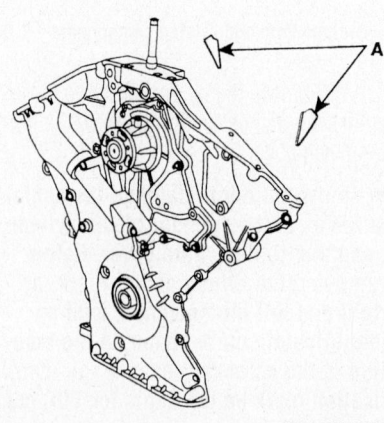

09474_SEDO_G0023

Timing cover gasket installation (A) gaskets—3.8L Engine

Tightening torque
B(17): 18.62 ~ 21.56Nm(1.9 ~ 2.2kgf.m, 13.74 ~ 15.91lb-ft)
C(4): 9.80 ~ 11.76Nm(1.0 ~ 1.2kgf.m, 7.23 ~ 8.68lb-ft)
D(1): 58.80 ~ 68.80Nm(6.0 ~ 7.0kgf.m, 43.40 ~ 50.63lb-ft)
E(1): 58.80 ~ 68.80Nm(6.0 ~ 7.0kgf.m, 43.40 ~ 50.63lb-ft)
F(2): 24.50 ~ 26.46Nm(2.5 ~ 2.7kgf.m, 18.08 ~ 19.53lb-ft)
G(4): 21.56 ~ 23.52Nm(2.2 ~ 2.4kgf.m, 15.91 ~ 17.36lb-ft)
H(1): 9.80 ~ 11.76Nm(1.0 ~ 1.2kgf.m, 7.23 ~ 8.68lb-ft)
I(1): 9.80 ~ 11.76Nm(1.0 ~ 1.2kgf.m, 7.23 ~ 8.68lb-ft)
J(1): 9.80 ~ 11.76Nm(1.0 ~ 1.2kgf.m, 7.23 ~ 8.68lb-ft)
K(4): 9.80 ~ 11.76Nm(1.0 ~ 1.2kgf.m, 7.23 ~ 8.68lb-ft)
L(1): 21.56 ~ 26.46Nm(2.2 ~ 2.7kgf.m, 15.91 ~ 19.53lb-ft) - New bolt

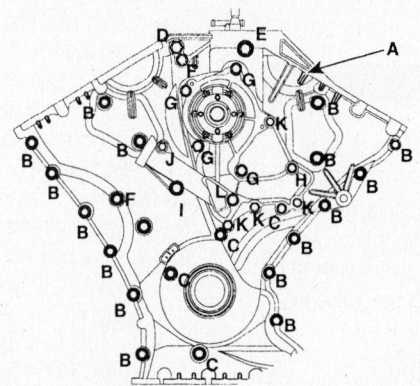

Timing cover bolt torque sequence and specification—3.8L Engine

36. Pull out the pins on both the left and right hydraulic tensioners. Install the oil pump chain cover.
37. Rotate the crankshaft two revolutions in the clockwise direction. Confirm the timing mark. Be sure to rotate the crankshaft in the clockwise direction.
38. Apply sealant on the chain cover and on the cylinder head, cylinder block, and lower oil pan. Be sure these parts are free of engine oil and dirt.

➡**Before assembling the timing chain cover, the liquid sealant TB1217H should be applied on the gap between the cylinder head and block. The part must be assembled with five minutes after the sealant is applied.**

39. Apply sealant on the chain cover.

➡**The liquid sealant TB1217H should be applied on the gap between the cylinder head and block. The part must be assembled with five minutes after the sealant is applied.**

40. Install a new gasket to the timing case cover.

➡**It is important that the dowel pins on the cylinder block and holes on the timing chain cover should be used as a reference in order to assemble the timing chain cover in the correct position.**

41. Install the timing chain cover. Torque the retaining bolts in the proper sequence and to specification. The engine should not be started for at least thirty minutes after timing chain cover assembly.
42. Install the timing case cover oil seal.
43. Continue the installation in the reverse order of the removal procedure.

Piston and Ring

POSITIONING

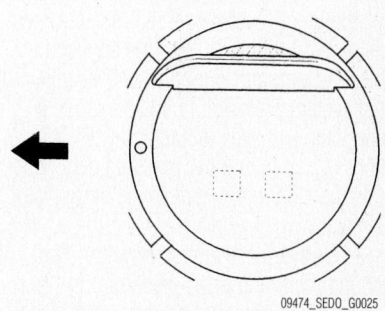

Piston ring end gap spacing—3.8L Engine

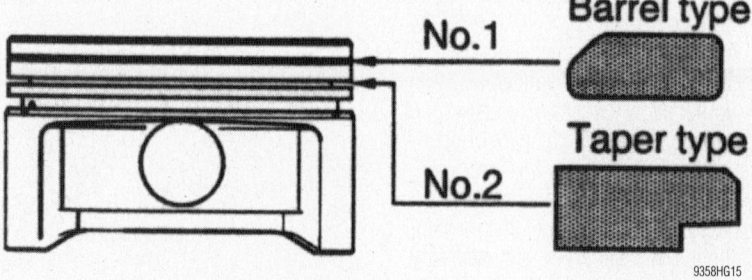

Compression ring identification—3.5L Engine

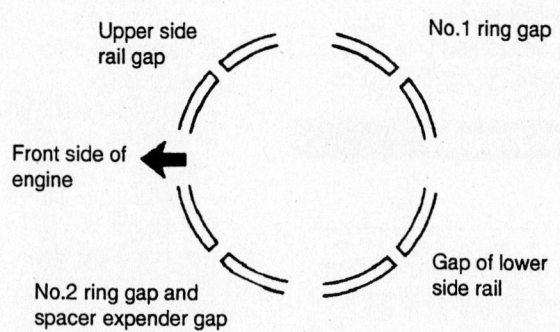

Piston ring end gap spacing—3.5L Engine

FUEL SYSTEM

Fuel System Service Precautions

Safety is the most important factor when performing not only fuel system maintenance, but any type of maintenance. Failure to conduct maintenance and repairs in a safe manner may result in serious personal injury or death. Maintenance and testing of the vehicle's fuel system components can be accomplished safely and effectively by adhering to the following rules and guidelines.

• To avoid the possibility of fire and personal injury, always disconnect the negative battery cable unless the repair or test procedure requires that battery voltage be applied.

• Always relieve the fuel system pressure prior to disconnecting any fuel system component (injector, fuel rail, pressure regulator, etc.), fitting or fuel line connection. Exercise extreme caution whenever relieving fuel system pressure, to avoid exposing skin, face and eyes to fuel spray. Please be advised that fuel under pressure may penetrate the skin or any part of the body that it contacts.

• Always place a shop towel or cloth around the fitting or connection prior to loosening to absorb any excess fuel due to spillage. Ensure that all fuel spillage (should it occur) is quickly removed from engine surfaces. Ensure that all fuel soaked cloths or towels are deposited into a suitable waste container.

• Always keep a dry chemical (Class B) fire extinguisher near the work area.

• Do not allow fuel spray or fuel vapors to come into contact with a spark or open flame.

• Always use a back-up wrench when loosening and tightening fuel line connection fittings. This will prevent unnecessary stress and torsion to fuel line piping. Always follow the proper torque specifications.

• Always replace worn fuel fitting O-rings with new. Do not substitute fuel hose or equivalent where fuel pipe is installed.

Fuel System Pressure

RELIEVING

1. Before servicing the vehicle, refer to the precautions at the beginning of this section.
2. Remove the rear seat.
3. Remove the access cover

4. Disconnect the fuel pump electrical connector.
5. Start the engine and allow it to run until it stalls.
6. Turn the ignition switch to the **OFF** position.
7. Disconnect the negative battery cable.
8. Restore the electrical connections after fuel system repairs are completed.

Fuel Filter

REMOVAL & INSTALLATION

The fuel filter is located inside the fuel tank and is replaced with the fuel pump as an assembly.

Fuel Pump

REMOVAL & INSTALLATION

1. Before servicing the vehicle, refer to the precautions at the beginning of this section.
2. Record the radio anti theft code data.
3. Remove or disconnect the following:
   • Rear seat
   • Access cover
   • Fuel pump electrical connector
4. Relieve the fuel system pressure.
5. Disconnect the negative battery cable.

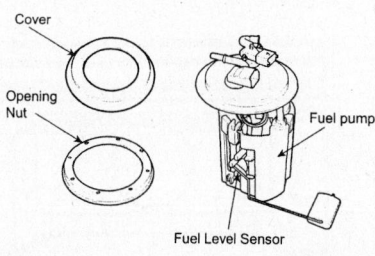

09474_SEDO_G0026

Fuel pump and related components—3.8L Engine

6. Remove or disconnect the following:
   • Fuel lines
   • Fuel pump and filter assembly
To install:
7. Install or connect the following:
   • Fuel pump and filter assembly
   • Fuel lines
   • Fuel pump electrical connector
   • Access cover
   • Rear seat
8. Start the engine and check for leaks.
9. Reprogram the radio anti theft codes.

Fuel Injector

REMOVAL & INSTALLATION

3.5L Engine

1. Before servicing the vehicle, refer to the precautions at the beginning of this section.
2. Record the radio anti theft code data.
3. Relieve the fuel system pressure.
4. Remove or disconnect the following:
   • Negative battery cable
   • Engine cover
   • Positive Crankcase Ventilation (PCV) hose
   • Brake booster vacuum hose
   • Surge tank stays
   • Fuel lines
   • Upper intake manifold (surge tank)
   • Fuel injector harness connectors
   • Fuel supply manifold with injectors attached
To install:
5. Replace all injector seals.
6. Install or connect the following:
   • Fuel supply manifold with injectors attached
   • Fuel injector harness connectors
   • Upper intake manifold (surge tank)
   • Fuel lines
   • Surge tank stays
   • Brake booster vacuum hose
   • Positive Crankcase Ventilation (PCV) hose
   • Engine cover
   • Negative battery cable
7. Start the engine and check for leaks.
8. Reprogram the radio anti theft codes.

3.8L Engine

1. Before servicing the vehicle, refer to the precautions at the beginning of this section.
2. Record the radio anti theft code data.
3. Relieve the fuel system pressure.
4. Disconnect the negative battery cable.
5. Remove the necessary components to gain access to the fuel injectors. Label all electrical connections to aid in reassembly.
6. Disconnect the electrical connector from the fuel injector. Remove the fuel injector from the engine.
To install:
7. Replace all injector seals.
8. Continue the installation in the reverse order of the removal procedure.
9. Start the engine and check for leaks.
10. Reprogram the radio anti theft codes.

KIA SEDONA 19-35

DRIVE TRAIN

Transaxle Assembly

REMOVAL & INSTALLATION

3.5L Engine

1. Before servicing the vehicle, refer to the precautions at the beginning of this section.
2. Record the radio anti theft code data.
3. Drain the cooling system.
4. Drain the transaxle fluid.
5. Remove or disconnect the following:
   - Battery and tray
   - Engine wiring harness
   - Transaxle wiring harness
   - Air filter assembly
   - Shift cable
   - Transaxle cooler lines
   - Radiator hoses
   - Radiator
   - Heater hoses
   - Steering intermediate shaft
   - Power steering pressure and return lines from the steering gear
   - Engine upper roll stopper and bracket
   - Front wheels
   - Heated Oxygen (HO2S) sensor connector
   - Front muffler
   - Outer tie rod ends
   - Lower ball joints
   - Axle halfshafts
   - Steering tube mounting bolt
6. Support the engine from below.
   - Front subframe
   - Starter motor
   - Transaxle housing cover
7. Support the transaxle with a transmission jack.
   - Torque converter bolts
   - Transaxle flange bolts
   - Transaxle

To install:

8. Install or connect the following:
   - Transaxle. Tighten the flange bolts to 29-38 ft. lbs. (42-54 Nm)
   - Torque converter bolts
   - Transaxle housing cover
   - Starter motor
   - Front subframe
   - Steering tube mounting bolt
   - Axle halfshafts
   - Lower ball joints
   - Outer tie rod ends
   - Front muffler
   - Heated Oxygen (HO2S) sensor connector

- Front wheels
- Engine upper roll stopper and bracket
- Power steering pressure and return lines from the steering gear
- Steering intermediate shaft
- Heater hoses
- Radiator
- Radiator hoses
- Transaxle cooler lines
- Shift cable
- Air filter assembly
- Transaxle wiring harness
- Engine wiring harness
- Battery and tray

9. Fill the transaxle to the correct level with the proper transmission fluid.
10. Fill the cooling system.
11. Start the engine and check for leaks.
12. Reprogram the radio anti theft codes.

3.8L Engine

1. Before servicing the vehicle, refer to the precautions at the beginning of this section.
2. Record the radio anti theft code data.
3. Disconnect the negative battery cable. Disconnect the positive battery cable. Remove the battery.
4. Disconnect the AFS connector. Remove the air cleaner upper cover. Remove the air cleaner assembly.
5. Disconnect the air cleaner hose. Remove the battery tray retaining bolts. Remove the battery tray.
6. Remove the inhibiter switch connector wire. Remove the solenoid valve connector wire. Remove the input speed sensor connector wire.
7. Remove the vehicle speed sensor connector wire. Remove the CKP sensor connector wire.
8. Disconnect and plug the transaxle

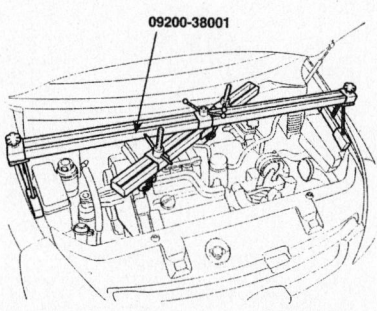

09474_SEDO_G0027

Engine/transaxle support tool installation—3.8L Engine

fluid cooler hoses from the transaxle lines. Remove the upper transaxle mounting bolts.
9. Using tool SST(09200-38001) or equivalent, hold the engine and transaxle assembly in position.
10. Remove the transaxle insulator mounting bolt.
11. Raise and safely support the vehicle. Remove the front tires. Remove the power steering column joint bolt.
12. Remove the engine/transaxle under cover. Drain the transaxle fluid. Drain the power steering fluid, through the return line. Disconnect the power steering pressure tube from the power steering oil pump.
13. Disconnect the lower arm, the tie rod end ball joint and the stabilizer bar link from the front knuckle.
14. Remove the roll stopper mounting bolts. Remove the mounting bolts from the subframe, by supporting the subframe with a jack. As necessary, remove the starter.
15. Remove the driveshaft from the transaxle. Position a jack, for support, under the transaxle assembly.
16. Remove the transaxle under mounting bolts. Remove the driveplate bolts.
17. Lifting the vehicle up and lowering the jack slowly, remove the transaxle assembly.

To install:

18. Installation is the reverse of the removal procedure.
19. Torque the lower transaxle to engine mounting bolts to 47-62 ft. lbs. (65-85 Nm).
20. Torque the upper transaxle to engine mounting bolts to 47-62 ft. lbs. (65-85 Nm).
21. Torque the roll stopper mounting bolts to 65-80 ft. lbs. (90-110 Nm).
22. Torque the transaxle insulator engine mounting bolt to 65-80 ft. lbs. (90-110 Nm).
23. Fill the transaxle with the proper grade and type transaxle fluid.
24. Run the engine and recheck the transaxle fluid. Roadtest the vehicle.
25. Reprogram the radio anti theft codes.

Halfshaft

REMOVAL & INSTALLATION

3.5L Engine

1. Before servicing the vehicle, refer to the precautions at the beginning of this section.

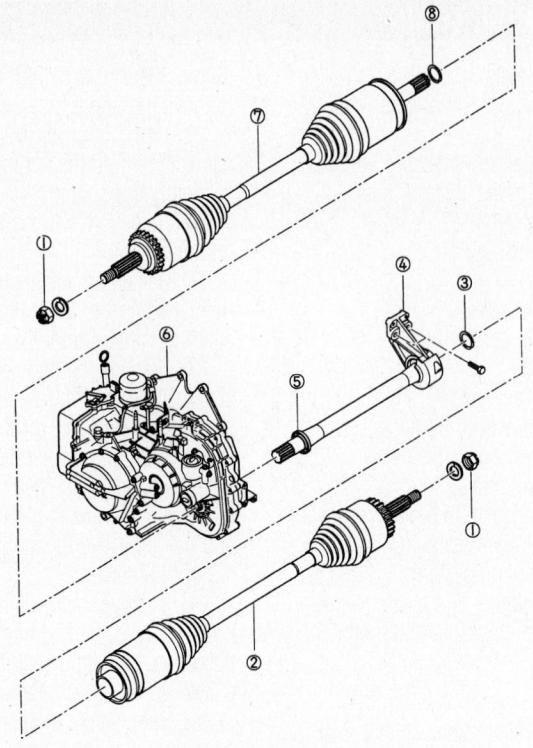

(1) Wheel nut
(2) Driveshaft(RH)
(3) Circlip
(4) Center bearing bracket

(5) constant velocity shaft
(6) Automatic transaxle
(7) Driveshaft(LH)
(8) Circle pin

9358HG17

Halfshaft and related components—3.5L Engine

2. Record the radio anti theft code data. Disconnect the negative battery cable.

3. Raise and support the vehicle safely. Drain the transaxle fluid.

4. Remove or disconnect the following:
- Front wheels
- Hub retaining nuts
- Sway bar links
- Outer tie rod ends
- Lower ball joints

5. Using a prybar, remove the left halfshaft from the transaxle. Separate the right halfshaft from the intermediate shaft.

6. Remove the axle halfshafts from the front hubs. It may be necessary to tap the stub shaft with a brass hammer to remove the axles.

To install:

➡**Use new circlips for assembly.**

7. Install the left halfshaft to the transaxle.

8. Install the right halfshaft to the intermediate shaft.

9. Install the halfshaft stub shafts to the wheel hubs.

10. Install or connect the following:

- Lower ball joints. Tighten the pinch bolts to 69-85 ft. lbs. (93-115 Nm).
- Outer tie rod ends. Tighten the nuts to 43-58 ft. lbs. (59-78 Nm).
- Sway bar links. Tighten nuts to 69-85 ft. lbs. (93-115 Nm).
- Hub retaining nuts. Tighten the hub nuts to 177-199 ft. lbs. (240-270 Nm).
- Front wheels. Tighten the lug nuts to 65-79 ft. lbs. (88-108 Nm).

11. Fill the transaxle to the correct level with the proper fluid.

12. Start the engine and check for leaks.

13. Reprogram the radio anti theft codes.

3.8L Engine

1. Before servicing the vehicle, refer to the precautions at the beginning of this section.

2. Record the radio anti theft code data.

3. Disconnect the negative battery cable. Raise and support the vehicle safely. Remove the front tires.

4. Drain the transaxle fluid.

5. Unstake the driveshaft lock nut. Remove the lock nut.

6. Remove the split pin and castle nut from the tie rod end ball joint. Disconnect the tie rod end from the knuckle, using tool SST (09568-4A000).

7. Remove the split pin and lower arm ball and nut. Using a plastic hammer, dis-

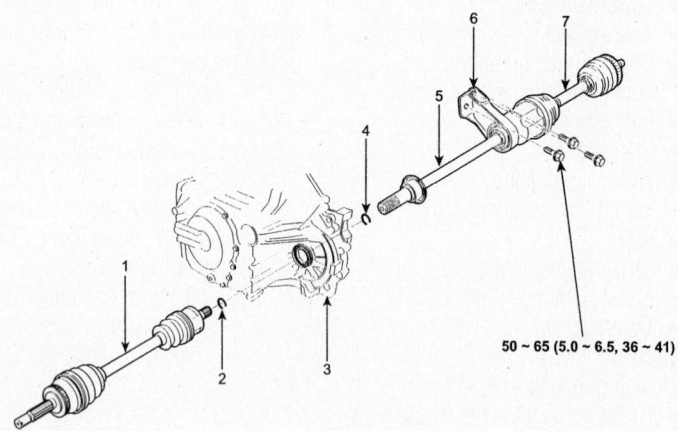

50 ~ 65 (5.0 ~ 6.5, 36 ~ 41)

TORQUE : Nm (kgf·m, lb-ft)

1. Driveshaft [LH]
2. Circlip
3. Transaxle
4. Circlip

5. Inner shaft
6. Bearing & bracket assembly
7. Driveshaft [RH]

09474_SEDO_G0028

Halfshaft and related components—3.8L Engine

connect the driveshaft from the front hub assembly.

8. To remove the right driveshaft, remove the right driveshaft heat protector. Remove the inner shaft bearing bracket assembly mounting bolts.

9. Insert a pry bar between the transaxle case and driveshaft joint; separate the driveshaft from the transaxle. Pull the driveshaft from the transaxle case.

To install:

➡**Use new circlips for assembly.**

10. Installation is the reverse of the removal procedure.

11. Torque the inner shaft bearing bracket assembly bolts to 36-47 ft. lbs. (50-65 Nm).

12. Torque the split pin and lower arm bolt and nut to 65-87 ft. lbs. (90-110 Nm).

13. Torque the tie rod end ball joint split pin and castle nut to 43-58 ft. lbs. (60-80 Nm).

14. Torque the driveshaft lock nut to 177-199 ft. lbs. (245-275 Nm).

15. Torque the tire and wheel assembly to 65-80 ft. lbs. (90-110 Nm).

16. Fill the transaxle with the proper grade and type transaxle fluid.

17. Run the engine and recheck the transaxle fluid. Roadtest the vehicle.

18. Reprogram the radio anti theft codes.

CV Joint

OVERHAUL

3.5L Engine

OUTER CV JOINT

The outer CV joint is serviced only with the axle halfshaft as an assembly. The outer CV joint boot can be serviced by first removing the inner Tripod joint.

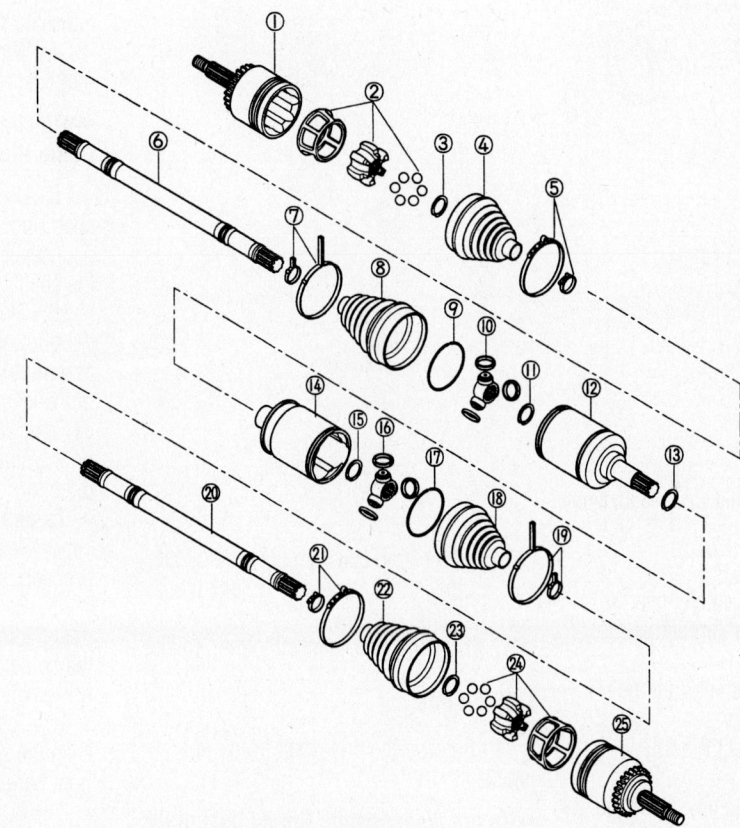

Driveshaft(LH)	(10) Spider assembly	(18) UTJ boot
(1) BJ assembly	(11) Snap ring	(19) UTJ boot band
(2) BJ inner race and ball	(12) UTJ assembly	(20) Driveshaft(RH)
(3) Snap ring	(13) Circlip	(21) BJ boot band
(4) BJ boot		(22) BJ boot
(5) BJ boot band	Driveshaft(RH)	(23) Snap ring
(6) Driveshaft(LH)	(14) UTJ assembly	(24) BJ inner race and ball
(7) UTJ boot band	(15) Snap ring	(25) BJ assembly
(8) UTJ boot	(16) Spider assembly	
(9) Circle pin	(17) Circle pin	

🖉 **Caution**
  a) **Install a protective material in a vice, and secure joint in the vice.**
  b) **Keep dust or foreign material from joint during procedure.**
  c) **Do not disassemble wheel side ball joint.**
  d) **Do not wash joint unless disassembling it.**

9358HG18

Inner and outer CV Joint exploded view and parts identification—3.5L Engine

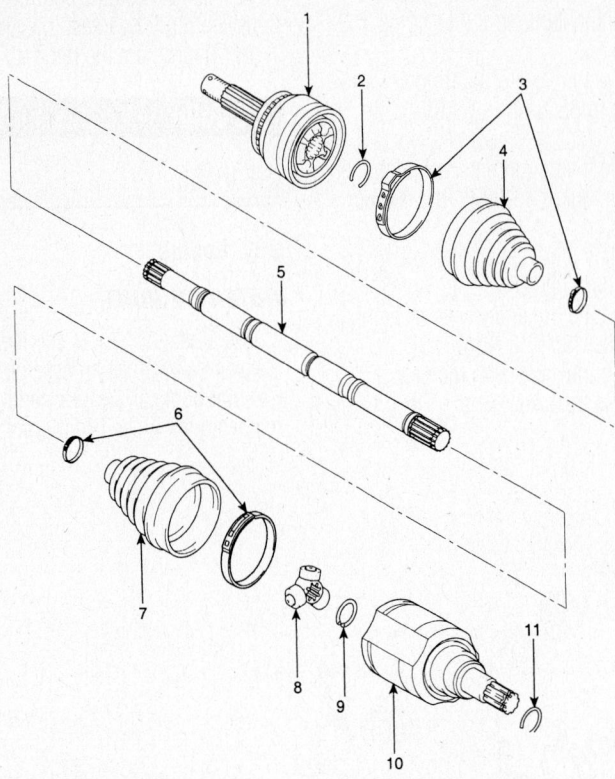

1. BJ assembly
2. Circlip
3. BJ boot bands
4. BJ boot
5. Shaft
6. UTJ boot bands
7. UTJ boot
8. Spider assembly
9. Circlip
10. UTJ case
11. Clip

09474_SEDO_G0029

Driveshaft assembly exploded view—3.8L Engine

INNER TRIPOT JOINT

1. Before servicing the vehicle, refer to the precautions at the beginning of this section.

2. Remove the axle halfshaft from the vehicle and place it in a vise.

3. Remove or disconnect the following:

- Tripod joint boot clamps. Slide the boot on the axle shaft to expose the joint.
- Circle pin
- Tripod joint housing
- Tripod joint snapring
- Tripod joint
- Inner Tripod joint boot

To assemble:

4. Install or connect the following:

- Inner Tripod joint boot
- Tripod joint
- Tripod joint snapring
- Tripod joint housing. Fill with 7.5 ounces of CV Joint grease.
- Circle pin
- Tripod joint boot clamps. Pull on

the clamp end with pliers and fold the locking tabs over to lock the clamp in place.

5. Install the axle halfshaft to the vehicle.

3.8L Engine

1. Remove the driveshaft from the vehicle.

➡ **Do not disassemble the BJ assembly.**

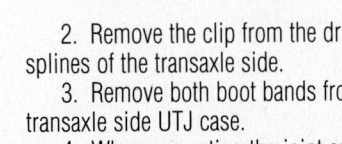

09474_SEDO_G0030

Put alignment marks across the spider roller assembly (A), UTJ case (B) and shaft splines (C)—3.8L Engine

2. Remove the clip from the driveshaft splines of the transaxle side.

3. Remove both boot bands from the transaxle side UTJ case.

4. When separating the joint and boot, remove the grease from the UTJ case.

➡ **Matchmark the assembly according to the illustration.**

5. Using the proper tool, remove the snap ring. Using tool (09495-3300), or equivalent remove the spider assembly from the shaft.

6. Clean the spider assembly. Remove the boot of the transaxle side joint (UTL assembly). Remove both bands on the side of the wheel.

7. Pull out the joint boot (BJ assembly) on the side of the wheel in the transaxle direction.

To assemble:

8. To assemble reverse the disassembly procedures.

9. Be sure to use the alignment marks made during disassembly when installing the spider assembly and snap ring on the driveshaft splines.

10. To control the air in the UTJ boot, keep the specified distance between the boot bands when they are tightened.

11. Be sure distance "L" is to specification.

STANDARD VALUE [L]

Items	Standard(L) mm(in)	
	LH	RH
3.8 Gasoline A/T	557.1 (21.93)	570.7 (22.47)

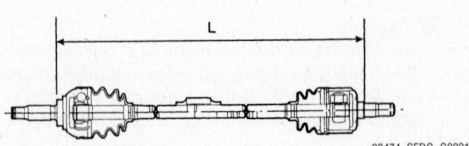

09474_SEDO_G0031

Distance L specification—3.8L Engine

STEERING

Air Bag

✳✳ CAUTION

Some vehicles are equipped with an air bag system. The system must be disarmed before performing service on, or around, system components, the steering column, instrument panel components, wiring and sensors. Failure to follow the safety precautions and the disarming procedure could result in accidental air bag deployment, possible injury and unnecessary system repairs.

PRECAUTIONS

Several precautions must be observed when handling the inflator module to avoid accidental deployment and possible personal injury.

• Never carry the inflator module by the wires or connector on the underside of the module.

• When carrying a live inflator module, hold securely with both hands, and ensure that the bag and trim cover are pointed away.

• Place the inflator module on a bench or other surface with the bag and trim cover facing up.

• With the inflator module on the bench, never place anything on or close to the module which may be thrown in the event of an accidental deployment.

DISARMING THE SYSTEM

2002–2005

1. Before servicing the vehicle, refer to the precautions at the beginning of this section.
2. Record the radio anti theft code data.
3. Position the vehicle with the front wheels in a straight-ahead position.
4. Disconnect both battery cables.
5. Wait at least 1 minute for the air bag back-up power supply to deplete its stored energy before continuing.
6. Proceed with the repair.
7. Reconnect both battery cables once the repair is complete.

2006

1. Before servicing the vehicle, refer to the precautions at the beginning of this section.

2. Record the radio anti theft code data. Remove the ignition key from the vehicle.
3. Disconnect the negative battery cable.
4. Wait at least 3 minute for the air bag back-up power supply to deplete its stored energy before continuing.
5. Proceed with the repair.
6. To confirm proper system operation, turn the ignition switch to the ON position. The SRS indicator light will be lit for at least six seconds and then go off.

Power Steering Gear

REMOVAL & INSTALLATION

2002–2005

1. Before servicing the vehicle, refer to the precautions at the beginning of this section.
2. Record the radio anti theft code data.
3. Drain the power steering fluid.
4. Remove or disconnect the following:
   • Negative battery cable
   • Front wheels
   • Outer tie rod ends
   • Power steering fluid pressure and return lines
   • Steering intermediate shaft
   • Steering rack brackets and fasteners
5. Remove the steering gear through the right wheel opening.

To install:

6. Install the steering gear through the right wheel opening.
7. Install or connect the following:
   • Steering rack brackets and fasteners. Tighten the fasteners to 55-69 ft. lbs. (74-93 Nm).
   • Steering intermediate shaft. Tighten the pinch bolt to 16-20 ft. lbs. (21-26 Nm).
   • Power steering fluid pressure and return lines. Tighten the bolts to 17-26 ft. lbs. (24-35 Nm).
   • Outer tie rod ends. Tighten the nuts to 43-58 ft. lbs. (59-78 Nm).
   • Front wheels
   • Negative battery cable

2006

1. Before servicing the vehicle, refer to the precautions at the beginning of this section.
2. Record the radio anti theft code data.
3. Disconnect the negative battery cable. Raise and support the vehicle safely. Drain the power steering fluid. Remove the front tires.
4. Remove the bolt connecting the steering column to the universal joint.
5. Disconnect the pressure line from the power steering oil pump. Disconnect the return hose.
6. Loosen the split pin and the castle nut, and remove the tie rod end from the knuckle using tool SST (09568-4A000).

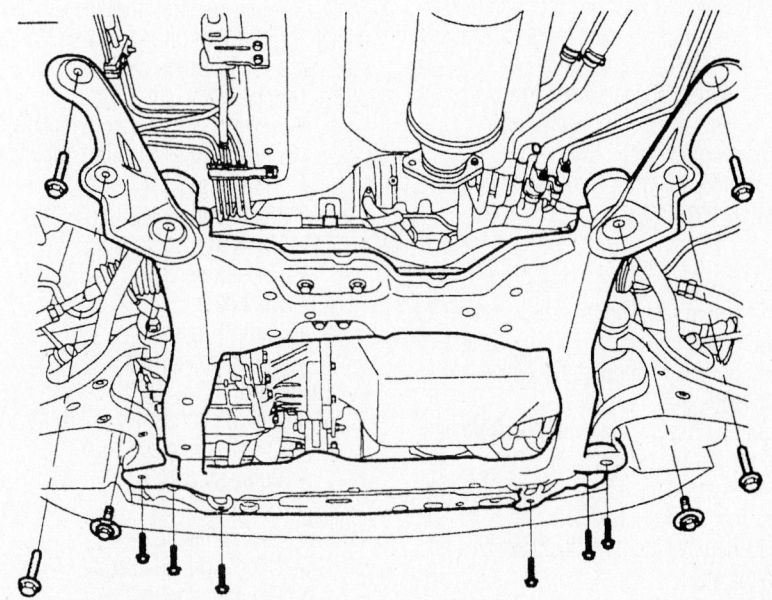

Subframe attaching bolts—2006 vehicles

09474_SEDO_G0032

7. Remove the split pin and lower arm bolts and nut. Disconnect the stabilizer link from the strut assembly. Repeat on the other side of the vehicle.

8. Remove the front and rear roll stopper bolts and nuts. Remove the subframe retaining bolts and nuts. Remove the rear roll stopper from the subframe.

9. Disconnect the pressure line and the return line from the steering gear valve body housing.

10. Remove the steering gear assembly from the subframe, after removing the retaining bolts.

To install:

11. Installation is the reverse of the removal procedure.

12. Torque the steering gear to subframe mounting bolts to 65-80 ft. lbs. (90-110 Nm).

13. Torque the split pin and lower arm bolt and nut to 65-87 ft. lbs. (90-110 Nm).

14. Fill the power steering pump with the proper grade and type fluid.

15. To air bleed the system, remove the fuel pump fuse, start the engine and allow it to stall.

16. While operating the starting motor intermittently (not more then 15 seconds at a time) turn the steering wheel all the way to the left and then all the way to the right. Repeat the procedure five or six times. Do not hold the steering wheel in the full turn position for more than ten seconds.

17. Check the fluid level to be sure it does not fall below the lower position on the filter.

➡ **If bleeding is done at engine idle, air will be broken up and absorbed into the fluid.**

18. Check the wheel alignment.

FRONT SUSPENSION

Strut

REMOVAL & INSTALLATION

2002–2005

1. Before servicing the vehicle, refer to the precautions at the beginning of this section.

2. Remove or disconnect the following:
   - Front wheel
   - Brake hose from the bracket
   - Wheel speed sensor cable
   - Steering knuckle mounting bolts
   - Upper strut mount nuts

3. Remove the strut from the vehicle.

To install:

4. Install the strut to the vehicle.

5. Install or connect the following:
   - Upper strut mount nuts and tighten them to 33-46 ft. lbs. (46-62 Nm).
   - Steering knuckle mounting bolts and tighten them to 88-101 ft. lbs. (119-137 Nm).
   - Wheel speed sensor cable
   - Brake hose to the bracket
   - Front wheel

6. Check the front end alignment and adjust as necessary.

2006

1. Before servicing the vehicle, refer to the precautions at the beginning of this section.

2. Raise and support the vehicle safely. Remove the front tires.

3. Remove the brake hose bracket bolt from the strut assembly.

4. Remove the speed sensor and wire bolts from the front knuckle.

5. Remove the front stabilizer link nut from the strut.

6. Remove the upper strut mounting nuts.

7. Remove the front strut mounting bolts from the knuckle.

8. Remove the strut from its mounting.

To install:

9. Install the strut to the vehicle.

10. Torque the upper retaining nuts to 32-43 ft. lbs. (45-60 Nm).

11. Torque the lower strut retaining bolts to 72-87 ft. lbs. (100-120 Nm).

12. Torque the front stabilizer link to the strut assembly nut to -87 ft. lbs. (100-120 Nm).

13. Continue the installation in the reverse order of the removal procedure.

DISASSEMBLY & ASSEMBLY

2002–2005

1. Before servicing the vehicle, refer to the precautions at the beginning of this section.

2. Remove the strut from the vehicle and attach Service Tool 0K2A1 341 001 or other suitable spring compressor.

3. Compress the coil spring.

4. Remove or disconnect the following:
   - Upper mount retaining nut
   - Mounting block
   - Upper spring seat
   - Upper spring isolator
   - Coil spring
   - Dust boot
   - Bump stopper
   - Lower spring seat

To install:

5. Install or connect the following:
   - Lower spring seat
   - Bump stopper
   - Dust boot
   - Coil spring
   - Upper spring isolator
   - Upper spring seat
   - Mounting block

   - Upper mount retaining nut and tighten it to 88-101 ft. lbs. (120-137 Nm).

6. Install the strut to the vehicle.

7. Check the front end alignment and adjust as necessary.

2006

1. Before servicing the vehicle, refer to the precautions at the beginning of this section.

2. Remove the strut from the vehicle and attach Service Tool 09546-2600 or other suitable spring compressor.

3. Compress the coil spring.

4. Remove the self locking nut from the strut. Remove the insulator, spring seat, coil spring and dust cover.

5. Inspect the insulator for wear and damage, replace as required.

6. Check the rubber parts for damage or deterioration, replace as required.

7. Install the spring lower pad so that the protrusions fit the holes in the spring lower seat.

8. Compress the spring, using the spring compressor tool. Install the compressed spring into the shock absorber.

➡ **There are two color identification marks on the coil spring, one indicates model option and the other indicates load classification. Install the coil spring with the identification mark directed toward the knuckle.**

9. After fully extending the piston rod, install the spring upper seat and insulator assembly.

10. After seating the upper and lower ends of the coil spring in the upper and lower spring grooves correctly, tighten the new self locking nut temporarily.

11. Remove the spring compression

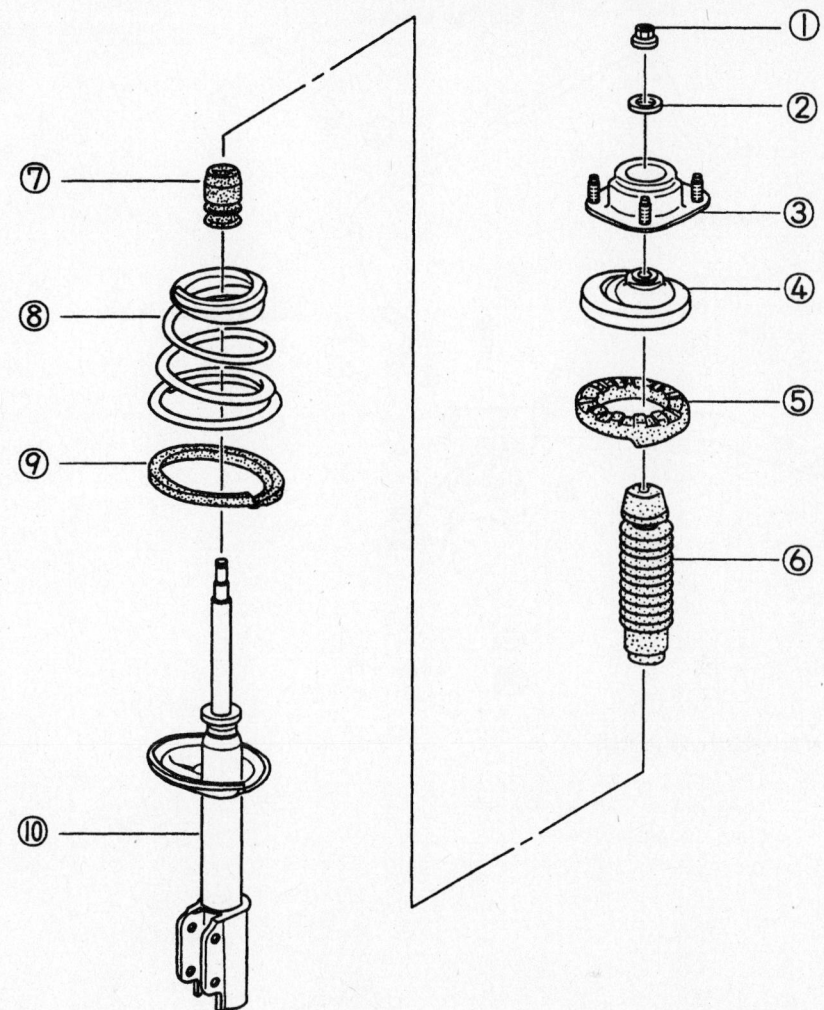

(1) Piston retaining nut
(2) Washer
(3) Mounting block
(4) Upper spring seat
(5) Upper spring isolator

(6) Dust boot
(7) Bump stopper
(8) Coil spring
(9) Lower spring isolator
(10) Shock absorber

9358HG19

Front strut and coil spring exploded view—2002—05 vehicles

tool. Tighten the self locking nut to 43-51 ft. lbs. (60-70 Nm).

12. Install the strut to the vehicle.

Stabilizer Bar

REMOVAL & INSTALLATION

2002–2005

1. Before servicing the vehicle, refer to the precautions at the beginning of this section.

2. Raise and support the vehicle safely. Remove the front tires.

3. Remove the stabilizer control link bolt.

4. Remove the control link lower control arm bolt.

5. Remove the exhaust pipe.

6. Position a suitable transaxle jack under the transaxle assembly.

7. Remove the engine number one and two mounting bolts from the subframe.

8. Remove the stabilizer mounting bolts from the subframe.

9. Lower the subframe enough to pull the stabilizer out from behind the subframe.

10. Remove the stabilizer bar from the vehicle.

To install:

11. Installation is the reverse of the removal procedure.

2006

1. Before servicing the vehicle, refer to the precautions at the beginning of this section.

2. Remove the connecting bolt between the steering universal joint assembly and the pinion assembly.

3. Raise and support the vehicle safely. Remove the front tires.

4. Remove the stabilizer link from the strut assembly.

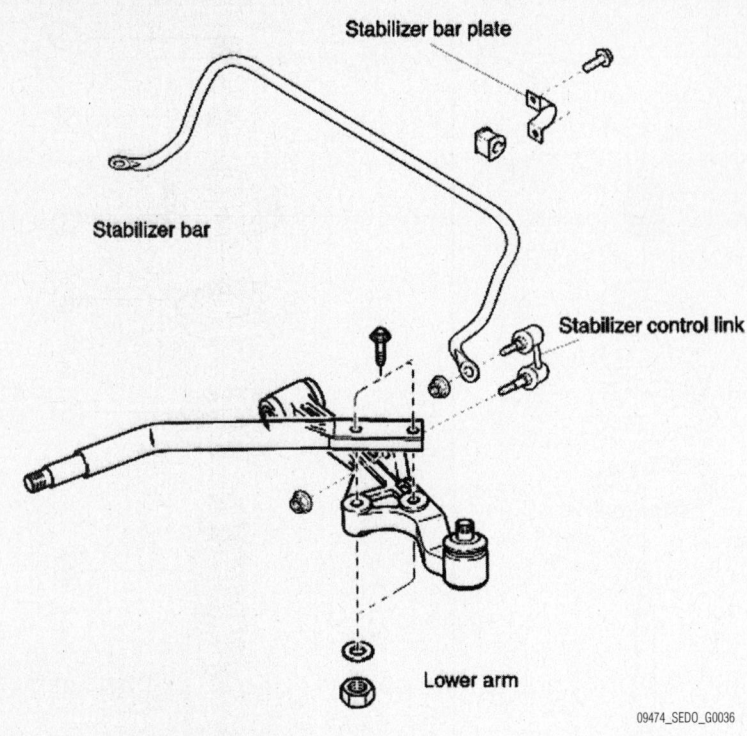

Stabilizer bar plate

Stabilizer bar

Stabilizer control link

Lower arm

09474_SEDO_G0036

Stabilizer bar—2002—05 vehicles

5. After removing the left and right tie rod end self locking nuts and cotter pins, remove the ball joint using a ball joint removal tool.

6. Remove the left and right lower arm mounting nuts and bolts.

7. Remove the engine mounting bolts. Remove the twelve bolts and nuts of the subframe, after supporting it with a jack.

8. After lowering the jack that is supporting the subframe, remove the stabilizer bar mounting bolts and stabilizer bar.

➡**Remove the stabilizer bar assembly through the gap between the body and rear side of the subframe. Be careful not to damage the power steering lines.**

9. Remove the brackets and bushings from the stabilizer bar, replace as required.

To install:

10. Installation is the reverse of the removal procedure.

11. Torque the stabilizer bar bracket mounting bolts to 28-43 ft. lbs. (39-60 Nm).

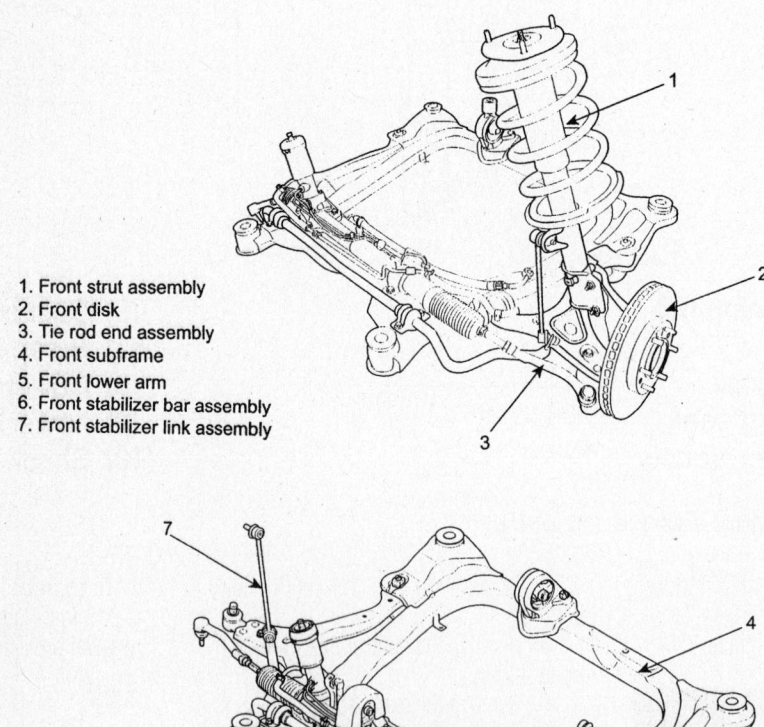

1. Front strut assembly
2. Front disk
3. Tie rod end assembly
4. Front subframe
5. Front lower arm
6. Front stabilizer bar assembly
7. Front stabilizer link assembly

09474_SEDO_G0034

Front suspension and related components—2006 vehicles

12. Torque the subframe bolts with washers to 116-130 ft. lbs. (160-180 Nm).

13. Torque the subframe bolts without washers to 32.5-43 ft. lbs. (45-60 Nm).

14. Torque the engine mounting bolts to 47-62 ft. lbs. (65-85 Nm).

15. Torque the lower arm mounting bolts to 65-87 ft. lbs. (160-180 Nm).

16. Torque the front stabilizer link nuts to 72-87 ft. lbs. (160-180 Nm).

17. Torque the steering universal joint bolt to 9-13 ft. lbs. (13-18 Nm).

18. Check the front end alignment and adjust as necessary.

Lower Ball Joint

REMOVAL & INSTALLATION

The ball joint is serviced with the lower control arm as an assembly.

Lower Control Arm

REMOVAL & INSTALLATION

2002–2005

1. Before servicing the vehicle, refer to the precautions at the beginning of this section.

2. Remove or disconnect the following:
   - Front wheel
   - Sway bar link
   - Tension rod bolts
   - Lower ball joint
   - Lower control arm subframe bolt
   - Lower control arm

To install:

3. Install or connect the following:
   - Lower control arm. Tighten the subframe bolt to 88-101 ft. lbs. (120-137 Nm)
   - Lower ball joint. Tighten the pinch bolt to 69-85 ft. lbs. (93-115 Nm)
   - Tension rod bolts and tighten them to 88-101 ft. lbs. (120-137 Nm)
   - Sway bar link. Tighten the nut to 69-85 ft. lbs. (93-115 Nm)
   - Front wheel

4. Check the front end alignment and adjust as necessary.

2006

1. Before servicing the vehicle, refer to the precautions at the beginning of this section.

2. Raise and support the vehicle safely. Remove the front tires.

3. Remove the lower arm mounting bolt from the knuckle.

4. Remove the lower arm mounting

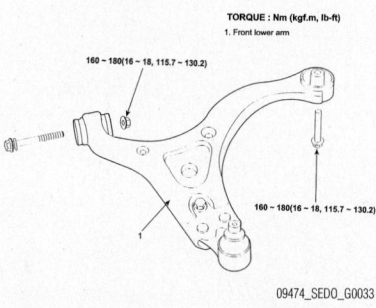

TORQUE : Nm (kgf.m, lb-ft)
1. Front lower arm

160 ~ 180(16 ~ 18, 115.7 ~ 130.2)

160 ~ 180(16 ~ 18, 115.7 ~ 130.2)

09474_SEDO_G0033

Lower control arm assembly—2006 vehicles

bolts. Remove the lower arm from the vehicle.

To install:

5. Installation is the reverse of the removal procedure.

6. Torque the lower arm mounting bolts to 116-130 ft. lbs. (160-180 Nm).

7. Torque the front lower arm ball joint mounting bolt to 65-87 ft. lbs. (90-120 Nm).

8. Check the front end alignment and adjust as necessary.

Wheel Bearings

ADJUSTMENT

2002–2005

The front wheel bearings are not adjustable. Replace the wheel bearings if play exceeds 0.002 inches (0.05 mm).

2006

The front wheel bearings are not adjustable. Replace the hub and bearing assembly as a complete unit.

REMOVAL & INSTALLATION

2002–2005

➡**The front wheel bearings are serviced with the steering knuckle as an assembly.**

1. Before servicing the vehicle, refer to the precautions at the beginning of this section.

2. Remove or disconnect the following:
   - Front wheel
   - Brake caliper and rotor
   - Wheel speed sensor harness
   - Hub nut
   - Lower ball joint
   - Strut mounting bolts
   - Steering knuckle

To install:

3. Install or connect the following:
   - Steering knuckle
   - Strut mounting bolts and tighten them to 88-101 ft. lbs. (119-137 Nm).
   - Lower ball joint. Tighten the pinch bolt to 69-85 ft. lbs. (93-115 Nm).
   - Hub nut and tighten it to 177-199 ft. lbs. (240-270 Nm).
   - Wheel speed sensor harness
   - Brake caliper and rotor
   - Front wheel

4. Check the alignment and adjust as necessary.

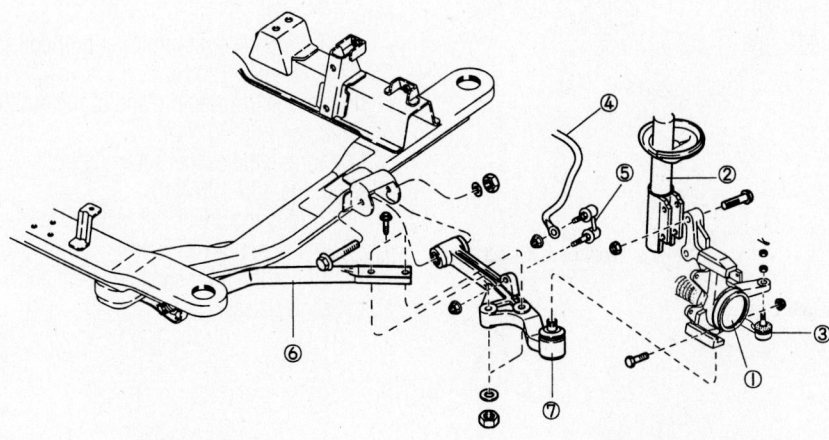

(1) Knuckle
(2) Shock absorber
(3) Tie rod end
(4) Stabilizer bar
(5) Stabilizer control link
(6) Tension rod
(7) Lower arm

9358HG21

Front suspension exploded view—2002—05 vehicles

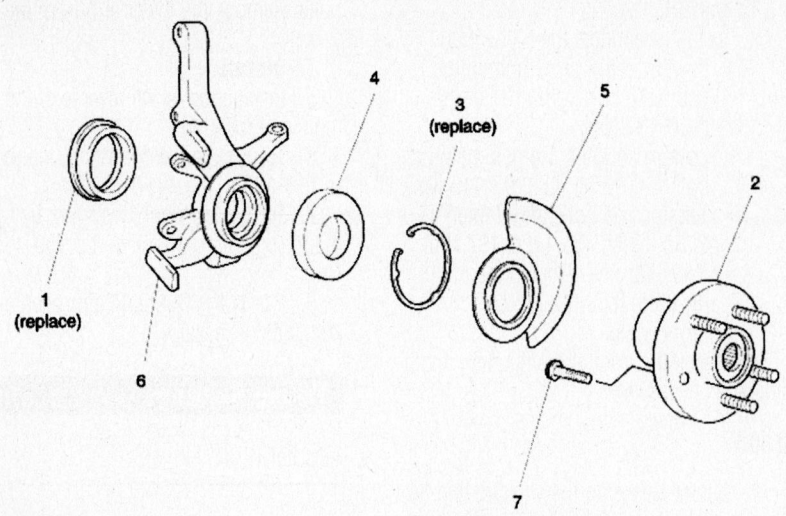

1. Oil seal
2. Front wheel hub assembly
3. Retaining ring
4. Wheel bearing
5. Dust cover
6. Steering knuckle
7. Wheel hub bolt

09474_SEDO_G0037

Front wheel bearing and related components—2002—05 vehicles

TORQUE : Nm (kgf·m, lb-ft)

160 ~ 180
(16.0 ~ 18.0, 157 ~ 177)

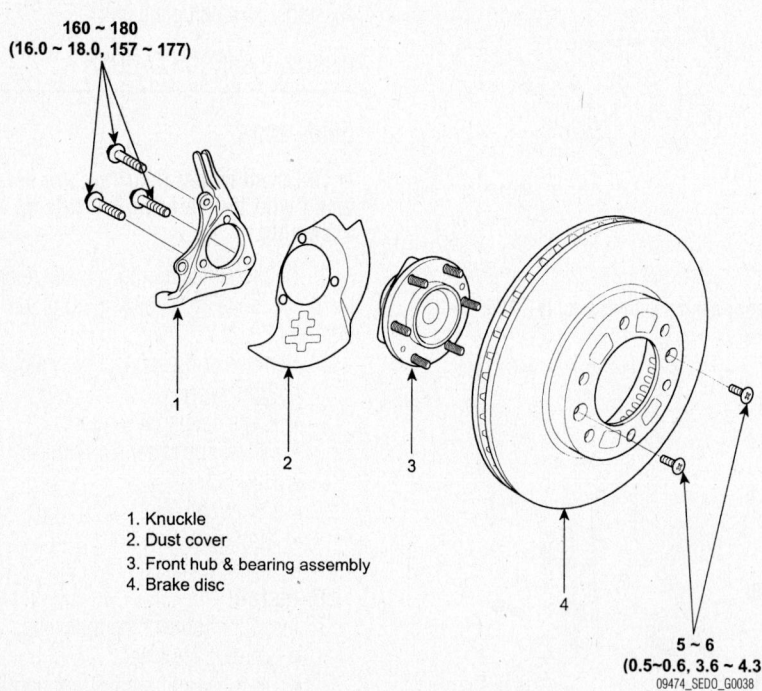

1. Knuckle
2. Dust cover
3. Front hub & bearing assembly
4. Brake disc

5 ~ 6
(0.5~0.6, 3.6 ~ 4.3
09474_SEDO_G0038

Front wheel bearing and related components —2006 vehicles

2006

➡ **The front wheel bearings are serviced with the steering knuckle as an assembly.**

1. Before servicing the vehicle, refer to the precautions at the beginning of this section.

2. Raise and support the vehicle safely. Remove the front tires.

3. Remove the wheel speed sensor and wire. Disconnect the brake hose from the strut assembly.

4. Unstake the driveshaft lock nut. Remove the driveshaft lock nut.

5. Remove the caliper mounting bolts. Secure it to the side.

6. Remove the split pin and castle nut from the tie rod end ball joint.

7. Disconnect the tie rod end from the knuckle, using tool SST(09568-4A000), or equivalent.

8. Remove the split pin and lower arm bolt and nut. Using a plastic hammer, disconnect the driveshaft from the hub assembly.

9. Remove the rotor from the hub assembly.

10. Remove the knuckle from the strut assembly.

11. Separate the hub and bearing assembly from the knuckle, by loosening the bolts.

To install:

12. Installation is the reverse of the removal procedure.

13. Torque the hub and bearing assembly bolts to 116-130 ft. lbs. (160-180 Nm).

14. Torque the knuckle to strut assembly mounting bolts to 72-87 ft. lbs. (100-120 Nm).

15. Torque the split pin lower arm bolt to 65-87 ft. lbs. (90-120 Nm).

16. Torque the split pin and castle nut to 43-58 ft. lbs. (68-80 Nm).

17. Torque the caliper retaining bolts to 61-72 ft. lbs. (85-100 Nm).

18. Torque the driveshaft lock nut to 177-199 ft. lbs. (245-275 Nm).

19. Check the front end alignment and adjust as necessary.

REAR SUSPENSION

Shock Absorber

REMOVAL & INSTALLATION

1. Before servicing the vehicle, refer to the precautions at the beginning of this section.

2. Raise and support the vehicle safely. Support the rear axle on jack stands.

3. Remove or disconnect the following:
   - Rear wheel
   - Shock absorber upper nut and washer
   - Shock absorber lower nut and washer
   - Shock absorber

To install:

4. Install or connect the following:
   - Shock absorber
   - Shock absorber lower nut and washer. On 2002–05 vehicles, tighten the nut to 55-69 ft. lbs. (74-93 Nm).
   - Shock absorber upper nut and washer. On 2002–05 vehicles, tighten the nut to 41-47 ft. lbs. (55-64 Nm).
   - Shock absorber bracket mounting bolts. On 2006 vehicles, tighten the bolts to 60-80 ft. lbs. (80-110 Nm) and the nuts to 65-87 ft. lbs. (90-120 Nm)
   - Shock absorber to knuckle nut. On 2006 vehicles, tighten the nut to 116-130 ft. lbs. (160-180 Nm)
   - Rear wheel

Coil Spring

REMOVAL & INSTALLATION

2002–2005

1. Before servicing the vehicle, refer to the precautions at the beginning of this section.

2. Support the vehicle with jackstands forward of the lower control arm mounting.

3. Support the rear axle with jackstands.

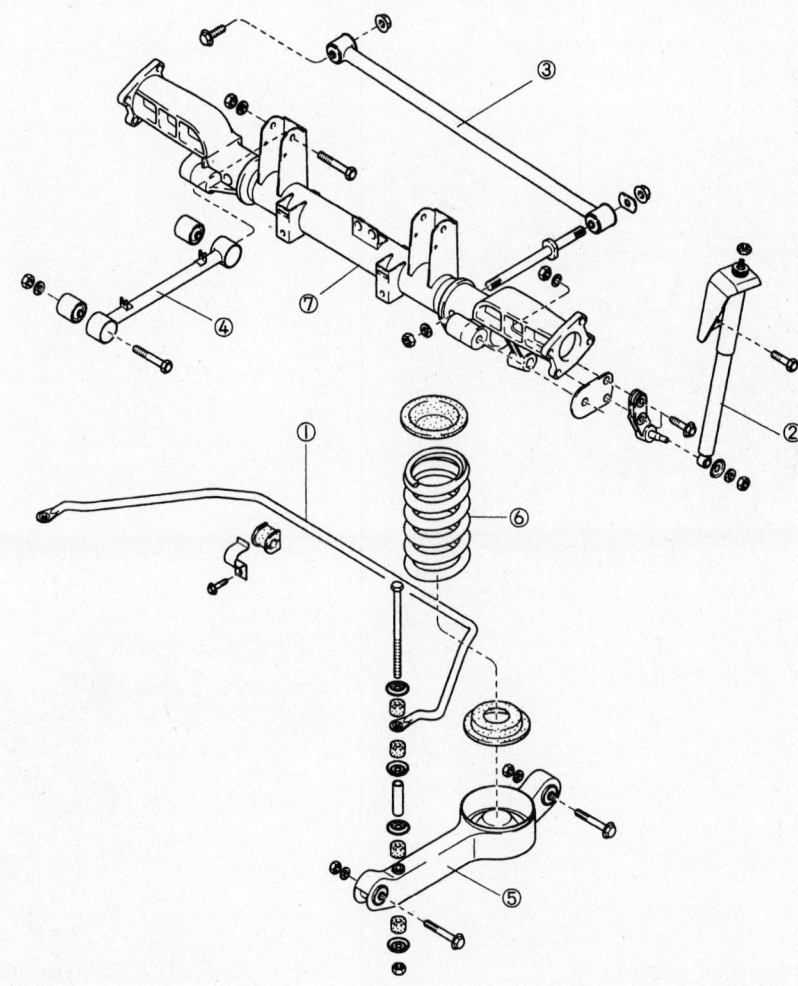

(1) Stabilizer bar
(2) Shock absorber
(3) Panhard rod
(4) Upper arm assembly

(5) Lower arm assembly
(6) Coil spring
(7) Rear casing

9358HG20

Rear suspension and related components—2002—05 vehicles

4. Support the lower control arm with a floor jack.

5. Remove or disconnect the following:

- Rear wheel
- Sway bar link
- Parking brake cable
- Lower control arm bolts

6. Carefully lower the floor jack and remove the lower control arm, coil spring, and spring seats from the vehicle.

To install:

7. Install or connect the following:

- Spring seats
- Coil spring
- Lower control arm. Tighten the bolts to 87-101 ft. lbs. (118-137 Nm)
- Parking brake cable. Tighten the bracket bolts to 14-19 ft. lbs. (16-23 Nm)

- Sway bar link. Tighten the locknut to 17-20 ft. lbs. (24-28 Nm)
- Rear wheel

2006

1. Before servicing the vehicle, refer to the precautions at the beginning of this section.

2. Raise and support the vehicle safely. Remove the rear tires.

3. Remove the lower arm bolt from the rear knuckle, while supporting the lower arm with a jack.

4. Loosen the lower arm bolt from the crossmember.

5. Remove the spring, lower seat and the upper pad.

To install:

6. Installation is the reverse of the removal procedure.

7. Torque the lower arm mounting bolt

to knuckle to 87-116 ft. lbs. (120-160 Nm).

8. Torque the lower arm to crossmember bolt to 145-195 ft. lbs. (200-270 Nm).

Stabilizer Bar

REMOVAL & INSTALLATION

2006

1. Before servicing the vehicle, refer to the precautions at the beginning of this section.

2. Raise and support the vehicle safely. Remove the rear tires.

3. Remove the left and right side stabilizer link retaining nuts from the trailing arm.

4. Remove the left and right side rear stabilizer mounting bracket nuts.

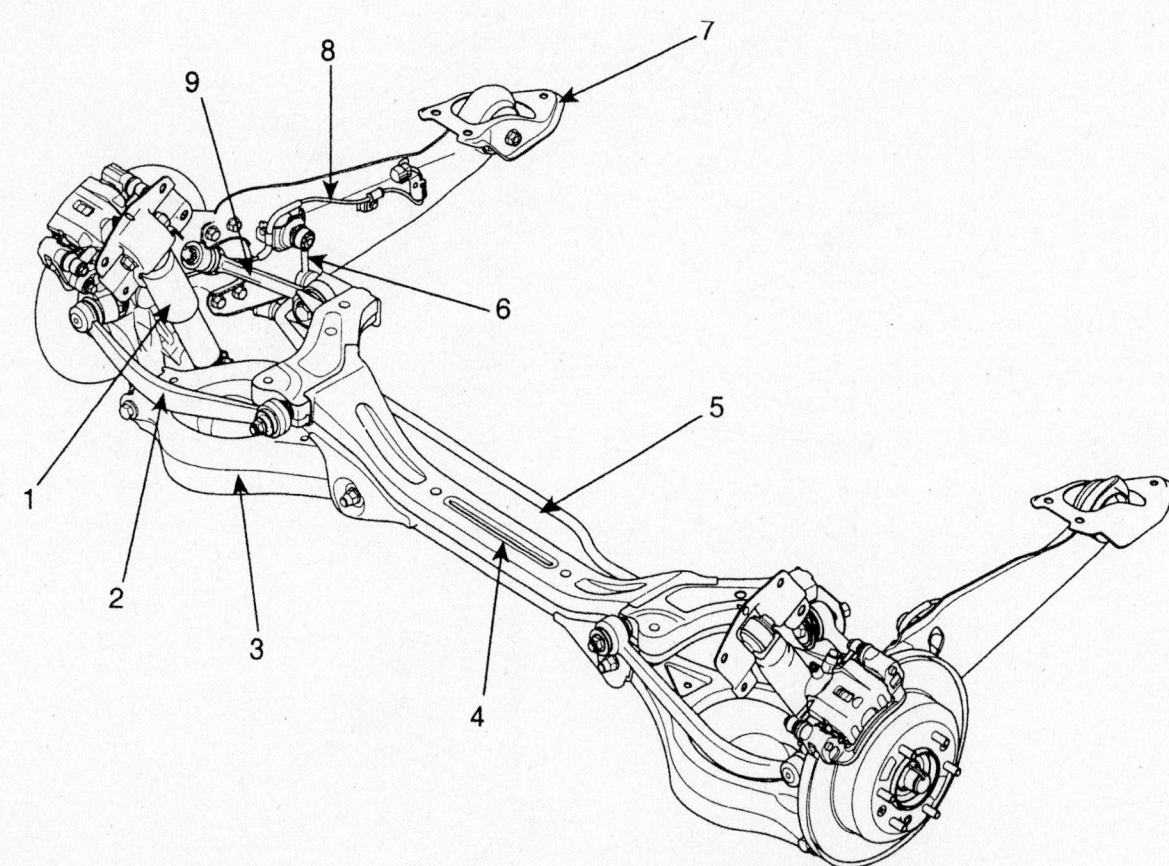

1. Rear shock absorber assembly
2. Rear upper arm
3. Rear lower arm
4. Rear cross member
5. Rear stabilizer bar assembly

6. Rear stabilizer link assembly
7. Trailing arm bracket
8. Trailing arm
9. Rear assist arm

Rear suspension and related components—2006 vehicles

5. Remove the stabilizer link nut from the stabilizer bar assembly.
   To install:
6. Installation is the reverse of the removal procedure.
7. Torque the stabilizer link nut to the stabilizer bar assembly to 36-47 ft. lbs. (50-65 Nm).
8. Install the bushing on the bar. Be sure the clamp of the stabilizer bar is in contact with the bushing.
9. Torque the stabilizer bracket bolt to the crossmember to 28-43 ft. lbs. (39-60 Nm).
10. Torque the stabilizer link nut to the trailing arm to 36-47 ft. lbs. (50-65 Nm).

Upper Control Arm

REMOVAL & INSTALLATION

2002–2005

1. Before servicing the vehicle, refer to the precautions at the beginning of this section.
2. Support the vehicle with jackstands forward of the lower control arm mounting.
3. Support the rear axle with jackstands.
4. Remove or disconnect the following:
   - Rear wheel
   - Upper control arm mounting bolts
   - Upper control arm
   To install:
5. Install or connect the following:
   - Upper control arm
   - Upper control arm mounting bolts. Tighten the bolts to 55-69 ft. lbs. (74-93 Nm)
   - Rear wheel

2006

1. Before servicing the vehicle, refer to the precautions at the beginning of this section.
2. Raise and support the vehicle safely. Remove the rear tires.
3. Remove the brake caliper mounting bolts. Position the caliper to the side, retaining it with wire.
4. Remove the rear upper arm ball joint, using tool 09568-4A000, or equivalent.
5. Remove the rear upper arm mounting nut from the crossmember.
6. Remove the assembly from the vehicle.
   To install:
7. Installation is the reverse of the removal procedure.
8. Torque the upper arm mounting nut to the crossmember to 116-130 ft. lbs. (160-180 Nm).

9. Torque the rear arm ball joint locking nut to 65-80 ft. lbs. (90-110 Nm).
10. Torque the brake caliper mounting bolts to 36-43 ft. lbs. (50-60 Nm).

Lower Control Arm

REMOVAL & INSTALLATION

2002–2005

1. Before servicing the vehicle, refer to the precautions at the beginning of this section.
2. Support the vehicle with jackstands forward of the lower control arm mounting.
3. Support the rear axle with jackstands.
4. Support the lower control arm with a floor jack.
5. Remove or disconnect the following:
   - Rear wheel
   - Sway bar link
   - Parking brake cable
   - Lower control arm bolts
6. Carefully lower the floor jack and remove the lower control arm, coil spring, and spring seats from the vehicle.
   To install:
7. Install or connect the following:
   - Spring seats
   - Coil spring
   - Lower control arm. Tighten the bolts to 87-101 ft. lbs. (118-137 Nm)
   - Parking brake cable. Tighten the bracket bolts to 14-19 ft. lbs. (16-23 Nm)
   - Sway bar link. Tighten the locknut to 17-20 ft. lbs. (24-28 Nm)
   - Rear wheel

2006

1. Before servicing the vehicle, refer to the precautions at the beginning of this section.
2. Raise and support the vehicle safely. Remove the rear tires.
3. Remove the lower arm bolt from the rear knuckle, while supporting the lower arm with a jack.
4. Loosen the lower arm bolt from the crossmember.
5. Remove the spring, lower seat and the upper pad.
6. Remove the lower arm mounting bolts from the crossmember. Remove the lower arm from the vehicle.
   To install:
7. Installation is the reverse of the removal procedure.
8. Torque the lower arm mounting bolt to knuckle to 87-116 ft. lbs. (120-160 Nm).
9. Torque the lower arm to crossmember bolt to 145-195 ft. lbs. (200-270 Nm).

Wheel Bearings

ADJUSTMENT

2002–2005

1. Before servicing the vehicle, refer to the precautions at the beginning of this section.
2. Remove or disconnect the following:
   - Rear wheel
   - Brake drum
   - Hub cover
   - Cotter pin
   - Lock nut cover
3. Adjust the locknut to achieve wheel bearing play of 0.001-0.006 inches (0.025-0.152 mm).

REMOVAL & INSTALLATION

2002–2005

1. Before servicing the vehicle, refer to the precautions at the beginning of this section.
2. Remove or disconnect the following:
   - Rear wheel
   - Brake drum
   - Hub cap
   - Cotter pin
   - Lock nut cover
   - Lock nut
   - Wheel bearing retainer washer and the outer wheel bearing
   - Hub assembly
   - Grease seal
   - Inner wheel bearing
3. Clean and inspect the wheel bearings and races for unusual wear or damage. Replace parts as necessary.
   To install:
4. Pack the wheel bearings with grease for assembly.
5. Install or connect the following:
   - Inner wheel bearing
   - Grease seal
   - Hub assembly
   - Lock nut. Adjust the locknut to achieve wheel bearing play of 0.001-0.006 inches (0.025-0.152 mm).
   - Lock nut cover
   - Cotter pin
   - Hub cap
   - Brake drum
   - Rear wheel

2006

1. Before servicing the vehicle, refer to the precautions at the beginning of this section.

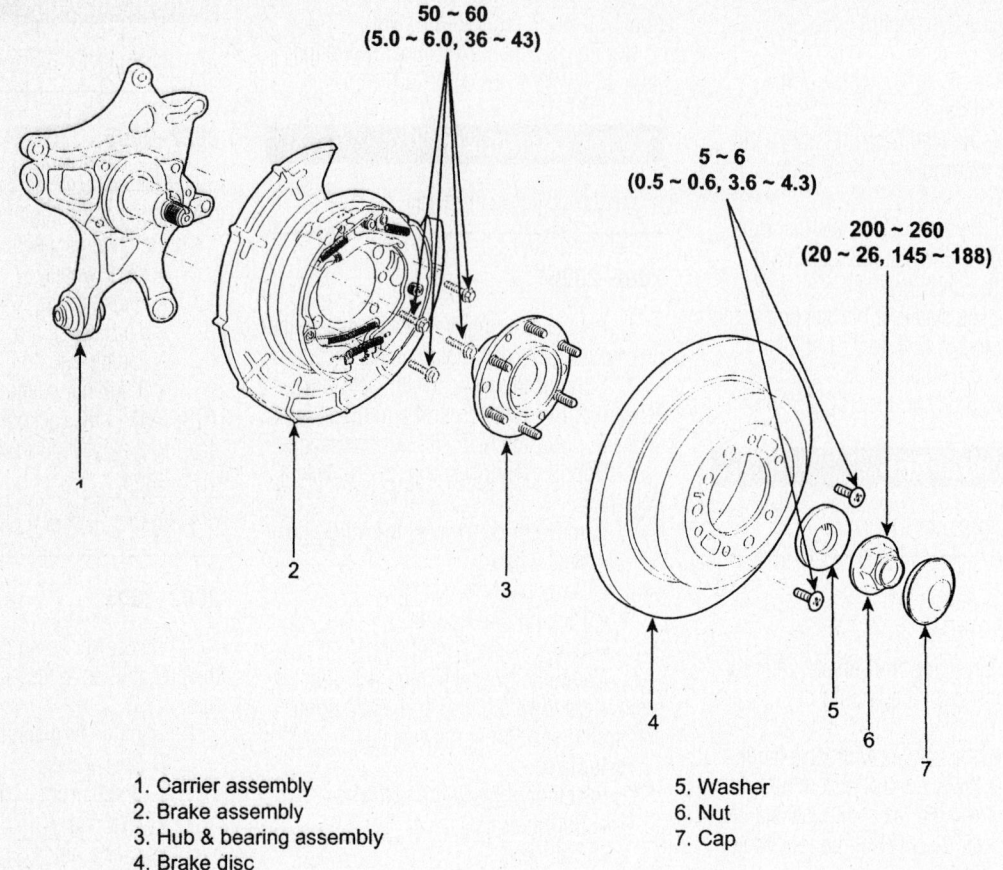

50 ~ 60
(5.0 ~ 6.0, 36 ~ 43)

5 ~ 6
(0.5 ~ 0.6, 3.6 ~ 4.3)

200 ~ 260
(20 ~ 26, 145 ~ 188)

1. Carrier assembly
2. Brake assembly
3. Hub & bearing assembly
4. Brake disc

5. Washer
6. Nut
7. Cap

TORQUE : Nm (kgf·m, lb-ft)

09474_SEDO_G0039

Rear hub and bearing assembly—2006 vehicles

2. Release the parking brake. Raise and support the vehicle safely. Remove the rear tires.

3. Support the lower part of the lower arm, using a jack. Remove the bolt and nut.

4. Remove the coil spring and upper pad. Remove the wheel speed sensor.

5. Remove the brake caliper, and position it to the side with wire.

6. Remove the brake rotor, after loosening the retaining screws.

7. Unstake the lock nut. Remove the lock nut and washer. Remove the hub and bearing assembly.

To install:

8. Installation is the reverse of the removal procedure.

9. Torque the lock nut to 145-188 ft. lbs. (200-260 Nm).

10. Torque the caliper retaining bolts to 36-43 ft. lbs. (50-60 Nm).

11. Torque the lower arm retaining nut and bolt to 116-130 ft. lbs. (160-180 Nm).

BRAKES

Brake Caliper

REMOVAL & INSTALLATION

Front

1. Before servicing the vehicle, refer to the precautions at the beginning of this section.

2. Remove or disconnect the following:

- Front wheel
- Brake fluid hose

- Caliper mounting bolts
- Brake caliper

To install:

3. Install or connect the following:

- Brake caliper
- Brake fluid hose
- Front wheel

4. Bleed the brake system.

5. Before attempting to move the vehicle, pump the brake pedal to seat the pads against the rotors. Make sure the vehicle has a firm brake pedal. Check the level of the brake fluid and add fluid if necessary.

Rear

1. Before servicing the vehicle, refer to the precautions at the beginning of this section.

2. Raise and support the vehicle safely. Remove the rear tires.

3. Disconnect and plug the brake fluid line.

4. Remove the caliper retaining bolts.

5. Remove the caliper from the vehicle.

To install:

6. Installation is the reverse of the removal procedure.

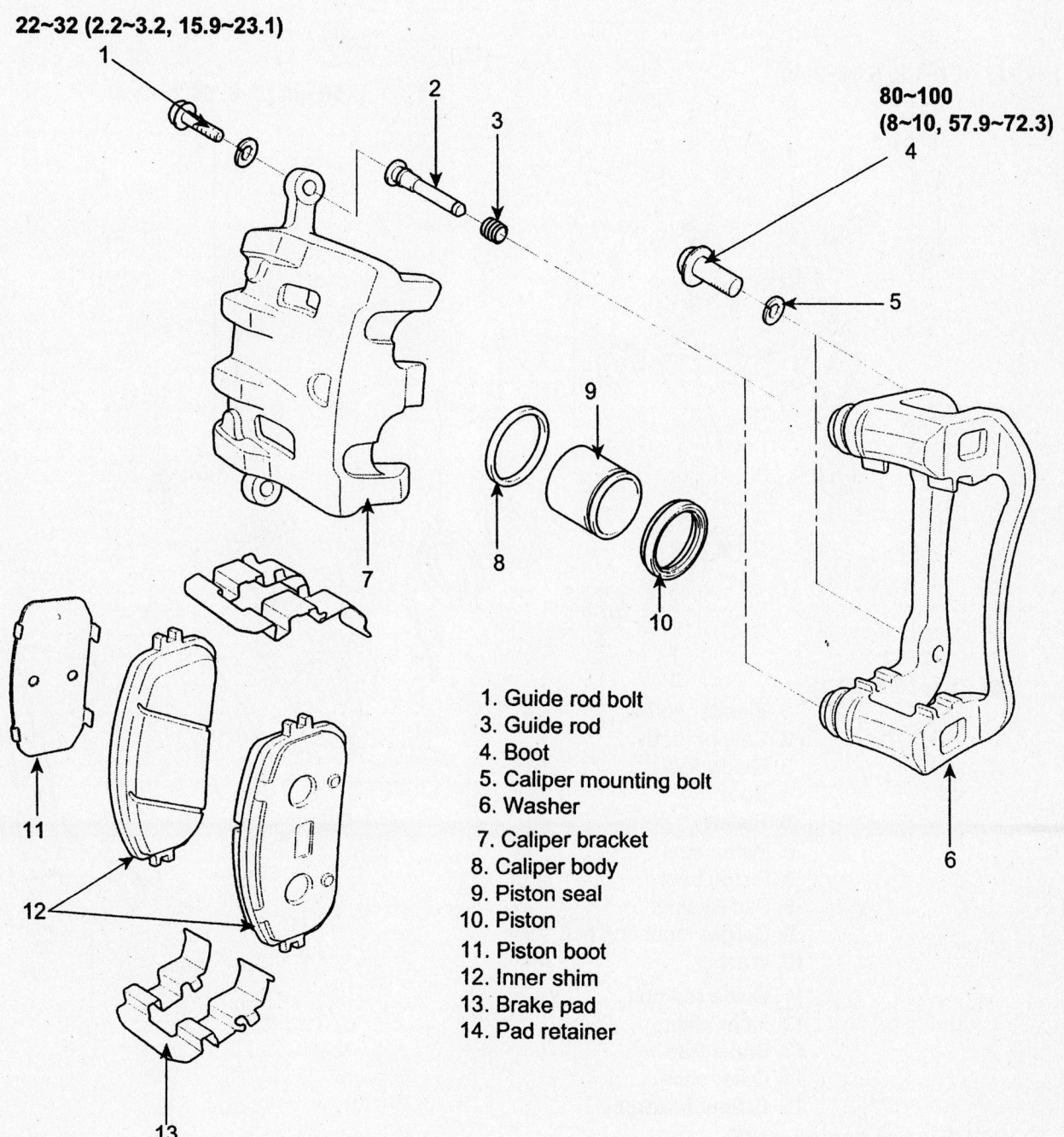

22~32 (2.2~3.2, 15.9~23.1)

**80~100
(8~10, 57.9~72.3)**

1. Guide rod bolt
3. Guide rod
4. Boot
5. Caliper mounting bolt
6. Washer
7. Caliper bracket
8. Caliper body
9. Piston seal
10. Piston
11. Piston boot
12. Inner shim
13. Brake pad
14. Pad retainer

TORQUE : Nm (kgf.m, lb-ft)

09474_SEDO_G0040

Front brake caliper and related components—2006 vehicles

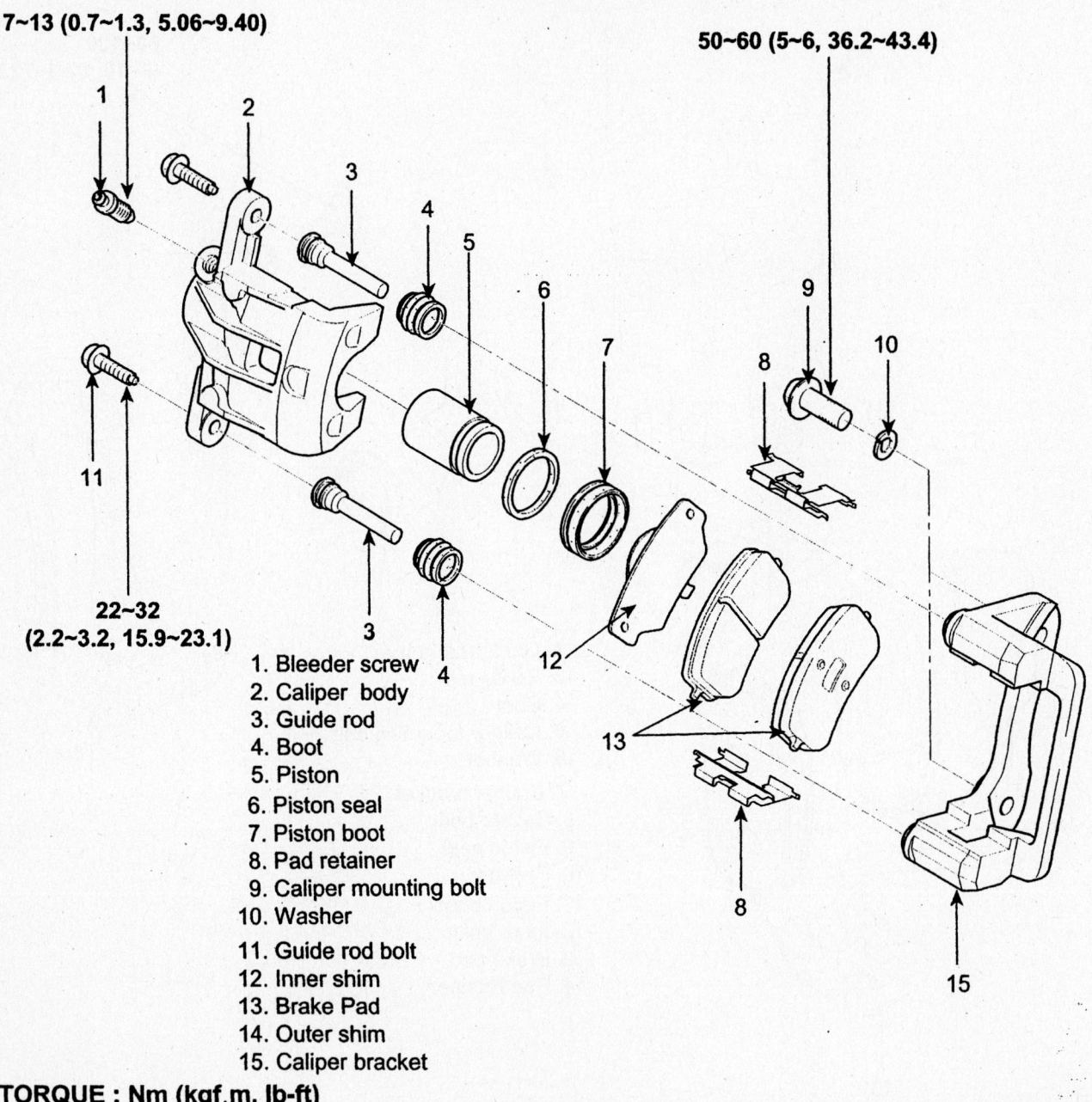

7~13 (0.7~1.3, 5.06~9.40)

50~60 (5~6, 36.2~43.4)

22~32
(2.2~3.2, 15.9~23.1)

1. Bleeder screw
2. Caliper body
3. Guide rod
4. Boot
5. Piston
6. Piston seal
7. Piston boot
8. Pad retainer
9. Caliper mounting bolt
10. Washer

11. Guide rod bolt
12. Inner shim
13. Brake Pad
14. Outer shim
15. Caliper bracket

TORQUE : Nm (kgf.m, lb-ft)

09474_SEDO_G0041

Rear brake caliper and related components—2006 vehicles

Disc Brake Pads

REMOVAL & INSTALLATION

Front

1. Before servicing the vehicle, refer to the precautions at the beginning of this section.
2. Remove or disconnect the following:
   - Front wheel
   - Brake hose from the support bracket
   - Outer brake pad retaining clip
   - Brake caliper
   - Inner and outer brake pads

To install:

3. Compress the caliper piston into the caliper bore.
4. Install or connect the following:
   - Inner and outer brake pads
   - Brake caliper
   - Brake hose to the support bracket
   - Front wheel

Rear

1. Before servicing the vehicle, refer to the precautions at the beginning of this section.
2. Raise and support the vehicle safely. Remove the rear tires.
3. Remove the caliper guide bolt, after raising the caliper assembly. Support it with wire.
4. Remove the shim pad, pad retainer and pad assembly in the caliper bracket.

To install:

5. Installation is the reverse of the removal procedure.

Brake Drums

REMOVAL & INSTALLATION

1. Before servicing the vehicle, refer to the precautions at the beginning of this section.
2. Remove or disconnect the following:
   - Rear wheel
   - Brake drum retaining screws
   - Brake drum

To install:

3. Install or connect the following:
   - Brake drum
   - Brake drum retaining screws
   - Rear wheel

Brake Shoes

REMOVAL & INSTALLATION

1. Before servicing the vehicle, refer to the precautions at the beginning of this section.

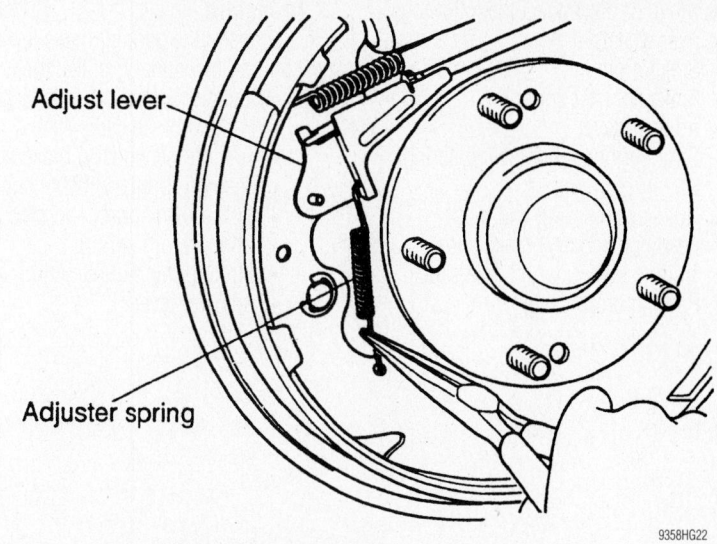

Brake adjuster spring removal—2002—05 vehicles

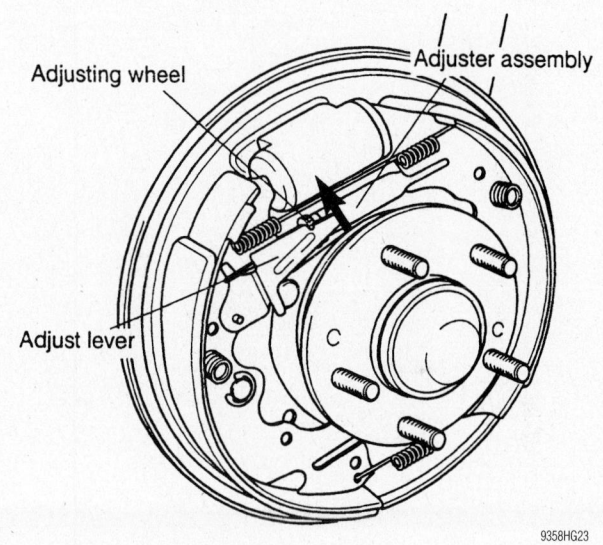

Brake adjuster assembly—2002—05 vehicles

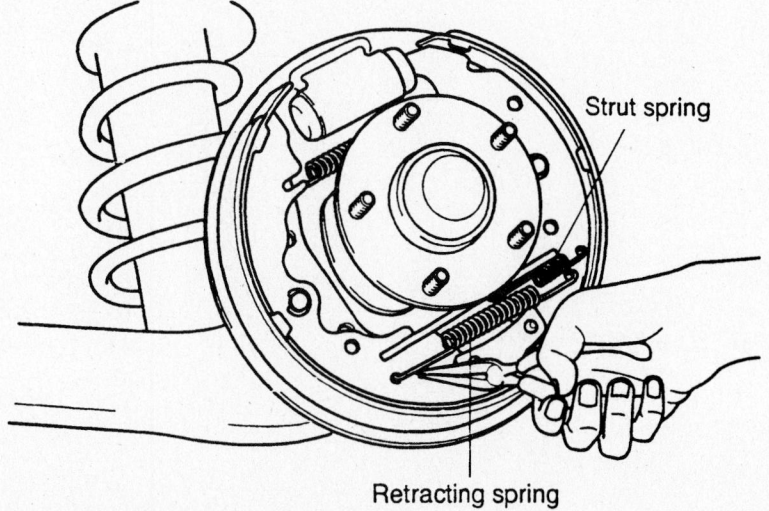

Brake strut spring and retracting spring—2002—05 vehicles

2. Remove or disconnect the following:
- Rear wheels
- Brake drum
- Brake adjuster spring
- Adjuster lever
- Strut spring and retracting spring
- Upper return spring
- Hold-down spring and pins
- Leading (primary) brake shoe
- Trailing (secondary) brake shoe
- Parking brake cable

To install:

3. Transfer the parking brake lever to the new trailing (secondary) brake shoe.

4. Install or connect the following:
- Parking brake cable
- Trailing (secondary) brake shoe
- Leading (primary) brake shoe
- Hold-down spring and pins
- Upper return spring
- Strut spring and retracting spring
- Adjuster lever
- Brake adjuster spring
- Brake drum
- Rear wheels

5. Work the parking brake control several times to complete the brake shoe adjustment and to check the parking brake adjustment as well.

6. Pump the brake pedal several times to assure a good pedal.

7. Road test the vehicle and check for proper brake system operation.

KIA

Sorento

20

SPECIFICATIONS AND MAINTENANCE CHARTS

ENGINE AND VEHICLE IDENTIFICATION

Code ①	Liters (cc)	Cu. In.	Cyl.	Fuel Sys.	Engine Type	Eng. Mfg.
3	3.5 (3497)	213	6	EGI	DOHC	KIA
4	3.5 (3497)	213	6	EGI	DOHC	KIA
5	3.5 (3497)	213	6	EGI	DOHC	KIA

Code ②	Year
3	2003
4	2004
5	2005

EGI: Electronic Gasoline Injection

DOHC: Double Overhead Camshafts

① 8th digit of VIN

② 10th digit of VIN

09474_SOREN_C0001

GENERAL ENGINE SPECIFICATIONS

Year	Model	Engine Displacement Liters	Engine VIN	Net Horsepower @ rpm	Net Torque @ rpm (ft. lbs.)	Bore x Stroke (in.)	Compression Ratio	Oil Pressure @ rpm
2003	Sorento	3.5	3	192@5500	217@3000	3.66x3.38	10:01	43-57@3000
2004	Sorento	3.5	3	192@5500	217@3000	3.66x3.38	10:01	43-57@3000
2005	Sorento	3.5	3	192@5500	217@3000	3.66x3.38	10:01	①

① 11.6psi at idle

09474_SOREN_C0002

ENGINE TUNE-UP SPECIFICATIONS

Year	Engine Displacement Liters	Engine VIN	Spark Plug Gap (in.)	Ignition Timing (deg.) MT	Ignition Timing (deg.) AT	Fuel Pump (psi)	Idle Speed (rpm) MT	Idle Speed (rpm) AT	Valve Clearance Intake	Valve Clearance Exhaust
2003	3.5	3	0.039-0.043	—	①	46-49	—	②	HYD	HYD
2004	3.5	3	0.039-0.043	①	①	46-49	②	②	HYD	HYD
2005	3.5	3	0.039-0.043	③	③	47-48	②	②	HYD	HYD

NOTE: The Vehicle Emission Control Information label often reflects specification changes made during production.

The label figures must be used if they differ from those in this chart

B: Before top dead center

HYD: Hydraulic

① Computer controlled, no adjustment possible

② N/P range with A/C OFF: 700-900 rpm

N/P range with A/C ON: 800-1000 rpm

D range: 650-850 rpm

③ 8-12 degrees

09474_SOREN_C0003

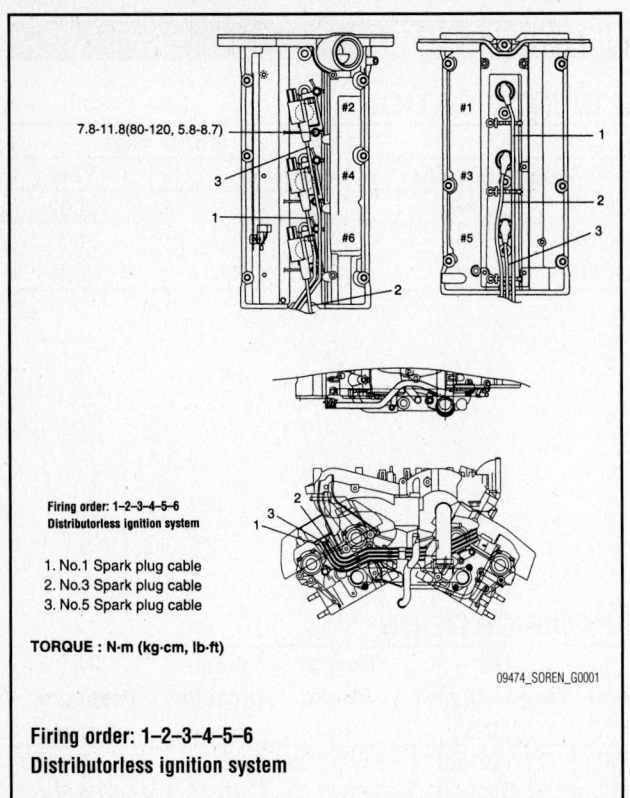

7.8-11.8(80-120, 5.8-8.7)

3

1

2

#2
#4
#6

#1
#3
#5

1
2
3

2
1
3

Firing order: 1-2-3-4-5-6
Distributorless ignition system

1. No.1 Spark plug cable
2. No.3 Spark plug cable
3. No.5 Spark plug cable

TORQUE : N·m (kg·cm, lb-ft)

09474_SOREN_G0001

Firing order: 1-2-3-4-5-6
Distributorless ignition system

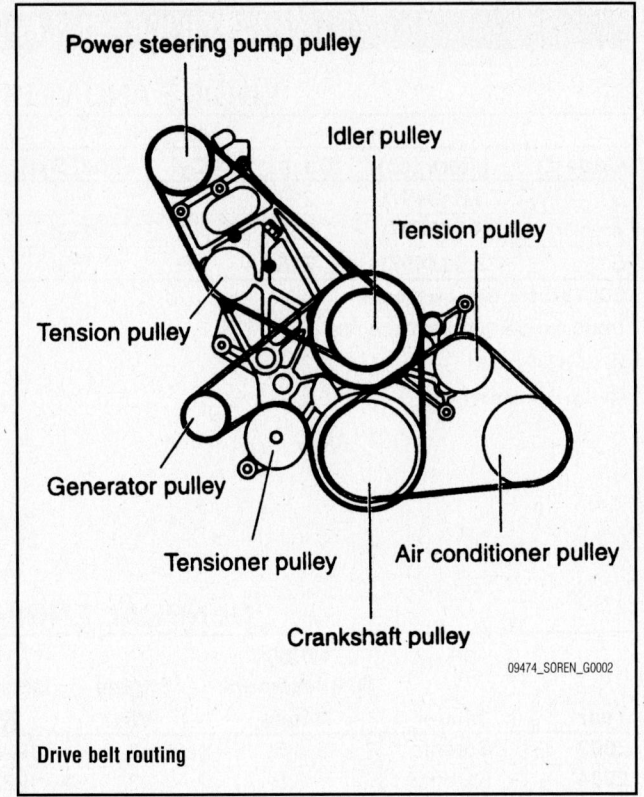

Power steering pump pulley

Idler pulley

Tension pulley

Tension pulley

Generator pulley

Tensioner pulley

Air conditioner pulley

Crankshaft pulley

09474_SOREN_G0002

Drive belt routing

CAPACITIES

Year	Model	Engine Displacement Liters	Engine VIN	Engine Oil with Filter	Transaxle (pts.) Manual	Transaxle (pts.) Auto.	Fuel Tank (gal.)	Cooling System (qts.)
2003	Sorento	3.5	3	4.6	—	21.0	21.1	11.6
2004	Sorento	3.5	3	4.6	①	21.0	21.1	11.6
2005	Sorento	3.5	3	4.6	①	②	21.1	11.6

NOTE: All capacities are approximate. Add fluid gradually and ensure a proper level is obtained.

① 2WD: 5.6 pts.
4WD: 4.75 pts.

② 30-40LE transaxle: 21 pts.
A5SR1 transaxle: 22 pts.

09474_SOREN_C0004

VALVE SPECIFICATIONS

Year	Engine Displacement Liters	Engine VIN	Seat Angle (deg.)	Face Angle (deg.)	Maximum out of Square (degrees)	Spring Free Length (in.)	Stem-to-Guide Clearance (in.)		Stem Diameter (in.)	
							Intake	Exhaust	Intake	Exhaust
2003	3.5	3	45	45-45.5	2	1.8268	0.0009-0.0020	0.0020-0.0039	0.258-0.259	0.257-0.258
2004	3.5	3	45	45-45.5	2	1.8268	0.0009-0.0020	0.0020-0.0033	0.258-0.259	0.257-0.258
2005	3.5	3	45	45-45.5	2	1.8268	0.0009-0.0020	0.0020-0.0033	0.258-0.259	0.257-0.258

NA: Not Available

09474_SOREN_C0005

CRANKSHAFT AND CONNECTING ROD SPECIFICATIONS

All measurements are given in inches.

Year	Engine Displacement Liters	Engine VIN	Crankshaft				Connecting Rod		
			Main Brg. Journal Dia.	Main Brg. Oil Clearance	Shaft End-play	Thrust on No.	Journal Diameter	Oil Clearance	Side Clearance
2003	3.5	3	2.5190-2.5197	0.0007-0.0014	0.002-0.0098	3	2.1650-2.1653	0.0009-0.0016	0.0039-0.0098
2004	3.5	3	2.5190-2.5197	0.0007-0.0014	0.002-0.0098	3	2.1650-2.1653	0.0009-0.0016	0.0039-0.0098
2005	3.5	3	2.5190-2.5197	0.0007-0.0014	0.002-0.0098	3	2.1650-2.1653	0.0009-0.0016	0.0039-0.0098

09474_SOREN_C0006

PISTON AND RING SPECIFICATIONS

All measurements are given in inches.

Year	Engine Displacement Liters	Engine VIN	Piston Clearance	Ring Gap			Ring Side Clearance		
				Top Compression	Bottom Compression	Oil Control	Top Compression	Bottom Compression	Oil Control
2003	3.5	3	0.0012-0.0020	0.0079-0.0118	0.0177-0.0236	0.0079-0.0276	0.0008-0.0031	0.0016-0.0024	SNUG
2004	3.5	3	0.0012-0.0020	0.0079-0.0118	0.0177-0.0236	0.0079-0.0276	0.0008-0.0031	0.0016-0.0024	SNUG
2005	3.5	3	0.0012-0.0020	0.0079-0.0118	0.0177-0.0236	0.0079-0.0276	0.0008-0.0031	0.0016-0.0024	SNUG

09474_SOREN_C0007

TORQUE SPECIFICATIONS
All readings in ft. lbs.

Year	Engine Displacement Liters	Engine VIN	Cylinder Head Bolts	Main Bearing Bolts	Rod Bearing Bolts	Crankshaft Damper Bolts	Flywheel Bolts	Manifold Intake	Manifold Exhaust	Spark Plugs	Oil Pan Drain Plug
2003	3.5	3	75-82	51-58	①	130-138	53-55	14-17	29-32	15-22	26-32
2004	3.5	3	75-82	51-58	①	130-138	53-55	14-17	29-32	15-22	26-32
2005	3.5	3	75-82	51-58	①	130-138	53-55	14-17	29-32	15-22	26-32

① Step 1: 26 ft. lbs.
  Step 2: Plus 92 degrees

09474_SOREN_C0008

WHEEL ALIGNMENT

Year	Model		Caster Range (+/-Deg.)	Caster Preferred Setting (Deg.)	Camber Range (+/-Deg.)	Camber Preferred Setting (Deg.)	Toe-in (in.)
2003	Sorento	F	0.50	3.30	0.50	0.39	0.1024
		R	—	—	—	—	—
2004	Sorento	F	0.50	3.30	0.50	0.39	0.1024
		R	—	—	—	—	—
2005	Sorento	F	0.50	3.30	0.50	0.39	0.1024
		R	—	—	—	—	—

09474_SOREN_C0009

TIRE, WHEEL AND BALL JOINT SPECIFICATIONS

Year	Model	OEM Tires		Tire Pressures (psi)		Wheel Size	Ball Joint Inspection	Wheel Lug Torque (ft. lbs.)
		Standard	Optional	Front	Rear			
2003	Sorento	P245/70R16	—	30	30	7JJ	①	65-86
2004	Sorento	P245/70R16	—	30	30	7JJ	①	65-86
2005	Sorento	P245/70R16	—	30	30	7JJ	①	65-86

OEM: Original Equipment Manufacturer

PSI: Pounds Per Square Inch

STD: Standard

OPT: Optional

① Replace if any measurable movement is found.

09474_SOREN_C0010

BRAKE SPECIFICATIONS

All measurements in inches unless noted

Year	Model		Brake Disc			Brake Drum			Minimum Lining Thickness	Brake Caliper	
			Original Thickness	Minimum Thickness	Maximum Run-out	Original Inside Diameter	Max. Wear Limit	Maximum Machine Diameter		Guide Bolts (ft. lbs.)	Mounting Bolts (ft. lbs.)
2003	Sorento	F	1.100	1.020	0.0012	—	—	—	0.079	16-24	47-54
		R	0.787	0.724	—	—	—	—	0.079	16-24	47-54
2004	Sorento	F	1.100	1.020	0.0012	—	—	—	0.079	16-24	47-54
		R	0.787	0.724	—	—	—	—	0.079	16-24	47-54
2005	Sorento	F	1.100	1.020	0.0012	—	—	—	0.079	16-24	47-54
		R	0.787	0.724	—	—	—	—	0.079	16-24	47-54

F: Front

R: Rear

09474_SOREN_C0011

SCHEDULED MAINTENANCE INTERVALS
Kia—Sorento

TO BE SERVICED	TYPE OF SERVICE	7.5	15	22.5	30	37.5	45	52.5	60	67.5	75	82.5	90	97.5	105	112.5	120
Accessory drive belts	S/I				✓				✓				✓				✓
Air cleaner element	I/R		✓		✓		✓		✓		✓		✓		✓		✓
Air conditioner system	S/I	Inspect the system operation and refrigerant amount annually.															
Brake lines, hoses and connections	S/I		✓		✓		✓		✓		✓		✓		✓		✓
Chassis and body fasteners	T				✓				✓				✓				✓
Cooling system hoses and coolant level	S/I				✓				✓				✓				✓
CV-joint boots	S/I		✓		✓		✓		✓		✓		✓		✓		✓
Engine coolant	R				✓				✓				✓				✓
Engine oil and filter	R	✓	✓	✓	✓	✓	✓	✓	✓	✓	✓	✓	✓	✓	✓	✓	✓
Exhaust system heat shields	S/I				✓				✓				✓				✓
Front and rear brakes	S/I		✓		✓		✓		✓		✓		✓		✓		✓
Front ball joints S/I	S/I				✓				✓				✓				✓
Fuel filter	R								✓								✓
Fuel lines and hoses	S/I				✓				✓				✓				✓
Rear differential fluid	S/I	✓	✓	✓	✓	✓	✓	✓	✓	✓	✓	✓	✓	✓	✓	✓	✓
Front differential fluid (if equipped)	S/I	✓	✓	✓	✓	✓	✓	✓	✓	✓	✓	✓	✓	✓	✓	✓	✓
Automatic transmission fluid	S/I		✓		✓		✓		✓		✓		✓		✓		✓
Manual transmission fluid (if equipped)	S/I	✓	✓	✓	✓	✓	✓	✓	✓	✓	✓	✓	✓	✓	✓	✓	✓
Transfer case oil (if equipped)	S/I	✓	✓	✓	✓	✓	✓	✓	✓	✓	✓	✓	✓	✓	✓	✓	✓
Ignition wires	S/I								✓								✓
Emission hoses	S/I				✓				✓				✓				✓
Driveshaft U-joints	L		✓		✓		✓		✓		✓		✓		✓		✓
Brake fluid	S/I		✓		✓		✓		✓		✓		✓		✓		✓
Clutch fluid (if equipped)	S/I		✓		✓		✓		✓		✓		✓		✓		✓
Idle speed	A				✓				✓				✓				✓
Locks and hinges	L	✓	✓	✓	✓	✓	✓	✓	✓	✓	✓	✓	✓	✓	✓	✓	✓
Spark plugs	R								✓								✓
Steering operation and linkage	S/I				✓				✓				✓				✓
Timing belt	R								✓								✓

R: Replace   S/I: Inspect and service, if needed   L: Lubricate   A: Adjust   T: Tighten

FREQUENT OPERATION MAINTENANCE (SEVERE SERVICE)

If a vehicle is operated under any of the following conditions it is considered severe service

- Towing a trailer or using a camper or car-top carrier.

- Repeated short trips of less than 5 miles in temperatures below freezing, or trips of less than 10 miles in any temperature.

- Extensive idling or low-speed driving for long distances as in heavy commercial use, such as delivery, taxi or police cars.

- Operating on rough, muddy or salt-covered roads.

- Operating on unpaved or dusty roads.

- Driving in extremely hot (over 90°F) conditions.

Engine oil and filter: replace every 3000 miles or 3 months, whichever occurs first.

Air cleaner element: inspect every 10,000 miles or 10 months and replace every 20,000 miles or 20 months, whichever occurs first.

Front differential fluid (if equipped): inspect every 5000 miles and replace every 15,000 miles.

Transfer case fluid (if equipped): inspect every 5000 miles.

Manual transmission fluid (if equipped): inspect every 5000 miles and replace every 60,000 miles.

Emission system hoses: inspect every 30,000 miles or 30 months, whichever occurs first.

Front and rear disc brakes: inspect every 15,000 miles or 15 months, whichever occurs first.

Chassis and body fasteners: tighten every 15,000 miles or 15 months, whichever occurs first.

Locks and hinges: lubricate every 5000 miles or 5 months, whichever occurs first.

GASOLINE ENGINE REPAIR

Alternator

REMOVAL & INSTALLATION

1. Before servicing the vehicle, refer to the Precautions Section.
2. Disconnect the negative battery cable.
3. Remove the accessory drive belt.
4. Disconnect the wiring to the alternator.
5. Remove the alternator mounting bolts.
6. Remove the alternator.

To install:

7. Install the alternator and insert a support bolt (Do not insert a nut this time).
8. After pushing the alternator forward, check to see if any spacers are required (a spacer thickness of 0.198 mm) is acceptable between the alternator front leg and front case.
9. If necessary, insert spacer(s), then insert and tighten nut securely.
10. Connect the alternator electrical connectors. Tighten the battery terminal connector nut to 60 inch lbs. (7 Nm).
11. Install the accessory drive belt.
12. To adjust the drive belt tension, hang the belt on the pulley of the tensioner and install the tensioner.

➡️**If the tensioner is already installed, loosen its mounting bolts to allow belt installation**

13. Install the drive belt.
14. Adjust the belt tension to specification by turning the adjusting bolt clockwise or counterclockwise.
15. Air conditioning belt tension should be 0.28-0.039 inch. Alternator belt tension should be 0.39-0.051 inch. Power steering belt tension should be 0.31-0.051 inch.
16. Torque the tensioner assembly bolt to 33-36 ft. lbs. (45-50 Nm).
17. Connect the negative battery cable.

Ignition Timing

This vehicle is equipped with a Distributorless Ignition System (DIS). No adjustment is necessary or possible.

Engine Assembly

REMOVAL & INSTALLATION

1. Before servicing the vehicle, refer to the Precautions Section.
2. Remove the battery and air cleaner assembly.
3. Remove the hood from the vehicle.

4. Drain the cooling system. Properly discharge the air conditioning system, as required.
5. Drain the engine oil.
6. Drain the transmission fluid.
7. Relieve fuel system pressure.
8. Remove the upper and lower radiator hoses.
9. Remove the heater hoses.
10. Remove the radiator.
11. Remove the breather hose and air-intake hose.
12. Disconnect the throttle and cruise control cables.
13. Disconnect the oil pressure switch connector.
14. Disconnect the oil pressure sensor connector.
15. Remove the engine ground cable.
16. Disconnect the EVAP canister hose.
17. Disconnect the brake booster vacuum hose.
18. Disconnect the fuel lines.
19. Remove the accessory drive belts.
20. Disconnect the power steering hoses from the pump.
21. Disconnect the alternator wiring harness connectors.
22. If equipped with an automatic transmission, disconnect the oil cooler lines.
23. Disconnect the starter motor wiring harness connectors.
24. Remove the starter motor.
25. Remove the exhaust front pipes from the exhaust manifolds.
26. Disconnect the transmission shift cable or clutch cable, if equipped.
27. Remove the halfshafts.
28. Support the transmission with a jack.
29. Make sure all cable, harness connectors and hoses are disconnected from the engine and transmission.
30. Remove the transmission from the crossmember.
31. Remove the crossmember.
32. Remove the driveshaft.
33. Remove the engine mounting bolts and remove the engine assembly.

To install:

34. Installation is the reverse of removal but please note the following steps:
- Tighten the transaxle mounting bracket bolts to 15–21 ft. lbs. (20–28 Nm)
- Tighten the engine mount bracket bolts to 51–65 ft. lbs. (69–88 Nm)

35. Make sure all cable, harness connectors and hoses are properly connected to the engine and transmission.

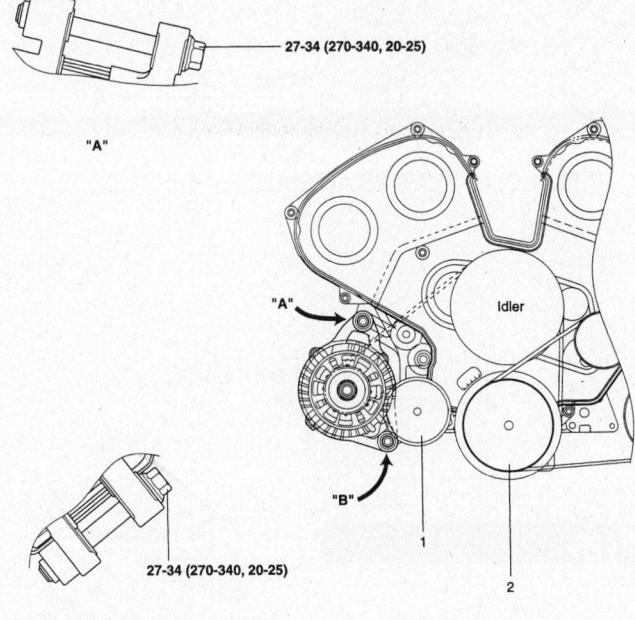

27-34 (270-340, 20-25)

"A"

"A"

Idler

"B"

1

2

27-34 (270-340, 20-25)

"B"

1. Generator tensioner
2. Crankshaft pulley

TORQUE : N·m (kg·cm, lb·ft)

09474_SOREN_G0003

Alternator mounting

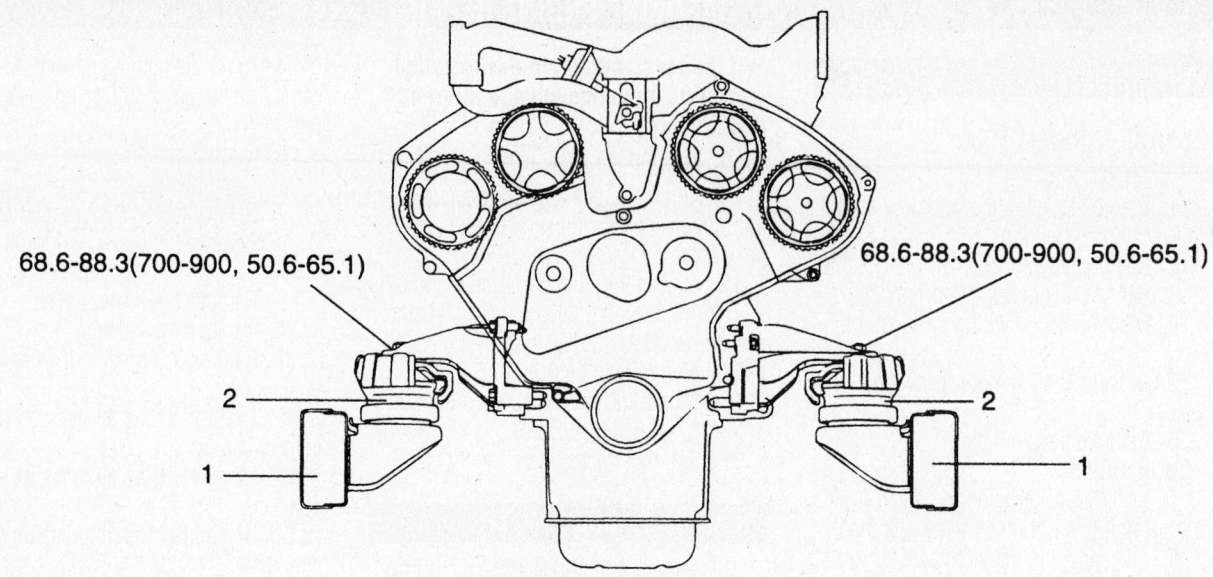

68.6-88.3(700-900, 50.6-65.1)

68.6-88.3(700-900, 50.6-65.1)

2

2

1

1

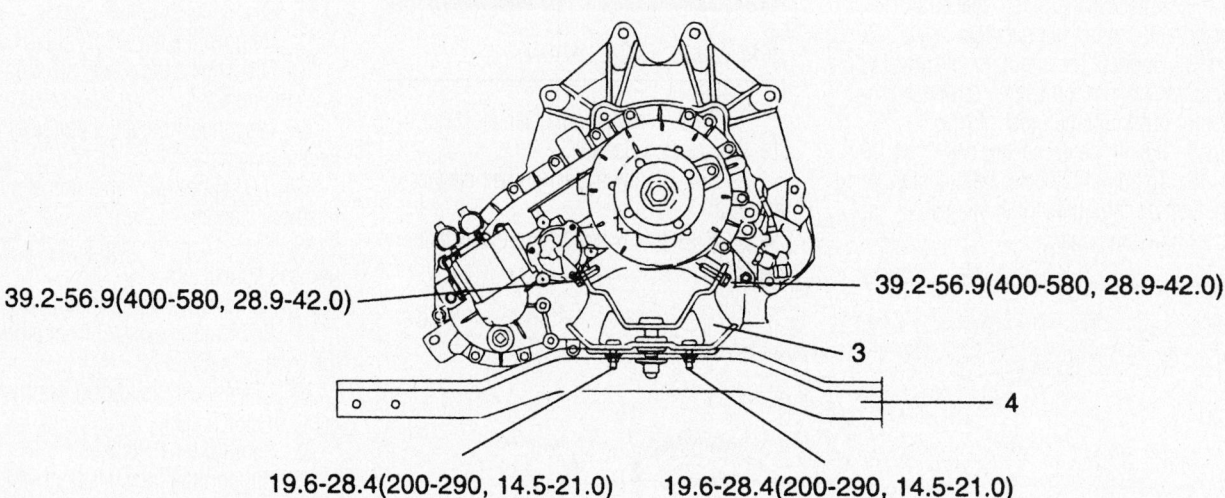

39.2-56.9(400-580, 28.9-42.0)

39.2-56.9(400-580, 28.9-42.0)

3

4

19.6-28.4(200-290, 14.5-21.0)   19.6-28.4(200-290, 14.5-21.0)

TORQUE : N·m (kg·cm, lb·ft)

1. Frame
2. Engine mounting

3. Transmission mounting
4. Cross member

09474_SOREN_G0004

Engine mount locations and torque specifications

36. Fill the engine crankcase to the correct level.
37. Fill the transmission to the correct level.
38. Fill the cooling system.
39. Fill the power steering system.
40. Start the engine and check for leaks.
41. Check the wheel alignment and adjust as necessary.

Water Pump

REMOVAL & INSTALLATION

1. Before servicing the vehicle, refer to the Precautions Section.
2. Drain the cooling system.
3. Disconnect the negative battery cable.

4. Remove the drive belt and the water pump pulley.
5. Remove the timing belt cover, timing belt, auto tensioner and idler pulley.
6. Remove the water outlet fitting, the thermostat case and the water pump fitting.
7. Remove the water pump mounting bolts.
8. Remove the water pump.

To install:

9. Clean the gasket surfaces of the water pump body and the cylinder block.

10. Install the new water pump gasket and pump assembly. Tighten the bolts to 12–16 ft. lbs. (17–22 Nm).

11. Install the water pump fitting, the thermostat case and the water outlet fitting.

12. Install the auto tensioner and timing belt. Adjust the timing belt tension, then install the timing belt cover.

13. Install the drive belt, water pump pulley and then adjust the auto tensioner.

14. To adjust the drive belt tension, hang the belt on the pulley of the tensioner and install the tensioner.

➥**If the tensioner is already installed, loosen its mounting bolts to allow belt installation**

15. Install the drive belt.

16. Adjust the belt tension to specification by turning the adjusting bolt clockwise or counterclockwise.

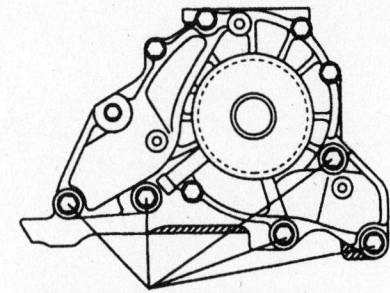

09474_SOREN_G0006

Water pump bolt locations

17. Air conditioning belt tension should be 0.28-0.039 inch. Alternator belt tension should be 0.39-0.051 inch. Power steering belt tension should be 0.31-0.051 inch.

18. Torque the tensioner assembly bolt to 33-36 ft. lbs. (45-50 Nm).

19. Refill the coolant.

20. Run the engine and check for leaks.

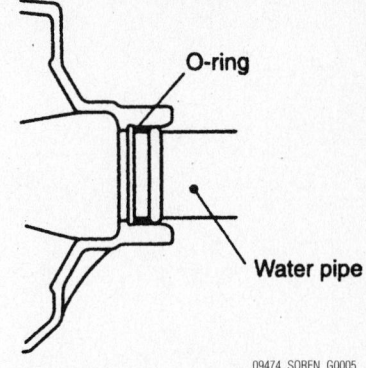

09474_SOREN_G0005

Water pump O-ring installation

Heater Core

REMOVAL & INSTALLATION

1. Before servicing the vehicle, refer to the Precautions Section.

2. Drain the cooling system.

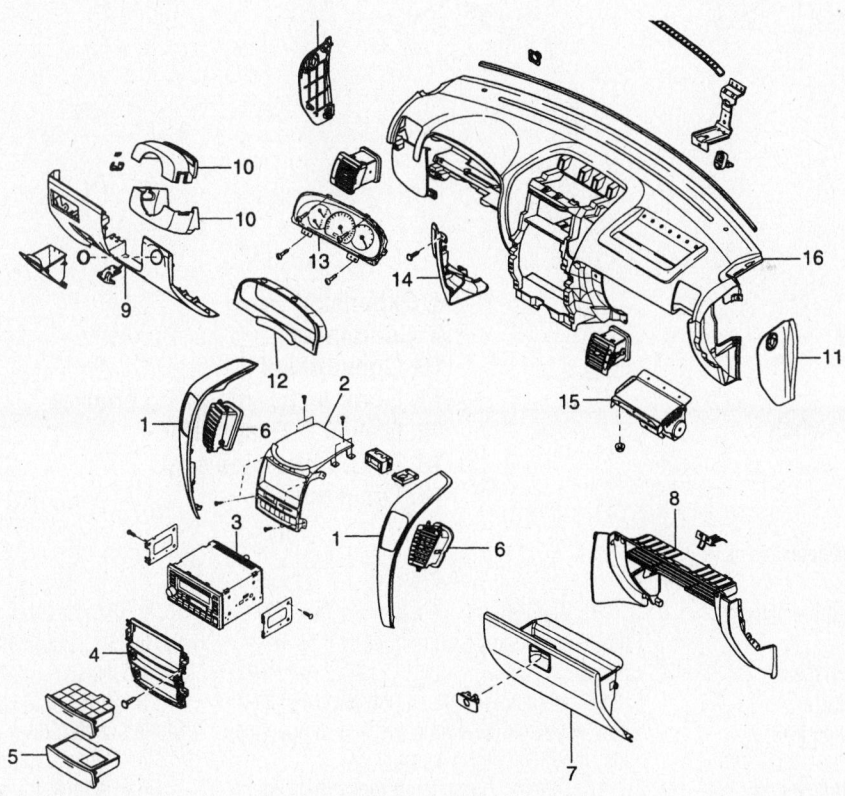

1. Center facia side panel assembly	7. Glove box assembly	12. Cluster facia panel assembly
2. Audio facia panel	8. Lower panel assembly (RH)	13. Instrument cluster assembly
3. Audio assembly	9. Lower panel assembly (LH)	14. Center lower panel assembly
4. Center facia panel assembly	10. Steering column shroud	15. Passenger airbag assembly
5. Ashtray	11. Crash pad side cover	16. Instrument panel assembly
6. Center air-vent assembly		

09474_SOREN_G0007

Instrument panel and related components

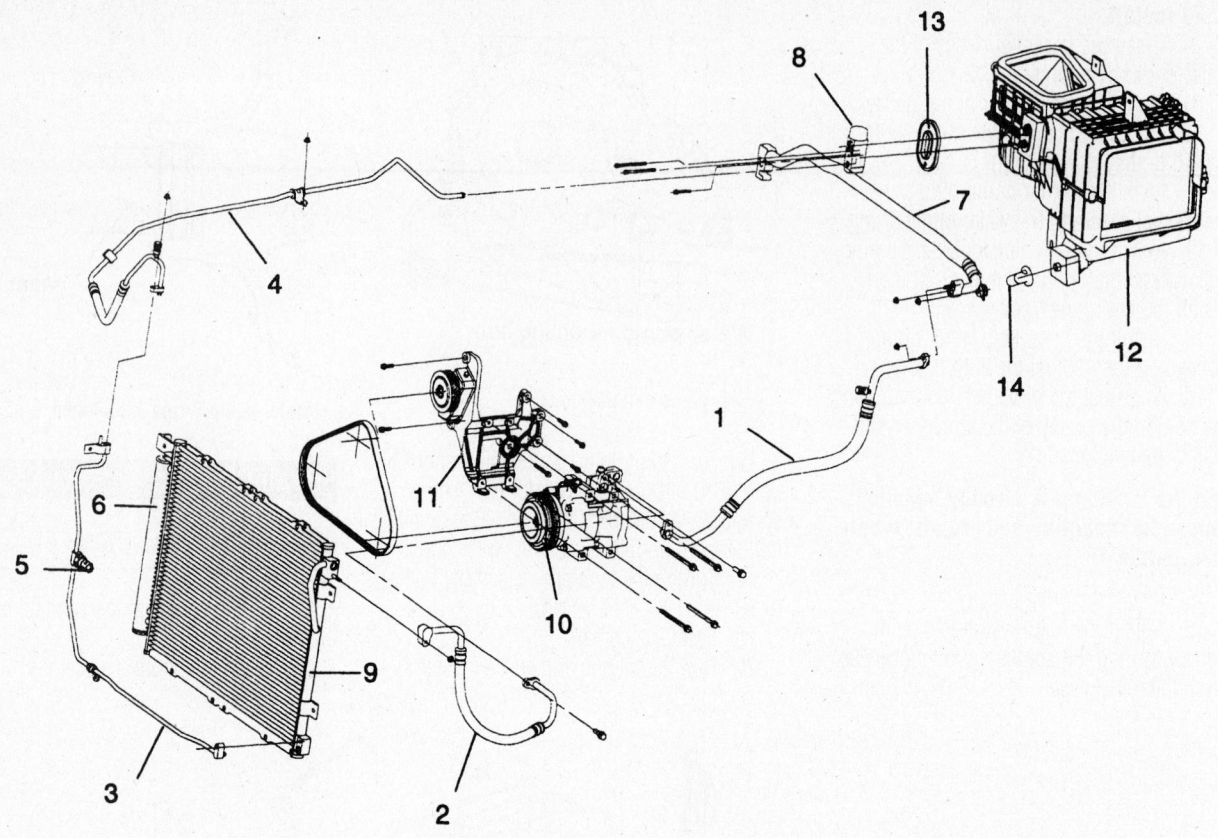

1. Suction hose
2. Discharge hose
3. Liquid pipe, A
4. Liquid pipe, B
5. Triple pressure switch
6. Receiver drier
7. Suction pipe

8. Expansion valve
9. Condenser
10. Compressor
11. Compressor mounting bracket
12. Blower & Evaporator
13. Evaporator pipe seal
14. Drain hose

09474_SOREN_G0008

Heater/AC unit and related components

3. Disconnect the negative battery cable.

4. Remove the heater hoses.

5. Remove the A pillar trim.

6. Remove the cowl side trim.

7. Remove the front door scuff trim.

8. Remove the shroud cover.

9. Remove the left side cover.

10. Remove the hood release cable.

11. Remove the lower panel screws (4), disconnect electrical connectors and remove the lower instrument panel.

12. Remove the driver air bag module.

13. Remove the steering wheel.

14. Remove the steering shroud cover.

15. Remove the lower instrument panel and steering shaft.

16. Remove the facia panel screws (2) and remove the instrument cluster facia panel.

17. Remove the cluster screws (4), disconnect electrical connector and remove the instrument cluster assembly.

18. Open ashtray, the remove instrument center facia side panel.

19. Remove audio facia panel screws (8), disconnect the electrical connector and remove audio facia panel.

20. Remove the radio mounting screws (4), disconnect the electrical connector, antenna cable and remove the radio unit.

21. Remove the screws (4), disconnect the electrical connector and remove heater controller.

22. Remove the screws (6), disconnect the electrical connector and remove center facia panel.

23. Remove the screws (4), then remove the left center lower panel.

24. Remove the glove box.

25. Remove the right side cover, instrument right lower panel screws (10), disconnect the electrical connector and remove the instrument right lower panel.

26. Disconnect the electrical connector and remove the main instrument panel assembly.

27. Remove crossbar side mounting bolts (4).

28. Remove crossbar lower mounting bolts (4).

29. Remove crossbar bracket bolts (2).
30. Remove the heater unit.

To install:

31. Install the heater unit.
32. Install the crossbar and tighten the mounting bolts to 12–19 ft. lbs. (17–26 Nm).
33. Install the main instrument panel assembly and reconnect the wiring harness connector.
34. Install the instrument right lower panel, reconnect the wiring harness connector and install the right side cover.
35. Install the glove box.
36. Install the left center lower panel.
37. Install the center facia panel.
38. Install the heater controller.
39. Install the radio.

40. Install the audio facia panel.
41. Install the instrument center facia side panel.
42. Install the instrument cluster assembly.
43. Install the instrument cluster facia panel.
44. Install the steering shaft and lower instrument panel.
45. Install the steering shroud cover.
46. Install the steering wheel.
47. Install the driver air bag module.
48. Install the lower instrument panel.
49. Install the hood release cable.
50. Install the left side cover.
51. Install the shroud cover.
52. Install the front door scuff trim.
53. Install the cowl side trim.

54. Install the A pillar trim.
55. Install the heater hoses.
56. Connect the negative battery cable.
57. Fill the cooling system.
58. Start the engine and check for leaks and proper heater operation.

Cylinder Head

REMOVAL & INSTALLATION

1. Before servicing the vehicle, refer to the Precautions Section.
2. Drain the cooling system.
3. Relieve the fuel system pressure.
4. Disconnect the negative battery cable.
5. Disconnect the upper radiator hose.

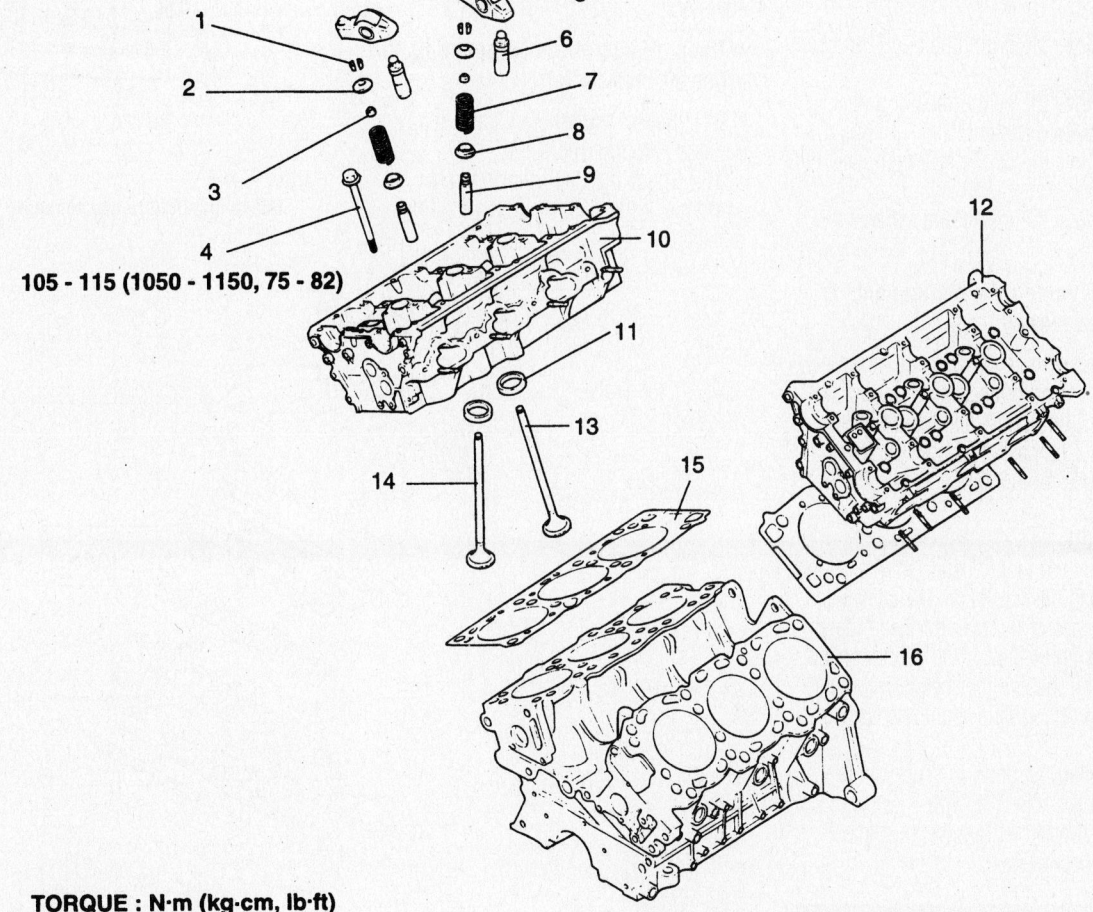

105 - 115 (1050 - 1150, 75 - 82)

TORQUE : N·m (kg·cm, lb·ft)

1. Retainer lock
2. Valve spring retainer
3. Valve stem seal
4. Cylinder head bolt
5. Rocker arm
6. Lash adjuster
7. Valve spring
8. Spring seat

9. Valve guide
10. Cylinder head (RH)
11. Exhaust valve seat ring
12. Cylinder head (LH)
13. Exhaust vlave
14. Intake valve
15. Gasket
16. Cylinder block

Cylinder head and related components

09474_SOREN_G0009

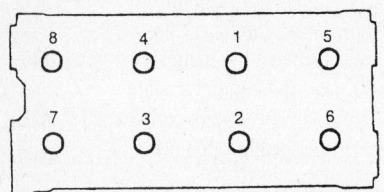

Cylinder head bolt torque sequence

09474_SOREN_G0010

6. Remove the breather hose and air-intake hose.

7. Remove the vacuum hose, fuel hose and coolant hose.

8. Remove the intake manifold.

9. Remove the cables from the spark plugs. The cables should be removed holding the boot portion.

10. Remove the ignition coil.

11. Remove the upper and lower timing belt cover.

12. Remove the timing belt and camshaft sprockets.

13. Remove the heat protector and exhaust manifold assembly.

14. Remove the coolant pump pulley and head cover.

15. Remove the intake and exhaust camshaft.

➡**Keep all valve train components in order for assembly.**

16. Remove the cylinder head assembly. The cylinder head bolts should be removed using the 12 mm socket, in two or three steps. Clean the gasket pieces from the cylinder block top surface and cylinder bottom surface.

To install:

17. Verify the identification marks on the cylinder head gasket. Install the gasket so that the surface of the cylinder head identification mark faces toward the cylinder head.

18. Install the cylinder heads with new gaskets. Do not apply sealant to these surfaces.

19. Tighten the bolts in sequence to 75–82 ft. lbs. (105–115 Nm).

20. Installation is the reverse of the removal procedure.

Rocker Arms

REMOVAL & INSTALLATION

1. Before servicing the vehicle, refer to the Precautions Section.

2. Drain the cooling system.

3. Relieve the fuel system pressure.

4. Disconnect the negative battery cable.

5. Disconnect the upper radiator hose.

6. Remove the breather hose and air-intake hose.

7. Remove the vacuum hose, fuel hose and coolant hose.

8. Remove the intake manifold.

9. Remove the cables from the spark plugs. The cables should be removed holding the boot portion.

10. Remove the ignition coil.

11. Remove the upper and lower timing belt cover.

12. Remove the timing belt and camshaft sprockets.

13. Remove the heat protector and exhaust manifold assembly.

14. Remove the coolant pump pulley and head cover.

15. Remove the camshafts.

16. Remove the rocker arms and lash adjusters.

➡**Keep all valvetrain components in order for installation.**

17. Inspect the roller visually. If any damage is found, replace it.

18. Check that the roller operates smoothly. If there is excessive clearance, replace it.

To install:

19. Installation is the reverse of the removal procedure.

Intake Manifold

REMOVAL & INSTALLATION

1. Before servicing the vehicle, refer to the Precautions Section.

2. Drain the cooling system.

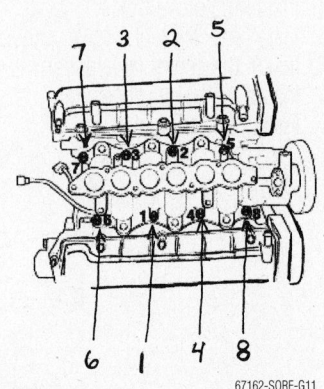

Intake manifold torque sequence

67162-SORE-G11

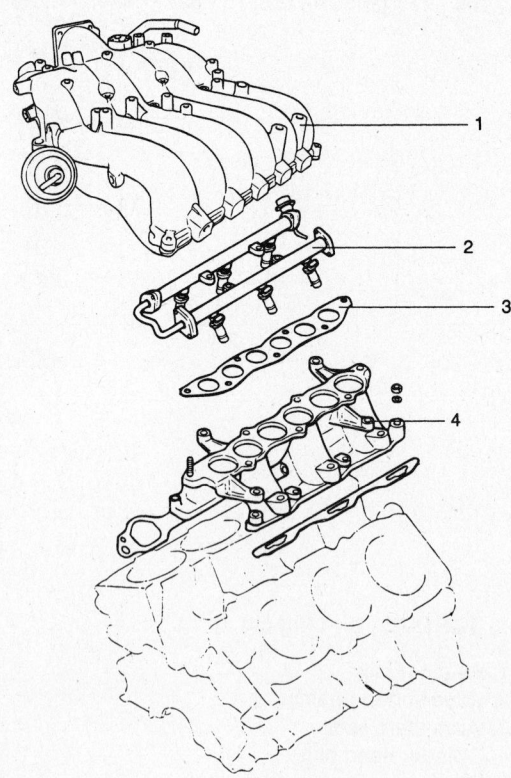

1. Surge tank
2. Delivery pipe

3. Gasket
4. Intake manifold

Intake manifold and related components

09474_SOREN_G0011

3. Relieve the fuel system pressure.

4. Disconnect the negative battery cable.

5. Remove the air intake hose connected to the throttle body.

6. Remove the accelerator and cruise control cables.

7. Remove the engine coolant hose and throttle body.

8. Remove the PCV hose and brake booster vacuum hoses.

9. Disconnect the vacuum hose connections.

10. Remove the surge tank stay.

11. Bleed off the pressure in the fuel pipe line to prevent the fuel from spilling.

12. Disconnect the connector from high pressure hose.

13. Remove the surge tank and gasket.

14. Disconnect the fuel injector harness connector.

15. Remove the delivery pipe with the fuel injector and the pressure regulator.

16. Disconnect the wiring harness of the coolant sensor assembly.

17. Remove the intake manifold.

To install:

18. Installation is the reverse of the removal procedure, while using the following torque values:
- Lower intake manifold nuts, in sequence to: 15-17 ft. lbs. (20-23 Nm)
- Surge tank stay bolts: 11-14 ft. lbs. (15-20 Nm)

Exhaust Manifold

REMOVAL & INSTALLATION

1. Before servicing the vehicle, refer to the Precautions Section.

2. Disconnect the negative battery cable.

3. Disconnect the Heated Oxygen (HO2S) sensor connectors.

4. Separate the exhaust Y pipe from the exhaust manifold(s).

5. Remove the heat protector.

6. Remove the exhaust manifold.

7. Remove the exhaust manifold gasket.

To install:

➡**Use only new gaskets and nuts for assembly.**

8. Install the exhaust manifolds with new gaskets. Tighten the fasteners to 20–24 ft. lbs. (27–32 Nm).

9. Install the exhaust manifold heat protectors. Tighten the bolts to 9–11 ft. lbs. (12-15 Nm).

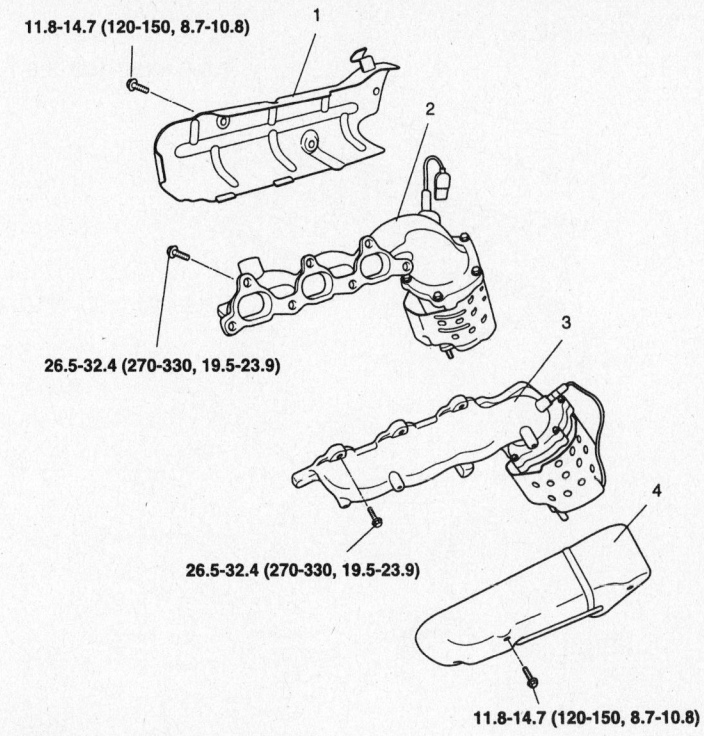

11.8-14.7 (120-150, 8.7-10.8)

26.5-32.4 (270-330, 19.5-23.9)

26.5-32.4 (270-330, 19.5-23.9)

11.8-14.7 (120-150, 8.7-10.8)

TORQUE : N·m (kg·cm, lb·ft)

1. Heat protector
2. Exhaust manifold

3. Exhaust manifold
4. Heat protector

09474_SOREN_G0012

Exhaust manifold and related components

10. Install the exhaust Y pipe.

11. Connect the Heated Oxygen (HO2S) sensor connectors.

12. Connect the negative battery cable.

Front Crankshaft Seal

REMOVAL & INSTALLATION

1. Before servicing the vehicle, refer to the Precautions Section.

2. Disconnect the negative battery cable.

3. Remove the accessory drive belts.

4. Remove the idler pulley.

5. Remove the crankshaft pulley.

6. Remove the power steering pump pulley.

7. Remove the belt tensioner pulley.

8. Remove the upper and lower timing belt covers.

9. Remove the timing belt.

10. Remove the timing belt crankshaft sprocket.

11. Remove the Crankshaft Position (CKP) sensor tone ring.

12. Remove the front crankshaft seal.

To install:

13. Install the front crankshaft seal. Use Seal Driver 09214-33000 or similar.

14. Install the CKP sensor tone ring.

15. Install the timing belt crankshaft sprocket.

16. Install the timing belt.

17. Install the upper and lower timing belt covers.

18. Install the belt tensioner pulley.

19. Install the power steering pump pulley.

20. Install the crankshaft pulley.

21. Install the idler pulley.

22. Install the accessory drive belts.

23. Connect the negative battery cable.

Camshaft and Valve Lifters

REMOVAL & INSTALLATION

1. Before servicing the vehicle, refer to the Precautions Section.

2. Disconnect the negative battery cable.

3. Remove the intake manifold.

4. Disconnect the breather hose and the engine harness.

5. Remove the power steering pulley, air conditioner pulley, crankshaft pulley, idler pulley and tensioner pulley.

6. Remove the timing belt cover.

7. Loosen the auto tensioner.

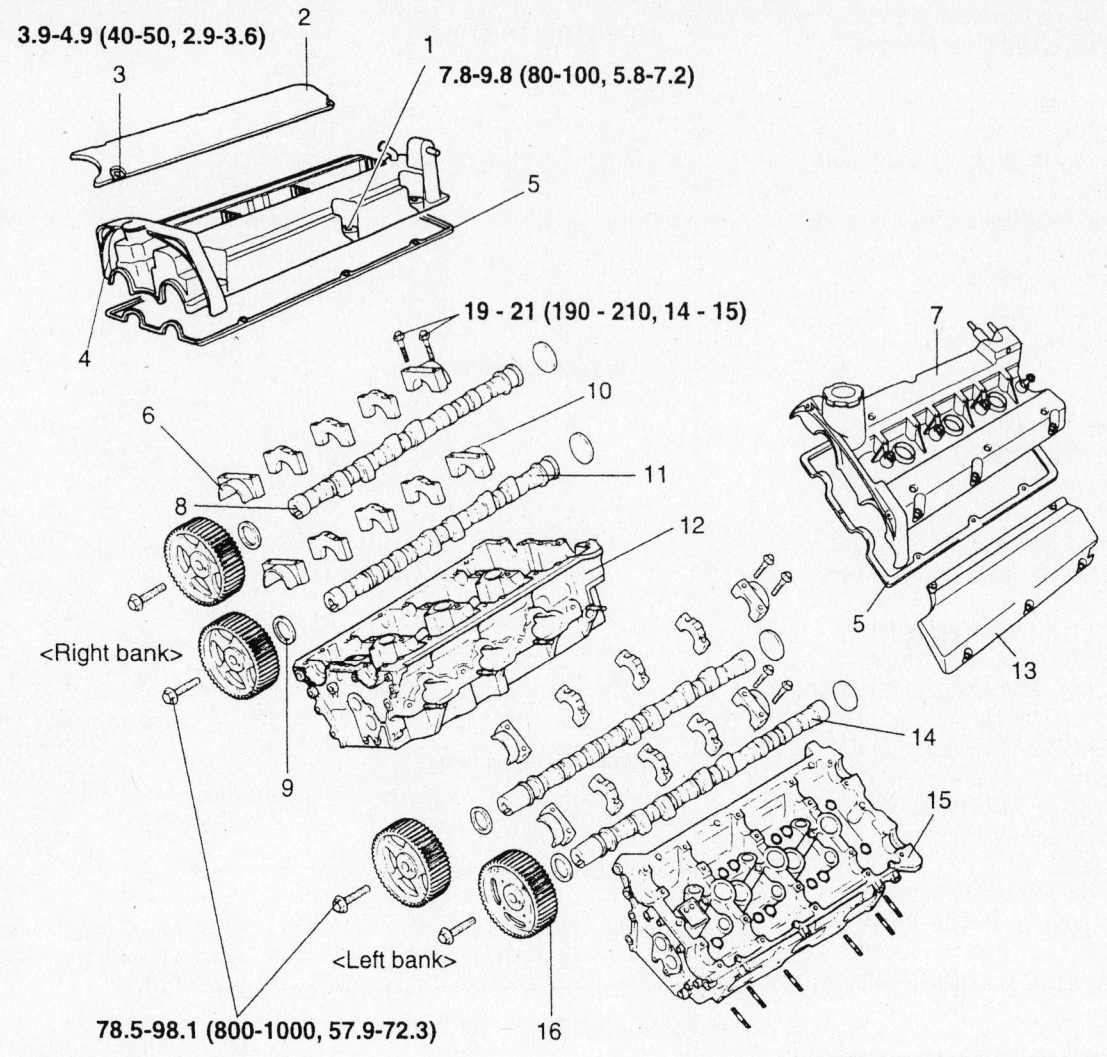

3.9-4.9 (40-50, 2.9-3.6)

7.8-9.8 (80-100, 5.8-7.2)

19 - 21 (190 - 210, 14 - 15)

78.5-98.1 (800-1000, 57.9-72.3)

\<Right bank\>

\<Left bank\>

09474_SOREN_G0013

TORQUE : N·m (kg·cm, lb·ft)

1. Cylinder head cover bolt
2. Center cover
3. Center cover bolt
4. Cylinder head cover
5. Gasket
6. Bearing cap (Front)
7. Cylinder head cover
8. Camshaft (EX)

9. Camshaft oil seal
10. Bearing cap (Rear)
11. Camshaft (IN)
12. Cylinder head (RH)
13. Center cover
14. Camshaft (EX)
15. Cylinder head (LH)
16. Camshaft sprocket

Camshaft and related components

8. Remove the timing belt from the camshaft sprocket.

9. Remove the spark plug cables.

10. Loosen the cylinder head cover bolts and then remove it.

11. Remove the camshaft sprockets.

➡**Keep all valve train components in order for assembly.**

12. Remove the camshaft bearing caps.

13. Remove the camshafts.

14. Remove the rocker arms.

15. Remove the hydraulic lifters.

To install:

16. Set the No. 1 cylinder to Top Dead Center of the compression stroke.

17. Install the lifters and rocker arms in their original positions.

18. Install the camshafts aligned according to the illustration.

19. The left and right banks of the

camshafts are different and you should be careful not to confuse them. Identification signals are as follows:

- Left bank—Intake (IN): I; Exhaust (Ex): E
- Right bank—Intake (IN): J; Exhaust (Ex): H

20. Confirm the identification mark and the number. Bearing caps of No.3, No.4, and No.5 have the front mark and arrange

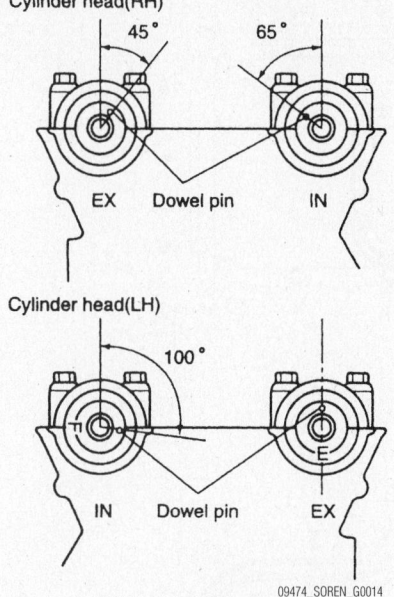

Camshaft installation alignment

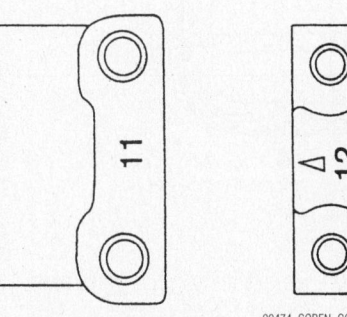

Bearing cap identification

the front mark upon the cylinder head while installing the bearing caps. Identification marks are as follows:

- Intake: I
- Exhaust: E

21. Install the bearing caps. Tighten the bolts evenly in 2 or 3 steps to the following torque specifications:

- Front and rear bearing caps: 14–15 ft. lbs. (19–21 Nm)
- Center bearing caps: 7–9 ft. lbs. (10–12 Nm)

22. Install the gasket to the cylinder head cover correctly.

23. Clean sealing surface on camshaft cap.

24. Apply sealant to the sealing surface on cylinder head cover and camshaft cap.

25. Install the cylinder head cover to the cylinder head.

26. Install the camshaft sprockets. Tighten the bolts to 58–72 ft. lbs. (79–98 Nm).

27. Install the timing belt.

28. Install the upper and lower timing belt covers.

29. Install the belt tensioner pulley.

30. Install the power steering pump pulley.

31. Install the crankshaft pulley.

32. Install the idler pulley.

33. Install the accessory drive belts.

34. Install the intake manifold.

35. Connect the breather hose and the engine harness.

36. Connect the spark plug cables.

37. Connect the negative battery cable.

Tightening Torque
Outer (*) 16 EA :
18.6-20.6 N·m(190-210 kg.cm, 13.7-15.2 lb.ft)
Inner 24 EA :
9.8-11.8 N·m(100-120 kg.cm, 7.2-8.7 lb.ft)

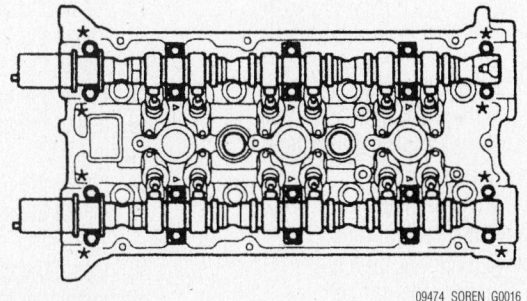

Camshaft bearing cap torque sequence

Sealant type : LT 5900

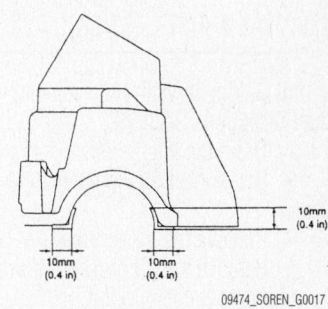

Cylinder head cover sealant application

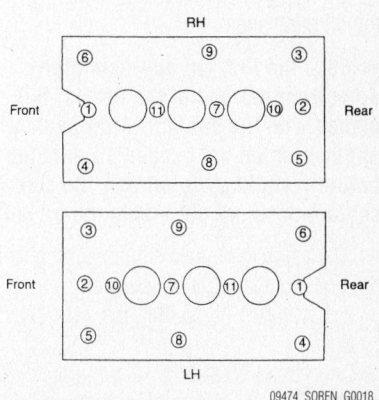

Cylinder head cover bolt torque sequence

Valve Lash

ADJUSTMENT

This vehicle is equipped with hydraulic valve lifters. No adjustment is necessary.

Starter Motor

REMOVAL & INSTALLATION

1. Before servicing the vehicle, refer to the Precautions Section.

2. Disconnect the negative battery cable.

3. Disconnect the starter motor electrical connectors.

4. Remove the starter heat shield.

5. Remove the starter motor.

To install:

6. Install the starter motor. Tighten the bolts to 7–9 ft. lbs. (10–12 Nm).

7. Install the starter heat shield.

8. Connect the starter motor electrical connectors. Tighten the battery terminal nut to 3–4 ft. lbs. (4–6 Nm).

9. Connect the negative battery cable.

Oil Pan

REMOVAL & INSTALLATION

1. Before servicing the vehicle, refer to the Precautions Section.
2. Drain the engine oil.
3. Disconnect the negative battery cable.
4. Remove the starter motor.
5. Remove the oil pressure switch.
6. Remove the oil filter.
7. Remove the lower oil pan.
8. Remove the upper oil pan.

To install:

9. Apply silicone sealant to the grove of the oil pan flange.

➡ **Make the first cut approximately 4mm from the end of the nozzle, furnished with the sealant. After sealant application do not exceed 15 minutes before installing the oil pan. Be sure sealant does not get inside the oil pan.**

10. Install the upper oil pan and tighten the bolts in sequence as follows:
   a. Bolts 1-14: 7–9 ft. lbs. (10–12 Nm).
   b. Bolts 15 and 16: 4–5 ft. lbs. (5–7 Nm).
   c. Bolts 17 and 18: Upper oil pan-to-transaxle mounting bolts: 22–30 ft. lbs. (29–41 Nm).
11. Apply sealant to the threads, then install the oil pressure switch and tighten to 6–9 ft. lbs. (8–12 Nm).
12. Install the lower oil pan and tighten the bolts in sequence to 7–9 ft. lbs. (10–12 Nm).
13. Install the oil filter.
14. Install the starter motor.
15. Connect the negative battery cable.
16. Fill the crankcase to the correct level with engine oil.
17. Start the engine and check for leaks.

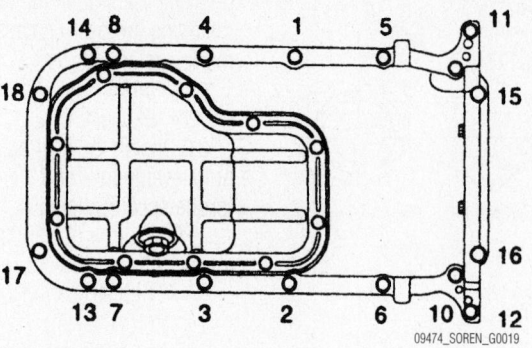

Upper oil pan torque sequence

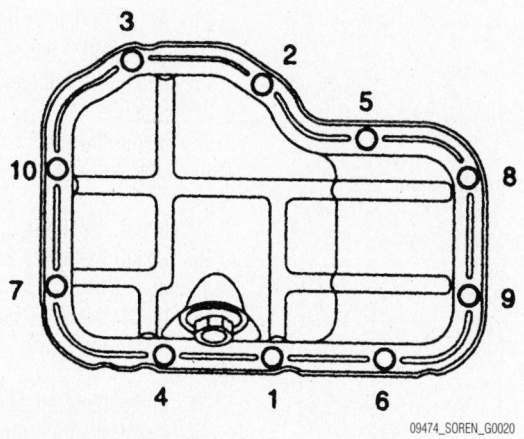

Lower oil pan torque sequence

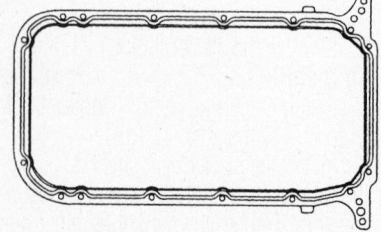

Oil pan flange sealant application

Oil Pump

REMOVAL & INSTALLATION

1. Before servicing the vehicle, refer to the Precautions Section.
2. Drain the engine oil.
3. Disconnect the negative battery cable.
4. Remove the cylinder head covers.
5. Remove the accessory drive belts.
6. Remove the idler pulley.
7. Remove the crankshaft pulley.
8. Remove the power steering pump pulley.
9. Remove the belt tensioner pulley.
10. Remove the upper and lower timing belt covers.
11. Remove the timing belt.
12. Remove the crankshaft sprocket.
13. Remove the Crankshaft Position (CKP) sensor tone ring.
14. Remove the oil filter.
15. Remove the starter motor.
16. Remove the lower oil pan.
17. Remove the upper oil pan.
18. Remove the oil pick-up tube.
19. Remove the oil filter bracket.
20. Remove the oil relief valve plug.
21. Remove the oil pump.

To install:

22. Install the oil pump with a new gasket. Tighten the oil pump case bolts to 9–10 ft. lbs. (12–15 Nm). Tighten the oil pump cover screws to 6–9 ft. lbs. (8–12 Nm).
23. Install the oil relief valve plug. Tighten the plug to 29–36 ft. lbs. (40–50 Nm).
24. Install the oil filter bracket.
25. Install the oil pick-up tube. Tighten the bolts to 11–15 ft. lbs. (15–22 Nm).
26. Install the upper oil pan.
27. Install the lower oil pan.
28. Install the starter motor.
29. Install the oil filter.
30. Install the CKP sensor tone ring.
31. Install the crankshaft sprocket.
32. Install the timing belt.
33. Install the upper and lower timing belt covers.
34. Install the belt tensioner pulley.
35. Install the power steering pump pulley.
36. Install the crankshaft pulley.
37. Install the idler pulley.
38. Install the accessory drive belts.
39. Install the cylinder head covers.

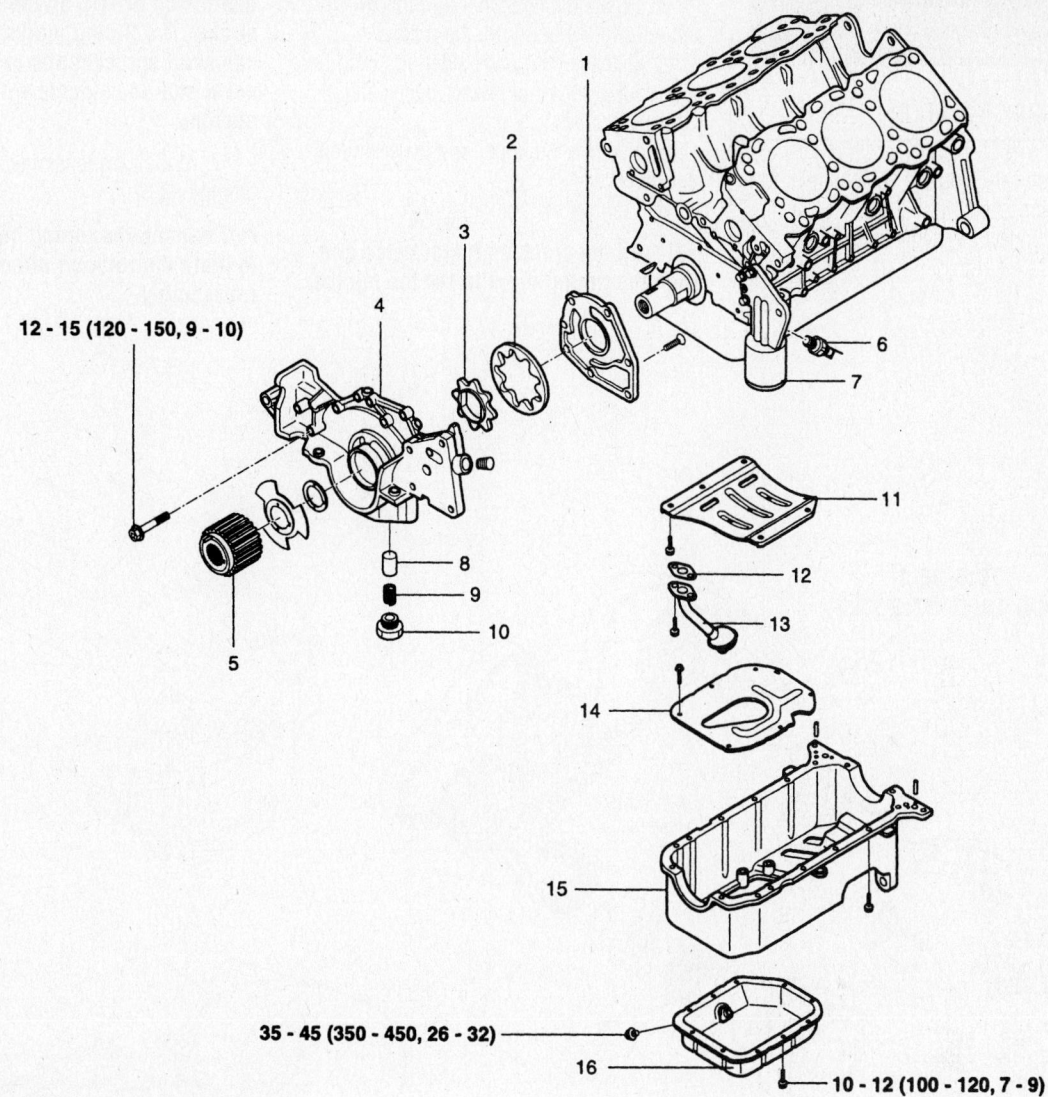

12 - 15 (120 - 150, 9 - 10)

35 - 45 (350 - 450, 26 - 32)

10 - 12 (100 - 120, 7 - 9)

TORQUE : N·m (kg·cm, lb·ft)

1. Oil pump cover
2. Oil pump outer rotor
3. Oil pump inner rotor
4. Oil pump case
5. Crankshaft sprocket
6. Oil pressure switch
7. Oil pump
8. Relief valve planger

9. Relief valve spring
10. Relief valve plug
11. Upper baffle plate
12. Gasket
13. Oil screen
14. Lower baffle plate
15. Upper oil pan
16. Lower oil pan

09474_SOREN_G0022

Oil pump and related components

40. Connect the negative battery cable.
41. Fill the crankcase to the correct level with engine oil.
42. Start the engine and check for leaks.

Rear Main Seal

REMOVAL & INSTALLATION

1. Before servicing the vehicle, refer to the Precautions Section.

2. Disconnect the negative battery cable.
3. Remove the starter motor.
4. Remove the transmission.
5. Remove the flexplate.
6. Remove the oil seal housing.
7. Remove the oil seal.

To install:

8. Install the oil seal to the seal housing using special tool 09231-33000 or similar seal driver.
9. Apply silicone sealant to the oil seal housing flange.

10. Install the seal housing and tighten the bolts to 7–9 ft. lbs. (10–12 Nm).
11. Install the flexplate. Tighten the bolts to 7–9 ft. lbs. (10–12 Nm).
12. Install the transmission.
13. Install the starter motor.
14. Connect the negative battery cable.
15. Fill the crankcase to the correct level with engine oil.
16. Start the engine and check for leaks.

Timing Belt, Cover and Crankshaft Seal

REMOVAL & INSTALLATION

1. Before servicing the vehicle, refer to the Precautions Section.
2. Disconnect the negative battery cable.

3. Rotate the tensioner arm clockwise and remove the belt from the pulley.
4. Remove the power steering pump pulley, idler pulley, tensioner pulley and crankshaft pulley.
5. Remove the upper and lower timing belt covers.
6. Remove the auto tensioner.

➡Rotate the crankshaft clockwise and align the timing mark to set the number one piston at TDC on the compression stroke. The timing marks on the camshaft sprocket and cylinder head cover should coincide with one another.

7. Unbolt the tensioner and remove the timing belt.

➡If reusing the timing belt make sure to mark the rotation direction for reassembly.

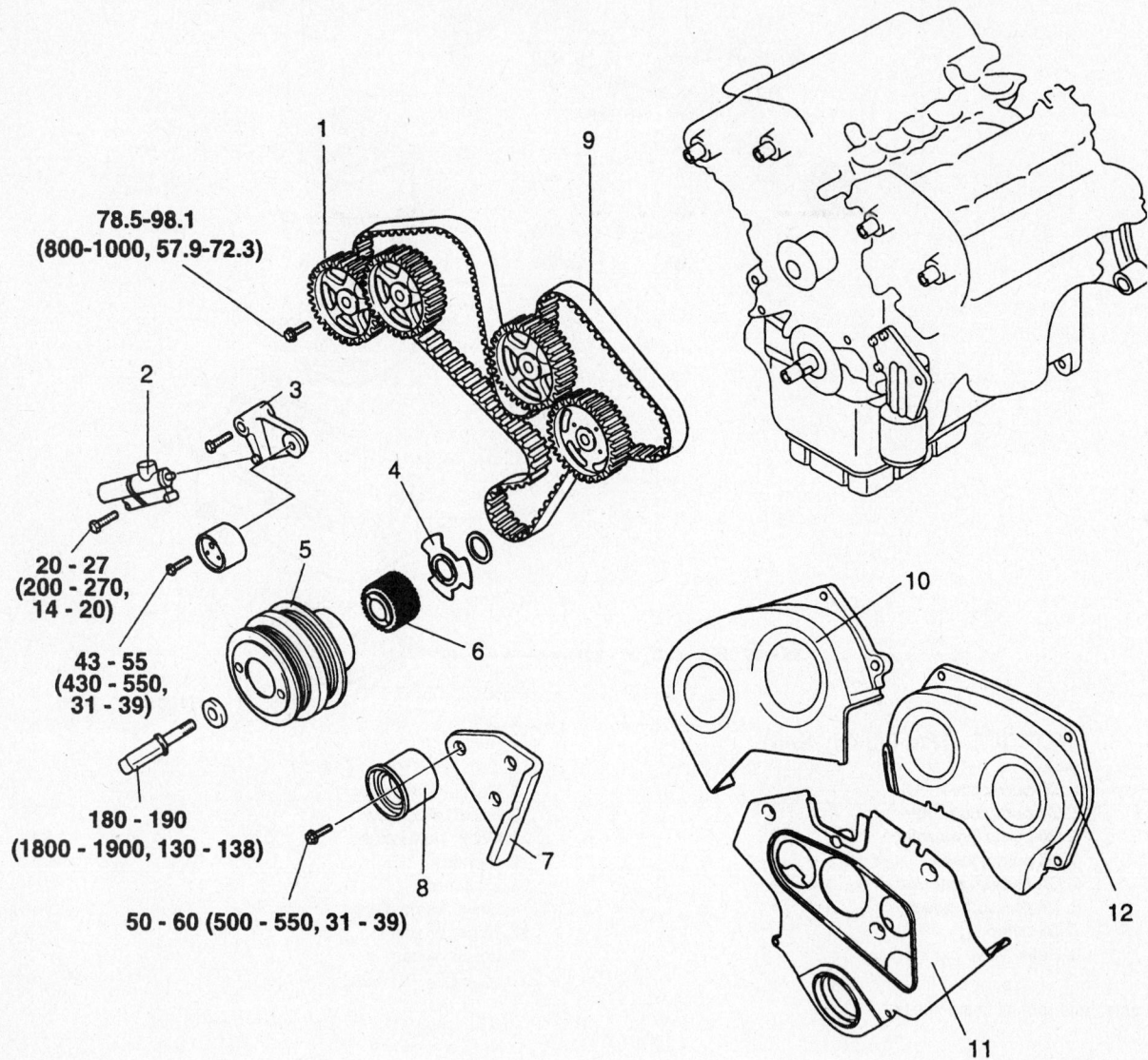

78.5-98.1 (800-1000, 57.9-72.3)

20 - 27 (200 - 270, 14 - 20)

43 - 55 (430 - 550, 31 - 39)

180 - 190 (1800 - 1900, 130 - 138)

50 - 60 (500 - 550, 31 - 39)

TORQUE : N·m (kg·cm, lb·ft)

1. Camshaft sprocket
2. Auto tensioner
3. Tensioner arm
4. Sensing blade
5. Crankshaft pulley
6. Crankshaft sprocket
7. Idler bracket
8. Idler
9. Timing belt
10. Timing belt upper cover (RH)
11. Timing blet lower cover
12. Timing belt upper cover (LH)

Timing belt and related components

09474_SOREN_G0023

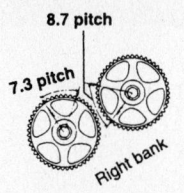
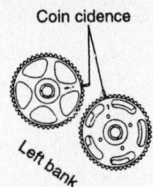

Camshaft sprocket initial installation position

To install:

8. Ensure that the engine is set to Top Dead Center (TDC).

9. Install the idler pulley to the idler bracket.

10. Install the tensioner arm, shaft and plane washer to the cylinder block. Torque the bolts to 25–40 ft. lbs. (35–55 Nm).

11. Install the crankshaft sprocket and align the timing mark. Do not bend the crankshaft sensing blade.

12. Install the camshaft sprocket and adjust the initial installation state according to the illustration.

13. Install the auto tensioner to oil pump case. At this time the auto tensioner's set pin should be completely assembled.

14. Align the timing marks of each sprocket and install the timing belt in the following order, crankshaft sprocket, idler

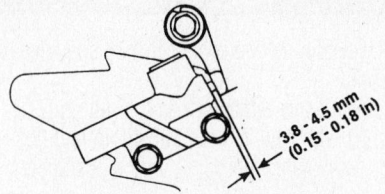

Measuring the auto tensioner rod

pulley, exhaust camshaft sprocket (LH), intake camshaft sprocket (LH), water pump pulley, intake camshaft sprocket (RH), exhaust camshaft sprocket (RH) and tensioner pulley.

15. After installing the timing belt, reconfirm the timing mark. Install the tensioner pulley. Pull the set pin out of the auto tensioner.

16. Rotate the crankshaft 2 revolutions **Clockwise**, then wait 5 minutes for the auto tensioner to adjust.

17. Measure the auto tensioner rod as shown. If the measurement is not 3.8–4.5 mm, then repeat the belt tensioning procedure.

18. After the auto tensioner measurement is correct, install the upper and lower timing belt covers.

19. Continue the installation in the reverse order of the removal procedure.

Piston and Ring

POSITIONING

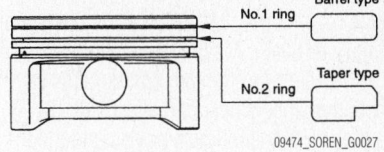

Compression ring identification

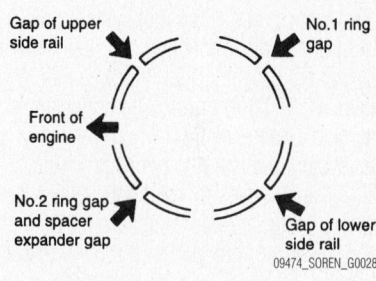

Piston ring end gap spacing

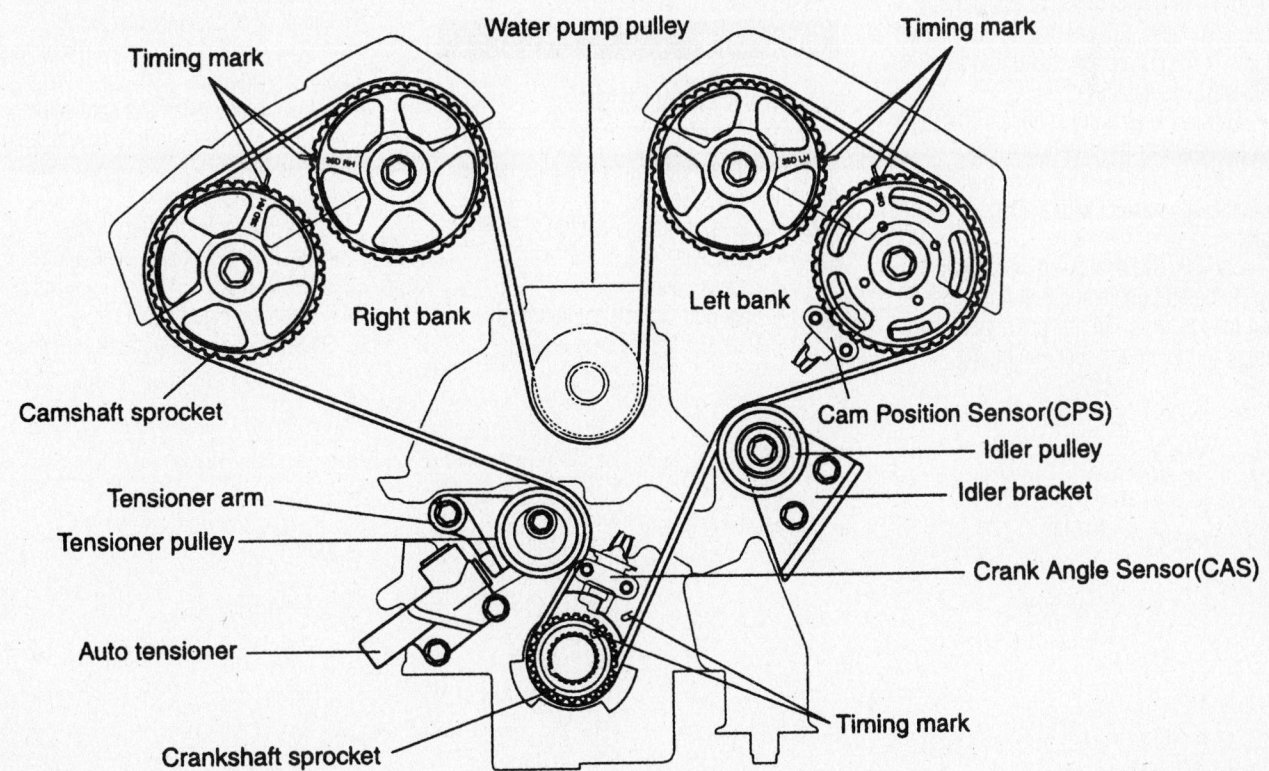

Timing belt routing and timing marks

FUEL SYSTEM

Fuel System Service Precautions

Safety is the most important factor when performing not only fuel system maintenance, but any type of maintenance. Failure to conduct maintenance and repairs in a safe manner may result in serious personal injury or death. Maintenance and testing of the vehicle's fuel system components can be accomplished safely and effectively by adhering to the following rules and guidelines.

• To avoid the possibility of fire and personal injury, always disconnect the negative battery cable unless the repair or test procedure requires that battery voltage be applied.

• Always relieve the fuel system pressure prior to disconnecting any fuel system component (injector, fuel rail, pressure regulator, etc.), fitting or fuel line connection. Exercise extreme caution whenever relieving fuel system pressure, to avoid exposing skin, face and eyes to fuel spray. Please be advised that fuel under pressure may penetrate the skin or any part of the body that it contacts.

• Always place a shop towel or cloth around the fitting or connection prior to loosening to absorb any excess fuel due to spillage. Ensure that all fuel spillage (should it occur) is quickly removed from engine surfaces. Ensure that all fuel soaked cloths or towels are deposited into a suitable waste container.

• Always keep a dry chemical (Class B) fire extinguisher near the work area.

• Do not allow fuel spray or fuel vapors to come into contact with a spark or open flame.

• Always use a back-up wrench when loosening and tightening fuel line connection fittings. This will prevent unnecessary stress and torsion to fuel line piping.

Always follow the proper torque specifications.

• Always replace worn fuel fitting O-rings with new. Do not substitute fuel hose or equivalent where fuel pipe is installed.

Fuel System Pressure

RELIEVING

1. Before servicing the vehicle, refer to the Precautions Section.
2. Remove the center floor carpet.
3. Remove the access cover.
4. Disconnect the fuel pump electrical connector.
5. Start the engine and allow it to run until it stalls.
6. Turn the ignition switch to the **OFF** position.
7. Restore the electrical connections after fuel system repairs are completed.

Fuel Filter

REMOVAL & INSTALLATION

The fuel filter is located inside the fuel tank, attached to the fuel pump and is replaced with the fuel pump as an assembly.

Fuel Pump

REMOVAL & INSTALLATION

1. Before servicing the vehicle, refer to the Precautions Section.
2. Remove or disconnect the following:
   • Rear seat
   • Access cover
   • Fuel pump electrical connector
3. Relieve the fuel system pressure.

   • Fuel lines
   • Fuel pump and filter assembly

To install:
4. Install or connect the following:
   • Fuel pump and filter assembly
   • Fuel lines
   • Fuel pump electrical connector
   • Access cover
   • Rear seat
5. Start the engine and check for leaks.

Fuel Injector

REMOVAL & INSTALLATION

1. Before servicing the vehicle, refer to the Precautions Section.
2. Relieve the fuel system pressure.
3. Remove or disconnect the following:
   • Negative battery cable
   • Positive Crankcase Ventilation (PCV) hose
   • Brake booster vacuum hose
   • Surge tank stays
   • Fuel lines
   • Upper intake manifold (surge tank)
   • Fuel injector harness connectors
   • Fuel supply manifold with injectors attached

To install:
4. Replace all injector seals.
5. Install or connect the following:
   • Fuel supply manifold with injectors attached
   • Fuel injector harness connectors
   • Upper intake manifold (surge tank)
   • Fuel lines
   • Surge tank stays
   • Brake booster vacuum hose
   • Positive Crankcase Ventilation (PCV) hose
   • Negative battery cable
6. Start the engine and check for leaks.

DRIVE TRAIN

Transmission Assembly

REMOVAL & INSTALLATION

Manual

1. Before servicing the vehicle, refer to the Precautions Section.
2. Disconnect the negative battery cable.
3. Remove the knob and the control lever.
4. Raise the vehicle.
5. Remove the transmission under cover.
6. Remove the clutch release cylinder.
7. Remove the front driveshaft (4WD vehicle).
8. Remove the front muffler and the heater protector.
9. Remove the transfer case connector (4WD vehicle).
10. Remove the rear driveshaft.
11. Support the transmission by the jack.
12. Remove the rear crossmember.
13. Remove the transmission with transfer case (4WD vehicle).

To install:

14. Install the transmission and transfer case (if equipped). Tighten the flange bolts to 31–40 ft. lbs. (43–55 Nm).
15. Install the rear crossmember.
16. Install the rear driveshaft.
17. Connect the transfer case connector (4WD).
18. Install the front muffler and the heater protector.
19. Install the front driveshaft (4WD).
20. Install the clutch release cylinder.
21. Install the transmission under cover.
22. Install the control lever and shift knob.
23. Connect the negative battery cable.
24. Start the engine and check for proper operation.

Automatic

1. Before servicing the vehicle, refer to the Precautions Section.
2. Disconnect the negative battery cable.
3. Drain the automatic transmission fluid.
4. Remove the control cable.
5. Remove the under cover.
6. Remove the front driveshaft (4WD).
7. Remove the front muffler and the heater protector.

8. Remove the transfer case connector (4WD).
9. Remove the rear driveshaft.
10. Remove the oil cooler pipe.
11. Remove the speed sensor connector.
12. Remove the back-up lamp switch connector.
13. Remove the starter motor.
14. Remove the transmission mounting bolt.
15. Support the transmission on the jack.
16. Remove the transmission with transfer case (4WD vehicle).

To install:

17. Install the transmission and transfer case (if equipped). Tighten the flange bolts to 29–38 ft. lbs. (42–54 Nm).
18. Install the starter motor.
19. Connect the back-up lamp switch connector.
20. Connect the speed sensor connector.
21. Install the oil cooler pipe.
22. Install the rear driveshaft.
23. Connect the transfer case connector. (4WD)
24. Install the front muffler and the heater protector.
25. Install the front driveshaft (4WD).
26. Install the under cover.
27. Install the control cable.
28. Connect the negative battery cable.
29. Fill the transaxle to the correct level with the proper transmission fluid.
30. Start the engine and check for leaks.

Clutch

REMOVAL AND INSTALLATION

1. Before servicing the vehicle, refer to the Precautions Section.
2. Disconnect the negative battery cable.
3. Remove the transmission from the vehicle.
4. Insert tool (09411-43000) or equivalent in the clutch dis. This prevents the disc from shifting.
5. Loosen the bolts that attach the clutch pressure plate to the flywheel in a star pattern. Loosen the bolts in succession, one or two turns at a time in order to prevent bending the pressure plate.

To install:

6. Installation is the reverse of the removal procedure.
7. Torque the pressure plate retaining bolts to 11–16 ft. lbs. (15–22 Nm).

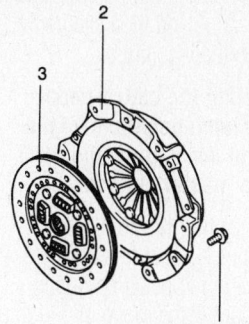

15 - 22 (150 - 220, 11 - 16)

2. Clutch cover
3. Clutch disc

09474_SOREN_G0029

Clutch and related components

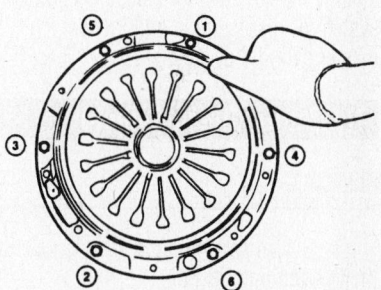

09474_SOREN_G0030

Clutch pressure plate bolt torque sequence

ADJUSTMENT

Clutch Pedal Height and Free-Play

1. Measure the clutch pedal height (From the face of the pedal pad to the floorboard) and the clutch pedal free-play (measured at the face of the pedal pad). The standard value is as follows:
   a. (A) 7.3–13.9mm (0.29–0.55 in)
   b. (A') 163.8mm (6.45 in)
2. If the clutch pedal free-play is not within the standard value range, adjust as follows :

Clutch pedal free-play (A) and Pedal height (A')

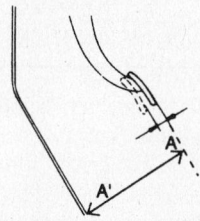

09474_SOREN_G0031

Clutch pedal height (A') and free-play (A) adjustment points

a. Turn and adjust the bolt, then secure it by tightening the lock nut.

b. Turn the push rod to coincide with the standard value and then secure the push rod with the lock nut.

➡ **When adjusting the clutch pedal height or the clutch pedal clevis pin play, be careful not to push the push rod toward the master cylinder.**

3. After completing the adjustments, check that the clutch pedal free play (measured at the face of the pedal pad) falls within the standard value range of 6–13mm (0.2–0.5 in.).

4. If the clutch pedal free play and the distance between the clutch pedal and the floor board when the clutch is disengaged do not meet the standard values, the cause may be either air in the hydraulic system or a faulty master cylinder clutch. Bleed the system or disassemble and inspect the master cylinder or clutch.

Hydraulic Clutch System

BLEEDING

1. Before servicing the vehicle, refer to the Precautions Section.

2. With an assistant in the vehicle, raise and safely support the vehicle.

3. Have your assistant pump the clutch pedal three times and hold the pedal to the floor.

4. Open the bleeder valve on the clutch slave cylinder until the air is purged from the cylinder.

5. Tighten the bleeder valve.

6. Have your assistant release the clutch pedal.

7. Fill the clutch master cylinder if below minimum.

8. Repeat Steps 2 through 6 until no air exits from the bleeder valve.

9. Lower the vehicle.

10. Fill the clutch master cylinder fluid reservoir.

Transfer Case Assembly

REMOVAL & INSTALLATION

1. Before servicing the vehicle, refer to the Precautions Section.

2. Drain the transfer case fluid.

3. Disconnect the negative battery cable.

4. Remove the front and rear console mounting screws. Slide the console forward to clear the parking brake handle and set aside. Open the shift boot.

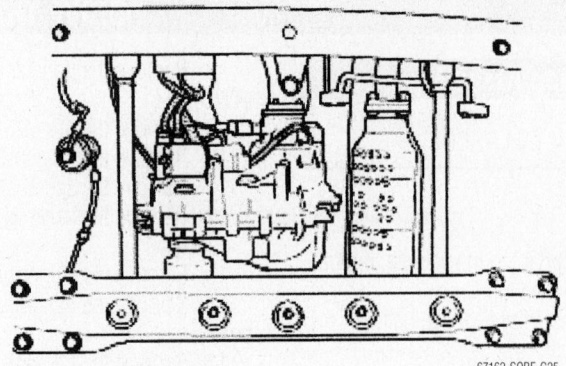

Transfer case and crossmember location

5. Loosen the transfer case shift lever locknut and remove the lever knob.

6. Pull the console up to access the wiring connector(s). Unplug the connector(s) and remove the console.

7. Shift the transfer lever to the 4L position.

8. Remove the cover plate.

9. Remove the retaining bolts from the transfer case and lift the shifter lever assembly straight out and properly support the transmission.

10. Matchmark the driveshaft at the flanges and remove the driveshaft.

11. Remove the crossmember bolts and support the transmission on the jack.

12. Disconnect the 4WD light switch connector.

13. Remove the transfer case mounting bolts.

14. Remove the crossmember.

15. Separate the transfer case from the transmission by striking the transfer case with a plastic mallet at the seal area.

16. Lower the transfer case from the vehicle.

To install:

17. Install the transfer case in position with a new gasket. Torque the bolts to 32 ft. lbs. (44 Nm).

18. Install the crossmember.

19. Connect the 4WD light switch connector.

20. Align the matchmarks on the driveshaft to the flanges. Torque the bolts to 27 ft. lbs. (36 Nm) and remove the transmission support

21. Install the retaining bolts to the transfer case and install the shift lever assembly.

22. Remove the transmission support jack.

23. Install the cover plate.

24. Install the floor console unit and connect or install the following:

- Switch wiring connector(s)
- Front console
- Lever knob
- Front console and tie the shift boot draw strings
- Slide the console over the parking brake handle

25. Connect the negative battery cable

26. Fill the transfer case to the proper level

27. Start the vehicle and check for leaks, repair if necessary.

Halfshaft

REMOVAL & INSTALLATION

1. Before servicing the vehicle, refer to the Precautions Section.

2. Remove the front wheels.

3. Remove the lock nut from front hub.

4. Remove the upper control arm link lock bolt, spring washer and nut.

5. Remove tie rod end cotter pin and using a ball joint puller, remove tie rod end from steering knuckle.

6. Mark driveshaft for identical installation position.

7. Using tool, pry the driveshaft from the differential housing. Separate the right halfshaft from the intermediate shaft.

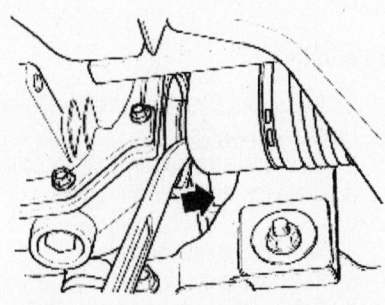

Pry the driveshaft from the differential housing

8. Remove the axle halfshafts from the front knuckle. It may be necessary to tap the stub shaft with a brass hammer to remove the axles.

To install:

9. Coincide the joining mark between the driveshaft and the differential and insert the shaft.

10. Install the knuckle assembly. Tighten the ball joint castle nut to 51–57 ft. lbs. (70–80 Nm).

11. Install the upper arm link lock bolt and tighten to 32–39 ft. lbs. (44–55 Nm).

12. Tighten the lock nut to 177–198 ft. lbs. (245–275 Nm) and then caulk the flange of lock nut on the end of driveshaft.

13. Install the front wheels.

CV Joint

OVERHAUL

Outer CV Joint (B.J.)

The outer CV joint (B.J.) is serviced only with the axle halfshaft as an assembly. The outer CV joint (B.J.) boot can be serviced by first removing the inner Tripod joint (T.S.J.).

Inner Tripod Joint (T.S.J.)

1. Before servicing the vehicle, refer to the Precautions Section.

2. Remove the axle halfshaft from the vehicle and place it in a vise.

3. Remove the T.S.J boot band and pull the boot away from T.S.J outer race.

4. Remove the circlip using a screwdriver.

5. Remove the driveshaft from the T.S.J outer race.

6. Remove the snap ring and disassemble the inner race and ball from the shaft.

7. Remove the B.J boot band and pull out the T.S.J boot and the B.J boot.

➡**If the boot is reused, wrap a tape around the driveshaft splines to protect the boot.**

To install:

8. Wrap tape around the driveshaft spline (T.S.J side) to avoid boot damage.

9. Install the outer and inner joint boots.

10. Install the inner race and ball to the shaft, then install the snap ring.

11. Install the driveshaft into the T.S.J outer race and install the circlip.

12. Apply CV joint grease to the driveshaft and position the boots over each joint.

13. Install the boot clamps. Pull on the clamp end with pliers and fold the locking tabs over to lock the clamp in place.

14. Install the axle halfshaft to the vehicle.

Rear Axle Shaft, Bearing and Seal

REMOVAL & INSTALLATION

1. Before servicing the vehicle, refer to the Precautions Section.

2. Disconnect the negative battery cable.

3. Remove the rear wheels.

4. Remove the disc brake and parking brake assembly.

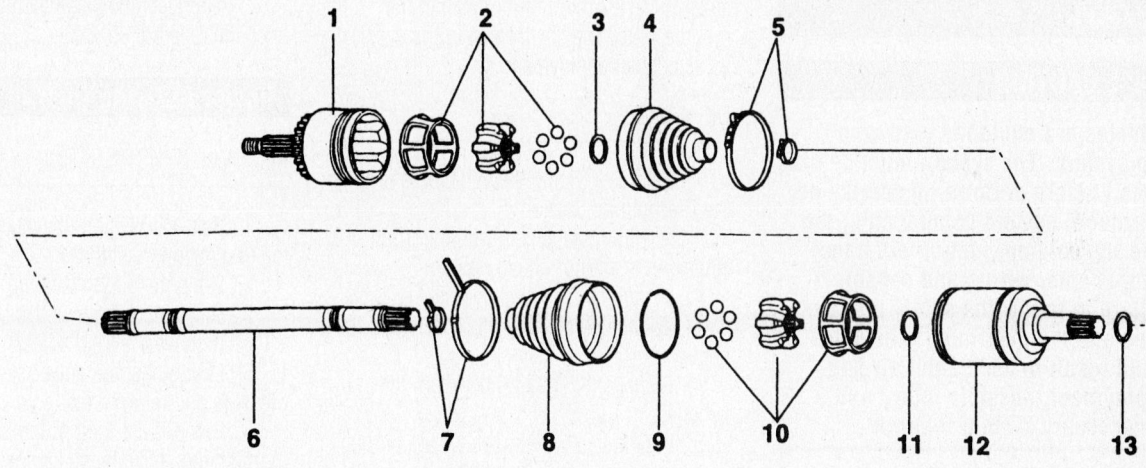

1. B.J assembly
2. B.J inner race and ball
3. Snap ring
4. B.J boot
5. B.J boot band
6. Drive shaft
7. T.S.J boot band
8. T.S.J boot
9. Circlip
10. T.S.J inner race and ball
11. Snap ring
12. T.S.J assembly
13. Clip

Inner and outer CV Joint and related components

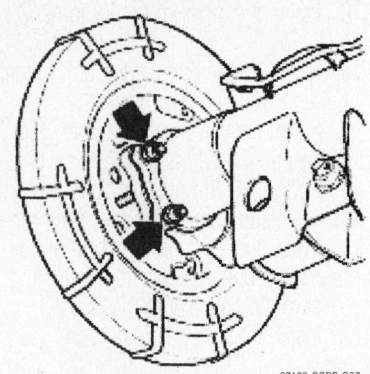

Rear axle shaft mounting bolt locations

67162-SORE-G27

5. Remove the parking brake cable and wheel speed sensor cable.

6. Remove the rear axle shaft mounting bolts.

7. Remove the rear axle shaft.

8. Remove the bearing collar and bearing from the axle.

9. Using a slide hammer, remove the oil seal.

To install:

10. Apply grease to the oil seal lip and using the appropriate seal driver, install the new axle seal into the differential.

11. Install the new wheel bearing and retainer collar to the rear axle shaft.

12. Install the rear axle shaft. Torque the axle shaft mounting bolts to 32–44 ft. lbs. (43–60 Nm).

13. Install the wheel speed sensor and parking brake cables.

14. Install the disc brake and parking brake assembly and the rear wheels.

15. Adjust the parking brake lever.

16. Connect the negative battery cable.

Pinion Seal

REMOVAL & INSTALLATION

1. Before servicing the vehicle, refer to the Precautions Section.

2. Drain the gear oil.

3. Disconnect the negative battery cable.

4. Remove the wheels.

5. Remove the brake drums (if removing the rear pinion seal).

6. Remove the driveshaft.

7. Remove the drive pinion.

8. Remove the pinion seal

To install:

9. Install a new pinion seal lightly coated with clean gear oil.

10. Install the drive pinion.

11. Rotate the pinion flange occasionally while tightening the flange nut and make certain that the pinion bearings are seated properly.

12. Install the driveshaft after aligning the matchmarks.

13. Install the brake drums.

14. Install the wheels.

15. Connect the negative battery cable.

16. Fill the gear oil to the proper level.

17. Start the vehicle and check for leaks, repair if necessary.

STEERING

Air Bag

✳✳ CAUTION

Vehicles are equipped with an air bag system. The system must be disarmed before performing service on, or around, system components, the steering column, instrument panel components, wiring and sensors. Failure to follow the safety precautions and the disarming procedure could result in accidental air bag deployment, possible injury and unnecessary system repairs.

PRECAUTIONS

Several precautions must be observed when handling the inflator module to avoid accidental deployment and possible personal injury.

• Never carry the inflator module by the wires or connector on the underside of the module.

• When carrying a live inflator module, hold securely with both hands, and ensure that the bag and trim cover are pointed away.

• Place the inflator module on a bench or other surface with the bag and trim cover facing up.

• With the inflator module on the bench, never place anything on or close to the module which may be thrown in the event of an accidental deployment.

DISARMING

1. Before servicing the vehicle, refer to the Precautions Section.

2. Position the vehicle with the front wheels in a straight-ahead position.

3. Disconnect the negative battery cable and keep the cable secure from touching the battery.

4. Remove the ignition key from the vehicle.

5. Wait at least 3 minutes for the air bag back-up power supply to deplete its stored energy before continuing.

6. Proceed with the repair.

7. Reconnect the negative battery cable once the repair is complete.

Power Steering Gear

REMOVAL & INSTALLATION

1. Before servicing the vehicle, refer to the Precautions Section.

2. Drain the power steering fluid.

3. Disconnect the negative battery cable.

4. Remove the front wheels.

5. Disconnect the power steering fluid pressure and return lines.

6. Remove the joint assembly connecting bolt and separate the steering intermediate shaft from the steering gear box.

7. Separate the outer tie rod ends from the steering knuckles.

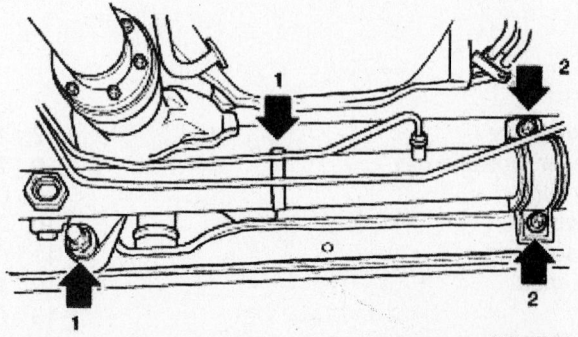

67162-SORE-G29

Power steering gear fastener (1) and mounting bracket fastener (2) locations

8. Remove the steering gear box mounting bolts and remove the steering gear box assembly together with mounting rubber.

To install:

9. Install the steering gear into position on the vehicle.

10. Install the steering gear mounting rubber, brackets and fasteners. Tighten the mounting bracket fasteners (2) to 54–68 ft. lbs. (75–95 Nm). Tighten the steering gear fasteners (1) to 88–114 ft. lbs. (122–158 Nm).

11. Connect the intermediate shaft to the steering gear. Tighten the pinch bolt to 15–19 ft. lbs. (22–27 Nm).

12. Connect the outer tie rod ends to the steering knuckles. Tighten the nuts to 50–57 ft. lbs. (70–80 Nm).

13. Connect the power steering fluid pressure and return lines. Tighten the bolts to 23–34 ft. lbs. (32–48 Nm).

14. Install the front wheels

15. Connect the negative battery cable.

16. Fill the power steering system with the correct amount of power steering fluid.

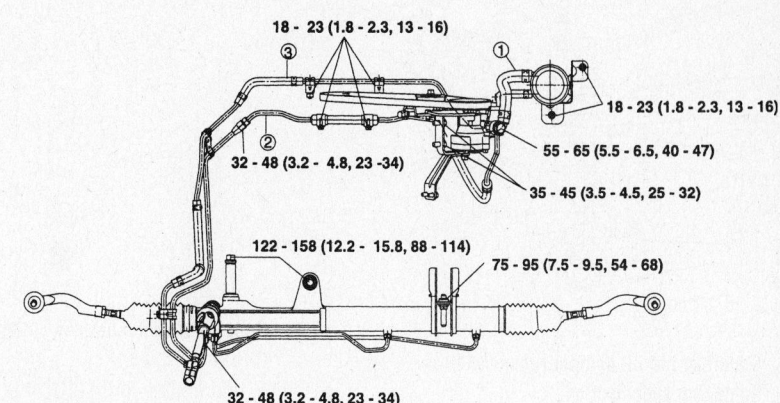

1. Suction hose
2. Pressure hose & tube
3. Return hose & tube

Torque : N·m (kg-m, lb-ft)

09474_SOREN_G0033

Power steering gear torque specifications and fluid line routing

FRONT SUSPENSION

Strut

REMOVAL & INSTALLATION

1. Before servicing the vehicle, refer to the Precautions Section.

2. Loosen battery cable and mounting bolt and then remove battery.

3. Remove three strut mounting block nuts from the mounting block.

4. Raise the front of the vehicle and support it with safety stands.

5. Remove the front wheels.

6. Remove the bolt on the steering knuckle side that secures the upper arm ball joint.

7. Remove the brake hose bracket and then the remove the upper arm bolts and nuts.

8. Remove the strut lower mounting nut.

9. Remove the strut from the vehicle.

To install:

10. After making sure identification mark on the spring seat. Position the strut assembly into the upper mounting block.

11. Install the mounting nuts by 3-4 threads only.

12. Insure the front of the vehicle is raised and supported with safety stands.

13. Tighten the lower nut of the strut assembly to 88–101 ft. lbs. (122–140 Nm).

14. Position the upper arm to the frame brackets, insert the bolts and hand tighten the nuts.

15. Install the upper arm ball joint into the top of the steering knuckle and tighten the side bolt and nut to 31–39 ft. lbs. (44–55 Nm).

16. Tighten the upper arm bolts and nuts to 54–68 ft. lbs. (76–95 Nm) and then install brake hose brackets.

17. Install the front wheels.

18. Lower the vehicle.

19. Tighten the mounting block nuts to 31–39 ft. lbs. (44–55 Nm).

20. Install the battery mounting bracket and the battery.

21. Check the front end alignment and adjust as necessary.

Stabilizer Bar

REMOVAL & INSTALLATION

1. Before servicing the vehicle, refer to the Precautions Section.

2. Raise and support the vehicle safely. Remove the front tires.

3. Remove the engine undercover (splash shield).

4. Remove the nuts and oil damper rubbers of the control link.

5. Remove the stabilizer bar bushing bracket retaining bolts.

6. Remove the stabilizer bar from the vehicle.

7. As required, remove the control link from its mounting.

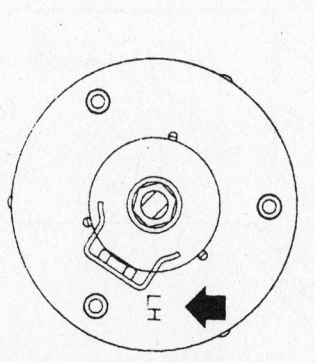

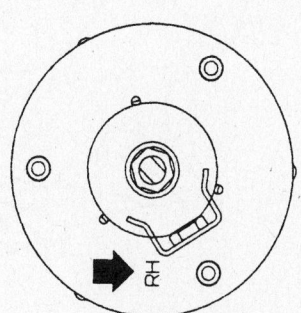

09474_SOREN_G0034

Correct strut mount placement

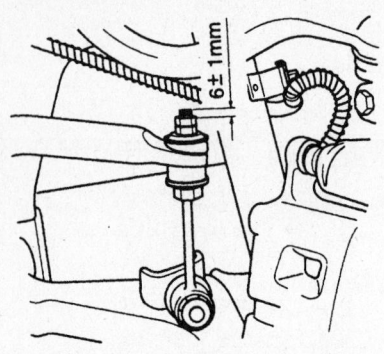

09474_SOREN_G0035

Stabilizer bar oil damper rubber and nut tightening specification

To install:

8. Installation is the reverse of the removal procedure.

9. Tighten the stabilizer bushing mounting bolts to 31–39 ft. lbs. (44–55 Nm).

10. Tighten the oil damper rubber and nut so that the nut protrudes about 6mm after it is tightened.

11. Tighten the lower nut of the control link to 68–84 ft. lbs. (95–117 Nm).

Upper Ball Joint

REMOVAL & INSTALLATION

The ball joint is serviced with the upper control arm as an assembly.

Lower Ball Joint

REMOVAL & INSTALLATION

1. Before servicing the vehicle, refer to the Precautions Section.

2. Raise the front of the vehicle and support it with safety stands.

3. Remove the front wheels.

4. Remove the lower nut of the control link of the stabilizer bar.

5. Remove the lower nut of the strut.

6. Remove the bolts and nuts that join the lower arm and lower arm ball joint.

7. Remove the cotter pin and castle nut from the lower arm ball joint.

8. Separate the lower arm ball joint from the steering knuckle.

To install:

9. Install the lower arm ball joint to the steering knuckle. Tighten the ball joint-to-control arm fasteners to 116–145 ft. lbs. (157–196 Nm).

10. Install a new castle nut and cotter pin through the castle nut.

11. Install the lower nut of the strut

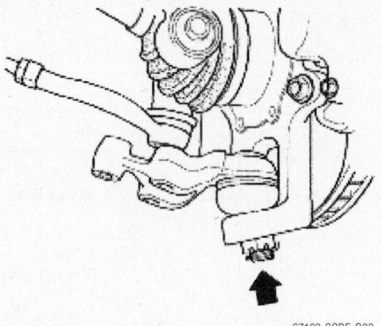

67162-SORE-G33

Lower ball joint, castle nut and cotter pin

assembly and tighten to 88–101 ft. lbs. (122–140 Nm).

12. Install the lower nut of control link of stabilizer bar and tighten the nut to 68–84 ft. lbs. (95–117 Nm).

13. Install the front wheels.

14. Remove the safety stands and lower the vehicle.

15. Check the front end alignment and adjust as necessary.

Upper Control Arm

REMOVAL & INSTALLATION

1. Before servicing the vehicle, refer to the Precautions Section.

2. Raise the front of the vehicle and support it with safety stands.

3. Remove the front wheels.

4. Remove the bolt on the steering knuckle side that secures the upper arm ball link.

5. Remove the brake hose bracket and then remove the upper arm bolts and nuts.

6. Remove the upper control arm from the vehicle.

To install:

7. Raise the front of the vehicle and support it with safety stands.

8. Position the upper arm to the frame

brackets, insert the bolts and hand tighten the nuts.

9. Install the upper arm ball joint into the top of the steering knuckle and tighten the side bolt and nut to 31–39 ft. lbs. (44–55 Nm).

10. Tighten the upper arm bolts and nuts to 54–68 ft. lbs. (76–95 Nm) and then install brake hose brackets.

11. Install the front wheels.

12. Check the front end alignment and adjust as necessary.

CONTROL ARM BUSHING REPLACEMENT

1. Secure the upper control arm in a suitable vise.

2. Using a standard bearing press, remove the old bushing.

3. Install the new bushing and then press it into the upper arm with a standard bearing press.

4. Apply lubricant to the new bushings to facilitate insertion into the upper control arm.

Lower Control Arm

REMOVAL & INSTALLATION

1. Before servicing the vehicle, refer to the Precautions Section.

2. Raise the front of the vehicle and support it with safety stands.

3. Remove the front wheels.

4. Remove the lower nut of the control link of the stabilizer bar.

5. Remove the lower nut of the strut.

6. Remove the bolts and nuts that join the lower arm and lower arm ball joint.

7. Remove the cotter pin and castle nut from the lower arm ball joint.

8. Separate the lower arm ball joint from the steering knuckle.

9. Remove the steering gear mounting bolts and nuts.

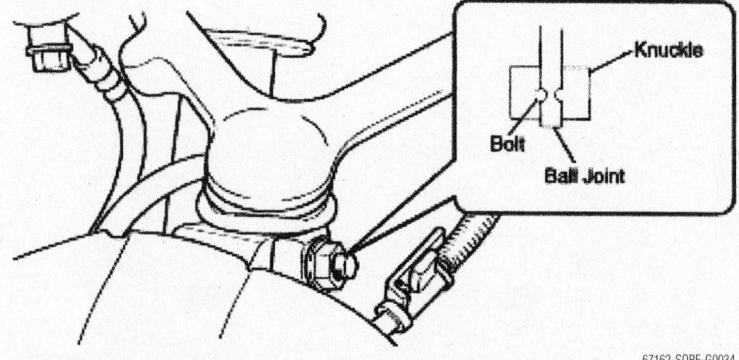

67162-SORE-G0034

Installation of upper ball joint to steering knuckle

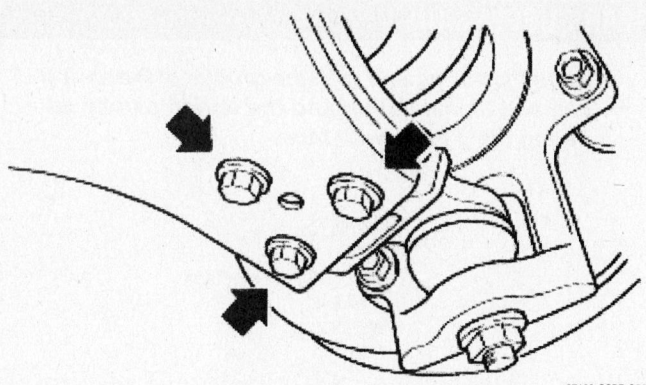

67162-SORE-G36

Lower ball joint-to-control arm fastener locations

10. Remove the spindle from the front frame crossmember brackets during raising the steering gear box by using a suitable bar.

➡**Before loosening the nuts of the spindles, make note of the numerical setting and mark the location on the frame bracket and plate so it can be re-installed to the same setting and location.**

11. Remove the lower control arm.

To install:

12. Install the lower arm ball joint to the steering knuckle. Tighten the ball joint-to-control arm fasteners to 116–145 ft. lbs. (157–196 Nm).

13. Install a new cotter pin through the castle nut.

14. Position the lower arm to the front frame crossmember brackets and then position the spindle up to the steering gear box by using a suitable bar.

15. Install the lower arm spindles and tighten the fasteners to 159–181 ft. lbs. (216–245 Nm).

16. Install the lower nut of the strut assembly and tighten to 88–101 ft. lbs. (122–140 Nm).

17. Install the lower nut of control link of stabilizer bar and tighten the nut to 68–84 ft. lbs. (95–117 Nm).

18. Install the front wheels.

19. Remove the safety stands and lower the vehicle.

20. Check the front end alignment and adjust as necessary.

Wheel Bearings

ADJUSTMENT

The front wheel bearings are not adjustable. Replace the hub/bearing assembly if there are any signs of excessive wear and/or damage.

REMOVAL & INSTALLATION

➡**The front wheel bearings are serviced with the steering knuckle as an assembly.**

1. Before servicing the vehicle, refer to the Precautions Section.

2. Raise and support the vehicle safely. Remove the front wheel.

3. Remove the wheel speed sensor.

4. Remove the brake caliper retaining bolts. Remove the caliper and position it to the side with wire.

5. Remove the brake rotor retaining screws. Remove the brake rotor.

6. Remove the lock nut and plain washer (2WD).

7. Remove the upper arm link lock bolt, spring washer and nut.

8. Remove the tie rod end cotter pin and separate the knuckle from the tie rod end.

9. Remove the ball joint cotter pin, and separate the knuckle from the lower ball joint.

10. Remove the steering knuckle.

11. Using a suitable tool, pry out oil seal from knuckle (4WD).

12. Press the wheel hub from the knuckle (4WD).

13. Press the knuckle and then remove wheel hub (2WD).

To install:

14. Install the dust cover to the knuckle and tighten the bolts to 12–16 ft. lbs. (16–23 Nm).

15. Install new oil seal and then install the wheel hub to the knuckle by pressing.

16. Apply grease to the wheel bearing and seal lip.

17. Put steering knuckle on the drive-shaft end with upper and lower ball joints in mounting holes.

18. Attach lower arm, tighten lock nut to 116–130 (160–180 Nm), and install cotter pin.

19. Attach tie rod end to knuckle, tighten nut to 51–57 ft. lbs. (70–80 Nm), and install cotter pin.

20. Insert upper arm link lock bolt with spring washer and tighten nut to 32–39 ft. lbs. (44–55 Nm).

21. Install the chamfer of plain washer toward the bearing (2WD).

22. Screw lock nut up against wheel hub assembly and using a lock nut wrench, tighten nut to tightening torque to set bearing preload. Use spring scale to measure:

- Bearing preload: 10 inch lbs.
- Tightening torque: 178–198 ft. lbs. (245–275 Nm)

23. Caulk the flange of lock nut on the end of driveshaft.

24. Install the wheel speed sensor harness.

25. Install the brake caliper and rotor.

26. Install the front wheel.

27. Check the alignment and adjust as necessary.

REAR SUSPENSION

Shock Absorber

REMOVAL & INSTALLATION

1. Before servicing the vehicle, refer to the Precautions Section.
2. Raise the rear of the vehicle and support it with safety stands.
3. Remove the rear wheels.
4. Raise the rear axle housing to facilitate removal of the shock absorbers.
5. Remove stabilizer link upper mounting nut.
6. Remove the rear shock absorber lower nut and washer. Remove the shock absorber upper bolt, and then remove the shock absorber.

To install:

7. Install the shock absorber and upper nut. Tighten the nut to 88–101 ft. lbs. (122–140 Nm).
8. Install the shock absorber lower bolt and tighten the bolt to 88–101 ft. lbs. (122–140 Nm).
9. Install the stabilizer link upper mounting nut.
10. Install the rear wheels.
11. Remove the safety stands and lower the vehicle.

Coil Spring

REMOVAL & INSTALLATION

1. Before servicing the vehicle, refer to the Precautions Section.
2. Raise the rear of the vehicle and support it with safety stands.
3. Remove the rear wheels.
4. Raise the rear axle housing to facilitate removal of the shock absorbers.
5. Remove stabilizer link upper mounting nut.
6. Remove the rear shock absorber lower nut and washer. Remove the shock absorber upper bolt, and then remove the shock absorber.
7. Lower the rear axle housing slowly to facilitate removal of the coil spring.
8. Remove the upper rubber seat.

To install:

9. Position the upper rubber seat to the coil spring. Align the spring end with the groove of the spring pad and fix the spring and the spring pad by adhering the 3 parts with tape.
10. Slowly raise the rear axle housing while installing the coil spring.

Align the spring end with the groove of the spring pad and fix the spring and the spring pad by adhering the 3 parts with tape.

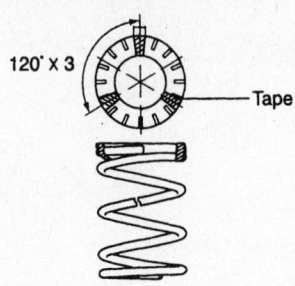

120° X 3 — Tape

09474_SOREN_G0036

Proper placement of coil spring upper rubber seat

11. Install the shock absorber and upper nut. Tighten the nut to 88–101 ft. lbs. (122–140 Nm).
12. Install the shock absorber lower bolt and tighten the bolt to 88–101 ft. lbs. (122–140 Nm).
13. Install the stabilizer link upper mounting nut.
14. Install the rear wheels.
15. Remove the safety stands and lower the vehicle.

Stabilizer Bar

REMOVAL & INSTALLATION

1. Before servicing the vehicle, refer to the Precautions Section.
2. Raise the rear of the vehicle and support it with safety stands.
3. Remove the rear wheels.
4. Support the bottom of the rear differential carrier with a floor jack.
5. Remove the stabilizer link mounting bolt. Remove the stabilizer bar bushing bracket.
6. Remove the stabilizer from the vehicle.

To install:

7. Installation is the reverse of the removal procedure.
8. Align the identification mark (white paint) on the stabilizer bar with the bushing, install the stabilizer bar bushing bracket.
9. Torque the retaining bolts to 13–16 ft. lbs. (19–23 Nm).
10. Install the joint cup. Tighten the lock nut until the thread of the bolt protrudes about 15mm.

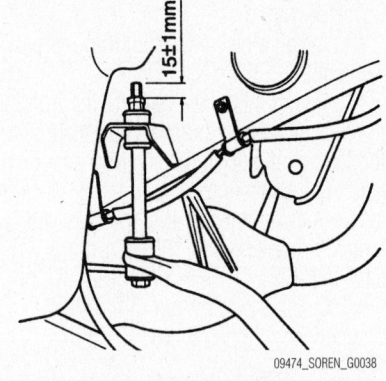

15±1mm

09474_SOREN_G0038

Rear stabilizer locknut tightening specification

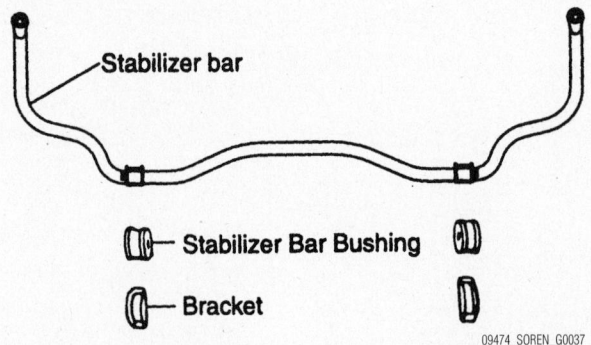

Stabilizer bar

Stabilizer Bar Bushing

Bracket

09474_SOREN_G0037

Rear stabilizer

Upper Control Arm

REMOVAL & INSTALLATION

1. Before servicing the vehicle, refer to the Precautions Section.

2. Raise the rear of the vehicle and support it with safety stands.

3. Remove the rear wheels.

4. Raise the rear axle housing to facilitate removal of the upper arm.

5. Remove shock absorber lower bolt.

6. Loosen the upper arm bolts and remove the upper arm.

7. Inspect the upper arm for bends, cracks and/or other damage. Inspect the upper arm bushings for wear and/or deterioration.

To install:

8. Install the upper arm and the bolts. Tighten the bolts to 88–101 ft. lbs. (122–140 Nm).

9. Install shock absorber lower bolt and tighten to 88–101 ft. lbs. (122–140 Nm).

10. Lower the rear axle housing.

11. Install the rear wheels.

12. Remove the safety stands and lower the vehicle.

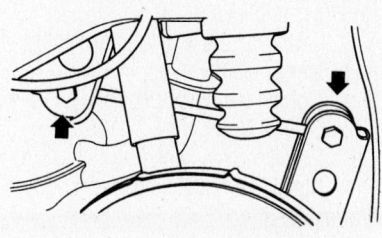

09474_SOREN_G0039

Rear upper control arm bolt locations

CONTROL ARM BUSHING REPLACEMENT

1. Secure the upper control arm in a suitable vise.

2. Using a standard bearing press, remove the old bushing.

3. Install the new bushing and then press it into the upper arm with a standard bearing press.

4. Apply lubricant to the new bushings to facilitate insertion into the upper control arm.

Lower Control Arm

REMOVAL & INSTALLATION

1. Before servicing the vehicle, refer to the Precautions Section.

2. Raise the rear of the vehicle and support it with safety stands.

3. Remove the rear wheels.

4. Raise the rear axle housing to facilitate removal of the lower arm.

5. Remove the shock absorber lower bolt.

6. Remove the wheel speed sensor cable from rear lower arm.

7. Loosen the lower arm bolts and remove the lower arm.

8. Inspect the lower arm for bends, cracks and/or other damage. Inspect the lower arm bushings for wear and/or deterioration.

To install:

9. Install the lower arm and the bolts. Tighten the bolts to 101–116 ft. lbs. (137–157 Nm).

10. Install wheel speed sensor cable to the rear lower arm.

11. Install shock absorber lower bolt. Tighten the bolt to 88–101 ft. lbs. (122–140 Nm).

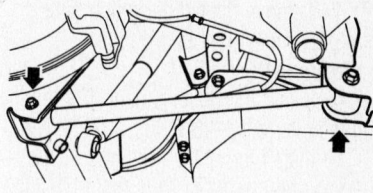

09474_SOREN_G0040

Rear lower control arm bolt locations

12. Lower the rear axle housing.

13. Install the rear wheels.

14. Remove the safety stands and lower the vehicle.

CONTROL ARM BUSHING REPLACEMENT

1. Secure the lower control arm in a suitable vise.

2. Using a standard bearing press, remove the old bushing.

3. Install the new bushing and then press it into the lower arm with a standard bearing press.

4. Apply lubricant to the new bushings to facilitate insertion into the lower control arm.

Wheel Bearings

ADJUSTMENT

The rear bearings are not adjustable. Replace the rear bearings if play exceeds 0.002 inches (0.05 mm).

REMOVAL & INSTALLATION

Refer to Axle Shaft, Bearing and Seal in the DRIVE TRAIN Section of this chapter.

BRAKES

Brake Caliper

REMOVAL AND INSTALLATION

Front

1. Before servicing the vehicle, refer to the Precautions Section.

2. Remove the front wheel.

3. Disconnect the brake fluid hose.

4. Remove the caliper mounting bolts.

5. Remove the brake caliper.

To install:

6. Install the brake caliper. Tighten the mounting bolts to 16–24 ft. lbs. (22–32 Nm).

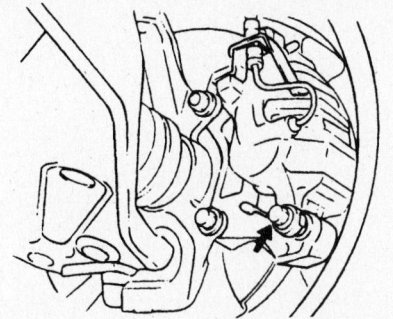

09474_SOREN_G0041

Front brake caliper mounting bolt locations

7. Install the brake fluid hose. Tighten the hose fitting to 12–14 ft. lbs. (17–20 Nm).

8. Bleed the brake system.

9. Install the front wheel.

10. Before attempting to move the vehicle, pump the brake pedal to seat the pads against the rotors. Make sure the vehicle has a firm brake pedal. Check the level of the brake fluid and add DOT 3 or 4 brake fluid if necessary.

Rear

1. Before servicing the vehicle, refer to the Precautions Section.

2. Remove the rear wheel.

3. Disconnect the brake fluid hose.
4. Remove the caliper guide bolts.
5. Remove the brake caliper.

To install:

6. Install the brake caliper. Tighten the guide bolts to 16–23 ft. lbs. (22–32 Nm).

7. Install the brake fluid hose. Tighten the hose fitting to 12–14 ft. lbs. (17–20 Nm).

8. Bleed the brake system.

9. Install the rear wheel.

10. Before attempting to move the vehicle, pump the brake pedal to seat the pads against the rotors. Make sure the vehicle has a firm brake pedal. Check the level of the brake fluid and add DOT 3 or 4 brake fluid if necessary.

Disc Brake Pads

REMOVAL AND INSTALLATION

Front

1. Before servicing the vehicle, refer to the Precautions Section.

2. Remove the front wheel.

3. Remove the guide pin, lift the caliper assembly up and suspend it with a wire.

4. Remove the following parts from the caliper support:
- Pad and wear sensor assembly
- Pad spring
- Outer shim

To install:

5. Compress the caliper piston into the caliper bore.

6. Install the pad clips.

7. Install the inner and outer pads on each pad clip.

1. Pad and wear sensor assembly
2. Pad spring
3. Outer shim

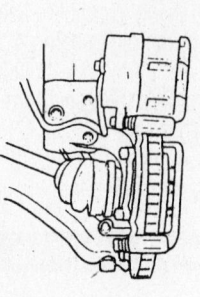 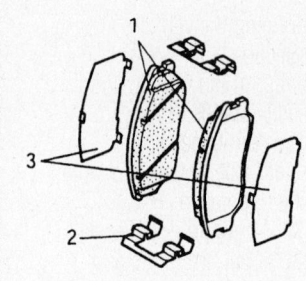

09474_SOREN_G0042

Front disc brake pad components

8. Lower the brake caliper carefully so as not to damage the boot.

9. Tighten the guide pin bolt to 16–24 ft. lbs. (22–32 Nm).

10. Install the front wheel.

Rear

1. Before servicing the vehicle, refer to the Precautions Section.

2. Remove the rear wheel.

3. Remove the guide pin bolts, lift the caliper assembly up and suspend it with a wire.

4. Before replacing the brake pads, drain brake fluid from the master cylinder reservoir until it remains half full.

5. Remove the brake pads by turning the piston in the housing assembly using special tool 09581-11000 to compress the piston.

6. Remove the inner and outer pads from the caliper.

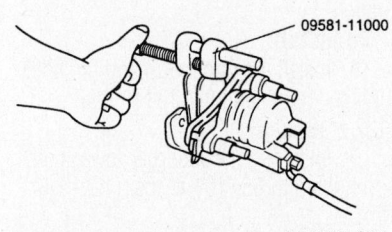

09581-11000

09474_SOREN_G0043

Rear disc brake pad removal

To install:

7. Install the inner and outer brake pads, engaging the clips securely onto the caliper assembly.

8. Lower the brake caliper assembly into proper position.

9. Tighten the guide pin bolts to 16–23 ft. lbs. (22–32 Nm).

10. Install the rear wheel.

11. Check the brake fluid level and top off, if necessary.

KIA

Sportage

21

SPECIFICATIONS AND MAINTENANCE CHARTS

ENGINE AND VEHICLE IDENTIFICATION

Engine							Model Year	
Code ①	Liters (cc)	Cu. In.	Cyl.	Fuel Sys.	Engine Type	Eng. Mfg.	Code ②	Year
2	2.0 (1975)	120.5	4	MFI	DOHC	KIA	2	2002
3 ③	2.0 (1975)	120.5	4	MFI	DOHC	KIA	5	2005
3 ④	2.0 (1998)	122	4	MFI	DOHC	KIA	6	2006
4	2.7 (2656)	162	6	MFI	DOHC	KIA		

MFI: Multi-port Fuel Injection

DOHC: Double Overhead Camshafts

① 8th digit of VIN

② 10th digit of VIN

③ 2005-06 California engine

④ 2002 engine

09474_KSPO_C0001

GENERAL ENGINE SPECIFICATIONS

Year	Model	Engine Displacement Liters	Engine VIN	Net Horsepower @ rpm	Net Torque @ rpm (ft. lbs.)	Bore x Stroke (in.)	Com-pression Ratio	Oil Pressure @ rpm
2002	Sportage	2.0	3	130@5500	127@4000	3.39x3.39	9.2:1	43-57 ①
2005	Sportage	2.0	2, 3	140@6000	136@4500	3.23x3.68	10:01	23 ②
		2.7	4	173@6000	178@4000	3.41x2.95	10:01	7.3 ③
2006	Sportage	2.0	2, 3	140@6000	136@4500	3.23x3.68	10:01	23 ②
		2.7	4	173@6000	178@4000	3.41x2.95	10:01	7.3 ③

MFI: Multi-port Fuel Injection

① The manufacturer does not provide an engine speed specification for oil pump pressure.

② At idle

③ Minimum

09474_KSPO_C0002

ENGINE TUNE-UP SPECIFICATIONS

Year	Engine Displacement Liters	Engine VIN	Spark Plug Gap (in.)	Ignition Timing (deg.) MT	AT	Fuel Pump (psi)	Idle Speed (rpm) MT	AT	Valve Clearance (in.) Intake	Exhaust
2002	2.0	3	0.039-0.043	4B	4B	38	750-850	750-850	HYD.	HYD.
2005	2.0	2, 3	0.039-0.043	NA	NA	50	NA	NA	0.0047-0.0110	0.0079-0.0150
	2.7	4	0.039-0.043	NA	NA	50	NA	NA	HYD.	HYD.
2006	2.0	2, 3	0.039-0.043	NA	NA	50	NA	NA	0.0047-0.0110	0.0079-0.0150
	2.7	4	0.039-0.043	NA	NA	50	NA	NA	HYD.	HYD.

B: Before Top Dead Center

HYD: Hydraulic lash adjusters

NA: Information not available

09474_KSPO_C0003

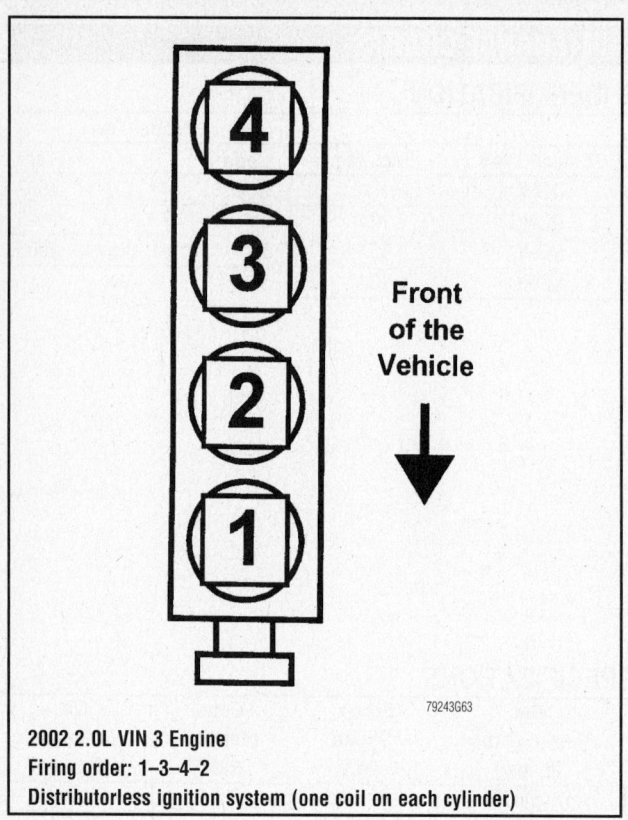

2002 2.0L VIN 3 Engine
Firing order: 1–3–4–2
Distributorless ignition system (one coil on each cylinder)

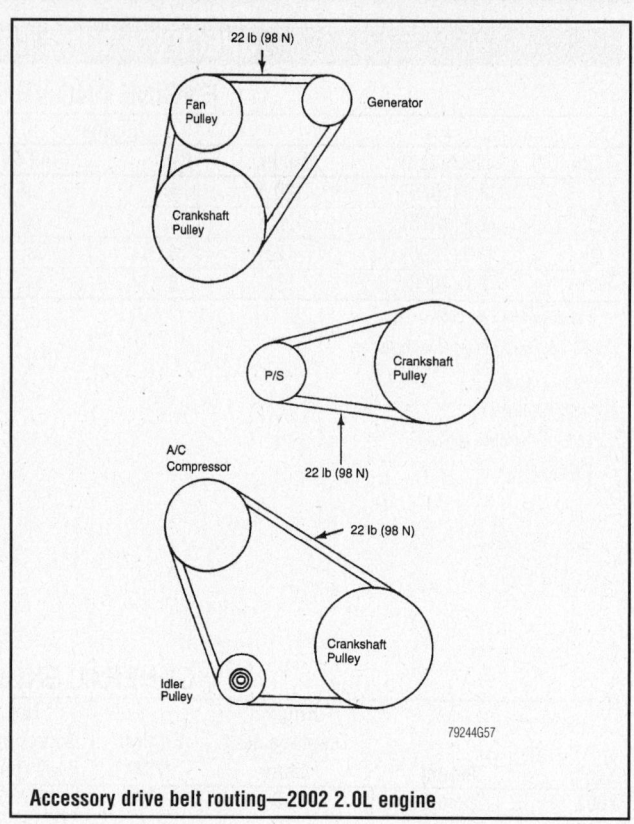

Accessory drive belt routing—2002 2.0L engine

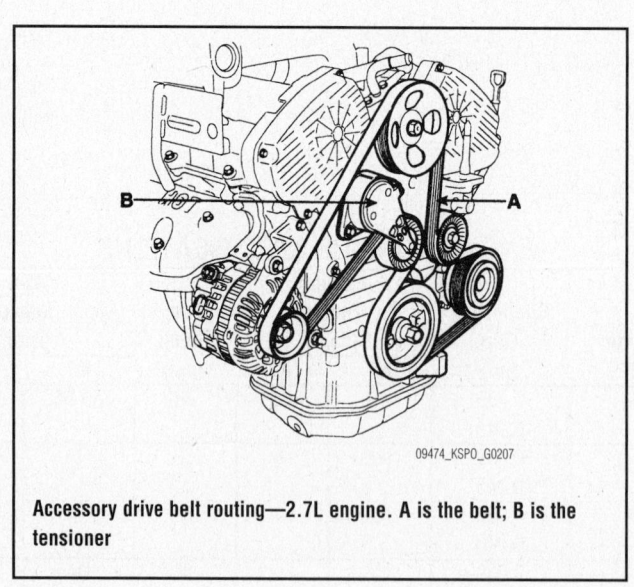

Accessory drive belt routing—2.7L engine. A is the belt; B is the tensioner

CAPACITIES

Year	Model	Engine Displacement Liters	Engine VIN	Engine Oil with Filter (qts.)	Transmission (pts.) Manual	Transmission (pts.) Auto.	Transfer Case (pts.)	Drive Axle Front (pts.)	Drive Axle Rear (pts.)	Fuel Tank (gal.)	Cooling System (qts.)
2002	Sportage	2.0	3	4.4	2.6	5.4	2.8	2.6	2.6	15.8	7.9
2005	Sportage	2.0	2, 3	4.2	4.6	16.4	1.6	①	1.8	15.3	6.4
		2.7	4	4.75	4.6	16.4	1.7	①	1.8	17.2	8.2
2006	Sportage	2.0	2, 3	4.2	4.6	16.4	1.6	①	1.8	15.3	6.4
		2.7	4	4.75	4.6	16.4	1.7	①	1.8	17.2	8.2

NOTE: All capacities are approximate. Add fluid gradually and check to be sure a proper fluid level is obtained.

① Included in transaxle capacity

09474_KSPO_C0004

VALVE SPECIFICATIONS

Year	Engine Displacement Liters	Engine VIN	Seat Angle (deg.)	Face Angle (deg.)	Spring Test Pressure (lbs. @ in.)	Spring Installed Height (in.)	Stem-to-Guide Clearance (in.) Intake	Stem-to-Guide Clearance (in.) Exhaust	Stem Diameter (in.) Intake	Stem Diameter (in.) Exhaust
2002	2.0	3	45	45	①	①	0.0010-0.0024	0.0012-0.0026	0.2350-0.2356	0.2348-0.2354
2005	2.0	2, 3	45-45.5	45-45.5	②	②	0.0008-0.0020	0.0014-0.0026	0.2348-0.2354	0.2343-0.2348
	2.7	4	45-45.5	45-45.5	48.4 @ 1.3780	1.3780	0.0008-0.0020	0.0012-0.0026	0.2350-0.2354	0.2340-0.2350
2006	2.0	2, 3	45-45.5	45-45.5	②	②	0.0008-0.0020	0.0014-0.0026	0.2348-0.2354	0.2343-0.2348
	2.7	4	45-45.5	45-45.5	48.4 @ 1.3780	1.3780	0.0008-0.0020	0.0012-0.0026	0.2350-0.2354	0.2340-0.2350

① Spring test pressure or installed height not provided by the manufacturer.
  Valve Spring Free Length:
  Outer spring: 1.524-1.539 in.
  Inner spring: 1.484-1.496 in.

② Intake: 41.4 lbs. @ 1.5354 in.
  Exhaust: 90.4 lbs. @ 1.2008 in.
  Lengths given are installed height.

09474_KSPO_C0005

CAMSHAFT SPECIFICATIONS CHART

All measurements are given in inches.

Year	Engine Displ. Liters	Engine VIN	Journal Dia.	Brg. Oil Clearance	Shaft End-play	Runout	Lobe Height Intake	Lobe Height Exhaust
2002	2.0	3	1.1787-1.1797	0.0014-0.0033	0.0030-0.0040	0.0012	1.7658-1.7377	1.7658-1.7377
2005	2.0	2, 3	1.1009-1.1016	0.0008-0.0024	0.0039-0.0079	0.0012	1.7527-1.7605	1.7487-1.7566
	2.7	4	1.0222-1.0226	0.0007-0.0024	0.0039-0.0059	0.0012	1.7303-1.7382	1.7303-1.7382
2006	2.0	2, 3	1.1009-1.1016	0.0008-0.0024	0.0039-0.0079	0.0012	1.7527-1.7605	1.7487-1.7566
	2.7	4	1.0222-1.0226	0.0007-0.0024	0.0039-0.0059	0.0012	1.7303-1.7382	1.7303-1.7382

09474_KSPO_C0015

CRANKSHAFT AND CONNECTING ROD SPECIFICATIONS

All measurements are given in inches.

Year	Engine Displacement Liters	Engine VIN	Crankshaft				Connecting Rod		
			Main Brg. Journal Dia.	Main Brg. Oil Clearance	Shaft End-play	Thrust on No.	Journal Diameter	Oil Clearance	Side Clearance
2002	2.0	3	2.3597-2.3604	①	0.0031-0.0071	3	2.0055-2.0061	0.0009-0.0021	0.0040-0.0103
2005	2.0	2, 3	2.2418-2.2426	0.0011-0.0019	0.0024-0.0102	3	1.8898-1.8905	0.0009-0.0017	0.0039-0.0098
	2.7	4	2.4402-2.4409	0.0002-0.0009	0.0028-0.0098	3	1.8891-1.8898	0.0007-0.0014	0.0039-0.0098
2006	2.0	2, 3	2.2418-2.2426	0.0011-0.0019	0.0024-0.0102	3	1.8898-1.8905	0.0009-0.0017	0.0039-0.0098
	2.7	4	2.4402-2.4409	0.0002-0.0009	0.0028-0.0098	3	1.8891-1.8898	0.0007-0.0014	0.0039-0.0098

① Journals 1, 2 and 4: 0.0010 - 0.0017 in.
  Journal 3: 0.0012 - 0.0019 in.

09474_KSPO_C0006

PISTON AND RING SPECIFICATIONS

All measurements are given in inches.

Year	Engine Displacement Liters	Engine VIN	Piston Clearance	Ring Gap			Ring Side Clearance		
				Top Compression	Bottom Compression	Oil Control	Top Compression	Bottom Compression	Oil Control
2002	2.0	3	0.0019-0.0024	0.006-0.012	0.008-0.014	0.008-0.028	0.001-0.003	0.001-0.003	SNUG
2005	2.0	2, 3	0.0008-0.0016	0.0091-0.0150	0.0130-0.0189	0.0024-0.0059	0.0016-0.0031	0.0012-0.0028	0.0024-0.0059
	2.7	4	0.0004-0.0012	0.0079-0.0138	0.0146-0.0205	0.0079-0.0276	0.0016-0.0031	0.0012-0.0028	SNUG
2006	2.0	2, 3	0.0008-0.0016	0.0091-0.0150	0.0130-0.0189	0.0024-0.0059	0.0016-0.0031	0.0012-0.0028	0.0024-0.0059
	2.7	4	0.0004-0.0012	0.0079-0.0138	0.0146-0.0205	0.0079-0.0276	0.0016-0.0031	0.0012-0.0028	SNUG

09474_KSPO_C0007

TORQUE SPECIFICATIONS
All readings in ft. lbs.

Year	Engine Displacement Liters	Engine VIN	Cylinder Head Bolts	Main Bearing Bolts	Rod Bearing Bolts	Crankshaft Damper Bolts	Flywheel Bolts	Manifold		Spark Plugs	Oil Pan Drain Plug
								Intake	Exhaust		
2002	2.0	3	62	63	50	11	73	16	31	11-17	26
2005	2.0	2, 3	①	②	36-38	123-130	87-94	16	16	15-22	30
	2.7	4	③	④	⑤	130-138	53-56	15	22-26	15-22	30
2006	2.0	2, 3	①	②	36-38	123-130	87-94	16	16	15-22	30
	2.7	4	③	④	⑤	130-138	53-56	15	22-26	15-22	30

① 10x99 bolts:
  Step 1: 16.6-19.5 ft. lbs.
  Step 2: + 60-65 degrees
  Step 3: +60-65 degrees
  12x151 bolts:
  Step 1: 20-23 ft. lbs.
  Step 2: + 60-65 degrees
  Step 3: +60-65 degrees

② Step 1: 20-23 ft. lbs.
  Step 2: + 60-65 degrees

③ Step 1: 18 ft. lbs.
  Step 2: +58-62 degrees
  Step 3: +43-47 degrees

④ M10 bolts:
  Step 1: 20-24 ft. lbs
  Step 2: +90-94 degrees
  M8 bolts:
  Step 1: 10-14 ft. lbs
  Step 2: +90-94 degrees

⑤ Step 1: 12-15 ft. lbs.
  Step 2: +90-94 degrees

09474_KSPO_C0008

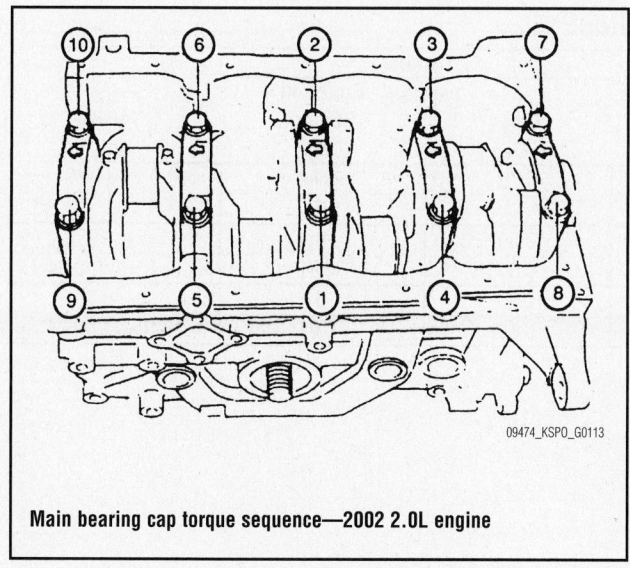

09474_KSPO_G0113

Main bearing cap torque sequence—2002 2.0L engine

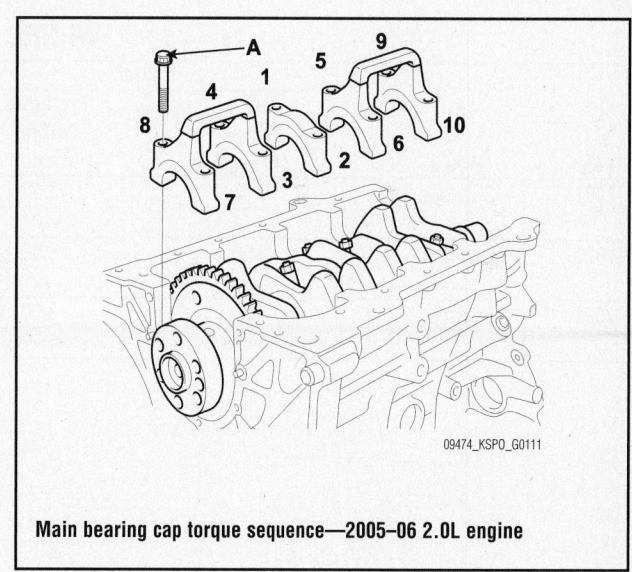

09474_KSPO_G0111

Main bearing cap torque sequence—2005–06 2.0L engine

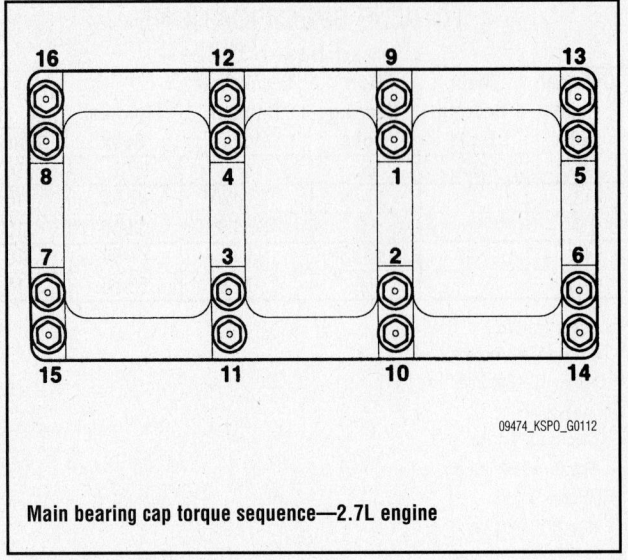

09474_KSPO_G0112

Main bearing cap torque sequence—2.7L engine

WHEEL ALIGNMENT

Year	Model		Caster Range (+/-Deg.)	Caster Preferred Setting (Deg.)	Camber Range (+/-Deg.)	Camber Preferred Setting (Deg.)	Toe-in (in.)
2002	Sportage	F	0.75	+3.58	0.75	+0.44	0.10+/-0.10
		R	—	—	—	0	0
2005	Sportage	F	0.50	+3.60	0.50	0	0+/-0.79
		R	—	—	0.50	-0.91	0.18+0.12-0.04
2006	Sportage	F	0.50	+3.60	0.50	0	0+/-0.79
		R	—	—	0.50	-0.91	0.18+0.12-0.04

09474_KSPO_C0009

TIRE, WHEEL AND BALL JOINT SPECIFICATIONS

| Year | Model | OEM Tires | | Tire Pressures (psi) | | Wheel Size | Ball Joint Inspection | Lug Nut Torque (ft. lbs.) |
		Standard	Optional	Front	Rear			
2002	Sportage	P205/75R15	none	26	26	6-JJ	①	73
2005	Sportage	P215/65R16	P235/60R16	30	30	6.5	②	67-80
2006	Sportage	P215/65R16	P235/60R16	30	30	6.5	②	67-80

OEM: Original Equipment Manufacturer

PSI: Pounds Per Square Inch

① Replace if any measurable movement is found.

② Starting torque: 0.36-1.78 ft. lbs.

09474_KSPO_C0010

BRAKE SPECIFICATIONS
All measurements in inches unless noted

| Year | Model | | Brake Disc | | | Brake Drum | | | Minimum Lining Thickness | Caliper Mounting Bolts (ft. lbs.) | Adapter Plate Bolts (ft. lbs.) |
			Original Thickness	Minimum Thickness	Maximum Run-out	Original Inside Diameter	Max. Wear Limit	Maximum Machine Diameter			
2002	Sportage		0.940	0.880	0.004	NA	9.89	NA	0.080	13	72
2005	Sportage	F	1.020	0.961	0.001	—	—	—	0.079	20-23	58-72
		R	0.390	0.315	0.001	9.000	NA	9.079	①	20-23	36-43
2006	Sportage	F	1.020	0.961	0.001	—	—	—	0.079	20-23	58-72
		R	0.390	0.315	0.001	9.000	NA	9.079	①	20-23	36-43

NA: Not Available

① Rear disc brakes: 0.079 in.; rear drum brakes: 0.039 in.

09474_KSPO_C0011

SCHEDULED MAINTENANCE INTERVALS
2002 Kia Sportage

TO BE SERVICED	TYPE OF SERVICE	7.5	15	22.5	30	37.5	45	52.5	60	67.5	75	82.5	90	97.5	105	112.5	120
Accessory drive belt	S/I				✓				✓				✓				✓
Air cleaner filter	R				✓				✓				✓				✓
Automatic transmission	R				✓		✓		✓		✓		✓		✓		✓
Ball joints	S/I				✓				✓				✓				✓
Brake lines & connections	S/I				✓				✓				✓				✓
Chassis/body fasteners	S/I				✓				✓				✓				✓
Cooling system	S/I				✓				✓				✓				✓
CV-joint boots	S/I		✓		✓		✓		✓		✓		✓		✓		✓
Disc brakes	S/I		✓		✓		✓		✓		✓		✓		✓		✓
Driveshaft U-joints	L		✓		✓		✓		✓		✓		✓		✓		✓
Drum brakes	S/I				✓				✓				✓				✓
Emission hoses & tubes	S/I								✓								
Emission hoses & tubes	R															✓	
Engine coolant	R				✓				✓				✓				✓
Engine oil & filter	R	✓	✓	✓	✓	✓	✓	✓	✓	✓	✓	✓	✓	✓	✓	✓	✓
Exhaust system heat	S/I				✓				✓				✓				✓
Front differential fluid	R				✓				✓				✓				✓
	S/I	✓	✓	✓	✓	✓	✓	✓	✓	✓	✓	✓	✓	✓	✓	✓	✓
Fuel filter	R				✓				✓				✓				✓
Fuel lines & hoses	S/I				✓				✓				✓				✓
Idle speed	S/I				✓				✓				✓				✓
Locks & hinges	L	✓	✓	✓	✓	✓	✓	✓	✓	✓	✓	✓	✓	✓	✓	✓	✓
Manual transmission fluid	R				✓				✓				✓				✓
PCV valve	S/I								✓								✓
Rear differential fluid	R				✓				✓				✓				✓
	S/I	✓	✓	✓	✓	✓	✓	✓	✓	✓	✓	✓	✓	✓	✓	✓	✓
Spark plug wires	S/I								✓								✓
Spark plugs	R				✓				✓				✓				✓
Steering operation &	S/I				✓				✓				✓				✓
Timing belt	R														✓		
	S/I								✓				✓				
Timing belt (non-California)	R								✓								✓
Transfer case fluid	R				✓				✓				✓				✓
Transfer case fluid	S/I		✓		✓		✓		✓		✓		✓		✓		
Transmission fluid	S/I	✓	✓	✓	✓	✓	✓	✓	✓	✓	✓	✓	✓	✓	✓	✓	✓

R: Replace   S/I: Inspect and service, if needed   L: Lubricate

FREQUENT OPERATION MAINTENANCE (SEVERE SERVICE)

If a vehicle is operated under any of the following conditions it is considered severe service:

- Towing a trailer or using a camper or car-top carrier.
- Repeated short trips of less than 5 miles in temperatures below freezing, or trips of less than 10 miles in any temperature.
- Extensive idling or low-speed driving for long distances as in heavy commercial use, such as delivery, taxi or police cars.
- Operating on rough, muddy or salt-covered roads.
- Operating on unpaved or dusty roads.
- Driving in extremely hot (over 90°) conditions.

Engine oil & filter: replace every 5000 miles or 5 months, whichever occurs first.

Air cleaner filter: inspect and replace if necessary, every 15,000 miles or 15 months, whichever occurs first.

Transfer case fluid: inspect the level every 5000 miles or 5 months, and replace every 15,000 miles or 15 months, whichever occurs first.

Transmission fluid: inspect the level every 5000 miles or 5 months, and replace every 15,000 miles or 15 months, whichever occurs first.

Front and rear differential fluid: inspect the level every 5000 miles or 5 months, and replace every 15,000 miles or 15 months, whichever occurs first.

09474_KSPO_C0012

SCHEDULED MAINTENANCE INTERVALS
2005-06 Kia Sportage Normal Maintenace

TO BE SERVICED	TYPE OF SERVICE	VEHICLE MILEAGE INTERVAL (x1000)													
		7.5	15	22.5	30	37.5	45	52.5	60	67.5	75	82.5	90	97.5	105
Accessory drive belt 2.0L ①	I	✓	✓	✓	✓	✓	✓	✓	✓	✓	✓	✓	✓	✓	✓
Accessory drive belt 2.7L ①	I				✓				✓				✓		
Air cleaner filter	R			✓			✓			✓			✓		
A/C filter	R	Every 10,000 miles													
Automatic transmission fluid	R		✓		✓		✓		✓		✓		✓		✓
Ball joints	S/I				✓				✓				✓		
Battery condition	S/I		✓		✓		✓		✓		✓		✓		✓
Brake/Clutch fluid	I		✓		✓		✓		✓		✓		✓		✓
Brake lines & connections	S/I		✓		✓		✓		✓		✓		✓		✓
Chassis/body fasteners	S/I				✓				✓				✓		
Disc brakes	S/I	✓	✓	✓	✓	✓	✓	✓	✓	✓	✓	✓	✓	✓	✓
Driveshaft	S/I	Clean shaft and retorque bolts every 15,000 miles													
Driveshaft U-joints	L		✓		✓		✓		✓		✓		✓		✓
Drum brakes	S/I	✓	✓	✓	✓	✓	✓	✓	✓	✓	✓	✓	✓	✓	✓
Engine coolant	R	Replace at 60,000 miles, then every 30,000 mile thereafter													
Engine oil & filter	R	Every 7,500 miles or 12 months													
EVAP canister filter	I				✓				✓				✓		
Exhaust system heat shields	S/I		✓		✓		✓		✓		✓		✓		✓
Fuel filter	R					✓					✓				
Fuel lines & hoses	S/I	✓	✓	✓	✓	✓	✓	✓	✓	✓	✓	✓	✓	✓	✓
Halfshafts and boots	S/I		✓		✓		✓		✓		✓		✓		✓
Locks & hinges	L	✓	✓	✓	✓	✓	✓	✓	✓	✓	✓	✓	✓	✓	✓
Manual transmission fluid	I		✓		✓		✓		✓		✓		✓		✓
Power steering fluid	I	✓	✓	✓	✓	✓	✓	✓	✓	✓	✓	✓	✓	✓	✓
Rear differential fluid ②	S/I	Check every 25,000 miles													
Spark plugs 2.0L platinum	R								✓						
Spark plugs 2.7L iridium	R	Every 100,000 miles or 10 years													
Steering operation & linkage	S/I	✓	✓	✓	✓	✓	✓	✓	✓	✓	✓	✓	✓	✓	✓
Throttle body	I	✓	✓	✓	✓	✓	✓	✓	✓	✓	✓	✓	✓	✓	✓
Timing belt	S/I				✓				✓				✓		
Tires	Rotate	✓	✓	✓	✓	✓	✓	✓	✓	✓	✓	✓	✓	✓	✓
Transfer case fluid ②	S/I	Check every 25,000 miles; replace every 60,000 miles													
Vacuum and crankcase hoses	S/I				✓				✓				✓		
Valve clearance	I	Every 60,000 miles or 4 years													
Water pump	I	Inspect when replacing drive belt													

R: Replace   S/I: Inspect and service, if needed   L: Lubricate

① Replace when excessive cracks occur or tension is not maintained

② Replace fluid whenever submerged in water

SCHEDULED MAINTENANCE INTERVALS
20005-06 Kia Sportage Severe Service

TO BE SERVICED	TYPE OF SERVICE	VEHICLE MILEAGE INTERVAL (x1000)													
		5	10	15	20	25	30	35	40	45	50	55	60	65	70
Accessory drive belt 2.0L ①	S/I	✓	✓	✓	✓	✓	✓	✓	✓	✓	✓	✓	✓	✓	✓
Accessory drive belt 2.7L ①	S/I						✓						✓		
A/C filter	R		✓		✓		✓		✓		✓		✓		✓
Air cleaner filter	R		✓		✓		✓		✓		✓		✓		✓
Automatic transmission fluid	I						✓						✓		
Automatic transmission fluid	R						✓						✓		
Ball joints	S/I				✓				✓				✓		
Battery condition	S/I		✓		✓		✓		✓		✓		✓		✓
Brake/Clutch fluid	I	✓	✓	✓	✓	✓	✓	✓	✓	✓	✓	✓	✓	✓	✓
Brake lines & connections	S/I	✓	✓	✓	✓	✓	✓	✓	✓	✓	✓	✓	✓	✓	✓
Chassis/body fasteners	S/I				✓				✓				✓		
Disc brakes	S/I	✓	✓	✓	✓	✓	✓	✓	✓	✓	✓	✓	✓	✓	✓
Driveshaft	SI	Clean driveshaft and torque bolts every 15,000 miles													
Driveshaft U-joints	L		✓		✓				✓				✓		✓
Drum brakes	S/I	✓	✓	✓	✓	✓	✓	✓	✓	✓	✓	✓	✓	✓	✓
Emission hoses & tubes	S/I	✓	✓	✓	✓	✓	✓	✓	✓	✓	✓	✓	✓	✓	✓
Engine coolant	R	Replace at 60,000 miles, then every 30,000 mile thereafter													
Engine oil & filter	R	Every 3,000 miles or 3 months													
EVAP canister filter	I						✓								
Exhaust system	S/I		✓		✓		✓		✓		✓		✓		✓
Fuel filter	R							✓							
Fuel lines & hoses	S/I	✓	✓	✓	✓	✓	✓	✓	✓	✓	✓	✓	✓	✓	✓
Halfshafts and boots	S/I	Every 7,500 miles													
Locks & hinges	L	✓	✓	✓	✓	✓	✓	✓	✓	✓	✓	✓	✓	✓	✓
Manual transmission fluid	I		✓		✓		✓		✓		✓		✓		✓
Manual transmission fluid	R												✓		
Parking brake	S/I	✓	✓	✓	✓	✓	✓	✓	✓	✓	✓	✓	✓	✓	✓
Power steering fluid	I	✓	✓	✓	✓	✓	✓	✓	✓	✓	✓	✓	✓	✓	✓
Rear differential fluid ②	R										✓				
Spark plugs	S/I	Inspect frequently; replace as necessary													
Steering operation & linkage	S/I	✓	✓	✓	✓	✓	✓	✓	✓	✓	✓	✓	✓	✓	✓
Throttle body	I	✓	✓	✓	✓	✓	✓	✓	✓	✓	✓	✓	✓	✓	✓
Timing belt	R								✓						
	S/I				✓			✓				✓			
Tires	Rotate	✓	✓	✓	✓	✓	✓	✓	✓	✓	✓	✓	✓	✓	✓
Transfer case fluid ②	R					✓					✓				
Valve clearance 2.0L	S/I	Every 60,000 miles or 48 months													
Water pump	I	Inspect when replacing drive belt													

R: Replace   S/I: Inspect and service, if needed   L: Lubricate

① Replace when excessive cracks occur or tesion is not maintained

② Replace fluid whenever submerged in water

09474_KSPO_C0014

ENGINE REPAIR

Alternator

REMOVAL

2002 Models

1. Before servicing the vehicle, refer to the Precautions Section.
2. Remove or disconnect the following:
   - Negative battery cable
   - Air cleaner inlet pipe front bolts
   - Top hose from the resonance chamber
   - Air cleaner inlet pipe
   - Alternator electrical connectors
   - Loosen the pivot and tensioner mounting bolts, do mot remove them
   - Drive belt from the alternator pulley
   - Drive belt tensioner
   - Alternator pivot bolt
   - Loosen the bolt at the base of the adjusting bracket and rotate the bracket up
   - Alternator

To install:

3. Install or connect the following:
   - Alternator
   - Pivot bolt and hand tighten
   - Rotate the bracket down on top of the alternator
   - Belt tensioner on the adjustment bracket
   - Tensioner mounting bolt and hand tighten it
   - Drive belt
   - Torque the tensioner bolt to 19 ft. lbs. (26 Nm).

- Torque the pivot bolt to 38 ft. lbs. (51 Nm).
- Alternator electrical connectors
- Air cleaner inlet pipe and tighten the clamp
- Hose to the resonance chamber
- Air inlet pipe bolts and tighten
- Negative battery cable

2005–06 Models

2.0L ENGINE

1. Before servicing the vehicle, refer to the Precautions Section.
2. Disconnect the battery negative terminal first, then the positive terminal.
3. Disconnect the alternator connector and "B" terminal cable from the alternator.
4. Remove the adjusting bolt (A) and mounting bolt (B), then remove the alternator belt.
5. Pull out the through bolt (C), then remove the alternator (D).

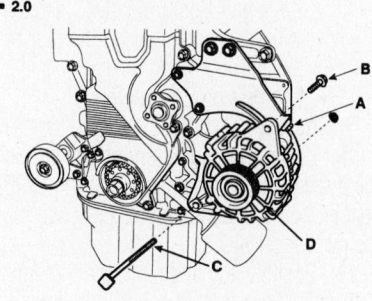

09474_KSPO_G0025

Alternator mounting—2005–06 2.0L engine

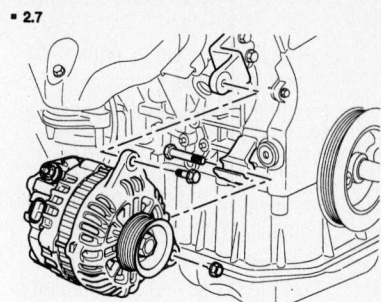

09474_KSPO_G0026

Alternator mounting—2005–06 2.7L engine

6. Installation is the reverse of removal. Adjust the alternator belt tension after installation

2.7L ENGINE

1. Before servicing the vehicle, refer to the Precautions Section.
2. Disconnect the battery negative terminal first, then the positive terminal.
3. Disconnect the alternator connector and "B" terminal cable from the alternator.
4. Remove the adjusting bolt and mounting bolt, then remove the alternator belt.
5. Pull out the through bolt, then remove the alternator.
6. Installation is the reverse of removal.

Ignition Timing

ADJUSTMENT

The engines in the Sportage are equipped with a distributorless ignition system. The ignition timing cannot be adjusted.

Engine Assembly

REMOVAL & INSTALLATION

2002 Models

1. Before servicing the vehicle, refer to the Precautions Section.
2. Properly relieve the fuel system pressure.
3. Drain the cooling system.
4. Drain the engine oil.
5. Drain the transmission fluid.
6. Remove or disconnect the following:
   - Both battery cables
   - Windshield washer hose from the hood
   - Hood

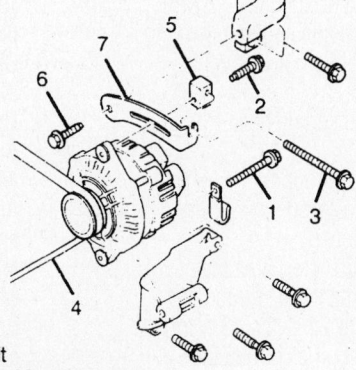

1. Pivot Bolt
2. Tensioner Mounting Bolt
3. Tensioner Adjusting Bolt
4. Drive Belt
5. Belt Tensioner
6. Bracket Bolt
7. Adjusting Bracket

09474_KSPO_G0001

Alternator mounting—2002 models

- 2 air duct mounting bolts from the top of the radiator. Loosen the clamp at the air intake housing and remove the duct
- Accelerator cable by pulling the throttle back and rotating the cable until it aligns with the slot in the pulley
- Transmission control cable
- Resonance chamber mounting bolt, chamber bolt and air silencer
- Idle Air Control (IAC) air hose, breather hose and vacuum line from the air intake tube
- Manifold Air Flow (MAF) sensor connector
- Loosen the air inlet hose clamp from the MAF sensor
- 3 bolts from the air intake tube to the throttle body
- Air intake hose and tube as an assembly
- Upper radiator hose
- Clutch fan nuts
- Cooling fan shroud bolts
- Fan and shroud at the same time
- Alternator drive belt
- Fan pulley
- Alternator electrical connectors
- Exhaust Gas Recirculation (EGR) solenoid valve connector on the intake manifold in front of the dynamic chamber
- Both heater hoses from the pipes
- Engine-to-body ground wire from the intake manifold and the harness bracket
- Brake booster vacuum hose from the dynamic chamber
- Fuel lines and fuel pressure regulator from the rear of the dynamic chamber
- Vacuum hose from the bottom of the EGR valve
- Purge solenoid valve vacuum hose from dynamic chamber
- Vacuum hoses from the top of the charcoal canister and slide the charcoal canister up and out of the bracket
- Lower radiator hose
- Cooling lines from the radiator, if equipped with an automatic transmission
- Radiator and raise and safely support the vehicle
- Lower splash panel
- Drive belt
- A/C pulley assembly
- A/C compressor mounting bolts and move the A/C compressor out of the way

- Power steering drive belt
- Intake manifold support bracket
- Starter wiring harness
- Starter
- Front exhaust pipe from the exhaust manifold
- Bracket bolt from the front exhaust pipe
- Exhaust-to-clutch (manual transmission) or converter (automatic transmission) housing bolts and the bracket
- Clutch housing (manual transmission) or converter (automatic transmission) housing bolts.
- Drive plate-to-torque converter bolts, if equipped

7. Support the transmission from underneath the vehicle.

8. Connect the engine hoist to the engine assembly.
- Left and right side engine mounting bolts

9. Lift the engine up and forward slightly to provide access to three electrical connectors on the rear of the cylinder head.
- Electrical connectors from the Camshaft Position (CMP) sensor, coil and condenser on the rear of the cylinder head
- Engine from the vehicle

To install:

10. Lower the engine enough to connect the three electrical connectors to the CMP sensor, coil and condenser on the rear of the cylinder head.

11. Position the engine to the transmission. Install the transmission bolts and tighten the bolts. Torque according to bolt size:
- 14mm bolts to 80 ft. lbs. (108 Nm)
- 10mm bolts to 28 ft. lbs. (38 Nm)
- 6mm bolts to 60 inch lbs. (7 Nm)
- Right and left side engine mounting bolts. Torque the bolt s to 38 ft. lbs. (52 Nm).

12. Disconnect the engine hoist from the engine assembly.

13. Raise and safely support the vehicle.
- Drive plate-to-torque converter bolts, if equipped.
- Connect the front exhaust pipe to the exhaust manifold. Torque the flange bolts to 24 ft. lbs. (31 Nm).
- Front exhaust pipe. Torque the bolts to 20 ft. lbs. (27 Nm).
- Starter
- Connect the starter wiring harness
- Intake manifold support bracket bolts and the bracket.

14. Install the power steering pump lock and mounting bolts. Install the power steer-

ing drive belt. Torque the bolts to 30 ft. lbs. (42 Nm).
- A/C compressor mounting bolts. Torque the bolts to 18 ft. lbs. (24 Nm).
- A/C belt pulley assembly and drive belt. Install the two A/C idler pulley bracket bolts and torque to 24 ft. lbs. (32 Nm).
- Lower splash panel
- Radiator
- Cooling lines to the radiator, if equipped
- Lower radiator hose
- Slide the charcoal canister in the bracket
- Vacuum hoses to the top of the charcoal canister
- Engine-to-body ground wire to the intake manifold and the harness bracket
- Brake booster vacuum hose to the dynamic chamber
- Fuel lines and fuel pressure regulator to the rear of the dynamic chamber
- Vacuum hose to the bottom of the EGR valve
- Purge solenoid valve vacuum hose to dynamic chamber
- EGR solenoid valve connector on the intake manifold in front of the dynamic chamber
- Heater hoses
- Electrical terminal connectors to the alternator
- Fan pulley
- Alternator drive belt. Torque the adjusting bolt 16 ft. lbs. (22 Nm) and the mounting bolt to 32 ft. lbs. (45 Nm).
- Fan and shroud as an assembly. Torque the five cooling fan shroud bolts to 72 inch lbs. (8 Nm)
- Clutch fan nuts. Torque the nuts to 27 ft. lbs. (37 Nm).
- Upper radiator hose
- Air intake hose and tube as an assembly
- Air intake tube to the throttle body and tighten the air inlet hose clamp to the MAF sensor
- MAF sensor electrical connector
- Resonance chamber mounting bolt, chamber bolt and air silencer
- IAC air hose, breather hose and vacuum line to the air intake tube
- Accelerator cable by pulling the throttle back and rotating the cable until it aligns with the slot in the pulley
- Transmission control cable

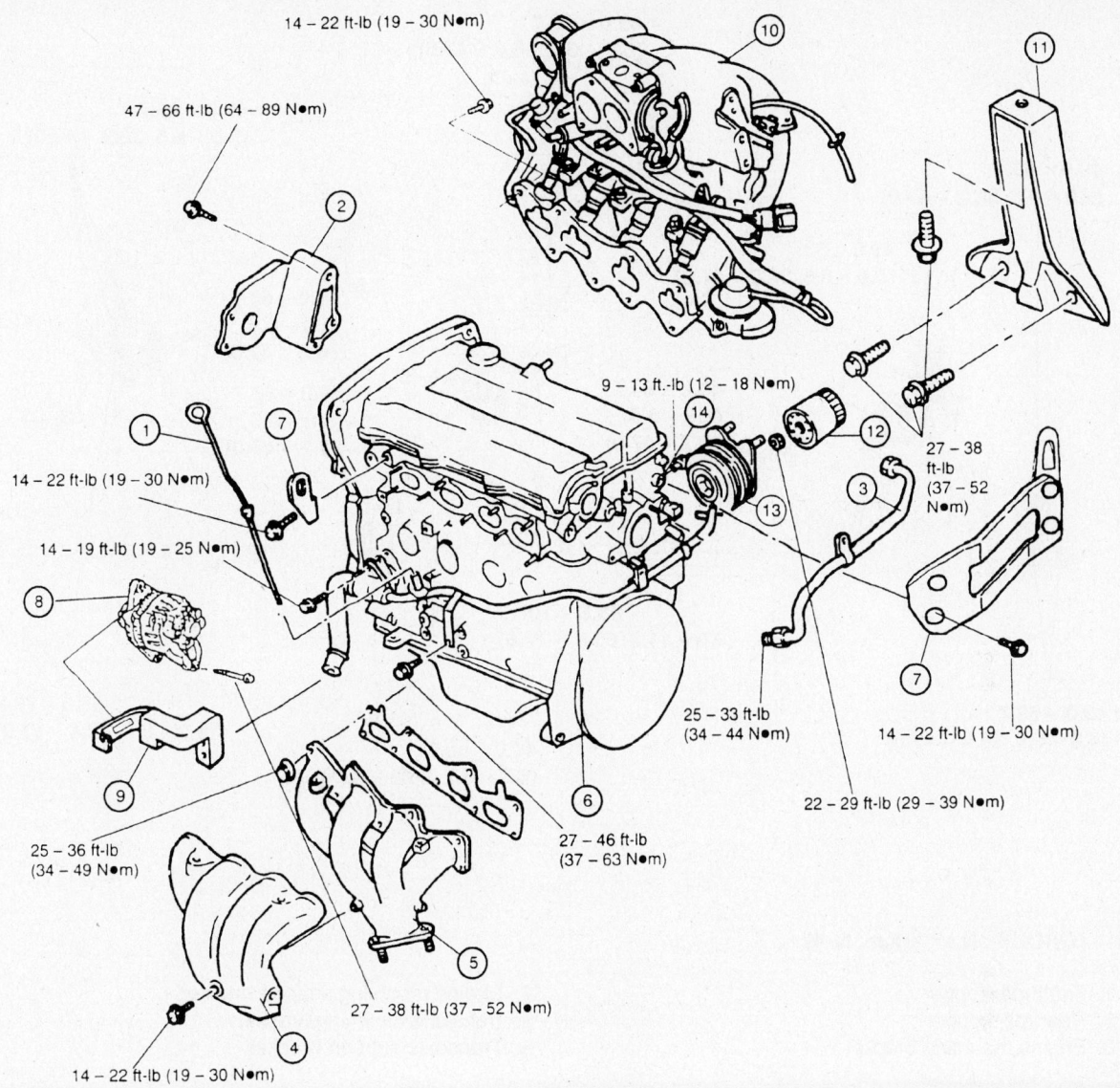

14 – 22 ft-lb (19 – 30 N•m)

47 – 66 ft-lb (64 – 89 N•m)

14 – 22 ft-lb (19 – 30 N•m)

14 – 19 ft-lb (19 – 25 N•m)

9 – 13 ft.-lb (12 – 18 N•m)

27 – 38 ft-lb (37 – 52 N•m)

25 – 33 ft-lb (34 – 44 N•m)

14 – 22 ft-lb (19 – 30 N•m)

22 – 29 ft-lb (29 – 39 N•m)

27 – 46 ft-lb (37 – 63 N•m)

25 – 36 ft-lb (34 – 49 N•m)

27 – 38 ft-lb (37 – 52 N•m)

14 – 22 ft-lb (19 – 30 N•m)

1. Oil Level Gauge
2. Thermo-Modulated Fan Bracket
3. EGR Pipe
4. Exhaust Manifold Heat Shield
5. Exhaust Manifold
6. Coolant Inlet Pipe and Bypass Pipe
7. Engine Hanger
8. Generator
9. Generator Strap and Bracket
10. Intake Manifold Assembly
11. Intake Manifold Support Bracket
12. Oil Filter
13. Oil Cooler
14. Oil Pressure Switch

79240G36

Exploded view of some peripheral engine component mountings—2002 models

- Air duct mounting bolts to the top of the radiator. Torque the clamp at the air intake housing
- Hood
- Windshield washer hose to the hood
15. Fill the engine with clean engine oil.
16. Connect the battery cables.
17. Fill the cooling system.
18. Fill the transmission fluid.
19. Recharge the A/C system.
20. Start the vehicle. Check for leaks, repair if necessary.

21. Road test the vehicle to check engine performance.

2005–05 Models

2.0L ENGINE

✳✳ WARNING

Use fender covers to avoid damaging painted surfaces. To avoid damage, unplug the wiring connectors carefully while holding the connector portion.

1. Before servicing the vehicle, refer to the Precautions Section.

➡**Mark all wiring and hoses to avoid misconnection.**

2. Remove the air duct.
3. Disconnect the battery terminals and remove the battery.
4. Remove the engine cover.
5. Drain the engine coolant. Remove the radiator cap to speed draining.
6. Remove the intake air hose and air cleaner assembly.

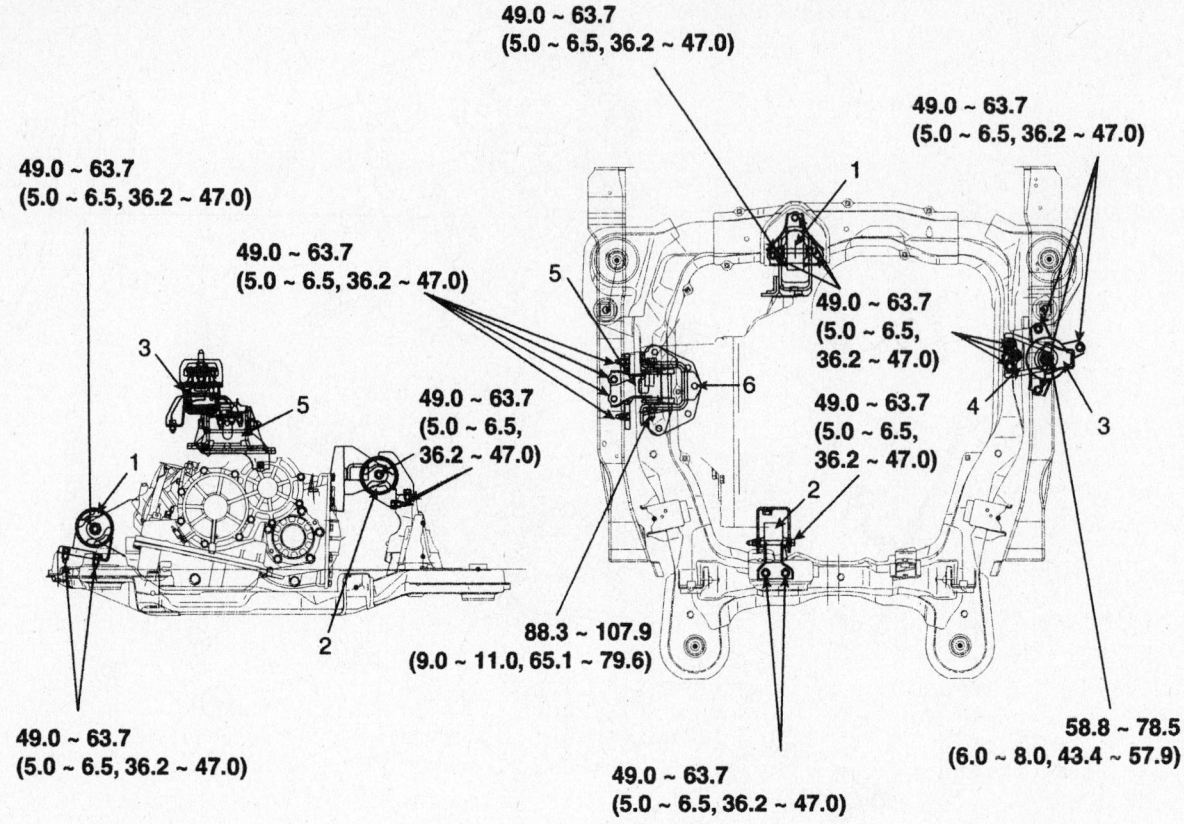

49.0 ~ 63.7
(5.0 ~ 6.5, 36.2 ~ 47.0)

49.0 ~ 63.7
(5.0 ~ 6.5, 36.2 ~ 47.0)

49.0 ~ 63.7
(5.0 ~ 6.5, 36.2 ~ 47.0)

49.0 ~ 63.7
(5.0 ~ 6.5, 36.2 ~ 47.0)

49.0 ~ 63.7
(5.0 ~ 6.5,
36.2 ~ 47.0)

49.0 ~ 63.7
(5.0 ~ 6.5,
36.2 ~ 47.0)

49.0 ~ 63.7
(5.0 ~ 6.5,
36.2 ~ 47.0)

49.0 ~ 63.7
(5.0 ~ 6.5, 36.2 ~ 47.0)

88.3 ~ 107.9
(9.0 ~ 11.0, 65.1 ~ 79.6)

49.0 ~ 63.7
(5.0 ~ 6.5, 36.2 ~ 47.0)

58.8 ~ 78.5
(6.0 ~ 8.0, 43.4 ~ 57.9)

TORQUE : N.m (kgf.m, lb-ft)

1. Front roll stopper
2. Rear roll stopper
3. Engine mounting bracket
4. Engine mounting support bracket
5. Transaxle mounting bracket
6. Transaxle support bracket

09474_KSPO_G0027

Engine mounting details—2005–06 2.0L engine

7. Remove the upper radiator hose and lower radiator hose.

8. Remove the heater hose.

9. Remove the accelerator cable.

10. Remove the engine wire harness connectors and wire harness clamps from the cylinder head and the intake manifold.

11. Disconnect the fuel inlet hose of the delivery pipe side.

12. Disconnect the hose of the PCSV (Purge Control Solenoid Valve) side.

13. Remove the brake booster vacuum hose.

14. Remove the power steering oil hose from the power steering pump.

15. Remove the power steering lower oil hose.

16. Remove the battery tray bracket.

17. Remove the transaxle wire harness connectors and control cable from transaxle (A/T).

18. Remove the transaxle oil cooler hose (AT).

19. Remove the under cover.

20. Remove the front exhaust pipe.

21. Disconnect the ABS wheel speed sensor from both front knuckles.

22. Remove the front strut lower mounting bolts and nuts.

23. Remove the caliper and hang the caliper assembly.

24. Remove the steering u-joint mounting bolt.

25. Install the jack for supporting engine and transaxle assembly.

26. Remove the engine mounting bracket.

27. Remove the transaxle mounting bracket.

28. Remove the sub frame mounting bolts and nuts.

29. Remove the engine and transaxle assembly by lifting vehicle.

⁕⁕ WARNING

When removing the engine and transaxle assembly, be careful not to damage any surrounding parts or body components.

30. Installation is in the reverse order of removal. Perform the following:

• Adjust the shift cable.
• Adjust the throttle cable.
• Refill the engine with engine oil.
• Refill the transaxle with fluid.
• Refill the radiator and reservoir tank with engine coolant.
• Place the heater control knob on "HOT" position.
• Start engine and let it run until it warms up. (the radiator fan operates 3 or 4 times.)
• Turn Off the engine. Check the level

in the radiator, add coolant if needed. This will allow trapped air to be removed from the cooling system.

- Put the radiator cap on tightly, then run the engine again and check for leaks.
- Clean the battery posts and cable terminals with sandpaper assemble them, then apply grease to prevent corrosion.
- Inspect for fuel leakage. After assembling the fuel line, turn on the ignition switch (do not operate the starter) so that the fuel pump runs for approximately two seconds and fuel line pressurizes. Repeat this operation two or three times, then check for fuel leakage at any point in the fuel line.

2.7L ENGINE

❊❊ WARNING

Use fender covers to avoid damaging painted surfaces. To avoid damage, unplug the wiring connectors carefully while holding the connector portion.

1. Before servicing the vehicle, refer to the Precautions Section.

➠**Mark all wiring and hoses to avoid misconnection. Inspect the timing belt before removing the cylinder head. Turn the crankshaft pulley so that the No. 1 piston is at top dead center.**

2. Remove the air duct.
3. Disconnect the negative terminal from the battery.
4. Drain the engine coolant. Remove the radiator cap to speed draining.
5. Remove the engine cover.
6. Remove the intake air hose and air cleaner assembly.
7. Disconnect the AFS connector.
8. Disconnect the breather hose from air cleaner hose.
9. Remove the intake air hose and air cleaner assembly.
10. Remove the upper radiator hose and lower radiator hose.
11. Remove the heater hoses.
12. Remove the engine wire harness connectors and wire harness clamps from the cylinder head and the intake manifold.
13. Disconnect front heated oxygen sensor (LH) connector, air pressure switch connector and oil pressure sensor connector.
14. Remove the fuel inlet from delivery pipe.

15. Remove the brake booster vacuum hose.
16. Remove the accelerator cable by loosening the locknut, then slip the cable end out of the throttle linkage.
17. Remove the power steering pump hose.
18. Remove the battery body bracket.
19. Disconnect the transaxle wire harness connector.
20. Remove the control cable transaxle range switch.
21. Remove the transaxle oil cooler hoses (A/T).
22. Remove the under cover.
23. Remove the front exhaust pipe.
24. Disconnect the ABS wheel speed sensor from both front knuckles.
25. Remove the front strut lower mounting bolts and nuts.
26. Remove the caliper and hang the caliper assembly.
27. Remove the steering u-joint mounting bolt.
28. Install the jack for supporting engine and transaxle assembly.
29. Remove the engine mounting bracket.
30. Remove the transaxle mounting bracket.
31. Remove the sub frame mounting bolts and nuts.
32. Raise and safely support the vehicle.
33. Installation is in the reverse order of removal. Perform the following:

- Adjust the shift cable.
- Adjust the throttle cable.
- Refill the engine with engine oil.
- Refill the transaxle with fluid.
- Refill the radiator with engine coolant.
- Bleed air from the cooling system with the heater valve open.
- Clean the battery posts and cable terminals with sandpaper assemble them, then apply grease to prevent corrosion.

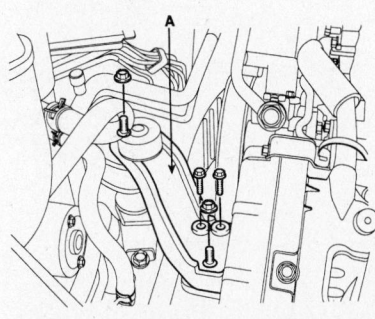

09474_KSPO_G0028

Engine mounting bracket—2.7L engine

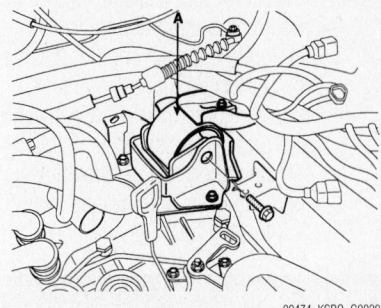

09474_KSPO_G0029

Transaxle mounting bracket—2.7L engine

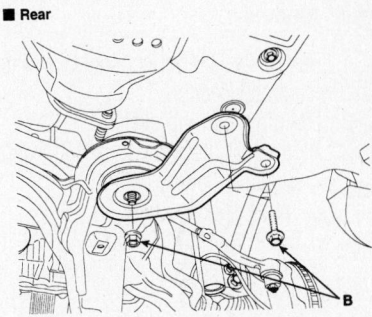

■ Rear

09474_KSPO_G0030

Subframe fasteners, part 1. "B" torque is 52–66 ft. lbs. (70–90 Nm) —2.7L engine

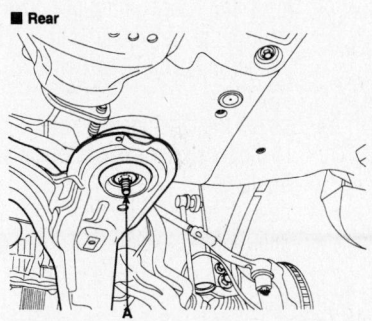

■ Rear

09474_KSPO_G0031

Subframe fasteners, part 2. "A" torque is 118–133 ft. lbs. (160–180 Nm) —2.7L engine

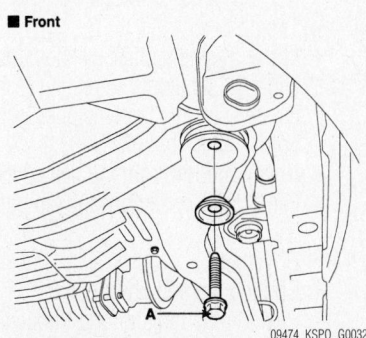

■ Front

09474_KSPO_G0032

Subframe fasteners, part 3. "A" torque is 118–133 ft. lbs. (160–180 Nm) —2.7L engine

- Inspect for fuel leakage. After assembling the fuel line, turn on the ignition switch (do not operate the starter) so that the fuel pump runs for approximately two seconds and fuel line pressurizes. Repeat this operation two or three times, then check for fuel leakage at any point in the fuel line.

Water Pump

REMOVAL & INSTALLATION

2002 Models

1. Before servicing the vehicle, refer to the Precautions Section.
2. Drain the cooling system.
3. Remove or disconnect the following:
   - Negative battery cable
   - Lower splash shield
   - Upper and lower radiator hoses
   - Coolant reservoir tank hose
   - Fresh air duct
   - Fan and shroud
   - Loosen the alternator mounting and adjusting bolts
   - Alternator belt
   - Fan pulley and bracket
   - Upper and lower timing belt covers and turn the crankshaft until No. 1 cylinder is at Top Dead Center (TDC).
   - Loosen the tensioner lockbolt and pry the tensioner away from the belt
   - Timing belt
   - Loosen the tensioner bolt
   - Water pump
   - Tensioners from the water pump

To install:

4. Clean the surface of any old gasket material.
5. Install or connect the following:
   - Tensioners on the water pump
   - Water pump and gasket. Torque the bolts to 19 ft. lbs. 25 Nm).
   - Timing belt
   - Loosen the tensioner lockbolt and allow the tensioner to rest against the belt. Torque the tensioner lockbolt to 32 ft. lbs. (43 Nm).
   - Upper and lower timing belt covers
   - Fan bracket assembly and fan pulley
   - Drive belt
   - Cooling fan and shroud
   - Position the radiator and torque the bracket bolts to 89 inch lbs. (10 Nm).
   - Torque the shroud bolts to 89 inch

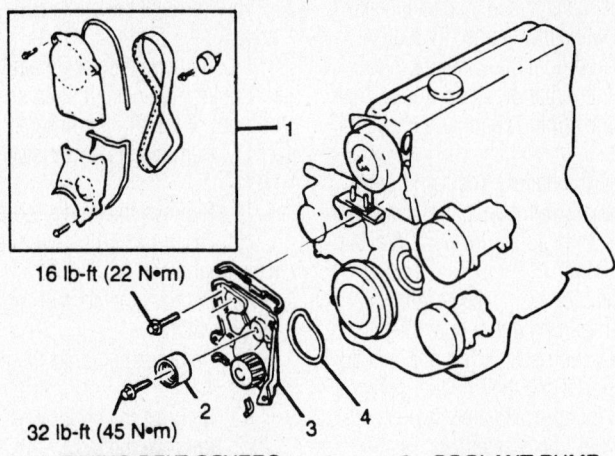

16 lb-ft (22 N•m)

32 lb-ft (45 N•m)

1. TIMING BELT COVERS, GASKETS AND TIMING BELT
2. IDLER PULLEY
3. COOLANT PUMP
4. GASKET

7924QG01

Exploded view of the water pump mounting—2002 models

lbs. (10 Nm) and torque the alternator adjusting and mounting bolts
   - Position the fresh air duct over the radiator and tighten the retaining bolt 89 inch lbs. (10 Nm)
   - Radiator hoses and tighten the clamps
   - Lower splash shield
   - Negative battery cable
6. Fill the cooling system.
7. Start the vehicle and bring the engine to operating temperature. Check for leaks and repair if necessary.

2005–05 Models

2.0L ENGINE

1. Before servicing the vehicle, refer to the Precautions Section.
2. Drain the engine coolant.

✴✴ CAUTION

System is under high pressure when the engine is hot. To avoid danger of releasing scalding engine coolant, remove the cap only when the engine is cool.

3. Remove the drive belts.
4. Remove the timing belt.
5. Remove the timing belt idler.
6. Remove the power steering pump and use a wire to secure the pump to the vehicle so that it is out of the way.
7. Remove the bolts and power steering pump bracket.
8. Remove the alternator.
9. Remove the 2 bolts and alternator brace.

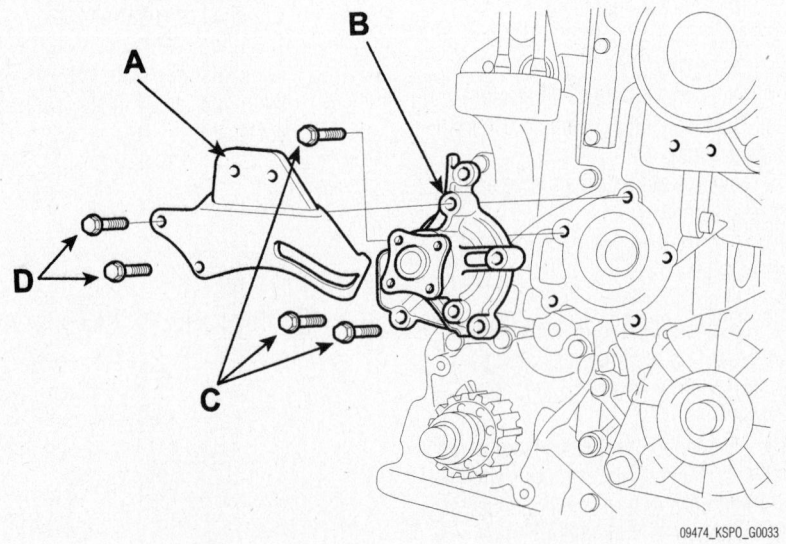

09474_KSPO_G0033

Water pump and related parts—2005–06 2.0L engine

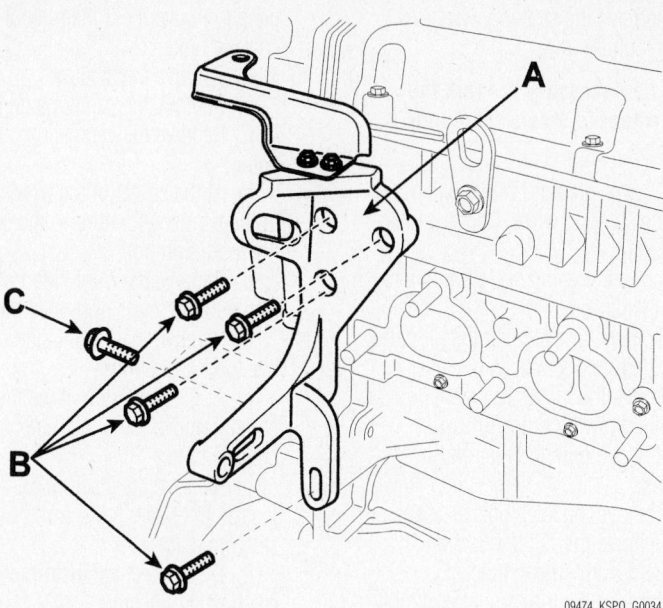

Power steering pump bracket—2005–06 2.0L engine

09474_KSPO_G0034

10. Remove the 3 bolts and remove the water pump and gasket.

To install:

11. Install the water pump (B) and a new gasket with the 3 bolts (C). Tighten to 11.8–14.7 Nm (8.7–10.8 ft. lbs.)

12. Install the alternator brace (A) with the 2 bolts (D). Tighten to 19.6–26.5 Nm (14.5–19.5 ft. lbs.)

13. Install the power steering pump bracket (A) and bolts (B, C). Tighten to:
- Bolts (B): 34.3–49.0 Nm (25.3–36.2 ft. lbs.)
- Bolts (C): 14.7–19.6 Nm (10.8–14.5 ft. lbs.)

14. Install the alternator.
15. Install the power steering pump.
16. Install the timing belt idler.
17. Install the timing belt.
18. Install the water pump pulley.
19. Install the drive belts.
20. Tighten the water pump pulley bolts to 7.8–9.8 Nm (5.8–7.2 ft. lbs.)
21. Fill with engine coolant.
22. Start engine and check for leaks.
23. Recheck engine coolant level.

2.7L ENGINE

1. Before servicing the vehicle, refer to the Precautions Section.
2. Drain the engine coolant.

❋❋ CAUTION

System is under high pressure when the engine is hot. To avoid danger of releasing scalding engine coolant, remove the cap only when the engine is cool.

3. Remove drive belts.
4. Remove the timing belt.
5. Remove the timing belt idler.
6. Remove the water pump.
7. Remove the water pump (A) and gasket (B).

To install:

8. Install the water pump (A) and a new gasket (B) with the 8 bolts. Tighten to 15–22 Nm (11–16 ft. lbs.).
- A=8x25
- B=8x30
- C=8x32
- D=8x40

9. Install the timing belt idler.
10. Install the timing belt.
11. Install drive belt.
12. Fill with engine coolant.
13. Start engine and check for leaks.
14. Recheck engine coolant level.

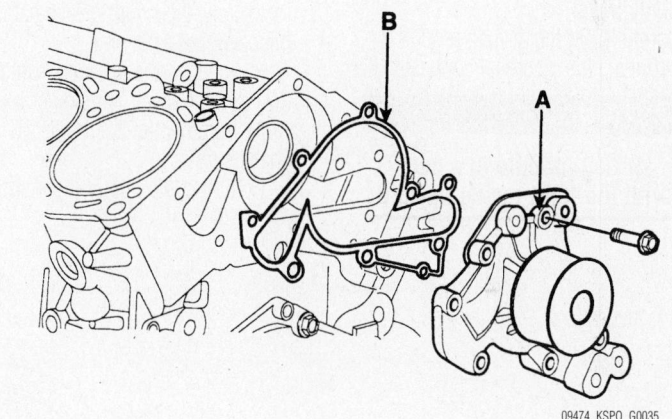

Water pump—2.7L engine

09474_KSPO_G0035

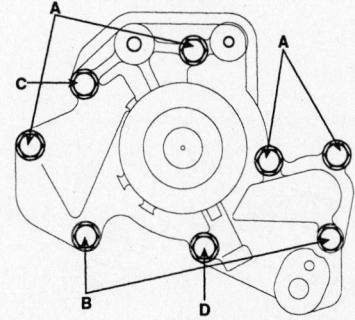

Bolt	Size	Number
A	8 × 25	4
B	8 × 30	2
C	8 × 32	1
D	8 × 40	1

09474_KSPO_G0036

Water pump bolt identification—2.7L engine

Heater Core

REMOVAL & INSTALLATION

2002 Models

1. Before servicing the vehicle, refer to the Precautions Section.
2. Disconnect the negative battery cable.

✳✳ CAUTION

After disconnecting the negative battery cable, wait for at least 10 minutes for the air bag module to deplete its stored energy.

3. Remove the driver's side air bag and steering wheel by performing the following procedure:
   a. Position the front wheels in the straight-ahead position.
   b. Remove the 4 steering wheel-to-air bag module bolts.
   c. Carefully, lift the air bag module and disconnect the electrical connector.

✳✳ CAUTION

Place the air bag module in a safe location with the front facing upward.

   d. Remove the steering wheel-to-steering column nut.

➡**If may be necessary to mark the steering wheel to steering column alignment.**

   e. Using a steering wheel puller, press the steering wheel from the steering column.
4. Drain the cooling system into a clean container for reuse.
5. Discharge and recover the air conditioning system refrigerant.
6. Remove the instrument panel by performing the following procedure:
   a. Remove both the rear and front consoles.
   b. Remove the knee bolster assembly.
   c. Remove the "T" bar section.
   d. Remove the relay bracket.
   e. Remove the turn signal assembly and the upper/lower steering column covers.
   f. Remove the hood release handle lockscrew, the hood release handle and the cable assembly nut.
   g. Remove the left side front pillar trim and the lower left side cover.
   h. Remove the 2 left side "T" bar-to-chassis bolts.
   i. At the left side of the instrument

panel, remove the 3 instrument panel-to-chassis bolts.
   j. Remove the ashtray.
   k. Remove the center panel trim.
   l. Remove the ventilation control panel.
   m. At the center of the windshield next to the windshield, remove the cap and the mounting bolt.
   n. Remove the right side front pillar trim and the lower right side cover.
   o. Remove the 2 right side of the "T" bar-to-chassis bolts.
   p. At the right side of the instrument panel, remove the 3 instrument panel-to-chassis bolts.
   q. Remove the steering column-to-instrument panel bolts and lower the steering column.
   r. Disconnect the instrument panel electrical connectors.
   s. Remove the instrument panel.
7. Remove the blower/evaporator housing by performing the following procedure:
   a. Disconnect the air conditioning refrigerant lines from the evaporator core and discard the gaskets. Plug the openings to prevent contamination.
   b. Disconnect the fresh air control cable from the blower/evaporator housing inlet duct.

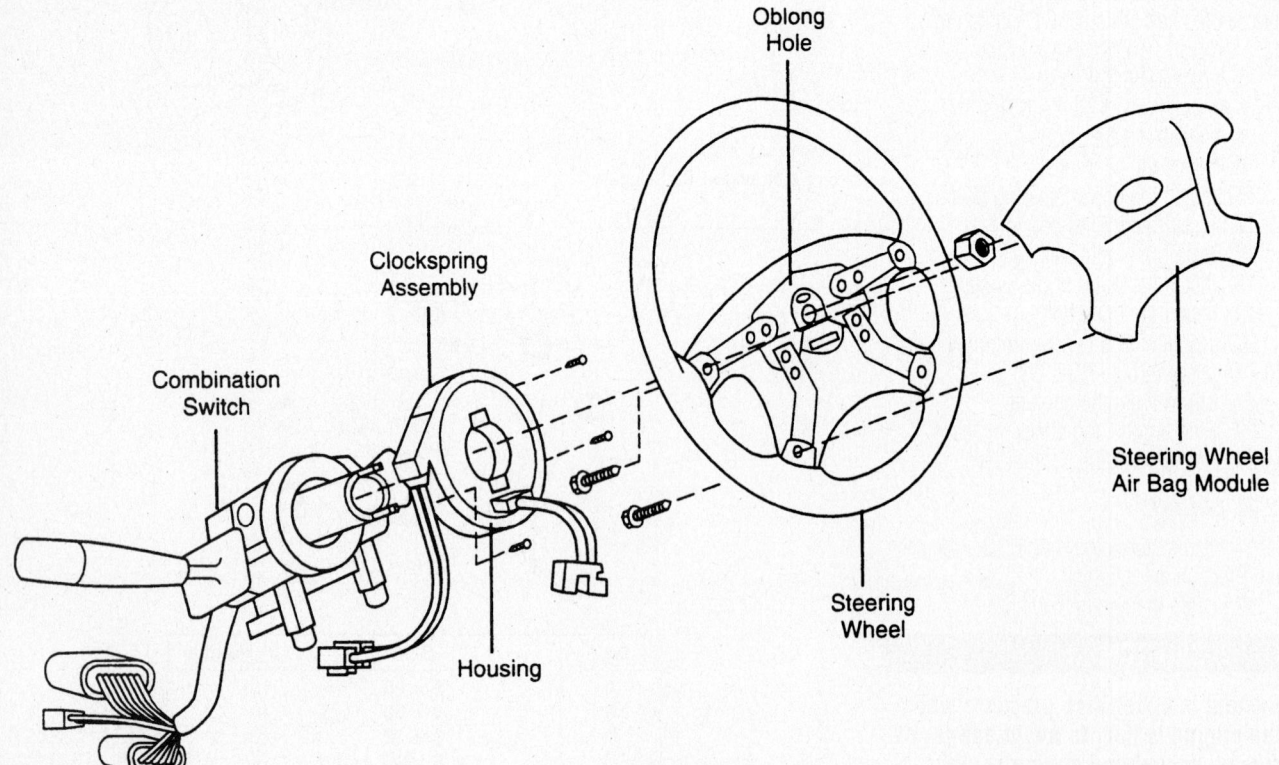

Exploded view of the steering wheel and air bag module assembly—2002 models

93113GI4

c. Disconnect the 5 connectors from the bottom of the blower/evaporator housing.

d. Move the carpeting from the bulkhead to gain access to the hole cover plate.

e. Remove the 4 hole cover plate nuts and the plate.

f. Remove the 2 upper blower/evaporator housing bolts.

g. Remove the 2 lower blower/evaporator housing-to-bulkhead nuts.

h. Remove the blower/evaporator housing.

8. Disconnect the heater hoses from the heater core.

9. Remove the temperature control cable from the heater housing.

10. Remove the 2 lower heater housing nuts and the upper heater housing-to-bulkhead nut.

11. Remove the heater housing.

12. Disassemble the heater housing by performing the following procedure:

a. Remove the seal from the heater core tube connections.

b. Remove the vent seal.

c. Remove the 2 wiring harness-to-heater servo screws.

d. Remove the 8 heater housing clips located on the servo side (left side).

e. Remove the left side of the heater housing.

f. Remove the 6 heater housing assembly clips.

g. Remove the 4 heater core tube mounting bracket screws and the bracket.

h. Remove the 8 remaining heater housing clips and disassemble the housings.

i. Remove the heater core from the housing.

To install:

13. Assemble the heater housing by performing the following procedure:

a. Install the heater core to the housing.

b. Assemble the housings and install the 8 remaining heater housing clips.

c. Install the heater core tube mounting bracket and the 4 bracket screws.

d. Install the 6 heater housing assembly clips.

e. Install the left side of the heater housing.

f. Install the 8 heater housing clips located on the servo side (left side).

g. Install the 2 wiring harness-to-heater servo screws.

h. Install the vent seal.

i. Install the seal to the heater core tube connections.

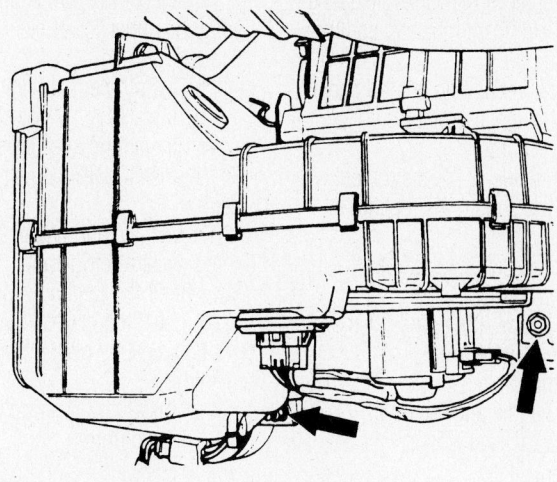

View of the blower/evaporator housing assembly—2002 models

93113GI5

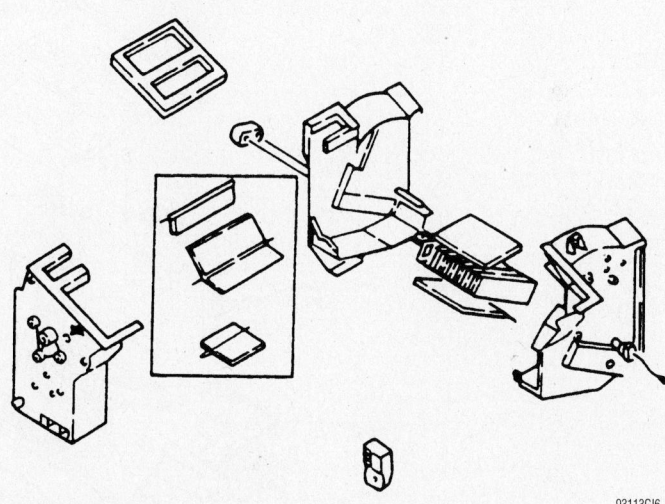

Exploded view of the heater core and heater housing assembly—2002 models

93113GI6

14. Install the heater housing.

15. Install the 2 lower heater housing nuts and the upper heater housing-to-bulkhead nut.

16. Install the temperature control cable to the heater housing.

17. Connect the heater hoses to the heater core.

18. Install the blower/evaporator housing by performing the following procedure:

a. Install the blower/evaporator housing.

b. Install the 2 lower blower/evaporator housing-to-bulkhead nuts.

c. Install the 2 upper blower/evaporator housing bolts.

d. Install the hole cover plate and the 4 plate nuts.

e. Move the carpeting over the bulkhead.

f. Connect the 5 connectors to the bottom of the blower/evaporator housing.

g. Connect the fresh air control cable to the blower/evaporator housing inlet duct.

h. Using new gaskets, connect the air conditioning refrigerant lines to the evaporator core.

19. Install the instrument panel by performing the following procedure:

a. Install the instrument panel.

b. Connect the instrument panel electrical connectors.

c. Install the steering column and lower the steering column-to-instrument panel bolts. Torque the bolts to 15 ft. lbs. (20 Nm).

d. At the right side of the instrument panel, install the 3 instrument panel-to-chassis bolts.

e. Install the 2 right side of the "T" bar-to-chassis bolts.

f. Install the right side front pillar trim and the lower right side cover.

g. At the center of the windshield next to the windshield, install the cap and the mounting bolt.

h. Install the ventilation control panel.

i. Install the center panel trim.

j. Install the ashtray.

k. At the left side of the instrument panel, install the 3 instrument panel-to-chassis bolts.

l. Install the 2 left side "T" bar-to-chassis bolts.

m. Install the lower left side cover and the left side front pillar trim.

n. Install the cable assembly nut, the hood release handle and the hood release handle lockscrew.

o. Install the turn signal assembly and the upper/lower steering column covers.

p. Install the relay bracket.

q. Install the "T" bar section.

r. Install the knee bolster assembly.

s. Install both the rear and front consoles.

20. Evacuate and charge the air conditioning system refrigerant.

21. Refill the cooling system.

22. Install the driver's side air bag and steering wheel by performing the following procedure:

a. Install the steering wheel to the steering column.

b. Install the steering wheel-to-steering column nut and torque the nut to 33 ft. lbs. (45 Nm).

c. Carefully, install the air bag module and connect the electrical connector.

d. Install the 4 steering wheel-to-air bag module bolts and torque to 72–106 inch lbs. (8–12 Nm).

23. Connect the negative battery cable.

24. Run the engine to normal operating temperatures; then, check the climate control operation and check for leaks.

2005–06 Models

1. Before servicing the vehicle, refer to the Precautions Section.

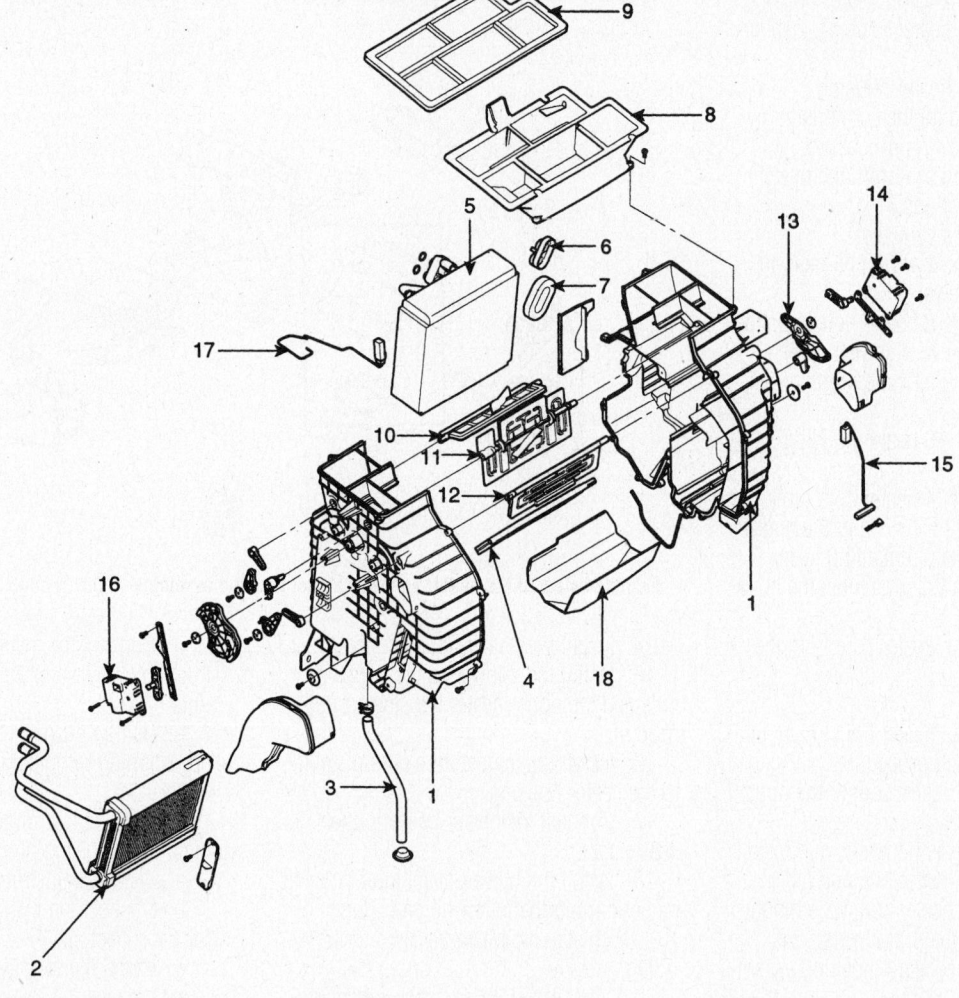

1. Heater & evaporator case	7. Seal	13. Cam
2. Heater core	8. Heater & evaporator upper case	14. Temp. control actuator
3. Drain hose	9. Upper case seal	15. Water temp. sensor (Full auto)
4. Temp. door	10. Defrost door	16. Mode control actuator
5. Evaporator core	11. Vent door	17. Evaporator temp. sensor
6. Bracket	12. Floor door	18. Insulation

09474_KSPO_G0037

Heater unit exploded view—2005–06 models

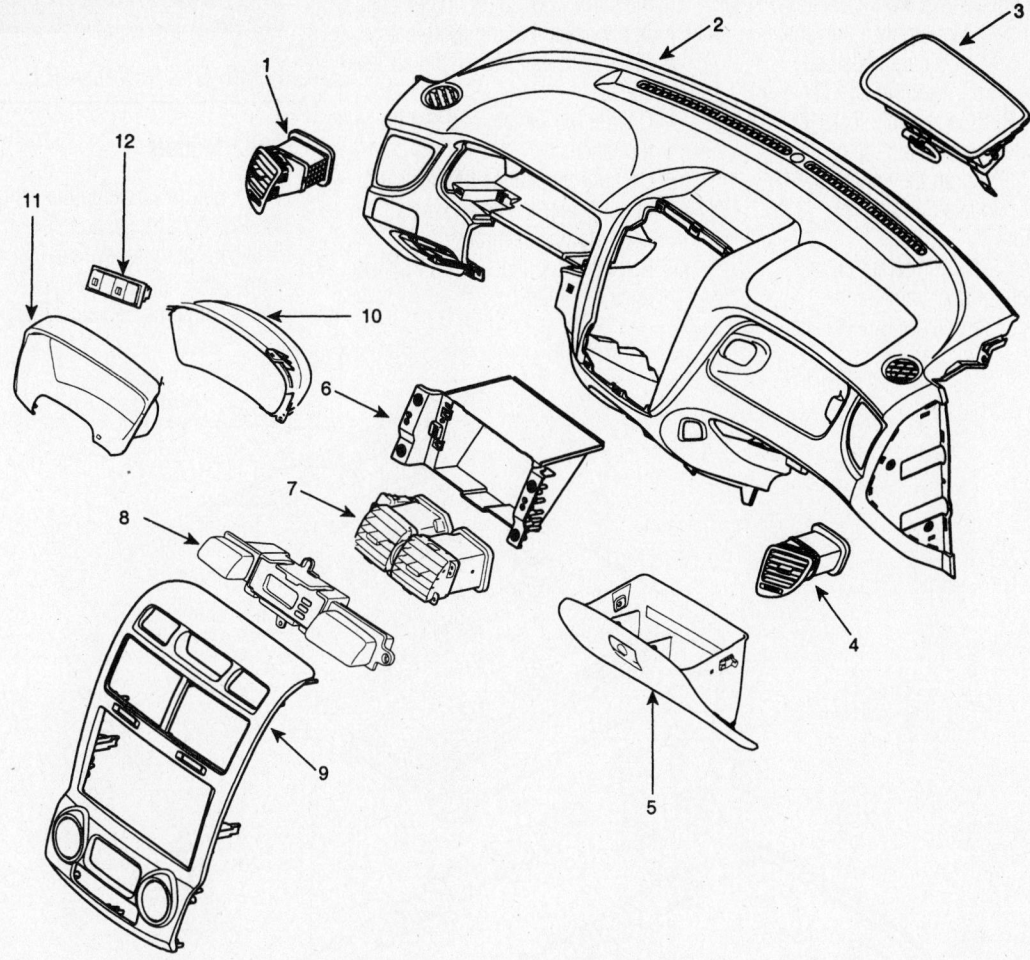

1. Air vent
2. Main crash pad
3. Air bag
4. Air vent
5. Glove box

6. Audio
7. Air vent
8. Clock assemble
9. Center facia panel

10. Cluster
11. Cluster facia panel
12. Switch assemble

09474_KSPO_G0039

Crash pad and related parts—2005–06 models

2. Disconnect the negative (-) battery terminal.

3. Recover the refrigerant with a recovery/recycling/charging station.

4. Remove the bolts (A) and the expansion valve (B) from the evaporator core. Plug or cap the lines immediately after disconnecting them to avoid moisture and dust contamination.

5. When the engine is cool, disconnect the inlet and outlet heater hoses from the heater unit. Engine coolant will run out when the hoses are disconnected; drain it into a clean drip pan. Be sure not to let coolant spill on electrical parts or painted surfaces. If any coolant spills, rinse it off immediately.

6. Remove the crash pad.

7. Remove the cross member.

8. Disconnect the connectors from the temperature control actuator, the mode control actuator and the evaporator temperature sensor, then remove the mounting nut (A) and the mounting bolts (B).

9. Remove the heater & evaporator unit after loosening the mounting screws (C).

10. Remove the self-tapping screws and the side bracket (A).

11. Be careful not to bend the inlet and outlet pipes during heater core (A) removal, and pull out the heater core.

12. Install the heater core in the reverse order of removal.

13. Install in the reverse order of removal, and note these items:
- If you're installing a new evaporator, add refrigerant oil (ND-OIL8).

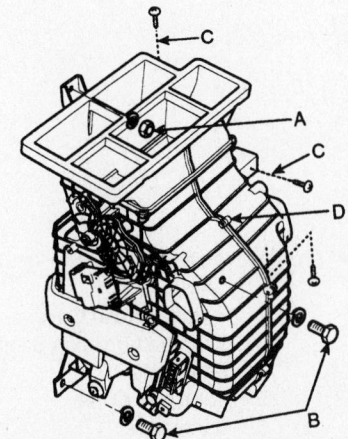

09474_KSPO_G0038

Heater unit fasteners—2005–06 models

- Replace the O-rings with new ones at each fitting, and apply a thin coat of refrigerant oil before installing them. Be sure to use the right O-rings for R-134a to avoid leakage.
- Immediately after using the oil, replace the cap on the container, and seal it to avoid moisture absorption.
- Do not spill the refrigerant oil on the vehicle; it may damage the paint; if the refrigerant oil contacts the paint, wash it off immediately.
- Apply sealant to the grommets.
- Make sure that there is no air leakage.
- Charge the system and test its performance.
- Do not interchange the inlet and outlet heater hoses and install the hose clamps securely.
- Refill the cooling system with engine coolant.

Cylinder Head

REMOVAL & INSTALLATION

2002 Models

1. Before servicing the vehicle, refer to the Precautions Section.
2. Properly relieve the fuel system pressure.
3. Drain the cooling system.

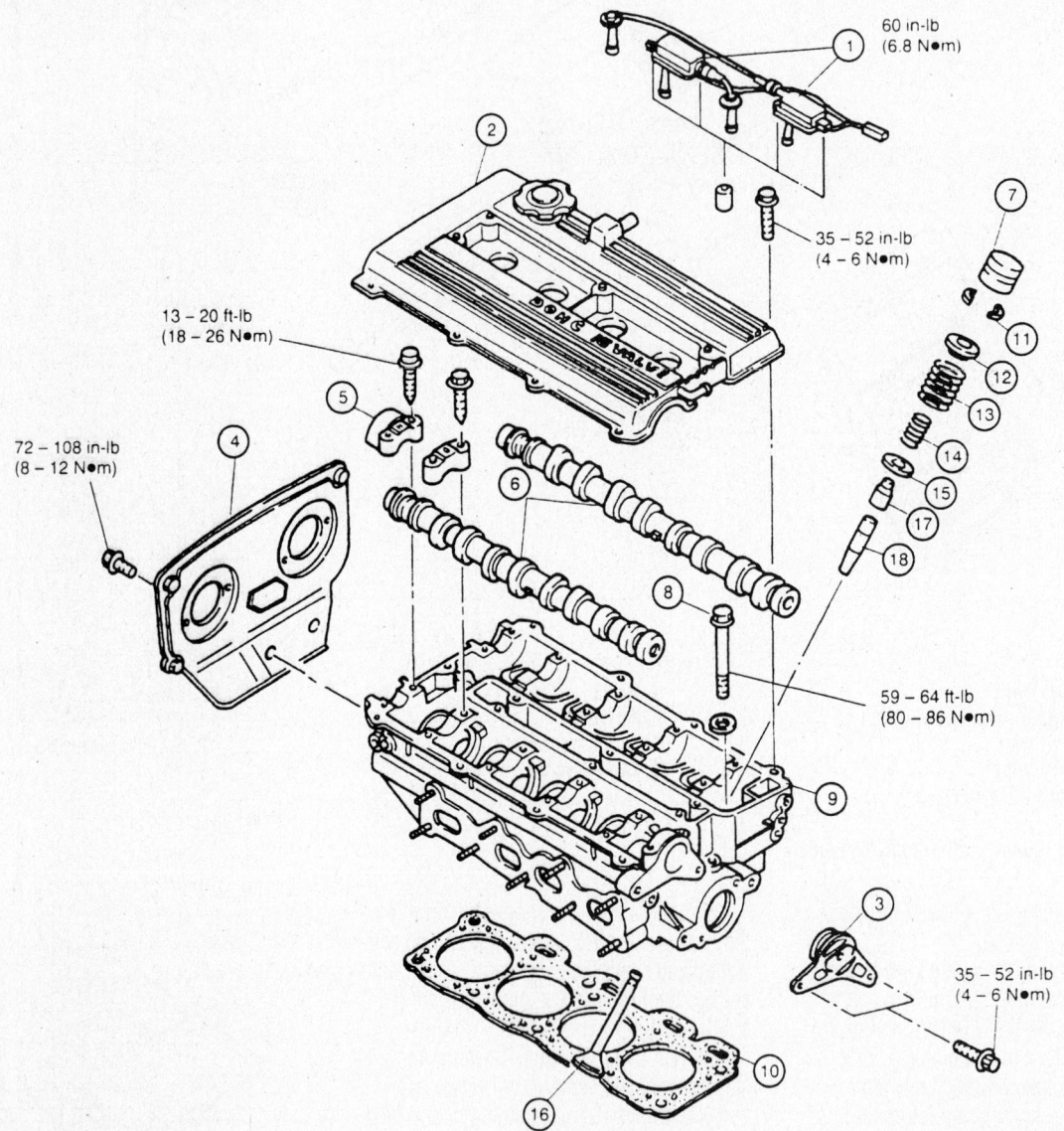

60 in-lb
(6.8 N•m)

35 – 52 in-lb
(4 – 6 N•m)

13 – 20 ft-lb
(18 – 26 N•m)

72 – 108 in-lb
(8 – 12 N•m)

59 – 64 ft-lb
(80 – 86 N•m)

35 – 52 in-lb
(4 – 6 N•m)

1. Ignition Coils and High Tension Leads
2. Cylinder Head Cover
3. Camshaft Position Sensor
4. Seal Plate
5. Camshaft Caps
6. Camshafts
7. Hydraulic Lash Adjuster
8. Cylinder Head Bolt
9. Cylinder Head
10. Cylinder Head Gasket
11. Valve Locks
12. Upper Spring Seat
13. Outer Valve Spring
14. Inner Valve Spring
15. Lower Spring Seat
16. Valve
17. Valve Stem Seal
18. Valve Guide

Exploded view of the cylinder head assembly—2002 models

7924QG37

4. Remove or disconnect the following:
- Negative battery cable
- Brake booster vacuum hose from the dynamic chamber
- Fuel line from the pressure regulator and the return line from the rear of the dynamic chamber
- Ground wire from the intake manifold
- Purge solenoid valve vacuum hose
- Upper radiator hose
- Intake manifold support bracket
- Converter flange inlet pipe
- Timing belt
- Cylinder head cover
- Cylinder head with the intake and exhaust manifolds attached
- 3 wire harness connectors at the rear of the cylinder head

To install:

5. Place the new head gasket on the engine block.

6. Install or connect the following:
- Cylinder head with the intake and exhaust manifolds attached
- 3 wiring connectors at the rear of the cylinder head
- Cylinder head bolts in the proper sequence. Torque the bolts to 64 ft. lbs. (87 Nm).
- Cylinder head cover
- Timing belt
- Converter inlet pipe flange nuts. Torque the nuts to 24 ft. lbs. (33 Nm).
- Upper radiator hose
- Vacuum hose from the intake manifold to the charcoal canister
- Purge solenoid vacuum hose
- Ground wire and harness bracket to the intake manifold. Torque the bolts to 18 ft. lbs. (25 Nm).

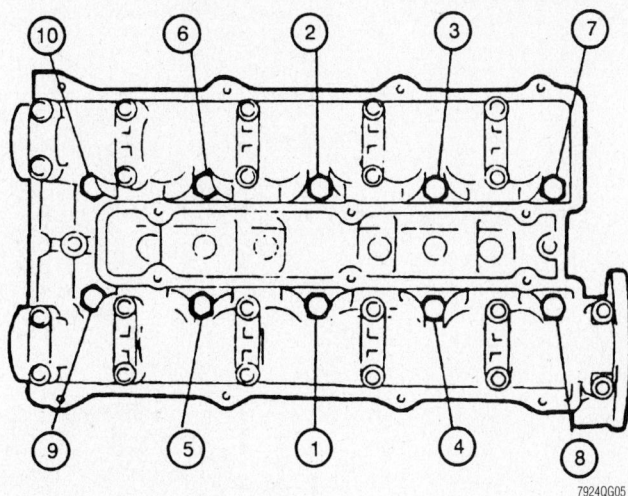

Cylinder head torque sequence—2002 models

- Fuel line to the pressure regulator and the return line to the fuel rail
- Brake booster vacuum hose
- Negative battery cable

7. Properly fill the cooling system.

8. Start the engine and check for leaks, repair if necessary.

2005–06 Models

2.0L ENGINE

1. Before servicing the vehicle, refer to the Precautions Section.

✳✳ CAUTION

Use Fender cover to avoid damaging painted surfaces. To avoid damaging the cylinder head, wait until the engine coolant temperature drops below normal temperature before removing it. When handling a metal gasket, take care not to fold the gas-

ket or damage the contact surface of the gasket. To avoid damage, unplug the wiring connectors carefully while holding the connector portion.

➡Mark all wiring and hoses to avoid misconnection. Inspect the timing belt before removing the cylinder head. Turn the crankshaft pulley so that the No. 1 piston is at top dead center.

2. Remove the air duct.

3. Disconnect the terminals from battery.

4. Remove the engine cover.

5. Drain the engine coolant. Remove the radiator cap to speed draining.

6. Remove the intake air hose and air cleaner assembly.

7. Remove the upper radiator hose and lower radiator hose.

8. Remove the heater hoses.

9. Remove the accelerator cable by loosening the lock-nut, then slip the cable end out of the throttle linkage.

10. Remove the engine wire harness connectors and wire harness clamps from cylinder head and the intake manifold.

11. Disconnect the fuel inlet hose of the delivery pipe side.

12. Disconnect the hose of the PCSV (Purge Control Solenoid Valve) side.

13. Remove the brake booster vacuum hose.

14. Remove the power steering pump drive belt.

15. Remove the power steering pump and fix the pump to vehicle with a wire.

16. Remove the bolts and power steering pump bracket.

17. Remove the spark plug cables.

18. Remove the exhaust manifold.

19. Remove the intake manifold.

20. Remove the timing belt.

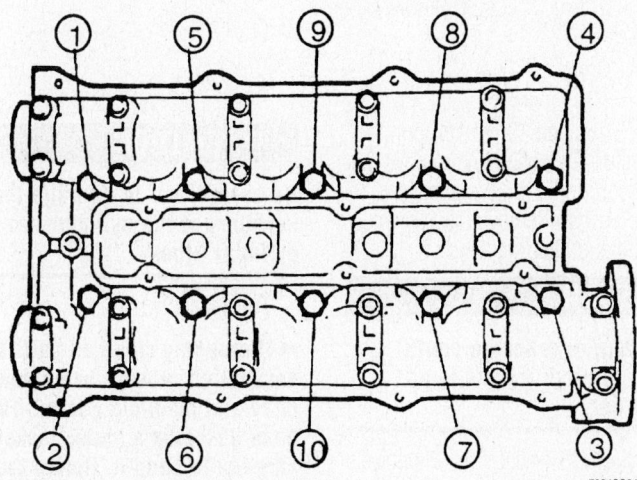

Cylinder head removal sequence—2002 models

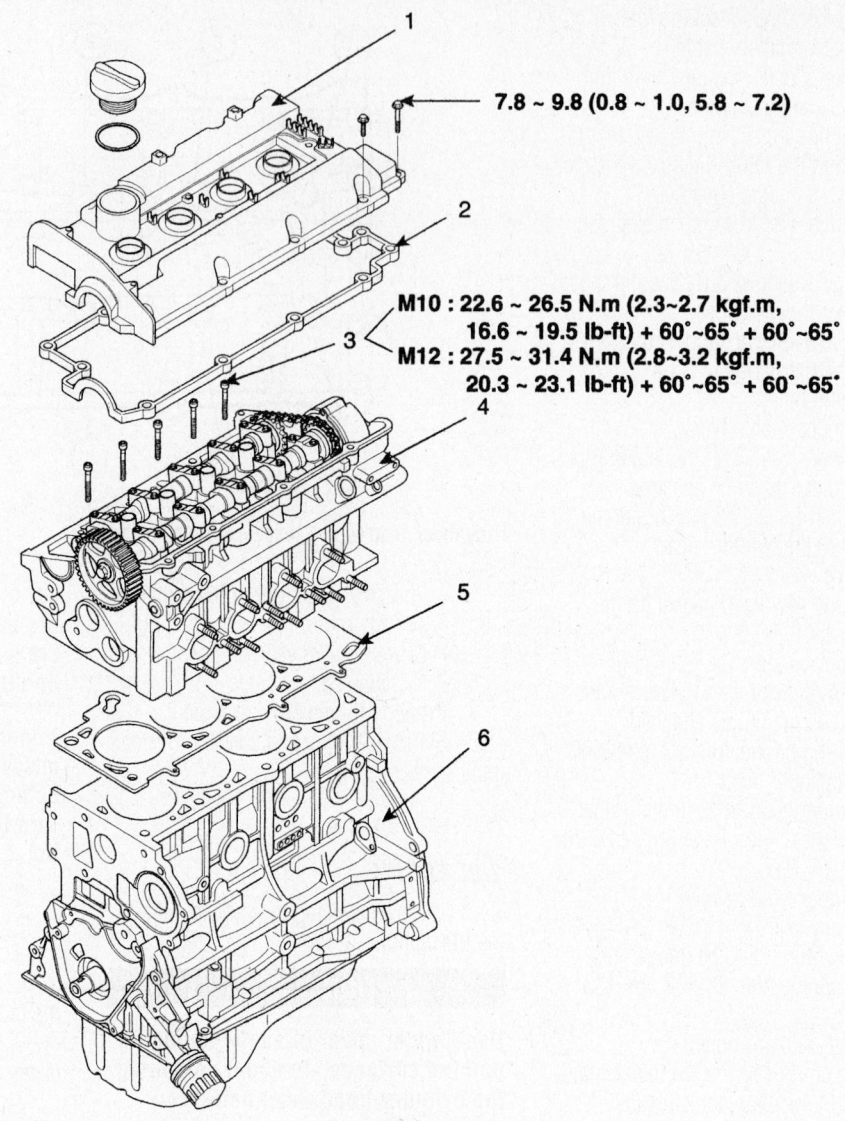

7.8 ~ 9.8 (0.8 ~ 1.0, 5.8 ~ 7.2)

M10 : 22.6 ~ 26.5 N.m (2.3~2.7 kgf.m,
16.6 ~ 19.5 lb-ft) + 60°~65° + 60°~65°
M12 : 27.5 ~ 31.4 N.m (2.8~3.2 kgf.m,
20.3 ~ 23.1 lb-ft) + 60°~65° + 60°~65°

TORQUE : N.m (kgf.m, lb-ft)

1. Cylinder head cover
2. Gasket
3. Cylinder head bolt
4. Cylinder head
5. Cylinder head gasket
6. Cylinder block

09474_KSPO_G0040

Cylinder head and related parts—2005–06 2.0L engine

21. Remove the PCV (Positive Crankcase Ventilation) hose.

22. Remove the cylinder head cover.

23. Remove the camshaft sprocket.

24. Insert a stopper pin or other device into timing chain auto tensioner and remove the auto tensioner.

25. Remove the camshaft bearing caps and camshafts.

26. Remove the OCV (Oil Control Valve).

27. Remove the OCV (Oil Control Valve) filter.

28. Remove the water hose from water pipe.

29. Using 8mm and 10mm hexagon wrenches, uniformly loosen and remove the 10 cylinder head bolts, in several passes, in the sequence shown. Remove the 10 cylinder head bolts and plate washers.

❋❋ CAUTION

Head warpage or cracking could result from removing bolts in an incorrect order.

30. Lift the cylinder head from the dowels on the cylinder block and replace the cylinder head on wooden blocks on a bench.

❋❋ CAUTION

Be careful not to damage the contact surfaces of the cylinder head and cylinder block.

To install:

➡Thoroughly clean all parts to be assembled. Always use a new cylinder head and manifold gasket. The cylinder head gasket is a metal gasket. Take care not to bend it. Rotate the crankshaft, set the No. 1 piston at TDC.

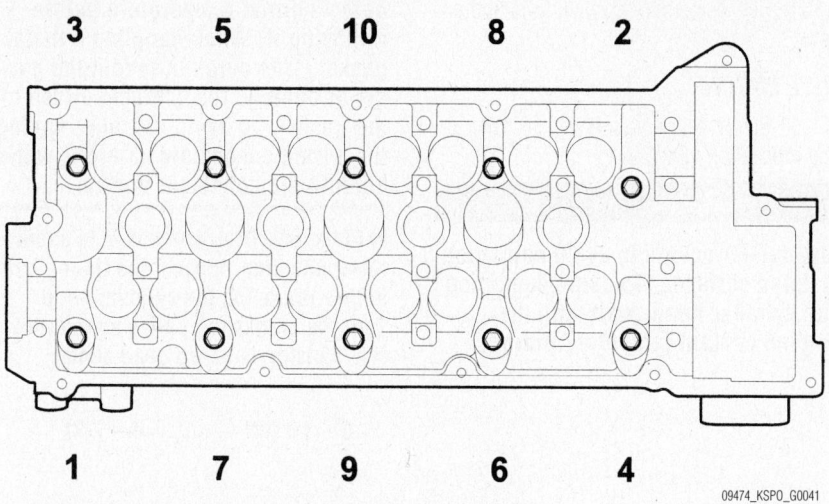

3   5   10   8   2

1   7   9   6   4

09474_KSPO_G0041

Cylinder head bolt removal sequence—2005–06 2.0L engine

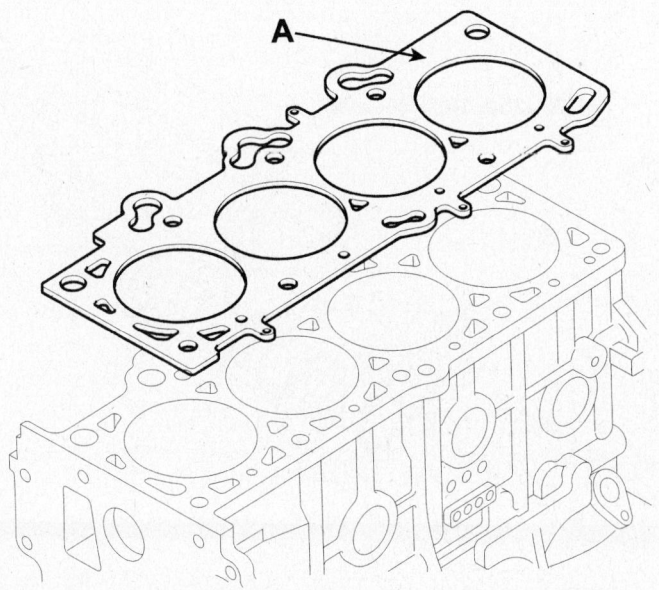

A

09474_KSPO_G0042

Cylinder head gasket installation—2005–06 2.0L engine

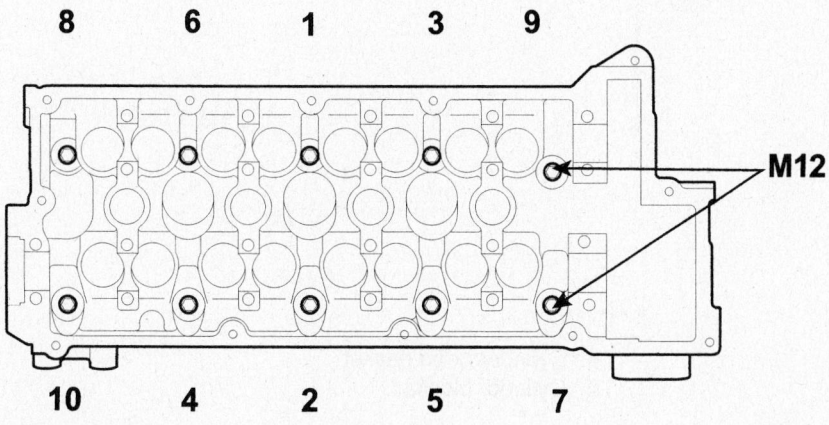

8   6   1   3   9

M12

10   4   2   5   7

09474_KSPO_G0043

Cylinder bolt torque sequence—2005–06 2.0L engine

31. Install the cylinder head gasket (A) on the cylinder block.

➥**Be careful of the installation direction.**

32. Place the cylinder head onto the block carefully in order to prevent damaging the gasket. If the gasket is damaged, fluid leakage could occur.

33. Apply a light coat if engine oil on the threads and under the heads of the cylinder head bolts. Using 8mm and 10mm hexagon wrenches, install and tighten the 10 cylinder head bolts and plate washers, in several passes, in the sequence shown:

- M10 bolts: 23–27 Nm (17–19 ft. lbs.) +60–65 degrees, + an additional 60–65 degrees
- M12 bolts: 28–31 Nm (20–23 ft. lbs.) +60–65 degrees, + an additional 60–65 degrees

34. Install the OCV (Oil Control Valve) filter. Tighten to 40–50 Nm (30–37 ft. lbs.)

➥**Always use a new OCV (Oil Control Valve) filter gasket. Keep the OCV filter clean.**

35. Install the OCV (Oil Control Valve). Tighten to 10–12 Nm (7–9 ft. lbs.)

✴✴ CAUTION

Do not reuse the OCV (Oil Control Valve) when dropped. Keep the OCV clean. Do not hold the OCV (Oil Control Valve) sleeve during servicing. When the OCV is installed on the engine, be careful not to rotate the engine while holding the OCV.

36. Install the camshafts.
37. Check and adjust valve clearance.
38. Install the camshaft bearing oil seal.
39. Install the camshaft sprocket.
40. Install the cylinder head cover.
41. Install the PCV (Positive Crankcase Ventilation) hose.
42. Install the timing belt.
43. Install the intake manifold.
44. Install the exhaust manifold.
45. Install the spark plug cables.
46. Install the power steering pump bracket and bolts. Tighten the mounting bolts to 34–49 Nm (25–36 ft. lbs.); the pivot bolt to 12–15 Nm (9–11 ft. lbs.)
47. Install the power steering pump.
48. Install the accelerator cable.
49. Install the brake booster hose.
50. Connect the hose of the PCSV (Purge Control Solenoid Valve) side.
51. Connect the fuel inlet hose of the delivery pipe side.
52. Install the engine wire harness con-

nectors and wire harness clamps to the cylinder head and the intake manifold.

53. Install the heater hose.

54. Install the upper radiator hose and lower radiator hose.

55. Install the intake air hose and air cleaner assembly.

56. Install the engine cover.

57. Reconnect the battery terminals.

58. Install the air duct.

59. Fill with engine coolant.

60. Start the engine and check for leaks.

61. Recheck engine coolant level and oil level.

2.7L ENGINE

1. Before servicing the vehicle, refer to the Precautions Section.

❊❊ WARNING

Use fender covers to avoid damaging painted surfaces. To avoid damaging the cylinder head, wait until the engine coolant temperature drops below normal temperature before removing it. When handling a metal gasket, take care not to fold the gasket or damage the contact surface of the gasket. To avoid damage, unplug the wiring connectors carefully while holding the connector portion.

➡Mark all wiring and hoses to avoid misconnection. Inspect the timing belt before removing the cylinder head. Turn the crankshaft pulley so that the No. 1 piston is at top dead center.

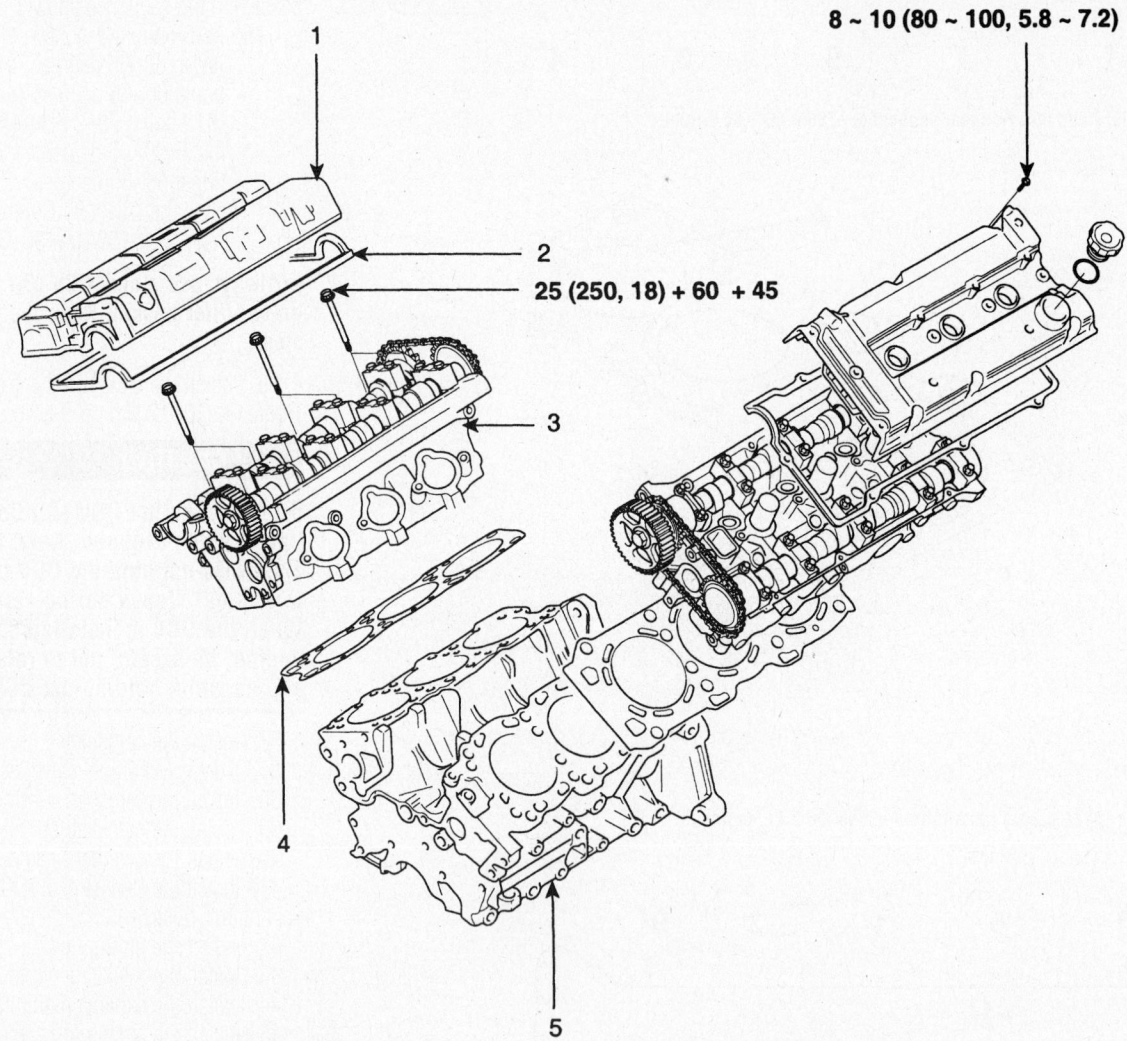

8 ~ 10 (80 ~ 100, 5.8 ~ 7.2)

25 (250, 18) + 60 + 45

TORQUE : N.m (kgf.cm, lb-ft)

1. Cylinder head cover
2. Gasket
3. Cylinder head
4. Cylinder head gasket
5. Cylinder blockad

Cylinder heads and gaskets—2.7L engine

09474_KSPO_G0045

2. Remove the air duct.

3. Remove the cowl grill and wiper motor.

4. Remove the strut bar.

5. Disconnect the negative terminal from the battery.

6. Drain the engine coolant.

7. Remove the radiator cap to speed draining.

8. Remove the engine cover.

9. Remove the intake air hose and air cleaner assembly.

10. Remove the coolant reservoir tank.

11. Remove the upper radiator hose and lower radiator hose.

12. Remove the heater hoses and throttle body heater hose.

13. Remove the engine wire harness connectors and wire harness clamps from the cylinder head and the intake manifold.

14. Remove the fuel inlet hose from delivery pipe.

15. Remove the brake booster vacuum hose.

16. Remove the accelerator cable by loosening the locknut, then slip the cable end out of the throttle linkage.

17. Remove the PCV hose.

18. Remove the intake manifold.

19. Remove the power steering pump.

20. Remove the exhaust manifold.

21. Remove the timing belt.

22. Remove the spark plug cable.

23. Remove the cylinder head covers.

24. Remove the camshaft sprocket.

25. Remove the camshaft bearing caps.

26. Remove the camshafts.

27. Remove the timing belt rear cover.

28. Remove the water temperature control assembly and water pipe.

29. Uniformly loosen and remove the 8 cylinder head bolts on each cylinder head in several passes and in the sequence shown, then repeat for the other side, as shown. Remove the 16 cylinder head bolts and plate washer.

To install:

➡**Thoroughly clean all parts to be assembled. Always use a new head and manifold gasket. The cylinder head gasket is a metal gasket. Take care not to bend it. Rotate the crankshaft, set the No.1 piston at TDC.**

30. Install the cylinder head gaskets on the cylinder block.

31. Place the cylinder head carefully, in order not to damage the gasket with the bottom part of the end.

32. Install cylinder head bolts. Tighten to:

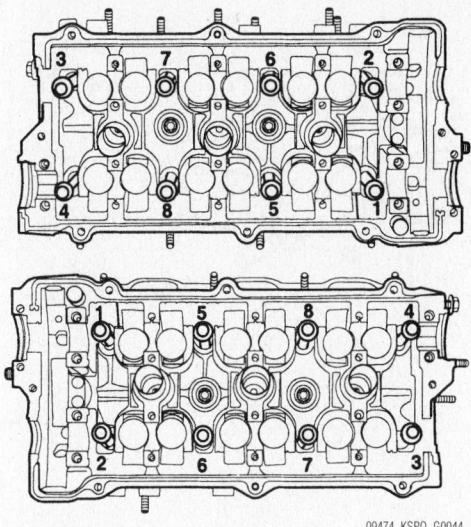

Cylinder head bolt loosening sequence—2.7L engine

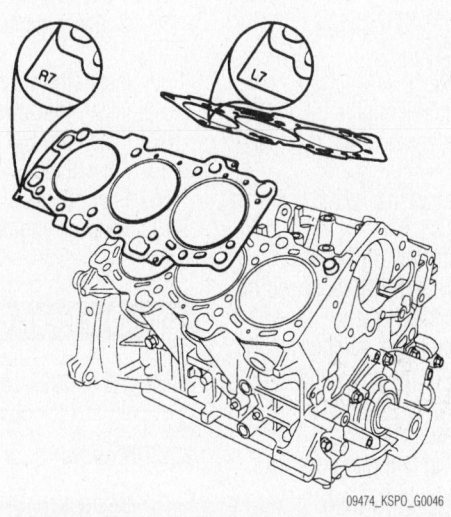

Head gasket installation—2.7L engine

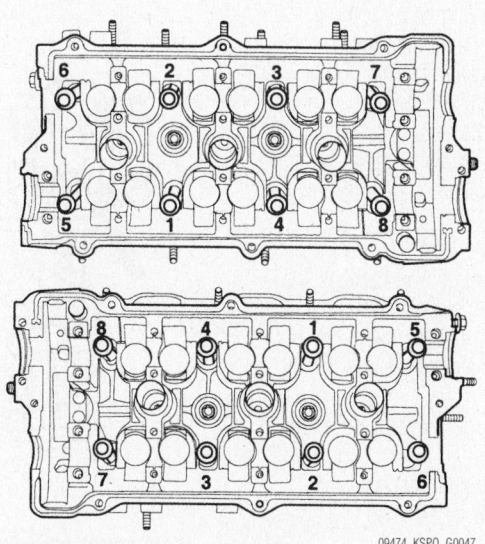

Cylinder head bolt tightening sequence—2.7L engine

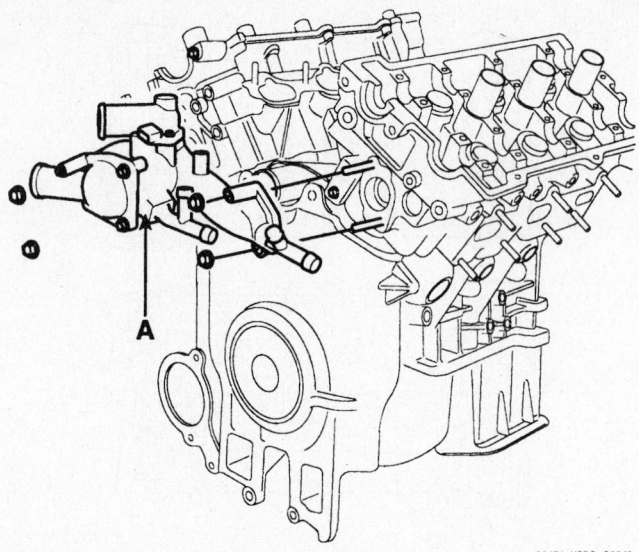

09474_KSPO_G0048

Water pipe and water temperature control assembly—2.7L engine

- Step 1: 25 Nm (18 ft. lbs.)
- Step 2: + 60 degrees
- Step 3: + 45 degrees

33. Apply a light coat if engine oil on the threads and under the heads of the cylinder head bolts.

34. Install the plate washer to the cylinder head bolt.

35. Install and uniformly tighten the cylinder head bolts on each cylinder head in several passes and in the sequence shown, then repeat for the other side, as shown.

36. Install the water pipe and water temperature control assembly (A). Tighten to 15–20 Nm (11–14 ft. lbs.)

37. Install the timing belt rear cover (A). Tighten to 10–12 Nm (7–9 ft. lbs.).

38. Install the camshafts.

39. Install the cylinder head covers.

40. Install the spark plug cables.

41. Install the timing belt.

42. Install the exhaust manifold.

43. Install the power steering pump.

44. Install the intake manifold.

45. Install the PCV hose.

46. Install the accelerator cable.

47. Install the brake booster vacuum hose.

48. Install the fuel inlet hose.

49. Install the engine wire harness connectors and wire harness clamps to the cylinder head and the intake manifold.

50. Install the heater hoses and throttle body heater hose.

51. Install the upper radiator hose and lower radiator hose.

52. Install the coolant reservoir tank.

53. Install the intake air hose and air cleaner assembly.

54. Install the engine cover.

55. Connect the negative terminal to the battery.

56. Install the strut bar.

57. Install the cowl grill and wiper motor.

58. Install the air duct.

59. Fill with engine coolant.

60. Start the engine and check for leaks.

61. Recheck engine coolant level and oil level.

Cylinder Head Cover(s)

REMOVAL & INSTALLATION

2002 Models

1. Before servicing the vehicle, refer to the Precautions Section.

2. Disconnect the negative battery terminal. Proceed to remove the hose at the resonance chamber.

3. Remove one bolt and the resonance chamber, then remove the two accelerator cable bracket bolts from the cylinder head cover.

4. Remove the two air intake tubes-to-cylinder head cover bolts.

5. Disconnect the accelerator cable by pulling back on the throttle shaft and rotating the accelerator cable until it lines up with the slot in the pulley.

6. Loosen the air hose clamp from the air intake hose to the MAF sensor.

7. Remove the IAC and breather hoses, and the vacuum line from the air intake tube.

8. Remove the three bolts from the air intake tube to the throttle body and remove the air intake tube and air intake hose as an assembly.

9. Remove the six coil cover bolts from the top of the cylinder head cover.

10. Disconnect the electrical connectors from the ignition coils.

11. Remove the four bolts (two each) from the ignition coils to the cylinder head cover.

12. Remove the spark plug wires from the spark plugs.

13. Remove the 15 cylinder head cover bolts.

14. Remove the cylinder head cover.

To install:

15. Place a small amount of sealer at the corners of the float camshaft caps and the distributor mounting cap.

16. Place cylinder head cover on top of cylinder head.

17. Secure cover with 15 bolts and tighten to 35–52 inch lbs. (4–6 Nm).

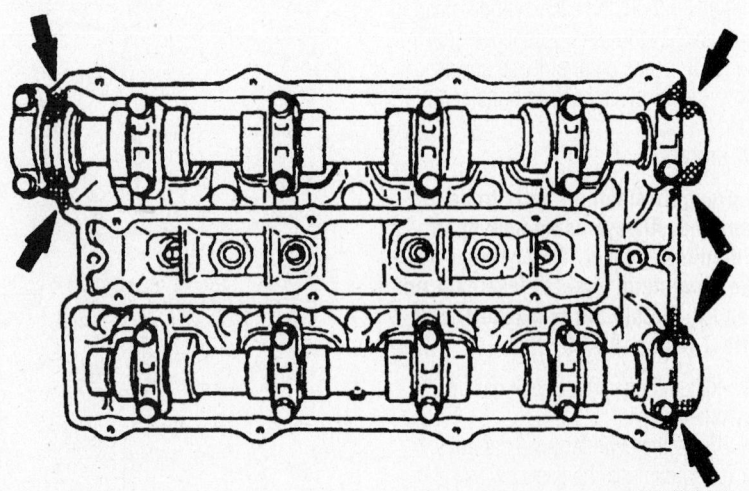

09474_KSPO_G0208

Place a small amount of sealer at the corners of the float camshaft caps and the distributor mounting cap

18. Install the ignition coils and connect the spark plug wires to the spark plugs.

19. Fasten the two coils to the cylinder head cover with the four mounting bolts and tighten to 60 inch lbs. (6.8 Nm).

20. Connect the two ignition coil connectors.

21. Install coil cover and secure with six bolts. Tighten to 30–40 inch lbs. (3–5 Nm).

22. Install air intake hose and air intake tube as an assembly. Secure air intake tube to throttle body with three bolts and tighten to 18 ft. lbs. (24 Nm).

23. Install the two air intake tube-to-cylinder head cover bolts and tighten to 16 ft. lbs. (22 Nm).

24. Connect the accelerator cable by pulling back on the throttle shaft pulley and inserting the cable in the slot. Release the throttle to the closed position.

25. Install the IAC and breather hoses, and the vacuum line to the air intake tube.

26. Connect the air intake hose to the MAF sensor and tighten the clamp.

27. Install the resonance chamber and secure with one bolt. Tighten the two clamps.

28. Connect the negative battery cable.

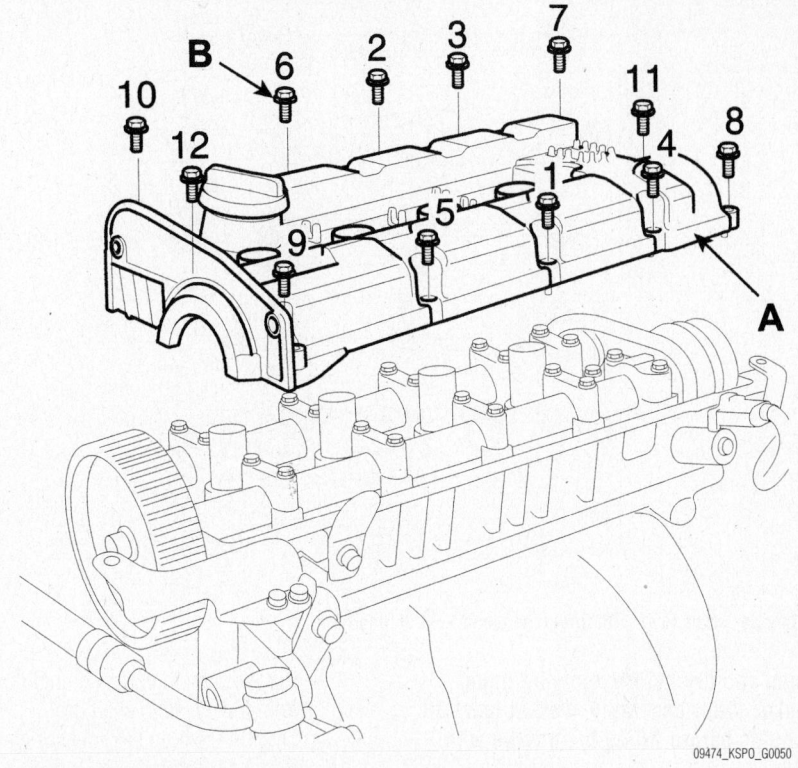

09474_KSPO_G0050

Cylinder head cover bolt torque sequence—2005–06 2.0L engine

2005–06

2.0L ENGINE

1. Before servicing the vehicle, refer to the Precautions Section.

2. Remove the air duct.

3. Disconnect the terminals from battery.

4. Tag and disconnect any hoses, wires or cables in the way of cylinder head cover removal.

5. Remove the bolts and lift off the cover. Discard the gasket.

To install:

6. Install the cylinder head cover gasket in the groove of the cylinder head cover.

➥**Before installing the cylinder head cover gasket, thoroughly clean the cylinder head cover and the groove. When installing, make sure the cylinder head cover gasket is seated securely in the corners of the recesses with no gap.**

➥**Use liquid gasket, Loctite® No. 5999. Check that the mating surfaces are clean and dry before applying liquid gasket. After assembly, wait at least 30 minutes before filling the engine with oil.**

7. Install the cylinder head cover (A) with the 12 bolts (B). Uniformly tighten the bolts in several passes. Tighten to 8–10 Nm (6–7 ft. lbs.)

8. Reconnect the battery terminals.

9. Install the air duct.

2.7L ENGINE

1. Before servicing the vehicle, refer to the Precautions Section.

2. Remove the air duct.

3. Disconnect the terminals from battery.

4. Tag and disconnect any hoses, wires or cables in the way of cylinder head cover removal.

5. Remove the bolts and lift off the cover. Discard the gasket.

To install:

6. Install the cylinder head cover gasket in the groove of the cylinder head cover.

➥**Before installing the cylinder head cover gasket, thoroughly clean the cylinder head cover and the groove. When installing, make sure the cylinder head cover gasket is seated securely in the corners of the recesses with no gap.**

➥**Use liquid gasket, Loctite® No. 5999. Check that the mating surfaces are**

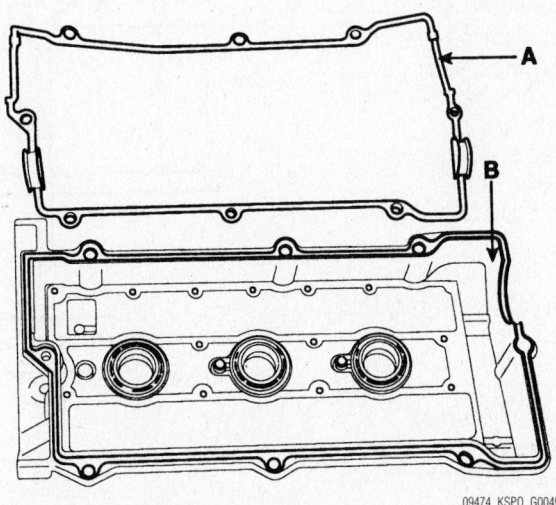

09474_KSPO_G0049

Cylinder head cover gasket application—2.7L engine

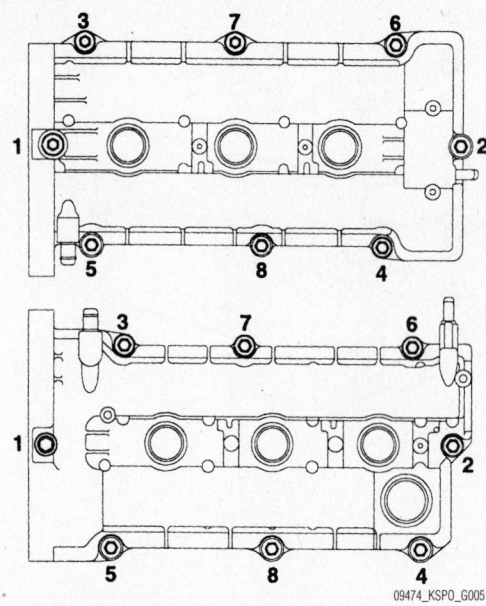

Cylinder head cover bolt torque sequence—2.7L engine

clean and dry before applying liquid gasket. After assembly, wait at least 30 minutes before filling the engine with oil.

7. Install the cylinder head cover. Tighten to 8–10 Nm (6–7 ft. lbs.)
8. Reconnect the battery terminals.
9. Install the air duct.

Intake Manifold

REMOVAL & INSTALLATION

2002 Models

1. Before servicing the vehicle, refer to the Precautions Section.
2. Properly relieve the fuel system pressure.
3. Drain the coolant.
4. Remove or disconnect the following:
   - Negative battery cable
   - Accelerator cable bracket bolts from the valve cover
   - Air intake tube to cylinder head cover bolts
   - Air intake tube to throttle body bolts
   - Loosen the clamp attaching the air tube to the Mass Air Flow (MAF) sensor
   - Idle Air Control (IAC) valve, breather hose and vacuum line from the air intake tube
   - Air intake tube
   - Positive Crankcase Ventilation hose (PCV) from the dynamic chamber
   - Purge solenoid valve vacuum hose from the dynamic chamber
   - Throttle Position (TP) sensor electrical connector
   - IAC valve electrical connector
   - Heater hoses
   - Engine-to-body ground strap from the intake manifold and the harness bracket below it
   - Brake booster vacuum line
   - Vacuum hose from the fuel pressure regulator hose
   - Dynamic chamber support bracket bolts
   - Fuel injector electrical connectors
   - Fuel pressure and return lines
   - Intake manifold support bracket
   - Oil filter
   - Intake manifold bolts
   - Bypass pipe from the heater hose
   - Intake manifold and discard the gasket

To install:
5. Install or connect the following:
   - Intake manifold with a new gasket to the cylinder head
   - Heater hose to the bypass pipe
   - Bolts and nuts attaching the intake manifold to the cylinder head. Torque the bolts to 14–22 ft. lbs. (19–30 Nm).
   - New oil filter
   - Intake manifold support bracket. Torque the bolts to 27–38 ft. lbs. (37–52 Nm)
   - Fuel lines
   - Fuel injector electrical connectors
   - Engine to body ground wire. Torque the bolt to 18 ft. lbs. (25 Nm).
   - Dynamic chamber support bracket. Torque the bolts to 18 ft. lbs. (25 Nm).
   - Purge solenoid valve vacuum hose to the dynamic chamber
   - Vacuum hose to the fuel pressure regulator
   - Coolant hoses to the throttle body
   - IAC valve electrical connector
   - Brake booster vacuum hose
   - Heater hoses
   - TP sensor electrical connector
   - Air intake tube and hose to the throttle body. Torque the bolts to 16 ft. lbs. (22 Nm).
   - PCV hose to the dynamic chamber

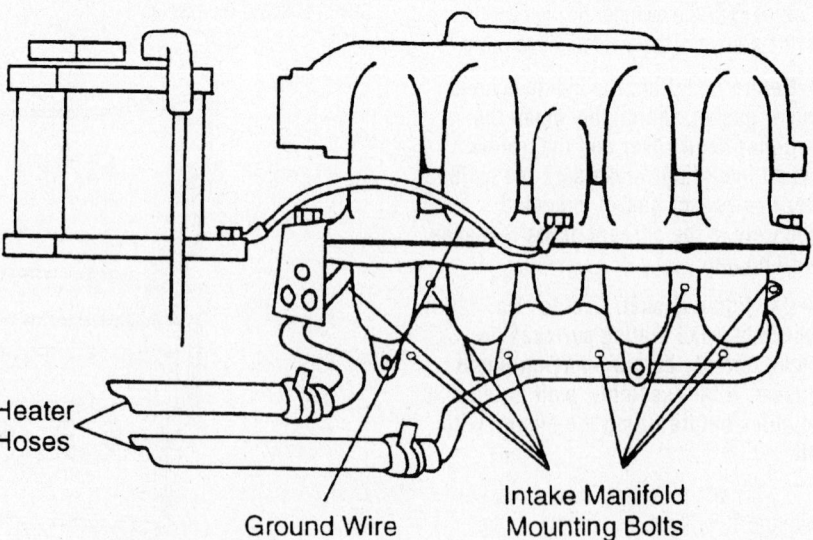

Intake manifold mounting bolt locations. Be sure to connect the ground cable—2002 models

Heater Hoses

Ground Wire

Intake Manifold Mounting Bolts

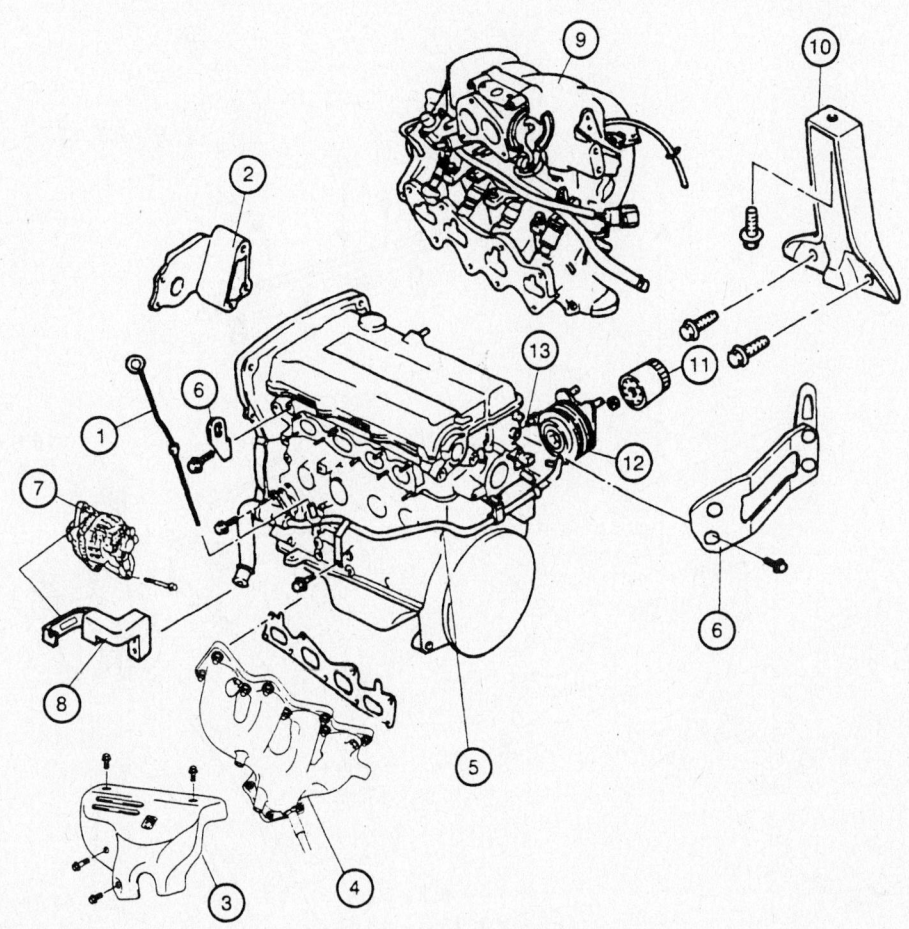

1. Oil Level Gauge
2. Thermo-Modulated Fan Bracket
3. Exhaust Manifold Heat Shield
4. Exhaust Manifold
5. Coolant Inlet Pipe and Bypass Pipe

6. Engine Hanger
7. Generator
8. Generator Strap and Bracket
9. Intake Manifold Assembly
10. Intake Support Bracket

11. Oil Filter
12. Oil Cooler
13. Oil Pressure Switch

9308QG01

Exploded view of the intake and exhaust manifolds and related components—2002 models

- Accelerator cable to the throttle body pulley
- Accelerator cable bracket. Torque the bolts to 10 ft. lbs. (15 Nm).
- Air intake hose to the MAF sensor
- Resonance chamber
- IAC hose, breather hose and vacuum line to the intake manifold
- Negative battery cable

6. Properly fill the cooling system.
7. Start the engine and check for leaks, repair if necessary.

2005–06 Models

2.0L ENGINE

1. Before servicing the vehicle, refer to the Precautions Section.
2. Remove the engine cover.
3. Disconnect the TPS (Throttle Position Sensor) connector and ISA (Idle Speed Actuator) connector.
4. Disconnect the PCV (Positive Crankcase Ventilation) hose and breather hose.

5. Remove the accelerator cable.
6. Remove the delivery pipe.
7. Remove the heater hose, PCSV (Purge Control Solenoid Valve) and the brake vacuum hose from throttle body and intake manifold.
8. Disconnect the negative cable from the battery.
9. Recover the refrigerant with a recovery/charging station.
10. Loosen the drive belt.
11. Remove the bolts. Then, disconnect

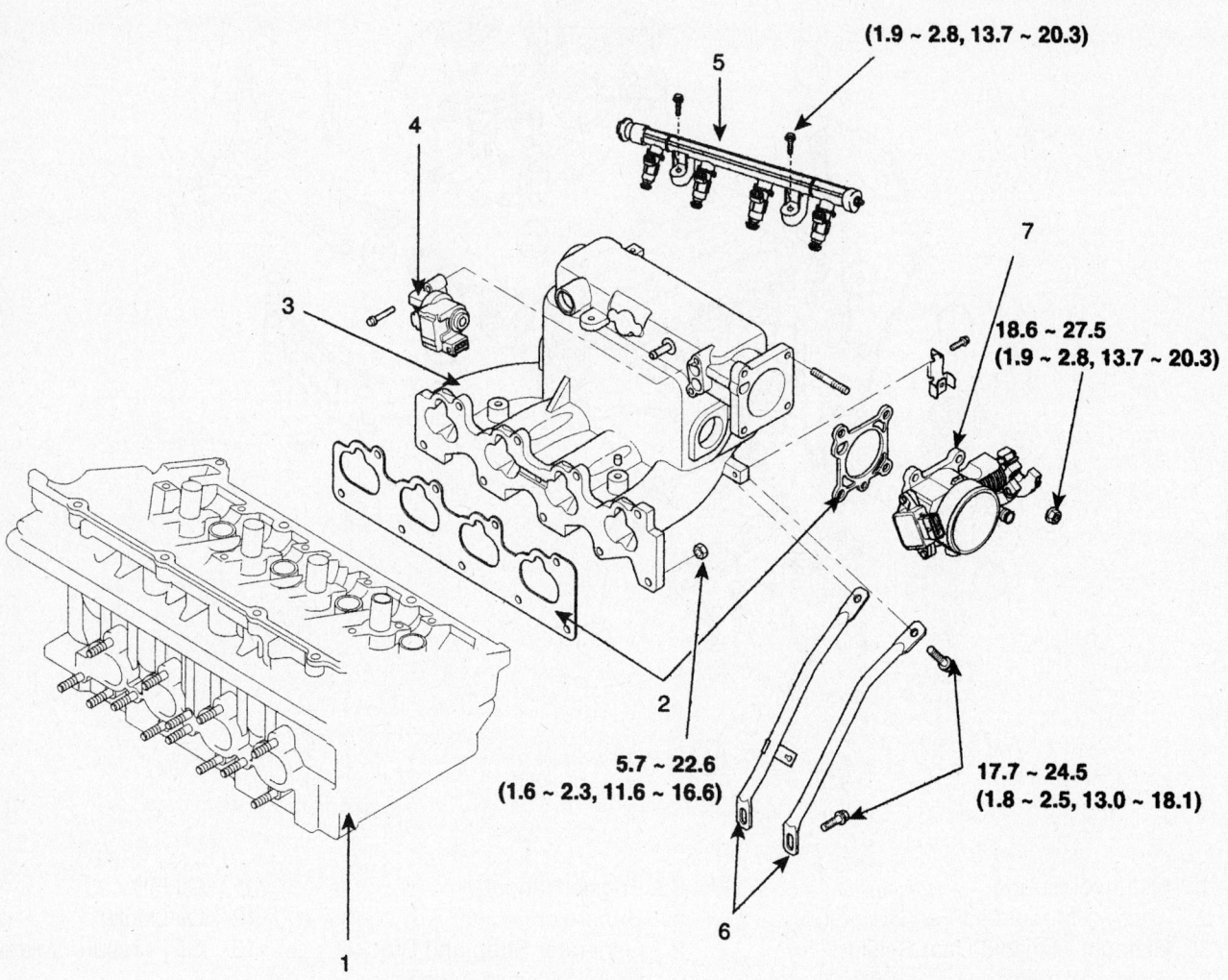

(1.9 ~ 2.8, 13.7 ~ 20.3)

18.6 ~ 27.5
(1.9 ~ 2.8, 13.7 ~ 20.3)

5.7 ~ 22.6
(1.6 ~ 2.3, 11.6 ~ 16.6)

17.7 ~ 24.5
(1.8 ~ 2.5, 13.0 ~ 18.1)

TORQUE : N.m (kgf.m, lb-ft)

1. Cylinder head
2. Gasket
3. Intake manifold
4. I.S.A (Idle Speed Actuator)

5. Delivery pipe assembly
6. Intake manifold stay
7. Throttle body

09474_KSPO_G0052

Intake manifold and related parts—2005–06 2.0L engine

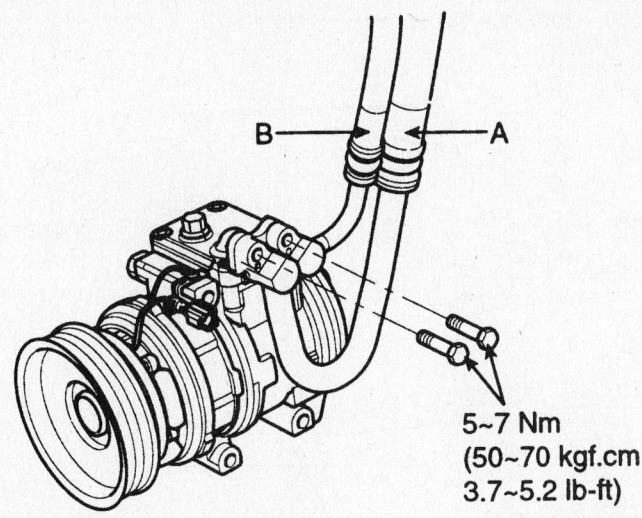

**5~7 Nm
(50~70 kgf.cm
3.7~5.2 lb-ft)**

09474_KSPO_G0053

Suction and discharge lines

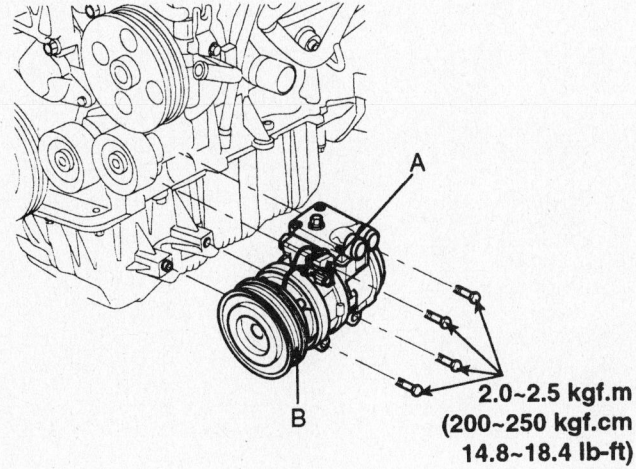

**2.0~2.5 kgf.m
(200~250 kgf.cm
14.8~18.4 lb-ft)**

09474_KSPO_G0054

Compressor mounting

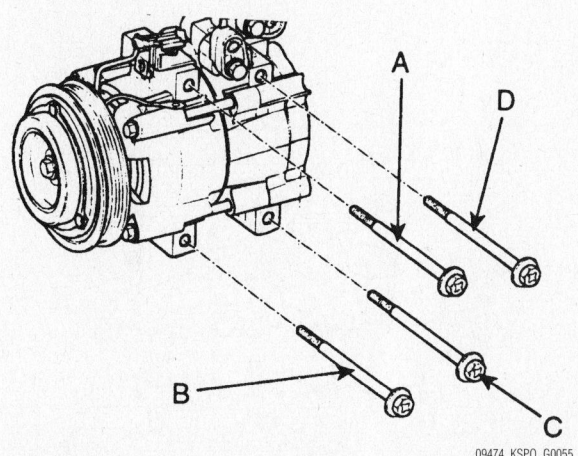

09474_KSPO_G0055

Compressor bolt identification—2.7L engine. A: 118mm (4.65 in.); B and C: 102mm (4.02 in.); D: 126mm (4.96 in.)

the suction line (A) and discharge line (B) from the compressor. Plug or cap the lines immediately after disconnecting them to avoid moisture and dust contamination.

12. Disconnect the compressor clutch connector (A), then remove the mounting bolts and the compressor (B).

13. Remove the intake manifold stay.

14. Remove the intake manifold.

15. Installation is in the reverse order of removal with new gasket. Observe the following torques:

- Intake manifold: 16–23 Nm (12–16 ft. lbs.)
- Intake manifold stay: 18–24 Nm (13–18 ft. lbs.)
- Fuel delivery pipe: 18–27 Nm (14–20 ft. lbs.)

2.7L ENGINE

1. Before servicing the vehicle, refer to the Precautions Section.

2. Remove the engine cover.

3. Remove air cleaner hose.

4. Disconnect the accelerator cable.

5. Disconnect the TPS connector (A).

6. Disconnect the ISA connector (B).

7. Disconnect the VIS actuator connector (C).

8. Disconnect the injector connector (D).

9. Disconnect the PCSV connector (E).

10. Disconnect the PCSV hose.

11. Disconnect the brake booster vacuum hose.

12. Disconnect the PCV hose.

13. Disconnect the IAT sensor connector.

14. Disconnect the VIS actuator connector.

15. Disconnect the ground cable from the surge tank assembly.

16. Remove the surge tank stay.

17. Remove the surge tank assembly.

18. Remove the injector assembly.

19. Remove the intake manifold and gasket.

To install:

20. Install the lower intake manifold and gasket. Tighten to 19–21 Nm (14–15 ft. lbs.)

21. Install the injector assembly.

22. Install the surge tank assembly.

23. Install the surge tank assembly. Tighten to 15–20 Nm (11–15 ft. lbs.)

24. Install the surge tank stay. Tighten to 15–20 Nm (11–15 ft. lbs.)

25. Install the ground cable.

26. Connect the VIS actuator connector.

27. Connect the IAT sensor connector.

28. Connect the PCV hose.

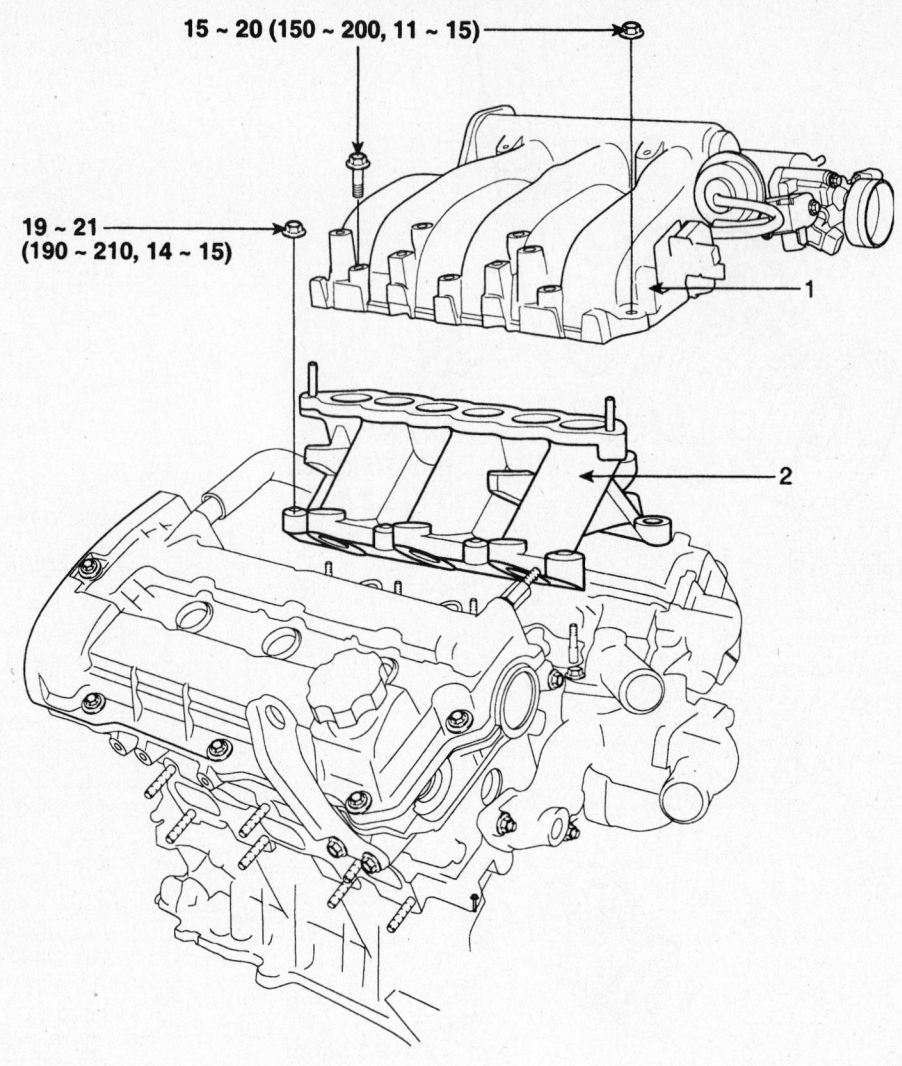

15 ~ 20 (150 ~ 200, 11 ~ 15)

19 ~ 21
(190 ~ 210, 14 ~ 15)

1

2

TORQUE : N.m (kgf.cm, lb-ft)

1. Surge tank assembly

2. Intake manifold

09474_KSPO_G0056

Upper (surge tank) and lower intake manifolds—2.7L engine

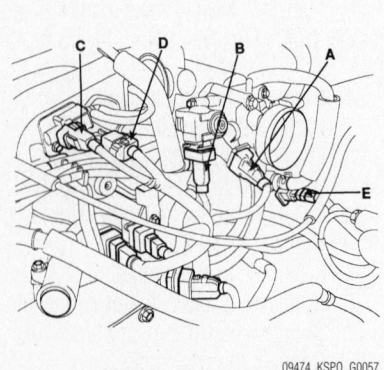

09474_KSPO_G0057

Connector identification—2.7L engine

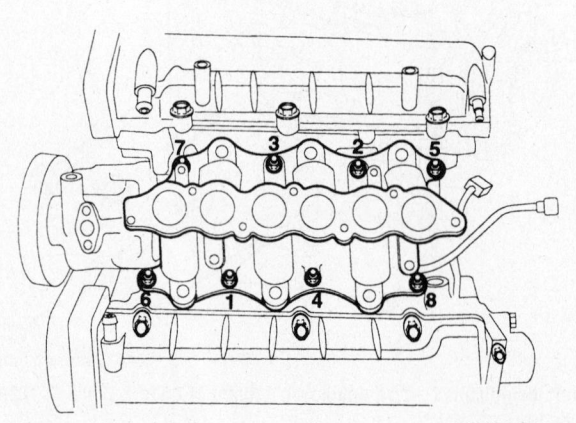

09474_KSPO_G0058

Lower intake manifold torque sequence—2.7L engine

29. Connect the brake booster vacuum hose.
30. Connect the PCSV hose.
31. Connect the PCSV connector.
32. Connect the injector connector.
33. Connector the VIS actuator connector.
34. Connector the ISA connector.
35. Connector the TPS connector.
36. Connector the actuator cable.
37. Install the air cleaner hose.
38. Install the engine cover.

Exhaust Manifold

REMOVAL & INSTALLATION

2002 Models

1. Before servicing the vehicle, refer to the Precautions Section.
2. Remove or disconnect the following:
   - Negative battery cable
   - Air intake hose
   - Exhaust manifold heat shield
   - Converter inlet pipe flange locknuts
   - Exhaust manifold and discard the gasket
3. Clean the mating surfaces.

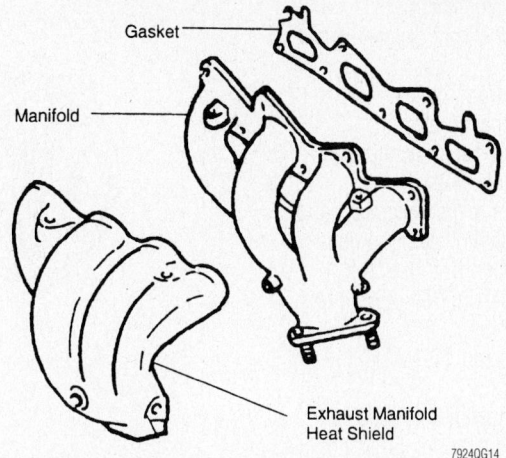

Exploded view of the exhaust manifold assembly—2002 models

To install:

4. Install or connect the following:
   - Exhaust manifold with a new gasket. Torque the bolts to 31 ft. lbs. (42 Nm).
   - New flange gasket and install the converter inlet pipe. Torque the bolts to 24 ft. lbs. (33 Nm).
   - Exhaust manifold heat shield. Torque the bolts to 18 ft. lbs. (25 Nm).
   - Air intake hose
   - Negative battery cable
5. Start the vehicle and check for leaks, repair if necessary.

2005–06 Models

2.0L ENGINE

1. Before servicing the vehicle, refer to the Precautions Section.

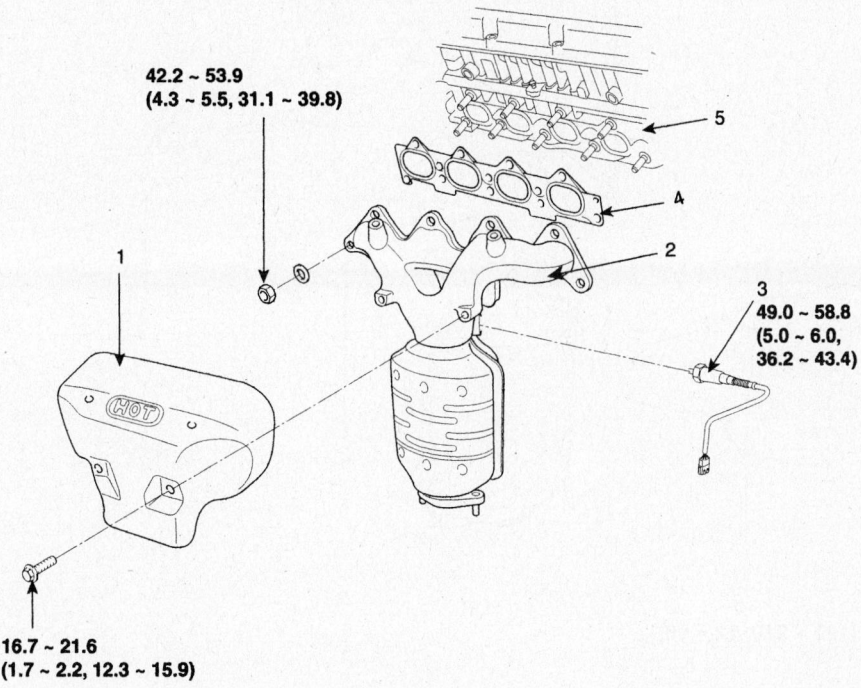

TORQUE : N.m (kgf.m, lb-ft)

1. Heat protector
2. Exhaust manifold
3. Front oxygen sensor
4. Gasket
5. Cylinder head

Exhaust manifold and related parts—2005-06 2.0L engine

2. Remove the engine cover.

3. Disconnect the front oxygen sensor connector.

4. Remove the front muffler.

5. Remove the heat protector.

6. Remove the exhaust manifold and catalytic converter assembly.

7. Installation is in the reverse order of removal. Observe the following torques:

- Front muffler: Nut: 39–59 Nm (29–43 ft. lbs.); Bolt: 29–39 Nm (22–29 ft. lbs.)
- Heat protector: 17–22 Nm (12–16 ft. lbs.)
- Exhaust manifold/catalytic converter assembly: 42–54 Nm (31–40 ft. lbs.)

2.7L ENGINE

1. Before servicing the vehicle, refer to the Precautions Section.

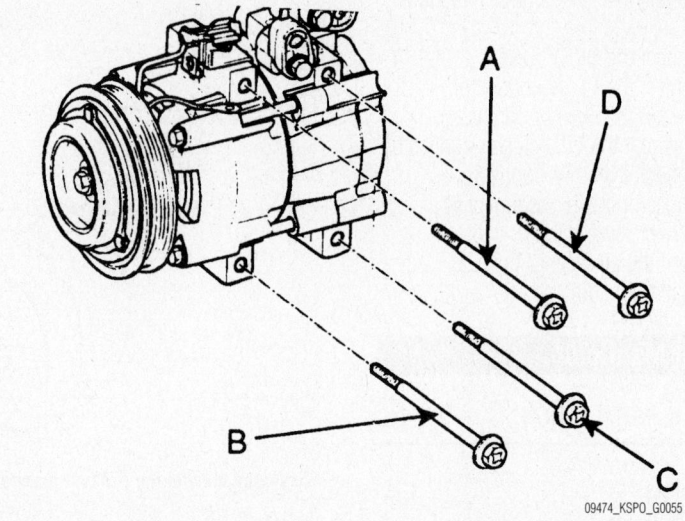

Compressor bolt identification—2.7L engine. A, B and C: 118mm (4.65 in.); D: 94mm (3.70 in.)

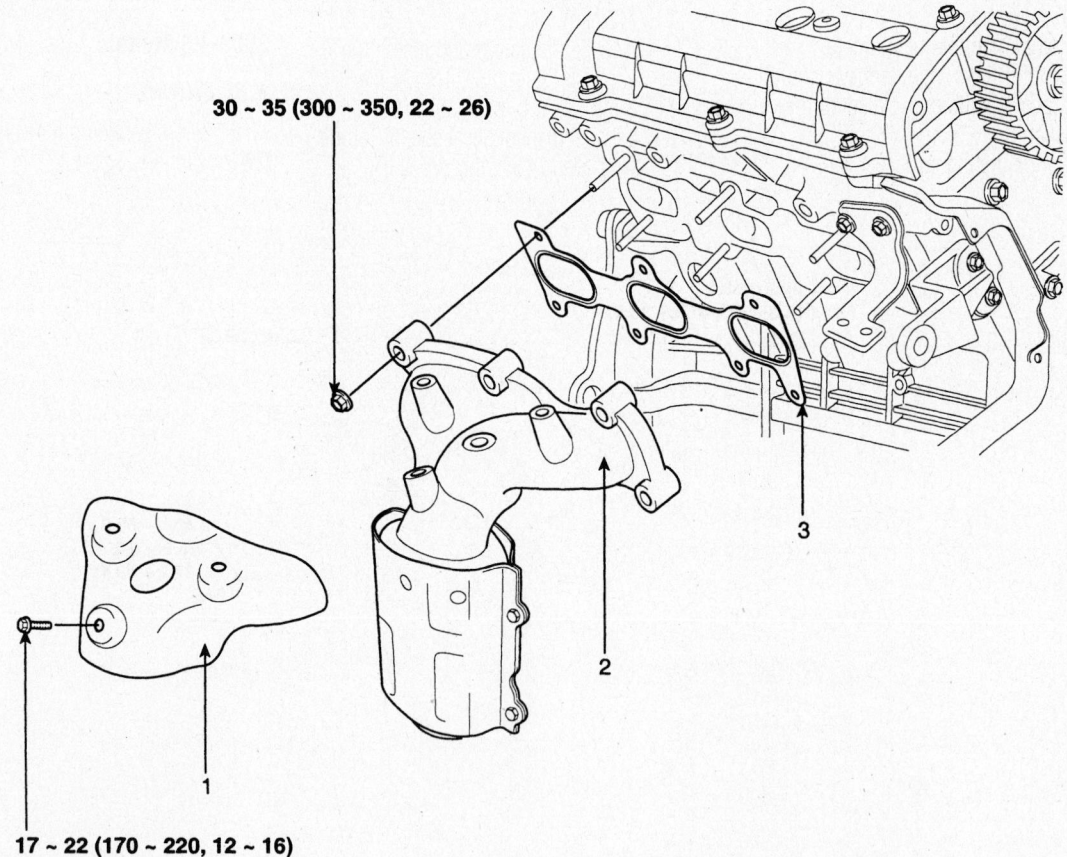

30 ~ 35 (300 ~ 350, 22 ~ 26)

17 ~ 22 (170 ~ 220, 12 ~ 16)

TORQUE : N.m (kgf.cm, lb-ft)

1. Heat protector
2. Exhaust manifold
3. Gasket

Exhaust manifold—2.7L engine

09474_KSPO_G0055

09474_KSPO_G0060

2. Remove the under cover.

3. Remove the front exhaust pipe.

4. Disconnect the oxygen sensor connector.

5. Remove the cooling fan motor assembly.

6. Disconnect the negative cable from the battery.

7. Recover the refrigerant with a recovery/charging station.

8. Loosen the drive belt.

9. Remove the bolts. Then, disconnect the suction line (A) and discharge line (B) from the compressor. Plug or cap the lines immediately after disconnecting them to avoid moisture and dust contamination.

10. Disconnect the compressor clutch connector (A), then remove the mounting bolts and the compressor (B).

11. Remove the left heat protector.

12. Remove the left exhaust manifold and gasket.

13. Remove the alternator.

14. Remove the right halfshaft.

15. Remove the right heat protector.

16. Remove the right exhaust manifold and gasket.

To install:

17. Install the right exhaust manifold and gasket. Tighten to 30–35 Nm (22–26 ft. lbs.).

18. Install the right heat protector. Tighten to 17–22 Nm (12–16 ft. lbs.).

19. Install the right halfshaft.

20. Install the alternator.

21. Install the exhaust manifold and gasket. Tighten to 30–35 Nm (22–26 ft. lbs.).

22. Install the left heat protector. Tighten to 12–15 Nm (9–11 ft. lbs.).

23. Install the air conditioner compressor.

24. Install the cooling fan motor assembly.

25. Connect the oxygen sensor connector.

26. Install the front exhaust pipe. Tighten to 30–40 Nm (22–30 ft. lbs.).

27. Install the under cover.

Camshaft

REMOVAL & INSTALLATION

2002 Models

1. Before servicing the vehicle, refer to the Precautions Section.

2. Properly relieve the fuel system pressure.

3. Drain the coolant into a suitable container.

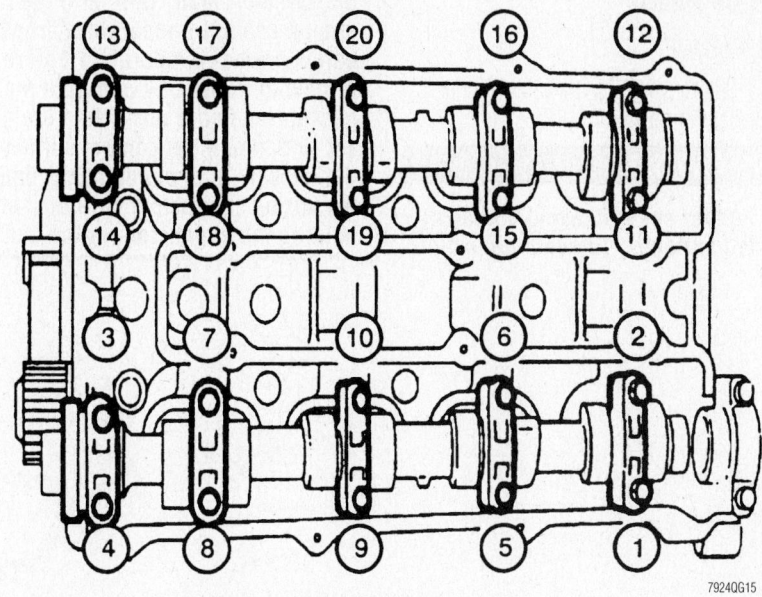

Camshaft cap bolt removal sequence—2002 models

7924QG15

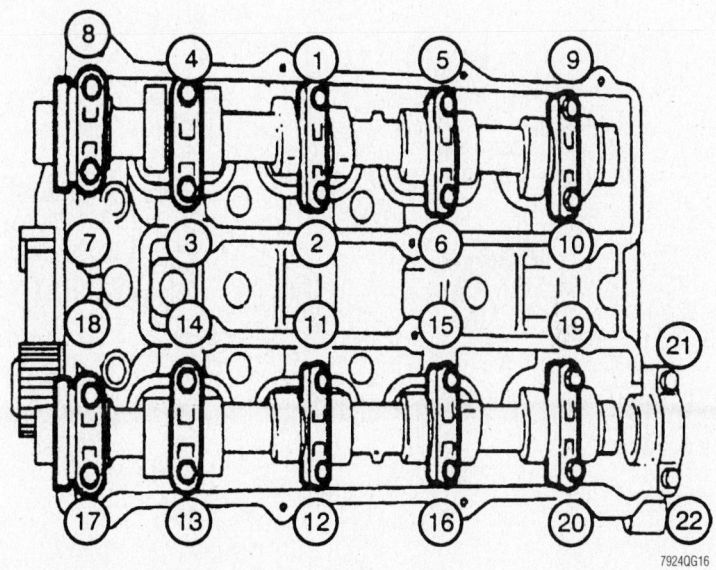

Camshaft journal bolt tightening sequence—2002 models

7924QG16

4. Remove or disconnect the following:

- Negative battery cable
- Upper timing belt cover
- Timing belt from the camshaft pulley
- Camshaft pulleys
- Camshaft cap bolts in the proper sequence
- Camshaft caps
- Camshafts

To install:

5. Install or connect the following:
- Camshafts into the cylinder head. The exhaust camshaft has a steel dowel pin at the rear for the camshaft position sensor
- Clean engine oil to the journals and bearings
- Camshaft oil seal
- Silicone sealant to the front camshaft cap and the camshaft position sensor mounting cap
- Camshaft caps in the proper sequence. Torque the bolts in three steps to 20 ft. lbs. (26 Nm).
- Camshaft pulleys
- Timing belt
- Timing belt cover
- Negative battery cable

2005–06 Models

2.0L ENGINE

1. Before servicing the vehicle, refer to the Precautions Section.

✳✳ CAUTION

Use Fender cover to avoid damaging painted surfaces. To avoid damaging the cylinder head, wait until the engine coolant temperature drops below normal temperature before removing it. When handling a metal gasket, take care not to fold the gasket or damage the contact surface of the gasket. To avoid damage, unplug the wiring connectors carefully while holding the connector portion.

→Mark all wiring and hoses to avoid misconnection. Inspect the timing belt before removing the cylinder head. Turn the crankshaft pulley so that the No. 1 piston is at top dead center.

2. Remove the air duct.
3. Disconnect the terminals from battery.

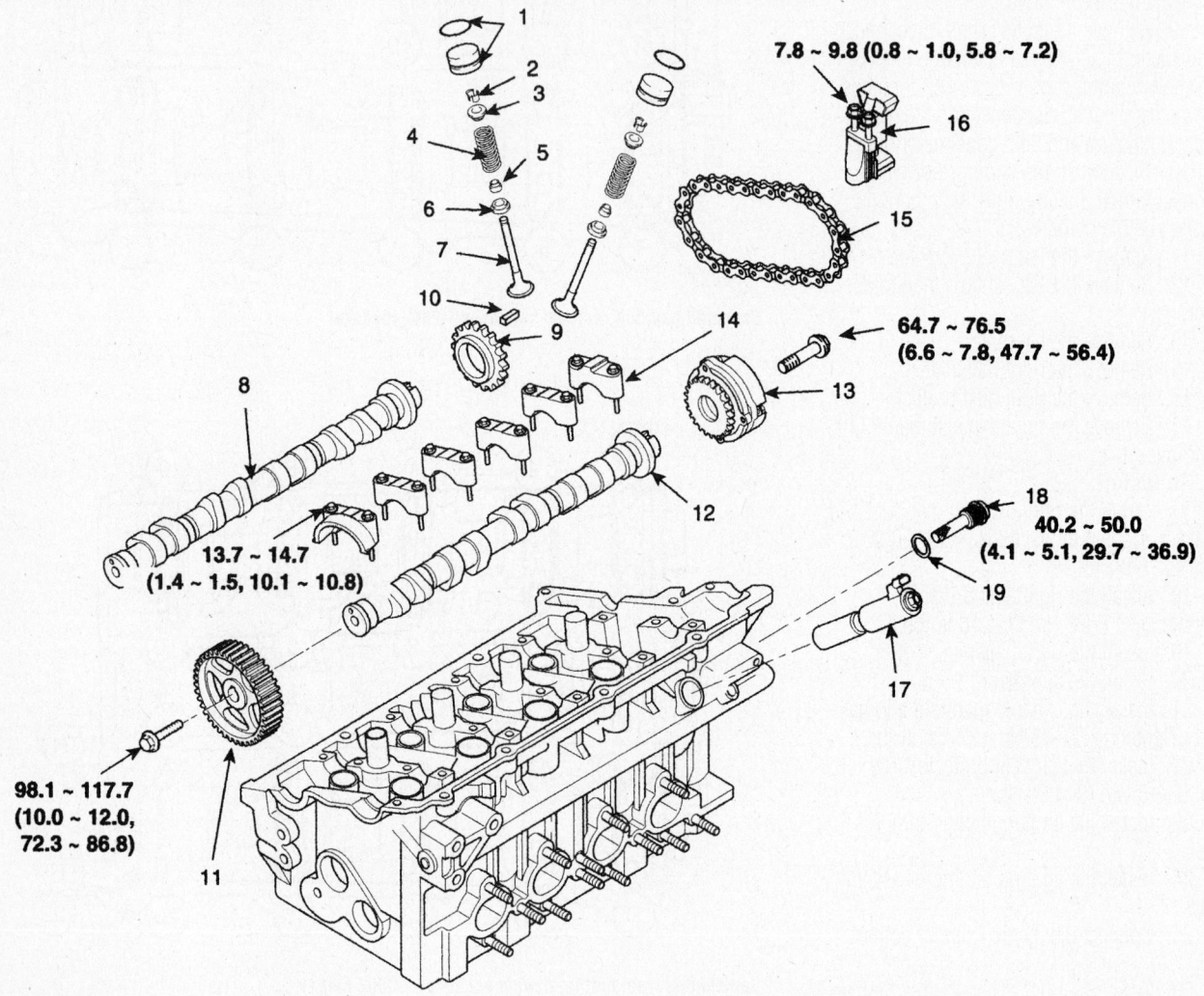

TORQUE : N.m (kgf.m, lb-ft)

1. MLA (Mechanical Lash Adjuster)
2. Retainer lock
3. Retainer
4. Valve spring
5. Stem seal
6. Spring seat
7. Valve
8. Intake camshaft
9. Chain sprocket
10. Key
11. Camshaft sprocket
12. Exhaust camshaft
13. CVVT (Continuously Variable Valve Timing) assembly
14. Camshaft bearing cap
15. Timing chain
16. Auto tentioner
17. OCV (Oil Control Valve)
18. OCV (Oil Control Valve) filter
19. Washer

Camshafts and related parts—2005–06 2.0L engine

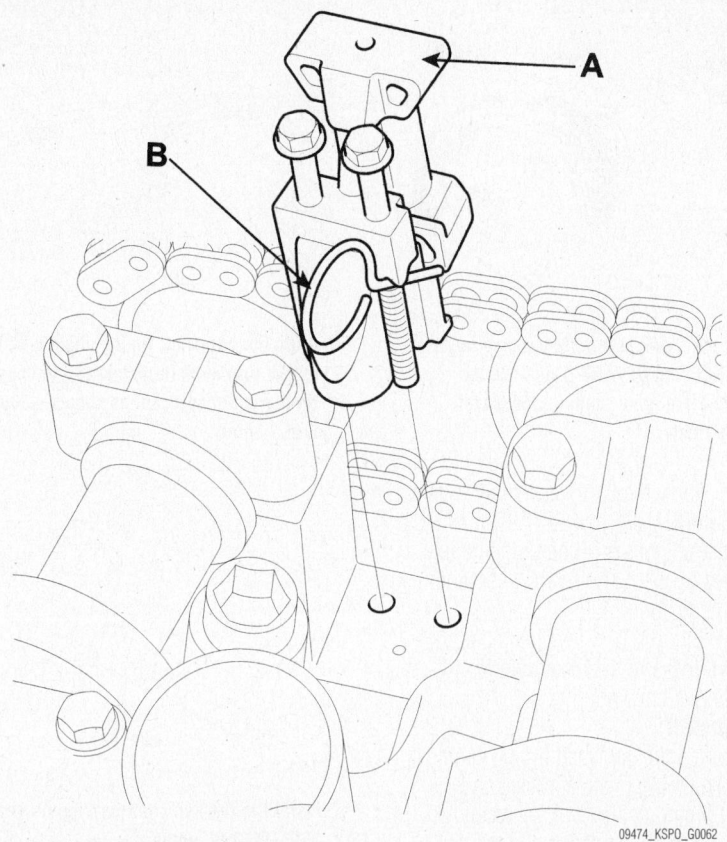

09474_KSPO_G0062

Insert a stopper pin or other device into timing chain auto tensioner—2005–06 2.0L engine

4. Remove the engine cover.

5. Drain the engine coolant. Remove the radiator cap to speed draining.

6. Remove the intake air hose and air cleaner assembly.

7. Remove the upper radiator hose and lower radiator hose.

8. Remove the heater hoses.

9. Remove the accelerator cable by loosening the lock-nut, then slip the cable end out of the throttle linkage.

10. Remove the engine wire harness connectors and wire harness clamps from cylinder head and the intake manifold.

11. Disconnect the fuel inlet hose of the delivery pipe side.

12. Disconnect the hose of the PCSV (Purge Control Solenoid Valve) side.

13. Remove the brake booster vacuum hose.

14. Remove the power steering pump drive belt.

15. Remove the power steering pump and fix the pump to vehicle with a wire.

16. Remove the bolts and power steering pump bracket.

17. Remove the spark plug cables.

18. Remove the timing belt.

19. Remove the PCV (Positive Crankcase Ventilation) hose.

20. Remove the cylinder head cover.

21. Remove the camshaft sprocket.

22. Insert a stopper pin or other device into timing chain auto tensioner and remove the auto tensioner.

23. Remove the camshaft bearing caps and camshafts.

24. Inspect the cam lobes.

   a. Using a micrometer, measure the cam lobe height.

   • Intake: 44.518–44.718mm (1.7527–1.7605in.)

   • Exhaust: 44.418–44.618mm (1.7487–1.7566in.)

   b. If the cam lobe height is less than minimum, replace the camshaft.

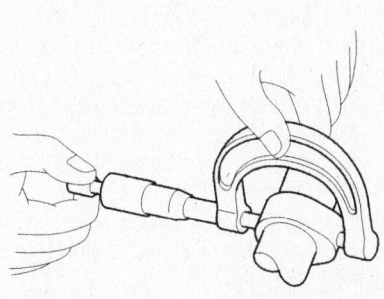

09474_KSPO_G0063

Using a micrometer, measure the cam lobe height

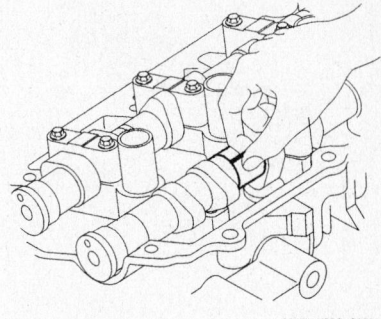

09474_KSPO_G0064

Lay a strip of Plastigage® across each of the camshaft journals

25. Inspect the camshaft journal clearance.

   a. Clean the bearing caps and camshaft journals.

   b. Place the camshafts on the cylinder head.

   c. Lay a strip of Plastigage® across each of the camshaft journals.

   d. Install the bearing caps and tighten the bolts to 10 ft. lbs. (14 Nm).

❋❋ WARNING

Do not turn the camshaft.

   e. Remove the bearing caps.

   f. Measure the Plastigage® at its widest point.

   g. Bearing oil clearance:

   • Standard: 0.020–0.061mm (0.0008–0.0024in)

   • Limit: 0.1mm (0.0039in.)

26. If the oil clearance is greater than maximum, replace the camshaft. If necessary, replace the bearing caps and cylinder head as a set.

27. Completely remove the Plastigage®.

28. Inspect the camshaft end play. Using a dial indicator, measure the end play while moving the camshaft back and forth. Camshaft end play should be 0.1–0.2mm

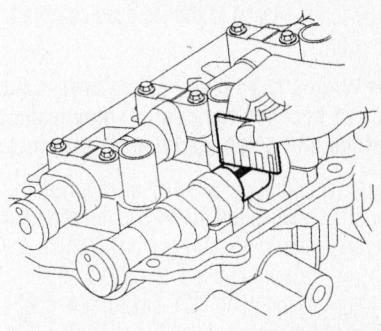

09474_KSPO_G0065

Measure the Plastigage® at its widest point

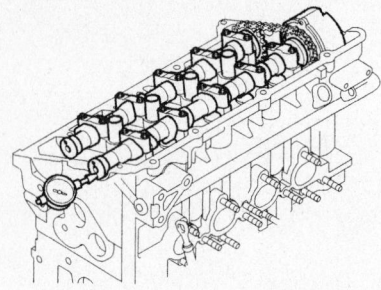

09474_KSPO_G0066

➡ Using a dial indicator, measure the end play while moving the camshaft back and forth

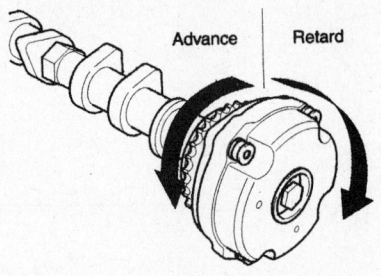

09474_KSPO_G0068

Allow the CVVT assembly to move in the ADVANCE and RETARD directions to ensure there is no binding and that it moves freely

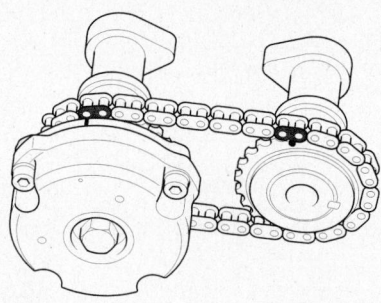

09474_KSPO_G0069

Align the camshaft timing chain with the intake timing chain sprocket and exhaust timing chain sprocket as shown—2005–06 2.0L engine

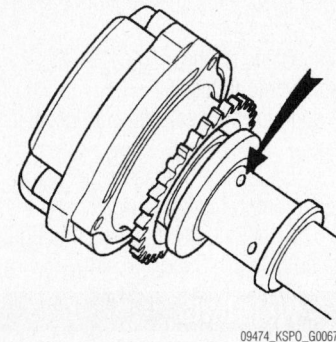

09474_KSPO_G0067

Apply vinyl tape to all the parts except the one indicated by the arrow

(0.0039–0.0079in.) If the end play is greater than maximum, replace the camshaft. If necessary, replace the bearing caps and cylinder head as a set.

29. Remove the camshafts.

30. Inspect the CVVT (Continuous Variable Valve Timing) assembly.

　a. Check that the CVVT (Continuous Variable Valve

　　Timing) assembly will not turn.

　b. Apply vinyl tape to all the parts except the one indicated by the arrow in the illustration.

　c. To release the CVVT lock pin, wrap some tape around the tip of an air pressure adapter and apply low air pressure (about 14 psi) to the exposed camshaft port.

➡**Wrap a shop towel or rag around the CVVT because residual oil may leak out of the unit when applying air pressure.**

　d. With low air pressure applied, turn the CVVT to the ADVANCE direction as indicated in the illustration. With the low air pressure applied, the CVVT should turn to the ADVANCE side. If too much air leaks when applying the low air pressure, the CVVT lock pin may not release and the CVVT may not turn.

　e. Allow the CVVT assembly to move in the ADVANCE and RETARD directions to ensure there is no binding and that it moves freely. It should move smoothly in the range about 20°.

　f. Turn the CVVT by hand and make sure it locks in the maximum retard angle position.

To install:

31. Align the camshaft timing chain with the intake timing chain sprocket and exhaust timing chain sprocket as shown.

32. Install the camshaft and bearing caps. Tighten to 13.7–14.7 Nm (10.1–10.8 ft. lbs.)

33. Install the timing chain auto tensioner. Tighten to 7.8–9.8 Nm (5.8–7.2 ft. lbs.)

34. Remove the auto tensioner stopper pin.

35. Check and adjust valve clearance.

36. Using seal driver, install the camshaft bearing oil seal.

37. Install the camshaft sprocket. Tighten to 72–87 ft. lbs. (98–117 Nm).

38. Install the cylinder head cover.

39. Install the PCV (Positive Crankcase Ventilation) hose.

40. Install the timing belt.

41. Install the intake manifold.

42. Install the exhaust manifold.

43. Install the spark plug cables.

44. Install the power steering pump bracket and bolts. Tighten the mounting bolts to 34–49 Nm (25–36 ft. lbs.); the pivot bolt to 12–15 Nm (9–11 ft. lbs.)

45. Install the power steering pump.

46. Install the accelerator cable.

47. Install the brake booster hose.

48. Connect the hose of the PCSV (Purge Control Solenoid Valve) side.

49. Connect the fuel inlet hose of the delivery pipe side.

50. Install the engine wire harness connectors and wire harness clamps to the cylinder head and the intake manifold.

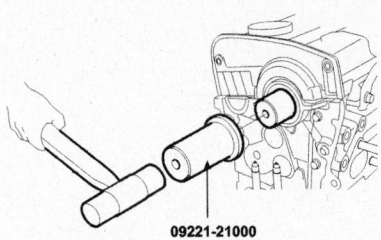

09221-21000

09474_KSPO_G0070

Installing the camshaft bearing oil seal—2005–06 2.0L engine

51. Install the heater hose.

52. Install the upper radiator hose and lower radiator hose.

53. Install the intake air hose and air cleaner assembly.

54. Install the engine cover.

55. Reconnect the battery terminals.

56. Install the air duct.

57. Fill with engine coolant.

58. Start the engine and check for leaks.

59. Recheck engine coolant level and oil level.

2.7L ENGINE

1. Before servicing the vehicle, refer to the Precautions Section.

❄❄ **WARNING**

Use fender covers to avoid damaging painted surfaces. To avoid damaging the cylinder head, wait until the engine coolant temperature drops below normal temperature before removing it. When handling a metal gasket, take care not to fold the gasket or damage the contact surface of the gasket. To avoid damage, unplug the wiring connectors carefully while holding the connector portion.

➡**Mark all wiring and hoses to avoid misconnection. Inspect the timing belt before removing the cylinder head. Turn the crankshaft pulley so that the No. 1 piston is at top dead center.**

2. Remove the air duct.

3. Remove the cowl grill and wiper motor.

4. Remove the strut bar.

5. Disconnect the negative terminal from the battery.

6. Drain the engine coolant. Remove the radiator cap to speed draining.

7. Remove the engine cover.

8. Remove the intake air hose and air cleaner assembly.

9. Remove the coolant reservoir tank.

10. Remove the upper radiator hose and lower radiator hose.

11. Remove the heater hoses and throttle body heater hose.

12. Remove the engine wire harness connectors and wire harness clamps from the cylinder head and the intake manifold.

13. Remove the fuel inlet hose from delivery pipe.

14. Remove the brake booster vacuum hose.

15. Remove the accelerator cable by loosening the locknut, then slip the cable end out of the throttle linkage.

16. Remove the PCV hose.

17. Remove the intake manifold.

18. Remove the power steering pump.

19. Remove the timing belt.

20. Remove the spark plug cable.

21. Remove the cylinder head covers.

22. Remove the camshaft sprocket.

23. Remove the camshaft bearing caps.

24. Remove the camshafts.

25. Using a micrometer, measure the cam lobe height. Cam height:
- Intake: 43.95–44.15 mm (1.7303–1.7382 in.).
- Exhaust: 43.95–44.15 mm (1.7303–1.7382 in.).

26. If the cam lobe height is less than minimum, replace the camshaft.

27. Using a micrometer, measure the journal diameter. Standard value: 25.964–25.980 mm (1.0222–1.0228 in.).

28. If the journal diameter is not as specified, check the oil clearance.

29. Check the bearings for flaking and scoring. If the bearings are damaged, replace the bearing caps and cylinder head as a set.

30. Clean the bearing caps and camshaft journals.

31. Place the camshafts on the cylinder head.

32. Lay a strip of Plastigage® across each of the camshaft journal.

33. Install the bearing caps. Tighten the bolts as described below. Remove the bearing caps.

34. Measure the Plastigage® at its widest point. Bearing oil clearance:
- Standard value: 0.02–0.061 mm (0.0008–0.0024 in.)
- Limit: 0.1 mm (0.0039 in.)

35. If the oil clearance is greater than maximum, replace the camshaft. If necessary, replace the bearing caps and cylinder head as a set.

36. Completely remove the Plastigage®.

37. Using a dial indicator, measure the end play while moving the camshaft back and forth. Standard value: 0.1–0.15 mm (0.004–0.0059 in.)

38. If the end play is greater than maximum, replace the camshaft. If necessary, replace the bearing caps and cylinder head as a set.

39. Remove the camshafts.

To install:

➡**Thoroughly clean all parts to be assembled. Always use a manifold gasket. Rotate the crankshaft, set the No.1 piston at TDC.**

40. Align the camshaft timing chain with the intake timing chain sprocket and exhaust timing chain sprocket as shown.

41. Apply new engine oil to the thrust portion and journal of the camshafts.

42. Apply a light coat of engine oil on the threads and under the heads of the bearing cap bolts.

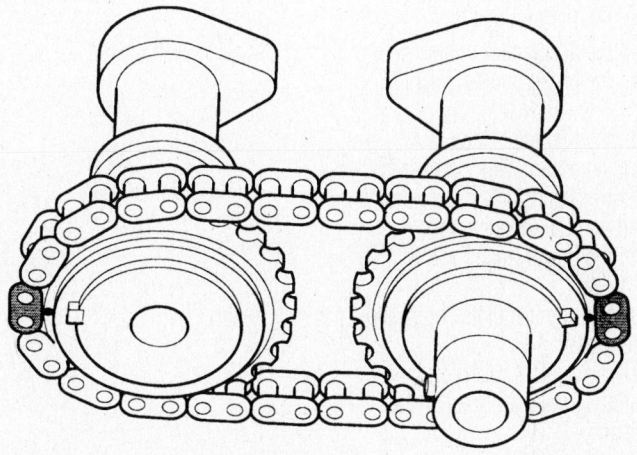

09474_KSPO_G0071

Align the camshaft timing chain with the intake timing chain sprocket and exhaust timing chain sprocket as shown—2.7L engine

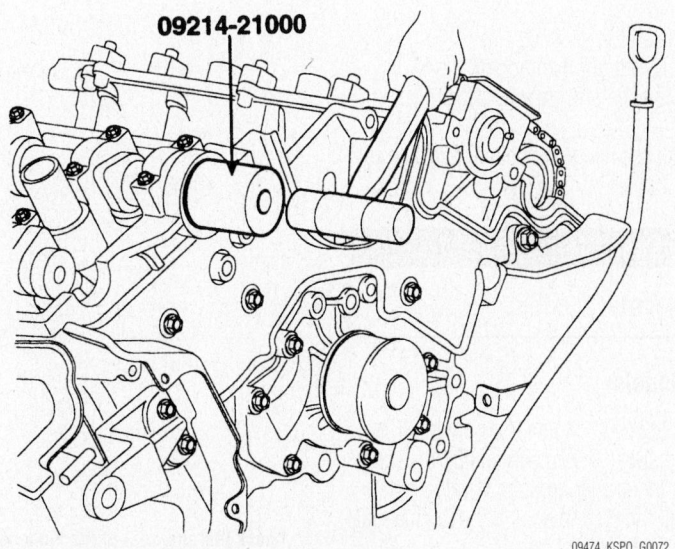

09214-21000

09474_KSPO_G0072

Installing the camshaft bearing oil seal—2.7L engine

43. Install the camshafts.
44. Install the camshaft bearing caps. Tightening torque:
   - M6 (38mm) (Mark 7): 10–12 Nm (7–9 ft. lbs.)
   - M6 (50 mm) (Mark 10): 14–16 Nm (10–12 ft. lbs.)
45. Using a driver, install the camshaft bearing oil seal.
46. Temporarily install the camshaft sprocket bolts.
47. Hold the hexagonal head wrench portion of the camshaft with a wrench, and tighten the camshaft sprocket bolts to 90–110 Nm (65–80 ft. lbs.).
48. Install the cylinder head covers.
49. Install the spark plug cables.
50. Install the timing belt.
51. Install the exhaust manifold.
52. Install the power steering pump.
53. Install the intake manifold.
54. Install the PCV hose.
55. Install the accelerator cable.
56. Install the brake booster vacuum hose.
57. Install the fuel inlet hose.
58. Install the engine wire harness connectors and wire harness clamps to the cylinder head and the intake manifold.
59. Install the heater hoses and throttle body heater hose.
60. Install the upper radiator hose and lower radiator hose.
61. Install the coolant reservoir tank.
62. Install the intake air hose and air cleaner assembly.
63. Install the engine cover.
64. Connect the negative terminal to the battery.
65. Install the strut bar.
66. Install the cowl grill and wiper motor.
67. Install the air duct.
68. Fill with engine coolant.
69. Start the engine and check for leaks.
70. Recheck engine coolant level and oil level.

Valve Lash

ADJUSTMENT

2002 Models

The 2002 2.0L engine uses Hydraulic Lash Adjusters (HLA's), which automatically maintain the proper amount of valve lash. Therefore, the engine does not need manual valve lash adjustment.

2005–06 Models

2.0L ENGINE

➡**Inspect and adjust the valve clearance when the engine is cold (engine coolant temperature 20C ± 5C [68 F ± 9 F]) and cylinder head is installed on cylinder block.**

1. Before servicing the vehicle, refer to the Precautions Section.
2. Remove the engine cover.
3. Remove the bolts (B) and timing belt upper cover (A).

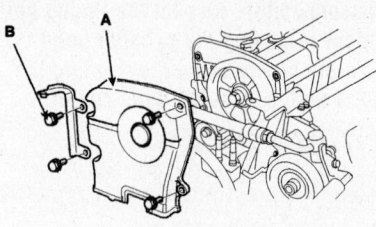

09474_KSPO_G0073

Timing belt upper cover—2005–06 2.0L engine

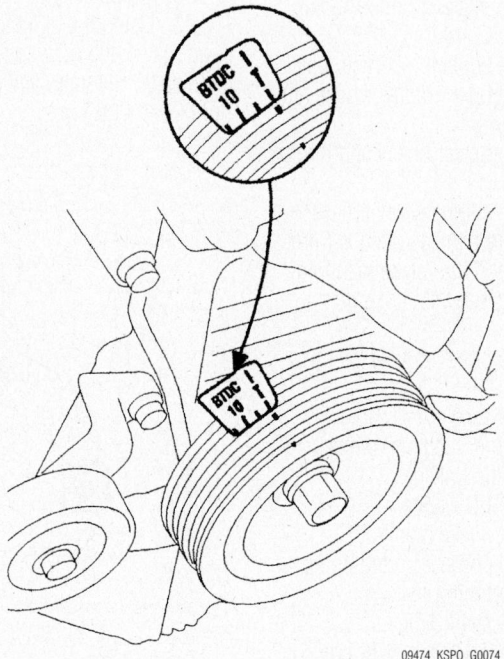

09474_KSPO_G0074

Turn the crankshaft pulley and align its groove with the timing mark "T" of the lower timing belt cover—2005–06 2.0L engine

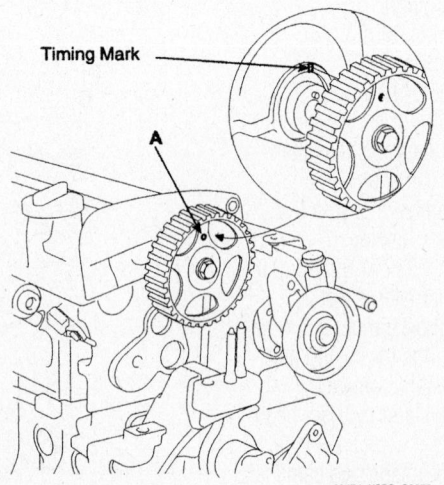

09474_KSPO_G0075

Check that the hole of the camshaft timing pulley (A) is aligned with the timing mark of the bearing cap—2005–06 2.0L engine

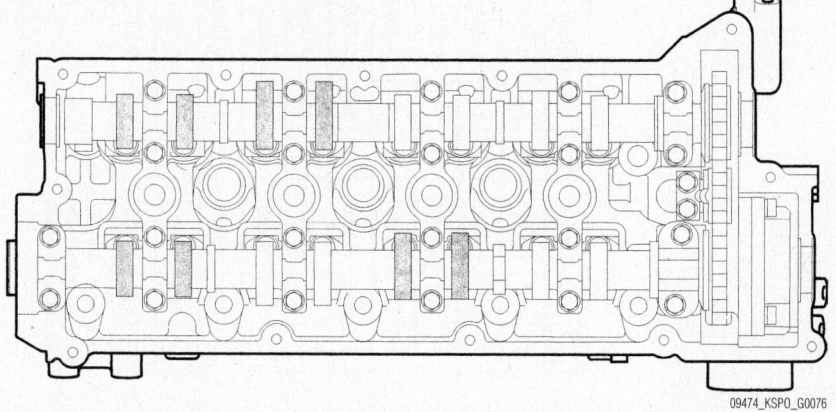

Check these valves (shaded) with No. 1 cylinder is at TDC/compression—2005–06 2.0L engine

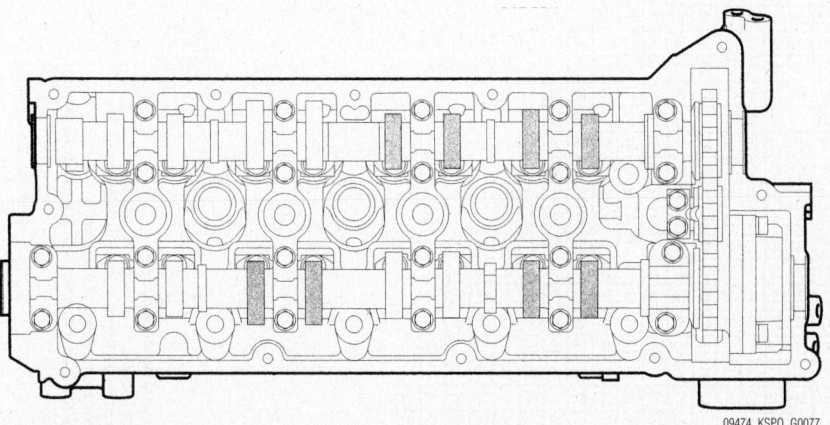

Check these valves (shaded) with No. 4 cylinder is at TDC/compression—2005–06 2.0L engine

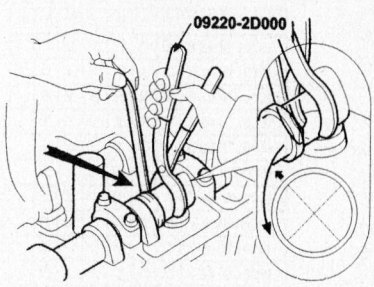

Using the SST (09220—2D000), press down the valve lifter and place the stopper between the camshaft and valve lifter—2005–06 2.0L engine

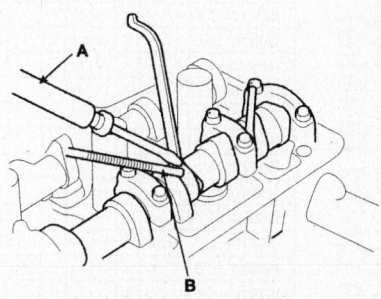

Remove the adjusting shim with a small screw driver (A) and magnet (B)—2005–06 2.0L engine

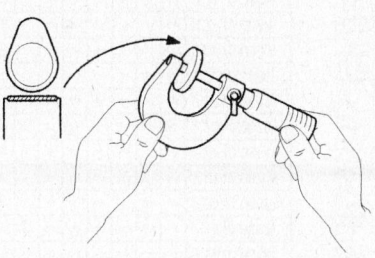

Measure the thickness of the removed shim using a micrometer—2005–06 2.0L engine

4. Remove the cylinder head cover.

5. Turn the crankshaft pulley and align its groove with the timing mark "T" of the lower timing belt cover.

6. Check that the hole of the camshaft timing pulley (A) is aligned with the timing mark of the bearing cap. If not, turn the crankshaft one revolution (360°).

7. Inspect the valve clearance:

   a. Check only the valves indicated as shown. No. 1 cylinder is at TDC/compression. Measure the valve clearance.

   b. Using a thickness gauge, measure the clearance between the tappet shim and the base circle of camshaft.

   c. Record the out-of-specification valve clearance measurements. They will be used later to determine the required replacement adjusting shim. Valve clearance at engine coolant temperature 20C ± 5C (68F ± 9F) should be:

   • Intake: 0.20mm (0.0079in)
   • Exhaust: 0.28mm (0.0110in)

   d. Turn the crankshaft pulley one revolution (360°) and align the groove with

the timing mark "T" of lower timing belt cover.

   e. Check only valves indicated as shown. No. 4 cylinder is at TDC/compression. Measure the valve clearance.

8. Adjust the intake and exhaust valve clearance:

   a. Turn the crankshaft so that the cam lobe of the camshaft on the adjusting valve is upward.

   b. Using the SST (09220-2D000), or equivalent, press down the valve lifter and place the stopper between the camshaft and valve lifter and remove the special tool.

   c. Remove the adjusting shim with a small screw driver (A) and magnet (B).

   d. Measure the thickness of the removed shim using a micrometer.

   e. Calculate the thickness of a new shim so that the valve clearance comes within the specified value at engine coolant temperature 20C ± 5C (68F ± 9F)
T: Thickness of removed shim
A: Measured valve clearance

N: Thickness of new shim
Intake: N = T + [A-0.20mm (0.0079 in.)]
Exhaust: N = T + [A-0.28mm (0.0110 in.)]

   a. Select a new shim with a thickness as close as possible to the calculated value. (Refer to the Adjusting shim selection chart)

➡ Shims are available in 20 size increments of 0.04mm (0.0016 in.) from 2.00mm (0.0787 in.) to 2.76mm (0.1087 in.)

   b. Place a new adjusting shim on the valve lifter.

Adjusting Shim Selection Chart (Intake)

New shim thickness mm (in.)

shim No.	Thickness		shim No.	Thickness	
1	2.00	(0.0787)	11	2.40	(0.0945)
2	2.04	(0.0803)	12	2.44	(0.0961)
3	2.08	(0.0819)	13	2.48	(0.0976)
4	2.12	(0.0835)	14	2.52	(0.0992)
5	2.16	(0.0850)	15	2.56	(0.1008)
6	2.20	(0.0866)	16	2.60	(0.1024)
7	2.24	(0.0882)	17	2.64	(0.1039)
8	2.28	(0.0898)	18	2.68	(0.1055)
9	2.32	(0.0913)	19	2.72	(0.1071)
10	2.36	(0.0929)	20	2.76	(0.1087)

HINT : New shims have the thickness in millimeters imprinted on the face

Intake valve clearance (cold) :
0.20 mm (Spec.), 0.12–0.28 mm (Limit)
Example : The 2.24 mm shim is installed, and the measured clearance is 0.450 mm.
Replace the 2.24 mm shim with a new No.13 shim.

09474_KSPO_G0081

Installed shim thickness mm (in): 2.00 (0.0787), 2.02 (0.0795), 2.04 (0.0803), 2.06 (0.0811), 2.08 (0.0819), 2.10 (0.0827), 2.11 (0.0831), 2.12 (0.0835), 2.13 (0.0839), 2.14 (0.0843), 2.15 (0.0846), 2.16 (0.0850), 2.17 (0.0854), 2.18 (0.0858), 2.19 (0.0862), 2.20 (0.0866), 2.21 (0.0870), 2.22 (0.0874), 2.23 (0.0878), 2.24 (0.0882), 2.25 (0.0886), 2.26 (0.0890), 2.27 (0.0894), 2.28 (0.0898), 2.29 (0.0902), 2.30 (0.0906), 2.31 (0.0909), 2.32 (0.0913), 2.33 (0.0917), 2.34 (0.0921), 2.35 (0.0925), 2.36 (0.0929), 2.37 (0.0933), 2.38 (0.0937), 2.39 (0.0941), 2.40 (0.0945), 2.41 (0.0949), 2.42 (0.0953), 2.43 (0.0957), 2.44 (0.0961), 2.45 (0.0965), 2.46 (0.0969), 2.47 (0.0972), 2.48 (0.0976), 2.49 (0.0980), 2.50 (0.0984), 2.51 (0.0988), 2.52 (0.0992), 2.53 (0.0996), 2.54 (0.1000), 2.56 (0.1008), 2.58 (0.1016), 2.6 (0.1024), 2.62 (0.1031), 2.64 (0.1039), 2.66 (0.1047), 2.68 (0.1055), 2.70 (0.1063), 2.72 (0.1071), 2.74 (0.1079), 2.76 (0.1087)

Measured clearance mm (in):
0.000 - 0.020 (0.0000 - 0.0008)
0.021 - 0.040 (0.0008 - 0.0016)
0.041 - 0.060 (0.0016 - 0.0024)
0.061 - 0.080 (0.0024 - 0.0031)
0.081 - 0.100 (0.0032 - 0.0039)
0.101 - 0.119 (0.0040 - 0.0047)
0.120 - 0.280 (0.0047 - 0.0110)
0.281 - 0.300 (0.0111 - 0.0118)
0.301 - 0.320 (0.0119 - 0.0126)
0.321 - 0.340 (0.0126 - 0.0134)
0.341 - 0.360 (0.0134 - 0.0142)
0.361 - 0.380 (0.0142 - 0.0150)
0.381 - 0.400 (0.0150 - 0.0157)
0.401 - 0.420 (0.0158 - 0.0165)
0.421 - 0.440 (0.0166 - 0.0173)
0.441 - 0.460 (0.0174 - 0.0181)
0.461 - 0.480 (0.0181 - 0.0189)
0.481 - 0.500 (0.0189 - 0.0197)
0.501 - 0.520 (0.0197 - 0.0205)
0.521 - 0.540 (0.0205 - 0.0213)
0.541 - 0.560 (0.0213 - 0.0220)
0.561 - 0.580 (0.0221 - 0.0228)
0.581 - 0.600 (0.0229 - 0.0236)
0.601 - 0.620 (0.0237 - 0.0244)
0.621 - 0.640 (0.0244 - 0.0252)
0.641 - 0.660 (0.0252 - 0.0260)
0.661 - 0.680 (0.0260 - 0.0268)
0.681 - 0.700 (0.0268 - 0.0276)
0.701 - 0.720 (0.0276 - 0.0283)
0.721 - 0.740 (0.0284 - 0.0291)
0.741 - 0.760 (0.0292 - 0.0299)
0.761 - 0.780 (0.0300 - 0.0307)
0.781 - 0.800 (0.0307 - 0.0315)
0.801 - 0.820 (0.0315 - 0.0323)
0.821 - 0.840 (0.0323 - 0.0331)
0.841 - 0.860 (0.0331 - 0.0339)
0.861 - 0.880 (0.0339 - 0.0346)
0.881 - 0.900 (0.0347 - 0.0354)
0.901 - 0.920 (0.0355 - 0.0362)
0.921 - 0.940 (0.0363 - 0.0370)
0.941 - 0.960 (0.0370 - 0.0378)
0.961 - 0.980 (0.0378 - 0.0386)

Adjusting Shim Selection Chart (Exhaust)

09474_KSPO_G0082

New shim thickness   mm (in.)

shim No.	Thickness	shim No.	Thickness
1	2.00 (0.0787)	11	2.40 (0.0945)
2	2.04 (0.0803)	12	2.44 (0.0961)
3	2.08 (0.0819)	13	2.48 (0.0976)
4	2.12 (0.0835)	14	2.52 (0.0992)
5	2.16 (0.0850)	15	2.56 (0.1008)
6	2.20 (0.0866)	16	2.60 (0.1024)
7	2.24 (0.0882)	17	2.64 (0.1039)
8	2.28 (0.0898)	18	2.68 (0.1055)
9	2.32 (0.0913)	19	2.72 (0.1071)
10	2.36 (0.0929)	20	2.76 (0.1087)

HINT : New shims have the thickness in millimeters imprinted on the face

Exhaust valve clearance (cold) :
0.28 mm (Spec.), 0.20-0.38 mm (Limit)
Example : The 2.24 mm shim is installed, and the measured clearance is 0.450 mm.
Replace the 2.24 mm shim with a new No.11 shim.

Installed shim thickness mm (in) / Measured clearance mm (in)

c. Using the SST (09220-2D000), press down the valve lifter and remove the stopper.

d. Recheck the valve clearance at engine coolant temperature 20°C ± 5°C (68°F ± 9°F)

Standard:
- Intake: 0.20mm (0.0079 in.)
- Exhaust: 0.28mm (0.0110 in.)

Limit (After adjusting valve clearance):
- Intake: 0.17–0.23mm (0.0067–0.0091 in.)
- Exhaust: 0.25–0.31mm (0.0098–0.0122 in.)

2.7L ENGINE

The 2.7L engine uses hydraulic lifters which don't require adjustment.

Starter

REMOVAL & INSTALLATION

2002 Models

1. Before servicing the vehicle, refer to the Precautions Section.
2. Drain the engine oil.
3. Remove or disconnect the following:
   - Negative battery cable
   - Intake manifold bracket upper bolts
   - Clutch release cylinder and move it aside, if equipped
   - Intake manifold bracket lower bolts and remove the bracket
   - Starter from the clutch, manual transmissions only
   - Starter from the torque converter, automatic transmissions only
   - Starter electrical connectors
   - Move the transmission wire harness aside
   - Starter

To install:
4. Install or connect the following:
   - Starter to the engine well
   - Starter electrical connectors
   - Lower intake manifold bracket and install the upper bolts
   - Starter into position. When aligned properly torque the bolts to 40 ft. lbs. (54 Nm).
   - Torque the intake manifold bracket bolts to 40 ft. lbs. (54 Nm).
   - Properly position the clutch release cylinder, if equipped. Torque the bolts to 40 ft. lbs. (54 Nm).
   - Torque the intake manifold bracket upper bolts to 40 ft. lbs. (54 Nm).
   - Negative battery cable

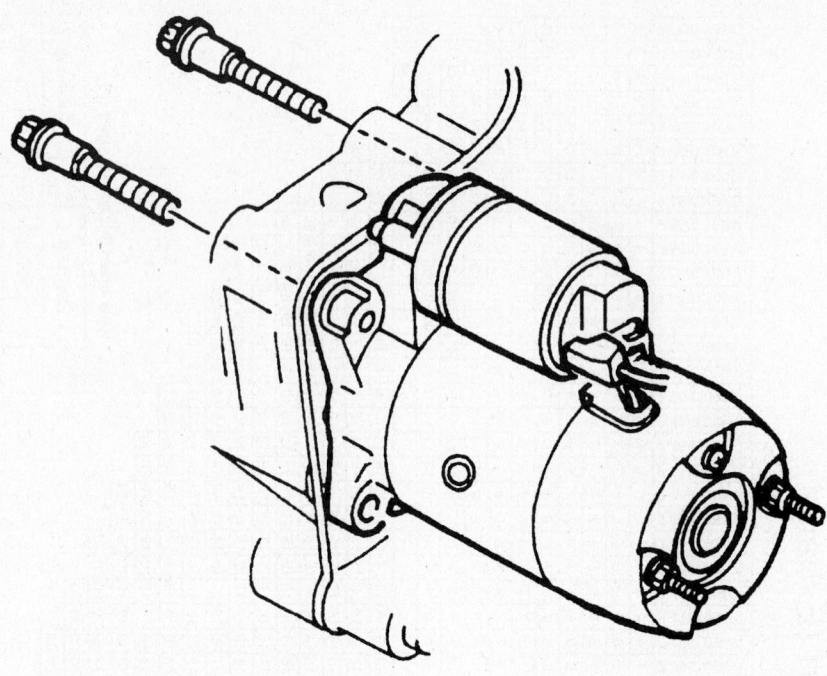

Exploded view of the starter—2002 models

9308QG02

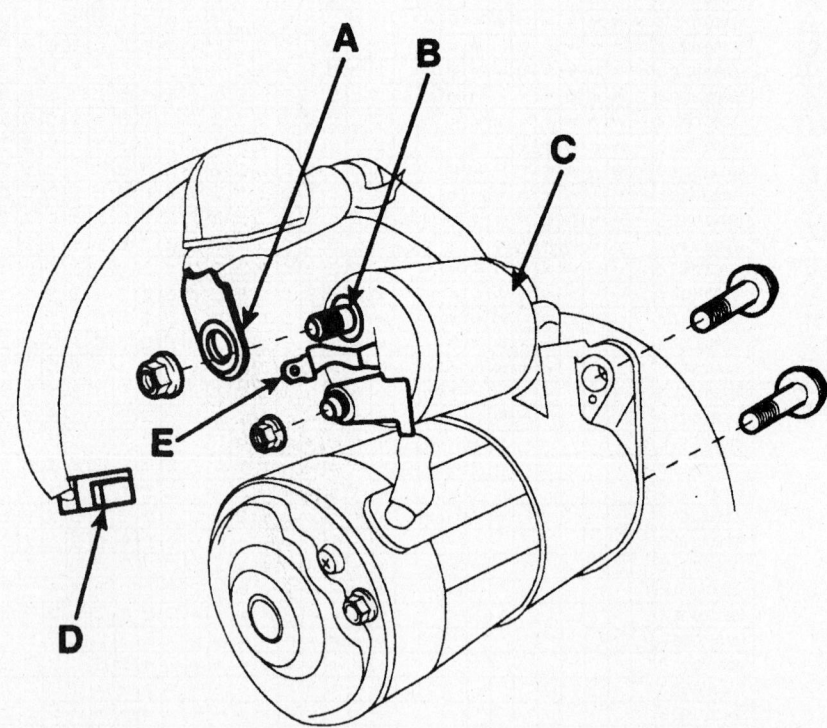

Starter mounting—typical of 2005–06 models

09474_KSPO_G0083

2005—06 Models

1. Before servicing the vehicle, refer to the Precautions Section.
2. Disconnect the battery negative cable.
3. Disconnect the starter cable (A) from the B terminal (B) on the solenoid (C), then disconnect the connector (D) from the "S" terminal (E).
4. Remove the 2 bolts holding the starter, then remove the starter.
5. Installation is the reverse of removal.

Torque the nuts and bolts to 20 ft. lbs. (27 Nm).

6. Connect the battery positive cable and negative cable to the battery.

Oil Pan

REMOVAL & INSTALLATION

2002 Models

1. Before servicing the vehicle, refer to the Precautions Section.
2. Drain the engine oil.
3. Remove or disconnect the following:
   - Negative battery cable
   - 2 Top intake manifold bracket bolts
   - Front 3 axle housing mounting bolts, 4WD only
   - Left front bushing from the axle housing mount and lower the front axle housing
   - Both gusset plates from the engine
   - Transmission under cover
   - Engine under cover
   - Oil pan mounting bolts and using a scrapper tool separate the oil pan
   - Oil pan
   - Oil strainer assembly
   - Oil baffle

To install:

4. Clean the engine block, oil pan and baffle pan surfaces of any gasket material.
5. Apply a continuous bead of Loctite Ultra Blue 587® silicone sealant around the baffle pan.
6. Install or connect the following:
   - Oil baffle. Torque the bolt to 84 inch lbs. (9.5 Nm).
   - Oil strainer. Torque the bolts to 84 inch lbs. (9.5 Nm).
7. Apply a continuous bead of Loctite Ultra Blue 587® silicone sealant around the oil pan.
   - Oil pan. Torque the bolts to 84 inch lbs. (9.5 Nm).
   - Transmission under cover. Torque the bolts to 84 inch lbs. (9.5 Nm).
   - Gusset plates to the engine. Torque the bolts to 33 ft. lbs. (45 Nm).
   - Engine under cover
   - Front axle housing into position. When properly aligned, torque the bolts to 48 ft. lbs. (65 Nm).
   - Intake manifold bracket bolts. Torque the bolts to 34 ft. lbs. (65 Nm).
   - Negative battery cable
8. Fill the engine with clean oil.
9. Start the vehicle and check for leaks, repair if necessary.

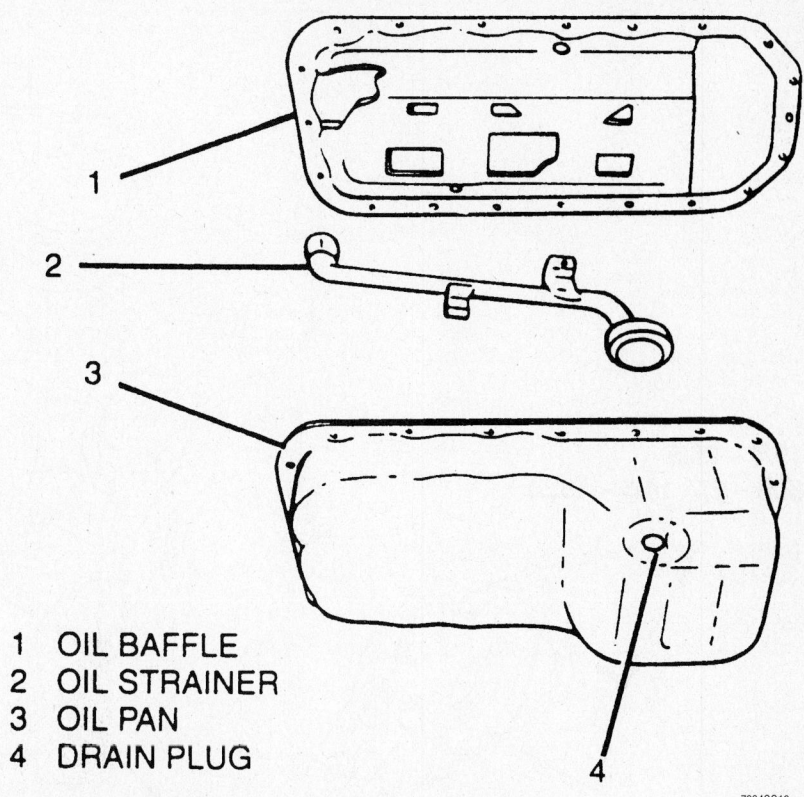

1. OIL BAFFLE
2. OIL STRAINER
3. OIL PAN
4. DRAIN PLUG

7924QG18

Exploded view of the oil pan assembly mounting—2002 models

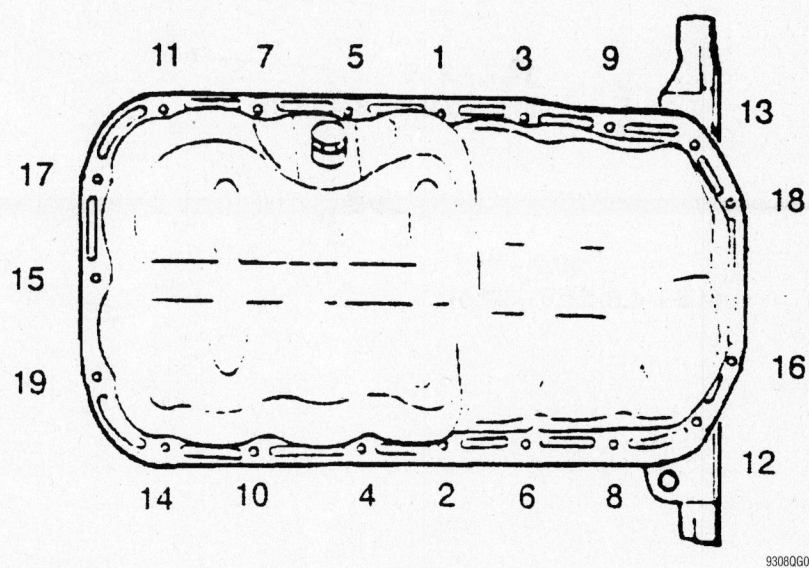

9308QG03

Tighten the oil pan bolts in sequence—2002 models

2005–06 Models

2.0L ENGINE

1. Before servicing the vehicle, refer to the Precautions Section.
2. Drain the engine oil.
3. Disconnect the rear oxygen sensor connector.
4. Remove the front muffler (A).
5. Remove the exhaust manifold.
6. Remove the front muffler bracket (A).
7. Remove the oil pan.

To install:

8. Using a razor blade and gasket scraper, remove all the old packing material from the gasket surfaces.

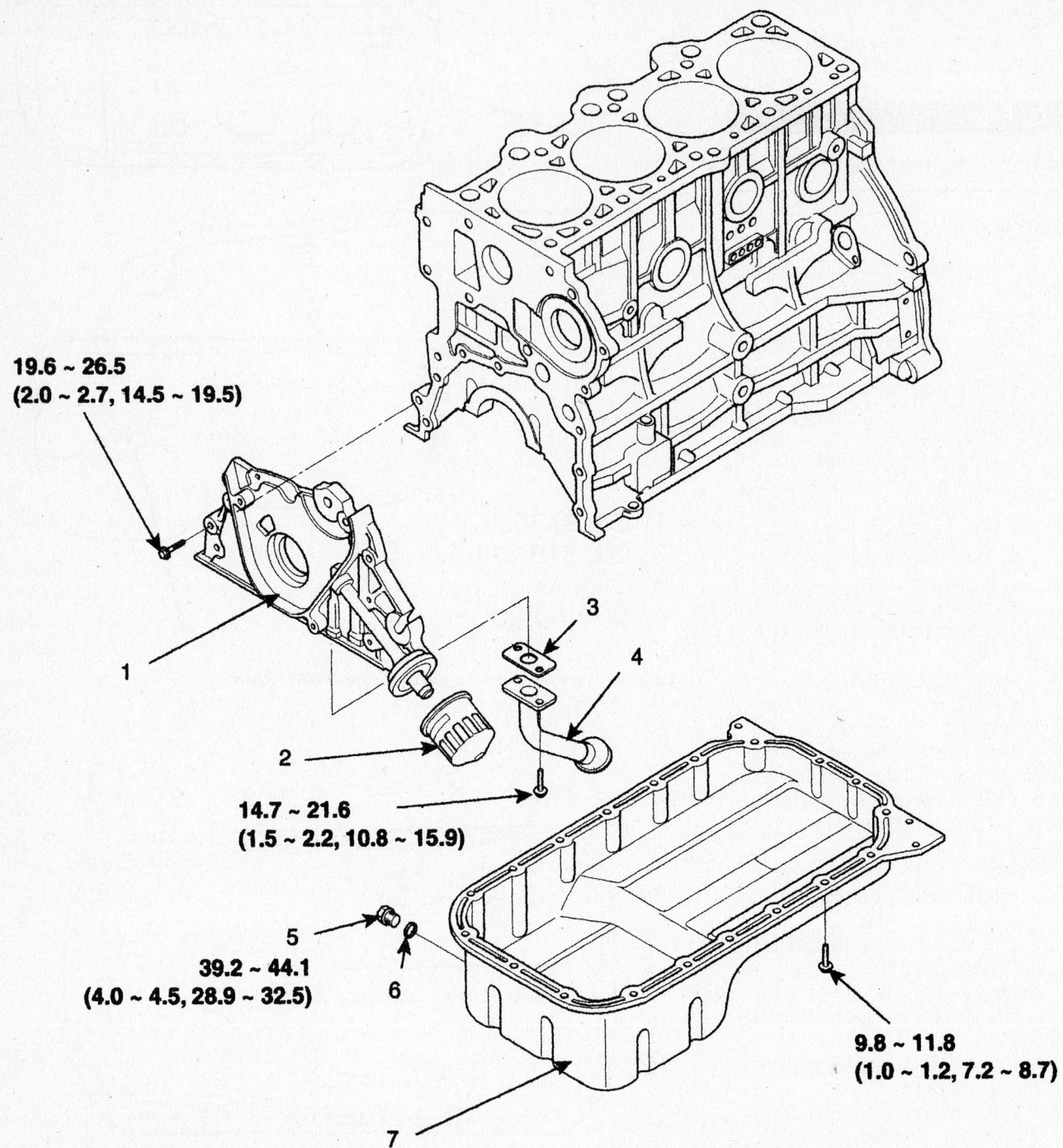

19.6 ~ 26.5
(2.0 ~ 2.7, 14.5 ~ 19.5)

14.7 ~ 21.6
(1.5 ~ 2.2, 10.8 ~ 15.9)

39.2 ~ 44.1
(4.0 ~ 4.5, 28.9 ~ 32.5)

9.8 ~ 11.8
(1.0 ~ 1.2, 7.2 ~ 8.7)

TORQUE : N.m (kgf.m, lb-ft)

1. Front case
2. Filter
3. Gasket
4. Oil screen

5. Drain plug
6. Gasket
7. Oil pan

09474_KSPO_G0084

Oil pan and related parts—2005-06 2.0L engine

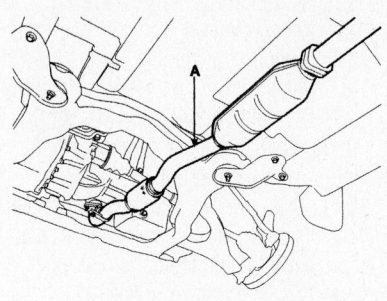

Front muffler—2005–06 2.0L engine

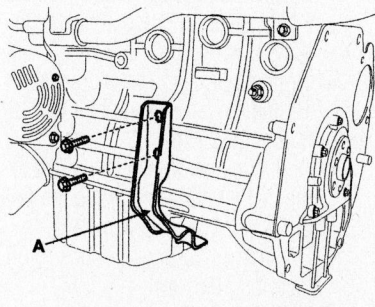

Front muffler bracket—2005–06 2.0L

➡**Check that the mating surfaces are clean and dry before applying liquid gasket.**

9. Apply liquid gasket MS 721-40A, or equivalent, as an even bead, centered between the edges of the mating surface.

➡**To prevent leakage of oil, apply liquid gasket to the inner threads of the bolt holes. Do not install the parts if five minutes or more have elapsed since applying the liquid gasket. Instead, reapply liquid gasket after removing the residue. After assembly, wait at least 30 minutes before filling the engine with oil.**

10. Install the oil pan with the bolts. Uniformly tighten the bolts in several passes to 10–12 Nm (7–9 ft. lbs.)
11. Install the front muffler bracket.
12. Install the exhaust manifold.
13. Install the front muffler.
14. Fill with engine oil

2.7L ENGNE

1. Before servicing the vehicle, refer to the Precautions Section.
2. Remove the engine and mount it on a work stand.
3. Remove the bolts and remove the lower pan.
4. Remove the 2 bolts and remove the pickup and gasket.

5. Remove the timing belt.
6. Remove the bolts and remove the upper oil pan.

To install:

7. Using a razor blade and gasket scraper, remove all the old packing material from the gasket surfaces. Check that the mating surfaces are clean and dry before applying liquid gasket.
8. Install the upper oil pan with the 17 bolts.
9. Uniformly tighten the bolts in several passes, in the sequence shown, to, bolts 1–15, 19–28 Nm (14–20 ft. lbs.); bolts 16 and 17, 5–7 Nm (4–5 ft. lbs.).

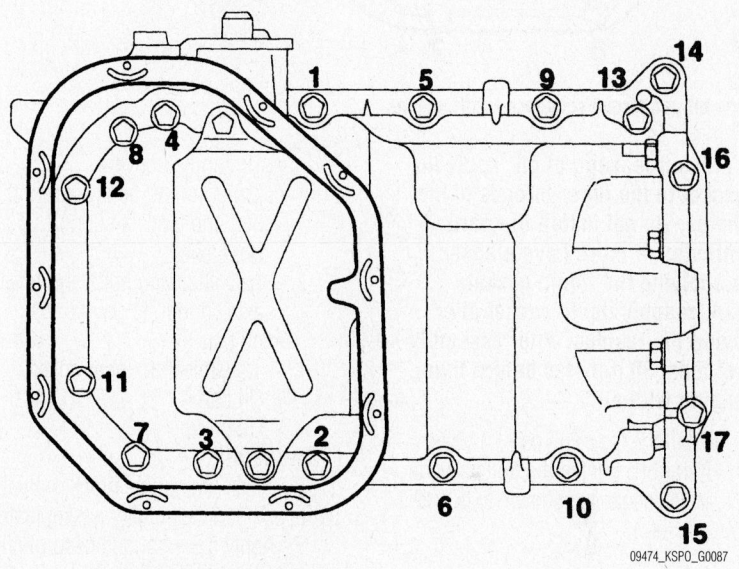

Upper pan torque sequence—2.7L engine

➡**To prevent leakage of oil, apply liquid gasket to the inner threads of the bolt holes. Do not install the parts if five minutes or more have elapsed since applying the liquid gasket. Instead, reapply liquid gasket after removing the residue. After assembly, wait at least 30 minutes before filling the engine with oil.**

10. Install a new gasket and oil screen (A) with 2 bolts (B) to 15–22 Nm (11–16 ft. lbs.)
11. Apply liquid gasket as an even bead, centered between the edges of the mating surface. Use liquid gasket MS 721-40 A, or equivalent.
12. Install the lower oil pan.

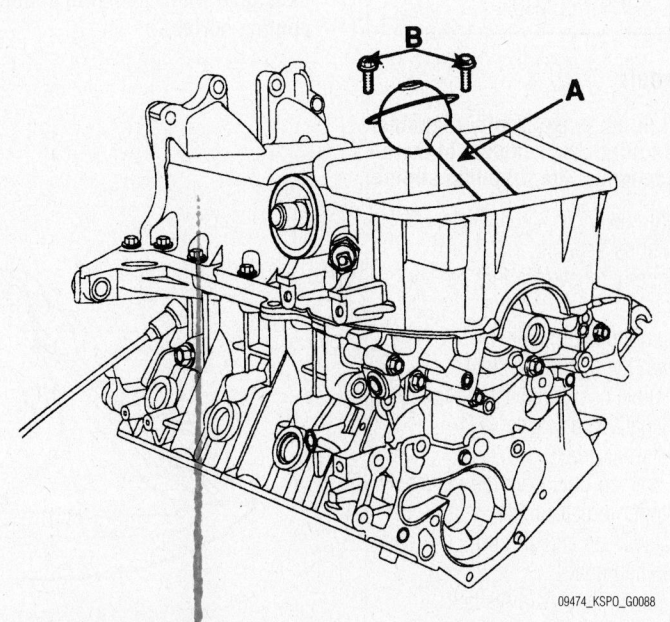

Oil pickup and screen installation—2.7L engine

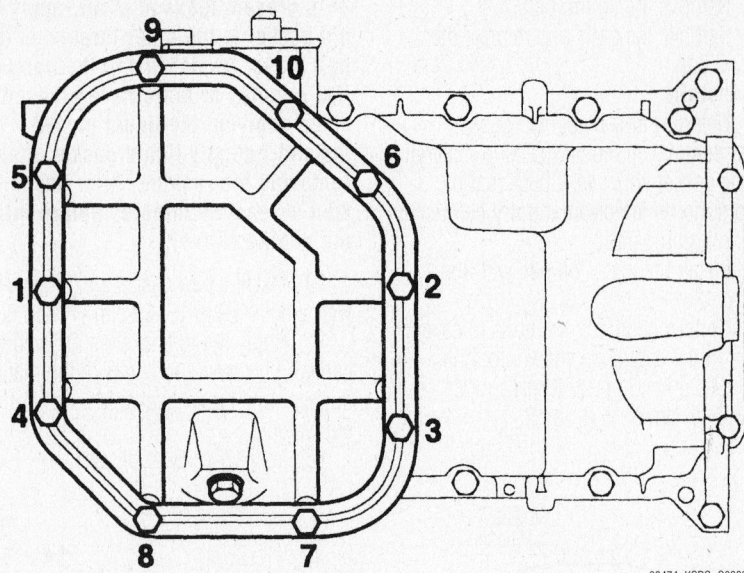

09474_KSPO_G0089

Lower oil pan torque sequence—2.7L engine

➡**To prevent leakage of oil, apply liquid gasket to the inner threads of the bolt holes. Do not install the parts if five minutes or more have elapsed since applying the liquid gasket. Instead, reapply liquid gasket after removing the residue. After assembly, wait at least 30 minutes before filling the engine with oil.**

13. Install the lower oil pan 10 bolts.

14. Uniformly, tighten the bolts several passes, in the sequence shown, to 10–12 Nm (7–9 ft. lbs.)

Oil Pump

REMOVAL & INSTALLATION

2002 Models

➡**The oil pump is externally-mounted, but still requires the removal of the oil pan to disconnect the oil pump strainer.**

1. Before servicing the vehicle, refer to the Precautions Section.

2. Properly relieve the fuel system pressure.

3. Drain the engine oil.

4. Drain the cooling system.

5. Remove or disconnect the following:
- Negative battery cable
- Alternator belt
- Fresh air duct from the radiator
- Upper radiator hose
- Clutch fan and shroud
- Splash guard
- Loosen the A/C drive belt
- Power steering belt

- Timing belt covers
- Lower timing belt pulley and lock bolt and place a support under the front axle
- Axle attaching bolts and lower the axle enough to gain access to the oil pan
- Transmission under cover
- Oil pan
- Oil pump

To install:

6. Clean the engine block, oil pan and baffle pan surfaces of any gasket material.

7. Apply a continuous bead of silicone sealant around the oil pump.

➡**Do not allow sealant to get in the oil passages when applying sealant to the contact surface.**

8. Install or connect the following:
- New O-ring and mount the oil pump to the engine. Torque the "A" bolts to 16 ft. lbs. (22 Nm) and the "B" bolts to 33 ft. lbs. (45 Nm).

9. Remove the upper and lower A/C compressor mounting bolts.

10. Loosen the A/C compressor bracket.
- Power steering pump bracket and hand tighten the bolts
- A/C compressor bracket
- A/C compressor. Torque the mounting bolts to 17 ft. lbs. (23 Nm).
- Torque the power steering pump bracket bolts to 24 ft. lbs. (33 Nm).
- Power steering pump. Torque the bolts to 43 ft. lbs. (58 Nm).
- Timing belt gear on the crankshaft. Torque the large crank bolt to 119 ft. lbs. (162 Nm).
- Oil baffle after applying sealant to the mating surface. Torque the bolt to 84 inch lbs. (9.5 Nm).
- Oil strainer. Torque the bolts to 84 inch lbs. (9.5 Nm).
- Oil pan
- Transmission under cover. Torque the bolts to 84 inch lbs. (9.5 Nm).
- Both gusset plates. Torque the bolts to 33 ft. lbs. (45 Nm).
- Raise the front axle into position. When aligned properly, torque the bolts to 123 ft. lbs. (167 Nm).
- Timing belt and cover
- Alternator belt
- A/C and power steering belt and adjust as needed
- Splash shield
- Upper radiator hose
- Clutch fan and shroud as an assembly

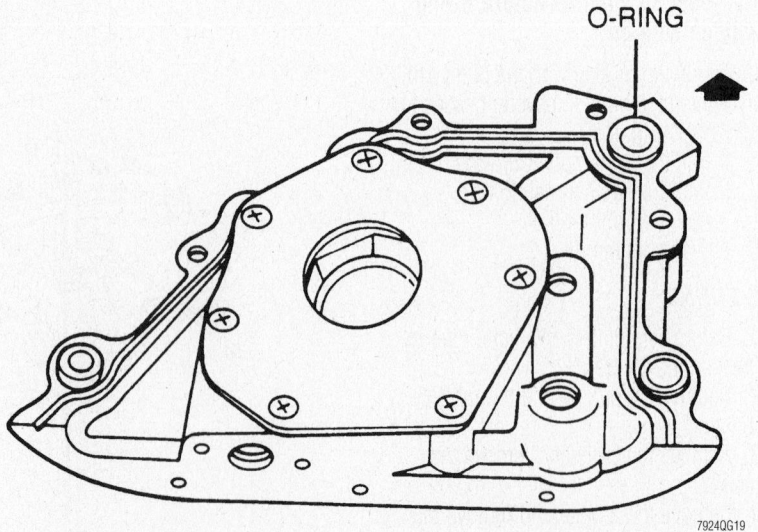

O-RING

7924QG19

Be sure the oil pump O-ring is in the proper location prior to installation—2002 models

- Air duct to the top of the radiator
- Engine under cover. Torque the bolts to 18 ft. lbs. (25 Nm).
- Negative battery cable
11. Properly fill the cooling system.
12. Fill the engine with clean oil.
13. Start the engine, check for leaks, and repair if necessary.

2005–06 Models

2.0L ENGINE

1. Before servicing the vehicle, refer to the Precautions Section.
2. Drain the engine oil.
3. Remove the drive belts.
4. Turn the crankshaft pulley, and align its groove with timing mark "T" of the timing belt cover.
5. Remove the timing belt.
6. Remove the bolt (B) and timing belt idler (A).
7. Remove the oil pan and oil screen.
8. Remove the alternator.
9. Remove the air conditioner compressor tensioner bracket.
10. Remove the bolts and front case.
11. Remove the screw from the pump housing, then separate the housing and cover.

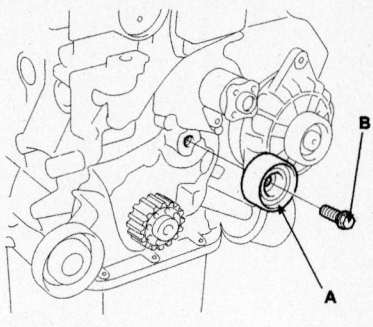

09474_KSPO_G0092

Remove the bolt (B) and timing belt idler (A)—2005–06 2.0L engine

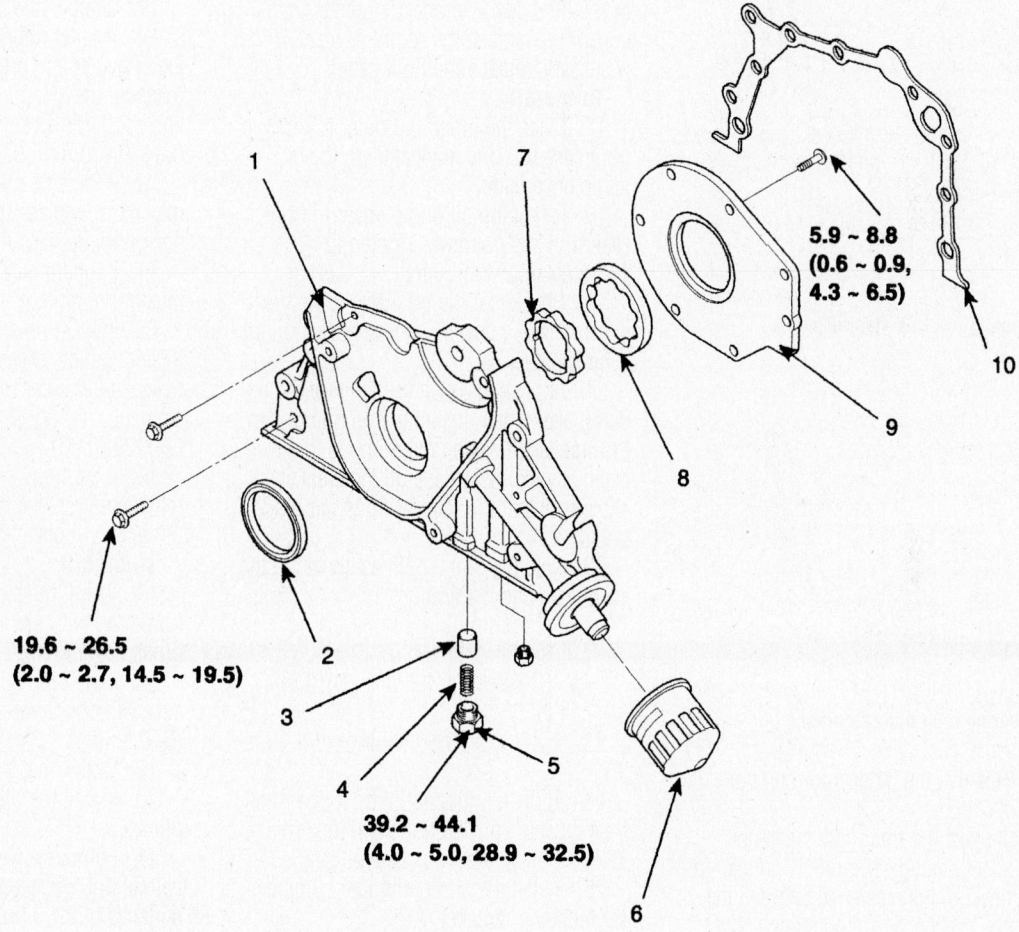

19.6 ~ 26.5 (2.0 ~ 2.7, 14.5 ~ 19.5)

39.2 ~ 44.1 (4.0 ~ 5.0, 28.9 ~ 32.5)

5.9 ~ 8.8 (0.6 ~ 0.9, 4.3 ~ 6.5)

TORQUE : N.m (kgf.m, lb-ft)

1. Front case
2. Oil seal
3. Relief plunger
4. Relief spring
5. Plug
6. Oil filter
7. Inner rotor
8. Outer rotor
9. Pump cover
10. Gasket

Oil pump—2005–06 2.0L engine

09474_KSPO_G0091

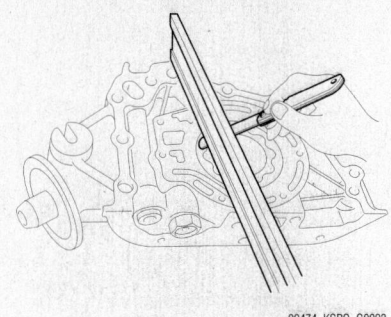

09474_KSPO_G0093

Measuring rotor side clearance

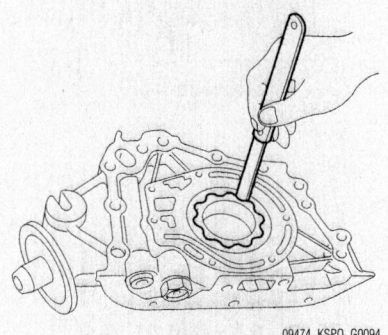

09474_KSPO_G0094

Measuring rotor tip clearance

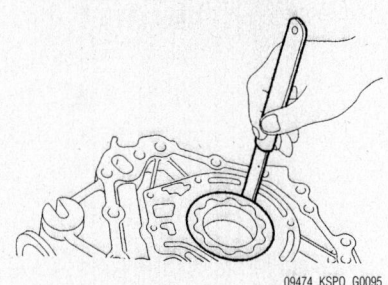

09474_KSPO_G0095

Measuring rotor body clearance

12. Remove the inner rotor and outer rotor.

13. Inspect the rotor side clearance. Using a feeler gauge and precision straight-edge, measure the clearance between the rotors and precision straight edge.

- Outer rotor: 0.04–0.09mm (0.0016–0.0035 in.)
- Inner rotor; 0.04–0.085mm (0.0016–0.0033 in.)

14. If the side clearance is greater than maximum, replace the rotors as a set. If necessary, replace the front case.

15. Inspect the rotor tip clearance. Using a feeler gauge, measure the tip clearance between the inner and outer rotor tips. Tip clearance: 0.025–0.069mm (0.0010–0.0027 in.). If the tip clearance is greater than maximum, replace the rotors as a set.

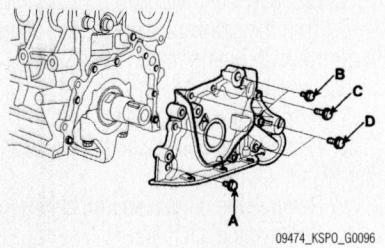

09474_KSPO_G0096

Oil pump bolt length identification— 2005–06 2.0L engine

16. Inspect the rotor body clearance. Using a feeler gauge, measure the clearance between the outer rotor and body. Body clearance: 0.120–0.185mm (0.0047–0.0073 in.). If the body clearance is greater than maximum, replace the rotors as a set. If necessary, replace the front case.

To install:

17. Place the inner and outer rotors into front case with the marks facing the oil pump cover side.

18. Install the oil pump cover to front case with the 7 screws. Tighten to 5.9–8.8N.m (4–6 ft. lbs.)

19. Check that the oil pump turns freely.

20. Place a new front case gasket on the cylinder block.

21. Apply engine oil to the lip of the oil pump seal. Then, install the oil pump onto the crankshaft. When the pump is in place, clean any excess grease off the crankshaft and check that the oil seal lip is not distorted.

22. Note the bolt length as shown in the accompanying illustration.

- A: 25mm (0.984 in.)
- B: 30mm (1.181 in.)
- C: 38mm (1.496 in.), (D): 45mm (1.772 in.)

23. Tighten to 19.6–26.5N.m (14–19 ft. lbs.).

24. Apply a light coat of oil to the front case oil seal lip. Using the SST(09214-32000), install the front case oil seal.

25. Install the air conditioner compressor tensioner bracket (A).

26. Install the alternator.

27. Install the oil screen, tighten to 14.7–21.6N.m (11–16 ft. lbs.).

28. Install the oil pan. Tighten to 9.8–11.8N-m (7–9 ft. lbs.).

➡Clean the oil pan gasket mating surfaces.

29. Install the timing belt idler. Tighten to 42.2–53.9N-m (31–40 ft. lbs.).

30. Install the timing belt.

31. Install the drive belts.

32. Fill with engine oil.

2.7L ENGINE

1. Before servicing the vehicle, refer to the Precautions Section.

2. Drain engine oil.

3. Remove right front wheel.

4. Remove right side cover.

5. Remove the front exhaust pipe.

6. Remove the alternator from engine.

7. Remove the drive belt.

8. Turn the crankshaft and align the white groove on the crankshaft pulley with the pointer on the lower cover.

9. Remove the timing belt.

10. Remove the oil pump case. Remove the screws from the pump housing, then separate the housing and cover.

11. Remove the inner and outer rotors.

12. Inspect rotor side clearance. Using a feeler gauge and precision straight edge, measure the clearance between the rotors and precision straight edge. Side clearance: 0.04–0.095 mm (0.0016–0.0037 in.).

13. If the side clearance is greater than maximum, replace the rotors as a set. If necessary, replace the front case.

14. If the tip clearance is greater than maximum, replace the rotor as a set.

15. Inspect rotor body clearance. Using a feeler gauge, measure the clearance between the outer rotor and body. Body clearance: 0.100–0.181 mm (0.0039–0.0017 in.).

16. If the body clearance is greater than maximum, replace the rotors as a set. If necessary, replace the front case.

To install:

17. Place the inner and outer rotors into front case with the marks facing the oil pump cover side.

18. Install the oil pump cover to front case with the 8 screws. Tighten to 8–12 Nm (6–9 ft. lbs.).

19. Check that the oil pump turns freely.

20. Install the oil pump on the cylinder block.

21. Remove any old liquid gasket and be careful not to drop any oil on the contact surfaces of the oil pump and cylinder block.

22. Using a razor blade and gasket scraper, remove all the old liquid gasket from the gasket surfaces and sealing grooves.

23. Using a non-residue solvent, clean both sealing surfaces.

24. Apply liquid gasket to the oil pump as shown in the illustration. Use liquid gasket MS 721-40A, or equivalent.

25. To prevent leakage of oil, apply liquid gasket to the inner threads of the bolt holes. Do not install the parts if five minutes or more have elapsed since applying the

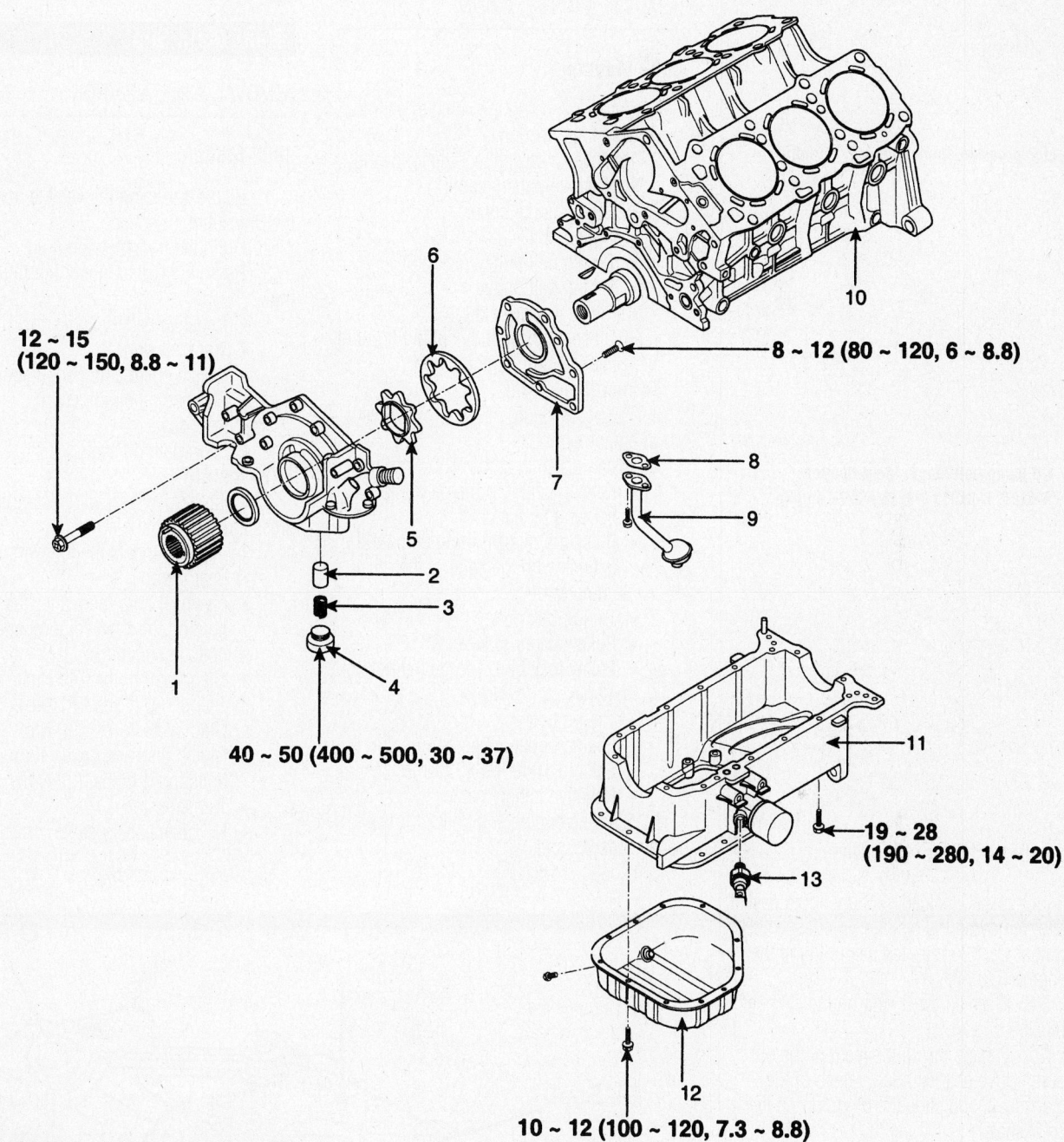

12 ~ 15
(120 ~ 150, 8.8 ~ 11)

8 ~ 12 (80 ~ 120, 6 ~ 8.8)

40 ~ 50 (400 ~ 500, 30 ~ 37)

19 ~ 28
(190 ~ 280, 14 ~ 20)

10 ~ 12 (100 ~ 120, 7.3 ~ 8.8)

TORQUE : N.m (kgf.cm, lb-ft)

1. Crankshaft sprocket
2. Relief plunger
3. Relief spring
4. Plug
5. Inner rotor
6. Outer rotor
7. Oil pump cover
8. Gasket
9. Oil screen
10. Cylinder block
11. Upper oil pan
12. Lower oil pan
13. Oil pressure switch

09474_KSPO_G0090

Lubrication system components—2.7L engine

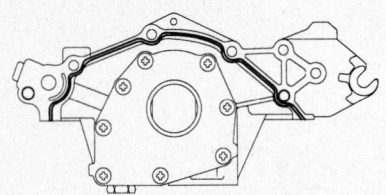

Liquid gasket application—2.7L engine

09474_KSPO_G0097

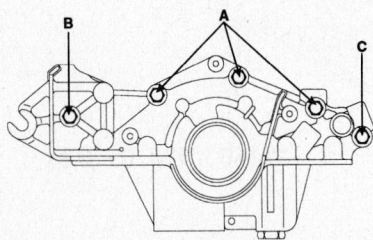

**Oil pump bolt length identification.
A-8x25, B-8x35, C-8x45—2.7L engine**

09474_KSPO_G0098

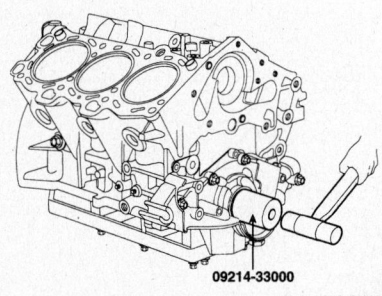

09214-33000

Seal installation—2.7L engine

09474_KSPO_G0099

liquid gasket. Instead, reapply liquid gasket after removing the residue. After assembly, wait at least 30 minutes before filling the engine with oil.

26. Place a new O-ring on the cylinder block.

27. Engage the spline teeth of the oil pump drive gear with large teeth of the crankshaft, and slide the oil pump on the crankshaft.

28. Install the oil pump with 5 bolts. Uniformly tighten the bolts in several passes. Tighten to 12–15 Nm (9–11 ft. lbs.).

29. Apply a light coat of oil to the seal lip. Using a seal driver, install the oil seal.

30. Install the oil pan and oil screen.
31. Install the timing belt.
32. Install the drive belt.
33. Install the alternator.
34. Install the front exhaust pipe.
35. Install the right front wheel.
36. Fill engine with oil.
37. Start engine and check for leaks.
38. Recheck engine oil level.

Front Crankshaft Seal

REMOVAL & INSTALLATION

2002 Models

1. Before servicing the vehicle, refer to the Precautions Section.
2. Remove or disconnect the following:
   - Negative battery cable
   - Engine under cover
   - Timing belt
   - Timing belt pulley lock bolt
   - Timing belt pulley
   - Pulley woodruff key
   - Oil seal by carefully cutting it out of the oil pump housing

To install:

3. Lubricate the lip of the new seal with clean engine oil.
4. Install or connect the following:
   - New oil seal into the oil pump housing by hand
   - Press the oil seal into pump until it is flush with the edge of the oil pump body
   - Timing belt pulley
   - Pulley woodruff key
   - Pulley lock bolt. Torque the bolt to 18 ft. lbs. (25 Nm).
   - Timing belt
   - Engine under cover. Torque the bolts to 18 ft. lbs. (25 Nm).
   - Negative battery cable
5. Start the engine and check for leaks, repair if necessary.

2005–06 Models

See the Oil Pump procedure.

Rear Main Seal

REMOVAL & INSTALLATION

2002 Models

1. Before servicing the vehicle, refer to the Precautions Section.
2. Drain the transmission fluid.
3. Remove or disconnect the following:
   - Negative battery cable
   - Transmission
   - Clutch cover and disc, if equipped
   - Flywheel, if equipped
   - Rear cover
   - Rear main oil seal

To install:

4. Coat the new seal with clean oil and press the seal into the cover.
5. Install or connect the following:
   - Rear cover
   - Flywheel onto the crankshaft. While holding the flywheel torque the bolts in sequence:
   a. Step 1: 30 ft. lbs. (41 Nm).
   b. Step 2: 60 ft. lbs. (81 Nm).
   c. Step 3: 73 ft. lbs. (99 Nm).
   - Clutch disc and cover. Torque the bolts to 16 ft. lbs. (22 Nm).
   - Transmission
   - Negative battery cable

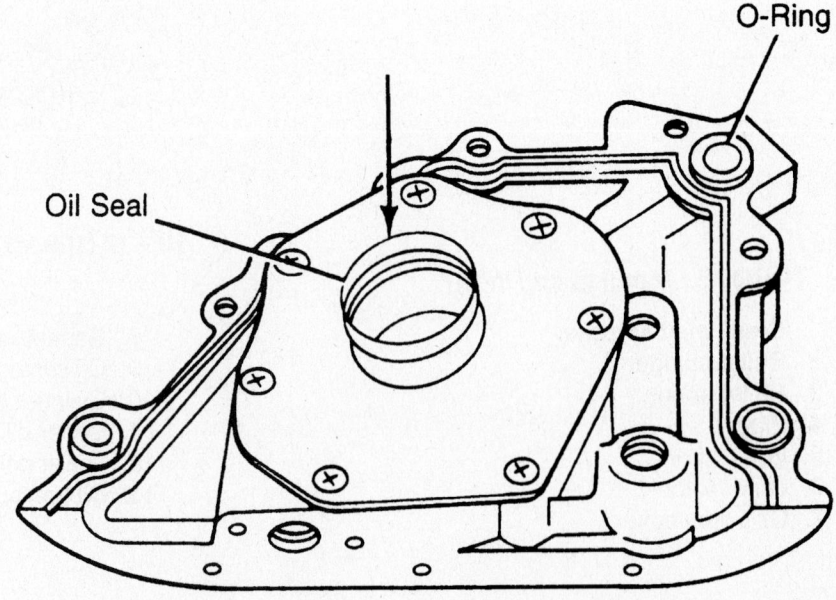

Install the oil seal into the oil pump housing—2002 models

O-Ring

Oil Seal

9308QG04

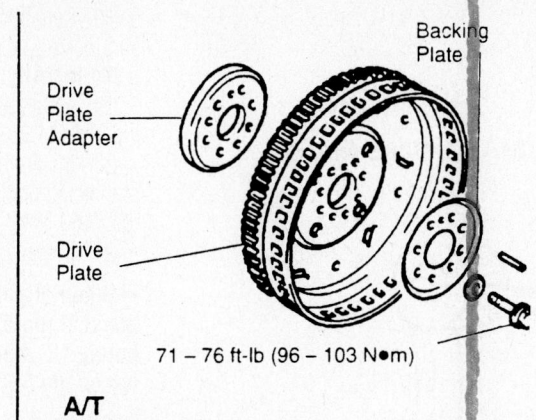

71 – 76 ft-lb (96 – 103 N•m)

A/T

Separator Plate

6 – 9 ft-lb
(8 – 12 N•m)

Flywheel (M/T)

Clutch Disc (M/T)

14 – 22 ft-lb
(19 – 30 N•m)

Rear Cover

71 – 76 ft-lb
(96 – 103 N•m)

Clutch
Cover (M/T)

16 – 24 ft-lb
(22 – 32 N•m)

7924QG38

M/T

Exploded view of the rear main seal and related components—2002 models

6. Fill the transmission to the proper level.

7. Start the vehicle and check for leaks, repair if necessary.

2005–06 Models

2.0L AND 2.7L ENGINES

1. Before servicing the vehicle, refer to the Precautions Section.
2. Remove the transaxle.
3. Remove the flywheel or flexplate.
4. Remove the rear main seal case.
5. Installation is the reverse of removal.

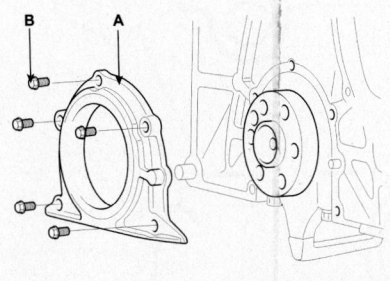

09474_KSPO_G0100

Rear main seal case—2005–06 2.0L engine

➡**Check that the mating surfaces are clean and dry.**

6. Apply engine oil to a new oil seal lip.
7. Using a seal driver and a hammer, tap in the oil seal until its surface is flush with the rear oil seal retainer edge. Torque the seal case bolts to 10–12 Nm (7–9 ft. lbs.).

Timing Belt

REMOVAL & INSTALLATION

2002 Models

1. Before servicing the vehicle, refer to the Precautions Section.
2. Disconnect the negative battery cable.
3. Properly relieve the fuel system pressure.
4. Remove the alternator drive belt.
5. Remove the fresh air duct from the top of the radiator.
6. Remove the upper radiator hose.
7. Remove the 4 attaching nuts to the clutch fan.
8. Remove the 5 fan shroud bolts. Remove the fan and shroud as an assembly.
9. Remove the 4 splash guard mounting bolts and the splash guard.
10. Loosen the lockbolts and loosen the air conditioning drive belt.
11. Loosen the power steering lock and mounting bolt. Remove the power steering belt.
12. Remove the 5 upper timing belt cover bolts and remove the cover.
13. Remove the 2 lower timing belt cover bolts and remove the cover.
14. Align the timing marks.

➡**When aligning the cam pulleys with the seal plate marks, align the left cam pulley I mark and the right cam pulley on the E mark.**

❄❄ WARNING

When aligning the timing marks, do not turn the timing gear counterclockwise. Damage to the engine will occur.

15. Loosen the tensioner bolt. Pry the tensioner away from the belt. Tighten the tensioner bolt to relieve the pressure against the timing belt.
16. Remove the timing belt.
17. Remove the camshaft pulley attaching bolts. Use a driver placed through one of the holes in the pulley to prevent it from moving when the attaching bolt is removed. Remove and mark the pulleys.

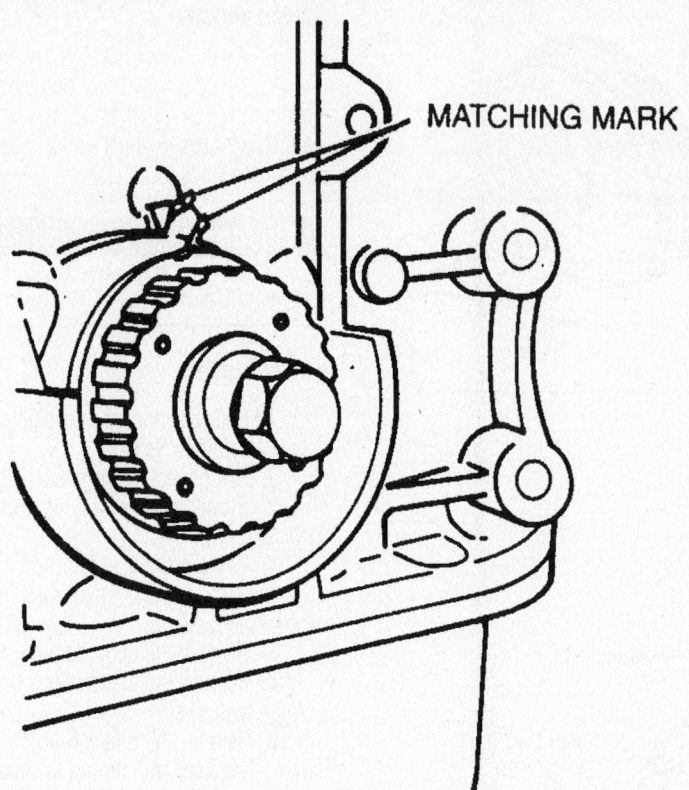

MATCHING MARK

79245G25

Align the crankshaft marks before removing the timing belt—2002 models

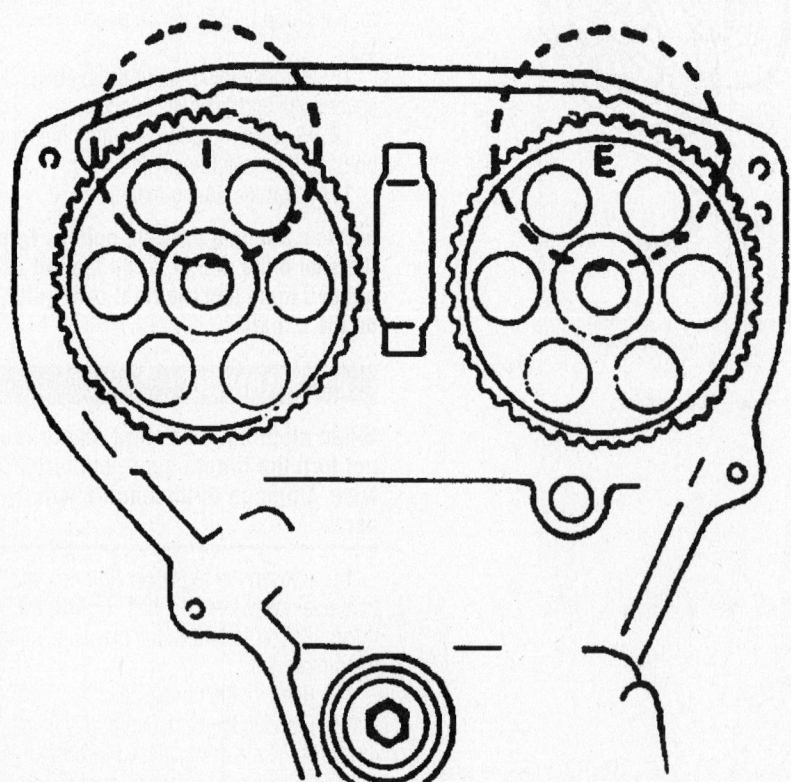

79245G28

Proper alignment of the intake and exhaust camshaft pulley timing marks—2002 models

18. Remove the lower timing belt pulley and locking bolt.

To install:

19. Install the camshaft pulleys. Tighten the bolts to 35–48 ft. lbs. (47–65 Nm).

20. Install the lower timing belt pulley and locking bolt. Tighten the bolt to 120 ft. lbs. (162 Nm).

21. If necessary, align the timing marks.

➡**When aligning the cam pulleys with the seal plate marks, align the left cam pulley "I" mark and the right cam pulley on the "E" mark.**

✳✳ WARNING

When aligning the timing marks, do not turn the timing gear counterclockwise. Damage to the engine will occur.

---

22. Loosen the tensioner bolt. Pry the tensioner away from the belt. Tighten tensioner bolt to relieve the pressure against the timing belt.

23. Install the timing belt.

✳✳ WARNING

If any binding is felt when adjusting the timing belt tension by turning the crankshaft, STOP turning the engine, because the pistons may be hitting the valves.

---

24. Loosen the tensioner bolt and allow the tensioner to tighten the timing belt. Tighten the tensioner bolt 27–38 ft. lbs. (37–52 Nm).

25. Check the timing belt deflection. If there is more than 0.30–0.33 in. (7.5–8.5mm) replace the tensioner spring.

26. Install the 2 lower timing belt cover bolts to the cover.

27. Install the 5 upper timing belt cover bolts to the cover.

28. Install and adjust the air conditioning and power steering drive belts.

29. Install the splash guard.

30. Install and tighten the alternator belt.

31. Install the upper radiator hose.

32. Install the fan and shroud as an assembly.

33. Install the 4 attaching nuts to the clutch fan.

34. Install the 5 fan shroud bolts.

35. Install the fresh air duct to the top of the radiator.

36. Properly fill the cooling system.

37. Connect the negative battery cable.

38. Start the engine and check for leaks.

39. Road test the vehicle.

2005–06 Models

2.0L ENGINE

1. Before servicing the vehicle, refer to the Precautions Section.
2. Remove the engine cover.
3. Remove the right front wheel.
4. Remove the 2 bolts and right side cover.
5. Set the jack to the engine oil pan.
6. Remove the bolt, nuts and engine mounting support bracket.
7. Remove the bolt and engine support bracket stay plate.
8. Temporarily loosen the water pump pulley bolts.
9. Remove the alternator drive belt.
10. Remove the air conditioner compressor drive belt.
11. Remove the power steering pump drive belt.

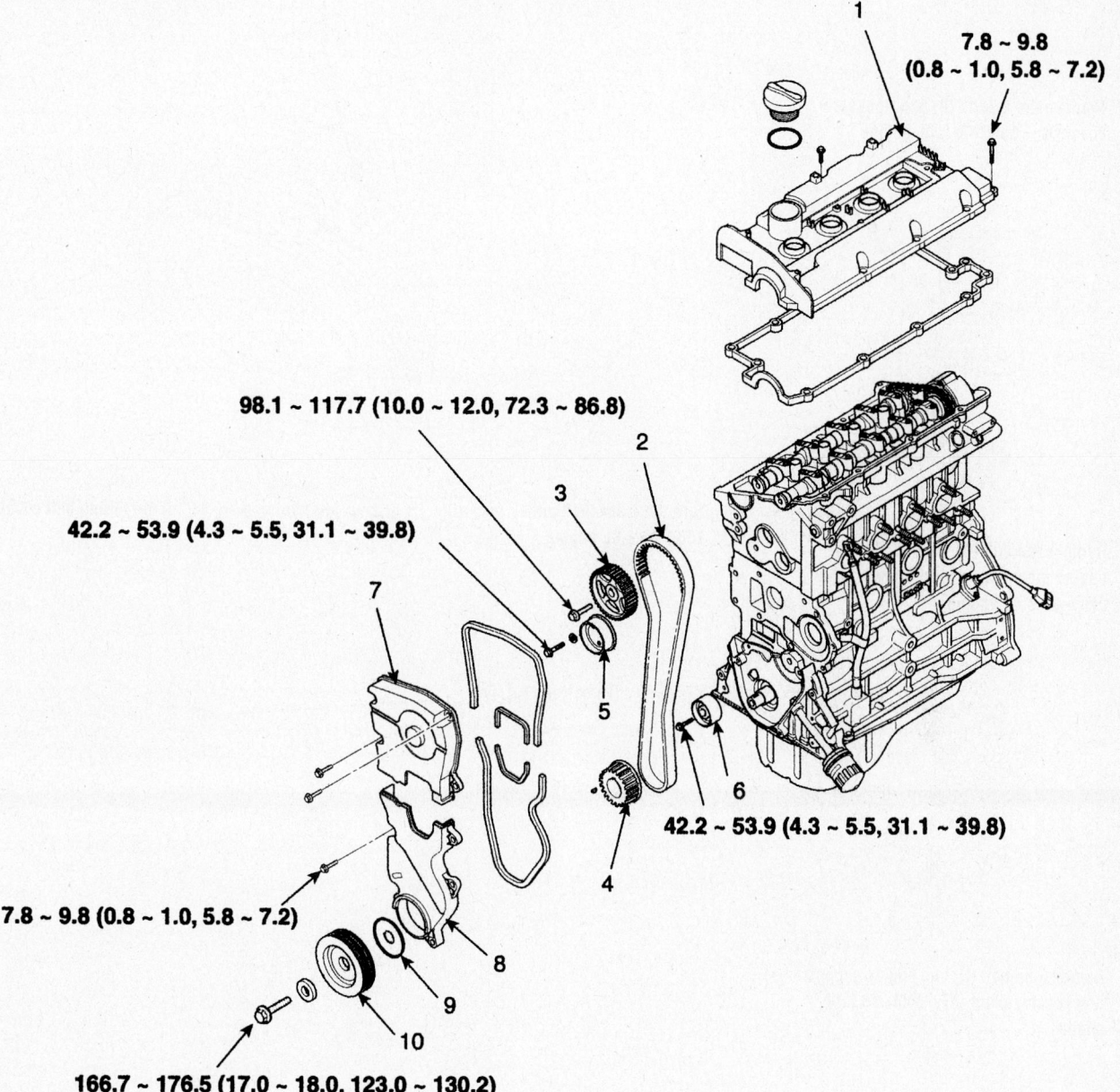

7.8 ~ 9.8 (0.8 ~ 1.0, 5.8 ~ 7.2)

98.1 ~ 117.7 (10.0 ~ 12.0, 72.3 ~ 86.8)

42.2 ~ 53.9 (4.3 ~ 5.5, 31.1 ~ 39.8)

42.2 ~ 53.9 (4.3 ~ 5.5, 31.1 ~ 39.8)

7.8 ~ 9.8 (0.8 ~ 1.0, 5.8 ~ 7.2)

166.7 ~ 176.5 (17.0 ~ 18.0, 123.0 ~ 130.2)

TORQUE : N.m (kgf.m, lb-ft)

1. Cylinder cover
2. Timing belt
3. Cam shaft sprocket
4. Crank shaft sprocket
5. Tensioner
6. Idler
7. Timing belt upper cover
8. Timing belt lower cover
9. Flange
10. Crank shaft pulley

09474_KSPO_G0101

Timing belt and related components—2005–06 2.0L engine

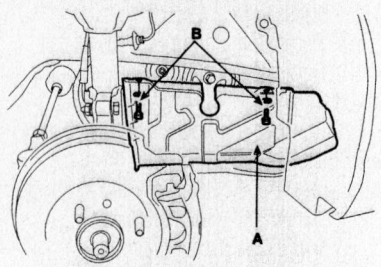

09474_KSPO_G0102

Remove the 2 bolts (B) and right side cover (A)—2005–06 2.0L engine

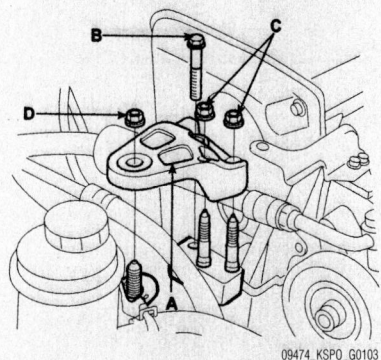

09474_KSPO_G0103

Remove the bolt (B), nuts (C, D) and engine mounting support bracket (A)—2005–06 2.0L engine

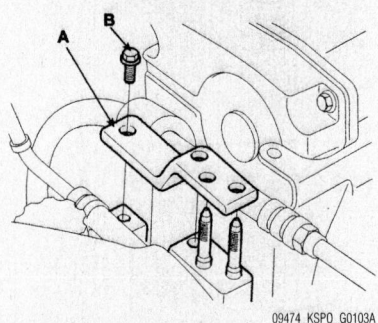

09474_KSPO_G0103A

Remove the bolt (B) and engine support bracket stay plate (A)—2005–06 2.0L engine

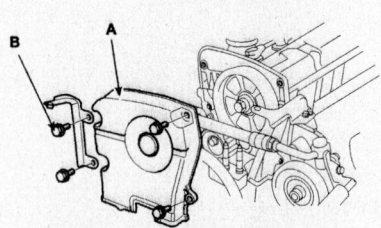

09474_KSPO_G01

Remove the 4 bolts (B) and timing belt upper cover (A)—2005–06 2.0L engine

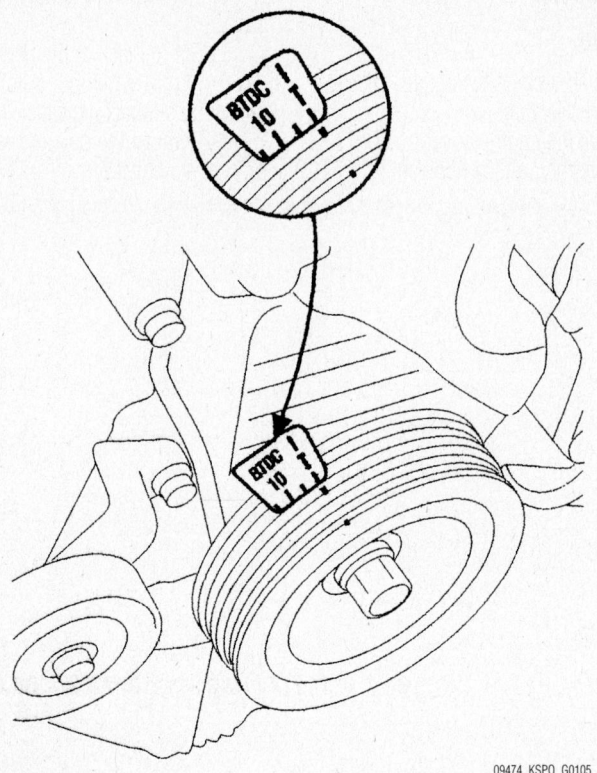

09474_KSPO_G0105

Turn the crankshaft pulley, and align its groove with timing mark "T" of the timing belt cover—2005–06 2.0L engine

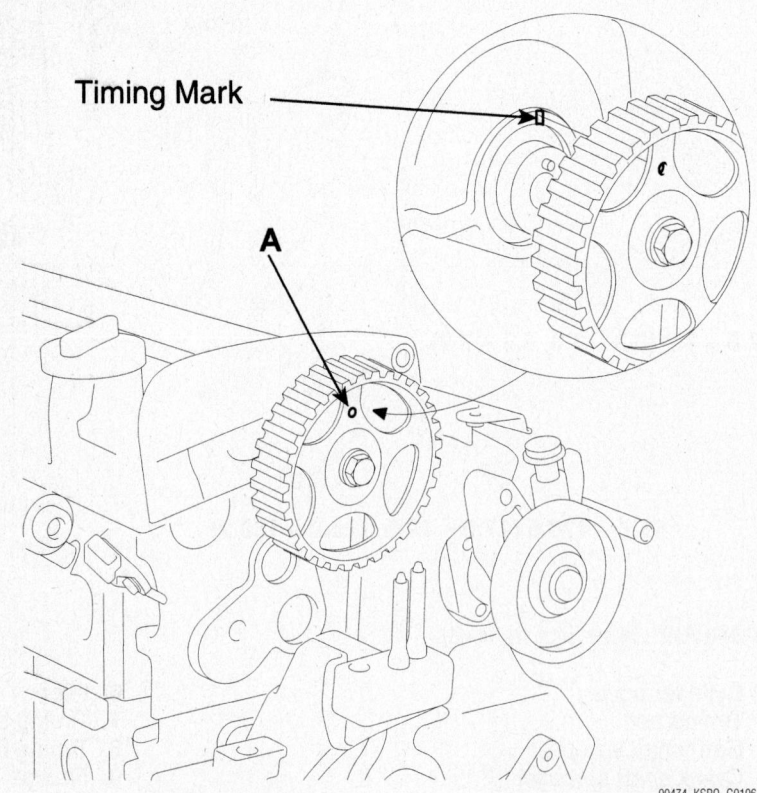

09474_KSPO_G0106

Check that the timing mark of camshaft sprocket is aligned with the timing mark of cylinder head cover. (No.1 cylinder compression TDC position)—2005–06 2.0L engine

12. Remove the 4 bolts and water pump pulley.

13. Remove the 4 bolts and timing belt upper cover.

14. Turn the crankshaft pulley, and align its groove with timing mark "T" of the timing belt cover. Check that the timing mark of camshaft sprocket is aligned with the timing mark of cylinder head cover. (No.1 cylinder compression TDC position)

15. Remove the crankshaft pulley bolt and crankshaft pulley.

16. Remove the crankshaft flange.

17. Remove the 5 bolts and timing belt lower cover

18. Remove the timing belt tensioner and timing belt.

➡ **If the timing belt is going to be reused, make an arrow indicating the turning direction to make sure that the belt is reinstalled in the same direction as before.**

19. Remove the bolt and timing belt idler.

20. Remove the crankshaft sprocket.

21. Disconnect the spark plug cables and do not pull on the cable by force.

➡ **Pulling on or bending the cables may damage the conductor inside.**

22. Remove the PCV (Positive Crankcase Ventilation) hose and the breather hose from the cylinder head cover.

23. Remove the accelerator cable from the cylinder head cover.

24. Loosen the cylinder head cover bolts and then remove the cover and gasket.

25. Remove the camshaft sprocket.

26. Hold the portion (A) of the camshaft with a hexagonal wrench, and remove the bolt (C) with a wrench (B) and remove the camshaft.

✳✳ WARNING

Be careful not to damage the cylinder head and valve lifter with the wrench.

27. Check the camshaft sprocket, crankshaft sprocket, tensioner pulley, and idler pulley for abnormal wear, cracks, or damage. Replace as necessary.

28. Inspect the tensioner pulley and the idler pulley for easy and smooth rotation and check for play or noise. Replace as necessary.

29. Replace the pulley if there is a grease leak from its bearing.

30. Check the belt for oil or dust deposits. Replace, if necessary. Small

deposits should be wiped away with a dry cloth or paper. Do not clean with solvent.

31. When the engine is overhauled or belt tension adjusted, check the carefully. If any damage is found, replace the belt.

➡ **Do not bend, twist or turn the timing belt inside out. Do not allow timing belt to come into contact with oil, water and steam.**

To install:

32. Temporarily install the camshaft sprocket bolt.

33. Hold the portion of the camshaft with a hexagonal wrench, and tighten the bolt with a wrench. Tighten to 98–118 Nm (72–87 ft. lbs.)

34. Install the cylinder head cover and bolts. Tighten to 8–10 Nm (6–7 ft. lbs.)

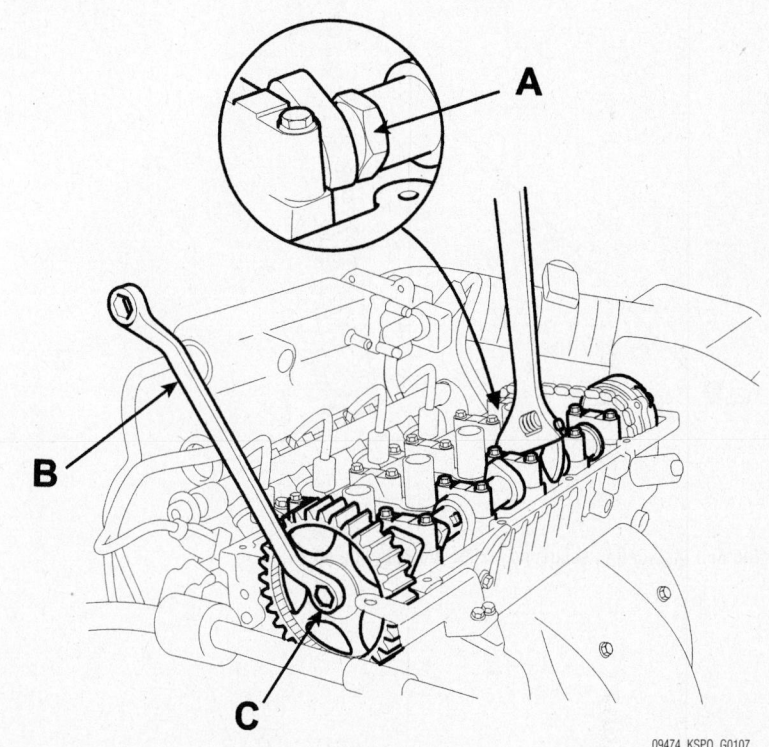

09474_KSPO_G0107

Hold the portion (A) of the camshaft with a hexagonal wrench, and remove the bolt (C) with a wrench (B) and remove the camshaft—2005–06 2.0L engine

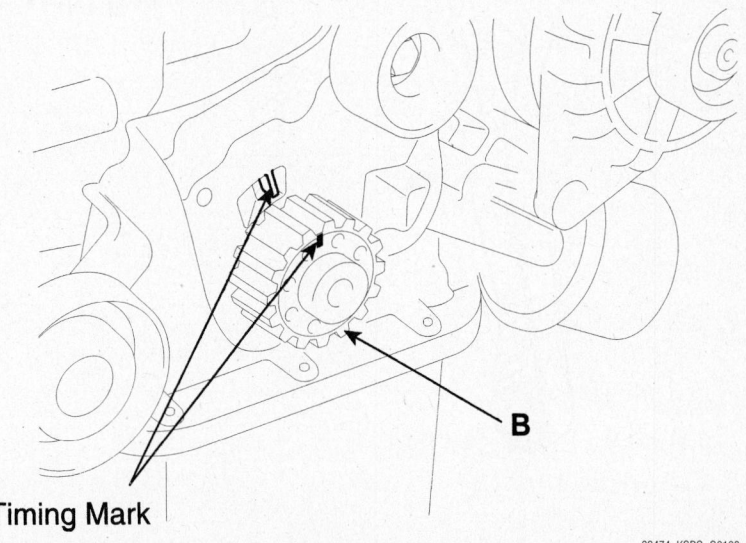

09474_KSPO_G0108

Crankshaft sprocket (B) timing marks aligned with the No.1 piston placed at top dead center and its compression stroke—2005–06 2.0L engine

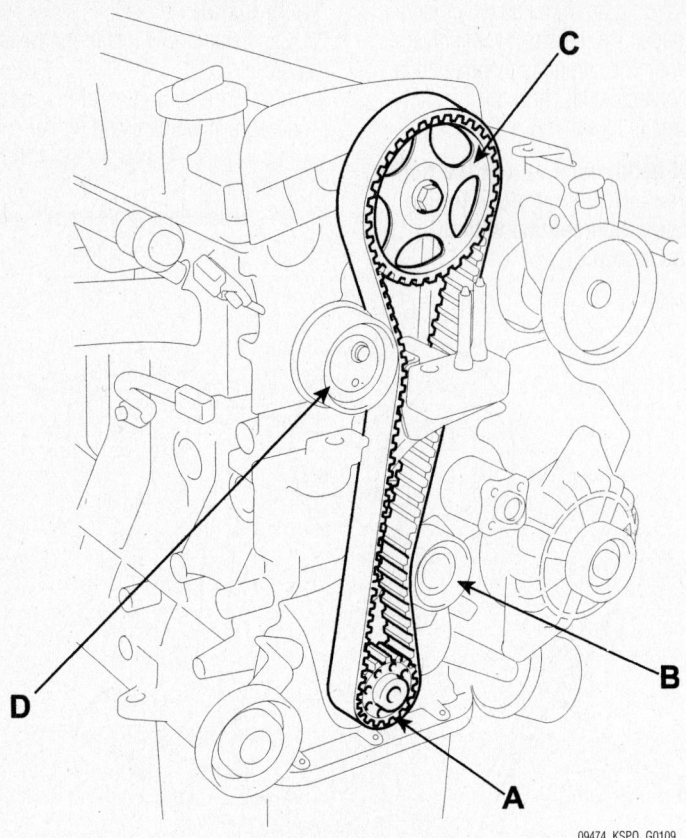

09474_KSPO_G0109

Timing belt installation sequence—2005–06 2.0L engine

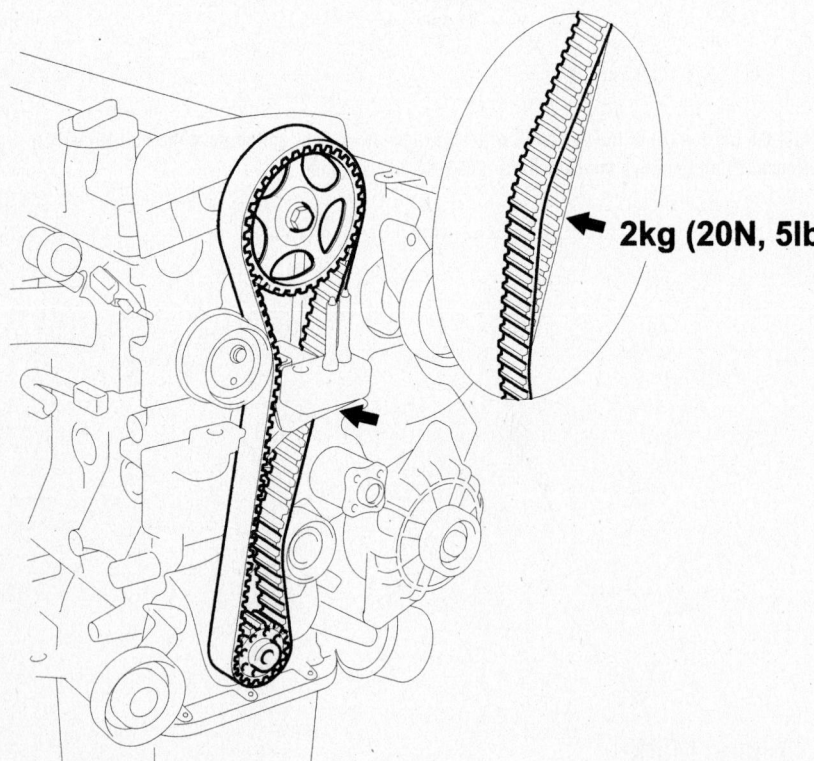

2kg (20N, 5lb)

09474_KSPO_G0110

Measuring timing belt deflection—2005–06 2.0L engine

35. Install the PCV (Positive Crankcase Ventilation) hose and breather hose to the cylinder head cover.

36. Install the accelerator cable to the cylinder head cover.

37. Install the spark plug cables.

38. Install the crankshaft sprocket.

39. Align the timing marks of the camshaft sprocket and crankshaft sprocket with the No.1 piston placed at top dead center and its compression stroke.

40. Install the idler pulley and tighten the bolt to 42–54 Nm (31–40 ft. lbs.)

41. Temporarily install the timing belt tensioner with plain washer.

42. Install the belt so as not give slack at each center of shaft. Use the following order when installing timing belt.
- Crankshaft sprocket (A)
- Idler pulley (B)
- Camshaft sprocket (C)
- Timing belt tensioner (D)

43. Temporarily install tensioner pulley using center bolt to add tension force to the belt.

44. Rotate the crankshaft clockwise (view from front) through angle equivalent to two teeth (18˚) of the camshaft sprocket.

45. Using a hex-wrench, apply tension to the timing belt in the clockwise direction so that there is no slack in the belt on the tension side. Tighten the tensioner bolt to 42–54 Nm (31–40 ft. lbs.)

46. Recheck the belt tension. When the tension side of timing belt is pushed horizontally with a moderate force, approx. 2kg (5 lbs.), the timing belt cog end deflects approx. 4–6mm (0.16–0.24in.).

47. Turn the crankshaft two turns in the operating direction (clockwise) and realign crankshaft sprocket and camshaft sprocket timing mark.

48. Install the timing belt lower cover with 5 bolts. Tighten to 8–10 Nm (6–7 ft. lbs.)

49. Install the flange and crankshaft pulley, and then tighten crankshaft pulley bolt. Make sure that the crankshaft sprocket pin fits into the small hole in the pulley. Tighten to 167–176 Nm (123–130 ft. lbs.)

50. Install the timing belt upper cover with bolts. Tighten to 8–10 Nm (6–7 ft. lbs.)

51. Install the water pump pulley and 4 bolts.

52. Install the power steering pump drive belt.

53. Install the air conditioner compressor drive belt.

54. Install the alternator drive belt.

55. Tighten the bolts of water pump pulley.

56. Install the engine mounting support bracket stay plate with bolt. Tighten to 42–54 Nm (31–40 ft. lbs.)

57. Install the engine mounting support bracket with nuts and bolt. Tightening torque:

- Nut (D): 59–78 Nm (43–589 ft. lbs.)
- Nut (C) and bolt (B): 49–64 Nm (36–47 ft. lbs.)

58. Install the right side cover with 2 bolts.

59. Install the right front wheel. Tighten to 88–98 Nm (65–72 ft. lbs.)

60. Install the engine cover with bolts. Tighten to 4–6 Nm (3–4 ft. lbs.)

2.7L ENGINE

1. Before servicing the vehicle, refer to the Precautions Section.
2. Remove the engine cover.
3. Remove right front wheel.
4. Remove 2 bolts and right side cover.
5. Turn the crankshaft pulley, and align its groove with timing mark "T" of the timing belt cover.

➡ **Always turn the crankshaft clockwise**

6. Remove drive belt and belt tensioner.
7. Remove the engine mount bracket.
8. Set the jack to the engine oil pan.

➡ **Place wooden block between the jack and engine oil pan.**

9. Remove the 2 bolts, 2 nuts and engine mount bracket.

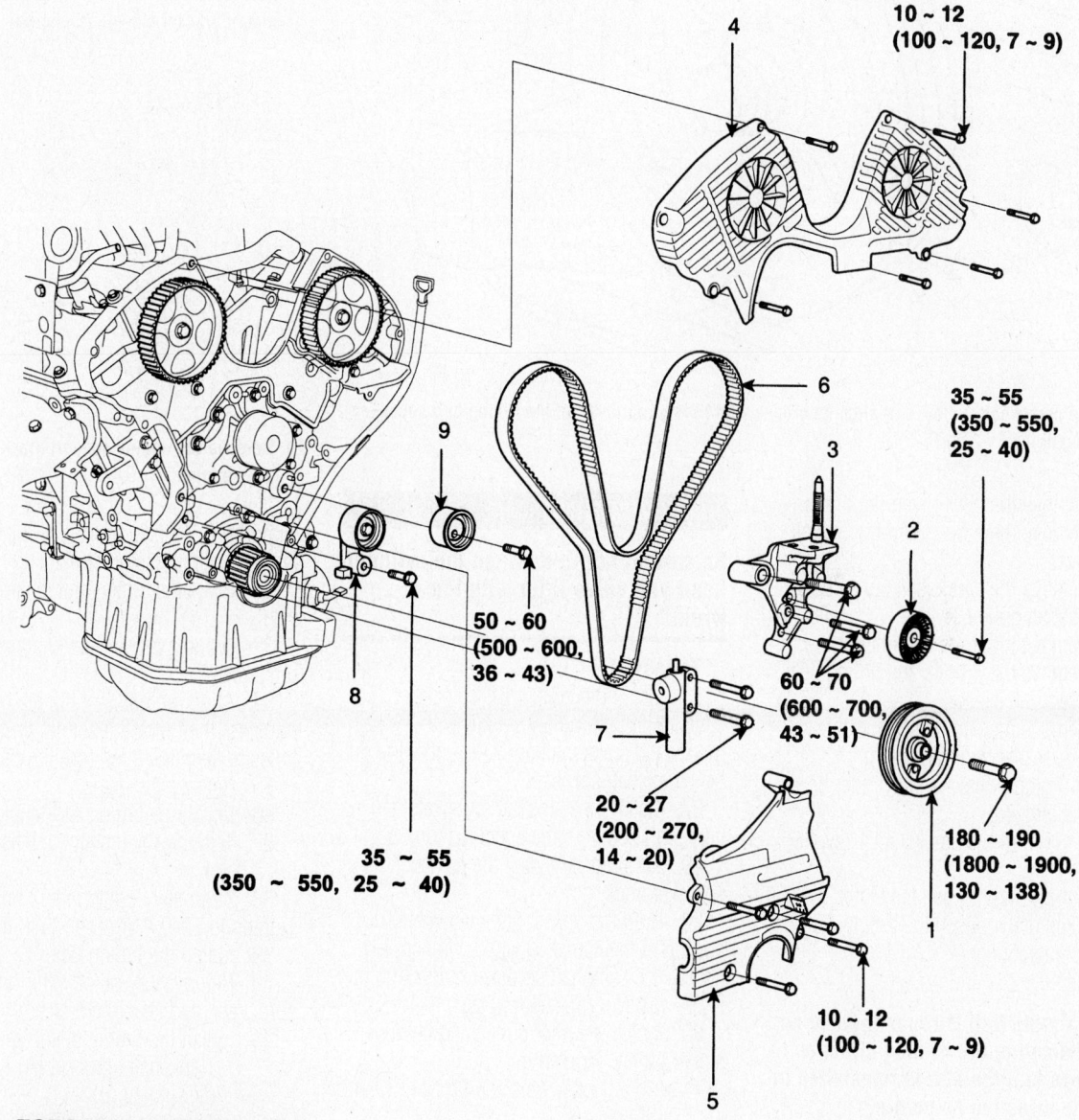

TORQUE : N.m (kgf.cm, lb-ft)

1. Crankshaft pulley
2. Drive belt idler pulley
3. Engine support bracket
4. Timing belt upper cover
5. Timing belt lower cover
6. Timing belt
7. Auto tensioner
8. Tensioner pulley
9. Idler pulleyy

Timing belt and related components—2.7L engine

09474_KSPO_G0114

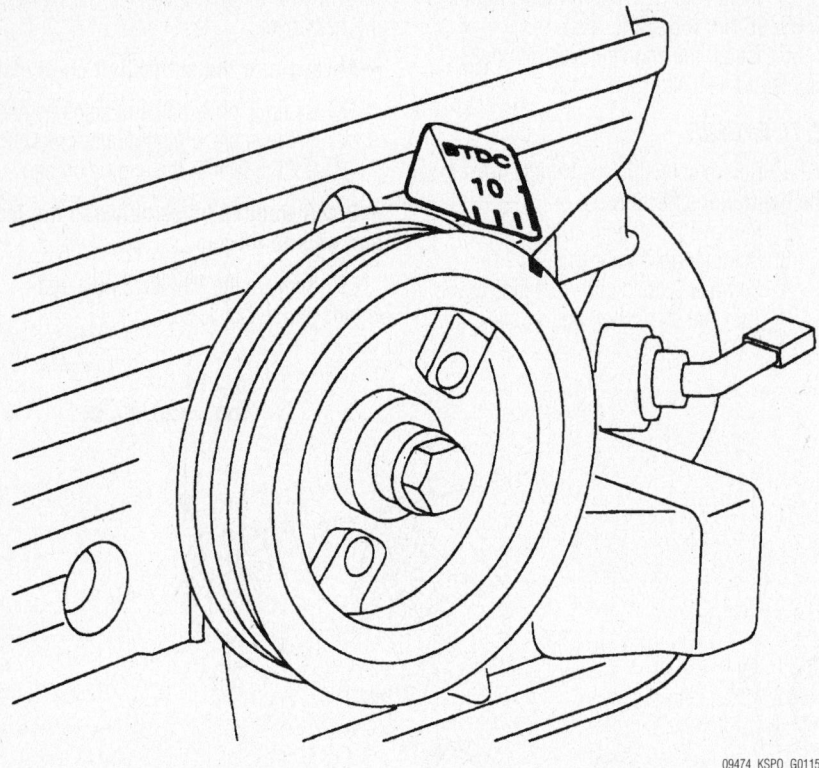

09474_KSPO_G0115

Turn the crankshaft pulley, and align its groove with timing mark "T" of the timing belt cover—2.7L engine

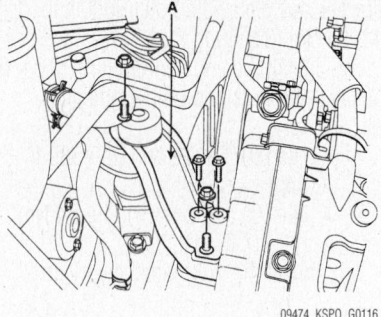

09474_KSPO_G0116

Remove the 2 bolts, 2 nuts and engine mount bracket (A)—2.7L engine

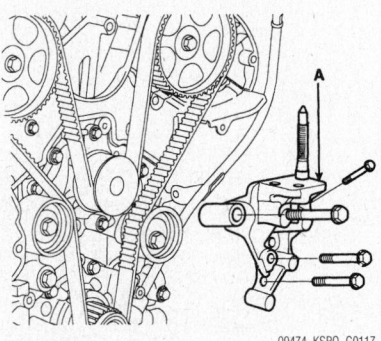

09474_KSPO_G0117

Remove the engine support bracket (A)—2.7L engine

10. Remove the power steering pump.

11. Remove the 7 bolts and timing belt upper cover.

12. Remove the crankshaft pulley bolt and crankshaft pulley.

13. Remove the drive belt idler pulley.

14. Remove the 4 bolts and timing belt lower cover.

15. Remove the engine support bracket.

16. Check that timing marks of the camshaft timing pulleys and cylinder head covers are aligned.

17. If not, turn the crankshaft 1 revolution (360°).

18. Remove timing belt tensioner.

19. Alternately loosen the 2 bolts, and remove the tensioner.

20. Remove the timing belt.

➡️If the timing belt is reused, make an arrow indicating the turning direction to make sure that the belt is reinstalled in the same direction as before.

21. Remove the tensioner pulley and timing belt idler pulley.

22. Remove the crankshaft sprocket.

23. Remove camshaft sprockets.

24. Hold the hexagonal head wrench portion of the camshaft with a wrench and remove the bolt and camshaft sprocket.

❊❊ WARNING

Be careful not to damage the cylinder head and valve lifter with the wrench.

To install:

25. Install the crankshaft sprocket.

26. Align the pulley set key with the key groove the crankshaft sprocket and slide on the crankshaft sprocket.

27. Install the camshaft sprockets and tighten the bolts to the specified torque.

28. Temporarily install the camshaft sprocket bolts.

29. Hold the hexagonal head wrench portion of the camshaft with a wrench, and tighten the camshaft sprocket bolts. Tighten to 90–110 Nm (65–80 ft. lbs.)

30. Install the idler pulley and the tensioner pulley. Tighten to:

- Idler pulley bolt: 50–60 Nm (36–43 ft. lbs.)
- Tensioner arm fixed bolt: 35–55 Nm (25–40 ft. lbs.)

➡️Insert and install the idler pulley to the roll pin that is pressed in the water pump boss.

31. Align the timing marks of the camshaft sprocket and crankshaft sprocket

with the No.1 piston placed at top dead center of its compression stroke.

32. Place the timing belt tensioner in a vise.

33. Using a press, slowly press in the push rod.

34. Align the holes of the push rod and housing, then install a set pin through the holes to keep the push rod, in position.

35. Release the press.

36. Install the timing belt tensioner.

37. Temporarily install the tensioner with the 2 bolts.

38. Alternately tighten the 2 bolts. Tighten to 20–27 Nm (14–20 ft. lbs.)

39. Install the timing belt.

40. Remove any oil or water on the sprockets, and keep them clean.

41. Install the timing belt in this order:

- Crankshaft sprocket (A)
- Idler pulley (B)
- Camshaft sprocket LH side (C)
- Water pump pulley (D)
- Camshaft sprocket right side (E)
- Tensioner pulley (F)

42. Remove the set pin from the tensioner.

43. Rotate the crankshaft 2 turns clockwise and measure the projected length of the auto tensioner at TDC (#1 compression

stroke) after 5 minutes. The projected length should be 7–9 mm (0.27–0.31 in.)

44. Install the engine support bracket. Tighten to:

- Front bolts: 60–70 Nm (43–51 ft. lbs.)
- Side bolt: 15–22 Nm (11–16 ft. lbs.)

45. Install the timing belt lower cover with 4 bolts. Tighten to 10–12 Nm (7–9 ft. lbs.)

46. Install the drive belt idler pulley. Tighten to 35–55 Nm (25–40 ft. lbs.)

47. Install the crankshaft pulley. Make sure that crankshaft sprocket pin fits the small hole in the pulley. Tighten to 180–190 Nm (130–138 ft. lbs.)

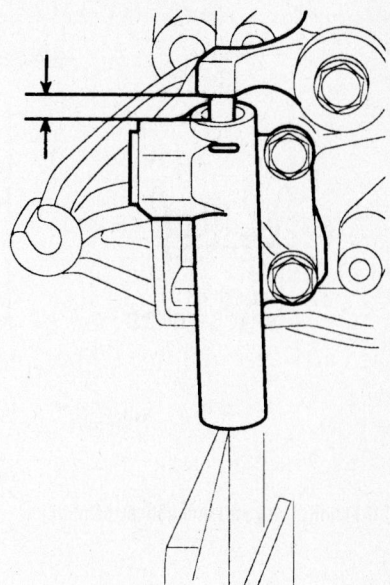

09474_KSPO_G0120

Measuring auto tensioner projected length—2.7L engine

48. Install the timing belt upper cover with 7 bolts.

49. Install the power steering pump.

50. Install the drive belt tensioner and drive belt.

51. Install the engine mount bracket.

52. Install engine mount bracket with 2 nuts and 2 bolts. Tighten to 60–80 Nm (44–59 ft. lbs.).

53. Install right side cover with 2 bolts.

54. Install right front wheel.

55. Install engine cover.

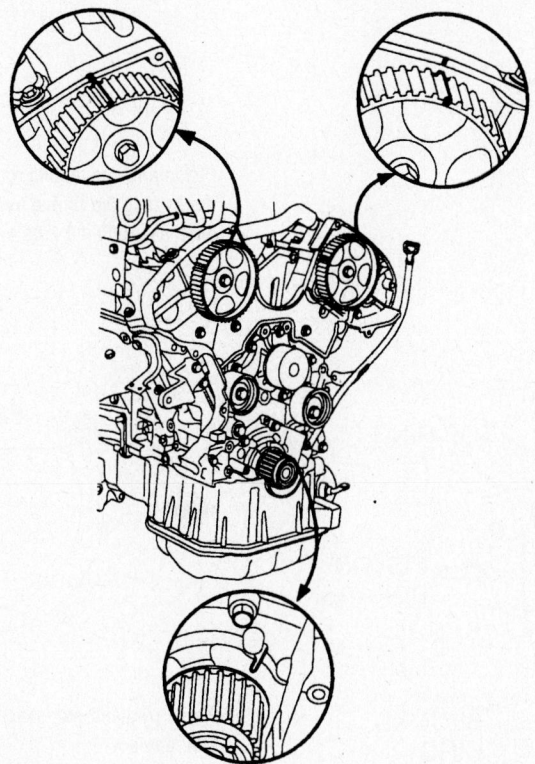

09474_KSPO_G0118

Timing marks aligned—2.7L engine

Piston and Ring

POSITIONING

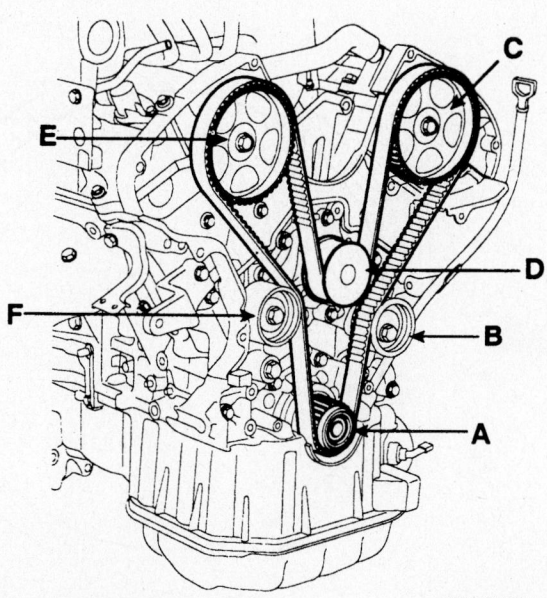

09474_KSPO_G0119

Timing belt installation sequence—2.7L engine

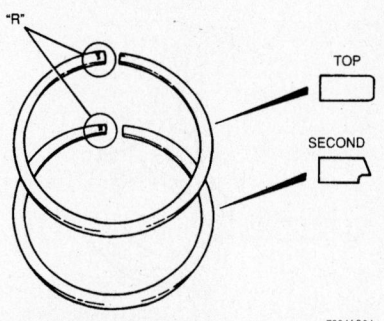

7924AG04

Compression ring positioning mark locations—2002 2.0L engine

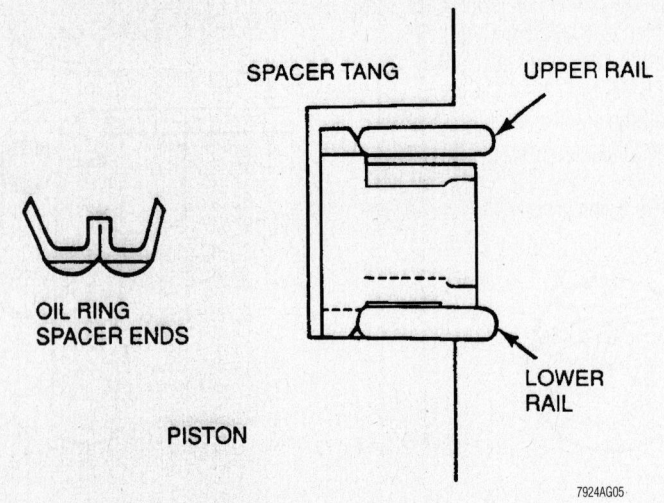

OIL RING
SPACER ENDS

PISTON

SPACER TANG

UPPER RAIL

LOWER RAIL

7924AG05

Oil control ring rail and spacer positioning—2002 2.0L engine

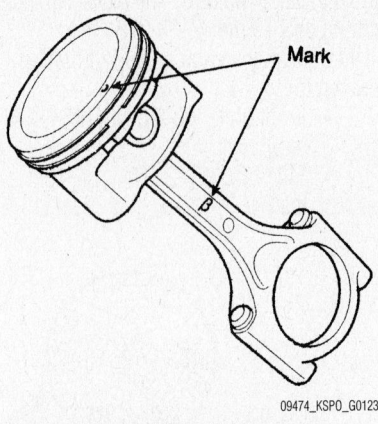

Mark

09474_KSPO_G0123

Piston and connecting rod orientation. The marks face the timing belt end of the engine—2005-06 2.0L engine

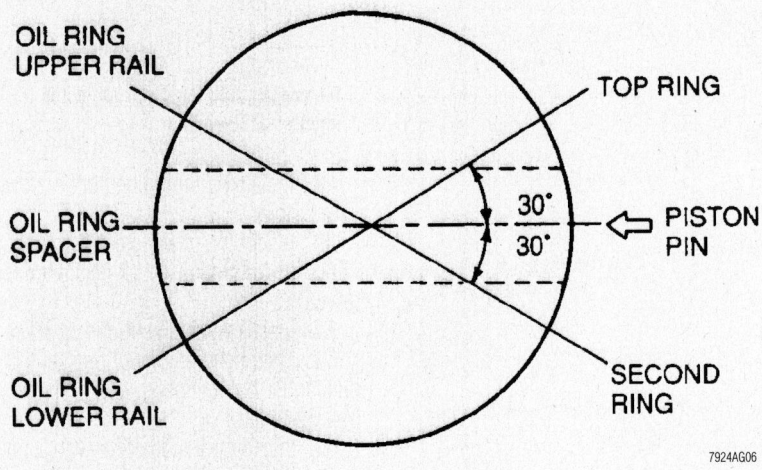

OIL RING UPPER RAIL

OIL RING SPACER

OIL RING LOWER RAIL

TOP RING

30°
30°

PISTON PIN

SECOND RING

7924AG06

Piston ring end-gap spacing—2002 2.0L engine

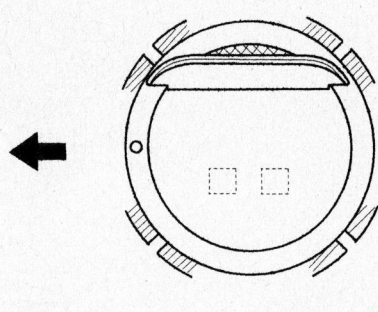

09474_KSPO_G0124

Piston ring end-gap spacing—2005-06 2.0L engine

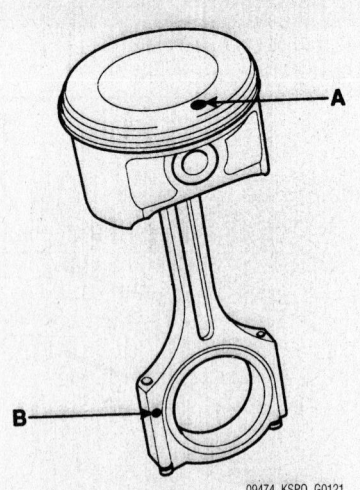

A

B

09474_KSPO_G0121

Piston and connecting rod orientation. The marks face the timing belt end of the engine—2.7L engine

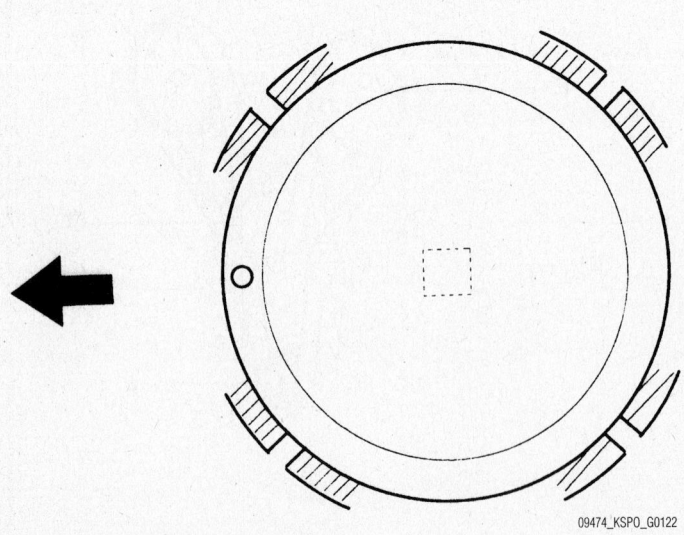

09474_KSPO_G0122

Piston ring end-gap spacing—2.7L engine

FUEL SYSTEM

Fuel System Service Precautions

Safety is the most important factor when performing not only fuel system maintenance but any type of maintenance. Failure to conduct maintenance and repairs in a safe manner may result in serious personal injury or death. Maintenance and testing of the vehicle's fuel system components can be accomplished safely and effectively by adhering to the following rules and guidelines.

• To avoid the possibility of fire and personal injury, always disconnect the negative battery cable unless the repair or test procedure requires that battery voltage be applied.

• Always relieve the fuel system pressure prior to disconnecting any fuel system component (injector, fuel rail, pressure regulator, etc.), fitting or fuel line connection. Exercise extreme caution whenever relieving fuel system pressure to avoid exposing skin, face and eyes to fuel spray. Please be advised that fuel under pressure may penetrate the skin or any part of the body that it contacts.

• Always place a shop towel or cloth around the fitting or connection prior to loosening to absorb any excess fuel due to spillage. Ensure that all fuel spillage (should it occur) is quickly removed from engine surfaces. Ensure that all fuel soaked cloths or towels are deposited into a suitable waste container.

• Always keep a dry chemical (Class B) fire extinguisher near the work area.

• Do not allow fuel spray or fuel vapors to come into contact with a spark or open flame.

• Always use a back-up wrench when loosening and tightening fuel line connection fittings. This will prevent unnecessary stress and torsion to fuel line piping.

• Always replace worn fuel fitting O-rings with new. Do not substitute fuel hose or equivalent where fuel pipe is installed.

Fuel System Pressure

RELIEVING

1. Before servicing the vehicle, refer to the Precautions Section.
2. Disconnect the fuel pump harness connector located behind the rear seat.
3. Start the engine and allow the engine to run out of fuel.

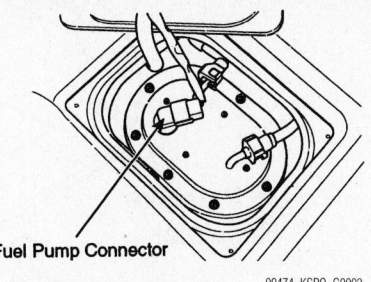

Fuel pump harness connector

09474_KSPO_G0002

4. Once the engine has stalled, turn the key to the **OFF** position and connect the electrical connector.
5. Disconnect the negative battery cable so pressure cannot build up until work has been completed.

Fuel Filter

REMOVAL & INSTALLATION

2002 Models

1. Before servicing the vehicle, refer to the Precautions Section.
2. Properly relieve the fuel system pressure.
3. Remove or disconnect the following:
   • Negative battery cable
   • Fuel pump connector
   • Fuel hoses from the fuel filter
   • Fuel filter from the bracket

To install:

4. Install or connect the following:
   • Fuel filter in the bracket
   • Fuel filter. Torque the bolts to 95 ft. lbs. (129 Nm).
   • Fuel hoses on the filter and make certain that the hoses are seated properly

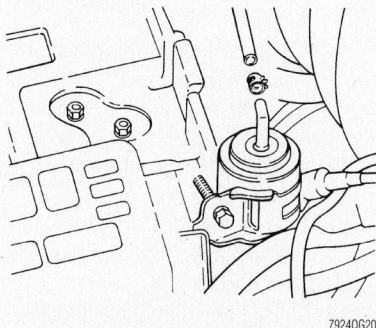

7924QG20

Fuel filter underhood mounting location— 2002 models

   • Fuel pump connector
   • Negative battery cable
5. Start the engine and check for fuel leaks, repair if necessary.

2005–06 Models

The fuel filter is part of the fuel pump module, located in the fuel tank.

Fuel Pump

REMOVAL & INSTALLATION

2002 Models

1. Before servicing the vehicle, refer to the Precautions Section.
2. Properly relieve the fuel system pressure.
3. Release the catch for the back seat and tilt the seat out of the way.
4. Move the carpet behind the seat that covers the fuel pump access panel.
5. Remove or disconnect the following:
   • Negative battery cable
   • Fuel pump electrical connectors
   • Bolt securing the ground wire
   • Fuel pump access panel
   • Hose clamps connecting the fuel hoses to the fuel pump
   • Hoses from the fuel pump
   • Screws securing the fuel pump to the fuel tank
   • Gradually lift the fuel pump from the tank
   • Plastic retaining bracket from the fuel pump assembly
   • Fuel hose from the fuel pump
   • Fuel tank pressure sensor
6. Wrap the fuel pump assembly in a rag before removing the assembly from the vehicle.
7. Cover or seal the fuel tank until installing the fuel pump assembly.

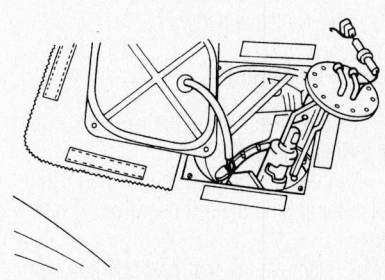

7924QG21

Removing the fuel pump through the access panel—2002 models

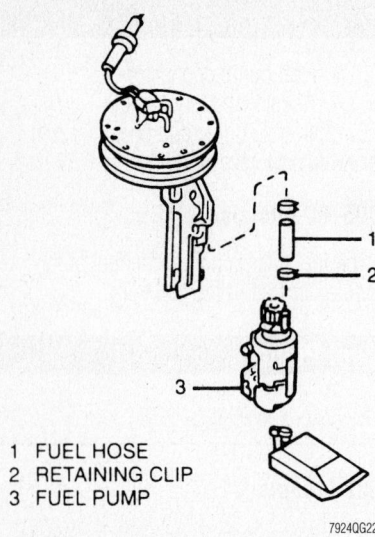

1 FUEL HOSE
2 RETAINING CLIP
3 FUEL PUMP

7924QG22

Exploded view of the fuel pump assembly—2002 models

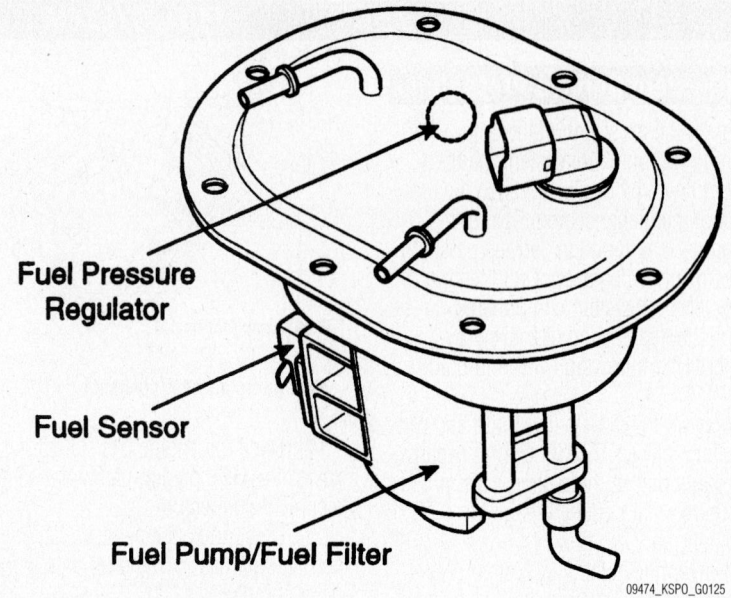

Fuel Pressure Regulator

Fuel Sensor

Fuel Pump/Fuel Filter

09474_KSPO_G0125

Fuel pump module—2005–06 models

To install:

➡**The fuel pump is part of the assembly and is replaced as a complete unit.**

8. Install or connect the following:
   - Plastic mounting bracket to the fuel pump and secure the bracket to the pump with 4 screws
   - Fuel hose to the pump and secure with a new clamp
   - Fuel pump into the access port on top of the fuel tank
   - Twist the fuel pump as necessary to properly position it in the fuel tank
   - 8 screws around the top surface of the fuel pump
   - Fuel hoses to the pump and secure with clamps
   - Fuel tank pressure sensor
   - Fuel pressure sensor electrical connector
   - Ground wire and fuel pump electrical connector
   - 2 halves of the fuel pump electrical connector
   - Access plate and reposition the carpet at seat
   - Negative battery cable

2005–06 Models

1. Before servicing the vehicle, refer to the Precautions Section.
2. Turn ignition switch to "OFF" and disconnect the battery negative (-) terminal.
3. Remove the rear seat cushion frame hinge bolts, and then lift the cushion.
4. Remove the service cover under the carpet.

5. Disconnect the fuel pump wiring connector.
6. Remove the fuel feed hose and suction hose.
7. Remove the fuel pump plate mounting bolts.
8. Remove the fuel pump assembly.
9. Installation is the reverse of removal.

Fuel Injector

REMOVAL & INSTALLATION

2002 Models

1. Before servicing the vehicle, refer to the Precautions Section.
2. Properly relieve the fuel system pressure.
3. Drain the cooling system.
4. Remove or disconnect the following:
   - Negative battery cable
   - Air intake hose assembly
   - Breather hoses from the air intake duct
   - Mass Air Flow (MAF) sensor electrical connector
   - Air intake hose bracket
   - Accelerator cable
   - Vacuum hose from the intake manifold to the vacuum pipe
   - Throttle Position (TP) sensor and Idle Air Control (IAC) valve electrical connectors
   - Coolant hoses from the throttle body
   - Clamp and hoses from the IAC valve

   - Brake booster vacuum hose from the dynamic chamber
   - Cruise control vacuum hose, if equipped
   - Bracket from the dynamic chamber
   - Manifold bracket
   - Heater inlet hose
   - Dynamic chamber
   - Fuel hose from the pressure regulator
   - Fuel injector rail clips
   - Fuel rail
   - Fuel injector insulators
   - Pressure regulator
   - Fuel injectors

To install:
5. Install or connect the following:
   - Fuel injectors to the fuel rail
   - New insulators
   - New injector clips
   - Fuel rail
   - Clamps and air hose to the fuel rail
   - Fuel hose to the pressure regulator
   - Dynamic chamber with a new gasket. Torque the bolts to 16 ft. lbs. (22 Nm).
   - IAC valve bracket bottom bolt
   - IAC valve and TP sensor electrical connectors
   - Heater inlet hose
   - Cruise control hose, if equipped
   - Vacuum hose to the pressure regulator
   - Air hose and clamp to the air rail
   - IAC valve vacuum hose
   - Manifold bracket bolts
   - Coolant hoses to the throttle body
   - Vacuum hose to the vacuum pipe

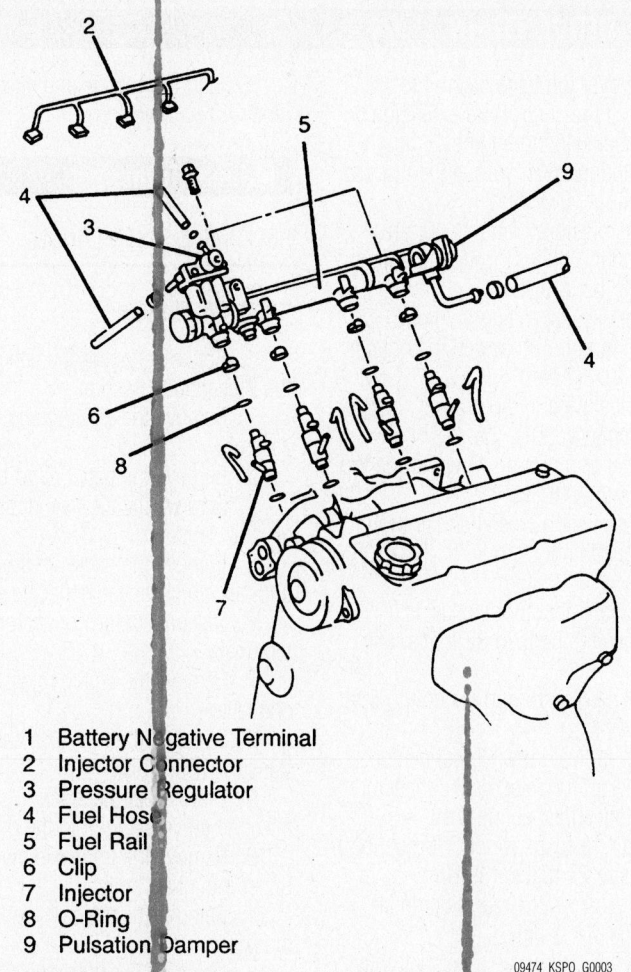

1 Battery Negative Terminal
2 Injector Connector
3 Pressure Regulator
4 Fuel Hose
5 Fuel Rail
6 Clip
7 Injector
8 O-Ring
9 Pulsation Damper

09474_KSPO_G0003

Fuel rail and injectors—2002 models

- MAF sensor bracket
- MAF sensor electrical connector
- Accelerator cable
- Breather hoses
- Negative battery cable
6. Fill the cooling system.
7. Start the vehicle and check for leaks, repair if necessary.

2005–06 Models

2.0L ENGINE

1. Before servicing the vehicle, refer to the Precautions Section.
2. Remove the engine cover.

3. Disconnect the TPS (Throttle Position Sensor) connector and ISA (Idle Speed Actuator) connector.
4. Disconnect the PCV (Positive Crankcase Ventilation) hose and breather hose.
5. Remove the accelerator cable.
6. Remove the delivery pipe.
7. Remove the clips and injectors.
8. Installation is in the reverse order of removal with new gasket. Coat the new O-rings with clean engine oil. Torque the delivery pipe bolts to 7–11 ft. lbs. (10–15 Nm).

2.7L ENGINE

1. Before servicing the vehicle, refer to the Precautions Section.
2. Remove the engine cover.
3. Remove air cleaner hose.
4. Disconnect the accelerator cable.
5. Disconnect the TPS connector (A).
6. Disconnect the ISA connector (B).
7. Disconnect the VIS actuator connector (C).
8. Disconnect the injector connector (D).
9. Disconnect the PCSV connector (E).
10. Disconnect the PCSV hose.
11. Disconnect the brake booster vacuum hose.
12. Disconnect the PCV hose.
13. Disconnect the IAT sensor connector.
14. Disconnect the VIS actuator connector.
15. Disconnect the ground cable from the surge tank assembly.
16. Remove the surge tank stay.
17. Remove the surge tank assembly.
18. Remove the clips and injectors.

To install:

19. Install the injector assembly. Coat the new O-rings with clean engine oil. Torque the delivery pipe bolts to 7–11 ft. lbs. (10–15 Nm).
20. Install the surge tank assembly.
21. Install the surge tank assembly. Tighten to 15–20 Nm (11–15 ft. lbs.)
22. Install the surge tank stay. Tighten to 15–20 Nm (11–15 ft. lbs.)
23. Install the ground cable.
24. Connect the VIS actuator connector.
25. Connect the IAT sensor connector.
26. Connect the PCV hose.
27. Connect the brake booster vacuum hose.
28. Connect the PCSV hose.
29. Connect the PCSV connector.
30. Connect the injector connector.
31. Connect the VIS actuator connector.
32. Connect the ISA connector.
33. Connect the TPS connector.
34. Connect the actuator cable.
35. Install the air cleaner hose.
36. Install the engine cover.

DRIVE TRAIN

Manual Transmission

REMOVAL & INSTALLATION

2002 Models

➡ **The removal of the manual transmission is virtually the same for 4WD and 2WD vehicles.**

1. Before servicing the vehicle, refer to the Precautions Section.
2. Drain the transmission fluid.
3. Drain the transfer case.
4. Remove or disconnect the following:
   - Negative battery cable
   - Rear portion of the center console
   - Shift lever and transfer lever knobs
   - Slide the boot cover over the shifters and remove the center console
   - Shift lever
   - Transfer lever
   - Front driveshaft by removing the bolts at the front differential and the bolts at the transfer case, if equipped
   - Bolts from the rear differential flange and the center support. Pull the driveshaft out of the tail shaft housing, if equipped with a 4 x 4
   - Bolts from the rear differential flange and center support. Pull the driveshaft out of the tail shaft housing, if equipped with a 4 x 2
   - Back-up light electrical connector and the Vehicle Speed Sensor (VSS) electrical connectors and move the wire harness aside
   - Crankshaft Position (CKP) sensor from the transmission housing
   - Clutch release cylinder and move it aside
   - Front exhaust pipe bracket
   - Front lower transmission housing bolts
   - Transfer case side mount, if equipped, and properly support the transmission
   - Transmission crossmember
   - transmission mount bolts
   - Starter bolts from the front housing
   - Transmission and transfer case, if equipped

To install:
5. Install or connect the following:
   - Transmission into position at the rear of the engine
   - Wire harness along the right side of the transmission and route the VSS

wire over the transmission to the rear of the control rod extension
   - Transmission to engine 14mm mounting bolts. Torque the bolts to 80 ft. lbs. (108 Nm).
   - Transmission to engine 10mm mounting bolts. Torque the bolts to 29 ft. lbs. (39 Nm).
   - Transmission to engine 6mm mounting bolts. Torque the bolts to 5 ft. lbs. (7 Nm).
   - Exhaust pipe to the bracket. Torque the bolts to 20 ft. lbs. (27 Nm).
   - Starter and ground wire. Torque the bolts to 29 ft. lbs. (39 Nm).
   - Transmission crossmember mount. Torque the bolts to 80 ft. lbs. (108 Nm).
   - Crossmember to the chassis. Torque the bolts to 32 ft. lbs. (44 Nm).
   - CKP sensor. Torque the bolt to 5 ft. lbs. (7 Nm).
   - 4WD indicator switch connector
   - Back-up light electrical connector
   - VSS electrical connector
   - Clutch release cylinder. Torque the bolts to 29 ft. lbs. (39 Nm).
   - Driveshaft to the rear differential flange, 4 x 4 only
   - Forward end of the driveshaft into the extension housing and attach the center support to the chassis
   - Shaft to the rear differential flange. Torque the bolts to 27 ft. lbs. (36 Nm).
   - Front driveshaft at the transfer case and install the bolts to the front differential, 4 x 4 only
   - Transfer case side mount, if equipped. Torque the bolts to 38 ft. lbs. (52 Nm).
   - Transfer case side mount to the chassis. Torque the bolts to 38 ft. lbs. (52 Nm).
   - Shift lever assembly. Torque the shift lever bracket bolts to 18 ft. lbs. (25 Nm).
   - Transfer lever assembly, if equipped. Torque the bolts to 18 ft. lbs. (25 Nm).
   - Dust cover plate over the shifter lever handles. Torque the bolts to 15 ft. lbs. (20 Nm).
   - Front console
   - Shifter lever knobs
   - Negative battery cable
   - Fill the transfer case.
6. Fill the transmission assembly.

7. Start the vehicle and check for leaks, repair if necessary.

Manual Transaxle

REMOVAL & INSTALLATION

2005–06 Models

1. Before servicing the vehicle, refer to the Precautions Section.
2. Remove the air duct and hose assembly.
3. Remove the battery and battery tray.
4. Remove the air cleaner assembly and air flow sensor.
5. Remove the back-up lamp connector and the vehicle speed sensor.
6. Disconnect the connectors and the terminals.
7. Remove the clutch release cylinder and lever.
8. Separate the transaxle cable from the transaxle assembly.
9. Disconnect the steering column shaft from the universal joint in the gear box.
10. Remove the transaxle clutch housing upper mounting bolts.
11. Remove the transaxle mounting brackets.
12. Using SST (09200-38001), or equivalent, support the engine assembly.
13. Disconnect the power steering oil pressure tube from the pump. Afterwards, stuff the hole with papers.
14. Remove the wheel and tire.
15. Disconnect the strut assembly, tie rod and stabilizer bar link from the knuckle.
16. Remove the wheel speed sensor.
17. Remove the brake caliper and suspend it with a string.
18. Lift up the vehicle and remove the undercover.
19. Remove the oil drain plug and drain the fluid.
20. Remove the front muffler.
21. Disconnect the steering tube on the cross member (sub-frame) and air condition switch.
22. Install a jack for the removal of a cross member (sub-frame).
23. After removing the cross member mounting bolts and nuts, remove the cross member with the steering gear box and the stabilizer bar.
24. Install a jack for the removal of the transaxle.
25. Remove the front and the rear roll stopper.

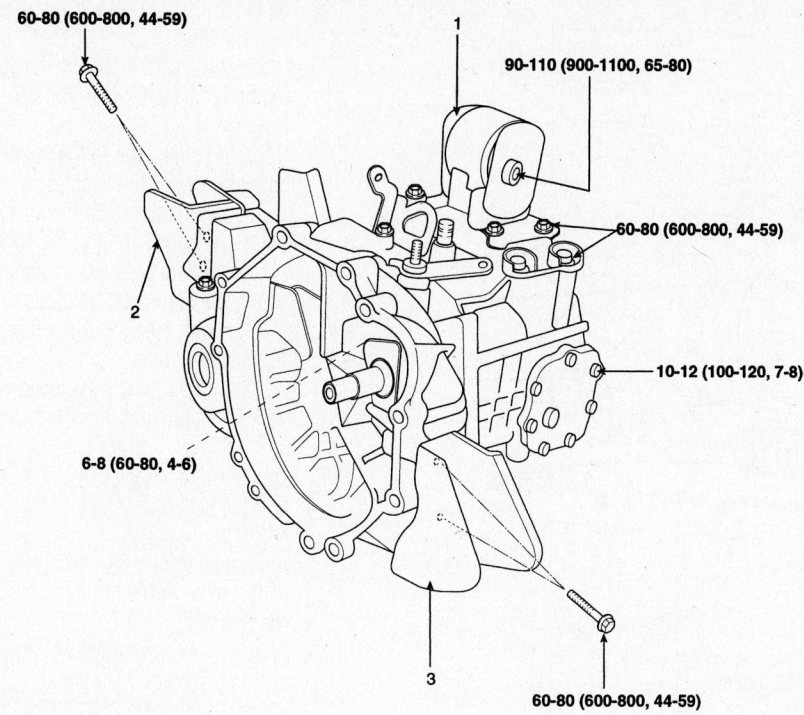

60-80 (600-800, 44-59)

90-110 (900-1100, 65-80)

60-80 (600-800, 44-59)

2

10-12 (100-120, 7-8)

6-8 (60-80, 4-6)

3

60-80 (600-800, 44-59)

TORQUE : Nm (kg.cm, lb-ft)

1. Transmission mounting bracket
2. Rear roll stopper bracket

3. Front roll stopper bracket

09474_KSPO_G0126

Manual transaxle mounting—2005–06 models

26. Remove the engine and the transaxle mounting bolts.

27. Lowering the jack slowly, remove the transaxle.

To install

28. Installation is the reverse of removal. Observe the tightening values in the accompanying illustration.

Automatic Transmission

REMOVAL & INSTALLATION

2002 Models

1. Before servicing the vehicle, refer to the Precautions Section.

2. Drain the transmission fluid.

3. Drain the cooling system.

4. Remove or disconnect the following:
   - Negative battery cable
   - Automatic transmission control cable from the throttle body

5. Slide the front seats forward and remove the 2 rear shift console mounting screws and set the parking brake.
   - Rear console and slide the front seats rearward

   - Front console mounting screws and untie the shifter boot draw strings
   - Loosen the transfer case lock nut and remove the transfer case shifter lever knob
   - Power/economy switch electrical connector
   - Front console and shift the transfer lever to the 4L position
   - transfer shift lever cover plate
   - transfer case shift lever assembly and place the transmission in the park position

- Selector lever nuts
- Split pin from the shift selector lever
- Shift selector rod and spring washers
- Electrical connectors from the base of the selector lever

➡**There is a fifth wire connection (Park Position). This wire is hard-wired, do not disconnect it.**

- Selector lever assembly
- Control cable from the throttle linkage and slide the cable from the bellcrank
- Upper dipstick tube from the lower dipstick tube
- Input/Turbine speed sensor from the top rear of the transmission
- Shift solenoids from the lower left side of the transmission
- Vehicle Speed Sensor (VSS) from the center of the transfer case
- Input speed sensor electrical connectors
- Matchmark the front and rear driveshafts. Remove the attaching bolts from both flanges and remove the driveshafts
- Oil cooler pipes at the transmission
- Starter
- Transfer case mounting bolts
- Transfer case nuts from the crossmember and support the transmission
- Crossmember
- Transmission to engine mounting bolts
- Front lower splash shield
- Left side lower gusset
- Torque converter inspection cover
- Torque converter to drive plate bolts
- Slide the transmission away from the engine and lower the transmission slightly

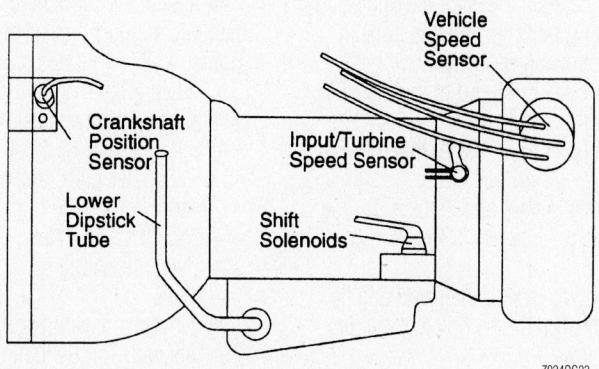

Automatic transmission wiring connections—2002 models

7924QG23

Radiator Side

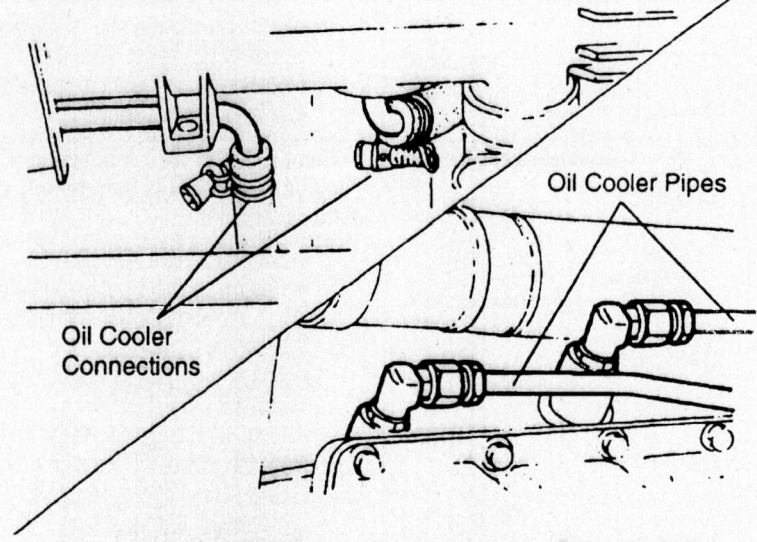

Oil Cooler Connections

Oil Cooler Pipes

Passenger Side

79240G24

Exploded view of the oil cooler pipe connections—2002 models

- Crankshaft Position (CKP) sensor attaching bolt and sensor
- 4WD, 4WD LOW indicator switches and the VSS from the transfer case

6. Lower the transmission. Be sure all wiring is clear and disconnected. Be sure the throttle cables come out without binding or attaching to anything.

To install:

7. Install or connect the following:
- Transmission into position to attach the sensor and indicator switch wiring. Be sure the throttle cables are guided into the engine compartment without binding or attaching to anything.
- 4WD, 4WD LOW indicator switches and the VSS to the transfer case
- Input/turbine speed sensor and CKP sensor. Torque the bolt to 12 ft. lbs. (16 Nm).
- Raise the transmission and position it to the engine. Install the upper housing bolts. Torque the bolts in the following sequence:

8. 10mm bolts to 38–60 ft. lbs. (57–81 Nm).

9. 12mm bolts to 51–65 ft. lbs. (69–88 Nm).
- Exhaust hanger and bracket. Torque the bolts to 38–60 ft. lbs. (57–81 Nm).
- Oil cooler pipe to the lower engine mount. Torque the bolt to 60 ft. lbs. (81 Nm).
- Crossmember. Torque the bolts to 23–34 ft. lbs. (31–46 Nm).

- Two transfer case mounting bolts located at the right center of the transfer case. Torque the bolts to 23–34 ft. lbs. (31–46 Nm).
- Four transfer case nuts to the crossmember. Torque the nuts to 23–34 ft. lbs. (31–46 Nm).
- Torque converter-to-drive plate bolts. Torque the bolts to 12–20 ft. lbs. (16–27 Nm).
- Torque converter inspection cover. Torque the bolts to 41–62 inch lbs. (5–7 Nm).
- Front splash guard. Torque the bolts to 41–62 inch lbs. (5–7 Nm).
- Starter. Torque bolts to 27–40 ft. lbs. (37–54 Nm).
- Left side lower gusset. Torque the bolts to 38–60 ft. lbs. (57–81 Nm).
- Right side lower gusset in 3 steps:

a. Install the bottom mounting bolts, but do not tighten.

b. Install the top bolt to the intake manifold support bracket and manifold. Tighten to 27–40 ft. lbs. (37–54 Nm).

c. Secure the manifold intake bracket by tightening the two attaching bolts to 38–60 ft. lbs. (57–81 Nm).
- Oil cooler pipes at the transmission. Torque the lines to 42–62 inch lbs. (5–7 Nm).
- Two oil cooling tube clamps to the lines
- Driveshafts with the matchmarks aligned. Torque the bolts to the differential flanges to 20–22 ft. lbs. (27–30 Nm) and the transfer case

flange bolts to 36–43 ft. lbs. (49–59 Nm).
- Undercover splash shield. Torque the bolts to 42–62 inch lbs. (5–7 Nm).
- Upper dipstick tube to the lower tube and lower the vehicle

10. Provide automatic transmission control cable slack by gently pulling the cable to the left until the cable pin has rotated sufficiently to line the automatic transmission control cable up with the slot in the rear of the throttle body bellcrank.

11. Slide the automatic transmission control cable and cable pin into the bellcrank.

12. Tighten the locknut.
- Throttle kickdown cable to the mounting bracket
- Automatic transmission control cable to the throttle body
- Air silencer
- Shifter lever assembly to the transfer case. Torque the bolts to 72–102 inch lbs. (22–28 Nm).
- Four wiring connectors under the shift selector lever
- Shift selector lever, do not exert any force when installing the shift selector lever
- Four shift selector lever nuts. Tighten to 72–102 inch lbs. (22–28 Nm).
- Shift rod and washers to the selector lever and install the split pin to the shift selector lever
- Power/Economy switch wiring connector
- Rear console
- Front console and tie the shift boot draw strings
- Negative battery cable

13. Fill the transmission to the proper level.

14. Fill the cooling system.

15. Start the vehicle, check for leaks, and repair if necessary.

Automatic Transaxle

REMOVAL & INSTALLATION

2005–06 Models

1. Before servicing the vehicle, refer to the Precautions Section.

2. Remove the air duct.

3. Remove the battery.

4. Remove the battery tray.

5. Remove the air cleaner assembly.

6. Remove the intercooler inlet pipe.

7. Disconnect the connectors relevant to the transaxle.

8. Disconnect the ground earth wire.

9. Detach the oil cooler hoses.

10. Using SST (09200-38001), or equivalent, support the engine.

11. Remove the transaxle mounting bracket (A) bolts.

12. Remove the transaxle-to-engine upper bolts.

13. Remove the bolts which mount the transaxle to the front sub frame.

14. Raise and safely support the vehicle.

15. Remove the oil drain plug and drain the fluid.

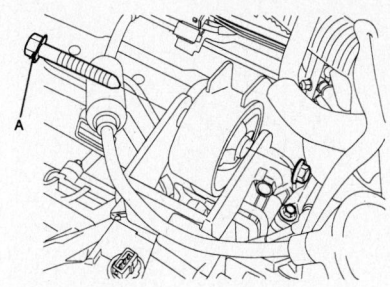

Remove the bolt (A) which mounts the transaxle to the rear sub-frame—automatic transaxle

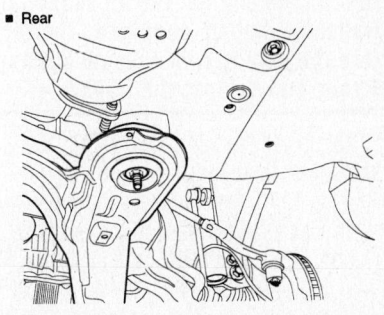

Rear sub-frame brace—automatic transaxle

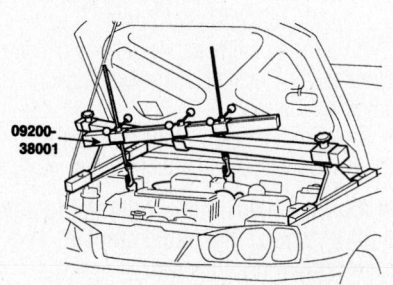
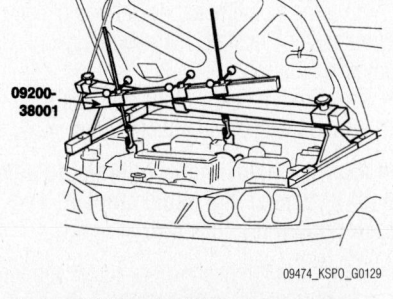

Using SST (09200-38001), or equivalent, support the engine—automatic transaxle

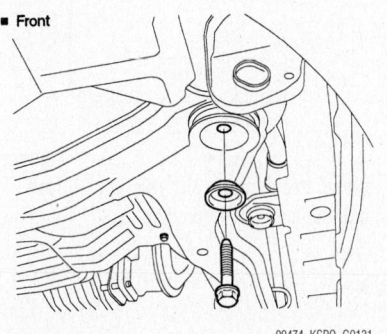

Front sub-frame bracket—automatic transaxle

Rear sub-frame bracket—automatic transaxle

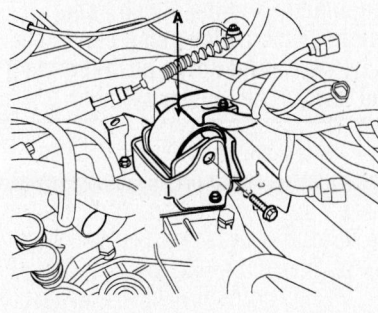

Transaxle upper mount—automatic transaxle

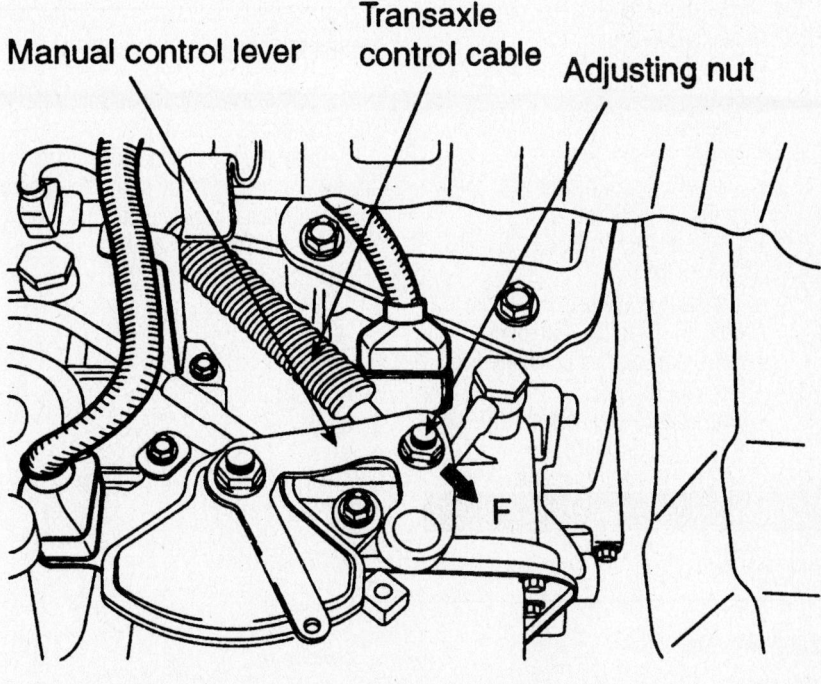

Manual control lever   Transaxle control cable   Adjusting nut

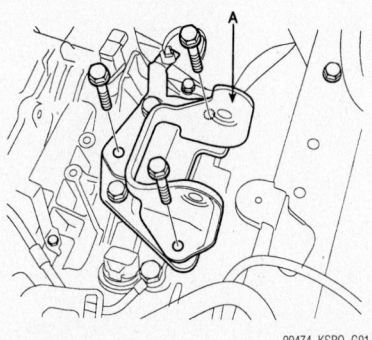

Transaxle mounting bracket—automatic transaxle

Automatic transaxle adjustment points—automatic transaxle

16. Support the transaxle with a jack.
17. Remove the steering column coupler bolt.
18. Remove the halfshafts.
19. Remove the bolt which mounts the transaxle to the rear sub-frame.
20. Remove the sub-frame. If it is a 4 wheel drive vehicle (4WD), remove the driveshaft
21. Remove the transaxle-to-engine lower bolts.
22. Remove the transaxle assembly.

To install:

❊❊ WARNING

If the torque converter is mounted first on the engine, the oil seal on the transaxle may be damaged. Therefore, be sure to first assemble the torque converter to the transaxle.

23. Install the transaxle control cable and adjust as follows:
  a. Move the shift lever and the transaxle range switch to the "N" position, and install the control cable.
  b. When connecting the control cable to the transaxle mounting bracket, install the clip until it contacts the control cable.
  c. Remove any free-play in the control cable by adjusting nut and then check to see that the selector lever moves smoothly.
  d. Check to see that the control cable has been adjusted correctly.
24. Installation is the reverse of removal. Observe the following torques:
  • Filler plug: 21–25 ft. lbs. (29–34 Nm)
  • Drain plug: 29–36 ft. lbs. (40–50 Nm)
  • Transaxle upper mount bolt: 36–47 ft. lbs. (50–65 Nm)
  • Upper bracket bolts: 43–58 ft. lbs. (60–80 Nm)
  • Transaxle to the rear sub-frame bolt: 36–47 ft. lbs. (50–65 Nm)
  • Sub-frame bracket bolts: 43–58 ft. lbs. (60–80 Nm)
  • Transaxle-to-engine bolts: 29–38 ft. lbs. (42–54 Nm)

Clutch

CLUTCH PEDAL HEIGHT ADJUSTMENT

2002 Models

1. Before servicing the vehicle, refer to the Precautions Section.
2. Pull back the carpet to measure the

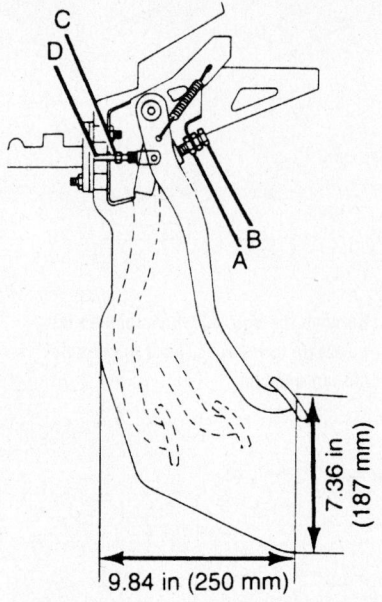

Clutch pedal height and free-play adjustment points—2002 models

distance from the firewall to the top of the pedal. The standard height is 9.84 in. (250 mm).
3. If adjustment is required, loosen the locknut and turn the stopper bolt.
4. After adjustment is made tighten the locknut to 12 ft. lbs. (16 Nm).

CLUTCH PEDAL CLEVIS PIN AND PEDAL HEIGHT ADJUSTMENT

2005–06 Models

1. Before servicing the vehicle, refer to the Precautions Section.
2. Measure the clutch pedal height (from the face of the pedal pad to the floorboard) and the distance between inner pad and clutch pedal. Standard values:

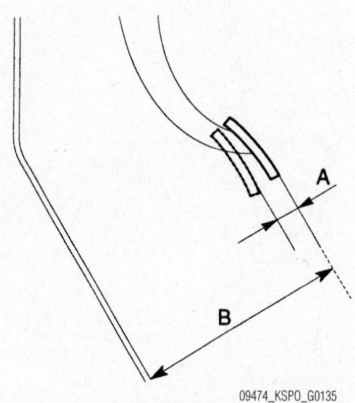

Clutch pedal clevis pin and pedal height measurement points—2005–06 models

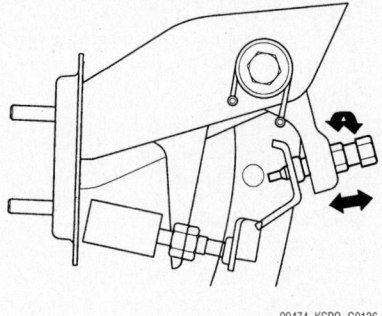

Pedal height or clevis pin adjustment—2005–06 models

  • A: 252 ± 2 mm (9.84–10.00 in.)
  • B: 189.5 mm (7.46 in.)
3. If the clutch pedal height and distance is not within the standard value range, adjust as follows:
  a. Turn and adjust the bolt, then secure by tightening the lock nut.

➡**After the adjustment, tighten the bolt until it reaches the pedal stopper, and then tighten the lock nut.**

  b. Turn the push rod to agree with the standard value and then secure the push rod with the lock nut.

❊❊ WARNING

When adjusting the clutch pedal height or the clutch pedal clevis pin play, be careful not to push the push rod toward the master cylinder.

4. After completing the adjustments, check that the clutch pedal free play (measured at the face of the pedal pad) is within the standard value ranges. Standard value: (C) 6–13mm (0.24–0.52 in.).

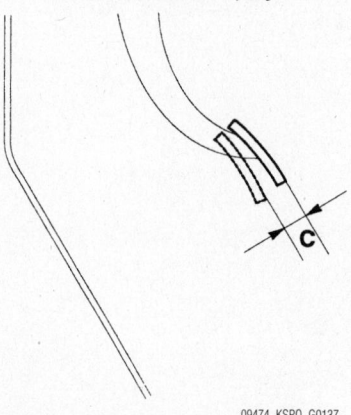

Final free-play adjustment—2005–06 models

5. If the clutch pedal free play and the distance between the clutch pedal and the floor board when the clutch is disengaged, do not meet with the standard values, it may be the result of either air in the hydraulic system or a faulty the clutch master cylinder. Bleed the air or disassemble and inspect the master cylinder or clutch.

CLUTCH PEDAL FREE-PLAY ADJUSTMENT

2002 Models

1. Before servicing the vehicle, refer to the Precautions Section.
2. Depress the clutch pedal gently by hand and measure the amount of free-play (distance the pedal travels before resistance is felt). The proper amount of free-play is 0.5 in. (12.7mm). If the free-play is not within the proper specifications, continue with the procedure.
3. Measure from the floor pan to the middle point of the clutch pedal when the pedal is in the fully released position. The proper clutch pedal height is 7.25 in. (184mm).
4. If the pedal height is incorrect, loosen locknut (A) and turn the pedal adjusting bolt (B) until the proper height is achieved, then retighten the locknut.
5. Remeasure the free-play. If it is still out of specification, loosen the clutch pushrod locknut (C) and turn the pushrod (D) until the proper free-play is achieved. Tighten locknut (C) securely.

REMOVAL & INSTALLATION

2002 Models

1. Before servicing the vehicle, refer to the Precautions Section.
2. Remove or disconnect the following:
   - Transmission
   - Pressure plate bolts and remove the clutch plate and disc

To install:
3. Install or connect the following:
   - Clutch disc and plate using a centering tool
   - Clutch cover. Torque the bolts to 73 ft. lbs. (99 Nm) and remove the centering tool
   - Check the release bearing condition and lubricate or replace as necessary
   - Transmission

2005–06 Models

1. Before servicing the vehicle, refer to the Precautions Section.
2. Remove the transaxle assembly.

3. Insert the special tool 09411-11000, or equivalent alignment tool, in the clutch disc to prevent the disc from falling.
4. Loosen the bolts, which attach the clutch cover to the flywheel in a star pattern. Loosen the bolts in succession, one or two turns at a time, to avoid bending the cover flange.

➡**Do not clean the clutch disc or the release bearing with cleaning solvent.**

5. Remove the release fork shaft and bushing.
6. Apply multipurpose grease to the splines of the disc.

✳✳ WARNING

When installing the clutch, apply grease to each part, but be careful not to apply excessive grease. It can cause clutch slippage and shudder.

7. Install the clutch disc assembly to the flywheel using the alignment.
8. Install the clutch cover assembly to the flywheel and temporarily tighten the bolts one or two steps at a time in the sequence shown. Tighten to 15–22 Nm (11–16 ft. lbs.)
9. Align the bearing (A) to the release fork (B) and then install it to the sleeve of the housing.

➡**Apply multipurpose grease, CAS-MOLY L9508, or equivalent, to the bearing sleeve, contact point of the release fork (B) and the bushing inner surface (C).**

10. Install the release lever to the release fork.
11. Install the transaxle assembly to the engine.

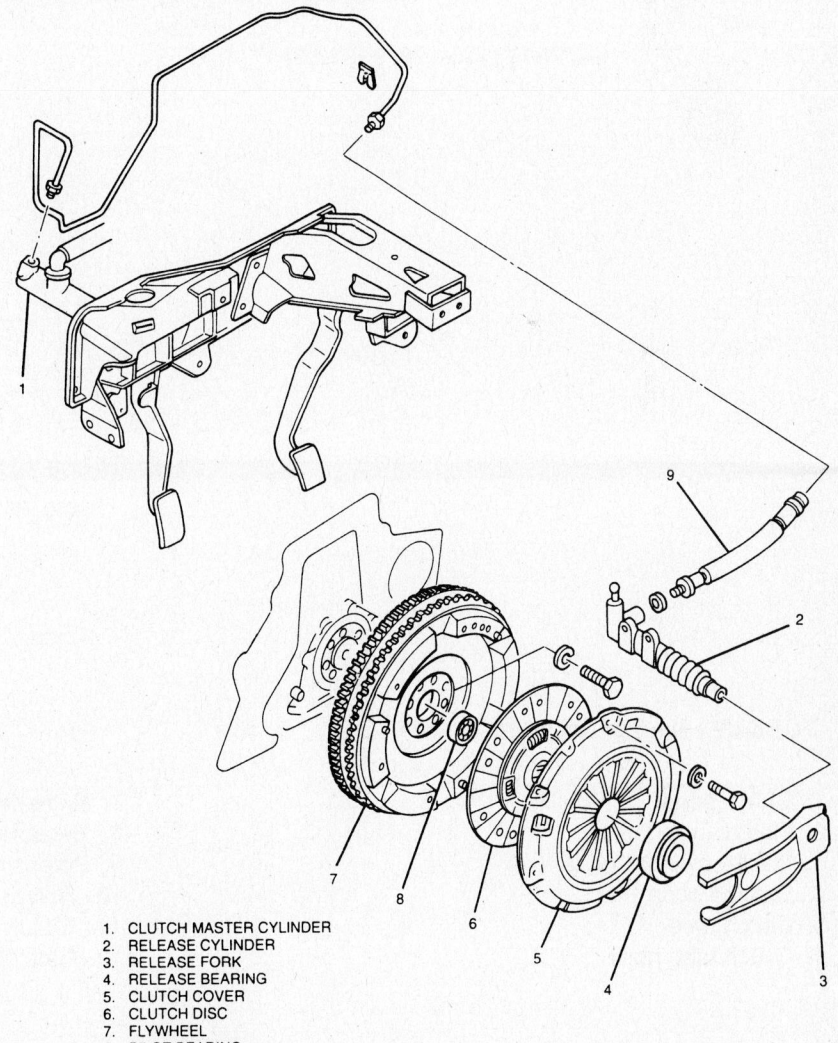

1. CLUTCH MASTER CYLINDER
2. RELEASE CYLINDER
3. RELEASE FORK
4. RELEASE BEARING
5. CLUTCH COVER
6. CLUTCH DISC
7. FLYWHEEL
8. PILOT BEARING
9. FLEXIBLE HOSE

7924QG26

Exploded view of the clutch assembly—2002 models

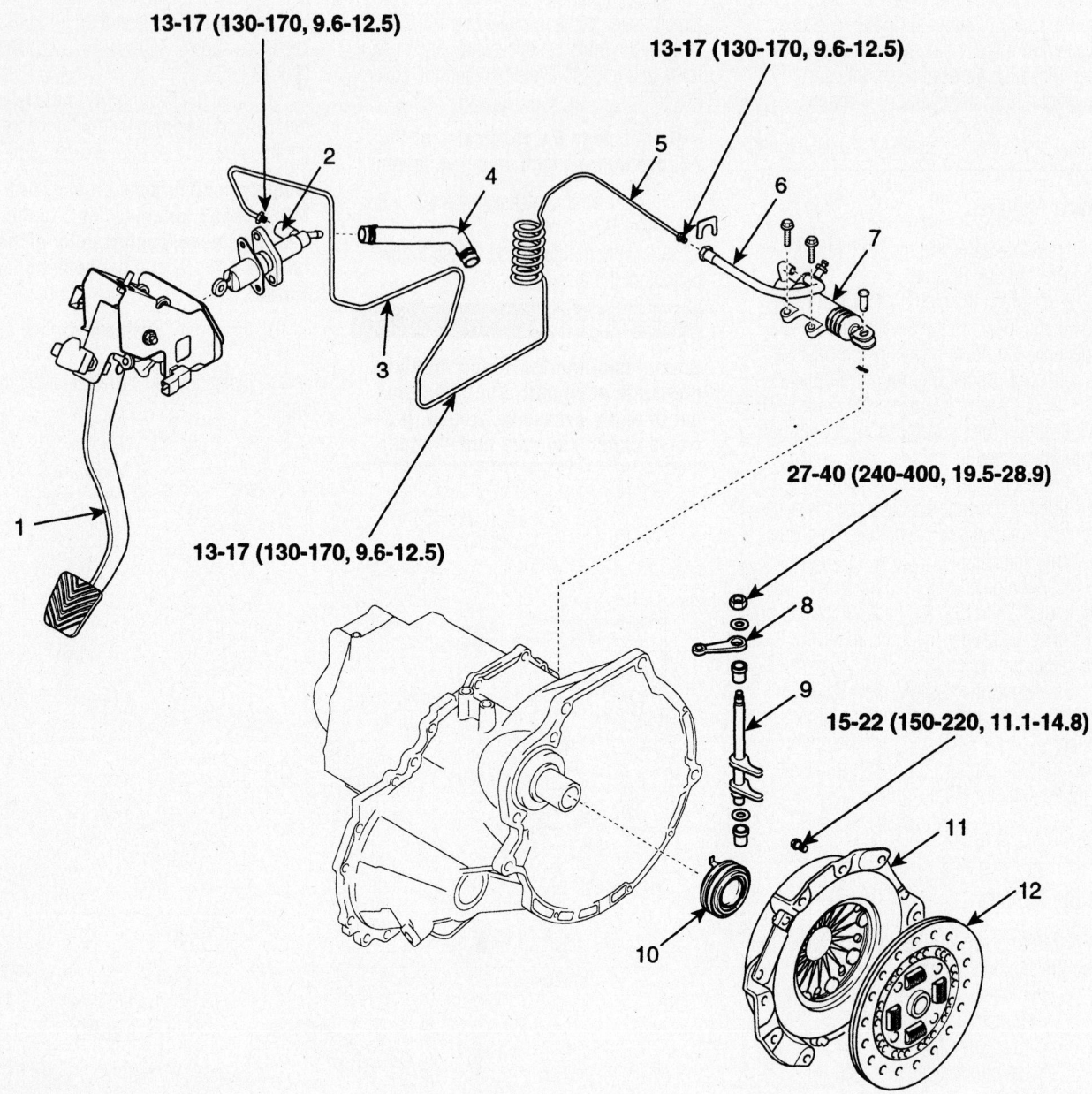

13-17 (130-170, 9.6-12.5)

13-17 (130-170, 9.6-12.5)

13-17 (130-170, 9.6-12.5)

27-40 (240-400, 19.5-28.9)

15-22 (150-220, 11.1-14.8)

TORQUE : Nm(kgf.cm, lb-ft)

1. Clutch pedal
2. Master cylinder
3. Master cylinder hose
4. Flexible hose
5. Clutch tube
6. Clutch tube hose
7. Release cylinder
8. Release lever
9. Release fork
10. Release bearing
11. Clutch disc cover
12. Clutch disc

09474_KSPO_G0138

Clutch system components—2005–06 models

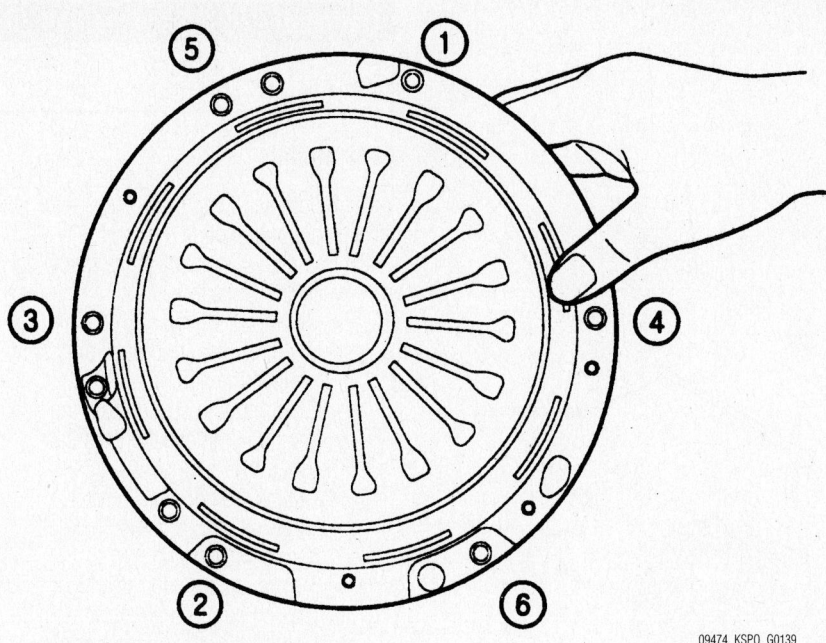

Clutch disc torque sequence—2005–06 models

09474_KSPO_G0139

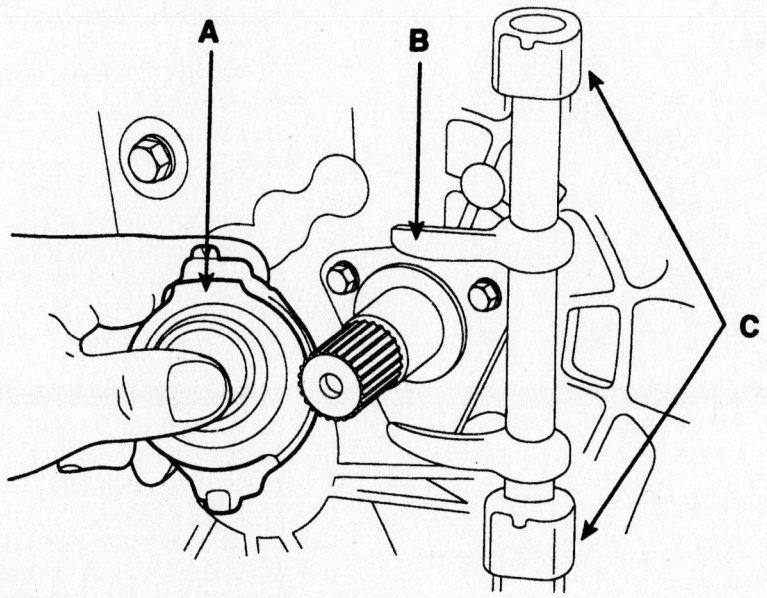

Bearing and release fork grease points—2005–06 models

09474_KSPO_G0140

Hydraulic Clutch System

BLEEDING

1. Before servicing the vehicle, refer to the Precautions Section.
2. With an assistant in the vehicle, raise and safely support the vehicle.
3. Have your assistant pump the clutch pedal three times and hold the pedal to the floor.
4. Open the bleeder valve on the clutch slave cylinder until the air is purged from the cylinder.
5. Tighten the bleeder valve.
6. Have your assistant release the clutch pedal.
7. Fill the clutch master cylinder if below minimum.
8. Repeat Steps 2 through 6 until no air exits from the bleeder valve.
9. Lower the vehicle.
10. Fill the clutch master cylinder fluid reservoir.

Transfer Case Assembly

REMOVAL & INSTALLATION

2002 Models

1. Before servicing the vehicle, refer to the Precautions Section.
2. Drain the transfer case.
3. Remove or disconnect the following:
   - Negative battery cable
   - Two rear console mounting screws. Slide the console forward to clear the parking brake handle and set aside
   - Three mounting screws from the front console. Untie the shift boot draw strings and open the boot
   - Loosen the transfer case shift lever locknut and remove the lever knob
   - Pull the console up to access the Power/Economy switch wiring connector. Unplug the connector and remove the console
4. Shift the transfer lever to the 4L position.
   - Cover plate
   - Retaining bolts from the transfer case and lift the shifter lever assembly straight out and properly support the transmission
   - Matchmark the driveshafts at the flanges and remove the driveshafts
   - Crossmember bolts
   - 4WD light switch connector
   - Transfer case mounting bolts located at the right center of the transfer case
   - Transfer case nuts from the crossmember
   - Separate the transfer case from the transmission by striking the transfer case with a plastic mallet at the seal area
5. Lower the transfer case from the vehicle.
   To install:
6. Install or connect the following:
   - Transfer case in position with a new gasket. Torque the bolts to 32 ft. lbs. (44 Nm).
   - Crossmember. Torque bolts to 32 ft. lbs. (44 Nm).
   - Transfer case mounting bolts located at the right center of the transfer case. Torque the bolts to 38 ft. lbs. (52 Nm).
   - Four transfer case nuts to the crossmember. Torque the nuts to 15 ft. lbs. (20 Nm).
   - Align the matchmarks on the drive-

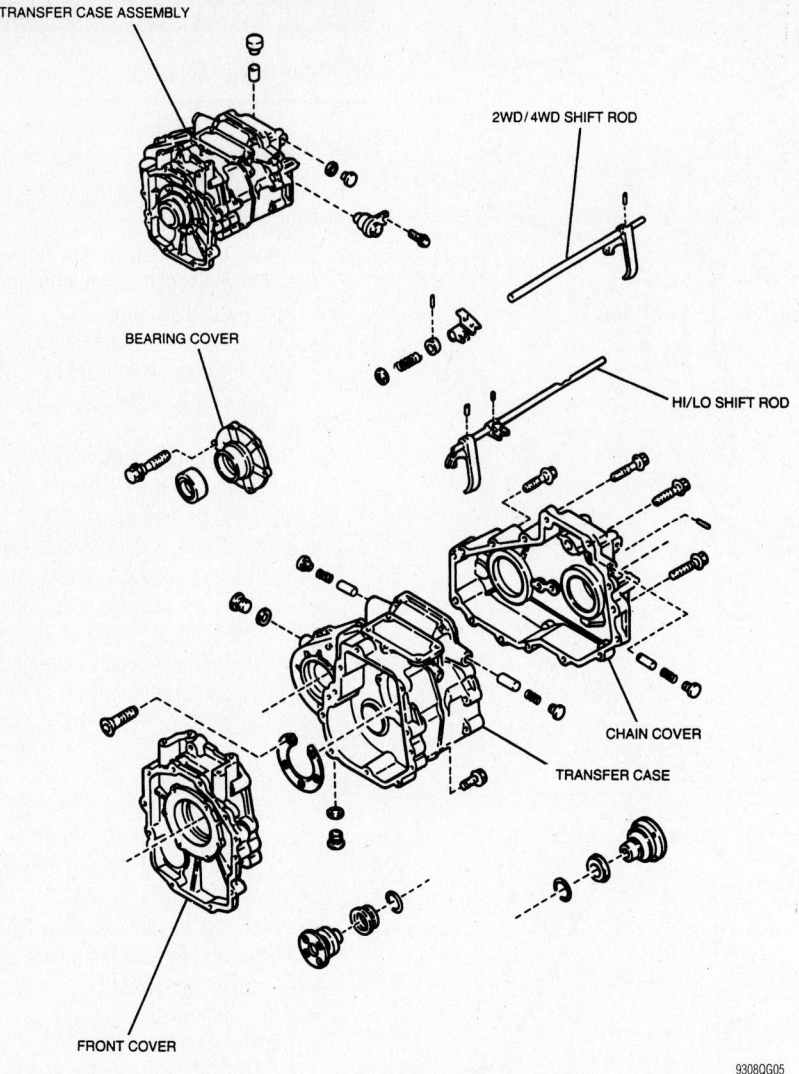

TRANSFER CASE ASSEMBLY

2WD/4WD SHIFT ROD

BEARING COVER

HI/LO SHIFT ROD

CHAIN COVER

TRANSFER CASE

FRONT COVER

9308QG05

Exploded view of the transfer case assembly—2002 models

shafts to the flanges. Torque the bolts to 27 ft. lbs. (36 Nm) and remove the transmission support
- 4WD light switch electrical connector
- VSS electrical connector
- Shifter lever assembly. Torque retaining bolts to 20 ft. lbs. (27 Nm).
- Cover plate. Torque the bolts to 15 ft. lbs. (20 Nm).
- Power/Economy switch wiring connector
- Front console
- Lever knobs
- Front console and tie the shift boot draw strings
- Slide the console over the parking brake handle
- Rear console
- Negative battery cable
7. Fill the transfer case to the proper level

8. Start the vehicle and check for leaks, repair if necessary.

2005–06 Models

1. Before servicing the vehicle, refer to the Precautions Section.
2. Remove the battery ground cable.
3. Raise and safely support the vehicle.
4. Remove the driveshaft.
5. Remove the front muffler.
6. Remove the right halfshaft.
7. Loosen the oil drain plug and drain the fluid.
8. After draining, re-tighten the oil drain plug. Tighten to 39–59 Nm (29–43 ft. lbs.)
9. Support the transfer assembly with a jack.
10. Remove the transfer assembly by loosening the mounting bolts.
11. Installation is the reverse of removal.
12. Refill the transfer case.

Halfshaft

REMOVAL & INSTALLATION

2002 Models

1. Before servicing the vehicle, refer to the Precautions Section.
2. Remove or disconnect the following:
- Negative battery cable
- Both front wheels
- Six free wheel hub bolts and remove the hub
- Snapring and spacer from the hub
- Carefully remove the fixed cam assembly
- Caliper from the brake rotor
- Upper control arm link lockbolt, spring washer and nut
- Tie rod end from the steering knuckle
- Loosen the drop link lower locknut
- Loosen the four upper drop link locknuts
- Spread open the drop link fork with a rubber mallet
- Matchmark the halfshaft and differential.
- Carefully pry the halfshaft from the differential
- Halfshaft

To install:

3. Install or connect the following:
- Halfshaft with the matchmarks aligned with the differential
- Torque the upper and lower drop link nuts to 36 ft. lbs. (49 Nm).
- Tie rod end to the steering knuckle. Torque the locknut to 27 ft. lbs. (36 Nm) and install a new cotter pin
- Upper control arm link lockbolt, spring washer and nut. Torque the bolt to 36 ft. lbs. (49 Nm).
- Fixed cam assembly
- Snapring and spacer in the hub
- Wheel hub and the six free wheel hub bolts. Torque the bolts in two

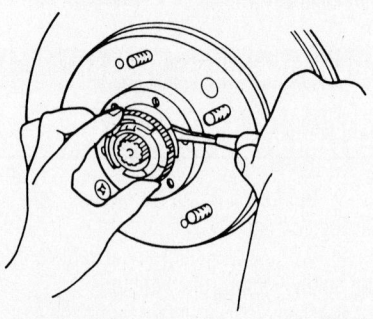

79240G27

Removing the 4WD fixed cam assembly— 2002 models

passes, 14 ft. lbs. (17 Nm), then to 23 ft. lbs. (31 Nm).
- Front wheel assemblies
- Negative battery cable

2005–06 Models

FRONT

1. Before servicing the vehicle, refer to the Precautions Section.

2. Loosen the wheel nuts slightly.

3. Raise the front of the vehicle, and make sure it is securely supported.

4. Remove the front wheel and tire from front hub.

5. Remove the split pin (A), then remove castle nut (B) and washer (C) from the front hub under applying the brake.

6. Remove the wheel speed sensor from the knuckle.

7. Disconnect the tie rod end ball joint (C) from the knuckle (D) using a ball joint separator.

8. Remove the split pin (A).

9. Remove the castle nut (B).

10. Disconnect the ball joint (C) from knuckle (D) using the special tool (09568-34000).

➥ **Apply a few drops of oil to the special tool. (Boot contact part)**

11. Remove the lower arm ball joint mounting bolts.

12. Using a plastic hammer, disconnect the halfshaft from the axle hub.

13. Push the axle hub outward and separate the halfshaft from the axle hub.

14. Insert a pry bar (A) between the transaxle case (B) and joint case (C), and

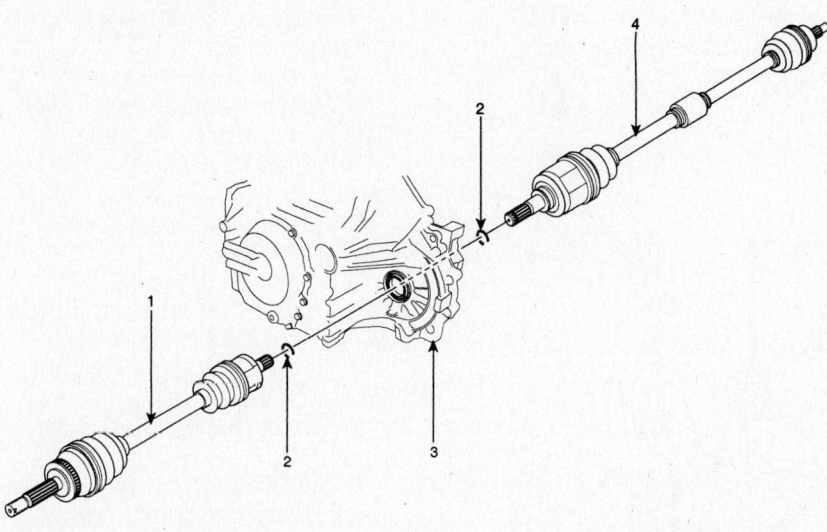

1. Driveshaft (LH)
2. Circlip

3. Transaxle
4. Driveshaft (RH)

09474_KSPO_G0141

Front halfshafts—2005–06 models

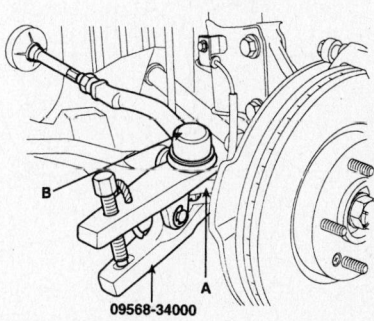

09568-34000

09474_KSPO_G0144

Disconnecting the ball joint from the knuckle—2005–06 models

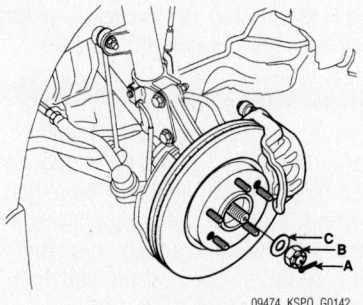

09474_KSPO_G0142

Removing the hub nut—2005–06 models

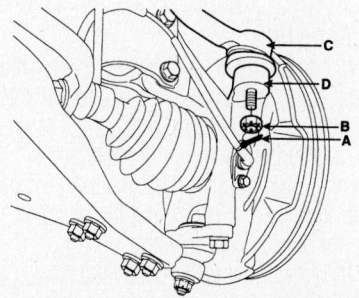

09474_KSPO_G0143

Disconnecting the tie rod end—2005–06 models

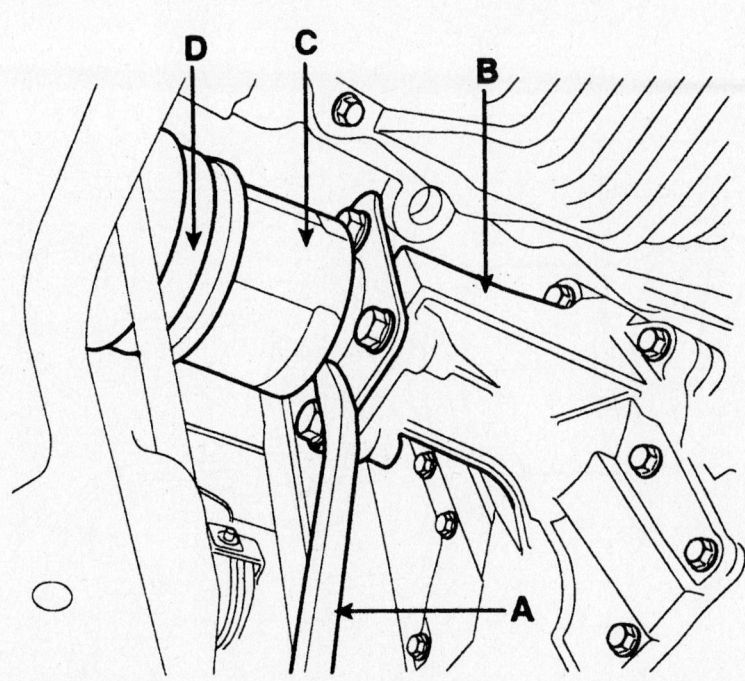

09474_KSPO_G0145

Separating the halfshaft from the transaxle—2005–06 models

separate the halfshaft (D) from the transaxle case.

➡ Use a pry bar (A) being careful not to damage the transaxle and joint. Do not insert the pry bar (A) too deeply, as this may cause damage to the oil seal. (max. depth: 7mm (0.28 in.). Do not pull the halfshaft by excessive force it may cause components inside the axle shaft joint kit to dislodge resulting in a torn boot or a damaged bearing. Plug the hole of the transaxle case with the oil seal cap to prevent contamination. Support the halfshaft properly.

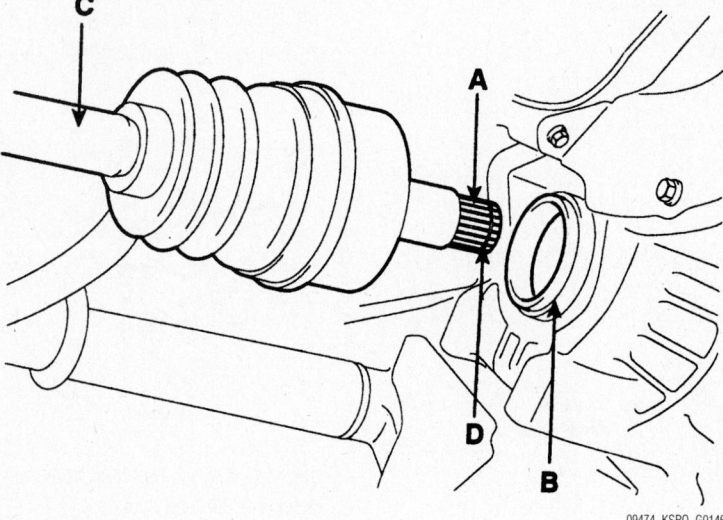

Halfshaft installation w/2wd—2005–06 models

09474_KSPO_G0146

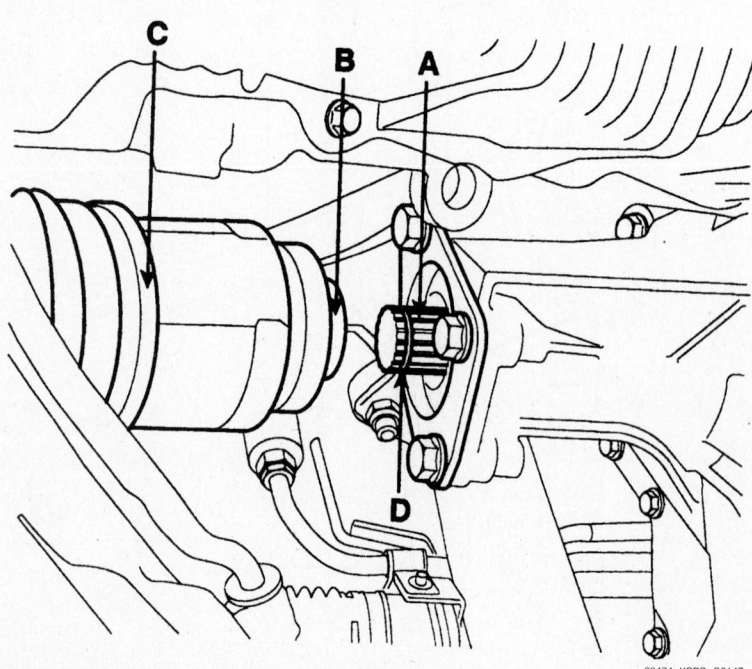

Halfshaft installation w/4wd—2005–06 models

09474_KSPO_G0147

15. Replace the retainer ring whenever the halfshaft is removed from the transaxle case.

To install:

16. Apply gear oil on the halfshaft oil seal case contacting surface (B) and transaxle case splines (A).

17. Before installing the halfshaft (C), set the opening side of the circlip (D) facing downward.

18. After installation, check that the halfshaft cannot be removed by hand.

19. Install the drive shaft into the knuckle.

20. Install the lower arm mounting bolts. Tighten to 100–120 Nm (74.8–88 ft. lbs.)

21. After installing the hub washer with convex surface outward, install the castle nut and the new split pin. Tighten to 200–280 Nm (147–206 ft. lbs.)

22. Install the front wheel and tire on the front hub. Tighten to 90–110 Nm (66–81 ft. lbs.)

REAR

1. Before servicing the vehicle, refer to the Precautions Section.

2. Loosen the wheel nuts slightly.

3. Raise the rear of the vehicle, and make sure it is securely supported.

4. Remove the rear wheel and tire from rear hub.

5. Remove the wheel speed sensor from the axle carrier.

6. Remove the split pin (A), then remove castle nut (B) and washer (C) from the rear hub under applying the brake.

7. Remove the trailing arm mounting bolt (B) from the knuckle (A).

8. Remove the suspension arm mounting nuts (C).

9. Push the axle hub outward and separate the halfshaft from the axle hub.

10. Insert a pry bar (A) between the differential case (B) and joint case (C), and separate the halfshaft (D) from the differential case.

➡ Use a pry bar (A) being careful not to damage the transaxle and joint. Do not insert the pry bar (A) too deep, as this may cause damage to the oil seal (max. depth: 7 mm (0.28in). Do not pull the halfshaft by excessive force because it may cause components inside the BJ or TJ joint kit to dislodge resulting in a torn boot or a damaged bearing. Plug the hole of the transaxle case with the oil seal cap to prevent contamination. Support the halfshaft properly.

11. Replace the retainer ring whenever the halfshaft is removed from the transaxle case.

To install:

12. Apply gear oil on the halfshaft differential case contacting surface and halfshaft splines.

13. Before installing the halfshaft, set the opening side of the circlip facing downward. After installation, check that the halfshaft cannot be removed by hand.

14. Install the BJ into the knuckle.

15. Install the suspension arm mounting nuts and trailing arm mounting bolt from the knuckle. Tighten to:

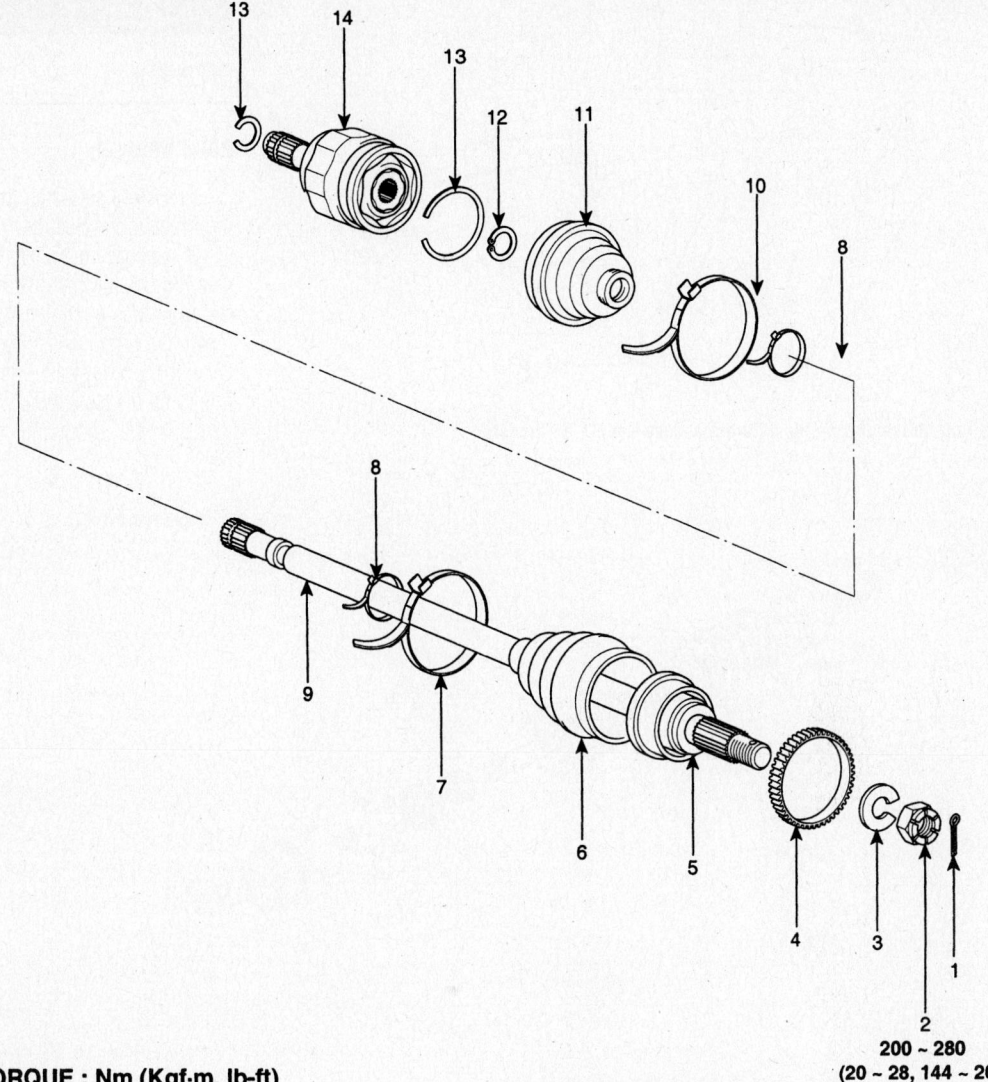

TORQUE : Nm (Kgf·m, lb-ft)

200 ~ 280
(20 ~ 28, 144 ~ 202)

1. Split pin
2. Castle nut
3. Washer
4. Dust cover & Tone wheel
5. BJ assembly
6. BJ boot
7. BJ boot big part band
8. Boot small part band
9. Shaft
10. TS boot big part band
11. TS boot
12. Snap ring
13. Circlip
14. TS assembly

09474_KSPO_G0148

Rear halfshaft components—2005–06 models

- Suspension arm mounting nuts (C): 140–160 Nm (104–118 ft. lbs.).
- Trailing arm mounting bolt (B): 100–120 Nm (74–88 ft. lbs.).

16. Install the washer, castle nut and split pin from the rear hub. Tighten to 200–280 Nm (147–206 ft. lbs.)

17. Install the wheel speed sensor on the knuckle.

18. Install the rear wheel and tire on the rear hub. Tighten to 90–110 Nm (66–81 ft. lbs.).

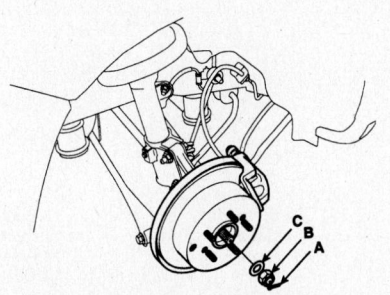

09474_KSPO_G0149

Rear hub nut removal—2005–06 models

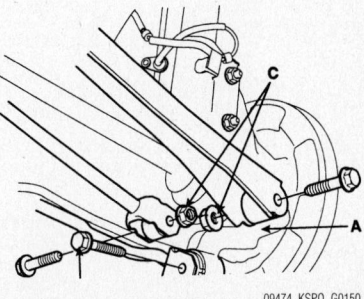

09474_KSPO_G0150

Trailing and suspension arms removal— 2005–06 models

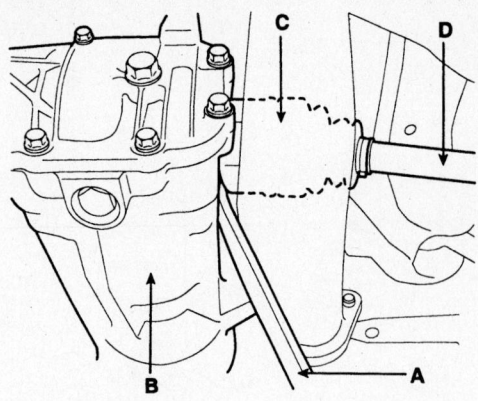

Separating the rear halfshaft from the differential case—2005–06 models

09474_KSPO_G0151

CV-Joints

OVERHAUL

2002 Models

1. Before servicing the vehicle, refer to the Precautions Section.

2. Secure the halfshaft in a soft-jawed vise. Be careful that dust or other foreign material does not enter the joint during repair. Do not disassemble the wheel side CV-joint. Do not wash joint insolvent unless it is to be disassembled.

3. Inspect boot and clamp for damage.

Right Hand (RH)

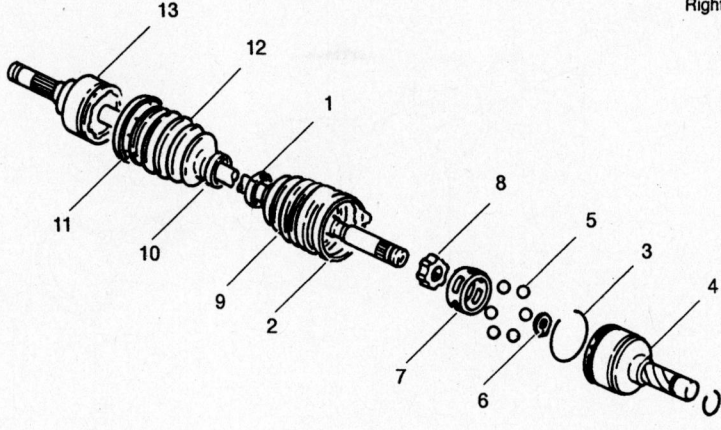

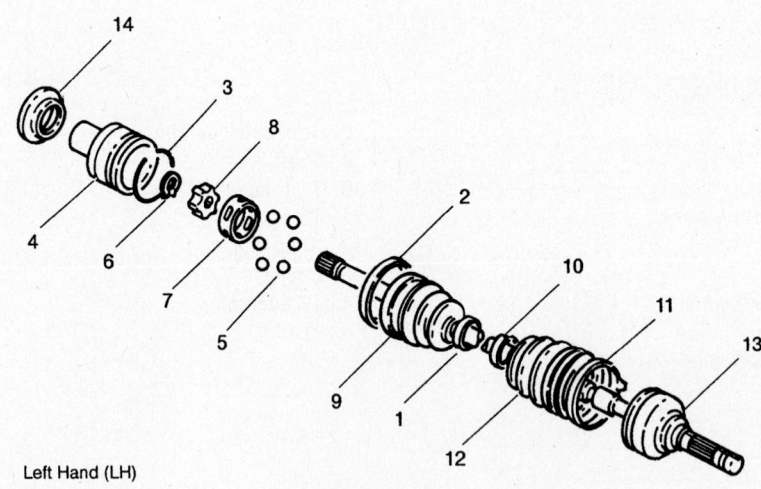

Left Hand (LH)

1	Boot Clamp	6	Snap Ring	11	Boot Clamp
2	Boot Clamp	7	Cage	12	Boot
3	Stopper Ring	8	Inner Ring	13	Axle Shaft And CV Joint
4	Outer Ring	9	Boot	14	Dust Cover
5	Ball	10	Boot Band		

Halfshaft components—2002 models

09474_KSPO_G0152

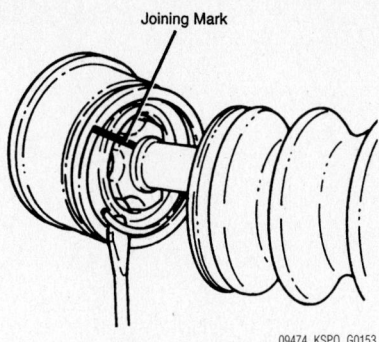

Remove the stopper ring using with a screwdriver—2002 models

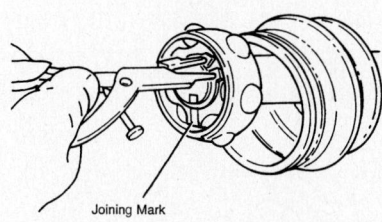

Remove the snapring—2002 models

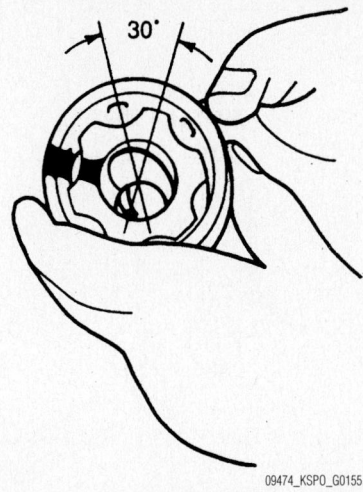

Turn the cage about 30 degrees and separate it from the inner ring—2002 models

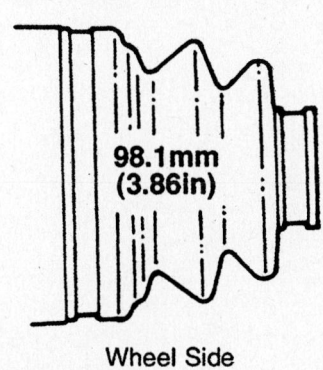
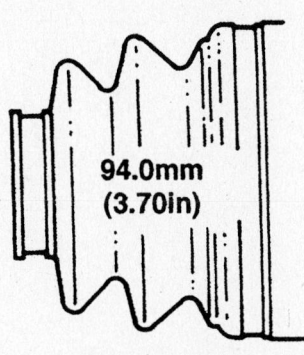

Wheel Side   **Differential Side**

98.1mm (3.86in)   94.0mm (3.70in)

Boot identification—2002 models

4. Pry up differential side boot locking clamps using a screwdriver and pliers.

5. Remove boot clamp and slide boot along shaft to expose the CV-joint.

6. Mark shaft and the outer ring for identical installation position.

7. Remove the stopper ring using with a screwdriver.

8. Mark the shaft, cage, and inner ring for identical installation position.

9. Remove the snapring with snapring pliers.

10. Remove the cage assembly.

11. Remove the ball bearings.

12. Mark the inner ring and cage for identical assembly position.

13. Turn the cage about 30 degrees and separate it from the inner ring.

14. Wrap the wheel side splines with tape and remove the boot. If necessary, remove the dust cover with a plastic hammer.

To assemble:

15. On the left side:

a. On differential end, install (if removed) a new dust cover with a plastic hammer.

b. If not already wrapped, wrap the wheel side splines with suitable tape and install a new boot with a new clamp.

➡**The wheel side and differential side boots are not identical.**

c. Align disassembly reference marks and put ball bearings into the inner ring.

d. Install the cage, inner ring and ball assembly to the halfshaft in the direction shown.

➡**Install the cage with the big end facing the snapring groove. If installed in reverse, the drive shaft may become disengaged.**

e. Using snapring pliers, install a new snapring into the halfshaft groove.

On the right side:

f. For wheel end, install the cage on the inner ring and turn about 30 degrees with respect to the inner ring.

g. Fit the ball bearings through the cage into the ball grooves of the inner ring.

h. Apply yellow joint grease to the cage, inner ring, and ball bearings.

i. Align reference installation marks and install a new snap and stopper ring.

j. Install boot with a new boot clamp. Using pliers, fold boot clamp. Bend the clamp clips to lock clamp in place.

➡**Always use new clamps and install clamp crimp-over in opposite direction of halfshaft forward rotation.**

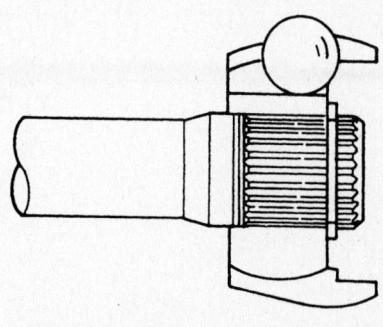

Install the cage, inner ring and ball assembly to the halfshaft in the direction shown—2002 models

k. Using pliers, install new clip onto differential halfshaft end groove.

2005–06 Models

FRONT

➡**Do not disassemble the outer end (BJ) joint. Special grease must be applied to the driveshaft joint. Do not substitute with another type of grease.**

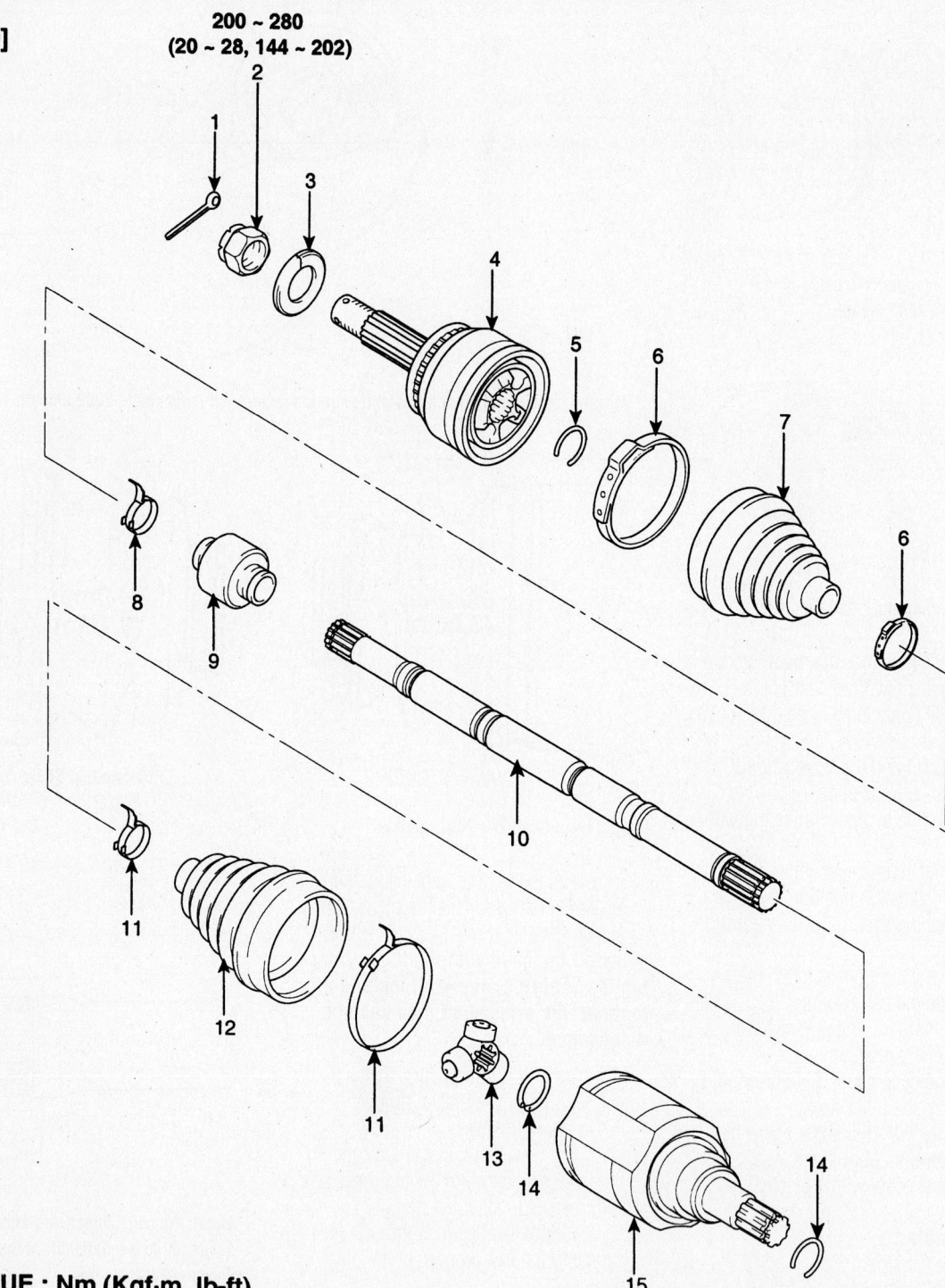

[2WD]

200 ~ 280
(20 ~ 28, 144 ~ 202)

TORQUE : Nm (Kgf·m, lb-ft)

1. Split pin
2. Castle nut
3. Washer
4. BJ assembly

5. Clip A
6. BJ boot band
7. BJ boot
8. Dynamic damper band

9. Dynamic damper
10. Shaft
11. UTJ boot band
12. UTJ boot

13. Trunion assembly
14. Circlip
15. UTJ assembly

09474_KSPO_G0158

Front halfshaft exploded view—2wd 2005–06 models

[4WD]

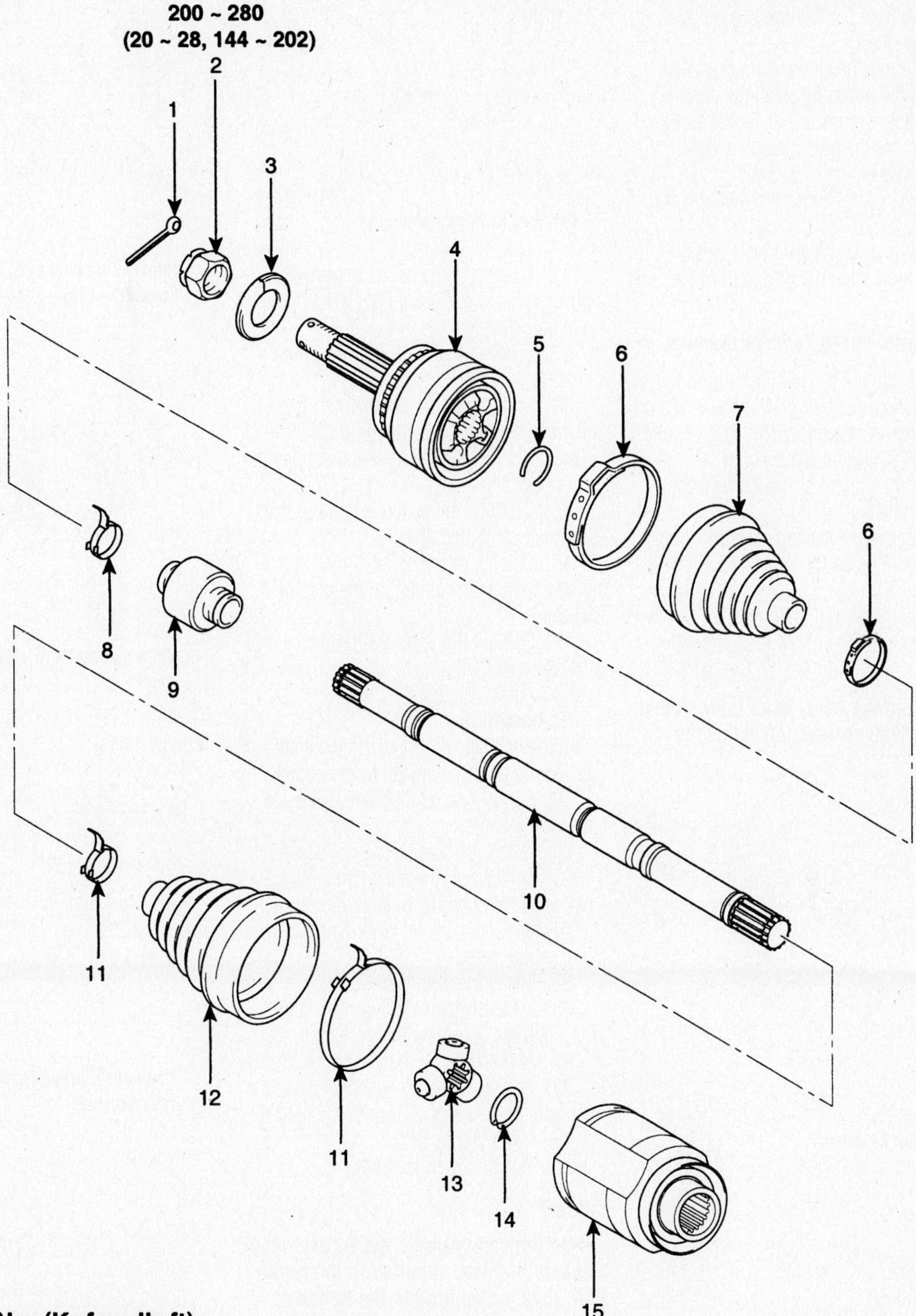

200 ~ 280
(20 ~ 28, 144 ~ 202)

TORQUE : Nm (Kgf·m, lb-ft)

1. Split pin
2. Castle nut
3. Washer
4. BJ assembly

5. Clip A
6. BJ boot band
7. BJ boot
8. Dynamic damper band

9. Dynamic damper
10. Shaft
11. UTJ boot band
12. UTJ boot

13. Trunion assembly
14. Circlip
15. UTJ assembly

09474_KSPO_G0209

Front halfshaft exploded view—2wd 2005–06 models

1. Before servicing the vehicle, refer to the Precautions Section.

2. The boot band should be replaced with a new one.

3. Remove the circlip from driveshaft splines of the transaxle side joint case.

4. Using a pliers or flat-tipped screwdriver, remove the both clamps on the transaxle side.

5. Pull out the boot from the transaxle side joint.

6. While dividing joint boot of the transaxle side, wipe the grease from the case.

➡ **Be careful not to damage the boot.**

7. Matchmark the joint according to the accompanying illustration. Put marks (D) on roller of trunion assembly (A), case (B) and spline part (C), for assembly.

8. Using a pliers or flat-tipped screwdriver, remove the circlip.

9. Remove the trunion assembly from the driveshaft using the special tool (09495-33000).

10. Clean the trunion assembly.

11. Remove the boot of the transaxle side joint.

➡ **If reusing the boot, wrap tape around the driveshaft splines to protect the boot.**

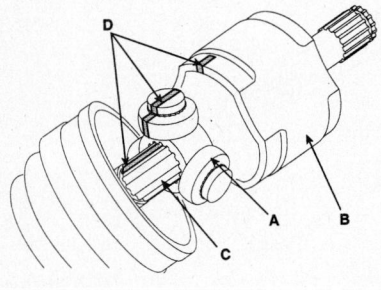

Matchmark locations

09474_KSPO_G0159

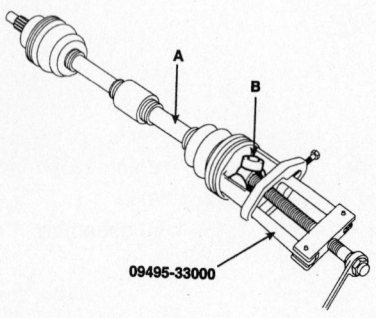

Trunion removal

09495-33000

09474_KSPO_G0160

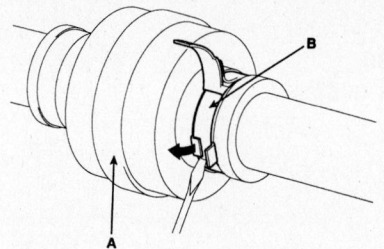

Dynamic damper installation

09474_KSPO_G0161

12. Using a pliers or flat-tipped screwdriver, remove both sides of clamp of the dynamic damper.

13. Fix the driveshaft in a soft-jawed a vise.

14. Apply soap on the shaft to prevent damage between the shaft spline and the dynamic damper when the dynamic damper is removed.

15. Separate the dynamic damper from the shaft carefully.

16. Using a pliers or flat-tipped screwdriver, remove the clamp on the side of wheel.

17. Pull out the joint (BJ) on the side of the wheel in the transaxle direction. Be careful not to damage the boot.

To assemble:

18. Wrap tape around the driveshaft splines to prevent damage to the boots.

19. Apply grease to the driveshaft and install the boots.

20. Install the clamps to both boots.

21. To reassemble the dynamic damper (A), keep the shaft (B) in straight, tighten the dynamic damper (A) with dynamic band (C), as illustrated.

22. Install the boot bands and boot.

23. Install the trunion assembly and the circlip to the spline on the driveshaft.

24. At this time align the matchmarks.

25. Add the as much specified grease to the joint as was wiped away at inspection.

26. Install the boots.

27. Tighten the boot bands.

REAR

➡ **Do not disassemble the BJ assembly. Special grease, supplied in the repair kit, must be applied to the halfshaft joint. Do not substitute with another type of grease. The boot bands should be replaced with new ones.**

1. Before servicing the vehicle, refer to the Precautions Section.

2. Using a pliers or flat-tipped screwdriver, remove the LH boot band and LH TJ boot band from the halfshaft.

3. Remove RH boot band and RH TJ

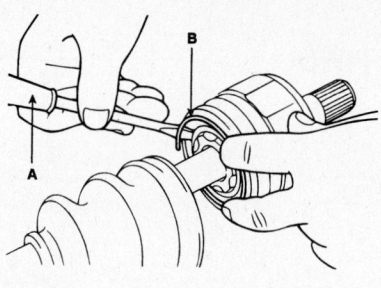

09474_KSPO_G0162

Remove the circlip (B) with a flat-tipped screwdriver (A)—2005–06 rear halfshaft

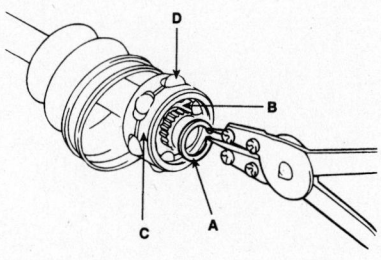

09474_KSPO_G0163

Joint disassembly—2005–06 rear halfshaft

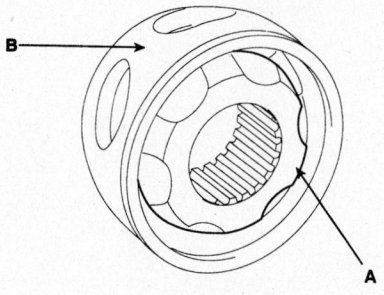

09474_KSPO_G0164

Cage offset properly positioned—2005–06 rear halfshaft

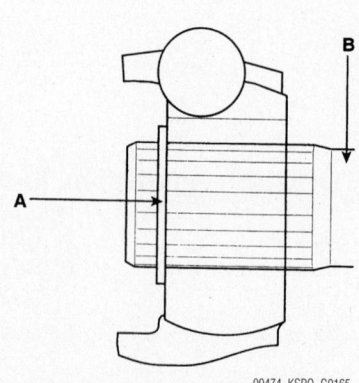

09474_KSPO_G0165

Proper inner race installation—2005–06 rear halfshaft

boot band in the same way of LH removal procedure.

➡ **Be careful not to damage the boot.**

4. Remove the circlip (B) with a flat-tipped screwdriver (A).

5. Pull out the halfshaft from the TJ outer race.

6. Remove the snapring (A) and take out the inner race (B), cage (C) and balls (D) as an assembly.

7. Clean the inner race, cage and balls without disassembling.

8. Remove the BJ boot bands and pull out the TJ boot and BJ boot.

➡ **If the boot is to be reused, wrap tape around the halfshaft splines to protect the boot.**

To assemble:

9. Wrap tape around the halfshaft splines (TJ side) to prevent damage to the boots.

10. Apply grease to the halfshaft and install the boots.

11. Apply the specified grease to the inner race (A) and cage. Install the cage (B) so that it is offset on the race as shown.

➡ **Use the grease included in the repair kit.**

12. Apply the specified grease to the cage and fit the balls into the cage.

13. Position the chamfered side (A) as shown in the illustration. Install the inner race on the halfshaft (B), and then the snapring.

14. Apply the specified grease to the outer race and install the BJ outer race onto the halfshaft.

15. Apply the specified grease into the TJ boot and install the boot with a clip.

16. Tighten the TJ boot bands.

17. Add the specified grease to the BJ as much as was wiped away at inspection.

18. Install the boots.

19. Tighten the BJ boot bands.

Driveshaft

REMOVAL & INSTALLATION

2002 Models

FRONT OR REAR

1. Before servicing the vehicle, refer to the Precautions Section.

2. Raise and support vehicle.

3. Place index marks (reference marks) on the driveshafts and their matching transfer case and differential input/output shafts.

4. Remove four nuts holding universal flange to transfer case.

5. Remove four nuts holding universal flange to differential.

6. Remove driveshaft.

7. Inspect driveshaft for dents or other damage. If shaft appears damaged, replace it.

8. Inspect universal joints for smooth pivoting. If joints bind or appear tight, replace them as assemblies.

9. Installation is the reverse of removal. Tighten the flange nuts to 40 ft. lbs. (54 Nm).

2005–06 Models

1. Before servicing the vehicle, refer to the Precautions Section.

2. After making a match mark on the rubber coupling and rear differential companion, remove the driveshaft mounting bolts.

3. Remove the center bearing bracket mounting bolts.

4. After making a match mark on the flange yoke and transaxle companion, remove the driveshaft mounting bolts.

5. Installation is the reverse of removal. Observe the following torques:
- Front flange bolts: 37–44 ft. lbs. (50–60 Nm)
- Center bearing bolts: 37–47 ft. lbs. (50–65 Nm)
- Rear flange bolts: 74–88 ft. lbs. (100–120 Nm)

Locking Hubs

REMOVAL & INSTALLATION

2002 Models

1. Before servicing the vehicle, refer to the Precautions Section.

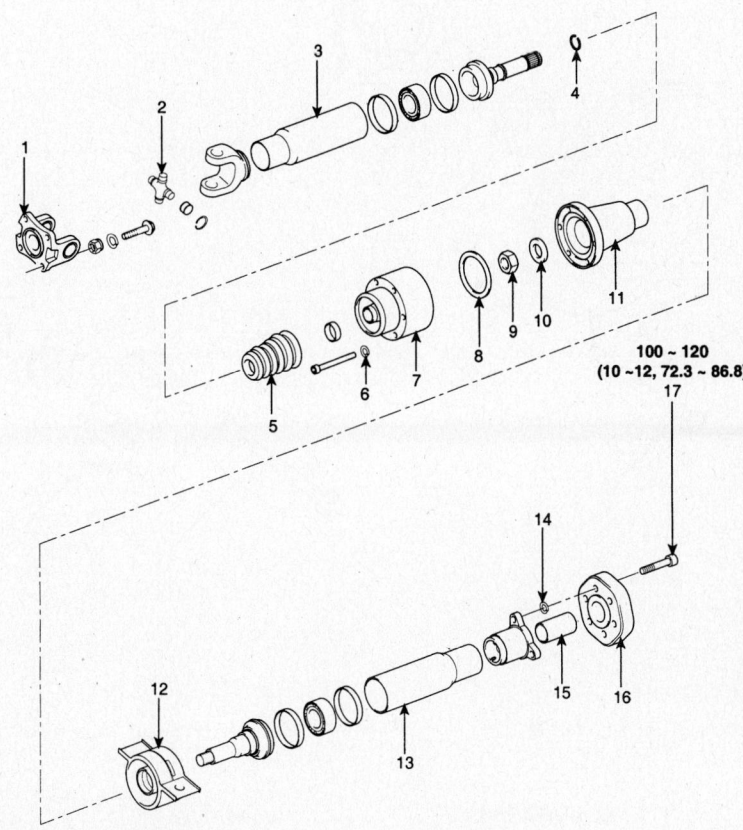

TORQUE : Nm (Kgf·m, lb-ft)

1. Flange yoke
2. Universal joint assembly
3. Front tube
4. Snap ring(VL)
5. LJ boot
6. Spring washer
7. VL joint
8. Sealing
9. Nut
10. Spring washer
11. Companion flange
12. Center bearing
13. Rear tube
14. Plain washer
15. Center device
16. Rubber coupling
17. Bolt

09474_KSPO_G0166

Driveshaft components—2005–06 models

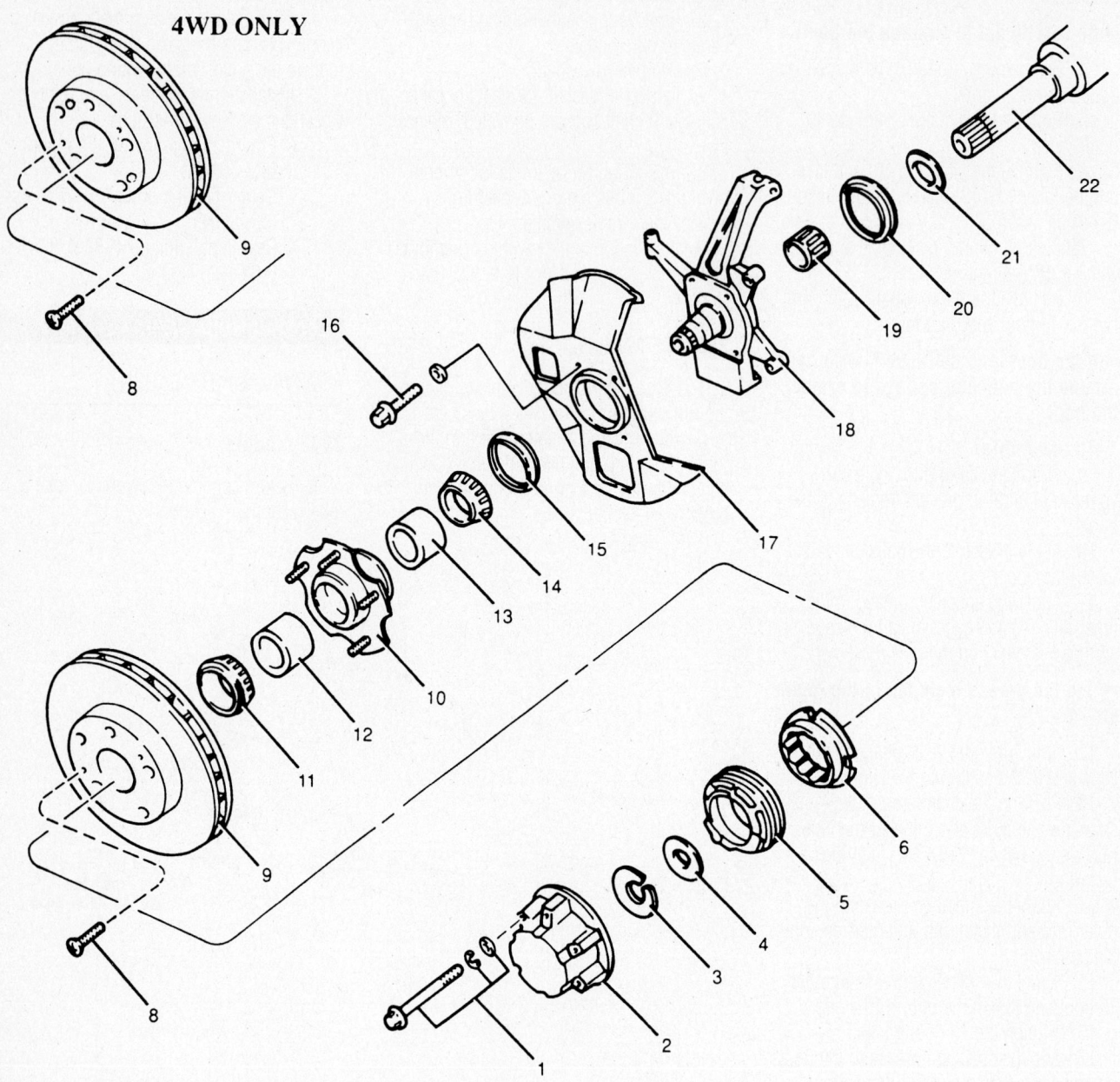

4WD ONLY

1	BOLT/WASHER	9	ROTOR	16	BOLT & SPRING WASHER
2	FREE WHEEL HUB BODY	10	WHEEL HUB	17	DUST COVER
3	SNAP RING	11	INNER BEARING INNER RACE	18	KNUCKLE
4	SPACER	12	INNER BEARING OUTER RACE	19	NEEDLE BEARING
5	FIXED CAM ASSEMBLY	13	OUTER BEARING OUTER RACE	20	OIL SEAL
6	LOCK NUT	14	OUTER BEARING INNER RACE	21	SPACER
8	SCREW	15	OIL SEAL	22	DRIVE SHAFT (LH)

7924QG28

Exploded view of the 4WD locking hub assembly—2002 models

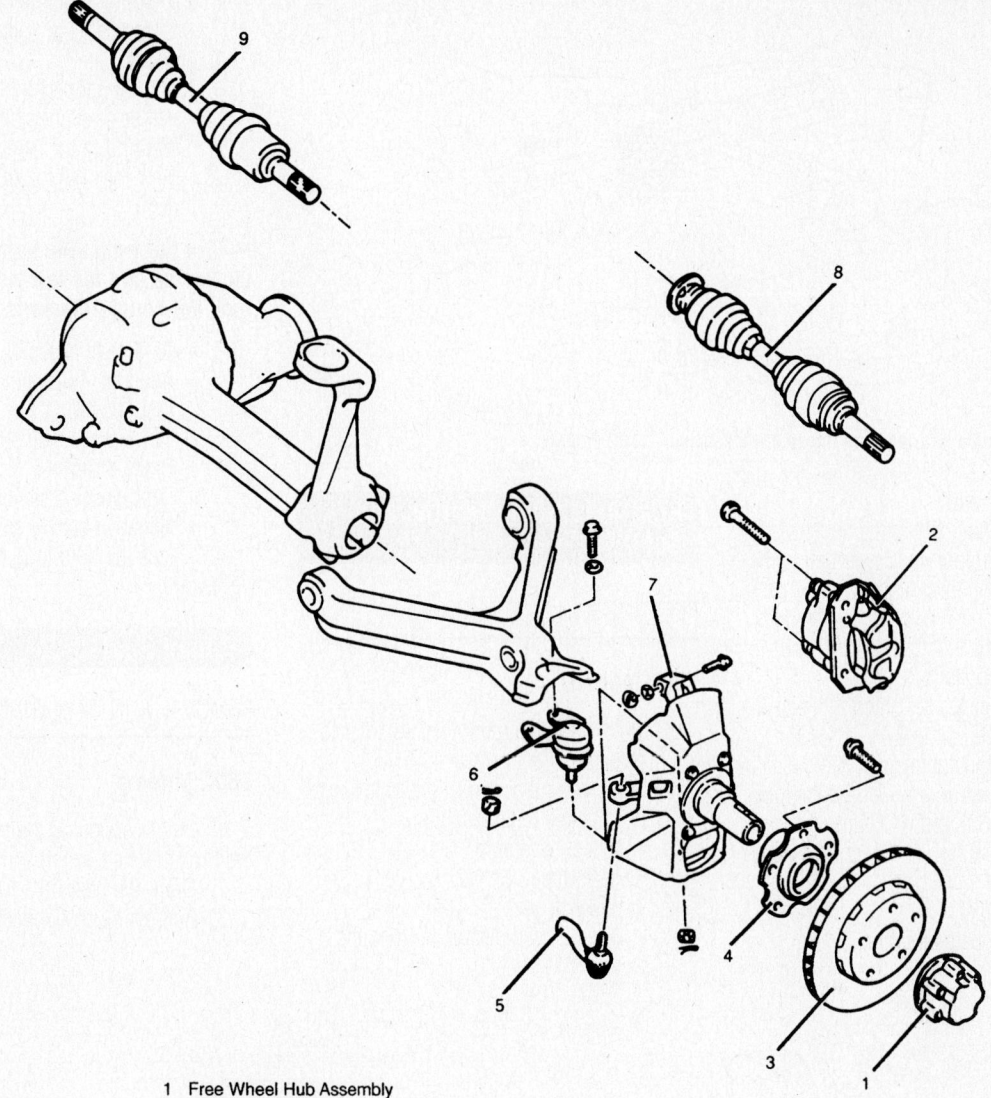

1 Free Wheel Hub Assembly
2 Caliper Assembly
3 Rotor
4 Wheel Hub Assembly
5 Tie - Rod End
6 Lower Ball Joint
7 Knuckle And Dust Cover
8 Drive Shaft(Half Shaft : LH)
9 Drive Shaft(Half Shaft : RH)

09474_KSPO_G0007

Exploded view of the front drive axle and related components—2002 models

2. Remove or disconnect the following:
    • Negative battery cable
    • Wheel assembly
    • Six free wheel hub bolts and remove the hub
    • Snapring and spacer from the hub
    • Carefully remove the fixed cam assembly

To install:
3. Install or connect the following:
    • Fixed cam assembly
    • Snapring and spacer in the hub
    • Wheel hub and the six free wheel hub bolts. Torque the bolts in two

passes, 14 ft. lbs. (17 Nm), then to 23 ft. lbs. (31 Nm).
    • Wheel assembly
    • Negative battery cable

Spindle Bearings

REMOVAL & INSTALLATION

2002 Models

1. Before servicing the vehicle, refer to the Precautions Section.
    2. Remove or disconnect the following:
    • Negative battery cable

    • Wheel assembly
    • Free wheel hub and bearing
    • Upper tie rod end from the steering knuckle
    • Lower control arm from the steering knuckle
    • Steering knuckle
    • Inner oil seal
    • Spindle bearing

To install:
3. Install or connect the following:
    • Spindle bearing using Bearing Installer tool K95B-5011-A
    • New oil seal and apply grease to the bearing and seal lip

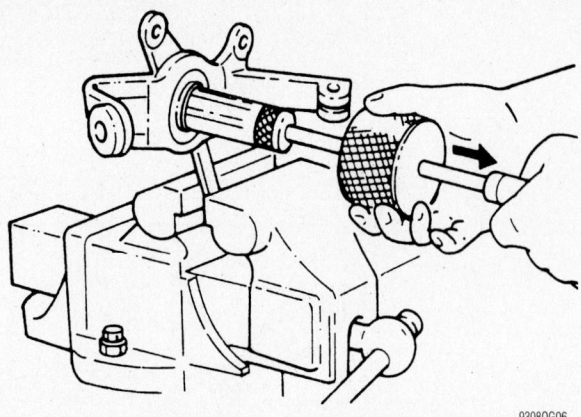

Remove the spindle bearing from the steering knuckle—2002 models

- Halfshaft end
- Steering knuckle to the halfshaft with the upper and lower ball joints in the mounting holes
- Lower control arm. Torque the lock nut to 110 ft. lbs. (148 Nm) and install a cotter pin
- Tie rod end to the steering knuckle. Torque the nut to 27 ft. lbs. (36 Nm) and install a cotter pin
- Upper control arm. Torque the lock bolt to 36 ft. lbs. (49 Nm).
- Free wheel hub and bearing assembly
- Front wheel
- Negative battery cable

Rear Axle Shaft Bearing and Seal

REMOVAL & INSTALLATION

2002 Models

1. Before servicing the vehicle, refer to the Precautions Section.
2. Remove or disconnect the following:
   - Negative battery cable
   - Wheel assembly
   - Rear wheel
   - Brake drum
   - wheel speed sensor, if equipped

- Bearing retaining nuts
- Axle shaft and bearing

To install:

3. Install or connect the following:
   - Oil seal retainer, oil seal and wheel bearing to the axle shaft. Torque the new lock nut to 220 ft. lbs. (300 Nm).

➡**Turn the right side halfshaft lock nut clockwise and the left side halfshaft lock nut counterclockwise.**

   - Axle shaft assembly into the axle housing. Torque the nuts to 74 ft. lbs. (100 Nm).
   - Wheel speed sensor, if equipped
   - Brake drum
   - Rear wheel assembly
   - Negative battery cable
4. Check the fluid level and top off if necessary.

Pinion Seal

REMOVAL & INSTALLATION

2002 Models

1. Before servicing the vehicle, refer to the Precautions Section.
2. Drain the gear oil.
3. Remove or disconnect the following:

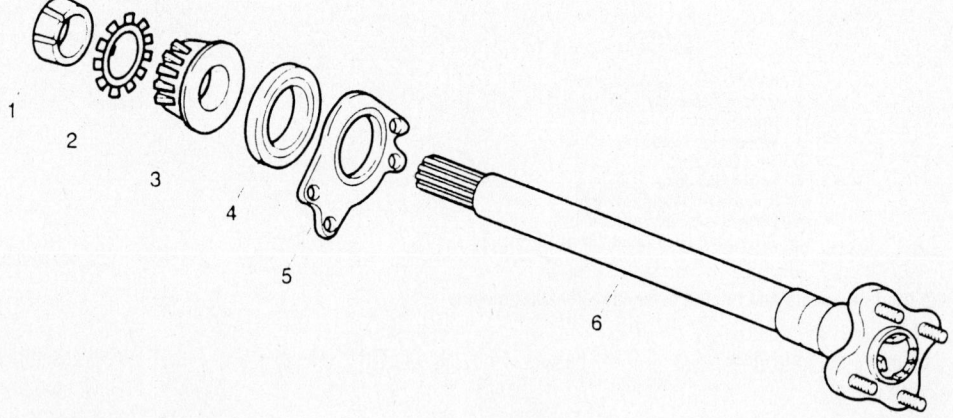

1. Lock Nut
2. Lock Washer
3. Bearing
4. Oil Seal
5. Oil Seal Retainer
6. Axle Shaft

Exploded view of the axle shaft, bearing and seal—2002 models

- Negative battery cable
- Wheel assemblies
- Brake drums
- driveshaft

➡**Using an inch lb. (Nm) torque wrench, measure and record the amount of torque required to maintain rotation of the pinion.**

- Pinion flange
- Pinion seal

To install:

4. Install or connect the following:
   - New pinion seal lightly coated with clean gear oil
   - Pinion flange
5. Rotate the pinion flange occasionally while tightening the flange nut and make certain that the pinion bearings are seated properly.
6. Take several bearing preload torque readings. Tighten the flange nut to achieve the preload torque reading. The maximum torque reading should not exceed 14 inch lbs. (1.6 Nm).
   - Driveshaft after aligning the match-marks
   - Brake drums
   - Wheel assemblies
   - Negative battery cable
7. Fill the gear oil to the proper level.
8. Start the vehicle and check for leaks, repair if necessary.

STEERING

Air Bag

PRECAUTIONS

Several precautions must be observed when handling the inflator module to avoid accidental deployment and possible personal injury.

1. Never carry the inflator module by the wires or connector on the underside of the module.
2. When carrying a live inflator module, hold securely with both hands, and ensure that the bag and trim cover are pointed away from you.
3. Place the inflator module on a bench or other surface with the bag and trim cover facing up.
4. With the inflator module on the bench, never place anything on or close to the module which may be thrown in the event of an accidental deployment.

DISARMING

2002 Models

1. Before servicing the vehicle, refer to the Precautions Section.
2. Turn the ignition switch to the **LOCK** position.
3. Disconnect the negative battery cable.
4. Wait 10 minutes for the back-up power to discharge.

2005–06 Models

1. Before servicing the vehicle, refer to the Precautions Section.
2. Turn the ignition switch to the **LOCK** position.
3. Disconnect the negative battery cable.
4. Wait 3 minutes for the back-up power to discharge.

ARMING

2002 and 2005–06 Models

1. Before servicing the vehicle, refer to the Precautions Section.

Assuming the system components (air bag control module, sensors, air bag, etc.) are installed correctly and are in good working order, the system is armed whenever the battery positive and negative battery cables are connected.

If you have disarmed the air bag system for any reason, to rearm, be sure no one is in the vehicle (as an added safety measure), then connect the negative battery cable.

Power Steering Gear

ADJUSTMENT

2002 Models

1. Before servicing the vehicle, refer to the Precautions Section.
2. Place the steering gear in a vise with protective jaws.
3. Place a torque wrench on the Pitman arm end of the shaft.
4. Loosen the locknut on the adjusting bolt.

5. Slowly turn the adjusting bolt to until the breakaway torque is 65 ft. lbs. (88 Nm).
6. Hold the adjusting bolt in position and tighten the locknut to 25 ft. lbs. (34 Nm).

REMOVAL & INSTALLATION

2002 Models

1. Before servicing the vehicle, refer to the Precautions Section.
2. Center the steering wheel.
3. Drain the power steering fluid.
4. Remove or disconnect the following:
   - Negative battery cable
   - Left front wheel

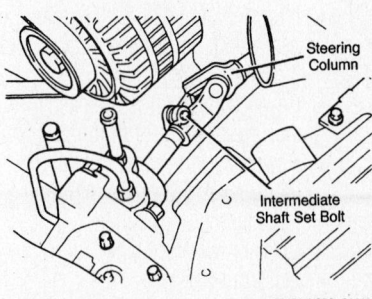

09474_KSPO_G0010

Intermediate shaft—2002 models

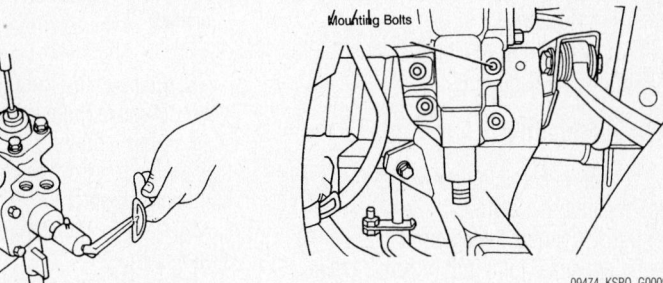

09474_KSPO_G0009

Steering gear mounting bolts—2002 models

09474_KSPO_G0008

Steering gear adjustment—2002 models

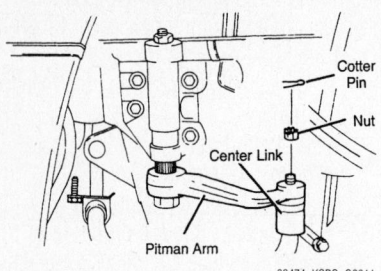

09474_KSPO_G0011

Center link and Pitman arm—2002 models

- Pitman arm-to-center link attaching nut
- Separate the Pitman arm from the center link with a ball joint puller
- Power steering hoses
- Coolant recovery tank and the power steering reserve tank
- Set bolt from the intermediate shaft
- Intermediate shaft from the steering gear
- Steering gear-to-frame bolts
- Steering gear

To install:

5. Install or connect the following:
- Position the steering gear to the frame. Torque the bolts to 159 ft. lbs. (215 Nm).
- Pitman arm to the center link. Torque the bolt to 36 ft. lbs. (49 Nm) and install a new cotter pin
- Intermediate shaft to the steering gear shaft. Torque the set bolt to 25 ft. lbs. (34 Nm).
- Power steering hoses/lines. Torque the fasteners to 29 ft. lbs. (34 Nm).
- Power steering reserve tank over the bracket and press until full engagement is reached
- Coolant recovery tank
- Left front wheel
- Negative battery cable

6. Fill the power steering fluid to the proper level and bleed the system.

7. Start the vehicle and check for leaks, repair if necessary.

8. Road test the vehicle to check that the steering wheel is straight.

2005–06 Models

1. Before servicing the vehicle, refer to the Precautions Section.

2. Disconnect the cover fixing clip (A) on the universal joint indoor driver inside, and loosen the noise covers (B).

3. Loosen the universal joint and the gear box mounting bolt (A) and disconnect the intermediate shaft (B) from the gear box.

4. Lift up the vehicle.

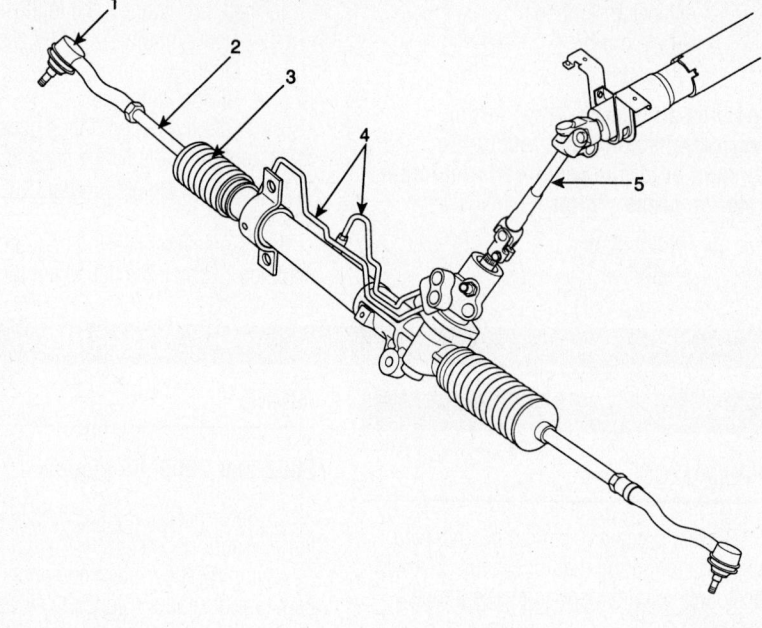

1. Tie rod end assembly
2. Tie rod assembly
3. Bellows
4. Feed tube
5. Joint assembly

09474_KSPO_G0167

Steering gear—2005–06 models

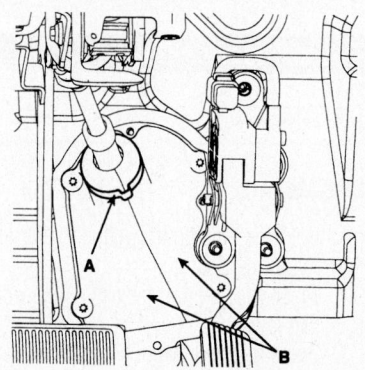

09474_KSPO_G0168

Disconnect the cover fixing clip (A) on the universal joint indoor driver inside, and loosen the noise covers (B)—2005–06 models

5. Remove the front tires.

6. Remove the engine under cover.

7. After removing the split pin, disconnect the tie rod from the knuckle using a ball joint separator.

8. Remove the stabilizer link from the strut assembly.

9. Remove the two bolts for lower arm ball joint.

10. Remove the propeller shaft and the front muffler assembly.

11. Drain oil from the transfer case (A). Then remove the rear flange assembly (B).

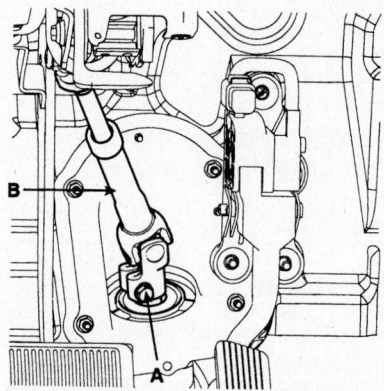

09474_KSPO_G0170

Loosen the universal joint and the gear box mounting bolt (A) and disconnect the intermediate shaft (B) from the gear box— 2005–06 models

12. Drain power steering oil.

13. Remove the connecting bolt for pressure tubes.

14. Remove two engine mounting bolts (B, C) and six subframe mounting bolts in order to remove the subframe (A).

15. Remove the power steering gearbox after removing four mounting bolts of the power steering gearbox.

16. Installation is the reserve of removal. Observe the following torques:

- Sub-frame bolts: 50–65 Nm (37–48 ft. lbs.)
- Steering gear mounting bolts: 44–59 ft. lbs. (60–80 Nm)
17. Add power steering fluid.
18. Air bleed the system.

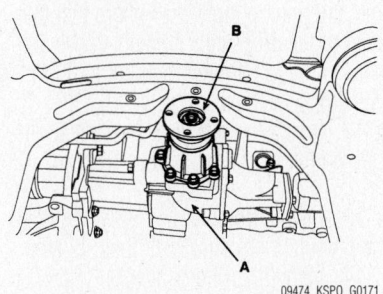

09474_KSPO_G0171

Transfer case rear flange—2005–06 models

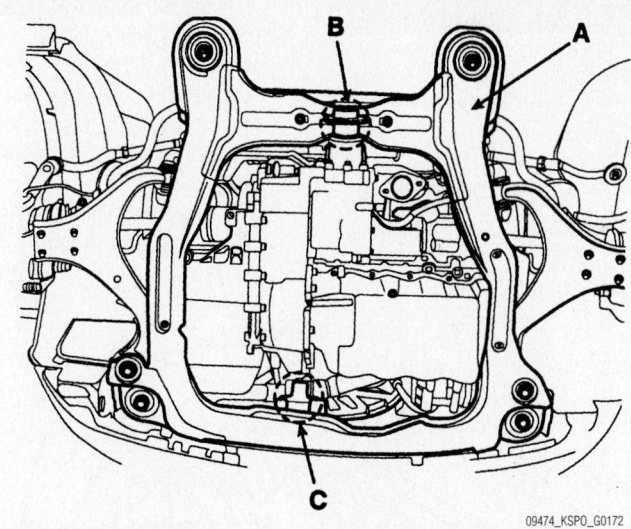

09474_KSPO_G0172

Sub-frame bolts—2005–06 models

FRONT SUSPENSION

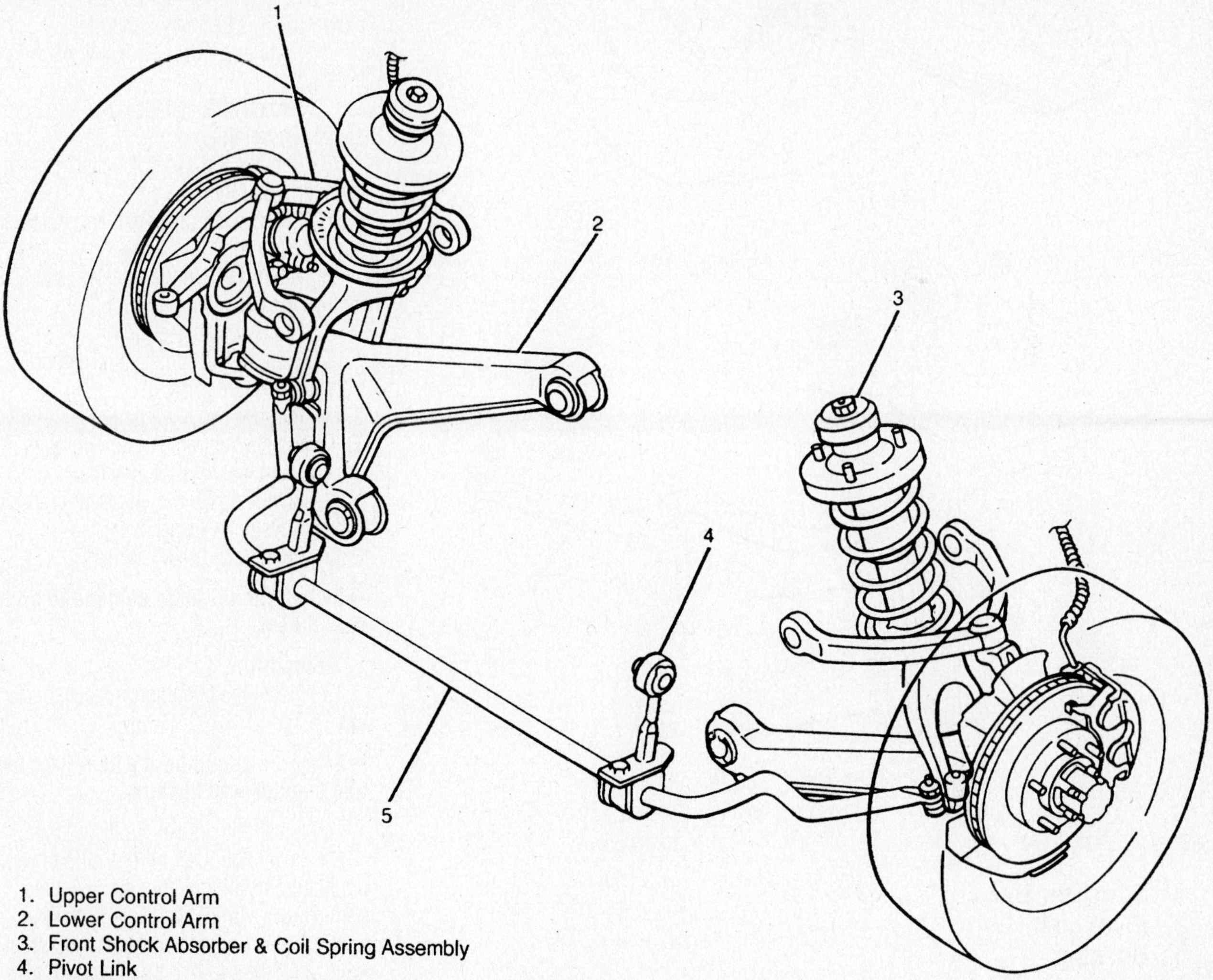

1. Upper Control Arm
2. Lower Control Arm
3. Front Shock Absorber & Coil Spring Assembly
4. Pivot Link
5. Stabilizer Bar

09474_KSPO_G0012

Front suspension components—2002 models

Stabilizer Bar

REMOVAL & INSTALLATION

2002 Models

1. Before servicing the vehicle, refer to the Precautions Section.

2. Raise the front of the vehicle and support it with safety stands.
3. Remove the wheels.
4. Remove the undercover.
5. Remove the pivot link bolts.
6. Remove the stabilizer bar nuts from the drop links.
7. Remove the stabilizer bar.

8. Remove the stabilizer bar plate bolts and remove the plate and pivot link.

To install:

9. Loosely tighten the pivot link bolts.
10. Insert the stabilizer bar plate into the pivot link.
11. Insert the stabilizer onto the drop link and tighten the nuts. Ensure that the head of the drop links are kept in the forward position. Tighten the nuts to 36 ft. lbs. (48 Nm)
12. Install the stabilizer bar plate bolts. Tighten the bar plate bolts to 18 ft. lbs. (25 Nm).
13. Install the pivot link bolts. Tighten the pivot link bolts to 31 ft. lbs. (42 Nm).
14. Install the undercover.
15. Replace the wheels; install lug nuts. Tighten lug nuts to 74 ft. lbs. (100Nm)

2005–06 Models

1. Before servicing the vehicle, refer to the Precautions Section.
2. Raise the front of the vehicle, and make sure it is securely supported.
3. Remove the front wheel and tire from front hub.
4. Remove the stabilizer bar link from the strut assembly.
5. Remove the two bolts attaching the lower arm ball joint.
6. Remove the driveshaft and the front muffler assembly.
7. Drain oil from the transfer case, then remove the rear flange assembly.
8. Drain power steering oil.
9. Remove the connecting bolt for steering pressure lines.
10. Remove two engine mounting bolts (B, C) and six subframe mounting bolts in order to remove the subframe (A).
11. Remove both two stabilizer brackets and two bushings respectively.
12. Remove the stabilizer bar.

➡ **Be careful not to do damage to pressure tubes.**

To install:

13. Install the bushing on the stabilizer bar.

➡ **Bring the clamp on the stabilizer bar into contact with bushing.**

14. Install the bracket on the bushing.
15. Install the bolts on one side finger-tight, then install the other side. Tighten both sides to 45–55 Nm (33–40 ft. lbs.)
16. Install the six sub-frame mounting bolts, then the two-engine mounting bolts (B, C). Tighten to 50–65 Nm (37–48 ft. lbs.)

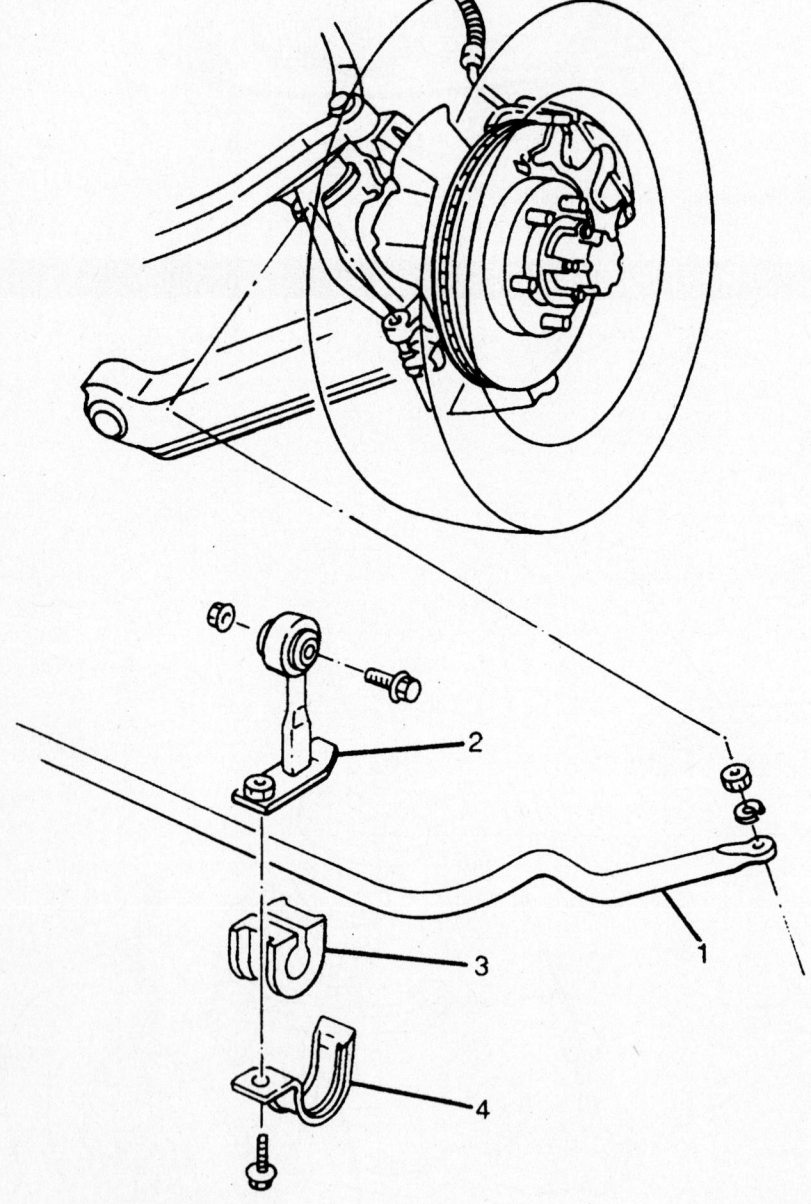

1   Stabilizer Bar
2   Pivot Link
3   Bushing
4   Stabilizer Plate

09474_KSPO_G0020

Front stabilizer bar and end link—2002 models

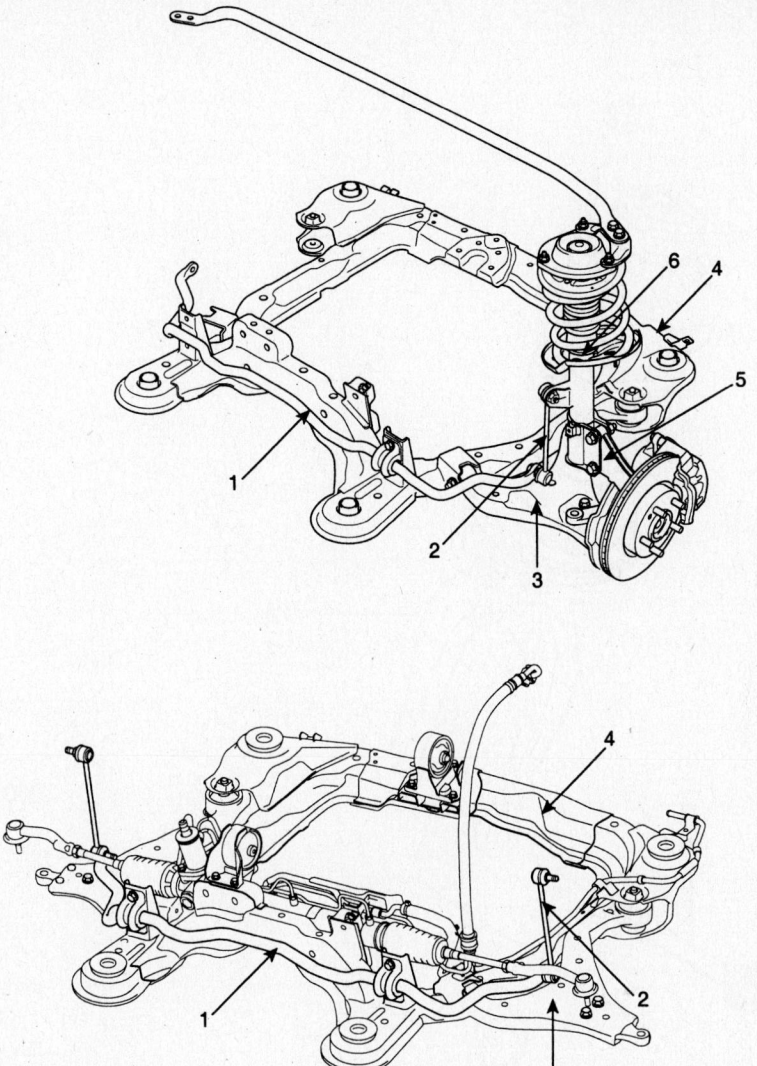

1. Stabilizer bar
2. Stabilizer bar link
3. Lower arm

4. Sub-frame
5. Knuckle
6. Strut assembly

09474_KSPO_G0173

Front stabilizer bar and related components—2005–06 models

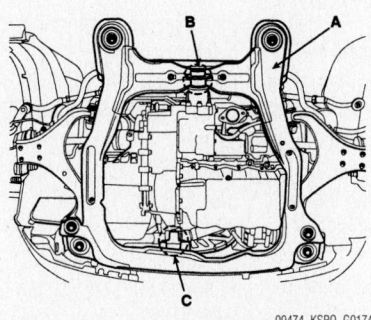

09474_KSPO_G0174

Remove two engine mounting bolts (B, C) and six subframe mounting bolts in order to remove the subframe (A)—2005–06 models

17. Install the connecting bolt for pressure lines.

➡**Be sure to align the white marks (A) on the tube and the hose.**

18. Install the rear flange assembly on the transfer case.

19. Install the driveshaft, then the front muffler assembly.

20. Install the two bolts for the lower arm ball joint. Tighten to 100–120 Nm (74–88 ft. lbs.)

21. Install the nut on the stabilizer bar link. Tighten to 100–120 Nm (74–88 ft. lbs.)

22. Install the front wheel and tire on the

front hub. Tighten to 90–110 Nm (66–81 ft. lbs.)

Strut (Shock Absorber)

REMOVAL & INSTALLATION

2002 Models

The front shock absorber and coil spring are removed as a single unit.

1. Before servicing the vehicle, refer to the Precautions Section.

2. Remove or disconnect the following:
   • Negative battery cable
   • Both front wheels

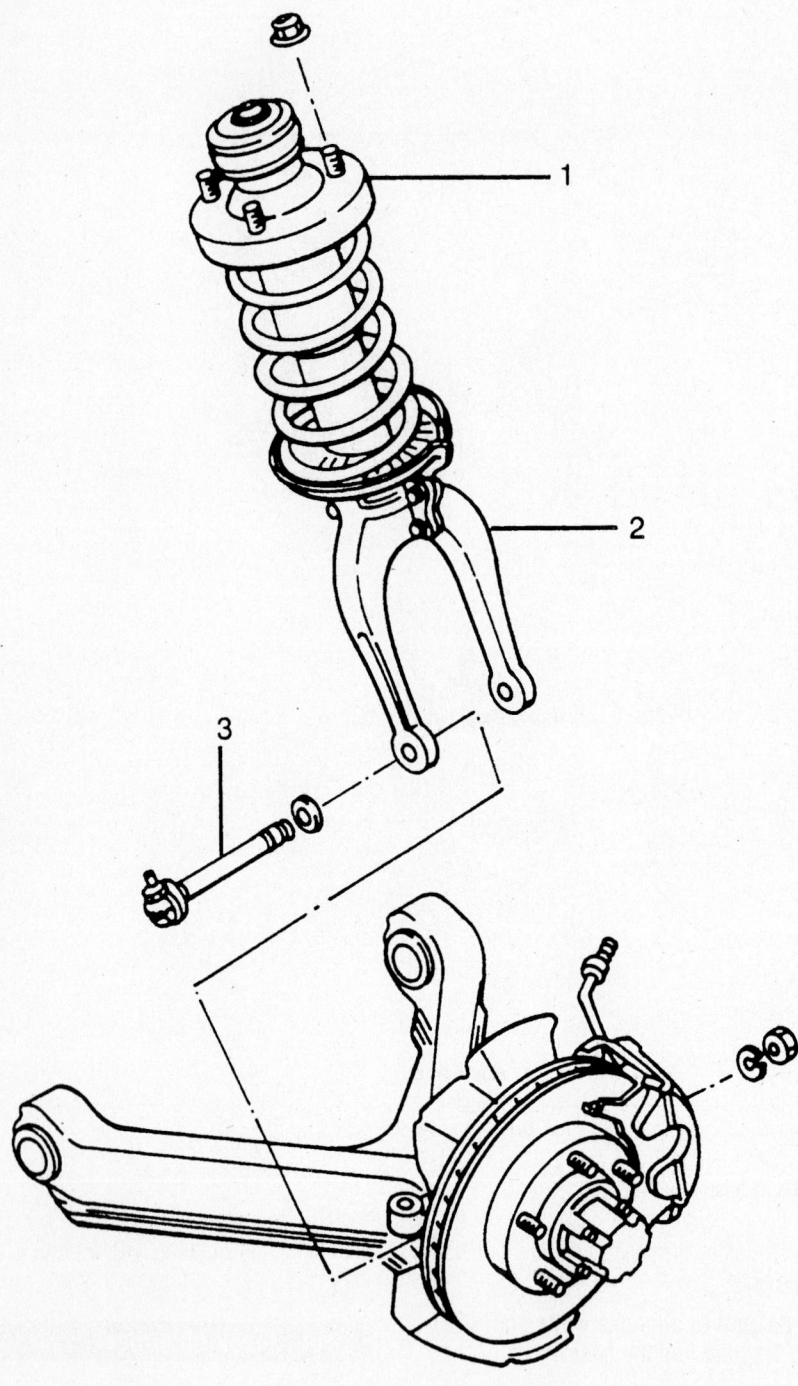

1 **Front Shock Absorber & Coil Spring Assembly**
2 **Front Fork**
3 **Drop Link**

9308QG09

Front strut mounting—2002 models

- Upper strut mounting block nuts
- Stabilizer bar
- Drop link nut and allow the drop link to remain in place
- Both halves of the front fork
- Drop link
- Strut and coil spring as an assembly

To install:

3. Install or connect the following:
- Coil spring to the shock absorber and position the assembly to the upper mounting block
- Upper mounting block nuts and hand tighten them
- Both front forks . Torque the bolts to 36 ft. lbs. (48 Nm).
- Drop link. Torque the nut to 145 ft. lbs. (197 Nm).
- Stabilizer bar to the drop link. Torque the nut to 36 ft. lbs. (48 Nm).
- Front wheels and torque the mounting block nuts to 18 ft. lbs. (25 Nm).
- Negative battery cable

2005–06 Models

1. Before servicing the vehicle, refer to the Precautions Section.

2. Raise the front of the vehicle, and make sure it is securely supported.

3. Remove the front wheel and tire from front hub.

4. Remove the brake hose bracket and speed sensor cable-mounting bolt from the strut assembly.

5. Remove the speed sensor cable-mounting bolt and speed sensor.

6. Remove the nut from the stabilizer bar link.

7. Remove the strut upper mounting nuts.

8. Remove the strut lower mounting bolts and then remove the strut assembly.

To install:

9. Install the strut assembly and then install the strut lower mounting bolts. Tighten to 140–160 Nm (103–118 ft. lbs.)

10. Install the strut upper mounting nuts. Tighten to 45–60 Nm (33–44 ft. lbs.)

11. Install the nut on the stabilizer bar link. Tighten to 100–120 Nm (74–88 ft. lbs.)

12. Install the speed sensor cable mounting bolt and speed sensor. Tighten to 7–11 Nm (5–8 ft. lbs.)

13. Install the brake hose bracket and speed sensor cable mounting bolt on the strut assembly. Tighten to 7–11 Nm (5–8 ft. lbs.)

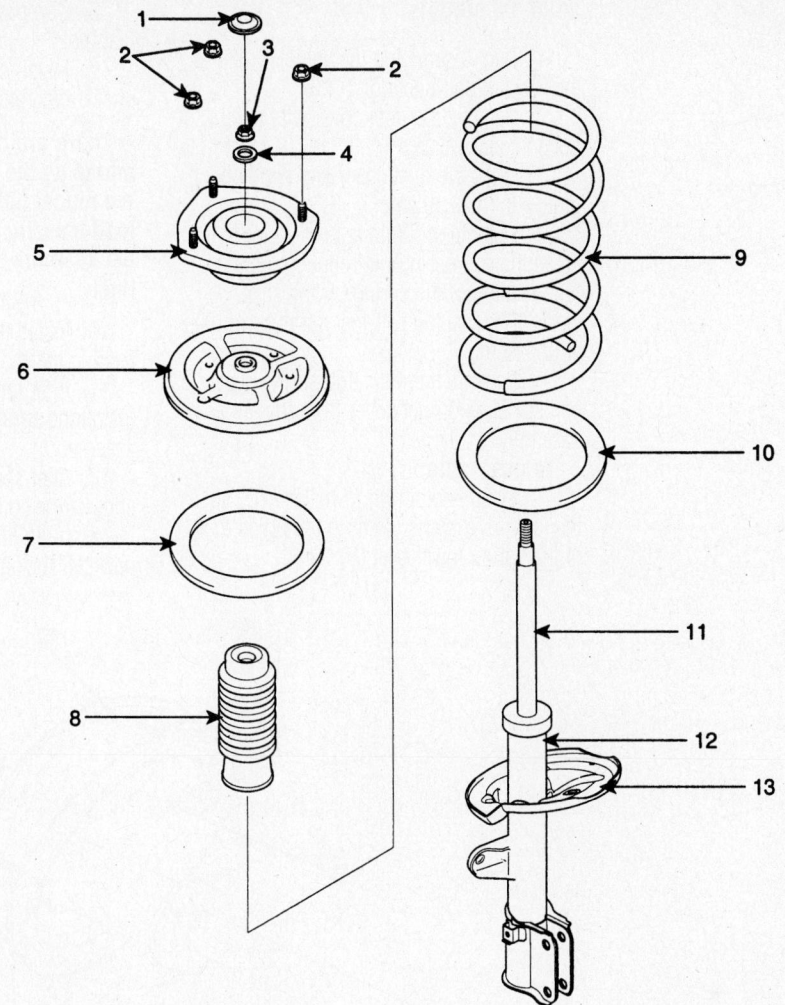

1. Insulator dust cover
2. Upper mounting nuts
3. Self-locking nut
4. Spacer
5. Insulator
6. Spring upper seat
7. Spring upper pad

8. Strut dust cover & bumper rubber
9. Coil spring
10. Spring lower pad
11. Piston rod
12. Strut assembly
13. Spring lower seat

09474_KSPO_G0175

Front strut exploded view—2005–06 models

14. Install the front wheel and tire on the front hub. Tighten to 90–110 Nm (66–81 ft. lbs.)

Coil Spring

REMOVAL & INSTALLATION

2002 Models

1. Before servicing the vehicle, refer to the Precautions Section.
2. Remove or disconnect the following:

- Negative battery cable
- Strut and coil spring as an assembly
- Place the strut in a vise
- Loosen the pivot rod nut several turns
- While still secured in a vise compress the coil spring
- Piston rod nut and disassemble the coil spring as needed

To install:
3. Install or connect the following:
- Bottom portion of the strut in a vise and compress the coil spring

- End of the coil spring to the rubber seat and install the spring
- Assemble the dust boot and lower retainer, lower insulator, spring seat, boss, center washer, upper insulator and install the coil spring
- Hand tighten the piston rod nut.
- Carefully loosen the spring compressor tool and remove the tool
- Torque the piston rod nut to 31 ft. lbs. (42 Nm).
- Coil spring and strut as an assembly

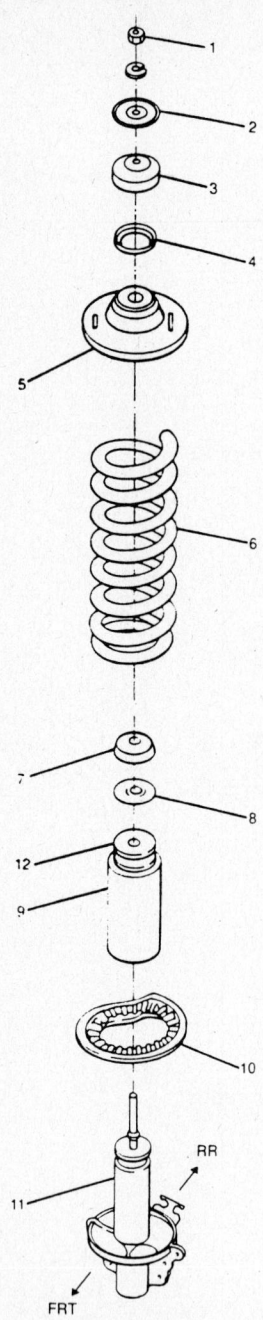

1. Nut
2. Upper retainer
3. Upper insulator
4. Centering washer
5. Spring seat
6. Coil spring
7. Lower insulator
8. Lower retainer
9. Dust boot
10. Rubber seat
11. Shock absorber
12. Front jounce stop

9308QG10

Exploded view of the front strut and coil spring assembly—2002 models

2005–06 Models

1. Before servicing the vehicle, refer to the Precautions Section.

2. Remove the dust cover with a flat-tipped screwdriver.

3. Open the dust cover and wipe off grease in the insulator.

4. Using a spring compressor, compress the coil spring until there is only a little tension of the spring on the strut.

5. Remove the self-locking nut from the strut assembly.

6. Remove the insulator, spring seat, coil spring and dust cover from the strut assembly.

To assemble:

7. Install the spring lower pad (D) so that the protrusions (A) fit in the holes (C) in the spring lower seat (B).

8. Compress coil spring using the special tool.

9. Install compressed coil spring into shock absorber.

➡**There are two identification color marks on the coil spring. One indicates the model option; the other indicates load classification. Distinguish between the two marks and then install them.**

10. Install the coil spring with the identification mark directed toward the knuckle.

11. After fully extending the piston rod, install the spring upper seat and insulator assembly.

12. After seating the upper and lower ends of the coil spring in the upper and lower spring seat grooves correctly, tighten new self-locking nut temporarily.

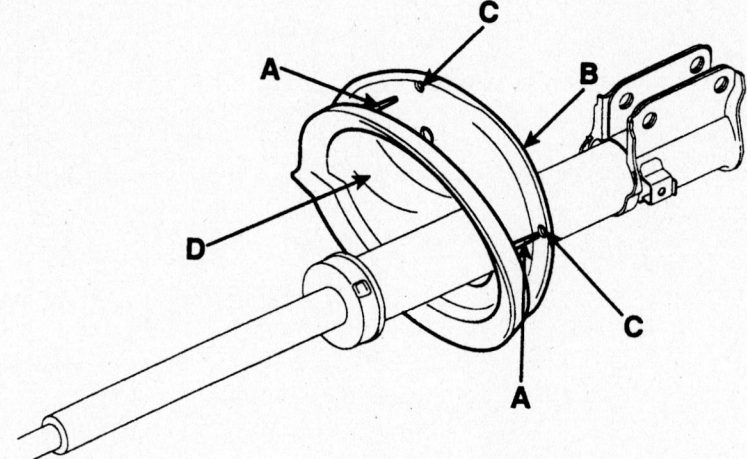

09474_KSPO_G0176

Install the spring lower pad (D) so that the protrusions (A) fit in the holes (C) in the spring lower seat (B)

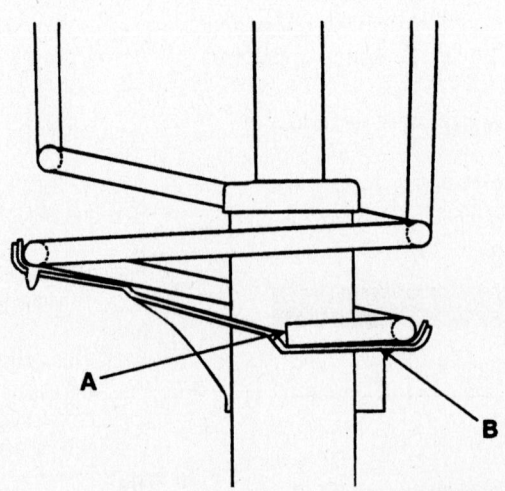

09474_KSPO_G0177

Install the spring lower pad (D) so that the protrusions (A) fit in the holes (C) in the spring lower seat (B)

13. Remove the compressor.
14. Tighten the self-locking nut to 60–70 Nm (44–51 ft. lbs.)
15. Apply grease to the strut upper bearing and install the insulator cap.

➡**When applying the grease, be careful so that it isn't smeared on the insulator rubber.**

Upper Ball Joint

REMOVAL & INSTALLATION

2002 Models

The upper ball joint is an integral part of the upper control arm. If the ball joint is worn, replacement of the upper control arm is necessary.

Lower Ball Joint

REMOVAL & INSTALLATION

2002 Models

1. Before servicing the vehicle, refer to the Precautions Section.
2. Remove or disconnect the following:
- Negative battery cable
- Front wheel assembly
- Cotter pin and lower ball joint nut
- Separate the lower ball joint from the spindle with a puller tool by prying down on the spindle to separate it from the lower ball joint
- Lower ball joint attaching bolts
- Lower ball joint

To install:
3. Install or connect the following:
- Position the lower ball joint and install the attaching nuts and bolts. Torque the fasteners to 36 ft. lbs. (48 Nm).
- Pry down and guide the spindle onto the lower ball joint. Torque the nut 87 ft. lbs. (118 Nm) and install a new cotter pin
- Front wheel assembly.
- Negative battery cable

2005–06 Models

1. Before servicing the vehicle, refer to the Precautions Section.
2. Raise and support the front end safely.
3. Support the lower control arm with a jack.
4. Remove the castellated nut and cotter pin from the ball stud. Discard the cotter pin.

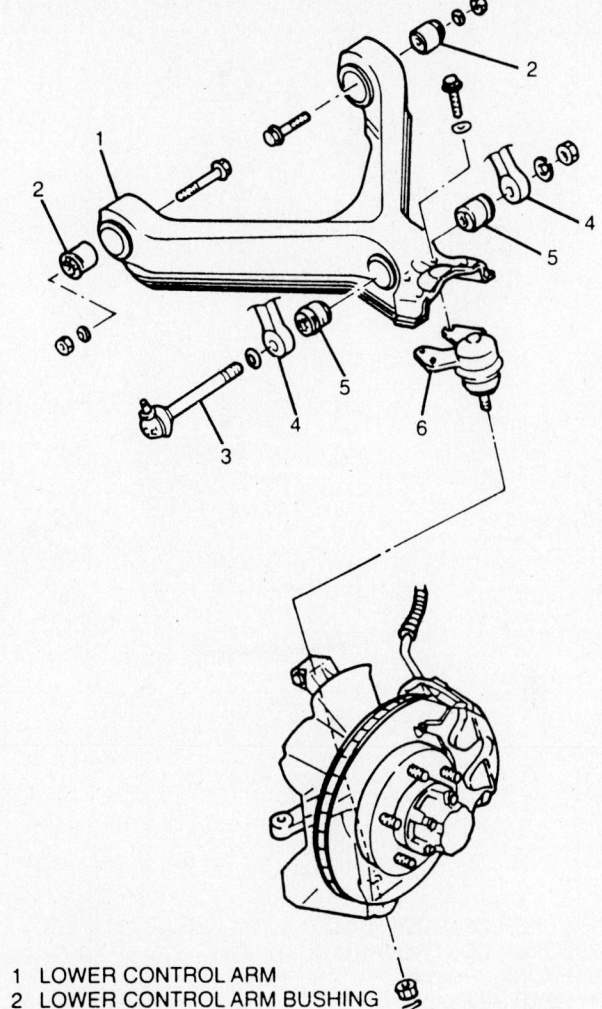

1 LOWER CONTROL ARM
2 LOWER CONTROL ARM BUSHING
3 DROP LINK
4 FRONT FORK
5 DROP LINK BUSHING
6 LOWER CONTROL ARM BALL JOINT

7924QG34

Exploded view of the lower control arm and ball joint assembly—2002 models

5. remove the ball joint attaching bolts and lower the arm. Remove the ball joint.
6. Installation is the reverse of removal. Torque the ball joint attaching bolts to 100–120 Nm (74–88 ft. lbs.). Torque the castellated nut to 200–280 Nm (147–206 ft. lbs.). Use a new cotter pin. Never back off the nut to align the cotter pin holes.

Upper Control Arm

REMOVAL & INSTALLATION

2002 Models

1. Before servicing the vehicle, refer to the Precautions Section.
2. Remove or disconnect the following:
- Negative battery cable
- Front wheel assembly

- Bolt securing the upper ball joint to the steering knuckle

➡**Note the matchmark setting on the upper control arm mounting bolts before removal.**

- Upper control arm mounting bolts
- Upper control arm from the vehicle.

To install:
3. Install or connect the following:
- Position the upper control arms in the frame mounting. Hand tighten the bolts
- Position the ball joint in the spindle. Torque the through bolt to 36 ft. lbs. (48 Nm).

➡**Be sure the slot in the ball joint aligns with the through-bolt during installation.**

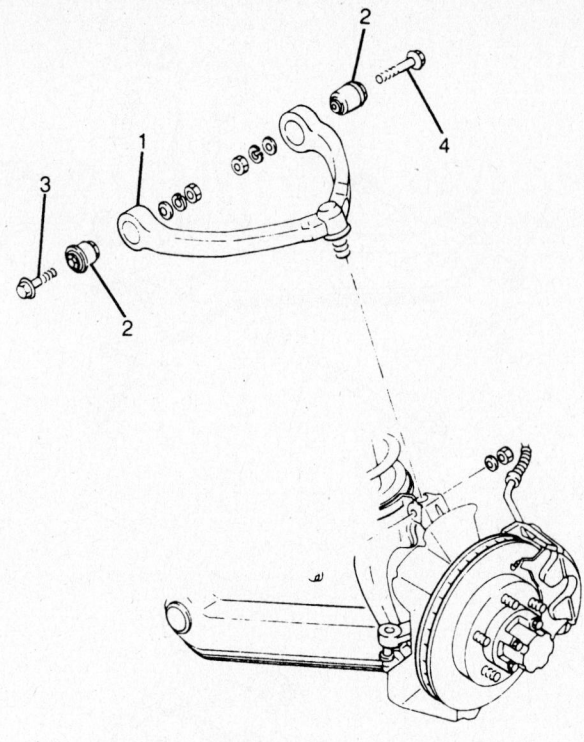

1 UPPER CONTROL ARM
2 UPPER CONTROL ARM BUSHING
3 FRONT SPINDLE
4 REAR SPINDLE

7924QG31

Exploded view of the upper control arm assembly—2002 models

- Align the upper control arm bolts to the previous settings. Torque bolts to 62 ft. lbs. (108 Nm).
- Front wheel assembly

4. Check and adjust the alignment, if necessary.

UPPER CONTROL ARM BUSHING REPLACEMENT

2002 Models

1. Before servicing the vehicle, refer to the Precautions Section.
2. Remove or disconnect the following:
   - Upper control arm assembly
   - Secure the control arm in a vise
   - Using a standard press, remove the bushing

To install:

3. Install or connect the following:
   - Lubricate the new bushing and press it into the upper control arm
   - Upper control arm to the vehicle

4. Check the wheel alignment and adjust if necessary.

Lower Control Arm

REMOVAL & INSTALLATION

2002 Models

1. Before servicing the vehicle, refer to the Precautions Section.
2. Remove or disconnect the following:
   - Negative battery cable
   - Front wheel assembly
   - Stabilizer bar
   - Driveshaft from the differential and steering knuckle
   - Shock absorber and coil spring from the front half of the fork
   - Cotter pin and castle nut from the lower control arm ball joint
   - Lower control arm ball joint from the steering knuckle

- Lower control arm bushing bolts from the front frame crossmember brackets
- Lower control arm

To install:

3. Install or connect the following:
   - Lower control arm to the front frame crossmember brackets and hand tighten the bushing bolts
   - Lower control arm ball joint and bolt to the steering knuckle. Torque the bolt to 87 ft. lbs. (118 Nm).
   - New cotter pin and castle nut
   - Torque the lower control arm bushing bolts to 206 ft. lbs. (280 Nm).
   - Front fork halves to the shock absorber and coil spring and position the lower portion over the lower control arm drop link holes
   - Drop link. Torque the nut to 145 ft. lbs. (197 Nm).
   - Driveshaft to the front differential and steering knuckle
   - Stabilizer bar
   - Front wheel assembly
   - Negative battery cable

4. Check and adjust the front wheel alignment if necessary.

2005–06 Models

1. Before servicing the vehicle, refer to the Precautions Section.
2. Raise the front of the vehicle, and make sure it is securely supported.
3. Remove the front wheel and tire from front hub.
4. Remove the lower arm ball joint mounting bolts.
5. Remove the lower arm mounting bolts.

To install:

6. Install the lower arm mounting bolts. Tighten to:
   - A bushing: 100–120 Nm (74–88 ft. lbs.)
   - G bushing: 140–160 Nm (103–118 ft. lbs.)

7. Install the lower arm ball joint mounting bolts. Tighten to 100–120 Nm (74–88 ft. lbs.)

8. Install the front wheel and tire on the front hub. Tighten to 90–110 Nm (66–81 ft. lbs.)

LOWER CONTROL ARM BUSHING REPLACEMENT

1. Before servicing the vehicle, refer to the Precautions Section.
2. Remove or disconnect the following:
   - Lower control arm assembly
   - Secure the control arm in a vise

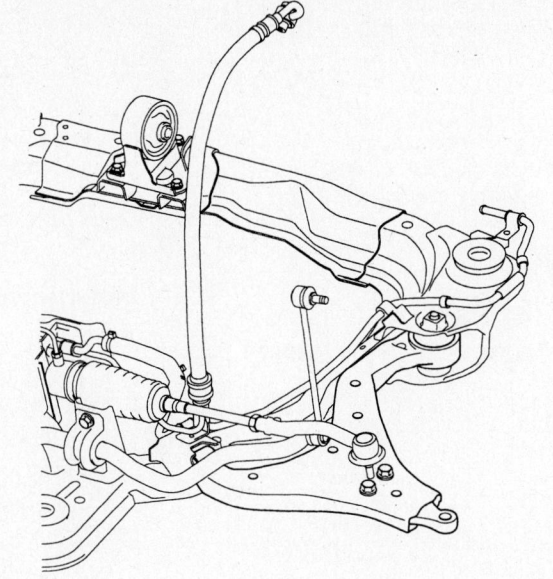

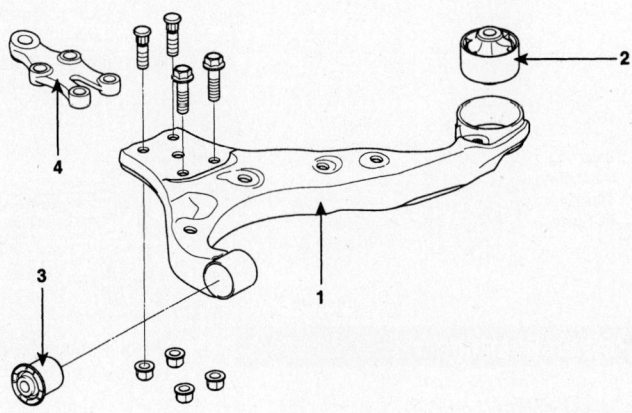

1. Lower arm
2. G bushing
3. A bushing
4. Connector

09474_KSPO_G0178

Lower control arm components—2005–06 models

- Using a standard press, remove the bushing

To install:

3. Install or connect the following:
- Lubricate the new bushing and press it into the lower control arm
- lower control arm to the vehicle

4. Check the wheel alignment and adjust if necessary.

Knuckle

REMOVAL & INSTALLATION

2002 Models

1. Before servicing the vehicle, refer to the Precautions Section.

2. Remove the locking hub.
3. Remove the wheel bearing hub.
4. Remove the dust cover.
5. Remove the upper control arm link lock bolt, spring washer and nut.
6. Remove the tie rod end cotter pin and using a ball joint puller, remove tie rod end from steering knuckle.
7. Remove lower control arm cotter pin using a ball joint puller, and remove control arm from steering knuckle.
8. Remove steering knuckle from vehicle and, using a screwdriver, pry out inner oil seal.
9. Using a needle bearing puller, remove the needle bearing.

To install:

10. Using SST K95B-5011-A, or equivalent, install the bearing.

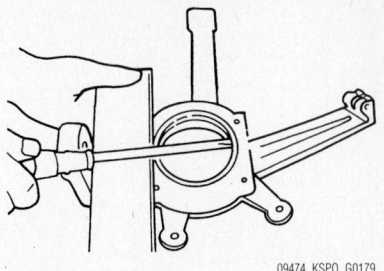

09474_KSPO_G0179

Prying out the inner knuckle seal—2002 models

11. Install a new oil seal.
12. Apply grease to the needle bearing and seal lip.
13. If the spacer was removed, apply grease to both sides and install on the half-shaft end.
14. Put the steering knuckle on the half-shaft end with the upper and lower ball joints in mounting holes.
Attach the lower control arm, tighten lock nut, and install a new cotter pin. Tighten the lock nut to 110 ft. lbs. (148 Nm). Never back off the nut to align the cotter pin holes.
15. Attach the tie rod end to the knuckle, tighten the lock nut to 27 ft. lbs. (36 Nm) and install a new cotter pin. Never back off the nut to align the cotter pin holes.
16. Insert the upper control arm link lock bolt with spring washer and tighten the nut to 36 ft. lbs. (49 Nm).
17. The remainder of installation is the reverse of removal.

Hub, Bearing and Knuckle

REMOVAL AND INSTALLATION

2005–06 Models

1. Before servicing the vehicle, refer to the Precautions Section.
2. Raise the front of the vehicle, and make sure it is securely supported.
3. Remove the front wheel and tire from front hub.
4. Remove the split pin, then remove castle nut and washer from the front hub.
5. Remove the caliper mounting bolts, and hang the caliper assembly to one side. To prevent damage to the caliper assembly or brake hose, use a short piece of wire to hang the caliper from the undercarriage.
6. Remove the wheel speed sensor from the knuckle.
7. Disconnect the tie rod end ball joint from the knuckle.

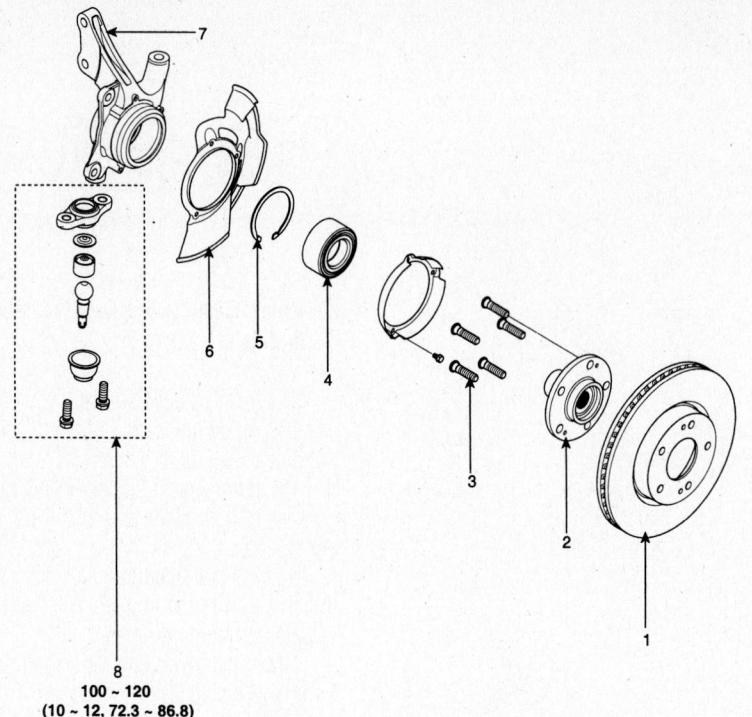

8
100 ~ 120
(10 ~ 12, 72.3 ~ 86.8)

TORQUE : Nm (Kgf·m, lb-ft)

1. Disc
2. Hub
3. Hub bolt
4. Hub bearing

5. Snap ring
6. Dust cover
7. Knuckle
8. Ball joint

09474_KSPO_G0180

Hub/knuckle components—2005–06 models

8. Remove the lower arm ball joint mounting bolts.

9. Remove the strut lower arm mounting bolts.

10. Remove the hub and the knuckle assembly.

To install:

11. Install the hub and the knuckle assembly.

12. Install the strut lower mounting bolts. Tighten to 140–160 Nm (103–118 ft. lbs.)

13. Install the lower arm ball joint mounting bolts. Tighten to 100–120 Nm (74–88 ft. lbs.)

14. Install the tie rod end ball joint in the knuckle. Tighten to 45–60 Nm (33–44 ft. lbs.). Use a new cotter pin.

15. Install the wheel speed sensor.

16. Install the brake caliper.

17. Install the washer, castle nut and split pin from the front hub. Tighten to 200–280 Nm (147–206 ft. lbs.)

18. Install the front wheel and tire on the front hub. Tighten to 90–110 Nm (66–81 ft. lbs.)

19. Be careful not to damage the hub bolts then install the front wheel and tire.

Wheel Bearings

ADJUSTMENT

2002 Models

1. Before servicing the vehicle, refer to the Precautions Section.

2. Remove or disconnect the following:
- Negative battery cable
- Front wheel
- Brake caliper from the rotor and hang it out of the way

3. Attach a dial indicator to the axle hub and measure the bearing play

4. If the play exceeds.004 inch (.10mm), check and adjust locknut torque. The bolts should be tightened to 23 ft. lbs. (31 Nm).

2005–06 Models

No adjustment is possible.

REMOVAL & INSTALLATION

2002 Models

1. Before servicing the vehicle, refer to the Precautions Section.

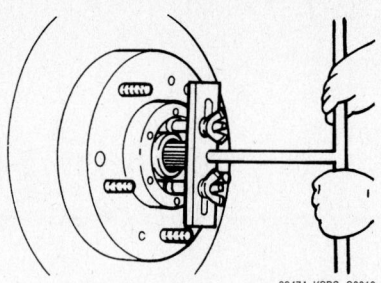

09474_KSPO_G0013

Locknut removal—2002 models

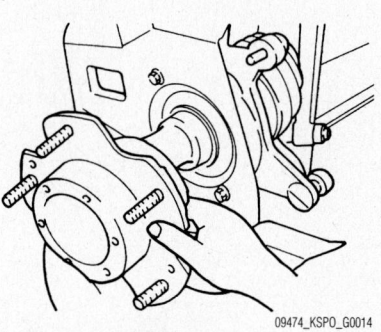

09474_KSPO_G0014

Hub removal—2002 models

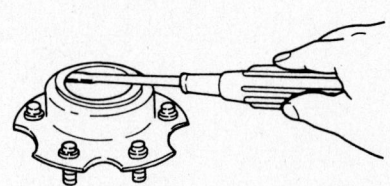

09474_KSPO_G0015

Oil seal removal—2002 models

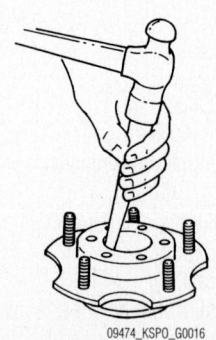

09474_KSPO_G0016

Inner and outer race removal—2002 models

2. Remove or disconnect the following:
- Negative battery cable
- Front wheel
- Free wheel hub body
- Brake caliper
- Brake rotor
- Wheel bearing and using a screw driver, pry out the oil seal
- Inner and outer bearings
- Using a drift punch, remove the inner and outer bearing race

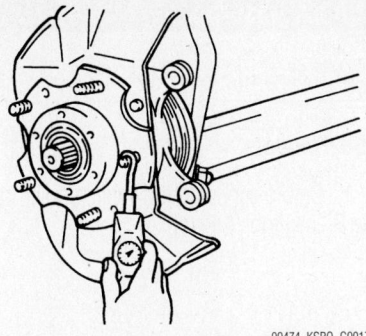

09474_KSPO_G0017

Measuring wheel bearing preload—2002 models

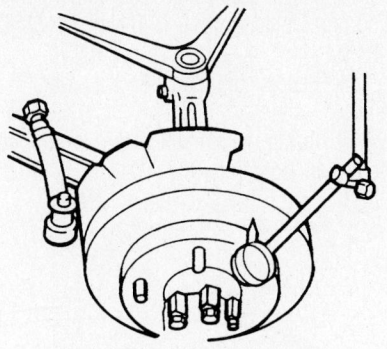

09474_KSPO_G0018

Checking rotor runout—2002 models

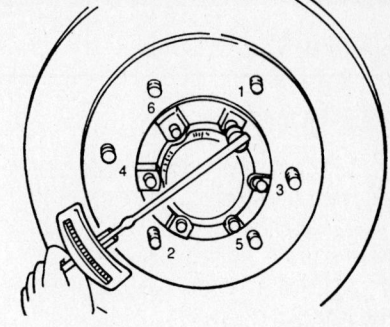

09474_KSPO_G0019

Hub bolt torque sequence—2002 models

To install:

3. Pack the new bearings with grease.
4. Install or connect the following:
   - Inner bearing and race and a new oil seal
   - Outer bearing and race and secure the dust cover with four screws
   - Apply grease to the new bearings and the lip of the oil seal
   - Hub assembly in the steering knuckle
   - Screw the locknut against the hub

assembly until there is 10 inch lbs. (1.3 Nm) of preload on the hub
- Brake rotor and retaining screws
- Attach a dial indicator to check the rotor run-out. The run-out should not exceed 0.004 inch (0.10mm)
- Brake caliper
- Free wheel hub fixed cam key with the locknut groove and push the on the fixed cam assembly
- Axle retainer snap ring
- Apply a light coat of sealant on the

free wheel hub body. Install the body on the hub. Torque the bolts in 2 passes. Tighten the bolts on the first pass to 18 ft. lbs. (25 Nm). Tighten the bolts on the second pass to 23 ft. lbs. (31 Nm).
- Negative battery cable

2005–06 Models

See the Hub, Knuckle and Bearing procedure.

REAR SUSPENSION

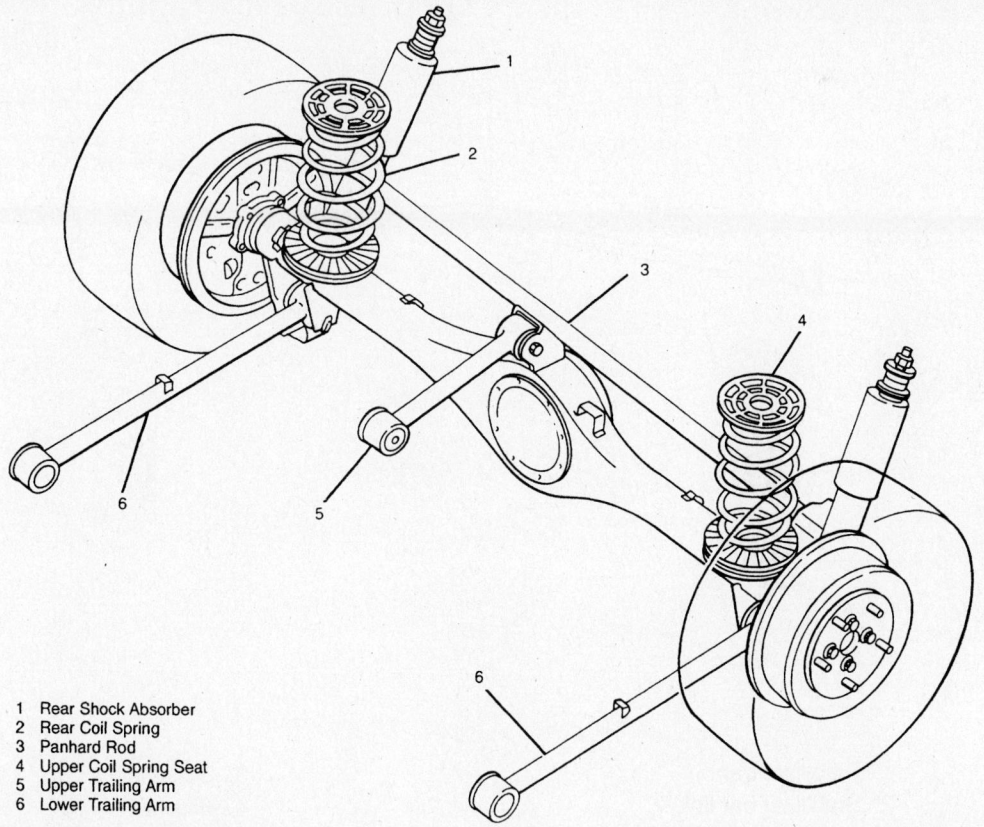

1 Rear Shock Absorber
2 Rear Coil Spring
3 Panhard Rod
4 Upper Coil Spring Seat
5 Upper Trailing Arm
6 Lower Trailing Arm

09474_KSPO_G0021

Rear suspension components—2002 models

Stabilizer Bar

REMOVAL & INSTALLATION

2005–06 Models

1. Before servicing the vehicle, refer to the Precautions Section.

2. Raise the rear of the vehicle, and make sure it is securely supported.

3. Remove the rear wheel and tire from rear hub.

4. Remove the stabilizer bar link mounting nuts.

5. Remove the stabilizer bar mounting bolts and then remove the stabilizer brackets.

6. Remove the stabilizer bar.

To install:

7. Install the bushing on the stabilizer bar. Bring clamp of stabilizer bar into contact with bushing.

8. Install the stabilizer bracket and then install the stabilizer bar mounting bolts.

9. One side bracket should be temporarily tightened, and then install the bushing on the opposite side. Tighten to 45–55 Nm (32–39 ft. lbs.).

10. Install the stabilizer bar link mounting nut. Tighten to 100–120 Nm (74–88 ft. lbs.).

11. Repeat step 3 and 4 for the other side.

12. Install the rear wheel and tire on the rear hub. Tighten to 90–110 Nm (66–81 ft. lbs.).

Shock Absorber

REMOVAL & INSTALLATION

2002 Models

1. Before servicing the vehicle, refer to the Precautions Section.

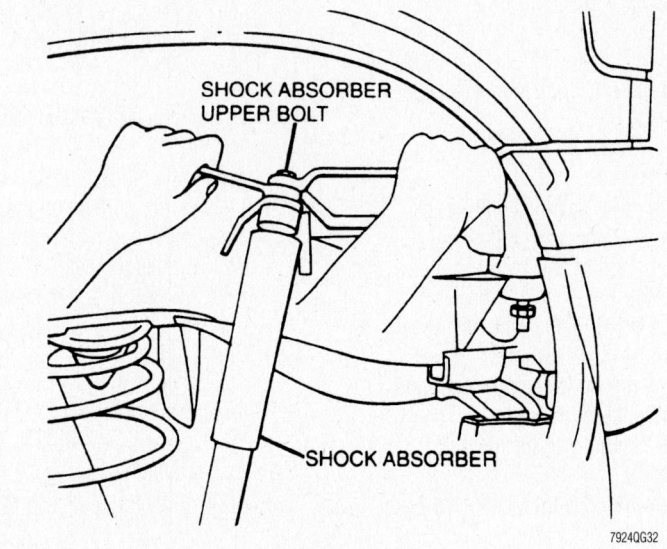

Upper shock absorber mounting nut—2002 models

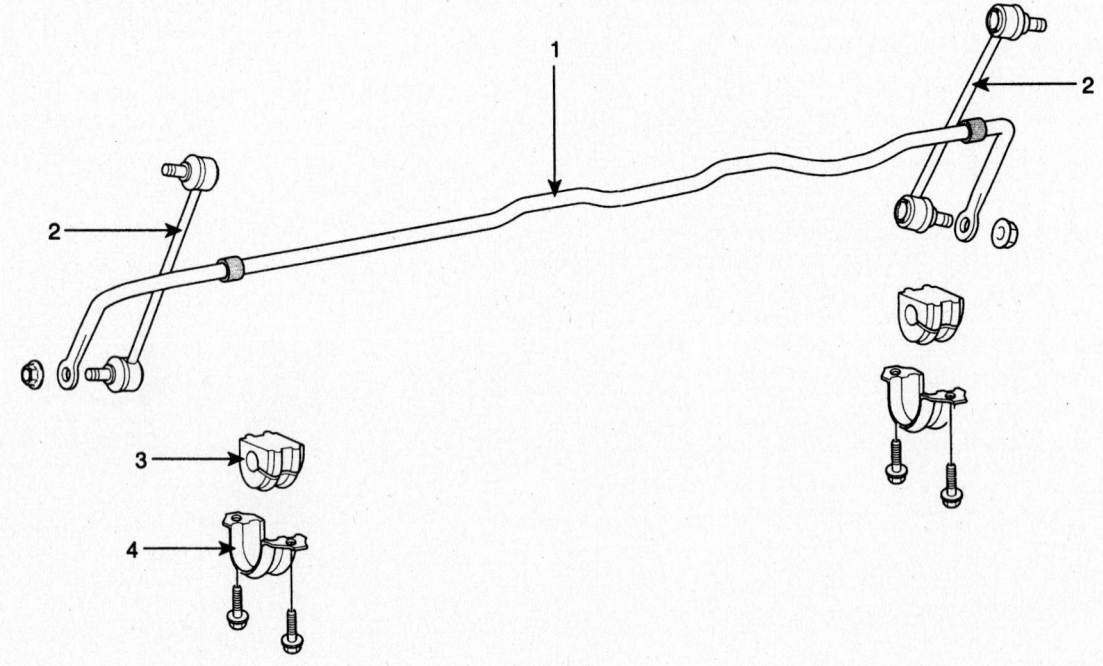

1. Stabilizer bar
2. Stabilizer bar link
3. Bushing
4. Bracket

Rear stabilizer bar components—2005–06 models

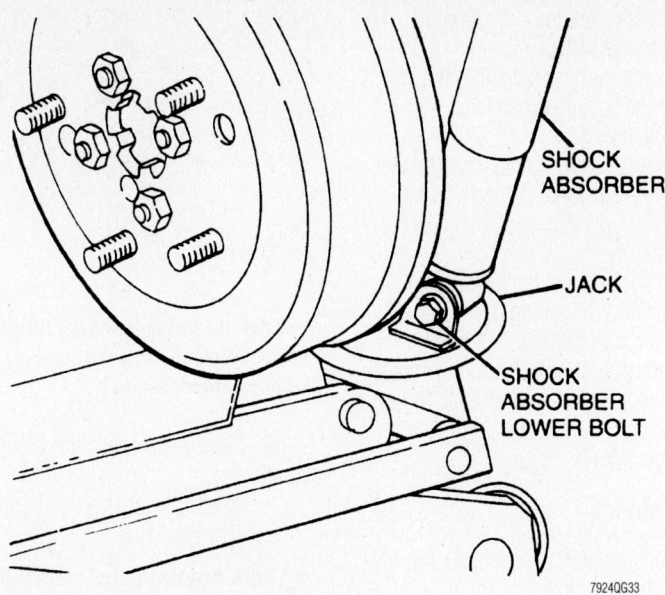

Lower shock absorber mounting bolt—2002 models

2. Remove or disconnect the following:
   - Negative battery cable
   - Rear wheels
   - Raise the rear axle with a floor jack to relax the shock absorbers and support the rear axle when the shock absorber is removed.
   - Rear safety nut, upper nut and washer
   - Upper rubber plate
   - Lower bolt from the shock absorber
   - Lower the rear axle housing
   - Shock absorber

3. Remove the lower mounting bolt and remove the shock absorber.

To install:

4. Install or connect the following:
   - Bottom washer and rubber cushion on the top of the shock absorber. Position the shock absorber on the vehicle
   - Lower bolt
   - Rubber cushion, washer and nut. Tighten to 53 ft. lbs. (72 Nm)
   - Safety nut. Torque the lower bolt to 62 ft. lbs. (84 Nm).
   - Rear wheels
   - Negative battery cable

Strut

REMOVAL & INSTALLATION

2005–06 Models

1. Before servicing the vehicle, refer to the Precautions Section.

2. Raise the rear of the vehicle, and make sure it is securely supported.

3. Remove the rear wheel and tire) from rear hub.

4. Remove the speed sensor cable mounting bolt.

5. With drum brakes, remove the speed sensor cable mounting bolts and the brake hose bracket.

6. With disc brakes, remove the speed sensor cable mounting bolt.

7. Remove the stabilizer bar link nut.

8. Remove the strut upper mounting nuts (A).

9. Remove the strut lower mounting bolts (A) and then remove the strut assembly (B).

To install:

10. Install the strut assembly and then install the strut lower mounting bolts. Tighten to 140–160 Nm (103–118 ft. lbs.).

11. Install the strut upper mounting nuts. Tighten to 30–40 Nm (22–29 ft. lbs.).

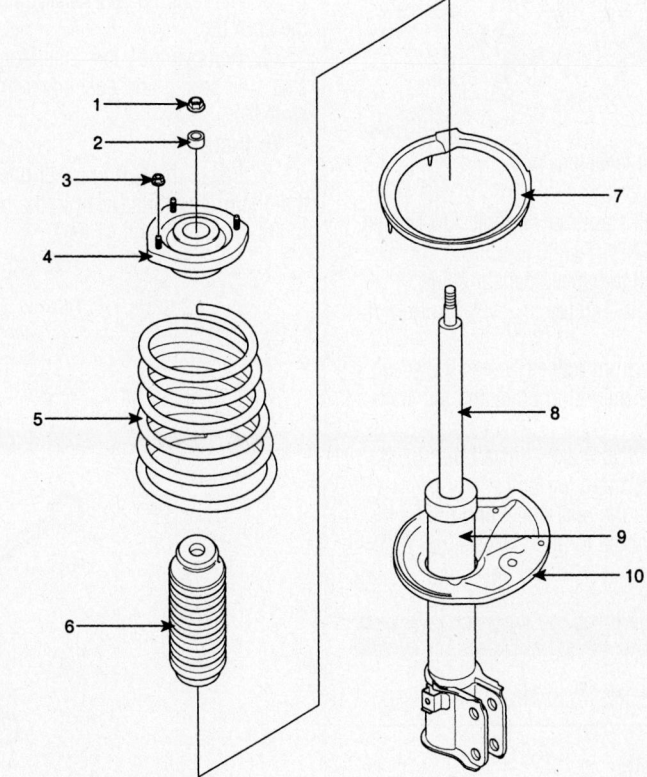

1. Self-locking nut
2. Spacer
3. Upper mounting nut
4. Insulator
5. Coil spring
6. Strut dust cover & bumper rubber
7. Spring lower pad
8. Piston rod
9. Strut assembly
10. Spring lower seat

Rear strut exploded view—2005–06 models

09474_KSPO_G0193

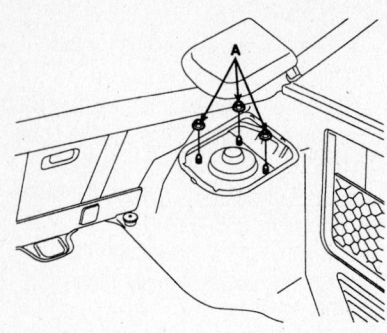

Upper strut mounting nuts

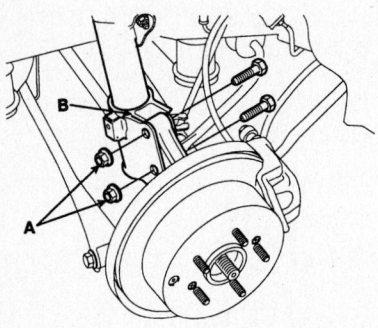

Lower strut mounting bolts

12. Install stabilizer bar link nut. Tighten to 100–120 Nm (74–88 ft. lbs.).

13. Install the speed sensor cable mounting bolt. Tighten to 7–11 Nm (5–8 ft. lbs.).

14. With drum brakes, install the speed sensor cable mounting bolts and the brake hose bracket.

15. With disc brakes, install the speed sensor cable mounting bolt.

16. Install the rear wheel and tire on the rear hub. Tighten to 90–110 Nm (66–81 ft. lbs.).

Coil Spring

REMOVAL & INSTALLATION

2002 Models

1. Before servicing the vehicle, refer to the Precautions Section.

2. Remove or disconnect the following:
   • Both rear wheels
   • Raise the rear axle with a floor jack to relax the shock absorbers and support the rear axle when the shock absorber is removed.

➡**For easier installation, complete one side at a time.**

   • Lower mounting bolt, then the shock absorber
   • Lower the floor jack until the coil spring is fully expanded
   • Coil spring
   • Inspect the upper and lower rubber spring seats and jounce stop for wear or damage, replace if necessary

To install:

3. Install or connect the following:
   • Position the spring in the upper and lower saddles
   • Raise the floor jack and connect the lower shock absorber bolt. Torque the bolt to 62 ft. lbs. (84 Nm).
   • Rear wheels

2005–06 Models

1. Before servicing the vehicle, refer to the Precautions Section.

2. Using the special tool (09545-26000), or equivalent spring compressor, compress the coil spring (A).

3. Remove the self-locking nut (C) from the strut (B).

4. Remove the pipe, insulator, spring seat, coil spring and dust cover from the strut (B).

To install:

5. Install the spring lower pad (D) so that the protrusions (A) fit in the holes (C) in the spring lower seat (B).

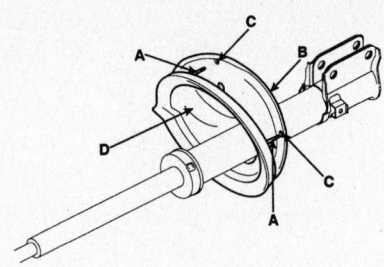

Install the spring lower pad (D) so that the protrusions (A) fit in the holes (C) in the spring lower seat (B)

6. Compress coil spring using the compressor.

7. Install compressed coil spring into shock absorber.

➡**There are two color marks on the coil spring. One corresponds to model option and the other corresponds to load classification. Ensure that the correct parts are being installed.**

8. Install the coil spring with the identification mark directed toward the knuckle.

9. After fully extending the piston rod, install the spring upper seat and insulator assembly.

10. After seating the upper and lower ends of the coil spring (A) in the upper and

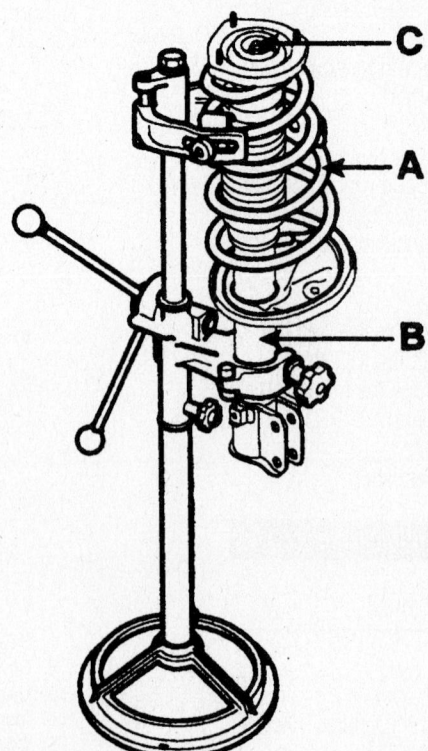

Rear strut mounted in a compressor

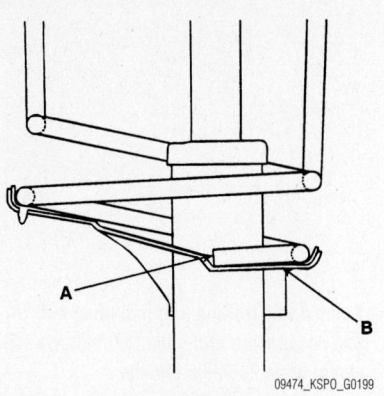

Proper spring seating on the strut

lower spring seat grooves (B) correctly, tighten new self-locking nut temporarily.

11. Remove the compressor.

12. Tighten the self-locking nut to 52–70 Nm (37–50 ft. lbs.)

Lateral Link (Panhard Rod)

REMOVAL AND INSTALLATION

2002 Models

1. Before servicing the vehicle, refer to the Precautions Section.

2. Raise the rear of the vehicle and support it with safety stands.

3. Remove the rear wheels.

4. Raise the rear axle housing to facilitate removal of the Panhard arm assembly.

5. Loosen the Panhard arm bolts and remove the Panhard arm.

6. Inspect the Panhard arm for bends, cracks and/or other damage.

7. Inspect the Panhard arm bushings for wear and/or deterioration.

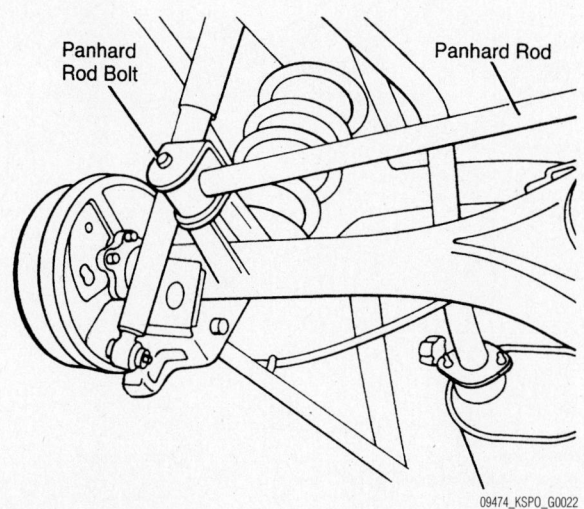

Panhard rod mounting—2002 models

8. Replace if damaged, deformed or cracked; replace bushings if worn or deteriorated.

To install:

9. Press out the bushing using a standard bearing press.

10. Apply lubricant to the bushing and press into place using a standard bearing press.

11. Install the Panhard arm and the bolts. Tighten the bolts to 62 ft. lbs. (84~Nm)

12. Lower the rear axle housing.

13. Replace the wheels and install the lug bolts. Tighten the bolts to 80 ft. lbs. (108Nm)

14. Remove the safety stands and lower the vehicle.

Upper Arm

REMOVAL AND INSTALLATION

2002 Models

1. Before servicing the vehicle, refer to the Precautions Section.

2. Raise the rear of the vehicle and support it with safety stands.

3. Remove the rear wheels.

4. Raise the rear axle housing to facilitate removal of the upper arm assembly.

5. Loosen the upper arm bolts and remove the upper arm.

6. Inspect the upper arm for bends, cracks and/or other damage.

7. Inspect the upper arm bushings for wear and/or deterioration.

8. Replace if damaged, deformed or cracked; replace bushings if worn or deteriorated.

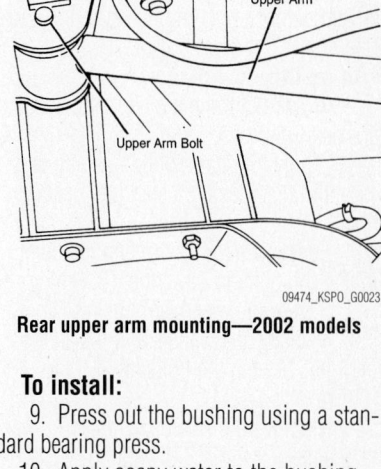

Rear upper arm mounting—2002 models

To install:

9. Press out the bushing using a standard bearing press.

10. Apply soapy water to the bushing and press into place using a standard bearing press.

11. Install the upper trailing arm and the bolts. Tighten the bolts to 62 ft. lbs. (84Nm)

12. Lower the rear axle housing.

13. Replace the wheels and install the lug bolts. Tighten the bolts to 74 ft. lbs. (100Nm)

14. Remove the safety stands and lower the vehicle.

Lower Arm

REMOVAL AND INSTALLATION

2002 Models

1. Before servicing the vehicle, refer to the Precautions Section.

2. Raise the rear of the vehicle and support it with safety stands.

3. Remove the rear wheels.

4. Raise the rear axle housing to facilitate removal of the lower arm assembly.

5. Loosen the lower arm bolts and remove the lower arm.

6. Inspect the lower arm for bends, cracks and/or other damage.

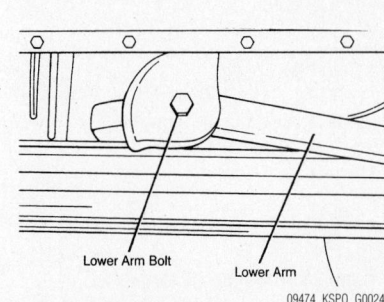

Rear lower arm mounting—2002 models

7. Inspect the lower arm bushings for wear and/or deterioration.

8. Replace if damaged, deformed or cracked; replace bushings if worn or deteriorated.

To install:

9. Press out the bushing using a standard bearing press.

10. Apply lubricant to the bushing and press into place using a standard bearing press.

11. Install the lower arm and the bolts. Tighten the bolts to 62 ft. lbs. (84Nm)

12. Lower the rear axle housing.

13. Replace the wheels and install the lug bolts. Tighten the bolts to 74 ft. lbs. (100Nm)

14. Remove the safety stands and lower the vehicle.

Trailing Arm

REMOVAL AND INSTALLATION

2005–06 Models

1. Before servicing the vehicle, refer to the Precautions Section.

2. Remove the trailing arm mounting bolts (A).

3. Remove the bracket mounting bolt, nut of the vehicle side.

4. Remove the trailing arm (B).

To install:

5. Install the trailing arm.

➡**The trailing arm mounting bolts/nuts should be installed snugly, but not fully torqued until the weight of the vehicle is on the suspension.**

6. Install the trailing arm mounting bolts/nuts. Tighten to 100–120 Nm (74–88 ft. lbs.).

Rear Suspension Arm

REMOVAL AND INSTALLATION

2005–06 Models

2-WHEEL DRIVE

1. Before servicing the vehicle, refer to the Precautions Section.

2. Remove the trailing arm mounting bolt (A) and suspension arm mounting bolt (B).

3. Remove the opposite side trailing arm mounting bolt and suspension arm mounting bolt.

4. After supporting the rear cross member assembly (B) with the jack (A), remove the cross member mounting bolts and nuts (C).

5. Remove the suspension arm bracket mounting bolts (A).

6. Remove the suspension arm (B).

To install:

7. Install the suspension arm bracket mounting bolts (A). Tighten to 160–180 Nm (118–133 ft. lbs.)

8. Make sure that the arrow mark (B) on the rear cross member (A) should place the front face of the vehicle.

9. Rear suspension arm (C)-to-rear carrier bolts should be temporarily tightened,

[2WD]

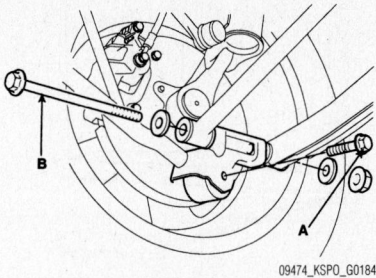

Remove the trailing arm mounting bolt (A) and suspension arm mounting bolt (B)—2-wheel drive 2005–06 Models

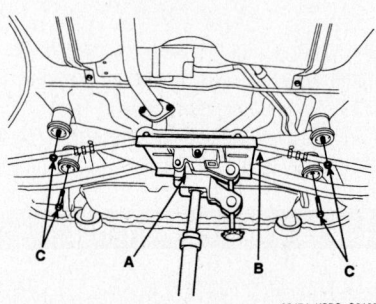

After supporting the rear cross member assembly (B) with the jack (A), remove the cross member mounting bolts and nuts (C)—2-wheel drive 2005–06 Models

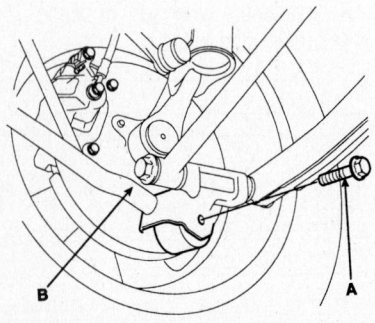

09474_KSPO_G0181

Trailing arm—2005–06 Models

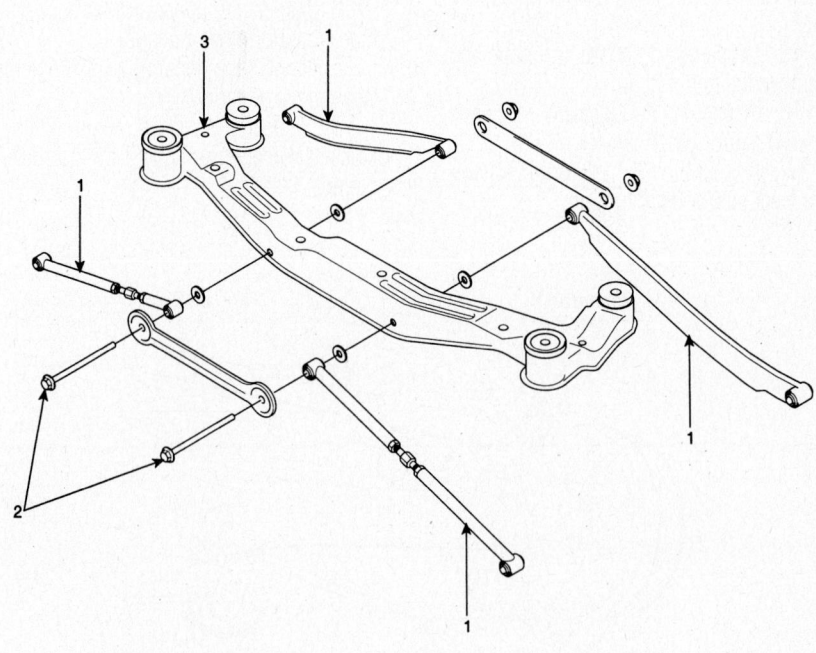

1. Suspension arm
2. Suspension arm bracket mounting bolt
3. Cross member

09474_KSPO_G0182

2WD rear suspension arms—2005–06 Models

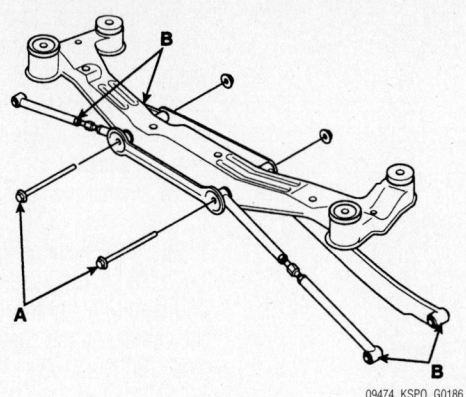

Suspension arm bracket bolts and suspension arm assembly—2-wheel drive 2005–06 Models

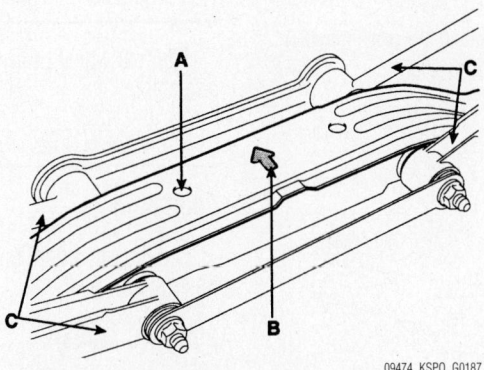

Rear suspension arm installation—2-wheel drive 2005–06 Models

[4WD]

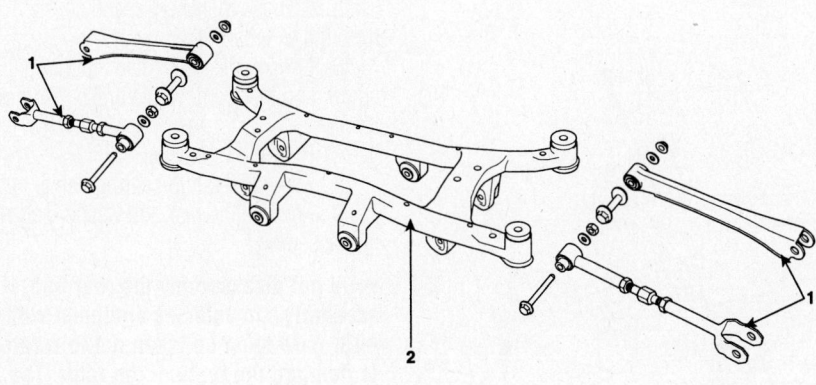

1. Suspension arm   2. Cross member

09474_KSPO_G0183

4WD rear suspension arms—2005–06 Models

and then fully tightened with the weight of the vehicle on the suspension. Tighten to 160–180 Nm (118–133 ft. lbs.)

4-WHEEL DRIVE

1. Before servicing the vehicle, refer to the Precautions Section.

2. Remove the muffler.

3. Remove the suspension arm mounting bolts (A).

4. Remove the opposite side suspension mounting bolts.

5. Remove the coupling control connector (A).

6. After supporting the rear cross member assembly (B) with a jack (A), remove the cross member mounting bolts and nuts (C).

7. Remove the driveshaft.

8. Remove the rear differential (A) from the cross member (B).

9. Remove the suspension arm bracket mounting bolts (A).

10. Remove the suspension arm (B).

To install:

11. Install the suspension arm bracket mounting bolts (A). Tighten to 140–160 Nm (103–118 ft. lbs.)

12. Install the rear differential (A) on the cross member (B). Tighten to 90–120 Nm (59–88 ft. lbs.)

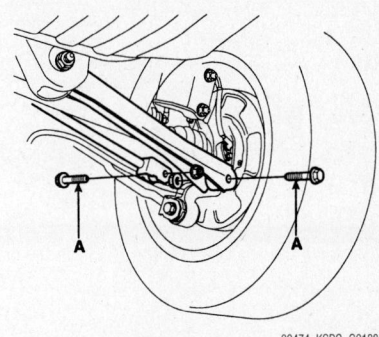

Remove the suspension arm mounting bolts (A)—4-wheel drive 2005–06 Models

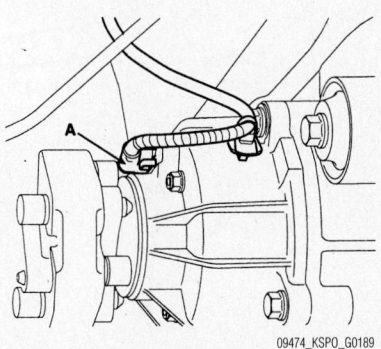

Remove the coupling control connector (A)—4-wheel drive 2005–06 Models

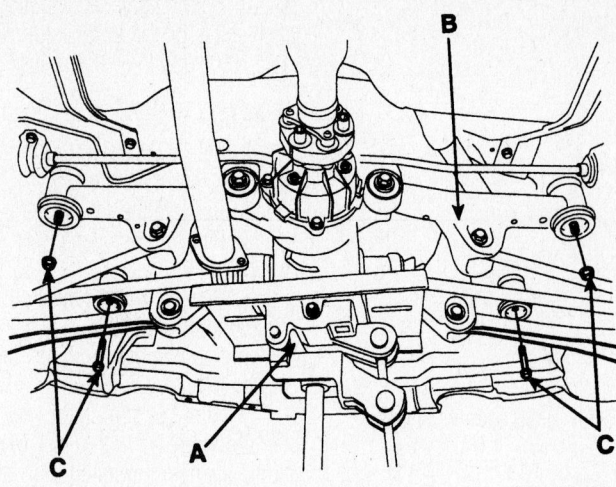

After supporting the rear cross member assembly (B) with a jack (A), remove the cross member mounting bolts and nuts (C)—4-wheel drive 2005–06 Models

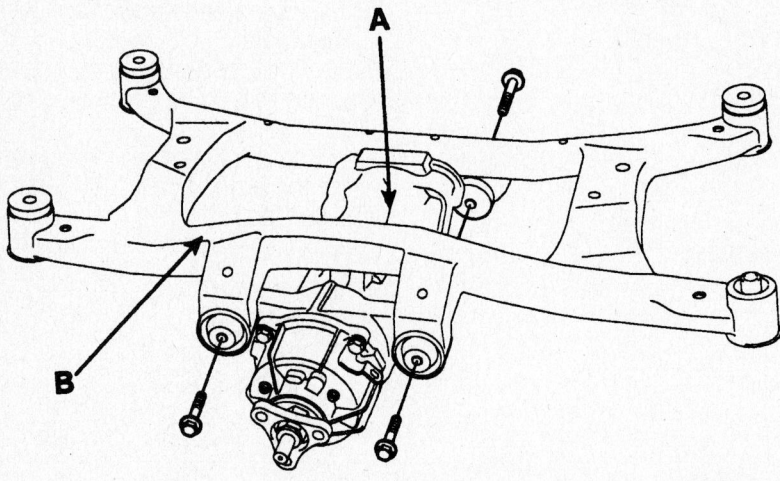

Remove the rear differential (A) from the cross member (B)—4-wheel drive 2005–06 Models

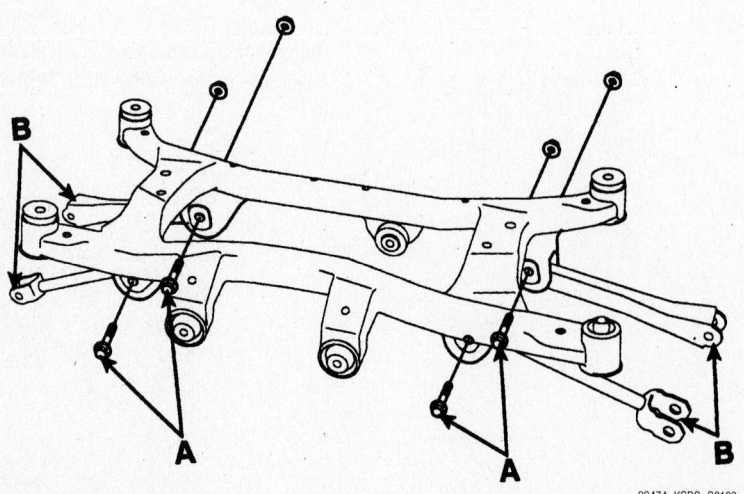

Remove the suspension arm bracket mounting bolts (A) and remove the suspension arm (B)—4-wheel drive 2005–06 Models

13. Install the driveshaft.

14. After supporting the rear cross member assembly (B) with the jack (A), install the cross member mounting bolts and nuts (C). Tighten to 100–120 Nm (74–88 ft. lbs.)

15. Install the coupling control connector (A).

16. Rear suspension arm-to-rear carrier bolts (A) should be temporarily tightened, and then fully tightened with the weight of the vehicle on the suspension. Tighten to 140–160 Nm (103–118 ft. lbs.)

Wheel Bearings

ADJUSTMENT

The rear wheel bearings are not adjustable.

REMOVAL & INSTALLATION

2002 Models

See the procedure under "Rear Axle Shaft Bearing and Seal".

2005–06 Models

2-WHEEL DRIVE

1. Before servicing the vehicle, refer to the Precautions Section.

2. Raise the rear of the vehicle, and make sure it is securely supported.

3. Remove the rear wheel and tire from rear hub.

4. Remove the caliper mounting bolts, and hang the caliper assembly to one side. To prevent damage to the caliper assembly or brake hose, use a short piece of wire to hang the caliper from the undercarriage.

5. Remove the wheel speed sensor from the axle carrier.

6. Loosen the brake disc mounting screw, and then remove the brake disc from the hub.

7. Pry off the hub cap.

8. Remove the hub bearing flange nut.

9. Remove the rear hub washer and rear hub assembly.

➡Do not disassembly the rear hub assembly. On vehicles equipped with ABS, care must be taken not to scratch or damage the teeth of the rotor. The rotor must never be dropped. If the teeth of the rotor are chipped, it results in deformation of the rotor. It will make it impossible to detect the wheel rotation speed accurately and to operate the system normally.

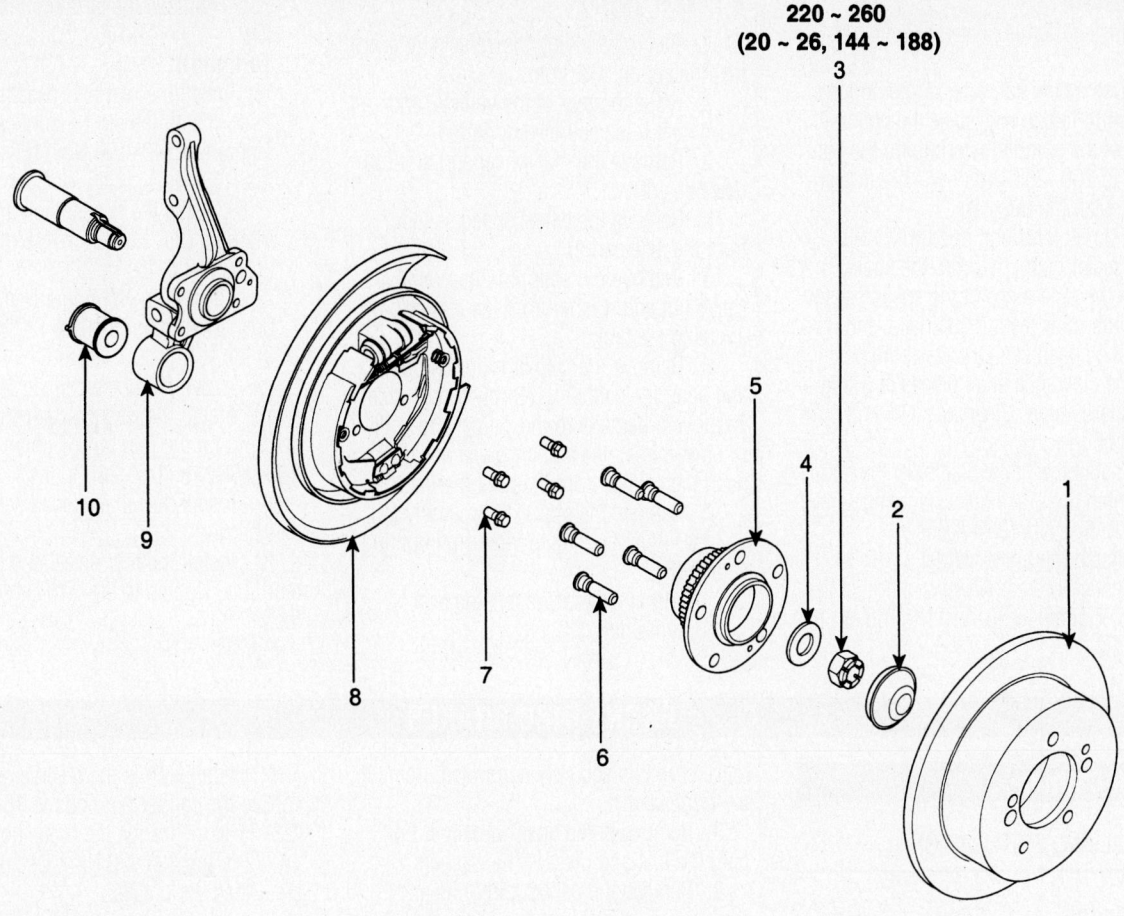

220 ~ 260
(20 ~ 26, 144 ~ 188)

TORQUE : Nm (Kgf·m, lb-ft)

1. Disc
2. Hub cap
3. Castle nut
4. Washer
5. Hub
6. Hub bolt
7. Dust cover mounting bolt
8. Rear parking brake assembly
9. Axle carrier
10. Bushing

09474_KSPO_G0200

Rear hub/bearing assembly—2-wheel drive 2005–06 models

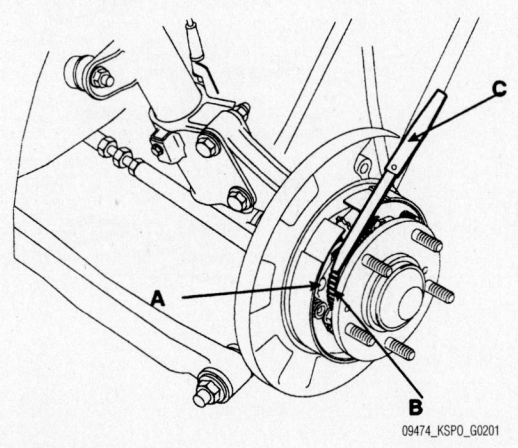

09474_KSPO_G0201

Speed sensor gap measurement

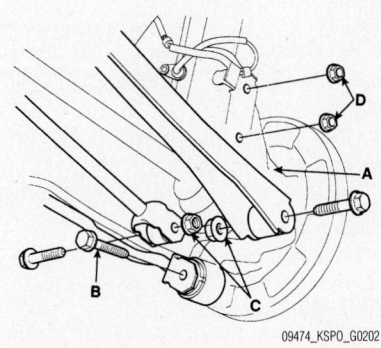

09474_KSPO_G0202

Rear hub/axle assembly—4-wheel drive
2005–06 models

To install:

10. Install the hub assembly and hub washer.

11. Use a new nut. After tightening the hub bearing flange nut, stake the concave portion of the spindle by crimping the nut. Tighten to 200–260 Nm (147–192 ft. lbs.)

12. Install the hub cap.

13. For vehicles equipped with ABS: Insert a feeler gauge (C) into the space between the pole piece of the speed sensor (A) and the rotor teeth (B) surface, and then tighten the speed sensors (A) at the position where the clearance at all places is within the standard value. Clearance: 0.5–1.5 mm (0.02–0.06 in.)

14. Install the brake disc from the hub, then tighten the brake disc mounting screw. Tighten to 5–6 Nm (3–4 ft. lbs.)

15. Install the brake caliper.

16. Install the rear wheel and tire on the rear hub. Tighten to 90–110 Nm (66–81 ft. lbs.)

4-WHEEL DRIVE

1. Before servicing the vehicle, refer to the Precautions Section.

2. Raise the rear of the vehicle, and make sure it is securely supported.

3. Remove the rear wheel and tire from rear hub.

4. Remove the wheel speed sensor from the axle carrier.

5. Remove the split pin, then remove castle nut and washer from the rear hub by applying the brake.

6. Remove the caliper mounting bolts, and hang the caliper assembly to one side. To prevent damage to the caliper assembly or brake hose, use a short piece of wire to hang the caliper from the undercarriage.

7. Remove the rear axle assembly (A).

  a. Remove the trailing arm mounting bolt (B).

  b. Remove the suspension arm mounting nuts (C).

  c. Remove the strut mounting nuts (D).

To install:

8. Install the rear axle assembly (A).

  a. Install the strut mounting nuts (D). Tighten to 140–160 Nm (103–118 ft. lbs.)

  b. Install the suspension arm mounting nuts (C). Tighten to 140–160 Nm (103–118 ft. lbs.)

  c. Install the trailing arm mounting bolt (B). Tighten to 100–120 Nm (74–88 ft. lbs.)

9. Install the brake caliper.

10. Install the washer, castle nut and new split pin on the rear hub. Tighten to 200–280 Nm (147–206 ft. lbs.)

11. Install the wheel speed sensor on the axle carrier.

12. Install the rear wheel and tire on the rear hub. Tighten to 90–110 Nm (66–81 ft. lbs.)

FRONT DISC BRAKES

Brake Caliper

REMOVAL AND INSTALLATION

2002 Models

1. Before servicing the vehicle, refer to the Precautions Section.

2. Raise and safely support the vehicle.

3. Remove the front wheels.

4. Remove the 2 caliper bolts and lift the caliper from the disc.

5. Disconnect the brake fluid flex line by removing the retaining bolt if the caliper is to be replaced.

To install:

6. Seat the caliper piston using a C-clamp.

7. Connect the brake fluid flex line bolt to the caliper. Tighten the bolt to 17 ft. lbs. (23.5 Nm).

8. Position the caliper over the disc assembly. Install the caliper bolts. Tighten to 13 ft. lbs. (17 Nm).

9. Install the front wheels. Tighten the lug nuts 77 ft. lbs. (99 Nm).

10. Bleed the hydraulic system if the flex hoses were removed.

11. Lower the vehicle.

2005–06 Models

1. Before servicing the vehicle, refer to the Precautions Section.

2. Raise the front of the vehicle, and make sure it is securely supported. Remove the front wheels.

3. Remove brake hose-to-caliper bolt, plug the line and discard the washers.

4. Remove the caliper pin bolts and remove the caliper from the anchor plate.

5. Installation is the reverse of removal. Torque the caliper pin bolts to 16–23 ft. lbs. (22–31 Nm). Torque the brake hose bolt to 18–22 ft. lbs. (24–29 Nm). Use new washers. Bleed the brakes.

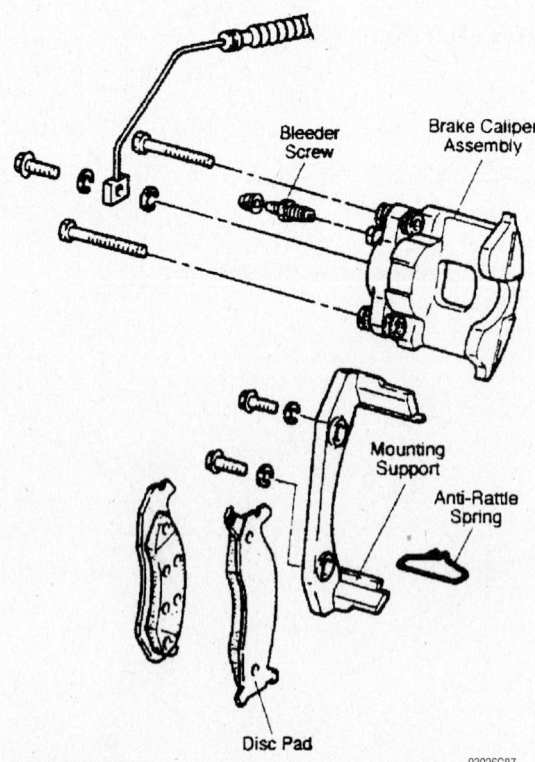

Exploded view of the front disc brake assembly—2002 models

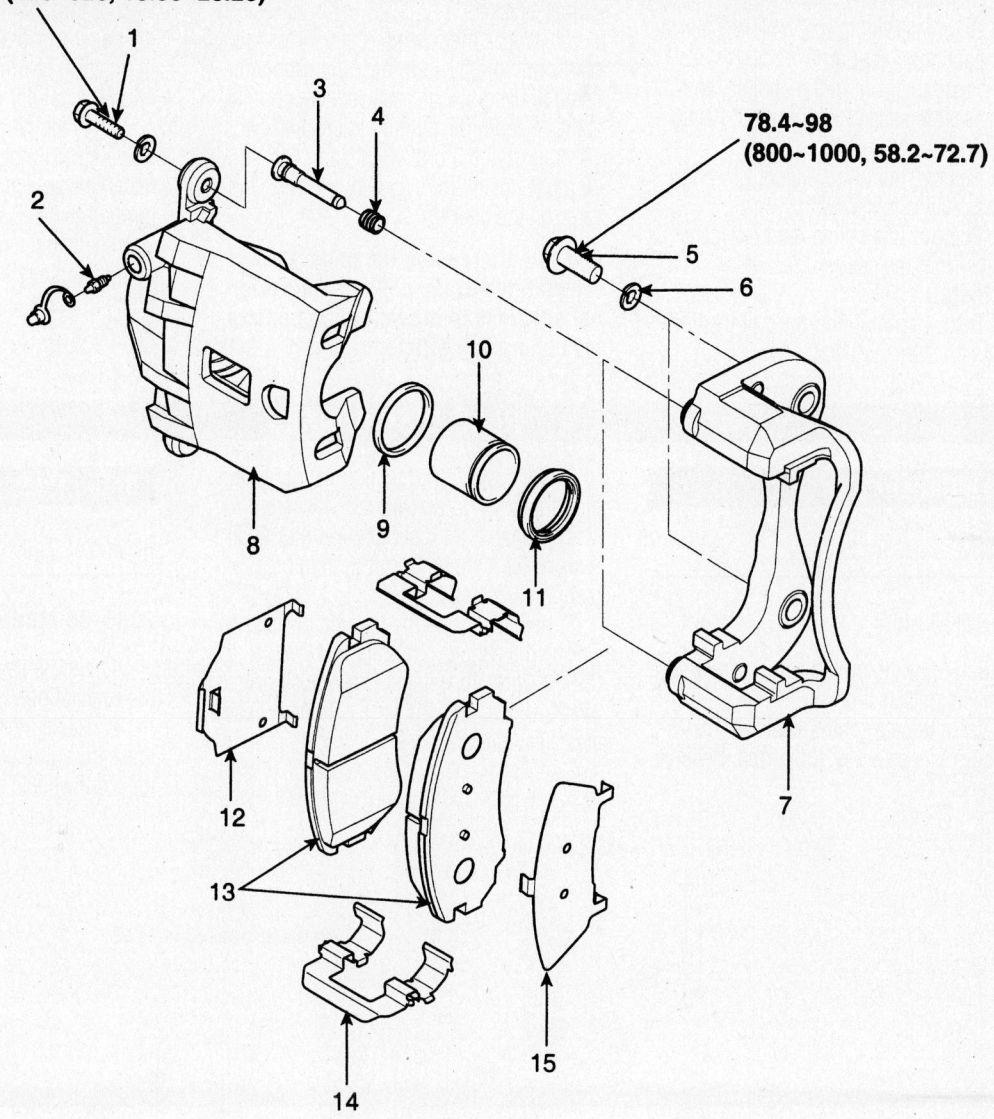

21.56~31.36 (220~320, 15.99~23.26)

78.4~98
(800~1000, 58.2~72.7)

TORQUE : Nm (kgf·cm, lb-ft)

1. Guide rod bolt
2. Bleeder screw
3. Guide rod
4. Boot
5. Caliper mounting bolt

6. Washer
7. Caliper bracket
8. Caliper body
9. Piston seal
10. Piston

11. Piston boot
12. Inner shim
13. Brake pad
14. Pad retainer
15. Outer shim

09474_KSPO_G0203

Front disc brake components—2005–06 models

Disc Brake Pads

REMOVAL AND INSTALLATION

2002 Models

1. Before servicing the vehicle, refer to the Precautions Section.
2. Raise and safely support the vehicle.

3. Remove the front wheels.
4. Remove the 2 caliper bolts and lift the caliper from the disc.
5. Slide the disc pads off the caliper bracket.

To install:

6. Clean the caliper bracket contact surface with a wire brush and lightly coat with assembly lube.

7. Position the disc pads.
8. Position the caliper over the disc assembly. Install the caliper bolts. Tighten to 13 ft. lbs. (17 Nm).
9. Install the front wheels. Tighten the lug nuts 77 ft. lbs. (99 Nm).
10. Bleed the hydraulic system if the flex hoses were removed.
11. Lower the vehicle.

2005–06 Models

1. Before servicing the vehicle, refer to the Precautions Section.
2. Raise the front of the vehicle, and make sure it is securely supported. Remove the front wheels.
3. Remove the lower caliper pin bolt and pivot the caliper upward.
4. Remove pad shims, pad retainers and pads from the caliper bracket.

To install:

5. Remove about one-third of the fluid from the master cylinder.

6. Install the pad retainers on the caliper bracket.
7. Check the foreign material at the pad shims and the back of the pads. Contaminated brake discs or pads reduce stopping ability. Keep grease off the discs and pads.
8. Install the brake pads and pad shims correctly. Install the pad with the wear indicator on the inside.

➡**If you are reusing the pads, always reinstall the brake pads in their original positions to prevent a momentary loss of braking efficiency.**

9. Push in the piston, using a piston forcing tool, so that the caliper will fit over the pads. Make sure that the piston boot is in position to prevent damaging it when pivoting the caliper down.
10. Pivot the caliper down into position. Being careful not to damage the pin boot, install the caliper pin bolt. Torque the caliper pin bolt to 16–23 ft. lbs. (22–31 Nm).
11. Refill the master cylinder as necessary.
12. Depress the brake pedal several times to make sure the brakes work, then test-drive.

REAR DISC BRAKES

Brake Caliper

REMOVAL & INSTALLATION

2005–06 Models

1. Before servicing the vehicle, refer to the Precautions Section.
2. Raise the rear of the vehicle, and make sure it is securely supported. Remove the wheels.

3. Remove brake hose-to-caliper bolt, plug the line and discard the washers.
4. Remove the caliper pin bolts and remove the caliper from the anchor plate.
5. Installation is the reverse of removal. Torque the caliper pin bolts to 16–23 ft. lbs. (22–31 Nm). Torque the brake hose bolt to 18–22 ft. lbs. (24–29 Nm). Use new washers. Bleed the brakes.

Brake Pads

REMOVAL & INSTALLATION

2005–06 Models

1. Before servicing the vehicle, refer to the Precautions Section.
2. Raise the front of the vehicle, and make sure it is securely supported. Remove the front wheels.

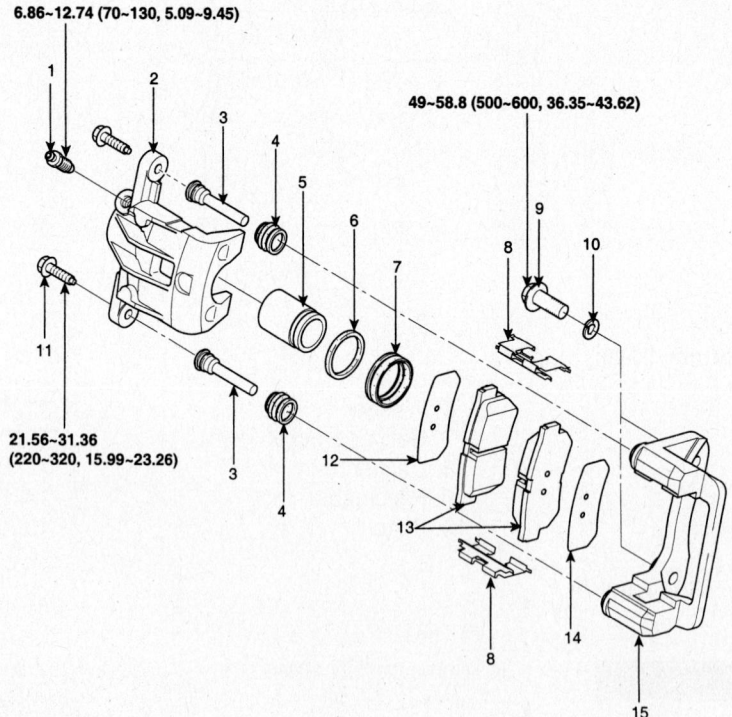

TORQUE : Nm (kgf·cm, lb-ft)

1. Bleeder screw	6. Piston seal	11. Guide rod bolt
2. Caliper body	7. Piston boot	12. Inner shim
3. Guide rod	8. Pad retainer	13. Brake Pad
4. Boot	9. Caliper mounting bolt	14. Outer shim
5. Piston	10. Washer	15. Caliper bracket

09474_KSPO_G0204

Rear disc brake components—2005–06 models

3. Remove the lower caliper pin bolt and pivot the caliper upward.

4. Remove pad shims, pad retainers and pads from the caliper bracket.

To install:

5. Remove about one-third of the fluid from the master cylinder.

6. Install the pad retainers on the caliper bracket.

7. Check the foreign material at the pad shims and the back of the pads. Contaminated brake discs or pads reduce stop-

ping ability. Keep grease off the discs and pads.

8. Install the brake pads and pad shims correctly. Install the pad with the wear indicator on the inside.

➡**If you are reusing the pads, always reinstall the brake pads in their original positions to prevent a momentary loss of braking efficiency.**

9. Push in the piston, using a piston forcing tool, so that the caliper will fit over

the pads. Make sure that the piston boot is in position to prevent damaging it when pivoting the caliper down.

10. Pivot the caliper down into position. Being careful not to damage the pin boot, install the caliper pin bolt. Torque the caliper pin bolt to 16–23 ft. lbs. (22–31 Nm).

11. Refill the master cylinder as necessary.

12. Depress the brake pedal several times to make sure the brakes work, then test-drive.

REAR DRUM BRAKES

Brake Drums

REMOVAL AND INSTALLATION

1. Before servicing the vehicle, refer to the Precautions Section.

2. Raise and safely support the vehicle.

3. Remove the rear wheels.

4. Apply the parking brake.

5. Remove the 4 attaching nuts.

6. Release the parking brake and remove the brake drum.

Installation is the reverse of the removal procedure.

Brake Shoes

REMOVAL AND INSTALLATION

1. Before servicing the vehicle, refer to the Precautions Section.

2. Raise and safely support the vehicle.

3. Remove the rear wheels.

4. Apply the parking brake.

5. Remove the 4 attaching nuts.

6. Release the parking brake and remove the brake drum.

7. Remove the spring from the secondary shoe to the adjusting lever.

8. Remove the adjusting lever.

9. Remove the return spring above the star adjusting wheel.

10. Turn the starwheel clockwise to relieve tension on the brake shoes.

11. Remove the starwheel.

12. Remove the hold-down pin clips.

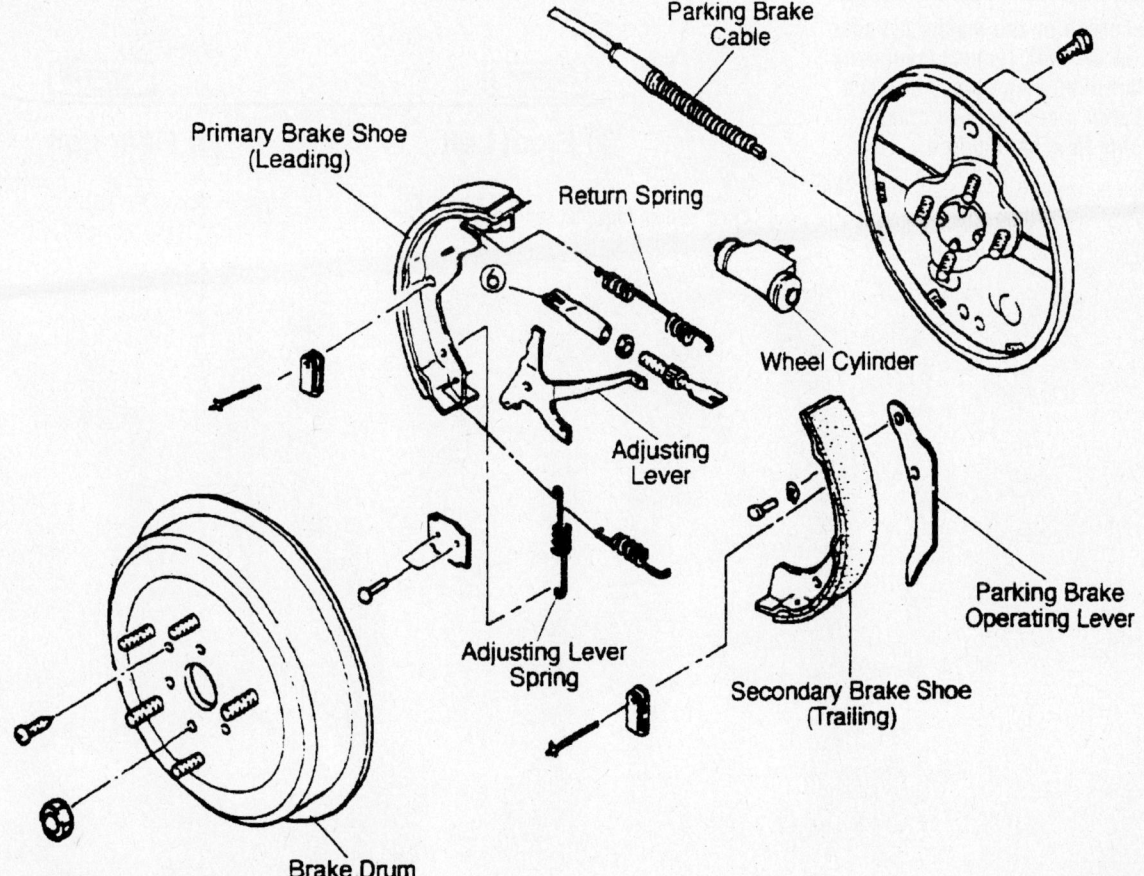

Exploded view of the rear drum brake assembly

93026G88

13. Remove the primary shoe.

14. Disconnect the C-clip and pin attaching the parking brake lever to the secondary brake shoe.

To install:

15. Lubricate the backing plate contact points.

16. Connect the parking brake lever to the secondary brake shoe.

17. Attach the primary and secondary brake shoes to the backing plate.

18. Install the starwheel.

19. Install the return spring above the starwheel.

20. Install the adjusting lever.

21. Install the spring from the secondary shoe to the adjusting lever. Be sure the lever contacts the starwheel.

22. Install the brake drum and retaining nuts.

23. Adjust the rear brakes through the slot in the rear of the backing plate.

24. Install the wheels. Tighten the wheel lugs to 77 ft. lbs. (99 Nm).

25. Lower the vehicle.

BRAKE HYDRAULIC SYSTEM

Brake System Bleeding

1. Before servicing the vehicle, refer to the Precautions Section.

✳✳ WARNING

Do not reuse the drained fluid. Always use Genuine DOT3 or DOT 4 Brake Fluid. Using unapproved brake fluid can cause corrosion and decrease the life of the system. Make sure no dirt of other foreign matter is allowed to contaminate the brake fluid. Do not spill brake fluid on the vehicle, it may damage the paint; if brake fluid does contact the paint, wash it off immediately with water.

➡ **The reservoir on the master cylinder must be at the MAX (upper) level mark at the start of bleeding procedure and checked after bleeding each brake caliper. Add fluid as required.**

2. Have someone slowly pump the brake pedal several times, then apply pressure.

3. Connect a length of clear plastic tube to the bleeder nipple and place the other end in a jar half full of clean brake fluid.

4. Loosen the right-rear brake bleed screw to allow air to escape from the system. Then tighten the bleed screw securely.

5. Repeat the procedure for each wheel in the sequence shown until air bubbles no longer appear in the fluid.

6. Refill the master cylinder reservoir to MAX (upper) level line.

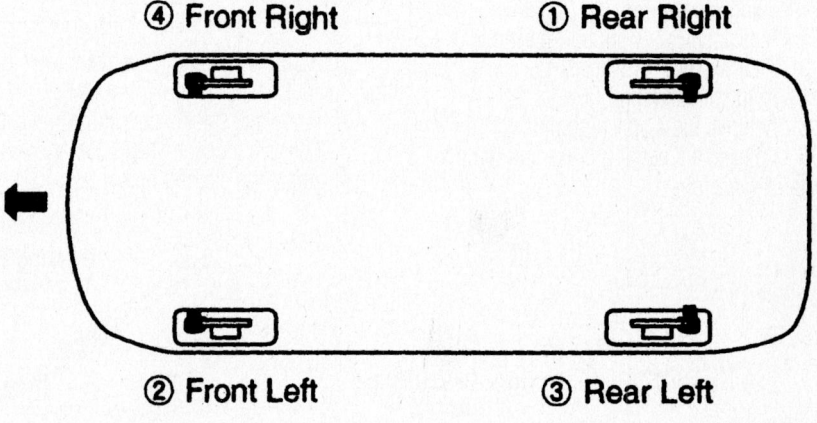

09474_KSPO_G0206

Brake system bleeding sequence